Preface

Microbiology continues to progress rapidly each year with the aid of new techniques, new research findings, and new insights into microorganisms and their many activities. Some of the most important scientific discoveries of this century have involved microbiologists. Since 1910, more than one-third of the Nobel Prizes in Medicine and Physiology have been awarded to microbiologists and individuals in related disciplines. From the results of studies conducted by these scientists and countless others, it has become increasingly evident that many exciting and significant discoveries in such areas as agriculture, biochemistry, genetics, medicine, molecular biology, pharmacology, and physiology rely on an understanding of microorganisms and applications of microbiological principles and associated techniques. For this reason, the fifth edition of *Microbiology* has been designed to emphasize carefully selected basic principles, concepts, techniques, and descriptive data. With this background, the reader can then incorporate information about new developments into a general framework of knowledge and can more readily understand how each new discovery resembles, differs from, extends, or disproves existing principles and concepts. A textbook for today's students must do justice to the newer principles and concepts without neglecting the older and proven ones that provide the foundation upon which the new body of knowledge rests. The fifth edition of *Microbiology* attempts to meet two important demands of a textbook: to stimulate interest and enhance learning, and to serve as a functional reference.

Organization of the Text

The present, fifth edition of *Microbiology* should be viewed as a new book, rather than as a revision, because the changes have been so major that many chapters of the text have been reorganized and rewritten, and more than half of the illustrations have been changed or added. We have tried to integrate the most recent advances in microbiology, molecular biology, immunology, biotechnology, and infectious diseases with the knowledge of the activities and structures of microorganisms. In presenting the subject matter the authors have taken into account the works of classical microbiologists—at times overlooked these days—that laid the foundation of our understanding not only of microorganisms, but the variety of other forms of life.

Microbiology is organized into eight parts. It focuses on a detailed introduction to the microbial world, starting with the basic principles and facts of microbiology, microscopy, and biochemistry and continuing through the properties, functions, activities, and applications of microorganisms. As the following description shows, the arrangement of the parts and the chapters within them allows for flexibility in the use of the text, so that instructors are free to design a course that meets the particular needs of their students.

Part I, "Introduction to Microbiology," begins with a chapter that draws the reader's attention to the impact of microorganisms and their activities on daily life. This chapter continues with an overview of microbiology as a discipline and includes a brief discussion of some of the challenges for microbiology in the near future, and a brief introduction to the historical development of microbiology. Particular attention is given not only to the events leading to the discovery of microorganisms but also to their numerous activities and potential value. Chapter 2 introduces the principles of classification and provides an introduction to microorganisms and their general properties. Chapter 3 discusses the broad range of microscopes used to detect and study microorganisms and other biological forms. Descriptions of specific techniques for specimen preparation also are included. It should be noted that a special Appendix (A) has been added near the end of the text to provide a brief treatment of basic chemistry and biochemical principles needed to become familiar with the various macromolecules associated with microbial cellular structures and functions, disease processes, and the immune system. This addition should serve as a ready reference for students who read other portions of the text.

Part II, "Microbial Growth, Cultivation, Metabolism, and Genetics," provides a comprehensive treatment of the general principles, processes, and techniques associated with microbial growth, cultivation, metabolism, and genetics. Chapter 4 introduces basic procedures used in bacterial cultivation and identification, while Chapter 5 covers fundamental as well as newer concepts of microbial metabolism and growth measurements. Chapter 6 discusses microbial genetics and includes topics such as genetic exchange mechanisms, mechanisms of cellular regulation, the features of plasmids, and genetic engineering, together with applications.

Part III, "The World of Microorganisms," provides a detailed introduction to the microbial world. The first chapter in this part describes the structures, organization, functions, and activities associated with procaryotes. The information here has been organized to provide an up-to-date frame of reference and basis for comparison with other microorganisms such as fungi, the protists (protozoa and certain algae), and the submicroscopic viruses, viroids, and prions. Chapters 8, 9, and 10 focus on the classification, cultivation, structure, organization, function, and ecological interaction of fungi, protozoa, algae, and viruses. Chapter 10 also provides a detailed survey of viruses found in all types of life and introduces the topic of the antiviral and anticancer protein known as interferon.

Part IV, "The Control of Microorganisms," discusses chemical and physical means of controlling microorganisms—topics of great practical significance. Both basic principles and particular techniques are presented. New developments are included—for example, chemotherapeutic control of fungi and viruses and techniques for determining the effectiveness of such chemotherapeutic agents. A detailed treatment of interferon drugs used in cases of AIDS and antibiotic resistance is provided in Chapter 12.

Part V, "Microbial Ecology and Interactions," focuses on the interactions and activities of microorganisms. Chapter 13 surveys the variety of procaryotes in the microbial world. Chapter 14 covers environmental microbiology, including the interrelationships among microbes in various environments and their involvement in the chemical cycles of life.

Part VI, "Microorganisms and Industrial Processes," begins with Chapter 15, which deals with the role of microorganisms in the preparation of various industrial products. Several applications of recombinant DNA technology to industry and agriculture are presented here. Chapter 16 discusses the microbial role in food preparation including wine, beer, and cheese making, and food spoilage. Attention also is given to the detection and identification of microorganisms in food. The chapter concludes with discussions of the microbiology of soil and water, and of waste treatment.

Part VII, "Principles of Immunology," presents the major features of the immune system, along with current concepts of immunology and several of their applications. Chapter 17 provides a general view of the host's defenses against disease agents. This chapter also discusses and characterizes antigens, and immunoglobulins. Chapter 18 presents a functional treatment of the states of immunity and discusses immunizing materials, their preparation, administration, and side effects. Chapter 19 deals with the techniques used for the isolation, identification, and applications of immunoglobulins and related substances. Special attention is given to monoclonal antibodies and the approaches used in the diagnosis of AIDS virus infection in this chapter. Chapter 20 discusses immunologic disorders, including acquired immune deficiency syndrome (AIDS), genetic factors, and autoimmune diseases.

Part VIII, "Microorganisms and Infectious Diseases," begins with Chapters 21 and 22, which set the stage for the following chapters by giving the student a thorough introduction to the principles of epidemiology and the many factors used by microorganisms to establish an infection or a disease process. Chapter 21 describes current approaches used to identify microorganisms in the laboratory. Chapters 23 through 29 apply these basic ideas to a survey of microbial diseases of specific human tissues and organ systems. Many of the currently important diseases such as AIDS, human T cell lymphotropic virus infections, Legionnaire's disease, hepatitis B, genital herpes, and toxic shock syndrome are included. Chapter 30 discusses the relationship between microorganisms, immunology, and cancer. Part VIII concludes with Chapter 31 which covers helminthic (worm) infections, their transmission, diagnosis, identification, and their impact on world health.

Distinctive Features of the Fifth Edition

The text includes a number of learning aids designed to help students absorb, review, apply, and retain its content. Some of these aids and distinctive features of the fifth edition are the following:

- Each chapter begins with an **overview** that briefly previews the topics to be covered. This overview is followed by a list of **learning objectives** organized to help the student master all important terms, concepts, techniques, and applications.
- Within the chapter, **summarizing tables** reinforce what the student has learned, and carefully designed and selected diagrams and photographs illustrate the ideas presented in the text.
- **Phonetic pronunciations** of most microorganisms are presented the first time mention of each organism is made in a chapter. A composite listing of pronunciations is included near the end of the text.
- Specially developed topics and/or applications of the subject matter known as **Microbiology Highlights,** have been prepared for each chapter. These boxed presentations also emphasize historical events, special applications and recent discoveries associated with microbiology.
- Much attention has also been given to assure that correct terms are clearly defined and accessible in order to help the student develop and retain a useful vocabulary. Key words are printed in bold type and other terms in italics.

MICROBIOLOGY

Overleaf: A microscopic view of the structures of a developing fungus. [From Azab, M. S., et al., Trans. Br. Mycol. Soc. *86:469–474, 1986.]*

MICRO

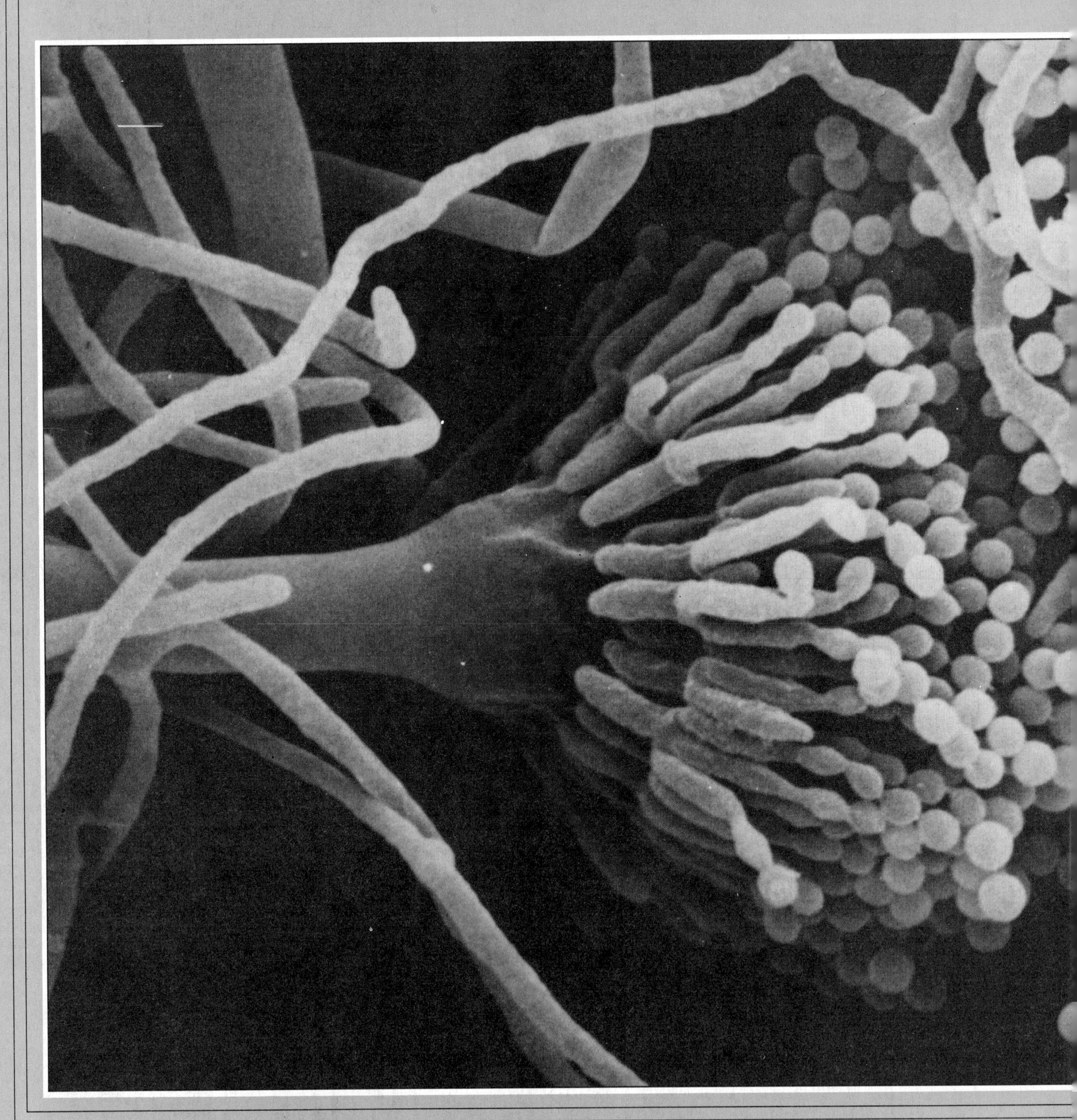

BIOLOGY

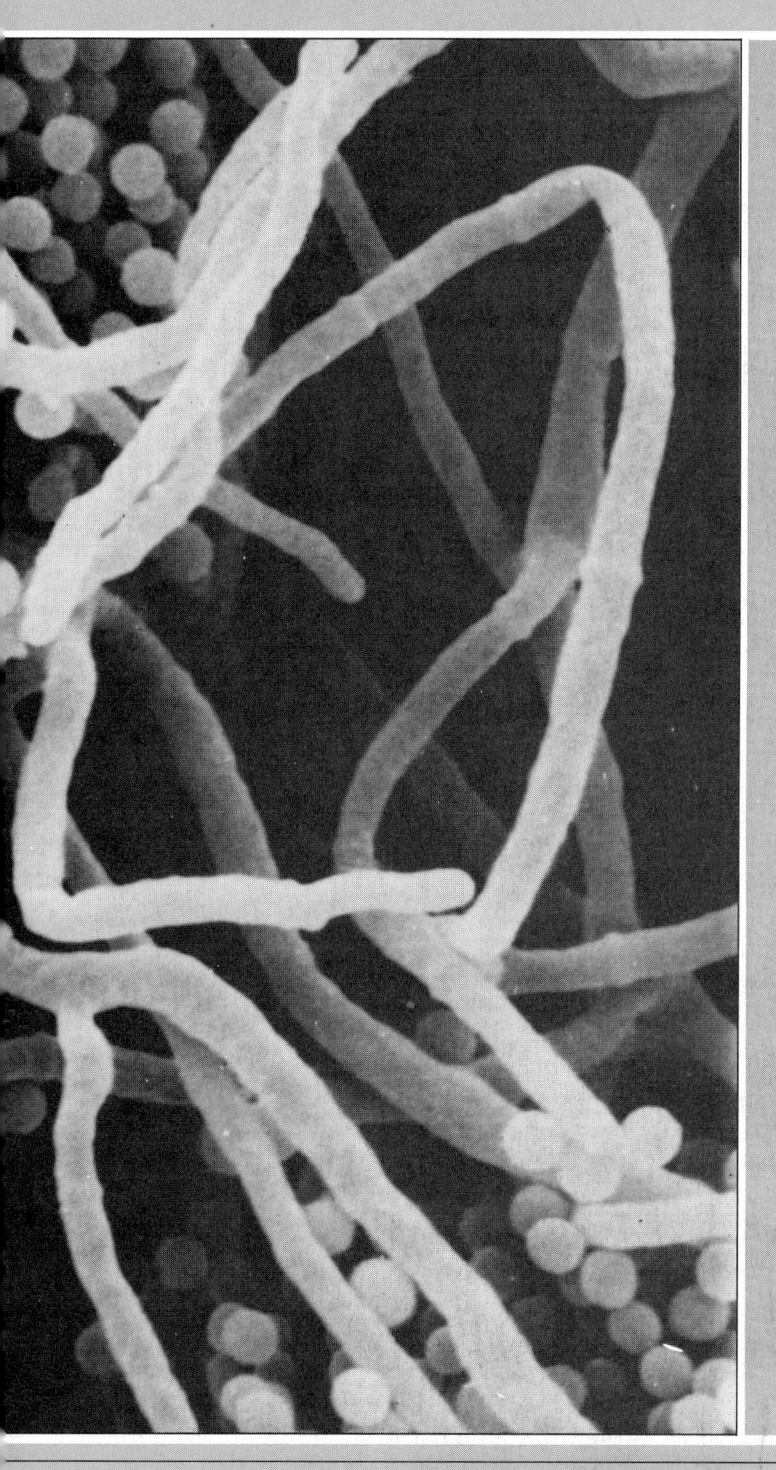

George A. Wistreich
East Los Angeles College

Max D. Lechtman
Golden West College

FIFTH EDITION

MACMILLAN PUBLISHING COMPANY
NEW YORK

Collier Macmillan Publishers
London

Microbiology is affectionately dedicated to our families, whose encouragement, patience, and sacrifices helped to make another edition possible.

Printed in the United States of America

Macmillan Publishing Company
866 Third Avenue, New York, New York 10022

Collier Macmillan Canada, Inc.

Library of Congress Cataloging in Publication Data

Wistreich, George A.
Microbiology / George A. Wistreich, Max D. Lechtman. — 5th ed.
p. cm.
Includes index.
ISBN 0-02-428950-7
1. Microbiology. I. Lechtman, Max D. II. Title.
QR41.2.W56 1988 87-24850
576—dc19 CIP

Printing: 4 5 6 7 8 Year: 0 1 2 3 4 5 6 7

- Each chapter concludes with a summary in outline form and Questions for Review to help students apply and draw out the important concepts and principles presented.

- The chapters dealing with infectious diseases now contain an **Illustrated Disease Challenge.** Some aspect of the disease being described or laboratory finding is illustrated and provides an opportunity to apply knowledge and critical thinking in problem-solving. These chapters also contain special **boxed sections presenting approaches to laboratory identification and acceptable treatment.**

- The **glossary** of the previous edition has been expanded not only to include up-to-date terminology, but also terms suggested by readers of the text.

- A new section has been incorporated dealing with the **make-up of biological and medical terms.** This section should enable the reader to analyze most biological and medical terms found in the current edition and most other texts.

- Many chapters contain in addition to general review questions, one or more photograph quizzes. These devices serve as opportunities for students to apply some of the information provided in the respective chapters.

- The **Suggested Readings** (reference) sections of each chapter contain annotated entries. These references are up-to-date, and should serve as foundations for students interested in obtaining more information and insight into the topics of the chapter.

- A total of 24 pages consisting of 159 **color photographs,** together with appropriate and functional captions have been included in the current edition. Much attention has been given to incorporate the types of subjects for which color is important for recognition and understanding. All color photos are coordinated and referenced with the text. Several of the photographs have never appeared in texts before.

- Considering the fact that many students undertaking an introductory microbiology course have had one or more chemistry courses, a special **Basic Chemistry and Biochemistry Appendix** has been included and designed to contain only essential and basic concepts.

- *Microbiology* also has six appendices that contain current classification schemes for bacteria, fungi, protozoa, algae, viruses, and helminths. The classification for bacteria follows that found in *Bergey's Manual of Systematic Bacteriology* and reported in the *International Journal of Systematic Bacteriology.*

Supplementary Materials

The *Study Guide to Accompany Microbiology,* by George A. Wistreich and David W. Smith, contains vocabulary lists, phonetic pronunciation of key terms, a review of important concepts, two self-tests for each chapter of the text, and enrichment sections that expand the coverage of specific text topics. Disease Challenges, which enable students to apply their diagnostic and observational skills, are included in chapters dealing with infectious diseases. These Disease Challenges are different from those in the text. Eight *Microbiology Trivia Pursuit* sections have been included to challenge the reader's powers of observation and attention to detail.

A *Slide Set* to complement both lecture and laboratory presentations has been carefully selected as an aid to adopting instructors. These slides are accompanied by descriptions, and suggestions for their use.

A *Transparency Set* to illustrate concepts in both lecture and laboratory presentations has been carefully selected and combined for effectiveness. These transparencies also are accompanied by descriptions, and suggestions for their use.

A test bank of questions appropriate to the text is also available to adopters.

The *Instructor's Manual to Accompany Microbiology,* available gratis to adopting instructors, contains annotated lists of current audiovisual materials with their sources, annotated references for each chapter and general references, discussion questions for each chapter, teaching suggestions, and answers to text questions.

Laboratory Exercises in Microbiology, sixth edition, contains 59 laboratory exercises and 11 special experimental exercises that can be used either in conjunction with the text or independently. Specific laboratory procedures are explained in step-by-step diagrams, and concepts learned in the laboratory are reinforced by postlab questions and photographic quizzes. Color photographs showing characteristic properties of microorganisms and experimental results are included in a separate section. Thirteen self-assessment sections have been added to help students determine their knowledge of basic principles and techniques. An accompanying *Instructor's Manual* offers the instructor resources, suggestions for effective use of the *Laboratory Exercises in Microbiology,* alternate experiments, annotated lists of current audiovisual materials, answers to questions in the laboratory manual, and sources of all laboratory-required equipment and supplies.

Acknowledgments

This edition of *Microbiology* was greatly influenced by the many enthusiastic comments of students and by the wise guidance and counsel offered by many instructors and professionals—in particular, the comments of the reviewers whose names follow this preface. The authors are greatly indebted to the large number of individuals responding to a detailed questionnaire prepared by Dr. Greg W. Payne.

The authors would also like to thank Madalyn Stone, J. Edward Neve, Leo Malek and other members of the editorial and production staffs of the Macmillan Publishing Company for their untiring and imaginative efforts expended in the preparation of the fifth edition of *Microbiology,* and accompanying supplementary materials. We are also indebted to Mr. Phillip Wisztreich and Ms. Rose Serber for the care and attention exercised in typing portions of the manuscript, and for their supportive services in preparing the manuscript.

The authors also would like to take this opportunity to acknowledge the contributions and support of Mr. Robert Rogers, Senior Biology Editor, who devoted a great deal of time, patience, and support to our text.

We would also like to thank the many contributors here and abroad who willingly provided us with photographs and diagrams. To these people we are especially indebted.

Finally, the authors would like to thank the following persons for reviewing the previous edition or the manuscript of the current edition and offering valuable suggestions:

Ted Johnson, *St Olaf College*
Thomas Jones, *Salisbury State College*
Gary E. Kaiser, *Catonsville Community College*
Alan Liss, *The State University of New York, Binghamton*
Bernard Marcus, *Genesee Community College*
Karen Messley, *Rock Valley College*
Christine Pootjes, *Pennsylvania State University*
Marion Price, *Onondaga Community College*
Rivers Singleton, *The University of Delaware*
Robert Sjogren, *The University of Vermont*
David W. Smith, *The University of Delaware*
Kenneth Thomulka, *Philadelphia College of Pharmacy*
Donald H. Walker, Jr., *The University of Iowa*

George A. Wistreich
Max D. Lechtman

Contents

PART VI
Microorganisms and Industrial Processes

PART VII
Principles of Immunology

PART VIII
Microorganisms and Infectious Diseases

APPENDICES

Detailed Contents

PART I Introduction to Microbiology

PART II Microbial Growth, Cultivation, Metabolism, and Genetics

PART III The World of Microorganisms

8 Fungi 223

9 The Protists 247

10 Viruses, Viroids, and Prions 275

PART IV The Control of Microorganisms

11 Chemicals and Physical Methods in Microbial Control 314

16 Food, Dairy, and Water Microbiology 455

PART VII Principles of Immunology

17 Introduction to Immune Responses 488

18 Immunization, States of Immunity and the Control of Infectious Diseases 531

19 Diagnostic Immunologic and Related Reactions 551

PART VIII Microorganisms and Infectious Diseases

25 Microbial Infections of the Respiratory Tract 716

26 Microbial Diseases of the Gastrointestinal Tract 749

27 Microbial Infections of the Cardiovascular and Lymphatic Systems 782

28 Microbial Diseases of the Reproductive and Urinary Systems 812

29 Microbial Diseases of the Central Nervous System and the Eye 842

30 Microbiology and Cancer 869

31 Helminths and Disease 889

Appendices A1

MICROBIOLOGY

PART I

Introduction to Microbiology

Louis Pasteur (1822–1896), working in the laboratory. [Courtesy Chas. Pfizer & Co., Inc.]

1

The Scope of the Microbial World

It is characteristic of Science and Progress that they continually open new fields to our vision. — Louis Pasteur

After reading this chapter, you should be able to:

1. Describe the major divisions of microbiology and the types of microorganisms and/or activities involved with each one.
2. Distinguish between the integrative and applied approaches to the study of microorganisms.
3. Briefly discuss the importance of microorganisms in everyday life.
4. Describe the contributions of individual scientists and the events leading to the discovery of microorganisms, their activities, and their functions.
5. Identify the problems of early investigators attempting to study, grow, and control microorganisms.
6. Explain the doctrine of spontaneous generation and the experiments used to disprove it.
7. List both Koch's and Rivers' postulates and relate their importance to the germ theory of disease.
8. Define chemotherapy and explain its importance in the control of microorganisms.

The first simple forms of life appeared on earth more than 3 billion years ago. Their descendants have changed and developed into the several million types of animals, plants, and microorganisms recognized today. No doubt, thousands more remain to be discovered and officially described.

Throughout the centuries, humans have used various forms of microorganisms to meet their needs and have been fascinated by the abundance and variety of microbial life around them. This chapter introduces the reader to the world of microorganisms, their activities and applications. It will soon become apparent that despite their small size, these microscopic forms of life play important roles in nature and in processes affecting the well-being of all other life forms.

Chapter 1 surveys the ideas, discoveries, problems, and successes of early scientists that contributed to establishing microbiology as a biological science. Among the most important of these events were the discoveries of microorganisms and their roles in natural processes such as fermentation and disease, the disproof of the doctrine of spontaneous generation, and the development of measures for controlling microorganisms.

Figure 1–1 Burying victims of the plague of London of 1664 in mass graves. In the late seventeenth century, one physician guessed that the annual number of deaths in London due to smallpox was about 3,000—an estimate that was probably low. *[The Bettmann Archive.]*

Microbiology is the study of microscopic forms of life. Like most other sciences, it had its origin deeply rooted in curiosity. At first, microorganisms were thought to be mysterious oddities of little practical importance. However, the work of many individuals, including Louis Pasteur, Robert Koch, and Joseph Lister, drastically changed this limited view of microbes during the late nineteenth century. For the first time, the world became aware of the desirable and undesirable effects of microorganisms on the environment—including spoilage, disease, and death (Figure 1–1). This realization ushered in a whole new era of research in microbiology.

In its early period, microbiology as a science was concerned with the isolation, identification, and control of microorganisms. Major advances in microscopy and biochemical techniques from the 1940s to the present showed microorganisms to be useful models for the study of various processes of living systems, particularly in the areas of genetics and metabolism. Moreover, these and numerous other studies have clearly shown that microorganisms perform many activities that are beneficial to humans. For example, these microscopic forms of life manufacture antibiotics, vitamins, and growth factors for humans, other animals, and plants; decompose sewage and solid and industrial wastes; and are essential to the formation of foods such as cheeses, yogurt, bread, pickles, and sauerkraut. Many of these microbial products are of commercial importance. Continued advances in microbiology and related branches of science are also providing many new and exciting ways to make microorganisms more useful. Through genetic engineering technology, it is possible to produce new forms of microorganisms by inserting into them specific pieces of foreign genetic material (genes). Such newly altered microorganisms can thus be directed to produce important proteins such as the hormones insulin (used for the treatment of diabetes) and human growth hormone (used to increase bone growth) and the antiviral agent interferon (used in the treatment of certain cancers and diseases caused by viruses); or they may be used to make an amino acid essential for good nutrition. This type of genetic tinkering has not only moved microbiology, and science in general, into a new stage of development but has also increased public concern with and interest in the advances of science and the impact of technology on the quality of human life.

Despite the established useful functions and the potentially valuable activities of microorganisms, these microscopic forms of life may be best known as agents of food spoilage (Figure 1–2) and causes of

Figure 1–2 An orange attacked by the Florida canker-causing bacterium. The oozing cankers (arrows) make the fruit unacceptable for market. *[Courtesy U.S. Dept. of Agriculture.]*

diseases, including acquired immune deficiency syndrome or AIDS (Figure 1–3), herpes (Color Photograph 1), Legionnaire's disease (Color Photograph 2), influenza, and listeriosis, the disease associated with contaminated Mexican-style cheeses. So far as is known, all primitive and civilized societies have experienced diseases caused by microbes, frequently with disastrous results. Moreover, microorganisms have played profound roles in warfare, religion, and the migration of populations (Table 1–1). An understanding of these and other activities of microorganisms can best begin with an examination of the microbial world.

The Microbial World

Microscopic forms of life are present in vast numbers in nearly every environment known. They are found in the soil, in bodies of water (Figure 1–4a), in the food and water we consume, and even in the air we breathe. Since the conditions that favor the survival and growth of many microorganisms are the same as those under which people normally live, it is not unusual to find these microscopic forms on the surfaces of our bodies and in the mouth (Figure 1–4b), nose, portions of the digestive tract, and other body regions. Fortunately, the majority of such microorganisms are not harmful to humans or to the various other forms of life.

The microbial world, including both microscopic life and the submicroscopic viruses, which are too small to be seen with an ordinary microscope, displays a wide variety of forms and activities. Many microorganisms, or microbes, occur as single cells; others are multicellular; and still others, such as viruses, do not have a true cellular appearance. Certain organisms, called **anaerobes,** are capable of carrying out their vital functions in the absence of free oxygen. However, the majority of organisms, the **aerobes,** require free oxygen. Some microbes can manufacture the essential compounds for their physiological needs from atmospheric sources of nitrogen and carbon dioxide. Other microorganisms, such as viruses and certain bacteria, are totally dependent for their existence on the cells of higher

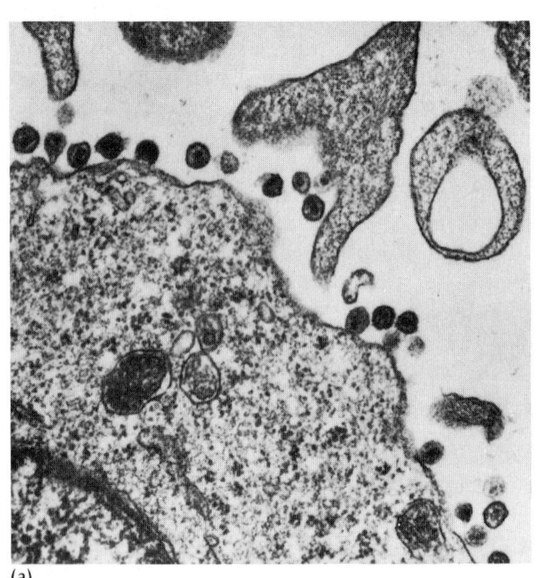

(a)

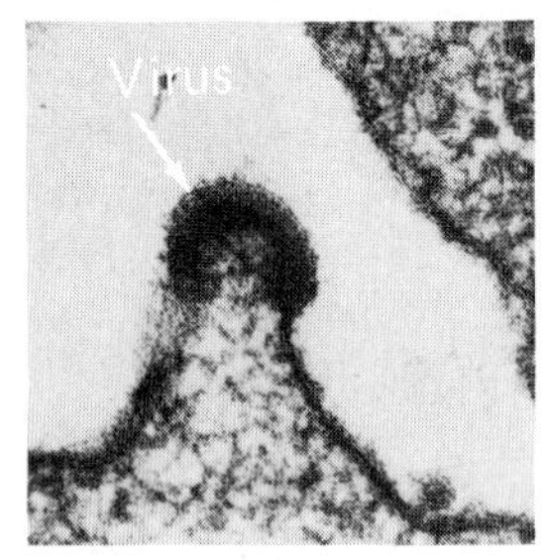

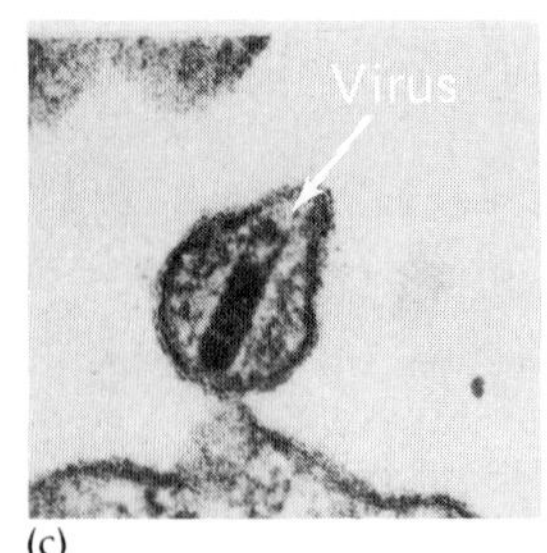

(c)

Figure 1–3 The AIDS virus (human immune deficiency virus, HIV). (a) New virus particles surrounding the cell in which they were formed. (b) An emerging virus. (c) A fully formed virus ready to attack another cell. *[From R. C. Gallo* et al., Science ***224:****500–503, 1984.]*

TABLE 1–1 Examples of the Involvement of Microorganisms in Human History

Historical Event	*Dates (A.D.)*	*Outcomes and Observations*
Decline of the Roman Empire during the time of Justinian	518–565	Repeated outbreaks of diseases such as plague greatly weakened the Empire and made the consolidation of power impossible.
Ethiopian siege of Mecca ("Elephant War")	570–571	Retreat of Ethiopians, probably due to smallpox, which prevented the destruction of Islam.
Spanish conquest of Mexico	1500s	More than 12 million Aztecs died of smallpox.
Great Fire of London	1666	End of the bubonic plague (Black Death) in London. Survivors used wood or stone rather than straw to rebuild roofs, thereby removing sites where rats and fleas might take refuge.
British versus North American Indians	1763	Blankets from smallpox victims were given to Indians. True case of biological warfare.
Proposed invasion of England by Louis XVI	1779	Smallpox, typhus fever, and scurvy prevented invasion attempt. Over 8,000 troops died at sea.
Invasion of Russia by Napoleon	1802	Russians suffered great loss of life due to typhus fever.
Irish diaspora	1845	Fungus blight devastated potato crop in Ireland and forced mass migrations.
U.S. Civil War, Battle for Vicksburg	1862–1863	Union forces suffered great losses from malaria.
World War I	1914–1917	Russia alone lost 3 million troops from typhus fever.

Figure 1–4 Some environments of microorganisms. (a) The clearly recognizable forms of diatoms, members of a phytoplankton algal community on the surfaces of one grain of sand. The bar marker (short line) on this and later illustrations serves as a size reference. The length of the marker represents the length of a metric unit at a particular magnification. The metric system is discussed in Chapter 3. *[From C. M. Pringle,* J. Phycol. ***21****:185–194, 1985.]*
(b) Bacteria in the mouth. Short, rodlike microorganisms can be seen on the tooth's surface. A few red blood cells are also present. *[A. L. Allen and J. B. Brady,* J. Peridontol. ***49****:415, 1978.]*

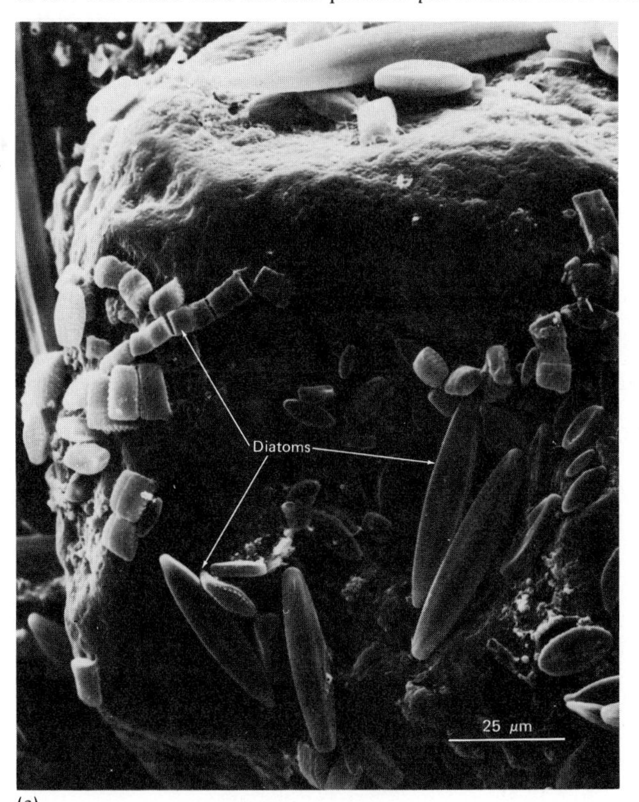

(a)

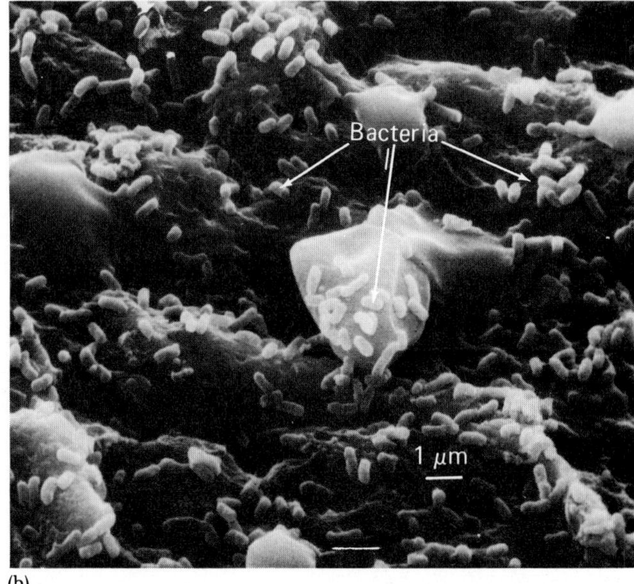

(b)

forms of life. The branch of science known as **microbiology** embraces all of these properties of microorganisms and many more. For the most part, microorganisms exhibit the characteristic features common to biological systems, such as reproduction, metabolism, growth, and mutation. Most microbial types can not only respond but also adjust to environmental stresses including temperature, acidity, radiation, pressure, and poisonous substances. It is clear from these processes performed by microbial forms of life that they possess a certain level of organization and extraordinary precision. Hence, it is quite appropriate to refer to microbes as small, organized units, or **microorganisms.**

Microbiology and Its Subdivisions

Taxonomic Arrangement

Microorganisms have offered and continue to provide important model systems for the study of basic biological events and processes. The microbiologist studies a wide range of these microscopic forms of life and their various activities in both natural and artificial environments. The field of microbiology includes the study of algae, bacteria, fungi (molds and yeasts), protozoa, viruses, and subviral agents of disease.

As a matter of convenience, the branches of microbiology are frequently named according to the topic on which they focus. This method of organization, referred to as the *taxonomic approach,* is briefly discussed in the following section.

BACTERIOLOGY

Bacteria represent a group of microscopic organisms that is large in terms of both variety and numbers. They are found in numerous types of environments (Color Photographs 3 and 66), ranging from soil and bodies of water to the external and internal surfaces of humans (Figure 1–5), lower animals, and plants. The activities in which bacteria are involved include the causation of disease (Color Photograph 4), the decomposition of decaying or dead organic matter, the digestion of food in humans and other organisms, and the production of various chemicals, foods, and other useful products—the subjects of **bacteriology.**

IMMUNOLOGY

The concepts of **immunology** are of ancient origin and are derived mainly from the study of resistance to infection. Centuries before the discovery of the germ theory of infectious disease, it was known that recovery from illness generally was accompanied by an ability to resist reinfection. In its infancy, immunology was devoted almost exclusively to the prevention of infectious diseases by vaccination (immunization).

One of the greatest triumphs of modern medicine has been the eradication of smallpox everywhere in the world. This dreaded viral disease, known since

Figure 1–5 Bacteria *(Staphylococcus epidermidis)* sticking to the curved cells surrounding the opening of a human sweat gland. The original magnification is 6500 ×. *[From B. E. Brooker and R. Fuller,* J. Appl. Bacteriol. ***57****:325–332, 1984.]*

Figure 1–6 Events in smallpox history. (a) The head of the mummy of Pharaoh Ramses V, who reportedly died in 1157 B.C. of an illness closely resembling smallpox. *[Courtesy of the World Health Organization, Geneva, Switzerland.]*

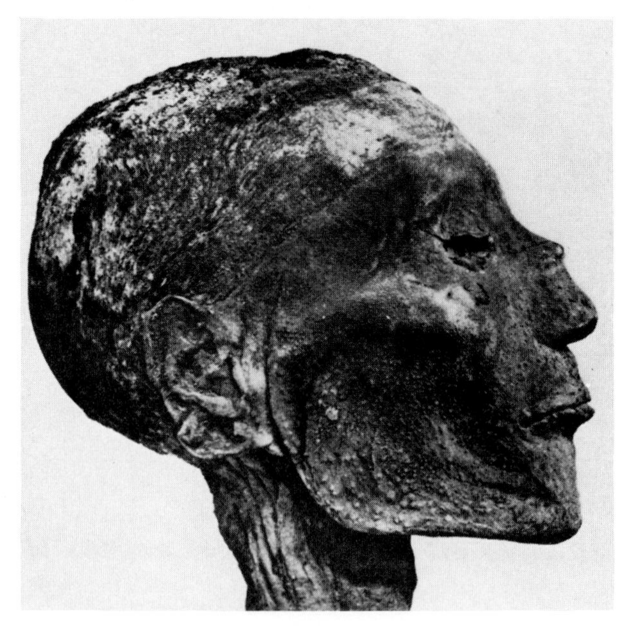

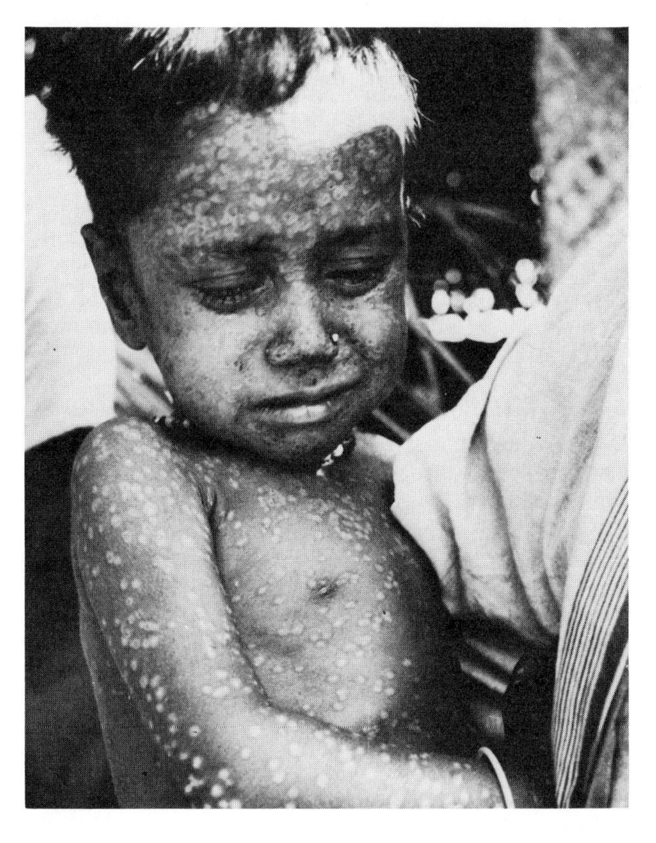

Figure 1–6 (b) The last known case of the more severe form of the virus disease smallpox in Bangladesh. *[Courtesy of the World Health Organization, Geneva, Switzerland.]*

ancient times (Figure 1–6a), has been responsible for the deaths of millions of people. However, not one case of smallpox has been reported anywhere in the world since 1978 (Figure 1–6b). This is largely the result of the concentrated vaccination campaign of the World Health Organization (WHO), a special agency of the United Nations. Because of WHO's success, similar campaigns are under way to eliminate other diseases caused by viruses, including measles and poliomyelitis. Considerable effort is also being made to improve older vaccines, develop new ones, and increase their use.

Classic immunology has expanded widely into numerous biological and nonbiological sciences. New applications of immunology include the use of rapid and more specific laboratory techniques for the detection, diagnosis, and treatment of many diseases, both infectious and noninfectious; uncovering the mechanisms of allergies and the immune response itself; and selecting tissues for organ transplants. **[Immunologic disorders are discussed in Chapter 20.]**

MYCOLOGY

Mycology is the study of the fungi, including molds, mushrooms, and yeasts (Figure 1–7). Fungi have a wide variety of sizes, colors, and shapes; they are nonphotosynthetic. Many fungi live in the soil and take an active part in the decomposition of organic matter (Color Photographs 5 and 6). In addition, fungi are involved in the production of several kinds

Figure 1–7 Mycology. (a) The microscopic appearance of a mold found in many different environments. Its reproductive structures, or *spores,* are in clusters. Molds of various types play a role in the production of antibiotics and in the decomposition of decaying matter. *[C. A. Shearer,* Mycologia ***56:****16–24, 1974.]* (b) Brewer's yeast, *Saccharomyces cerevisiae,* is a fungus. This microorganism is frequently used in baking and in the production of beer and wine. Yeasts may reproduce asexually by forming buds. *[Courtesy of Standard Brands, Fleischmann Laboratories, Stamford, Conn.]*

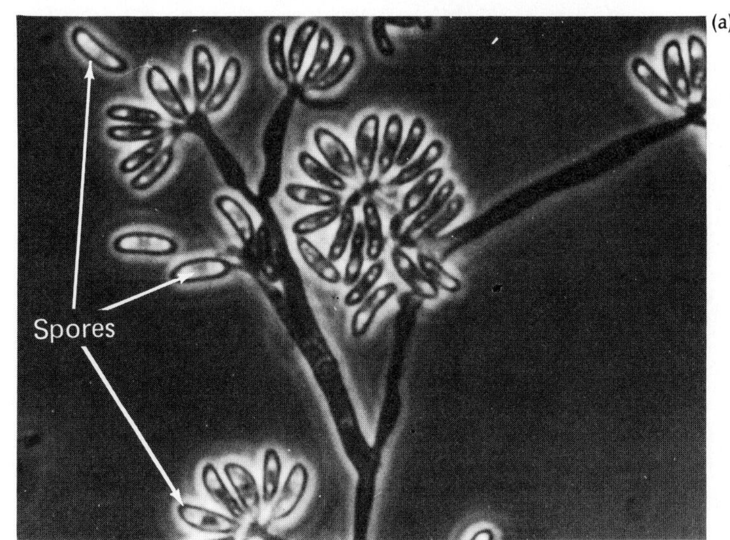

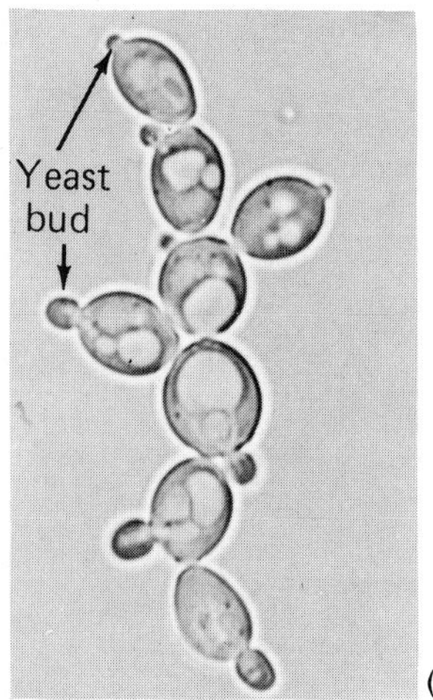

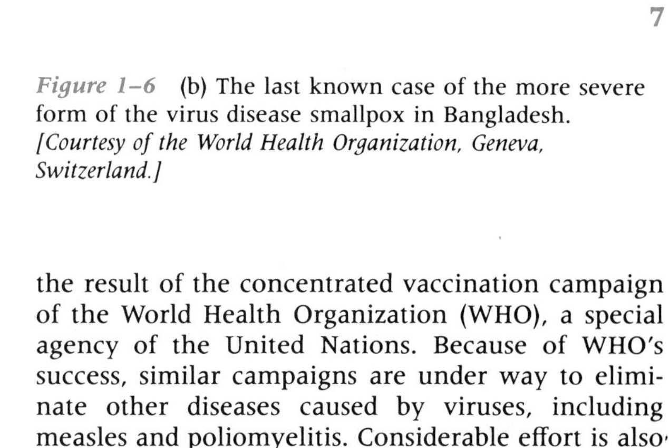

of food and antibiotics and in causing disease in humans (Color Photograph 41), other animals, and plants. All of these aspects of fungi are the subject matter of **mycology.**

PHYCOLOGY

Phycology, the study of algae, is sometimes also called *algology.* Algae range in size from microscopic unicellular forms (Figure 1–8 and Color Photograph 7b) to the multicellular giant kelp (seaweed), which can reach lengths of 50 meters or more. In this text only certain microscopic forms will be discussed.

The pigments in algae, which include the colors brown, green, red, and yellow, give these organisms the ability to carry out photosynthesis. Algae increase the level of dissolved oxygen in their immediate environment and serve as food sources for humans and livestock.

Not all of the activities of algae are beneficial to higher life forms, however. For example, algae have caused problems ranging from formation of algal blooms (Color Photograph 7a) to vast fish kills.

PROTOZOOLOGY

Protozoa, the subject of **protozoology,** are larger than most other microorganisms and are complex in structure and activities. They are found in sewage, various bodies of water, the intestinal tracts of insects (Figure 1–9), and damp soil. Most protozoa feed on other microorganisms and rotting or dead organic matter. These are harmless. But some types are responsible for severe diseases in humans and other animals. Among these, malaria, African sleeping sickness (Figure 1–10), and gastrointestinal infections have caused and continue to produce widespread suffering and death among peoples of the tropical world and elsewhere. Certain protozoa have also caused the deaths of many AIDS victims.

VIROLOGY

Before the discovery in the nineteenth century that bacteria cause disease, the term *virus,* a Latin word for poison, was used by Louis Pasteur to describe the poisonous and then invisible cause of rabies. In the twentieth century, "virus" acquired a new and more specific meaning. Today, the term designates a large group of submicroscopic disease-causing agents that are basically quite different from all other forms of life. An individual virus generally consists of nucleic acid surrounded by a protein covering of some type (Figure 1–11).

Virology is the subdivision of microbiology that

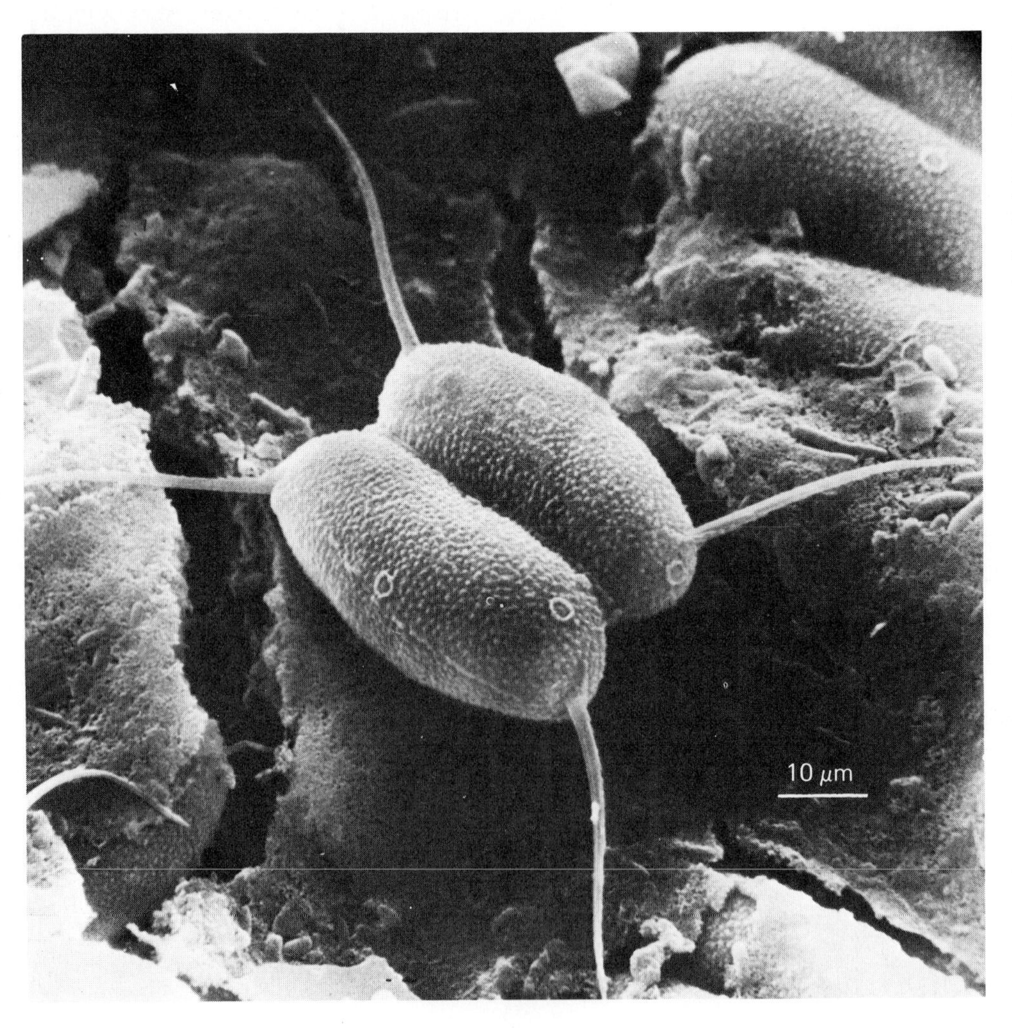

Figure 1–8
Phycology. Microscopic algae are remarkable microorganisms, showing amazing variety in shape, surface appearance, and the way they group together. *[From L. A. Staehelen and J. D. Pickett-Heaps,* J. Phycol. ***11**:163, 1975.]*

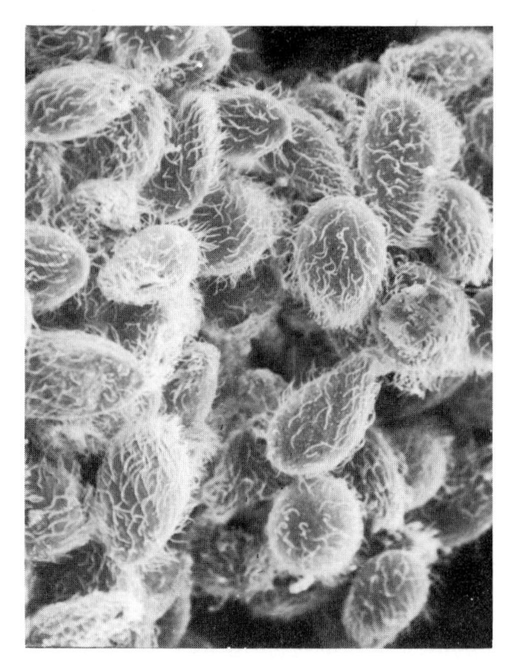

Figure 1–9 The *Tetrahymena*, covered with cilia (little hairs). This protozoon is a common inhabitant of the intestinal tracts of termites and other related forms. *[Courtesy G. A. Thompson, Jr.]*

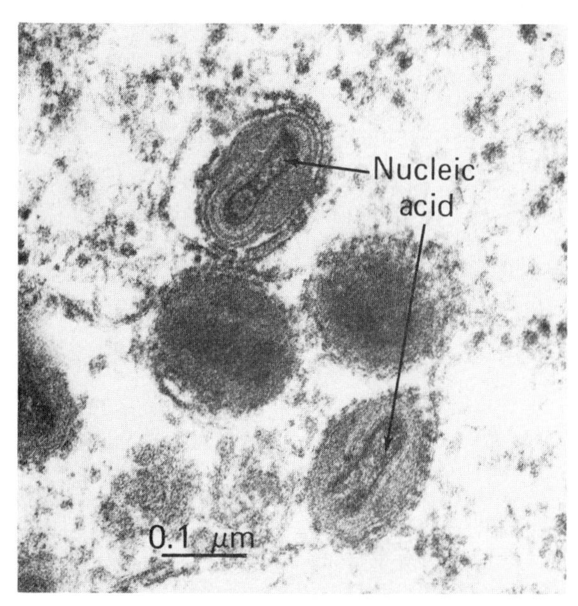

Figure 1–11 An electron micrograph of vaccinia virus. This virus was used for vaccination against smallpox, a disease that has been virtually eliminated by effective immunization. *[From F. I. Holowaczak, L. Flores, and V. Thomas,* J. Virol. ***61****:376–396, 1975.]*

is concerned with these submicroscopic organisms, which have an organization and patterns of growth and multiplication different from those of other microorganisms. Because viruses depend on the cells they invade (their hosts) for growth and reproduction, they are referred to as *obligate intracellular parasites.* Virtually every form of life, including microorganisms (Figure 1–12), can be infected by viruses.

Because viruses are smaller than other microorganisms, they could not be viewed clearly until the late 1940s. They were surrounded by mystery because of their apparent invisibility and their disease-producing capabilities. Combating the effects of these invisible disease agents was for many years limited to prevention through vaccine development. As mentioned earlier, smallpox, a disfiguring and deadly human viral infection, was the first to yield to such preventive measures.

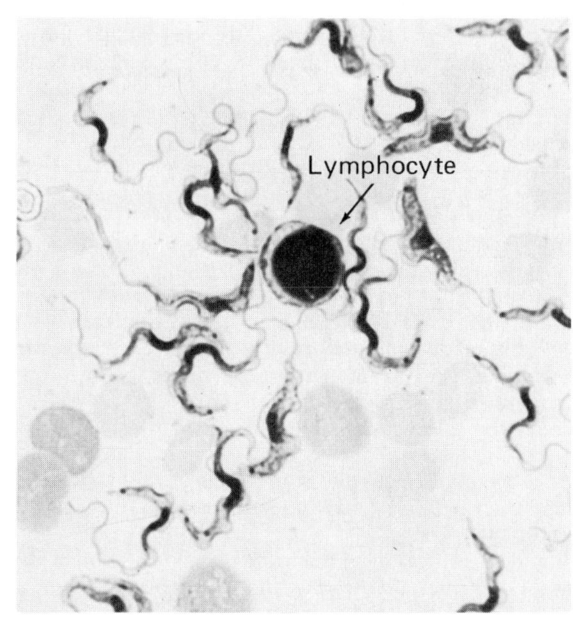

Figure 1–10 The cause of African sleeping sickness, a disease of the central nervous system. *[From I. Cunningham,* J. Protozool. ***33****:226–231, 1986.]*

Figure 1–12 A view through an electron microscope of bacterial viruses capable of destroying selected bacteria found in milk. *[Courtesy of Prof. Dr. Teuber.]*

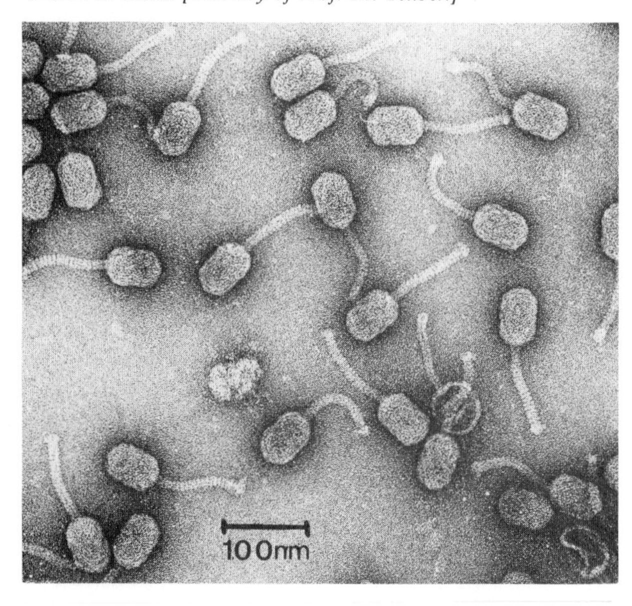

Figure 1–13 A leaf of a tobacco plant infected with tobacco mosaic virus. Note the variation in color shown on the infected leaf (arrows). *[Courtesy USDA.]*

THE DISCOVERY OF VIRUSES. The recognition of a nonbacterial infectious agent was made by Dmitrii Ivanowski in 1892 during his investigations with a plant disease known as the tobacco mosaic disease (Figure 1–13). He demonstrated the infectious nature of plant sap from infected plants by injecting healthy plants with the collected material and producing the disease. Because Ivanowski could neither observe any microscopic agent nor prevent its activity by passing infectious plant sap material through bacterial retaining filters, he concluded that here was a previously unknown form of life.

In 1898, Löffler and Frosch discovered the virus that causes foot-and-mouth disease (FMD), and three years later Walter Reed and his associates discovered that mosquitoes transmit the virus that causes yellow fever. With the discovery of several viruses responsible for diseases in animals and the demonstration of viruses capable of attacking and causing the destruction of bacteria in 1915, interest in these agents began to grow. Then in 1935, W. Stanley reported the crystallization of a pure virus, the tobacco mosaic virus, to the scientific world.

This discovery, in the words of W. Hayes, "gave birth to the romantic idea that viruses are a kind of missing link between living and nonliving material," a borderline form of life. The debate as to whether or not viruses are alive continues today. Nevertheless, viruses have gained a prominent position in the scientific world, not only because of their obvious relationship to disease but also because of their value as research tools in unlocking the intricate mechanisms of life's processes. Today, most scientists consider viruses a distinct category of living things. **[Smaller agents of disease are discussed in Chapter 10.]**

Integrative Arrangement

The need for specialization has led to the development of microbiological subdisciplines, each of which is defined by the types of activities, functions, applications, and related features of microorganisms. There are various ways of deriving subject areas in microbiology. One approach is the **integrative arrangement.** Here the boundaries between the subdivisions of microbiology are drawn in a way that emphasizes the common properties of microbes and their many interactions. The knowledge gained from the areas of study listed in Table 1–2 increases

TABLE 1–2 Integrative Arrangement of Microbiological Subdisciplines

Subdiscipline	*Brief Description*
Microbial ecology	The study of the relationships between microorganisms and their environments. An investigation of the way microorganisms respond to unfavorable situations is an example of this area of specialization.
Microbial genetics	The study of genes, how they produce characteristics and inheritance patterns that are passed to further generations of microorganisms. This subdiscipline deals with genetic exchanges between microbes, how the information in genes is expressed within organisms, control of cellular activities, and the effects of permanent genetic changes or mutations.
Microbial morphology and ultrastructure	The study of the shapes and the microscopic and submicroscopic details of microbial cells.
Microbial physiology	The study of how microorganisms function. Metabolic activities, the effects of the environment on microbial synthetic pathways, and the nutritional requirements of different groups of microbes are among the subjects of microbial physiology.
Microbial taxonomy	This subdiscipline deals with the naming and classification of microorganisms. It involves determining the similarities and differences among microbial species and using these data to formulate a classification system that shows the relationship of microorganisms to one another.

the understanding of microorganisms. Such insight is of great importance to scientists working in applied areas of microbiology, which include preventing and treating diseases caused by microorganisms; preventing the microbial spoilage of foods; and maintaining the quality of air, water, and soil.

Applied Microbiology

Microbiology can also be organized into subdisciplines by using an applied approach. Several areas of microbiology are oriented along lines of problem solving or involve the study of locations where organisms occur (habitat). Table 1–3 lists and describes a number of these disciplines. It should be pointed out that the approaches used to define the subdisciplines are quite arbitrary. Variations and overlapping are quite common. What is important here is that the individuals working in these specialty areas ultimately share and relate their findings to the general field of microbiology.

Since many animal parasites, such as hookworms and tapeworms, have microscopic stages in their life cycles, courses concerned with the applied aspects of medical microbiology may include the study of these forms (Color Photograph 8a). Ordinarily, however, these *metazoan* (multicellular) *parasites* are considered separately in the specialized branch of *parasitology.* In this textbook the parasitic worms (helminths) will be considered in a separate chapter.

From this brief introduction of microbiology and descriptions of its many divisions and subdisciplines, it is apparent that microorganisms play a major role in the advancement of human health and welfare. As we proceed to discussions of the basic properties and applications of microorganisms, it will become quite evident, as the French scientist Louis Pasteur expressed it, that "The role of the infinitely small is infinitely large." Pasteur was one of the first scientists to recognize the true biological function of microbes (Figure 1–14). He was born on December 27, 1822, in the little French village of Dôle and eventually became a major figure in the development of biology and medicine. His discoveries brought to light dramatic new concepts as well as new approaches to age-old problems. Before beginning a detailed consideration of microorganisms, let us consider some of the events that have contributed to our understanding of the microbial world.

TABLE 1–3 Microbiological Subdisciplines Arranged According to Applications

Subdiscipline	*Brief Description*
Aquatic microbiology	The study of microorganisms in aquatic environments such as lakes, ponds, and rivers and the effects of environmental changes on microbial processes.
Food and dairy microbiology	Food and dairy microbiology involves the efforts of chemists, engineers, and microbiologists to gain efficient control of fermentation (the conversion of raw materials into desirable end products by carefully selected microorganisms). Food microbiology is also concerned with preventing spoilage and making improvements in the nutritional value, aroma, flavor, and general quality of foods. Dairy microbiology deals with the manufacture of cheeses and other fermented milk products, such as yogurt. The production of foods free of disease-producing agents is the goal of both applied areas.
Industrial microbiology and biotechnology	These subdisciplines, wherever applicable, are concerned with the use of microorganisms to produce economically important products and with the development of techniques used to prevent microbial destruction of economically important products formed by other means.
Marine microbiology	The study of microorganisms in marine or oceanic environments and the effects of environmental changes on microbial populations and processes.
Medical microbiology	This subdiscipline is concerned with the harmful effects of microorganisms **(pathogens).** Additional investigations in this area include determining the properties and disease-producing capabilities of microorganisms, developing procedures to detect the presence of pathogens (diagnosis), developing treatments and/or vaccines against pathogens, and using both chemical and physical methods to manage and control infectious diseases in the population.
Soil microbiology	The study of microorganisms that use soil as a terrestrial habitat, the various microbial processes associated with biodegradation (the biological breakdown of various compounds in soil), and the cycling of minerals (biogeochemical cycles).
Veterinary microbiology	This subdiscipline deals with harmful effects of infectious disease agents (microorganisms and worms) on domestic livestock and agriculturally important plants. The spread and control of such pathogens are of major concern.

Figure 1–14 Louis Pasteur (1822–1885), the Freelance of Science. *[National Library of Medicine, Bethesda, Md.]*

Early Development of Microbiology

Microbiology came of age about 100 years ago. The discoveries and achievements of the early microbiologists are not only of great importance and interest for the history of science but also the basis for the many and varied roles that microorganisms have come to play in the most practical applications of technology, medicine, and the biological sciences in the twentieth century. The modern uses of microorganisms have simplified the unraveling of biological mysteries that have puzzled scientists for generations.

Here are but a few of the individuals and events that played important rules in the history of microbiology.

Leeuwenhoek and the First Views of Microorganisms

In 1673, in Delft, Holland, Anton van Leeuwenhoek, a successful linen merchant, was the first to observe the mysterious and exciting world of microorganisms. For the next 50 years, until his death in 1723, Leeuwenhoek continued to watch microorganisms with the aid of small, *simple (one-lens) microscopes* (Figure 1–15a). Although the *compound microscope* had already been invented by Hans and Zacharias Janssen, Leeuwenhoek found his device more suitable for observing specimens with transmitted light. This was because his lenses provided greater detail, or resolving power. The magnifying power of his early instruments ranged from approximately 50 to 300 times the diameter of a particular specimen. **[The compound microscope, Chapter 3.]**

Leeuwenhoek's position in the development of microbiology has been firmly established because of his remarkable observations and descriptions of microscopic forms of life. His unending curiosity led him to spend hours upon hours examining specimens collected from lakes and rain barrels and even from his own teeth and those of others. Among his first observations were descriptions of protozoa and of the basic shapes of bacteria, yeasts, and algae.

Leeuwenhoek recognized the value of recording his observations (Figure 1–15b). He sent more than 200 handwritten, occasionally illustrated letters to the Royal Society in London. Many of these were translated into English and published.

Because of his numerous observations and careful

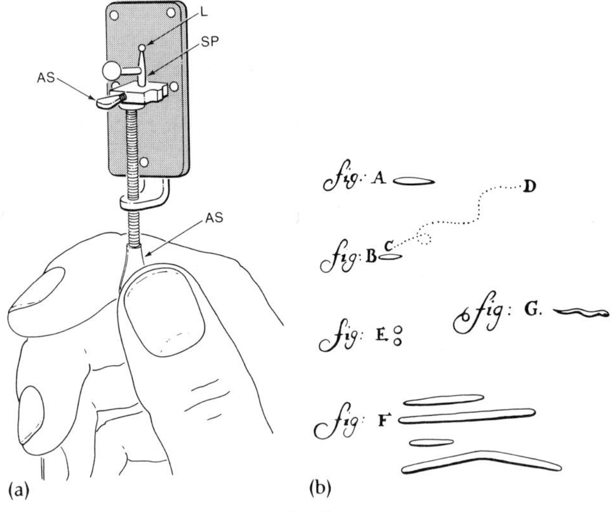

Figure 1–15 (a) Leeuwenhoek's simple microscope (rear view). The small size of the instrument is apparent from its relationship to the fingers grasping it. A single magnifying lens (L) was held between two thin metal plates. Objects for examination were mounted on a specimen pin (SP) that could be moved by adjusting screws (AS). Specimens were viewed by holding the microscope close to the eye and placing a source of illumination in back of the lens. (b) Leeuwenhoek's drawings of bacteria. The recognizable forms include rodlike, spherical (coccus), and spiral forms. *[From J. O. Corliss,* J. Protozool. ***22**:3–7, 1975.]*

measurement and recording of specimens, Leeuwenhoek is considered the father of bacteriology, hematology, histology, protozoology, and other sciences for which the microscope is the main investigative tool. After Leeuwenhoek's death, the study of microorganisms was neglected for some time. This was because of the difficulty in constructing better microscopes and because many scientists still considered microorganisms to be nothing more than oddities and were firm believers in the theory of spontaneous generation.

The Spontaneous Generation, or Abiogenesis, Controversy

The theory of **spontaneous generation,** stated by Aristotle in 346 B.C., expressed a belief widely held as late as the nineteenth century—that life could and did appear spontaneously from nonliving or decomposing matter. People were constantly seeing what they thought were examples of such spontaneous generation: snakes, frogs, and related forms of life apparently developing from the mud of river banks and maggots and flies appearing in decaying food. Aristotle taught that insects develop from morning dew and rotting manure and that tapeworms arise from animal wastes. These beliefs, held by all Greek scholars, were accepted and expanded throughout the Middle Ages and until comparatively recent times.

Today the theory of spontaneous generation seems absurd. Certainly it was the product of inadequate observation and faulty deduction. Nevertheless, it figured importantly in scientific thought throughout the centuries—especially in the study of disease and of various natural processes such as fermentation. Before any relationship could be shown between microorganisms and disease, the concept of spontaneous generation had to be disproved.

The Macroscopic Level of Attack

FRANCESCO REDI AND THE FLY EXPERIMENT

One of the first to refute the doctrine of spontaneous generation was the Italian naturalist and physician Francesco Redi. Around 1665 he showed that maggots did not emerge spontaneously from putrefied (decayed) meat. Redi put meat in three separate containers. One of these was closed with a paper cover, another was left uncovered, and the third was covered with fine gauze (Figure 1–16). Naturally, the meat readily putrefied and attracted flies. Redi made the following observations:

1. The paper-covered container showed no evidence of any flies or maggots.
2. Flies laid their eggs on the meat in the uncovered container, and within a short period of time maggots and newly emerging adult flies appeared.
3. Although no maggots were present in the meat in the gauze-covered container, they did appear on the covering. Apparently, the smell of putrefying meat attracted the flies. Unable to reach it, they laid their eggs on the gauze.

The conclusion seemed inescapable. Maggots, and the flies into which they develop, come not from the meat itself but from the eggs left on the meat by other flies. Thus Redi dealt a crushing blow to the myth that flies were spontaneously generated from meat.

The Microscopic Level of Attack: The "War of Infusions"

Leeuwenhoek's later discovery of bacteria—*animalcules,* as they were called—revived the arguments over spontaneous generation on a microscopic level.

Figure 1–16 Francesco Redi's experiment showing that flies were not spontaneously generated in meat. Among the systems used were: (a) those with no direct environmental contact, (b) those open to the environment, and (c) those covered with gauze.

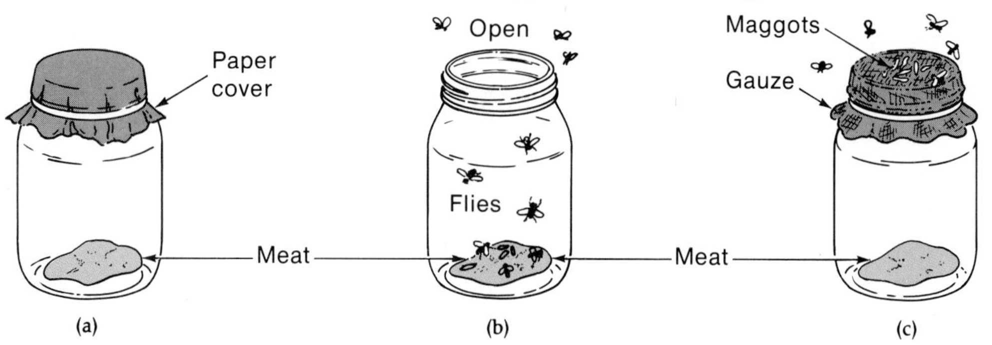

Microbiology Highlights

THE SCIENTIFIC METHOD: FROM HYPOTHESIS TO LAW

Collecting the Information

A sense of mystery together with the challenge of discovering an underlying order of events excite the active curiosity of most scientists. One goal of microbiology and every other branch of the biological sciences is to find explanations for observed phenomena and to show interrelationships between them and related events. To achieve this aim, the approach known as the *scientific method* is frequently used. Like so many truly great ideas, it is basically simple and is applied to some degree by almost everyone every day. The English biologist T. H. Huxley described the scientific method as "nothing but trained and organized common sense." While all of the steps of this procedure may not be applied to every aspect of microbiology, the essence of the method does direct the investigator to pose pertinent questions and to look for testable answers. This is not as simple as it sounds. The scientific method involves making *careful observations* of objects or events that may occur naturally or may be the result of planned and controlled situations.

Forming a Hypothesis

The information collected must be analyzed and arranged into some logical pattern or generalization that could explain the observed phenomena. This particular step is called forming a *hypothesis* or simply making a well-calculated guess.

Shaping a hypothesis is not the end of the process. It must next be tested for accuracy. This phase of the scientific method is called *experimentation*. The simplicity or complexity of the hypothesis will determine the type and number of experiments needed. Such experiments must be designed to (1) test the pertinent point of the hypothesis, (2) determine if predictions based on the hypothesis are correct, (3) include adequate controls for purposes of comparison, and (4) avoid the subjectivity or bias of the scientist. Experiments that cannot be repeated by other competent investigators are discarded.

The discovery of the microbial cause of tooth decay demonstrates the forming of a hypothesis and its testing by experimentation. For years hypotheses were developed and attempts made to find the cause of tooth decay, or dental caries. It was shown that the bacterium *Streptococcus mutans* (strep-toe-KOK-us MEW-tans) was a potent producer of caries in animals reared in a germ-free environment. This finding led to the isolation of these bacteria from humans and other animals and to attempts to reproduce tooth decay experimentally in laboratory animals. Results of these experiments are illustrated in these photographs.

Establishing a Theory

The next stage is the *theory*, which is a hypothesis that has been repeatedly and extensively tested. It is supported by various types of observations and controlled experiments. Despite the fact that no theory in science is ever absolutely and finally proven, a good theory can be used to predict new facts, show relationships between phenomena, and relate new information as it is uncovered. The term *law* is often used interchangeably with "theory." A distinction must be made, however, for a law is a theory that has reached universal acceptance. Many theories do not achieve this distinction.

There is no doubt that the scientific method is a functional tool for inquiry, but it is not without limitations and sources of frustration. The brief description given here cannot adequately demonstrate the trials and errors experienced by investigators. Moreover, the individuals using the scientific method also have limitations. They are human and, being so,

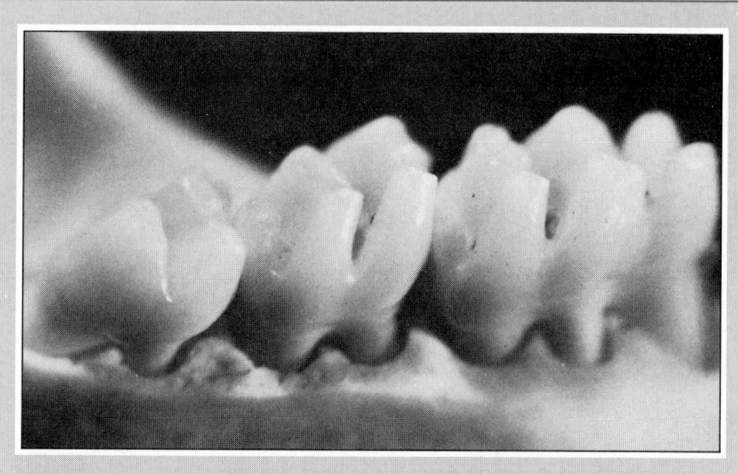

The teeth of a control noninfected animal.

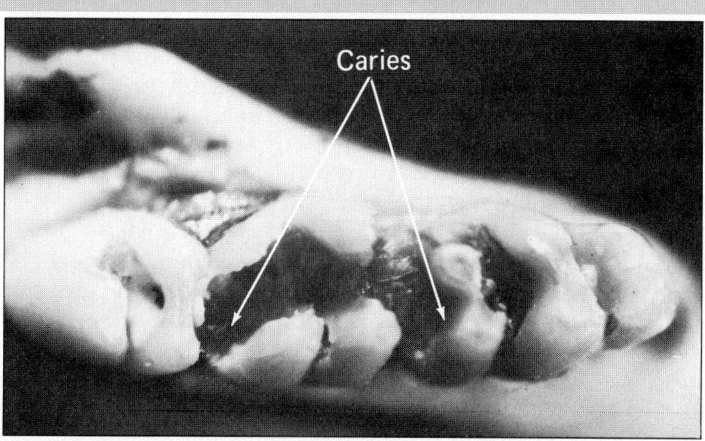

Here the caries-producing (tooth-decay) effects of the suspected bacterium are quite obvious. *[From Hamada, S. T. Ooshima, M. Torii, H. I. Imanishi, N. Masuada, S. Sobue, and Kotanti,* Microbiol. Immunol. *22:301–314, 1978.]*

Bacteria such as these *(S. mutans)* produce sugar substances that help them stick to the surfaces of teeth. Diets containing large amounts of sugar (sucrose) support an increase in the number of bacteria. *[From Dr. Z. Skobe, Forsyth Dental Center.]*

have feelings, failings, and even prejudices that may, from time to time, interfere with their reasoning and competence. Scientific investigation depends on a combination of subjective judgments and objective tests and a delicate mixture of intuition and logic. Done well, scientific research is an art. But it should be noted that in the final analysis, the basic rules are the same for all.

Although he believed that his newly discovered forms came from the surrounding air, Leeuwenhoek performed no systematic study to prove it. Another view shortly developed. Many individuals believed that nonliving material of animal or vegetable origin held a "vital," or life-generating, force that could give rise to animalcules. As proof, they cited the fact that microorganisms eventually appeared in boiled extracts *(infusions)* of hay and meat. In 1711 Louis Joblot found contradictory evidence when he observed that infusions stoppered tightly immediately after boiling remained free of microorganisms. However, if these stoppered preparations were later opened and exposed to the air, animalcules soon appeared. Joblot's findings were challenged, and thus the "war of infusions" began.

JOHN NEEDHAM AND A VITAL FORCE

In 1749, John Needham, a Roman Catholic priest, reported the results of his experiments, which he believed proved that bacteria arose spontaneously where no such living forms existed before. Needham's studies consisted of tightly corking flasks of boiled mutton broth and observing them periodically for cloudiness as an indication of microbial growth. Some containers remained clear, but most eventually became turbid. Examining a few drops of these cloudy preparations with a microscope, Needham found them to be teeming with microorganisms. Since boiling was known to destroy microorganisms as well as any other living cells, Needham believed that his experiments not only provided a clear demonstration of spontaneous generation but also showed that the organic matter in his flasks possessed a "vital or vegetative force" that could confer the properties of life on the nonliving elements present.

LAZZARO SPALLANZANI AND THE HEATING CHALLENGE

Sixteen years later, the Abbé Lazzaro Spallanzani, an Italian naturalist, reinvestigated Needham's findings and conclusions. He questioned the heating procedure used by Needham. Spallanzani found that sealed flasks containing infusions heated for one hour showed no cloudiness after a reasonable period. This experiment was repeated several times, with the same results every time. Needham argued that the prolonged boiling procedure destroyed the life-rendering "vegetative force." Spallanzani responded by breaking the seal on his heated, closed flasks, allowing exposure to air. Within a short time, the contents of these flasks became turbid, showing that the long-heated organic matter was still capable of supporting life.

The effect of Spallanzani's experiments was short-lived. Soon after, the discovery of oxygen by Joseph Priestley, a Unitarian minister, and the demonstration of oxygen's importance to life by Antoine-Laurent Lavoisier in 1775 rekindled arguments for spontaneous generation. Spallanzani's findings were criticized on the grounds that sufficient oxygen was not present in his sealed flasks to suport microbial growth.

Filtration Approaches

SCHWANN AND SCHULZE AND HOT FILTERS

It was now necessary to show that bacterial growth in nutrient-containing flasks was brought about by exposure to air containing these organisms and not by spontaneous generation. In 1836 Theodor Schwann set up two separate flasks, both of which held an infusion of some type. In one such experimental system (Figure 1–17), air passed through a red-hot tube before entering the flask containing the nutrient material. The other system, which received unheated air, was the experimental control. Soon growth developed in the control, while the system receiving heated air remained **sterile** (free of any living organisms).

Similar experiments were performed by Franz Schulze in the same year. However, his approach involved a different treatment of entering air. Air was allowed to enter nutrient flasks only after it had passed through solutions of strong chemicals such as sulfuric acid and sodium hydroxide. Schulze's results were the same as Schwann's. No growth developed in the flask receiving the treated air.

SCHRÖDER, VON DUSCH, AND COTTON FILTERS

Upon learning of the experiments of Schwann and Schulze, the supporters of spontaneous generation insisted that the drastic treatments of the air systems destroyed all possible "life-rendering power." Life could not possibly be spontaneously generated. This objection was countered in 1854 by Heinrich Georg Friedrich Schröder and Theodor von Dusch, who introduced the use of cotton plugs for bacteriological

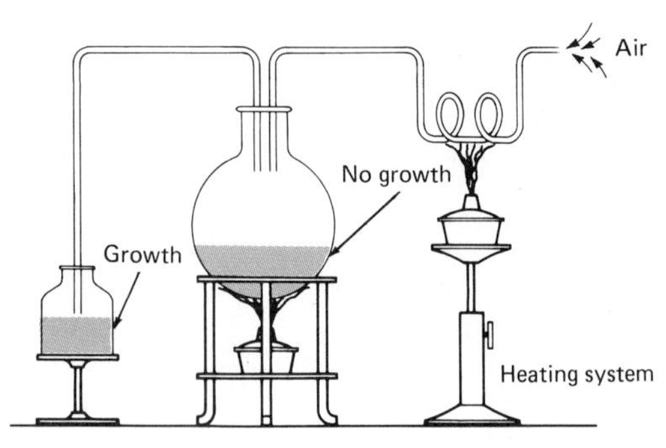

Figure 1–17 Schwann's experimental system. When the center flask containing an infusion or other nutrient material was exposed to heated air, growth did not occur. Note the coiled glass tubing and the heating device on the right-hand side of the drawing.

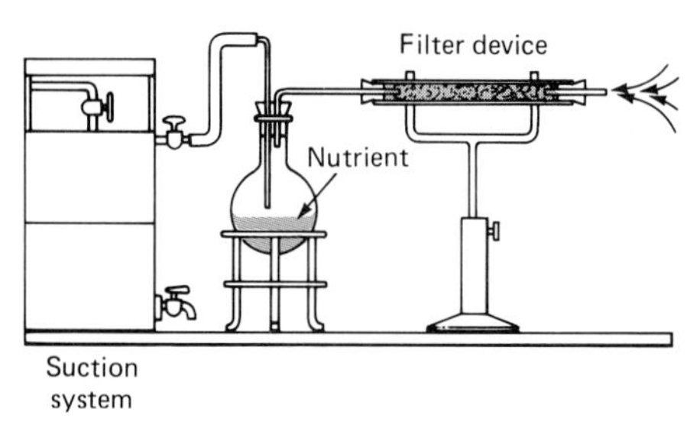

Figure 1–18 The experimental system used by Schröder and von Dusch in 1854 to demonstrate the removal of living organisms from air by filtration. When liquid was drained from the bottle on the left, suction was produced, which drew air through the long tube containing cool cotton (at the right). This air then flowed into the flask of nutrient material.

culture flasks and tubes. Using a system similar to that of Schulze and Schwann, these scientists allowed air to enter untreated in any way except by being filtered through cotton wool that had been previously heated in an oven (Figure 1–18).

The results of these experiments were the same as those of Schwann and Schulze. Flasks that received filtered air showed no signs of life, while those systems exposed to unfiltered air clearly contained microorganisms. These studies demonstrated that treatment of air with chemicals or heat was unnecessary to prevent growth in nutrient-containing flasks and that the living forms in air could be removed from air by filtration through cotton wool. The French chemist and physicist Louis Pasteur later recovered some of the bacteria trapped in the cotton wool used for filtration, completing the proof of its filtering action. **[See Chapter 11 for modern filtration methods.]**

PASTEUR AND THE CRUCIAL BLOW

Although these various experiments might seem conclusive, the issue was far from settled. In 1859, French naturalist Félix Pouchet claimed to have carried out experiments showing clearly that microbial growth could occur without contamination by air, thereby providing renewed hope for the supporters of spontaneous generation. About this time the studies of Pasteur were becoming known, and several other scientists began to recognize the roles of microorganisms in wine and vinegar production (fermentation) and food spoilage (putrefaction) processes. However, the acceptance of Pasteur's findings on the biological functions of microbes was threatened by the claims of Pouchet. Irritated by these arguments, Pasteur set out to disprove spontaneous generation once and for all.

Changing the procedure of Schröder and von Dusch slightly, Pasteur passed large volumes of air through a tube with a plug of gun cotton as a filter. A piece of the gun cotton was then dissolved in an alcohol-ether mixture, and the remaining sediment was examined under the microscope. Pasteur found small round and oval bodies that resembled the *spores* (reproductive structures) of plants. To show that the gun cotton not only stopped the passage of microorganisms but contained these forms of life as well, Pasteur simply added a little of the used filter to sterile meat infusions. Soon microbial growth appeared. Thus Pasteur confirmed the ways in which microbes gained access to nutrients and how they could be prevented from doing so.

Despite his obvious success, Pasteur was not fully satisfied. He had to perform what some regard as his most elegant experiment on the subject, namely, to show that microbe-free air did not create life in organic infusions. To prove his point, Pasteur designed and used swan-necked or S-shaped flasks (Figure 1–19). Nutrient liquid media were placed in the flasks and both the container and liquid contents were sterilized by boiling. Once this had been done, no plugs of any type were used to prevent the passage of microorganisms into these systems. Although these systems were open to the environment, allowing the flow of air through the system, no growth occurred. Because of the length and the

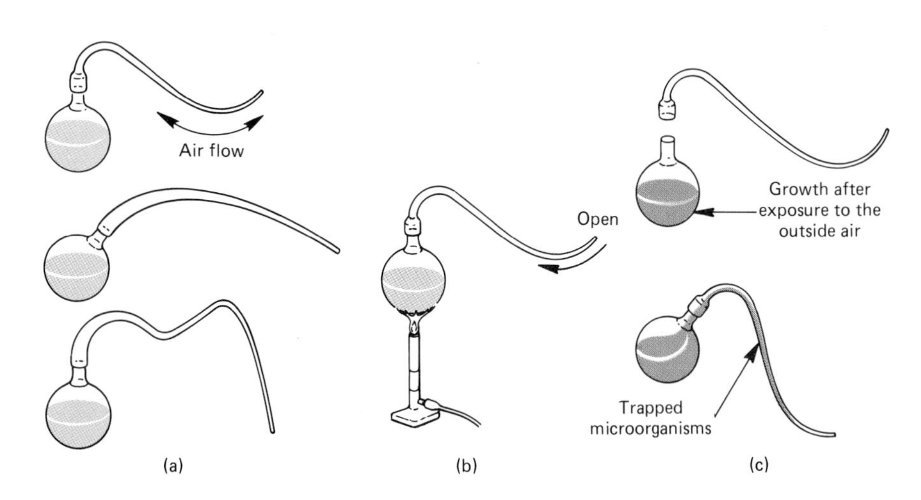

Figure 1–19 Pasteur's experiment disproving spontaneous generation. Steps were as follows: (a) before boiling, (b) boiling, and (c) after boiling. These flasks used by Pasteur permitted the entrance of oxygen; but because of their long, curving necks, they trapped bacteria, fungal spores, and other microbial life, thereby protecting the sterile contents from contamination. Only if the curved neck of a flask was broken off, allowing microbes from the environment to enter the flask, did microbial growth occur.

bend of the flask's swan-neck, microorganisms present in the air could not reach the main part of the flask. However, if the top of a system were broken off, or if a flask were tilted so that the sterile liquid nutrient ran into the exposed part of the neck and then returned, microbes soon appeared in the fluids.

TYNDALL AND THE FINAL BLOW

Most authorities agree that the final blow to spontaneous generation was delivered by the Irish physicist John Tyndall in 1877. In the course of his work on the optical properties of atmospheric dust, he observed that a beam of light passing through air without any dust particles was invisible. On the other hand, a light beam passing through a dust-containing environment was clearly visible. Moreover, the dust particles in such an atmosphere could also be seen. Aware of Pasteur's filter experiment conclusions regarding the presence of microorganisms on dust and the greater likelihood of microbial contamination in a dusty environment, Tyndall devised a system (Figure 1–20) to determine if air lacking dust particles contained microorganisms. He built a chamber equipped with side windows and curved tubular vents through which bacteria could not enter. The inside surfaces of this box were coated with *glycerol* to trap the dust particles that sooner or later would come to settle on the surfaces. This chamber was fitted with a rack of test tubes. When the chamber was found to be optically empty of floating matter, simply by shining a beam of light through its windows, the test tubes were filled with a broth, which was then sterilized by immersing the tubes in boiling brine. Tyndall found that the broth remained sterile even though it was in direct contact with the air of the chamber. When dust-containing air was introduced, microbial growth appeared after a brief time. Thus, with his specialized chamber and techniques, Tyndall demonstrated that bacterial life occurred in sterile broth only after it was introduced from an outside source. His results also showed a correlation between the presence of particles in the air and the fermentation or putrefaction of the material exposed to it.

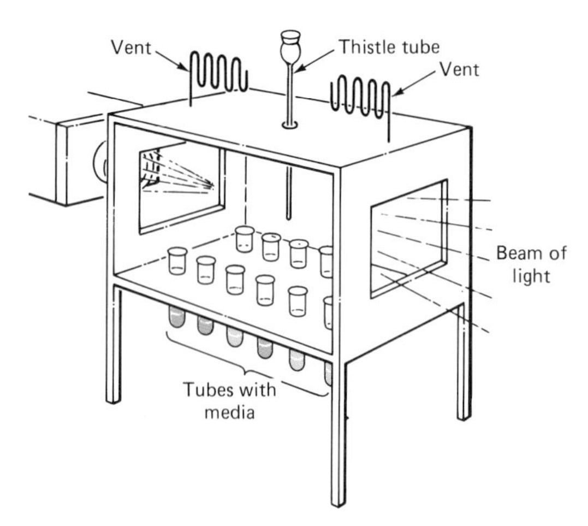

Figure 1–20 The culture chamber designed by John Tyndall to investigate the relationship of bacteria and dust particles. Note the side windows, the specialized vent system to allow airflow, and the thistle tube for the introduction of nutrient medium into the test tubes located in the bottom of the chamber.

During the course of his studies, Tyndall concluded that bacteria existed in two forms. The *thermolabile* form was killed easily by a few minutes of exposure to boiling temperatures. However, the *thermostable* form was resistant to boiling for several hours. These incredibly heat-resistant bacterial structures, which are now known as *spores,* were independently found and named in 1877 by the German botanist-bacteriologist Ferdinand Cohn. While attempting to repeat his experiments with dust-containing environments, Tyndall was unable to obtain similar results after a bale of hay used in broth preparations was brought into his laboratory. The hay harbored too many spores and spore-forming bacteria. Only when tests were performed in different rooms could Tyndall duplicate his earlier results. In an effort to eliminate the spores, Tyndall developed an intermittent sterilization procedure, which later became known as *tyndallization.* His procedure consisted of boiling the nutrient solutions for short periods of time on each of three successive days and incubating these preparations between heatings to allow microbial growth to occur. This was an effective procedure because growing bacteria are easily killed by boiling, and the time between heatings allowed the spores to germinate into growing bacteria that had lost their heat resistance. Today, spores are destroyed in the laboratory by more rapid means with the apparatus called the **autoclave** (Figure 11–11). This device incorporates steam under pressure, usually at a temperature of 121.5°C, a higher temperature than Tyndall could attain without pressurized vessels.

The Biological Functions of Microbes

Except for Leeuwenhoek, French and German scientists dominated early microbiology. Notable among these individuals was Louis Pasteur, the so-called freelance of science. His contributions led to new, more effective measures for disease prevention (Figure 1–21), to the improvement of health in general, and to the understanding of basic aspects of microbial life. Among the practical results of this understanding are the control of fermentation processes, pasteurization, and the development of vaccines against such dread diseases as rabies and anthrax.

Figure 1–21 Pasteur in his laboratory during work on the viral disease rabies in 1884. Pasteur's idea for weakening the rabies virus led to an effective control measure against the disease. *[The Bettmann Archive.]*

The Germ Theory of Fermentation

Nonvitalist versus Vitalist Views

Fermentation is a natural process in which alcohols and organic acids such as vinegar and lactic acid are formed from dissolved sugar in the presence of microorganisms and in the absence of air. The results of fermentation reactions, including the souring of milk and the preparation of alcoholic beverages, have been known and used all over the world throughout history. Yet, despite the recorded descriptions of microorganisms by Leeuwenhoek, the biological basis of fermentation was not formulated until well into the nineteenth century. Basically two viewpoints evolved to explain these processes—the *nonvital* (nonbiological) theory and the *vital* (biological) theory.

According to the nonvital view, yeasts seen in fermenting materials are by-products rather than causes of fermentation. During the period from approximately 1839 to 1869, supporters of the nonvital theory, including the three influential chemists Jöns Jakob Berzelius, Justus von Liebig, and Friedrich Wöhler, believed that essential, unstable chemical entities called *ferments* produced the reactions by acting as **catalysts** or **enzymes,** simply activating the chemical reactions. These unstable ferments were formed by the action of air on sugar-containing fluids. The resulting ferments passed their instability to sugar molecules, which in turn decomposed to form the products of fermentation. Liebig used as support for the nonvital theory the absence of any yeasts in acetic and lactic acid fermentations. The nonvitalists neglected to consider the possibility that other microorganisms could produce these "essential ferments." Several years later these reactions were shown to be caused by bacteria.

The biological theory of fermentation was strengthened by the work of French and German scientists in the 1830s. The German physiologist Theodor Schwann clearly demonstrated the role of yeasts in alcoholic fermentation. He showed that exposing these microorganisms to heat and chemical agents stopped all fermentation. In addition, Schwann described the **asexual** form of reproduction (budding) of yeasts and showed that this "sugar fungus," *Saccharomyces cerevisiae* (SAK-uh-row-MY-seez sair-a-VIS-e-eye)* was needed in large numbers for fermentation to proceed. These observations were not readily accepted by the nonvitalists, and the stage was set for a bitter controversy that was not settled until Louis Pasteur provided experimental proof for the biological theory in 1857.

Pasteur's Contributions to Fermentation Research

Pasteur's fermentation studies occupied a major portion of his scientific career. They began as a search for the cause of souring and spoilage of beer and wines. In 1854 Pasteur was a professor at the University of Lille, France, and in that city the production of wine and beer was a very important industry. Pasteur found that the spoilage was caused by a different fermentation process in which sugar is converted to lactic acid rather than to alcohol. The microscopic examination of sediments from wine vats in which this lactic acid fermentation occurred revealed microorganisms that were eventually recognized as bacteria and unwanted "wild" yeasts.

*The phonetic pronunciation of an organism will be given the first time it appears in a chapter.

The classification of such organisms was difficult, since no formal rules of identification had yet been developed.

On further experimentation, Pasteur found a way to destroy these microbes without changing the quality of the wine. He discovered that wine could be heated and held for a specific period of time at a temperature between 50° and 60°C. This procedure, which came to be known as **pasteurization,** prevented the wine from spoiling.

Pasteur also showed that the souring of milk is caused by microorganisms. Today, milk and certain other foods are routinely pasteurized by heating at 63°C for 30 minutes or at 71°C for 15 seconds. These temperatures are adequate to destroy or to reduce in number food-spoiling organisms and many **pathogenic** (disease-producing) organisms.

In investigating many other fermentative processes, Pasteur discovered some interesting things:

1. Each type of fermentation, as defined by its particular end products, is caused by a specific microbial type.
2. Each microorganism requires specific conditions, such as a definite degree of **acidity** or **alkalinity,** in order for fermentation to occur.

While studying the fermentation that produces butyric acid, a substance present in rancid butter, Pasteur discovered microorganisms that could live only in the absence of free oxygen. Normally these organisms are *motile,* but he noticed that those in close contact with air ceased to move. Pasteur quickly realized the possibility that air might inhibit these bacteria. He confirmed his suspicion by introducing a stream of air into fermentating systems. The effect was striking. The process either stopped totally or slowed down considerably. The terms **aerobic** and **anaerobic** were coined by Pasteur to distinguish microbes capable of living only in the presence of free oxygen from those capable of living only in the absence of free oxygen, or air.

Pasteur's discoveries contributed enormously to the control of microorganisms and established the bases with which to show the relationship of numerous organisms to disease and to food spoilage.

Antiseptic Surgery and Surgical Infections

When anesthetics were introduced into surgery and obstetrics during the 1840s, surgeons began performing longer, more complex procedures than ever before. Unfortunately, the number of surgical wound infections increased at the same time, often causing the death of patients. The young English surgeon Joseph Lister (Figure 1–22) undertook the task of preventing wound infection.

Figure 1–22 Dr. Joseph Lister (1827–1912). *[National Library of Medicine, Bethesda, Md.]*

Impressed with Pasteur's studies showing the involvement of microorganisms in fermentation and putrefaction, Lister reasoned that surgical infections, *sepsis,* might actually be caused by microbes. He devised procedures designed to prevent access of microorganisms to wounds (Figure 1–23). Lister's system, which came to be known as *antiseptic* surgery, included the heat sterilization of instruments and the application of carbolic acid *(phenol)* to wounds by means of dressings. These procedures, received critically at first by the medical community, ultimately proved to be effective in preventing surgical sepsis. Although Lister was not aware of the exact nature of the microorganisms involved, antiseptic surgery did provide an indirect source of evidence in support of the germ theory of disease.

The Germ Theory of Disease

In developed countries today, most major diseases, including cholera, plague, diphtheria, typhoid, typhus, and yellow fever, are controlled by means of immunization (vaccination), sanitation, and the de-

Figure 1–23 A surgical operation in the early days of antiseptic surgery. Steam and the anesthetic chloroform are being administered to the patient. Lister's steam spray device is used here for the administration of phenol. *[The Bettmann Archive.]*

struction of **arthropod** carriers such as fleas, lice, mosquitoes, and ticks, which serve to transmit specific infections. Unfortunately, in some parts of the world, many of these diseases still take a heavy toll. Because of the great suffering and disability caused by these diseases, humanity has benefited greatly from the contributions of Robert Koch (Figure 1–24), Louis Pasteur, and others, who established the relationship between a specific disease agent and a disease and who developed methods for the control of infections. **[The major arthropods involved are discussed in Chapter 21.]**

From earliest times, disease was associated with natural phenomena such as earthquakes and floods, mysterious and supernatural forces, and poisonous vapors called miasmas. Although ancient Greek and

Figure 1–24 (a) Robert Koch (1843–1910), one of the trailblazers of microbiology. This German-born bacteriologist identified human pathogens, developed bacteriological techniques, and discovered old tuberculin, a tuberculosis skin-testing material. *[National Library of Medicine, Bethesda, Md.]* (b) A view of Robert Koch's study. Note the various laboratory-related items: microscope, test tubes and rack, staining supplies, a bell jar under which microorganisms were incubated, and a photomicrographic camera (at left horizontal position). *[The Bettmann Archive.]*

(a)

(b)

Roman physicians suspected that certain types of disease were caused by invisible, minute agents, no direct proof for this view was found until the nineteenth century. The concept of *contagion*—the spread of infectious disease—preceded the proof of the existence of pathogenic agents by many centuries.

Fungi were the first microorganisms shown to be pathogenic. Agostino Bassi proved experimentally that an agent of this type caused an infection in silkworms. This discovery was followed in 1839 by the first isolation of a fungus from a human skin disease by Johann Schönlein. As we will see in Chapter 8, many fungi are serious threats not only to humans but also to plants and other forms of animal life. Protozoa also were among the first microorganisms shown to have an association with disease. This relationship is credited to Pasteur, who in 1865 discovered that an infection of silkworms, which were vital to the silk industry in Europe, was protozoan in nature. The disease was called pebrine.

Koch's Postulates

A direct demonstration of the role of bacteria as agents of disease was given by Koch in 1876 and confirmed by Pasteur and Jules Joubert. The organism used was *Bacillus anthracis* (bah-SIL-lus ann-THRAY-sis) (Color Photograph 94), the cause of anthrax. Although rod-shaped structures had earlier been observed in the blood and organs of sheep dying of anthrax, there was no clear-cut proof at that time that these bodies caused anthrax.

Koch established a definite sequence of experimental steps to show the causal relationships between a specific organism and a disease beyond a shadow of a doubt. Although this procedure is known as **Koch's postulates,** it is based upon the earlier theoretical work of the German scientist Jacob Henle. In showing the causal relationship of *B. anthracis* to anthrax, Koch first isolated the organism from an animal with the disease and ultimately completed the cycle by obtaining similar cultures from laboratory-inoculated animals with symptoms of anthrax (Figure 1–25). The steps of Koch's postulates are as follows:

1. The suspected causative agent must be found in every case of the disease.
2. This microorganism must be isolated from the infected individual and grown in a culture containing no other kinds of microorganisms. (Koch utilized the fluid filling the eyeballs of cattle for this purpose.)
3. Upon inoculation into a normal, healthy, susceptible animal, a pure culture of the agent must reproduce the specific disease.
4. The same microorganism must be recovered again from the experimentally infected host.

Microbiology Highlights

THE BATTLE AGAINST HOSPITAL INFECTIONS

Surgery before 1800 was done primarily for life-threatening conditions. Speed rather than technique was stressed, since there was no anesthesia. When anesthesia was introduced by W. G. Morton, a dentist, in 1846, it was of major importance to surgeons and their patients. Surgeons could operate more thoroughly and slowly and patients were no longer overcome by the pain. Yet surgery remained unacceptable for most patients for a very simple reason: as many as 80 percent of all hospital operations resulted in gangrene (presumably streptococcal and related infections), and almost half of all patients died after a major operation.

Several medical practices and hospital conditions added significantly to the problems of surgical patients. For example, from 1850 to 1880, in Bellevue Hospital in New York City, it was common practice to keep a surgeon's instruments sharp and free of rust, but they were never thoroughly cleaned. Instruments were frequently returned to their cases after use with little more than a wipe, even if they had been dropped on the floor or used to amputate an infected leg that had pus oozing from the cut surface. The surgical wards of the late eighteenth and early nineteenth centuries were described as having mixtures of human feces, urine, blood, and pus on the floors and the patients' saliva clinging to the walls. Wounds were washed with the same water and sponges passed from one patient to another. Even bandages were reused without being washed.

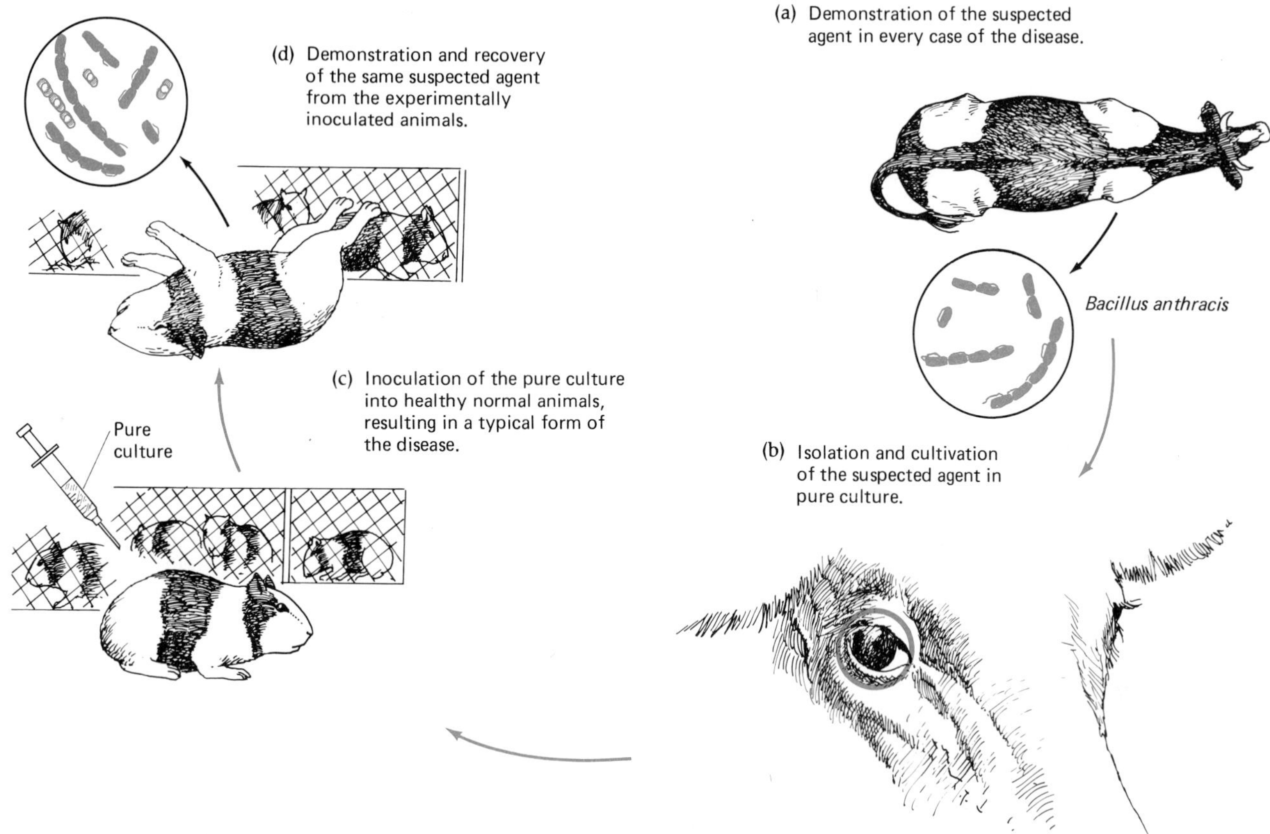

Figure 1–25 Koch's postulates. Shown is the sequence of experimental steps by which Robert Koch, using material from a cow's eye, established the bacterial cause of anthrax. (b) Koch was able to obtain a pure culture of the suspected agent. (c) By inoculating normal, healthy animals with the pure culture, one can produce a disease state similar to the one from which the suspected pathogen was isolated. (d) Microscopic supporting evidence of the disease agent.

With relatively few exceptions, the causal relationship between pathogenic bacteria and a particular disease has been shown according to the dictates of Koch's postulates. However, if an appropriate animal model is not available, Koch's postulates cannot be fulfilled.

Rivers' Postulates

At the time when Koch's postulates were formulated, true viral pathogens were unknown. In 1937, T. M. Rivers created a similar group of rules to establish the causal role of viruses in disease. **Rivers' postulates,** applicable to both animal and plant viruses, are as follows:

1. The viral agent must be found either in the host's body fluids at the time of the disease or in the cells showing specific lesions.
2. The viral agent obtained from the infected host must produce the specific disease in a suitable healthy animal or plant or provide evidence of infection in the form of **antibodies** (substances produced in response to a virus) against the viral agent. (It is important to note that all host material used for inoculation must be free of any bacteria or other microorganisms.)
3. Similar material from such newly infected animals or plants must, in turn, be capable of transmitting the disease in question to other hosts.

Early Technical Achievements

Microbial Cultivation

With the growing interest in microorganisms, improved techniques for their handling and study developed rapidly. The advances included the preparation of several nutrient combinations, known as **media,** for the growth, or **culture,** of bacteria. Typical media include sources of carbon, nitrogen, vitamins, and trace chemical elements. All of these early preparations were in broth or liquid form. Although these media provided excellent growth

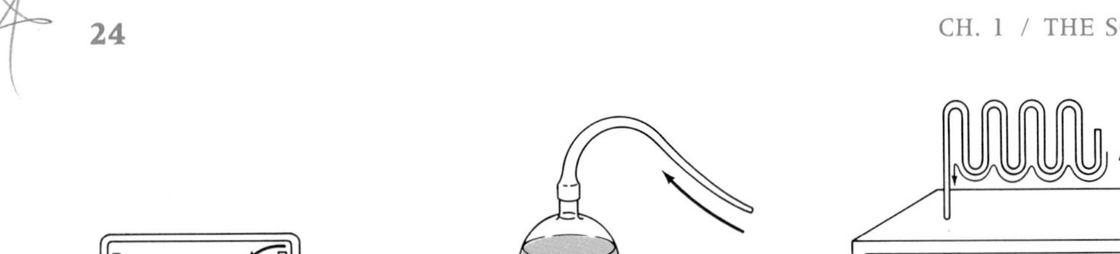

Figure 1–26 The flow of air (arrows) into the Petri dish and the systems used by L. Pasteur and J. Tyndall. (a) The Petri dish, (b) Pasteur's swan-necked flask, (c) the specialized system for Tyndall's culture chamber. Refer to Figure 2–19 for additional details.

conditions, they posed problems in the isolation and separation of bacteria from the mixed populations usually found in natural specimens. A general method of separating bacteria in mixtures was needed, especially in studies seeking the causative agents of infectious diseases.

The early approaches to isolating bacteria focused on the development of solid natural media. Freshly cut surfaces of potato were used to culture bacteria. After a suitable incubation period, distinctively colored bacterial growths, or *colonies,* formed (Color Photograph 3). Each isolated colony contained a single kind of organism. Other such media were carrot slices, freshly baked bread, meat, and coagulated egg white. Although each one proved satisfactory for bacterial culture, some were either difficult to work with or did not support the growth of all organisms of interest.

In 1881, Koch reported a relatively simple procedure for the surface isolation of bacteria from contaminated materials using a medium of coagulated blood plasma or gelatin-solidified broth. In the latter case, gelatin, a simple protein, would be added to a liquid medium that could support the growth of the desired organisms. While this preparation was still warm and liquid, it was poured on sterilized glass plates and allowed to harden. Organisms could then be introduced by spreading one drop of specimen solution across the surface of the gelatin. This process is known as *streaking* (Figure 4–13). The inoculated glass plate was covered by a bell jar and left to incubate. Some six years later, an assistant of Koch's, Richard J. Petri, introduced the **Petri dish** as a more effective medium container than the glass plate. The flow of air into the dish was controlled by its design, which was based on the systems used by Pasteur and Tyndall (Figure 1–26). [Cultivation techniques are described in Chapter 4.]

Using his gelatin medium, Koch could obtain isolated colonies of bacteria that could not be grown on potato. However, the gelatin had several limitations, one of which was that it melted when warmed to more than 28°C. This made it impossible to obtain isolated colonies of human pathogenic bacteria, since the most favorable growing temperatures of these organisms range from 35° to 37°C. Other problems soon became apparent. Certain microorganisms were found to digest gelatin, leaving a liquefied broth. Others would not grow well at low temperatures. Both limitations had to be overcome before pure cultures of human pathogens could be obtained.

Around this time, another associate of Koch's, Walter Hesse, learned of a far better solidifying agent than gelatin. The real credit for this new agent belongs to Hesse's wife. Upon learning of her husband's difficulty in finding a better way to grow pathogens, Frau Hesse told him of a substance used by her grandmother in the tropics to keep jams and jellies solid at warm temperatures. This material proved to be agar-agar (or simply **agar**), which is extracted from algae found along the coasts of Sri Lanka, China, Japan, Malaysia, and southern California. Agar is liquid at temperatures of nearly 100°C but becomes a firm gel at approximately 42°C. Most agar media now contain approximately 1.5% agar. As few bacteria are capable of digesting agar, it has proved to be indispensable in microbiology.

Chemotherapy

By 1900 the microbial causes of many important human diseases were known. These included cholera, diphtheria, leprosy, plague, tetanus, tuberculosis, and typhoid fever. During the ten years that followed, the bacterial agents of syphilis and whooping cough were added to the list. Despite the relative success in uncovering the cause of bacterial disease, advances in treatment were disappointing. Up to this time, infectious diseases were controlled mainly through the use of vaccines, such as the vaccine against smallpox. Furthermore, such approaches were largely preventive; little could be done to cure already infected individuals.

The modern era of control and treatment began with the use of chemicals that would kill or interfere with the growth of the disease agent without

Figure 1–27 Paul Ehrlich (1845–1915), the famous physician-chemist who provided one of the first effective drugs to combat syphilis. In 1908, he and E. Metchnikoff were jointly awarded the Nobel Prize in immunology. *[National Library of Medicine, Bethesda, Md.]*

damaging the infected individual. This approach, known as **chemotherapy,** was introduced by Paul Ehrlich (Figure 1–27) during his search for "magic bullets" to combat African sleeping sickness and syphilis. Ehrlich's efforts led to the discovery of an arsenic-containing compound that proved effective against syphilis (Color Photograph 62) and related diseases but produced undesirable side effects in patients.

Of all the advances made in the early part of this century, the one that is regarded as the major modern triumph of microbiology is the discovery and application of the wonder drugs known as **antibiotics.** In 1929, Alexander Fleming (Figure 1–28a) observed, as had many microbiologists before him, the presence of the mold *Penicillium* (penn-uh-SIL-ee-um) in a Petri dish culture of the bacterium *Staphylococcus* (staff-ill-oh-KOK-us). Strangely enough, the area in the immediate vicinity of the mold was free of

(a)

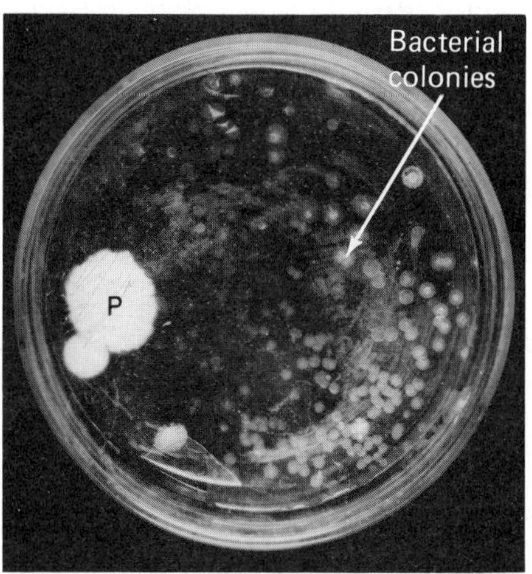

(b)

Figure 1–28 (a) Sir Alexander Fleming working in his laboratory in 1954. In conjunction with Ernest B. Chain and Howard W. Florey, he received the Nobel Prize in 1945 for the discovery and purification of penicillin. (b) The original culture plate on which Fleming observed the effect of penicillin on microorganisms. Note the limited number of bacterial colonies immediately around the fungus *Penicillium* shown on the left. *[St. Mary's Hospital Medical School, London.]*

bacteria (Figure 1–28b). On further study, Fleming isolated a mold-produced substance that inhibited bacteria but was nontoxic to laboratory animals. Fleming named his newly isolated antibacterial material **penicillin.** Years later, penicillin was purified sufficiently for human use. Today this antibiotic remains one of the most effective chemotherapeutic agents against bacterial disease.

Since Fleming's discovery, hundreds of new antibiotics have been developed for the control and treatment of infectious diseases. Many of the human bacterial infectious diseases that were once major causes of death have been brought under control through the use of these miracle drugs.

The Growth of Organized Microbiology

In approximately three-quarters of a century, microbiology has become a significant influence in our society and one of the most dynamic and important branches of biology. Like any growing field, it requires a formal means of communication by which to exchange ideas and experimental findings. It is not surprising, therefore, to find that there are numerous journals, as well as several organizations, representing the specialties of microbiology. In the United States, for example, the American Society for Microbiology publishes several monographs, laboratory aids, and journals, including *Applied and Environmental Microbiology, Microbiological Reviews, Infection and Immunity, The Journal of Bacteriology, The Journal of Clinical Microbiology,* and *The International Journal of Systematic Bacteriology.* Comparable organizations are well established in other countries. Because these organizations are concerned with the advancement of microbiology or one of its specialties, particular attention is given to maintaining the highest professional and ethical standards.

Some Challenges for Microbiology

During the late nineteenth century, microbiology developed into an established discipline with a distinctive core of concepts and techniques. For almost 50 years after the death of Pasteur in 1895, the major activities of microbiology were isolated from the mainstream of biological thinking and directed toward the isolation and identification of pathogens, the study of mechanisms of immunity during times of health and disease, the search for chemotherapeutic agents, and the effort to unravel the many environmental activities of microorganisms. While these areas continue to be of major interest and importance, microbiology has also ventured out in new directions.

A number of major discoveries involving microorganisms between 1920 and 1935 contributed significantly to the development of *biochemistry.* This discipline is concerned with the chemical basis of living matter and the reactions associated with it. The discoveries that the energy-yielding processes in animal muscle and alcoholic fermentation by yeasts were basically the same and that the vitamins required by animals in trace amounts were chemically identical to the growth factors needed by bacteria and yeasts demonstrated the fundamental similarities of all living systems at the metabolic level. This concept, frequently referred to as *the unity of biochemistry,* set the stage for a major revolution in biology: the arrival of *molecular biology.* This is a challenging field concerned with explaining life processes in molecular terms.

The invaluable role of microorganisms in unraveling the molecular mysteries of biochemistry and genetics was clearly defined by several important advances in the 1940s. The first of these was made in 1941 by G. Beadle and E. Tatum, who succeeded in isolating a number of genetically different forms (mutants) of the fungus *Neurospora* (noo-RAHS-pawr-uh). This discovery opened the way to analysis of the consequences of permanent genetic change (mutations) in biochemical terms.

In 1944, O. Avery, C. MacLeod, and M. McCarty, studying the transfer of hereditary factors in bacteria, found that deoxyribonucleic acid (DNA) was responsible for the process. Thus, the chemical nature of heredity was revealed. With this discovery, the isolation of microbiology from the mainstream of biological thought came to an end; the stage was set for the formation of a close alliance of microbiology, biochemistry, and genetics.

In 1953, J. D. Watson and F. H. Crick proposed a model for the molecular organization of DNA. Acceptance and knowledge of the model led to numerous discoveries related not only to the structures and functions of deoxyribonucleic acid and ribonucleic acid (RNA) but also to the mechanism by which genetic information in DNA controls the synthesis of proteins. In fact, how genetic information flows from DNA to RNA and finally results in the formation of a protein molecule is referred to as the "central dogma" of molecular biology, which is common to all life.

Microbiology Highlights

RESOURCES OF MICROBIOLOGY

Examples of scientific publications associated with microbiology.

Within a relatively few decades an enormous body of experimental data has accumulated and been recorded in the journals for the various divisions of microbiology. Discussions and interpretations of these data are also published. Obviously, the need to communicate knowledge and ideas is vital to any phase of scientific endeavor. Knowledge of important developments forms a most powerful and necessary tool for scientific methodology.

In beginning the study of any new area of specialization, understanding of and information about publications in the field are essential. Scientific periodicals are produced by scientific societies, individuals, institutions, private companies, and commercial publishing firms. They range in frequency from weeklies such as *Science* to annuals such as the *Annual Review of Microbiology.* The content of periodicals also may vary. Certain journals, such as *Applied Microbiology* and the *Journal of Phycology,* are devoted to providing reports of recent research developments in the fields of pure and applied microbiology. Other publications, such as *Advances in Immunology* and *Microbiological Reviews,* provide review articles that draw together information from many separate sources in a particular area. Such reviews are especially important to individuals approaching a new field of interest or a specialized topic for the first time. Still other periodicals, including *The American Scientist* and *The Sciences,* combine both of these approaches and may include additional items such as book reviews, society developments, and announcements. Some periodicals contain advertisements of products useful to the microbiologist.

Most publications provide an annual index of subjects and authors, either in a separate issue or as part of the last issue of a volume. There are also reference publications devoted solely to providing brief summaries or article titles for specific subjects, (*Biological Abstracts* and *Index Medicus*). Others list the titles of all articles contained in selected periodicals (*Current Contents* and *Science Citation Index*). Either type of reference work helps the microbiologist keep abreast of publications in his or her particular specialization.

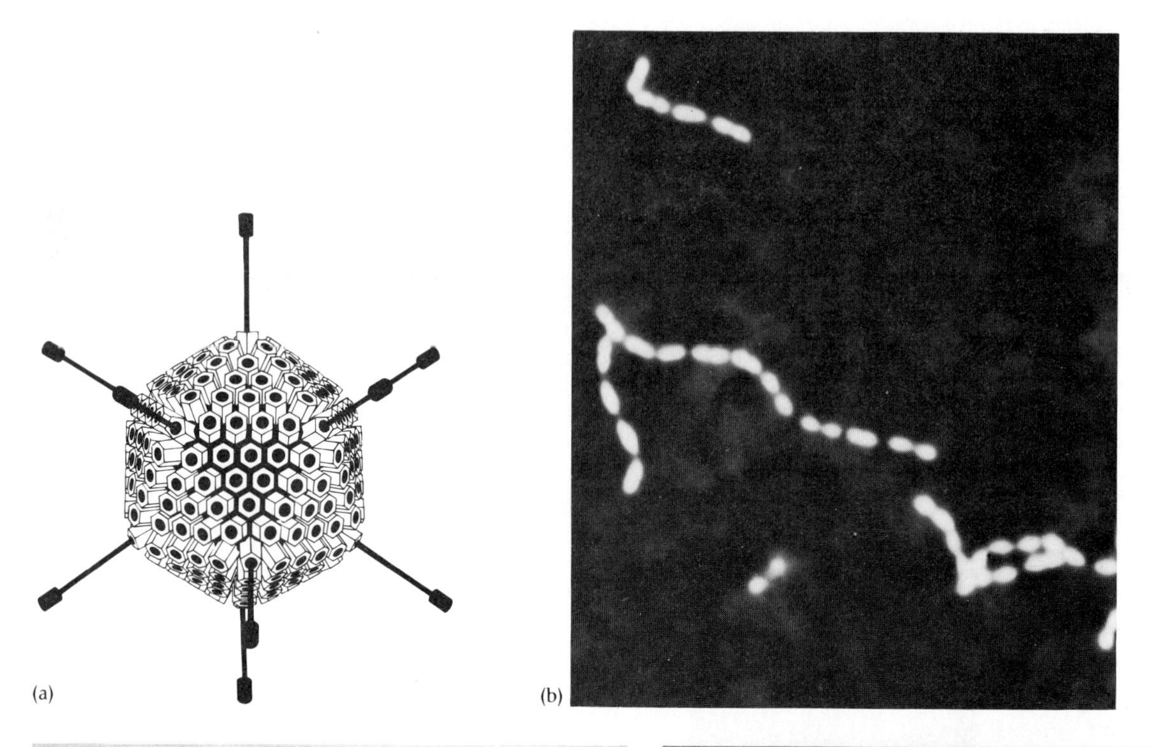

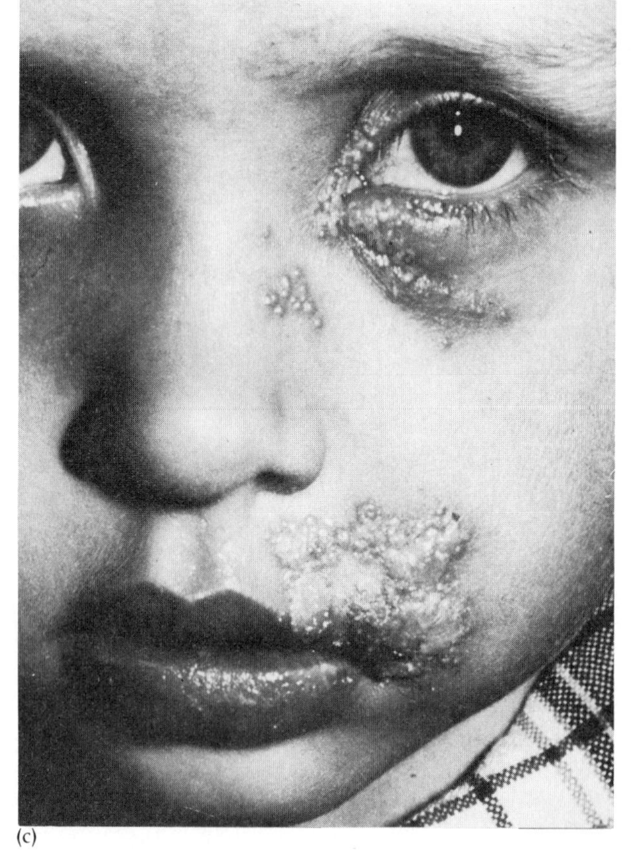

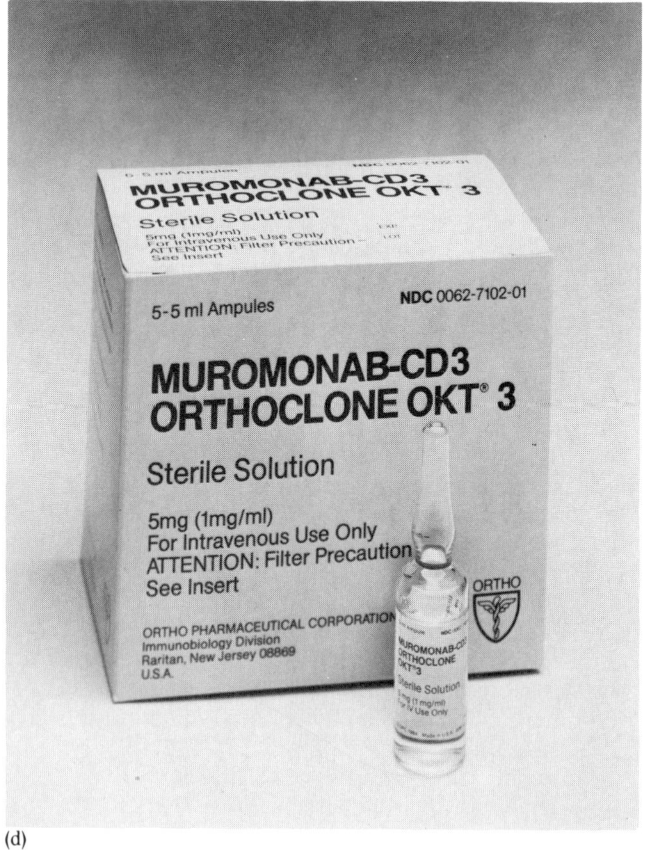

Figure 1–29 A glimpse of topics to come. (a) A diagrammatic representation of an adenovirus and its parts. (b) The clear arrangement of bacterial cocci in a chain (streptococcus) shown by fluorescence microscopy. (c) A child with the bacterial disease impetigo and herpes simplex virus infection. *[Photo kindly provided by Dr. H. Zachariae.]* (d) A container of the first specific antibody (monoclonal) for kidney transplants approved for therapeutic use in the United States. *[Courtesy of Johnson & Johnson.]*

A Glimpse of Things to Come

One of the attractive features of microbiology as well as other areas of science is the amount of investigation and work remaining to be done. Despite the decades of intense study of microorganisms and their activities, key questions remain unanswered. Microbiology has assumed a position of great importance in modern society. Nowhere is this more evident than in the area of genetic engineering, also known as **recombinant DNA technology.** This area of investigation, in which the genetic machinery of microorganisms and other forms of life can be changed, is only a small part of an accelerating revolution that promises to change the way we think about our world and the way we live. Researchers, with the aid of genetic technology, are continuously trying to provide better solutions or better methods to deal with the problems of food production, pollution control, energy production, and the control and treatment of diseases not only of humans but of domestic animals and plants as well.

This chapter has presented a brief overview of the organization and significance of microbiology as well as selected insights into the historical events responsible for its establishment as a science. The chapters that follow will take you, the reader, on a journey into the *microbial world.* What are microorganisms? How are they constructed and studied (Figures 1–29a and 1–29b)? Can they be controlled (Figure 1–29c)? These are but a few of the questions to be answered. Attention will also be directed to the cause, diagnosis, and treatment of both old and newly discovered infectious diseases (Figure 1–29d), microbial environmental interactions (Figures 1–30), and biotechnological applications of microorganisms. As you will see from reading about these topics, any decisions affecting the future of the world may depend upon and involve the activities of microorganisms.

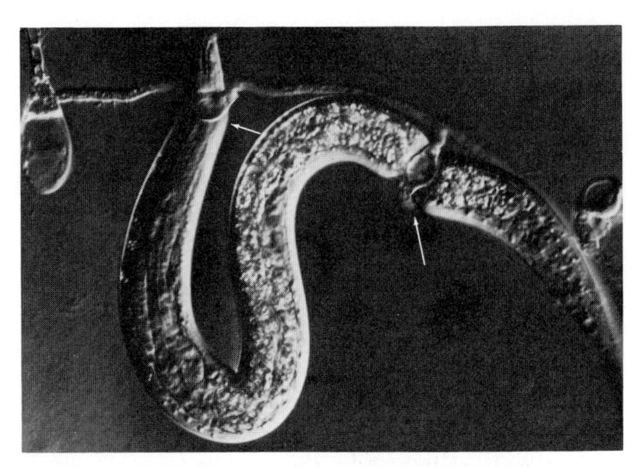

Figure 1–30 A worm in a choke-hold of fungus (arrows). The fungus traps and eventually destroys the worm. Original magnification 400 ×. *[G. L. Barron,* The Nematode Destroying Fungi, *1977.]*

Summary

The Microbial World

1. Microscopic forms of life can be found in abundance in nearly every environment.
2. The majority of microorganisms are not harmful.
3. Most microorganisms are single cells and exhibit the characteristics common to all biological systems, including *reproduction, metabolism, growth, irritability, adaptability, mutation,* and *organization.*
4. Viruses and subviral disease agents do not have a true cellular form of organization.

Microbiology and Its Subdivisions

TAXONOMIC ARRANGEMENT

1. The general branches of microbiology include *bacteriology,* the study of bacteria; *immunology,* the study of a host's resistance against disease and the detection and prevention of disease; *mycology,* the study of fungi; *phycology,* the study of algae; *protozoology,* the study of protozoa; and *virology,* the study of viruses and subviral agents of diseases.
2. The first demonstration of the existence of a nonbacterial infective agent, the virus, was made by Ivanowski in 1892.
3. Demonstration that viral diseases infect plants, humans, and other animals made viruses an important subject of study.

INTEGRATIVE ARRANGEMENT

1. Special areas of microbiology focus on the study of common properties of microorganisms and their interactions.
2. Such areas of study include cellular features of microorganisms *(microbial morphology and ultrastructure),* interactions between microorganisms and their environments *(microbial ecology),* hereditary mechanisms *(microbial genetics),* metabolic activities *(microbial physiology),* the relationship between chemical structure and genetic makeup *(molecular biology),* and naming and classification of microorganisms *(microbial taxonomy).*

Applied Microbiology

1. Applied areas of microbiology utilize the basic principles, information, and techniques of microbiology in various ways.
2. Such applied areas deal with the microbiology of foods and other products formed by microorganisms and with the environments in which microorganisms may be found.

Early Development of Microbiology

LEEUWENHOEK AND THE FIRST VIEWS OF MICROORGANISMS

1. Leeuwenhoek designed and constructed simple microscopes with lenses that provided greater viewing detail *(resolving power)* than was previously possible.
2. He made the first accurate descriptions of most of the major types of single-celled microorganisms known today—*algae, bacteria, protozoa,* and *yeasts.*

The Spontaneous Generation, or Abiogenesis, Controversy

According to the Aristotelian doctrine of *spontaneous generation,* lower forms of animal life arose spontaneously from inanimate or decomposing organic matter.

THE MACROSCOPIC LEVEL OF ATTACK

Francesco Redi demonstrated that spontaneous generation did not apply to animals by showing that flies did not develop spontaneously from putrefied meat.

THE MICROSCOPIC LEVEL OF ATTACK: THE "WAR OF INFUSIONS"

1. The eventual appearance of bacteria and protozoa in boiled hay or meat preparations *(infusions)* was offered as proof of spontaneous generation on a microscopic level.
2. Louis Joblot and Lazzaro Spallanzani independently showed that heating infusions under controlled conditions prevents the appearance of microscopic life.

FILTRATION APPROACHES

1. Various filtration methods were found to be effective in removing microorganisms from air.
2. Pasteur demonstrated that air free of microbes could not create life in organic infusions.
3. By means of a specially constructed chamber, John Tyndall showed that bacterial life occurred in sterile broth only after the broth was exposed to dust-bearing bacteria from outside sources.
4. Tyndall and Ferdinand Cohn independently demonstrated the existence of heat-resistant bacterial bodies called *spores.*

The Biological Functions of Microbes

1. Louis Pasteur demonstrated the biological functions of microorganisms.
2. He developed effective vaccines against microbial diseases such as anthrax and rabies.

The Germ Theory of Fermentation

NONVITALIST VERSUS VITALIST VIEWS

1. *Fermentation* is a process in which alcohols and organic acids such as vinegar and lactic acid are formed from sugar-containing substances. Examples of the process include the souring of milk and the production of wine and beer.
2. According to the nonvital, or nonbiological, theory of fermentation, yeasts are by-products of fermentation. Supporters of this theory held that unstable chemicals called ferments cause fermentation.
3. The vital, or biological, theory of alcoholic fermentation, eventually proved true, held that yeasts are responsible for the reaction.

PASTEUR'S CONTRIBUTIONS TO FERMENTATION RESEARCH

1. Pasteur obtained experimental proof for the microbial nature of fermentation and for the specificity of fermentation reactions.
2. He developed the heating process, *pasteurization,* that kills or inactivates most disease- and spoilage-causing organisms.
3. He discovered *anaerobes,* microorganisms that can live only in the absence of free oxygen.

Antiseptic Surgery and Surgical Infections

1. The introduction of anesthetics into surgery during the 1840s enabled surgeons to perform longer and more complex operations, with a corresponding increase in the incidence of wound contamination and infection.
2. Joseph Lister developed methods to prevent the access of microorganisms to surgical wounds, instruments, and operating rooms.

The Germ Theory of Disease

1. From earliest times, diseases have been associated with natural phenomena, supernatural forces, and poisonous vapors.
2. Robert Koch, one of several individuals who established the causal relationship between a disease agent and a disease state, suggested a method of demonstrating this connection.

KOCH'S POSTULATES

Koch's postulates are a definite sequence of experimental steps by which to prove the causal relationship between a disease agent and a specific disease state:

1. Identification of the causative agent in all cases of the disease.
2. Isolation of the agent from the disease state and preparation of a pure culture.
3. Reproduction of the disease in a susceptible host by inoculation with the pure culture.
4. Recovery of the same microorganism from the experimentally infected host. (Koch's postulates cannot be achieved with bacteria without an appropriate animal host.)

RIVERS' POSTULATES

Rivers' postulates are a definite sequence of experimental steps similar to those developed by Koch, but they are applicable to both animal and plant viruses.

Early Technical Achievements

MICROBIAL CULTIVATION

1. An important development in the study and handling of microorganisms was the development of nutrient combinations called *media.*
2. Richard J. Petri developed a culture-medium container known as the *Petri dish.* The paths of air in the dish were based on the systems of Pasteur and Tyndall.
3. The introduction of the solidifying agent *agar-agar* into media made possible the study of microorganisms that grow over a wide temperature range.

CHEMOTHERAPY

1. Paul Ehrlich was the first to develop and apply chemicals for the treatment of disease and the control of disease-causing microorganisms. This approach is known as *chemotherapy.*

2. Alexander Fleming is known for the discovery of *antibiotics*. The subsequent development and application of penicillin and other antibiotics have been important in the control of bacterial infectious disease agents.

The Growth of Organized Microbiology

1. Both national and international organizations exist that are directed toward the advancement of general microbiology and associated specialties. One such organization in the United States is the American Society for Microbiology.
2. Journals, laboratory aids, and special monographs are published by professional organizations as a formal means of communication for the scientific community.

Some Challenges for Microbiology

1. Microorganisms have an invaluable role in determining the molecular basis of life processes.
2. The close alliance between microbiology, biochemistry, and genetics resulted from genetic studies with the fungus *Neurospora*, the discovery of DNA as the major chemical in genes, and the Watson and Crick molecular model of DNA.

A Glimpse of Things to Come

Microbiology has a position of importance in the modern world. Many decisions affecting the future of the world may depend on the activities of microorganisms.

Questions for Review

1. a. List the characteristic features of a *biological system?*
b. Must all of these properties be present before a microorganism can be considered living? Explain.

2. Which of the properties of a biological system would you consider essential to the well-being of microorganisms in the following situations?
a. pathogenic bacteria in the bloodstream
b. a microorganism stranded on the moon's surface
c. bacteria in the human small intestine

3. a. List three ways in which viruses differ from other microorganisms.
b. Give two examples of viral diseases.

4. Briefly describe the activities associated with the following areas of microbiology:
a. food microbiology
b. medical microbiology
c. mycology
d. phycology
e. dairy microbiology
f. microbial ecology
g. microbial genetics
h. biochemistry

5. Describe Leeuwenhoek's contributions to the development of microbiology. Did his discoveries go beyond the microbial world? Explain.

6. Why was the study of microbiology neglected after Leeuwenhoek's death?

7. What is the "doctrine of spontaneous generation"? How would you attempt to disprove it? Why did the discovery of oxygen in 1775 provide new support for the proponents of this concept?

8. Describe the experiments of Pasteur and Tyndall, and explain how they dealt the final blow to the doctrine of spontaneous generation.

9. Why was it important to disprove the doctrine of spontaneous generation?

10. Who first demonstrated the biological significance of microorganisms?

11. What were the major arguments used to support the nonvital and vital theories of fermentation? What are "ferments"? Describe the contribution of Schwann in demonstrating the role of yeasts in alcoholic fermentation.

12. What consistent patterns did Pasteur find in studying different fermentative processes? How did he show the microbial basis for fermentation?

13. What is pasteurization? What types of microorganisms are destroyed by this process?

14. Describe the early concepts of the cause of diseases. What were some of the first indications of the role of microorganisms as a cause of disease?

15. What purpose do Koch's postulates serve? What significant contributions did Koch make to the germ theory of disease? What are Rivers' postulates?

16. What factors or situations made the early study of microorganisms difficult?

17. In what ways are the contributions of Paul Ehrlich and Alexander Fleming related? Describe their most important contributions.

18. Define or explain the following:
a. antiseptic surgery
b. pure culture
c. spore
d. agar
e. fermentation
f. antibiotic
g. unity of biochemistry
h. central dogma of molecular biology

19. Of what value are bacteria to molecular biology?

Suggested Readings

BENDINER, E., "Liberator of Surgery from Shackles of Sepsis," *Hospital Practice* **21**:1265–1266, 1986. *A well-written presentation of the life of Joseph Lister, the individual who brought the vital refinements of antisepsis and asepsis into the hospital.*

BEVERIDGE, W. I. B., *The Art of Scientific Investigation.* New York: Vintage Books, A Division of Random House, 1957. *A short, thought-provoking book dealing with the basis of scientific investigation, thought, and ethics.*

BROCK, T. D., *Milestones in Microbiology.* Reprint. Washington, D.C.: American Society for Microbiology, 1975. *An enjoyable publication containing the translated and edited papers of Koch, Pasteur, and many others involved in the history of microbiology.*

BURNET, M., and D. O. WHITE, *Natural History of Infectious Disease.* Great Britain: University Press, 1972. *An interesting account of the discovery of various infectious disease agents and how diseases are spread and treated.*

DOBELL, C. (ed. and trans.), *Anton van Leeuwenhoek and His "Little Animals."* Reprint. New York: Dover Publications, 1960. *A well-written introduction to the life and times of Anton van Leeuwenhoek.*

DOYLE, R. J. and N. C. LEE, "Microbes, Warfare, Religion and Human Institutions," *Canadian Journal of Microbiology* **32:**193–200, 1986. *An interesting article emphasizing the significant role played by microorganisms in warfare, religion, migration of populations, art, and diplomacy.*

DUBOS, R. J., *Louis Pasteur: Free Lancer of Science.* Boston: Little, Brown and Co., 1950. *Pasteur's life and work are presented in an enjoyable and interesting manner.*

FULLER, J. G., *Fever.* New York: Ballantine Books, A Division of Random House, 1974. *A true medical detective thriller about the events leading to the discovery of the viral cause of the deadly disease, Lassa fever.*

MCGEOCH, D. J., I. W. HALLIBURTON, V. TER MEULEN, and C. R. PRINGLE, "Some Highlights of Animal Research in 1985." *Journal of General Virology* **61:**813–830, 1986. *A presentation of the most important and interesting advances in virus research during 1985.*

PORTUGAL, F. H., and J. S. COHEN, *A Century of DNA.* Cambridge, Mass.: The MIT Press, 1977. *A timely book describing the discovery of, and advances made with, the genetic substance DNA. The immediate and overwhelming impact of this genetic substance on the nature and basic properties of all forms of life, as well as the nature of scientific investigation, is presented in a readable and interesting manner.*

READINGS FROM SCIENTIFIC AMERICAN: INDUSTRIAL MICROBIOLOGY, vol. **245,** no. 3. San Francisco: W. H. Freeman, September 1981. *A collection of articles dealing with the uses and applications of microorganisms in the making of food, beverages, pharmaceuticals, and industrial chemicals. Particular attention is given to the genetic programming of microorganisms for their tasks.*

2

Survey of the Microbial World and Introduction to Classification

There are those who consider questions in science that have no unequivocal, experimentally determined answer scarcely worth discussing. Such feelings, coupled with a conservative stance, may have been responsible for the long and almost unchallenged dominance of the system of two kingdoms—plants and animals. . . . The unchallenged position of these kingdoms has ended, however, alternative systems are being widely considered. — R. H. Whittaker, 1969

After reading this chapter, you should be able to:

1. State the importance of and need for classification.
2. List the taxonomic ranks.
3. Describe both the early and modern approaches to classification, and distinguish between natural and artificial systems.
4. Explain the organization of the five kingdoms that make up the biological world.
5. Distinguish clearly between eucaryotic and procaryotic cellular organization.
6. Identify the positions of microbial groups in the biological world.
7. List the major characteristics of each kingdom in the biological world.
8. Describe how viruses differ from other microorganisms.
9. Describe modern concepts of the history of life and procaryote and eucaryote evolution.
10. Describe current trends in microbial classification.

Where do algae, bacteria, fungi, protozoa, and even viruses fit in the scheme of life? To understand microorganisms—their functions and activities—and even to solve the mystery of how life originated requires a logical system by which to show differences, similarities, and relationships among organisms—a system of classification. In this chapter we shall see how a classification scheme and modern approaches to show relationships among microorganisms developed. In addition, we shall consider differences in cellular organization and structure as criteria for classification.

The simplest cellular forms of life consist of individual, small cells that always originate from preexisting cells through the processes of growth and cell division. Such microscopic units are typical of the microbial groups known as the bacteria, algae, protozoa, and fungi. Even though these microorganisms share this common property, they differ from one another in terms of size, shape, energy-yielding and other types of biochemical reactions, and internal organization. This chapter considers how these and other differences are used to separate microorganisms and place them into specific groups. The general properties of cellular forms of microbial life and the noncellular forms known as viruses also are described.

The Importance of Biological Classification

Classification, an orderly arrangement of organisms with similar physical properties or biochemical and genetic characteristics into groups, is an extremely important aspect of the biological sciences. The most obvious reasons for classification are to establish the criteria for identifying organisms, to arrange similar organisms into groups, to provide important information on how organisms evolved, and to exchange ideas among research scientists.

Taxonomy is the science of classification. It involves naming different forms of life according to an international code of principles, rules, and recommendations as well as identifying unknown organisms by comparing their characteristics with those of known ones. Taxonomists arrange organisms into a ranked series of categories that reflect their interrelationships. As a science, taxonomy is dynamic and subject to change as new and more precise information becomes available or as old conflicts are resolved. Changes affect existing classification, names of organisms, bases for identification, and even the recognition of new forms of life.

Natural versus Artificial Classification

The classification of organisms is based on their similarities. Although there are several approaches to determining degrees of similarity, no agreement exists as to which one is best. The ideal goal is a *natural*, or *phylogenetic*, classification system, which identifies relationships between organisms on the basis of their probable origin. In such a system, organisms with a common origin are grouped more closely than those with dissimilar origins. The final outcome of a developed natural system resembles the structure of a tree, with the trunk representing the main course of evolution from its origin at the base, the branches and twigs representing later stages in evolutionary development, and the outermost leaves indicating biological forms now in existence.

The alternative to a natural classification system is an *artificial* one, based on easily recognizable properties of known organisms. This approach provides a practical and useful guide for the identification of unknown organisms.

Naming Organisms: The Binomial System of Nomenclature

For centuries, biologists have classified the forms of life visible to the naked eye as either animal or plant. This practice was eventually adopted as a scientific basis for separating the living world into the two kingdoms (groupings) **Animalia** and **Plantae.** In 1735, the Swedish naturalist Carolus Linnaeus published *Systema Naturae*, which together with his later works served to organize much of the current knowledge about living things. Since Linnaeus knew nothing of evolution, he based his classification system on structural and physiological properties of organisms.

Although the modern system of classification is based on evolutionary relationships, it makes use of one of Linnaeus' innovations, the **binomial system of nomenclature.** Under this system, all organisms have two-word names. For example, the name *Mycobacterium tuberculosis* (my-koh-back-TI-ree-um too-BUR-kyuh-LOW-sis) first gives the genus *(Mycobacterium)* to which the organism belongs. The genus name always begins with a capital letter. The species name follows and begins with a lowercase letter. The **genus** is a group of related organisms; a simple definition of a species is more difficult to write. For higher animals and plants, a **species** is often defined as a group whose members have a limited geographic distribution and interbreed. Because this definition assumes sexual reproduction, it does not readily apply to microorganisms such as bacteria. For such biological forms, a species may be defined as a population of cells that are descended from a single cell and that therefore have the same genetic makeup.

Usually the genus and species names appear printed in either **boldface** or *italics*. They are Latin terms, or words from other languages to which Latin endings are added. The name of a microorganism frequently refers to a distinctive property of the organism, such as its color (*citreus*, yellow), the disease it causes *(pneumoniae)*, or its habitat (*coli*, colon). The name may also indicate its discoverer—*Escherichia* (esh-er-IK-ee-ah), by Escherich—or some other individual for whom it was named—*Yersinia* (yur-SIN-ee-ah), for Yersin.

In bacterial taxonomy, when a new species is identified and named, a particular representative or **strain** is designated as the **type strain.** Type strains are preserved in national culture collections so that continued use and the study of specific organisms can be ensured. If a type strain is lost, a closely resembling substitute, or **neotype** strain, is utilized.

Groupings from the Linnaean System

In the Linnaean system, classification begins by dividing all life forms into **kingdoms.** Today five kingdoms are recognized. Each kingdom, in turn, is divided into general groupings, or taxonomic ranks, called **phyla** (plural of **phylum**) for animals and **divisions** for plants and bacteria. For all kingdoms these ranks are subdivided into **classes,** classes into **orders,** and orders into **families.** Families are subdivided into **genera** (plural of **genus**), and finally the genera are composed of **species.** A species may be further divided into two or more **subspecies.** The subspecies is the lowest taxonomic rank having official standing in the naming of microorganisms. Table 2–1 shows a comparative classification of human beings and *Treponema pallidum* (TREP-uh-NEE-mah PAL-leh-dum), the bacterial causative agent of syphilis (Figure 2–1).

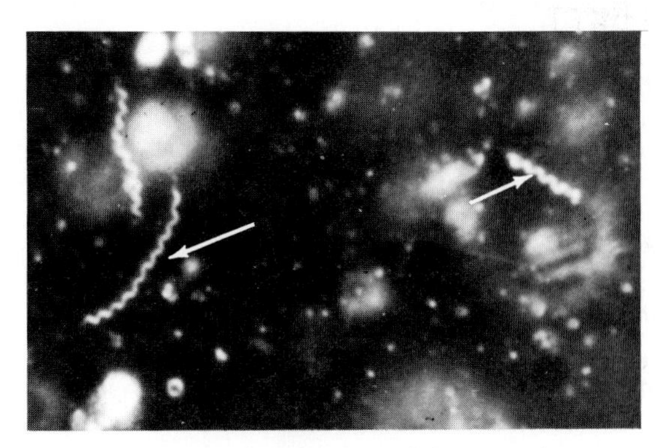

Figure 2–1 A microscopic view of the spirochete (corkscrew-shaped bacterium) that causes the sexually transmitted bacterial disease syphilis.

The Position of Microorganisms in the Living World

In general, while animals have readily identifiable structures of their own, they are characterized as lacking the typical structures of plants, such as leaves, stems, and roots. In addition, they are noted to be actively motile, nonphotosynthetic, and quite complex. Plants are regarded as the opposite of animals in all of these respects.

TABLE 2–1 A Comparative Classification of the Human Species and *Treponema pallidum*, the Cause of Syphilis

Taxonomic Rank	*Designation*	*Taxonomic Rank*	*Designation*
Kingdom	Animalia (animals)	Kingdom	Procaryotae
Phylum	Chordata (chordates)	Division	Gracilicutes
Subphylum	Vertebrate (vertebrates)		
Class	Mammalia (mammals)	Class	Scotobacteria
Order	Primates (primates)	Order	Spirochaetales
Family	Hominidae (humans and closely related forms)	Family	Spirochaetaceae
Genus	*Homo* (the human and precursors)	Genus	*Treponema*
Species	*sapiens* (modern human)	Species	*pallidum*

Until about 1830, the taxonomic status of most forms of life remained fairly constant. However, the discovery of a great variety of microorganisms presented biologists with several problems. While some microorganisms could be categorized as either plant or animal, many could not. Additional properties had to be considered. One of these was the presence of an outer structure called a **cell wall,** which would define a cell as a plant. Animallike cells, such as protozoa, did not have this feature. Moreover, such cells were able to capture and ingest solid foods, such as smaller protozoa and cell fragments, something plant cells could not do. Thus microscopic algae, bacteria, and fungi (molds and yeasts) were grouped into the Plantae, while the protozoa were considered to be members of the Animalia.

Unfortunately, other problems soon became apparent. Microorganisms were discovered that had properties of both animals and plants. Biologists continued to assign these microscopic forms arbitrarily to one of the kingdoms. It was clear, however, that a suitable classification of microorganisms could not be based on the properties of larger animals and plants.

Early Bacterial Classification Schemes

Up to the mid-nineteenth century, bacteria were regarded as animal largely because of Leeuwenhoek's term *animalcule* for the microscopic forms he observed. The assumption that all such forms of life were of a similar type persisted until 1857, when the Swiss botanist Carl von Nägeli proposed that bacteria should have a class of their own in the plant kingdom. He designated this class Schizomycetes, or "splitting fungi." Somewhat later, Ferdinand Cohn published a new system of classification for bacteria, which, like previous systems, was based mainly on their **morphology** (form and structure).

The work of Pasteur and Koch, including the development of solid media and pure culture techniques, provided the tools with which to uncover a variety of characteristics for the classification and identification of bacteria. (These contributions are discussed in Chapter 1.) Thus, by the start of the twentieth century, an immense body of information on various organisms had been accumulated. A notable contribution was made by Karl Lehmann and R. O. Neumann, who in 1903 established the beginnings of a formal manual for classification. For the first time, bacterial staining reactions and the ability of organisms to form the highly heat-resistant endospores were considered to be formal diagnostic features.

In the years that followed, a number of significant traits were incorporated into classification, including pathogenicity and metabolic activities. These and related developments led to the preparation in 1923 of the first edition of *Bergey's Manual of Determinative Bacteriology* by the Society of American Bacteriologists (now the American Society for Microbiology). Over the years, this most famous of modern texts on bacterial classification has incorporated all significant advances as they have been made. It records the discoveries of new species, new criteria of classification, and improved schemes for classification. Through the years this manual has become a widely used international reference work for bacterial taxonomy. Because of its importance and the need to keep pace with the rapidly accumulating volumes of significant information, a new approach to its organization was introduced with the first edition of *Bergey's Manual of Systematic Bacteriology.* This publication consists of four subvolumes, each of which focuses on specific bacterial groups and which can be prepared, published, and revised independently. This new approach provides a greater degree of coverage on topics that are related. Eventually, information that is useful for the identification of specific groups, in addition to detailed descriptions of bacterial species, will be incorporated into smaller but improved editions of *Bergey's Manual of Determinative Bacteriology.*

Unfortunately, no classification scheme entirely satisfactory to everyone has ever been developed, although several are in use. Moreover, it is not unusual to find the same organisms classified by specialist groups in a different manner or assigned to a different taxonomic level. In this text, we will use the most widely accepted terminology whenever possible.

The Third Kingdom: The Protista

The difficulties in applying the two-kingdom system to the classification of microorganisms are obvious. There is an enormous variety of microorganisms differing widely in metabolic and structural properties. Some microorganisms are plantlike, others are animallike, and still others are totally different from all other forms of life.

In 1866, one of Charles Darwin's students, Ernst Haekel, proposed the establishment of a third kingdom to eliminate the existing confused status of microorganisms and to provide a logical position for them in the living world. This kingdom was to be called **Protista,** from the Greek word meaning "primitive" or "first." The new kingdom consisted of single-celled microbes—such as algae, bacteria, fungi, and protozoa—and multicellular organisms that were not differentiated (organized) into distinct tissues and organs, as are higher animals and plants. The concept of protists gained in popularity over the years but was never universally accepted. In 1957, Roger Stanier and his associates gave new life to the term *protist.* They distinguished two subgroups based on cellular characteristics: the lower protists **(procaryotes),** microorganisms having a primitive nucleus, and the higher protists **(eucaryotes),** those having a well-defined nucleus.

Figure 2–2 A comparison of (a) eucaryotic animal and (b) plant cells, showing the better-known individual components of each. Compare these diagrams with the electron micrographs shown in Figures 2–3 and 2–4.

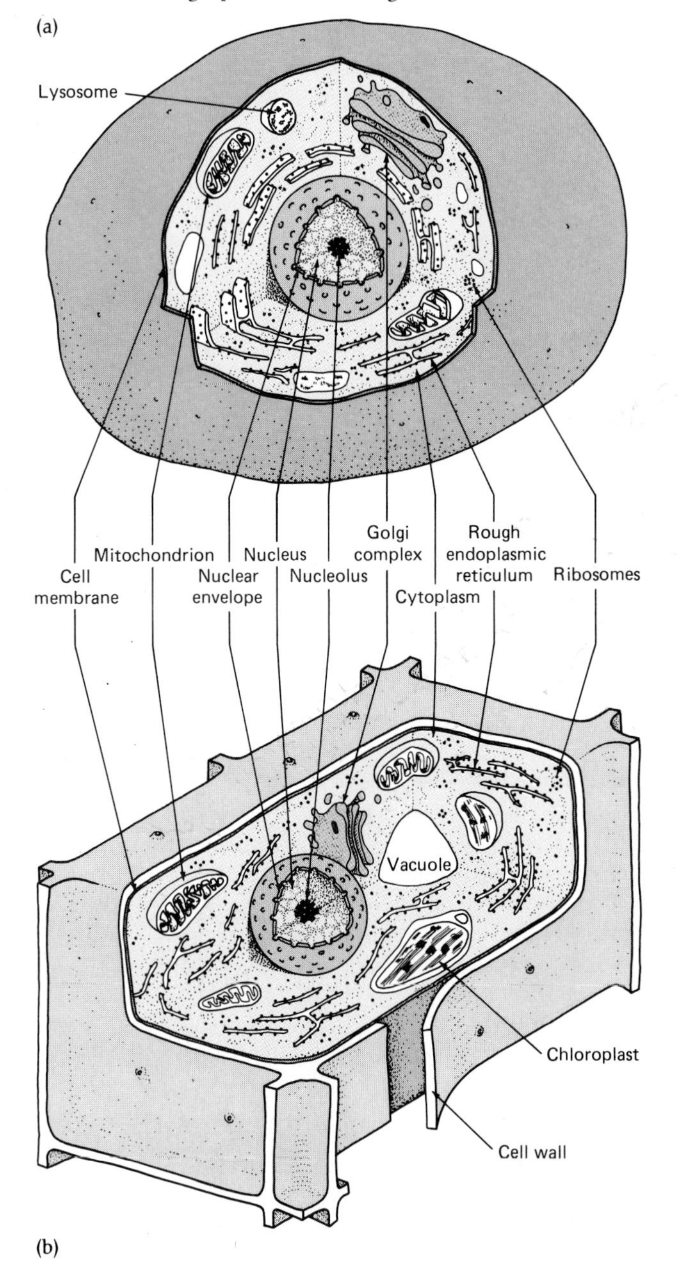

Eucaryotic and Procaryotic Cellular Organization

Cells are considered the basic biological units of structure and function. However, they can live only under rather limited conditions of pH, temperature, water content, and so forth. If these conditions change too greatly, the cells cannot readily adjust to them and die. Therefore, cells contain mechanisms not only to control their internal activities but also to adjust to various external events and forces.

With the development of the electron microscope and new techniques for the preparation of biological specimens, we have learned a great deal about the organization of cells. Later we will discuss the structures (together with their known functions) that contribute to the distinctive architecture of representative microorganisms. Here we shall study characteristic features of two fundamental cellular organizations, namely, the *procaryotic* and the *eucaryotic.*

The more complex eucaryotic organization is found in fungi, protozoa, and algae as well as in typical animals and plants (Figures 2–2, 2–3, and 2–4). Eucaryotic cells contain (1) a nuclear envelope and structures within the **nucleus** called **nucleoli;** (2) organized structures in the **cytoplasm** outside the nucleus that are bound by membranes (for example, **endoplasmic reticulum, mitochondria, chloroplasts, Golgi apparatus,** and **lysosomes);** (3) **cell membranes** with special components of some lipids called *sterols;* (4) 80 S **ribosomes** consisting of 60 S and 40 S subunits **[S = Svedberg unit, an indirect measure of molecular size];** (5) whiplike **flagella** or short **cilia** structures by which

Figure 2–3 An electron micrograph of antibody-producing cells exhibiting eucaryotic organelles. Explanation of symbols: cell membrane (C), mitochondrion (M), nucleus (N), nucleolus (Nu), rough endoplasmic reticulum (RER). *[From L. V. Chalifoux, N. W. King, M. D. Daniel, M. Kannagi, R. C. Desrosiers, P. K. Sehgal, L. M. Waldron, R. D. Hunt and N. L. Letvin,* Lab Invest. ***55****:43, 1986.]*

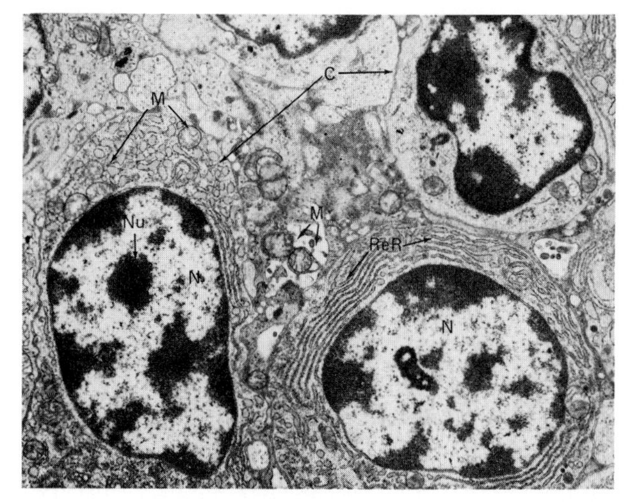

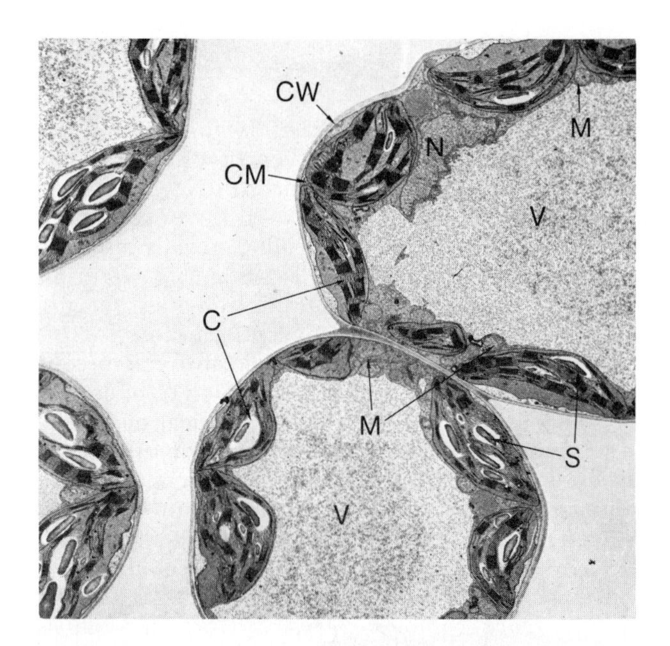

Figure 2–4 A plant cell. This ultrathin section shows various cellular components including the cell membrane (CM), chloroplasts (C), cell wall (CW), nucleus (N), starch granules (S), mitochondria (M), and a central vacuole (V). *[From R. R. Camp, and W. F. Whittingham,* Amer. J. Bot. ***59:****1057–1067, 1972.]*

certain cells move; (6) a *mitotic spindle* during nuclear division; (7) more than one *chromosome;* (8) a basic group of chromosomal proteins called **histones;** and (9) *nucleosomes,* elementary subunits of chromosomes. Most eucaryotic plantlike cells have a rigid outer covering known as a **cell wall** (Figure 2–2). The functions and activities of eucaryotic structures (organelles) are listed in Table 2–2.

Procaryotic organization is typical of bacteria (Figure 2–5). Procaryotic cells lack most of the structures present in the cytoplasm of eucaryotic cells (Figure 2–2). Because of this relative simplicity, the procaryotic cell is generally considered more

TABLE 2–2 **Eucaryotic Organelles and Their Functions and Activities**

Organelle	*Associated Functions and Activities*
Plasma (cell) membrane	1. Transport of substances into and out of cells 2. In some cells, engulfment of foreign material (phagocytosis) 3. Pinocytosis
Cell wall	1. Found only in plants, algae and fungi; imparts shape and strength to the cell 2. Protection against certain osmotic imbalances
Centrioles	Mitotic spindles form between these structures in animal cells
Chloroplast	Photosynthesis
Chromosomes	Contain the genes (heredity units) that control the structure and activities of the cell
Cilium (plural: cilia)	Motion, or movement of substances past the ciliated cell
Endoplasmic reticulum	
Rough form	Protein synthesis and transport
Smooth form	Production of hormonelike compounds (steroids)
Flagellum (plural: flagella)	Propulsion (movement)
Golgi apparatus (complex)	1. Transfer of proteins and other cellular components to a secretory cell's exterior 2. Storage and packing structure for cellular products
Lysosome	Localized intracellular digestion
Microbody, or peroxisome	Enzymatic activities
Microtubule	1. Cell transport of materials 2. Development and maintenance of cell shape 3. Cell division 4. Ciliary and flagellar movement
Mitochondrion	Synthesis of the energy-rich compound adenosine triphosphate (ATP)
Nucleolus	Major site for the formation of ribosomal components
Nucleus	1. Control of cellular physiological processes 2. Transfer of hereditary factors to subsequent generations
Ribosome	Protein synthesis
Vacuoles	1. Locations of water 2. Storage site for certain amino acids, carbohydrates, and proteins 3. Dumping ground for cellular wastes

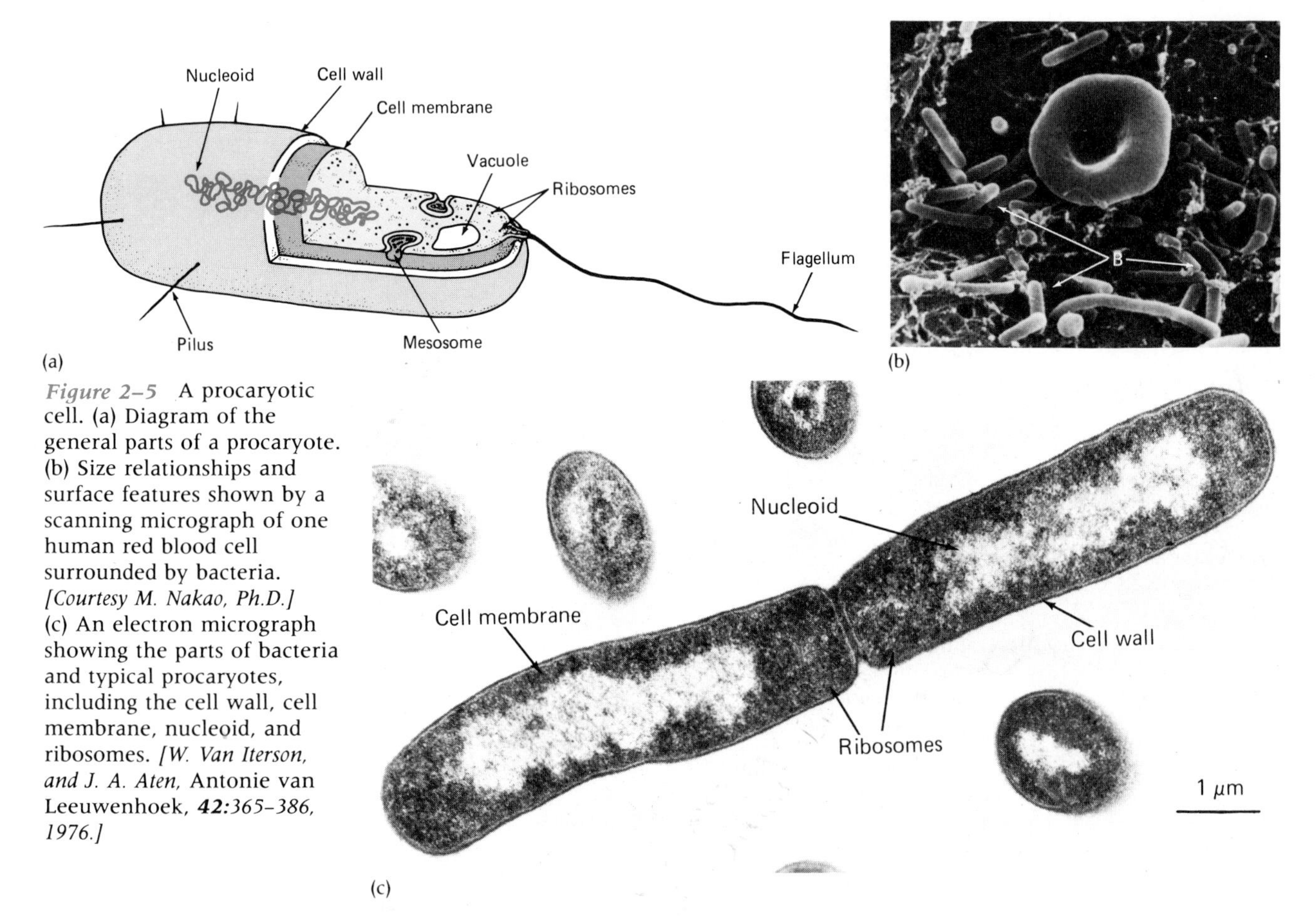

Figure 2–5 A procaryotic cell. (a) Diagram of the general parts of a procaryote. (b) Size relationships and surface features shown by a scanning micrograph of one human red blood cell surrounded by bacteria. *[Courtesy M. Nakao, Ph.D.]* (c) An electron micrograph showing the parts of bacteria and typical procaryotes, including the cell wall, cell membrane, nucleoid, and ribosomes. *[W. Van Iterson, and J. A. Aten,* Antonie van Leeuwenhoek, ***42**:365–386, 1976.]*

primitive from an evolutionary standpoint than the eucaryotic cell.

Procaryotic organisms, however, have some unique structural features. Among them are a nuclear region, or **nucleoid,** containing a single large molecule of DNA, which exists as a closed circular molecule and is referred to as a *chromosome.* Histones found in eucaryotic cells are completely lacking in this DNA structure. All blue-green or cyanobacteria and a few photosynthetic bacteria have layers of membrane, or **lamellae,** that are involved in photosynthesis utilizing bacterial chlorophylls. Table 2–3 compares the general properties of eucaryotic and procaryotic cells.

The Five-Kingdom Approach

The Components of the Five-Kingdom System

Of all the classification schemes used in recent years, the one that now has gained widest acceptance is the **five-kingdom system** proposed by Robert H. Whittaker in 1969. Originally this scheme divided the living world into the five kingdoms of **Plantae, Animalia, Protista, Fungi,** and **Monera** (procaryotes). Currently, depending on the authority, procaryotic organisms are placed either in the kingdom of Monera, and under the superkingdom of Procaryotae, or in the kingdom of Procaryotae. In this text the latter arrangement will be used, since it is the one followed in *Bergey's Manual of Systematic Bacteriology.* The relationships between the five kingdoms are shown in Figure 2–6, which suggests the descent of all organisms from a common ancestor, the first living cell. Whittaker's system is based on three levels of cellular organization: procaryotic **(Procaryotae);** eucaryotic, unicellular **(Protista);** and eucaryotic, multicellular and multinucleate **(Fungi, Animalia,** and **Plantae).** The divergence at each of these levels depends upon three possible forms of nutrition: photosynthesis, absorption, and ingestion (Figure 2–7). The Procaryotae lack an ingestive mode of nutrition, but at the highest level (multicellular-multinucleate), the nutritional modes distinguish the three higher kingdoms, Plantae, Fungi, and Animalia. Brief descriptions of the kingdoms follow.

PLANTAE

Many of the species in the kingdom Plantae are well known and easily recognized green plants—garden

TABLE 2–3 A Comparison of Eucaryotic and Procaryotic Cells

Structure	*Eucaryotic Cell*	*Procaryotic Cell*
Nuclear envelope	Present	Absent
Nucleolus	Present	Absent
Chromosomes	Multiple and generally linear	Single and generally circular
Mitochondria	Present	Absent
Photosynthetic system	Chlorophyll, when present, and contained in chloroplasts	May contain chlorophyll, and other pigments, but not in chloroplasts
Golgi apparatus, endoplasmic reticulum, lysosomes, etc.	Present	Absent
Histones	Present	Absent
Nucleosomes	Present	Absent
Ribosomes	Large (80 S = 60 S + 40 S)[a]	Small (70 S = 50 S + 30 S)[a]
Membrane sterols	Present	Absent[b]
Mitotic apparatus	Present	Absent
Flagella	Present	Present but simpler structurally
Representative microorganisms	Algae, fungi, and protozoa	All bacteria (including the archaeobacteria, cyanobacteria, rickettsia, and chlamydia)

[a] S = Svedberg units, a measure of molecular size.
[b] With the exception of mycoplasma.

Microbiology Highlights

SUBDIVIDING SUBSPECIES

Several taxonomic ranks below subspecies are established and often used to indicate particular bacterial strains that can be distinguished by certain property characteristics. Such ranks, while of great practical value in the identification of microorganisms, have no official standing in the naming of microorganisms. Some of the more common "infrasubspecific" ranks include those listed below.

Infrasubspecific Ranks

Preferred Term	*Common Term*	*Distinguishing Property*
Biovar	Biotype	Specific biochemical or physiological reaction or property
Morphovar	Morphotype	Special shape or morphological feature
Pathovar	Pathovar	Destructive property for certain hosts
Phagovar	Phagotype	Susceptibility to the destructive effects (lysis) of certain bacterial viruses
Serovar	Serotype	Distinctive antigenic properties (ability to promote antibody production)

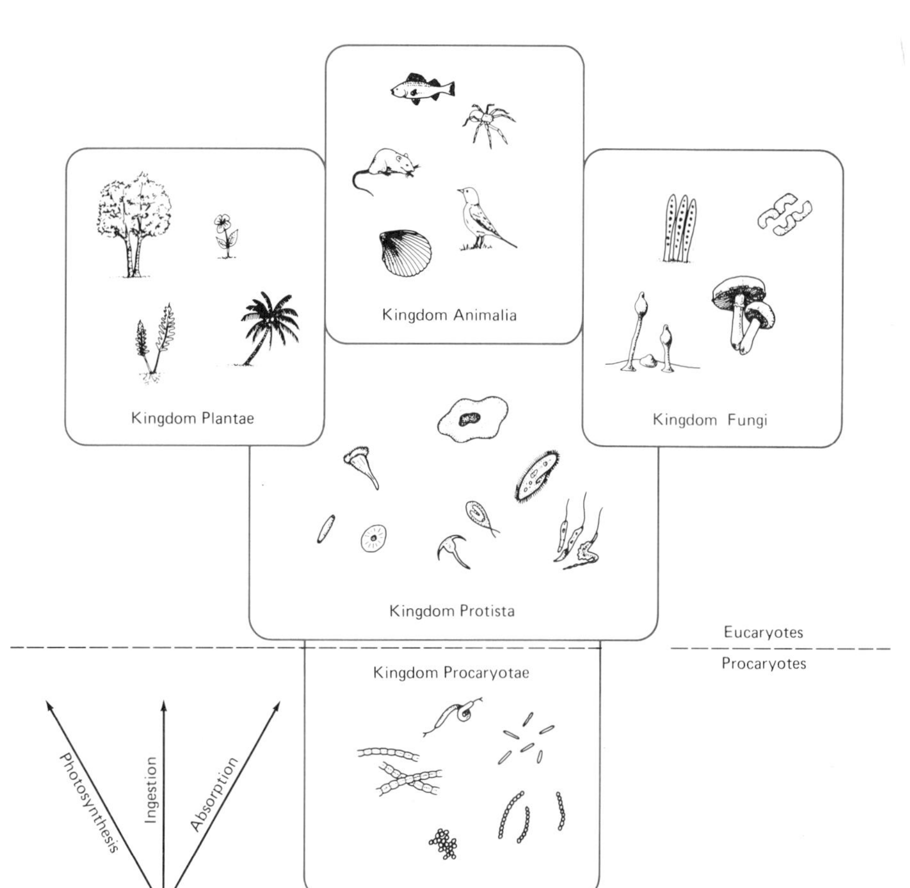

Figure 2–6 The five-kingdom system of classification is based on three levels of cellular organization: (1) procaryotic (Procaryotae); (2) eucaryotic, unicellular (Protista); and (3) eucaryotic, multicellular, and multinucleate (Fungi, Animalia, and Plantae).

vegetables, grasses, shrubs, mosses, and most algae. In general, plants are multicellular, eucaryotic, and have rigid cell walls. Most plants are photosynthetic, containing one or more types of chlorophyll, and, with few exceptions, they are nonmotile. Other properties of plant kingdom members include differentiation of their tissues (Color Photograph 9) into organized structures such as roots, stems, and leaves, and complex life cycles, generally involving both asexual and sexual reproduction. Table 2–4 summarizes the characteristics of Plantae as well as those of the other kingdoms.

ANIMALIA

The animal life included in the kingdom Animalia ranges from organisms without backbones, such as sponges and worms, to highly developed forms with backbones, such as mammals. Animal cells are eucaryotic and enclosed only by a flexible membrane. They lack cell walls. Animals are multicellular and motile most of their lives, and they exhibit ingestive nutrition. Some show tissue differentiation (Color Photograph 10). Animals may reproduce sexually and/or asexually.

PROTISTA

Various protozoa (Figure 2–7 and Color Photographs 46 and 47) and unicellular and colonial forms of algae (Figure 2–8 and Color Photographs 7

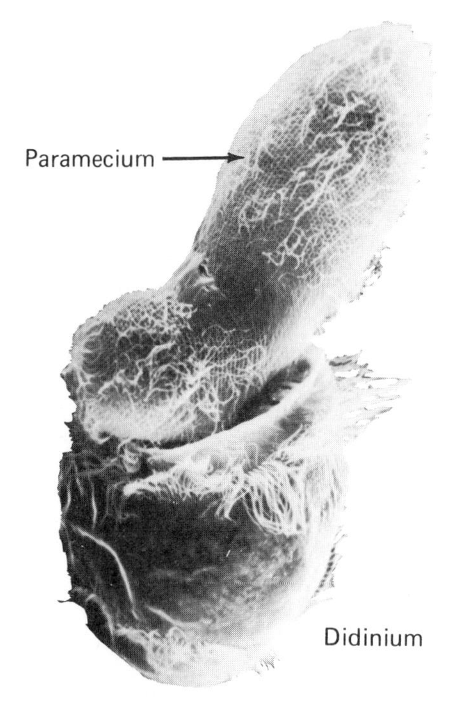

Figure 2–7 The carnivorous nature of certain protozoa is evident in this scanning micrograph. If the ciliate *Didinium* (die-din-EE-um) is hungry during a chance collision with the *Paramecium* (par-ah-MEE-see-um), capture and ingestion of the larger protozoon almost always occur. *[H. Wessenberg, and G. Antipa,* J. Protozool. ***17****:250–270, 1970.]*

TABLE 2–4 Characteristics of the Five Kingdoms

	Kingdoms				
Characteristics	*Plantae*	*Animalia*	*Protista*	*Fungi*	*Procaryotae*[a]
Cellular organization	Eucaryotic and multicellular	Eucaryotic and multicellular	Eucaryotic, unicellular, and some colonial forms	Eucaryotic, multicellular, and unicellular	Procaryotic and unicellular
Cell wall	Present	Absent	Present with algae	Present	Present
Differentiation of tissues	Present	May be present or absent	Absent	Absent	Absent
Mode or type of nutrition	Primarily photosynthetic	Ingestive and some absorptive	Absorptive, ingestive, photosynthetic, and combinations	Absorptive	Absorptive; few photosynthetic
Reproduction	Generally both asexual and sexual	Generally both asexual and sexual	Asexual and sexual	Asexual and sexual	Asexual and rarely sexual
Motility (ability to move)	Mostly nonmotile	Motile	Both motile and nonmotile	Generally nonmotile	Both motile and nonmotile
Microbial representatives	None	None	Microscopic algae and protozoa	Molds and yeasts	Bacteria including cyanobacteria

[a] Also known as the kingdom of Monera.

and 50) make up the Protista. Unlike typical animals and plants, protists are biologically and biochemically independent. There is no dependence upon other cells, such as exists in the tissues and organs making up higher plants and animals. The Protista is considered by some to be a transitional kingdom; in an evolutionary sense it falls between procaryotic organisms and the more structurally sophisticated members of the animal and plant kingdoms. Others view Protista as a catchall kingdom for any organism that does not readily fit into another kingdom.

Two phyla of algae are primarily multicellular: the brown algae and the red algae. A third phylum, the green algae, includes several kinds of multicellular organisms and an even larger number of unicellular forms. Some authorities place these phyla in the plant kingdom and not under the kingdom of Protista.

Protists are eucaryotic and obtain their nutrients in various ways, including absorption, ingestion (Figure 2–7), photosynthesis, and combinations of these modes. They exhibit both asexual and sexual reproduction.

Figure 2–8 A scanning micrograph of a colony forming green algae *Coelastrum* (SEE-lass-trum). *[H. J. Marchant,* J. Phycol. ***13****:102–110, 1977.]*

FUNGI

The Fungi includes the familiar mushrooms (Color Photograph 5), molds (Figure 2–9 and Color Photo-

Figure 2–9 Fungi are important sources of antibiotics. Here, droplets of antibiotics (arrows) can be seen on the surface of the mold *Penicillium* (pen-ee-SIL-lee-um). *[Courtesy of Lederle Laboratories.]*

graphs 37 and 98) and yeasts (Color Photograph 38), as well as the less familiar slime molds (Color Photograph 43), rusts, and smuts. An individual fungus organism may be microscopic or weigh several pounds (Color Photograph 36). Fungi are eucaryotic, possess cell walls, and may be unicellular or multicellular. Like animals, they are not photosynthetic. Fungi obtain their nutrition from the environment by absorption.

PROCARYOTAE

The kingdom of Procaryotae consists of unicellular and colonial microorganisms with procaryotic cellular organization; these are designated as the true bacteria, or **eubacteria.** All of the functions and activities of life are carried out within the confines of these extremely small but far from simple cells. The procaryotes include a wide variety of bacteria (Figure 2–10), such as the blue-greens and other photosynthetic organisms, agents of disease such as the rickettsiae and chlamydiae, and numerous beneficial and harmless forms. Most organisms in this group have absorptive nutrition, although some procaryotes are photosynthetic. Reproduction is mainly asexual. **[Chapter 13 provides a survey of the procaryotes.]**

Recent evidence obtained from research techniques used to trace the evolutionary relationship between currently known bacteria and the common ancestor of all existing life suggests that there are two fundamentally different types of procaryotes, the **eubacteria** and the more primitive **archaeobacteria.** This division is generally accepted by the bacterial taxonomists. The number of traits that show the archaeobacteria as a distinct group different from the eubacteria is now quite impressive (Table 2–5).

A Third Line of Evolution

In 1977, C. R. Woese and his co-workers found that the RNA composition and genetic structure of specific bacteria known as the **methanogens** (methane-producing bacteria, Figure 2–11) were distinctly different from those of other microorganisms or animals and plants. This discovery has far-reaching im-

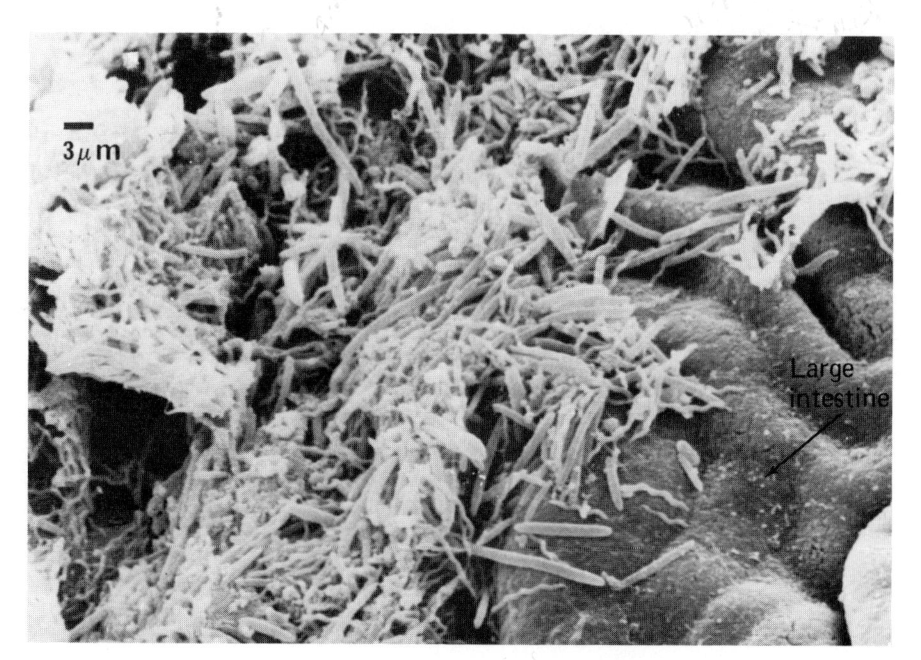

Figure 2–10 The presence of large numbers and varieties of bacteria on the surfaces of the large intestine. *[From A. Lee, J. L. O'Rourke, P. J. Barrington, and T. J. Trust,* Inf. Immun. ***51****:536–546, 1986.]*

TABLE 2–5 General Cellular Comparison of Eucaryotes, Eubacteria, and Archaeobacteria

	Cell Group		
	Eucaryotes	*Eubacteria*	*Archaeobacteria*
CELLULAR PROPERTY			
Cellular organization	Eucaryotic	Procaryotic	Procaryotic
Cell wall	Present (plants, fungi, algae)	Present (except in mycoplasma)	Present[b]
Murein in cell wall	Absent	Present	Absent
CYTOPLASMIC COMPONENTS			
Chloroplasts	Present (plants, algae)	Absent	Absent
Nuclear membrane	Present	Absent	Absent
Endoplasmic reticulum	Present	Absent	Absent
Golgi apparatus	Present	Absent	Absent
Mitochondria	Present	Absent	Absent
Ribosomes	Present (80 S)[a]	Present (70 S)	Present (70 S)
GENETIC SYSTEM			
Number of chromosomes	>1	1	1
Chromosome form	Linear	Circular	?

[a] S = Svedberg units, a measure of molecular size. Chapter 7 provides additional details.
[b] Differs in structure from the cell walls found in the eubacteria.

plications, since it suggests that there may be a third line of evolution. Older views have long assumed that all terrestrial life evolved along two lines, one of which gave rise to animals and plants and the other to bacteria, fungi, and protists (Figure 2–6). On the basis of the genetic studies with methanogens, it appears that a very ancient point of divergence occurred along these evolutionary lines.

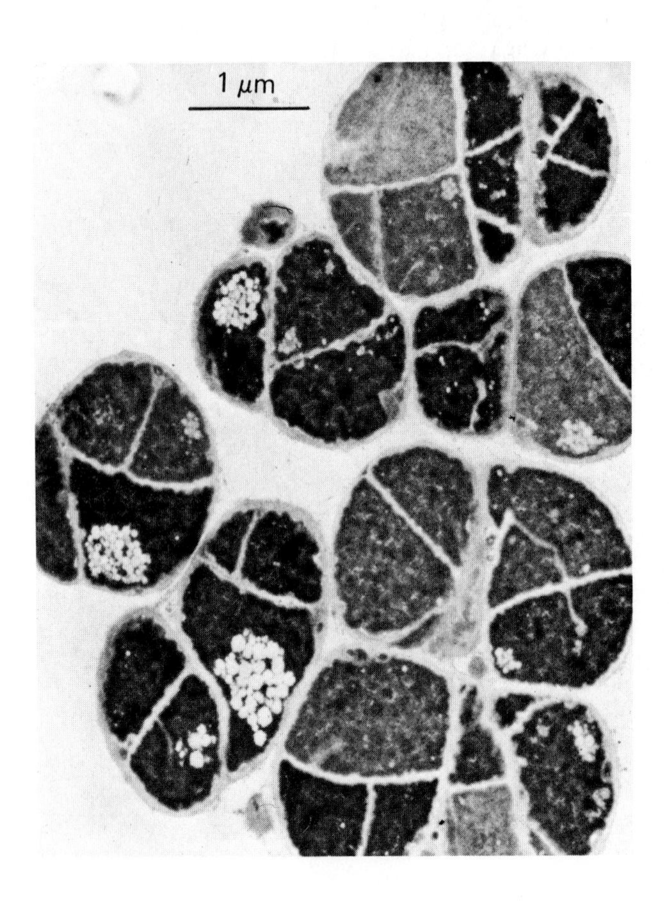

Woese's findings also provide an important clue to the earth's early environment. Scientists have long believed that for about the first billion years after the formation of the earth, the planet was very warm and enveloped in clouds consisting largely of hydrogen, carbon dioxide, and other gases but virtually no free oxygen. It was during this period that methanogens are believed to have evolved. In his studies of methanogens, Woese discovered the following properties: they thrive at high temperatures, between 65° and 70°C (Color Photograph 66), such as in the superheated springs at Yellowstone Park; they are anaerobic; and they utilize carbon dioxide and hydrogen and give off methane gas.

True Bacteria and the Archaeobacteria

Based on the results of comparative ribosomal RNA analyses with methanogen-related organisms, G. E. Fox and C. R. Woese in 1977 suggested that procaryotes are of two different kinds, the eubacteria and the archaeobacteria. As their name suggests, the archaeobacteria are thought to have descended from ancient organisms. The archaeobacteria, truly unusual microorganisms, include three very different kinds of bacteria: **methanogens, extreme halophiles,** and **thermoacidophiles.**

Figure 2–11 An electron micrograph showing the methane-producing bacterium *Methanosarcina barkeri* (meh-THAN-oh-back-terr-ee-um BAR-kerr-eye), which has been identified as a descendant of a form of life that possibly dates back to the earth's first billion years. *[J. G. Zeikus and V. G. Bowen,* Can. J. Microbiol. ***21****:121–129, 1975.]*

Microbiology Highlights

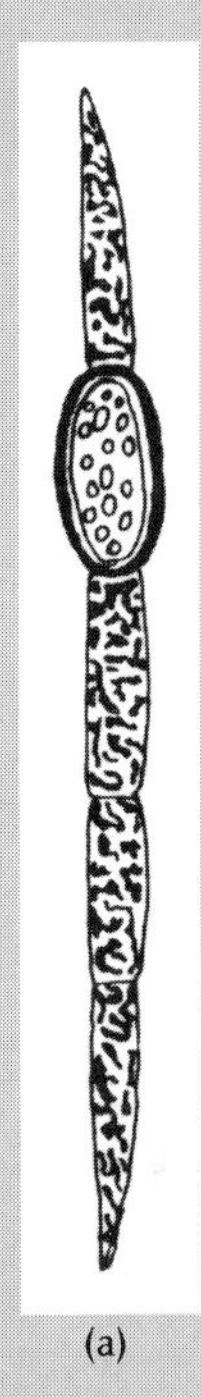

(a)

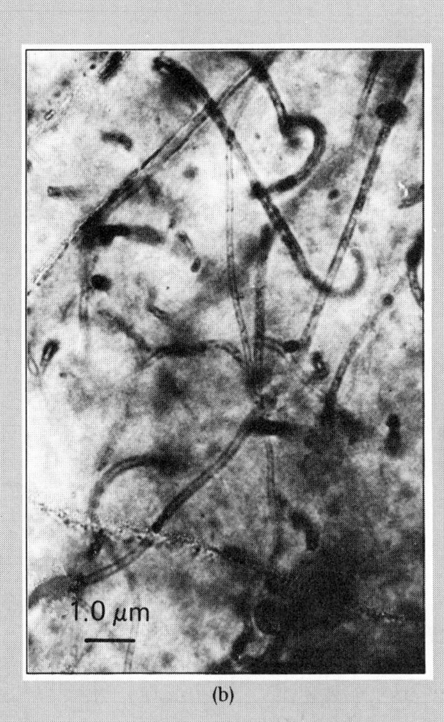

(b)

(a) Microfossils. The microstructures shown are from the oldest known sediments on earth. These round or ellipsoidal microstructures appear to have double walls and are found individually or in chains or clusters. *[L. A. Nagy,* Science *138:514, 1974.]* (b) A photomicrograph of well-preserved microfossil filaments and spherical forms probably related to blue-green bacteria found in the late Precambrian Chuar group from the eastern Grand Canyon. This area has long been considered a promising locale for the detection of early life. *[J. W. Schopf, T. D. Fort, and W. J. Breed,* Science *170:1319–1321, 1973.]*

Before Pasteur's brilliant experiments refuting the spontaneous generation of life, it seemed obvious that the multitude of worms, snakes, and insects simply arose out of dirt, soil, water, decaying material, and so forth. When spontaneous generation was discredited, many looked to other planets for the origin of life. Such notable scientists as the German naturalist Hermann von Helmholtz and the Nobel Prize-winning physical chemist Svante Arhennius proposed that life traveled to earth from outer space in the form of spores. Until and unless space missions find some proof of this hypothesis, the consensus is that we must look for the origins of life here on earth.

Paleomicrobiology

The earth is considered to be 4.5 billion years old. Evidence about the history or evolution of various forms of life has been deduced largely from the study of *fossils,* the remains of the past. A fossil may be in a petrified state, and therefore observed as a solid object, or it may be in the form of an imprint on a rock, left by a structure that has long since disappeared.

By studying fossils, we can observe changes in the body structure and appendages of animals and in the vascular systems of plants that have occurred since the time of fossil formation. Although it is easy to visualize the fossils of leaves, footprints, and bones, it is exceedingly difficult to imagine the appearance of different forms of life billions of years ago. Some of the oldest pieces of evidence left by biological communities in the fossil record are those of blue-green bacteria, which take the form of columnlike masses or **stromatolites.** Such structures are fossilized bacterial colonies embedded with minerals.

Examination of well-preserved microfossils in rocks from the Precambrian period (about 3.5 billion years ago), revealed the presence of procaryotes and algae. Interpretation of the findings suggest that these microorganisms flourished during this period. If these fossil data are correct, oxygen-producing microorganisms were in existence 3.5 billion years ago. Presumably, the first organisms were anaerobic, since molecular oxygen did not accumulate until later.

The extreme halophiles require high concentrations of salt in order to survive. They grow in salty locations along ocean borders and in inland bodies of water such as the Dead Sea and the Great Salt Lake. These bacteria have a photosynthetic mechanism based not on green chlorophyll but on the pigment bacterial rhodopsin, which is remarkably like one of the visual pigments found in mammals.

Thermoacidophiles represent another group of bacteria that also are notable for the environments they inhabit. These organisms normally grow in hot, acidic locations. Growth above 90° Celsius (C) has been observed with some of them.

The number of properties of archaeobacteria that not only support their evolutionary origin but distinguish them from the true bacteria is steadily increasing. These properties include differences in cell wall, cell membrane, and ribosomal chemical composition and organization; the presence of unique proteins; and antibiotic sensitivities of certain metabolic reactions that differ from those found with other bacteria (Table 2–5). Clearly, the evidence from various studies establishes the archaeobacteria as a distinct and separate group of procaryotes. Thus, all cellular forms of life can be placed into one of three categories: eubacteria, eucaryotes, and the archaeobacteria. **[Chapter 13 describes additional properties of the archaeobacteria.]**

Procaryote-Eucaryote Evolutionary Relationships

The discovery of archaeobacteria suggests some changes in current views concerning the evolutionary relation between procaryotes and eucaryotes. Research results have tended to support the idea that mitochondria and chloroplasts, two eucaryotic structures, are descended from procaryotes that became trapped in a larger cell and established a close internal relationship **(endosymbiosis)** with it. The mitochondrion is thought to have been an actively metabolizing bacterium and the chloroplast to have been a photosynthesizing relative of blue-green bacteria. It thus appears that at least two lines of procaryotic descent may have occurred in eucaryotic cells. **[The blue-greens or cyanobacteria are described in Chapter 13.]**

The origin of the remaining portion of the eucaryotic cell is not clear. Many investigators have tended to favor the view that the ancestor of this cell line may have been a bacterium (Figure 2–12). While this view is satisfying in several respects, it fails to explain many biochemical and metabolic differences between the two cells. To settle some of these differences, the **urcaryote** was proposed to represent a eucaryotic ancestral cell line distinct from the procaryotes. This proposal establishes three lines of descent equidistant from one another and provides a much better perspective to judge which properties are ancestral and which have evolved recently. With the discovery of the archaeobacteria, the nature of the common ancestor of all forms of life and the evolution of the eucaryotic cell become approachable problems (Figure 2–12).

A Consideration of Viruses

Like other microorganisms, viruses have definite properties that set them apart from all other forms of biological life. Since viruses are neither procaryotic nor eucaryotic, determining their taxonomic position in the biological world and establishing an acceptable classification system has been difficult. The classification of the submicroscopic microbes known as **viruses** has long been disputed by biologists. The debate has centered on whether these forms are living or not. Viruses are not cells (Figure 2–13). They do not possess the typical structures of procaryotic or eucaryotic cells, such as cell membrane, mitochondria, nucleus, and ribosomes. Individual viruses consist of one type of nucleic acid, either deoxyribonucleic acid (DNA) or ribonucleic acid (RNA), surrounded by a protein coat. After invading living cells, viruses use the metabolic and genetic machinery of the host cell to produce hundreds of new virus particles. Recent studies have also shown that some viruses can transform normal cells in laboratory animals into cancer cells. Viruses infect all types of life (Color Photograph 56). Even microorganisms such as bacteria, microscopic algae, protozoa, and fungi are not free from attack by these submicroscopic organisms.

Perhaps viruses exist at the boundary between living and nonliving matter. Although they resemble higher forms of life in that they undergo permanent genetic changes **(mutation)** and have a means of replicating and increasing their number, they do not readily lend themselves to classification by the rules and characteristics used in the classification of animals, plants, and other microorganisms. A further discussion of viruses and smaller infectious agents is presented in Chapter 10.

Trends in Microbial Classification

Numerical Taxonomy

Some taxonomists consider certain properties of microorganisms to be more significant than others. For example, staining reactions or morphologic properties might be weighted more heavily than fermentative capability in assigning a microorganism to a

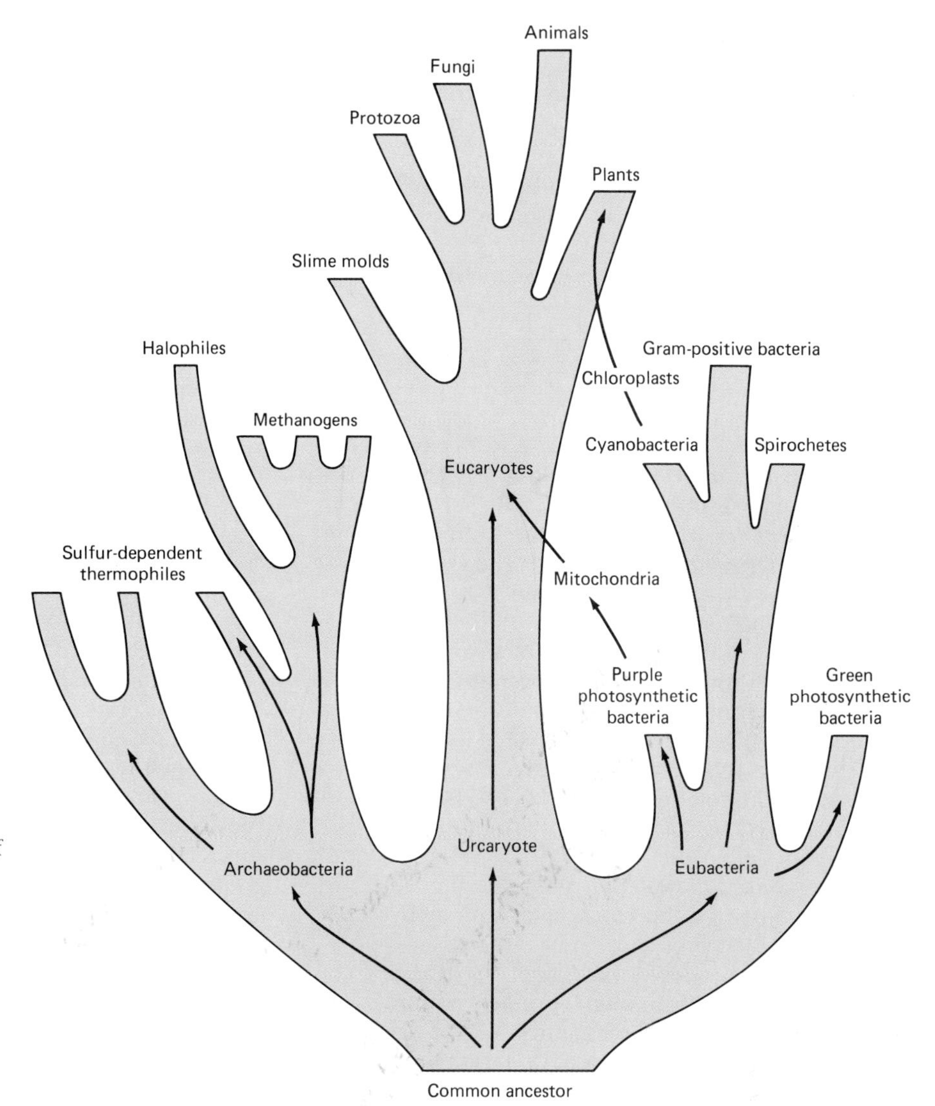

Figure 2–12 A representation of the three lines of descent emerging from a common ancestor based on ribosomal ribonucleic acid *(rRNA)* comparisons. The eubacteria lead to a variety of procaryotes. Chapter 13 surveys several of the procaryotes indicated in this figure.

group. Unfortunately, classification under such conditions tends to yield results that are biased.

Since 1957, the use of numerical techniques and computer methods has provided a more objective approach to bacterial taxonomy. The purely mathematical approach of numerical taxonomy or Adansonian analysis eliminates the need to establish different values or weights for different characteristics. This method takes into account many tests, each with equal weight, in determining the similarity between microorganisms. Taxonomic groups are generally defined on the basis of the overall similarity of the observed properties of the organisms rather than on their ancestry (family tree).

Using computers, the properties of many organisms can be compared. This produces the information needed to arrange similar organisms in taxonomic groupings, or *clusters.* In this approach, each organism is classified as an *operational taxonomic unit,* or *OTU.* The computer determines the percentage of

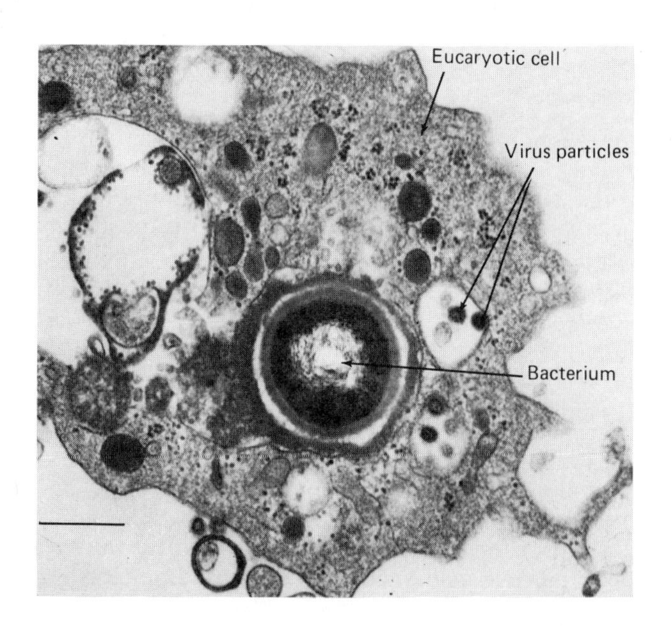

Figure 2–13 An electron micrograph showing size relationships among virus particles, a bacterium, and a eucaryotic cell. The bacterium is enclosed in a white blood cell vacuole. The bar marker equals 0.5 μm. *[Courtesy of J. S. Abramson, E. L. Mills, G. S. Giebink, and P. G. Quie,* Inf. Immun. ***35****:350–355, 1982.]*

TABLE 2–6 The Determination of the Similarity Coefficient for Two Bacterial Strains

Explanation of Symbols

a	Number of characteristics positive (present) in both strains.
b	Number of characteristics positive in strain 1 and negative (absent) in strain 2.
c	Number of characteristics negative in strain 1 and positive in strain 2.

Calculation

$$\text{Similarity coefficient } (S_J) = \frac{a}{a + b + c}$$

similarity between OTUs by finding the ratio of the number of characteristics they have in common to the total number of characteristics compared. This result is expressed as a percentage and is referred to as the *similarity coefficient* (S_J). Table 2–6 shows how the S_J can be determined.

Based on S_J calculations, clusters of organisms are arranged according to the highest mutual similarity in terms of numerical relationships. Such similar clusters are called *phenons.* Branched diagrams, or **dendrograms,** can be constructed using similar values to show the relationships between clusters or organisms. Based on their percentage of similarity, a dendrogram joins one group with others. The numerical taxonomic approach provides the objective and stable basis for the construction of such groupings (Figure 2–14).

In the field of clinical medicine, the use of computers in classification offers greater potential for the identification of disease agents. With a pure culture, instruments can perform a large number of biochemical, morphological, and antibiotic sensitivity tests that, when linked to a special computer program, can not only identify a pathogen but also suggest appropriate chemotherapy.

Molecular Approaches to Taxonomy

DNA or genetic relatedness is an attractive addition to taxonomy, since it evaluates the relationship among organisms at a very fundamental level by comparing their genomes (genetic material) or their products, which may be either protein or one of several kinds of RNA molecules. There are three major types of RNA, all of which have specific roles in the production of proteins: *messenger RNA (mRNA), ribosomal RNA (rRNA),* and *transfer RNA (tRNA).* **[Chapter 6 describes the nucleic acids.]**

Modern approaches to tracing evolutionary relationships include (1) determination of the mating capabilities of organisms; (2) comparisons of the overall chemical composition of DNA; (3) DNA homology (degree of evolutionary similarity), a comparison of the entire linear arrangement of nucleotides (nucleic acid subunits); (4) RNA homology, a comparison of those portions of DNA that contain the information for the formation of certain types of RNA; and (5) comparisons of specific proteins or their components from different organisms. DNA homology values are average measurements and demonstrate similarities between closely related organisms. RNA homology values, on the other hand, are specific for each type of RNA and are used to detect similarities between distantly related organisms. For example the DNA-rRNA hybridization technique uses the ribosome, since it is the only cellular organelle found in all living organisms. The ribosome is composed of ribosomal RNA (rRNA) and protein. Since ribosomal RNA molecules are direct transcripts from portions of the cell's DNA, some degree of similarity exists between these two macromolecules. The rRNA from one organism is labeled (combined) with radioactive material and mixed with the DNA of another organism to determine the

Figure 2–14 The construction of a dendogram. Hypothetical similarity values are used to show relationships between clusters of organisms. Group 1 contains organisms that are about 95% similar, it is joined with a fourth organism at the level of 80% similarity. Group 2 consists of three organisms that are 90% similar. Group 3 represents two organisms that are 98% interrelated to a third organism, to which they are about 85% similar. Similarity between Groups 1 and 2 is at the 60% level, and Group 3 is about 58% similar to Groups 1 and 2.

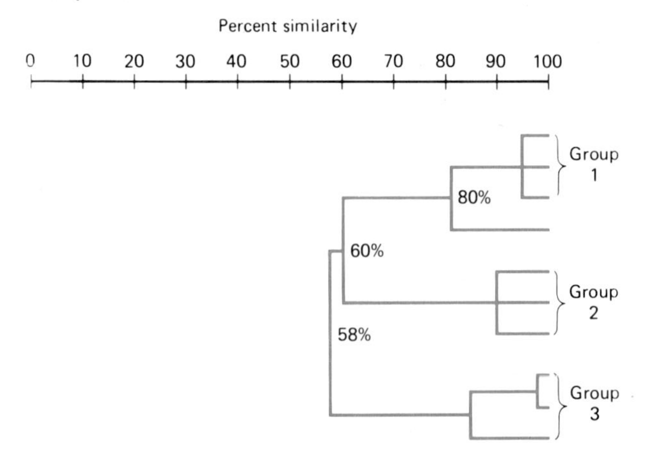

degree of similarity between the two forms of life. Details of this and related approaches, which are presented in later chapters, are also being used to study evolutionary (ancestral) relationships.

This introduction to classification and the thumbnail descriptions of microorganisms and their relative positions in the biological world provide a foundation for further study. In the next chapter we shall survey several methods and a broad range of instruments used to view microorganisms.

Summary

The Importance of Biological Classification

1. Classification of biological forms has several purposes: establishing the guidelines and properties necessary for identification; arranging organisms with similar properties into groups; determining evolutionary relationships; and reducing confusion.
2. *Taxonomy* is the science of classification. It involves the naming of different forms of life according to an international code of principles, rules, and recommendations and the identification of unknown organisms.

Natural versus Artificial Classification

1. A *natural classification system* indicates relationships between organisms on the basis of their probable origin.
2. An *artificial classification system* is based on easily recognizable properties of known organisms. This approach is useful for the identification of unknown organisms.

Naming Organisms: The Binomial System of Nomenclature

1. All forms of life have two-word names, a system introduced by the Swedish naturalist Carolus Linnaeus.
2. The first name, the *genus* designation, always begins with a capital letter. The *species* name, which follows, is not capitalized. Both names are printed either in boldface or italic (italic in this text).
3. The scientific names of a microorganism frequently refer to a distinctive property or to its discoverer.
4. A species of higher animals or plants often is defined as a group of interbreeding organisms having a limited geographic distribution.
5. For microorganisms such as bacteria, a species may be defined in terms of a population of cells. All descendants of a single parent cell are all genetically identical.
6. When a new bacterial species is identified and named, a representative strain is designated as the type strain and preserved in national culture collections.
7. Neotype strains are used as substitutes for lost type strains.

Groupings from the Linnaean System

1. All forms of life are classified into *kingdoms*.
2. Kingdoms are subdivided into smaller and more specific groupings or taxonomic ranks in the following sequence: *phyla* (singular, *phylum*) for animals, and *divisions* for plants; *classes, orders, families, genera* (singular, *genus*), and *species* (singular, *species*). A species may be divided into two or more subspecies.

The Position of Microorganisms in the Living World

1. Until about 1830, all known forms of life, even microorganisms, were considered either animals or plants.
2. Newly discovered microorganisms with properties of both animals and plants presented taxonomic problems.

EARLY BACTERIAL CLASSIFICATION SCHEMES

1. Early systems were based on *morphological* (shape) properties.
2. Staining reactions, spore formation, physiological reactions, and pathogenicity were incorporated into later classification schemes.
3. Since 1923, editions of *Bergey's Manual of Determinative Bacteriology* have served as significant sources of information on bacterial classification. *Bergey's Manual of Systematic Bacteriology* represents the newest form of this reference.

The Third Kingdom: The Protista

1. In 1866 Ernst Haekel proposed the establishment of the kingdom Protista for all single-celled microbes (such as algae, bacteria, fungi, and protozoa) and multicellular forms that were not organized into tissues and organs.
2. In 1957, Stanier and his associates suggested that two subgroups be established: the *lower protists* (procaryotes), or those with primitive nuclei, and the *higher protists* (eucaryotes), or those with well-defined nuclei.

EUCARYOTIC AND PROCARYOTIC CELLULAR ORGANIZATION

1. Eucaryotic cells contain intracellular organized structures bounded internally by sterol-containing membranes including nuclei, large ribosomes, an endoplasmic reticulum, mitochondria, a mitotic spindle during nuclear division, linear chromosomes, the basic group of chromosome proteins called *histones,* and nucleosomes. Certain plantlike eucaryotic cells have an outer covering, a cell wall.
2. Eucaryotic cellular organization is found in microorganisms such as fungi, protozoa, and algae, as well as in typical animals and plants.
3. Procaryotic cells lack most of the cytoplasmic structures of eucaryotic cells. However, procaryotic cells have unique properties that include single, circular chromosomes and bacterial chlorophylls associated with membrane layers.
4. Procaryotic organization is typical of bacteria.

The Five-Kingdom Approach

THE COMPONENTS OF THE FIVE-KINGDOM SYSTEM

1. The living world is divided into the kingdoms *Plantae, Animalia, Protista, Fungi,* and *Procaryotae (Monera).*
2. The system is based on the differences among eucaryotic and procaryotic forms, cellular organization, and modes of nutrition—absorptive, ingestive, photosynthetic, or combinations of these.

A Third Line of Evolution

Genetic studies of methane-producing bacteria suggest that there may be a third line of evolution. Currently existing *methanogens* may differ little from their ancestors that evolved in the anaerobic primeval atmosphere.

TRUE BACTERIA AND THE ARCHAEOBACTERIA

1. Results of comparative ribosomal RNA analyses suggest that there are two different kinds of procaryotes, the true bacteria and the archaeobacteria.
2. The archaeobacteria are believed to have descended from ancient organisms, and include the methanogens, extreme halophiles, and thermoacidophiles.
3. Several properties separate the archaeobacteria from true bacteria. These include differences in cell wall, cell membrane, and ribosomal chemical composition and organization, the presence of unique proteins, and antibiotic sensitivities of certain metabolic reactions.

PROCARYOTE-EUCARYOTE EVOLUTIONARY RELATIONSHIPS

1. The discovery of archaeobacteria suggests some changes in current views concerning the evolutionary relation between procaryotes and eucaryotes and the consideration of a common ancestor for all forms of life.
2. The urcaryote is proposed to represent a eucaryote ancestral cell line.

A Consideration of Viruses

1. *Viruses* are not cells. They do not have typical parts of procaryotic or eucaryotic cells.
2. Individual virus particles contain either DNA or RNA and require living cells to produce new viruses.

Trends in Microbial Classification

NUMERICAL TAXONOMY

1. This method utilizes a large number of equally weighted properties for microbial classification.
2. Taxonomic groups generally are defined by the overall similarity of observed properties rather than by ancestral relationships.
3. The degree of similarity between organisms is established by the ratio of the number of characteristics they have in common to the total number of characteristics compared. This result, expressed as a percentage, is the *similarity coefficient* (S_J).
4. Branched diagrams, or dendrograms, based on similarity values are used to show relationships between clusters of organisms.

MOLECULAR APPROACHES TO TAXONOMY

1. DNA relatedness determines the relationships among organisms by comparing the genetic content or genomes of their products.
2. Modern approaches to tracing evolutionary relationships include: (a) mating capabilities; (b) comparisons of the overall composition of DNA; (c) DNA homology (evolutionary similarities); (d) RNA homology; and (e) comparisons of specific proteins or their parts from different organisms.
3. DNA homology values detect similarities between closely related organisms, whereas RNA homology values show similarities between distantly related organisms.

Questions for Review

1. What is the significance of classification to the biological sciences?
2. Distinguish between classification and taxonomy.
3. What contribution(s) of Carolus Linnaeus are still used today?
4. Distinguish between natural and artificial classification systems.
5. What are taxonomic ranks?
6. Explain the binomial system of nomenclature.
7. Define "species."
8. What are the advantages of the currently accepted five-kingdom system of classification?
9. Name at least five characteristics of each kingdom in the biological world.
10. What are the major differences between fungi and plants?
11. What important characteristics do the Procaryotae and Protista have in common?

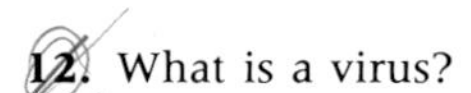

12. What is a virus?
13. Why is the classification of viruses in the biological world a problem?
14. Define, describe, or explain:
    a. paleomicrobiology
    b. type strain and neotype strain
    c. urcaryote
    d. archaeobacteria
    e. methanogen
15. What kinds of advances in microbiology have been critical to improvements in classification?
16. Of what value are computers to microbial classification? List at least three applications.
17. How does the fossil evidence for ancient microorganisms support the chemical theory of microbial evolution?
18. a. Identify the type of cellular organization shown in Figure 2-15.
    b. Identify the labeled structures and give the general function of each one.

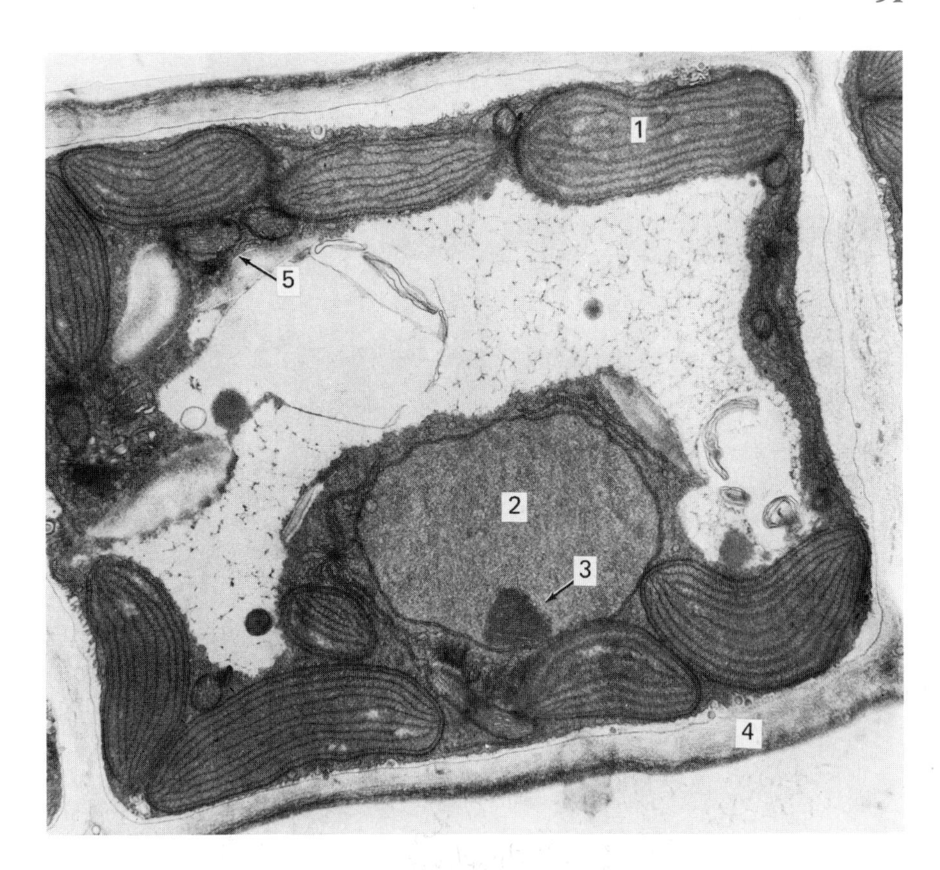

Figure 2–15 [*From D. N. Young,* J. Phycol. ***15****:42–48, 1979.*]

Suggested Readings

ALBERTS, B., D. BRAY, J. LEWIS, M. RAFF, K. ROBERTS, and J. D. WATSON, *Molecular Biology of the Cell.* New York: Garland Publishing, Inc., 1983. *A complete and functional examination of cells in light of results from current studies.*

KEETON, W. T., and J. L. GOULD, *Biological Science,* 4th ed. New York: W. W. Norton & Company, 1986. *An up-to-date, functional general biology reference.*

KRIEG, N. R. (ed.), *Bergey's Manual of Systematic Bacteriology,* vol. 1. Baltimore: Williams & Wilkins, 1984. *The first of four volumes covering the descriptions of bacterial species and their distinctive properties. The classification scheme presented is the latest of its kind and adopts* Procaryotae *as the kingdom for all bacteria.*

LEE, C. P., et al., *Mitochondria and Microsomes: In Honor of Lars Ernster.* Reading, Mass.: Addison-Wesley Publishing Co., 1981. *An interesting up-to-date, concise summation of important advances in mitochondrial and microsomal research.*

MARGULIS, L., *Symbiosis in Cell Evolution.* San Francisco: W. H. Freeman and Company, 1981. *A well-written book including molecular genetics and the taxonomy of all major groups of lower organisms.*

MARGULIS, L., and K. V. SCHWARTZ, *Five Kingdoms, An Illustrated Guide to the Phyla of Life on Earth.* San Francisco: W. H. Freeman and Company, 1981. *A unique catalog of earth's life forms, complete with descriptions and illustrations of all major groups of organisms.*

SCHOPF, J. W., "The Evolution of the Earliest Cell," *Scientific American* **239:**111–138, 1978. *An interesting account of the earliest cells and the events that led to the appearance of biochemical systems and the oxygen-enriched atmosphere on which life depends today.*

WOESE, C. R., "Archaeobacteria," *Scientific American* **244:** 98–122, 1981. *A well-organized description and discussion of a group of bacteria that do not seem to belong to either the procaryotes or the eucaryotes.*

3

Techniques Used in the Observation of Microorganisms

Accordingly, I took (with the help of a magnifying mirror) the stuff off and from between my teeth further back in my mouth where the heat of the coffee couldn't get at it. This stuff I mixed with a little spit out of my mouth (in which there were no air bubbles) . . . then I saw with as great a wonderment as ever before, an inconceivable great number of little animalcules . . . the whole stuff seemed to be alive and moving. . . .
— A. van Leeuwenhoek, 1692

After reading this chapter, you should be able to:

1. Apply the metric system of measurement to the structures of typical animal and plant cells and microorganisms.
2. Describe the general properties of light as they apply to microscopy.
3. Name and give the functions of the components of a bright-field microscope.
4. List the uses and relative advantages of microscope objectives in the examination of cells.
5. Describe the different types of microscopes, explaining associated techniques used in the study of microorganisms and other biological materials.
6. Outline the major differential staining techniques used in the identification of microorganisms and their parts.
7. List the advantages and limitations of staining procedures.
8. Describe how specimens and preparation techniques used with electron microscopes differ from those used with other microscopes.
9. Summarize the general advantages, limitations, and areas of application for different types of microscopy.
10. Define or explain the following terms: micrometer, resolving power, dyes, gram-positive, acid-fast, freeze-fracture, sectioning, high-voltage electron microscopy, and image enhancement.

The microscope is the key investigative tool of microbiology. The microscopic observation of organisms involves staining procedures for identification and the use of the metric system to record the dimensions of organisms. These and the advantages and limitations of major types of microscopes are the subject of this chapter.

Microscopes and Microscopy

Microorganisms cannot be seen with the naked eye. Although several microbial forms—algae, bacteria, protozoa, and yeasts—had been observed by Anton van Leeuwenhoek as early as 1674, not until the development of modern compound microscopes did biologists the world over become aware of the tremendous number and variety of microorganisms.

The microscopes used today have evolved significantly from the first microscopes (Figure 3–1). Depending upon the magnification principle involved, microscopes are known as either *light microscopes* or *electron microscopes.* Light, or optical, microscopes may be bright-field, dark-field, fluorescence, or phase-contrast instruments. Electron microscopes are of two kinds—*transmission* and *scanning electron microscopes.* Modern instruments are equipped with special devices for the analysis and more precise viewing of specimens. Before considering the various forms of microscopy, we shall review the units of measurement used in describing the morphological characteristics of microorganisms and some fundamentals of optics.

Units of Measurement

For most of our everyday activities, we find it convenient to use familiar units of measurement, such as the inch, foot, yard, ounce, or pound. In addition, we usually use numbers that are neither very large nor very small to specify the quantities of these units. However, the exacting nature of science requires that experimenters be able to repeat and verify results using the most precise measurements. Thus the quantitative nature of science has given rise to a variety of units of measurement and to procedures for writing both very small and very large numbers.

In the biological sciences, the two major systems of units used to express measurements are the *English system* (pounds, feet, and inches) and the **metric system.** The latter is far more widely used internationally and for scientific work. Since about 1960, an International System of Units based upon the metric system has quickly been gaining acceptance. It is referred to as the **SI system,** from the French *Système International d'Unités.*

The metric system has the great advantage of being simple, since using it requires remembering just a few basic terms. The basic unit of length is the *meter (m);* the basic unit of volume is the *liter (l);* the basic unit of mass (weight) is the *gram (g);* and the basic unit of temperature is the *Celsius degree (°C).* Converting to English units,

1 meter = 39.4 inches

1 liter = 1.06 quarts

1 gram = 0.0352 ounce

An additional feature of the metric system is that it involves decimal counting. Units are defined so that every unit is some power of 10 (10, 100, 0.01, and so on) times the basic unit. A meter, for example, consists of 100 centimeters (cm) or 1,000 millimeters (mm); 1 centimeter can then be seen to consist of 10 millimeters. Converting from one unit to another in the metric system merely involves shifting the decimal point.

Instead of having a number of different root words for length, as in the English system (league, mile, rod, yard, foot, inch, mil), the metric system has just one, the meter. With suitable prefixes, the meter can express a useful unit for even very small and very large distances (see Table 3–1). The same

Figure 3–1 (a) A modern-day laboratory compound microscope. Note the following components: ocular, mechanical stage, condenser, iris diaphragm, and source of illumination.

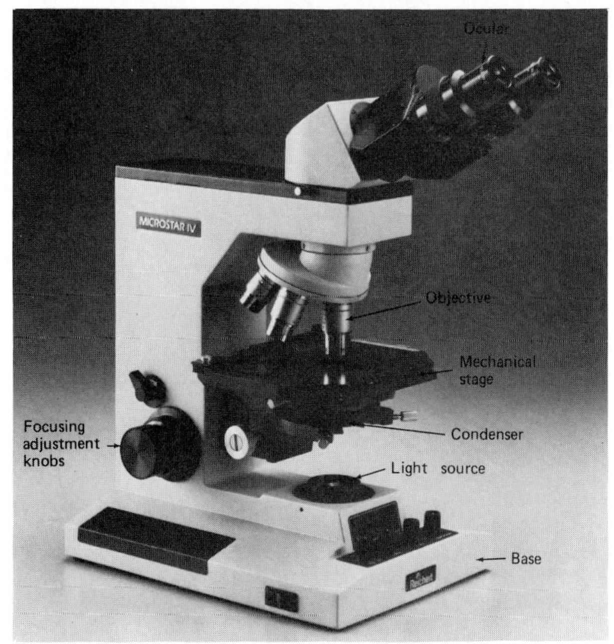

(a)

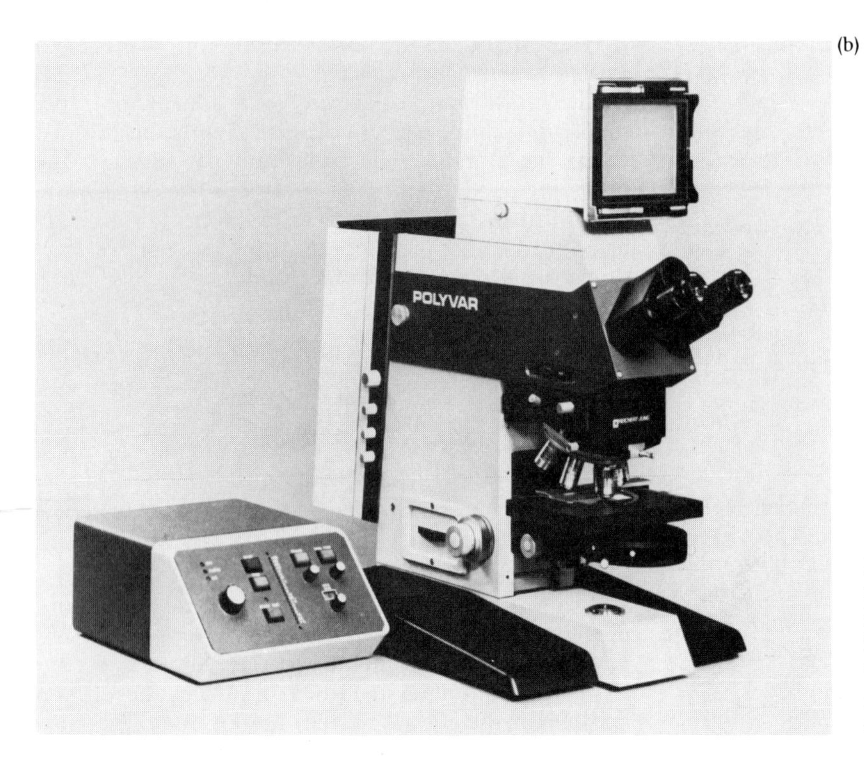

(b)

Figure 3–1 (b) An example of a research microscope, the Polyvar. The instrument includes a special illumination control system together with a photography component. *[Courtesy of Reichert-Jung, Inc.—A Cambridge Instrument Company.]*

prefixes are applied to the measurement units of mass and volume as well.

In the study of biological materials, the most commonly used measurements of length are the **centimeter (cm** = 10^{-2} m), the **millimeter (mm** = 10^{-3} m), the **micrometer** (**μm** = 10^{-6} m), the **nanometer (nm** = 10^{-9} m), and the **Angstrom unit** (**Å** = 10^{-10} m). Most animal, plant, and microbial cells are measured in micrometers; cellular parts and viruses in nanometers. Table 3–2 gives some common metric units and their equivalents.

To estimate distances from a map, one usually looks for a scale or bar marker as an aid. The same approach is used in the biological sciences to indi-

TABLE 3–1 International System of Units (SI) and Some Common Prefixes

Quantity	*SI Base Unit*	*Symbol*
length	meter	m
mass	kilogram	kg
time	second	s
temperature	kelvin	K
PREFIX	MEANS MULTIPLY BY	SYMBOL
giga	10^9 = 1,000,000,000	G
mega	10^6 = 1,000,000	M
kilo	10^3 = 1,000	k
hecto	10^2 = 100	h
deca	10^1 = 10	da
deci	10^{-1} = 0.1	d
centi	10^{-2} = 0.01	c
milli	10^{-3} = 0.001	m
micro	10^{-6} = 0.000,001	μ
nano	10^{-9} = 0.000,000,001	n
pico	10^{-12} = 0.000,000,000,001	p

FOR EXAMPLE:
1 millimeter = 10^{-3} meter, or 0.001 meter
1 nanometer = 10^{-9} meter, or 0.000,000,001 meter
1 kilogram = 10^3 grams, or 1,000 grams
In symbols, 1 mm = 10^{-3} m, 1 nm = 10^{-9} m, 1 kg = 10^3 g

Microbiology Highlights

DISCOVERING THE LITTLE BEASTIES

Anton van Leeuwenhoek (1632–1723), the father of bacteriology and protozoology. *[Courtesy of Fisher Scientific Company, Chicago.]*

Between 1653 and 1673, Anton van Leeuwenhoek developed the curious hobby of constructing simple (one-lens) microscopes. Although he was not the first to build a microscope, his were the finest of that time. Leeuwenhoek actually improved his microscopes and developed his techniques of observation for 20 years before reporting any of his findings. Even though his greatest recognition comes from the observation of microorganisms, Leeuwenhoek's discoveries went beyond the world of microbes. He made many other contributions of biological significance. For example, he provided confirming evidence for William Harvey's theory of blood circulation by constructing an aquatic microscope *(aalkijer)*. This instrument enabled scientists to observe the flow of red blood cells through the capillaries of a fish's tail fin. Leeuwenhoek also observed muscle fiber striations (1682), nuclei of fish blood cells (1682), and the coverings (myelin sheaths) of nerve fibers (1717).

Leeuwenhoek continued making better microscope lenses and observing microorganisms or, as he referred to them, the "little beasties," until he was 91 years old. It is interesting to note that Leeuwenhoek had little formal education, did not study at any university, and knew no language but his native tongue. Yet because of his almost childlike curiosity and great skills as an objective observer, he discovered some of the greatest secrets of nature.

cate the dimensions of cellular structures and associated components in a drawing or photograph. The *bar marker,* a short, straight line usually found at the bottom corner of a photograph, is placed there as a size reference. The length of the marker represents the length of a metric unit at that particular magnification. Typical units include 1 μm, 1 nm, and fractions or multiples of these units.

Each type of microscope has definite limitations. Each instrument can be used to observe cells or other objects that are at least a certain size and no smaller. For example, an ordinary light *(bright-field)*

TABLE 3–2 Some Metric Units and Useful Equivalents

Unit	*Symbol*	*Equivalents*[a]
meter	m	1 m = 39.37 in.
centimeter	cm	1 cm = 10^{-2} m = 0.39 in.
micrometer	μm	1 μm = 10^{-6} m = 0.39×10^{-4} in.
nanometer	nm	1 nm = 10^{-9} m = 0.39×10^{-7} in.
Angstrom	Å	1 Å = 10^{-10} m = 0.39×10^{-8} in.
liter	ℓ or L	1 ℓ = 1 dm^3 = 0.001 m^3 = 1.06 qt
milliliter	ml or mL	1 mL = 0.001 ℓ = 0.001 qt
degree Celsius	°C	1°C = 1 K = 1.8°F
		0°C = 273.15 K = 32°F

[a] English equivalents are approximate.

microscope can be used to observe objects with dimensions down to about 0.2 μm. An electron microscope can be used with specimens less than one-tenth that size—15 nm (or 0.015 μm). The limitations of these and other instruments are compared in Figure 3–2. The advantages, disadvantages, and special features of each form of microscopy are given in the sections that follow.

To understand microscopy, one must be aware of certain fundamental properties of light and optics. With this foundation, many of the problems encountered in using a microscope will be more readily understood.

Properties of Light

Light travels with a velocity of approximately 186,300 miles, or 3×10^{10} centimeters, per second. According to the wave theory, light is propagated through space as waves of varying electric and magnetic fields.

Figure 3–2 A diagrammatic representation of the types of specimens that can be viewed by different microscopes commonly used in microbiology and related areas. The operational ranges of these instruments are also indicated.

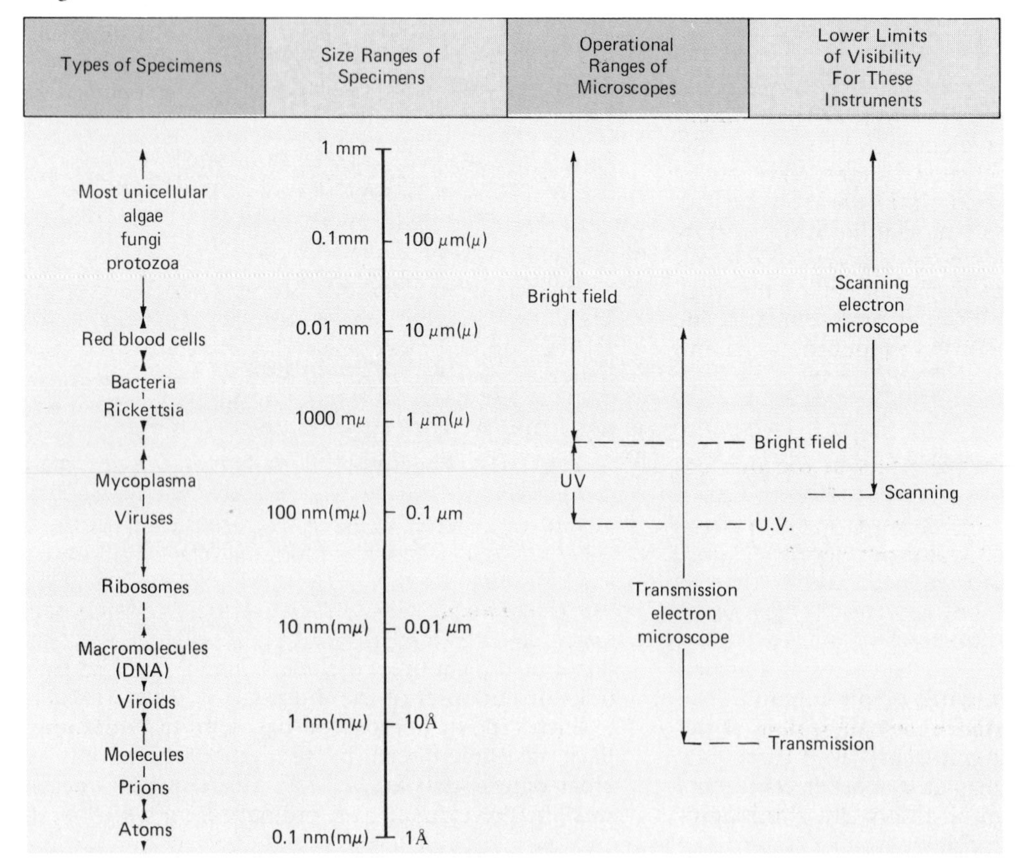

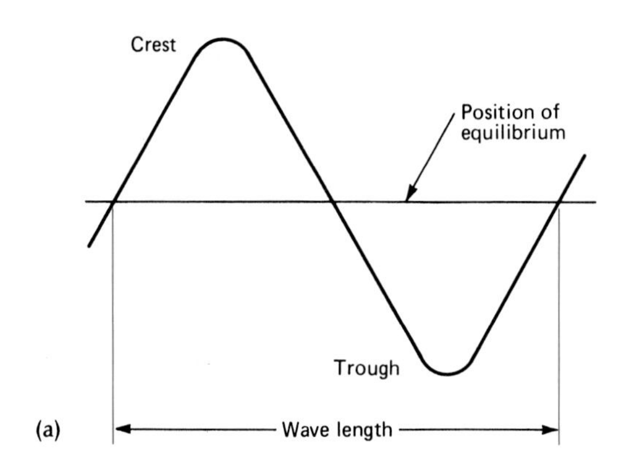

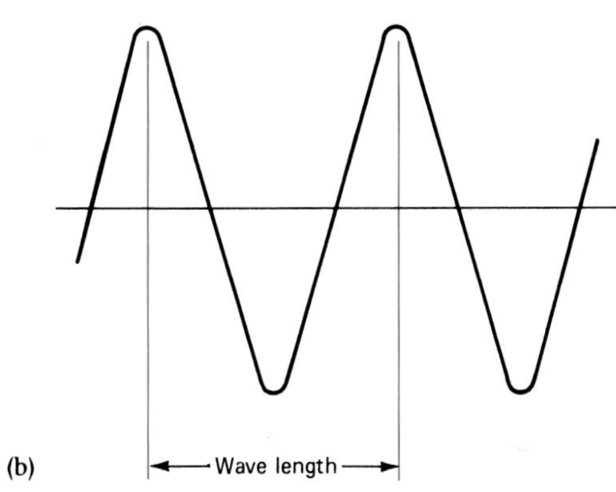

Figure 3–3 Properties of light waves. (a) The anatomy of a wavelength. Note the position of equilibrium. (b) A representation of frequency.

AMPLITUDE

Light waves are described in terms of their amplitude, frequency, and wavelength (Figure 3–3). An analogy can be made between a light wave and a jump rope. When two people pull on the rope and stretch it tight, a position of equilibrium is established. Now if the rope is swung, a wavelike effect is produced. The maximum displacement of the rope from the equilibrium position, as represented by the crests and troughs of the curves produced, is the amplitude of the wave.

FREQUENCY

This property of a light wave refers to the number of vibrations that occur in one second. Specifically, *frequency* is the number of times a wave crest or trough passes a particular point per second (Figure 3–3). If two people take a rope and shake it, the frequency of the wave can be regulated by the speed with which the rope is shaken.

WAVELENGTH AND FREQUENCY

A *wavelength* is the distance between two corresponding points on a wave, such as the distance between two successive peaks or crests (Figure 3–3a). Because the frequency of light is the number of wave crests or troughs that move past a point in one second, and because the speed of light in a medium is constant, frequency is inversely related to wavelength:

$$\text{frequency} = \frac{\text{velocity}}{\text{wavelength}}$$

The wavelengths of light rays that make up the visible spectrum range from approximately 4,000 to 7,000 Å (400 to 700 nm) (Figure 3–4). The colors we see result from a combination of factors, including amplitude and wavelength. Wavelengths either less than or greater than the limits of the visible spectrum also exist. *Ultraviolet* light rays, for example, have wavelengths ranging from approximately 1,000 to 3,850 Å (100 to 385 nm). Infrared light has wavelengths greater than those of visible light.

The **resolving power (RP)** of a microscope—its ability to reveal the fine details of a specimen—depends on wavelength. As a general rule, the shorter the wavelength of illuminating light, the greater the RP. Thus, with ultraviolet light as illumination, finer details can be seen than with visible light.

REFRACTIVE INDEX AND REFRACTION

The nature of the medium through which light passes during the operation of a microscope affects the image seen. The rate at which light moves is not the same for all transparent media. Denser materials exert a slowing effect on light rays. This difference in velocity is expressed in the form of a *refractive index*, or index of refraction. The refractive index, represented by the Greek letter η (eta), is defined by the formula

$$\eta = \frac{\text{speed of light in a vacuum}}{\text{speed of light in the medium being tested}}$$

In making determinations of light velocity, it is important to keep the temperature constant during the testing period. The indexes of some commonly used materials are 1.00003 for air, 1.33 for water, an average of 1.6 for various glasses, and 1.55 for immersion oil.

Light rays traveling in a single medium generally move along a straight path. However, if these rays pass at an oblique angle from one material into another material having a different refractive index, the light wave changes direction. This phenomenon, called *refraction*, takes place at the boundary between the two media. The importance of refraction and the refractive indexes of materials is clearly demonstrated in the use of the oil-immersion objective (described in the following section) and other forms of microscopy.

Successful operation of a microscope depends upon understanding the principles involved in its operation, the functions of the instrument's compo-

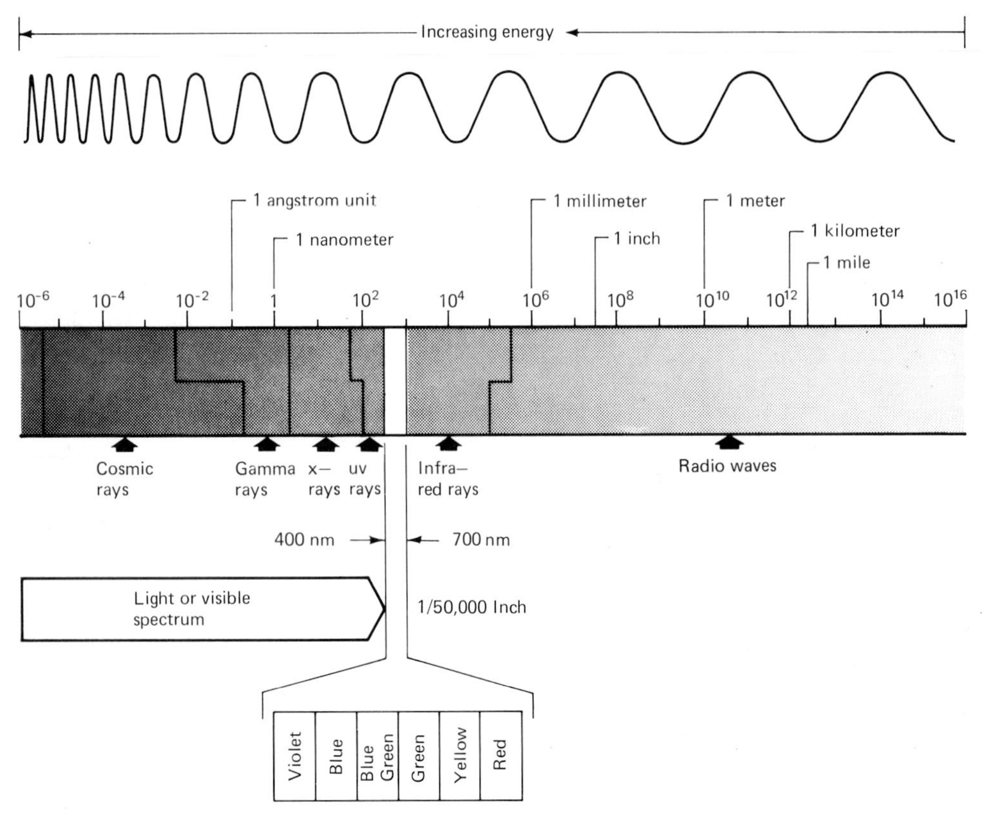

Figure 3–4 Relative wavelengths of representative forms of radiation. Which of these are important in studying microorganisms? Would any type of radiation be harmful to these and other forms of life? See Chapter 11 for discussions of the use of ultraviolet (UV) rays.

nents, and the procedures for proper maintenance and use. We shall now review the most important types of optical microscopes and their uses in microbiology.

Light Microscopes

Bright-Field Microscopy

Leeuwenhoek's early microscopes were simple in the sense that they used a single lens—rather like a magnifying glass. There are fundamental limitations to the magnifying power of a simple microscope, no matter how carefully it is constructed. For this reason, the development of the compound microscope was a welcome advance in microbiology. Credit for the invention and early development of the compound microscope is still debated, but most experts acknowledge the basic contributions of Hans and Zacharias Janssen. Around 1590, the Janssens introduced a second lens to magnify the image formed by the primary lens. This is the principle upon which today's compound microscopes are based.

A *compound microscope* consists of at least two lens systems: the *objective,* which magnifies the specimen and is close to it, and the *ocular,* or *eyepiece,* which magnifies the image produced by the objective lens. The total magnification thus obtained is equal to the product of the magnifying powers of the two sets of lenses. Under optimal conditions, maximum magnifications can range from approximately 1,000 to 2,000 times (×) the diameter of a specimen being observed.

The compound microscopes used today consist of a series of optical lenses (systems), mechanical adjustment parts, and supportive structures for these components. The optical lenses include the ocular, the objectives (usually three with different magnifying powers), and the substage condenser. The coarse, fine, and condenser adjustment knobs, together with the iris diaphragm lever, comprise the major mechanical parts. The various components of the scope are held in position by supportive structures such as the base, arm, pillar, body tube (barrel), and revolving nosepiece. Several of these microscope components are shown in Figure 3–5 and discussed in the following sections.

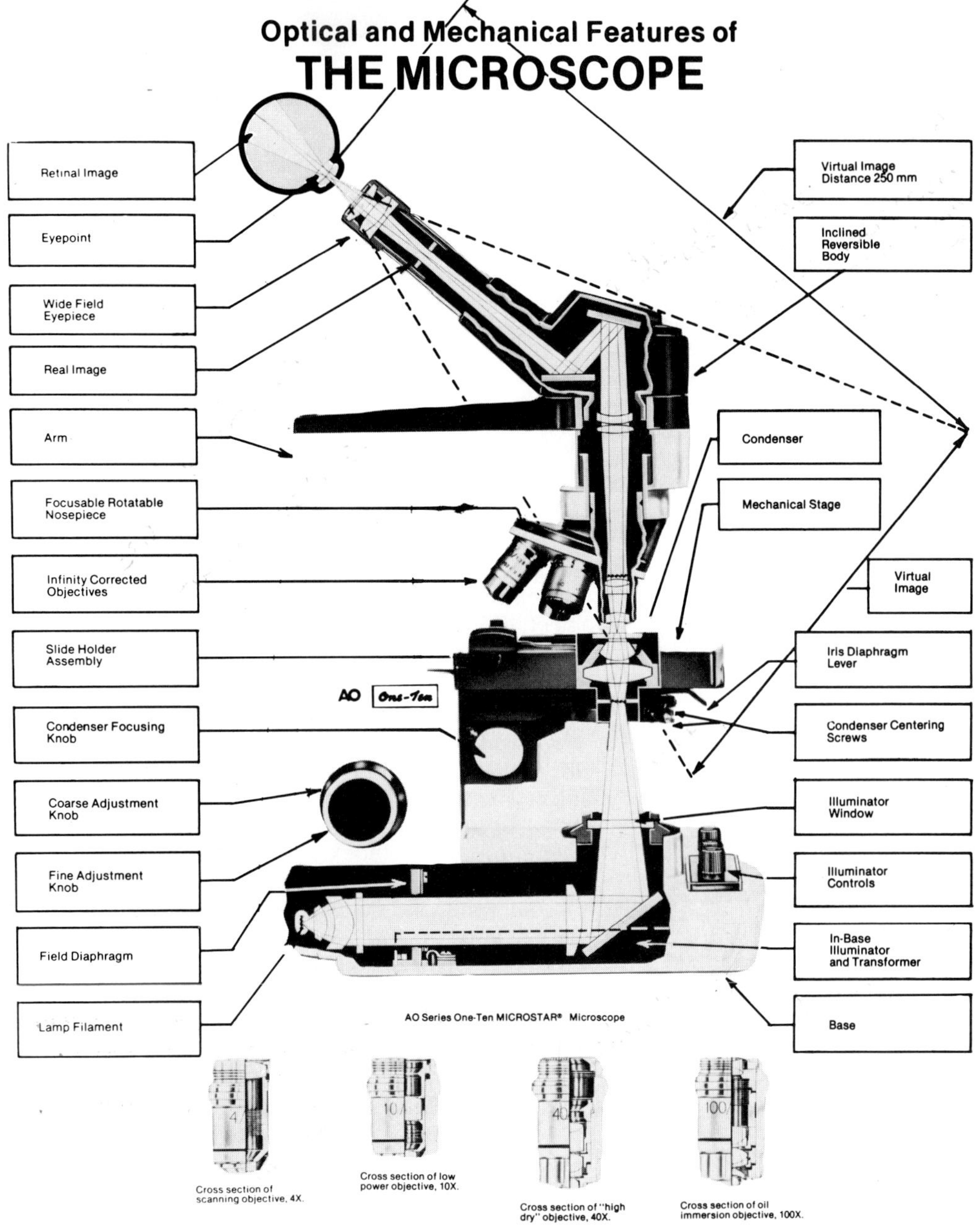

Figure 3–5 A cutaway diagram of a representative compound monocular microscope. The various parts of the instrument and the pathways followed by light waves through the microscope are shown. *[Courtesy, Reichert Instrument Division, Warner Lambert Technologies, Inc.]*

OCULAR (EYEPIECE)

A short tube generally containing two lenses, the ocular fits into the upper portion of the microscope's body tube. Several different types of eyepieces can be used to examine specimens. The specific type used generally depends upon the objective lenses on the instrument. The magnifying power of the ocular is usually engraved on it. Common magnifications for oculars include 1×, 2×, 5×, 10×, and 15×. Eyepieces are used to magnify the image of the specimen produced by the objective and to correct certain distortions produced by the objective.

OBJECTIVES

Objectives are considered the most important of the microscope's optical parts, primarily because they affect the quality of the image seen by the observer. Most microscopes are equipped with three objectives with different magnifying powers: the low-power, high-power (or high-dry), and oil-immersion lenses.

Objective lenses are used to gather, or concentrate, the light rays coming from the specimen being viewed, form the image of the specimen, and magnify this image. Several important properties of a microscope are directly associated with the objectives. One of these is *resolving power,* or *resolution,* which is specifically defined as the ability to distinguish clearly two points that are close together within the specimen. This feature is largely determined by the wavelength of the light source (with a shorter wavelength providing finer detail) and the angle formed by the outer edges of the lens and a point on the object. The resolution is also affected by the refractive index of the medium through which light passes before entering the microscope objective. The relationship of these factors is expressed in the combined formula

$$\text{Numerical aperture (NA)} = \eta \sin \theta$$

where η represents the refractive index of the medium through which light passes before entering the objective lens and θ (Greek letter theta) is the angle formed by light rays (in the shape of the cone) coming from the condenser and passing through the specimen. Light in this form is frequently referred to as "a pencil of rays." Sin θ is the trigonometric sine of this angle. Figure 3–6 depicts NA. Values for NA are engraved on the barrels of objectives and indicate the maximum obtainable resolution.

Another important feature of modern-day instruments is the property of *parfocal.* Stated simply, this means the changing of objectives without making major focusing adjustments. Thus, if a higher magnification is needed during the course of examining a specimen, one would just rotate the desired higher objective into place and make some minor focusing adjustment to bring the specimen into view.

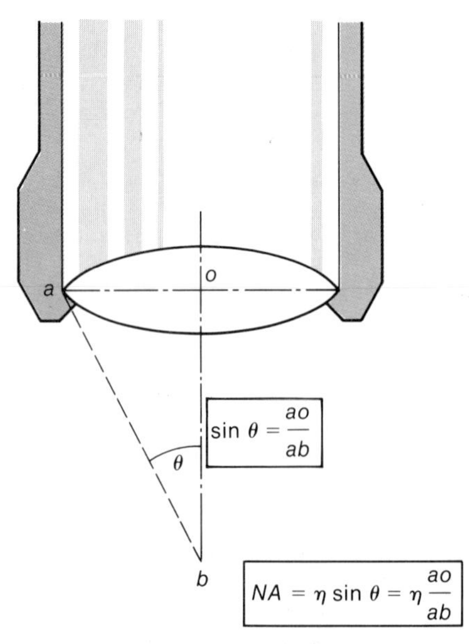

Figure 3–6 A diagrammatic representation of numerical aperture, (NA).

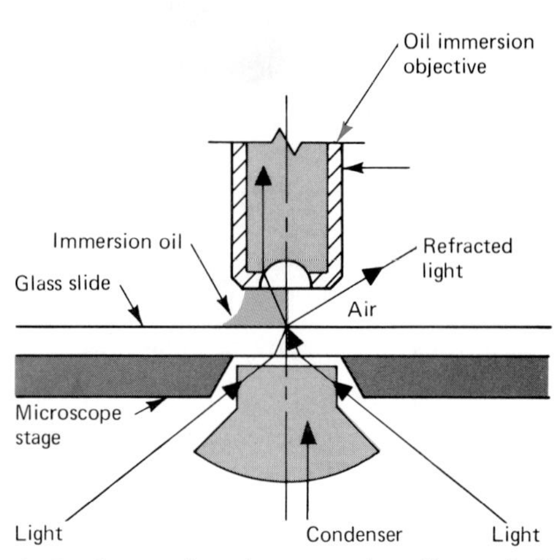

Figure 3–7 Comparison between the effects of oil and air on the passage of light rays from a specimen to the front lens of the oil-immersion objective.

The oil-immersion lens is an especially important tool for studying microorganisms. The highest-magnification objective used in general microbiology courses, it can magnify specimens about 100 times. To obtain the best possible results, the objective is immersed in a medium that has approximately the same index of refraction as glass, 1.6. One medium commonly employed for this purpose is cedarwood oil. Other materials containing mineral oil also are in common use. Oils have the advantage of not evaporating when exposed to air for long periods of time. Further, since its index of refraction is the same as that of glass, oil does not bend the light rays entering the front lens of the oil-immersion objective (Figure 3–7). If air rather than oil is present between the specimen and the objective, some light is lost. The image observed is usually fuzzy, and the finer details cannot be seen (Figure 3–8). The RP of the oil-immersion objective is definitely enhanced by the oil medium. The oil acts as an additional lens in the system and prevents the loss of necessary light rays.

THE CONDENSER AND IRIS DIAPHRAGM

A condenser is found under the microscope stage between the source of illumination and the specimen or object to be viewed. This component frequently is called a *substage condenser.* One of the most commonly used is the Abbé condenser (Figure 3–11). It consists of two lenses that illuminate specimens with transmitted light. The condenser is important to high-resolution microscopy. Microscopic examinations of specimens with either high-power or oil-immersion objectives require adequate illumination. When an objective is changed to increase magnification, the amount of light passing through the lens system decreases, since the size of the lens

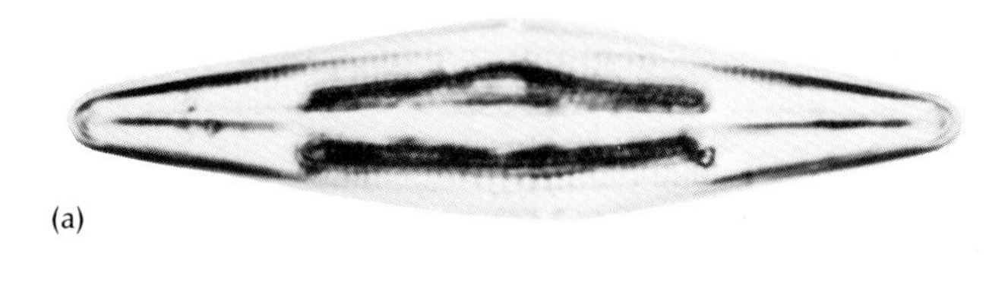

(a)

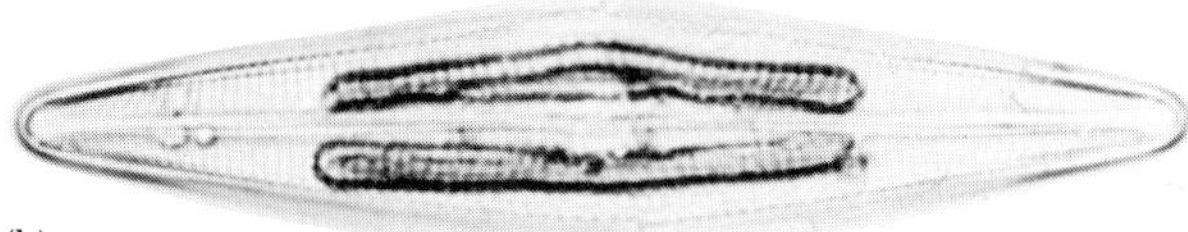

(b)

Figure 3–8 A comparison of an observed image of a diatom (alga) when viewed with the oil-immersion objective. (a) No oil was used here. Air was present between the objective and the specimen. (b) The effect with immersion oil. Note the clearer image and the greater detail.

opening is smaller. The condenser serves to concentrate light rays before they reach the specimen. Abbé, variable-focus and achromatic condensers are also commonly used.

Occasionally, too much light may pass through the specimen and into the objective lens, significantly decreasing the contrast of the specimen and causing a loss of resolution. Microscope condensers are generally equipped with an *iris diaphragm* to control light intensity. This component functions to control the amount of light entering the condenser. When unstained material, such as living protozoa or hanging-drop preparations of bacteria, is to be examined, the opening of the iris diaphragm generally is reduced. This component is regulated by the iris diaphragm lever. Many newer microscopes are equipped with fixed condensers and iris diaphragms regulated for general use.

Micrometry

In certain aspects of microbiology, it is necessary to measure the dimensions of cells or, if possible, of their components. With light microscopy, this type of procedure can usually be performed with the aid of an *ocular micrometer,* a special ocular containing a graduated scale, and a *stage micrometer.* The latter device is a glass slide on which a millimeter scale is usually imprinted. Graduation of this scale is in hundredths of a millimeter.

The measurement procedure requires that the ocular micrometer first be calibrated with the stage device. This is done by replacing the normal eyepiece of the microscope with the ocular micrometer and determining the exact number of divisions, that is, hundredths of a millimeter, that correspond to those of the millimeter scale on the stage micrometer. The particular objective used should be noted, as this component is important to the accuracy of the measurements taken. The manipulation or exchange of these parts will change calibration values. With the calibration of the ocular micrometer complete, the stage micrometer is removed and the specimen to be measured is placed in position.

Photographic variations of this procedure are also used. With photography, permanent records of the dimensions of organisms can readily be made.

Specimen Preparation for Bright-Field Microscopy

The procedures used to prepare microorganisms for microscopic examination are of two types. The *hanging-drop* and *temporary wet-mount* techniques are used with living organisms. The second type of procedure, *staining,* employs **smears,** which are thin films of microorganisms spread on the surface of a clean glass slide that has been air-dried. The smear is then heat-fixed by passing the slide through the flame of a Bunsen burner. This step not only kills but coagulates the protein of cells and thereby fixes the organisms to the slide.

HANGING-DROP AND TEMPORARY WET-MOUNT TECHNIQUES

The direct examination of microorganisms in the normal living state can be extremely useful in determining size and shape relationships, motility, and reactions to various chemicals or immune sera. The hanging-drop and temporary wet-mount techniques both maintain the natural shape of organisms and reduce the distorted effects that can occur when specimens are dried and fixed. Because the majority of microorganisms are not very different from the fluid in which they are suspended in terms of color and refractive index, a low-intensity light source is used for viewing them.

Hanging-drop preparations (Figure 3–9) are made by placing a drop of a microbial suspension on the center of a cover slip, which is usually ringed with petroleum jelly or a similar material. This sealing material is used primarily to eliminate air currents and reduce evaporation. A depression slide (hollow-ground slide with a concave central area) in an inverted position is lowered onto the prepared cover slip. Slight pressure is applied to the slide in order to ensure contact between the cover slip and the depression slide. The finished preparation then is turned right side up and is ready for examination.

Temporary wet-mounts of specimens are prepared by placing a drop of microbial suspension on the center of a clean glass slide and placing a cover slip over it.

STAINING

The microscopic study of live cells is limited, in that usually only the outline and structural arrangement of cells are revealed by bright-field microscopy. Stained preparations of fixed cells permit greater visualization of cells, observation of internal cellular

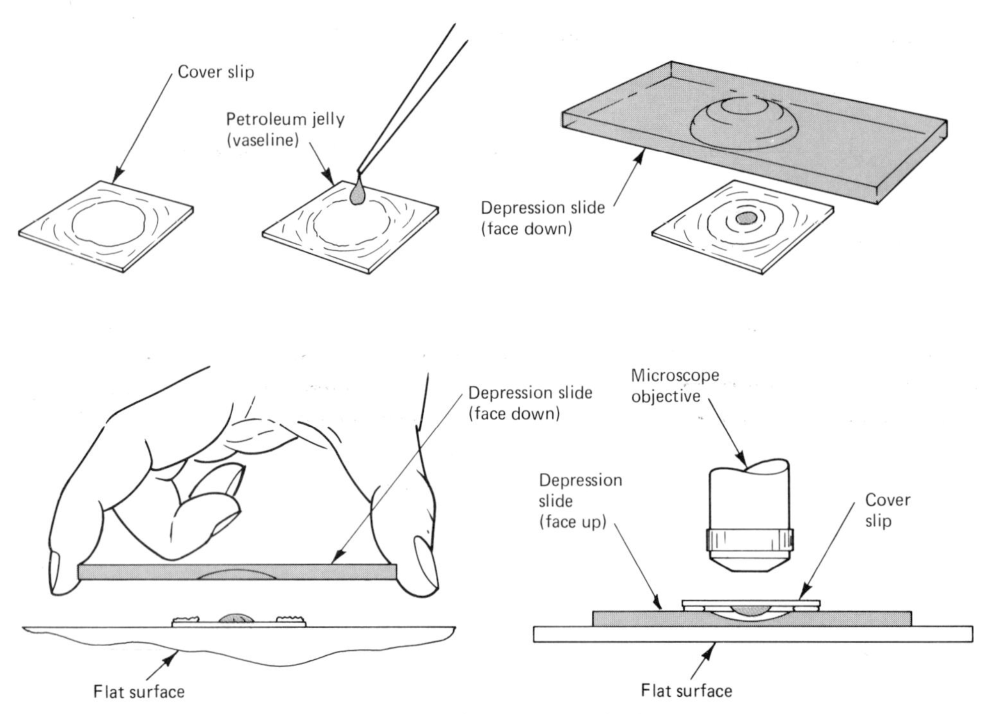

Figure 3–9 A schematic representation of the steps in the hanging-drop preparation technique. The specimen is placed in the center of a prepared cover slip by means of a dropper or inoculating loop.

components, and, to some extent, differentiation of microbial species.

THE NATURE OF DYES. Dyes used for staining are usually in the form of salts. They are of two general types, basic and acidic. The electric charge on the dye ion determines the type of dye it is. Basic dyes are positively charged. Substances of this kind stain or react with nuclear components. Acid dyes are negatively charged. Compounds of this type stain cytoplasmic material and certain kinds of granules and other related materials.

SIMPLE STAINING. Various bacteriological procedures utilize staining solutions that contain one and only one dye dissolved in either a dilute alcoholic solution or water. Such preparations are referred to as *simple stains*. The concentrations of the commonly used dyes are quite low, approximately 1% to 2%. Simple staining solutions include carbolfuchsin, crystal violet, methylene blue, and safranin (Color Photographs 11a, b, and c).

This type of staining procedure involves applying the simple stain to a fixed bacterial smear for a period of time that may range from a few seconds to several minutes. Such stains should never be allowed to dry on the smear. Before the microscopic examination, preparations are rinsed to remove excess stain and dried by blotting between layers of filter paper or other appropriate material. Bacterial cells to which these simple staining solutions are applied take the color of the dye preparation (Color Photograph 11).

Simple stains can be used to demonstrate the shapes, arrangements, and sizes of microorganisms, to differentiate or distinguish bacterial cells from nonliving structures, and to show the possible presence of bacterial spores in certain cases.

DIFFERENTIAL STAINING. The preliminary grouping of bacteria is usually based upon their general shape and the manner in which they react to certain staining techniques called **differential staining methods**. Procedures of this type use more than one dye preparation. When properly carried out, these techniques will divide nearly all bacteria into major groups. The two most widely used differential staining methods in bacteriology are the **Gram** and **acid-fast staining** techniques. The differentiation principle is also used to view bacterial structures inside or outside the cell—spores and capsules, for example.

GRAM STAIN. The Gram-staining reaction was developed about 1883 by Hans Christian Gram, a Danish physician. He discovered this technique while trying to stain biopsy (pathologic) specimens so that microorganisms could be distinguished from surrounding tissue. Gram noted that some bacterial cells exhibit an unusual resistance to decolorization. He used this observation as the basis for a differential staining technique.

The Gram differentiation is based upon the color reactions of bacteria in a fixed smear when they are treated with the primary dye crystal violet followed by a mordant such as an iodine-potassium iodide solution. Certain organisms lose the violet color rapidly when a decolorizing agent such as ethyl alcohol or a mixture of acetone and alcohol is applied. Others lose their color more slowly. After this decoloration step, a counterstain, usually the red dye safranin, is used. A standardized procedure listing the specific reagents and respective staining times is given in Table 3–3. Bacterial cells resistant to decolorization will retain the primary dye and exhibit a blue or purple color. They will not take the counterstain. Such organisms are referred to as *gram-positive* (Color Photograph 12a). Those microorganisms unable to retain the crystal violet stain, the decolorized cells, will take the counterstain, and consequently exhibit a pink or red color (Color Photograph 12b). The term *gram-negative* is used to describe these organisms. Several characteristics of the gram-positive and gram-negative groups appear to be correlated with their staining reactions. These include chemical composition, sensitivity to penicillin, and relative cell-wall thickness. When properly used, the gram stain can be useful in the diagnosis of many infectious diseases. **[Bacterial diseases are discussed in Part VIII.]**

In addition to the two major categories of bacteria, gram variables and gram-nonreactives have also been recognized. The *gram-nonreactive* category includes those microorganisms that do not stain, or

TABLE 3–3 A Representative Standardized Gram-Staining Procedure

Reagents in Their Order of Application[a]	*Length of Time Applied*	*Reactions and Appearance of Cells*[b] *Gram-positives*	*Gram-negatives*	*Appearance*
Crystal violet (CV)	1 min	1. Dye is taken up by cells in two forms, bound and unbound 2. Cells appear violet	1. Same 2. Same	Appearance G+ G−
Iodine solution (I) (a mordant)	1 min	1. Iodine reacts, i.e., fixes probably both the unbound and bound crystal violet[c] 2. A CV-I precipitate (CV-I complex) is formed 3. Cells remain violet in color	1. Same 2. Same 3. Same	G+ G−
Decolorizer (ethanol or an acetone-ethanol mixture)	Applied cautiously drop by drop until a purple color no longer comes from the smear	1. The decolorizer causes the dissociation of the CV-I precipitate, CV-I = CV + I 2. The components of the complex are now soluble 3. Dehydration of the thick cell wall occurs 4. Diffusion of dye proceeds slowly	1. Same 2. Same 3. Dehydration of the thin cell envelope occurs 4. Diffusion of dye proceeds faster	G+ G−
Counterstain (safranin or a dilute solution of carbolfuchsin)	½–1 min	1. Some displacement of CV may occur, but in general cells are not affected 2. Cells appear purple or dark blue	1. Displacement of any CV left occurs 2. Cells take up counterstain 3. Cells appear red or pink	G+ G− (purple) (red)

[a] Note that wash steps are used after the application of each reagent.

[b] Based on information provided by J. W. Bartholomew, T. Cromwell, and R. Gan, "Analysis of the Mechanism of Gram Differentiation by Use of Filter Paper Chromatographic Technique," *J. Bacteriol.* **90:**766, 1965.

[c] Iodine is called a *mordant* in this capacity.

that stain poorly. Various spirochetes fall into this group.

Gram-variables are bacteria that under ordinary conditions may appear gram-positive and gram-negative on the same slide. This variation in staining may also be due, in large part, to improper technique, such as lack of attention to decolorization, thick smears, or use of old cultures of gram-positive bacteria. The Gram-staining reactions given most commonly are those of 24-hour-old cultures. With certain organisms, gram positivity disappears in older cultures—possibly leading to this kind of gram variability.

Customarily, each reagent (the primary dye, the iodine solution, the decolorizer, and the counterstain) is removed after its period of application by rinsing with water. Excessive washing, however, can remove the dye or dye-iodine complexes within the cells and consequently greatly affect the overall staining reaction. As the decolorization step is probably the most critical, it should be performed with special care. A common control is the use of a mixed smear of cultures of a gram-positive coccus and a gram-negative rod on an area of the same slide containing an unknown specimen. Appropriate positive and negative reactions of the known cultures indicate proper technique. The results of the unknown specimen can then also be considered accurate. The final step in the procedure, application of the counterstain, must be performed very carefully. Given too much exposure time, the counterstain will replace the primary dye in gram-positive organisms, thus affecting the stained appearance of these cells.

Since Gram's original work, many investigators have tried to find the mechanism involved in Gram differentiation. The explanations offered for the gram-positive state can be grouped in at least three categories: (1) the existence of a specific chemical material that reacts gram-positively; (2) different affinities for the primary dye crystal violet by gram-positive and gram-negative cells; and (3) permeability differences between gram-positive and gram-negative microorganisms.

At the present time, Gram differentiation appears to be a permeation phenomenon. Both the thickness of an organism's cell wall and the size of the spaces in these structures, the pore size, are important to the final outcome of the Gram-stain procedure.

Acid-fast Stain Procedure. This staining technique was developed by Paul Ehrlich in 1882. Acid-fastness is a resistance of cells stained with a basic dye to decolorize in a 3% solution of acid alcohol—a hydrochloric acid (HCl) and ethyl alcohol mixture. Acid-fastness is the most important property differentiating mycobacteria from other bacterial species. Among the better-known members of this acid-fast group are *Mycobacterium leprae*, (my-koh-back-TI-ree-um LEP-ray), which causes leprosy, *M. tuberculosis*, the cause of tuberculosis, and *M. avium-intracellulare* (M. a-VEE-um-in-tra-sell-YOU-larr), a cause of death in some AIDS victims. **[See tuberculosis in Chapter 25, leprosy in Chapter 23, and AIDS in Chapter 27.]**

Most bacterial cells can be easily stained by general techniques. However, certain microorganisms have cell walls with substantial quantities of fatty or waxy substances. Staining or decolorizing them by the usual methods can prove to be difficult. This difficulty prompted Ehrlich to develop the acid-fast staining procedure. The Ziehl-Neelsen procedure, a modification of Ehrlich's method, is the one now used in laboratories to identify acid-fast organisms.

The acid-fast staining procedure is performed by applying the primary dye, carbolfuchsin, to a heat-fixed bacterial smear. Next, the preparation is steamed for approximately five minutes. More dye is added as needed to prevent drying. After heating, the slide is allowed to cool; then it is rinsed gently in running tap water. Acid alcohol, the decolorizing agent, is applied to the smear drop by drop until a red color (the primary dye) no longer runs off from the smear. The preparation is rinsed again immediately and then is covered with the counterstain methylene blue. After one minute, this reagent is rinsed from the slide. On microscopic examination, acid-fast cells are red (Color Photograph 13a) and non-acid-fast cells (Color Photograph 13b) are blue.

A modified version of the procedure known as the Kinyoun acid-fast technique also is in use. Steaming is not required and Kinyoun carbolfuchsin and malachite green serve as the primary and counterstain, respectively.

The mechanism of the acid-fast staining reaction is unclear at the present time. Some researchers believe that mycolic acid (a fatty acid) alone is responsible for the property of acid-fastness. However, there is no conclusive evidence to support this view. One general requirement for the reaction to occur appears to be intact cells. Disrupted cells of normally acid-fast organisms become non-acid-fast.

Spore Stain. Bacterial spores are known for their resistance to high temperature, radiation, desiccation (drying), chemical disinfection, and staining. These structures, which are formed within the cell **(endospores)**, cannot be stained by ordinary methods, such as those described earlier. The dyes do not penetrate the spore's wall. Stained smears of spore-containing cells appear to have oval holes, or colorless spheres, within them. Special procedures are needed to demonstrate the presence of spores. A modified Ziehl-Neelsen method (Color Photograph 14b) can be used for this purpose, but the Schaeffer-Fulton procedure is more commonly employed (Color Photograph 14a). In this procedure the primary stain, malachite green, is applied to a heat-fixed smear and heated to steaming for approximately five minutes. The steaming is done to enhance the penetration of the dye into the relatively impermeable spore coats. Next, the preparation is washed for approximately 30 seconds in running water. The wash step mainly removes the malachite green from cellular parts other than the spores. A counterstain, safranin, is then applied to the smear. This compound displaces any residual primary dye in the vegetative, or nonsporulating, cells but not in the spores. In adequately prepared smears, one can readily observe green spores within red or pink cells, in addition to free green spores (Color Photograph 14a).

Spore-staining techniques are of taxonomic importance in that they can be used to help identify spore-producing bacteria such as those belonging to the genera *Bacillus* and *Clostridium* (klos-TRI-dee-um). The genus *Bacillus* includes the causative agent of anthrax, *B. anthracis*, while the genus *Clostridium* contains a wealth of human pathogens, including *C. botulinum* (bo-tyou-LIE-num), *C. perfringens* (per-FRIN-jens), and *C. tetani* (TEH-tan-ee). These organisms cause botulism (fatal food poisoning), gas gangrene, and tetanus (lockjaw), respectively. The intracellular location, shape, and size of endospores are relatively constant and characteristic for a given bacterial species (Color Photograph 14b). Consequently, they can be used to aid in identifying a newly isolated, unknown organism. **[Anthrax and tetanus are described in Chapter 23. The different forms of botulism are discussed in Chapter 26.]**

Dark-Field Microscopy

Specimens examined by *dark-field microscopy* are usually seen as bright objects against a black or dark background (Figure 3–10). This effect is opposite to the one obtained with bright-field microscopy, in which specimens usually appear darker than the light background. The dark-field procedure is commonly performed by fitting an Abbé condenser with an opaque disk or "dark-field stop" (Figure 3–11). This dark-field stop is placed below the condenser, thus eliminating all light from the central portion of the condenser. A thin cone of illumination reaches the specimen at an angle to the objective lens. Light is scattered by the specimen, which thus acts as a light source. It appears as a glowing object against a dark background.

Figure 3–10 A photomicrograph showing the dark-field image produced with bacteria. Note the bright appearance of rods (B) and spirochetes (A) against the dark background. *[H. Konishi, and Z. Yoshii,* J. Gen. Microbiol. ***131****:3131–3134, 1985.]*

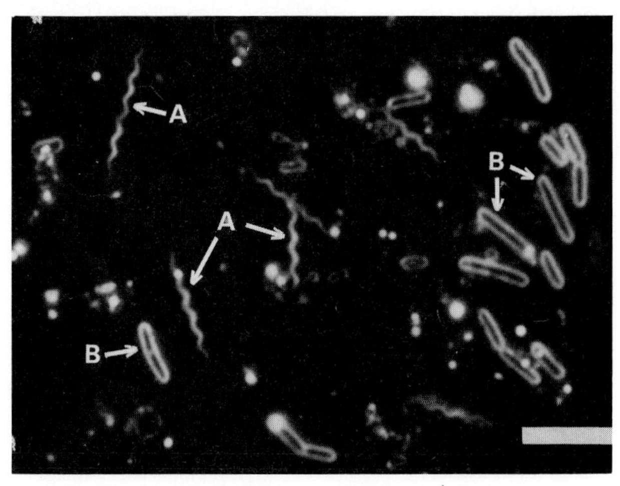

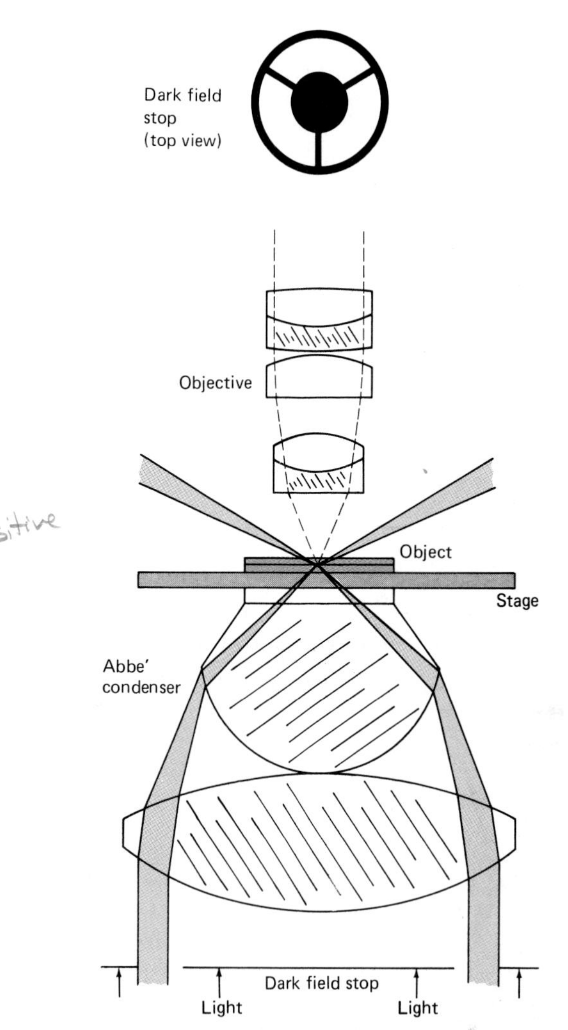

Figure 3–11 Dark-field microscopy. Because of the application of the dark-field stop, the specimen is illuminated only by oblique rays of light. An Abbé condenser is also shown, as well as a typical dark-field stop (top). The light path can be seen at the sides of the dark-field stop. *[After Bausch and Lomb, Inc., Rochester, N.Y.]*

The microscope slides and cover slips used in dark-field examinations must be absolutely free from dirt, dust, and scratches. Unwanted objects and marks can reflect light and could easily brighten the background.

Dark-field microscopy is used for examination of unstained microorganisms, the contents of hanging-

drop preparations, and colloidal solutions. This procedure had diagnostic importance, especially in the case of syphilis (Figure 2–1b).

Fluorescence Microscopy

An early-twentieth-century advance, *fluorescence microscopy* is used for examining various cell types and cellular structures (Figure 3–12a and Color Photograph 15), for locating chemical components in cells and tissues, and for the rapid immunofluorescence diagnosis of otherwise difficult disease states. This type of microscopy provides a means of studying structural details and other properties of a wide variety of specimens. Instruments used in fluorescence microscopy do not differ optically or mechanically from conventional microscopes, but they do require special filter systems to protect the eyes of viewers from the ultraviolet light. The different types of specimens differ in "fluorescing power" from their surroundings. This property of **fluorescence** is obvious when a substance becomes luminous upon exposure to ultraviolet light. With certain substances, including some dyes, fats, oil droplets, and uranium ores, exposure to this form of radiation causes them to absorb the energy of the invisible ultraviolet light waves and re-emit the energy as visible light waves. Many living tissues fluoresce naturally, absorbing ultraviolet radiation and emitting green, yellow, or red light (Color Photograph 15). Selective combinations of natural fluorescence with fluorescent dyes makes possible identification, location, and counting of cell types and differentiation of cell parts (Figure 3–12). Fluorescence of cells can be measured and used to compute the concentration of chemical components present.

Fluorescent dyes used in this way include acridine orange R (Color Photograph 15), auramine O, primulin, and thiazo-yellow G. These substances apparently have a selective action for microorganisms and their components. For example, the fluorescent dye auramine O is used in a detection procedure for *M. tuberculosis.* The dye, which glows yellow when exposed to ultraviolet light, has a strong selective action for the waxlike substances present in this organism. Auramine O is applied to a smear of a sputum specimen suspected of containing *M. tuberculosis.* Excess dye is removed by washing. Then the stained preparation is examined with the aid of the fluorescence microscope. The presence of the tubercle bacilli is indicated by the bright yellow organisms against a dark background. (Although the effect produced is similar to that observed with a dark-field microscope, the principles involved differ significantly.)

Immunofluorescence is another adaptation of this type of microscopy. Fluorescent dyes, such as fluorescein isothiocyanate and lissamine rhodamine B, are employed to react chemically with blood serum proteins, called *antibodies,* and thereby "label" or tag them. The former compound produces an apple-green color and the latter an orange one. Antibodies are noted for their ability to react with protein or protein-polysaccharide components or products of various types of cells, including bacteria (Color Photograph 16). Such chemical substances that react with antibodies are known as **antigens.**

Antibodies can be obtained by injecting antigens into a suitable animal. After a sufficient length of time, the injected animal is bled to obtain the antibodies that it produced. A fluorescent dye such as fluorescein is combined with these antibodies, thus creating a fluorescein-labeled antibody preparation. When this material is applied to a specimen smear containing the antigen responsible for the initial production of the antibody, a specific antigen-antibody reaction occurs. The antibody fluoresces when a treated smear is examined by fluorescent microscopy. The details of antibody production and fluorescent antibody methods can be found in Chapters 17 and 19, respectively.

Immunofluorescence techniques have several important uses, including the detection of disease agents in tissues, cells, and other specimens and the detection of the products of various types of microorganisms. The chemical structure of cells can be effectively investigated by such procedures.

Figure 3–12 A comparison of light microscopy techniques. Note the presence and location of cross walls (C) in the different views. (a) Living cells, unstained. (b) Killed cells, stained. (c) Living cells stained with a fluorescent dye. *[From M. Miyata, H. Miyata, and B. F. Johnson,* J. Gen. Microbiol. ***132****:883–891, 1986.]*

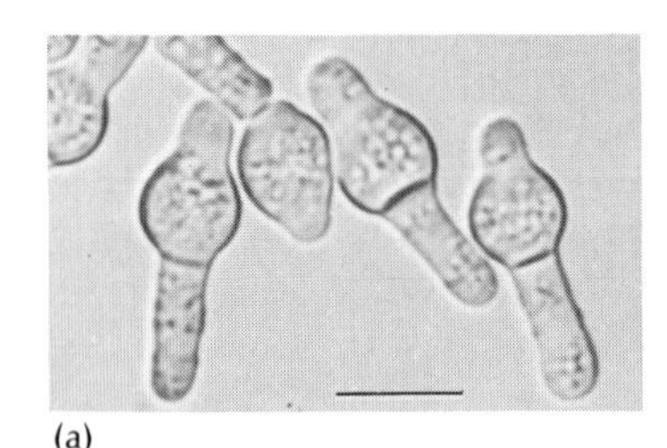

(a)

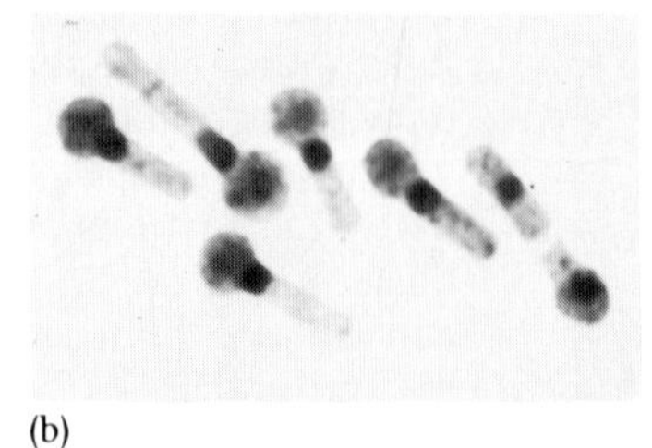

(b)

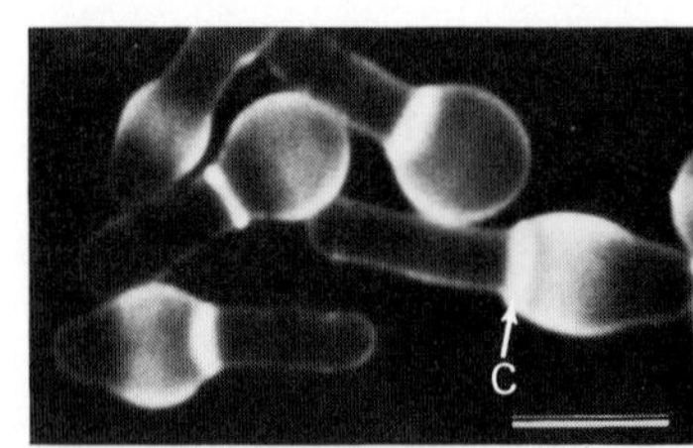

(c)

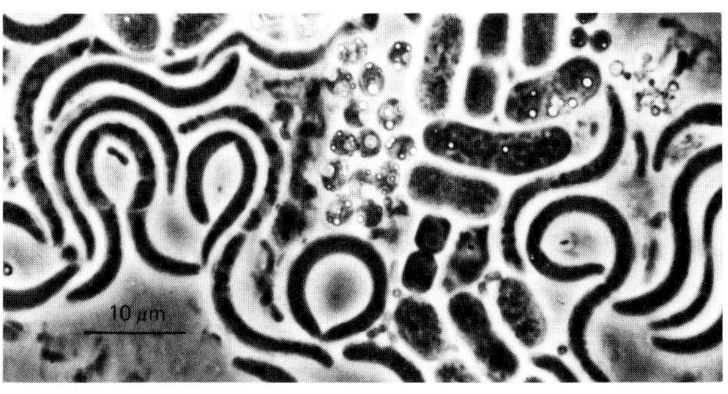

Figure 3–13 A phase-contrast micrograph showing a variety of bacterial cells comprising a microbial community in a Michigan Lake. Cocci (spherical forms), rods (somewhat rectangular and oval), and spirals (wormlike) are evident. *[Reproduced by permission of the National Research Council of Canada from D. E. Caldwell, and J. M. Tiedje,* Can. J. Microbiol. ***21****:377–385, 1975.]*

Phase Microscopy

Phase microscopy is sometimes useful in examining the internal structures of transparent, living cells or in demonstrating their presence in certain fluid environments (Figure 3–13). The phase-contrast microscope can also be used to estimate concentrations of substances within cells or cellular regions. Phase microscopy takes advantage of the fact that different parts of cells have different densities. Cells are also denser than the material surrounding them. However, in the living state cells are difficult to see because of their transparent quality. Use of the phase-contrast principle distinguishes cells and their structures from the background so that their shapes, sizes, and other features can be studied.

The phase-contrast principle was discovered by Frits Zernike, who was recognized for his achievement by being awarded the Nobel Prize for Physics in 1953. As discussed earlier, light waves have several variable characteristics, including frequency and amplitude. Two light waves with the same amplitude and frequency may be traveling so that their troughs and crests pass a given point at the same time. In this case they are said to be *in phase* (Figure 3–14a). They are *out of phase* when their crests pass a given point at different instants (Figure 3–14b). A special *condenser* makes these phase differences visible to the human eye (Figure 3–15).

In a phase-contrast instrument, and sometimes in a regular compound microscope, the condenser has a special *annular diaphragm.* This component allows only a ring of light to pass through the condenser and strike the specimen being viewed. The microscope objective also contains a transparent disk, called the *phase-shifting plate* or *phase-shifting element.* A ring on this disk is used for focusing the light from the annular diaphragm. Depending on its composition, this ring can alter the phase of light waves passing through it by either delaying or advancing them. By briefly slowing one wave, the disk forces it out of phase with one that has not been delayed.

Figure 3–14 The properties of light-wave amplitude and frequency with respect to phase. (a) Two light waves of equal frequency and amplitude in phase. (b) Light waves showing the same properties of amplitude and frequency but out of phase. The phase-contrast microscope transforms this difference into a visible phenomenon. *[After L. A. Wren,* Understanding and Using the Phase Microscope. *Unitron Instrument Company, Inc., Newton Highlands, Mass., 1963.]*

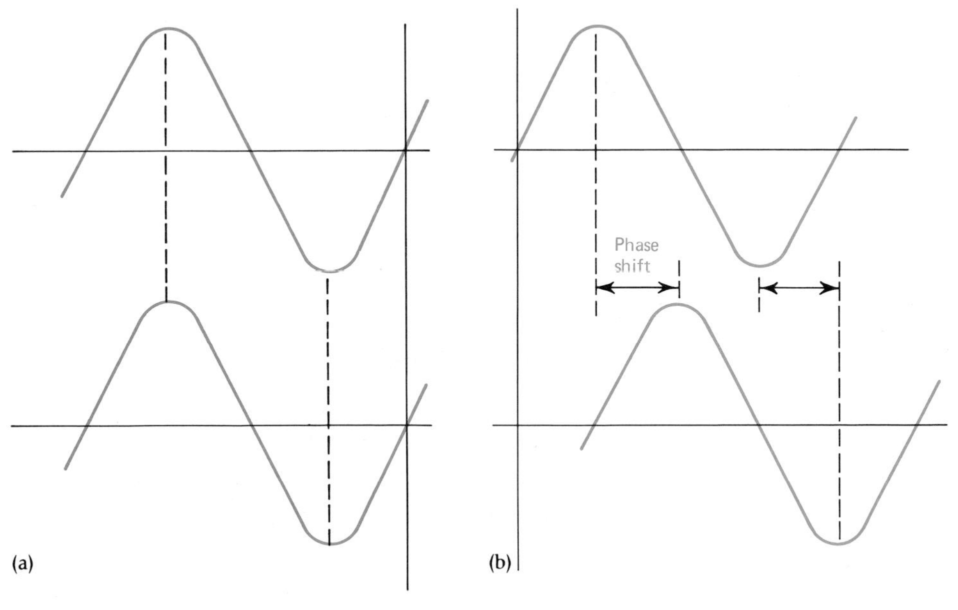

Figure 3–15 The operation of the phase-contrast principle.

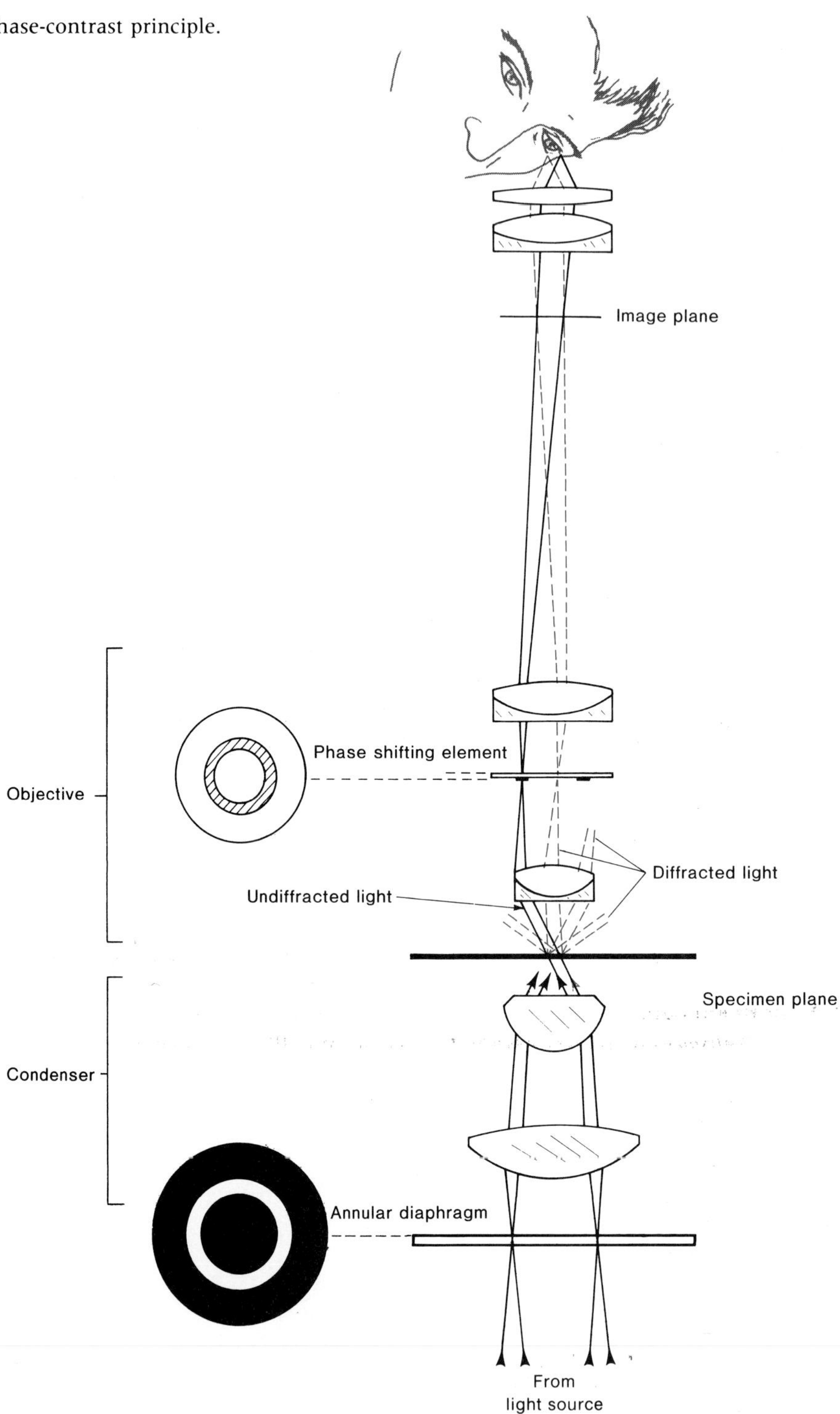

Electron Microscopes

Transmission Electron Microscopy

The realization that the light microscope had reached its limit of resolution (0.2 μm), with further improvements unlikely, prompted the development of the electron microscope. "It was inevitable," says physicist R. Wyckoff, "that the realization of the exceedingly short wavelength of electrons would give microscope research quite a new direction." Electrons are very small, negatively charged particles that behave in many respects like light waves. Accelerating electrons up to 100,000 volts, for example, produce a wavelength of approximately 0.05 Å, much shorter than even ultraviolet light. In actual

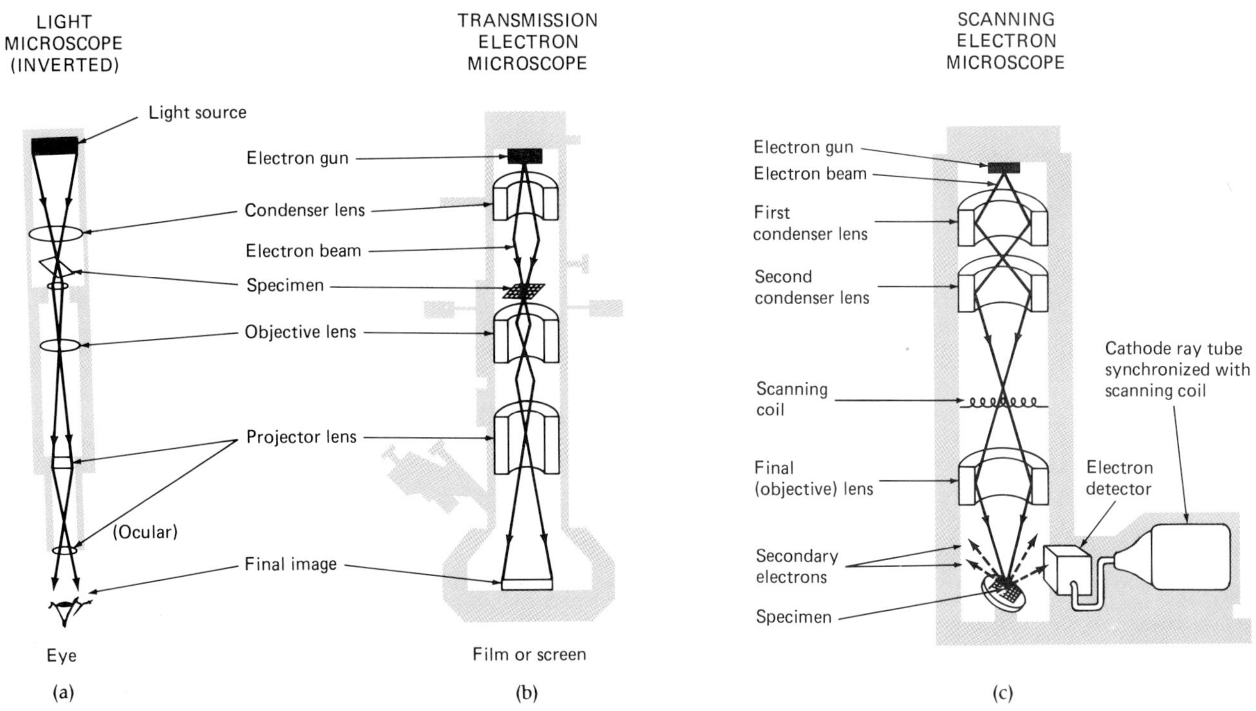

Figure 3–16 A comparison of three different microscopes. (a) The light microscope, which is shown upside down (inverted), so that the parts correspond with those of the two electron microscopes (b) and (c). (b) The transmission electron microscope (TM). This instrument also uses electrons and electromagnets. (c) The scanning coils deflect a fine beam of electrons so that it travels rapidly across the specimen, in synchrony with the spot on a cathode ray tube. The specimen is seen as a "television" picture on the screen of the cathode ray tube.

practice, the resolving power achieved in the electron microscope for biological specimens is about 4 Å, or 0.0004 μm.

Most *transmission electron microscopes* are similar in basic design to the light microscope (Figures 3–16 and 3–17). However, certain distinct differences exist (Table 3–4). The basic components of the transmission electron microscope include a source of

Figure 3–17 One of the large number of currently available transmission electron microscopes. *[Courtesy of Hitachi Scientific Instruments.]*

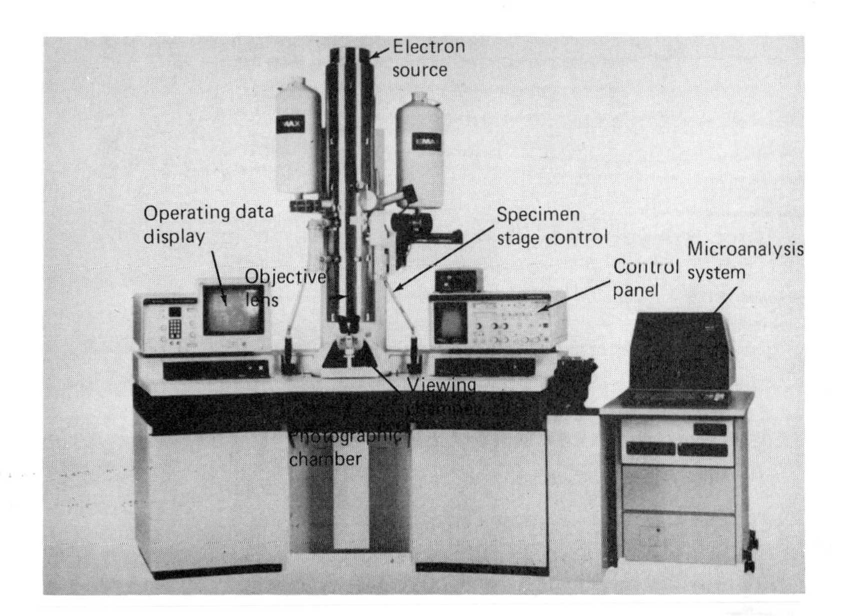

TABLE 3–4 Differences Between the Ordinary Light Microscope and a Typical Transmission Electron Microscope

Property or Procedure	*Light Microscope*	*Transmission Electron Microscope*[a]
Source of radiation for image formation	Visible light	Electrons
Medium through which radiation travels	Air	Vacuum (approx. 10^{-4} mm mercury)
Specimen mounting	Glass slides	Thin films of collodion or other supporting material on copper grids
Nature of lenses	Glass	Magnetic fields or electrostatic lenses
Focusing	Mechanical (i.e., raising or lowering objectives)	Electrical, i.e., the current of the objective lens coil is changed
Magnification adjustments	Changing objectives	Electrical, i.e., changing current of the projector lens coil
Major means of providing specimen contrast	Light absorption	Electron scattering

[a] Most of these properties are also true for scanning and related instruments. Specimens, however, are supported on the round, solid surfaces of stubs (Figure 3–18).

illumination, optical, viewing, and vacuum systems, and electronic systems designed to hold the electron voltage (wavelength) stable.

The source of illumination for the transmission electron microscope, the electron gun, generates an electron beam from a thin, V-shaped tungsten filament. This metal wire must be heated enough to release a sufficient quantity of electrons from the tip of the filament by thermionic emission. These electrons are concentrated and accelerated by other components of the electron gun, producing a fast-moving narrow beam of electrons. In order for the transmission electron microscope to function properly, the components of the electron gun and the lenses must be aligned with one another.

The lens system for the electron beam is electromagnetic rather than glass. The lens of an electron microscope consists of a lens coil formed by several thousand turns of wire encased in a soft iron casing. A magnetic field is created by passing a current through the coil. The electron beam is concentrated within the soft iron casing and other lens components also made of soft iron, called *polepieces,* located in the lens coil.

Three general types of lenses are found in an electron microscope: condenser, objective, and projector lenses. The projector serves to project the final image of the specimen onto a viewing screen, in place of the ocular, or eyepiece, of a conventional light microscope. The screen is coated with a phosphorus compound that fluoresces upon being irradiated by electrons. Permanent records of an image are made photographically. Specimens are generally viewed directly through clear, lead-treated thick glass with the aid of a binocular microscope.

The transmission electron microscope will function only if a vacuum is maintained in the microscope column. Otherwise molecules in the path of the electrons can deflect them and interfere with the image.

Specimen Preparation for Electron Microscopy

SPECIMEN VIEWING AND CONTRAST

As electrons pass through a specially treated specimen being studied, the denser portions of the specimen scatter electrons more readily than those that are less dense. The resulting contrast creates an image of the specimen. The image is focused on a fluorescent screen or photographic emulsion.

When an untreated biological specimen is placed on the usual type of metal grid (Figure 3–18), coated with a plastic support film, and viewed in an

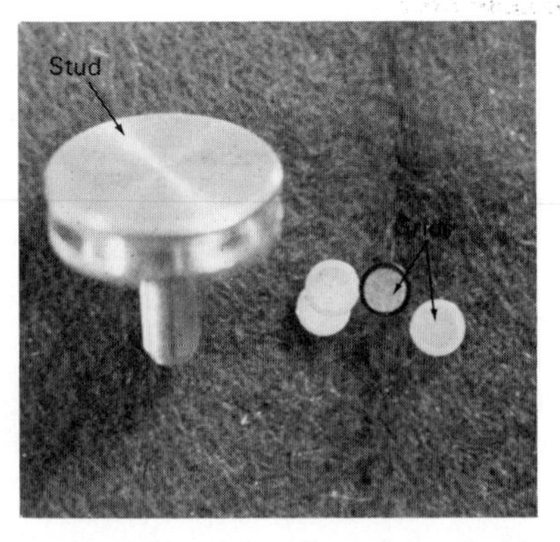

Figure 3–18 A comparison of specimen holders used in electron microscopy. A specimen stub used for scanning electron microscopy (SEM) and a number of metal specimen grids used in transmission electron microscopy (TEM).

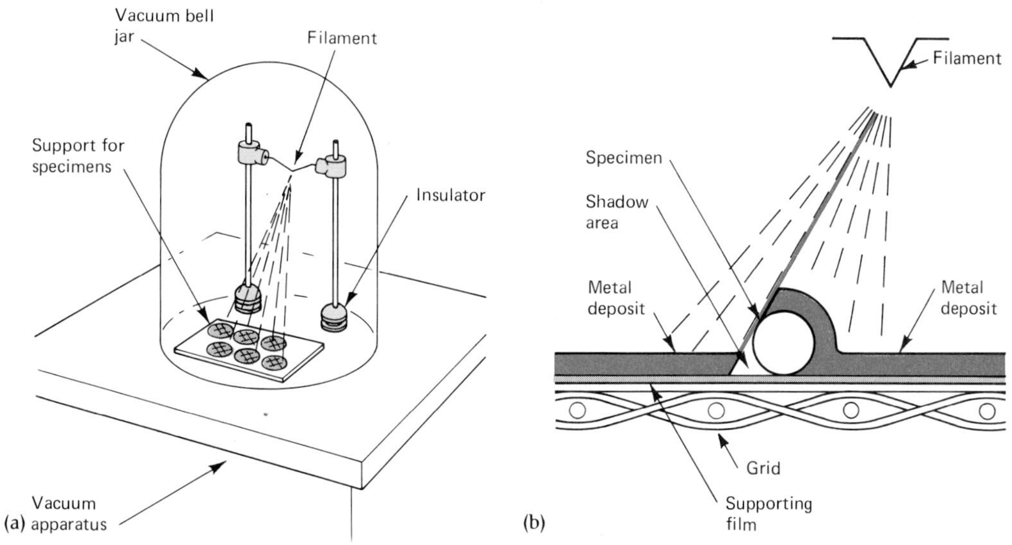

Figure 3–19 (a) A diagrammatic representation of the shadow-casting apparatus. (b) The distribution of metal evaporated onto a specimen. The shadow area is transparent and consequently appears light in viewing the specimen. Because most electron micrographs are printed as negatives, this area registers as a dark region in electron micrographs. (See Figure 3–20.)

electron microscope, the contrast between the specimen's electron image and its supporting film is very poor. This is principally because biological materials are mainly composed of the elements carbon, hydrogen, nitrogen, and oxygen. Since these elements have low atomic weights, their ability to scatter, or deflect, electrons is poor. Thus, in order to see specimens such as isolated macromolecules of protein or nucleic acid, or to study the structural features of cells, special techniques are necessary to make specimens stand out against the supporting film (see Figure 3–20). Three such specimen preparation techniques are *shadow casting, replicas,* and *electron staining.*

SHADOW CASTING

In the *shadow-casting* procedure, a thin film of an electron-dense material is deposited at an angle on a specimen (Figure 3–19). Substances employed for this purpose are heavy metals (chromium, nickel, platinum, uranium, or alloys of gold and palladium or platinum and palladium). The metal to be used is placed on a tungsten filament or other device, which is heated to cause the electron-dense material to evaporate from the surface.

The shadow-casting technique decidedly increases the specimen's contrast. This is illustrated by the electron micrograph shown in Figure 3–20.

SURFACE REPLICAS

Surface replicas have been widely used in areas of microbiology involving the surfaces of specimens such as algae, bacterial and fungal spores, and viruses. Surface replicas are generally prepared by a single-stage technique (Figure 3–21a). In this technique, a thin layer of a low-molecular-weight material such as carbon is deposited on the surface of a specimen in a vacuum. The newly coated specimen is then floated onto a water surface, from which it is transferred to a strong acid or alkali solution capable of dissolving away the specimen without damaging the replica. The resulting cleaned replica is washed in water and placed on a specimen grid for viewing in the electron microscope (Figure 3–21b). Replicas are

Figure 3–20 An electron micrograph demonstrating the effect of shadow casting. The specimen here is vaccinia (cowpox) virus. This microorganism was used in vaccination against smallpox. *[Courtesy of Dr. Robley C. Williams and the Virus Laboratory, University of California, Berkeley.]*

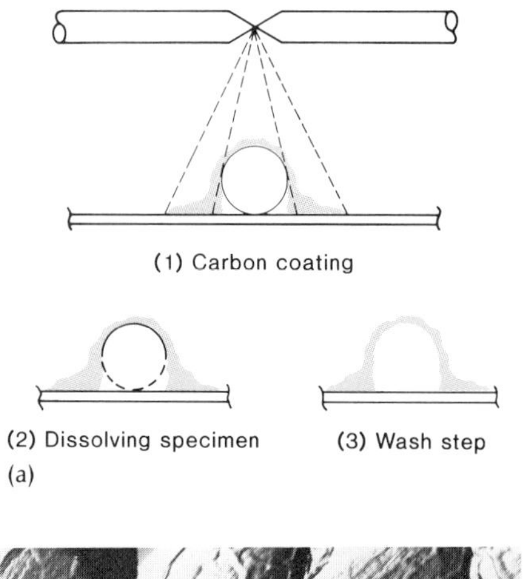

Figure 3–21 (a) Surface-replica preparation technique. The final step leaves only the replica or surface impression of the specimen. (b) The result of carbon replication showing the surface of resting spores of *Bacillus megaterium.* Note the clear surface detail. *[Unpublished electron micrographs by L. J. Rode and Leodocia Pope, Department of Microbiology. The University of Texas at Austin, Austin, Tex.]*

generally shadow-cast in order to emphasize certain aspccts of surface detail.

ELECTRON STAINING

Electron stains are solutions that contain heavy metal elements, for example, osmium tetroxide, phosphotungstic acid, and uranyl acetate. These preparations are used to increase the contrast of specimens. Most electron stains function either by being absorbed on the surface or by combining with specific chemically reactive groups of the specimen. Two general types of staining techniques are known—positive and negative staining. *Positive staining* makes electron-transparent particles visible against a relatively transparent background, while *negative staining* sets specimens against an electron-dense background. In 1959, while studying preparations of turnip yellow mosaic virus, S. Brenner and R. W. Horne introduced this simple negative-contrast technique into general use for electron microscopy. The procedure provides a high resolution for the examination of cell parts, viruses (Figure 3–22), and other specimens. One advantage of this technique is that it does not require any specialized vacuum equipment. Negative staining in electron microscopy is often used as a quantitative device to determine numbers of viruses, as well as being used as a diagnostic tool.

THIN SECTIONING (MICROTOMY)

Many biological specimens are too thick for direct examination in the electron microscope. To make them suitable for investigation, either surface replication or thin sectioning (slicing) is employed (Figure 3–23). *Thin sectioning,* or *microtomy,* seems applicable to a larger variety of specimens. By this technique, structural arrangements of tissue (cellular interconnections), the internal organization of cells, developmental cycles of microorganisms, and many other items of biological interest can be effectively studied. Figure 3–24 compares the information obtained by thin sectioning with that obtained from electron staining.

Before sections or slices of biological material can be prepared, the specimen must be treated in some manner to preserve a specific structure or structures. Preservation of biological structures involves chemical fixation, dehydration, and embedding in plastic. The reliability of the final appearance of a preparation depends upon these steps, as the reagents used

Figure 3–22 Negative staining of sowthistle yellow vein virus. The appearance of the virus when stained with phosphotungstic acid. The surface proteins of the virus are quite evident. *[Courtesy of D. Peters and E. W. Kitajima.]*

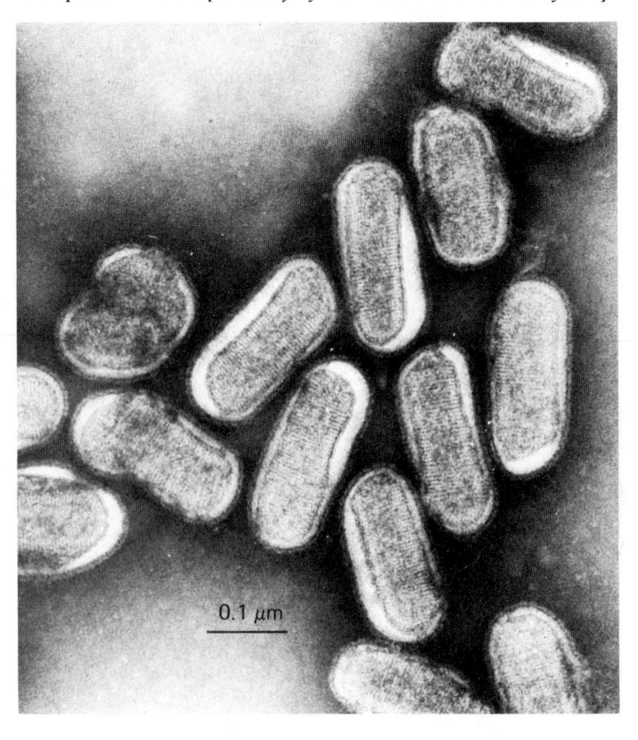

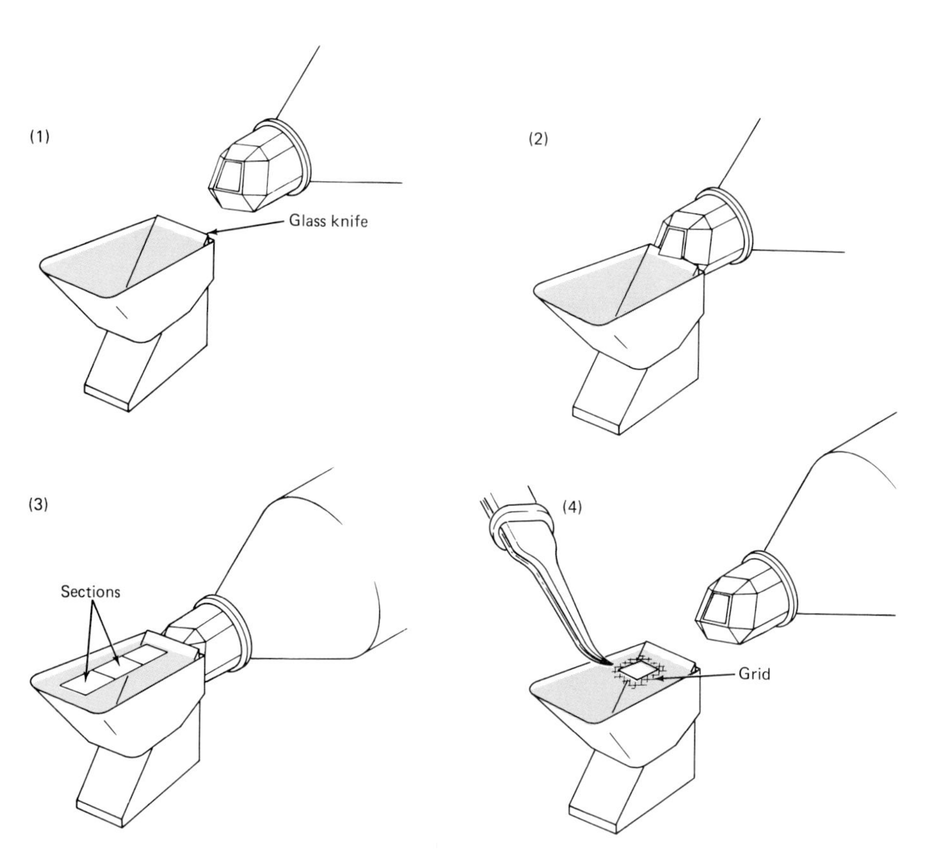

Figure 3–23 A representation of how ultrathin sectioning is performed: (1) The specimen approaches the surface of the glass knife. (2) The beginning of a section. (3) The appearance of several sections. (4) The removal of a specimen for viewing on a coated grid.

Figure 3–24 Electron micrographs of *Veillonella* sp. comparing two techniques used in electron microscopy. (a) A negatively stained whole-cell preparation. Note the convoluted surfaces of the diplococci. (b) An ultrathin section. Note the internal appearance of the organisms, as well as their convoluted outer regions. The cell on the top has developed a cross wall that will eventually result in division. *[From S. E. Mergenhagen, H. A. Bladen, and K. C. Hsu,* Ann. N.Y. Acad. Sci. ***133****:279, 1966. © The New York Academy of Sciences; 1966; reprinted by permission.]*

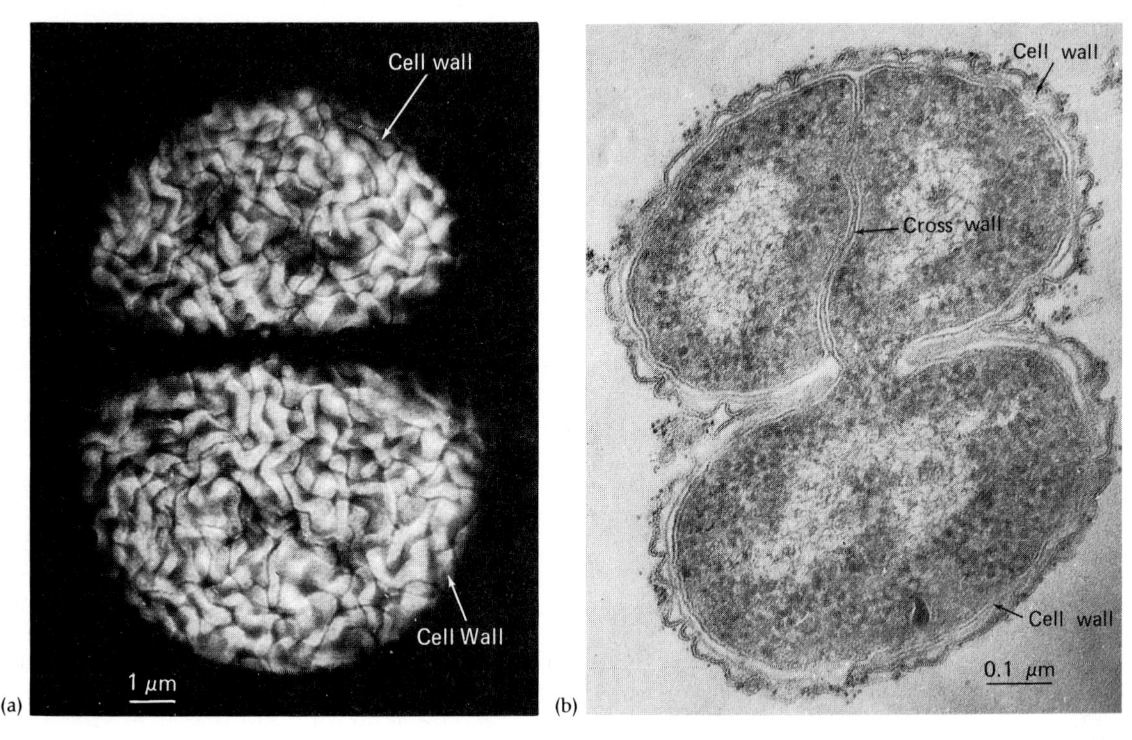

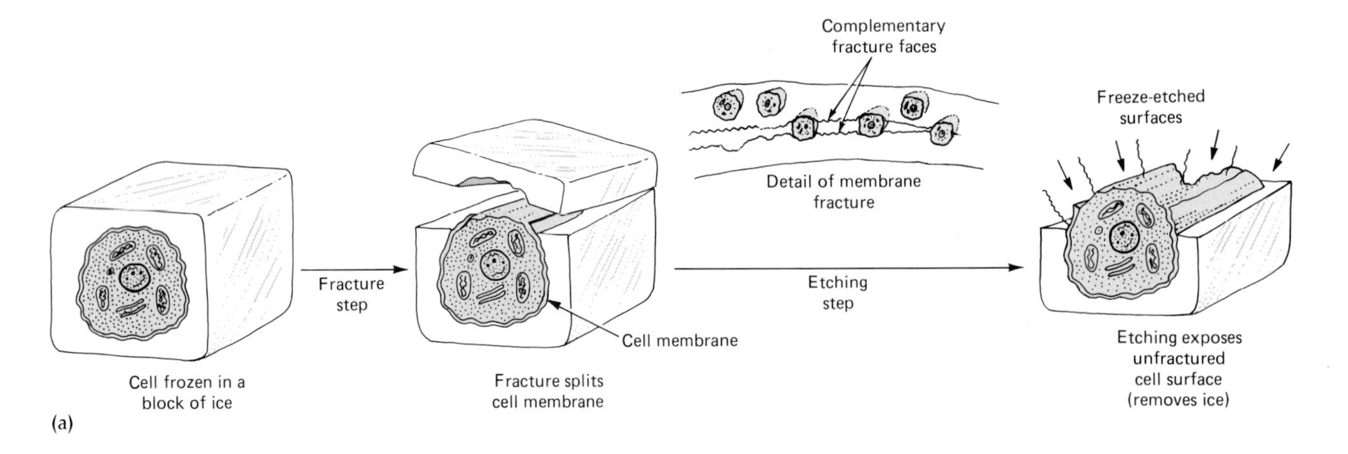

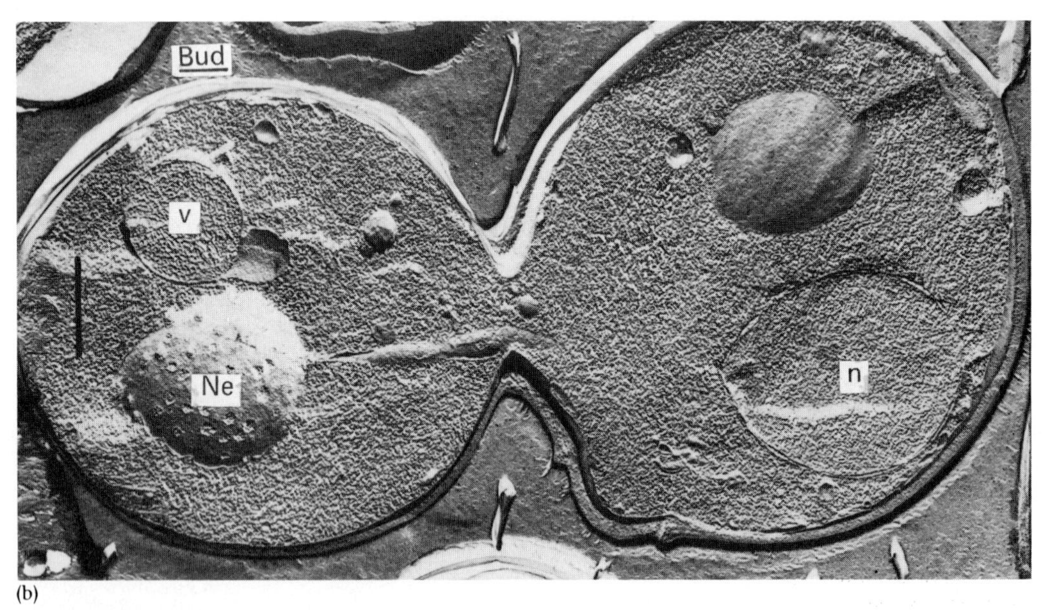

(b)

Figure 3–25 The freeze-fracture and freeze-etching technique. (a) The frozen cell is fractured with a sharp blow of a blade or knife. The fracture line very often runs through the interior of the membranes, separating the two layers there and revealing the membrane proteins that would otherwise remain buried. The last step allows for the etching of the membrane through the sublimation of surface ice. This exposes the unfractured membrane surface, thereby providing detail of the complementary fracture faces of the membrane. (b) Typical results of a free-fracture technique. The surfaces of several structures of a yeast cell and its bud are shown. Note the convex surface nuclear envelop (ne), and the concave appearance of the nucleus (n), and the vacuole (v). *[From J. H. M. Willison, and G. C. Johnston,* Can. J. Microbiol. ***31****:109–118, 1985.]*

may easily affect the physical and chemical properties of the specimen. **[See Chapters 7 through 10 for the internal features of microorganisms.]**

FREEZE-FRACTURE

The technique of **freeze-fracture** is quite valuable to the biological sciences, since it avoids fixation, embedding, and sectioning of specimens. With this technique, unfixed specimens can be examined without the production of artifacts. The cleaving, or fracture, of a specimen may show both the outer and inner surfaces of a membrane or may even split the membrane at such an angle as to reveal structures or materials passing through it. This technique is especially valuable in the classification of various forms of life.

In freeze-fracture, cells are quick-frozen in water, split, or cleaved, and treated to expose their interior structures. Exposed surfaces are shadowed to provide contrast, and a replica is prepared. The procedure and results are shown in Figure 3–25.

LOCALIZATION TECHNIQUES

To know the cellular site of enzymes and their activities, or to be able to follow the formation or incorporation of specific structures within a cell, can be essential to understanding both the organization and function of cells. Sometimes the identity and position of specific antigens, either within or on the surface of a cell, can be of considerable importance. Several techniques and procedures have been developed for the precise localization of enzyme reaction products and surface antigens.

In enzyme studies, the reaction product of an enzyme is precipitated and made electron-dense so that the position of the enzyme can be identified. The most widely used method for the detection of antigens involves the use of antibody preparations tagged with an electron-opaque, iron-containing protein such as ferritin. These tagged materials permit precise discrimination of the distribution of antigens both on a cell surface and within a cell. **[See Chapter 19 for electron microscopic diagnostic tests.]**

Electron Microscopic Image Enhancement

The image of a specimen seen in an electron microscope generally includes information that is unrelated to the structure or other features of the specimen. This information, called *background noise,* has various sources, including the electron microscope itself, photographic materials, radiation damage, and the specimen-supporting film. Three general techniques are used to reduce or separate the noise from the picture obtained and to bring about *image enhancement.* These are *photographic averaging, optical diffraction,* and *computer processing.* A specific requirement for all techniques is that the specimen under study must exhibit regularity in its organization and its structural appearance. This requirement limits the application of these techniques to subjects such as viruses and various bacterial structures.

With photographic averaging, a composite image is produced on a single photographic plate by accurately superimposing a number of images of similar objects. The common features of the objects are enhanced, while at the same time, randomly distributed noise that contributes nonspecific detail is reduced.

In the optical diffraction technique, the image on an electron micrograph is used as a diffraction device by changing the path of light waves passing through it. Light waves, all in the same phase, are passed through the image and focused on a screen to produce a diffraction pattern. The pattern is directly related to the structural detail of the image. This technique not only provides distinct information, free of noise, concerning the shape and form of specimen parts but also makes possible the regeneration of an image without noise and interfering structures.

Computer processing is the most recent advance to be used for the perfection of image detail. In this technique, the computer changes the visual image into an electrical image (signal) it can understand. Using this electrical image, the computer performs a mathematical analysis by obtaining multiple images that can be compared and analyzed for similar detail and used to produce an average image. The computer is programmed to detect all corresponding parts and their specific orientation from all images. All nonessential information therefore is eliminated from the final averaged image by the mathematical limitations of the computer program. Figure 3–26 shows the process and the type of information produced by the application of computer processing to the study of cellular structures. **[See bacterial ribosomes in Chapter 7.]**

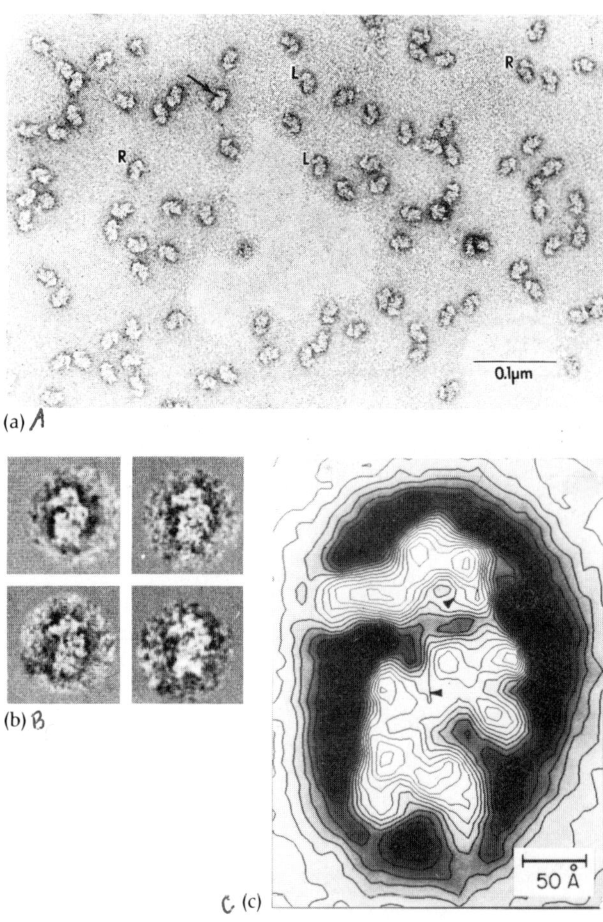

Figure 3–26 An example of computer-image averaging. A computer is used to align and average large numbers of individual electron-microscopic images of a ribosomal particle. (a) An electron micrograph showing left-oriented (L) and right-oriented (R) side views of ribosomal subunits taken from cells. Look closely for differences. The L views were selected for computer processing. (b) A gallery showing several L views. One of these is used as a reference view. (c) The computer image representing the total average from all views considered. *[J. Frank, A. Verschoor, and M. Boublik,* Science ***214****:1353–1355, 1981.]*

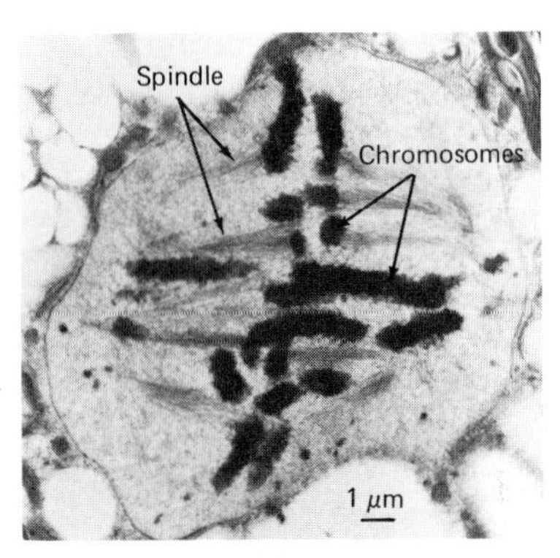

Figure 3–27 A high-voltage electron micrograph of the early anaphase stage of cell division (mitosis) in the green algae *Oedogonium cardiacum.* The chromosomes and remains of the mitotic spindle are well preserved. What advantages does this form of microscopy offer to the microbiologists? *[From R. A. Coss and J. D. Pickett-Heaps,* J. Cell Biol. ***63****:84–98, 1974.]*

High-Voltage Electron Microscopy

The capability of transmission electron microscopes to show minute details and to produce functional images of preparations depends on the specimen's ability to scatter electrons passed through it. This passage of electrons is influenced by two factors, the thickness of the specimen and the energy of the electrons. The latter depends on the accelerating voltage used to generate the electrons. A new instrument, the *high-voltage electron microscope,* is similar to the general-transmission electron microscope, except that its electron voltage is considerably greater. This feature of the microscope makes it possible to penetrate thicker specimens and to observe the relationships of cell components (see Figure 3–27).

Scanning Electron Microscopy

The *scanning electron microscope* (Figure 3–28) is comparatively new and quite different in principle as well as in application from the conventional transmission electron microscope. The scanning electron microscope gives a three-dimensional quality to specimen images. It has been used a great deal to study the surfaces of specimens too thick for conventional transmission electron microscopy. Specimens for the scanning electron microscope (SEM) require less preparation than those for transmission instruments. Many specimens are first fixed and then coated with a thin film of heavy metal. Large portions of surfaces can be seen in detail with excellent contrast (Figure 3–29). Normally, the scanning microscope is operated by scanning, or sweeping, a very narrow beam of electrons (an electron probe) back and forth across a metal-coated specimen. Secondary or backscattered electrons leaving the specimen surface are collected, the current is amplified, and the resulting image is displayed on the screen of a cathode-ray tube. The three-dimensional view pro-

Figure 3–28 The Hitachi Perkin-Elmer Model HHS-2R scanning electron microscope. A television monitor, camera, and related equipment are shown. *[Courtesy the Perkin-Elmer Corporation, Scientific Instruments Department.]*

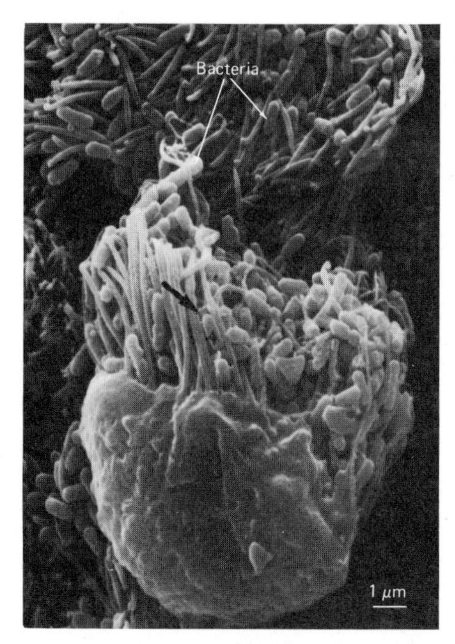

Figure 3–29 The presence of bacteria (arrows) on the cilia of a respiratory tract epithelial cell. Original magnification, 9,900 ×. *[T. Matsuyuma, and T. Takino,* J. Med. Microbiol. ***13****:159–161, 1980.]*

duced reveals the specimen's surface features rather than its internal organization. The most valuable feature of the scanning electron microscope is its great depth of focus. In the technique known as *inverted backscatter scanning,* a specimen section (slice) is first exposed to one of several heavy-metal electron stains and then examined with a backscatter detector. This device aids in the formation of a sharp image because of its ability to detect electrons and to distinguish between the atoms of heavy-metal stains, which penetrate into deeper portions of the specimen. Polarity of the SEM is reversed to give a reverse image of the specimen, which is comparable to the images seen under the light microscope (Figure 3–30).

In microbiology, the scanning electron microscope is especially useful in studies of bacterial, fungal, and protozoan morphology. It is also becoming an important tool in microbial classification in studies of morphological changes in tissues infected with microorganisms and in the direct examination of cells in microbial communities in various environments (Figure 3–29). The current protozoan classification is based in part on electron microscopy findings (see Appendix D).

Figure 3–30 Inverted backscatter-SEM. (a) A view of the intestinal lining of a normal mouse containing large numbers of bacteria (boxed area). (b) A higher magnification of the boxed area showing the curved shapes of the bacteria present. *[From A. Lee, J. L. O'Rourke, P. Barrington, and T. J. Trust,* Inf. Immun. ***51****:536–546, 1986.]*

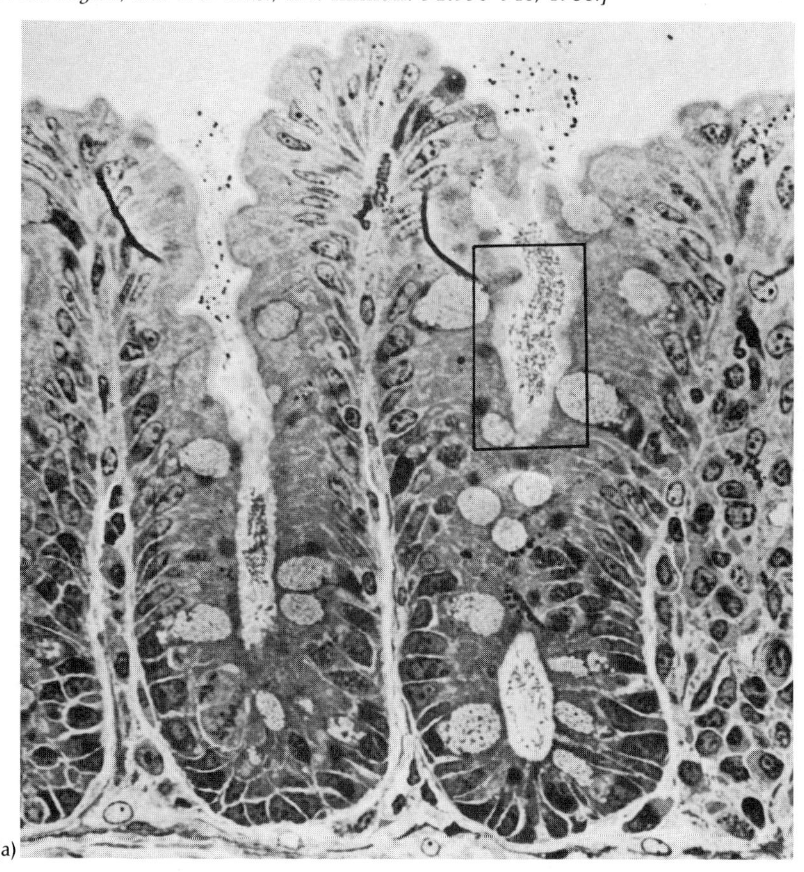

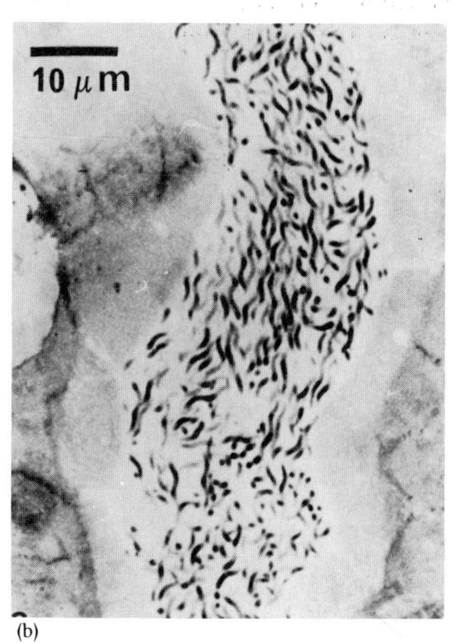

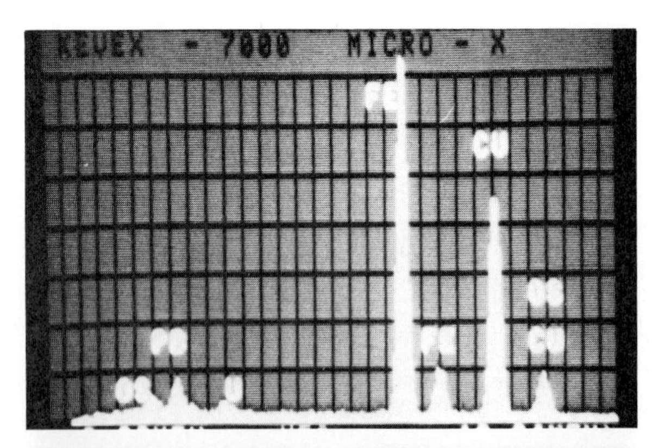

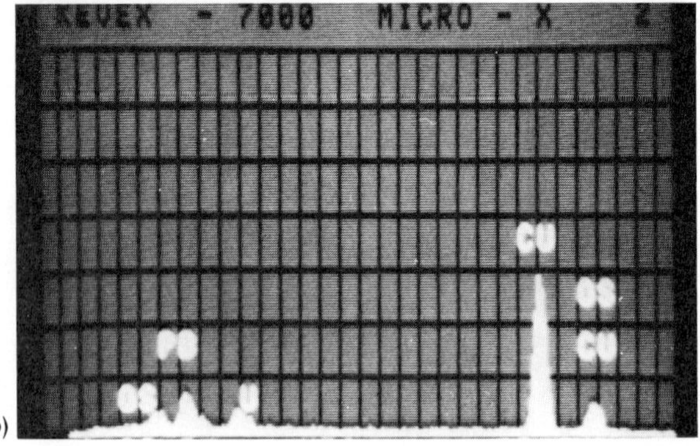

Figure 3–31 Energy-dispersive x-ray spectrum of an ultrathin-sectioned magnetotactic bacterial cell. Cells such as these contain chains of iron-rich particles. (a) Transmission micrograph showing points (arrows) where cells were analyzed to obtain a spectrum (display) shown in (b). (b) The spectrums obtained from electron-dense particles in a chain. The major elements detected are iron (Fe) and copper (Cu). *[D. C. Balkwill,* et al., J. Bacteriol. ***141****:1399, 1980.]*

Scanning Transmission Electron Microscopy

The *scanning transmission electron microscope* is a combined conventional transmission and scanning electron microscope high-resolution system. The instrument also serves as an analytical device with which the chemical composition of an observed structure can be determined. This elementary microanalysis of selected regions of a specimen is done by energy-dispersive x-ray spectroscopy (EDX). In x-ray spectroscopy, an x-ray detector is used to monitor the distinct x-ray pattern produced by the interaction between the electron beam of the microscope and the chemical elements in specific areas of the specimen. The resulting x-ray pattern is typically displayed as peak heights for quantitative measurement (Figure 3–31). Scanning transmission electron microscopes are commonly equipped with small computers to control data acquisition and to make the necessary measurements and calculations.

Microorganisms and Life Processes

New properties and uses of microorganisms are revealed by a growing number of advances in microscopy, molecular biology, biochemistry and DNA-recombinant technology. Such discoveries not only provide a greater understanding of microbial life and the factors that affect it but also offer a deeper insight into the properties and workings of higher forms of life.

The chapters in the next section will describe the major and minor growth requirements of microorganisms and how different nutrients and techniques are used for the isolation and cultivation of various microbes. Attention also will be given to the organized chemical activities occurring within microbial cells. These include energy-generation reactions and those reactions that meet the growth and reproductive needs of microorganisms. The functions of microbial genetic material and how such material can be manipulated for the synthesis of important chemicals such as insulin, human growth hormone, and the antiviral substance interferon are among the main topics covered in the last chapter of the section. Appendix A contains a basic review of chemistry and biochemistry as an aid for the section.

Summary

Microscopes and Microscopy

1. Depending upon the magnification principle involved, microscopes fall in one of two categories: *light* (optical) or *electron*.
2. Light microscopes include *bright-field, dark-field, fluorescence,* and *phase-contrast* instruments.
3. Electron microscopes include the *transmission* and *scanning* instruments.

UNITS OF MEASUREMENT

The basis of scientific concepts, laws, and theories requires careful measurement of lengths, volumes, weights, and other quantities.

1. *Le Système International d'Unités,* or *SI system,* is a widely accepted system for measurement.
2. The units used for length, volume, mass, and temperature are the meter, liter, gram, and Celsius degree, respectively.
3. Commonly used metric units for the measurement of lengths include the meter (m), centimeter (cm), millimeter (mm), micrometer (μm), nanometer (nm), and Angstrom (Å).
4. The *bar marker* on photographs serves as a size reference and represents the length of a metric unit at a particular magnification.

PROPERTIES OF LIGHT

1. *Light waves* are described in terms of their amplitude, frequency, and wavelength.
2. *Amplitude* refers to the maximum displacement of a light wave from its equilibrium position.
3. *Frequency* is the number of vibrations that occur in one second.
4. *Wavelength* is defined as the distance between two corresponding points on a wave.
5. The *resolving power (RP)* of a microscope refers to its ability to reveal detail. It depends on the wavelength of light used.
6. The type of material or medium through which light passes during the operation of a microscope plays a major role in the image seen.

Light Microscopes

BRIGHT-FIELD MICROSCOPY

1. Two types of bright-field microscopes are known, the *simple* (one-lens) microscope and the *compound* microscope.
2. A compound microscope consists of at least two lens systems: the *objective,* which forms and magnifies the image of the specimen, and the *eyepiece,* or *ocular,* which magnifies the image produced by the objective lens.
3. *Resolving power (RP)* (the ability to show details of specimens) and *parfocal* (changing objectives without major focusing adjustments) are two important microscope features.
4. The highest-magnification objective used in general courses is the oil-immersion objective, which for best results should be immersed in a medium having the same *index of refraction* as glass.
5. Condensers serve to concentrate light for specimen viewing. Most condensers are equipped with a shutter-like device, the *iris diaphragm,* to control light intensity.

MICROMETRY

Micrometry refers to procedures used to measure the dimensions of cells or, if possible, their parts.

SPECIMEN PREPARATION FOR BRIGHT-FIELD MICROSCOPY

1. *Hanging-drop* and temporary *wet-mount* techniques are used for the microscopic examination of living organisms. These procedures maintain the natural shape of the specimen and are useful in determining sizes, shapes, detection of movement, and reactions to various chemicals and immune sera.
2. General staining procedures involve the applications of dye-containing solutions to smears (thin films of microorganisms on glass slides, which have been air-dried and heat-fixed). These procedures permit greater observation of cells and, to some extent, a differentiation of microbial species.
3. Two general types of dyes are used for staining: basic dyes, which react with nuclear components, and acid dyes, which stain cytoplasmic and related materials.
4. *Simple staining* methods involve the application of only one dye solution. Such procedures provide information on microbial size, shape, arrangement, and the possible presence or absence of spores.
5. *Differential staining* methods involve the application of more than one dye solution. Such procedures make possible the separation of bacteria into major groups and the identification of certain bacterial parts.
6. The two most widely used differential staining methods are the *Gram* and *acid-fast* staining techniques.
7. Based on resulting color reactions of the Gram stain, microorganism are classified as either *gram-positive* (purple) or *gram-negative* (red).
8. Several bacterial properties appear to be correlated with gram-staining reactions; these include chemical composition, sensitivity to penicillin, and relative cell wall thickness.
9. The acid-fast procedure is used mainly to differentiate mycobacteria from other bacteria. The decolorizing agent, acid alcohol, differs from the one used in the Gram stain.
10. *Spore-staining* techniques are of taxonomic importance in that they can be used to help identify spore-forming bacteria.

DARK-FIELD MICROSCOPY

1. Specimens examined by *dark-field microscopy* usually appear as bright objects against a dark background.
2. Dark-field microscopy is useful with unstained preparations and may have diagnostic importance.

FLUORESCENCE MICROSCOPY

1. *Fluorescence microscopy* is used for examination of various cell types and cellular structures, the location of chemical components in cells, and the rapid diagnosis of certain diseases.
2. Ultraviolet light, fluorescent chemicals, and special microscope filter systems are required for this type of microscopy.

PHASE MICROSCOPY

1. *Phase microscopy* is of value in the examination or detection of living cells in certain fluid environments.
2. The *phase-contrast* principle makes cells and their parts stand out from the background.

Electron Microscopes

TRANSMISSION ELECTRON MICROSCOPY

1. Most *transmission electron microscopes* are similar in basic design to the light microscope.
2. The fundamental difference is that electron microscopes use electrons as the source of illumination. Therefore, they have different lens construction and specimen preparation procedures. They can produce higher magnification with better resolving power.

SPECIMEN PREPARATION FOR ELECTRON MICROSCOPY

1. Because of their chemical composition, untreated biological specimens are difficult to see in the transmission electron microscope.
2. Special treatment procedures are used to increase the contrast of such specimens. These include shadow casting, replicas, and electron staining.
3. *Shadow casting* involves depositing, at an angle, a film of heavy metal on a specimen.
4. *Surface replicas* are surface impressions of specimens.
5. *Electron stains* are solutions that contain heavy metal elements that may combine with certain components of cells.
6. *Thin sectioning,* or *slicing,* is used with specimens that are too thick for direct examination in the electron microscope.
7. Thin sectioning can be used to study the internal organization, structural arrangement, and related properties of cells.
8. *Freeze-fracture* can be used to study biological specimens without the preparation associated with other techniques.
9. *Localization techniques* can be used to find the precise site of enzymes and their activities, or to determine the exact distribution of antigens on a cell surface or within a cell.

ELECTRON MICROSCOPIC IMAGE ENHANCEMENT

1. Image enhancement removes or separates information that does not contribute to the visible structural properties of a specimen.
2. Techniques used for image enhancement are of three general types: photographic averaging, optical diffraction, and computer processing.
3. In microbiology, image enhancement is applied to microorganisms or their parts displaying regularity in structure.

HIGH-VOLTAGE ELECTRON MICROSCOPY

This form of microscopy makes it possible to penetrate thicker specimens and to observe three-dimensional relationships among cell components.

SCANNING ELECTRON MICROSCOPY

1. The *scanning electron microscope* gives a three-dimensional quality to the images of the specimens viewed.
2. Large portions of specimen surfaces can be seen in detail with excellent contrast.
3. Specimens are metal-coated for examination.
4. Inverted backscatter scanning provides greater detail of thin specimen slices and images comparable to those obtained by light microscopy.

SCANNING TRANSMISSION ELECTRON MICROSCOPY

1. The *scanning transmission electron microscope* combines conventional transmission electron microscopy with scanning electron microscopy to form a high-resolution system.
2. The system serves as an analytical device with which the chemical (element) composition of a specimen can be determined.
3. Energy-dispersive x-ray spectroscopy, equipped with a computer system, is used to perform the chemical analysis.

Questions for Review

1. Define or explain the following:
   a. simple microscope
   b. gram-variable
   c. compound microscope
   d. wet-mount
   e. micrometer
   f. scanning electron microscope
   g. nanometer
   h. contrast
   i. acid alcohol
2. Distinguish between RP and magnification.
3. What is parfocal?
4. What are the functions of the following microscope components?
   a. condenser
   b. iris diaphragm
   c. low-power objective
   d. ocular
5. Compare the types of microscopy listed below with respect to (1) source of illumination, (2) specimen preparation, (3) limits of magnification, and (4) general uses. The construction of a table with these categories would be quite helpful.

a. bright-field microscopy
b. fluorescence microscopy
c. dark-field microscopy
d. transmission electron microscopy
e. phase-contrast microscopy
f. scanning electron microscopy
g. high-voltage electron microscopy
h. scanning transmission electron microscopy

6. Compare the results that can be obtained using the hanging-drop procedure with those associated with simple stains.

7. Give the equivalent metric units specified for the following:
a. 10 μm = ________ nm
b. 10 mm = ________ μm
c. 1 m = ________ nm
d. 100 nm = ________ Å

8. What is the purpose of differential staining procedures? What are common examples of this type of technique?

9. What are the functions of the different reagents used in the Gram stain?

10. What factors play prominent roles in determining gram-positivity?

11. What are the specific color characteristics of the cell types or microbial structures listed below after the performance of the standard gram, acid-fast, and spore stains?
a. gram-positive bacteria
b. gram-negative bacteria
c. acid-fast bacteria
d. spores and associated vegetative cells
e. gram-variable organisms
f. non-acid-fast cells
g. gram-nonreactives

12. What factors contribute to the resolving power of a microscope?

13. What is the purpose of using immersion oil? How is this material applied?

14. Indicate the general range of sizes of the following microorganisms:
a. bacteria
b. protozoa
c. rickettsiae
d. viruses
e. fungi

15. What is an acid dye?

16. What types of cellular structures are stained by a basic dye?

17. Why are localization techniques used in electron microscopy?
a. What does noise represent in an electron micrograph?
b. How can it be reduced or eliminated?

18. What is inverted backscatter scanning?

Suggested Readings

BRADBURY, S., *The Optical Microscope in Biology.* London: Edward Arnold Ltd., 1976. *A short, simple text on the theory of microscopy and various techniques employed.*

BURRELLS, W., *Microscope Technique, A Comprehensive Handbook for General and Applied Microscopy.* New York: John Wiley & Sons, 1977. *In-depth discussions of light microscopy are provided by this general reference. Techniques available to light microscopists and an introduction to electron microscopy are also included.*

HAYAT, M. A., *Basic Techniques for Electron Microscopy.* Orlando, Florida: Academic Press, Inc., 1986. *An up-to-date presentation not only of methods but also of the concepts behind them.*

HAYAT, M. A., *Introduction to Biological Scanning Electron Microscopy.* Baltimore: University Park Press, 1978. *Provides a functional treatment of the operation, design, and uses of the scanning electron microscope. It is well written and contains many references and illustrations.*

HOLT, S. C., and T. J. BEVERIDGE, "Electron Microscopy: Its Development and Application to Microbiology," *Canadian Journal of Microbiology* **28**:1–53, 1982. *A comprehensive and readable coverage of the instrumentation and techniques associated with electron microscopy. Topics range from conventional transmission electron microscopy to computer enhancement of images.*

LILLIE, R. D., *H. J. Conn's Biological Stains: A Handbook on the Nature and Uses of the Dyes Employed in the Biological Laboratory,* 9th ed. Baltimore: Williams & Wilkins, 1977. *An internationally accepted publication that contains the most complete and physical information on biological dyes and their application.*

SPENCER, M., *Fundamentals of Light Microscopy.* New York: Cambridge University Press, 1982. *A functional reference work.*

WREN, L. A., *Understanding and Using the Phase Microscope.* Newton Highlands, Mass.: Unitron Instrument Company, 1963. *Principles and the use of phase-contrast microscopy are presented in an understandable manner.*

PART II

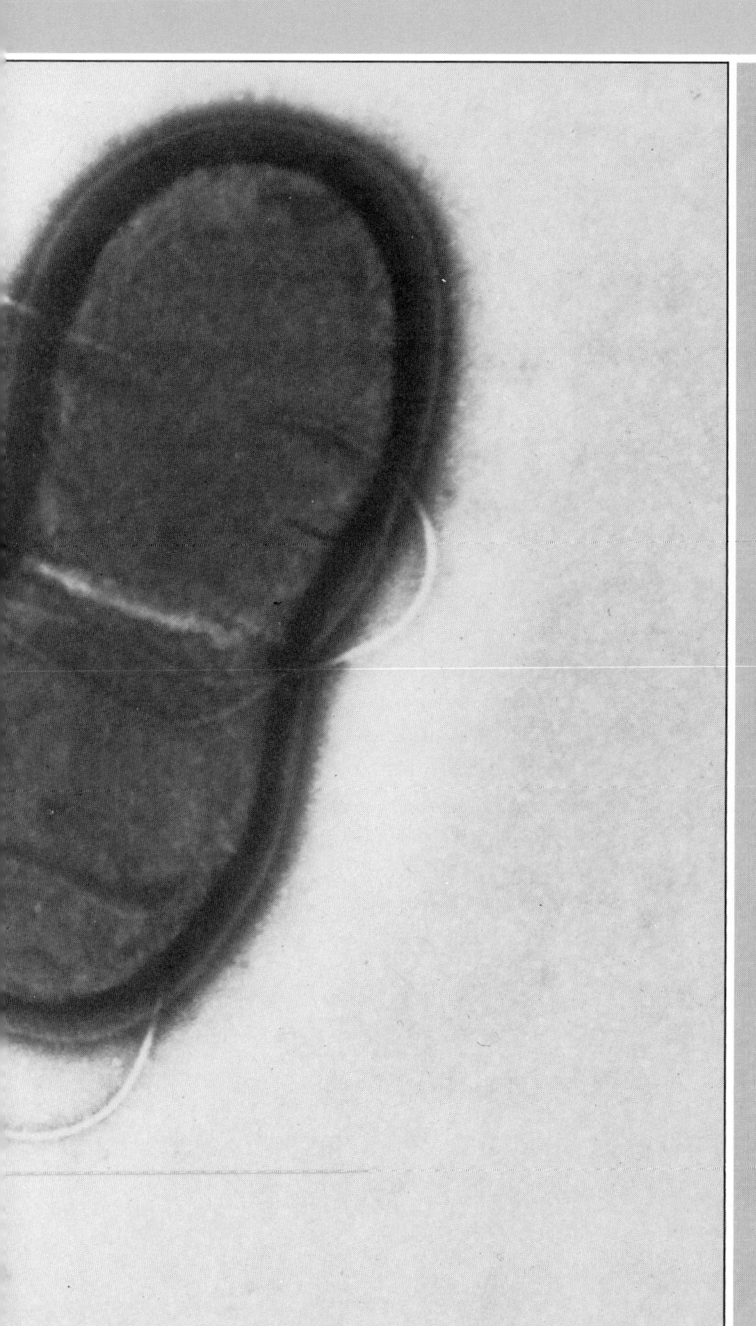

Microbial Growth, Cultivation, Metabolism, and Genetics

Electron micrograph of Eubacterium desmolans, *an aerobic steroid digesting bacterium isolated from cat feces. [From G. N. Morris, J. Winter, E. P. Cato, A. E. Ritchie, and V. D. Bokkenheuser,* Internat. J. System Bacteriol, *36:183–186, 1986.]*

4

Bacterial Growth and Cultivation Techniques

The stability of the **internal medium** *is a primary condition for the freedom and independence of certain living bodies in relation to the environment surrounding them. — Claude Bernard*

After reading this chapter, you should be able to:

1. Describe the forms of nutrition known as heterotrophy, autotrophy, and hypotrophy.
2. Identify microbial categories according to gaseous requirements.
3. Describe what is meant by cardinal temperatures and classify bacteria according to temperature requirements.
4. Identify the role of pH in bacterial growth and list selected organisms with unusual pH requirements.
5. Identify the role and common sources of nitrogen, carbon, vitamins, growth factors, mineral salts, and water in bacteriological media.
6. Describe the basic elements of media preparation, the requirements for storage, and selected uses of bacteriological media.
7. Identify the role of aseptic technique in bacterial inoculation and transfer procedures.
8. Describe the pour-plate and streak-plate techniques and their use in the isolation of pure cultures.
9. Explain the need for pure cultures.
10. Describe general methods for establishing capneic and anaerobic incubation conditions and the concept of biological indicators.
11. Describe what occurs during each phase of the bacterial growth curve.
12. Describe differences in how bacteria grow in liquid and solid media.
13. Identify the basic aspects and uses of continuous culture techniques.
14. Describe techniques used to determine bacterial growth by estimation of cell mass and cell numbers.

Chapter 4 examines various aspects of bacterial growth, including specific nutritional categories and physical requirements. This chapter also describes the preparation and handling of media, the conditions needed for growth of organisms with different requirements for oxygen and carbon dioxide, and the phenomenon of growth itself.

Bacteria are grown for many purposes, and each species has particular requirements for its nutrition and growth. These requirements reflect an organism's metabolic and physiological needs as well as its chemical composition; they are also related to its environmental location. Because of such precise growth needs, pure cultures must be obtained and verified periodically. This is particularly true when an organism is to be identified, since so many bacteria appear the same morphologically.

An Introduction to Bacterial Cultivation

A wide variety of procedures and nutrient preparations are used to induce microorganisms to grow and reproduce. Different microbes require different environments and combinations of nutrients, called *culture media* (singular, *medium*). Thus procedures and media for culturing anaerobic bacteria, for instance, are of little value in inducing viruses to grow.

Microorganisms are cultivated in containers ranging in size from test tubes, flasks, and Petri dishes (Figure 4–1) to huge steel tanks. The tanks are used commercially to obtain large quantities of the desired organisms or their products, as discussed in Chapter 15. Whatever the actual container material may be, the procedures are referred to as *in vitro*—literally, "in glass," even though such containers are often made of plastic.

Figure 4-1 An example of a three-compartment Petri dish. Dishes are also available without compartments or with two or more compartments. Different media or specimens can be introduced into each separate area. (See Color Photograph 00.) *[Courtesy of Falcon Plastics, Division of BioQuest, Los Angeles.]*

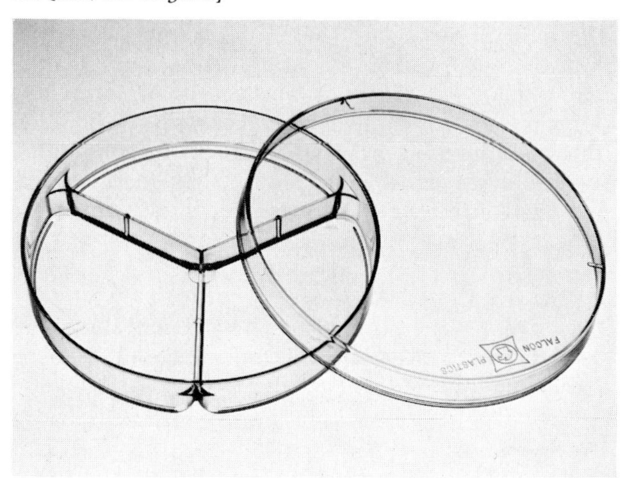

Some microorganisms cannot be grown *in vitro*, but only in living animals. Cultivation methods using live animals are called *in vivo* techniques (Color Photograph 53). When animal tissues are removed and used as a culture medium for microbes, the procedure is an *in vitro* one because a culture vessel is used, not the living (*vivo*) animal. Some microbes can be cultured both *in vivo* and *in vitro*.

In order for certain organisms to cause disease, they must be able not only to survive but also to grow and reproduce in or on the body of their hosts. These pathogens—as well as nonpathogenic microbes that populate different body regions—carry on their life processes in environments containing different organic and inorganic substances, different oxygen and carbon dioxide concentrations, and different degrees of alkalinity or acidity (pH). Cultivation techniques and media used to provide suitable conditions for the growth of bacteria will be discussed from the viewpoint of these different conditions.

General Nutritional Requirements

Heterotrophy, Autotrophy, and Hypotrophy

Heterotrophic organisms require certain preformed, carbon-rich organic compounds for their nutrition. These materials may be sources of carbon (sugars), nitrogen (amino acids), vitamins, or other growth factors. Heterotrophs grow well in any environment with a source of adequate organic and other nutrients. **Autotrophic** microorganisms, on the other hand, can synthesize all or nearly all of their essential organic materials from inorganic carbon compounds. They usually thrive best in soils and bodies of water (Color Photograph 7a).

Both heterotrophic and autotrophic organisms are free-living and can, as a rule, be cultivated on artificial media—if the particular nutrients and environmental conditions for growth are known and provided.

A third group of microorganisms, the **hypotrophs,** are obligate intracellular parasites and thus will grow only within living host cells (Figure 4–2). The hypotrophs include the viruses of animals, plants, and some bacteria such as all rickettsiae and chlamydiae. [See the rickettsias, Chapter 13.]

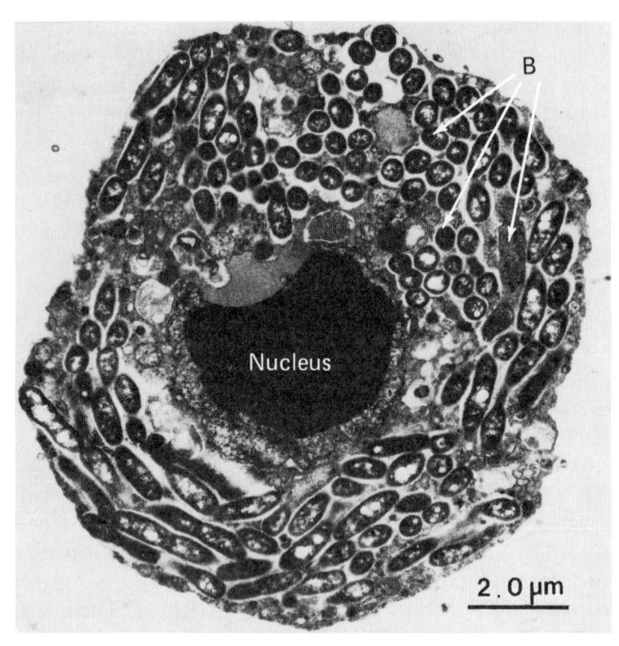

Figure 4–2 A large white blood cell filled with cells of the causative agent of Legionnaires disease. These bacteria (B) have nearly completely occupied the cell's cytoplasm. *[From R. A. Kashimoto, M. D. Kastello, J. D. White, F. G. Shirley, V. G. McGann, E. W. Larson and K. W. Hedlund,* Inf. Immun. ***25****:761–763, 1979.]*

Physical and Chemical Requirements for Bacterial Growth

GASEOUS REQUIREMENTS

Microorganisms are also categorized by their response to available gases such as oxygen and carbon dioxide. Oxygen is involved as an electron receptor in cellular chemical activities (metabolism), and carbon dioxide is an essential nutrient for some organisms.

Aerobic organisms require oxygen, either at atmospheric levels or at some lower concentration.

Microaerophilic organisms require an oxygen concentration lower than that found in the atmosphere (about 20%). Consequently, they do not grow in the upper portions of liquid (broth) cultures where oxygen is abundant. If this avoidance of oxygen is due to a need for more carbon dioxide rather than less oxygen, the microaerophilic organism is called *capneic.* Capneic organisms require an atmosphere with a carbon dioxide level of approximately 3% to 10%.

Certain medically important bacteria, such as *Neisseria gonorrhoeae* (nye-SE-ree-ah go-nor-REE-ah), must be provided with carbon dioxide for growth purposes. With other pathogens such as *Brucella* (broo-SEL-la) spp., causative agents of brucellosis or undulant fever, carbon dioxide is used only for primary isolations. Since many microorganisms of medical importance grow better in the presence of carbon dioxide, it is likely that capneic incubation will become the preferred technique in the laboratory. Details of the conditions of incubation are discussed later in this chapter.

Anaerobes are organisms that do not require free oxygen but possess varying degrees of oxygen tolerance. *Strict,* or *obligate, anaerobes* cannot tolerate any free oxygen in their environment. The human pathogenic bacteria *Bacteroides* (back-teh-ROI-deez) spp. are examples of such organisms. They must be collected from the patient, transported, cultured, and grown under strict anaerobic conditions. The slightest exposure to free oxygen will kill them. *Aerotolerant anaerobes,* on the other hand, can grow in the presence of free oxygen, although it is not required for their metabolism. Streptococci are aerotolerant anaerobes. *Facultative anaerobes* do not require oxygen but will grow better in its presence. The colon bacterium *Escherichia coli* belongs to this category, as do certain yeasts.

One explanation for the sensitivity of obligate anaerobes to an oxygen environment is related to the formation of a highly reactive form of oxygen during microbial metabolic processes. This product, the superoxide radical (O_2^-), alone can be quite destructive, since it has been shown to react with and damage many biochemical compounds. In addition, the superoxide radical is toxic because it can be converted to hydrogen peroxide (H_2O_2):

$$2O_2^- + 2H^+ \rightarrow H_2O_2 + O_2$$

The enzyme in this reaction is superoxide dismutase. Hydrogen peroxide is commonly used as an antiseptic to prevent bacterial growth. Some bacteria produce an enzyme called catalase that breaks down the peroxide to oxygen and water. Another enzyme, peroxidase, protects some aerobic and facultative anaerobic organisms by converting the peroxide to water. Strict anaerobic bacteria do not produce this enzyme and are therefore poisoned by environments containing peroxide.

An additional explanation for the inability of anaerobes to grow in the presence of oxygen concerns the oxidation-reduction potential of the medium. The oxidation-reduction potential is a concept that indicates the electron-accepting or electron-yielding potential of substances. When the oxidation-reduction potential of a medium is lowered by the addition of a reducing or oxygen-absorbing compound, such as cysteine or thioglycollic acid, many anaerobes will grow. Because some reducing agents are toxic, they must be selected with care.

THERMAL CONDITIONS

As a group, microorganisms have been shown to grow between the temperatures of about freezing and above 90°C. Organisms can be classified as **psychrophiles** (cold-loving, 0° to 20°C), **mesophiles** (moderate temperature-loving, 20° to 45°C), or **thermophiles** (heat-loving, 45° to above 90°C),

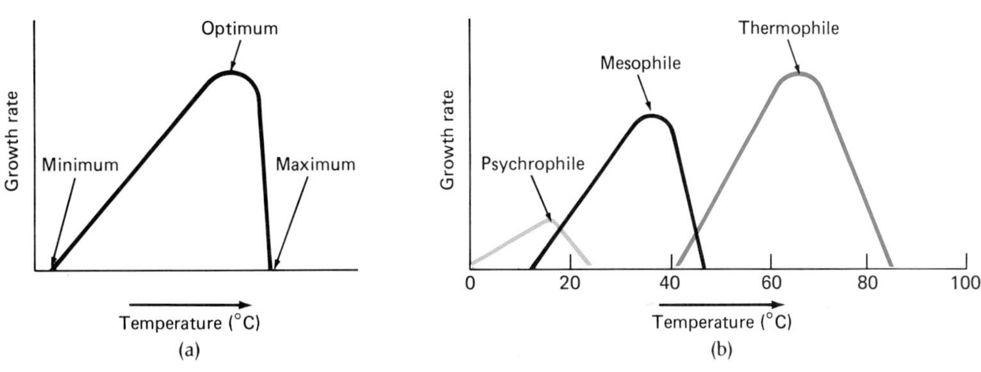

Figure 4–3 (a) The growth responses of a bacterial species over a range of temperatures. Minimum, maximum, and optimum growth temperatures are shown. (b) A comparison of the growth responses over a range of temperatures. Note that there is some overlapping.

based upon their favorable growth temperature range.

Psychrophiles can be found growing in most soil and water environments. They have been found in the Antarctic ice and in the home freezer unit. Mesophiles are more familiar to us, since their temperature range includes our own. These organisms are also found in soil and water. Many are of interest primarily because of their disease-producing capabilities. Thermophiles also are found in soil and water. Unusual thermophilic organisms were reported in 1983, when biologists trying to recreate environmental conditions such as those in certain deep-sea environments found archaeobacteria capable of thriving under pressure and at water temperatures of 250°C and higher.

An organism, although not actively growing at unfavorable temperatures, may still endure them. For example, many organisms can survive in freezing environments; these forms may be called *psychroduric* or *cryoduric*. Organisms that survive at high temperatures are known as *thermoduric* types. *Bacillus* (bah-SIL-lus) spp. are good examples of the latter, due to their formation of heat-resistant spores.

Each organism has a restricted range of temperatures within which it will grow. These are the minimum, maximum, and optimum temperatures and are known as an organism's *cardinal temperatures* (Figure 4–3). The cardinal temperatures for any organism depend upon several factors, including the enzymes of the organism, the age of the culture, the supply of nutrients, and various physical and chemical conditions.

ACIDITY OR ALKALINITY (pH)

Microorganisms also have certain pH (hydrogen ion concentration) needs, as reflected by their growth responses in various media. **Acids** supply hydrogen ions, while **bases** accept them and neutralize solutions containing hydrogen ions. Basic substances also supply hydroxyl ions (OH^-).

Reactions that take place in cells, tissues, and many organ systems, particularly those involving **enzymes,** are quite sensitive to changes in acidity. The acidity of various compounds is determined by the number of hydrogen ions produced (Table 4–1). In the same sense, the alkalinity of a substance depends upon the concentration of hydroxyl (OH^-) ions it produces. The degree of acidity or alkalinity of a substance is expressed as its *pH.* The pH scale (Figure 4–4) is a convenient way of expressing the hydrogen ion concentration of a substance. Techni-

Figure 4–4 The pH scale. A solution with a pH of 7 is neutral, since the concentrations of H^+ and OH^- are equal. The lower the pH below 7, the more H^+ ions are present, and the more acidic the solution. With pH increases above 7, the concentration of H^+ ions decreases and the concentration of OH^- increases, thus making the solution more basic (alkaline).

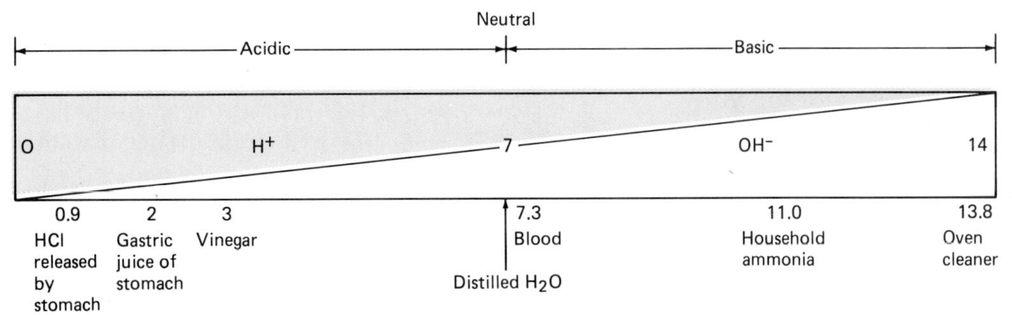

TABLE 4–1 pH Values and Corresponding Hydrogen Ion Concentrations

pH Scale	*pH Value*	*Normal Concentrations*	*Common pH of Substances*	
Highly acid	0	10^{0}	Hydrochloric acid from human gastric juice	0.9
	1	10^{-1}		
	2	10^{-2}	Orange juice	2.0
	3	10^{-3}		
	4	10^{-4}	Tomato juice	4.2
	5	10^{-5}		
	6	10^{-6}	Media for fungi	5.6
Neutral	7	10^{-7}	Milk	6.6
			Distilled water	7.0
			Blood (approx.)	7.3
			Bile	7.8
Alkaline	8	10^{-8}	Some media	8.6
	9	10^{-9}		
	10	10^{-10}		
	11	10^{-11}		
	12	10^{-12}	Lime water	12.3
	13	10^{-13}		
Highly alkaline	14	10^{-14}	A 1-N sodium hydroxide solution	14.0

cally, the pH value is the negative logarithm of the hydrogen ion concentration. To understand the operation of this scale, consider that distilled water at room temperature is slightly dissociated into an equal number of H^+ and OH^- ions and is therefore neither acidic nor alkaline. The actual concentration of H^+ ions in distilled water is *0.0000001* mole/liter. The logarithm of this number is -7, and the negative logarithm is $-(-7)$, or 7. Distilled water, with a pH of 7, is a standard against which other substances can be measured. If a substance contains considerably more hydrogen ions, perhaps *0.0001* mole/liter, its pH would be 4. Conversely, a substance with a smaller hydrogen ion concentration, *0.000000001* mole/liter, would have a pH of 9. The full range of the pH scale goes from 0 (pure hydrogen ion) through neutrality at pH 7, to 14 (pure hydroxyl ion). Note that a pH change of only one unit, say from 6 to 5, involves a tenfold change in H^+ concentration, from 0.000001 to 0.00001 mole/liter.

Acids and bases also are produced as waste products of metabolism. Certain bacteria can withstand very high acid concentrations, and some of those that form acetic acid (vinegar) are actually able to live in extremely strong acid. However, too much acid or base in a cell can be harmful. Fortunately, cells have mechanisms to control pH changes. One such mechanism is the **buffer** system. A buffer is a chemical or combination of chemicals that resists changes in pH when acids or bases are added. The buffer either accepts or donates H^+ ions. Buffers also work to maintain pH when OH^- ions are added. One of the most common buffering systems, and one that is important in human blood, is the carbonic acid-bicarbonate system (H_2CO_3—HCO_3^-). Bicarbonate ions are formed in the body as follows:

$$\underset{\text{Carbon dioxide}}{CO_2} + \underset{\text{Water}}{H_2O} \rightleftharpoons \underset{\text{Carbonic acid}}{H_2CO_3} \rightleftharpoons H^+ + \underset{\text{Bicarbonate ion}}{HCO_3^-}$$

As indicated by the arrows, the reactions are reversible.

When excess H^+ ions are present in blood or other body fluids, bicarbonate ions combine with them to form carbonic acid, a weak acid. The carbonic acid is unstable and rapidly breaks down into carbon dioxide and water. A buffer may release H^+, which combines with OH^- to form water.

Various indicators and electronic pH meters are commonly used to measure the hydrogen concentrations of preparations. Examples of indicators incorporated into bacteriological media are bromcresol purple, litmus, and phenol red (Color Photographs 27 and 28).

Some organisms can be found growing in sulfur springs containing sulfuric acid with a pH of less than 2, others in ammoniated solutions at a pH greater than 8. Fungi, as a rule, grow well at an acid pH range of 5.5 or 6. This property is used in the preparation of media selective for these organisms, since many contaminating bacteria cannot grow under these conditions.

Necessary adjustments to obtain the desired pH usually can be made by the careful addition of standard acids such as hydrochloric acid (HCl) or of bases such as potassium hydroxide (KOH) or sodium hydroxide (NaOH).

The causative agent of Asiatic cholera, *Vibrio cholerae* (VIB-ree-oh KOL-er-ee), can tolerate a pH of 8. This fact can be used in the preparation of media for the isolation of the organism, since it must be

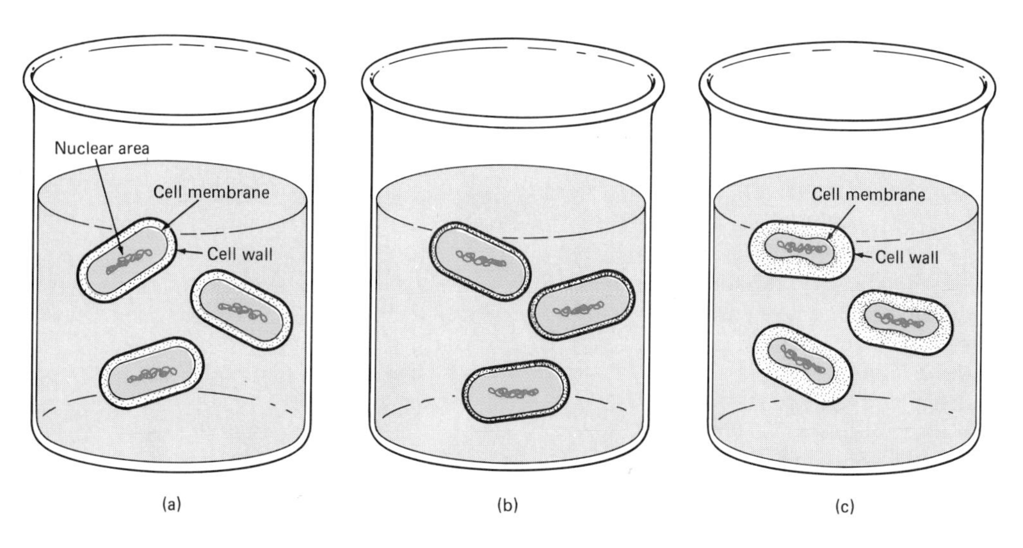

Figure 4–5 Osmotic pressure effects on cell wall-containing microorganisms. (a) The **isotonic** environment. No changes in cell size occur. (b) The **hypotonic** environment. Water molecules enter cells causing them to swell. The presence of a cell wall enables the cell to withstand the increase in osmotic pressure. (c) The **hypertonic** environment. Despite the presence of a cell wall, water loss and shrinkage of cellular content occurs. This is called process plasmolysis.

separated from other typical bacteria comprising the enteric flora found in feces. As a rule, microorganisms prefer a more neutral pH, between 6 and 7.5. [See Chapter 8 for fungal cultivation.]

OSMOTIC PRESSURE

Bacterial growth may also be influenced by the force or tension built up when water diffuses through cell membranes. The force that drives the water is known as osmotic pressure and results from the tendency of water molecules to pass through a membrane and to equalize their concentration on both sides of the membrane. The fluid compartments of all living cells contain dissolved salts, sugars, and other substances that give a certain osmotic pressure to the fluid. When such a cell is placed into a fluid environment that has exactly the same osmotic pressure as the cell, there is no net movement of water molecules either in or out of the cell. Thus, the cell neither swells nor shrinks (Figure 4–5a). The osmotic pressure of the surrounding fluid is equal to that within the cell and is said to be *isotonic.*

When bacterial cells are taken from an isotonic solution and placed into a fluid having a lower osmotic pressure, water will enter these cells and cause them to swell (Figure 4–5b). The surrounding fluid is said to be *hypotonic.* In most cases the rigid cell walls of bacteria, fungi, algae, and plants enable these cells to withstand, without bursting, an external fluid environment that is hypotonic (contains low concentrations of dissolved substances). The cell wall can be stretched only slightly, and a steady state is reached when the resistance of the cell wall to stretching prevents any further increase in cell size. Cells without walls, however, will burst, resulting in a condition called *plasmoptysis.* This movement is caused by the greater concentration of dissolved substances within the cell in comparison to the surrounding hypotonic environment. Since the dissolved solids cannot distribute themselves because of the microbial membrane, the water molecules are pulled into the cells.

When cells with walls are placed into a (*hypertonic*) solution having a greater concentration of dissolved materials, water passes out to the surrounding environment and the contents of the cells shrink (Figure 4–5c). This process is called *plasmolysis.* In this situation, the cytoplasm becomes concentrated and usually pulls away from the cell wall. Erythrocytes, cells without walls, become crinkled or crenated due to the loss of water (Figure 4–6).

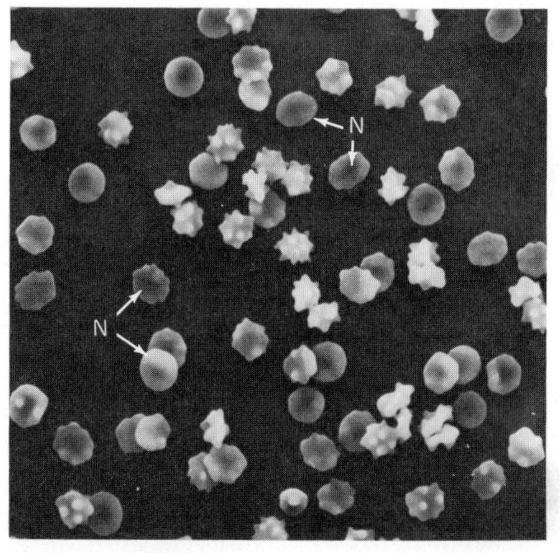

Figure 4–6 The appearance of crenated blood cells (erythrocytes). Normal cells (N) also are shown. *[Courtesy Dr. J. R. Warren.]*

Components of Bacteriological Media

The process of combining various substances into nutritive combinations that are used for the isolation and identification of medically and environmentally important microorganisms has long been an integral part of microbiology. A great many such combinations, or *media,* are available for sterility testing, food and water analysis, and the preparation of biological materials such as antibiotics, vaccines, and genetically engineered products.

Culture media can be prepared either in broth (liquid) or gel (solid) form. Agar, a complex carbohydrate, is the most commonly used solidifying agent. Since it is not a nutrient for most microorganisms, agar can be used in culture media without changing the quality of any preparation.

As living cells, microorganisms require sources of nitrogen, carbon, vitamins, and minerals. These media components and their sources are briefly described in the following section.

Nitrogen Sources

Nitrogen is a component of cellular proteins, nucleic acids, and vitamins (see Appendix A). Microorganisms must therefore be supplied with this element in some form. Many can use ammonium salts such as ammonium chloride (NH_4Cl) as inorganic nitrogen; others require the breakdown products of proteins, such as peptones (partially hydrolyzed proteins), peptides, and amino acids. Some bacteria (for example, *Bacillus* spp.) produce extracellular protein-digesting enzymes (proteases) that break down gelatin and other proteins into the smaller components of peptides and amino acids. These components can then be brought into the cells for further metabolic action. An example of this ability to digest materials in the environment is shown in Figure 4–7. A bacterial culture is shown growing on a nutrient agar that contains the milk protein casein. The presence of the casein causes the agar to appear white. The zones around the bacterial growth demonstrate the ability of the bacterial enzymes to digest the protein for use by the cells.

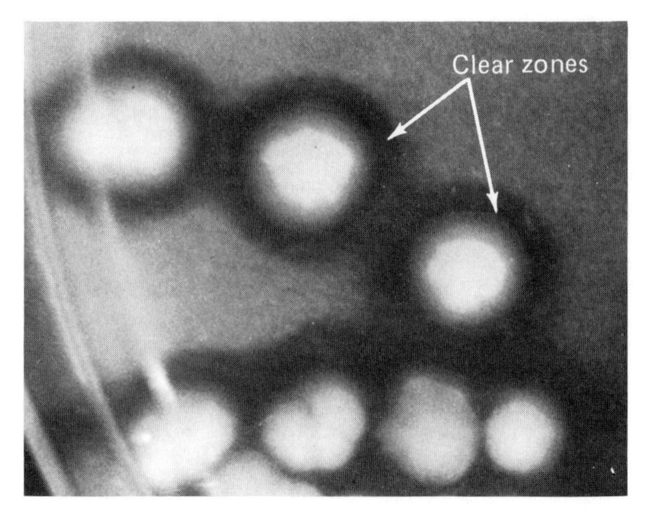

Figure 4–7 Bacteria growing on casein (milk) agar. The clear zone demonstrates the ability of the bacterial enzymes to digest materials in the environment of the cells.

Carbon and Energy Sources

Carbon is the most basic structural element of all living forms. Organisms obtain it from organic nutrients (carbohydrates, lipids, and proteins) and from carbon dioxide, which is produced through a large number of metabolic (chemical) reactions. *Metabolism* refers to the sum total of combined chemical and energy changes that take place in an organism. It consists of *catabolism* (energy-producing reactions) and *anabolism* (energy-requiring synthetic reactions). **[Chapter 5 describes metabolism in more detail.]**

Catabolic reactions produce amino acids, sugar, fatty acids, and other related compounds. Such materials may function in anabolic reactions, thus producing the enzymatic and structural proteins, nucleic acids, carbohydrates, and other biochemical compounds required by the organism. The same compounds may be involved in energy metabolism, producing the impetus for growth and reproduction, or may be used as storage products rich in energy-yielding chemical bonds.

Carbohydrate sources of energy and carbon include starch, glycogen, various pentose (five-carbon) monosaccharides and hexose (six-carbon) monosaccharides, and disaccharides such as lactose, sucrose, and maltose. To utilize the polysaccharides (starch and glycogen), the organism must produce extracellular enzymes to break the complex compound down into smaller molecules that can enter the cell. The enzyme amylase, for example, degrades starch into maltose units, which can then be transported into the cell. Subsequently, the enzyme maltase splits maltose into two glucose units for use in further metabolic activities. Thus catabolism not only yields building blocks for protein, polysaccharide, lipid, and nucleic acid biosynthesis but results in the production of energy with which cells can perform anabolic reactions. **[Refer to Appendix A for descriptions of biochemical compounds.]**

Vitamins and Growth Factors

Several of the vitamins important in the treatment of human nutritional deficiency diseases are also required by microorganisms. Certain microorganisms are capable of synthesizing their own required vitamins; others must obtain them from their nutrient medium. Vitamin compounds that have been shown to be effective in microbial nutrition include thiamine chloride, riboflavin, nicotinic acid, pantothenic acid, pyridoxine, biotin, para-aminobenzoic acid, folic acid, cyanocobalamin, and inositol.

Vitamins can function as portions of coenzymes or as integral components of other biologically active materials. A **coenzyme** (a nonprotein organic molecule) may be thought of as the active portion of an enzyme when it is associated with the protein component of that enzyme.

Various microorganisms require other growth factors as well. These vitamin precursors include purines and pyrimidines for nucleic acid synthesis.

Some microorganisms appear to have peculiar requirements. For example, *Haemophilus influenzae* (hee-MAH-fil-us in-flew-EN-zee) and certain other bacterial species must have a certain component of hemoglobin for growth, in addition to a complete coenzyme—either nicotinamide adenine dinucleotide (NAD) or the related compound NADP (the P indicates a phosphate in the molecule). The interrelationships of the various vitamins and growth factors will be more meaningful when seen in the context of metabolic events, presented in Chapter 5.

Essential Mineral Salts

Various inorganic compounds are also essential for microbial nutrition. These include phosphates, which are required in nucleic acid synthesis, and sulfates, which may be needed in the formation of certain amino acids. Potassium, magnesium, manganese, iron, and calcium serve as inorganic **cofactors** for particular enzymes or may be incorporated into several biochemical reactions. For example, iron is a constituent of cytochromes, which are important in energy metabolism. Calcium is a major component of bacterial spores. Still other inorganic nutrients, including cobalt, copper, zinc, and molybdenum, are required in very small, or trace, amounts. Such inorganics are components of special enzymes and are believed to be required by most, if not all, life. Some organisms have special mineral requirements. The diatoms (Color Photograph 50) require silicon for formation of their cell walls if they are to grow.

Water

The bacterial cell is approximately 80% water and must be in intimate contact with a water supply for its survival, growth, and reproduction. Such contact may involve the presence of a mass of cells on a moist, solid surface or a cell suspension in a liquid medium. When nutrient media are prepared, the various ingredients are dissolved in distilled water, not tap water. Distilled water is used in order to minimize the presence of excess inorganic salts or extraneous organic compounds that may be in tap water. Since tap water may vary in its mineral and dissolved solids composition from day to day, more consistent preparations are obtained with distilled water.

Preparation and Storage of Media

Laboratory media have been devised for the cultivation of specific microorganisms to suit their particular growth requirements. Several preparations of this type can be obtained in dehydrated form from commercial supply companies. Reconstitution is simple. The required weight of the powdered medium is added to distilled water and heated gently to dissolve and mix the ingredients of the preparation. Adjustment in pH may be necessary before the medium is dispensed into smaller vessels. Once this has been done, the medium is sterilized with a high-pressure steam system (autoclaved) according to the directions of the supplier.

Certain components of media, such as serum, plasma, some carbohydrates, and vitamins, are destroyed or inactivated by heat. These components are usually sterilized by filtration and then introduced aseptically (free of organisms) into the remaining portion of the medium. It is also possible to filter-sterilize the entire medium. **[Sterilization and procedures are described in Chapter 11.]**

Media Storage

Most media can be stored under conditions that prevent dehydration. Media are usually kept refrigerated or in airtight packaging. In either case, the media should first be incubated overnight to determine if sterilization was adequate. All media found to be contaminated should be sterilized and then discarded.

Media Usage and Categories

The cultivation of microorganisms may be necessary for any of the following reasons: the isolation and identification of organisms; the determination of antibiotic sensitivities of pathogens isolated from patients; sterility testing of products destined for hu-

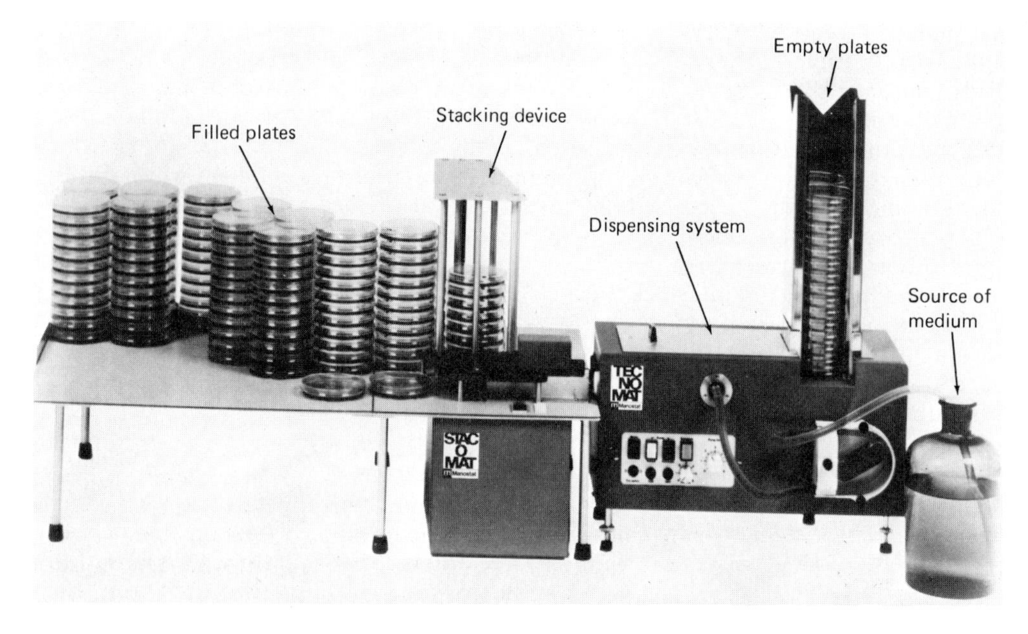

Figure 4–8 Automation in microbiology. Various routine procedures and manual methods in microbiology can be automated. This includes the preparation of plate media. The device shown pours and stacks filled Petri plates for immediate use in a much shorter time than manual preparation takes.

man use; food and water analyses; environmental control; antibiotic and vitamin assays; industrial testing; and the preparation of biological products, such as materials used for immunizations. The choice of which medium to use for a specific purpose is important and can be a problem. Hundreds of formulations in dehydrated or completed form exist. Because of the personnel time required in making media and pouring plates, many laboratories use automated devices to pour media and stack Petri plates rapidly (Figure 4–8). A wide range of sterile, prepackaged, and ready-to-use broth and agar media also are commercially available. The selection of media can be influenced by such factors as availability and cost, personal habit and experience, preferences of instructors or of chief laboratory personnel, and reported research findings. It should be noted that the most efficient laboratory is not necessarily the one with the greatest variety of media. Efficient performance and results depend on carefully chosen media. The following sections will present the different categories of media used for the cultivation of microorganisms and representative examples of each. **[See Chapter 15 for industrial application.]**

Chemically Defined and Complex Media

Media are generally divided into two categories: (1) *chemically defined* or *synthetic* and (2) *complex.* In a chemically defined preparation, the composition of each essential nutrient is known, thus making it possible to duplicate the medium exactly when needed. Table 4–2 lists the ingredients of such a medium together with their associated functions.

TABLE 4–2 Ingredients of a Standard Chemically Defined Culture Medium

Ingredient	*Amount/Liter*	*Associated Functions*[a]
$(NH_4)_2SO_4$	1.0 g	Nitrogen and sulfur source
$MgSO_4$	0.1 g	Magnesium and sulfur source
K_2HPO_4	7.0 g	Potassium and phosphorus source and pH buffer
KH_2PO_4	2.0 g	Potassium and phosphorus source and pH buffer
Glucose	5.0 g	Carbon and energy source
Agar	15.0 g	Solidifying agent[b]
Distilled H_2O	1000 mL	Solvent

[a] The medium's pH is generally adjusted to 7.0.
[b] Not used in the broth form of this medium. In addition, silica gel can be used as an alternate solidifying agent.

TABLE 4–3 **Ingredients of a Standard Solid Complex Culture Medium**

Ingredient	*Amount/Liter*	*Associated Functions*
Glucose	10.0 g	Source of energy and carbon
Tryptone	10.0 g	Source of amino acids and some minerals
Yeast extract	3.0 g	Source of B complex and other vitamins as well as trace elements
Agar[a]	1.5–2.0 g	Solidifying agent
Distilled water	1000 mL	Solvent

[a] Broth media of the same type contain all ingredients except the agar.

Certain ingredients used in the preparation of complex media are not as chemically pure as those used in defined media. The essential nutrients of complex media are generally supplied by extracts from beef, yeasts, and plants (barley) and from the milk protein casein. Nutrient broth, which is a commonly used preparation, incorporates beef extract together with NaCl and distilled water. Table 4–3 lists the ingredients of some standard complex culture media and the functions of each.

The isolation and identification of unknown bacteria and fungi that are of medical and public health importance often require the use of specially prepared complex media including *differential, selective,* and *both selective and differential* plate media. Table 4–4 describes several examples.

Differential Media

Several combinations of nutrients and pH indicators can be used to produce a visual differentiation of several microorganisms growing on the same medium. A *differential medium* is one that supports the growth of various species while providing an environment that makes it easier to distinguish among different organisms. For example, an *enriched* medium such as blood agar can be used to differentiate among many bacterial species belonging to the genus *Streptococcus.* For this group of organisms, approximately 5% to 10% sheep blood is added to a base medium (for example, blood-agar base), which is sterilized separately and cooled before the sheep blood is added. Because of the enzymes and other compounds they produce, colonies of different streptococci exhibit different visual signs on the medium. Based on these effects, a limited identification or classification can be established (see Table 4–4).

In addition to different sources of mammalian blood, carbohydrates, proteins, and other nutrients are used for differentiation purposes. The pH indicators may also be included in media. Those most commonly employed are bromothymol blue, neutral red, and phenol red (Color Photographs 27 and 28).

Selective Media

A preparation that can interfere with or prevent the growth of certain microorganisms while permitting that of others is a *selective medium.* Selective media provide a means for isolating a particular species or category of microorganisms (Figure 4–9). Among the substances that can be used as selective agents are dyes such as crystal violet, eosin Y, methylene blue, and brilliant green, all of which inhibit gram-positive organisms. Bile salts and high concentrations of sodium chloride may also be used. A medium can be made selective by adjusting its pH to a very high or very low level, thus permitting the growth of some organisms and inhibiting others. A well-known example of a selective medium is Sabouraud's glucose agar with a pH of 5.6, used for fungus cultivation (Color Photographs 37 and 98b). Other substances employed to make culture media selective include antibiotics.

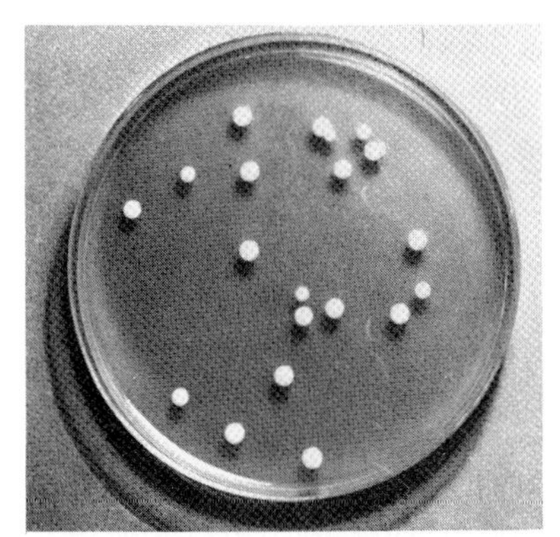

Figure 4–9 The effects of a selective medium. A 100:1 ratio of *Micrococcus luteus* (my-kroh-KOK-kuss LOO-tee-uss) (smaller colonies) and *Staphylococcus aureus* (staff-il-oh-KOK-kuss OH-ree-us) (larger colonies) was plated on a nonselective medium (left) and a selective medium (right). The larger *S. aureus* colonies were favored by the selective medium. *[K-H. Schleifer, and E. Kramer,* Zbl. Bakt. Hyg. I. Abt. Orig. C. 1, *270–280, 1980.]*

TABLE 4–4 Examples of Differential (D), Selective (S), and Both Selective and Differential (DS) Plate Media

Medium[a]	*Type*	*Properties and Reactions*
Blood agar	D	Used to distinguish among species hemolyzing (disrupting) blood. Hemolytic reactions around colonies include alpha (α) hemolysis (green zones); beta (β) hemolysis (clear zones); and gamma (γ) hemolysis (no changes in the medium) (Color Photographs 23a, 23b, and 23c.)
Eosin methylene blue agar	DS	Used in the isolation and preliminary identification of certain gram-negative bacteria. Lactose fermenters form dark-purple colonies, nonfermenters form pink ones. Some lactose fermenters produce a green metallic sheen on colonies. (Color Photographs 24a and 24b.)
Hektoen enteric agar	DS	Used in the isolation and preliminary identification of certain gram-negative bacteria. Reactions include pink to orange zones (lactose fermenters), green zones (nonlactose fermenters), and red zones (fermenters of the carbohydrates salicin and sucrose); nonfermenters leave the medium unchanged and hydrogen sulfide producers cause colony centers to turn black. Bile salts are the inhibitory agents in the medium. (Color Photographs 25a and 25b.)
MacConkey agar	DS	Used in the isolation and preliminary identification of gram-negative enteric bacteria. The carbohydrate lactose is the differentiating substance. Lactose fermenters form red colonies, while colonies of nonlactose fermenters are colorless and clear. Bile salts and crystal violet serve as inhibitors to most gram-positive bacteria.
Mannitol salt agar	DS	Used for the isolation of potentially pathogenic *Staphylococcus aureus* (staff-il-oh-KOK-kus OH-ree-us), which ferments mannitol, turning the medium from pink to yellow (Color Photograph 26). Salt in a concentration of 7.5% is the inhibitory agent.
Salmonella-Shigella agar	DS	Used for the isolation and selection of *Shigella* (shi-GEL-la) and *Salmonella* (sal-mon-EL-la) species. Lactose fermenters form red colonies, while nonlactose fermenters form clear, generally nonpigmented colonies.
Sabouraud's dextrose agar	S	Used for the isolation of yeasts and molds. A pH of 5.6 serves to inhibit most bacteria.
Thayer-Martin medium	DS	Used for the isolation of *Neisseria gonorrhoeae* (nye-SE-ree-a go-nor-REE-a), *N. meningitidis* (nye-SE-ree-a meh-nin-JI-tee-diss), and certain related species. The medium contains several antibiotics to inhibit the growth of contaminating organisms.

[a] The applications of these media are described in later chapters concerned with infectious diseases.

Selective and Differential Media

Many media are both selective and differential. The properties of both types of media are combined for the purpose of isolating and identifying organisms in one general procedure (Color Photographs 25 and 26). Table 4–4 lists several examples.

The ingredients and applications of many differential, selective, and selective and differential media are discussed in the chapters concerned with particular disease-causing agents.

Identification Tube and Plate Media

These media are differential in nature and contain ingredients to detect specific enzymatic reactions and any variations in growth. Although these preparations are not inhibitory, only one or a limited number of organisms produce a specific reaction. *Pure cultures* are required for these differential media, which are widely used for the identification of unknown microorganisms (Color Photographs 27 and 28).

Media for the Cultivation of Anaerobes

Fortunately, most pathogenic anaerobes encountered are heterotrophic and do not require unusual growth factors (are not fastidious). Thus, most of these organisms can be cultivated routinely and identified without the use of complicated equipment or complex media. Quite often, the same media may be used for both aerobes and anaerobes.

Basically, procedures for cultivation and/or identification of anaerobic bacteria are much like those for aerobes. Many of the specific techniques are identical except for the incubation atmosphere.

Broth media vary in composition. Some contain meat and partially hydrolyzed proteins (peptones), sodium, and various phosphates and chlorides (Table 4–2). The sodium serves primarily to ensure a suitable osmotic pressure in the medium; the phosphates provide a buffering action against the acids produced by the metabolic activities of cultured organisms. A reducing agent, which will convert oxygen to water, may also be added to a medium for the cultivation of microaerophilic and certain anaerobic bacteria.

With a preparation such as thioglycollate medium, anaerobic conditions are maintained by two reducing agents, L-cystine and agar, in addition to the thioglycollic acid. Although agar is used as a solidifying agent for many media, its 0.075% concentration here is insufficient for this purpose. However, here agar functions by slowing the diffusion of oxygen.

Resazurin serves as an oxidation-reduction indicator. This compound is colorless or nearly colorless in its reduced state, thereby indicating the existence of anaerobic conditions. When oxidized, it turns pink or red, indicating an aerobic state. A sterile tube of this medium suitable for the growth of aerobes, microaerophiles, and several but not all anaerobes will have an upper pink layer and a lower yellow to light brown layer (see Color Photograph 29). Aerobic organisms grow well in the upper region, anaerobes in the lower zone.

Prereduced Anaerobically Sterilized Media

Prereduced anaerobically sterilized media, when used properly, greatly increase the chances of isolating many fastidious anaerobic bacteria. Prereduced anaerobically sterilized media are prepared, dispensed, sterilized, and stoppered in an oxygen-free nitrogen environment, preventing any exposure to oxygen. In order to maintain the exclusion of air during inoculations, a stream of nitrogen or carbon dioxide without any free oxygen is directed into the tube as it is inoculated. Various media of this kind are commercially available.

Inoculation and Transfer Techniques

Once the medium is selected and prepared, the next step in culturing a microorganism is inoculation of the medium with the specimen. This is accomplished by use of an inoculating loop (Figure 4–10) or its modified form, the inoculating needle.

The inoculating or transfer tool basically consists of a thin piece of heat-resistant wire, usually platinum or stainless steel, measuring 5 to 7.5 cm (2 to 3

Figure 4–10 Inoculation and transfer techniques. (a) Flaming the inoculating tool. Note that the entire wire portion must turn red hot in order to eliminate (incinerate) undesirable microorganisms. (b) Transferring the inoculum to two sterile tubes of broth. The mouths (openings) of both tubes must be flamed before and after the transfer, as must the inoculating tool. Note how the tube cap is grasped. (c) Streaking an agar plate. This procedure involves spreading an inoculum over the surface of the medium.

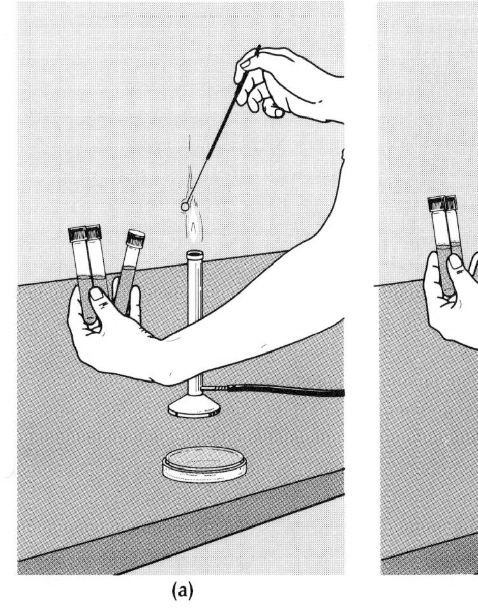

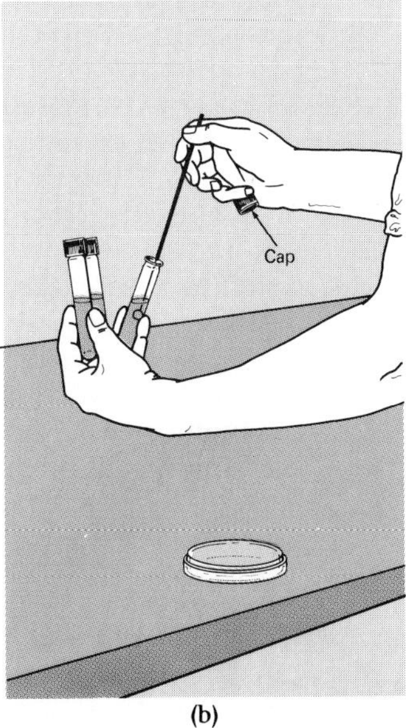

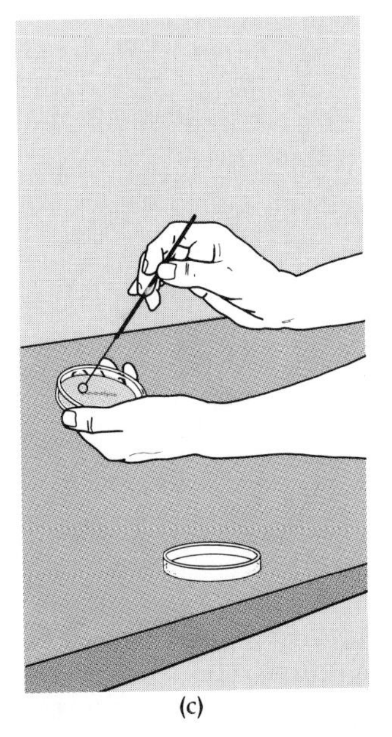

in.) in length and usually provided with a short handle.

Most types of media, such as broth, agar plates, and slants, may be inoculated with either a transfer loop or a needle. However, the loop is generally used with liquid inocula sources of cultures.

The inoculation and transfer techniques are important to master because they are routinely part of many bacteriological procedures. In addition to inoculation, they may be used for isolation and characterization of pure cultures; transfer of microbial growth from one medium to another; preparation of samples for microscopic examination—smears, hanging-drop, or temporary wet-mounts; and maintenance of stock cultures of organisms.

The steps in a typical tube inoculation are shown in Figures 4–10a and b. A tube containing the inoculum and one or more tubes with sterile media are held together in one hand, supported by the three middle fingers. The thumb functions to keep the tubes in proper position. The inoculating tool, which is in the other hand, is passed through a flame (flamed)—heated to redness to incinerate any organism on its surface. Next, the plugs of the tubes are grasped by the little and ring fingers of the hand holding the inoculating tool. The mouths of these tubes are passed through the flame. This flaming creates convection currents that prevent airborne contaminants from falling into the tubes. The heating may also kill some organisms on the tubes and prevent them from contaminating any other media or the individual holding the cultures.

When an inoculum is obtained and introduced into the proper tubes, the lips of the tubes are again passed through the flame, and each plug is inserted into the same tube from which it was removed. Once again the inoculating loop is flamed to redness.

This technique is carried out to prevent the introduction of contaminating organisms, collectively referred to as *sepsis.* Hence the term **aseptic technique** is used for procedures of this type.

Other means for the transfer of microorganisms include sterile pipettes, cotton applicator swabs, and syringes. The device used is determined by the needs of the particular situation.

Techniques for the Isolation of Pure Bacterial Cultures

The isolation and subsequent identification of a bacterial species are important facets of microbiology. Much of our present knowledge of the properties of bacteria and the interactions among them is based on studies with pure cultures. The successful isolation of a given organism into a pure culture is extremely important. *Enrichment techniques* that make it possible to obtain pure cultures are designed to increase the numbers of a particular bacterial species by favoring its growth, survival, or spatial separation from other members of the microbial population. Several techniques can be effectively used to isolate, detect, and enumerate different microorganisms present in specimens by spatially separating them in or on a solid medium and allowing them to develop into colonies. Among these procedures, two methods, the *pour-plate* and *streak-plate techniques* (Color Photographs 30 and 31), have become indispensable to the bacteriologist. Automation of these and related procedures is possible (Color Photograph 86a).

Whatever isolation technique is used, its effectiveness can be greatly increased by using differential or selective and differential culture media, such as eosin methylene blue agar and Hektoen enteric media (Table 4–4 and Color Photographs 24 and 25).

The Pour-Plate Technique

The forerunner of the present pour-plate method (Color Photograph 30) was developed in the laboratory of the famous German bacteriologist Robert Koch (1843–1910). Today this technique consists of cooling a melted agar-containing medium to approximately 45° to 50°C, inoculating the medium with a specimen, and immediately pouring it into a sterile Petri plate, allowing the freshly poured medium to solidify, and incubating the preparation at the desired temperature (Figure 4–11 and Color Photograph 30). When the number of bacteria in the specimen is not known beforehand, suitable dilutions must be made in order to ensure that the colonies will be isolated (Figure 4–11). By means of the pour-plate method, bacteria are distributed throughout the agar and trapped in position as the medium solidifies. Although the medium restricts bacterial movement from one area to another, it is soft enough to permit growth, which occurs both on the surface and in the depths of the inoculated medium.

In addition to isolating and detecting bacterial species in a mixed culture, this procedure is useful in quantitative measurements of bacterial growth. Unfortunately, however, it has several drawbacks. Colonies of several species may present a similar appearance in the agar environment, thus making differentiation difficult. Also, certain species of bacteria may not grow under the cultural conditions of the method. And, finally, removing colonies for further study may be difficult.

The Streak-Plate Technique

The streak-plate technique (Color Photograph 31) was also developed in the laboratory of Robert Koch—this time by bacteriologists Friedrich Loeffler

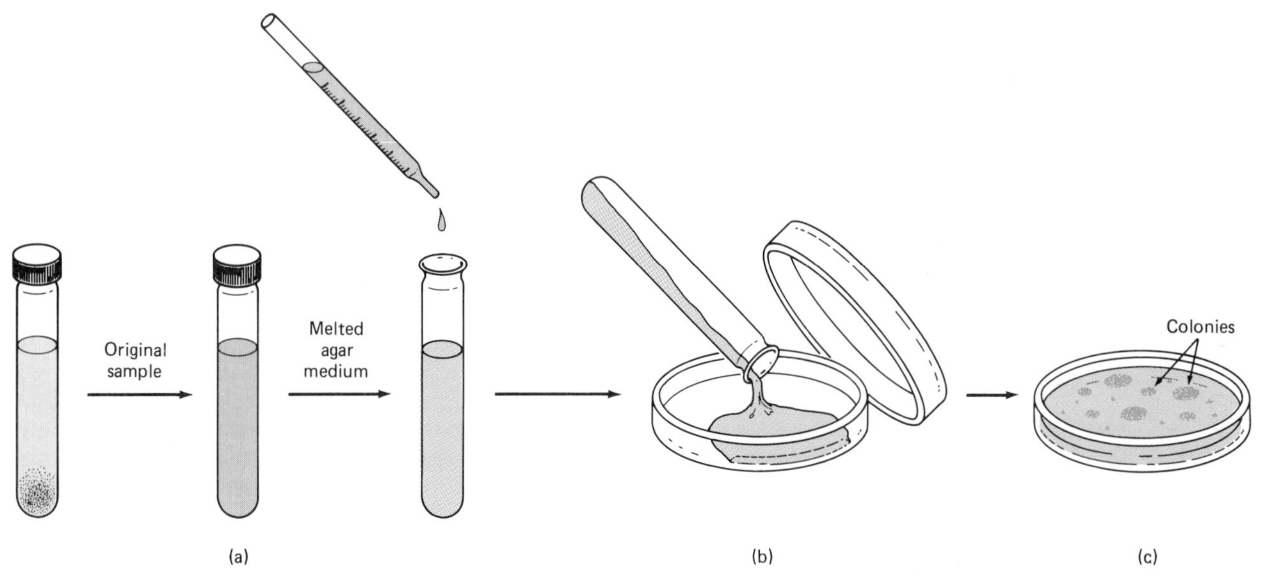

Figure 4–11 The pour-plate technique. Depending on the concentration of microorganisms (microbial load) in an original sample, one or more dilutions in a liquid medium may be necessary (a). Once dilutions are made, they are used to inoculate an agar-containing medium, which is then poured into a sterile Petri plate (b). After incubation, colonies appear on and within the agar medium (c). (See Color Photograph 30.)

and Georg Gaffky. The modern method for the preparation of a streak-plate involves spreading a single loopful of material containing microorganisms over the surface of a solidified agar medium. Figure 4–12 shows streaking methods that have been used.

The clock-plate technique is one of the more common forms of this procedure (Figures 4–13 and 4–14). First, the inoculum is spread over a small portion of the medium's surface. Then the inoculating loop is flamed to destroy any residual bacteria, and the plate containing the medium is rotated approximately one-quarter of a turn. The flamed loop is next used to make a second set of streaks, thus diluting (spreading) the bacterial population in the original set of streaks. As Figure 4–13 shows, the original surface streaks are crossed only once. Further dilutions of the specimen are carried out by repeating this sequence—flaming the loop, rotating the medium, and making additional streaks.

If this technique is properly performed, well-

Figure 4–12 Representative streaking patterns used in the isolation of bacteria.

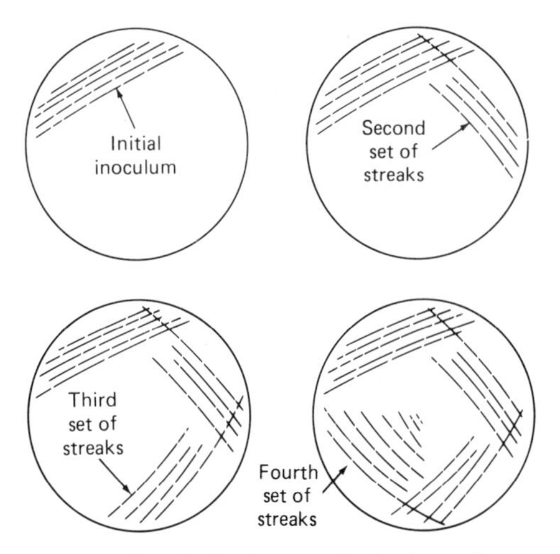

Figure 4–13 The clock-plate method of streaking. The initial inoculum is spread (streaked) over a small portion of an agar plate surface (top left plate). A small amount of the inoculum is then spread over a second portion of the plate (top right plate). Two additional sets of streaks are made in a similar way (two plates on the bottom). It should be noted that the inoculation tool must be flamed between each of the streak sets.

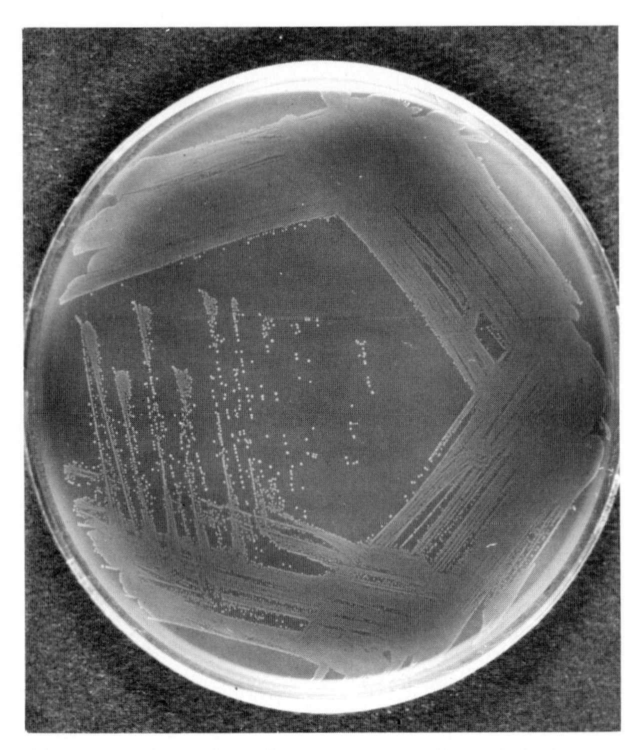

Figure 4–14 A clear demonstration of the clock-plate technique. Note the isolated colonies of *Moraxella phenylpyruvica.* (See Color Photograph 31.) *[P. S. Riley, et al., Appl. Microbiol. **28**:355, 1974.]*

isolated colonies should grow after incubation at an appropriate temperature (Color Photograph 31). One colony is supposed to develop from a single cell, thereby producing a pure culture. It is customary to "pick" (transfer) a small portion of a desired colony to a tube of medium, such as broth or agar slant, and utilize the culture as a source of organisms for additional studies. Where large numbers of specimens and a variety of media are to be streaked, computerized inoculating systems are available, as shown in Figure 4–15.

Another technique, the spread-plate procedure, is used in certain types of investigations. Here a dilution of a bacterial specimen is placed on an agar medium and spread over its surface with the aid of a sterile bent glass rod. The agar plate can be placed on a rotating wheel device to aid the spreading out of the bacterial specimen.

Because of the high concentration of water in agar preparations, condensation forms in most Petri plates. Water of condensation on media surfaces may cause bacterial colonies to run together. Plates are therefore incubated in an inverted position.

Anaerobic Transfer

The lack of growth of many clinical isolates may not be due to the oxidation-reduction potential of the properly prepared anaerobic medium, but instead to brief exposure to air during the specimen collection or transportation to the laboratory, or, after growth, during the transfer of organisms to subculture media. In order to prevent inactivation of anaerobes, transfer should be performed under a flowing stream of nitrogen for cultures on solid media or by the use of a pipette for liquid media. For example, in the transfer of cultures from thioglycollate broth to a fresh tube of medium, a pipette should be introduced into the bottom of the tube with the finger on the mouthpiece. Carefully lifting the finger allows bacterial inoculum from the most anaerobic area to flow into the pipette. Next, with the finger on the mouthpiece, the pipette is removed and inserted into the bottom of a fresh tube of thioglycollate medium. The finger is then quickly removed, which causes the release of the inoculum—again into the most anaerobic portion of the medium. The pipette is then removed, and the newly inoculated preparation is incubated at the desired temperature.

THE ANAEROBIC GLOVE BOX

The methods used to isolate and identify anaerobic bacteria generally require an anaerobic glove box. The equipment consists of a gas-tight chamber with glove portals and an entry lock for the transfer of materials in and out of the chamber (Figure 4–16). An individual using the chamber places his or her hands in the gloves to work with the materials inside. The atmosphere within the chamber may consist of nitrogen, or nitrogen and carbon dioxide, or nitrogen with small amounts of hydrogen and carbon dioxide. Several commercially designed chambers are available.

Figure 4–15 An autostreaker automatically streaks a specimen on the required type of agar plate according to a computerized program, either for isolation or for determining the numbers of bacteria present. *[Courtesy Tomtec, Orange County.]*

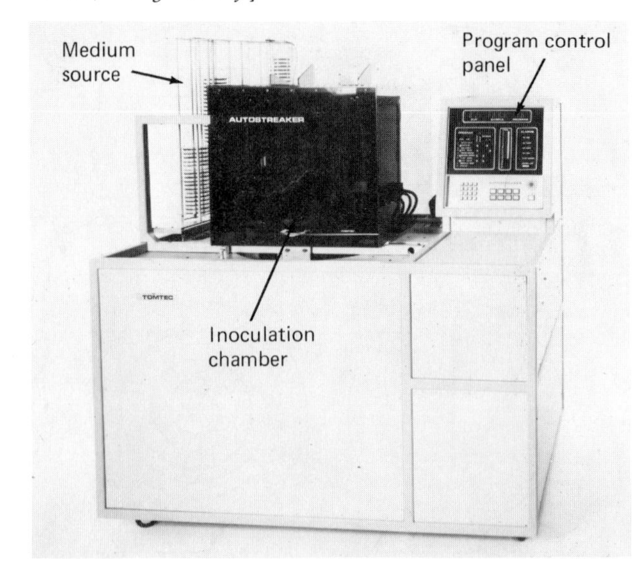

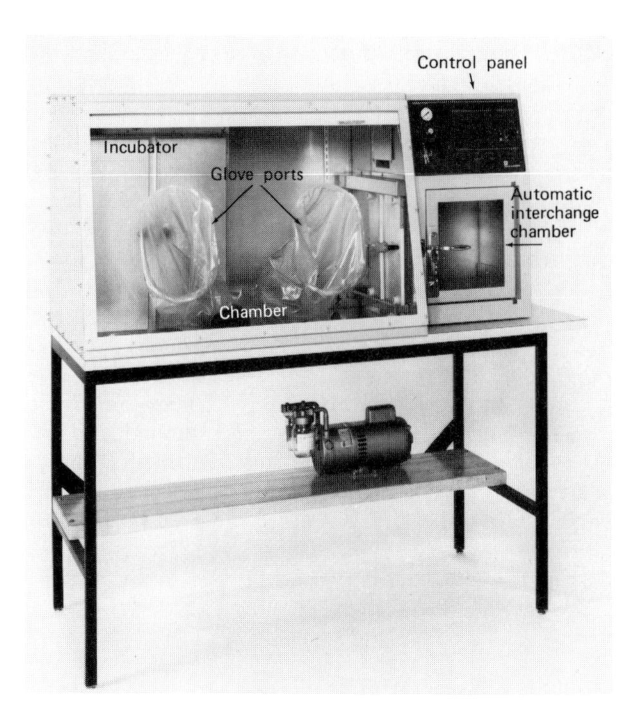

Figure 4–16 An anaerobic glove box. The device consists of a gastight chamber with glove ports and an entry lock for the transfer of materials into or out of the chamber. *[Courtesy of Forma Scientific, a Division of Mallenickrodt, Inc.]*

Conditions of Incubation

Aerobic Incubation

Microbial growth, as noted earlier, can depend upon suitable levels of oxygen and carbon dioxide. While aerobic organisms may be able to grow under anaerobic conditions, the obligate aerobes must be in intimate contact with air or approximately 20% oxygen. Many organisms have less demanding oxygen requirements and will grow well in the depths of liquid or solid media. Therefore, cultivation of aerobic or facultative anaerobic bacteria is relatively simple under usual laboratory conditions.

Once an appropriate solid or liquid medium has been inoculated and protected from contamination, all that remains to be controlled is the temperature and occasionally the level of humidity. Standard laboratory incubators are usually sufficient to maintain such environmental factors.

With any incubator, it is advisable to employ a maximum-minimum thermometer to determine the temperature range provided by the thermostatic control. The thermometer registers the highest and lowest temperature values in any given time and ensures that the incubator is functioning properly.

Capneic Incubation

Air enriched with carbon dioxide or a capneic environment is required by some bacteria, including *Neisseria gonorrhoeae, N. meningitidis* (nye-SE-ree-a meh-nin-jit-EE-diss), and *Brucella* (broo-SELL-la) spp., for primary isolation from clinical specimens. Many other microorganisms, such as *Mycobacterium tuberculosis* (my-koh-back-TI-ree-um too-ber-koo-LOW-sis), appear to grow better under capneic conditions. This type of environment can be provided in a variety of ways including commercially available, disposable CO^2 gas-generating envelopes (GasPak) and the candle jar. The CO_2-generating envelope contains two tablets, one citric acid and the other sodium bicarbonate. When water is added to this system, CO_2 is released. With the candle-jar technique, Petri plates or tubes with inoculated media are placed in a jar with a candle. The lighted candle continues to burn after the lid has been placed on the jar until the carbon dioxide concentration increases enough to stop combustion, usually at about 3% to 5% carbon dioxide. This does not represent complete combustion of oxygen, as aerobic organisms will grow and strict anaerobes will not grow in such a candle jar.

Anaerobic Incubation

Standard laboratory incubators are available for anaerobic conditions and temperature control. These incubators are designed for vacuum attachments by which air is removed and inert gases are added.

OTHER METHODS OF ANAEROBIC INCUBATION

Before the advent of the anaerobic incubator, several techniques were used for the culture and incubation of anaerobes. Today, certain useful devices are used to remove all traces of oxygen and to provide a suitable atmosphere for anaerobes. These include heavy-walled anaerobic jars with gas-tight lids (Figure 4–17a) that can hold a number of tubes, plates, or other containers as well as disposable units such as plastic bags or pouches (Figure 4–17b) that have a limited holding capacity for plates and other items.

With the anaerobic jar, plates or tubes are placed in the unit with an envelope containing the anaerobic generating system. The envelope is opened, water is added to the reagents, and the jar is sealed immediately. The generation of hydrogen in the presence of air and the catalyst causes reduction of the oxygen to water, which condenses on the side of the unit. This system eliminates the need for hydrogen and nitrogen tanks and vacuum pumps, which aids the ability of a small clinical laboratory to carry out anaerobic cultivation. Disposable systems work much the same way.

(a)

(b)

Figure 4–17 Commercial anaerobic devices. (a) The components of the GasPak® Anaerobic system include: GasPak anaerobic jar (J), lid (L), clamp (C), and disposable hydrogen GasPak carbon dioxide generator envelope (E). (b) The GasPak Pouch anaerobic system. This system utilizes a reagent that contains all the reactants necessary to create an anaerobic environment. The reagent also contains a built-in indicator to show that anaerobiosis has been achieved. *[Courtesy BBL Microbiology Systems. Division of Becton, Dickinson and Company.]*

ANAEROBIC INDICATORS

Anaerobic incubation systems should be checked whenever possible to ensure that adequate conditions have been obtained and maintained during incubation. One system used for this purpose incorporates methylene blue as an indicator of oxidation-reduction potential in the medium. It is usually dispensed into small screw-capped vials and autoclaved. This reduces the methylene blue to a colorless state, leukomethylene blue. The procedures for testing involves placing an opened vial of medium in the incubator and initiating the anaerobiosis environment. After a suitable incubation period, the vial is removed and examined. If anaerobic conditions were maintained during the test period, the methylene blue should still be in the colorless reduced state. However, if anaerobiosis was not achieved, the medium will be oxidized and appear blue. (Note that in the first case, the brief exposure to air may cause a slight oxidation of the leukomethylene blue at the surface of the medium.)

Another system for checking an anaerobic incubator involves subculturing an obligate anaerobic bacterium in the same incubator as other specimens. If the indicator organism grows, satisfactory conditions were met. Organisms that may be used for this purpose include certain *Clostridium* (klos-TREH-dee-um) and *Bacteroides* spp. However, the use of chemical indicators is recommended because of their simplicity and reliability.

Patterns of Growth

The ways in which organisms grow and respond to various media provides information helpful to their identification and to detecting culture contamination (unwanted organisms).

After incubation, certain cultural characteristics, such as pigmentation, type of growth on agar slants, pellicle formation, and colonial appearance of organisms can be observed (Figures 4–18 and 4–19 and Color Photographs 19 and 20). Such properties are often useful in describing a particular bacterial species.

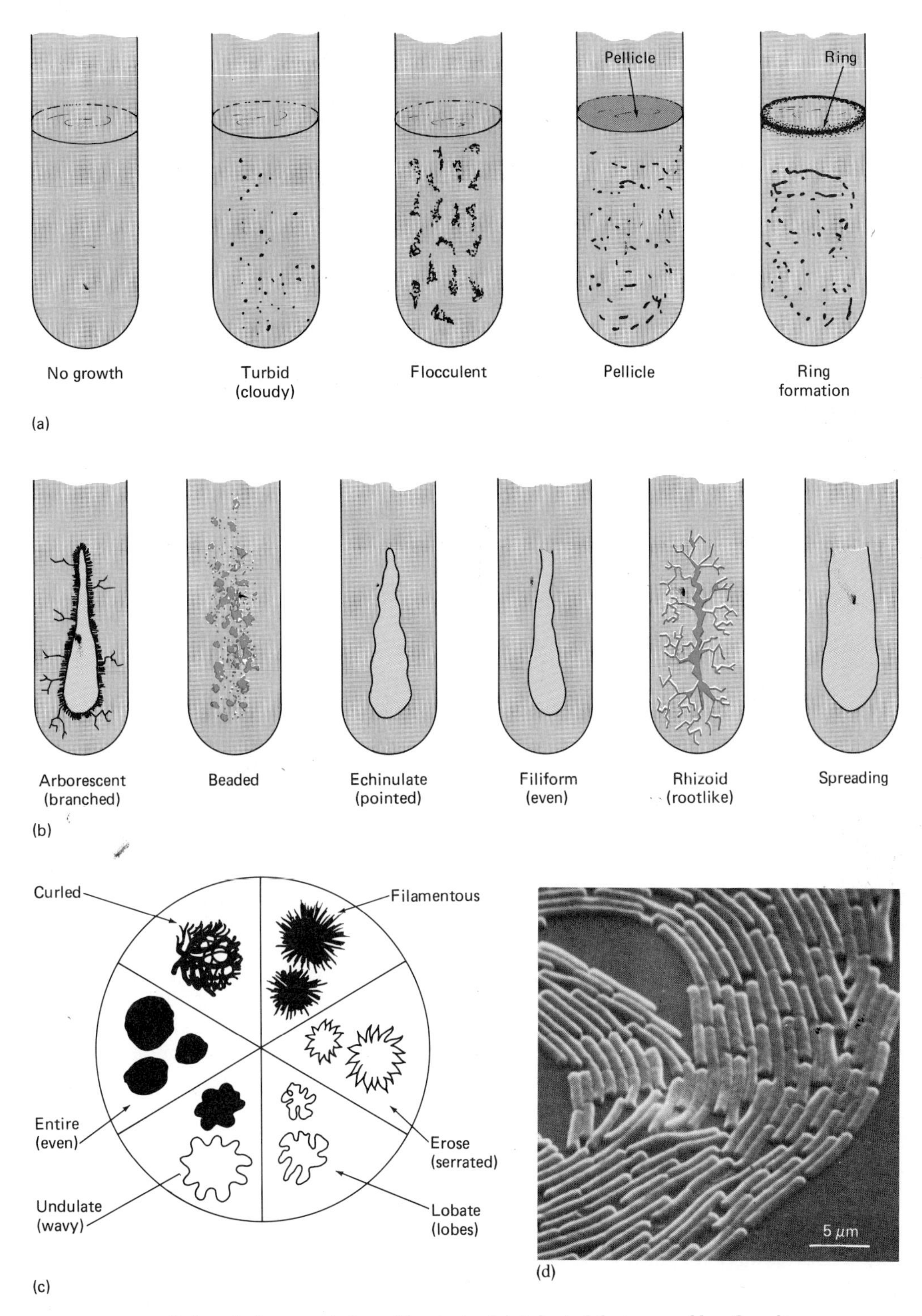

Figure 4–18 Cultural characteristics of bacteria. (a) Selected features of broth cultures. It should be noted that the pellicle covers the broth surface. (b) Agar slant strokes. In this case the inoculum is introduced to the base of the slant and down along the surface from this region to the end of the slant. (c) Some margin characteristics of bacterial colonies. (d) A scanning micrograph of colonial growth showing the arrangement of bacterial rods at the edge of a colony. *[E. G. Afrikian,* et al., J. Appl. Microbiol. ***26****:934–937, 1973.]*

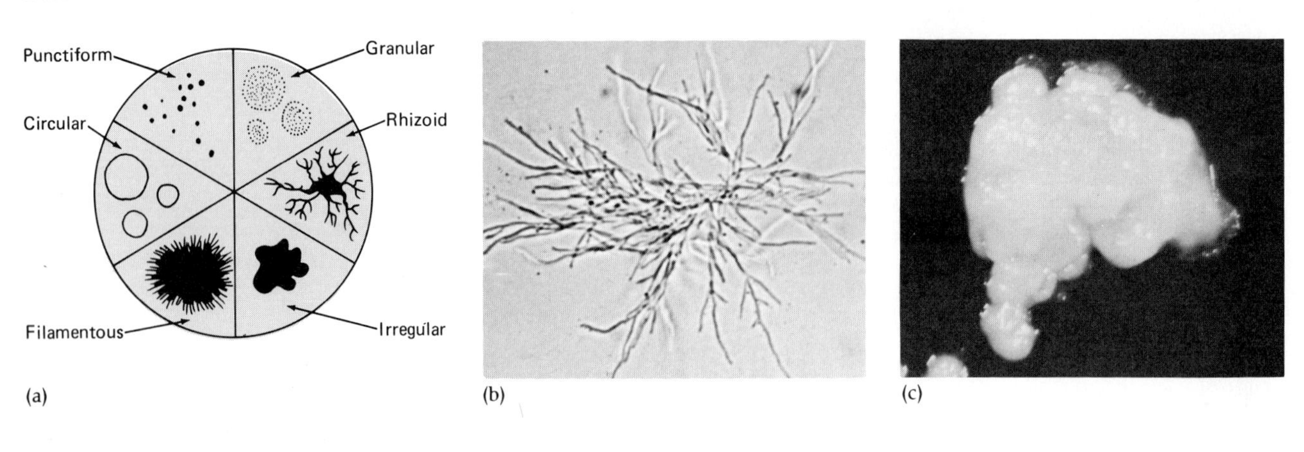

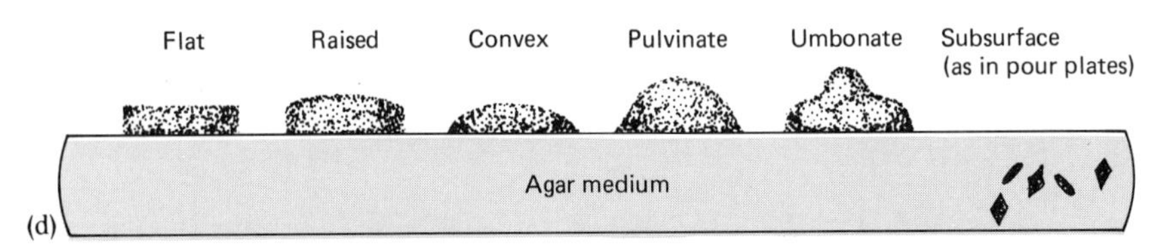

Figure 4–19 Selected characteristics of bacterial colonies. (a) Diagrammatic representation of colonial forms of growth. (b) A developing microcolony of *Actinomyces propionicus* (ak-tin-oh-MY-sees PROH-pee-on-ee-kus). Note the filamentous nature of the organism. (c) "Molar" tooth colonies of the same organism shown in its mature form. (d) Elevation characteristics of bacterial colonies. *[M. A. Gerencser, and J. M. Slack,* J. Bacteriol. ***94:****109, 1967.]*

Maintaining Pure Cultures

Once pure cultures are obtained, they must be kept free from contamination. For long-term storage, cultures may be frozen in liquid nitrogen. Cultures may also be preserved in a dried state by the freeze-drying process known as *lyophilization.* Most pure cultures remain true to type and they remain viable under these conditions almost indefinitely.

Bacterial Growth

Most bacteria reproduce by binary fission, the production of two new cells from one parent. This can be seen quite clearly in Figure 4–20. However, some bacterial species, bacterial variants lacking cell walls, and mycoplasma increase their cell numbers asexually by budding. This budding process can be seen quite clearly with *Hyphomicrobium* (high-fo-my-KROW-bee-um) spp. **[See budding bacteria, Chapter 13.]**

Bacterial Growth Curve

When bacteria are transferred to a fresh broth medium that is otherwise unchanged, subsequent growth over time generally follows a curve, as shown in Figure 4–21. The actual times and shape of each portion of the curve and the numbers of viable living organisms obtained will vary between species and different types of media used.

THE LAG PHASE OF GROWTH

The *lag phase* represents what appears to be a transition period for bacteria transferred to new conditions. At this time, the bacteria are producing the necessary enzymes so that they can grow in the new environment. While there is no increase in cell number, there is a considerable increase in the size of individual cells, with increases in cell protein, DNA, dry weight, and overall metabolic activity. The increased activity appears to be essential to the adjustment process, which may involve altering the pH or oxidation-reduction potential of the medium. These cells are said to be in a state of physiological youth. Although this state of activity is critical to the development of the new culture, it creates conditions potentially harmful to the cells. During physiological youth, cells are more permeable to materials in their environment and contain a higher percentage of free water. These properties appear to increase the susceptibility of the cells to various toxic chemicals as well as to heat.

Microbiology Highlights

BIOLOGICAL OR "STERILITY TEST" CABINETS

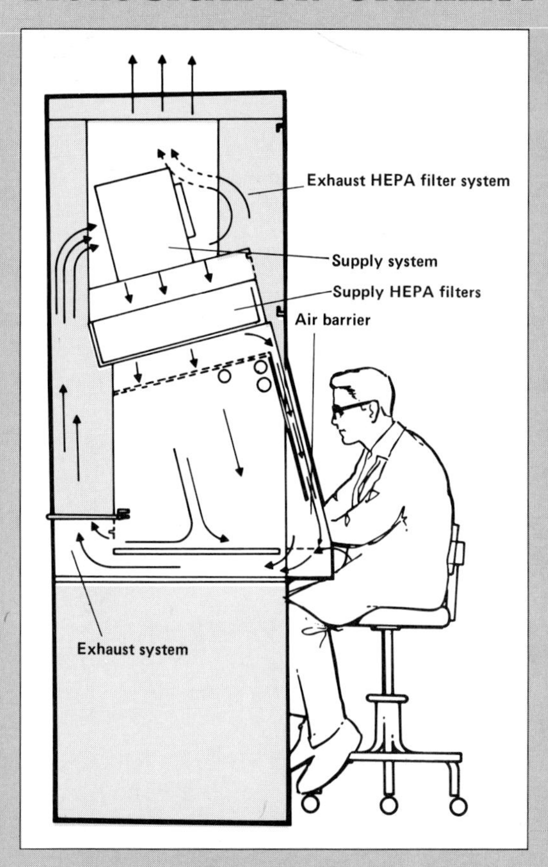

A biological cabinet. The air flow patterns create a barrier to contaminants passing either into or out of the work area. [Courtesy of BioQuest, Cockeysville, Md.]

Protection from airborne contamination is important both for obtaining and maintaining pure cultures and for the safety of those working where microbes are being cultured. An example of a compact cabinet designed for this purpose is shown in the accompanying figure. Such units are valuable in laboratories that handle biohazardous substances or that prepare bacteriological media for sterility testing and for inoculation of a variety of media.

Cabinets of this type are equipped with a double transparent viewscreen. Air circulating between the layers creates a "front air barrier" at the opening of the work area. This barrier protects both the laboratory personnel and materials from contamination by preventing airborne particles from leaving or entering the cabinets. Air entering and leaving the working area passes through HEPA (high-efficiency particulate air) filters, which remove particles 0.3 μm and greater. Although contamination of personnel and materials is prevented in a unit such as this one, workers must still observe aseptic precautions and techniques.

The actual length of the lag phase depends on several factors, including the status of the transferred cells (normal or damaged), the previous environment, and the number of viable (living) organisms in the inoculum. When conditions are suitable, division begins, and after an acceleration in the rate of growth, the cells enter the logarithmic (log) or exponential phase.

THE LOGARITHMIC PHASE

During the *logarithmic (log) phase,* the number of cells increases in a geometric progression—one splits to make two, two to make four, four to make eight, and so on. Division occurs at a constant and maximum rate for the existing cultural conditions. When the number of cells is plotted against time on a logarithmic scale (Figure 4–21), a straight line is

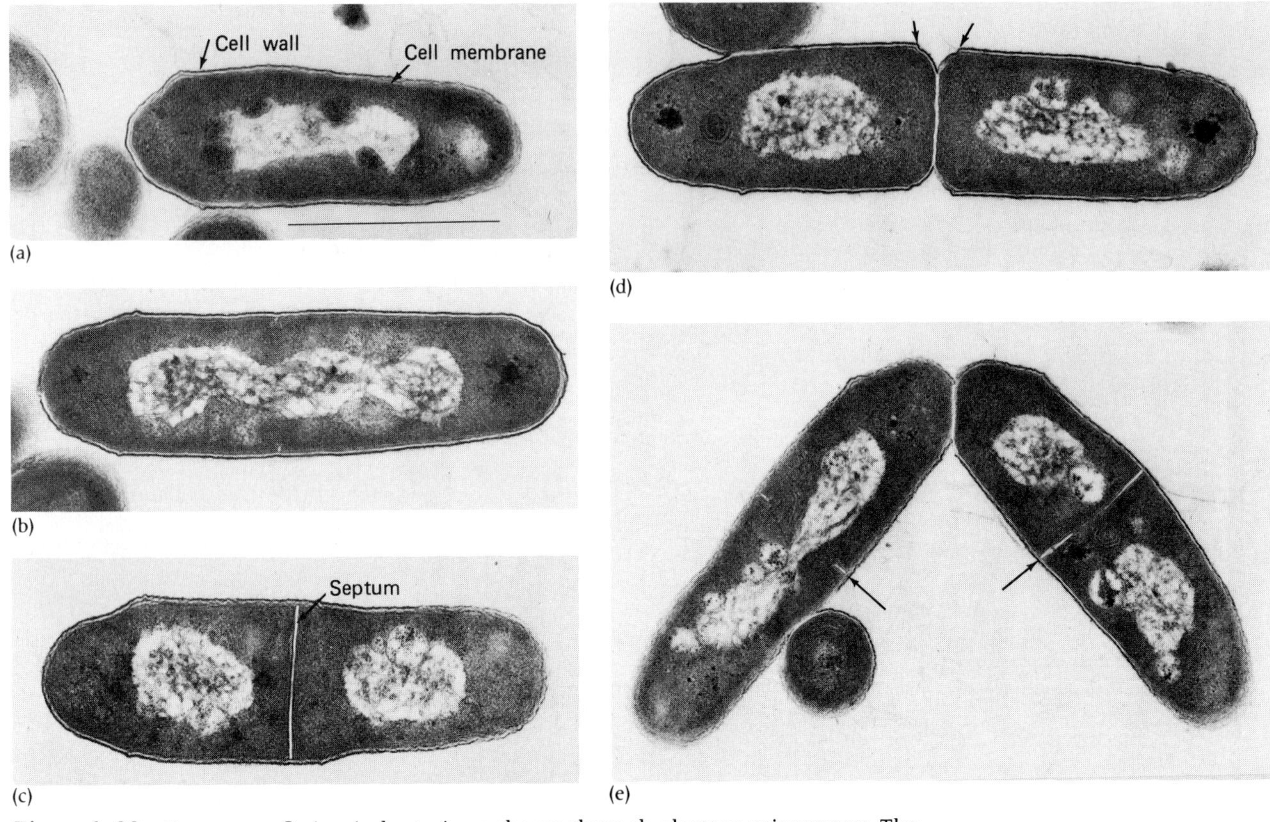

Figure 4–20 Transverse fission in bacteria as shown through electron microscopy. The sequence begins with a parent cell (a) and continues with (b) cell elongation and cross-wall or septum formation. Later steps in the process include (c) complete formation of a cross wall, (d) rupture of outer portion of the cell wall, and (e) two daughter cells bending to separate from each other. Note the new septum formation already starting (arrows). The bar marker represents 1μm. *[From A. Takade, K. Takeya, H. Taniguchi, and Y. Mizguchi,* J. Gen Microbiol. ***129**:2315, 1983.]*

obtained. This graphic presentation emphasizes the geometric change in cell populations over time. It is from this logarithmic increase in cell number that one is able to calculate the average time for a cell to divide, the *generation time (g)*. The generation time is equal to the period of time required for the number of cells present to double.

The generation time of a bacterium can be determined directly from a graph in which the logarithm of the number of cells is plotted against time. It is

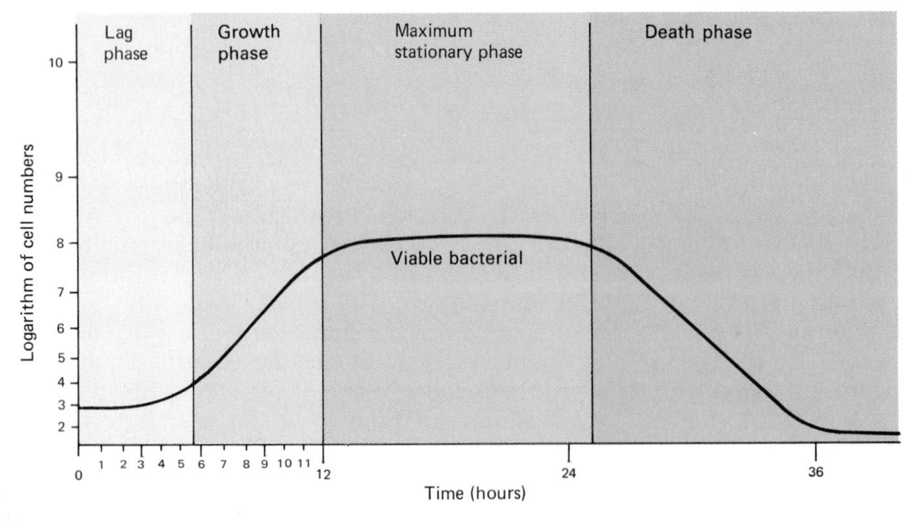

Figure 4–21 A representation of a generalized bacterial growth curve showing numbers of bacteria and times of incubation.

also possible to calculate generation time mathematically by means of the following formula:

$$\text{Generation time} = \frac{t}{3.3 \log (B_0/B_1)}$$

where t = the time interval between the measurement of cell numbers at one point in the log phase (B_0) and then again at a later point in time (B_1)

B_0 = initial bacterial population

B_1 = bacterial population after time t

log = logarithms to the base 10

3.3 = log 10 to log 2 conversion factor

It is important to note that as cell numbers increase, nutrients are consumed, and metabolic waste products accumulate. These events inevitably lead to the end of the log phase.

STATIONARY GROWTH PHASE

At some point, the growth rate begins to taper off and the *stationary phase* starts. Here the growth and death rates are more nearly identical and a fairly constant population of viable cells is achieved. While some cells die, releasing nutrients, others reuse

Microbiology Highlights

CELL CYCLES

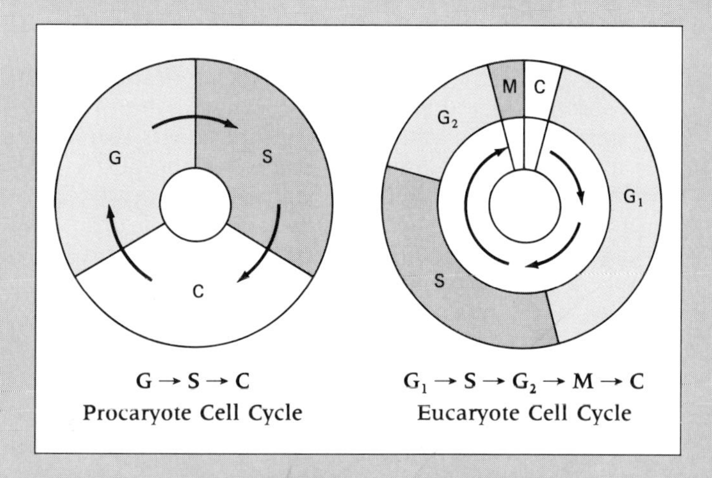

Procaryote Cell Cycle

Eucaryote Cell Cycle

Cell division is a continuous process, which together with cell growth forms the basis of a *cell cycle*. This is the interval between one cell generation and the next. Quite frequently the progressive events or phases of such a cycle may be represented as a *cell-cycle diagram* and the various stages of division indicated by symbols (see the accompanying figures). For a procaryote, three general phases or periods are recognized: G or growth phase, during which active protein synthesis takes place; S or synthesis phase, in which DNA duplication occurs; and the C or division (fission) phase, in which the cell splits into two new cells, each of which normally would be capable of repeating the cycle.

For the eucaryote, the cell cycle may be divided into five phases: G_1 (gap), in which active protein synthesis and the formation of new organelles such as mitochondria and the endoplasmic reticulum occurs; S, in which DNA and chromosome duplication takes place; G_2, in which the structures directly related to mitosis are produced; M (mitosis), in which even distribution of genetic material takes place; and C or the cytokinesis phase, in which the duplicated cytoplasmic parts are divided between newly formed cells. The S period as described is preceded and succeeded by the G_1 and G_2 phases. It should be noted that no DNA is formed during these gap periods. The actual duration of cell cycles and the different phases varies considerably in different cell types.

these nutrients and continue to divide. During this phase, energy metabolism continues; with some organisms, commercially important biosynthetic products such as enzymes and antibiotics are formed.

LOGARITHMIC DEATH RATE

In the stationary phase, conditions develop that accelerate the rate of death. These conditions include the accumulation of toxic wastes such as acids and the decreasing concentration of essential nutrients. After a short period, a *logarithmic death rate* is observed. During this period, the number of viable cells decreases in a geometric progression that is the reverse of the one in the log growth phase. This situation will continue until the number of cells is very low and remains almost constant for a time. The culture now has entered the final phase of the growth curve.

Viable cells may persist for some time in this phase. Some may tolerate the ever-increasing accumulation of wastes, and some may even reproduce, although with an extremely long generation time. During this period, dead cells will lyse from naturally occurring *autolytic* (self-digesting) *enzymes* and release cellular contents, which may serve as nutrients for the remaining viable organisms. At some point, after days, weeks, or even months, conditions will cause even the hardiest of organisms to perish. This usually occurs with the drying out of the medium.

Continuous Culture

The batch processes for growing bacteria are often adequate for various biochemical studies and for antibiotic, antigen, and vaccine production. However, there is usually considerable variation in the ages of the resulting cells and their metabolic activities. Techniques of continuous culture were developed to avoid these variations. Devices like that shown in Figure 4–22 have been used for large-scale cell production and the controlled production of many individual biochemical compounds.

In this system, the rate of growth is controlled by providing a liquid medium with some essential nutrient at a concentration that is less than optimal and with a flow rate less than the growth rate. This *external control* keeps all cells in the log phase of growth. For example, in a simple synthetic medium for *Escherichia coli* containing glucose, ammonium chloride, and phosphate, the nitrogen source is usually chosen as the *limiting growth factor.* By keeping the glucose and phosphate at high levels and determining the growth rate or generation time at varying concentrations of ammonium chloride, one is able to determine the suboptimal concentration for the limiting growth factor. With external control, the system is self-stabilizing, continuously providing cells with whatever growth rate is needed to yield the best product.

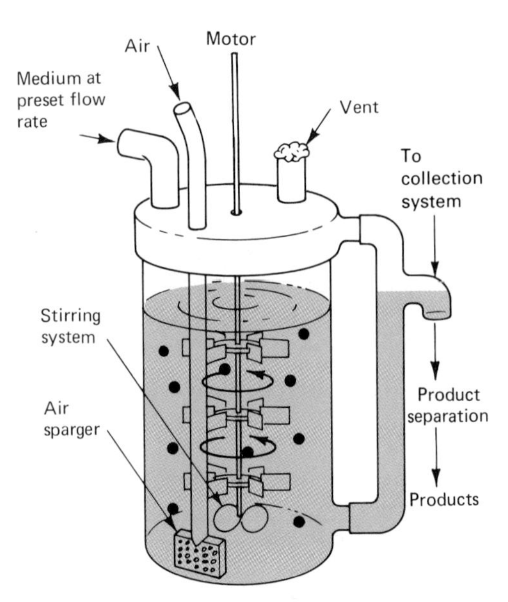

Figure 4–22 A device for large-scale cell production. The example shown is of a continuous culture system utilizing a growth-limiting medium.

Growth on Solid Media

Mass culture techniques and biochemical studies generally use either batch or continuous broth cultures. However, colony and slant growth cultures can be extremely useful in evaluating the purity of cultures and in isolating organisms from a mixture and beginning their identification (Color Photographs 19, 20, and 21). Common growth patterns and terminology used for their descriptions are given in Figures 4–19 and 4–20. The growth of bacteria into colonies is illustrated by *Brochothrix (Microbacterium) thermosphactum* (broh-CHOH-tricks thermoh-SVAK-tum) in Figure 4–23.

Measurement of Growth

Microbial growth can be determined by observing an increase in mass or numbers. The selection of techniques for this purpose depends upon the particular organism involved or the requirements of a particular problem. Several different analytical procedures are performed for comparative purposes, taking into account dry weight, protein or nitrogen concentration, and turbidity (cloudiness).

Cell Mass Determination

DRY WEIGHT

Cell mass can be measured from the dry weight of microbial cells in culture. This technique is commonly used to determine the growth of fungi and bacteria. With fungi the mycelial mat (fungal mass)

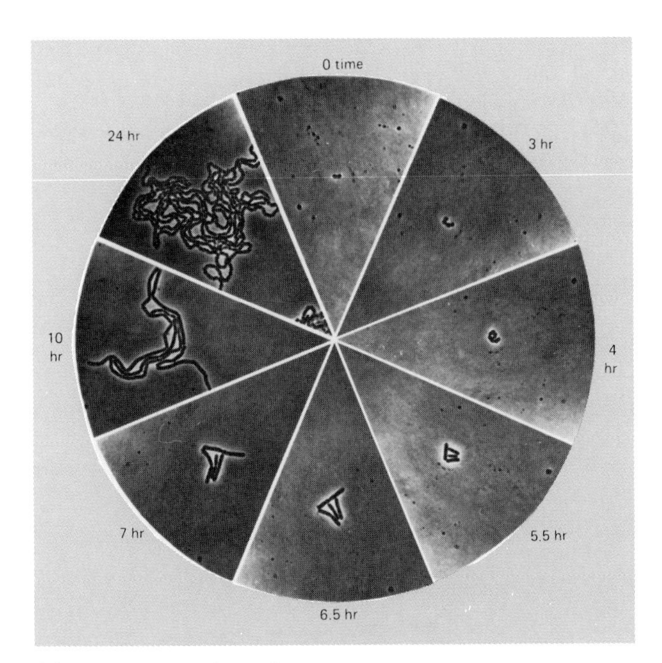

Figure 4–23 Colony formation in *Brochothrix (Microbacterium) thermosphactum* (BRO-jhoh-thrix (my-KROW-back-teer-ee-um) ther-MOH-sphak-tum). [*C. M. Davidson,* et al., J. Appl. Microbiol. ***31****:551–559, 1968.*]

is removed from the growth medium, possibly washed briefly to remove extraneous solids, placed in a weighing bottle, and dried in a heated desiccator. When the microbial remains appear dry, the contents of the bottle are accurately weighed. Usually this procedure—heating followed by weighing—is repeated until a constant weight is obtained. The weight is used to estimate the fungal mass produced under particular growth conditions. Such data may be used to compare the rates of growth of different antibiotic-producing molds or to determine the relative effects of antifungal agents.

Cell mass values of bacteria can be obtained in essentially the same manner. However, to avoid including growth medium along with the bacteria, it is customary to remove the bacteria by centrifugation from a quantity of the medium and then to determine the weight of solids present. The weight of a comparable volume of medium treated in the same manner is then subtracted from the total weight of the bacterial suspension to determine the dry weight of the bacterial cells.

Growth can often be differentiated from reproduction in terms of an increase in cell mass or growth rather than cell numbers. This is not always the case, however, since cellular reserve material may be accumulating without the production of major biological compounds such as nucleic acids or proteins. Thus, in some situations, an increase in mass is not a reflection of growth.

CHEMICAL ANALYSIS

Mass can also be determined by chemical analyses for protein or deoxyribonucleic acid, since the concentration of the biochemicals in a growing bacterial culture can be correlated with the increase in cell mass. In this case, samples of the culture are obtained together with cell-free specimens of the medium for control purposes. In this manner, the effect of various nutrients or antimetabolites and related substances upon the protein synthesis of a growing culture can be determined.

TURBIDITY

Dry weight and protein or nitrogen concentrations are not difficult to determine, but a more rapid measure of cell mass is often necessary. One visible and measurable characteristic of a growing culture is the increasing cloudiness (turbidity) of the medium. A cell suspension looks cloudy because each cell scatters light. The more cell material present, the greater the scattering of light, thus making the culture more turbid. Changes in turbidity are used in various studies and correlate well with other growth characteristics of a particular organism grown under a particular set of conditions.

Turbidity measurements can be performed with a variety of instruments, including *colorimeters* and *spectrophotometers.* With these two instruments, the amount of light lost (by absorption or scattering) in passing through the sample is measured. The intensity of the transmitted light is compared with the intensity of light transmitted by a sample of the clear medium to give a measure of turbidity called *optical density* (Figure 4–24).

Comparison of the culture medium containing bacteria with the bacteria-free medium, as shown in Figure 4–24, makes it possible to estimate the turbidity produced by organisms even in a medium that is itself colored or slightly cloudy.

Viable Counts

Viable counts are carried out to estimate the number of living and/or infective microorganisms in a given material. Cultures of organisms are diluted in a liquid medium and placed in an environment that allows them to produce some type of visual effect. Agar media or membrane filters are used for certain algae, bacteria, and fungi. Tissue cultures are employed for the chlamydiae, rickettsiae, and viruses (see Chapter 10). The actual count is of colonies for bacteria (Figure 4–25). Estimates thus obtained are expressed in *colony-forming units* (*CFU*).

To obtain a colony count for bacteria, 1 mL of a well-mixed bacterial suspension (contaminated drinking water in Figure 4–25) is placed in a 99-mL sterile dilution liquid (water blank). This 99 mL of sterile fluid may be nutrient broth, physiological saline, or distilled water. This produces a dilution of 1:100, which can also be expressed as 1/100 or 1×10^{-2}. After the dilution step, one or more 1-mL samples are removed and introduced into separate sterile Petri dishes, followed by the addition of

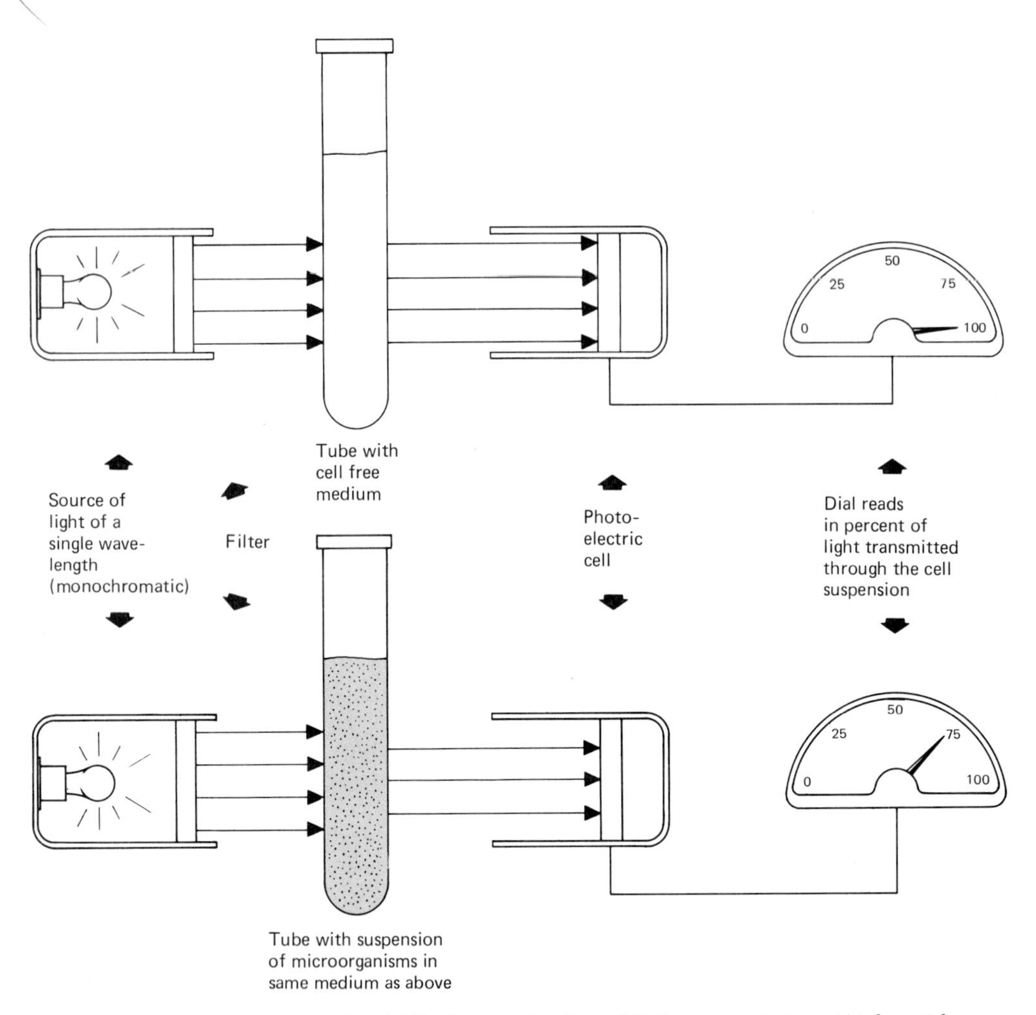

Figure 4–24 Measurement of turbidity by monitoring of light transmission. A tube with a cell-free medium (top). A tube with a microbial suspension (bottom).

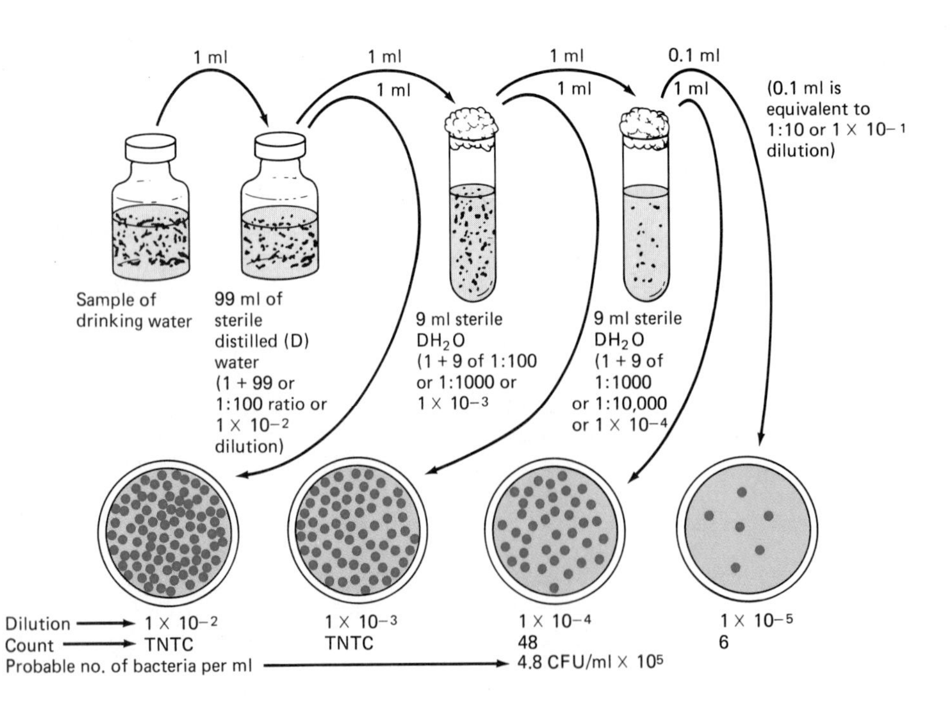

Figure 4–25 The procedure used to determine the viable population in a bacterial culture. The pour-plate technique is used here.

melted and cooled nutrient agar. (This step is not indicated in Figure 4–25, and only one Petri dish per dilution is shown.) The plates are rotated to ensure proper mixing, and the agar is allowed to solidify. The finished plates are then incubated at a desired temperature. After incubation, it can be seen that confluent growth occurred with the 1×10^{-2} dilution (Figure 4–25). This result is referred to as being *too numerous to count (TNTC)*. Thus, the initial 1/100 dilution was not adequate to obtain the necessary separation of colonies for counting. Unless some additional information is available to give an indication of the expected number of bacteria in the original sample, a series of dilutions must be made. Each of these dilutions, in turn, must be plated out in a manner similar to the one described.

The remainder of Figure 4–25 shows the following serial dilution process. A 1-mL sample of the 1/100 dilution is added to a 9-mL dilution blank, which makes a dilution ratio of 1/10 or 1/100, or 1/1,000 (also written as 10^{-3}). This, too, produces colonies too numerous to count. This procedure is carried still further, to 1/10,000 (or 10^{-4}) and 1/100,000 (or 10^{-5}) dilutions, which are plated, incubated, and counted. Only the 1/10,000 dilution results in separating the organisms sufficiently to yield an easily countable plate. This dilution produced a colony count of 48 CFU/mL.

A good rule of thumb for accuracy is not to count plates with fewer than 30 or more than 300 colonies. The lower values are likely to be affected by the mixing of the dilution blank, and with the higher figures a large proportion of colonies may be formed by multiple groups of organisms.

To calculate the probable number of bacteria per milliliter in the original sample, it is necessary only to multiply the bacterial colony count by the reciprocals of the dilution and of the volume used. Thus

$$\text{CFU} = \frac{48}{\underset{(10{,}000)}{1\ \text{mL} \times 10^{-4}}} = 48 \times 1 \times 10{,}000$$

$$= 480{,}000 \text{ or } 4.8 \times 10^{5}$$

The result is the estimated number of bacteria in terms of CFU. This designation is used rather than absolute numbers, since more than one organism may produce a single colony. Several laboratories use an automatic colony counter (Figure 4–26), when large numbers of Petri plates must be counted.

Determining Total Counts of Microorganisms

The term *total count* refers to all organisms, living and dead. Since viable counts require incubation time, it may be necessary to determine the total

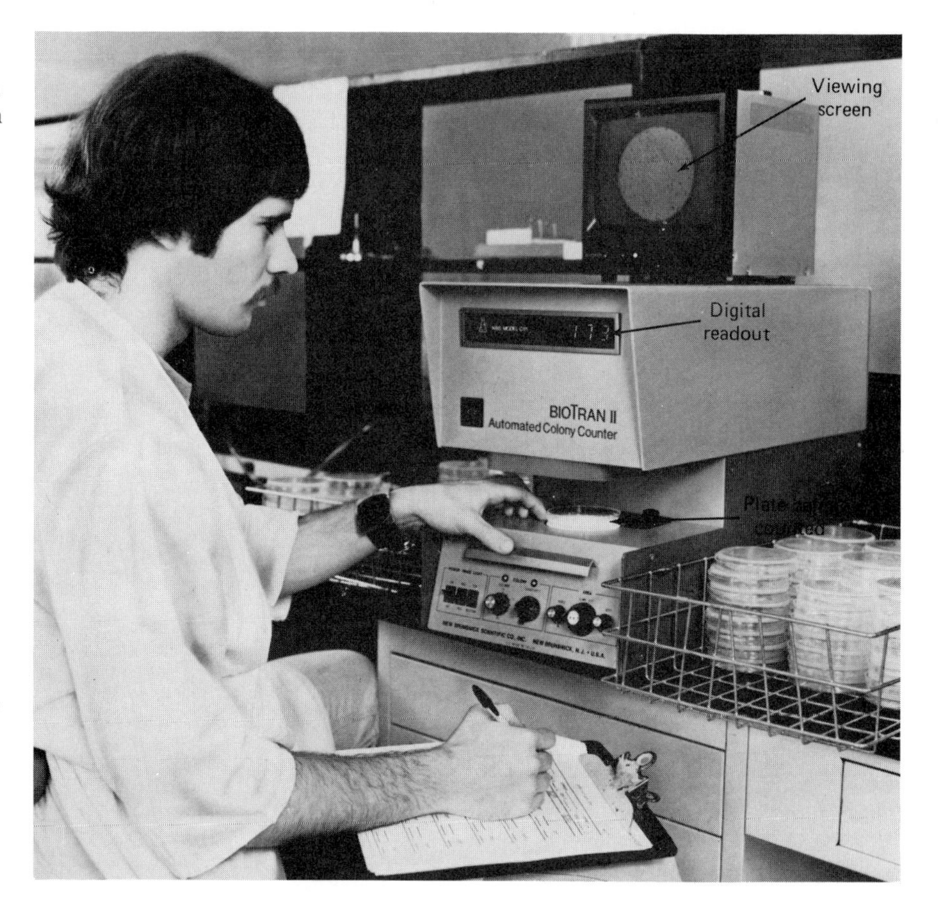

Figure 4–26 An automatic bacterial colony computer, the BioTrain II. This device can obtain a colony count from any Petri dish of standard size in 1 second. The count is flashed on a digital readout visual display. *[Courtesy of New Brunswick Scientific Company, Inc. Edison, N.J.]*

count to estimate the potential viable count. The total count also gives an estimate of the total number of microorganisms (bioburden) to which a substance has been exposed.

The method used to obtain total counts of microorganisms in specimens is determined by the particular type of microorganism involved. Techniques and equipment that can be used include a direct microscopic count, proportional counting, a counting chamber, and electronic counting.

DIRECT MICROSCOPIC COUNT (BREED SMEAR TECHNIQUE)

A direct microscopic count involves spreading 0.01 mL of a well-mixed specimen over a square centimeter area of a glass slide. After drying, any substance in the sample that might affect the accuracy of the determination (fat in milk, for example) can be removed by an appropriate solvent, such as xylene. The average number of microorganisms per field can be determined by using the high-dry or oil-immersion objectives. The number obtained is then multiplied by a microscope correction factor (MCF), which takes into account the calibration of the instrument.

This correction factor is calculated according to the following formula:

$$\text{MCF} = \frac{\text{area of the smear}}{\text{area of the microscopic field}} \times 100$$

Commercially available stage micrometers are used to measure the diameter of the microscopic field. From this value the area can be calculated according to the formula $A = \pi r^2$, where $\pi = 3.142$ and r is the radius of the field. The dilution factor, which takes into account the volume of material used from the sample and adjusts the number of microorganisms to that found in 1 mL, is represented by 100 in the MCF calculation. In other words, 100 times the number of organisms found in a 0.01-mL sample will give the number of such microbes expected to be present in 1 mL of the sample studied. The final count reported is usually the average of counts on approximately 50 different fields.

Direct microscopic counts have been used in approximating the quality of various grades of raw and pasteurized milk and in computing the growth curves of microorganisms.

Unfortunately, direct microscopic counts are not particularly accurate when only a few microorganisms are present in a specimen—less than 20,000 to 50,000 per milliliter.

PROPORTIONAL COUNTING

Since the turn of this century, microbiologists have used the technique of proportional counting for estimating microbial populations. In this method a standardized suspension of yeast cells, or of inert particles such as polystyrene latex spheres, is prepared. A measured volume of this suspension is then mixed with a measured volume of the unknown specimen. By examining the mixture, one can determine the relative numbers of specimen organisms and particles or yeast cells present. If the ratio turns out to be 1:1, then the original unknown sample contained a number of organisms equal to that in the standardized preparation. This method is useful for determining the number of virus particles in suspensions. Ratios are determined with electron microscopy and suitably prepared shadowed specimens. **[See Chapter 3.]**

COUNTING CHAMBER

In a counting chamber, a known volume of sample suspension is placed over a calibrated, etched grid contained in a special chamber. In the Petroff-Hauser device (Figure 4–27), the etched lines can be

Figure 4–27 (a) A Petroff-Hauser bacterial counter. (b) A vertical view. (c) An enlarged view of the ruled chambers in the center of the platform.

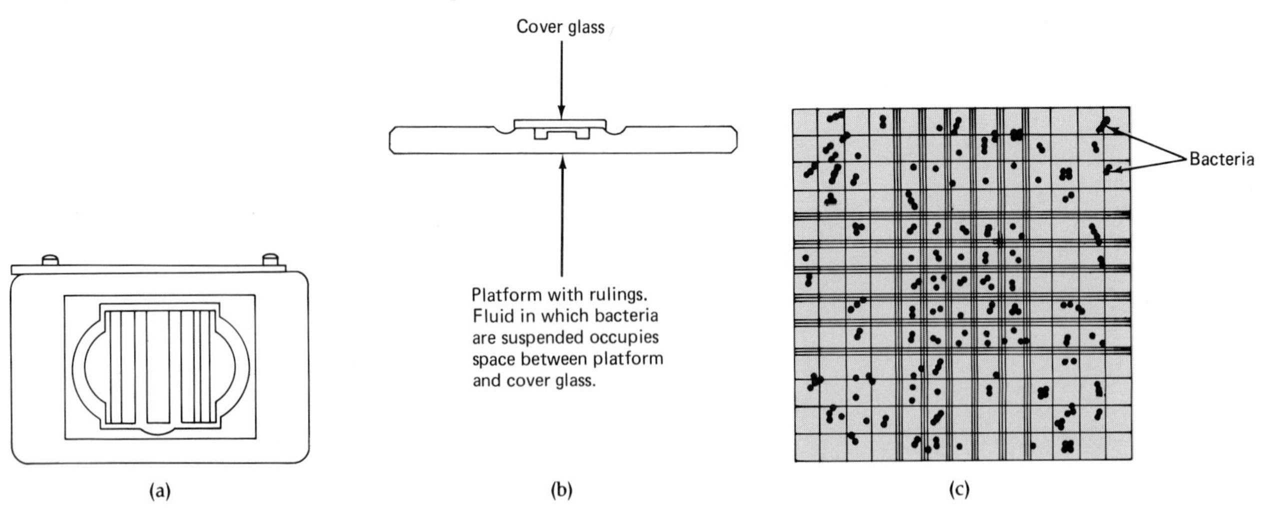

seen on the central surface. This surface is purposely depressed slightly, so that a space of 0.02 mm is formed when the calibrated region is covered with a special cover glass. The counting chamber consists of regular, partitioned, cubical chambers of known volume.

A bacterial suspension is introduced into this space with the aid of a calibrated pipette. After the bacteria settle and the liquid currents have slowed, the microorganisms are counted and their number per unit volume of the suspension is calculated, using an appropriate formula.

Counting chambers can also be used in estimating the total and viable numbers of yeast cells in a suspension. In this case, methylene blue is added to the yeast suspension to act as a *vital dye,* a dye that can distinguish between living and dead cells. Viable yeast cells absorb the dye and reduce it to the leuko or colorless form. Thus, living cells are colorless, and dead or dying cells are stained blue. As with the Breed smear, it is essential to have large populations in order to obtain a reasonably accurate count.

This method has not proved practical with bacterial cells. Methylene blue or other oxidation-reduction (redox) indicators are limited in this respect.

ELECTRONIC COUNTING

Electronic counting devices rely on changes in the electrical conductivity of the medium to signal the presence of a cell. As the suspension passes through a narrow opening, each cell is detected by a fluctuating conductivity across the orifice, and passage of the cell registers on the counter. Counters of this type are used routinely in clinical hematology to count white and red blood cells. Excellent reproducibility and accuracy of results are usually obtained.

Automation in the Laboratory

The use of automated and computerized instruments to improve the efficiency of both clinical and research laboratories is steadily increasing. With such improvements, laboratories can generally obtain results of microbiological detection and identification tests more rapidly and lower operating costs without reducing accuracy. Instruments used in these types of activities can prepare dilutions of specimens (inocula), inoculate, determine cell numbers, identify unknown organisms, and determine antibiotic susceptibilities of disease agents. Aspects of this topic as they relate to the identification of disease agents are explored further in Chapters 16 and 21.

Summary

An Introduction to Bacterial Cultivation

1. A wide variety of procedures and nutrient preparations called *culture media* are used for the cultivation of microorganisms.
2. Living animals, plants, and/or their cells are used for those microorganisms that require living cells for growth and reproduction.
3. Procedures using nonliving materials contained in culture vessels are called *in vitro* techniques, while techniques using living cells or entire animals or plants are called *in vivo* techniques.

General Nutritional Requirements

HETEROTROPHY, AUTOTROPHY, AND HYPOTROPHY

1. *Heterotrophic* organisms need preformed organic compounds for their growth, reproduction, and other activities. Examples of such compounds include amino acids and carbohydrates.
2. *Autotrophic* organisms utilize inorganic compounds such as carbon dioxide and minerals.
3. *Hypotrophs* need living cells for their activities. These cell forms include viruses, the rickettsias, and chlamydias.

Physical and Chemical Requirements for Bacterial Growth

GASEOUS REQUIREMENTS

1. Microorganisms can also be grouped by their response to available gases in the environment, such as oxygen and carbon dioxide.
2. Organisms requiring oxygen are called *aerobic.* Those needing low concentrations are *microaerophilic.* Those having no requirement for free oxygen are called *anaerobes.* Strict, or *obligate,* anaerobes cannot tolerate any free oxygen in their environment.
3. Facultative anaerobes can grow with and without oxygen.
4. Capneic organisms require a carbon dioxide concentration greater than that normally found in the atmosphere.

THERMAL CONDITIONS

1. Microorganisms need specific temperatures for their growth and related activities.
2. The *minimum, maximum,* and ideal, or *optimum,* requirements are known as an organism's *cardinal temperatures.*
3. *Psychrophiles* have a temperature range of 0° to 20°C; *mesophiles,* 20° to 45°C; and *thermophiles,* 45° to more than 90°C.

ACIDITY OR ALKALINITY (pH)

1. Microorganisms have certain pH needs for growth.
2. Various indicators and electron devices can be used to determine the hydrogen ion concentrations of preparations. Examples of pH indicators used in bacteriological media are bromcresol purple, litmus, and phenol red.
3. Buffers are chemical combinations that resist changes in pH when acids or bases are added.

OSMOTIC PRESSURE

1. Osmotic pressure results from the tendency of water molecules to pass through a membrane and to equalize their concentration on both sides of the membrane. Osmotic conditions can be isotonic, hypotonic, or hypertonic.
2. Cells with walls (algae, fungi, and bacteria) are able to withstand certain osmotic imbalances.

Components of Bacteriological Media

Media are used for the isolation and identification of microorganisms from materials such as dairy products, foods, soil, water, and specimens associated with microbial diseases. Culture media may be liquid (broth) or solid (gel). Agar, a complex carbohydrate, is the most commonly used solidifying agent.

NITROGEN SOURCES

Nitrogen-containing compounds—proteins and their products—are an important ingredient of media.

CARBON AND ENERGY SOURCES

1. Carbon and energy sources are also important ingredients of media. Organisms obtain carbon from organic nutrients such as carbohydrates, lipids, and proteins, as well as from carbon dioxide.
2. Carbohydrate sources of energy and carbon include starch, glycogen, various pentose monosaccharides and hexose monosaccharides, and various disaccharides.

VITAMINS AND GROWTH FACTORS

Several of the vitamins important in the treatment of human nutritional deficiency diseases are also required by microorganisms. Other growth factors may be needed as well.

ESSENTIAL MINERAL SALTS

Various inorganic compounds are also essential for microbial nutrition. These include phosphates and sulfates. Potassium, magnesium, manganese, iron, and calcium are also used.

WATER

The bacterial cell is 80% water and must be in intimate contact with a water supply for its survival, growth, and reproduction.

Preparation and Storage of Media

1. Many types of media may be obtained in dehydrated (powder) form from commercial establishments. Such preparations are then reconstituted with distilled water, adjusted in pH, dispensed in appropriate containers, and sterilized for use.
2. Appropriate storage conditions are necessary to maintain media quality.

Media Usage and Categories

Media can be used for different purposes, such as the isolation and identification of microorganisms, the determination of microbial antibiotic sensitivity, sterility testing of medically important products, analyses of food and water, and the preparation of vaccines.

CHEMICALLY DEFINED AND COMPLEX MEDIA

Chemically defined or synthetic media contain essential chemical ingredients in exactly known amounts. Complex media contain ingredients obtained from extracts of animals, plants, or microorganisms. The exact chemical composition of these and related nutrients are not known at this time.

DIFFERENTIAL MEDIA

Differential media are used to distinguish among growing microorganisms based on the visible reactions produced. Blood agar is an example of this type of medium.

SELECTIVE MEDIA

Selective media favor the growth of certain microorganisms and interfere or prevent the growth of others. Various substances such as dyes, high concentrations of salt, and antibiotics are used to make media selective. Sabouraud's dextrose agar is an example of this medium type.

SELECTIVE AND DIFFERENTIAL MEDIA

Selective and differential media combine the properties of both selective and differential types of media. Examples of this category include mannitol salt agar and MacConkey agar.

IDENTIFICATION TUBE AND PLATE MEDIA

These media are differential in nature and contain ingredients and indicators to detect specific enzymatic reactions and variations in growth.

MEDIA FOR THE CULTIVATION OF ANAEROBES

Media for anaerobes incorporate reducing agents to create anaerobic conditions.

PREREDUCED ANAEROBICALLY STERILIZED MEDIA

Prereduced anaerobically sterilized media greatly increase the chance of isolating many fastidious anaerobic bacteria.

Inoculation and Transfer Techniques

1. Inoculating instruments, loop or needle, are used for introducing specimens into or onto nutrient media, isolation of pure cultures, culture transfer, and preparation of specimens for microscopic study.
2. Before use, the inoculating instrument must be heated (flamed) to redness to sterilize it. In addition, the lips of tubes from which cultures are removed or into which cultures are introduced must be carefully flamed.
3. *Aseptic technique* refers to the prevention of microbial contamination.

Techniques for the Isolation of Pure Bacterial Cultures

1. Enrichment techniques make it possible to obtain pure cultures, and are designed to increase the relative numbers of a particular bacterial species.
2. The pour-plate and streak-plate techniques are two effective procedures used for the isolation and subsequent identification of bacteria.

THE POUR-PLATE TECHNIQUE

The *pour-plate technique* consists of cooling melted agar-containing medium, inoculating the medium with a specimen, and immediately pouring it into a sterile Petri plate, allowing the freshly poured medium to solidify and incubating the preparation at the desired temperature.

THE STREAK-PLATE TECHNIQUE

The *streak-plate technique* involves spreading a single loopful of material containing microorganisms over the surface of a solidified agar medium.

ANAEROBIC TRANSFER

Transfer should be performed under a flowing stream of nitrogen for cultures on solid media or by the use of a pipette for liquid media.

Conditions of Incubation

AEROBIC INCUBATION

Once an appropriate solid or liquid medium has been inoculated and protected from contamination, cultures must be incubated under controlled conditions of temperature and humidity.

CAPNEIC INCUBATION

Various devices and procedures are used to provide the suitable concentration of carbon dioxide for capneic incubation.

ANAEROBIC INCUBATION

Incubation under anaerobic conditions can be accomplished with vacuum devices such as the anaerobic incubator, heavy-walled anaerobic jars with gas-tight lids, and disposable units of various types.

Patterns of Growth

Colony, slant growth, and broth cultures can be useful in evaluating the purity of cultures and in the identification of bacterial cultures.

Maintaining Pure Cultures

Pure cultures may be maintained in a frozen state, under liquid nitrogen, or in the frozen-dried condition produced by lyophilization.

Bacterial Growth

Most bacteria reproduce by binary fission, the production of two new cells from one parent cell.

BACTERIAL GROWTH CURVE

1. Upon transfer of bacteria to a fresh broth medium and under controlled conditions, a typical growth curve will result.
2. The actual phases and shape of such a curve vary with the species and media used.
3. A bacterial growth curve consists of a *lag phase, logarithmic phase, stationary growth phase,* and *logarithmic death rate.*
4. *Generation time* is the time it takes for a population to double.
5. The generation time of a bacterium can be calculated mathematically by the following formula:

$$\text{Generation time} = \frac{t}{3.3 \log (B_0/B_1)}$$

where t = time interval between the measurement of cell numbers of one point in the log phase (B_0) and then again at a later point in time (B_1)

B_0 = initial bacterial population

B_1 = bacterial population after time t

log = logarithms to the base 10

3.3 = log 10 to log 2 conversion factor

CONTINUOUS CULTURE

Continuous culture techniques were developed to avoid variation in the ages of the bacterial cells that are grown.

GROWTH ON SOLID MEDIA

Colony and slant growth cultures can be useful in evaluating the purity of cultures and in isolating organisms from a mixture.

Measurement of Growth

Microbial growth can be determined by observing an increase in mass or numbers.

CELL MASS DETERMINATION

Analytical procedures used to determine cell mass include determination of dry weight, chemical analysis for protein or nitrogen concentration, and turbidity, or cloudiness, measurements.

VIABLE COUNTS

1. Viable counts are estimates of the numbers of living microorganisms in a given material. For bacteria these estimates can be expressed in *colony-forming units* (CFU). Only plate counts with 30 to 300 colonies can be used.
2. CFU can be calculated according to the formula

$$\text{CFU} = \frac{\text{number of colonies}}{\text{volume of sample used}} \times \text{reciprocal of dilution used}$$

DETERMINING TOTAL COUNTS OF MICROORGANISMS

Techniques and equipment used to obtain total counts of microorganisms in materials include the *direct microscopic count, proportional counting, counting chamber,* and *electronic counting devices.*

Automation in the Laboratory

An increasing number of cultivation-associated procedures are being automated.

Questions for Review

1. Explain the differences between *in vivo* and *in vitro* techniques.
2. What does pH mean? How can the pH of a medium be altered?
3. Compare the nutritional requirements of autotrophic, heterotrophic, and hypotrophic organisms and give examples of each organism.
4. Explain the following terms:

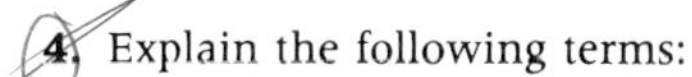

a. aerobe
b. microaerophilic
c. anaerobe
d. facultative anaerobe
e. capneic
f. obligate aerobe

5. What is a hypotroph?

6. Discuss the grouping of microorganisms according to their temperature growth requirements.

7. a. What is a growth medium?
b. What substances serve as sources of carbon, mineral salts, nitrogen, and vitamins?
c. Distinguish between broth and agar media and give common examples of each.
d. Distinguish between a chemically defined medium and a complex one.

8. Discuss anaerobic cultivation methods. Include in your answer a treatment of media and equipment used in general procedures.

9. What is an anaerobic jar?

10. Describe the pour-plate and streak-plate techniques. What disadvantages does each procedure possess? What advantages?

11. What is aseptic technique?

12. What are synthetic media? Can both bacteria and fungi be grown with them?

13. List three cultural characteristics of bacteria that can be noted from broth cultures. Do the same for agar slant and agar plate cultures.

14. a. Distinguish between selective and differential media.
b. What is a selective and differential medium?
c. Give examples of each of these medium categories.
d. What is an enrichment technique?

15. Describe three methods by which you can measure growth in microorganisms.

16. Define:
a. growth
b. lag phase of growth
c. reproduction
d. Breed smear technique
e. bacterial growth curve
f. stationary growth phase
g. proportional counting
h. continuous cultivation
i. generation time

17. Calculate the colony-forming units of a sample if 1 mL of a 10^{-6} dilution yielded 66 colonies.

18. Identify the broth patterns of growth shown in Figure 4–28.

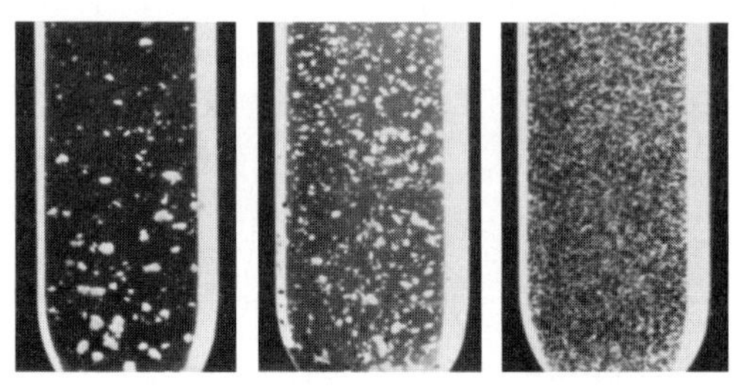

Figure 4–28 *[From P. A. Scherer, H. P. Bochem, J. D. Davis and D. C. White,* Can. J. Microbiol. ***32****:137–144, 1986.]*

Suggested Readings

GERHARDT, P., R. G. E. MURRAY, R. N. COSTILOW, E. W. NESTER, W. A. WOOD, N. R. KRIEG, and G. B. PHILLIPS (eds.), *Manual of Methods for General Bacteriology.* Washington, D.C.: American Society for Microbiology, 1981. *A comprehensive manual describing currently used techniques for the study of bacterial growth, morphology, genetics, and metabolism. Attention also is given to laboratory safety and bacterial systematics.*

GRAY, T. R. G., and G. R. POSTGATE (eds.), *The Survival of Vegetative Microorganisms.* New York: Cambridge University Press, 1976. *This book has several articles dealing with the many ways vegetative microorganisms manage to survive in spite of starvation, cold, heat, osmotic shock, ultraviolet light, and other factors.*

INNISS, W. E., "Interaction of Temperature and Psychrophilic Microorganisms," *Annual Review of Microbiology* **29**:445 (1975). *The effects of temperature on psychrophilic microorganisms and their activities, whether in the laboratory or in their natural environments, are discussed.*

LENNETTE, E. H., A. BALOWS, W. J. HAUSLER, JR., and H. J. SHADOMY (eds.), *Manual of Clinical Microbiology* (4th ed.). Washington D.C.: American Society of Microbiology, 1985. *A highly comprehensive publication which includes chapters describing techniques in media used in the cultivation and the study of microorganisms.*

PIRT, S. J., *Principles of Microbe and Cell Cultivation.* New York: John Wiley & Sons, 1975. *A slightly advanced text describing specific topics associated with methods of cultivation of microorganisms, such as the estimation of biomass and chemostat cultures, as well as the effects of oxygen, temperature, pH, and water on growth.*

WASHINGTON, J. A. (ed.), *Laboratory Procedures in Clinical Microbiology.* New York: Springer-Verlag, 1981. *A highly functional publication emphasizing proper techniques of specimen collection, processing, and the cultivation of microorganisms. An appendix describing currently used media and reagents for the isolation and identification of microorganisms is included.*

5

Microbial Metabolism and Cellular Regulation

... Man, woman, horse, horsefly that bit the horse, bacteria and viruses that infected the horsefly, toxins manufactured in the bacteria and the viruses, toxins coming out and killing a physician or a politician, all are metabolism. The infinitude of change forever taking place in us, and those who inhabit air and earth and sea with us, is metabolism.
— *Gustav Eckstein,* **The Body Has a Head**

After reading this chapter, you should be able to:

1. Discuss the concept of metabolism and define anabolism and catabolism in terms of general reactions and energy changes.
2. Define enzymes and describe their function in cellular reactions.
3. Describe the concept of oxidation-reduction.
4. Describe the general mechanisms for metabolism exhibited by photosynthetic and chemosynthetic autotrophs and heterotrophs.
5. Outline the general steps in glycolysis in terms of the initial chemical activation process and the end products.
6. Show the general steps in the citric acid cycle by which acetyl CoA is metabolized to CO_2, NADH, FADH, and ATP.
7. Discuss the involvement of ATP in energy production in the cell and the mechanisms for making ATP by photophosphorylation, substrate phosphorylation, and oxidative phosphorylation.
8. Define fermentation and outline typical reactions that use pyruvic acid.
9. Differentiate between the types of potential energy (number of ATP molecules) produced as a result of glycolysis under anaerobic versus aerobic conditions.
10. Indicate the metabolic steps by which proteins and lipids can be used for energy in glycolysis and the citric acid cycle.
11. Distinguish between glycolysis, the hexose monophosphate shunt, and the Entner-Doudoroff pathway as catabolic reactions of glucose.
12. Outline the steps in photosynthetic carbon dioxide fixation as an example of anabolism.
13. Discuss feedback inhibition as a mechanism for controlling metabolic activities in the cell.
14. Describe each of the following methods of studying metabolic reactions: simple fermentation tests, oxidation-reduction activity, radioisotopes, and chromatography.

Chapter 5 describes how life forms use available biochemicals for energy, growth, and reproduction. Carbohydrate metabolism is discussed briefly in order to provide a foundation for understanding the interrelationship of carbohydrate, protein, and lipid metabolism. This chapter also covers the regulation of metabolic activities and offers brief discussions of methods for analyzing metabolism and various aspects of microbial metabolism. Emphasis is placed on the similarities and differences between the metabolic reactions of microbes and those of other organisms.

While all the activities of living organisms use energy, the ways in which energy is acquired are numerous and varied. Many people find it difficult to imagine that a microscopic bacterial cell performs many of the same biochemical reactions that occur in human or other forms of life. The chemical activities of microorganisms have been studied for several reasons and found to have numerous applications in areas such as the commercial production of dairy products, antibiotics, and industrially important chemicals as well as in the approaches used for the control and treatment of infectious diseases. This chapter describes the nutrients available to bacteria and the processes used by them to acquire energy and drive the reactions needed to form proteins, lipids, and nucleic acids. **[Refer to Appendix A for descriptions of these macromolecules.]**

General Metabolism

The word **metabolism** refers to the entire set of chemical reactions by which cells maintain life. Since such reactions are involved in either the production or use of energy, metabolic activities can be grouped into two general phases: **catabolism** and **anabolism.** Catabolic or degradative reactions are those by which complex organic nutrients are broken down into simpler organic and inorganic compounds. During these reactions, energy is obtained and stored in the *energy-rich bonds* of the molecules that are formed during catabolism. This energy is used in part for the active transport of molecules across biological membranes and for the series of reactions that synthesize macromolecules, such as nucleic acids, lipids, polysaccharides, and proteins as well as the subunits (monomers) of which macromolecules are formed (Table 5–1). These synthetic processes form the basis of anabolism. Thus, within a cell, nutrient molecules are being degraded while others are being synthesized for the formation of cell parts. This balanced flow of biologically important chemicals and energy maintains the essential processes of cells. Conceptually, all living organisms use the same mechanism: they couple (link) the reactions that require energy to the reactions that yield energy (Figure 5–1).

TABLE 5–1 Macromolecules and Their Subunits

Macromolecule[a]	*Subunits*
Proteins	Amino acids
Polysaccharides	Sugars
Simple lipids	Glycerol and fatty acids
Nucleic acids	Sugar (pentose, five-carbon sugars) Purines and pyrimidines Phosphate

[a] Refer to Appendix A.

ATP

The most common energy carrier in biological systems is *adenosine triphosphate,* better known as ATP. This compound is referred to as a *nucleoside triphosphate.* Each molecule contains one five-carbon sugar, ribose, three phosphate groups linked in a chain (triphosphate group), and one adenine molecule (a purine). The relationship of these components is shown in Figure 5–2. ATP is formed from one adenosine diphosphate (ADP) molecule and one inorganic phosphate (P_i). The cycle of ATP $\rightleftharpoons$ ADP (Figure 5–3) is the major means by which energy is exchanged in biological systems.

Figure 5–1 An overview of metabolism showing the relationship between anabolism and catabolism. Note the distribution of energy. A major portion of the energy released in catabolism is lost to the environment. Therefore a cell has continuous need of new sources of energy to drive anabolic reactions.

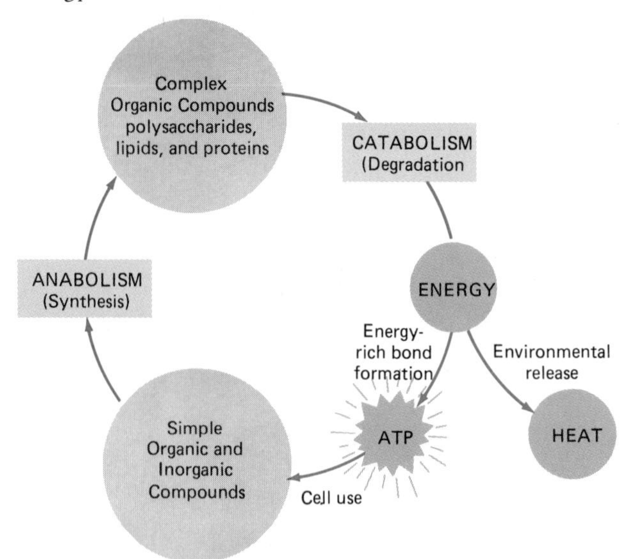

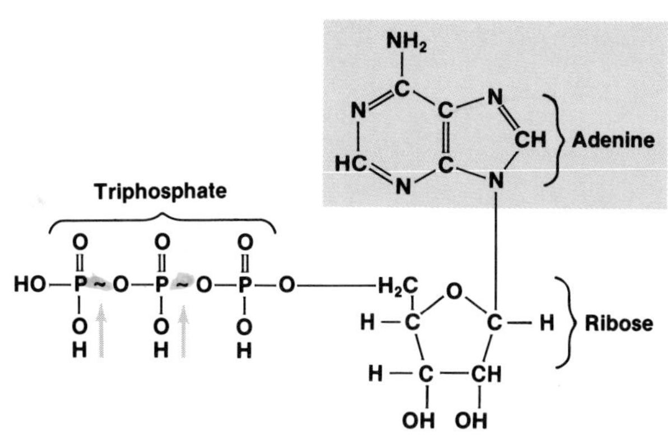

Figure 5–2 Adenosine triphosphate (ATP). The components of this compound are adenine, ribose, and phosphate groups. High-energy bonds are indicated by arrows.

ATP is known as an energy-rich molecule because it has high-energy-yielding bonds between its first and second phosphates and between its second and third phosphate groups (Figure 5–2). These bonds enable ATP to hold much greater quantities of energy than compounds having ordinary linkages. It is customary to indicate high-energy phosphate bonds by the symbol ~ and ordinary bonds by —.

Sometimes ATP is directly hydrolyzed to ADP plus phosphate-releasing energy for different types of cellular activities (Figure 5–3). A variety of enzymes (organic catalysts) known as *adenosine triphosphases* control the reaction. Usually, however, the end or terminal group of ATP is transferred to another molecule. This addition of a phosphate group is called *phosphorylation,* and the enzymes involved with such transfers are known as *kinases.* In phosphorylation reactions, some of the energy of the phosphate group in the ATP molecule is transferred to the phosphorylated compound, which, in its newly energized state, can now participate in a subsequent reaction. Reactions of this type, emphasizing the ATP/ADP cycle as a universal energy-exchange system, are described later in this chapter.

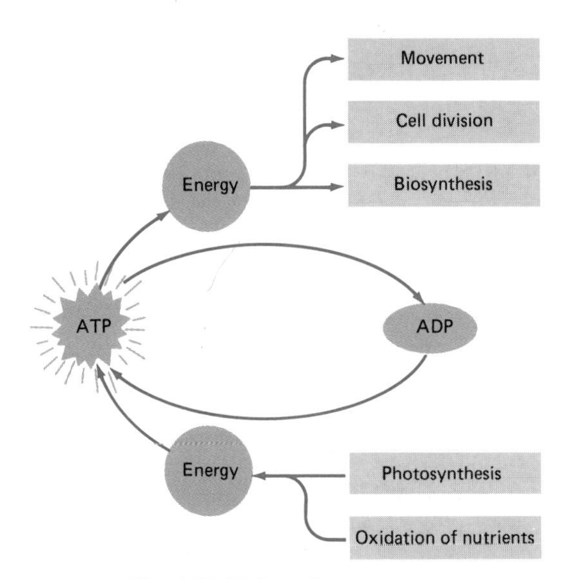

Figure 5–3 The ATP-ADP cycle.

Unity of Biochemistry

Microorganisms have evolved many different processes for obtaining energy and building blocks to permit growth, repair, reproduction, and movement. Even with such diversity, a striking similarity, referred to as the **unity of biochemistry,** can be observed repeatedly in different organisms. Although many differences do occur in biochemical pathways, the similarities in the metabolic reactions of bacterial and human cells are remarkable. Among the various factors contributing to keeping the reactions of metabolism operating smoothly are the *enzymes.* What they are and how they function will be described in the following section.

Enzymes

Enzymes are protein biological or organic catalysts. They are associated with metabolic reactions, respiration, the conversion and transfer of energy within living systems, and the formation of various macromolecules and cellular components. At least 2,000 different enzymes are now known, each of them capable of catalyzing a specific reaction. Moreover, different cell types synthesize different types of enzymes—no cell contains all of the known enzymes. The specific enzymes produced by a cell are the major factors determining the biological activities and functions performed by that cell.

It is well known that chemical reactions between compounds occur when molecules that have sufficient energy collide. Therefore, any condition that increases the collision frequency between molecules also increases the rate at which chemical reactions occur. Among the conditions that will do this are increases in pressure, temperature, and the concentrations of the reacting substances. Catalysts are substances that can speed a reaction by increasing the contact between molecules or lowering the energy requirement for a reaction, the activation energy. Enzymes are the catalysts in living systems. They lower the activation energy (Figure 5–4) by forming a temporary association with the reacting molecules (reactants). While this association brings the reactants close to one another, it may also weaken exist-

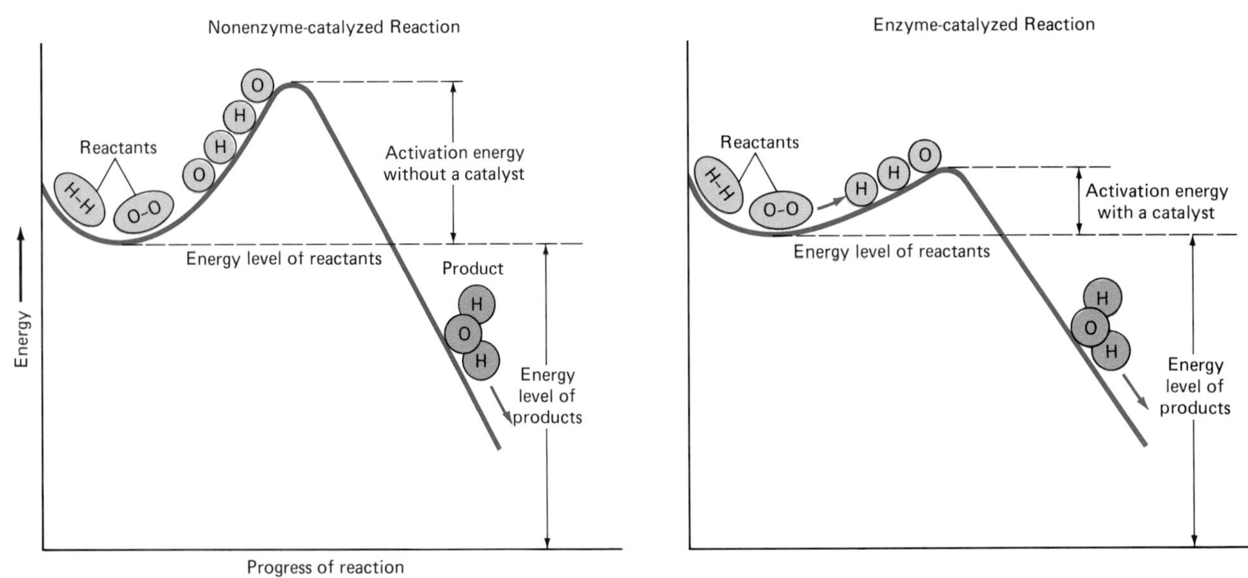

Figure 5–4 Activation energy and lowering the energy barrier. If collisions between molecules are to be increased to the point where breaking occurs and a chemical reaction begins, the reacting substances must be activated. Once activation has taken place, the reaction proceeds spontaneously. As shown here, the activation levels are lowered with the help of catalysts such as enzymes.

ing chemical bonds, making it easier for new ones to form. As a result of enzyme involvement, the reaction goes more rapidly than it would without it. Furthermore, since the enzyme itself is not altered permanently in the process, it can be used over and over again.

Biological systems continually show evidence of very high rates of chemical reactions: the rapid growth of seeds, the almost visible growth of young animals, and the tremendous increases in the number of bacterial cells growing in a suitable medium within a relatively short period of time. These changes require rapid chemical reactions for energy production and cell formation. Such reactions do not proceed at a sufficient speed under normal environmental conditions. For reactions to occur fast enough to be suitable for biological systems, they would have to occur at temperatures too high or at a pH too extreme for life. Biological systems, therefore, require catalysts that can function under conditions compatible with life. By allowing selective controlled acceleration of specific reactions at appropriate times, these biological catalysts function effectively without causing changes that could disrupt or kill living cells.

The Parts of an Enzyme

Many enzymes are entirely protein and are called **simple enzymes.** Those that contain a nonprotein group are called **conjugated enzymes.** The protein portion is called an *apoenzyme,* meaning that it is incomplete and is inactive without the nonprotein portion, known as the *cofactor.* Together the apoenzyme and the cofactor form an entire enzyme. When the cofactor is a metal ion—for example, magnesium, zinc, iron, copper, or manganese—it is referred to as an *activator.* When the cofactor is a small organic substance and easily removed, it is called a *coenzyme.* Many coenzymes are derived from dietary vitamins. Vitamin deficiency diseases are thus often related to the decreased ability of certain enzymes to function in normal metabolic activities. *Prosthetic* groups differ from coenzymes and cofactors in that they are bound more tightly and are therefore more difficult to remove from the enzyme. No matter what the accessory group is called, its function is related to the activity of the particular enzyme involved.

FACTORS AFFECTING ENZYME ACTIVITY

Earlier we discussed the fact that enzymes function under conditions compatible with life. Without enzymes, life as we know it would not be possible, since most biochemical reactions have significant activation energies and will not proceed spontaneously (Figure 5–3). Thus conditions that affect life often do so by exerting some effect on enzyme activity. Being protein in composition, enzymes are sensitive to pH, temperature, ionic strength, heavy metals—that is, to anything that will change (denature) their structure and arrangement. When such changes occur, enzymes lose both their structural properties and biological functions.

Enzymes have particular conditions of pH and temperature under which they function best. Each enzyme has a functional or optimal pH and tem-

TABLE 5–2 Selected Enzymes and Their Catalytic Actions

Enzyme Type[a]	*Catalytic Action*	*Specific Example*
Deaminase	Removes amino (NH_2) groups	Alanine deaminase
Decarboxylase	Causes the release of carbon dioxide (CO_2)	Pyruvate decarboxylating enzyme complex
Dehydrogenase	Causes removal and transfer of hydrogen to acceptor compound	Isocitric dehydrogenase
Ligase	Forms a bond between two molecules using energy obtained from the breakage (cleavage) of ATP bonds	DNA ligase
Oxidoreductase	Transfers electrons or hydrogen atoms (electron-transfer reactions)	Cytochrome oxidase
Phosphorylase	Causes the incorporation of phosphate (PO_4) groups	Phosphorylase[b]
Transferase	Moves functional groups such as amino (NH_2) groups from one molecule to another	Transaminase

[a] Reactions involving several of these enzymes are described in this chapter and in Chapter 6.
[b] This enzyme is involved in the breakdown of starch.

perature, since the rate of enzymatic reactions is affected by these factors. Temperature and pH affect not only the attractions among the amino acids of the enzyme molecule but also those between the location on the enzyme where the reaction takes place (*active site*) and the substance (*substrate*) on which the enzyme acts.

Because various denaturing chemicals (including acids), radiation (ultraviolet light), and extremes of temperature inhibit or destroy enzymes, these conditions are of value in the control of microorganisms. Although denaturation is partially or completely reversible in certain cases, denaturation that is allowed to continue prevents an enzyme from regaining those structural properties responsible for the formation of the active site and the enzyme's activity. Thus, denaturation and enzyme inactivity become irreversible. Chapters 11 and 12 describe several chemical and physical factors affecting enzyme activity in greater detail.

Enzymes are usually unaltered by the process they promote. Thus, they can be used over and over again by the cell. Some enzymes are quite specific, often catalyzing only one particular reaction (Table 5–2). An enzyme is frequently named after the material, or **substrate,** on which it acts. For example, amylose, a major component of starch, is the substrate for the enzyme amylase. Enzymes are also named for their associated actions. Thus the enzymes classified as dehydrogenases remove hydrogen, and those called decarboxylases removes carbon dioxide. Note the ending *ase,* which identifies almost all enzyme names.

MECHANISM OF ACTION

When an enzyme, which is a three-dimensional circular (globular) protein, combines with a substrate, an *enzyme-substrate* (*ES*) *complex* results (Figure 5–5). With the formation of the complex, the substrate is activated and a chemical reaction can take place. The rate at which the reaction proceeds is related to both the speed of the catalytic reaction at the active site and the persistence of the ES complex. Figure 5–6 shows how an enzyme catalyzes a chemical reaction using the sugar maltose. The enzyme *maltase* catalyzes the reaction by combining with the substrate maltose to produce an ES complex. This ES complex proceeds to break and form the appropriate chemical bonds, resulting in the formation of two glucose molecules and the regenerated enzyme. Since this chemical conversion involves the addition of water, it is referred to as a *hydrolytic reaction* (*hydro,* meaning water, and *lytic,* referring to disruption).

It is important to emphasize that the catalytic activity of an enzyme depends on its chemical composition and spatial organization, of which the major determining factors are the sequence of amino acids in the polypeptide chains of the enzyme and the polypeptide folding (spatial) arrangement. Moreover, the types of enzymes found in a cell are established by its genetic makeup.

With this brief introduction to metabolism we can now examine the various metabolic strategies and nutritional patterns developed by microorganisms and other forms of life to generate ATP and change carbon-containing molecules into the important macromolecules needed by cells.

Energy Metabolism

Chemical reactions are essentially energy transformations in which the energy stored in chemical bonds is transferred to other newly formed chemical bonds. Living systems derive energy from nutrients by a series of chemical reactions, some of which are oxidations. During oxidation, energy is released and

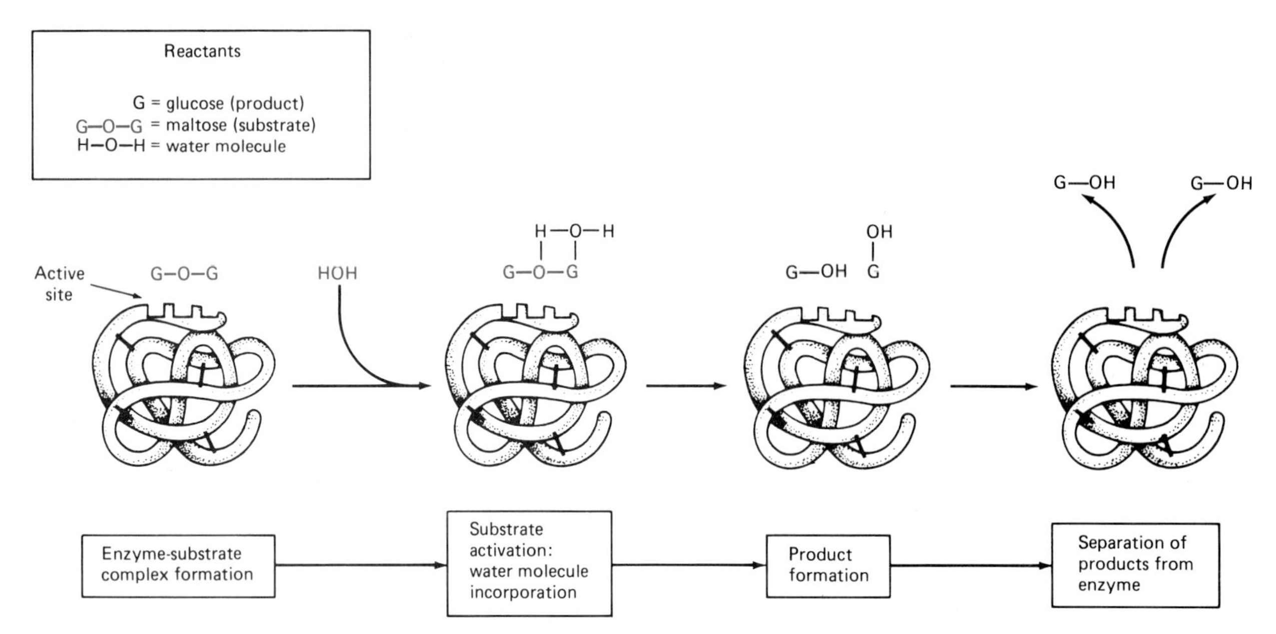

Figure 5–5 The mechanism of enzyme action. A diagrammatic representation of the position in which the carbohydrate maltose would be held by the enzyme maltase to allow the breaking and forming of the appropriate bonds for the production of two molecules of glucose.

energy-rich chemical bonds like those of ATP may be formed to store the released energy. Mechanisms also exist in living systems to release small amounts of the energy stored in particular bonds as a cell needs it. These mechanisms generally involve sequences of reactions, some of which are oxidation-reduction reactions.

Figure 5–6 A photomicrograph of *Chromatium violascens* (kroh-MA-tee-um vye-oh-LACE-ens), a representative purple sulfur bacterium. Note the presence of the bright refractile sulfur globules. Original magnification, 2,000X. *[Reproduced with the permission of the National Research Council of Canada from P. Caumette,* Can. J. Microbiol. ***30****:273–284, 1984.]*

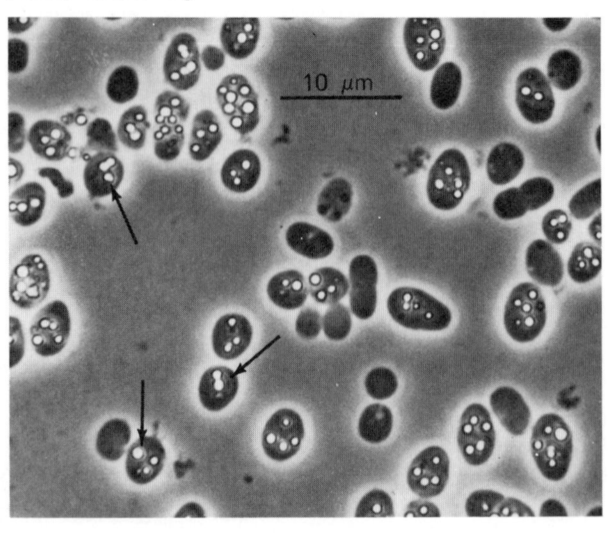

Oxidation-Reduction Reactions

Whenever an atom, molecule, or ion loses one or more electrons, (e^-) in a reaction, the process is called *oxidation* and the particle is said to have been *oxidized.* The electrons lost by an oxidized molecule do not float randomly. They are reactive and are readily picked up by another molecule. The resulting gain of one or more electrons is referred to as *reduction,* and the second molecule is said to have been *reduced.* In oxidation-reduction reactions, therefore, electron transfer takes place. Many oxidations in biological systems, in which organic molecules are involved, result in the removal of two electrons and two hydrogen ions or protons (H^+) at the same time. For this reason, biological oxidations are referred to as *dehydrogenations.* This reaction can be shown with the oxidation of glyceraldehyde-3-phosphate to form 1,3 diphosphoglyceric acid. Both of these molecules represent important steps in the oxidation of carbohydrates, such as glucose. (The P_i designation in the reaction indicates inorganic phosphate.)

$$\begin{array}{c} CH_2\text{—O—P} \\ | \\ CHOH \\ | \\ C\text{=O} \\ | \\ H \end{array} \xrightarrow[P_i]{2H^+,\ 2e^-} \begin{array}{c} CH_2\text{—O—P} \\ | \\ CHOH \\ | \\ C\text{=O} \\ | \\ \text{O—P} \end{array}$$

Glyceraldehyde-3-phosphate → 1,3 diphosphoglyceric acid

In biological systems, oxidations and reductions are always paired with one another (coupled) and are referred to as *oxidation-reduction reactions.* This coupling of reactions can be illustrated with glyceraldehyde-3-phosphate oxidation and the reduction of the coenzyme nicotinamide adenine dinucleotide (NAD^+). This coenzyme, together with nicotinamide adenine dinucleotide phosphate (NADP), is used in biological systems to carry 2 e^- and 1 H^+. In this oxidation, NAD^+ accepts a hydrogen ion (H^+) and two electrons (e^-) to form NADH. The remaining proton is released as free H^+.

$$\underset{\text{Glyceraldehyde-3-phosphate}}{\begin{array}{l} CH_2{-}O{-}P \\ | \\ CHOH \\ | \\ C{=}O \\ | \\ H \end{array}} + NAD^+ \xrightarrow{2\,H^+ \rightarrow NADH} \underset{\text{1,3 diphosphoglyceric acid}}{\begin{array}{l} CH_2{-}O{-}P \\ | \\ CHOH \\ | \\ C{=}O \\ | \\ O{-}P \end{array}} + H^+$$

Additional applications of oxidation-reduction reactions are described later in this chapter.

Metabolic Differences Among Microorganisms

Microorganisms vary not only in biosynthetic capabilities as indicated by their respective collections of enzymes, but also in their nutrient requirements. The sources of chemical elements and energy needed by microorganisms and other forms of life to synthesize biochemical compounds must be available in the environment and ready for use. Organic sources include a large number of compounds ranging from two-carbon molecules to more complex ones such as starch, which consists of thousands of carbon atoms. Inorganic sources include carbon dioxide, hydrogen sulfide, and other molecules.

The simplest type of nutrition is that of the *autotroph* or self-feeder. Such organisms are able to obtain all of their carbon needs from carbon dioxide. In contrast to the autotroph is the *heterotroph.* This type of organism uses organic compounds as carbon sources and may also use the same compounds to satisfy its energy needs. Both groups are further separated into more specific nutritional types according to the energy and carbon sources they use for growth and metabolism (Table 5–3).

Autotrophic Microorganisms

PHOTOAUTOTROPHS

Photoautotrophic (photosynthetic) organisms use carbon dioxide to obtain their carbon, and light as their energy source. Photosynthetic bacteria possess chlorophylls chemically similar to those found in plants. Photosynthetic pigments in bacteria are designated *bacteriochlorophyll-a, -b, -c,* and *-d,* to distinguish them from plant *chlorophyll a.* In addition to the bacteriochlorophylls, photosynthetic organisms also have various carotenoid pigments, which give color to their colonies. The colors include yellow to orange-brown, red, pink, reddish purple, and violet. Such pigmentation is largely dependent upon the type and concentration of carotenoid present. Green bacteria also contain carotenoids, but these compounds, which are light yellow, do not mask the green bacteriochlorophyll. The carotenoids in photosynthetic bacteria absorb light energy, which is

TABLE 5–3 Classification of Microorganisms by Nutritional Type

Type	*Energy Source*	*Primary Carbon Sources*	*Selected Electron Sources*	*Representative Groups*
Photoautotroph	Light	CO_2	H_2S	*Chlorobium* (kloh-ROH-bee-um) *Chromatium* (krow-MAH-tee-um)
Photoheterotroph	Light	CO_2 and organic substances	Organic substances	*Rhodospirillum* (row-doh-spy-RILL-um) *Rhodopseudomonas* (row-doh-soo-doh-MOH-nass)
Chemoautotroph	Oxidation of inorganic compounds	CO_2	H_2S S^0 H_2 Fe^{2+} CH_4 CO	*Beggiatoa* (beg-ee-a-TOE-ah) *Thiobacillus* (thigh-oh-ba-SILL-us) *Hydrogenomonas* (high-droh-gen-oh-MOH-nass) *Thiobacillus* *Pseudomonas* (soo-doh-MOH-nass) *Carboxydomonas* (kar-BOX-ee-doh-MOH-nass)
Chemoheterotroph	Oxidation of organic compounds	Organic substances	Organic substances	Most microorganisms

then transferred to the bacteriochlorophyll molecules for use in photosynthesis. Organisms such as cyanobacteria and green plants use water as the source of electrons and are said to carry out *oxygenic photosynthesis.* In the process, the energy absorbed from light by chlorophyll is used to split water molecules, yielding molecular oxygen (O_2), hydrogen ions, and electrons.

$$2H_2O \xrightarrow{\text{oxidation}} 4H + O_2\uparrow + 4e^-$$

The hydrogen and electrons serve to reduce coenzymes and are thereby made available for reducing additional substrates. The energy (ATP) and reducing power (NADPH) provided from the light reaction are required for the conversion (fixation) of carbon dioxide (CO_2) into carbohydrate and cell material. The sequence of reactions involved here is independent of light and forms the **Calvin cycle,** the means by which most autotrophs capture CO_2. Additional features of carbon dioxide fixation are presented later.

In *anoxygenic* or *bacterial photosynthesis,* organisms such as the green sulfur bacteria and purple sulfur bacteria do not oxidize water and thereby cannot produce molecular oxygen. Other electron sources, including hydrogen sulfide (H_2S) and hydrogen (H_2), are oxidized. With sulfide oxidation, an accumulation of elemental sulfur (S^0) is generally deposited within cells as refractile (light-deflecting) small round bodies or globules (Figure 5–6). When the sulfide is depleted, the sulfur is oxidized to sulfuric acid. It is important to note that the oxidation of environmental sources of inorganic compounds supplies the electrons to reduce coenzymes, and light supplies the energy needed for ATP synthesis. The energy and reducing power provided by these molecules are used by anoxygenic photosynthesizing organisms.

CHEMOAUTOTROPHS

Chemoautotrophs (chemosynthesizing autotrophs) are nonphotosynthesizing organisms that rely on the oxidation of inorganic compounds for the energy to fix CO_2 as the sole source of carbon. They may oxidize hydrogen or hydrogen sulfide (as do photosynthetic bacteria), ammonia, nitrites, and iron-containing substances. Representative reactions and bacteria capable of performing them are listed in Table 5–4. The Calvin cycle, named after its discoverer Melvin Calvin, is a complex series of reactions used by autotrophic microorganisms for the fixation of CO_2. It represents three slightly different but integrated metabolic sequences for converting CO_2 into glyceraldehyde-3-phosphate (Figure 5–11). Since this molecule contains three carbon atoms, the Calvin cycle sometimes is referred to as the C_3 pathway. A substantial amount of energy (ATP) and reducing power (NADPH) is needed to perform the reactions. Photoautotrophs obtain the energy and reducing power from the light reactions of photosynthesis (presented later), while *chemolithotrophs* obtain theirs from the oxidation of inorganic compounds. Chemolithotrophs are able to grow in a strictly mineral environment in the absence of light. Additional features of CO_2 fixation and the Calvin cycle are presented later.

Heterotrophic Microorganisms

PHOTOHETEROTROPHS

Photosynthesizing heterotrophs use light as their energy source, and organic compounds serve as primary sources of carbon. These organisms are adaptable in their oxygen requirements. Anaerobic conditions must be present when the photosynthetic process is being used to obtain energy. This is also true for photosynthetic autotrophs. However, unlike autotrophs, heterotrophs receive electrons from alcohols, fatty acids, and other organic acids. Photosynthesizing heterotrophs can grow aerobically only in the dark. Under these conditions they are, in essence, the same as chemoheterotrophs, a class that includes most microorganisms and members of the animal kingdom.

CHEMOHETEROTROPHS

These microorganisms are quite varied in their metabolism and can obtain both their carbon and energy needs from the metabolism of a single organic

TABLE 5–4 Energy Production by the Oxidation of Inorganic Compounds

Representative Bacterial Genera	*Energy Source*	*Energy-yielding Reactions*
Hydrogenomonas (high-droh-jen-OHM-oh-nass)	H_2	$2H_2 + O_2 \rightarrow 2H_2O$
Nitrobacter (nye-troh-BACK-ter)	Nitrite	$2NO_2^- + O_2 \rightarrow 2NO_3^{2-}$
Nitrosomonas (nye-troh-so-MOH-nass)	Ammonia	$2NH_3 + 3O_2 \rightarrow 2NO_2^- + 2H^+ + 2H_2O$
Thiobacillus (thigh-oh-ba-SIL-lus)	Sulfur compounds sulfur	$2H_2S + O_2 \rightarrow 2S + 2H_2O$ $2S + 3O_2 + 2H_2O \rightarrow 2SO_4 + 4H^+$
	ferrous (iron) ion	$4Fe^{2+} + 4H^+ + O_2 \rightarrow 4Fe^{3+} + 2H_2O$

compound. All animals, fungi, and protozoa as well as most bacteria fall into this group.

Chapter 13 describes several examples of the four major nutritional categories described here. The next section closely examines some of the principal reactions of catabolism.

Energy Production

The catabolic release and storage of energy in useful forms, such as ATP, is often described in terms of **respiration** and **fermentation.** These are two major mechanisms for generating energy from the oxidation of chemical compounds. In general, respiration refers to energy-producing oxidative sequences or pathways in which inorganic compounds act as the last electron acceptors in a series of reactions. Specifically, *aerobic respiration* concerns those pathways in which oxygen is the last electron acceptor—being converted, as a rule, to water.

In the oxidative pathways of *anaerobic respiration,* inorganic electron acceptors other than oxygen are used; these include nitrate, sulfate, and carbonate. *Fermentations* involve those oxidative pathways in which an organic compound serves as both electron donor and acceptor. Thus, oxidation occurs in the absence of any outside electron acceptors. Many fermentation products are of considerable practical importance. These include vinegar, beer, wine, and yogurt. **[Chapter 16 describes several industrial applications of fermentation.]**

Generating ATP

PHOSPHORYLATION

As indicated earlier, the process of making ATP involves combining adenosine diphosphate (ADP) and inorganic phosphate. This reaction requires 7 to 8 kilocalories per ATP molecule. In metabolism, ATP is generated by three fundamentally different biochemical mechanisms: **substrate-level phosphorylation, oxidative phosphorylation,** and **photophosphorylation** (Figure 5–7).

SUBSTRATE-LEVEL PHOSPHORYLATION. In this reaction, the energy released through the oxidation of a substrate is used to phosphorylate ADP and to gen-

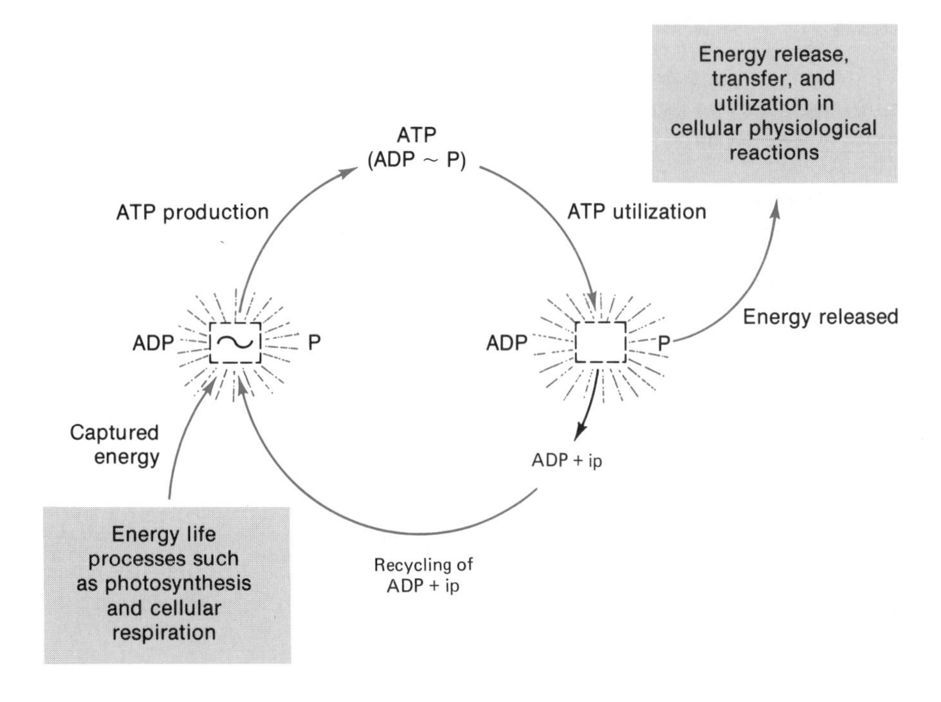

Figure 5–7 The formation of adenosine triphosphate (ATP), a fuel that serves all living organisms. By means of energy-yielding processes such as photophosphorylation, energy-rich bonds (~) are formed and used to combine ADP and inorganic phosphate (iP) into ATP. ATP, in turn, is used for various life processes, resulting in ADP and released inorganic phosphate. The inorganic phosphate and ADP are used again to continue the cycle.

erate an ATP molecule. The formation of ATP molecules by substrate phosphorylation occurs in the cytoplasm of both procaryotes and eucaryotes as well as in the fluid environment (matrix) of mitochondria. **[See Chapter 2 for the cellular organization of procaryotes and eucaryotes.]**

Oxidative Phosphorylation. ATP is generated in different metabolic mechanisms, including respiration and photosynthesis, by transporting electrons along a chain of carrier molecules with fixed positions within procaryotic plasma membranes and eucaryotic inner mitochondrial membranes. Each component of the chain can be reduced by reacting with the carrier molecule that precedes it and oxidized by the carrier that follows it (Figure 5–8).

In oxidative phosphorylation, the substance being oxidized is referred to as the *electron donor,* while the substrate receiving the electrons removed from the donor is called an *electron acceptor.* The series of electron acceptors is called the *electron transport chain* (see Figure 5–14). Those transport chains involved in the oxidation of organic compounds consist of a sequence of electron-transport coenzymes that pass electrons from one to the next in a defined order to generate ATP from ADP. The most common of these coenzymes is nicotinamide adenine dinucleotide (NAD), which exists in both an oxidized state (NAD) and a reduced state (NADH). Two other coenzymes that participate in oxidation-reduction reactions are nicotinamide adenine dinucleotide phosphate (NADP) and flavin adenine dinucleotide (FAD), which are $NADPH_2$ and $FADH_2$ in their reduced forms.

The electron transport system is a common pathway for the utilization of electrons formed during a variety of metabolic reactions. It is primarily concerned with aerobic metabolism; however, certain components of the system have been identified with anaerobic organisms. These organisms use inorganic electron acceptors other than oxygen in electron transport processes. Because these processes are similar to respiration but occur in the absence of oxygen, they are called anaerobic respiration. Other features of the electron chain are described in the section dealing with respiration.

Figure 5–8 The flow of electrons along a segment of the electron transport chain. Electrons e⁻ received from an oxidized substrate are passed from one coenzyme to another in a defined sequence through a series of oxidation and reduction reactions. Figure 5–14 shows an example of a complete system.

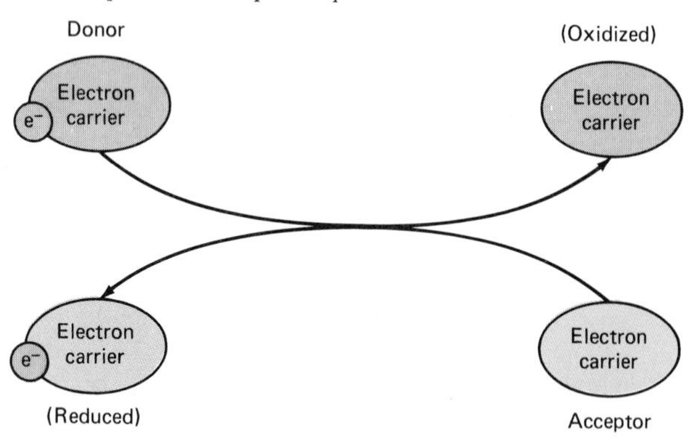

Photophosphorylation. The third phosphorylation mechanism occurs in photosynthesizing cells. Light energy trapped by pigments (such as the chlorophylls in photoautotrophs, described earlier) liberates an electron, which is then passed along a series of electron acceptors. With each transfer, energy is released for the generation of ATP through a chemiosmotic mechanism. *Chemiosmosis* is a process in which energy produced by proton movement across a membrane is used for the formation of ATP molecules.

These three biochemical mechanisms are of major importance to energy production. Later sections will show their respective roles in catabolic and anabolic pathways.

CHEMIOSMOSIS

Many organisms have transport membrane (transmembrane) channels that function in pumping protons (H^+) out of cells or organelles such as mitochondria and chloroplasts. The proton-pumping channels—using a flow of excited electrons obtained from carbohydrate oxidation or from light-excited pigments in photosynthesis—cause a structural change in transmembrane proteins. The conformational change, in turn, forces protons (H^+) to pass outward (Figure 5–9).

As the proton concentration outside the membrane rises higher than that on the inside, an electrical *gradient* (positive charge outside versus negative charge inside) is created. This imbalance drives the outer protons back across the membrane toward a region of lower proton concentration through the only channels available to them, namely, those coupled to ATP production (Figure 5–9). The return flow of protons (H^+) resulting from this proton gradient drives the generation of ATP. Thus, the net result is an expenditure of excited electron energy produced by photosynthesis, or metabolism to power the synthesis of ATP.

Since the chemical formation of ATP is driven by a diffusion force similar to osmosis, the process is referred to as *chemiosmosis.* In 1961 Peter Mitchell proposed the basis of this process, namely, that electron transport and ATP synthesis are coupled by a proton gradient across membranes. The chemiosmotic mechanism to generate energy for ATP synthesis is found in both oxidative phosphorylation and photophosphorylation and used by both procaryotes and eucaryotes.

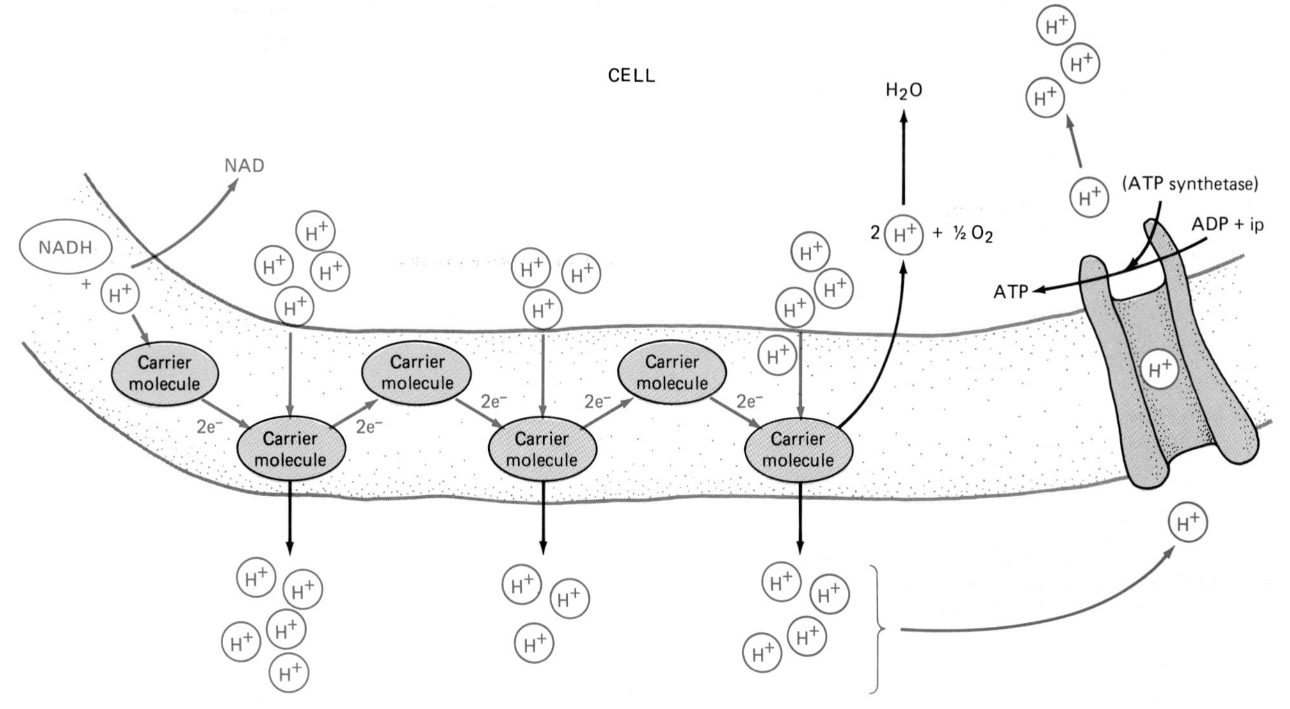

Figure 5–9 The chemiosmotic mechanism of ATP production. This model shows the use of energy released through electron passage to pump protons (H^+) out of a cell. As the outer proton concentration becomes greater than that in a cell, the outer protons are driven inward by a diffusion force similar to osmosis to generate ATP formation.

Photosynthetic Carbon Dioxide Fixation

Photosynthesis is, along with glycolysis (presented in the next section), one of the oldest and most fundamental life processes. It is ultimately responsible for the formation of all of the organic molecules that heterotrophs use to drive their various activities. All the major forms of photosynthesis developed among the photoautotrophic bacteria. As described earlier, in anoxygenic photosynthesis (the earliest form), hydrogen sulfide is enzymatically split, generating elemental sulfur as a by-product. The more advanced version, oxygenic photosynthesis, contains the additional and necessary stage to provide the power with which to split water and thereby generate molecular oxygen as a by-product. The overall reaction of oxygenic photosynthesis is as follows:

$$6CO_2 + 6H_2O \rightarrow C_6H_{12}O_6 + 6O_2 \uparrow$$

Photosynthesis consists of a complex set of events involving three different types of chemical processes: *light reactions, dark reactions,* and the *regeneration of photosynthetic pigment* (Figure 5–10). In the first photosynthetic process, sunlight is used to eject electrons from a donor molecule (pigment) to drive the chemiosmotic generation of ATP. The light reactions involve photosynthetic membranes and take place only in the presence of light. These reactions are followed by a series of enzyme-catalyzed reactions that use the newly generated ATP and NADPH to bring about the formation of organic molecules by the incorporation of atmospheric CO_2 (carbon fixation). The reactions here occur readily in the dark as long as sufficient ATP and reducing power in the form of NADPH (produced by the light reactions) are available.

Carbon fixation, in the dark reactions of photosynthesis, depends on the presence of a molecule to which CO_2 can be attached. The five-carbon sugar ribulose diphosphate (RuDP) serves this purpose and is a major component in the Calvin cycle, briefly described earlier.

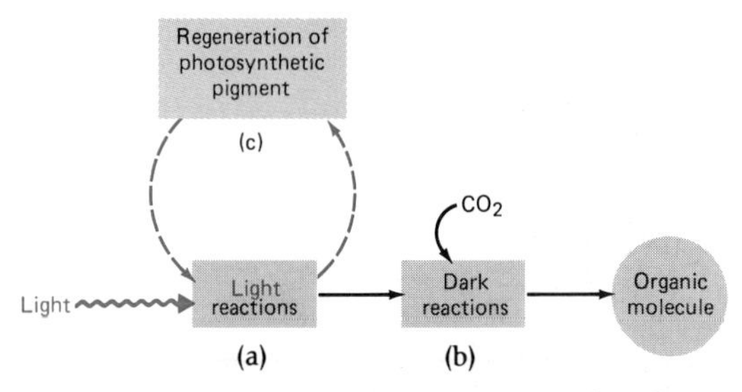

Figure 5–10 The three processes associated with photosynthesis: (a) light reactions, (b) dark reactions, and (c) regeneration of photosynthetic pigment.

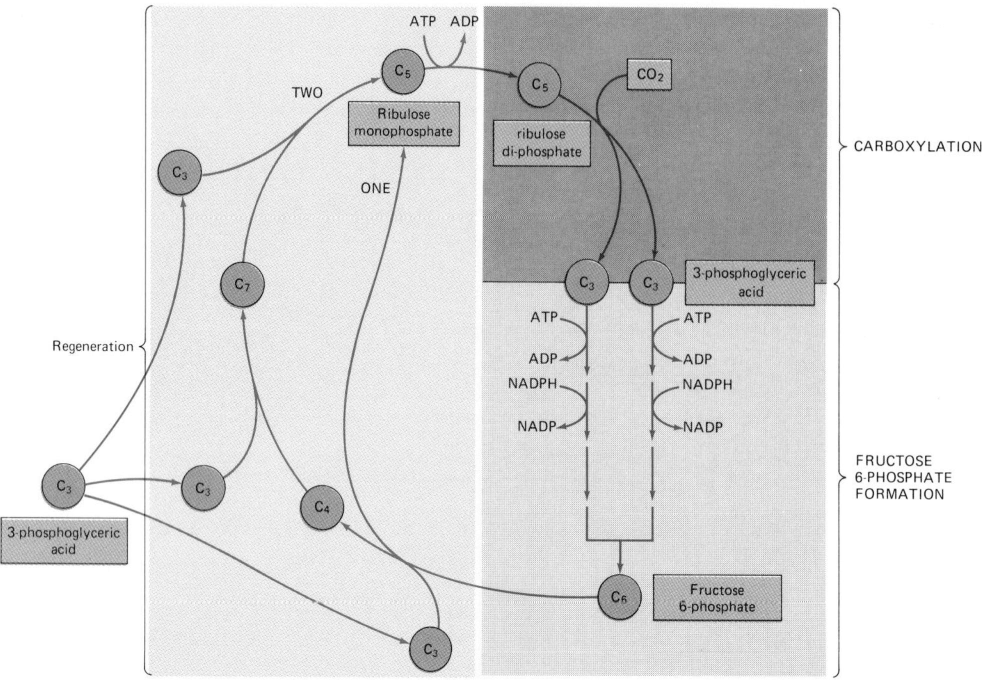

Figure 5–11 A summary of the dark reactions. The energy cycle that drives the reactions of this cycle is in the form of ATP and reduced NADP. The number of carbon atoms in each compound is indicated.

In the Calvin cycle (Figure 5–11), RuDP combines with CO_2 (by carboxylation) to form a temporary six-carbon intermediate compound that is immediately cleaved into two molecules of three-carbon phosphoglyceric acid (PGA). This reaction is catalyzed by the enzyme RuDP carboxylase. In a series of five reactions, two molecules of PGA form one molecule of fructose 6-phosphate (F6P). In still another series of reactions, some of the F6P molecules formed during the cycle are used to regenerate the cycle's starting material, RuDP, and others enter the cell's metabolism. Each turn of the Calvin cycle to fix one CO_2 molecule depends on the light reaction to supply two ATPs and two NADPHs.

Finally, the last event in photosynthesis involves rejuvenating the pigment that absorbed the light energy to start the process.

Central Pathways for Energy Production

The first step in catabolism of organic molecules is to degrade complex macromolecules into simple ones (depolymerization). For example, starches are dismantled into simple sugar molecules such as glucose, proteins into amino acids, fats into fatty acids and glycerol, and nucleic acids into nucleotides. (Refer to Appendix A for descriptions of these macromolecules.) Although these initial steps do not yield usable energy, they still serve to provide a number of small molecules such as the starting material for carbohydrate metabolism, the six-carbon sugar glucose ($C_6H_{12}O_6$). Through the oxidation of such organic molecules, organisms produce energy by *aerobic respiration, anaerobic respiration,* and *fermentation.* While a major portion of this energy is lost as heat, cells use the remaining energy in the form of ATP to perform their numerous activities. In this section we will examine the degradation of organic compounds resulting in energy release through a series of enzymatically catalyzed reactions known as **biochemical pathways.**

Glycolysis

Glycolysis, which literally means the breakdown of sugar, represents the most common process by which a fermentable substrate can be degraded. The Embden-Meyerhof-Parnas (EMP) pathway is the best-known series of catabolic reactions converting glucose into two three-carbon molecules of the important compound pyruvate (Figure 5–12). Two other pathways, the hexose monophosphate shunt (also known as the pentose phosphate pathway) and

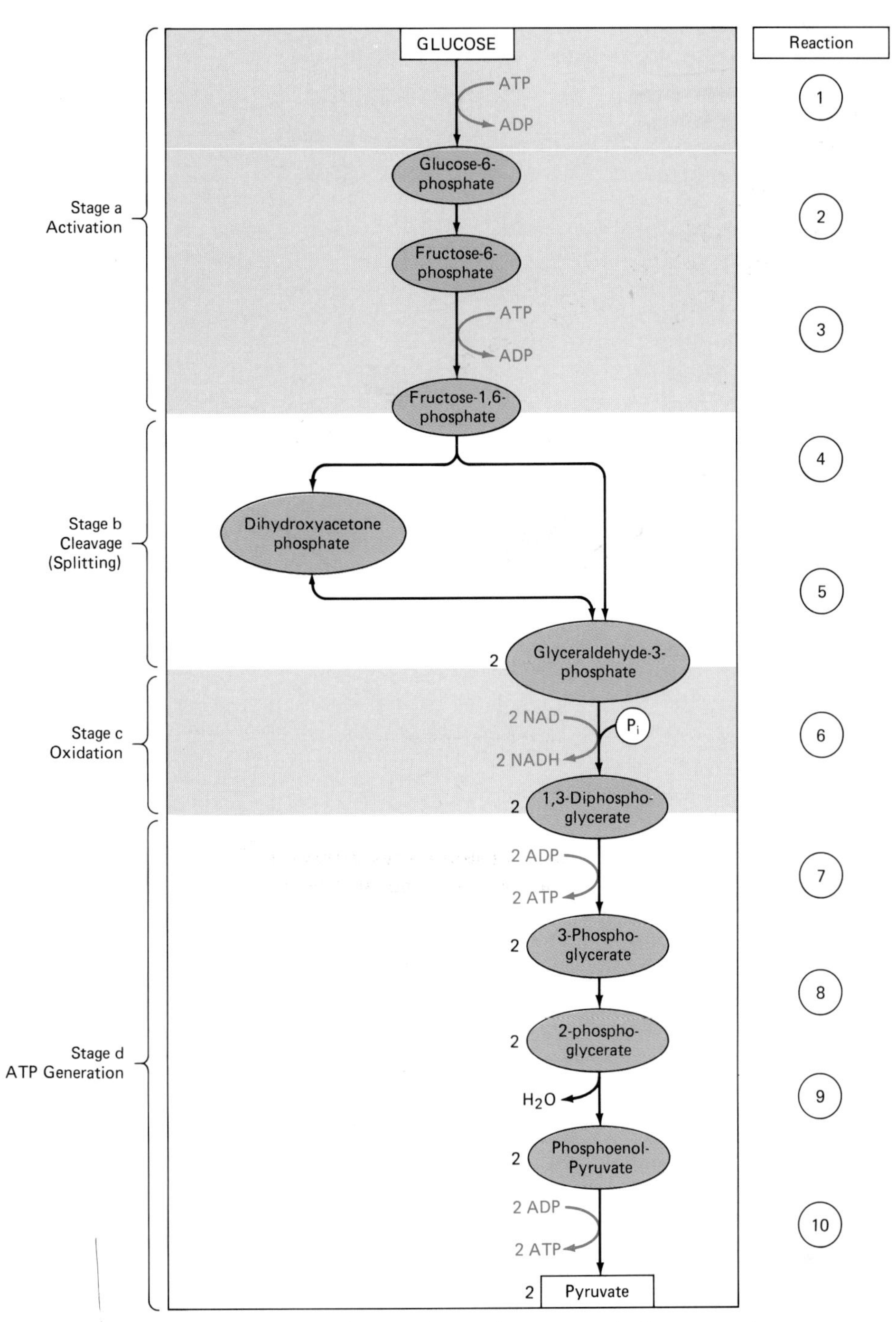

Figure 5–12 Four stages in glycolysis by the Embden-Meyerhof-Parnas pathway. (a) Activation of glucose by ATP. (b) Splitting of glucose into two 3-carbon molecules. (c) Oxidation involving the removal of electrons donated to nicotinamide adenine dinucleotides. (d) ATP generation and formation of pyruvate. Ten specific reactions in the pathway are also indicated.

the Entner-Doudoroff pathway lead to the catabolic conversion of sugars to pyruvate.

Figure 5–12 shows the ten reactions of the EMP pathway organized into four stages: *activation, cleavage, oxidation,* and *ATP generation.* The *activation* stage includes the phosphorylation of a glucose molecule by the transfer of a phosphate from an ATP molecule. This reaction activates the glucose. Two other reactions change the glucose molecule into a compound (fructose 1,6-diphosphate) that can readily be split into two three-carbon phosphorylated units. An additional ATP molecule is needed in the third reaction of this stage.

In the *cleavage stage,* fructose 1,6-diphosphate is

split into two three-carbon molecules. One of these is glyceraldehyde 3-phosphate (G3P), and the second is dihydroxy-acetone-phosphate (DHAP). The DHAP is enzymatically converted to a second G3P molecule.

The oxidation stage involves the removal of two electrons and two protons (H^+). Two electrons and one proton are taken up by the coenzyme nicotinamide adenine dinucleotide (NAD^+) to form NADH, while the remaining proton is dispersed into the cellular environment. The compound 1,3-diphosphoglyceric acid is formed from the oxidation.

The final stage, *ATP generation*, consists of four reactions leading to the conversion of the oxidation product of the previous stage into the three-carbon molecule *pyruvate*. Two ATP molecules are generated in the process.

In summarizing the overall net result of the pathway in terms of ATP production, the gain amounts to two ATP molecules. This is because each glucose molecule is split into two G3P molecules. The following accounting shows the results according to ATP expenditures and gains.

$$\begin{array}{r} -2\text{ATP stage a} \\ \underline{2(+2\text{ATP}) \text{ stage d}} \\ +2\text{ATP} \end{array}$$

Although the glycolytic reaction sequence is quite inefficient, since it generates a small number of ATP molecules for cellular activities, it does provide smaller molecules for use as building blocks by procaryotes and eucaryotes. Most of the remaining energy is associated with pyruvate, the key important compound. Upon further oxidation of compounds derived from pyruvate, the energy can be recovered.

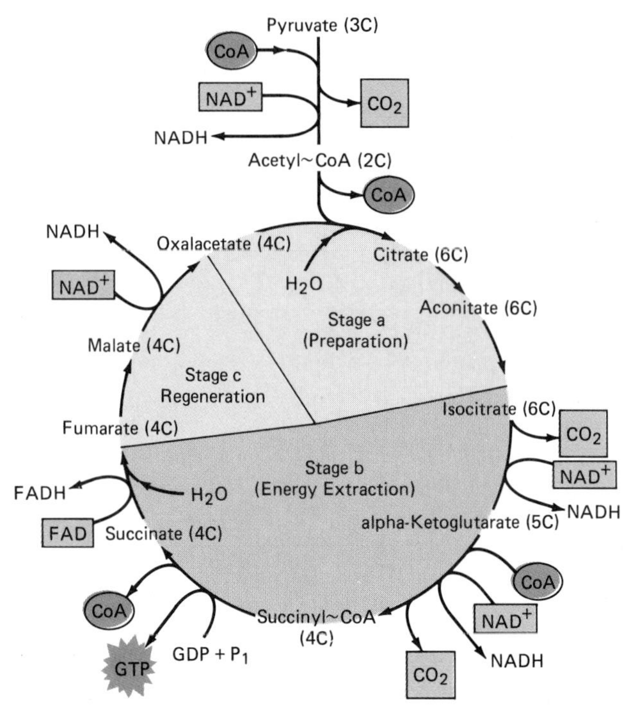

Figure 5–13 The reactions of the Krebs or citric cycle divided into three stages: (a) preparation, (b) energy extraction, (c) regeneration of the starting molecule. Acetyl~COA enters at every turn of the cycle and is oxidized to carbon dioxide (CO_2) and water (H_2O). ATP is generated by the proton pumps driven by the electrons removed at various points along the cycle. Each reaction is numbered to serve as a guide through the cycle.

Aerobic Respiration

KREBS CYCLE (CITRIC ACID CYCLE)

The Krebs cycle, named after Hans A. Krebs to acknowledge his discovery of several participating compounds in this cycle, consists of a sequence of enzyme-catalyzed chemical reactions. The series of reactions is referred to as a cycle since the compound oxaloacetate, formed at the end of the sequence, is identical to the substance at the starting point (Figure 5–13). Each time a so-called spin of the cycle occurs, pyruvic acid molecules from the glycolytic or related pathway must be incorporated. However, before pyruvic acid enters the Krebs cycle (also known as the citric acid cycle), it must undergo an enzymatic removal of CO_2. This transition step between the two cycles then leaves a two-carbon compound to combine with coenzyme A (Co-A) to form acetyl~coenzyme A (acetyl~CoA) to start the next reaction in the sequence.

The nine reactions of the citric acid cycle can be arranged into three stages: *preparation, energy liberation,* and *starting material regeneration.* As indicated earlier, such rotation of the cycle begins and ends with oxaloacetate and incorporates acetyl~CoA, which is oxidized to CO_2 and H_2O. The CO_2 is released into the environment as a gaseous by-product of aerobic respiration. This cycle may be referred to as "the energy wheel of cellular metabolism" because it is crucial to supplying the energy needs of cells. Through its chemical reactions, the cycle releases the large amount of potential chemical energy stored in intermediate compounds derived from pyruvate in a step-by-step fashion.

The *preparation* stage consists of three reactions. Acetyl~CoA joins the four-carbon oxaloacetate to form the six-carbon citric acid. The remaining two reactions cause a repositioning of citric acid's hydroxyl (OH) group, which results in the formation of isocitric acid (Figure 5–13).

The *energy liberation* stage consists of four reactions. The first reaction yields two electrons that reduce NAD^+ to NADH, causing the oxidative removal of CO_2 (decarboxylation) from isocitric acid and the formation of the five-carbon alpha-ketoglutaric acid. The alpha-ketoglutaric acid is, in turn, oxidatively decarboxylated to produce a molecule that combines with CoA to form the four-carbon succinyl~CoA. The reaction also yields a pair of electrons that reduce NAD^+ to NADH. The bond connecting the succinyl group to CoA is a high-energy one similar to those found in ATP. In the sixth reaction, the high-energy bond is cleaved with its released energy, causing the phosphorylation of guanosine diphosphate to guanosine triphosphate (GTP). The four-carbon succinic acid is the remaining compound at the end of this reaction. In the seventh reaction, succinic acid is oxidized to form the four-carbon fumaric acid. A different electron acceptor, flavin adenine dinucleotide (FAD^+), is used here.

The last stage of the citric acid cycle consists of two reactions in which a water molecule is added to fumaric acid, resulting in the four-carbon malic acid. This molecule is then oxidized, yielding the four-carbon oxaloacetate. The oxidation reaction also produces two electrons that reduce NAD^+ to NADH. With the regeneration of oxaloacetate, the cycle can begin again by the incorporation of another acetyl~CoA.

Briefly summarizing the cycle (Figure 5–13), citric acid is decarboxylated twice, once for each carbon in the original acetyl compound. Each of these reactions yields CO_2 and reduced coenzymes. The citric acid cycle is considered complete and begins again with additional acetyl CoA as soon as the formation of oxaloacetic acid has occurred. The course of the cycle involves a sequence of carrier molecules known as the *electron transport system (ETS)* or *chain*. These molecules are capable of oxidation and reduction and include the flavoproteins, cytochromes, quinones, and iron-sulfur proteins. (The features of these carriers are described in the next section.) As electrons are transferred through the ETS, a stepwise release of energy occurs and ATP is generated from ADP in the process.

Each pyruvic acid molecule oxidized via this cycle results in the formation of 12 ATP molecules. Because each molecule of glucose can result in the formation of two pyruvic acid molecules, a total of 24 ATP molecules can be formed from a single molecule of glucose.

ELECTRON TRANSPORT CHAIN (SYSTEM)

The major purpose of the electron flow along the electron transport chain or system (Figure 5–14) is to pump protons across the inner membranes of eucaryotic organelles, such as mitochondria and chloroplasts, and procaryotic cytoplasmic membranes (Figure 5–9). This flow is tightly linked to the phosphorylation processes described earlier.

Electrons entering the electron transport system from NADH possess a relatively high energy content. As the electrons pass from one electron carrier to the next, some of their energy is conserved in the form of ATP and the remainder is released as heat.

As mentioned earlier, the ETS includes three classes of carrier molecules involved in the respiration of organic compounds. The *flavoproteins,* such as flavin adenine dinucleotide (FAD), are proteins that contain a coenzyme derived from riboflavin (vitamin B_2); they participate as proton (H^+) carriers within the ETS. The *cytochromes* belong to an iron-containing (heme) group of proteins. Four types of heme groups are known, *a, b, c,* and *d;* they are the basis on which one cytochrome is distinguished from another. The central iron atom of the cytochromes can be cycled between the oxidized ferric state (Fe^{3+}) and the reduced state (Fe^{2+}). Cytochromes are electron carriers. The cytochrome content of bacteria varies among species and environments and therefore has diagnostic importance in the identification of certain aerobic bacteria. For example, the presence of cytochrome *c* can be shown by the oxidase test. This test is performed by applying oxidase reagents, such as dichlorophenol or indophenol, to a small amount of bacterial growth.

Figure 5–14 The electron transport system. The flow of electrons between NADH and the final electron acceptor oxygen (½O_2) is shown in this figure. Note that sufficient energy is generated at three points along the chain, which is then coupled to oxidative phosphorylation and ATP formation.

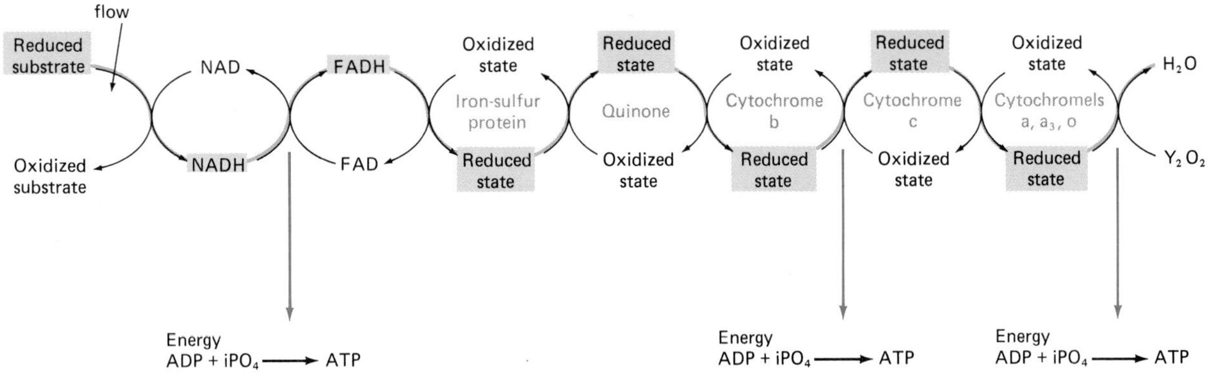

These dyes, which are colorless in the reduced state, are rapidly oxidized to colored forms by cytochrome *c*-containing organisms (Color Photograph 116b). *Quinones* are nonprotein substances that function as proton carriers in the ETS. The iron-sulfur proteins are electron carriers in the ETS. They contain two, four, or eight atoms of unstable sulfur, which are so called because strong acids cause their release as hydrogen sulfide. In addition, since the iron in these proteins is not in a cytochrome, they are also referred to as nonheme iron proteins.

All members of the ETS are arranged in precise order (Figure 5–14), and must be capable of being reduced by the reduced form of a previous carrier and oxidized by the oxidized form of a carrier next in line. The sequence of reactions is started by the oxidation of NADH, which, in turn, reduces FAD. This process of oxidizing NADH and reducing FAD results in the formation of FADH. What is more significant, it releases enough energy to form ATP. At the end of the series of oxidations and reductions in this system, electrons freed of their excess energy react with oxygen, which then combines with hydrogen ions to form water.

Summary of ATP Generation During Aerobic Respiration of Glucose

The relationship of ATP production (net gain) to the various pathways in glucose catabolism is shown in Figure 5–15. An overall reaction for aerobic respiration can be summed up as follows:

$$\underset{C_6H_{12}O_6}{\text{Glucose}} + \underset{6O_2}{\text{Oxygen}} + \underset{38\ ADP}{\text{Adenosine diphosphate}} + \underset{38\ P_i}{\text{Phosphate}}$$

$$\downarrow$$

$$\underset{6CO_2}{\text{Carbon dioxide}} + \underset{6H_2O}{\text{Water}} + \underset{38\ ATP}{\text{Adenosine triphosphate}}$$

Aerobic procaryotic microorganisms can use the pathways shown in Figure 5–15 to generate a maximum of 38 ATP from glucose. Glycolysis yields 2 ATP by substrate phosphorylation, and by the same process an additional 2 ATP are obtained in the Krebs cycle. The total of 10 NADH can yield 30 ATP and the 2 FADH can yield 4 ATP, all by oxidative phosphorylation. This totals 38 ATP.

Aerobic eucaryotic microorganisms, however, can produce a total of only 36 ATP. The transition and citric acid cycle reactions occur within the fluid matrix and inner membranes of the mitochondria in close proximity to the sites of oxidative phosphorylation activity. However, glycolysis occurs in the cytoplasm, and the 2 NADH produced in glycolysis must reduce FADH in the process of entering the mitochondria, thus losing 2 potential ATP.

Figure 5–15 A summary of ATP usage and generation during the aerobic respiration of glucose. The total net yield is 38 ATPs. In fermentation, only the glycolysis segment (upper right) is used. The ATP molecules generated in fermentation processes are much fewer in number.

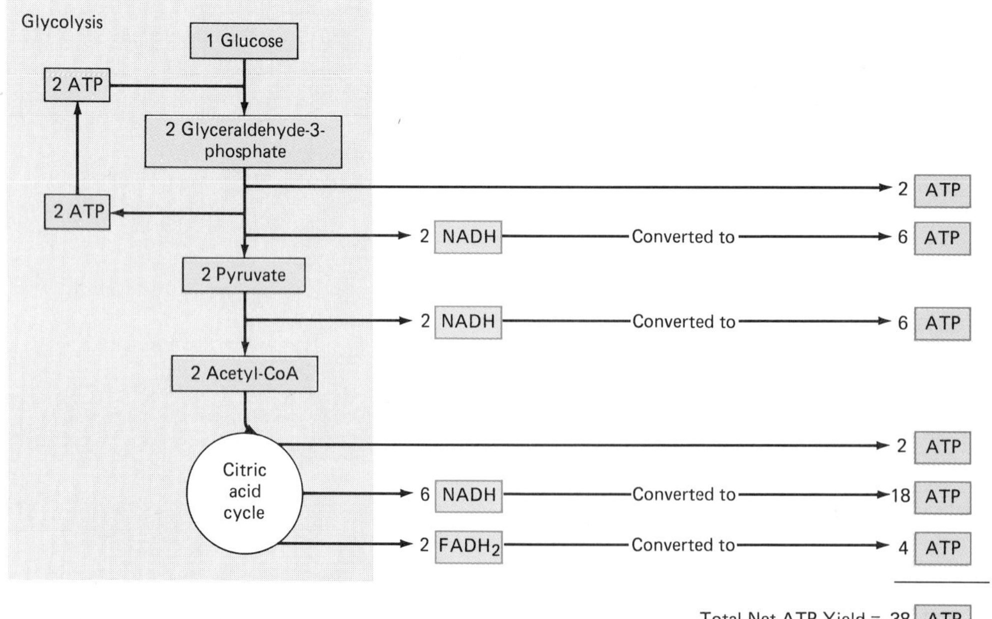

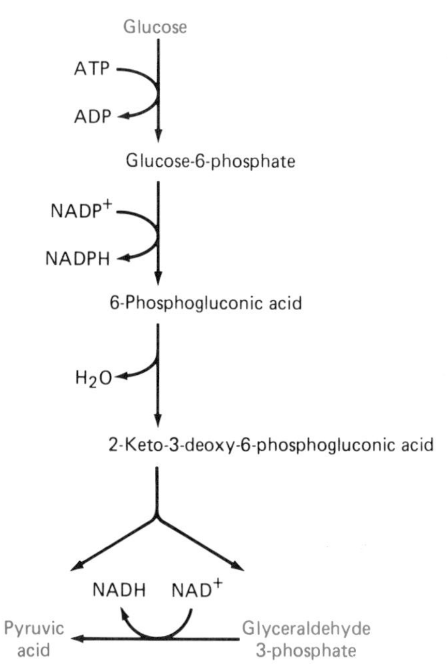

Figure 5–16 The Entner-Doudoroff (ED) pathway.

Selected Alternative Pathways in Carbohydrate Catabolism

As indicated earlier other significant catabolic reactions also occur. Two representative examples are the *Entner-Doudoroff (ED) pathway,* in which glucose is converted to pyruvic acid in fewer steps than it is in the pathways of glycolysis, and the *hexose monophosphate shunt* (HMS), in which glucose is converted to five-carbon carbohydrates (pentose units) for nucleic acid synthesis.

ENTNER-DOUDOROFF (ED) PATHWAY

Like glycolysis, the ED pathway produces pyruvic acid from glucose. However, it is thought to occur in a representative evolutionary line of bacteria, separating them from most organisms that utilize glycolysis. The pathway is shown in Figure 5–16. The ED pathway involves an initial phosphorylation, as in glycolysis, but then is followed by an oxidative step of the compound to an acid form (phosphogluconic acid). Subsequently, dehydration occurs, with the formation of ketodeoxyphosphogluconic acid. The last reaction in the pathway involves an enzymatic splitting that produces pyruvic acid and glyceraldehyde phosphate, which can be converted to pyruvic acid, as indicated by the dotted line in Figure 5–16.

The ED pathway is found in some gram-negative bacterial genera, including those of *Pseudomonas, Rhizobium* (rye-ZOH-bee-um), and *Agrobacterium* (a-grow-back-TIR-ee-um). It is generally not found in gram-positive bacteria.

THE HEXOSE MONOPHOSPHATE SHUNT (HMS)

The HMS, or pentose phosphate pathway, operates in conjunction with glycolysis in many bacteria. This set of reactions generates ribose for nucleic acid synthesis and produces NADPH for other synthetic reactions, including carbon dioxide fixation in photosynthesis. The HMS is shown in Figure 5–17. The first reaction, the phosphorylation of glucose to glucose 6-phosphate, is the same as that found in glycolysis and the ED pathway. The next reaction results in the conversion of glucose 6-phosphate to 6-phosphogluconic acid, as in the ED pathway, followed by the conversion to ribulose 5-phosphate and then to ribose 5-phosphate. Ribulose 5-phosphate also combines with carbon dioxide in the dark reaction of photosynthesis (Figure 5–10). Here $NADP^+$ is reduced at two reaction sites, rather than the more commonly encountered NAD^+. The pentose units can subsequently be converted to two different intermediates in glycolysis and into pyruvic acid. Thus, the HMS serves as a loop in glycolysis for the production of pentose units and NADPH.

ANAEROBIC RESPIRATION

The emphasis on oxygen in the electron transport system suggests that corresponding anaerobic systems do not occur. This is not true. The comparable anaerobic system of biological oxidations that does not use oxygen as the final acceptor of electrons is

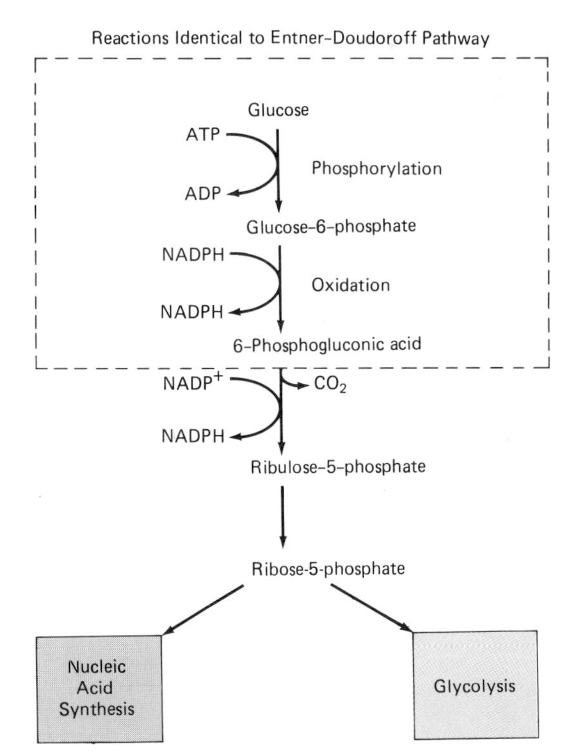

Figure 5–17 The hexose monophosphate shunt (HMS).

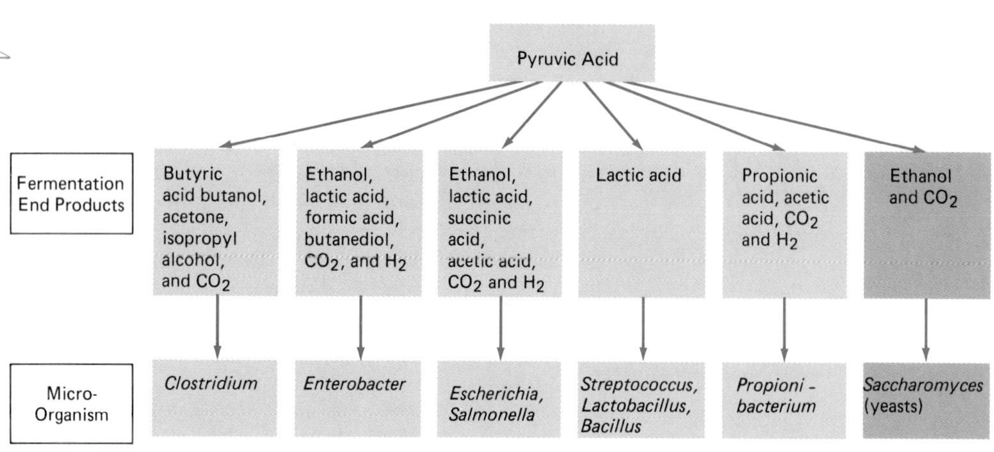

Figure 5–18 Fermentation end products formed by various microorganisms from pyruvic acid.

called *anaerobic respiration.* In anaerobic respiration, compounds such as carbonates, nitrates, and sulfates ultimately are reduced. Many facultative anaerobic bacteria can reduce nitrate to nitrite under anaerobic conditions. This type of reaction permits some continued growth when free oxygen is absent, but the accumulation of nitrite, which is produced by the reduction of nitrate, is eventually toxic to the organism. Certain species of *Bacillus* and *Pseudomonas* (soo-doh-MOH-nass) are able to continue the reduction of nitrite to gaseous nitrogen. This process, called **denitrification,** can occur only when these aerobic organisms are grown under anaerobic conditions. The organisms that reduce sulfate and carbonate are strictly anaerobic. *Desulfovibrio desulfuricans* (dee-sul-foh-VIB-ree-oh dee-sul-furr-EE-kans) reduces sulfate to hydrogen sulfide as it oxidizes carbohydrate to acetic acid. *Methanobacterium bryantii* (meh-THAN-oh-back-teer-ee-um br-eye-AN-tee-eye) is able to couple the reduction of carbon dioxide to methane with a similar oxidation of carbohydrate to acetic acid. These reactions occur in nature in processes that recycle these elements and are presented in more detail in Chapter 14.

In anaerobic respiration, the reduction of nitrate, sulfate, and carbon dioxide involves cytochrome-containing electron transport systems comparable to that illustrated for aerobic respiration.

FERMENTATION

In glycolysis, the same reactions occur whether oxygen is present or not. The products are primarily pyruvic acid, NADH, and ATP. The essential difference between what occurs aerobically and anaerobically is what happens to pyruvic acid and NADH. In the case of fermentation reactions, pyruvic acid is converted to a variety of organic compounds (Figure 5–18). These reactions involve the transfer of electrons and hydrogen from NADH to organic compounds. The end products of fermentations and the pathway by which they are formed are specific for microbial species or groups.

In fermentation, organic compounds serve as both ultimate electron donors and acceptors. Thus, a

TABLE 5–5 A Comparison of Fermentation and Respiration

Metabolic Process	*Conditions of Growth*	*Electron Acceptor*	*Reduced Product Formed*	*Type of Phosphorylation Used for ATP Generation*	*Net ATP Gain*
		Inorganic compound, such as			
Respiration	Aerobic	O_2	H_2O	Oxidative and substrate-level	38
	Anaerobic	NO_3^-	NO_2^-, N_2O, N_2	Oxidative and substrate-level	Varible
		SO_4^{2-}	H_2S		
		CO_2	CH_4		
Fermentation	Aerobic or anaerobic	Organic compounds, such as pyruvic acid	Ethanol[a]	Substrate-level	2

[a] A product commonly found with yeast.

fermentable substance such as glucose often yields both oxidizable and reducible products of metabolism (metabolites).

Fermentation is a major source of energy for those organisms that can survive only in the absence of air (obligate anaerobes). Other fermentative organisms that can grow in the presence or absence of air (facultative anaerobes) use fermentation as a source of energy only when oxygen is absent. An important property of all fermentation reactions is that most of the energy of substrates remains untapped. Energy gain is very low and occurs as a result of substrate-level phosphorylation. The synthesis of ATP in fermentation is largely restricted to the amount formed during glycolysis. Table 5–5 compares several features of fermentation to respiration.

During glycolysis glucose is oxidized to pyruvic acid (Figure 5–19). Some fermenting organisms reduce the pyruvic acid to lactic acid as the sole or primary end product (Figure 5–19). Such organisms are said to be *homofermentative.* They include species of *Bacillus, Lactobacillus* (lack-toh-ba-SILL-lus), and *Streptococcus.* Microorganisms that produce lactic acid, as well as acetic, formic, and other acids, possibly ethanol, other alcohols, and acetone, are referred to as *heterofermentative.* This group includes yeasts, other fungi; several gram-positive bacteria belonging to the genera of *Clostridium* and *Streptococcus;* and several gram-negative species. Heterofermenters often use the hexose monophosphate shunt. **[See Chapter 15 for industrial uses of microbes.]**

Noncarbohydrate fermentation can occur with amino acids, organic acids, purines, and pyrimidines. A reaction of this type is illustrated by the action of *Clostridium* on amino acids. In one kind of amino acid fermentation, the *Stickland reaction,* two different amino acids participate, neither of which can be fermented singly by most clostridia. In this coupled fermentation, one amino acid, alanine, is oxidized while the other is reduced. The alanine is also deaminated to yield pyruvic acid and ammonia. The pyruvic acid is further oxidized and decarboxylated to yield acetic acid and carbon dioxide. During this time, glycine is reduced and deaminated to produce acetic acid and ammonia. The overall reaction can be represented as follows:

$$\text{alanine} + \underset{\text{glycine}}{2} \rightarrow \underset{\text{ammonia}}{3} + \overset{3}{\underset{\text{acid}}{\text{acetic}}} + CO_2$$

Figure 5–19 Selected reactions involving pyruvic acid.

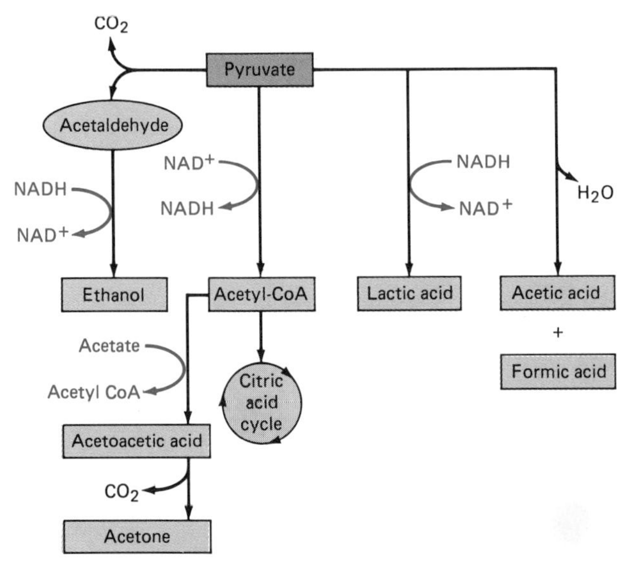

Metabolic Interrelationship of Carbohydrates, Fats, and Proteins

Although the metabolic pathways of carbohydrates, lipids, proteins, and other significant biochemical compounds are often described separately, they are actually closely related. A simplified version of the network formed by these pathways is shown in Figure 5–20.

In the catabolism of glucose, acetyl CoA can be produced (Figure 5–13), as previously discussed. Acetyl CoA is also a catabolic product of fat metabolism. Moreover, it is an important precursor in the synthesis of fatty acids. The interrelationship of fatty acid synthesis and glucose catabolism is well illustrated by the dependence of the fatty acid synthesis process on glycerol, an alternate product of glycolysis (Figure 5–12).

Acetyl CoA can also be obtained from proteins. For example, deamination, loss of the amino ($-NH_2$) group, converts the amino acid alanine to pyruvic acid and subsequently to acetyl CoA (Figure 5–20). An additional example of the interrelationship of protein synthesis and glucose catabolism is the conversion of proteins to amino acids and then to pyruvic acid and the citric acid cycle intermediates, alpha-keto glutaric and oxaloacetic acids. The addition of an amino ($-NH_2$) group converts these compounds into the building blocks of protein. Conversely, deamination converts glutamic and aspartic acids to the citric acid cycle intermediates, alpha-keto glutaric and oxaloacetic acids. Many of the steps in glycolysis are reversible, so that glucose units are available for combination into starch, glycogen, or cellulose. Thus, metabolic intermediates of carbohydrates, lipids, and proteins are used to create energy and building blocks for more carbohydrates, lipids, nucleic acids, and proteins. In this manner, the food digested by living beings becomes distinctively assimilated.

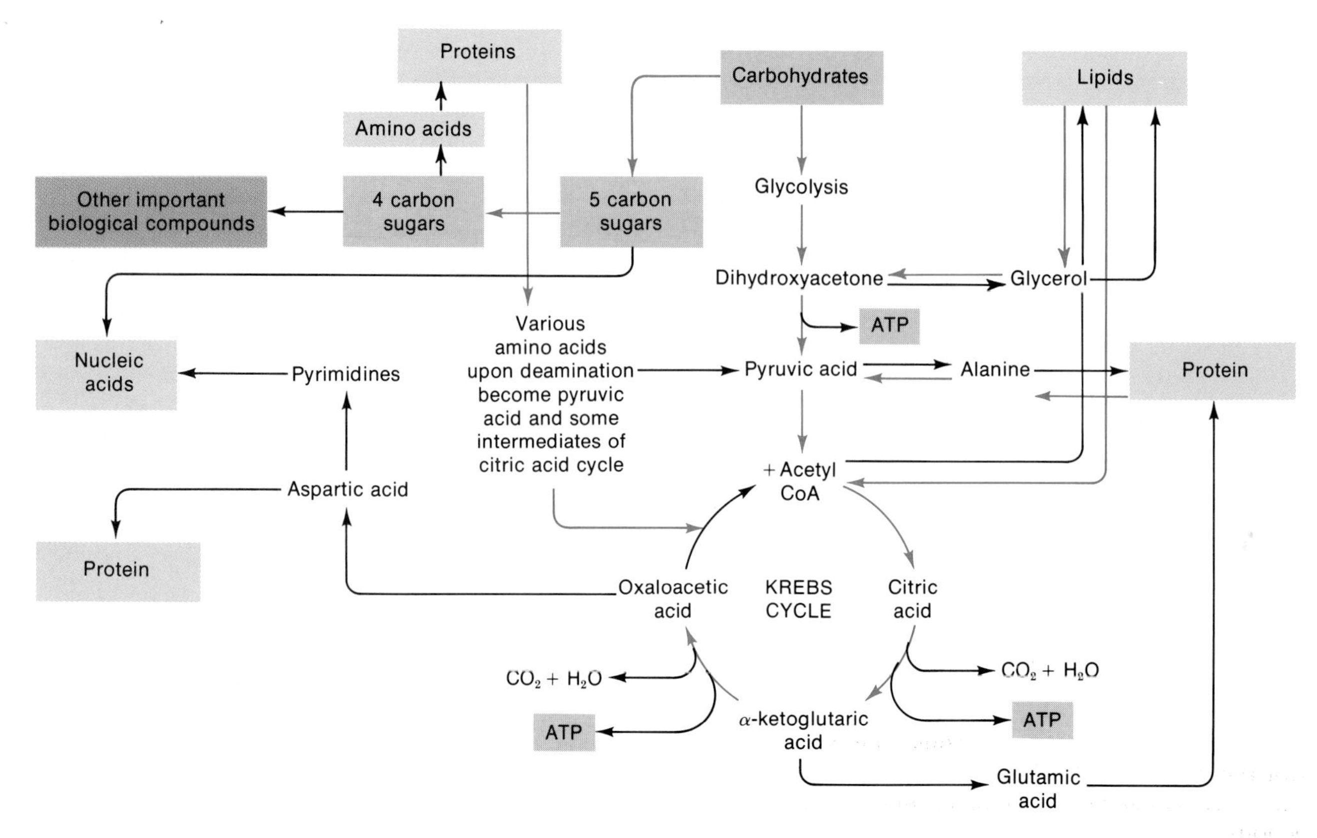

Figure 5–20 The relationship between the metabolic pathways of carbohydrates, lipids, and proteins. Reactions involved in the degradation (breakdown) of compounds (carbohydrates, amino acids, etc.) are indicated by colored lines, and reactions leading to the formation of important macromolecules are shown by dark lines.

Biosynthetic (Anabolic) Reactions

In general, individual cells synthesize their own proteins, nucleic acids, lipids, polysaccharides, and other complex compounds. Such macromolecules are not received, preformed, from other cells. Biosynthetic or anabolic processes include not only the formation of various macromolecular components but also their assembly into the different membranes that form organelles. It is important to note that synthetic processes are, in general, not simply the reverse of the catabolic processes by which mol-

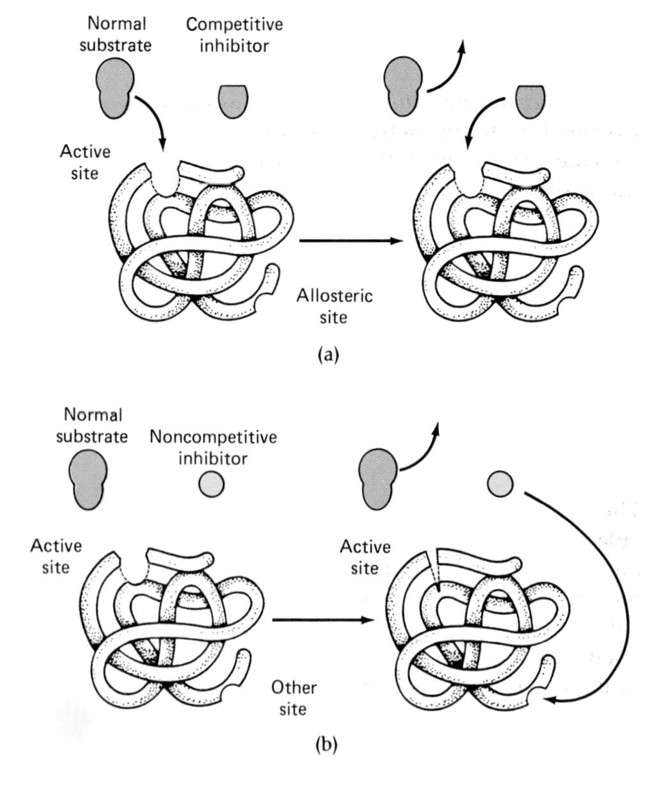

Figure 5–21 Competitive and noncompetitive enzyme inhibitors. (a) A competitive inhibitor competes with the substrate for the active site of the enzyme. Once bound to the active site, the competitive inhibitor prevents the enzyme's binding of the substrate. Refer to Figure 5–5. (b) A noncompetitive inhibitor binds to an allosteric site on an enzyme, which is different from its active site. The inhibitor causes a change in shape, thereby inactivating the enzyme.

ecules are degraded. In addition, the individual steps of biosynthetic pathways are catalyzed by different enzymes, permitting a separate control mechanism. Figure 5–20 presents a good overview of some of these processes. Chapter 6 provides insights into related topics, including protein synthesis and the influence of genetics on biosynthetic processes.

Control and Regulation of Metabolism

If a cell is to make the best use of the energy available to it, it must function efficiently. Cell products are produced as needed. When a substance is either produced in the cell or enters it, a specific enzyme required to catalyze a reaction for that substance is formed by protein synthesis. Thus, one way to control the growth and other activities of microorganisms such as bacteria is to affect their enzymatic reactions. As described earlier, most enzymes can be inhibited or even destroyed by certain chemical or physical agents. Enzymatic activities can also be affected by substances chemically similar to normal substrates competing for an enzyme's active site. This type of interference is called *competitive inhibition* (Figure 5–21a). Since the competitive inhibitor is not similar enough to fully serve as the normal substrate, the enzyme cannot form end products. **[Chapter 12 describes application of this form of inhibition with sulfa drugs.]**

In *noncompetitive inhibition,* the inhibitor may bind irreversibly at the active site, or it may distort the enzyme's shape by binding at a location other than the active site. The attachment at a site other than the active site is specific for the inhibitor and prevents substrate binding by causing a change in the shape of the active site (Figure 5–21b). There is no competition between the substrate and the noncompetitive inhibitor. In addition, the change in enzyme shape may be reversible or irreversible.

Feedback or End-Product Inhibition

The ability to change the shape of an enzyme molecule provides a basis for microorganisms to regulate enzymatic activities. *Allosteric interaction* is an example of an ingenious mechanism by which an enzyme may be temporarily activated or inactivated. Such interactions occur among enzymes that have at least two binding sites, one the *active site* and the other the *allosteric effector site,* into which molecules known as allosteric effectors fit. The binding of an effector changes the shape of the enzyme and either activates or inactivates it. Allosteric interactions are often involved in *feedback inhibition,* a common means of biological control.

In many enzymatic reactions, the final product inhibits, or slows, the series of reactions that produced it. This control system is known as *feedback inhibition.* Synthesis of the amino acid isoleucine in *Escherichia coli* is one such set of reactions. When extra quantities of an amino acid are added to a bacterial culture actively synthesizing that particular amino acid, synthesis of the compound ceases. Thus, by controlling the amount of the isoleucine made available to a culture, the pathway for this compound can be either started or stopped.

As long as isoleucine is required for protein synthesis or for other metabolic reactions, synthesis continues. When the reactions utilizing isoleucine stop and the compound begins to accumulate, it interferes with the particular enzyme responsible for starting the pathway. Studies of this control mechanism using isolated enzymes show that the end product inhibits the activity of the first enzyme involved in the pathway leading to its synthesis.

Evidence indicates that the inhibition is caused by isoleucine, which binds to a site on the first enzyme of the pathway, the *allosteric* site; this site is different from the catalytic site at which the substrate, threonine, would attach. The attachment by isoleucine changes the shape of the enzyme so that it cannot combine with its substrate (Figure 5–22). Thus, feedback inhibition occurs with enzymes having two types of binding sites. The process is reversible if the end product is not present. The enzyme then can assume its original shape and combine with its specific substrate.

Isoleucine synthesis offers one of many examples of regulatory mechanisms involved in metabolic reactions. The presence of such regulation makes one think of a cell as a finely tuned automated system for the performance of life processes.

Determining Metabolic Activities

Metabolic studies have shown how antibiotics interfere with microbial growth and also how microorganisms can alter their metabolism to become resistant to these drugs. They have also demonstrated some basic differences between pathogenic and nonpathogenic bacteria. The results of such investiga-

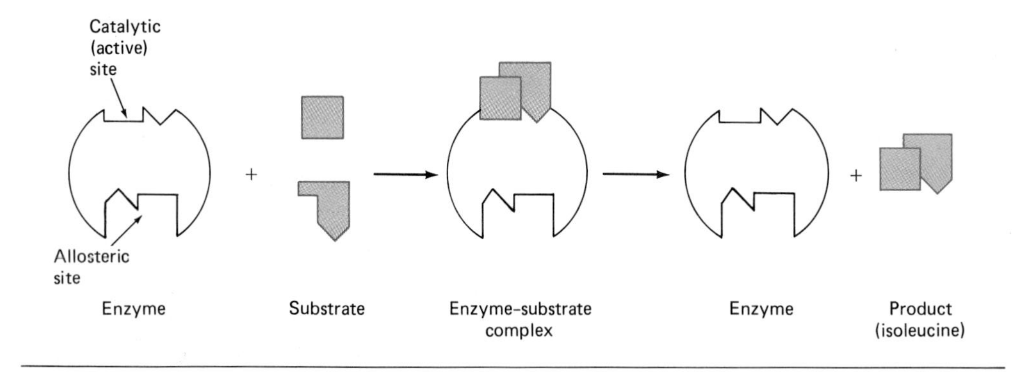

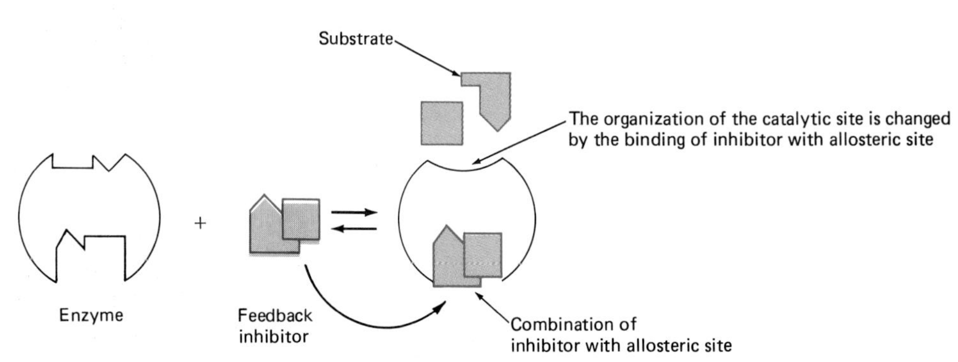

Figure 5–22 Feedback inhibition using the isoleucine biosynthetic pathway. The top view shows the normal pathway. Isoleucine is the end product. In the lower view, the end product causes a change in the shape of the enzyme involved in the reaction and thereby prevents substrate attachment. The end product binds to the allosteric site of the enzyme.

tions are being used to develop faster methods of identifying medically important microorganisms. These studies have also provided procedures with which to evaluate, improve, or uncover the mechanisms operating in industrial processes such as the production of antibiotics, various other drugs, and nutrient supplements such as amino acids and vitamins. Among the methods and materials used are (1) simple fermentation tests, (2) oxidation-reduction activity, (3) radioisotopes, and (4) chromatography. **[See Chapter 16 for industrial uses of microorganisms.]**

Simple Fermentation Tests

When an organism has the necessary enzymes to ferment a particular sugar, fermentation can generally be detected by the production of organic acids with or without gas. This can be accomplished either by incorporating some nontoxic pH indicator in the medium or by adding it after growth. A decrease in pH indicates that a fermentation reaction or the production of organic acids (Color Photograph 27) took place. In liquid media, gas production can be observed by the presence of obvious bubbling or by trapping the gas in an inverted small glass tube (Figure 5–23a). This tube, known as a *Durham tube,* is placed upside down in the fermentation medium prior to autoclaving. In a fermentation, the formation of gas within the system displaces the liquid in the inverted vial.

Generally, the Durham tube indicates only the presence or absence of gas. The *Smith fermentation tube* (Figure 5–23b) can be used to measure the relative proportions of different gases produced. Because microorganisms often produce more gas than the Durham tube is capable of holding, the Smith tube is preferable for obtaining a reasonably accurate measurement of the volume of gas produced. This can be accomplished by measuring the portion of tube containing gas before (Figure 5–23b) and after (Figure 5–23c) the addition of alkali to absorb CO_2. Thus, one can determine the relative amounts of carbon dioxide produced. Because the remaining gas is primarily hydrogen, the measurements indicate the ratio of hydrogen to carbon dioxide formed.

Oxidation-Reduction Activity

Under anaerobic conditions, the metabolic activity of cells can be monitored in several ways. One such procedure involves adding an electron acceptor such as methylene blue to a resting microbial suspension and then exposing it to a vacuum to remove air from the medium. Metabolic activity is then monitored by noting the rate of decolorization of the

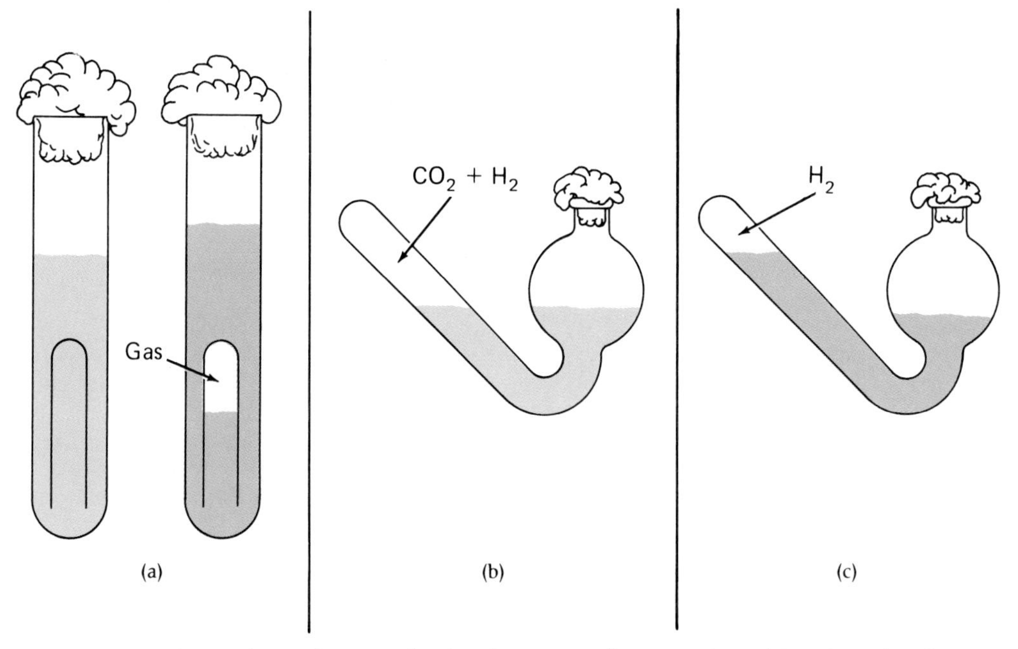

Figure 5–23 Comparison of gas production from sugar fermentations. (a) Uninoculated Durham fermentation tube and the results of active fermentation. Note the displacement of the fluid from the inverted vial. (b) A Smith tube before the addition of an alkaline compound. (c) The results after addition of an alkaline compound.

dye. During the operation of metabolic pathways, electrons are transported by various intermediates and dehydrogenases in place of NAD^+, which can use methylene blue as the electron acceptor, thus causing it to form leukomethylene blue, which is colorless.

A common source of electrons in biological systems is the Krebs cycle intermediate, succinic acid. If succinic acid dehydrogenase is present in a particular unknown system, the resting suspension will remove electrons from the substrate to methylene blue with some intermediate steps. The loss of color indicates the presence of that dehydrogenase. Using properly controlled studies, one can determine: (1) possible metabolic pathways, (2) the effect of various respiratory inhibitors, and (3) chemical blocking agents in metabolic pathways.

Radioisotopes

Some of the elements that make up biochemical compounds, such as carbon, nitrogen, oxygen, and phosphorus, exist in several forms, called *isotopes.* Isotopes of an element have the same atomic number but different atomic masses. Certain isotopes are unstable; as deterioration occurs, they release various radioactive particles—alpha, beta, and gamma rays. These can be detected by radiation counters or by their effect on photographic emulsions. Table 5–6 lists some of the common isotopes used in biochemical research and their respective *half-lives* (the time required to lose half of their radioactivity).

Melvin Calvin began his study on the pathway of carbon in photosynthesis in 1946, using radioactive carbon dioxide, $^{14}CO_2$. This isotope was introduced into the atmosphere of growing algae. Periodically the photosynthetic process was interrupted briefly to remove small samples of the compounds formed. The various carbon compounds of the algae were isolated and tested for radioactivity, an indication of ^{14}C. In this manner, the pathway of carbon in the photosynthesis cycle could be traced step by step through the presence of ^{14}C in successive metabolic products.

Chromatography

The term **chromatography** refers to procedures used for separating the substances found in body fluids or cells or separating the components of a single substance from one another. The many forms of chromatography all share the same basic mechanism, namely, the simultaneous exposure of the

TABLE 5–6 Common Isotopes Used in Biochemistry

Element	*Radioisotope*	*Half-life*
Carbon (^{12}C)	^{14}C	5,570 years
Hydrogen (^{1}H)	^{3}H	12.2 years
Sulfur (^{32}S)	^{35}S	87.1 days
Iron ($^{55.8}Fe$)	^{59}Fe	45 days
Phosphorus ($^{30.9}P$)	^{32}P	14.3 days
Iodine ($^{126.9}I$)	^{131}I	8 days

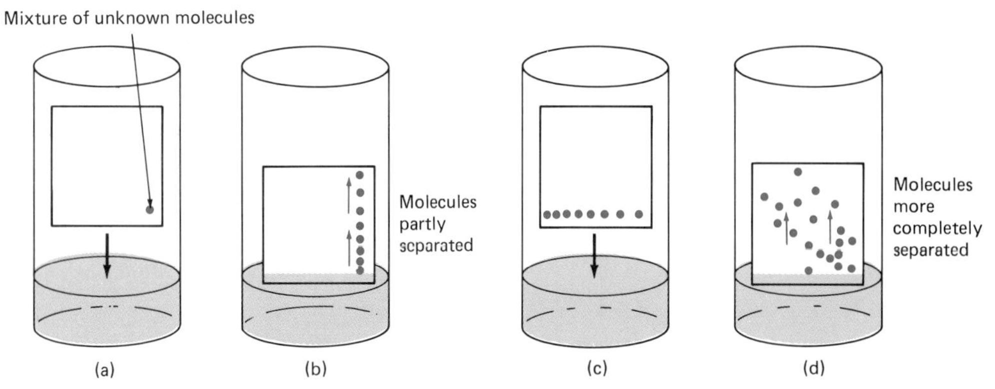

Figure 5–24 (a) The procedure for the separation and identification of the amino acids in a single protein by two-axis chromatography. Refer to the text for details of steps a through d.

mixture under study to two different solvents that will not mix or to a solvent mixture and an absorbent solid to which molecules of gas, dissolved substances, or a liquid will adhere.

One of the simplest of chromatographic techniques is paper chromatography. Several drops of an unknown sample are placed near one end of an absorbent material such as a filter paper strip moistened with water (Figure 5–24a). The bottom edge of the paper is then placed into a nonaqueous solvent such as phenol, which moves up the paper. Those substances (solutes) in the mixture that have a much higher attraction for the phenol than for the water in the filter paper migrate up the paper along with the phenol. Those solutes having a much greater attraction for the water in the filter paper do not travel far; they transfer from the moving phenol and adhere to the water on the filter paper. In this manner separation of the various substances in the original mixture is achieved. At a specified time the movement of the solvent is stopped and the results of the separation are examined (Figure 5–24b).

Depending on the type of material under study, further separation may be needed. Such separation is possible by using a second solvent and allowing the solutes to travel across the filter paper in a new direction (Figures 5–24c, d). The technique is known as *two-axis chromatography.* An example of the results obtained by this method is shown in Figure 5–24e.

The first major study involving chromatography was made by Michael Tswett in 1906. By running an ether solution of chlorophyll through a column of calcium carbonate, he was able to separate the various colored pigments.

Chromatography can be performed with various solvent mixtures on filter paper, silica, or cellulose-coated glass, or in columns containing various solid matrices such as powdered calcium carbonate or alumina.

Gas chromatography is a form of column chromatography used for the separation of gases or vaporized chemicals, with an inert carrier gas as the mobile solvent. As the carrier gas passes from the column, it enters a detector, which often uses the differences in thermal properties of the separated material to detect it. Generally, the detector plots this information on a graph. Materials can be identified by comparing their times of detection under particular conditions with tabulations of times for known materials. Gas chromatography has been used in the identification of microorganisms, particularly the obligate anaerobes that produce many easily detectable fatty acids.

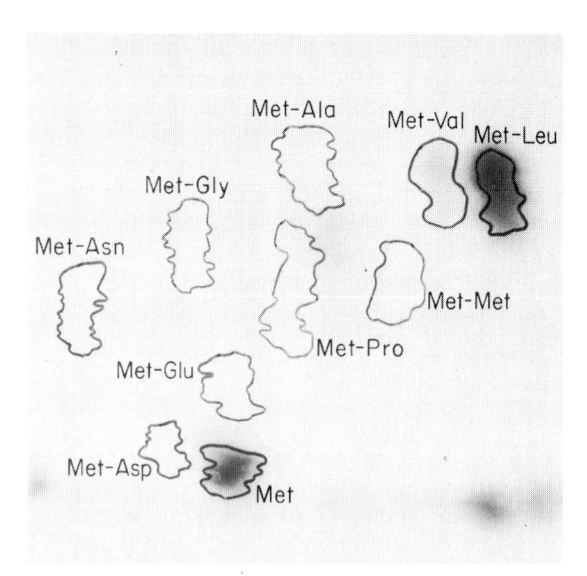

Figure 5–24 (b) The positions that would be occupied by specific amino acids are indicated. Only the darkened areas show the amino acids present. Symbols used include Ala (alanine), Asp (aspartic acid), Gly (glycine), His (histidine), Leu (leucine), Met (methionine), Pro (proline), and Val (valine).

The metabolic transformations that cells perform increase the quantity of cell components (biosynthesis). The formation of these components requires energy, which involves the conversion of chemical bond energy into the concentrated high-energy bonds of ATP or, in the case of phototrophs, the transformation of light energy into the chemical bond energy of ATP. Chapter 6 presents other important aspects of microbial life, such as the respective roles of DNA, RNA, and proteins in cellular activities and in the exchange of genetic information among microorganisms.

Summary

General Metabolism

1. *Metabolism* refers to the entire set of chemical reactions by which a cell produces and forms the various macromolecules it needs to maintain itself. The two general categories of metabolism are *catabolism* and *anabolism*.
2. Catabolic reactions through which complex compounds are broken down with the release of energy are linked to anabolic reactions that result in the formation of important macromolecules.

ATP

1. In biological systems, the energy to drive biosynthetic reactions is provided by a compound such as adenosine triphosphate, or ATP.
2. One ATP molecule contains ribose, three phosphate groups, and one adenine molecule.
3. ATP is formed from ADP and inorganic phosphate; it contains two high-energy phosphate bonds shown as ~P.

Unity of Biochemistry

Similarities can be observed in the metabolic reactions of biological systems, thereby suggesting the existence of the *unity of biochemistry*.

Enzymes

1. Enzymes are biological catalysts.
2. They lower the amount of energy (*activation energy*) necessary to start a chemical reaction.
3. Enzymes are highly specific for the reactions they catalyze.

The Parts of an Enzyme

1. Some enzymes consist only of protein.
2. Certain enzymes consist of two parts, a protein component, the *apoenzyme*, and an additional chemical part, the *cofactor*.
3. Metal cofactors are called *activators*, while organic cofactors are referred to as *coenzymes*.

Factors Affecting Enzyme Activity

1. Extremes of temperature, pH, radiation, and heavy metal concentration can affect enzyme activity.
2. Enzymes exposed to such factors may undergo temporary or permanent changes in structure (denaturation) and become nonfunctional.
3. The molecule with which an enzyme reacts is called the *substrate*.
4. Specific portions of an enzyme that interact with substrates are known as *active sites*.

Mechanism of Action

1. An *enzyme-substrate* (ES) *complex* is formed by the physical joining of a specific enzyme with the substrate molecule.
2. In the ES complex, the substrate is changed into end products (new molecules).
3. At the end of the reaction, the enzyme returns to its original form.

Energy Metabolism

The major role of catabolic reactions is to bring about the release of energy concentrated in high-energy bonds for use in energy-requiring biosynthetic cellular reactions.

Oxidation-Reduction Reactions

1. Organisms that use chemical energy do so by performing oxidation-reduction reactions.
2. Oxidation is the removal of electrons from a substance, while reduction is the gain of electrons by a substance.
3. In biological systems, oxidation-reduction reactions are always paired (coupled).
4. The most common electron carrier is nicotinamide adenine dinucleotide (NAD^+). Upon accepting two electrons, NAD^+ becomes reduced to NADH.

Metabolic Differences Among Microorganisms

1. Microorganisms vary in biosynthetic capabilities and nutrient requirements.
2. Autotrophs obtain all of their carbon needs from CO_2.
3. Heterotrophs use organic compounds as carbon sources and many use the same compounds for energy.

Autotrophic Microorganisms

1. Photoautotrophic (photosynthetic) organisms use carbon dioxide to obtain carbon and light as their energy source.
2. The bacteriochlorophylls found among photosynthetic autotrophic bacteria include *a, b, c,* and *d.* Such chlorophylls are different from those found in plants.
3. The photosynthetic process carried out by green or purple bacteria (anoxygenic photosynthesis) does not result in the formation of oxygen. However, cyanobacterial and plant photosynthesis (oxygenic) does produce oxygen.

4. Nonphotosynthetic organisms such as the *chemoautotrophs* rely on the oxidation of inorganic compounds, including hydrogen or hydrogen sulfide, for the energy to fix carbon dioxide in the Calvin cycle.

HETEROTROPHIC MICROORGANISMS

1. Heterotrophic organisms that are capable of photosynthesis (photoheterotrophs) can adjust to varying levels of oxygen.
2. Chemoheterotrophs obtain both their carbon and energy needs from organic compounds.

Energy Production

1. *Respiration* refers to those energy-producing oxidative pathways in which inorganic compounds act as the last electron acceptor. In aerobic respiration, gaseous oxygen serves this purpose, while in anaerobic respiration, compounds such as nitrate, carbonate, or sulfate are involved.
2. *Fermentation* refers to those oxidative pathways in which organic compounds serve as both electron donors and acceptors.

Generating ATP

PHOSPHORYLATION

1. The basic process for producing adenosine triphosphate (ATP) involves combining adenosine diphosphate (ADP) with inorganic phosphate (P_i). This reaction requires energy, which can be obtained through *photosynthetic phosphorylation, substrate phosphorylation,* or *oxidative phosphorylation.*
2. In photosynthetic phosphorylation, the required energy is obtained from the absorption of light by chlorophyll.
3. During substrate phosphorylation reactions, energy is released from substances (substrates) acted upon by enzymes. This energy can be used to form ATP.
4. Oxidative phosphorylation requires coenzymes such as nicotinamide adenine dinucleotide (NAD), nicotinamide adenine dinucleotide phosphate (NADP), and flavin adenine dinucleotide (FAD). During a series of oxidation and reduction reactions, enough energy is produced to form ATP. This series of reactions constitutes the *electron transport chain.*
5. Oxidative phosphorylation occurs along the cristae of eucaryotic mitochondria and mesosome membranes of procaryotes.

CHEMIOSMOSIS

1. According to the chemiosmotic mechanism, electron transport and ATP synthesis are connected by a proton (H^+) gradient across cell membranes and membranes of organelles.
2. A diffusion force similar to osmosis drives protons into the cell through membrane channels associated with ATP production.
3. Chemiosmosis operates in both oxidative phosphorylation and photophosphorylation and is used by both procaryotes and eucaryotes.

Photosynthetic Carbon Dioxide Fixation

1. Photosynthesis is the conversion of light energy to chemical energy; chlorophyll and other pigments help to collect the light energy. It consists of three different chemical processes: light reactions, dark reactions, and regeneration of photosynthetic pigment.
2. Light reactions are used to collect electrons to drive the chemiosmotic generation of ATP.
3. Dark reactions use the products of light reactions to fix carbon by the reduction of CO_2 into carbohydrate. The Calvin cycle is the major metabolic pathway to fix carbon.
4. The last event in photosynthesis is restoring the photosynthetic pigment to start the process again.

Central Pathways for Energy Production

1. Through the oxidation of complex molecules, organisms produce energy by *aerobic respiration, anaerobic respiration,* and *fermentation.*
2. Biochemical pathways consist of a series of enzymatically catalyzed reactions.

GLYCOLYSIS

1. Glycolysis is the process resulting in the breakdown of sugar.
2. The reactions of the Embden-Meyerhof-Parnas pathway result in the conversion of one glucose molecule into two molecules of pyruvate.
3. The pathway yields a net gain of two ATP molecules.
4. The Krebs (citric acid) cycle consists of nine sequential reactions that the cell uses to liberate electrons to drive the synthesis of 38 ATP molecules, (a high yield of energy).
5. The Krebs (citric acid) cycle involves the production of citric acid by the combination of acetyl CoA and oxaloacetic acid. The citric acid is then converted through a series of reactions to oxaloacetic acid, and the cycle can then begin once more.
6. A summary of glycolysis and respiration is shown in Figure 5–25.
7. Two selected alternate pathways are the Entner-Doudoroff (ED) pathway and the hexose monophosphate shunt.
8. In the ED pathway, glucose is converted to pyruvic acid in fewer steps than in glycolysis. This pathway appears to be found in relatively few bacteria and may indicate a basis for a separate evolutionary line.
9. The hexose monophosphate shunt (HMS) serves as a side loop of glycolysis in many bacteria for the production of pentose units for nucleic acid synthesis. It also serves as a source of NADPH for other biosynthetic reactions, such as photosynthetic carbon dioxide fixation.
10. Glycolysis functions the same under anaerobic as under aerobic conditions. The difference involves the use of NADH for reduction of organic substrates, for example, the conversion of pyruvic acid to lactic acid.
11. In fermentation, pyruvic acid is converted to a variety of organic compounds and involves the transfer of electrons and hydrogen from NADH to organic compounds.

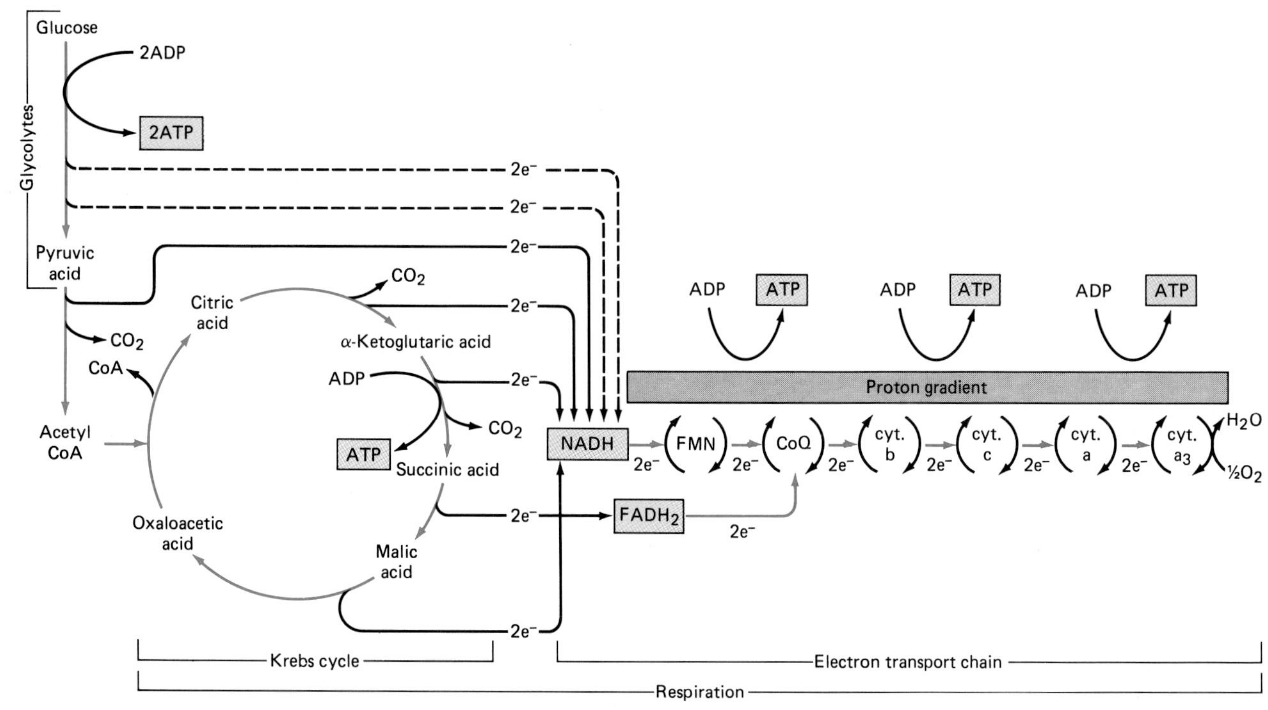

Figure 5–25 A summary of glycolysis and respiration.

12. The energy gain in fermentation is low and is a result of substrate-level phosphorylation.

Metabolic Interrelationship of Carbohydrates, Fats, and Proteins

1. Several aspects of the metabolism of carbohydrates, lipids, proteins, and other compounds are interrelated.
2. Various metabolic intermediate compounds of these pathways are used for energy and the construction of other macromolecules.

Biosynthetic (Anabolic) Reactions

1. Living cells exist in a dynamic state and are continuously involved with reactions used to construct and degrade many cell components.
2. Each cell generally synthesizes its own macromolecules. Each step in a biosynthetic pathway is catalyzed by a separate enzyme and requires ATP.

Control and Regulation of Metabolism

1. In competitive inhibition, an inhibiting substance structurally similar to the normal substrate competes for an enzyme's active site.
2. In noncompetitive inhibition, an inhibitor binds to an enzyme at a location other than the active site. Binding to this site changes the enzyme's shape and thereby inactivates it.
3. The inhibition of enzyme activity by the accumulation of an end product is called *feedback inhibition.* This type of inhibition is an example of an allosteric interaction and can occur only with enzymes that have both active and allosteric (other) sites.

Analysis of Metabolism

1. Metabolic studies have shown how antibiotics may interfere with microbial growth and cause changes in metabolic activities of microorganisms.
2. Specific methods used to study microbial metabolism include simple fermentation tests, determination of oxidation-reduction activity, the use of radioisotopes in pathways, and chromatography.

Questions for Review

1. Differentiate between catabolic and anabolic reactions in a cell's metabolism.
2. What is an enzyme?
3. Explain the following properties and/or activities associated with enzymes:
   a. parts of an enzyme
   b. active site
   c. formation of an ES complex
   d. factors affecting enzyme action
4. Briefly explain what occurs in oxidation-reduction reactions.
5. Compare the general features of the metabolic patterns of photosynthetic and chemosynthetic autotrophs and heterotrophs.

6. Briefly describe three mechanisms for making ATP.
7. Outline the general steps in glycolysis and indicate the primary end products of this process.
8. Outline the general process of the citric acid cycle, indicating the starting materials and products of each reaction.
9. Summarize aerobic respiration in terms of each major phase and the numbers of ATP molecules possible from substrate and oxidative phosphorylation.
10. Compare the total numbers of ATP molecules that can be formed by glycolysis in terms of substrate and oxidative phosphorylation under anaerobic and aerobic conditions.
11. What is meant by fermentation? List typical products that may be produced by bacteria from pyruvic acid under anaerobic conditions.
12. What is anaerobic respiration?
13. Indicate how proteins and lipids can be catabolized and utilized to make ATP in aerobic metabolism.
14. Describe the events occurring in photosynthesis.
15. Discuss three mechanisms by which enzyme activity in the cell is regulated.
16. Define or explain the following terms:
    a. electron acceptor
    b. chemiosmosis
    c. Calvin cycle
    d. light receptor
    e. chemolithotroph
    f. electron transport chain

Suggested Readings

CECH, T. R., "RNA as an enzyme." *Scientific American* **255:** 64–75, 1986. *The discovery that RNA can cut, splice, and assemble itself and that every cellular reaction is not catalyzed by a protein enzyme is described in this article.*

DANKS, S. M., E. H. EVANS, and P. A. WHITTAKER, *Photosynthetic Systems: Structure, Function and Assembly.* New York: John Wiley & Sons, Inc., 1983. *A concise, up-to-date review of photosynthesis in both bacteria and plants.*

GOTTSCHALK, G., *Bacterial Metabolism,* 2nd ed. New York: Springer-Verlag, 1986. *A detailed, up-to-date coverage of bacterial energy metabolism. Includes topics such as nitrogen fixation, biosynthesis of cell components, and regulation of bacterial metabolism.*

HINKLE, P., and R. MCCARTY, "How Cells Make ATP." *Scientific American* **238:**104–125, 1978. *A functional summary of oxidative respiration, with an understandable account of the events occurring at the mitochondrial membrane.*

MARGULIS, L., *Early Life.* Boston: Science Books International, 1982. *While the book emphasizes the probable evolution of microorganisms, it does so with clear and simple discussions of cellular chemistry.*

MILLER, K., "The Photosynthetic Membrane." *Scientific American* **239:**100–113, 1979. *A current description of the mechanism by which the major components of the photosynthetic apparatus are embedded in membranes.*

STRYER, L., *Biochemistry,* 2nd ed. San Francisco: W. H. Freeman and Company, 1981. *A well-written biochemistry text providing easy-to-understand and up-to-date descriptions of molecular mechanisms of metabolism.*

6

Microbial Genetics and Its Applications

Everything in nature contains all the powers of nature. Everything is made of one hidden stuff. — Ralph Waldo Emerson

After reading this chapter, you should be able to:

1. Explain how experiments performed by Griffith and by Hershey and Chase helped to prove that DNA is the genetic material.
2. Describe the organization of DNA and RNA and their respective roles in protein synthesis.
3. Outline the steps involved in protein synthesis and distinguish between the transcription step in procaryotes and eucaryotes.
4. Describe the general features of the genetic code and how mutations may affect it.
5. Distinguish among the general types of mutation and their causes.
6. Describe the operon model of protein synthesis.
7. Explain the one-gene, one-enzyme hypothesis.
8. Distinguish between transformation, conjugation, and transduction as means of recombination in asexually reproducing organisms.
9. Define and/or describe the roles of plasmids, bacteriocins, R plasmids, metabolic plasmids, and transposons in a bacterium's life.
10. Explain how a scientist might develop a biological assay for a particular amino acid.
11. Explain how genetic studies can be used to determine taxonomic relationships between asexual organisms.
12. Explain what is meant by gene manipulation. Contrast its potential benefits and hazards and describe one technique for gene manipulation with the bacterium *Escherichia coli.*

What chemical substance is responsible for an organism's genetic heritage and behavior? The beginnings of an answer were provided by several pioneers attempting to unravel this mystery. During the period from 1869 to 1953, both the structure of the substance and its chemical properties responsible for genetic phenomena were identified. This chapter will deal with the genetics of microorganisms and their role as test systems for understanding the mechanisms of heredity at the molecular level. Considerable attention will also be given to protein synthesis and some applications of microbial genetics, including the use of microorganisms as precise molecular instruments in the areas of genetic engineering and biotechnology.

Like biochemical principles, genetic principles are universal. The study of the inheritance and variability of the characteristics of microorganisms (microbial genetics) has contributed greatly to what we now know about the genetics of all organisms. Research in genetics at the molecular level has identified deoxyribonucleic acid (DNA) as the chemical substance making up genes, the discrete units that physically carry the information of heredity. Much has been discovered about the molecular structure of DNA in studies dealing with its formation and how the information it contains is transmitted to control the various activities of cells. Microorganisms, especially bacteria, have been extremely valuable test subjects, since they require very little laboratory space and grow rapidly; thereby they allow many experiments to be performed quickly. Cells numbering in the billions can be screened for various properties by techniques requiring little time and money. In addition, all offspring (progeny) of a single microbial cell are essentially identical. This chapter will describe how research with bacteria such as *Streptococcus pneumoniae* (strep-toh-KOK-kuss new-MOH-nee-ah) and *Escherichia coli* raised, and essentially resolved, the modern dilemma of the chemical nature of the genetic material and of mutations, and how studies with *Neurospora crassa* (new-RAH-spor-ah KRASS-ah), a common bread mold, led to the careful mapping of metabolic pathways and genetic fine structure.

While the study of microbial genetics has made fundamental contributions to our understanding of gene structure and function, it has also opened the door to one of the most exciting areas of the biological sciences today, *genetic engineering.* This area, which involves the manipulation of DNA outside an organism for the purpose of constructing a new form of life with different or altered properties, promises to have far-reaching effects on our everyday life. This is especially evident from the series of discoveries in the early 1970s that led to the development of *DNA recombinant techniques.* By means of these techniques genes isolated from an unrelated organism can be introduced into a bacterium or yeast so that the progeny of that microorganism can make medically important drugs, such as human insulin to treat diabetes, growth hormone to correct bone-growth defects, and interferon to treat certain cancers and virus diseases. Other applications of genetic engineering to the commercial uses of organisms, called *biotechnology,* are described in Chapter 15.

This chapter presents both the established and newer concepts and principles that gave rise to and will continue to underlie the remarkable advances being made at the molecular level.

Evolution and Inheritance

It was the nineteenth-century naturalist Charles Darwin who wrote, in his book *The Origin of Species,* that the evolution of new species occurred because of natural selection. Darwin proposed that natural differences or variations appear spontaneously among individuals of a species. The conditions of the environment exert selective pressures on these varying individuals, and the resulting struggle for existence in the environment determines which individuals survive to pass on their characteristics to future generations. Thus, according to Darwin's concept of evolution, in a changing environment and over a period of time, only those organisms that could adapt genetically to the changing environment would survive. Therefore, the ability to acquire new genetic properties confers a survival advantage to such organisms. These organisms would reproduce and continue to survive by natural selection and thereby implement evolution.

Modern biology recognizes that variations in cells and organisms can arise as a result of two phenomena. These are permanent changes in the chemical composition of an organism's DNA, known as *mutations,* and the influence that different environments have on an organism. Before dealing with these important topics, we will briefly describe some of the major historical events that led to the discovery of DNA as the genetic material and the various roles played by microorganisms in this discovery.

Heredity and Mutation

DNA as the Material of Heredity

The basic genetic material was first isolated by Frederick Miescher in 1868. His isolate, which is called *nuclein* because it came from the nuclei of white blood cells, had an interesting chemical composition: 14% nitrogen and 2.5% phosphorus. We now know that Miescher's nuclein was mainly deoxyribonucleic acid (DNA).

In 1881, nuclein was shown to be associated with chromosomes. Findings of this kind suggested to some biologists that nuclein was responsible for the transmission of hereditary characteristics. Until the mid-twentieth century the predominant view, however, was that the protein of the nucleus was the material responsible for heredity. It was argued that nuclein could not be the genetic substance because it appeared to vary in quantity during asexual cell division, or mitosis.

Proteins, on the other hand, appeared to be more constant, and were favored as possible carriers of genetic information because they were known to contain large numbers of amino acid subunits. **[See Appendix A for a description of amino acids.]** The belief that nuclear protein was the basis of heredity remained widespread until 1949, when quantitative chemical analyses of DNA from cell nuclei proved that DNA, not protein, remained constant throughout the mitotic cycle of cell division. This finding suggested that DNA was probably the critical substance in chromosomes.

Meanwhile, a series of experiments on the biological role of DNA, begun in the 1920s, were destined to lead us to our current understanding of the cellular and molecular mechanisms of heredity.

Griffith and the Pneumococcus

In 1928, F. J. Griffith observed some startling results while investigating the destructive effects of pneumococci in mice. The ability of a pneumococcus to produce disease is largely dependent upon the polysaccharide capsule that prevents its destruction by phagocytes (white blood cells) in the host. These encapsulated organisms can mutate to unencapsulated forms that are *avirulent,* or unable to produce disease. If the virulent pneumococcus culture is heat-killed and then injected into mice, virulence is lost, as one might expect (Figure 6–1a). Griffith inoculated mice with a mixture of live, unencapsulated, avirulent bacteria and heat-killed encapsulated pneumococci of a virulent type. To his surprise, some of the mice died (Figure 6–2d). Analyzing his results, Griffith concluded that neither the avirulent live bacteria nor the heat-killed virulent ones could have killed the mice, because neither had that effect when injected alone. He isolated living pneumococci from the dead mice and identified them as encapsulated organisms of the type represented by the heat-killed bacteria. It therefore appeared that either the dead bacteria were rejuvenated by the living avirulent ones or the dead bacteria somehow transformed the avirulent organisms into virulent ones. (Additional details of transformation will be presented later on in this chapter.)

In 1933, J. L. Alloway reported the next major step in explaining this transformation phenomenon. He performed *in vitro* the same type of experiment that Griffith had performed *in vivo.* In place of the killed cells, these studies used a sterile, cell-free extract of the virulent pneumococci. This extract did not cause the death of mice when injected alone, but when mixed with avirulent bacteria of a different type, the result was the production of virulent pneumococci, indistinguishable from the ones that were used to make the extract. A substance called a *transforming principle* (TP) had been prepared in crude form. During the 1940s, O. Avery and his co-workers proved absolutely that the TP was DNA.

Thus began the era of intensive investigations into the molecular basis of genetics and of life itself. The crude extracts of TP were separated into protein, lipid, polysaccharides, ribonucleic acid (RNA), and DNA, and it was shown that only the DNA fraction caused transformation.

The Hershey-Chase Experiment

The findings of Griffith, Alloway, Avery, and others were not widely appreciated at first. Many biologists still preferred to believe that proteins were the storehouse of hereditary information. In 1952, A. Hershey and M. Chase reported decisive results clearly demonstrating that only DNA and not protein was the genetic material. These investigators used a strain of bacterial virus (bacteriophage) that contained DNA and was known to attack the bacterium *Escherichia coli.* The viral nucleic acid is contained in a protein cover or coat. **[Refer to Chapter 10 for additional details on viruses.]**

When a bacteriophage infects a bacterium, it first attaches to the cell wall of the organism by means of a taillike structure (Figure 6–2a). This step is followed by the injection of the viral nucleic acid (hereditary material) into the bacterial cell (Figure 6–2b). The hereditary material, which is in the viral head, takes over the metabolism of the infected bacterium and directs it to produce large numbers of new viruses (Figure 6–2c). The production of the new viruses eventually leads to the disruption and death of the host bacterium.

Hershey and Chase took advantage of the fact that DNA contains phosphorus (P), whereas protein does not, and that protein contains some sulfur (S), whereas DNA does not. By culturing bacteriophage on bacteria grown in a medium containing radioactive phosphorus (^{32}P) and/or radioactive sulfur (^{35}S), they were able to obtain viruses that incorporated ^{32}P into their DNA and ^{35}S into their protein. Hershey and Chase then infected nonradioactive bacteria with these new, radioactively labeled bacterial viruses. After the attachment and infection pro-

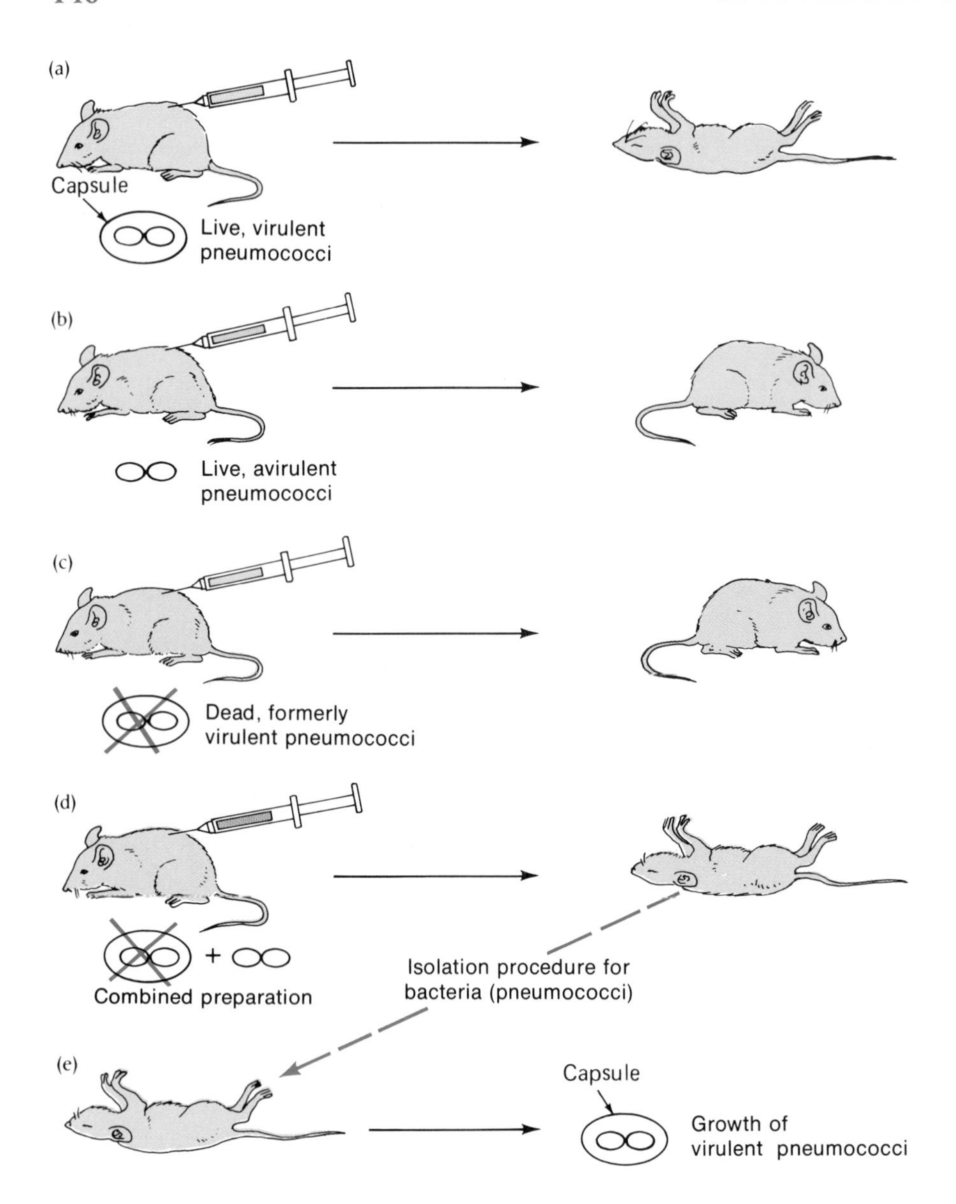

Figure 6–1 Griffith's experiment contributed greatly to the understanding of transformation and the nature of DNA. See the text for details of this experiment.

cesses had begun, the bacteria were agitated in a blender to detach what remained of the viruses from their surfaces (Figure 6–2d). Analysis of the viral remains showed the presence of a substantial amount of ^{35}S but little ^{32}P. This finding indicated that only the empty protein viral coat had been left on the bacterial cell surfaces. Analysis of the infected bacteria showed the presence of large amounts of ^{32}P but little ^{35}S. This demonstrated that only DNA had been injected in these bacteria. Thus it was clear that since the only link between the virus and its progeny was viral DNA, this viral DNA alone must have contained all the genetic information needed for the production of new viruses (Figure 6–2e). This experiment strongly supported the earlier findings on transformation. The identity of DNA had now been proved by a wide variety of established chemical and physical analytical techniques.

The organization and respective functions of nucleic acids are described in the next section.

Nucleic Acids

All forms of life store the information specifying the structure of their proteins in the two general types of *nucleic acids* referred to as DNA and RNA. These macromolecules consist of repeating units called *nucleotides*. A nucleotide is constructed by attaching a phosphate group to the nonring carbon of a pentose (five-carbon) sugar and either a purine or a pyrimidine (nitrogen-containing bases) to the opposite end of the sugar molecule (Figure 6–3a).

In the formation of a nucleic acid chain, individual nucleotides are connected in a line by phosphate groups (Figure 6–3b). The phosphate group of one nucleotide is linked to the hydroxyl group of another, forming a *phosphodiester* (O-P-O) *bond.* A nucleic acid, then, can be seen simply as an arrangement of pentoses linked together by phosphodi-

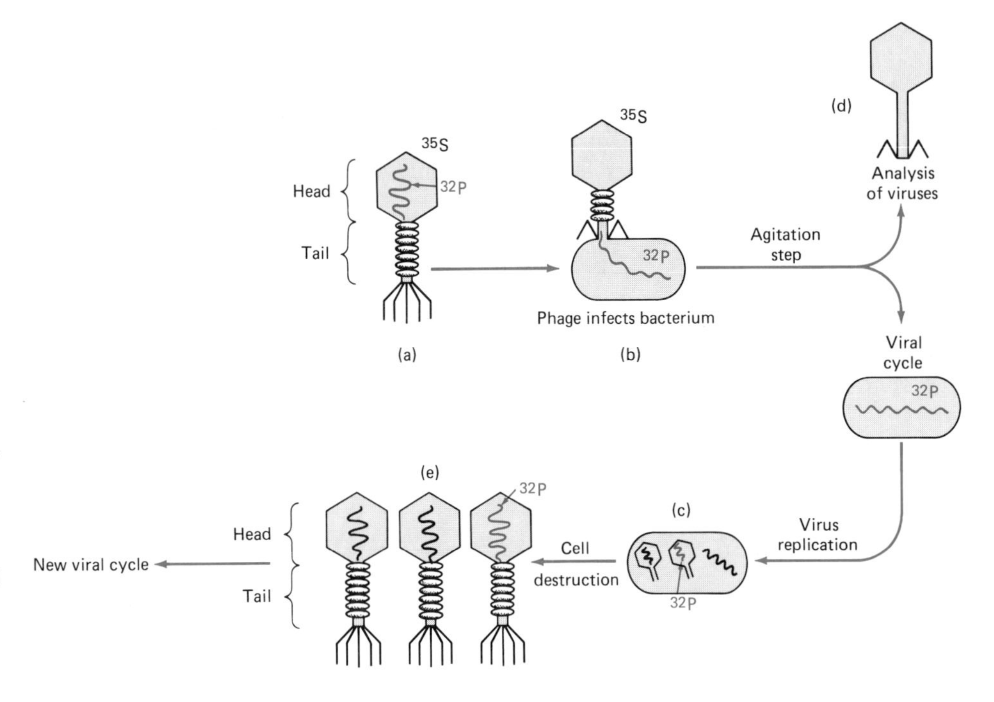

Figure 6–2 Steps in the production of viral particles and the Hershey and Chase experiment. (a) A bacterial virus (bacteriophage) showing the locations of radioactive ^{32}P and ^{35}S; (b) viral attachment and injection of genetic material; (c) production of new viruses, some containing radioactively labeled nucleic acid; (d) analysis of bacterial viruses after the agitation step; (e) new viruses, released after the destruction of a bacterium, ready to start the infection cycle.

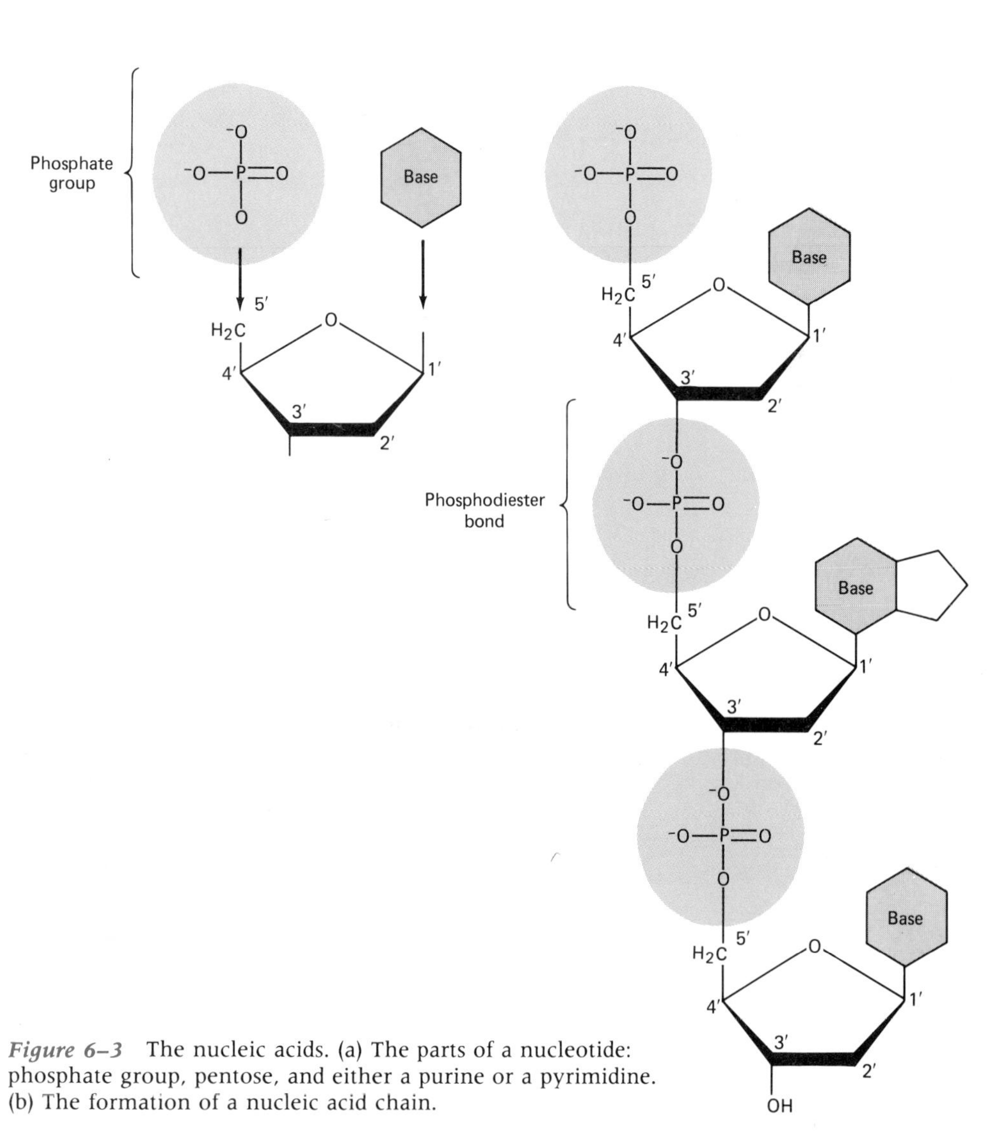

Figure 6–3 The nucleic acids. (a) The parts of a nucleotide: phosphate group, pentose, and either a purine or a pyrimidine. (b) The formation of a nucleic acid chain.

ester bonds, with either a purine or pyrimidine extending from each sugar. Now we will consider DNA, the form of nucleic acid that serves as the basic storage house of genetic information. This will be followed by a description of the different forms of RNA and their respective roles in the formation of proteins.

Deoxyribonucleic Acid (DNA)

DNA consists in most organisms of two nucleotide chains or strands. These two strands exist in the form of a double helix (spiral), in which one strand is wound around the other in a definite pattern (Figure 6–4). The nucleotide (nitrogen-containing) bases

Figure 6–4 A model of the DNA molecule showing the helical nature and the relationship of the two nucleotide strands. Note the complementary combinations of the purine and pyrimidine nucleotides. (a) A diagrammatic representation of the molecule. The ribbonlike strands represent the sugar-phosphate backbones of the DNA. Pairing sequences, hydrogen bonding, and the physical dimensions of the molecule are indicated. Note that normally there are 10 base pairs per turn of the helix. (b) A small portion of the DNA molecule showing the relationship of the sugar and phosphate to their respective nucleotide base pairs. Note that the sugar groups on the right-hand strand appear to be upside down. Actually the chains run in opposite directions and are referred to as being antiparallel.

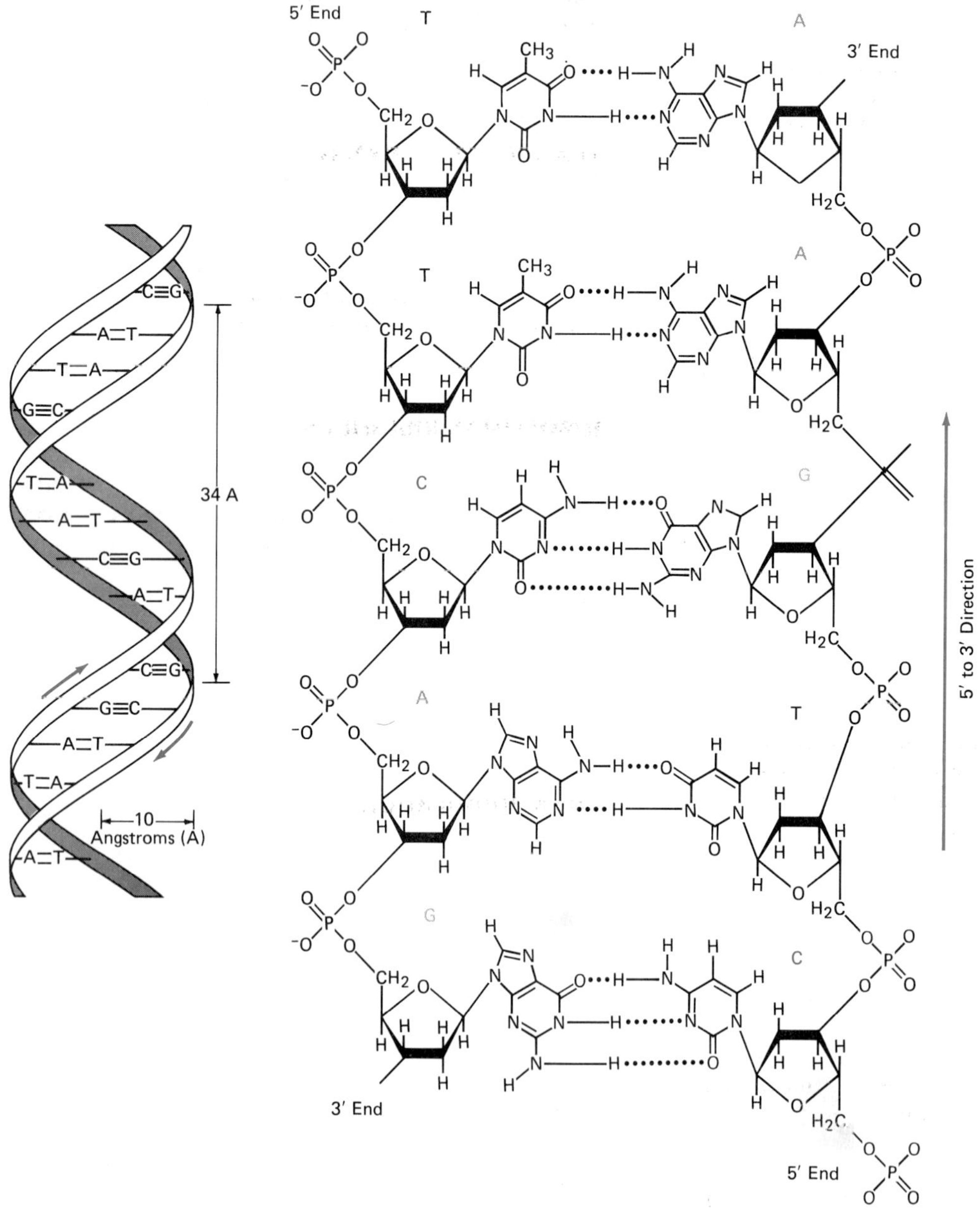

follow a specific form of pairing, so that *adenine* and *thymine* are always joined together, as are *guanine* and *cytosine*. In addition, the two strands of the molecule are complementary to one another; that is, the nucleotide sequence in one chain dictates a complementary nucleotide sequence in the other. Two hydrogen bonds join the adenine of one strand to the thymine of the other, while three hydrogen bonds join cytosine to guanine (Figure 6–4b). Thus thymine is linked to adenine and cytosine is linked to guanine. As a consequence, the ratio of thymine and adenine or of guanine to cytosine in a double-stranded DNA molecule is always 1:1.

In examining a DNA molecule, it also becomes apparent that the two strands extend in opposite or antiparallel directions, and their terminal phosphate (PO_4) groups are located at opposite ends of the double helix. In addition, at one end of a single DNA chain, the sugar has a free hydroxyl (OH) group at the 3-carbon or 3′ position; whereas at the opposite end of the chain, the sugar has a phosphate at the 5-carbon or 5′ position. Thus one chain runs from the 3′ end to the 5′ end, and the complementary strand goes from the 5′ end to the 3′ end (Figure 6–4b).

In procaryotes, the genetic information is contained within a single double-stranded DNA molecule, the ends of which join together to form a closed circle (Figure 6–8). While no other macromolecules are associated with procaryotic DNA, the genetic material of eucaryotic cells is surrounded by proteins called *histones*.

GENOTYPE AND PHENOTYPE

The characteristics exhibited by an organism at any one time reflect the genetic information coded within its DNA, or *genome*, and expressed through interactions with the chemical and physical factors of a particular environment. Specific nucleotide sequences contain the hereditary information in *genes*. The specific structure that carries the genes is the *chromosome*. The genetic makeup of any organism—that is, the numbers and kinds of its genes—is called its *genotype*. The properties of an organism that are expressed within a particular environment are referred to as the *phenotype* of the organism.

Ribonucleic Acid (RNA)

Like DNA, RNA also consists of a sequence of nucleotides; but unlike DNA, it usually exists as a single-stranded molecule. There are two other basic differences: in RNA the pentose ribose is substituted for deoxyribose, and the pyrimidine base, uracil, is substituted for thymine.

RNA participates in several different ways to bring about the expression of the genetic information enclosed within DNA molecules. The genetic or hereditary information specifies the particular type and position of amino acids in a protein molecule (amino acid sequence). Based on their respective functions and roles, three different forms of RNA are recognized: *messenger RNA (mRNA)*, *transfer RNA (tRNA)*, and *ribosomal RNA (rRNA)*. All RNA is synthesized as a complementary copy of a DNA sequence. We shall discuss some of their general properties here.

MESSENGER RNA (mRNA)

This form of RNA is synthesized as a complementary copy of a DNA nucleotide sequence. The function of mRNA is to obtain and carry the genetic information encoded in the DNA to the location where proteins are to be synthesized. Messenger RNA molecules are very unstable and survive for only a short time in microbial cells.

TRANSFER RNA (tRNA)

Transfer RNA is the shortest of the ribonucleic acid molecules. It assumes a three-dimensional structure, which is usually shown in the shape of a cloverleaf (Figure 6–5). Each tRNA molecule, of which there are at least 20, is equipped to transport a single amino acid to the site of protein synthesis. Transfer RNA makes up approximately 15 percent of a cell's total RNA.

RIBOSOMAL RNA (rRNA)

Ribosomal RNA is combined with specific ribosomal proteins to form the structural units known as *ribosomes*. Ribosomes serve as the sites or workbenches where the information in mRNA is read and proteins are formed. Ribosomal RNA comprises the bulk of the RNA in cells. (Additional properties of the RNAs are presented in the section on protein synthesis and in later chapters describing the features of microorganisms.)

DNA Replication

As in other biological reactions, special enzymes are needed for the duplication of identical DNA molecules carrying genetic information. This process of *DNA replication* involves enzymes that specialize in performing various operations on DNA. Some unwind the molecule for replication, others assemble short fragments consisting of nucleotides, and still others join these DNA fragments together.

When DNA replicates, the double strands of the molecule unwind and separate, forming a Y-shaped replication fork at the site of DNA synthesis (Figure 6–6). Each individual strand serves as a template or pattern for the formation of a new complementary

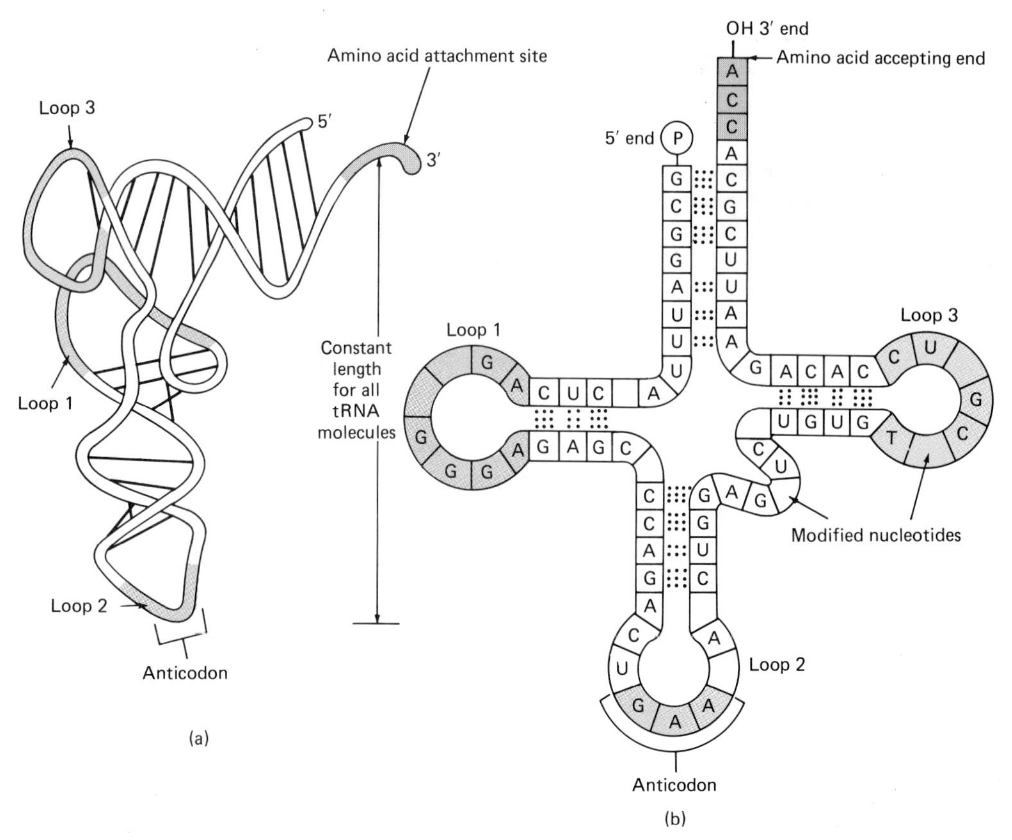

Figure 6–5 Two views of a typical tRNA molecule. (a) The actual shape of the molecule. One loop of the molecule contains the triplet anticodon that attaches to the specific base pairs of an mRNA transcript. The *amino acid-accepting end* is located at a constant distance from the anticodon region. (b) The two-dimensional cloverleaf form of tRNA.

strand. The nucleotides in the new strands are assembled in a specific order, because each purine or pyrimidine in the original (parental) chain forms hydrogen bonds only with complementary pyrimidine or purine nucleotides (thymine opposite adenine and guanine opposite cytosine). As the replication fork continues to move, two new DNA strands are synthesized, with their respective nucleotides arranged in a complementary order. The new and original strands then wind around each other, resulting in the formation of two complete double-strand DNA molecules. Since each of the original strands serves as a template or pattern against which a new strand is formed, the method of information copying is referred to as *semiconservative replication* (Figure 6–7).

THE ENZYMES AND THE PROCESS

The initiation of the replication process requires a *primer*, a short RNA molecule that is complementary to one of the DNA strands. Once the RNA molecule is formed by the enzyme RNA polymerase, DNA synthesis begins. Next, the RNA primer is removed and replaced by newly formed DNA. The replication process continues with the enzyme DNA polymerases adding nucleotides only to the 3′ hydroxyl (OH) end of the replicating DNA strand or strands in a 5′ to 3′ direction (Figure 6–6). The fact that these polymerases can only add nucleotides to the 3′ hydroxyl group of the strand raises the question of how DNA synthesis can proceed in proper order (sequentially) from the single starting point along *both* DNA strands. As mentioned earlier, the two strands of DNA are antiparallel, since one strand runs from the 5′ OH to the 3′ OH free end and the complementary strand runs from the 3′ OH to the 5′ OH free end (Figure 6–6). Therefore, the formation of the complementary strands must proceed in opposite directions. One of the new strands is formed *continuously* as the replication fork moves, since it runs in the appropriate direction for the addition of nucleotides to the free 3′ OH of the primer. The other new DNA strand must be formed *discontinuously* in the 5′ to 3′ direction. Since the formation of the discontinuous strand can start only after unwinding of the DNA molecule has begun, it will lag behind the continuous strand. Short segments of DNA known as *Okazaki fragments*, named after their discoverer, Reijii Okazaki, are used in the formation of the lagging strand. These fragments are joined together by the action of the enzymes called *DNA ligases* (Figure 6–6). Thus new complementary strands of DNA are formed by the coordinated activities of DNA polymerases and ligases.

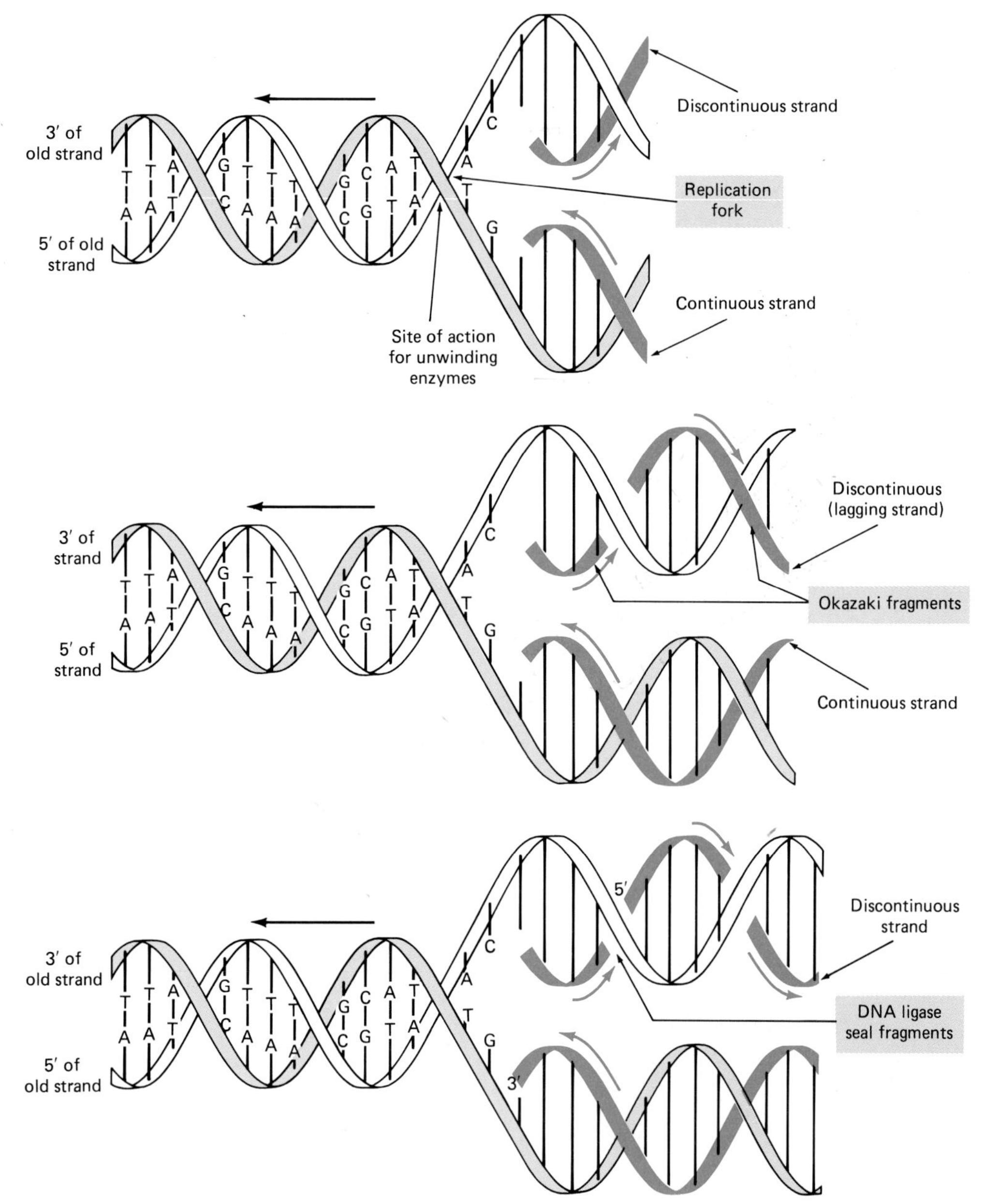

Figure 6–6 A DNA replication fork (Y) and DNA strand replication. DNA polymerase functions in the replication of both new strands. The upper-left-hand strand is being copied by the addition of nucleotides to the 3′ end, while the lower-left-hand strand is being copied by the addition of nucleotides to the 3′ end. When the DNA polymerase reaches the 5′ end of an earlier fragment (arrow), the enzyme attaches the new fragment to it.

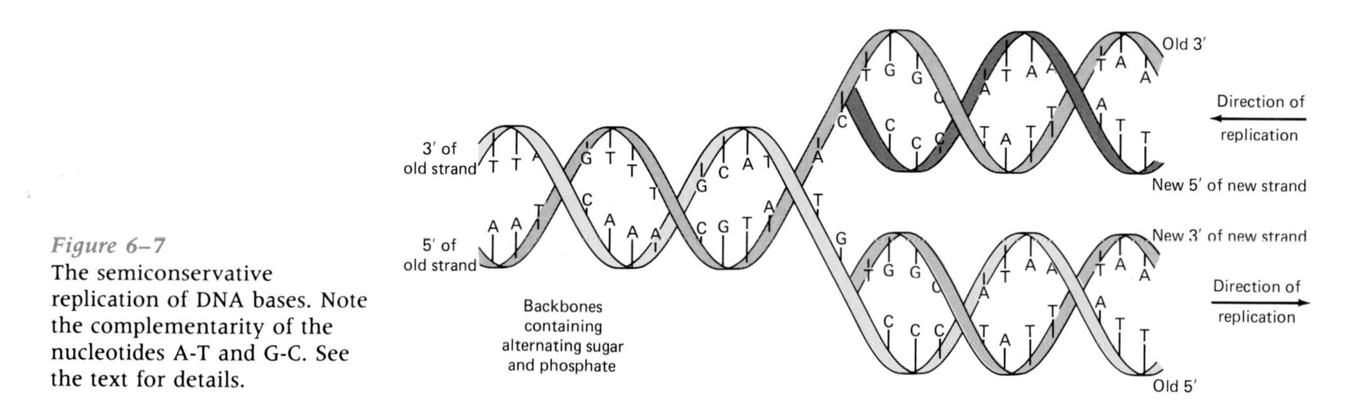

Figure 6–7
The semiconservative replication of DNA bases. Note the complementarity of the nucleotides A-T and G-C. See the text for details.

How Bacteria Replicate Their DNA

The genetic material of a bacterium is contained within a single double-stranded molecule that is generally organized into one circular chromosome. Unlike eucaryotic chromosomes, the bacterial chromosome is not enclosed by a nuclear envelope (Figure 6–8a) and may be attached to one or several points along a bacterium's cell membrane. While the replication process in bacteria is complex, and involves the enzymes described earlier, it is not difficult to visualize overall (Figure 6–9).

The DNA replication process begins at a single point of origin, or nick site, and moves in two opposite directions *(bidirectional)* around the chromosome, resulting in the formation of two replicating forks. The newly created replication forks move away from the original starting site and around the circular chromosome to form a newly replicated DNA molecule (Figure 6–8b). Since the chromosome is a circular structure, the forks eventually meet when the process is completed. DNA replication is a relatively rapid process in bacteria growing at their respective optimal temperatures. Bacterial cells regulate the initiation of the process to conserve energy and limit the formation of unwanted products.

RNA Synthesis

Although the functions of the three major kinds of RNA are different, the mechanism for their formation is identical. The enzyme *DNA-dependent RNA polymerase* (also called *RNA polymerase*) catalyzes their polymerization from the ribonucleotides: adenosine triphosphate (ATP); cytodine triphosphate (CTP); guanosine triphosphate (GTP); or uridine triphosphate (UTP). One strand of DNA serves as a template or pattern. An example of a single step in RNA formation can be written as shown at top of next page.

In this step the RNA molecule is lengthened by the addition of one ribonucleotide. The specific nucleotide added is directed by the hydrogen bonding specified by the DNA template. The same hydrogen bondings occur in RNA formation as in the case of DNA replication *with the exception that thymine is replaced by uracil.* A pyrophosphate (P-P) molecule that

Figure 6–8 Bacterial DNA. (a) The bacterial nucleoid—the structure containing the cell's DNA. *[From J. A. Hobot, W. Villiger, J. Escaig, M. Maeder, A. Ryter, and E. Kellenberger,* J. Bacteriol. ***162**:960–971, 1985.]* (b) An autoradiograph of the *Escherichia coli* chromosome, emphasizing its circular form and mode of replication. (c) A diagram showing replication to be approximately one-third completed. *[J. Cairns,* Cold Spring Harbor Symp. Quant. Biol. ***28**:43–46, 1963.]*

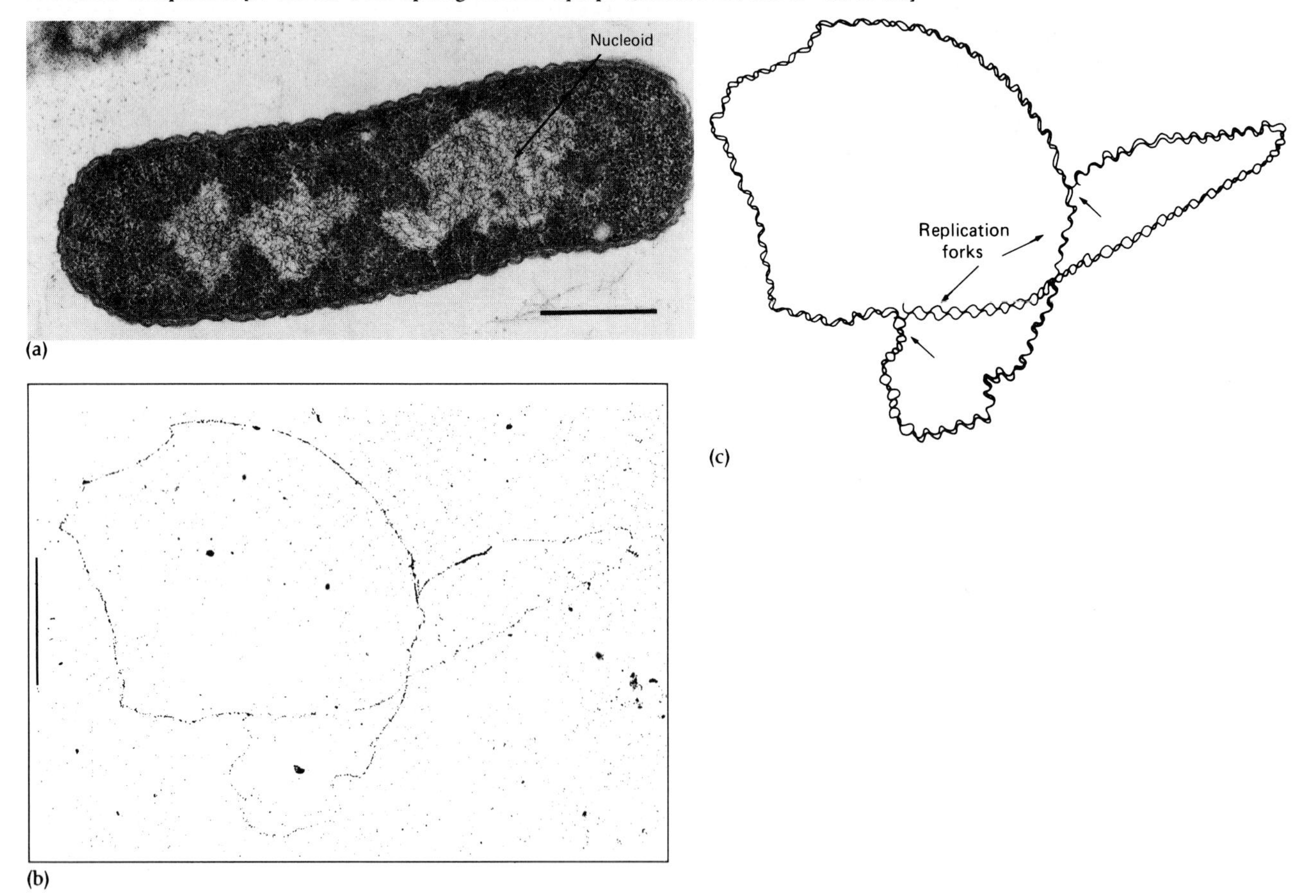

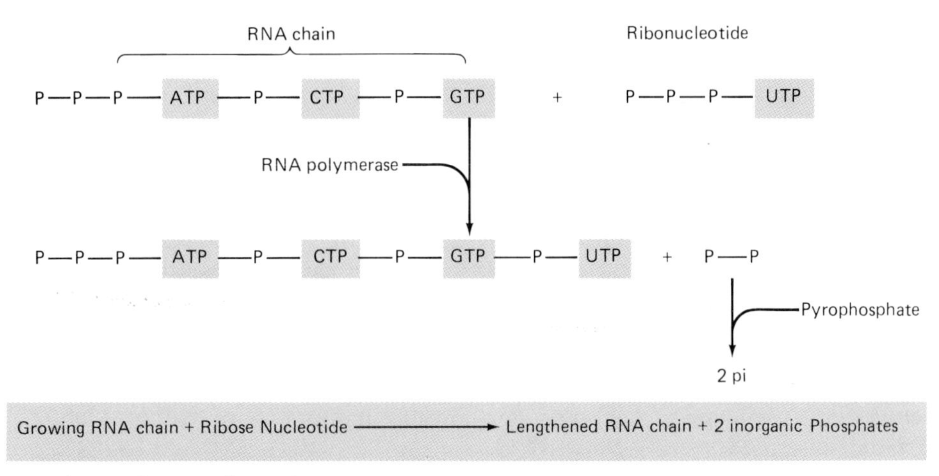

Single step in RNA formation

is formed with each step subsequently is split into two inorganic phosphate molecules.

Protein Synthesis

Probably one of the greatest scientific accomplishments of the 1960s was the delineation of the steps in protein synthesis. Chief among the critical aspects of this work was the initial breaking of the genetic code, or nucleotide language, reported by Marshall W. Nirenberg in 1961.

The Genetic Code

When it had been determined that DNA was indeed the genetic substance, investigators wondered what combinations of the four nucleotides containing the nitrogenous bases guanine (G), adenine (A), cytosine (C), and thymine (T) might code for each of the 20 amino acids found in proteins. The first question to be answered was how many nucleotides were required to specify an amino acid. The four nucleotides could be used to write out only 16 two-letter "code words"—GA, GC, GT, AC, AT, etc.—and could, therefore, specify only 16 of the 20 amino acids. A code using three nucleotides (GTA, GCA, CTG, etc.) would allow 64 combinations, more than

Figure 6–9 The circular chromosome of a procaryote starts to replicate at a single site, moving out in both directions (arrows). The two moving replication forks meet on the far side of the chromosome, thus duplicating the circular chromosome.

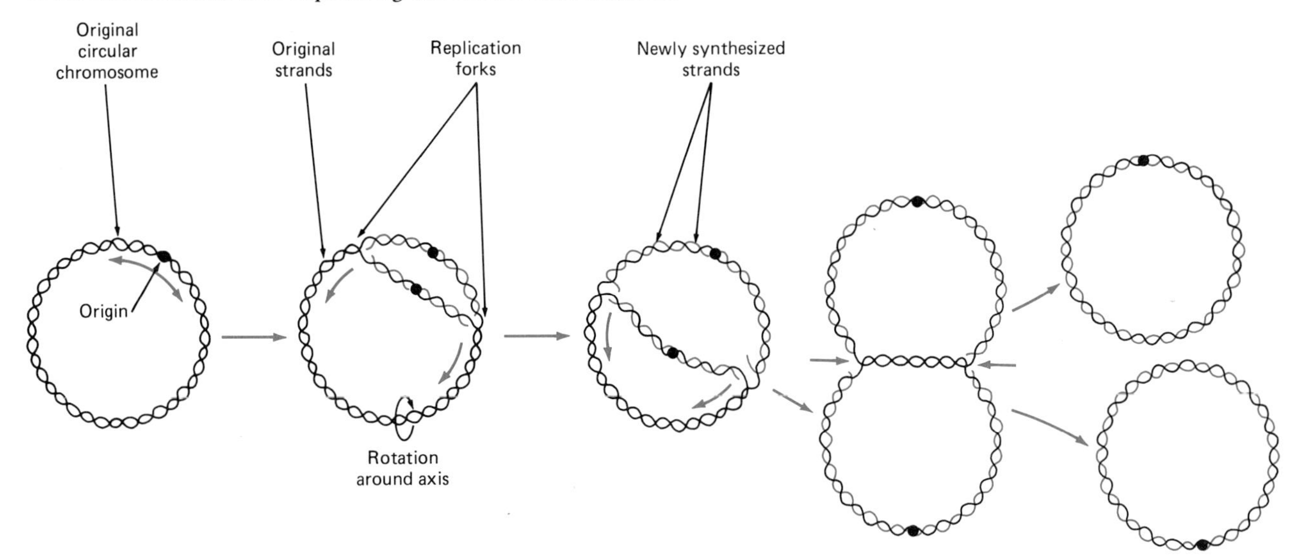

enough to specify all the natural amino acids. Four-letter codes would permit 256 combinations.

The triplet code with its 64 combinations would permit the 20 needed combinations, with 44 extra for alternate combinations to code for certain amino acids and to act as spacers or regulatory areas—punctuation marks of a sort. Nirenberg and others have demonstrated that the triplet code is, indeed, the operating genetic system.

From the results of various investigations, the features of the genetic code were uncovered. This system, which appears to be found in all organisms, is used to read the genetic information derived from DNA and contained within the nucleotide sequence of mRNA molecules *(transcripts)*. The DNA code information in the nucleotide sequence of mRNA consists of sets of three nucleotides or triplets called *codons*. One end of individual transfer RNA (tRNA) molecules contain complementary triplet nucleotide sequences, known as *anticodons*, that attach to specific codons during the process of protein synthesis. Another end of the tRNA molecule attaches to one particular amino acid; this attachment is catalyzed by one of 20 tRNA-activating enzymes. There is one activating enzyme for each of the 20 amino acids found in proteins. **[See Appendix A for descriptions of amino acids.]**

The **genetic code** represents the precise relationship between the sequence of pyrimidine and purine bases in the DNA and RNA molecules and the sequence of amino acids in the proteins these molecules encode. Table 6–1 shows the codons for the complete code. The codons are read from left to right. With few exceptions, a particular codon always corresponds to the same amino acid in all organisms. It should be noted that 3 of the 64 codons do not code for any of the known amino acids. These three, known as *nonsense codons*, function as "stop" signals in a mRNA nucleotide sequence, marking the end of a particular protein molecule. The codon AUG serves as the "starting" signal for the beginning of a protein molecule. This same triplet also encodes for the amino acid methionine (Table 6–1).

The Genetic Code in Operation

As mentioned earlier, the fundamental unit of hereditary information is the gene. When such information is translated, a protein with an amino acid sequence specified by the gene is formed. In all living organisms, this type of gene expression occurs in two distinct phases: *transcription* and *translation*.

TRANSCRIPTION

The first step in gene expression is transcription. It is the enzymatic process by which genetic informa-

TABLE 6–1 The Genetic Code Dictionary. The RNA nucleotide sequences, or *codons*, are shown together with the specific amino acids they represent. Two different codons may specify the same amino acid. Nonsense codons, which do not stand for any of the amino acids but serve as punctuation signals in the protein synthesis process, appear in "Stop" boxes.

First Position		Second Position U (1 2 3)		Second Position C (1 2 3)		Second Position A (1 2 3)		Second Position G (1 2 3)		Third Position
U (uracil)		U U U	Phe[a]	U C U	Ser	U A U	Tyr	U G U	Cys	U
		U U C		U C C		U A C		U G C		C
		U U A	Leu	U C A		U A A	Stop	U G A	Stop	A
		U U G		U C G		U A G	Stop	U G G	—Trp	G
C (cytosine)		C U U	Leu	C C U	Pro	C A U	His	C G U	Arg	U
		C U C		C C C		C A C		C G C		C
		C U A		C C A		C A A	Gln	C G A		A
		C U G		C C G		C A G		C G G		G
A (adenine)		A U U	Ile	A C U	Thr	A A U	Asn	A G U	Ser	U
		A U C		A C C		A A C		A G C		C
		A U A		A C A		A A A	Lys	A G A	Arg	A
		A U G	—Met	A C G		A A G		A G G		G
G (guanine)		G U U	Val	G C U	Ala	G A U	Asp	G G U	Gly	U
		G U C		G C C		G A C		G G C		C
		G U A		G C A		G A A	Glu	G G A		A
		G U G		G C G		G A G		G G G		G

[a] The amino acid represented by the three-base sequence.

Phe = phenylalanine; Ser = serine; Tyr = tyrosine; Cys = cysteine; Leu = leucine; Trp = tryptophan; Pro = proline; His = histidine; Arg = arginine; Gln = glutamine; Ile = isoleucine; Thr = threonine; Asn = asparagine; Lys = lysine; Met = methionine; Val = valine; Ala = alanine; Asp = aspartic acid; Gly = glycine; Glu = glutamic acid.

Microbiology Highlights

A GENETIC MELODY

What's the name of that tune? Or what's the DNA coding sequence in that melody? These may not be totally unrelated questions. Dr. Susumu Ohno, a geneticist at the Beckman Research Institute of the City of Hope in Duarte, California, believes that a certain harmony exists between DNA and music. According to Dr. Ohno, similar principles regulate the construction of gene coding sequences and musical composition. The repetition of cycles or periodicity provides the common link. For biologically related situations, additional examples of such repeated cycles include nights and days, the four seasons in the year, and even the mating calls of songbirds. In musical pieces, the main melody and some variation of it are first presented and then repeated in many ways.

Dr. Ohno's approach involves assigning two consecutive positions (notes) in the octave scale to each of the four bases: adenine (A), guanine (G), thymine (T), and cytosine (C). Several melodies that correspond to gene coding sequences have been found among the compositions of the music masters Bach, Mozart, and Beethoven.

This notation system used by Dr. Ohno shows the relative positions of A, G, T, and C.

Transcribing gene base sequences into music—as well as applying the technique to transcribe music into such sequences—produces interesting and intriguing results. For example, the musical transformation of 51 codons for the enzyme phosphoglycerate kinase resulted in a pleasing melody when played on a violin. On the other hand, the conversion of Chopin's Funeral March produced a coding sequence very much like that of tyrosine kinase, the heart of many cancer-causing genes.

A portion of Chopin's Nocturne, Opus 55 No. 1, transcribed into base sequences.

tion in one DNA strand is used to specify a complementary sequence of bases in an RNA molecule. There are three aspects to the transcription process: *initiation, elongation,* and *termination.*

INITIATION. All the final products of transcription—mRNA, tRNA, and rRNA—participate in protein synthesis. The initiation of transcription occurs at specific locations on the DNA molecule that are defined by short nucleotide sequences called *promoters.* These sequences also determine which DNA strand (the *sense strand*) will be transcribed. The transcription process begins with the help of a protein subunit of RNA polymerase, known as the *sigma factor,* which binds to the promoter. This binding causes the strands of the DNA double helix to separate and begin the formation of mRNA, with the sense strand functioning as the template (Figure 6–10). It appears that the sigma factor's function is mainly to recognize the promoter site. Once transcription has been started, the sigma factor is released.

ELONGATION. During this transcription phase the RNA polymerase alone continues to catalyze the formation and elongation of mRNA, the *transcriptional unit.*

TERMINATION. Just as the initiation of transcription is under the exact control of specific DNA nucleotide sequences, so too is its termination. The DNA template contains nucleotide sequences that serve as "stop" signals for RNA polymerase (Table 6–1). The *rho protein,* a specific molecule, sometimes aids the polymerase in recognizing certain base pairs as the signal to end the transcription process. Thus, a DNA molecule carries, in addition to the information for protein structure, a series of regula-

Figure 6–10 The transcription of DNA to form mRNA. (a) An overall view showing several locations (genes) involved in transcription. Note that only one of the two DNA strands is transcribed for a given DNA segment (gene). An opposite strand (arrow) may be used to transcribe a neighboring gene. (b) The double helix of DNA being unwound by RNA polymerase and its subunit, the **sigma factor.** The enzyme now has access to the nucleotide sequence (genetic code). After transcription, DNA is rewound behind the moving process. The newly formed mRNA molecule (transcript) is shown. The **rho protein** aids in stopping or terminating the process. The ribonucleotides used in mRNA formation are found in the area of the transcription.

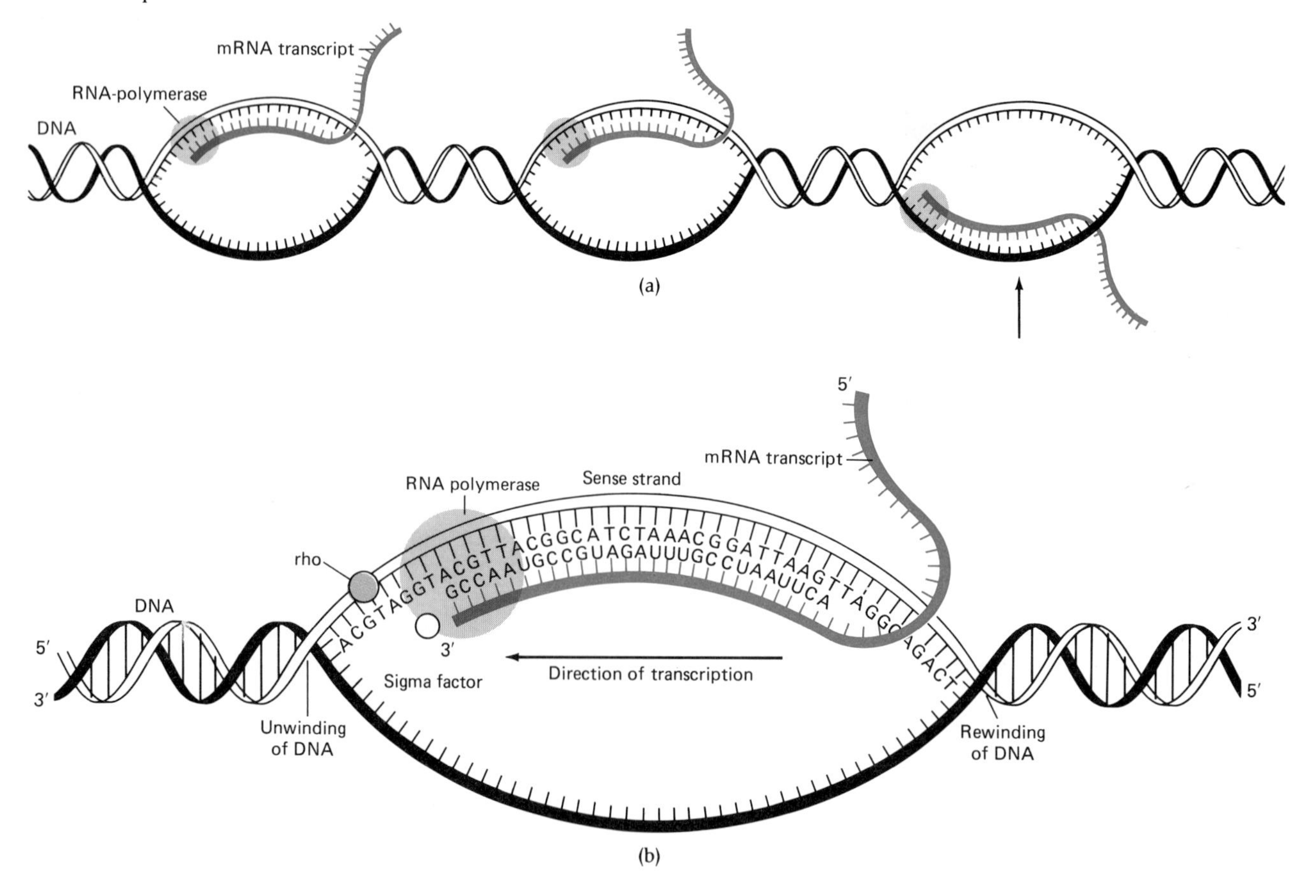

tory or control commands. Such commands also occur in the second or translation stage of protein synthesis. When the mRNA molecule is completed, it passes from its site of formation to the ribosomes.

The flow of information generally goes from DNA to RNA. However, in the AIDS virus, which contains only viral RNA, and related disease agents such as other retroviruses, an enzyme, reverse transcriptase, as its name suggests, reverses the information flow. In such cases the viral RNA first must be converted to DNA. **[See Chapters 10 and 30 for more details of this enzyme and the associated viruses.]**

TRANSLATION

Decoding the information contained in the mRNA transcript is known as *translation.* The mRNA directs the sequence in which specific amino acids are linked together to form proteins and related molecules. It is important to keep in mind that protein synthesis, like all other types of biosynthetic processes, requires energy.

Amino Acid Activation. Before amino acids can be incorporated into protein molecules, they must first be activated. As indicated earlier, a group of 20 enzymes, called aminoacyl-tRNA synthetases, are responsible for amino acid activation. Each of these enzymes is specifically designed for a particular amino acid. The activation process occurs in two steps. In the first step, an amino acid reacts with adenosine triphosphate (ATP) to form an enzyme-linked intermediate form, aminoacyl adenylic acid (Figure 6–11a). The aminoacyl group is then transferred to the amino acid-accepting site of a specific tRNA molecule (Figure 6–11b). The activated amino acid is now ready to participate in the translation process.

Translation of the mRNA transcript occurs on the ribosome surface (Figure 6–12). In order for the codon sequence on the mRNA to have meaning, the translation process must start at the correct nucleotide base of the message. Each DNA sequence has a correct *reading frame.* If translation were to be initiated at one or two bases beyond the correct frame, information in the mRNA would be interpreted incorrectly, changing the meaning of the encoded message. A situation of this type could lead to the formation of a nonfunctional or faulty protein. The next phase of translation is dependent on the ribosomes.

Ribosomes are among the more complex components of living systems. Each one is composed of a small and a large subunit. The smaller component has a binding site for mRNA, while the larger subunit has two tRNA binding sites, the **A** (aminoacyl or amino acid acceptor) site, and the **P** (polypeptide) site. Only two tRNAs can be held by a ribosome at one time. The ribosome-dependent process described in the following sections can be divided into the three phases of *initiation, elongation,* and *termination.* **[See Chapter 7 for other properties of ribosomes.]**

Initiation. To begin this phase of protein synthesis, an *initiation complex* is formed. This step involves the binding of a methionine-carrying tRNA to the small ribosomal subunit (the tRNA that corresponds to the start codon *AUG*). Special proteins, called *initiation factors,* position the tRNA to ensure that the proper reading frame (the groups of three nucleotides) will be translated into protein. The complex, guided next by another initiation factor, binds to mRNA. After this binding step, the large ribosomal subunit becomes attached and completes the formation of the initiation complex.

Upon the completion of the initiation complex, the mRNA codon next to the *initiating* or *start codon, AUG,* interacts with the large ribosomal unit. This step positions the codon so that it can interact with an incoming tRNA. The initial methionine-carrying tRNA binds to the **P** (polypeptide) *site* of the ribosome mentioned earlier (Figure 6–12). This site also carries the growing peptide molecule. Any new incoming amino acid-carrying tRNAs bind to the **A** (amino acid) *site* (Figure 6–12). A new incoming

Figure 6–11 The activation of amino acids in the process of translation. (a) The reaction between an amino acid and ATP, resulting in the formation of the aminoacyl adenylic acid. (b) The transfer of the aminoacyl group to the tRNA. (Explanation of other symbols used: AMP, adenosine monophosphate; P-P, pyrophosphate.)

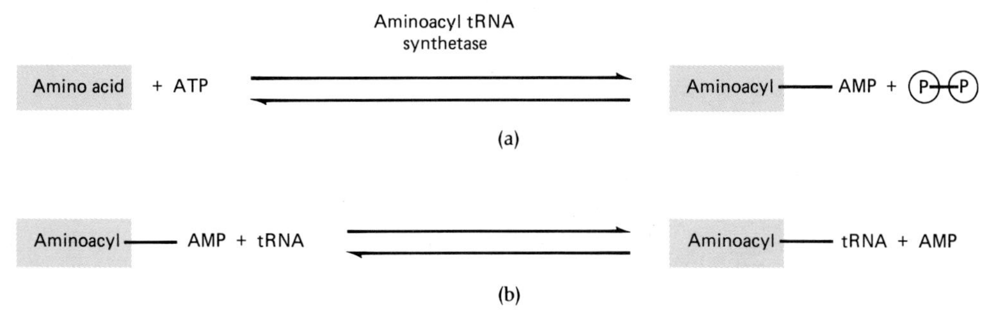

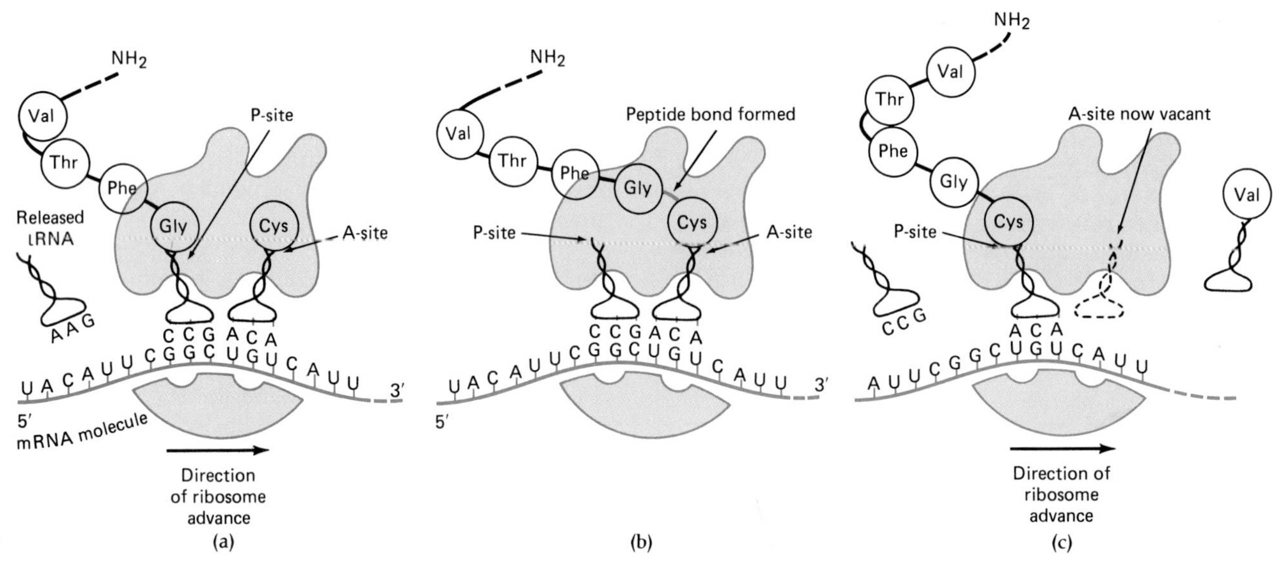

Figure 6–12 The process of translation involves the addition of amino acids to a growing polypeptide chain according to the sequence of codons in the mRNA transcript. (a) An amino acid-tRNA combination is bound to an **empty** ribosome A (aminoacyl) site next to an **occupied** P (polypeptide) site. (b) The growing polypeptide chain detaches from the tRNA molecule in the P site and its amino acid is connected to the amino acid at the A site by a peptide bond. The released tRNA molecule will be ejected from the ribosome and will be free to combine with another amino acid, as needed, in the formation of another protein. (c) As the ribosome moves along the mRNA transcript, the growing peptide molecule (still attached to the tRNA) is transferred to the P site. The A site is vacant again and ready to receive another amino acid-tRNA combination.

tRNA is thus able to attach to the mRNA molecule at the exposed codon position. Special proteins, called *elongation factors,* aid in positioning the incoming tRNAs. As the mRNA transcript is translated, other tRNA molecules carry the code-specified activated amino acids to the complementary codons (positions) on the mRNA. Peptide bonds form between the amino acids as an amino acid is transferred by enzymes from the **A** site to the **P** site of the ribosome (Figure 6–12). The resulting abandoned tRNA falls from its site on the ribosome, thus providing a vacant site for the next amino acid-carrying tRNA.

ELONGATION. The ribosome in this phase moves along the mRNA transcript a distance corresponding to three nucleotides. The move repositions the growing chain of amino acids and exposes the next mRNA codon. An incoming tRNA brings and places a new amino acid on the growing chain. This is how the peptide or protein molecule elongates. **[See Appendix A for a description of proteins.]**

TERMINATION. Protein formation stops when an amino acid-terminating stop signal or nonsense codon is encountered. Special enzymes, *release factors,* bind to the nonsense codons and break the bond holding the newly formed peptide or protein molecule to the ribosome. The released protein may be changed by the cell to form the final product.

The mRNA is unstable and must be replaced fairly often. This liability may be a mechanism to ensure that one particular protein is not overproduced. With many mRNAs being synthesized and released to the cytoplasm, the degradation of existing mRNA makes ribosomes available for the synthesis of whatever structural protein or enzyme the cell may need.

Procaryotic and Eucaryotic Transcription Differences

Certain differences exist between procaryotic and eucaryotic protein synthetic processes. These differences are primarily related to how mRNA is formed and whether or not additional processing is necessary after transcription to form functional mRNA molecules. In eucaryotic (but not procaryotic) cells, messenger RNA molecules are generally and extensively processed in the nucleus, before translation. Eucaryotic genes frequently contain noncoding regions, called *introns,* separating protein coding regions, known as *exons.* Both introns and exons are transcribed into precursor mRNA. The precursor or primary mRNA transcript, also referred to as *heterogenous nuclear RNA,* is several times larger than the mature or final form of the molecule. It contains regions that code for amino acid sequences, or *transcribed exons,* spatially separated by segments that do

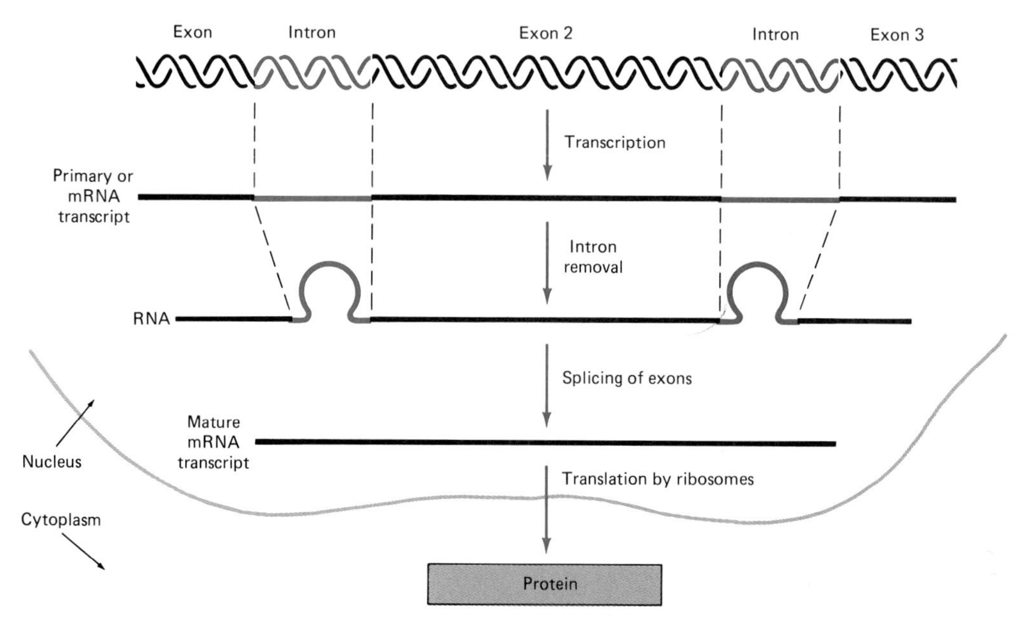

Figure 6–13 Mechanism of mRNA processing in eucaryotes. The genetic sequence on the DNA is transcribed into a mRNA molecule, some of which **(exons)** contains transcribed genetic information and some which **(introns)** does not. Excision (removal) of transcribed introns and the splicing (joining) of exons results in a functional mRNA transcript.

not code for amino acid sequences, or *transcribed introns* (Figure 6–13). Processing of the primary transcript includes the removal of the introns and the splicing together of the exons or sensible segments of the RNA molecule to form a mature mRNA molecule that contains the regulatory sequences for the *start* and *stop* signals needed for translation. As a result of mRNA processing, the sequence of eucaryotic mRNA nucleotides is not complementary to the specific sequences of DNA nucleotides used for the formation in the primary transcript.

In procaryotes, mRNA molecules are not modified from the time of their formation through the translation step in protein synthesis. In other words the mRNA molecule is transcribed directly and no form of processing comparable to that found with eucaryotic cells occurs. Procaryotes do not have introns. In addition, the attachment of the procaryotic mRNA molecule to ribosomes often takes place before transcription is complete, thus emphasizing the close relationship between the processes of transcription and translation in procaryotes (Figure 6–14).

Control and Regulation of Metabolism

If a cell is to make the best use of the energy available to it, it must function efficiently. Cell products are produced as needed. When a substance is either produced in the cell or enters it, a specific enzyme required to catalyze a reaction for that substance is formed by protein synthesis. In certain systems, the cell's DNA controls the production of the particular mRNA molecules needed for the synthesis of specific proteins. In other cases products of enzymatic reactions influence cellular metabolism. The cell coordinates its genetic capabilities with the available nutrients to manufacture the enzymes it needs for its various activities. Therefore, protein synthesis

Figure 6–14 Transcription and translation in a procaryotic cell. (a) An electron micrograph showing two DNA strands, one inactive and the other actively forming mRNA (transcription). Polyribosomes attach to and move along the mRNA, reading the codon sequence it contains. *[After B. Hamkalo and O. Miller, Jr.,* Ann. Rev. Biochem. ***42:****379, 1973.]*

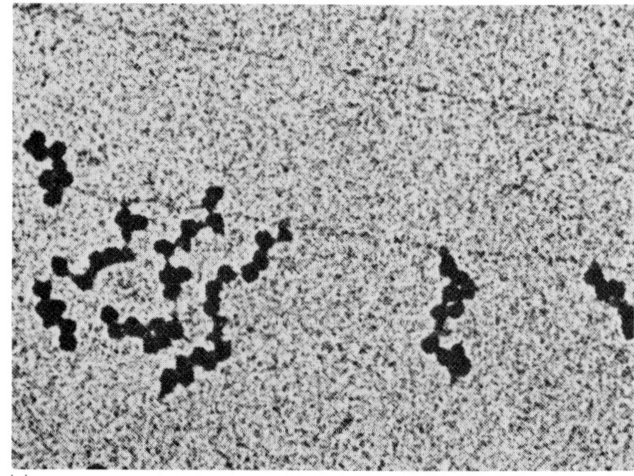
(a)

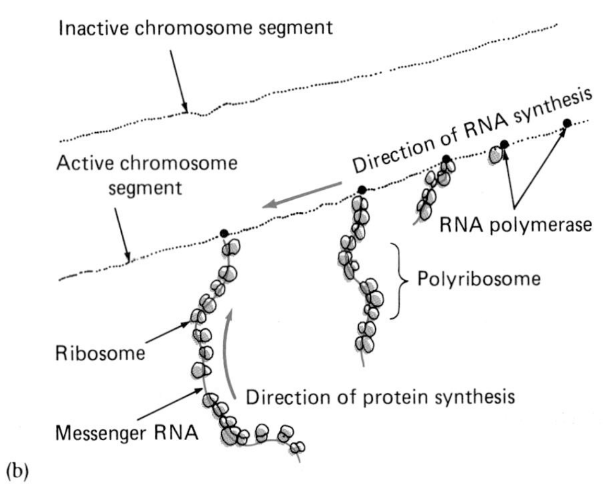

Figure 6–14 (b) A diagrammatic representation and interpretation of transcription and translation.

is not an unrelated process but rather one that is under genetic control. Two types of metabolic regulation are known: one affects enzyme formation and the other affects enzyme activity.

Some enzymes, called *constitutive,* are formed continuously in a cell, whereas others, called *adaptive,* are synthesized only at certain times and as determined by environmental factors, such as the presence of certain nutrients or various chemicals. Those proteins formed only in response to the presence of environmental factors are referred to as *inducible,* while those whose formation is inhibited by such factors are considered to be *repressible.*

Regulation of Protein Synthesis

Feedback or end-product inhibition, as described in Chapter 5, depends on control of metabolism at the enzyme level. Here a given enzyme beginning a pathway is turned off when a particular metabolite accumulates and is then turned on when the concentration decreases. This is a clever system, but apparently it does not prevent the synthesis of the enzyme or enzymes of a pathway. The DNA of a particular microorganism contains the information, or code, for all the protein molecules it can synthesize. However, not all of those proteins are being formed at all times. Some enzymes, as indicated earlier, are *constitutive,* and others are *adaptive.* Therefore, in addition to the direct control of feedback or end-product inhibition, enzymes also are regulated at the genetic level; genetic mechanisms involve the *induction* and *repression* of enzyme synthesis. Much of the present knowledge of such genetic control and expression resulted from the work of two French microbiologists, François Jacob and Jacques Monod. On the basis of their studies of lactose catabolism in *E. coli,* these investigators proposed the *operon hypothesis of genetic control* (Figure 6–15).

Figure 6–15 The lactose *(lac)* operon, an example of a generalized operon model in a catabolic system. The **top view** shows the components of the operon: the promotor region (P); the operator (O); structural genes Z, Y, and A involved with the production of specific enzymes; and the unlinked regulatory gene (i). The events involved with inducing the operon also are shown at the top. The **inducer,** lactose, binds to the repressor protein product of the regulatory gene. The **structural genes** are then transcribed (mRNA formation) and the lactose catabolic enzymes are synthesized. The **bottom view** shows the events involved with repressing the operon. Here the repressor protein product binds to the operator, thereby preventing transcription of the operon's structural genes.

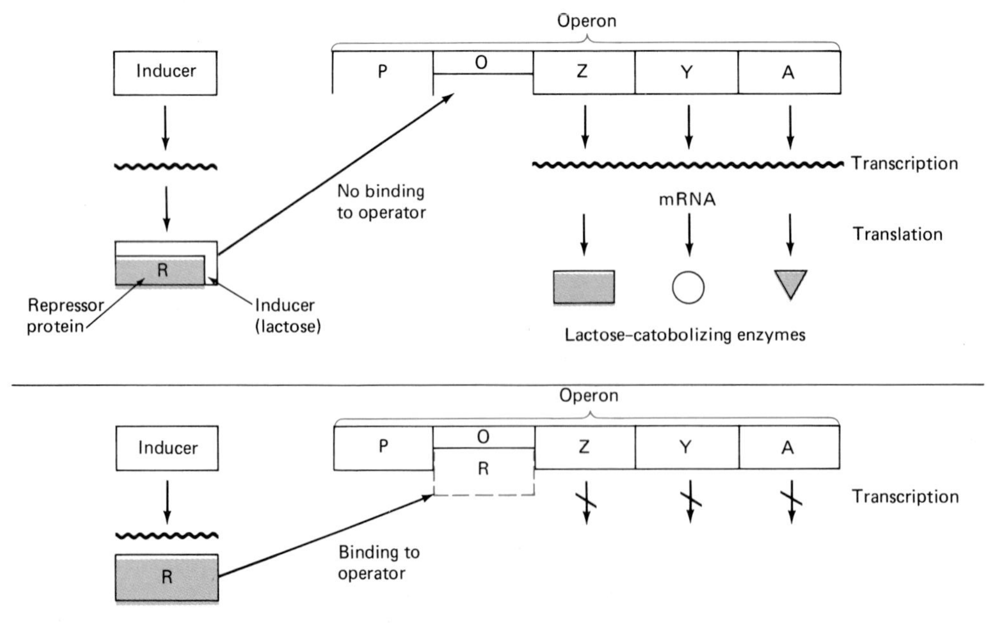

Operons

An *operon* is a region of DNA that codes for a group of structural or adaptive genes responsible for the synthesis of enzymes in a specific metabolic pathway. These structural genes are controlled by an *operator,* a *promoter region,* and a *regulatory gene.* The operator controls the transcription of structural genes, turning on or off mRNA formation by the structural genes. RNA polymerase starts transcription by binding to the promoter region. The efficient binding of the polymerase to this region seems to occur only if another protein, called *catabolite activator protein* (CAP), has bound first. The CAP, in turn, cannot bind to the promoter region unless it has first bound a low-molecular-weight substance called *cyclic adenosine monophosphate (cAMP).* Cyclic AMP has been shown to be a major factor in a variety of control systems not only in procaryotes but in higher organisms as well. The regulatory gene, which lies outside of the operator region, is responsible for turning the operator on and off. This gene codes for the synthesis of a protein repressor molecule that binds to the operator region and blocks protein synthesis by structural genes (Figure 6–15).

In catabolic systems with an inducible operon system, an appropriate inducer, when present, interacts with the repressor. This interaction causes a conformational change in the molecule and thus prevents it from binding to the operator and turning on protein synthesis. An appropriate inducer can be a substrate such as lactose (Figure 6–15), which would be catalyzed by the enzymes to be formed. In the absence of an inducer, the repressor protein binds to the operator and thereby turns off protein synthesis.

The regulation of a *repressible operon* system is the opposite of that of an inducible system. In biosynthetic systems, the repressor protein normally cannot bind to the operator, and the protein-synthesizing activity of the operon is not turned off. Only when a specific inhibitor is present can the repressor bind to the operator and prevent the transcription of structural genes. The inhibitor combines with the repressor to form the binding unit. Specific inhibitors often include biosynthetic metabolic end products, the production of which exceeds the level required by the cell. Biosynthetic end products are called *corepressors.*

Enzyme induction and repression come into play as cells adapt to changing environments. Induction is found in many catabolic pathways involving amino acids, sugars, and other carbohydrates. End-product repression is found in many anabolic pathways involving amino acids, purines, and pyrimidines.

Mutation and Protein Synthesis

Mutations (permanent hereditary changes) may develop as a consequence of natural or spontaneous events, or they may be induced by various physical or chemical agents (Table 6–2). The result of a mutation depends upon the change it causes in the transcription of the DNA code into messenger RNA (mRNA) and on protein synthesis.

The change in one base causes the substitution of a complementary base in mRNA. The resulting codon is read by a different tRNA, thereby causing the placement of an amino acid in the forming protein molecule different from the one specified by the code. Changes in a single DNA base are called *point mutations.* Several types of these mutations are known; they include *substitution mutations, addition mutations,* and *deletion mutations* (Figure 6–16).

The insertion of an incorrect base in place of the normal one within a base pair is a substitution mutation (Figure 6–16a). This type of error can cause a codon change, which is significant if it results in the coding for a different amino acid. Another possible effect is the creation of a stop signal. Here the change is referred to as a *nonsense* mutation and results in the formation of an incomplete or abbreviated protein.

Mutations involving either the deletion or addition of one or more purine or pyrimidine bases cause a misreading of the genetic code (Figure 6–16b). These events are called *reading-frame shifts.* Both deletion and addition mutations occur from abnormal stretching or twisting of the DNA strands and may cause the synthesis of a nonfunctional protein.

TABLE 6–2 Some Sources and Types of Mutations

Source	*Type of Mutation and/or Effect(s)*
Chemical mutagens	Single nucleotide substitution.
Ionizing radiation	Deletions of one or more nucleotides from the DNA sequence; attachment of part of one chromosome to another (translocation).
Spontaneous mutation	Single nucleotide substitution.
Transposition	Insertion of a transposon (a mobile gene) into a random location on a gene; results in activation.
Ultraviolet (UV) radiation	Formation of pyrimidine dimers; causes errors in DNA replication. (See Chapter 11 for UV effects.)

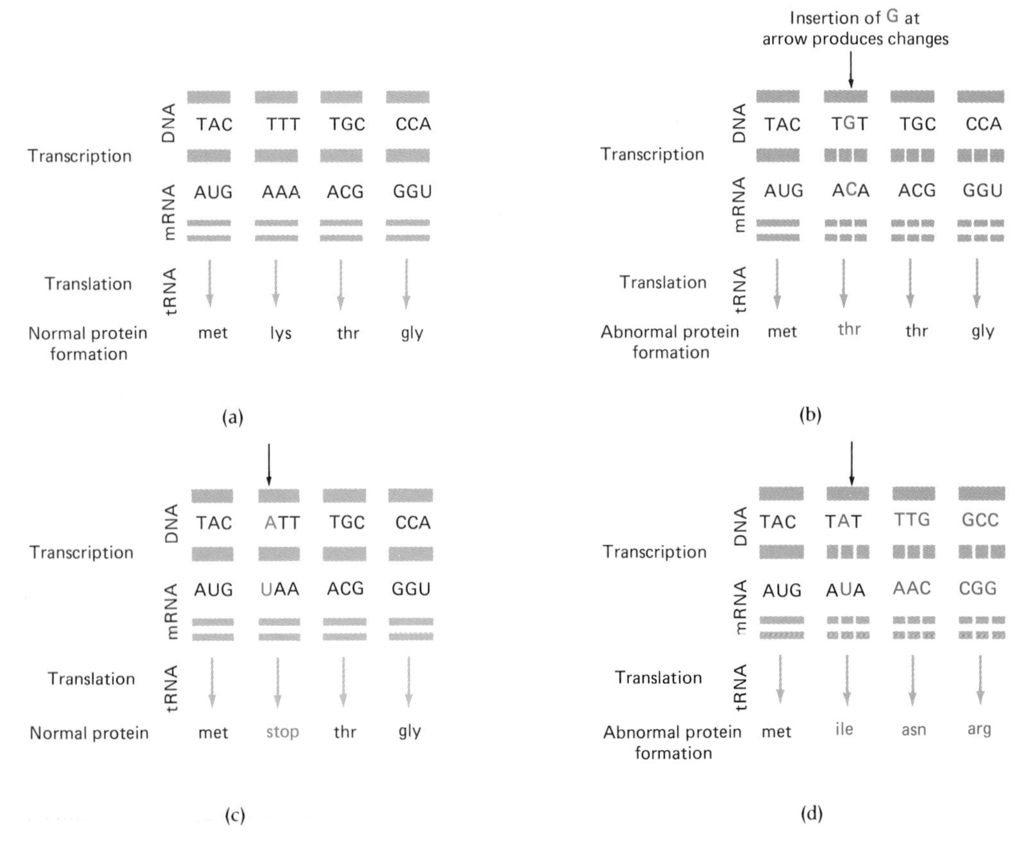

Figure 6–16 Types of mutations. (a) A **normal DNA segment** being transcribed into an mRNA transcript coding for four amino acids. (b) A **substitution mutation.** Here the nucleotide base thymine (T) is replaced by guanine (G), causing the insertion of the amino acid threonine for lysine. (c) A **nonsense mutation.** Replacement of thymine (T) with adenine (A) results in a stop signal nucleotide sequence UAA. The synthesis process shuts down as a consequence of this substitution. (d) **Addition-Frameshift** mutation. The addition of adenine (A) between the bases of thymine (T) and thymine (T) results in a shift in the reading frame, thus causing an incorrect reading of the code and the insertion of different amino acids in the developing protein or related molecule.

Mutations can also be caused by the rearrangement of chromosome pieces on the same or separate chromosomes. Situations of this nature are referred to as *translocation* forms of chromosomal aberration (abnormality). Through translocations, parts of an operon, such as the regulatory genes, may be lost. Depending on the control system affected, enzymes may become completely repressed or constitutive.

MUTAGENIC AGENTS

Early work on mutations with *Drosophila* was performed with x-ray irradiation because x-rays cause chromosomal breakage. Consequently, one of the first chemical mutagens studied, mustard gas, was chosen because its action is similar to that of x-rays. Mustard gas represents a group of mutagenic chemicals known as *alkylating agents,* which replace a hydrogen with another group—for example, methyl ($-CH_3$) or ethyl ($-CH_2CH_3$), on a molecule such as guanine. The substituted guanine appears to cause defects by pairing with a thymine molecule rather than with cytosine. Thus, in the next generation of DNA, the original guanine is replaced by adenine, which will pair with the incorrect thymine. Alkylating agents also produce deletions and chromosome rearrangements.

Many other chemical and physical agents have been tested for mutagenic activity. Ultraviolet (UV) rays appear to act as a mutagen because UV wavelengths are strongly absorbed by DNA. Some of the chemicals studied include molecules similar in structure to the components of DNA. These base analogs are purines and pyrimidines not normally found in DNA (Figure 6–17a). Their incorporation into DNA during nucleic acid synthesis results in incorrect pairing during the replication process. For example, 5-bromouracil can be incorporated into DNA in place of thymine. Because of stereochemical differences, 5-bromouracil can pair with guanine rather than adenine, thus changing the code. 5-Bromouracil also makes the DNA more sensitive to UV radiation and other mutagenic agents.

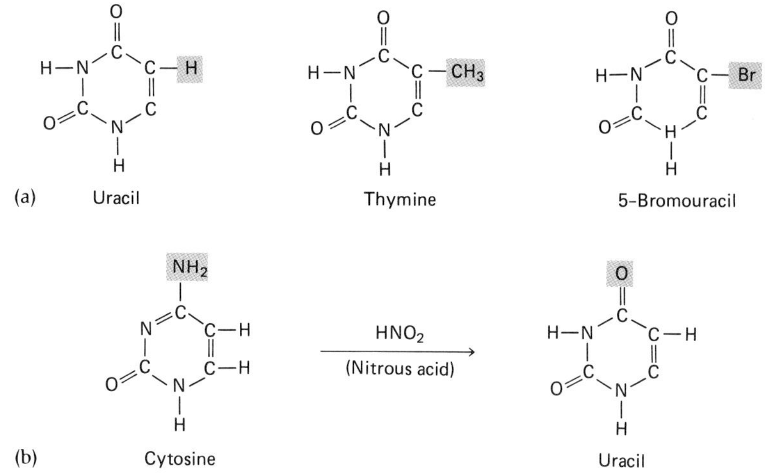

Figure 6–17 Mutagenic agents: (a) 5-bromouracil compared to uracil and thymine. (b) The effect of nitrous acid on cytosine.

Another group of chemical mutagens, the acridines, appear to penetrate the DNA in a way that separates the bases. This separation may lead to the unnatural insertion or deletion of a base, which causes reading-frame shifts. The mutagen nitrous acid was originally tested because of its known interaction with proteins, which were thought to be the genetic material. However, experiments have shown that nitrous acid does cause mutations by removing an amino group (NH_2) found in adenine, guanine, and cytosine. When these bases are **deaminated** by removal of the amino group, they are converted to hypoxanthine, xanthine, and uracil, respectively (Figure 6–17b). The newly formed compounds will not pair with the appropriate purine or pyrimidine, thus creating spot changes in the genetic code.

Genetic Repair

Several forms of DNA polymerase, the enzyme responsible for DNA replication, are able to repair damaged DNA. Such repair enzymes can detect DNA sequences that contain missing or unnatural purines or pyrimidines and substitute the appropriate bases.

With certain mutations—such as those produced by carcinogenic chemicals—ultraviolet light or nuclear radiation causes significant defects in one of the DNA strands. A typical defect of this type takes place when neighboring thymine bases on one DNA strand combine with one another to form a double pyrimidine or dimer (Figure 6–18). Dimer formation has the effect of stopping protein or DNA formation and often causes cell death. When such a defect appears, the cell emits an emergency SOS signal that activates a group of repair enzymes and cellular mechanisms associated with DNA repair. This repair system removes the damaged bases and replaces them all with the appropriate bases in the proper sequence specified by the undamaged complementary strand. Occasionally, errors resulting in mutations are made during the repair process. Visible light, through a mechanism referred to as *photoreactivation,* greatly enhances the speed and effectiveness of the repair process. Repair does not occur in the dark.

Mutations in Microorganisms

The theory that mutations are actually defects in enzymes preventing specific metabolic activities has been tested, primarily with microorganisms. The first significant work in this area was reported by

Figure 6–18 Thymine dimer formation caused by action of ultraviolet light on adjacent thymine [T]-containing nucleotides.

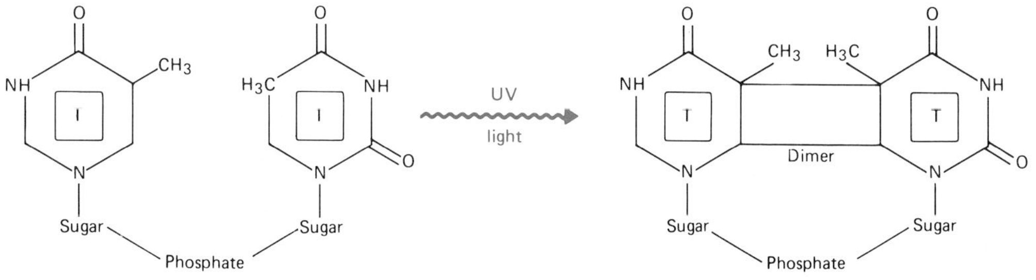

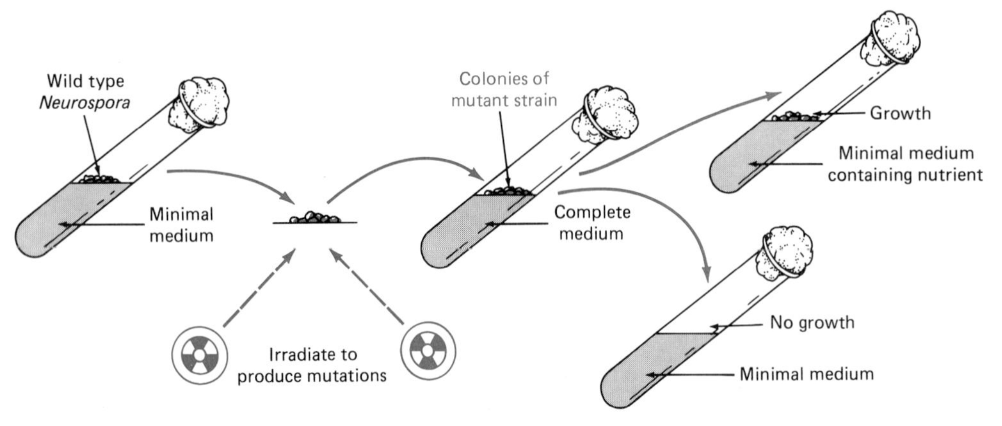

Figure 6–19 The features of the Beadle and Tatum one gene–one enzyme process. Mutant *Neurospora* (new-RAH-spor-a) strains produced by irradiation can grow on a complete medium but not on a minimal one that lacks a particular nutrient. Growth can occur if the minimal medium is supplemented with the needed nutrient.

George W. Beadle and Edward L. Tatum in 1941. Their research organism was *Neurospora crassa,* an ascomycete fungus often found as a pink-red mold on bread. The important feature of the microorganism was that it grew on a simple, defined medium called *minimal medium,* in which the primary, or parent, culture did not have complex nutritional requirements (Figure 6–19). For example, the minimal medium for *N. crassa* includes ammonium chloride for nitrogen, glucose for energy and carbon, a few mineral salts, one or more vitamins, water, and agar (when used for solid media). The parent culture was called *wild-type,* referring to the form usually found in nature. After irradiation, mutants of *N. crassa* were found that could not grow on the minimal medium but would grow on a complex medium containing added malt and yeast extracts. Beadle and Tatum found that each mutant actually had one particular nutritional requirement rather than a general requirement for different nutrients. These growth requirements were for a specific amino acid, vitamin, purine, or pyrimidine, and each mutant grew on the minimal medium supplemented with its particular requirement. Such mutants are called **auxotrophs,** in contrast to the parent or wild-type culture referred to as a **prototroph.** (An auxotroph is defined as a mutant having nutritional requirements in addition to those of its parental culture.) The fact that each mutant tended to have a specific nutritional requirement served as the basis for the *one-gene, one-enzyme* hypothesis proposed by Beadle and Tatum, that is, that each gene is responsible for the production of a single enzyme. These scientists correlated the development of the nutritional requirement with the absence of a particular enzyme. This hypothesis has proven helpful in determining metabolic pathways in both procaryotic and eucaryotic cells. Since the early 1940s, mutational research has included most if not all microorganisms, a favorite being *E. coli* and its associated viruses.

Microbial Experiments and the Random Nature of Mutation

In the introduction to this chapter we briefly described the controversy surrounding the theories of evolution advanced by Darwin. That controversy has carried over into microbiology. The concept of evolution or development of a species was viewed by Jean-Baptiste de Lamarck (1744–1829) and others as a continual process of gradual changes in plant and animal species in response to environmental conditions. According to this theory of adaptation or acquired characteristics, the properties developed in response to the environment became a part of an individual's genetic makeup. Darwin's concept opposed the Lamarckian theory.

The view that changes are caused by adaptation or adjustment to conditions is certainly logical. An excellent example of this type of event in microbiology could involve the development of resistance to antibiotics by microorganisms.

Consider a penicillin-sensitive *Staphylococcus aureus* culture obtained from a patient. Giving penicillin should greatly reduce if not eliminate the infection. However, the patient may subsequently return to the hospital with a severe uncontrolled infection, and the *Staphylococcus* (staff-il-oh-KOK-kuss) culture obtained this time is found to be penicillin-resistant. According to the Lamarckian adaptation theory, some of the bacteria that were in contact with the antibiotic managed to survive by adaptation. On the other hand, supporters of modern Darwinism would theorize that those bacteria that happened to be resistant to the antibiotic were the only survivors. In either case, the time between apparent recovery and relapse is the time required for the resistant bacteria to increase to sufficient numbers. The issue that remains is to differentiate between adaptation and random variation.

In 1943, S. E. Luria and M. Delbrück reported an

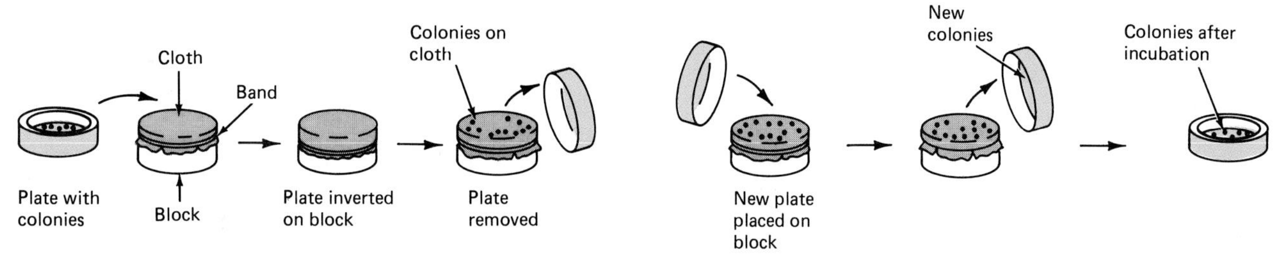

Figure 6–20 The replica-plating technique. Among other applications, this procedure can be used to demonstrate that spontaneous mutations arise in microorganisms.

indirect approach to finding the nature of microbial mutation. The procedure used was a "fluctuation test," designed to determine statistically whether mutation to bacterial virus resistance in *E. coli* was random or directed. Their data presented strong evidence for the spontaneous, random nature of mutation. In other words, the environment does not direct the mutation but rather selects for those mutations that have occurred.

The Indirect Selection Procedure

After the demonstrations by Luria and Delbrück and others that mutation is spontaneous in microorganisms, J. and E. W. Lederberg developed a relatively simple method by which virus-resistant mutants could be isolated with no exposure to the bacterial virus. In this replica-plating technique (Figure 6–20), a velveteen nap was used to transfer bacteria from colonies on one plate to several other plates containing fresh media. When this is properly done, the transferred bacteria develop into colonies at locations on the new plates corresponding to those occupied by the same colonies on the original inoculum plate (Figure 6–20). The relative locations of such colonies on all plates inoculated with the same velveteen pad should be the same. Using the bacterial colonies on one nutrient agar plate (1) as their inoculum sources, the Lederbergs placed these organisms on nutrient agar and another plate with nutrient agar plus a bacterial virus (2) that could infect and kill the bacteria, thus preventing colonies from developing. Figure 6–21 shows a hypothetical experiment that illustrates the main points of the procedure and the interpretation of the results.

The sequence diagrammed in Figure 6–21 has been greatly simplified for clarity and brevity. Only three colonies of the original 12—numbers 2, 7, and 12—were resistant to virus and developed on the medium containing the virus. Because the two plates were oriented in the same way as the original

Figure 6–21 Spontaneous mutation in bacteria as shown through replica plating. The selective agent here is a virus to which the bacteria used are sensitive.

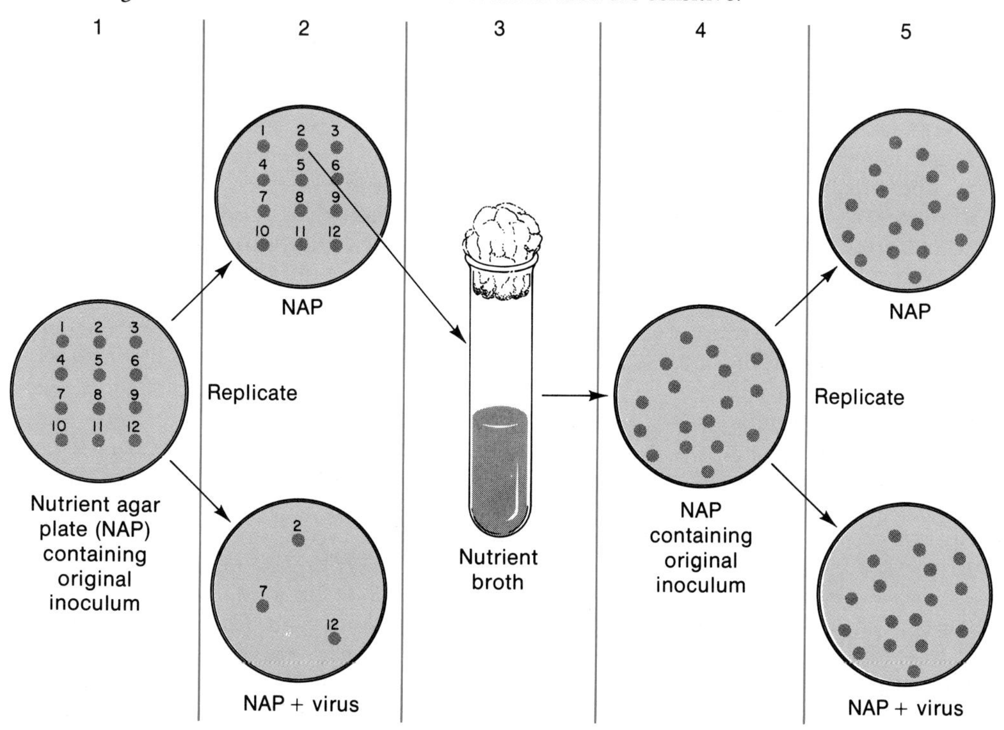

one, colonies that appeared on the medium without virus could be identified and related by position to those found to be virus-resistant. Colony 2 from the plate without virus was moved by an inoculating needle from the nutrient agar plate to nutrient broth and allowed to grow (3). When growth from the nutrient broth was streaked on a nutrient agar plate, 17 colonies developed (4) and were used as the original inoculum to repeat the replication technique as before. This time, the resultant colonies appeared on both types of media (5). We would not have known that virus-resistant bacteria had arisen and were present unless such colonies had been identified by replicating onto the second plate containing the virus. This procedure leaves little doubt that the virus-resistant cells were in the original colony *before* any exposure to virus had occurred.

The techniques developed by Luria and Delbrück and by the Lederbergs have been used to study mutations of a wide variety of characteristics with many different microorganisms, including antibiotic resistance in *S. aureus, E. coli, Pseudomonas aeruginosa* (sue-doh-MOH-nass a-roo-jin-OH-sah), *Shigella* (shi-GEL-la) spp., and *Salmonella typhi* (sal-mon-EL-la TIE-fee).

DNA Transfer in Procaryotes

The generally accepted view that the production of offspring different from their parents by sexual cross-breeding (hybridization) is important to the survival of a species (hybrid vigor) does not seem to apply to bacteria because they are haploid. That is, they contain a single chromosome. Although plant and animal species are classified according to their ability to combine gametes with a resultant diploid (paired-chromosome) cell, microbial species are classified by common structural and biochemical characteristics and genetic relatedness. Bacteria are known to exchange genetic material between genera and between species. This exchange can occur through the transfer of entire chromosomes or their fragments, or through the transfer of extrachromosomal DNA molecules called *plasmids* from one bacterium to another. Plasmids exist and replicate independently of the host cell's chromosome and genetically code for nonessential properties. However, these molecules of DNA may contain genes that might prove to be especially useful for some organisms in certain environments. Examples of valuable and adaptive features associated with plasmids include antibiotic resistance, pili formation, nitrogen fixation, oil degradation, synthesis of metabolic enzymes for unusual substrates, and increased virulence.

The process by which new combinations of chromosomal genes are obtained from two different types of cells is called *genetic recombination.* On a molecular level, a new chromosome is formed from the DNA contributed by the two different parental cells. In bacteria, three processes by which such recombinant chromosomes can form are **transformation, conjugation,** and **transduction.** Genetic material is transferred from a donor cell to a recipient and does not involve cell fusion, which is a common feature of the sexual process as it occurs with eucaryotes.

Chromosomal Transfer

Transformation

As described previously in this chapter, the early studies on DNA and bacterial transformation (Figure 6–1) by Griffith and others had a profound impact on the biological sciences. In the process of *transformation,* DNA is released from donor cells into the surrounding environment, where recipient cells incorporate it (Figure 6–26a). Transformation was the first mechanism of bacterial genetic exchange to be uncovered. A recipient cell that is able to absorb donor DNA and subsequently undergo transformation is a *competent cell.* Competence can be defined as the ability of the recipient strain to transport DNA from the culture medium into the cell. As competence appears in a culture during the growth phase, a protein is produced that can change incompetent cells into competent ones. It is believed that this *competence factor* influences some change in cell surfaces that causes development of receptor sites or increases permeability to DNA molecules. It has been shown with *Streptococcus pneumoniae* that new surface antigens develop and are released into the medium. These materials can attach to new cells and induce competence. The cell wall becomes more porous, and there is an increase in the number of mesosomes. Mesosomes are believed to be involved in the internal transport of the DNA to the chromosome for possible integration.

Since Griffith's pioneering work with *S. pneumoniae,* transformation has been demonstrated with many genera, including *Haemophilus, Bacillus, Rhizobium* (rye-ZOH-be-um), *Neisseria, Acinetobacter* (a-sin-et-OH-back-ter), *Pseudomonas,* and *Escherichia.* The research with the pneumococci and *Haemophilus* was complicated by their complex nutritional requirements, making it difficult to study more than a few characteristics at any one time. Recall that the study of mutation with *N. crassa* was aided by that organism's relatively simple nutritional needs. This property allowed careful study of metabolic pathways and the formulation of the one-gene, one-enzyme hypothesis. The discovery of transformation in *Bacillus subtilis* (ba-SILL-us SUT-ill-is) in 1961 made such studies in transformation possible because this species has relatively simple nutritional requirements.

Transformation experiments can yield much information on the transfer of genetic information *per se* and the location of genes on the chromosomes or genomes of several bacterial species.

TRANSFECTION

In transfection, transformationlike situations occur in which viral DNA is taken up by the bacterial host cell. The term also applies to situations in which an artificial process is used to induce plasmids and sources of naked DNA molecules to be taken up by host cells. In transfected cells, the absorbed DNA is transported to the nuclear area, where it is assembled into many copies of the transfecting DNA called *transgenomes*. These transgenomes are unstable in most cells.

However, they can be incorporated into the chromosomes of a small percentage of cells and actually transcribed into mRNA. (See the discussion of gene manipulation later in this chapter.) An exciting study in transfection with *B. subtilis* showed that DNA extracted from an animal virus such as vaccinia (cowpox) could be used in transformation to lead to the development of animal viruses in infected *B. subtilis*.

Bacterial Conjugation

Conjugation is the transfer of genetic material between two living bacteria that are in physical contact with one another (Figure 6–26b). In all known cases, the mechanism is encoded by plasmid genes. Conjugation, or recombination, the major means of genetic exchange and variability in sexually reproducing higher organisms, had been investigated with bacteria as early as 1908. However, bacterial conjugation was not proven until Lederberg and Tatum obtained multiple mutants *(polyauxotrophs)* of *E. coli* in 1946.

The bacterial strains used were biochemically deficient mutants. Each had two or more different genetic defects. One parent organism had defects that caused it to require the vitamin biotin (B) and the amino acid methionine (M) for survival; the other parent strain required the amino acids threonine (T) and leucine (L) but not biotin or methionine. These parent strains were symbolized as $B^-M^-T^+L^+$ and $B^+M^+T^-L^-$, respectively. The minus signs represented a requirement for the particular biochemical substance for growth on minimal medium, and the positive signs indicated no deficiency requirement. Neither strain was able to grow on minimal medium. Only a transfer of B^+M^+ to one parent or T^+L^+ to the other would permit growth on the minimal medium of the test conditions (Figure 6–22).

When the two strains were mixed and placed on minimal medium, some bacteria were able to sur-

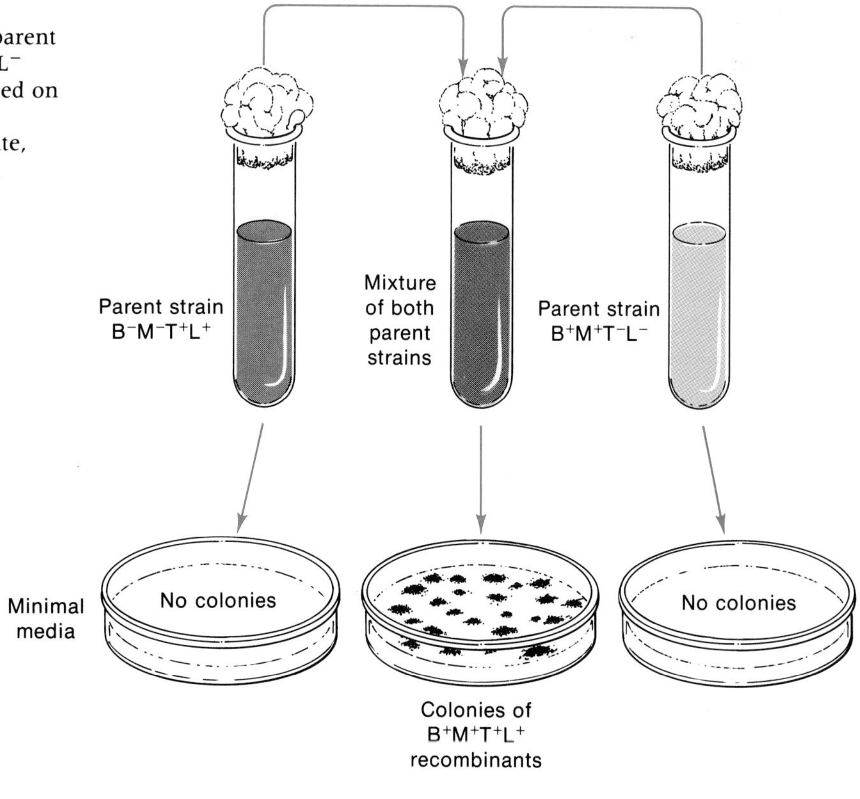

Figure 6–22 Lederberg and Tatum's original conjugation experiment. The parent strains, $B^-M^-T^+L^+$ (left) and $B^+M^+T^-L^-$ (right), were unable to grow when plated on minimal media. However, when these strains were mixed, allowed to conjugate, and then plated, colonies developed on minimal media that represented the presence of $B^+M^+T^+L^+$ recombinants.

vive and grow. These organisms represented the natural type of genetic makeup, $B^+M^+T^+L^+$. The statistical probability that the double spontaneous mutations would yield the same result is about one in 10^{14}. The number of resulting colonies obtained was far greater than this. These findings suggest that recombination by crossing over was occurring, indicating a linear arrangement of genes *(linkage)* in bacteria, as in higher organisms. The crossing over of the T^+L^+ portion of a chromosome would produce a wild-type *E. coli.* This phenomenon can be represented as follows:

$$\frac{B^+M^+T^-L^-}{B^-M^-T^+L^+} = B^+M^+T^+L^+$$

To eliminate the possibility that transformation, transfer of DNA without contact between cells, might be taking place, B. Davis in 1950 carried out a series of experiments using a U-tube device (Figure 6–23). The U tube was constructed with two arms, separated by a porous glass filter that would prevent passage of bacteria between the arms but would not block DNA. When two biochemically deficient parental strains such as $B^+M^+T^-L^-$ and $B^-M^-T^+L^+$ were placed in the same arm, recombinants were obtained at expected frequencies. However, if one mutant strain was placed in one arm, the other in the second arm, and the medium flushed between the arms, no recombinants were obtained. Thus, transformation was eliminated as an explanation of this phenomenon, and the need for intimate contact between bacterial cells was firmly established.

Through subsequent research in this area, sexuality in bacteria was discovered. In addition, certain strains were shown to be donors of genetic material and others to be recipients in any bacterial population in which conjugation occurs. Donor (male) cells are designated $\mathbf{F^+}$, while recipient (female) cells are indicated as $\mathbf{F^-}$. Only donor cells contain extra pieces of DNA, known as *F (fertility) particles,* which are not part of the chromosomes of such cells (Figure 6–24). They are plasmids, however.

Upon mixing of F^+ and F^- bacteria, the donor F^+ cells attach to recipient F^- cells by sex pili and, within minutes, F particles are transferred. F^- cells become F^+ cells through this process. Bacterial chromosomes are rarely transferred.

The F particle is an extrachromosomal genetic complex plasmid composed of circular DNA that possesses genes for the following: (1) the regulation of its own replication; (2) the synthesis of sex or F pili (Figure 6–25) necessary for conjugation so that donor DNA can pass to a recipient cell; and (3) the formation of a particular surface component that may serve to lower negative electrical surface charges of donor cells, so that intimate contact with recipient cells can occur after random collision.

Some streptococci have been observed to conjugate, although pili were not readily seen. Such organisms also secrete chemicals, "sex attractants," which cause mating pairs to aggregate.

Hfr CELLS

Bacterial chromosomes are transferred to F^- cells by a cell type known as *Hfr,* an abbreviation for *high frequency of recombination* (Figure 6–24). Such cells arise from integration of the F particle into the donor cell chromosome. This integration appears to

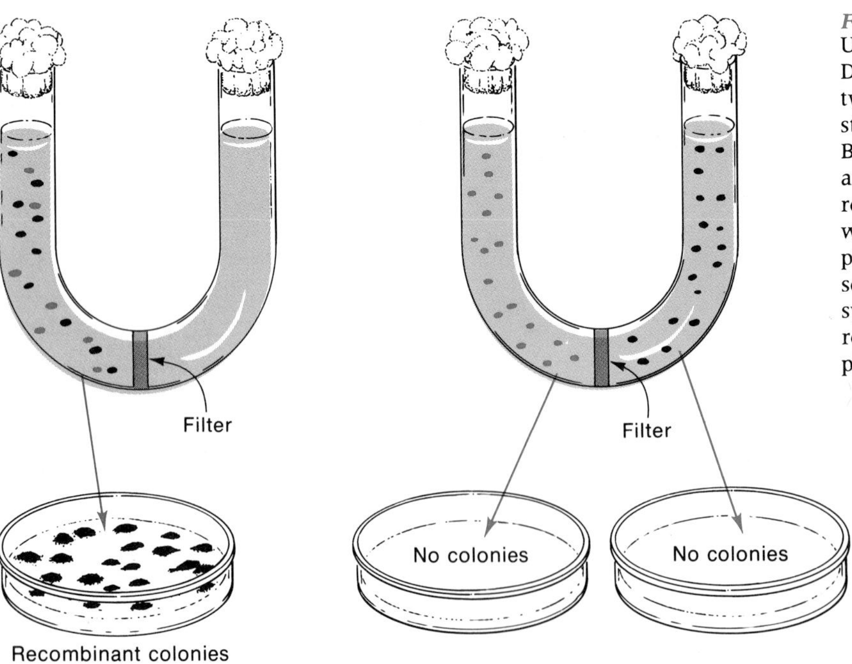

Figure 6–23 A diagram of the U-tube experiments performed by Davis. In the system on the left, two biochemically deficient parent strains such as $B^+M^+T^-L^-$ and $B^-M^-T^+L^+$ are placed in the same arm. When a sample is plated, recombinants develop. However, when the individual strains are placed in different arms and separated by a filter, as in the system on the right, no recombinants develop upon plating.

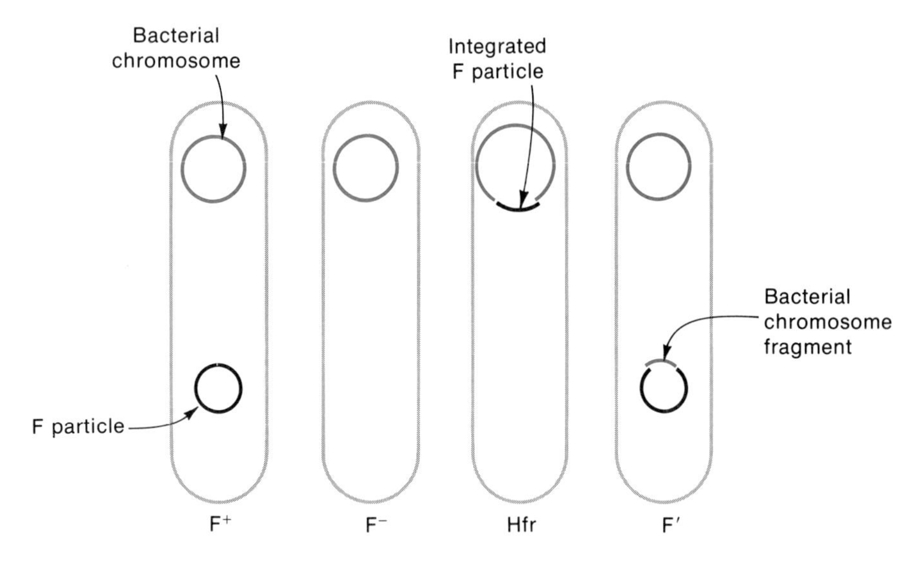

Figure 6–24 The structure of F^+, F^-, Hfr, and F′ bacterial cells. The diagram shows the possible relationships between the bacterial chromosome and the F particle.

mobilize donor chromosomes for the transfer process. Since given *Hfr* strains always donate their respective genes in the same order, they have been used to determine the arrangement and orientation of a large number of chromosomal genes and to develop a genetic map for the bacterial chromosome. A later section in this chapter describes this approach to mapping the bacterial genome.

Several studies have shown that fusing bacterial spheroplasts (cells with most of their cell walls removed) for the purpose of genetic transfer is also effective (Color Photograph 32). With such spheroplast fusion, larger quantities of genetic material are transferred and genetic recombination occurs with greater frequency.

Figure 6–25 An electron micrograph of a presumed specific pairing between an Hfr cell (bottom) and an F^- cell. The sex pili (S) can be clearly differentiated from other pili (P). Small RNA containing viruses are attached to the sex pili. *[From R. Curtiss, III,* et al., J. Bacteriol. ***100:****1091–1104, 1969.]*

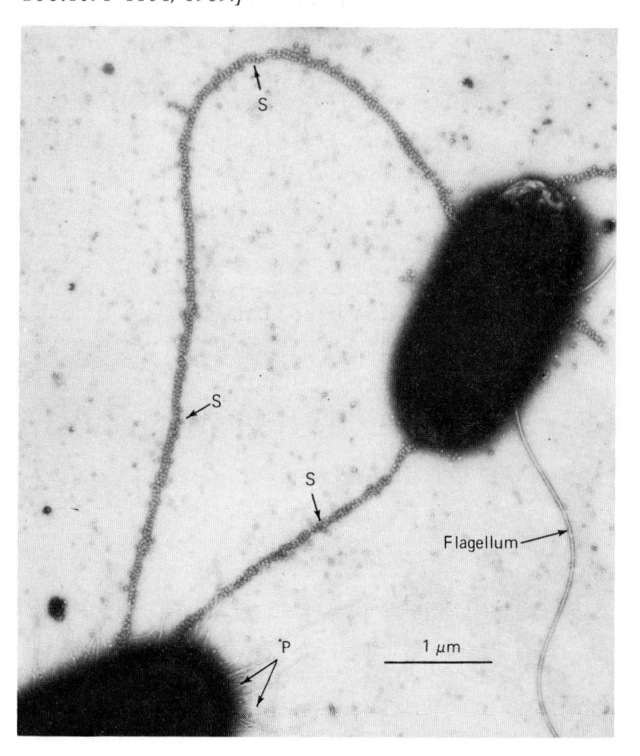

THE F-PRIME (F′) PARTICLE

There is another type of extrachromosomal particle that some bacterial strains are capable of transferring. Such particles carry several chromosomal genes of a donor cell (Figure 6–24). F particles can change their extrachromosomal status by integrating into a bacterial cell chromosome. However, in rare instances when they detach themselves, several bacterial chromosome genes remain attached to the particle. The resulting F particle carrying bacterial genes is referred to as an *F-prime (F′) particle.* It behaves in a manner similar to an F^+ particle. As a consequence of an F′ × F^- mating, the recipient bacterial cell acquires the additional genes carried by the F′ particle upon integration into the recipient's chromosomes.

Transduction

When Lederberg extended his recombination research to *Salmonella* species, the unidirectional transfer of genetic material occurred. This phenomenon resembled conjugation as observed with *E. coli;* but when performing these studies in a Davis U tube, Lederberg observed that physical contact was not required. This situation appeared similar to the transformation observed with pneumococci, *Haemophilus* spp., and *B. subtilis.* To confirm the transformation in *Salmonella,* DNase, the enzyme that breaks down free DNA, was added to the culture to prevent the genetic transfer of free DNA. Contrary to expectation, the DNase did not prevent the transfer of ge-

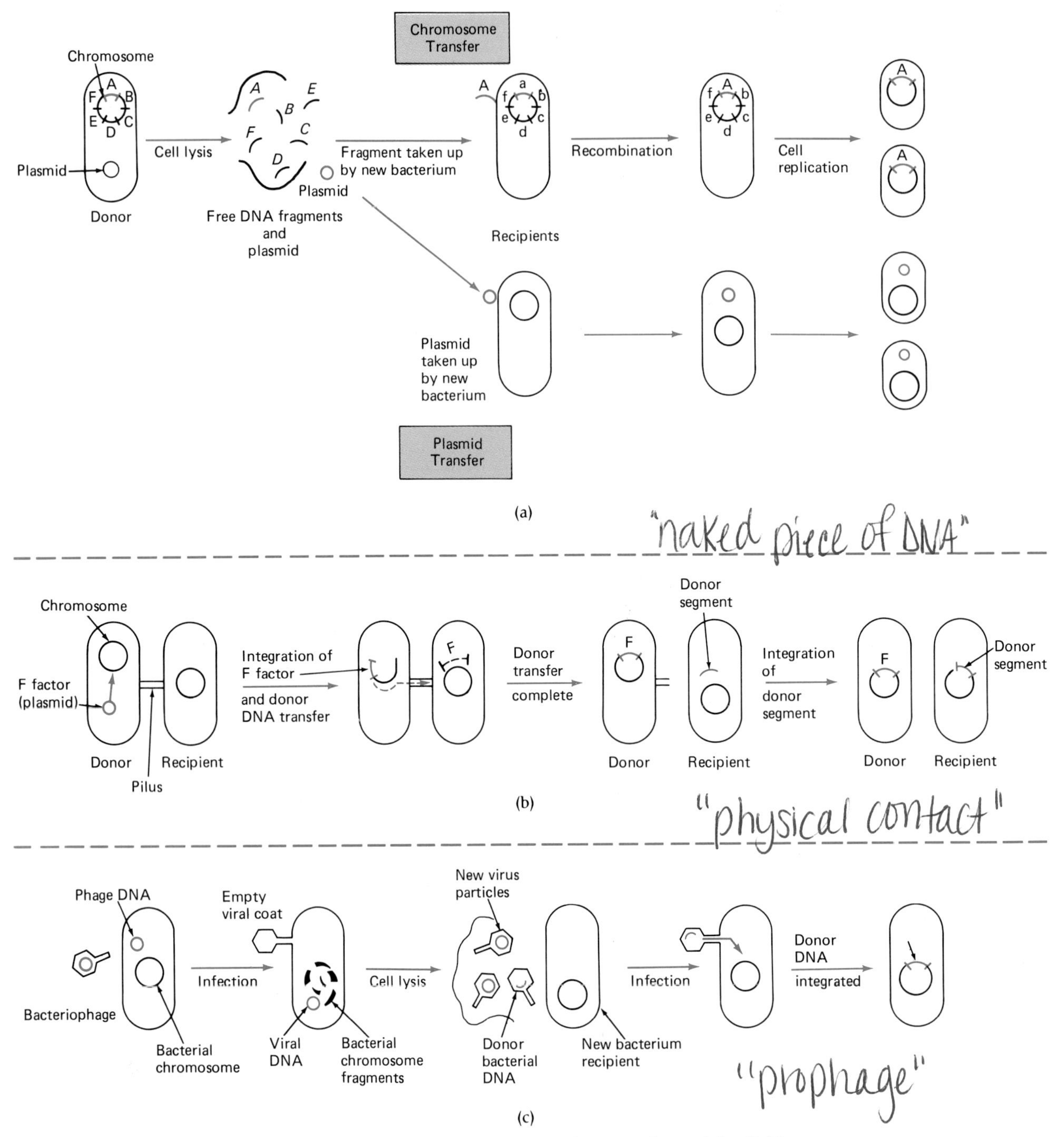

Figure 6–26 Three mechanisms by which DNA transfer occurs from one bacterial cell (the donor) to another (the recipient). (a) **Transformation,** a process for the transfer of a "naked" piece of DNA. (b) **Conjugation,** a process of DNA transfer that requires physical contact between the participants. Pili usually function in this capacity. (c) **Transduction,** a process of DNA transfer by means of a bacterial virus (bacteriophage). Donor bacterial DNA is enclosed within the protein coat (capsid) of the virus.

netic information. This surprising result spurred research on the transduction phenomenon.

In 1952, N. D. Zinder and J. Lederberg reported finding bacterial genetic transfer via a filterable agent identified as a bacterial virus. Certain bacteriophages are able to infect bacterial cells and, through the process of *lysogeny,* become an integral part of the bacterial genome. In this condition the viral DNA actually becomes part of the bacterial chromosome. However, the virus is eventually activated, either spontaneously or by UV light, replicates at the expense of the cell, and eventually causes the host cell to disintegrate. Thus, many viruses can be released into the culture to infect other bacteria. **[See Chapter 10 for additional details of bacteriophages.]**

At times during the intracellular development of some bacteriophages, errors occur and a defective virus, called a *transducing particle,* is formed. Such a particle may carry one or more bacterial genes—small pieces of the bacterial DNA that remained with the maturing bacteriophage (prophage) as it broke loose from the host cell chromosome. It is these small pieces of genetic material that are transferred to other bacterial cells by bacterial viruses in *transduction* (Figure 6–26c).

Compared with bacterial conjugation, transduction permits the transfer of relatively small segments of DNA because of limited space within the virus. Usually this transfer consists of the virus DNA (prophage) and a segment of DNA taken from its attachment site on the bacterial chromosome.

There are two forms of transduction that are recognized, *generalized* (nonspecialized) and *specialized* (restricted). In the generalized type, virtually any gene (genetic marker) of the host bacterium can be transferred and does not require lysogeny. In the specialized form of genetic transfer, only specific genes near the attachment site of the viral DNA on the chromosome of the host cell are involved. This phenomenon has been used to study the fine structure of bacterial DNA using short linkage patterns.

Summary of Genetic Exchange Mechanisms

Figure 6–26 compares the different mechanisms by which DNA is transferred in bacteria. Table 6–3 lists the various features of each transfer process.

Lysogenic Conversion

During the 1950s, a more medically significant phenomenon was discovered: *lysogenic,* or *viral, conversion.* In lysogenic conversion, unlike transduction, a particular genetic change occurs in all infected microorganisms as a consequence of a virus integrating itself into the host cell's DNA (lysogeny). Such conversions, which are recognized as physical changes, include the development of smooth colonial types in mycobacteria and alterations in the antigenic types of salmonellae.

Probably the most significant alteration of a microbial characteristic by lysogeny was reported by V. J. Freeman in 1951. This investigation showed that the bacterium causing diphtheria, *Corynebacterium diphtheriae* (ko-ri-nee-back-TI-ree-um dif-THI-ree-a), could produce toxin only when infected by a specific bacteriophage. There are other requirements as well, primarily the presence of a certain concentration of iron, but if the phage is absent, lysogenic-state toxin production does not occur. **[See bacterial toxins in Chapter 22.]**

Various small, extrachromosomal DNA particles

TABLE 6–3 A Comparison of Bacterial Transformation, Conjugation, and Transduction

	Mechanism of DNA Transfer		
Property	*Transformation*	*Conjugation*	*Transduction*
Bacteria involved	Gram-positive and gram-negative	Generally gram-negative	Both gram-positive and gram-negative
Mechanism of transfer	Through cell membrane of recipient cell	Through sex pili after cell-to-cell contact	Bacterial virus infection
Plasmid-dependent transfer	Occurs	Yes	None
Transfer in one direction (unidirectional)	No	Yes	Yes
Extent of DNA transferred	About 20 genes	From 20 genes to complete chromosome	About 20 genes
Susceptible to DNA degrading enzymes	Yes	No	No

capable of independent replication have been found in bacteria. These genetic elements are called **plasmids.** They replicate either autonomously or as part of the host (usually a bacterial cell) chromosome. Plasmids are able to integrate only at specific points on a chromosome where a nucleotide sequence corresponds to the one on the plasmid DNA.

Plasmids occur primarily in procaryotes and share the properties of being self-replicating, dispensible, circular molecules of double-stranded DNA. Differences among plasmids include the types of functions they encode, the number of copies per cell, the host bacterium used for replication, compatibility with other plasmids in a host cell, and the mechanism used for their transmission among bacterial species. Size is another distinguishing property. Both large and small plasmids are known.

Plasmids have been studied in most bacterial genera. These DNA-containing structures are categorized as nongenetic factors and are associated with properties that can be transferred between bacteria by conjugation, transformation, and transfection. The following cellular features have been found to be encoded by plasmids: resistance to antibiotics (R factors), toxin production, bacteriophage sensitivity and resistance, pigment production, plant tumor formation, and catabolic (degradation) activities. A number of these functions will be described.

BACTERIOCINS

The bacteriocins were first described in 1932 as antibiotic-like material from *E. coli.* Bacteriocins are a diverse group of substances, usually proteins, that inhibit or kill sensitive members of strains closely related to the one that produced the material. Colicins, bacteriocins produced by *E. coli,* act by adsorbing to specific molecular groupings, or receptors, on the surfaces of sensitive organisms, causing modification or loss of these receptor sites.

Antibioticlike substances similar to those from *E. coli* have been found in *Pseudomonas* (pyocin) and *Bacillus megaterium* (megacin). Collectively, these substances are called *bacteriocins.* The ability to form bacteriocins requires a bacteriocinogenic plasmid. Chemically, bacteriocins may be simple proteins, proteins and carbohydrates, or comparable to incomplete bacteriophage particles (phage tails). The mode of action differs among bacteriocins. Some appear to damage the cell membrane of susceptible hosts; others interfere with the synthesis of nucleic acid or protein. Colicin I, for example, is able to pass into the region between the outer and rigid layers of the cell wall (periplasmic space) of gram-negative bacteria. When the colicin comes into contact with the cell membrane, channels appear, allowing potassium ions to leave. This results in decreased membrane function and a loss of ATP production.

R FACTORS

Genes that render host cells resistant to antibiotics and other toxic substances are referred to as *R factors.* The number of these factors carried by plasmids has increased dramatically and includes resistance to almost all antibiotics and a variety of heavy metals such as mercury, cobalt, and cadmium. Many of the pathogenic bacteria encountered today were once controllable by the actions of a broad group of chemotherapeutic agents. However, they are now highly resistant to these substances because of R factors. Unfortunately, R factors have spread rapidly through bacterial populations by conjugative transfer and are now found in almost all pathogenic bacteria. Different combinations of resistance genes may be carried by plasmids. The number of such genes may range from one to eight. **[See Chapter 12 for additional details of antibiotic resistance.]**

METABOLIC PLASMIDS

Pseudomonas spp., commonly found in soil, have unique metabolic activities. They can break down xylene, hexane, phenol, and camphor. The complete set of enzymes in each catabolic pathway is synthesized under the control of genes on a plasmid. These plasmids may be transferred by conjugation or along with the sex-factor plasmid. When they are transferred to a recipient cell, that cell acquires an entire metabolic pathway.

TRANSPOSONS

Not all genes are fixed in their locations on chromosomes. Some can move about on the chromosome or on plasmids (Figure 6–27). Such movable genetic elements are called *transposons* (Tn) and are found in both procaryotes and eucaryotes. The movement of a transposon is possible due to the presence on the chromosome of a short segment that holds the genetic information for an enzyme system capable of cutting and reforming DNA fragments. Transposons appear to have no autonomous existence and can cause DNA deletions and inversions (changes in position). The combination of gene movement within bacteria and plasmid transfer between bacteria has resulted in the wide distribution of such genes (Figure 6–27).

Transposable elements are extremely important to the movement of genes that determine antibiotic resistance. R genes are examples of such transposons, and include Tn 1 and Tn 10, which individually code for ampicillin and tetracycline resistance. **[See Chapter 12 for a description of these antibiotics and antibiotic resistance.]**

Mapping the Bacterial Genome

Transformation, transduction, and conjugation techniques are all used to determine the location of individual genes on the bacterial chromosome. The

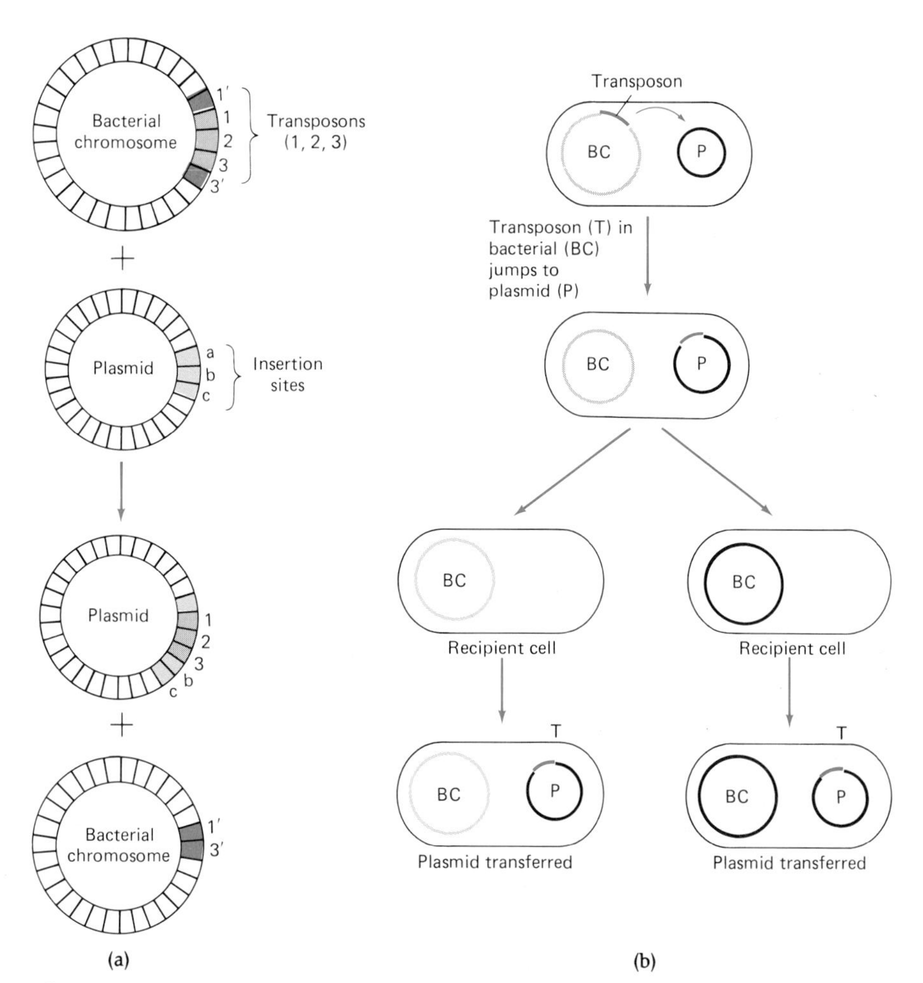

Figure 6–27 Transposons, the jumping genes. (a) An individual gene can move from one DNA molecule to another within a bacterial cell. (b) These jumping genes can be transferred among several members of a bacterial population.

characteristics transferred by transformation or transduction are usually very small linkage groups. Many very complex analyses of recombinant cells are, therefore, necessary to piece together enough information to map the genes. Although the process of mapping by conjugation is probably equally complex, it is much more easily described.

Linear transfer of DNA between the F^+ and F^- bacteria yielded information on genetic mapping in *E. coli.* Before extensive work was begun, it was known that with a particular F^+ strain, rates of recombination of approximately 1 in 1 million occurred. Unfortunately, this was not often enough to permit large-scale studies of conjugation. However, during these studies, mutants of F^+ strains, the Hfrs, were found that recombined at a rate of approximately 1 in 100. As indicated earlier, investigations with these mutant strains have shown that fertility in *E. coli* depends upon the presence of the F particle in the bacterium. In F^+ cells, the sex particle is present as an individual cytoplasmic unit. When an F^+ cell is mated with an F^-, the F^- cell receives this F particle at a high rate without the bacterial genome. Only in the case of Hfr is this F particle integrated with the genome, allowing other genes to be transferred to the recipient (Figure 6–29). Thus, it was found that the small rate of recombination observed in F^+ populations was caused by the presence of a small number of Hfr mutants in the population.

In 1955 a relatively simple procedure for genetic mapping analyses in *E. coli* was reported. After Hfr and F^- strains were mixed, conjugation could be interrupted by whirling the bacteria in a kitchen blender. Analysis of recombinants in the F^- bacteria indicated that the genome was indeed linear. This was demonstrated by breaking the mating pairs periodically and observing that, after certain intervals, additional genetic characteristics were expressed by the original F^- bacteria.

Figure 6–28 summarizes the conjugation process in *E. coli.* In the hypothetical experiment presented, Hfr and F^- strains of *E. coli* are mixed, and samples of the mixture are removed to a blender after selected time intervals. After blending to rupture the conjugation tube, the recombinants are cultured on media suitable for analyzing nutritional characteris-

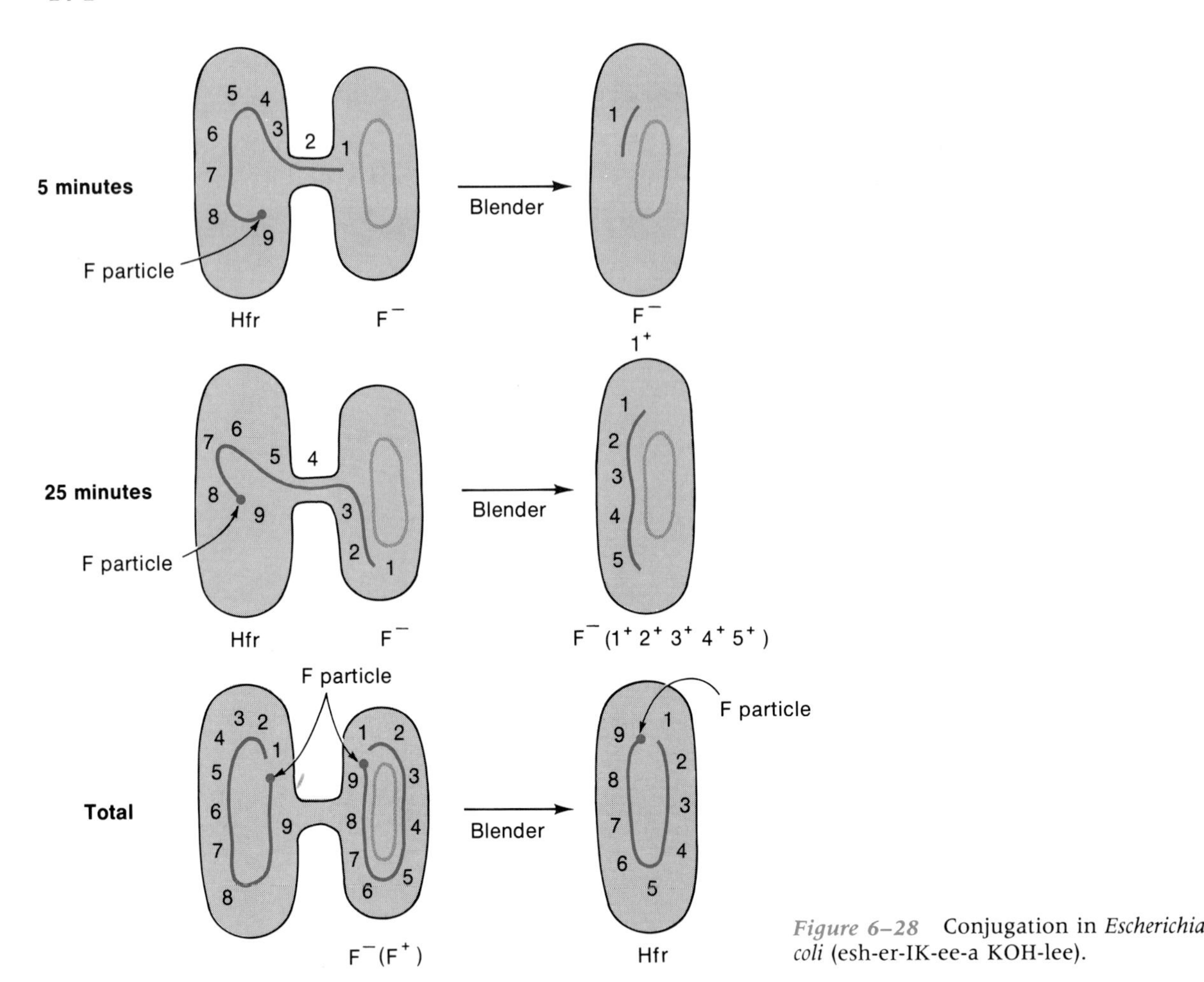

Figure 6–28 Conjugation in *Escherichia coli* (esh-er-IK-ee-a KOH-lee).

tics and characteristics of antibiotic and bacterial virus resistance. These genetic traits, or *markers,* are indicated in Figure 6–28 by numbers 1 through 9. The sexuality factor, or F, is always the last marker transferred. When it is transferred, the former F^- cell becomes either an Hfr cell or an F^+ cell.

The time intervals at which the mating bacteria were separated can be used to locate the genes on the chromosome. Suppose, for example, that in a hypothetical experiment using interrupted mating, one additional characteristic was transferred to the recipient cell during each five minutes of mating. By the end of the experiment, the F^- cells contained all the characteristics under study. The order of transfer would indicate the order in which the characteristics are located along the length of the chromosome.

Actual research has shown that the entire chromosome of *E. coli* is transferred in 90 minutes. Resistance to penicillin, for example, is transferred at 83 minutes; resistance to streptomycin at two points, 7 and 64 minutes; and formation of pili at 88 minutes.

Other types of analysis have proved that the bacterial chromosome is truly one circular DNA molecule. This circular cellular component of genetic information, which controls the destiny of *E. coli,* always breaks at the point of F-particle integration.

Gene Manipulation

Transformation, transduction, and conjugation are natural means by which asexual organisms produce DNA recombinations similar to those of sexual organisms. When these processes are performed in the laboratory rather than in nature, they are examples of human-directed *gene manipulation.* This section deals primarily with processes geared to produce recombinant DNA situations that might be considered unnatural, such as producing microorganisms with DNA of entirely unrelated microorganisms or of higher forms of life incorporated in their genetic makeup, most commonly as plasmids or bacteriophage DNA. These methods are referred to as **genetic engineering** or **recombinant DNA technology.** **[See Chapter 10 for descriptions of bacteriophages.]**

TABLE 6–5 Examples of Proteins Produced by Recombinant-DNA Engineering That Are Currently Approved for Therapeutic Use or for Testing

Protein	*Source of Genetic Material*	*Function(s)*
Growth hormone	Mammalian	Correction of growth defects
Insulin	Mammalian	Treatment of diabetes mellitus
Interferon (alpha)	Mammalian	Antiviral and anticancer agent
Interferon (gamma)	Mammalian	Anticancer and antiviral agent
Interleukin-2	Mammalian	Anticancer agent
Parathyroid hormone	Mammalian	Regulation of calcium metabolism
Serum albumin	Mammalian	Replacement fluids in transfusions
Tissue-plasminogen factor	Mammalian	Dissolving blood clots
Tumor necrosis factor	Mammalian	Anticancer agent
Urokinase	Mammalian	Correction of blood-clotting disorders
Virus protein coats for cytomegalovirus, influenza virus, and hepatitis B virus	Viral	Vaccine production for respective human diseases
Foot-and-mouth virus protein coat	Viral	Vaccine production

Potential Benefits

Various pharmaceutical products are being produced by recombinant DNA technology (Table 6–5). The recombinant process involves the construction of a hybrid plasmid *in vitro* (Figure 6–29) by inserting a fragment of human DNA into a plasmid capable of replicating in the bacterium *E. coli* or the yeast *Saccharomyces cerevisiae* (sak-a-roh-MY-sees se-ri-VISS-ee-eye). Since these microorganisms do not distinguish between foreign human genes and their own, they can manufacture the proteins or related substances specified by the foreign DNA. Because of the early successes of genetic engineering and its great poten-

Figure 6–29 Genetically engineered bacteria. (a) Scanning micrograph of *Escherichia coli* cells containing plasmids for the biosynthesis of human insulin proteins. Note the swollen regions. The insert shows normal cells not containing plasmids. Original magnification, 8,500×. (b) Transmission electron micrographs showing the involved regions of individual cells. Original magnification, 55,000×. *[D. C. Williams, R. M. Van Frank, W. L. Muth, and J. P. Burnett,* Science ***215****:687, 1982.]*

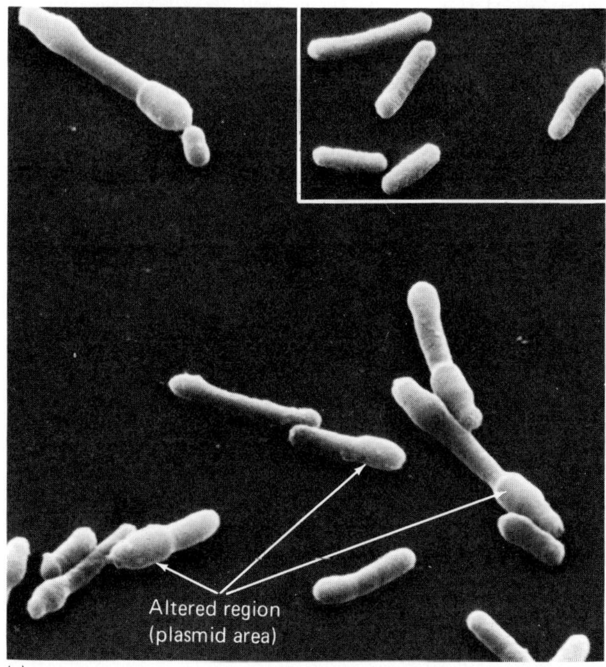

(a)

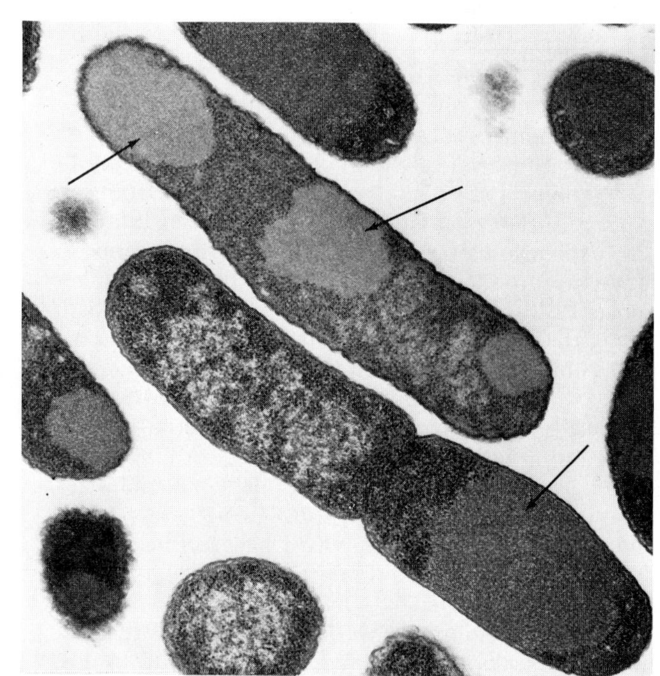

(b)

tial for improving the quality of life, applications of DNA-recombinant technology have been extended into areas such as synthetic vaccine production (Chapter 19) and agriculture. One agricultural application involves the use of recombinant organisms (chimeras) to place nitrogen-fixing enzymes into plants, which can then use nitrogen gas directly. Currently, nitrogen fixation is limited to some free-living bacteria in the soil and water and to some mutualistic bacteria in association with legumes and a few other types of plants. In nitrogen fixation, nitrogen gas (N_2) in the atmosphere is incorporated into protein of bacteria and plants and subsequently into the soil upon decomposition (see Chapter 14). Additional aspects of commercial and agricultural uses of genetic engineering are presented in Chapter 15.

Techniques

The basis of gene manipulations is the isolation of a specific gene from one cell, its enzymatic incorporation into a suitable carrier *(vector),* the introduction of the specific gene-carrying vector into an appropriate host cell, and the production of large quantities of the specific gene. This process is referred to as gene cloning (production of gene copies) and consists of four major steps:

1. Breaking and joining DNA from different sources.
2. Obtaining a suitable gene carrier that can replicate with attached foreign DNA, (a cloning vector).
3. Introducing the composite DNA into a bacterium or yeast cell.
4. Selecting the recombinant organism (chimera) produced from the large number of bacterial yeast cells exposed to the gene carriers.

Among the general goals of gene cloning is the establishment of a *clone library,* or collection of bacterial strains, each of which contains a specific separate gene or a cluster (group) of such genes from an organism of particular interest such as the human. The expression of cloned genes is of major importance to producing large amounts of the cloned gene's DNA or large quantities of the product coded for by the gene such as antibiotics or specific industrially important enzymes.

Two major biochemical factors are required for gene cloning. They are restriction endonucleases and cloning vectors such as plasmids, certain bacteriophages, and cosmids (bacteriophage-plasmid artificial combinations, or hybrids). Only restriction endonucleases and plasmids will be described here.

RESTRICTION ENDONUCLEASES

A restriction endonuclease is a unique DNase that not only breaks DNA at particular points but, in doing so, creates a condition that permits attachment of portions of foreign DNA cleaved by the same enzyme. The enzyme produces regions of unpaired bases at the ends of the DNA strands ("sticky ends"), as shown in Figure 6–30a. The *E. coli* restriction endonuclease called *EcoRI* is capable of splitting the DNA complementary sequences at the G–A and A–G

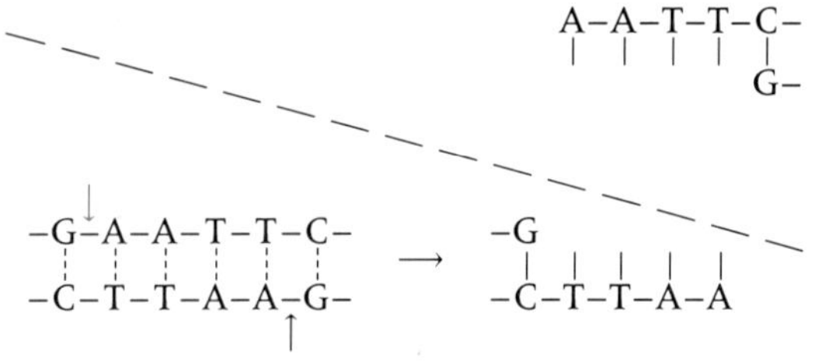

```
                                        A-A-T-T-C-
                                        | | | | |
                                                G-
  ↓
-G-A-A-T-T-C-         -G
 | | | | | |    →      | | | | |
-C-T-T-A-A-G-         -C-T-T-A-A
           ↑
```

bonds, thereby creating the sticky ends. The restriction endonuclease apparently functions by recognizing the $\begin{matrix} A-A-T-T \\ T-T-A-A \end{matrix}$ pattern, which has been called a *palindrome.* By definition, a palindrome is a word, verse, or sentence that is the same when read backward or forward.

The enzyme does not distinguish between different species in which that sequence is found and therefore produces the complementary single strands wherever the sequence is present (Figure 6–32b). This technique, or a similar one, has been used to produce composite DNA structures between *E. coli* and *S. aureus* as well as between *E. coli* and DNA from the frog *Xenopus laevis* (zee-NO-pus LEE-viss). When the pieces of foreign DNA are mixed with the opened plasmid, complementarity rules, and the appropriate single-stranded ends are joined to form the composite DNA, Figure 6–30c.

Over 200 restriction endonucleases have been characterized. They are designated according to the bacterial species in which they occur by a capital initial of the genus followed by the first two letters of the species expressed in lowercase. Occasionally letters or numbers are added to distinguish between various enzymes produced by the same species. Examples of this restriction-endonuclease naming system are *Eco*R1 *(Escherichia coli)* and *Hae*III *(Haemophilus aegyptus;* he-MA-fi-lus ee-GIP-tus*).*

PLASMIDS

Although, as described earlier in this chapter, plasmids are generally transferred by cell-to-cell contact (Figure 6–27b), transfer can be achieved in the laboratory by transformation procedures (Figure 6–31). Plasmids can replicate in appropriate host bacteria, thereby making additional copies of the foreign gene or genes. Some strains of bacteria may produce as many as 20 copies of the plasmid or composite plasmid and make 20 times as much gene product. This process has been called *gene amplification.* Another significant feature of plasmids is that the property they pass to the host cells enables the investigator to recognize and isolate the bacteria containing the desired gene or genes and eliminate the cells not containing the plasmid. An example of a suitable

Microbiology Highlights

GENETIC MACHINERY—MAKING DESIGNER GENES

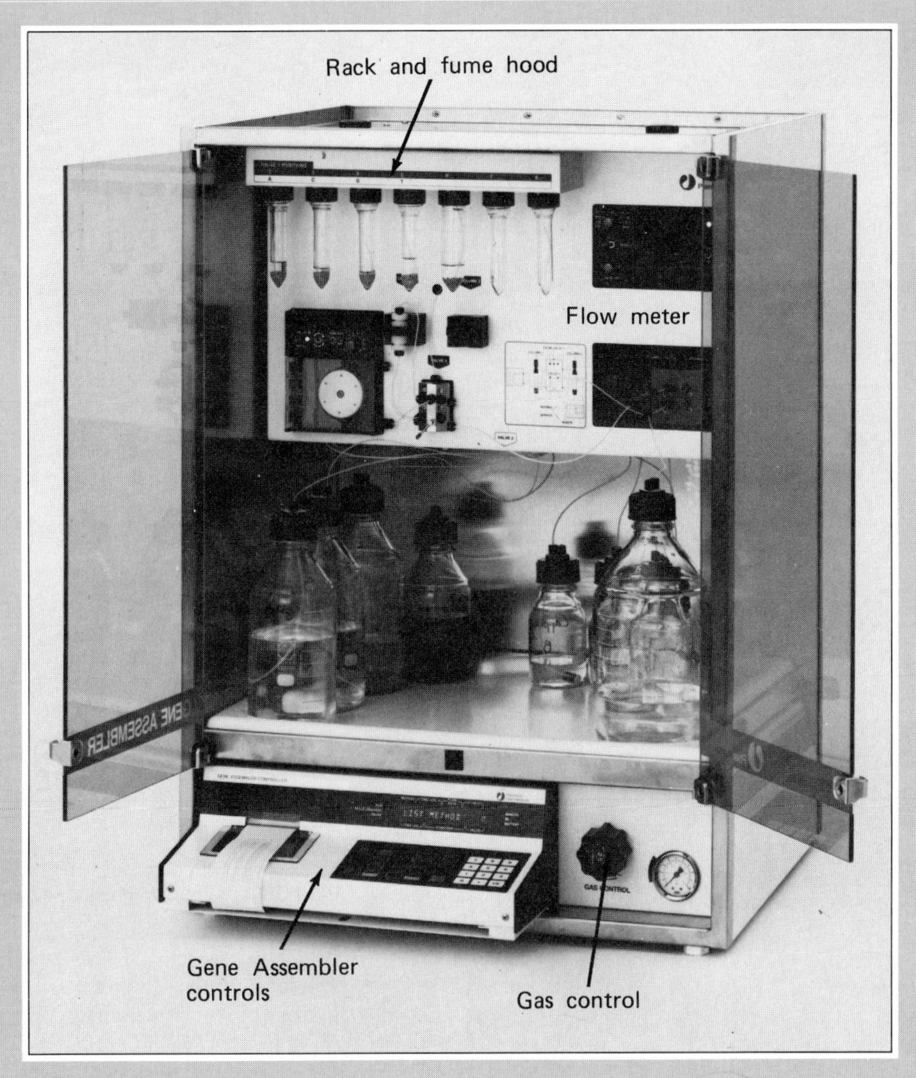

A fully automated DNA synthesizer, The Gene Assembler. *[Courtesy Pharmacia, Molecular Biology Division.]*

A number of automated genetic engineering instruments are on the market to synthesize DNA and to determine the sequence of amino acids making up any protein. All these devices are electronically controlled, miniature biochemistry sets capable of completing tasks in one day or less that had previously taken technicians or scientists four to eight months. Gene machines for DNA synthesis can modify naturally occurring genes or design new ones. Used in combination with a protein sequencer, information on a protein's amino acid sequence can be used to program a gene machine to synthesize a specific gene. When such a gene is inserted into bacteria, unlimited quantities of the associated protein can be manufactured for additional study or, possibly, therapeutic use. Two proteins, insulin and the antiviral interferon, are currently being synthesized in this manner (refer to Table 6–5). The production of hundreds of other medically and industrially important proteins no doubt will follow a similar path.

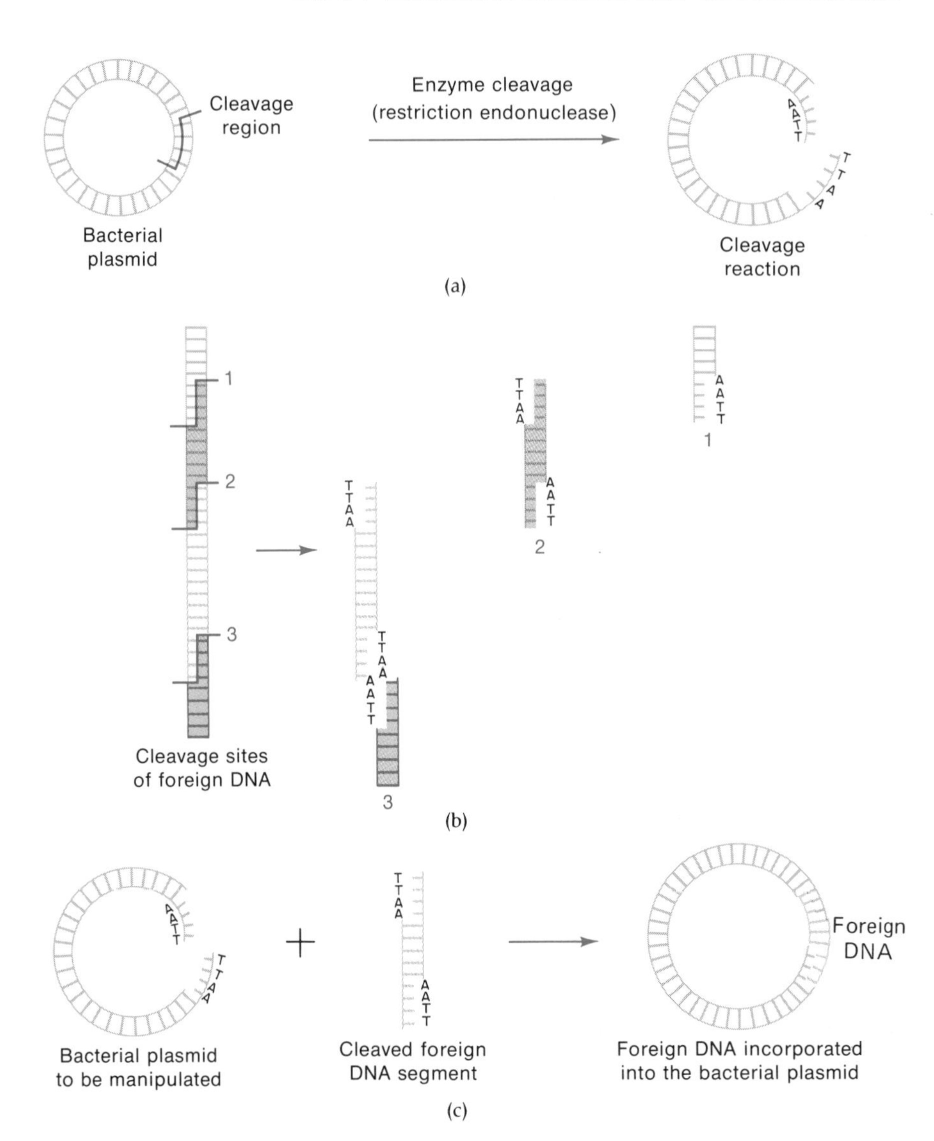

Figure 6–30 Incorporation of foreign DNA segments into a bacterial plasmid. This is an example of gene manipulation or gene splicing.

cloning plasmid for this purpose is pBR322 (Figure 6–32). This small plasmid, which replicates in *E. coli,* was artificially constructed as a cloning vector. It contains genes for ampicillin and tetracycline resistance and a number of sites susceptible to specific restriction endonuclease attack. Since the complete nucleotide base sequence of this plasmid is known, the location of a large number of restriction sites can be determined. It is important for a plasmid to contain only a single recognition site within a gene for at least one restriction enzyme. Thus treatment with a specific enzyme will open the plasmid and not break it into pieces. The plasmid pBR322 has several restriction sites for enzymes, including BamHI and PstI. The BamHI site is located within the gene for tetracycline resistance, and the PstI site is within the gene for ampicillin resistance. The insertion of a foreign DNA fragment into one of these sites will result in the loss of the antibiotic resistance determined by the site. This phenomenon, called *insertional inactivation,* is used to detect the presence of foreign DNA within a plasmid. Thus preparing pBR322 plasmids with foreign DNA inserted into the gene for tetracycline resistance (Figure 6–32b) for use in transformation procedures will result in the isolation of transformed bacterial recombinants (clones) that are either ampicillin-resistant and tetracycline-resistant or ampicillin-resistant and tetracycline-sensitive. The reason for the two types of clones is that not all plasmids will incorporate the foreign DNA. Thus only those cells showing tetracycline sensitivity contained the plasmids with inserted foreign DNA. Isolation of the desired recombinant bacteria can be achieved using agar plates containing the appropriate antibiotics.

E. coli has been used to produce clones of foreign DNA from *S. aureus* and *X. laevis,* as mentioned earlier. In addition it has been the recipient of DNA from the sea urchin, the fruit fly *Drosophila melanogaster,* mouse mitochondria, and yeast. This transformation capability is not limited to *E. coli.* It has been reported that plasmids from *E. coli* have been inserted into cowpea protoplasts, where they were later found to be stable extranuclear particles. A very exciting experiment showed that the fungus *Pseudosaccharomyces* (soo-doh-sak-a-roh-MY-sees) Tc-

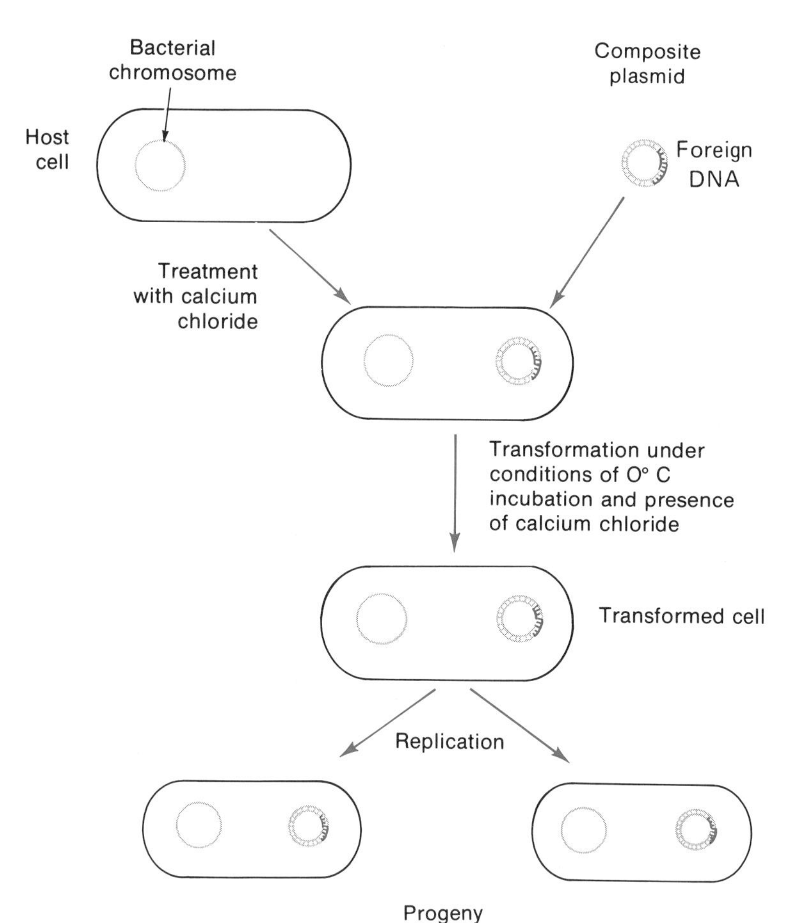

Figure 6–31 Transformation process for gene manipulation in *Escherichia coli.* Here a plasmid (foreign DNA) is introduced into a receptive host cell. The newly received DNA is replicated along with the host cell's chromosome and is passed on to future offspring (progeny).

Figure 6–32 The preparation of plasmid pBR322 as a cloning vector. (a) The structure of plasmid pBR322. (b) The scheme for the insertion of foreign DNA into the site within the gene for tetracycline resistance. Since the foreign DNA is not incorporated by all plasmids, two forms pBR322 are shown. Refer to the text for additional details.

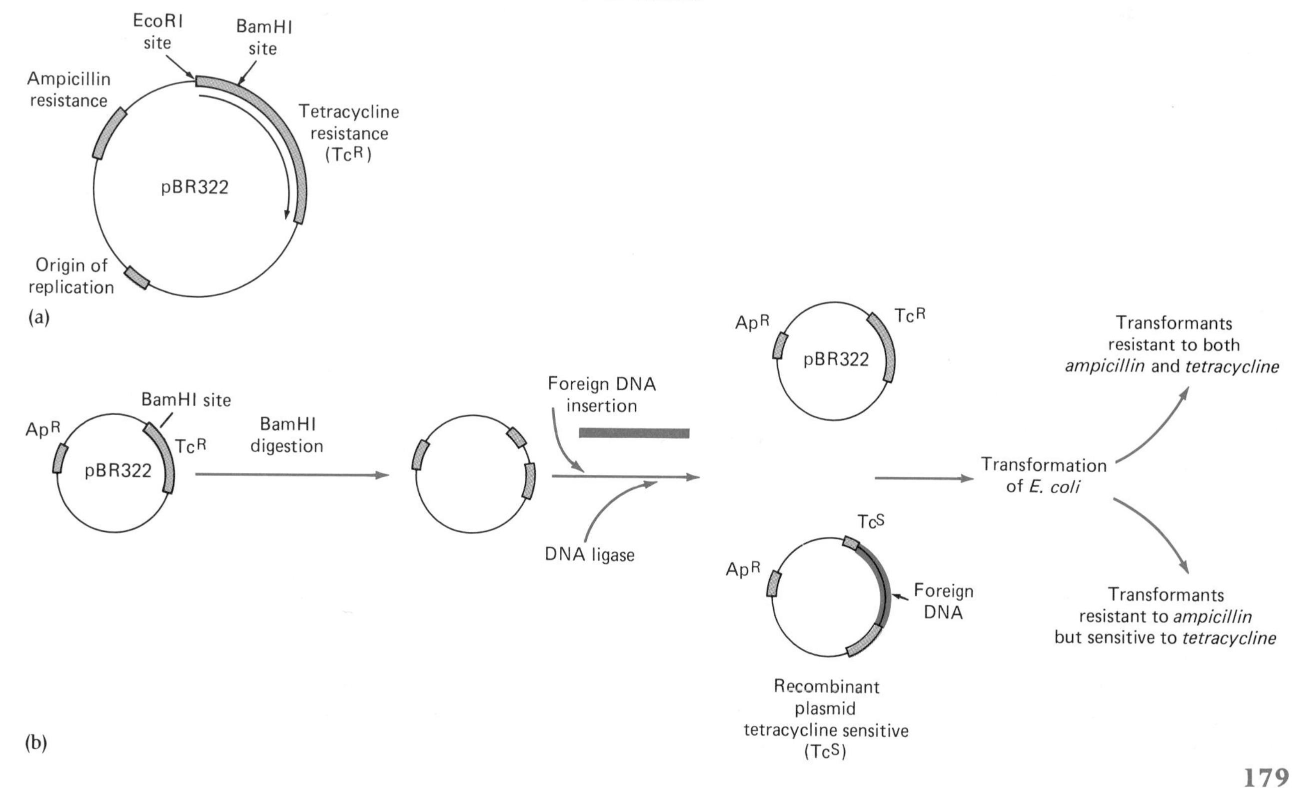

1176 could be transformed with extracted DNA from pancreatic beta cells and that these fungi were able to synthesize the biologically active insulin normally made by the beta cells. Thinumalachar, Narasimhan, and Anderson were awarded a patent in 1978 to make insulin by this process. **[See Chapter for other aspects of genetic engineering.]**

Potential Hazards

A number of potential hazards have been suggested; for example, it is feared that mixing genes between widely separated species of life forms might create some deadly monster. Recombinant organisms are called *chimeras* after the fire-breathing mythological monster with the head of a lion, the body of a goat, and the tail of a dragon. It might be possible to create a chimera that produces toxins and has a greater potential for spreading disease than the original bacterium. Let us assume that an irresponsible scientist were to create an *E. coli* chimera with the capability of making botulism toxin. Normally, *Clostridium botulinum*, an anaerobic soil organism, gains only occasional access to conditions in which its toxin can be produced, ingested, and ultimately cause food poisoning. However, a strain of *E. coli* with that capability normally living in the human intestine might more easily gain access to a suitable environment and might produce an epidemic. Some concern has also been voiced over creating bacterial chimeras with antibiotic resistance patterns not found in nature and chimeras of tumor or other animal viruses and bacteria. Imagine *E. coli* with the genetic information to cause cancer, rabies, viral encephalitis, or even severe birth defects such as those caused by the virus of measles (rubella).

The concern over such potential hazards led to the development of strict research guidelines in 1976 by the National Institutes of Health Recombinant DNA Molecule Program Advisory Committee. The guidelines include handling all bacterial chimeras as highly virulent pathogens, with appropriate protective clothing and laboratory isolation procedures. These strict guidelines have since been relaxed for the following reasons: (1) no problems, health or otherwise, have been associated with chimeras, and (2) the bacteria used, such as *E. coli* K12, appear to be unable to grow normally in the intestine or transfer plasmids to normal enteric bacteria. It is generally agreed that the chimera as a source of foreign DNA is at least a million times less hazardous than the original source of the DNA.

Viewing the Members of the Microbial World

Microorganisms, whether procaryotic or eucaryotic, are microscopic. Their respective sizes are frequently limited by requirements for minimal yet sufficient space in which to hold all genetic information and the biochemical apparatus such as ribosomes and enzymes. Despite the great variations in size and internal organization that exists among the cellular members of the microbial world, they all share in common the fact that many of their cellular functions are in or on associated structures, the *cell organelles*.

The next series of chapters will examine structures, functions, and chemical compositions of bacteria, fungi, and the protists composed of the protozoa and selected groups of algae. The last chapter in Part III deals with similar features of the noncellular viruses and the smaller infectious agents, viroids, and prions.

Summary

Evolution and Inheritance

1. Darwin proposed that spontaneously occurring variations and natural selection were the causes of evolutionary change.

2. Modern biology recognizes that variations in cells and organisms arise by permanent genetic changes in the chemical composition of an organism's DNA (mutations) and the influence that different environments have on an organism.

Heredity and Mutation

1. In 1868, F. Miescher isolated nuclein from white blood cell nuclei. Later, in 1881, several individuals proposed nuclein to be responsible for the transmission of hereditary traits.

2. The results of experiments conducted by F. J. Griffith in 1928 and J. L. Alloway in 1933 suggested that the genetic substance in bacteria was deoxyribonucleic acid (DNA).

3. A. D. Hershey and M. Chase, using bacteriophages containing radioactive ^{32}P in their DNA and ^{35}S in protein components, showed that DNA is the storehouse of genetic (hereditary) information.

Nucleic Acids

1. Two general kinds of nucleic acids are recognized, deoxyribonucleic acid (DNA) and ribonucleic acid (RNA).

2. Both types of nucleic acids contain the basic repeating units, called *nucleotides*.

3. Nucleotides consist of a pentose, purine, or pyrimidine and a phosphate group.

DEOXYRIBONUCLEIC ACID (DNA)

1. The DNA molecule consists, in most organisms, of a double helix (spiral structure).
2. The nucleotide bases in the molecule exhibit a specific form of pairing: adenine is always linked to thymine, as is guanine to cytosine.
3. Each strand extends in an opposite direction and has terminal phosphate groups at opposite ends of the double helix.
4. One strand of the DNA molecule has either a free hydroxyl at the 3′ or the 5′ carbon of the sugar molecule.
5. Specific nucleotide sequences contain hereditary information in genes. Chromosomes carry genes.
6. The word "genotype" refers to the numbers and kinds of genes, while "phenotype" refers to the physical expression of genes.

RIBONUCLEIC ACID

1. RNA is constructed of nucleotides similar to those found in DNA.
2. RNA differs from DNA by containing ribose instead of deoxyribose and the substitution of uracil for thymine.
3. The three major types of RNA are messenger RNA (mRNA), transfer RNA (tRNA), and ribosomal RNA (rRNA). All of these molecules play specific and different roles in protein synthesis.
4. All RNA molecules are enzymatically copied from DNA.

DNA REPLICATION

1. Replication of the DNA molecule involves the separation of its two chains, with each one bringing about the formation of a new complementary chain. DNA polymerase catalyzes the process.
2. The new and original chains then wind around each other, resulting in the formation of two new DNA molecules.
3. The method of information copying with each new molecule containing one of the original chains is called semiconservative replication.

HOW BACTERIA REPLICATE THEIR DNA

1. The circular DNA molecule of bacteria starts replication at a single site and moves out in two directions, thus creating two replication forks.
2. The replication forks move around the circular DNA, duplicating the molecule as they go.

RNA SYNTHESIS

1. Although the functions of the three major kinds of RNA are different, the mechanism of their formation is identical.
2. One strand of DNA functions as a template and DNA-dependent RNA polymerases catalyze RNA formation.

Protein Synthesis

The expression of hereditary information in all organisms takes place in two stages: *transcription* and *translation.*

THE GENETIC CODE

1. The DNA molecule contains hereditary information in the form of a coding sequence of three nucleotides.
2. The genetic code of DNA is transcribed into nucleotide sequences of mRNA. These sequences each consist of three nucleotides arranged in a triplet called *codon.*
3. A complementary three-nucleotide sequence on tRNA molecules, called an *anticodon,* attaches to a codon during protein synthesis.
4. The sequence code specifies the specific type and location of amino acids in a protein molecule.
5. Nonsense codons serve as stop signals in the process.

THE GENETIC CODE IN OPERATION

1. *Transcription* is the process by which the specific information necessary for the formation of a protein molecule is incorporated into the newly formed molecule.
2. The mRNA travels to a site within the cell where ribosomes decode the information contained within the molecule. This process is called *translation.*
3. After activation of specific amino acids, these building blocks of proteins are carried to the site of protein production by transfer RNA (tRNA).
4. Finished proteins are used by cells for enzymes or structures.

PROCARYOTIC AND EUCARYOTIC TRANSCRIPTION DIFFERENCES

1. Intervening segments, called *introns,* are found scattered within eucaryotic primary RNA transcripts. These introns do not code for amino acids in a protein molecule; they separate segments known as *exons* that contain codes for amino acids.
2. mRNA processing in eucaryotic cells removes the introns and joins exons in the mature form of mRNA used in the synthesis of proteins.

Control and Regulation of Metabolism

1. The cell coordinates its genetic capabilities with the availability of nutrients in its environment to form proteins.
2. Continuously formed enzymes are called *constitutive,* while those formed in response to environmental factors are referred to as *adaptive.*
3. Proteins formed in response to environmental factors are *inducible;* and those proteins inhibited by such factors are considered *repressible.*

OPERONS

1. An *operon* represents a mechanism for the regulation of enzyme production. Two types of enzyme systems are recognized, *inducible* and *repressible.*
2. An operon is a region of DNA that codes for a group of structural or adaptive genes responsible for enzyme synthesis in a metabolic pathway.
3. The genes of an operon are controlled by an operator, a promoter region, and a regulatory gene.
4. The operator controls structural gene transcription.
5. The promotor region initiates transcription by the entire operon; the regulatory gene controls the operator.

Mutation and Protein Synthesis

1. Mutations result from changes in the sequences of purines and pyrimidines in DNA strands.
2. Examples of mutations include substitutions, additions, and deletions.

3. Nonsense mutations result in the formation of incomplete proteins.
4. Various chemical and physical factors can cause mutations. Such mutagenic agents include mustard gas, nitrous acid, and UV light.
5. DNA polymerases repair damaged DNA. Photoreactivation also aids in the repair process.

Mutations in Microorganisms

1. Mutations are expressed as defects in enzymes that prevent specific normal metabolic activities from occurring.
2. Experiments involving bacterial resistance to viruses have provided strong evidence for the spontaneous, random nature of mutation and the Darwinian theory of natural selection.

DNA Transfer in Procaryotes

1. Genetic material can be exchanged between members of different bacterial genera and species. Such exchanges can occur by means of transformation, bacterial conjugation, and transduction.
2. *Transformation* results in a genetic change caused by soluble extracts of DNA.
3. *Bacterial conjugation* refers to the transfer of genetic material between two living bacteria that are in intimate physical contact.
4. *Transduction* is a process in which genetic material transfer is accomplished by a bacterial virus.
5. Notable changes occur in microorganisms as the result of virus infection. Such *lysogenic*, or *viral, conversions* are of medical importance and include toxin production and antigenic changes.
6. *Plasmids* are small extrachromosomal DNA particles capable of independent replication. Plasmids control nonessential properties of bacteria, including antibiotic resistance, the synthesis of bacteria-killing substances called *bacteriocins*, and certain metabolic activities.
7. Transposons are genes capable of changing their positions on both chromosomes and plasmids. These jumping genetic elements are found in both procaryotes and eucaryotes.

Mapping the Bacterial Genome

Transformation, transduction, and conjugation techniques can be used to determine the location of individual genes on the bacterial chromosome.

Gene Manipulation

1. Laboratory-controlled DNA recombination of bacteria is known as *gene manipulation*, also called *genetic engineering* or *recombinant DNA technology*.
2. The possible benefits of gene manipulation include the potential for improving sources of antibiotics, producing important hormones, and making animal protein for food by fermentation. One possible hazard is that of creating disease agents that are difficult or impossible to control, although experience has shown this possibility to be unlikely.

Questions for Review

1. Why were the investigations of Miescher and of Griffith significant to genetics?

2. What was the basis of the Hershey and Chase experiment with bacterial viruses that showed that DNA was the genetic material?

3. Indicate which of these structures and/or factors are directly associated with transcription or translation:
   a. activating enzymes
   b. sigma factor
   c. nonsense codons
   d. rho factor
   e. promoter
   f. P site
   g. RNA polymerase
   h. tRNA

4. Distinguish among the different types of RNA in relation to their respective locations in cells and functions.

5. Describe how DNA replicates.

6. What is the genetic code?

7. Describe the components and functioning of an operon.

8. Using the chemistry involved in transcription and translation, describe how UV light and nitrous acid can cause mutations.

9. List and distinguish among the different types of mutations.

10. What is the SOS signal? How does a bacterial cell respond to the signal?

11. What is the one-gene, one-enzyme hypothesis of Beadle and Tatum?

12. Define "minimal medium," "prototroph," and "auxotroph."

13. Explain Lederberg's indirect selection procedure for proving the spontaneous nature of mutation in microorganisms.

14. Define and distinguish among "transformation," "conjugation," and "transduction."

15. Compare transformation with conjugation and lysogenic conversion with transduction.

16. How can the bacterial chromosome be mapped using conjugation?

17. Define "plasmid," "transposon," "bacteriocin," "R plasmid," and "metabolic plasmid."

18. What is the function of a restriction endonuclease in gene manipulation?

19. Discuss the potential benefits and hazards of gene manipulation (genetic engineering).

20. Describe the general process of gene cloning.

21. Outline a procedure to obtain a plasmid with the property of tetracycline sensitivity.

Suggested Readings

ALBERTS, B., et al., *Molecular Biology of the Cell.* New York: Garland, 1983. *An in-depth discussion of the molecular basis of genetics.*

BECKWITH, J., J. DAVIES, and J. A. GALLANT (eds.), *Gene Function in Prokaryotes.* Cold Spring Harbor, N.Y.: Cold Spring Harbor Laboratory, 1984. *A series of articles covering research areas such as gene regulation, protein secretion, and transcription-translation.*

CHAMBON, P., "Split Genes." *Scientific American* **244:**60–71, 1981. *The differences between bacteria and higher forms of life in the processing of mRNA for protein synthesis are clearly presented in this article.*

COHEN, S. N., and J. A. SHAPIRO, "Transposable Genetic Elements," *Scientific American* **242:**40–49, 1980. *A well-illustrated and clear presentation of the mechanisms by which transposons join unrelated DNA segments and transfer groups of genes among plasmids, viruses, and chromosomes in living cells.*

EIGEN, M., W. GARDINER, P. SCHUSTER, and R. WINKLER-OSWATITSCH, "The Origin of Genetic Information." *Scientific American* **244:**88–118, 1981. *This article discusses possible ways in which early RNA genes interacted with proteins and how the genetic code developed.*

MOTULSKY, A. G., "Impact of Genetic Manipulation on Society and Medicine." *Science* **219:**135–140, 1983. *Recent discussions of the social and ethical issues raised by recombinant DNA techniques are presented.*

NOMURA, M., "The Control of Ribosome Synthesis." *Scientific American* **250:**102–114, 1984. *A discussion of the structure and function of the ribosome and how ribosome assembly is linked to the needs of the cell.*

NOVICK, R. P., "Plasmids." *Scientific American* **243:**102–127, 1980. *Broad coverage is given to these accessory genetic factors in bacteria, which are best known as carriers of antibiotic resistance and as vehicles for genetic engineering.*

WATSON, J. D., J. TOOZE, and D. T. KURTZ, *Recombinant DNA: A Short Course.* San Francisco: W. H. Freeman, 1983. *A good introduction to the theory and practice of genetic engineering.*

WEINBERG, R. A., "A Molecular Basis of Cancer." *Scientific American* **249:**126–144, 1983. *A discussion of cancer genes (oncogenes) and the molecular changes associated with them.*

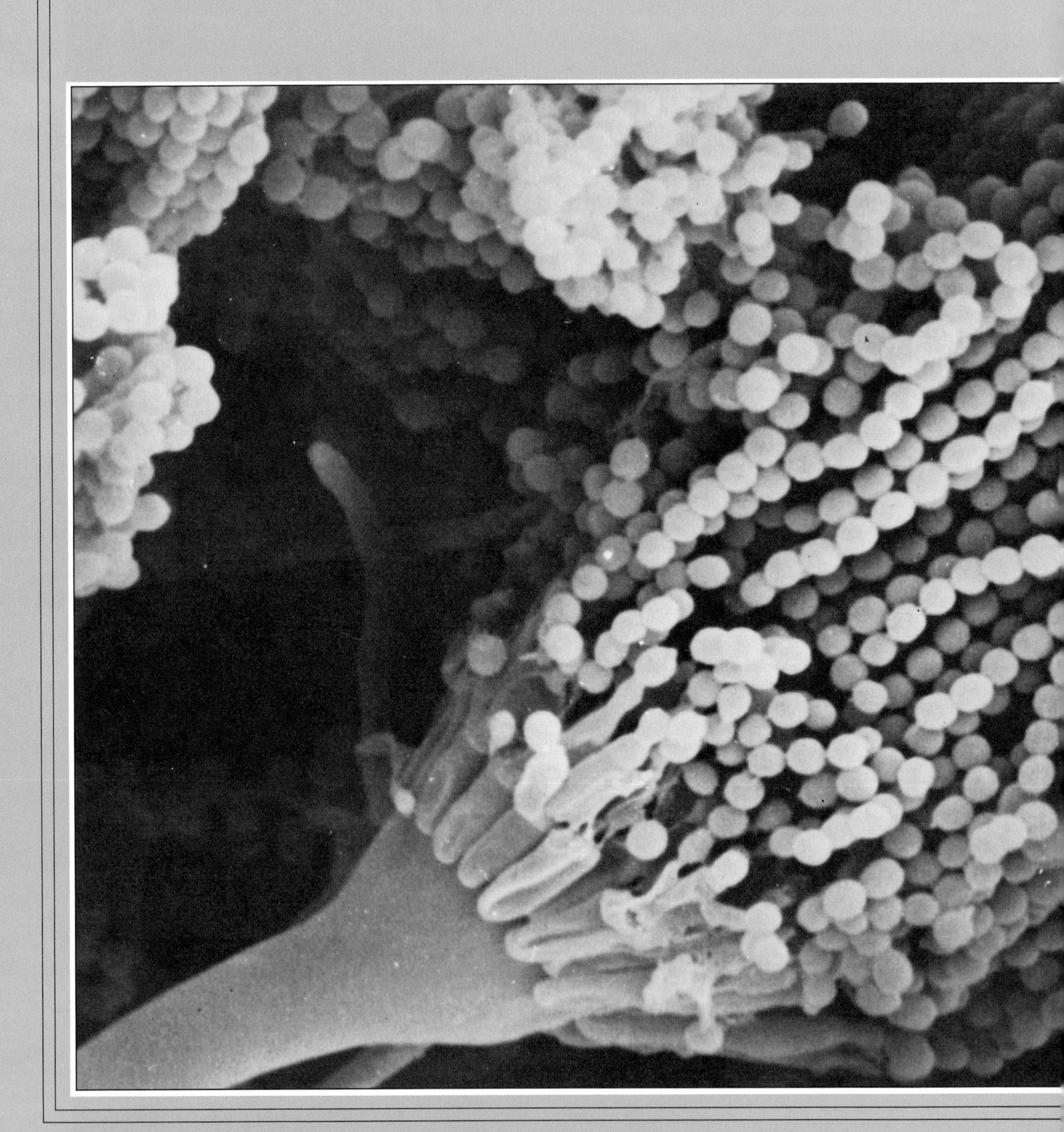

PART III

The World of Microorganisms

A microscopic view of the fungus **Aspergillus terrus** *showing spores and other components. [Courtesy M. S. Azab, P. T. Peterson, and T. W. K. Young.]*

7

The Procaryotes: Their Structure and Organization

The numerous and fundamental differences between eucaryotic and procaryotic organisms . . . have been fully recognized only in the past few years. In fact, this basic divergence in cellular structure which separates the bacteria and the blue-green bacteria from all other cellular organisms probably represents the greatest single evolutionary discontinuity to be found in the present-day living world. — R. Y. Stanier, E. Adelberg, and M. Douderoff

After reading this chapter, you should be able to:

1. Describe the size range, organization, cellular arrangements, and distinctive structures and associated functions of procaryotes.
2. Compare the biochemical, structural, and functional properties of procaryotes with those of other microorganisms such as the protists, fungi, and viruses.
3. Identify the cellular structures of bacteria in different types of micrographs.
4. Identify the photosynthetic machinery of procaryotes.
5. Compare the features of a gram-positive bacterium with those of a gram-negative bacterium.
6. List and explain the different possible arrangements of bacterial flagella.
7. Distinguish among the resting structures found in the procaryotes.
8. Outline the processes of bacterial sporulation and germination.

This chapter presents a detailed account of procaryotes, including their shapes, organization and arrangement, and distinctive structures. Throughout, emphasis is on those features that distinguish bacteria from other types of microbial life.

Bacteria, first discovered by Anton van Leeuwenhoek, are among the most widely distributed forms of life. They are found in air, water, the upper layers of soil, and internal and external regions of the human body and lower animals and plants. Well over 1,700 species are known at the present time. As later chapters will show, these procaryotes are almost completely at the mercy of their environment. Temperature, the availability of suitable nutrients, and the presence of toxic substances greatly affect the activities and survival of bacteria and other microorganisms. This chapter describes the structural organization of procaryotes. In this text, the procaryotes include the cyanobacteria. (Chapter 14 presents the features of these bacteria and another somewhat related group, the archaeobacteria.)

Bacterial cells are distinguished by morphological features such as size, shape, patterns of cell arrangement, and ultrastructure, (the fine structural detail of the cell). Many of these properties are important in identifying particular bacterial species and in correlating various structures with the overall functioning of an organism and its responses to different environments.

Figure 7–1 The relative size of a bacterium compared to a eucaryotic cell (left) and its parts. Structures are drawn to approximate size. (a) A portion of a generalized plant cell. (b) Gram-negative bacterium; a bacterial virus (bacteriophage) shown close to the bacterial cell wall is approximately five times its normal size. *[After W. Van Iterson,* Bact. Rev. ***29****:299–325, 1965.]*

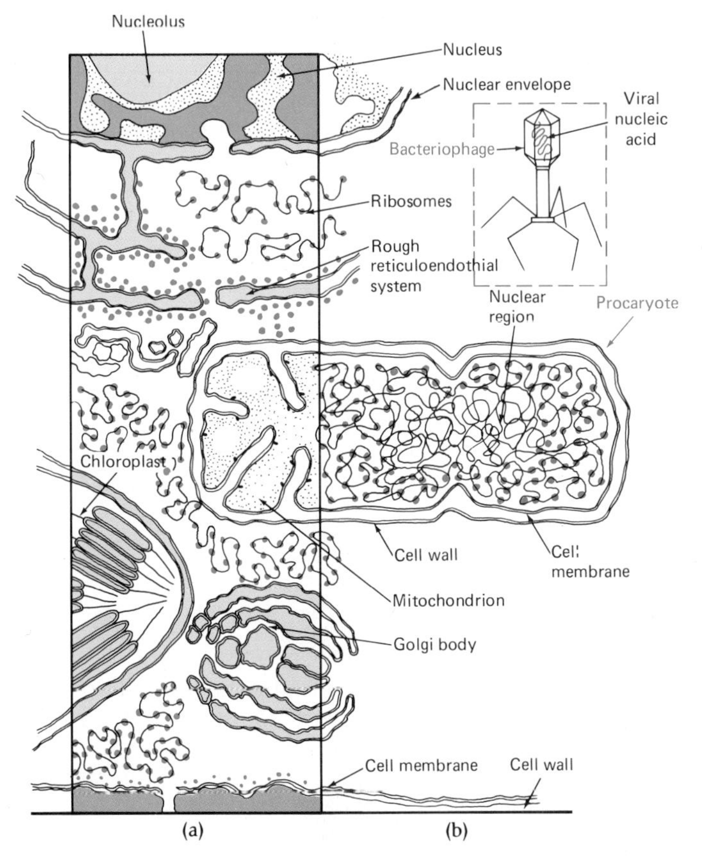

Bacterial Size

The small size of bacterial cells is obvious from microscopic examination. The dimensions of the smallest procaryote species border on the limits of resolution of the bright-field microscope. Most disease-causing bacteria range in size from approximately 0.2 to 1.2 μm in diameter and 0.4 to 14 μm in length.

The relative size of bacteria compared to the cells of higher forms of life and to a bacterial virus (bacteriophage) are shown in Figure 7–1. Note that bacterial cells are about the size of mitochondria and chloroplasts, organelles found with eucaryotic cells.

Shapes and Patterns of Arrangement

Most individual bacteria have one of three shapes: spherical, rodlike (cylindrical), or spiral. A new morphological type of procaryote, the box or square, was described in 1980 by A. E. Walsby. Square cells appear as flat rectangular boxes with perfectly straight edges (Figures 7–2 and 7–3d). Small cells measure 2 by 2 μm and are 0.25 μm thick, and may occur in pairs or associations containing as many as 64 cells.

Spherical bacteria, called **cocci** (singular, **coccus**), can have a wide variety of arrangements (Figure 7–2 and Color Photograph 12a). Among the most common are the following:

1. Diplococci: pairs of cells
2. Streptococci: chains of four or more cells
3. Tetrads: four cocci in boxlike, or square, arrangements
4. Sarcinae: cubical packet consisting of eight cells
5. Staphylococci: irregular, grapelike clusters of cocci (Figures 7–2 and 7–3a and Color Photograph 11a).

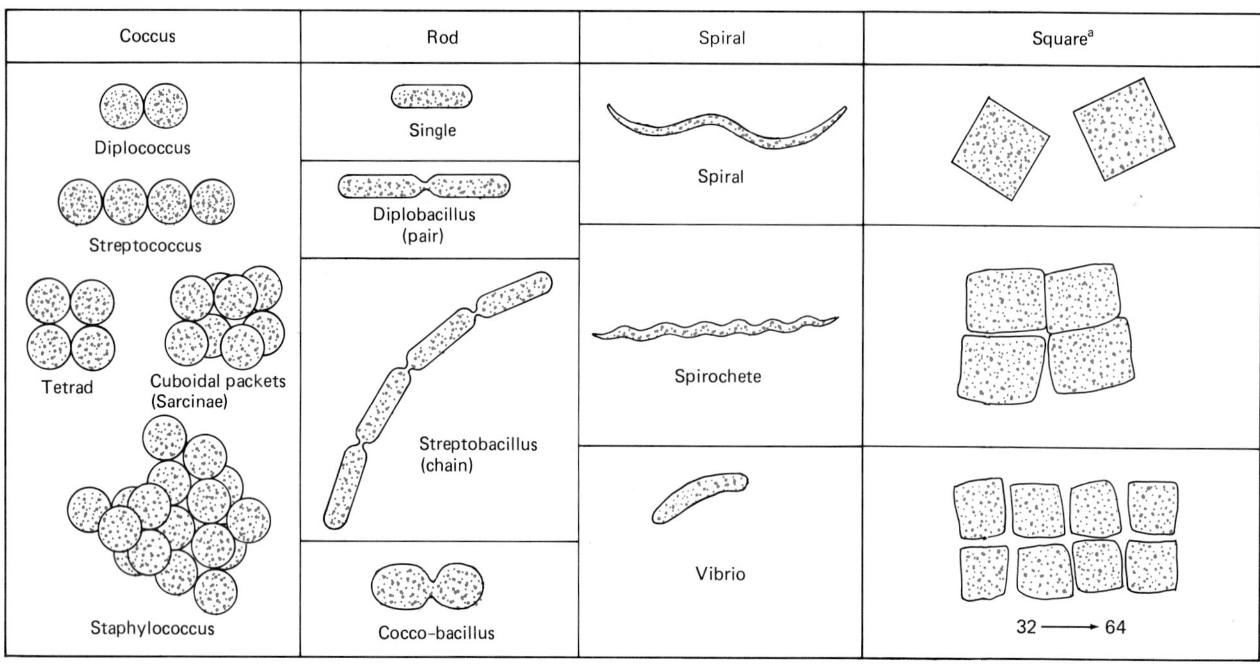

[a]The term *arcula* (from the Latin), meaning small box, has been proposed for these bacteria.

Figure 7–2 Characteristic bacterial cell shapes and morphological arrangements.

Rodlike bacteria, or *bacilli* (singular, *bacillus*), are cylindrical or relatively long ellipsoids (Figures 7–2 and 7–3b and Color Photograph 11a). These cells may occur singly, in pairs (diplobacilli), or in chains (streptobacilli). The cells of certain species appear to be small, rounded rods difficult to distinguish clearly from cocci. Such cells are referred to as *coccobacilli* (Figure 7–2).

Certain bacterial species may produce unique groupings of their cells; *Corynebacterium diphtheriae* (ko-ri-nee-back-TI-ree-um dif-THI-ree-ah) is one example. The arrangement of its cells, resembling a picket fence (palisading), is well known. Spiral organisms, known as *spirilla* (singular, *spirillum*), exhibit significant differences in the number and fullness of spirals and the length and rigidity of the spiral turns, or coils, depending on the species. The *Vibrio* species (Figure 7–2) are bacteria that consist of only a portion of a spiral. Other organisms possess several loose turns (Figure 7–3c). The agent that causes syphilis, *Treponema pallidum* (tre-poh-NEE-ma PAL-li-dum) (Color Photograph 62), has the coiled appearance of a corkscrew.

Structures and Functions

Before microbiologists could begin to investigate the molecular architecture of cells and the functional interrelationships of cellular components (Figure 7–4), they needed to know the chemical composition of the cell and the arrangement of its parts. This information came from extremely sensitive microanalytical chemical procedures and from electron microscopic examinations of intact cells and isolated intracellular components. **[See Chapter 3 for electron microscopic techniques.]**

Several procedures are used to disrupt microorganisms and obtain the cell components to be studied. In general, mechanical means of cell disintegration are successful for such purposes. Techniques of this type include (1) grinding or violent agitation with abrasive materials, such as alumina, glass beads, or sand; (2) pressure cell disintegration by forcing cells under pressure through a cooled needle valve; and (3) sonic and ultrasonic disintegration. The centrifugation and washing of resulting preparations remove unwanted substances. The separation and isolation of particular cellular components—such as cell membranes, cell walls, enzymes, and ribosomes—can be achieved through the use of different centrifugation speeds *(differential centrifugation)*.

The Bacterial Surface

In recent years, the surfaces and structures of bacterial cells (Figure 7–4) have been the focus of much attention. Many intriguing studies have furthered our understanding of transport across surface barriers, formation and function of surface structures, effects of antibiotics, and the roles of bacterial surface components in the causation, diagnosis, and prevention of infectious diseases. Certain surface

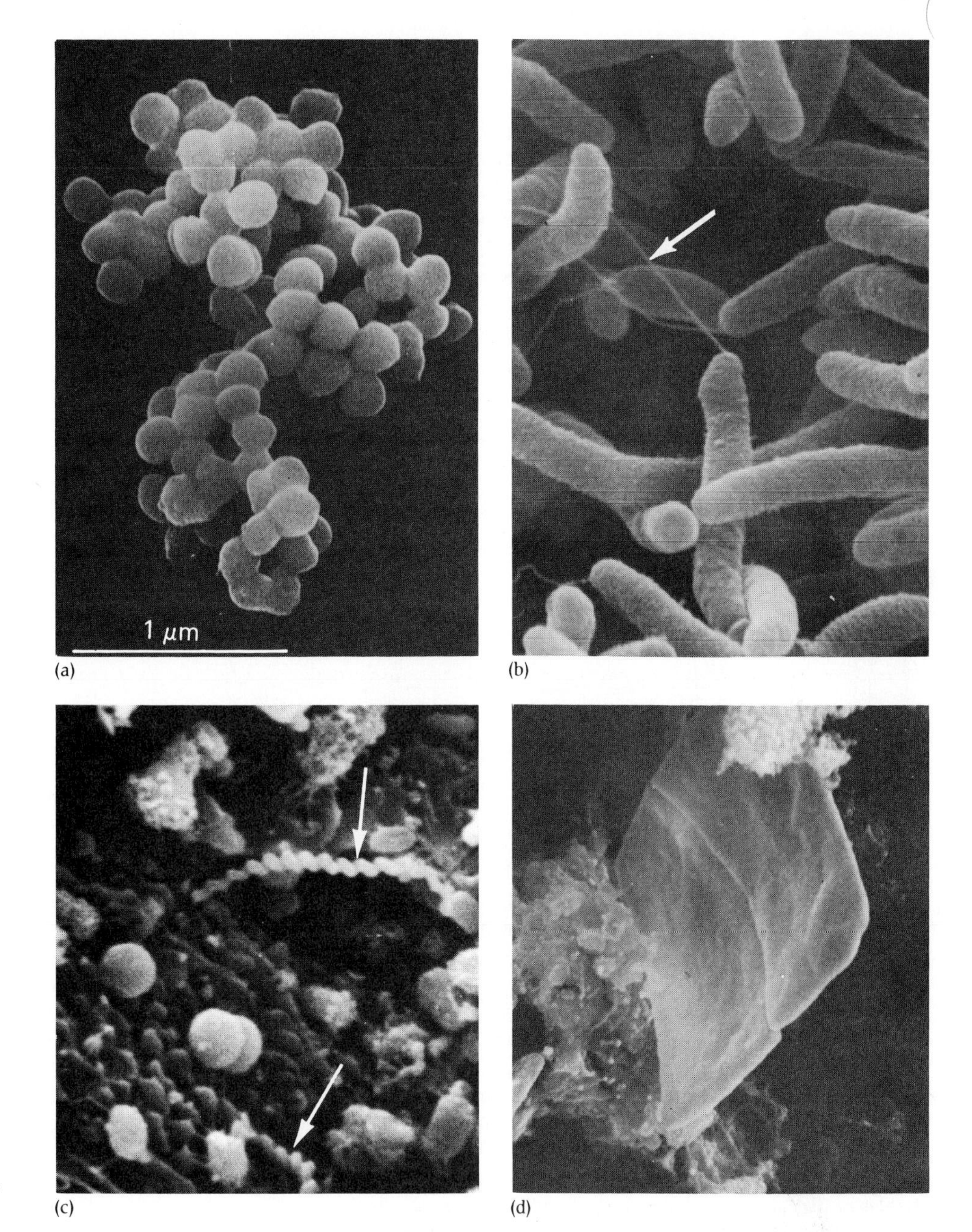

Figure 7–3
Scanning micrographs of morphological arrangements. (a) The staphylococcal pattern. *[M. Yamada,* et al., J. Bacteriol. ***123:****678–686, 1975.]* (b) Rods. Note the flagella on some cells (arrow). *[Y. Tago, and K. Aida.* Appl. Environ. Microbiol. ***34:****308–314, 1977.]* (c) The corkscrew spirochete, (arrows). *[T. Banchop,* et al., Appl. Microbiol. ***30:****668–675, 1975.]* (d) The newly discovered square bacteria. *[W. Stoeckenius,* J. Bacteriol. ***148:****352–360, 1981.]*

structures also play important roles in enabling bacteria to remain in favorable environments or to migrate from unfavorable ones. This section describes the currently recognized surface and closely related structures. They are (1) surface appendages, namely, the **flagella, axial filaments, pili, spines,** and **spirae;** (2) the **glycocalyx,** which includes S layers and capsules; (3) **cell walls;** and (4) **protoplasts** and **spheroplasts.**

Surface Appendages

The bacterial surface-associated structures—the flagella, pili, spines, and spirae (Figures 7–5 to 7–10)—differ both in function and in overall appearance. However, they do share certain common properties and participate in the organism's response to the environment.

FLAGELLA

The surface filaments known as **flagella** (singular, **flagellum;** Figure 7–5) are responsible for the motility of most bacteria. Some other bacteria use different structures for movement, including axial filaments. In addition to being observed under the microscope, motility can be demonstrated in various semisolid agar-containing media. Motile organisms are recognized by the visible spread of their growth pattern throughout the medium.

Motility can be significant in identifying a bacterial species. However, care must be taken to distin-

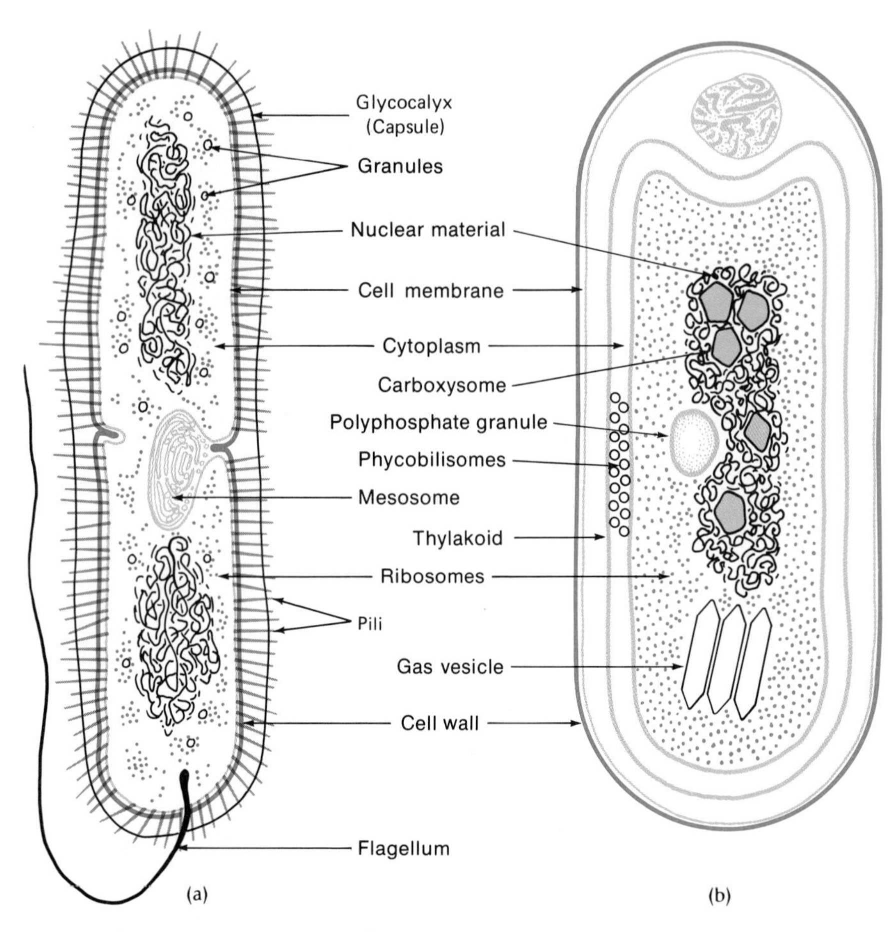

Figure 7–4 Two procaryotic cell types. (a) The ultrastructural features of a general bacterial cell. (b) A blue-green procaryote (cyanobacterium). Certain structures—such as capsules, flagella, mesosomes, pili, and photosynthetic apparatus—have not been found with all procaryotes.

guish true movement from the quivering to-and-fro motion known as **Brownian movement.** The latter is caused by a bombardment of the bacteria by molecules of the fluid in which they are suspended.

The presence of flagella and their associated activity do not necessarily correlate with other bacterial physiological properties. However, it appears that flagellation does bear a direct relationship to growth rate. Several factors may affect flagellation, among them the chemical composition of the medium, the pH, and the liquid or solid state of the medium. More flagellation occurs in liquid preparations.

The existence of locomotor organelles such as flagella had been suspected since the mid-1800s. Yet it was not possible to see them without the aid of electron microscopy and the special flagella stain developed by Einar Leifson that is used in light microscopy (Color Photograph 16).

Flagellar Shape and Arrangement. Flagella are extremely delicate structures that are readily detached from their bacterial cells. In a stained preparation (Color Photograph 17) they are long and slender, with an undulating shape. Their shape is fairly uniform for most bacterial species.

The thickness of flagella varies from species to species, ranging from 12 to 15 nm, much thinner than those of eucaryotic structures (Figure 7–6).

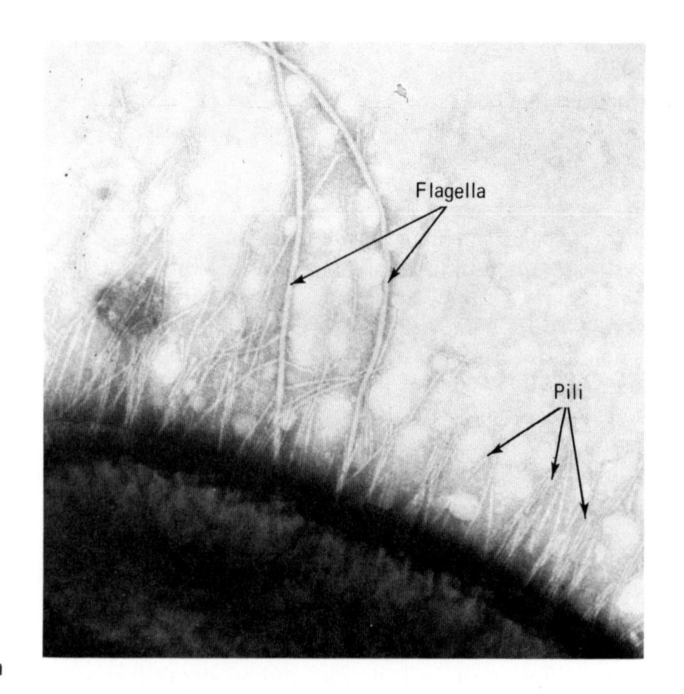

Figure 7–5 A comparison of surface appendages. A negatively stained preparation showing a clear distinction between the thin pili and the thicker flagella of *Proteus* (PRO-tee-us) spp.

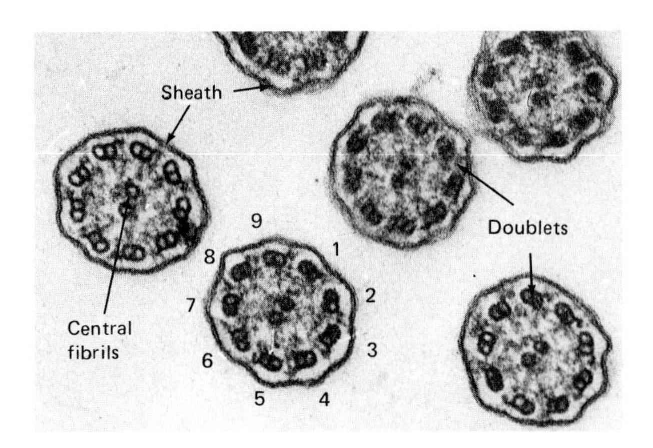

Figure 7–6 An electron micrograph of a cross section through the cilia of a eucaryotic ciliated cell. Note the presence in each cilium of an outer ring of nine pairs of fibrils surrounding two centrally located fibrils, a characteristic of higher animals and plants. Procaryotic flagella have dimensions similar to those of the two central fibrils. Original magnification approximately 100,000×. *[Courtesy Drs. B. A. Afezelius and R. Eliasson.]*

They vary in number and arrangement (Figure 7–7). Some organisms possess no flagella, a condition referred to as *atrichous;* others may have one *(monotrichous)* or several *(multitrichous).* The flagella may also be arranged in several ways. The tufted arrangement of Figure 7–7b is *lophotrichous* flagellation; distribution of flagella all around the cell is *peritrichous* flagellation.

Organisms having peritrichous flagella can spread in large numbers over media surfaces. Such spreading zones involve the movement of bullet-shaped microcolonies, or rafts (Figure 7–8).

FLAGELLAR ULTRASTRUCTURE. Bacterial flagella are much thinner than the cilia of vertebrates or the flagella of protozoa. A typical flagellum is relatively uniform in diameter along its length. However, where it attaches to a *basal granule, or body,* just be-

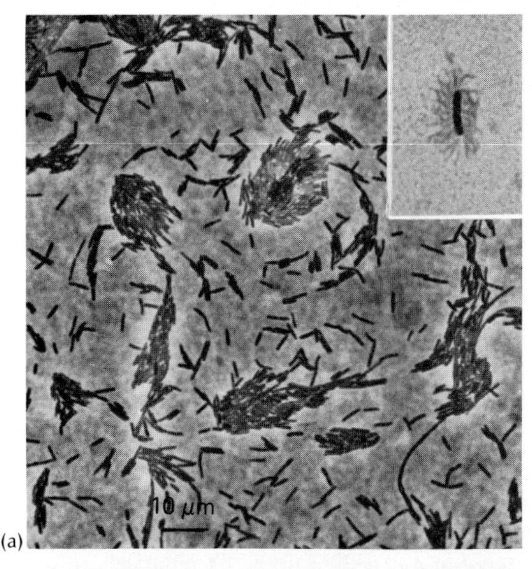

(a)

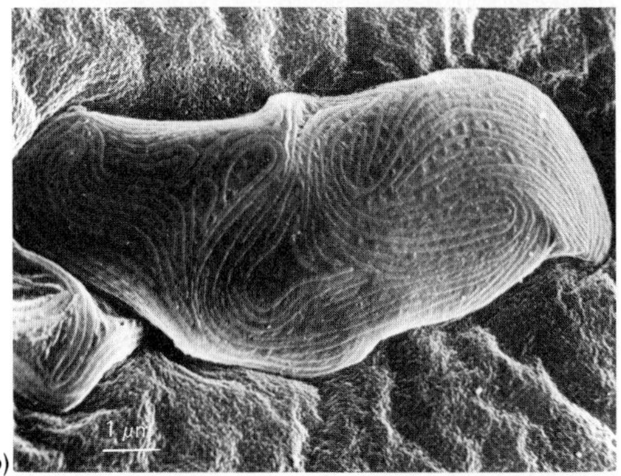

(b)

Figure 7–8 The swarming of bacteria. (a) A light micrograph showing the outer swarming zone. The inset is of a heavily flagellated cell from this zone. *[J. Henrichsen,* Bacteriol. Revs. ***36****:478–503, 1972.]* (b) A scanning micrograph of swarming organisms. Note the large number of cells. *[Reproduced by permission National Research Council of Canada from F. D. Williams, and G. E. Vandermolen,* Can. J. Microbiol. ***23****:107–112, 1976.]*

Figure 7–7 Bacterial flagella. (a) *Pseudomonas diminuta* (soo-doh-MOH-nass dim-EE-noo-ta) has a single flagellum at one end (polar monotrichous flagellation). The wavelength of the flagellum is quite short and very unusual. (b) A freshwater isolate, *Spirillum* (spy-RIL-lum) sp. representing the tufted, or lophotrichous, form of flagellation. (c) *Flavobacterium* (flay-voh-back-teer-EE-um) sp. showing flagellation all around the cell (peritrichous). The magnification is the same for all preparations. Preparations were stained by the Leifson flagella-staining procedure. *[Micrographs courtesy of E. Leifson.]*

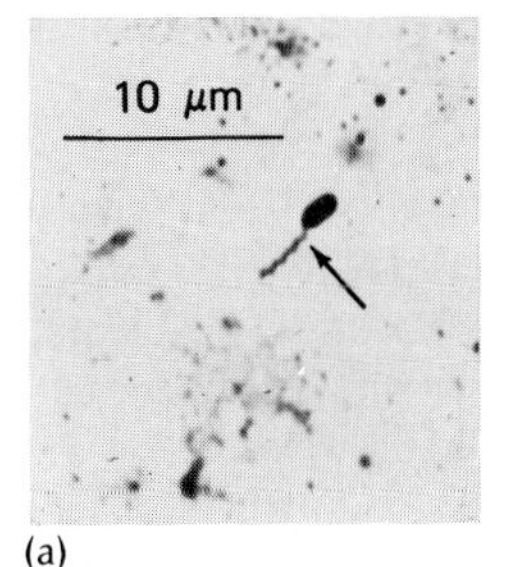

(a)

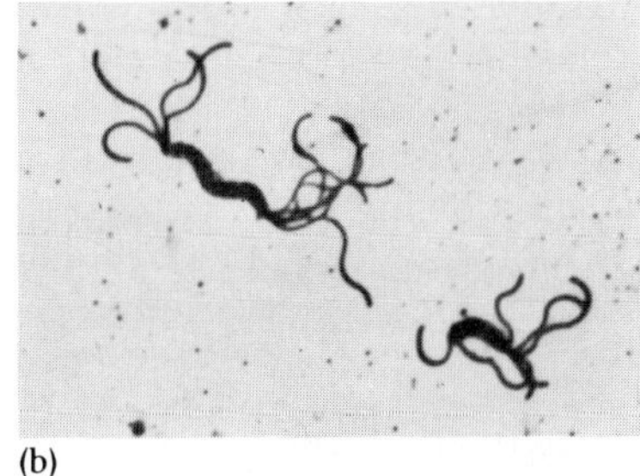

(b)

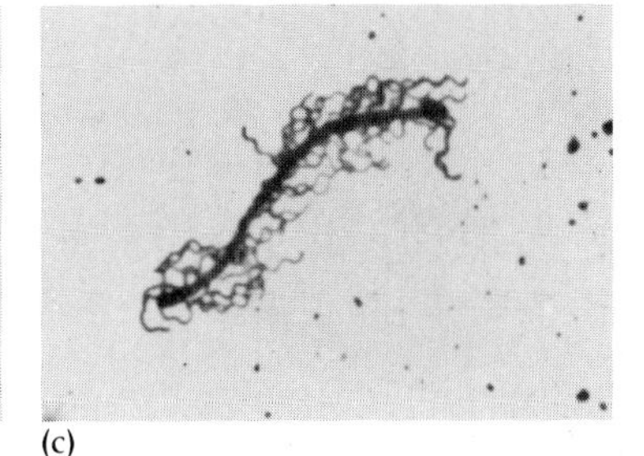

(c)

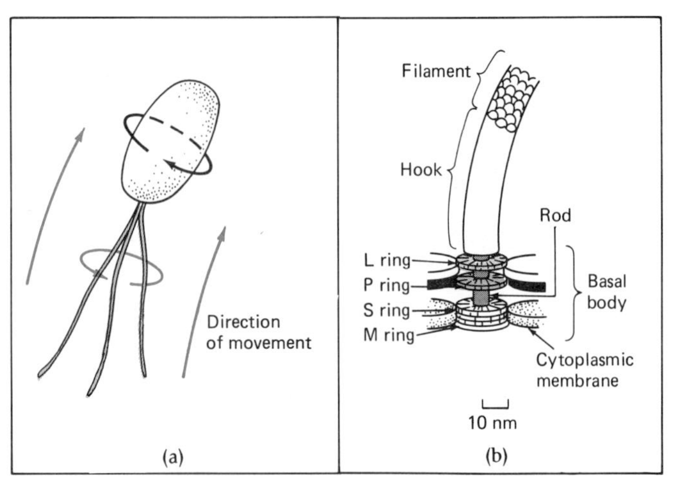

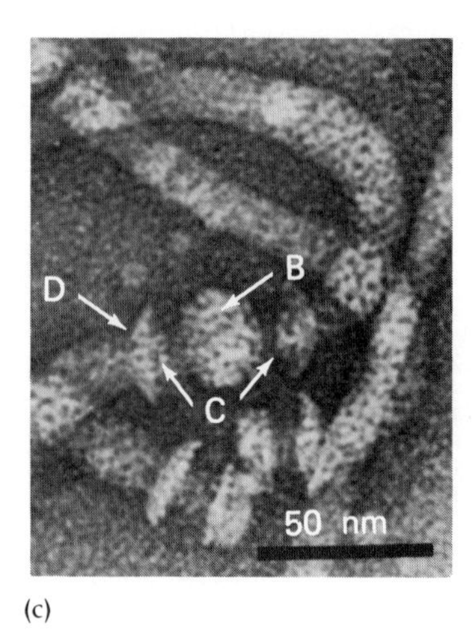

(c)

Figure 7–9 Bacterial flagella. (a) The mechanism of flagellar motion. The flagellum pushes the organism. The flagellum and the bacterium rotate in opposite directions. (b) The parts and attachment of a flagellum to the bacterial cell. In gram-positive bacteria, only the lower (S and M) rings of a basal body are present. Apparently, the upper pair is not required to support the rod portion of the flagella as it passes through the relatively thick cell wall. This difference is significant, because it implies that only the S and M rings are essential for flagellar activity. (c) A micrograph showing the relationship of a basal body to its flagellum. Note the presence of a collarlike structure (C) on the flagellum and a narrowing region where the flagellum is attached to the basal body (B). A disc-shaped structure can be seen there (D). Bar marker = 50 nm. *[From L. S. Thomashow, and S. C. Rittenberg,* J. Bacteriol. ***163:****1038–1046, 1985.]*

neath the cytoplasmic membrane, a thickened, hook-shaped portion, the *basal hook,* can be observed (Figure 7–9). Flagella are believed to originate in the basal body. In negatively stained preparations of some species, basal flagellar ends are found to be connected to a broadened body, called a *collar,* which in turn is connected by a constricted region, or neck, to a disc- or cup-shaped part (Figure 7–9b). The latter structure, which may have a paired disc appearance, is believed by some microbiologists to be a detached section of the basal body. The collar is believed to be a cell wall fragment. It should be noted here that a flagellum originates in the cytoplasmic region and pierces the cell wall as it emerges from the bacterium. It is *not* part of the cell wall.

Electron micrographs show that the flagella of bacteria consist of three parallel protein fibers intertwining in a triple helical structure. These fibers are composed of a protein called *flagellin.* The molecular weight of flagellin is relatively low, approximately 20,000 to 40,000. An amino acid not found elsewhere has been identified in this protein compound. It is ϵ (epsilon)-*N*-methyl-lysine. Although there is a similarity in the amino acid composition and molecular weight among the flagellins of different bacterial species, these compounds are by no means identical. This fact is demonstrated by the immunological specificity, that is, the production of different antibodies in response to flagella preparations from different bacterial species, subspecies, and strains. Such differences are important in the identification of certain pathogens.

The movement of flagellated bacteria is believed to be associated with mechanical changes in the basal body. Each flagellum rotates in a counterclockwise direction and pushes the bacterium (Figure 7–9a). Bacteria have a primitive sensory system that allows them to detect the presence of nutrients and poisonous or toxic substances.

The movement of cells toward *(positive chemotaxis)* or away from *(negative chemotaxis)* chemicals depends on the ability of the bacteria to detect these substances and the transmission of such information to flagella. When nutrients are detected, the rotation of a single flagellum, or flagellar bundle, propels the cell in a coordinated way in the desired direction. Wandering off course causes a loss of coordination, with flagellar bundles separating and cells tumbling about until the nutrients are sensed again. Toxic substances also cause flagellar bundles to come apart, so that bacteria move away from potential danger. The exact mechanism for such coordinated movement is not completely understood.

FUNCTION. Bacterial cells benefit from flagella in the following ways: (1) they can migrate toward environments favorable for growth and away from those that might be harmful; (2) they can increase the concentration of nutrients or decrease the concentration of poisonous materials near the bacterial surfaces by causing a change in the flow rate of en-

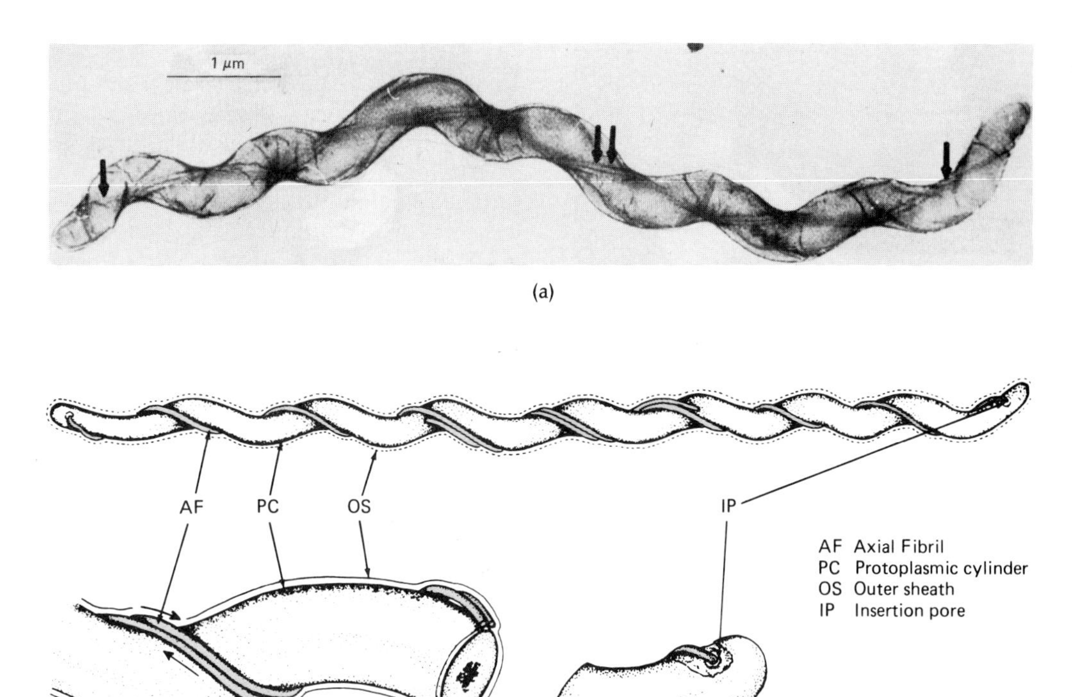

Figure 7–10 Axial fibrils. (a) An electron micrograph showing axial filaments extending for most of the spirochete's length. *[S. Holt,* Microbiol. Revs. ***42**:114–160, 1978.]* (b) The anatomical parts of spirochetes showing axial fibrils (AF) and outer sheath (OS). *[After S. Holt,* Microbiol. Revs. ***42**:114–160, 1978.]*

vironmental fluids; and (3) they can move flagellated organisms to uninhabited areas for colony formation. It has also been suggested that flagellated pathogens may more easily penetrate certain host defense barriers, such as mucous secretions.

AXIAL FILAMENTS

Spirochetes, which have flexible cell walls, move by unique structures called *axial filaments* (Figure 7–10). Electron micrographs show that each filament is composed of two fibrils morphologically similar to the procaryotic flagellum. Each fibril has two basic parts: a shaft and its covering sheath and an insertion apparatus that is differentiated into a terminal knob composed of a proximal hook and insertion discs (Figure 7–11b).

PILI (FIMBRIAE)

Pili (singular, **pilus**), or *fimbriae,* are surface filaments of varying diameters and lengths. Their presence on bacterial cells can be detected directly by such electron microscopy procedures as shadow casting (Figure 7–12), negative staining (Figure 7–5), or directly, in certain cases, by the clumping (agglutination) of red blood cells from humans and a variety of laboratory animals.

In general, pili differ from bacterial flagella in several properties, including (1) their smaller diameter, (2) the absence of the wavelike appearance so characteristic of flagella, and (3) the apparent lack of association with an organism's true motility.

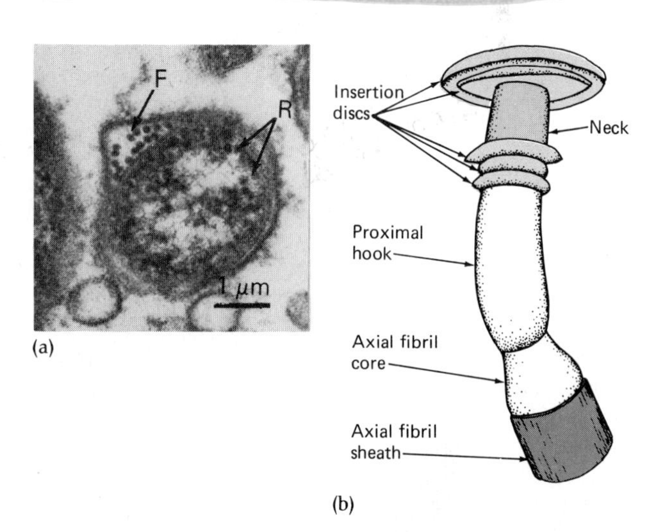

Figure 7–11 (a) This thin section shows the location of fibrils (F) between the outer sheath and the plasma membrane. Ribosomes (R) within the cell are also evident. *[K. H. Hougen,* Acta Path. Microbiol. Scand. Sect B. ***82**:799–809, 1974.]* (b) The parts of the axial fibrils involved with its insertion. *[After S. C. Holt,* Microbiol. Revs. ***42**:114–160, 1978.]*

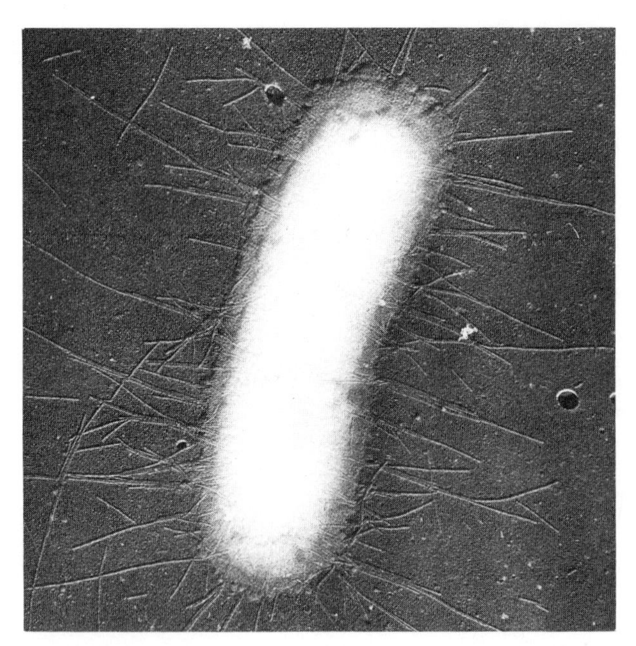

Figure 7–12 Bacterial pili (fimbriae). Type 1 piliated *Escherichia coli* (esh-er-IK-ee-ah KOH-lee), strain K-12 F. *[Courtesy of Drs. C. C. Brinton, Jr., and J. Carnahan, University of Pittsburgh.]*

Chemical analyses of pili show them to be mainly protein. Specific homogeneous protein subunits called *pilin* interlock and form the rigid, helical, tubelike pilus. The production of pili is under genetic control.

These filamentous surface structures have been found primarily in gram-negative bacteria. Included in this group are members of the genera *Branhamella* (bran-ham-EL-a), *Escherichia, Klebsiella* (kleb-see-EL-la), *Neisseria, Pseudomonas, Shigella,* and *Vibrio* (VIB-ree-oh). There have been reports of pili in numerous strains of the gram-positive organism *Corynebacterium renale* (ko-ri-nee-back-TI-ree-um REE-nal).

FUNCTION. The functions of type I pili include (1) attachment to most surfaces, cellular or otherwise, and (2) formation of surface films of organisms *(pellicles),* which could enhance microbial growth in still-culture situations when the oxygen supply is limited. Pili can function as important aids to establishing a disease process. In this case certain piliated bacteria attach to mammalian cell surfaces where they reproduce and produce toxins. **[See Chapter 21 for a discussion of virulence.]** Pili are also used as receptor sites by some bacteriophages (bacterial viruses) to inject their genetic material into a susceptible bacterial cell.

Another type of pilus, called the *F* or *sex pilus,* while similar in form to other pili, is different chemically. The sex pili are formed by donor bacteria containing plasmids for conjugative transfer of genetic material. Conjugative pili have been found among several bacterial species (Figure 7–13).

SPINES (SPINAE) AND SPIRAE

Certain marine gram-negative bacteria have been found to produce unusual rigid protein appendages known as *spines* or *spinae* (Figure 7–14). These structures may be several micrometers in length and consist of hollow shafts with expanded cone-shaped bases used for attachment to the outer cell wall surface. As many as ten spines may be carried by one bacterium. Flagella and pili, if present, are lost when spinae are produced. The chemical nature of these structures, their rigidity, and their capacity to increase the effective size of individual cells may protect spined bacteria against ingestion by larger microorganisms such as predatory protozoa in natural environments.

When grown in liquid cultures, nitrogen-fixing strains of *Azospirillum lipoferum* (a-zoh-spy-RIL-lum lip-oh-FER-um) produce spiral structures, called *spirae* (Figure 7–15). Each spira is a single coiled thread with a diameter varying from 40 to 55 nm, and may be either wound around a polar flagellum or separate from it. The function of spirae is not known.

The next sections describe bacterial structures

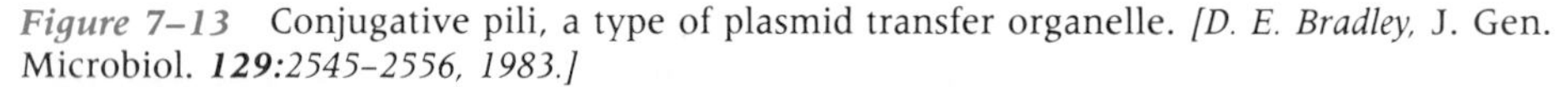

Figure 7–13 Conjugative pili, a type of plasmid transfer organelle. *[D. E. Bradley,* J. Gen. Microbiol. ***129:****2545–2556, 1983.]*

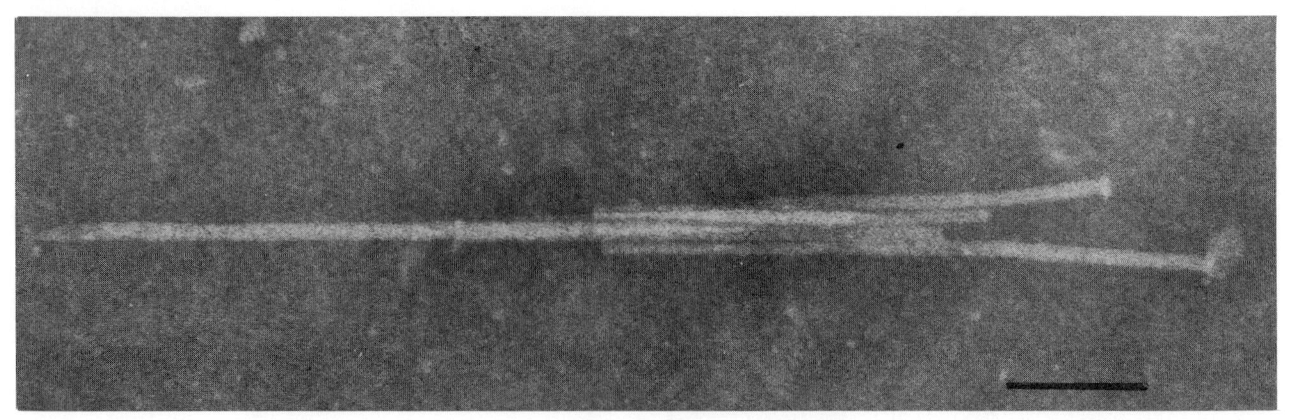

Microbiology Highlights

PILI BY VISIBLE LIGHT

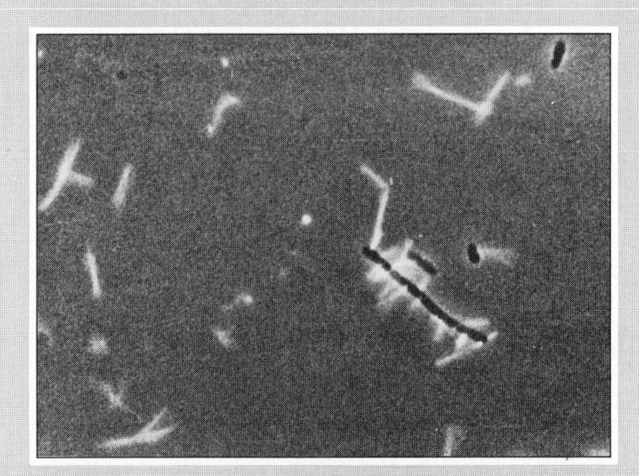

Pili by visible light.

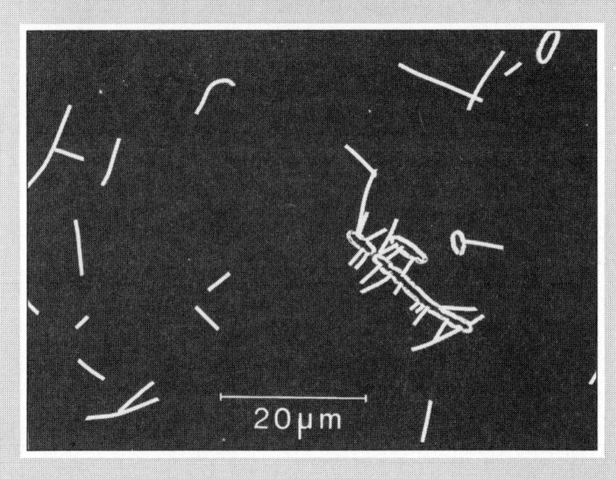

Diagrammatic sketch. (From C. K. Biebricher, and E.-M. Duker, *J. Gen. Microbiol.*, **130**:941–949, 1984).

F (sex) pili and several other bacterial surface filamentous appendages have diameters far below the resolution of light microscopy. But by using an electron microscope, F pili, to which RNA bacteriophages (phages) are attached, can easily be seen and distinguished from other filamentous structures. Labeling phages with an appropriate fluorescent dye can effectively demonstrate the presence and structure of F pili. About 700 RNA phages per micrometer (μm) of F pili are necessary to cover the pilus surface for visualization.

that may play important roles in establishing a disease process or protecting organisms from unfavorable environments.

THE GLYCOCALYX

Bacterial cells in many natural environments are surrounded by an extracellular component known as a *glycocalyx.* This bacterial part is defined as any polysaccharide-containing structure outside of the cell wall that may be composed of fibrous polysaccharides or spherical glycoproteins produced by the cell itself. The extent of glycocalyx formation depends on nutritional and other environmental conditions. The glycocalyx has several important functions and activities. With this structure, bacteria can attach to other bacteria and to various solid surfaces. In this way bacteria form *microcolonies,* the major type of bacterial growth in nature and several diseases. Attachment to surfaces in these environments also is a protective function, since the glycocalyx prevents the removal of cells from the system and anchors them in areas where nutrients tend to concentrate. The glycocalyx provides still another measure of protection to cells from antibacterial agents such as antibiotics, bacterial viruses, immunoglobulins, and phagocytes.

Glycocalices are subdivided into two types: *S (regular surface) layers* and *capsules.*

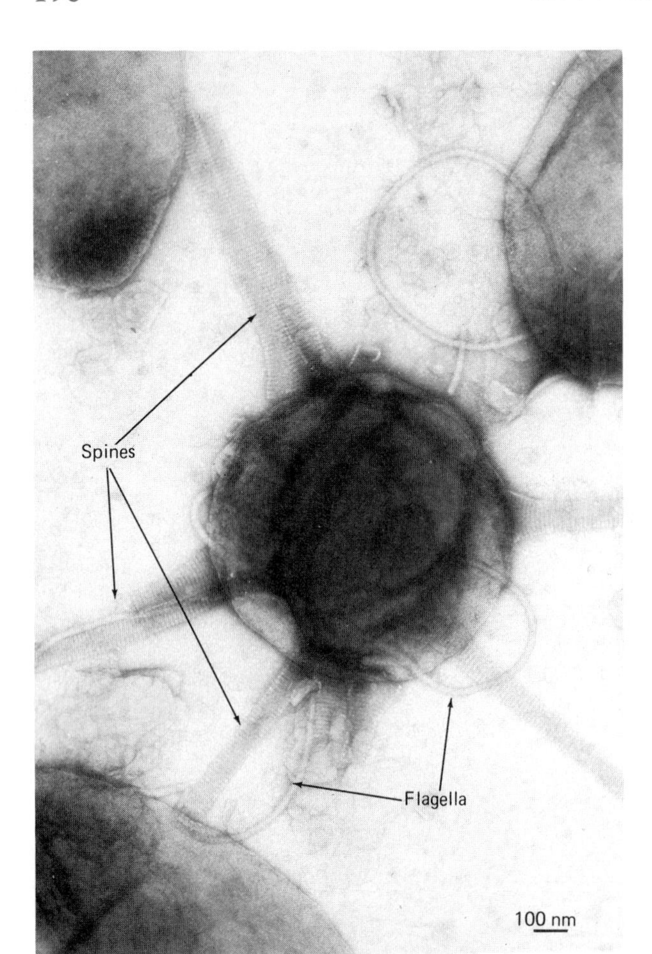

Figure 7–14 An electron micrograph showing spines with cone-shaped bases attached to bacterial surfaces. Flagella are also present. *[J. H. M. Willison,* et al., Can. J. Microbiol. ***23****:258–266, 1977.]*

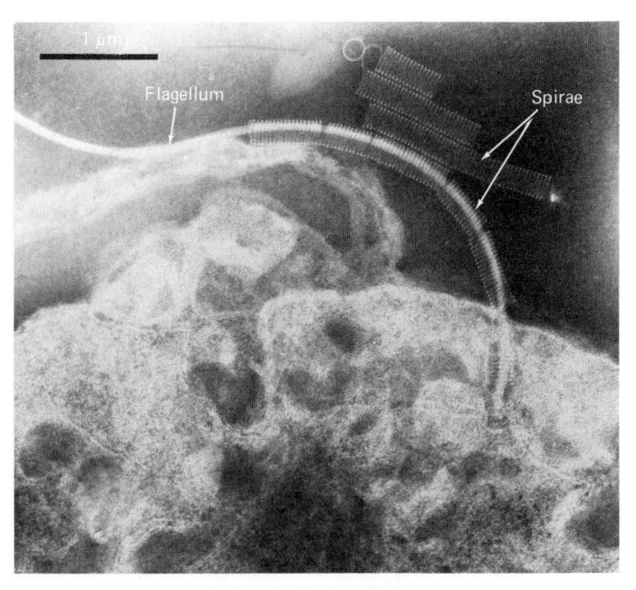

Figure 7–15 Spirae around a flagellum and unattached. *[K. B. Easterbrook, and S. Sperker,* Can. J. Microbiol. ***28****:130–136, 1982. Reproduced by permission of the National Research Council of Canada.]*

S LAYERS. S layers are generally constructed from a single type of protein that forms either hexagonal or tetragonal patterns covering the entire cell surface (Figure 7–16a). A principal function of these layers appears to be related to protection of the bacterium from hostile environmental agents. According to this view, the gaps or holes generally present in the layers are sufficiently large to allow the free exchange of small molecules, such as nutrients and waste products, between the cell and its environment, while being sufficiently small to screen the cell from large molecules, such as cell-dissolving (lytic) enzymes and extrachromosomal pieces of DNA plasmids. S layers also have been found to be excellent vehicles for the study of macromolecular formation and arrangement when electron microscopy and computer image processing techniques are used (Figure 7–16c). **[Chapter 3 describes these techniques.]**

CAPSULES. Bacterial capsules are described as organized accumulations of gelatinous material on cell walls, in contrast to *slime layers,* which in structural terms are generally considered to be a loose, unorganized network of similar material extending from the cell surface. The production of capsules is determined largely by genetic as well as environmental conditions, such as the presence or absence of capsule-degrading enzymes and various growth factors. Consequently, capsules will vary in thickness and rigidity and may or may not be closely associated with the bacterial cell surface. Electron micrographs of capsules generally do not show any specific structural features; however, many of the true capsules, as opposed to slime layers, have been shown to possess definite borders (Figure 7–17a).

Capsules vary in chemical composition. Complex polysaccharides, either alone or in combination with nitrogen-containing mucinlike substances and polypeptides, are the most common constituents of bacterial capsules. Uronic acid is one type of compound commonly found in capsules but not in cell walls.

In the laboratory, the presence of capsules usually is indicated when colonies are mucoid (Figure 7–17b), that is, when they exhibit a stringy consistency when touched with an inoculation loop. With the aid of India ink or negative stains and brightfield microscopy, capsules can usually be demonstrated as uncolored halos (clear zones) between the opaque background and the individual bacterial cells (Color Photograph 15). Capsule formation may be lost during repeated transfer of bacteria on laboratory media. It is thought that this loss occurs because bacteria growing in pure cultures are no longer challenged by the same unfavorable factors that exist in their natural environment. Genetically altered organisms (mutants) lacking a capsule will

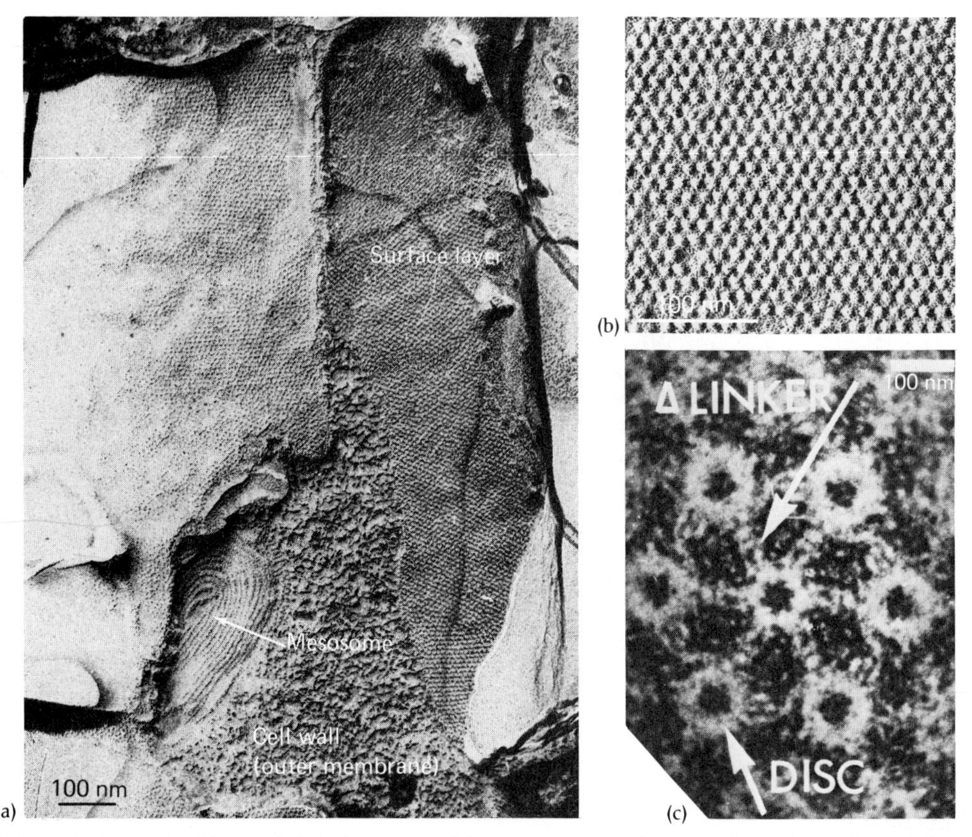

Figure 7–16 The S layer. (a) A freezing-etching preparation showing the relationship of the regular surface layer that covers the outer portion of the cell wall (pitted layer). A mesosome also is present. (b) A portion of the hexagonal array or pattern of the layer. (c) A computer-enhanced image of a hexagonal unit. *[Courtesy of T. J. Beverage and R. G. E. Murray, Dept. of Microbiology, University of Guelph, and Dept. of Microbiology and Immunology, University of Western Ontario, Canada.]*

Figure 7–17 Bacterial capsules. (a) An ultrathin section of *Klebsiella pneumoniae* (kleb-see-EL-la noo-MOH-nee-ah), a capsule (Ca)-forming bacterium. Other bacterial components shown include the cell wall (Cw), ribosomes (R), and the nuclear area (Nu). Note the constricted area (arrows) of cell separation where the capsular material is scanty. *[A. Cassone, and E. Garaci,* Can. J. Microbiol. ***23****:684–689, 1977.]* (b) The left side shows rather sticky mucoid colonies of capsule-forming organisms; on the right, the rough or coarse colonies of non-capsule formers are evident. *[T. J. Chai, and R. E. Levin,* Appl. Microbiol. ***30****:450–455, 1975.]*

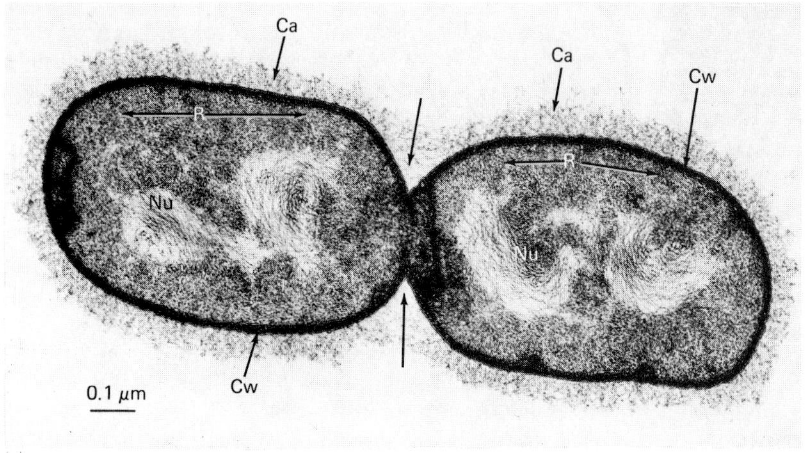

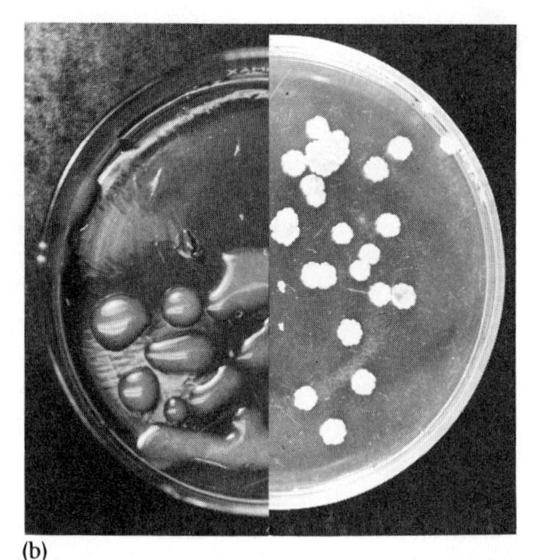

outgrow other cells because there is no competitive advantage to expending the energy required to form such a protective structure.

Encapsulation (capsule production) is important to certain pathogenic bacteria, influencing their disease-producing capabilities. With the exception of bacteria that produce large amounts of toxins (poisons), pathogens either attach to a tissue and cause disease by the production of toxins and/or enzymes or stick to a tissue, multiply, and persist until they interfere with organ function and produce an obvious disease condition.

Capsules protect pathogens against phagocytosis and bactericidal factors in the body fluids of the host. The loss of the capsule-producing property often results in a reduction or disappearance of the disease-causing capacity. Many pathogens will form capsules on their initial laboratory isolation from a diseased individual. Examples of bacteria whose virulence is associated with encapsulation include *Bacillus anthracis* (ba-SIL-lus an-THRAY-sis) (the cause of anthrax), *Clostridium perfringens* (klos-TRE-dee-um per-FRIN-jens) (gas gangrene), and *Streptococcus pneumoniae* (strep-toh-KOK-kus noo-MOH-nee-eye) (lobar pneumonia) (Color Photograph 15). **[Capsules and disease are discussed in Chapter 21.]**

CELL WALLS

The presence of cell walls was implied in Leeuwenhoek's descriptions of bacteria. Apparently, he recognized the need for some type of structure that would not only preserve a bacterium's shape but would also hold its cellular contents together.

The rigid cell wall is the main structural component of most procaryotes. Its presence was first demonstrated by placing bacterial cells in a very concentrated sucrose solution. The cellular membrane and its contents were seen to shrink away from an outer, enclosing rigid envelope as water from the inside of the cell diffused out into the sucrose under the influence of osmotic pressure. Thus, the bacterial cell's outer structural limit was defined.

Some of the functions of the cell wall are (1) to prevent rupture of bacteria caused by osmotic pressure differences between intracellular and extracellular environments, (2) to provide a solid support for flagella, (3) to maintain the characteristic shape of the microorganism, and (4) to regulate, to a certain degree, the passage of molecules into and out of the cell *(molecular sieving)*. The sites of attachment of most bacterial viruses (bacteriophages) are on the cell wall. At distinct locations, the wall is also modified to accommodate the surface appendages such as pili and spines. **[See viral attachment in Chapter 10.]**

The cell wall accounts for 20% to 40% of a bacterium's dry weight. Several factors affect this percentage, including the organism's stage of growth and its nutritional deficiencies. The strength of the cell wall is based on the physical and chemical properties of a complex molecular layer found only in bacteria, the **peptidoglycan layer** (Figure 7–18). However, not all bacterial cell walls contain peptidoglycan.

Various types of equipment and techniques for isolating and characterizing bacterial cell walls have provided much information on their chemical composition and ultrastructure. The first chemical analysis of bacterial cell walls was carried out in 1887 by L. Vincenzi. His studies showed the presence of substantial amounts of nitrogen, indicating that bacterial cell walls were not made of cellulose, as are plant cell walls. Beginning in the 1950s, M. R. J. Salton and others performed analyses that determined the following properties for bacterial cell walls:

1. Cell walls contain two simple sugars related to glucose, *N*-acetylglucosamine (NAG) and *N*-acetylmuramic acid (NAM) (Figure 7–18b). NAM occurs only in bacteria, including the cyanobacteria and the rickettsiae. NAG and NAM are connected by bridges of amino acids to form the mucopeptide or peptidoglycan layer of the wall (Figure 7–18a). This "backbone layer" imparts mechanical strength to both gram-positive and gram-negative cell walls. It is the innermost part of the cell wall next to the cell membrane, and has been named the *murein sacculus*. **[See Chapter 13 for features of cyanobacteria and the rickettsias.]**
2. Cell walls may contain a variety of natural or common amino acids that are used to build the necessary cross-linkages or bridges binding the peptidoglycan polymers (units) of the cell wall together. Included in this group are alanine, glycine, glutamic acid, and lysine. Differences exist between the amino acid composition of gram-positive and gram-negative cell walls.
3. Bacterial cell walls also contain certain unique amino acids, the atoms of which are arranged differently in space than those of most natural amino acids.
4. Certain organisms contain a molecule called *diaminopimelic acid* (Figure 7–19). It is found only in these bacteria in walls lacking lysine.

Certain structural and chemical differences exist between the walls of gram-positive and gram-negative bacteria. The walls of gram-positive bacteria are relatively thick (15 to 18 nm). Moreover, they are uniformly dense (Figure 7–20a and 7–21) and exhibit a delicate ordering of chemicals throughout their structure. In several gram-positive species special polysaccharides are present in significant amounts, are bound to peptidoglycan layers, and contribute to cell wall structure. These substances include teichoic and teichuronic acids, which also contribute to the wall's ability to bind heavy metals and control the autolytic (cell-digesting) actions of enzymes in certain cell walls.

The walls of gram-negative cells are multilayered membranous structures and are chemically and structurally more complex than those of gram-

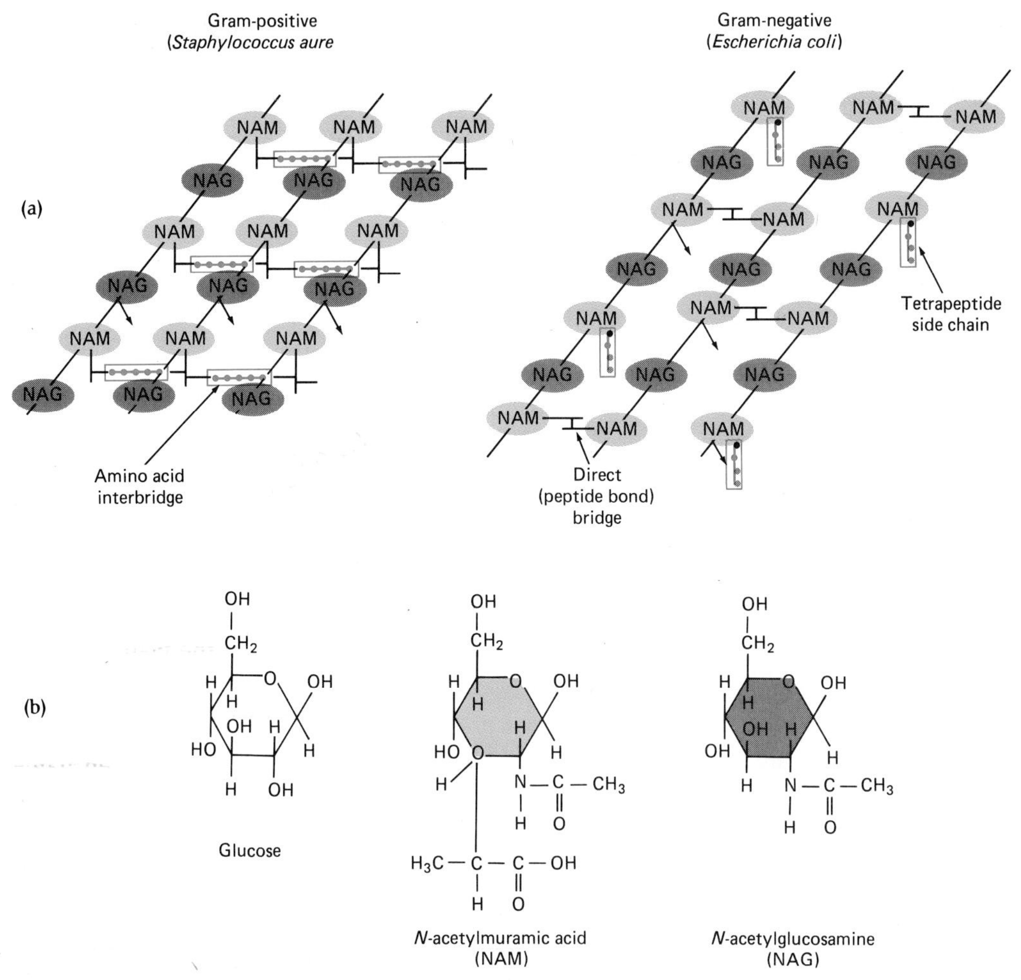

Figure 7–18 The chemical organization of the peptidoglycan or murein layer of gram-positive and gram-negative bacterial cell walls. (a) A three-dimensional view showing the components of the peptidoglycan layer, the amino sugars *N*-acetyl glucosamine (NAG), *N*-acetyl muramic acid (NAM), and the amino acids that form peptide connecting bridges. The number of peptide bridges and the number of amino acids in these bridges vary with the bacterial species. Arrows show the sites of enzyme (lysozyme) action. (b) The structural formulas of the amino sugars NAM and NAG. For comparison, glucose is included to show where these compounds differ from glucose.

positive cells. They contain a wider range of amino acids and significant amounts of lipid, polysaccharide, and protein constituents. Combinations of these components form an outer layer or envelope that surrounds a thin peptidoglycan layer (Figure 7–21b). The envelope is firmly attached to the murein layer and is connected to it by lipoproteins. Lipopolysaccharides (LPS) prevent the penetration of antibacterial substances, such as penicillin, that interfere with the formation of the peptidoglycan and

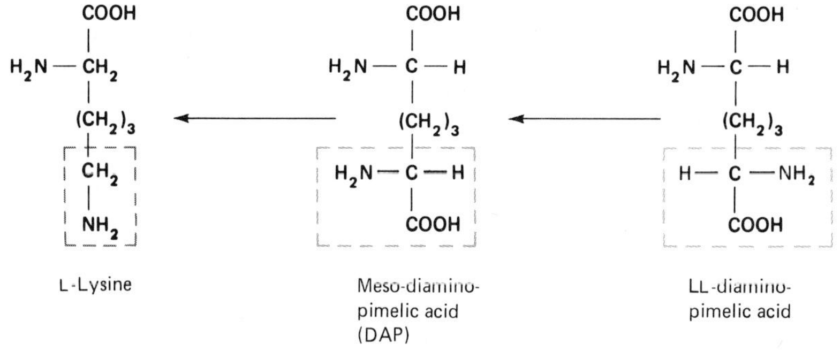

L-Lysine
Meso-diamino-
pimelic acid
(DAP)
LL-diamino-
pimelic acid

Figure 7–19 L-lysine and meso- and LL-diamino-pimelic acids. Certain organisms utilized LL-DAP in their cell walls; others convert its derivative meso-DAP into lysine for wall formation. Differences among the molecules are shown in boxed form.

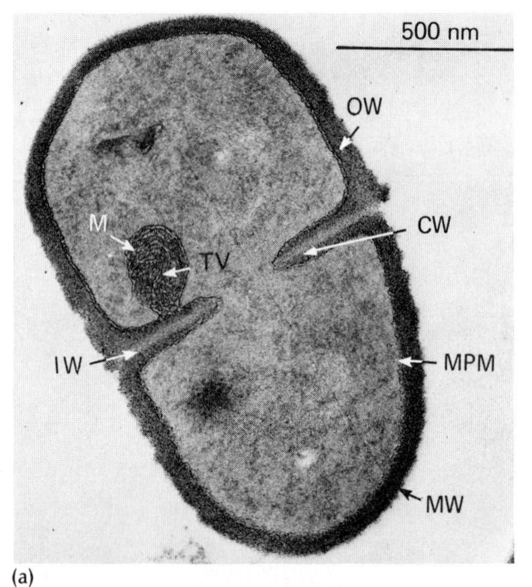

(a)

(b)

Figure 7–20 (a) An electron micrograph of dividing cells of *Lactobacillus casei* (lack-toh-ba-SILL-lus KAY-see-eye), showing thick outer walls (OW) and middle cell walls (MW) extending into the area of cross wall or septum formation (CW). The darkly stained inner cell wall (IW) of the cell on the lower right is seen in the peripheral region but not in the cross wall. The cell membrane (MPM) of the cell on the upper left is closely applied to the wall and cross wall and can be seen to extend into and around the mesosome (TV) boundary. Mesosomes, their tubular-vesicular (TV) components, and ribosomes (R) are quite evident. *[Courtesy of K. J. I. Thorne, and D. C. Barker,* J. Gen. Microbiol. ***70****:87–98, 1972.]* (b) An electron micrograph of a normal rod-shaped murein sacculus or peptidoglycan layer. *[Courtesy of U. Schwartz, and W. Leutgeb,* J. Bacteriol. ***106****:588–595, 1971.]*

are also known for their toxic (harmful) and antigenic (antibody forming) properties. The lipid portion of the LPS, referred to as *lipid A,* (Figure 7–21b), is an endotoxin that is known to cause the destruction of blood cells and fever during the course of certain infections. The polysaccharide portion of the LPS form envelope-associated antigens, which are of value in the immunological identification of some disease-causing gram-negative bacteria. **[See Chapter 19 for a description of bacterial agglutination tests.]**

Several specialized functions and/or activities of the outer membrane of gram-negative cells are reflected in its chemical composition. The structure contains a number of characteristic and unusual major proteins. Among these are the *porins,* which act as pores by forming watery channels to allow the passive passage of various-sized molecules.

Although most bacteria have rigid cell walls, one important group of organisms, mycoplasma, lack them. Because of the absence of a cell wall, these procaryotes exhibit a wide variety of shapes (Figure 7–22). **[See Chapter 13 for a description of the *Mycoplasma*.]**

PROTOPLASTS AND SPHEROPLASTS

Under certain conditions, bacterial cells may lose all or a portion of their walls. Three general conditions can bring about this state: (1) the presence of an inhibitory substance, such as penicillin; (2) the lack of essential nutrients needed for formation of cell walls; and (3) treatment with enzymes such as lysozyme capable of hydrolyzing linkages in the murein sacculus of bacterial walls. When the cytoplasmic membrane is found either to contain only trace amounts of cell wall or to be completely free of such material (Figure 7–23), the term *protoplast* is usually used to denote the remaining unit. This designation customarily refers to gram-positive organisms. If the cell wall material is not completely removed from the bacterium, the structure is called a *spheroplast.* This term is usually applied to rounded gram-negative organisms that have developed as a consequence of the treatment mentioned above.

Penicillin, when used in proper concentrations and under appropriate culture conditions, can transform *growing cells* of several species into spheroplasts. The antibiotic inhibits cell wall synthesis at a particular point in the process.

The enzyme lysozyme has been used by numerous investigators to produce cells either completely or almost completely free of wall material. Rod-shaped cells are converted by such treatment into osmotically sensitive spherical structures. To prevent the destruction of such cells, procedures are performed in stabilizing solutions, such as polyethylene glycol or sucrose. These precautionary measures are necessary regardless of the method used for protoplast or spheroplast production.

The cytoplasm of isolated protoplasts does not contain the membranous organelles called *mesosomes* that are often present at or near the site of division in whole cells.

L Forms. Certain bacteria can spontaneously give rise to variants that can replicate in the form of small cells with defective or absent cell walls. These **L forms,** named after the Lister Institute in London, where they were discovered, are small enough to pass through most filters. They can also be formed by several other species when their cell wall synthesis is inhibited or impaired. Penicillin and other antibiotics, specific immunoglobulins, and lysosomal enzymes that degrade cell walls can create the environment necessary for L-form produc-

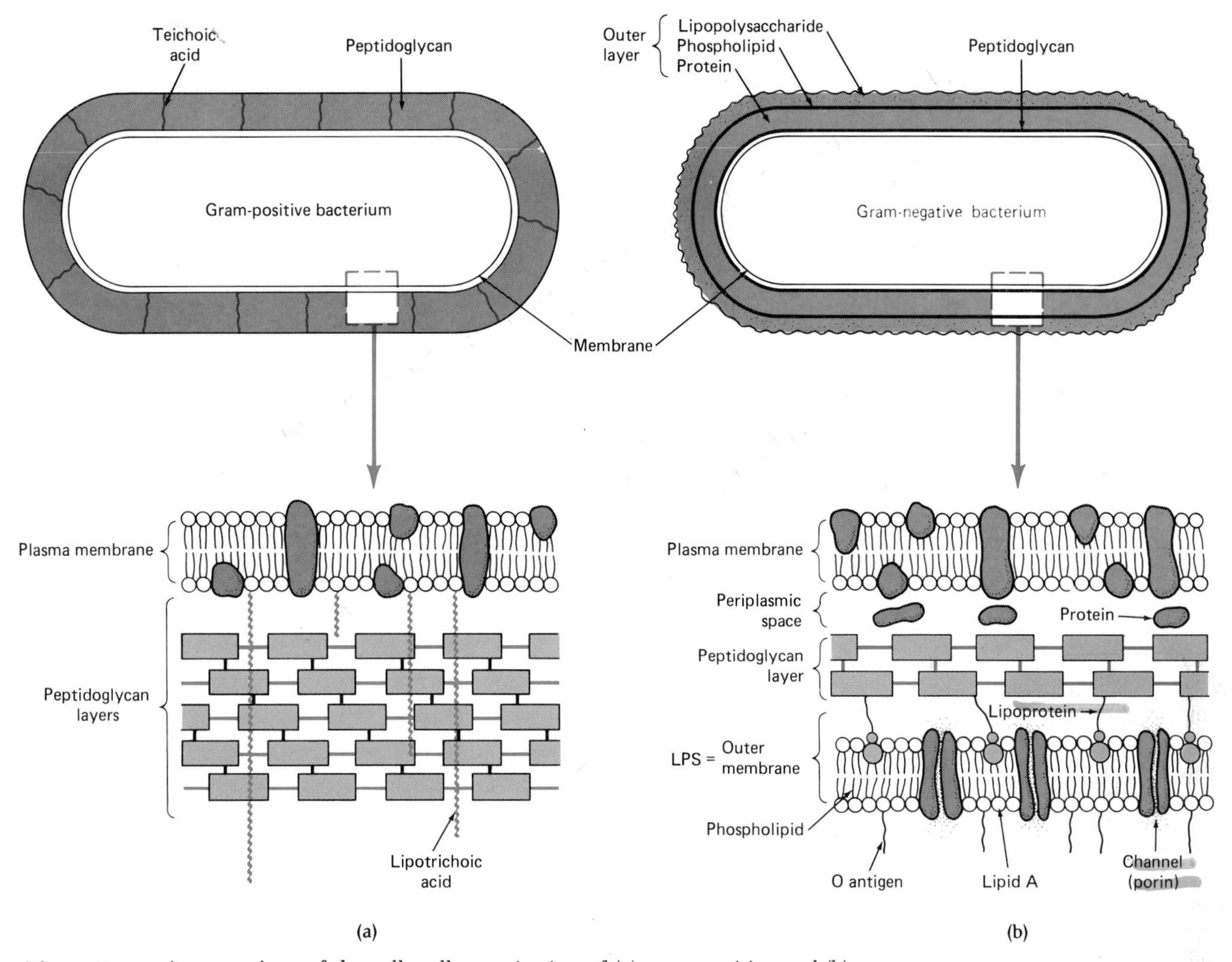

Figure 7–21 A comparison of the cell wall organization of (a) gram-positive and (b) gram-negative bacteria. The relationship of the cell wall to the cell membrane is also shown in each case. In addition, a schematic representation of the molecular organization of the major chemicals in these cell walls is presented.

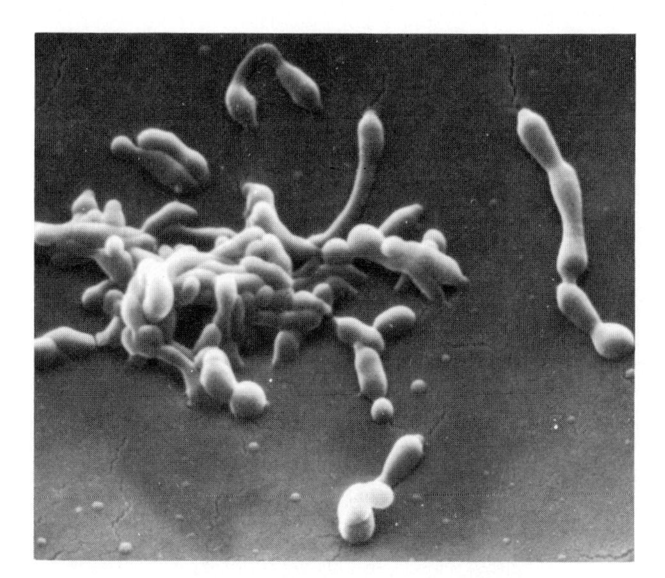

Figure 7–22 The various shapes of mycoplasma. *[S. Razin,* et al., Inf. Immun. ***30****:538–546, 1980.]*

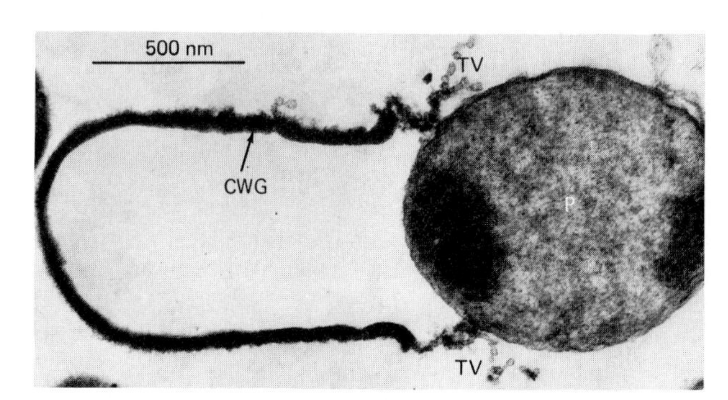

Figure 7–23 A protoplast (P) of *Lactobacillus casei.* Note the absence of a cell wall and mesosome. The tubular-vesicular (TV) components of the mesosomes are attached to the cell wall ghost (CWG) with protoplast formation. *[Courtesy of K. J. I. Thorne, and D. C. Barker,* J. Gen. Microbiol. ***70****:87–98, 1972.]*

Microbiology Highlights

CELL WALL ARCHITECTURE

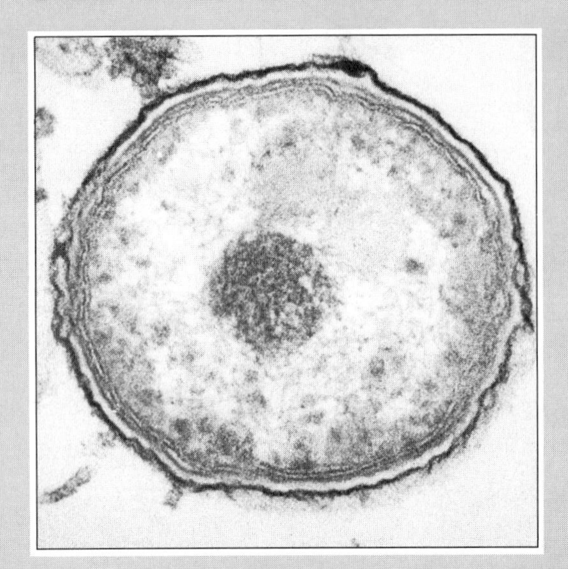

The triple-layered cell wall of mycobacteria. (From Rastogi, N., et al., *Current Microbiol.* **13**:237–242, 1986.)

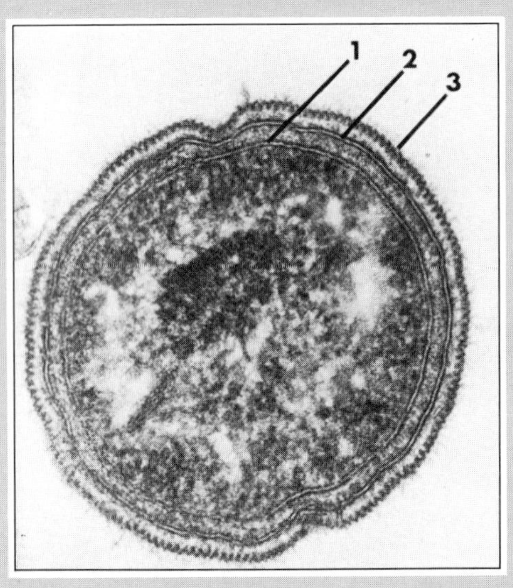

The unique outer cell wall layer showing archlike units. (From Tanner, A. C. R., et al., *Internat. J. System. Bacteriol.*, **36**:213–221, 1986.)

As described in the section dealing with the bacterial surface, one of the most important structures of procaryotes is their cell wall. The procaryotic cell wall is not only chemically quite different from any comparable eucaryotic cell structure but is also one of the features distinguishing procaryotic from eucaryotic organisms. In addition to the ultrastructural differences between gram-positive and gram-negative cells, electron microscopic studies have shown some unique properties of cell walls of bacterial genera as well as of species. N. Rastogi and associates uncovered a triple-layered structure in 18 species of mycobacteria. This cell wall may play an important role in the survival of the species by providing protection against ingesting white blood cells (phagocytes). In another study, A. C. R. Tanner and associates found an outer layer of a cell wall belonging to a newly isolated *Bacteriodes* (back-te-ROY-deez) species to have unusual archlike units (as shown in the accompanying micrograph). This cell wall feature distinguished the species from all other members of the genus. The function of the sculptured cell wall is not known at this time.

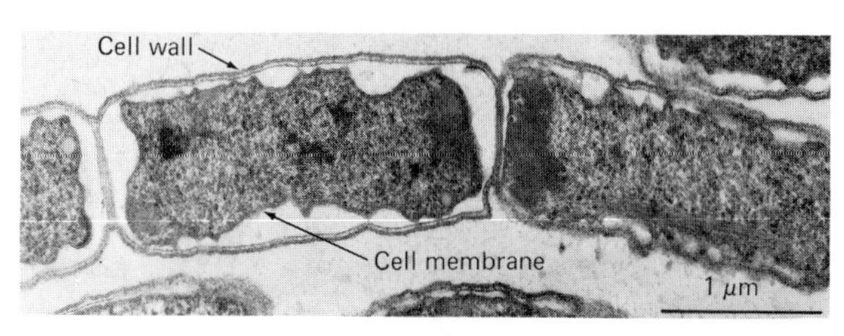

Figure 7–24 An ultrathin section of the gram-negative *Thermothrix thioparus* (ther-MOH-tricks tigh-oh-PAR-us). This electron micrograph clearly shows the organism's cell membrane, which was separated from the cell wall. *[Reproduced with permission of the National Research Council of Canada from D. E. Caldwell,* et al., Can. J. Microbiol. ***22**:1509–1517, 1976.]*

tion in the tissues of a suitable host. L forms have been isolated from several disease states in humans, lower animals, and plants.

L forms are structurally equivalent to protoplasts (Figure 7–23) and spheroplasts. However, the designation "L form" refers only to cells that can multiply. Some L forms revert to normal cell wall-bearing cells in a suitable host or a favorable medium, whereas others maintain their morphological property. Special media must be used for the isolation and growth of all L forms.

Structures Within the Cell Wall

THE PLASMA MEMBRANE

The plasma membrane lies just beneath the cell wall and separates it from the bacterium's cytoplasm. This structure, which is also called the *cytoplasmic membrane* or *cell membrane,* is an integral and indispensable functional part of all types of cells (Figure 7–24).

In eucaryotic cells, membranes partition the cellular space into compartments in which biochemical reactions occur. Structures similar, if not identical, to the cytoplasmic membrane also surround intracellular eucaryotic organelles. The plasma membrane of a eucaryote is continuous or interconnected with the cell's internal membrane systems.

The membrane is semipermeable, regulating the passage of molecules into and out of cells in a selective or differential manner. It sometimes acts as a *passive barrier* and at other times as an *active barrier.* When a cell membrane acts as a passive barrier, substances pass through it by simple diffusion, moving freely from an area of greater to one of lesser concentration until an equilibrium is reached. Studies have shown that the membrane contains certain regions capable of performing work to "pump" certain molecules into or out of the cell against a difference in concentration. This so-called osmotic work involves the expenditure of energy and is generally referred to as *active transport,* shown in detail in Figure 7–25. Generally, a bacterium brings more molecules in than out, so that specific molecules accumulate inside the cell.

Active transport provides bacteria with certain advantages, including the ability to maintain a fairly constant intracellular ionic state in the presence of varying external ionic concentrations and the means with which to capture nutrients present in low concentrations in media. Membranes contain several types of transport systems for such substances as amino acids, mineral ions, sugars, and related compounds. The transport systems, called **permeases,** consist of several enzymes that are situated in the membrane and function as carriers. In certain situations, bacteria encounter potential nutrient molecules too large to penetrate the cell wall or membrane. Most bacteria deal with this problem by secreting enzymes *(exoenzymes)* into the surrounding environment to break down the large molecules into smaller units, which can then be transported by the permeases into the cells.

The bacterial plasma membrane also participates in the outward transport of molecular waste products and substances necessary for the formation of the cell wall. The membrane and membrane-associated structures, such as mesosomes, are important to the energy-producing reactions of the cell. **[Mesosomes are described later in this chapter.]**

Figure 7–25 Active processes involving the movement of substances (molecules) across the cell membrane. In active transport, substances are carried across membranes from an area of lower concentration to one of higher concentration with the expenditure of energy. Energy is involved at various stages of the process. A carrier located in the membrane picks up a molecule to be transported (1), carries it through the membrane (2), releases the molecule into the cell's cytoplasm (3), and then becomes available to perform the transporting activity again (4).

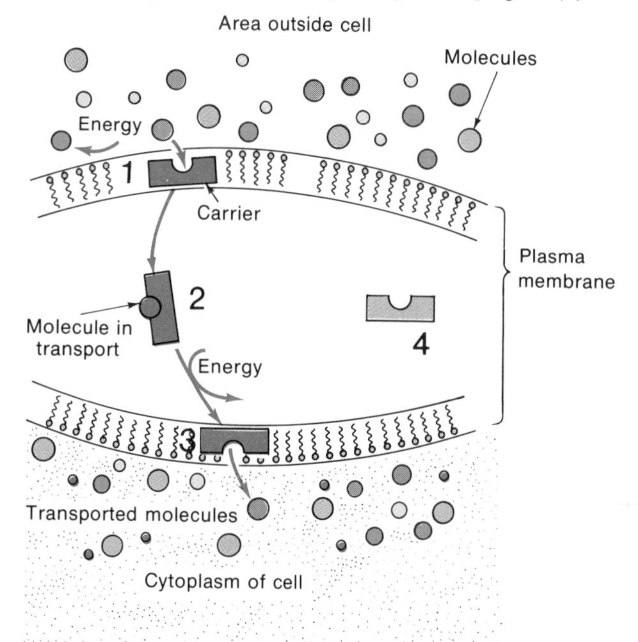

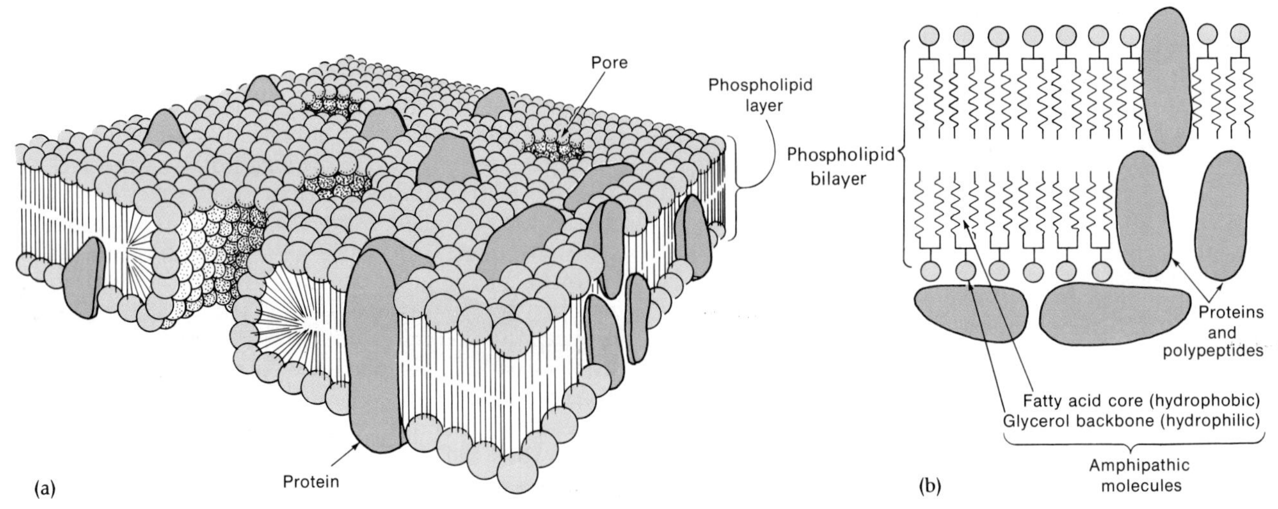

Figure 7–26 (a) An accepted model of the cell membrane, showing the relationship of the phospholipid layer and protein molecules. (b) A view of the cell (plasma) membrane showing the relationship between the phospholipid layer and protein molecules.

Chemically, the plasma membrane of bacteria consists of both proteins and lipids. Sterols, high-molecular-weight lipids, are not found in the bacterial membranes with the exception of mycoplasma, but they are common constituents of eucaryotic structures. The cell membrane contains several types of protein such as the cytochromes, iron-sulfur proteins, and the components of the electron transport chain (see Chapter 5). In addition, several types of enzymes are localized in the cell membrane and include the permeases, mentioned earlier, and the biosynthetic enzymes that regulate the last steps in the formation of various cell wall macromolecules (peptidoglycans, teichoic acids, simple polysaccharides, and lipopolysaccharides), and membrane lipids. The lipids provide strength and other structural properties for the membrane. Membrane lipids are *phospholipids.* These molecules are amphipathic; they consist of regions that are spatially separated, with one end repelled by water *(hydrophobic)* and the opposite end attracted to it *(hydrophilic).* The hydrophilic end, which carries an ionic charge and is polar, consists of glycerol attached to phosphate and other chemical groups. The nonpolar, hydrophobic end consists of hydrocarbon chains of fatty acids (Figure 7–26). With respect to the position and arrangement of proteins on the phospholipid bilayer, it appears that protein molecules are not arranged in an orderly fashion (Figure 7–26), but actually penetrate the phospholipid region and may extend through it completely. Further, the protein molecules continually move and change their shape. The dynamic arrangement of the membrane molecules is referred to as the **fluid mosaic model.**

Membranes differ chemically in different cell types. Analyses have clearly shown differences in the types of lipids and in the protein:lipid ratios of membranes obtained from different kinds of cells. The enzyme compositions also vary considerably. In general, bacterial cell membranes comprise 10% to 20% of a cell's dry weight.

In electron micrographs, eubacterial and eucaryotic membranes appear to consist of several layers—two dark layers, each about 2 to 3 nm thick, surrounding a lighter layer that is 4 to 5 nm thick. This arrangement represents the typical fine structure of the so-called **unit membrane.**

Periplasmic Space. The region that separates the cell wall from the plasma membrane (inner membrane) is called the *periplasmic space* (Figure 7–27) and contains the *periplasm.* Both are important parts of the wall of gram-negative bacteria and are subject to osmotic changes in the environment surrounding the cell. The periplasm exhibits definite

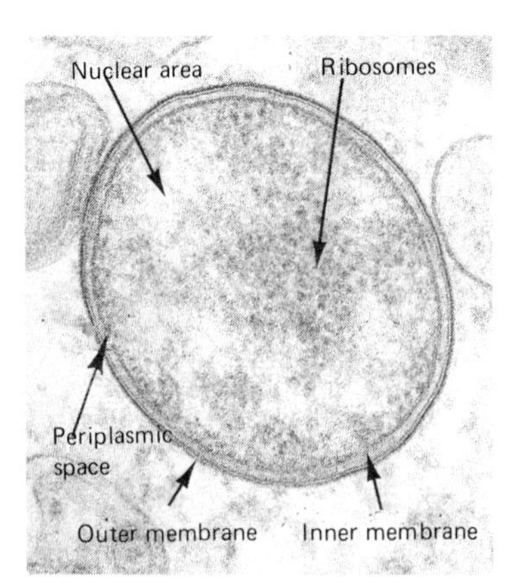

Figure 7–27 An electron micrograph showing the relationship of the outer and inner membranes. The periplasmic space is between the peptidoglycan layer and the cytoplasmic membrane. It also separates the cell wall (outer membrane) from the inner or plasma membrane. *[Courtesy Y. Rikihisa.]*

enzymatic activity that is important in the maintenance of the peptidoglycan layer and other parts of the cell wall. It also provides a microenvironment for the preprocessing of nutrients and the postprocessing of waste products secreted by cells. **[The peptidoglycan layer is described earlier in this chapter.]**

MESOSOMES

The bacterial membranous structures called **mesosomes** were observed in *Mycobacterium avium* (my-koh-back-TI-ree-um a-VEE-um) (the cause of tuberculosis in birds) by Shinohara, Fukushi, and Suzuki in 1957. Although some observers consider mesosomes to be artifacts created during specimen preparation, many other investigators have found similar organelles in various bacterial species. Mesosomes occur primarily in species of various gram-positive bacteria. In this group of organisms are species of *Bacillus (ba-SIL-lus), Lactobacillus* (lack-toh-ba-SIL-lus) (Figure 7–20a), *Staphylococcus,* and *Streptococcus.* Membranous structures similar to the mesosomes of gram-positives are seldom seen in gram-negative bacteria. The appearance of their organelles differs substantially. Mesosomes in gram-positive bacteria generally appear as pocketlike structures that contain tubules, vesicles (Figure 7–20a), or lamellae (folded-membrane arrangements). Gram-negative organisms primarily exhibit the lamellar form.

Mesosomes are apparently involved in several different bacterial processes. They function essentially to increase the cell's membrane surface, which, in turn, increases enzymatic content. A relationship seems to exist between the enzymatic needs of cells and the formation of new mesosomes.

The activities with which mesosomes have been associated include (1) cell wall synthesis, (2) division of nuclear material, (3) respiration, and (4) spore formation.

PROCARYOTIC NUCLEOIDS (GENOME, NUCLEOPLASM REGION) AND PLASMIDS

Procaryotic cells lack the distinct nucleus of eucaryotic cells. They also lack other features and structures characteristic of higher forms, such as a mitotic apparatus, nuclear envelope, and nucleolus. The essential genetic information, or *genome,* of a procaryotic cell is contained within a single chromosome composed of a single DNA molecule. This structure is located in the *nucleoplasm region,* or *nucleoid,* of the cell (Figure 7–28). The DNA is not surrounded by a nuclear membrane and exists in the form of a closed loop. In electron micrographs, nucleoids may appear as coarse, dense areas or as bundles of relatively fine fibrils in ribosome-free spaces (Figure 7–28). The DNA in the center of the nucleoid is thought to be tightly packaged and to have a high degree of supercoiling. In both gram-negative and gram-positive cells, the size of the nucleoid will increase during division. **[Chapter 6 describes the properties of DNA.]**

Many bacteria contain additional genetic information in extrachromosomal circular DNA molecules, known as **plasmids** (Figure 7–29). These extrachromosomal components are capable of independent replication and carry genetic information for a variety of different functions, such as drug resistance. As a general rule, extrachromosomal DNA is not essential to the life of an organism. **[Plasmids are discussed in Chapter 6.]**

RIBOSOMES

In procaryotes as in eucaryotes, the *ribosome* is important to the protein-synthesizing process. This cellular structure receives, processes, and translates genetic instructions for the formation of specific proteins. The ribosome captures amino acids in a properly coded sequence and joins them in a chain to form a new protein. Clusters of ribosomes, or *polysomes,* participate in this assembly-line process together with various molecules of ribonucleic acid

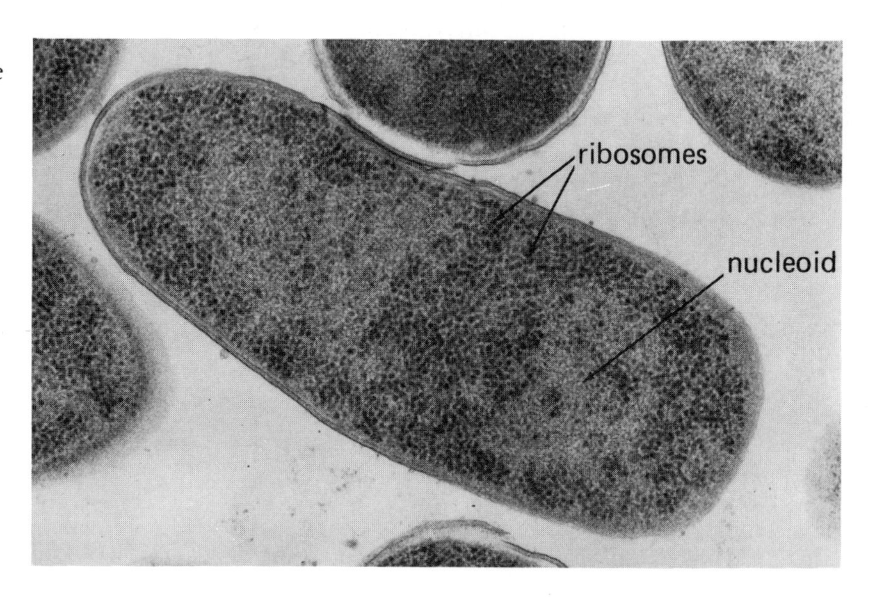

Figure 7–28 The procaryotic nucleoid. The fibrillar structure occupies areas free of ribosomes within the cell. Note the large numbers of ribosomes. *[From J. A. Hobot,* et al., J. Bacteriol. ***162:****960–971, 1985.]*

Figure 7–29 A plasmid DNA from *Vibrio parahaemolyticus* (VIB-ree-oh par-a-hee-moh-LIT-ee-cuss). These extrachromosomal structures contain genes that can provide additional capabilities to the bacterium. *[H. Guerry and R. R. Codwell,* Inf. Imm. ***16****:328, 1977.]*

(RNA) and the cell's chromosome. Procaryotic ribosomes are smaller than and differ in protein composition from eucaryotic ribosomes. Therefore these differences provide selective sites for the action of several protein-synthesis-inhibiting antibiotics. **[Refer to Chapter 6 for a discussion of protein synthesis.]**

Ultrathin sections show these submicroscopic structures as fine granules within cells (Figure 7–28). A ribosome consists of two subunits, one of which is about half the size of the other. Each subunit consists of both ribosomal RNA and protein. Specific models for the structure, arrangement, and location of these ribosomal components have been developed (Figure 7–30). A typical bacterium may hold as many as 15,000 ribosomes. These structures

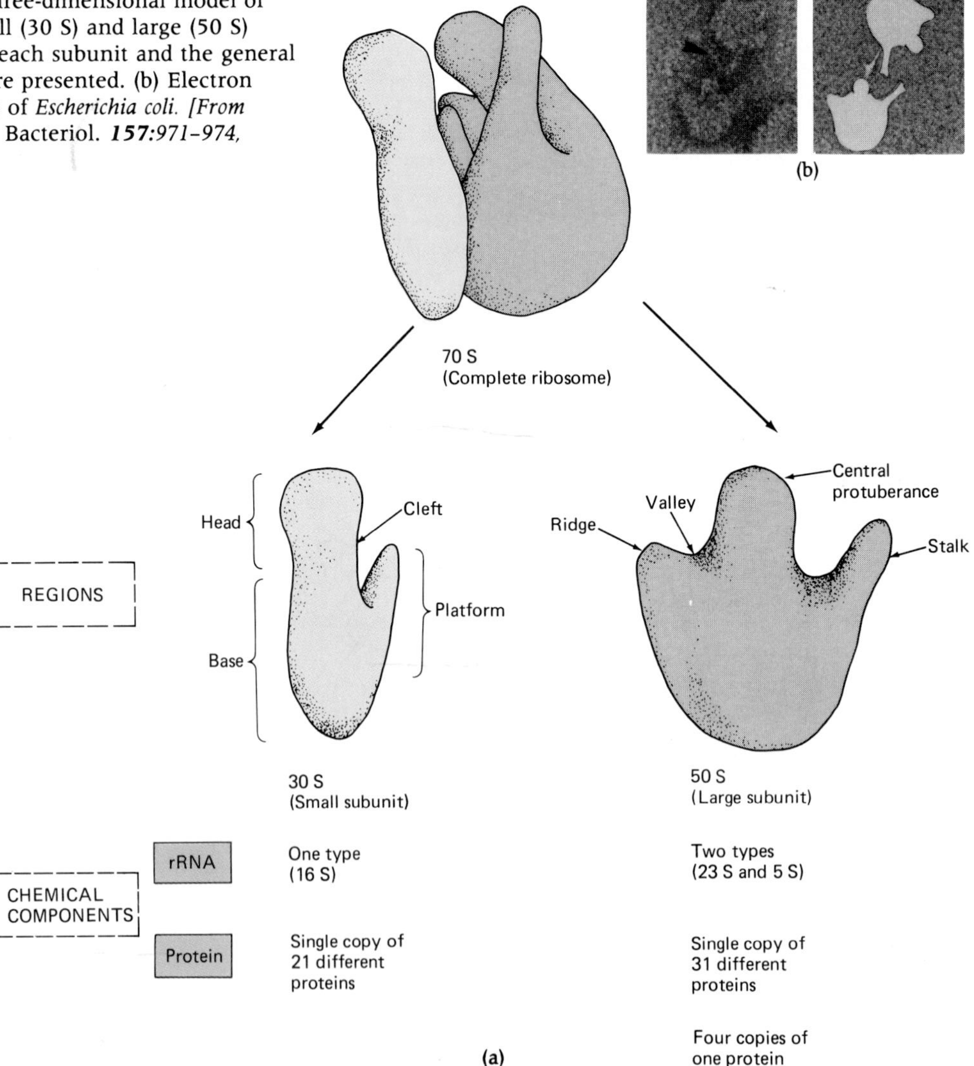

Figure 7–30 (right) (a) A three-dimensional model of a ribosome showing the small (30 S) and large (50 S) subunits. Specific regions of each subunit and the general chemical composition also are presented. (b) Electron micrograph of the ribosomes of *Escherichia coli. [From M. W. Clark, and J. A. Lake,* J. Bacteriol. ***157****:971–974, 1984.]*

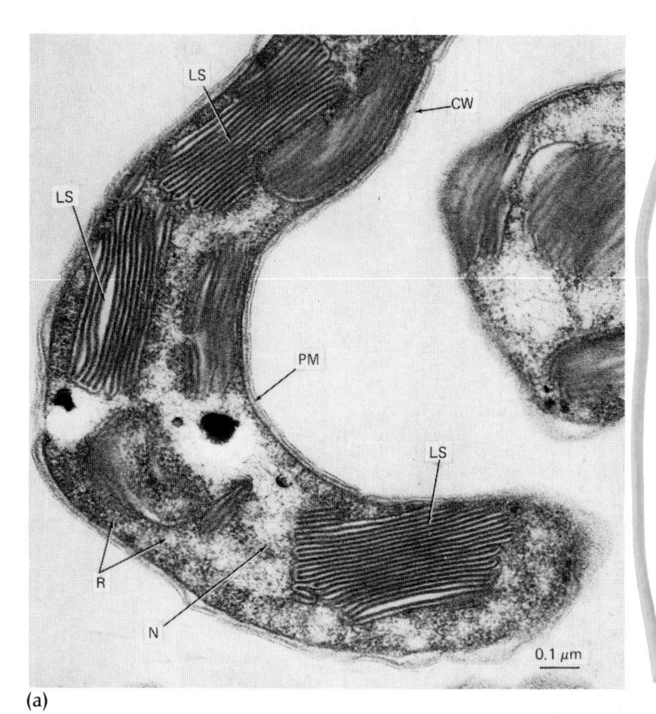

Figure 7–31 The photosynthetic machinery of procaryotes. An electron micrograph of a thin section of *Ectothiorhodospira mobilis* (ek-toh-thigh-oh-roh-doh-SPY-rah MOH-bill-iss) showing its general internal arrangement. Included are a multilayered cell wall (CW), photosynthesizing membrane, stacks (LS), nucleoplasm (N), plasma membrane (PM), and ribosomes (R). *[C. C. Remsen,* et al., J. Bacteriol. ***95****:2374–2392, 1968.]*

are estimated to account for approximately 40% of a bacterial cell's dry weight.

PHOTOSYNTHETIC MACHINERY

Purple and green bacteria as well as cyanobacteria (Color Photograph 64) carry out the essential steps of photosynthesis using specific light-harvesting pigments and cellular parts different from those of the complex eucaryotic organelle, the chloroplast (Figure 7–1b). The photosynthetic pigments of purple bacteria are incorporated in a complex cell membrane system (Figure 7–31a). In green bacteria, such pigments are housed in special organelles called *chlorobium vesicles* (Figure 7–31b). In cyanobacteria, the major photosynthetic pigments, phycobiliproteins, are localized in specialized structures known as *phycobilisomes.* These components are attached to the outer surfaces of membranes known as *thylakoids* (Figure 7–31c). Other distinguishing properties of these photosynthesizing procaryotes are given in Table 13–5.

SELECTED CYTOPLASMIC INCLUSIONS

During growth cycles, bacteria can accumulate several kinds of reserve materials, both water-insoluble and water-soluble. Such accumulations, which are nonliving bodies, are called *inclusions* and consist of lipids, polysaccharides, and certain inorganic substances. Some of these cellular inclusions are common to a wide variety of microorganisms, while others appear to be limited to a few and can there-

Figure 7–31
(b) An electron micrograph of *Rhodopseudomonas capsulata* (roh-doh-soo-doh-MOH-nass cap-SOOL-a-tah), a phototrophic bacterium exhibiting a large number of photosynthetic units (vesicles). *[N. Kaufman,* et al., Arch. Microbiol. ***131****:313–322, 1982.]* (c) Ultrastructure of the cyanobacterium *Plectonema boryanum* (plek-TOH-nee-ma borr-ee-AN-um). Note the following structures: cell wall (CW); nucleoid (N), containing DNA; phycobilisomes (PH); polyhedral carboxysome (PB); ribosomes (R); developing septum (S); and thylakoids (TH). *[M. Kessel,* et al., Can. J. Microbiol. ***19****:831–836, 1973.]*

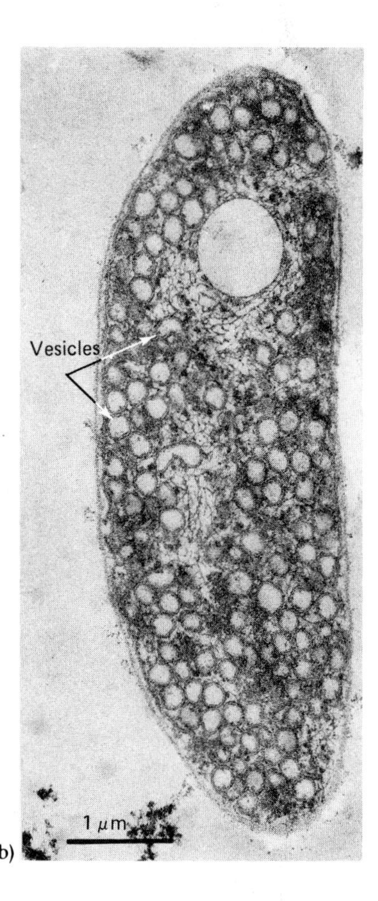

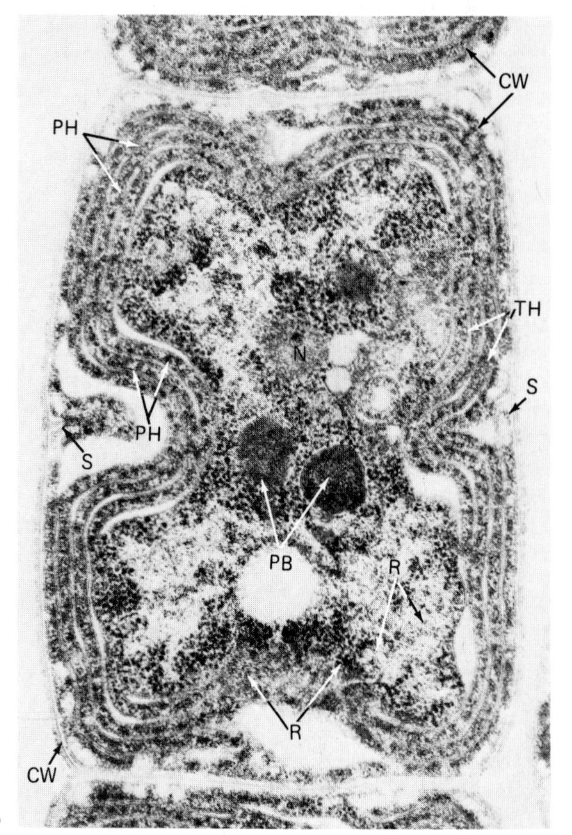

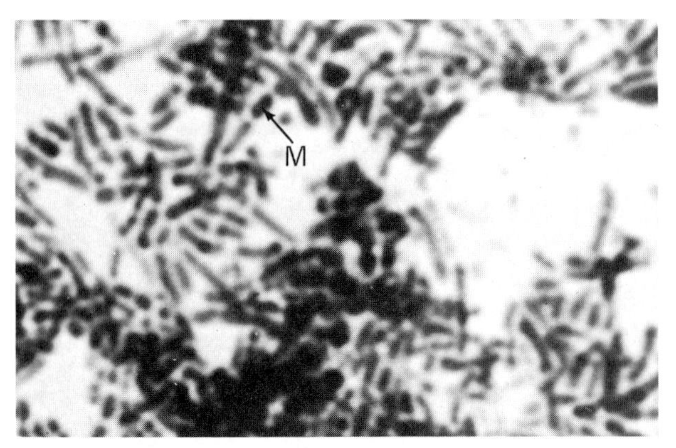

Figure 7–32 A preparation of *Corynebacterium diphtheriae* (ko-ri-nee-back-TI-ree-um dif-THI-ree-a) stained by an aged methylene blue solution, showing metachromatic granules (M). The characteristic arrangements of the bacterial cells shown here are associated with members of the genus *Corynebacterium*. *[Courtesy of J. Mosley, East Los Angeles College.]*

fore be extremely useful in the identification of certain bacterial species. For example, structures called *metachromatic granules* are conspicuous in *Corynebacterium diphtheriae* (Figure 7–32), the cause of diphtheria. These inclusions can be shown by several methods. The choice of procedure is generally determined by the chemical nature of the inclusion in question.

Cytoplasmic inclusions are divided into two major groups based on the presence or absence of a surrounding membrane.

NON-MEMBRANE-ENCLOSED INCLUSIONS

Non-membrane-enclosed inclusions are found in several types of microorganisms, including algae, bacteria, fungi, and protozoa.

METACHROMATIC GRANULES. These structures derive their name (*meta*, meaning "change," and *chromatic*, meaning "color") from the fact that they become red on staining with aged methylene blue solution. The actual chemical composition of metachromatic granules (also called *Babes-Ernst granules* or *volutin*) has been disputed for some time. Certain investigators hold that they are polymerized inorganic phosphate. Others contend that the granules are composed of nucleic acid, lipid, and protein. Metachromatic granules are believed to serve as temporary storage areas for reserve food. Unstained, the inclusions appear refractile in the light microscope and opaque in the electron microscope (see Figures 7–32 and 7–33).

POLYSACCHARIDE GRANULES. Glycogen and starch are two common types of *polysaccharide granules*. Iodine solutions have been used to distinguish between them. The areas which produce a bluish color contain starch. Those areas which produce a reddish-brown color have glycogen.

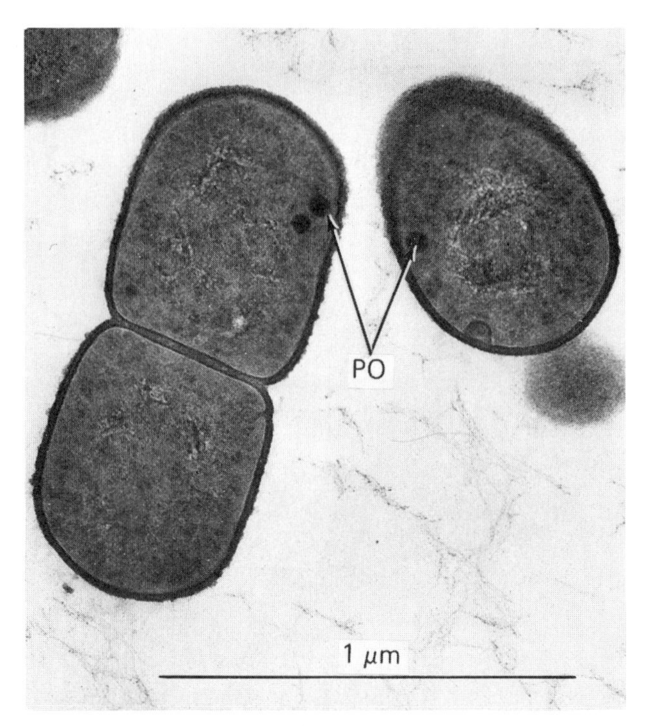

Figure 7–33 A section of *Microbacterium thermosphactum* (my-kroh-back-TI-ree-um ther-moh-SFAC-tum) showing polymetaphosphate inclusions (PO)—large reserves of inorganic phosphate. *[C. M. Davidson,* et al., J. Appl. Bact. ***31****:551–559, 1968.]*

MEMBRANE-ENCLOSED INCLUSIONS

CARBOXYSOMES. Nonunit membrane organelles called *carboxysomes* contain the enzyme responsible for carbon dioxide fixation in photosynthesis. The enzyme appears always to be present in polyhedral (many-sided) inclusions of several CO_2-fixing cyanobacteria (Figure 7–34).

LIPID INCLUSIONS. Lipid droplets have been reported in many bacteria, including species of the genera *Azotobacter* (a-zoh-toh-BAK-ter), *Bacillus, Corynebacterium, Microbacterium* (my-kroh-back-TI-ree-

Figure 7–34 The carboxysomes, inclusions found in many CO_2-fixing microorganisms. (a) The polyhedral-shaped inclusions. (b) The enzyme contents of a carboxysome. The large subunits of ribulosebisphosphate carboxylase. *[Courtesy W. N. Konigs.]*

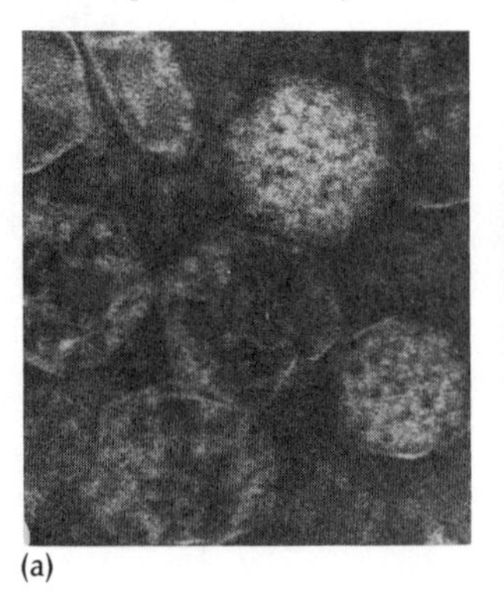
(a)

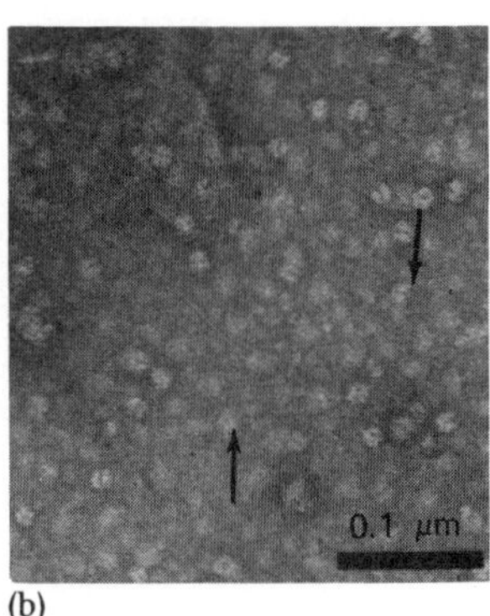

(b)

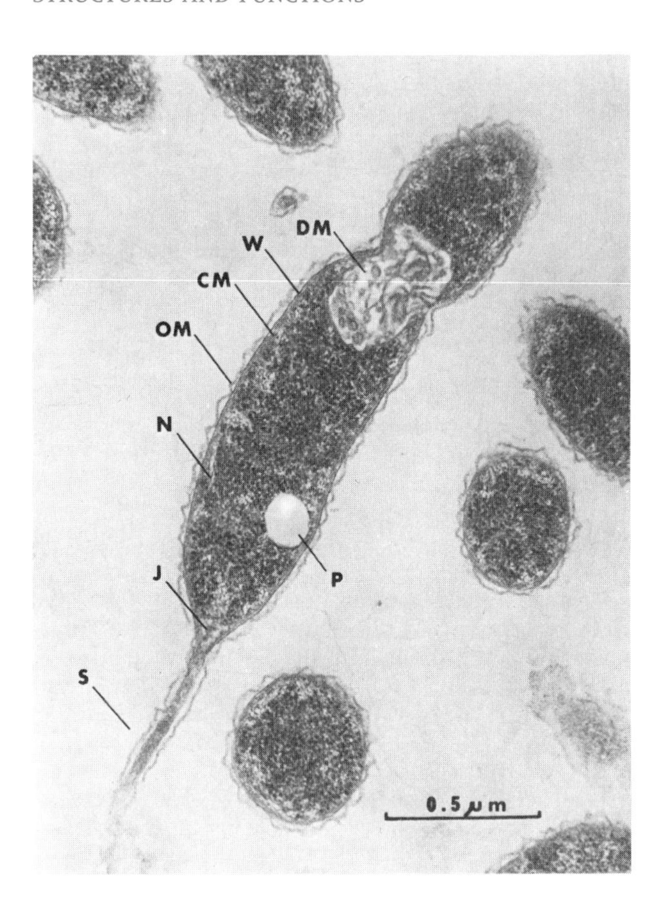

Figure 7–35 A dividing *Caulobacter crescentus.* These stalked (S) bacteria are found in practically every type of aquatic community and in many soils. The electron micrograph shows several procaryotic structures discussed in the text; cytoplasmic membrane (CM), division site of the mesosome (DM), outer membrane of the cell envelope (OM), nucleoplasm (N), granule of poly-hydroxybutyric acid (P), and peptidoglycan layer of the cell wall (W). *[J. S. Poindexter,* Microbiol. Revs. ***45:*** *123–179, 1981.]*

um), *Mycobacterium,* and *Spirillum.* Fat granules appear most regularly in gram-positive species and become more prominent as the cell ages. Lipids may take the form of either neutral fats or granules of poly-hydroxybutyric acid (Figure 7–35). The latter is the only food storage lipid for certain organisms.

For light microscopy, lipid or poly-hydroxybutyrate granules can be stained easily with fat-soluble dyes such as the Sudan series. In simple-stain (one-stain) preparations, these lipid inclusions appear as colorless areas.

GAS VACUOLES. Many aquatic procaryotes—including the photosynthetic cyano-, green, and purple sulfur bacteria—contain gas-filled structures known as *gas vacuoles* (Figure 7–36a). Electron microscopic examination shows that each gas vacuole consists of several individual gas vesicles (Figure 7–

Figure 7–36 Gas vesicles. (a) An electron micrograph showing typical cyanobacterial cell inclusions, gas vacuoles (G), and polyhederal (many-sided) bodies (PB). Note the great number of gas vacuoles. *[M. J. Kessel,* Ultrastruct. Res. ***62:****203–212, 1978.]* (b) Collapsed gas vesicles of the cyanobacterium *Anabaena flos-aquae* (an-a-BEE-nah floss-AK-quah), negatively stained. Magnification 200,000×. *[Courtesy of D. Branton,* Bact. Revs. ***36:****1–32, 1972.]*

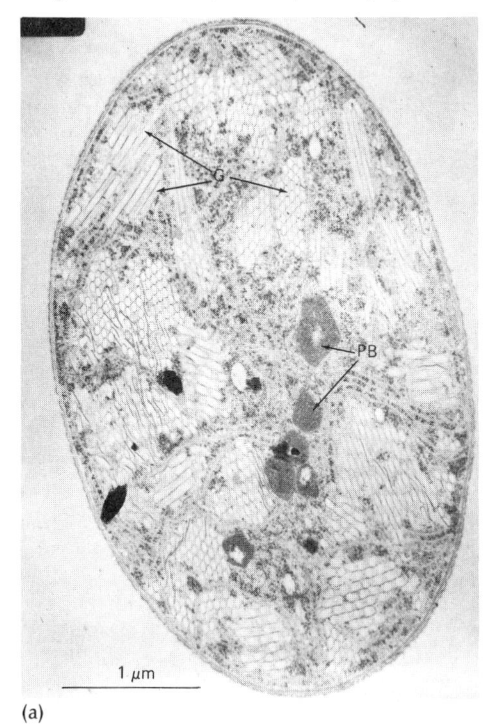

(a)

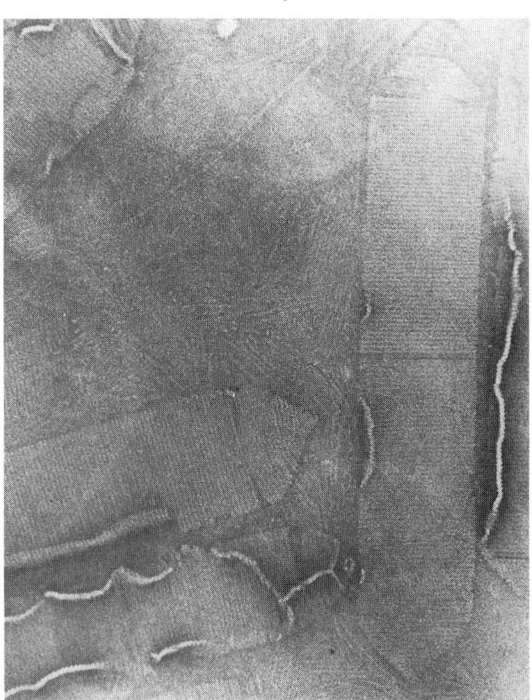

(b)

TABLE 7–1 Selected Characteristics of Bacterial Dormant Structures

	Structures			
Property	*Endospores*	*Exospores*	*Cysts*	*Conidia*
Heat resistance	Characteristically present	Present	Absent	To a limited degree
Cortex	Present	Absent	Absent	Absent
Dipicolinic acid (DPA)	Present	Absent	Absent	Absent
Number formed per cell	1	1–4	1	Formed in chains

36b). These vesicles are hollow cylinders with conical ends bounded by a layer of protein. They are arranged in regular rows in the vacuoles. Among the functions attributed to vacuoles are provision and regulation of cell buoyancy, light shielding, surface-to-volume regulation, and various combinations of these functions.

Specialized Cells

The cellular events in the cycles of certain procaryotes may change and lead to the formation of new cell types. This type of activity is *differentiation* at a primitive level. In bacteria, *dormant,* or resting, structures of four kinds can be produced: heat-resistant **endospores** and **exospores, cysts,** and heat-susceptible **conidia** (Table 7–1). Endospores (Figure 7–37 through 7–39), exospores (Figure 7–40), and cysts (Figure 7–41) are formed asexually, that is, without the union of nuclear material from two different types of cells. Usually one endospore or cyst is produced per cell. In certain bacterial species, up to four exospores are formed. Conidia present a different situation, as several are formed and used for purposes of reproduction (Figure 7–42). Table 7–2 (page 214) lists certain bacteria that form each type of dormant structure.

Cyanobacteria can produce cystlike cells called **heterocysts,** which are involved in nitrogen fixation, and sporelike cells called **akinetes** (Figure 7–43). The differentiation event may be permanent, as in the case of heterocysts, whereas differentiated cells such as the spore may revert to the original cell type.

Several species of bacterial genera are capable of forming heat-resistant endospores (Table 7–2). This property is most common among the members of *Bacillus* and *Clostridium* (klos-TRE-dee-um). Some cyanobacteria form small reproductive cells that are also called *endospores.* Heat-resistant exospores also can be found among species of *Methylosinus* (met-el-OH-sigh-nus) and *Rhodomicrobium* (roh-doh-my-KROW-bee-um) (Table 7–2). Both cyanobacterial endospores and heat-resistant exospores differ from bacterial endospores in chemical composition, structure, mode of formation, and development.

Figure 7–37 The bacterial endospore. An ultrathin section of *Bacillus sphaericus* (ba-SILL-lus spheer-EE-cuss), showing several components of the cell and its spore. Note the relationships of the outer coat (OC), lamellar midcoat (LMC), inner coat (IM), and cortex (CX), Low-density material (LDM) and the exosporium (E) also are shown. A large crystalline inclusion (C), which can be toxic to young forms of insects, is located between the exosporium and the spore. The bar marker represents 1 μm. *[A. Kalfon, J.-F. Charles, C. Bourgoin, and H. De Barjac,* J. Gen. Microbiol. ***130**:893–900, 1984.]*

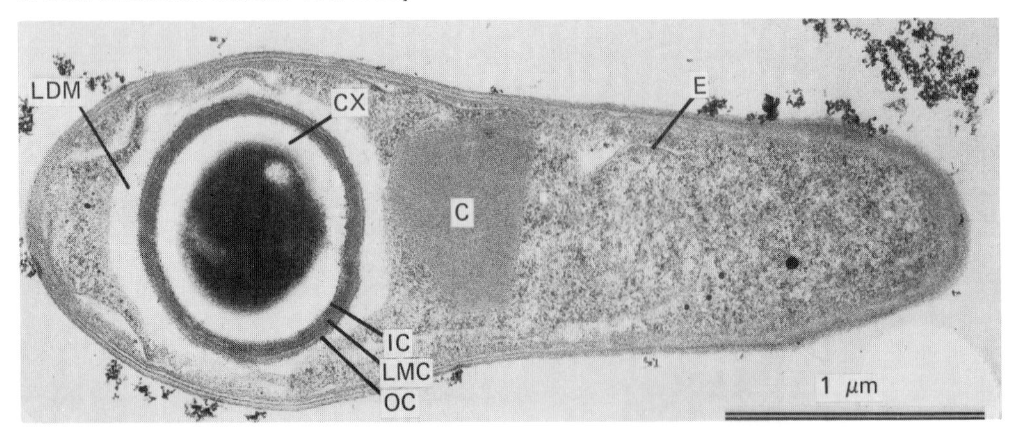

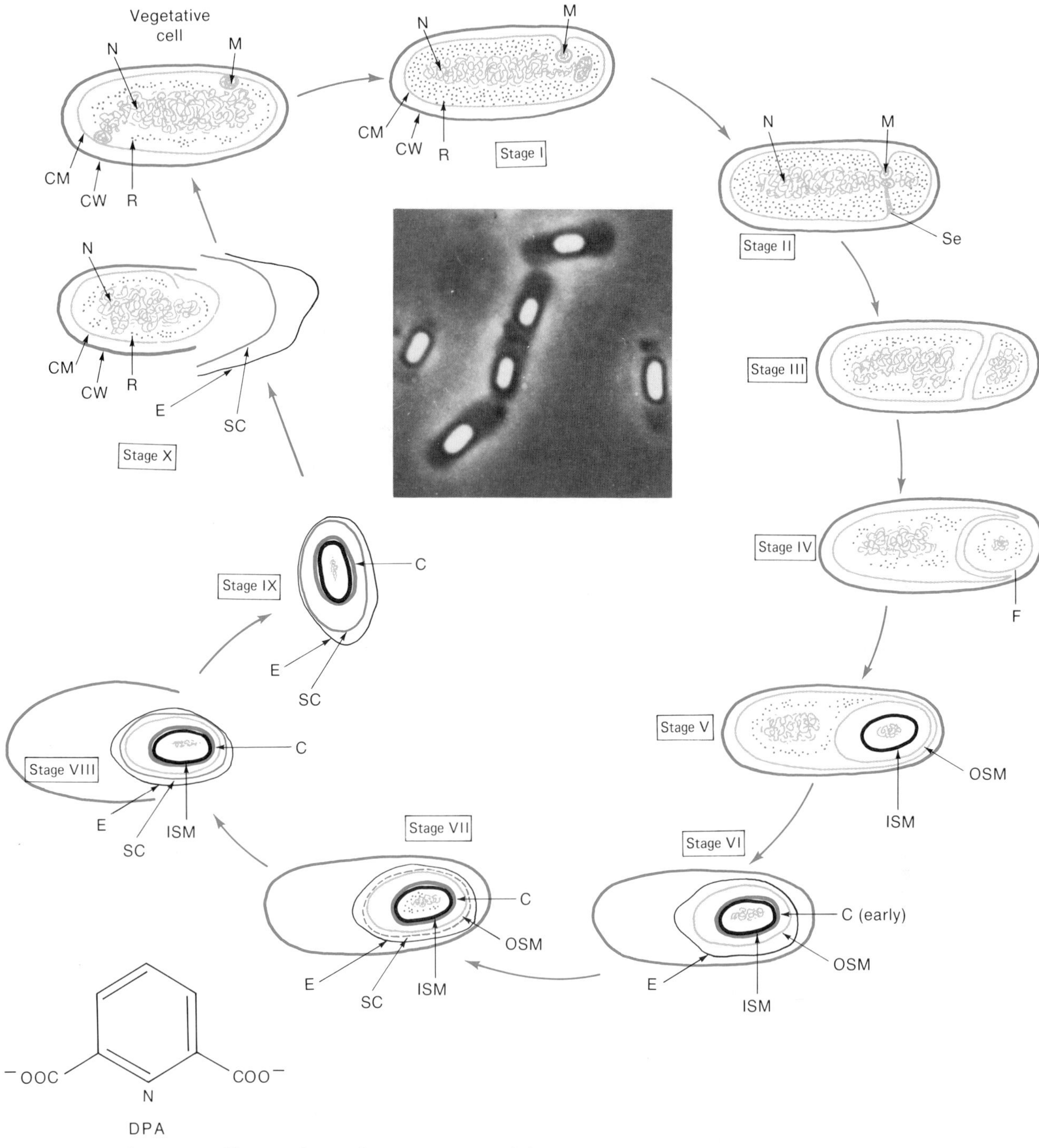

Figure 7–38 The stages of bacterial sporulation. Portions of the vegetative cell and spore are shown, including the cell membrane (CM), cell wall (CW), cortex (C), exosporium (E), forespore (F), inner spore membrane (ISM), mesosome (M), nuclear material (NM), outer spore membrane (OSM), septum (Se), and spore coat (SC). Note the formula for dipicolinic acid (DPA), a factor in the heat resistance of spores. Refer to the text for an explanation of the sporulation stages. The center micrograph shows the appearance of light deflecting bacterial spores. *[Courtesy of Drs. P. Fitz-James and A. Tam.]*

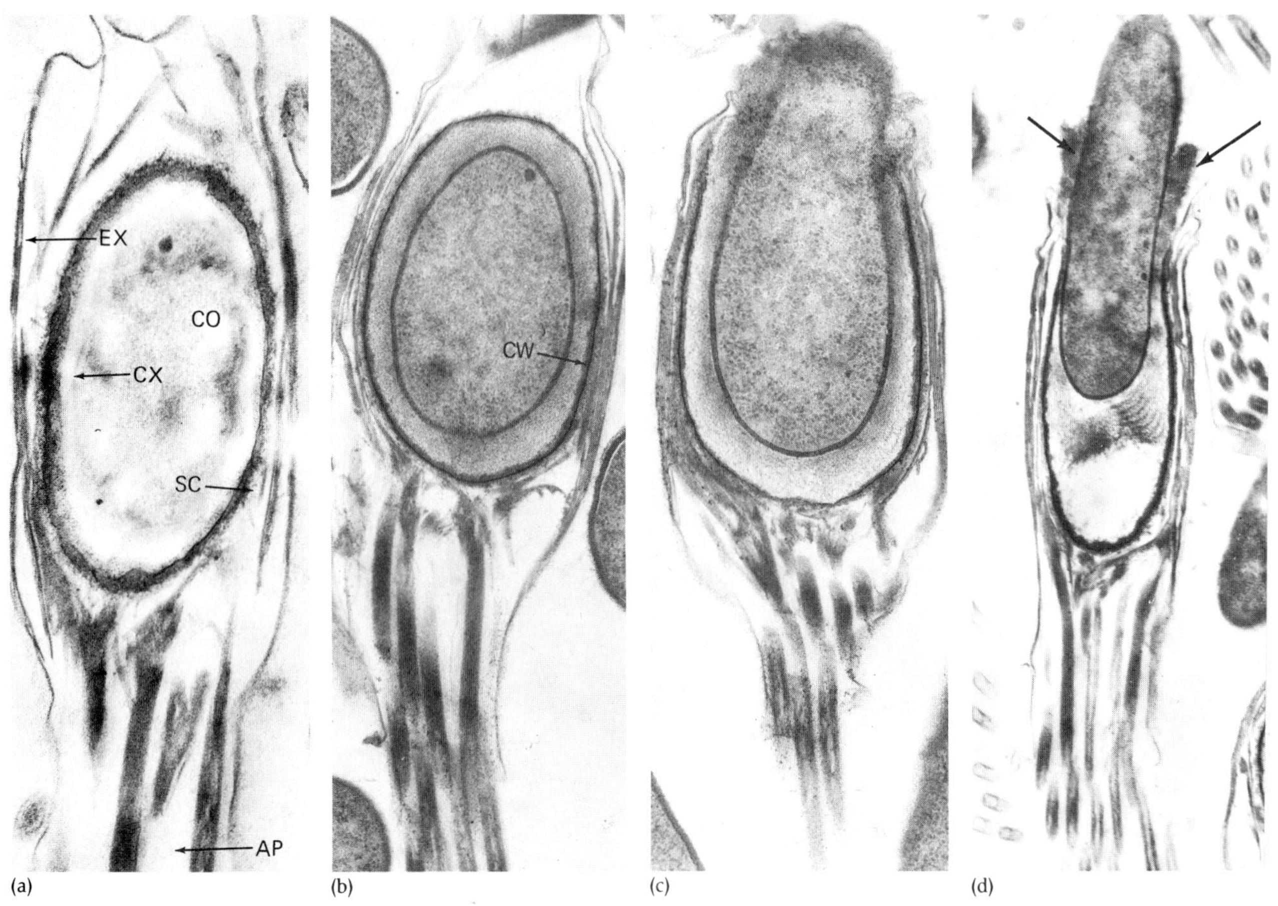

Figure 7–39 Spore germination bacterial cell outgrowth. (a) A thin section of a dormant spore of *Clostridium bifermentans* (klos-TREH-dee-um by-fur-MEN-tens). The components include appendages (AP), core (CO), cortex (CX), spore coat layers (SC), and exosporium (EX). (b) The early stage of germination. The cortex has been changed into a region without any visible ultrastructure inside the core wall (CW). (c) and (d) are the stages of elongation and outgrowth. In (c), the bacterial cell is constricted by the spore coat as it emerges. In (d), cortex material (arrows) is squeezed from the remains of the spore. *[W. A. Samsonoff,* et al., J. Bacteriol. ***101:****1038–1045, 1970.]*

Figure 7–40 Stages in bacterial exospore formation. (a) Vegetative cell showing the grove and surface structures. (b) Formation of the enlarged bud structure at the end of the vegetative cell. (c) Constriction of the bud, which leads to an invaginated exospore that is connected to the vegetative cell. The entire structure shows bilateral symmetry. The developing exospore from the vegetative cell. (e) A mature exospore showing a nearly spherical shape and thick exospore wall (EW). The bar equals 0.5 μm. *[After P. R. Duggan,* et al., J. Bacteriol. ***149:*** *354–360, 1982.]*

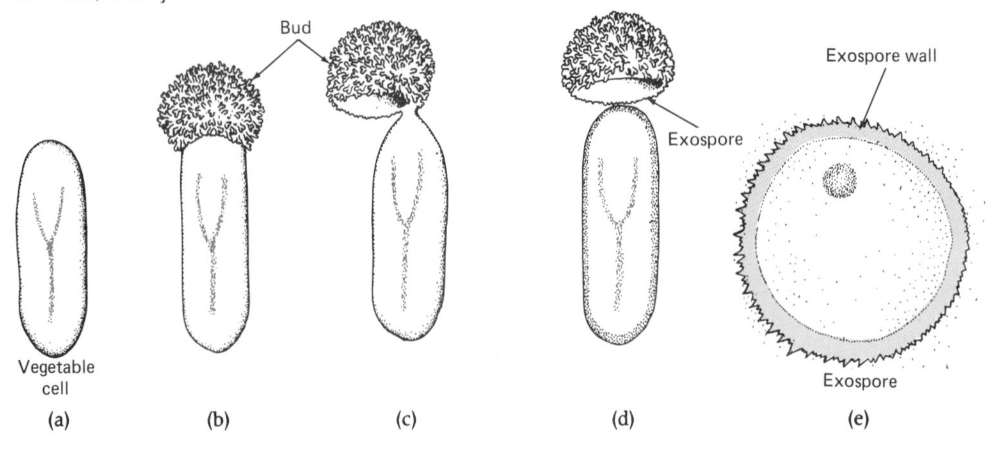

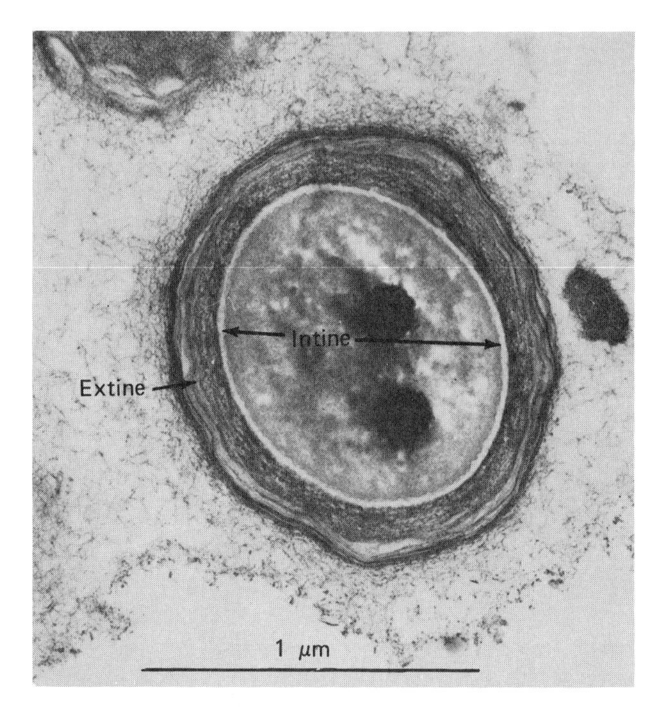

Figure 7–41 An ultrathin section through a cyst of *Azotobacter vinelandii* (a-zoh-toh-BAK-ter vin-LAN-dee-eye). This preparation has been stained with ruthenium red, which demonstrates capsular material. The cyst coat shown clearly demonstrates the outer exine and inner intine layers. Note the lamination visible in the exine layer. *[L. M. Pope, and O. Wyss,* J. Bacteriol. ***102****:234–239, 1970.]*

Bacterial Endospores

An **endospore** is a dormant structure formed inside an individual bacterial cell, or *sporangium*, during the spore formation, or **sporulation,** period (Figure 7–38). Disintegration of the parent cell releases the endospore (Color Photograph 14). The term *free spore* is used to describe such released structures.

A bacterial dehydrated spore resembles dried protein in its density and its ability to bend light rays. Spores have characteristic resistance to the effects of heat, drying, chemical disinfection, and radiation and impermeability to common stains. These characteristics can be used not only to detect the presence of spores but also to determine when these structures develop into bacterial cells—**germination.**

Dormancy, Sporulation, and Germination

DORMANCY

The **dormancy** of bacterial spores is well documented. Spores of *B. anthracis*, for example, have been found to survive for 60 years in soil at room temperature. Meat that was canned for 118 years was found to contain spores of a thermophilic bacillus. In addition, examination of ancient materials

Figure 7–42 Bacterial conidia. (a) The single heat-sensitive spores of the actinomycete *Thermomonospora formosensis* (ther-moh-moan-OS-por-ah for-moss-EN-sis). *[T. Hasegawa, S. Tanida, and H. Ono,* Intern. J. System. Bacteriol. ***36****:20–23, 1986.]* (b) Spiral chains of conidia of the actinomycete *Actinomadura yumaensis* (ak-tin-oh-MA-durr-a you-MAH-en-sis), which was isolated from a Yuma County, Arizona, soil sample. *[From D. P. Labeda, R. T. Testa, M. P. Lechevalier, and H. A. Lechavalier,* Intern. J. System. Bacteriol. ***35****:833–336, 1985.]*

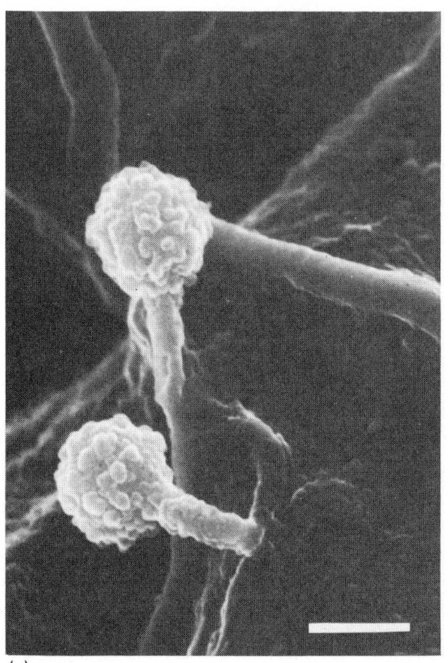

(a)

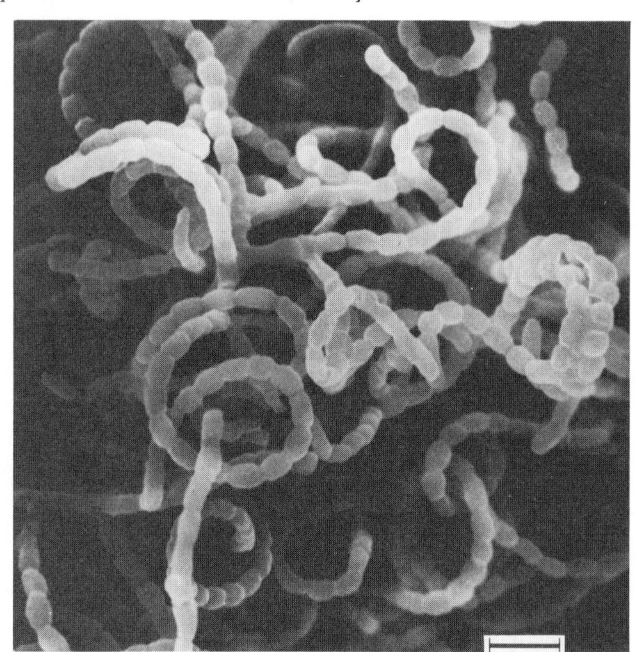

(b)

TABLE 7–2 **The Occurrence of Dormant Structures in Bacterial Genera**

Structures			
Heat-resistant Endospores	*Heat-resistant Exospores*	*Cysts*	*Conidia (Heat-susceptible Spores)*
Bacillus	*Methylosinus*	*Azotobacter*	*Actinomyces*
Clostridium	*Rhodomicrobium*	*Myxococcus*	*Micromonospora*
Desulfotomaculum		*Sporocytophaga*	*Nocardia*
Sporosarcina			*Streptomyces*
Thermoactinomyces			*Streptosporangium*

has uncovered the presence of certain spore formers, such as *Bacillus circulans* (B. sir-KOO-lans). The abdominal regions of mummies in Bohemia 180 to 250 years old contained several species of *Clostridium*—species not found on the surface of the mummies, on their coffins, or in the ground.

SPORULATION

Sporulation, or spore formation, is a survival mechanism and may also be a primitive mechanism by which a portion of a cell's cytoplasm and genetic material can be separated from the cell's contents and segregated into a distinct package, the spore.

During the 1880s, Behring viewed sporulation as an intermediate stage in the normal development of the bacterial cell, a process that may be partially or completely inhibited by physiological injury short of total prevention of growth. This concept still serves as a good description of sporulation, since it does not attempt to explain the purpose of sporulation beyond defining it as a natural step in the life cycle of certain bacteria. Sporulation represents an extremely complex process of differentiation. In most vegetative cells, the process leads to the formation of a new cell type that is totally different from the parent cell in chemical composition, fine structure (Figure 7–37), and physiological properties.

Specific genetic information and a suitable physiological environment, both internal and external, are necessary for sporulation. These physical factors are needed: (1) a narrow temperature range that approximates the optimum for vegetative growth, (2) a narrow pH range, about the same optimum level for vegetative growth, (3) increased oxygen when cells such as those of *Bacillus* spp. begin to sporulate.

These chemical substances appear to be required for sporulation: (1) glucose, (2) particular amino acids, (3) growth factors such as vitamins and minerals, including folic acid (for *B. coagulans*), phosphate, calcium, manganese, and bicarbonate.

Spore Characteristics. Several stages can be observed in sporulation (Figure 7–38). A variety of biochemical and physical activities are associated with each stage.

The first definite indication of the beginning of sporulation is formation of a cross wall, or septum, near one end of the cell (stages I, II, and III in Figure 7–38). This structure separates the cytoplasm and the DNA of the smaller cell from the rest of the cell contents. The larger cell engulfs the smaller one to produce a *forespore* (stage IV). This particular sporulation phase is called the *point of commitment;* the organism has reached a point from which it has no alternative but to complete the process.

Following the development of the forespore, rapid formation of the new spore-associated structures takes place (stages V and VI in Figure 7–38). These structures include the *cortex,* which develops between the two concentric sets of membrane of the forespore (the inner and outer spore membranes), and the *spore coat,* which forms outside the outermost membrane (stage VII in Figure 7–38). In certain species an additional thin layer, the *exosporium,* forms outside the spore coat (Figure 7–37 and stage VI in Figure 7–38). After endospore formation, the parent cell disintegrates, and the dormant structure becomes a released or free spore (stage IX in Figure 7–38). A mature spore is shown in stage VIII of Figure 7–38. The central area is the *core,* or *spore cytoplasm.* The core wall (CW) and cell membrane (CM) become the cell wall and cytoplasmic membrane of the cell that appears upon germination (stage X in Figure 7–38). The appearance of the spore under light microscopy and its imperviousness are attributed to the cortex and spore coats.

Chemical Makeup and Thermal Resistance. Chemically, the spore has very little free water. Calcium dipicolinate makes up about 10% of the structure's dry weight. The core region contains a characteristic quantity of DNA but relatively small amounts of enzymes and RNA. No messenger RNA (mRNA) is present. The spore wall and cortex contain glycopeptides. The coats are mostly proteins with a high concentration of cystine (amino acid) that permits cross-linkages. These bridges may be partly responsible for the spore's resistance to heat and to the actions of enzymes or chemical agents.

Onset of sporulation in many procaryotes is accompanied by the formation of protein-related antibiotic substances. In certain organisms such as *Bacillus thuringiensis* (ba-SIL-lus thur-in-jee-EN-sis), a protein crystal inclusion is formed. Only one of these *parasporal bodies* or crystalline inclusions (Figure 7–37) is produced per cell during sporulation.

Parasporal bodies are used to control certain insects. The production of these in some bacteria is genetically determined by high-molecular-weight plasmids.

The thermal resistance of spores appears to be caused by a variety of factors. Some of these are (1) the presence of specific heat-resistant components, such as thermostable enzymes, (2) an absence of free water, (3) a high content of various minerals, particularly calcium, and (4) the presence of dipicolinic acid (DPA) (Figure 7–38). A good correlation exists between the massive intake of calcium ions (Ca^{2+}), the production of DPA, and heat resistance. If spore-forming bacteria are grown in media deficient in calcium, the level of heat resistance appears to be correspondingly low. A similar phenomenon has been observed with spores grown in a manner that results in a DPA deficiency. Now, evidence seems to indicate that a high mineral and DPA content is closely related to the heat resistance of bacterial spores.

GERMINATION OF SPORES

The transition of the resting spore to an actively dividing vegetative bacterial cell is *germination* (stage X in Figure 7–38). The state of dormancy is broken by an overall process consisting of activation, germination, and outgrowth.

PHASE I: ACTIVATION. Dormant structures may fail to develop even when placed in an environment that allows vegetative growth. Generally, a spore will germinate if it is exposed to a triggering or germination agent, either physical or chemical. Physical triggering agents include a few minutes of heat treatment at 60° to 70°C, mechanical treatment, and incubation at 42°C for some time. Chemical factors that act as triggering agents include surface-wetting agents; inorganic materials such as chloride, cobalt, manganese, phosphate, and zinc; and various normal metabolic compounds, such as adenosine, alanine, calcium dipicolinate, carbon dioxide, glucose, lactic acid, and tyrosine.

Regardless of the agent involved, activation appears to involve (1) a breaking down of permeability barriers by activation of lytic enzymes, (2) physical disruption of the spore coat and/or the cortex material, and (3) a subsequent activation of carbohydrate metabolism. Thus *activation,* or phase I, represents a lag period during which the spore emerges from dormancy (Figure 7–39a).

PHASE II: GERMINATION. *Germination* is the transition of a heat-resistant, refractile, impervious structure into one that has lost these characteristics. Phase II appears to be initiated by the formation of a germination groove in the spore coat. This groove may serve as the means by which water and nutrients enter the spore. Several events occur with water uptake: (1) a significant increase in oxygen consumption and glucose oxidation, (2) swelling of the spore (Figure 7–39b), and (3) the excretion of approximately 30% of the spore's dry weight. The excreted solids are about equal amounts of calcium dipicolinate, various proteins, amino acids, and glycopeptide (Figure 7–39c). At this point, the spore is considered to have germinated, based on four major criteria: (1) sensitivity to heat, (2) stainability with simple dyes, (3) loss of refractility, as determined by its phase-contrast appearance, and (4) decrease in the optical density of the spore suspension.

PHASE III: OUTGROWTH. Once the spore has been activated and germinated, *outgrowth* may occur (Figure 7–39d). A complete growth medium is now required, or the germinated cell will not be able to reproduce. Initial stages of outgrowth include visible swelling of the spore within its spore coat and a rapid formation of a vegetative cell wall and a plasma membrane. The newly formed vegetative cell emerges from the spore coat and elongates. If an antibiotic such as chloramphenicol, a potent inhibitor of protein synthesis, is added during phase II, germination will occur without subsequent outgrowth. Figure 7–39 shows spore germination and outgrowth in *Clostridium bifermentans* (klos-TRE-dee-um by-fer-MEN-tens), as shown by electron microscopy.

Bacterial Exospores

Members of the genus *Methylosinus* (meh-thel-OH-sye-nus) and strains of the photosynthetic bacterium *Rhodomicrobium vannielii* (row-doh-my-KROW-bee-um van-NEEL-ee-eye) produce exospores and are among the few genera of true bacteria apart from *Bacillus* and *Clostridium* that produce bodies resistant to heat and drying. Unlike bacterial endospores, exospores do not contain DPA (Figure 7–38). *Methylosinus* exospore formation is initiated by the appearance of a budlike enlargement and a surrounding capsule at one end of a vegetative cell (Figure 7–40a). A constriction begins in the area where the bud is attached and continues until the developing immature, cup-shaped spore becomes detached. After separation, the exospore matures by forming a thick exospore wall and becoming spherical (Figure 7–40e). Exospores of *Methylosinus* are structurally similar to those of *R. vannielii* and lack a well-defined cortex.

Bacterial Cysts

The cysts produced by *Azotobacter* spp. and various myxobacteria are usually spherical, with contracted cytoplasm and a thick wall. The cytoplasm of *Azoto-*

bacter spp. usually contains evident nuclear material, lipid globules, and electron-dense bodies (Figure 7–41). The thick wall of this organism consists of an inner layer, or **intine,** and an outer layer, or **exine.** Cysts of *A. agilis* (A. aj-IL-iss) are not heat resistant, while those of another species, *A. chroococcum,* have considerable heat resistance. Cysts are resistant to drying and are formed singly within vegetative cells.

Heat-susceptible Spores, Heterocysts, and Akinetes

The dormant structure of the funguslike actinomycetes is an asexual spore that is formed at the end of special surface (aerial) cells by a process of fragmentation. These resting bodies are called *conidia* (Figure 7–42). Mild heat resistance seems to be the only characteristic shared by some conidia and bacterial spores. Among the differences between these two types of dormant bodies are absence of a cortex, absence of DPA, and a lack of refractility.

Cyanobacteria grow in either filamentous (threadlike) or nonfilamentous form. The filamentous forms produce chains of vegetative cells (Figure 7–43a) known as **trichomes** in an enclosing gelatinous sheath. The filamentous cyanobacteria reproduce by the breakup of filaments into short segments called **hormogonia.** Fragmentation, which may be caused by the wave action of water or the feeding habits of aquatic animals, results in a population increase of cells. Cyanobacteria can also reproduce by forming differentiated cells called *heterocysts* and *akinetes,* which occur singly at intervals along the filament (Figure 7–43a). The heterocyst is similar in size to vegetative cells but differs from them in having a thickened wall (Figure 7–43b) and two swollen points called *papillae.* The major function of heterocysts is nitrogen fixation. Akinetes (Figure 7–43c) are usually larger than most vegetative cells.

Table 7–3 summarizes the properties, functions, and chemical composition of procaryotic structures.

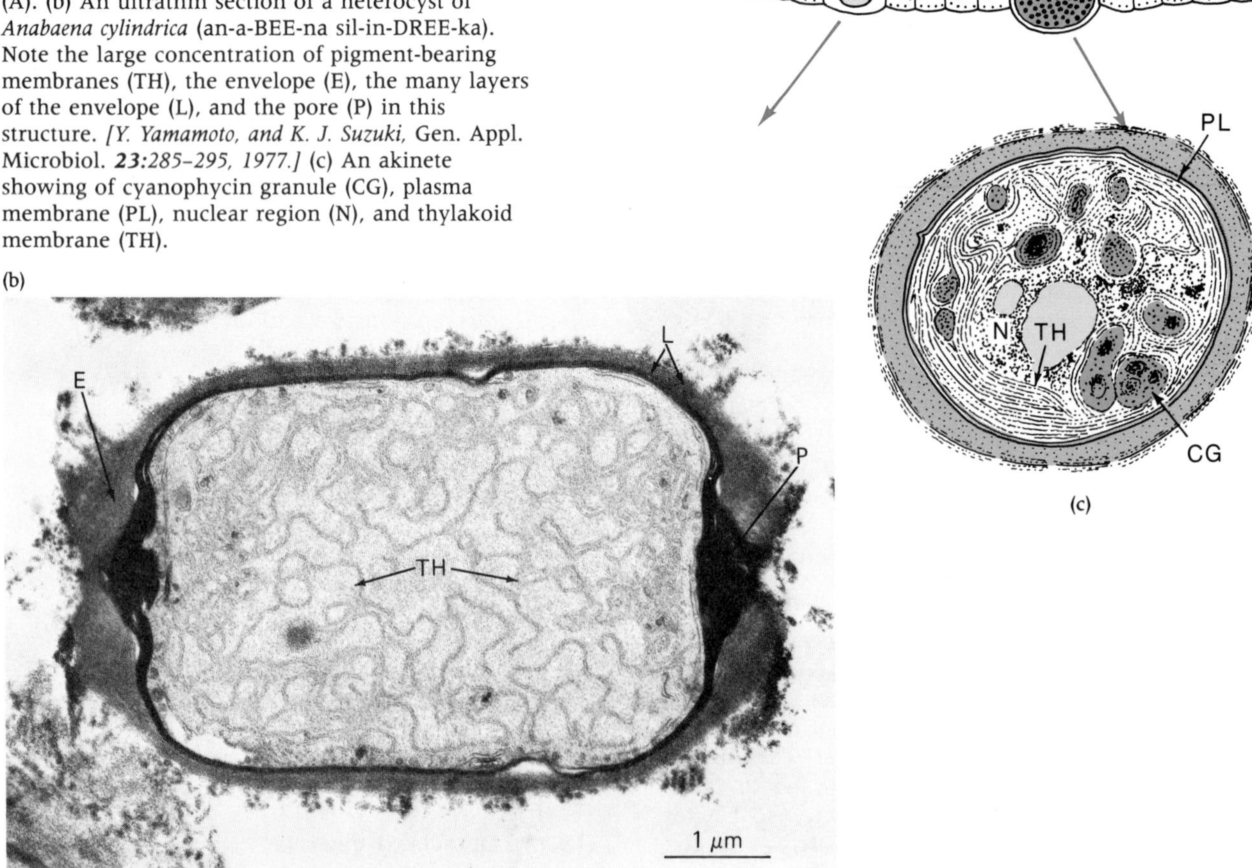

Figure 7–43 Heterocysts and akinetes. (a) The filament of a cyanobacterium. A filament showing vegetative cells (V), heterocysts (H), and akinetes (A). (b) An ultrathin section of a heterocyst of *Anabaena cylindrica* (an-a-BEE-na sil-in-DREE-ka). Note the large concentration of pigment-bearing membranes (TH), the envelope (E), the many layers of the envelope (L), and the pore (P) in this structure. *[Y. Yamamoto, and K. J. Suzuki,* Gen. Appl. Microbiol. ***23****:285–295, 1977.]* (c) An akinete showing of cyanophycin granule (CG), plasma membrane (PL), nuclear region (N), and thylakoid membrane (TH).

TABLE 7–3 **Properties, Functions, Activities, and Chemical Composition of Procaryotic Structures**

Structure	*Properties, Functions, Activities*	*Major Chemical Components*
Akinete	1. Limited protection? 2. Resting cell (spore) 3. Nitrogen fixation?	General components of a cyanobacterial procaryotic cell
Axial filament	Movement in spirochetes	Protein
Carboxysome	Utilization of carbon dioxide	Protein
Cell membrane	1. Selective barrier between the cell's interior and exterior 2. Biosynthesis 3. Chromosome separation	Protein, no sterols, phospholipids
Cell wall	1. Encloses procaryotic cell 2. Provides shape and mechanical protection 3. Contains bacterial virus receptor sites	Amino sugars (*N*-acetylglucosamine and *N*-acetylmuramic acid), protein, lipopolysaccharides
Chlorobium vesicle	Photosynthesis	Protein, lipid, photosynthetic pigment
Cyst	1. Limited protection? 2. Resting stage	General components of procaryotic cell
Endospore	1. Protection against physical heat, pH changes, and drying 2. Cellular differentiation 3. Reproduction, for some cyanobacteria[a]	General components of procaryotic cell plus calcium and dipicolinic acid (DPA)
Exospore	1. Protection against physical heat and drying 2. Cellular differentiation	General components of procaryotic cells; lacks DPA
Flagellum	Movement	Protein
Gas vesicle	1. Regulates buoyancy 2. Light shielding	Protein, common gases
Genome (nuclear region, or nucleoplasm)	Contains all of the genetic information of the procaryote	DNA
Glycocalyx (includes capsules and S layers)	1. Protection against antibiotics, bacteriocins, bacterial viruses, immunoglobulins, and phagocytosis 2. Enables bacteria to stick to other cells and to surfaces of inert materials (soil, sand, etc.) 3. Increases virulence	Glycoprotein, polysaccharide
Heterocyst	Nitrogen fixation	Protein, lipid
Mesosome	1. Nucleoplasm division 2. Sporulation 3. Biosynthesis 4. Cell wall formation	Protein, lipid
Metachromatic granules	Storage of reserve nutrients	Nucleic acids, lipid, protein, phosphate

[a] Endospores of cyanobacteria differ both in chemical composition and function from those of other procaryotes.

TABLE 7–3 (continued)

Structure	*Properties, Functions, Activities*	*Major Chemical Components*
Plasmid	Carries genetic factors associated with drug resistance and certain metabolic enzymes	Extrachromosomal DNA
Pilus	1. Attachment 2. Formation of surface films 3. Receptor sites for viruses	Protein
Pilus (sex)	Transfer of genetic material	Protein
Ribosome	Protein synthesis	Protein, ribosomal RNA
Spinae	Protection against ingestion by predatory microorganisms	Protein
Spirae	Unknown	Protein
Thylakoid	Photosynthesis	Protein, lipid, photosynthetic pigment

Summary

Morphological properties of bacteria are important because they can be used to identify bacterial species, locate specific structures and their functions, and show the relationships of structures to the overall function of the organism.

Bacterial Size

The dimensions of bacteria border on the limits of the resolution of the bright-field microscope.

Shapes and Patterns of Arrangement

1. Most bacteria have one of four principal shapes: *coccus (spherical), rodlike (cylindrical), spiral,* or *square.*
2. Square cells appear as flat rectangular boxes that may occur in pairs or in associations containing as many as 64 cells.
3. Cocci can appear as pairs *(diplococci),* chains *(streptococci),* fours in a square arrangement *(tetrads),* cubical packets *(sarcinae),* and irregular clusters *(staphylococci).*
4. Rods may occur singly *(bacilli),* in pairs *(diplobacilli),* or in chains *(streptobacilli).*
5. Small, rounded rods are known as *coccobacilli.*
6. Spiral bacteria vary in the number and amplitude of spirals and the length and rigidity of their coils.
7. *Vibrios* are bacteria that consist of only a portion of a spiral.

Structures and Functions

1. Procedures used for cell disruption include (1) grinding or violent shaking with abrasives, (2) pressure cell disintegration, and (3) sonic and ultrasonic disintegration.
2. The separation and isolation of cellular parts can be obtained with different centrifugation speeds.

The Bacterial Surface

Recognized structures associated with the bacterial cell surface include (1) surface appendages *(flagella, axial filaments, pili, spines,* and *spirae),* (2) the *glycocalyx,* which includes *capsules* and *S layers,* (3) cell walls, and (4) protoplasts and spheroplasts.

Surface Appendages

Bacterial surface-associated structures differ both in function and overall appearance. However, they share certain common properties and participate in an organism's environmental responses.

FLAGELLA

1. Flagella are responsible for bacterial motility.
2. Differences in the thickness, number, and arrangement of flagella exist among bacterial species.
3. Organisms without flagella are referred to as *atrichous,* those with one as *monotrichous,* and those with several as *multitrichous.*
4. Spreading rapidly or swarming over media surfaces involves microcolonies, or rafts, of bacteria with flagella surrounding the cell.
5. Flagella originate in the cell's cytoplasm from a *basal granule* or *body.*
6. Bacterial flagella are thinner than those of eucaryotic cells and are composed of a protein called *flagellin.*
7. Flagella give bacteria the ability to migrate toward favorable environments *(chemotaxis)* and a means by which to increase nutrients or decrease poisonous substances near the cell surface.
8. The movement of flagellated bacteria is believed to be associated with mechanical changes in the basal body and the continuous generation of ATP. The rotation of each flagellum pushes a bacterial cell in a specific direction.

AXIAL FILAMENTS

1. Spirochetes move by structures called *axial filaments.*

2. These filaments are composed of two fibrils identical in structure to flagella.

PILI

1. Pili (fimbriae) can be seen only with the aid of special techniques and electron microscopy.
2. Pili differ from bacterial flagella in several ways: smaller diameter, general appearance, and a general lack of association with an organism's true motility.
3. Pili are composed of a specific protein called *pilin*.
4. Pili enable bacterial cells to stick to surfaces, attach to other bacteria prior to the transfer of DNA, and form surface films.
5. Pili provide receptor sites for some bacterial viruses (bacteriophages).
6. Sex (F) pili are used for the transfer of genetic material.

SPINES AND SPIRAE

Unusual rigid protein appendages, *spines* or *spinae*, are carried by certain marine gram-negative bacteria. These structures may protect against ingestion by larger microorganisms such as predatory protozoa.

THE GLYCOCALYX

1. A *glycocalyx* is any polysaccharide-containing structure outside of the bacterial cell wall. It may be composed of fibrous polysaccharides or spherical glycoproteins produced by the cell itself.
2. Functions and activities of the glycocalyx include (a) enabling bacteria to attach to solid surfaces in marine and freshwater systems, soil, animal and plant tissues, and other microbial habitats, and (b) providing protection from antibacterial agents such as antibiotics, bacterial viruses, immunoglobulins, and phagocytes.
3. Glycocalices are subdivided into two types: S (regular surface) layers and capsules.
4. *S layers* consist of a single type of protein units that forms definite patterns covering the entire cell surface. The major function of such layers is protective.
5. *Capsules* are organized accumulations of polysaccharides and/or polypeptides on cell walls. These structures vary in thickness and rigidity. Formation is determined by genetic and environmental factors, and may be lost on repeated transfer in the laboratory.
6. Colonies of capsule-producing organisms appear as sticky, mucoid growths.
7. Encapsulation protects pathogenic organisms against certain drugs, phagocytosis, and bactericidal factors in a host's body.
8. *Slimes* are unorganized accumulations of material similar to that found in capsules.

Cell Walls

1. The functions of cell walls include (a) protection against rupture caused by osmotic pressure differences between intracellular and extracellular environments, (b) support for flagella, (c) maintenance of characteristic shape of organisms, and (d) control of the passage of certain molecules.
2. Bacterial cell walls contain (a) two simple amino sugars, *N-acetylglucosamine (NAG)* and *N-acetylmuramic acid (NAM)*, which are interconnected by amino acids to form the rigid mucopeptide or peptidoglycan layer of the wall, (b) a variety of naturally occurring amino acids, and (c) unnatural forms of amino acids.
3. Some bacterial walls contain *diaminopimelic acid*.
4. Certain chemical and physical differences exist between the walls of gram-positive and gram-negative organisms.
5. Gram-positive walls are relatively thicker than those of gram-negatives, exhibit a delicate ordering of chemicals, and contain special polysaccharides, such as teichoic and teichuronic acids.
6. Gram-negative cell walls are multilayered membranous structures and are more complex than those of gram-positive cells. The peptidoglycan layer is surrounded by a lipopolysaccharide outer membrane that interferes with the penetration of antibiotics such as penicillin.
7. The murein sacculus or peptidoglycan layer is the backbone layer of bacterial cell walls.

Protoplasts and Spheroplasts

1. Bacterial cells treated with or cultured in the presence of certain antibiotics or enzymes may lose all or a portion of their walls, and form protoplasts or spheroplasts.
2. *Protoplasts* are bacterial cells with little or no cell wall material.
3. *Spheroplasts* are cells having some cell wall material.
4. *L forms* are morphologically equivalent to protoplasts and spheroplasts. They occur naturally and revert to cells with normal cell walls in a suitable environment.

Structures Within the Cell Wall

THE PLASMA MEMBRANE

1. The *plasma membrane* is a semipermeable structure separating the cell wall from a bacterium's cytoplasm.
2. It functions as an osmotic barrier to regulate and control the internal environment of a cell, and as an aid for the outward transport of wastes.
3. Permeases in the membrane transport various molecules into the cell.
4. Chemically, the membrane consists of proteins and phospholipids, and ranges in thickness from 5 to 8 nanometers (nm).
5. The *periplasmic space* and its contents, the *periplasm*, separate the cell wall from the plasma membrane in gram-negatives. The periplasm is important to the maintenance of the wall and the processing of nutrients and waste products by the cell.

MESOSOMES

1. *Mesosomes* have been reported primarily for gram-positive bacterial species.
2. The activities with which mesosomes have been associated include (a) cell wall formation, (b) division of nuclear material, (c) cellular respiration, and (d) spore formation.

PROCARYOTIC NUCLEOIDS (GENOME, NUCLEOPLASM REGION) AND PLASMIDS

1. Procaryotes lack the distinct nucleus of eucaryotes and the associated structures and features, such as mitotic apparatus, nuclear membrane, and nucleolus.

2. Essential genetic information, the *genome,* of a procaryote is contained within a single chromosome, which is composed of a single DNA molecule. The nucleic acid is contained within a centrally located nucleoid and in a ribosome-free space.
3. Many bacteria contain additional genetic information in the form of extrachromosomal circular DNA molecules, known as *plasmids.*
4. Plasmids can replicate independently and carry genetic information for functions that are not essential to the life of an organism.

RIBOSOMES
1. Procaryotic ribosomes are important to the formation of proteins. They differ from eucaryotic ribosomes in chemical composition and smaller size.
2. Ribosomes are submicroscopic and composed of protein and ribosomal RNA. Each ribosome consists of two subunits.

PHOTOSYNTHETIC MACHINERY
1. In purple bacteria, *photosynthetic* pigments are incorporated into a complex cell membrane system.
2. Green bacterial pigments are housed in *chlorobium vesicles.*
3. Major cyanobacterial photosynthetic pigments, phycobiliproteins are localized in *phycobilisomes,* which are attached to outer surfaces of *thylakoids.*

SELECTED CYTOPLASMIC INCLUSIONS
1. *Inclusions* are accumulations of reserve materials.
2. Cytoplasmic inclusions can be divided into two major groups, non-membrane-enclosed and membrane-enclosed. The former include *metachromatic granules* and *polysaccharide granules.*
3. Membrane-enclosed inclusions include *carboxysomes, lipid inclusions,* and *gas vacuoles.*

Specialized Cells
1. Examples of differentiated cells include *heat-resistant endospores* and *exospores, cysts,* heat-susceptible *conidia,* and the *heterocysts* and *akinetes* of cyanobacteria.
2. These structures can be distinguished from one another by their respective structure, chemical composition, mode of formation, and development.
3. Differentiation may be permanent, as with heterocysts, or temporary, as with the spore.

Bacterial Endospores
1. An endospore is a *dormant* structure formed within an individual bacterial cell. Endospores appear within a parent cell as spore formation occurs.
2. Spores are able to exist outside of the cells that formed them.
3. Bacterial spores exhibit unusual resistance to heat, drying, chemical disinfection, and radiation and an impermeability to common stains.
4. The spore formation process is called *sporulation.* The development of spores into vegetative bacteria is known as *germination.*

DORMANCY, SPORULATION, AND GERMINATION
1. A spore is a dormant, or resting, cell. It can remain in this state for many years.
2. Specific physical factors, including temperature and pH, as well as nutrients such as glucose, particular amino acids, vitamins, and minerals, are needed for sporulation to occur.
3. A mature spore has a central area, or *core,* a *cortex,* a spore wall, and a cell membrane.
4. Chemically, the heat-resistant spore has very little water but does have a high content of dipicolinic acid and calcium, which seems to correlate with thermal resistance.
5. The change of a resting spore into an actively dividing vegetative cell involves stages of *activation, germination,* and *outgrowth.*

Bacterial Exospores
1. The photosynthetic bacterium *Rhodomicrobium vannielii* and members of the genus *Methylosinus* form exospores resistant to heat and drying.
2. Spore formation starts at one end of a vegetative cell. Mature spores do not contain DPA.

Bacterial Cysts
Members of the genus *Azotobacter* produce distinctive resting cells called *cysts.* These structures are resistant to drying but not to heat.

Heat-susceptible Spores, Heterocysts, and Akinetes
1. The dormant structure of the actinomycetes is an asexual spore (sometimes called the *conidium*) formed at the end of cells by fragmentation.
2. These spores are not heat resistant.
3. Specific conditions are necessary for germination.
4. Some cyanobacteria form rounded specialized cells known as *heterocysts.* These cells arise from *vegetative* cells and are the major sites for the utilization of atmospheric nitrogen (nitrogen fixation).
5. *Akinetes,* or spores, protect cyanobacteria that form them against drying and freezing.

Questions for Review

1. Distinguish between procaryotic and eucaryotic cells and give representative examples of each.
2. Compare structural similarities and differences between a typical plant cell and a photosynthetic procaryote.
3. Construct a table listing the functions and chemical composition of the following organelles:

a. bacterial cell wall
b. capsule
c. mesosome
d. pilus
e. bacterial cell membrane
f. cilium
g. bacterial flagellum
h. thylakoid
i. phycobilisome
j. endoplasmic reticulum
k. ribosome
l. nucleolus
m. polyribosome
n. mitochondrion
o. gas vacuole
p. sex pilus

4. Compare the cell walls of gram-positive and gram-negative bacteria.

5. Describe the different cell group arrangements found among bacteria. Can these arrangements be utilized for purposes of classification?

6. Identify the morphological arrangements shown in Figure 7–44.

7. List the basic shapes of bacteria.

8. List and give the general properties of spores formed by bacteria.

9. What procedure is necessary to observe the following cellular parts?
a. bacterial spores
b. cell membranes
c. flagella
d. ribosomes
e. capsules
f. mesosomes
g. periplasm
h. pili
i. murein sacculus
j. axial filaments

10. What is a bacterial glycocalyx?

11. a. Briefly describe the general organization of a gram-negative cell wall.
b. How does it differ from that of a gram-positive cell wall?
c. Why is penicillin less effective with gram-negative bacteria than with gram-positives?

12. How does the structure containing the genetic information of bacteria differ from that of a typical animal cell?

Figure 7–44 *[D. L. Shungu, J. B. Cornett, and G. D. Shockman,* J. Bacteriol. ***138****:598–608, 1979.]*

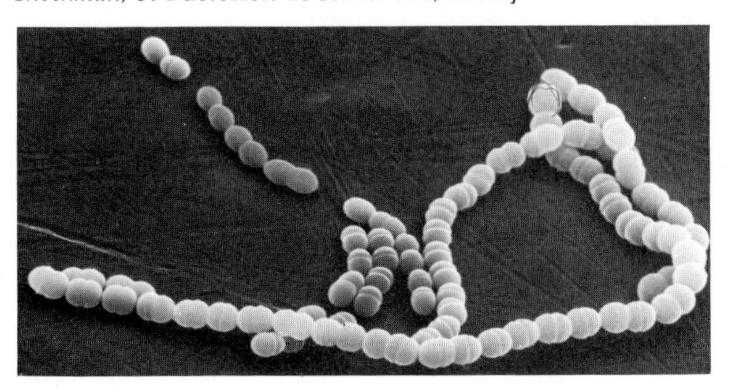

13. a. Compare a bacterial flagellum to one of a eucaryotic cell.
b. Compare a bacterial ribosome to one of a eucaryotic cell.

14. a. What is a bacterial spore?
b. Describe the process of sporulation.
c. How can bacterial spores be destroyed?
d. Differentiate between endospores and exospores.

15. How does a protoplast differ from a spheroplast?

16. What is the importance of each of the following structures to the virulence (disease producing capability) of a microorganism, disease transmission, or diagnosis?
a. bacterial flagella
b. capsules
c. pili
d. endospores
e. gram-negative cell walls
f. exospores
g. heterocysts
h. S layer
i. akinete

17. How could one obtain a preparation of cell walls from a bacterial culture?

18. What are L forms?

19. Identify the bacterial structures indicated in the electron micrograph shown in Figure 7–45.

20. What type of cell wall does the procaryote in Figure 7–45 possess?

Figure 7–45 An ultrathin section of *Staphylococcus aureus* (staffy-low-KOK-kuss O-ree-us). *[T. J. Popkin,* et al., J. Bacteriol. ***107****:907, 1971.]*

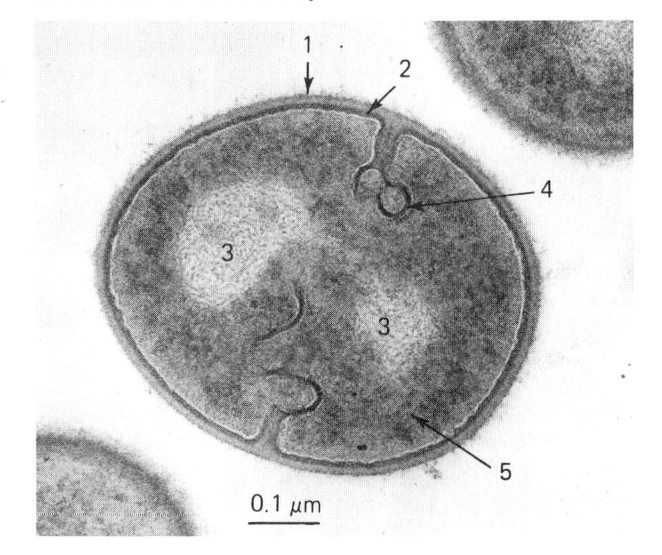

Suggested Readings

BERG, H. C., "How Bacteria Swim," *Scientific American* **234:** 36, 1975. *Describes how the bacterial flagellum, a thin helical filament, moves the bacterium, not by waving or beating, but by rotating like a propeller driven by a reversible rotary action at its base.*

BEVERIDGE, T. J., "Ultrastructure, Chemistry, and Function of the Bacterial Wall," *International Review of Cytology* **72:**229, 1981. *A well-organized article that defines the properties of a bacterial cell wall.*

COSTERTON, J. W., and R. T. IRVINE, "The Bacterial Glycocalyx in Nature and Disease," *Annual Review of Microbiology* **35**:299–324, 1981. *The chemical and physical properties of a newly defined bacterial component are presented.*

FERRIS, F. G., and T. J. BEVERIDGE, "Functions of Bacterial Cell Surface Structures." *BioScience* **35**:172–177, 1985. *A functional coverage of procaryotic ultrastructures and associated functions.*

HOLT, J. G. (ed.), *Bergey's Manual of Systematic Bacteriology, vol. 2.* Baltimore, Md.: Williams and Wilkins, 1986. *A highly detailed coverage of gram-positives, actinomycetes, and related forms.*

HURST, A., and G. W. GOULD, *The Bacterial Spore.* New York: Academic Press, 1984. *An update of information concerning the mechanisms of sporulation and germination as related endospore structures.*

KRIEG, N. R. (ed.), *Bergey's Manual of Systematic Bacteriology, vol. 1.* Baltimore, Md.: Williams and Wilkins, 1984. *A detailed coverage of the properties of spirochetes, gram-negatives, mycoplasmas, and endosymbionts.*

UNWIN, N., and R. HENDERSON, "The Structure of Proteins in Biological Membranes." *Scientific American* **250**: 78–95, 1984. *A well-illustrated description of how proteins embedded within cell membranes function to transport molecules.*

WALSBY, A. E., "The Gas Vacuoles of Blue-green Algae," *Scientific American* **237**:90, 1977. *Discussion of how gas vacuoles of cyanobacteria help to regulate their buoyancy. The structure and function of these vacuoles are discussed.*

8

Fungi

After reading this chapter, you should be able to:

1. Describe the general characteristics of fungi and the position they occupy in the biological world.
2. List the distribution and features of the major classes of fungi.
3. Explain how fungi differ from other types of microorganisms.
4. Outline the features of fungal reproduction.
5. Identify and give the function(s) of the structures of fungi.
6. Describe the methods used to grow fungi.
7. List and describe the major beneficial and destructive activities of fungi.
8. Distinguish between the following:
   a. a sporocarp and a basidiocarp
   b. an anamorph and a telomorph
9. Explain what a lichen is.

For 300 years after the birth of Christ rainfall was abnormally heavy. . . . The crops fluorished and so did the host of tiny living organisms—the rusts, rots, mildews, molds, smuts, and blights that were to ultimately destroy the grain. . . . Field after field was laid to waste. History is made by a multiplicity of little things, as well as major revolutions, wars, earthquakes and floods. And so it was that the fungi, by bringing hunger and unrest, contributed to the decline of the great Roman Empire. — Lucy Kavaler,
Mushrooms, Molds and Miracles

The fungi are a distinct group of organisms consisting of over 100,000 species. They are an ancient form of life, no less than 400 million years old and possibly much older. Together with bacteria, many fungi serve as major decomposers of rotting and decaying matter in the environment. Some fungi, because of their biochemical activities, are of major commercial value for the production of beer, wine, fermented dairy products, and antibiotics. Still others spoil many different materials as they obtain their nutrients and cause serious diseases of plants, humans, and other animals. This chapter describes the distribution and organization of fungi as well as their interactions with other forms of life and their commercial importance.

Fungi (singular, **fungus**) are larger than bacteria and structurally more complex. Like animals, protozoa, and most bacteria, they require organic nutrients as sources of energy. Since fungi contain no photosynthetic pigments, they depend on other structures and enzyme systems for energy to carry out their various activities.

The fungi are a distinctive life form of great practical and ecological importance. This unique group of eucaryotic, nonphotosynthetic microorganisms contains more than 80,000 species and includes the large, conspicuous mushrooms, puffballs, and woody bracket forms (Color Photographs 33–37), as well as the smaller molds and yeasts (Color Photographs 38, and 39).

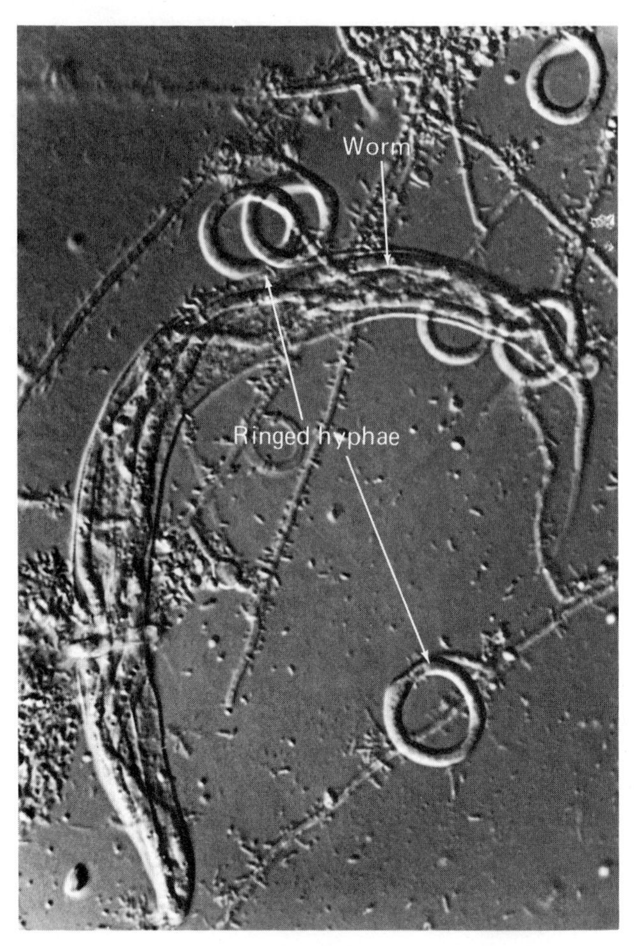

Figure 8–1 A nematode-trapping fungus. Here the ringed hyphae of *Arthrobotrys anchonic* (are-throw-BOTT-ris ann-COE-nick) grasp and eventually immobilize the unsuspecting roundworm. Original magnification 400×. *[G. L. Barron,* The Nematode Destroying Fungi, *1977.]*

Distribution and Activities

Fungi are widespread in nature (Figures 8–1 and 8–2). They grow well in dark, moist environments and in habitats where organic material is available. Several fungi grow at high temperatures (are thermophilic) in natural and human-made locations. Others have been isolated from various antarctic and subantarctic soils. Thus, fungi are found in terrestrial and aquatic environments with a broad range of temperatures.

Since fungi do not contain cholorophyll, they are dependent upon the organic products of other organisms, either living or dead, as sources of energy. Therefore, fungi are heterotrophic and are usually referred to as *saprobes* or *saprophytes.*

Many of these organisms are active producers of enzymes that enable them to break down complex substances for their use. The production of such digestive enzymes has several beneficial effects, such as the recycling of elements to the soil, making it more suitable for plant growth and enabling certain plants to obtain minerals. However, these enzyme systems can also have undesirable effects. Some fungi are known for their parasitic and destructive effects on humans, plants, lower animals, and even fungi (Figure 8–3 and Color Photographs 6 and 41). Mold spoilage of food is a familiar problem to the marketer of agriculture produce. Each year millions of dollars worth of fruits and vegetables are lost because of damage caused by fungi (Color Photograph 6). Fortunately, most fungi that spoil foods do not invade the tissues of humans and lower animals.

Fungi are of considerable interest to humans because of their value as food and their role in the processing and production of many foods and drugs such as antibiotics. Fungi collected for food include morels, certain mushrooms, truffles, and nonwoody polypores (Color Photographs 33 and 36). Commercially grown mushrooms are higher in protein than many fruits and other vegetables and are good sources of minerals such as iron and phosphorus, and the vitamins niacin, riboflavin, and thiamine.

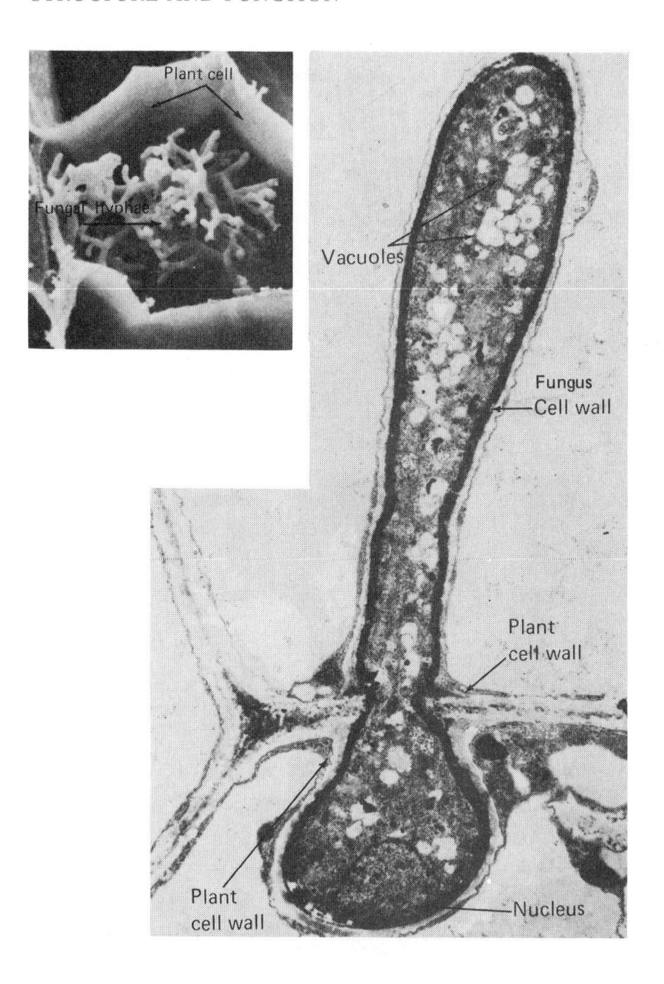

Figure 8–2 An infected root system of mycorrhiza is generally beneficial to both the fungus and the host plant. The extent of invasion by fungus is usually kept in check to prevent injury to the plant rootlet. (a) A scanning micrograph showing the fungus cell or hypha penetrating into the plant cell. (b) A transmission micrograph of the association. Note the eucaryotic organization of the fungus. The main component of the host plant cell shown is the cell wall. Hyphal structures shown include the cell wall, nucleus, and vacuoles. *[Reproduced by permission of the National Research Council of Canada from D. A. Kinden, and N. F. Brown,* Can. J. Microbiol. ***21****:1768–1780, 1975.]*

Foods produced worldwide that utilize fungi include bread, cheese, beer, and wine.

Structure and Function

Fungi differ from bacteria in several ways, including their size (fungi are larger), structural development, cellular organization, and methods of reproduction. In addition, although fungi may exist as single-celled forms, the vast majority are multicellular. It is important to note that the term *fungus* is a general one that includes the forms **molds** and **yeasts.**

Figure 8–3 (a) The destructive effects of the yeast pathogen *Cryptococcus neoformans* (kryp-toe-KOK-kus nee-oh-FOR-manz), or cryptococcosis of the face. *[Photograph given to Dr. Zimmerman by Bilgisi Sheseti, Istanbul, Turkey. Courtesy of the Armed Forces Institute of Pathology, Neg. No. 55-8225.]* (b) The fungus disease of dry beans, called *bean rust.* The causative agent, *Uromyces phaseola typica* (you-row-MY-sees fay-ZEE-oh-la tip-EE-ka), produces destructive lesions on the plant's leaves. *[Courtesy of the USDA's Bureau of Plant Industry.]*

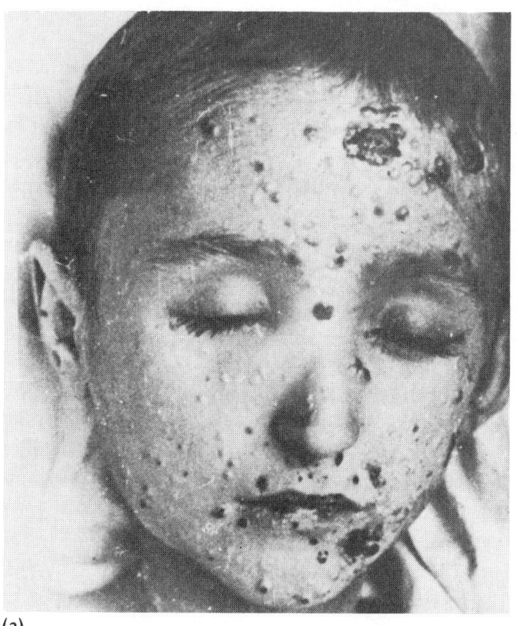

(a)

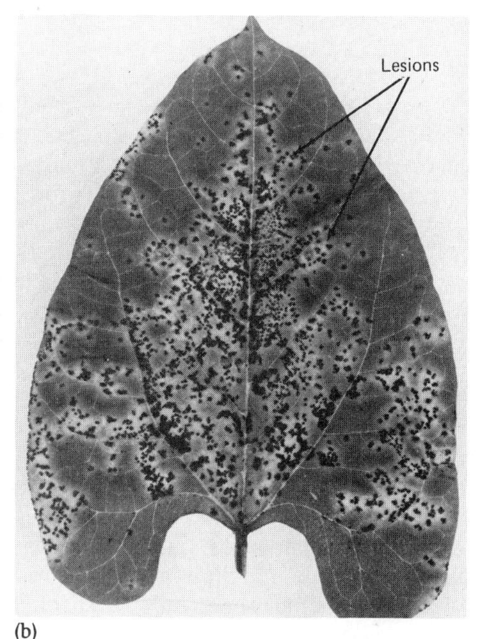

(b)

Microbiology Highlights

THE INVASION OF THE FUNGUS (MYCOPARASITISM)

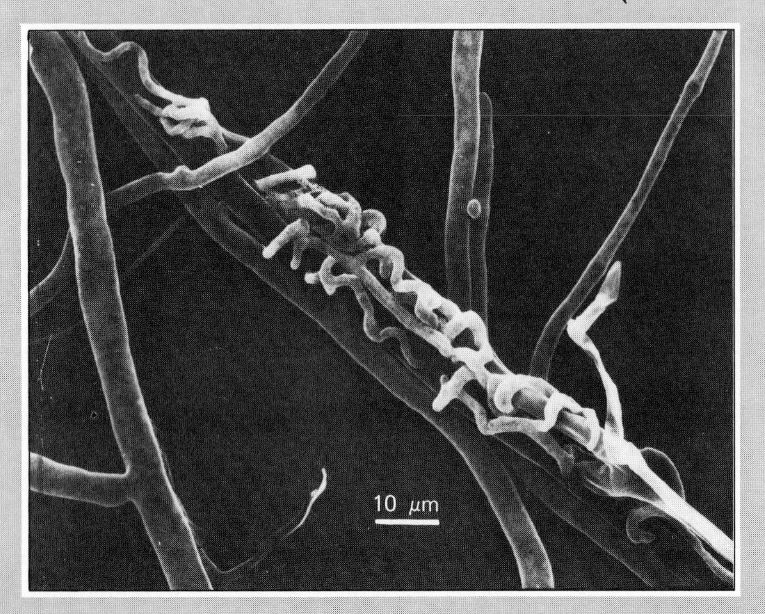

Mycoparasitism in action. The hyphal coils of a parasitic fungus surrounding its host. *[From Y. Elad, R. Barak, and I. Chet,* J. Bacteriol. ***154****:1431, 1983.]*

Since several fungi are antagonistic and parasitic to other fungi, they can serve as potential biocontrol agents against soil-borne plant pathogens. When a parasitic fungus reaches a host's hypha, it usually coils around it and may produce hook-shaped contact branches. By manufacturing enzymes that digest cell walls and other related enzymes, the attached fungus proceeds with its mechanical and enzymatic penetration of the host fungus. Penetration occurs at different locations even when the host fungi have thick, melanin-containing walls. The pigment melanin generally protects against the biological and chemical digestion of fungi, but not under these conditions. After it is penetrated, the host fungus forms a common layer with the invader. Soon after this, the cytoplasm of the host rapidly decomposes and the space it occupied is filled by the invading fungus.

Molds

Although molds do not have true roots, stems, or leaves, they do show differentiation (Figure 8–4). Most molds, for example, consist of tubular, branching eucaryotic cells called **hyphae** (Figure 8–5c). Filaments of these cells are often subdivided by crosswalls (septa) into multicellular hyphae (Figure 8–5d). Septa in many fungi are perforated by one or more pores, which are small but allow nuclei and cytoplasm to pass from cell to cell (Figure 8–5d). In species that do not contain septa between the two or more nuclei of the hyphae (the coenocytic state), the contents of the hyphae can move freely throughout the filament.

As molds continue to grow, the hyphae branch, intermingle, and often fuse, eventually forming a visible cobweblike aggregation like that seen on moldy bread or fruit. These structures, which are analogous to bacterial colonies, are called **mycelia** (singular, **mycelium**) (Figure 8–5a). Hyphal masses appear dry and powdery (Color Photograph 37 and Figure 8–5c). This is often the result of the formation of various types of reproductive units, called **propagules** or **spores.** Typically, there is a division of roles for different structures.

In some instances, the mycelium is made up of two general regions (Figure 8– 5a). One of these regions extends below the surface of the medium on which the mycelium is growing for the purpose of food collection. This part of the mycelium is called the *vegetative mycelium.* Certain species have rootlike structures called *rhizoids* that obtain food and also serve as anchoring devices. The part of the fungus that is above the substrate is the *aerial mycelium.* It is the reproductive portion of the microorganism, with specialized branches that produce spores (Figures 8–4 and 8–5b).

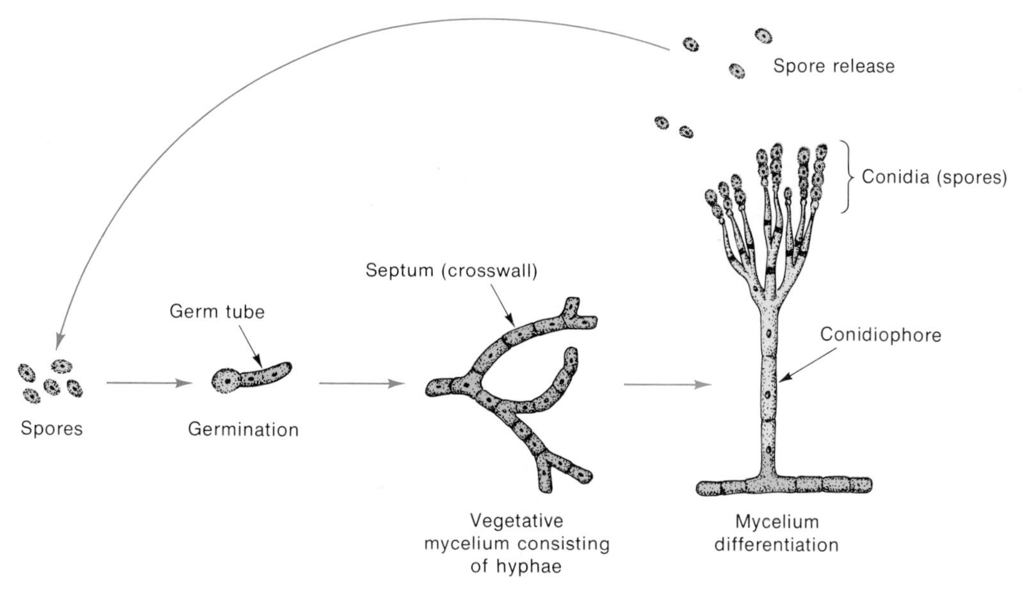

Figure 8–4 The asexual life cycle of the common fungus *Penicillium* (pen-ee-SILL-lee-um) sp. showing several typical structures and their arrangements. Spores, hyphae, vegetative mycelium, and cross walls (septa) are indicated. The germination of the spore into a hypha starts the development of the fungus. Refer to Color Photograph 37 for the coloration of this organism's mycelium.

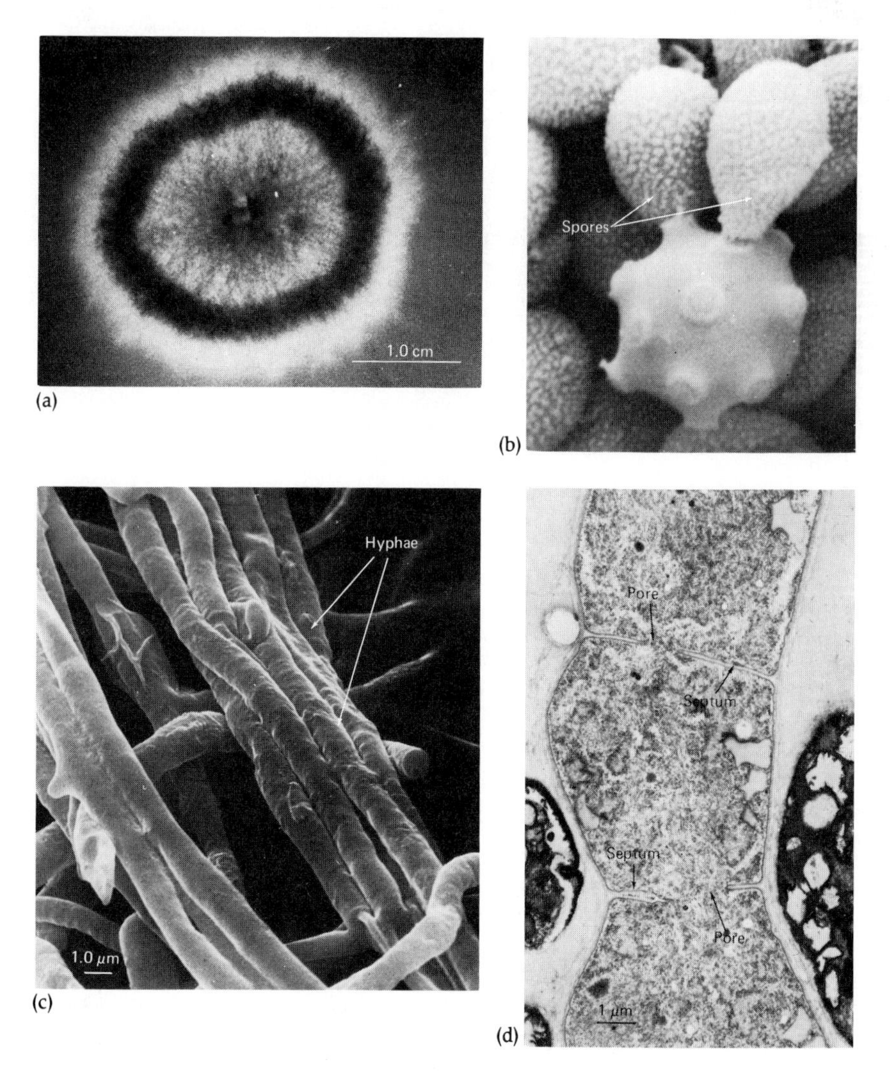

Figure 8–5 Components of molds. (a) The cottony mycelium. (b) A scanning micrograph of spores and their attachment to the fungal structure from which they arose. *[Courtesy of P. Jeffries and J. W. K. Young.]* (c) The tubular hyphae, or threads, that form most of the fungal mycelium. *[(a) and (c) reproduced by permission of the National Research Council of Canada from D. H. Ellis, and D. A. Griffiths,* Can. J. Microbiol. ***21****:442–452, 1975.]* (d) An ultrathin section of hyphal cells showing the cross walls, or septa, characteristic of specific fungi. *[Reproduced by permission of the National Research Council of Canada from D. H. Ellis, and D. A. Griffiths,* Can. J. Microbiol. ***21****:442–452, 1975.]*

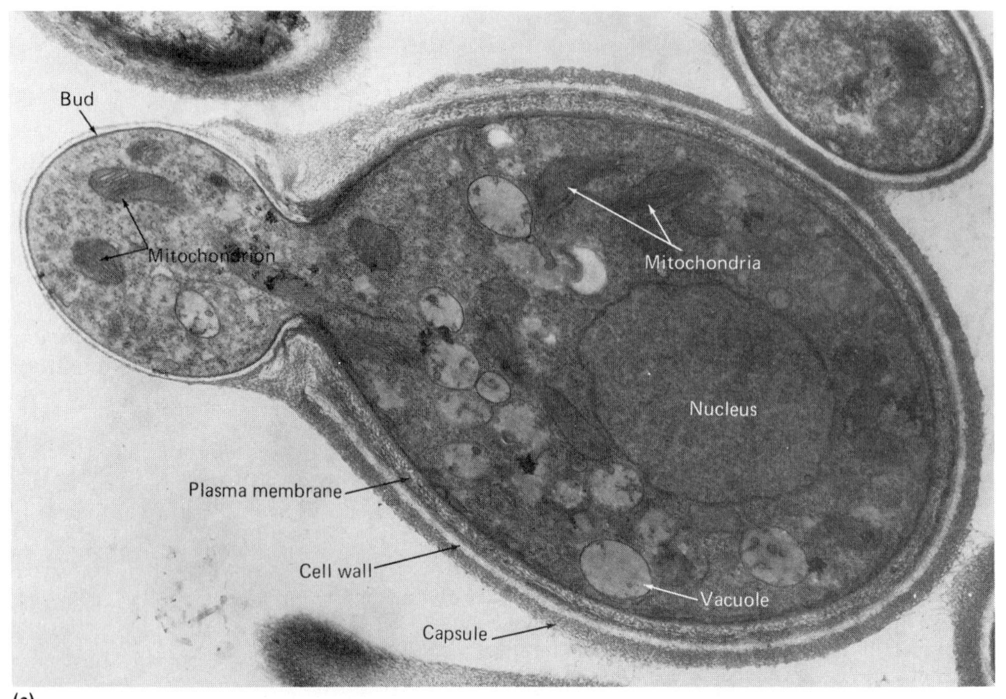

(a)

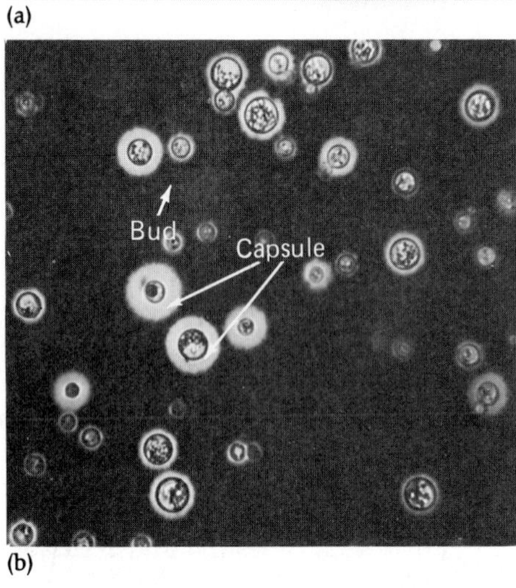

(b)

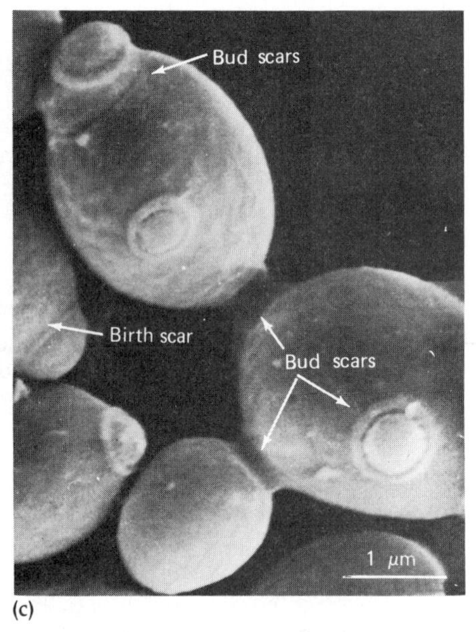

(c)

Figure 8–6 (a) An electron micrograph showing the eucaryotic organization of a yeast cell and its bud. Note the presence of the bud, capsule, cell wall, mitochondria, nucleus, plasma membrane, and vacuole. *[Reproduced by permission of the National Research Council of Canada, from E. M. Peterson, R. J. Hawley, and R. A. Calderone,* Can. J. Microbiol. ***22:****1518–1521, 1976.]* (b) *Cryptococcus neoformans* stained with nigrosin clearly shows the capsular regions surrounding the cells. Note also the newly separated bud. *[Courtesy of Dr. M. Silva-Hutner, Columbia University, College of Physicians and Surgeons, N.Y.]* (c) Various stages in the formation of buds on yeast cells. Birth scars, sites left by buds on the main cells, and the corresponding scars on buds can also be seen. *[K. Watson, and H. Arthur,* J. Bacteriol. ***130:****312–317, 1977.]*

Yeasts

The fungi known as *yeasts* are oval, spherical, or elongated cells (Figure 8–6) that form moist, shining colonies (Color Photographs 38 and 39). Yeasts also have a eucaryotic organization and a thick cell wall. In some yeasts, a capsule may form around the cell wall (Figure 8–6b).

Certain yeasts reproduce asexually by a process of division (fission) that results in the formation of a new *bud,* or *daughter cell* (Figure 8–6a). The mother cell bears a bud scar at the region where separation took place; the bud has a birth scar (Figure 8–6c). In some species the newly formed cells do not separate. These connected yeasts cells are called *pseudomycelia.*

Dimorphism

Many fungi are **dimorphic,** exhibiting two different forms under two different environmental conditions. Under certain environmental conditions these organisms exhibit their normal type of saprobic form, but in animal tissues or when grown on rich nutrient preparation at higher temperatures, they appear as yeasts. The term **saprophytic form** has been suggested for the mold phase, **parasitic form** for the yeast stage.

Reproduction and Spores

Both sexual and asexual reproduction occur in fungi. However, not all of these organisms reproduce sexually; some must depend entirely upon asexual means. Mitosis in fungi differs from that found in other eucaryotes. The main differences are that the nuclear envelope does not break down and reform and the mitotic spindle forms within the nucleus. In addition, centrioles are absent from all fungi. **[Refer to Chapter 2 for a comparison of eucaryotes and procaryotes.]**

The sexual form of reproduction may involve the union of hyphae, sex cells, or differentiated, multinucleated sex organs (Figure 8–12). All of these structures have a haploid or single set of chromosomes in one nucleus. When male and female cells unite, such haploid nuclei fuse and a diploid fertilized zygote is formed.

A typical reproductive unit of fungi is the **spore,** which may have one or more nuclei obtained through sexual or asexual reproduction. Spores may be produced upon or within hyphae or within isolated cells. In several complex fungi, multicellular reproductive structures are formed that bear spores. Large spore-bearing structures are often called **sporocarps.** The familiar mushroom is an example of a sporocarp. (Color Photograph 35). The type of spore produced and the manner of sporulation are both important in the identification and classification of fungi. These specialized structures function as reproductive cells, but usually they are no more heat-resistant than any other type of fungal cell.

Asexual Spores

Depending on the species, both asexual and sexual spores or only asexual spores are produced. Asexual spores (Figure 8–7) include the *arthroconidium (arthrospore), blastoconidium (blastospore), chlamydoconi-*

Figure 8–7 Asexual spores of fungi. (a) A scanning micrograph showing isolated, barrel-shaped arthrospores of *Trichophyton mentagrophytes* (trik-oh-FI-ton men-tag-row-FI-tease), a ringworm-causing fungus. These spores break off from the hyphae. (b) Microconidia (small spores) developing from hyphae. These spores emerge directly from the tubular hyphae of the fungus. The bar marker measures 1 μm. *[(a) and (b) from D. J. Bibel, D. A. Crumrine, K. Yee, and R. D. King,* Infect. Immunol. ***15****:958–971, 1977.]*

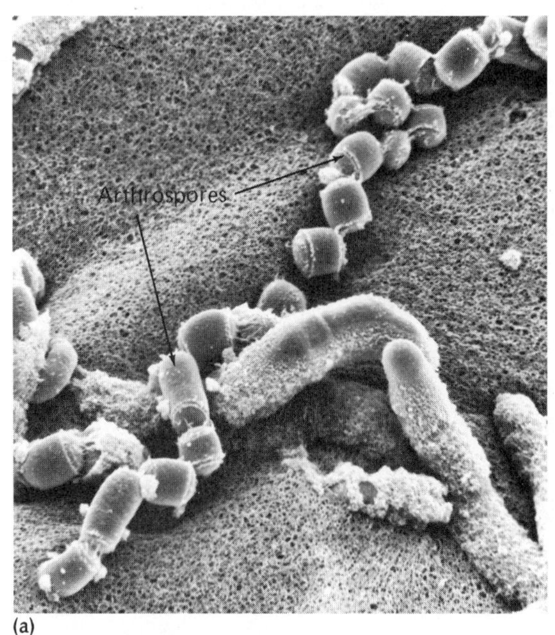

(a)

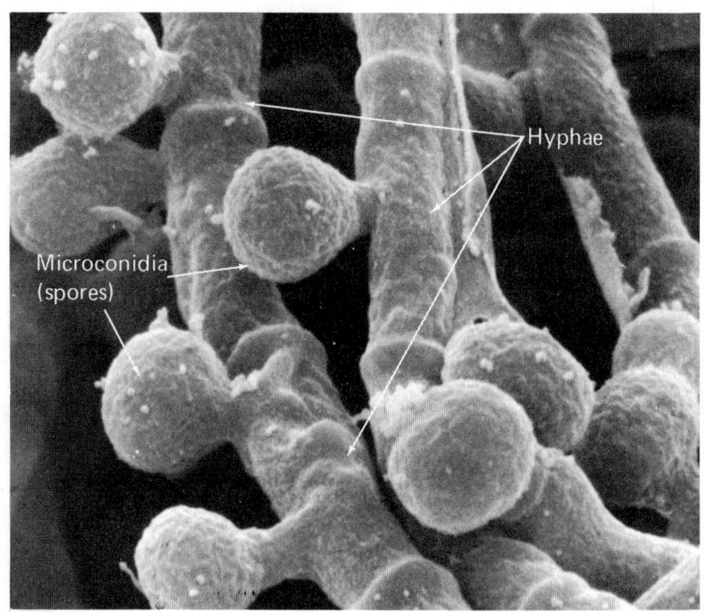

(b)

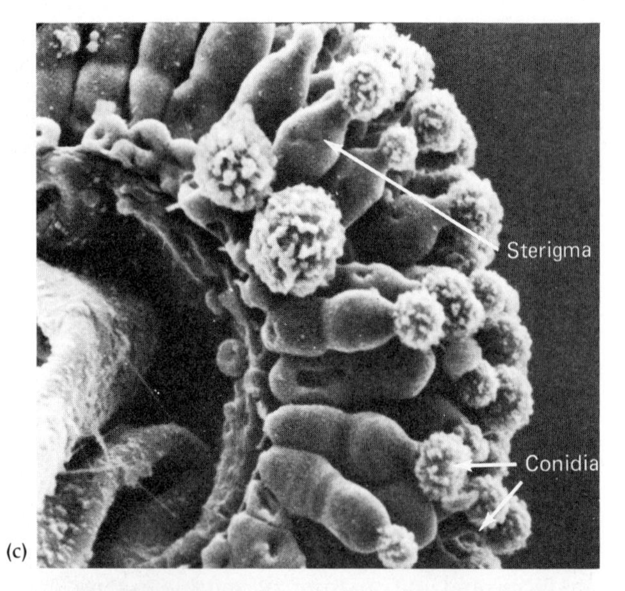

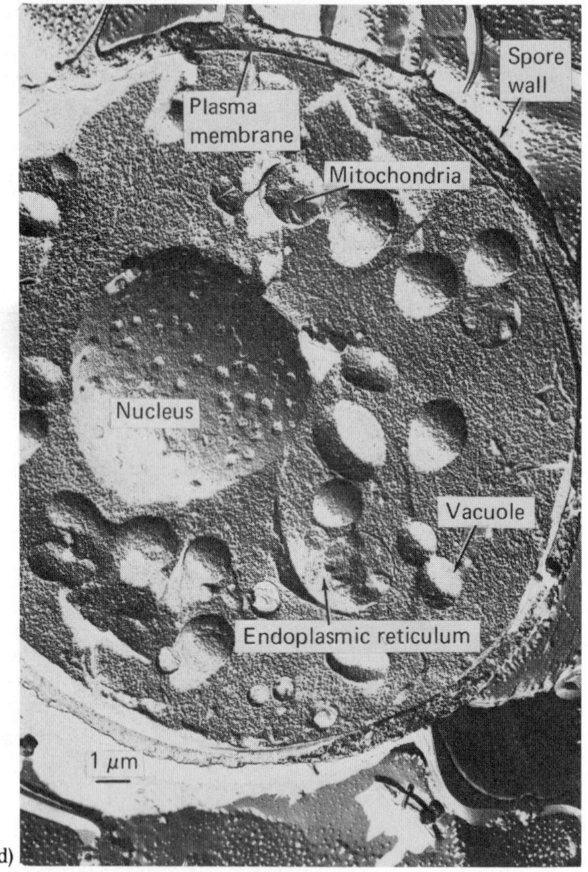

Figure 8–7 (c) Conidia of *Aspergillus* (a-sper-JILL-lus). [*M. Tokunaga, J. Tokunaga, and K. Harada,* J. Elect. Micro. ***22****:27–38, 1973.*] (d) An electron micrograph of a freeze-etched fungal spore (conidium) showing its internal organization. Note the thickness of the spore wall, the typical eucaryotic organelles, the association of the endoplasmic reticulum with the plasma membrane, many vesicles, mitochondria, and the nuclear envelope. *[Reproduced by permission of the National Research Council of Canada from J. Sekiguchi, G. M. Gaucher, and J. W. Costerton,* Can. J. Microbiol. ***21****:2048–2058, 1975.]*

dium (chlamydospore), conidiospore (conidium), sporangiospore, and *zoospore.* Some fungi have the ability to form two types of asexual spores. Asexual spores vary in color, shape, surface appearance, and size and may have one or several nuclei (Figure 8–8). Many of the properties of these reproductive cells are summarized in Table 8–1.

Under suitable conditions of nutrition, moisture, pH, and temperature, fungal spores germinate and produce one to several filamentous structures called *germ tubes* (Figures 8–4 and 8–9). These usually penetrate through thin or weakened portions of the spore *(germ pores).* The fungal germ tubes develop by elongating and branching to form the hyphae.

Figure 8–8 With the aid of scanning electron microscopy, the distinctive surface appearance of fungal spores has been uncovered. The sporangiospores of two different species are shown. *[Courtesy D. H. Ellis.]*

TABLE 8–1 Properties of Fungal Spores

Spore Type	*Site and/or Type of Formation*	*Single or Multicellular*	*Shape*	*Resistance to Environment*	*Examples of Genera that Form Spore Type*[a]
ASEXUAL SPORES					
Arthroconidium (Arthrospore)	Fragmentation of hyphae (Figure 8–7a)	Single	Cylindrical to round	Usually none	*Coccidioides*[b] (kok-sid-ee-OY-deez), *Geotrichum* (gee-oh-TRICK-um), *Trichophyton*[b] (trik-oh-FYE-ton)
Blastoconidium (Blastospore) (buds)	Formed on main cell	Single	Round to oval	Usually none	*Candida*[b] (KAN-did-a), *Saccharomyces* (sak-a-row-MY-sees)
Chlamydoconidium (chlamydospore)	Enlargement of terminal hyphal cells	Single	Considerable variation but usually round	These thick-walled cells exhibit unusual resistance to drying and heat	*Candida*[b], *Mucor* (MYOO-kore)
Conidium	Borne on specialized hyphal branches, conidiophores	Single (microconidia)	Round to oval	Usually none	*Aspergillus*[b] (a-sper-JILL-lus), *Cephalosporium* (seff-a-low-SPOR-ee-um), *Penicillium* (pen-ee-SILL-lee-um)
		Multicellular (macroconidia)	Long and tapering	None	*Alternaria*[b] (al-ter-NARE-ee-a), *Microsporum*[b] (my-krow-SPOH-rum), *Trichophyton*[b]
Phialospore (modified conidium)	Borne on specialized hyphal branches, conidiophores, phialides	Single	Round to oval	None	*Phialophora* (fil-a-LOW-foh-ra)
Sporangiospore	Formed within sacs, sporangia, at end of hyphal cells	Single	Round	None	*Absidia* (ab-sid-EE-a), *Coccidioides*[b], *Mucor, Rhizopus* (rye-ZOH-pus)
Zoospore	Formed within sacs, sporangia, at end of hyphal branches	Single, flagellated	Round	None	*Saprolegnia*[b] (sap-row-LEG-nee-a)
SEXUAL SPORES					
Ascospore	Formed within saclike cells, asci, after cellular and nuclear union	Single (usually 8 per ascus)	Round to oval	None	*Allescheria* (al-lesh-ER-ee-a), *Neurospora* (new-RAH-spor-a)
Basidiospore	Formed at end of club-shaped structures, *basidia* (Figure 8–16)	Usually single (usually 4 in number)	Round to oval	None	*Amanita*[b] (am-an-EE-tah), *Agaricus* (a-GARE-i-kuss), *Coprinus* (koh-PRIN-us)
Oospore	Developed within a fertilized egg cell, *oogonia*	Single (usually 1 to 20 per oogonium)	Round	More resistant than most asexual spores	*Saprolegnia*[b]
Zygospore	Formed after cellular and nuclear fusion (Figure 8–10)	Large, thick-walled, single structure	Round to oval	None	*Rhizopus*

[a] Certain fungi can form more than one type of spore.
[b] Causes disease and/or allergies.

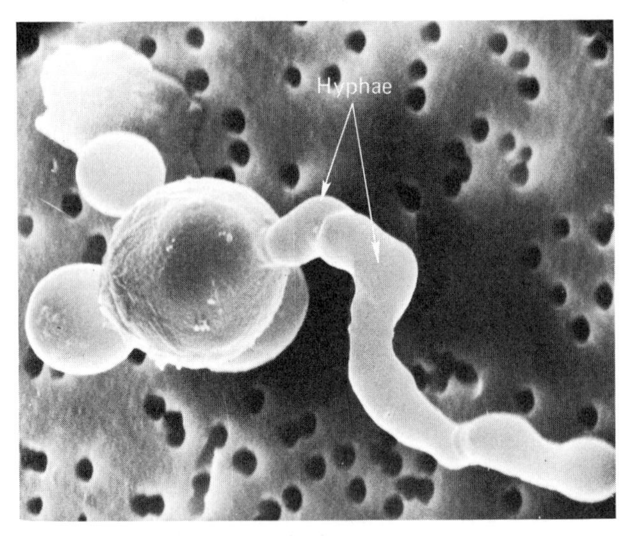

Figure 8–9 A scanning micrograph showing germ tube formation by arthrospores. Separate, newly formed hyphal cells can also be seen. *[S. W. Queener, and L. F. Ellis,* Can. J. Microbiol. ***21****:1981–1966, 1975.]*

Sexual Spores

Sexual spores (Figure 8–10) are produced by nuclear fusion. Reproductive cells of this type contain the products of meiosis and include *ascospores, basidiospores, oospores,* and *zygospores.* In general, sexual spores are observed less often than asexual ones. Certain environmental conditions must be provided before sexual sporulation can be induced. Table 8–1 lists several properties of sexual spores.

Figure 8–10 Representative zygospores of *Rhizopus* (rye-ZOH-puss) species. Note the two hyphal branches (arrows).

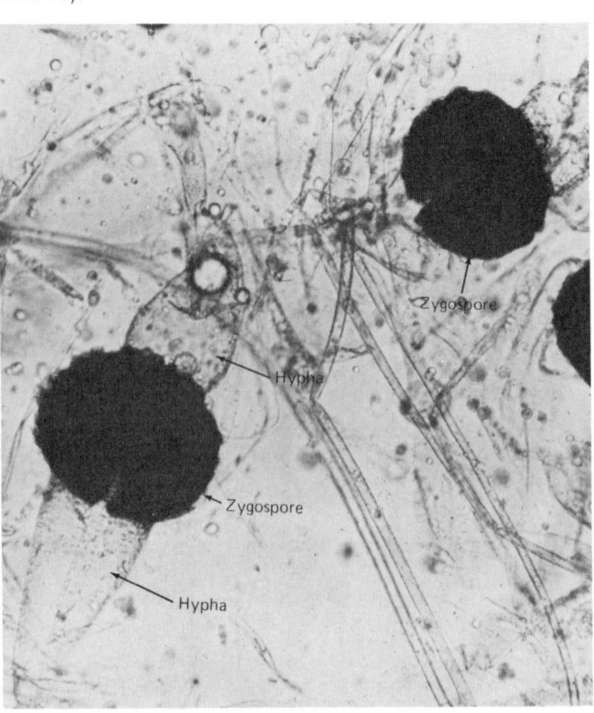

Ultrastructure

Both yeasts and molds are eucaryotic (Figures 8–2b, 8–5d, 8–6b, and 8–7d). Their cells usually contain membrane-bound organelles, including endoplasmic reticula, mitochondria, and well-defined nuclei. The membranes surrounding the nuclear material of the cells contain sterols such as cholesterollike compounds. Sterols, with the exception of bacteria such as the mycoplasma, are not found in procaryotes but only in the structures of higher forms of life. Since the fungi resemble higher plants and animals both in their cellular complexity and in their organelles, only the cell wall and fimbriae will be discussed here.

As in bacteria, the cell wall lies immediately external to the fungal cell's cytoplasmic membrane. Under the electron microscope the walls of filamentous fungi appear to be composed of thin, threadlike structures called **microfibrils** (Figure 8–11). These measure approximately 15 to 25 nanometers in diameter and are arranged in a type of thatchwork. Chemically, the microfibrils in many fungi are composed of **chitin,** a polymer of *N*-acetylglucosamine (Figure 7–18) that is also the principal structural substance in the exoskeletons of crayfish, crabs, and related forms. Cellulose, the major component of plant cell walls, has also been found in the hyphal walls of filamentous fungi. **[Chapter 7 describes bacterial cell walls.]**

The chemical composition of yeast cell walls is quite different. They contain approximately equal quantities of the highly branched, insoluble polysaccharide glucan and the soluble polysaccharide mannan. Other compounds found in the cell wall include lipids, proteins, and the amino sugar glucosamine.

Figure 8–11 An electron micrograph showing the appearance of interlacing microfibrils. The relationship of the cell wall to the plasma (cell) membrane is also shown. *[R. Caputo, M. Innocenti, and M. Shimono,* J. Ultrastruct. ***59****:149–157, 1977.]*

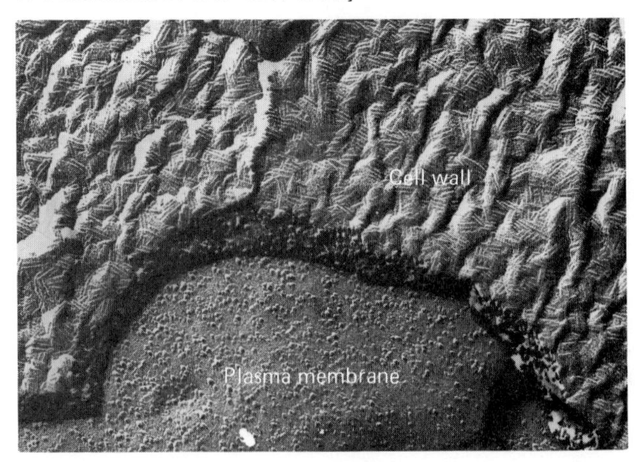

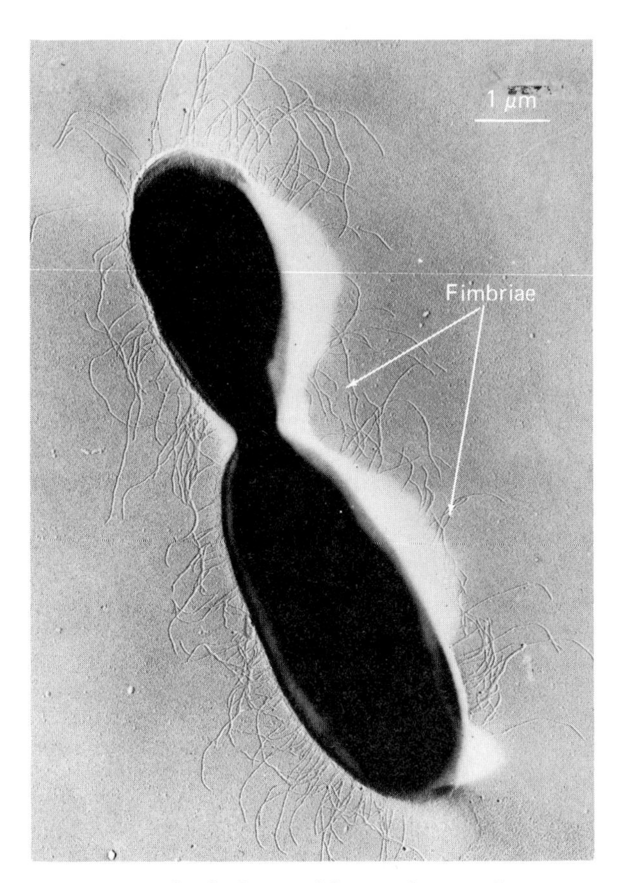

Figure 8–12 The fimbriae of fungi. These surface hairlike appendages have been found on several yeasts and are believed to be involved in sexual reproduction. *[Reproduced by permission of the National Research Council of Canada from N. H. Poon, and A. W. Day,* Can. J. Microbiol. ***21**:537–546, 1975.]*

The surfaces of fungi are important to several interactions, including the joining of cells for sexual reproduction **(conjugation)** and the uptake and excretion of molecules. Recent ultrastructural studies have revealed the presence of pili, or fimbriae (Figure 8–12), on the surfaces of various yeasts. These structures, which are remarkably similar in several respects to those of bacteria (Figure 7–10), originate below the cell wall. They may be necessary to complete conjugation. **[See bacterial pili in Chapter 7.]**

Cultivation and Identification

Molds and yeasts can be grown and studied by cultural methods similar to those used for many bacteria. Most fungi are able to grow under aerobic conditions, but many grow more slowly than bacteria. Consequently, media that can support both bacteria and fungi are apt to become overgrown with bacteria. In working with fungi, therefore, it is advisable to use culture media that limit the growth of other microbial types. Ingredients that can be added to media for this purpose include antibiotics (such as actidione, chloramphenicol, gentamicin, penicillin, and streptomycin), dyes (such as crystal violet), and high concentrations of sugar. Another important property of preparations for fungal cultivation is their acidity, pH 5.6 to 6.

Nearly all of the culture media used for fungi can be obtained in the dehydrated state from commercial sources. Most can be used in either the liquid or solid state. **[See Chapter 4, "Media Categories."]**

Types of Media

In general, three basic types of media can be used in fungus cultivation: natural, dehydrated or complex, and synthetic (chemically defined) preparations. Natural media are not widely used in the laboratory. Examples of such cultivation materials include slices or infusions of animal tissues, fruits, and vegetables.

Complex media contain, in addition to dextrose (glucose) and/or sodium chloride, a wide assortment of organic ingredients such as peptones, beef extract, and corn meal, which have variable compositions. One of the most commonly used of these preparations is Sabouraud's (Sab) dextrose or maltose medium, which also contains peptone, agar, and distilled water. After sterilization, it can be used in the form of slants or plates for the isolation and identification of a wide variety of molds (Color Photographs 37 and 98b), and yeasts (Color Photograph 38). Incorporation of the dye trypan blue in Sabouraud's agar permits rapid detection and tentative identification of several species of yeast (Color Photograph 39). Other examples of basic media include blood agar, brain agar, corn meal agar, and thioglycollate broth.

Synthetic media—those for which the exact composition is known—are used for the isolation of saprophytic and pathogenic fungi. Most saprophytic fungi grow at room temperature and are unable to grow at 37°C, but most pathogens grow readily at 37°C.

Distinctive Mycelia

Several fungi can be identified by the appearance of their mycelia. Color, diffusion of pigment, and texture are but a few of the important features. Cultures of these fungi are generally prepared by inoculating the center of a particular medium. After a suitable incubation period at the appropriate temperature, the distinctive mycelium should have developed (Color Photographs 98b and 98c). The undersurfaces of certain pathogenic fungi also exhibit a characteristic appearance. A trained worker usually has little difficulty in identifying typical fungal species.

Classification

Several properties of fungi are important to their classification and identification. These include method of reproduction, mycelial formation, and cellular structure and formation (Table 8–2). Additional approaches based on biochemistry, physiology, electron microscopy, and mathematical analyses are used to determine relationships, compare properties, and analyze the similarities and differences that exist among fungi. The information gained from such studies not only increases the knowledge and understanding of fungi but also provides the bases with which to modify and improve the classification of fungi.

As a group, the fungi have been classified with different kingdoms. We shall consider them as belonging to the kingdom Mycetae (Fungi), a classification proposed by G. C. Ainsworth in 1973, and discuss characteristics such as the methods of reproduction and the presence or absence of motile spores *(zoospores)*. The fungi in the division Amastigomycota, which do not produce zoospores, include the **ascomycetes** (Ascomycotina), **basidiomycetes** (Basidiomycotina), and **deuteromycetes** (Deuteromycotina). Members of the division Mastigomycota form zoospores and include oomycetes. The general properties, ecological roles, and economic importance of representatives from these divisions and of the slime molds (Color Photograph 43), a group that exhibits features of both fungi and protozoa, are presented in the following sections. **[See Appendix C for fungal classification.]**

Ascomycotina (Ascomycetes)

Several species of the Ascomycotina, often called *ascomycetes,* are biologically and economically important. This group, also known as *sac fungi,* is the largest, containing almost 2,000 genera. Some members of the group live in aquatic or moist land environments. Others are parasitic or live with other microorganisms or plants in a mutually beneficial association. The asexual spores by which ascomycetes reproduce are produced at the ends of hyphae. The spores germinate to form new mycelia.

Ascospores are the characteristic sexual spores of the ascomycetes, formed following the fusion of tubelike structures (bridges) from two neighboring sex-cell producing cells, the *gametangia.* The female cell is called an *ascogonium,* and the male is known as the *antheridium* (Figure 8–13). The two nuclei of these cells fuse and form a single nucleus. Daughter nuclei are then produced through meiotic division. As many as eight nuclei may form. Each becomes surrounded by a dense protoplasmic layer and a spore coat, or wall, thus forming an ascospore. These reproductive structures are contained in the original union of the two neighboring cells. The enclosing sac is called an *ascus.* Asci may or may not be produced within or upon multicellular fruiting bodies known as *ascocarps* (Figure 8–14). Four major types of such structures are known: *apothecium, cleistothecium, locule,* and *perithecium.* The organization and morphological differences among ascocarps are of value in fungal classification. The life cycle of an ascomycete is shown in Figure 8–15.

The yeasts, such as *Saccharomyces cerevisiae* (sak-a-ROW-my-sees se-ri-VISS-ee-eye), which leaven breads and ferment beer and wine, are also ascomy-

TABLE 8–2 Selected Characteristics of the Major Groups of Fungi

Subdivisions	*Type of Mycelium*	*Site of Formation (Asexual Spores)*	*Site of Formation (Sexual Spores)*	*Representative Groups*
Ascomycotina (ascomycetes)	Septate	At the tips of hyphae	Within sacs	*Claviceps purpurea* (KLA-vee-seps pur-poo-REE-a), morels, truffles, and yeasts
Basidiomycotina (basidiomycetes)	Septate	At the tips of hyphae	On a surface of a basidium	Poisonous mushrooms (*Amanita* spp.), mushrooms, rusts, smuts
Deuteromycotina (deuteromycetes or fungi imperfecti)	Septate	At the tips of hyphae	None present	Most human pathogens
Zygomycotina (zygomycetes)[a]	Almost completely aseptate (coenocytic)	In sacs	In mycelium	Bread mold *(Rhizopus nigricans),* aquatic species
Oomycetes[b]	Aseptate	In sacs	Within a unicellular female sex organ (oogonium)	Some aquatic forms, pathogens responsible for powdery mildew, blights of plants, and fish infections

[a] The designation *Phycomycetes* used previously for this group is no longer considered valid in modern classification systems.
[b] This is a class of fungi belonging to the division Mastigomycota.

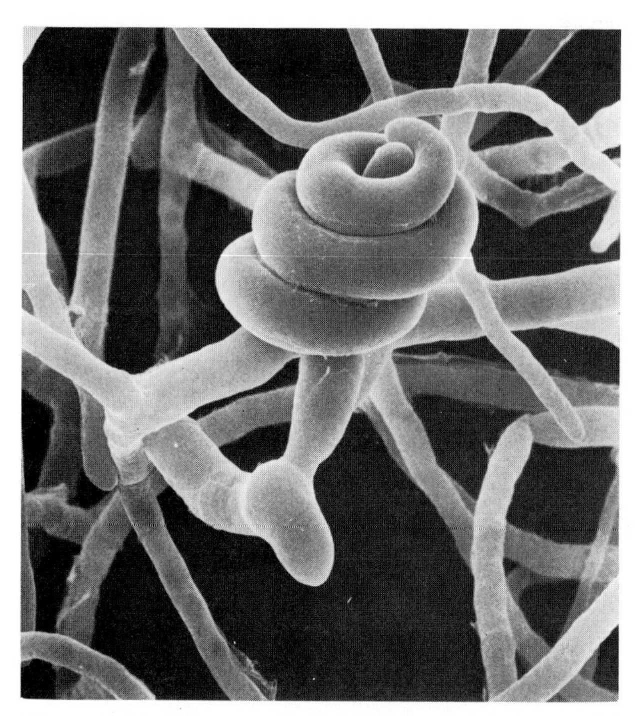

Figure 8–13 The sex cell-producing system (gametangia) of the ascomycetes. The antheridium (male gametangium) surrounded by a coiling ascogonium (female gametangium). *[From M. N. Takashio, and M. Osumi,* Mycologia, **77**:*166–168, 1985.]*

Figure 8–14 One of four major types of ascocarps, the perithecium. Ascospores and sterile, hairlike processes, the paraphyses, can be seen. Original magnification 560×. *[Courtesy Dr. David H. Ellis, The Adelaide Children's Hospital, Inc.]*

Figure 8–15 The life cycle of an ascomycete showing both asexual and sexual reproduction portions.

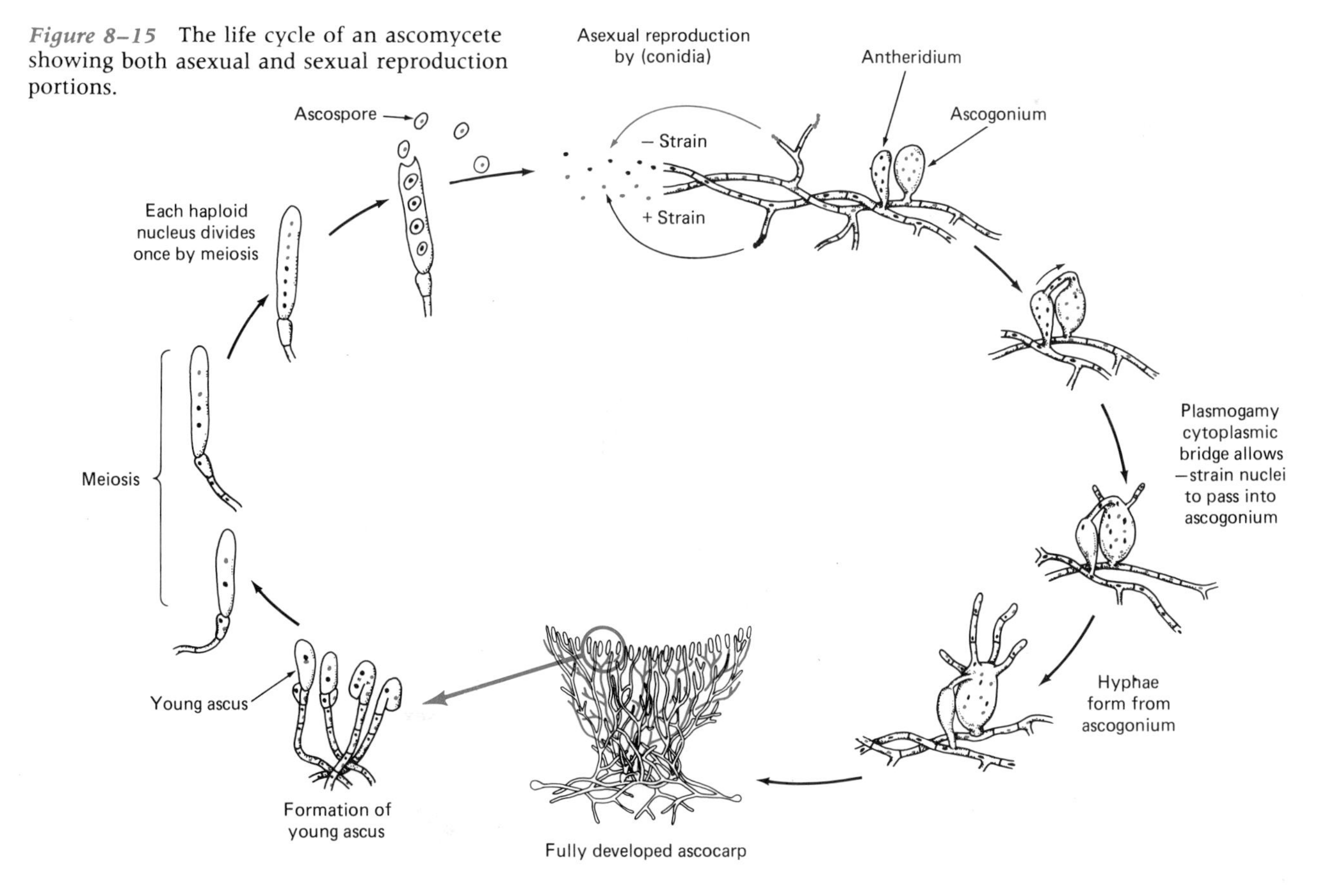

Microbiology Highlights

DEATH BY YEAST: A BLIGHT AGAINST THE COCKROACH

The German cockroach is an age-old scourge in North America and elsewhere. This arthropod has been known to carry *Salmonella* (sal-mon-EL-la) species (a group of bacteria known for their food-poisoning activities) and to spread the causative agents of diseases such as amebic dysentery, infectious hepatitis, and poliomyelitis. The roach also is a source of allergy-causing substances and thereby serves as a major health hazard to many allergy sufferers. Because of their obvious impact on public health, numerous efforts are being made to find effective and safe means with which to eliminate roaches. These include both chemical and natural biological controls. Despite the inroads humans have made, cockroaches seem to find ways of overcoming or resisting each new control weapon.

A rather curious event took place recently at the German cockroach breeding laboratory at the University of California (Riverside campus). A yeast somewhat similar to the soil fungus *Metarhizium* infected and killed 3 million roaches. While the loss was costly, amounting to about $75,000, it pointed to a possible new and effective natural biological control. The yeast reproduces in the circulatory system of the roach over a period of weeks, competes with it for food, and eventually causes the arthropod's death by starvation. Taking into consideration the delayed effect of the yeast and the cockroach's ability to spread microorganisms, infected specimens could reach a colony and share their infection before dying of it themselves. Since infected cockroaches don't reproduce well, the roach colony could eventually die out. This accidentally discovered yeast blight provides one more potential weapon to overcome the arthropod that some think is "virtually impossible to kill."

cetes. In the production of alcoholic beverages, carbon dioxide is a by-product of this fermentation. In baking, however, carbon dioxide is the essential ingredient. Yeasts are mixed with dough, where they multiply and produce sugar-fermenting enzymes. The fermentative activity of the yeasts produces alcohol and carbon dioxide. Tiny bubbles of carbon dioxide form in the dough and cause it to rise. In baking, the gas expands and the alcohol escapes.

Certain yeasts also serve as important sources of the B complex vitamins and ergosterol, which is used in the production of vitamin D. Yeast cells reproduce asexually by *budding*. In this process the cell nucleus divides by mitosis, and one of the two nuclei becomes enclosed in a small projection, or *bud*, formed on the side of the parental cell (Figure 8–6b). Yeasts are also capable of sexual reproduction.

Two edible fungi are also ascomycetes. One of these is the mushroomlike morel (Color Photograph 33). The other is the truffle, a subterranean spherical fungus found around certain species of oak trees. The truffle mycelium, and possibly that of morels, grows in association with tree roots. The unique flavor of truffles and morels is highly valued by many cooks and gastronomes. The rarity and difficulty of harvesting truffles, which mainly grow in certain regions of France and Italy, combined with their flavor, place them among the most expensive of all foods, even though they have little nutritive value.

Several ascomycetes are known for their harmful activities. These include a variety of plant diseases such as apple scab, Dutch elm disease, and powdery mildew of roses and related plants (Color Photograph 40); the production of the poisonous substance (toxin) *ergot* by *Claviceps purpurea* (KLA-vee-seps pur-poo-REE-a) growing on rye, other cereals and grasses; and the production of substances related to ergot, such as the well-known hallucinogen lysergic acid diethylamide (LSD). On the other hand, substances derived from ergot, the ergotamine drugs, are used medically to treat vascular diseases such as migraine headaches.

Basidiomycotina (Basidiomycetes)

The Basidiomycotina, or basidiomycetes, include the bracket fungi of trees (Color Photograph 36), jelly fungi, mushrooms (Color Photographs 34 and 35), puffballs, rusts, smuts, and toadstools. Of these various species, the most familiar is the often edible mushroom. The visible part of such fungi is the reproductive body known as the *basidiocarp* (Figures 8–16 and 8–17).

Commercial mushrooms, *Agaricus* (a-GAR-i-kuss) species, are the only cultivated food crop of this group. Mushrooms are grown by "planting" the primary mycelia, known as *spawn*, in soil enriched with animal manure in a dark, cool place, such as a cave, basement, or similar environment.

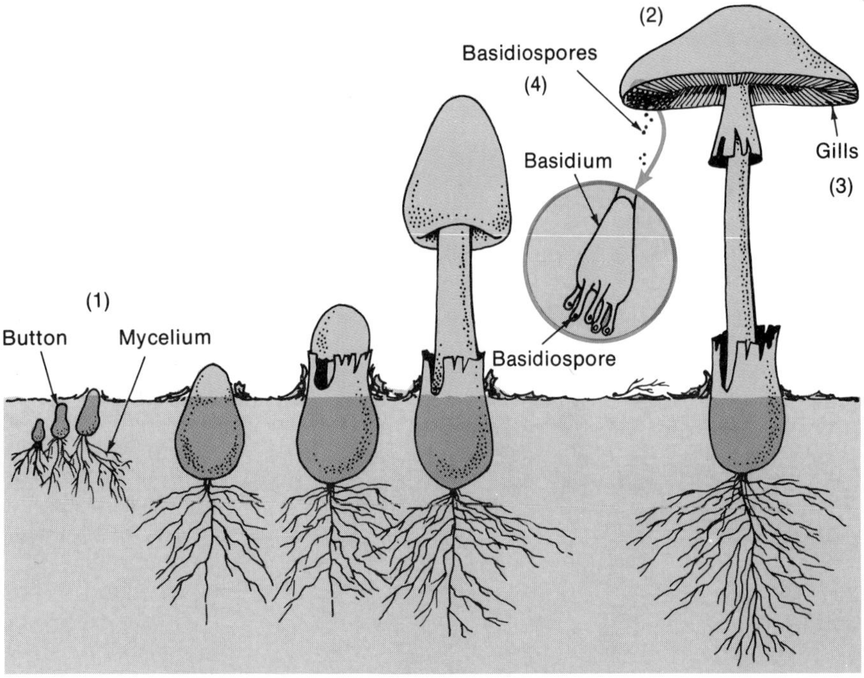

Figure 8–16 Mushroom development stages from the hyphae masses (mycelium) to the completed mushroom. A compact button appears underground (1) and grows into the complete fruiting body, or mushroom (2). On the undersurface of the mushroom, thin perpendicular gills appear (3), on which specialized hyphae (basidia) develop and produce basidiospores (4). The spores are released, and in a favorable environment they germinate and give rise to new hyphae and mycelia.

Mushroom poisoning occurs frequently. When not caused by bacteria (botulism) in marinated mushroom preparations, it is most often attributable to species of the mushroom genus *Amanita* (am-an-EE-tah) (Color Photograph 35). The species of *Amanita* can be identified by three characteristics: (1) the veil that covers the emerging mushroom, (2) white spores when the mushroom is mature, and (3) a distinctive cup at the base. However, mushrooms lacking these characteristics are not necessarily safe to eat. Studies with a so-called sacred mushroom, another *Amanita* species, which has been used for thousands of years in tribal religious ceremonies, have led to the development of drugs that eventually may treat central nervous system diseases such as epilepsy and schizophrenia.

Figure 8–17 A scanning micrograph of a small cross section of a gill from a common mushroom (basidiocarp). Typical basidia in various developmental stages and basidiospores are visible. Original magnification 1,050×. *[Courtesy of Dr. David H. Ellis, The Adelaide Children's Hospital, Inc.]*

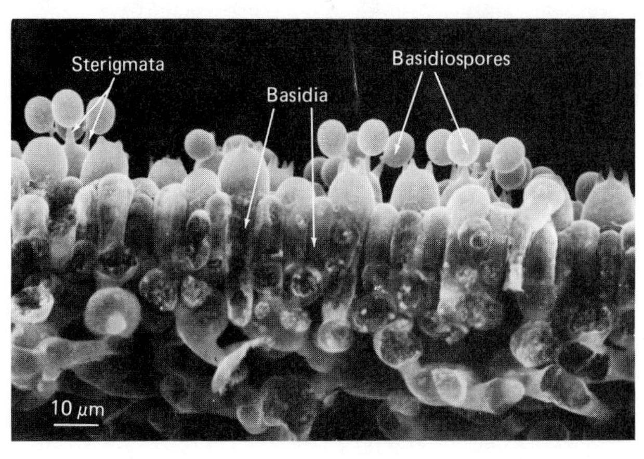

The rusts and smuts, which are plant parasites, cause considerable damage and economic loss. This is true for grain plants such as corn, oats, and wheat. Each species of smut is restricted to a single host species. Some of these plant pathogens, such as the stem rust of wheat and white pine blister rust, have complicated life cycles, passed in two or more different plants and involving the production of several kinds of spores.

The sexual spore-bearing structure of the basidiomycete is known as a *basidium* and grows out of an extensive underground mycelium (Figure 8–16). Each basidium is an enlarged, club-shaped hypha, at the tip of which four basidiospores develop. These spores, which develop outside of the basidium, are eventually released to develop into new mycelia under appropriate environmental conditions. The life cycle and other distinguishing properties of a mushroom are shown in Figures 8–16 and 8–17 and described in Table 8–2.

Deuteromycotina (Deuteromycetes)

The Deuteromycotina, or deuteromycetes, include all fungi that apparently lack a sexual stage; they are known as imperfect fungi. It is believed that the majority of these organisms are the nonsexual stages *(anamorphs)* of sexually reproducing fungi *(telomorphs)* that belong to the Ascomycotina or to the Basidiomycotina. Frequently, the sexual stage is found later, and then both stages of the fungus are taxonomically transferred to the group to which the

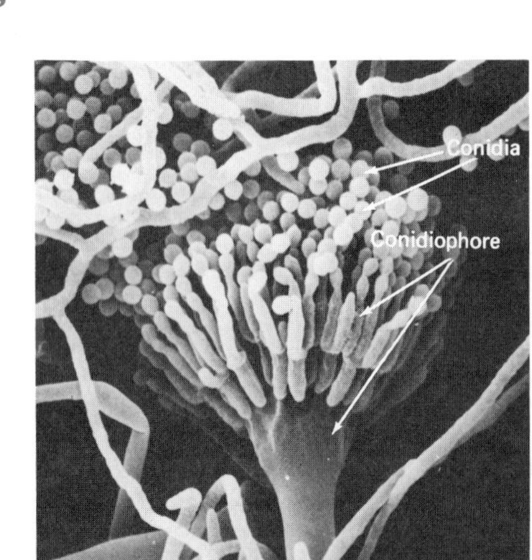

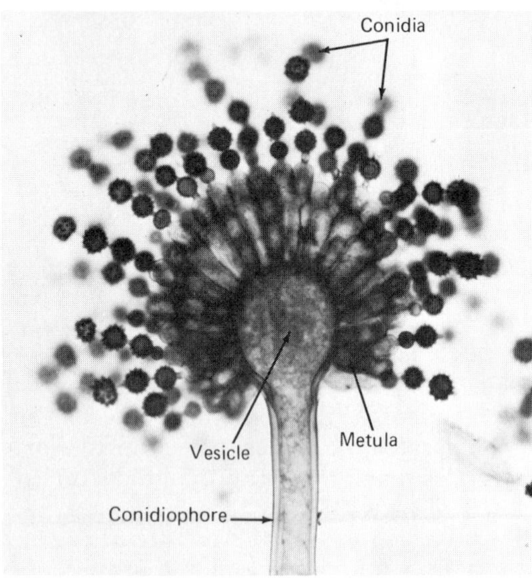

Figure 8–18 The structures of common fungi. The typical microscopic appearances of fungal species are shown. (a) *Penicillium* species. The characteristic brushlike feature of this mold is quite apparent. *[From M. S. Azab,* et al., Trans. Br. Mycol. Soc. ***86****:469–474, 1986.]*
(b) *Aspergillus niger* (a-sper-JILL-lus NYE-jerr). Note the free spores. *[A. D. Ciegler, I. Fennel, G. A. Sansing, R. W. Detroy, and G. A. Bennett,* Appl. Microbiol. ***26****:271–278, 1973.]*

sexual stage belongs. Such a transfer is consistent with the concept that the asexual and sexual stages together form the complete fungus life cycle *(holomorph)*. As a group, the members generally do not have a common origin or relationship.

Several deuteromycete species are commercially and medically important. Among the most important are members of the genus *Penicillium* (pen-ee-SILL-lee-um) (Figure 8–18a and Color Photograph 37). The distinctive flavor and soft consistency of Camembert and several other cheeses are caused by the enzymatic activities of specific fungi. The mycelium forms the white coat that covers the surface of the cheese. Eventually, the hyphae penetrate into the cheese, softening it by enzymatic action and imparting a unique flavor. The characteristic flavor and appearance of Roquefort and related blue cheeses are caused by the activities of another *Penicillium* species.

Several *Penicillium* species are important sources of the antibiotic penicillin. Other commercially important fungi are discussed later in the text.

Pathogenic fungi belonging to the deuteromycetes cause two types of human infections, namely, those that involve the superficial tissues, such as hair, nails, and skin, and those that affect the deeper tissues and organs. Examples of the former include the variety of infections called *ringworm* (Figure 8–19 and Color Photographs 41 and 98a). Some deuteromycetes cause only one of the general disease states; others can produce both types of infection (Figures 23–8 and 25–5). Many of these mycotic infections are described in later chapters.

Certain deuteromycetes produce poisons known as **mycotoxins** (Figure 8–20). Although most molds growing on foods are harmless, some are not. *Aspergillus flavus* (a-sper-JILL-lus FLAY-vus) and other fungi, which often grow on stored cereal products and nuts, produce one type of mycotoxin, **aflatoxins** (from the phrase A*spergillus fla*vus *toxin*), which are quite toxic to many animal species. The real and potential danger of these toxins was dramatically shown in 1960 by a large-scale trout poisoning in commercial fish hatcheries. The fish had been fed rations later shown to be contaminated by fungi. Concurrent outbreaks of turkey X disease in England were traced to similar contamination.

The presence of mycotoxins in food is not too surprising considering the wide distribution of fungi and their growth during storage and handling of food and crops. The recognition of aflatoxins stimulated research that led to isolation and identification of many additional mycotoxins. Other mycotoxins

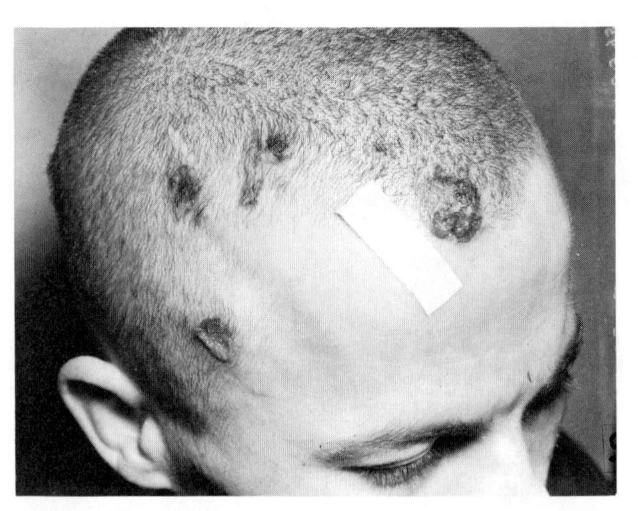

Figure 8–19 Favus, a severe form of ringworm of the scalp. *[Courtesy of the Armed Forces Institute of Pathology, Washington, D.C., Neg. No. B-535-1.]*

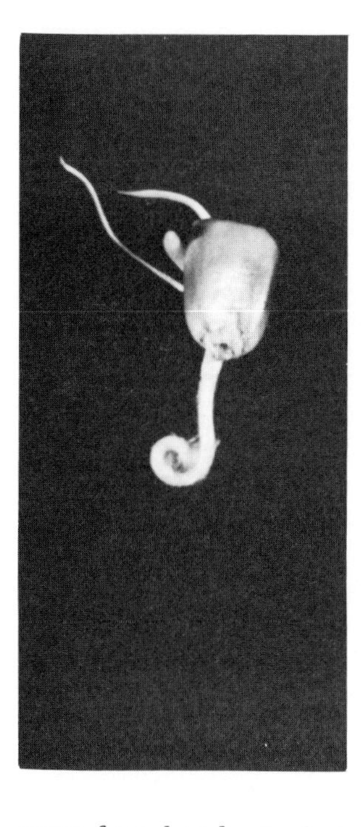

Figure 8–20 Fungi produce a variety of compounds. Certain of these substances have antibiotic action for bacteria, but they may exert serious disturbances in other forms of life. The *malformins* are a group of compounds that affect plant growth. The corn plant on the left has been exposed to malformin from *Aspergillus niger.* The plant on the right has been exposed only to water. *[R. W. Curtis, W. R. Stevenson, and J. Tuite,* Appl. Microbiol. ***28:****362–365, 1974.]*

were found to be toxic to many animal species and also to cause cancer in the liver of humans.

Poisoning by such fungal toxins assumes worldwide significance in view of population groups suffering from protein deficiency diseases. While being treated for malnutrition, affected individuals may suffer severe liver injury from mycotoxins. This poisoning occurs because the protein sources often used to correct malnutrition, peanuts and cereals, may be contaminated by mycotoxin-producing fungi. The World Health Organization (WHO) has led in directing attention to the serious health hazards associated with fungus-contaminated foods.

Zygomycotina (Zygomycetes)

The smallest class of fungi are the Zygomycotina, or Zygomycetes. The hyphae of these fungi have few or no crosswalls; they form coenocytic mycelia. Their asexual spores include chlamydospores, conidia, and sporangiospores. Sexual reproduction is by fusion of the hyphal tips of two different mating cell types, resulting in a zygospore (Figures 8–10 and 8–21).

Figure 8–21 The common bread mold *Rhizopus nigricans* (rye-ZOE-puss NYE-gree-kans). The life cycle of this mold, shows the various structures of the mycelium and both the asexual and sexual stages.

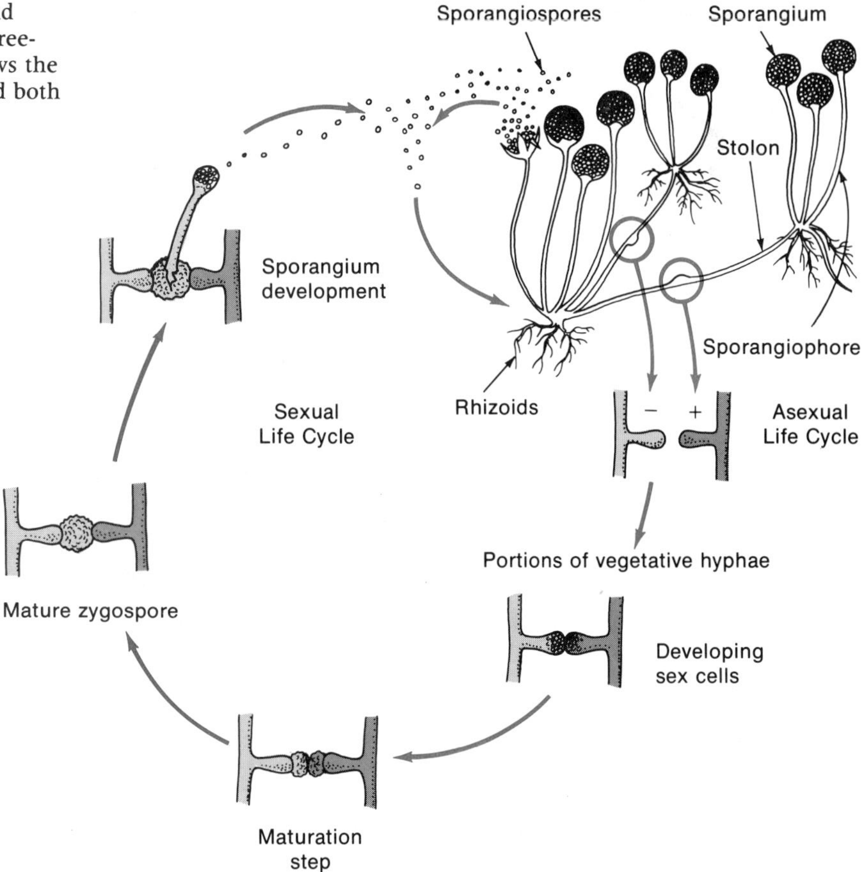

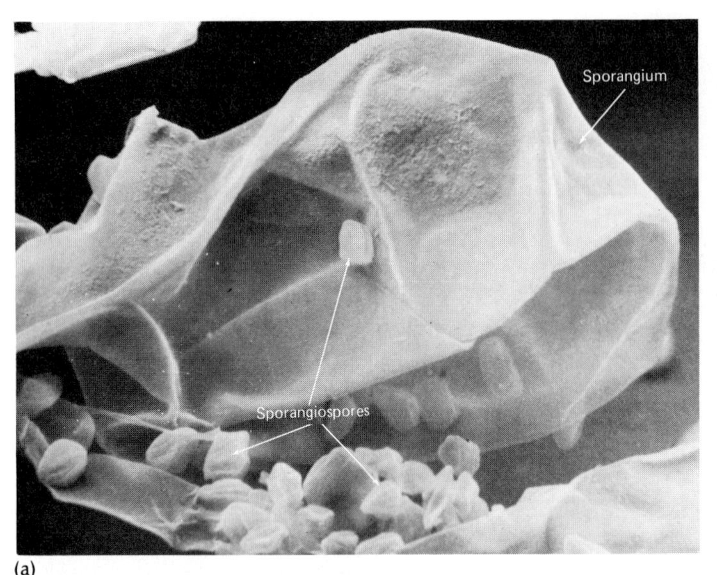

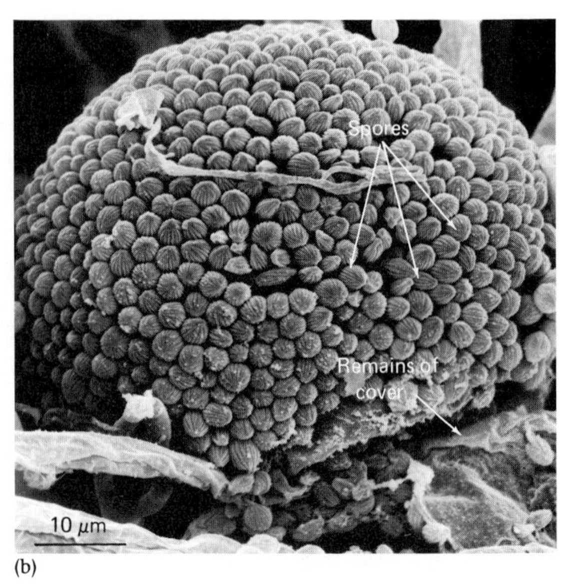

Figure 8–22 *Rhizopus nigricans.* (a) A scanning micrograph showing a collapsed sporangium and numerous sporangiospores. Note the thin texture of the saclike structure. The asexual cycle is shown in the upper right portion of this figure. (b) The mass of sporangiospores remaining after complete removal of the sporangial sac covering. Original magnification 1,875×. *[Courtesy Dr. David H. Ellis, The Adelaide Children's Hospital, Inc.]*

A familiar but unwelcome zygomycete is the common bread mold, *Rhizopus nigricans* (rye-ZOH-pus NYE-gri-kans) (Figure 8–22). Bread becomes moldy when a spore of this organism falls on it, germinates, and grows to form the mycelium that covers the bread surface. Some of the hyphae, the *rhizoids,* penetrate the bread to obtain nutrients and anchor the mold. Others, the *stolons,* spread horizontally with amazing speed. Eventually, certain hyphae grow upward and develop sacs, or *sporangia,* at their tips. Clusters of brown to black spherical asexual spores develop within each sac (Figure 8–21) and are released when the delicate spore sac ruptures. Growth of molds such as this is usually prevented in commercial bakeries by the addition of preservatives to the bread dough before baking.

In the sexual reproduction phase of zygomycetes, equal-sized sex-cell-forming structures called *gametangia* fuse. These gametangia are multinucleated; they form on a single hypha or on hyphae of different mating types, usually indicated as + and −. Once fusion takes place, a massive zygospore often forms between the two gametangia (Figure 8–21). A sporangium forms next to complete the reproduction cycle.

Certain zygomycetes cause fungus infections in humans and other animals. Others parasitize roundworms (Figure 8–23a and b) and arthropods.

Oomycetes

Oomycetes are noted for the production of single-celled, motile, asexual zoospores that can move only in water or moist environments. The life cycle of oomycetes includes two reproductive stages, one asexual and the other sexual, involving oospores (Figure 8–24).

Species of this fungal class cause some of the most destructive plant diseases known. These include downy mildew of grapes, which at one time nearly ruined the French wine industry. It was finally brought under control by the invention of Bordeaux mixture, a combination of copper sulfate and lime, which is still a common fungicide. A far more serious disease caused by oomycetes is late blight of potatoes, which almost totally destroyed the potato crop in Ireland in 1845, 1846, and 1848. This agricultural disaster resulted in widespread famine and death in Ireland during those years. Thousands fled the country and emigrated to other lands, especially the United States.

Certain oomycetes also are parasitic on mosquito larvae and other insects. Specific chemicals, pheromones, in such hosts appear to attract the fungi. The potential value of this attraction in mosquito control is under study.

Gymnomyxa: Slime Molds and Associates

The Gymnomyxa contain a wide variety of organisms that are grouped together by several properties, many of which are unlike those of true fungi. For example, these forms of life have a naked cell mass of protoplasm, called *slugs* or *plasmodia* (singular *plasmodium*), that has animallike movement and may creep about (Figure 8–25). Their actively feeding vegetative structures are composed of masses of

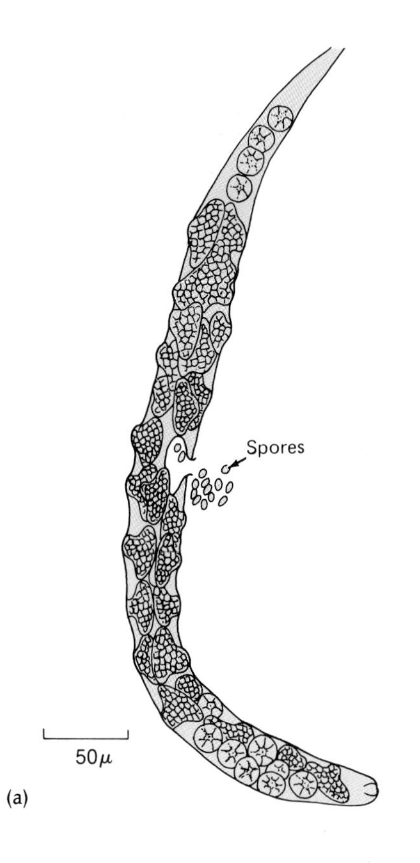

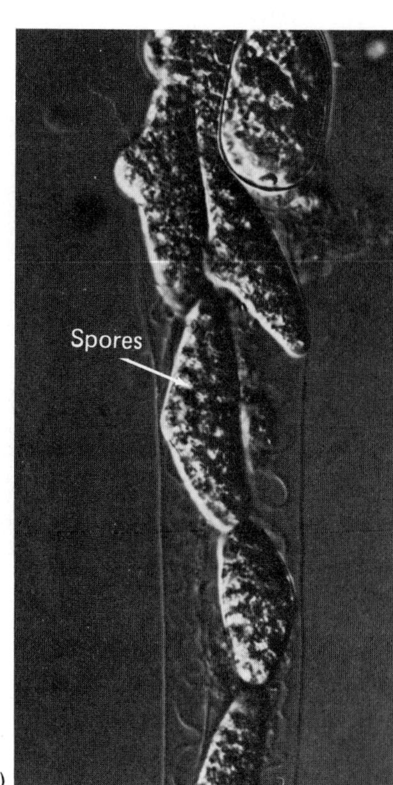

Figure 8–23 An animal's body is used for the production of spores. (a) The overall view. (b) The result of penetration and infection of a fungus parasitic on roundworms. The spore sacs can be seen within the worm's body. *[Reproduced by permission of the National Research Council of Canada, from G. L. Barron,* Can. J. Microbiol. ***22:****752–762, 1976.]*

amoebalike cells without cell walls, and their fruiting (reproductive) structures resemble those of fungi (Figure 8–26).

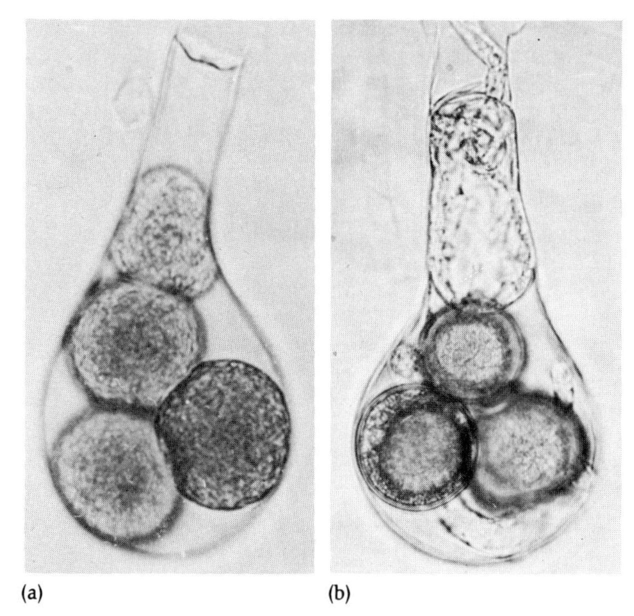

Figure 8–24 Sexual spores of oomycetes. (a) Immature oospores. (b) A germinating oospore with a germ tube and several mature oospores inside a fertilized egg cell (the oogonium). Original magnification 1000×. *[Courtesy of G. W. Beakes, University of Newcastle Upon Tyne.]*

The slime molds, so called because of the smooth, glossy appearance of their plasmodia, are included in the Gymnomyxa. Approximately 500 species of slime molds are recognized. Usually they are saprophytic, feeding on decaying plant life, and appear in various colors on decaying logs; on dead leaves in dense, shaded forests; or in damp soil (Color Photograph 43). Certain parasitic slime molds, such as *Plasmodiophora brassicae* (plaz-MOH-dee-oh-fore-a brass-IK-ee), cause significant injury to food crops such as cauliflower, radish, rutabaga, and turnip. This organism causes club-root disease of plants, in which the roots of the plant increase in size and provide an environment for slime mold growth and development.

The slime molds can be divided into two groups, *cellular* and *acellular.* The vegetative forms of the cellular mold consist of single amoebalike cells. The acellular slime molds are masses of protoplasm of indefinite size and shape (plasmodia) in their vegetative forms.

One of the outstanding features of this group of microorganisms is their rather unusual life cycle, consisting of a series of processes through which an organism goes as it develops from one form to another. The cycle of a cellular slime mold provides an example. (Sexual reproduction apparently does not occur in these forms, whereas it does in the acellular slime molds.) Cellular slime molds (Figure 8–26) live and multiply in soil habitats as amoebae with one nucleus. They use the bacteria in these envi-

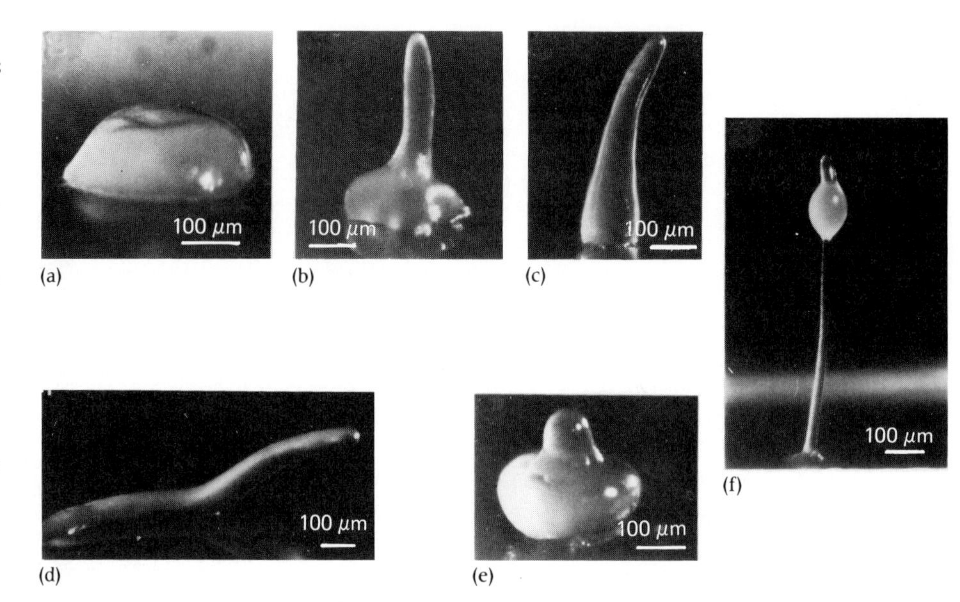

Figure 8–25 Major stages in the shaping of cells and tissues (morphogenesis) of the slime mold *Dictyostelium discoideum* (dic-tee-oh-STEL-lee-um dis-KOY-dee-um). An initial accumulation of amoebae that occurs in response to starvation, elongates into a migratory slug, and eventually completes a developmental cycle by forming a fruiting body. (a) A mound, (b) a mound with a short column, (c) a vertical column, (d) a slug, (e) a sombrero, and (f) a fruiting body. The bar markers represent 100 μm. *[R. L. Clark, G. S. Retzinger, and J. L. Steck,* J. Gen. Microbiol. ***121****:319–331, 1980; and photographer, K. Tanchik.]*

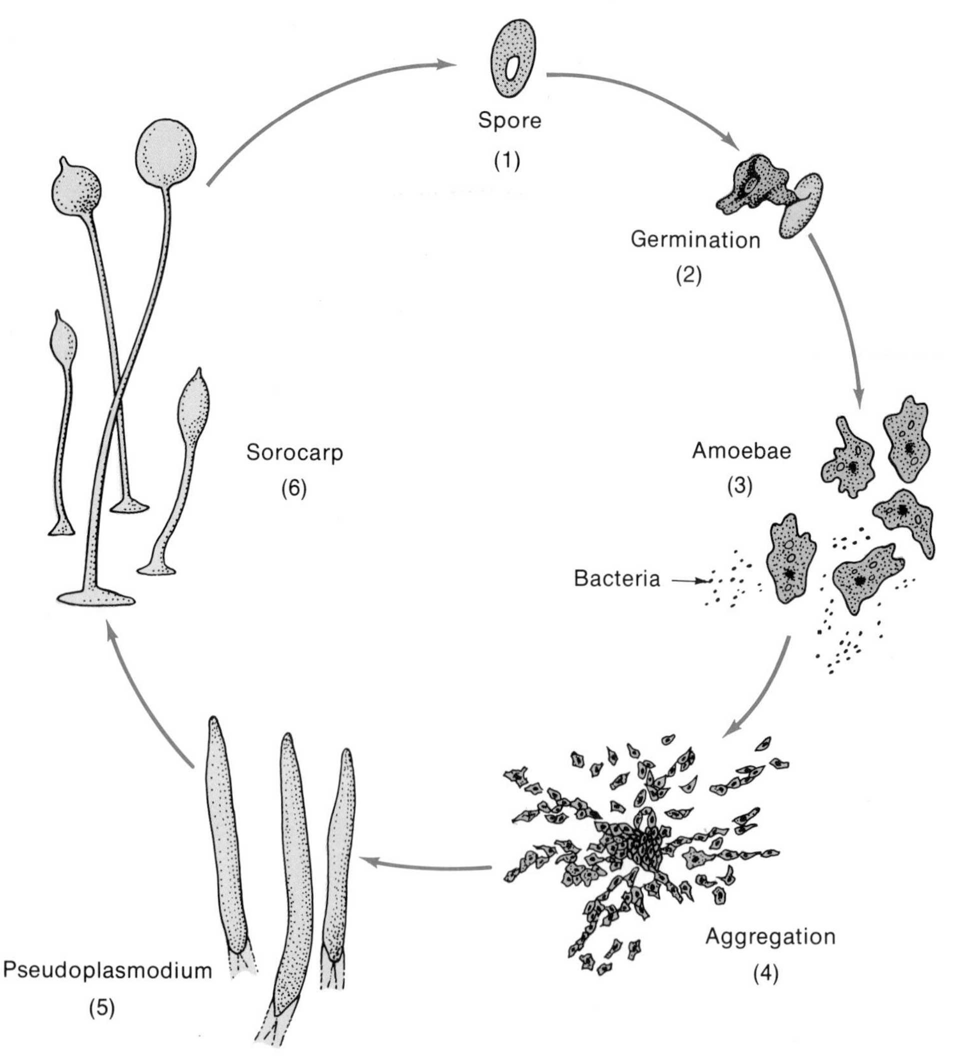

Figure 8–26 The life cycle of the cellular slime mold *Dictyostelium discoideum.* In a suitable environment, (1) the spores of this organism germinate (2) and give rise to amoebalike cells (3) that feed on bacteria, grow, and divide. When food sources decrease, some cells produce a hormonelike substance, a crasin, which causes other cells to stream toward a central area, forming an aggregation (4). Next, aggregations form a pseudoplasmodium, or slug (5). At a particular stage of this organism's development, the cells of the slug differentiate into a spore-producing sorocarp (6), and the cycle starts again.

ronments as food. However, when the supply of bacteria is exhausted, the amoebae combine their efforts and collect into multicellular aggregates (pseudoplasmodia, or slugs) of 100,000 or more. The cells comprising this structure lose some of their individuality but do not fuse. The aggregation process is initiated by the production of acrasin, a chemical identified as cyclic adenosine monophosphate (cAMP). Each aggregate undergoes a complex development cycle resulting in the formation of a fruiting body. The latter structure consists of a base, a slender stalk, and a cluster of spores in a capsule at the head. The fruiting body is composed of two specialized types of cells. *Stalk cells* lift the spore mass from the substratum, the surface on which the slime mold is situated. The *spores* are reproductive cells. Environmental factors apparently play an important role in the development of the fruiting structure.

When moistened, the spores of the slime mold germinate and give rise to the single cells called *myxamoebae.* These forms, which closely resemble typical protozoan amoebae, divide or reproduce by simple fission (splitting). Eventually, under the conditions mentioned earlier, pseudoplasmodia again are formed, and the development cycle begins once more.

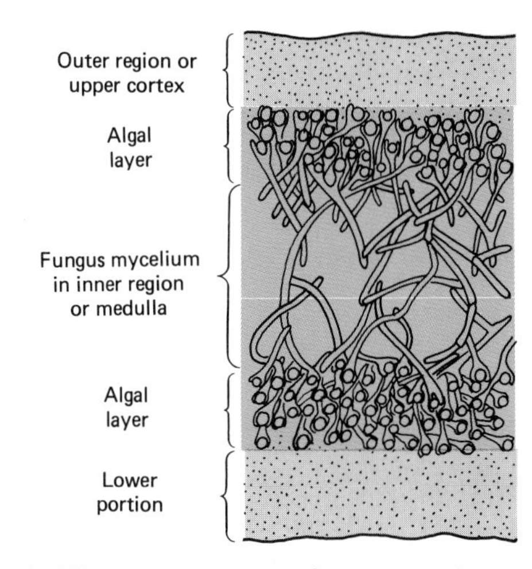

Figure 8–27 A cross section of one type of lichen, fruticose. Note the algal layer and the fungus-containing medulla and cortex layers. (See Color Photograph 44.)

Lichens

Many interesting relationships exist between various microorganisms. One of these is the symbiotic association between certain blue-green bacteria or green algae and fungi, the combination of which is called a *lichen* (Color Photographs 44 and 45). In the resulting plant body, called a **thallus,** the hypha of the fungus produces a tightly interwoven mycelium (Figure 8–27). Algal cells are usually housed and protected within this basic structure. The function of algae, according to one view, is to provide nutrients through the photosynthetic process. Fungi absorb from their environment water and minerals that are utilized by the companion algae. Since lichens are capable of living in environments that cannot support most other forms of life (for example, on rocks and in the arctic), it may be that this association is formed as a matter of utmost necessity for survival of the microorganisms involved. Lichens are able to survive in these environments partly by being able to dry or freeze and reaching a condition we may consider to be a form of suspended animation. Once conditions improve, lichens recover and continue with their normal metabolic activities.

Although lichens are remarkably resistant to drying, cold, heat, and other environmental conditions, they are extremely sensitive to air pollutants and quickly disappear from heavily polluted urban regions. Apparently, lichens absorb the pollutants from rain water. Because they have no means of excreting them, lethal concentrations of toxic air pollutants gradually build up.

At least 25,000 different lichen species have been identified. The fungal member of the lichen is usually a member of the class Ascomycotina. The other member may be one of many blue-green bacteria or green algae. Fungi are usually the dominant organisms and therefore determine the shape and size of the basic structure.

The reproduction of lichens may take place through a combination of normal sexual processes. Ascospore formation usually occurs with the fungal partner, and asexual division takes place with the photosynthetic partner.

Many different arrangements and colors are displayed by lichens. These differences largely determine their type, or category. Three major types of lichens are recognized. *Crustose* (crustlike) lichens are usually found on rocks or bark as irregular, flat patches. The colors include black, gray, green, yellow, and brown (Color Photograph 45). *Foliose* (leaflike) lichens are curled, leafy, and usually greenish-gray, possessing rootlike structures for attachment and the absorption of minerals. *Fruticose* (shrublike) lichens are highly branched and either hang from different tree parts or originate in the soil (Color Photograph 44).

Lichens are ecologically and economically important. They serve as food for arctic caribou and reindeer and are used in the preparation of litmus paper, a well-known indicator in chemistry and related sciences. Recently, certain lichens have been shown to be capable of producing antibiotics. From such reports, it appears that lichens may have great potential in the treatment of certain bacterial and fungal diseases. Another role of lichens, one that is sometimes overlooked, is their contribution to the organic content of soil. Quite often the activity of their rootlike structures aids in the decomposition of rocks. The decaying remains of dead lichens becomes intermixed with the small rock particles, providing nutrients and a foundation for plant development.

Summary

1. Fungi are eucaryotic and nonphotosynthetic.
2. Fungi are of great practical and ecological importance; they include mushrooms, puffballs, woody bracket fungi, molds, and yeasts.

Distribution and Activities

1. Fungi are widespread in nature, growing well in dark, moist environments where organic material is available, and over a broad range of temperatures.
2. These organisms are heterotrophic and are referred to as *saprobes,* since they utilize preexisting organic products.
3. Fungi produce digestive enzymes that, depending on the species, may be beneficial or destructive.
4. Fungi are important as sources of foods and in the processing and production of other foods and drugs.

Structure and Function

1. Fungi differ from bacteria in size, structure, cellular organization, and methods of reproduction.
2. The general term *fungus* includes the forms molds and yeasts.

Molds

1. Molds do not have true roots, stems, or leaves, but do exhibit a differentiation of parts.
2. The structural unit of most molds is the *hypha.*
3. Strands, or filaments, of hyphae can be subdivided into multicellular forms by crosswalls, or septa.
4. Hyphae without septa are called *coenocytic.*
5. Mold growth resulting in a visible cobweblike aggregation of hyphae is called a *mycelium.*
6. Spores are the specialized reproductive cells of molds.

Yeasts

1. Yeasts appear as oval to spherical cells that form moist, shining colonies.
2. Some yeasts may produce capsules.
3. These microorganisms reproduce sexually by producing new *buds,* or *daughter cells.* Birth scars are carried by new cells, while main cells exhibit bud scars.
4. Newly formed cells that do not separate form *pseudomycelia.*

Dimorphism

Under certain environmental conditions some fungi exhibit two different forms, appearing as either molds or yeasts. This phenomenon is called *dimorphism.*

Reproduction and Spores

1. The type of spore and the sporulation process are both important to fungal identification and classification.
2. Fungal spores function as reproductive cells and are not more resistant to environmental factors than other fungal cell types.
3. Both sexual and asexual reproduction occur in fungi. Some organisms are entirely dependent on asexual reproduction. Mitosis in fungi differs from that found in other eucaryotes.
4. Sexual reproduction may involve the union of hyphae, sex cells, or differentiated, multinucleated sex organs.

Asexual Spores

1. Spores are formed either through asexual or sexual processes. Depending on the species, both asexual and sexual spores or only asexual spores are formed.
2. Asexual spores include the *arthrospore, blastospore, chlamydospore, conidium, sporangiospore,* and *zoospore.*
3. Under appropriate conditions of nutrition, moisture, pH, and temperature, fungal spores germinate and produce one or more long structures called *germ tubes.* Germ tubes subsequently develop into *hyphae.*

Sexual Spores

Sexual spores are the result of nuclear fusion between two different cell types and include *ascospores, basidiospores, oospores,* and *zygospores.*

Ultrastructure

1. Both molds and yeasts exhibit eucaryotic cellular organization.
2. Cellular membranes contain sterols, a property that separates fungi from procaryotes.
3. The cell walls of filamentous fungi are composed of thin, threadlike structures called *microfibrils* (which are composed of *chitin*) and cellulose. Yeast cell walls contain the complex polysaccharides glucan and mannan, as well as lipids, proteins, and the amino sugar glucosamine.
4. Pili appear on the cell walls of various yeasts. These structures are similar to those of bacteria and may be involved with the sexual reproduction of yeasts.

Cultivation and Identification

1. Molds and yeasts can be grown and studied by cultural methods similar to those used for many bacteria.
2. Media used for fungus cultivation are modified to limit the growth of other microbes. Ingredients used for this

purpose include antibiotics, dyes, high concentrations of sugars, and compounds that lower the pH of media.

TYPES OF MEDIA
Three basic types of media are used: *natural* (carrot plugs, potato slices), complex *dehydrated* (Sabouraud's dextrose agar), and *synthetic.*

DISTINCTIVE MYCELIA
The appearance of pigment and its diffusion from mycelia are used for identification purposes.

Classification

1. Several properties of fungi are used in fungus classification. These include methods of reproduction, myceial formation, cellular structure and formation, biochemical and physical properties, and mathematical analyses.
2. Fungi also may be classified on the basis of properties such as methods of reproduction and the presence of swimming spores (zoospores).
3. Members of the division Amastigomycota do not produce zoospores, while those of the Mastigomycota do.
4. The Amastigomycota includes the zygomycetes, ascomycetes, and deuteromycetes.
5. The Mastigomycota includes the oomycetes.

ASCOMYCOTINA (ASCOMYCETES)
The ascomycetes, or sac fungi, include a variety of economic and biologically important organisms. Members of this group can be found in a variety of environments, including aquatic, land, and other forms of life. The ascomycetes reproduce both asexually and sexually. Sexual spores, asci, may or may not be produced within fruiting bodies known as *ascocarps.* Four types of ascocarps are known: apothecium, cleistothecium, locule, and perithecium.

BASIDIOMYCOTINA (BASIDIOMYCETES)
The basidiomycetes include the bracket fungi of trees, mushrooms, puffballs, rusts, smuts, and toadstools. These fungi are also economically and biologically important. Some members of the group, such as the species of *Amanita,* are highly poisonous. The basidiomycetes reproduce asexually and sexually.

DEUTEROMYCOTINA (DEUTEROMYCETES)
The deuteromycetes represent a group of fungi in which sexual reproduction is unknown; they are known as imperfect fungi. Several members of the group are medically and commercially important. Some species are noted for their production of highly poisonous *mycotoxins* such as *aflatoxins.*

ZYGOMYCOTINA (ZYGOMYCETES)
The zygomycetes contain the smallest number of species. They reproduce both asexually and sexually. Some members are parasitic.

OOMYCETES
The oomycetes are noted for their production of single-celled, motile, asexual zoospores, which can move about only in water or moist environments. These fungi can also reproduce sexually. Certain species cause some of the most destructive plant diseases known. They also are parasitic for certain insects.

GYMNOMYXA: SLIME MOLDS AND ASSOCIATES

1. Slime molds have at times been classified as both fungi and protozoa because of the cell types produced in their unique life cycle.
2. The slime molds can be divided into two groups, the *cellular* and *acellular* forms.
3. While most slime molds are saprophytic, feeding on decaying plant life, some can be parasitic and cause the destruction of various vegetable plants.

Lichens

1. Lichens consist of the symbiotic association between certain blue-green bacteria or green algae and fungi.
2. The fungus component provides a tightly woven foundation of water and minerals for the association, while the algae provide nutrients by means of photosynthesis.
3. Three major types of lichens are recognized: *crustose* (crustlike), *foliose* (leaflike), and *fruticose* (shrublike).
4. Lichens serve as sources of food for arctic animals, contribute to the organic content of soil, and are used in litmus indicator production. They are also extremely sensitive to air pollution.

Questions for Review

1. List three ways in which fungi differ from bacteria.

2. a. What type of cellular organization do fungi have?
b. List at least six organelles found in fungi and give their functions.

3. a. Differentiate between asexual and sexual spores.
b. List and describe three examples of each type of fungal spore.
c. What type of spore do yeasts produce?

4. Distinguish between aerial and reproductive mycelia.

5. Define or explain the following terms:
a. hypha
b. mycelium
c. septum
d. coenocytic
e. bud
f. basidium
g. mold
h. yeast
i. basidiocarp
j. anamorph

6. What is dimorphism?

7. Do the techniques used for the cultivation of fungi differ from those used with bacteria? Explain.

8. a. What criteria are used in the identification and classification of fungi?
b. Do such criteria differ from the ones used for bacterial classification? Explain.

9. List the major subdivisions of fungi and give two beneficial and two harmful activities of each.

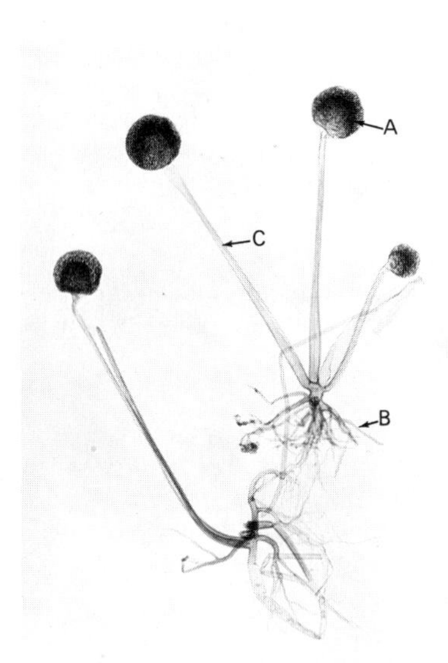

Figure 8–28 *[Courtesy Carolina Biological Supply Company.]*

10. Identify and give the function of each of the labeled structures in Figure 8–28.

11. A mycelium was found growing on an old melon. Microscopic examination reveals the view shown in Figure 8–29.
   a. Identify the fungus.
   b. Identify the labeled parts.

12. a. What are slime molds?
   b. How do slime molds differ from members of the ascomycetes?

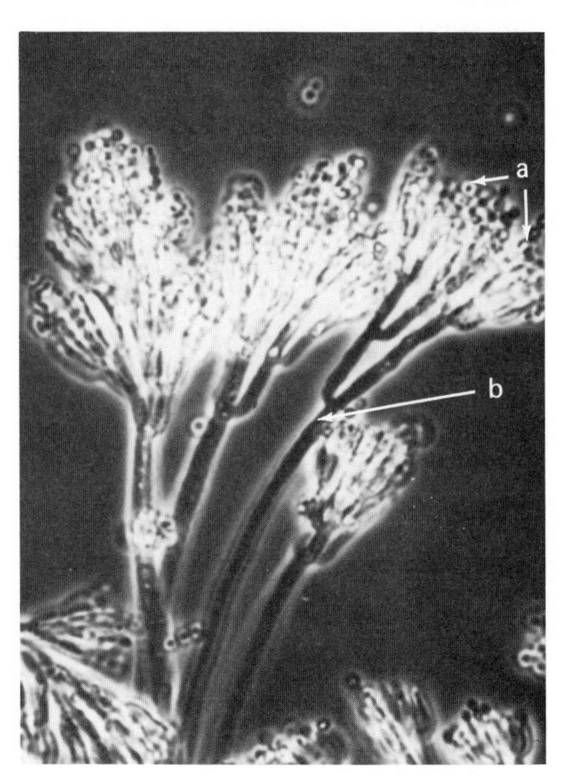

Figure 8–29 *[A. D. Ciegler, I. Fennel, G. A. Sansing, R. W. Detroy, and G. A. Bennett,* Appl. Microbiol. ***26****:271–278, 1973.]*

13. a. What are lichens?
   b. What beneficial functions do they have?

Suggested Readings

CHANG, S. T., and P. G. MILES, "A New Look at Cultivated Mushrooms" *BioScience* **34:**358–362, 1984. *An in-depth, fascinating presentation of this rapidly growing industry.*

COLE, G. T., "Models of Cell Differentiation in Conidial Fungi." *Microbiological Reviews,* **50:**95–132, 1986. *A review of the development aspects of the large group of economically, medically and ecologically important fungi that reproduce asexually by means of conidia.*

COOKE, R. C., *The Biology of Symbiotic Fungi.* New York: John Wiley & Sons, 1977. *This book attempts to outline all major symbiotic associations between fungi and either animals or plants and show how these various associations function.*

COURTENAY, B., and H. H. BURDSALL, JR., *A Field Guide to Mushrooms and Their Relatives.* New York: Van Nostrand Reinhold Co., Inc., 1982. *An excellent identification manual containing more than 350 descriptions of ascomycetes and basidiomycetes species.*

DEMAIN, A. L., "Industrial Microbiology." *Science* **214:** 987–995, 1981. *A highly readable review of the current and future potential contributions by fungi and bacteria to industrial processes.*

HALE, M. E., *The Biology of Lichens. 3rd ed.* Baltimore, Md.: University Park Press, 1983. *A concise account of lichen biology.*

HESSELTINE, C. W., "Fungi, People, and Soybeans," *Mycologia* **77:**505–525, 1985. *An excellent article describing U.S. Department of Agriculture research studies dealing with food fermentation, mainly in Asia.*

HOWARD, D. H. (ed.), *Fungi Pathogenic for Humans and Animals-B. Pathogenicity and Detection: II.* New York: Marcel Dekker Publishers, 1985. *An excellent collection of general reviews and practical guidelines dealing with culture techniques, microscopy detection, epidemiology, and pathogenesis.*

WASSON, R. G., *Soma, Divine Mushroom of Immortality.* New York: Harcourt Brace Jovanovich, 1967. *A fascinating discussion of the long history of the poisonous mushroom Amanita. Numerous illustrations, some in color.*

WEBB, A. D., "The Science of Making Wine." *American Scientist* **72:**360–367, 1984. *A short article that traces the ways in which an ancient art has become a modern-day science technology.*

9

The Protists

Much of our economic life is under the control of microbial activities over which we have little if any control. . . . Except in a few situations microorganisms are today as undisciplined a force of nature as they were centuries ago. — René Dubos, **Mirage of Health**

After reading this chapter, you should be able to:

1. Describe the distinctive features of protists and their position in the biological world.
2. Explain the basis of protozoan classification.
3. Identify protozoan structures and give their functions.
4. List and describe the distinguishing properties of the major groups of protozoa.
5. List and describe the major beneficial and destructive activities and functions of protozoa.
6. Distinguish between trophozoites and cysts.
7. Describe the methods of reproduction found among the protozoa.
8. List the distinguishing features of the divisions of algae.
9. Identify and give the associated functions of algal structures.
10. Compare algae with other types of microorganisms and explain how algae differ from them.
11. Describe the various types of spores found among algal protists.
12. List and describe the major beneficial and destructive activities of algae.

Of the five kingdoms of organisms, the protists are far and away the most diverse in structure and life cycles. They include some of the simplest as well as many of the most complex forms of life. This chapter surveys the protozoa and algae, two forms of protists. Their distribution, structure, life cycles, as well as their beneficial and harmful activities will be discussed.

The kingdom of Protista includes a variety of microbial groups whose members are predominantly unicellular. Some associations of individual cells, referred to as colonies, and some rare multicellular forms also are known. Protists—like animals, plants and fungi but unlike bacteria—are eucaryotic. Some organisms included in this kingdom tend to have animallike features, other have plantlike qualities and still others have features befitting fungi. However, they also share many properties. This chapter presents the two major members of the Protista kingdom, the **protozoa** and the **algae.**

Protozoa, the Animallike Protists

Since their discovery by Anton van Leeuwenhoek in 1674, protozoa have been extensively studied and even put to work. Today, protozoa are of interest to biologists working in various fields: the dating of oil beds by means of the fossil remains of protozoa; chemotherapy for protozoan diseases such as amebic dysentery, malaria, and toxoplasmosis; electron microscopy of protozoan structures; and use of protozoa as biologic indicators of pollution.

Protozoa are unicellular, eucaryotic microorganisms. Many are motile, and many require organic food and obtain it from their environments. Because of properties such as these, protozoa have traditionally been considered animals. Modern systems that divide living organisms classify protozoa as belonging to the Protista (Figure 2–6). At least 45,000 protozoan species have been described to date. While these microorganisms vary widely in shape, size, structure, and physiological properties, in most cases they are completely independent.

Distribution and Activities

Protozoa (singular **protozoon**) are found in many different environments. Some are present in bodies of water, where they play an important role in the food chains of natural communities. Others have mutually beneficial *(symbiotic)* relationships with higher forms of animal life (Figure 9–1) or with other microorganisms (Figure 9–2). Protozoa contribute to soil fertility through their role in the decomposition of organic matter by enhancing bacterial metabolism and excreting simple chemicals that return to plants. They function as a natural control

Figure 9–1 Beneficial associations. (a) A scanning micrograph showing the ciliate *Epidininium crawley* (ep-ee-DIN-in-ee-um KRAW-lee) (arrows) attached to the surface of plant fragments being digested in the intestinal tract of sheep. *Epidininium* are found in the digestive tracts of sheep and cattle, where they ingest whole or damaged chloroplasts and starch grains. *[T. Bauchop, and R. T. J. Clark,* Appl. Environ. Microbiol. ***32****:417–422, 1976.]* (b) An enlarged view of *Epidininium.* The functions of the various structures are explained later in the chapter.

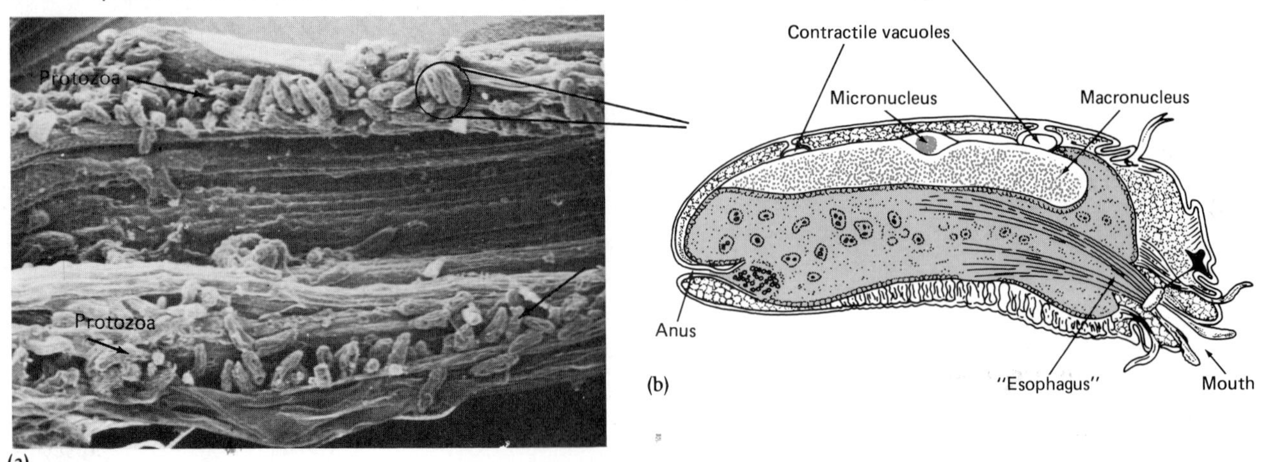

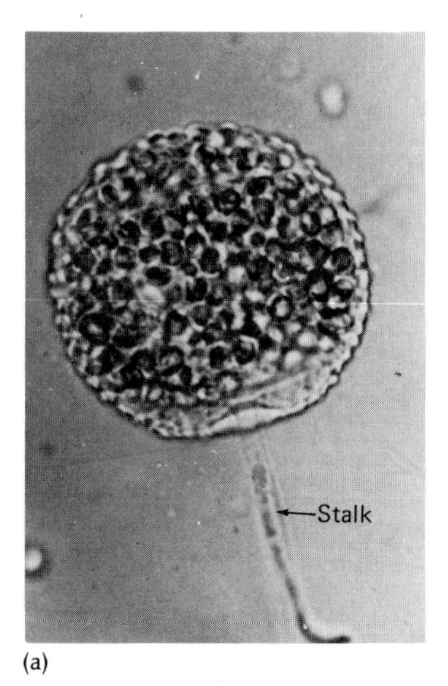

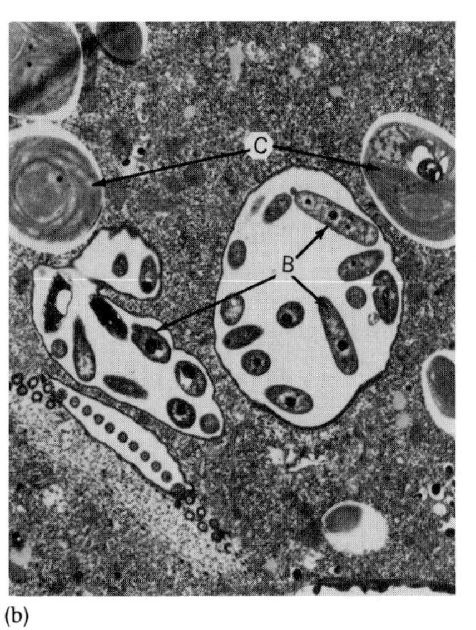

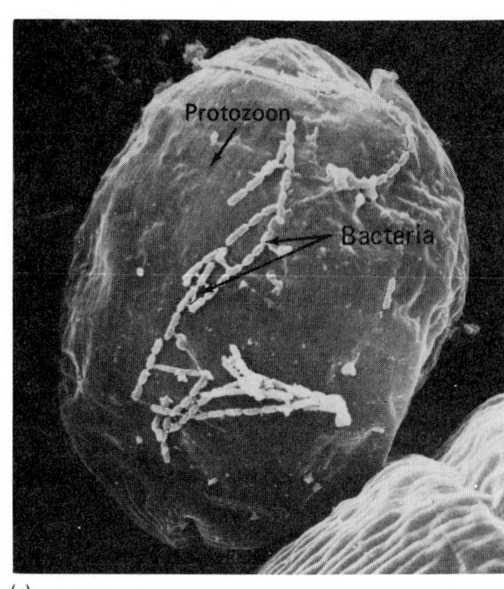

(a) (b) (c)

Figure 9–2 Associations of protozoa with other forms of life. (a) A photomicrograph of a single green *Vorticella* (vor-tee-SELL-la) with its stalk. The green coloration in *Vorticella* is usually caused by the presence of endosymbiotic (living within) algae. *[L. E. Graham, and J. M. Graham,* J. Protozool. ***25****:207–210, 1978.]* (b) An ultrathin view of this protozoon shows the presence not only of endosymbiotic green algae, *Chlorella* (klor-ELL-a) (C), but also intact bacteria (B). (c) Methanobacteria on the surface of a protozoon. These bacteria are believed to be ectosymbionts, organisms that have close external association with another form of life. *[G. D. Vogels, W. F. Hoppe, and C. K. Stumm,* Appl. Environ. Microbiol. ***40****:608–612, 1980.]*

on microbial populations by feeding on various types of microorganisms.

Protozoa are also known for their harmful activities. African sleeping sickness, amebic dysentery, malaria, and toxoplasmosis are but a few of the human diseases associated with these microorganisms. Several protozoa also infect wild and domestically important animals (Figure 9–3). In severe cases, infected hosts are crippled and disfigured and eventually die. **[See Part VIII for protozoan diseases.]**

Structure and Function

Even in protozoa, the complexity of essential functions requires division of labor among cellular parts (organelles), which are variously involved with movement, obtaining and utilizing nutrients, excretion and osmoregulation, reproduction, and protection.

Locomotion

Three types of locomotor organelles occur among the protozoa: **pseudopodia, flagella,** and **cilia.** These structures are important to the classification of protozoa. Pseudopodia ("false feet") are temporary cellular processes commonly associated with amoebae. The cell moves as a whole when the organism extends an area of its protoplasm, and then the rest of the cell flows into the extension (Figure

Figure 9–3 The appearance of rainbow trout with "black tail," more commonly known as *whirling disease.* This protozoan infection is recognized as a major communicable disease in fish hatcheries and in waters stocked with hatchery-infected fishes. *[K. Wolf, and M. E. Markiw,* J. Protozool. ***23****:425–427, 1982.]*

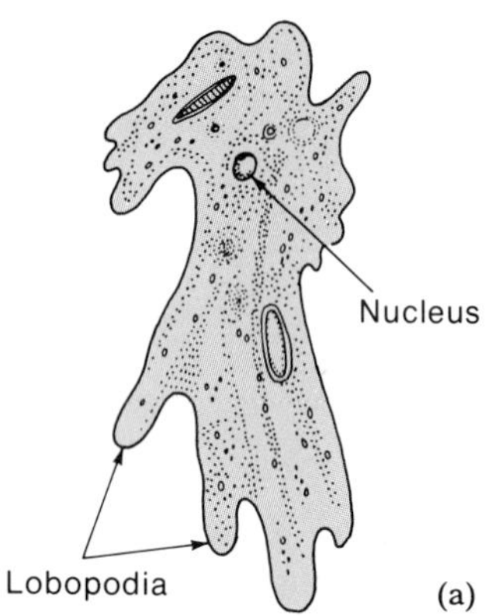

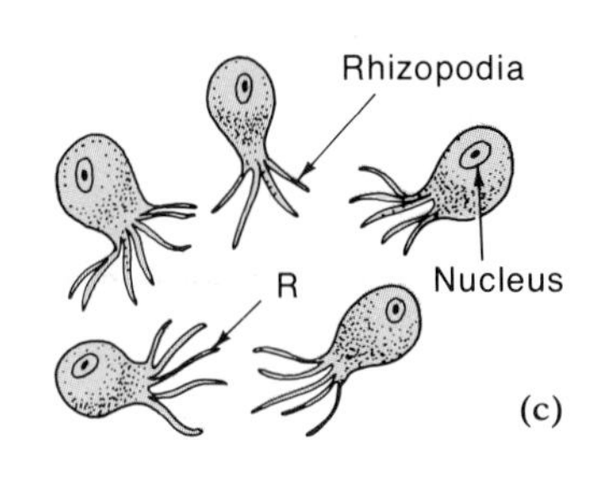

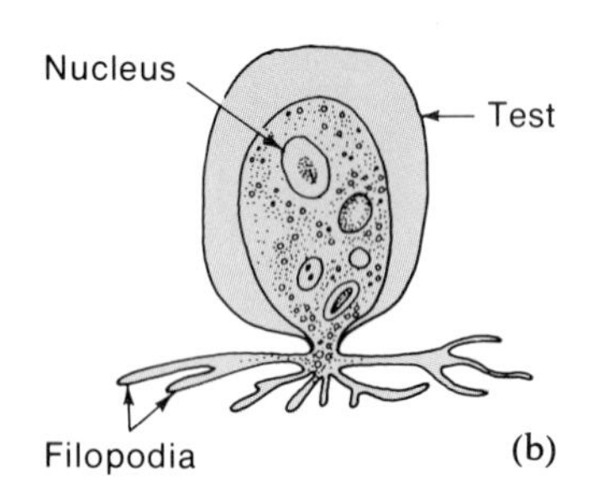

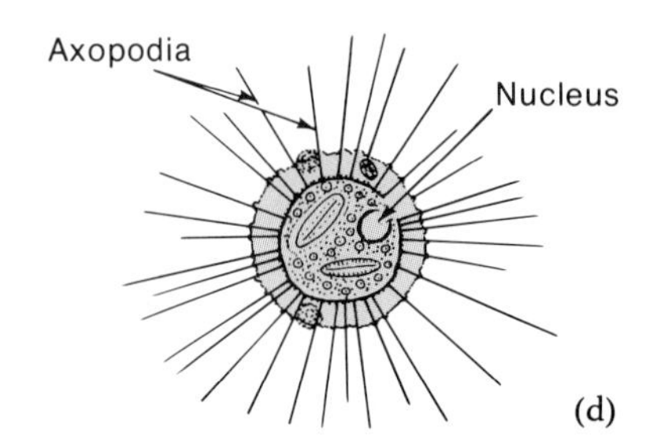

Figure 9–4 Types of pseudopodia. (a) Lobopodia. (b) Filopodia. (c) Rhizopodia. (d) Axopodia. *[After P. A. Meglitsch,* Invertebrate Zoology. *London: Oxford University Press, 1967.]*

9–7). Pseudopodia also aid in capturing food. Several types of pseudopodia are found among amebalike organisms (Figure 9–4). These include finger-shaped, round-tipped *lobopodia;* thin, pointed *filopodia;* branching, slender, pointed *rhizopodia;* and slender *axopodia,* with several fibers forming an axial filament.

Protozoan flagella are delicate whiplike structures that beat to propel the organism. These organelles appear to respond to stimuli such as chemicals and touch. The structure and organization of flagella are approximately the same in all types of eucaryotic cells and organisms. Internally, an individual flagellum consists of two central microtubules, surrounded by nine double tubules. This 9 + 2 arrangement is characteristic of most cilia and flagella associated with all eucaryotic cells, from paramecia to humans (Figure 9–17). The outer covering, or sheath, of a flagellum is a continuation of the cell membrane. A deoxyribonucleic acid (DNA)-containing granule known as the *kinetoplast* is situated at the base of each flagellum (Figure 9–5). The extranuclear DNA of this structure *(kDNA)* is contained

Figure 9–4 (e) An amoeba with several lobopodia (L) and a long filopodium (F). Original magnification 3000×. *[From W. B. Lushbaugh, and F. E. Pittman,* J. Protozool. ***26:****186–195, 1979.]*

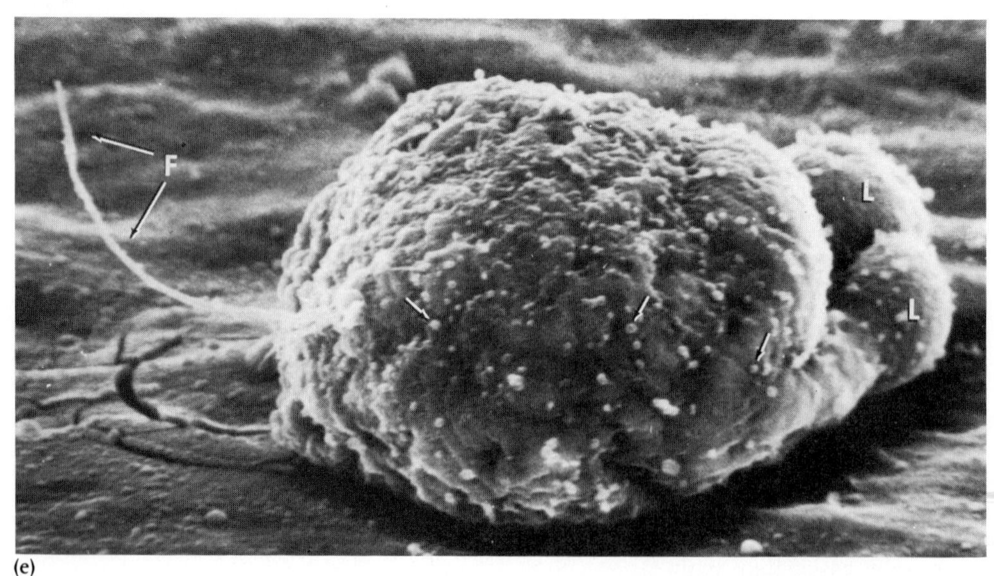

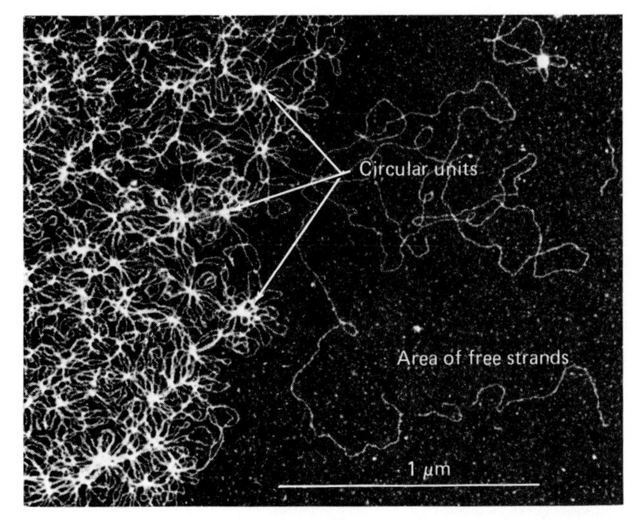

Figure 9–5 A portion of the circular units of kinetoplast DNA making up a characteristic network. Long DNA strands free of the network can also be seen. *[D. C. Barker,* Micron ***11****:21–62, 1980.]*

within a modified region of the protozoon's mitochondrion and consists of several circular molecules arranged in a characteristic network (Figure 9–5). Biochemical analyses of this DNA are of value in tracing evolutionary changes occurring among related flagellates and in the classification and identification of unknown organisms isolated from cases of disease. In some flagellates, the flagellum may be buried in the cell membrane along much of its length, forming a finlike undulating membrane (Figure 9–16 and Color Photograph 90).

Cilia are shorter forms of flagella. Like flagella, they are composed of two central and nine peripheral microtubules (Figure 9–6) enclosed by a covering that is continuous with the cell membrane. These organelles may cover the surface of the protozoon, or they may be restricted to a particular region, such as the oral region. In some organisms, cilia are fused in stiffened tufts called *cirri* (Figure 9–6), which may function as tiny legs. Ciliary structures have a coordinating network of fibers that lies just beneath the protozoan covering (Figures 9–6c).

Feeding and Digestion

Three major methods of obtaining nutrients are found in the protozoa. Some organisms are **autotrophic,** capable of forming organic compounds from inorganic ones. Others are heterotrophic, requiring preformed organic substances from their environment. Some heterotrophs are **saprobic,** obtaining nutrients by absorbing what is needed through the cell surface, or **holozoic,** ingesting entire organisms or smaller particles of food. Holozoic protozoa must have mechanisms for food capture. These include food cups (Figure 9–7), cytosomes (mouths) (Figure 9–8), and tentacles with which to trap prey (Figure 9–16b). Ingested food particles pass into intracellular digestive cavities known as *food vacuoles.* Indigestible material is voided either through a temporary opening or through a permanent opening, the *cytopyge.*

Excretion and Osmoregulation

The excretion of wastes occurs at the cell surface. Most protozoa excrete most of the nitrogen from protein degradation as ammonia. Other waste products created by intracellular parasitic forms such as the causative agents of malaria (Color Photograph 110) are secreted and accumulated in the host cell. With the destruction of infected cells in the human, the release of waste products produces chills and high body temperatures.

In many protozoa, excretion and osmoregulation are the function of the **contractile vacuoles.** The contractile vacuole system separates a dilute *(hypotonic)* solution of water and electrolytes from the cy-

Figure 9–6 Ciliary structures of the protozoon *Gastrostyla steinii* (gas-TROH-sty-la styn-EYE). (a) Diagrammatic representation of a cirrus-bearing protozoon. (b) A view showing the rows of cirri. (c) Network of fibrils (A) and individual fibrils (B) just beneath the network. The numbering of triplet instead of double microtubules is also shown in (A). *[J. N. Grim,* J. Protozool. ***19****:113, 1972.]*

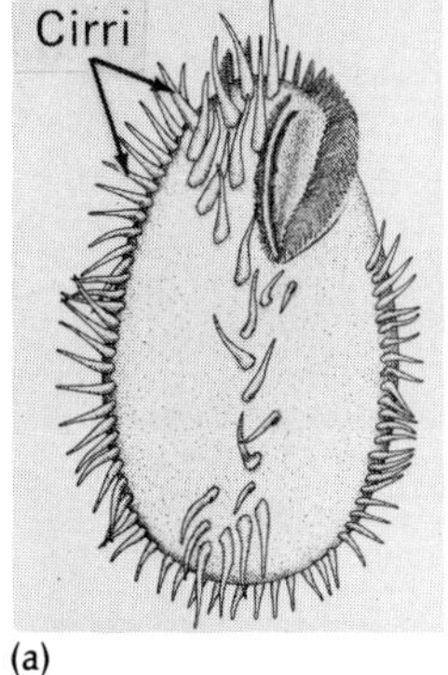

(a)

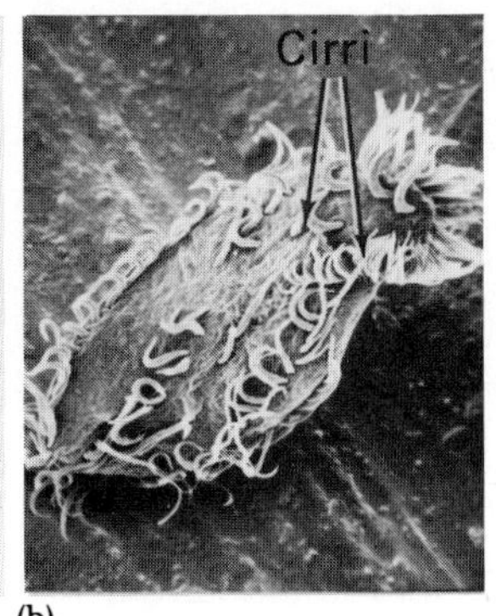

(b)

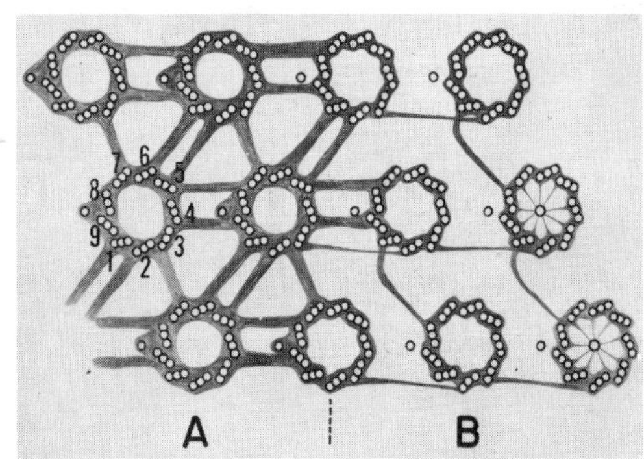

(c)

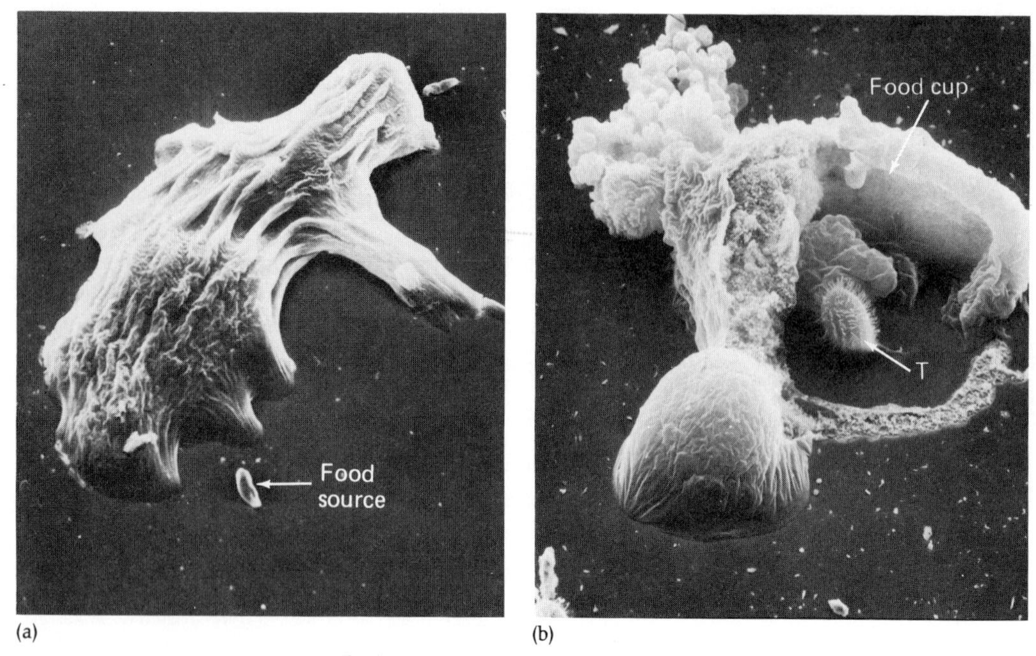

Figure 9–7 The feeding action of an amoeba. (a) A nonfeeding amoeba beginning to extend itself toward a potential food source. (b) An amoeba showing food cups, one of which is open. The inside surface of a food cup enclosing the ciliated protozoon *Tetrahymena* (tet-rah-HIGH-men-a) (T) is evident. *[K. W. Joan, and M. S. Jeon,* J. Protozool. ***23**:83–86, 1976.]*

Figure 9–8 (a) A vegetative protozoon showing a prominent feeding apparatus, the cytosome (C), fringed by a row of cilia. *[J. A. Kloetzel,* J. Protozool. ***22**:385–392, 1975.]* (b) A close-up view of a ciliate ingesting a blue-green bacterium (cyanobacterium) (A) through its cytosome. Note the many rows of cilia. *[R. K. Peck, and K. Hausman,* J. Protozool. ***27**:401–409, 1980.]*

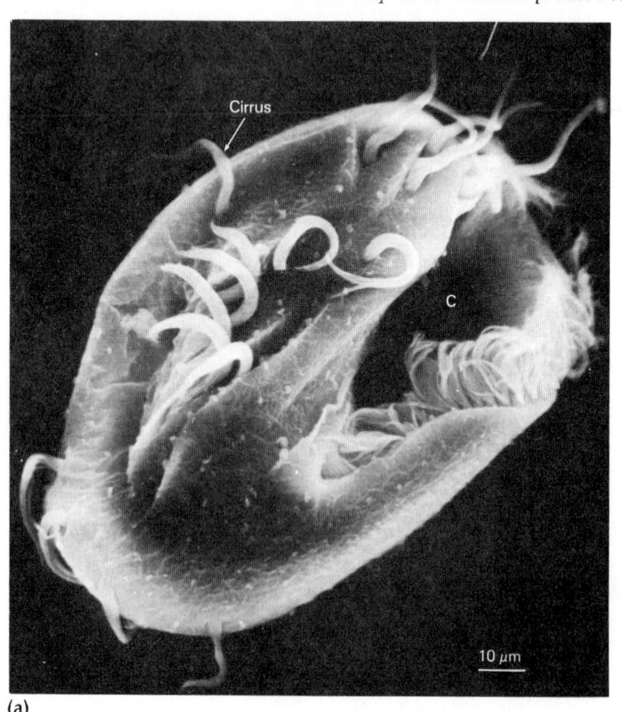

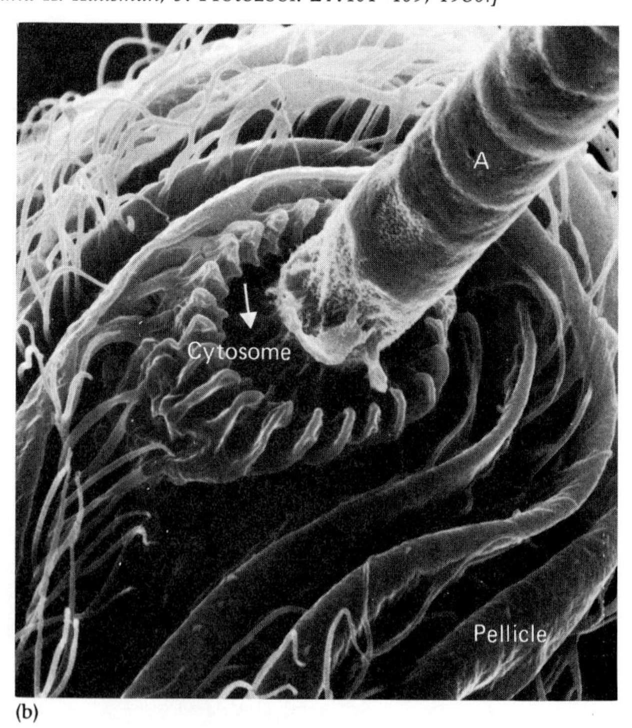

toplasm. The protozoan vacuole is filled from several small tubes, and pumps its collected contents from the cell at a specialized opening such as a pore in the cell surface (Figure 9–9a). In protozoa such as amebas, the pore is a temporary structure formed at the onset of each contraction *(systole)*. In *Paramecium* and other ciliates, the pore is a permanent opening in the cellular covering *(pellicle)*. Certain flagellated protozoa discharge through a temporary pore into a pocket surrounding a flagellum (Figure 9–9b).

Protective Structures

Protozoa are exposed to many environmental hazards. Much of their behavior is in response to toxic, life-threatening conditions. Therefore, the development and use of special protective devices help them to survive. Most of the protective structures of protozoa prevent mechanical injury or protect against drying, excessive water intake, and predators. They include several types of surface coverings and trichocysts. Other protozoa resort to encystment.

Several flagellates, some amebas, and all ciliates have an outermost surface envelope known as a **pellicle.** This covering is stronger than the cell membrane to which it is attached. Pellicles provide protection against chemicals, drying, and mechanical injury.

Many stationary and some free-living protozoa build coverings called **tests** or *shells* (Figure 9–10). Some structures consist of sand grains or other foreign particles cemented together; others are composed of secretions of calcium carbonate or silica.

Trichocysts (Figure 9–11) are specialized intracellular organelles used for the capture of food and possibly for defense. Some ciliates discharge these structures in response to toxic environments.

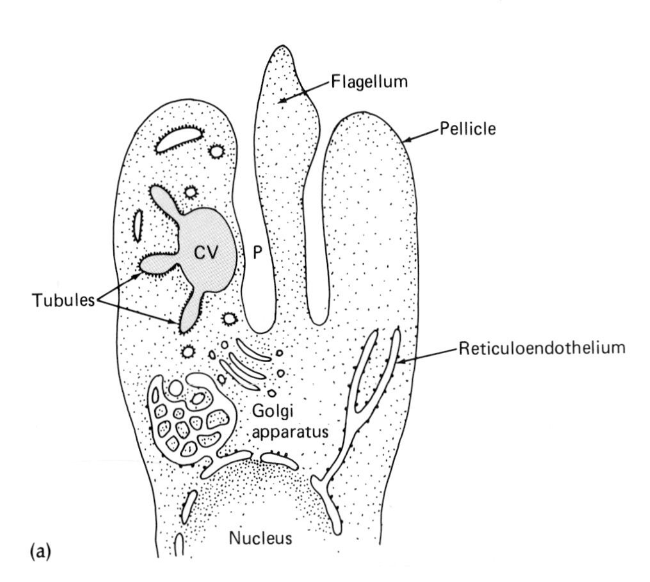

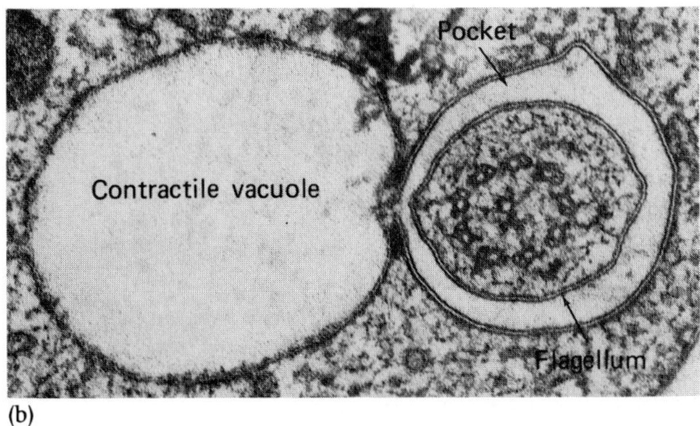

Figure 9–9 The contractile vacuole. (a) A general view of the organization of a contractile vacuole system found in some flagellates. The tubules that fill the contractile vacuole (CV) and the receiving pocket (P) surrounding the flagellum are emphasized. *[From J. C. Linder, and L. A. Staehelin,* J. Cell. Biol. ***83****:371–382, 1979.]* (b) An electron micrograph showing the relationship between the contractile vacuole and the pocket surrounding the flagellum. *[From J. C. Linder, and L. A. Staehelin,* J. Cell. Biol. ***83****:371–382, 1979.]*

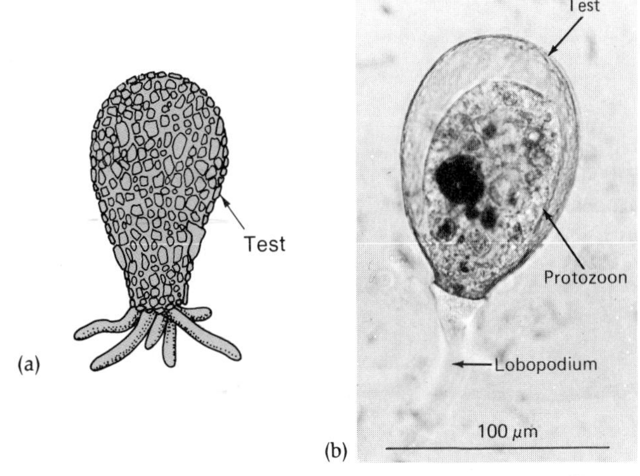

Figure 9–10 Protective devices of protozoa. (a) *Difflugia* (dif-floo-GEE-a), a test-forming organism. (b) A light micrograph of another test-forming protozoon, *Nebela tincta* (neh-BELL-a TINK-ta), showing the relationship between the protective structure and the organism. *[Courtesy Von Mark Andreas Gnekow.]*

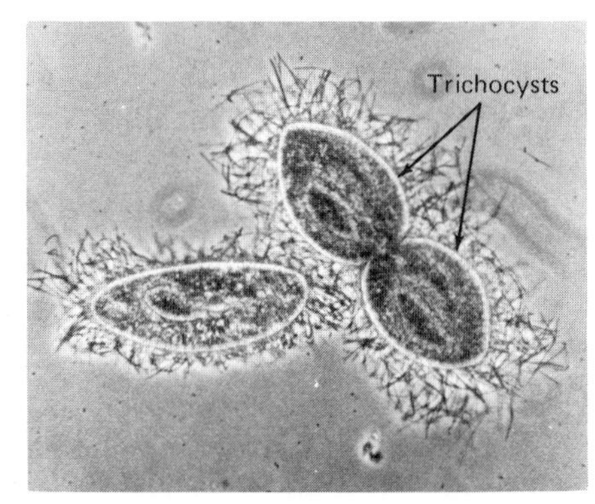

Figure 9–11 Paramecia with trichocysts discharged in response to an unfavorable environmental situation. *[D. Nyberg,* J. Protozool. ***25****:107–112, 1978.]*

Trophozoites and Cysts

In several parasitic species and free-living protozoa found in temporary bodies of water, the normal, active feeding form, known as the **trophozoite** (Figure 9–12 and Color Photograph 46), often cannot withstand the effects of various chemicals, food deficiencies, temperature or pH changes, and other harsh factors in the environment. To overcome such conditions, many protozoa can secrete a thick, resistant covering and develop into a resting stage called a **cyst** (Figure 9–12).

The conditions necessary for such encystment are not fully known, but the process definitely represents a type of cellular differentiation. In addition to providing protection, cysts may serve as sites for cellular reorganization and nuclear division followed by multiplication after the organism leaves the cyst. In the case of certain pathogens such as *Entamoeba histolytica* (en-tah-ME-ba his-toe-LEH-tee-ka), which causes amebic dysentery, cysts also aid in spreading the disease agent. The importance of the trophozoite and cyst stages in the diagnosis of disease is discussed in later chapters. **[See Chapter 25 for gastrointestinal infections.]**

Methods of Reproduction

Ciliated protozoa are multinucleate. They possess at least one large *macronucleus* and one smaller *micronucleus* (Figure 9–17). The macronucleus, which varies in shape, regulates metabolic and developmental functions and maintains all visible traits. The micronucleus exerts overall control over a cell's macronucleus and regulates cellular sexual and reproductive processes.

Asexual Reproduction

Both asexual and sexual reproduction occur among protozoa. Like other microbial types, some protozoa are capable of reproducing only asexually. In this process, the parent cell divides, either equally or unequally, to produce one or more offspring that eventually develop into mature forms. **Binary fission** (division in two parts) is the most common type of asexual reproduction (Figure 9–13). Division may be nuclear or cytoplasmic. Neither meiosis (chromosome reduction division) nor fertilization (sexual reproduction) takes place during binary fission.

In addition to binary fission, protozoa reproduce

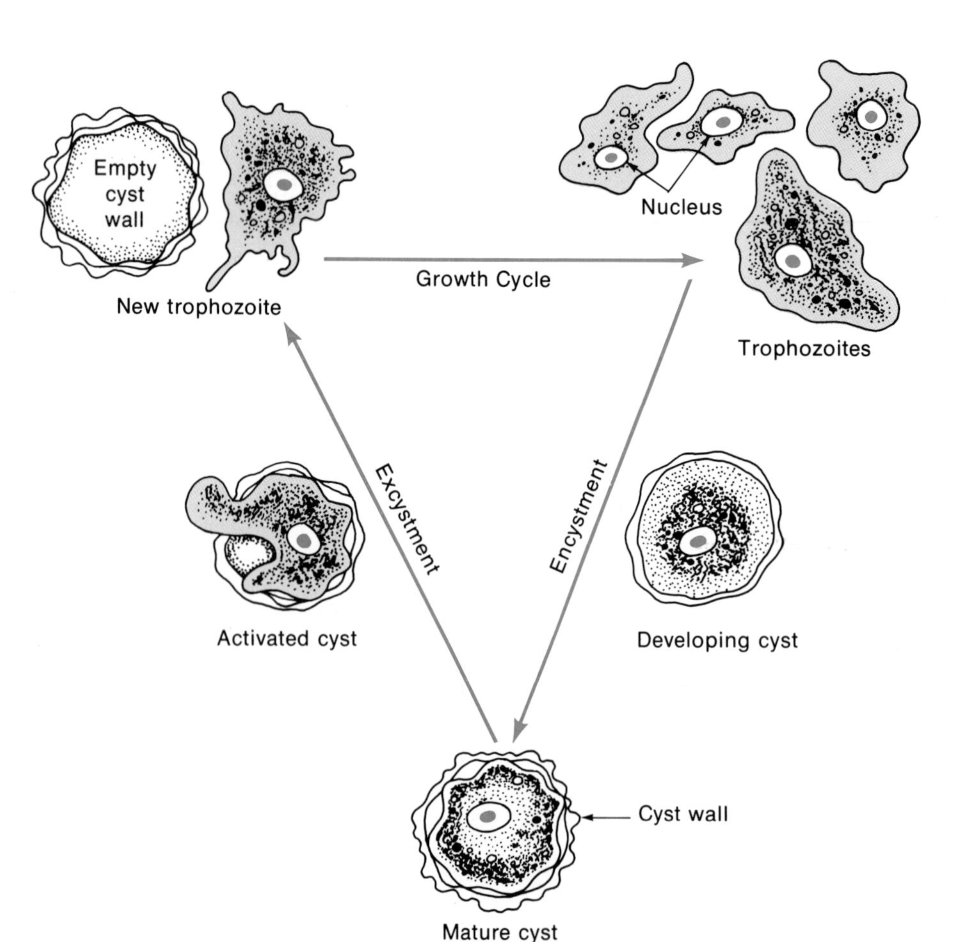

Figure 9–12 Trophozoites and cysts. The growth and differentiation cycle of protozoa having both trophozoite and cyst stages. In response to stimuli, the vegetative stage, the trophozoite, undergoes metabolic changes to form the resting stage, or cyst. Such cells can withstand prolonged starvation and drying.

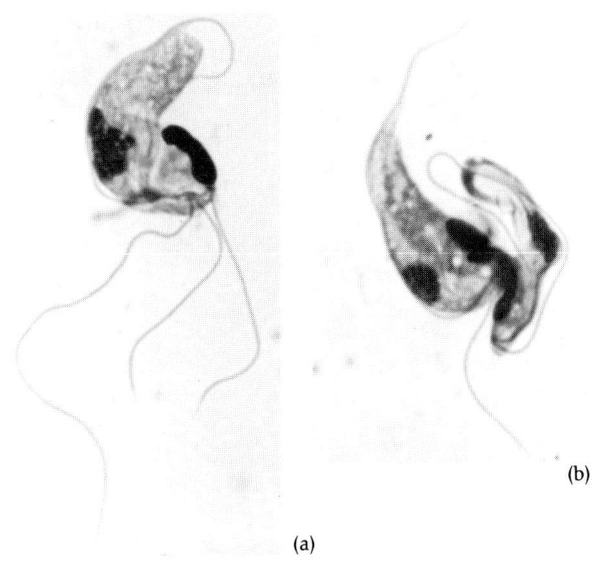

Figure 9–13 Stages in binary fission of the flagellate *Cryptobia salmostitica* (crip-toe-BEE-a sal-mos-TIT-ee-ka). This protozoon is commonly found in the blood of spawning salmon on the west coast of North America. A leech transmits it from fish to fish. (a) Before division. (b) After division. *[Reprinted with permission of the National Research Council of Canada from P. T. K. Woo,* Can. J. Zool. ***56:****1514–1518, 1978.]*

asexually by budding, multiple fission, and plasmotomy. In **budding,** a new individual is formed either at the protozoon's surface or in an internal cavity. **Multiple fission,** or *schizogony,* involves the formation of a multinucleate organism that undergoes division. Division produces a large number of single-nucleus-containing cells almost simultaneously. The nucleus and other essential organelles divide repeatedly before the cytoplasm divides (cytokinesis). Multiple fission is characteristic of sporozoan parasites (Color Photograph 47). Multinucleated protozoa sometimes divide into two or more multinucleated daughter cells. This type of asexual reproduction is known as **plasmotomy.**

Sexual Reproduction

Sexual reproduction may be relatively simple or complex. Although the production of sex cells (meiosis) apparently takes place, the details of this phenomenon are not well understood. Types of sexual reproduction processes include syngamy, conjugation, and autogamy.

In **syngamy,** the union of two different sex cells *(gametes)* results in the formation of a fertilized cell, or **zygote.** The zygote may undergo additional development. This type of reproduction is found in sporozoa such as malarial parasites. **[See Chapter 27 for a description of malaria.]**

Conjugation characteristically takes place in ciliated protozoa (Figure 9–14). The process involves the partial union of two different mating types of ciliates for the exchange of a pair of haploid micronuclei. These organelles fuse in both cells to form new diploid micronuclei. After their fusion, the new micronuclei divide by mitosis, thus giving rise to two new identical diploid organelles in each mating ciliate. In each cell, one micronucleus becomes the precursor of future micronuclei while the other undergoes multiple DNA replication and becomes the new macronucleus. (See Figure 9–17 for the location of these organelles.)

Autogamy is a modified version of conjugation that occurs only within one protozoon. The micronucleus divides into two parts that then reunite to form a zygote nucleus. The protozoon then divides to yield two cells, each with its full complement of nuclear structures.

Regeneration

The regeneration of lost or damaged parts is a characteristic property of most protozoa. When a cell is cut in two, only the portion containing the nucleus regenerates.

Figure 9–14 Conjugation shown by scanning electron microscopy. The conjugating partners are joined by a bridge of cytoplasm (not seen here). The dorsal ridges and bristle rows are prominent in the protozoon *Euplotes* (you-PLOY-teez) at the right. *[J. A. Kloetzel,* J. Protozool. ***22:*** *385–392, 1975.]*

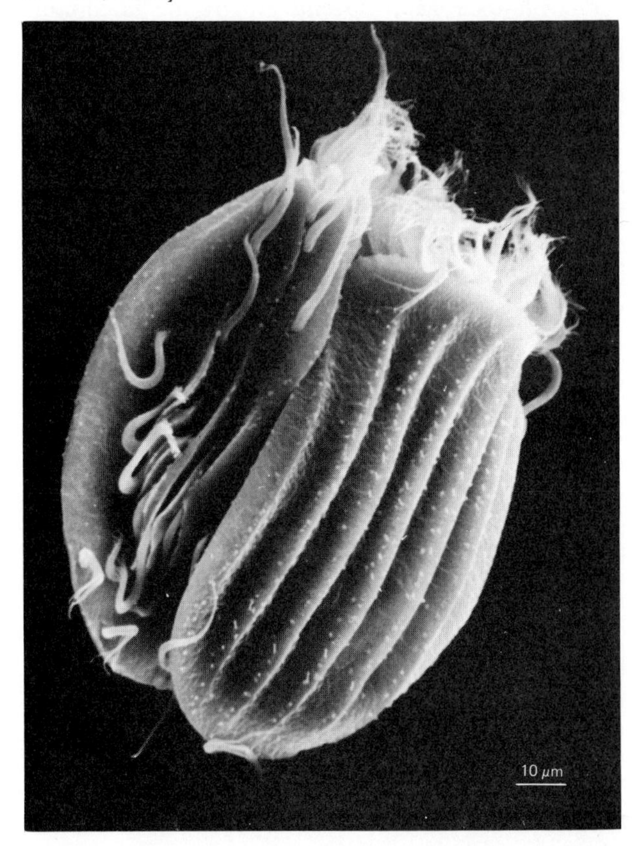

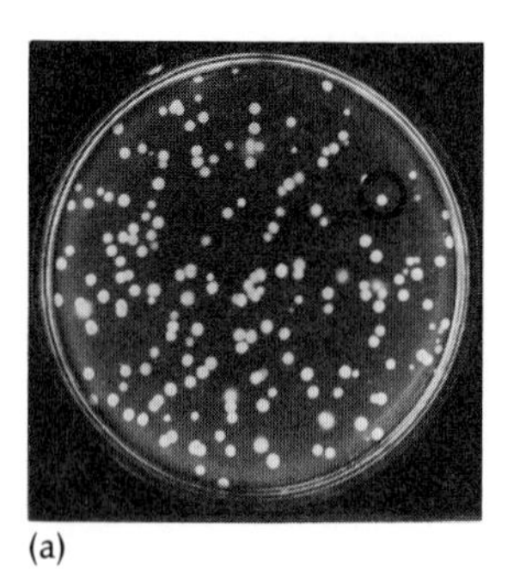

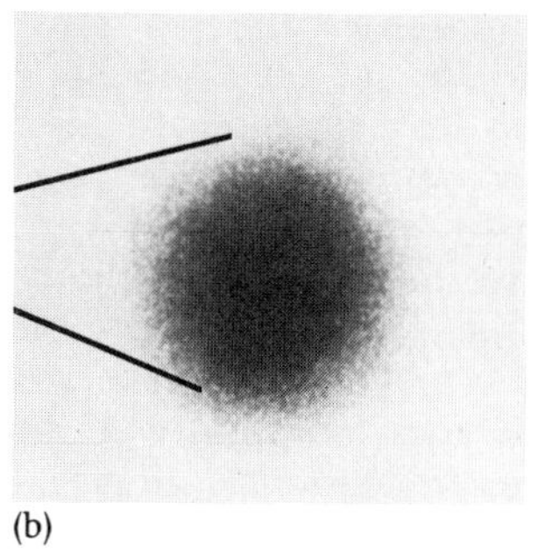

(a) (b)

Figure 9–15 Protozoan cultivation. (a) The appearance of *Trichomonas vaginalis* (trik-oh-MON-us va-jin-AL-iss) colonies in agar. (b) An enlarged view of an isolated colony. *[D. H. Hollander,* J. Parasitol. ***62****:826–828, 1976.]*

Cultivation

The protozoa are a large and varied group. Their requirements for cultivation—especially those of parasites—are also quite diverse. Here we shall present some basic considerations.

In general, free-living protozoa are best cultured under conditions of moderate light, temperatures ranging from 15° to 21°C, and a neutral to slightly alkaline pH. Artificial media used for general cultivation may contain rice, wheat grains, skim milk, hay, or lettuce. More specific growth media contain glucose, proteins and related substances, minerals, and yeast extract. Certain protozoa such as *Amoeba* (a-ME-ba) species and ciliates, including *Didinium* (dye-din-EE-um), require the addition of other protozoa or microbes as food sources (Figures 9–7). Solid or semisolid preparations can also be used successfully (Figures 9–15a and b) with some species. Commercially prepared media are available for protozoan cultivation.

Parasitic protozoa have been cultured successfully in tissue culture preparations. In tissue culture, living cells of various animals serve as sites for development and reproduction (Figure 9–16). Such cultures are of major importance to the development of vaccines and the testing of chemotherapeutic agents. Depending on the species, liquid or solid growth media can also be used.

Classification

The principal subgroups of protozoa are distinctive and easily defined, but their classification is not without problems. This is largely because as new facts are discovered and new ideas grow out of them, the details of classification change. Various properties of protozoa are used in their classification. Among them are method of obtaining nutrients; method of reproduction; cellular organization, structure, and function; biochemical analyses of nucleic acids and proteins from specific cellular structures; and organelles of locomotion. In 1980, the Society of Protozoologists introduced a new classification scheme in which the protozoa were raised to subkingdom rank. As the scheme stands now, all parasitic and either medically or agriculturally important protozoa belong to one of six phyla: *Ciliophora, Sarcomastigophora, Apicomplexa, Myxozoa, Microspora,* and *Acetospora* (Table 9–1). The slime molds are grouped under a seventh phylum, the *Labyrinthomorpha.* In this text, the slime molds are treated as fungi in the overall approach to the classification of microorganisms. The general features of three principal phyla (Ciliophora, Sarcomastigophora, and Apicomplexa) and selected subgroups are described in the following sections and summarized in Table 9–1. **[See Appendix D for protozoan classification.]**

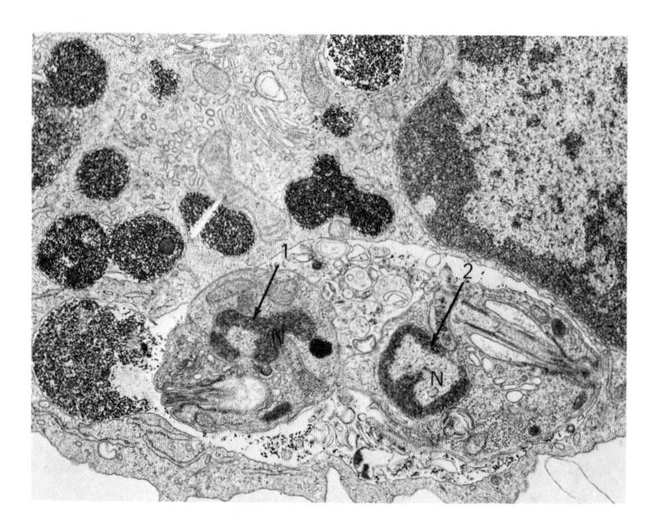

Figure 9–16 The intracellular appearance of the human parasite *Leishmania donovani* (lysh-may-NEE-a don-oh-VAY-nee). The two protozoa shown (1 and 2) were grown in a tissue culture using guinea pig cells. Note the nucleus (N) of each protozoon. *[K. P. Chang, and D. M. Dwyer,* Science ***193****:678–680, 1976.]*

Ciliophora

Members of the phylum Ciliophora, typified by *Paramecium,* are characterized by the presence of several thousand fine cilia on their surfaces (Figure 9–17). These cilia function both in moving the organism and in obtaining food. The beating strokes of the cilia cause the cell to revolve as it swims. Cili-

TABLE 9–1 **Description of Protozoa**

		Selected Differentiating Properties			
			Method of Reproduction		
Phylum	*Subphylum or Class*	*Means of Movement*	*Asexual*	*Sexual*	*Representatives*
Ciliophora	Kinetofragminophorea[c]	Cilia	Transverse fission	Conjugation	*Balantidium coli*[a] (bal-an-TID-ee-um KOH-lee), *Didinium, Tokophyra*[d] (toe-KA-phra)
	Oligohymenophorea[c]		Transverse fission	Conjugation	*Paramecium* (par-a-MEE-see-um), *Tetrahymena* (tet-rah-high-MEN-a), *Vorticella* (vor-tee-SELL-la)
	Polymenophorea[c]		Transverse fission	Conjugation	*Euplotes* (you-PLOH-teez), *Stentor* (sten-TORR)
Sarcomastigophora	Opalinata	Cilia	Binary fission	Syngamy	*Opalina* (opah-LINE-a), *Protoopalina* (PROH-opah-line-a)
	Sarcodina	Pseudopodia (false feet)	Binary fission	When present, involves flagellated sex cells	*Amoeba* spp., *Difflugia* (dif-floo-GEE-a), *Entamoeba histolytica*[a]
	Mastigophora	Flagella	Binary fission	None	*Chlamydomonas* (clam-id-oh-MOH-nass), *Giardia intestinalis* (jee-AR-dee-a in-tes-tin-AL-iss), *Trichomonas* (trik-oh-MOH-nass) spp.,[a] *Trichonympha* (trik-oh-NIM-pha), *Trypanosoma brucei gambiense*[a] (tri-pan-oh-SOH-ma brew-SEE gam-bee-ENS-ee), *T. cruzi*[a] (T. KRUZ-ee)
Apicomplexa	Sporozoa[c]	Generally nonmotile except for certain sex cells	Multiple fission	Involves flagellated sex cells	*Eimeria*[b] (EYE-meer-ee-a), *Plasmodium* (plaz-MOH-dee-um) spp.,[a] *Toxoplasma gondii*[a] (toks-oh-PLAZ-ma GON-dee-eye)

[a] Human parasite.
[b] Small animal pathogen.
[c] Class.
[d] A member of the subclass suctoria.

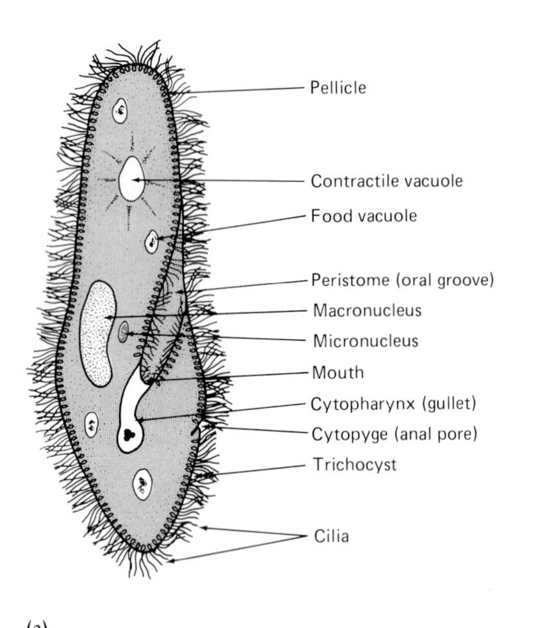

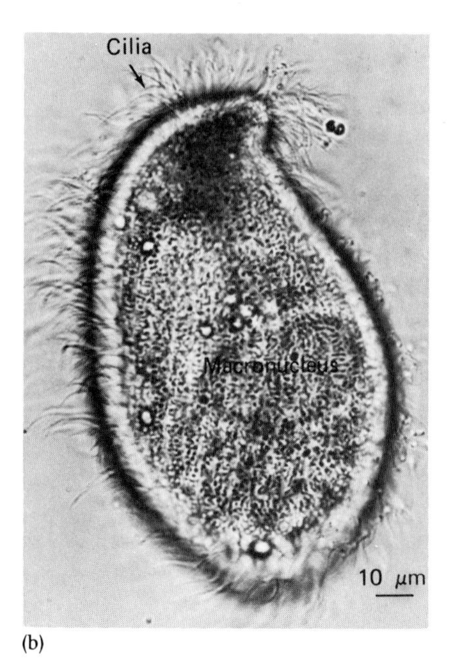

Figure 9–17 The ciliates. (a) A *Paramecium* (par-ah-ME-see-um) and its parts. (b) A bright-field view of a ciliate showing several characteristic features. *[From J. J. A. Van Bruggen,* et al., Arch. Microbiol. ***144**:367–374, 1986.]*

ates also have a definite shape due to the presence of a sturdy, flexible outer covering, the *pellicle* (Figure 9–17).

Of all the protozoa, the members of Ciliophora are the most specialized because they have organelles that carry out particular vital processes. One such organelle is the bottle- or rod-shaped trichocyst (Figure 9–11). Depending on the species, trichocysts may be discharged to serve as anchoring devices, weapons of defense, or tools with which to capture prey. Ciliates and suctorians (discussed in the following section) also differ from other protozoa in having at least two nuclei per cell, a *micronucleus* and a *macronucleus* (Figure 9–17).

Ciliates are found in both fresh and salt water. Some are free-living; others are either parasitic (Color Photograph 48) or *commensal,* the term used when two or more different organisms live together and one benefits while the other remains unaffected.

Suctoria

The Suctoria, a subclass of the Kinetofragminophorea, are a particularly interesting group of ciliated protozoa. Young suctorians are free-swimming. As they develop into adults, they lose their cilia and attach to an object by means of a stalk or disc (Figure 9–18a).

Suctoria, which live in both fresh and salt water and on aquatic plants and animals, obtain food with delicate protoplasmic tentacles. Some of these are pointed and can spear unsuspecting prey. Others are rounded and function as suckers in catching food (Figure 9–18b). These protozoa reproduce by asexual budding.

Sarcomastigophora

OPALINATA

The Opalinata occur in the large intestines of frogs and toads. These protozoa have many cilialike organelles arranged in rows over their body surfaces. Some organisms have two or more nuclei. However, there is no differentiation of these structures into micro- and macronuclei. The Opalinata reproduce by syngamy (the union of sex cells).

SARCODINA

For the Sarcodina, locomotion and food capture are both accomplished by means of pseudopodia (Figures 9–4 and 9–19). The amebas, well known for this mode of action, are members of the Sarcodina. The Sarcodina are simpler in structure than the ciliates and flagellates, with fewer organelles and no definite body shape (Figure 9–19).

The many kinds of protozoa that comprise the Sarcodina are found in all bodies of water. One of the most interesting is the large subgroup called the *foraminifera* (approximately 18,000 different species, found mostly in salt water). These microorganisms form shells made of lime or of substances such as sand from the surrounding waters. Since foraminifera have inhabited the seas for millions of years,

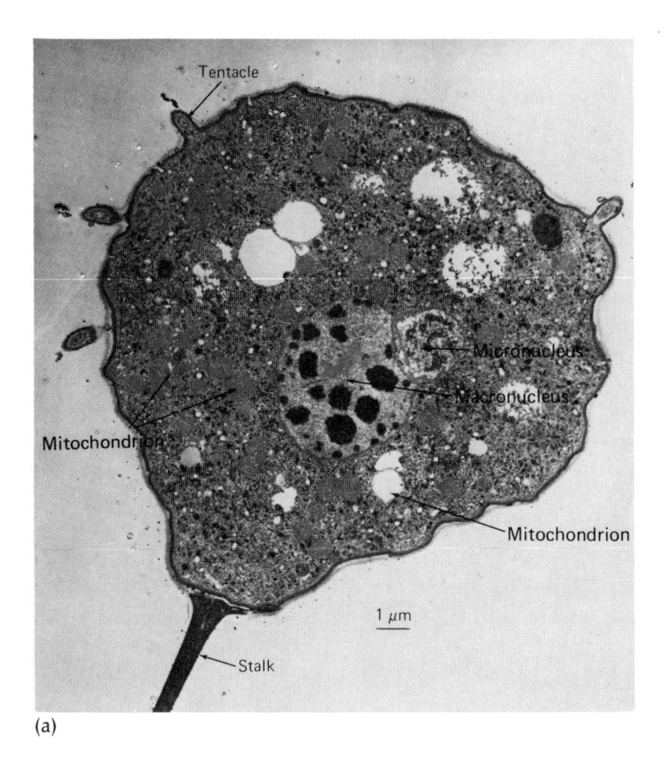

(a)

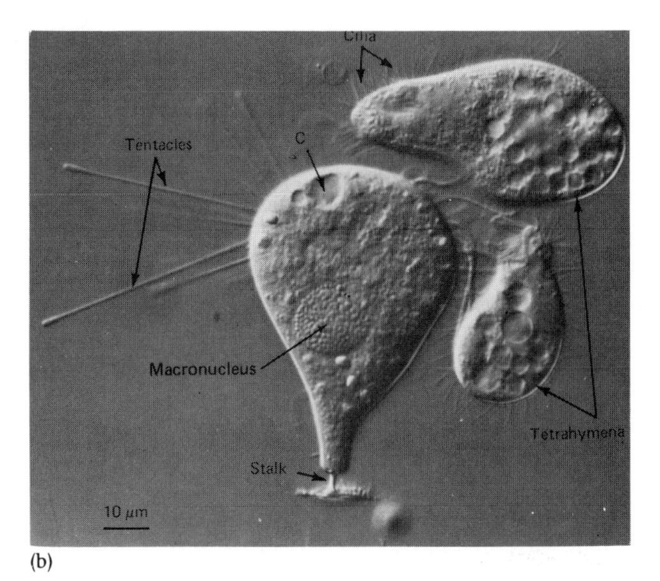

(b)

Figure 9–18 The *Suctoria* (suck-TOR-ee-a). (a) An electron micrograph of the adult suctorian *Tokophyra* (toe-KA-phrah), showing eucaryotic organization: macronucleus, micronucleus, mitochondria, tentacles, and a rodlike stalk. (b) *Tokophyra* feeding on two *Tetrahymena* (ciliates). The contractile vacuole (C), macronucleus (Ma), and feeding tentacles are clearly shown. *[Courtesy J. B. Tucker.]*

their shells have accumulated on the ocean floors and form a large portion of certain layers of sedimentary rock. Deposits of these organisms form a grayish ooze that can be transformed into chalk under proper geologic conditions. In England, the White Cliffs of Dover represent one such accumula-

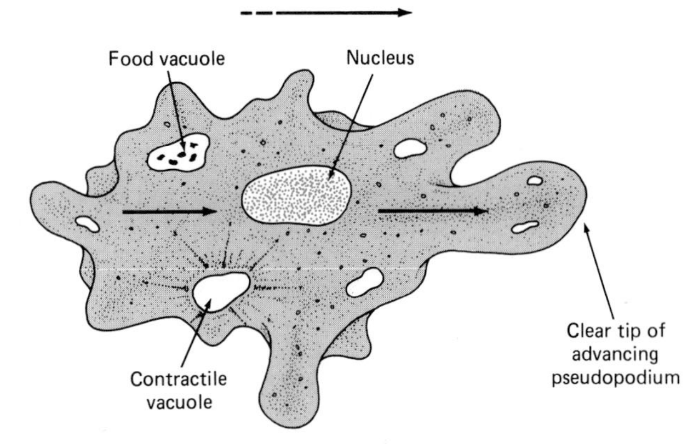

Figure 9–19 A diagram of the parts of *Amoeba* (a-ME-ba) spp. Arrows indicate the direction of movement.

tion, and the pyramids near Cairo, Egypt, were carved from limestone deposits of the same origin. The presence of foraminifera has also proved to be a guide to the petroleum geologist seeking new oil fields.

Other members of the Sarcodina, the parasitic amoeba, may be found in most kinds of animals. The most important form to attack humans is *Entamoeba histolytica* (Color Photograph 46), which causes amebic dysentery. This disease presents a serious medical problem in tropical and subtropical regions and in occasional severe outbreaks in temperate regions. Amebic dysentery is spread by ingestion of cysts (Figure 9–12) in contaminated food or water. **[Protozoan diseases are described in Part VIII.]**

MASTIGOPHORA

The Mastigophora, or flagellates, are mostly unicellular and usually possess at least one flagellum at some stage of their life cycle (Figure 9–20). These flagella are used for locomotion, for obtaining food, and as sense receptors.

The flagellates include more than half of the living species of protozoa and are an extremely variable group. They are believed to be the oldest of the eucaryotic organisms and the ancestors of the other major forms of life. In some classifications these protozoa are divided into two groups: the phytoflagellates (plantlike) and the zooflagellates (animallike). The phytoflagellates contain chlorophyll and are photosynthetic. The zooflagellates lack photosynthetic pigments and are heterotrophic.

Free-living Mastigophora are common in both fresh and salt water. Many others inhabit the soil or the intestinal tracts of some animals. Some Mastigophora are free-living, commensal, mutualistic, or parasitic. Certain parasitic flagellates are medically important. One disease caused by Mastigophora is African sleeping sickness (Figure 9–21). *Trypanosoma brucei gambiense* (tri-pan-oh-SOH-mah broo-SEE gam-bee-ENS-ee) (Color Photograph 90) and *T. b. rhodesiense* (T. b. row-dee-zee-ENS-ee) are transmitted by

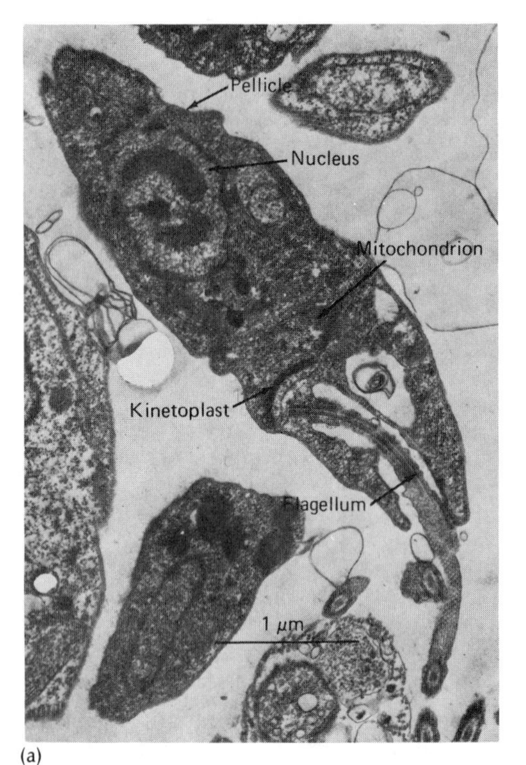

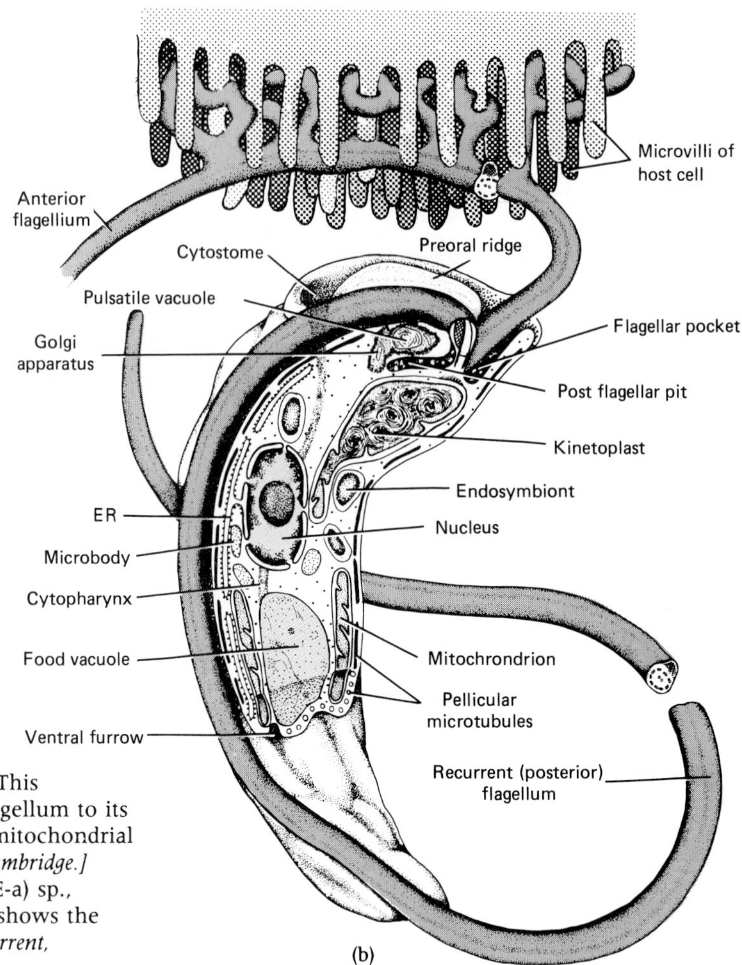

Figure 9–20 Structural features of flagellates. (a) This electron micrograph shows the relationship of a flagellum to its kinetoplast. The kinetoplast is contained within a mitochondrial membrane. *[Courtesy Dr. D. C. Barker, University of Cambridge.]* (b) A drawing of a flagellate, *Cryptobia* (crip-toe-BEE-a) sp., attached to the border of a host cell. This diagram shows the eucaryotic organization of the protozoon. *[W. L. Current, J. Protozool. **27**:278–287, 1980.]*

the bite of the tsetse fly. In an untreated case of the disease, the victim becomes drowsy and passes into a coma, which is followed by death. **[See African sleeping sickness in Chapter 29.]**

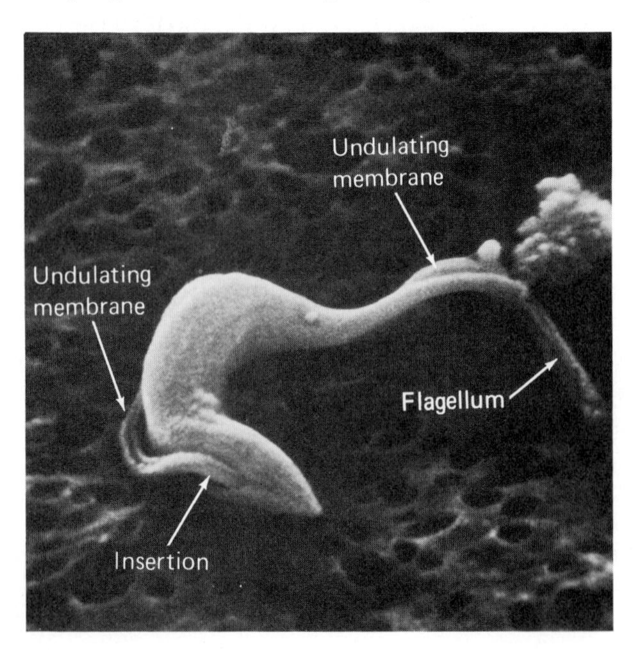

Figure 9–21 A scanning micrograph of a trypanosome, showing the point of insertion of the undulating membrane and the flagellum. *[P. Gardiner, J. Protozool. **27**:183–185, 1980.]*

Another genus of this group that is parasitic to humans and other animals is *Leishmania* (lie-sh-may-NEE-a). The human skin disease known as *cutaneous leishmaniasis* is especially common in the Mediterranean region. Uncomplicated cases do not present any problems, and infected persons usually recover within a year. However, when bacteria complicate this condition, a skin ulcer develops (Color Photograph 49 and Figure 9–22). The details of this disease, as well as others caused by mastigophorans, are discussed more fully in later chapters. **[See kala azar in Chapter 27.]**

An interesting relationship exists between certain members of Mastigophora and termites (Figure 14–1). This association is one in which the two organisms live together for mutual benefit (mutualism). The termite ingests wood but cannot digest the cellulose in it. Protozoa living in the intestine of the termite digest the cellulose, providing sugar for the insect and for themselves.

Microbiology Highlights

CATTLE TRYPANOSOMIASIS: A MAJOR SOURCE OF ECONOMIC LOSS

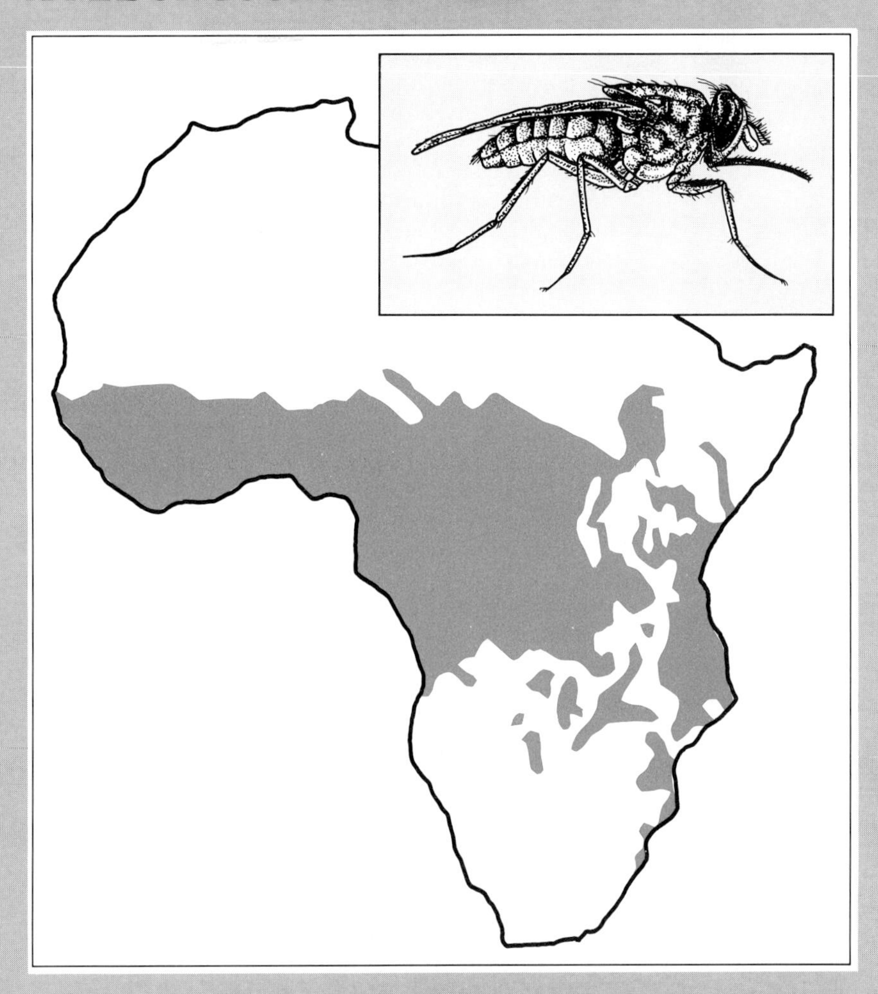

The flying range and distribution of tsetse flies in Africa.

African trypanosomiasis, known as sleeping sickness in humans and nagana in livestock, is found throughout the tropical areas of Africa where tsetse flies are present. The disease is caused by several trypanosome species that affect humans, cattle, and a variety of other domestic as well as wild animals. Typical symptoms include a reduction in the number of red blood cells (anemia), fever, irritation of the inner lining of the eyelid (conjunctivitis), nervous symptoms, paralysis, and death.

The parasitic protozoa are spread from animals that serve as reservoirs of the disease agents to humans and domesticated livestock by several tsetse fly species. The total flying range of tsetse flies includes 37 countries and extends over 10 million square kilometers. About 37 percent of Africa is "tsetse fly territory." Since the 1950s, the situation has become steadily worse and has greatly affected Africa's economy.

It is estimated that at least 50 million people and a similar number of cattle are exposed to trypanosomiasis yearly. As a consequence of such exposure, Africa produces 70 times less animal protein per unit area than Europe. Theoretically, the area involved could support an additional 120 million cattle and other domesticated animals if trypanosomiasis were brought under control. This projected increase in production could provide an additional 1.5 million tons of meat per year.

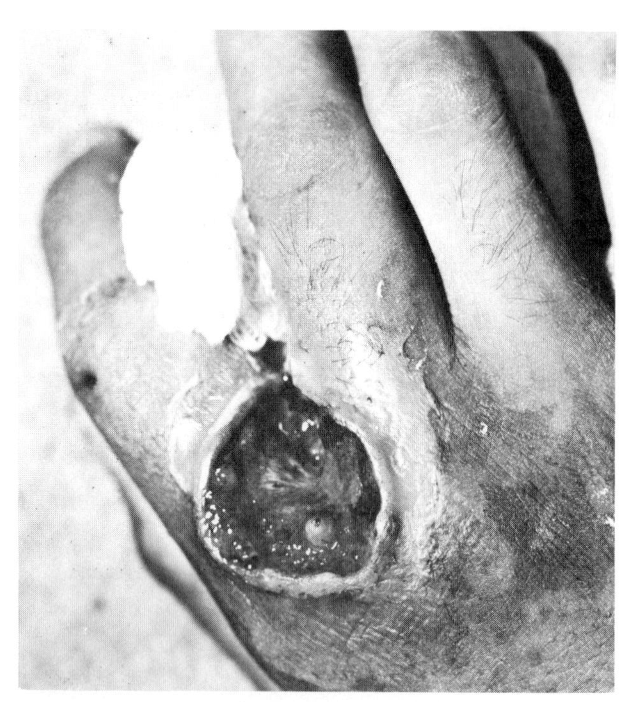

Figure 9–22 Tropical ulcer of the hand. *[Courtesy of the Armed Forces Institute of Pathology, Neg. No. A-43127-1.]*

Apicomplexa

SPOROZOA

All sporozoa are parasitic, absorbing nutrients from their hosts (Figure 9–23). Some are intracellular; that is, they live within the host's cells (Color Photographs 47 and 110). Others live in body fluids or various body organs. Adult sporozoa have no locomotor organelles.

Figure 9–23 A scanning micrograph showing the parasitic sporozoon *Eperythrozoon wenyoni* (ep-erith-ROW-zoon wen-YON-ee) on red blood cells. This protozoon, which is found in cattle severely ill from other diseases, can cause significant red blood cell destruction. *[K. S. Deeton, and N. C. Jain,* J. Parasitol. ***50:****867–873, 1973.]*

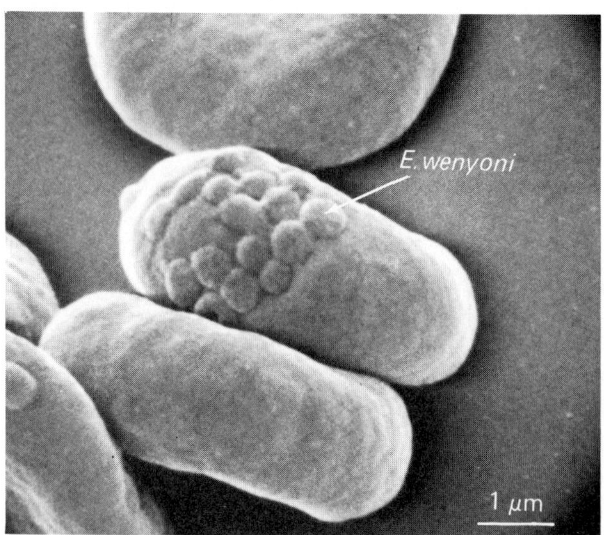

Both asexual and sexual reproduction occur among sporozoa. A number of sporozoa undergo sporulation, producing numerous small, infective spores called *oocysts.* Infective spores reach a susceptible host by way of food, water, or arthropod bites. Spores typically contain one or more smaller individual organisms called **sporozoites.** Many sporozoa have complicated life cycles, certain stages taking place in one host and other stages in a different host. An example of such a cycle involves mosquitoes and the different species of *Plasmodium,* which cause malaria not only in humans but also in several other animals, including canaries, chickens, ducks, lizards, pigeons, mice, monkeys, rats, and snakes. **[The general life cycle of the malarial parasite together with disease symptoms, treatment, and related topics are presented in Chapter 27.]**

A related protozoon, *Toxoplasma gondii* (Color Photograph 47), has generated considerable interest in recent years. The disease caused by this organism, toxoplasmosis, has been found in cats, cattle, dogs, humans, and sheep. This organism may enter a host by way of the nose or mouth. Originally toxoplasmosis gained public notice when it was revealed that this disease in a pregnant woman might also infect the fetus, causing serious defects or death. Currently the disease has achieved additional importance as one of the major causes of death in cases of acquired immune deficiency syndrome (AIDS). **[See Chapter 29 for toxoplasmosis.]**

Several other sporozoan species affect human well-being, not only because they cause infections in humans but also because they infect many domestic animals, causing losses of millions of dollars to agriculture every year.

Algae, the Plantlike Protists

In the early twentieth century, most aquatic plant life was lumped together and collectively referred to as *algae.* By the 1920s it became clear that the algae contained several related yet distinct groups of microscopic and massive forms of life. The five-kingdom system, by establishing the kingdom Protista, raised the problem of where to place several groups of algae traditionally regarded by botanists as plants. According to most current classification schemes, the green, red, and brown algae are placed in the plant kingdom. This approach will generally be followed here, thus leaving only the golden algae, diatoms, euglenoids, and dinoflagellates as members of the kingdom Protista.

Algae serve as the basic source of food in the sea upon which other marine life depends. Algae fix, or capture, more carbon by photosynthesis than all land plants. They are useful in illustrating the relationships between biological structure and function. These and other properties make the algae both fascinating and important to study.

"The Grass of the Waters"

Algae are photosynthetic aquatic protists. Most algal forms are free-floating and free-living. Certain species, however, participate in symbiotic associations, living together with other organisms. **Lichens** are a primary example of this type of mutualism. [See lichens in Chapter 8.]

Algae are frequently found in bodies of water used by humans—lakes, ponds, reservoirs, rivers, streams, and swimming pools. Unfortunately, in certain cases, algae may be quite troublesome, giving drinking water a disagreeable taste and odor and clogging water-filtering systems. The occasional spurts of growth known as **algal blooms** (Color Photograph 7a) may disrupt aquatic communities by the increased accumulations of their waste products. Such blooms may also be formed by cyanobacteria. [See cyanobacteria in Chapter 14.]

Algae constitute an important part of the **plankton** population, the microscopic forms of life floating near the surfaces of most bodies of water. The term *phytoplankton* is used to designate the algae of the group, while *zooplankton* is applied to the animallike organisms. Phytoplankton are the basis of several food chains. These miniscule organisms, along with bacteria, are eaten by progressively larger forms of life, which in turn may be consumed by humans. A simplified food chain for an open-water aquatic zone is shown in Figure 9–24.

Structure and Organization

Algae are eucaryotic organisms having a wide range of sizes and shapes (Color Photographs 7b and 50). They range in size from microscopic dimensions to lengths of 60 meters (m). Some algae are unicellular; others are highly differentiated and multicellular. The unicellular species may be curved, rodlike, or spherical (Figure 9–25). Multicellular algae (now considered plants) assume many filamentous or colonial arrangements with shapes of varying complexity. Some algae may even have rootlike organs, stems, and leaves. All photosynthesizing algae have developed specialized means of regulating their exposure to sunlight. One form this adaptation has taken among the flagellated algae is their *phototactic response,* the ability to swim toward or away from a source of light. The response is found among repre-

Figure 9–24 An example of a food chain beginning with plankton and showing the position of algae in this most important chain of events. Additional information on such environmental situations is presented in Chapter 14. The zooplankton feed on the phytoplankton and then, in turn, serve as food for small fish.

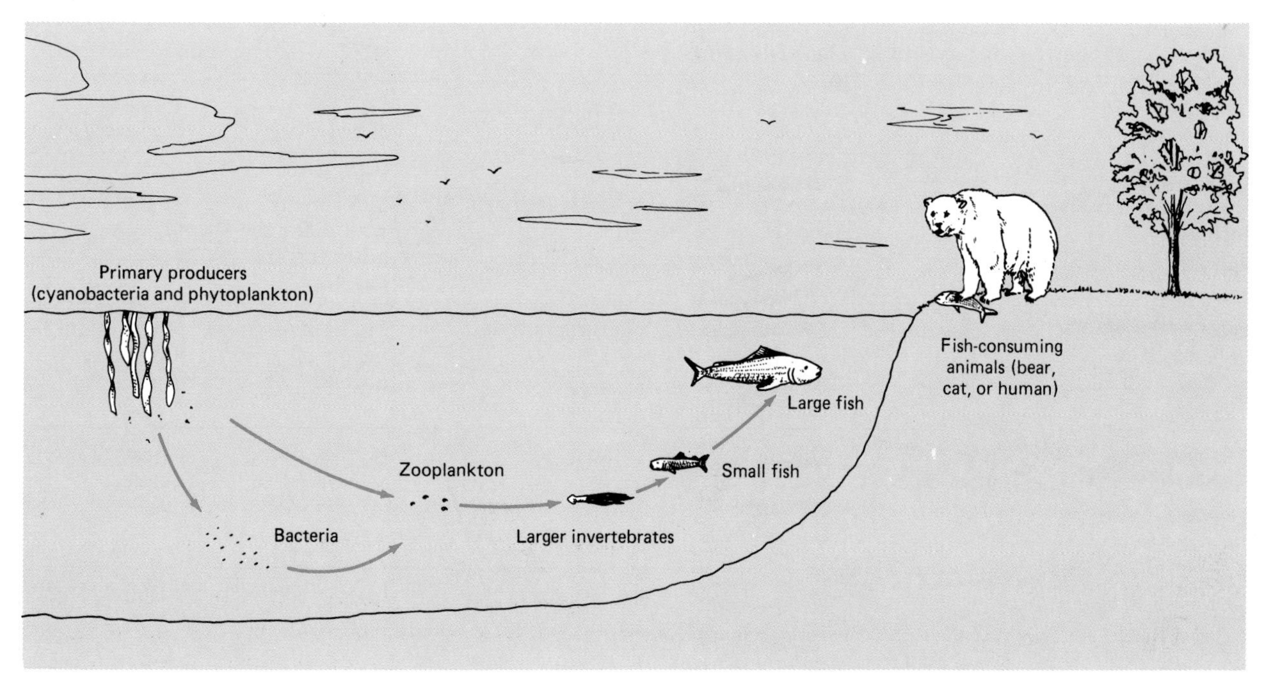

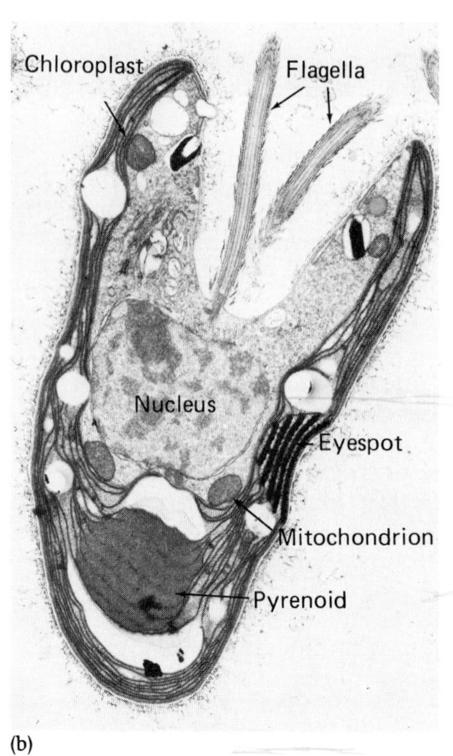

(a) (b)

Figure 9–25 The interesting appearance of a green alga found in both fresh and marine waters. (a) A scanning micrograph showing the external features of the single-celled alga. (b) An internal view of the same organism. *[From R. N. Pienaar, and M. E. Aken,* J. Phycol. ***21****:428–447, 1985.]*

sentatives of all the major algal protists (Table 9–2). Specialized structures, referred to as *directional light-wave antennae,* are used for phototaxis. They take several forms and are independent of a cell's photosynthetic apparatus. The geometry of both the alga and its swimming path are important for phototaxis. When the alga changes its position relative to the light, the distribution of light within the cell changes. Since cellular shape and structures vary so much among the different divisions of algal protists, further consideration will be given to these topics in later sections dealing with specific groups.

TABLE 9–2 Properties of the Protist Algal Divisions

Division and Common Name	*Habitat*	*Pigments Contained*	*Selected Reserve Materials*	*Motility*	*Method of Reproduction*
Chrysophycophyta Golden algae (includes diatoms)	Fresh- and saltwater	Chlorophylls a and c, special carotenoids, xanthophylls	Oils, leucosin, chrysolaminarin	Unique movement with diatoms; others utilize flagella	Asexual and sexual
Euglenophycophyta Euglenoids[a]	Freshwater	Chlorophylls a and b, carotenes, xanthophylls	Fats, paramylum	Motile by means of flagella	Asexual only by binary fission
Pyrrophycophyta Dinoflagellates[b]	Fresh- and saltwater	Chlorophylls a and c, carotenes, xanthophylls	Starch, oils	Motile	Asexual; rarely sexual

[a] These microorganisms possess characteristics of both animals and plants. Euglenoids seem to be intermediate between algae and protozoa.
[b] Species of the genera *Gonyaulax* and *Gymnodinium* occur in algal blooms referred to as the "red tide."

Means of Reproduction

Algae may reproduce either asexually or sexually. At the cellular level, sexual reproduction can involve the union of cells *(plasmogamy)* (Figure 9–26), the union of nuclei *(karyogamy)*, and the reduction of chromosome number **(meiosis).** As later sections will show, some algal species have complicated life cycles.

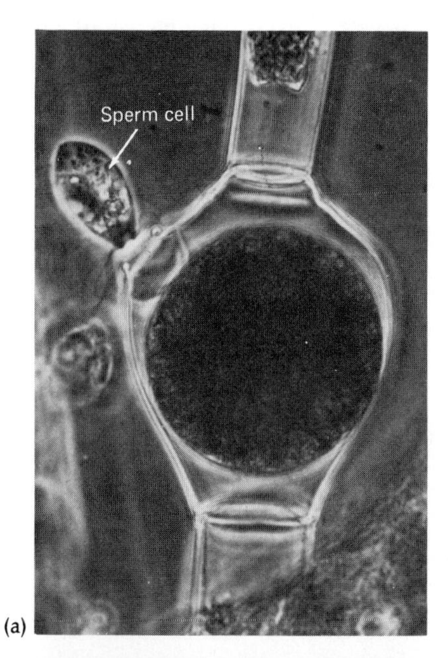

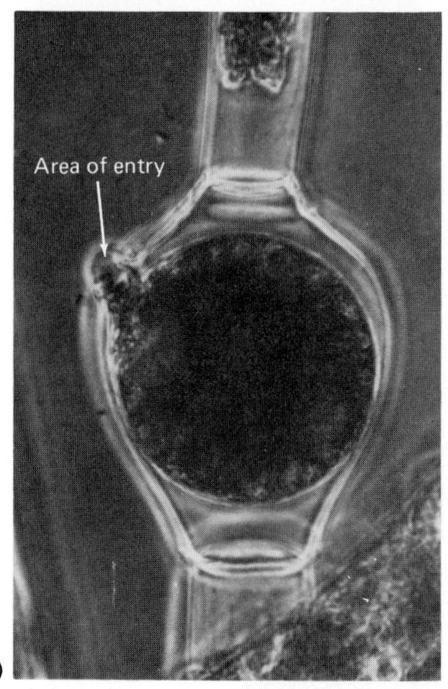

Figure 9–26 A view of the fertilization process with an alga. (a) The small sperm cell is shown making its initial contact with larger egg cell. *[L. R. Hoffman,* J. Phycol. ***9****:62–81, 1973.]* (b) A later view showing the fusion of nuclear material of both cells. *[L. R. Hoffman,* J. Phycol. ***9****:296–301, 1973.]*

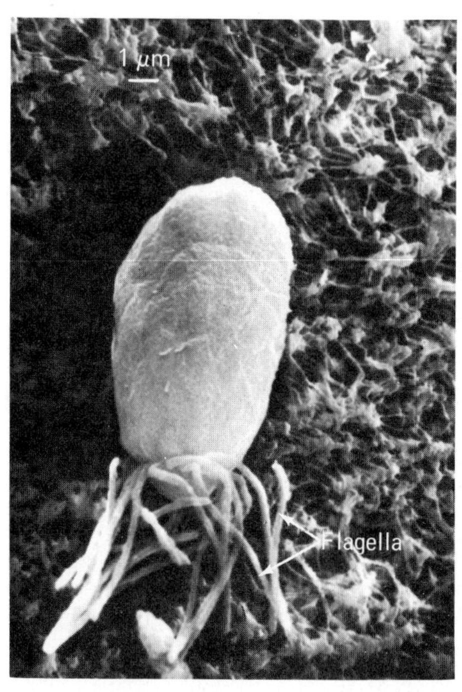

Figure 9–27 A scanning micrograph of a motile algal zoospore. These flagellated cells exhibit a typical eucaryotic organization. Zoospores are usually produced from the contents of a vegetative algal cell. *[M. M. Markowitz,* J. Phycol. ***14****:289–302, 1978.]*

Means of asexual reproduction among the algae include production of unicellular spores that germinate without fusing with other cells; fragmentation of filamentous forms; and cell division by splitting to form new individuals like the parent cell. Many algal spores, especially those of aquatic forms, have flagella (Figure 9–27), are motile, and lack a cell wall. These motile forms are called **zoospores.** Nonmotile spores known as **aplanospores** and other types are produced by various algae and will be discussed later.

Cultivation

Many years ago, the study of algae in laboratories depended on collecting fresh samples or preserving specimens. New approaches to the cultivation of algae have led to the development of various types of liquid and solid (agar-based) nutrient media. Many media contain exact amounts of substances required for growth, such as vitamins, and inorganic as well as organic sources of nitrogen, phosphorus, and other elements. Other preparations used may be composed mostly of undefined ingredients including soil, rice grains, and split peas. Depending on the algae, either marine or fresh water is used to sup-

port cultures. Proper pH and illumination also must be provided.

Temperatures used for many freshwater algal cultures range from 19° to 21°C. Algae from colder ocean habitats require lower temperatures. Some of the references at the end of this chapter pertain to additional aspects of cultivation.

Classification

Classification has undergone many changes since the time of Carl von Linnaeus, when only 14 genera were recognized. At the present time, three algal protist divisions are generally recognized. To draw attention to the algal level of organization, the names of these divisions include the root *phyco,* from the Greek *phykos,* meaning seaweed.

While there is no universally accepted classification of algae, taxonomic divisions may be distinguished by several characteristics, including cellular organization, cell wall chemistry and physical properties, flagellation or its absence, pigmentation, reserve storage products, and reproductive structures and methods. Table 9–2 summarizes these and other properties of the three protist divisions. Selected features of these unicellular groups are presented in the following sections.

The Golden Algae and the Diatoms (Chrysophycophyta)

The Chrysophycophyta group is highly diverse in pigmentation, cell wall chemistry, and flagellation (Table 9–2). The golden algae, the yellow-green algae, and the diatoms are included in this division. Because of space limitations, we shall concentrate here on the diatoms.

The large floating populations in fresh and salt water referred to as *plankton* are in large part diatoms. This accumulation of organisms is composed of trillions of algae that as a group produce more food through photosynthesis than all the rest of the plant world combined. Diatoms are found abundantly even in arctic regions (Color Photograph 50).

Structure and Organization

An obvious characteristic of diatoms is the intricately sculptured bilateral (two-sided) and radial (wheel) patterns of their cell walls (Figures 9–28a and b). Diatoms are usually classified on the basis of the shape, symmetry, and structure of their cell walls, called **frustules.** The walls of diatoms consist of two halves, referred to as *valves,* which fit together much like the parts of a pillbox (Figure

Figure 9–28 Representative diatoms from the orders Centrales and Pennales.
(a) *Arachnoidiscus* (a-rack-NOY-diss-kuss), a marine centric diatom. (b) *Navicula* (na-VIK-oo-la), a marine pennate diatom. *[Courtesy of Dr. Paul E. Hargraves, Narragansett Marine Laboratory, Kingston, R.I.]*

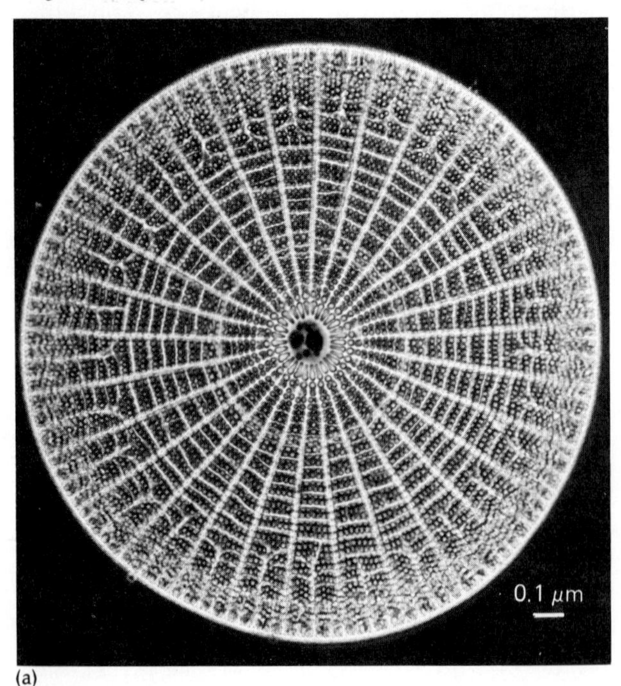

(a)

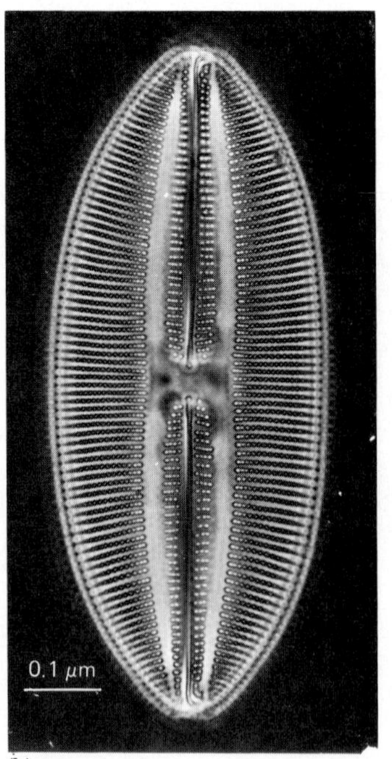

(b)

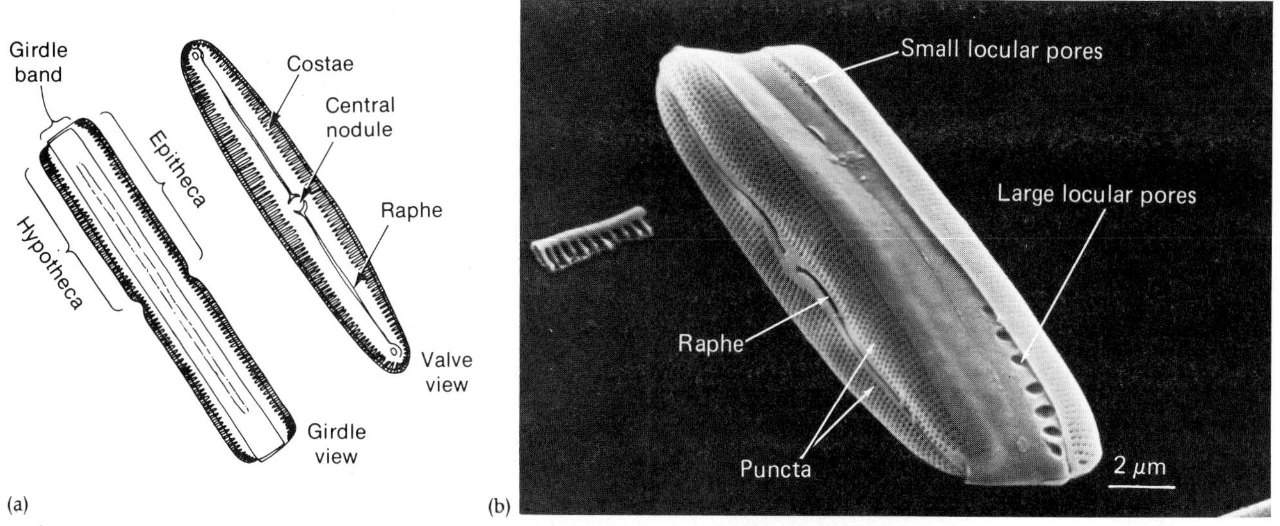

Figure 9–29 The structure of a pennate diatom. (a) A drawing of a pennate diatom, *Pinnularia* (pin-YOU-lair-ee-a) sp. (b) A scanning micrograph showing the external girdle view of the entire frustle, or cell wall, of *Mastogloia* (mass-toe-glow-EE-a). District puncta, the raphe, and large locular pores can be seen. Such ornamentation of the cell wall is extremely important to the classification of diatoms.

9–29). The larger upper half is the *epivalve;* the lower one is the *hypovalve.* The organisms with circular valves (Figure 9–28a) are called *centric* diatoms, while those characterized by boat-shaped structures like that in Figure 9–28b are called *pennate* diatoms. Various-sized chambers *(loculi)* are located internally on the margins of valves (Figure 9–29b). The structural details of these skeletons provide a basis for identification and more detailed classification.

Cell wall markings help distinguish diatoms from certain green algae. These may be holes *(puncta)* arranged in rows *(striae)* (Figure 9–29b) or ridges. The valves of some diatoms also have a long, narrow opening called a *raphe,* which are usually supported by riblike structures (Figure 9–30).

Figure 9–30 A tilted internal view of a valve. The marginal chambers, or loculi, large and small, and supportive riblike structures are shown. *[F. C. Stephens, and R. A. Gibson,* J. Phycol. ***16****:354–363, 1980.]*

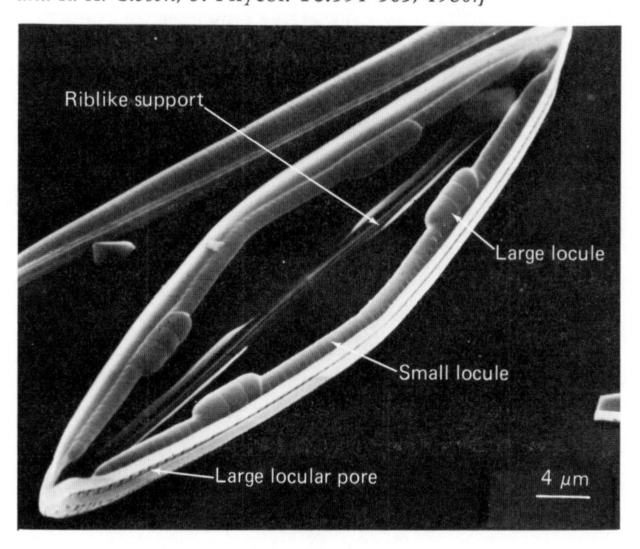

Diatoms are chemically unique in that their cell walls contain large concentrations of silicates. These substances are the basic components of glass, granite, and sand. Diatoms continually absorb silicates and deposit them in their cell walls. When a diatom dies, the silica in the cell wall begins to dissolve rapidly. However, under favorable conditions, these glassy structures accumulate and form deposits of fossil diatoms called **diatomaceous earth.** Such deposits have been gathering for thousands of years, and in some parts of the world deposits 900 meters thick have been found. Locations of such deposits include the city of Lompoc, California, and various sites in Maryland and Virginia.

Centric diatoms are capable of forming resting spores (Figure 9–31). These spores serve as the major means by which diatoms survive unfavorable growth conditions. Resting spores can be produced singly, in pairs, or in groups of four. They usually consist of two valves.

Reproduction

Asexual cell division is the usual method of reproduction in diatoms (Figure 9–32). During the process, the two valves of the parent cell separate and serve as epivalves for the two newly formed products of division. One of the new cells is always slightly smaller than the parent. Figure 9–32 shows this feature as well as other aspects of the process.

Sexual reproduction does not result in diatom size reduction. Instead it is the means by which

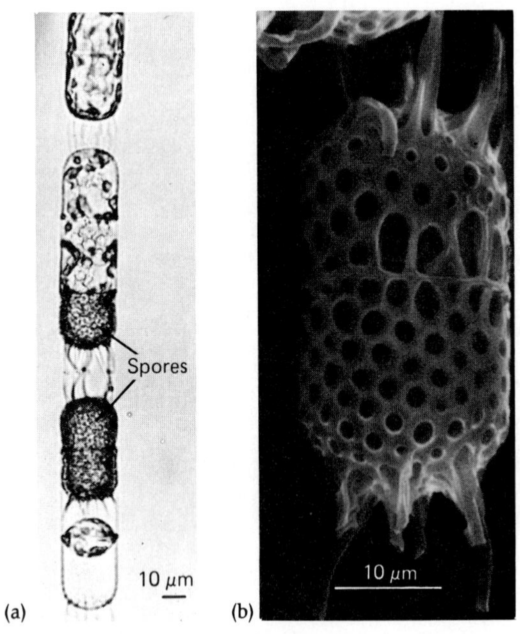

Figure 9–31 Resting spores of the centric diatom *Stephanopyxis turris* (steff-an-oh-PICKS-iss TURR-iss). (a) Resting spore formation. (b) A scanning micrograph of an entire resting spore. Note the extensions on each end of the spore. *[P. E. Hargraves,* J. Phycol. ***12****:18–128, 1976.]*

maximum cell size can be attained. The types of sexual reproduction found among other algal groups are also found among diatoms. That is, they may reproduce sexually by several means. Certain diatoms produce a unique sexual spore known as the *auxospore.* Such spores can be produced by a number of mechanisms, including conjugation and the fusion of two gametes.

Economic and Ecological Importance

Diatomaceous earth or the fossil remains of diatoms has many important applications. For example, because of its abrasive quality, it is used as a polishing agent in toothpastes and metal polishes. It is also used in the manufacture of insulating materials and dynamite sticks. Other industries use diatomaceous earth for the filtration of beer, oil, and other fluids.

Since diatoms possess photosynthetic pigments, these organisms, as well as the golden algae and the yellow-green algae, produce and store the reserve food substances, chrysolaminarin, and oils. It is believed that these algae have been significant sources of petroleum.

Aside from their present commercial value, diatoms are particularly interesting to scientists as members of food chains and indicators of various geological changes. Diatoms have been used to trace the pattern of glacier formation in northern Europe and as indicators of industrial water pollution.

The Euglenoids (Euglenophycophyta)

Euglenoids are a small group of eucaryotic, unicellular microorganisms (Figure 9–33) that have a curious combination of animal and plant properties. For example, like animal cells, some lack chloroplasts and have the ability to ingest food particles. Like plants, many euglenoids possess photosynthetic pigments apparently identical to those found in the

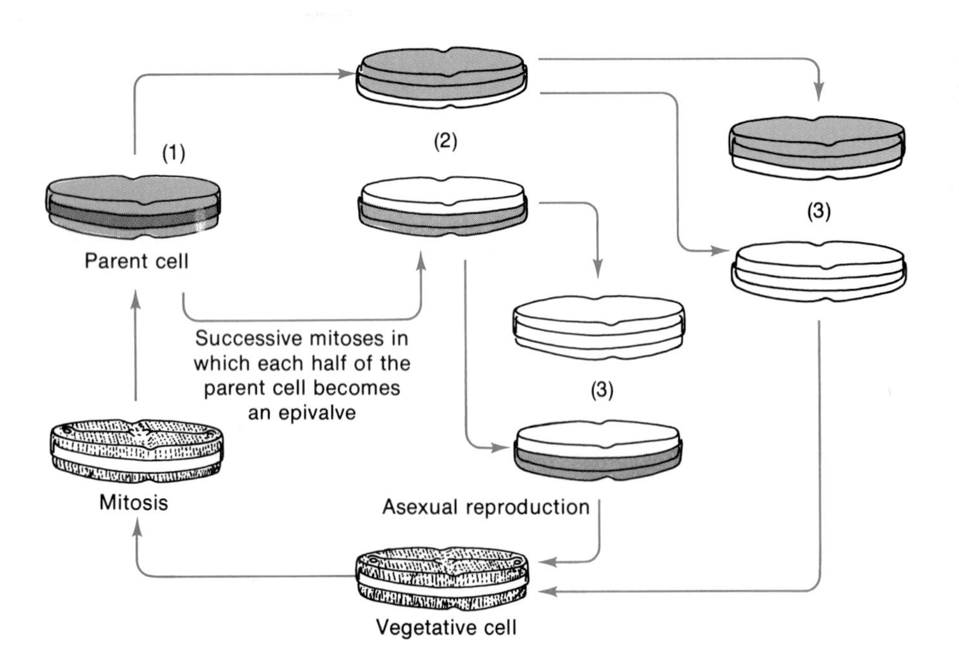

Figure 9–32 Asexual reproduction of diatoms. During the process, the two parent valves (1) separate (2) and serve as respective epivalves for the daughter diatoms. The process continues with the newly formed pennate diatoms (3).

Microbiology Highlights

CIQUATERIC FISH

Ciquatera (see-gwa-TAY-ra) poisoning is caused when the potent toxin ciquatoxin, produced by the algal dinoflagellate *Gambierodiscus* (gam-beer-OH-dis-kuss) contaminates the flesh of fish. One of the most powerful poisons known, it remains active even after the fish is cooked. Its detection in fish is presently not possible. The fish's general appearance and activity are not affected by the toxin. It causes approximately 1,200 cases of disease and death annually in the Pacific region, where fish is a major staple of the diet.

Ciquatoxin is transmitted via the marine food chain, appearing and disappearing unpredictably in a wide range of edible fish. Certain fish species are more commonly affected than others. These include red bass, moray eels, and Spanish or gray mackerel.

The toxin generally causes gastrointestinal symptoms. However, potentially fatal cases show central nervous system involvement, which includes convulsions and respiratory failure. Without appropriate resuscitation and medical assistance, death may occur from oxygen deficiency (hypoxia).

green algae (Table 9–2). All euglenoids lack cell walls and are bounded by an outer thickened cell membrane, a *pellicle*, composed of a system and grooves that run along the cell (Figure 9–33c).

Some euglenoid cells move about by means of a flowing contracting and expanding motion known as *euglenoid movement*. Others are flagellated. The basic number of flagella per cell is two. One of these, however, may not emerge and often is reduced to a short stump. Both flagella are inserted at the base of

Figure 9–33 The euglenoids. (a) A scanning micrograph of stalked cells of *Colacium mucronatum* (coh-lay-SEE-um myoo-KROW-nay-tum). Species of *Colacium* grow on a variety of aquatic animals, such as free-living worms and protozoa, as well as on mud and in fresh water, attached to aquatic plants. *Colacium* cells attach to surfaces by means of a sticky stalk. *[J. R. Rosowski, and R. L. Willey,* J. Phycol. ***13****:16–21, 1977.]* (b) The structures found in euglenoids. (c) Features of *Euglena* (yon-GLEE-nah) as seen through transmission electron microscopy. This ultrathin section shows the following structures: ridged pellicle, nucleus, and chloroplasts. *[J. R. Cook, and T. C. Li,* J. Protozool. ***20****:652–653, 1973.]*

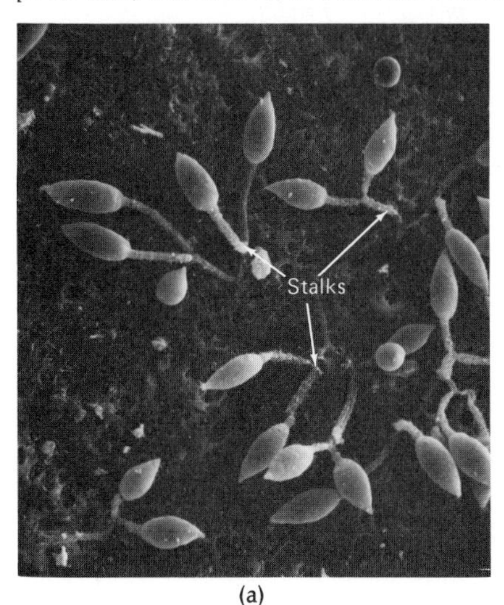

(a)

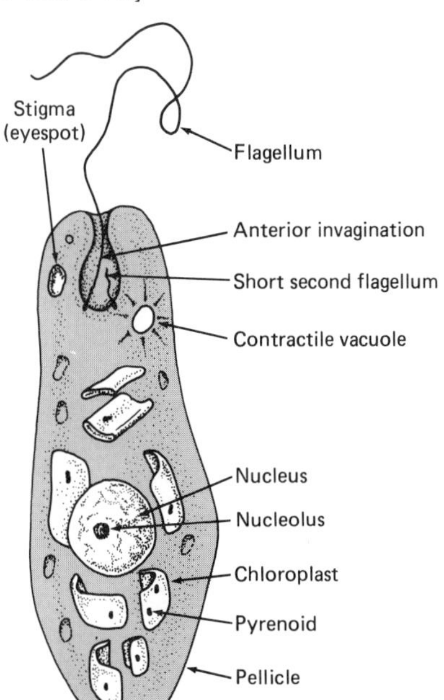

(b)

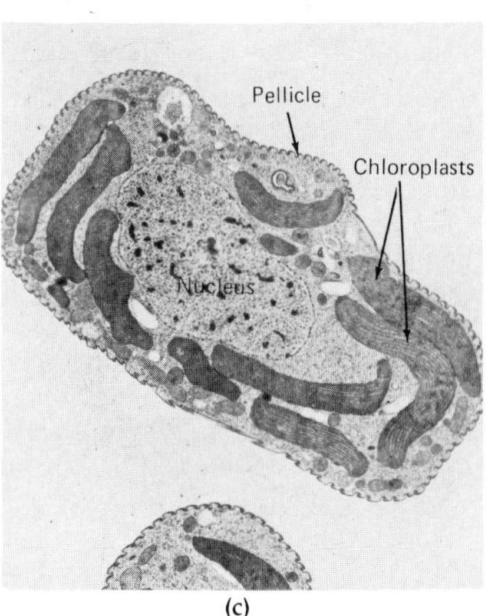

(c)

a deep invagination called the *reservoir* at the anterior end of a cell. Some euglenoid flagella have thin, small, delicate hairs that project outward. They are believed to wrap themselves around the organelle and increase the efficiency with which the organism moves.

The chloroplasts of euglenoids are scattered throughout the cellular interior; each chloroplast contains a pyrenoid. Pyrenoids produce a special type of starch called *paramylum*, which is stored as a reserve food and also is used for the resynthesis of chloroplast constituents.

Green euglenoid cells have a prominent orange-red pigmented eyespot, or stigma. It is located near the reservoir where flagella originate. The eyespot consists of many pigment-containing granules and enables cells to respond to light stimuli. Some euglenoids have a contractile vacuole that is used to discharge excess water and wastes. Some species are even able to form resistant structures, *cysts*, against unfavorable environments. Euglenoids reproduce asexually by simple cell division.

These microorganisms, with their unique mixture of animal, plant, and microbial features, are used as experimental subjects in photosynthesis studies and assays for vitamin B_{12}. They also present an interesting challenge to taxonomists.

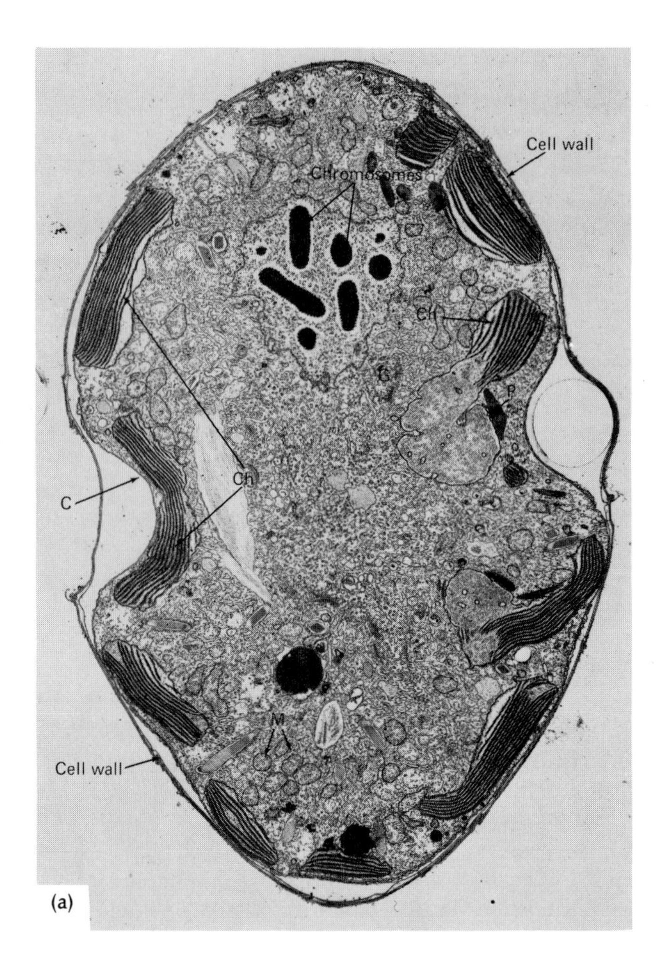

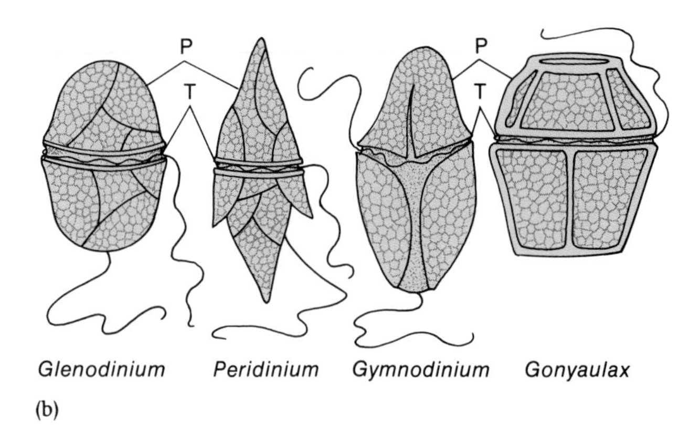

Figure 9–34 (a) An internal view of the dinoflagellate *Cachonina illdefina* (kack-OH-nin-a ill-DE-fin-a). This cell shows its eucaryotic organization and the cell membrane (C), cell wall, chloroplasts (Ch), Golgi bodies (G), and pyrenoids (P). *[E. M. Herman, and B. L. Sweeney,* J. Phycol. ***12**:198–205, 1976.]* (b) Four dinoflagellate species. Note the plates (P) that enclose the single-celled organism. The presence of two flagella, one of which is located in the transverse groove (T), is another characteristic feature of these organisms. A dinoflagellate. One type of algal protist. *[From J. T. Berdach,* J. Phycol. ***13**:243–251, 1977.]*

The Dinoflagellates (Pyrrophycophyta)

The dinoflagellates comprise a diverse group of biflagellated (Figure 9–34) and nonflagellated eucaryotic, unicellular organisms. For many years they were considered protozoa because of their motility and, in certain cases, their ability to ingest solid food particles. Further study of the group identified nonmotile and photosynthetic representatives, thereby establishing the dinoflagellates' relationship to algae.

The dinoflagellates occupy a variety of aquatic environments. They are found in both marine and freshwater habitats, where they exhibit parasitic, saprobic, and symbiotic relationships.

Structure and Organization

Most individual dinoflagellates have a heavy cell wall, or *theca*, composed of cellulose-containing plates. Both the armored and unarmored (naked) organisms are circled by a transverse groove (Figure 9–34b). In motile cells, two unequal flagella extend from a pore at a point along this groove.

Reproduction

The most common form of reproduction in dinoflagellates is by cell division. Some dinoflagellates reproduce by fragmentation. Others form zoospores and nonmotile spores called *aplanospores.* Sexual reproduction has also been observed in a number of dinoflagellates.

Economic and Ecological Importance

Dinoflagellates are widely recognized as important members of food webs in the oceans and in some freshwater environments. They provide a basic food source. Aside from this beneficial function, most of the activities of dinoflagellates are unpleasant or harmful.

Many dinoflagellates impart an offensive odor and taste to water. Several members of this division also produce "red tides," a type of algal bloom. In this situation, environmental conditions are right for a sudden increase in the concentration of cells, which color the ocean in the immediate area red, brown, or yellow. Warm water, large concentrations of iron and phosphate, and other factors contribute to algal reproduction.

Most blooms of dinoflagellates produce toxic effects that may kill fish or invertebrates or contribute to paralytic shellfish poisoning in humans. Paralytic shellfish poisoning results from eating clams, mussels, oysters, scallops, or other filter-feeding molluscs that have concentrated toxins of *Gonyaulax catenella* and related species (Figure 9–34b). The toxin of *G. catenella,* saxitoxin, affects the nervous system and is said to be 100,000 times as potent as cocaine.

Summary

The kingdom Protista includes the protozoa and certain unicellular algae.

Protozoa, the Animallike Protists

1. *Protozoa* are unicellular, eucaryotic, animallike protists.
2. Many protozoa are motile and require organic food sources.

Distribution and Activities

Protozoa are widely distributed and exhibit both beneficial and harmful symbiotic relationships.

Structure and Function

LOCOMOTION

1. There are three basic types of organelles for movement among the protozoa: *pseudopodia, flagella,* and *cilia.* These structures are important to classification.
2. Individual flagella and cilia consist of two centrally located microtubules surrounded by nine double tubules. These organelles originate from a cytoplasmic DNA-containing granule, or *kinetoplast.* Biochemical analyses of kinetoplast DNA are of value in tracing evolutionary relationships and in the classification and identification of unknown flagellates. Various modifications of cilia and flagella occur among protozoa.

FEEDING AND DIGESTION

1. Three major methods of obtaining nutrients are found in protozoa, classifying them as *autotrophic, saprobic,* or *holozoic* (the last two being heterotrophic forms).
2. Holozoic protozoa have various mechanisms for food capture, including food cups, cytosomes (mouths), and tentacles.

EXCRETION AND OSMOREGULATION

1. Disposal of excretory wastes occurs at the cell surface.
2. Most nitrogen wastes from protein digestion are excreted as ammonia.
3. In the case of parasitic protozoa, waste products are released with the destruction of the host cells.
4. The *contractile vacuole* system is involved with osmoregulation.

PROTECTIVE STRUCTURES

1. Most of the protective structures of protozoa function to prevent mechanical injury or to guard against drying, excessive water intake, and predators.
2. Examples of protective structures include the *pellicle,* a strong outer covering of ciliates and some amebas; *tests,* or *shells,* coverings of calcium carbonate or silica; and *trichocysts,* specialized defense organelles.

Trophozoites and Cysts

1. The normal, active feeding form of free-living as well as several parasitic protozoa is the *trophozoite.*
2. Many protozoa can secrete a resistant, thick covering that develops into a resting stage, a form known as a *cyst.*
3. Cysts also serve as sites for division and as a means for spreading pathogenic protozoa.

Methods of Reproduction

1. Ciliated organisms are multinucleated. Usually one macronucleus and one micronucleus are present in the organism.
2. The macronucleus regulates metabolism and developmental functions and maintains all visible traits.
3. The micronucleus controls the entire cell and its sexual and reproductive processes.

ASEXUAL REPRODUCTION

1. Both asexual and sexual reproduction occur among protozoa.
2. Some protozoa reproduce only asexually.

3. The forms of asexual reproduction include splitting, either equally or unequally, into two new cells *(binary fission)*, *budding*, splitting into several new cells *(multiple fission)*, and forming two or more multinucleated cells *(plasmotomy)*.

SEXUAL REPRODUCTION
The types of sexual reproduction include the union of two different types of sex cells *(syngamy)*, the transfer of nuclear material from one cell to another through a temporary connection *(conjugation)*, and a modified version of conjugation known as *autogamy*.

REGENERATION
Regeneration of lost or damaged parts is characteristic of protozoa. When a cell is cut in two, only the portion containing the nucleus regenerates.

Cultivation
Several protozoa can be cultured under a variety of natural or laboratory conditions. Both liquid and solid media also can be used for their cultivation.

Classification
Protozoa can be classified according to mode of nutrition, structures and associated functions, methods of reproduction, and means of movement.

CILIOPHORA
1. The ciliophorans are protozoa that characteristically have short hairs called *cilia* that function in moving the organism and obtaining food.
2. Ciliates can be free-living, *commensal*, or parasitic.

SUCTORIA
1. Suctorians are ciliated during the early stages of their life cycles, but adults lose these organelles.
2. Adult forms develop a stalk for attachment to surfaces and delicate tentacles to obtain food.

SARCOMASTIGOPHORA
Opalinata
1. The opalinates are found in the large intestines of frogs and toads.
2. These protozoa have cilialike structures arranged in rows on their body surfaces and two or more nuclei. They reproduce sexually by syngamy.

Sarcodina
1. Protozoa belonging to the group Sarcodina move and capture food by means of pseudopodia and are simpler in structure than ciliates.
2. Members of the Sarcodina are widely distributed in nature and can be free-living as well as parasitic.

Mastigophora
1. The organisms of Mastigophora are mostly unicellular and usually have one or more flagella at some or all stages of their life cycle.
2. Flagellates are widely distributed in natural environments. Flagellates are believed to be the oldest eucaryotic organisms and the ancestors of other major forms of life.
3. These protozoa can be free-living, commensal, mutualistic, or parasitic.

APICOMPLEXA
Sporozoa
1. All of these protozoa are parasitic and obtain essential nutrients from their hosts.
2. Adult sporozoa have no organelles for movement.
3. The causative agents of malaria and toxoplasmosis belong to this group of protozoa.

Algae, the Plantlike Protists
The golden algae, diatoms, euglenoids, and dinoflagellates are recognized members of the kingdom Protista.

"The Grass of the Waters"
1. *Algae* are photosynthetic aquatic organisms. Many are important members of several food chains. Others, such as algal blooms, pose environmental problems.
2. Many algae are free-floating and free-living. Some species live together with other organisms.

Structure and Organization
1. Algae are eucaryotic forms that exhibit a wide range of sizes and shapes.
2. Algae range in size from microscopic dimensions to lengths of 60 m.
3. Cellular shape, organization, and arrangement vary considerably among algae divisions.

Means of Reproduction
1. Algae may reproduce either sexually or asexually. Some have complicated life cycles.
2. Asexual reproduction includes production of a variety of unicellular spores, fragmentation of filaments, and division by cellular splitting.

Cultivation
A variety of liquid or solid nutrient preparations are available for cultivation of fresh algal specimens.

Classification
1. At least three eucaryotic algal Protist divisions are recognized.
2. The names of the divisions include the root *phyco* ("seaweed") to draw attention to the algal level of organization.
3. Algal divisions differ from one another in several respects, including cellular organization, cell wall chemistry, flagellation, pigmentation, reserve storage products, and means of reproduction.

The Golden Algae and the Diatoms (Chrysophycophyta)
1. Golden algae and diatoms contain a highly diverse group of algae.
2. Diatoms form a major part of floating algal populations worldwide.

STRUCTURE AND ORGANIZATION
Obvious distinguishing characteristics of diatoms are their intricately sculptured cell walls, or *frustules*, which contain large concentrations of silicates, the basic components of sand and glass.

REPRODUCTION
Asexual cell division is the usual method of reproduction in diatoms. It results in two cells, with one of the new cells always slightly smaller than the parent. Sexual reproduction does not result in size reduction.

ECONOMIC AND ECOLOGICAL IMPORTANCE
Diatoms have several important commercial applications, such as in polishes and filters. They can also serve as indicators of industrial water pollution.

The Euglenoids (Euglenophycophyta)

1. Euglenoid, which are eucaryotic unicellular forms, exhibit a curious combination of animal and plant properties.
2. Euglenoids have at least one of each of the basic organelles necessary for life in an aquatic environment.
3. A number of cellular features, flagella structure and insertion, and cell membrane, or *pellicle,* unite and distinguish this group from other organisms.
4. Euglenoids are important experimental subjects and present a taxonomic challenge.

The Dinoflagellates (Pyrrophycophyta)
The dinoflagellates include biflagellated as well as nonflagellated forms.

STRUCTURE AND ORGANIZATION
The dinoflagellates have a characteristic transverse groove that divides cells into halves.

REPRODUCTION
The most common form of reproduction in dinoflagellates is by cell division. Other methods of reproduction are fragmentation, formation of spores, and sexual reproduction.

ECONOMIC AND ECOLOGICAL IMPORTANCE
Dinoflagellates are important members of food webs in aquatic environments. Some of these algae are also known for harmful effects such as red tides and toxin production.

Questions for Review

1. List three characteristics that all protozoa have in common.

2. List at least three ways in which protozoa differ from fungi.

3. a. What type of cellular organization do protozoa exhibit?
b. List at least six organelles found in protozoa, and give the functions of each.

4. a. Differentiate between a trophozoite and a cyst.
b. What are the respective functions or activities associated with trophozoites and cysts?
c. Do all protozoa have both trophozoites and cysts? Explain.

5. a. What criteria are used to identify or classify protozoa?
b. Do such criteria differ from the ones used for bacterial and fungus classification? Explain. (Refer to Chapters 7 and 8.)

6. a. How do protozoa reproduce?
b. List and describe three reproductive mechanisms.

7. List the major groups of protozoa, and indicate at least one human disease agent from each group.

8. How are protozoa cultured?

9. Define or explain the following terms:

a. pseudopodium	e. cirri
b. contractile vacuole	f. macronucleus
c. pellicle	g. micronucleus
d. kinetoplast	h. tentacle

10. Do protozoa have any ecological value?

11. Figure 9–35 shows a phase-contrast micrograph of the ciliate *Euplotes eurystomus.* What organelles of this protozoon do you see?

12. Do algae resemble any other microbial type? Explain.

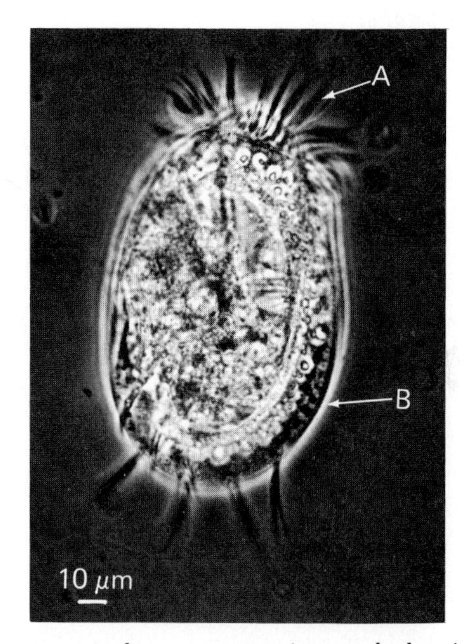

Figure 9–35 A phase-contrast micrograph showing the ciliate *Euplotes eurystomus. [Y. Shigenaka, K. Wantanabe, and M. Hareda,* J. Protozool. ***20****:414–428, 1973.]*

13. a. How are algae classified?
b. What properties or features do algae have that are not found in any other microbial group? List at least three.
c. List six criteria used in algal classification.

14. How do algae reproduce? Explain.

15. List three algal protist divisions and describe the distinguishing features of each one.

16. What is unique about the cell walls of diatoms?

17. Of what commercial value are algae?

18. Compare the cellular organization and photosynthetic process of algae with that of photosynthetic bacteria.

19. List some of the beneficial and detrimental effects of algae.

20. Define or explain the following:
a. food chain
b. paramylum
c. phytoplankton
d. algal bloom
e. zoospore
f. frustule

21. Photographic Quiz
a. What type of alga is shown in Figure 9–36?

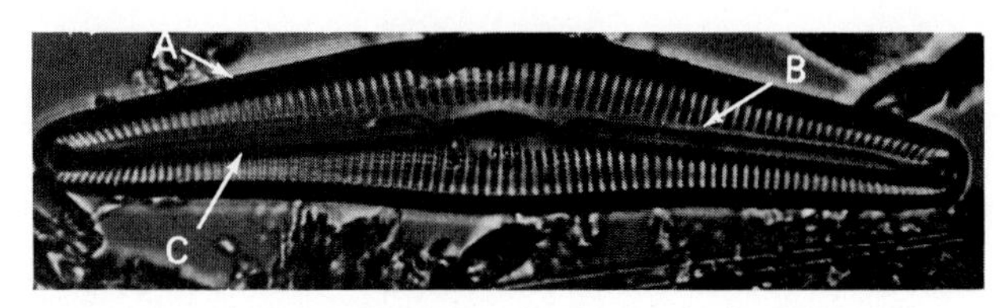

Figure 9–36 [*T. L. Hufford, and G. B. Collins,* J. Phycol. **8**:*192–195, 1972.*]

b. Identify the labeled structures of the alga in Figure 9–36.

Suggested Readings

BAMFORTH, S. S., "Terrestrial Protozoa," *Journal of Protozoology* **27**:33–36, 1980. *An excellent description of the effects of conditions and various forms of life on activities of soil protozoa.*

BOLD, H. C., and M. J. WYNNE, *Introduction to the Algae: Structure and Reproduction.* Englewood Cliffs, N.J.: Prentice-Hall, Inc., 1985. *A well-written and functionally illustrated coverage of algae.*

BUETOW, D. E. (ed.), *The Biology of Euglena,* vol. 3: *Physiology.* New York: Academic Press, 1982. *A comprehensive review of the structures and associated functions of* Euglena.

COLLINS, M., "Algal Toxins," *Microbiological Reviews* **42**:725–746, 1978. *A review of the toxins produced by algae, with specific reference only to those that are harmful to multicellular organisms.*

DAWES, C. J., *Marine Botany.* New York: John Wiley & Sons, Inc., 1981. *A comprehensive treatment and an ecological approach to photosynthesizing marine plants.*

FARMER, J. N., *The Protozoa: Introduction to Protozoology.* St. Louis: The C. V. Mosby Company, 1980. *A basic comprehensive introduction to this protist group.*

LEVINE, N. D., et al., "A Newly Revised Classification of the Protozoa." *Journal of Protozoology* **27**:37–58, 1980. *The original article presenting the description of the currently accepted protozoan classification system.*

RUDZINSKA, M. A., "Do Suctoria Really Feed by Suction?," *BioScience* **23**:87–94, 1973. *A discussion of the feeding habits of this most interesting group of protozoa.*

SCHOLTYSECK, E., *Fine Structure of Parasitic Protozoa.* New York: Springer-Verlag, 1979. *An excellent combination of electron micrographs and diagrams showing the fine structural features of a number of representative protozoa.*

TRAGER, W., "Some Aspects of Intracellular Parasitism," *Science* **183**:269–273, 1974. *An interesting short account of intracellular parasitism, with an emphasis on nutrition.*

10

Viruses, Viroids, and Prions

"So, Naturalists observe a flea
hath smaller fleas that on him prey
and these have smaller still to bite 'em
and so proceed ad infintitum."
— Jonathan Swift

After reading this chapter, you should be able to:

1. Indicate the differences between viruses and other types of microorganisms.
2. Describe the size, host range, organization, and functions of viruses.
3. Explain representative replication cycles of viruses.
4. Identify viral structures in electron micrographs.
5. List and describe the methods used for the cultivation of viruses.
6. Discuss the bases for virus classification and how they differ from those of other microbial types.
7. List the differences that exist among viruses, viroids, and prions with respect to structure, replication cycle, and host range.
8. Describe the general consequences of viral infection in tissue cultures.
9. What is lysogeny?
10. Indicate the respective roles of plus and minus strands in viral nucleic acid formation.
11. Distinguish between transcriptase and reverse transcriptase.
12. Describe the differences separating viruses from virusoids, viroids, and prions.

True microorganisms, however small and simple, are cells. Unicellular microorganisms such as certain algae, bacteria, fungi, and protozoa always contain deoxyribonucleic acid as their storehouse of genetic information and have their own machinery for the production of energy and macromolecules, such as carbohydrates, lipids, proteins, and nucleic acids. Viruses are totally different. They are representatives of the ultimate in parasitism. This chapter surveys the structure and function of these most unusual submicroscopic organisms. Attention also is given to even smaller (subviral) forms, the viroids and prions.

Virtually every kind of life can be infected by viruses—vertebrate and invertebrate animals, plants, procaryotes, and eucaryotic microorganisms such as fungi, protozoa, and certain algae (Figure 10–1). There are even some "satellite" viruses, which are considered in a sense parasites on other viruses. Until about 1972, it was generally believed that the smallest infectious disease agents were viruses. This view changed in the 1980s, with the discovery of smaller and less complex agents called *viroids.* A number of plant diseases are known to be caused by viroids, which consist of small uncovered (naked) molecules of ribonucleic acids. The discovery of subviral agents of diseases apparently did not end with viroids. Another smaller and quite different disease agent, the *prion,* has been shown to be the cause of scrapie, a neurological disease of sheep and goats. This chapter discusses what is known about the physical, chemical, and biological properties of viruses and subviral agents of disease. Descriptions of the life (replication) cycles, which include cell parts and how new viruses are formed (replicated), also are presented. Various aspects of human diseases caused by viruses are described in later chapters.

What Is a Virus?

Viruses are unlike any other form of microorganism. This is obvious not only from their submicroscopic size but also from other differences related to the way they function (Table 10–1). Viruses cannot be cultivated outside of a living system; they require a living host cell in order to replicate (Figure 10–2). Because viruses are so different from other microorganisms, investigators debated for years whether or not they were actually alive.

A completely satisfactory definition of a virus has yet to be formulated. However, there are a number of characteristic properties that distinguish viruses from all other biological forms and provide a genu-

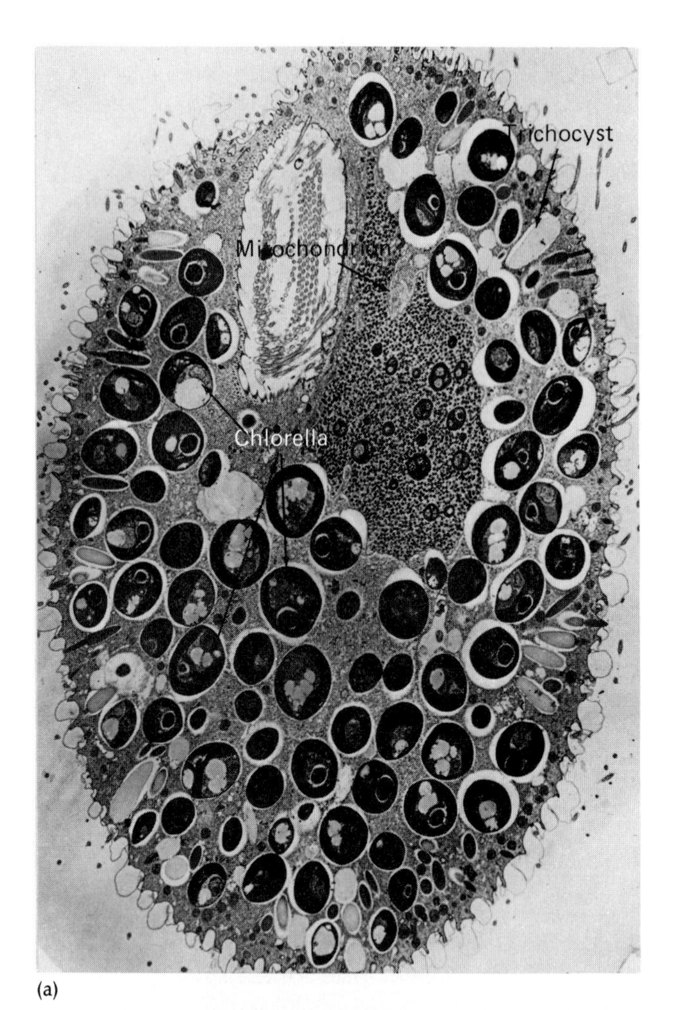

(a)

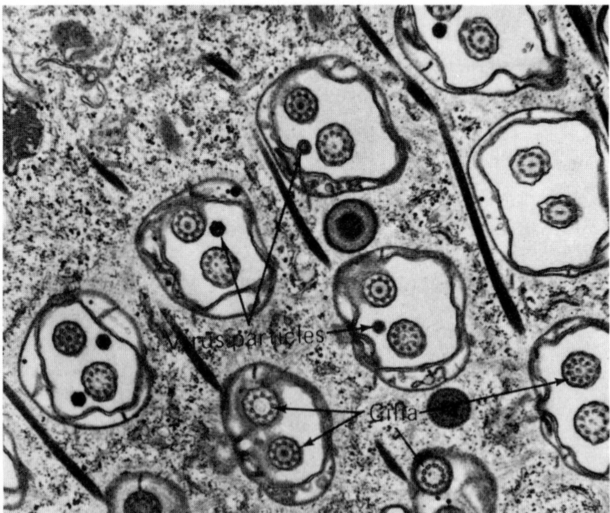

(b)

Figure 10–1 An interesting interrelationship involving algae, viruses, and protozoa. (a) An ultrathin preparation of *Paramecium* (par-a-ME-see-um) containing many green algae of the genus *Chlorella* (klor-ELL-ah) (arrows). *[H. Kawakami, and N. Kawakami,* J. Protozool. ***25:****217–225, 1978.]* (b) On closer examination, several infective virus particles can be seen in the intracellular symbiotic algae. *[H. Kawakami, and N. Kawakami,* J. Protozool. ***25:****217–225, 1978.]*

TABLE 10–1 Properties of Viruses, Viroids, and Other Microorganisms

	Microbial Components				Growth Requirements		
Microbial Group	*Cell Wall*	*Internal Membrane Parts*	*Ribosomes*	*Nucleic Acid Content*	*Cultivation in or on Artificial Media*	*Require Living Cells*	*Sensitivity to Antibiotics*
Algae	Present	Present	Present	DNA, RNA	Yes	No	Variable
Bacteria	Present	Absent	Present	DNA, RNA	Yes	Some	Present
Fungi	Present	Present	Present	DNA, RNA	Yes	No	Variable
Protozoa	Absent	Present	Present	DNA, RNA	Yes	Some	Variable
Viruses	Absent	Absent	Absent[a]	DNA or RNA[b]	No	Yes	Absent
Viroids[c]	Absent	Absent	Absent	RNA	No	Yes	Absent
Prions[c]	Absent	Absent	Absent	None	No	?	Absent

[a] One group of viruses, the *arenaviruses,* which cause natural infections in rodents, contain host cell ribosomes.
[b] Individual virus particles contain either DNA or RNA, never both.
[c] Described later in this chapter.

ine basis for their definition. A virus can be defined as an infectious agent having the following specific properties:

1. The possession of one or several molecules of either deoxyribonucleic acid (DNA) or ribonucleic acid (RNA), usually but not necessarily covered by a coat of one or several proteins.
2. The ability to transfer its nucleic acid from one host cell to another.
3. The ability to use and direct a host cell's enzyme systems for its intracellular replication.
4. The absence of the binary fission characteristic of bacteria and related microorganisms.
5. The lack of any energy-harnessing metabolic cycle like those in other life forms.

Some viruses are able to integrate their nucleic acid into a host's DNA. By such mechanisms, certain viruses become dormant or continually active, while others are able to change the genetic properties of a cell, affecting its growth and, unfortunately, leading to abnormal and cancerous formations. The structures and activities of organisms that meet this definition and description are detailed in the following sections.

Figure 10–2 An electron micrograph of a cell showing the result of infection with immunodeficiency virus (HIV). Numerous newly formed virus particles surround the cell (arrows). Human immunodeficiency virus is the cause of acquired immune deficiency syndrome, or AIDS *[From R. C. Gallo,* et al., Science *224:500–503, 1984.]*

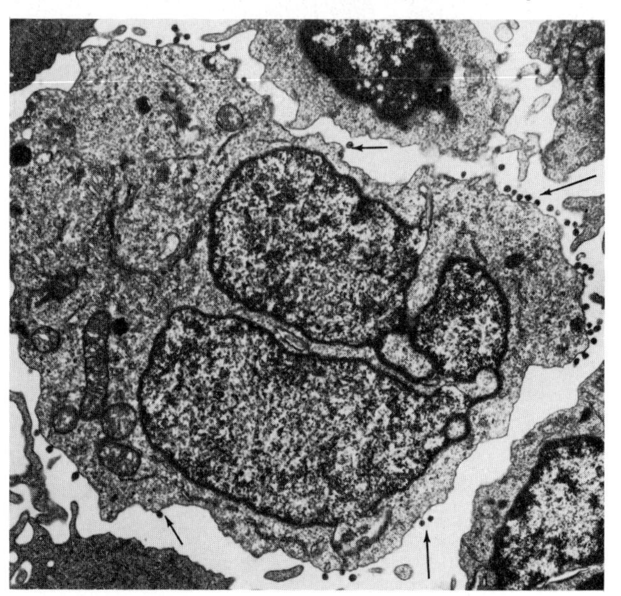

Basic Structure of Extracellular Viruses

Each mature virus particle, or **virion,** has a characteristic structure and organization. Electron microscopy, x-ray diffraction analysis, and computer imaging show virions to consist of a nucleic acid inner core surrounded by a protein outer coat called a **capsid** (Figure 10–3). Each capsid is constructed from a definite number of protein subunits called **capsomeres.** These structural subunits are grouped together to form a characteristic shape. In electron micrographs, the capsomeres appear in recognizable patterns and are the visible portions of viruses (Figure 10–4). The capsid covering the viral nucleic acid core comprises the **nucleocapsid** (Figure 10–3b and 10–3c). [See electron microscopy in Chapter 3.]

The Nucleic Acid of the Mature Virion

The viral nucleic acid, whether DNA or RNA, contains the viral genetic material, or **genome.** The nucleic acid may be either single-stranded or double-stranded and assumes a long, filamentous, folded or

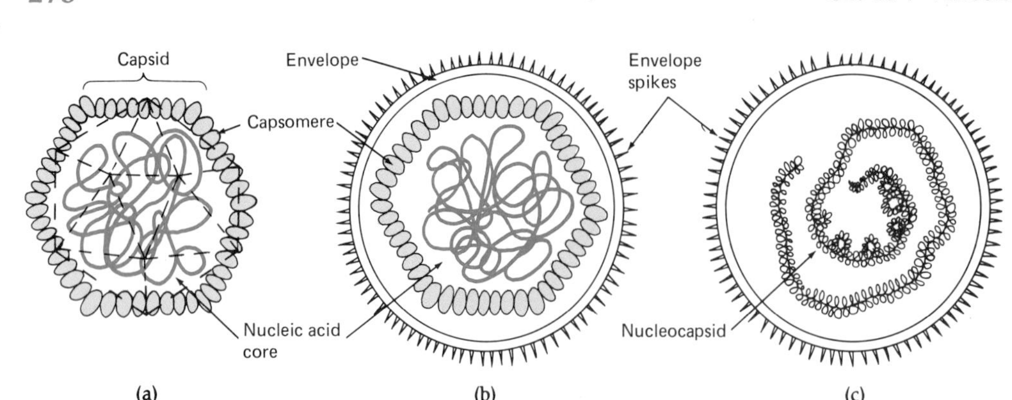

Figure 10–3 Viral architecture. (a) A "naked" icosahedral (20-sided) virus particle. Note the folding of the nucleic acid core and the capsomeres. The capsid is composed of capsomeres. (b) An enveloped, icosahedral (20-sided) nucleic acid-containing capsid. The envelope itself has spike structures. (c) A enveloped helical virus. Here the capsid is closely and tightly bound to the nucleic acid.

coiled form (Figure 10–5). The specific form and arrangement of the nucleic acid vary according to the virus. While most viruses have either single-stranded (ss) RNA or double-stranded (ds) DNA, some of these agents may possess double-stranded RNA or single-stranded DNA. In most cases the nucleic acid is found to be a single molecule. Some large RNA-containing viruses are exceptions to this finding. Additional RNA molecules in a virion can be regarded as extra chromosomes. The genomes (genetic material) of certain RNA viruses serve as viral messenger RNA and are referred to as *plus strands.* This means that upon infection, a viral nucleic acid strand acts as a blueprint of information in the form of messenger RNA (mRNA) and is translated directly into new viral protein (Figure 10–6). In other viruses, transcribed RNA, which is complementary to the viral RNA genome, functions as mRNA and is termed *minus-strand* RNA.

The length of the nucleic acid molecule varies among different viruses, but it is constant for a particular type. The DNA of several viruses is circular or cyclic, but in some DNA-containing agents and in all RNA viruses, the nucleic acid of the mature virus particle is linear, or noncyclic.

Figure 10–4 (a) An electron micrograph showing the capsomere structure of adenovirus type 5 in a negatively stained preparation. (b) A capsomere model of an adenovirus. *[Courtesy of Dr. R. W. Horne, The John Innes Institute; Electron Microscope Laboratory, Norwich, England.]*

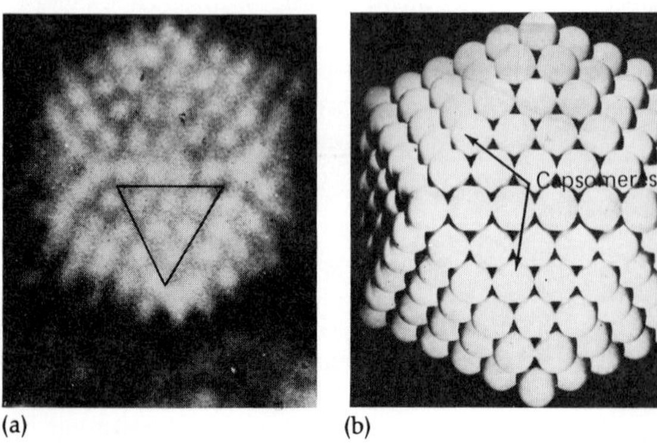

Enveloped Viruses

Several animal virus particles, as a consequence of their intracellular development pattern, acquire an outer coat, or covering, from the cytoplasmic and/or nuclear membranes of infected cells as they pass through or are released from them (Figure 10–7).

Figure 10–5 A negatively stained preparation of the RNA-containing influenza virus. The virion's shape is quite irregular. Note the coiled appearance of the nucleic acid and the envelope covered with short spikes. Original magnification 217,500×. *[Courtesy of D. Hockley and J. Wood.]*

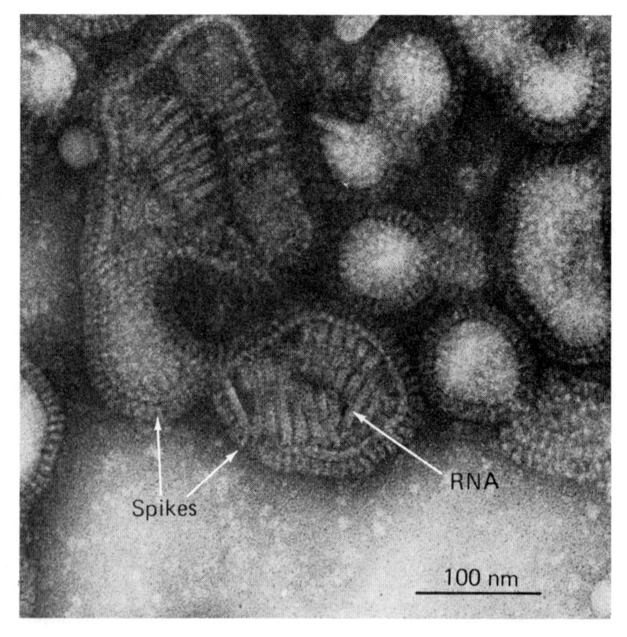

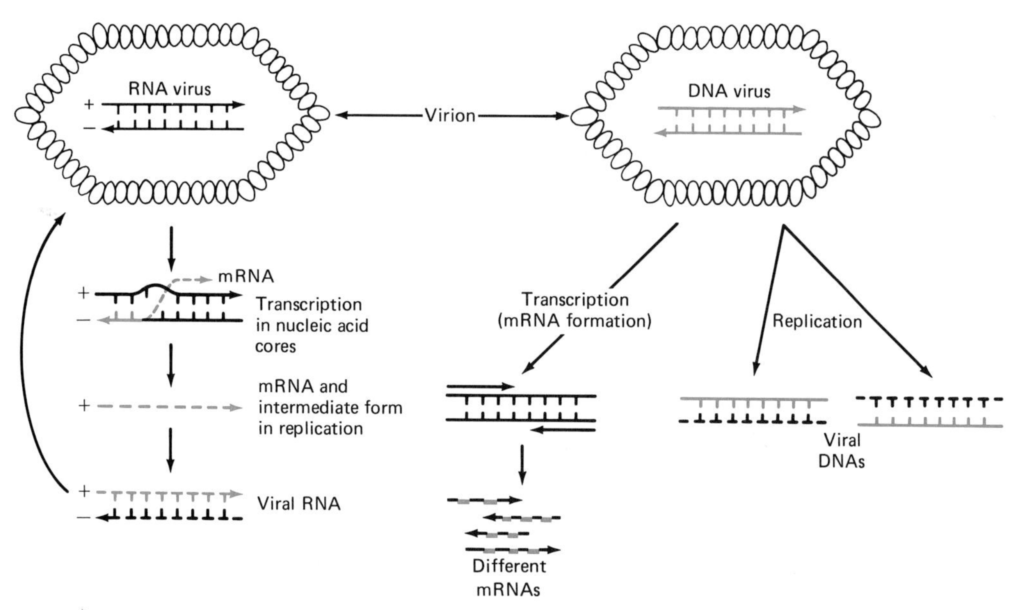

Figure 10–6 A comparison of the replication of virion double-stranded (ds) RNA and DNA molecules. In the replication of double-stranded RNA (left), the pattern or information for new strand information comes from only one strand, the plus (+) strand. Minus (−) strands for RNA viruses are then enzymatically made from newly liberated plus strands (indicated by arrow on the left side of the figure). This transcription process is different with double-stranded DNA viruses (right). Here both strands of the nucleic acid function as a pattern or source of information for the formation of new strands. New nucleic acid strands are shown as broken lines. (Refer to Chapter 6 for a description of protein synthesis.)

Figure 10–7 The development of an enveloped herpes simplex virion. (a) A diagrammatic summary. In the intranuclear development, virus particles are budding through the nuclear membrane (1) or being released into the space directly around the nucleus (2). In both cases they acquire an envelope composed of host membranes. Resulting enveloped viruses travel through the cytoplasm within smooth vacuoles (3). At the cell membrane, enveloped virions are released by a process of reverse phagocytosis (4), followed by the folding of the membrane around released virions into vesicles (5). Virus budding (6); extracellular viruses with significant amounts of intramembranous particles. *[After M. Rodriguez, and M. J. Dubois-Dalcq,* Virology ***26****:435–447, 1978.]* (b) Developing viral particles in the nucleus. (c) Budding viruses acquiring envelopes.

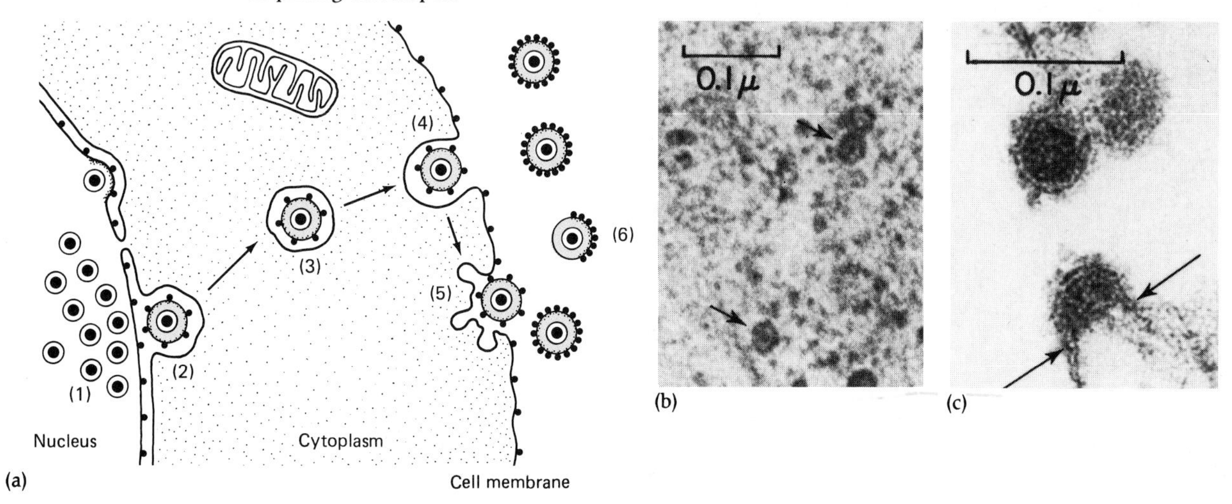

The term **envelope** has been given to this outer coat (Figure 10–3b). Viruses lacking such envelopes are commonly referred to as "naked." Envelopes are chemically composed of carbohydrates, phospholipids, and proteins. The proteins in these envelopes are encoded by viral genes, while the phospholipids are derived from host cell membranes in most cases. Envelopes usually range in thickness from 10 to 15 nanometers (nm). Depending on the viral agent, the envelope may be covered with projecting spikes called **peplomers** (Figures 10–3b and 10–7), which in profile appear as knobs or a fringe (Figure 10–8a). The physical characteristics of such fringes differ among viral groups and can be used for identification purposes.

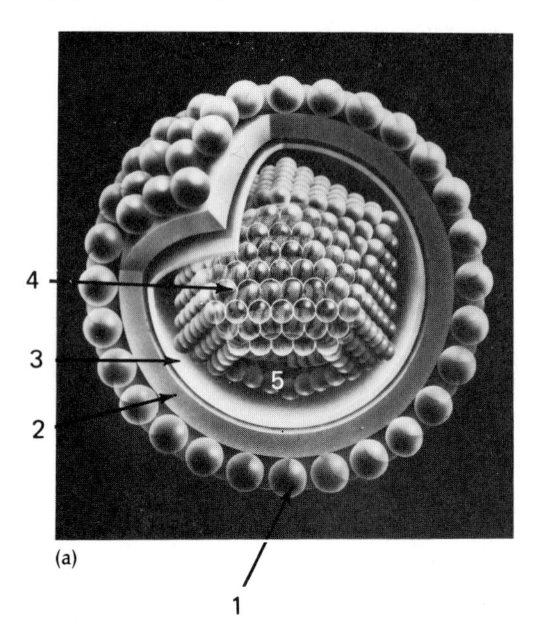

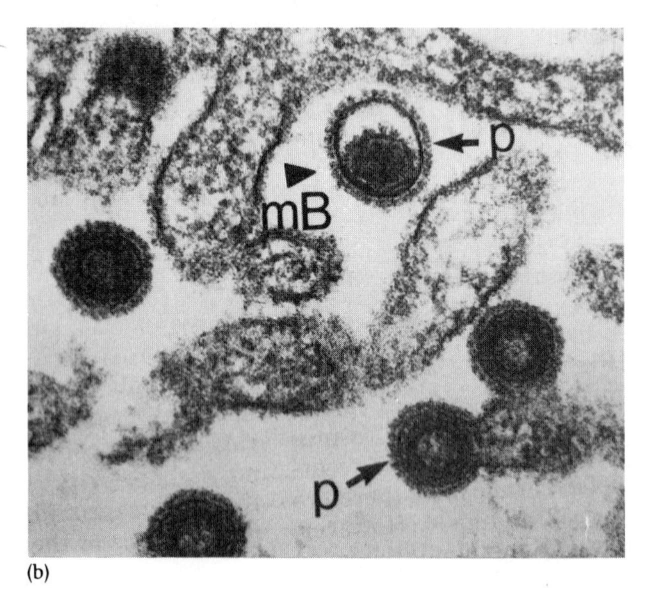

Figure 10–8 The structure of a mouse tumor virus. (a) A sketch of virus structure and organization showing surface knobs or projections (1), viral membrane (2) and (3), capsid composed of individual capsomeres (4), and the nucleic acid core (5). The knobs and viral membrane form the viral envelope. (b) An electron micrograph of viruses (p) in various stages of maturation (mB). *[N. H. Sarkar,* Virology ***150:****419–438, 1986.]*

VIRAL ENZYMES

Characteristically, viruses lack most of the enzymes and metabolic processes found with other forms of life. However, several have enzymes associated with the formation of their nucleic acids and, in certain situations, an enzyme that helps viruses attach to host cells. In the case of influenza, envelope spikes contain *neuraminidases,* which bind these agents to respiratory linings. These and other enzymes are discussed in later sections.

Shape and Size

Virions vary not only in size but also in shape. The virus particle has one of several forms (Figure 10–9). Among the possible basic shapes are the polyhedron, the helix, and combinations of shapes known as binal, bullet-shaped, and filamentous. Many virus particles lacking a basic shape or containing accessory structures are called *complex* (Figure 10–9c).

The capsomeres of different viruses show variations and are arranged in definite geometric patterns. In the case of tobacco mosaic virus, the morphological units form a helical structure. This arrangement is shown diagrammatically in Figure 10–9f. The virus particle itself is a long rod and is referred to as a *helical virion.*

Several animal, plant, and bacterial viruses have polyhedral shapes; that is, they are many-sided (Figures 10–9b and 10–9h). Their capsids commonly have 20 triangular faces (Figure 10–9b). Such viral particles are called *icosahedral virions.* The size of this type of virus is determined by the number of capsomeres it contains (Figure 10–4).

When either helical or icosahedral virions are enclosed by envelopes, they are described as *enveloped helical* or *enveloped icosahedral.* An example of the latter is herpes simplex virus, the causative agent of fever blisters (Figure 10–9a). The viruses that cause influenza (Figure 10–5) and mumps are typical enveloped helical forms.

Classification

Viral classification is still in its infancy compared to the classifications of procaryotes or higher forms of life. Some of the newer and more sensitive ap-

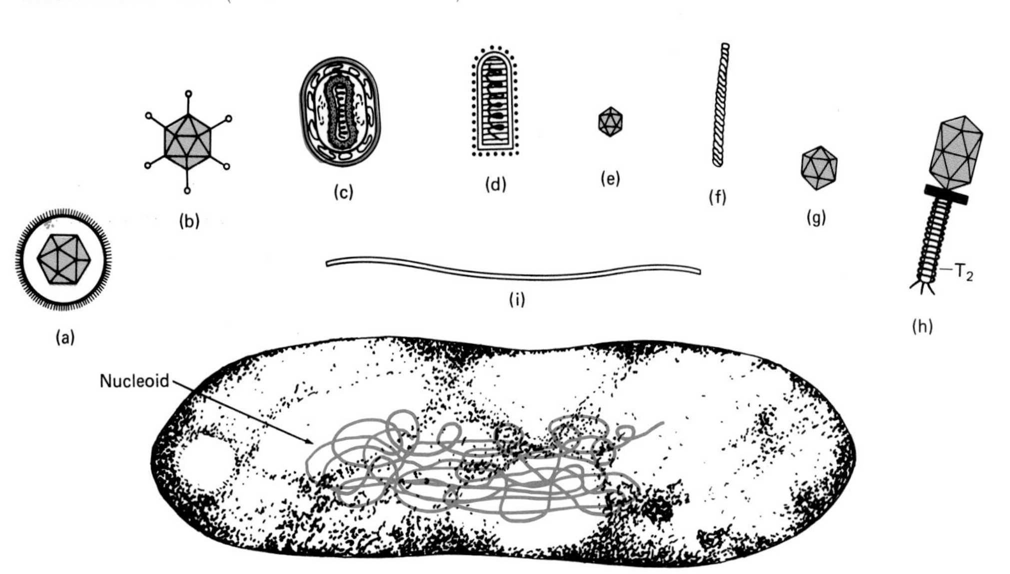

Figure 10–9 Comparative sizes and shapes of animal, plant, and bacterial viruses and the much larger cell of the bacterium *Escherichia coli* (esh-er-IK-ee-a KOH-lee). Representative viral shapes and viruses shown are (a) polyhedral and enveloped (herpes viruses), (b) icosahedral (adenoviruses), (c) complex (cowpox or vaccina), (d) bullet-shaped (rabies), (e) icosahedral animal (polio), (f) helical (tobacco mosaic), (g) icosahedral plant naked (cauliflower mosaic), (h) binal (T_2, or type 2 bacteriophage), and (i) filamentous (inoviruses).

proaches to the classification of viruses, however, show a new refinement in determining the relationship between different forms of life.

Viruses traditionally have been grouped in several ways—according to the host normally infected, the particular tissue attacked, and even the general symptoms associated with a disease state. Although such approaches were useful, they were not very scientific or reliable. Not only could several different viruses cause similar disease symptoms, but one type of virus could produce quite different diseases depending on the host and the environment. Thus, it became apparent that a useful system of classification must be based on the basic properties of the virus itself. The constant chemical, structural, immunologic, and genetic properties of viruses form an appropriate basis for such a precise system.

Several classification systems for viruses have been proposed. One of the most commonly accepted systems uses Latinized binomial names like those used for other forms of life and is based on the characteristics of viruses as defined by Lwoff and Tournier in 1962. Under this system and the principles published by the International Committee on Taxonomy of Viruses, viruses are classified according to the properties of virions, such as the following:

1. Capsid organization
   a. Shape and general size of the virus particle
   b. Number of capsomeres
   c. Presence or absence of an envelope
   d. General symmetry of the nucleocapsid
2. Nucleic acid chemistry
   a. Type of nucleic acid (whether DNA or RNA)
   b. Number of strands (whether single or double)
   c. Molecular weight of nucleic acid
   d. Manner in which genetic information is translated into proteins
   e. Presence of a **transcriptase** (an enzyme involved in the formation of nucleic acids)
   f. Nucleic acid hybridization information

Special features—such as cellular location for viral development, synthesis of viral parts, additional enzymes or unique proteins, and host-parasite interactions—are used for further subdivision. Representative examples of this system are shown in Table 10–2. Animal, plant, and bacterial viruses are classified separately.

Bacteriophages (Bacterial Viruses)

Almost all members of the large group of readily cultivable bacteria serve as hosts for *bacteriophages* (from the Greek word *phagein,* meaning "to eat") or bacterial viruses. The host range of a bacteriophage

TABLE 10–2 Representative Characteristics Used in the Classification of Viruses

Nucleic Acid	Double- (ds) or Single-Stranded (ss)	Capsid Symmetry (Shape)	Number of Capsomeres	Capsid Size in Nanometers	Enveloped or Naked	Host Range and Examples of Viruses: Animal	Invertebrate	Microbial	Plant
DNA	ss	Polyhedral		50	Naked	Parvoviruses			
	ss	Filamentous		50 × 800	Naked			Coliphage fd[a]	
	ss	Polyhedral	12	22	Naked			ϕX–174[b]	
	ss	Polyhedral			Naked				Geminiviruses
	ds	Polyhedral	72	45 × 55	Naked	Polyoma			
	ds	Polyhedral	252	60 × 90	Naked	Adenoviruses			
	ds	Polyhedral	162	100	Enveloped	Herpesviruses			
	ds	Helical		40 × 30	Enveloped		Nuclear polyhedrosis		
	ds	Helical		9–10	Enveloped	Poxviruses			
	ds	Binal		Polyhedral head, 95 × 65, and helical tail, 17 × 115	Naked			Coliphages $T_2T_4T_6$[a]	
	ds	Polyhedral	812	140	Naked		Triple iridescent virus		
	ds	Polyhedral			Naked				Cauliflower mosaic
RNA	ss	Helical		17.5 × 300	Naked				Tobacco mosaic virus
	ss	Polyhedral		31	Naked				Tomato bushy stunt virus
	ss	Polyhedral		20–25	Naked			Colliphage f_2[a]	
	ss	Polyhedral		50	Enveloped	Alpha viruses (encephalitis viruses)			
	ss	Bullet-shaped		170 × 70	Naked	Rabies virus			
	ds	Polyhedral		25	Naked			*Penicillium* virus (PcV)[c]	
	ds	Polyhedral	32	28	Naked	Picornaviruses (polio)			
	ds	Polyhedral		100	Enveloped			ϕ6[b]	
	ds	Polyhedral	92	70	Naked				Wound tumor virus

[a] Bacterial viruses that attack the bacterium *Escherichia coli*.
[b] Other bacterial viruses.
[c] Virus of molds.

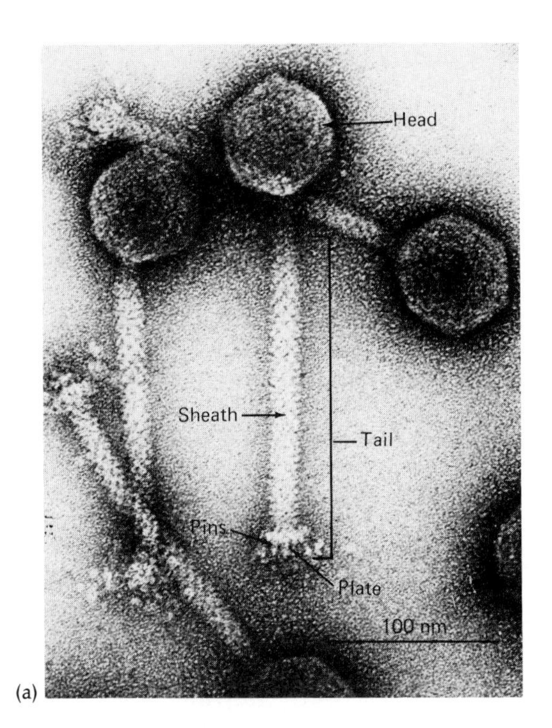

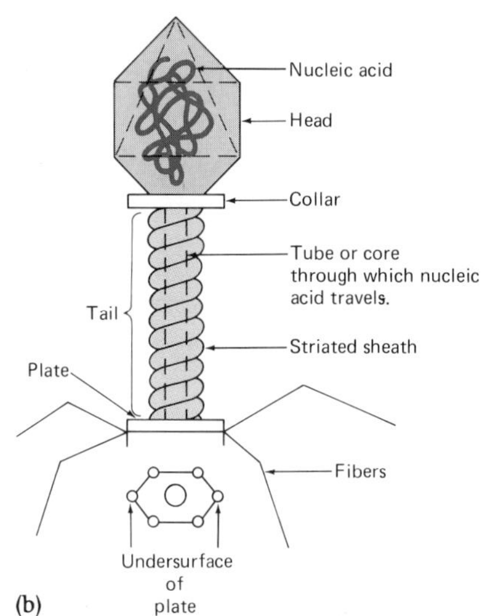

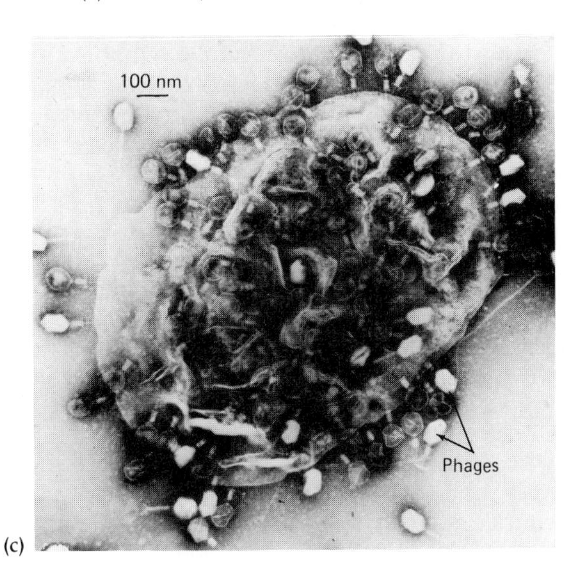

Figure 10–10 Bacterial viruses. (a) A negatively stained bacteriophage preparation active against the streptococci involved with cheese production. *[Courtesy J. Lembke and M. Teuber, Kiel, Federal Dairy Research Center.]* (b) A diagram showing the various components of a T_2 (type 2) phage. The plate of this virus is hexagonal and contains a pin at each of its corners, with long, thin fibers connected to it. (c) An electron micrograph showing T_{4r} phages normally adsorbing onto *Escherichia coli.* Contracted viral sheaths can be seen with some viruses. Numerous phage heads are clear, but others are opaque. Which ones still have their nucleic acid? *[L. D. Simon,* et al., Virology ***41****:77, 1970.]*

(also called *phage*) may involve a single specific bacterial species, strain, or several bacterial genera. Viruses have been found to attack several cyanobacterial species. Their morphological features greatly resemble those of other bacterial viruses (Figure 10–10).

Many bacterial viruses have proved useful in research on the mechanisms of cellular infections. Other phages have been valuable in the identification of bacterial pathogens and epidemiological investigations. For this reason, the bacterial viruses and their cultivation have been intensively studied.

The Basic Structure of Bacteriophages

Bacterial viruses may contain either DNA or RNA. In most DNA phages, the nucleic acid is double-stranded (Figure 10–11). In RNA bacteriophages, the nucleic acid is either single-stranded or double-stranded. The nucleic acid of most bacterial viruses except for one phage group is in a polyhedral capsid, frequently referred to as the "head" (Figure 10–10b). In many cases this capsid is attached to a helical protein structure called a "tail," which, with its additional parts, aids the phage in adsorbing onto a susceptible bacterial host (Figure 10–10c). The head and tail combination is referred to as a binal form.

At the present time, many basic structures have been recognized among bacteriophages, with no particular form of phage being restricted to a specific bacterial genus. Bacteriophages are thought to show greater variation in form than any other viral group. They vary greatly in size, shape, and nature of tail, if any, as shown by electron microscopy. The most common basic morphological types are illustrated in Figure 10–12 (see page 285).

The first quantitative and critical studies of viral replication were carried out in large part with the bacterium *Escherichia coli.* Many of the principles of bacterial virus replication were later found to be valid for other viruses as well. Representatives of several main virus groups will be described.

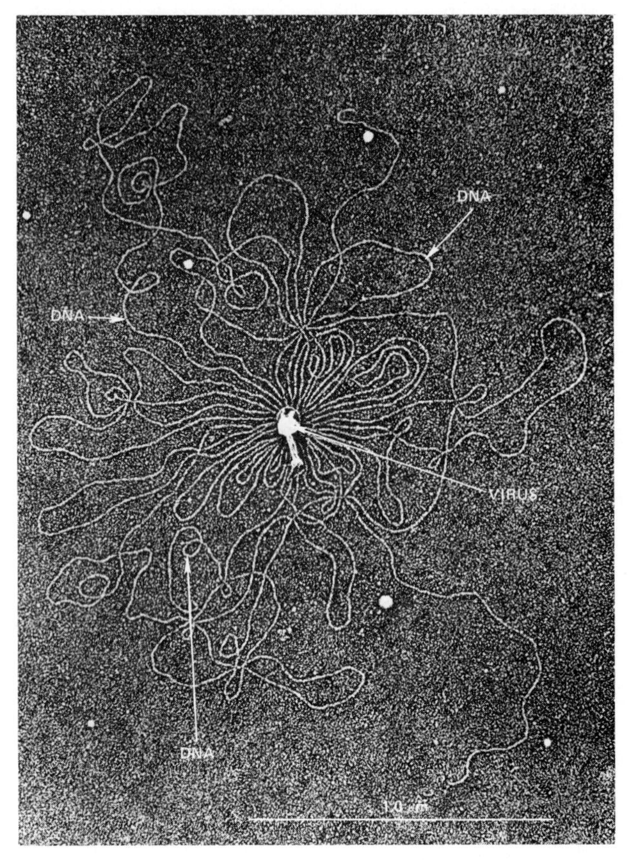

Figure 10–11 A special preparation of a bacteriophage showing its DNA (arrows), which contains the genetic information of the virus. The virus particle is located approximately within the center of the nucleic acid. The bar marker equals 1.0 micrometers. *[Courtesy of Dr. A. K. Kleinschmidt; from A. K. Kleinschmidt, D. Lang, D. Jackerts, and R. K. Zahn,* Biochem. Biophys. Acta ***61****:857, 1962.]*

Filamentous Bacteriophages

The filamentous, or threadlike, viruses, discovered in 1963, are among the smallest viruses known to date. They are the lone exception to the polyhedral form of bacteriophages. Filamentous viruses are long deoxyribonucleoproteins measuring approximately 5.5 nm in diameter. Purified preparations of DNA from these microorganisms have been found to exhibit single-stranded form. Two different lengths have been reported for filamentous viruses: approximately 870 nm and about 1,300 nm.

Gram-negative bacteria, including *E. coli, Pseudomonas aeruginosa* (soo-doh-MOH-nass air-roo-jin-OH-sa), *Salmonella typhimurium* (sal-mon-EL-la tie-fee-MUR-ee-um), *Vibrio parahaemolyticus* (VIB-ree-oh par-a-hee-moh-LIT-ee-cuss), and *Xanthomonas oryzae* (zan-tho-MOAN-us or-EE-zay), are hosts for filamentous viruses. Adsorption to bacterial cells takes place at the ends of threadlike bacterial sex *pili.* Binal-shaped, pilus-dependent phages also are known. **[Pili are described in Chapter 7.]**

Filamentous viruses have focused attention on a form of symbiotic behavior previously unrecognized in bacteria. Unlike other bacterial viruses whose mode of replication destroys their respective hosts, filamentous viruses are released from dividing and growing bacteria without any apparent marked injury to the host cells. In short, this is a nondestructive or *nonlytic form of viral release.* Research suggests that filamentous viruses are assembled during their release from the host. This resembles the mode of replication found in certain animal viruses. Some virologists believe that filamentous viruses may serve as excellent models in studies concerning the development and effects of *oncogenic* (cancer-inducing) viruses. **[See Chapter 30 for viruses and cancer.]**

Cultivation of Bacteriophages

In general, the cultivation of bacteriophages is relatively uncomplicated and neither expensive nor time-consuming. Phages are commonly propagated on appropriate, actively growing young bacterial cells in either broth or agar cultures. In liquid cultures, the destruction of enough susceptible bacteria by the replicating viruses will cause the nutrient medium to clear. When agar plates are used, the presence of bacterial viruses is indicated by the development of transparent **plaques** against the dense background of bacterial growth on the medium's surface (Figure 10–13). Plaques also may develop within colonies (Color Photograph 51).

Various types of bacterial viruses can be isolated from natural sources such as dairy products, diseased tissues, feces, sewage, soil, and water. In principle, the presence of such viruses and their bacteriolytic action can be demonstrated by a fairly straightforward procedure. Fluid samples of any of the above materials are passed through a filter that retains bacteria but not viruses. The resultant filtrate material is then tested for the presence of phage. This usually involves introducing a small amount of the test material into an appropriate fresh culture of a test host organism. Another tube, without the addition of the suspected phage-containing sample, is used as a control. Both tubes are then incubated under optimum environmental conditions. If the tube to which the test material was added (the inoculated tube) shows clearing, or at least less turbidity than the control tube, then the presence of phages in the original sample is indicated.

Unfortunately, in practice, the demonstration of bacterial viruses from natural sources is not always as easy as this description suggests. In many studies, more involved procedures are required. Moreover, many samples of natural substances may not yield bacterial viruses.

Once a new phage is found, its characterization is usually undertaken. This involves determining its chemical and physical properties, uncovering its

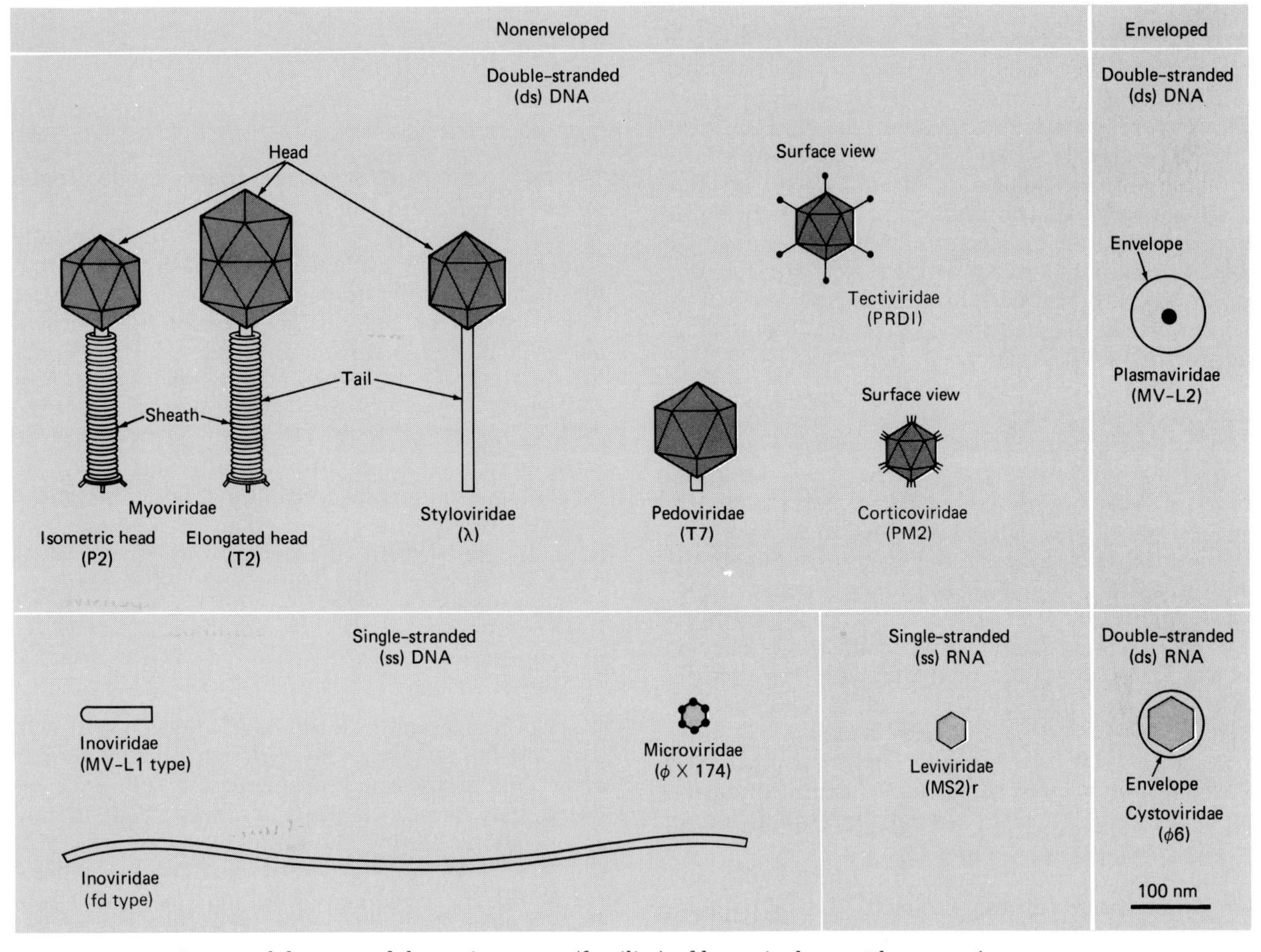

Figure 10–12 Structural features of the main groups (families) of bacteriophages. The general position and type of nucleic acid are indicated for each virus. *[Modified from R. E. F. Mathews,* Intervirology ***12**:156, 1979.]*

mechanism of action, and demonstrating its basic differences from and similarities to other bacterial viruses.

Enumeration of Phages

Several methods can be used to estimate accurately the number of virus particles in a sample, including electron microscopy. However, one of the most useful and most commonly employed procedures is the *plaque count assay,* which is simple and accurate and yields highly reproducible results. A *plaque* is a clear area on the surface of the cell culture; it results from the destruction of host cells by lytic viruses (Figure 10–13).

The phage suspension to be assayed is added to a tube containing a small quantity of melted soft agar. One drop of an appropriate fresh bacterial culture is also introduced. The contents of the tube are mixed and quickly poured onto the surface of a dried, hard agar layer contained in a Petri dish. This plate is

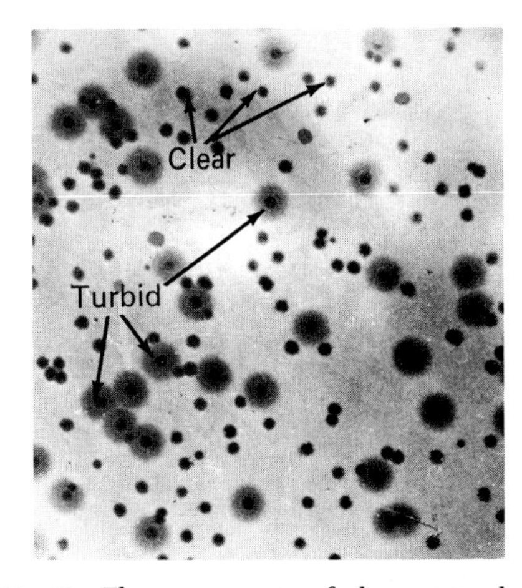

Figure 10–13 The appearance of plaques produced by a mixture of bacteriophages. Smaller, sharp-edged, clear plaques produced by one type of lytic virus can be distinguished from the larger, turbid ones produced by others. Plaque appearance can be used in classification. *[Courtesy of S. E. Follansbee,* et al. Virology ***58**:180–198, 1974.]*

rocked gently back and forth to distribute both the bacteria and the phage mixture evenly over the agar surface before the soft agar hardens. Viral particles diffuse through the medium, infect, multiply, and lyse susceptible bacteria, leaving plaques on the agar. Those bacterial cells not infected grow, multiply, and form a cloudy, or turbid, layer over the hard agar surface. The number of viral particles in the original sample is determined by the number of plaques multiplied by a correction factor for the specimen dilution used. This product is expressed as the number of plaque-forming units (PFU) per milliliter of the initial sample.

Phage Typing

Bacterial viruses continually propagated in a specific bacterial species can become adapted to that particular bacterial host. Such viruses exhibit the phenomenon called *host-controlled modification.* They will primarily infect and lyse a specific bacterial species or cells of associated strains. This adaptation is the basis for phage typing (Color Photograph 52), used to identify strains in certain bacterial species such as *Salmonella typhi* and *Staphylococcus aureus.*

Phage typing is an extremely important procedure in tracing the sources of epidemics and distinguishing between pathogenic bacterial strains that cannot be differentiated by other means.

Phage typing is performed by growing the unknown bacterial culture—isolated from a patient, for example—on an agar plate. Known phages are systematically spotted onto the "lawn" of bacteria. After a suitable incubation period, zones of lysis appear if the appropriate phage is present. From the phage or phages that produce plaques, the bacterial host can be identified. Plastic plates with grid patterns embossed on their bottom surfaces (Figure 10–14) may be used to simplify the procedure.

Figure 10–14 "Integrid" Petri plate, an example of a Petri dish that can be used to conduct systematic phage-typing procedures. The grids are used to locate a particular phage or host. *[Courtesy of Falcon Plastics, Division of BioQuest, Los Angeles.]*

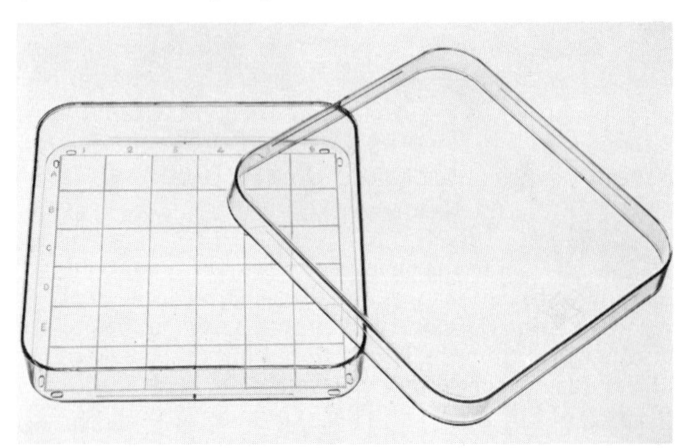

Replication Cycle of Bacteriophages

The fact that viruses do not reproduce by division is an essential feature common to the replication of all viruses. Phages that regularly infect, replicate, and complete their life cycles in bacterial hosts, and that ultimately cause the destruction of bacterial cells are called *virulent* or *lytic* viruses. Not all infections of bacteria proceed in this manner; some viruses establish an entirely different relationship with a host cell. These viruses may replicate and bring about the destruction of the host cell, or their DNA may be incorporated into that of the host (host DNA), so that it is passed on to succeeding generations of the host cell. Such viruses are said to be *temperate,* and their integrated form is referred to as a *prophage.* A bacterial cell containing a prophage is **lysogenic.**

Lysogeny

When a temperate bacteriophage infects a host cell, two possibilities exist. The lytic cycle characteristic of virulent viruses may occur, or the cell, once infected, may harbor the virus in a noninfectious state as a *prophage.* During the prophage phase, the host cell may undergo significant changes in colonial morphology, antigenicity, and toxin production. These effects, called *lysogenic* or *viral conversion,* are discussed in Chapter 6.

Occasionally, a prophage can be activated to enter into a lytic cycle. Exposure of a lysogenic culture to ultraviolet light increases the rate of such prophage activation significantly. This is one means of determining if a given culture is lysogenic. The interrelationship of lytic and temperate viral cycles is represented in Figure 10–15.

Lytic and temperate viral activities can usually be differentiated on the basis of the appearance of plaques that develop when bacterial hosts are plated. When relatively few viruses are mixed with many sensitive bacteria and placed on the surface of solid media, the plaques that develop after a suitable incubation period (Figure 10–12) will be clear for a lytic virus and turbid for a temperate one.

The clear plaque represents virulent viral activity in all infected cells, resulting in the death and subsequent lysis of such cells. In the turbid plaque, only some of the infected bacteria have undergone a lytic cycle; the surviving bacterial cells can and will reproduce, unaffected by the virus under the existing conditions.

During the lysogenic cycle, occasional bursting of cells liberates free viruses into the general culture containing lysogenic bacteria. Even though additional viruses may be adsorbed and infection occa-

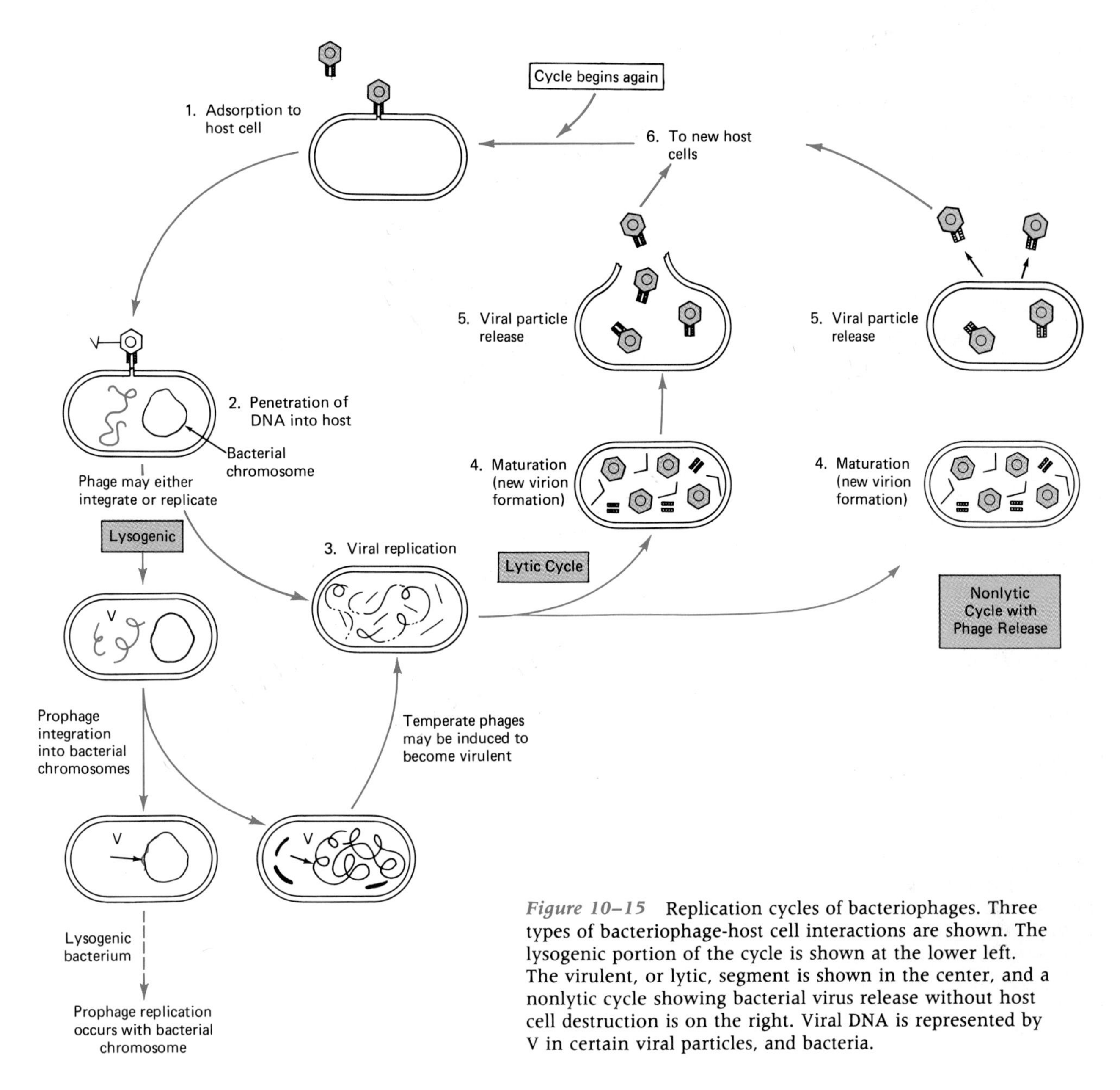

Figure 10–15 Replication cycles of bacteriophages. Three types of bacteriophage-host cell interactions are shown. The lysogenic portion of the cycle is shown at the lower left. The virulent, or lytic, segment is shown in the center, and a nonlytic cycle showing bacterial virus release without host cell destruction is on the right. Viral DNA is represented by V in certain viral particles, and bacteria.

sionally occurs with some bacteria in such lysogenic cultures, a lytic cycle usually cannot occur. Bacteria harboring a virus as a prophage (i.e., lysogenic bacteria), are immune to infection by other phages of the same or similar type.

Another form of bacterial cell immunity occurs when a mutation alters the chemistry or structure of a receptor site in such a way as to prevent further viral adsorption. This phenomenon is called *resistance.*

The Lytic Cycle

The lytic cycle has been studied extensively in certain phages for which *E. coli* and other bacteria are the hosts. Details of the structural and molecular properties of these viruses, designated T_2, T_4, and T_6 (or collectively, Type-even), are known well enough to allow construction of a general model of the processes of phage infection and host cell destruction. The steps in such a T-even cycle (Figure 10–16) include *adsorption* (attachment), *penetration, replication* of new viral DNA, *maturation,* and release of mature viruses. Other bacteriophages that multiply are known, and the release of their progeny occurs without destruction of the host cell (Figure 10–16).

ADSORPTION AND PENETRATION

Adsorption is a specific event, because only certain viruses are able to infect particular bacterial host cells. The phage tail fibers, shown in Figure 10–10b, function as *adsorption sites* that bind to specific *receptor* or attachment *sites* on the host bacterium's cell wall.

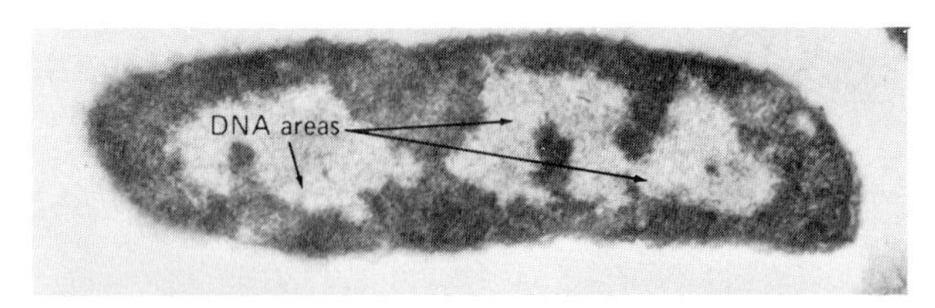

(a) Time of infection shows *E. coli* with normal DNA. Adsorbed viruses are not evident.

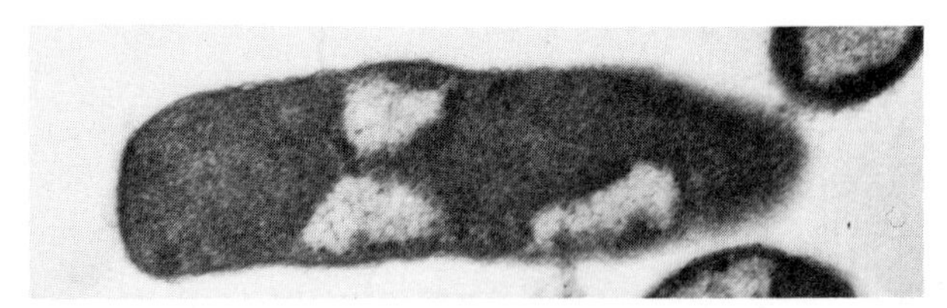

(b) Two minutes after infection. The appearance of the DNA has changed.

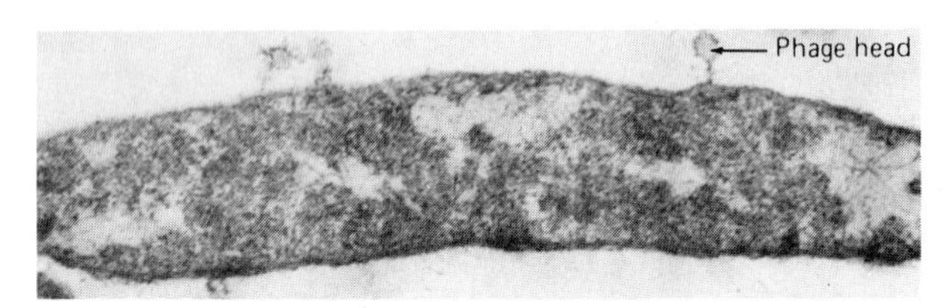

(c) Ten minutes after infection. DNA has become more diffuse. Empty viral heads are evident on the cell wall.

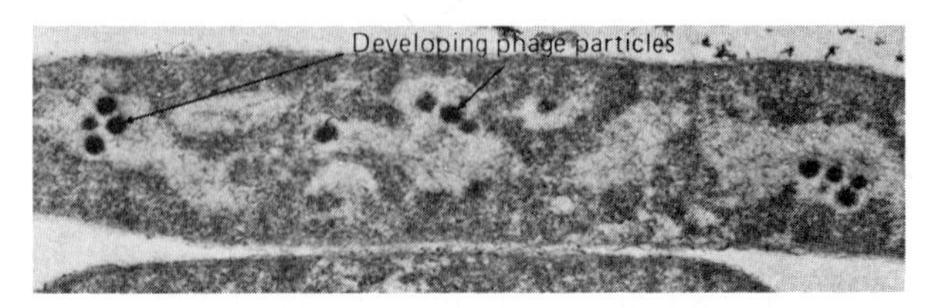

(d) Twelve minutes after infection. First phages appear.

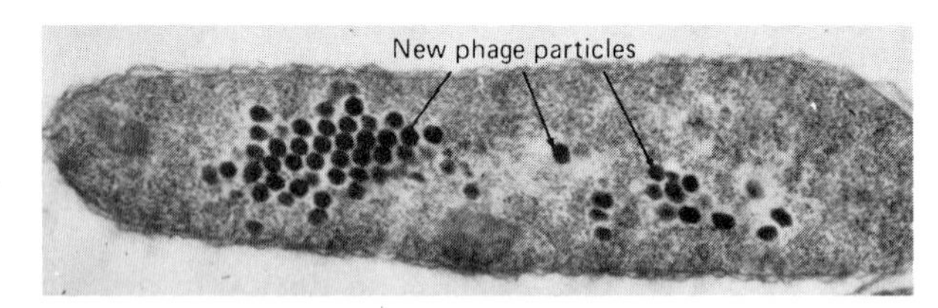

(e) Thirty minutes after infection. The cell is nearly ready to burst.

Figure 10–16 The intracellular development of virus T_2 in *Escherichia coli. [F. Jacob, and E. L. Wollman, "Viruses and Genes,"* Sci. Amer. ***204****:93, June 1961. Micrographs by E. Kellenberger and A. Ryter.]*

Once the bacteriophage has attached, an enzyme, *lysozyme,* is released from within the phage tail and digests a small portion of the host's cell wall. Following this step, the virus's tail sheath contracts, and the tail core, or central tube, mechanically penetrates the cell wall, preparing the way for a viral DNA injection. The phage DNA, located in the head component (Figure 10–10), is introduced from the open tip of the tail through the bacterial cell wall, across the cell membrane, and into the host cell's interior. The exact mechanism of this injection process is unknown.

For phages of the T-even viruses, the protein capsid remains outside the host. For filamentous, single-stranded DNA phages, the capsid enters the host. Although a small amount of cytoplasmic leakage can occur after perforation of the cell wall, the bacterial host does not appear to experience any great difficulty at this stage. If, however, the ratio of virus particles to bacterial cells, the *multiplicity of infection,* is great, sufficient injury to the cell wall results to cause lysis. This phenomenon is generally called *lysis from without,* to differentiate it from lysis that would normally occur upon release of viruses at the end of the lytic cycle.

VIRAL NUCLEIC ACID REPLICATION

Following the injection of phage DNA, there is an *eclipse,* or latent period, during which no whole infective viruses can be observed. Once infection begins, the virus assumes command and blocks the cellular synthesis of the host's own nucleic acids and proteins. Certain components of the host cell, such as ribosomes and enzymes, still function, but they are utilized for the replication of viral components. During this period, a portion of the viral DNA, upon reaching the host's cytoplasmic region, is immediately used for the formation of "early" viral messenger RNA (vmRNA). This nucleic acid contains the necessary information for the formation of viral proteins. The information of the newly formed vmRNA is translated by the host's ribosomes, resulting in the synthesis of specific viral enzymes, including those necessary for phage nucleic acid replication, protein capsid formation, and virus particle assembly.

MATURATION

During viral maturation, viral DNA is used to form other essential viral parts, including the capsomeres of the capsid. As these capsid subunits accumulate, viral DNA molecules combine with a specific protein and become tightly packed into polyhedral units. The capsomeres crystallize on the surfaces of the polyhedrons to form mature bacteriophage heads. The activity continues with the remaining viral subunits being synthesized and assembled into complete *virions.* The stepwise assembly process is controlled by viral genes described as *morphopoietic* (from the Greek *morphe,* meaning "form," and *poiein,* meaning "to make").

LIBERATION OF VIRUS PARTICLES

As in other phases of the lytic cycle, the liberation process differs among bacteriophages. Basically, however, it includes the following sequence of events. As the maturation period of virus particles comes to an end, another viral protein product appears and steadily increases in concentration. This substance, known as *bacteriophage lysozyme,* disrupts the chemical bonds holding together the components of the cell wall's rigid layer. The wall becomes progressively thinner. Eventually it ruptures from an osmotic pressure imbalance that causes water to flow into the cell from the surroundings *(plasmoptysis).* The virus particles and the remaining contents of the cell are thus released into the immediate environment. The lytic cycle is complete, and infectious viruses are once more available to begin the cycle in new host cells.

Growth Curve

From events of the lytic bacteriophage cycle, it is possible to obtain experimentally a growth curve for bacterial viruses. This type of experiment is used to determine the time lapse between injection and release of mature virus particles and to estimate the approximate number of these particles produced per cell, called the *burst size.* This *one-step growth curve* experiment shows how viral development differs from the growth cycle in cellular systems.

In performing a growth curve, a virus suspension is added to bacteria at a low multiplicity of approximately 0.1 to 0.01 (one viral particle for each 10 to 100 bacterial cells). The low multiplicity of infection is needed to reduce the chance of several viruses infecting a single bacterium and producing abnormal results. After a short period of incubation to allow adsorption, nonadsorbed viruses are removed. Thus, the opportunity for subsequent infection of additional bacteria and the possibility of more than one virus attacking a bacterium are substantially reduced. Next, the bacterial suspension is diluted to a desired concentration in fresh media. Samples are removed periodically, and the number of free virus particles and infected bacteria is determined by plaque count. The virus-containing samples are mixed with a suspension of sensitive bacteria. The number of plaques produced by each sample is plotted and results in the formation of a curve such as the one shown in Figure 10–17.

The plaque count usually remains constant for a period of time. This phase of the growth curve is referred to as the *latent stage,* or *latent period.* The designation is misleading in the sense that significant viral activity is going on even though the plaque count does not indicate it. The lysis of infected bacteria and the subsequent release of viral particles are indicated by an abrupt increase in plaque number. This stage, called the *burst period,* continues until all infected cells have lysed. Again, the plaque count reaches a more or less constant level, even if uninfected bacteria are present. In the diluted bacterial suspension, used cells are so widely separated that newly liberated phages cannot spread to uninfected

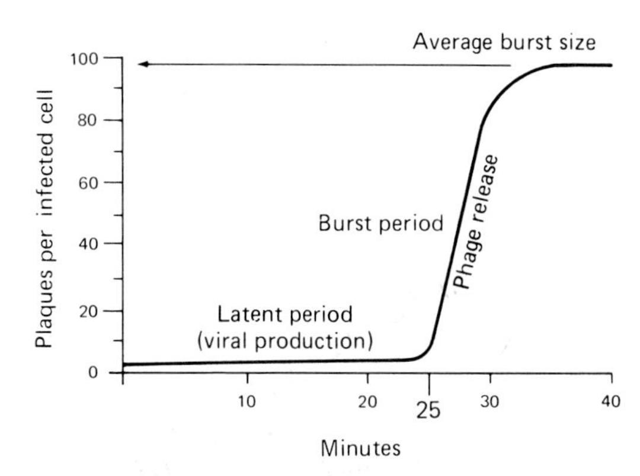

Figure 10–17 A typical one-step viral growth curve. This representation shows the length of the latent period, burst period, and average burst size.

bacteria. By knowing the number of infected bacteria at time zero, the number of plaques obtained at various intervals can be calculated in terms of PFU per infected cell by dividing the number of plaques present after the burst period by the number present before. In Figure 10–17, the burst size was found to be 100 PFU. This is comparable to the burst size that would be found with virus T_2 shown in Figure 10–16e.

Cyanophages

In 1963, R. S. Safferman and M. E. Morris discovered a virus that attacks and lyses several species of blue-green bacteria. These viruses are the **cyanophages** now known to be present in nearly all bodies of fresh water.

Several of the cyanophages isolated thus far have been given names that correspond to their hosts. Phage groups have been designated by the initials of the generic names of the hosts, to which Arabic numerals are added to signify specific subgroups. For example, LPP phages attack three different filamentous cyanobacterial genera, *Lyngbya* (ling-BY-a), *Phormidium* (for-meh-DEE-um), and *Plectonema* (plek-TOH-nee-ma; Figure 10–18). Because the use of host specificity as a criterion for classification poses some problems, morphological properties and antigenic specificities are preferred for classification of cyanophages.

Cyanophages are very similar to other bacterial viruses both in structure (Figures 10–19a and 19b) and in infection cycle. LPP phages, one of the most extensively studied cyanophages, have virions of the head-tail type. All the nucleic acids of these viruses isolated to date are found to be linear, double-stranded DNAs.

The hosts of cyanophages, the cyanobacteria, are widespread in the aquatic environment and often

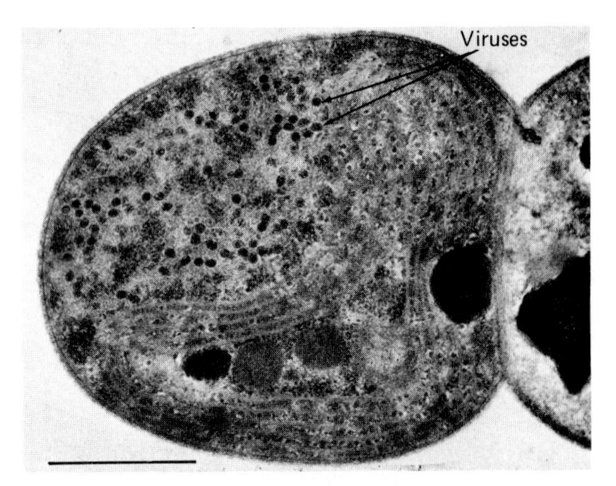

Figure 10–18 The cyanophages. The blue-green bacterium *Plectonema* (plek-TOH-nee-ma) infected with LPP-1G cyanophage. Note the photosynthetic lamellae (layers) and the numerous viral heads.

occur in great numbers in the form of blue-green bacterial or algal blooms (Color Photograph 7a), which sometimes result in fish poisonings. The discovery of cyanophages has led several investigators to consider these viruses as possible agents for the biological control of algal blooms. **[See Chapter 13 for cyanobacteria.]**

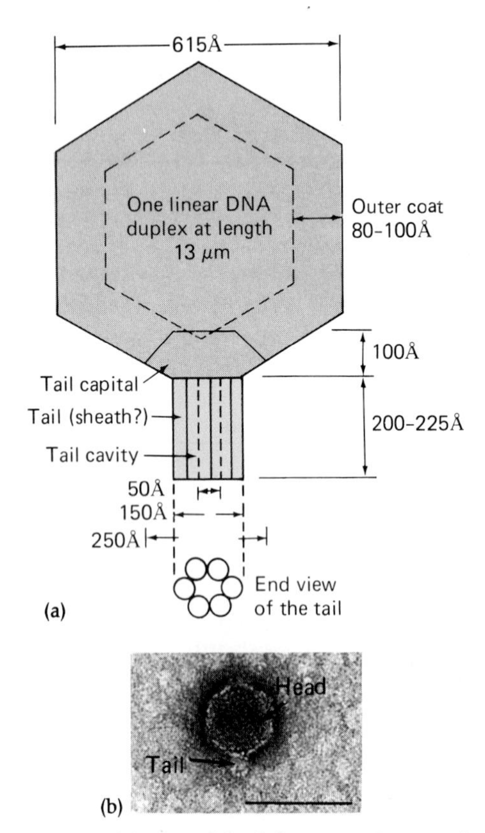

Figure 10–19 (a) A model of the LPP-1G cyanophage. (b) A mature virus particle. *[Courtesy of E. Paden, and M. Shilo,* Bact. Revs. *37:343–370, 1973.]*

Cyanophage Cultivation and Enumeration

The conditions, media, and procedures used for the preparation of suitable hosts (and subsequently for cyanophages) differ significantly from those for bacteriophages. For example, most bacterial hosts and phages are stable from pH 5 to 8, whereas many blue-green bacteria and cyanophages are stable from pH 7 to 11. As in bacteriophage culture, evidence of infection is indicated by plaque formation or using electron microscopy. Many of the studies involving cyanophages are complicated by the fact that their hosts are filamentous.

Animal Viruses

Disease

Many human diseases known for centuries are now recognized as caused by viruses. These include smallpox, poliomyelitis, and yellow fever. Several additional major illnesses of humans and other animals are also viral in nature (Table 10–3). Among this group are diseases such as AIDS, canine distemper, chickenpox, foot-and-mouth disease, influenza, measles, mumps, rabies, and various types of encephalitis. In 1911, P. Rous discovered the ability of viruses to produce malignant tumors (Rous sarcoma) in chickens. This finding was the first of many to recognize the viral nature of such tumors in both animals (Figure 10–20) and plants.

The effects of animal viral agents can range from the production of mild skin rashes and upper respiratory symptoms to tissue destruction and death. Several of the later chapters discuss the various viral infections. The organization and nucleic acid composition of major animal virus groups are indi-

Figure 10–20 The effects of tumor-causing viruses (Rous sarcoma). A greatly enlarged diseased liver (left) shows many tumorous growths. The liver on the right is from an uninffected chicken of the same age. *[USDA photograph by Madeleine Osborne.]*

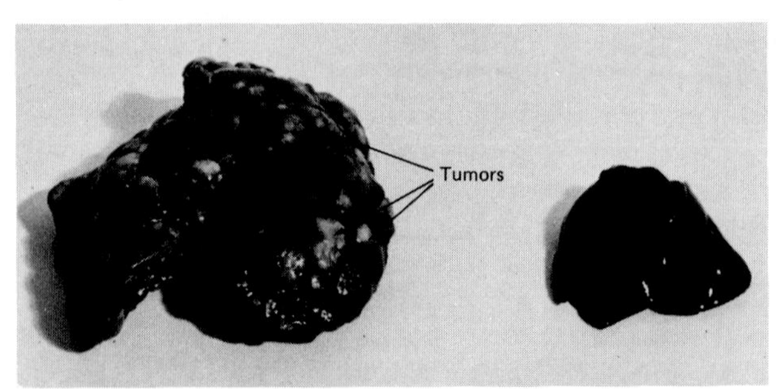

TABLE 10–3 Properties of Major Animal Virus Groups

Virus Group[a]	*Type of Nucleic Acid*	*Single- (ss) or Double-Stranded (ds)*	*Enveloped (E) or Naked (N)*	*General Properties*
Adenovirus	DNA	ds, linear	N	Found in several animal species, including humans; associated with respiratory infections and with tumors in laboratory animals
Baculovirus	DNA	ds, circular	E	Group contains insect viruses that are located in a protein-crystalline substance (matrix)
Hepadnavirus	DNA	ds, circular	E	Causative agent of hepatitis B infection
Herpesvirus	DNA	ds, linear	E	Important causative agents of human disease such as chickenpox, infectious mononucleosis, and infections of the skin and mucous membranes
Iridovirus	DNA	ds, segmented	N	Group includes viruses having the widest range of insect hosts
Papovavirus	DNA	ds, circular	N	Causes warts; used in the study of tumor development
Parvovirus	DNA	ss, linear	N	Group includes satellite viruses that are incapable of replication except in the presence of a helper virus; also insect viruses
Poxvirus	DNA	ds, linear	E	Found in several animal species, including humans; examples of infections include monkey pox, smallpox, and molluscum contagiosum
Arenavirus	RNA	ss, circular	E	Particles contain cellular ribosomes; viruses cause natural inapparent infections of rodents
Bunyavirus	RNA	ss, circular	E	Viruses (arboviruses) are spread by a variety of arthropods
Calcivirus	RNA	ss, plus strand	N	Group contains viruses causing gastrointestinal infections
Coronavirus	RNA	ss, linear, plus strand	N	Includes several agents of the common cold
Flavivirus	RNA	ss, circular	E	Includes the viruses causing yellow fever and several nervous system diseases
Orthomyxovirus	RNA	ss, segmented, plus strand	E	Includes the influenza viruses
Paramyxovirus	RNA	ss, minus strand	E	Many produce human childhood diseases such as measles and mumps and localized respiratory infections
Picornavirus	RNA	ss, plus strand	N	Includes several agents of diseases such as poliomyelitis, rashes, meningitis, mild upper respiratory infections, and gastrointestinal infections
Reovirus	RNA	ds, segmented	N	Most are spread by fecal-contaminated water and food; several have yet to be associated with specific diseases
Retrovirus	RNA	ss, plus strand	N	Group includes tumor- and cancer-causing agents and the AIDS virus (human immunodeficiency virus, HIV)
Rhabdovirus	RNA	ss, minus strand	E	Group includes large bullet-shaped viruses such as the agents for rabies and vesicular stomatitis; as well as insect viruses
Togavirus	RNA	ss, plus strand	E	Arthropod-spread diseases, including several nervous system diseases, as well as insect viruses

[a] Several new viruses, as well as some unclassified viruses, may require the formation of new groups.

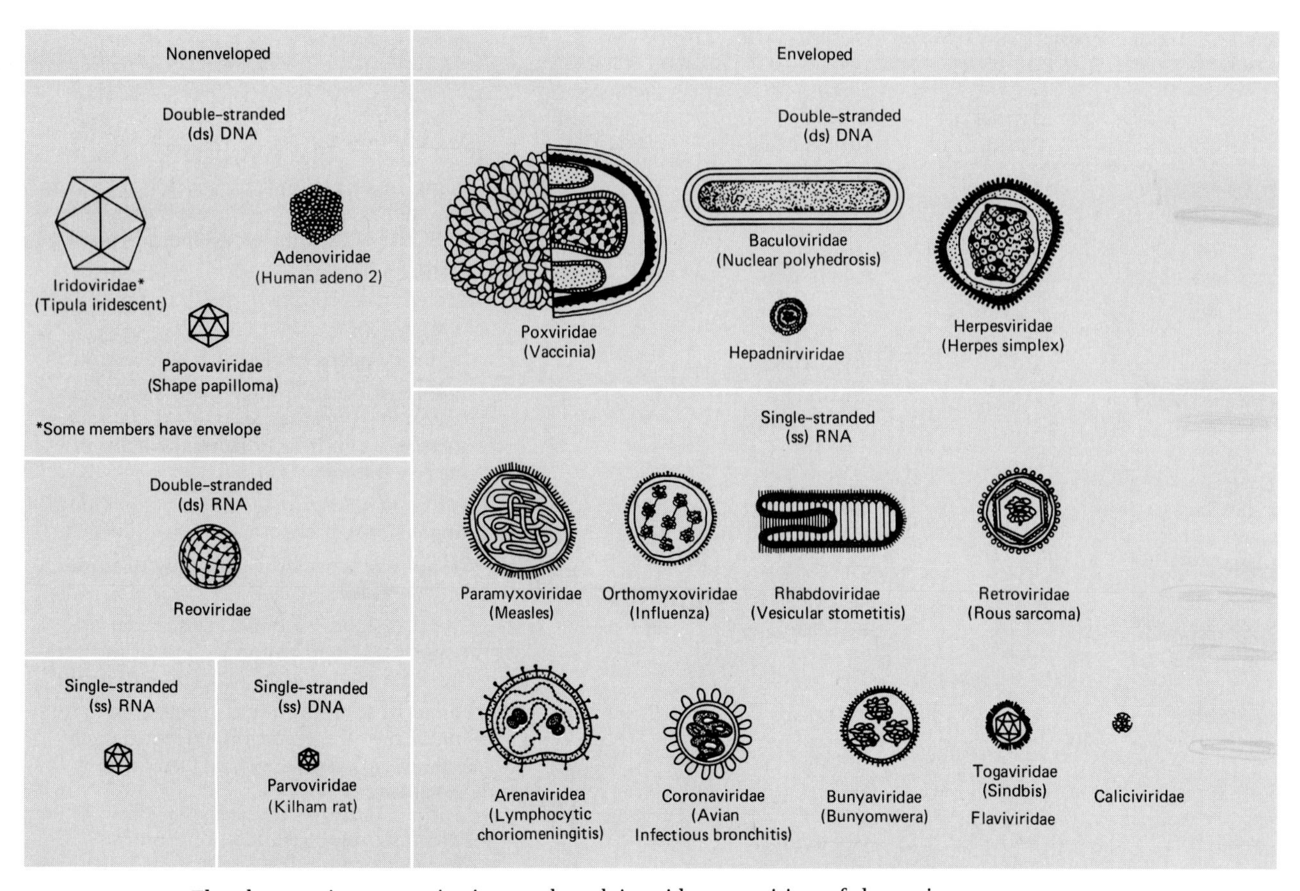

Figure 10–21 The shapes, sizes, organization, and nucleic acid composition of the major animal virus groups. Locations of nucleic acid cores are also shown in some cases. *[Modified from R. E. F. Mathews,* Intervirology ***12**:156, 1979.]*

cated in Figure 10–21. **[See cancer viruses in Chapter 30.]**

Insect Viruses

It has been found that many insects are attacked by viruses (Figure 10–22). A wide variety of such diseases are known. Some are of economic importance, involving insects such as honeybees and silkworms. Still other viruses attack some of the most serious insect pests of agricultural crops and forests. Many of these destructive insects become extensively diseased when they are infected by certain viruses. This finding, together with the rapid development of insects resistant to chemical insecticides and the concern about the harmful effects of chemical residues on the environment, has been largely responsible for the consideration of insect viruses as biological insecticides.

In many insects, viral infections cause the formation of specific intracellular protein products, called *inclusions.* These inclusions may be either granular or polyhedral (Figure 10–23a). Diseases associated with them are referred to as *granuloses* or *polyhedroses,* respectively. A polyhedral inclusion may be formed in either the cytoplasmic or nuclear region of the cells of infected animals. Granular inclusions, which may be either capsules or oval grains of protein, are generally found in the cytoplasm.

The baculovirus, picornavirus, and reovirus groups contain virus particles that are uniquely embedded into a crystalline protein material or matrix (Figure 10–23b). The term *occlusion* refers to enclosing protein material. The protein is formed and deposited around virions during virus replication and is thought to contribute to the stability of the viruses in the physical environment outside the insect host.

Animal Virus Replication

Studies on the replication of animal, bacterial, and plant viruses clearly show that the essential sequence of events in replication processes is common

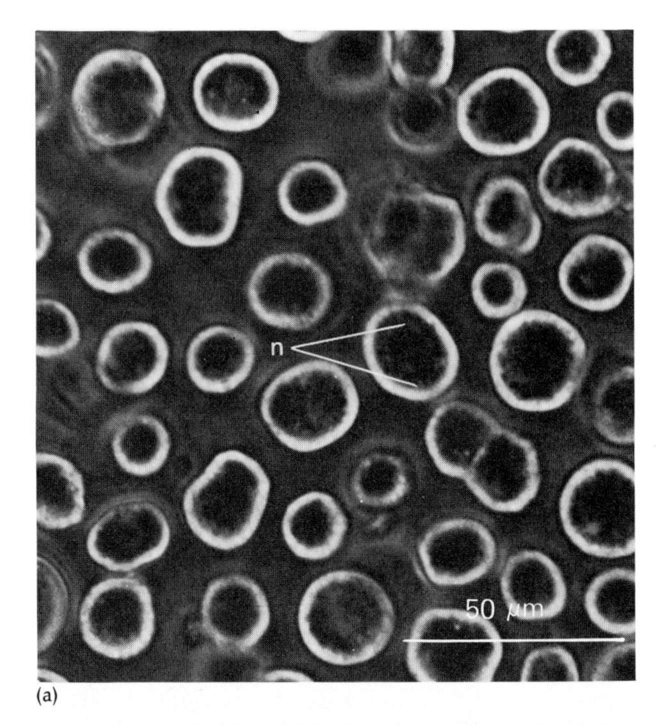

(a)

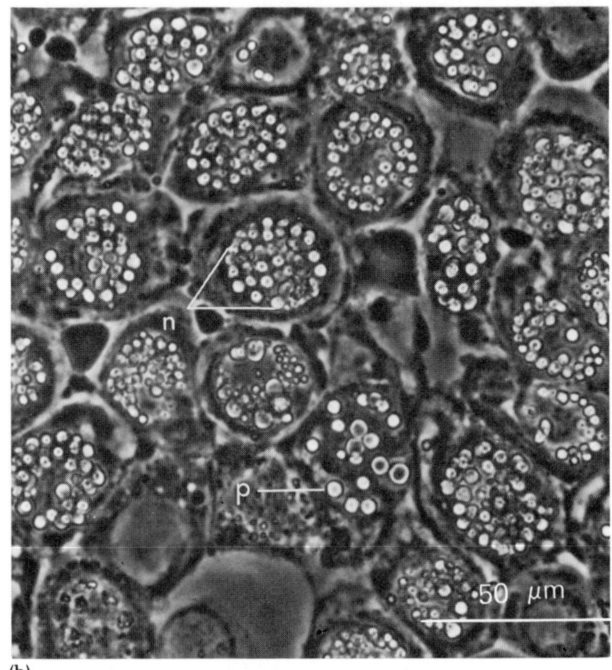

(b)

Figure 10–22 Nuclear polyhedrosis insect virus infection. (a) Normal insect tissue cells shown with intact and noninfected nuclei (n). (b) The appearance of infected cell nuclei packed with polyhedral inclusion bodies (p). *[From S. S. Sohi, J. Percy, and J. C. Cunningham,* Intervirology ***21****:50–60, 1984.]*

to all viruses. These events include (1) entry into a susceptible host cell, (2) reproduction within specific cellular locations to produce new virions (progeny), and (3) release of mature virus particles from the host cell.

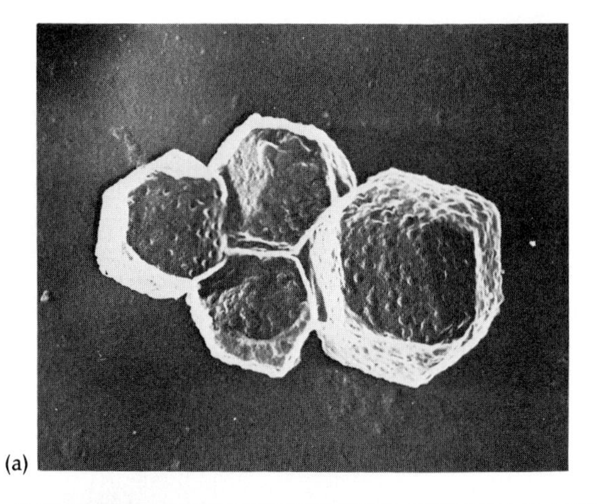

(a)

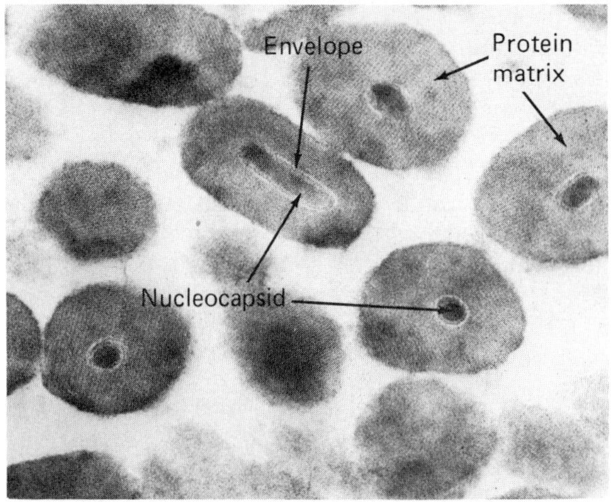

(b)

Figure 10–23 Insect viruses. (a) An electron micrograph of a cytoplasmic polyhedrosis virus. Spherical viral particles can be seen in the polyhedra shown. *[Courtesy of Dr. R. Markham, Agricultural Research Council, Cambridge, England.]* (b) The appearance of granular viruses showing the viral envelope, nucleocapsid, and occlusion or protein matrix. *[Courtesy of K. A. Tweeten, L. A. Bulla, Jr., and R. A. Consigli,* Microbiol. Revs. ***45****:379–408, 1981.]*

ATTACHMENT AND ENTRY

While most viruses lack specialized attachment structures, their surfaces contain a number of chemical attachment sites. The first encounter between a virus particle and a cell may involve only one viral attachment protein molecule and one receptor molecule on the host cell.

Naked Virions. In the case of naked virions, an individual virus may easily fall off since it may not be tightly bound to the cell's surface. Irreversible binding probably occurs when several sites on one virus particle bind to several host cell receptors (Figure 10–24). The nucleocapsid of such nonenveloped viruses does not enter intact into the host cell. The host cell membrane causes a rearrangement of capsid proteins, which frees the nucleic acid and allows it to pass through the membrane and into the cytoplasm.

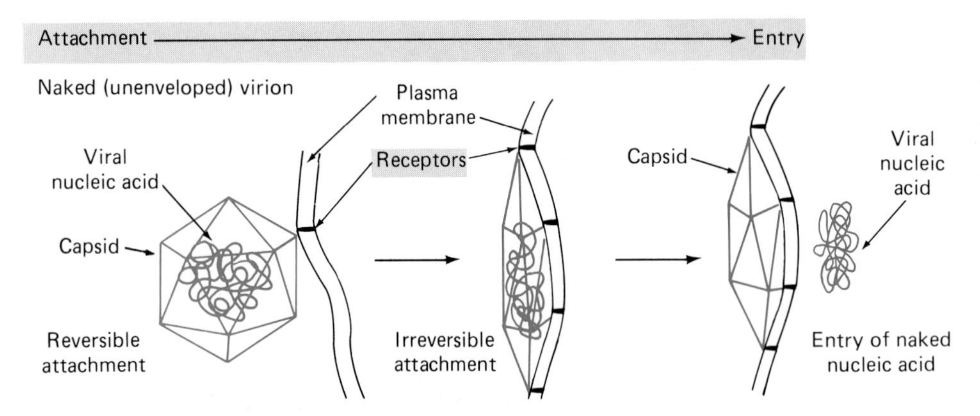

Figure 10–24 Early events in the infection processes of naked animal viruses. Naked (unenveloped) virus particles permanently attach to a host's cell membrane only after membrane receptors migrate to the site to bind the virus. Once this occurs, the virus particle is reoriented to allow its nucleic acid to pass through the membrane.

The nucleic acid of most DNA viruses must also pass through the host cell's nuclear membrane to the site of DNA replication (Figure 10–24). The operating mechanisms are not known.

Enveloped Viruses. The projections or spikes of enveloped virions serve as attachment devices. Most enveloped viruses enter the host cell cytoplasm by fusion of the envelope with the plasma membrane (Figure 10–25). As a result of this fusion, the nucleocapsid is deposited more or less intact into the cytoplasm of the host cell (Figure 10–25a).

With structurally more complex virus particles, virions may be actively ingested by the host cell by the same process used to take in other particles, namely, phagocytosis. Virions are often observed inside membranous vesicles that form by an infolding of the plasma membrane and engulf the particle. Within such vesicles, cellular digestive enzymes released from lysosomes may digest outer layers of complex nucleocapsids, thereby releasing the viral nucleic acid. This process of removing the capsid and freeing the nucleic acid is called *uncoating* (Figure 10–25b).

Figure 10–25 Early events in the infection process of various types of enveloped animal viruses. (a) With enveloped viruses, attachment to several receptors occurs immediately, resulting in a fusion of the virions with the membrane and the release of the nucleocapsid into the cell's interior. (b) In the case of virions with a complex structure, the entire particle is taken into the cell by a phagocytic (engulfing) process. Once inside the host cell the enzymes of lysomes digest the portions of the nucleocapsid, thereby freeing the viral nucleic acid (uncoating process).

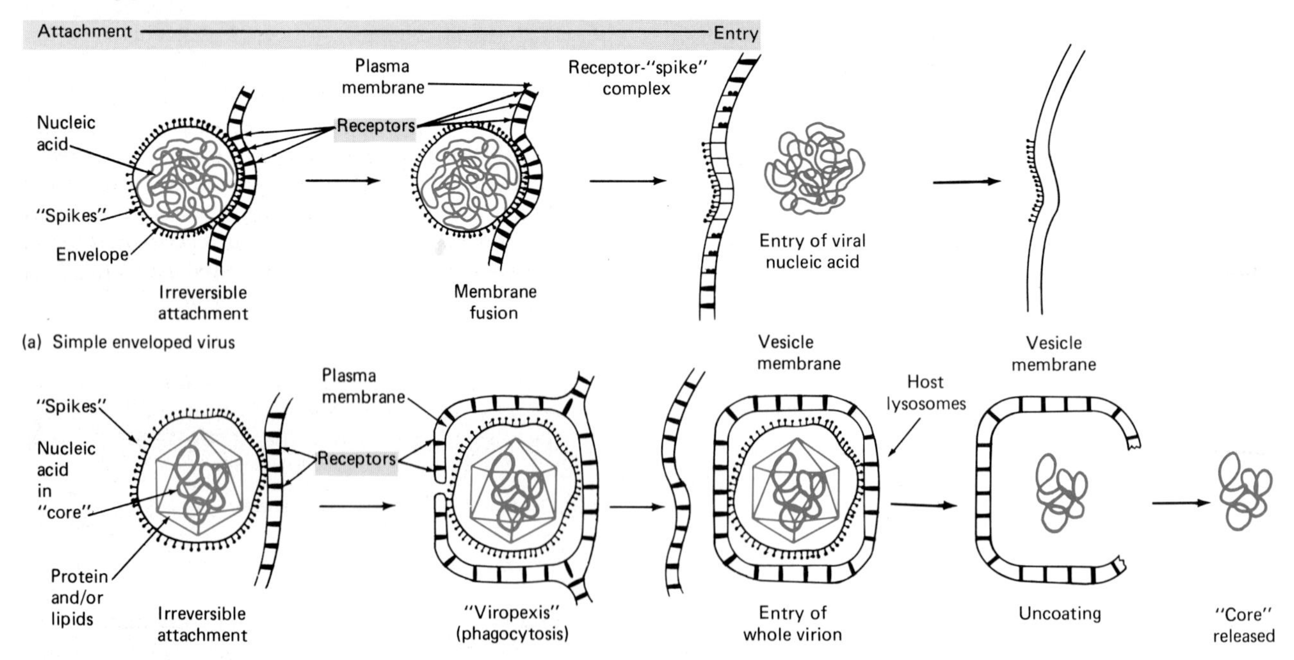

REPLICATION, MATURATION, AND VIRION RELEASE

The replication or biosynthesis of new viral components occurs some time after the viral nucleic acid is released from an invading viral particle. Although different viral groups are replicated in different locations (Figure 10–26), and by different strategies, the basic process is like an assembly line. Various parts of the virus are made separately and then efficiently assembled into complete virions in the maturation stage.

The nucleic acid of viruses codes (contains the information) only for the synthesis of various viral proteins. These codes include viral enzymes that regulate the replication of nucleic acid and are synthesized early in the infectious process, other enzymes that may be incorporated into virions, and viral structural proteins. All replication cycles have complex control mechanisms for both cellular and viral activities leading to the formation of viral proteins. Various cellular organelles are used for the replication process (Figure 10–26).

DNA Viruses. Replication of viral DNA can take place by a variety of mechanisms. For example, poxviruses, which contain their own *transcriptase* required for the formation of viral DNA, virtually establish a nucleic acid factory in the cytoplasm of the host cell. The formation of additional viral parts and virion assembly also take place in the cytoplasm. Other DNA viruses, including adenoviruses (Figure 10–27), herpesviruses, and papovaviruses, replicate in the nucleus and use the host cell's enzymes (DNA polymerases) for viral nucleic acid replication.

Viral structural proteins formed in the cytoplasm are transported back to the nucleus for the construction of viral nucleocapsids. Nucleocapsids of viruses such as those of the herpesvirus group acquire envelopes by passing through the nuclear membrane. Such envelopes contain virus-coded proteins. [See protein synthesis in Chapter 6.]

Enveloped viruses lcavc a host by pushing through the cell membrane, using a budding process (Figure 10–28). This means of exiting does not necessarily lead to cell death. Other viruses, such as those consisting of naked virions, may be released through host cell disruption. Figure 10–27 shows the sequence of events involved with the replication of a nonenveloped (naked) DNA virus.

RNA Viruses. RNA viruses are a remarkable biological phenomenon in that their genetic information is coded in RNA. The flow of information in all biological systems except RNA viruses follows the pattern starting from DNA to RNA to protein. Also interesting is the fact that RNA viruses, instead of relying on a single strategy, have developed several mechanisms for their replication.

Viruses such as coronaviruses, picornaviruses, and togaviruses (Table 10–2) rely on a strategy

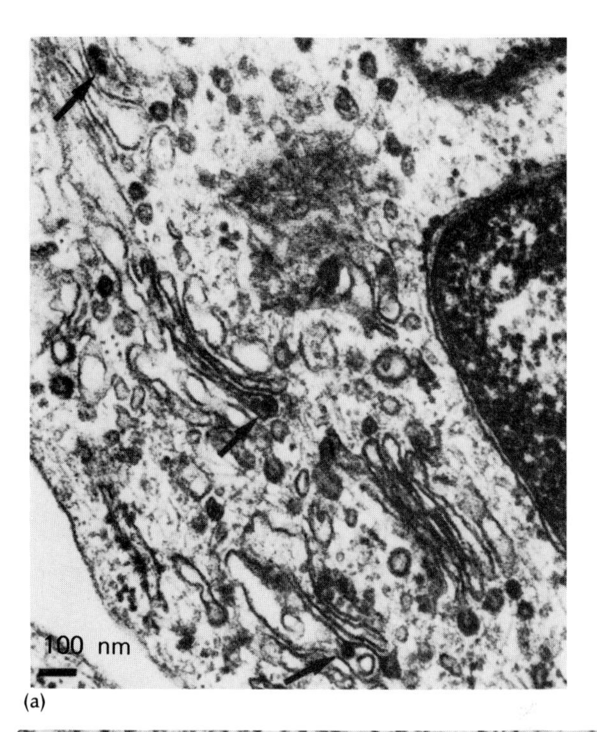

(a)

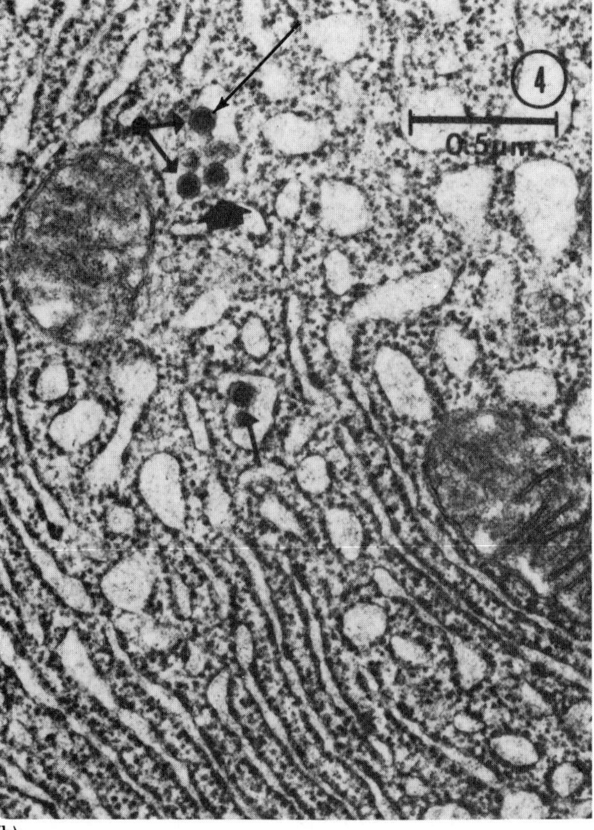

(b)

Figure 10–26 Cellular locations of viral replication (arrows). (a) The presence of virions within the Golgi system of the infected cell. *[From M. Weiss, and M. C. Horzinek,* J. Gen. Virol. ***67**:1305–1314, 1986.]* (b) Immature virus particles within portions of the rough endoplasmic reticulum. *[From P. C. Freeback, R. Macleod, and T. O'Brien,* J. Gen. Virol. ***65**:65–71, 1984.]*

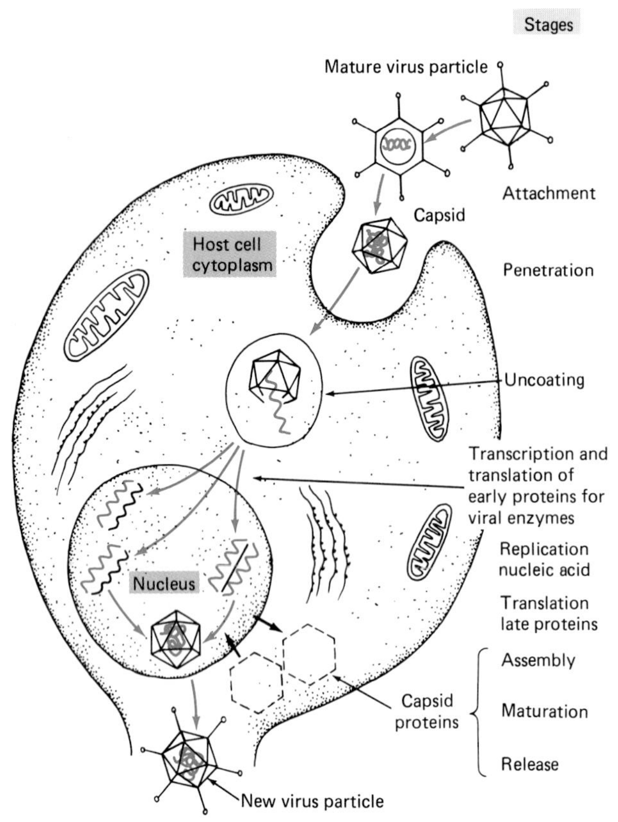

Figure 10–27 Replication of adenovirus, a non-enveloped DNA virus. Specific events such as the early transcription and translation of viral DNA required for the replication of viral nucleic acid and the late transcription and translation event associated with new viral protein synthesis are emphasized. Details of the attachment process are shown in Figures 10–24 and 10–25.

wherein single strands of RNA function as giant mRNA molecules, referred to as *positive* or *sense strands* (Figure 10–6). These large molecules direct the formation of long multiprotein units that are divided into several polypeptides by host cell enzymes (Figure 10–29a).

A second strategy for RNA replication, found with viruses such as orthomyxoviruses, paramyxoviruses, and rhabdoviruses, involves the enzyme *RNA-dependent RNA polymerase.* Here a noninfectious single strand of RNA, referred to as the *negative strand* or *no sense* (Figure 10–6), is used for the synthesis of an effective positive strand. The two strands form a double-stranded replicative RNA, with the positive strand serving as a template for the synthesis of new viral structural proteins or for new viral RNA (Figure 10–29b).

Enveloped RNA viruses acquire their envelopes by budding through altered cell membranes of cell organelles such as the endoplasmic reticulum, mitochondria, or from the cytoplasmic membrane by budding (Figure 10–28). With most enveloped polyhedral viruses, the shape of the nucleocapsid determines the size of the envelope. This type of control leads to the formation of uniform virions. In the case of helical nucleocapsids, such uniformity may not occur. Virions sometimes containing more than one genome or exhibiting a variety of sizes and shapes develop. The mechanisms for viral particle release are similar to those noted for DNA viruses. Figure 10–28 shows the sequence of events in the replication of influenza virus.

A third and quite unusual strategy of viral RNA replication is found with the retroviruses (animal tumor viruses, including the AIDS virus). These viruses contain an RNA-dependent DNA polymerase *(reverse transcriptase)* that reverses the normal direction of information flow (DNA to RNA to protein). This enzyme permits the formation of a complementary strand of DNA, which in turn makes another complementary DNA strand, thus forming a double-stranded DNA. The newly synthesized DNA is integrated into the DNA of a host cell chromosome and assumes a provirus, or integrated, state. In this position, it functions as a template for the transcription of mRNA for viral protein synthesis or single-

Figure 10–28 A summary of the stages in the development of the enveloped RNA influenza virus. The steps shown extend from adsorption through the budding of new virus particles. A complete virion is also shown in Figure 10–5.

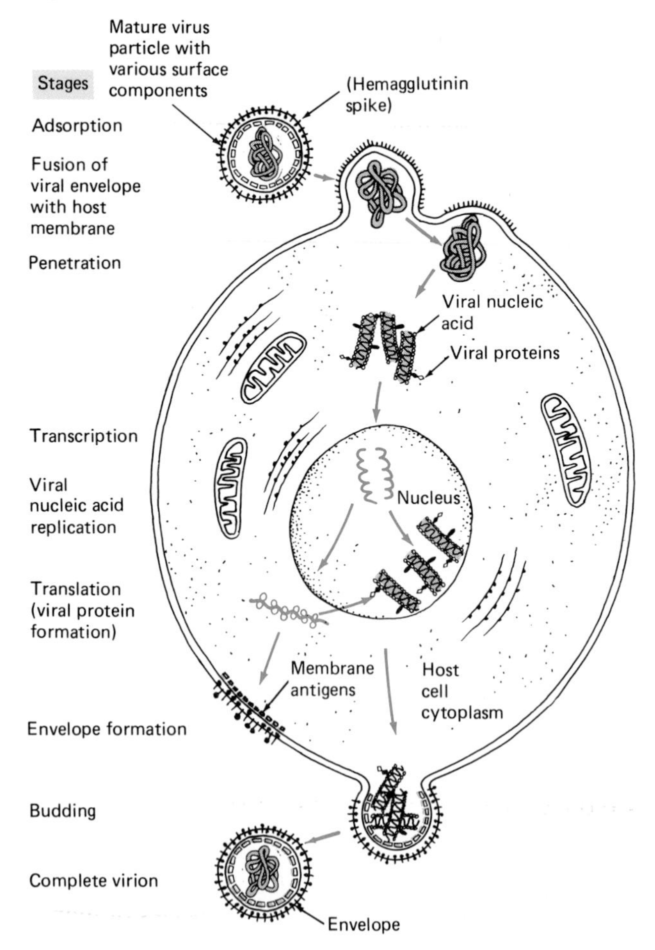

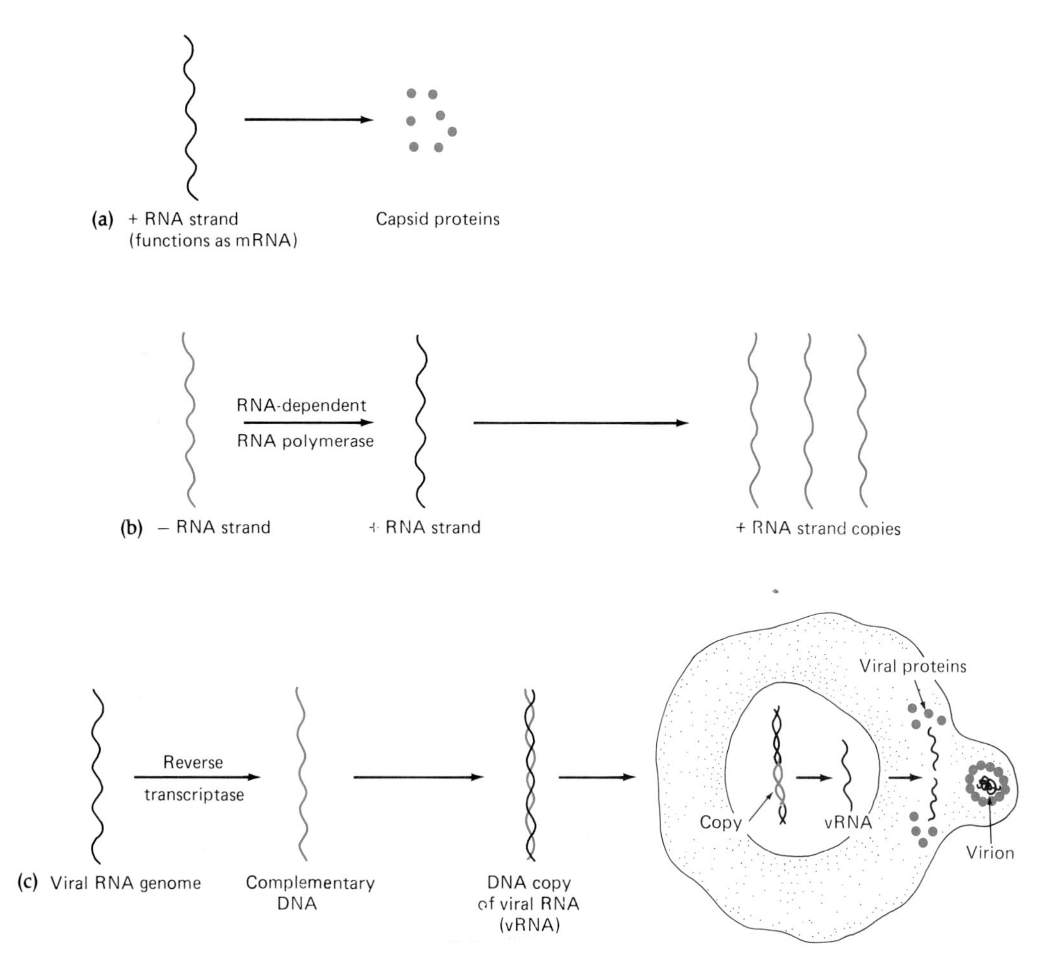

Figure 10–29 Mechanisms for RNA replication. (a) A plus (+) or sense RNA strand serves as mRNA for protein synthesis. (b) A minus (−) or no-sense strand of RNA is used for the synthesis of a plus strand, which in turn makes new viral RNA copies. (c) Reverse transcriptase is used to form a complementary DNA molecule. The new DNA molecule may be integrated into the host cell's chromosome. Upon activation, the viral molecule controls the replication of new virions.

stranded viral RNA, both of which are used in new virion replication. Figure 10–29 compares the different RNA virus strategies.

Defective Interfering Particles

The production of defective interfering virus particles was first observed with influenza viruses. Although such particles have capsids, they contain only a portion of the viral nucleic acid. Defective interfering virus particles are associated with various virus infections, can interfere with normal viral replication, and may require coinfection with a specific virus for their replication.

Virus Cultivation in Animals

The early attempts to grow animal viruses involved susceptible laboratory animals. Included among these animals were rhesus and other species of monkeys, hamsters, mice, and rats. Although these efforts were quite successful, other means had to be found to meet the needs of scientists working with viruses and yet keep laboratory expenses within working limits. We will now describe some current procedures for cultivating animal viruses.

EMBRYONATED EGGS

The introduction of *embryonated eggs* as a suitable medium for the growth of viruses provided an invaluable means of obtaining large quantities of viruses for the preparation of vaccines and diagnostic reagents, as well as for viral studies.

There are many advantages in using embryonated eggs: their availability in virtually unlimited quantities; their relative ease of handling; the presence of a naturally constant environment within the confines of the egg's components (the embryo as it ordinarily comes from the hen is sterile); the general inability of the embryo to produce antibodies against the viruses used as **inocula;** and the availability of eggs with a relatively uniform genetic constitution from flocks that have been inbred for several generations.

The embryonated eggs used for virus cultivation are fertilized chicken eggs that have undergone normal embryonic development in the hen.

Several portions of the embryonated egg are used for viral cultivation, including the allantoic and amniotic cavities, the yolk sac, the chorioallantoic membrane (Figure 10–30), and the embryo itself. The particular region used depends on the virus to be cultured, as certain viral agents are capable of proliferating only in some parts of the embryo. Mumps and Newcastle disease viruses, for example, replicate particularly well in the allantoic cavity.

SIGNS OF VIRUS INFECTION. In addition to being used to produce large numbers of viruses, embryo inoculation techniques are used for studying the morphological features of viruses, isolating viruses from specimens, determining the effectiveness of drugs on viruses (chemotherapy), investigating the mechanism of viral infection, and preventing viral infection. **[See viral chemotherapy in Chapter 12.]**

Virus infection in chick embryos may appear in several ways. Certain viral agents produce local lesions called *pocks* that vary in size, shape, and opacity (Color Photograph 53). However, it is difficult to identify a particular agent on the basis of this effect alone. Additional signs of infection are the death of the embryo and the demonstration of a blood clumping reaction, or **hemagglutination,** associated with the allantoic or amniotic fluid. Virus infection can sometimes be shown by detection of the virus or related agents by light or electron microscopy. However, it is also possible that there will be *no* obvious signs of viral infection.

Tissue Culture Cultivation of Viruses

Although the technique of growing tissues outside of an animal or plant (tissue culture) is almost as old as the study of viruses, early virologists could not use this technique because of problems of contamination by bacteria and fungi. It was only after three advances in virology that tissue culture gained routine acceptance for the cultivation of viruses and other agents requiring a living cell. The three advances were the introduction of antibiotics, which greatly reduced the contamination problem; development of an excellent, defined growth medium for cells; and introduction of the enzyme trypsin to free cells from fragments of tissue so that they could be grown in single-cell layers.

The work that ushered in the era of tissue culture was the *in vitro* cultivation of poliomyelitis virus in tissues other than nerve cells. This was achieved in 1949 by J. F. Enders, T. H. Weller, and F. C. Robbins using human embryonic tissue. Before then, polio virus had been grown only in lower animals and nervous tissue. The destructive action, or *cytopathic effect,* of the virus (Figure 10–31) was clearly demonstrated. The three scientists received a Nobel Prize for their work in 1954.

PREPARATION OF ANIMAL TISSUE CULTURES

Although the tissues of plants and cold-blooded animals can be cultivated, we shall concentrate here on the cultivation of mammalian tissue. Many types of cells can be used for tissue cultures.

Two types of cell cultures are in common use: *primary* and *continuous.* Primary cultures consist of cells obtained from normal tissue that cannot divide or survive indefinitely. Continuous cultures are derived from malignant (cancerous) tissues, which can undergo an unlimited number of cellular divisions.

Figure 10–30 The anatomy of the chick embryo and the locations of possible inoculation sites. (a) Yolk sac inoculation. (b) Chorioallantoic membrane inoculation.

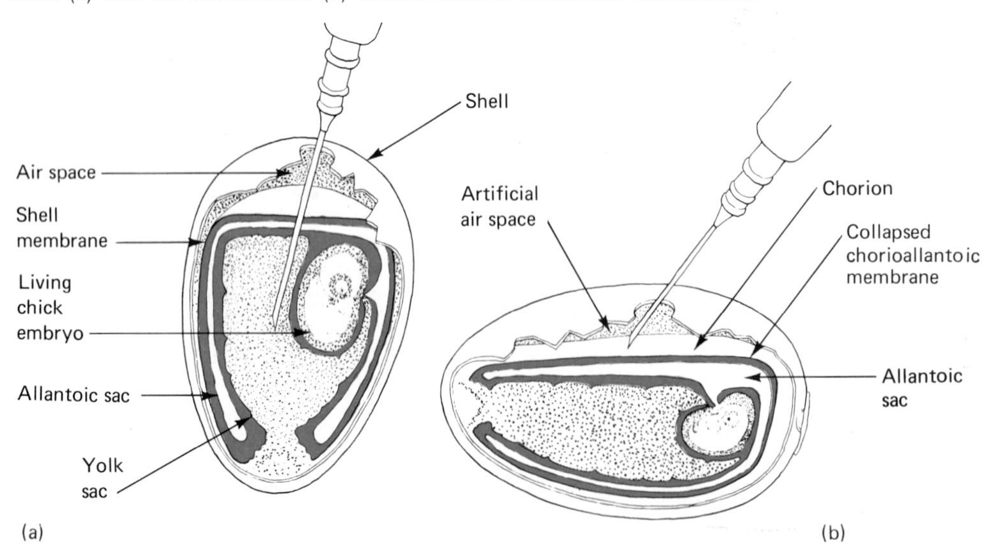

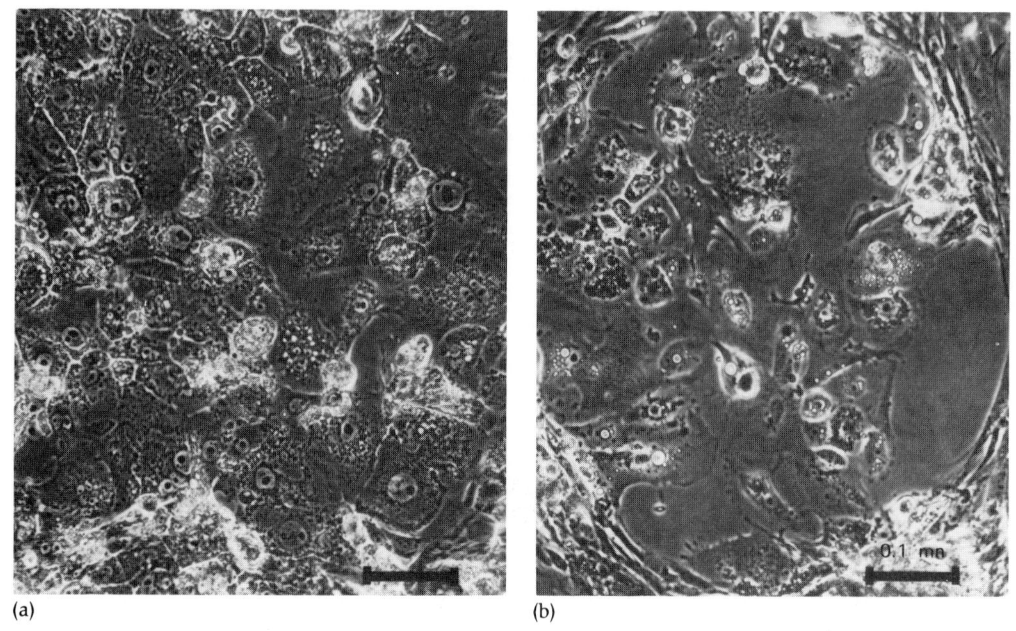

(a) (b)

Figure 10–31 Demonstration of *in vitro* virus infection. (a) Uninfected tissue culture cells. Note the fairly even distribution of cells. (b) The cytopathic effect (CPE) of viruses is shown here. Cells have a granular appearance and show loss of nuclei and separation from other cells. *[W. Kurz, H. Gelderblom, R. M. Flugel, and G. Durai,* Intervirology ***25****:88–96, 1986.]*

In the preparation of animal tissue cultures, cells are removed from animals and maintained under suitable conditions *in vitro,* so that they can serve as hosts for viruses or as sources of nucleic acids. The growth medium must provide ample nutrients to keep the metabolism of the animal tissues functioning well, and it must be kept free from contamination by unwanted microorganisms such as mycoplasma and latent viruses. An unexpected problem that confronts tissue culture workers is the presence of viruses other than those desired. The animal that is the source of tissue may be harboring a latent viral infection. When these tissues are grown *in vitro,* the viral agent may use its new environment to unleash some destructive effect, thus producing an extremely confusing picture for the virologist.

DETECTION OF VIRAL MULTIPLICATION IN TISSUE CULTURE SYSTEMS

There are several tissue culture indications of viral infection, including cytopathic effects (Figure 10–31b) and alterations in cellular metabolic reactions.

Cytopathic Changes. The gross cytopathic changes in virus-infected cultures can readily be seen by standard light microscopy. Although fixation and staining of infected cells are not necessary to detect virus multiplication, they may be used for making permanent records of its stages (Color Photographs 54 and 55).

Characteristic cytopathic effects (CPEs) brought about by several viruses include cytoplasmic granulation (Figure 10–31) or vacuolation, and condensation of nucleic acid and protein material in the nuclear membrane of cells (Figure 10–32). Viral infection can also cause an extreme form of binding of surface membranes called *cell fusion.* The resulting fused cells, which contain several nuclei within one common cytoplasm (Figure 10–33), are called *syncytia.* Cell fusion is generally limited to certain enveloped viruses, such as measles virus and some herpesviruses. The particular cytopathic effect produced can sometimes aid in diagnostic virology, the identification of a particular viral group. Certain viruses, however, characteristically do not cause any observable structural change in cells.

Efforts to detect viral infection by observing cytopathic effects are complicated by the fact that newly isolated viruses occasionally must be transferred several times before the characteristic cytopathic effects are produced. Furthermore, changes in the fine structure of cells may result from the effects of a temporary lack of nutrients, changes in pH, or other related factors. These effects can usually be distinguished from cytopathic effects, however, because they develop within 24 hours and are not neutralized by antiviral substances such as antibodies.

Plaquing Techniques. In the plaquing method, a single or monolayer tissue culture system is inoculated with a viral suspension. After the virus particles are adsorbed by the tissue cells, the system is overlaid with nutrient agar and incubated at the desired temperature. Viruses that infect the cells multi-

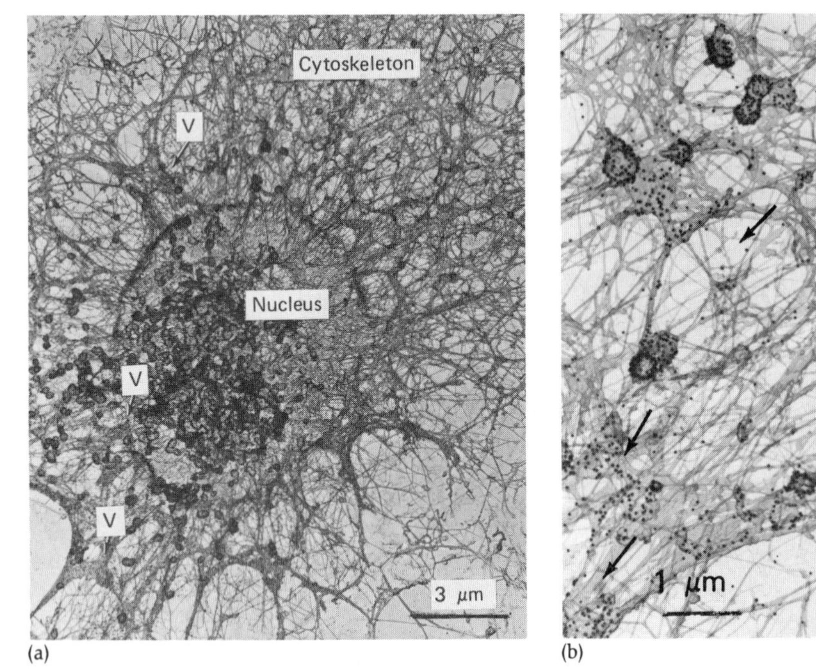

Figure 10–32 The use of a cell's cytoskeletal framework for animal virus growth. (a) The normal appearance of the cytoskeleton after the removal of soluble proteins. (b) The massive reorganization of the cytoskeleton during infection by poliovirus, resulting in cytopathic effects. Viral nucleocapsids are formed on and remained attached to this framework. *[From H. G. Weed, G. Krochamalnic, and S. Penman,* J. Virol. ***56****:549–557, 1985.]*

ply within their hosts and produce cellular damage. Because of the agar overlay and the single layer of cells, viruses can spread only along the agar surface, producing clearly defined areas of cellular degeneration (Figure 10–34). Such sites of destruction, called **plaques,** are detected by staining either the remaining living cells (with neutral red) or the dead cells (with trypan blue). The characteristics of the plaques, such as their diameter and margin properties, in a particular type of tissue culture system facilitate the identification of viruses.

The plaque technique can also be used to obtain pure viral lines. This procedure, called *plaque purification,* is based on the presumption that a single

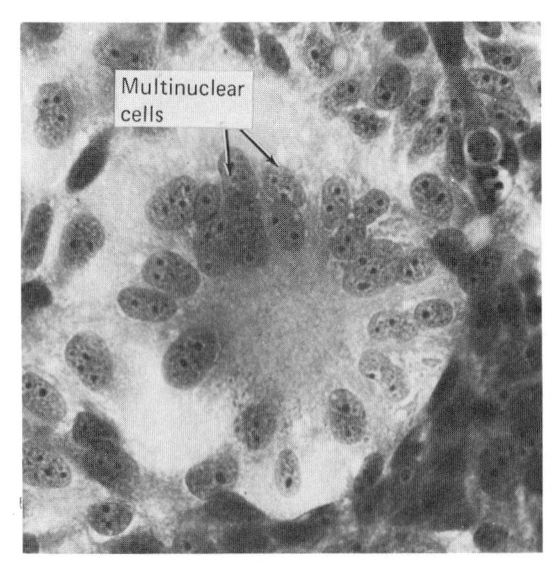

Figure 10–33 Photograph of measles virus-induced multinucleated cell formation with monkey kidney cells. *[Courtesy of Dr. F. Rapp, Baylor University College of Medicine, Houston.]*

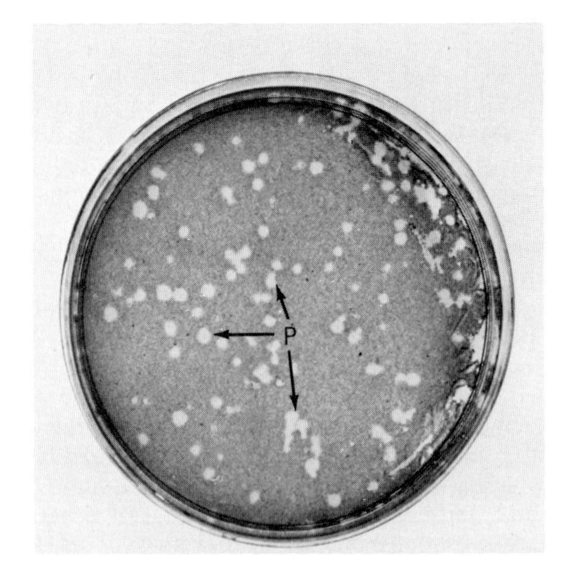

Figure 10–34 Plaque (P) formation by virulent measles virus in BSC-1 cell culture. *[Courtesy of Dr. F. Rapp, Baylor University College of Medicine, Houston.]*

virus particle produces each plaque, just as a single bacterium produces a bacterial colony.

Plaque Assays. *Plaque assay* is used to determine the number of infectious virus particles in a particular specimen. The measurement of infectivity is accomplished by carrying out the plaque technique with specific dilutions of virus-containing suspensions. The infectivity of the original virus-containing suspension is computed from the number of resultant plaques, each of which is referred to as a *plaque-forming unit (PFU),* together with a correction factor for the specimen dilution employed. Modifications of this technique are necessary at times, depending upon the virus being assayed.

Cocultivation. In this approach to virus detection, tissue thought to be infected is cultivated and grown in the presence of another tissue cell culture known to be susceptible to the suspected virus. If any virus is released from the cells of the infected tissue, the susceptible cocultivated virus cell line becomes infected and usually produces enough virus to permit detection and identification.

Transformation Assays. Methods have been developed for detecting and measuring the infectivity of viruses that do not cause cell death. One of these methods is the *transformation assay.* Certain cancer-producing viruses can destroy some cultured cells but can also transform others into malignant systems. Such transformed cells grow in an uninhibited manner to produce a small clump of tumor cells that stands out among the normal cells in a tissue culture. For some viruses, such as the agent of Rous sarcoma, the transformation assay is the basic method for determining viral infectivity.

Interferon

Interferon was detected as an antiviral agent by A. Isaacs and J. Lindemann in 1957. The ability to inhibit viral growth was the property for which this protein was named. Later studies have shown that interferons are a family of proteins produced by various cells when exposed to interferon inducers such as viruses or double-stranded ribonucleic acids. Originally three distinct types of human interferon were found: fibroblast (IFF), immune (IFI), and leukocyte (IFL). They differed biologically, chemically, and physically from one another. Interferons have several effects other than antiviral activity (Figure 10–35), including cell growth inhibition and the suppression of immune responses. Because of their properties, interferons are being studied for a therapeutic agent in cases of severe incapacitating virus infections, virus-caused diseases in individuals with lowered resistance, and different cancerous states. Increased production of interferon is a major use of recombinant DNA technology. **[Chapter 12 describes interferon in greater detail.]**

Figure 10–35 The effects of genetically engineered (cloned) interferon on viral replication. (a) Human cells infected with herpes simplex virus-1 (HSV-1). Normal replication and release of virus particles are evident (arrows). (b) Cells pretreated with interferon before infection, showing a marked inhibitory effect on viral replication and release. *[From S. Chatterjee, E. Hunter, and R. Whitley,* J. Virol. ***56****:419–425, 1985.]*

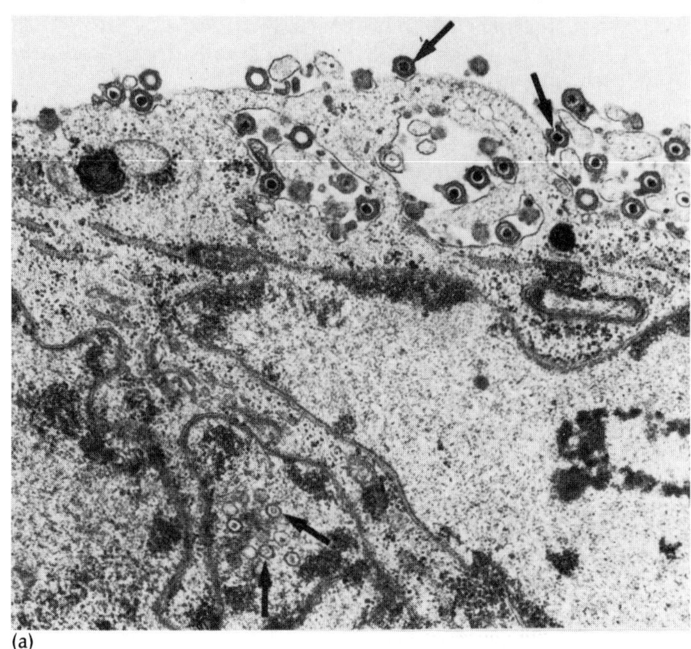
(a)

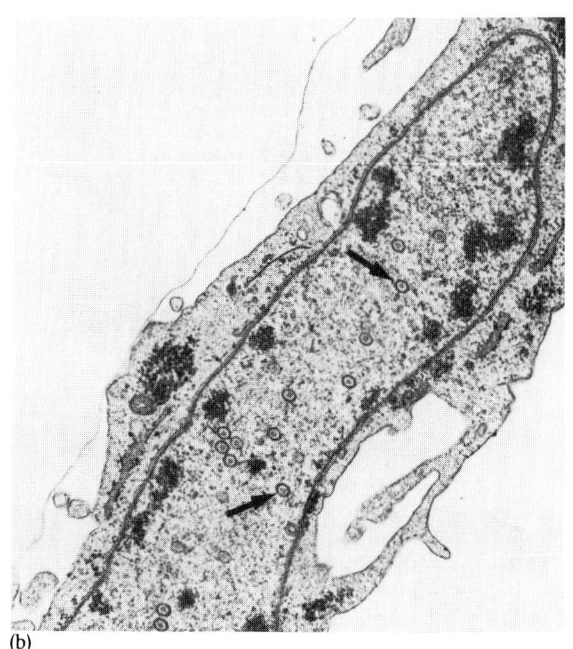
(b)

Plant Viruses

Virus diseases of plants, although not recognized as such, were known long before the discovery of bacteria. Among the disease states first recorded (in 1576) was a variegation in the color of tulips, which is now called *tulip break* (Figure 10–36). Since that time, more than 300 different viruses have been found that affect plants. Currently, because of the economic losses caused by plant viruses, there is significant interest in these agents and the responses of host plants to infection.

Virions

Individual plant virions generally consist of RNA surrounded by a protein capsid. Plant viruses are also known to contain single-stranded DNA, including the *geminiviruses* and double-stranded DNA, the *caulimoviruses.* The capsomere pattern of virions and the position and orientation of the nucleic acid components to their capsids differ among plant viruses (Figure 10–37).

The shapes of plant virus particles fall into two general categories—polyhedral and helical (Figure 10–37). The latter group includes rigid rodlike particles (Figure 10–38); bullet-shaped viruses; and long, flexible threads.

Figure 10–36 Virus-caused tulip break. A drawing showing the streaking or "break" of tulips caused by virus infection from a book entitled *The Clergy-Man's Recreation,* by J. Lawrence, published in 1714. *[Courtesy of Dr. R. Markham, Agricultural Research Council, Cambridge, England.]*

Plant Virus Cultivation

Plant virus cultivation serves many other purposes besides ultrastructural studies. These include the study of virus-infected cells, the determination of viral environmental and replication requirements, and the evaluation of effects of chemicals and radiation on virus replication and other activities.

Tissue culture, cell culture, and protoplast culture are all uses in cultivation. In tissue culture, pieces of various plant parts such as roots, stems, and seeds are used. Diseased tissues, including tumors, have also been used. Cell culture techniques use cells isolated from plant tissues. Such cells can be obtained from the vigorous shaking of cut plant parts or from the enzymatic breakdown of the intercellular connecting substances in plant tissues. The methods used to produce protoplast cultures involve the enzymatic digestion of plant cell components.

The presence of viruses in plant cells can be demonstrated in various ways, including staining with fluorescent antibody procedures and electron microscopy (Figure 10–39).

Plant Virus Infections

Plant viruses apparently have no specific mechanisms to ensure their penetration into host cells. They can enter their hosts through breaks or abrasions in the plant or seed, or they can be introduced by insects, parasitic worms, or other plants such as dodder.

The response of plants to viruses is extremely varied. Because of the effect of virus infection on particular cells or cell types, abnormalities develop in all parts of infected plants. Such cellular responses, in turn, depend on the changes in organelles that result from biochemical and molecular biological events that occur during the establishment and replication of the virus.

External signs of virus infection include mottling of leaves, with dark green, light green, or yellow areas often accompanied by raised blisterlike spots (Color Photograph 56); changes in flower color (Color Photograph 57); formation of unusually small, misshapen, or bumpy fruits; cracked bark; and abnormal growths (tumors) on roots. Of the various internal changes that occur, degeneration of conductive tissues, increase in cell number, and the presence of cellular inclusions are the most common. The cellular inclusions frequently contain virus particles.

Influence of Host Genotype

The genetic makeup, or genotype, of the host plant has a profound effect on the outcome of infection with a particular virus. This genetic influence of the host plant is important in the following situations:

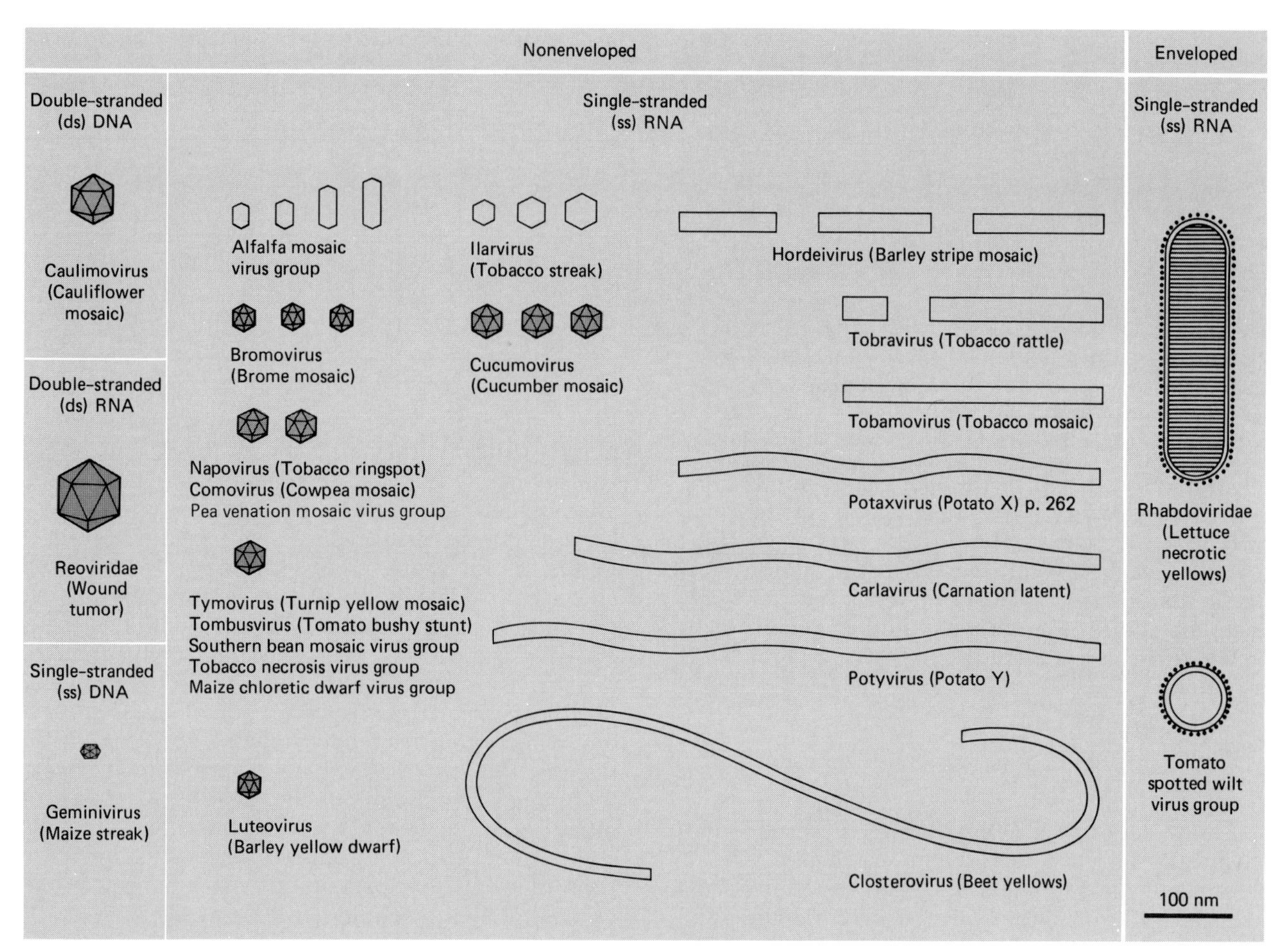

Figure 10–37 Typical shapes and approximate size relationships of plant virus groups. The general position and nucleic acid type are indicated in most cases. *[Modified from R. E. F. Mathews,* Intervirology ***12**:157, 1979.]*

Figure 10–38 (a) An electron micrograph of negatively stained tobacco mosaic virus. *[Courtesy of Dr. R. Markham, Agricultural Research Council, Cambridge, England.]* (b) A diagrammatic representation of a tobacco mosaic virion showing how its RNA chain is arranged within the supporting framework of the capsid. *[D. L. D. Caspar, and A. Klug,* Adv. Virus Res. ***7**:225, 1960.]*

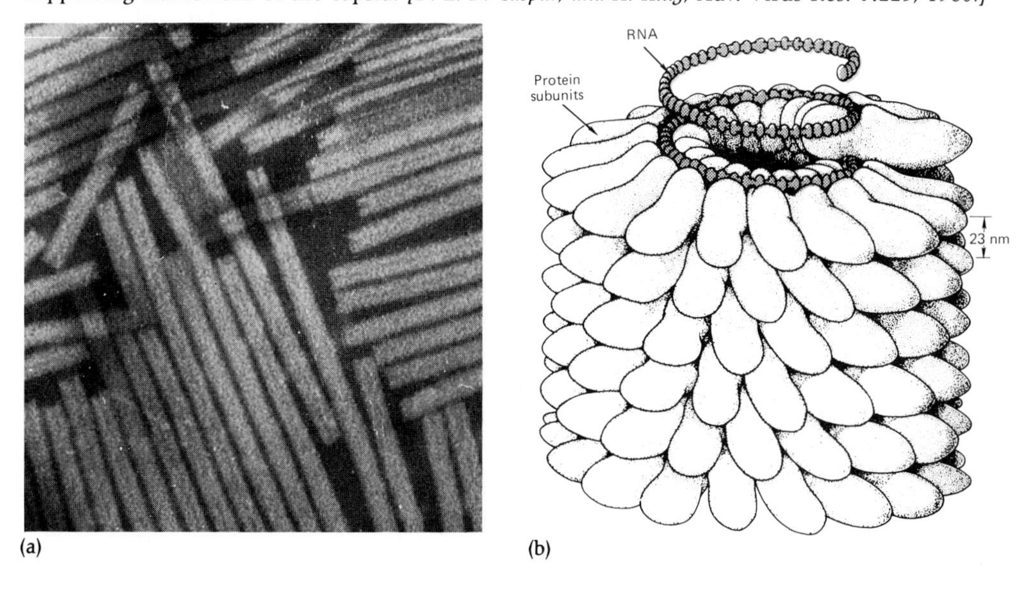

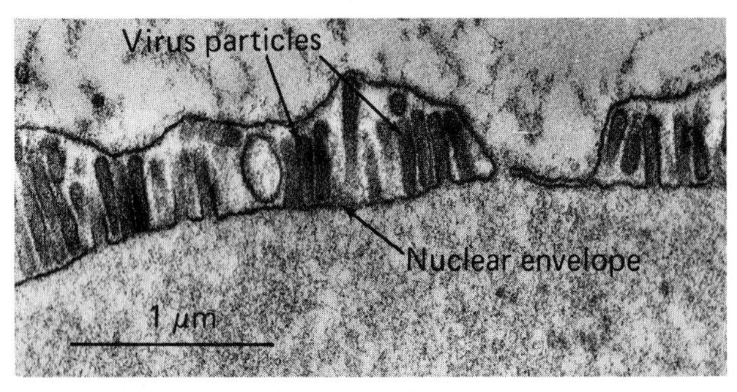

Figure 10–39 Plant virus particles forming within the nuclear envelope of a host cell. *[M. Russo, and G. P. Martelli,* Virology ***56****:39–48, 1973.]*

(1) immunity, in which the plant does not become infected under any circumstances; (2) resistance to infection; (3) hypersensitivity, in which the host reacts by localized death of cells at the site of infection without further spread of virus; and (4) tolerance, in which the virus multiplies and spreads widely through the plant but the disease produced is very mild.

Satellite Viruses

Special relationships exist between certain biochemically unrelated viruses in which one virus is an obligate parasite of another. This situation occurs with the fully infectious tobacco necrotic virus and its noninfectious satellite virus. The small satellite agent is totally dependent on the larger helper virus, which contains information only for the formation of a protein capsid. While the exact nature of this dependence is unknown, no replication of such defective or satellite viruses takes place in the absence of a helper virus. Similar satellite situations occur with some bacteriophages and DNA animal viruses.

Viruses of Eucaryotic Microorganisms

More and more virus associations with eucaryotic microorganisms, including algae, fungi, and protozoa, are being described (Figure 10–40). Such associations take one of two forms. In one type, the eucaryotic microorganism serves as a **vector,** or transmitter, of the virus. This virus-vector relationship appears to be highly specific. At present there is no evidence that any of these viruses multiplies in its eucaryotic microbial vector. Interestingly enough, certain viruses present in these eucaryotes actually may be associated with procaryotic or other eucaryotic microorganisms (endosymbionts) present within the cell (Figure 10–40). Those eucaryotes that exhibit phagocytosis are likely to acquire viruses of other organisms simply by feeding from a nonsterile environment. Once introduced, such viruses may be kept intact. Various parasitic protozoa have acquired viruses following contact with host cells.

In the second situation, the virus infects the eucaryote, which then functions as a host. Since relatively little is known about virus infections of eucaryotic microbes other than fungi, we shall concentrate here on these fungal viruses, or *mycoviruses.*

Figure 10–40 Viruses of eucaryotic microorganisms. (a) Polyhedral viruslike particles in a spore of the brown algae *Chorda tomentosa* (KORD-a toe-men-TOES-a). *[R. Toth, and R. T. Wilce,* J. Phycol. ***8****:126–130, 1972.]* (b) A virus particle invading a green alga through its cell wall. Note the taillike (T) structure of the invading particle. *[H. Kawakami, and N. Kawakami,* J. Protozool. ***25****:217–225, 1978.]*

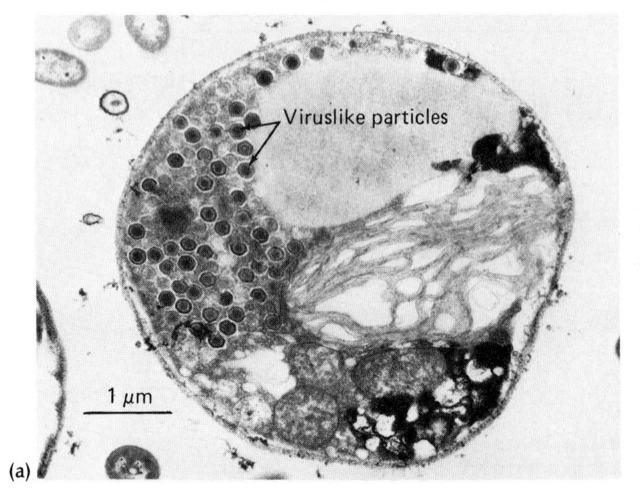

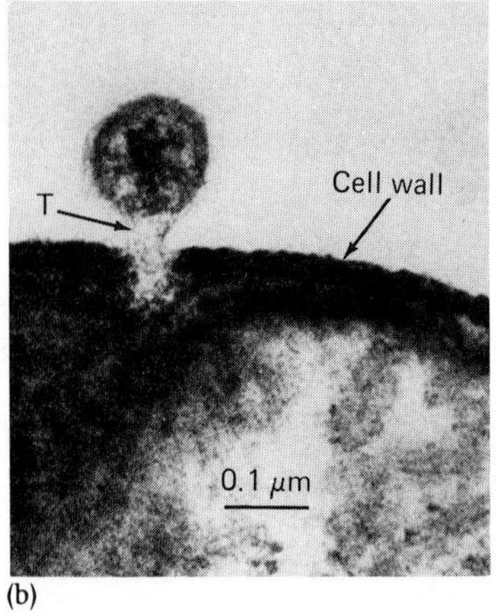

In 1967 a virus was found in an ascomycete. Since then, different viruses have been found in more than 100 species of fungi, including genera from all the main taxonomic groups of fungi. A number of these mycoviruses have now been characterized *in vitro* (in test tubes or similar containers), and their replication has been demonstrated *in vivo* (in living organisms). **[See Ascomycotina in Chapter 8.]**

The viruses studied are primarily nonenveloped, polyhedral particles that contain double-stranded RNA. An enveloped, double-stranded, herpesvirus-type virus particle has also been isolated from fungi. Transmission of fungal viruses can occur through hyphal connections and through asexual or sexual spores. Although there may be a total absence of harmful effects, mycoviruses may cause the death of the host or the initiation of abnormal mycelial growth activities. From the evidence, it appears that there is no consistent relationship between the presence or absence of mycoviruses and the production of antibiotics and mycotoxins such as aflatoxins.

The biological significance of viruses in fungi is still unclear. Under normal conditions, most of these viruses seem to be latent. However, as experience has shown with bacteria and other forms of life, latent infections can be triggered into virulent ones.

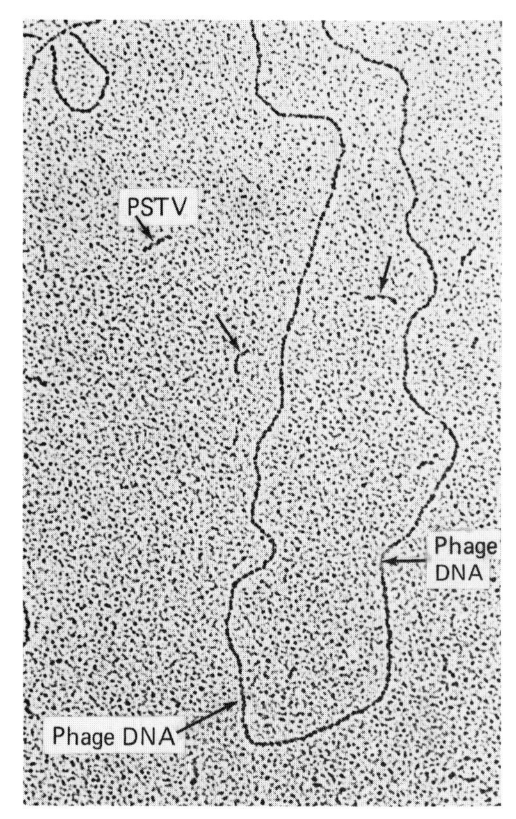

Figure 10–41 An electron micrograph of the potato spindle tuber viroid (PSTV) in comparison with a conventional viral nucleic acid from bacteriophage T_7. Note the great size difference between the viroids (arrows) and the viral DNA molecule. *[Courtesy of T. O. Diener, Research Plant Pathologist, USDA.]*

Viroids and Other Disease-Associated Small RNAs

Several additional small RNA molecules associated with disease states are known. These include *viroids, satellite RNA molecules,* and *virusoids.* Properties of these molecules are compared in Table 10–4.

Viroids

Until about sixteen years ago, it was generally believed that all infectious diseases of plants and animals were caused by bacteria, fungi, or viruses. However, in 1971, T. O. Diener introduced the term *viroid* to represent a newly discovered category of subviral plant pathogens (Color Photograph 58). Diener and W. B. Raymer found that the first member of this unique class of disease agents causes potato spindle tuber disease. To date, about one dozen viroid diseases of other crop plants have been identified. Undoubtedly additional pathogens will be discovered.

All currently known viroids consist of low-molecular-weight (75,000 to 120,000 daltons) RNA, making the viroids about one-tenth the size of the smallest known plant virus (Figure 10–41). They exist both intracellularly and extracellularly as circular, single-stranded RNA molecules and have an average

TABLE 10–4 Comparison of Disease-Associated Small RNAs

Property	*Viroids*	*Virusoids*	*Satellite RNA Viruses*
Cause of infections	Yes	No	No
Possession of a specific capsid	No	Yes	No
Contained in helper capsid	No	No	Yes
Replicates only in the presence of a helper capsid	No	?	Yes
Interferes with helper capsid replication	No	?	Yes
In vivo and *in vitro* stability of RNA	No	High	High

length of about 50 nanometers (nm). Viroids are smaller than any known viral chromosome.

Although many details of viroid replication are still unknown, a general concept of the process has emerged. Viroids are transcribed from RNA patterns or templates. These templates, as well as new viroids, are synthesized by enzymes already present in healthy host cells. The new viroids are formed into circles by a host cell enzyme.

Viroids are located mainly in the nuclei of infected cells. Their location, together with their ability to serve as mRNA templates, suggests that disease symptoms may result from viroid interference with host genetic mechanisms and metabolic functions. Such interference could cause the production of faulty proteins.

The existence of viroids poses several intriguing and interesting questions in relation to causing disease, especially since they introduce a far smaller amount of genetic information into host cells than do viruses yet, in most cases, are able to replicate and cause disease. Viroids are clearly independent genetic systems, with properties determined by the nucleotide sequence of their respective RNAs. **[Refer to Chapter 6 and Appendix A for descriptions of RNA.]**

VIRUSOIDS

A new dimension has been added to viroid investigations with the finding of encapsidated, circular, linearlike RNA closely associated with much larger viral RNA molecules in certain virus particles. Apparently these viroidlike RNAs or virusoids need the viral RNA to aid in their replication. While these new submicroscopic agents also differ from viroids in other ways, their relationship to viroids, if any, remains to be determined.

Satellite RNAs

Specific virus-dependent replicating RNA molecules, *satellite RNAs*, are present in varying numbers in the protein coats of certain helper or satellite viruses. These molecules are similar in size to viroids, replicate only in the presence of a specific virus, and may or may not produce devastating effects on infected plants. The mechanism of action and the activities of these molecules are not fully understood.

Prions

One group of animal and human diseases of the central nervous system (CNS) is caused by agents having unconventional properties compared to those of known viroids and animal viruses. Since these diseases require a long time to develop, they have been referred to as slow infections. One agent causes scrapie, a progressively destructive and fatal disease of sheep and goats. Kuru (Figure 10–42) and Creutzfeldt-Jakob disease are related human diseases.

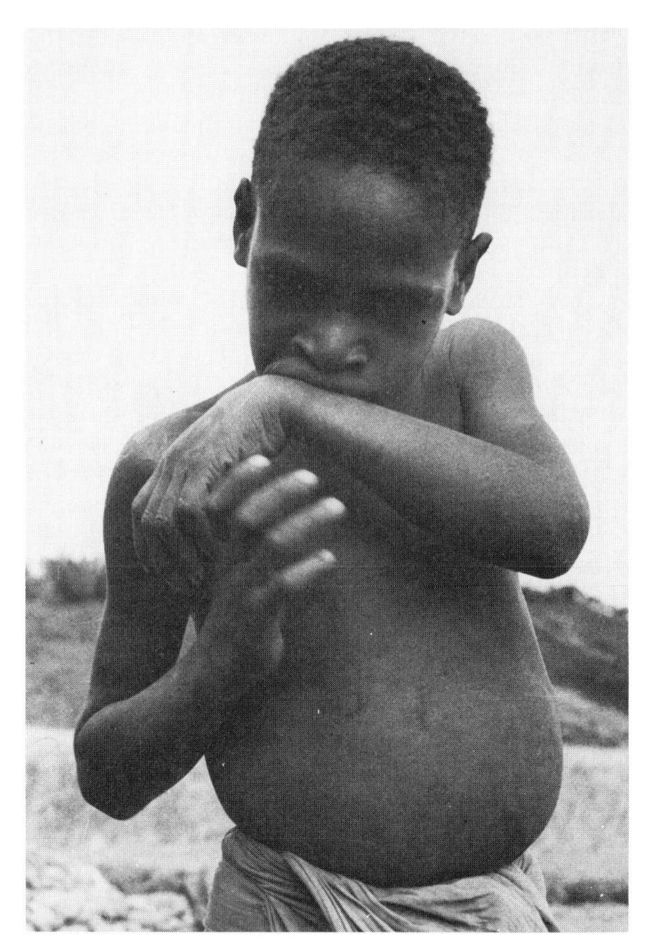

Figure 10–42 A victim of kuru. This child is condemned. He has lost the use of his limbs and can no longer eat. Kuru prevents him from swallowing. *[Courtesy World Health Organization.]*

The scrapie agent consists exclusively of protein and exhibits unusual physical, chemical, and biological properties. These include a molecular weight of 30,000 daltons, a filamentous form (Figure 10–43), a size smaller than that of any known viroid, and a remarkable resistance to various types of radiation and enzymes known to inactivate nucleic acids. Since infectious activity appears to depend primarily on a specific protein, S. B. Prusiner in 1981 proposed the name *prion* (proteinaceous infectious particle) for the scrapie agent. While the method of replication for the prion still remains a mystery, three hypotheses have been proposed (Table 10–5).

The scrapie agent has been a source of both fascination and frustration for many years. As more information is gathered, it appears that prions may well represent a new group of infectious agents. **[Additional details concerning slow infections and prions can be found in Chapter 29.]**

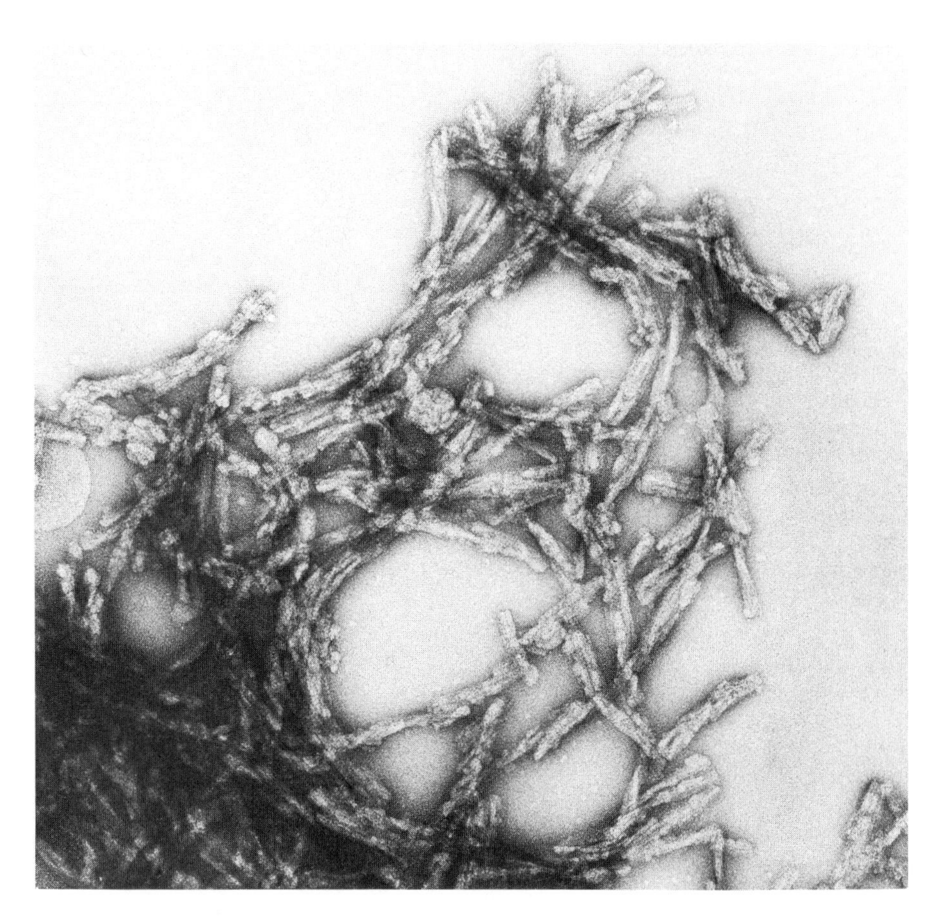

Figure 10–43 Scrapie prions isolated from an infected hamster brain. Original magnification 100,000×. *[Courtesy of Dr. Stanley B. Prusiner.]*

Approaching the Control of Microorganisms

Microorganisms exert most of their effects through growth and/or an increase in number. Knowledge of the structures and processes involved in microbial growth and reproduction is essential if we are to predict their activities and to control them. A characteristic property of microbial growth and reproduction is the short time needed for a population to increase. Very large numbers may not only occur quickly but can also have harmful consequences. The chapters in the next section will describe many of the currently used chemical and physical agents and approaches used to control microorganisms and the infectious diseases they cause.

TABLE 10–5 Scrapie Agent Hypotheses

Hypothesis	*Chemical Composition of Agent*	*Brief Description of Replication Process*
Prion	Protein only	Protein-directed protein synthesis (protein → protein), or reverse translation (protein → mRNA), or induction of host cell transcription.
Virino	Protein and nucleic acid; host protein with regulating nucleic acid (virino)	A scrapie-specific nucleic acid disrupts host cell functions and regulates the replication of the scrapie agent using host cell enzymes
Filamentous virus	Protein and nucleic acid; protein encoded by virus-specific nucleic acid	Normal viral replication process takes place.

Summary

Every kind of life can be parasitized by viruses.

1. A *virus* can be considered to be an extraordinarily complex organization of nonliving chemicals, an exceptionally simple form of microorganism, or both.
2. A mature virus particle, or *virion,* has several distinguishing properties, including the absence of cellular structures, the possession of either DNA or RNA, viral nucleic acid regulation of new virus production (replication), dependence on the host cell and its parts for energy and protein needs, and an ability to transfer its nucleic acid from one host cell to another.

Basic Structure of Extracellular Viruses

1. A typical viral life cycle consists of both intracellular and extracellular phases. Intracellular phases involve the production of viral parts used in the formation of complete viruses.
2. Most virions exhibit a characteristic structure.
3. A complete virus consists of a nucleic acid core surrounded by a protein *capsid* (coat), referred to as a *nucleocapsid.*
4. The capsids of viruses consist of subunits known as *capsomeres.*
5. Virus particles can exhibit several forms. These include *polyhedral, icosahedral, helical* (spiral), *binal* (a combination of polyhedral and helical), *bullet-shaped,* and *filamentous.*

The Nucleic Acid of the Mature Virion

1. The viral nucleic acid, whether DNA or RNA, contains the viral genetic material, or *genome.*
2. The specific type and form of nucleic acid vary according to the virus.
3. Individual virus particles contain either DNA or RNA, never both.
4. Length also varies, but it is constant for a specific virus.
5. The genomes of certain RNA viruses (plus strands) function as mRNA and are translated into new viral proteins.

Enveloped Viruses

1. Viruses that through their intracellular developmental cycle acquire an outer coat, or covering, from the cytoplasmic and/or nuclear membranes of infected host cells are referred to as *enveloped.*
2. The envelopes of some viruses are covered with projecting spikes that may appear as a fringe. These projections serve to attach viruses to various cells.
3. Characteristically viruses lack metabolic enzymes. However, some have enzymes for attachment to host cells.

Shape and Size

Virions vary not only in size but in shape as well.

Classification

Current classification schemes for viruses are mainly based on capsid organization, nucleic acid chemistry, and immunologic and genetic properties.

Bacteriophages (Bacterial Viruses)

1. Almost all large groups of cultivable bacteria contain viruses.
2. Many bacterial viruses are useful in identification of bacterial pathogens and in research on the mechanisms of cellular infections.

The Basic Structure of Bacteriophages

1. These viruses contain either RNA or DNA.
2. A characteristic feature of some bacteriophages is their taillike structure, which serves as an attachment device as well as a tube through which viral nucleic acid can be injected into a host cell.
3. Bacterial viruses exhibit a great variety of forms.

Filamentous Bacteriophages

1. The *filamentous viruses* are among the smallest viruses known to date.
2. Filamentous viruses have focused attention on a form of symbiotic behavior previously unrecognized in bacteria.

Cultivation of Bacteriophages

1. Actively growing young bacterial cells are used for phage cultivation. Either broth or agar cultures are used for this purpose.
2. When agar plates are used, the presence of bacterial viruses is indicated by the development of *plaques.*

Enumeration of Phages

Accurate estimations of the number of virus particles in a sample can be done by several techniques including the plaque count assay and electron microscopy.

Phage Typing

Phage typing is used to identify certain specific bacterial species.

Replication Cycle of Bacteriophages

1. Bacterial viruses that regularly infect and complete their life cycles in bacterial hosts and ultimately produce more virus particles with the accompanying destruction of the host are called *virulent* or *lytic* viruses.
2. Not all virus infections of bacteria end in host cell destruction. Some viruses integrate their DNA into that of the host. Such viruses are said to be *temperate,* and their integrated form is referred to as a *prophage.* The bacterial cell containing a prophage is *lysogenic.*

Lysogeny

1. When a temperate phage infects a host cell, it can either cause the destruction of the host (lytic cycle) or enter a noninfectious state.
2. The presence of a virus in a noninfectious state can result in changes of the host cell, including changes in colonial morphology and toxin production.
3. Bacteria harboring a virus as a prophage are immune to infection by other phages of the same type. This is a form of *resistance.*

The Lytic Cycle

The lytic cycle of bacterial viruses such as the T (type)-even phages includes the stages of adsorption, penetration, and replication of new viral DNA, maturation, and new virus particle release.

GROWTH CURVE
A *one-step growth curve* utilizes the events of the lytic cycle to determine the time sequence from injection to the release of mature viruses and the number of virus particles released per cell.

Cyanophages
Cyanophages, viruses of blue-green bacteria, are similar to other bacterial viruses in both structure and infection cycle.

CYANOPHAGE CULTIVATION AND ENUMERATION
The conditions, media, and procedures used for preparation of suitable hosts for cyanophages are different from those used for bacteriophages.

Animal Viruses
DISEASE
1. Examples of human viral diseases include AIDS, chickenpox, and yellow fever.
2. Lower animal diseases include canine distemper, foot-and-mouth disease, and various tumors.
3. The effects of animal viruses can range from mild skin rashes and upper respiratory symptoms to tissue destruction and death.
4. Viruses that attack insects such as honeybees and silkworms can cause severe economic losses.

ANIMAL VIRUS REPLICATION
Although differences exist among viruses, the general sequence of events involved in replication are attachment to the host cell, penetration, *uncoating* (removal of viral envelope and capsid from the nucleic acid), replication of viral nucleic acid, manufacture of other viral parts, assembly of viral parts to form new virus particles, and release of new viruses.

DEFECTIVE INTERFERING PARTICLES
Defective interfering virus particles may interfere with the normal replication of certain viruses.

VIRUS CULTIVATION IN ANIMALS
1. Fertilized eggs are useful for virus cultivation, especially when large numbers of viruses are needed.
2. Egg cultivation of viruses is useful in vaccine production, virus isolation, virus morphology studies, determinations of the effectiveness of drugs on viruses, and in studies dealing with the mechanism of infection.
3. Signs of infection include death of the embryo, demonstration of red blood cell clumping by virus-containing fluids, and evidence of virus particles in embryo material by various types of microscopy.

TISSUE CULTURE CULTIVATION OF VIRUSES
1. Viruses can be grown in a variety of tissues maintained in various types of containers. This type of cultivation is known as tissue culture.
2. Two types of cell culture in common use are the *primary* and the *continuous.*
3. The destructive effects of viruses in tissue systems are called *cytopathic effects* and include nuclear changes, formation of multinucleated and giant cells, cell fusion, and vacuole formation. The particular type of cytopathic effect is used, at times, for virus identification.
4. A special technique used in determining the number of virus particles in specimens and in purifying and establishing virus lines is the plaque method. A *plaque* is a clear area in a single tissue culture layer that contains numerous virus particles.
5. Cocultivation involves the growth of tissues suspected of containing a viral pathogen in a cell culture known to be susceptible to the virus. If virus is present, its existence will be shown by the susceptible cell line.
6. Transformation assays are special tissue culture methods used to detect viruses.

Plant Viruses
Virus diseases of plants were known long before the discovery of bacteria.

VIRIONS
1. Most individual plant virus particles consist of RNA surrounded by a protein capsid.
2. The shapes of plant viruses fall into two general categories, polyhedral (many-sided) and helical.

PLANT VIRUS CULTIVATION
Plant viruses can be cultivated in tissue culture, cell culture, and protoplast culture systems.

PLANT VIRUS INFECTIONS
1. Plant viruses produce both internal and external effects in infected hosts. Abnormalities can be noted on different plant parts including flowers, roots, and bark.
2. The genotype of the host influences the effect of viral infection.
3. *Satellite viruses* are dependent on helper viruses for their replication.

Viruses of Eucaryotic Microorganisms
Viruses may be found in association with eucaryotic microorganisms such as algae, fungi, and protozoa.

Viroids and Other Disease-Associated Small RNAs
1. Several small RNA molecules associated with disease states are known: viroids, satellite viruses, satellite RNA molecules, and virusoids.
2. *Viroids* are among the smallest infectious agents known and consist only of a low-molecular-weight RNA. They are mainly found in plant disease states.
3. *Virusoids* are encapsidated circular and linearlike RNA closely associated with and dependent on larger viral RNA molecules for their replication.
4. *Satellite RNA molecules* are present in the protein coats of certain helper or satellite viruses and may influence the outcome of a disease process.

Prions
1. The term *prion* (proteinaceous infectious particle) has been proposed for the scrapie agent and similar disease agents.
2. The scrapie agent is remarkably resistant to various types of radiation, contains a hydrophobic protein that is esssential for infecting, and appears to be even smaller than the smallest viroid. The mechanism for replication is not known.

Questions for Review

1. a. What is a virus?
   b. How does the organization of a virus particle differ from that of a procaryote?
   c. How does the organization of a virus particle differ from that of a eucaryotic microorganism?
2. Explain or define the following terms:
   a. capsomere
   b. virion
   c. capsid
   d. icosahedron
   e. nucleic acid core
   f. envelope
   g. binal arrangement
   h. spike
3. Distinguish between a virion and a viroid.
4. Compare the structural features of:
   a. a typical animal virus
   b. a typical plant virus
   c. a bacteriophage
   d. a cyanophage
   e. viroid
5. Do all viruses contain DNA? Explain.
6. Which of the structures shown in the electron micrograph of Figure 10–44 would you *not* expect to find in the following:
   a. bacteriophage
   b. typical diatom
   c. typical flagellated protozoon
   d. yeast cell
7. Examine the electron micrograph in Figure 10–45.
   a. What type of virus is shown?
   b. Identify and give the function(s) of each labeled part.
8. a. How are viruses classified?
   b. Does virus classification differ from the system used for bacteria? If so, how?
   c. List six properties of viruses used for their classification.
9. a. List three different techniques used for virus cultivation.
   b. How do such techniques differ from the ones used for bacterial cultivation?
10. a. How is the replication of viruses detected in the laboratory?
    b. What does CPE mean?
11. a. Describe, in order, the events that occur in the replication of one animal DNA and one RNA virus.
    b. Describe, in order, the events that occur in the replication of a binal-shaped bacteriophage.
12. What is *lysogeny?*
13. a. What types of microorganisms harbor viruses?
    b. Do these viruses differ from either animal or plant virus particles? If so, how?
    c. What is a satellite virus?
14. Do animal viruses infect plant cells? Explain.
15. What forms of life do not have viruses?
16. List eight specific structures of procaryotes not found in a virus particle.
17. a. Define the statement that viruses are living.
    b. Present arguments against the above statement.
    c. What do you think?
18. The electron micrograph in Figure 10–46 shows enveloped virus particles emerging from an infected cell. Briefly outline the events that were necessary for the development of such viruses.

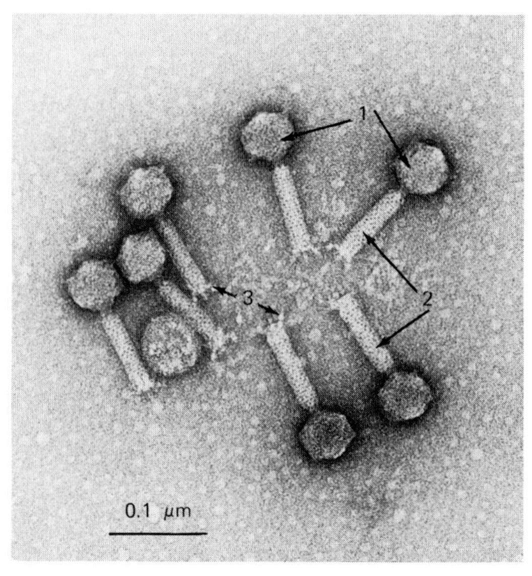

Figure 10–45 [From J. C. Gerdes, and E. R. Romig, J. Virol. *15*:1231–1238, 1975.]

Figure 10–44 An electron micrograph of an ultrathin section of *Bacillus subtilis*. [Courtesy of W. Van Iterson, University of Amsterdam, and the North-Holland Publishing Company, Amsterdam.]

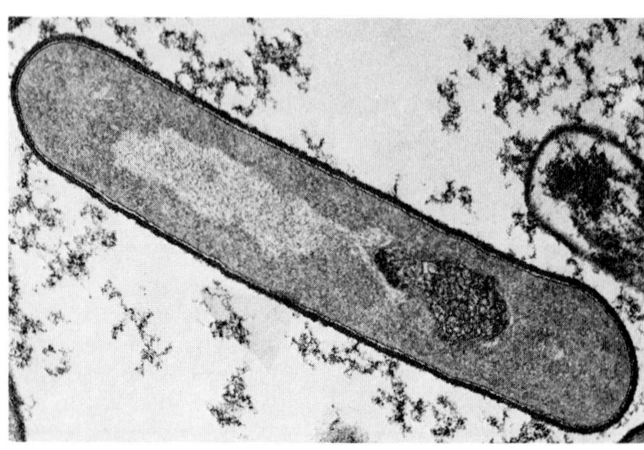

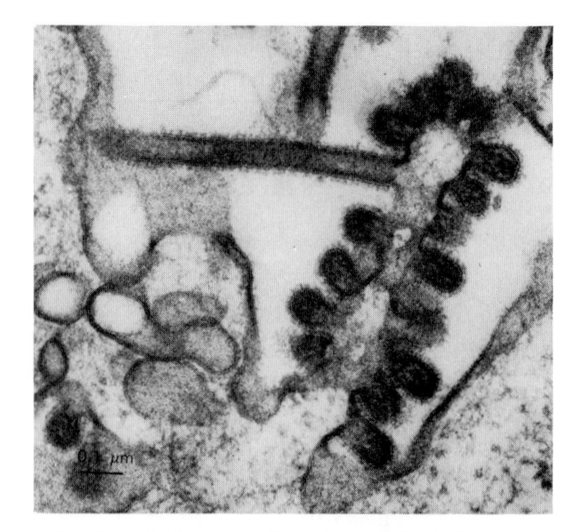

Figure 10–46 Influenza virus formation. [Courtesy L. Herrero-Uribe and D. Hockley.]

Suggested Readings

CAMPBELL, A. M., "How Viruses Insert Their DNA into the DNA of the Host Cell." *Scientific American* **237**:102, 1976. *Details of the insertion of a virus into the host cell are presented. How viruses can coexist peacefully with host cells without causing cell death is explained.*

CARP, R. I., P. A. MERZ, R. J. KASCSAK, G. S. MERZ, and H. M. WISNIEWSKI, "Nature of the Scrapie: Current Status of Facts and Hypothesis." *Journal of General Virology* **66**: 1357–1368, 1985. *An up-to-date presentation of this slow infection affecting the central nervous system of various animals.*

COOPER, J. I., and F. O. MACCALLUM, *Viruses and the Environment.* London: Chapman & Hall, 1984. *A well written work giving an up-to-date presentation of the major aspects of the biology of viruses.*

DIENER, T. O., "Viroids: Minimal Biological Systems." *BioScience* **32**:38–44, 1982. *An account of the research studies that established the existence of viroids.*

GALLO, R. C., "The AIDS Virus." *Scientific American* **256**: 46–56, 1987. *The second of a two-part series on retroviruses that describes the features of human T-lymphotropic virus III* (the human AIDS virus) *and compares it to the virus found in monkeys.*

HUGHES, S. S., *The Virus, A History of the Concept.* New York: Science History Publications, 1977. *This is a good short book on the history of the virus from the germ theory of infectious disease to concepts of the virus in the twentieth century.*

PRUSINER, S. B., "Prions." *Scientific American* **251**:50–59, 1984. *A fascinating introduction to protein infectious disease agents that contain no nucleic acids yet are able to multiply within cells.*

SPECTOR, S., and G. J. LANCZ, *Clinical Virology Manual.* New York: Elsevier Science Publishing Co., Inc., 1986. *A comprehensive collection of data on basic properties and pathogenic mechanisms of major human virus groups, with intensive descriptions of isolation and identification methods.*

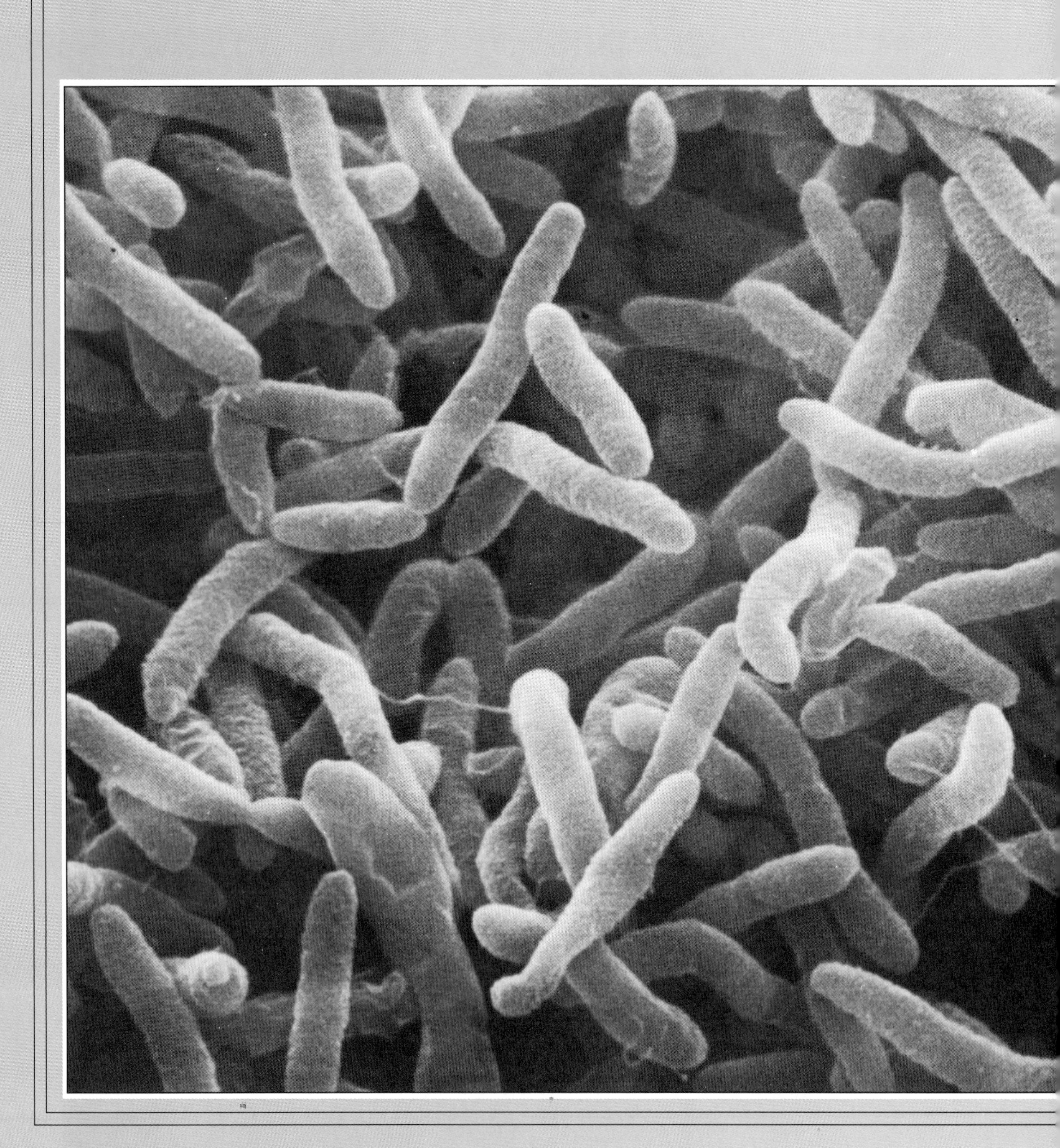

PART IV

The Control of Micro-organisms

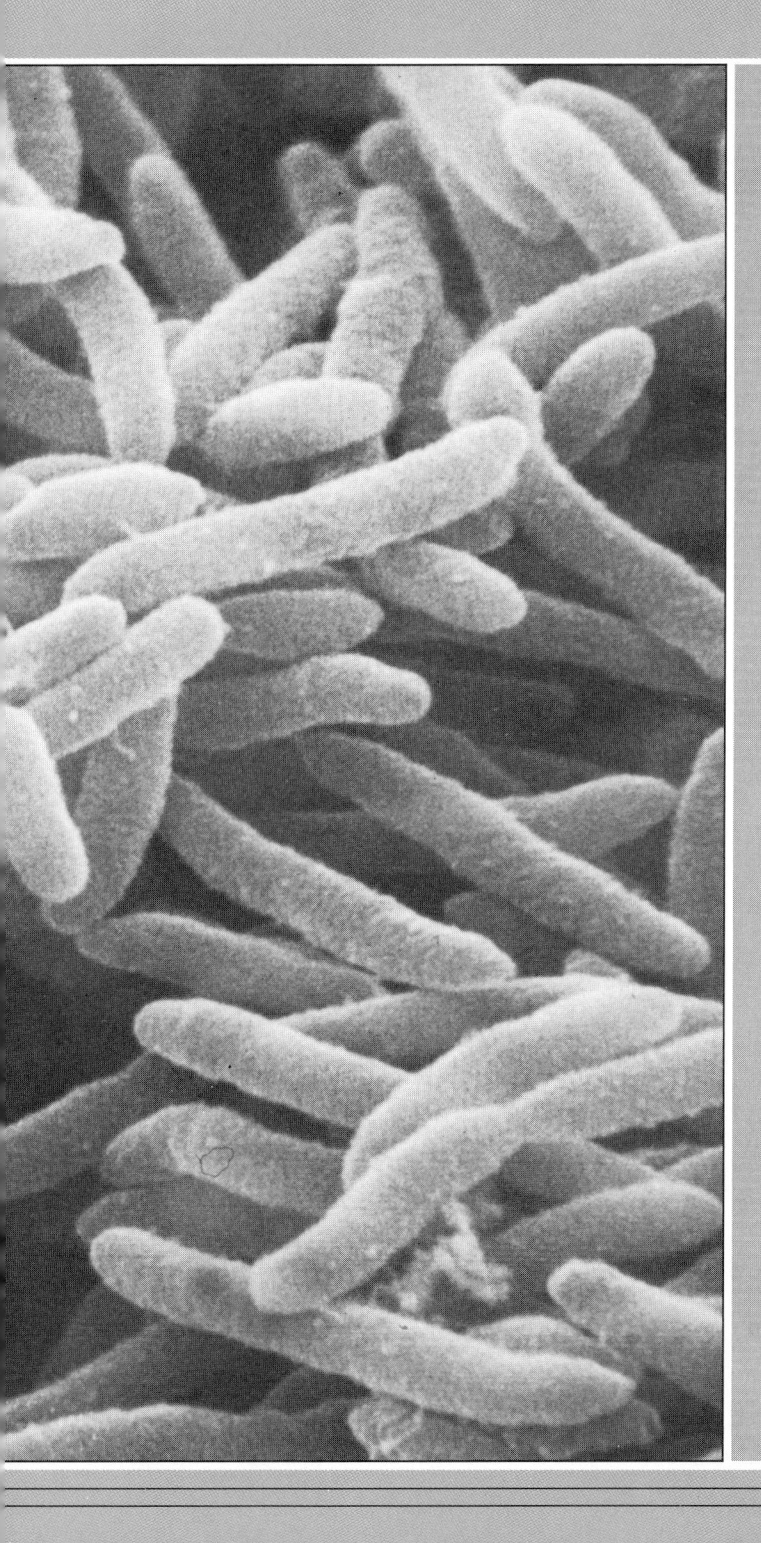

Uncontrolled growth of bacterial cells. [Courtesy of Y. Tago, N. Kuraishi, and K. Aida.]

11

Chemicals and Physical Methods in Microbial Control

Each organic being is striving to multiply . . .
each has to struggle for life . . .
the vigorous survive and multiply.
— Charles Darwin

After reading this chapter, you should be able to:

1. Define the following terms: disinfection, antisepsis, sterilization, disinfectant, antiseptic, bactericidal, bacteriostatic, and sanitizer.
2. Discuss the general principles of preparing materials for heat sterilization.
3. Explain the mechanisms by which moist heat, dry heat, and radiation kill microorganisms.
4. Discuss the monitoring of heat-sterilization procedures and give examples of physical, chemical, and biological devices for accomplishing this.
5. Discuss the essential points in using an autoclave properly.
6. Differentiate between the use of boiling and pasteurization as means of heat disinfection.
7. Discuss what is meant by the target theory in connection with the effectiveness of ionizing radiation as a method of killing microorganisms.
8. Explain the processes of photoreactivation and dark reactivation in protecting microorganisms against ultraviolet light.
9. Explain how filters can sterilize fluids and give several examples of materials used for filtration.
10. Identify the contributions of Paré, Semmelweis, Lister, Koch, Neuber, and Krönig and Paul.
11. List and describe seven general directions for using a chemical in antisepsis and disinfection.
12. Discuss the representative chemical agents from each of the following groups in terms of mechanism of killing microorganisms, general effectiveness, and advantages and disadvantages: halogens, alcohols, phenols, peroxides, detergents, heavy metals, aldehydes, and gaseous sterilants.
13. Discuss the use of gaseous chemical sterilants.
14. Describe and discuss the following methods in determining the effectiveness of chemicals as antiseptics or disinfectants: the phenol coefficient test, the use-dilution method, the direct spray method, and the toxicity index.
15. Describe a testing procedure to determine if a chemical is bacteriostatic or bactericidal.

Many chemicals and physical methods are available for use in the control of microorganisms. Chapter 11 presents basic concepts in chemical and physical control, discussions of selected chemicals, and an overview of the means by which they can be tested and compared.

Micoorganisms essentially occupy all parts of the earth where life exists. Thus one may reasonably expect to find them in almost any material and on most surfaces. Some of these microscopic forms of life infect animals (including humans) and plants and cause disease. Others contaminate foods and, by producing chemical changes, make them inedible or even poisonous. Still others are responsible for the deterioration of a variety of materials such as fabrics, leather goods, and wooden structures. Thus the need for practical procedures to control microbial growth or prevent contamination is obvious. Currently available control measures are used to remove microorganisms from an environment, to inhibit microbial growth, or to kill microorganisms.

This chapter begins by presenting important terms and a brief coverage of historical events associated with the chemical and physical approaches to microbial control. The next portion of the chapter describes the chemicals and physical methods used with inanimate materials to destroy actively growing, undesirable microorganisms but not necessarily microbial spores *(disinfection)* as well as methods used on body surfaces to reduce the numbers of normally present microorganisms (microbiota) and disease-causing contaminants. The remaining part of the chapter emphasizes the physical techniques and chemicals used for the complete destruction or removal of all forms of life (sterilization). Chapter 12 describes the various chemicals (such as antibiotics) used in the treatment of infectious diseases and their mode of action.

The problem of controlling disease-causing microorganisms has troubled scientists for as long as the origin of disease has been known. Many substances have been tested in attempts to find the perfect one to eliminate microbial contamination of living and nonliving surfaces. In the process, many different chemicals, physical methods, and disease-control concepts have been developed.

The antimicrobial agents developed have a wide range of effectiveness and are proper to use in many circumstances. The terms by which they are described reflect these differences. An **antiseptic** and a **disinfectant,** for example, differ mainly in the way they are used. An antiseptic is applied to living tissue; a disinfectant is used for contaminated inanimate objects (**fomites**). When these agents are used, it should be understood that both pathogens and nonpathogens may be affected. Table 11–1 lists and defines the most important processes, conditions, or states and agents associated with the chemical and physical control of microorganisms. **[See means of disease transmission in Chapter 21.]**

TABLE 11–1 Important Terms in Chemical and Physical Control

Process, Condition, or Agent	*Explanation*
Antisepsis	Prevention of the growth or activity of microorganisms by inhibition or killing; applies to the use of chemicals on living tissue.
Antiseptic	A chemical agent used for the purpose of antisepsis.
Bactericide	An agent that kills the vegetative forms of bacteria. The suffix "cide" is used to denote agents, usually chemical, that kill. Commonly used terms are "biocide," "bactericide," "fungicide," "virucide," and "algicide." The term "germicide" is used if the agent kills pathogens but not necessarily spores. An agent that kills bacterial spores is a "sporicide."
Bacteriostasis	A condition in which bacterial growth is inhibited. The suffix "static" is used to denote agents, usually chemical, that prevent growth but do not necessarily kill the organism or bacterial spores. Commonly used terms include "bacteriostatic" and "fungistatic."
Disinfection	The elimination of microorganisms to prevent their transmission, disinfection is directed against their metabolism or structure and is used directly on inanimate objects.
Disinfectant	An agent, usually chemical, used for the purpose of disinfection.
Sanitization	A process that reduces the number of microorganisms on inanimate objects and in environments to safe levels as specified by public health standards.
Sanitizer	Any agent that reduces the numbers of microbial contaminants to acceptable levels; applies to the use of agents on inanimate objects and is usually associated with the cleaning of eating and drinking utensils and the cleaning operations for dairy equipment.
Sterilization	Any process, chemical or physical, that kills or removes *all* forms of life, especially microorganisms.
Sterilant	A chemical used for sterilization.

[a] The terms for respective agents are in bold print.

Microbiology Highlights

FROM SEWAGE TO SUTURES, IN THE WORDS OF JOSEPH LISTER*

In the course of the year 1864 I was struck with an account of the remarkable effects produced by carbolic acid [phenol] upon the sewage of the town of Carlisle, the admixture of a very small proportion not only preventing all odor from the lands irrigated with the refuse material, but, as it was stated, destroying the entozoa which usually infest cattle fed upon such pastures.

My first attempt [to use carbolic acid] was made in the Glasgow Royal Infirmary in March, 1865, in a case of compound fracture of the leg. It proved unsuccessful, in consequence, as I now believe, of improper management, but subsequent trials have more than realized my most sanguine anticipations.

Carbolic acid proved in various ways well adapted to the purpose. It exercises a local sedative influence upon the sensory nerves; and hence is not only painless in its immediate action on a raw surface, but speedily renders a wound previously painful entirely free from uneasiness.

* From H. A. Lechevalier and M. Solotorovsky, *Three Centuries of Microbiology.* New York: Dover Publications, 1974, p. 46.

Disinfection and Antisepsis

Historical Background

The Arabs learned several hundred years ago that the burning of a wound with hot metal *(cauterization)* prevented infection. This was a common procedure, despite the fact that the patient would be scarred for life. Even though cauterization was traumatic, it gave victims a better chance of overcoming the effects of disease agents. In 1537, the French surgeon Ambroise Paré treated gunshot wounds with bandages soaked in egg yolk, turpentine, and other materials. The turpentine caused a kind of chemical cauterization, and the egg material supplied the antibacterial enzyme lysozyme. **[See lysozyme in Chapter 17.]**

The concepts of antisepsis and disinfection were introduced largely by Ignatz Semmelweis (1816–1865) and Joseph Lister (1827–1912).

Ignatz Semmelweis

Ignatz Semmelweis, a Hungarian physician working in Vienna, observed that the incidence of *childbed fever,* otherwise known as *puerperal fever,* was much higher in the obstetrics ward run by physicians in the Vienna General Hospital than in a similar ward run by midwives. In comparing the procedures in both locations, Semmelweis observed that the midwives washed their hands frequently, whereas the physicians came directly from performing autopsies to treat patients, without changing their blood-splattered clothes or washing their hands. Semmelweis hypothesized that these procedures were important factors accounting for the difference in infection rates between the two wards. In 1847, to lower the rate of infection in his maternity clinics, he required the attendants to wash their hands with chlorinated lime. The infection rate dropped significantly. Unfortunately, Semmelweis's efforts to persuade other physicians of the necessity for cleanliness and disinfection ended in failure and his ideas were not accepted. Recognition of Semmelweis's contributions to the control of hospital-acquired (nosocomial) infections was not generally appreciated until after his death.

In the early nineteenth century, infections were still believed to be caused by some magical power in the air or by an imbalance of body fluids. Certainly, contaminated hands were not involved. It took the incredible tenacity of Pasteur, who showed that microorganisms could not only ferment fruit juice to wine but could also cause spoilage of the wine, to put across the idea that microorganisms could cause disease.

Surgical Antisepsis

Aware of Pasteur's work, Joseph Lister in England sought to prevent surgical infections. His efforts with carbolic acid (phenol) clearly established modern surgical procedures.

In 1881, Koch and his associates evaluated 70 different chemicals for use in disinfection and antisepsis. Among these chemicals were various phenols and mercuric chloride ($HgCl_2$). In 1886, Neuber used the latter compound as a surgical antiseptic as well as for the disinfection of operating rooms. He also insisted that patients wear clean gowns and that surgeons wear clean apparel in the operating room.

In 1897, Krönig and Paul published standardized procedures for the evaluation and comparison of chemical disinfectants. Their reports specified the concentrations of chemical compounds, specific test bacteria used, viable plate counts, the temperatures at which the tests were performed, and the culture media used. Effectiveness was reported in terms of the number of surviving bacteria in relation to the period of exposure.

Use of Disinfectants and Antiseptics

The diversity of microbial populations on an object to be disinfected is seldom known. The general assumption is that the most resistant forms of microorganisms are present, namely, bacterial spores. **[See Chapter 7 for bacterial spores.]** Today an almost limitless number of chemical agents are used for controlling microorganisms, and new ones appear on the market regularly. A common problem confronting those who use disinfectants or antiseptics is which to select and how to use it. Since there is no ideal or all-purpose agent, the compound of choice is the one that will kill the organisms present in the shortest time without any damage to the contaminated material. It should be noted that the selective toxicity of antimicrobial agents may vary. Some agents exert their action in a nonselective manner and have similar effects on all cell types. Other agents, such as antiseptics, are much more selective in their effects and are more harmful to microbes than to animal tissues. **[Table 11–2 presents selected chemicals and physical methods used for disinfection and antisepsis.]**

General Directions for Chemical Disinfection

All surfaces of the contaminated material that are to be treated must be exposed to the chemical agent. Therefore, before chemical treatment is begun, the material must be thoroughly cleaned. Furthermore, there must be enough space left between items so that all surfaces of each item are fully exposed to the solution.

If possible, the cleaning agent should be germicidal, so that viable microorganisms will not remain in the solution and perhaps contaminate floors or attendants by splashing.

The time at which a particular article is immersed in solution and the time of its removal should be noted. This information prevents others from interrupting the process before it is completed.

Because solutions used to kill spores are highly volatile, the room in which they are used should be well ventilated.

Certain chemicals alter the composition of the materials to be treated. Usually, the labels of commercial containers with such compounds indicate the materials on which they should not be used. Never use a chemical without first checking the label.

Disinfectants should always be diluted in the proportions suggested by the manufacturer.

Solutions should be changed often, especially if cloudiness or sedimentation appears.

It is generally a good practice to have hand lotions in the vicinity for proper hand care after using disinfectants. A safety eyewash bottle should also be available in the event of an accident.

Chemical Antiseptics

Antiseptics are usually applied and allowed to evaporate, as in the case of alcohol. Some are rinsed off with alcohol after drying, as in the case of iodine-alcohol; or rinsed off with sterile water for surgical preparations, as in the case of iodophors; or rinsed with water, leaving an active residue, as with hexachlorophene preparations.

In general, 70% to 90% isopropyl alcohol is the least expensive antiseptic yet a very effective one. The addition of iodine to alcohol greatly increases its disinfecting properties. With or without iodine, isopropyl alcohol solutions are not active against spores. The least expensive and best sporicide appears to be a combination of formaldehyde and alcohol, but this solution is too toxic for antiseptic use. Since disinfectant solutions or gases do not have to come into contact with human skin or mucous membranes, greater toxicity is acceptable, making them more generally applicable as antimicrobial agents.

The choice of antiseptic depends largely upon the needs of the particular operation and the desired effects. It should be noted that some compounds are extremely irritating, and skin sensitivity varies greatly.

Selected Disinfecting Chemicals

Chemical disinfectants are used in what are often termed "cold sterilization" methods. In such procedures, contaminated objects are submerged in a

TABLE 11–2 Selected Chemicals and Physical Methods for Disinfection and Antisepsis

Chemicals/ Physical Methods	*Effectiveness*	*Advantages*	*Disadvantages*	*Preferred Use*	*Recommended Exposure Time*
CHEMICALS					
Halogens:					
Chlorine	Kills most vegetative cells, some viruses and fungi; spores usually resistant; hypochlorite in a 1:10 dilution inactivates the AIDS virus[a]	Excellent deodorant	Activity reduced by organic material and some metallic catalysts; irritating odor and residue; solutions are somewhat unstable	Purification of drinking water	Immediate effect (from seconds to minutes, depending upon compound and concentration)
Iodine	Kills vegetative cells, some spores and viruses when used in high concentration	Extremely useful as a skin disinfectant; can be used over wide pH range	Irritating odor and residue, except with iodophors	Wound dressing, preoperative preparation	At least 60 sec to kill vegetative bacterial cells
Alcohols	Kill vegetative cells and many viruses; spores unaffected	Unaffected by organic compounds; no residue left; stable and easily handled	No major disadvantages as a disinfectant	Skin and surfaces	10 to 15 min in 70% to 80% solutions
Phenols and related compounds	Kill vegetative cells and some fungi; only moderately effective against spores	Stable to heating and drying; unaffected by organic compounds	Pure phenol is harmful to tissues and has disagreeable odor	In combination with halogens and detergents, they make excellent disinfectants	Effect is immediate
Detergents, quaternary ammonium compounds	Kill bacteria (including staphylococci; *M. tuberculosis,* TB) some viruses and spores are unaffected	Stable in the presence of organic compounds; easy to handle; no irritating residue left	Hard water, detergents, and fibrous materials interfere with activity; can rust metals	Small metal instruments	Depends upon type and concentration, generally 10 to 30 min
Heavy metals	Kill some vegetative cells and viruses; TB and spores are unaffected	Fast and inexpensive; no special equipment required	Inactivated by organic compounds and chemical antagonists	Rarely used except as preservatives and in fungal and protozoan infections	Effective as long as in contact
Aldehydes	Kill staphylococci, TB,[b] and viruses readily; spores killed only upon prolonged exposure	Glutaraldehyde is nontoxic and nonirritating to most tissues	Prolonged exposure required	Instrument disinfection	A 2% solution for 3 to 18 hours
PHYSICAL METHODS					
Boiling	Kill all microorganisms; spores vary in resistance	Inexpensive	Not practical for objects that must remain dry	Liquids	100°C for 15–20 min
Pasteurization	Effective; denatures protein and kills all pathogens and some nonpathogens	Prolongs shelf life of products	All exposed microorganisms are not killed	For milk, vinegar, and certain alcoholic beverages	Varies with properties of products, such as acidity; for milk, 70°C for 15 sec

[a] Reported by Hospital Infections Program, Division of Viral Diseases, Division of Host Factors, Division of Hepatitis and Viral Enteritis, AIDS Activity. Center for Infectious Diseases, Office of Biosafety, Centers For Disease Control; Division of Safety, National Institutes of Health.
[b] Recent evidence suggests that aldehydes may not be as effective against *Mycobacterium tuberculosis* as once thought.

disinfectant for a specified time period at room temperature. Choosing among the many agents available involves considering the conditions under which the material will be used. Then the product or formulation (combination of substances) that best suits the particular requirements should be selected. Table 11–2 summarizes many of the factors involved in making the right choice of a chemical. It is important to note that several chemicals can and do sterilize. However, the rinsing, handling, and/or time of exposure associated with their application often results in recontamination and thereby makes sterility difficult to achieve.

HALOGENS

Halogens include the compounds of chlorine and iodine, both organic and inorganic. Most inorganic halogen compounds are deadly to living cells. They kill cells by oxidizing protein, thus disrupting membranes and inactivating enzymes.

IODINE. Iodine solution, either in water or alcohol, is highly antiseptic and has been used for years as a preoperative preparation of the skin, usually applied to the skin's surface immediately before a surgical procedure. It is also effective against many protozoa, such as the ameba that causes dysentery. In proper concentration, iodine does not seriously harm human tissues. However, the clinical use of tincture (alcohol solution) of iodine stains tissue and may cause local skin irritation and occasional allergic reactions. Preparations of iodine compounded with nonionic detergents and polymers such as polyvinylpyrrolidone have been developed for disinfection and antiseptic uses, respectively. The iodine binds loosely with the organic compound and is released slowly to produce effective disinfection. Antiseptic iodophors are used routinely for preoperative skin cleansing and disinfection.

CHLORINE. Free chlorine has a characteristic color (green) and a pungent odor. Chlorine in any of its various forms has long been recognized as an excellent deodorant and disinfectant. It has been a standard treatment for drinking water in many communities. Unfortunately, most compounds of chlorine are inactivated in the presence of organic material and some metallic catalysts. Currently chlorine and chlorine-containing compounds are being replaced in water treatment by the application of ozone and other methods.

Hypochlorite solutions are most commonly used in disinfecting and deodorizing procedures, as they are relatively harmless to human tissues, easy to handle, colorless, and nonstaining, although they do bleach. They are widely employed in hospitals to disinfect rooms, surfaces, and nonsurgical instruments. They tend to leave a residue that can be irritating to skin and other tissues. A 1:10 dilution of ordinary household bleach is recommended as a disinfectant against the AIDS virus by the Centers for Disease Control in Atlanta, Georgia. Household bleach contains 5.25% sodium hypochlorite.

Several organic chlorine derivatives are also used to disinfect water. This is particularly useful for campers and others who must use water that is likely to be contaminated. The most common of the compounds employed is halazone, or parasulfone dichloramidobenzoic acid (Figure 11–1a). A halazone concentration of 4 to 8 mg/liter safely disinfects even fairly hard water containing typhoid bacilli in approximately 30 minutes. Another compound used for the same purpose is succinchlorimide (Figure 11–1b). A concentration of 11.6 mg will disinfect 1 liter of water in 20 minutes. These organic chlorides are quite stable in tablet form, becoming active when placed in water. However, these compounds have been shown to be inadequate against cysts of the protozoon *Entamoeba histolytica* (en-tah-ME-ba his-toh-LI-tee-ka), the etiological agent of amebic dysentery. Iodine preparations are available for this purpose. However, boiling for 15 to 20 minutes is often more desirable, if practical at the time. [See amebic dysentery in Chapter 26.]

ALCOHOLS

Alcohols are among the most effective and frequently used agents for sterilization and disinfection. Alcohols denature proteins, possibly by dehydration, and act as solvents for lipids. Thus, membranes are likely to be disrupted and enzymes inactivated in the presence of alcohols. Three alcohols are used: methanol, CH_3OH; ethanol, CH_3CH_2OH; and isopropanol, $(CH_3)_2CHOH$. As a general rule, the bactericidal value increases with the molecular weight; isopropyl alcohol is therefore the most widely used of the three. In practice, a solution of 70% to 80% alcohol in water is employed. Percentages above 90 and below 50 are usually less effective except for isopropyl alcohol, which is effective in percentages up to 99.

Although alcohols are somewhat affected by organic material, they leave no residue on surfaces. Large pieces of office equipment and furniture are commonly wiped with alcohol. A 10-minute exposure is sufficient to kill vegetative cells but not spores. A quick wipe serves only to cut down the bacterial population and thus reduce the chance of infection. It has been common practice to dip instruments into alcohol and then flame them. The effectiveness of this procedure is questionable, and it should not be substituted for better sterilization

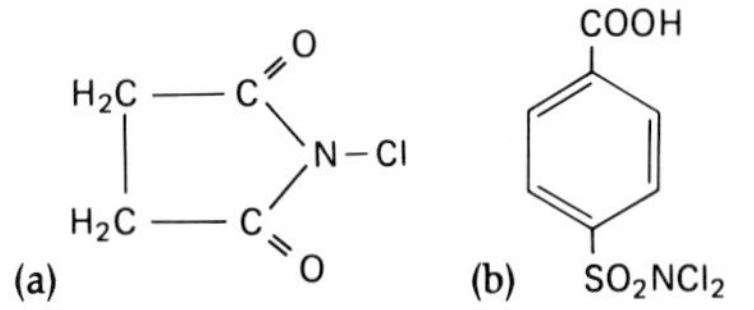

(a) H_2C—C(=O), H_2C—C(=O), N–Cl   (b) COOH, SO_2NCl_2

Figure 11–1 (a) Halazone (parasulfone dichloramidobenzoic acid). (b) Succinchlorimide.

methods. The alcohols, alone or in combination, are often used as antiseptics.

PHENOLS

Phenol (carbolic acid) is probably the oldest recognized disinfectant. It was used by Lister in 1867 as a germicide in the operating room. In low concentration, its deadly effect is caused by the fact that it precipitates proteins actively. Like the alcohols, the phenols denature protein. In addition, they disrupt membranes by lowering surface tension. Phenol is the standard of comparison for determining the activities of other disinfectants. Phenol and cresol (methylated phenol; Figure 11–2) have a typical irritating odor and are corrosive to tissues. However, they are very stable to heating and drying and retain activity in the presence of organic material. Unfortunately, they are only moderately effective against bacterial spores. The addition of a halogen such as chlorine or an organic compound enhances the activity of the phenols.

Hexachlorophene is one of the most useful phenol derivatives. Combined with a soap, it is a commonly used, highly effective skin disinfectant, although slow-acting. Unlike most phenolic compounds, hexachlorophene has no irritating odor and has a high residual action. It is also a good deodorant, a property put to extensive use by commercial deodorant and soap makers before such widespread use in over-the-counter preparations was banned by the Food and Drug Administration (FDA). After hexachlorophene was declared a controlled substance because of possible neurotoxicity in infants, chlorhexidene was made available. This compound has proven to be a highly effective substitute for hexachlorophene.

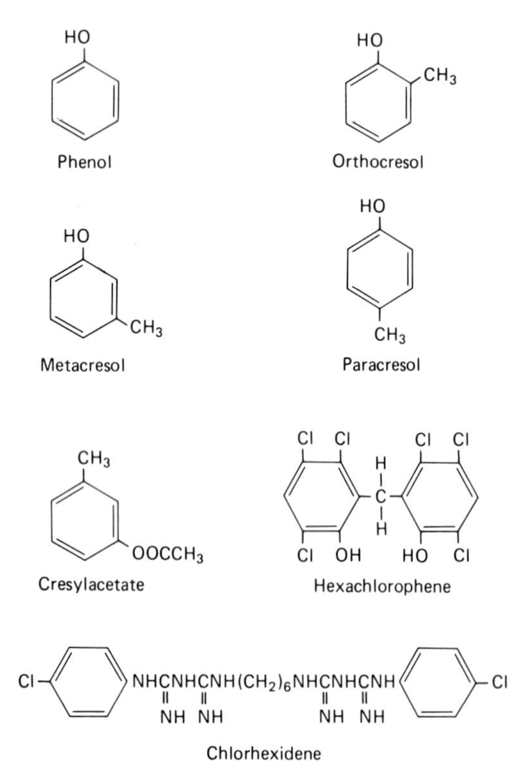

Figure 11–2 Structures of the phenolic group of antimicrobial agents.

An interesting feature of phenol and the cresols is their pain-killing properties. They can only be used externally because they are highly toxic. Slight modification of cresol yields cresylacetate (Figure 11–2), which has been used in spray form for antisepsis and as an analgesic on the mucous membranes of the ear, nose, and throat.

PEROXIDES

Hydrogen peroxide (H_2O_2) is an effective and nontoxic antiseptic. The molecule is unstable and, when warmed, degrades into water and oxygen:

$$2H_2O_2 \longrightarrow H_2O + O_2\uparrow$$

During the generation of oxygen gas, the superoxide radical (O_2^-) is formed in the presence of metal ions usually present in the cytoplasm. This highly reactive species reacts with negatively charged groups in proteins and leads to inactivation of vital enzyme systems.

At concentrations of 0.3 to 6.0%, H_2O_2 is used in disinfection, and concentrations of 6.0% to 25.0% have been used in sterilization. It is particularly recommended for materials such as surgical implants and hydrophilic soft contact lenses, as it leaves no residual toxicity after a few minutes of exposure. There is good evidence that H_2O_2 at 10% is actively virucidal and sporicidal. A 3% solution is often used to cleanse and disinfect wounds, since anaerobic bacteria are particularly sensitive to oxygen. Sodium peroxide (Na_2O_2) paste has been used for the treatment of acne. Zinc peroxide (ZnO_2) is used medically in a creamy suspension with zinc oxide (ZnO) and zinc hydroxide [$Zn(OH_2)$] for skin infections caused by microaerophilic and anaerobic organisms. Benzoyl peroxide has recently been added as an effective ingredient in skin scrubs; it is used in the treatment of acne.

ANTISEPTIC DYES

There are a variety of dyes that have a growth-inhibiting, or **bacteriostatic,** activity. We shall present two of these varieties, the *acridine* derivatives and *rosaniline* dyes. *Acriflavine,* a mixture of acridine derivatives (Figure 11–3), has low toxicity and is relatively free from skin-sensitizing properties. It has a broad spectrum of activity and has been used for treatment of urinary tract infections. The mechanism of action appears to be caused by the ability of acridines to react with deoxyribonucleic acid (DNA).

One methyl derivative of a rosaniline dye, *crystal violet,* is a potent bacteriostatic agent for gram-positive bacteria, besides being the primary stain in the

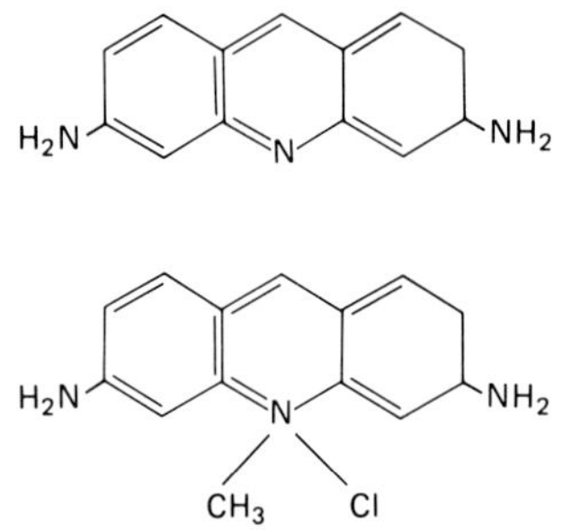

H_2N ... NH_2 (acridine structure)

Figure 11–3 Diaminoacridine (top) and diamino-methyl-acridinium-chloride (bottom).

gram reaction (Figure 11–4). Crystal violet has been used for the treatment of candidiasis, a yeast infection, and vaginitis caused by the protozoon *Trichomonas* (trik-oh-MON-as). *Candida albicans* (KAN-did-ah al-BEH-kans) is particularly sensitive to the dye. The mechanism of action of this compound against gram-positive bacteria appears to be blockage of a step in the synthesis of peptidoglycan cell wall material. [See inhibition of cell wall formation in Chapter 12.]

DETERGENTS

Detergents are organic compounds that, because of their structure, bond both to water and to nonpolar organic molecules. The molecules of a detergent have one *hydrophilic* end, which mixes well with water, and one *hydrophobic* end, which does not. Therefore, the detergent molecules orient themselves on the surface of lipid material with their hydrophilic ends toward the water (Figure 11–5). Detergents are referred to as *surfactants* because of their surface activity.

Detergents may or not not be *ionic* (electrically charged; Figure 11–6). The nonionic ones do not usually qualify as good disinfectants and may in some cases even support the growth of bacteria and fungi. Of the ionic surface-active agents, the *anionic,* or negatively charged, ones are only mildly **bactericidal.** The *cationic,* or positively charged, detergents are extremely bactericidal, especially for the bacterium *Staphylococcus* (staff-il-oh-KOK-kuss) and some viruses, although they do not affect spores.

The most widely used cationic detergents are quaternary ammonium compounds (quaternaries), which include benzalkonium chloride and acetylpyridium chloride. These compounds are highly bactericidal against gram-positive bacteria and only slightly

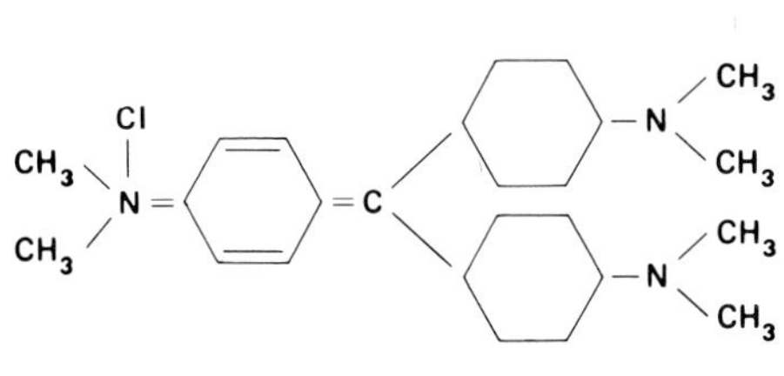

Figure 11–4 Crystal violet (hexamethylpararosaniline).

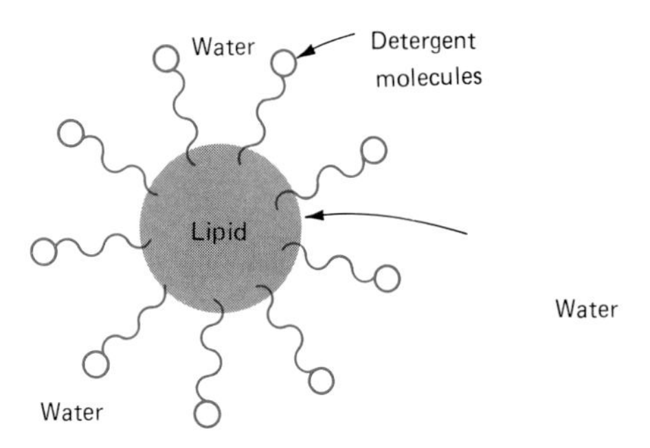

Figure 11–5 Detergent molecules bond both to nonpolar material (lipid) and to water.

less effective against gram-negative organisms. Quaternaries are also active against fungi and protozoa but inactive against bacterial spores and viruses. These compounds react with the lipid of the microbial cell membrane, altering the surface features and permeability of the structure and causing a leakage of essential cellular components. Since quaternaries have little toxicity for the skin, they are extensively used as skin antiseptics. Higher concentrations are applied as sanitizers in eating establishments, dairies, and food processing plants.

Unfortunately, there are a few problems with cationic detergents. They are absorbed by porous or fibrous materials such as cotton and cork, which may reduce their efficiency. Hard water (containing calcium or magnesium), soaps, anionic detergents, and organic matter all interfere with the action of cationic detergents. These compounds also rust metal objects unless an antirust agent such as nitrite is added. Even with drawbacks, cationic detergents are among the most widely used disinfecting chemicals as they are easily handled and are not irritating to tissues in the concentrations ordinarily used.

HEAVY METALS

Heavy metals in antimicrobials usually act by precipitating enzymes or other essential proteins of the cell. The most commonly used heavy metals are mercury, silver, and zinc.

MERCURY. Mercuric bichloride ($HgCl_2$), once a popular disinfectant, is largely deactivated by the

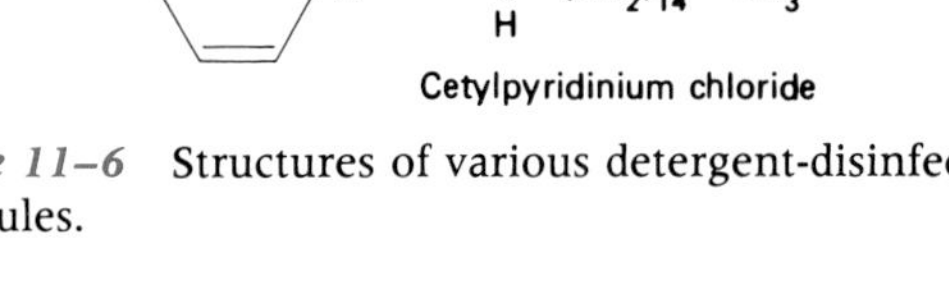

Nonionic detergent: $CH_2O.OC(CH_2)_{16}CH_3$ / $CHOH$ / CH_2OH — Stearic acid monoglyceride

Anionic detergent: $CH_3(CH_2)_{10}COO^-(Na^+)$ — Sodium laurate

Cationic detergent: $N^+ \, (Cl^-) \, CH_2-(CH_2)_{14}-CH_3$ — Cetylpyridinium chloride

Figure 11–6 Structures of various detergent-disinfectant molecules.

presence of organic material and is now considered obsolete. Organic mercury compounds are effective in the treatment of minor wounds and as a preservative in serums and vaccines.

SILVER. In a 1% concentration, silver nitrate can be used to prevent possible gonococcal and certain other bacterial infections of the eyes of newborns. The chemical is placed in the eyes of the infant immediately after birth, because if the pathogens are present in the birth canal, the resulting infection can cause blindness of the newborn (ophthalmia neonatorum). Since silver nitrate has been found not to be effective against a number of potential bacterial pathogens, it is being replaced by the antibiotic erythromycin. [See Chapter 12 for the properties of this antibiotic.]

ZINC. A mixture of a long-chained fatty acid and the zinc salt of the acid is commonly used as an antifungal powder or ointment. It is particularly effective for the treatment of athlete's foot. The zinc salt also acts as an astringent and aids in healing any superficial lesions, as does zinc oxide paste, which is commonly recommended for treating diaper rash and concurrent bacterial or fungal infections.

ALDEHYDES

The aldehydes, too, kill cells by protein denaturation. A 20% solution of formaldehyde (Figure 11–7) in 65% to 70% alcohol makes an excellent decontaminating bath if instruments are suspended in it for 18 hours. However, because the residue it leaves can cause skin irritation, the instruments must be rinsed before use. A related compound, glutaraldehyde (Figure 11–7) in solution is as effective as formaldehyde, especially if the pH is 7.5 or above. Staphylococci and other vegetative cells are killed within 5 minutes, *Mycobacterium tuberculosis* and viruses in 10 minutes, and spores sometimes in 3 hours, although as many as 12 hours may be required. The solution is nontoxic and practically nonirritating to patients.

GASEOUS METHODS

FORMALDEHYDE VAPOR. In addition to formaldehyde's usefulness in liquid disinfection, it can also be very effective as a gas. When formalin (37% aqueous formaldehyde) or paraformaldehyde (polymerized HCHO) is warmed, it releases formaldehyde vapor, an extremely effective disinfectant for instruments and various materials that have been contaminated with spores or *Mycobacterium tuberculosis.* Formalin and paraformaldehyde have certain disadvantages. The vapor has poor penetration, reacts with extraneous organic materials, and has a tendency to polymerize as a thin white film on the surface of the objects being treated.

$HCHO$
Formaldehyde

$OCH\,(CH_2)_3\,CHO$
Glutaraldehyde

CH_3Br
Methyl bromide

CH_3CH_2OH
Ethyl alcohol

Figure 11–7 Chemical structures of aldehyde and gaseous disinfectants.

METHYL BROMIDE VAPOR. Methyl bromide (Figure 11–7) vapor requires humidity for activity. Although it penetrates well, its microbicidal activity, which is by alkylation, is weak. It has been used primarily as a disinfectant for fungi and non-spore-forming bacteria.

FRAGRANCE MATERIALS

Fragrance materials or essential oils often used in perfumes have been reported to be antimicrobial because of their content of acids, aldehydes, alcohols, ketones, esters, ethers, acetals, and lactones. Through the use of standard testing methods, these oils have been shown to be highly effective against many microorganisms. They are used in Europe as the sole antimicrobial agents in some toiletries such as body deodorants. Thus, antimicrobial or deodorant action can be obtained without the more typical chlorinated phenols, benzethonium chloride (quaternary), or aluminum chlorhydrate, which can be sensitizing or allergenic for some people.

Examples of selected fragrance materials and their natural sources are presented in Table 11–3. Their mechanism of action appears to affect enzyme activity and/or inhibit glycolysis. Organisms that have been shown to be controlled by the materials in Table 11–3 include the bacteria *Staphylococcus aureus, Escherichia coli, Pseudomonas aeruginosa,* the mold *Aspergillus niger* (a-sper-JIL-lus nye-JER), and the yeast *Candida albicans.*

Testing Methods for Chemical Antiseptics and Disinfectants

The same standards of effectiveness are applied to antimicrobials whether they are intended as disinfectants or antiseptics. Until the early 1950s, the

TABLE 11–3 Selected Natural Fragrance Materials with Antimicrobial Activity

Material	*Natural Sources*
Anethol	Anise seed and bitter fennel oils
Cinnamaldehyde	Cinnamon bark and cassia oils
Eugenol	Clove and cinnamon leaf oils
Limonene	Citrus fruit oils

only accepted procedure for proving sterilizing power was the *phenol coefficient test.*

The antimicrobial efficiency of a disinfectant can be examined using a three-stage testing approach. The first stage involves laboratory screening tests that verify whether a chemical preparation or substance has antimicrobial activity. The *phenol coefficient* procedure can be used for this purpose. The second stage also is carried out in the laboratory, but under conditions simulating real-life situations. Conditions are determined at which a specific dilution or concentration of the preparation is active. The *Use-dilution test* is the one of choice in evaluating disinfectants at this stage. The last stage is conducted in the field, under actual working conditions and determines, after a normal period of use, whether microorganisms exposed to the disinfectant solution are still killed.

Phenol Coefficient Test

The phenol coefficient test compares the activity of a given product with the killing power of phenol under the same test conditions. Various dilutions of phenol and the test product are mixed with a specified volume of a broth culture of *Staphylococcus aureus* or of a certain species of *Salmonella.*

At intervals of 5, 10, and 15 minutes, a specified volume from each diluent tube is removed, added to a nutrient broth medium, and incubated for at least two days. The broth medium is selected to suit the product being tested. For example, oxidizing chemicals or mercurials must be tested with fluid thioglycollate medium, while a broth made up of nutrient ingredients plus lecithin and sorbitan monoleate (Tween 80) is necessary for testing phenolics and quaternary ammonium compounds. However, if the product is bacteriostatic rather than bactericidal, it may be necessary to incubate the system for as long as 10 to 14 days to determine the agent's effectiveness.

After the incubation period, the broth subcultures from the disinfectant dilutions are examined for visible evidence of growth. The *phenol coefficient* is defined as the ratio of the highest dilution of a test germicide that shows a killing effect in 10 minutes but not in 5 minutes to the dilution of phenol that has the same effect. For example, if the greatest dilution of a test disinfectant producing a killing effect was 1:200, and the greatest dilution of phenol showing the same result was 1:90, the phenol coefficient (PC) value would be 2.2. The value is calculated by dividing the phenol dilution into the dilution of the test substance used:

$$PC = \frac{\text{dilution of test disinfectant}}{\text{dilution of phenol}} = \frac{200}{90} = 2.2$$

Use-dilution Test

Certain questions are often raised concerning products for floor and wall disinfection. What concentration should be used for disinfecting surfaces? Is a given disinfectant compatible with soaps and detergents, or might it be deactivated by them? Is the disinfectant active in hard water or at low or high pH levels? What is the spectrum of activity against various microorganisms? Until the development of standard *use-dilution* methods by the Association of Official Analytical Chemists (AOAC), the appropriate dilution of a germicide was computed by multiplying the phenol coefficient by 20. While this practice appeared reasonable for some products, its applicability was not universal.

The AOAC use-dilution method has now been adopted to establish appropriate dilutions of a germicide for actual conditions. In this procedure, three bacterial species are tested against the product: *Staphylococcus aureus* (ATCC 6538), *Salmonella choleraesuis* (sal-mon-EL-la KOL-er-ah-soo-us) (ATCC 10708); and *Pseudomonas aeruginosa* (ATCC 15442). (The ATCC designation refers to the catalog number for the particular organism at the American Type Culture Collection in Rockville, Maryland. Cultures of these bacterial species can be obtained from this agency.) The cultures and dilutions of disinfectant are prepared according to the specific instructions of the AOAC. The bacterial species are used to contaminate small stainless steel cylinders, which are dried briefly and then placed in specified volumes of the test product. The cylinders are exposed for 10 minutes, allowed to drain on the side of the test product tube, transferred to appropriate subculture media, and incubated for two days. The results are read simply as "growth" or "no growth," using at least 10 replicates of each organism at the test dilution of the product. A satisfactory *use-dilution* is one that kills all test organisms, producing at least a 95% level of confidence. Occasionally it may be necessary to perform a 30- or even 60-cylinder test per organism in order to achieve this level of effectiveness.

The AOAC use-dilution method, although superior to the phenol coefficient procedure, should not be considered infallible. It is essential that each institution modify this AOAC procedure to take into account the use conditions and pathogens of that particular locale.

Direct Spray Method

Antimicrobial substances that have relatively poor solubility in water are difficult to test by the usual methods. They usually show very little inhibition and thus are difficult to compare in terms of relative effectiveness. Because of this problem as well as the desire for a test method more comparable to one

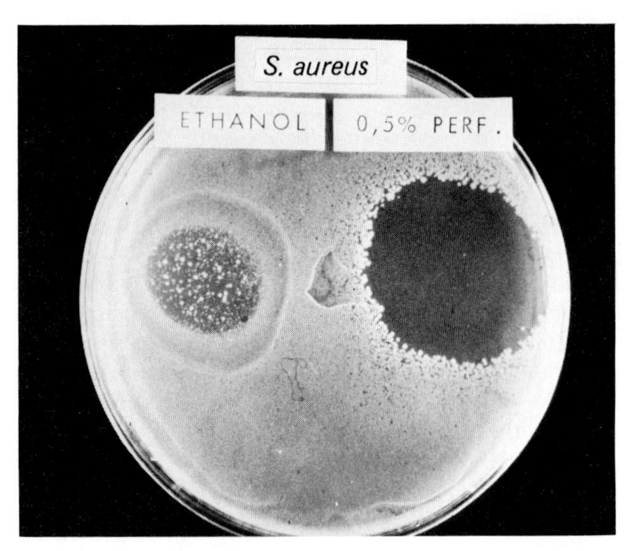

Figure 11–8 The relative effectiveness of pure ethanol and ethanol with 0.5% fragrance mixture against *Staphylococcus aureus* (staff-il-oh-KOK-kuss O-ree-us) by the direct spray method. *[Courtesy of Fermenich, Inc., New Jersey.]*

that simulated a direct application to the skin, Fermenich, Inc., developed the direct spray method to test the fragrance materials discussed earlier.

In this method, an accurate measured sample of material is sprayed with a metered aerosol valve system directly onto a previously inoculated agar plate surface. The spray is controlled so that the same-sized area of the plate is always covered with aerosolized material. Figure 11–8 shows the results of a typical experiment testing the effect of ethanol alone versus ethanol with a proprietary fragrance mixture against *Staphylococcus aureus*.

Bacteriostatic-Bactericidal Test

Many compounds may be either bacteriostatic or bactericidal, depending on the concentration at which they are used. This point can be tested by inoculating serially diluted disinfectant solutions in growth media. After a two-day incubation period, the dilutions that show no growth are subcultured to fresh media. If the growth was prevented by bacteriostatic rather than bactericidal action, the organisms will grow on subculture. In this manner, the concentration of a particular product that is bacteriostatic can be compared with results of the bactericidal tests discussed previously.

Tissue Toxicity Test

One approach to comparing antiseptics has been the *tissue toxicity test.* Germicides were tested for their killing effect with bacteria and their toxicity for chick heart tissue cells. A *toxicity index* was formulated, defined as the ratio of the greatest dilution of the product that can kill the animal cells in 10 minutes to the dilution that can kill the bacterial cells in the same period of time and under identical conditions. For example, a tincture of iodine solution was found to be toxic for chick heart tissue at a 1:4,000 dilution and bactericidal for *Staphylococcus aureus* at a 1:20,000 dilution, giving a toxicity index of 1/5, or 0.2. In contrast, a tincture of merthiolate solution was found to have a toxicity index of 3.3, and tincture of metaphen an index of 10.0. Theoretically, an antiseptic should have an index less than 1.0, indicating a greater toxicity to bacteria than to tissue cells. But it is very difficult to assess exactly how an index based upon toxicity to chick heart *in vitro* relates to human skin *in vivo.*

Physical Methods for Disinfection

Public health and welfare in many situations depend on the control of microbial populations. The previous section described many of the chemical agents available for the removal, inhibition, and killing of microorganisms. In this section attention will focus on certain physical processes used to prevent the spread of pathogens and to reduce the numbers of spoilage-causing organisms.

Boiling

The least expensive and most readily available disinfection technique is boiling. The recommended time for this procedure is 15 minutes once the water has reached a rolling boil. Vegetative cells are killed with 5 to 10 minutes' exposure. However, most spores and viruses can survive many hours of this treatment. The addition of certain substances and chemicals to the boiling bath may add to the killing power of this method. However, for critical items, a better and more reliable disinfection technique is advised.

Pasteurization

Pasteurization is a method of heat disinfection commonly applied to milk, wine, and cider. The process prolongs the shelf life of such products by decreasing the number of organisms that can cause spoilage. Although this was the original intent of pasteurization, the process has attained greater significance as a means of preventing milk-borne diseases such as tuberculosis, brucellosis (undulant fever), Q fever, certain streptococcal infections, staphylococcal food poisoning, salmonellosis, shigellosis, and diphtheria. Pathogens may gain access to milk from infected cows or handlers or by contamination of the product before pasteurization. Adequate sanitation is crucial in preventing contamination after pasteurization. **[See microbial food spoilage in Chapter 16.]**

The microorganisms that cause the milk-borne diseases are killed by exposure to 62.9°C for 30 minutes or 71.6°C for 15 seconds. The causative agents of tuberculosis, *Mycobacterium tuberculosis* and *M. bovis,* were once thought to be the most heat-resistant of the pathogenic microorganisms encountered in milk. Consequently, they were used to test the pasteurization method for its general applicability in controlling milk-borne disease. However, as improved techniques for studying rickettsiae were developed, it was found that *Coxiella burnetti* (kocks-ee-EL-la bur-NEH-tee), the causative agent of Q fever, could survive pasteurization under certain conditions that the tuberculosis organisms could not. During the batch process most commonly used for pasteurization, milk is placed in kettles and heated at 62.9°C for 30 minutes with some mixing. This method does not produce the desired temperature at the surface of the dairy product—an inadequacy that became apparent only after the development of several Q fever cases traced to pasteurized milk. With improvement in mixing, the danger of contracting Q fever was eliminated. **[See tuberculosis in Chapter 25; and rickettsial diseases in Chapter 27.]**

The *high-temperature, short-time (HTST)* or *flash pasteurization method* (71.6°C for 15 seconds) uses coiled tubing or thin sheets of milk flowing between metal plates. Proper mixing is not a problem, but adequate disinfection depends upon the cleanliness of the milk, as dirt and debris protect organisms, resulting in failure of the pasteurization process.

Milk that has been properly pasteurized has good flavor and food value comparable to that of raw milk. However, excessive heating causes a flavor change that is objectionable to some people. Moreover, the vitamin content may be substantially reduced.

The temperatures used for pasteurization are adequate to control what appear to be the significant disease agents, but they do not inactivate staphylococcal enterotoxin, a bacterial product that causes severe gastroenteritis. The organisms present in milk from cows with infected udders (mastitis) are killed by pasteurization; however, their toxic products are not inactivated. For this reason, dairy herds are frequently inspected for the presence of infected cows.

Sterilization

In biblical times, the clothes and other belongings of lepers were burned because it was believed that purification by fire was the only way to prevent contamination. The use of heat to sterilize and disinfect is still important in public health. However, during this past century, other physical means have been developed to cleanse and sterilize. **Table 11–4 summarizes the commonly used physical and chemical sterilization methods** with the advantages, disadvantages, and applications of each (see page 326).

High-Temperature Killing of Microorganisms

Thermal procedures are usually simple, reliable, and relatively inexpensive. Much of the knowledge concerning the heat destruction ("thermal kill") of microorganisms comes from studies by the food-processing industry.

Terminology of Thermal Kill

Although this chapter does not compare various procedures in depth, three important terms concerning the thermal kill of microorganisms should be noted and defined:

1. *Thermal death point:* That temperature at which a suspension of organisms is sterilized after a 10-minute exposure.
2. *Thermal death time:* The length of time required for a particular temperature to sterilize a suspension of organisms.
3. *Decimal reduction time* or *D value:* The time that will be required to kill 90% of the organisms in a suspension at a specified temperature. The temperature is generally expressed as a subscript, as in $D_{100°C}$ or $D_{59°F}$.

The thermal death point and the thermal death time take into consideration the interaction of time and temperature without any particular guidelines. The *D* value is the actual number of minutes required to reduce the viable count of microbes according to certain established guidelines. Figure 11–9 shows and explains the procedure used to calculate the decimal reduction time for a bacterial population at a specific temperature.

TABLE 11–4 Selected Physical and Chemical Means of Sterilization

Procedure	*Effectiveness*	*Disadvantages*	*Preferred Use*	*Recommended Exposure Time*
PHYSICAL METHODS				
Autoclaving	Liquids can be sterilized; good penetration; 100% effective	Equipment is expensive; dampens fabrics; corrodes metals; new high-pressure autoclaves improved but costly	All critical materials that can withstand temperature and pressure	Generally 15 to 30 min at 121°C and 15 pounds pressure
Dry heat, oven	Easily handled; 100% effective	May char fabric and melt rubber; poor penetration, so procedure is slow	Glassware, wax, oils, powders	2 hr at 160°C or 1 hr at 170°C
Dry heat, heat transfer	Can disinfect complex mechanical equipment that cannot be handled by other methods	Items must be dried after treatment; residue left; effective only with small items	Dental handpieces and small instruments	Oils and beads: 125°C for 20 to 30 min; spores: 160°C for 1 hr
Dry heat, incineration	Extremely effective	Not applicable for reusable items; public ordinances may prohibit incineration	In laboratory, inoculation loops and needles, contaminated disposable items, infectious wastes, dressings, and animal remains	Until material is completely burned to ashes
Ultrapasteurization	100% effective if managed properly	Heat transfer problems; requires thin packaging; equipment is expensive	Liquid and semiliquid foods	141°C for 2 seconds
Radiation, ultraviolet	Kills bacteria, some viruses, and some fungi; leaves no residue	Poor penetration (surface sterilization only); effects may be reversible; may cause burns to human tissue	Air and surface disinfectant	Prolonged exposure
Radiation, ionizing	Effective for vegetative bacteria and some fungi; viruses are relatively resistant	Limited effectiveness against microbial spores, toxins, and enzymes	Large-scale commercial use, although can be used in laboratory experiments, mainly used for medical devices	A radiation dose equivalent to standard heat effects produced by 121°C for 15 min with saturated steam
Filtration	100% effective for bacteria and larger organisms; may be used for removal of viruses (ultrafiltration)	Some filter media are adsorptive, fragile, or electrically charged	For thermolabile liquids such as enzymes, some sugars, and certain antibiotics	Not applicable
CHEMICAL METHOD				
Ethylene oxide	100% effective if properly managed	Slow; equipment is expensive; leaves irritating residue; materials must be aired before use	Plastic, rubber, and instruments that are sensitive to heat or chemicals	1 to 2 hours at 60°C for small loads; 10 to 12 hours necessary for large loads

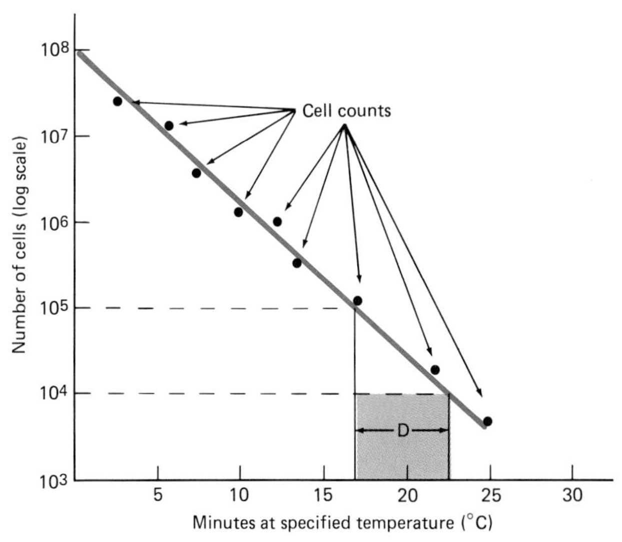

Figure 11–9 Finding the decimal reduction time (D value). Measurements at various time periods to determine the number of microorganisms killed at a specified temperature provides the data. The "•" represent the cell counts at various times. The D value can be determined by (1) connecting the cell counts to produce a straight line or curve, (2) drawing lines (dotted) from the ordinate to the curve corresponding to one log unit for the number of cells, (3) extrapolating to the abscissa (time scale). The D value in this case is approximately six minutes. Thus it takes six minutes to reduce the microbial population by a factor of 10.

How Heat Works

Moist heat appears to kill by denaturing proteins, chiefly enzymes and cell membranes. Another theory of the killing action of moist heat involves a change in the physical state of cell lipids. This view is supported by the observation that *Clostridium botulinum* (klos-TRE-dee-um bot-you-LYE-num) spores are more heat-resistant when grown in the presence of higher-molecular-weight fatty acids, which would liquefy at a higher temperature. When grown in lipid-free media, the spores exhibit less heat resistance. There are probably several factors involved in the effects of moist heat.

Bacterial death caused by dry heat appears to be due largely to the oxidation of the cell components. Studies have shown that the drier the preparation, the greater the heat resistance. If dried organisms are heated in a vacuum or under nitrogen, the killing effect is slower and the organisms appear to be more resistant. Freeze-dried or lyophilized pellets (small tablets) of *Escherichia coli* prepared by rapidly freezing and dehydrating cell suspensions in a vacuum have shown levels of resistance to dry heat close to those of spore-forming bacteria. If, however, pellets are dropped in boiling water, the *E. coli* cells are killed rapidly. Thus, the rapid denaturation of proteins and possibly the physical change in lipids appear to occur faster in moist-heat sterilization than through oxidative effects alone.

Sterilization Considerations

Preparing Materials for Physical Sterilization

With all sterilization procedures, it is absolutely essential that everything to be sterilized be scrupulously clean. This means the complete removal of all debris, particularly organic material such as blood or serum. Reducing a bacterial population through the physical action of wiping organisms off surfaces aided by the bactericidal effect of a good detergent enhances the effect of any sterilizing technique. Instruments and other metal objects should be placed in hot sodium triphosphate solution to remove organic debris before being disinfected or sterilized.

For items that must be wrapped before sterilization, either paper or muslin is recommended. A loosely woven material such as muslin allows for better penetration and circulation of heat than more tightly woven fabrics. Items should be wrapped with several thicknesses and then sealed with a heat-sensitive tape. Hinged instruments should be wrapped in open position to ensure proper sterilization. The date must be written on wrapped packs when they are sterilized, since they must be resterilized if not used within four weeks.

Monitoring Sterilization

Sterilization requires the killing or removal of *all* forms of life, especially microorganisms. Table 11–4 summarizes the conditions of time and temperature needed to ensure sterilization. Should these conditions not be met at any time, the items being treated will probably not be sterile.

It is imperative to use some means of monitoring sterilizers. There are physical, chemical, and biological approaches to use in following the sterilization procedure. With either moist or dry heat, a thermometer is very important to indicate whether adequate heat has been applied to the materials. A recording thermometer is particularly helpful, since an actual record can be filed for future reference. With the pressurized steam sterilizer *(autoclave)*, the use of a pressure gauge is also useful. It should be used with the thermometer, not in place of it, as we shall explain in detail later.

Chemicals that darken on exposure to specified heat treatments are also used to monitor moist- and dry-heat sterilization. These substances are often impregnated in paper strips or wrapping tape in such a way that heating causes the chemical to spell out STERILE. This is useful both in identifying sterilized materials and in monitoring the sterilizing device.

Biological indicators (BI) offer the surest means of monitoring sterilizers. Two bacterial spore form-

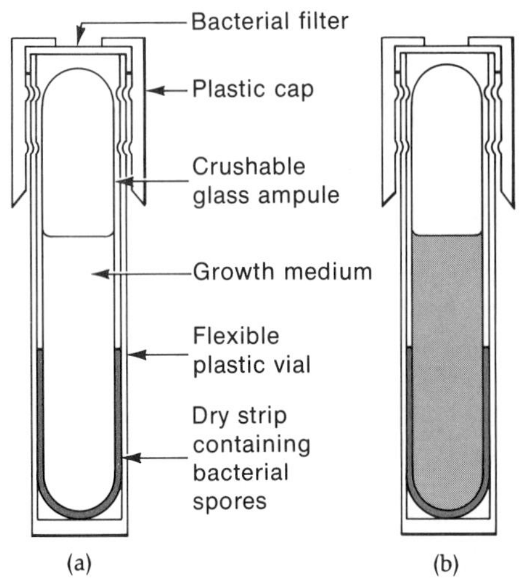

Figure 11–10 (a) A diagram of one type of biological monitoring system (the Attest Indicator) that can be used to test the effectiveness of either steam or gas sterilizers. After the sterilizing period, the indicator system is removed and allowed to cool; the ampule is crushed between the thumb and forefinger, thereby mixing the broth medium with the dry spore strip. The preparation is then incubated. (b) If growth appears after incubation, as indicated by a cloudiness or a color change of the medium, the sterilizing procedure and/or sterilizer were not effective.

ers have been selected for use (Figure 11–10). *Bacillus stearothermophilus* (bah-SIL-lus ste-roh-ther-MAH-fil-us) spores are extremely resistant to steam; *B. subtilis* var. *niger* (bah-SIL-lus SAH-til-us variety NYE-jer) spores are more resistant to dry heat. After undergoing the sterilization procedure, the control spore materials are incubated at an appropriate temperature. Growth after 24 to 48 hours indicates a failure of the sterilizer (see Color Photograph 59).

Physical Sterilization Methods

Moist Heat

AUTOCLAVING

The surest and most preferred technique for sterilization is the application of steam under pressure, or *autoclaving.* The autoclave (Figure 11–10a) consists of a steel chamber capable of withstanding more than 1 atm, or 15 pounds per square inch (psi), of pressure. Items to be sterilized are placed in the autoclave. As steam vapor enters the chamber, the air inside is forced out a vent. When the temperature inside the chamber reaches 100°C, or boiling, and all the air is removed, the vents are closed, but the steam continues to enter. This process causes an increase in the internal pressure to 15 psi above atmospheric pressure. Most autoclaves have a control valve that can be set for any desired pressure. A temperature of 121.5°C at a pressure of 15 psi must be maintained for 15 minutes for sterilization. In theory, all living material—including bacteria, fungi, spores, and viruses—is destroyed in 10 to 12 minutes. The extra time is a margin of safety. Thick packs or large liquid volumes may be held for longer times to ensure that adequate sterilizing temperatures have been reached at their centers. The various components of an autoclave and the path of proper gas flow for sterilization are shown in Figures 11–11 and 11–12.

In an autoclave the moist heat does the sterilizing, not the pressure. It is essential that *all air in the autoclave be forced out* by the steam in order to achieve the appropriate temperature. Residual air will contribute to the pressure in the system, but it will effectively lower the maximum attainable temperature. Adequate *circulation of steam* within the chamber is also essential. Improper loading of items can create "dead spaces" that are not heated to the sterilizing temperature. This, of course, may cause conditions within the chamber that prevent sterilization. When the chamber is being loaded, vessels or beakers must be arranged so that all the air may be freely replaced with steam. Packs or wrapped goods must be positioned so that the steam can reach the center of each. That is, these items must usually be tipped or laid on their sides and dry goods set far enough apart so that steam can circulate freely between them. The type of load is a factor in determining the length of exposure.

Condensation of the steam on cooler items causes the release of latent heat from the water to the items' surface, raising their temperature to a level incompatible with life. Unfortunately, the collection of water upon cooler surfaces during autoclaving dampens fibrous materials, dulls the edges of instruments, and causes metals to rust. It is recommended that a *corrosion inhibitor* such as sodium benzoate or other commercially available product be used to protect metallic items before autoclaving.

If operating instructions are followed accurately and the loading of the chamber is done correctly, the use of the autoclave for sterilization is highly effective. However, the equipment requires special installation and regular maintenance.

Faster autoclaves are commercially available, and others are being developed. These instruments operate at higher temperatures, produced by steam at higher pressures. Table 11–5 presents steam pressures, temperatures, and suggested sterilization times that are possible with various kinds of autoclaves. It is essential to repeat that *the temperature is the critical variable and is dependent upon proper displacement of the air.*

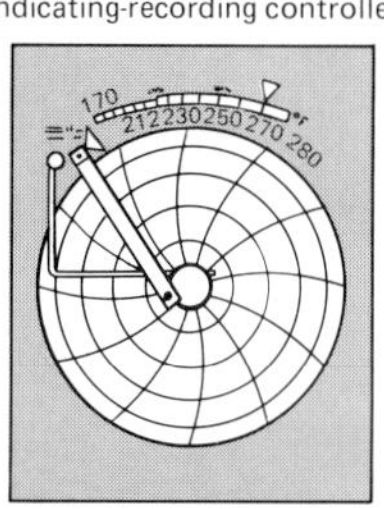

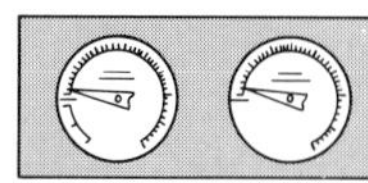

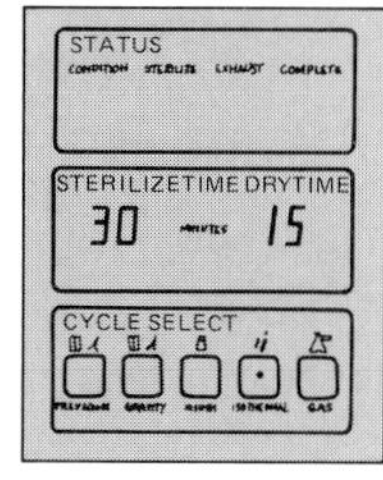

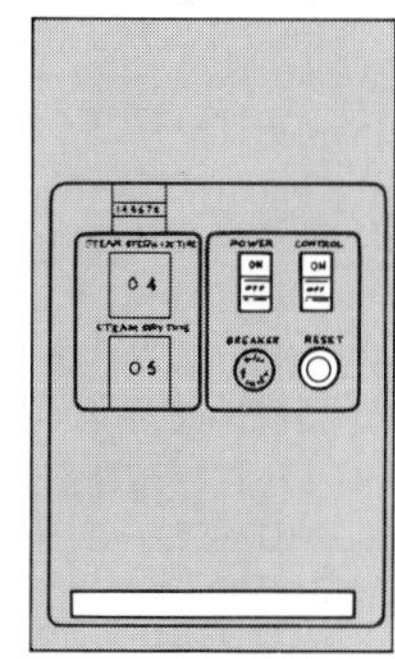

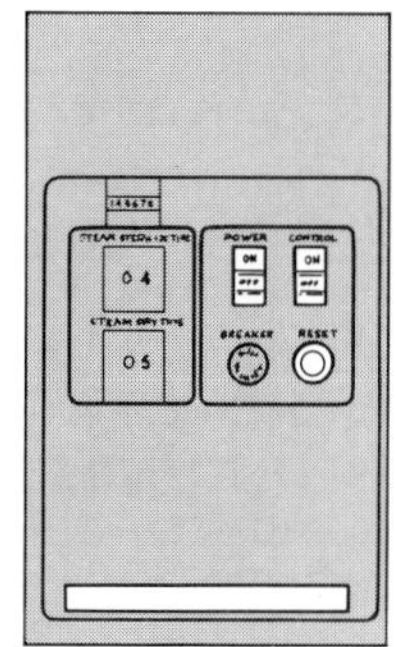

Figure 11–11
The autoclave operating system. A modern autoclave with its control column showing. At the top of the control column, the following parts are shown: *an indicating recording controller* (controls and records temperature); pressure gauges of the system; a *primary control panel,* which provides information as to what the sterilizer is doing at a given moment, timing controls, and appropriate cycle selection for load; a *secondary control panel,* which includes the power switch; and parts with which to reset the sterilization cycle and adjust the cycle as to exposure and drying times. *[Courtesy AMSCO, Erie, Pennsylvania.]*

Figure 11–12 A diagram of a downward displacement (gravity) sterilizer, showing various components and the path of airflow (light arrows) and steam flow (dark arrows). One safety valve is shown.

TABLE 11–5 General Autoclaving Pressure-Temperature-Time Relationships

Steam Pressure in Pounds per Square Inch (psi)	*Temperature* (°C)	*Temperature* (°F)	*Sterilization Time (min)*
15	121	250	15
20	126	259	10
30	135	273	3

STERILIZING SOFT ITEMS

Surgical packs, dressings, various cloth, paper, and other materials usually require drying after exposure to the steam condensate during autoclaving. This can be accomplished by evacuation of the chamber. For sterilizing soft items, the standard gravity autoclave is being replaced by high-vacuum systems. Reasons for this change include shortening of overall exposure to heat, more complete drying, and shortening of the overall process time.

In a high-vacuum system, the load is subjected to a prevacuum of 15 mm Hg absolute pressure for a few minutes prior to the entrance of steam. This procedure allows the steam to penetrate all parts of the load more rapidly than would be possible otherwise. The load is then processed at 121.5°C for 15 minutes or at 135°C for 3 minutes. The steam is removed until the pressure comes down to about 40 mm Hg absolute pressure to remove excess heat and moisture. Then sterile air is let in to bring the internal pressure to 760 mm Hg (15 psi). This procedure produces a dry, sterile load.

ULTRA-HIGH-TEMPERATURE (UHT) STERILIZATION (ULTRAPASTEURIZATION)

UHT sterilization, also known as *ultrapasteurization,* is applicable to liquid and semiliquid food materials. The development of autoclavable, flexible packaging has allowed in-package sterilization of milk, fruit juices, and drinks, and other foods. For this type of processing, the packages containing food are held between heating plates and heated to 141°C for two seconds. In another version of the method, the material is first sprayed into a pressurized steam chamber at the desired temperature; then the sterilized product is packaged aseptically. A food that can be processed and UHT-sterilized does not require refrigeration until the package is opened. The fact that refrigeration is not needed during shipping or storage offers a considerable saving in energy costs.

Dry Heat

DIRECT FLAMING OR INCINERATION

One of the simplest sterilization procedures is direct flaming. No special equipment other than a hot flame is needed, and the method is 100% effective. It merely requires that the material to be sterilized be heated to a red glow. No known living organism can withstand such treatment. As a practical means of sterilization, however, direct flaming has little application. Obviously, it cannot be used on flammable materials such as cloth, rubber, and plastic or on liquids. Moreover, instruments and other metal objects cannot withstand repeated exposure to the high temperature. Nevertheless, bacteriologists find this technique of great value in the flaming of wire transfer loops for the inoculation of sterile tubes and flasks. **[See inoculation techniques in Chapter 4.]**

HOT-AIR STERILIZATION

Items to be sterilized by the hot-air method are placed in an ovenlike apparatus capable of reaching 160° to 170°C. Materials are placed well inside the oven to allow good circulation. Hot-air sterilization procedures require a two-hour exposure at 160°C or one hour at 170°C. Because dry heat has less power of penetration than moist heat, a longer exposure time is necessary to kill all forms of life (see Table 11–6).

This method offers two advantages over autoclaving. First, there is no water present to dampen materials or to corrode instruments. Second, the hot-air equipment is relatively economical. Little installation is necessary, and hot-air ovens are easy to use.

There are, however, several disadvantages. Many smaller offices cannot afford to have equipment tied up with procedures of washing, sterilizing in dry heat, and cooling. The entire sterilization process may require a total of several hours. Another limitation is that not all items can be sterilized by this method. Fibrous materials are often scorched or charred at the prolonged high temperatures necessary for sterilization. Plastics and rubber, unless they are the more expensive heat-resistant varieties, do not fare well in dry heat. Even certain kinds of solder will melt at 170°C.

But for the sterilization of glassware such as Petri dishes and pipettes, dry heat is the method to use. After Petri plates have been sterilized by this means, the oven is often used as a convenient storage facility. Dry-heat sterilization is also the method of

TABLE 11–6 Hot-Air Sterilization Time-Temperature Relationships

Temperature (°C)	*Temperature* (°F)	*Sterilization Time (min)*
121	250	1080–1440 (18–24 hr)
140	285	180
150	300	150
160	320	120
170	340	60

choice for powders, waxes, mineral oil, petroleum jelly, and other materials that must be kept dry or that do not allow moisture to penetrate.

A newer technique for dry-heat sterilization of instruments using 200°C has been proposed. Stainless steel and metal items have been sterilized in 38 minutes, including a 10-minute cooling time. Because of the higher heat necessary, this type of procedure may be of only limited use.

Radiation

Ultraviolet Radiation

Microorganisms in the air are killed by exposure to ultraviolet (UV) radiation. The wavelengths for killing microorganisms are found in the range of 290 to 220 nanometers (nm; Figure 11–13). The most effective radiation is 253.7 nm. UV lamps are normally employed at this wavelength. Compounds such as purines and pyrimidines absorb UV at approximately 260 nm. The aromatic amino acids—such as tryptophan, phenylalanine, and tyrosine—absorb UV at 280 nm.

It appears that absorption of UV radiation produces chemical modifications of the nucleoproteins, creating cross-linkages between pairs of thymine molecules (Figure 11–15a, page 332). These abnormal linkages may cause a misreading of the genetic code, resulting in mutations that impair vital functions of the organism and consequently causing its death (Figure 11–14). Actively multiplying organisms are the most easily killed, while bacterial spores are the most resistant.

REACTIVATION OF UV-TREATED ORGANISMS

In experiments on ultraviolet killing of *E. coli*, certain suspensions, although treated in the same manner as others, were observed to exhibit a lower incidence of mutation or death. Further investigation showed that the length of exposure to visible light to which the suspensions were subjected after UV exposure but prior to their placement in the dark incubators correlated well with the decrease in mutation or kill efficiency. This *photoreactivation* was found to be caused by an enzyme active in the presence of visible light (540 to 420 nm) but not in a dark environment. The enzyme involved in photoreactivation is capable of breaking the thymine cross-linkages formed in deoxyribonucleic acid (DNA) by UV exposure. It thus repairs the UV-induced effect (Figure 11–15b). Studies have shown a wide range among bacteria in photoreactivation capability. Bacterial spores, for example, cannot be photoreactivated, whereas vegetative cells of sporulating cultures can. **[See Chapter 6 for additional details.]**

Figure 11–13 Radiation. (a) The radiation energy spectrum. (b) The germicidal effectiveness of ultraviolet radiation, a form of nonionizing radiation. Effective microbial wavelengths are found in the range of 290–220 nanometers.

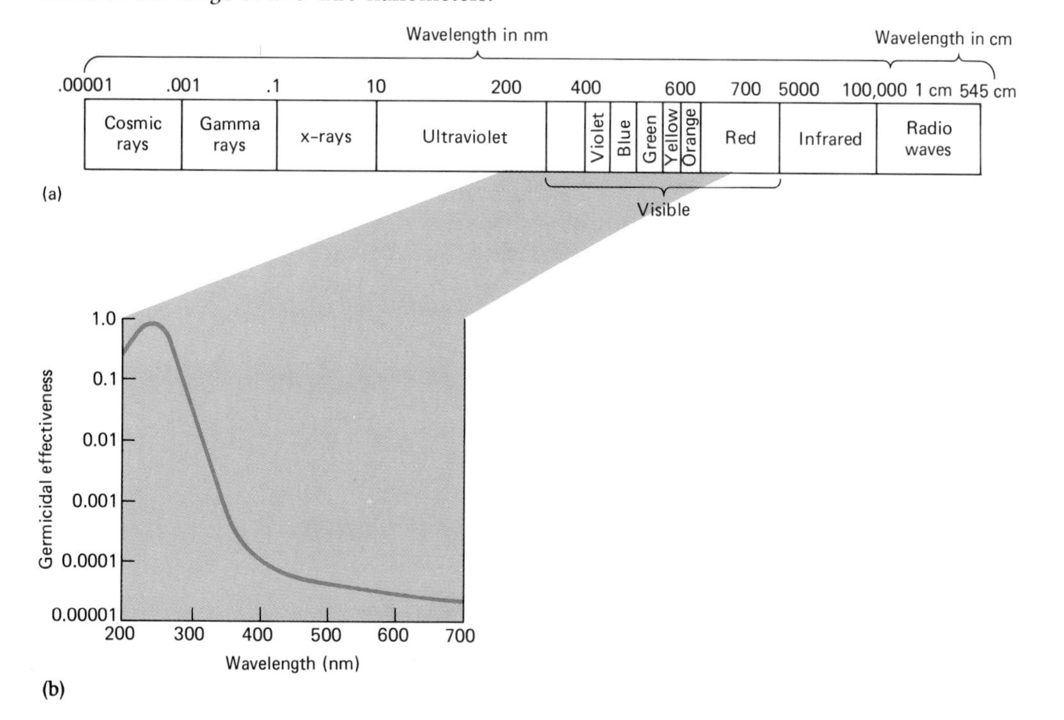

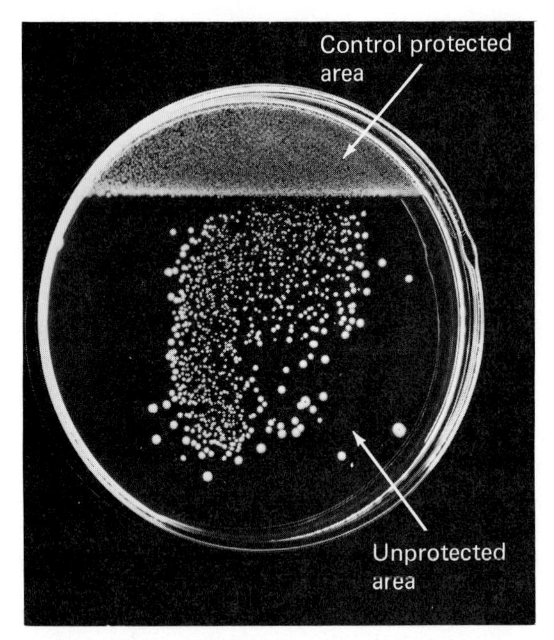

Figure 11–14 A demonstration of ultraviolet killing of *Escherichia coli.* As a control, a portion of the plate was covered and protected from UV exposure. The effectiveness of the surface exposure is indicated by the reduced number of bacterial colonies formed in the unprotected area. Longer exposure would reduce the colony number even further. The entire plate was seeded with *E. coli* before UV exposure.

Dark reactivation may also occur by a mechanism that appears to be quite different from photoreactivation. The thymine-thymine linkage is not simply broken but removed. The defect is then corrected by a DNA polymerase that replaces the thymines, using the complementary DNA strand for the necessary information.

To avoid dark reactivation, the irradiated suspension must be stored cold or in an inadequate growth medium prior to the completion of the experiment.

UV APPLICATIONS

UV radiation has been applied with success to air and water sterilization. However, poor penetration of the rays is a limiting factor. Material to be sterilized, whether liquid, gas, or aerosol, must be passed over or under the surface of a suitable lamp in thin layers if the treatment is to be effective. Any person working at or near a UV source must always wear glasses to protect the retina from severe irritation or possible permanent damage.

Commercial UV units have been developed for use in water systems where chlorination is to be avoided. Such units may have a capacity of up to 20,000 gallons/hour and have been used with wells, swimming pools, aquarium systems, and shipboard drinking water as well as in the treatment of sea water for shellfish aquaculture. In the last case, chlorine was found to inhibit the feeding of oysters. Water UV systems have been shown to kill coliforms, three types of polio virus, several types of ECHO virus, several coxsackie viruses, and some reoviruses. Similar UV units have been used in the preparation of vaccines and in combination with beta-propiolactone to sterilize plasma.

The potential of UV radiation as a relatively inexpensive treatment of waste water is receiving considerable study. The findings suggest that UV irradiation may be an effective tool in the disinfection of sewage effluent in estuaries. Such treatment may protect shellfish and their consumers from microbial contamination and infection.

Ionizing Radiation

Ionizing electromagnetic radiation includes alpha, beta, gamma, and x-rays; cathode rays; and high-energy protons and neutrons. On absorbing such radiation, an atom emits high-energy electrons, thus ionizing its molecule. The ejected electron is absorbed by another atom, creating a chain of ionizations, or an ionization path, in the irradiated substance. This activity excites chemical groups in DNA, causing the production of highly reactive, short-lived chemical radicals. Such radicals may alter chemical groups in DNA or actually break DNA strands, causing mutations.

According to the *target theory,* a cell will be killed when an ionization path occurs in a significant portion of its "sensitive volume." It is safe to assume that the target, or sensitive volume, is none other than DNA. When one wades through the complicated mathematics of the target theory concerning

Figure 11–15 The effects of ultraviolet radiation. (a) The thymine dimer ($\widehat{\text{T-T}}$) in ultraviolet-irradiated DNA. (b) Photoreactivation by a photoreactivating enzyme (PR). This reaction results in a repair of the damage to DNA. (See Chapter 6 for additional details.)

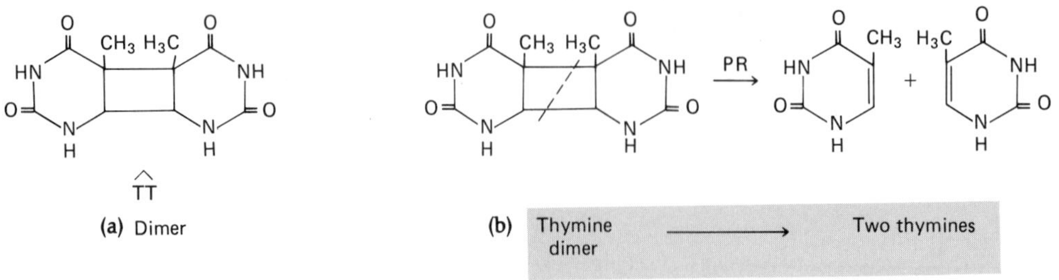

$\widehat{\text{TT}}$

(a) Dimer

(b) Thymine dimer $\xrightarrow{\text{PR}}$ Two thymines

the relative sensitivities of different organisms to ionizing radiation, it becomes evident that the sensitivity of such cells varies inversely with the size of their DNA volume. Thus, the smallest cells with the smallest targets are the most resistant.

In general, multicellular forms of life are more sensitive to ionizing radiation than are unicellular organisms. Gram-negative bacteria are more sensitive than gram-positive ones, and bacterial spores are more resistant than vegetative forms. Yeasts and molds tend to be intermediate in their resistance. Viruses, in general, are more resistant than bacteria. Some viruses are among the living systems with the highest known resistance to radiation. One conspicuous exception is the extremely radiation-resistant, gram-positive coccus *Micrococcus radiodurans* (my-krow-KOK-kuss ray-dee-oh-DUR-anz). No harmful effects are associated with this bacterium apart from its ability to spoil irradiated food.

The enzymes, toxins, and antigens of microbial origin are generally very resistant to radiation compared to living cells. Therefore, the number of microorganisms present before the application of ionizing radiation is important when dealing with medical and consumable products.

Organisms can be protected to some degree against ionizing radiation by reducing agents such as sulfhydryl-containing compounds and various chemicals that bind metal ions.

The use of ionizing rays has been considered for disinfection and sterilization procedures. This is a low-temperature form of sterilization that can be used when heat-sterilization methods would cause unacceptable damage to products. For economic reasons, ionizing radiation procedures are suitable for large-scale sterilization only. Radiation sources powerful enough to be used commercially have been available since World War II. Such sources are either a radioactive isotope or a machine that accelerates subatomic particles such as electrons to high energy levels. The radioactive isotope cobalt 60 is produced by exposing the naturally occurring form of the element to neutrons in the core of a nuclear reactor.

The main industrial application of ionizing radiation is for the sterilization of plastic hypodermic syringes and sutures. High levels of radiation have also been used to prevent spoilage in packaged meats without causing significant changes in the appearance, taste, texture, or nutritive value of the product. Ionizing radiation has been used for the radiopasteurization of fruits, seafood, eggs, milk, and poultry products as an adjunct to refrigeration, thus increasing the market life of these foods. Ionizing radiation has also been used to sterilize insect pests in stored grain. Since 1963, the Food and Drug Administration (FDA) has permitted unrestricted public consumption of fresh bacon "radiosterilized" with cobalt 60.

Radiation sterilization can be used to produce vaccines. A 1970 report by Reitman and his associates indicates that gamma irradiation was used to produce an effective vaccine against Venezuelan equine encephalitis virus infection. The vaccine appeared to be superior to both live and formalin-inactivated vaccines. Irradiated vaccines are also under study for protection against a variety of worm infections. [See vaccine production in Chapter 18.]

Microwave Irradiation

One of the more modern approaches to food cooking and preparation techniques involves microwave irradiation. Microwave ovens are quite common in the consumer market, and microwave cooking is steadily becoming more popular. Microwaves are a form of energy, not heat. When they interact with such materials as water or oils in food products, heat is generated.

The FDA, which is responsible for verifying oven safety and radiation processes, certifies only 915 and 2,450 megahertz (MHz) as usable and safe power frequencies for commercial and home ovens. (A hertz is a unit of frequency equal to one cycle per second.) Both frequencies contain energy sufficient to cause heating without breaking chemical bonds. Heat is generally produced when microwaves either cause negatively charged particles (ions) to accelerate and collide with other molecules or cause dipoles (one pair of electric poles with different charges and separated by a small distance) to attempt to rotate and line up with the rapidly alternating electrical field that is created by the 915 or 2,450 MHz power frequencies.

Several studies concerned with the effects of microwave irradiation on bacteria and fungi generally have shown cell death to be caused mainly by the heat produced. In addition, the presence or absence of water was found to control effectiveness. For example, suspensions of *Clostridium sporogenes* (klos-TRE-dee-um spo-RAH-jen-eez) spores were more susceptible to microwave heating than to conventional heating. Dried spores or freeze-dried (lyophilized) cultures of *E. coli*, *Staphylococcus aureus*, and *Salmonella* species were unaffected by such treatment.

Filtration

Filtration involves the passage of a liquid or gas through a screenlike material that has pores small enough to retain microorganisms of a certain size. The screen or filter medium becomes contaminated, while the liquid or gas that passes through it is sterilized. Certain filters also utilize materials that adsorb microorganisms. Most commonly used filters do not remove viruses.

Microbiology Highlights

A PROPHECY IN THE MAKING, FROM CHARLES CHAMBERLAND, 1884*

I have noted that even the most impure water, filtered through these vases [filtering units], was deprived of microbes and germs.

The apparatus that I have the honor of presenting to the Academy can be directly attached to a water tap and water passes through the filter under the pressure of the water system. Under a water pressure of about 2 atmospheres, as found in M. Pasteur's laboratory, one obtains, with a single porous tube or filtrating candle of 0.20 m of length by 0.025 m of diameter, about twenty liters of water a day. This seems to me sufficient for the ordinary consumption of a family. By increasing the number of candles, by grouping them in batteries, one can obtain the necessary water flow for satisfying the need of a school, a hospital or barracks. This filter literally permits one to have his own individual spring at home. M. Pasteur has demonstrated that spring waters, collected at their origin, are deprived of microbes.

* From H. A. Lechevalier and M. Solotorovsky, *Three Centuries of Microbiology*. New York: Dover Publications, 1974, p. 281.

Filtration is used for sterilizing substances that are sensitive to heat. Included in this group are enzyme solutions, bacterial toxins, cell extracts, and some sugars.

Liquid Filtration

Liquids can be filtered through any of a variety of materials, such as clay, paper, asbestos, glass, diatomaceous earth, and cellulose acetate. The earliest filters of this type, developed by Chamberland in Pasteur's laboratory, were made of unglazed porcelain (Figure 11–16). Three other types of so-called *bacteriological depth filters* were used for many years. These were: (1) Berkefeld filters, primarily made from a mixture of diatomaceous earth, asbestos, plaster of Paris, and water; (2) fritted or sintered glass filters made of fine, uniformly ground glass particles fused together at a high temperature in the form of disks; and (3) asbestos filters made of compacted layers of asbestos and paper in the form of pads. These filter types have largely been replaced by highly versatile, paper-thin membrane filters made from cellulose acetate and other related materials (Figure 11–17a).

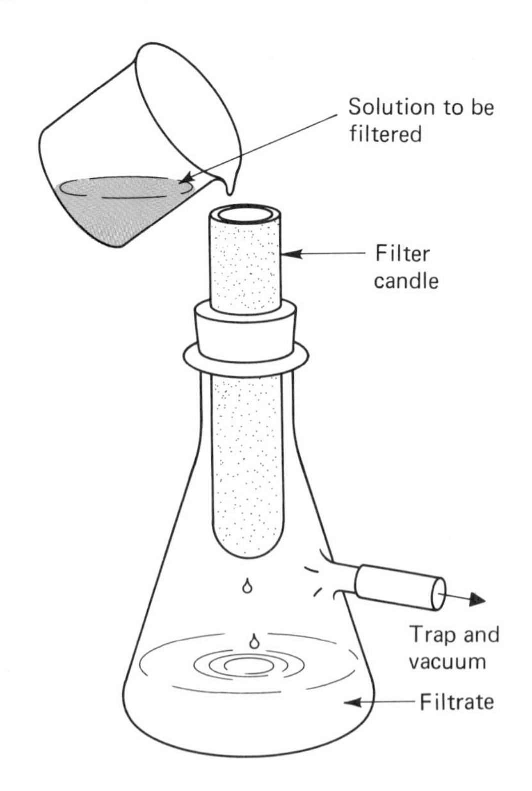

Figure 11–16 The Chamberland porcelain filter. The candle or cylinder was closed at one end, and made of hydrous aluminum silicate, or kaolin (china clay) and quartz. This cylinder was heated just enough to bind the particles in the mixture but not to glaze the porcelain formed. Varying the proportions of the ingredients produced cylinders of varying sized pores.

MEMBRANE FILTERS

Membrane filters of cellulose acetate are also made in disk form. However, the filter pad is usually much thinner than those mentioned previously. Membrane filters are 0.1 mm thick, compared to thicknesses of approximately 5 mm for the asbestos and fritted glass filters. The membrane is held in place on a supporting screen by a suitable holder (Figure 11–17b).

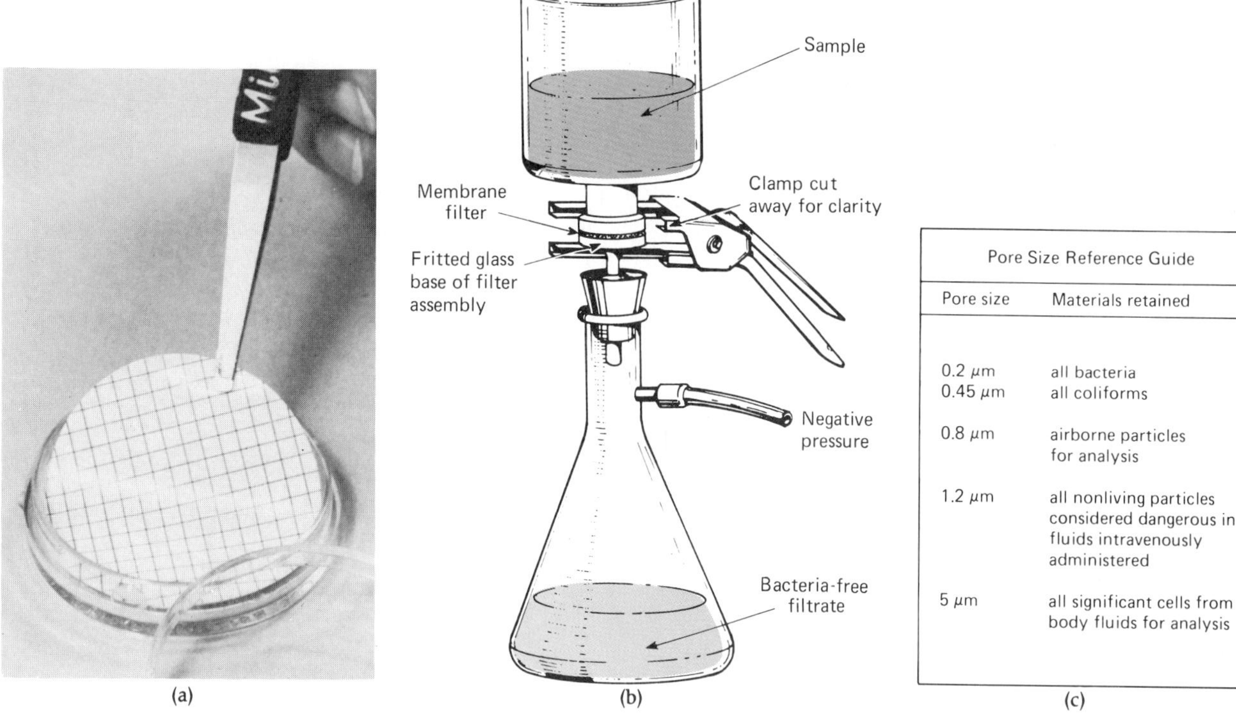

Pore Size Reference Guide	
Pore size	Materials retained
0.2 μm	all bacteria
0.45 μm	all coliforms
0.8 μm	airborne particles for analysis
1.2 μm	all nonliving particles considered dangerous in fluids intravenously administered
5 μm	all significant cells from body fluids for analysis

Figure 11–17 Membrane filtration. (a) After filtration of a sample, the thin membrane can be placed directly on a selective and differential medium. After incubation, distinct colonial types, which can be counted and identified, form on the membrane filter's surface. *[Courtesy Millipore Corporation, Bedford, Mass.]* (b) The basic features of a membrane filter apparatus. The sample is pulled through the membrane filter by negative pressure. Bacteria and related organisms unable to pass through the membrane's pores remain on its surface. (c) A guide to the pore sizes of membrane filters and the types of materials accordingly retained. Coliforms are bacteria found in the human large intestine.

All the other filters described earlier combine sieving with adsorption based on opposite charge effects. Thus, filtration of various organic compounds through porcelain, diatomaceous earth, glass, or asbestos devices may yield a filtrate of lower concentration than the original liquid. In addition to charge effects, the adsorptive nature of such filters removes certain components from solutions. The membrane filter, on the other hand, works by sieve action alone. Membrane pores are generally uniform in size. In addition, the rate of flow is superior to that of depth filters because of the number of pores. Pore sizes cover a broad range. Some membranes may retain most bacteria as well as larger protein molecules (Figure 11–17c). It should be noted that membrane filtrations remove only materials larger than the pore size of the filter used. Membrane filters are disposable, autoclavable, and compatible with many chemicals (Figure 11–18).

Some membranes can be made transparent, so that materials collected on their surfaces can be stained and examined microscopically. Various fluids can be passed through a filter and the filter then aseptically placed on an agar medium (Figure 11–17a). Organisms present in such fluids grow on the filter surface and can be identified and counted (Color Photograph 60).

Membrane filters are being used more frequently because of their wide range of applications. Such applications include the examination of body fluids, the testing of water, and the processing of various pharmaceutical products and certain alcoholic beverages such as beer and wine. **[See Chapter 16 for the use of membrane filters in the detection of water contamination.]**

Air Filtration

Filtering air to reduce microbial contaminants has found significant application in hospital operating rooms and in assembly rooms for space vehicles, among other uses. Air filtration has been practiced by microbiologists for nearly a century to prevent contamination of culture media. Sterile nonabsorptive cotton is one of the oldest known materials used for this purpose. When a cotton plug is properly inserted to plug a test tube or flask of medium (Figure 11–19), the sterilized contents will remain sterile unless the cotton becomes charred or moist-

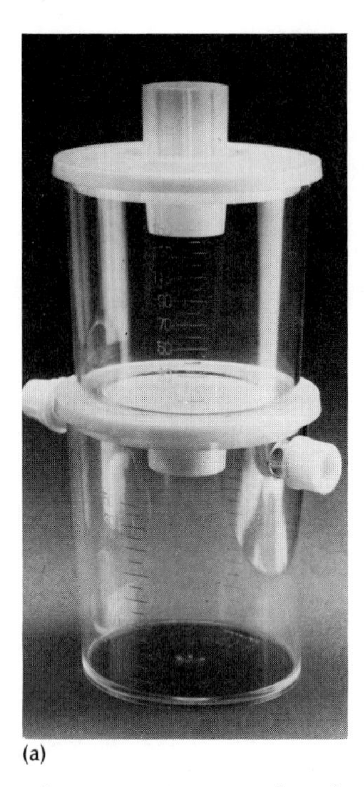
(a)

(b)

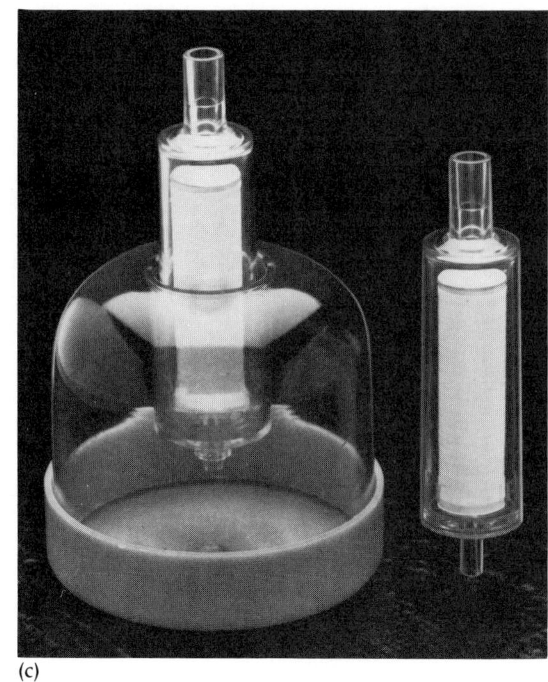
(c)

Figure 11–18 Examples of commercially available disposable and/or autoclavable membrane filter systems. (a) A single-use Sterifil system; (b) The sterile aseptic system; (c) The Sterivex system. *[Courtesy of Millipore Corporation, Bedford, Mass.]*

ened. A wet cotton plug allows bacteria to penetrate.

Surgical masks, which are made of cloth, paper, or fiberglass, behave in a similar manner. Because bacteria penetrate wet masks, masks must not be worn longer than 20 to 30 minutes and should be changed more often if necessary.

The access of bacteria to culture media can also be prevented by the use of plastic or stainless steel caps, which fit over the mouth of the tube or flask (Figure 11–19). The enclosure forces the air stream to reverse its direction in order to enter the tube. Solid particles continue in the original direction because of velocity and gravity. This was the basic principle used by Pasteur to disprove the idea that air contained some vital force responsible for the spontaneous generation of life. The Pasteur flask is shown in Figure 1–18.

HIGH-EFFICIENCY PARTICULATE AIR (HEPA) FILTERS

Commercial air filtration systems for air conditioning perform with a wide range of efficiency. The application of such devices to the extensive decontamination of air requires the use of high-efficiency particulate air (HEPA) filters. These systems have an efficiency of 99.97% for the removal of particles that are 0.3 μm or more in diameter.

HEPA filters are constructed of cellulose acetate pleated around aluminum foil and are manufactured in various sizes for different applications. One such application is the formation of nonturbulent or laminar air flow (LAF) for maximum contamination control in a given area. LAF is described as air flow moving in one direction within a confined area with uniform velocity and minimum turbulence. Originally, LAF was designed to remove dust particles from air by filtration. It was used initially in the electronics and aerospace industries to produce air with low particle levels, which was necessary to prevent electron circuitry and instrument malfunction.

Figure 11–19 Air filtration by means of a plug made of nonabsorbent material (left) and a metal cap (right).

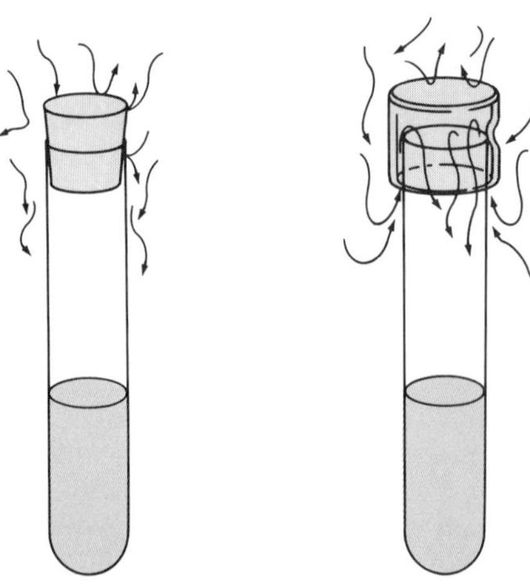

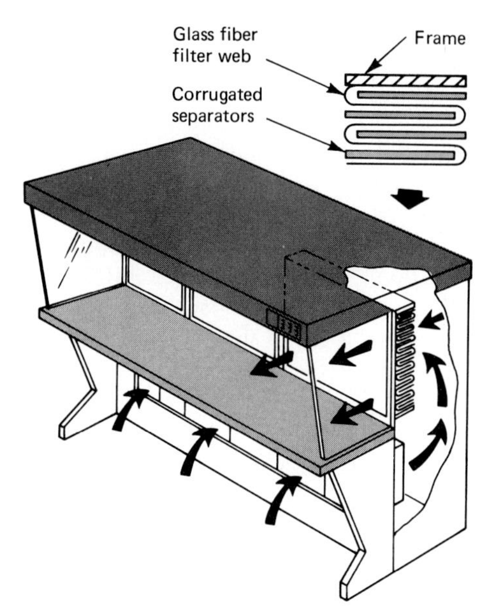

Figure 11–20 A laminar airflow bench for the prevention of sterile media contamination.

Today LAF is widely used in the pharmaceutical and cosmetics industries. HEPA filters are used in LAF systems (Figure 11–20).

LAF can be used in several forms, including rooms with wall or ceiling units, where the air flow originates through one wall or at the ceiling and exits at the opposite end, thus displacing air. LAF rooms have been used in a wide range of activities, such as providing protection to patients undergoing bone marrow transplants, protecting leukemic patients from microbial contamination by the environment during various types of treatment, preparing media and tissue cultures, and protecting personnel processing infectious material. Smaller versions of LAF units are also in use; these include the LAF bench (Figure 11–20), which is used for sterility testing and media preparation.

While LAF is an effective method to carry airborne contamination away from the work area, it is not without limitations. For example, sterilization of a contaminated object or area is not possible. Thus, proper aseptic techniques must be followed to maintain high efficiency in LAF systems.

Chemical Sterilants

Ethylene Oxide

Perhaps the best-known and most often used sterilizing gas is ethylene oxide (EtO; Figure 11–21). EtO is a cyclic ether and kills cells by acting as an alkylating agent. This means that the $CH_2CH_2O^-$ portion of the gas molecule replaces the $-H$ of

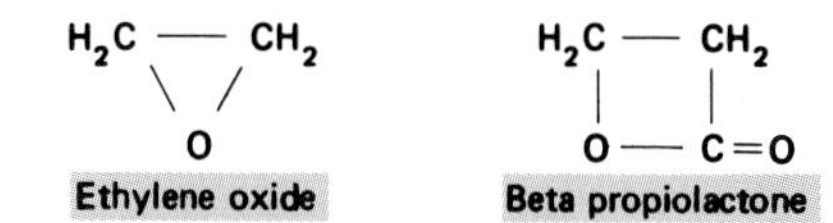

Figure 11–21 Chemical sterilants.

functional groups (CH_3, COOH) of proteins, nucleic acids, and probably other biochemicals. This addition of the EtO molecule is called *aklylation* and results in the blocking of reactive groups. In the case of proteins, this process causes denaturation.

Introduced in 1940, EtO is a highly explosive gas that is soluble in water. A special autoclave-type sterilizer is used with EtO. The chamber should be humidified for at least one hour before actual sterilizing for highest EtO activity. To ensure sterility of critical objects, an overnight exposure to 12% EtO at 60°C is recommended. Other combinations include four hours' exposure at 50° to 56°C, or 6 to 12 hours' exposure at room temperature. Because the gas is explosive in air, it must be used with caution. EtO is often diluted with inert gases such as carbon dioxide, nitrogen, or, more rarely, methyl bromide.

The maximum allowable concentration for prolonged human exposure to EtO has been set at 50 mg/liter. One study showed that acute human exposure resulted in nausea, vomiting, and mental disorientation. No deaths were reported in that study. However, other studies have implicated EtO as a cause of an increased rate of spontaneous abortions in exposed pregnant women and as a carcinogen. Because of the significant toxicity of EtO, recommendations have been made to license or to certify personnel who must operate EtO sterilizers as part of their job.

Unfortunately, EtO leaves a residue that is irritating to tissue, and all exposed items must be well aired before use. The procedure is slow and time-consuming, and the equipment is expensive. The true advantage of EtO lies in the ease with which it penetrates plastic to sterilize contents of wrapped or sealed packages. Materials that are heat- or moisture-sensitive and *not damaged by EtO* are readily sterilized. These materials include optical equipment, catheters, heart-lung machine components, artificial heart valves, respiratory therapy equipment, and often difficult-to-decontaminate items such as pillows, mattresses, and shoes. In situations when ethylene oxide sterilizers are not available or not recommended for the sterilization of heat-sensitive devices, *cold sterilization* with glutaraldehyde is used. Glutaraldehyde, a chemical relative of formaldehyde, was discussed earlier in this chapter. Items for sterilization are immersed in alkaline glutaraldehyde-containing solutions such as Cidex for 10 hours at room temperature or in acid glutaraldehyde containing preparations such as Sonacide for 1 hour at 60°C.

As with other methods of physical sterilization, the use of biological indicators (BI) or monitors for sterility testing is very important. The organism that has been chosen as the BI for EtO is *Bacillus subtilis* var. *niger* (morphotype *globigii*) and is usually provided in dried form on filter paper or sealed in ampules. The BI units may be placed in and around the material to be sterilized. After exposure, the BIs are allowed to air for 48 hours to eliminate the residual EtO before culturing and incubation to test the sterility of the process.

Beta-Propiolactone

Beta-propiolactone (BPL) (Figure 11–21) is stable at temperatures below freezing, but when vaporized in a humid environment at room temperature, it becomes a powerful sterilant. BPL is also a powerful alkylating agent. In this case the active group would be $-CH_2CH_2COO^-$. As a liquid, BPL has been used to sterilize vaccines, tissues, sera, and surgical ligatures. BPL decomposes rapidly on exposure to moisture and disappears within a few hours. When used properly as a vapor, it can be relatively nontoxic. However, *liquid BPL* has been shown to be carcinogenic and is currently rarely used.

Virus Disinfection

Disinfection of objects contaminated with viruses poses a particularly difficult problem. While some viruses, such as influenza virus, are as sensitive as vegetative bacteria to disinfectants, others such as the hepatitis B, non-A/non-B hepatitis, and poliomyelitis viruses are more resistant and therefore comparable to *Mycobacterium tuberculosis.*

The true resistance of various viruses is as yet unknown, because of the difficulties encountered in determining the effectiveness of disinfectants, the time of exposure, and the concentrations needed to inactivate them. When hepatitis, AIDS, or related viruses are suspected, heat sterilization is the preferred method, even if it degrades the contaminated material.

In both dental and medical offices, the greatest potential danger to the patient, excluding obvious disease, lies in the use of contaminated needles and the transmission of hepatitis B infection. Infected patients often carry the living virus in their bloodstream for several weeks before symptoms arise and for several years after recovery from the disease. In either case, the organism is highly infective and can cause severe illness and even death. Its transmission has been traced directly to injection of the virus by the use of contaminated needles and syringes. Hepatitis viruses are extremely stable and resistant to considerable heating, drying, and most chemicals. For this reason, any items that come into contact with contaminated serum, blood, or other specimen material must be processed rigorously to ensure sterilization. The easiest and surest way to protect a patient is to use the many disposable presterilized products available commercially. **[See viral hepatitis and AIDS in Chapters 26 and 27, respectively.]**

Summary

Disinfection and Antisepsis

1. Preventing the growth or activity of microorganisms by inhibition or killing by means of chemicals applied to living tissues is known as *antisepsis.*

2. The concepts of disinfection and antisepsis were introduced in the mid-1800s by Ignatz Semmelweis and Joseph Lister.

Use of Disinfectants and Antiseptics

An ideal disinfectant is a chemical that kills pathogens rapidly without damaging the contaminated material.

GENERAL DIRECTIONS FOR CHEMICAL DISINFECTION

1. All surfaces of contaminated material to be disinfected must be exposed to the chemical agent.

2. For efficiency and safety, all procedures should (a) be timed; (b) use properly diluted, fresh disinfectant solution; and (c) be performed in well-ventilated rooms.

CHEMICAL ANTISEPTICS

The choice of antiseptic depends largely upon the demands of the particular operation. Consideration should be given to the fact that some compounds are irritating.

SELECTED DISINFECTING CHEMICALS

1. Chemical disinfectants are used in "cold sterilization" methods.

2. A large number of such agents are used, including the following:
   a. *Halogens:* compounds of chlorine and iodine that disrupt membranes and inactivate enzymes by the oxidation of protein.
   b. *Alcohols:* three types of alcohols—ethanol, methanol, and isopropanol—that disrupt membranes and inactive enzymes by denaturizing proteins and removing of lipids.
   c. *Phenols:* cresol and other phenolic derivatives, which disrupt membranes, usually by protein denaturation.

d. *Peroxides:* hydrogen peroxide and related compounds generate oxygen gas, which inactivates essential enzyme systems.
e. *Antiseptic dyes:* acriflavine and crystal violet, which react with DNA and interfere with cell wall formation, respectively.
f. *Detergents:* molecules with nonpolar and polar regions, which generally act by dissolving the fatlike substances or lipids on or in cell walls or membranes.
g. *Heavy metals:* compounds containing mercury, silver, arsenic, zinc, and copper inactivate and/or disrupt the functions of enzymes and cellular parts by precipitating proteins.
h. *Aldehydes:* formaldehyde and glutaraldehyde kill cells by protein denaturation.
i. *Gases:* vapors of formaldehyde, methyl bromide, and ethyl alcohol denature proteins.
j. *Fragrance materials,* which are used in some toiletries for their pleasant odors, as well as antimicrobial and deodorant activities.

Testing Methods for Chemical Antiseptics and Disinfectants

PHENOL COEFFICIENT TEST

1. The *phenol coefficient test* is a standard accepted procedure for determining the antimicrobial activity of various compounds.
2. This test compares the relative activity of a given product with the killing power of phenol under the same conditions.

USE-DILUTION TEST

1. Developed by the Association of Official Analytical Chemists (AOAC), the *use-dilution test* is an accepted procedure to establish appropriate germicide dilutions to use for disinfection.
2. A satisfactory use-dilution is one that kills all test organisms, producing at least a 95% level of confidence.

DIRECT SPRAY METHOD

The testing of materials such as essential oils and fragrance materials that are not readily water-soluble can be done by this method.

BACTERIOSTATIC-BACTERICIDAL TEST

Many compounds may be either *bacteriostatic* or *bactericidal,* depending on the concentration used.

TISSUE TOXICITY TEST

The toxicity of antiseptics can be determined by exposing tissue culture systems to different dilutions of the test solution.

Physical Methods of Disinfection

1. *Boiling* is the least expensive and most readily available disinfection technique. Fifteen minutes after the water has reached a rolling boil, this method is generally effective.
2. *Pasteurization* decreases the number of spoilage-causing organisms and pathogens that gain access to certain products. Temperatures of 62.9°C for 30 minutes or 71.6°C for 15 seconds are used for the procedure.

Sterilization

HIGH-TEMPERATURE KILLING OF MICROORGANISMS

1. Sterilization by heat is usually simple, reliable, and relatively inexpensive.
2. *Thermal death point, thermal death time,* and *decimal reduction time* are important terms related to the heat killing of microbes.
3. Moist heat kills by causing the denaturation of proteins in cell membranes and enzymes.
4. Dry heat causes the oxidation of cell parts.

Sterilization Considerations

PREPARING MATERIALS FOR STERILIZATION

1. *Sterilization* refers to any act or process that kills or removes all forms of life, especially microorganisms. Actively multiplying organisms are the most easily killed, while bacterial spores are the most resistant.
2. Items to be sterilized must be cleaned to remove all organic material, such as blood or serum.
3. Wrapped sterilized materials should be labeled with the date of sterilization.

MONITORING STERILIZATION

1. Several different devices and materials are used to determine whether sterilization is effective.
2. Examples of such sterilizing monitors include recording thermometers, paper strips or wrapping tape impregnated with chemicals that darken on exposure to specific heat treatments, and biological indicators containing bacterial spores.

Physical Sterilization Methods

MOIST HEAT

1. The most effective sterilization technique is the application of steam under pressure, or *autoclaving.*
2. A temperature of 121.5°C at a pressure of 15 psi maintained for 15 minutes is commonly used in this technique.

ULTRA HIGH-TEMPERATURE (UHT) STERILIZATION (ULTRAPASTEURIZATION)

1. Selected food materials such as milk can be sterilized at 141°C for two seconds in special packaging.
2. These materials do not require refrigeration until they are opened.

DRY HEAT

1. One of the simplest means of sterilization is direct flaming, or incineration. The technique is of great value in the flaming of transfer tools during inoculation procedures.
2. Hot-air sterilization is carried out in an ovenlike apparatus capable of producing temperatures ranging from 160° to 170°C. This technique is the method of choice for the sterilization of various types of glassware.

Radiation

ULTRAVIOLET RADIATION

1. The most effective wavelength for killing microorganisms is 253.7 nm.
2. Absorption of UV radiation produces chemical changes in the nucleoproteins of cells, creating cross-linkages between pairs of the pyrimidine thymine. Such abnormal linkages may cause mutations.
3. Commercial UV light units are used in vaccine preparation and water treatment procedures.

IONIZING RADIATION

1. Ionizing electromagnetic radiations include alpha, beta, gamma, and x-rays; cathode rays; and high-energy protons and neutrons.
2. Application of ionizing radiation causes changes in DNA molecules and leads to mutations.
3. Gram-negative bacteria are more sensitive than gram-positive ones, and bacterial spores are more resistant than vegetative forms. Viruses are more resistant than bacteria, and fungi tend to be intermediate.
4. The enzymes, toxins, and antigens of microbes are more resistant to radiation than living cells.
5. Industrially, ionizing radiation is used for the sterilization of plastic items and the preservation of certain food products.

MICROWAVE IRRADIATION

1. Microwaves are a form of energy, not heat. Heat is generated when microwaves interact with water or oils.
2. Safe usable power frequencies used for commercial and home ovens are 915 and 2,450 MHz.
3. Microwave irradiation works best when water is present in the material to be heated.

Filtration

1. Filtration involves passing liquids or gases through porous material in which the pores are small enough to retain microorganisms of certain sizes.
2. Filtration is used for sterilizing substances that are sensitive to heat.
3. Some materials used as filters for liquids are diatomaceous earth, ground glass, asbestos, and cellulose acetate (membrane filters).
4. Laminar air flow (LAF) is air flow moving in one direction within a confined area with uniform velocity and minimum turbulence. High-efficiency particulate air (HEPA) filters are used in LAF systems.
5. LAF systems are used in a variety of situations, including protecting leukemic, transplant, or burn patients from microorganisms.

Chemical Sterilants

1. Ethylene oxide (EtO) is used in equipment that is similar to an autoclave for the sterilization of heat-sensitive materials such as artificial heart valves and pillows and shoes.
2. Beta-propiolactone (BPL) can be used in liquid form for the sterilization of certain vaccines, surgical ligatures, and other heat-sensitive materials.

Virus Disinfection

1. Disinfection of objects contaminated with viruses poses a difficult problem because of the resistance of some species.
2. Hepatitis and certain other viruses are extremely stable and resist considerable heating, drying, and most chemicals.
3. The use of disposable, presterilized instruments has reduced the potential transmission of such diseases by contaminated needles and syringes.

Questions for Review

1. Define the following:
   a. disinfection
   b. antisepsis
   c. sterilization
   d. disinfectant
   e. antiseptic
   f. bactericidal
   g. bacteriostatic
   h. sanitizer

2. Describe the important contributions of the following scientists:
   a. Paré
   b. Semmelweis
   c. Lister
   d. Koch
   e. Neuber
   f. Krönig and Paul

3. List and describe seven general directions for using a chemical as a disinfectant or antiseptic.

4. Describe each of the following chemicals in terms of the mechanism by which it kills microorganisms, its general effectiveness, and its advantages and disadvantages: chlorine, isopropyl alcohol, hexachlorophene, hydrogen peroxide, quaternary ammonium compounds, silver nitrate, glutaraldehyde, and ethylene oxide.

5. How is the phenol coefficient test performed and interpreted?

6. How is the use-dilution test performed and interpreted?

7. What are the advantages of the direct spray method of testing antimicrobial agents?

8. A chemical has been shown to prevent the growth of *S. aureus*. How can you determine if that chemical is bactericidal or bacteriostatic?

9. What is the purpose of the toxicity index and how is it interpreted?

10. Which disinfectant or particular method might be considered in each of the following cases?
   a. oral thermometer for use with tuberculous patients
   b. blood pressure cuff
   c. a child's rectal thermometer
   d. disposable needles and syringes
   e. kitchen floor
   f. dishes from an individual with the flu
   g. a glass contaminated with saliva from an AIDS patient

11. How would you prepare to sterilize a surgical device that must be wrapped?

12. Define:
   a. thermal death point
   b. thermal death time
   c. *D* value

13. Explain why boiling and pasteurization are means of disinfection rather than sterilization.
14. Differentiate between the mechanisms of kill for moist and dry heat.
15. How can physical, chemical, and biological monitoring be used to test heat sterilizers?
16. What are the key points to remember when using an autoclave?
17. What are the required times of exposure for sterilization and the expected temperature for pressurized steam at 20 psi?
18. Indicate the required times of exposure for sterilization at hot-air temperatures of 121°C and 160°C.
19. What is microwave irradiation?
20. What is meant by photoreactivation and dark reactivation in terms of UV radiation treatment of microorganisms?
21. Explain the target theory.
22. What types of materials are used for filtration?
23. Do filters sterilize fluids? Explain.
24. Of what medical importance is LAF?

Suggested Readings

AMERICAN HOSPITAL ASSOCIATION. *Infection Control in the Hospital,* 4th ed. Chicago: American Hospital Association, 1979. *This text contains a description of and approaches to infection control.*

BLOCK, S. S. (ed.), *Disinfection, Sterilization and Preservation.* 3rd ed. Philadelphia: Lea & Febiger, 1983. *This text contains functional coverage of current methods used for disinfection and sterilization of a wide variety of materials and environments.*

BORICK, P. M., *Chemical Sterilization.* Stroudsburg, Pa.: Dowden, Hutchinson, and Ross, 1973. *Various methods of chemical sterilization are described, including antibacterial agents, ethylene oxide, formaldehyde, and detergents.*

CASTLE, M., *Hospital Infection Control.* New York: John Wiley & Sons, 1980. *A functional reference text on practical surveillance and maintenance of asepsis in a health care facility.*

CHEREMISINOFF, N. P., P. N. CHEREMISINOFF, and R. B. TRATTNER, *Chemical and Nonchemical Disinfection.* Ann Arbor, Mich.: Ann Arbor Science Publishers, Inc., 1981. *An overview of waste-water disinfection emphasizing nonchemical techniques: radiation, sound, filtration, and heat. Typical chemical techniques are also included for a comparison.*

COLLINS, C. H., et al. (eds.), *Disinfectants: Their Use and Evaluation of Effectiveness.* New York: Academic Press, 1981. *A detailed presentation of selected disinfectants and the methods used to determine their effectiveness.*

RUSSELL, A. D., W. B. HUGO, and G. A. J. AYLIFFE (eds.), *Principles and Practice of Disinfection, Preservation and Sterilization.* Oxford: Blackwell Scientific Publications, 1982. *A comprehensive publication dealing with different types of antimicrobial agents, their properties, mechanisms of action, and applications as disinfectants, antiseptics, and preservatives in pharmaceuticals, cosmetics, foods, and related areas.*

12

Antimicrobial Chemotherapy

If it is terrifying to think that life may be at the mercy of the multiplication of those infinitesimally small creatures, it is also consoling to hope that science will not always remain powerless before such enemies.
— Louis Pasteur

After reading this chapter, you should be able to:

1. Define the terms "antibiotic" and "chemotherapy."
2. List and describe six factors influencing drug selection.
3. Distinguish between bacteriostatic and bactericidal antibiotics.
4. List at least six commonly used antibiotic drugs and indicate their respective range of effectiveness.
5. Describe the general mechanisms of antibiotic action.
6. Discuss the interactions of penicillin-binding proteins (PBPs).
7. Outline drug sensitivity testing methods for bacteria and viruses.
8. Discuss the importance and basis of microbial drug resistance.
9. List and explain interactions that may occur between antimicrobial agents.
10. Summarize problems and limitations of drugs used in the treatment of fungus, protozoan, and viral diseases.
11. Define or explain the following terms: "aminoglycoside," "beta-lactamase," "MIC," "antibiotic abuse," "plasmid," and "interferons."

Chemotherapeutic agents are chemicals used in the treatment of infectious diseases or, in certain cases, for their prevention. Some of these agents are obtained from microorganisms or plants, others are synthesized in commercial pharmaceutical laboratories, and still others, such as interferon, are produced by means of recombinant DNA procedures. For a chemotherapeutic agent to be useful, it must inhibit or kill the disease agent while causing little or no injury to the host. This chapter will describe the features of various types of chemotherapeutic agents, their applications, and the methods used to evaluate their antimicrobial activity.

Earlier in this century the leading causes of death included bacterial diseases such as diphtheria, dysentery, pneumonia, and tuberculosis. Moreover, treatment for the sexually transmitted diseases syphilis and gonorrhea not only required long periods but at times was toxic to patients. Today these and many other dreaded diseases can be effectively treated with one of several antibiotic chemotherapeutic agents. Antibiotics are among the most frequently prescribed drugs for treatment and control of microbial infections. In recent years, however, problems that threaten the effectiveness of chemotherapeutic agents have appeared with increasing frequency. These include antibiotic-resistant bacteria, the transfer of drug resistance among organisms, the multiple effects of the host environment on antibiotic action, the limitations of antiviral preparations, and drug toxicity. An understanding of the mechanisms, limitations, and effectiveness range of antibiotics is critical to dealing with such problems.

Historical Background

The discoveries of Salvarsan by Ehrlich in 1909, sulfa drugs by Domagk in 1935, and penicillin by Fleming in 1929 ushered in the modern era of chemotherapy as a means of controlling infectious diseases. Various drugs for the treatment of such diseases have been in use since ancient times. However, they were largely in the form of extracts of plants and their parts. During the sixteenth century, for example, extracts of cinchona bark (quinine) were used to treat malaria, and extracts of ipecacuanha roots (emetine) were used in the treatment of amebic dysentery. Until Paul Ehrlich (Figure 1–26) developed the "magic bullet"—Salvarsan (Figure 12–1)—for the treatment of syphilis, the selection and study of synthetic compounds or natural extracts for therapeutic purposes were largely haphazard. Ehrlich's approach to chemotherapy laid the groundwork for all modern drug development. He stressed that the effectiveness of a chemotherapeutic drug was dependent on the degree of its *selective toxicity*. Thus, a functional drug would kill or inhibit the growth of a parasite without causing serious damage to the host.

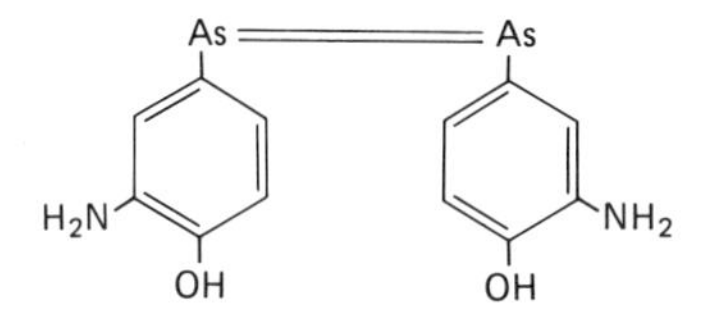

Figure 12–1 Salvarsan, Ehrlich's "magic bullet." This arsenic-containing compound was useful in the treatment of a form of sleeping sickness caused by trypanosomes in horses as well as for treatment of syphilis.

Chemotherapeutic Agents

Most chemotherapeutic agents originate in one of two ways: (1) as natural products of microorganisms or (2) as antimicrobial substances or agents synthesized in the laboratory. The term **antibiotic** originally referred only to chemicals produced by one microorganism that could kill or inhibit the growth of other microbial forms. Today, the term is applied to a variety of antimicrobial drugs, including the totally synthesized laboratory products and the chemically modified (semisynthetic) forms of natural antibiotics.

After Sir Alexander Fleming's discovery that microorganisms produce substances that inhibit or kill other microorganisms, the search for antibiotics blossomed. Table 12–1 lists the microbial sources of more than 20 antibiotics. Several of these drugs are antibacterial, antifungal, and antiprotozoan. The organisms producing them are largely soil and water species that must constantly compete for food and room to live. Thus, the production of antibiotics, which is a natural means of controlling competing microbe populations *in situ*, has been put to many uses in addition to the treatment of human infections. In recent years, attention has been given to agricultural uses of antibiotics, such as for feed additives to protect plants and livestock against infectious diseases and to accelerate their growth. They have also been used as food additives to retain product freshness for an extended period.

Antibiotics are also of great interest in research. They offer remarkable experimental devices for biochemistry—novel biochemical tools that can make a significant contribution to progress in this and related fields. The continuing search for synthetic antimicrobial agents involves efforts to isolate new natural antibiotics from novel microorganisms as well as synthetic research.

TABLE 12–1 Common Microbial Sources of Some Antimicrobial Drugs

BACTERIA			
Micromonosporaceae[a]	**Streptomycetaceae**[a]	**Bacillaceae**[a]	**Vibrionaceae**[a]
Micromonospora spp.	*Streptomyces* spp.	*Bacillus* spp.	*Chromobacterium violaceum*
Eveminomicin	Amphotericin B[b]	Bacitracin	Monobactams
Gentamicin	Chloramphenicol	Colistin	
Megalomicin	Erythromycin	Gramicidin	
Microcin	Kanamycin	Polymyxins	
Micromonosporin	Lincomycin		
Rosaramicin	Neomycin		
Sisomicin	Nystatin[b]		
Verdamicin	Rifampin		
	Streptomycin		
	Tetracyclines		
	Vidarabine		

FUNGI Deuteromycotina (Fungi Imperfecti) **Moniliaceae**[a]		
Cephalosporium spp.	*Penicillium* spp.	*Aspergillus* sp.
Cephalosporins	Griseofulvin[b]	Fumigillin[c]
	Penicillin	
	Statalon[d]	

Note: Drugs are primarily antibacterial unless otherwise designated.
[a] Family taxonomic rank
[b] Antifungal drugs; have also been used against certain protozoan pathogens.
[c] Antiamoebic drug
[d] Antiviral drug

Principles of Chemotherapy

When a pathogenic microorganism has been isolated from a patient, its sensitivity to a variety of antimicrobial agents is monitored. With the results of such tests, a physician can choose the drug best suited for the patient. The administration of an antibiotic depends upon several factors, including the patient's general physical condition, existence of drug allergies, the pathogen, and the site of infection. The site of infection is particularly important. For example, certain orally administered drugs may reach high levels in urine and fair levels in blood. However, they may fail to cross the blood-brain barrier or may not penetrate well into the tissue that contains the organisms causing the particular problem.

Another factor to consider is the dosage of the antibiotic to be given. When the laboratory performs drug susceptibility tests and reports that a certain organism is susceptible to a particular antibiotic, the result should mean that the organism is sensitive to the level of drug that can be achieved in the body. Unfortunately, it is not always possible to correlate laboratory findings with results in the body. Part of the problem is the selective concentration of chemotherapeutic agents in certain tissues, which produces drug concentrations either greater or lower than those used in laboratory testing. It is important, therefore, that the obtainable levels of the drug in the various parts of the body be known, as well as the relative susceptibilities of the pathogen. This relative susceptibility is called the **minimal inhibitory concentration (MIC),** meaning the lowest concentration of a drug that will prevent growth of a standardized suspension of the organism. Additional discussion dealing with antibiotic testing methods is presented later in this chapter.

Minimal Antibacterial (Active) Concentrations

Minute amounts of antibacterial agents have been shown to produce structural changes in bacteria and to inhibit their growth (Figure 12–2). The effects of sublethal or minimal antibiotic concentrations (MACs) as well as minimal bactericidal (lethal) concentrations (MBCs) are of great importance to successful treatment of disease states and to determining the mechanism of antibiotic action.

Properties of an Effective Antibiotic

Before any chemotherapeutic agent can be considered for use, it must meet two important requirements. First, the drug must be shown to be relatively nontoxic to the host. Second, it must exhibit antimicrobial activity at low concentrations when introduced into the body of an infected individual.

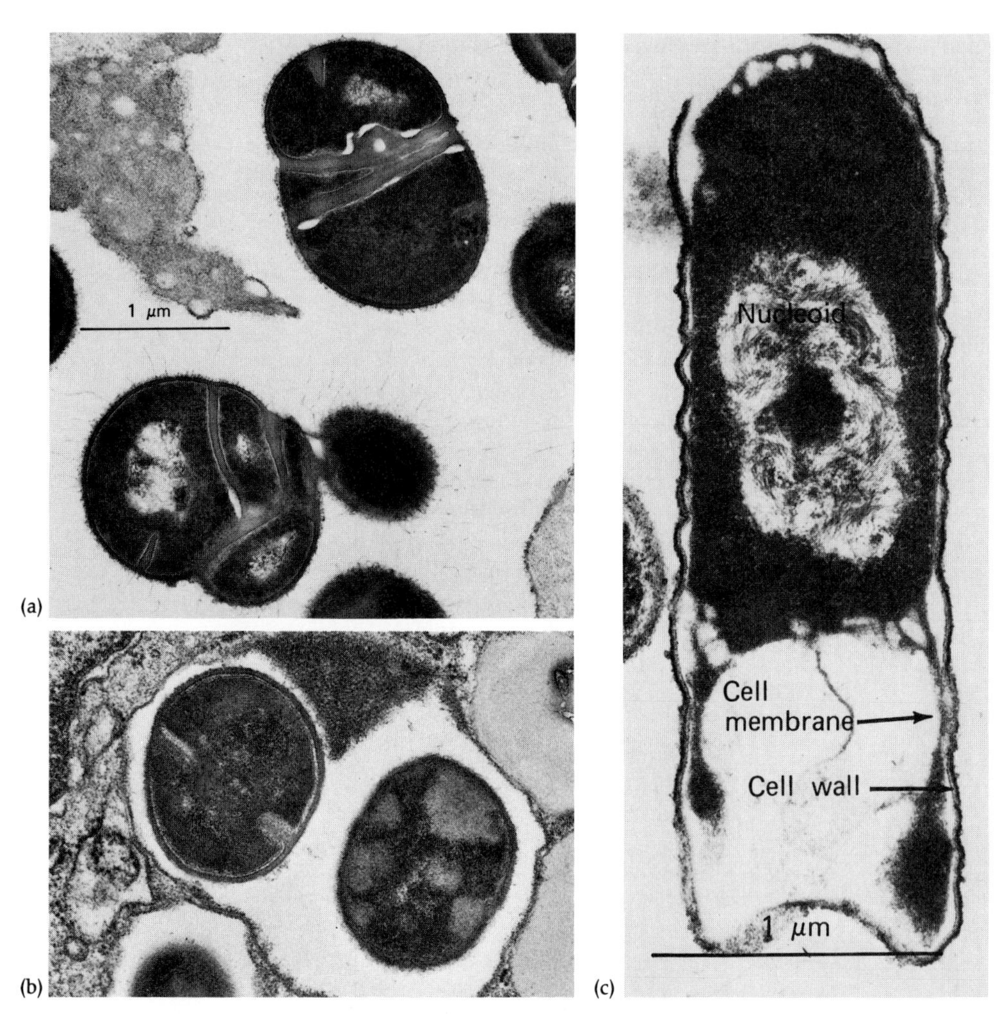

Figure 12–2 The effects of low concentrations of beta lactam antibiotics. (a) A gram-positive bacterium, such as *Staphylococcus aureus* (staff-il-oh-KOK-kuss OH-ree-us), enlarges and exhibits abnormal structures (arrows). Normal cells are shown in (b). (c) A gram-negative bacterium showing cell membrane in the middle of the cell. Original magnification, 36,000 ×. *[From V. Lorian,* J. Antimicrob. Chemother. ***15****:15–26, 1985.]*

Physicians are frequently faced with the problem of balancing the usefulness of a chemotherapeutic agent for a particular pathogen against its potential undesirable side effects, which may include nerve damage, irritations of the gastrointestinal tract or kidneys; interference with natural defense mechanisms of the host, such as phagocytosis; elimination of the host's normal microbial flora (this may create an imbalance of microbial populations within the body that can lead to unchecked growth and reproduction of pathogenic microorganisms); and the development of a patient's sensitivity or allergy to the drug itself. Table 12–2 lists a broad range of antibiotics and other preparations used to treat various bacterial infectious diseases, as well as their possible side effects. Most of the drugs listed are routinely used. The association of a single drug with a particular microorganism should not be considered an endorsement. Furthermore, it should be noted that under certain conditions, some medications may not be effective. Determining the usefulness of antibiotics for treatment takes into consideration factors such as the patient's history and clinical condition and the results of laboratory isolations and antibiotic sensitivity tests. Of particular importance is the *therapeutic index,* the ratio between the minimum toxic dose of the drug for the host and the minimum effective microbial lethal dose. Antibiotics with a high therapeutic index (a greater selective toxicity for the microorganism than for the host), are the most useful. Penicillin is an example of such a drug.

Mechanisms of Action of Antimicrobial Agents

The use of several chemotherapeutic agents leads to irreversible injury of susceptible microorganisms and ultimately cell death. Effects of this type are re-

TABLE 12–2 Mechanisms of Action and Effects of Chemotherapeutic Agents

Drug of Choice or Multiple Therapy	*Mechanism of Action*		*Possible Side Effects*
	Bacteriostatic[a]	*Bactericidal*[b]	
Aminoglycosides (gentamicin, kanamycin, streptomycin, etc.)		+	Dermatitis, fever, kidney damage, nerve injury, hearing loss
Ampicillin		+	Allergy, bone marrow depression, diarrhea, yeast infection (candidiasis)
Bacitracin		+	Kidney damage
Cephalosporins (cephalothin, cephalexin, moxalactam, etc.)		+	Generally low toxicity, allergy (hypersensitivity) may develop
Chloramphenicol	+		Drug-associated fever, interference with blood cell formation
Clindamycin		+	Diarrhea, nausea
Diaminodiphenyl sulfone (DDS or dapsone)	+(?)		Vertigo, malaise, headache, fever, reduction in white blood cells, hematuria (bloody urine)
Gentamicin, carbenicillin (alone or in combination with gentamicin)		+	Kidney damage, hearing loss
Erythromycin	+		Somewhat toxic
Isonicotinic hydrazide	+		Renal toxicity, constipation, gastritis, nerve injury, drowsiness
Methicillin		+	Allergy, renal toxicity, neutropenia (a reduction in the number of neutrophils)
Para-aminosalicylic acid	+		Generally low toxicity
Penicillins		+	Allergy
Rifampin		+	Some disturbance of liver activity
Sulfonamides	+		Vertigo, malaise, headache, fever, dermatitis, hematuria
Tetracyclines	+		Nausea, vomiting, diarrhea, skin irritation (dermatitis), vaginitis, allergy
Vancomycin		+	Toxic to the ear and kidneys

[a] Growth inhibiting
[b] Lethal

ferred to as *cidal.* A drug that causes the death of bacteria is termed **bactericidal** (Table 12–2). Those chemotherapeutic agents that do not cause cell death but inhibit growth produce a *static* effect. Dilution or removal of such drugs enables microorganisms to resume growth and reproductive activities. A chemotherapeutic agent that inhibits bacteria is referred to as **bacteriostatic.**

Antibiotics that inhibit a single bacterial group or only a few species are considered to have a narrow, or limited, spectrum of activity. Other drugs active against several gram-positive and gram-negative bacteria are said to have a *broad spectrum.*

Whether a drug exerts a static or cidal effect can be an important factor in the outcome of certain diseases. Static chemotherapeutic agents are dependent on the host's defense mechanisms for the eventual elimination of pathogenic microorganisms. If such mechanisms are ineffective, relapses or a worsening of the host's condition can result. Cidal drugs are usually independent in their actions and cause their effects directly on disease agents.

Specific Mechanisms of Action

There are several major mechanisms of action by which antimicrobial agents can inhibit or kill microorganisms (Table 12–3). These include (1) inhibition of the formation of a specific product of metabolism (metabolite); (2) inhibition of cell wall formation (Figure 12–7); (3) inhibition of protein synthesis; (4) irreversible damage to the cell membrane (Figure 12–2); and (5) inhibition of nucleic acid synthesis.

TABLE 12–3 Summary of Mechanisms of Action for Some Commonly Used Antibiotics[a]

Process or Structure Affected	*Antibiotic*
Metabolism	Isonicotinic hydrazide (INH), para-aminosalicylic acid (PAS), sulfonamides, trimethoprim
Cell wall synthesis	Bacitracin, cephalosporins, cycloserine, fosfomycin, penicillins, ristocetin, vancomycin
Protein synthesis	Aminoglycosides, chloramphenicol, clindamycin, cycloheximide, erythromycin, fucidin, lincomycin, puromycin, rifampicin, tetracyclines
Cell membrane	Amphotericin B, colistin, imidazole compounds, nystatin, permeability polymyxins, vancomycin
DNA synthesis	Mitomycins
RNA synthesis	Actinomycin, ethambutol, 5-fluorocytosine, griseofulvin, rifamycins

[a] Antibiotics such as amphotericin B and nystatin are used in the treatment of mycotic and protozoan diseases.

The following sections will describe the specific effects of some commonly used antimicrobial drugs.

Combination Chemotherapy

At times, the effectiveness of a particular antibiotic can be increased when it is administered in combination with another antimicrobial agent. Combination antimicrobial therapy has become an accepted approach to treatment, since it can (1) provide broad coverage in coping with infections due to unidentified organisms; (2) be used for mixed infections, since a single drug may not be active against all of the infecting organisms; (3) increase bactericidal activity; and (4) prevent or delay the emergence of strains resistant to either one of the combined drugs. The combined effect of two drugs acting together may be *synergistic, additive,* or *antagonistic.* If the resulting antimicrobial activity is greater than when each agent is given separately, the effect is referred to as a *synergistic* one (Figure 12–3). For example, the use of an antifungal agent such as amphotericin B, which alters the permeability of cell membranes, enhances the antifungal effects of 5-fluorocytosine and other compounds by increasing their penetration into fungal cells. If the combined activity of these drugs is equal to the sum of their independent activities measured separately, the effect is considered additive. Some drug combinations may be *antagonistic,* a condition resulting when one drug interferes with the antimicrobial activity of another. Situations of this type may occur when an antibiotic such as tetracycline, which inhibits bacterial growth, is administered along with penicillin, which interferes with actively growing cells.

The use of antibiotic combinations for treatment prevents or reduces the development of antibiotic-resistant microorganisms. Toxic effects of individual drugs can also be reduced by lowering the concentrations of each one in combination. Moreover, combinations can provide broad-spectrum coverage for life-threatening disease states before a diagnosis has been established with a great degree of certainty.

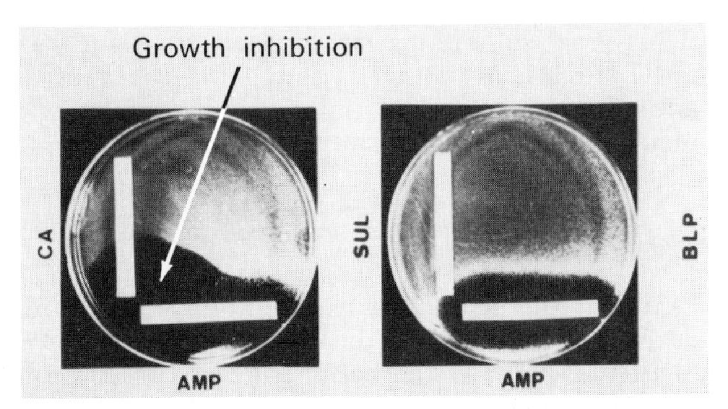

Figure 12–3 The *in vitro* effects of ampicillin used in combination with inhibitors of penicillin inactivation (lactamase inhibitors). (a) Ampicillin and clavulanic acid used together show a synergistic result against the test organism. (b) Ampicillin and sulbactam do not increase the effectiveness of the antibiotic *[From M. D. Kitzis, L. Gutmann, and J. F. Acar,* J. Antimicrob. Chemother. *15:23–30, 1985.]*

Some Commonly Used Antimicrobial Drugs

Sulfa Drugs

Gerhard Domagk reported in 1935 that a red dye compound, prontosil, was chemotherapeutic for streptococcal infections in mice. This drug was not active *in vitro* and supported Ehrlich's assumption that a chemotherapeutic agent must be modified in the body in order to be effective. By 1936, scientists at the Pasteur Institute in Paris had discovered that prontosil was converted to sulfanilamide (Figure 12–4), which was active both *in vitro* and *in vivo.*

Sulfanilamide is but one of many types of sulfa drugs. Each of these compounds has certain peculiarities related to its solubility and relative toxicity.

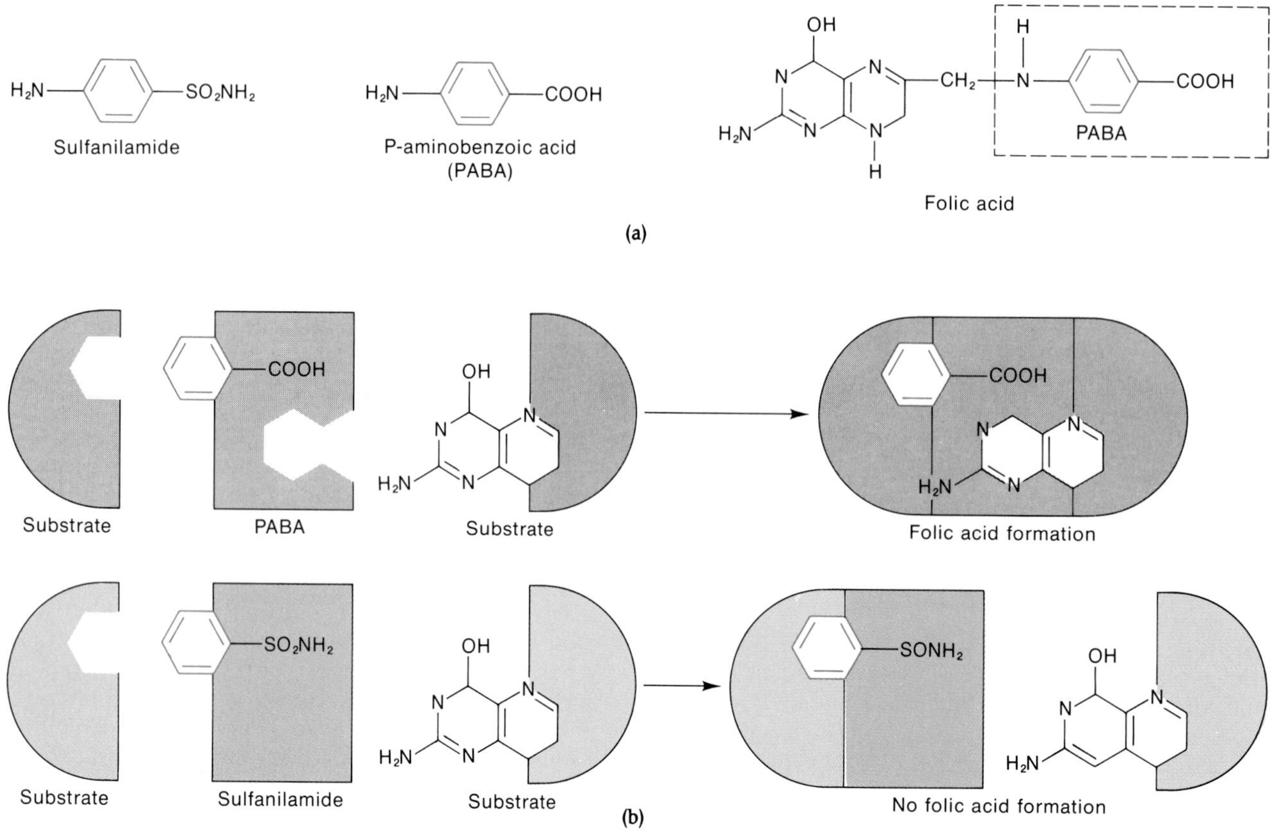

Figure 12–4 Sulfa drugs and their mode of action. (a) The formulas of the simplest sulfa drug, sulfanilamide, and para-aminobenzoic acid (PABA), an important compound needed by many bacteria for the formation of the essential growth factor, or coenzyme, folic acid. Note the position of PABA in folic acid. (b) PABA normally combines with other reacting molecules (substrates) to form the pure product folic acid (top). Sulfonamides inhibit the growth of susceptible organisms by replacing or competing with PABA in this type of reaction, thus preventing the formation of folic acid (bottom).

RANGE OF ACTIVITY

Sulfa drugs are used for treatment of *Escherichia coli* urinary tract infections, meningococcal infections, certain protozoan diseases, chancroid, trachoma, and infections caused by *Nocardia* spp. They are not particularly effective in the presence of dead tissue or pus and therefore are used only for mild to moderate infections. **[See urinary tract infections in Chapter 28.]**

MECHANISM OF ACTION

Sulfa drugs are bacteriostatic, acting upon bacteria that are growing and actively metabolizing. The mechanism of action is associated with the similarity of their structures to paraminobenzoic acid (PABA), which is a part of the essential coenzyme and growth factor folic acid (Figure 12–4a). Thus, a sulfonamide may enter a metabolic reaction in place of the PABA and interfere with the synthesis of folic acid. A lack of this coenzyme will disrupt normal cellular activities. Sulfonamides are active against bacteria that synthesize their own folic acid and cannot differentiate between these compounds and PABA. The mechanism here is an example of competitive inhibition of enzymes. Sulfa drugs do not usually inhibit growth of cells that require preformed folic acid (Figure 12–4b). The level of sensitivity of the sulfa drugs varies widely from one bacterial species to another. Trimethoprim is another antimicrobial agent that interferes with folic acid metabolism.

Penicillins

Penicillin, the first widely used antibiotic, was introduced to clinical use for the general public in 1945. Figure 12–5 shows the basic structure of penicillin as it appears in (1) the natural product, penicillin G; (2) a semisynthetic penicillin, ampicillin; (3) another semisynthetic penicillin, carbenicillin; and (4) cephalothin, one of a group of natural and semisynthetic "penicillins" called *cephalosporins.*

The cephalosporin antibiotics are included with penicillin because of the great similarities in basic structure. The portions of the molecules shown to

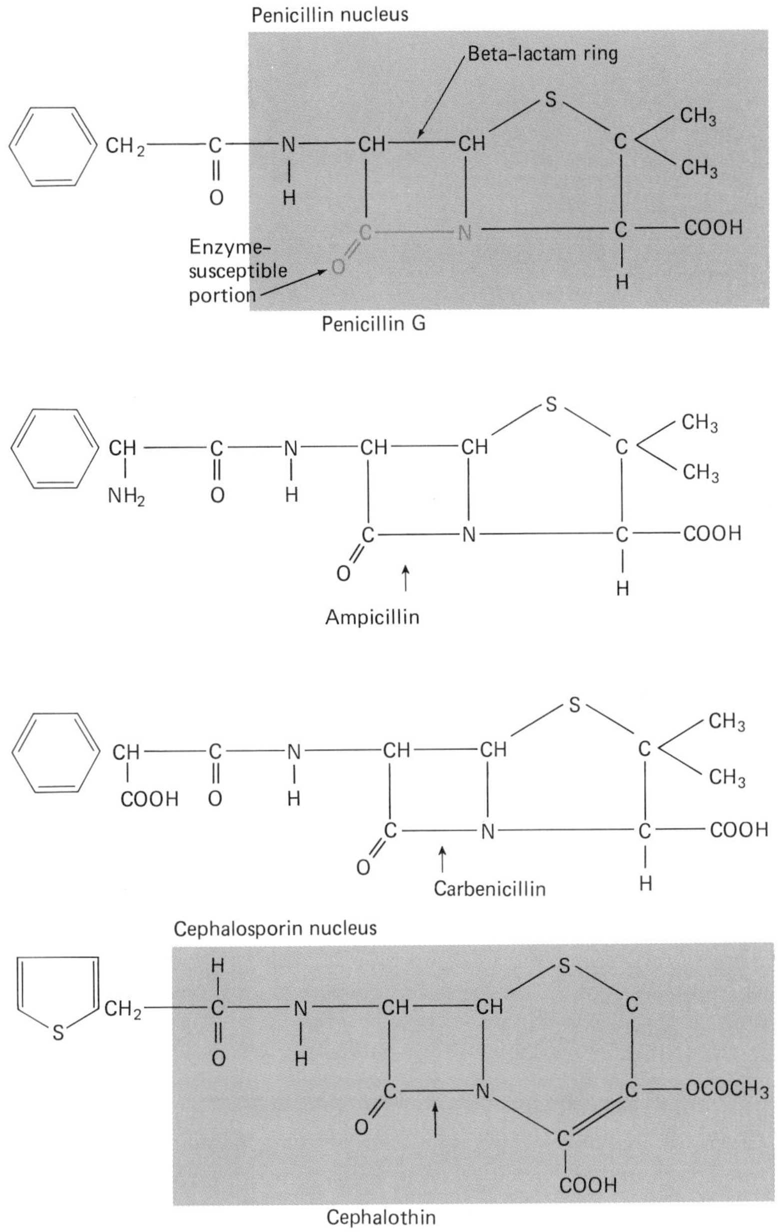

Penicillin nucleus
Beta-lactam ring
Enzyme-susceptible portion
Penicillin G

Ampicillin

Carbenicillin

Cephalosporin nucleus
Cephalothin

Figure 12–5 Four antibiotics of the penicillin family. The penicillin nucleus or core is shown on one side for each of the first three antibiotics. The beta-lactam ring and the bond that is susceptible to the action of beta-lactamase enzymes (arrows) are also indicated. A cephalosporin nucleus is indicated in the cephalothin molecule. This site (arrow), susceptible to attack by cephalosporinase, is also shown.

the left of the *penicillin nucleus* in Figure 12–5 differ from the natural penicillin G, while the portions to the right are identical. Cephalosporins have become very important in the treatment of bacterial infections because of their low toxicity, broad antibacterial activity, and stability to beta-lactamases. Beta-lactamases are bacterial enzymes that attack the beta-lactam ring component of penicillin and related antibiotics (Figure 12–6).

Ampicillin was developed in an effort to obtain improved drugs. The slight modifications of the molecule by the addition of an amino (NH_2) group converted penicillin into a broader-spectrum chemotherapeutic agent.

Disodium carbenicillin was the first semisynthetic penicillin to attain extensive clinical use specifically because of its pronounced activity against selected gram-negative organisms, primarily *Pseudomonas* and certain strains of *Proteus.* The main application of this antibiotic, alone or in combination with gentamicin, has been for serious *Pseudomonas* infections such as severe burns and infections of the pulmonary and urinary systems. Microbial resistance to carbenicillin appears to develop primarily through inappropriate use of the drug. **[See skin infections in Chapter 23.]**

The original cephalosporin-producing molds, *Cephalosporium* spp., were isolated from salt water by Brotzu in 1945. As with penicillin, additional work showed that other antibiotics obtained from these molds were more effective than the original. By making chemical changes in their basic formula, several new broad-spectrum antibiotics were developed. The cephalosporins are divided into three

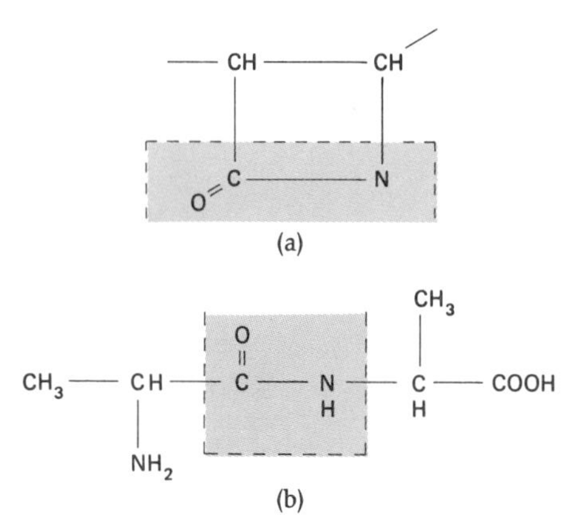

Figure 12–6 The portion of the lactam ring of penicillin (a) that is comparable to a region of the dipeptide (alanylalanine). (b) It prevents alanylalanine from being incorporated into the developing bacterial cell wall.

groups or generations, primarily on the basis of their antibacterial spectrum. Among those currently available for use are cephalothin (Figure 12–5), cephalexin, cefazolin, cephaloridine, and moxalactam.

RANGE OF ACTIVITY

Penicillin G and closely related drugs are highly active against sensitive strains of gram-positive and gram-negative cocci, gram-positive bacilli, and gram-positive and gram-negative anaerobic bacteria. At high drug concentrations, which can be produced in the urinary tract, penicillin is known to be effective against *E. coli* and *Proteus mirabilis.* However, ampicillin is the drug preferred because it can be used in smaller concentrations against *E. coli, P. mirabilis, Haemophilus influenzae, Salmonella,* and *Shigella* spp., as well as the other organisms that respond to penicillin. The spectrum of the cephalosporins is similar to that of ampicillin, but they are active also against pneumococci, *H. influenzae,* and most anaerobic bacteria. Unlike penicillin and ampicillin, cephalosporins are not inactivated by beta-lactamase, an enzyme produced by penicillin-resistant staphylococci, but they are susceptible to cephalosporinases produced by various bacteria.

MECHANISM OF ACTION

All four of these beta-lactam chemotherapeutic agents are bactericidal, and all act by interfering with cell wall synthesis. The portion of their structure associated with the four-member lactam ring is comparable to a region of the dipeptide alanylalanine (Figure 12–6). **[See cell wall structure in Chapter 7.]**

Apparently, the lethal target of beta-lactam antibiotics is a *transpeptidase,* the enzyme that connects the peptides of the peptidoglycan layer of cell walls. Penicillin is effective only during the growth stages of sensitive organisms, since fully formed cell walls are not sensitive to its action. Sensitive bacterial cells grown in the presence of penicillin or modified forms of the antibiotic have unusual shapes and abnormal internal organization (Figure 12–7). These effects are caused by the interactions of the antibiotic and three to eight distinct types of penicillin-binding proteins (PBPs). PBPs have been found in the cytoplasmic membranes of both gram-positive and gram-negative bacteria and are thought to be involved in cell wall formation, cell elongation, maintenance of cell shape, and cross-wall (septum) formation. The binding of beta-lactams to different PBPs results in certain structural changes.

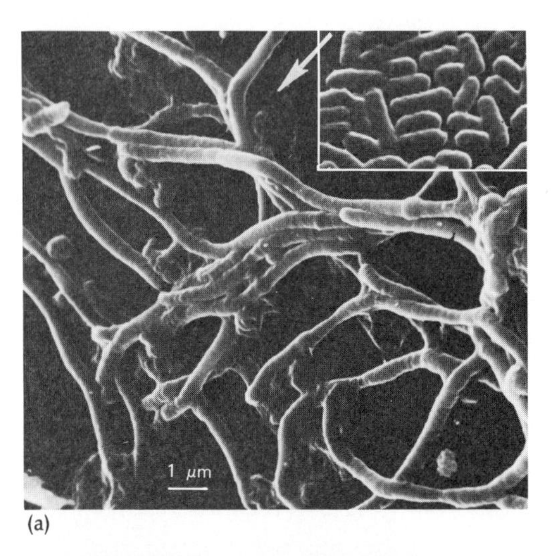

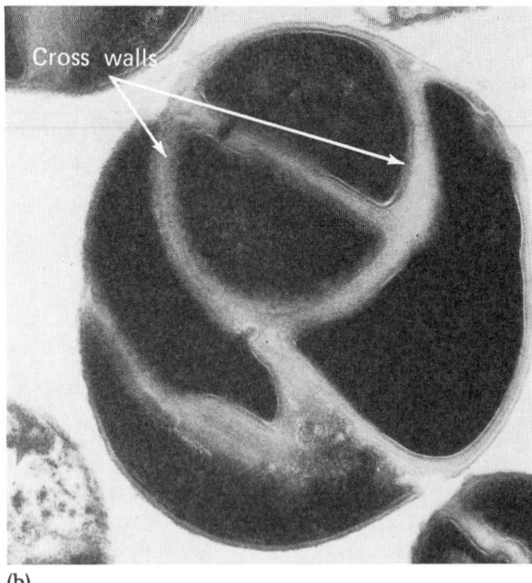

Figure 12–7 The effects of penicillin and its modified forms on bacteria. Even small amounts of these antibiotics are associated with changes in morphology and/or growth inhibition. (a) Antibiotic treatment with ampicillin converts *Salmonella typhimurium* from the form shown in the inset to long filaments. (b) The formation of several crosswalls in *Staphylococcus aureus* after exposure to a sublethal inhibitory concentration of penicillin. *[V. Lorian, et al.,* Proceedings of the 10th International Congress of Chemotherapy, *American Society for Microbiology, 1978, pp. 72–78.]*

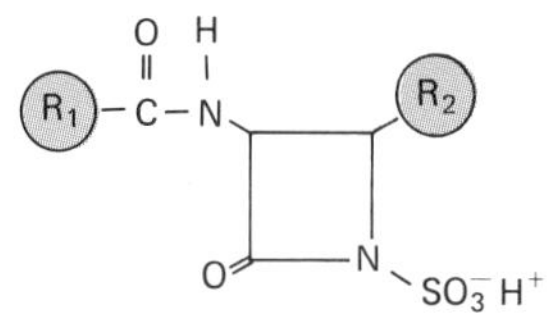

Figure 12–8 The structure of a monobactam. The Rs represent side chains of additional portions of the molecule.

Monobactams

Monobactams represent a group of monocyclic, bacterially produced beta-lactam antibiotics (Figure 12–8). The first isolation of these compounds from a variety of soil-dwelling bacteria—including *Acetobacter, Agrobacterium, Chromobacterium* (krow-moh-back-TIR-ee-um), *Gluconobacter* (gloo-KONO-back-ter), and *Pseudomonas*—came 50 years after the momentous discovery of penicillin by Fleming in 1929. Monobactams function like other beta-lactam antibiotics and interfere with bacterial cell wall formation. While monobactams have no activity against gram-positive bacteria or anaerobes, they exhibit remarkable activity against certain aerobic gram-negative bacteria, including species of *Enterobacter* (en-te-roh-BACK-ter), *Haemophilus, Klebsiella, Neisseria, Pseudomonas,* and *Serratia* (sir-RAY-sha). One example of monobactams under study is a synthetic compound known as aztreonam. It is stable to most plasmid- or chromosome-directed (mediated) beta-lactamases and thus is of considerable interest as a means of controlling beta-lactamase-producing pathogens and mixed infections. Aztreonam is also able to penetrate the outer membrane of gram-negative bacteria. Another particular value of the compound is its minimal effect and selective antibacterial action on the microbial inhabitants (microbiota) of the host. The microbiota in various body regions present a defensive barrier to invasion by potential pathogens. The use of broad-spectrum antibiotics can disturb or destroy this defense capacity of the host.

Clindamycin: A Useful Alternative to Penicillin-Cephalosporin Antibiotics

Clindamycin is a semisynthetic antibiotic active mainly against gram-positive bacteria. The antibiotic inhibits pneumococci, streptococci, and most *Staphylococcus aureus* isolates. It is active against gram-negative anaerobic bacteria, as well as the protozoan causative agents of malaria and toxoplasmosis. Clindamycin acts by inhibiting protein synthesis in the bacterial cell.

Chloramphenicol and the Tetracyclines

Chloramphenicol and naturally occurring tetracyclines (Figure 12–9) are produced by species of *Streptomyces* (Table 12–1). These compounds are bacteriostatic, broad-spectrum antimicrobials having similar mechanisms of action. Both are thought to disrupt protein synthesis by blocking the transfer of activated amino acids from transfer RNA (tRNA) to the growing polypeptide chain. Two semisynthetic tetracyclines, doxycycline and minocycline, are efficiently absorbed and differ only in minor respects from other tetracyclines. Minocycline is active against certain tetracycline-resistant bacterial species. **[See protein synthesis in Chapter 6.]**

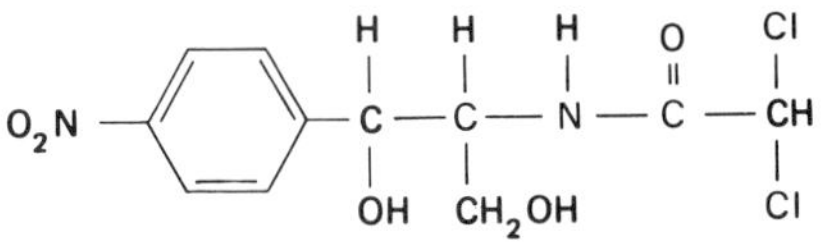

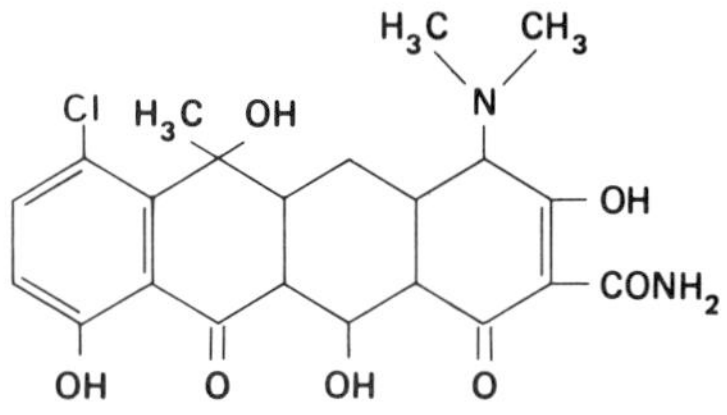

Chloramphenicol

Chlortetracycline

Figure 12–9 Chloramphenicol and chlortetracycline.

RANGE OF ACTIVITY

Chloramphenicol is particularly useful for infections caused by many gram-positive and gram-negative bacteria, rickettsiae, and chlamydiae. It is the drug of choice for typhoid fever. However, because of its highly toxic effect on blood-cell-forming tissues, chloramphenicol must be reserved for cases resistant to other forms of treatment.

Tetracyclines are also active against a variety of gram-positive and gram-negative bacteria, as well as the rickettsiae and chlamydiae. They are much less toxic than chloramphenicol and can be used more freely. However, they are inferior to chloramphenicol for *Salmonella* infections. **[See *Salmonella* infections in Chapter 26.]**

Aminoglycosides

Aminoglycoside antibiotics are one of the oldest and most functional groups of broad-spectrum antibiotics. They include streptomycin, kanamycin, and gentamicin (Figure 12–10).

The name of this group of antibiotics is derived from its complex structure, which includes the connection of two or three components by glycosidic bonds (Figure 12–10). Streptomycin was discovered by Waksman and Schatz in 1944 as a product of *Streptomyces griseus.* Kanamycin was isolated from *S. kanamyceticus.* Gentamicin is produced by species of *Micromonospora,* which are closely related to those of the *Streptomyces.*

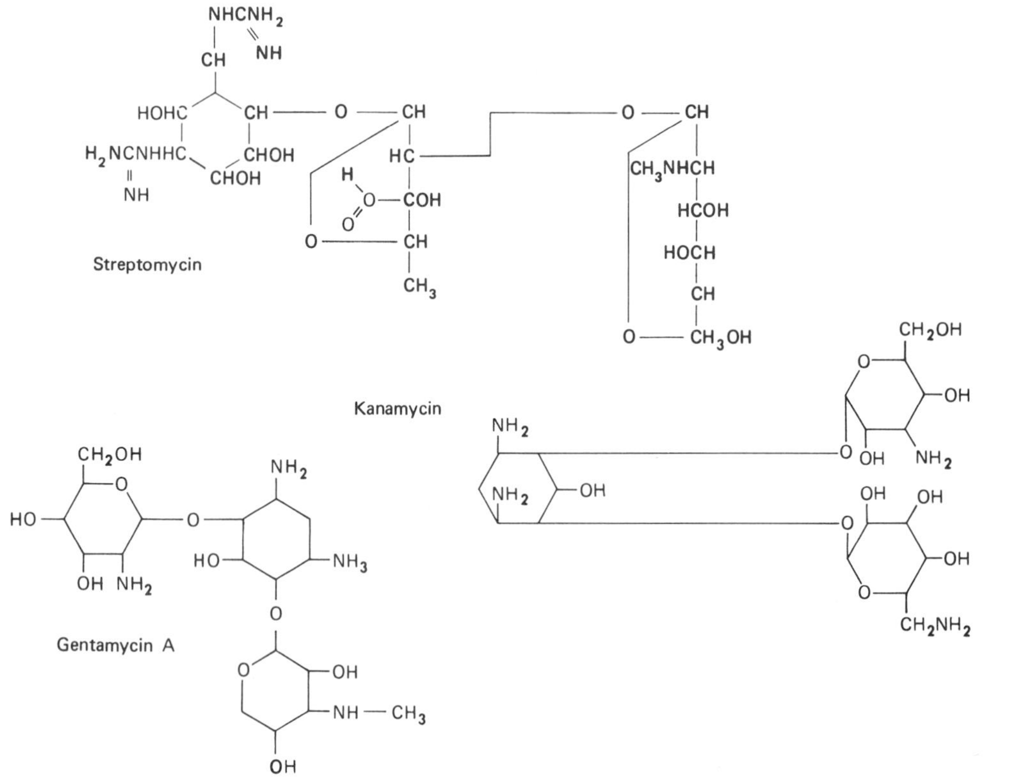

Streptomycin

Kanamycin

Gentamycin A

Figure 12–10 Representative antibiotics of the aminoglycoside group. The parts of these and other antibiotics are connected by glycosidic bonds, which are links between the hydroxyl (OH^-) group of one molecule and the aldehyde group ($\overset{O}{\overset{\|}{C}}$—H) of another molecule. These glycosidic linkages are shown for these antibiotics.

RANGE OF ACTIVITY

Streptomycin's primary activity is against gram-negative bacteria, enterococci, and *Mycobacterium tuberculosis.* Since organisms rapidly develop resistance to this drug, it must be used in combined therapy. When combined with penicillin or ampicillin, it is particularly effective against enterococcus infections, especially endocarditis. In the treatment of tuberculosis, it can be combined with isonicotinic hydrazide (INH) and para-aminosalicylic acid (PAS). **[See tuberculosis in Chapter 25.]**

Kanamycin is active against a wide variety of gram-positive and gram-negative bacteria and *M. tuberculosis.* It is not particularly effective against *Pseudomonas* spp., various streptococci, or anaerobes.

Although gentamicin is a broad-spectrum drug, it is primarily active against infections from gram-negative bacteria and is a drug of choice for *Pseudomonas* infections. It is effective against most *Proteus* spp., but kanamycin is usually the more effective drug. Most aminoglycosides can be toxic to kidneys and auditory nerves (Table 12–2). However, they do not cause allergies or interfere with immunologic processes.

Since the isolation of streptomycin, many newer aminoglycosides have been developed for use in the treatment of serious infections caused by aerobic gram-negatives, including those resistant to several currently available antibiotics. Examples of such newer aminoglycosides are sisomicin, tobramycin, amikacin, and the derivatives of sisomicin, netilmicin, and 5-episisomicin.

MECHANISM OF ACTION

All aminoglycosides, although somewhat different in spectrum, are bactericidal and interfere with protein synthesis. They appear to act by combining with a subunit of the ribosome, causing a misreading of the genetic code.

Polypeptides

Two of the more common polypeptides are polymyxin B and colistin (polymyxin E). The members of this group can be isolated from *Bacillus polymyxa* (ba-SIL-lus pol-EE-mix-a). Colistin was also found in a *Bacillus colistinus* (ba-SIL-lus col-is-TIN-us) culture from a Japanese soil sample.

RANGE OF ACTIVITY

These drugs are effective against most gram-negative bacteria, with the exception of *Proteus* spp. They are used with gentamicin for *Pseudomonas* infections.

MECHANISM OF ACTION
Polymyxin B and colistin act as detergents on the microbial membranes, causing leakage of essential cytoplasmic components. They may be bacteriostatic or bactericidal, depending upon the dosage used and the relative number of organisms to be treated. Both compounds are somewhat toxic. However, kidney damage and nerve injury are usually reversible.

Antimycobacterial Drugs

Among the drugs commonly used in combination with streptomycin are isonicotinic hydrazide (INH) and para-aminosalicylic acid (PAS) (Figure 12–11).

RANGE OF ACTIVITY
After the discovery that the activity of sulfa drugs is based on competition with PABA in microbial metabolism, investigators set out to apply this type of mechanism to other organisms. Salicylic acid was found to stimulate the metabolism of *M. tuberculosis.* On the basis of this finding, PAS was shown to be active against bovine tuberculosis in 1946.

In acting against tuberculosis, INH penetrates into the tissues so well that it can act against bacilli located in tubercles and inside phagocytes—unlike streptomycin and PAS. INH is fairly nontoxic, but it may cause renal (kidney) complications, usually in patients with renal tuberculosis.

MECHANISM OF ACTION
INH is bacteriostatic initially, becoming bactericidal later. The activity appears to be due to its incorporation into nicotinamide adenine dinucleotide (NAD) or nicotinamide adenine dinucleotide phosphate (NADP), both of which are coenzymes. INH also resembles vitamin B_6 and probably interferes with those enzymes that incorporate vitamin B_6 into a coenzyme molecule. Thus, INH appears to act by blocking essential enzyme activity due to its incorporation into coenzymes. PAS acts as sulfa drugs do, by interference with PABA metabolism (Figure 12–4b). PAS is a relatively ineffective bacteriostatic drug. The prime value of this antitubercular agent is its activity in delaying the emergence of resistance to streptomycin and INH. The usual side effects of PAS are nausea and vomiting, which are commonly prevented by the simultaneous administration of an antacid compound.

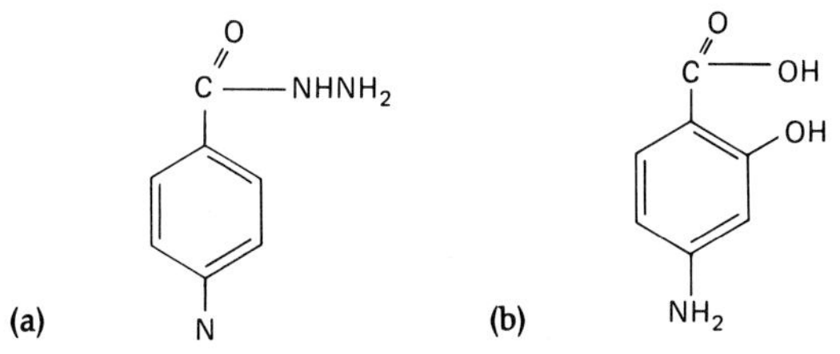

(a) O, C, $NHNH_2$, N (b) O, C, OH, OH, NH_2

Figure 12–11 Two antimycobacterial drugs. (a) Isonicotinic hydrazide. (b) Para-aminosalicylic acid.

Ethambutol

Another drug used in the treatment of tuberculosis is ethambutol. It is often used in combination with INH. This antibiotic functions as a competitive inhibitor that interferes with ribonucleic acid (RNA) synthesis. In combination with other drugs, it helps delay the emergence of drug-resistant strains in tuberculosis patients.

Rifampin

Rifampin is a relatively new semisynthetic derivation of rifamycin B produced by *Streptomyces mediterranei.* This is a broad-spectrum drug that is chemically unrelated to other antibiotics. It is active against gram-positive and gram-negative organisms and is highly effective in the treatment of tuberculosis and leprosy. The development of rifampin is a major advance in antituberculosis chemotherapy, since it is of great value in patients with drug-resistant organisms. Microbial resistance to rifampin can also develop, however. Therefore, in antituberculosis therapy it is used in combination with other drugs. Rifampin interferes with nucleic acid synthesis by inhibiting deoxyribonucleic acid (DNA) transcription. Rifampin blocks the initiation of the process. **[See leprosy in Chapter 23.]**

Mechanisms of Drug Resistance

When many antimicrobial agents became available for chemotherapy in the early 1950s, these "wonder drugs" were thought to be the final answer to the control of infections. Penicillin, in particular, was added to items such as chewing gum, mouthwash, and toothpaste. In addition to this indiscriminate use of drugs, many physicians routinely prescribed them for minor infections. Partly as a result of drug misuse, microorganisms such as *S. aureus* became resistant to these agents and consequently became more difficult to eliminate. Even with the development of more and more specific and broad-spectrum drugs, resistance remains a problem that requires constant consideration. Antibiotic resistance has been demonstrated in all pathogens, and virtually no known antibiotic is exempt.

Drug resistance may either be a natural (intrinsic) property of the microorganism or it may be acquired. Intrinsic patterns of resistance are a common property of an entire species, while acquired resistance affects only one strain.

Acquired antimicrobial drug resistance may result either from chromosomal mutation or from the extrachromosomal genetic exchange involving plasmids and transposons (jumping genes). While the development of chromosome-directed resistance is relatively uncommon, it does occur. The resistance to antibiotics such as the new cephalosporins, nalidixic acid, and rifampin is chromosome-directed.

The most common method of transfer of resistance is by plasmids and transposons. The plasmids encoding genes that are responsible for the mechanisms of antimicrobial drug resistance are called R plasmids or R factors. Such factors exist in most bacteria around the world and are responsible for the multiple antibiotic resistance of most clinically important gram-negative pathogens. Transposons—movable segments of DNA that can change positions on DNA molecules—are also found in gram-negative bacteria. They are believed to account for the widespread distribution of drug-resistant genes among unrelated bacteria. **[See Chapter 6 for additional details of plasmids and transposons.]**

Mechanisms by which organisms develop resistance to antimicrobial agents include (1) an enzymatic alteration of the drug, (2) an increase in the permeability of the cell walls and membranes of organisms, (3) a change in the sensitivity of affected enzymes, (4) an increased production of a competitive substrate, (5) the formation of a more resistant target site, and (6) bypassing the drug-blocked reaction.

Beta-Lactamases

A major mechanism by which bacteria become resistant to an antibiotic is the inactivation or destruction of the drug by enzymes produced by resistant cells. One example of this phenomenon is the production of *beta-lactamase* enzymes by resistant strains of *Haemophilus influenzae, Neisseria gonorrhoeae, S. aureus,* and species of *Enterobacter* and *Pseudomonas.* These enzymes attack the beta-lactam portion of penicillin molecules, thus destroying the antibiotic. The first penicillin-destroying enzyme of this type was detected by E. P. Abraham and E. Chain in 1940 and was named *penicillinase.* The synthesis of beta-lactamases may be regulated by genes on the bacterial chromosomes or on *plasmids.* These enzymes may be produced continuously or only upon an organism's exposure to a beta-lactam antibiotic, the inducer. The amount of enzyme synthesized is often variable and depends on the regulatory gene activity and the availability of the inducer.

Since most bacteria have the ability to synthesize beta-lactamases, the location of the enzyme is important in terms of the function of the organisms that form them and the laboratory that tests for them. All beta-lactamases are present at the cell wall. The enzymes of gram-positive bacteria are excreted into the surrounding environment, while most of those produced by gram-negatives are located in the periplasmic space. (See Chapter 7 for a discussion of bacterial cell wall organization.) Several mechanisms of action for the beta-lactamases conferring antibiotic resistance have been proposed

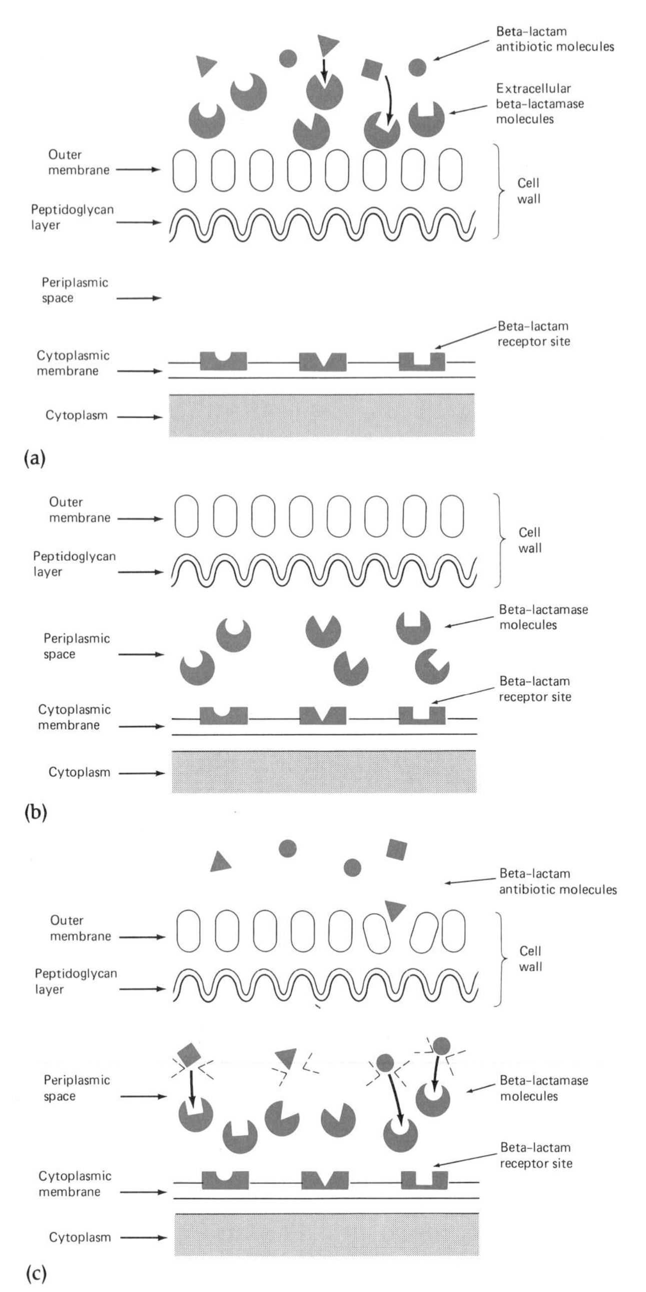

Figure 12–12 Mechanisms of action proposed for beta-lactamases. (a) Extracellular inactivation of beta-lactam molecules; (b) strategic placement of a limited number of different types of beta-lactamase molecules in the periplasmic space; and (c) induction of large amounts of beta-lactamase molecules forming a physical barrier. Refer to Chapter 7 for descriptions of cell wall and cytoplasmic organization. *[After C. W. Stratton,* Beta-Lactamases *(monograph). Kansas City, Mo.: Marion Scientific, 1981.]*

(Figure 12–12). These include (1) the secretion of large amounts of the enzymes extracellularly by the organism, thereby destroying the beta-lactam antibiotic; (2) the strategic placement of enzyme molecules in the periplasmic space, which could inactivate beta-lactam antibiotic molecules before the latter reach receptor sites on the bacterial cytoplasmic membrane; and (3) the induction of large amounts of beta-lactamase to saturate the periplasmic space, thereby not only inactivating antibiotic molecules entering the cell but also creating a physical barrier to block access to receptor sites on the cytoplasmic membrane.

The pharmaceutical industry has attacked the problem of penicillin resistance by altering the structure of the drug, specifically the side chains (Figure 12–13). Such changes have produced most of the newer penicillins and cephalosporins used today. The antibiotics methicillin, oxacillin (Figure 12–13), and nafcillin are effective against beta-lactamase-producing bacteria.

Detecting Beta-Lactamases

Several different techniques are available to screen isolated pathogens for beta-lactamase production. Because of the need for practical and rapid results, standardized antibiotic-containing paper disk methods are generally used. A change in the color of dye in the disk indicates that an acid, such as penicilloic acid, is being produced as the microorganism destroys (by hydrolysis) the antibiotic.

Other Mechanisms

One example of drug resistance caused by changes in cellular permeability and sensitivity involves streptomycin. Apparently, streptomycin interferes with translation of genetic information involving messenger RNA (mRNA) and ribosomes. Resistance can occur by the development of a decreased sensitivity of the enzymes concerned with attaching mRNA to ribosomes. Streptomycin also appears to affect cell membrane permeability. Resistance, therefore, can occur because of changes in the selective permeability of the cell.

Sulfonamides also seem to act both at an enzymic level (the pathway from para-aminobenzoic acid to folic acid) and at membrane sites. Resistance to sulfonamides may be caused by changes in enzyme sensitivity and/or selective permeability. With sulfonamides, the bacterium can also develop resistance by overproducing PABA, as occurs in *S. aureus.*

The Dangers of Antibiotic Abuse

As more antibiotics become known, both medical and nonmedical uses increase. Very few of these compounds have become obsolete or been withdrawn from the market. Several antibiotics are overused and overprescribed. For example, market research data show that almost two-thirds of the prescriptions given to patients for the common cold are for antibiotics, yet most colds and sore throats are caused by viruses, microorganisms that are not affected by most currently available antibiotics or antimicrobials. Antibiotics can be obtained without prescription in various foreign countries. The resulting increased unnecessary exposure to these drugs has produced a significant increase in antibiotic-resistant organisms. The use of antibiotics has undoubtedly increased the number of plasmid-containing bacteria. Future application of chemotherapeutic agents must be directed toward reducing the incidence of such plasmid carriers if the usefulness of antibiotics is to be not only retained but also improved. Furthermore, during the early 1980s, as many as 1.5 million people were hospitalized annually in the United States for adverse drug reactions, and approximately 130,000 died from reactions associated with antibiotics and antimicrobials.

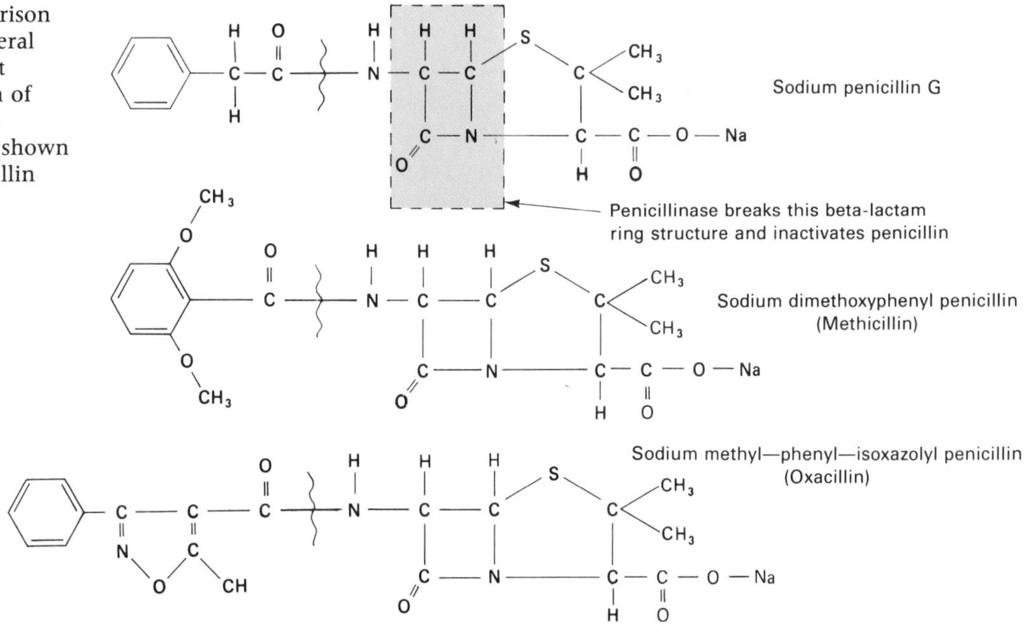

Figure 12–13 Comparison of penicillin G and several beta-lactamase-resistant penicillins. The portion of the penicillin molecule susceptible to attack is shown with the sodium penicillin G molecule.

Sodium penicillin G

Penicillinase breaks this beta-lactam ring structure and inactivates penicillin

Sodium dimethoxyphenyl penicillin (Methicillin)

Sodium methyl—phenyl—isoxazolyl penicillin (Oxacillin)

Antibiotic Sensitivity Testing Methods

Minimum Inhibitory and Minimum Bactericidal Concentration Determinations

The best methods for determining the antibiotic susceptibility of microorganisms involve careful estimation of an antimicrobial agent's minimal inhibitory concentration (**MIC**) and minimal bactericidal concentration (**MBC**). These determinations can be performed with liquid or solid media.

BROTH DILUTION METHODS

Two types of broth dilution tests are in use: *tube dilution* and *microdilution.* In the tube method, tubes of sterile broth containing decreasing concentrations of a specific antibiotic are inoculated with a standard concentration of the test organism (Figure 12–14). After 24 hours of incubation, the MIC is determined by the lowest drug concentration inhibiting growth. The MBC may be found by inoculating a second set of tubes of broth, or agar plates without antibiotics, with specimens from the original tubes showing no visible growth (Figure 12–14). The lowest drug concentrations in the original tubes that produces no growth in the second set of media is the MBC. Microdilution tests are carried out in special small disposable trays with large numbers of in-

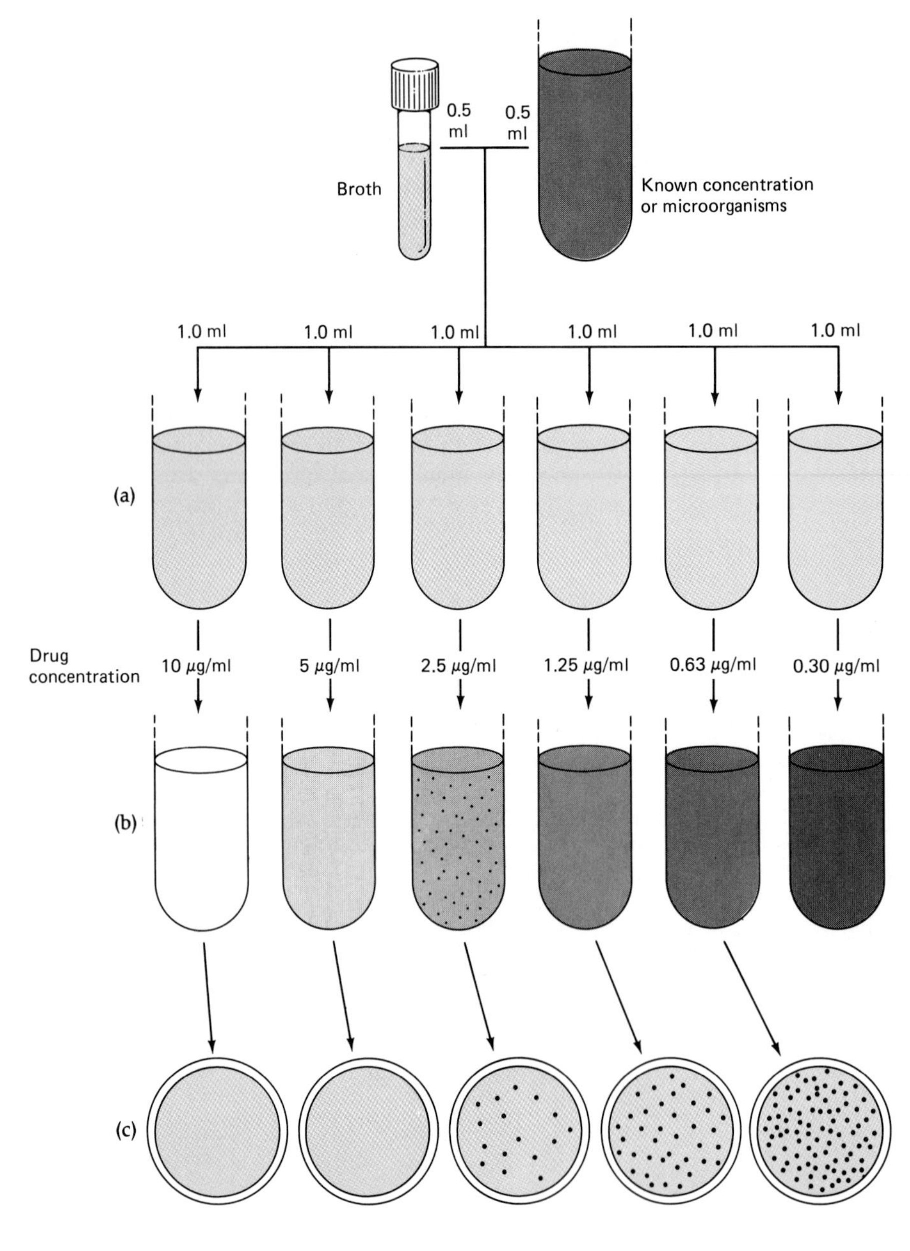

Figure 12–14 Minimal inhibitory concentration (MIC) and minimal bactericidal concentration (MBC) of antimicrobial agents as determined by broth dilution testing. After inoculation with a standard and known concentration of test microorganisms, the tubes are incubated. (a) After incubation all tubes are examined for growth. (b) The lowest concentration of drug inhibiting growth is the MIC. Here the finding is 1.25 mg. Determining the MBC requires inoculating agar plates with inocula taken from tubes without visible growth. After incubation, plates without growth represent drug concentrations that kill the test microorganisms.

Microbiology Highlights

CONTAMINATION BY ANTIBIOTICS

There is a continual and substantial level of concern about the contamination of the environment, particularly with industrial wastes. There appears, however, to be a more subtle and widespread contamination with antibiotics. As a result of indiscriminate antibiotic usage and dispersal, a major environmental and ecological change has occurred through the years: the increase in antibiotic-resistant bacteria. How do antibiotics get into the environment? Direct antibiotic contamination of the environment can take place in numerous ways. For example, in 1983 in the United States alone, over 18 billion tons of antibiotics were produced for use in the treatment of human, domestic animal, and plant diseases. In most situations, the antibiotics were not inactivated after use but rather introduced into the environment in the form of human and animal fecal material. Such wastes are disposed of in a variety of ways. Depending on the country, fecal material may be dumped into landfills or offshore waters or sprayed onto planted fields as fertilizer. Other sources of environmental contamination include the spraying of antibiotics on bacteria-infected fruit trees and the use of antibiotics in the feed of farm animals in order to increase their growth and size.

In 1986, in response to the growing evidence of environmental contamination by antibiotics and the increase in resistant bacteria, an international organization was formed called the *Alliance for the Prudent Use of Antibiotics.* The goal of this group is to increase awareness of antibiotic overuse and its consequences for public health on a worldwide basis.

dividual wells (chambers) containing a number of antibiotics in varying concentrations.

AGAR DILUTION METHOD

In the agar dilution method, different antibiotic concentrations are incorporated into an agar medium for both aerobes and anaerobes. A replicator device (Figure 12–15a) may be used to inoculate multiple specimens onto a series of plates with varying concentrations of antibiotics. After incubation, the minimum inhibitory concentration may be determined electronically (Figure 12–15b).

Figure 12–15 (a) The Cathra Replicator multipoint device designed to simultaneously inoculate up to 36 unknown isolates onto the surface of a plate medium (Replicator plates). This system also is applicable to determining antibiotic susceptibilities. (b) The components of the Cathra Replianalyzer. This system accepts, stores, and interprets the results of antibiotic susceptibilities and biochemical identification tests produced on Replicator plates. *[Courtesy MCT Medical.]*

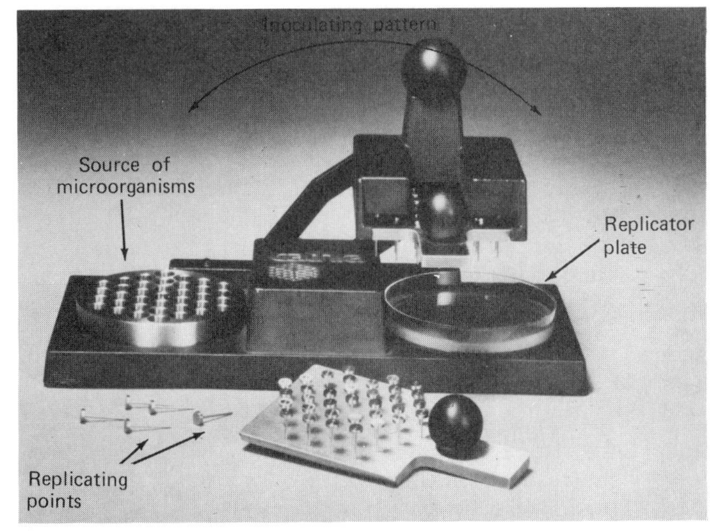

(a)

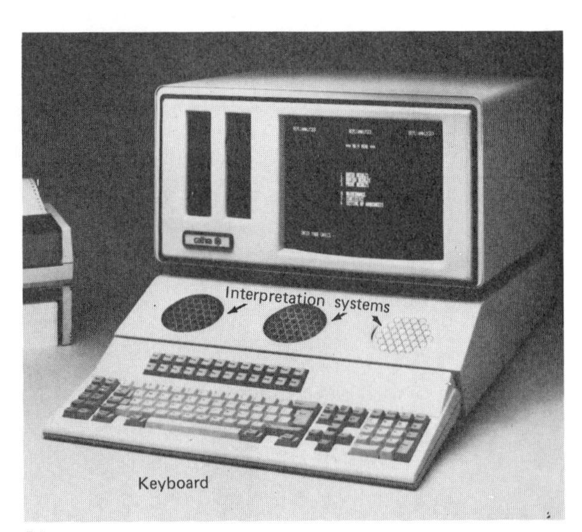

(b)

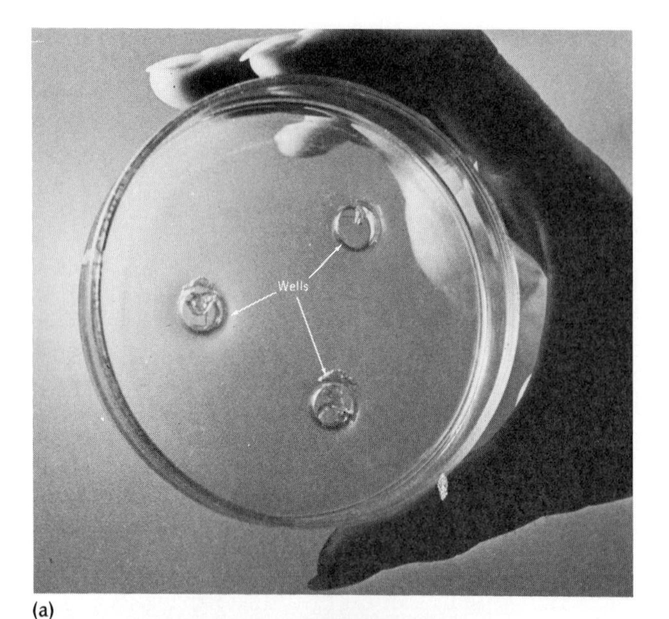

(a)

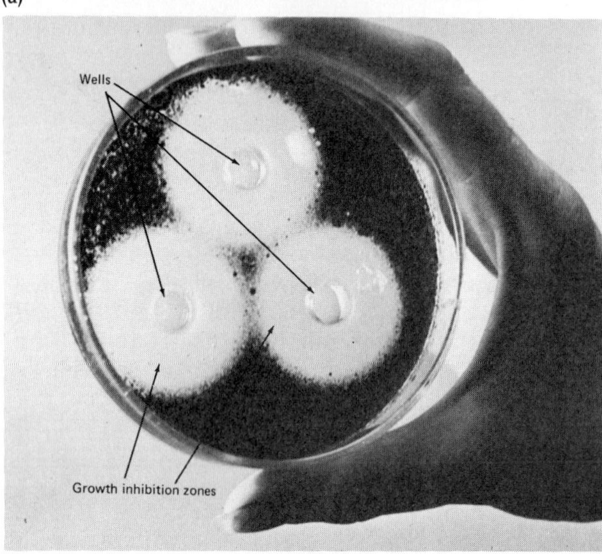

(b)

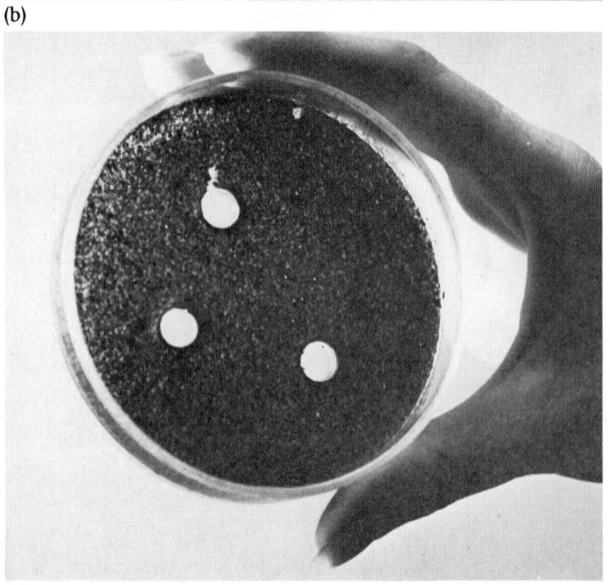
(c)

Figure 12–16 (a) A Petri dish containing a medium seeded with a test organism. Three wells have been cut for the placement of an experimental drug. (b) The results of a test in which the test organisms are sensitive to the drug. (c) Drug resistance on the part of the test organisms. *[Courtesy of Lederle Laboratories, Pearl River, N.Y.]*

Drug Diffusion Methods

CYLINDER AND WELL METHODS

The screening of large numbers of bacteria with various antibiotics requires simple techniques that can be used with several samples at the same time. For example, small cylinders can be placed into the agar plates or wells can be cut into the agar for the purpose of holding a specified quantity of a particular antimicrobial agent. This type of procedure is shown in Figure 12–16a. The Petri dish here contains an agar medium seeded with a test organism. The three wells have been cut out and filled with dilutions of a drug. If the antimicrobial agent is effective against the test organism, three zones of inhibition will develop (Figure 12–16b). If the drug is ineffective, then the results shown in Figure 12–16c are found.

FILTER-PAPER DISK

Because the well method is still a bit awkward to perform as a routine laboratory procedure, *impregnated paper disks* have received wide acceptance (Figure 12–17). A. Bondi in 1947 reported using filter-paper disks containing specified concentrations of antibiotics (Color Photograph 61). The standardization of variables to minimize difficulties in interpretation was a critical aspect of this work.

Several factors can affect the size of the zone of antibacterial activity. These include (1) the depth of the medium used, (2) the choice of medium, (3) the size of the inoculum, and (4) the diffusion rate of a particular antibiotic. The last factor, in particular, has resulted in unfortunate misinterpretations of results.

Some laboratories used single- or double-disk methods. The single-disk methods use one disk of either a high- or low-antibiotic concentration. Determining the relative sensitivity of the organism to the drug requires interpretation of zone sizes. With the double-disk method, the interpretation is simpler. Here both high- and low-strength disks are applied for each antibiotic to be tested. The organism is reported as being sensitive if a clear zone appears around both disks. If a zone appears around the high-concentration disk alone, the organism is called *moderately susceptible.* If zones are lacking in both disks, the organism is considered resistant to the drug. Although interpretation is simpler here, the accuracy of the double-disk method does not approach that of the Kirby-Bauer procedure.

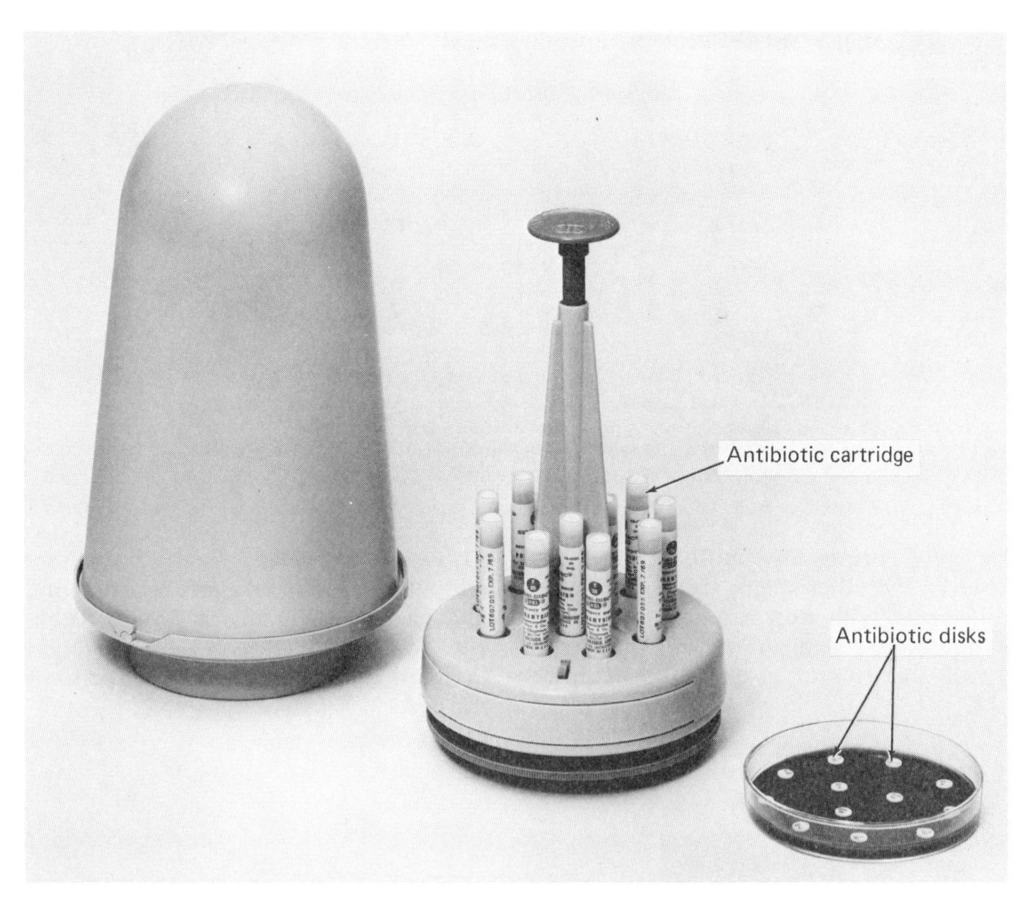

Figure 12–17 Commercially available paper disks can be obtained in cartridges and applied to the medium in a Petri plate. This is usually done with a multiple applicator device of the type shown. *[Courtesy of BioQuest, Division of Becton Dickinson and Company, Cockeysville, Md.]*

KIRBY-BAUER (K-B) STANDARD SINGLE-DISK METHOD

First reported in 1966, the Kirby-Bauer method uses a single high-strength antibiotic disk with Mueller-Hinton agar dispensed in 150 × 15 mm Petri plates. The depth of the medium is 5 to 6 mm (approximately 80 mL of medium). Standardization of the test organisms is accomplished by (1) introducing the growth from five isolated colonies into 4 mL of brain-heart infusion broth; (2) incubating the preparation for 2 to 5 hours in a water bath or thermal block or until adequate turbidity is evident; and (3) adjusting the turbidity of the bacterial suspension according to a standard made from barium sulfate (McFarland Standard 0.5).

A sample is taken by a sterile cotton swab. Excess fluid is removed by rolling the swab against the side of the tube containing the bacterial suspension. Then it is used to spread the organisms onto the surface of the agar medium. After a few minutes' wait for the surface moisture to be absorbed by the agar, disks are applied. This procedure may be done with the large-size applicator like the one shown in Figure 12–17. After incubation, the size of the zones can be measured with the aid of calipers, or they can be compared with a template consisting of various zones, the sizes of which are based on the data published by Ryan and his associates in 1970.

The zone sizes for selected drugs are presented in Table 12–4. These sizes were evaluated according to MICs for pathogens and obtainable levels of antimicrobial agents in the human body. Extensive published tables are available from pharmaceutical companies. With this system, when an organism is reported to be susceptible (S) or resistant (R), the level of confidence in such a report is very high. An intermediate (I) classification indicates that the organism is probably resistant to obtainable body levels of the drug in question. The drug should not be used without first performing a tube dilution test confirming the MIC obtained. In practice, an I designation is considered to be the same as an R rating—meaning the drug is not effective. The drug is not used if the microorganism has been found to be sensitive to several other drugs.

The K-B method and the agar-overlay modification techniques presented next are suitable only for rapidly growing pathogenic bacteria, such as staph-

TABLE 12–4 Zone Size Comparison of Selected Chemotherapeutic Drugs

		Inhibition Zone Diameter (mm)		
Chemotherapeutic Drug	*Concentration in Disk*	*R*[a] *less than*	*I between*	*S more than*
Ampicillin[b]	10 μg	21	21–28	28
Chloramphenicol	30 μg	13	13–17	17
Colistin	10 μg	9	9–10	10
Kanamycin	30 μg	14	14–17	17
Penicillin[b]	10 units	21	21–28	28
Sulfonamides	300 μg	13	13–16	16
Tetracyclines	30 μg	15	15–18	18

[a] R = resistant, I = intermediate resistance, S = susceptible. See text for additional information concerning the differences between R and I and reasons for the various zone sizes for different drugs.
[b] Interpretation of zone sizes with ampicillin and penicillin varies with different organisms. These values are primarily for staphylococci.

ylococci and *Pseudomonas* spp. Fortunately, most pathogens are in this category. The other single- or double-disk procedures can be used with most bacterial species. However, without some form of standardization, the accuracy of the results is questionable.

THE AGAR-OVERLAY METHOD

The agar-overlay modification of Barry, Garcia, and Thrupp (1970) has greatly simplified the Kirby-Bauer procedure without affecting the interpretation of zone sizes. Figure 12–18 shows some of the steps involved in this test.

Figure 12–18 Agar-overlay modification of the Kirby-Bauer standardized antibiotic disk sensitivity test. *[Courtesy of St. Mary's Long Beach Hospital, Long Beach, Calif.]*

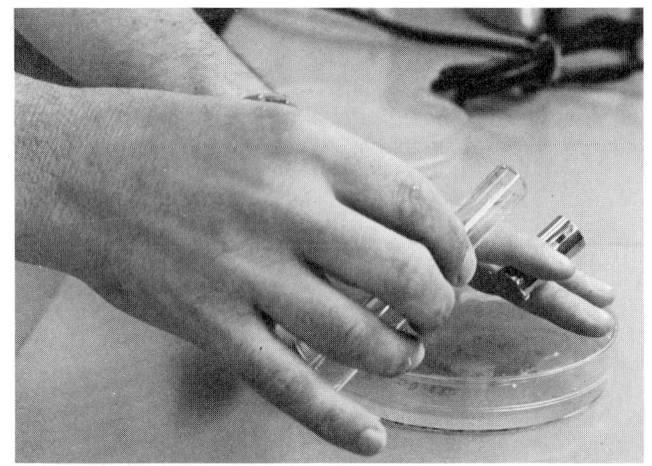

(a) Pour plate preparation with organism

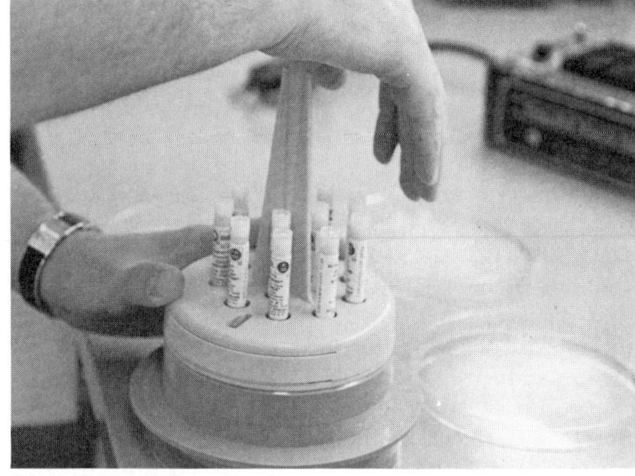

(b) Antibiotic disk application

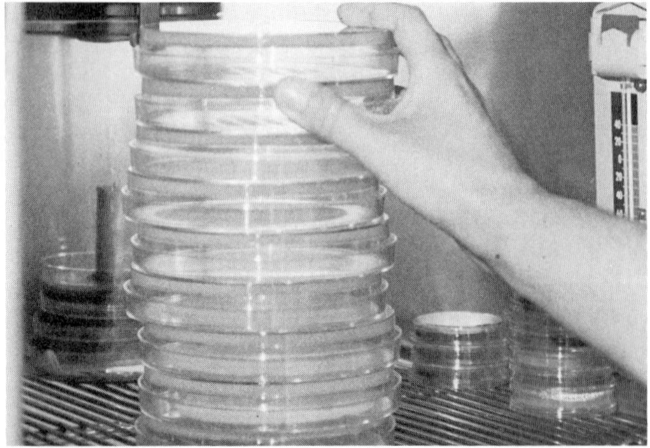

(c) Incubation

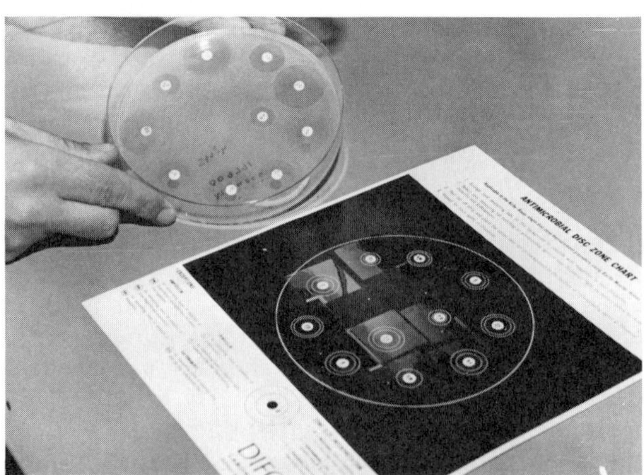

(d) Test results interpretation

Figure 12–19 The Biogram System. An electronic computing system used to determine MICs on a Kirby-Bauer disk diffusion plate. An electronic caliper is used to measure precise zone sizes. *[Courtesy Giles Scientific, Inc.]*

In this procedure, 0.5 mL of brain-heart infusion broth is inoculated with the test colonies to make the solution slightly cloudy. The tube is then held at 37°C for 4 to 8 hours. During this incubation period, tubes containing 8 mL of melted 1.5% agar are placed in a constant-temperature bath or block maintained at approximately 52°C.

After the incubation period, 0.001 mL of the broth culture is introduced by means of a calibrated loop into a tube of melted agar. The contents of the tube are mixed and then poured onto the surface of 70 mL of solidified Mueller-Hinton agar in 150 × 15 mm plastic Petri plates (Figure 12–18a). It is imperative that the plates be warmed to at least room temperature before attempting this last step. Tilting and/or rotating the plate is usually necessary to produce a uniform layer of the seeded agar. The colder the plate, the more difficult this step is to perform.

The required antibiotic disks are applied 3 to 5 minutes after the agar layer has solidified (Figure 12–18b). These disks are placed firmly on the agar surface. Then the Petri plate cover is replaced and the entire system is incubated overnight at 35°C (Figure 12–18c). The zones of inhibition that develop are compared with standards to determine relative susceptibilities (Figure 12–18d). Measurements can be made electronically (Figure 12–19).

SUSCEPTIBILITY OF ANAEROBES

Anaerobes are associated with virtually all types of infections. These organisms are responsible for 10% to 15% of blood infections and are isolated in the majority of many common infections involving the respiratory, gastrointestinal, and urogenital systems.

Selection of a susceptibility testing method for anaerobes is determined by the technical capabilities of the laboratory, costs involved, and accuracy. Both qualitative and quanitative tests are available. These include the *agar dilution* and *broth dilution* methods described earlier as well as *disk elution* and *disk diffusion.*

The disk elution procedures are modifications of the dilution method in which antibiotics are incorporated into an appropriate test medium by adding a paper disk containing specific antibiotics. The disk diffusion techniques are similar to those described earlier.

Blood Levels of Antimicrobial Agents

Antimicrobial therapy is becoming increasingly complex as more resistant organisms emerge as important pathogens. *In vitro* susceptibility tests performed in most laboratories indicate whether an infection caused by the organism tested is likely to respond to the antibiotic concentration recommended for treatment (Color Photograph 61). In cases of life-threatening diseases, proper treatment often requires close monitoring of blood levels of antibiotics in the patient.

A number of tests are available to perform such serum assays, but they have not been standardized to the same extent as the better-known bacterial susceptibilty methods. In these tests, specimens are obtained from patients who have received the drug for at least 24 hours. The blood sample is usually drawn just before administration of the antibiotic (trough level) and also at the time the highest concentration of antibiotic is expected in the bloodstream (peak level). Biological or enzymatic methods are then used to measure antibiotic concentrations.

In biological assays, the concentration of antibiotic in the patient's bloodstream is measured by comparing growth inhibition of a test organism by the patient's serum to inhibition of the same organism by a known amount of the antibiotic administered. A disk diffusion procedure is widely used for this purpose (Figure 12–20).

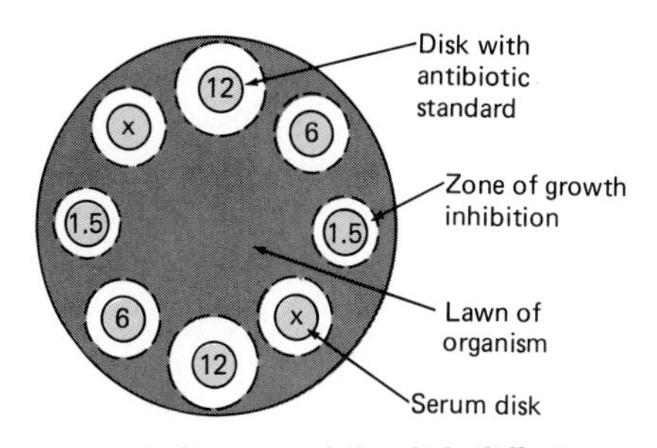

Figure 12–20 A diagram of the disk diffusion assay plate and its components.

Two plates are normally used for each assay. Specified amounts of the patient's serum and of three standard concentrations of antibiotic are incorporated onto ¼-inch filter-paper disks. Two sets of disks are placed clockwise on each agar plate, with the disk containing the most concentrated antibiotic standard first and the serum disk last. After two to three hours of incubation at 35° to 37°C, the zones of inhibition caused by the antibiotic and by the patient's serum are measured, compared, and used to calculate the inhibitory concentrations of the antibiotic being monitored.

Enzymatic assay is more rapid, accurate, and specific than biological assays. The test uses radioactive material and an appropriate counter to detect levels of radioactivity of reaction mixtures.

Antimycotic Agents

Unlike chemotherapy for bacterial infections, treatment of fungus infections or mycoses by chemical agents is still quite limited. There are several reasons for this. Infections caused by fungi are much less common than those produced by bacteria or other microorganisms; the most frequent mycotic infections, such as the various ringworm conditions (Color Photographs 41 and 42), are not life-threatening and in most cases trivial. Also, it has proven especially difficult to develop antifungal agents that have a specificity for fungal cellular structures and for cellular macromolecular synthetic processes such as protein and RNA synthesis.

The most important antifungal agents are those that affect cell membranes. Of these, the polyene antibiotics (Figure 12–21) are the most important and most useful in the treatment of systemic infections. These antibiotics interact with the membranes of susceptible cells and distort their selective permeability. This results in the leakage of potassium and magnesium ions followed by decreased protein and RNA synthesis. All organisms susceptible to polyenes contain sterols. These include algae, flatworms, mammalian cells, protozoa, and yeasts. Because of their effect on membrane permeability, some polyenes have been found to increase the effectiveness of other drugs. Some problems have been encountered with the polyenes, however, especially with respect to absorption, solubility, stability, and toxicity. In addition, various studies indicate that microbial resistance to amphotericin B and other polyene antifungal agents may develop during the course of treatment, especially in patients with immune system defects. Most episodes of drug resistance have involved yeasts belonging to the genus *Candida*. **[See yeast infections in Chapter 24.]**

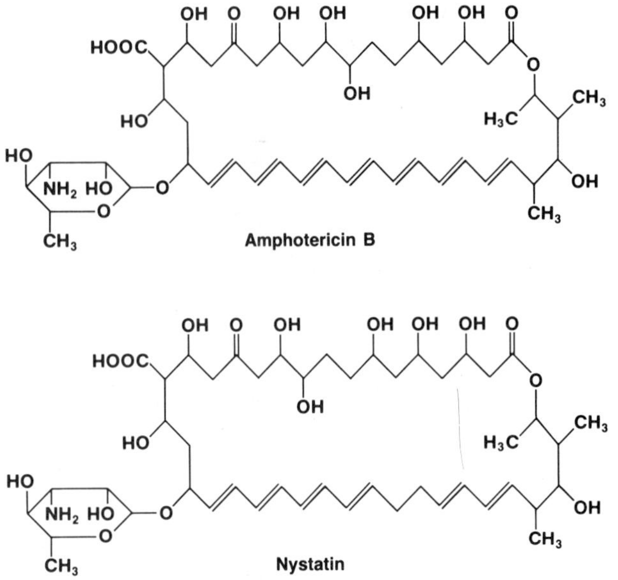

Amphotericin B

Nystatin

Figure 12–21 The chemical structures of selected antifungal agents. Two polyene antibiotics are shown, amphotericin B and nystatin. Interactions between these antibiotics and the membranes of susceptible cells causes leakage of potassium and magnesium ions followed by decreased protein and ribonucleic acid synthesis.

The imidazoles (Figure 12–22) represent another group of antimicrobial agents that are active against fungi and certain pathogenic protozoa. The imidazoles are believed to act by interfering with the synthesis of sterols. Three of these compounds, clotri-

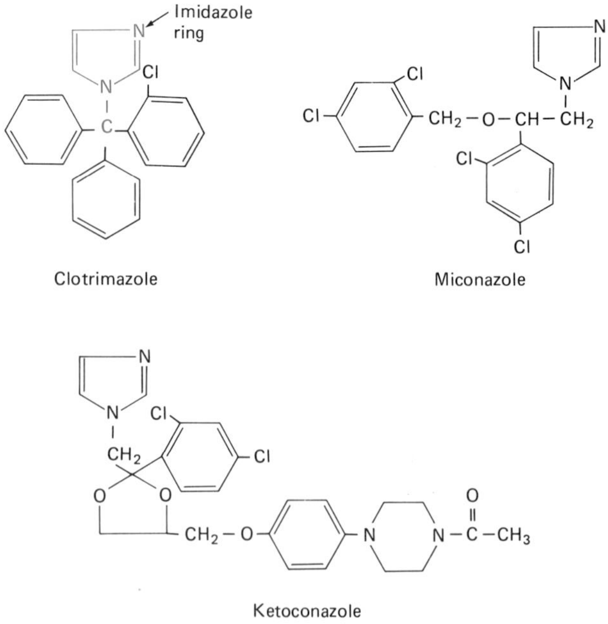

Clotrimazole

Miconazole

Ketoconazole

Figure 12–22 Structures of antifungal imidazoles. The imidazole ring also is shown.

TABLE 12–5 Characteristics of Antimycotic Drugs

Drug	*Fungus Diseases Affected*	*Side Effects*	*Mechanism of Action*
Amphotericin B (Fungizone)	Superficial candidiasis	Essentially none	Interferes with membrane function; increases cell permeability
	Systemic candidiasis, not involving the heart (endocarditis)	Kidney damage, convulsions, hypotension, nausea, vomiting, abdominal pain, metallic taste, cardiac arrest, and anemia[a]	
	Deep-seated fungus infections, including aspergillosis, coccidioidomycosis, histoplasmosis, systematic sporothricosis, blastomycosis, and cryptococcosis	Same as for systemic candidiasis	
Clotrimazole	Ringworm of the skin	Generally none	Changes cell permeability
Fluorocytosin (Ancobon)	Systemic candidiasis, not involving the heart, and cryptococcosis	Gastrointestinal distress and reduction of white blood cells (leukopenia)	Inhibits RNA synthesis
Griseofulvin (Fulvicin, Grifulvin, Grisactin)	Dermatophytoses (ringworm of the hair, nails, and/or skin)	Fatigue, skin eruptions, nausea, vomiting, and diarrhea	Damages cell membranes; inhibits cell division
Ketoconazole	Disseminated fungus infections	Generally well tolerated	Interferes with membrane function; increases cell permeability
Miconazole	Dermatophytoses (ringworm of the hair, nails, and/or skin), candidiasis, candidal vulvovaginitis, and deep-seated fungus infections such as aspergillosis and coccidioidomycosis	Bleeding and ulcerations at sites where the drug is introduced	Increases cell permeability
Nystatin (Mycostatin)	Superficial yeast infections of the oral cavity, skin, and vagina	Basically none when applied topically (on surfaces)	Damages cell membrane

[a] The side effects listed occur more readily with the rapid introduction of the drug.

mazole, ketoconazole, and miconazole, are effective in the treatment of fungal skin diseases (Color Photographs 4 and 98a). Miconazole and ketoconazole also appear to be useful in the management of fungal diseases involving body organs (deep-seated mycoses). [See deep-seated mycoses in Chapter 25.]

Other important antifungal, or antimycotic, agents are 5-fluorocytosine, which affects RNA synthesis, and griseofulvin, which inhibits cell division in a variety of different types of cells. Both of these agents are taken orally. The characteristics of several traditional as well as new antimycotic agents are given in Table 12–5.

Treatment of Protozoan Diseases

At the present time, control of most of the protozoan diseases of humans and domestic animals relies largely on sanitation rather than on drugs. Even though several medications have been developed for most of these disease states, they are not always readily available. Chemotherapeutic agents used for protozoan diseases have other limitations. Some preparations, especially those containing arsenic or antimony, are highly toxic. Others are not effective

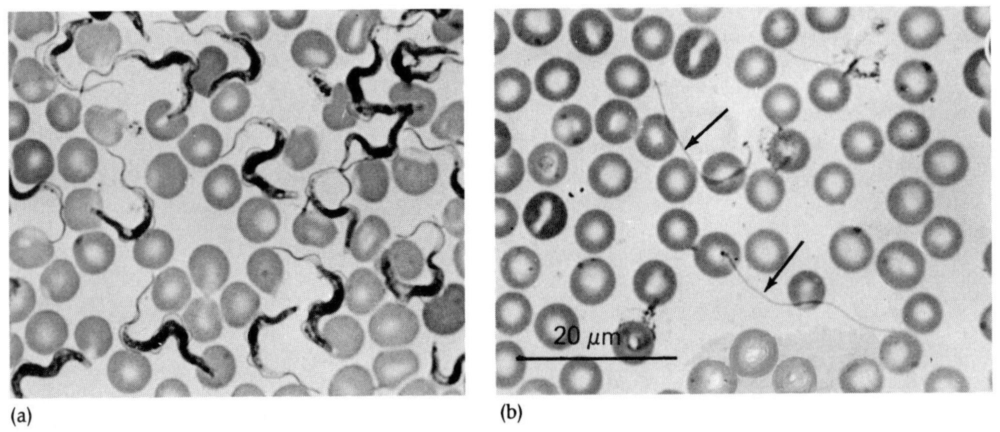

Figure 12–23 The dramatic effect of chemotherapeutic agents. (a) The protozoon *Trypanosoma brucei,* one of the causative agents of African sleeping sickness, before treatment. (b) After treatment with salicyl hydroxamic acid and glycerol. Note the absence of trypanosomes and the remaining two flagella (arrows). *[A. B. Clarkson, Jr., and F. H. Brohr,* Science ***194:*** *204–206, 1976.]*

against all stages of the parasite. Resistance to certain drugs is also becoming a greater problem.

Antiprotozoan drugs are thought to exert their antimicrobial action by several mechanisms. These include interference with energy metabolism (Figure 12–23); disruption of membrane function; and interference with nucleic acid production, protein synthesis, or other biosynthetic reactions of the parasite. Later chapters will discuss specific drugs for treatment of protozoan diseases.

Antiviral Agents

The history of human viral chemotherapy is quite short. Although numerous attempts to develop antiviral drugs had been made prior to the 1960s, the first such compound licensed for medical use in the United States was amantadine in 1966. A lack of information concerning specific drug-sensitive virus sites, and the limited number of well-controlled clinical trials in the past, were major stumbling blocks to finding effective antiviral drugs. In addition, considerable prophylactic and therapeutic success had been achieved with a large number of viral agents through the use of vaccines, but limited success had been obtained with chemotherapeutic drugs. During the last few years, with a better understanding of virus-cell interactions and the viral mechanisms of replication, several promising antiviral drugs have been developed. Examples of these include acyclovir, amantadine, azido-deoxythymidine (AZT), cytosine arabinoside (Cytarabin), 5-iodo-2′-deoxyuridine (IDU), methisazone, ribavirin, and vidarabine. Table 12–6 provides a comparative summary of these agents, including the viral diseases affected, mechanism of action, and possible side effects. Figure 12–24 shows some of their chemical structures. Among the various antiviral substances that has received considerable attention in recent years is interferon. This antiviral substance is now recognized as a group of agents with broad biological effects, including cancer therapy.

TABLE 12–6 A Comparative Summary of Antiviral Drugs

Drug	*Representative Viral Disease(s) Affected*[a]	*Possible Side Effects*	*Mechanism of Action*
Acyclovir	Genital herpes, fever blister, chickenpox	Local pain of short duration at site of application	Selective inhibition of viral DNA synthesis
Amantadine	Influenza A_2	General irritability, insomnia, confusion, hallucinations, inability to concentrate	Prevents penetration of certain viruses into host cells
3′-azido-3′-deoxythymidine (AZT)	AIDS	Inhibits normal blood cell formation; headache	Interferes with viral DNA synthesis

TABLE 12–6 (continued)

Drug	*Representative Viral Disease(s) Affected*[a]	*Possible Side Effects*	*Mechanism of Action*
5-iodo-2′-deoxyuridine	Severe herpes simplex virus infections	Nausea, vomiting, hair and fingernail loss, lowering of white blood cells (leukopenia) and platelets (thrombocytopenia)	Blocks synthesis of nucleic acids
Cytosine arabinoside (Cytarabin)	Progressive varicella (chickenpox) and zoster (shingles) infections	Nausea, vomiting, loss of appetite, chromosomal changes, lowering of white blood cells and platelets, anemia	Inhibits DNA synthesis
Methisazone	Progressive vaccinia	Nausea, vomiting, loss of appetite, liver toxicity (hepatotoxicity)	Interferes with protein synthesis at the level of translation
Ribavarin	Respiratory syncytial virus infection, AIDS	Difficulty in breathing, chest soreness, anemia, rash and inflammation of eyelid lining (conjunctivitis)	Inhibits early replication step leading to viral nucleic acid synthesis
Vidarabine	Genital herpes, fever blister, cytomegalovirus infection, infectious mononucleosis	Nausea, vomiting, diarrhea; tremors, pain, and seizures in chronic hepatitis patients and kidney transplant recipients	Selective inhibition of viral DNA synthesis

[a] Most of these diseases are described in the chapters of Part VIII.

The majority of antiviral agents are nucleic acid derivatives. They function by interfering with DNA or RNA synthesis. Unfortunately, certain agents such as cytosine arabinoside block DNA formation in both DNA viruses and human cells. Thus they have limited application in treatment.

Interferons

Interferons (IFNs) are a family of proteins that occur in a variety of vertebrates ranging from fish to humans and are biological regulators of cell function. The first activity of IFN to be discovered was

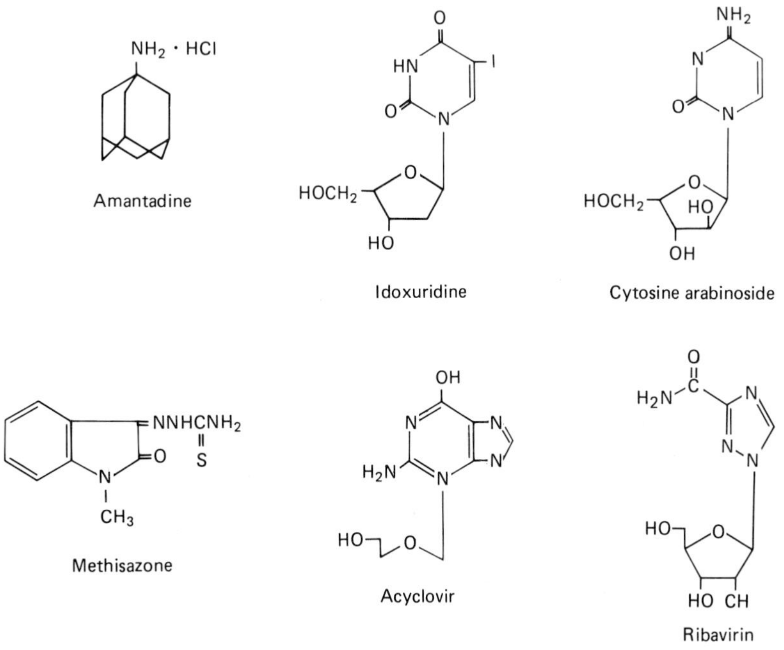

Figure 12–24 The chemical structures of antiviral agents.

Amantadine

Idoxuridine

Cytosine arabinoside

Methisazone

Acyclovir

Ribavirin

interference with the replication of various viruses. This is the property for which these agents were named. [See other properties of IFN in Chapter 10.]

IFNs are synthesized by cells *of different animal species* and appear to be host-specific; the IFN produced by a given species is the same regardless of the viral agent that causes its formation. IFNs differ, however, among animal species with respect to antigenicity and molecular weight. Interferons produced from natural sources and, more recently, by recombinant DNA technology have been used to treat skin infections and certain cancers often resistant to other types of treatment.

Human IFNs are classified into three distinct antigenic types: α (alpha), β (beta), and γ (gamma). Alpha and beta interferons also are designated as type 1 interferons while the gamma form is called type 2 interferon. The different interferons can be distinguished from one another on the basis of their sensitivity to acids.

The production of α and β IFNs can be induced in macrophages and fibroblasts, respectively, by members of most major virus groups and many other microorganisms, particularly those with an intracellular phase in their growth cycle. Included in this group are the causative agents of malaria, the rickettsiae, mycoplasma, and other bacteria, such as *Brucella abortus* (broo-SEL-la a-BOR-tus) and *Francisella tularensis* (fran-sis-EL-la too-la-REN-sis). Many chemical substances will also induce IFN. These include bacterial endotoxins, natural or synthetic *double-stranded ribonucleic acid* (dsRNA), synthetic RNA, and complex polysaccharides. The IFNs can be induced in lymphocytes by mitogens, proteins that stimulate cell division, and by antigens to which cells have been sensitized. The first intermediate substances formed during IFN production are mRNAs. The fact that these compounds can be translated into IFNs in cell-free systems has facilitated efforts to isolate the genes specifying human interferons. The secreted IFNs bind to surface receptors of responsive cells and affect numerous biological activities.

PROPERTIES

In addition to the properties mentioned earlier, IFNs exhibit unusual stability at low pH and a general resistance to temperatures of 50°C and slightly higher in certain instances. They are susceptible to various protein-digesting enzymes.

In addition to their antiviral actions, IFNs affect cell multiplication, cell motility, cell membrane rigidity, and various immunological processes, including immunoglobulin responses, rejection of foreign tissue grafts, skin test reactions as in the tuberculosis skin test, the activation of macrophages, and the selection of natural killer T cells. [See Chapters 17 and 20 for T cells and macrophages.]

ACTION

The most extensively studied effect of IFNs is the conversion of cells into an antiviral state in which they are poor hosts for virus replication. This conversion requires an interaction between IFN and the surface of active cells, and the presence of a nucleus, RNA synthesis, and protein synthesis (Figure 12–25). The antiviral state is accompanied by changes in the level of mRNAs and proteins, including enzymes that have the ability to inhibit the replication of DNA and RNA viruses of most virus groups to a greater or lesser extent. A number of new enzymatic activities are induced in an IFN-treated cell. Specific enzymes prevent the translation of viral mRNA (protein kinase) by at least inactivating a substance necessary to start translation and disrupting viral mRNA (mRNA endonuclease). As inhibition of viral protein synthesis becomes more complete, the quantities of virus synthesized are greatly reduced.

THERAPEUTIC VALUE

Given their range of activity and the nature of their effects on cellular processes, it is not surprising that IFNs sometimes inhibit cellular replication. What is unusual is that this inhibition often appears to be selective for cancer-associated cells. Thus, the IFNs appear to have great potential as therapeutic agents not only against viruses but also against certain cancer cells.

Progress in this direction has been hindered by difficulty in obtaining large quantities of IFNs and determining the duration of their activity. In addition, side effects such as fever, nausea, temporary interference with blood cell production, and vomiting develop with injections.

A large portion of the human IFNs used in small-scale clinical trials now underway have been produced from leukocytes at great expense. A breakthrough in the technology of IFN production, which occurred in early 1980, may provide a cheaper source and greater availability of these therapeutic agents.

In this process, recombinant DNA techniques were used first to isolate and then to insert human IFN genes into a bacterial strain of *E. coli*. The bacteria, in turn, synthesized biologically active human IFNs. If the process can be perfected to maintain high levels of IFN production inexpensively, the control of several virus-caused diseases and cancers can be achieved.

The use of interferons in treatment has not been without side effects. The most frequently found symptoms include fever, chills, loss of appetite, headache, and muscle and joint pain. Fortunately, the severity of the side effects does not appear to be influenced by the dose given, but in several cases the symptoms are decreased with repeated exposure to interferon.

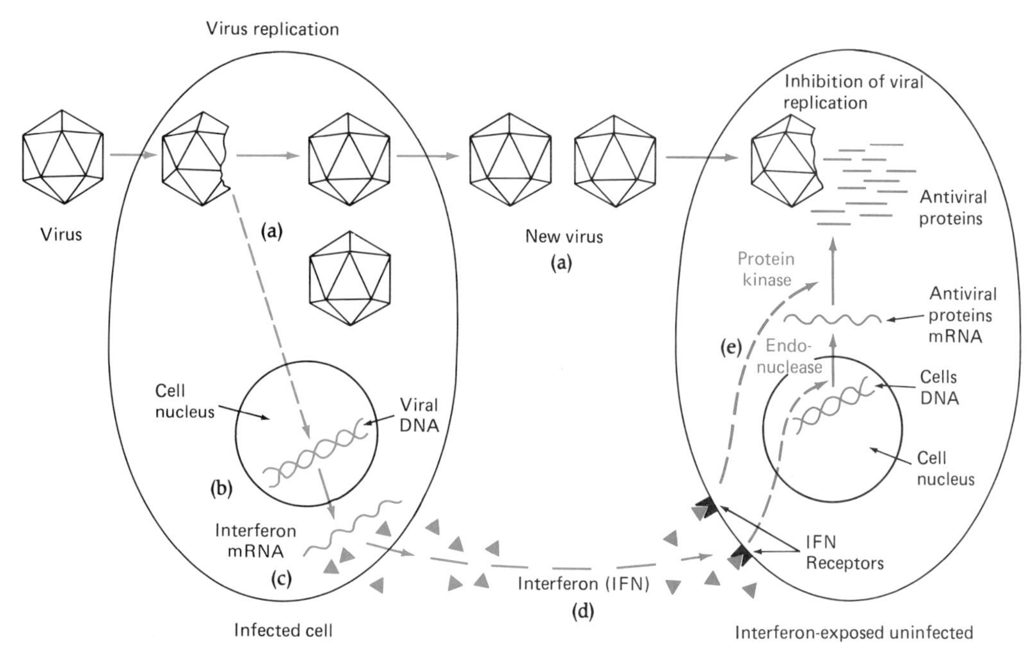

Figure 12–25 Proposed action of interferon (IFN). Infecting virus brings about the production of new viruses (a), but also induces the infected cell to produce interferon mRNA (b). The messenger RNA is translated to form interferon (c). The newly formed IFN may inhibit viral replication in the infected cell (not shown), or it may be released to bind to a neighboring cell (d). The neighboring, uninfected cell (right) is induced by the IFN to form several viral inhibiting enzymes or antiviral proteins (e).

Antiviral Agent Sensitivity Testing

With the recent unraveling of many of the biochemical, biological, and biophysical properties of disease-causing viruses has come great anticipation of an era of effective viral chemotherapy. Despite the wealth of knowledge that has been gathered, there remain many problems peculiar to the field of viral chemotherapy. Among these obstacles is the fact that chemical agents effective against viruses are often also toxic to mammalian cells. Also, antiviral agents are affected by a multiplicity effect. An agent may be effective against viruses in low concentration but ineffective against high concentrations. Demonstrating the clinical effectiveness of antiviral agents is difficult, and drug-resistant viral strains may well emerge.

Microbes and the Environment

The activities of microorganisms are greatly affected by the physical condition of their environments. Understanding environmental influences and the associations between microorganisms and other forms of life helps to explain the distribution of microorganisms in nature and makes possible the development of methods with which to control microbial activities and destroy undesirable organisms. Not all organisms respond similarly to a given environmental factor. As a matter of fact, a particular condition may be beneficial to one organism yet harmful to others. It is also quite possible for certain organisms to tolerate some unfavorable environmental conditions under which they cannot grow but still survive. The chapters in Part V will provide a brief survey of bacteria and their environments as well as discussions of representative microbial associations and environmental activities in which a wide variety of microorganisms are involved.

TABLE 12–7 Current Clinical Status of Antiviral Agents

	Type of Application			
	Topical		*Systemic*	
Agent	*Prophylaxis (preventative measure)*	*Therapy*	*Prophylaxis*	*Therapy*
Acyclovir	+[a]	+	+	−
Amantadine	−	+	+	−
Azido-deoxythymidine	−	−	−	+
Cytosine arabinoside	−	−	−	−
Idoxuridine	−	+	−	−
Interferon	+	−	−	−
Methisazone	−	−	+	−
Ribavirin	−	−	−	+[b]
Vidarabine	−	+	−	+

[a] +, effective; − not effective.
[b] Administered by aerosol.

Of all these important problems, the evaluation of clinical effectiveness is especially critical, because a drug must be thoroughly tested before it can be made available for general use. The clinical value of a few antiviral drugs has been established (Table 12–7). Before the clinical effectiveness of an antiviral agent can be determined, the drug in question must be evaluated for antiviral activity and toxicity both in cell culture and in animal models. In the second phase of drug evaluation, human experimentation is performed, with the objective of determining tolerable dosages and general effects of the antiviral agent on the body. The third phase is the formal therapeutic trial, which is directed toward determining the true clinical effect of the drug on actual cases of virus infections. Certain preparations satisfy some of these drug evaluation requirements but for various reasons fall short of satisfying all of them. Clearly, the problems peculiar to viral chemotherapy must be overcome before a proven treatment for viral diseases will be at hand.

Side Effects of Chemotherapy

Reasonable expectations for the outcome of treatment with antibiotics are quite clear. However, at times, unfavorable side effects or responses develop that limit the usefulness of antibiotics in arresting or curing certain diseases. Examples of side effects and complications include an exaggerated allergic response (hypersensitivity), direct toxic effects on cells and/or organs such as the kidney and liver, destruction of blood-forming tissues in the bone marrow, and changes in an individual's normal bacterial flora, which may contribute to the establishment of and overgrowth by opportunistic microorganisms. Depending on the severity and duration of such responses, death of the patient may result. Tables 12–2 and 12–5 list the general signs and symptoms associated with the side effects of chemotherapy.

Summary

1. Antibiotics are among the most frequently prescribed drugs for treatment and control of microbial infections.
2. Certain problems have appeared that can limit their effectiveness. These include antibiotic-resistant microorganisms, transfer of drug resistance among organisms, and the effects of the host on antibiotic action.

Historical Background

The discoveries of Salvarsan by Ehrlich in 1909, sulfa drugs by Domagk in 1935, and penicillin by Fleming in 1929 ushered in the modern era of chemotherapy.

Chemotherapeutic Agents

1. Chemotherapeutic agents are either natural products of microorganisms or antimicrobial agents produced in the laboratory. Today the term *antibiotic* applies to both categories of substances.
2. Antibiotics can be used as feed additives to protect plants and livestock against infectious diseases as well as to keep agricultural products fresh for extended periods.

Principles of Chemotherapy

1. The administration of an antibiotic depends on several factors, including the physical condition of the recipient, the existence of drug allergies, the pathogen, the site of infection, and the dosage of the antibiotic to be given.
2. Laboratory findings of effective antibiotic levels do not always correlate with similar levels in the body.

3. Laboratory testing is used to determine the lowest concentration of a drug that will prevent the growth of a standardized microbial suspension. This relative sensitivity is expressed as *minimal inhibitory concentration (MIC).*

MINIMAL ANTIBACTERIAL (ACTIVE) CONCENTRATIONS

1. Minute amounts of antibacterial agents can produce structural changes as well as growth, and therefore are referred to as minimal antibiotic concentration (MACs).
2. Minimal bactericidal (lethal) concentrations are called MBCs.

PROPERTIES OF AN EFFECTIVE ANTIBIOTIC

1. An effective chemotherapeutic agent must be relatively nontoxic to the host and exhibit antimicrobial activity at low concentrations in the body of an infected individual.
2. The ration between the minimum toxic dose of a drug for the host and the minimum effective microbial lethal dose is called the *therapeutic index.*

MECHANISMS OF ACTION OF ANTIMICROBIAL AGENTS

1. Chemotherapeutic agents causing irreversible damage or death to microbial cells are called *cidal* drugs. These preparations are independent in their actions on microorganisms.
2. Drugs inhibiting growth and reproduction of microbial cells exert a *static* effect. Static drugs are dependent on the host's immune system for the elimination of pathogens.
3. Drugs active against several gram-positive and gram-negative bacteria are called *broad-spectrum.* Those active against a single group or only a few species have a *narrow,* or limited, *spectrum* of activity.

SPECIFIC MECHANISMS OF ACTION

Major mechanisms by which antimicrobial agents can inhibit growth or kill microorganisms include (1) inhibition of specific products of metabolism; (2) inhibition of cell wall formation; (3) inhibition of protein synthesis; (4) irreversible damage to the cell membrane; and (5) inhibition of nucleic acid synthesis.

COMBINATION CHEMOTHERAPY

1. The effectiveness of a particular antibiotic can be increased when it is given in combination with another drug. The result is called a *synergistic effect.*
2. Some antibiotics administered in combination produce an *antagonistic* or an *additive* result.

Some Commonly Used Antimicrobial Drugs

SULFA DRUGS

1. Sulfa drugs are bacteriostatic and act upon growing and metabolizing cells.
2. The mechanism of action is an example of competitive inhibition of enzymes, which interferes with cellular metabolism.

PENICILLINS

1. Cephalosporins are a group of antibiotics known for stability to beta-lactamases.
2. Penicillins are bactericidal and act by interfering with cell wall formation.
3. The penicillins are effective only during the growth stages of sensitive organisms.
4. Effects of penicillin are caused by its interaction with penicillin-binding proteins (PBPs) in the cytoplasmic membranes of bacteria.

MONOBACTAMS

1. Monobactams function like other beta-lactam antibiotics and interfere with bacterial cell wall formation.
2. The antibiotics are effective against gram-positive and certain gram-negative bacteria. They are stable to most plasmid- and chromosome-mediated beta-lactamases.

CLINDAMYCIN: A USEFUL ALTERNATIVE TO PENICILLIN-CEPHALOSPORIN ANTIBIOTICS

This semisynthetic antibiotic inhibits protein synthesis.

CHLORAMPHENICOL AND THE TETRACYCLINES

1. These agents are bacteriostatic, broad-spectrum antibiotics.
2. Both antibiotics inhibit protein synthesis.

AMINOGLYCOSIDES

The aminoglycosides are all bactericidal and interfere with protein synthesis.

POLYPEPTIDES

Polypeptides include polymyxin B and colistin and cause a leakage of essential cytoplasmic parts.

ANTIMYCOBACTERIAL DRUGS

Drugs commonly used against tuberculosis include isonicotinic hydrazide (INH), para-aminosalicylic acid (PAS), and ethambutol.

RIFAMPIN

Rifampin is a broad-spectrum drug used against tuberculosis.

Mechanisms of Drug Resistance

1. Antibiotic resistance in certain bacteria is determined and controlled by extrachromosomal particles called *plasmids,* or chromosomal transfer.
2. The ability of organisms to transfer resistance limits the effectiveness of antibiotics.
3. Mechanisms by which organisms develop resistance to antimicrobial agents include enzymatic alteration of the drug; changes in selective permeability of cell walls and membranes; changes in the sensitivity of affected enzymes; and increased production of a competitive substrate.

BETA-LACTAMASES

1. Beta-lactamases are important enzymes responsible for resistance to many beta-lactam antibiotics, such as the penicillins and cephalosporins.
2. Among the bacteria known for their beta-lactamase production are *Haemophilus influenzae, Neisseria gonorrhoeae, Staphylococcus aureus,* and species of *Enterobacter* and *Pseudomonas.*
3. Detection of beta-lactamase producers (antibiotic-resistant organisms) in the laboratory can be achieved using special antibiotic paper disks.

THE DANGERS OF ANTIBIOTIC ABUSE

Increased unnecessary exposure to antibiotics has resulted in a significant increase in antibiotic-resistant organisms.

Antibiotic Sensitivity Testing Methods

Methods for determining the antibiotic susceptibility of microorganisms include drug diffusion techniques, performed with cylinders and filter paper disks, and tests monitoring blood levels of antibiotics in patients.

Minimal Inhibitory and Minimal Bactericidal Concentration Determinations

Both qualitative and quantitative antibiotic susceptibility tests are available. These include agar dilution, broth dilution, disk elution, and disk diffusion.

Blood Levels of Antimicrobial Agents

In cases of life-threatening diseases, it is often necessary to monitor closely blood levels of antibiotics in the patient.

Antimycotic Agents

1. Fewer agents exist for the treatment of fungus infections than for bacterial diseases.
2. Antifungal drugs affect cell membrane permeability and protein and RNA synthesis and inhibit cell division.
3. The imidazoles are a group of antimicrobial agents active against fungi and certain pathogenic protozoa.

Treatment of Protozoan Diseases

1. Chemotherapeutic agents for protozoan diseases are toxic and not always readily available.
2. Their antimicrobial action is similar to those of drugs used against other microbes.

Antiviral Agents

1. Some promising antiviral agents have been developed.
2. Before such drugs can be used, they must be shown to have antiviral activity and be relatively nontoxic.
3. Interferons (IFNs) are a family of proteins with antiviral and antitumor activities. Three distinct types are known: α (alpha), β (beta), and γ (gamma).
4. IFNs also affect cell multiplication, cell motility, cell membrane rigidity, and various immunological processes such as immunoglobulin production, rejection of foreign tissue grafts, and the activation of certain white blood cells.
5. IFN inhibits viral replication by interfering with the viral protein translation process.
6. Recombinant DNA technology may be the commercial means by which high levels of therapeutically valuable IFNs can be produced inexpensively.

Side Effects of Chemotherapy

Various side effects develop in patients with the use of certain chemotherapeutic drugs. Such effects include allergic responses, direct toxic effects on body organs, destruction of blood-forming tissues, changes in an individual's normal bacterial flora, and death.

Questions for Review

1. a. What are antibiotics?
   b. Of what value are such chemicals?
   c. What are the sources of antibiotics?

2. Differentiate between bacteriostatic and bactericidal drugs.

3. List the properties of a functional antibiotic.

4. a. List and explain four ways in which antimicrobial drugs interfere with microbial cells and/or activities.
   b. Give an example of one antibiotic that acts by each mechanism that you listed for 4a.

5. What is a broad-spectrum antibiotic?

6. Why are antibiotics such as penicillin and streptomycin ineffective against viruses?

7. a. What is an *in vitro* antibiotic sensitivity test?
   b. What factors affect the accuracy of such procedures?

8. a. Of what value is the determination of the blood or serum levels of an antibiotic in the treatment of a disease?
   b. What methods are used for this type of determination?

9. a. Explain microbial drug resistance.
   b. How is such resistance transferred among bacteria?
   c. What mechanisms are responsible for drug resistance?
   d. Is such resistance increasing or decreasing? Explain.

10. What is the value of semisynthetic or synthetic antibiotics?

11. Why are antibiotics, which are effective against bacterial pathogens, given to individuals with viral infections?

12. a. How do antimycotics function?
   b. List three antimycotic agents.

13. a. How are antiviral agents tested and selected?
   b. Are there any limitations to the use of such agents?

14. a. What are IFNs?
   b. How do they work?
   c. Do IFNs have any chemotherapeutic value?

15. Does the host of an infectious agent exert any influence on the effectiveness of an antibiotic?

16. Define or explain the following:
   a. MIC
   b. *Streptomyces*
   c. beta-lactamase
   d. competitive inhibition
   e. plasmid
   f. MAC
   g. MBC
   h. AZT
   i. synergy
   j. monobactams

Suggested Readings

ABRAHAM, E. P., "The Beta-Lactam Antibiotics," *Scientific American* **244:**76 (1981). *A clear presentation of the structure and activity of the important chemotherapeutic drugs.*

AINSWORTH, G. C., *Introduction to the History of Mycology.* Cambridge: Cambridge University Press, 1976. *An interesting and scholarly work containing many anecdotes associated with various breakthroughs in microbiology. The account of Sir Alexander Fleming's discovery of penicillin and how it revolutionized the practice of contemporary medicine is particularly worthwhile.*

FRIEDMAN, R. M., *Interferons, A Primer.* New York: Academic Press, 1981. *A straightforward description in relatively nontechnical terms of the biology, mechanisms of production and action, and clinical aspects of IFNs. A greater emphasis is placed on principles than on details of experiments.*

HIRSCH, M. S., and S. M. HAMMER, "Nucleoside Derivatives and Interferons as Antiviral Agents," in *Current Clinical Topics in Infectious Diseases,* Remington, J. S., and M. N. Swartz (eds.). New York, McGraw-Hill Book Company, 1982, pp. 30–55. *A summary of the status of the most promising antiviral agents.*

HIRSCH, M. S. and J. C. KAPLAN, "Antiviral Therapy," *Scientific American.* **256:**76 (1987). *A functional description of how antiviral drugs such as interferon and acyclovir distinguish between the host and the virus.*

NEU, H. C., "The In Vitro Activity, Human Pharmacology, and Clinical Effectiveness of New β-Lactam Antibiotics," *Annual Review of Pharmacology and Toxicology.* **22:**599 (1982). *A review of a number of new β-lactam antibiotics that are currently being evaluated for clinical use.*

PETERSON, P. K. and J. VERHOEF, *The Antimicrobial Agents Annual I.* New York: Elsevier Science Publishing Company, Inc. 1986. *A collection of authoritative reviews of new antimicrobial agents as well as recently reported information on established agents.*

PRUSOFF, W. H., "Some Thoughts on the Present and Future Approaches to the Development of Antiviral Agents," in *Human Viruses: An Interdisciplinary Perspective,* Nahmias, A. J., W. Dowdle, and R. F. Schinazi (eds.). New York: Elsevier North-Holland, 1981, p. 507. *A general discussion of basic challenges in antiviral chemotherapy emphasizing the importance of understanding the molecular basis of the enzyme interactions controlling virus replication.*

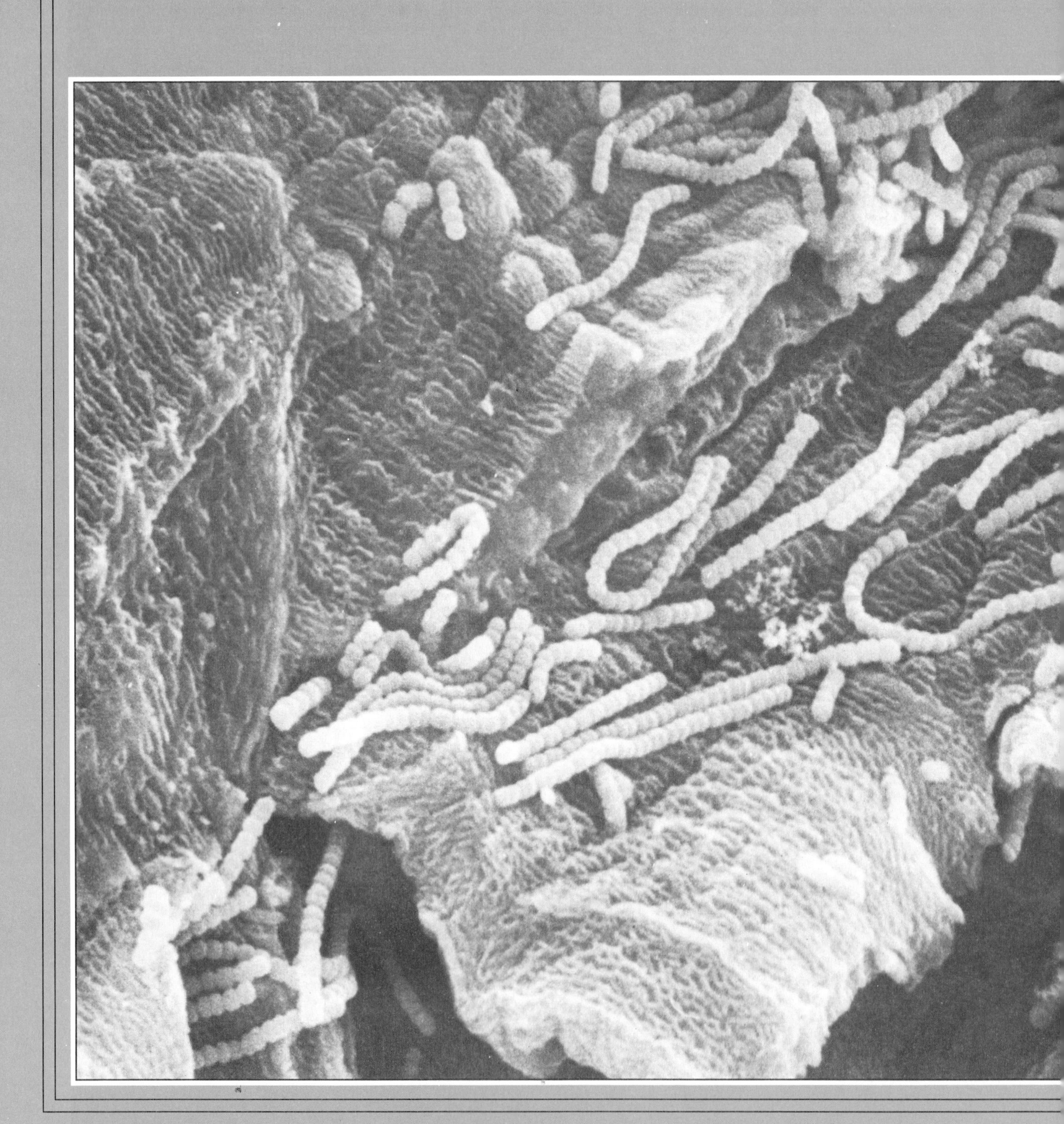

PART V

Microbial Ecology and Interactions

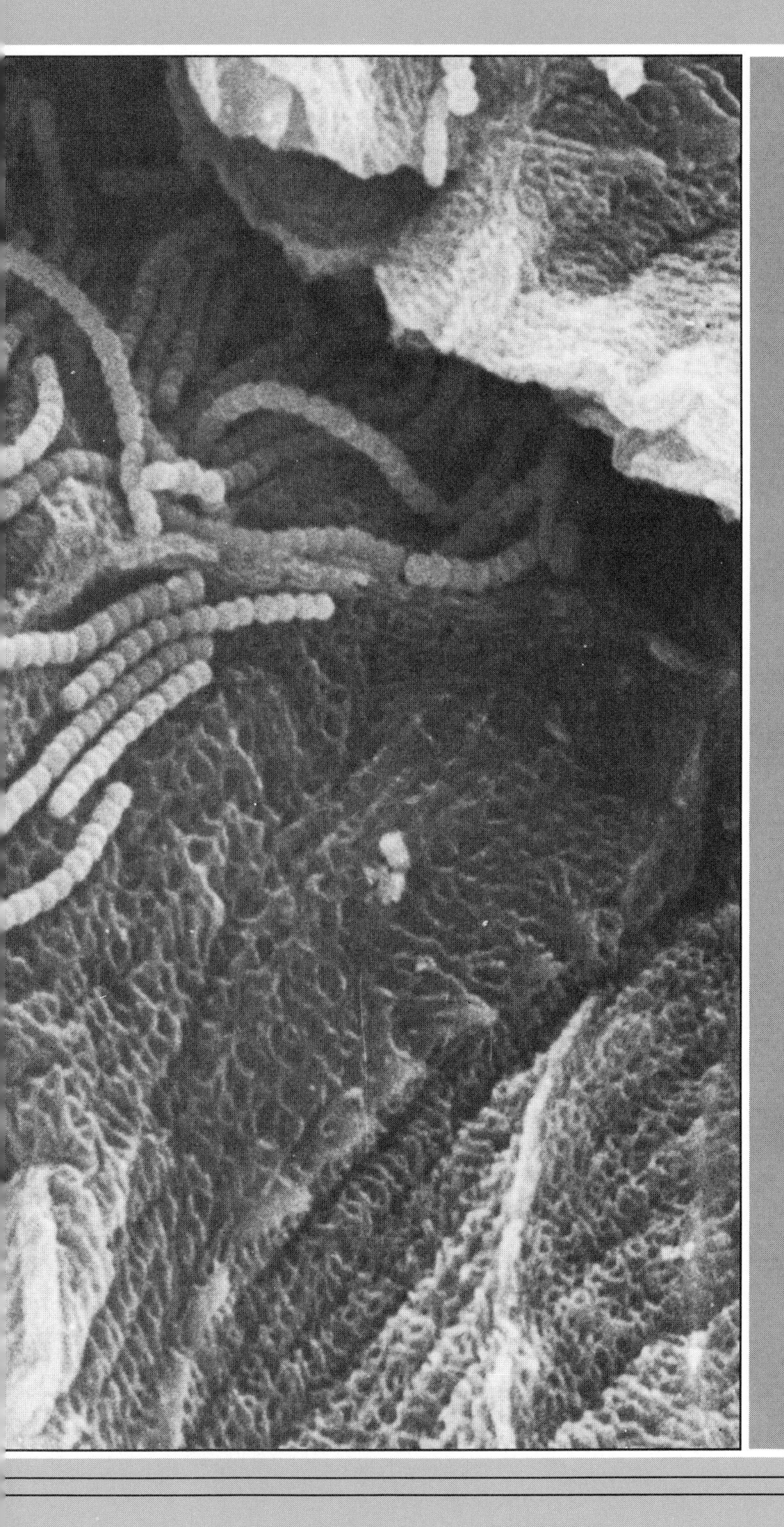

Simonsiella *filaments on the surface of a tongue sample taken from cattle. [From R. P. McCowan, K-J. Cheng, and J. W. Costerton,* Appl. Environ. Microbiol. *37:1224–1229, 1979.]*

13

A Survey of Procaryotes

Microbes possess a wider range of physiological and biochemical potentialities than do all other organisms combined. Microbes represent forms of life that can persist in nature because they fill particular ecological niches. — C. B. Van Niel

After reading this chapter, you should be able to:

1. Define and distinguish among the following terms: "classification," "taxonomy," and "identification."
2. Describe two techniques used by molecular biologists to show taxonomic relationships among microorganisms.
3. Briefly describe the features of the divisions in the kingdom Procaryotae.
4. List and describe the unusual features of the gliding bacteria, sheathed bacteria, spiral and curved bacteria, and mycoplasma.
5. Describe the life cycles of the prosthecate bacteria *Caulobacter* and *Hyphomicrobium*.
6. Discuss obligate intracellular parasitic procaryotes and their effects on host cells.
7. Select and describe the life cycle of one obligate intracellular parasitic procaryote.
8. Describe the general features and activities of the cyanobacteria.
9. Compare and describe those properties of the archaeobacteria that distinguish them from other procaryotic organisms.

It is estimated that the procaryotes preceded eucaryotic forms of life on earth by as much as a billion years. During this time, certain procaryotes became specialized as photosynthetic cells, and molecular oxygen, a by-product of photosynthesis, began to accumulate. The accumulation of oxygen led to further development of cellular life and quite likely to the evolution of eucaryotes. In this chapter, we shall consider some of the modern techniques used to find the natural relationships among the procaryotes and then briefly survey this most interesting and diverse group of microorganisms.

The earliest evidence of life dates back some 3 billion years. And until about half a billion years ago, microorganisms dominated the zones of air, land, and water occupied by living forms on earth (the biosphere). The success of the procaryotes, biologically speaking, is due to their rapid rate of cell division and their great metabolic diversity. They can survive in many environments that support no other form of life. Some procaryotes have been found in the dark depths of the ocean, others in the icy waters of Antartica, and still others in the near-boiling waters of natural hot springs. Some unusual and distinctive members of the kingdom Procaryotae will be presented in this chapter. But, first we will describe some of the molecular biological techniques used to show the relatedness between organisms and to expand on the introduction to classification presented in Chapter 2.

Applications of Molecular Biology to Taxonomy

As indicated in Chapter 2, the purpose of **classification** is to group organisms with similar properties and to separate those that are different. **Taxonomy,** the science of classification, serves to identify and describe as completely as possible the basic taxonomic units, the species. Taxonomists arrange organisms in a ranked series of categories that reflect interrelationships. The process by which an organism is assigned to an established genus and species is called **identification.** As new and more precise information becomes available, classification schemes, the names of organisms (nomenclature), the bases for identification, and even the recognition of new forms of life are subject to change. Figure 13–1 shows the relationship of sources of information and the flow of such information to classification.

Molecular biology is one of the specialty areas providing more precise measures of evolutionary relatedness among organisms. A select number of techniques are presented in this section. These methods are based on attempts to compare more directly the information contained in the DNA of different groups of organisms. **[Refer to Chapter 6 for a discussion of the genetic code and protein synthesis.]**

Figure 13–1 Sources and flow of information used in the description and classification of bacteria. *[After H. G. Trüper, and J. Kramer,* Principles of Characterization and Identification of Prokaryotes, *in M. P. Starr* et al., *eds.,* The Prokaryotes, *Vol. 1. New York: Springer-Verlag, 1981, pp. 176–193.]*

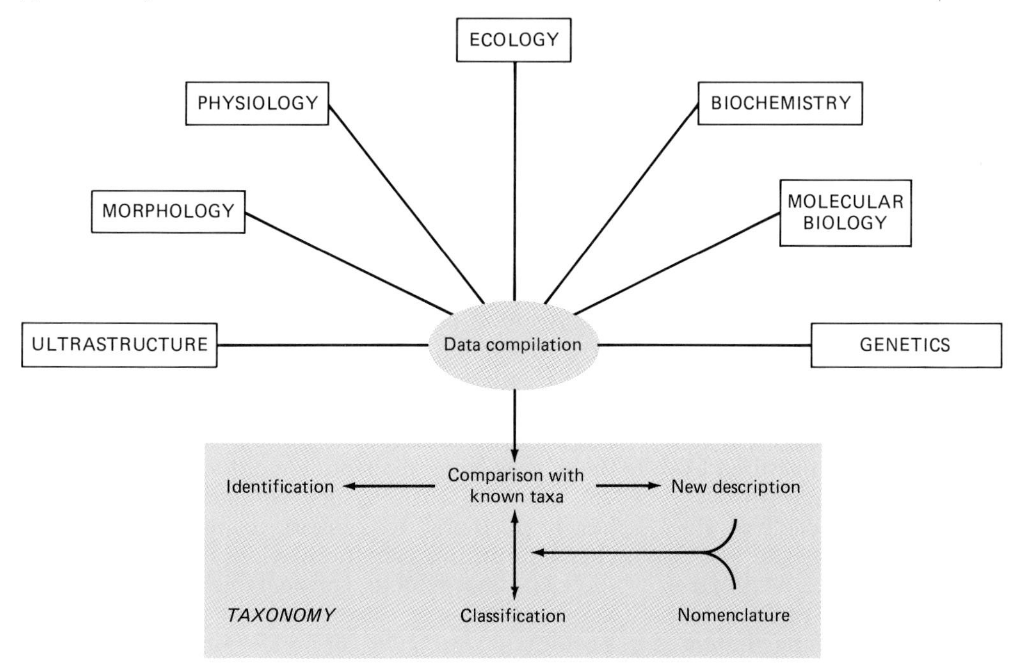

TABLE 13–1 Selected Bacterial Genera and Their % (G + C) Ratios

% (G + C)	*Representative Genera*[a]
30–36	*Bacillus, Clostridium, Fusobacterium, Staphylococcus, Streptococcus*
38–44	*Bacillus, Coxiella, Haemophilus, Lactobacillus, Neisseria, Proteus, Streptococcus*
46–52	*Clostridium, Corynebacterium, Enterobacter, Escherichia, Klebsiella, Neisseria, Pasteurella, Proteus, Salmonella, Vibrio*
54–60	*Alcaligenes, Corynebacterium, Enterobacter, Klebsiella, Lactobacillus, Pseudomonas, Serratia, Spirillum*
62–68	*Bacillus, Micrococcus, Mycobacterium, Pseudomonas, Rhizobium, Vibrio*
70–80	*Mycobacterium, Nocardia, Sarcina, Streptomyces*

[a] For pronunciations, see the Organism Pronunciation Guide at the end of the text.

Percent (G + C) Comparison

One means of studying the relatedness of two organisms is directly connected to the complete composition and sequence of purines and pyrimidines in their respective DNA molecules. This molecular approach to taxonomy involves determining the percentage of guanine and cytosine, % (G + C), in the DNA of an organism, as compared to the total base composition of adenine (A), thymine (T), guanine, and cytosine. Organisms with similar % (G + C) values have been shown to be related genetically and to be very similar in terms of numerical taxonomy. Selected bacteria and their % (G + C) values are presented in Table 13–1. These figures can be calculated from the formula:

$$\%\,(G + C) = \frac{G + C}{A + T + G + C} \times 100$$

In many instances, genera are well grouped by their % (G + C) value. However, certain significant discrepancies exist, either because of incomplete data or because of true unrelatedness. Values vary from 23% to 75%.

DNA Hybridization

When the DNA double-stranded helix is heated, the hydrogen bonds holding the base pairs together are weakened and the strands can separate (denaturation). Upon cooling, the strands reassociate and the DNA appears to be identical to the original. As would be expected, when DNA from *Escherichia coli* (esh-er-IK-ee-ah KOH-lee), for example, is studied in this manner, there is complete reassociation with more DNA from *E. coli.* When two different strains, or species, are mixed and heated, the degree of relatedness of the two organisms can be measured by determining the ability of their respective DNAs to reassociate to form a "hybrid" DNA molecule or undergo DNA–DNA hybridization (Figure 13–2).

The degree of hybridization is referred to as *percent homology* and provides an average measure of the similarity of DNA base sequences along the entire genome (genetic composition) of an organism. With the help of information from DNA–DNA hybridization, misclassified strains can be detected or a newly discovered organism can be assigned to a specific genus. Figure 13–2 shows the features of a DNA–DNA hybridization technique.

DNA–Ribosomal RNA Hybridization

The ribosome, as described in earlier chapters, consists of protein and ribosomal RNA (rRNA), both of which have been shown to have highly stable chemical sequences. rRNAs are direct transcriptions from portions of the cell's DNA. Therefore, there is some homology between the two molecules. DNA–rRNA hybridization involves the radioactive labeling of rRNA of one organism and mixing it with an excess of a single-stranded DNA preparation from another organism. After the mixture is incubated, unpaired rRNA segments are removed by enzymatic treatment and the radioactivity is expressed in terms of the heat stability of the DNA–rRNA hybrids.

Sequencing of 16S Ribosomal RNA

In this procedure the 16S rRNA component from various microbial species is digested with the enzyme ribonuclease. The fragments produced by the enzymatic treatment vary in length from 1 to 20 nucleotides. The purine and pyrimidine base sequence of each fragment is determined, and fragments consisting of six or more nucleotides are placed into a so-called nucleotide-word dictionary or catalogue, to be used for comparisons with the 16S rRNA fragments of other species. Each fragment ends in guanine (Figure 13–3). The degree of similarity or relatedness between two microbial species can then be determined through a comparison of their respective nucleotide-word dictionaries. The 16S rRNA is used in such comparisons because it has been found to remain relatively stable (unchanged) during evolution, while the other part of the ribosome exhibits considerable variability. **[See Chapter 7 for a description of ribosome components.]**

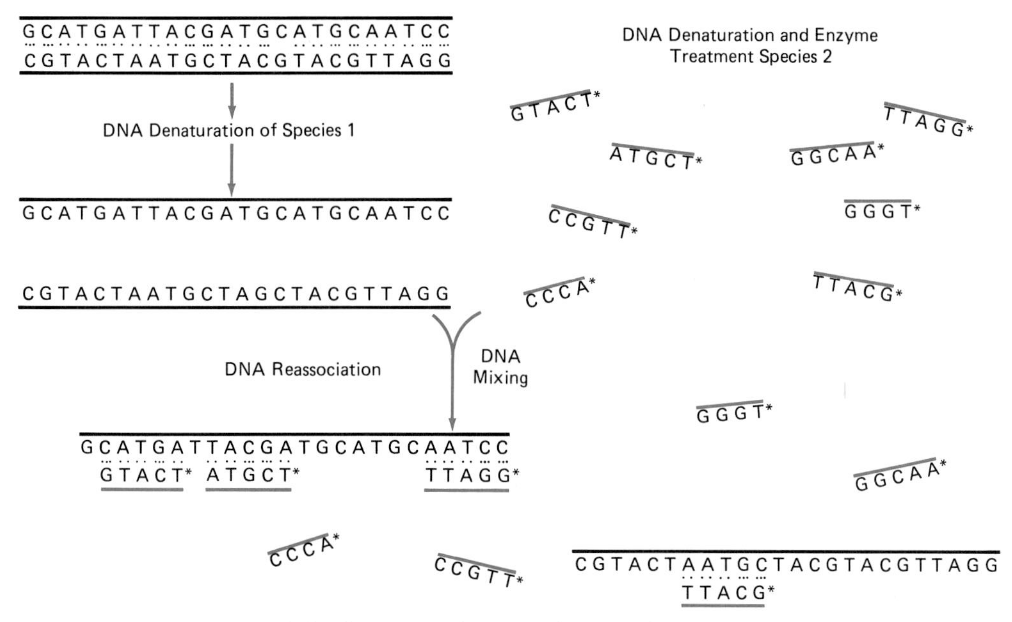

Figure 13–2 DNA-DNA hybridization technique. The DNA from species 1 is denatured to produce single strands, which are trapped on some form of inactive material. The DNA from species 2 is made radioactive(*), then denatured and fragmented by enzymatic action. Mixing, heating, and combining (reassociation) of the DNAs from the two species allow complementary segments of the bases to form double-stranded DNA molecules. The relative amount of these molecules formed indicates the degree of relatedness between the two species.

Methods Used to Study the Molecular Architecture of the Procaryotic Cell

Several additional techniques are useful in showing natural or phylogenetic relationships among the procaryotes. These include **chemotaxonomy** and **serology.** Chemotaxonomy involves the application of chemical and physical methods to analyze the chemical composition of entire bacterial cells or their parts. For example, the chemical composition of procaryotic cell walls was found to be one of the earliest properties to be of great chemotaxonomic value. The characteristic cell wall component of many procaryotes (also present in gram-negative cells, gram-positive cells, and the cyanobacteria) is the peptidoglycan layer. **[Refer to Chapter 7 for a description of this cell wall component.]** This complex molecular layer is not found in the mycoplasmas and the archaeobacteria (see Table 13–3). Further study has shown that the amino acid and/or sugar composition of the peptidoglycan layer varies among gram-positive species. Such findings provide additional useful chemotaxonomic information.

The chemical analyses of other cell components have also provided information that is useful for classification and for shedding light on possible natural relationships. These include the lipids of procaryotic cytoplasmic membranes and the cytochromes, the specialized hemeproteins described in Chapter 5. Several distinguishing properties of procaryotes are described later in this chapter.

Serological studies depend on the ability of chemical components of procaryotic cells to cause the production of highly specific protein molecules known as *antibodies* or *immunoglobulins* when introduced into vertebrates (animals with backbones). Antibodies are found in the blood of such animals

Figure 13–3 The nucleotide word dictionary resulting from ribonuclease digestion. The fragments of 16S rRNA are of value in determining the degree of relatedness between microbial species. Note that only fragments consisting of six or more nucleotides (arrows) are of use and that each fragment, regardless of the number of nucleotides it contains, ends in a guanine (G).

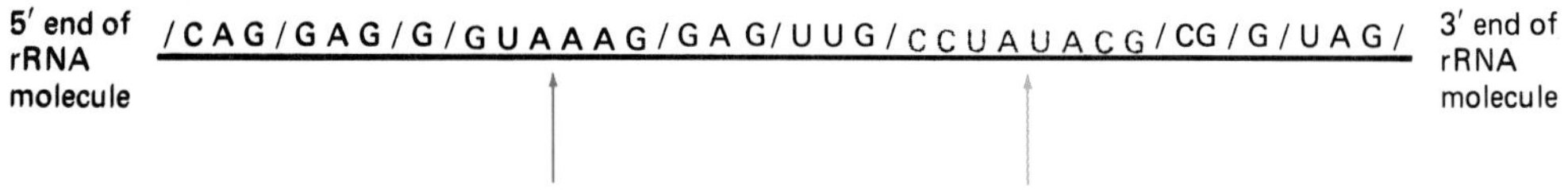

5′ end of rRNA molecule /CAG/GAG/G/GUAAAG/GAG/UUG/CCUAUACG/CG/G/UAG/ 3′ end of rRNA molecule

and can be demonstrated by specific laboratory techniques. [Chapter 19 describes a large number of such procedures.] The chemicals provoking antibody production are called *antigens*.

The serological techniques used for bacterial classification fall into two categories: (1) those concerned with showing chemical differences or similarities associated with the surface structures of procaryotic cells such as pili, flagella, capsules, cell walls, and cytoplasmic membranes and (2) those concerned with finding similarities or differences in similar proteins from different bacterial species.

Diagnostic Microbiology

As Chapter 22 will describe and emphasize, the major goal of approaches used in clinical diagnostic laboratories is to accurately identify bacterial agents isolated from diseased patients as quickly as possible. The emphasis is placed on rapid and practical procedures.

The identification of bacterial species is based on schemes or keys that take into consideration a variety of characteristics, such as differential staining reactions (e.g., Gram's stain, acid-fast stain, etc.), colonial properties, various biochemical tests, and—in certain cases—the results of specific serological tests. The elements of a diagnostic key are organized into a flowchart pattern. The results obtained from tests and/or properties of isolated organisms determine the path followed to their identification. Established reference and identification schemes are available in laboratories to describe the expected results of major reactions obtained with known bacterial species. Such schemes are used as standards or comparisons in the identification of isolated bacteria. Additional details of diagnostic microbiology can be found in Chapter 21. The remaining portion of this chapter will survey the kingdom of the Procaryotae.

The Kingdom of the Procaryotae

Chapter 7 described the differences in structure and organization between bacteria and other forms of life. Here we shall discuss a number of species and

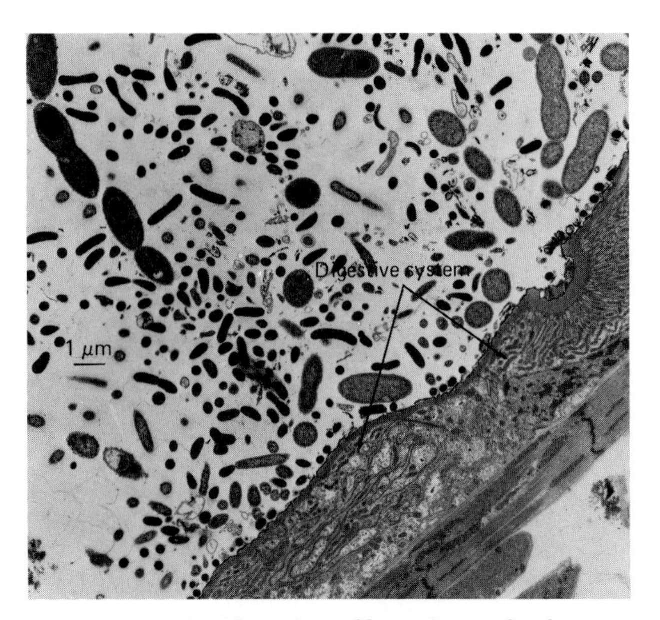

Figure 13–4 A wide variety of bacteria are closely associated with the digestive system of a termite. Such an association is believed to enable the termite and the bacteria to exchange nutrients to the benefit of both. [*J. Breznak, and H. S. Pankrantz,* Appl. Environ. Microbiol. ***33***:*406–426, 1977.]*

their activities to show the widespread distribution and major contributions of this microbial group. The term "bacteria" will be used interchangeably with "procaryotes."

In their natural habitats, bacteria associate with other bacteria, different microbial types (Figure 13–4) or higher forms of life. Such relationships may be either beneficial or harmful to the forms of life involved. Fortunately, most bacteria are harmless, and many perform functions that are favorable to humans and other forms of life. These include:

1. Aiding the digestive processes of animals
2. Decomposing organic material
3. Producing and flavoring foods
4. Returning chemical elements to the soil for use by plants.

The functions of bacteria are so important that if they were to stop, all animals and plants would soon become extinct.

There are several detailed classifications of bacteria. While none of these is official, the approach to the classification of procaryotic organisms presented in the 1984 edition of *Bergey's Manual of Systematic Bacteriology* is widely accepted. The goals of the manual are to assist in bacterial identification and to show the relationships that exist among various types of bacteria. [Refer to Chapter 2 for additional details of this manual.]

According to the classification system proposed in *Bergey's Manual,* the kingdom Procaryotae is subdivided into the following four divisions: (I) Gracilicutes (GRAS-il-ee-cue-tees), (II) Firmicutes (FUR-me-cue-tees), (III) Tenericutes (teh-NER-ee-cue-tees), and (IV) Mendosicutes (MEN-doss-ee-cue-tees). General divisions of these properties are given in Table 13–2. Each division is organized into sections composed of several taxa (taxonomic groupings) of bacteria, including classes, orders, families, genera, and species. [See Chapter 2 for a description of taxonomic ranks.] Traditional microbial characteristics based on information from studies dealing with biochemistry, physiology, molecular biology, and genetics form the basis for the arrangement of the taxa. Table 13–3 (on pages 380–381) shows the arrangement of the sections organized according to the specific divisions, together with descriptions of the bacteria belonging to each section. (A more detailed classification can be found in Appendix B and in some of the texts listed in the Suggested Readings for this chapter.)

The remaining portion of this chapter presents descriptions of a number of representatives from the divisions of Gracilicutes, Tenericutes, and Mendosicutes. Many organisms from division II, Firmicutes, are described in detail in relation to their disease-causing activities in Chapters 21 through 29. Some members of this division are also discussed in Chapter 14.

TABLE 13–2 General Features of the Four Divisions the Kingdom Procaryotae[a]

	Divisions			
Features	*I. Gracilicutes*	*II. Firmicutes*	*III. Tenericutes*	*IV. Mendosicutes*
Cell wall	Complex, typical gram-negative structure[b]	Typical gram-positive cell wall; thick[b]	Absent	Most members have some form of cell wall
Peptidoglycan layer	Present (thin)	Present (thick)	Absent	Absent
Gram stain	Usually gram-negative	Generally gram-positive	Gram-negative	Some gram-positive, others gram-negative
Morphology (cell shape)	Cocci, ovals, rods (straight and curved), springlike (helices), and filaments (threadlike)	Cocci, rods, filaments and branching forms	Highly pleomorphic (variable in shape), filamentous with branching	Cocci, spheres, rods, filamentous forms, and pleomorphic
Distinctive cell structures include:	Sheaths, capsules	None	Cell membrane	Cell walls
Endospores	Not formed	Formed by some	Not formed	Not observed
Fruiting bodies and myxospore formation	With certain species only	None formed	None formed	Not observed
Motility	Motile and non-motile forms are present; some exhibit gliding and swimming motility	Motile forms present	Usually nonmotile forms are present; some exhibit gliding motility	Many are motile
Reproduction	Binary fission; budding, and multiple fission rare	Binary fission	Budding, fragmentation, and/or binary fission	Constriction, budding and binary fission[c]
Growth conditions required	Aerobic, anaerobic, facultative aerobic	Aerobic, anaerobic, facultative anaerobic	Aerobic	Most are strictly anaerobic; some are aerobic
Type of metabolism	Photoautotrophic, lithotrophic, or chemoheterotrophic[d]	Chemoheterotrophic	Chemoheterotrophic	Chemolithotrophic, chemoheterotrophic

[a] Adapted in part from Krieg, N. R. (ed.) *Bergey's Manual of Systematic Bacteriology,* vol. 1. Baltimore: Williams & Wilkins, 1984.
[b] Refer to Chapter 7 for descriptions of cell walls.
[c] The mode of reproduction has not been determined for all members of the Division.
[d] Refer to Chapter 5 for descriptions of metabolism.

TABLE 13–3 The Various Sections of the Four Divisions of the Kingdom Procaryotae

Division[a]	*Category*	*Special Properties*	*Representative Genera*[e]
I	Spirochetes	Slender, flexible, coiled cells; they may occur in chains and exhibit transverse fission; motile, gram-negative	*Borrelia,*[b] *Christispira, Leptospira,*[b,c] *Treponema*[b,c]
I	Aerobic/microaerophilic, motile, helical/vibrioid, gram-negative bacteria	Rigid, helically curved rods with less than one complete turn to many turns	*Aquaspirillum, Bdellovibrio,*[c] *Campylobacter,*[c] *Microcyclus, Spirillum*[c]
I	Nonmotile (or rarely motile), gram-negative, curved bacteria	Straight to curved or C-shaped rods; may form rings; obligate aerobes	*Meniscus, Spirosoma*
I	Gram-negative aerobic rods and cocci	Rods that are usually motile, with polar flagella; bluntly rod-shaped to oval cells, some of which are motile by polar or peritrichous flagella and some of which are cyst formers; some rods and cocci that require high concentrations of sodium chloride for growth; certain species utilize methane	*Acetobacter, Alcaligenes,*[c] *Agrobacterium, Azotobacter,*[b] *Bordetella,*[b,c] *Brucella,*[b,c] *Francisella,*[b,c] *Legionella, Methylococcus, Pseudomonas*[b,c]
I	Facultatively anaerobic gram-negative rods	Straight and curved rods; some are nonmotile; others are motile by polar or peritrichate flagella; all members are non-spore-formers; some have special growth requirements	*Citrobacter,*[c] *Edwardsiella, Enterobacter,*[c] *Escherichia,*[b,c] *Haemophilus,*[b,c] *Klebsiella,*[b,c] *Pasteurella,*[b,c] *Salmonella,*[b,c] *Serratia,*[b,c] *Shigella,*[b,c] *Streptobacillus,*[c] *Vibrio,*[b,c] *Yersinia*[b,c]
I	Anaerobic gram-negative rods, straight, curved, or helical	Strict (obligate) anaerobic, non-spore-forming organisms; some members are motile; pleomorphism (variation in shape) occurs	*Bacteroides,*[b,c] *Fusobacterium,*[b] *Leptotrichia*
I	Dissimilatory sulfate- or sulfur-reducing bacteria	Gram-negative; use sulfate or other oxidized sulfur compounds or elemental sulfur as electron acceptor	*Desulfobacter, Desulfuromonas, Desulfovibrio*
I	Anaerobic gram-negative cocci	Cocci of variable size and characteristically in pairs; they are not flagellated	*Acidaminococcus, Veillonella*[c]
I	Chemolithotrophic bacteria	Gram-negative pleomorphic rods; these organisms use inorganic materials for energy	*Nitrobacter,*[b] *Nitrococcus,*[b] *Thiobacillus*[b]
I	Anoxygenic phototrophic bacteria	Gram-negative spherical or rod-shaped bacteria; multiplication is by binary fission and/or budding; they are photosynthetic without producing oxygen; pigments are purple, purple-violet, red, orange-brown, brown, or green	*Chromatium, Rhodomicrobium*
I	Cyanobacteria	Gram-negative rods or cocci; unicellular or filamentous; motility, when it occurs, is by gliding; contain chlorophyll and phycobiliproteins; produce O_2 during photosynthesis; several species perform N_2 fixation	*Anabaena, Chroococcus*
I	Budding and/or appendaged bacteria	Bacteria with rod-, oval-, egg-, or bean-shaped filamentous growth; multiplication is by budding or binary fission; these bacteria sometimes have a holdfast cell	*Caulobacter, Hyphomicrobium*
I	Sheathed bacteria	Gram-negative rods that occur in chains within a thin sheath; they sometimes have a holdfast cell for attachment to surfaces	*Crenothrix, Leptothrix, Sphaerotilus*

TABLE 13–3 (continued)

Division[a]	*Category*	*Special Properties*	*Representative Genera*[e]
I	Gliding, fruiting bacteria	Gram-negative rods typically embedded in a tough slime coat; they are capable of a slow gliding movement; reproduction is by binary fission; gliding bacteria sometimes form colorful fruiting bodies	*Myxococcus, Stigmatella*
I	Gliding, nonfruiting bacteria	Gram-negative; appear as rods or helical filaments; some species are pathogenic to fish	*Beggiatoa, Cytophaga*
I	Rickettsias and chlamydias	The majority of cells are gram-negative coccoid or pleomorphic rods; most are obligate intracellular parasites transmitted by arthropods	*Chlamydia,*[b,c] *Cowdria, Coxiella,*[b,c] *Ehrlichia, Neorickettsia, Rickettsia,*[b,c] *Rickettsiella, Rochalimaea,*[c] *Symbiotes*
II[d]	Gram-positive cocci	Various arrangements of cocci that are aerobic, facultative (adaptable), or anaerobic; includes some pathogens	*Aerococcus,*[c] *Micrococcus, Peptococcus, Sarcina, Staphylococcus,*[b,c] *Streptococcus*[b,c]
II	Endospore-forming gram-positive rods and cocci	Members are aerobic, facultatively anaerobic, or anaerobic	*Bacillus,*[b] *Clostridium*[b,c] *Desulfotomaculum, Sporosarcina*
II	Regular non-spore-forming gram-positive rods	Members may be aerobic, facultatively anaerobic, or anaerobic	*Erysipelothrix,*[c] *Lactobacillus,*[b] *Listeria*[b,c]
II	Irregular non-spore-forming gram-positive rods	Rods or pleomorphic rods, with filamentous and branching filaments; included are aerobic, facultatively anaerobic, and anaerobic rods	*Actinomyces,*[c] *Arachnia, Arthrobacter, Bifidobacterium,*[b,c] *Corynebacterium,*[b,c] *Propionibacterium*
II	Mycobacteria	Rods or pleomorphic rods, these organisms are usually gram-positive and acid-alcohol-fast (acid-fast); some species are pathogenic	*Mycobacterium*
II	Nocardiforms	Form filamentous and branching filaments; reproduce by fragmentation; often acid-fast; some species are pathogenic	*Nocardia*
II	Sporangiate actinomycetes[d]	Rods or pleomorphic rods, with filamentous and branching filaments; included are aerobic, facultatively anaerobic, and anaerobic rods	*Actinoplanes* (including *Amorphosporangium*), *Streptosporangium, Ampullariella, Spirillospora*
III	Mycoplasma	Highly pleomorphic, gram-negative organisms that contain no cell wall; they reproduce by fission, by production of many small bodies, or by budding; members may be aerobic, facultatively anaerobic, or anaerobic	*Acholeplasma, Mycoplasma,*[b,c] *Spiroplasma, Anaeroplasma, Ureaplasma*
IV	Archaeobacteria	Rods, cocci, and variable forms; includes aerobes, anaerobes, autotrophs, and heterotrophs; three major types are halophiles, methanogens, and thermophiles; no peptidoglycans in cell walls; distinctive lipids in cell membrane	*Methanobacterium, Halobacterium, Sulfolobus*

[a] Divisions I. Gracilicutes; II. Firmicutes; III. Tenericutes; IV. Mendosicutes based on descriptions in *Bergey's Manual of Systematic Bacteriology.* N. R. Krieg, ed. (Baltimore: Williams and Wilkins, 1984).
[b] Genera discussed in various chapters of this text.
[c] Medically important species are contained in this genus.
[d] The sections: actinomycetes dividing in more than one plane, streptomycetes and their allies, and other conidiate (conidia forming) genera are not described here. See Appendix B.
[e] For pronunciation, see the Organism Pronunciation Guide at the end of the text.

Microbiology Highlights

WHERE DOES A MICROBIOLOGIST GET CULTURES?

The American Type Culture Collection, commonly referred to as ATCC, is a nonprofit scientific organization established in 1925 and charged with the task of collecting, preserving, and distributing authenticated cultures of microorganisms. Through the ensuing years the ATCC has grown to become the world's largest and most diverse depository of organisms. Over 34,000 strains of organisms representing more than 9,500 different species of algae; bacteria; animal, bacterial, plant, and fungal viruses; protozoa; plasmids; oncogenes (cancer genes); and DNA probes are maintained by the current organization, located in Rockville, Maryland. Some of the microbial strains in the collection were isolated before the turn of the century. Actually the lineage of the ATCC can be traced to some extent to well-known collections such as those of Louis Pasteur in France and the National Collection of Type Cultures in England.

Periodically the strains in the collection are tested to ensure their authenticity for the many users. The ATCC packages more than 60,000 cultures each year for worldwide distribution.

While there are many users of the strains in the collection, the ATCC is of special importance to the mycologist, since it houses the only major depository of living fungi in the United States. Several of these cultures are used in studies dealing with infectious diseases, environmental problems, and industrial applications. The original strain of *Penicillium notatum* (pen-ee-SIL-lee-um no-TAY-tum), the mold used by Alexander Fleming in his discovery of the antibiotic penicillin, as well as newer, industrially important fungi are contained within the ATCC.

Many of the important research studies dealing with the classification, identification, and industrial uses of microorganisms would be seriously hampered without the services of the ATCC.

Division I—Gracilicutes

The Spirochetes

Spiral bacteria are heterotrophic organisms with a unique cellular anatomy and distinctive forms of movement. Spirochetes (Figure 13–5) and spirilla are such procaryotes. These organisms grow in a wide range of natural habitats, including fresh and marine bodies of water, mud, and internal and external body surfaces of humans and other animals (Figure 13–5). A small number of spiral bacteria cause diseases such as leptospirosis (Figure 13–6), syphilis (Color Photograph 62), and relapsing fever.

A spirochete is a flexible helical bacterium with a coiled protoplasmic cylinder that consists of the cytoplasmic and nuclear regions surrounded by a plasma-cell wall complex. Wound around the cylinder are axial fibrils (Figure 13–5). Because of their particular structure, spirochetes can move in liquid environments without being in contact with solid surfaces. Some spirochetes also exhibit a creeping or crawling movement on solid surfaces similar to the locomotion of gliding bacteria. [See axial filaments in Chapter 7.]

Despite basic similarity of structure, many differences exist among spirochetes. They can be aerobic, facultatively anaerobic, or strictly anaerobic. The general properties of five genera of spirochetes are given in Table 13–4.

Spirilla have rigid cell walls and one or more fla-

Figure 13–5 *Cristispira,* a genus of large spirochetes colonizing selected structures of a variety of molluscs such as oysters. (a) Spirochetes on the surface and within the digestive tract of a Chesapeake Bay oyster.

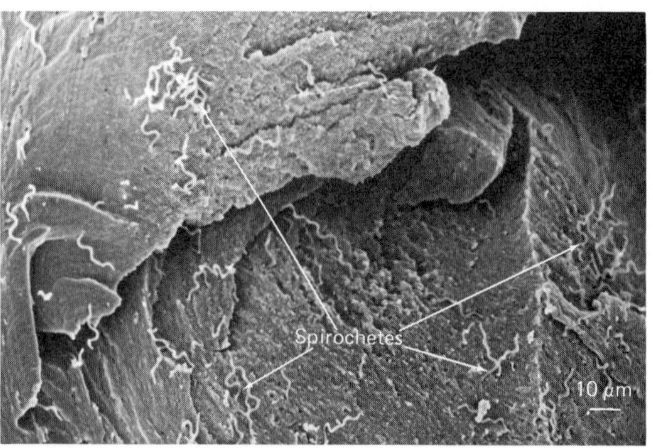

(a)

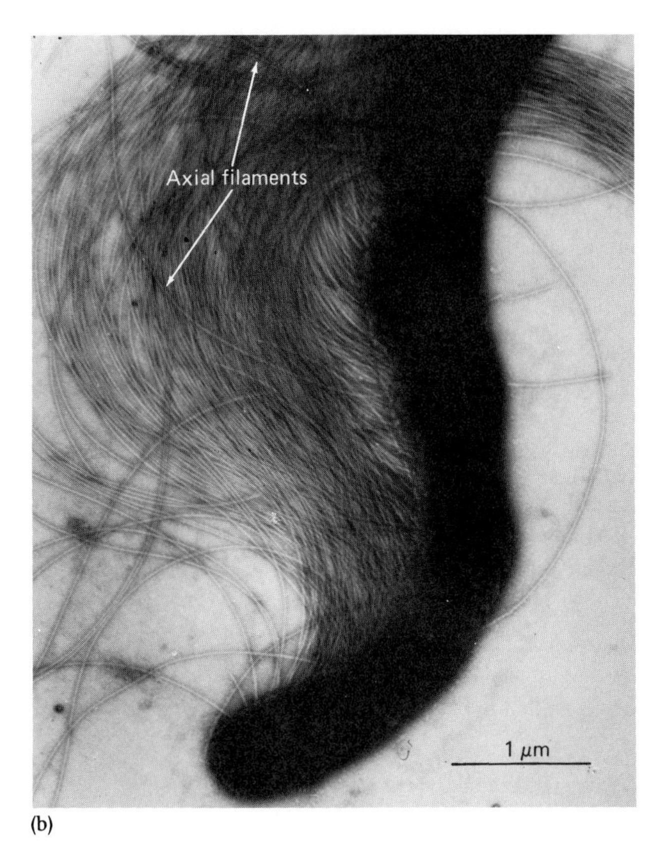

(b)

Figure 13–5 (b) A negatively stained portion of *Cristispira* showing numerous axial filaments. *[Courtesy B. D. Tall, and R. K. Nauman,* Appl. Environ. Microbiol. ***42:****336–343, 1981.]*

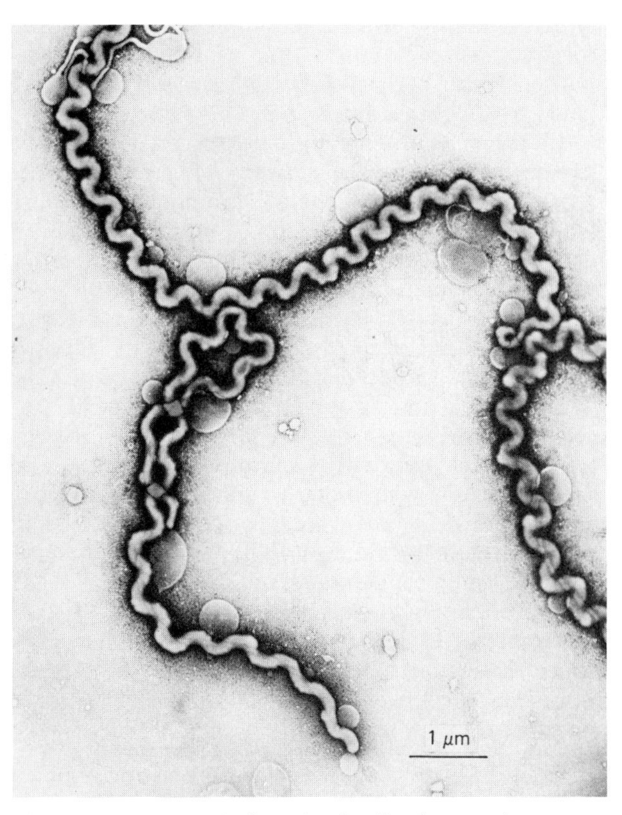

Figure 13–6 Negatively stained cells of *Leptospira canicola.* Note the smooth, coiled shape of these organisms. *[D. L. Anderson, and R. C. Johnson,* J. Bacteriol. ***95:****2293–2309, 1968.]*

gella at one or both ends (Color Photographs 16 and 17). Most spirilla are free-living, but one species is a human pathogen.

Aerobic Microaerophilic, Motile, Helical/Vibrioid Gram-Negative Bacteria

Members of *Campylobacter* (kam-pee-loh-BACK-ter) are curved, spiral rods. Some species form chains and appear ribbon-shaped. These organisms are gram-negative, require small amounts of oxygen *(microaerophilic),* have a single flagellum at one or both ends of their cells, and move with a characteristic corkscrew motion. They are found in the mouths and intestines of humans and other animals. One species, *C. fetus,* is known to cause abortion in cattle and a variety of diseases in humans. **[See *Campylobacter* infections in Chapter 26.]**

Many bacteria require other living cells for their development and reproduction. Some species re-

TABLE 13–4 General Properties of Spirochete Genera

Genus[a]	*Properties*
Borrelia	Generally anaerobic or microaerophilic; species cause relapsing fever in humans and other animals; lice or ticks spread the disease agents
Cristispira	Large spirochetes, usually found in the digestive tracts of many freshwater and marine molluscs[b]
Leptospira	Strictly (obligate) aerobic; usually found free-living in soil or surface waters or in association with humans and other animals; some species cause disease
Spirochaeta	Most are anaerobic; usually found as free-living forms in aquatic environments
Treponema	Anaerobes; found in genital areas, the intestinal tract, or mouths of humans and other animals; many species are normally present in healthy individuals; others produce disease

[a] For pronunciations, see the Organism Pronunciation Guide at the end of the text.
[b] The genus *Pillotina* is a proposed designation for the large spirochetes normally present in termite intestinal tracts.

quire eucaryotic animal and plant cells, while other species use procaryotes. One well-known intracellular parasite of bacteria is *Bdellovibrio bacteriovorus* (dell-oh-VIB-ree-oh back-tir-ee-OH-vore-us).

In 1962, an interesting parasitic relationship between bacteria was observed. *Bdellovibrio bacteriovorus,* an aerobic, flagellated, gram-negative vibrio (comma-shaped), was found to attack certain host bacteria for its own benefit. The generic name of this particular attacking bacterium indicates the organism's behavior. *Bdello* comes from the Greek word meaning "leech." These organisms are now known to be widely distributed in soil and sewage.

Bdellovibrio attacks its prey by striking the host cell surface at a high velocity (Figure 13–7). It penetrates the cell wall with a spinning motion and with the aid of cell-wall digestive enzymes. In the process, the parasite's flagellum is left behind. The invading organism situates itself between its host's cell wall and plasma membrane. Here the *Bdellovibrio* uses partially degraded host components to grow and reproduce. Newly formed offspring leave the host cell to find new susceptible bacteria in which to repeat this parasitic cycle. In cultures containing hosts and parasites, *Bdellovibrio* destroys the host cells. This results in clear circular areas known as *plaques* (Color Photograph 63).

Anoxygenic Photosynthetic Bacteria and the Cyanobacteria

The phototrophic procaryotes include those bacteria that are capable of using environments that contain inorganic sources of carbon. Three distinct and well-defined groups of gram-negative phototrophs are recognized: the purple bacteria, the green bacteria, and the cyanobacteria. The photosynthetic pigments of these three groups distinguish them.

Photosynthesis by green and purple bacteria (Figure 13–8) or other related anaerobic organisms does not produce oxygen **(anoxygenic photosynthesis).** Moreover, aerobic conditions generally prevent the expression of photosynthetic activity in these microorganisms. The cyanobacteria, on the other hand, do produce oxygen as a by-product of photosynthe-

Figure 13–7 *Bdellovibrio.* (a) The stages in the life cycle of this parasitic bacterium. The stages of the cycle show the approach to the susceptible host (1), *Bdellovibrio* attachment (2), penetration of host cell by the invading *Bdellovibrio* (3), the situation of the invader in the host bacterium between the host's cell wall and plasma membrane (4 and 5), the use of the host for the production of new *Bdellovibrio* cells (6 and 7), and the release of newly formed *Bdellovibrio* cells (8). (b) After the parasite has penetrated its host, it will use the nutrients there for reproductive purposes. Eventually the host will break open, releasing newly formed Bdellovibrios. *[Abram, et al.,* J. Bacteriol. ***118****:663, 1974.]*

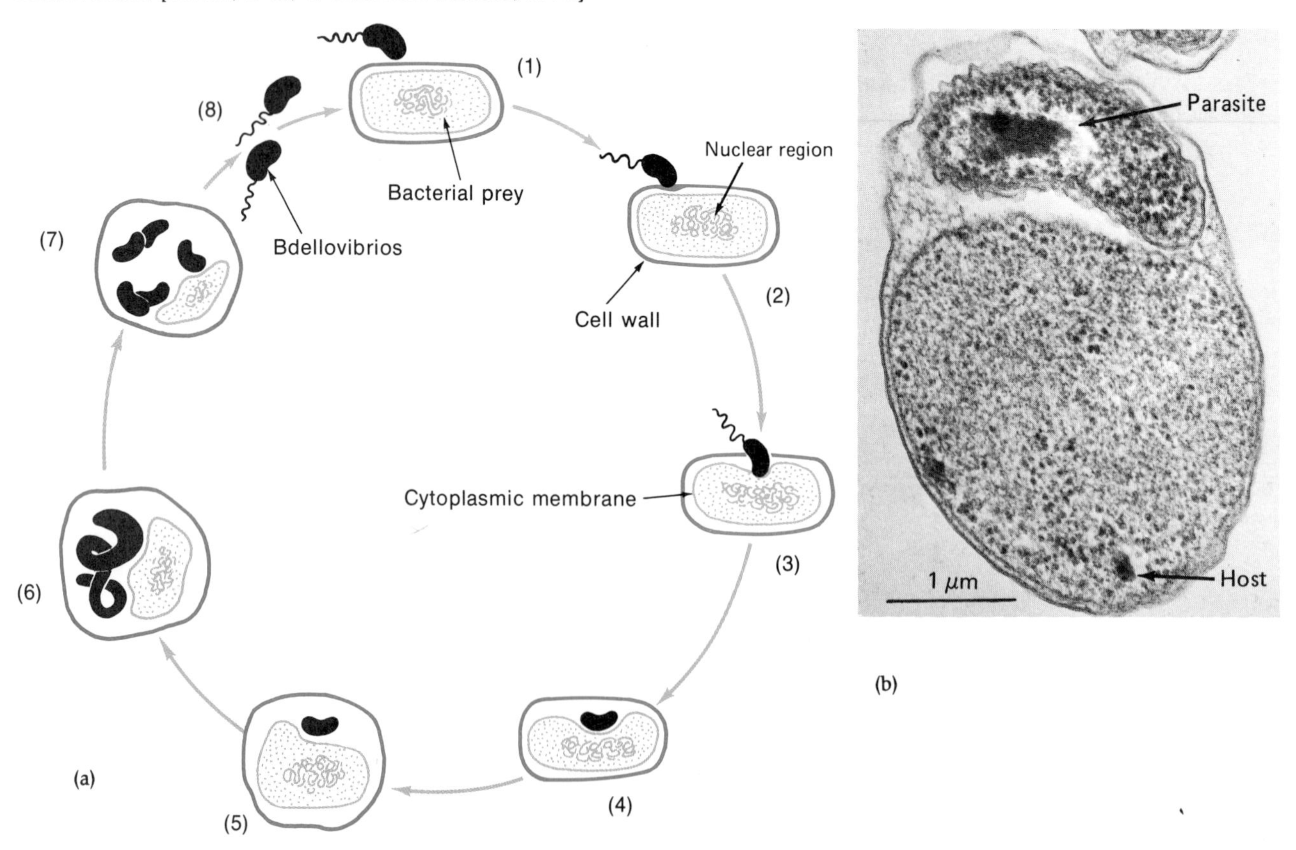

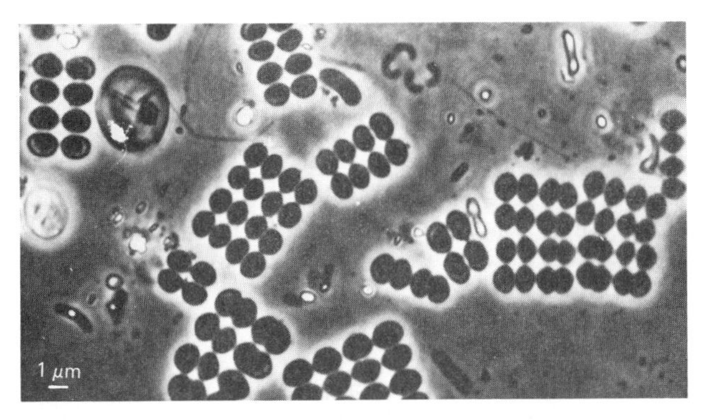

Figure 13–8 A green sulfur bacterial community. *[Reproduced with the permission of the National Research Council of Canada from D. E. Caldwell, and J. M. Tiedje,* Can. J. Microbiol. ***21****:377–385, 1975.]*

sis *(oxygenic photosynthesis).* Certain of these organisms also have the capacity to carry out the anoxygenic process. Table 13–5 lists other differences between the phototrophic groups.

PURPLE BACTERIA

This small group of gram-negative bacteria consist of only about 30 species. The purple bacteria are unicellular and reproduce by binary fission or, in a few species, by budding. All members of the group are capable of anaerobic growth in the presence of light, with CO_2 as the carbon source and reduced inorganic compounds as electron donors.

Two subgroups are recognized among these purple procaryotes: the *purple sulfur* (Figure 13–9) and the *purple nonsulfur bacteria.* Table 13–6 (on page 386) presents several distinguishing properties of these procaryotes.

GREEN BACTERIA

The green bacteria consist of the sulfur and nonsulfur types. Together they form a smaller taxonomic group than the purple bacteria. Both forms of green bacteria resemble their respective purple bacteria counterparts to a large extent. However, they can be distinguished from one another on the bases of structure, type of nutrition, metabolism, and environmental interactions. The green sulfur bacteria contain bacteriochlorophyll *c* or *d* and *a* as their photosynthetic pigment.

Figure 13–9 The appearance of the purple sulfur bacterium. (a) *Chromatium gracile* when grown in the presence of several substrates. Note the intracellular granules (arrows). *[Reproduced with the permission of the National Research Council of Canada from Caumette, P.,* Can. J. Microbiol. ***30****:273–284, 1984.]*

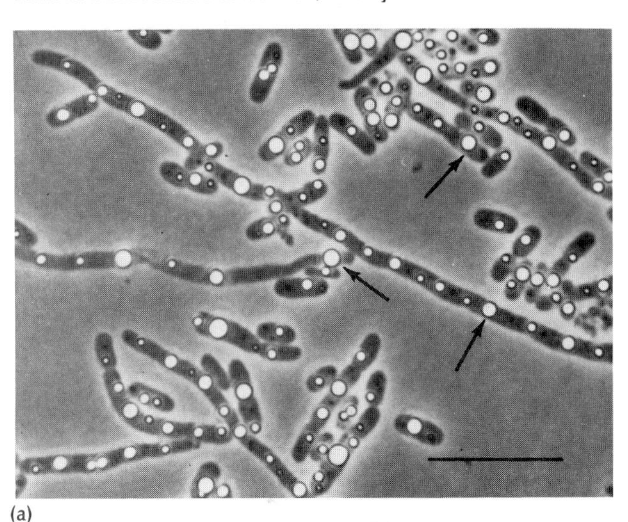

(a)

TABLE 13–5 Basic Differences Between Purple Bacteria, Green Bacteria, and the Cyanobacteria

	Phototrophic Procaryotes		
Property	*Purple Bacteria*	*Green Bacteria*	*Cyanobacteria*
Motility	Motile	Nonmotile and gliding forms	Gliding
Nitrogen fixation	Occurs in some species	Absent	Occurs in all major groups
Oxygenic photosynthesis[a,b]	Absent	Absent	Present
Major photosynthetic pigment	Bacteriochlorophylls *a, b, c, d,* or *e,* carotenoids	Bacteriochlorophylls *a, b, c, d,* or *e,* carotenoids	Chlorophyll *a* and phycobiliproteins
Electron donors	H_2, H_2S, S	H_2, H_2S, S	H_2O
Carbon source	Organic C or CO_2	CO_2	CO_2
Reserve non-nitrogenous organic material	Glycogen and poly-β-hydroxybutyrate	Absent	Glycogen

[a] Oxygen produced as a product of photosynthesis.
[b] Under certain conditions, photosynthesis is anoxygenic and H_2S functions as the electron donor.

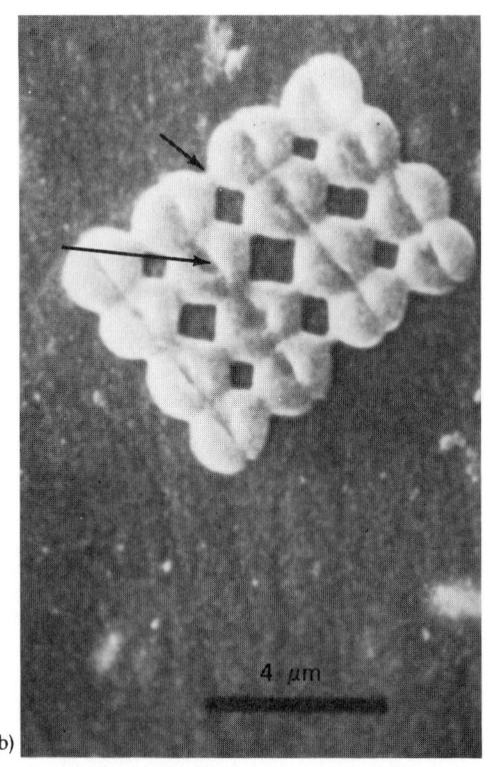

Figure 13–9 (b) *Thiopedia rosea,* a purple sulfur bacterium commonly found in the anaerobic regions below the surface of sulfide-containing lakes, ponds, ditches, and waste-treatment facilities. *[Reproduced with the permission of the National Research Council of Canada from R. Scherrer, and V. E. Shull,* Can. J. Microbiol., ***32****:607–610, 1986.]*

CYANOBACTERIA

The cyanobacteria (blue-green bacteria) represent the largest, most diverse, and most widely distributed group of photosynthetic bacteria. They bear a resemblance to gliding bacteria (described later) in general morphology, to some photosynthetic bacteria in their ability to use both carbon dioxide and gaseous nitrogen with light energy, and to eucaryotic algae and higher plants in having chlorophyll *a* and in being able to split water enzymatically *(photolysis)*. The cyanobacteria are found in a wide variety of environments. Some grow freely in snow on high mountain tops. Others thrive in thermal springs, such as those found in Yellowstone National Park, where the water temperature may be as high as 85°C (185°F). Still others are found in marine and fresh waters, in soil, and even in wet flower pots. Certain blue-green bacteria can grow on volcanic rock where most plant life fails to develop. The explanation lies in the ability of these microorganisms to utilize gaseous nitrogen, carbon dioxide, and water vapor from the air for their nutritional needs. This use of elemental nitrogen to form nitrogen-containing compounds is called **nitrogen fixation.** It is a process that introduces nitrates into soil and maintains soil fertility. Occasionally, when water temperature, nutrients (usually pollutants), and other factors reach a favorable level, certain blue-green bacteria multiply very rapidly, resulting in a microbial bloom (Color Photograph 7a). When this occurs, the waste products of such organisms accumulate and may seriously affect the other forms of water life and even humans.

The cellular properties of blue-green bacteria are clearly unlike those of any eucaryotic algal group. Cell wall composition, ribosome structure, features of protein synthesis, and certain nucleic acid properties in cyanobacteria all indicate that cyanobacteria are procaryotes.

Many subgroups of cyanobacteria are unicellular, but new cells produced by division may remain connected to form aggregates (Figure 13–10). Cyanobacteria may be spherical, rod-shaped, or spiral (Figure 13–10c). Their cells are 1.5 to 2 μm in diameter, gram-negative, with a multilayered peptidoglycan cell wall, plasmids, and pili. Small reproductive cells, a type of endospore, called *baeocytes* (Figure 13–10a), are formed from the multiple splitting of vegetative cells.

TABLE 13–6 A Comparison of Purple Sulfur Bacteria and Purple Nonsulfur Bacteria

Property	*Purple Sulfur Bacteria*	*Purple Nonsulfur Bacteria*
Typical habitat	Sulfide-rich waters	Organically rich freshwater lakes or ponds with low sulfide concentrations
Type of photosynthesis[a]	Photoautotrophic	Photoheterotrophic
Capable of aerobic growth	No[b]	Yes[b]
Accumulation of elemental sulfur as an intermediate in H_2S oxidation	Yes	No[b]
Sensitivity to growth inhibition by H_2S	Generally low	Generally high

[a] Refer to Chapter 5 for a discussion of types of nutrition and photosynthesis.
[b] Some exceptions exist.

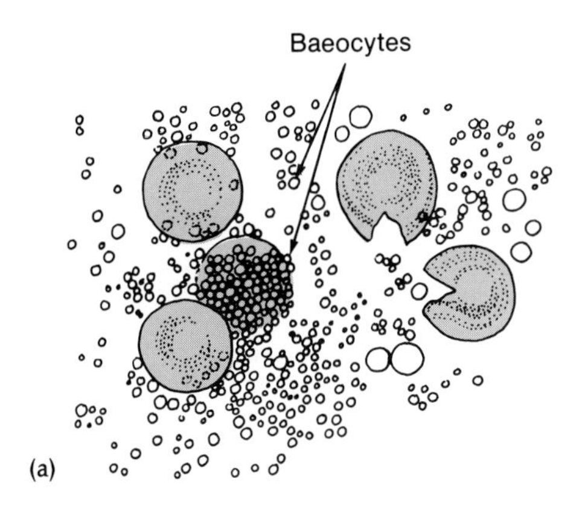

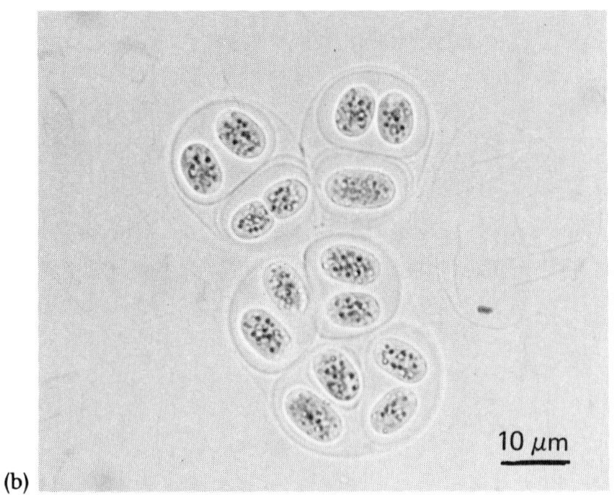

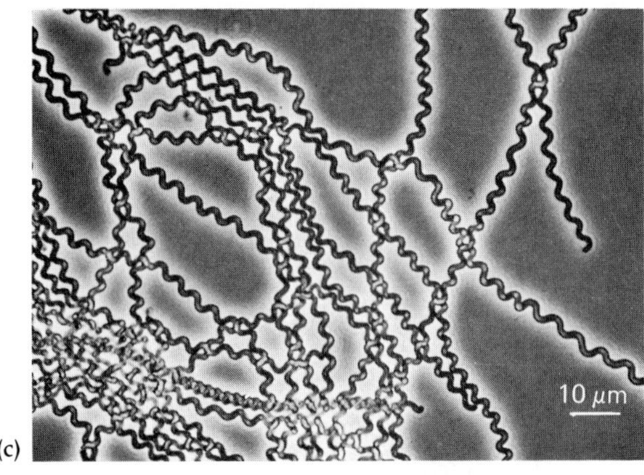

Figure 13–10 Cyanobacteria. (a) A drawing of *Dermocarpa.* Spherical cells of varying size are filled with reproductive spores known as *baeocytes.* The formation of spores distinguishes this organism from other cyanobacteria. (b) Young aggregates of the coccal cyanobacterium *Myxosarcina.* Note the packets of cells. (c) The spiral, loosely coiled cells of *Spirulina. [Courtesy of R. Rippka.]*

In other subgroups, the unit of structure is the filament, in which cells are bound together in a common sheath *(trichome).* Reproduction occurs by breakage of these trichomes into shorter chains of cells called *hormogonia* (Figure 13–11).

Heterocysts and akinetes are produced by certain filamentous cyanobacteria (Figures 7–42a, b, and c).

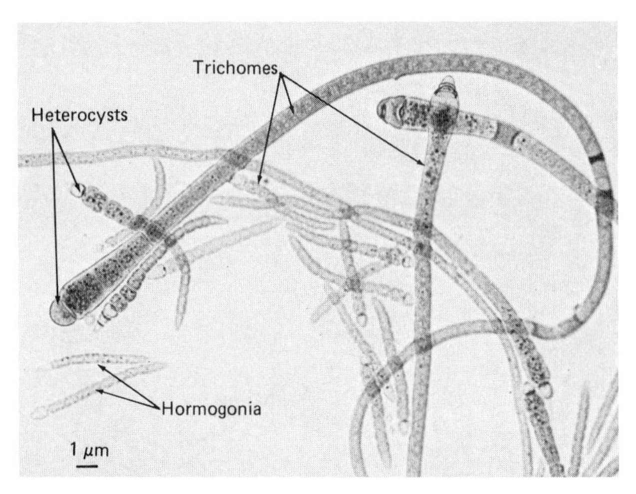

Figure 13–11 Trichomes and hormogonia. This photo of the cyanobacterium *Calothrix* shows the long trichomes, containing many cells in a common sheath, and the shorter cell chains, the hormogonia. Heterocysts can be seen at the end of the trichomes. *[Courtesy of R. Rippka.]*

These cells can also be distinguished structurally from vegetative cells (Figure 7–42b). By its involvement in nitrogen fixation, the heterocyst enables blue-green bacteria to survive and develop in nitrogen-deficient environments.

Cyanobacteria may actually be green (Color Photograph 64), purple, red, yellow, or even colorless. These variations are brought about by different kinds and amounts of pigments (Table 13–5). The photosynthetic pigments of cyanobacteria are contained in structures known as *phycobilisomes,* which are attached to the outer surfaces of thylakoids (Figure 7–42b).

The cyanobacteria constitute one of the largest and most diverse subgroups of gram-negative procaryotes known today.

Budding and/or Appendaged Bacteria

Several procaryotes with complex structures and morphologically distinctive cell cycles are found in freshwater and saltwater environments. Among the most interesting of these bacteria are those with extensions derived in part from the cell envelope. These projections (Figure 13–12), which appear during cell cycles, are called *prosthecae.* In organisms such as *Caulobacter* (ko-loh-BACK-ter) species, the prosthecae help to attach the bacteria to cells or to nonliving substances and surfaces. Sticky material on the prosthecate structure, or *stalk,* aids in the cellular attachment. When these projections are not sticky, they are called *pseudostalks.*

The cell cycles of several prosthecate bacteria result in the formation of different cell types.

Microbiology Highlights

SOLVING THE WORLD'S FOOD PROBLEM

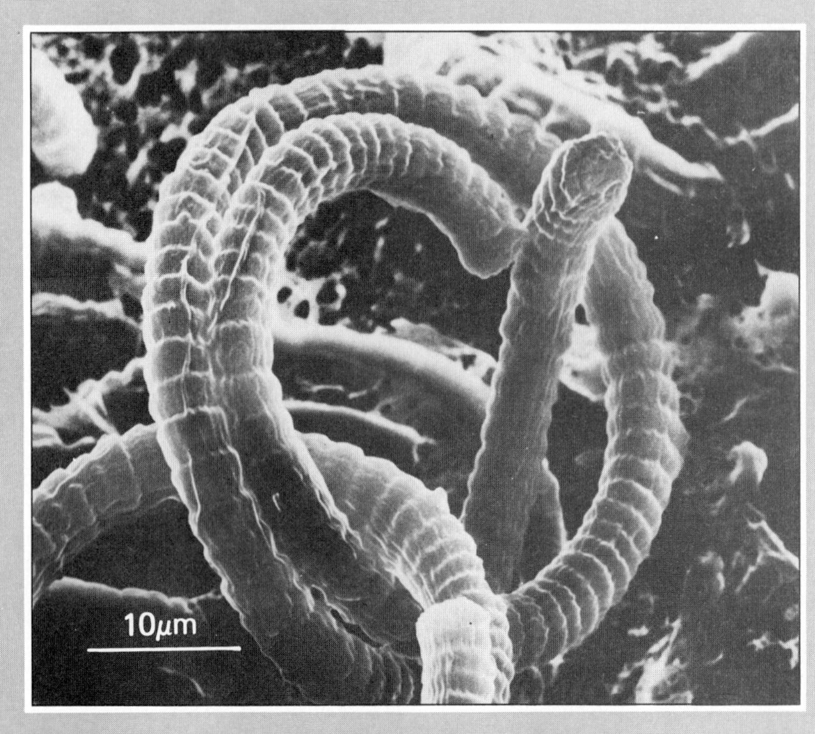

A scanning micrograph of filaments and the helical shape of *Spirulina platensis*. The bar marker equals 10 μm. *[O. Ciferri,* Microbiol. Rev. *47:551, 1983.]*

Two thirds of the world's people suffer from inadequate nutrition. Furthermore, the production of traditional agricultural foodstuffs by existing methods cannot possibly meet the protein demands of a human population that continues to grow in an uncontrolled manner. Attempts to solve problems of protein nutrition are usually directed at increasing productivity of agriculture, animal breeding, and fishery. However, attention is shifting to the wider use of nonconventional proteins such as protein concentrates of low-quality fish; protein products of oil-producing crops such as cotton seeds, soybean, and sunflowers; and protein provided by microorganisms. As difficult as it may be to believe, milligram for milligram, bacteria and many other microorganisms contain as much protein as steak. The protein content of microbes is high, as much as 40% to 60% or more of their weight.

In view of the insufficient world food supply and the high protein content of microbial cells, the use of the total cell weight (biomass) would appear to be an ideal supplement to the conventional food supply. One source of such nutrient material is "single-cell protein," or SCP. This term, coined in 1966 at the Massachusetts Institute of Technology, refers to the microbial biomass used as food and cattle feed additives. Either the isolated cell protein or the total cell material may be called SCP.

While animals and plants have always provided the main food sources, microorganisms have also been consumed in small quantities in such products or forms as cheese, vinegar, mushrooms, yeast, and even cyanobacteria. *Spirulina platensis* (spy-ROO-line-ah plah-TEN-sis), shown in the accompanying figure, is a cyanobacterium and a common member of aquatic populations that has been collected, sun-dried, and used as a food by the local populations in the Chad area and the Rift Valley in Africa. Because of their gas vacuoles, cyanobacterial filaments float on water surfaces, forming large mats that can easily be harvested. Historical research has shown that at the time of the Spanish conquest of Mexico, *Spirulina maxima* (S. max-EH-mah) was similarly gathered from Lake Texcoco and used for food. Thus it appears that *Spirulina* species have been regular components of human diets in two different parts of the world.

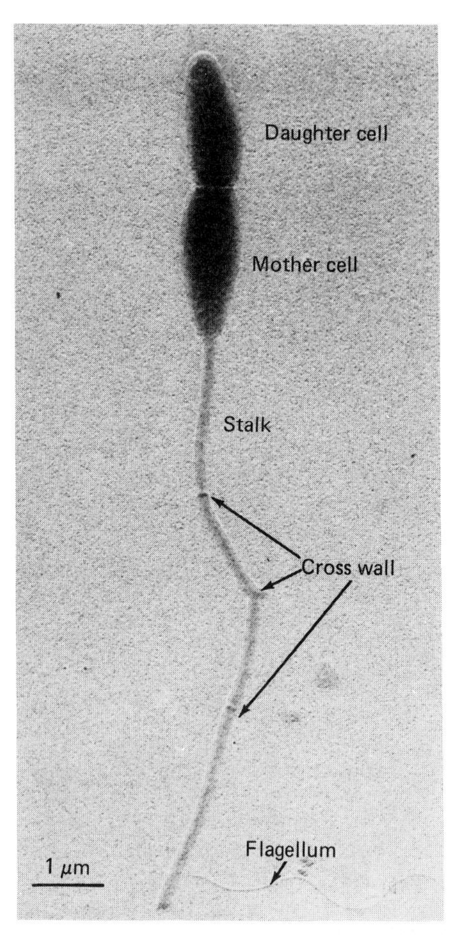

Figure 13–12 *Caulobacter* and related species. An electron micrograph showing the various components of a *Caulobacter* species. *[R. Whittenbury, and C. S. Dow,* Bacteriol. Rev. ***41**:754–808, 1977.]*

CAULOBACTER

The division cycle of *Caulobacter* differs significantly from those typical of most bacteria. Stalked cells divide to produce two structurally different cells (Figure 13–13). One cell keeps the stalk; the other becomes a flagellated swarmer cell without a stalk (Figure 13–13b). After separation, the swarmer cell does not divide until it, too, has developed a stalk.

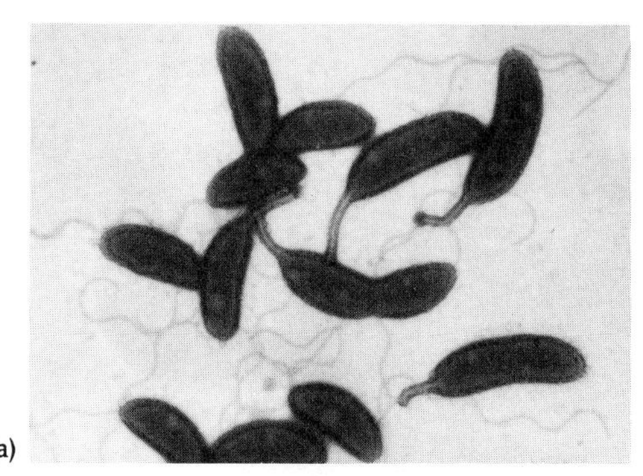

(a)

Figure 13–13 The stalked bacterium *Caulobacter.* (a) This bacterium is noted for its division into one swarmer and one stalked cell. Compare this electron micrograph with the life cycle shown in Figure 13–13b. *[Courtesy S. Koyasu, The Tokyo Metropolitan Institute of Medican Sciences.]*

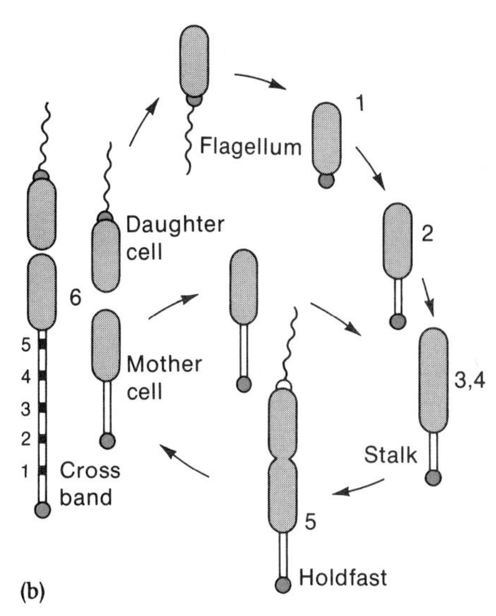

(b)

Figure 13–13 (b) The cell cycle of *Caulobacter,* starting with the motile swarm cell stage. Note the loss of the flagellum (1), stalk formation (2), swarm cell production phases (3, 4), flagellum and holdfast synthesis (5), and cell division to produce a stalked cell and a daughter swarm cell (6). At the far left a swarm cell and a mother cell, showing five cross bands (walls), can be seen. *[R. Whittenbury, and C. S. Dow,* Bacteriol. Rev. ***41**:754–808, 1977.]*

HYPHOMICROBIUM (A BUDDING PROCARYOTE)

Species of *Hyphomicrobium* (high-foh-my-KROW-bee-um) and of the related photosynthetic genus *Rhodomicrobium* (roh-doh-my-KROW-bee-um) have a division cycle different from that of *Caulobacter.* Reproduction of these organisms occurs by the formation of a bud either directly from the main, or mother, cell or at the tip of a filament (Figure 13–14a). Buds are flagellated and normally detach from the main cell. Upon separation, only one cell has a filament (Figure 13–14b).

Figure 13–14 A budding bacterium *Rhodomicrobium vannielii.* (a) An electron micrograph showing the components of a budding organism. *[R. Whittenbury, and C. S. Dow,* Bacteriol. Rev. ***41**:754–808, 1977.]*

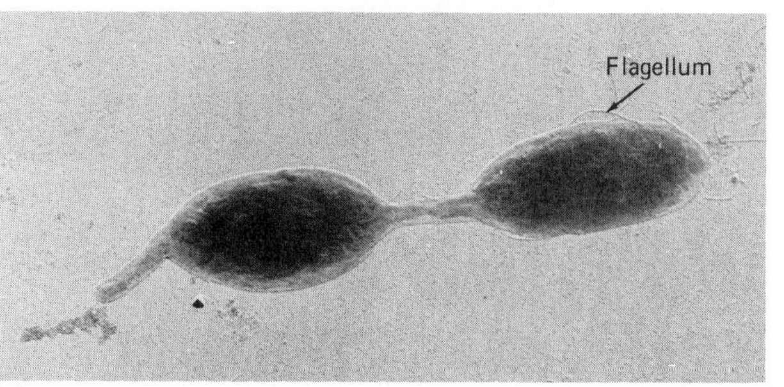

(a)

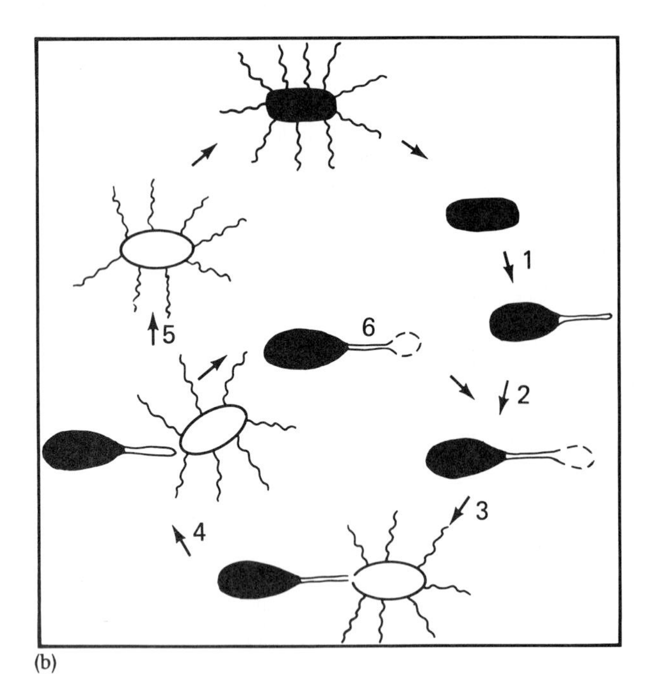

Figure 13–14 (b) A cell cycle starting with a swarm cell. Note that a filament forms before every daughter cell. Specific stages show the formation of a thin outgrowth (1), the enlargement of the outgrowth or the formation of a bud (2), the formation of flagella by the bud (3), the breaking away of the flagellated bud (4), the swimming away of the flagellated bud to start cycle again (5), and a new bud formation on the original cell that began the cycle (6). *[R. Whittenbury, and C. S. Dow,* Bacteriol. Rev. ***41****:754–808, 1977.]*

Sheathed Bacteria

A limited number of aquatic procaryotes form a specialized covering known as a **sheath** (Figure 13–15). The presence of a sheath enables organisms to attach themselves to solid surfaces and offers protection against predators. Several bacterial genera are noted for sheath-forming ability. These include *Crenothrix* (kren-OH-thricks), *Sphaerotilus* (sphere-ot-IL-lis), and the *Sphaerotilus-Leptothrix* (lep-TOH-thricks) group (Figure 13–15b). The bacteria of this last group often form sheaths surrounded by a slime layer. This outer layer is closely connected with their iron-accumulating and storing capacity (Figure 13–16a) and with their use of manganese for energy and growth. When environmental conditions become unfavorable, the sheathed bacteria form rounded cells known as *gonidia* (Figure 13–16b). These are released from the end of cellular filaments.

Sheathed bacteria are widely distributed in nature. They are found in sewage, soil, and water. Some of these organisms are noted for their contamination of water pipes and the formation of slimy masses of cell filaments.

Gliding Bacteria

Gliding motility is slow, occurs only when organisms are in contact with a solid surface, and usually involves secretion of a slime track. A number of filament-forming bacteria (organisms consisting of individual cells in a common outer cell wall) exhibit this type of movement. These include members of the genus *Flexibacter* (flex-ee-BACK-ter) (Figure 13–17).

Figure 13–15 Sheathed bacteria. (a) The relationship of the sheath (S), cell wall (CW), and cell membrane (CM) are clearly shown from different views of these organisms. *[M. H. Deinema, S. Henstra, and E. W. von Elgg,* Antonie van Leeuwenhoek ***43****:19–29, 1977.]* (b) Many trichomes of *Leptothrix lopholea* radiating from common holdfasts. *[W. L. van Veen, et al.,* Microbiol. Revs. ***42****:329–356, 1978.]*

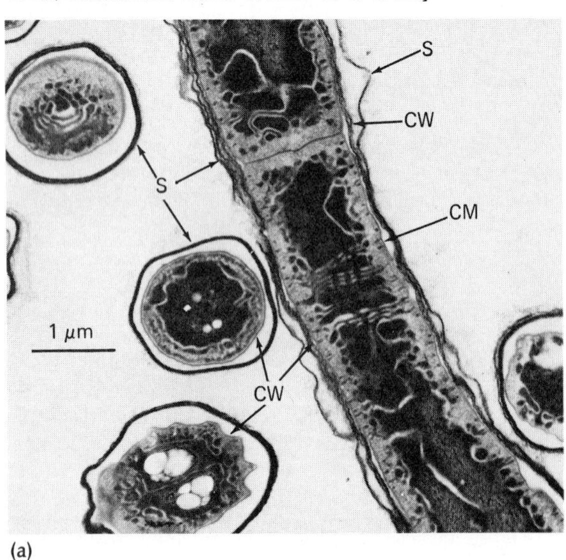

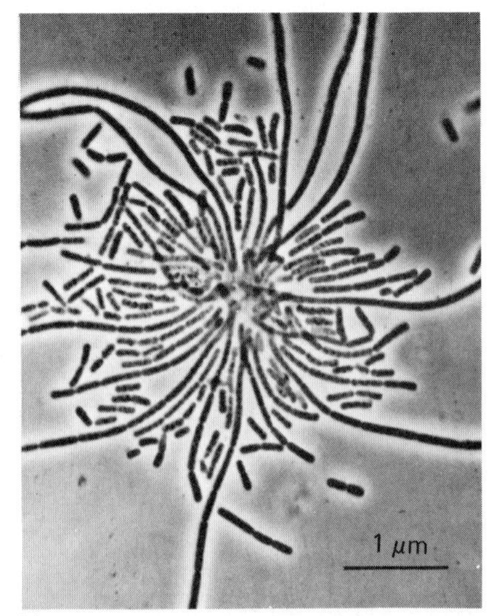

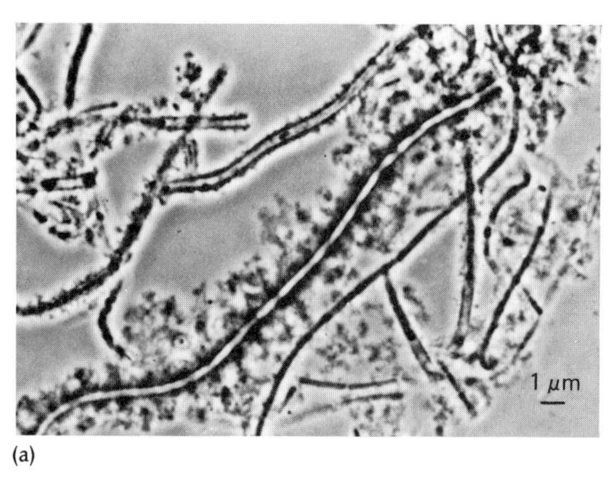

(a)

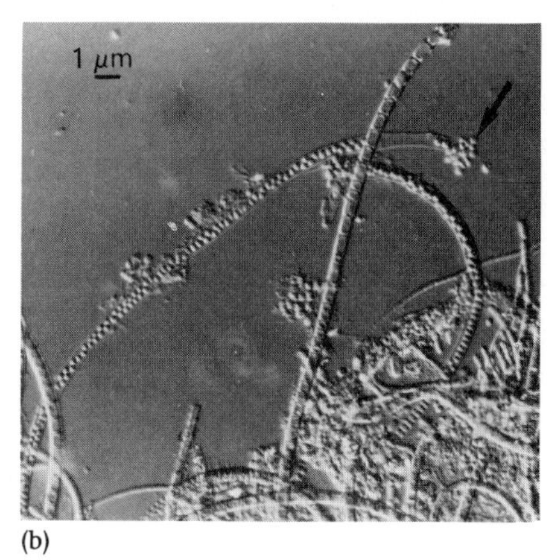

(b)

Figure 13–16 (a) *Leptothrix cholodnii* sheaths covered with ferric chloride. *[W. L. van Veen, et al.,* Microbiol. Revs. ***42****:329–356, 1978.]* (b) The masses of cell filaments of this bacterium *Crenothrix polyspora* can block water pipes and wells. The sheaths surrounding the cells (gonida) of the filaments are visible here (near arrow). *[H. Volker, et al.,* J. Bacteriol. ***131****:306–313, 1977.]*

Although the mechanism of gliding motility is not known, a number of hypotheses have been proposed. Recent studies have shown the presence of long fibers with goblet-shaped units *(goblets)* (Figure 13–17b) on the surfaces of gliding *Flexibacter.* These may be important to the movement of these organisms or to the sticking of cells to solid surfaces.

One group of gliding bacteria, the myxobacteria, have the interesting property of forming multicellular, complex structures called *fruiting bodies* (Figure 13–18, on page 392). This process involves cell-to-cell signaling and the binding of competent cells to one another.

Fruiting body formation can be viewed as occurring in three sequential stages and does not begin so long as adequate nutrients for vegetative growth are available. The first stage consists of localized accumulations of cells into discrete multicellular aggregates resulting from movement into numerous collection centers. Shortly after each aggregate forms, the cells move upward and construct a *stalk.* In the third stage, a partitioning of the cell population occurs at the tip of the stalk and several saclike structures, *sporangia,* form. The majority of cells in a sporangium undergo differentiation and become resting cells called *myxospores.* A myxospore enclosed in a capsule is a *myxocyst.*

Certain myxobacteria secrete *pheromones* during fruiting body formation. Such chemical secretions are found with other bacteria and higher forms of life, and are known for their ability to bring about specific behavioral or developmental responses in other organisms of the same species.

Subunits
Outer membrane
Cytoplasmic membrane
1 μm

(a)

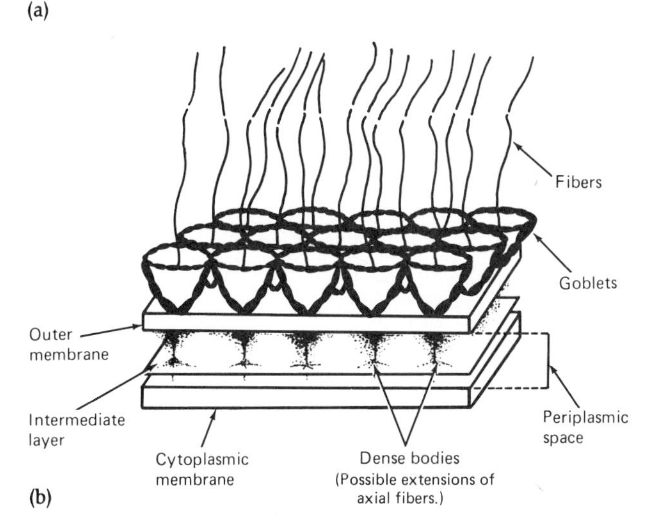

(b)

Figure 13–17 Gliding bacteria. (a) An outer array of goblet-shaped subunits on *Flexibacter,* which serve as a continuous source of sticky fibers. These units and their associated secretions play a role in bacterial attachment and possibly in the gliding movement. The cytoplasmic membrane of the organism is also shown. (b) A diagrammatic representation of the goblets and long fibers. *[Reproduced with the permission of the National Research Council of Canada from H. F. Ridgeway, et al.,* Can. J. Microbiol. ***21****:1733–1750, 1975.]*

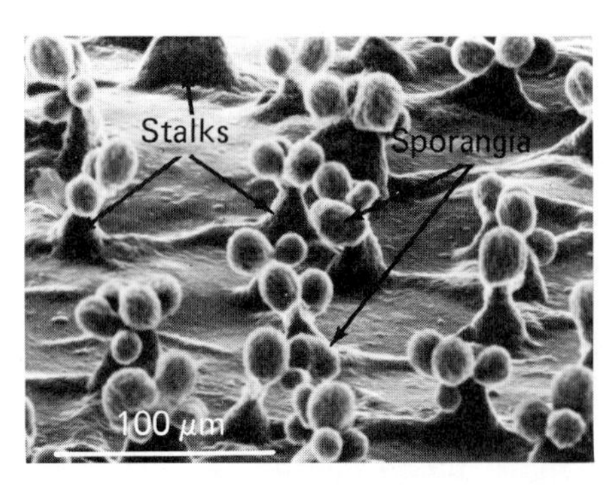

Figure 13–18 The multicellular fruiting bodies of the gliding myxobacterium *Stigmatella aurantiaca.* Specific parts of a typical fruiting body include the characteristic myxospore-containing sac (sporangium) and the supporting stalk. *[D. White, J. A. Johnson, and K. Stephens,* J. Bacteriol. ***144****:400–405, 1980.]*

Fruiting body formation distinguishes the myxobacteria from all other procaryotes and provides experimental material for studying the molecular bases for cell interactions and directed cell movement.

The Rickettsias and Chlamydias

RICKETTSIAS

Rickettsias have been responsible for the death of millions of people during times of war and natural disasters. It has been claimed that Napoleon's defeat in Russia in 1812 was brought about by typhus fever (a well-known rickettsial disease). This infection struck again during World War I, killing more than 3 million soldiers and civilians. Even during World War II and the Korean conflict, epidemics threatened the outcome of several military operations.

These microscopic forms were named after Howard Taylor Ricketts, who first isolated the agent causing Rocky Mountain spotted fever. He died in Mexico City in 1910 while studying typhus fever. Three genera are recognized among the rickettsias: *Rickettsia* (re-KET-see-ah), *Rochalimaea* (rosh-ah-LIM-ee-ah), and *Coxiella* (KOKS-ee-el-la). At one time, the rickettsias were classified somewhere between bacteria and viruses. However, it is now quite clear that they are a special form of bacteria with definite metabolic requirements. All rickettsias—with the exception of the causative agent of trench fever, *Rochalimaea quintana*—are obligate intracellular parasites; that is, they must have living, susceptible cells for growth and multiplication (Figure 13–19). This property distinguishes them from most other bacteria.

Coxiella, unlike the other two rickettsial genera, form endospores that are resistant to drying. The resting cells lack dipicolinic acid, the chemical associated with the heat-resistance of gram-positive bacterial spores described in Chapter 7. *Coxiella* spores may play an important role in the transmission of the human respiratory disease known as Q fever. **[See Chapter 25 for a description of Q fever.]**

Rickettsias can infect a wide range of natural hosts including arthropods (fleas, mites, and ticks), birds, and mammals. Members of the genera *Cowdria* (CO-dree-ah), *Ehrlichia,* and *Neorickettsia* (nee-oh-re-KET-see-ah) are pathogenic for some mammals. For example, *E. canis* causes a serious disease in dogs. *Cowdria* species cause heartwater, a blood disease of cattle, goats, and sheep. *Neorickettsia* are involved in a complicated worm-borne disease of canines. Characteristically, most of these microorganisms are transmitted from vertebrate host to vertebrate host by an arthropod, referred to as a **vector.** Some rickettsias may be passed on from one generation of arthropods to the next by introduction into eggs. The public health dangers posed by rickettsias have stimulated research on the production of antibiotics, insecticides, and vaccines.

Although many rickettsias are pathogenic, there are also several species that are apparently nonpathogenic. Rickettsias belonging to the genera *Rickettsiella* (rick-ett-see-EL-lah) and *Symbiotes* (sim-buy-O-teez) do not harm their insect hosts and appear to

Figure 13–19 A tissue culture cell infected with *Rickettsia prowazeki* (R), the cause of the rickettsial disease typhus fever. A nucleus, cell membrane, and mitochondrion are shown. *[Courtesy of Drs. David J. Silverman and Charles L. Wisseman, Jr., Department of Microbiology, University of Maryland School of Medicine.]*

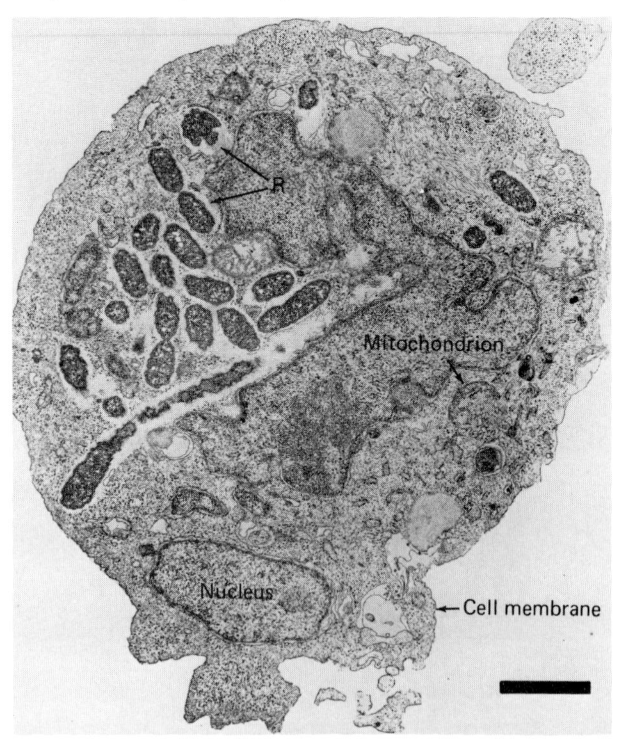

be essential for development and reproduction. This situation may be due to the establishment of a symbiotic, mutually beneficial relationship known as **mutualism.** Relatively little is known about these nonpathogenic rickettsias.

CHLAMYDIAS

The genus *Chlamydia* (clam-ID-ee-ah) consists of procaryotic microorganisms that invade and use eucaryotic cells for their own survival and reproduction (Figure 13–20). Because of this parasitic property, these microbes were once considered viruses. However, they are different from viruses in that chlamydias use their own ribosomes and enzymes for the formation of protein and nucleic acids; they depend on their host cells only for certain growth factors. Viruses, on the other hand, are totally dependent on host cells for their development. One of the distinctive properties of the chlamydias is their apparent complete lack of ATP-generating biochemical pathways.

Because they can survive only by parasitism and because of their ability to infect humans and other animals, chlamydias cause a number of different diseases. These include the respiratory infection psittacosis, or parrot fever; trachoma, a leading cause of preventable blindness that occurs worldwide; and lymphogranuloma venereum and several other sexually transmitted diseases. *C. psittaci* (C. SIT-a-sigh) strains appear to be well adapted to birds and lower mammals, and are the causative agents of respiratory infections. Different strains of *C. trachomatis* (C. trah-KOM-ah-tis) cause the other diseases mentioned. [See chlamydial infections in Chapters 25 and 28.]

Chemical and morphological analyses show that chlamydias are gram-negative coccoid forms; they are inhibited by a variety of antibiotics, including the tetracyclines. In addition, the chlamydias have little or no muramic acid in their cell walls and exhibit a complicated and unique growth cycle. This cycle starts with the attachment to susceptible cells of extracellular chlamydial particles known as *elementary bodies.* After entering the host cell, usually through phagocytosis, the elementary body develops into a second form, known as the *reticulate particle* or *initial body.* These larger growing particles are metabolically active, produce proteins, nucleic acids, and other macromolecules, and multiply by splitting *(binary fission).* Some initial bodies become infectious elementary bodies and are released from the host cell to infect other cells. These agents reinitiate the infection. In the laboratory, tissue cultures are used for their cultivation.

Division III—Tenericutes

Mycoplasmas and Related Organisms

The term **mycoplasma** refers to a group of microorganisms previously known as pleuropneumonia-like organisms. The many different isolates of this group have been separated from other bacteria in the class Mollicutes and catalogued into several genera (Table 13–7). Mycoplasmas are medically and economically important microorganisms, including the causative agents of a variety of animal and plant

Figure 13–20 *Chlamydia trachomatis.* (a) An electron micrograph of *C. trachomatis* elementary body (EB). The elementary body is the infectious extracellular form of the organism. Original magnification 100,000×. (b) A typical inclusion in an infected cell showing EBs. Original magnification 15,000×. *[From E. Manor, and I. Sarov,* Infect. Immun. ***54****:90–95, 1986.]*

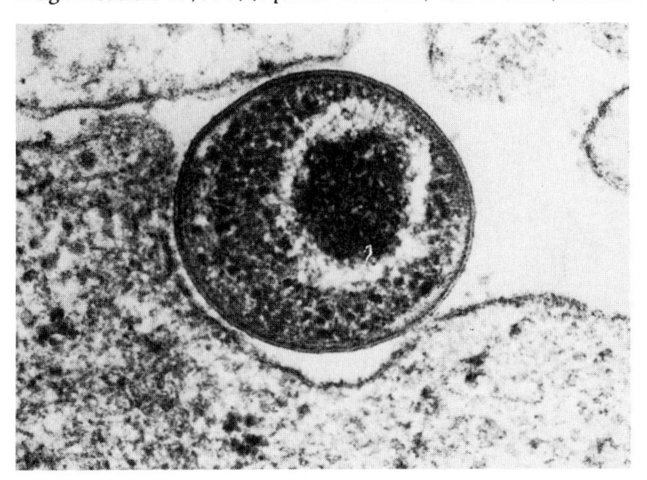

(a)

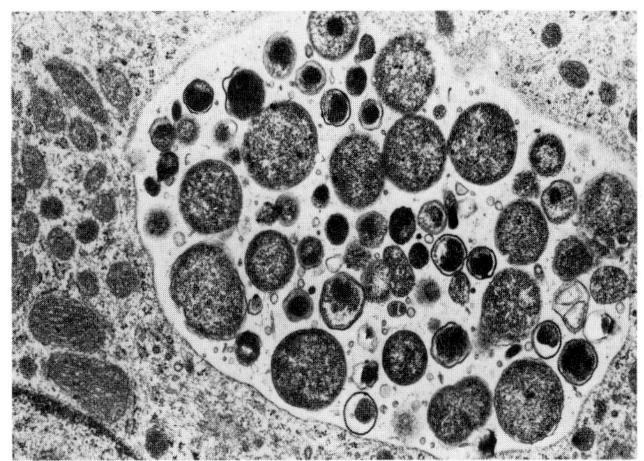

(b)

TABLE 13–7 **Genera of the Class Mollicutes**

Genus[a]	*Habitat*	*Sterol Required for Growth*
Mycoplasma	Animals	Yes
Acholesplasma	Saprophyte[b]	No
Ureaplasma (T strains)	Animals	Yes
Spiroplasma	Plants, insects	Yes
Thermoplasma	Burning coal refuse piles	No
Anaeroplasma	Animals	Yes

[a] For pronunciations, see the Organism Pronunciation Guide at the end of the text.
[b] Obtains its nutrients from rotting and decaying organic matter.

diseases. *Mycoplasma pneumoniae* (my-koh-PLAZ-mah new-MOH-nee-eye), for example, is the cause of one form of human pneumonia. *Spiroplasma* (spy-roh-PLAZ-mah) species are known to infect several insects and cause a large group of plant diseases including those known as the "yellows" group (Figure 13–21). For many years, viruses were assumed to be responsible for the yellows diseases. These plant diseases can be found in most parts of the world. However, they tend to be most serious in long-term crops in tropical and subtropical regions.

The mycoplasmas, among the smallest and simplest self-replicating procaryotes, are distinguished from all other bacteria by the absence of a cell wall (Figure 13–21b). The mycoplasma cell is bounded by a single lipoprotein cell membrane and contains only the minimum set of structures essential for growth and replication. Different species in this group have characteristic shapes (Color Photograph 65). [See procaryote cell membranes in Chapter 7.]

These microorganisms use amino acids, carbohydrates, and other growth factors. Species of *Anaeroplasma* (ah-NARE-oh-plaz-mah), *Mycoplasma, Spiroplasma,* and *Ureaplasma* (you-REE-ah-plaz-mah) require blood serum, a source of sterols such as cholesterol, which they cannot manufacture themselves (Table 13–7). The sterols help stabilize cell membranes and protect the organisms against osmotic destruction. With agar media, the mycoplasma grow down into the preparation, often producing colonies that look like fried eggs (Figure 13–22).

Division IV—Mendosicutes

Archaeobacteria

Volcanic activity on earth has occurred for about 4 billion years. Although few volcanic areas are still active today, such extremely hot regions have re-

Figure 13–21 The spiroplasmas. (a) Coconut palms infected with lethal yellowing, one example of "yellows" diseases. The leaves on the palm in the foreground are yellow and beginning to fall. The palms in the background (arrows) have completely lost their leaves, leaving bare trunks. (b) A scanning micrograph showing the helically coiled spiroplasmas. The bar marker represents 1.0 μm. *[Courtesy M. J. Daniels,* Zbl. Bact. Hyg. *I. Abt. Orig. A.* ***245:*** *184–199, 1974.]*

(a)

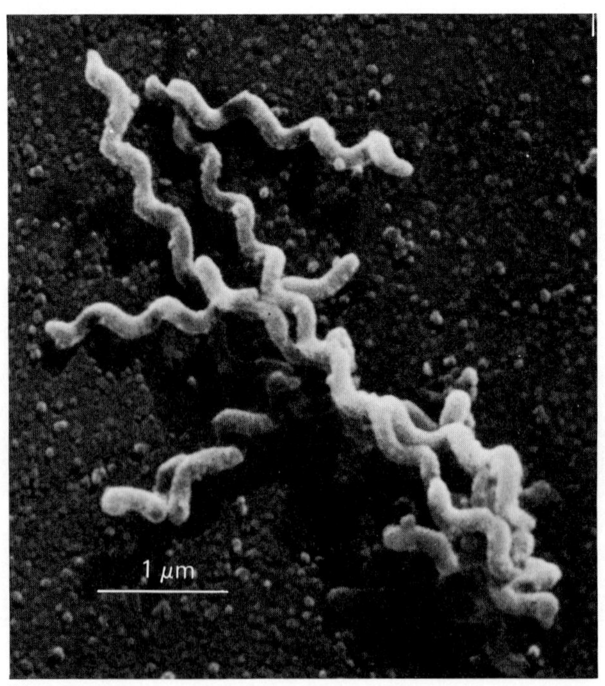

(b)

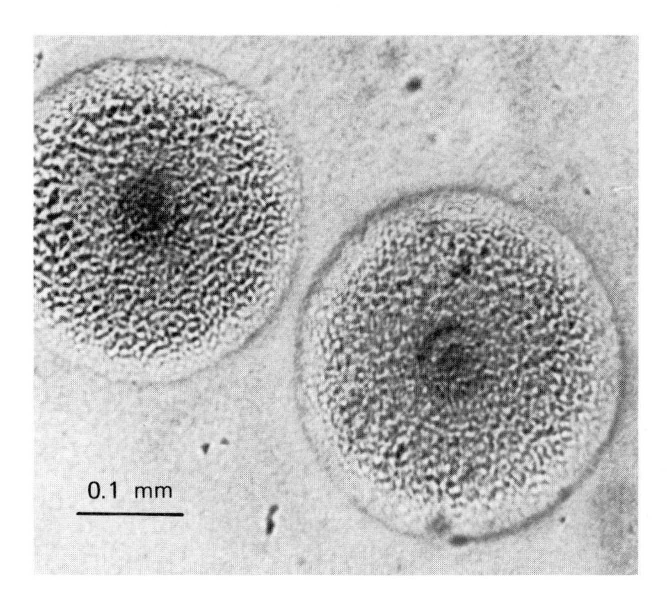

Figure 13–22 Two unstained mycoplasma colonies. Note the fried-egg appearance. *[From R. G. Cluss, and N. C. Somerson,* J. Clin. Microbiol. ***19****:543–545, 1984.]*

mained almost unchanged. Conditions in several of these volcanic areas have limited the evolution or development of the organisms that flourish there (Figure 13–23). This suggests that recently isolated procaryotes may possibly be quite similar to their ancient ancestors.

The archaeobacteria are a diversified group of procaryotes that are evolutionarily quite distinct from the eubacteria (true bacteria). Archaeobacteria are single-celled and have several features separating them from other procaryotes. Their major distinguishing properties include a different sequence of nucleotides in the rRNA, a universal absence of muramic acid (a typical peptidoglycan compound) as a component of cell walls, differences in the translation process of protein synthesis and the chemical composition of membrane lipids, and resistance to antibiotics such as penicillin.

Since the archaeobacteria are known for their ability to grow in extremely harsh environments, they can be placed in three major groups based on metabolic or ecological properties: **methanogens, halophiles,** and **thermoacidophiles.**

Methanogens live only in anaerobic environments such as stagnant pools, sewage sludge, and volcanic areas (Figure 13–24). These archaeobacte-

Figure 13–23 *Methanothermus fervidus,* an extremely thermophilic archaeobacterium capable of growing at temperatures up to 97°C. *[Courtesy K. S. Stetter, University of Regensburg.]*

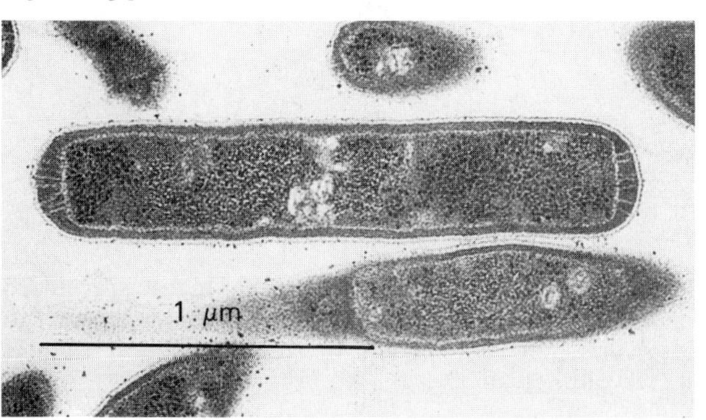

(a)

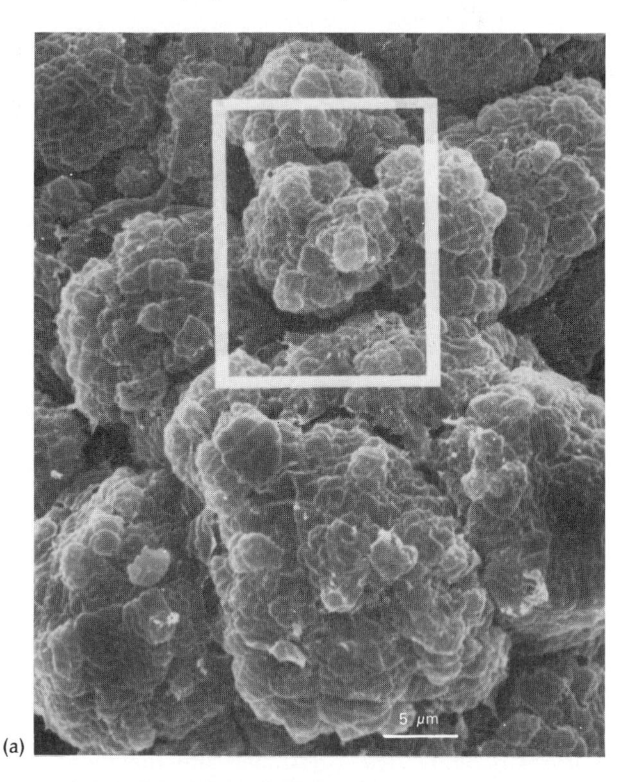

(b)

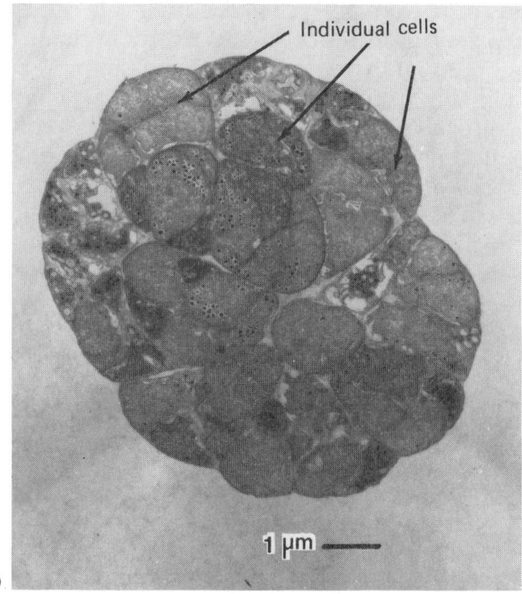

Figure 13–24 *Methanosarcina mazei.* (a) A scanning micrograph of the archaeobacterium growing in small, irregular aggregates (sarcinae formation). (b) A transmission micrograph showing the packaging of individual cells in these aggregates. *[Courtesy R. W. Robinson et al., Department of Microbiology and Cell Science, University of Florida.]*

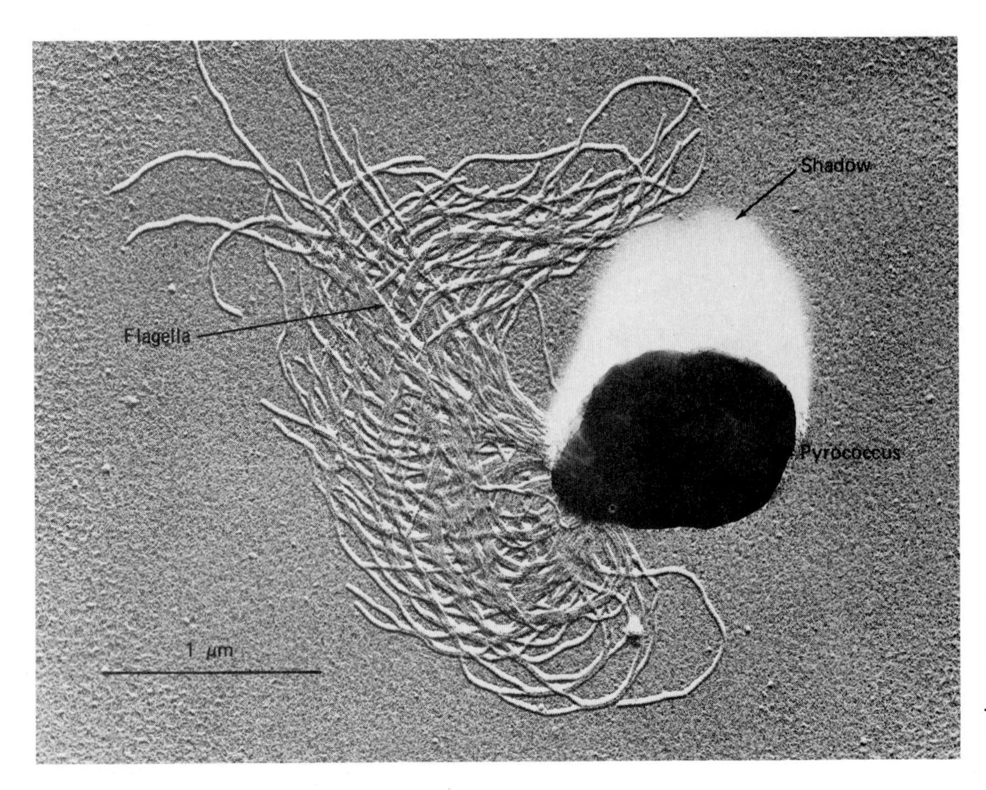

Figure 13–25
A shadow-cast preparation of the marine heterotrophic archaeobacterium *Pyrococcus furiosus* (the fireball). *[From G. Fiala, and K. O. Setter,* Arch. Microbiol., ***145****:56–61, 1986.]*

ria are distinguished by their unique energy metabolism and are responsible for the production of "marsh gas" or methane (Color Photograph 66).

The halophilic (salt-loving) bacteria are a distinctive group of procaryotes clearly able to adapt to the high salt concentrations and light intensities of their natural habitats. The ribosomes and metabolic enzymes of these bacteria require high salt concentrations in order to function effectively. A characteristic property of extreme halophiles is the presence of red pigment in their cell membranes. These bacteria are known to produce a typical brick-red color in seawater ponds, from which salt is recovered by evaporation, and on salted dry fish or hides treated with salt containing the halophiles. Apparently the red pigment protects cells against photochemical damage from the high light intensities.

The thermoacidophiles are a mixed group of archaeobacteria having the ability to grow at high temperature and low pH. These organisms are commonly found in hot acid springs and soils around the world (Figure 13–25). The group includes facultative as well as strict anaerobes.

Finding a Place in the Biosphere

Throughout its history, microbiology has focused on the identification and control of microorganisms and the discovery of their functions. Through the years, it has become apparent that many of the metabolic processes of microorganisms such as procaryotes are similar to those of cells in higher forms of life.

This chapter has been concerned mainly with current approaches to bacterial classification and a brief survey of the kingdom Procaryotae. The next chapter will describe the interactions of procaryotes, a selected number of factors that control their growth and development in the biosphere, and the role of microorganisms in the cycling of biologically important chemical elements.

Summary

Applications of Molecular Biology to Taxonomy

1. The purpose of classification is to group organisms with similar properties and to separate those that are different.
2. The development, in recent years, of highly specialized techniques in molecular biology has given rise to several new approaches for the description of procaryotes.

3. Specific techniques in molecular biology include % (G + C) comparisons, DNA hybridization, DNA–rRNA hybridization, and 16S rRNA sequencing.

Methods Used to Study the Molecular Architecture of the Procaryotic Cell
Several additional techniques are used in showing natural relationships among procaryotes. These include chemotaxonomy and serological tests.

Diagnostic Microbiology
The identification of bacterial species generally involves the use of established identification keys. Such keys resemble a flowchart consisting of a series of specific tests.

The Kingdom of the Procaryotae
1. *Bergey's Manual of Systematic Bacteriology,* published in 1984, is a widely used classification reference.
2. According to the classification scheme proposed in *Bergey's Manual,* the kingdom of Procaryotae is divided into four divisions: Gracilicutes, Firmicutes, Tenericutes, and Mendosicutes.
3. Each division is organized into sections that are composed of several taxonomic ranks—that is, classes, orders, genera, and species.

Division I—Gracilicutes
THE SPIROCHETES
1. *Spiral bacteria* are heterotrophs with unique cellular anatomy and motion.
2. Examples of these procaryotes include spirochetes and spirals.
3. Spirals occur in a wide range of natural habitats. A small number cause diseases of humans and lower animals.
4. Despite similarities in morphology, spirochetes differ in their physiology, activities, and distribution.
5. Most spirilla are free-living bacteria with rigid cell walls and flagella at one or both ends.

AEROBIC/MICROAEROPHILIC, MOTILE, HELICAL/VIBRIOID GRAM-NEGATIVE BACTERIA
Curved, spiral rods of the genus *Campylobacter* are *microaerophilic.* They have a single flagellum at one or both ends, and they move with a characteristic corkscrew motion. One species, *C. fetus,* causes disease in lower animals and humans.

ANOXYGENIC PHOTOSYNTHETIC BACTERIA AND THE CYANOBACTERIA
1. Phototrophic procaryotes use inorganic sources of carbon.
2. Three distinct groups of gram-negative organisms are recognized as phototrophs: *cyanobacteria (blue-green bacteria)* and the *green* and *purple bacteria.*
3. The green and purple bacteria do not produce oxygen as a product of their photosynthetic activity (anoxygenic photosynthesis) and contain pigments different from those of the blue-green bacteria.
4. Cyanobacteria are the largest, most diverse, and most widely distributed group of photosynthetic bacteria.
5. Certain cyanobacteria form heterocysts where *nitrogen fixation* occurs.
6. Cyanobacteria, in the presence of various nutrients, favorable temperature, and other factors, can multiply rapidly to form an algal bloom.
7. Morphologically, many subgroups of cyanobacteria are unicellular; others may appear as long filaments of cells bound together in a common sheath *(trichome).*
8. Reproduction in cyanobacteria, depending on the species, can occur by the breakage of trichomes or by the formation of reproductive cells, *baeocytes* or *endospores.*
9. Heterocysts and akinetes are produced by certain filamentous cyanobacteria.
10. The photosynthetic pigments of cyanobacteria are contained in structures known as phycobilisomes that are attached to outer surfaces of *thylakoids.*

BUDDING AND/OR APPENDAGED BACTERIA
1. Several procaryotes found in aquatic environments have complex fine structures and morphologically distinct cell cycles.
2. Cell wall extensions containing cytoplasm are called *prosthecae.*
3. Some organisms, such as species of *Caulobacter,* use prosthecae called *stalks* for attachment purposes. These structures contain sticky material at the tip, known as a *holdfast.*
4. The prosthecate bacteria have cell cycles that involve formation of new cell types and are significantly different from those of typical bacteria.
5. *Hyphomicrobium* and the related photosynthetic *Rhodomicrobium* are budding bacteria that have extensions known as *filaments.* Bud formation can occur at the tip of such structures.

SHEATHED BACTERIA
1. A limited number of aquatic bacteria form specialized coverings known as *sheaths.*
2. The sheath has both ecological and nutritional consequences for the organisms that form it.
3. The sheathed bacteria include *Crenothrix, Sphaerotilus,* and the *Sphaerotilus-Leptothrix* group.
4. When environmental conditions become unfavorable, rounded cells known as *gonidia* are formed.
5. Sheathed bacteria are found in sewage, soil, and water.

GLIDING BACTERIA
1. *Gliding* motility is found among both heterotrophs and autotrophs.
2. This form of movement is slow, occurs only when organisms are in contact with a solid surface, and involves secretion of a slime track.
3. A number of filamentous bacteria exhibit gliding motility.
4. The mechanism of gliding motility is not known.
5. One group of gliding bacteria, the myxobacteria, form distinctive, multicellular structures that are called *fruiting bodies.*

6. Fruiting body formation occurs in three stages and results in a structure with a stalk and saclike *sporangia.* Resting cells known as *myxospores* are found in sporangia.

THE RICKETTSIAS AND CHLAMYDIAS

1. Several bacterial species require living eucaryotic animal, plant, or microbial cells for development and reproduction.
2. Examples of such parasitic procaryotes include *Coxiella, Chlamydia, Rickettsia* and *Rochalimeae.*
3. Each genus of obligate intracellular bacteria has distinctive features of its cycle, including transmission, penetration of host cells, reproduction and release of progeny, and cultivation.

Division III—Tenericutes

MYCOPLASMAS AND RELATED ORGANISMS

1. The mycoplasma are procaryotes that normally do not have cell walls.
2. Blood serum, which contains sterols, is required for the growth and reproduction of these organisms.
3. Several mycoplasmas are the causative agents of a variety of animal and plant diseases.

Division IV—Mendosicutes

ARCHAEOBACTERIA

1. The archaeobacteria are evolutionarily quite distinct from the true bacteria (eubacteria).
2. Three major groups of archaeobacteria are recognized on the basis of metabolic and ecological properties: halophiles, methanogens, and thermoacidophiles.
3. The major distinguishing features of the archaeobacteria include an absence of peptidoglycan, a different form of translation process in protein synthesis, a unique chemical composition of membrane lipids, resistance to penicillin, and an ability to survive extremely harsh environments.

Questions for Review

1. Distinguish between classification and taxonomy.

2. Briefly describe the following molecular biological techniques used in microbial classification:
   a. % (G + C) comparison
   b. DNA-DNA hybridization
   c. DNA-rRNA hybridization

3. What techniques are used to investigate the molecular architecture (organization) of the bacterial cell for purposes of classification?

4. List and give the distinguishing features of the four divisions introduced in the first edition of *Bergey's Manual of Systematic Bacteriology.*

5. Compare the steps involved in the life cycles of the following procaryotes:
   a. *Chlamydia* and budding bacteria
   b. *Bdellovibrio* and stalked bacteria

6. List and explain at least two distinguishing properties of the following procaryotes:
   a. sheathed bacteria
   b. gliding bacteria
   c. mycoplasma
   d. archaeobacteria
   e. spirochetes
   f. green bacteria
   g. purple bacteria
   h. cyanobacteria

7. What are the beneficial functions or activities that are performed by the various procaryotes discussed in this chapter?

8. What is an algal or microbial bloom, and how does it form?

9. How do the mycoplasma differ from the archaeobacteria?

10. How do the archaeobacteria differ from the cyanobacteria?

Suggested Readings

BERKELEY, R. C. W., and M. GOODFELLOW (eds.), *The Aerobic Endospore-Forming Bacteria: Classification and Identification* (Special Publications of the Society for General Microbiology, Vol. 4). London: Academic Press, 1981. *An important contribution to an understanding of the aerobic, endospore-forming genera* Bacillus, Sporolactobacillus, Sporosarcina, *and* Thermoactinomyces.

CARR, N. G., and B. A. WHITTON (eds.), *The Biology of Cyanobacteria.* Oxford: Blackwell Scientific Publications, 1982. *This successor to the 1973 edition provides a thorough treatment of blue-green bacteria. Specific areas covered include structure, function, carbon metabolism, nitrogen fixation, and the origin and early evolution of the cyanobacteria.*

DE ROSA, M., A. GAMBACORTA, and A. GLIOZZI, "Structure, Biosynthesis, and Physicochemical Properties of Archaebacterial Lipids." *Microbiological Reviews* **50**:70–80, 1986. *A short article describing the specific features of the unusual and distinctive membrane lipids of archaeobacteria.*

HOVIND-HOUGEN, K., "Determination by Means of Electron Microscopy of Morphological Criteria of Value for Classification of Some Spirochetes in Particular Treponemes." *ACTA Pathologica et Microbiologica Scandinavica,* Section B, Supplement **255**:1–41, 1976. *A thorough study emphasizing the ultrastructural properties of members of the genera* Borrelia, Leptospira, *and* Treponema. *This paper also clearly shows the major importance of electron microscopy to microbial classification.*

KANDLER, O., *Archaeobacteria.* Deerfield Beach, Fla.: Verlag Chemie International, 1982. *A compact, up-to-date account of the molecular, biological, and biochemical aspects of archaeobacteria.*

KRIEG, N. R., "Identification of Bacteria," in Sneath, P. H. A. (ed.), *Bergey's Manual of Systematic Bacteriology,* Baltimore: Williams and Wilkins, 1986, pp. 988–990. *A concise and specific discussion of the bacterial identification process.*

KROGMANN, D. W., "Cyanobacteria (Blue-Green Algae)—Their Evolution and Relation to Other Photosynthetic Organisms," *BioScience* **31**:121–123, 1981. *A clearly written article describing the general properties of cyanobacteria, their distribution, and their taxonomic relationship to other forms of life.*

MURRAY, R. G. E., "The Higher Taxa, or, A Place for Everything . . . ?" in Sneath, P. H. A. (ed.), *Bergey's Manual of Systematic Bacteriology,* Baltimore: Williams and Wilkins, 1986, pp. 995–998. *An excellent discourse on the classification of bacteria by one of the primary thinkers in the field.*

POINDEXTER, J. S., "The Caulobacters: Ubiquitous Unusual Bacteria." *Microbiological Reviews* **45**:123–179, 1981. *A thorough and updated coverage of this widely distributed and distinctively shaped group of procaryotes. Many topics are covered, including taxonomy, cultivation and nutrition, ultrastructure, caulophages, and ecological implications.*

RIPPKA, R., J. DERUELLES, J. B. WATERBURY, M. HERDMAN, and R. Y. STAINER, "Generic Assignments, Strain Histories and Properties of Pure Cultures of Cyanobacteria." *Journal of General Microbiology* **111**:1–61, 1979. *An excellent, well-illustrated study of 178 strains of cyanobacteria. Revised definitions of several genera also are proposed.*

14

Microbial Ecology and Environmental Activities

A thing . . . never returns to nothing, but all things after disruption go back into the first bodies of matter. . . . None of the things, therefore, which seem to be lost is utterly lost, since nature replenishes one thing out of another and does not suffer any thing to be begotten. — Lucretius

After reading this chapter, you should be able to:

1. Define the term *ecosystem.*
2. Distinguish between the abiotic and biotic factors of an ecosystem.
3. Explain the role of producers, consumers, and decomposers in an ecosystem.
4. Describe the types of natural habitats in which microbes are found.
5. Define and discuss action, reaction, and coaction as ecological concepts.
6. List microbial genera known to grow at extremes of temperature, pH, osmotic and hydrostatic pressure; specify the associated abiotic conditions.
7. Describe the carbon, nitrogen, sulfur, and oxygen cycles in terms of their significance in ecology.
8. Interrelate the carbon, nitrogen, and sulfur cycles with the oxygen cycle to show that these cycles are not independent events.
9. Explain the phrase "microbial infallibility."
10. Discuss the role of bioconversion in nature in association with methanogenesis, petroleum degradation, pesticides, and mercury.
11. Discuss the role of bioconcentration in nature in relation to manganese deposits and the toxic accumulations of pesticides and radioisotopes.
12. Discuss what is meant by the following types of coaction, giving examples for each type: syntrophism, competition, and predation.
13. Describe and discuss examples of commensalism, parasitism, and mutualism as types of symbiosis.
14. Distinguish between the following combinations: autotroph and heterotroph; chemosynthesis and methanogenesis; nitrification and ammonification; population and community; habitat and niche; and legume and lichen.

The soils and waters of planet Earth abound with microorganisms. The influence of the environment on these organisms and on the relationships among microbes and other life forms is critical to the survival of life as we know it. The vast majority of microorganisms do not cause spoilage or disease; they do influence the environment and perform a variety of natural processes that affect the well-being of life on this planet. This chapter describes many of these microorganisms, their environments, and their interaction with different life forms.

The word "ecology" was coined by Ernst Haeckel in 1869. He wrote: "By ecology we mean the body of knowledge concerning the economy of nature—the investigation of the total relations of the animal both to its inorganic and its organic environment"* Haeckel's use of the word "economy" in the definition is very apt, for this term is derived from two Greek words, *oikos* and *nemain*, meaning "to manage a home." Thus, ecology is the study of the relationships among different forms of life and their interactions with their physical environment. It includes all the activities and relationships by which living and nonliving components of the environment manage to sustain life on this planet.

Basic Ecological Principles

Life and the physical environment go together, for the environment supplies both the nutrients and the conditions for the existence of life. This combination of living things and the environment in which life exists is the **biosphere.** The biosphere includes the rocks and other solid portions of the **lithosphere,** extends up into the gaseous **atmosphere** to more than 10,000 m, and penetrates into the aquatic regions of the earth, or **hydrosphere,** to a depth of about 8,000 m.

Most forms of life are adapted to a particular environment that is restricted in general living conditions and food resources. Moreover, the nutrition of any species must be in balance with the biosphere as a whole. Thus, the biosphere is a global system in which materials are cycled and energy is transformed and ultimately consumed.

Despite their small size, microorganisms have important roles in many natural processes that contribute to the survival of animals, plants, and microbes. Microorganisms as well as other forms of life function as catalysts in numerous chemical reactions.

The purpose of this chapter is to survey environmental microbiology, including descriptions of representative microorganisms in their natural environments and discussions of microbial nutrition in relation to the requirements of other forms of life.

* Quoted in R. Brewer, *Principles of Ecology* (Philadelphia: W. B. Saunders Company, 1979).

Organization of the Biosphere

At the present time, Earth, among all the planets of our solar system, is the most favorable for the production and survival of life as we know it. Its distance from the sun, neither too close nor too far, creates an environment that permits the various biochemical and biophysical reactions on which life depends. Environment includes all factors external to the individual organism that in some way affect it.

The biosphere is made up of ecological units called **ecosystems.** Groups of individuals having similar properties and living in the same area are called **populations** (Figure 14–1). When various

Figure 14–1 Some microbial members of the biosphere. An impressive number of the diatom *Cocroneis scutellum* forming an algal population. *[J. McN. Sieburth, and C. D. Thomas,* J. Phycol. ***9****:46–50, 1973.]*

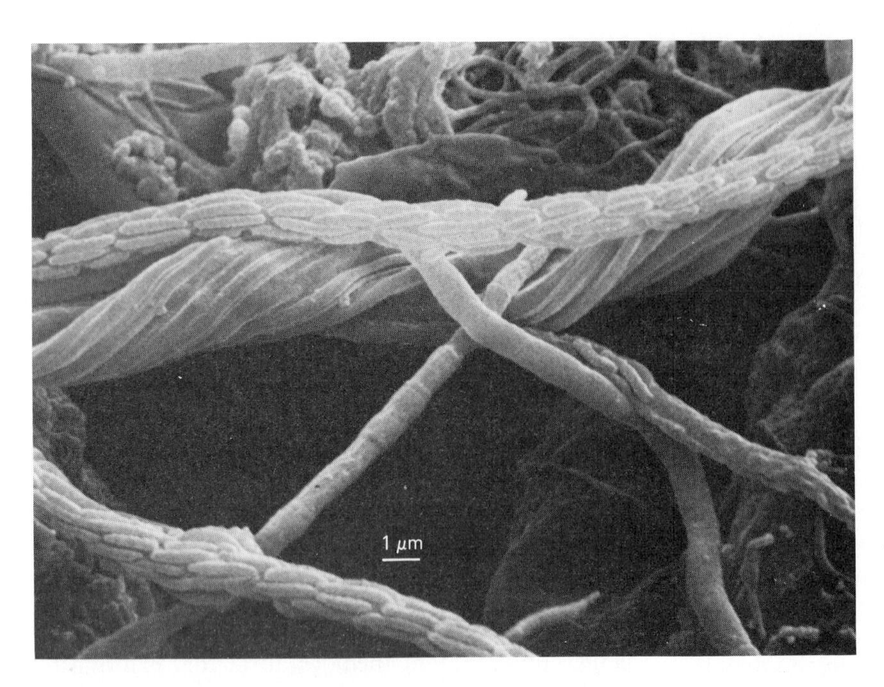

Figure 14–2 Microbial communities in the intestinal tract of a termite. What bacterial shapes are present? *[J. A. Breznak, and Pankratz, H. S.,* Appl. Environ. Microbiol. ***33****:406–426, 1977.]*

populations interact with one another, they form a **community** (Figure 14–2). The interactions of such a community and its physical and chemical environment constitute an *ecosystem.*

The living members of an ecosystem are divided into three categories, depending on their role in maintaining the stable ecosystem: *producers, consumers,* and certain types of consumers referred to as *decomposers.*

Photosynthesizers—such as the cyanobacteria and purple and green bacteria, microscopic algae, larger algae, and plants—function as the producers of organic compounds from inorganic materials. Such forms of life, as noted in Chapter 5, are called *autotrophs.* The consumer and decomposer categories are made up of *heterotrophs,* species that cannot produce organic substances from inorganic ones. Microbial consumers include a number of protozoa (Figure 14–3) as well as other microbial forms of life. The microbial decomposers are various bacteria and fungi that break down complex organic products—such as the remains of animals, plants, and other microbes—into smaller compounds or into inorganic materials that can be used by plants for their producer activity. In short, these microorganisms are microscopic "recycling centers." It would be difficult for life to continue normally without them.

Decomposers function in the ecosystem by using dead or rotting organic matter as nutrients. Like consumers, these other organisms are heterotrophic. They secrete a variety of extracellular enzymes that render the waste materials on which they act soluble, so that they and other organisms can utilize them for food (Figure 14–4). Subsequent metabolic activities by these and other organisms further decompose the wastes and turn the essential elements

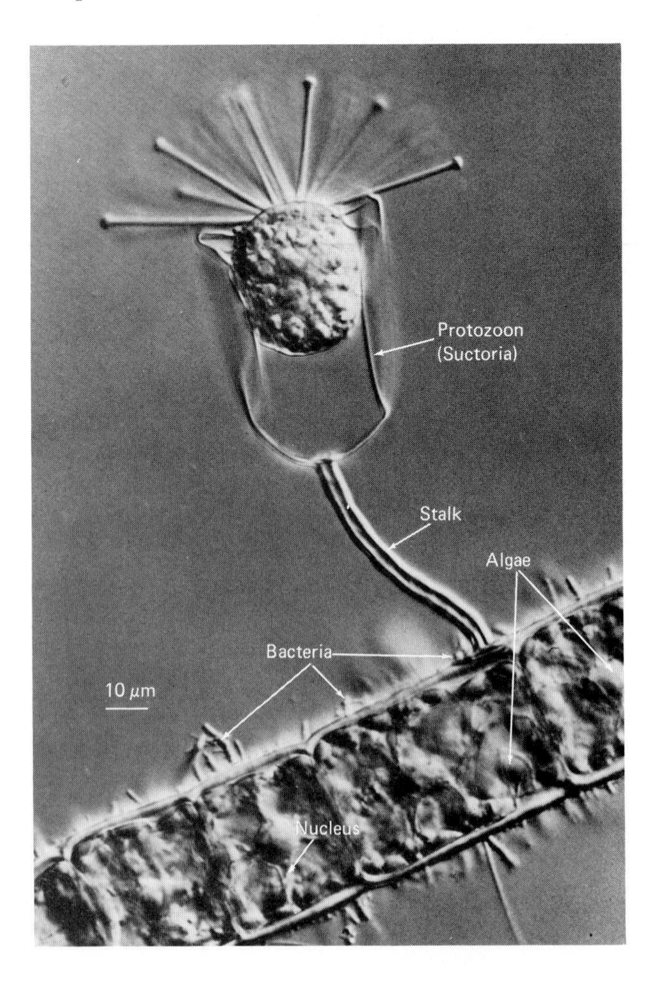

Figure 14–3 A filamentous green alga, containing chloroplasts and nuclei, serving as food not only for a feeding (phagotrophic) protozoon but also for a large number of bacteria. *[John McN. Sieburth,* Sea Microbes. *New York: Oxford University Press, 1979.]*

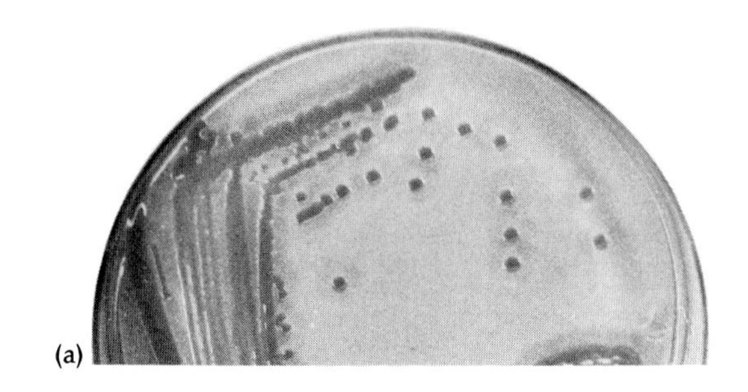

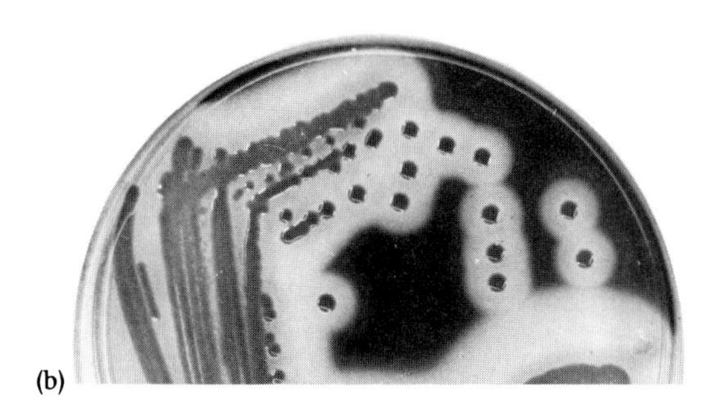

Figure 14–4 An example of microbial enzymatic activity. Agar-degrading bacteria play an important role in the recycling of organic material at the edges of the seas and oceans, where agar forms a significant component of the polysaccharides of red algal cells and tissues. (a) Colonies of the agar-degrading (agarolytic) bacterium *Cytophaga saccharophila.* Depressions caused by the organism's enzymatic activity on the agar-containing medium can be seen. (b) The extent of the agar-degrading action is more apparent after the plate has been flooded with Lugol's iodine. *[Courtesy J. A. C. Agbo, and M. O. Moss,* J. Gen. Microbiol. ***135****:355–368, 1979.]*

into chemicals that can be used in the nutrition of plants and other organisms.

The activities of producers, consumers, and decomposers together constitute food chains. These are interrelated processes that promote the exchange of nutrients and result in the consumption of energy in the ecosystem (Figure 14–5). The biologic world is organized into a large number of interacting **food chains.** Each of these includes a number of species, the populations of which generally have reached a steady state in terms of their reproduction and destruction.

The Components of an Ecosystem

The total community of organisms together with the activities and interactions of living and nonliving components in a natural physical location form an **ecosystem,** with the ability to respond to and modify its environment. Ecosystems include lakes, forests, ponds, and even smaller regions such as a fishbowl or a handful of soil. The various parts of the human body also are excellent examples of ecosystems. The specific location or place of residence of an organism is called its **habitat;** its specific role or function in a community is referred to as its **niche.** The niche includes an organism's habits, relationships to other forms of life, food-related reactions, and ability to change or be changed by its environment. No two species can occupy the same niche at the same time. Thus, certain plants produce food for a community, while other forms of life have the role of decomposer and, at times, predator. An invasion of an organism's niche by another form

Figure 14–5 Interactions of producers, consumers, and a form of consumer, the decomposers (decay producing organisms).

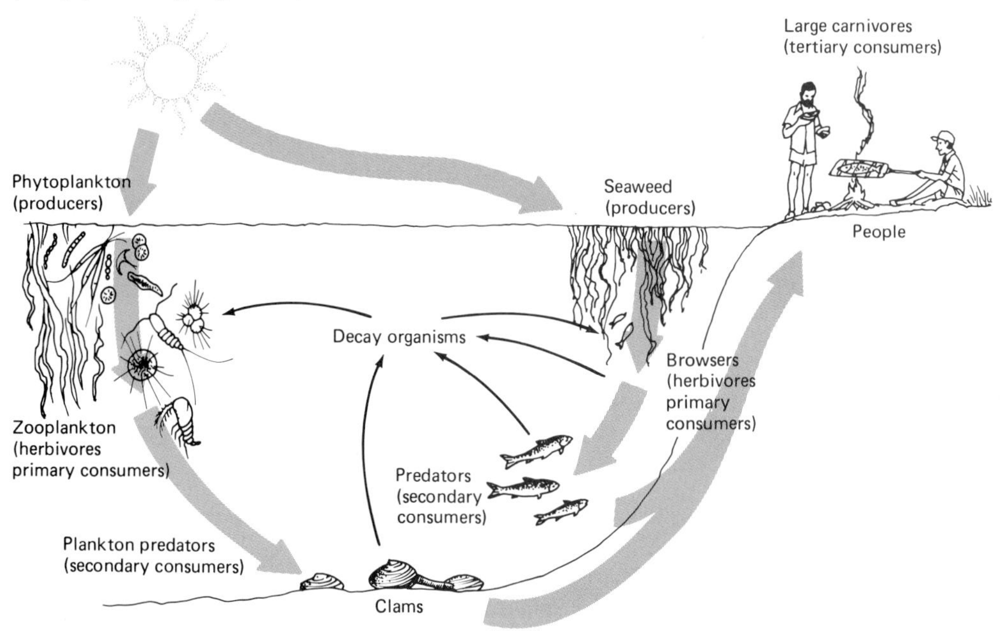

of life results in *competition* and ends in the elimination of one species. Competition is also a frequent occurrence among members of the same species.

BIOTIC AND ABIOTIC FACTORS

The possible habitats of any living organism are limited by both *biotic* (living) and *abiotic* (nonliving) factors in the biosphere. Biotic factors include other animals, plants, and microbes that may compete or otherwise interfere with essential processes. Examples of abiotic components are hydrostatic pressure, osmotic pressure, pH, light, and temperature. The nonliving components also include a variety of inorganic substances such as water, carbon dioxide, oxygen, and minerals as well as organic substances.

Natural Habitats of Microorganisms

As indicated earlier, the biosphere can be divided into three zones. These are the *atmosphere;* the *hydrosphere,* or aquatic environments; and the *lithosphere,* the solid portions of the Earth.

Terrestrial Habitats

The great variety of terrestrial habitats extends from the rocky slopes of mountains to the lush tropical jungles to the hot desert sands. All terrestrial environments are composed of either rock or rock materials. Life forms associated with rocks are algae (Color Photograph 67), lichens (Color Photograph 45), and higher plants such as mosses.

Disintegrating rock material forms, in order of decreasing particle size, gravel, sand, silt, and clay. These particles also form the basis for the loose surface material that covers the earth's crust, the *soil.* Particles of soil often bear a film made of moisture, organic materials, and associated microbes. It is estimated that fertile soils commonly contain as many as 100,000 to 500,000 organisms per gram.

SOIL AND STEPS IN ITS FORMATION

The general ability of soil to support life varies considerably, as does its chemical and microbial composition. Soil consists of solids, minerals and organic compounds, water, and air. The minerals in soil range from the larger particles of gravel and sand to the smaller silt and finest of clay. The variety and proportion of such materials affect the suitability of soil for agricultural and other purposes.

Soil, which is formed by the breakdown of rock, usually contains a combination of minerals and organic compounds in various stages of decomposition, water, and dissolved organic and inorganic materials and gases. The disintegration of rocks is caused by a variety of mechanical, chemical, and biological processes collectively called *weathering* (Figure 14–6).

Mechanical weathering includes the breaking of rock by forces such as glaciers and avalanches, as well as the effects of wind, blown sand, and cracking by freezing and thawing. *Chemical weathering* involves the dissolving of rock materials by rain, streams, or groundwater. *Biological processes* include the cracking of rock by root growth and the action of metabolic by-products of microbial growth, such as acids, to dissolve and disintegrate rocks.

Many diverse microorganisms and other forms of life are found in soil and contribute to its formation and fertility. Microbial contributions in addition to those mentioned earlier include the decomposition of organic matter to produce better soil texture and water-binding capacity, release of minerals important for plant growth from soil particles, and the transformation of various chemical compounds into substances useful for plants and other forms of life. Soil also serves as the habitat for antibiotic-producing microbes and pathogens of lower animals, humans, and plants.

Aquatic Habitats

More than 70% of the earth's surface is water. The aquatic habitats that this water represents fall into two general categories: *fresh water* and *salt (marine) water.* Freshwater habitats, which make up about 2% of the hydrosphere, include ponds, streams, lakes, and rivers. Marine habitats contain salt concentrations of 3.5% and higher and include the oceans, seas, and certain inland lakes such as the Great Salt Lake in Utah. *Estuaries* are coastal regions where sea water mixes with fresh water under the influence of the daily tidal cycle.

Two unique features make water habitats suitable for microorganisms: (1) there is less temperature variation in bodies of water than in terrestrial environments, and (2) sunlight penetrates the water, allowing photosynthetic activities well below the surface. Many aquatic environments are rich in nutrients in the form of waste organic material.

Microhabitats

The habitats of interest in microbiology are often much smaller than a forest or lake. These small, inhabited regions are called *microhabitats.* Every environment consists of many different microhabitats. A decaying piece of wood, a soil particle, the moist undersurface of a fallen leaf, and the carcass of a fish or bird are just a few of the microhabitats to be found in a forest. These few examples represent a

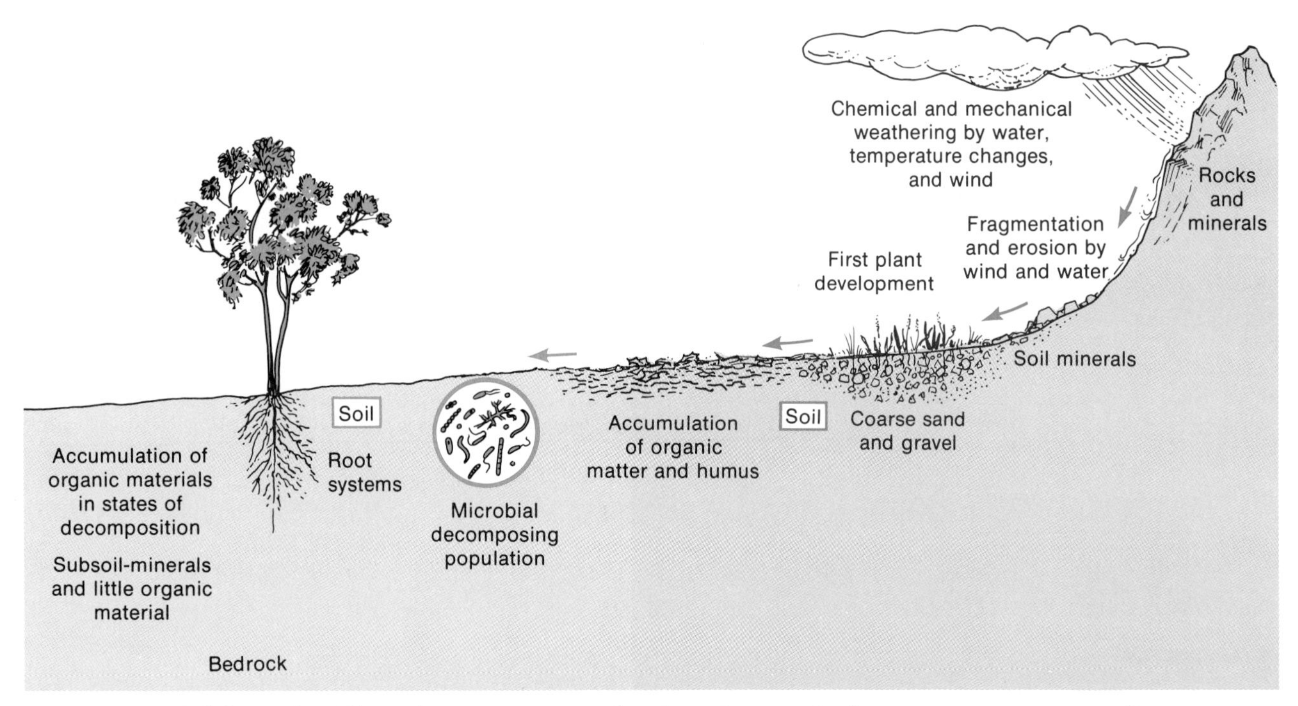

Figure 14–6 Soil formation. The various components of rocks and mountains become disintegrated into soil particles. Soil particles are produced by erosion, water, and wind.

wide range of conditions in terms of pH, temperature, gaseous oxygen, and moisture content. Each is dynamic in that its conditions change constantly because of abiotic factors and the microbial activities of various producers, consumers, and decomposers.

The human body contains a variety of microhabitats. Many microorganisms normally share our bodies and form various ecosystems on the skin and in the upper respiratory tract, mouth, lower gastrointestinal tract, and various body openings. These habitats and niches are discussed in later chapters in terms of their potential microbial benefits and hazards.

Microbial Habitats

Microorganisms have been found in nearly all regions of the earth. In many cases the organisms are only transients in the environment, but some are definitely native microorganisms. These indigenous microorganisms are able to grow during those periods when environmental conditions are appropriate and to survive during periods when conditions for growth are unsuitable. Survival in some cases results from the formation of bacterial endospores or other dormant microbial structures. [See procaryotic structures in Chapter 7.]

Some microorganisms have evolved unusual mechanisms to remain in their optimum environments. One such organism is *Aquaspirillum* (ah-qua-spy-RIL-lum) sp. Members of this genus can orient themselves to the earth's magnetic field and migrate into anaerobic sediments when disturbed by turbulence in the water and sediment. Cellular orientation is produced by a magnetic pull that causes individual organisms to move downward. This phenomenon, called *magnetotaxis,* involves structures called *magnetosomes,* which contain iron in the form of magnetite (Figure 14–7). A mechanism similar to the one observed in bacteria also appears to be responsible for the migration patterns of some animals able to detect and use geomagnetism as a cue for orientation. Magnetite has been detected in honeybees as well as in birds and fish.

Microbial Activities

The pioneer American ecologist F. E. Clements has classified the activities in ecosystems as one of three kinds: *action, reaction,* or *coaction.* Action deals with the way abiotic forces affect organisms. Reaction deals with the biological activities in the ecosystem: how organisms react to the environment and perform their functions in shaping the environment. Coaction is concerned with the effect organisms have upon one another. The last portion of this chapter will describe coaction involving microorganisms, plants, and animals. Five kinds of coactional relationships—**syntrophism, competition, antagonism, predation,** and **symbiosis**—will be presented.

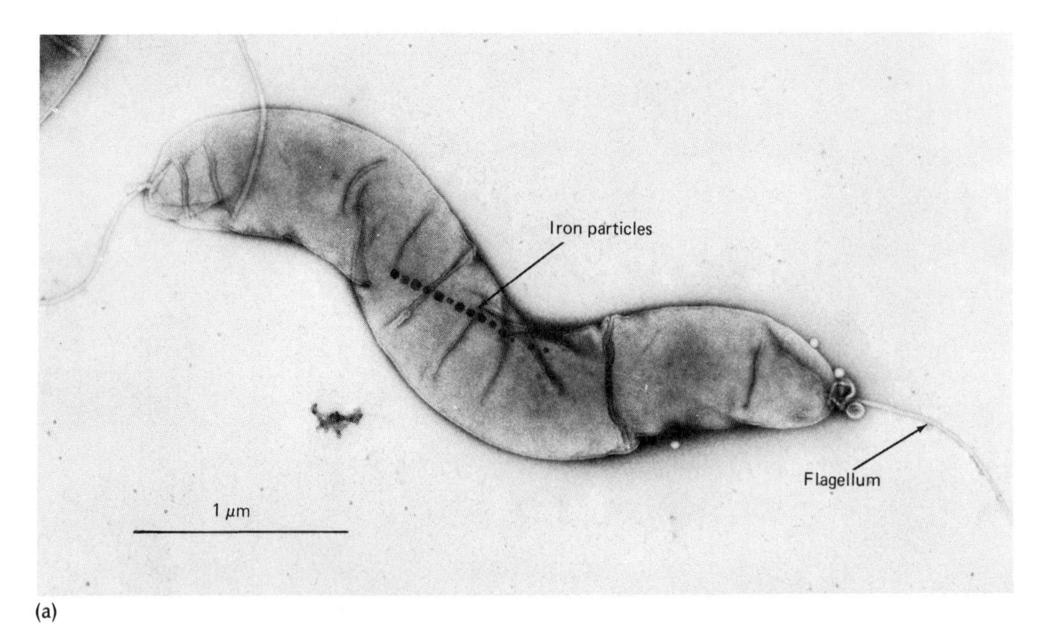

(a)

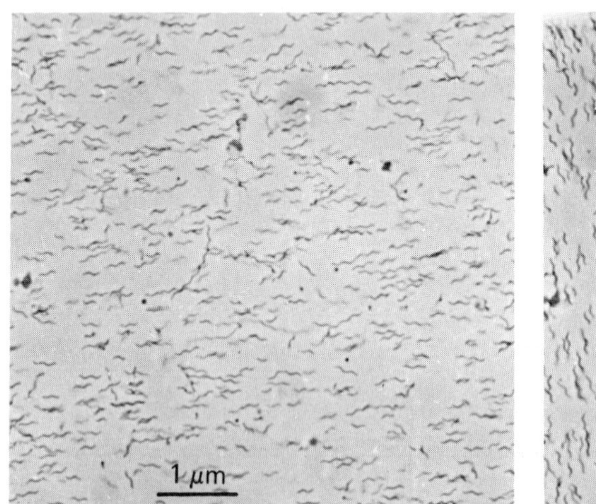

(b)

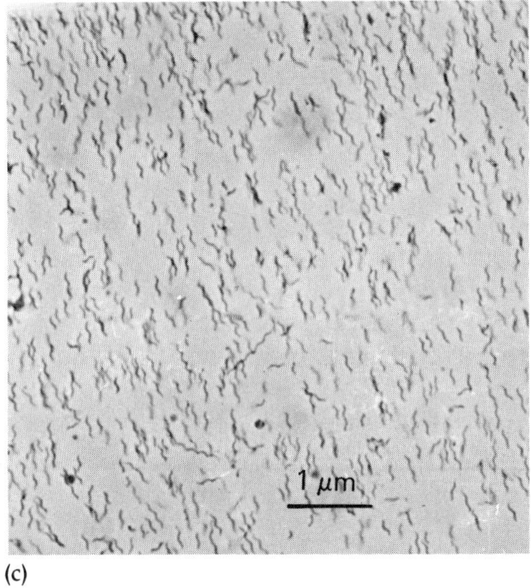

(c)

Figure 14–7 Magnetically responsive cells. (a) A transmission micrograph of a magnetotactic cell of *Aquaspirillum magnetotacticum.* The intracellular chain of electron-dense, iron-rich particles, *magnetosomes,* are arranged in a reasonably straight line. *[D. Maratea, and R. P. Blakemore,* Internat. J. System. Bacteriol. ***31****:452–455, 1981.]* (b) and (c) Phase-contrast micrographs showing the response of living magnetic cells in artificial magnetic fields. In (b), cells moving in a local field of a small magnet held near the microscope with its north-south magnetic axis directed right to left. In (c), with the position of the magnetic axis changed and directed top to bottom, the cells reorient themselves and swim in a new direction. *[R. P. Blakemore, et al.,* J. Bacteriol. ***140****:720–729, 1979.]*

Action Activities

Abiotic factors include those physical and chemical conditions in the environment that may not necessarily result from biological activity. For example, a low pH can be a result of mining wastes. A high temperature may be caused by volcanic activity or by microbial catabolism.

Table 14–1 presents selected microorganisms and the extreme conditions under which they grow.

Temperature

Microorganisms have been observed growing in extreme temperatures of −10° to 110°C in their natural habitats. *Psychrophilic* (cold-loving) yeasts have been isolated from glacial ice and soil from Antarc-

TABLE 14–1 Three Abiotic Factors and Selected Microorganisms that Grow Under Extreme Conditions in Nature

Abiotic Factor	*Microorganism*[d]	*Conditions for Growth*
Temperature	*Pyrodictium*	82°C and above
	Thermoactinomyces spp.	68°C
	Rhodotorula sp.[a]	14°C
	Flavobacterium spp.	4°C
	Bacillus globisporus	−10°C
pH	*Agrobacterium* sp.	12.0
	Vibrio cholerae	9.0
	Bacillus pasteurii	8.5
	Sulfolobus sp.	0.5
	Thiobacillus thiooxidans	0
Osmotic Pressure	*Candida* spp.[a]	60% sugar[b]
	Hansenula spp.[a]	60% sugar[b]
	Saccharomyces spp.[a]	60% sugar[b]
	Halobacterium salinarum	27–30% NaCl[c]
	Sarcina morrhuae	27–30% NaCl[c]
	Pseudomonas cepacia	Distilled water

[a] These organisms are yeasts; all others are bacteria.
[b] Honey is one example of such a high sugar concentration in nature.
[c] This concentration of NaCl can be found in Great Salt Lake, Utah, and the Dead Sea.
[d] For pronunciations, see the Organism Pronunciation Guide at the end of the text.

tica. A *Flavobacterium* sp. (flay-voh-back-teer-EE-um) isolated from the Great Lakes demonstrated higher respiration rates at 4°C than at 20°C. Since this organism grows best at 4°C, it is considered an obligate psychrophile. Various genera of algae, bacteria, fungi, and even protozoa contain members that are psychrophilic. Some psychrophiles that grow and metabolize optimally at 4°C or less can change the quality of refrigerated foods and eventually cause spoilage.

Thermophilic (heat-loving) microorganisms grow in volcanic springs and soils and heated water ecosystems, as well as in *compost.* Compost is the natural decomposition product of microbial activity in mixtures of vegetation supplemented with manure. This decomposition of plant and animal wastes is usually an anaerobic process, but it can be made aerobic by frequent mixing of the materials. Heat generated by composting can reach 80°C, which is usually adequate to disinfect the compost. The final product is used for soil enrichment. *Thermoactinomyces* (thermo-ak-tin-oh-MY-sees) spp. have been observed growing optimally at 65° to 68°C and have been isolated with other thermophilic bacteria from compost piles.

Probably the most remarkable organisms are species of the genus *Pyrodictium* (pie-ro-DIK-tee-um, Figure 14–8). These archaeobacteria are anaerobic, sulfur-reducing thermophiles found in volcanically heated marine waters near Vulcano, Italy. The cardinal temperatures for this genus are a minimum of 85°C, an optimum of 105°C, and a maximum of 110°C. **[See Chapter 5 for discussions of growth and temperature.]**

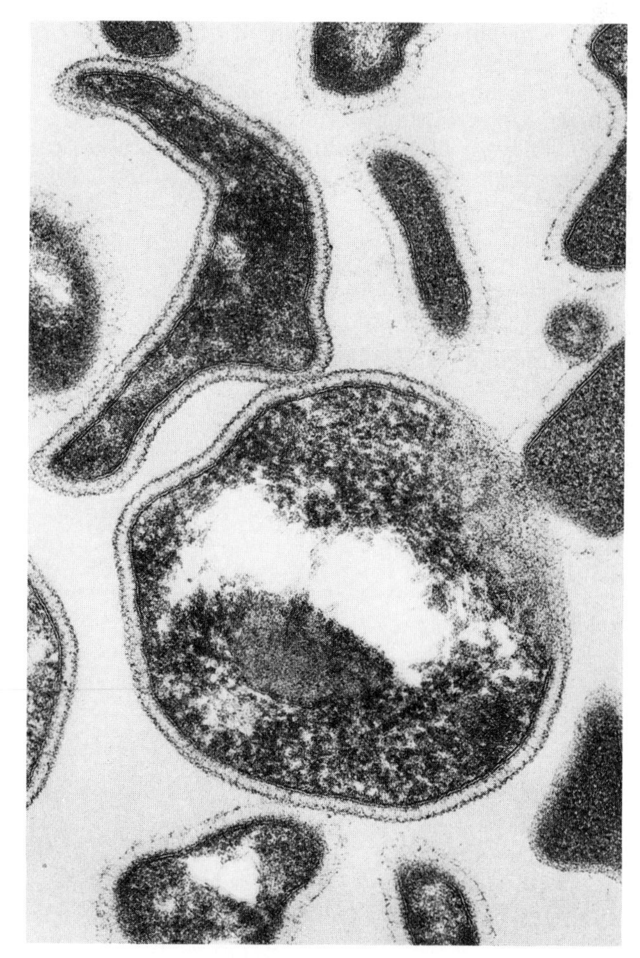

Figure 14–8 An electron micrograph of *Pyrodictium*. *[Courtesy K. O. Stetter, University of Regensburg.]*

pH

Terrestrial and aquatic habitats vary in acidity and alkalinity. The pH of such environments is determined by the types and amounts of various minerals and microbial activities. For example, bacteria can oxidize sulfur to sulfuric acid, creating aquatic environments with pH values as low as 0 to 2.0. A pH of 0 means that the acidity is equivalent to that of a 1-*N* solution of a strong acid, which is a great deal of acid for a living creature to tolerate. Bacteria are also known to decompose urea and proteins to produce large quantities of ammonia. This compound forms ammonium hydroxide in both water and soil and can produce pH values as high as 12.0.

Acidophilic (acid-loving) microorganisms include *Thiobacillus thiooxidans* (thigh-oh-bah-SIL-lus thigh-oh-OX-ee-dans), a sulfur oxidizer that will grow at pH 0, although its optimum pH for growth is 2.0 to 3.5. *T. ferrooxidans* (T. fer-oh-OX-ee-dans), an iron oxidizer, has been isolated from acid soils such as peat bogs and from acid mine waters. Some sulfur-containing waters of volcanic habitats are also hot, and as indicated earlier, the bacterium *Sulfolobus* (sul-foh-LOW-bus) has been isolated and shown to be an acid-loving thermophile. Fungi in general prefer a somewhat acid environment of pH 5.0 to 5.5.

Alkaliphilic (alkaline-loving) microorganisms include *Bacillus pasteurii* (bah-SIL-lus pas-TOOR-eye), which will grow at pH levels of 8.5 and higher, *Vibrio cholerae* (VIB-ree-oh KOL-er-eye) at 9.0, and *Agrobacterium* (a-grow-back-TIR-ee-um) sp. at 12.0.

Osmotic Pressure

As the amount of dissolved materials increases in the immediate environment of living cells, a condition of hypertonicity results and water tends to escape from the cells. This drying out of cells with hypertonic solutions is used in meat and fish preservation, since common spoilage microorganisms cannot tolerate pickling brines of 30% sodium chloride. It is also used in the preparation and preservation of fruits by the addition of 40% to 60% sugar to make jams, jellies, preserves, and syrups.

High salt conditions also occur in nature. The Dead Sea and Great Salt Lake in Utah have salt concentrations of 27%. Some areas can be found with salt concentrations as high as 30%. *Halophilic* (salt-loving) microorganisms such as *Halobacterium salinarium* (hail-oh-back-TI-ree-um sal-ee-NARE-ee-um) and *Sarcina morrhuae* (sar-SEE-nah mor-YOU-eye) will grow in 36% salt. As a rule, these organisms will not grow in salt concentrations of less than 12% to 15%. In such situations active growth will occur only when the salt concentration increases, usually by evaporation. Obligate halophilic bacteria require the sodium and magnesium ions in the saline solutions to maintain the integrity of their cell envelope, since their cell walls have no rigid layer. Some slightly halophilic bacteria may even concentrate sodium and chlorine as granular inclusions, possibly in an attempt to survive suboptimal environments (Figure 14–9).

Saccharophilic (sugar-loving) microorganisms can be found naturally in honey and other high-sugar-content materials. These microorganisms are primarily fungi. Saccharophilic yeasts include species of *Candida* (KAN-did-ah), *Hansenula* (han-sen-YOU-la), *Pickia* (pik-EE-ah), *Saccharomyces* (sak-ah-roh-MY-sees), and *Torulopsis* (tor-you-LOP-sis). Species of the molds *Rhizopus* (rye-ZO-puss) and *Neurospora* (neu-RAH-spor-ah) may also be saccharophilic.

At the other extreme of osmotic environments are three strains of *Pseudomonas cepacia* (soo-doh-

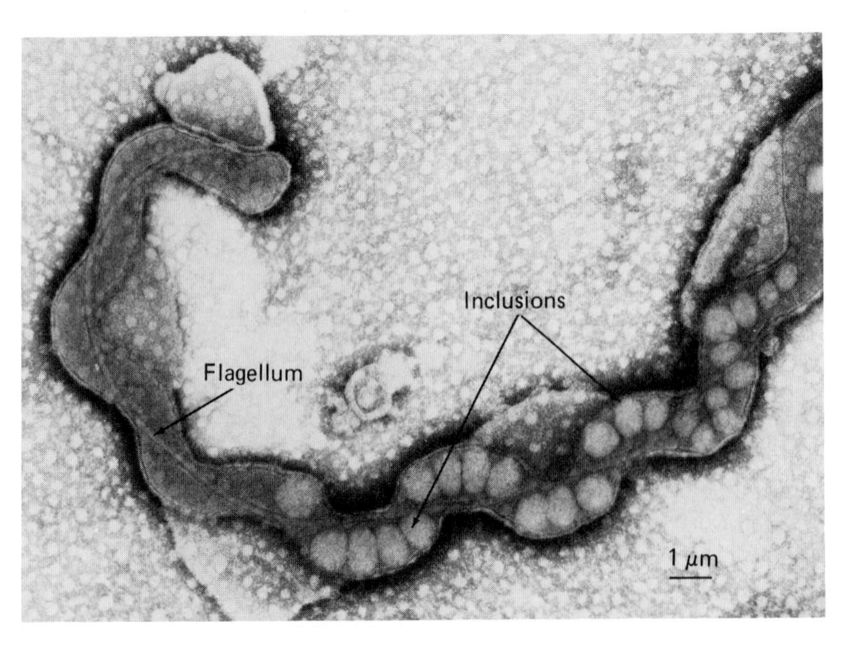

Figure 14–9 A halophilic leptospire. The cell's cytoplasm is packed with inclusions that were found by x-ray microanalysis to contain sodium and chlorine. It is hypothesized that these granules are of importance to the organism's survival. *[K. Hovind-Hougen, et al.,* Arch. Microbiol. ***130****:339–343, 1981.]*

MOH-nuss sep-a-SHE-ah) capable of growing in distilled water. Such organisms can achieve a population density of 10^6 to 10^7 with only the trace amounts of volatile organic materials, salts, and metal ions usually present in distilled water. Under these conditions, the organism is able to grow within a wider temperature range and has a very small cell size. Certain microorganisms can survive for up to 5 years in distilled water. Examples of such organisms include non-spore-forming bacteria such as *Nocardia* (no-KAHR-dee-ah) spp., species of yeasts belonging to the genera *Candida, Cryptococcus* (kryp-toh-KOK-kus) and *Geotrichum* (gee-OTT-ree-kum) and species of molds, such as *Alternaria* (al-ter-NARE-ee-ah) and *Aspergillus* (a-sper-JIL-lus).

Hydrostatic Pressure

The pressure exerted by the weight of water is referred to as hydrostatic pressure. In nature, high hydrostatic pressures are found mainly at ocean depths. For every 10 m of depth, the hydrostatic pressure increased by 1 atmosphere (atm); that is, 15 pounds per square inch (psi). Organisms that grow better at higher-than-normal atmospheric pressure are called *barophiles* (Figure 14–10). Hydrostatic pressure is known to affect the activity of most enzymes, with protein synthesis and the transport of materials across membranes being the most sensitive. Most bacteria isolated from shallow-water or soil habitats are completely inhibited or killed by hydrostatic pressures ranging from 200 to 600 atm. Despite the harmful effects of hydrostatic pressure, many microorganisms that can tolerate high hydrostatic pressures to a variable degree have been isolated from deep water. These organisms, referred to as being *barotolerant,* are unique in their ability to grow well at pressures corresponding to that of the deep-water habitat.

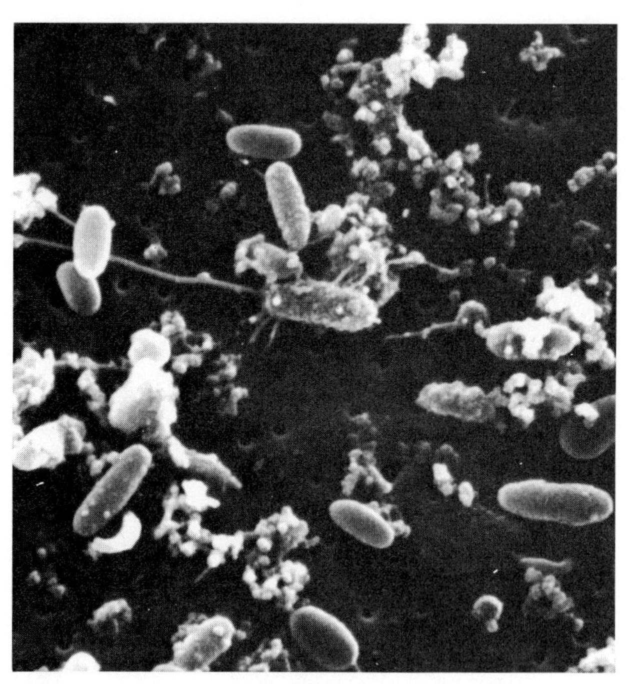

Figure 14–10 A scanning micrograph of bacteria from material collected on a membrane filter at an ocean depth of 4,400 m and fixed prior to decompression. *[Courtesy of Dr. H. W. Jannasch, Woods Hole Oceanographic Institution.]*

Reaction Activities

The Life-supporting Biogeochemical Cycles

The various types of microbial, animal, and plant life on this planet have used and reused inorganic materials since life began. For an estimated 3.5 billion years, water, carbon, oxygen, nitrogen, phosphorus, sulfur, iron, manganese, and many other materials have been cycled through the biosphere. Oxidized and reduced inorganic and organic forms are used repeatedly to transform energy and make cell substance. Some metabolic reactions cause insoluble stored compounds to go into solution and become available. The reverse process stores excess materials to prevent their loss from the immediate environment. The participation of organisms in this recycling is one aspect of *reaction,* the response of organisms to abiotic factors and their function in shaping nature. The reactions that cycle critical materials are often called collectively *biogeochemistry.* This term refers to the important role of life *(bio)* in the chemistry of the earth *(geo).* In this portion of the chapter, we shall discuss the role of microorganisms in biogeochemistry and in *bioconversion, biodegradation,* and *bioconcentration.*

The various elements that are vital to the structures and activities of living systems, including carbon, hydrogen, nitrogen, oxygen, phosphorus, and sulfur, are drawn from the environment, incorporated into the cells, and eventually returned to the environment to be used over again. The earth gains little matter from other portions of the universe and loses little to outer space. Interrelated series of natural biogeochemical cycles ensure the flow of essential elements between the living and nonliving parts of ecosystems. In these cyclic transformations of chemicals from one form to another, microbes such as bacteria, fungi, and protozoa are intricately associated with animals and plants (Figures 14–11 through 14–18). It is important to emphasize here that no one cycle can function in the absence of the others. Moreover, all cycles are working at the same time.

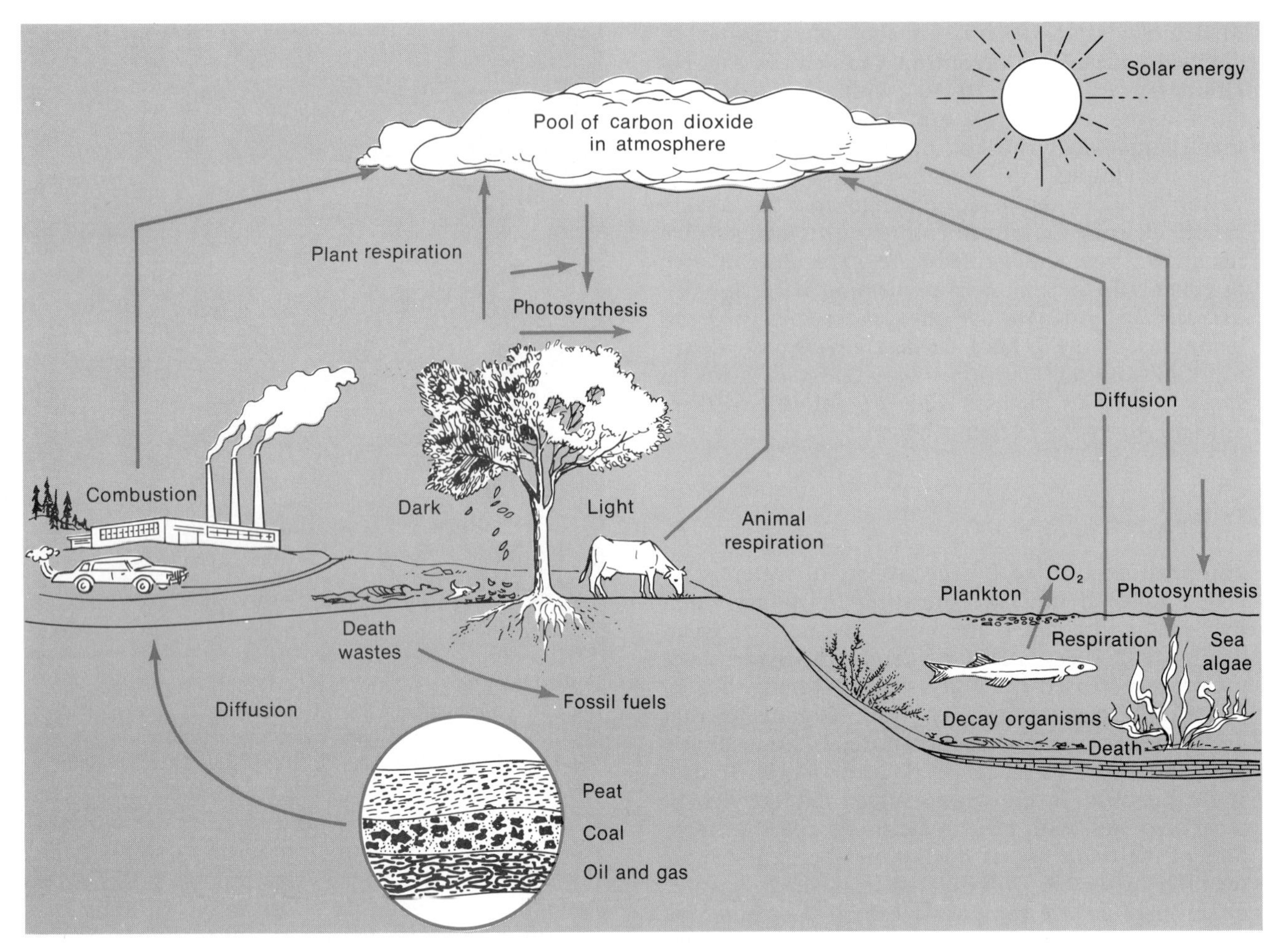

Figure 14–11 The biogeochemical cycle of carbon involves a chemical element needed by all forms of life to synthesize a full range of organic compounds and cellular components. Energy from the sun is used by higher plant life, algae, and various microorganisms to convert carbon dioxide and water into organic compounds. Eventually these compounds are decomposed to again form carbon dioxide and release it into the atmosphere. A variety of microorganisms are involved in this decomposition phase of the cycle.

The Carbon Cycle

Carbon is one of the most common and important of the elements involved with living systems. It circulates among a large number of biosphere components by means of the **carbon cycle** (Figure 14–11). This cycle is of particular importance because it is the route by which useful energy from the sun is stored in the biosphere. Through photosynthesis, carbon dioxide is first absorbed by many different plant and microbial producers to form cellular substances; later it is released to the atmosphere by consumers. Most photosynthetic organisms use the energy from the sun to generate oxygen, energy-rich organic molecules, and some heat. The consumers, which are mainly animals, utilize the organic molecules and, by respiration, generate heat and return some of the carbon dioxide to the air or water. A small amount of carbon dioxide dissolves in water, where it can be trapped in ice caps and groundwater or may form bicarbonate ions. Bicarbonate ions react with calcium ions to form limestone (calcium carbonate). The carbon in this substance may not be returned to the atmosphere for millions of years. Carbonates formed by living organisms as shells or other protective structures of animals and protozoa can also accumulate on lake and sea bottoms and eventually form limestone.

When burned, wood and fossil fuels yield large quantities of carbon dioxide to the atmosphere. Under conditions of limited oxygen, incomplete combustion occurs, resulting in carbon monoxide (CO). Interestingly, certain bacteria metabolize this toxic gas, converting it to carbon dioxide with the release of energy.

Under anaerobic conditions, carbon dioxide is reduced to methane (CH_4) by a select group of microorganisms. This particular metabolic activity is discussed later in this chapter because of its significance as a means of producing fuel. Other aerobic organisms such as *Methylococcus* (meth-el-o-KOK-kus) spp. are able to oxidize methane as a source of energy and produce carbon dioxide.

Ultimately, various decomposer microorganisms, mainly bacteria, oxidize animal and plant remains, with the production of additional heat and the return of carbon dioxide to the environment. The bulk of plant materials is made up of cellulose and lignin, the structural units of vascular plants. These substances usually decompose slowly.

Some cellulose decomposition occurs in digestion by ruminants. Lignin digestion is performed by various bacteria and fungi. These organisms apparently require cellulose as an energy source in order to digest lignin. Under certain conditions, such as those that exist in acid soils and bogs, lignin and cellulosic wastes do not degrade and may be converted to peat. This type of process appears to be the origin of most, if not all, of our fossil fuels, and requires anaerobic conditions and thousands of years.

The Nitrogen Cycle

Organisms need nitrogen to synthesize essential compounds such as amino acids, proteins, and nucleic acids. Although 79% of the atmosphere is nitrogen gas (N_2), this elemental form of nitrogen is useless to most organisms. Producer organisms such as plants must obtain their nitrogen from the soil in the form of inorganic nitrogen compounds known as the *nitrates.* Animals secure their nitrogen from the compounds produced by plants. By means of the **nitrogen cycle** (Figure 14–12), the essential compounds are converted from one type to another by various organisms, often with the release of energy that is captured during metabolic activity. The cycle consists of two phases: *nitrogen fixation,* and *nitrogen assimilation.*

NITROGEN FIXATION

Nitrogen fixation is the reduction of atmospheric molecular nitrogen to ammonia (NH_3) and other

Figure 14–12 The nitrogen cycle. One of the most important microbial processes in the nitrogen cycle is nitrogen fixation, the conversion of nitrogen gas, N_2, into nitrogen compounds. As this diagram shows, the process is carried out by microorganisms. Higher plants are dependent on microorganisms and this process for their nitrogen requirements. By far the most important nitrogen-fixing organisms in soils are those that are symbiotic.

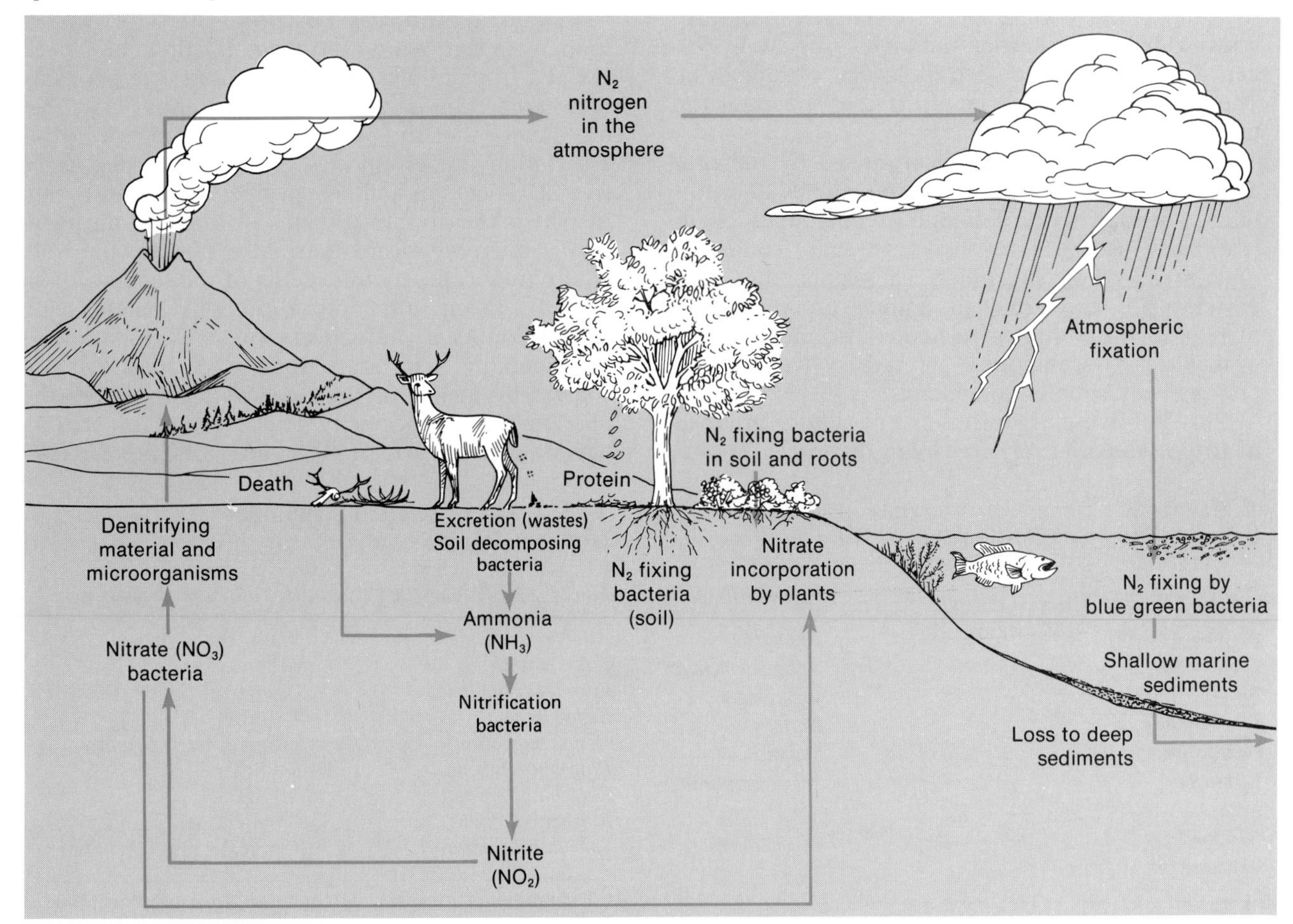

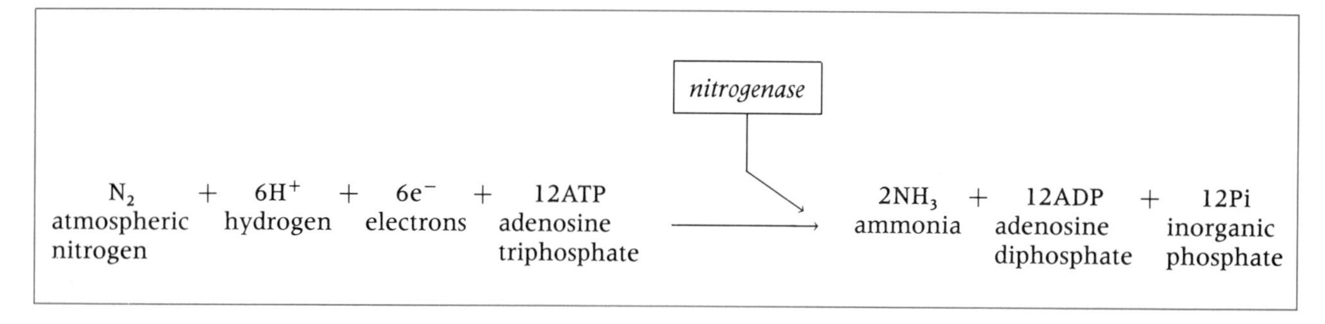

biologically essential nitrogen-containing compounds (Figure 14–12). Under natural conditions, nitrogen fixation can occur in two ways: (1) by inorganic chemical processes in the atmosphere, such as photochemical and electrical (lightning) reactions, and (2) by biological processes involving cyanobacteria and various other bacterial species.

Certain microorganisms can reduce nitrogen by living in association with the roots of higher plants (symbiotically) or living freely without such an association (nonsymbiotically), producing intracellular ammonia. The nitrogen-fixation reaction requires a bacterial enzyme, *nitrogenase,* and adenosine triphosphate (ATP). The active component of the enzyme is formed from two protein molecules and iron and molybdenum. The nitrogenase binds the nitrogen (N_2) and reacts with a special electron carrier protein, oxidizing the carrier and picking up six hydrogen ions (H^+) to form two molecules of ammonia. The overall process can be pictured as shown at the top of the page.

The ammonia (NH_3) is incorporated by the producers of the system to their amino acids and proteins. These nitrogen-containing compounds circulate from the primary fixers through consumer organisms. When consumers eat plants, the plant proteins are converted to animal proteins. Ultimately, such proteins or related compounds become available to decomposers and are converted into progressively simpler compounds.

Nitrogen-fixing organisms can usually be grouped in three categories: (1) free-living bacteria, (2) cyanobacteria, and (3) leguminous plants and their symbiotic bacteria. However, nitrogen fixation also occurs with some nonleguminous root-nodule plants, plants with leaf nodules, and lichens (Color Photograph 44).

Nonsymbiotic Nitrogen Fixation. Many free-living cyanobacteria and other bacteria that exist in both soil and water habitats are known for their nonsymbiotic nitrogen fixation. These microorganisms can be found in environments as extreme as the soil and rocks of Antarctica and the hot springs of Yellowstone National Park. Representatives of this group are listed in Table 14–2.

Symbiotic Nitrogen Fixation. Most of what is known today about nitrogen fixation has been learned from studies of the mutualistic relationship between certain bacteria and leguminous plants that are members of the pea family. The bacteria in this association are all species of the genus *Rhizobium* (rye-ZOH-bee-um) and are specific to each host plant in which they occur (Table 14–3). These nitrogen-fixing microorganisms normally live in the soil, where they can penetrate roots of the legumes (Figure 14–13) and enter into a mutually beneficial relationship. As a prerequisite for the formation of this symbiotic association, the two partners come in contact by means of their cell surfaces, where the phenomena of specificity and recognition are believed to take place. Some plants will interact only

TABLE 14–2 Genera of Nonsymbiotic, or Free-living, Microorganisms Capable of Nitrogen Fixation

Free-living Bacteria[a]	*Cyanobacteria*
Azotobacter	*Anabaena*
Azotomonas	*Aphanizomenon*
Bacillus	*Chlorogloea*
Clostridium	*Nostoc*
Desulfovibrio	*Stigonema*
Klebsiella	*Trichodesmium*
Nocardia	
Pseudomonas	
Rhodospirillum	

[a] For pronunciations, see the Organism Pronunciation Guide at the end of the text.

TABLE 14–3 Selected Symbiotic Nitrogen-fixing Systems

Mutualistic Microorganisms[a]	*Host Plant or Animal*
Anabaena azollae[b]	Water fern *(Azolla)*
Frankia alni	Alder tree
Frankia ceanothi	Chaparral shrub *(Ceanothus)*
Klebsiella sp.	Termite
Nostoc muscorum[a]	Tropical herb *(Gunnera)*
Rhizobium japonicum	Soybean
Rhizobium meliloti	Alfalfa
Rhizobium trifolii	Clover
Spirillum lipoferum	Tropical grass *(Digitaria)*

[a] For pronunciations, see the Organism Pronunciation Guide at the end of the text.
[b] Cyanobacterium.

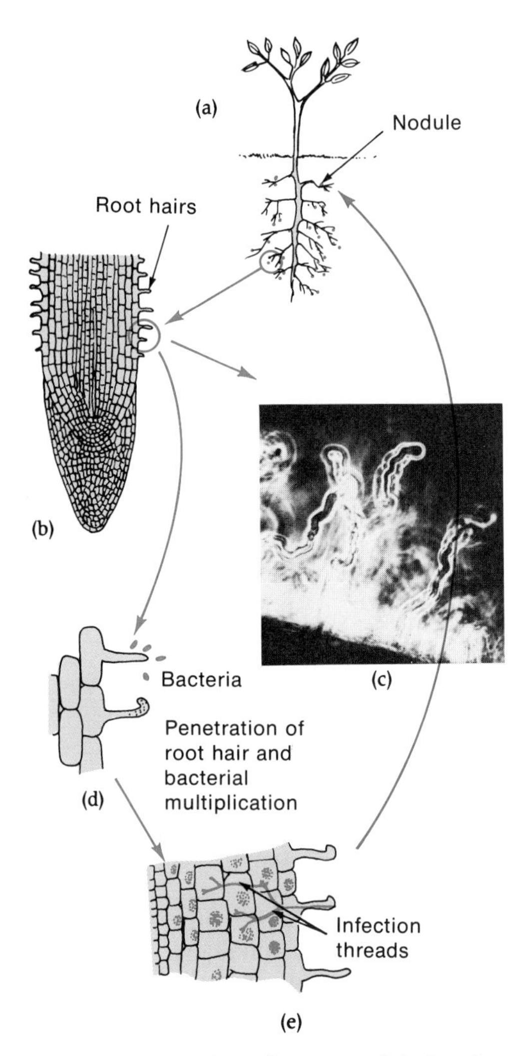

Figure 14–13 Formation of a root nodule in a legume infected by species of *Rhizobium*. (a) The appearance and location of root nodules. (b) Root hairs, extensions of specialized surface root cells. (c) Close-up of root hairs on a clover plant. (d) Penetration and multiplication of bacteria. (e) Infection thread formation and the spreading of infection thread to nearby cells leading to nodule formation. *[F. B. Dazzo, and D. H. Hubbell,* Appl. Microbiol. ***30**:1017–1033, 1975.]*

with specific bacterial strains. The invading bacteria then establish themselves through a series of complex interactions that stimulate plant host responses, including curling of the root hair tip and the formation of so-called infection threads (Figure 14–14). Following these events, additional plant cells in the interior of the root are penetrated, and eventually enlarge to form localized swellings, or nodules. Free nitrogen picked up by these nodule systems from the soil is converted into ammonia. Although various other nitrogen-fixing systems exist (Figure 14–12), the microbe-legume association is extremely efficient in trapping nitrogen and is thought to be one of the most important sources of fixed nitrogen.

NITROGEN ASSIMILATION

The proteins produced by plants are consumed and eventually returned to the environment, mainly in excretions. Through the enzymatic activities of the microorganisms associated with decay, the organic nitrogen molecules in such excretions and in decaying matter are broken down to ammonia. Ammonia or ammonium ion (NH_4^+) formation is known as *ammonification.* Most of the ammonia produced through decay is converted into nitrates (NO_3^-) in two steps

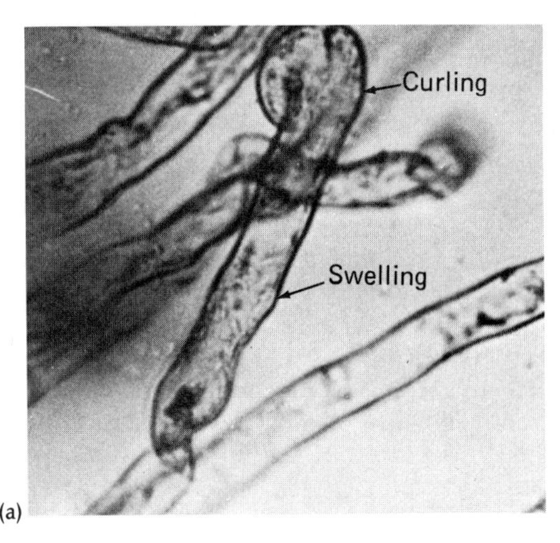

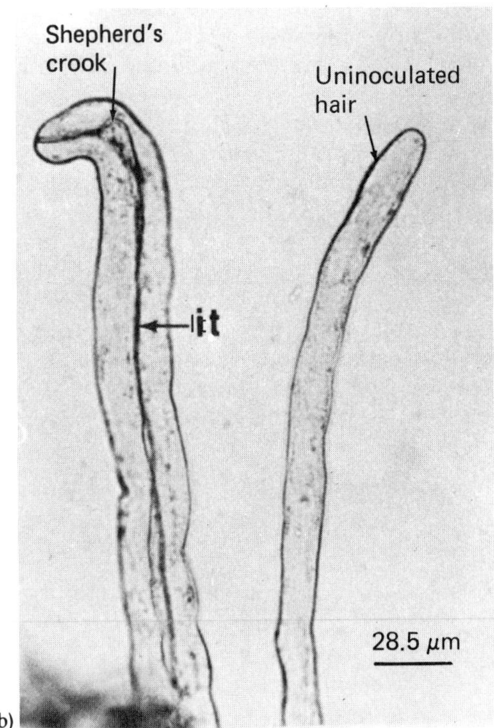

Figure 14–14 The early host responses to infection leading to the formation of N_2-fixing root nodules. (a) A root hair two hours after inoculation, showing curling and swelling. (b) A root hair (left) two days after inoculation, showing a shepherd's crook and an infection thread (it). A thin uninoculated root hair is shown on the right. *[S. Shantharam, and P. P. Wong,* Appl. Environ. Microbiol. ***43**:677–685, 1982.]*

TABLE 14–4 Selected Nitrifying and Denitrifying Microorganisms

Nitrification Step 1: $NH_3 \rightarrow NO_2$	Step 2: $NO_2 \rightarrow NO_3$	*Denitrification* $NO_3 \rightarrow NO_2 \rightarrow N_2O \rightarrow N_2$
Bacteria[a]	**Bacteria**	**Bacteria**
Halobacterium	*Nitrobacter*	*Alcaligenes*
Nitrococcus	*Nitrococcus*	*Bacillus*
Nitrosococcus	*Nitrospina*	*Corynebacterium*
Nitrosolobus		*Flavobacterium*
Nitrosomonas		*Halobacterium*
Nitrosospira		*Pseudomonas*
Nocardia		
Streptomyces		
Molds	**Molds**	
Aspergillus	*Aspergillus*	
Cephalosporium	*Cephalosporium*	
Penicillium	*Penicillium*	

[a] For pronunciations, see the Organism Pronunciation Guide at the end of the text.

(Table 14–4). Chemosynthetic bacteria belonging to the genus *Nitrosomonas* (nye-troh-soh-MOH-nass) oxidize NH_3 to nitrites (NO_2^-). Nitrites, in turn, are oxidized to nitrates (NO_3^-) by bacteria of the genus *Nitrobacter* (nye-troh-BACK-ter). The end products of these reactions are not only made readily available to plants through their roots, but are also used (assimilated) by other members of the microbial community (Table 14–5). Some bacteria, such as *Pseudomonas,* and a few fungi use the nitrogen atom in the nitrate as the electron acceptor and reduce it, usually to gaseous N_2.

The conversion of nitrate to gaseous nitrogenous compounds, or *denitrification,* completes the cycle. By means of certain reactions, nitrates are converted to nitrogen gas, which reenters the atmosphere or water (Table 14–4). The process of denitrification is strictly anaerobic. Once again, microorganisms in terrestrial and aquatic environments are the agents involved (Table 14–4). The major steps in the nitrogen cycle are shown in Figure 14–15. The aerobic and anaerobic phases of the cycle also are indicated.

The Phosphorus Cycle

Phosphorus is essential for the growth and development of all forms of life. Without phosphorus there could be no organic phosphorus-containing compounds such as adenosine triphosphate (ATP), deoxyribonucleic acid (DNA), and ribonucleic acid (RNA). [See ATP, DNA, and RNA in Appendix A.]

Producer organisms acquire phosphorus in the form of inorganic phosphate (PO_4^{3-}) and convert it into organic phosphates important in the metabolism of carbohydrates, fats, and nucleic acids. Lower animals obtain their phosphorus as inorganic phosphate in water or as inorganic and organic phosphate compounds in the food they consume.

The **phosphorus cycle** (Figure 14–16) is not as completely balanced as the others described. As water runs over rocks, gradually wearing away their surfaces, various minerals, including phosphates, are carried as sediments to the bottom of the sea faster than they can be returned to land. Sea birds have an

TABLE 14–5 Microorganisms that Can Assimilate Nitrate

Bacteria	*Molds*	*Yeasts*
Aeromonas	*Alternaria*	*Candida*
Bacillus	*Aspergillus*	*Hansenula*
Clostridium	*Fusarium*	*Rhodotorula*
Nocardia	*Penicillium*	*Trichospora*

Figure 14–15 The major steps in the nitrogen cycle: ammonification (NH_3 or NH_4^+ formation), nitrification, denitrification, and nitrogen fixation.

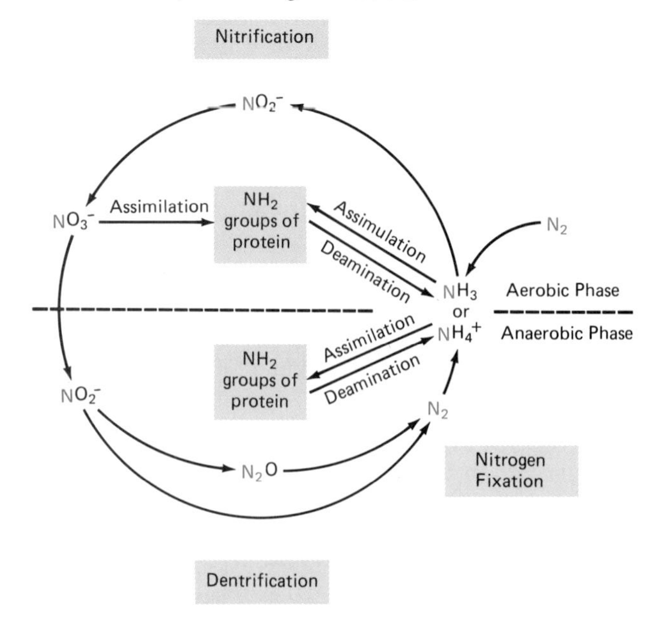

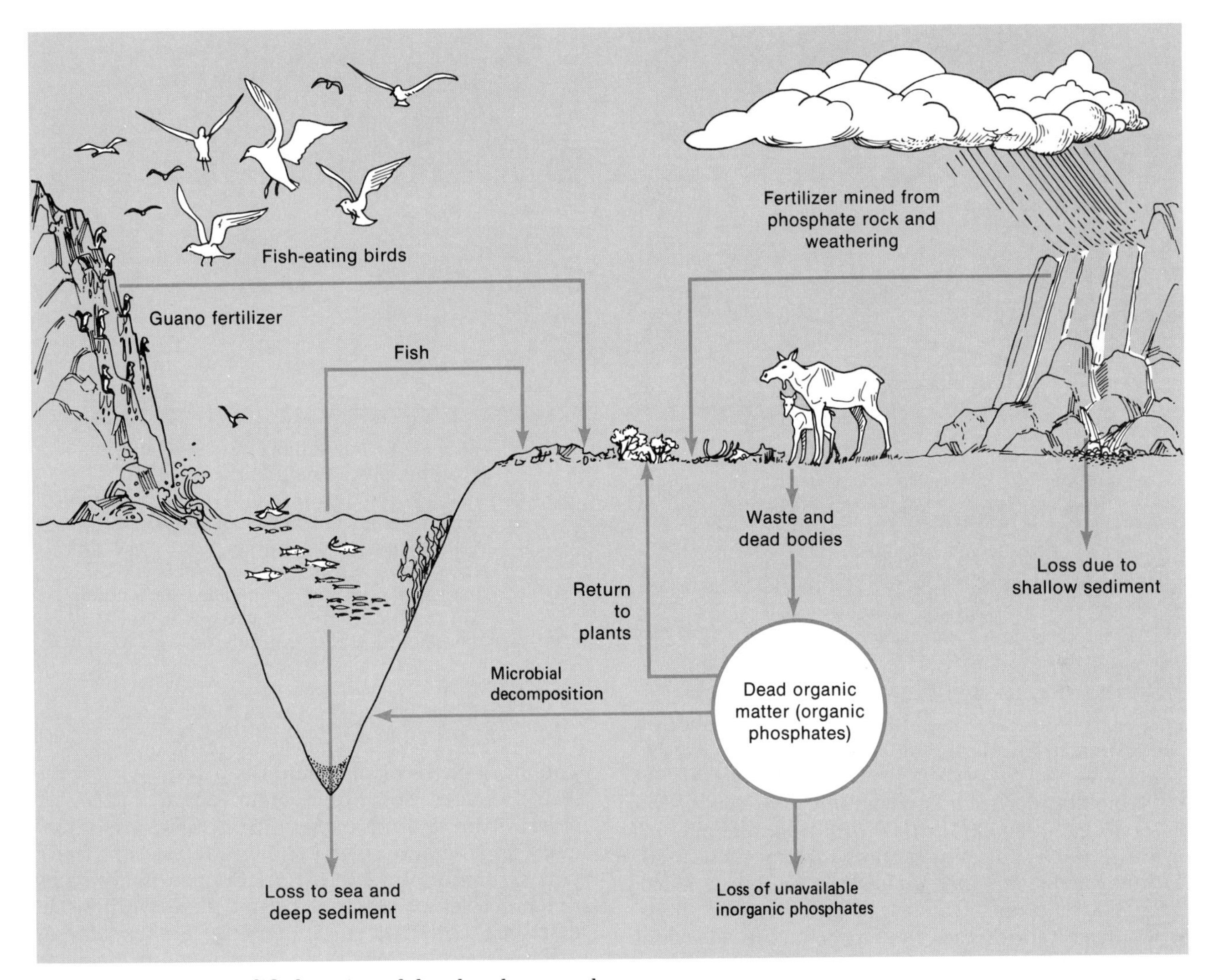

Figure 14–16 A simplified version of the phosphorus cycle.

important role in this process, since they return phosphorus to the cycle in the form of phosphate-rich droppings. Fish also recover phosphates from the sea and thereby serve as a source of the mineral.

Under natural conditions, much less phosphorus than nitrogen is available to organisms. However, various human activities have substantially increased the concentration of this mineral. Phosphates in fertilizers, detergents, and sewage are being added to ocean sediments faster than phosphates can be recycled in the aquatic ecosystem. One outcome of this process has been a dramatic increase in algal and cyanobacterial populations (Color Photograph 7a), for which phosphorus had been a limiting nutrient. This proliferation of microorganisms, in turn, has brought about additional changes in the ecology of aquatic environments.

The Sulfur Cycle

Sulfur is another essential component of proteins and is therefore required by all life forms. In nature, sulfur exists in the form of elemental sulfur (S^0) and in oxidation states including hydrogen sulfide (H_2S), sulfite (SO_2^{2-}), and sulfate (SO_4^{2-}). Hydrogen sulfide can also be produced by decomposition of protein by a wide variety of other microorganisms. Species of *Salmonella* (sal-moan-EL-la) and *Proteus* (PRO-tee-us) are noted for this ability, and it is used in their identification. Natural deposits of elemental sulfur are due to volcanic action and microbial activity.

The sulfur cycle (Figure 14–17) bears several similarities to the nitrogen cycle and includes the following sequence of events. Animal and plant protein is first broken down into its constituent amino acids by the protein digesting (proteolytic) enzymes excreted by a variety of organisms. Two of these amino acids, cysteine and methionine, contain sulfur. **[See Appendix A for a description of amino acids.]** These sulfur-containing amino acids may be used directly, or the sulfur from the compounds can be completely reduced and converted to hydrogen sulfide (H_2S) by a process similar to ammonification. This breakdown process is anaerobic and occurs in sewage, polluted waters, fresh and marine water

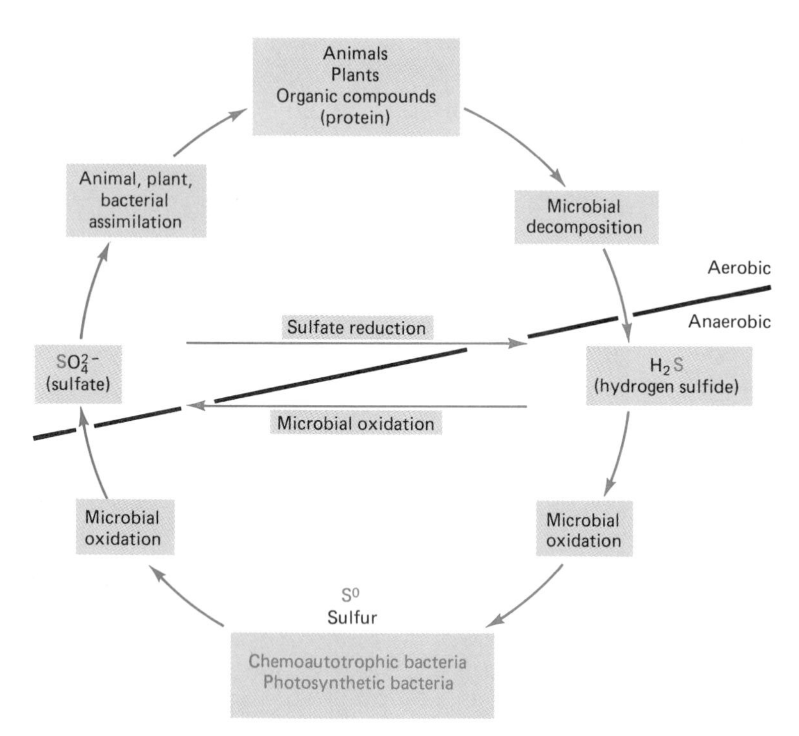

Figure 14–17 Sulfur cycle. Several forms of sulfur exist in nature; the three forms that are of practical significance are sulfate (SO_4^{2-}), sulfide (S^{2-}) and elemental sulfur S^0. Only the biological aspects of the cycle are shown. The burning of fossil fuels and other industrial processes have added substantially to the concentration of gaseous sulfur compounds in the atmosphere.

and muds, and the rumen (the large first chamber of the stomach) of ruminant animals such as cattle, goats, and sheep. The H_2S formed is highly toxic to most biological systems. In addition, H_2S reacts with several heavy metals such as iron, lead, and zinc to form insoluble precipitates, thus making these metals unavailable in their usable soluble forms. Since iron is a required element for all organisms, high H_2S concentrations can limit the amounts of available iron.

Under aerobic conditions (Figure 14–17), H_2S is first oxidized to sulfur (S^0) and then oxidized to its most usable form, (SO_4^{2-}), by chemoautotrophic bacteria. (The process is similar to nitrification.) Under anaerobic conditions (Figure 14–17), the oxidation of H_2S to sulfate SO_4^{2-} mainly is performed by two major groups of bacteria. These include the nonphotosynthetic chemoautotrophs of the genera *Thiobacillus, Thiomicrospira* (thigh-oh-my-kroh-SPY-rah), *Sulfolobus,* and *Beggiatoa* (beg-ee-ah-TOE-ah) and, in the presence of light, by the photosynthetic autotrophs, the green and purple sulfur bacteria, and certain cyanobacteria. This process occurs in fresh and marine water and muds, sewage, bogs, coal mine drainage, sulfur springs, and bodies of stagnant water. The oxidized sulfur is assimilated by most forms of life and incorporated into the sulfur-containing amino acids.

In this cycle, sulfate and S^0 are reduced to H_2S under anaerobic conditions by a small group of anaerobic bacteria that are able to use sulfate as an electron acceptor. These organisms include species of the genus *Desulfovibrio* (dee-sul-foh-VIB-ree-oh) and some species of *Desulfuromonas* (dee-sul-fur-oh-MOH-nas) that can reduce S^0 to H_2S.

The Oxygen Cycle

The atmosphere of our planet is a reservoir of molecular oxygen. Some microorganisms and all higher forms of life require oxygen for cellular respiration and the generation of energy by the electron transport system. [See Chapter 5.] Oxygen is cycled by the processes of respiration and photosynthesis as described in Chapter 5. During photosynthesis, when water functions as an electron donor, molecular oxygen (O_2) is released and hydrogen and electrons are transferred to form reduced coenzymes for use in biosynthetic reactions. The oxygen released by the splitting of water in photosynthesis is the major source of O_2 upon which aerobic respiration depends. During respiration, oxygen is recycled with the formation of water and carbon dioxide. The current level of O_2 in the atmosphere is kept in balance by the relative global rates of chemical reactions in the upper atmosphere, the burning of fossil fuels, photosynthesis, and aerobic respiration. From Figure 14–11 it can be seen that the supply of oxygen is closely linked to the carbon cycle.

Additional Reactions of Ecological Concern

There is a common belief among microbiologists that given sufficient time, soil microorganisms can decompose any organic molecule. This concept has been called *microbial infallibility* by the noted soil microbiologist Martin Alexander. Alexander admits, however, that certain materials are quite resistant to decomposition. These substances include certain pesticides, unique detergents, various plastics, and lignin.

Mineralization, or the complete biodegradation of organic compounds in waters or soils, is almost always a consequence of microbial activity. As microorganisms convert the organic substrate to inorganic products, the responsible microbial populations use some of the carbon in the substrate and convert it to cell components. At the same time, energy is released, and the populations increase in number and biomass as they utilize some of the carbon and acquire energy for biosynthesis. As a consequence, mineralization is typically a growth-linked process. Destruction of the harmful effects of toxic substances, *detoxication,* is a common outcome of mineralization processes except when one of the products itself is of environmental concern, as is the case with nitrate in certain waters or sulfide under anaerobic conditions.

The biodegradation of synthetic organic molecules polluting terrestrial and aquatic ecosystems is of great concern. It is apparent from several studies that an enormous number of synthetic chemicals used in industry, farming, and the home are deliberately or accidentally released into waters and soils and thereby present a major threat to the environment. A particularly striking effect of such toxicity on the protozoon *Tetrahymena pyriformis* (tet-rah-high-MEN-ah pie-ree-FOR-mis) is shown in Figure 14–18. Normal cells have the typical pear shape and obvious cilia. Exposure to an environmental chemical pollutant, phenol, sometimes used as a disinfectant, produces the damaging effects shown in Figures 14–18b and c.

With many chemicals, a microbial conversion quite different from mineralization takes place. Although compounds that include the pesticides DDT, aldrin-heptachlor, and many other chlorinated and nonchlorinated molecules are acted upon biologically, no microorganisms able to use these chemicals as sources of nutrients or energy are present. Microbial populations presumably are growing on other materials while performing a type of bioconversion referred to by some soil scientists as *cometabolism.*

During natural cycles, some reactions lead to storage deposits of particular materials, for example, limestone storage of carbon, sulfur deposits in or near bacterial cells, and rare nitrate deposits probably produced by nitrification. These types of reactions can be called **bioconcentrations.**

The next portion of this chapter will describe additional examples of bioconversion and bioconcentration, indicating both beneficial and harmful reactions of ecological concern.

Bioconversion

As used here, the term "bioconversion" refers to the biological or enzymatic production of a useful product such as methane, the degradation of petroleum

Figure 14–18 Dangerous environmental pollutants resulting from various industrial processes are being discovered. One of these is phenol. This series of micrographs shows the effect of this organic pollutant on ciliated protozoa, which are important links in the food chain and occur worldwide in a variety of freshwater habitats. (a) *Tetrahymena pyriformis,* control organisms. (b) After a three-minute exposure to phenol (100 mg/liter). (c) After a 60-minute exposure. *[T. W. Schultz, and J. N. Dumont,* J. Protozool. ***24:****164–172, 1977.]*

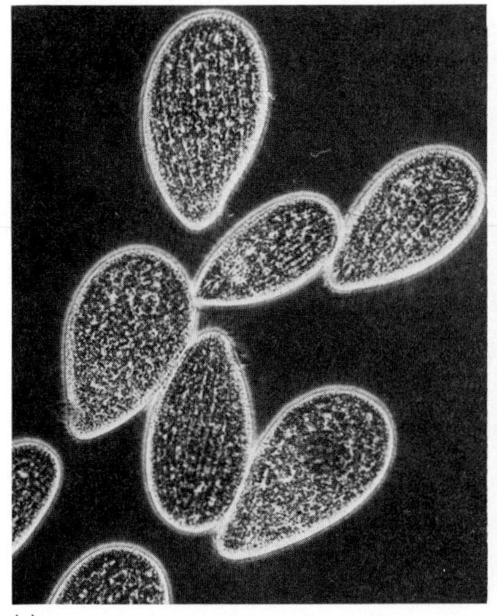
(a)

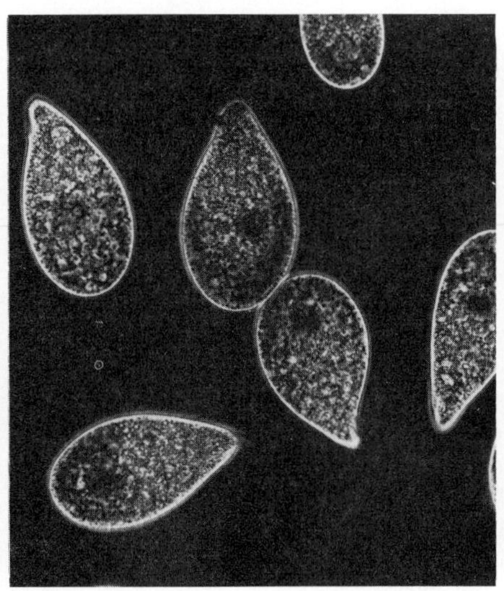
(b)

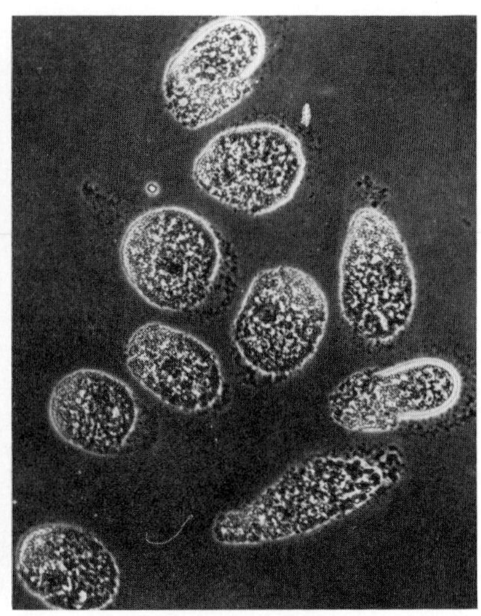
(c)

products such as pollutants, and the detoxification of organophosphate pesticides. It should be noted that the microbial processes associated with bioconversion may lead to the formation of new toxic substances (toxicants) or persistent products.

METHANOGENESIS

Methane production occurs as a normal part of the carbon cycle. The worldwide concern over dwindling petroleum reserves has focused attention on methane as an important fuel. Methane is especially interesting because it is, in essence, a recyclable commodity. Various plant and animal wastes and vegetation forms have been investigated as raw materials for commercial methane production. For example, an important by-product of sewage treatment is methane. A recent research report detailed a process of cattle waste digestion that yields nearly 20 times the methane produced during normal sewage treatment. There are, however, a number of problems with the collection and distribution of sufficient quantities of biologically produced methane to make the process practical at the present time.

Methanogenesis is an anaerobic process in which methanol, organic acids, or carbon dioxide are reduced to methane. This process occurs in aquatic sediments, black muds, marshes, swamps, and the rumen of appropriate animals. The methane-producing bacteria depend upon the syntrophic production of the necessary substrates, including hydrogen, in these habitats.

In freshwater sediments, acetate is a major substrate, and nearly pure methane has been collected at a slow but consistent rate in Lake Erie. In salt marshes along the southeastern coast of the United States, hydrogen and formic acid were found to stimulate methanogenesis. Ruminants are also known to have a system in which carbon dioxide and hydrogen are the important substrates in methanogenesis, since the organic acids produced by fermentation in the rumen are absorbed by the animal for its own nutrition. Typical methanobacteria include a number of the archaeobacteria described in Chapters 2 and 13.

PETROLEUM, OR HYDROCARBON, DEGRADATION

The degradation of crude oil materials is a major ecological concern. It was estimated in 1971 that 12 million metric tons of petroleum pollute marine environments each year as a result of spills by tankers or wells or as a result of the cleaning of tankers while they return to load more cargo. Crude oil is a complex mixture of hydrocarbons containing paraffins, kerosene, octane, petroleum oils, and many other components. The fact that oil spills eventually disappear from waters, soil around refineries, leaky pipelines, and polluted beaches is clear evidence that microorganisms can decompose these hydrocarbons.

A species of *Corynebacterium* (ko-re-nee-back-TI-ree-um) has been isolated that can oxidize paraffin in a manner comparable to fatty acid catabolism to acetyl coenzyme A (CoA). *Pseudomonas putida* (soo-doh-MO-nass poo-TEE-dah) strains from terrestrial and aquatic habitats have been shown to oxidize benzene, toluene, ethylbenzene, octane, naphthalene, camphor, and salicylates as sole sources of carbon. Metabolic plasmids are associated with many of the hydrocarbon-degrading enzymes in these organisms. **[See plasmids in Chapter 6.]**

Petroleum-degrading microorganisms adsorb to oil droplets. The smaller the droplets, the greater the ability of the organisms to oxidize the substrates and reproduce. These organisms also have the ability to produce emulsifying agents that help form small droplets (Figure 14–19a). Yeasts (Figure 14–19b) such as *Candida tropicalis* (CAN-did-ah trop-ik-AIL-iss) produce cell-wall material of mannan and fatty acids, which assists in droplet formation. Most organisms excrete some form of extracellular material for this process.

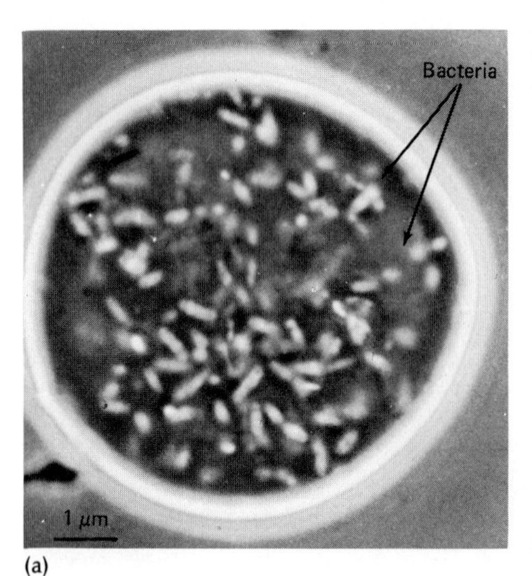

(a)

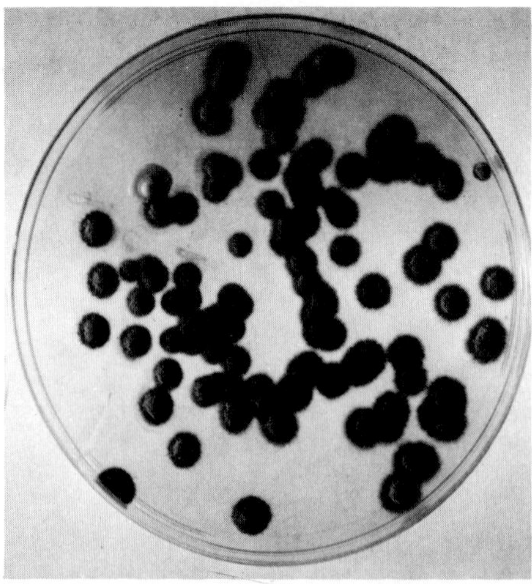
(b)

Figure 14–19 Petroleum-degrading microorganisms. (a) Bacteria in an oil globule from an Eastern Bay sediment. (b) Colonies of petroleum-degrading yeasts. *[J. D. Walker, and R. R. Colwell,* Microbial Ecol. *1:63–95, 1974.]*

Although oil-degrading microorganisms have been found in varied habitats, including arctic coastal water, their use in inoculating oil spills has proven ineffective. The major problem appears to be the low levels of nitrogen and phosphorus compounds in sea water. One solution that has been tried is to treat urea and octyl phosphate with paraffin to produce a slow-release fertilizer. This material, distributed with the organisms in powder form, appears to promote microbial action.

Although the prospect of using microorganisms to clean up oil spills is encouraging, it must be kept in mind that microorganisms can also cause spoilage and deterioration of petroleum products. Species from the bacterial genera *Pseudomonas, Chromobacterium* (kroh-moh-back-TI-ree-um), *Alcaligenes* (al-ka-LIJ-en-eez), *Mycobacterium* (my-ko-back-TIR-ee-um), and *Sarcina* (sar-SI-nah), as well as certain fungal species of *Aspergillus* and *Candida,* can decompose gasoline. In addition, the deterioration of asphalt highways and pipe coatings has been caused by species of *Mycobacterium* and *Nocardia* (no-KAR-dee-ah).

A common consequence of microbial action on a toxic chemical (toxicant) introduced into waters or soils is **detoxication.** In modifying the chemical, the detoxifying microorganisms destroy its actual or potential harmful influence on one or more susceptible animal, plant, or microbial species. Although mineralization of toxicants characteristically produces detoxication, harmful intermediates formed in the sequence of reactions may remain for some time. This occurs, for example, in the conversion of phenoxyherbicides in soil to yield compounds toxic to plants.

The pesticide parathion belongs to a large group of chemicals known as *organophosphates.* These compounds are extremely toxic to insects and all other animals because they prevent nerve cell conduction across synapses. In a study on the concept of microbial infallibility, various soil and sewage microorganisms were tested for their ability to detoxify parathion. Mixed bacterial cultures were found to be effective in preliminary studies. Some of the organisms involved were species of *Brevibacterium* (brev-ee-back-TI-ree-um), *Pseudomonas,* and *Azotomonas* (a-zoh-to-MOH-nas).

Unfortunately, the experience with another group of pesticides has not been so promising. The chlorinated hydrocarbons, which include dichlorodiphenyl trichloroethane (DDT), aldrin, dichlorodiphenyl dichloroethane (DDD), lindane, and endrin, can be "degraded" under anaerobic conditions by various facultative anaerobic bacteria and yeasts, but the end products are still extremely toxic. In one study, it was observed that DDT was dechlorinated to form DDD, another very toxic pesticide. The problems associated with the buildup of chlorinated hydrocarbons in various aquatic and terrestrial environments appear to be related to their relative invulnerability to attack by microorganisms. Organic chemicals that remain for long periods in natural ecosystems because of the inability of microorganisms to degrade them rapidly, if at all, are known as *recalcitrant molecules.* Although some of these substances are toxic, a few, such as many plastics and other synthetic products, are not hazardous but are landscape eyesores.

Certain chemicals that are themselves harmless are converted enzymatically to products that are hazardous to some form of life. Toxicants may be generated from harmless substances (by a process that is often called *activation*) in a mineralization sequence.

Microbial activation is evident in the case of mercury, another problem pollutant in terrestrial and aquatic environments. Mercury is toxic to humans and other life forms, as demonstrated by the mercury poisoning tragedy that occurred in Minamata Bay, Japan, where industrial pollution produced exceptionally high mercury levels in fish. The form of mercury was methyl mercury, which is 50 to 100 times more toxic than inorganic mercury compounds. This methylation of mercury can occur abiotically, but considerable evidence points to a significant role of many bacteria and fungi in this reaction. Some organisms, such as *Clostridium cochlearium* (klos-TRI-dee-um kok-LEER-ee-um) introduce a methyl group on the molecule cobalamin by using vitamin B_{12} and cysteine. Subsequently, the methyl group can be transferred to mercury nonenzymatically. *C. cochlearium* is anaerobic, performing methylation in aquatic bottom sediments and sludge. The ability to methylate mercury has also been shown to exist in indigenous microorganisms such as species of *Pseudomonas* on the gills and in the intestines of fish. Aerobic methylation also occurs. Soil organisms with this capacity include a species of the bacterial genus *Pseudomonas* and the yeast *Neurospora* crassa (new-RAH-spor-ah KRAS-ah). One source of mercury in soils has been the application of organomercurial fungicides. Research in Sweden since the introduction of these fungicides has shown considerable increases of methyl mercury in plant and animal tissues.

Bioconcentration

Bioconcentration is the ability of microorganisms to store materials in the environment. The storage may be within cells, as in the case of sulfur, or outside the cells, as in the case of nitrate or sulfur. The metabolic activities of sulfur oxidizers have been used to concentrate low-grade mineral ores. Sulfur oxidizers, such as *Thiobacillus,* can grow in slag from copper mines; the material remains after ore is roasted to remove most of the copper. The sulfuric acid produced by the oxidizing activity of *Thiobacillus* dissolves the copper, and copper sulfate accumulates at the base of the slag pile. A similar process is used to concentrate uranium from ores in the form of uranyl sulfate.

Bioconcentration has caused the formation of commercially attractive deposits of manganese, but also hazardous accumulations of pesticides and radioisotopes in nature.

MANGANESE DEPOSITS

Manganese, an important trace element in plant nutrition, is found widely distributed in terrestrial and aquatic environments. It is commercially important in alloys of iron, aluminum, and copper. In the soil, bacteria such as *Leptothrix (Sphaerotilus) discophorus* (lep-TOH-thricks dis-KOFF-oh-rus) oxidize manganese compounds and accumulate manganese dioxide (MnO_2) in or on their cells (Figure 14–20). This oxidation may be a primary energy source for these organisms. The accumulation depletes the manganese available to algae and plants until reducer organisms render MnO_2 deposits soluble and available once more.

In the oceans, both of these reactions have also been noted. A species of *Arthrobacter* has been found in association with ferromanganese nodules. These nodules contain as much as 63% MnO_2 deposited by the oxidative capacity of the bacteria. A species of *Bacillus* isolated from nodules and deep-sea sediments can reduce MnO_2 as a stage of its electron transport system. The reduction process must be significantly slower than that of oxidation or the nodules would not have formed to the extent that has been observed. International discussions have been held on cooperative ventures for the mining of these nodules.

Figure 14–20 *Metallogenium* species having radiating filaments encrusted with manganese oxides (MnO_2). *[E. Gregory, R. S. Perry, and J. T. Staley,* Microb. Ecol. *6:125–140, 1980.]*

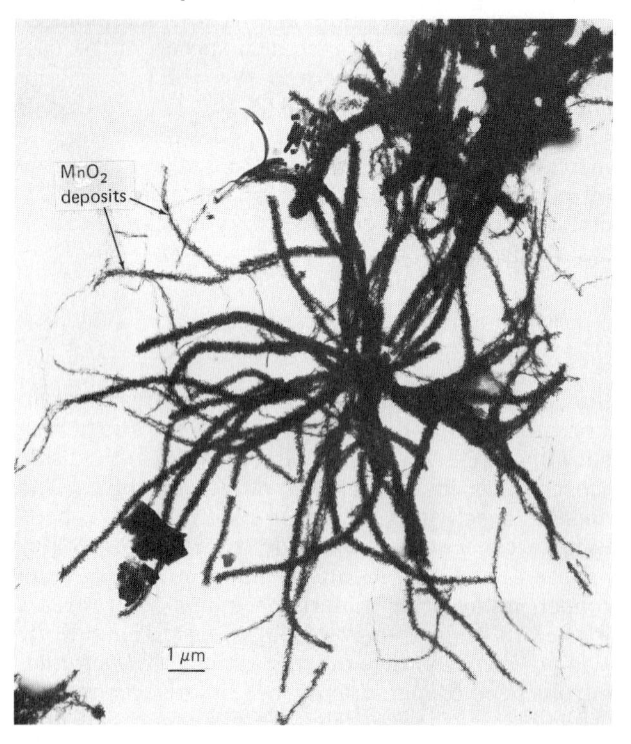

Coaction Activities

The previous portion of this chapter emphasized the interactions of microorganisms with abiotic factors in the ecosystem or reaction. The last and remaining section concerns interactions between various organisms in the ecosystem, or the ecological concept of coaction. Attention will be given to five kinds of coactional relationships: **syntrophism, competition, antagonism, predation,** and **symbiosis.**

Syntrophism

Organisms exhibiting *syntrophism* are not intimately associated with one another, but benefit from one another. For example, a major activity found in soil is the decomposition of the polysaccharide cellulose in decaying plant matter. This involves bacteria from the genera *Cytophaga* (sigh-TOH-fa-ja) and *Spirillum* (spy-RIL-lum) and a number of fungi, all of which produce the cellulose-digesting enzyme cellulase. The digestion produces the disaccharide cellobiose, which is used not only by the microorganisms that help produce it but also by others in the general area. Cellobiose is subsequently degraded to glucose, which is used by many organisms in the immediate environment. Thus, several different organisms feed together. Other examples of this basic activity include processes involved with yogurt production, waste digestion in sewage disposal, and the actions of different microorganisms in the digestive tracts of various animals.

Competition

Ecologists define "competition" as the interaction between organisms resulting from a demand for nutrients, space, and energy that are in short supply in a habitat. The species capable of using such resources more efficiently will usually eliminate its competition.

Antagonism

Some organisms can control the inhibitants of their environments by secreting toxic substances. Molds such as *Penicillium* (pen-ee-SIL-lee-um) and other an-

Microbiology Highlights

THE MOBILITY OF HAZARDOUS PRODUCTS AND WASTES

A characteristic of the 1970s in many Western industrialized nations, including the United States, was the expansion of government intervention and regulation in the areas of occupational and environmental health and consumer protection. This expansion was an outcome of a demand by the majority of such countries' populations for better protection of the environment and the health and safety of workers and consumers.

In 1976 the U.S. Council on Environmental Quality reported an incident involving the chlorinated hydrocarbon kepone. A Virginia company had been making kepone for 16 months, and cases of poisoning in their personnel had been linked to this substance. When the U.S. Cancer Institute reported animal studies showing kepone to be carcinogenic, the company was required to stop making the pesticide. Evaluation of the environment, including the James River, showed that water and shellfish contained traces of the chemical as far as 40 miles away. It is suspected that the kepone accumulated in microorganisms upon which the shellfish feed.

More definitive examples of the accumulation of toxic substances have been reported for radioisotopes, which have also been shown to be carcinogenic. In "The Myth of the Peaceful Atom," by Curtis and Hogan, two pertinent examples are presented.* The Savannah River Nuclear Power Plant in Aiken, South Carolina, released "insignificant" amounts of radioactivity in its cooling water. An examination of algae in the water, however, showed a concentration of up to 6,000 times that in the water itself. These algae serve as food for the bluegill fish; the bones of this fish had a concentration 8,200 times that in the water.

The Hanford Nuclear Power Plant in Hanford, Washington used the Columbia River for cooling water. As suspected, the water there, too, had insignificant levels of radioactivity and was considered safe. However, algae and other microorganisms in the water concentrated this safe level of radioactivity as much as 2,000 times. Fish that fed on the plankton had levels 15,000 times that in the water. Ducks that fed on vegetation in the water had levels 40,000 times that in the water, and their egg yolks had concentrations 1 million times that of the safe water. Clearly, one aftermath of the failure of the Three Mile Island nuclear power plant in 1979 that will be closely followed is the bioconcentration of radioactivity in the plant and animal life in and around the Susquehanna River.

These few examples of bioconversion and bioconcentration of toxic materials in nature should clearly serve to emphasize the potential and real dangers in using aquatic and terrestrial habitats for the dumping of hazardous products and waste materials. While some countries have established an extensive body of law to protect the public from occupational, environmental, and consumer hazards, other countries have not done so. For example, in Mexico an estimated 16.5 metric tons of hazardous wastes are generated each year. As of 1985, the country still had no law for the regulation of hazardous waste disposal. This major health problem is further compounded by the fact that hazardous wastes, pesticides, and drugs formerly shipped *within* countries, to locations of low public resistance or feeling against dumping, now also cross national borders and are dumped in other countries or in open seas.

* R. Curtis and E. Hogan, "The Myth of the Peaceful Atom," in C. E. Johnson, *Eco-crisis* (New York: John Wiley & Sons, 1970).

tibiotic-producing organisms in the soil may be able to control the growth of bacteria that are sensitive to their secretions. Microorganisms produce various substances such as bacteriocins (described in Chapter 6) that have an inhibitory effect on other organisms (Figure 14–21).

Predation

Predation plays an important role in controlling the populations of higher forms of life, as well as those of microorganisms. The *predator* is the species that ingests other species, the *prey.* The paths of two dif-

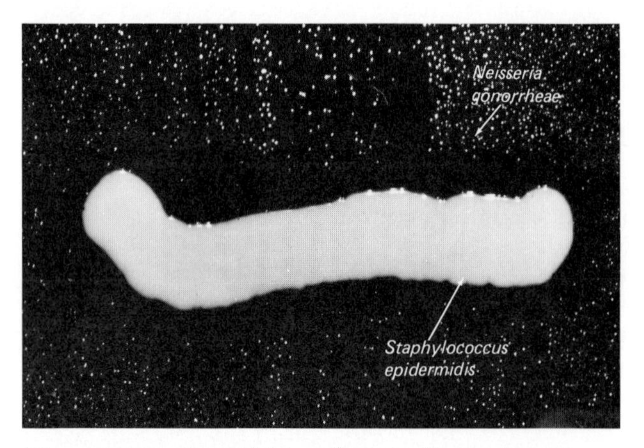

Figure 14–21 Bacterial interference. Various bacteria produce substances that can inhibit the growth of other bacteria in their vicinity. Here the growth of *Neisseria gonorrheae* is inhibited by another bacterium, *Staphylococcus epidermidis. [Reproduced with the permission of the National Research Council of Canada, from R. Shtibel,* Can. J. Microbiol. ***22****:1430–1436, 1976.]*

ferent forms of life cross and interact to the benefit of one and to detriment of the other. The predator is usually larger than the prey, but as Figure 14–22 shows, the smaller protozoon, *Didinium* (dye-din-EE-um), is able to ingest the larger *Paramecium* (par-a-ME-see-um) with little trouble. Microbial predators may show little or no preference for one prey or another.

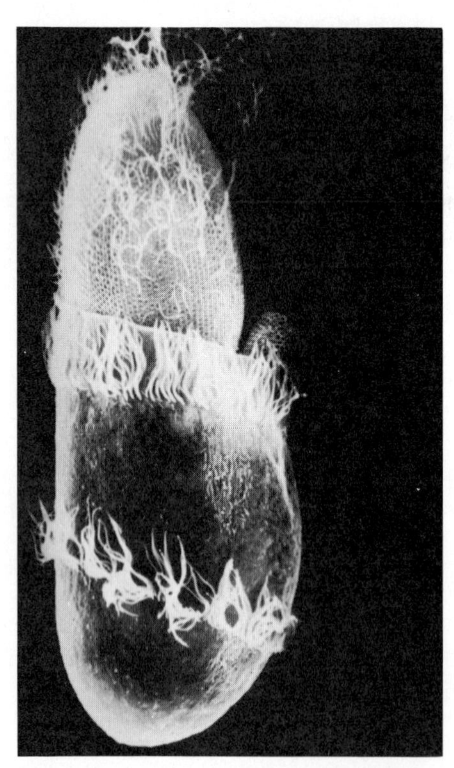

Figure 14–22 The ingestion of the protozoon *Paramecium* by another ciliated form, *Didinium*. The half-swallowed *Paramecium* is shown in the feeding apparatus of *Didinium. [H. Wessenberg, and G. Antipa,* J. Protozool. ***17****:250–270, 1970.]*

Symbiosis

In *symbiosis* two different forms of life coexist in an intimate ecological relationship. This relationship, which may be of long or short duration, requires close physical contact. Symbiotic associations in which one partner is actually inside a cell or tissue of the other partner are called **endosymbiotic** (Figure 14–23 and Color Photograph 68); those in which one member is external to the other are called **ectosymbiotic** (Figure 14–24). The participants in these associations are called *endosymbionts* and *ectosymbionts,* respectively.

Symbiosis is classified as of one of three forms, depending on the benefits it provides the symbionts. In **mutualism,** both members benefit from the association; in **commensalism,** only one organism benefits while the other partner neither benefits nor is harmed; and in **parasitism,** one organism benefits at the expense of the other.

MUTUALISM

Mutualism, like the other forms of symbiosis, occurs widely among most of the principal groups of animals, plants, and microorganisms and includes an amazing range of physiological and behavioral adjustments. Mutualism is usually essential to the survival of the organisms involved because of the dependence they have developed for one another. In many situations, one partner cannot survive without the other.

INTERMICROBIAL MUTUALISM. Among microorganisms, the lichens represent the ultimate mutualistic association. Lichens consist of highly specialized fungi that house green algae or cyanobacteria

Figure 14–23 The presence of intracellular symbionts (endosymbionts, arrow) in the small nucleus (micronucleus) of the ciliated protozoon *Paramecium caudatum. [Courtesy Dr. Hans-Dieter Görtz, Zoological Institute, University of Münster.]*

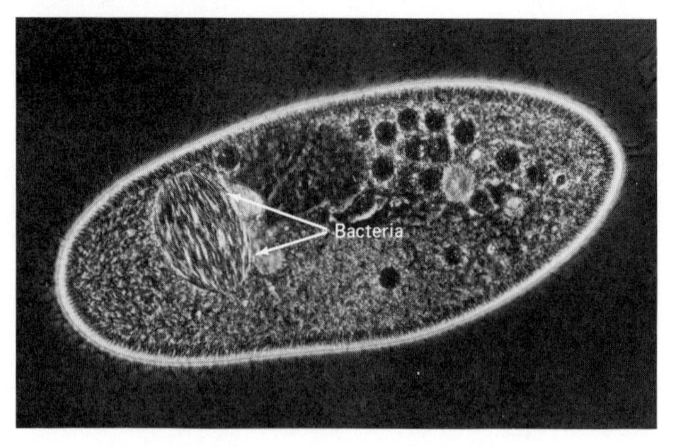

Figure 14–24 Symbiotic relationships require close contact. In the ectosymbiotic association shown, physical attachment of the bacteria (arrows) to the plant part is essential for the microorganisms to infect and cause developmental changes. *[L. D. Spiess, et al.,* Amer. J. Bot. ***64****:1200–1208, 1977.]*

among their hyphae. The fungus absorbs moisture, minerals, and carbon dioxide from the environment. The algae or cyanobacteria then photosynthesize and supply themselves and the partner fungus with organic nutrients and oxygen (Color Photographs 44 and 45).

MICROBE–PLANT MUTUALISM. Some of the most advanced and ecologically important examples of mutualism exist among plants. Nitrogen fixation, described earlier, is the conversion of gaseous nitrogen (N_2) into a form that can be used by plants or other soil or water organisms. In nature this is usually accomplished by microorganisms. Nitrogen-fixing bacteria of the genus *Rhizobium* live in specially formed nodules in the roots of legumes such as alfalfa, clover, and soybeans (Figure 14–25). Here, in exchange for the protection and a constant environment provided by the plant, these bacteria provide the legume with substantial amounts of nitrogen essential to its growth (Figure 14–26).

Figure 14–25 Nitrogen fixation. (a) Nodules (small tumors) that house nitrogen-fixing bacteria. *[Courtesy USDA.]* (b) A view through a nodule showing the bacterial endosymbionts within large cells of the nodule. *[J. Kummerow, et al.,* Amer. J. Bot. ***65****:63–69, 1978.]*

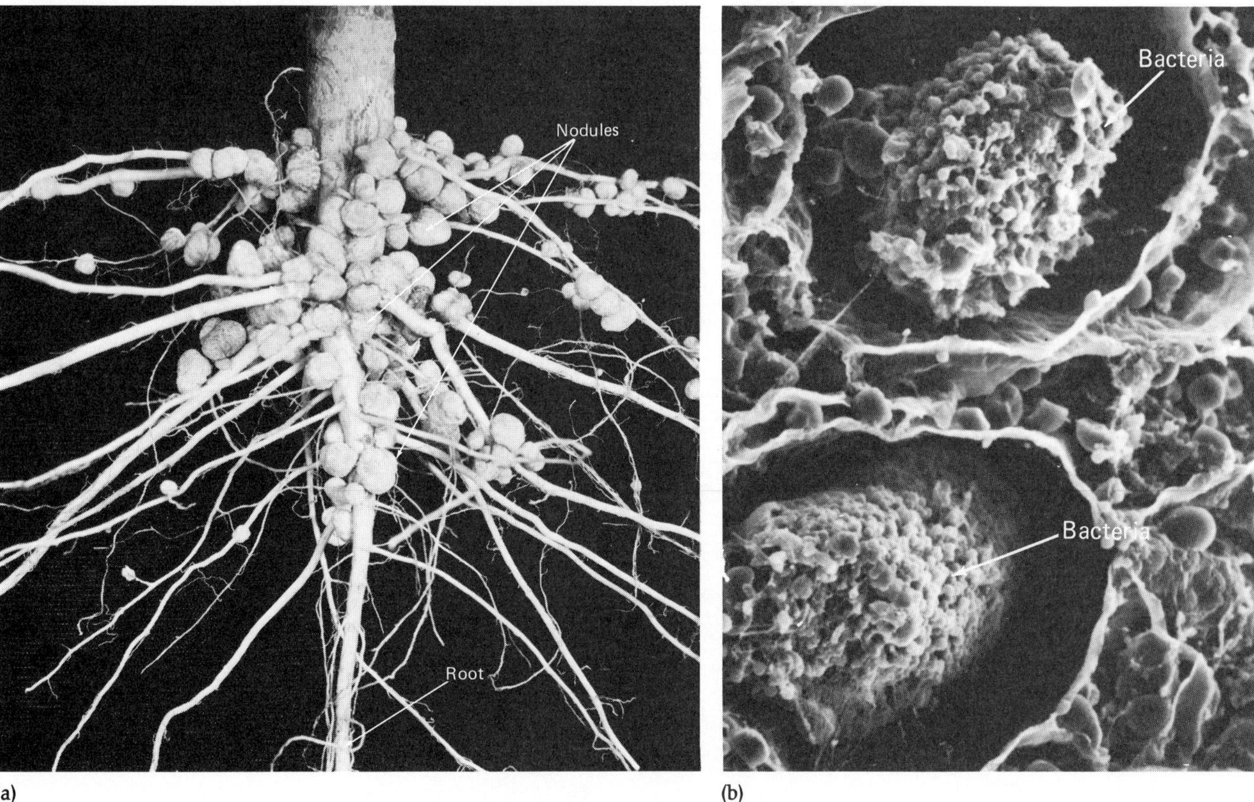

Figure 14–26 The effects of symbiotic nitrogen fixation. The alfalfa plant on the left does not have nodules and is growing in nitrogen-poor soil. The plant on the right is nodulated and is flourishing despite the nitrogen-poor soil. *[Courtesy Nitagin Company.]*

Another example of microbe–plant mutualism is the mycorrhyzae; fungi *(myco)* in a symbiotic relationship with plant roots *(rhyza)*. The fungal mycelium attaches to the small plant roots and serves to greatly expand the ability of the plant to absorb nutrients from the environment. In return, the fungus is benefited by having a ready source of organic nutrients. Additional benefits to the plant appear to include (1) protection from root pathogens, (2) detoxification of soil substances, and (3) production of growth-stimulating hormones.

MICROBE–ANIMAL MUTUALISM. Mutualistic partnerships between microorganisms and animals are quite common. One of the most famous pairs is the termite and its intestinal protozoan populations, discussed in Chapter 9. Another interesting and unusual example is the fungus-garden cultivating ant (Figure 14–27). Several ant species have mutually beneficial relationships with fungi that parallel those between humans and crop plants. Depending on the ant species, the fungi are provided with bits of fresh leaves, plant debris (mulch), or ant excrement as nourishment. Workers among the ants tend the fungus beds inside the ant colony and remove certain fungi that the colony does not eat (that is, the weeds). The fungi may be carried to a new ant colony by the young queen, sometimes in a special body pouch and usually as fungus (hyphal) fragments. The participating fungi that have been identified are mostly mushrooms (or related forms) and yeast. The arrangement is clearly beneficial to both parties. The hyphae or spores provide food for the ants, while the fungi receive nutrients and are freed from fungal competitors. The growing fungi also function to keep a constant level of moisture in the colony.

Other insects, such as termites and beetles, are also known to cultivate particular fungi as foods. The insects involved usually construct their nests in decaying plant materials, which are invariably inhabited by fungi. Probably the insects originally fed indiscriminately on fungi, but the nutritional advantages of utilizing certain fungi resulted in the evolution of lines that fed on and maintained only one kind of fungus.

Another interesting mutualistic relationship is the one between species of luminescent bacteria such as *Photobacterium fischeri* (foto-back-TIR-ee-um fish-ER-eye) (Figure 14–28c) and various ocean fish. Much of the light in ocean depths originates in bacteria specially nurtured in fish structures called *light organs*. For example, the flashlight fish (Figures 14–

Figure 14–27 Fungus-garden cultivating ants. The queen ant of a young *Atta sexdens* colony on her fungus garden. The smaller workers that constitute the queen's brood are also evident. *[N. A. Weber,* Science ***153**:587–604, 1966.]*

(a)

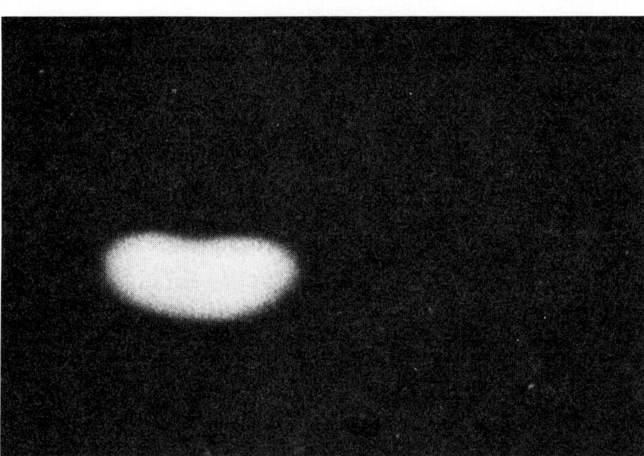
(b)

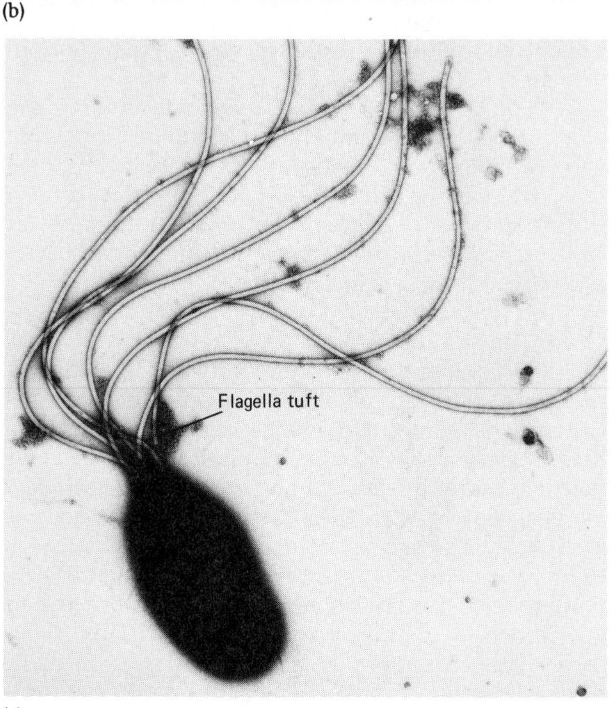

(c)

Figure 14–28 Luminous bacteria and the flashlight fish, an interesting symbiotic association. The flashlight fish, *Photoblepharon palpebratus,* has a microbiological light-generating organ that can be used for several functions. It can frighten off a would-be predator or assist the flashlight fish in capturing prey and communicating with members of its own species. (a) The location of the luminous region. (b) The light emitted by the fish's luminous organ at night. *[J. G. Morin, et al.,* Science ***190:****74–76, 1975.]* (c) *Photobacterium fischeri,* one of several species known to colonize the luminous organs of fish and other aquatic life. Note the tuft of flagella. (Refer to Color Photograph 69 for a demonstration of the light emitted by these bacteria in culture.) *[J. L. Reichelt, and P. Bauman,* Arch. Microbiol. ***94:****283, 1973.]*

28a and b) swims above reefs in the Indian Ocean and Red Sea, shining its beam of light much like a car on a dark country road. These luminous organs are used by fish to attract and capture prey, as well as to communicate with other members of its species.

Why this type of mutualistic relationship exists is unknown. Some investigators believe that it is a matter of nutrient exchange. The bacteria obtain glucose from the fish; the fish utilize pyruvic acid excreted by the luminous bacteria. What *is* known is that the light organs are surprisingly specific as to which bacteria they will host. The light-producing reaction in these bacteria (Color Photograph 69) is similar to that in fireflies. An enzyme called *luciferase* combines hydrogen from a component in the electron transport system of the bacteria with oxygen to make water and transform an aldehyde compound into an acid. The energy released by this reaction is given off as light.

Rumen Symbiosis. Rumen symbiosis is a mutualistic relationship that involves a higher form of animal life. The ruminants are a group of plant-eating mammals including cows, goats, camels, and sheep that have a special structure known as the *rumen* in their digestive tracts (Figure 14–29). Within this structure, microbial digestion of cellulose and other plant polysaccharides takes place. Like other mammals, ruminants are incapable of manufacturing the enzymes necessary to digest cellulose. The rumen, which actually comprises the first two of four stomachs, serves as large incubation chambers in which anaerobic bacteria and protozoa break down cellulose. The types of reactions associated with a typical ruminant are shown in Figure 14–29. The ingested grasses and leafy plant material are mixed with large amounts of saliva and then moved into the rumen. There cellulolytic bacteria such as a *Bacteroides* (back-teh-ROY-deez) sp. and ciliated protozoa hydrolyze cellulose to the disaccharide cellobiose and the monosaccharide glucose. Microbial fermentation of these sugars by various bacterial species produces acetic, butyric, and propionic

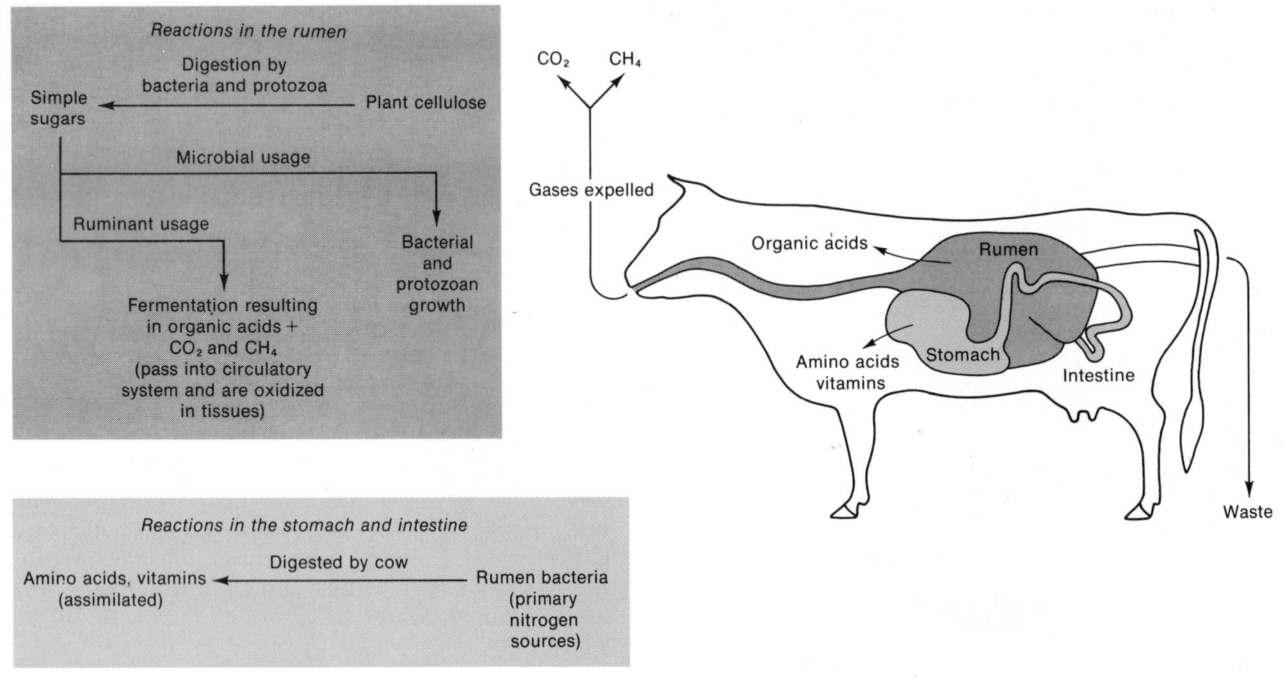

Figure 14–29 Ruminants, like other mammals, cannot digest the cellulose of the plants they ingest because of their inability to make cellulase. With the mutualistic relationship they have with microorganisms, ruminants such as cows can eat and make use of cellulose, their major source of carbon. The digestive tract of a typical ruminant and the microbial reactions and processes that occur in specific locations are shown in this diagram.

acids, as well as the gases carbon dioxide and methane. The ruminant absorbs and uses these organic acids as its main source of energy.

Figure 14–30 Potato spindle tuber disease, one effect of viroid infection. Healthy potatoes are shown at the top, and diseased specimens at the bottom. *[J. S. Semancik, et al.,* Virology ***52**:292–294, 1973.]*

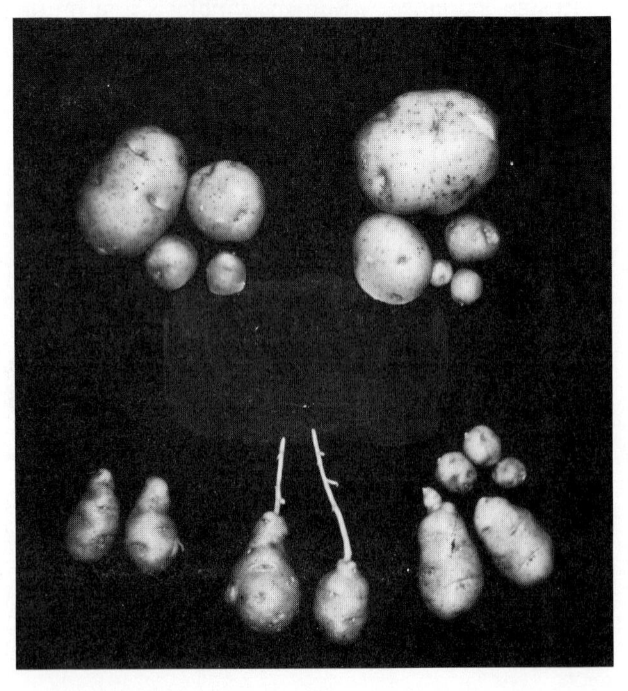

The microorganisms in the rumen also provide other specific functions, such as the synthesis of amino acids and vitamins. Some microorganisms leave the rumen and are digested in other parts of the gastrointestinal system in order to serve as a major supply of proteins and vitamins for the ruminant. This is particularly important for the nutrition of ruminants because grasses are deficient in protein.

The biochemical reactions occurring among the participants in the rumen are complex, involving vast numbers and types of microorganisms. We have emphasized the bacteria, but the protozoa in the rumen are also important. They, too, serve as sources of protein and appear to control the bacterial populations within the system.

COMMENSALISM

The literal meaning of "commensalism" is "eating at the same table." Commensalism is an association between two organisms in which one partner is benefited and the other partner is neither benefited nor harmed. Many intestinal bacteria such as *Escherichia coli* (esh-er-IK-ee-ah KOH-lee) are normally commensals, as are many intestinal protozoa such as *Entamoeba coli* (en-tah-MEE-bah KOH-lee) and trichomonads. It is not always easy to distinguish between a commensal organism and a parasite. This is because many commensals living harmlessly on or in body surfaces also have a capacity for disease production.

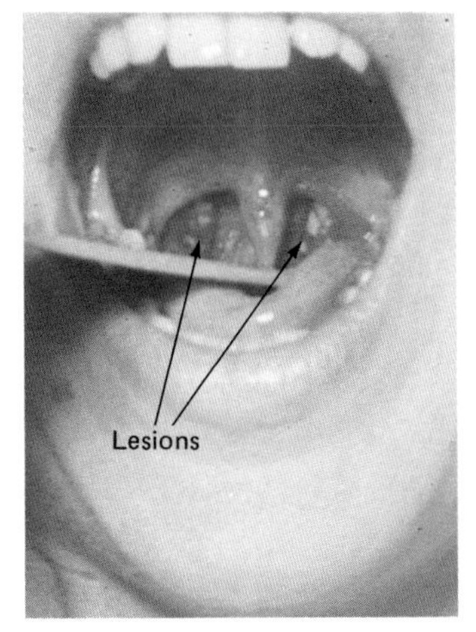

[1] Herpes infection.

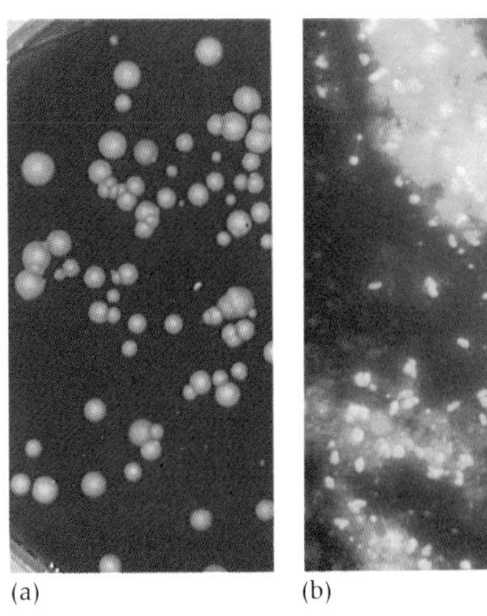

[2] Legionnaire's disease agent.

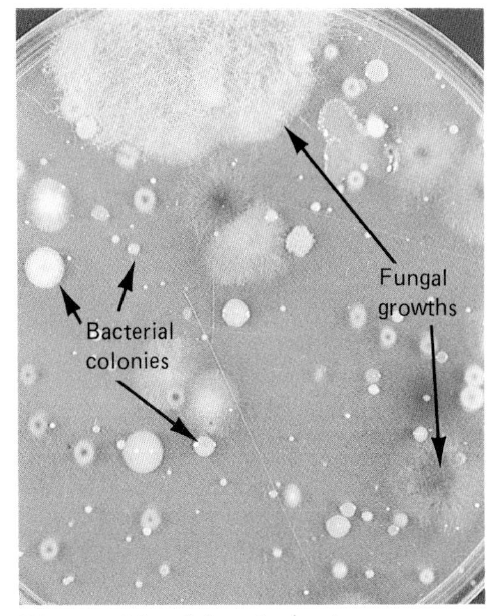

[3] Bacterial and fungal growth.

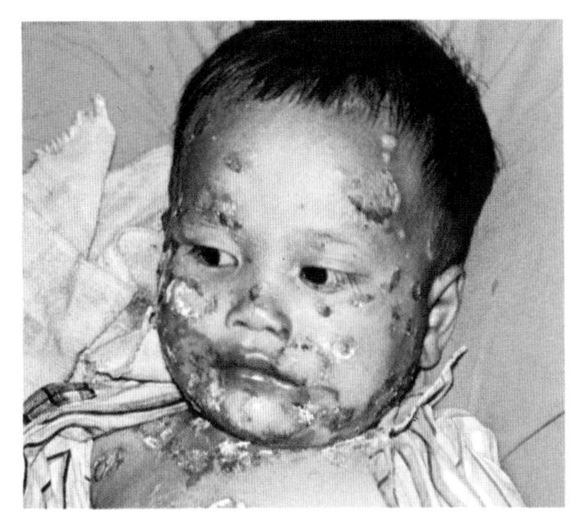

[4] Impetigo.

[5] Common mushroom.

[6] Brown rot of peach.

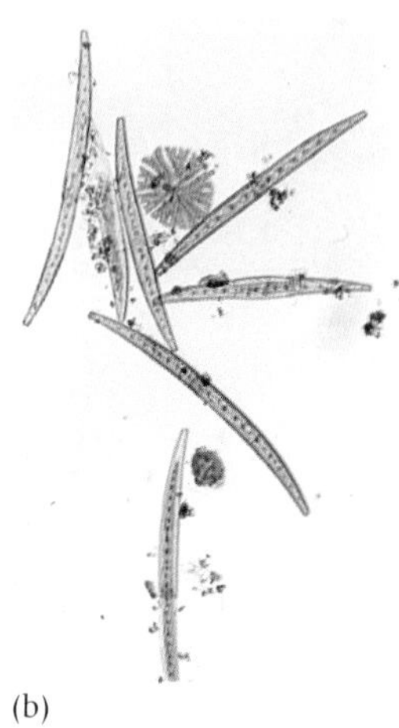

[7] Algae and their effects.

Detailed Color Atlas Legends follow Color Atlas

[8] Animal worm infections.

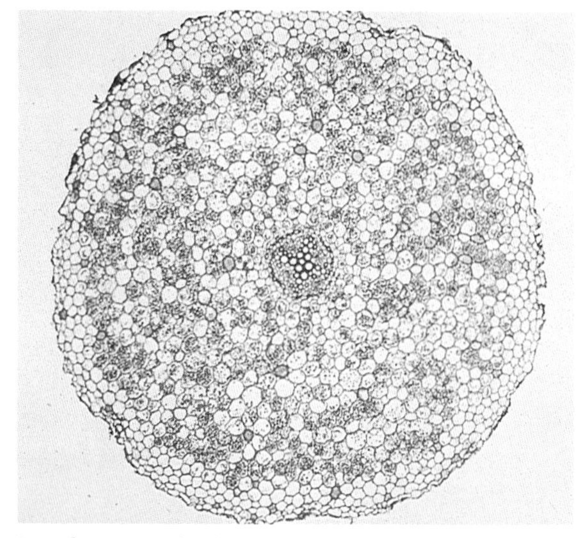

[9] Plant organization.

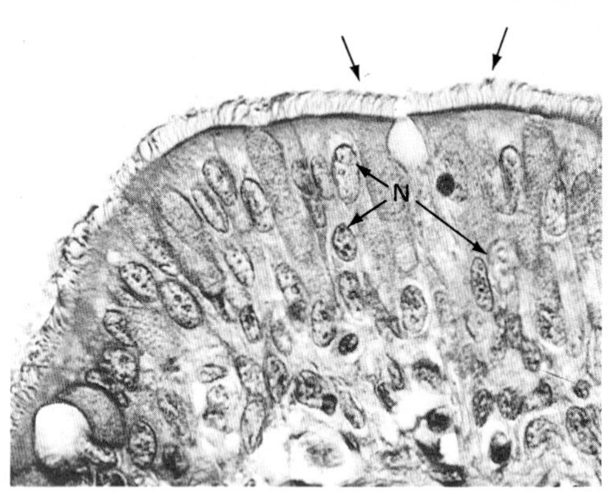

[10] Animal organization.

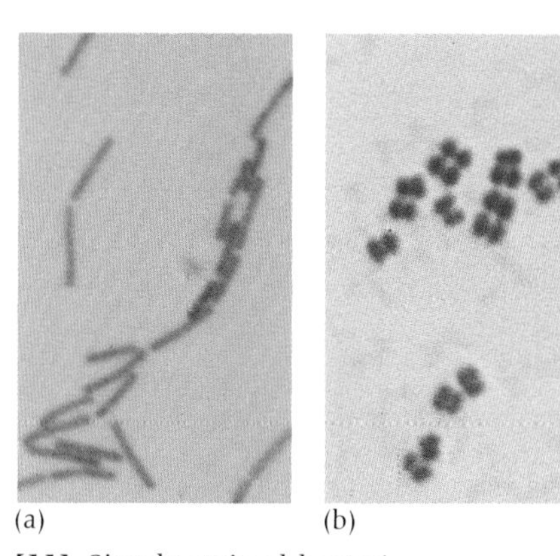
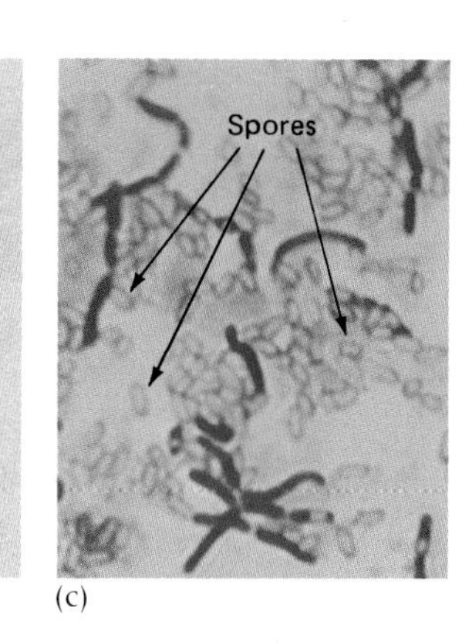

(a) (b) (c)

[11] Simple stained bacteria.

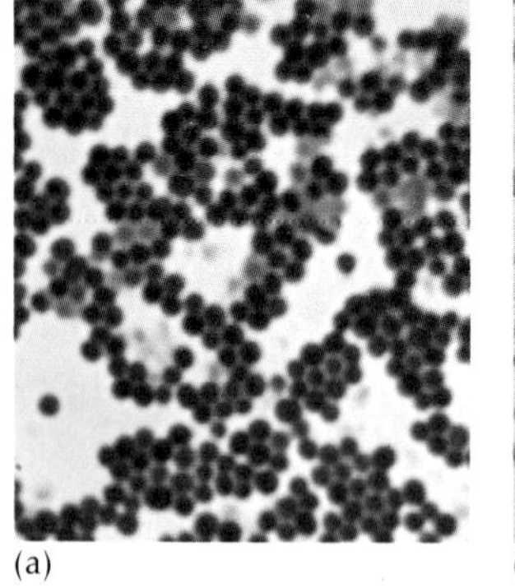
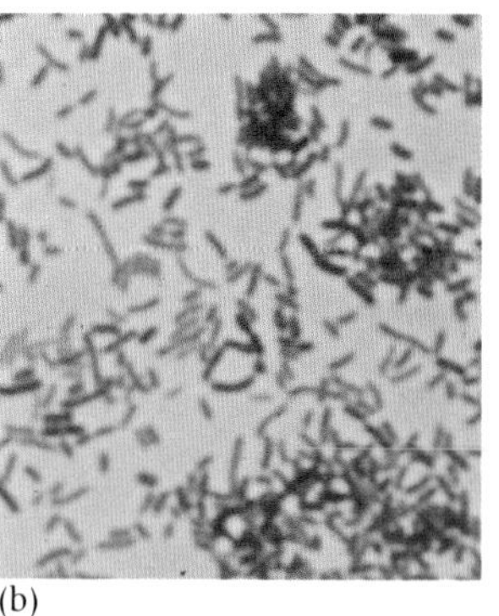

(a) (b)

[12] Gram stain reactions.

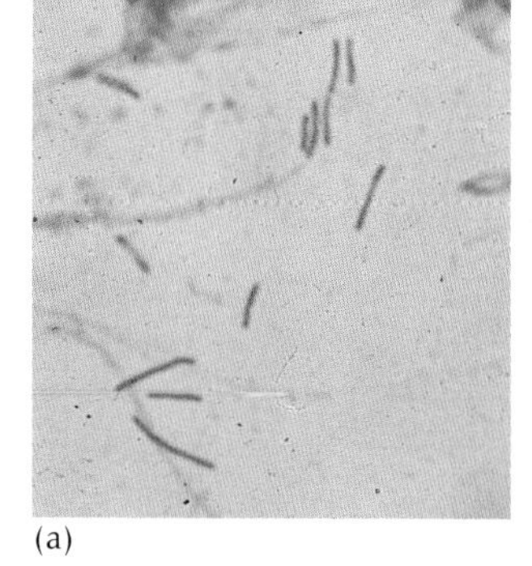
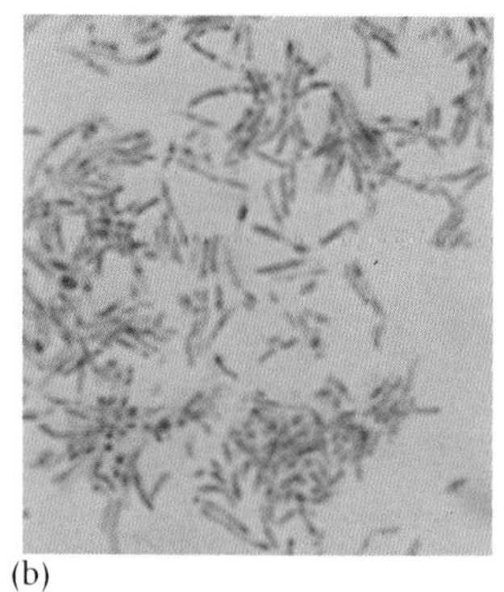

(a) (b)

[13] Acid-fast reactions.

Detailed Color Atlas Legends follow Color Atlas

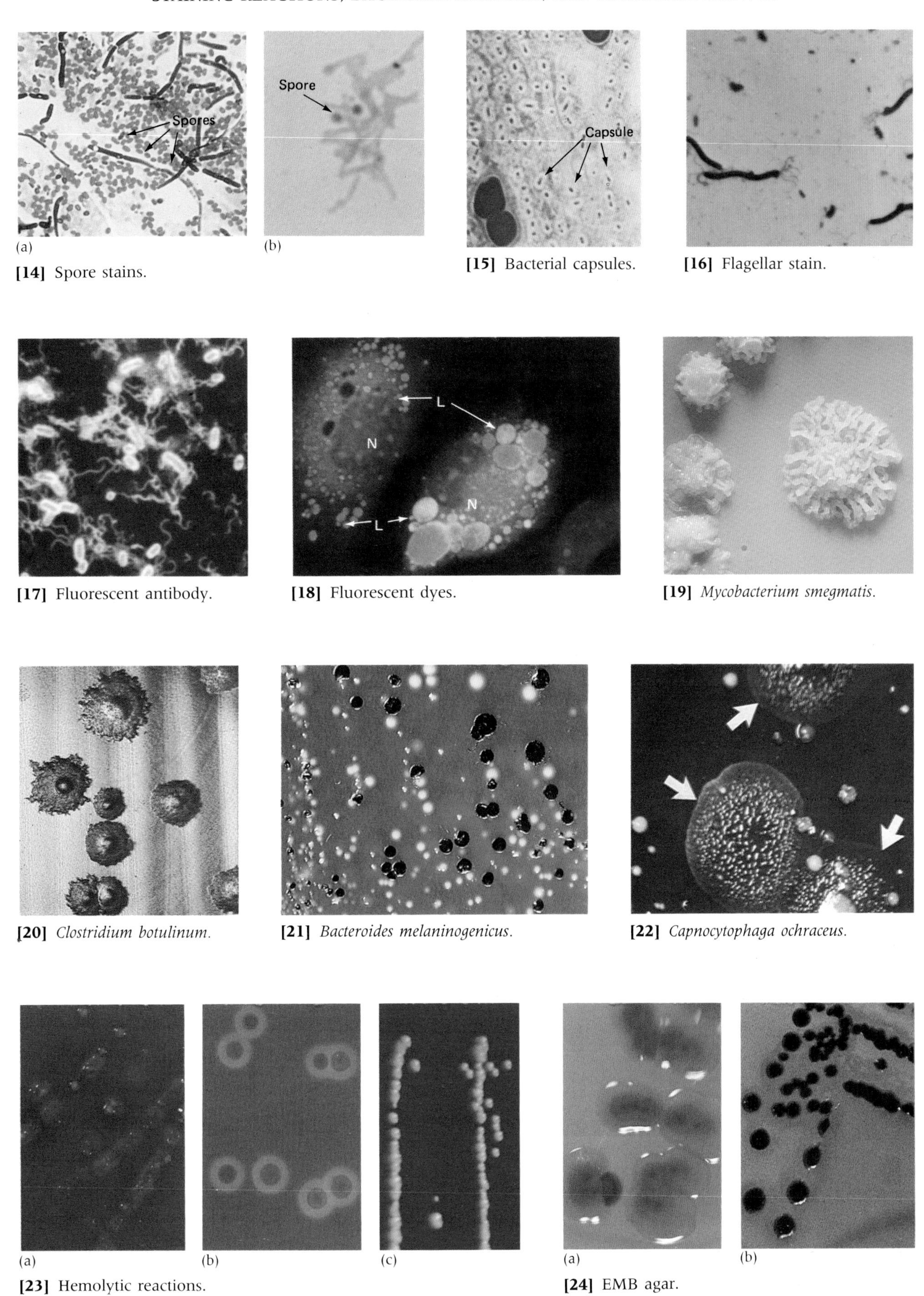

(a) (b)

[14] Spore stains.

[15] Bacterial capsules.

[16] Flagellar stain.

[17] Fluorescent antibody.

[18] Fluorescent dyes.

[19] *Mycobacterium smegmatis.*

[20] *Clostridium botulinum.*

[21] *Bacteroides melaninogenicus.*

[22] *Capnocytophaga ochraceus.*

(a) (b) (c)

[23] Hemolytic reactions.

(a) (b)

[24] EMB agar.

Detailed Color Atlas Legends follow Color Atlas

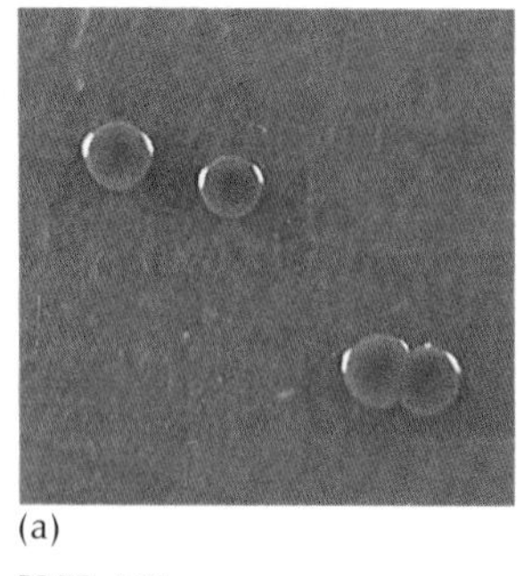
(a)

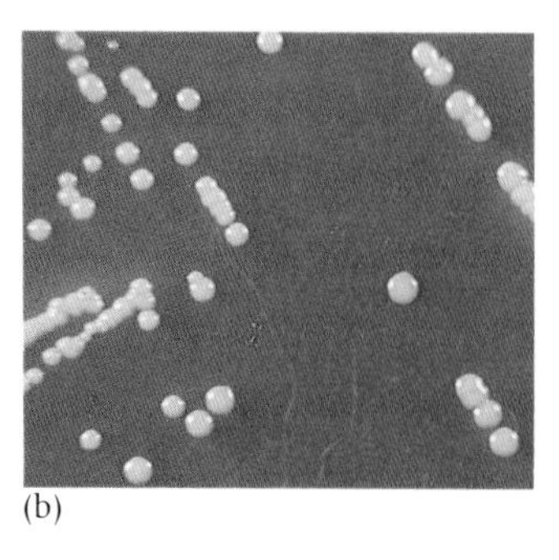
(b)

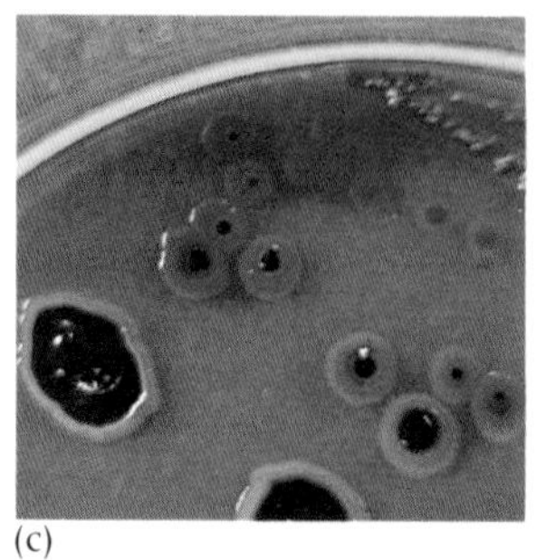
(c)

[25] HE agar.

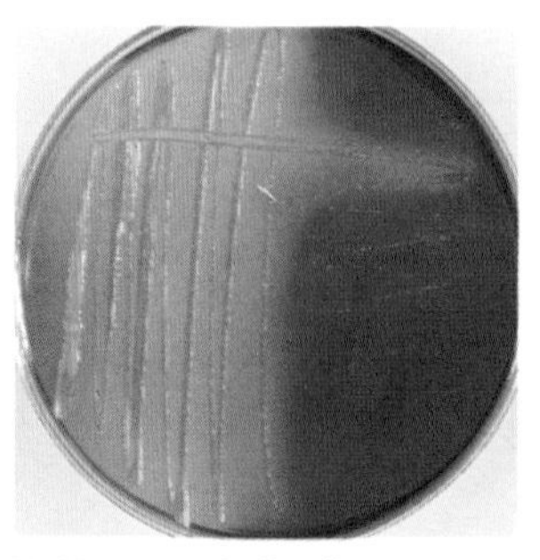

[26] Mannitol-salt agar.

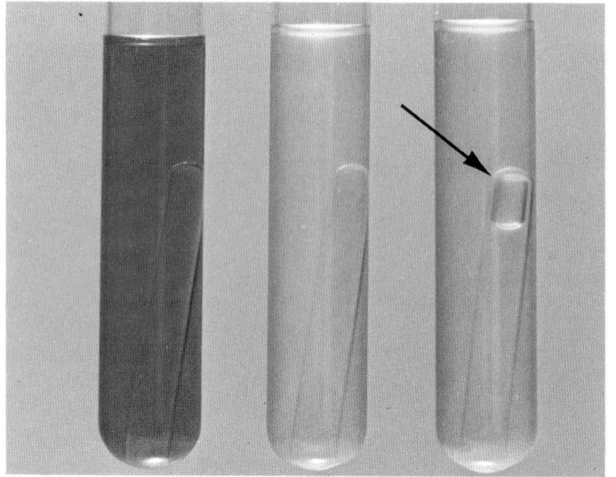

[27] Durham tube fermentations.

[28] Triple sugar iron agar.

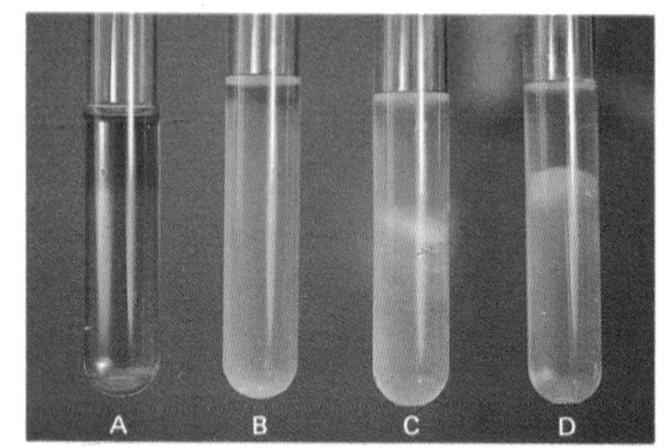

[29] Thioglycollate broth.

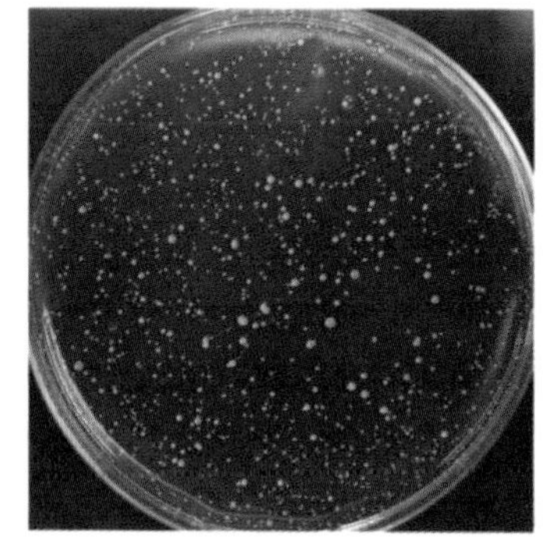

[30] Pour plate.

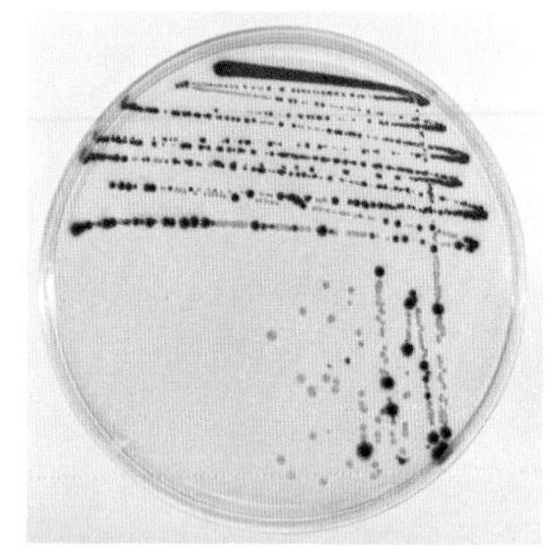

[31] Streak plate.

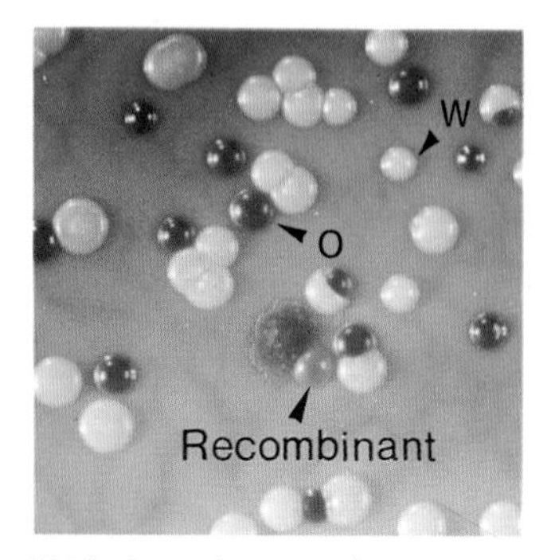

[32] Genetic recombination.

[33] Morels.

[34] Poisonous puffball.

Detailed Color Atlas Legends follow Color Atlas

[35] *Amanita.*

[36] Polypore.

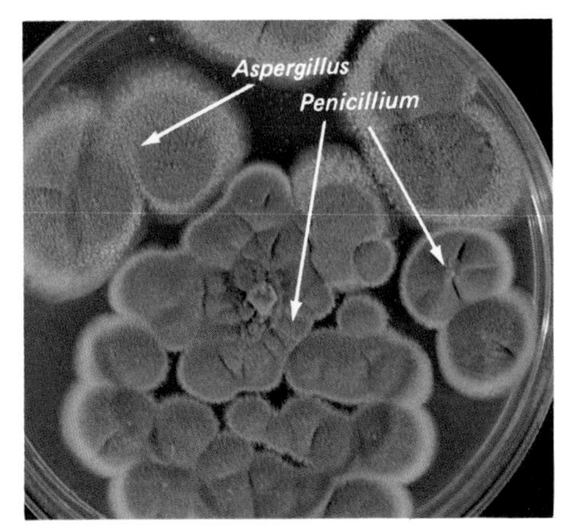

[37] *Penicillium* and *Aspergillus.*

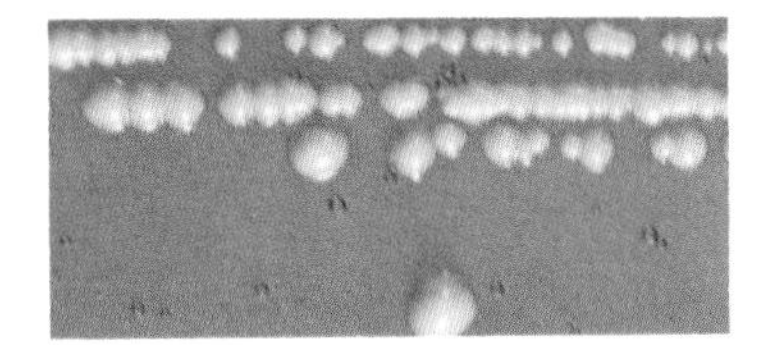

[38] Brewer's yeast.

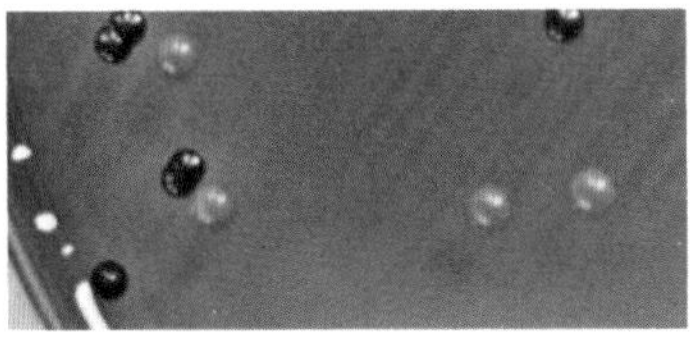

[39] Pathogenic yeasts.

[40] Fairy ring disease.

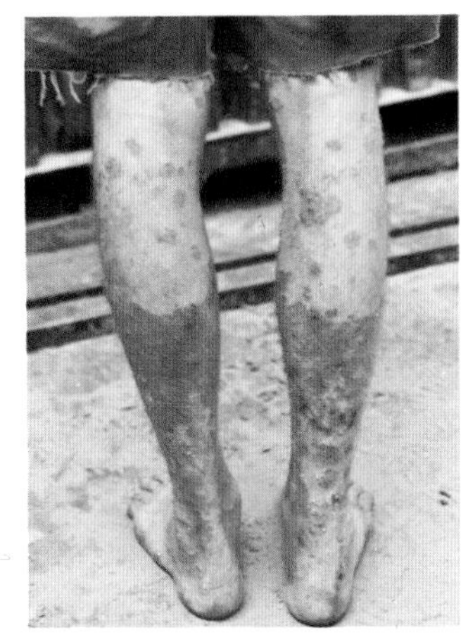

[41] Athlete's foot.

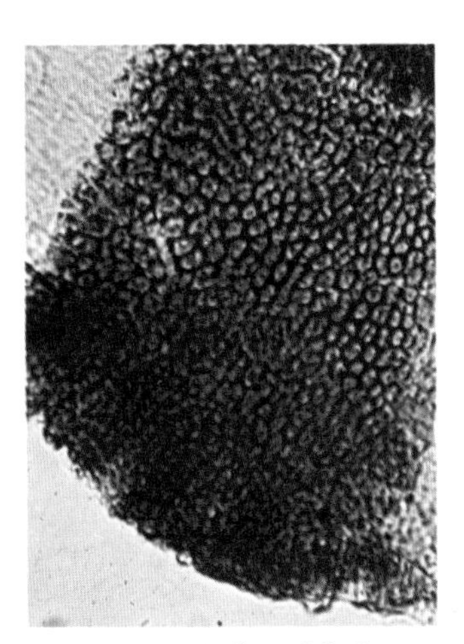

[42] Fungus infected hair.

[43] Slime mold.

[44] Fruiticose lichens.

Detailed Color Atlas Legends follow Color Atlas

[45] Fibrous and crustose lichens.

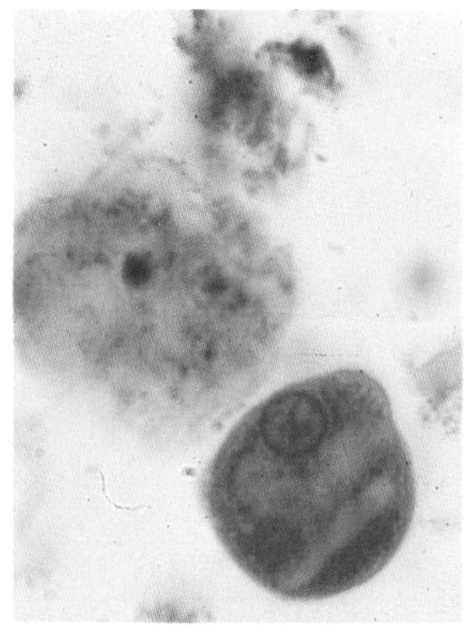

[46] *Entamoeba histolytica.*

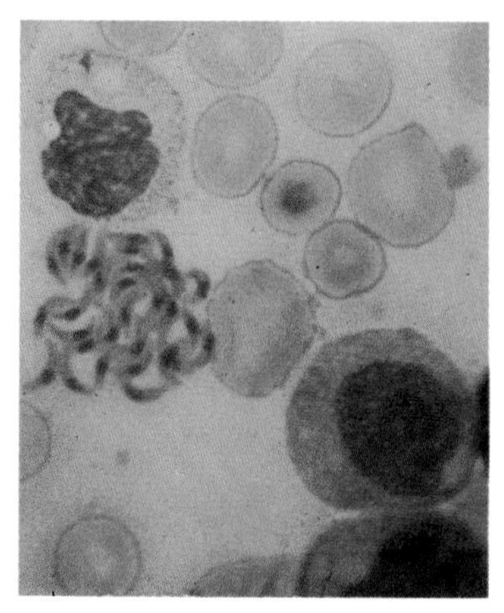

[47] *Toxoplasma gondii.*

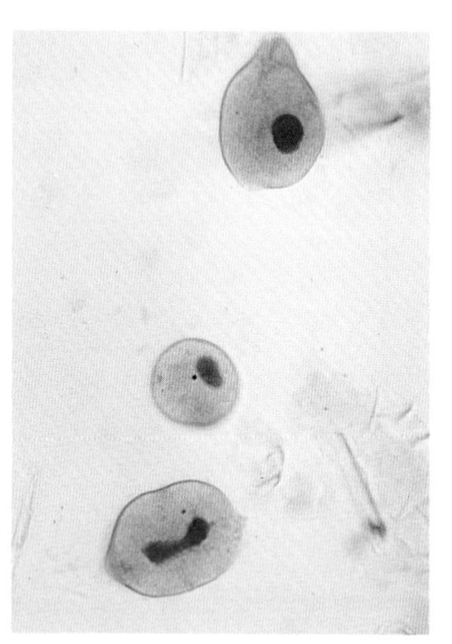

[48] *Balantidium coli.*

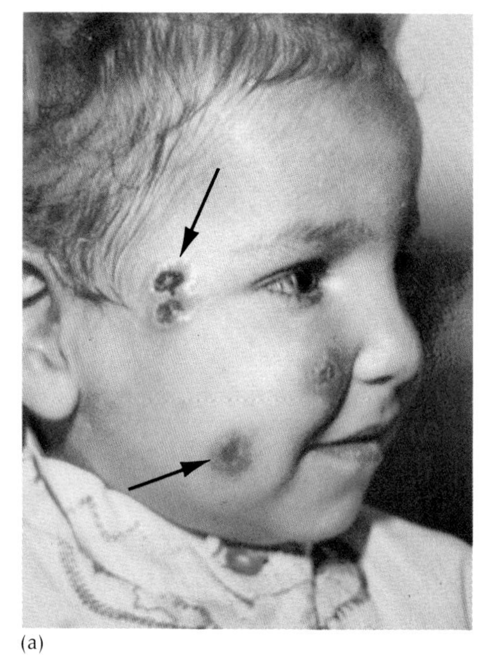

(a)

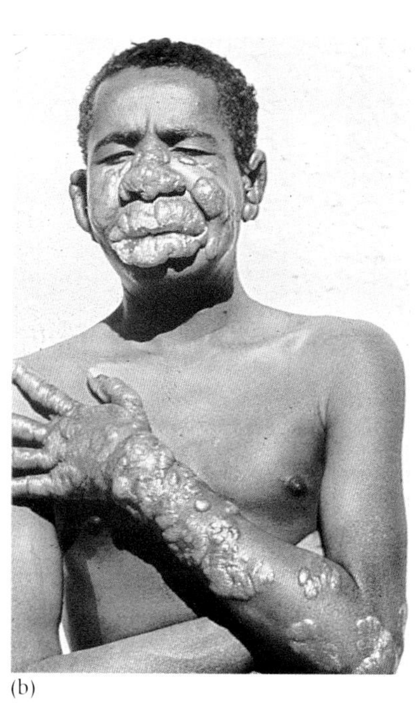

(b)

[49] Leishmaniasis.

[50] Diatoms.

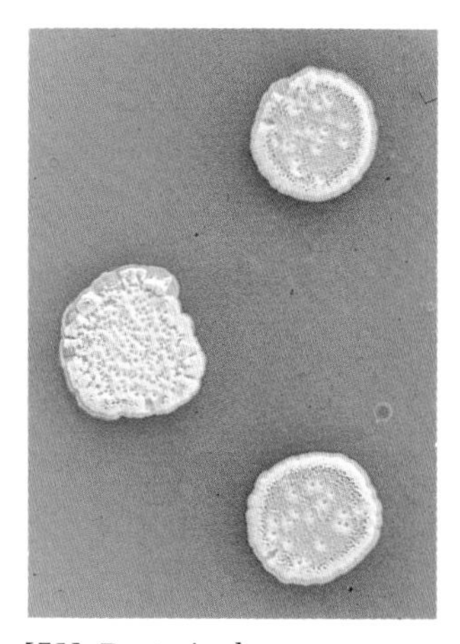

[51] Bacteriophage plaques.

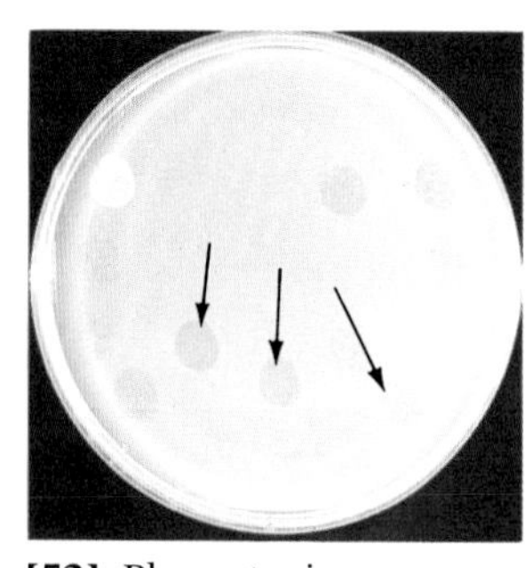

[52] Phage typing.

Detailed Color Atlas Legends follow Color Atlas

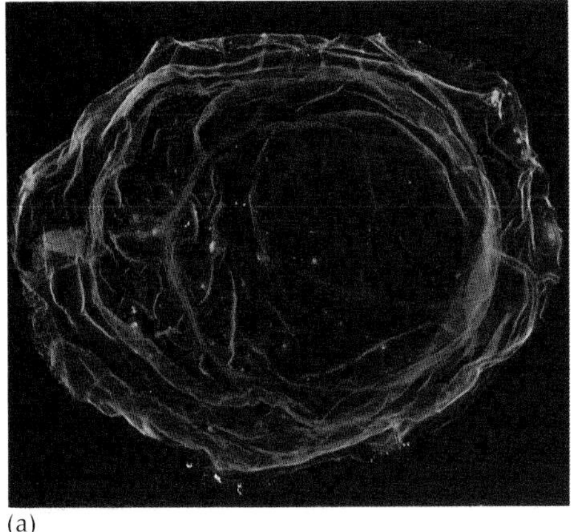

(a)

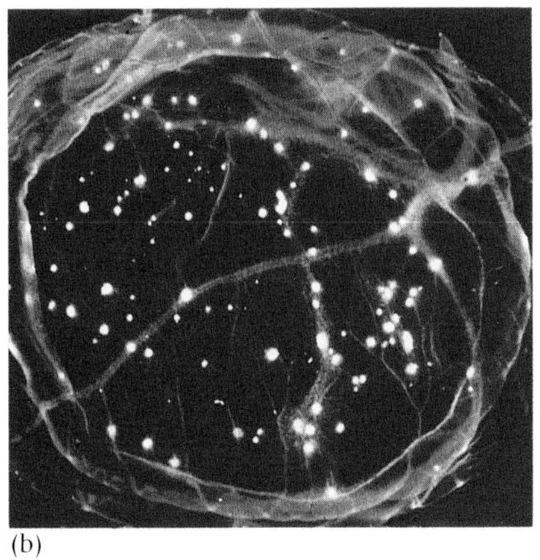

(b)

[53] Virus infection in chick embryos.

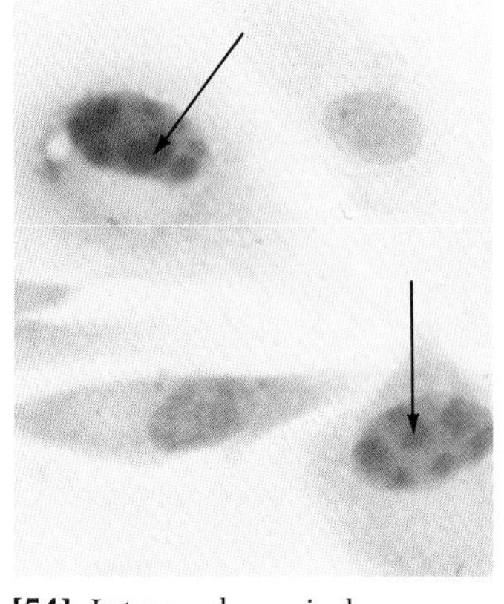

[54] Intranuclear viral inclusions.

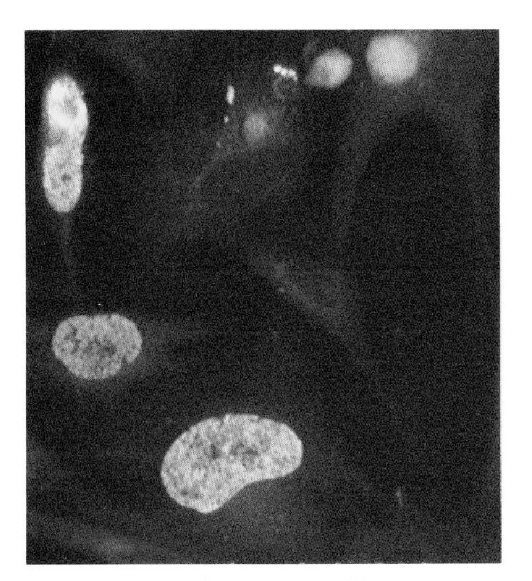

[55] Virus infected lung cells.

[56] Plant virus infection.

[57] Virus infected plant.

[58] Viroid infection.

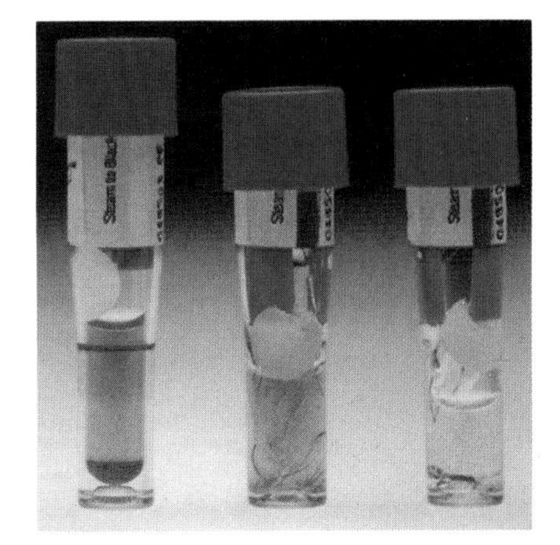

[59] Sterility monitor.

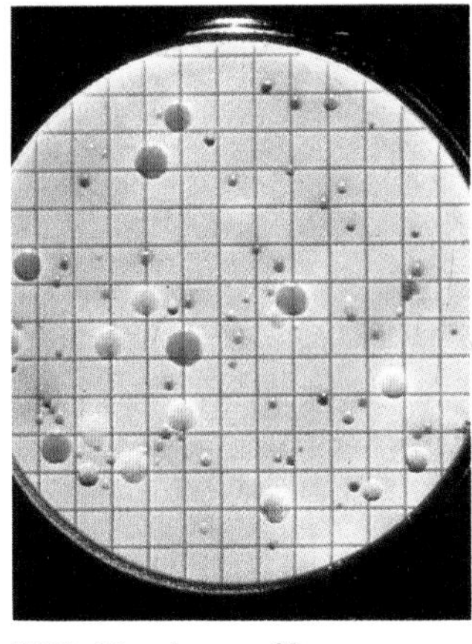

[60] Membrane filter.

Detailed Color Atlas Legends follow Color Atlas

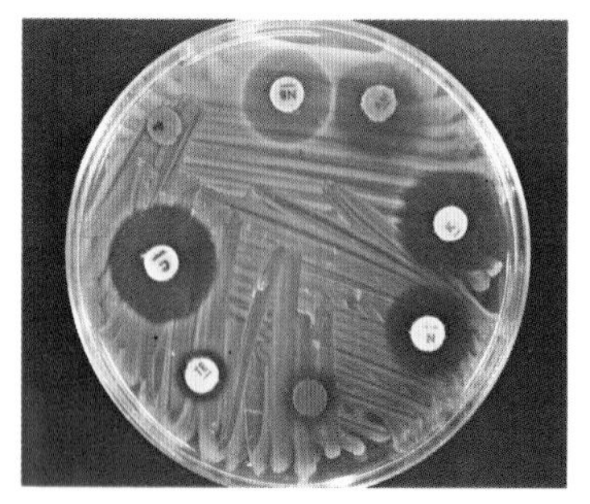

[61] Antibiotic sensitivity test.

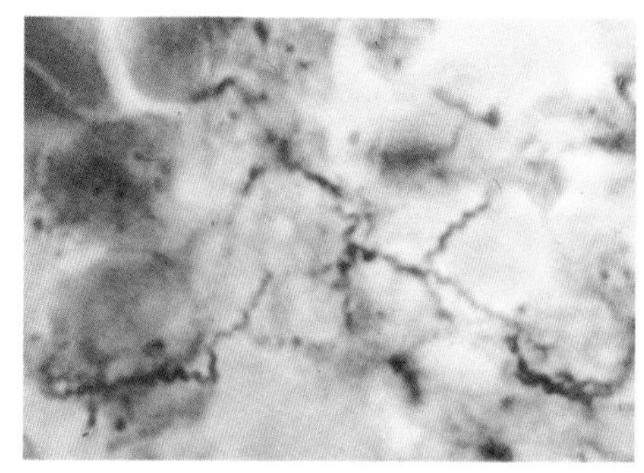

[62] *Treponema pallidum.*

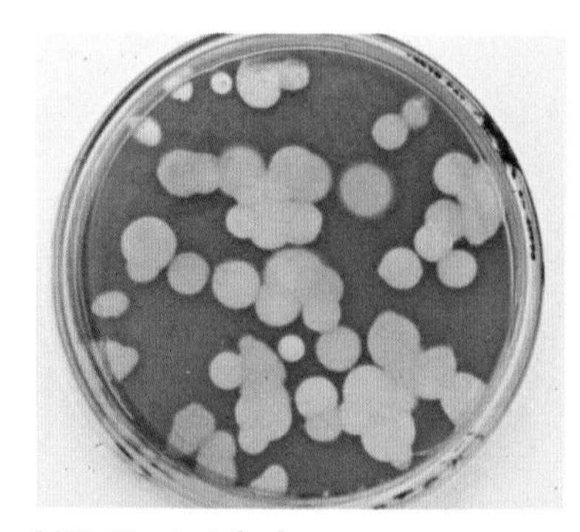

[63] Bacterial plaques.

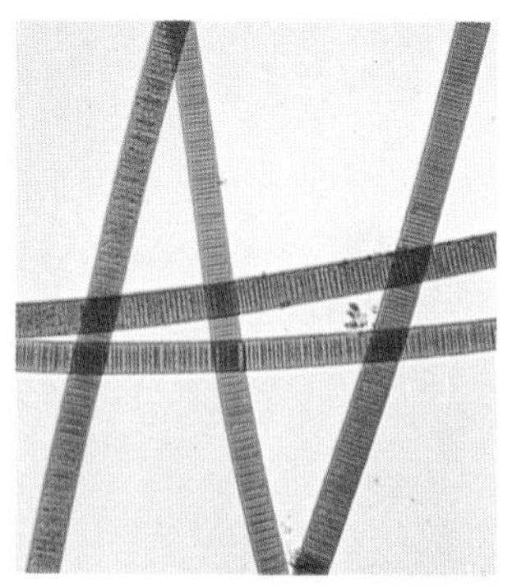

[64] Cyanobacteria.

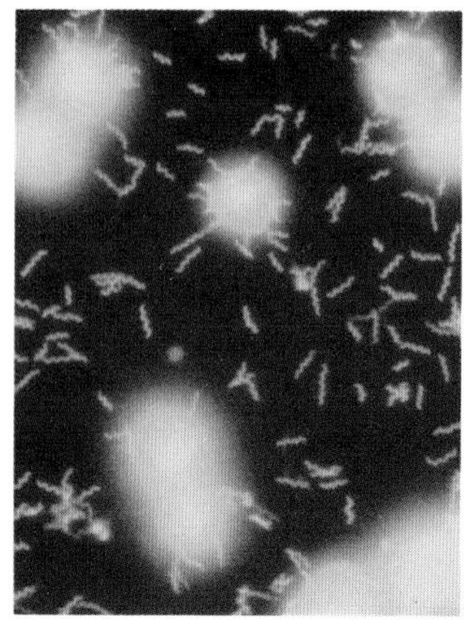

[65] Spiroplasma.

[66] Methanogen environment.

(a)

(b)

[67] Green algae colonization.

[68] Endosymbiosis.

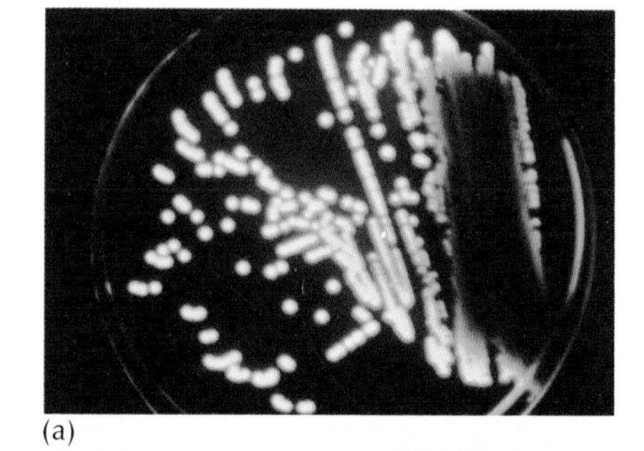

(a)

(b)

[69] Bioluminescent bacteria.

Detailed Color Atlas Legends follow Color Atlas

[70] Plant galls.

[71] Products of fermentation.

[72] A brewing vessel.

[73] Wine testing.

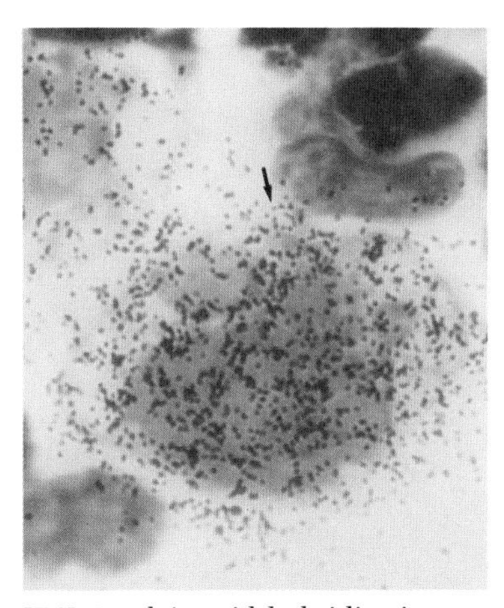

[74] Nucleic acid hybridization.

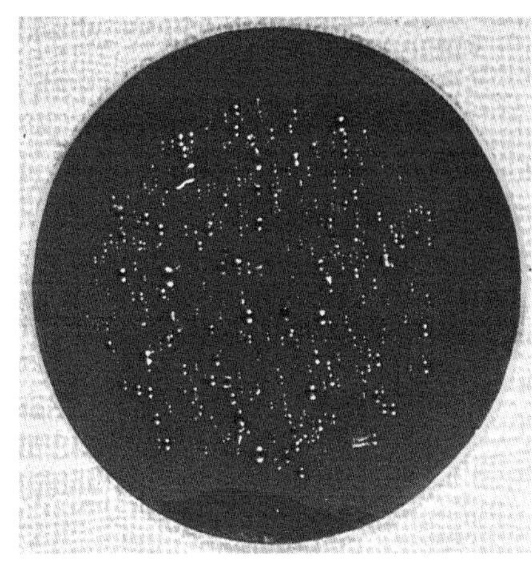

[75] Coliforms on a membrane filter.

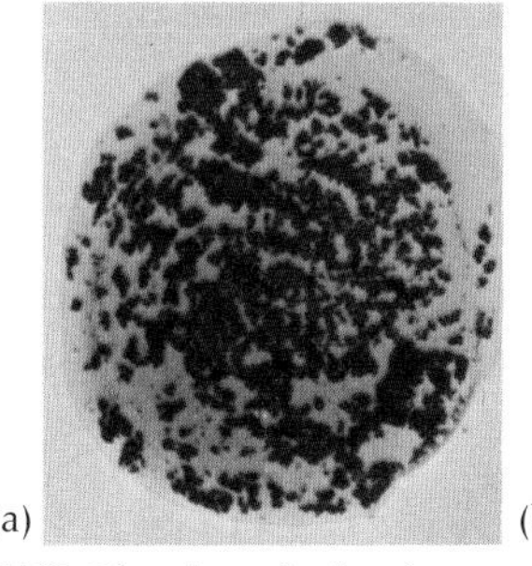

(a)

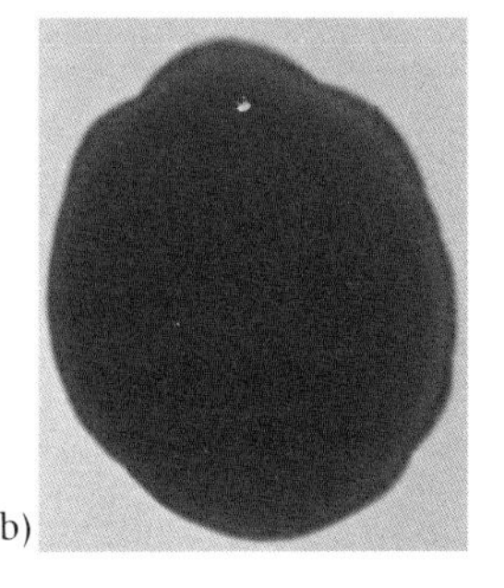

(b)

[76] Blood agglutination.

[77] Latex agglutination.

[78] Complement fixation.

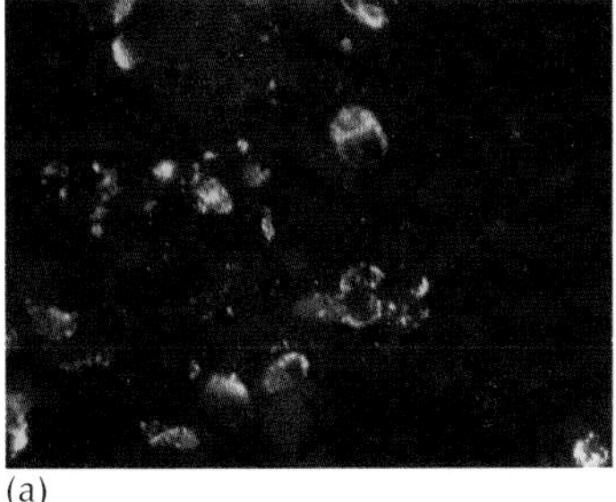

(a)

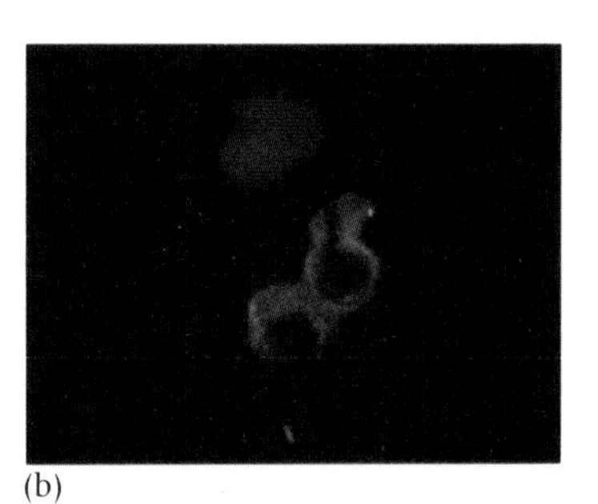

(b)

[79] Polyclonal and monoclonal antibodies.

Detailed Color Atlas Legends follow Color Atlas

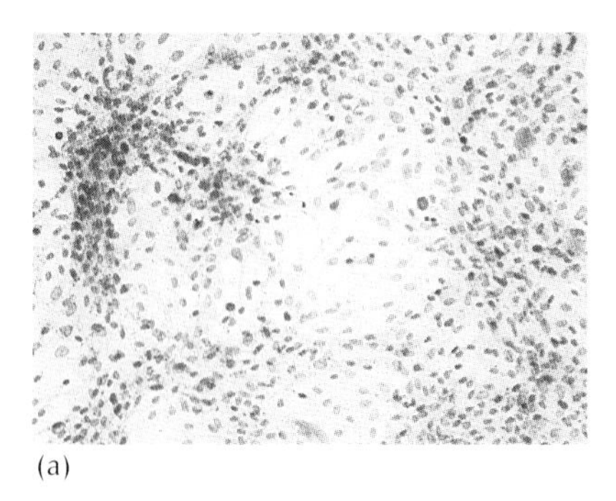
(a)

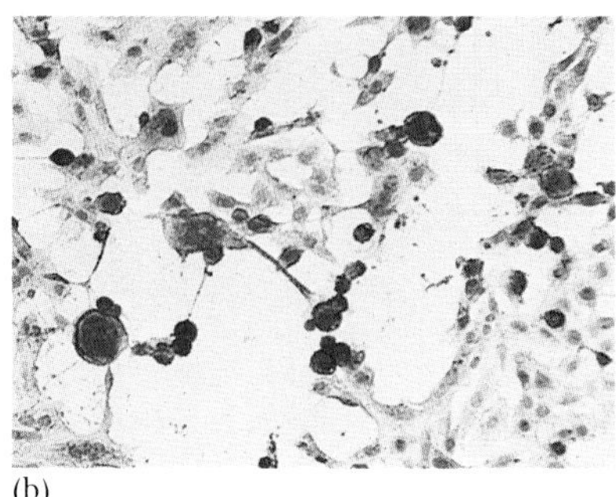
(b)

[80] Peroxidase-antiperoxidase test.

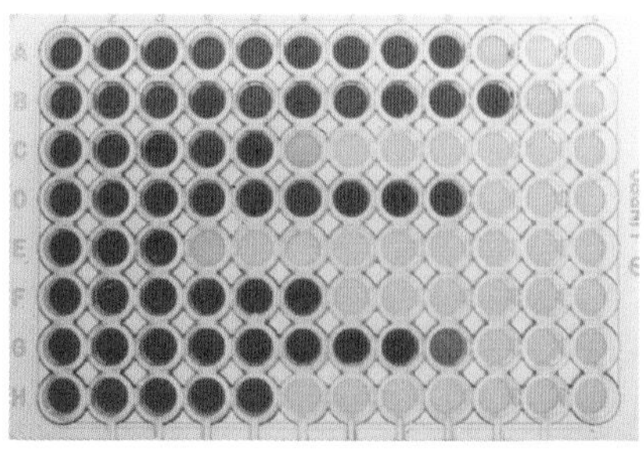

[81] ELISA.

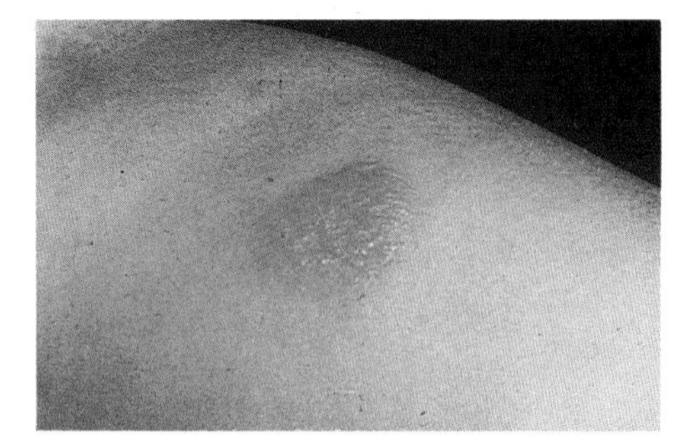

[82] Human skin test.

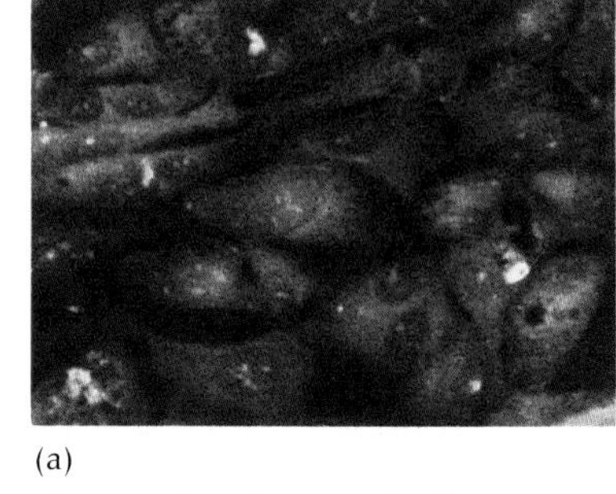
(a)

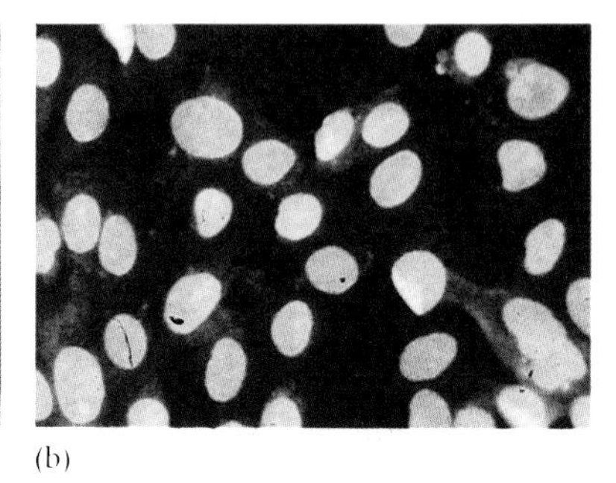
(b)

[83] Antinuclear antibody test.

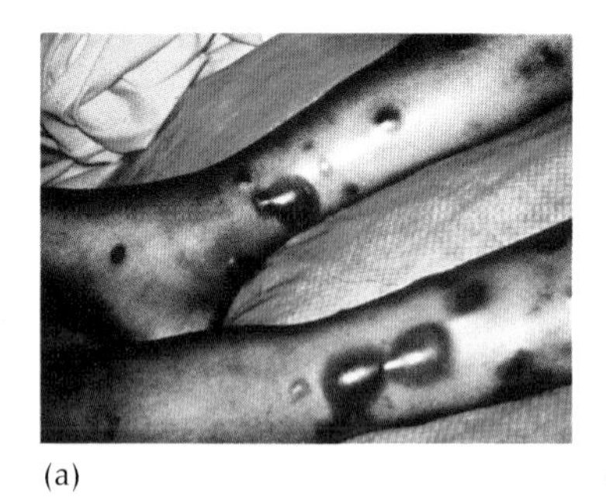
(a)

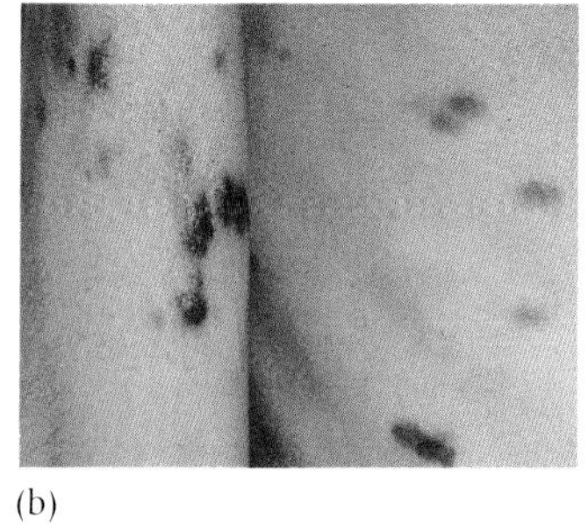
(b)

[84] Immune deficiencies.

[85] Reduviidae bugs.

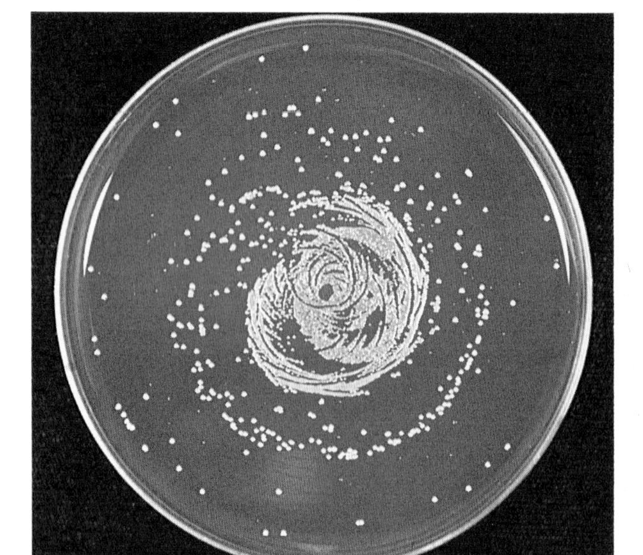

[86a] Automated streaking.

[86b] API enteric system.

Detailed Color Atlas Legends follow Color Atlas

[87] Bacitracin sensitivity.

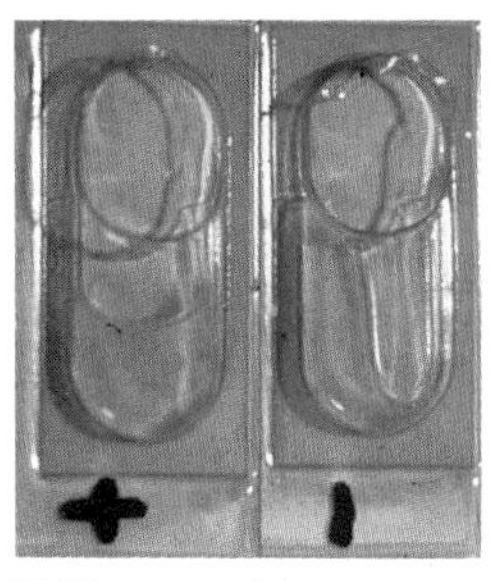

[88] Coagulase test.

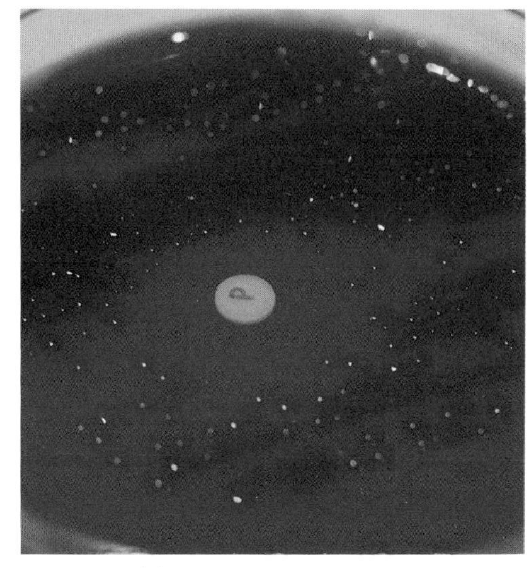

[89] Optochin test.

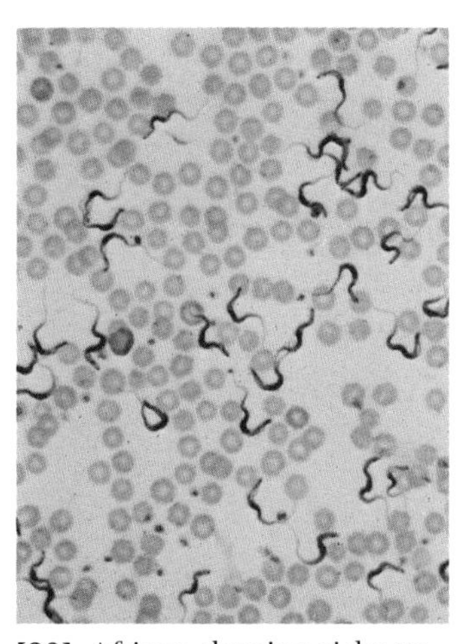

[90] African sleeping sickness.

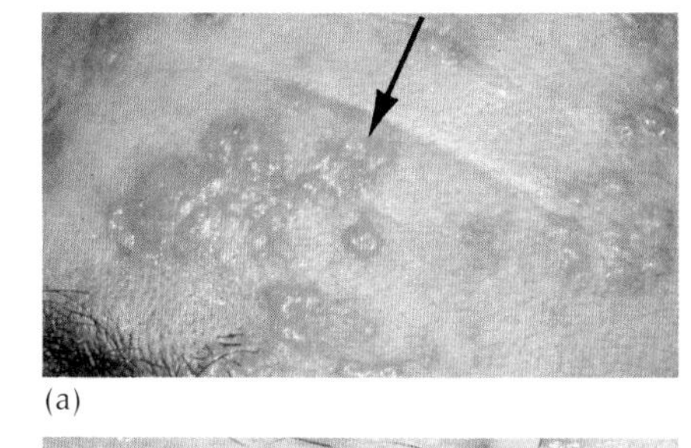

(a)

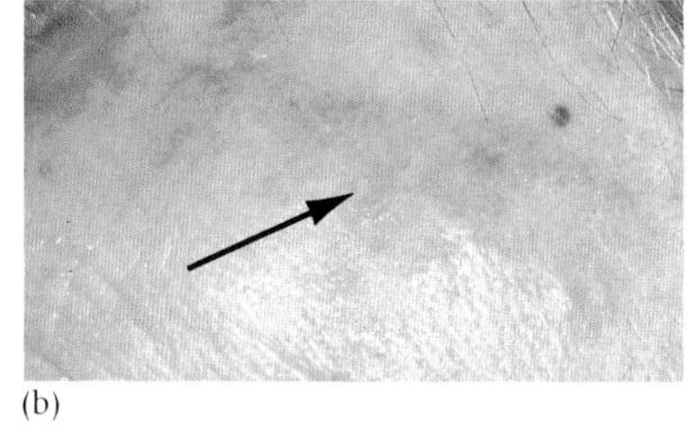

(b)

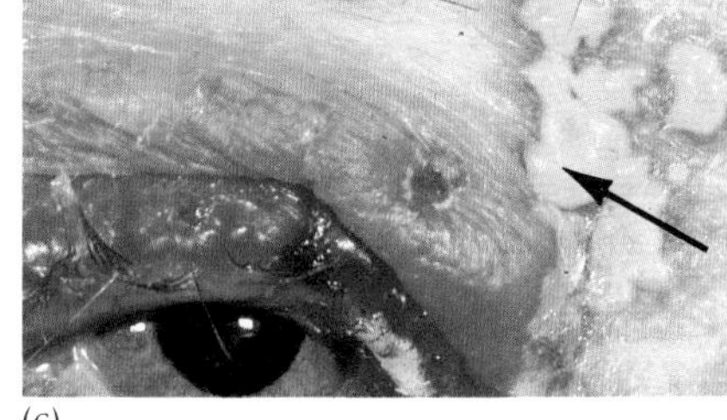

(c)

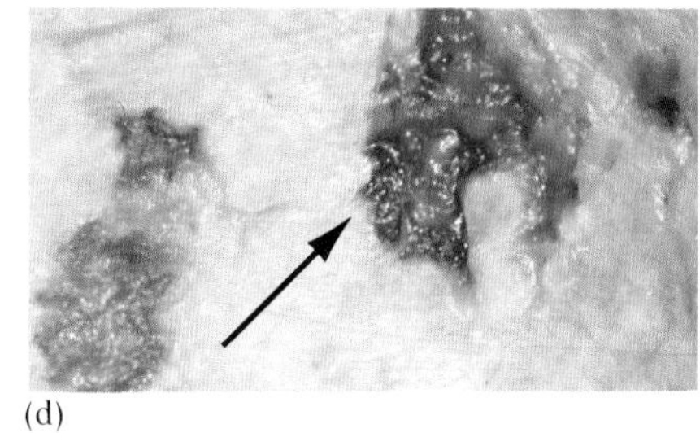

(d)

[91] Skin lesions.

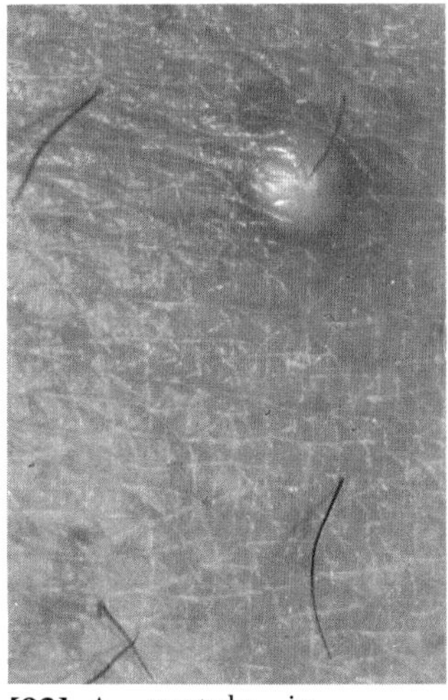

[92] A pustule in folliculitis.

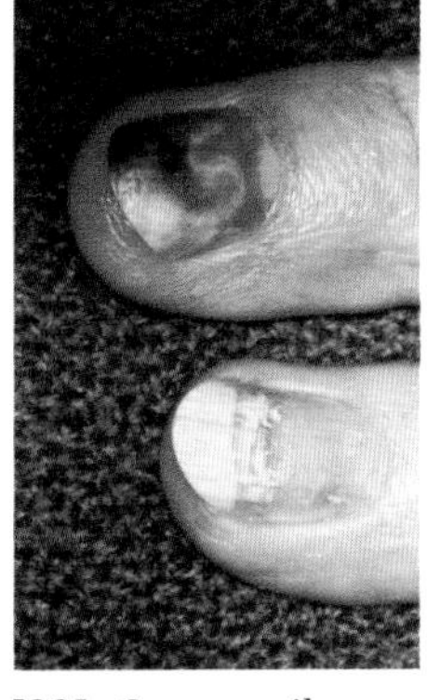

[93] Green nail syndrome.

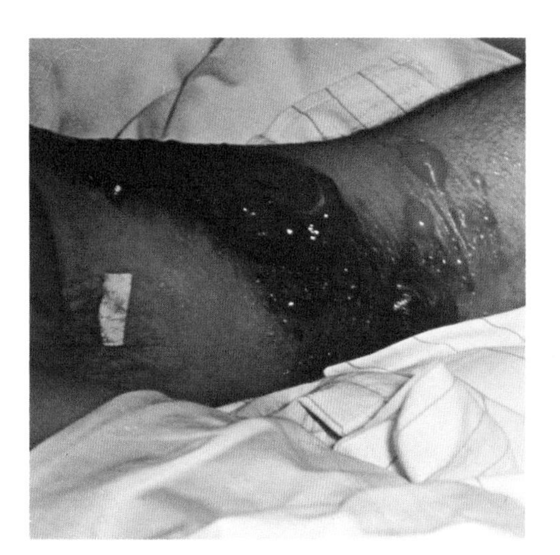

[94] Cutaneous anthrax.

Detailed Color Atlas Legends follow Color Atlas

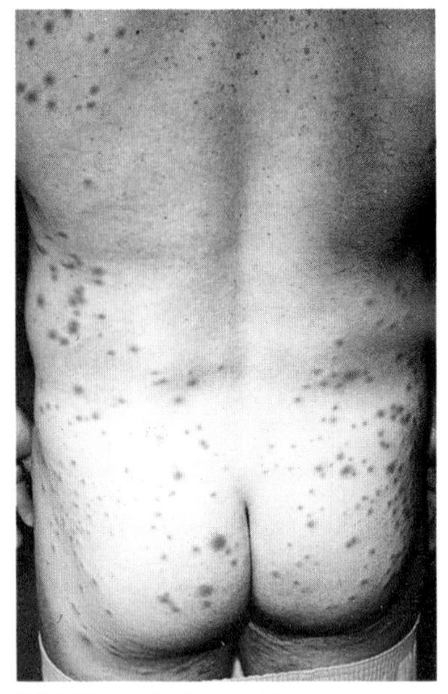

[95] Hot tub dermatitis.

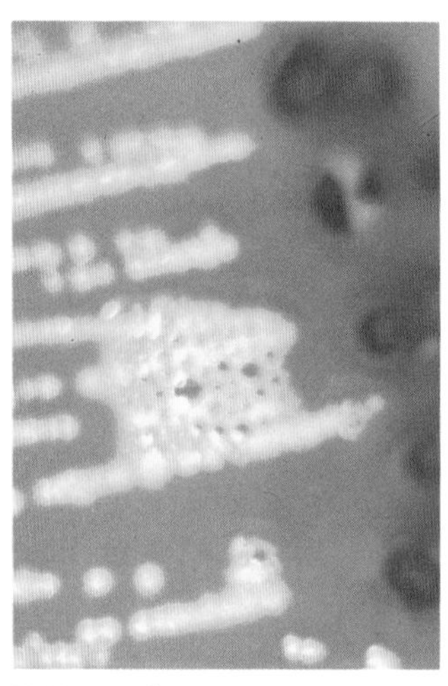

[96] Catalase reaction.

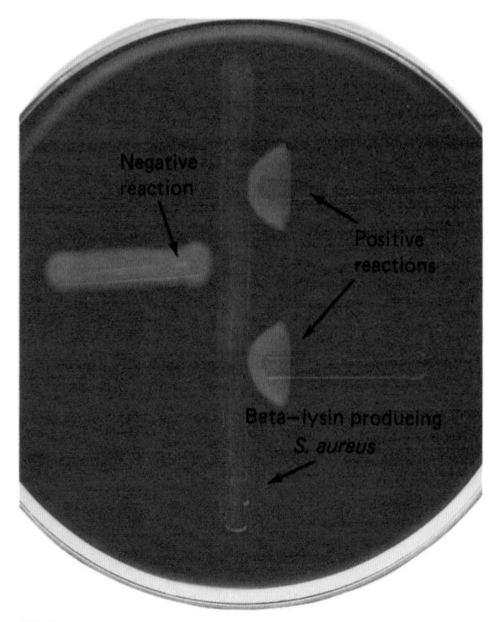

[97] CAMP reactions.

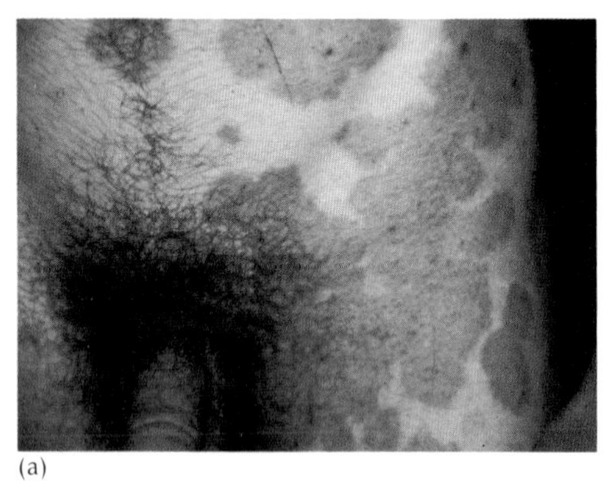

(a)

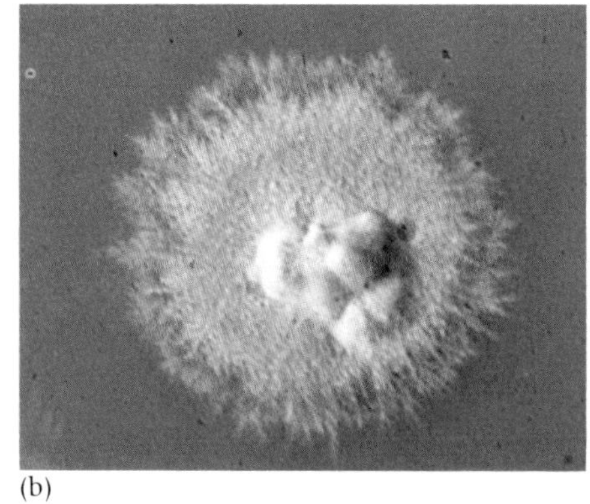

(b)

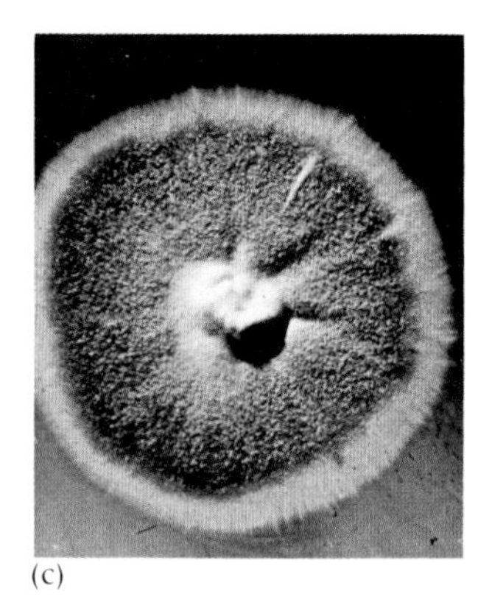

(c)

[98] Ringworm and causative agents.

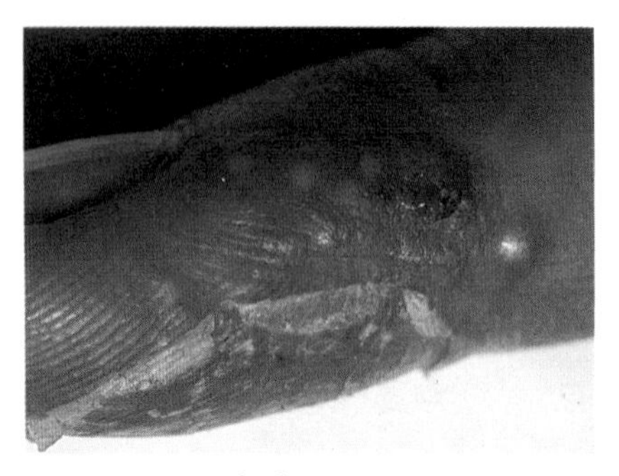

[99] Herpes whitlow.

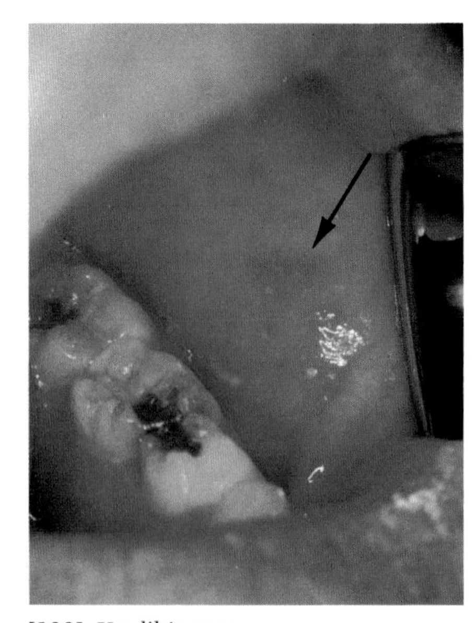

[100] Koplik's spots.

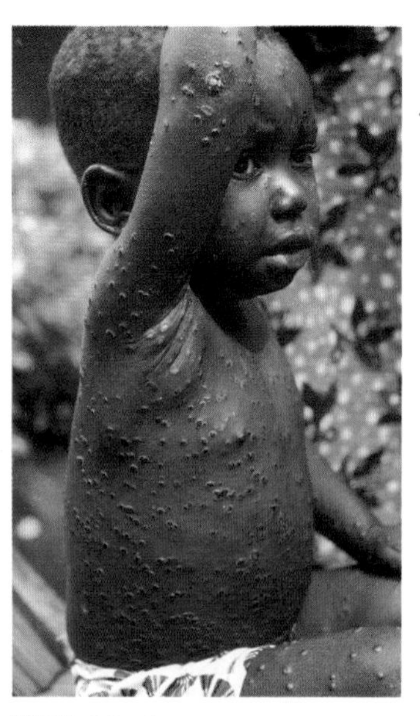

[101] Monkeypox.

Detailed Color Atlas Legends follow Color Atlas

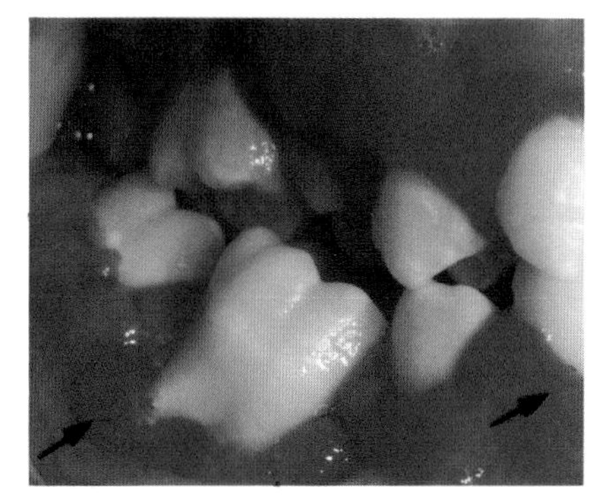

[102] Gingivitis.

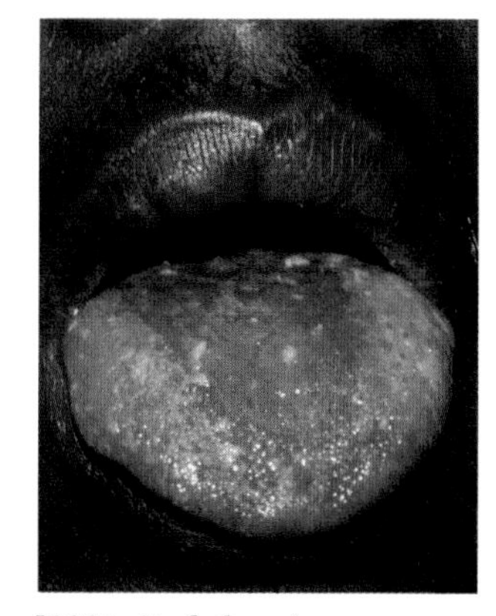

[103] Oral thrush.

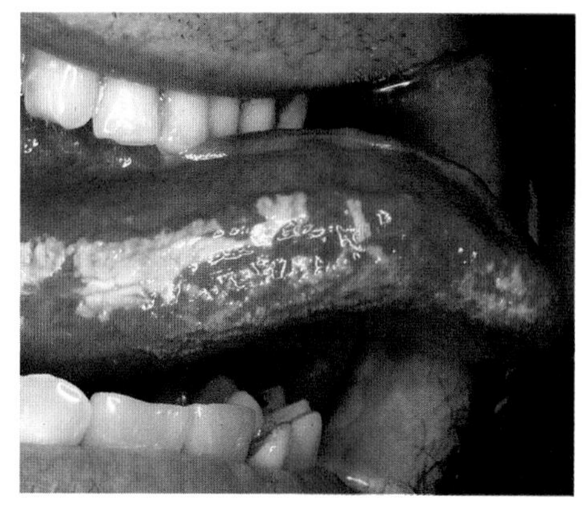

[104] Hairy leukoplakia.

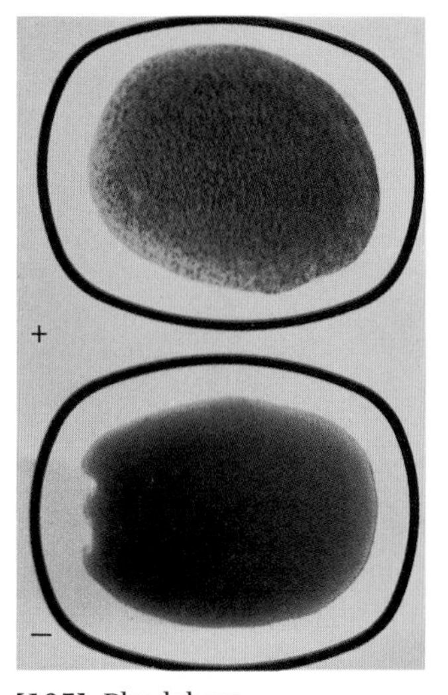

[105] Phadebact test.

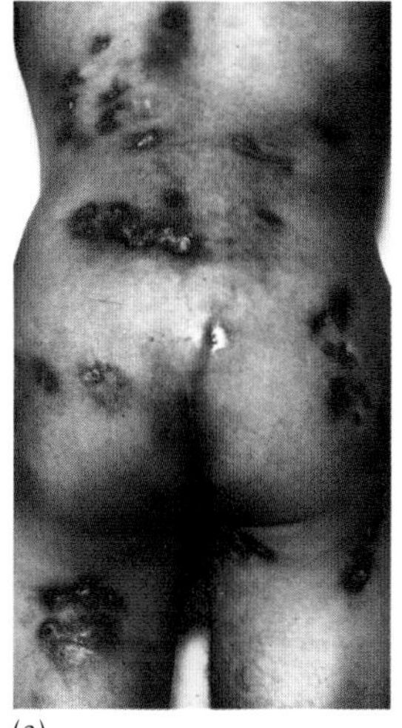

(a)

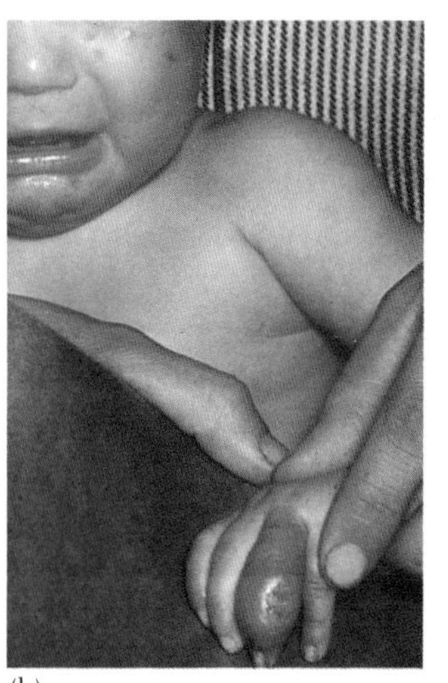

(b)

(c)

[106] Coccidioidomycosis and *C. immitis*.

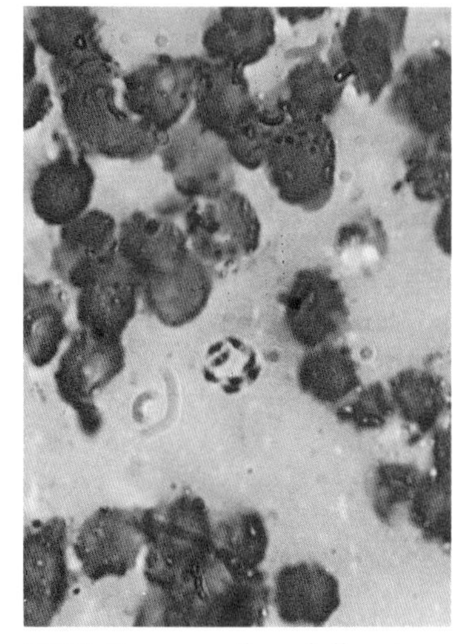

[107] *Pneumocystis carinii* cyst.

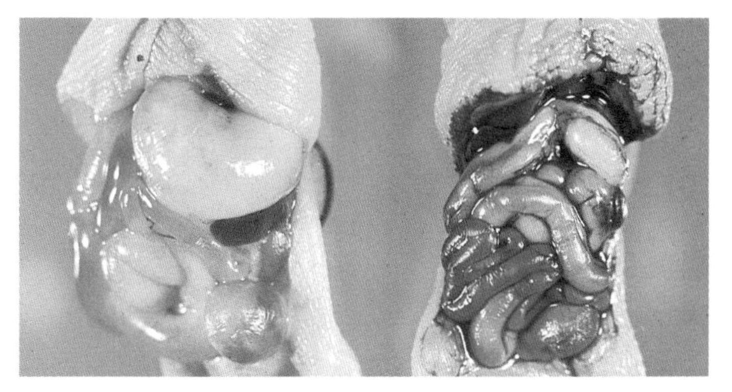

[108] Microbial toxin detection.

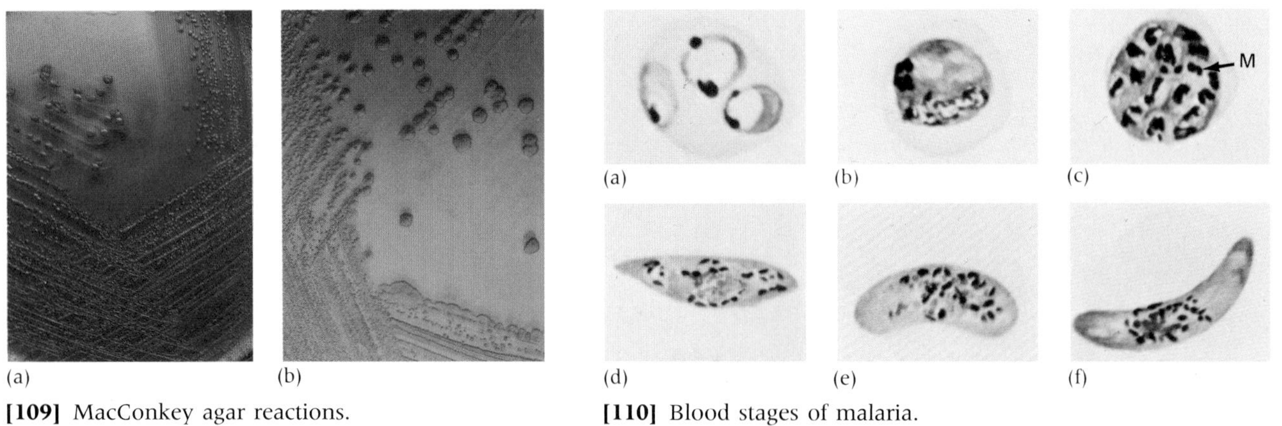

(a) (b)

[109] MacConkey agar reactions.

(a) (b) (c) (d) (e) (f)

[110] Blood stages of malaria.

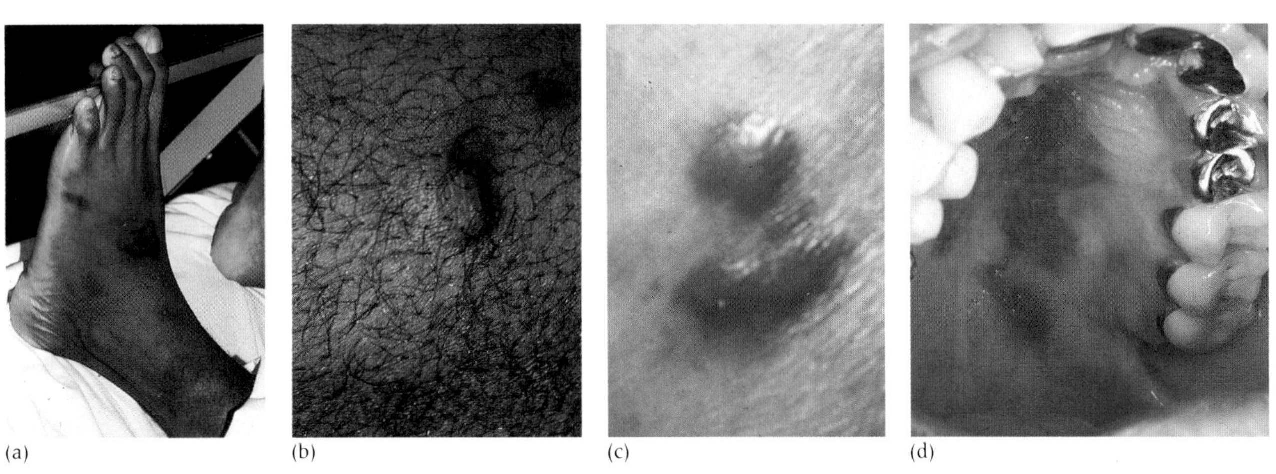

(a) (b) (c) (d)

[111] Forms of Kaposi's sarcoma.

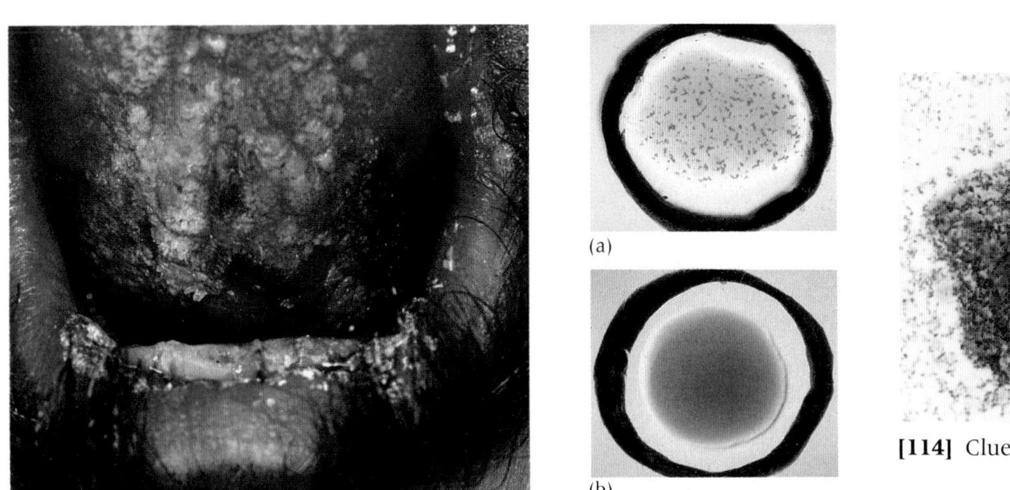

[112] Chronic oral candidiasis.

(a) (b)

[113] Staphyloside test.

[114] Clue cells.

Detailed Color Atlas Legends follow Color Atlas

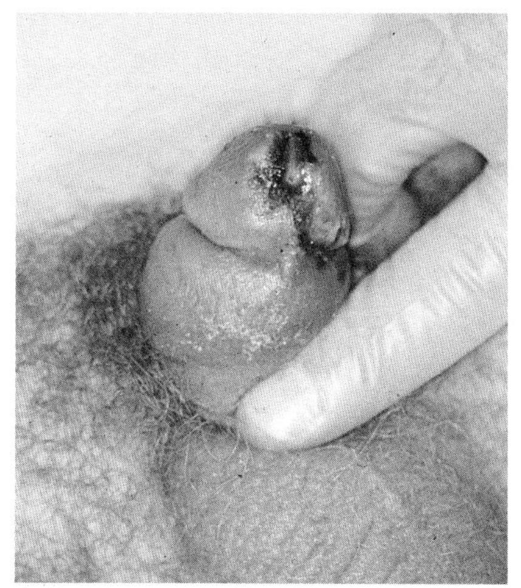

[115] Herpes balanitis.

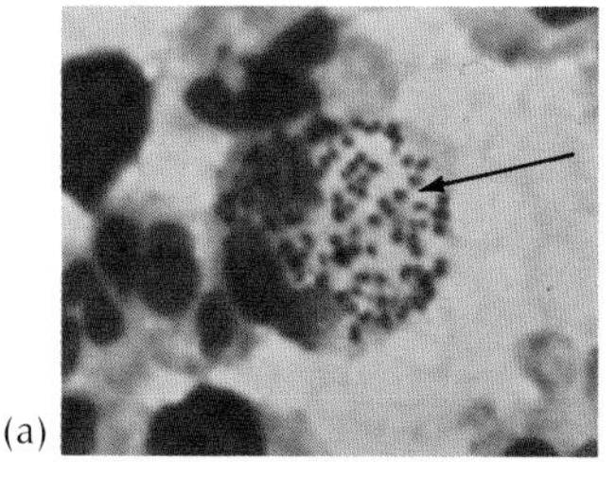

(a)

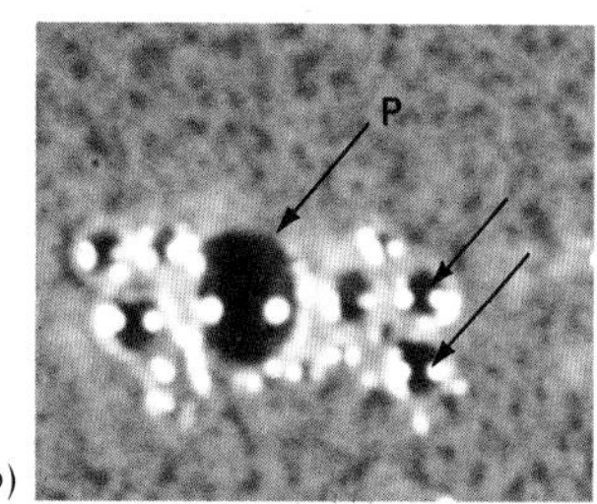

(b)

[116] Features of *N. gonorrhoeae.*

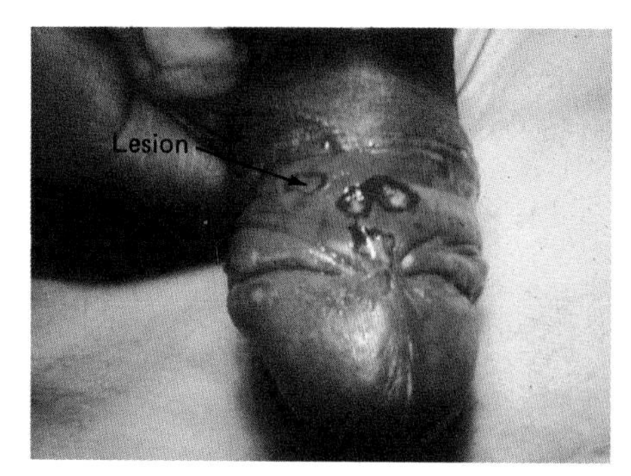

[117] Chancroid.

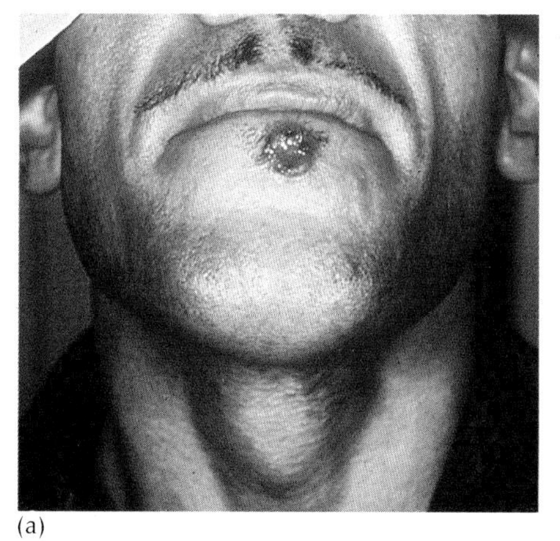

(a)

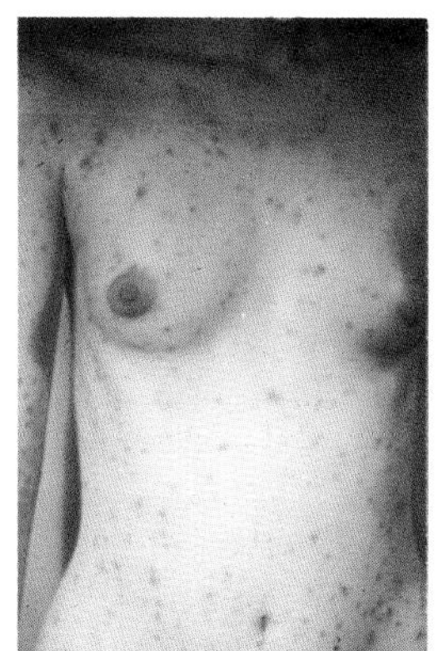

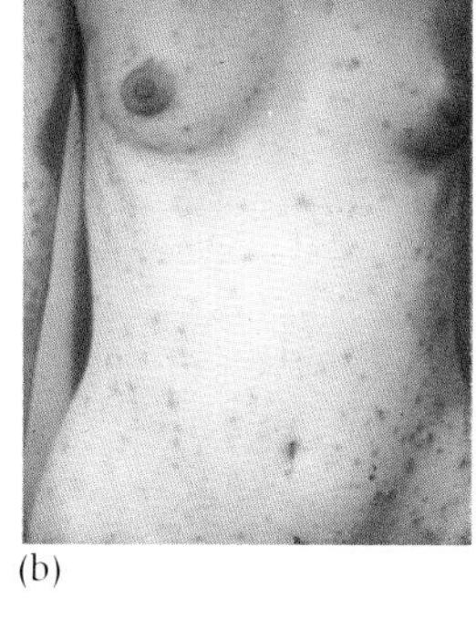

(b)

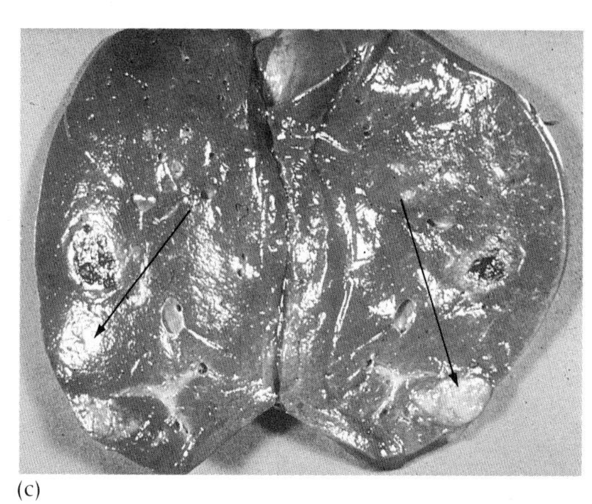

(c)

[118] Stages of syphilis.

Pustule

[119] Herpetic skin lesion.

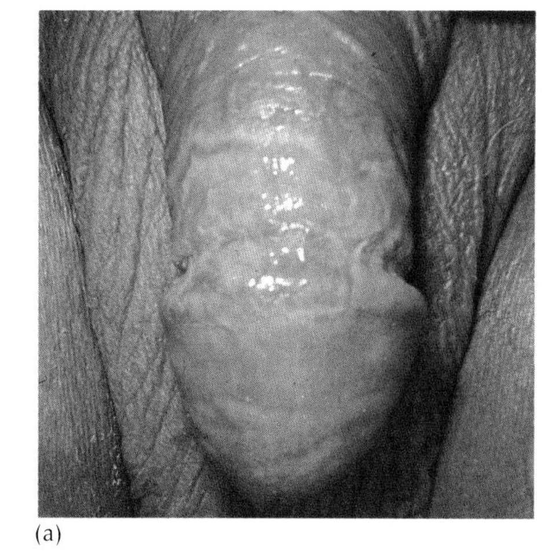

(a)

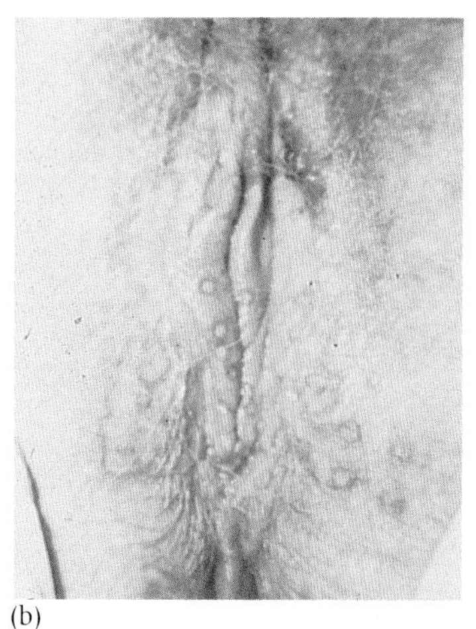

(b)

[120] Herpes simplex type 2 infections.

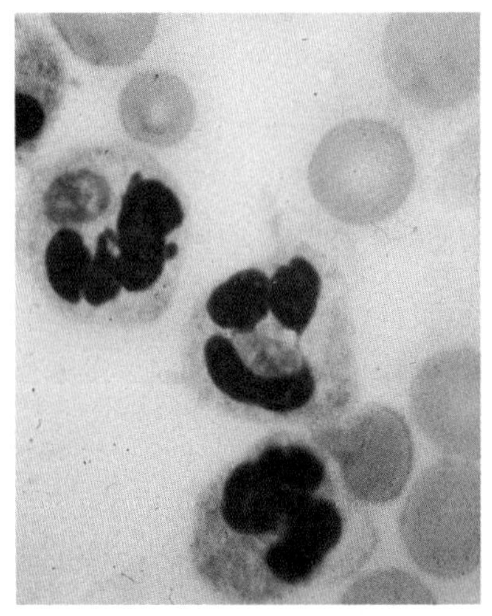

[121] Intracellular *T. gondii.*

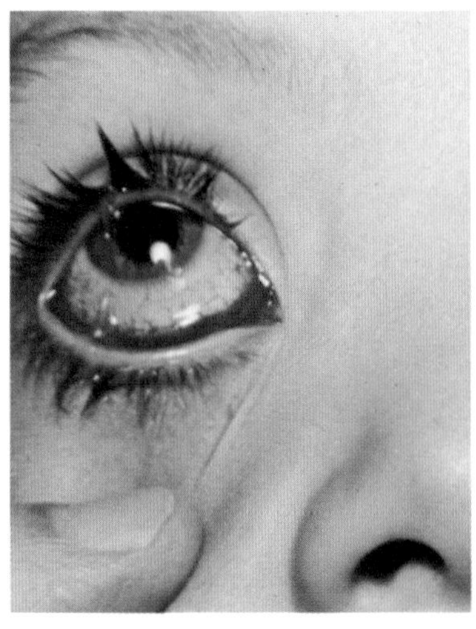

[122] Conjunctivitis.

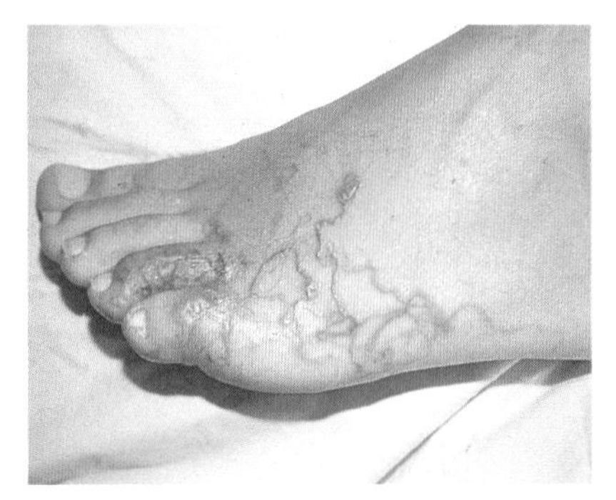

[123] Cutaneous larva migrans.

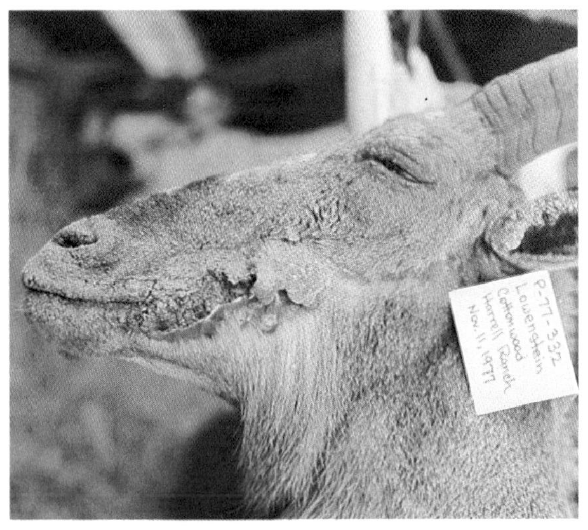

[124] Barbary sheep infection.

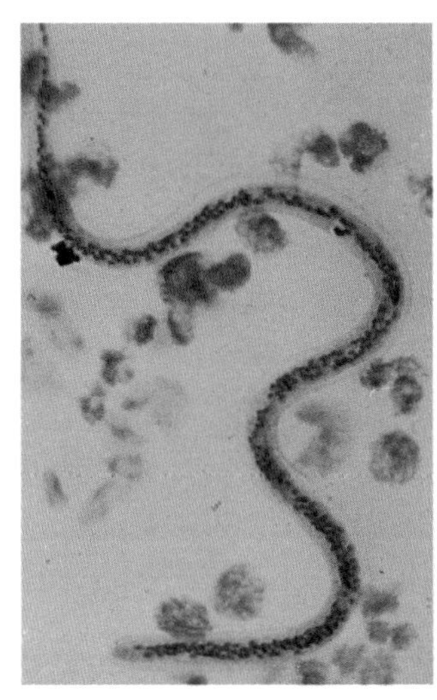

[125] *Wuchereria bancrofti.*

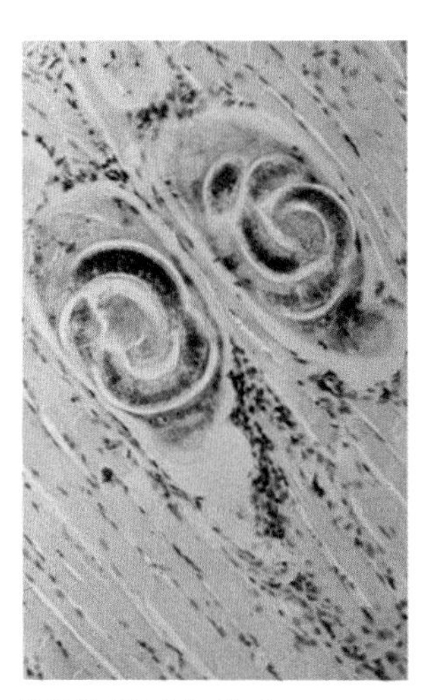

[126] *Trichinella* in muscle.

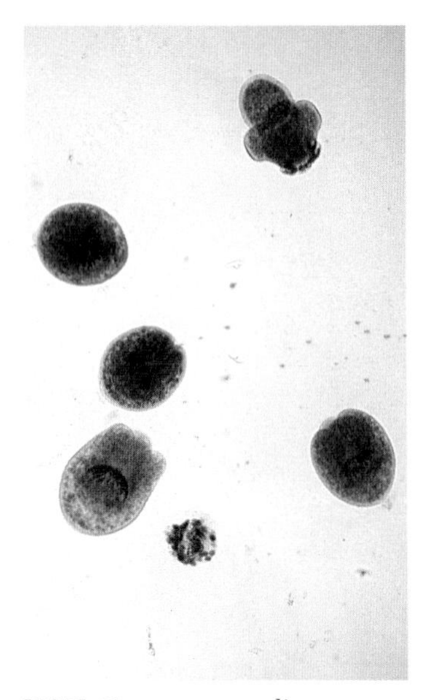

[127] Tapeworm scolices.

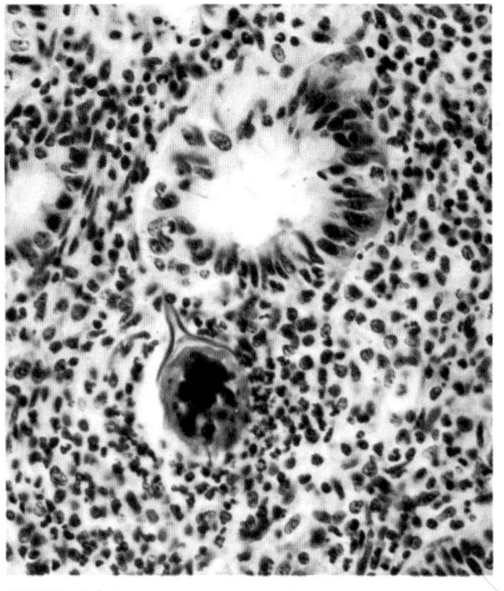

[128] Schistosome ova in tissue.

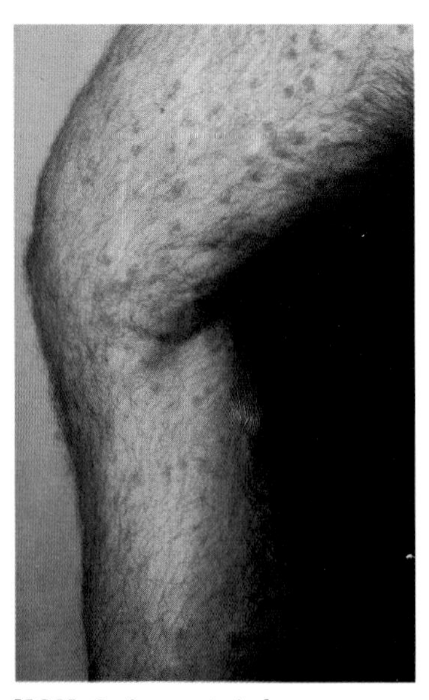

[129] Swimmer's itch.

Detailed Color Atlas Legends follow Color Atlas

Color Atlas Legends

[1] Herpes simplex viruses have made their mark on the human population. Type 1 is primarily responsible for fever blisters or the common cold sore, whereas type 2 affects the genital area and causes a sexually transmitted disease. An infection involving the throat is shown. [Courtesy Dr. W. Lawrence Drew, Mt. Zion Hospital Center and Medical Center.]

[2] The cause of Legionnaire's disease (a) Accumulations (colonies) of the Legionnaire's disease bacterium *(Legionella pneumophila)* 6 days after its placement on the surface of a special nutrient preparation. (b) The microscopic appearance of a related bacterium discovered after *L. pneumophila.* A special stain and procedure, known as the *fluorescent antibody technique,* makes the bacteria glow when exposed to ultraviolet light. [Courtesy Centers for Disease Control, Atlanta.]

[3] The appearance of bacterial and fungal (cottony) growth. After exposing a Petri dish containing a nutrient mixture to room air, colonies of different microorganisms developed within 24 to 48 hours.

[4] Impetigo, one of many different types of infection caused by bacteria. [A. M. Allen, D. Taplin, and L. Twigg, *Arch. Derm.,* **104**:271 (1971).]

[5] One form of fungus. *Mycena* species are common mushrooms in areas of rotting and decaying wood and leaves. More than 200 different types of the fungus are found in North America.

[6] Brown rot of peach. This fungus disease eventually completely penetrates the fruit, causing it to rot. The peach becomes a dry, distorted object called a *mummy.*

[7] (a) An algal bloom in a small pond near Williamsburg, Virginia. Note the few small clear patches in the massive accumulation of algae on the surface of the water. (b) Types of algae commonly encountered in natural bodies of water. Note the characteristic symmetry exhibited by many of these microorganisms.

[8] The effects of worms. Many of the techniques used to detect microorganisms and to treat diseases caused by them are applicable to worms. Adult heartworm infection in dogs. These roundworms are being removed from the right ventricle of the heart. Canine dirofilariasis, or dog heartworm disease, can be treated in many situations by drug therapy (chemotherapy) or surgical procedures. Although humans also can become infected with the causative agent, *Dirofilaria immitis,* the outcome is rarely fatal. [H. Shiang, et al., *JAVMA.* **163**:981 (1973).]

[9] Higher plants have their specialized cells organized into groups known as *tissues* that combine to make up organs. This stained cross-sectional view of a buttercup root shows an outer epidermis and an extensive region of cell—the cortex—used for storage of such substances as starch. The inner region contains vascular tissue that is used for the transport of various substances.

[10] Animals contain cells and tissue specialized for some of the same functions as higher plants. This stained section shows epithelial cells found in the mucosal lining of the human trachea. The ciliated nature of the surface epithelial cells (arrows) and the cylindrical features of the columnar epithelial cells are readily apparent. Note the appearance of the stained nuclei (N) in these eucaryotic cells.

[11] Examples of simple stained bacterial smears observed under oil immersion. (a) The rod *Bacillus subtilis* stained with safranin. (b) A young culture of the coccus *Micrococcus luteus* stained with crystal violet. Note the appearance of these cocci in pairs and other arrangements. (c) *B. subtilis* stained with crystal violet. This bacterium can form heat-resistant structures known as *spores* that do not stain with this procedure. The spores appear either as clear areas within the cells in which they were formed or as free structures.

[12] Differential staining reactions—the Gram stain procedure. (a) A typical gram-positive reaction exhibited by *Staphylococcus aureus.* These bacteria characteristically retain the primary dye, usually crystal violet, and appear dark purple in color. This photomicrograph also demonstrates the possible variations in the arrangement of cells (e.g., diplococci, single cocci) that may be observed during the examination of a microscope field. (b) The appearance of gram-negative rods. In this case, the primary dye was not retained following the decolorization step. Consequently, the organisms take the counterstain (usually safranin).

[13] Differential staining reactions—the acid-fast reaction. (a) A typical acid-fast reaction. Microorganisms exhibiting this reaction show a red coloration. (b) Bacterial cells exhibiting a non-acid-fast (blue) reaction.

[14] Differential staining reactions—the spore stain procedure. (a) The spores and vegetative cells of *Bacillus megaterium.* The preparation shown was stained by the Schaeffer-Fulton technique. Spores appear green, while vegetative portions of cells are red. Note the presence of spores (endospores) inside their respective vegetative cells. (b) A photomicrograph of the anaerobic spore former *Clostridium tetani.* This preparation was stained by a combination of carbol fuchsin and methylene blue. In this case, spores are red, while the vegetative cells are blue. Note the end position taken by these structures. This type of appearance is referred to as a "drumstick."

[15] A specially stained sputum specimen from a patient infected with *Streptococcus pneumoniae.* Note the capsules (clear areas) surrounding the individual diplococci.

[16] A photomicrograph of the flagellar arrangement of *Spirillum volutans.* In addition to showing the spiral morphology of this organism, the arrangement of two or more flagella at one or both ends of a cell is evident. Note that special staining procedures are needed to demonstrate these organelles.

[17] The direct fluorescent antibody technique showing the presence of bacterial flagella. The cells and their wavy flagella are quite evident. [J. G. Elliot, et al., *Appl. Microbiol.* **28**:1063–1065 (1974).]

[18] The use of fluorescence microscopy. With the aid of the dye acridine orange, the distribution of enzyme-containing structures in eucaryotic cells, lysosomes (L) can be shown. Both small and giant orange lysosome clusters are evident around the nucleus (N). [J. M. Oliver, et al., *J. Cell Biol.* **69**:205–210 (1976).]

[19] The curled colonies of *Mycobacterium smegmatis,* a rapidly growing bacterial species. [A. L. Vestal, *Procedures for the Isolation and Identification of Mycobacteria,* DHEW No. (HSM) 73–8230 (1973).]

[20] The appearance of colonies of *Clostridium botulinum,* an anaerobic bacterium, and the cause of the fatal food poisoning, botulism. [Courtesy C. L. Hatheway, Ph.D., Anaerobe Section, Centers for Disease Control.]

[21] These black-pigmented colonies contain cells of the anaerobe *Bacteroides melaninogenicus.* [Courtesy Dr. M. C. Newman.]

[22] Colonies of *Capnocytophaga (Bacteroides) ochraceus.* These organisms are involved in oral infections. [Courtesy of Dr. M. C. Newman.]

[23] Representative hemolytic reactions on blood agar media. (a) Alpha (α) hemolysis. Note the greenish discoloration of the medium surrounding the bacterial colonies. (b) Beta (β) hemolysis. Here clear zones surround the individual bacterial colonies. (c) Gamma (γ) hemolysis. Neither discoloration nor clear zones are present.

[24] Reactions on the selective and differential medium, eosin-methylene blue agar (EMB). This medium is used to detect and isolate gram-negative intestinal pathogens. The preparation contains two sugars, lactose and sucrose, and two indicator dyes, eosin Y and methylene blue. The dyes differentiate between lactose-fermenting and non-lactose-fermenting organisms. (a) Non-lactose-fermenting bacteria (light colonies). (b) Lactose-fermenting bacteria (dark colonies).

[25] The characteristics of bacterial colonies grown on Hektoen enteric (HE) agar, another selective and differential medium. The medium incorporates several ingredients, including the three carbohydrates lactose, salicin, and sucrose. HE agar favors the detection and isolation of enteric pathogens and inhibits many nonpathogenic bacteria. In addition to demonstrating sugar fermentation, the medium serves as an indicator of protein-digesting reactions through H_2S production. Examples of obtainable reactions include (a) the moist green colonies of a nonfermenting organism; (b) the yellow colonies of a lactose fermenting organism; and (c) the characteristic black centered colonies of an H_2S producer.

[26] The appearance of *Staphylococcus aureus* on bacto-mannitol salt agar. The yellow zones around the bacterial colonies indicate that acid has been formed from mannitol. This medium does not differentiate between coagulase-positive and coagulase-negative staphylococci.

[27] Carbohydrate fermentation using Durham fermentation tubes. The pH indicator here is phenol red. When enzymatic action occurs, the color of the medium changes. The tube on the left is an uninoculated control. Note its color. The tube in the center has undergone a definite color change, indicating that enzymatic breakdown of the carbohydrate has occurred. Such reactions result in the formation of various acids, some of which have commercial value, while others can cause spoilage under certain conditions. Certain organisms can further decompose these acids to gas. In the tube on the right, gas resulting from this type of reaction has been trapped in an inverted vial.

[28] Selected triple sugar iron agar (TSIA) medium reactions. TSIA is an especially valuable aid to identification when it is used with additional media, such as Endo, EMB, or MacConkey agar. The ability of an organism to ferment dextrose (glucose), lactose, and saccharose (sucrose), producing acid and gas, can be determined with TSIA. The indicator here is phenol red. Also, the medium can be used to detect hydrogen sulfide (H_2S) formation. Alkaline reactions (Alk) are indicated by a red coloration, acid (A) production by a yellowing of the medium, gas formation by the presence of air pockets in the preparation, and the presence of hydrogen sulfide (H_2S) by a blackening of the medium.

[29] Representative growth characteristics of bacteria in fluid thioglycollate broth. Tube A is uninoculated. Note the presence of the upper, pink zone (for aerobes) and the lower, yellow zone (for anaerobes). Tube B shows the growth of a typical anaerobic organism, and tubes C and D contain facultative (adjustable) organisms.

[30] Pour-plate preparation with *Micrococcus luteus.* Note the random distribution and varied shapes of the bacterial colonies. Compare the appearance of *M. luteus* in this preparation to that in Color Photograph 31.

[31] Streak plate preparation with a mixed culture of *Micrococcus luteus* and the red *Serratia marcescens.* Note the well-separated bacterial colonies that have formed on the plate.

[32] Genetic recombination through the fusion of spheroplasts. The parental strains, one white (W) and one orange (O), produced yellow-salmon colored recombinants. [Courtesy H. David and N. Rastogi, Institut Pasteur, Paris, France.]

[33] *Morchella esculenta,* one of several well-known and the most sought-after of all edible fungi. This true morel is an ascomycete (ascomycotina).

[34] An example of a poisonous puffball, *Scleroderma (aurantium) citrinium,* has a round to oval shape and a yellow-brown outer skin with dark warts. Cut in half, the flesh of young forms is dark purple to black.

[35] The poisonous mushroom *Amanita muscaria.* Note the characteristic scales on the cap of this fungus. [Courtesy Dr. H. Bigelow, University of Massachusetts.]

[36] An interesting example of the most obvious group of fungi in North America, the polypores. The characteristic shelflike fruiting bodies of *Polyporus sulphureus* are commonly found season after season on old or dying trees.

[37] The mycelia of *Penicillium* species and *Aspergillus flavus.*

[38] The glistening colonies of brewer's yeast. *Saccharomyces cerevisiae,* growing on Sabouraud's agar.

[39] The colonies of two additional yeasts. *Cryptococcus neoformans* (dark blue) and *Candida albicans* (light blue), on Sabouraud glucose agar containing trypan blue. This agar is a selective and differential medium for yeasts. [R. M. Vickers, et al., *Appl. Microbiol.* **27**:38 (1974).]

[40] Fairy ring disease is destructive to grasslands, including lawns, pastures, and parks. The destructive action is caused by a poisonous fungus excretion that damages the

roots of the host plant, thereby weakening it and permitting a massive invasion by the fungus.

[41] Disabling ringworm infection on areas of the feet. The patient had been wearing wet boots and socks. This fungus infection, also known as *athlete's foot,* is a fairly common condition. It generally is much milder in its effects. [A. M. Allen and D. Taplin. *JAMA.* **226**:864 (1973).]

[42] The appearance of spores on the surface of a fungus-infected hair. [Courtesy Dr. E. S. Beneke, Department of Botany and Plant Pathology, Michigan State University.]

[43] The scrambled egg slime, *Fuligo septica.* Both a well-formed yellow plasmodium and the filamentous newly forming plasmodium are shown.

[44] Lichens are combinations of fungi and either cyanobacteria or algae functioning together in a mutually beneficial relationship. Lichens are found in an extremely wide range of locations. A fruiticose lichen is shown.

[45] Several species of lichens on a tree branch, the fibrous form called "the old man's beard" and orange and white crustose lichens.

[46] The trophozoite of *Entamoeba histolytica,* the cause of amebic dysentery. This organism moves by means of pseudopods.

[47] *Toxoplasma gondii* in a stained bone marrow smear. White and red blood cells are also shown. This protozoon can develop intracellularly as well as extracellularly; note the crescent shape of this microorganism. [C. Abell and P. Holland, *Amer. J. Dis. Child.* **118**:782–787 (1969).]

[48] The trophozoite of the ciliated protozoon, *Balantidium coli,* a cause of gastrointestinal infections.

[49] Humans are subject to a wide variety of protozoan infections, many of which are most frequent in tropical areas. The effects of the skin disease leishmaniasis are shown. (a) A mild case of cutaneous leishmaniasis on the face of a young boy. [Courtesy of Dr. Muna Al-Jaqi.] (b) A severe infection. The appearance of this disease can be mistaken for the bacterial infection leprosy. [Courtesy Dr. Med. K. F. Schaller, Lomé, Togo.]

[50] This marine phytoplankton community of various diatom genera, including *Asterionella* (A), *Navicula* (N), *Rhizosolenia* (R), *Skeletonema* (S), *Thalassionema* (T), and *Thalassiosira* (Th), flourishes during the winter months. [Courtesy Dr. Paul Hargraves, University of Rhode Island.]

[51] The colonies of the bacterium *Streptomyces azureus* with plaques produced by temperate bacteriophages mutated to virulent forms. [Courtesy S. Ogata, Kyushu University.]

[52] An example of phage typing (arrows). [Courtesy of the National Medical Audiovisual Center, Atlanta.]

[53] Signs of virus infection in chick embryos. (a) The appearance of a normal, uninfected chorioallantoic membrane. (b) The development of pocks (small, whitish, dense spots) as a result of virus infection. [Courtesy Centers for Disease Control, Atlanta.]

[54] Intranuclear inclusion bodies (arrows) formed in virus-infected cells. Note the appearance of the nuclei of uninfected cells. [Y. Matsunaga, S. Matsuno, and J. Mukoyama, *Inf. Immun.* **18**:495–500 (1977).]

[55] Immunofluorescent staining of virus-infected lung cells. The cells were exposed to antibodies against a specific virus. The infected cell nuclei can be recognized by their bright yellow-green color. [K. V. Shah, R. W. Daniel, and T. J. Kelley, Jr., *Infect. Immun.* **18**:558–560 (1977).]

[56] Symptoms of plant virus infections. A virus-infected plant showing discoloration and yellowing of leaves. [Reproduced with permission of the National Research Council of Canada from L. N. Chiykowski, *Can J. Bot.* **43**:373–378 (1965).]

[57] The flower on the top shows normal coloration and appearance. The flowers on the bottom exhibit signs of infection, including reduction in size, pointed petals, and lighter than normal coloration. [Reproduced with permission of the National Research Council of Canada from L. N. Chiykowski, *Can. J. Bot.* **54**:1171–1179 (1976).]

[58] The stunting of these tomato plants is a result of potato spindle tuber viroid (PSTV) infection. [Courtesy T. O. Diener, Research Plant Pathologist, USDA.]

[59] The appearance of an AMSCO sterility "Proof Plus" monitoring system. No color change from the control after incubation indicates proper sterilization, whereas a yellow color indicates sterilization failure.

[60] The appearance of various bacteria isolated from raw milk by membrane filtration. [Courtesy Nalge Company, div. Sybron Corporation, Rochester, New York.]

[61] Antibiotic-sensitivity testing is used to determine the antibiotic or antibiotics that may be effective against specific bacterial infections. Although the results of this type of test generally provide some insight into the specific chemotherapy to be used, the *in vitro* effectiveness of antibiotics is not always evident in the patient. In the test shown, the large, clear areas surrounding the individual antibiotic-containing discs indicate that the organism tested is sensitive to several chemotherapeutic agents. The green disc contains penicillin.

[62] The microscopic appearance of the spirochete *Treponema pallidum.* This organism is the causative agent of syphilis.

[63] Plaques, or local zones of destruction formed by lytic bacteria on lawns of the cyanobacterium *Anabaena cylindrica.* The vegetative cells of cyanobacteria are easily attacked and disintegrated by certain parasitic bacteria. [Courtesy Y. Yamamoto and K. Suzuki.]

[64] The cyanobacterium (blue-green bacterium) *Oscillatoria.*

[65] *Spiroplasma,* a helical mycoplasma demonstrated by DNA fluorescent staining. [From T. G. Steiner, J. McGarrity, and D. M. Phillips, *Inf. Immun.* **35**:296–304 (1982).]

[66] Methanogens were isolated from this hot spring in Yellowstone National Park. The gas bubbles on the surface contain methane and carbon dioxide. [J. C. Zeikus, *Bacteriol. Revs.* **41**:514–541 (1977).]

[67] Algae can be found in many varied environments. (a) In hot deserts, certain green algae colonize the microscopic air-space system a few millimeters below the surface of certain porous rocks. (b) For the quartz shown here, the main source of water is dew, which is absorbed and retained by the porous rock. The cooling of the desert at night causes the dew to condense and provide a source of moisture for the algae. With the appearance of the sun, the algae carry out photosynthesis, providing that various minerals and other necessary factors are present.

[68] Endosymbiosis. Certain sea anemones have formed an endosymbiotic relationship with green algae, as evidenced by their green coloration. Nonendosymbiotic anemones are white.

[69] Bioluminescent bacteria growing on agar plants. The colonies were photographed using only their own light. [Courtesy Kenneth Nealson.]

[70] Tumors or galls on plants are sometimes caused by the bacterium *Agrobacterium tumefaciens.* Crown gall, which occurs at the border between the root and stem, is common in several plant species. Some evidence suggests that large plasmids found in strains of *A. tumefaciens* are responsible for such tumors. [H. N. Miller, *Phytopathology.* **65**:850–851 (1975).]

[71] Each of the foods and beverages shown is a product of microbial fermentation. [Courtesy Wine Institute, San Francisco.]

[72] A copper brewing vessel used for boiling wort. Hops are added at this stage of beer making.

[73] The removal of a wine sample to determine alcohol content and other properties. [Courtesy Sebastiani Vineyards.]

[74] Detection of human B lymphocyte virus-infected cells. A ^{35}S-labeled probe derived from the viral genetic material (genome) was used in the preparation of the probe. Detection of viral nucleic acids is shown by the presence of large numbers of black grains. [S. Z. Salahuddin, et al., *Science* **234**:596–601 (1986).]

[75] Coliforms on the surface of a millipore filter. [Courtesy Millipore Filter Corporation.]

[76] Typical blood agglutination reactions. The left side of the slide contains anti-A blood-typing serum, and the right side contains anti-B blood-typing serum. Since clumping, or agglutination, occurs only on the left, the blood type shown is A. Had the reaction occurred on the right, the blood type would have been B, and if the reaction had occurred on both sides, the blood type would have been AB. The complete absence of clumping indicates blood type O.

[77] An example of a latex slide agglutination reaction. The test detects bound coagulase and protein A of *Staphylococcus aureus.* A clumping result is a positive reaction. [Courtesy Scott Laboratories, Inc.]

[78] Complement fixation—tube reaction. The left tube shows the cloudy, strongly positive result (4+), the center tube a weakly positive reaction (2+), and the right tube a clear, negative result. [Courtesy S. Stanley Schneierson, M.D., and Abbott Laboratories, Chicago.]

[79] A comparison of polyclonal and monoclonal antibodies using immunofluorescence. (a) Green fluorescence of cells from a nasopharyageal specimen, indicating the presence of influenza virus by polyclonal antibodies. (b) The specific action of monoclonal antibodies, showing a higher sensitivity for a specific virus. Only green fluorescing cells are infected. [From I. Shalit, P. A. McKee, H. Beauchamp, and J. L. Waner, *J. Clin. Microbiol.* **22**:877–879 (1985).]

[80] Peroxidase-antiperoxidase staining technique. Diagnostic procedures incorporating immunological principles are valuable in the rapid identification of herpesviruses and several other disease agents. (a) A negative result. (b) A positive staining reaction, specific in this case for herpesvirus. [Courtesy Immulok, Carpenteria, California.]

[81] The results of an enzyme-linked immunoabsorbent assay (ELISA) procedure. The assay is based on the ability of antigen or immunoglobulin to absorb to a solid-phase support surface and to be linked to an enzyme, forming a complex showing both detectable immunological and enzyme activity.

[82] Positive human skin test. Indurations of 5 mm or more are considered positive. [Courtesy Mycology Section, Laboratory Division, Centers for Disease Control.]

[83] An indirect immunofluorescent test for antinuclear antibodies. The antinuclear antibody (ANA) test for the detection of the autoimmune condition systemic lupus erythematosus (SLE). (a) A negative reaction. (b) A positive result. Here, anti-DNA antibodies from an SLE patient are bound to the DNA in human tissue culture cell nuclei [Courtesy Smith Kline Instruments, Inc., a subsidiary of Smith Kline & French Laboratories.]

[84] Immunological disorders. (a) Large, blisterlike swellings caused by the bacterium *Pseudomonas aeruginosa* in an individual with agammaglobulinemia. [Courtesy D. T. Lim, M.D., Head, Section Allergy/Clinical Immunology, Cook County Hospital.] (b) The appearance of Kaposi's sarcoma. This condition is found with some acquired immune deficiency syndrome (AIDS) patients. [Courtesy Dr. W. Lawrence Drew, Mt. Zion Hospital and Medical Center.]

[85] Cone-nosed bugs of the family Reduviidae. Species of the genus *Panstrongylus* transmit the causative agent of Chagas' disease, a protozoan infection that damages the heart tissue of newborn infants. [Courtesy Bureau of Vector Control, Berkeley, California.]

[86] (a) The results of an automated streaking procedure. Note the circular pattern of colonies. [Courtesy Tomtec, Orange County, California.] (b) A ready-to-use system manufactured by Analytab Products, Inc. (API). The combination of substrates in this system allows the user to perform 22 standard biochemical tests. The tests in the top row show the appearance of uninoculated systems. The tests in the bottom row demonstrate examples of the color changes indicating positive reactions.

[87] The characteristic growth inhibitory reaction of bacitracin. (The discs contain two units of bacitracin). [Courtesy BBL Microbiology Systems, div. Becton, Dickinson and Co.]

[88] The coagulase test. A positive reaction is generally considered to be the best single indicator of potential pathogenicity. The formation of a coagulated plasma clot is a positive result, while the absence of coagulation is a negative one. [Courtesy Analytab Products, Inc.]

[89] A positive optochin (ethylhydrocupreine) reaction with *Streptococcus pneumoniae.* Note the absence of growth around the disc.

[90] A blood smear containing the protozoon causative agent of African sleeping sickness, *Trypanosoma brucei gambiense.* This disease is spread by tsetse flies.

[91] The appearance of skin lesions found with infections. (a) Blister-like vesicles. (b) Flat macules and erythema (reddening). (c) Pustules, yellow cloudy vesicles, and some dried scabs. (d) Ulcers. [From T. J. Liesegang, *J. Amer. Acad. Dermatol.* **11**:165–192 (1984).]

[92] A pustule from a staphylococcal folliculitis. A small hair can be seen projecting from the center of the lesion. [From L. G. Wickboldt and N. A. Fenske, *Hospital Practice* **21**:41–47 (1986).]

[93] An effect of *Pseudomonas aeruginosa,* the green nail syndrome. The green color is due to pigments produced by *Pseudomonas.* [J. H. Hall, et al., *Arch. Derm.* **93**:312–324 (1968). Copyright 1968, American Medical Association.]

[94] Cutaneous anthrax. [Courtesy Drs. M. H. Matz and H. G. Brugsch, *JAMA* **188**:115 (1964).]

[95] A maculopapular, pustular rash caused by *Pseudomo-*

nas aeruginosa developing after a bath in a hot tub. Note the distribution pattern of the rash. [From P. H. Chandrasekar, et al., *Arch. Dermatol.*, **120**:1337 (1984).]

[96] The catalase reaction. Catalase is an enzyme that catalyzes the breakdown of hydrogen peroxide (H_2O_2), thereby releasing free oxygen gas. The formation of a white froth when a few drops of 3% H_2O_2 are added to a microbial colony or to a broth culture is a positive reaction.

[97] The CAMP test. A positive CAMP reaction seen as a triangle perpendicular to the streak growth of a beta-lysin producing *Staphylococcus aureus* strain. [Courtesy BBL Microbiology Systems, div. Becton, Dickinson and Co.]

[98] (a) Confluent rings of an inflammatory ringworm infection involving the lower abdomen, groin, and hip areas. [A. M. Allen, et al., *Mil. Med.* **137**:295 (1972).] (b) The mycelium of *Epidermophyton floccosum*, a causative agent of ringworm of the groin, *tinea cruris*. (c) The mycelium of *Trichophyton rubrum*. The fungus is a common cause of athlete's foot. [Courtesy Dr. E. S. Beneke, Department of Botany and Plant Pathology, Michigan State University.]

[99] Herpes whitlow. Note the vesicle formation and bleeding. [Courtesy Hugh Zachariae, M.D.]

[100] Koplik's spots, a typical sign of measles infection. [Courtesy Centers for Disease Control, Atlanta.]

[101] A case of monkeypox in a three-year-old child. [Courtesy K. Ruti, I. Arita, Z. Jezek and L. Khodavekich and Chief, Smallpox Eradication Unit, World Health Organization.]

[102] A clinical case of severe periodontitis showing gingivitis. The gums are extremely inflamed and receding, thereby exposing more of the individual teeth. [From D. C. Anderson, et al., *J. Infect. Dis.* **152**:668 (1985).]

[103] Oral thrush. Note the whitish surface film on the sides of the tongue. The film consists of large numbers of yeast cells, which are visible on microscopic examination.

[104] Hairy leukoplakia on the side of the tongue. [Courtesy Sol Silverman, Jr., D.D.S., University of California, San Francisco, and Centers for Disease Control, Atlanta.]

[105] An example of a rapid coagglutination Phadebact test. The presence of a blue precipitate (blue snow) or clumping indicates a positive result.

[106] (a) Lesions of disseminated coccidioidomycosis on the buttocks and back. (b) Coccidioidomycosis involving the patient's finger. In addition to being a respiratory system disease agent, *Coccidioides immitis* can attack other regions of the body. In this case, the infection resulted from a contaminated cut. (c) The appearance of *C. immitis* arthrospores (1,200X). These reproductive units can be found in dust. They serve to spread the disease agent. [Courtesy Mycology Section, Laboratory Division, Centers for Disease Control, Atlanta.]

[107] A cyst of *Pneumocystis carinii* showing eight elongated sporozoites (arrow). The other structures shown are red blood cells. This pulmonary specimen was stained with polychrome methylene blue. [H. K. Kim and W. T. Hughes, *Amer. J. Clin. Path.* **60**:462–466 (1973).]

[108] The detection of microbial toxins. These intestines and stomachs of mice were treated with toxic (left) and nontoxic (right) preparations. The swelling of the organs on the left is apparent. [S. Stavric and D. Jeffrey, *Can. J. Microbiol.* **23**:331–336 (1977).]

[109] MacConkey agar reactions: (a) The colonies of *Salmonella typhi*, a lactose nonfermenting organism. (b) The colonies of a lactose fermenter such as *Escherichia coli*. [Courtesy BBL Microbiology Systems, div. Becton, Dickinson and Co.]

[110] Selected blood stages of *Plasmodium falciparum*. (a) Young trophozoites. The three parasites show a form called a "signet-ring" form. (b) The parasite is undergoing initial nuclear division (red area). (c) A mature schizont with merozoites (M). Note the number of the merozoites. (d) One of the successive events that take place in gametocyte (sex-cell) development. Gametocytes are generally not found in the peripheral circulation. (e) A mature macrogametocyte (female sex cell). (f) A mature microgametocyte (male sex cell). [These descriptions are adapted from U.S. Department of Health, Education, and Welfare, *Manual for the Microscopical Diagnosis of Malaria in Man.* Washington, D.C. (1960).]

[111] Forms of Kaposi's sarcoma. (a) The violet nodules (masses) of Kaposi's sarcoma on the foot of an AIDS patient. (b) A closer view of a violet nodule seen with AIDS patients. (c) Brown papules seen with Kaposi's sarcoma. (d) Kaposi's sarcoma in the mouth of an AIDS patient. [Color Photographs 111a–c courtesy Centers for Disease Control, Atlanta; Color Photograph 111d courtesy Sol Silverman, Jr., D.D.S., University of San Francisco, and the Dental Disease Prevention Activity, Centers for Disease Control, Atlanta.]

[112] A severe form of chronic oral candidiasis in an AIDS patient. [Courtesy of John Molinari, Ph.D., University of Detroit, and the Dental Disease Prevention Activity, Centers for Disease Control, Atlanta.]

[113] (a) Positive (clumping) and (b) negative (no clumping) Staphyloside test. [Courtesy BBL Microbiology Systems, div. Becton, Dickinson and Co.]

[114] A vaginal smear showing a clue cell and mixed bacterial flora (microbiota): p, gram-positive cocci; b, small gram-negative rods; c, curved rods; and g, *Gardnerella*. [From C. A. Spiegel, R. Amsel, and K. K. Holmes, *J. Clin. Microbiol.* **18**:170–177 (1983).]

[115] Herpes simplex virus, type 1, penile ulceration (balanitis). [R. D. Powers, M. F. Rein, and F. G. Hayden, *JAMA* **248**:215 (1982).]

[116] Gonorrhea. (a) A Gram stain preparation of a clinical smear from a patient suspected of having gonorrhea. The organisms are located intracellularly within leukocytes and exhibit the characteristic "coffee bean" appearance (paired cocci with adjacent sides flattened). During the early stages of gonorrhea, or in cases of longstanding infection, *Neisseria gonorrhoeae* can be found extracellularly. [Courtesy Dr. D. S. Kellogg, Jr., and the Venereal Disease Research Laboratory, National Communicable Disease Center, Atlanta.] (b) The oxidase test. Members of the genus *Neisseria* produce a positive oxidase reaction when tetramethyl-p-phenylenediamine (oxidase reagent) is applied to their colonies. Colonies demonstrating a positive result are first pink and later dark red (P). Negative reactions are indicated by colorless colonies (arrows). This test does not differentiate one *Neisseria* species from another. [Courtesy S. Stanley Schneierson, M.D., and Abbott Laboratories, Chicago.]

[117] Chancroid. The lesions here differ from those found in syphilis. Chancroid lesions are usually soft and tender.

[118] Stages of syphilis. (a) Secondary syphilis: typical Hunterian chancre on the lower lip. (b) Secondary syphilis: extensive papulosquamous rash on the body. (c) The gumma is the hallmark of lesions of late benign syphilis and is a classic example of granulomatous inflammation. The

gumma, from an anatomic viewpoint, is a firm, white lesion that may vary from microscopic size to 10 centimeters or more in diameter. This photograph shows two of these lesions in a liver specimen. A firm, white, bleeding, and largely necrotic gumma can be seen at the lower end (arrow). [Courtesy U.S. Department of Health, Education, and Welfare, Centers for Disease Control, *Syphilis, a Synopsis,* Public Health Service Publication No. 1660 (1968). Reproduced by permission.]

[119] Characteristic herpetic skin lesion (pustule) in an inguinal area. [R. D. Powers, M. F. Rein, and F. G. Hayden, *JAMA* **248**:215 (1982).]

[120] Herpes simplex virus, type 2 infections. (a) Penile infection. [Courtesy Dr. W. Lawrence Drew, Mount Zion Hospital Center and Medical Center.] (b) Herpes genitalis in a female patient. Note the large number of blisters. [Courtesy Dr. Lawrence Corey, Head, Virology Division, University of Washington.]

[121] Intracellular *Toxoplasma gondii* in a bone marrow smear. The preparation was stained by the Giemsa method. [C. Abell and P. Holland, *Amer. J. Dis. Child.,* **118**:782–787 (1969).]

[122] The general appearance of conjunctivitis (left eye). The right eye is normal. [Courtesy Drs. G. Cobbs and I. C. Ryden.]

[123] The condition known as cutaneous larva migrans caused by the invasion of the foot and subsequent movement of young forms of a dog hookworm. [Courtesy of Dr. Christopher Lee.]

[124] The effects of the roundworm *Elaeophora schneideri* on a Barbary sheep. These skin lesions extend from the area surrounding the eyes to the ears and muzzle of the animal. [Courtesy D. B. Pence and G. G. Gray.]

[125] A blood smear showing the presence of the microfiliarian parasite *Wuchereria bancrofti.* This helminth is the causative agent of filiariasis. *Culex* spp. are frequently the mosquito vectors for the disease agent.

[126] A microscopic view of *Trichinella spiralis* larvae in muscle tissue.

[127] A large number of sheep tapeworm scolices. A fleshy elevated region or rostellum can be seen on one scolex.

[128] A microscopic view of a spined schistosome ovum (egg) in host tissue.

[129] The appearance of the papules in a case of swimmer's itch caused by the penetration of bird cercariae. [From P. Kimmig, *Zbl. Bakt. Hyg. I. Abt. Orig. B.* **181**:390–408 (1985).]

(a)

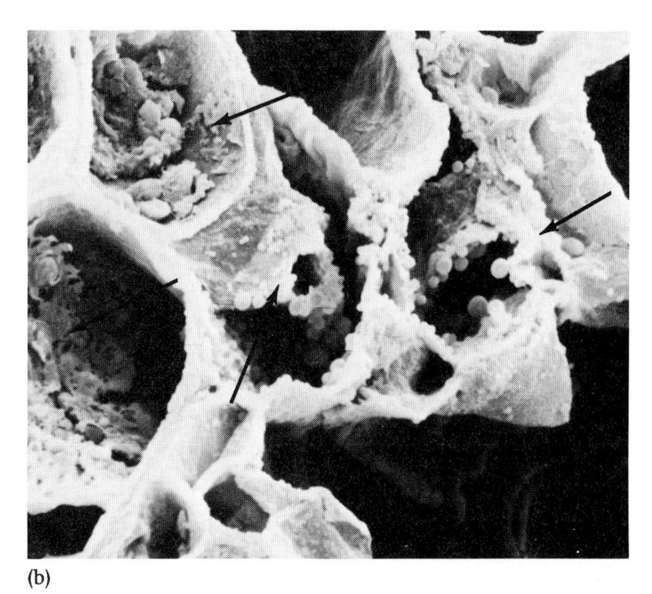

(b)

Figure 14–31 Parasitism in plants. (a) A fungus-caused canker on the mark of a pine tree. (b) Plant (parenchyma) cells, showing the presence of the fungal components (arrows) that cause the canker. *[B. L. Welch, and N. E. Martin,* Phytopathology ***64****:1541, 1974.]*

PARASITISM

Parasitism is an association between two specifically distinct organisms in which one, the **parasite,** lives on or within the other in order to obtain essential nutrients. The organism that harbors a parasite is called the **host.** One of the more familiar parasites is the tapeworm, but parasitism is not limited to multicellular animals. Parasites may also be plants, such as mistletoe and dodder, or microorganisms (Color Photograph 70). All disease-producing agents can be considered parasites. This category includes viroids (Figure 14–30), viruses, bacteria, fungi (Figure 14–31), and protozoa. The effects of parasites will be described in later chapters.

Microbes in Industry and Biotechnology

The descriptions and discussions in Chapters 13 and 14 clearly emphasize the many significant roles microorganisms play in environmental activities and their interrelations with and among a wide variety of forms of life. Industry has also recognized the economic benefit to be gained from the proper control of several of these microbial processes. The large-scale growth of microorganisms for the production of food, animal feed, antibiotics, vitamins, and industrial chemicals involves ever-changing industrial activities that are responsible, in the United States alone, for the generation of tens of billions of dollars annually.

Added to the industrial importance of microbes is their association with the handling and purification of water. While water is essential for life, it can also be a hazard, since it can carry microbial pathogens and toxic chemicals. The chapters in the next section examine some of the microorganisms and microbiological principles associated with and applied to water treatment, industry, and biotechnology.

Summary

The influence of the environment on all forms of life and the various relationships among organisms is critical to their survival.

Basic Ecological Principles

1. The *biosphere* includes the general environment of all living things. It extends up into the atmosphere to more than 10,000 m, down into the ocean to about 8,000 m, and more than 250 m below land surfaces.
2. The biosphere is a global biological system that is based on a continuous, or cyclic, flow of energy and nutrients.

Organization of the Biosphere

1. *Ecosystems* are ecological units of the biosphere.
2. Groups of individuals having similar properties and living in the same area are called *populations.*
3. The different forms of life of an ecosystem fall into three categories: *producers, consumers,* and *decomposers.* These organisms are responsible for maintaining the stability of the system.
4. Producers—which include photosynthesizing algae, bacteria, and plants—form organic compounds from inorganic materials.
5. Consumers utilize these organic compounds for their essential activities.
6. Decomposers break down complex decaying organic matter, which can then be used by plants and other organisms for their producer activities.
7. Producers, consumers, and decomposers together constitute circular processes that promote the exchange of nutrients and energy flow in the ecosystem.

The Components of an Ecosystem

1. The specific location or place of residence of an organism is called its *habitat;* its specific role in a community is its *niche.*
2. Habitats for any living organism are limited by both *biotic* (living) and *abiotic* (nonliving) factors.
3. Biotic factors include other animals, plants, and microbes; examples of abiotic components are hydrostatic pressure, light, moisture, minerals, osmotic pressure, pH, temperature, and chemical pollutants. The balance of an ecosystem can be disrupted by disturbing a factor of either type.

Natural Habitats of Microorganisms

1. The earth can be divided into three zones: the *atmosphere* (gaseous regions), the *hydrosphere* (aquatic environments), and the *lithosphere* (solid portions).
2. Most habitats are *terrestrial* and *aquatic* environments.
3. *Terrestrial habitats* include the rocky slopes of mountains, the lush tropical jungle, and the hot desert sands.
4. *Aquatic* environments fall into two general categories: fresh water and marine water.
5. *Estuaries* are coastal regions where sea water mixes with fresh water.

Microbial Activities

1. The activities in ecosystems can be one of three types: *action, reaction,* or *coaction.*
2. Action is concerned with how abiotic factors affect organisms.
3. Reaction is concerned with the biological activities in an ecosystem.
4. Coaction is concerned with the interactions among organisms.

Action Activities

Microorganisms have been found growing under severe environmental conditions including extremes of temperature, acidity and alkalinity, osmotic pressure, and hydrostatic pressure.

Reaction Activities The Life-supporting Biogeochemical Cycles

1. Biologically important elements such as carbon, hydrogen, oxygen, nitrogen, phosphorus, and sulfur are removed from the environment, incorporated into cellular structures, and eventually returned to the environment to be used over again.
2. The natural biogeochemical cycles are interrelated to ensure the flow of essential chemical elements between the living and nonliving parts of ecosystems. No one cycle can function without the others.
3. The microbial processes associated with bioconversion may also lead to the formation of new toxic substances *(toxicants)* or persistent products.
4. The biodegradation of synthetic organic polluting substances may involve the process known as *cometabolism.* Here toxic chemicals, although subject to microbial action, do not support microbial growth. Other nutrient sources are used by responsible microorganisms.

The Carbon Cycle

1. Carbon is one of the most common and important elements involved with living systems.
2. The *carbon cycle* is the route by which useful solar energy enters the biosphere through the process of photosynthesis.

The Nitrogen Cycle

1. Organisms need nitrogen to synthesize amino acids, proteins, and nucleic acids.
2. The *nitrogen cycle* serves to convert essential compounds from one type to another, often with the release of energy that is captured during metabolic activity.
3. The cycle consists of two phases, *nitrogen fixation* and *nitrogen assimilation.*
4. Nitrogen fixation is the reduction of atmospheric molecular nitrogen to ammonia (NH_3) and other biologically essential nitrogen-containing compounds. The process involves cyanobacteria and other bacteria.
5. In nitrogen fixation, the enzyme *nitrogenase* binds N_2 and the following reaction occurs:
   $N_2 + 6H^+ + 6e^- + 12ATP \rightarrow 2NH_3 + 12ADP + 12Pi$

6. Nitrogen fixation can occur symbiotically between legumes and bacteria of the genus *Rhizobium,* and nonsymbiotically by some cyanobacteria and various other bacteria, or various nonleguminous root-nodule plants.
7. Microbial enzymatic activities involving decaying organic matter break down organic nitrogen molecules to NH_3. The process is known as *ammonification.*
8. Chemosynthetic bacteria of the genus *Nitrosomonas* oxidize NH_3 to nitrites (NO_2^-), which in turn are oxidized to nitrates (NO_3^-) by *Nitrobacter* species. This is nitrification.
9. Denitrification completes the nitrogen cycle and results in the conversion of nitrates into nitrogen gas.

THE PHOSPHORUS CYCLE

1. Without phosphorus, compounds such as ATP, DNA, and RNA would not be formed.
2. Through the *phosphorus cycle,* producer organisms convert inorganic phosphate (PO_4^{3-}) into organic phosphates.

THE SULFUR CYCLE

1. Sulfur is another essential component of proteins.
2. Through the *sulfur cycle,* oxidized sulfur is obtained and assimilated by living organisms into sulfur-containing amino acids, cysteine, cystine, and methionine.

THE OXYGEN CYCLE

By means of the *oxygen cycle,* an oxygen pool is generated to meet the needs of all land and aquatic respiring organisms.

Additional Reactions of Ecological Concern

Most metabolic and recycling reactions in terrestrial and aquatic environments are beneficial.

BIOCONVERSION

1. The biological or enzymatic formation of a useful product is referred to as *bioconversion.*
2. Examples include methane production, degradation of polluting petroleum products, and the detoxication of certain pesticides.

BIOCONCENTRATION

Bioconcentration refers to the ability of microorganisms to store materials in the environment. Such materials include sulfur, manganese deposits, pesticides, and radioactive materials.

Coaction Activities

Five types of coactional relationships are recognized: *syntrophism, competition, antagonism, predation,* and *symbiosis.*

SYNTROPHISM

1. While organisms are not intimately associated with one another in syntrophism, they benefit from one another's activities.
2. Examples of syntrophism include the feeding of organisms in soil where the digestion and decaying plant material occurs; yogurt production; and waste digestion in sewage disposal.

COMPETITION

1. Competition is the interaction between organisms resulting from a demand for nutrients and energy that exceeds the immediate supply.
2. Some microbes eliminate competitors by using a limited supply of nutrients more efficiently.

ANTAGONISM

Some organisms can control the inhabitants of their environments by secreting toxic substances that interfere with microbial growth.

PREDATION

Predation plays an important role by controlling various populations of life. The *predator* ingests other species, called the *prey.*

SYMBIOSIS

1. In symbiosis two different forms of life coexist in an intimate ecological relationship. This type of relationship requires close physical contact.
2. *Endosymbiotic* relationships refer to situations in which one partner is actually inside a cell or tissue of the other partner; in *ectosymbiotic* relationships, one member is external to the other.
3. Symbiosis can take one of three basic forms: in *mutualism* both members benefit from the association; in *commensalism,* only one organism benefits from the association, while the other partner neither benefits nor is harmed; in *parasitism,* one organism benefits at the expense of the other.
4. Mutualism can be found between microorganisms (lichens), between microorganisms and plants (nitrogen-fixing legumes), and between microorganisms and animals (termites, fungus-cultivating ants, fish with luminous organs, and ruminants).
5. An example of commensalism is the relationship between higher forms of animal life and their intestinal microorganisms.
6. In parasitism, the organism living on or within the other partner is called the *parasite,* while the supplier of nutrients is the *host.* Parasites can be animal, plant, or any disease-producing microorganism.

Questions for Review

1. a. What is the biosphere?
   b. How is it organized?
   c. What is the difference between a population and a community?

2. Describe the functions and importance of producers, consumers, and decomposers in an ecosystem.

3. a. Distinguish between abiotic and biotic factors in the biosphere.
   b. How do abiotic and biotic factors affect one another?

4. a. What is an ecosystem?
   b. Give two examples.

5. Explain the concept of a food chain.

6. Distinguish between a habitat and a niche.

7. Give the major features of terrestrial and aquatic habitats.

8. In what types of habitats are microorganisms found?

9. What is meant by action, reaction, and coaction activities? Explain these terms in relation to ecological concepts.

10. What is a barophile?

11. Briefly outline the carbon, nitrogen, sulfur, and oxygen cycles.

12. Of what value is nitrogen fixation?

13. Of what biological value is phosphorus to living systems?

14. Explain or define each of the following terms:
a. photosynthesis
b. chemosynthesis
c. nitrogen fixation
d. legume
e. nitrification
f. denitrification

15. Define and discuss *microbial infallibility.*

16. Discuss methanogenesis as a bioconversion process of considerable importance.

17. What is the involvement of microorganisms in the degradation of petroleum in nature?

18. Relate microbial activities to the detoxification of organophosphate pesticides and to their role in modifying chlorinated hydrocarbons and inorganic mercury.

19. What is the involvement of microorganisms in the bioconcentration of manganese in the oceans?

20. Relate microbial activities to the toxic accumulation of chlorinated hydrocarbons and radioisotopes in aquatic and terrestrial habitats.

21. Define each of the following relationships and give at least one example of each type of association involving microorganisms:
a. syntrophism
b. predation
c. mutualism
d. commensalism
e. competition
f. parasitism

22. Define or explain each of the following terms:
a. lichen
b. autotroph
c. ectosymbiont
d. host
e. prey
f. rumen
g. nodule
h. viroid
i. magnetotaxis

Suggested Readings

ALEXANDER, M., "Biodegradation of Chemicals of Environmental Concern," *Science* **211**:132 (1981). *An up-to-date presentation of the role that microorganisms play in conversion of a variety of synthetic organic compounds to inorganic products.*

BLAKEMORE, R. P., and R. B. FRANKEL, "Magnetic Navigation in Bacteria," *Scientific American* **245**:58, 1981. *An updated coverage of certain aquatic bacteria that tend to swim along magnetic field lines and are influenced by the earth's magnetic field.*

BOTKIN,, D. B., and E. A. KELLER, *Environmental Studies.* Columbus, Ohio: Charles E. Merrill, 1982. *A readable general textbook on ecology that concentrates on fundamental principles of environmental studies, the physical and biological processes, resources, and environmental planning.*

BRILL, W. J., "Biological Nitrogen Fixation," *Scientific American* **236**:68–81, 1977. *A nicely illustrated article on the few bacteria and algae that can "fix" atmospheric nitrogen into ammonia.*

BROCK, T. D. (ed.), *Thermophiles: General Molecular, and Applied Microbiology.* New York, N.Y.: John Wiley & Sons, A Wiley-Interscience Publication, 1986. *Describes many thermophilic microorganisms in their various habitats as well as methodologies for study, physiology, genetics, and industrial applications.*

COOKE, R., *The Biology of Symbiotic Fungi.* New York: John Wiley & Sons, 1977. *Attempts to outline all major symbiotic relationships that exist between fungi and either animals or plants, describing how these various associations function.*

CURDS, C. R., "The Ecology and Role of Protozoa in Aerobic Sewage Treatment Processes," *Annual Review of Microbiology* **36**:27, 1982. *An article describing the major role of ciliated protozoa in aerobic waste treatment processes.*

EHRLICH, H. L., *Geomicrobiology.* New York: Marcel Dekker, 1981. *Describes the important role of microorganisms in geological processes. There are excellent discussions of methods and transformations of many organic and inorganic substances.*

GUTNICK, D. L., and E. ROSENBERG, "Oil Tankers and Pollution: A Microbiological Approach," *Annual Review of Microbiology* **31**:379, 1977. *This review stresses the special character of hydrocarbon-utilizing bacteria and the use of these organisms in some of the problems arising from oil tanker operations.*

JANNASCH, H. W., "Experiments in Deep-Sea Microbiology," *Oceanus* **31**:50, 1978. *An interesting article dealing with the study of microorganisms in deep-sea environments and how decomposition of organic materials occurs under conditions of high pressure and low temperature.*

KLUG, M. J., and C. A. REDDY (eds.), *Current Perspectives in Microbial Ecology,* Proceedings of the Third International Symposium on Microbial Ecology, Washington, D.C.: American Society for Microbiology, 1984. *An excellent, if somewhat detailed, look at the broad aspects of microorganisms in nature, partly in terms of their applications and the future directions foreseen for microbial ecology.*

MAH, R. A., D. M. WARD, L. BARESI, and T. L. GLASS, "Biogenesis of Methane," *Annual Review of Microbiology* **31**:309, 1977. *The methane bacteria are discussed as regards their role in interspecies hydrogen transfer, methane from carbohydrates, acetate, and the bacteria isolated and grown in pure culture thus far.*

RANGASWAMI, G., *Agricultural Microbiology.* New York: Asia Publishing House, 1966. *A good general book on historical aspects of agricultural microbiology, the organisms involved, and the methods used to study them. Pathogenic organisms in the soil and on plants and organisms in food and dairy microbiology are also discussed.*

WEINBERG, E. D. (ed.), *Microorganisms and Minerals.* New York: Marcel Dekker, 1977. *Describes the way microbial cells acquire and store selected mineral elements and the functions of manganese, calcium, magnesium, and iron in microbial cells. Microbial roles in cycling of mineral elements and antimicrobial action of mineral elements are also discussed.*

WOLIN, M. J., "Fermentation in the Rumen and Human Large Intestine," *Science* **213**:1463, 1981. *A comprehensive treatment of the microbial ecosystem of the complex stomach (rumen) of domestic ruminants. These animals rely on the digestion of food by microorganisms for essential macro- and micronutrients.*

PART VI

Micro-organisms and Industrial Processes

The aging of wine in oak barrels. [Courtesy of Sebastiani Vineyards.]

15

Industrial Microbiology and Biotechnology Applications

Science discovers, genius invents, industry applies, and man adapts himself to, or is molded by new things. — Lowell Tozer

After reading this chapter, you should be able to:

1. Describe at least three microbial processes that are of commercial interest.
2. Discuss the basic similarities and differences in the making of beers, wines, and spirits, and ethanol as a fuel source.
3. Outline the processes by which the following microbial products are made and then used: antibiotics, vinegar, amino acids, vitamins, lactic acid, citric acid, butanol, isopropanol, and polysaccharides.
4. List six enzymes, their microbial origins, and their uses.
5. Discuss the significance of immobilized enzymes and cell systems in industrial processes.
6. Describe the basic difference between most microbial processes and steroid transformation.
7. Discuss the use of microorganisms in the biological control of insects and weeds.
8. List at least three activities of microorganisms that can be monitored in order to control industrial processes.
9. Discuss why the patenting of microorganisms was not possible before 1980.
10. Discuss the role of genetic engineering in the development of new industrial processes.

Human beings have used a great variety of microorganisms for economic gain or for prevention of economic loss. The use of microbes has been an indisputable factor in the economic livelihood of the world, in meeting individual needs, as an environmental consideration, and for food production. Modern industrial microbiology not only seeks to improve traditional processes but to manipulate and control microorganisms in the large-scale production of foods and antibiotics, and to develop entirely new microbial-related industries. This chapter presents several aspects of genetic engineering applications to industrial microbiology.

Many household substances are products of microbial activity. Several of these substances have been used since antiquity. Through the combined efforts of chemists, engineers, and microbiologists, modern technology can direct microorganisms to do the bidding of human beings.

Introduction to Industrial Microbiology

Industrial microbiology is the use of microorganisms or a microbiological technique in a commercial enterprise. Beneficial microbial activities range from the production of foods, antibiotics, enzymes, and various organic chemicals to the preparation of alcoholic beverages, blood plasma substitutes, and **steroids** such as **hormones.** Important microbiological techniques are used for microbial contamination control, sterilization, and vitamin assays. Clinical microbiology, which includes the identification and diagnosis of infectious diseases, traditionally has not been considered part of industrial microbiology. However, the advent of many highly profitable diagnostic laboratories concerned with medical microbiology and chemistry suggests that clinical testing laboratories should be added to the roster. Automation and associated approaches have also influenced several of these industries.

This chapter deals with a group of important and representative industrial processes: the manufacture of alcoholic beverages, antibiotics, selected organic acids, alcohols, enzymes, and steroids; the use of microorganisms in biological control; the application of DNA recombinant (genetic engineering) techniques; and the patenting of microorganisms for use in industrial processes.

Aerobic Versus Anaerobic Processes

Microorganisms are well suited to serve as chemical factories to perform industrial processes. Various microbes not only have a broad range of enzymes to chemically convert a large number of organic substances but also are able to perform such metabolic conversions rapidly.

At times references to fermentation have been incorrectly applied or misunderstood. In a strict biochemical sense the term **fermentation** applies only to reduction of organic substrates performed under anaerobic conditions for the generation of energy. A classic example of fermentation is the reduction of pyruvic acid to lactic acid, a reaction particularly important in the food and dairy industries. Many industrial processes are performed with microorganisms growing within a medium. While such *submerged cultures* resemble fermentation systems, they are aerobic and require large quantities of air to be maintained. A major example of an industrial submerged culture process involves the production of antibiotics. Thus, in industrial microbiology circles, "fermentation" may refer to any large-scale process regardless of whether it is aerobic or anaerobic.

Alcoholic Beverages and Related Products

The production of alcoholic beverages is an industry of enormous economic significance in many countries (Color Photograph 71). The manufacture of all three major types of these products—beer, wine, and spirits (distilled alcoholic beverages)—is an application of industrial microbiology.

Historically, there is evidence that professional wine and beer making was well established by 3000 B.C. An Assyrian tablet dating back to 2000 B.C. states that Noah took beer with him aboard the ark. Egyptian and Chinese documents describe the production and use of beer in 2500 and 2300 B.C., respectively.

Beer

Beer is the product of a yeast fermentation process that uses barley or other grains as the source of sugars and various nitrogen-containing compounds (Figure 15–1). Traditionally, beer results from yeast fermentations of barley. Because yeasts cannot ferment starch (the complex form of sugars contained in grains), a preliminary step, called *malting,* is required to break down the starches. In this process, the grain is kept moist, causing it to sprout and initiating the production of the enzyme amylase. The enzyme breaks starch down into various simpler sugars that can be fermented by yeast. The sprouting malt grains are readied for fermentation by drying and then mixing or mashing with water to achieve further starch digestion. Before the introduction of yeast, the liquid fermentation medium or

Microbiology Highlights

BATCH AND CONTINUOUS FERMENTERS

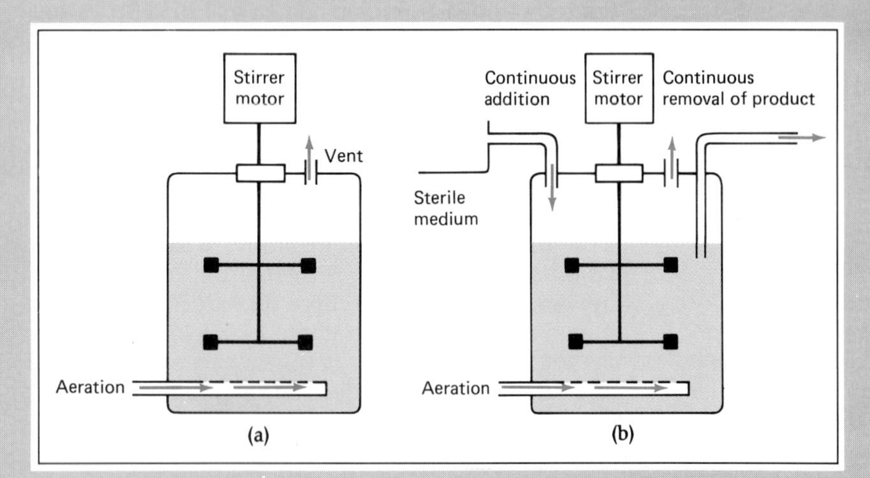

Types of fermenters. (a) Batch system. (b) Continuous system.

Large-scale industrial fermentations are performed in vessels called fermenters, of which there are two types, *batch* and *continuous*. In a batch system, the course of the fermentation process is similar to that of a typical microbial growth cycle. Cell numbers increase rapidly during exponential growth and reach stationary levels when the nutrients in the growth medium are exhausted. Desired fermentation products generally appear only after the stationary phase begins. Harvesting is carried out when maximum levels of such products are reached.

Continuous fermentation differs from the batch system in several ways. In the continuous process, fresh growth medium is slowly and continuously pumped into an already filled fermenter. This addition of the new medium causes a steady overflow of fluid containing the desired fermentation product. The process may be carried out for several days or weeks and left at the maximum stage for long periods of time, thus making continuous collection of the product possible.

Fermentation processes of any type must be carefully monitored and controlled to ensure maximum yield of products. Many factors including heat and pH can be regulated automatically.

wort is heated to eliminate undesirable microorganisms. Hops, dried petals of the vine *Humulus lupulus,* are added to enhance the color and flavor of the final product and to stabilize it. Hops contain, among other chemical compounds, two antibacterial substances, *humulon* and *lupulon,* which prevent bacterial contaminants from spoiling the beer. The fermentation process is complete when the desired alcohol concentration is reached. The beer is then filtered and pasteurized, or filtered through a very fine screen to remove bacteria. Finished beer may become cloudy on being refrigerated due to the precipitation of proteins by the cold temperature. Beer can be protected from this reaction or *chillproofed* by the addition of proteolytic enzymes to degrade the proteins responsible.

There are two general types of beer, *ales* and *lagers.* Ales are fermented by *Saccharomyces cerevisiae* (sak-ah-roh-MY-sees se-re-VISS-ee-eye) strains, which are called *top yeasts.* This terminology refers to the fact that the yeast cells produce uniform cloudiness (turbidity) and are carried up in containers to form foams by carbon dioxide. Ales are incubated at 14° to 23°C for five to seven days and have higher alcohol concentrations than lagers. Lagers are fermented by strains of *Saccharomyces carlsbergensis* (sak-ah-roh-MY-sees karls-BER-jen-sis), also known as *bottom yeasts* because they tend to form sediments. Lagers are incubated at lower temperatures, 6° to 12°C, for a slightly longer time, 8 to 10 days. They contain more unfermented carbohydrates that are not converted into ethanol.

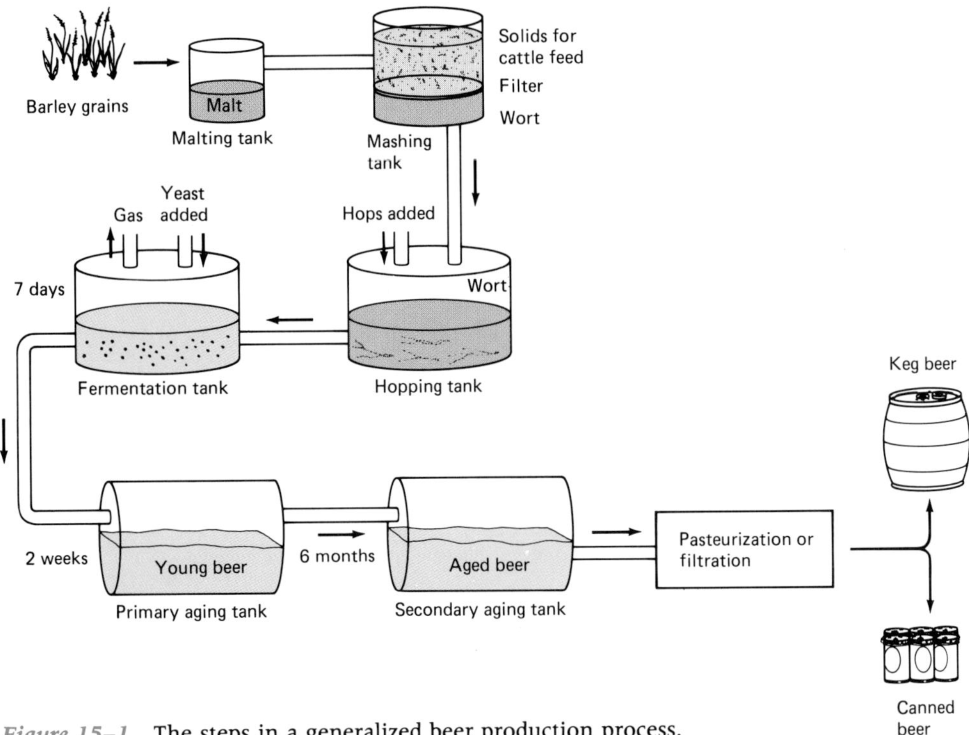

Figure 15–1 The steps in a generalized beer production process.

Wine

Wine is made by the fermentation of almost any ripe fruit juice or from extracts from certain vegetable products, such as dandelions. Because these starting materials contain a high concentration of sugar (up to 30%), as much as 15% alcohol may be obtained with certain yeast strains. The grapes used for wine usually range in sugar content from 12% to 30% and in water content from 70% to 85%. The other constituents in grapes, including acids and minerals, account for the different color, taste, and sometimes bouquet of various wines.

As soon as the skin of the grape is broken, the fermentation begins. Commercially, the fruit is crushed and pressed mechanically (Figure 15–2). The resulting pressed grape juice, called **must,** can be fermented naturally, as is done in Europe, by the enzymatic activity of wild yeasts that grow on the grapes. In the United States, on the other hand, grape juice and other fermentable substances are usually either sterilized or treated with sulfur dioxide to prevent the growth of undesirable microorganisms. Specific yeast cultures, strains of *Saccharomyces ellipsoideus* (sak-ah-roh-MY-seez el-lip-SOY-dee-us), are then introduced, and the mixture of fruit juice and microorganisms is aerated in order to promote yeast growth. Anaerobic conditions result when yeasts are present in sufficient numbers to carry out the fermentation.

When the fermentation is complete, as determined by the alcohol content of a sample, the wine is placed in vats to clarify and age (Color Photograph 73). Aging involves continuing the enzymatic activities under anaerobic conditions until the flavor and aroma for that particular wine have developed. While aging takes place, the wine slowly clears and the suspended solid material settles to the bottom of the container in the form of a sediment. Normally, the product is separated from its sediment and transferred several times to smaller containers. This procedure, called *racking,* is continued until the aging process reaches its limit in the tanks. To prevent spoilage, the resulting product is pasteurized or filtered and then bottled.

The wide range of wines currently available is the result of changes in the basic fermentation process. Must from both red and white wine grapes is white and is used to produce white wine. Red wines are made by allowing the fermentation to proceed in the presence of red grape skins. The alcohol, formed during the process, extracts the coloring into the wine. Rosé wines are prepared by lightly pressing the skins of purple grapes and allowing fermentation to occur rapidly in the presence of the grape skins. Certain wines are also distinguished from others on the basis of sweetness. In dry wines, for example, most or all of the sugar is metabolized during fermentation. In sweet wines, fermentation is stopped while some sugar still remains in the product.

Some types of wines are known for a particular

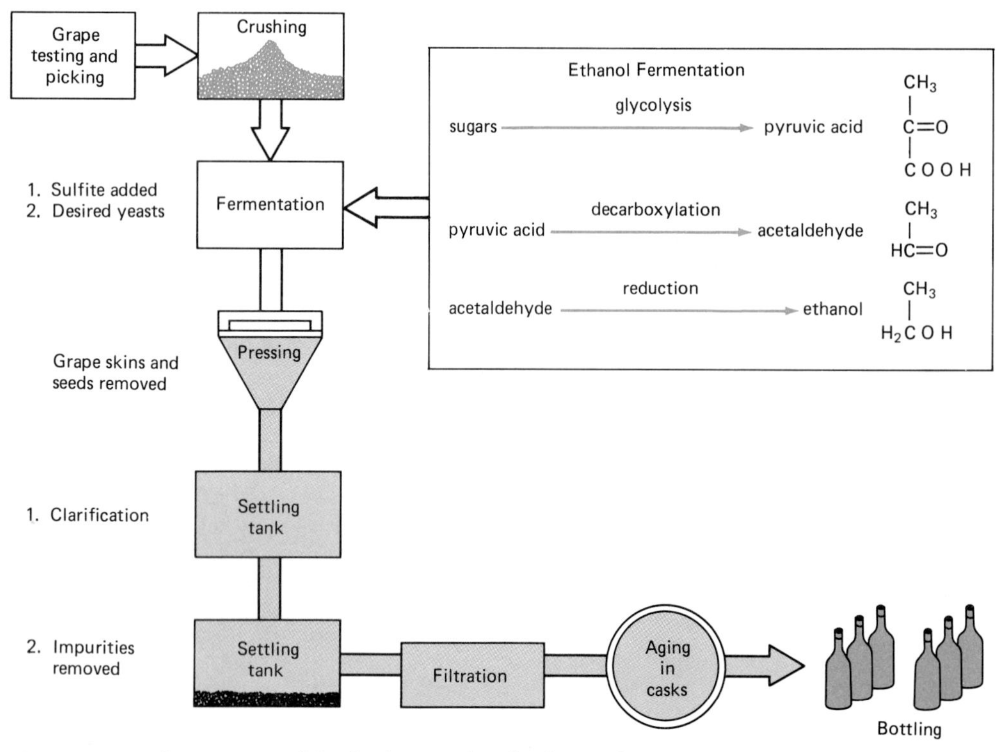

Figure 15–2 A summary of the basic steps involved in making red wine. The sulfite is added to control spoilage microorganisms. For white wines, pressing precedes fermentation. The general steps involved in ethanol fermentation are also shown.

flavor, effervescence (sparkling property), or higher alcohol content. These include *sparkling wines* and *fortified wines.* Sparkling wines, such as champagne, undergo a second alcoholic fermentation inside the bottle under pressure. Sugar and yeast added to an already fermented product results in CO_2 production, which carbonates the wine.

Since yeasts cannot survive in wines that are more than 15% alcohol, most table wines contain 10% to 12%. Alcohol concentrations of wines may be increased through the addition of brandy or other spirits. Such products are called fortified wines and include Madeira, port, and sherry.

While wines are usually considered to result from yeast fermentation, at least one exception is known: palm wine. This alcoholic product is formed by the activity of the bacterium *Zymomonas mobilis* (ZYE-moh-moh-nass moh-BIL-is). The rate of production and yield of ethanol are both higher with *Z. mobilis* than with yeast.

Sake

Although sake is called a rice wine, it is actually manufactured more like beer. A mixed culture of the mold *Aspergillus oryzae* (a-sper-JIL-lus oh-REE-zay) and a species of *Saccharomyces* is used to inoculate steamed rice. The mold grows throughout the rice, producing a partially fermented material called *koji,* which is then mixed with water and more steamed rice and incubated. As the *Aspergillus* digests the starch in the rice to form maltose and glucose, the *Saccharomyces* uses the sugars to produce ethyl alcohol to the extent of about 14%.

Distilled Beverages

Beverages obtained through distillation include whiskey, brandy, and rum. The product is dependent upon the material used in the fermentation. The whiskeys are prepared from different types of grains, brandy from fruit juices, and rum from molasses. Rye whiskey is produced from distilled fermented rye grains, bourbon from corn, and scotch from barley malt. The various flavors of whiskeys result from the presence of minor ingredients known as *congeners.* These substances include aldehydes, esters, ethers, higher alcohols, and acids that evaporate easily.

Distillation can be performed by the use of a vacuum and heat. This separates the volatile products from most of the water and solids. The resulting distillate has an alcohol content of 90% to 96%. Distilled products are labeled according to their alcohol content in terms of *proof levels.* Proof is twice the value of the percent alcohol. Thus, in beverages containing 90% to 96% alcohol, the proof is between 180 and 192. The raw product is then aged, or matured, in wooden casks or barrels. The distillate is also diluted with water to reduce the alcohol content to approximately 51.5%, or 103 proof. During storage, the concentration will vary because of evaporation. When ready for bottling, the product is usually standardized to a particular value, commonly 80 proof.

Fuel Alcohol

The rising cost of petroleum fuel has renewed interest in ethanol as fuel. In contrast to petroleum, which is a fossil fuel and nonrenewable, ethanol can be produced from renewable resources such as plant carbohydrates. These carbohydrates include sucrose from beets or beet molasses, starch from grains or potatoes, and cellulose and hemicellulose from various plant materials. In terms of complexity, sucrose can be fermented directly, while starches require mashing and cellulose requires extensive digestion. Because of the alcoholic beverage industry, the conversion of sucrose and starch to ethanol has received more attention. However, cellulose conversion is potentially the most significant process, since it would result in the conversion of waste cellulose such as paper products, leaves, and tree trimmings to ethanol.

In addition to cellulose and hemicellulose, which make up 50% to 75% of the total plant weight or biomass, lignin, a cell-wall polysaccharide, contributes 15% to 25%. Both cellulase and ligninase-producing microorganisms have been studied for their potential use in fuel alcohol production. These organisms have included the bacterium *Arthrobacter* (ar-thro-BACK-ter) species KB-1 and the fungus *Phanaerochaete chrysosporum* (fan-air-OH-kete cry-SOH-spor-um). The biodegrading of lignin yields the 5-carbon sugar xylose, which must then be fermented to ethanol. Four yeasts capable of this fermentation are *Candida shehatae* (KAN-did-ah shee-HA-tee), *C.* species xF217, *Kluyvermyces marxianus* (KLY-ver-my-sees marks-EE-ann-us), *Pachysolen tannophilus* (pack-y-SOH-len tan-OFH-il-lus), and *Pichia stipitis* (pick-EE-ah stip-EE-tiss).

Brazil has developed extensive fuel alcohol production facilities using beet molasses as the source of sugar. Conversion of the sucrose to ethanol can be accomplished in less than a day, yielding a product having 8% ethanol. One project in the United States has achieved a product with a yield of 16% ethanol with a three-day incubation period. Considering the overall cost of materials and distillation, the ethanol fuel can be produced for approximately $1.50 per gallon in the United States, according to recent estimates.

Butanol and Isopropanol

The industrial need for large amounts of alcohol for solvents can be met by such compounds as butanol and isopropanol. Both compounds are produced primarily as by-products of the nonbiological process for making gasoline from crude petroleum. However, as petroleum becomes more scarce, it is entirely likely that the microbial fermentation processes will replace this source. Butanol and isopropanol can be produced together by certain subspecies of *Clostridium butylicum* (clos-TRID-ee-um boo-TILL-eh-kum). Butanol alone can be produced by *C. acetobutylicum* (C. a-set-ah-boo-TILL-eh-kum). The products are produced from fruit cannery wastes or molasses.

In addition to their use as solvents, both of these alcohols are used in perfume manufacturing. Isopropanol is also well known as rubbing alcohol and as an excellent antiseptic and disinfectant.

Antibiotics

The discovery of penicillin was reported by Alexander Fleming in 1929, and the subsequent development of this compound in the early 1940s by Ernst Boris Chain and Sir Howard Walter Florey was the basis for today's multi-billion-dollar antibiotics industry. **Antibiotics** are usually produced by growing a specific bacterial or fungal species in a submerged culture in large, well-aerated tanks (Figure 15–3). In the case of penicillin production, for example, a suitable medium (which usually contains by-products of the distilling industry) is aseptically inoculated with fungi such as *Penicillium chrysogenum* (pen-ee-SIL-lee-um cry-SOH-jen-um) or *P. notatum* (P. no-TAY-tum) (Color Photograph 37). After one or two weeks of growth, conditions develop that interfere with the production of penicillin. At this time the fungal growth is removed by centrifugation, sometimes with filtration, followed by the complex extraction and purification of the antibiotic. [See antibiotics in Chapter 12.]

Other commercial antibiotics have been obtained from similar activities of bacteria (such as *Bacillus polymyxa* [bah-SIL-lus poly-MIX-ah], *B. subtilis* [B. SAH-til-us], *Actinomyces* [ak-tin-oh-MY-sees] spp. and *Streptomyces* [strep-toh-MY-sees] spp.) and fungi (such as *Aspergillus fumigatus*). The antibiotics produced by these and other microorganisms are largely responsible for the successful treatment of many dread bacterial diseases, including anthrax, gonorrhea, meningitis, strep throat, syphilis, and tuberculosis. Moreover, research by the pharmaceutical industry

Figure 15–3 A portion of the fermentation unit used for streptomycin production. *Streptomyces griseus,* the microorganism from which the antibiotic is produced, is grown from test tube quantities in increasingly larger tanks. When it has multiplied itself billions of times, the living material is transferred to the huge tanks shown for final fermentation. *[Courtesy of Merck and Co., Inc.]*

is constantly developing new and modified antibiotics that have already broadened the application of these "wonder drugs" (Table 12–1).

Organic Acids, Amino Acids, and Vitamins

A number of useful organic acids, amino acids, and vitamins are produced commercially with the use of microorganisms. Representative chemicals from each of these categories together with organisms responsible for their production are listed in Table 15–1.

Organic Acids

VINEGAR

Commercial vinegar is usually made from wine or cider. It contains 3% to 5% acetic acid, which is responsible for its sour taste. In one process, fruit juice is fermented by brewer's yeast, *Saccharomyces cerevisiae,* until it contains 10% to 12% alcohol. This alcoholic solution is then sprayed into a tank that contains aerated wood shavings, coke, or gravel. These materials provide the surfaces on which species of *Acetobacter* can grow and oxidize the alcohol to acetic acid. The overall reaction is

$$\underset{\text{ethanol}}{CH_3CH_2OH} \rightarrow \underset{\text{acetaldehyde}}{CH_3CHO} \rightarrow \underset{\text{acetic acid}}{CH_3COOH}$$

Acetobacter aceti (ah-see-toh-BACK-ter a-SEE-tee), which was first studied by Pasteur in 1862, is one of the more common species involved in the reaction. Another organism, *A. xylinum* (A. ZYE-lin-um) has been observed to produce its own supporting material by forming a meshwork of cellulose fibers. This meshwork enables the organisms to have intimate contact with the alcohol solution and permits air to circulate around participating cells. The entire process must be carefully controlled, especially in regard to aeration and temperature. The oxidation of the alcoholic fluid produces so much heat that bacterial growth may be stopped if the system is not cooled. The vinegar is collected at the bottom of the tank and may be recirculated to produce more oxidation and a higher-strength product. Residual alcohol evaporates in the process, leaving none to be bottled with the vinegar.

CITRIC ACID

Citric acid is produced as a surplus, intermediate compound of the citric acid cycle. The molds *Aspergillus niger* and *A. wentii* are grown in cornstarch, corn meal, beets, molasses, sugar-cane juice, and raw sugar. The process is an aerobic one carried out in large, aerated fermentation tanks or large shallow pans. The *Aspergillus* species growing within the nutrient medium enzymatically break down glucose to eventually form citric acid (Figure 15–4). The process does not go beyond this compound because the mold lacks the next enzyme in the cycle. Citric acid accumulates and is collected for use in various commercial products. **[See Chapter 5 for details of the Krebs (citric acid) cycle.]**

LACTIC ACID

Lactic acid is produced by a fermentation process carried out by *Lactobacillus delbrueckeii* (lack-toh-ba-SILL-us del-BROOK-ee) or *L. bulgaricus* (L. bull-GAR-i-kus). These bacteria are grown in acid-hydrolized corn or potato starch, molasses, or whey from cheese making. In the process, the lactic acid is usually neutralized with lime (CaO) to form calcium lactate. This compound can be used as a food supplement, or it can be converted back to lactic acid for other uses.

EICOSAPENTAENOIC ACID (EPA)

This organic acid, also known as one of the omega-3 fatty acids, is found in high concentrations in salt-water fish, particularly salmon and albacore. The value of a diet rich in fish for lowering blood cholesterol has been traced to omega-3 fatty acids. However, the fish do not produce the fatty acids, they obtain them from the diatoms that they consume.

TABLE 15–1 Selected Organic Acids, Amino Acids, and Vitamins Produced by Microorganisms

Product	*Microorganisms*[a]	Uses
ORGANIC ACIDS		
Acetic acid (vinegar)	*Acetobacter* species	Flavoring, food preservation
Citric acid	*Aspergillus niger,*[b] *A. wentii*[a]	Anticoagulant; flavoring of foods, etc.; manufacture of ink and dyes; engraving
Eicosapentaenoic acid	Marine diatoms[c]	Treatment for high blood cholesterol levels
Lactic acid	*Lactobacillus delbrueckii, L. bulgaricus*	Calcium lactate and iron lactate for nutritional deficiency conditions; solvent in lacquers; leather manufacture
AMINO ACIDS		
Glutamic acid	*Micrococcus glutamicus, Arthrobacter* sp., *Brevibacterium* sp.	Medicinals, food supplements
Lysine	*Escherichia coli,*[d] *Enterobacter aerogenes*[d]	Food supplements
Valine	*Escherichia coli*	
VITAMINS		
Cyanocobalamine (B_{12})	*Streptomyces olivaceus, Propionibacterium freundenreichii*	Food supplement, treatment of anemia caused by vitamin B_{12} deficiency
Beta-carotene	*Dunaliella salina*[c]	Food coloring, vitamin A precursor, potential cancer treatment
Riboflavin (B_2)	*Ashbya gossypii*[b]	Food supplement

[a] For pronunciations, see the Organism Pronunciation Guide at the end of the text.
[b] These organisms are molds.
[c] This organism is an alga.
[d] Both of these organisms are required for the process.

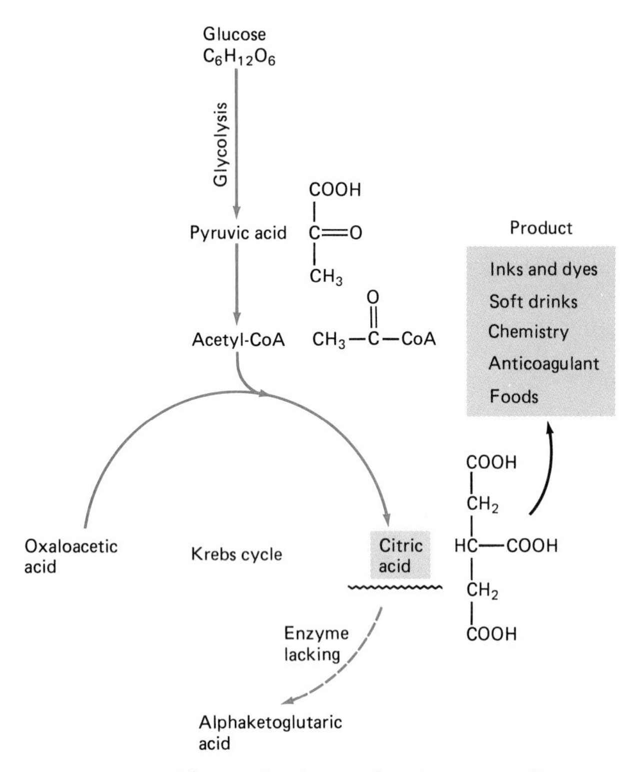

Figure 15–4 The production and various uses of citric acid. Molds such as *Aspergillus niger* are able to breakdown glucose to pyruvic acid and bring about the formation of citric acid. The product accumulates in the system and is removed for use in various commercial products.

Companies are attempting to cultivate the diatoms and obtain the EPA directly as a food supplement and as a treatment for individuals with high cholesterol levels.

Amino Acids

Although amino acids are obtained through the consumption of foods, some plant proteins lack amino acids essential for human nutrition and well-being. If foodstuffs prepared from plant products are supplemented with the amino acids for which they are deficient, their nutritional value is substantially improved. A variety of microorganisms can be used to synthesize amino acids for use as supplements and food flavorings for humans and other animals. Table 15–2 lists several products and the microorganisms involved with their formation.

In the glutamic acid process, bacteria such as *Micrococcus glutamicus* (my-kroh-KOK-kuss gloo-tam-EE-kuss) or species of *Arthrobacter* or *Brevibacterium* (brev-ee-back-TIR-ee-um) are able to add ammonia to alpha-ketoglutaric acid (cf. Chapter 5), one of the intermediate compounds of the citric acid cycle. This process produces glutamic acid in yields much greater than needed by the bacteria. The surplus can be recovered from the medium. The usual substrates in the glutamic acid process are molasses, fish meal, soybean cake, and ammonium compounds. In addition to being a potential food supplement, glutamic acid is used in flavoring as monosodium glutamate (MSG).

The lysine process requires two different bacteria, each for a different step. *Escherichia coli* (eh-sher-IK-ee-ah KOH-lee) is inoculated into glycerol, corn steep liquor, and ammonium compounds to produce diaminopimelic acid. This chemical is then modified by *Enterobacter aerogenes* (en-te-roh-BACK-ter a-RAH-jen-eez) to produce lysine.

Other amino acids produced by microorganisms and fermentation include methionine, threonine, tryptophan, and valine. **[See amino acids in Appendix A.]**

Vitamin Production

Vitamins can be made commercially by synthetic processes. However, some vitamins are rather complicated and expensive to produce by such means and can be more economically made by microbial fermentation. Vitamin B_2 (riboflavin) and vitamin B_{12} (cyanocobalamin) are two such products of microbial activity. To produce riboflavin, the mold *Ashbya gossypii* (ash-BYE-ah goss-EE-pie) is grown in corn steep liquors. The process for cyanocobalamin utilizes either *Streptomyces olivaceus* (strep-toh-MY-sees ol-eh-VA-see-us) or *Propionibacterium freundenreichii* (proh-pee-on-ee-back-TER-ee-um froy-den-REICH-eye), grown in malt extract and corn steep liquor and supplemented with cobalt, an essential part of the vitamin's structure.

Beta-carotene, a vitamin A precursor, is produced in considerable quantities by the saltwater-tolerant alga *Dunaliella salina.* This organism produces up to 20% of its cell weight as beta-carotene and grows in environments as salty as the Dead Sea. Although the carotene is largely used as a food coloring agent and vitamin A precursor, its potential as an anticancer agent is possibly more significant.

Polysaccharides

Selected microorganisms, their polysaccharides, and some of the uses of these polysaccharides are presented in Table 15–2. These materials are often formed in the slime layer or capsule of bacteria and are easily extracted because of their ready solubility in water. Many applications are based on their ability to increase the viscosity of solutions and to emulsify fats. Applications have increased consider-

TABLE 15–2 Selected Microorganisms, Their Polysaccharides, and Applications

Microorganism[a]	*Polysaccharide Type*	*Applications*
Aureobasidium pullulans[b]	Glucan	Styrene-type plastic, adhesives, fibers, packing materials
Xanthomonas campestris	Xanthan gum	Thickener for inks, adhesives, ceramic glazes; food products such as sauces, gravies, salad dressing, and ice cream; pharmaceuticals
Leuconostoc mesenteroides, *L. citrovorum,* *L. dextranicum*	Dextran	Blood plasma extender; complexing agent with iron for food supplement; photographic emulsions

[a] For pronunciations, see the Organism Pronunciation Guide at the end of the text.
[b] This organism is a mold; all the others are bacteria.

ably in the past few years and involve plastics, components of inks, adhesives, sauces, gravies, and dairy products, as well as food supplements and artificial blood plasma.

Steroid Transformations

Steroid hormones are extremely important in contraception and in the treatment and management of various conditions, including arthritis and shock. Corticosterone, an adrenal cortex hormone, has been very useful in the treatment of shock. Supplies of this compound were obtained from cattle and were in limited supply until studies showed that microorganisms could transform other steroids into corticosterone. Investigations also showed that a combination of chemical and microbial reactions could be used to produce cortisone, another steroid hormone widely applied in the treatment of **arthritis** (joint inflammations) and in control of other types of inflammatory states. **[See infectious arthritis in Chapter 23.]**

Industrial fermentative processes, such as the production of beer, wine, and rum, involve many enzymes in the conversion of the sugar substrate to the alcohol product desired. Steroid transformations are quite different: a single specific enzyme changes one particular chemical component on a steroid molecule, thus creating a new compound. Such an enzyme may be associated with only one microbial species. Table 15–3 gives several examples of microorganisms and the particular steroid each can produce from progesterone (Figure 15–5 on page 444). These and related microorganisms are the only practical means available for the large-scale production of selected steroids. These compounds are used to correct or lessen the effects of hormonal and related disorders.

Biological Control

Among the large number of prepared chemicals released into the environment each year, some of the most harmful are pesticides and herbicides. Several compounds originally used to kill food-crop destroying insects have proven to produce decreases in crop yields and undesirable effects in other forms of life. Biological control, in contrast to chemical control, includes a number of approaches to producing food crops without the harmful accumulations of chemicals. One of these methods involves the use of microorganisms to naturally control pest population.

TABLE 15–3 Microbial Steroid Transformations

Microorganism	*Transformation*
Aspergillus ochraceus	Progesterone → 11-α-hydroxyprogesterone
Curvuloria lunata	Progesterone → 4-pregnene-11β,21-diol-3,20-dione
Dactylium dendroides	Progesterone → 11-α-hydroxyprogesterone
Streptomyces lavendulae[a]	Progesterone → 20-β-hydroxyprogesterone

[a] This organism is a bacterium; all the others are fungi.

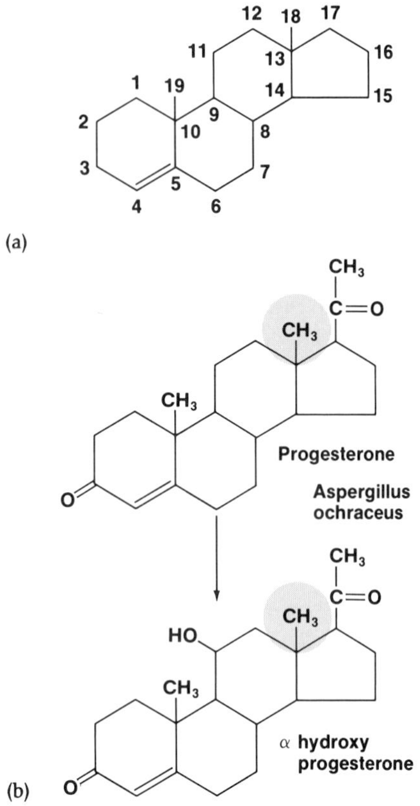

(a)

(b)

Progesterone

Aspergillus ochraceus

α hydroxy progesterone

Figure 15–5 The microbial transformation of progesterone. (a) The basic steroid structure. The numbers designate the location of particular carbon atoms. (b) The microbial transformation by the mold *Aspergillus ochraceus* of progesterone into 11 α (alpha) hydroxyprogesterone. This product can be modified further to produce cortisone.

Microbial Insecticides

Many insects associated with the destruction of agricultural products (Figure 15–6) or with the transmission of various disease agents to humans, other animals, and plants appear to be susceptible to the effects of certain microorganisms. Agostino Bassi in 1838 first proposed the use of microorganisms to control insect pests. In order for a prospective insect pathogen to be an effective microbial insecticide, it should meet certain requirements, including (1) the ability to injure the specific insect host enough so that its competitive activities are inhibited; (2) fast action; (3) specificity for a particular insect pest (the insect pathogen should be harmless to useful insects and invertebrates); (4) relative stability to environmental factors, such as drying and sunlight, and to the manner in which it is dispensed in the field (dust or spray); and (5) economically feasible manufacture. Unfortunately, the majority of microorganisms isolated from diseased insects do not seem to fulfill these requirements, making microbial control of agricultural pests a difficult task. The microorganisms that have proven to be particularly effective as pesticides are listed in Table 15–4 with the pests they control. Unfortunately, these organisms either infect the insect pests and produce a disease that spreads to other insects (*Bacillus popillae* [bah-SIL-lus pop-ILL-eye], *B. lentimorbus* [B. lent-ee-MOR-bus], *Beuveria bassiana* [boo-VER-ee-ah bass-EE-ana] for example) or produce a toxin that is effective when applied to the insects but do not reproduce in the field.

B. popillae and *B. lentimorbus* produce milky white disease. The name refers to the massive whitish growth of the bacteria in the insect's body fluids. *B. popillae* and *B. thuringiensis* (B. thur-in-jee-EN-sis) produce crystalline structures during their sporulation. Such crystals are called *parasporal bodies* because of their location. In the case of *B. popillae,* they do not appear to be related to the disease process. On the other hand, for *B. thuringiensis,* the significance of these crystals lies in the fact that the protein of which they are made functions as a toxin capable of injuring the midgut cells of susceptible insects. The activity of these crystals may inhibit the feeding of the pest or may cause other harmful effects. *B. thuringiensis* variety *israelensis* is extensively used in various areas as a biological pesticide to control mosquito populations.

Another bacterium, *Pseudomonas aeruginosa,* has also been shown to be effective in controlling adult grasshoppers, locusts, some caterpillars, and the wax moth. However, because of its known ability to

Figure 15–6 Feeding injury to developing apples caused by green fruitworms *(Orthosia hibisci)*. Fruitworms are the larval stages of certain insects (lepidoptera). These larvae have been given this name because of their habit of eating holes in the fruits of apple, cherry, peach, pear, and plum trees. *[Courtesy of Dr. R. W. Rings and the Ohio Agricultural Research and Development Center. From* J. Econ. Entomol. ***63****:1562–1568, 1970.]*

TABLE 15–4 Microorganisms Used as Pesticides

Microorganism[a]	*Insect and Other Types of Pests Affected*
Bacillus popillae[b]	Japanese beetle European chafer
B. lentimorbus[b]	European chafer
B. thuringiensis[b]	Chicken louse Tobacco hornworm Fall and spring cauberworm
B. thuringiensis[b] var. *israelensis*	*Aëdes, Culex, Anopheles* mosquitoes Blackfly
B. thuringiensis[b] and nuclear polyhedrosis virus[c]	Cotton bollworm Cabbage looper Alfalfa caterpillar Western tent caterpillar
B. thuringiensis[b] and granulosus virus[c]	Cabbage worm
B. thuringiensis[b] and *Beuveria bassiana*[d]	European cabbage worm
B. sphaericus[b]	*Culex, Anopheles,* and *Aëdes* mosquitoes
Hirsutella thompsonii[d]	Soil roundworms
Nosemia algerae[e]	Mosquitoes of various species
Vairimorpha necatrix[e]	Soybean caterpillar
Verticillium species[d]	Aphids Whitefly

[a] For pronunciations, see the Organism Pronunciation Guide at the end of the text.
[b] These organisms are bacteria.
[c] These combinations are required for effectiveness. See polyhedrosis viruses in Chapter 10.
[d] These organisms are fungi.
[e] These organisms are protozoa.

cause infections in humans and other animals, its potential as a microbial insecticide is questionable.

Microbial Herbicides

A number of fungi are used to control weeds in Florida. These include the mold *Phytophthora citrophthora* (fight-OFF-thor-ah sit-ROFF-thor-ah), which is very effective for the control of the milkweed vine, a weed that competes with citrus trees for water and nutrients. The mold *Cercospora rodmanii* (ser-KOH-spor-a rod-MAN-eye) is being used to inhibit the growth of the water hyacinth, which clogs waterways and poses a considerable hazard to inland boating.

Microbial Activity Against Plant Pathogens

As described in Chapter 14, antagonism and competition play significant roles in ecosystem associations. These roles have also been exploited in agricultural settings to prevent a number of plant diseases.

Techniques have been developed that make use of various microorganisms to occupy plant wounds and exclude pathogens either by direct competition or through the production of an antimicrobial agent. Examples include the application of the bacterium *Erwinia herbicola* (er-WI-nee-ah her-BIK-oh-la) to apple trees to exclude the plant pathogen *E. amylovora* (E. am-ee-LOH-vor-ah) and the use of an avirulent strain of *Agrobacterium tumefaciens* (a-grow-back-TIR-ee-um too-mee-FAYSH-ee-enz) to prevent crown gall disease, caused by a virulent strain. Agrocin, a growth-inhibiting substance, is produced by the avirulent *Agrobacterium.*

Microbial Enzymes

Enzymes of fungi and bacteria have a wide variety of industrial applications, including alcoholic beverages, food, detergents, and pharmaceuticals. Table 15–5 lists some of these enzymes, along with their microbial origins and selected uses.

TABLE 15–5 Microbial Enzymes and Selected Uses

Enzyme	*Microorganism*[a]	*Uses*
Amylase	*Aspergillus oryzae*[b]	Adhesives; baking; brewing; laundry presoak products; spot remover; syrups
Asparaginase	*Azotobacter vinelandii, Bacillus coagulans, Bacterium cadaveris, Escherichia coli B, Pseudomonas* sp., *Serratia marcescens*	Antitumor activity
Lipase	*Bacillus licheniformis*	Leather manufacturing
Pectinase	*Aspergillus aureus,*[b] *A. wentii*[b]	Clarifying fruit juices and wines; retting flax for linen
Protease	*Aspergillus oryzae,*[b] *Bacillus* spp.	Baking and brewing; chill-proofing beer; degum silk goods; glue; laundry presoak; leather manufacturing; meat tenderizer; cheese ripener; spot remover; wound cleaning
Streptodornase	*Streptococcus hemolyticus*	Wound cleaning
Streptokinase	*Streptococcus hemolyticus*	Dissolving blood clots

[a] For pronunciations, see the Organism Pronunciation Guide at the end of the text.
[b] These organisms are molds; all the others are bacteria.

In the production of alcoholic beverages, microbial amylase may be used instead of the enzyme from barley malt to prepare grains for fermentation. Protease can also chillproof beer by degrading the proteins that cause cloudiness in such beverages upon refrigeration.

Proteases and lipases are used in cleaning animal hides, a necessary step in leather production. Proteases are also used commercially for meat tenderizing, as they are capable of breaking down some of the tough meat protein.

One commercial application of microbial enzymes has been the removal of stains from clothing. Many substances that stain clothes either are made of lipids, proteins, or starches or are held in a fabric by these types of compounds. Certain enzymes, under suitable conditions, can decompose stains or the cementing substances associated with them, thereby helping to clean the soiled fabric. Needless to say, stains can be caused by a large variety of substances, and appropriate enzymes may not be available, or, if available, may not be incorporated into detergent or presoak compounds used in laundering. Thus, it is not surprising that all stains are not removed by commercial products. Moreover, such enzyme-containing products have been associated with the development of **allergies.** **[See allergy in Chapter 20.]**

Microbial enzymes are obtained by processes carried out either in shallow pans for optimum aeration by diffusion or in tanks where aeration is produced by bubbling air through the growth medium. Usually rich organic wastes from dairy or canning plants are used as the growth medium. When microbial growth is judged to be complete on the basis of an analysis for the desired enzyme, the microorganisms are removed by filtration, leaving the culture filtrate, which can be processed for the desired enzyme.

Immobilized Enzymes

Immobilized (stationary) enzymes, often within polyacrylamide gel columns, are used in certain commercial processes, such as the conversion of starch to simpler sugars (sucrose to glucose and fructose). Other applications of enzymes are being developed: the use of lactase for the processing of whey during the manufacture of cheese; urease for treatment of certain waste waters; phenol oxidase for the processing of phenolic wastes; and pesticide-hydrolyzing enzymes for various pesticides. The advantage of immobilizing the enzymes wherever possible is that it provides a means of removing them once the process is complete.

Immobilized Cell Systems

In the previous section, it was noted that microbial enzymes could be immobilized and perform their

TABLE 15–6 Selected Industrial Immobilized Microbial Cell Systems

Microorganism	*Products and/or Applications*
Curvularia lunata[a]	Cortisol (steroid)
Erwinia herbicola	Tyrosine, L-DOPA (amino acid and related products)
Escherichia coli	Tryptophan (amino acid)
Candida tropicalis[b]	Degrades phenols
Micrococcus denitrificans	Denitrification

[a] Mold
[b] Yeast

specific activities without having to be removed once the process was completed. Many microbial enzymes are not readily extractable and/or do not function well outside of the organisms. Therefore, immobilized systems have been developed whereby the organisms, rather than extracted enzymes, are trapped in supporting materials. These materials include collagen, gelatin, agar, alginate, carrageenan, polystyrene, and the most commonly used polyacrylamide. Examples of microorganisms used in these systems and their applications are presented in Table 15–6.

Controlling Industrial Processes

Basic research in microbial physiology and cultivation has led to the development of instrumentation to monitor industrial processes. Such instrumentation includes monitoring of environmental factors such as temperature, pH, oxygen, and carbon dioxide. These instruments also monitor more direct physiological activities, such as nicotinamide adenine dinucleotide (NAD) reduction, adenosine triphosphate (ATP) concentration, and the development of chemicals with fluorescent properties. In addition, monitoring of cell growth can be correlated with changes in pH, ATP, turbidity, and electrical conductivity. This monitoring allows engineers and scientists to follow an industrial process and know either immediately or in a relatively short time when problems develop.

Biosensors (enzyme-associated sensing devices) are used to monitor the concentrations of various chemicals in raw materials, either as intermediates in reactions or as the products themselves. Table 15–7 presents some commercially available biosensors, together with the microbial sources of the enzymes and selected applications.

One device developed to monitor industrial processes is the Bactometer (Bactomatic, Inc., Princeton, N.J.), which detects impedance changes in the microbial environment. *Impedance* is defined as opposition to the flow of alternating current through a conducting medium.

Microbial metabolism usually produces ionized by-products that influence the impedance of the medium. The sample material to be tested is placed in special chambers containing electrodes. These chambers are then placed into a temperature-controlled area of the device, and readings are obtained on a printer. The concentration of organisms in the material determines the relative detection time, which takes only a few hours in contrast to the usual one to two days needed with the traditional plate count technique. The Bactometer is shown in Figure 15–7.

TABLE 15–7 Some Commercially Available Biosensors

Chemical Detected and Enzyme	*Microbial Source of Enzyme*	*Applications*
Cholesterol-cholesterol oxidase	*Nocardia erythropolis*[a]	Food processing; blood tests
Ethanol-alcohol oxidase	*Candida boidinii*[b]	Ethanol fermentation
Glucose-glucose oxidase	*Aspergillus niger*[c]	Food processing; automatic insulin pump
Pyruvic acid—pyruvate oxidase	*Lactobacillus delbrueckii*[a]	Fermentation

[a] Bacteria
[b] Yeast
[c] Mold

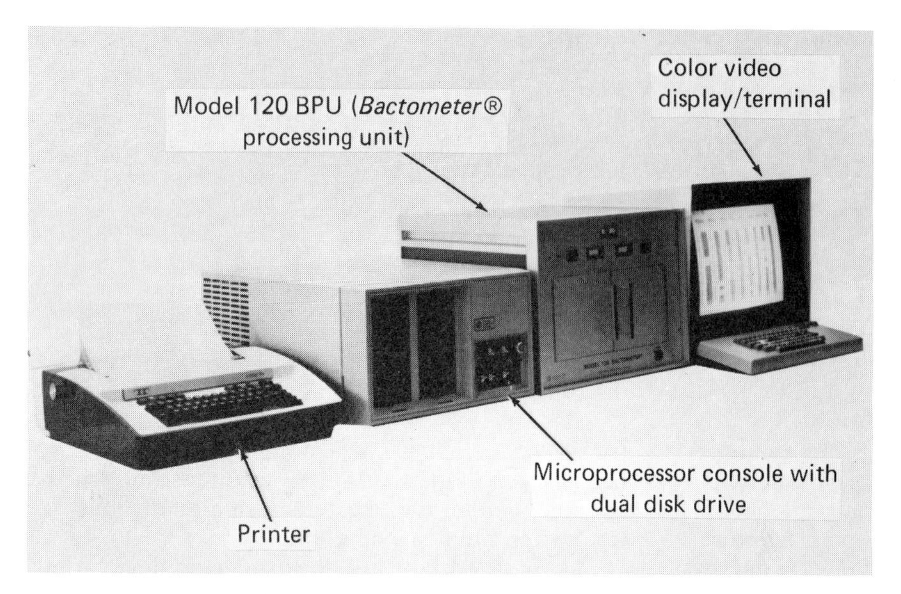

Figure 15–7 Instrument package of the Bactometer growth-monitoring device. *[Courtesy Bactomatic, Inc., Princeton, N.J.]*

Advances in Microbial Identification

The isolation, characterization, and identification of microorganisms are essential aspects of industrial microbiology and clinical microbiology. Industrial processes generally are concerned with finding new and more efficient organisms to make pharmaceuticals, various acids, alcohols, foods, and food supplements.

In clinical microbiology, where the rapid identification of disease agents is a major concern, emphasis is placed on the development of more rapid and accurate techniques. Several approaches have been developed using applications of highly specific antibody preparations (monoclonals), DNA hybridization, and fatty acid analysis. Attention here will focus on DNA probes and computerized microbial fatty acid analysis. Chapter 19 describes the features of monoclonal antibodies and their applications.

DNA Probes

A DNA probe is a highly specific tool of genetic engineering that has assumed great significance in studies dealing with the genetic similarities among microorganisms and among higher forms of life as well as for the detection of diseases such as AIDS, various cancers, and genetic disorders (Table 15–8). DNA probes are single-stranded segments of DNA

TABLE 15–8 DNA Probes for Medical Diagnosis

Probe	*Target Population (number of individuals actually affected in the U.S. alone)*
Cancers	
Leukemia/lymphoma	25,000 new cases in 1985
Oncogenes (cancer genes)	The number of individuals with cells containing these cancer genes is undetermined.
Genetic Disorders	
Cystic fibrosis	33,000
Huntington's disease	25,000
Sickle-cell anemia	50,000; 2 million Afro-Americans carry the gene
Susceptibility to heart disease	62 million
Infectious Diseases[a]	
AIDS	70–80 million tests per year
Chlamydial infections	3 million women per year
Cytomegalovirus infection	0.5 million cases per year
Hepatitis B virus	3 million new cases per year
Herpesvirus I and II infections	2 to 4 million cases per year
Periodontal disease	23 million

[a] Descriptions of these diseases can be found in Chapters 23 through 29.

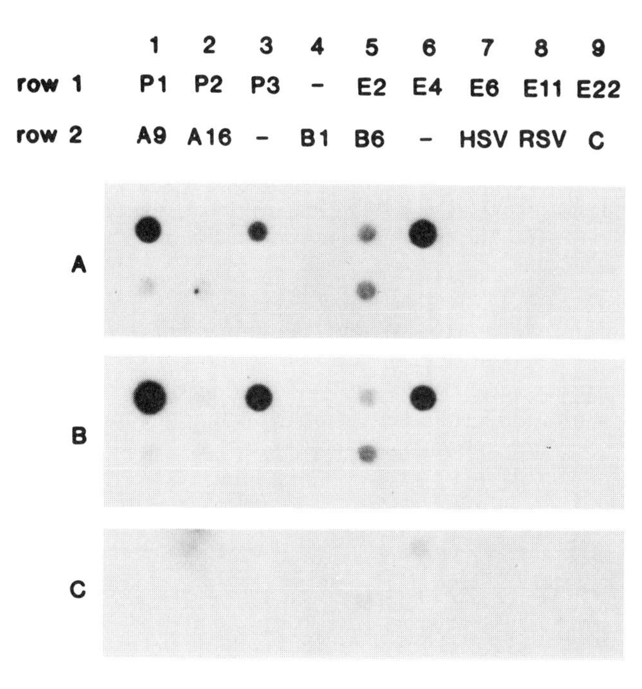

Figure 15–8 Hybridization (dot) patterns obtained with the nucleic acids of poliovirus and several different types of enteroviruses. The degree of blackening obtained with specific probes can be measured quantitatively and is generally proportional to the amount of DNA or RNA that has been hybridized. Abbreviations: P1, poliovirus 1; P2, poliovirus 2; P3, poliovirus 3; E2, echovirus 2; E4, echovirus 4; E6, echovirus 6; E11, echovirus 11; E22, echovirus 22; A9, coxsackievirus A9; A16, coxsackievirus A16; B1, coxsackievirus B1; B6, coxsackievirus B6; C, control. *[From H. A. Rotbart, et al.,* J. Clin. Microbiol. ***22****:220–224, 1985.]*

synthesized by so-called gene machines. **[See Microbiology Highlights in Chapter 6.]** Individual probes are constructed so that their respective gene (nucleic acid) sequences will match and combine with the gene sequence of genetic material extracted from specific microorganisms (Figure 15–8), cancer cells, or cells from individuals with genetic disorders. Probes permit quicker and more accurate laboratory analyses and thereby allow for more effective and rapid treatment. Table 15–8 lists a number of probes currently available or under development, together with targeted diseases and the numbers of individuals estimated to be afflicted with the disease or disorder toward which the probes directed. The world market for these probes is estimated to grow to more than $1 billion by 1990. **[Refer to Appendix A for a description of nucleic acids.]**

DNA probes have the property of being able to chemically recognize and specifically interact with nucleic acid sequences present only in members of the target organism or group of target organisms (Color Photograph 74). A properly designed DNA probe has an absolute specificity for the target sequence present only in the target cell or related material. The process of recognition and interaction is called *nucleic acid hybridization.*

To be useful as probes, DNA fragments must be marked or labeled in some way. The most commonly used markers in research laboratories are radioactive isotopes, which can be detected by autoradiography. **[See Chapter 5 for a description of autoradiography.]**

A nucleic acid sequence specific for any given cell or virus can always be uncovered, and a probe can be developed to detect that sequence. Figure 15–8 shows the behavior of a DNA probe for poliomyelitis virus, type 1 RNA. The probe created was completely representative of the entire polio 1 virus RNA sequence. It was produced by copying polio 1 virus RNA to produce radioactive DNA complementary to the viral RNA. Since polio virus is a member of a general group known as the enteroviruses, its DNA probe will also hybridize to other related enteroviruses but not to the nucleic acid of any other viral group.

SPECIFIC TECHNIQUES

Several applications and variations of hybridizing probe techniques are in use. These include Southern blotting and Northern blotting.

The Southern blotting procedure, developed by E. Southern, is a hybridization method that can be applied to a large number of particular unknown DNA segments without the need to purify individual DNA fragments. A variant of the Southern procedure, called Northern blotting, is used when it is more convenient to separate RNA molecules migrating in an electrical field. **[See Chapter 17 for a description of electrophoresis.]** Bound RNA can be hybridized with probes of radioactive RNA or single-stranded DNA.

While the use of radioactivity provides the basis for very sensitive detection systems, it is not without disadvantages. These include high costs, radiation hazards, and a short half-life of useful isotopes. Consequently, a significant amount of money has been directed toward developing nonradioactive detection systems. Figure 15–9 shows the general features of a probe system using a nonradioactive detection substance that can be observed visually or with the aid of electronic devices.

Fatty Acid Analysis

The recent development of a computer-controlled gas chromatography system has given rise to a highly specific and sensitive microbial identification instrument (Figure 15–10). The application of this instrument involves an analysis of an unknown organism's fatty acid composition (*profile*) and a computer comparison of the results with a memory-bank collection of known microbial fatty acid compositions. Upon finding a match of fatty acid com-

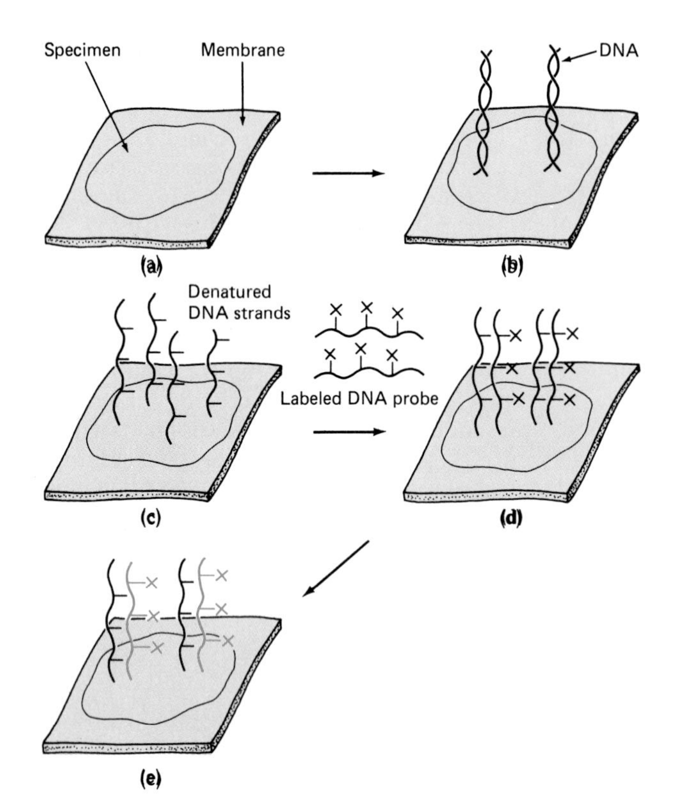

Figure 15–9 A DNA-probe identification test.
(a) A specimen is applied to the membrane. (b) The organisms' DNA is released from the cells and bound to the membrane. (c) The DNA strands are denatured.
(d) The enzyme-labeled probes are added and allowed to hybridize (combine) to complementary segments. The specimen is then washed to remove unbound probes (not shown). (e) The color-developing system is examined for identifying color.

parisons, the computer reports the identification and its level of accuracy. Since pure cultures are required, the unknown must be isolated and grown before performing the analysis. Studies to date show that fatty acid analyses are very specific and correspond well to the genetic relatedness of organisms. The applications of the computer-controlled gas chromatography system will undoubtedly increase as more microbial fatty acid profiles are added to the memory bank.

Industrial and Agricultural Applications of Genetic Engineering

An increasing number of industries are utilizing the advances resulting from genetic engineering, or recombinant DNA (rDNA) technology. Among these are the pharmaceutical, chemical, and food-processing industries. Regardless of the industry, certain criteria must be met before genetic technologies can make the production of a product commercially fea-

Figure 15–10 The Hewlett Packard HP5898A Microbial Identification System using a computer controlled and comparison system based on fatty acid analysis by gas chromatography.

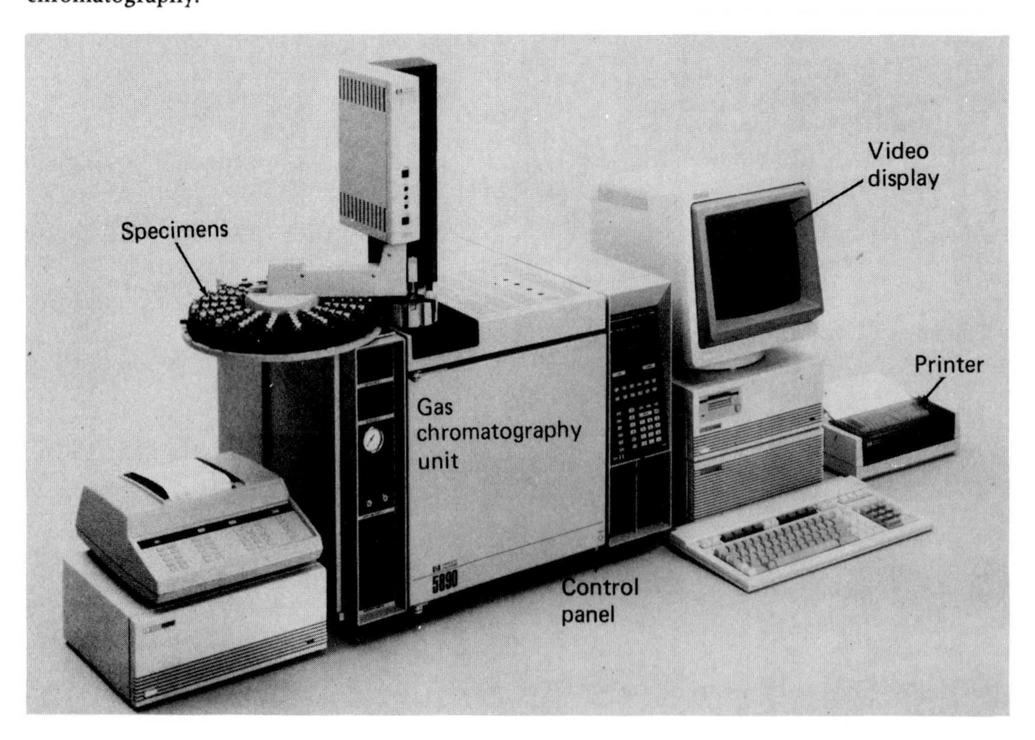

Microbiology Highlights

SLAYING MODERN MONSTERS

The Chimera of Greek mythology was a fire-breathing monster with the forequarters of a lion, the body of a goat, and the hindquarters of a dragon. This Chimera was slain to protect the people.

Of the many chimeras (hybrids) formed by gene splicing, none is more remarkable than the recombinant bacterial vaccine consisting of *Salmonella typhi* (sal-mon-EL-lah TIE-fee) cells, the cause of typhoid fever, containing the genes of *Vibrio cholerae* (VIB-ree-oh KOL-er-ee), the cause of Asiatic cholera. Typhoid fever and cholera are two modern monsters about to be slain by this genetically engineered chimera. These diseases, seen less frequently in developed countries, are still ravaging populations in the Third World.

D. Rowley and co-workers at the Department of Microbiology at the University of Adelaide, Australia, are developing an oral vaccine with a non-disease-causing strain of *S. typhi.* Although this was found to be harmless, it proved capable of stimulating antibody formation in field trials conducted in Chile and Egypt. Six genes for antibody-stimulating polysaccharides from *V. cholerae* were spliced into *S. typhi,* thus creating a microbial chimera with the general properties of *S. typhi* and the combined surface characteristics of both bacterial species.

Preliminary tests are planned in collaboration with the Center for Vaccine Development in Baltimore, Md., with subsequent extensive field trials to take place in China, India, and Bangladesh. The availability of a marketable vaccine is anticipated in 1990.

sible and profitable. These requirements include establishing the need for a biochemical product and demonstrating the usefulness of a biological approach to commercial production as well as the efficiency of a genetic approach to increased production of the product. Once these criteria have been met, the application of genetic engineering can be undertaken. The pharmaceutical industry was the first to make widespread use of genetic technologies and has applied them to increase the production efficiency of various pharmaceuticals, such as antibiotics, amino acids, and certain hormones, including insulin and human growth hormone. Figure 15–11 outlines the development process used to produce a genetically engineered pharmaceutical product. Several other human-associated hormones currently used in medical treatment are potentially attractive for manufacture through rDNA techniques. These include corticotropin, calcitonin, secretin, and glucagon. [Refer to Table 6–5 for currently approved rDNA products.]

Recombinant DNA technology has a number of current and potential agricultural applications. The most immediate applications in agriculture will involve the large-scale manufacture of such products as amino acids and vitamins for animal feed, as well as pesticides, herbicides, and antibiotics. Benefits of genetic engineering include the development of a foot-and-mouth virus vaccine in 1981. Potential applications of the new technology are highly varied, ranging from breeding sheep that produce more wool to boosting the world's food supply, which includes stimulating animal production. Several rDNA experiments presently in progress are concerned with the microbial production of specific hormones; these include bovine growth hormone for increased production of milk and meat, porcine hormone for increased meat production, and ovine hormone for increased wool growth in sheep.

The potential benefits of plant genetics are definite but less immediate. The rDNA techniques are aimed at the manipulation of plant cells and genes to increase plant resistance to disease and unfavorable weather and soil conditions, and to improve the yield, quality, and variety of feed, food, and fiber crops.

One interesting application of genetic engineering along these lines involves the splicing of *Escherichia coli* genes into *Methylophilus methylotrophus,* an organism that can utilize methanol and ammonia for its nutrients. The resulting chimera is much more efficient in producing nutrient proteins than the original culture and has been developed for a process to convert waste fossil fuels into animal feed.

Patenting Microbial Processes

In a 1948 judgment of the U.S. Supreme Court, microorganisms could not be patented for a use that was consistent with their activities in nature. In this

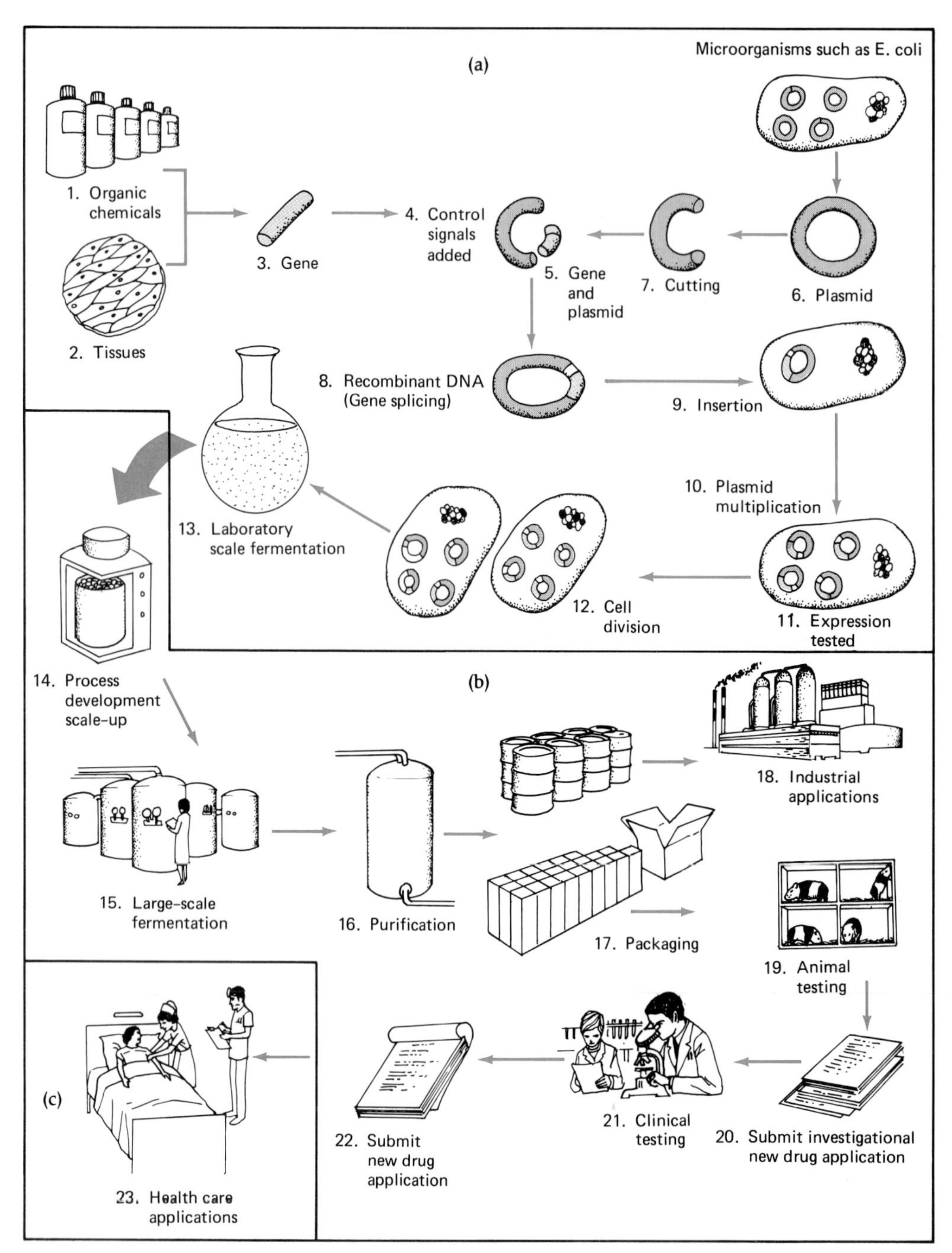

Figure 15–11 A representation of the product-development process involving genetic engineering. (a) The main features of recombinant DNA techniques. (b) The commercial phase includes large-scale fermentation, product purification, animal safety testing, and packaging. (c) The resulting product can be either directed for industrial (non-health-care) or health-care applications. For health-care use, additional studies for clinical safety and effectiveness and the filing of a new drug application with the Food and Drug Administration (FDA) are required. Once the drug product is considered safe and effective by the FDA, it is ready for human use.

particular case, a company was attempting to patent mixtures of the bacterium *Rhizobium* spp. for an industrial application involving a variety of leguminous plants. In order to be patentable, an invention must be new, useful, and not obvious. The usual types of inventions that have been patented in microbiology include changes in industrial process conditions, media, and chemical recovery steps.

During the 1970s, recombinant DNA research began to challenge the then current interpretation of whether microorganisms or only processes could be patented. The U.S. Patent and Trademark Office

(USPTO) could consider only patent applications based on current law and would not patent microorganisms, even though genetic engineering was providing organisms with uses not apparently consistent with their activities in nature.

A USPTO publication in 1977, however, stated: "It has been suggested, for example, that research in this field [recombinant DNA] might lead to ways of controlling or treating cancer and hereditary defects. The technology also has possible applications in agriculture and industry. . . . In view of the exceptional importance of recombinant DNA . . . Assistant Secretary of Commerce for Science and Technology has requested . . . 'special' status to patent applications . . . and . . . provided for the accelerated processing of patent applications for inventions relating to recombinant DNA."

A Supreme Court decision in 1980 has allowed microorganisms to be patented as long as they are on deposit with an approved culture collection as part of the disclosure requirements of the patent application. The number of such applications has steadily increased. In 1985 alone, over 3,000 were filed with the USPTO.

Summary

Introduction to Industrial Microbiology

1. "Industrial microbiology" refers to the use of microbes, or microbiological techniques, in a commercial enterprise.
2. Examples of beneficial industrial applications include the production of antibiotics, enzymes, and certain steroid hormones.

AEROBIC VERSUS ANAEROBIC PROCESSES

Fermentation is the anaerobic reduction of organic substrates for the generation of energy. Products of fermentation reactions include ethyl alcohol, butanol, isopropyl alcohol, acetone, and glycerol. Some industrial processes include aerobic phases.

Alcoholic Beverages and Related Products

1. The production of beer, wine, and spirits is a significant division of industrial microbiology in many countries.
2. Beer is produced by a yeast fermentation process that uses barley or other grains as sources of sugar and nitrogen-containing compounds. Beer production also incorporates the addition of hops for color, flavor, and antibacterial substances.
3. Wine is made by the fermentation of almost any ripe fruit juice or of certain extracts from vegetable products. The presence of other substances such as acids and minerals accounts for the different color, taste, and sometimes bouquet of wines. The general wine-making process includes grape testing and picking, crushing, fermentation, pressing, clarification, aging, filtration, and bottling.
4. Distilled beverages include whiskey, brandy, and rum. The type of product depends upon the material used in the fermentation.
5. Specialized fermentation and distillations have been designed to produce ethanol for fuel.
6. Additional alcohols produced by fermentation include butanol and isopropanol.

Antibiotics

Antibiotics are usually produced by growing a specific bacterial or fungal species in a submerged culture in large, well-aerated tanks.

Organic Acids, Amino Acids, and Vitamins

1. Vinegar, or acetic acid, is usually made from wine or cider. Species of the bacterial genus *Acetobacter* perform the overall reaction.
2. Additional organic acids that are produced by microbial processes include citric, lactic, and eicosapentaenoic acids.
3. A variety of microorganisms can be used to synthesize amino acids and vitamins. Such products are used as food supplements, as flavorings, and for medical purposes.

Polysaccharides

Microbial polysaccharides such as glucans, dextrans, and xanthan gums are produced for use as thickeners in food products, in inks, and as blood plasma substitutes.

Steroid Transformations

A combination of chemical and microbial reactions can be used to produce various steroid hormones such as cortisone.

Biological Control

1. Many insect and weed pests are susceptible to microbial agents that can be used in their control.
2. Microorganisms function as *microbial insecticides, microbial herbicides,* and as *plant pathogen antagonists.*

Microbial Enzymes

Commercially produced microbial enzymes have applications in a variety of industrial areas, including the manufacture of alcoholic beverages, foods, detergents, and pharmaceuticals. Some of these enzymes have been immobilized to improve the efficiency of recovery processes.

IMMOBILIZED CELL SYSTEMS

Cultures of organisms can be immobilized in substances such as polyacrylamide to permit improved process efficiency in situations in which it is impractical to extract and immobilize enzymes.

CONTROLLING INDUSTRIAL PROCESSES

1. The normal physiological activities of microorganisms can be monitored to permit reasonable evaluation of process performance by the use of specialized instrumentation.

2. Microbial enzymes may also be employed as biosensors in the detection, monitoring, and control of chemical reactions or processes.

Advances in Microbial Identification

1. The accurate and rapid identification of microorganisms is of major importance in industrial and clinical activities.
2. Processes and devices for identification are becoming a major commercial enterprise. Two important recent approaches are the use of DNA probes and computerized fatty acid analyses.
3. DNA probes are single-stranded segments of DNA synthesized by so-called gene machines. Individual probes are constructed so that their gene sequences match specific gene sequences of genetic material from specific cancer cells, viruses, bacteria, or cells from individuals with genetic disorders.

PATENTING MICROBIAL PROCESSES

Until 1980, microorganisms could not be patented in the United States, because patents were limited to improvements in process methodology rather than improvements in the organisms involved in the process.

Industrial Applications of Genetic Engineering

Gene-splicing techniques are being used to improve industrial processes and to produce mammalian hormones, agriculturally important vaccines, and other commercially important biological products.

Questions for Review

1. What is industrial microbiology?

2. a. Describe the processes involved in making beer, wine, and spirits.
b. What microorganisms are involved?
c. What is a top yeast?
d. What is a bottom yeast?

3. For each of the following, give a brief outline of the manufacturing process and/or industrial uses, including the types of microorganisms involved:
a. antibiotics
b. lactic acid
c. isopropanol
d. vinegar
e. citric acid
f. amino acids
g. vitamins
h. butanol
i. eicosapentaenoic acid
j. polysaccharides

4. For each of the following, give the microbial origin and uses:
a. amylase
b. asparaginase
c. lipase
d. pectinase
e. protease
f. biosensor

5. Compare and contrast immobilized enzymes and cell systems in industrial applications.

6. a. How do steroid transformations differ from most other microbial processes?
b. Describe one transformation.

7. a. What two categories of microbial insecticides are recognized?
b. What are the advantages and disadvantages of each?

8. What types of physiological activities can be monitored to control industrial processes utilizing microorganisms?

9. What was the difficulty in patenting microorganisms prior to 1980?

10. List three industrial applications of genetic engineering.

Suggested Readings

AHARONOWITZ, Y., and G. COHEN (eds.), "Microbial Production of Pharmaceuticals," *Scientific American* (September 1981). *An entire issue devoted to various aspects of industrial microbiology.*

DICKSON, D., "Patenting Living Organisms—How to Beat the Bug-rustlers," *Nature* **283**:128–129, 1980. *A short article dealing with the legal aspects of patenting industrially important microorganisms.*

FREIFELDER, D., *Molecular Biology*, 2nd ed. Boston: Jones and Bartlett Publishers, Inc., 1987. *An excellent and well-designed overview of molecular biology. The text covers the various subject areas essential to understanding the many and varied aspects of biotechnology.*

GRANADOS, R. B., and B. A. FEDERICI (eds.), *The Biology of Baculoviruses*, Vol. III. Boca Raton, Fla.: CRC Press, Inc., 1986. A comprehensive reference to the use of baculoviruses for insect control and detailed descriptions of associated methods and procedures.

KENNY, M., *Biotechnology: The University-Industrial Complex.* New Haven, Conn.: Yale University Press, 1986. *A discussion of the biotechnology industry, its structure, potential impact, and how it has altered the character of university-based research.*

PORTER, J., *All About Beer.* Garden City, N.Y.: Doubleday & Company, 1975. *A rather complete, easy-to-read book about all aspects of beer, from ingredients and the brewing process to the unusual ways beer can be used with food.*

SICHEL, P. M. F., and J. L, *Which Wine? The Wine Drinker's Buying Guide.* New York: Harper & Row, 1975. *A clearly indexed, at-a-glance guide to wines. While the emphasis appears to be on wine purchases, this book contains interesting and functional information about wine making and features of wines known the world over.*

UNDERKOFLER, L. A., and M. L. WULF (eds.), *Developments in Industrial Microbiology*, Vol. 22. Arlington, Va: Society for Industrial Microbiology, 1981. *This volume includes well-written articles on microbial control of mosquitoes, conversion of biological materials to fuel alcohol and animal feed, and many other articles of general interest in applied microbiology.*

16

Food, Dairy, and Water Microbiology

"Eat, drink, and be leary." — *O. Henry (William Sidney Porter)*

After reading this chapter, you should be able to:

1. Discuss the types, sources, and importance of microorganisms in foods.
2. List five fermented foods or related products of fermentation and name the microorganisms and the starting material involved in their production.
3. Discuss three factors that influence the fermenting activities of microorganisms used in food and dairy products.
4. Describe the general sequence of events involved in the production of a fermented food such as pickles or sauerkraut.
5. Describe the processes by which cheese and yogurt are produced.
6. List four microbial disease agents and/or their products that can be spread by food.
7. Discuss the sanitary quality of food, food preservation, and the detection of microorganisms in food.
8. List the types of microorganisms involved in food spoilage and explain how microorganisms cause it.
9. List and describe at least four methods of preventing microbial food spoilage.
10. Define the term "single-cell protein" and discuss its potential role in solving the world's food shortage.
11. Describe the general processes for water treatment.
12. List the types of materials found in domestic wastewater.
13. Describe the basic features of the following procedures used to determine water quality: most probable number (MPN), membrane filter technique (MF), and standard plate count (SPC).
14. Discuss biological oxygen demand (BOD) and the way it is measured.
15. Outline the steps involved in wastewater treatment and list potential products and problems associated with waste disposal.
16. List and describe two general procedures for solid waste disposal.
17. Discuss the applications of prepared microbial cultures for waste treatment.
18. Define or explain the following: humus, trickling filter, compost, floc, reclamation of wastewater, and indicator.

The availability and safety of foods, including dairy products and water supplies, are of primary concern to developed and developing countries alike. Throughout the centuries, microbial fermentations of various foods including milk have provided a means to decrease spoilage and usually eliminate potential disease agents. This chapter describes several types of food and dairy products and the different fermenting organisms involved with their production. Attention will also be given to the role of genetic engineering in food production and to the detection of microorganisms in food spoilage.

Another major public concern is the incredible accumulations of wastes in the environment. Frequently, microorganisms are intricately involved with the processing and ultimate decomposition of such wastes. This chapter also considers microbial involvement in water purification and waste treatment.

Many foods and dairy products owe their production and characteristics to the activities of microorganisms. Fermented milk products; innumerable varieties of cheese, pickles, sauerkraut; and fermented sausages all have unique aromas and flavors that result directly or indirectly from the enzymatic actions of specific fermenting organisms. In some cases, microbial activity even increases the vitamin content of the fermented foods as well as their digestibility. No other single group of foods and food products has been so important for human well-being throughout the world and throughout history.

While many microorganisms have beneficial roles in food preparation, several also are responsible for some of the most serious types of food poisoning and the spoilage of a wide variety of food and dairy products. The general term "food poisoning" is used to describe the conditions developing in persons who have consumed foods containing harmful chemicals, microorganisms, or microbial toxins (poisons). The major sources of microbial contamination in foods are numerous and may include the ingredients of the food being prepared, the equipment, the water and air in a food processing plant, and even the food handlers. These major sources are briefly discussed in this chapter. The actual food-associated diseases and their respective causative agents are described in Chapters 26 and 31.

In this chapter we will first describe various aspects of the microbial ecology of food and dairy products and then continue with water treatment and analysis and waste disposal.

Fermented Foods

The microorganisms that grow in or on a food and the changes they cause in flavor, texture, and general appearance are influenced by factors such as acidity, available carbohydrates, oxygen, temperature, and water. Fermented foods and food products used throughout the world are the result of microbial activities (Table 16–1). The commercial manufacture and sometimes the domestic or "home" production of many of these fermented foods are begun by appropriate microbial **starter cultures.**

TABLE 16–1 Examples of Fermented Foods and Products Produced by Microorganisms[a]

		Fermenting and for Flavor Contributing Microorganisms	
Food or Product	*Raw Starting Material*	*Bacteria*	*Fungi*
BREADS			
Cakes, rolls, etc.	Wheat flours		*Saccharomyces cerevisiae*
San Francisco sourdough bread	Wheat flours	*Lactobacillus sanfrancisco*	*Saccharomyces exiguus*
DAIRY PRODUCTS			
Acidophilus milk	Milk	*Lactobacillus acidophilus*	
Cheeses (ripened)	Milk curd	*Brevibacterium* spp. Lactic acid bacteria *Micrococcus caseolyticus* Propionibacteria *Streptococcus cremoris* *Streptococcus lactis* *Streptococcus thermophilus*	

[a] Alcoholic beverages and related products, including beer, vinegar, and wines, are described in Chapter 15.

TABLE 16–1 (continued)

		Fermenting and for Flavor Contributing Microorganisms	
Food or Product	*Raw Starting Material*	*Bacteria*	*Fungi*
Cultured buttermilk	Milk	*Leuconostoc cremoris* *Streptococcus cremoris* *Streptococcus lactis*	*Geotrichum* spp., *Penicillium camemberti,* *Penicillium roqueforti*
Yogurt	Milk and milk solids	*Lactobacillus bulgaricus* *Streptococcus thermophilus*	
MEAT PRODUCTS			
Country cured hams	Pork hams		*Aspergillus* spp. *Penicillium* spp.
Dry sausages (salami, etc.)	Pork, beef	*Pedicoccus cerevisiae*	
NONBEVERAGE PLANT PRODUCTS			
Olives	Green olives	*Leuconostoc mesenteroides* *Lactobacillus plantarum*	
Pickles	Cucumbers	*Lactobacillus plantarum* *Pedicoccus cerevisiae*	
Poi	Taro roots	Lactic acid bacilli	
Sauerkraut	Cabbage	*Leuconostoc mesenteroides* *Lactobacillus plantarum*	
Soy sauce (shoyu)	Soybeans, wheat, rice	*Lactobacillus delbrueckii*	*Aspergillus oryzae* or *A. soyae,* *Saccharomyces rouxii*

Fermented Vegetables

Foods such as sauerkraut and pickles result from the microbial fermentation of carbohydrates in plant tissues into acids. The addition of salt controls undesirable microbial activities, establishing a favorable environment for the desired fermentation. The organic acids and related compounds formed during the process act as preservatives and contribute to the characteristic flavor and aroma of fermented products. Foods of this type are less likely to spoil or to harbor pathogenic microbes. In addition, they lose very little of their nutritive value during fermentation. Most green vegetables and fruits can be preserved by this type of *pickling* process (Table 16–1).

SAUERKRAUT

The origins of sauerkraut are not known, but they are believed to derive from central Europe, where this fermented product of cabbage is still used as a main-course vegetable. Sauerkraut is the product of lactic acid fermentation, which is usually carried out by the normal mixed bacterial flora of cabbage and lactobacilli. Its manufacture is simple and inexpensive and requires very little equipment (Figure 16–1). Cabbage is shredded and layered in large vats. The layers are pressed down to produce anaerobic conditions. Salt is added between the layers to restrict the actions of undesirable bacteria while favoring the two most desirable lactic acid bacteria, *Leuconostoc mesenteroides* (loo-koh-NOS-tok mes-ter-

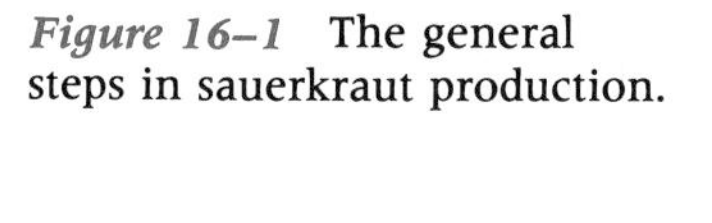

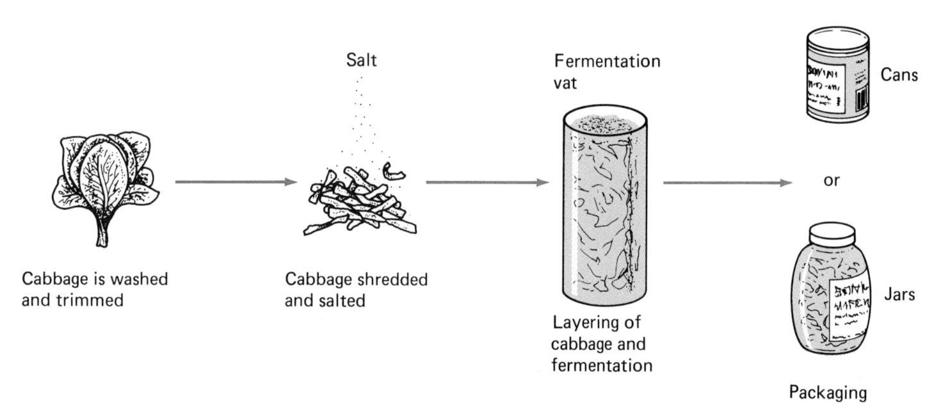

Figure 16–1 The general steps in sauerkraut production.

OYD-ees) and *Lactobacillus plantarum* (lack-toh-bah-SIL-lus plan-TARE-um). Kept at room temperature, these organisms produce large quantities of lactic acid from the cabbage juices released within the first week of the fermentation. As acidity increases, the *L. mesenteroides* decrease in number and are replaced by various species belonging to the genus *Lactobacillus.* In most cases, the extremely acid-tolerant *L. brevis* (L. BREV-iss) predominates and carries on the fermentation, producing acetic and lactic acids and alcohol. A variety of other organic acids and related compounds are also formed, giving sauerkraut its distinctive flavor. The entire process requires about three to six weeks for completion.

PICKLES

Pickles are the fermentation products of fresh cucumbers. The microorganisms in the normal mixed flora of cucumbers are responsible for the process. *L. plantarum* is the most essential microbial species in pickle production. In general, the fermentations are the same for most types of pickles—dill, sour, and sweet. The differences among the pickles result from differences in the spices and other ingredients used in pickling or in the postpickling procedure. The keeping quality of various pickles is assured by heat pasteurization or the addition of vinegar. Other details of pickle production as well as related fermented foods can be found in the "Suggested Readings" given at the end of this chapter.

Dairy Products

Milk acquires microorganisms at the time it is drawn from the cow or other milk-producing animals, and it may be further contaminated as it is handled and processed. Since milk contains a variety of nutrients including fat, minerals, protein, the sugar lactose, vitamins, and water, it can serve as an excellent medium for numerous microorganisms. Unless measures are taken to eliminate spoilage organisms, souring occurs. To prevent such spoilage and to make milk commercially marketable in the United States, it is usually pasteurized. Samples of products are routinely checked for bacterial content by direct microscopic count and standard plate count techniques. Recommended grading standards for dairy products have been developed and maintained by the U.S. Public Health Service based on the results of such procedures.

Milk is the starting material for a variety of products that both increase its appeal and serve to preserve it (Tables 16–1 and 16–2). The fermentation of milk sugar (lactose) is the basis for a variety of these food products. Milk usually contains a wide variety and number of microorganisms. In order to ensure a high-quality dairy food, such organisms must be eliminated. After this step, pure cultures of reliable microorganisms, called lactic *starter cultures,* are introduced into the starting material. Lactic

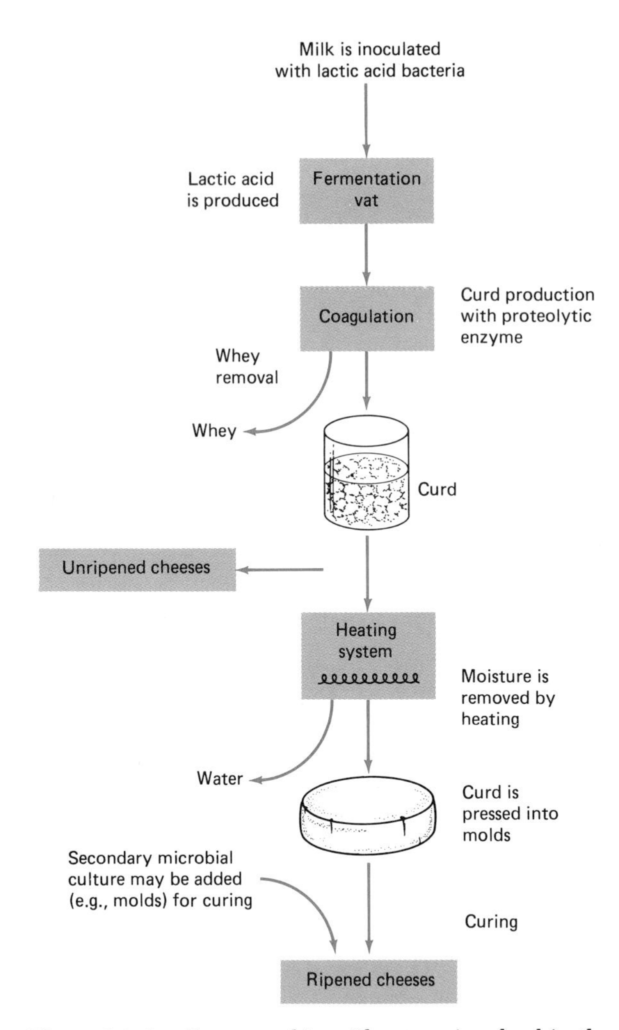

Figure 16–2 Cheese making. The steps involved in the making of unripened and ripened cheese products are outlined.

starters always include bacteria such as *Streptococcus cremoris* (strep-toh-KOK-kuss kree-MORE-is), *S. diacetylactis* (S. dye-ah-see-TIL-lak-tis), or *S. lactis,* which convert lactose to lactic acid. The characteristic flavors and textures of butter, buttermilk, cheeses (Figures 16–2 and 16–3), yogurt, and related foods are produced by the types of microorganisms (bacteria or fungi) used in fermentation and subsequent processes. A number of species belonging to the bacterial genera *Lactobacillus, Propionibacterium* (proh-pee-on-ee-back-TEER-ee-um), and *Streptococcus* are used in cultured dairy foods for acid production, flavor development, and flavor production.

While starter organisms may remain active and maintain their characteristics for some time, changes in normal fermentation activity occur. Such changes, which are considered to be defects, include insufficient acid development, abnormal flavors, accumulation of gas, and bitterness. Defects may be caused by the presence of antibiotics and chemicals, which can inhibit starter organisms; mutations; and destruction of starter organisms by bacterial viruses.

BUTTER AND SOUR CREAM

Butter and sour cream are produced by inoculating pasteurized cream or milk with a specific lactic starter culture and holding it until the necessary amount of acidity is obtained. In the case of butter, acidified cream is churned until the thick butter forms. It is then worked to remove excess liquid, washed, salted, and packaged. The characteristic aroma and taste of butter result from the compound diacetyl, which is formed by bacteria such as *Streptococcus diacetylactis* and other organisms. In addition to butter and other dairy products, diacetyl is found in red and white wines, brandy, roasted coffee, and many other fermented foods. The compound also exhibits antibacterial activity.

Cultured sour cream is prepared by fermenting pasteurized light cream with a lactic starter.

CHEESE PRODUCTION

The production of most cheeses is the result of microbial activity (Color Photograph 71). A great number and variety of cheeses are manufactured throughout the world (Table 16–2). Basically, they can be divided into three texture types: *soft* (such as cottage cheese or cream cheese); *hard* (such as cheddar or Swiss); and *semisoft* (such as Camembert).

Most cheese-making processes start with cow's milk, either whole or skimmed. The customary first step is to curdle the milk (Figure 16–2). Adding an appropriate bacterial culture to the starter material causes a firm *curd* and a watery fluid portion (called *whey*) to form. Addition of the enzyme rennin (which is obtained from butchered calves) hastens the process. Rennin alone also can be used with good results and is often recommended for home cheese making. The best curds for cheese making are obtained with the combined actions of lactic acid bacteria and rennin. Cheese made directly from milk, or in some cases from whey, is known as *natural cheese.* Pasteurized process cheese and cheese spreads are made by blending and heating one or more varieties of natural cheese.

Moisture removal is the second step in cheese production (Figure 16–3a). The extent to which moisture is reduced depends upon the type of cheese to be produced. Heat, pressure, or cutting and compression of the curd are used for this purpose. When suitable amounts of whey have been removed, the curds are molded into characteristic shapes. Salt is sometimes added during cheese production to reduce moisture, prevent the growth of unwanted microbes, and contribute to the flavor of particular cheeses (Figure 16–3b).

Another important step in cheese production is ripening. Certain cheeses are obtained after milk curdles, without any further curing. These are called *unripened* cheeses. A good example of this type of product is cottage cheese. When cheeses undergo *ripening* by bacteria or fungi, protein or fat is usually degraded, depending upon the dairy product desired. For instance, in the case of Camembert and Limburger cheeses, proteins are broken down during the ripening process. In the case of Roquefort and blue cheese, fat is degraded. Swiss cheese presents another situation. The activity of bacteria in Swiss cheese causes significant gas production during lactose fermentation, which results in the characteristic holes in the cheese.

Ripening can be conducted in one of two ways. The particular procedure depends, again, on the type of cheese desired. In the case of hard cheeses, such as cheddar and Swiss, the microorganisms are introduced and distributed into the interior of the cheese. In the case of soft cheeses (Camembert and Lim-

TABLE 16–2 Categories, Examples, and Origin[a] of Natural Cheese Varieties

Texture Category	*Unripened*	*Ripened by Bacterial Action*	*Ripened by Mold Action*
Soft	Cottage cheese (uncertain) Cream cheese (United States) Neufchatel (France) Ricotta (Italy)	Bel Paese (Italy) Limburger (Belgium)	Brie
Semisoft	Primost (Norway)[b]	Brick (United States) Gouda (Holland) Monterey Jack (United States) Port du Salut (France and Canada)	Blue (France) Camembert (France) Gorgonzola (Italy) Roquefort (France)
Hard	Gjetost (Norway)[b] Sapsago (Switzerland)	Asiago (Italy) Cheddar (England)[c] Edam (Holland)[c] Gruyere (Switzerland)[d] Parmesan and Provolone (Italy)[c] Swiss (Switzerland)[c]	

[a] The country of origin of each cheese is shown in parentheses.
[b] Produced by acid and high heat coagulation of milk and whey.
[c] The final product does not have holes.
[d] The final product characteristically has holes.

(a)

(b)

Figure 16–3 Selected aspects of cheese making. (a) Curd is poured into containers or forms to allow for drainage and to separate it from whey. *[Courtesy Fisher Cheese Co., Wapukoneta, Ohio.]* (b) To give flavor and start the formation of a crust, a product such as blue cheese is taken from its cold, dry storage room and punctured so that the mold can get air in order to grow and develop. Next, it is placed in curing cellars where the temperature and humidity encourage mold growth. *[Courtesy Borden, Inc.]*

burger, for example), the microorganisms are encouraged to grow on the surface and not in the interior of the product. These ripening methods also affect the sizes and shapes of cheese bricks or other units, such as wheels. Hard cheeses can be prepared in rather large dimensions, while soft ones are usually found in small sizes with a relatively large surface area so that microbial products can spread through the cheese rapidly.

Other Fermented and Cultured Dairy Foods

In addition to butter, sour cream, and cheeses, peoples of the world consume fermented milk in the form of buttermilk, yogurt, and related products. The historical, geographical, ecological, and dietary patterns in various regions of the world are reflected in the diversity, variety, and types of fermented milks in vogue today (Table 16–3). These products are generally produced by the intense activity of lactic acid bacteria added as pure cultures. In addition, certain yeasts may be a part of the fermenting microflora producing low levels of alcohol in the product. Furthermore, the use of milk of various animals adds another dimension to the variety of flavor, body, and texture of fermented milk foods. For example, kefir, a fermented milk originating in the Caucasus Mountains, can be made from cow's, sheep's, or goat's milk. Moreover, the microbial flora is mixed and complex, including lactobacilli (Figure 16–4) and various yeast species. Table 16–3 lists other origins of fermented milk products.

BUTTERMILK

Buttermilk is the liquid remaining after cream is churned for butter production. Commercially, it is usually prepared without churning by inoculating skim milk with a starter culture and holding until sufficient souring typical of the final product occurs. The various types of buttermilk available differ both in the source of milk and the microorganisms used in the fermentation process.

YOGURT

Yogurt is another well-known fermented dairy product that forms an important part of the human diet in many parts of the world. In the United States alone, 1979 sales figures indicated that more than 565 million pounds of this fermented dairy product were sold. Two lactose-fermenting bacteria, *Lactobacillus bulgaricus* (L. bul-GA-ri-kuss) and *Streptococcus thermophilus* (S. ther-MOH-fil-lus), are used as the starter culture to make yogurt. Equal numbers of *L. bulgaricus* and *S. thermophilus* are desirable for flavor and texture production. The lactobacilli grow first, liberating the amino acids glycine and histidine and stimulating the growth of streptococci. The production of the characteristic flavor by the afore-

TABLE 16–3 Categories, Examples, Predominating Microorganisms, and the Origins of Selected Cultured Dairy Products

Categories	*Examples*	*Countries of Origin*	*Predominant Microogranisms*
Cultured buttermilks	Filmjolk	Norway	*Streptococcus* and *Leuconostoc* species
	Lattfil	Sweden	
	Langfil		
	Ymer	Denmark	
	Villi	Finland	
	Skyr	Iceland	
Buttermilk-type products	Bulgarian buttermilk	Bulgaria	*Lactobacillus* species
	Acidophilus milk	Bulgaria	
	Yakult	Japan	
Yogurts	Yogurt	Egypt, Iraq, Lebanon, Syria, Turkey	*Streptococcus* and *Lactobacillus* species
	Dahi	India	
Mixed lactic acid and ethanol products	Kefir	USSR, Finland	*Streptococcus, Lactobacillus,* and yeast species
	Koumiss	USSR	
	Laban	Lebanon	

mentioned cultures is a function of time as well as the sugar content of the starting yogurt material.

Commercially, yogurt is made from a mix of whole, partially defatted milk; condensed skim milk; cream; and nonfat dry milk. Milk fat levels in such preparations range from 1.0% to 3.25%. Only products containing a minimum of 3.25% milk fat are labeled yogurt. Those with 0.5% to 2.0% or less than 0.5% milk fat are labeled low-fat and nonfat yogurts, respectively. In addition to milk fat, other milk ingredients are found in yogurt. These include casein, sodium and calcium caseinates, whey, and whey protein concentrates. Additives of several kinds are also permitted in commercially produced yogurt, such as nutritive carbohydrate sweeteners, coloring, stabilizers (for smooth texture and longer shelf life), and fruit preparations for flavoring. Fruit preparations for blending in yogurt are specially designed to meet the marketing requirements for different types of yogurt. The types most commonly marketed include the sundae style (fruit preserves are layered on the bottom of containers) and the Swiss style (fruit preparations are thoroughly blended in yogurt after culturing).

Yogurt and other cultured dairy foods are excellent sources of high-quality protein and certain B-complex vitamins and minerals. In addition to the nutritional importance of yogurt, it has been found to be of value in maintaining a balance in the intestinal microbial flora. **[See indigenous microbiota in Chapter 17.]**

Figure 16–4 A scanning micrograph showing the long, curved lactobacilli from a kefir sample. *[From V. M. Marshall, W. M. Cole, and B. E. Brooker,* J. Appl. Bacteriol. **57***:491–497, 1984.]*

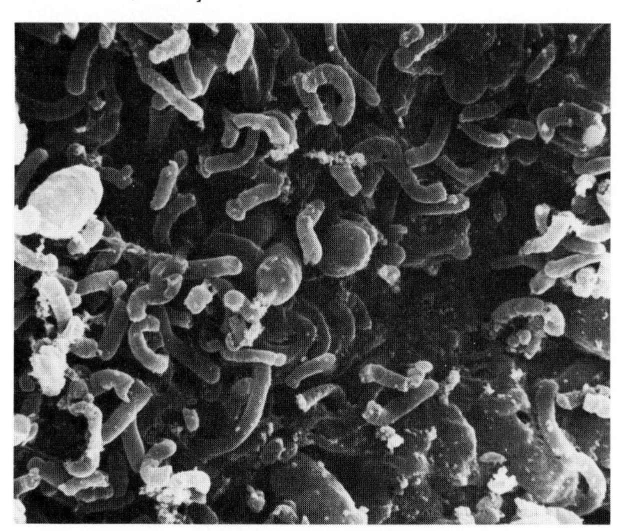

Direct Acidification

Direct acidification as an alternative method for the production of fermented dairy products has been receiving increasing attention. This process eliminates the need for acid production by lactic acid starters. An acidic condition is created in the product by the simple addition of a food-acceptable acid. Stabilizers and artificial flavors are added to produce the texture and flavor characteristic of a cultured dairy product. Imitation fermented dairy foods can be readily developed with this technique. Several advantages are claimed for the use of direct acidification in place of the conventional bacteriological methods. These advantages include a significant reduction in equipment and personnel; fewer controls;

an elimination of defects resulting from the use of microbial cultures; and the production of foods with reproducible flavors.

Single-Cell Protein: Microorganisms as Food

Two thirds of the world's people suffer from inadequate nutrition. The production of traditional agricultural foodstuffs by existing methods cannot possibly meet the protein demands of a human population that continues to grow in an uncontrolled manner. Attempts to solve problems of protein nutrition are usually directed at increasing productivity of agriculture, animal breeding, and fishery. However, attention is shifting to the wider use of nonconventional proteins such as protein concentrates of low-quality fish; protein products of oil-producing crops such as cotton seeds, soybeans, and sunflowers; and protein provided by single-cell microbes.

As difficult as it may be to believe, milligram for milligram, bacteria and many other microorganisms contain as much protein as steak. The protein content of these one-celled forms of life is high, as much as 40% to 60% or more of their weight. Although it may be some time before this source of protein, called *single-cell protein (SCP),* is used to replace conventional human foods, more and more attention is being directed to the ability of microorganisms to grow rapidly on materials previously not considered food sources or on substances that are edible but not nutritious, such as cellulose. Bacteria and fungi can be utilized to exploit nuisance plants

TABLE 16–4 Examples of Microorganisms and Materials Used in the Production of SCP

Materials Utilized (Substrates)	*Microorganisms*[e]		
	Algae	*Bacteria*	*Fungi*
Cellulose		*Cellulomonas* sp.	*Trichoderma viride* *Chaetomium cellulolyticum*[a]
CO_2 (and photosynthesis)	*Chlorella pyrenoidosa*[b] *Scenedesmus quadricauda* *Chlorella stigmatophora*[c]		
Ethanol		*Acinetobacter calcoaceticus*	*Candida utilis*[d]
Food processing wastes			
Banana peels			*Pichia spartinae*[b]
Beef fat (tallow)			*Saccharomycopsis lipolytica*[d] and *Candida utilis*[d]
Potato peels			*Saccharomycopsis fibuliger*[d] and *C. utilis*[d]
Gas oil		*Acinetobacter calcoaceticus*	
H_2 and CO_2		*Alcaligenes eutrophus*	
Kerosene and related substances		*Nocardia* sp.	*Candida intermedia,*[d] *C. lipolytica,* *C. tropicalis*
Methane		*Methylococcus capsulatus,* *Methylomonas* sp.	*Trichoderma* sp.
Methanol		*Methylomonas methanica* *Methylophilus methylotrophus*	
Simple sugars			*Candida utilis,*[d] *Kluyvermyces fragilis,*[d] *Fusarium graminearum,* *Saccharomyces cerevisiae*
Starches			*Endomycopsis fibuligera*

[a] Mold
[b] Freshwater alga
[c] Marine alga
[d] Yeast
[e] For pronunciations, see the Organism Pronunciation Guide at the end of the text.

Microbiology Highlight

"MEAT" OF THE FUTURE?

Normally when one thinks of mold and meat, the vision is one of the gray or green furry coating that can sometimes be found on leftovers. In one of the more recent exploitations of microorganisms for high-protein food, two British companies have joined forces to manufacture and market mycoprotein-based meatlike products. The fungus *Fusarium graminearum* (foo-SAIR-ee-um gram-ee-NEAR-um) produces hyphae that are comparable in texture to meat fibers. The organism is grown in a continuous system using glucose from hydrolyzed cornstarch, mineral salts, trace metals, choline, and biotin. The mycelium is continuously harvested, treated to reduce the high RNA content, mixed with egg albumin (to improve texture after cooking) and with colorings and flavorings that are specific to different products. These products include "fish cakes," "veal patties," "meat pies," "chicken and ham patties," and chicken-flavored pieces. Mycoprotein has about 75% of the protein content of lean beef with less than half the fat and no cholesterol. In addition, it contains a significant amount of dietary fiber. One of the companies, Ranks Hovis McDougall, has sold considerable quantities of these mycofoods in its company restaurants and reports no unfavorable responses.

such as various weeds, urban solid waste, oil, and almost anything containing carbon (Table 16–4). The substances used by microorganisms in the production of SCP are supplemented with other nutrients, such as inorganic nitrogen compounds. By microbial enzymatic activity the energy in the carbon source is converted into biomass, that is, the total weight of living matter. This resulting SCP is harvested, usually dried, and subsequently used for animal feed. With further processing, most SCP can be used as a source of food for humans. The by-products of the generation of SCP are largely carbon dioxide, heat, and some waste water.

In an overpopulated world, there is not enough animal protein to go around. Animal protein has the best nutritional value because it contains more of the essential amino acids than does plant protein. The burgeoning demand for animal protein in the United States and other affluent nations will require some type of livestock population explosion. It will also require gigantic quantities of feed to fatten all of these animals.

Various animals, such as cattle and hogs, although excellent sources of amino acids, must consume an extremely large amount of plant mass to produce their own smaller mass. The fattening of livestock requires unbelievable quantities of human-edible plant material. This process often involves feeding such animals one of the best protein sources known, namely soybean meal, while the people in poor and developing countries may have to subsist on a protein-deficient plant diet. SCP has the potential to correct this situation by replacing the human-edible feeds used in the livestock industry, thereby making these feeds available for human consumption.

A safe SCP product can also be a direct source of protein for humans. It would probably first have to be fortified with certain essential amino acids. SCP preparations are often deficient in lysine or methionine, and very high in RNA. The general processing of SCP involves reducing the RNA content.

SCP may sound unappealing to people who associate microorganisms only with disease and spoilage. Nonetheless, as we have seen, microbes already exist in several types of food and beverages consumed by humans.

Genetic Engineering

Traditionally, microorganisms have been used to stabilize, flavor, or change various properties of food. In addition, continuous efforts are made to control microbial spoilage and to ensure that foods are free from microorganisms that may be hazardous to public health. Now, with the availability of genetic engineering techniques, improvements in SCP production can be expected. This improvement will largely be accomplished by designing microorganisms capable of transforming inedible organic waste, known as *biomass,* into food that is safe for human consumption or for animal feed. **[See genetic engineering techniques in Chapter 6.]**

Microbial Food Spoilage

Major Sources of Microorganisms

Most food for humans is of animal or plant origin. Most of these foods, including market fruits and vegetables, may normally be expected to contain a wide variety and a varying number of bacteria and fungi. Most of these organisms are occupying their

Figure 16–5 The nutrient content of fruits and vegetables can support the growth of bacteria, mold, and yeasts. Their high water content also contributes to the growth of many spoilage bacteria. These tomatoes have bacterial spot disease. *[Courtesy USDA.]*

normal ecological habitats and do not cause any problems, but some can be destructive and produce significant food spoilage (Figure 16–5 and Color Photograph 6). The sources of microorganisms in foods include air, dust, soil, water, utensils, food handlers, and the intestinal tracts of humans and other animals.

Various factors influence the numbers and types of microorganisms in fruits and vegetables or a finished food product. Among these factors are the properties of the food itself—its pH, moisture, and nutrient content—as well as the general environment from which the food was obtained and the conditions of handling, processing, and storage to which it was exposed. The number of microorganisms should be kept as low as possible. Excessively high numbers of microbes, some of which may cause food spoilage (Table 16–5), or the presence of pathogens or their products (Table 16–6) are cause for alarm.

TABLE 16–5 Examples of Microorganisms and Effects Associated with Food Spoilage

		Microorganisms Involved	
Food	*Spoilage Effects*	*Bacteria*	*Fungi*
CANNED FOOD			
Fruits	Mold growth		*Byssochlamys fulva*
CEREAL PRODUCTS			
Bread	Ropy texture	*Bacillus subtilis* var. *mesentericus*	
	Mold growth		*Aspergillus niger, Mucor* sp., *Penicillium* spp., *Rhizopus nigricans*
DAIRY PRODUCTS			
Pasteurized milk	Gas formation	*Bacillus* spp., *Clostridium* spp.	
	Lactic acid souring	*Lactobacillus thermophilus, Microbacterium lacticum, Streptococcus thermophilus*	
	Proteolysis (protein breakdown)	*Bacillus* spp., *Clostridium* spp., *Micrococcus* spp.	
FISH			
Fresh fish	Discoloration	*Micrococcus* spp., *Pseudomonas fluorescens, Sarcina* spp.	Various fungi (both molds and yeast)

TABLE 16–5 (continued)

		Microorganisms Involved	
Food	*Spoilage Effects*	*Bacteria*	*Fungi*
Fresh fish	Pronounced fishy flavor	*Flavobacterium* spp., *Micrococcus* spp., *Pseudomonas* spp., *Serratia* spp.	
JELLIES AND JAMS	Mold growth		*Aspergillus* spp. *Penicillium* spp.
MEAT			
Fresh meats	Slime formation in cold environments	*Flavobacterium* spp., *Micrococcus* spp., *Pseudomonas* spp.	Yeasts
Cured and smoked bacon, sausage	Greening	*Lactobacillus* spp. *Leuconostoc* spp.	
	Mold growth		*Alternaria* spp., *Aspergillus* spp., *Mucor* spp., *Rhizopus stolonifer*
Eggs	Green discoloration and rot	*Pseudomonas fluorescens*, *Proteus* spp.	

How Microorganisms Affect Food

Not only can foods spoil, they can also contain a variety of other biological hazards (Tables 16–5 and 16–6). Microbial pathogens, their products and the diseases they can cause are presented in several later chapters in Part VIII. Here, we will deal with microbial food spoilage.

Spoiled food can be defined as food that has been so damaged or altered as to make it unfit for human consumption. The principal causes of such spoilage are dehydration, enzymatic action, microbial growth, and oxidation. The type of spoilage is influenced by such factors as the composition of the food, the types of microorganisms present and the degree of microbial contamination, the presence of microbial growth inhibitors, and conditions of storage, including temperature (Figure 16–6) and moisture. Spoilage microorganisms are affected by these same factors. Examples of foods, spoilage effects and the microorganisms causing these effects are listed in Table 16–5. Control of one or several of the factors mentioned usually reduces microbial spoilage of foods.

TABLE 16–6 Some Biological Hazards Associated with Contaminated Foods[a]

Food-Poisoning Bacteria	*Other Bacteria*	*Fungal and Algal Toxins*
Bacillus cereus, *Campylobacter jejuni*, *Clostridium botulinum*, *C. perfringens*, *Escherichia coli* (EPEC), *Salmonella* spp., *Staphylococcus aureus*, *Streptococcus faecalis*, *Vibrio parahaemolyticus*	*Brucella* spp., *Leptospira*, *Listeria monocytogenes*, *Mycobacterium tuberculosis*, *Vibrio cholerae*, *Yersinia enterocolitica*	Aflatoxins, other mycotoxins, saxitoxin (red tide product)
Protozoa	*Viruses*	*Worms*
Entamoeba histolytica, *Toxoplasma gondii*	Enteroviruses, hepatitis A virus, Delta virus, Newcastle disease virus	Chinese liver fluke; beef, fish, pork, and sheep tapeworms; pork roundworm

[a] Refer to Part VIII chapters for details concerning these and other microorganisms and the disease states associated with them.

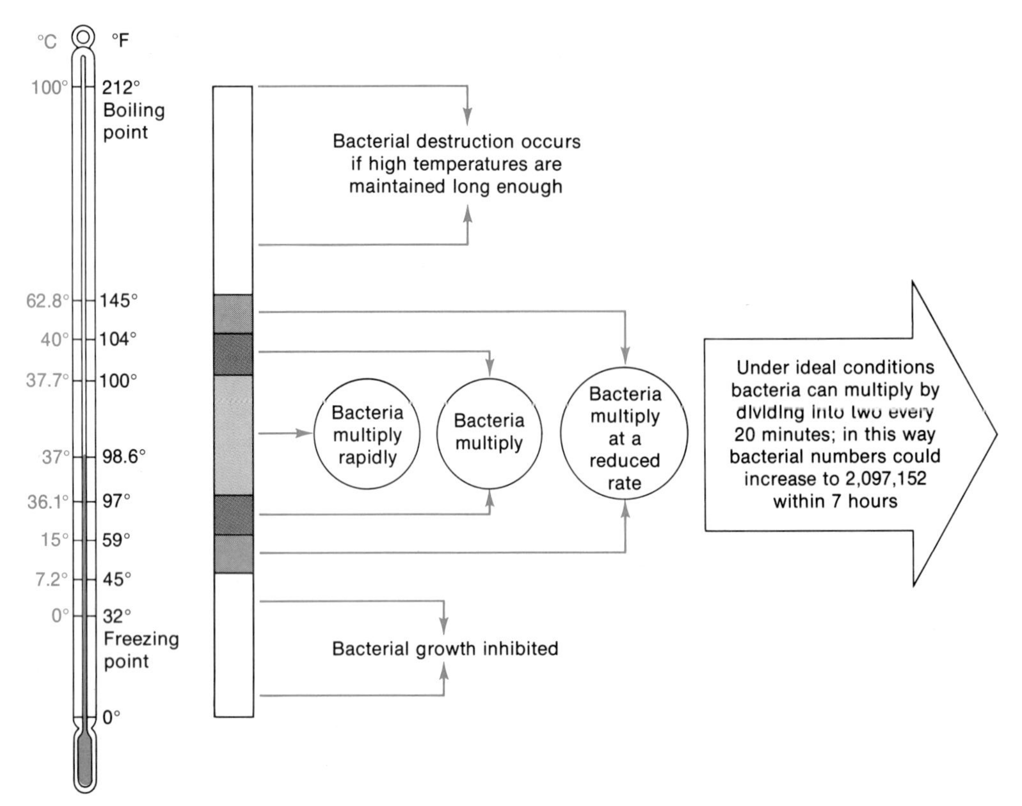

Figure 16–6 Microorganisms grow over a wide range of temperatures. Knowing the temperature growth ranges for organisms such as those capable of causing spoilage or disease is important to selecting the proper temperature for the storage of different foods. This diagram shows the effect of temperature on the growth of bacteria in foods. The safe and growth-supportive temperatures for bacteria are indicated. *[After B. C. Hobbs,* Food Poisoning and Food Hygiene. *London: Edward Arnold Publishers, Ltd., 1968.]*

Detection of Microorganisms in Food

Various techniques are used to determine the numbers and kinds of microorganisms and their products in foods. These methods include direct microscopic counts for both living and dead cells, standard plate counts, specific staining procedures, immunological tests for enzymes and other microbial products, isolation and biochemical identification, and statistical determinations for living cells. Table 16–7 lists bacteria that are considered to be indicators of potential human or fecal contamination and also bacteria that are considered to be indicators of post-heat-processing contamination.

TABLE 16–7 Bacteriological Indicators for Food Contamination

Indicators of Potential Human or Fecal Contaminants	
Escherichia coli	*Streptococcus faecalis*
Pseudomonas aeruginosa	*S. faecium*
Staphylococcus aureus	
Indicators of Post-Heat-Processing Contamination	
Aeromonas spp.	*Escherichia coli*
Enterobacter spp.	*Klebsiella* spp.
Erwinia spp.	*Serratia* spp.

Membrane filtration has become the method of choice for determining bacterial counts in water and beverages because this technique offers several advantages over other counting procedures. In spite of these advantages, membrane filtration suffers from drawbacks that have discouraged its application to food microbiology. Included are the tendency of many foods, particularly dairy products, to clog the pores of a membrane during filtration, limitations in determining the number of a specific group of microorganisms in the presence of a high background microbial flora, and the retention of food particles on the surface of the membrane, thus interfering with accurate bacterial colony counts.

Relatively inexpensive solutions to these problems can be achieved by prefiltration of food samples to prevent the deposit of food particles on a membrane surface and the use of an appropriate enzyme treatment with foods to be tested.

A reliable membrane filtration system for food microbiology is of major importance. It offers many advantages over other techniques, such as the ability to detect low numbers of organisms by filtering larger sample volumes, the removal of water-soluble materials such as substances that could interfere with microbial growth, the absence of temperature stress from melted agar (as in the pour-plate technique), and the availability of more rapid results.

Food Preservation

Drying

Among the oldest methods to prevent or control food spoilage are fermentation, refrigeration, and drying (Table 16–8). The preservation of foods by drying is based upon the fact that microorganisms and enzymes need water to be active. The term **water activity** (a_w) is used to indicate the water content of foods. In preserving foods by drying, the water activity is lowered to a point where the activities of food-spoiling and food-poisoning microbes are inhibited. Examples of foods so preserved include dried fruits (apricots, figs, prunes, raisins), powdered eggs and milk, and frozen-dried foods. Organisms including molds, yeasts, and salt-tolerant bacteria can grow in foods with a low water activity.

Canning

The developing of canning as a process to prevent food spoilage began in Napoleonic France in 1795, when the French government offered a cash prize for the development of a practical method of food preservation. In 1809 a French confectioner, Francois Appert, succeeded in preserving meat in stoppered glass bottles that were kept in boiling water for varying periods of time. Appert was probably unaware not only of microorganisms but also of the long-term significance of his heating process. Interestingly enough, this beginning of canning as it is known and used today preceded by some 50 years Pasteur's demonstration of the role of microbes in the spoilage of wine.

Both canning and pasteurization are high-temperature treatments used in the processing of foods. Pasteurization, however, destroys only non-spore-forming pathogens. It is not a method for sterilization. In the canning process, foods are sterilized by heating in cans or jars that can be sealed in such a way that no microorganisms can enter. Commercial cans are made of 98.5% sheet steel with a thin coating of tin. The time and temperature for heating vary with the type of food, the acid content, and the ease with which sterilization can be achieved. Acid foods such as lemon juice, sauerkraut, and tomatoes, for example, require shorter heat-processing times than do low-acid foods such as milk, ripe olives, and sardines. Commercial canning operations are adjusted to prevent the destruction of vitamins and other important food components.

A wide variety of foods are canned, including fruits, vegetables, meats, and fish. There are many steps involved in the commercial canning operations for such foods. Preparatory operations include cleaning, washing, and *blanching.* Blanching involves treatment of raw food products such as fruits or vegetables with hot water or live steam to soften the plant tissues, making the filling of cans easier and destroying any enzymes that may change the color, flavor, or texture of the final product. Depending on the food, peeling, slicing, or dicing may also be necessary during the early operation. Cans are then filled, heated for a short time, exposed to a vacuum system to remove any excess air *(thermal exhaustion),* and sealed. The finished, sealed cans are subjected to heat sterilization for thermal processing. Established time and temperature exposure schedules are followed to ensure the destruction of spoilage and pathogenic microorganisms that may

TABLE 16–8 **Food Preservation Approaches and Examples of Methods**

Method	*Examples*
Killing of microorganisms	Heating, e.g., canning Ionizing radiation, e.g., gamma radiation Mechanical disruption Nonionizing radiation, e.g., microwave and ultraviolet light
Growth inhibition	Drying by dehydration and freeze-drying Chemical preservatives Refrigeration or freezing High osmotic pressure by use of salt brines and sugar syrups Low pH with acidulants Anaerobic conditions
Removal of microorganisms	Centrifugation Filtration
Prevention of contamination	Observing and maintaining aseptic conditions

be in or on the raw food product. The heat-treated cans are usually cooled in cold-water tanks or by cold-water sprays. Home canning follows the same general steps except that the glass jars most often used do not undergo a separate evacuation and sealing process. They are sealed by cooling after heat processing. Because it is not usually feasible to check for sterilization of homecanned foods, the prescribed times and temperatures should be followed scrupulously.

SPOILAGE OF COMMERCIAL CANNED FOODS

The microbial spoilage of various commercial canned foods can be divided into five categories: *incipient spoilage, gross underprocessing, leaker spoilage, thermophilic spoilage,* and *insufficient heat treatment.*

Incipient spoilage is caused by keeping products, prior to canning, under conditions favoring microbial growth and using highly contaminated foods or ingredients. Gross underprocessing results from ineffective sterilization procedures. Poorly constructed containers, defects in the canning operation, or contamination of foods after processing result in leaker spoilage. Storage of foods at temperatures favorable for microbial growth or contamination by thermophilic spore formers is responsible for thermophilic spoilage. Reduced exposure to heat (insufficent heat treatment) will favor the growth of spoilage-causing and toxin-producing microorganisms.

Chemical Preservatives

Several chemical additives are used to prevent microbial spoilage of foods. Examples are citric acid, used in jams, jellies, and soft drinks; the mold inhibitor sodium propionate, used in bakery goods and processed cheese; and the bacterial inhibitors sodium nitrate and sodium nitrite, used in the curing of meat and fish (Table 16–9). These substances are classified as food additives under the Federal Food, Drug, and Cosmetic Act, indicating that they have been tested and found safe for human consumption by the Food and Drug Administration. This agency is currently reviewing all food additives as to levels of safety and effectiveness for food preservation. The levels of substances such as nitrates and nitrites may be reduced considerably in the near future as a result of such studies.

Chemical preservatives include both inorganic and organic compounds. These compounds either kill microorganisms or inhibit their growth, enzymatic activities, or reproduction. Some compounds are also used to destroy or reduce the number of microorganisms on food-processing equipment and food utensils or to treat water used in the processing of foods.

Like food, water is essential for life. However, water can also be hazardous, since it can contain disease agents and poisonous materials. The next section will consider the treatment of water for human consumption and approaches to the treatment of waste water and other wastes generated by humans and their activities.

Water Microbiology

The significance of microorganisms in aquatic environments is well recognized. As described in Chapter 14, some of these microorganisms play vital roles

TABLE 16–9 Examples of Antimicrobial Chemical Agents Used as Preservatives

Chemical Agent	*Uses Include:*
Acetic acid	To achieve desired acidity levels (acidulant) in bakery goods, beverages, confections, cottage cheese, pasteurized cheese, sherbets, and syrups
Benzoic acid (also sodium benzoate)	Common preservatives in carbonated soft drinks, confections, fruit juices, jellies, jams and preserves, and oleomargarine
Carbon dioxide (also carbonic acid)	Common preservatives in beer, controlled gas storage of fruits, fruit juices, soft drinks, and vegetables
Citric acid	To achieve desired acidity levels in various products, such as jams, jellies, and soft drinks
Lactic acid	Acidulant and preservative in carbonated soft drinks and cottage cheese
Sodium propionate	Mold inhibitor in bakery goods and processed cheese
Sodium nitrate	Bacterial inhibitor, and curing of meat and fish[a]
Sodium nitrite	Bacterial inhibitor, and curing of meat and fish[a]
Sorbic acid (also potassium sorbate)	Fungus inhibitor in bakery goods, fresh fruit, salads, and syrups
Sulfur dioxide and sodium sulfite	Preservative and antibrowning agent in dried fruits, fruit juices, syrups, and wine

[a] Allowable levels.

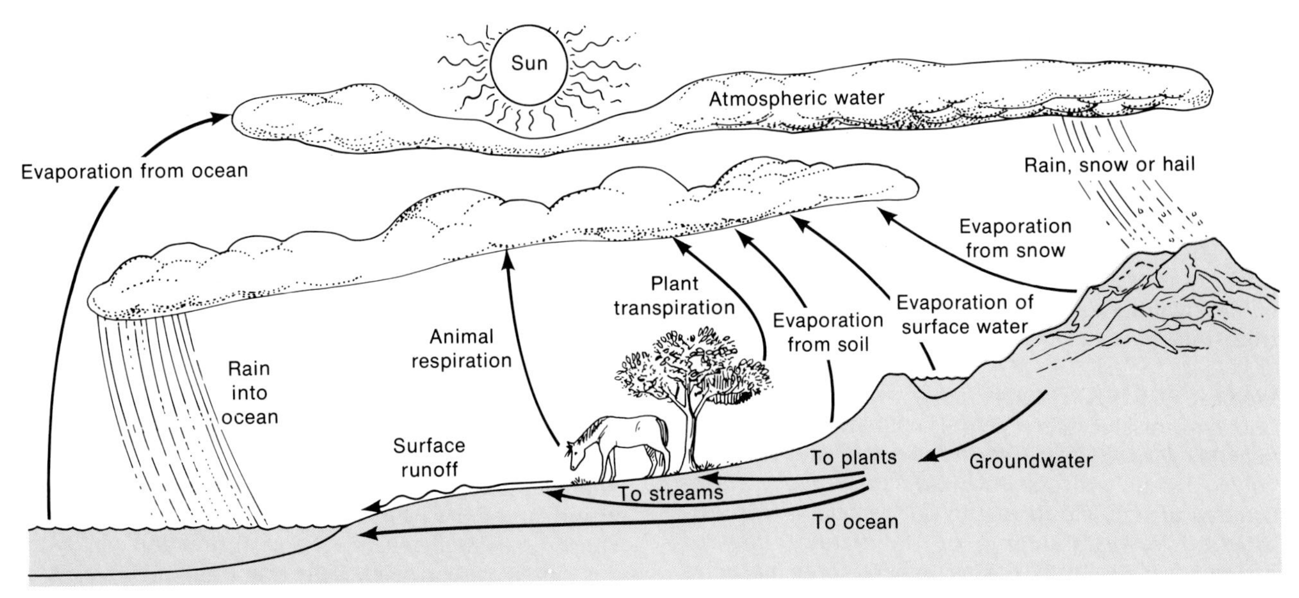

Figure 16–7 The water cycle provides the mechanisms by which constant interchanges of water take place among air, sea, and land and between different forms of life and their environments. Such interchanges produce daily and seasonal changes in the environments of animals, plants, and microorganisms.

in aquatic ecosystems and contribute to the operation of biogeochemical cycles. Others are of interest and importance because of their association with waste treatment and their potential dangers to safe water supplies.

Environmental reservoirs of infectious disease agents for humans include soil, water supplies, sewage, and some foods. Thousands of other substances used or produced by human society are introduced into various bodies of water through natural means as well as by industrial and domestic processes. Through the years several areas of the world have been declared unsafe because of potentially toxic levels of biological and chemical pollutants. Raw sewage, with its vast numbers of potentially pathogenic bacteria and viruses, is the most dangerous threat to water supplies. Indeed, it may have disastrous consequences for fish, wildlife, and humans. Thus it is quite obvious that not all water is potable, or fit for human consumption. Let us consider the source of our water supply.

The Water (Hydrologic) Cycle

Because of the *water cycle* (Figure 16–7), there is a constant interchange of water among air, land, and sea and between different forms of life and their environments. Moreover, water greatly influences such environments. The water cycle involves *evaporation, cloud formation, precipitation, surface water runoff,* and *percolation through the soil.*

Evaporation from bodies of water provides some of the moisture that forms clouds. Additional atmospheric moisture results from a plant process known as *transpiration,* through which water vapor is released from leaves as water and minerals are drawn up stems from the roots. Under the proper atmospheric conditions, the clouds lose their water as precipitation in the form of rain, snow, or hail. Some of this water seeps into the ground and filters, or percolates, through dirt and sand to enter the underground water, known as an *aquifer.* This *groundwater,* which is one of two main sources of water, is usually of good overall quality. Although it may be high in dissolved minerals, such water has little color and turbidity (cloudiness) and only low numbers of microorganisms. Groundwater generally has the lowest microbial levels. Surface waters, the other main water source, often contain large numbers of microorganisms and come from streams, ponds, rivers, lakes, and shallow wells.

AQUATIC MICROORGANISMS

The microbial populations of water environments are influenced by several chemical and physical factors, including hydrostatic pressure, light, pH, salinity, and temperature. Various microorganisms are found naturally in fresh and salt waters. Others enter these waters from natural sources such as air or soil or from industrial or domestic processes. Most microorganisms that are native to natural waters are nonpathogenic. Disease-causing organisms found in ground- or surface waters usually are unable to grow in these environments and gain access to these waters as a result of contamination from domestic or industrial activities.

The most widely used sources of water supplies

are generally surface waters. Since these sources are often subject to pollution, they must be treated before being used for domestic or industrial purposes. After treatment, the water may be pumped to reservoirs for storage and then distributed to homes and industrial facilities by means of extensive networks of adequately maintained underground pipes. The basic steps used to make water microbiologically safe and to improve its general quality according to public health standards forms the basis of water purification processes.

WATER PURIFICATION

The type or necessary degree of water purification depends on the water source. Several steps are involved in the process, including *sedimentation, aeration, coagulation, filtration,* and *chlorination.* **Sedimentation** involves pumping or the natural flow of incoming water into basins, where large particles such as sand or gravel can be deposited by settling. Fast-flowing rivers or streams usually require sedimentation. Slower-flowing surface waters, lakes, or reservoirs may not need this process. Waters containing significant levels of organic materials, such as plant material and animal wastes often require **aeration,** or spraying of the water into the air. This step provides the water with a high level of dissolved oxygen, thereby speeding up the microbial oxidation of organic materials. Most water sources are exposed to a **chemical coagulation** step, in which aluminum- and iron-containing substances are added to the water being processed. These chemicals form insoluble precipitates that trap and remove most of the remaining organic and inorganic materials as well as microorganisms, including viruses. These precipitates settle out of the water and most of the remaining suspended contaminants are removed by **filtration.** The filtration process incorporates sand filters or pressure filters through which water is simply pumped under pressure. Pressure filters are used in facilities such as swimming pools and small industrial plants. The last step in the process is **chlorination,** the most common method of ensuring the microbiological safety of drinking water. The addition of chlorine to water accomplishes several goals: disinfection, oxidation of taste- and odor-containing chemicals, and oxidation and subsequent removal of iron and manganese compounds. [See Chapter 11 for the properties of chlorine.]

While the most common disinfectant chemical is still chlorine, ozone treatment of water is becoming more available. It has been found that chlorine may react with the chemicals in the water and yield potentially cancer-causing (carcinogenic) compounds such as chloroform and chlorophenol. Chlorine is also harmful to various forms of aquatic life. In addition to its disinfecting properties, the ozone treatment of water also promotes aerobic biological activity and decreases the concentration of suspended solids, thereby improving the efficiency of filtration. Ozone has been found to be safer and, under proper operating conditions, as economical as chlorine compounds.

Wastewater Treatment

Each village, town, municipality, or region must deal with its liquid and solid waste materials. Very small communities can run their wastewaters into shallow ponds or streams for biological decomposition and bury the solids. Natural cycles such as those associated with carbon, sulfur, and phosphorus can process considerable amounts of these materials and do so with or without human assistance. The breakdown of much litter results in the formation of slowly decaying animal and plant remains (humus) and the decomposition of small pieces of organic matter (detritus) in fresh- and marine-water sediments. Large communities produce far too much waste for environmental processes to handle. Thus, specialized ponds, or lagoons (Figure 16–8), have been developed for handling some wastes, and treatment facilities are built and maintained for many domestic and industrial wastes (Figure 16–9). Solid waste materials, if not properly disposed of, can and usually do contribute many toxic chemicals as well as depleting the dissolved oxygen in water. These wastes are still, for the most part, buried, but in a manner that should not prove harmful to the environment. Consolidating treatment processes in

Figure 16–8 One of two lagoons for a standard rendering plant. Waste material is processed before being released into a neighboring creek. Here waste material from poultry preparation facilities (which includes intestines, head, feet, and dead animals) is processed. Three agitator aerators can be seen in this lagoon. A major portion of the resulting clear water is recycled into the waste treatment plant and reused. *[Courtesy USDA-Soil Conservation Service.]*

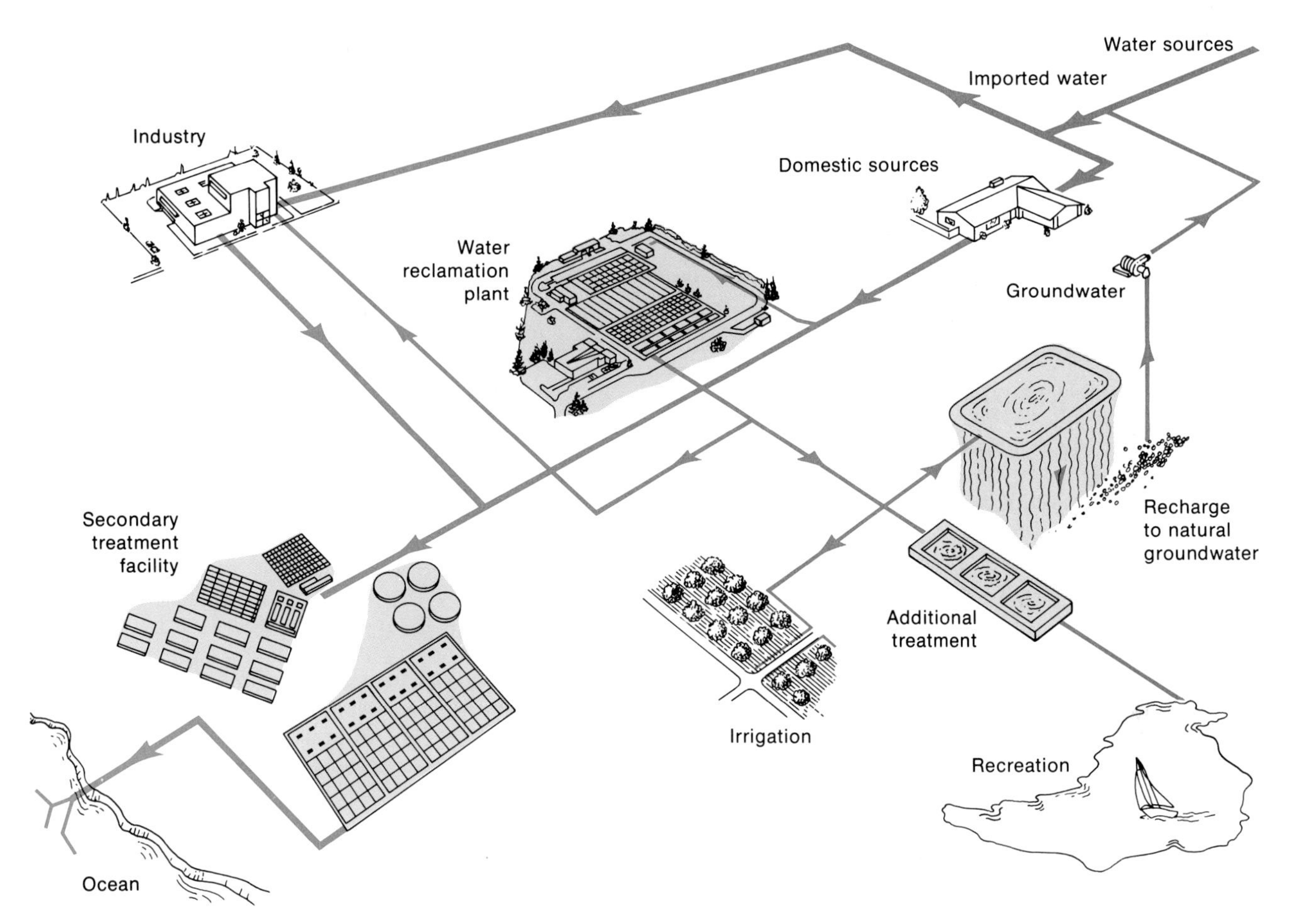

Figure 16–9 A collection system for waste water, showing the treatment of water to render it reusable. Note that treated water can be used for recreational activities, but additional treatment is needed first. *[Courtesy of the Sanitation Districts of Los Angeles County.]*

larger facilities reclaims valuable by-products in a way that might otherwise not be practical. Some materials are converted into fertilizer, compost, and methane. Certain residual substances can be partially pyrolized (decomposed by heating) to produce activated carbon for water purification or more completely pyrolized to ashes for the recovery of various heavy metals.

Most waste-treatment processes can be considered natural, in the sense that the microorganisms are present in the soil, water, and/or materials to be treated. Subsequently, specialized communities of organisms develop and perform the treatment more efficiently. If toxic materials enter the process, the killing of important populations in the community often causes the treatment to slow and even stop until the proper balance is restored. In contrast to this natural category of processes are the applied treatment processes, in which prepared microbial populations and/or communities are used. In applied treatment processes, the organisms are grown under suitable conditions and formulated for addition to materials that require specialized treatment, such as petroleum wastes and greases.

Natural Processes

Wastewater

Wastewater, or sewage, contains about 99% water and is therefore easily pumped through sewage pipes to a treatment plant (Figure 16–9). The remainder of the material consists of suspended, relatively insoluble solids such as starch granules, cellulosic (plant) remains, lignins, and proteins. Dissolved substances in sewage include sugars, fatty acids, alcohols, amino acids, and many inorganic compounds. Sewage also contains greases, oils, litter, paper, gravel, and many varied materials from garbage disposal units and community industrial processing facilities (Figure 16–9).

Domestic wastewater is largely human fecal matter and urine diluted with laundry and wash water. One estimate is that each person in the United States produces more than 100 gallons of wastewater per day. A treatment facility serving an area of 1 million people must therefore process 100 million or more gallons of domestic water daily. Industrial and commercial wastewaters include materials from food canning processes, dairies, slaughterhouses, and cattle or hog feedlots, as well as from many chemical and pharmaceutical industrial plants. The food- or animal-related enterprises and pharmaceutical companies produce wastewaters of considerably higher organic content than that of domestic wastes. The chemical industries vary considerably in their organic contributions to wastewater. Unfortunately, chemical wastes often contain toxic materials, such as heavy metals or pesticides, that may reduce the effectiveness of a treatment facility.

Microorganisms commonly found in wastewaters include bacteria of intestinal origin such as *Bacteroides* (back-te-ROY-deez) spp., *Bifidobacterium* (bye-feh-doh-back-TIR-ee-um) spp., *Clostridium perfringens* (klos-TRE-dee-um per-FRIN-jens), *Enterobacter aerogenes* (en-te-roh-BACK-ter a-RAH-jen-eez), *Escherichia coli* (esh-er-IK-ee-ah KOH-lee), *Lactobacillus* spp., and *Streptococcus faecalis* (S. fee-KAL-is). Because of the origin of some of the wastes, human pathogens are sometimes found in sewage. Table 16–10 lists some of these microorganisms. One of the major purposes of wastewater treatment is to eliminate such potential hazards to public health. [See intestinal flora in Chapter 26.]

Some microorganisms found in wastewater are primarily from soil and water. These include species of bacteria including *Sphaerotilus* (sphere-OT-ill-us), *Crenothrix* (kren-OH-thricks), and *Beggiatoa* (beg-ee-ah-TOH-ah) and fungi such as *Saprolegnia* (sap-roh-LEG-nee-ah) and *Leptomitus* (lep-TOH-my-tis). All of these organisms can form slimy growths in pipes and ditches. Some can also produce iron, sulfur, or manganese deposits, which clog pipes, or hydrogen sulfide, which is noted for its offensive rotten-egg odor. Other bacteria found in wastewater include species of *Cytophaga* (sigh-TOH-fah-jah), *Micrococcus* (my-kroh-KOK-kuss), *Pseudomonas* (soo-doh-MOH-nass), *Bdellovibrio* (dell-oh-VIB-ree-oh), *Chromobacterium, Aeromonas* (air-OH-moh-nass), *Rhodospirillum* (roh-doh-spy-RILL-um), and various methanogenic organisms.

Bacteriological Water Testing

A wide variety of microorganisms may be present in drinking water. If water supplies are contaminated by wastewaters, a number of infectious disease agents could be transmitted, many of which cause severe gastrointestinal diseases (Table 16–10). Asiatic cholera, bacillary and amebic dysenteries, leptospirosis, and several viral and protozoan diseases are some of the diseases transmitted by contaminated drinking water. Even water that looks clear and pure may be sufficiently contaminated with pathogens to pose a considerable health hazard. [See microbially-caused gastrointestinal diseases in Chapter 26.]

Contamination Detection

It is neither practical nor possible to test drinking water for all the possible microbial pathogens that may be present. Fortunately, it is also unnecessary. There are several microorganisms commonly found in human and other animal intestinal tracts that can be used as *indicators* of fecal pollution. Their presence in water supplies in sufficient numbers signals contamination of an intestinal source. The most common of these indicator organisms are the *coliforms*. These microbes are facultatively anaerobic, gram-negative, non-spore-forming rods that ferment lactose with the production of acid and gas in 24 to 48 hours when grown at 35°C. The group includes various strains of *Escherichia coli, Enterobacter aerogenes,* and *Klebsiella pneumoniae* (kleb-see-EL-lah new-MOH-nee-ah). These microbes populate the intestine in large numbers and survive longer in water than do the pathogenic bacteria. *Salmonella typhi* (sal-moan-EL-la TIE-fee), the causative agent of typhoid fever, has been shown to be outnumbered in feces by the coliforms by a ratio of one million to one. *S. typhi* survives in water for only about one week, but the coliforms survive for weeks and even

TABLE 16–10 Examples of Principal Human Pathogens Associated with Wastewaters[a,b]

Bacteria	*Protozoa*	*Viruses*
Campylobacter species	*Entamoeba histolytica*	Adenoviruses
Leptospira spp.		
Salmonella paratyphi	*Giardia lamblia*	Enteroviruses (coxsackieviruses, echoviruses, and polioviruses)
Salmonella typhi		
Salmonella typhimurium		Parvoviruses (hepatitis A virus)
Shigella dysenteriae		Reoviruses
Vibrio cholerae		

[a] Most of these pathogens are described in more detail in Chapter 26.
[b] For pronunciations, see the Organism Pronunciation Guide at the end of the text.

months. The presence of a large number of coliforms in water usually indicates recent pollution; relatively low numbers reflect a past contamination problem.

Procedures for testing water for coliforms and other indicator organisms are published regularly by the American Public Health Association in *Standard Methods for the Examination of Water and Wastewater.* Two specific procedures used for coliforms are the *most probable number (MPN) method* and *membrane filtration (MF).* A general count of microorganisms in the water is also performed.

MOST PROBABLE NUMBER

One means of performing the MPN procedure uses fifteen Durham lactose fermentation tubes, five of which are inoculated with 10 mL of the water sample. A second set of five tubes receives 1 mL each, and a third set of five tubes is inoculated with 0.1 mL of the sample. The lactose tubes are incubated for 24 to 48 hours at 37°C. The number of coliform organisms per 100 mL of the water is determined by recording the number of fermentation tubes showing the presence of gas for each sample size and comparing these data with a statistical table, part of which is presented in Table 16–11.

Consider a complete procedure for bacteriological examination of water. A water sample that is estimated to have 2.2 or more coliforms per 100 mL is presumed to be contaminated. This, then, is known as the *presumptive test* for total coliforms (Figure 16–10). The *confirmed test* involves plating an inoculum from the lactose tubes positive for gas production onto an eosin-methylene blue (EMB) agar plate or other comparable selective and differential medium (Color Photographs 24 and 25). On EMB plates, coliform bacteria may produce deep purple colonies, often with a green metallic surface or pink colonies. The dark color indicates lactose fermentation. Appearance of such colonies confirms the presumptive test. Isolated positive cultures are then subjected to the *completed test,* which involves their inoculation into lactose broth fermentation tubes and nutrient agar slants. If the medium is fermented and stained smears of the organisms from the slant are gram-negative, non-spore-forming rods, the proof for coliform contamination is complete. By incubating cultures at 44.5°C ± 0.2°C, coliforms of enteric or fecal origin can be distinguished from those normally present in soil or water. Only fecal coliforms grow at this temperature. [See selective and differential media in Chapters 4 and 21.]

TABLE 16–11 Application of the MPN Technique to Water Bacteriology

Number of Positive Lactose Broth Tubes			*An Estimate of the Number of Coliform Organisms per 100 mL of Water*
10 mL	*1 mL*	*0.1 mL*	
0	0	0	0
0	0	1	2
0	1	0	2
0	1	1	4
1	0	0	2.2
1	0	1	4.4
1	1	0	4.4
1	1	1	6.7

MEMBRANE FILTER TECHNIQUE

Membrane filters can be used in a more rapid procedure for the isolation and identification of various types of microorganisms including coliforms (Figure 16–10). Total and fecal coliform tests can be performed using these filters (Color Photograph 75). In this method, a specified volume of water is filtered to trap any bacteria present. The filter is then removed and incubated at 35°C or 44.5°C on a pad soaked with an appropriate selective and differential medium, such as EMB or Endo agar. Suspect colonies of coliforms appear on the filter within 24 hours.

A cloudy water sample cannot be tested by this method, since the suspended materials will clog the pores of the filter. Even with this limitation, the membrane filter method has significant advantages in terms of the amount of suspect water that can be tested and the time involved in obtaining completed tests for water quality.

Water Standards and Other Indicator Microorganisms

Standards for water quality usually refer to coliform counts such as the ones described. They are based on the general assumption that the possible presence of disease agents increases with the pollution of water by feces and urine. When the results of bacteriologic examinations of water show unacceptable levels of fecal contamination, the search is made for pollution sources or for failures in the water purification system.

Other than coliforms, the most common indicator organisms for fecal contamination in the United States are strains of *Streptococcus faecalis.* Membrane filtration and MPN procedures are performed for these organisms using selective media with sodium azide as the selective ingredient. *S. faecalis* does not survive well in water, so its presence indicates recent pollution. Two other test organisms are *Bifidobacterium bifidus* and *Clostridium perfringens. B. bifidus* is unable to reproduce in wastewater and is therefore an indicator of recent pollution. *C. perfringens* may survive indefinitely due to its spore-forming capability. The presence of *C. perfringens* therefore indicates any contamination by feces or wastewater over an extended period of time. Results of recent studies have implicated water-borne mycobacteria as causes of infection in individuals with low immunity or resistance. This problem has become so se-

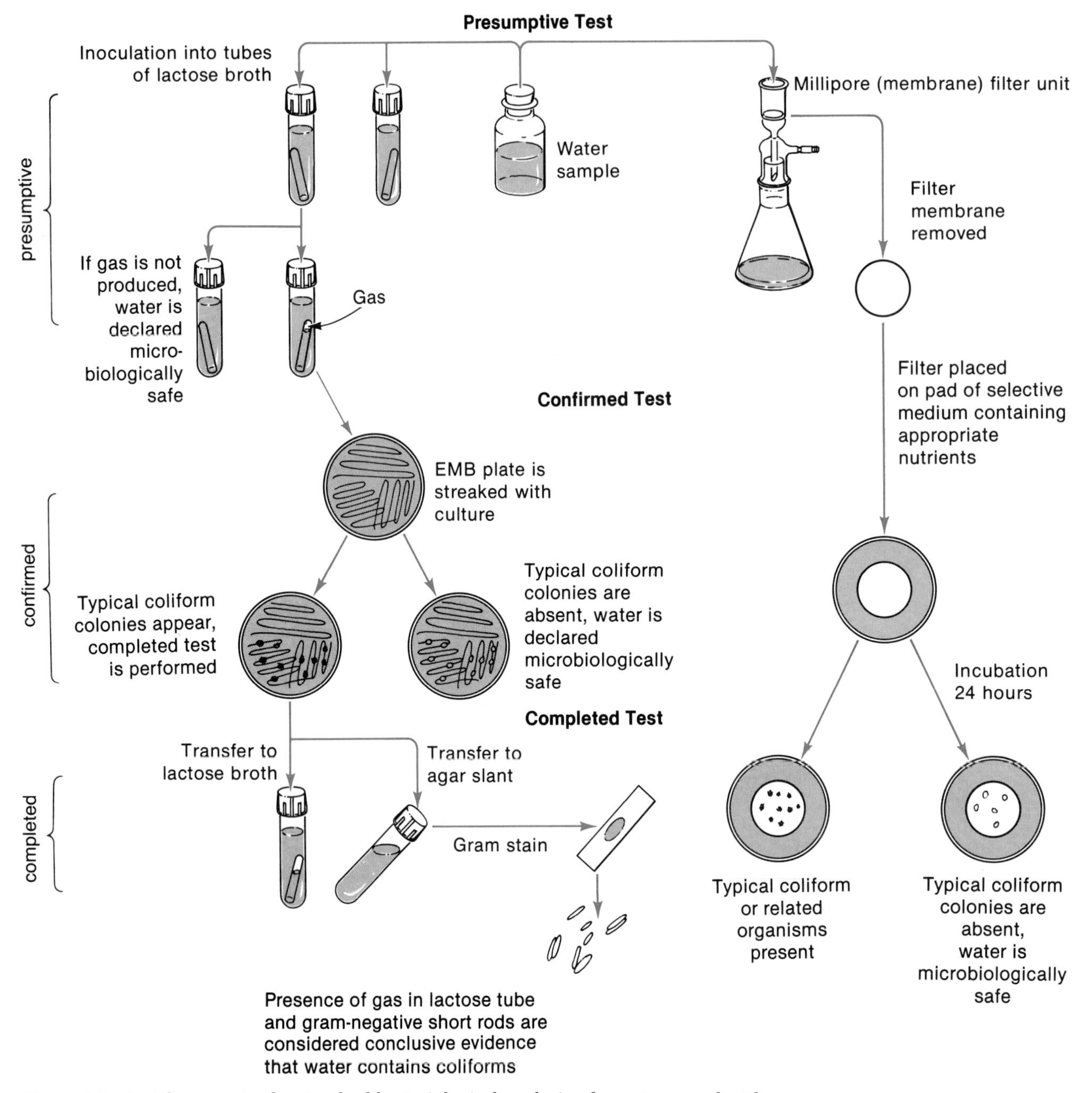

Figure 16–10 The steps in the standard bacteriological analysis of a water sample. The presumptive, confirmed, and completed tests are represented. The use of a membrane filter is also shown.

vere in the Boston area that several scientists have recommended that mycobacteria be included as indicators of unsafe water.

STANDARD PLATE COUNT

Although there are no specific regulations concerning water quality in terms of total aerobic colony count, the periodic use of the assay procedure known as the *standard plate count* can be used to find unusual changes in bacterial counts of water samples. The procedure involves the preparation of nutrient agar pour plates, incorporating 1-mL samples of water or dilutions of samples. After incubation at 35°C for 24 hours, colony counts are made. A finding of 100 or fewer colony-forming units per milliliter of sample is considered to be a reasonably safe standard for drinking water.

The standard plate count can be used to detect the presence of noncoliform organisms that may affect the results obtained with assays of drinking water. The procedure is also used in evaluating the efficiency of water treatment processes. Plate counts on samples of raw water and on samples of water at various purification steps indicate how effectively

Figure 16–11 One consequence of the development of anaerobic conditions in a river. This fish kill was caused by excessive pollution. The remains of such animals putrefy, not only producing foul odors but adding significant amounts of organic matter to natural water systems. *[Courtesy USDA.]*

the bacterial contamination is being reduced. Different incubation temperatures are frequently necessary to allow for the growth of a greater number of organisms present in the water samples.

Nutrient Contamination

The quantity of organic and inorganic materials in the wastewaters is the second critical problem in water treatment. Organisms in the soil and water often cannot metabolize such materials fast enough, so they accumulate and establish anaerobic conditions. The depletion of dissolved oxygen in water, for example, is caused by the bacteria oxidizing the organic matter and consuming the oxygen in the process. The decrease in oxygen content has several important and usually obvious effects (Figure 16–11).

The *biological* or **biochemical oxygen demand (BOD)** is the amount of oxygen consumed by microorganisms during their decomposition of organic materials. BOD is used as an indication of the extent of organic pollution in bodies of water. From the BOD, the level of pollution or nutrient load, as well as the efficiency in correcting unfavorable situations, can be determined. To make this determination, several water samples are obtained in sterile, glass-stoppered bottles. In one sample, the initial concentration of dissolved oxygen is measured in milligrams per liter. The other samples are incubated at 20°C for five days, at which time the final dissolved oxygen in these samples is measured. The difference between the initial value and the later ones represents the BOD.

Wastewater Treatment

PRELIMINARY AND PRIMARY TREATMENT

Most wastewaters are processed for disposal by one or more of the steps outlined for wastewater treatment in Figure 16–12. Step A represents what is called *preliminary treatment* in which wastewater enters a tank or settling chamber for the physical removal of some wastes.

Figure 16–12 A general procedure used in waste-water treatment.

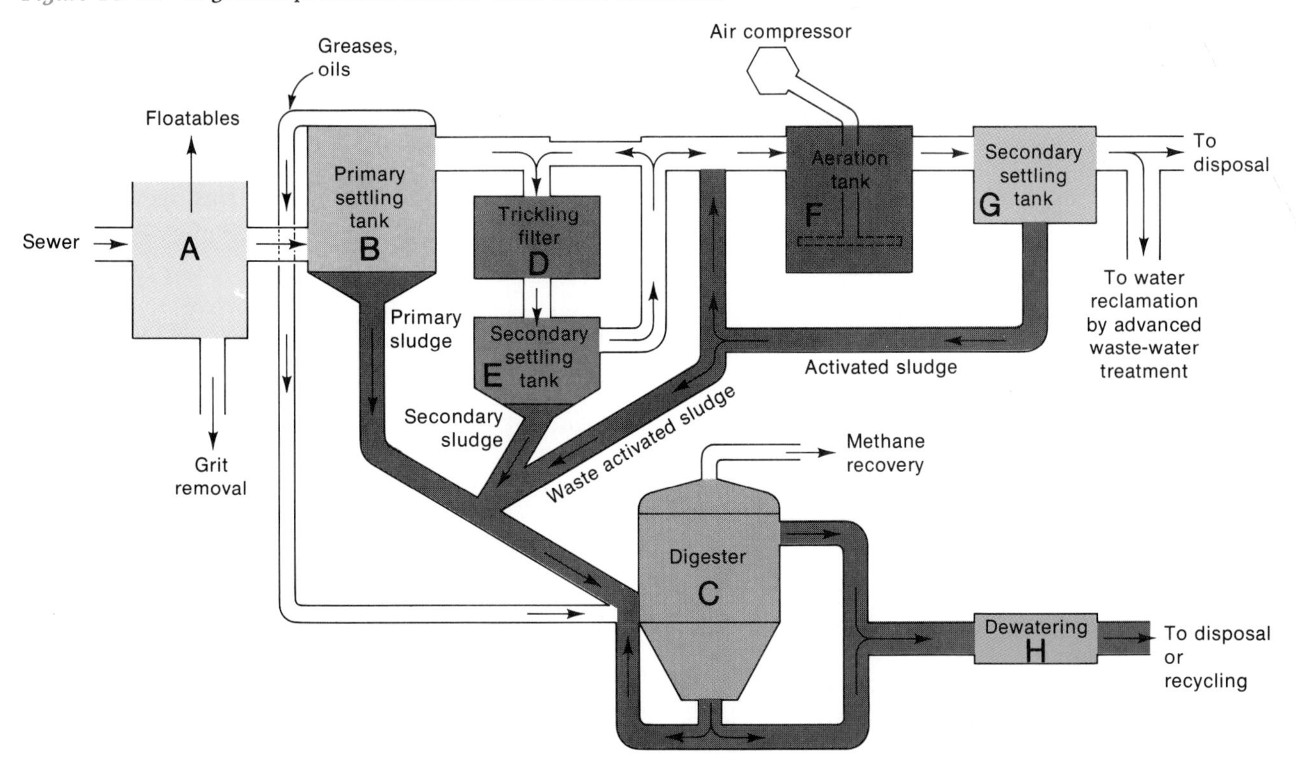

This stage of sewage treatment relies on the force of gravity and a few screens to take out the heavier materials; thus, it begins with the removal of solids, including cardboard, plastics, glass, wood, and gravel, from the more readily digestible components. The large solids, grease, and scum caught in this manner are raked off. Some plants use a shredding device in place of screens to chop up the heavier items, which then stay in the sewage water until the next stage of treatment.

Screened or shredded sewage passes next through a grit chamber, which removes cinders, sand, and small stones. This process serves a protective function in that pumps and other equipment used in later stages are protected from damage. The effluent (liquid material) from the preliminary treatment is then pumped to the primary settling tank (step B). In this *primary treatment,* suspended solids of plant, animal, and mineral material settle slowly. In some systems such settling requires 1½ to 2 hours. The settled solids are collected by a wiper blade and pushed to a pit in the center of the tank. There solids collectively form *sludge,* which is pumped to the next station in the process, the *digester* (step C). Any material, such as grease or oil, that floats in primary settling is removed and also pumped to the digesters. This ends primary treatment. Like preliminary treatment, primary treatment involves only physical activities: floating, settling, scraping, and pumping. If treatment ends at the primary stage, the settled material, sludge, may be removed either manually or by mechanical means. Sometimes sludge is dried in beds and disposed of on land. However, because sludge itself can be a pollutant, it is preferable to initiate microbial decomposition in sludge digestion tanks before drying. This process is an aspect of secondary treatment.

The effluent in the sedimentation tank is usually treated with chlorine to kill microorganisms and reduce objectionable odors. Once this step occurs, primary treatment is complete and the effluent can be released into some natural water environment. Ozone also is replacing chlorine in waste treatment. As of July 1983, 35 ozone facilities were in operation in the United States for sewage disinfection.

SECONDARY TREATMENT

Because of the objectionable properties of the effluent from primary treatment, secondary treatment is employed. This operation involves the biological degradation of organic material by microorganisms under controlled conditions. It is performed in two phases, anaerobic and aerobic digestion.

ANAEROBIC DIGESTION. The digester labeled step C in Figure 16–12 represents anaerobic decomposition of the wastes. The primary sludge, greases, and oils are pumped to the digester. Anaerobic digestion is much slower than the aerobic processes, but it can tolerate materials of much higher BOD values. A 90% BOD reduction may require several weeks to a month. The main purpose of this process is to convert some organic carbon to carbon dioxide and methane and convert some organic nitrogen to nitrogen gas. Note the gas removal pipes at the top of the digester in Figure 16–13. The methanogenesis process often produces a gas mixture of 70% to 77% methane and 23% to 30% carbon dioxide. The mixture can fire boilers or run motors, or it can be stored in a gas sphere for future use. Methanogenic bacteria isolated from digesters include *Methanococcus vannielli* (me-THAN-oh-kok-kuss van-NEEL-ee), *Methanobacterium ruminantium* (me-THAN-oh-back-ter-ee-um room-in-ANN-ti-um), *M. formicicum* (M. for-ME-see-kum), and a *Methanospirillum* (me-THAN-oh-spy-ril-lum) sp. Many other anaerobic and facultatively anaerobic heterotrophic and denitrifying bacteria have been isolated from digesters. The few fungi found show little or no apparent activity. Some protozoa have been observed and apparently are predators. Some of the processed sludge is returned to the inlet of the digester to ensure that the correct inoculum is present for digestion of the new material.

The most common sludge systems operate in the range of 35° to 40°C. Recent research studies and

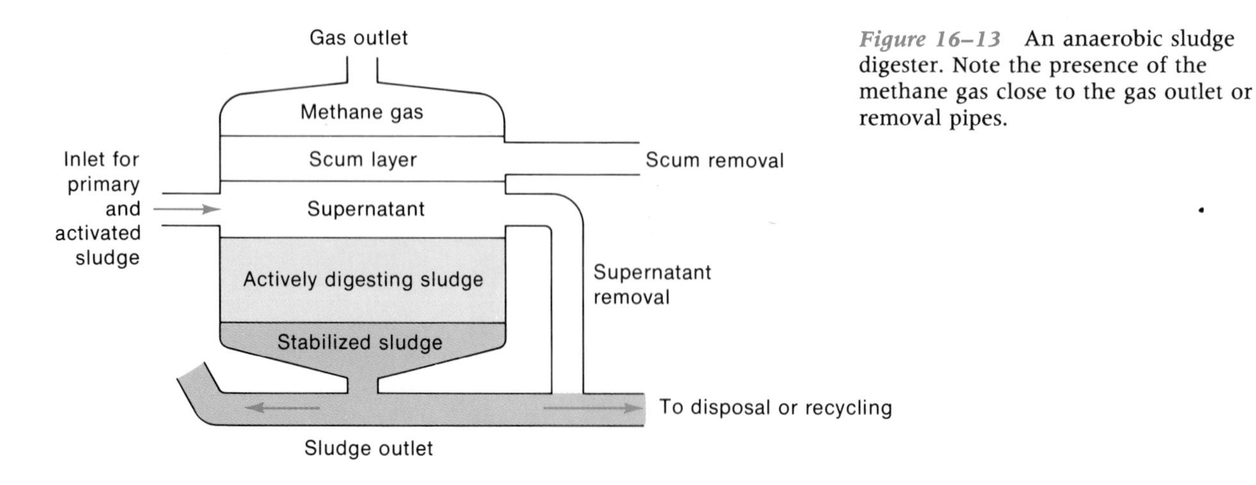

Figure 16–13 An anaerobic sludge digester. Note the presence of the methane gas close to the gas outlet or removal pipes.

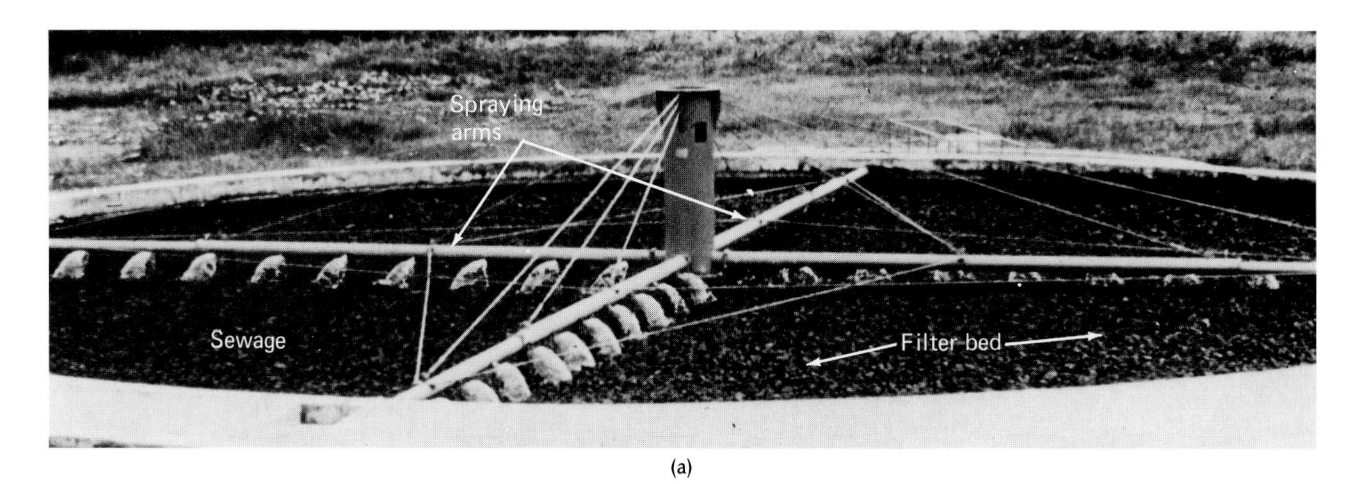

(a)

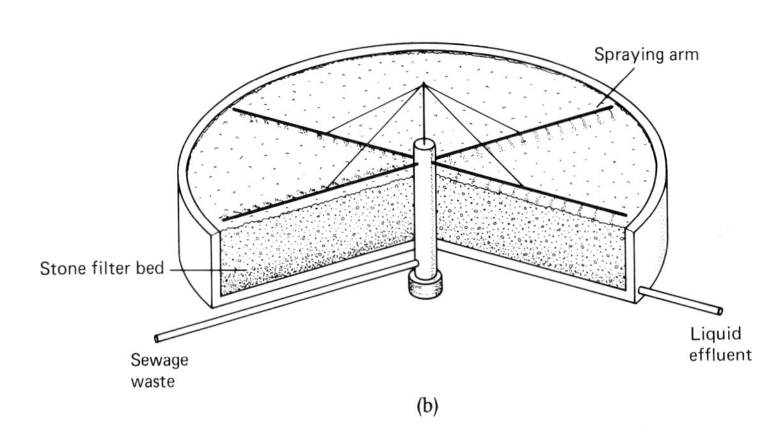

(b)

Figure 16–14 A trickling filter.
(a) This type of filter is commonly used in sewage treatment plants. *[Courtesy of the National Medical Audiovisual Center, Atlanta.]*
(b) A diagrammatic view. Sewage wastes are distributed by revolving, spraying arms onto a gravel-and-rock filter bed. Microorganisms coating the rocks stabilize the sewage as it trickles through the filter bed. This activity greatly reduces the load (content) of degradable organic materials in the effluent fluid.

practice have shown that operating anaerobic digesters at 55°C decreases both thickness (thus requiring less energy for mixing) and retention time and increases methane production and disinfection. In addition, the higher temperature reduces oxygen solubility and sensitivity to toxic chemicals in wastes. As of 1987, the Hyperion Sewage Treatment Plant in Los Angeles County was operating the only full-scale, high-temperature (thermophilic) anaerobic digester with very good results.

Aerobic Stabilization. Waste decomposition by aerobic processes is often called *aerobic stabilization* because of the greater degree of BOD removal. About 85% to 95% of the BOD can be removed from the primary effluent by two common processes in a matter of hours. The same result might require up to a month in the anaerobic digester. These aerobic processes are the *trickling filter* (step D) and *activated sludge* (step F) shown in Figure 16–12.

Trickling Filter. Suspended solids may be sprayed onto the surfaces of artificial beds of a trickling filter made up of broken stone, cork, plastic balls, or another supporting matrix contained in tanks. Microorganisms form a living film on the components of the matrix and digest the organic matter of the sewage as it filters through the bed. Figure 16–14 shows a typical trickling filter equipped with spraying arms. Fungi and algae appear to be the primary agents in the aerobic stabilization process, but a number of bacteria are also involved. The microorganisms commonly isolated from trickling filters are listed in Table 16–12. The film averages about 2 mm in thickness and consists of three distinct layers. The outer layer is primarily fungi. The middle layer is primarily fungi and algae (Figure 16–15). The inner layer consists of fungi, algae, and bacteria. In addition to being quite efficient for BOD removal, the process also removes 90% to 98% of viruses present in the initial wastewater. The effluent of the trickling filter is usually pumped to a settling tank (step E in Figure 16–12), where the fluid is separated from any portions of the film that may have dislodged from the supporting materials. Such settled solids are pumped to digesters. The ef-

TABLE 16–12 Common Microorganisms Found in Films on Rock Supports in Trickling Filter Systems

Algae	*Bacteria*	*Fungi*
Anacystis	*Alcaligenes* sp.	*Aureobasidium pullulans*
Euglena	*Flavobacterium*	*Fusarium* sp.
Ulothrix	*Pseudomonas* sp.	*Subbromyces splendeus*
	Zoogloea	

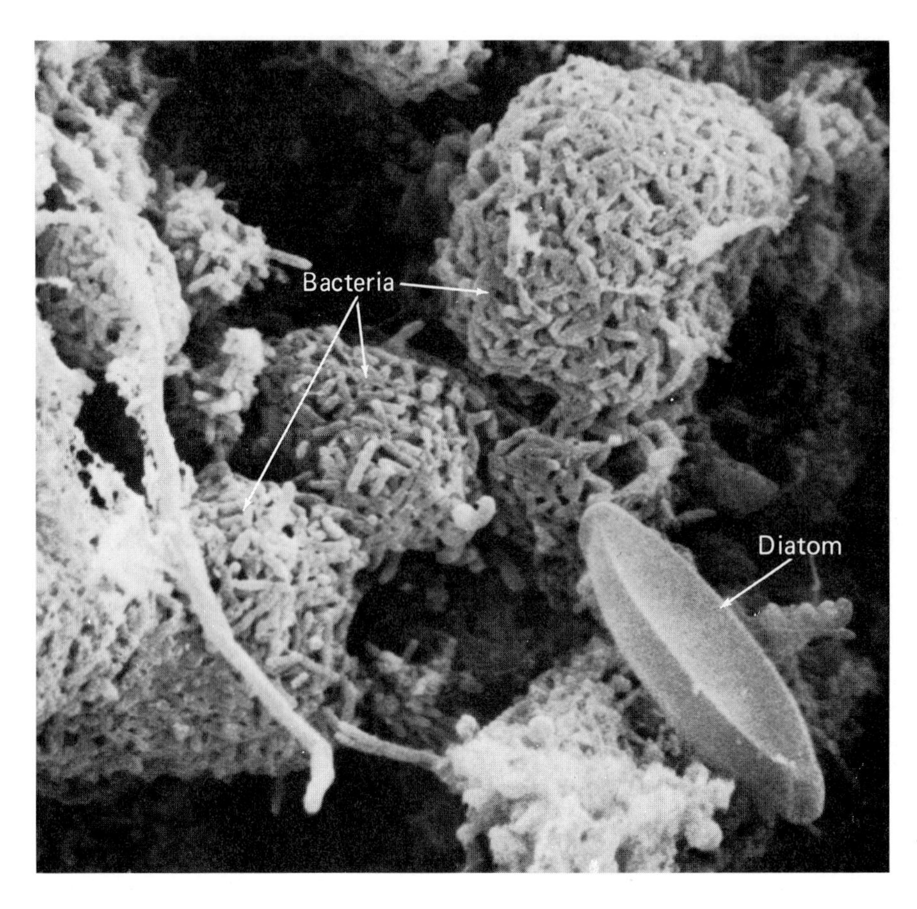

Figure 16–15 Microbial films develop on all objects submerged in natural waters. Microbial growth is especially plentiful when the water environment is rich in organic material and oxygen. Algae, bacteria, fungi, and protozoa are among the forms of life in such growth. These bacterial colonies were found 30.48 cm below the surface of a trickling filter. A diatom can also be seen. *[From W. N. Mack, J. P. Mack, and A. O. Ackerson,* Microbial Ecol. ***2:****215–226, 1975.]*

fluent of the secondary settling process can be diverted to the second aerobic stabilization process.

The effluent from primary or secondary settling tanks is pumped into aeration tanks or rectangular chambers for the activated sludge process (Figure 16–16). Previously digested material is mixed in with the effluents to ensure the proper inoculum (Figure 16–12), and air compressors provide oxygen for stabilization and mix the digesting materials. Some foamy material is caused by the formation of masses of microbial growth known as *floc* (Figure 16–17) or *zoogloeal masses.* This floc exists as a branched system or as a formless gel containing a celluloselike polymer and a polymer of hydroxybutyric acid. This mass is thought to be due to the

Figure 16–16 An activated sludge system demonstrating the foaming and agitation caused by the compressed air. *[Courtesy of the Orange County Sanitation District, Fountain Valley, Calif.]*

Figure 16–17 Bacteria in an activated sludge floc. *[Y. Tago, and K. Aida,* Appl. Environ. Microbiol. ***34:*** *308-314, 1977.]*

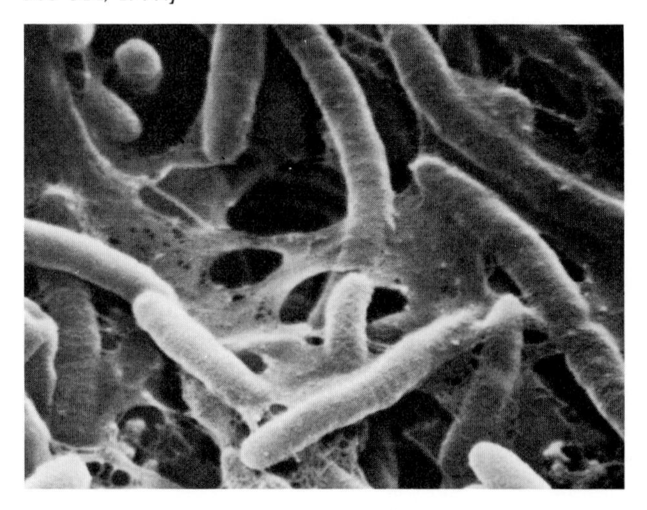

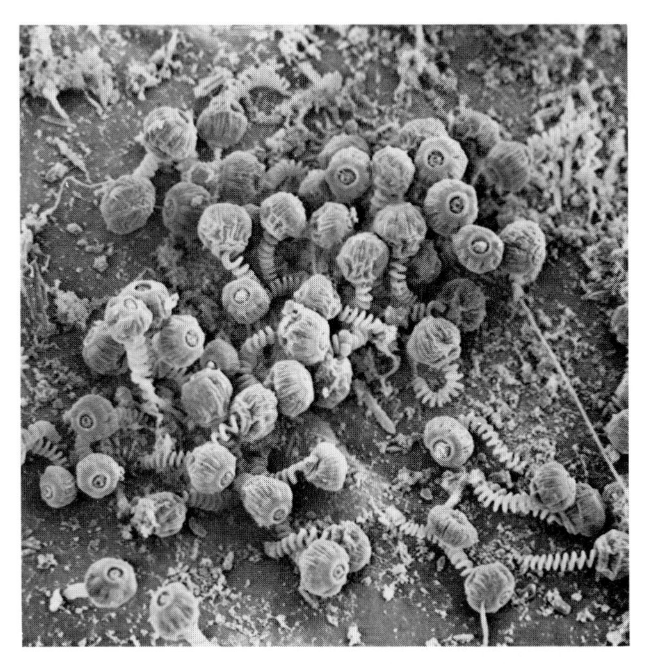

Figure 16–18 One of the many types of protozoa (*Vorticella* sp.) found in sludge systems as well as trickling systems. *[Courtesy W. N. Mack and J. P. Mack.]*

activities of the bacterium *Zoogloea ramigera* (zoh-GLEE-ah rah-MIG-err-ah), which forms fingerlike colonies almost resembling multiple pseudopods on an amoeba. The name *zoogloea* means "living glue" and is quite appropriate. Other bacteria identified in floc include species of *Pseudomonas, Alcaligenes* (al-ka-LIJ-en-eez), *Achromobacter* (a-KROME-oh-back-ter), *Microbacterium* (my-kroh-back-TE-ree-um), and *Brevibacterium* (brev-ee-back-TIR-ee-um). Slime and capsule formation by these organisms may contribute to the floc formation. Ciliated protozoa, such as the *Vorticella* sp. shown in Figure 16–18, control the overgrowth of bacteria by predation and contribute to floc formation. As many as 67 species of protozoa have been reported in activated sludge. The formation of floc is important for two reasons: (1) it speeds the adsorption and absorption of nutrients and (2) the floc settles out in the secondary settling tank (step G, Figure 16–12), leaving a clear supernatant liquid. Some of the floc is returned to inoculate incoming wastewater, and most of it is pumped to the digesters. The effluent of the activated sludge system has been reduced in its BOD by about 85% to 95%. This fluid may be chlorinated and disposed of in streams and lakes, or it may be pumped for reclamation by an advanced wastewater treatment as shown in Figure 16–18. It may also be used to water lawns and trees as well as for irrigation. The final step in secondary treatment is the disposal or reuse of digested sludge from the digester (step H, Figure 16–12). In this step, much of the water is removed and the thickened sludge is dried. This material is dumped as landfill or processed for sale as fertilizers and soil conditioners.

ADVANCED WASTEWATER TREATMENT

Figure 16–19 represents one approach to the further purification of the treated wastewater to provide a high-quality product. In advanced wastewater treatment, lime is mixed with the effluent to produce accumulations of impurities and settling of the limed sludge. The lime is recovered by pyrolysis and the liquid is aerated to remove ammonia and other gases. The liquid is then filtered first in a separation bed and finally in activated carbon. The resultant water has been used by some communities for the development of recreational lakes or for injection into groundwater for storage as well as to prevent saltwater intrusion in coastal communities.

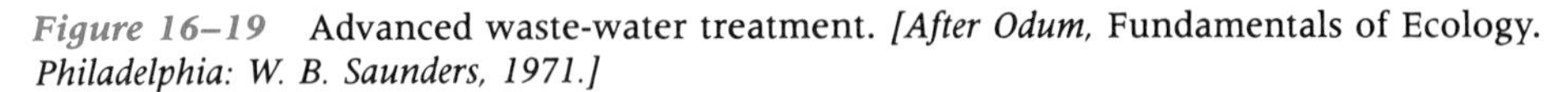

Figure 16–19 Advanced waste-water treatment. *[After Odum,* Fundamentals of Ecology. *Philadelphia: W. B. Saunders, 1971.]*

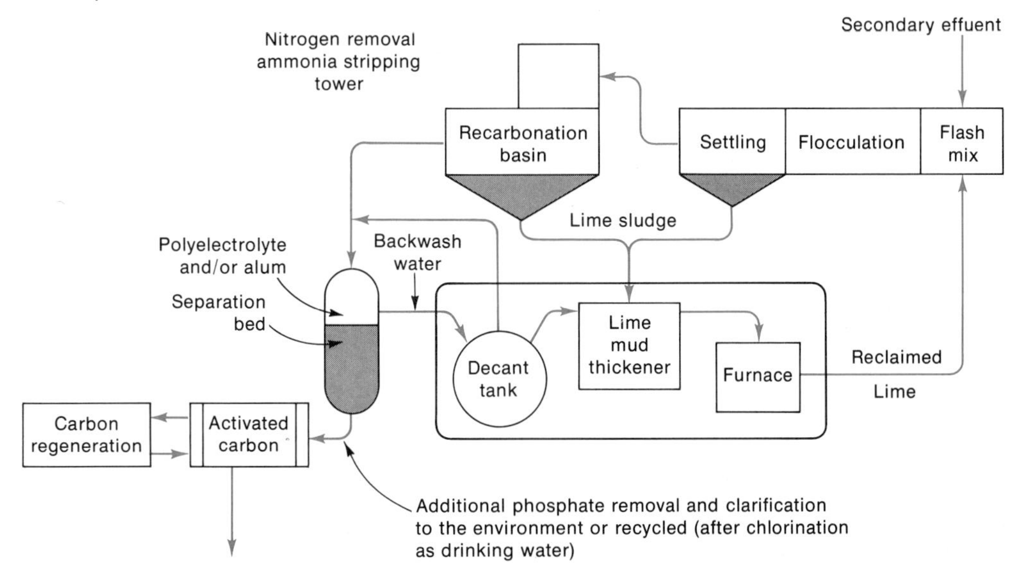

Lagooning. Aerobic waste stabilization does not have to be performed in large buildings and concrete reservoirs, but is also being done in so-called algal lagoons. These are actually very shallow ponds into which wastes are pumped for treatment. If the proper slime- or floc-forming bacteria are present, along with other microbes necessary for aerobic stabilization, one requirement for suitable treatment is present, namely, suitable flora. Since mixing and spraying are impractical in a situation of this type, algae are used as a source of oxygen for the organisms. Together these components form a sound ecological system, with groups of different organisms working for mutual benefit. The algae require carbon dioxide, water, and other compounds that the bacteria supply by the digestion of the sewage. The bacteria require the oxygen that the algae now are able to produce. This concept of combining algae and other microorganisms in waste treatment has been considered both for the production of food and oxygen for astronauts and as a means of disposing of their waste materials. It has also been applied to the problem of disposing of unusual industrial waste (Figure 16–18).

Algae and bacteria appear to predominate in these lagoons. The algae must usually be harvested to prevent the development of anaerobic conditions, but in some systems brine shrimp are used to graze on the algae. These shrimp subsequently are used as food by fish. Wastewaters of particularly high BOD may be forcibly aerated to speed up the process.

Anaerobic waste digestion can also be performed in lagoons when the BOD loading is too great for an aerobic process. The lagoons are made anaerobic by their being at least 4 ft deep and containing the high-BOD wastewater. As the aerobic and facultatively anaerobic microorganisms grow in the system, the oxygen is depleted. Subsequently, the bubbling of generated carbon dioxide and methane helps carry more oxygen out from the water. Photosynthetic bacteria, such as *Rhodospirillum* (roh-doh-spy-RILL-um) spp., play an important role in odor control in these anaerobic lagoons. Hydrogen sulfide is produced during anaerobic decomposition of proteins containing the amino acids methionine, cysteine, and cystine. If allowed to accumulate, the H_2S would become offensive. Since the bacteria use the hydrogen sulfide as a hydrogen donor, they prevent the condition from happening. **[See structures of amino acids in Appendix A.]**

Solid-Waste Disposal

Solid waste materials probably pose a greater and more complex problem than does wastewater because of their higher BOD loading. In addition, solid wastes cannot be pumped and may contain considerable material that either is not biodegradable or that degrades very slowly. Solid wastes from urban areas include garbage, glass, metals, ceramics, ashes, paper, dried sewage sludge, plant and tree trimmings, plastics, and other miscellaneous ingredients. The processes used for solid-waste disposal are directed toward the reduction of its volume and weight and its conversion into a less troublesome form.

Solid wastes have traditionally been dumped into pits or gullies and allowed to contaminate the air and water in the environment. Burning dumps were sources of pollution. More modern approaches involve controlled dumping and the establishment of sanitary landfills. In such systems, solid wastes are spread and compacted by machines into layers. After 2 to 3 meters of such material has been deposited, a layer of clean earth is applied and compacted. When the physical capacity for solid wastes of a sanitary landfill has been reached, it is sealed with an appropriate amount of compacted earth. This approach to solid waste disposal controls the usual problems of odor, pests, and air and water pollution.

Since a large portion of a city's wastes is organic in nature, microorganisms are also involved in the decomposition of such material. Microbial decomposition activities in landfills are *primarily anaerobic digestion.*

Initially, aerobic and facultative anaerobic microorganisms are involved with the decomposition process. Because of the compacted layers in a sanitary landfill, the free movement of air into the system is prevented. Soon the oxygen is depleted and the activities of anaerobes dominate the digestion of the organic matter in the landfill. The heat that accompanies decomposition may raise the temperature to more than 65°C within the landfill.

The decomposition processes in the landfill are similar to those in anaerobic sludge digestion and in the digestive systems of ruminants. Cellulose and other insoluble organic substances undergo digestion, and the resulting products are fermented. Methane and carbon dioxide are slowly released by these activities. While the release of methane is slow and does not appear to cause problems, accidental ignition of the escaping gas can occur. Studies are under way in Los Angeles and San Francisco to determine the commercial value of collecting this natural gas to help meet energy needs. The carbon dioxide produced dissolves in water, which percolates through the fill. In general, it does not cause any environmental problem. The landfill areas are usually made into parks and other types of recreational areas not affected by small amounts of escaping methane or substances produced through decomposition and compaction. Studies are also underway to develop digestion facilities for solid wastes comparable to those used in secondary wastewater treatment. This would permit better control of the production and collection of methane. **[See organisms involved with cellulose digestion in Chapter 14.]**

COMPOSTING

Another approach to the disposal of solid wastes is the aerobic process of *composting*. In this microbial process, waste materials are mixed with soil for an inoculum. With an appropriate mixture of nutrients for microorganisms, the organic wastes are converted to humus. This material is rich in nitrogen, has good water-holding capacity, and is used for planting as a soil conditioner. Building a compost heap on a small individual basis is quite feasible.

Applied Waste-Treatment Processes

As indicated earlier, waste-treatment processes involve specially grown and formulated cultures for mixing with the materials to be treated. The organisms used may be mixtures of pure cultures or communities adapted from natural sources, such as rumen bacteria for cellulolytic activity, soil, or treated wastes. The preparations are usually grown under optimum conditions and then freeze-dried, air-dried, or stabilized. Stabilized cultures are microbial suspensions containing additives to promote growth or to emulsify wastes, wetting agents, and/or flocculating agents. All preparations are formulated to cause a rapid breakdown of the materials to be treated. Selected organisms that have been identified in some of these preparations are listed in Table 16–13.

Important application of these prepared cultures include the clearing of grease traps in restaurants and the removal of accumulated greases in sewer systems. The accumulation of greases causes odors and blockage of pipes. Treatment with grease-digesting microorganisms has been shown to eliminate odors and clear blockages. Ultimately, in the wastewater treatment facilities, the processed greases have contributed to improved BOD removal, increased overall efficiency, and increased methane production.

These cultures can also be applied to the cleaning of heavy and crude oils from oil tankers and other ships. Such was the case with the cleaning of oily wastewater remaining in the *Queen Mary* when it docked in Long Beach, California, to be refitted as a museum and convention center. There were 800,000 gallons of such waste to be treated before it could be released into the ocean. In only six weeks, 150 pounds of dried culture yielded water of adequate quality to be dumped in the ocean. The cost of any other treatment process would have been prohibitive.

Some Products and Problems of Waste Treatment

Throughout this discussion on wastewater and solid-waste treatment, a number of potentially valuable products have been discussed—among them methane, humus, fertilizer, and good-quality water. Other beneficial substances have been found in waste treatment. For example, activated sludge may contain from 1 to 10 mg/kg dry weight of vitamin B_{12}, cyanocobalamin. Sludge can also be pyrolized to produce activated carbon necessary for the advanced water treatment process described.

Unfortunately, problems that may limit the use of some of these products have surfaced. The use of treated water is regulated because it can harbor viruses. In areas where digested sludge has been used as fertilizer, vegetables have been found to be contaminated with viruses originally found in raw sewage. The composting process can also pose problems. Humus has been found to contain the fungal pathogen *Aspergillus fumigatus* (a-sper-JIL-lus few-me-GAY-tus), which is often associated with composting of sewage sludge, municipal refuse, and wood chips. This organism can cause severe infection of individuals who are otherwise ill or who have diminished immunological capacity due to age, physical condition, or use of drugs such as steroids. Many people are allergic to this mold, and asthma can be a common complication of exposure.

The Interaction Between Disease-causing Agents and the Human Host

This chapter has highlighted several of the positive roles of microorganisms in the production of foods and dairy products. The benefits derived from mi-

TABLE 16–13 Selected Bacteria in Applied Waste-Treatment Processes

Bacteria	*Waste Materials*
Bacillus subtilis KJ	Starches, proteins
Bacillus thuringiensis	Fats, greases
Nocardia corallina	Phenolics
Micrococcus sp., *Pseudomonas* sp., *Achromobacter* sp., *Flavobacterium* sp.	Hydroquinone monosulfate from photographic processing

crobes do not stop here, especially when one considers the environmental roles of certain microorganisms and their effect on the recovery of water and the disposal of wastes. In the next section, we will turn our attention to the interactions between disease-causing microorganisms and hosts such as the human.

An essential part of the defense against disease-causing agents in higher animals is the immune response. This type of resistance develops during the course of an infection. Foreign substances known as **immunogens** (antigens) stimulate the immune system to produce specific proteins called **immunoglobulins** or antibodies. How do these and related components of the immune system interact to neutralize the activities of disease agents? How are some of these interactions used to detect and determine the outcome of diseases? And how might immune factors cause undesirale effects? These are but a few of the major questions considered in Part VII.

Summary

1. Microorganisms have both beneficial and harmful roles in food preparation.
2. Certain foods, milk products, and water are potential sources of chemicals and disease agents.
3. *Fermentation* is a valuable process for the preparation and preservation of various foods and dairy products.
4. The unique aromas and flavors of fermented foods result directly or indirectly from the enzymatic action of specific microorganisms.

Fermented Foods

1. Microorganisms used in the fermentation of foods are influenced by factors such as acidity, available carbohydrates, oxygen, temperature, and water.
2. Microbial *starter cultures* contain the specific organisms necessary to begin fermentation processes.

Fermented Vegetables

Fermented vegetables such as sauerkraut and pickles result from the fermentation of carbohydrates present in the plant tissues. Such fermentations are started by the microorganisms in the normal mixed flora of the vegetables.

Dairy Products

1. Milk contains a variety of nutrients, which makes it an excellent medium for many microorganisms.
2. Unless measures are taken to eliminate unwanted spoilage organisms, milk quickly sours.
3. In the United States, milk and related products are tested by procedures such as standard plate counts and direct microscopic examinations to determine their safety for human consumption.
4. The fermentation of lactose (milk sugar) is the basis for the production of various dairy products, including butter, buttermilk, sour cream, yogurt, and numerous cheeses.
5. The particular microorganisms and their enzymatic actions in starter cultures are responsible for the characteristic flavors, texture, and aromas of dairy products.
6. The first step in most cheese-making processes is the curdling of milk, which results in the formation of a solid portion, or *curd,* and the liquid *whey.* Other steps include moisture removal and further enzymatic action by microorganisms, called *ripening.* Based on their texture and method of production, cheeses are classified as *soft, semisoft,* and *hard.*

Direct Acidification

Direct acidification is being used to replace lactic acid starters in the production of fermented dairy products. The process involves the addition of a food-acceptable acid, stabilizers, and artificial flavors.

Single-Cell Protein: Microorganisms as Food

The product obtained through the use of microorganisms as a source of protein is known as *single-cell protein* (SCP). In their ability to produce protein, microorganisms offer a potential solution to meeting the world's demands for protein. They have several advantages as sources of protein: they can utilize a variety of substrates as carbon sources, they are easily harvested, and milligram for milligram they contain as much protein as steak. For human consumption, SCP preparations must be fortified with certain essential amino acids, and processed to remove potentially toxic RNA.

Genetic Engineering

Genetic engineering techniques are expected to improve SCP products.

Microbial Food Spoilage

Major Sources of Microorganisms

1. Most foods used for human consumption contain a variety and a varying number of bacteria and fungi. Some of these microorganisms can be destructive and cause food spoilage.
2. Sources of microorganisms in foods include air, dust, soil, water, utensils, food handlers, and the intestinal tracts of humans and other animals.
3. Factors such as pH, moisture, nutrient content, and conditions of processing and storage influence the numbers and types of microorganisms in finished food products.

Detection of Microorganisms in Food

1. Techniques used to determine the numbers and kinds of microorganisms in food are similar to those used with various types of medically important specimens. They include direct microscopic counts for both living and dead cells, standard plant counts, staining procedures, immunological tests for enzymes and other microbial products, and isolation and biochemical identification.
2. Different microorganisms are used as indicators of potential human or fecal contamination in food materials before as well as after heat processing.

3. The application of conventional membrane filtration techniques to the microbial analysis of foods has several drawbacks, including the tendency of various foods to clog membrane pores, limitations in determining the actual number of specific microorganisms in samples, and the retention of food on membranes, which interferes with accurate bacterial colony counts.
4. Prefiltration and the use of appropriate enzyme treatment of foods to be tested provide increased reliability and accuracy to bacterial counts in filter membrane techniques.
5. A reliable membrane filter system is of major importance especially one that can process food samples in a way that eliminates most difficulties and problems found with most current membrane filter applications to food samples.

Food Preservation

1. Food preservation methods are used to prevent and control food spoilage.
2. Examples of such methods include fermentation, refrigeration, drying, canning, and the use of chemical preservatives.
3. Canning and pasteurization are examples of high-temperature methods used in the processing of foods to control microbial numbers. Only canning is a method for sterilization.
4. Chemicals added to foods to prevent microbial spoilage are classified as food additives. Such substances must be determined as safe for human well-being by the Food and Drug Administration. Chemical preservatives can also be used to destroy or reduce the number of microorganisms on food-processing equipment and utensils and to treat water to be used in the processing of foods.

Water Microbiology

Water-borne microbial disease agents pose a major health hazard. Natural sources of water and wastewaters therefore must be treated to make them safe for consumption and/ or disposal into the environment.

THE WATER (HYDROLOGIC) CYCLE

1. The water cycle provides for a constant interchange of water among air, land, sea, and different life forms.
2. The water cycle involves evaporation, cloud formation, precipitation, surface water runoff, and percolation through the soil.

AQUATIC MICROORGANISMS

Various microorganisms are found regularly in natural waters. Their presence in such environments is influenced by both chemical and physical factors including hydrostatic pressure, light, pH, salinity and temperature.

WATER PURIFICATION

Common water treatment steps include the following: sedimentation, aeration, coagulation, filtration, and chlorination. Ozone is beginning to be substituted for chlorine for disinfection.

Wastewater Treatment

Large communities produce too much waste for environmental processes to handle. They have developed a variety of methods for dealing with disposal of wastewater and solid wastes.

Natural Processes

WASTEWATER

1. Wastewater, or sewage, contains about 99% water and is easily pumped through sewage pipes to treatment plants. The remainder of the material consists of suspended insoluble solids.
2. Domestic waste is largely human excrement diluted with laundry and wash water.
3. Industrial and commercial wastewaters include materials from canning processes, dairies, slaughterhouses, cattle feedlots, and chemical and pharmaceutical plants. Chemical wastes often contain toxic components, such as heavy metals or pesticides.
4. One of the major purposes of wastewater treatment is to eliminate such potential hazards to public health as human pathogens.
5. Some microorganisms found in wastewater are from soil and water sources. Many of them form slimy growths in pipes and ditches. Others produce metal deposits, which clog pipes, or the highly odorous hydrogen sulfide.

Bacteriological Water Testing

1. A wide variety of microorganisms may be present in drinking water, some of which may be pathogenic, causing severe gastrointestinal diseases.
2. The presence of coliforms, which include *Escherichia coli*, *Enterobacter aerogenes*, and *Klebsiella pneumoniae*, serves as an indicator of fecal pollution.
3. Two procedures used to detect coliforms in water are the *most probable number (MPN)* method and *membrane filtration* (MF). *Presumptive, confirmed,* and *completed tests* are used.
4. Indicator microorganisms for fecal contamination other than coliforms include *Bifidobacterium bifidus, Clostridium perfringens,* and *Streptococcus faecalis.*
5. Noncoliform organisms can be detected by the standard plate-count procedure.
6. The depletion of dissolved oxygen in water through the addition of considerable quantities of organic and inorganic materials creates severe environmental conditions.
7. The consumption of oxygen by microbes is known as the *biological* or *biochemical oxygen demand (BOD)*. This term refers to the amount of oxygen consumed by microorganisms during their decomposition of organic materials.

WASTEWATER TREATMENT

1. The vast bulk of wastewaters are processed for disposal by one or more steps, which include *preliminary and primary treatment, secondary treatment,* and *advanced treatment.*
2. Preliminary treatment removes solids from plant, animal, and mineral materials. In primary treatment, solid material known as *sludge* is removed. The liquid portion obtained is usually exposed to chlorine to kill microbes and to reduce objectionable odors.
3. Secondary treatment involves the biological degradation of organic material by microorganisms, and is performed in two phases: *anaerobic digestion* and *aerobic digestion.*

4. The main purposes of anaerobic digestion are to convert some organic carbon compounds to carbon dioxide and methane and to convert some organic nitrogen compounds to nitrogen gas.
5. Aerobic digestion results in a greater degree of BOD removal. Two processes used are the *trickling filter* and *activated sludge.*
6. Advanced wastewater treatment represents one approach to the further purification of treated wastewater to provide a high-quality product.

SOLID-WASTE DISPOSAL

1. Solid waste materials probably pose a greater and more complex problem than does wastewater. This is because of their higher BOD loading, inability to be pumped, and inclusion of materials that either are not biodegradable or degrade very slowly.
2. Solid-waste disposal is directed toward reduction in volume and weight and conversion into a less troublesome and offensive form.
3. Traditionally, solid wastes have been dumped in pits or similar depressions. Modern approaches involve controlled dumping and the establishment of sanitary landfills, where solid wastes are spread and compacted by machines into layers. Microbial activities in landfills primarily consist of *anaerobic digestion.*

APPLIED WASTE-TREATMENT PROCESSES

1. Specially grown cultures of microorganisms, often in combinations, are freeze-dried, air-dried, or stabilized with additives for mixing with materials to be treated.
2. Applications of these prepared cultures include grease removal, the breakdown of heavy and crude oils.

SOME PRODUCTS AND PROBLEMS OF WASTE TREATMENT

1. Potential products obtainable from waste treatment include methane, humus, fertilizer, potable water, vitamins, and activated carbon for filtration.
2. Potential problems with waste treatment include the presence of virus pathogens and the development of mold allergies.

Questions for Review

1. List the types of microorganisms and starting materials needed for the production of the following foods:
a. butter c. yogurt
b. pickles d. ripened cheese

2. What is lactic acid fermentation?

3. a. Outline the general procedure for sauerkraut production.
b. What is the source of the microorganisms responsible for the reaction?

4. Do fermented foods harbor pathogens? Explain.

5. List four factors that influence the activities of microbes involved with fermented food production.

6. a. Describe the general process of cheese making.
b. Distinguish between curd and whey.

7. Give the categories of cheeses and an example of each one.

8. a. What is single-cell protein (SCP)?
b. What types of microorganisms can be used for its production?
c. Why are microorganisms good sources for SCP?

9. Why is milk an extremely good medium for microbes?

10. Distinguish between pasteurization and sterilization.

11. a. Is milk sterile when it is drawn from a cow? Explain.
b. How is the safety of milk and related products for human consumption determined?

12. Are fermented foods less nutritious than other types of food products?

13. What techniques are used to determine the numbers and kinds of microorganisms in foods? List three.

14. List four principal causes of food spoilage.

15. a. What is a spoiled food?
b. What factors influence the growth and effects of spoilage microorganisms?

16. List four different types of biological hazards associated with contaminated foods.

17. List four general approaches to food preservation and two methods for each of them.

18. a. List two chemical preservatives and the types of food for which each is used.
b. Examine the labels of two food products and find the food preservatives they contain.
c. Do you know of any food preservatives that are unsafe for human consumption? Explain.

19. List the various types of materials found in wastewater.

20. Name five water-borne infectious diseases and their specific causative agents. (Refer to Part VIII.)

21. a. Describe a water-quality testing procedure for indicator microorganisms.
b. What are the common indicator organisms?

22. How does the nutrient load in wastewater affect the environment?

23. Define BOD and describe how it is determined.

24. What is the role of the water cycle in water purification?

25. Describe each of the following wastewater treatment processes:
a. preliminary c. secondary anaerobic
b. primary d. secondary aerobic

26. What is the basis for efficient aerobic digestion of wastes in the trickling filter process and in the activated sludge process?

27. Compare the effectiveness of aerobic and anaerobic wastewater processes.

28. How does advanced water treatment work?

29. How can solid wastes be processed?

30. Distinguish between natural and applied waste treatment processes.

31. List the potential products and problems associated with waste treatment.

Suggested Readings

AMERICAN PUBLIC HEALTH ASSOCIATION, *Standard Methods for the Examination of Water and Waste Water,* 16th ed. American Water Works, Water Pollution Control Federation, 1985. *An up-to-date and complete reference source for water analysis.*

BEUCHAT, L. R., *Food and Beverage Mycology.* Westport, Conn.: Avi Publishing Co., 1978. *Molds and yeasts and their relationships to food and beverage spoilage and processing are the subjects of this excellent book.*

CURDS, C. R., "The Ecology and Role of Protozoa in Aerobic Sewage Treatment Processes." *Annual Review of Microbiology* **36**:27, 1982. *An interesting article describing the major role of ciliated protozoa in aerobic waste treatment processes.*

DAVIES, F. L., and B. A. LAW (eds.), *Advances in the Microbiology and Biochemistry of Cheese and Fermented Milk.* New York: Elsevier Applied Science Publishers, 1984. *An up-to-date discussion of the microorganisms in the dairy industry—their interrelationships, physiology, genetics, and identification.*

DU MOULIN, G. C., and K. D. STOTTMEIR, "Waterborne mycobacteria: An increasing threat to health." *ASM News* **52**(10):525, 1986. *A well-written and interesting article outlining the causes of the increased numbers of immunosuppressed patients and subsequent infections by a wide variety of mycobacteria with low virulence.*

JANNASCH, H. W., "Experiments in Deep-sea Microbiology." *Oceanus* **21**:50, 1978. *An interesting article dealing with the study of microorganisms in deep-sea environments and how decomposition of organic materials occurs under conditions of high pressure and low temperature.*

KHARATYAN, S. G., "Microbes as Food for Humans." *Annual Review of Microbiology* **32**:301, 1978. *The use of biomass of various microorganisms for human consumption and SCP substrates and their nutritional value and safety are discussed.*

NATIONAL RESEARCH COUNCIL (U.S.), FOOD PRODUCTION COMMITTEE. SUBCOMMITTEE ON MICROBIOLOGICAL CRITERIA. *An Evaluation of the Role of Microbiological Criteria for Foods and Food Ingredients.* Washington, D.C.: National Academy Press, 1985. *A reasonably complete and well-written presentation of the microbiology of foods, with particular attention to the selection of pathogens, indicator organisms, and other means for assessment of food quality and safety.*

ROSSMORE, H. W., *The Microbes, Our Unseen Friends.* Detroit: Wayne State University Press, 1976. *A well-written and enjoyable book that emphasizes the beneficial importance of microorganisms, especially in food production. The author combines technical information with touches of humor.*

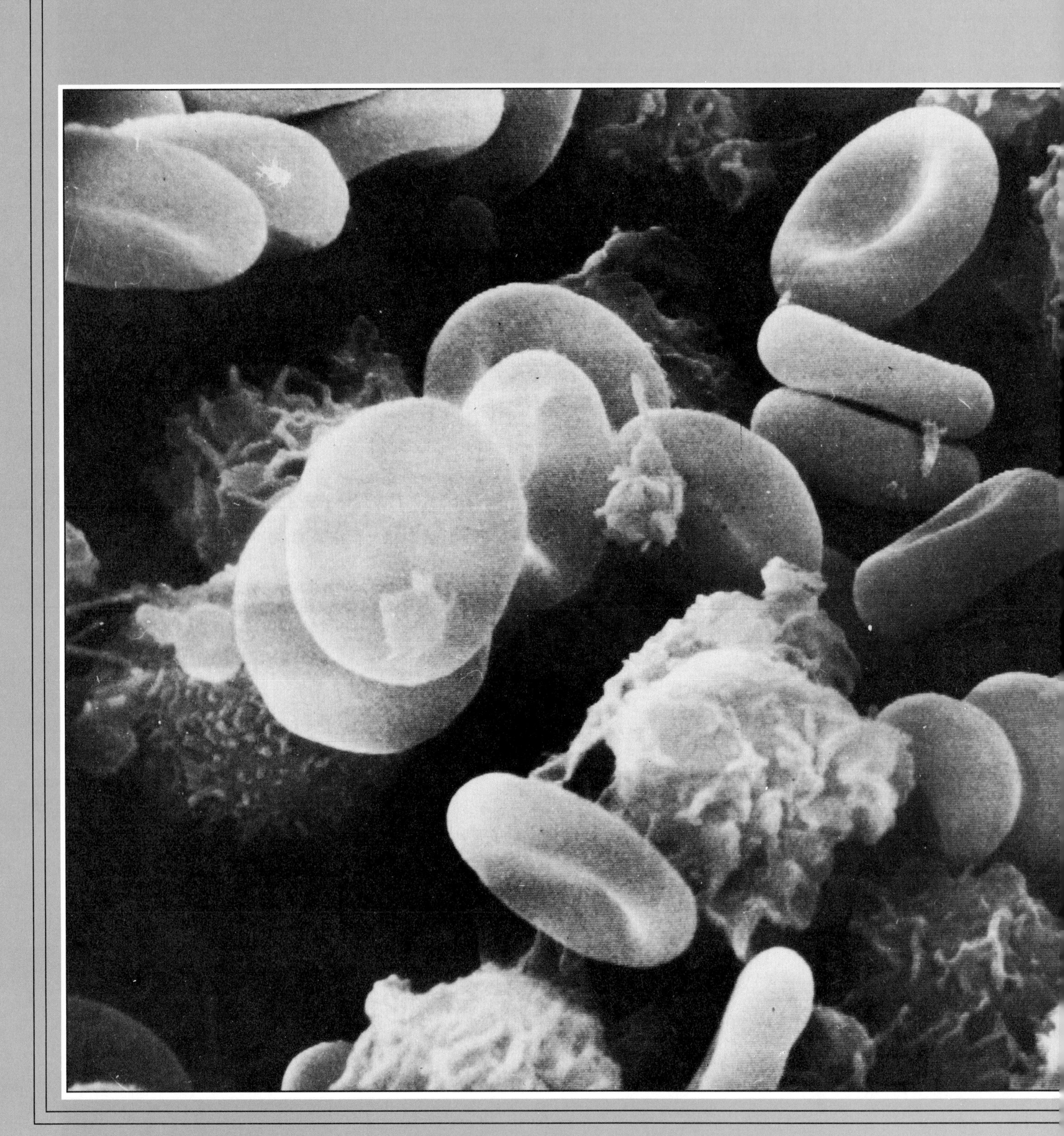

PART VII

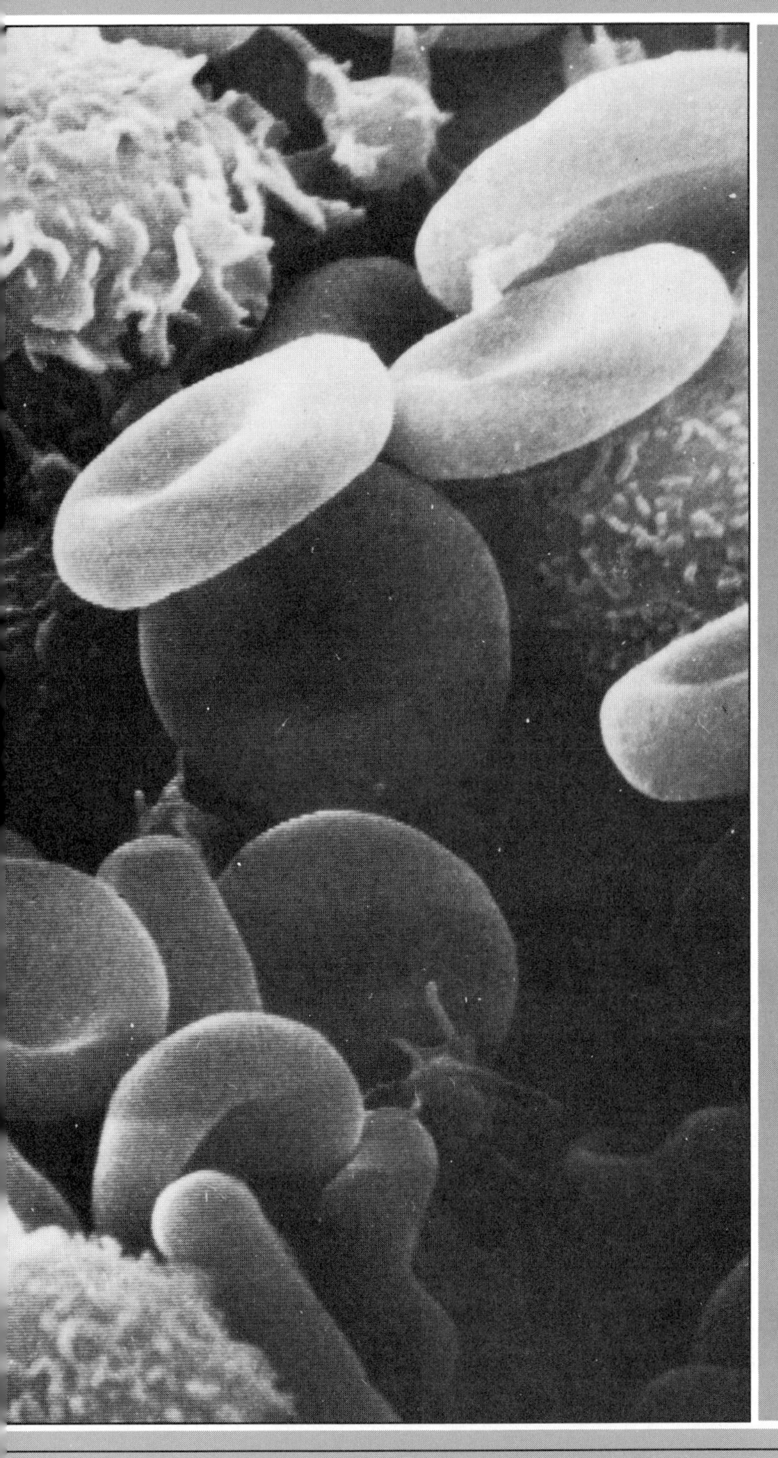

Principles of Immunology

A scanning electron micrograph showing the surface features of the disk-like red blood cells and an assortment of lymphocytes.

17

Introduction to Immune Responses

The development of immunity, in its broadest sense, is a process by which the body learns from experience of past infections to deal more efficiently with subsequent ones.
— Sir Macfarlane Burnet

After reading this chapter, you should be able to:

1. Distinguish between nonspecific and specific immunological responses.
2. Differentiate among the components of blood and the lymphatic system with respect to their roles in health and disease.
3. List and describe five body systems that pose barriers to potential disease-causing agents.
4. Explain the benefits of microbiota (indigenous microbes) to resistance.
5. List and describe the antimicrobial substances produced in the human body.
6. Explain the functions and importance of phagocytosis, inflammation, and fever in host resistance.
7. Explain the general properties of antigens and antibodies.
8. Distinguish among the major classes of immunoglobulins.
9. Describe the general structural features of immunoglobulin molecules.
10. Explain the primary and secondary responses to antigenic stimuli.
11. Define the functions of the thymus gland, macrophages, B cells, and T cells in the immune response.
12. List the sources and locations of normal immunoglobulins in the body.
13. Name and give the main features of current hypotheses regarding antibody formation.
14. Distinguish between humoral and cell-mediated immunity.
15. Briefly describe the gene systems that have important roles in the immune system.
16. Identify the significant factors and conditions that lower host resistance.

Chapter 17 deals with the various means and mechanisms by which the body normally combats infectious disease agents (pathogens) and other factors recognized by the body's unique defensive immune system as being foreign. We shall see what happens when such protection fails and what factors and conditions lower an individual's resistance. We also discuss the features of immunological responses, the properties of immunogens (foreign matter), and the properties of specific protein substances—immunoglobulins—that are produced by the body in response to antigens (immunogens).

Humans and other vertebrates are protected in varying degrees from disease-causing microorganisms and cancer cells by a surveillance mechanism referred to as the *immune system.* Collectively, the various components of this system provide protection or *immunity* by imposing barriers to invasion by microorganisms or other disease agents or by selectively neutralizing or eliminating materials recognized by the immune system as being foreign. Immunological responses, which may be either *nonspecific* or *specific,* serve three functions: defense against invasion by microorganisms, maintenance of a stable internal environment (homeostasis), and surveillance or recognition of abnormal cell types.

Nonspecific immunological responses serve as the foundation of nonspecific resistance (immunity). They are quite general and represent the first lines of defense used to protect the body against foreign cells or substances. Some contributing factors to nonspecific resistance are species factors, mechanical and chemical barriers, phagocytosis, inflammation, and various antimicrobial chemical products of the body.

Specific immunological responses (the production of immunoglobulins) provide resistance to particular microorganisms and their products. Components of the immune system are responsible for the ability of vertebrates (animals with a backbone) not only to detect but to eliminate cells and related factors that have gained access to their bodies. The associated immunologic responses are remarkably specific and involve a number of key cell types of the immune system.

The various components of the immune system are combined in an exquisitely complex communications network that functions as an effective defense against foreign microorganisms and against body cells that have become abnormal (cancerous). Thus, defective functioning of the system results in abnormalities or disease. We shall begin this chapter by studying this most important body defense; we shall then discuss those factors that contribute to both nonspecific and specific resistance.

The Immune System

The invasion of the body by foreign microorganisms or chemical agents may pose a threat to health. Defending against the effects of such foreign invaders or antigens is the function of the various components of the *immune system,* a series of protective mechanisms, specialized cells, and molecules operating in the body. The immune system has several means of coping with disease agents. Among these mechanisms are (1) removing them from the body; (2) neutralizing infectious organisms and biologically active molecules; and (3) destroying foreign cells. Some of the components of this important system and their associated activities are described in the following sections and in later chapters.

The Components of Normal Blood and Their Roles in Health and Disease

Blood is the body's transportation system and the means of intercommunication among the various tissue cells of the body. It transports food and hormones, removes cellular waste products, assists in the regulation of body temperature, and aids in the removal and, in certain situations, the destruction of foreign substances and invading microorganisms. As several of the following chapters will show, blood can play an important role in the transmission, production, diagnosis, cure, and prevention of many conditions caused by microorganisms.

As a liquid, blood is a somewhat atypical form of connective tissue. It consists of cellular elements in a fluid substance called **plasma** (Figure 17–1). The structural components of mammalian blood are not all considered to be true cells. They are often referred to as the *formed elements* and include *erythrocytes* (red cells), *leukocytes* (white cells; Figure 17–2), and *platelets* (thrombocytes). *Chylomicrons,* which are visible minute fat globules, are also suspended in the plasma portion of blood. The cellular elements comprise approximately 45% of the blood (Table 17–1), and plasma forms the remaining portion. A special type of calibrated tube called a *hematocrit* can be used to determine the relative proportions of cells and the liquid portion, or plasma, in a blood sample. Table 17–1 provides a summary of the properties of the formed elements.

PLASMA AND SERUM

When blood is removed from the body by means of a sterile syringe and needle and introduced into a test tube, the specimen normally clots (coagulates) within two to six minutes. The soluble protein substance *fibrinogen,* which normally circulates in the plasma, is converted into the insoluble protein *fi-*

TABLE 17–1 Properties of the Formed Elements of Blood

Formed Element	*Function*	*Diameter (in μm)*	*Life Span*
Erythrocyte	Oxygen and carbon dioxide transport	7.5–7.7	100–120 days
GRANULOCYTES			
Eosinophil	Phagocytosis; destruction of parasitic worms and eggs; counters the effects of certain chemicals in allergies	10–12	No longer than two weeks
Basophil	Releases various chemicals during certain allergy reactions[a]; phagocytosis	10–12	No longer than two weeks
Neutrophil	Phagocytosis	10–12	No longer than two weeks
AGRANULOCYTES			
Lymphocyte (B cell)	Gives rise to antibody-producing cells	7–15	Two or more days
Lymphocyte (T cell)	Participates in antibody production; destruction of foreign cells	7–15	Two or more days
Monocyte	Phagocytosis; assists in antibody production	14–19	Some for 200 days
Thrombocyte	Blood clotting	2–4	Five to nine days

[a] See Chapter 20.

brin, which forms the fiber framework of the clot. Most of the blood cells in the specimen become enmeshed in the fibrin. Within a few hours, the clot shrinks and expels a clear, yellow fluid called *serum* (Figure 17–1). The serum contains several types of proteins, including albumins and globulins.

Albumins are important to the maintenance of adequate cellular nutrition and a normal osmotic pressure. These proteins are mainly responsible for the thickness of blood.

Serum globulins occur in three major groups, known as *alpha, beta,* and *gamma* globulins. Alpha and beta globulins are involved with the transport of other proteins in the body, while the gamma globulins are most important to immunity. Most immunoglobulins or antibodies are found in the gamma globulin fraction. These proteins circulate in the blood and other body fluids to provide *humoral immunity.* They are produced by the body in response to (1) infectious agents; (2) vaccine preparations of killed, weakened, or attenuated organisms or their products; or (3) other foreign substances, such as pollens, which are associated with allergies.

Figure 17–1 The distinction between serum (left) and plasma (right). Serum does not contain fibrinogen.

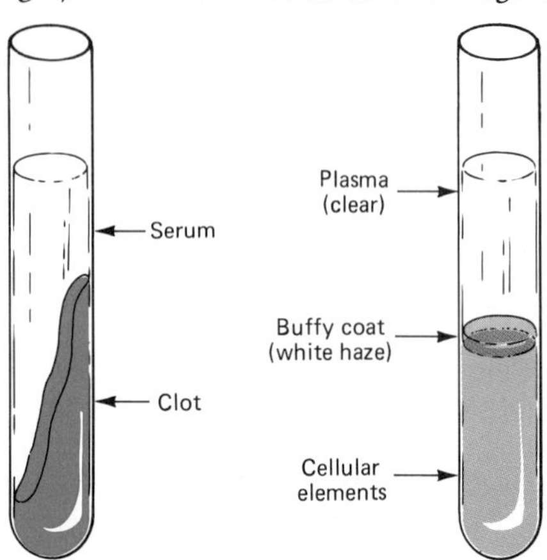

In the laboratory, plasma is obtained by mixing a blood sample with an anticoagulant such as potassium, or sodium oxalate, or heparin. These substances interfere with the formation of *thrombin,* which is a vital factor in the conversion of fibrinogen into fibrin. The cellular elements will settle to the bottom of the tube containing the specimen, either on standing or on centrifugation. The clear fluid left above is called *plasma.* Thus, plasma is blood without cells but with fibrinogen; serum is *defibrinated* plasma.

Plasma is a complex mixture of substances, including carbohydrates, lipids, proteins (including albumins and globulins), gases, inorganic salts, hormones, and water. The pH of plasma normally is slightly alkaline, approximately 7.4, or near neutral.

ERYTHROCYTES

Red cells, or *erythrocytes,* are formed in the bone marrow and measure about 7.5 to 7.7 μm in diameter and 1.9 to 2 μm in thickness. In mammals, they are not nucleated when mature and appear as biconcave disks. The red cells are composed of a membrane that is closely associated with the iron-containing protein compound *hemoglobin.* Hemoglobin has great attraction for oxygen, and the red cell is specialized for the transport of this gas. In addition, these cells possess the major, minor, and Rh blood factors. An erythrocyte's life span generally ranges from 100 to 120 days. Production of insufficient or hemoglobin-deficient red cells by the body is known as *anemia.*

LEUKOCYTES

White cells, or *leukocytes,* as shown in stained blood smears, are classified into two groups based upon (1) the presence and type of cytoplasmic granules, (2) the shape of the nucleus, (3) the appearance of the cytoplasm, and (4) the size. The groups are called **granulocytes** and **agranulocytes.** Granulocytes contain distinct cytoplasmic granules and have irregular multilobed nuclei. Like erythrocytes, they are formed in the bone marrow. The life span of these cells is no longer than two weeks. Upon staining (as in the case of blood-smear preparations), these cytoplasmic components react with dyes to yield characteristic colors. Based upon the resulting chemical reaction, three types of granulocytes are defined: (1) *Eosinophils* contain granules that react with acid dyes and become red. They are phagocytic and help combat certain parasitic worm infections. Eosinophils are present in greatest number in individuals with parasitic infections or allergies. (2) *Basophils* contain granules that react with basic dyes and become blue. Basophils are phagocytic, capable of engulfing and digesting invading foreign particles. They are also known to play a role in immunity to virus infection and in rejection of grafts. These cells also contain chemical substances that have powerful effects on the body's blood vessels and pulmonary system (Figure 17–2a). Mast cells (Figure 17–3), which may develop from basophils, are also found in large numbers along blood vessels. As Chapter 20 will describe, the mast cell contains a variety of chemicals associated with certain allergic reactions. (3) *Neutrophils* or *polymorphonuclear leukocytes (PMNLs)* contain two types of granules that react with neutral dyes or acid and basic dye mixtures and become neutral or orange-colored. The primary granules contain enzymes such as acid hydrolase and myeloperoxidase, as well as lysozyme, while the secondary or specific granules have lysozyme and the iron-containing protein lactoferrin. Mature neutrophils are primarily phagocytic cells (Figure 17–4) and serve as important defenses against microbial disease agents and foreign particles. These cells also contain modified amino acids, called *defensins,* which have antimicrobial activity against viruses, bacteria, and fungi. The sequence of cellular and biochemical details involved with phagocytosis is described later in this chapter.

The *agranulocytes* do not possess granules, and their nuclei are rounded rather than lobed. The two general cell types in this group are the *lymphocyte* and the *monocyte.*

Lymphocytes have individual rounded nuclei that occupy most of the cell. Some lymphocytes survive

Figure 17–2 Components of blood involved with the immune system. (a) An electron micrograph of a basophil. Note the characteristic granules of this white blood cell type. A lymphocyte is shown at the lower right. *[J. M. Clark, G. Altman, and F. B. Fromowitz,* Infect. Immun. ***15****:305–312, 1977.]* (b) An electron micrograph showing the ultrastructural features of a monocyte. Original magnification 11,000×. *[W. Deimann, and H. D. Fahimi,* J. Exp. Med. ***15****:883, 1979.]*

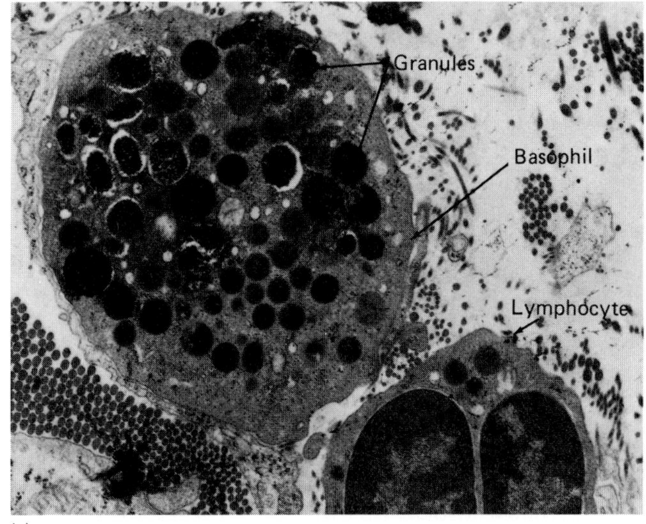

(a)

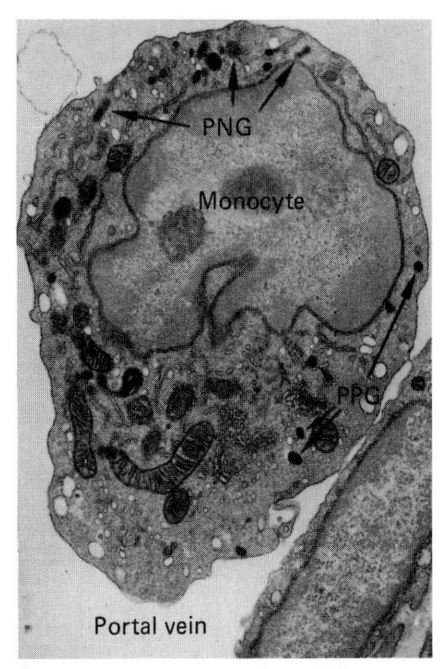

(b)

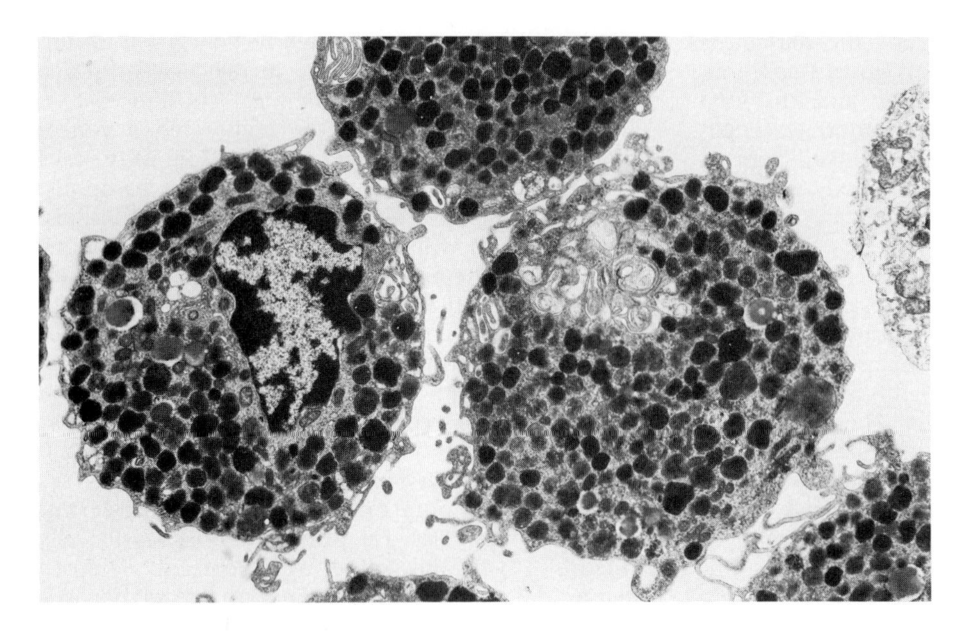

Figure 17–3 Mast cells, important participants in certain allergic reactions. *[From A. M. Dvorak,* et al., Lab. Invest. ***53**:45–56, 1985.]*

for as long as 200 days. Lymphocytes are principally produced in the appendix, lymph nodes, spleen, thymus, and other lymphoid tissues. Two types of lymphocytes, B and T cells, are important components of the immune system. B cells give rise to antibody-producing plasma cells, while the various forms of T cells contribute to an individual's resistance by participating in antibody production and the destruction of foreign cells. Additional features of these cell types are given later in this chapter.

Monocytes are larger than lymphocytes, and their nuclei are generally kidney-shaped (Figure 17–2b). They are formed in the bone marrow. As to function, certain monocytes are also involved with specific antibody formation. In tissues, monocytes also participate in phagocytosis.

Disease states can cause changes in the proportions of the different types of leukocytes. Also, certain pathogens stimulate an increase in white cells, called *leukocytosis.* These include *Streptococcus pneumo-*

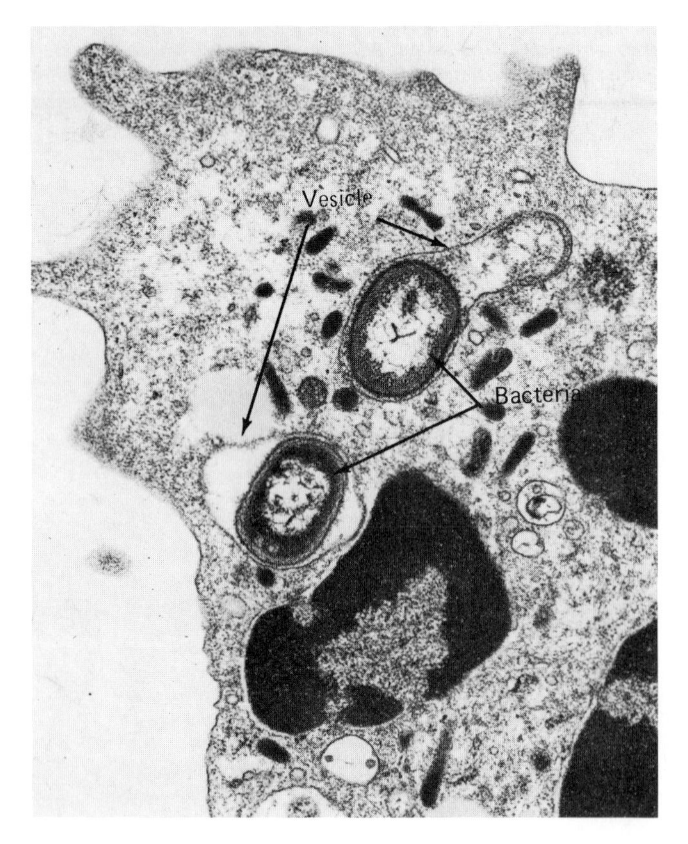

Figure 17–4 Two bacterial cells phagocytized by a neutrophil. Note the position of the bacterial cells within phagocytic vesicles (arrows). *[From T. A. Bertram,* et. al., Infect. Immun. ***37**:1241–1247, 1982.]*

niae (strep-toe-KOK-kuss new-MOH-nee-eye), which causes lobar pneumonia; *Neisseria gonorrhoeae* (Nye-SE-ree-ah go-norr-REE-ah), the cause of gonorrhea; and *Staphylococcus aureus* (staff-ill-oh-KOK-kuss AW-ree-us), the cause of boils, carbuncles, pneumonia, etc. Other microorganisms cause leukopenia, a reduction in the number of leukocytes. The causative agents of influenza, measles, tuberculosis, and typhoid fever are included in this group.

Differential Count. A differential count is performed to determine the proportions of the various white cells. The respective numbers as well as the increases and decreases in certain cells are important clues to the diagnosis of several disease states. For example, the number of neutrophils increases rapidly in the early stages of many bacterial infections. The differential count is made from a stained blood smear. The normal values, based on an approximate total of 100 cells, are as follows:

Granulocytes	
Basophils	0–1
Eosinophils	1–4
Neutrophils	60–70
Agranulocytes	
Large lymphocytes	0–3
Small lymphocytes	25–30
Monocytes	4–8
	90–116

The Lymphatic System

The cells and molecules of the immune system are carried to most tissue by the bloodstream. They enter the tissues by penetrating the walls of the capillaries (Figure 17–5). These cells eventually make their way to a return vascular system of their own, the *lymphatic system.* Components of this system include lymphatic vessels, lymph fluid, lymph nodes, and lymphocytes. The lymphatic vessels exist throughout most of thc body (Figure 17–6). They may pass through lymph nodes, which are nodular accumulations of lymphocytes and macrophages. Macrophages are derived from circulating blood monocytes. They are capable of several immunologically related functions, including clearing and degrading foreign substances within the body (phagocytosis) and preparing foreign substances in the antibody formation process. In addition to being of particular value in the transport of fatty acids, proteins, and white blood cells, the lymphatics are important during times of infection or other tissue injury. This system serves to remove material that accumulates at the site of tissue damage, such as foreign cells and their products, white cells, and tissue debris. Pathogens and their products are carried by the lymphatics to the lymph nodes, where the immune responses by the host's macrophages and lymphocytes are initiated. During times of infection, lymphatic vessels may become inflamed, and involved lymph nodes may be considerably enlarged.

Figure 17–5 A granulocyte in transit. (a) This transmission electron micrograph shows the white blood cell passing through the capillary wall. (b) A similar cell shown by scanning electron microscopy. *[J. K. Chamberlain, P. F. Leblond, and R. I. Weed,* Blood Cells *1:655–674, 1975.]*

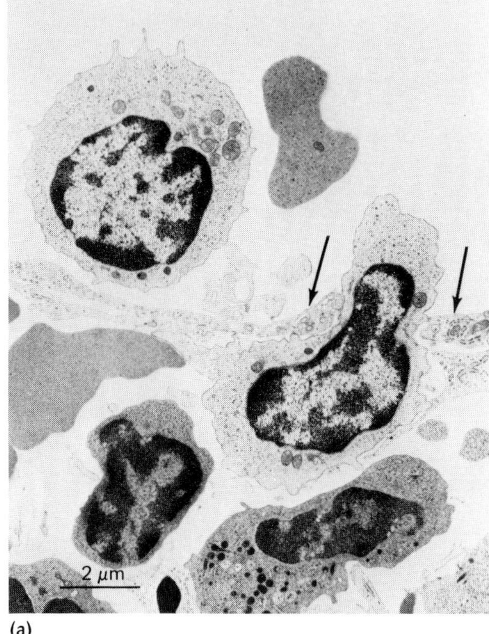

(a)

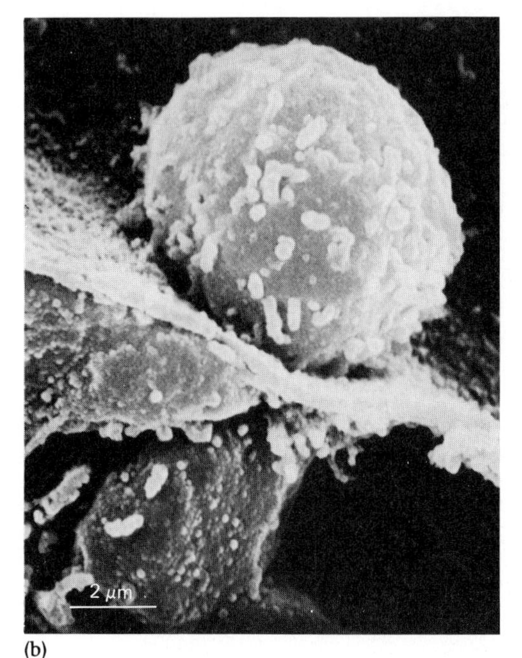

(b)

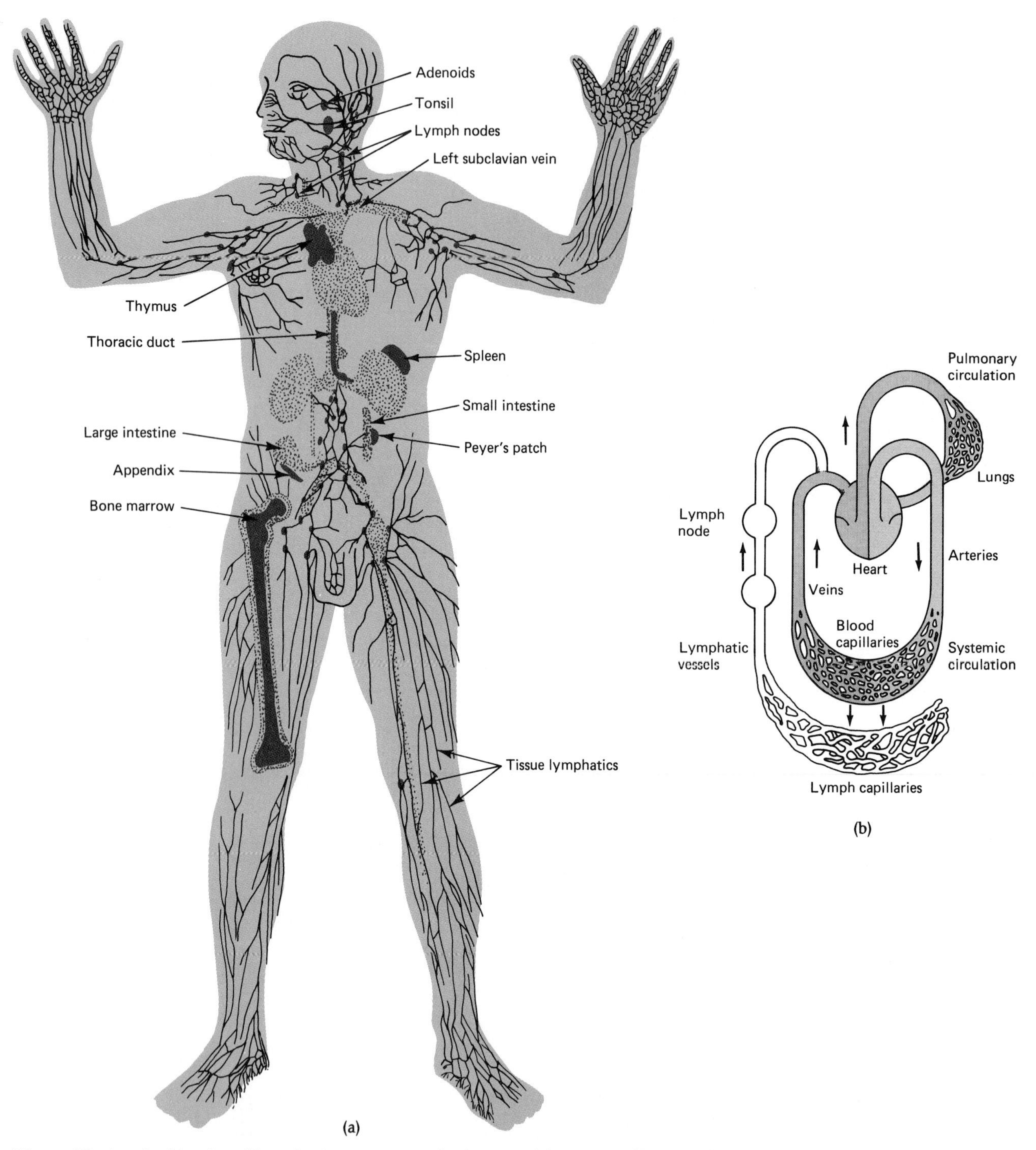

Figure 17–6 The blood and lymphatic systems in the human. (a) An overall view of the lymphatic system, showing major locations of lymph nodes (the parts known to remove bacteria and foreign particles and cells). Note the positions of major body organs in relation to the lymph nodes. (b) A diagrammatic representation of the relationship between the circulatory and lymphatic systems.

Microbiology Highlights

CELLULAR COMPONENTS OF THE IMMUNE SYSTEM AND AIDS

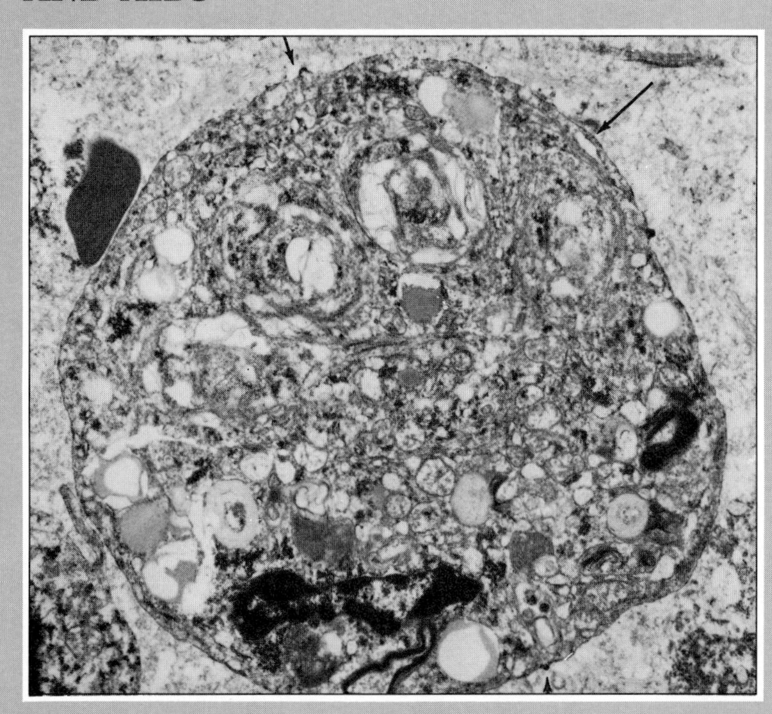

A transmission electron micrograph of a mononuclear macrophage with virus particles in varying stages of maturation. Budding viruses are shown (arrows). *[From Koenig, S., et al., Science, 233: 1084–1093, 1983.]*

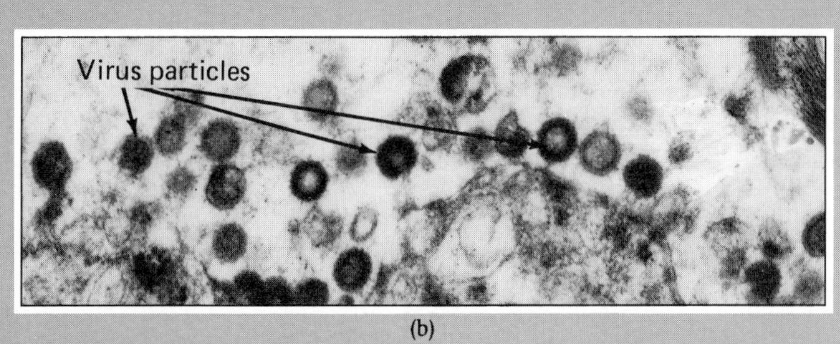

(b)

Acquired immune deficiency syndrome (AIDS) is associated with a variety of medical disorders involving the central and peripheral nervous systems. Destructive nervous system changes are present in over three-fourths of autopsied victims of the disease. The AIDS virus is well known for its use of certain T lymphocytes that replicate and cause destruction. However, the type of cell or cells involved with the nervous system damage was unknown until 1985, when L. G. Epstein and his associates found viral particles in multinucleated giant cells of AIDS patients with impaired brain function. In 1986, S. Koenig—using the combined techniques of viral isolation, transmission electron microscopy, and recombinant-DNA technology—identified mononuclear and multinucleated macrophages in brain tissue as the sources of new AIDS viruses. Such infected cells may serve not only as reservoirs for these viruses but also as the vehicles by which the viruses are spread in the infected patient.

THE MONONUCLEAR PHAGOCYTIC (RETICULOENDOTHELIAL) SYSTEM
The mononuclear phagocytic or reticuloendothelial system consists of a variety of cells that ingest and digest foreign and host substances. Such phagocytic activity is carried out by cells called *macrophages* that may be either attached to tissue or free to wander into tissues. The fixed cells are found in the body's network of loose connective tissue (reticulum), lining the capillaries and the sinuses, and in regions such as the liver (Figure 17–7), spleen, bone marrow, and lymph nodes. The wandering phagocytes are found in air sacs of the lungs (Figure 17–8a), blood, and the cavity-holding portions of the digestive system.

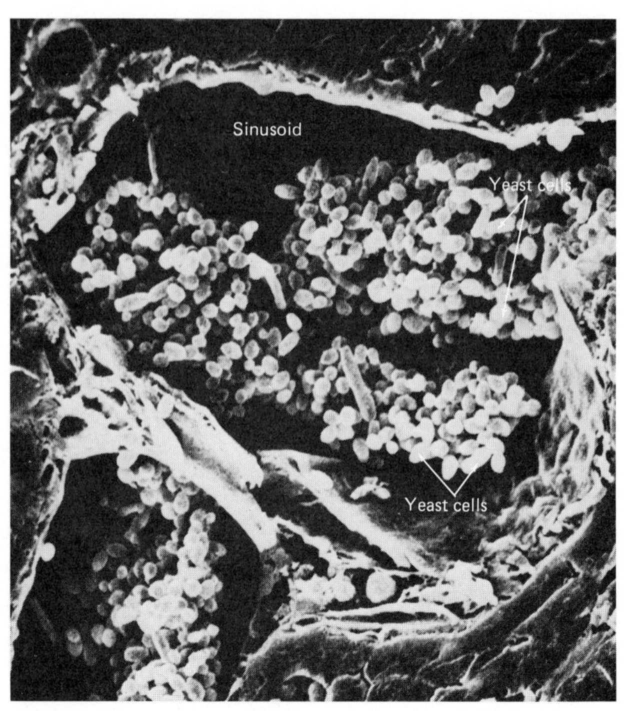

Figure 17–7 The removal of the yeast *Candida albicans* (KAN-did-ah AL-beh-kans) from the bloodstream by the liver. This organ, part of the mononuclear phagocytic (reticuloendothelial) system, helps remove foreign microorganisms by trapping cells (arrows) in its spaces (sinusoids). *[R. T. Sawyer, R. J. Moon, and E. S. Beneke,* Infect. Immun. ***14:****1348–1355, 1976.]*

Nonspecific Resistance

Species Resistance

It is well known that some animal species are normally resistant to diseases that can have disastrous effects on other animals. Humans exhibit such *nonsusceptibility* toward a variety of infectious diseases of other animals, including canine distemper, cattle plague, chicken cholera, and hog cholera. On the other hand, lower animal species are similarly resistant to human-associated bacterial infections such as dysentery, gonorrhea, typhoid fever, and whooping cough, and to viral diseases such as measles and mumps. This kind of nonsusceptibility, or *species resistance,* is determined by physiologic and anatomic properties of the particular animal species and is inheritable. Demonstrable antibodies (protective protein molecules) are not associated with this state of resistance.

Changes in body temperature, diet, and stress can affect species resistance, as is shown by several classic experiments. Chickens and frogs are normally not susceptible to the bacterial disease anthrax. However, the cold-blooded frog will succumb to inoculation with the infectious agent if its temperature is artificially raised to approximately 35°C, and the warm-blooded chicken will succumb if its body temperature is lowered to that level. The multiplication of various pathogens depends upon the availability of growth factors in usable form. Some disease agents depend on the food of their host. For example, dogs are normally resistant to anthrax; however, when meat is omitted from their diets, they become susceptible to the infection.

Several instances of resistance have been found within and among various races of the world. Most forms of racial resistance are reflections of evolutionary changes. For example, in black Africans, two different mechanisms contribute to their resistance to the protozoan disease malaria. One of these is the absence of the Duffy factor, a specific chemical on the surfaces of red blood cells of some individuals. This blood factor is used by *Plasmodium vivax* (plaz-MOH-dee-um VYE-vacks) to attach to red blood cells and initiate an infection. **[Chapter 19 describes blood factors in more detail.]** The second instance of racial resistance involves the genetic disease sickle-cell anemia. In individuals with this condition, red blood cells assume a sickle or crescent shape and contain a faulty form of hemoglobin, which interferes with the body's use of oxygen. While malarial disease agents can attach to the distorted red blood cells, they cannot develop within them and are thereby prevented from establishing an infection. **[Chapter 27 describes the features of malaria.]**

Species or racial resistance depend on the interplay of many factors, not all of which are known. Moreover, most of them appear to be genetically determined and are specific for certain disease agents. The body also uses a number of other mechanisms and factors that are of a more nonspecific nature. These are described in the following section.

Mechanical, Microbial, and Chemical Barriers: The Body's First Lines of Defense

Several systems of the body are barriers to potential disease-causing agents. Their effectiveness depends on the normal physiologic or disease state of the host. Conditions such as alcoholism, poor nutrition, and the debilitating effects of aging, fatigue, and prolonged exposure to extreme temperatures and to immunosuppressive therapy contribute to establishment of a disease process. The chapters in Part VIII contain general descriptions of the various parts of the body's defense system. Some pathogenic agents seem to initiate their infectious process only when they gain access through particular portals of entry, each of which has its own defense barriers.

INTACT SKIN

Unbroken skin serves as an excellent mechanical barrier that most microorganisms cannot penetrate. In addition, certain secretions formed by skin-associated glands provide protection against various bacteria and fungi. For example, secretions from the sebaceous glands (oil glands associated with hair follicles) contain both saturated and unsaturated fatty acids that are both bactericidal and fungicidal. Injuries to the skin, such as abrasions, lacerations, bites, or burns, provide the opportunity for microorganisms to pass this first line of defense. However, mere penetration of the skin does not establish an infection. Once organisms enter the body by this means, they may or may not encounter conditions favorable for their growth and multiplication; nevertheless, accidental injuries, regardless of how minor they appear to be, should not be neglected. Given the right set of circumstances, any type of wound can result in a serious infection. The microorganisms most likely to cause such infections include bacteria such as staphylococci, which normally inhabit hair follicles and sweat glands. [See skin diseases in Chapter 23.]

MUCOUS TISSUES

The mucosal tissues of mammals are composed of cells (epithelial), many of which form the outer surface of their bodies and line body cavities and passageways leading to the external environment. These tissues form an enormous surface area and include the entire gastrointestinal and respiratory tracts, salivary and tear (lacrimal) glands, bile system, and portions of the genitourinary system. It is with the mucosal tissues of a host that microbial pathogens and toxic agents frequently make their first contact and establish a disease process. Several immune and nonimmune mechanisms and substances function at mucosal surfaces to prevent microbial invasion or damage. Predominant among immune factors are specific antibodies, the secretory immunoglobulins (sIgA) found in fluids that bathe mucous membranes. Functions of sIgA include inhibition of bacterial attachment, neutralization of toxins and viruses, and prevention of the uptake of foreign substances by epithelial cells. Additional properties of this class of immunoglobulins and others are described later in this chapter.

Nonimmune mechanisms and factors associated with mucous membranes include the secretion of several biologically active substances, such as lysozyme in tears, and the presence of the body's own microorganisms (indigenous microbiota) necessary to maintain a normal balance in the region.

RESPIRATORY SYSTEM

The respiratory tract, and in particular the upper portion, is exposed continually and directly to numerous potentially damaging agents. These agents include microorganisms, gases, and particles that are foreign to the system. Protection against such factors is provided by a number of coordinated nonspecific defense mechanisms and specific immunologic responses. [See immunoglobulins later in this chapter.]

Among the nonspecific defenses of the respiratory tract are the nasal hairs coated with a thick, slimy secretion known as *mucus,* which filter and trap particles from inhaled air, and the mucous membranes of this region, which are also covered with mucus that serves to trap dust, foreign particles, and various microorganisms. Respiratory tract fluids also contain various antibacterial proteins such as the enzyme lysozyme and the iron-collecting lactoferrin, discussed later. Parts of the respiratory passages are also lined with cilia, microscopic hairlike projections, which beat rhythmically in such a way as to move particles trapped by mucus upward toward the back of the throat, where they are swallowed. These barriers, plus the coughing and sneezing reflexes, help to eliminate foreign particles. Microorganisms penetrating the defenses of the respiratory system may activate immunologic responses, including circulating phagocytes (Figure 17–8) and antibodies or immunoglobulins that participate in the neutralization, inactivation, or destruction of microorganisms or their toxic products.

GENITOURINARY SYSTEM

The mucous membranes of the female genital tract are protected against several pathogens by a thick vaginal secretion that tends to trap certain invading organisms and eliminate them from the body. In addition, the acidity of the vaginal environment discourages some infectious agents. The outward flow of urine and its acidity contribute to the defense of the urinary tract. However, various pathogens, including those causing the bacterial diseases gonorrhea and syphilis and the virus infection genital herpes (Color Photograph 120) are able to invade the

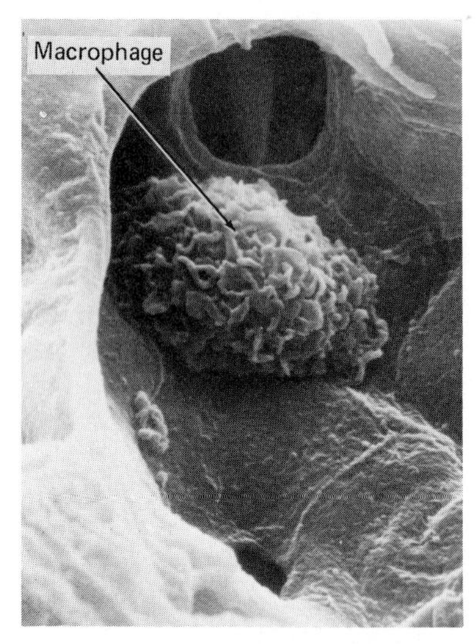

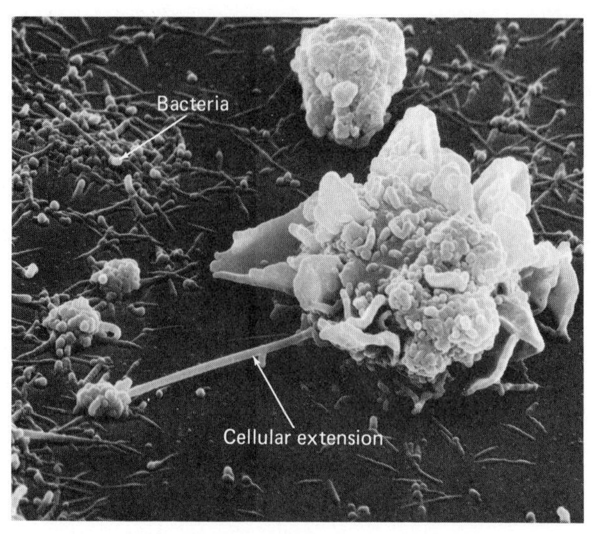

Figure 17–8 Phagocytic cells in action. (a) A wandering macrophage on the surface of an air sac lying just beneath the entrance to the lung. *[M. F. Greenwood, and P. Holland,* Lab. Invest. ***27****:296, 1972.]* (b) Phagocytosis of bacteria. Note the cellular extension of the phagocyte. *[D. A. Powell, and K. A. Muse,* Lab. Invest. ***37****:535, 1977.]*

body through this portal. **[See sexually transmitted diseases in Chapter 28.]**

EYES

Several factors function to prevent disease agents from entering and attacking the inner lining of the eyelid (the conjunctivae). These factors include the mechanical motion of the eyelids, eyelashes, and eyebrows and the washing effects of tears, which contain the bactericidal substance lysozyme, discussed later.

GASTROINTESTINAL SYSTEM

The composition and acidity of gastric juice provide considerable protection to the stomach; however, some organisms are shielded by the presence of food.

In the small intestine, mucus, certain enzymes, bile, and phagocytosis are important factors contributing to the body's defense. The large intestine usually harbors many microorganisms (indigenous or native microbiota) that are important in maintaining a normal balance.

Indigenous Microbiota or Normal Flora

CONDITIONS THAT POSE PROBLEMS: AN INTRODUCTION

The microorganisms associated with particular regions or tissues are often referred to as the *indigenous (native) microbiota* or *normal flora.* In the latter context, "flora" denotes all microscopic life, and "normal" is a statistical term. The reader must not equate "normal" with "nonpathogenic," for many organisms found on and in the body can pose problems under conditions such as the following:

1. Deterioration of the host's defense mechanisms
2. Relocation of microorganisms, as when an organism finds its way to an area of the body previously uninhabited by it
3. A disturbance of the balance of the normal flora

The microbiota include beneficial as well as pathogenic microorganisms. Moreover, some of these indigenous microscopic forms of life may flourish in the general region of tissue damage and contribute to the disease state as **opportunists** rather than as primary causative agents of disease. An opportunist is an organism that is able to produce disease only in an individual (host) with weakened defense mechanisms. Many present-day infections result from an interaction between the host and certain microorganisms that form the host's microbiota rather than from external (outside of the body) sources of microorganisms. This type of situation is in sharp contrast to the infectious epidemics of the past, such as cholera, plague, and smallpox. Those diseases involved pathogenic microorganisms from external sources, capable of attacking any susceptible host.

SITES OF MICROBIOTAL OCCUPATION

As a rule, few or no microorganisms are found in the following anatomical locations: blood, larynx, trachea, nasal sinuses, bronchi, esophagus, stomach, upper intestinal tract, upper urinary tract (including the posterior urethra), and posterior genital tract (including the passage above the cervix).

The major habitats for indigenous microorganisms include the skin and contiguous mucous membranes (Figure 17–9), conjunctivae, portions of the upper respiratory tract, mouth, lower intestine, and several of the external and internal parts of the reproductive system. It will become apparent later that certain characteristics of each region allow a different range of microorganisms to thrive. These differences can be categorized into the following three types of environment:

1. Extremely high levels of both moisture and nutrients, as in the lower intestines and the mouth
2. A high level of moisture and a low level of nutrients, as with mucous membranes
3. A low level of moisture and a moderate level of nutrients, as on the skin

Other variables include availability of oxygen, pH and temperature levels, and relative exposure to contaminants and ventilation.

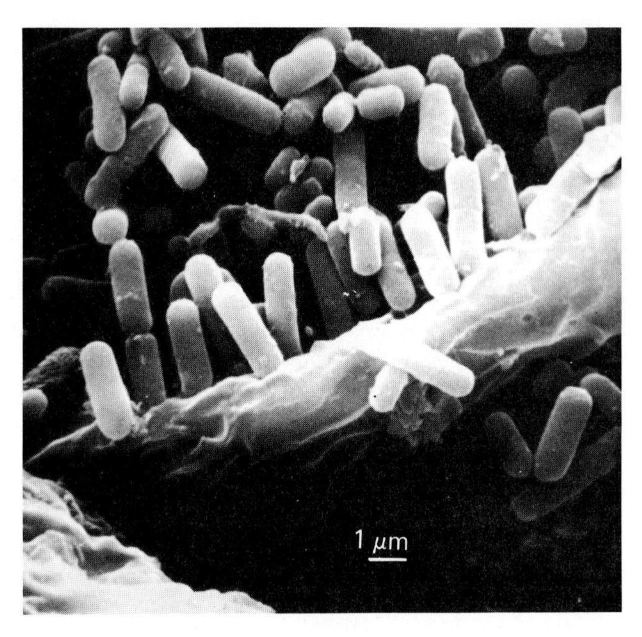

Figure 17–9 Attachment of short, rod-shaped bacteria to the epithelial surface of an adult mouse's gastrointestinal system. *[D. C. Savage, and R. V. H. Blumershine,* Infect. Immun. ***10****:240–250, 1974.]*

DEVELOPMENT

Because the infant is bathed during pregnancy in a sterile amniotic fluid, development of the indigenous microbiota begins with the normal birth process. As the baby passes through the birth canal, it picks up organisms, many of which may remain with it throughout its lifetime. Additional microorganisms are acquired by contact with the air, with hospital personnel, and with the mother. Such organisms may be transient (temporary) or may become permanent members of the microbiota. Cultures from the mouths of infants 6 to 10 hours old show appreciable numbers of bacteria. Bacteria appear in the feces 10 to 20 hours after birth.

Because each anatomical area varies in pH, oxygen content, nutrients, moisture, and bactericidal factors, different organisms will predominate. While members of the microbiota persist in their respective locations, saprophytic organisms and many parasitic microorganisms are destroyed or excreted. Conditions in these locations can change as a result of maturation of the individual, alteration in dietary habits, or chemotherapy. Thus, microorganisms may be temporary or permanent, depending upon the conditions that exist in the body. In addition, many potential pathogens may be present in various body habitats in the absence of disease. A slight imbalance of the host's defense mechanisms or ecological shifts as a consequence of chemotherapy or other factors could result in disease.

BENEFITS FROM THE INDIGENOUS MICROBIOTA

Probably the main benefit derived by humans from their microbial inhabitants is protection from disease. This statement may seem contradictory, because we have noted that some members of the microbiota can cause disease under certain circumstances. The key phrase is *under certain circumstances.* As a rule, the normal flora occupy their own niches and thus inhibit foreign organisms invading from other portions of the body or from the external environment. Such inhibition is brought about by competition for food, by the production of antibiotics or other inhibitory substances such as bacteriocins (antibacterial proteins produced by bacteria), or by changes in environmental conditions such as oxygen content or pH.

This ecological balance apparently prevents indigenous pathogens, such as the yeast *Candida albicans* (KAN-did-ah AL-beh-kans) and the bacteria *Streptococcus pneumoniae, Haemophilus influenzae* (he-MAH-feh-lus in-flew-EN-zee), and *Staphylococcus aureus* from causing severe disease. When the balance is upset by chemotherapy, for example, one or more pathogens may grow unchecked. A frequent consequence of antibiotic therapy is the appearance of *C. albicans* infections. These infections may occur in the mouth or in the perianal region. Untreated candidiasis can result in serious involvement of the lungs, infection of the coverings of the brain and spinal cord (meningitis), and blood poisoning (septicemia). Indigenous microorganisms that have been shown to inhibit the growth of *C. albicans* include *Enterobacter aerogenes* (en-te-row-BACK-ter a-RAH-jen-eez), *Escherichia coli* (esh-err-IK-ee-ah KOH-lee), *Pseudomonas aeruginosa,* and streptococci.

Another significant role played by indigenous organisms is helping to maintain mechanisms for antibody production. Studies with germ-free animals show that such animals generally have very low levels of immunoglobulins, the protective proteins formed by the body in response to various foreign substances. Because indigenous flora are lacking and their immune responses are weak, these animals are particularly susceptible to infection. The available evidence indicates that the microbiota acts as a constant source of antigens or irritants to the antibody-producing systems of the body and thereby permit a more rapid immunological response when it is needed.

The role of the microbiota in nutrition is the subject of considerable research. Some of these organisms synthesize a variety of vitamins in excess of their own needs, which are thus made available for the host. These vitamins include biotin, pyridoxin, pantothenic acid, and vitamins K and B_{12}. This function of the microbiota must be of a supplementary rather than an indispensable nature. If not, certain vitamin-deficiency diseases would not have been so readily discovered. However, it is likely that some individuals with deficient diets are benefited by this function. One interesting side effect of chemotherapy is the occasional development of symptoms of vitamin B_{12} deficiency, perhaps as the result of a reduction in the population of normal flora.

Antimicrobial Substances

Chemical substances capable of *in vitro* and in several instances *in vivo* antimicrobial activity have been isolated from various animal fluids, tissues, and cells. The full extent of their *in vivo* effectiveness is not known. Several antimicrobial substances and the types of microorganisms affected by them are listed in Table 17–2. The better-known ones are discussed more fully in the following sections.

CATIONIC PROTEINS

These proteins were the first antimicrobial substances extracted and identified from neutrophil granules. Cationic proteins, which are also called leukins and phagocytin, are heat-stable, most active at neutral pH, rich in the basic amino acid arginine, and function in conjunction with other tissue and fluid products such as lysozyme and histones. Different cationic proteins are active against specific

TABLE 17–2 Representative Antibacterial Substances in Animal Tissues and/or Fluids

Substance	*Common Sources*	*General Chemical Composition*	*Types of Microorganisms Affected*
Cationic proteins	Neutrophils	Protein	Gram-negative and gram-positive bacteria
Complement	Sera of most warm-blooded animals	Believed to be a protein-carbohydrate-lipoprotein complex	Gram-negative bacteria
Histones	Components of the lymphatic system	Protein	Gram-positive bacteria
Interferons	Virus-infected cells[a]	Protein	Various viruses and certain protozoa
Lactoferrin	Various body secretions, including tears, breast milk, bile, etc.	Glycoprotein	Bacteria and fungi
Lysozyme	Include leukocytes, saliva, perspiration, tears, egg whites	Protein	Mainly gram-positive bacteria
Properdin	Serum	Protein	Gram-negative bacteria and certain viruses
Protamine	Spermatozoa	Protein	Gram-positive bacteria
Spermidine, spermine	Prostate and pancreas	Basic polyamines	Gram-positive bacteria
Tissue polypeptides	Components of the lymphatic system	Basic peptides	Gram-positive bacteria
Transferrin	Serum, and tissue and organ spaces	Glycoproteins	Bacteria, fungi, and protozoa

[a] Refer to Chapter 12 for other sources and details of interferon production.

bacterial groups and rapidly interfere with microbial reproduction without destroying the microbes' structural organization.

COMPLEMENT

The bactericidal property of serum, as well as that of whole blood, has been recognized since approximately 1888. These antibacterial substances were found to be inactivated by heating to 56°C for 30 minutes and to function as bactericides only in the presence of specific antibody. Paul Ehrlich discovered that antibody was also required for other activities, such as the lysis of red blood cells (hemolysis) by this thermolabile (heat-sensitive) component of serum, which he named **complement.** Unlike the immunoglobulins and other serum factors, complement concentrations do not increase in response to immunization.

Complement, commonly designated C, is a complex group of at least 11 chemically and immunologically distinct plasma proteins capable of interacting with each other, with immunoglobulins, and with cell membranes. The individual proteins of a complement are normally present in the sera of most vertebrates and are functionally inactive molecules. In addition, some of the complement components are themselves protein complexes. For example, the first component of complement, C 1, is a macromolecule that consists of three individual subunits, C 1q, C 1r, and C 1s.

When complement is activated, a series of sequential reactions known as the *complement cascade* is set into motion. Through such reactions a number of biological reactions are generated, ranging from the destruction (lysis) of different eucaryotic cells, bacteria, and viruses to regulation of the events in inflammation. Complement can also enlist the participation of other host cells, biologically active enzymes, and other substances in various host immune responses, including the release of histamine from mast cells, the migration of leukocytes, and phagocytosis. Details of the complement cascade and certain favorable and unfavorable effects when this important system is activated are described in Chapter 20. The cytotoxic activity of complement has also been used in the diagnosis of several infectious diseases, as in the complement fixation test discussed in detail in Chapter 19.

Only limited information exists on the synthesis of complement components. Macrophages have been shown to manufacture C 4 and C 2. C 1 synthesis has been associated with epithelial cells of the gastrointestinal mucous membranes, and the liver has been suggested as the site for the formation of other components.

INTERFERONS

Originally, interferon, a natural antiviral protein, was discovered in the late 1950s. During studies on viral interference (the ability of one virus to inhibit infection by another), interferon was shown to be a cellular product or cytokine formed in response to infection with a variety of RNA and DNA viruses. Interest in this glycoprotein has intensified because of its natural antitumor activity and its potential as an effective treatment for cancer. The biological and physical properties of the currently known three types of interferon (alpha, beta, and gamma) together with their mechanism of action and approaches to commercial production, are described in Chapter 12.

LACTOFERRIN AND TRANSFERRIN

Iron is an essential nutrient for almost all organisms. The lack of free iron in a host may serve as a barrier against invasion by some bacteria, fungi, and protozoa. In the human, two iron-binding proteins, *lactoferrin* and *transferrin,* function in the absorption, transport, and metabolic exchange of iron. In addition, these proteins inhibit microbial growth by binding available ferric ions, thereby depriving bacteria and fungi of a nutrient essential for their growth.

Lactoferrin and transferrin are chemically closely related but immunologically distinct glycoproteins. Lactoferrin is synthesized by glandular epithelial cells and is present in relatively high concentrations in tears, semen, breast milk, bile, and nasopharyngeal, bronchial, cervical, and intestinal mucosal secretions. In addition, lactoferrin is found in PMNLs and pus. Transferrin is present primarily in serum and in spaces within tissues and organs.

LYSOZYME

Lysozyme, a thermostable enzyme, is present in several different body fluids and tissues, including leukocytes, perspiration, saliva, and tears. Its bactericidal action is associated with the hydrolysis of bacterial cell walls, especially where repeating units of *N*-acetylglucosamine and *N*-acetylmuramic acid are exposed. Lysozyme is effective against gram-positive bacteria and, under certain conditions, against several gram-negative species. Recent studies have shown lysozyme's effectiveness against several microorganisms to be increased by the presence of very low antibody concentrations. **[See bacterial cell walls in Chapter 7.]**

PROPERDIN

Properdin was first reported in 1954 as a relatively heat-sensitive protein found in normal serum. The substance was described as having bactericidal activity against gram-negative bacteria in the presence of magnesium ions (Mg^{2+}) and complement. Later studies have shown properdin to be one of several proteins involved in the destruction of certain bacteria, the neutralization of certain viruses, and the destruction of red blood cells of individuals with the rare blood disease paroxysmal nocturnal hemoglobinuria (recurring episodes of blood in the urine).

The properdin system consists of at least four specific proteins that participate in the activation of an important defense mechanism, the complement system (see Chapter 20).

SPERMINE

In 1953, Dubos and Hirsh isolated a basic peptide that is active *in vitro* against tubercle bacilli. This substance, called *spermine,* is known to be present in human tissues and those of various other animal species. The effectiveness of the peptide in these tissues is, at present, unknown. Chemically, spermine is low in the amino acid lysine but high in arginine. Its *in vitro* activity against tubercle bacilli depends upon activation by the enzyme spermine oxidase. Dubos found that tissue containing this enzyme was more resistant to tubercle bacilli than those tissues apparently lacking spermine oxidase.

Other Lines of Defense in Nonspecific Resistance

Phagocytosis

After the mechanical barriers provided by the skin and mucous membranes, phagocytosis is one of the organism's most important defenses against invading matter (Figure 17–10). **Phagocytosis** is the ingestion and subsequent digestion of particles by single cells. This process is carried out by circulating granulocytes (Figure 17–5), by wandering mononuclear cells (Figure 17–8), and by fixed macrophages such as Küpffer cells. Foreign matter is ingested by extensions of phagocytic cells (Figure 17–10). In 1882, Elie Metchnikoff observed "ameboid cells" (leukocytes) ingest cells of the yeast *Monospora bicuspidata* (mon-OSS-pore-ah bye-KUSS-peh-datta) within the water flea *Daphnia* (DAFF-nee-ah). Apparently, he was among the first investigators to recognize the important role played by leukocytic ingestion in protecting a host from disease.

Figure 17–10 The major stages in the phagocytosis of bacteria. The ingestion, intracellular digestion, and elimination of indigestible materials are emphasized.

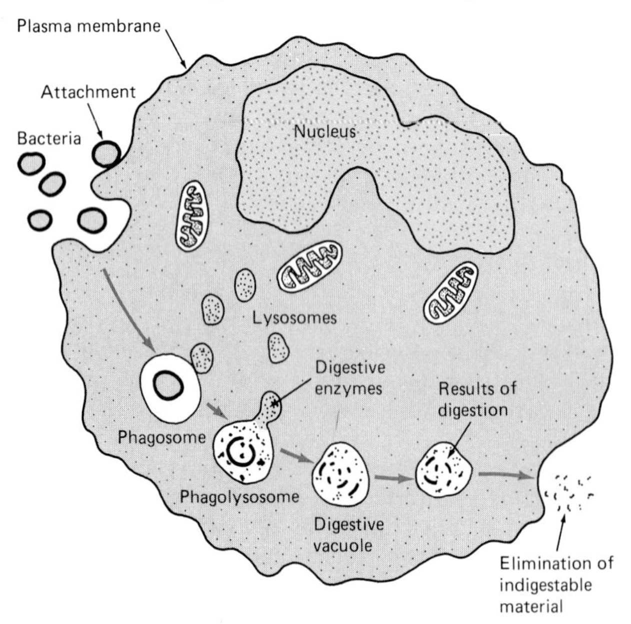

Noncellular components of blood have been shown to be important in phagocytic ingestion. By the early twentieth century, immune serum was known to enhance active phagocytosis. When a specific antibody combines with microbial cells in the presence of a complex group of proteins known as *complement,* cells are made susceptible to phagocytosis. This process is called **opsonization.** Opsonizing antibodies combine with the surface antigens of a cell (antigen-antibody complex) and prepare it for ingestion by the phagocyte. Ingestion of bacteria can occur in the absence of detectable antibody. In this phenomenon, called *nonimmune* or *surface phagocytosis,* the pathogen is trapped against tissue surfaces.

If phagocytosis of the invading organism is complete, the disease state is either averted or cured. However, in the event that pathogens manage to escape ingestion and intracellular destruction, they can reproduce and cause serious infections. This tends to happen with *Mycobacterium tuberculosis, Neisseria gonorrhoeae, N. meningitidis,* and pathogenic staphylococci. Exposure to certain anesthetics, significant increases or reductions of body temperature, and various drugs can depress phagocytic activity, thus seriously lowering an individual's resistance.

Phagocytic Cell Types

It was Metchnikoff who first identified two major kinds of cells having phagocytic capabilities. He named them *microphages* and *macrophages.* The microphage category includes polymorphonuclear leukocytes (PMNLs), of which neutrophils appear to have the most pronounced phagocytic activity. PMNLs are the principal cells involved in the inflammatory response to invading microorganisms. Inflammation, as described later, is the result of a complex series of reactions that are coordinated to isolate and destroy harmful agents and to prepare the damaged area for healing and repair.

The *fixed macrophages* of the mononuclear phagocytic system function in disposing of old and fragmented red blood cells, bacteria, and the remains of other foreign materials. The *wandering macrophages,* including monocytes, also aid in the disposal of various blood cells. This is especially true of macrophages that have passed through and out of blood vessels. In addition, these macrophages assist in the repair of damaged tissue by destroying and absorbing cellular debris.

Stages in Phagocytosis

Destruction of microorganisms by phagocytosis occurs in three stages: (1) contact between the phagocyte and the particle to be ingested; (2) ingestion; and (3) intracellular killing (microbicidal activity) and destruction (digestion).

CONTACT

The contact stage may be either random or directed. Random contact depends upon chance collision between ingesting cells and particles to be ingested. In the more specific response, phagocytic cells migrate toward bacteria or particulate matter, drawn by *chemotaxis,* a reaction to chemical stimuli. Several bacterial extracts or whole organisms have been reported to attract leukocytes *(positive chemotaxis).* The exact mechanism of chemotactic responses is unknown.

INGESTION

The ingestion stage of phagocytosis is similar to food intake by amoebae. Bacteria are ingested through an invagination of a leukocyte's cytoplasmic membrane. In general, bacteria or particles coated, or opsonized, with antibody molecules are more readily ingested. In the ingestion process, a phagocytic vacuole or *phagosome* (Figure 17–11) forms, engulfs the bacterial cell. This activity requires the expenditure of energy by the leukocyte, energy derived for the most part from glucose metabolism. The vacuole migrates into the cytoplasm. There it collides with lysosomelike granules, which explosively discharge their antibacterial contents into the bacterium-containing vacuole. The membranes of the vacuole and granule fuse, forming a digestive vacuole, or *phagolysosome.*

The granules of phagocytic cells appear to decrease in number as the organisms or particles are ingested. This event is called the *degranulation phenomenon* and is directly related to the number of particles ingested. Chemical analysis has shown the granules to contain a large number of degradative enzymes. Degranulation and the burst of metabolic activity are closely connected events that serve to deliver toxic substances into the phagocytic vacuole. The phagosome concentrates these toxic substances around an ingested bacterial cell and thereby reduces the harmful effects on the phagocyte and surrounding cells.

INTRACELLULAR DESTRUCTION

Both lysosomal components and metabolic products contribute to the bactericidal activity in the phagocyte. As a result of the burst of metabolic activity *(respiratory burst)* following ingestion, lactic acid (which lowers the pH in the vacuole) and the strong oxidizing agent hydrogen peroxide are produced. These substances, together with histones, lysozyme, and various digestive enzymes released from lysosome granules, degrade the dead bacterium. Several of these antimicrobial agents are discussed in more detail elsewhere in this chapter.

Two types of lysosome granules are formed as neutrophils mature: primary and secondary granules (Figure 17–11). The first to form is the primary granule, which contains typical lysosomal enzymes. The secondary granule is the predominant granule in a mature cell and contains various bactericidal substances. Both can discharge their contents into phagocytic vacuoles.

Microbicidal Mechanisms. Since neutrophils can ingest and kill organisms under both aerobic and anaerobic conditions, their killing mechanisms are divided into two broad categories: *oxygen-dependent* and *oxygen-independent* systems. Oxygen-dependent killing can further be divided into myeloperoxidase-dependent and myeloperoxidase-independent reactions. The combination of hydrogen peroxide with

Streptococci

(a)

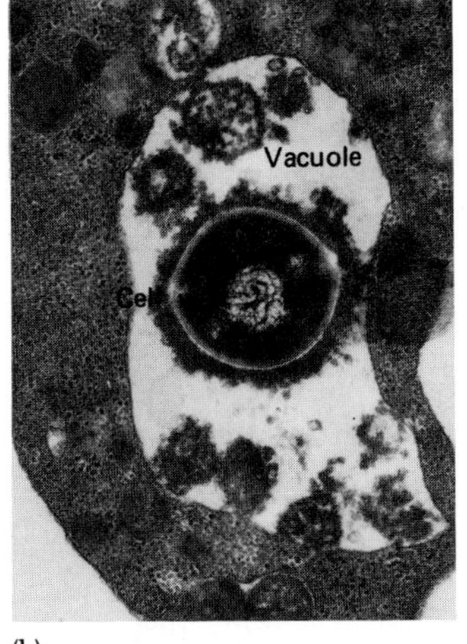

(b)

Figure 17–11 Intraphagocytic degradation of group A streptococci as shown by electron microscopy. (a) Streptococci contained within polymorphonuclear leukocytes (PMNLs) 45 min after *in vitro* phagocytosis. Early evidence of degradation can be seen with certain bacteria (arrows). (b) A streptococcal cell undergoing the initial process of intraphagocytic degradation in the vacuole. Changes in the bacterial cell wall and internal regions are evident at this point. *[E. M. Ayoub, and J. G. White,* J. Bacteriol. ***98:****728–736, 1969.]*

myeloperoxidase (MPO), a protein released from primary granules, and a halide ion such as chloride produces a very effective bactericidal system. While the hydrogen peroxide released in the phagocyte by itself is a powerful bacterial killing agent, the described myeloperoxidase combination has at least 50 times more activity. Although myeloperoxidase is required for the best microbicidal activity, its absence from phagocytes is not necessarily disabling. Oxygen-dependent killing not involving MPO obviously takes place, but the mechanisms of action are not clearly defined. The microbicidal activity of oxygen-independent systems involves the actions of various proteins, including cationic proteins, lactoferrin, lysozyme, complement, proteases (protein-splitting enzymes) and histones as well as a low pH within the phagosome (see Table 17–2).

Failure of Intracellular Destruction. Once ingested by phagocytes, several gram-positive cocci are killed and degraded. This set of events is well documented for pneumococci (the cause of lobar pneumonia) and streptococci (the cause of strep throat and several other disease states). However, toxin-producing staphylococci are apparently not killed after ingestion. Tubercle bacilli are readily ingested, but they remain intact inside the phagocyte because of their resistance to digestion. Such tubercle bacilli are provided not only with an environment in which to multiply but also with a means for spreading to other regions and establishing new infections.

Several types of phagocytic dysfunction are known. Many of them are genetic disease states. One currently under study is the rare genetic disease Chediak-Higashi syndrome (C-HS). This condition has been found in cattle, humans, mice, and killer whales. Its characteristics include increased susceptibility to bacterial infections and the presence of large, abnormal granules (C-HS granules) in several different cell types. Large primary granules that fail to degranulate after phagocytic ingestion appear to be a major factor in the increased susceptibility to infection (Figure 17–12). In humans this change in granule activity is accompanied by an impaired capacity to kill certain gram-negative and gram-positive bacteria, although the ingestion of organisms generally is normal.

Inflammation

Inflammation is the body's second line of defense against infection. Inflammation can be produced by infectious disease agents and by irritants such as chemicals, heat, and mechanical injury. The single most important event in this process is the accumulation of large numbers of phagocytic cells at the site of inflammation.

SIGNS

The characteristic or *cardinal signs* of inflammation are heat, pain, redness, swelling, and loss of normal function. The redness is the result of increased blood in the involved area. The dilation of local blood vessels causes a slowing of blood flow. This dilation is accompanied by an increased permeability, which causes swelling, or *edema*, as tissue fluid accumulates in the spaces surrounding tissue cells. The increased diameter of the blood vessels speeds the flow of blood to the injured area, thereby raising the temperature. Clots can form in small vessels in the general area of injury, which may prevent infectious agents or their products from entering the circulatory system. The pain experienced in an inflammatory reaction is believed to be caused by the pressure accumulating tissue fluid exerts on sensory nerves. The blood-dilating substances may also irritate these nerve endings. The loss of normal function is associated with the pain.

As the inflammatory response develops, a noticeable change occurs in the behavior of granulocytic cells. These cells first attach themselves to the inner lining of small blood vessels, the capillaries, and then push their way between the cells of these blood vessels into the areas of tissue injury at a rate faster than normal. In the later stages, these granulocytes are replaced by monocytes.

Pus formation may also be associated with inflammation. After the phagocytes have destroyed the microbial cells and engulfed the tissue debris, they become degranulated and die. In the involved area, a central mass of fluid is formed by the remains of damaged tissue cells, dead phagocytes, and microbial casualties. This fluid is pus.

MECHANISM

Inflammation is a complex mechanism. Its characteristic symptoms are thought to be caused by substances released from damaged cells. Included in this group of suspected factors are the compounds *histamine* and *serotonin*. **[See histamine in Chapter 20.]**

INFLAMMATORY EXUDATE

The process of inflammation brings several mechanisms of the host's defense into play. With the increased blood flow to the injured area, the concentration of white blood cells and various antimicrobial factors—including complement, chemotactic substances, specific antibody—in the fluid or exudate associated with the inflammatory reaction are greatly increased. The chemotactic substances promote the attraction and accumulation of phagocytic leukocytes, while the complement and specific antibody increase the phagocytosis of microorganisms. The injuries and the increasing number of dead host cells cause the release of still more antimicrobial substances, which make the involved area increasingly unfavorable for the several types of microor-

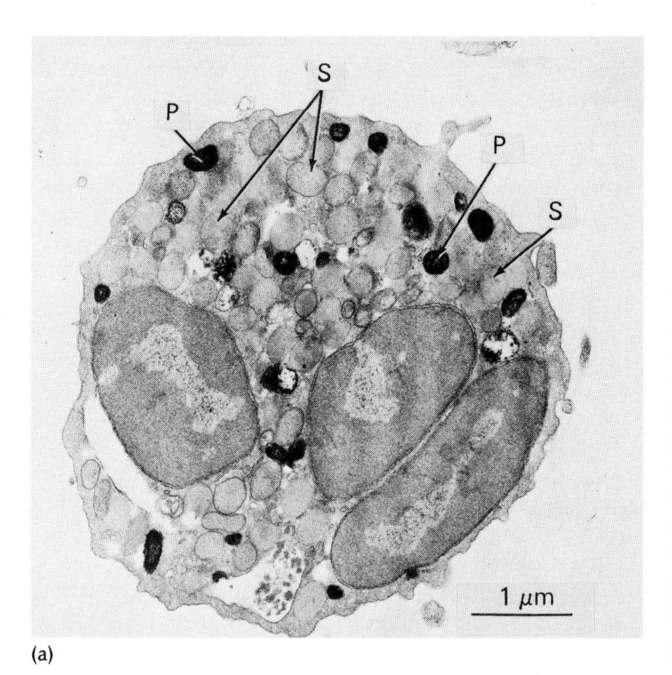

(a)

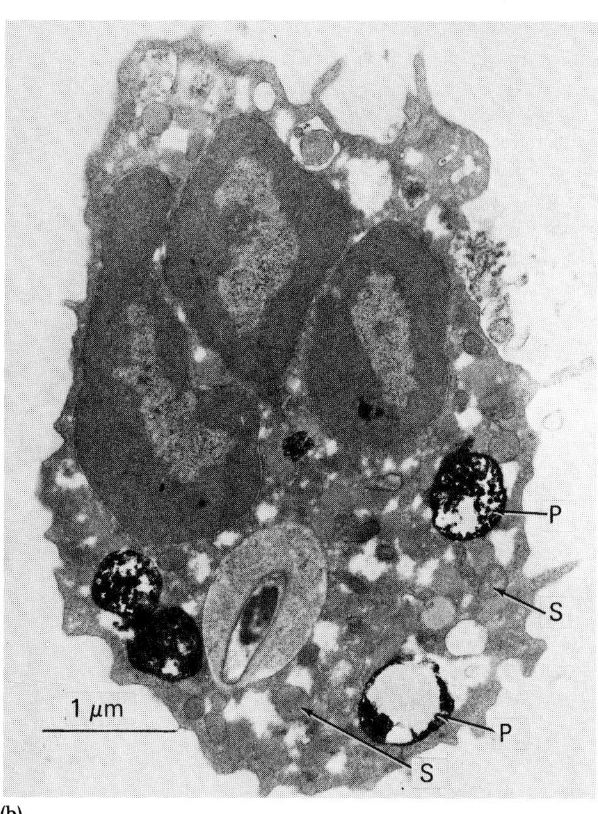

(b)

Figure 17–12 Leukocyte dysfunction in Chediak-Higashi syndrome (C-HS). (a) An electron micrograph of a normal PMNL incubated without bacteria. The cell contains primary (P) and secondary (S) granules characteristic of mature cells of this type. (b) A C-HS PMNL incubated in the presence of *Bacillus subtilis.* Phagocytic vacuole formation has occurred with an intact, apparently undigested bacterium in the lower part of the cell. However, the appearance of the cell indicates that the degranulation of primary granules (P) was delayed. *[H. W. Renshaw, W. C. Davis, H. H. Fudenberg, and G. A. Padgett,* Infect. Immun. ***10:****928–937, 1974.]*

ganisms. A beta globulin, C-reactive protein, normally not found in serum appears when there is inflammation anywhere in the body. A laboratory test for the protein is used as a sensitive indicator of inflammation.

The exudate associated with inflammation also contains the elements needed for blood coagulation. These elements may wall off the site of activity, preventing its spread to other areas. A closed region or sac of this type, containing pus and microorganisms, is referred to as an **abscess.** *Pimple, boil,* and *furuncle* are common terms for this kind of lesion. Abscesses that are not isolated but intercommunicating are known as *carbuncles.* [See skin infections in Chapter 23.]

FEVER

Fever is an elevation of body temperature above normal. It is a frequently treated symptom of many disease states. Attempts to lower body temperature during a febrile (feverish) episode are based on the assumption that fever is a harmful by-product of infection. Although the effects of fever may be disagreeable, they are beneficial if the body temperature does not rise too high. Temperatures of 38.5° to 39.0°C speed the destruction of disease agents by increasing immunoglobulin production and phagocytic activity. Fever is thought to be caused by the action of small proteins, known as *pyrogens,* entering the bloodstream. These pyrogens are synthesized and secreted in response to stimuli by cells such as PMNLs, monocytes, tissue macrophages, and phagocytic cells of the mononuclear phagocytic (reticuloendothelial) system. Many of these cells also have major roles in other aspects of the immune response to foreign stimuli.

Prolonged fever can be associated with unfavorable effects on the host. These effects include immune reactions; an increased heart rate, which, in a patient with compromised cardiovascular function, may result in convulsions; metabolic changes leading to dehydration and electrolyte losses; interference with the activities of chemotherapeutic agents; and an increased susceptibility to the effects of some microbial toxins. [See microbial toxins in Chapter 22.]

Specific Immune Responses

From our earlier discussion, it is apparent that the immune system is a highly active, multicompartmental network of elements with ever-changing parts and functions. As with other body (physiologic) mechanisms, immunologic responses should be looked upon as adaptive processes used by the body to achieve and maintain stability between internal and external environments. In this section we will examine the specific immune response of a host to foreign materials.

In general, individuals who successfully recover from an infectious disease acquire some degree of resistance toward the inciting cause. The resistance acquired may be toward a specific microorganism or toward certain microbial products that are recognized by the body's immune system as foreign. Molecules that are viewed as foreign by this system and which stimulate an immune response are called **antigens.** The immune system has several means of defending against such foreign substances. These include the **humoral response,** the formation of antibodies or **immunoglobulins,** which function in the specific recognition of antigens, and the **cell-mediated response,** regulated by a group of differing antigen-reactive cells, the **T lymphocytes.** While the T lymphocytes do not synthesize specific antibody, they can still recognize specific antigens and respond by changing into cells capable of killing any cells bearing such antigens on these surfaces or by releasing biologically active substances. Under certain conditions, complement, the complex protein discussed in this chapter and in Chapter 20, must interact with antigens and antibodies, and even lymphocytes, for the immune system to respond effectively to foreign substances.

The nature of antigens and antibodies, selected theories concerning antibody formation, and acquired states of immunity are among the major topics considered in the following sections of this chapter. The complement pathways in the immune response are described in Chapter 20.

Antigens and Immunogens

Macromolecules that can react with antibodies but do not necessarily stimulate the production of antibodies are generally referred to as *antigens.* Substances that provoke antibody formation and can combine with them are increasingly being referred to as **immunogens.** The stimulation of antibody production has been termed **immunogenicity.** In the interest of maintaining some uniformity of expression with other chapters of this text, the terms *antigen* and *antigenicity* generally will be used for both types of responses; however, *immunogen* will be used to emphasize antibody-stimulating activity. Because many microorganisms and various types of cells are effective immunogens, they can be detected and identified by laboratory tests involving the antibodies produced in response to them. Immunogenic substances also play a most important role in the prevention of disease. They are used in the preparation of vaccines and related materials, which in turn are used to produce active states of resistance to disease agents. [See vaccines in Chapter 18.]

Properties of Immunogens

Numerous substances can stimulate antibody production (specific immune response). Included in this group of both natural and synthetic macromolecules are (1) most free proteins; (2) combinations of proteins and other substances, including nucleoproteins (nucleic acid plus protein), lipoproteins (lipid and protein), and glycoproteins (carbohydrate and protein); and (3) certain polysaccharides. The majority of lipids are not considered immunogenic. Various bacterial components, including flagella, capsules, cell walls, and pili, and of course the entire microorganism, are immunogenic.

In addition to these naturally occurring immunogens, synthetic ones are also possible. Such preparations result from chemical modifications of nonantigenic substances. Reactions of this sort can occur both *in vitro* and *in vivo.* Several drugs and small reactive molecules that, by themselves, do not stimulate an immune response can be chemically connected *(coupled)* to a larger molecule, called the *carrier,* and thus become immunogenic.

Immunogens provoke their particular effects because antibody-forming *(immunopoietic)* tissues of an animal recognize them as foreign matter. The greater the incompatibility between the immunogen and the recipient's tissues, the greater the *immune response.* This is generally true except when the toxicity (poisonous nature) of the foreign substance overwhelms the animal's recognition mechanism.

Factors That Determine Immunogenicity and Immune Responses

Immunogens generally have molecular weights greater than 5,000 and a large molecular surface with room for many *antigenic determinant sites* (Figure 17–13). These determinants are specific groups of atoms on the surface of the antigen that both stimulate the formation of antibodies and react with them. An antibody molecule recognizes and responds to an antigen by binding closely to the antigenic determinant, or *epitope.* The number of determinant antigenic sites on the surface of a molecule

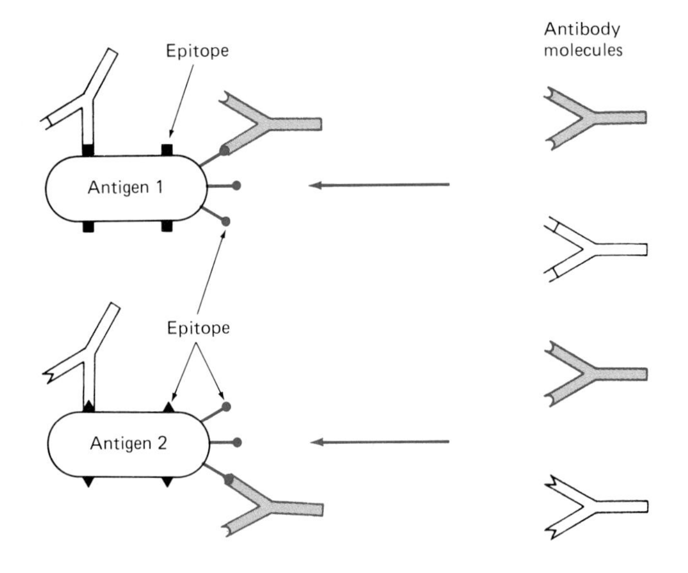

Figure 17–13 Examples of different types of antigenic determinants or epitopes. Most antigens have several determinants that are bound by different antibody molecules or T cells involved with cell-mediated immunity. This figure shows two antigens containing different as well as similar epitopes.

is known as its *valence.* In general, an immunogen's valence is proportional to its molecular weight.

Other factors that affect the immune response include the species of animal receiving the antigen, the animal's degree of *immunological maturity* (the degree to which its immune mechanisms are functioning), the route of inoculation, and the use of an *adjuvant.* An adjuvant is a preparation that consists of material such as mineral oil mixed with an immunogen to prolong and intensify the antibody-provoking stimulus.

Classes of Antigens

HAPTENS

In 1921, Karl Landsteiner and his collaborators, studying the specificity of an antibody response to its antigenic determinant, found small molecules of known chemical composition that were not immunogenic unless coupled to large proteins called *carriers.* These small, chemically defined substances have been given the name *hapten,* from the Greek word *haptein,* meaning "to fasten." Thus, a hapten is a molecule that cannot bring about an immune response by itself but can do so when it is conjugated (linked) to a larger protein carrier. In such situations, the hapten becomes one of the antigenic determinants with which antibodies react. Virtually any chemical, either small or large molecules, may serve as an antigenic determinant if linked to a suitable immunogenic carrier. This property is characteristic of certain lipids and polysaccharides of animal and bacterial cells. **[See macromolecules in Appendix A.]**

AUTOANTIGENS ("SELF" VERSUS "NOT-SELF")

Generally, an animal does not produce antibodies against its own body substances or against cells from which they can be derived. In other words, the immunopoietic tissues recognize the individual's cells as belonging to the "self," and not as foreign matter, "not-self." In exceptional situations, however, antibodies are produced against body components. These substances are referred to as **autoantibodies** and the antigens as **autoantigens.** The resulting state represents the process known as *autoimmunization,* or *autoallergy.*

Under normal conditions, autoantigens are limited to particular cells and tissues and do not gain access to immunopoietic tissues. However, under certain conditions, disease states associated with the production of antibodies toward various normal cellular components are possible. Representative autoimmune disease and the immunogenic substances believed to be responsible for each are shown in Table 17–3. **[Autoimmune diseases are discussed in greater detail in Chapter 20.]**

TABLE 17–3 Some Autoimmune Disease States and Their Incriminated Immunogenic Substances

Autoimmune Disease	*Immunogenic Substance*
Acquired hemolytic anemia	Red blood cells
Allergic encephalomyelitis	Myelin (fatlike material) from the central nervous system
Aspermatogenesis	Spermatozoa
Idiopathic thrombocytopenic purpura[a]	Blood platelets
Rheumatoid arthritis	Immunoglobulins[b] (IgG)
Systemic lupus erythematosus (LE)	Deoxyribonucleic acid
Thyroiditis (Hashimoto's disease)	Thyroglobulin (one of the hormones from the thyroid gland)

[a] This disease state is characterized by bleeding in various tissues and the presence of a rash and purpura (little areas of hemorrhaging) in the skin. It is also called *purpura hemorrhagica.*
[b] Other causes are believed to be operative in this disease condition.

ISOANTIGENS

The erythrocytes of individuals within the same species are known to contain different antigens. In addition, antibodies capable of specifically reacting with these blood cell antigens also differ among individuals. These different factors are referred to as **isoantigens** and **isoantibodies.**

The agglutination (clumping reaction) that results from mixing antigenically different red blood cells, such as blood type A, B, or AB from one person with the normal serum of another individual of a different blood type, is called *isohemagglutination* (Color Photograph 76). Blood-typing procedures are based on this phenomenon.

TOLEROGENS

There are some antigens that interact with the host's immune system but, instead of producing a positive immune response, they cause a state of unresponsiveness. These are referred to as *tolerogens.* These antigens may work in several ways, for example, by blocking cells of the immune system from performing their normal functions or by causing a suppression of naturally occurring immune responses. The end result is that the immune system will not respond to later antigen exposures.

HETEROPHILE ANTIGENS

Various cells and tissues cause production of antibodies that react with other tissues derived from some mammals, fish, and even plants of completely unrelated species. In 1911, Forssmann reported that the injection of emulsions containing guinea pig tissues into rabbits caused the rabbits to produce antibodies that caused the clumping or lysis of sheep erythrocytes in the presence of complement. Antigens that stimulate production of antibodies effective against material unrelated to the original antigens are called **heterophile antigens.** Antibodies to these antigens will cross-react with the cells of various animal species and microorganisms. The best-known example of the heterophile antigens is the *Forssmann antigen.* It has been found in other animals, including birds, cats, dogs, mice, and tortoises, and has been associated with certain bacterial species e.g., *Bacillus anthracis* (ba-SILL-us ann-THRAY-sis), *Streptococcus pneumoniae, Salmonella* (salmon-EL-la) spp., and *Shigella dysenteriae* (shi-GEL-la dis-en-TE-ree-eye). In these situations, however, the antigen might have become implanted in some microorganisms because of the intimate contact between pathogens and their respective hosts.

Forssmann antibodies—those that will react with sheep red cells—are present in the sera of individuals with infectious mononucleosis, a fact that has been used for diagnostic purposes. [See infectious mononucleosis in Chapter 27.]

Antibodies

Blood serum contains many proteins, some of which can be distinguished on the basis of their physiochemical and immunogenic features. These properties include (1) chemical composition, (2) chromatographic features, (3) electrical charge and migration in an electrical field, (4) molecular weight, (5) relative solubilities in alcohol, electrolytes, and water, and (6) sedimentation coefficients.

Most antibodies are members of the protein group known as *gamma globulins.* Their presence in the blood of an immune animal was demonstrated by von Behring and Kitasato in 1890. Serum of a laboratory animal that had been injected with several small doses of diphtheria toxin was shown to have protective powers. Two experimental animals were injected with lethal doses of diphtheria toxin. In one case, the toxin was mixed with antitoxin from the immune animal. Only the recipient of the antiserum-toxin mixture survived.

Additional experiments showed the reaction to be specific: serum from animals receiving diphtheria toxin produced antibodies specific only for diphtheria toxin; serum from animals given another toxin did not produce a protective effect against diphtheria. The results clearly demonstrated the phenomenon of *passive immunization,* the transfer of antibodies from an immunized individual to a nonimmune recipient. [See passive immunity in Chapter 18.]

Electrophoresis and Immunoelectrophoresis (IEP)

Immunoelectrophoresis is an extremely valuable tool that enables the immunologist to separate and identify antibodies and antigens in a mixture. In the first step, the mixture is placed on a gel or gel-like material. This gel is placed in an electric field that draws any charged molecules toward either the positive or the negative electrode depending on its charge. The rate at which different proteins migrate separates them into bands or spots on the gel. This separation by an electric field is the basis of electrophoresis. Next, a thin trough is hollowed out in the agar, parallel to the migration lines of the individual protein components. These steps and the appearance of a typical IEP system are shown in Figure 17–14. Antisera against one or several proteins in the original mixtures are placed in these troughs. In areas where homologous antigen from the original mixture and antibody meet by diffusion through the gel, precipitation lines develop, usually in arcs.

IEP can be particularly valuable in analyzing and identifying a large number of different antigens present in a solution. In addition, the presence of

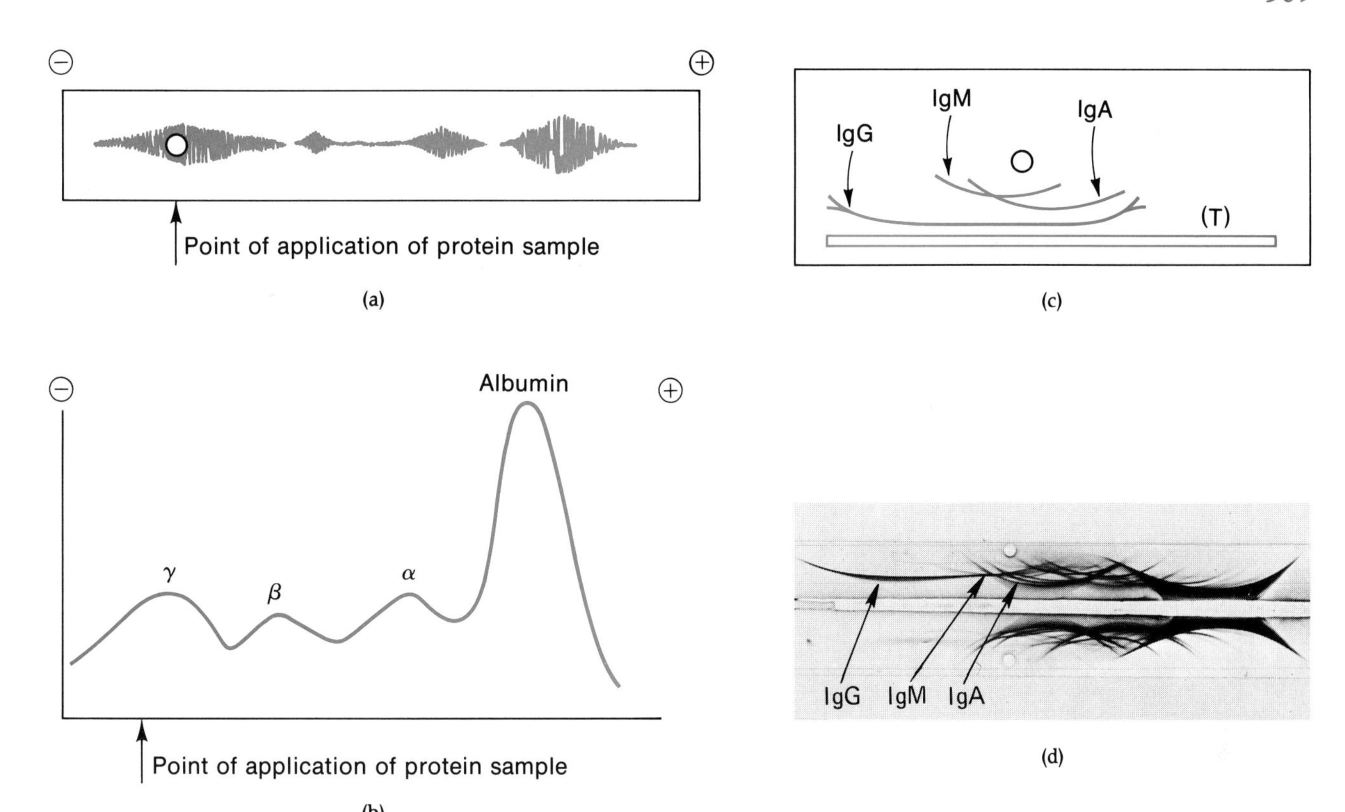

Figure 17–14 Approaches to demonstrating the presence of various serum proteins. (a) A diagrammatic representation of the electrophoresis procedure. A protein sample is placed in a prepared hole in the gel and then allowed to migrate in an electric field. (b) A graph of the locations on the gel of different proteins that migrated in the electrophoresis procedure. The four separated protein categories shown are albumin, alpha (α) globulins, beta (β) globulins, and gamma (γ) globulins. (c) Immunoelectrophoresis. Here antibodies (immunoglobulins) to serum proteins are introduced into a horizontal central trough (T). These antibodies and the separated serum proteins (which now function as antigens) diffuse toward each other to form precipitin lines, as shown in (d). The diagrammatic representation in (c) shows an idealized immunoelectrophoretic pattern with only the relative positions of IgG, IgM, and IgA. (d) Immunoelectrophoretic patterns of normal serum. Although numerous serum fractions are shown, only a select few are specifically indicated. *[Courtesy of Dr. M. D. Poulik, Wayne State University School of Medicine, and the Child Research Center of Michigan.]*

abnormal proteins in the serum of individuals with malignancies and certain blood abnormalities can be shown with this procedure.

IEP can be applied to a variety of body fluids, including amniotic fluid, cerebrospinal fluid, human plasma and serum, respiratory secretions, saliva, animal and plant tissue antigens, and microbial antigens.

Immunoglobulins: Activities and Occurrence

CLASSES

Antibodies are primarily in the slowest-moving electrophoretic fraction. Such substances are designated as gamma (γ) globulin (Figure 17–14b). Any protein exhibiting antibody activity or having antibody reaction sites in common with antibody molecules is called an *immunoglobulin (Ig)*. The known immunoglobulins are grouped in the five classes IgG, IgA, IgM, IgD, and IgE, based on differences in their antigenicity to other animals and their physicochemical properties, such as carbohydrate content, charge, and molecular size and structure. Table 17–3 lists the classes of immunoglobins present in serum of every normal individual and some of their properties.

On the basis of antigenic reactivity and subsequent immunochemical analysis, several subclasses of immunoglobulins have been found.

IgG. In the normal adult IgG is the most plentiful of the immunoglobulins. It forms approximately 75% of the total serum immunoglobulin content. Four subclasses, IgG1 to IgG4, are known to occur within the class. The capacity of a given individual to form these immunoglobulins may be under genetic control.

TABLE 17–3 The Immunoglobulins and Selected Properties

	Immunoglobulins					
Properties	*IgA*	*sIgA*[c]	*IgG*	*IgM*	*IgE*	*IgD*
H chain class	α (alpha)	α (alpha)	γ (gamma)	μ (mu)	ϵ (epsilon)	δ (delta)
L chain class	κ and λ (kappa and lambda)	κ and λ	κ and λ	κ and λ	κ and λ	κ and λ
J chain	Absent	Present	Absent	Absent	Absent	Absent
Secretory component	Absent	Present	Absent	Absent	Absent	Absent
Molecular weight (approximate in daltons)	160,000	400,000	150,000	900,000	190,000	180,000
Sedimentation coefficient[a]	75	105	6–7	19	8	7–8
Percentage of total serum antibody	15	—	75	10	0.004	0.2
Participation in classic complement fixation	None	None	+	+++	None	None
Placental transfer	None	None	Yes	None	None	None
Antibacterial destruction (lysis)[b]	+	+	+	+++	Questionable	Questionable
Viral neutralization[b]	+++	+++	+	+	Questionable	Questionable

[a] Refers to the speed with which a particle settles in an ultracentrifuge system, and is influenced by both particle size and molecular weight.
[b] + = low level of activity; +++ = high level of activity.
[c] Secretory IgA.

IgG is the only immunoglobulin class that can cross the human placenta from the maternal circulation to that of the fetus. These maternal IgG molecules are responsible for the protection of the newborn during the first months of life. The number of IgG molecules decreases sharply by approximately the fourth month after birth. It is about this time that the child's own IgG-synthesizing machinery comes into play. Other maternal immunoglobulins, such as IgA and IgM, do not appear in the newborn's serum because they cannot pass through the human placenta (Table 17–3). Synthesis of the newborn's IgM occurs at about the same time as the synthesis of IgG. IgA levels increase later, generally between the ages of four and ten.

Other activities of the IgG class include neutralizing bacterial toxins, combining with (fixing) serum complement (a complex protein in serum), and attaching to macrophages, thereby arming these cells to function effectively in phagocytosis. **[See phagocytosis earlier in this chapter.]**

IgA. IgA is next in order of abundance. Two types of this immunoglobulin class are recognized (Table 17–3). One of them is found in serum, while the other occurs in fluids that bathe the mucous tissues of the body and is called *secretory IgA (sIgA)*. In these body secretions, sIgA provides the primary defense against certain microbial toxins and viruses (Table 17–3). Such secretions include colostrum (the early form of breast milk), saliva, tears, the nasal mucosa, bronchial secretions, mucous secretions of the small intestine and vagina, and prostatic fluid. Serum IgA normally constitutes about 15% of serum immunoglobulins. The concentration of sIgA in the body fluids listed above is substantially higher than that of other immunoglobulins.

Several differences exist between the two types of IgA. For example, sIgA is much larger and contains a secretory component (SC) and J (joining) chain that function to bind together the parts of the immunoglobulin (Figure 17–15). The SC also transports sIgA molecules to mucosal surfaces and protects them from bacterial enzyme (IgA proteases) attack.

IgM. IgM is the largest of the immunoglobulins and comprises about 10% of the serum immunoglobulins. Generally, after exposure to an immunogen, IgM is the main immunoglobulin appearing in the early immune response. Its presence is followed by a more permanent and larger number of IgG molecules.

Structurally, IgM consists of five Y-shaped units similar to IgG that form a pentamer. This immunoglobulin class fixes complement and aids phagocytosis by macrophages.

IgD. IgD is found in serum in trace amounts, comprising about 0.2% of the total serum immunoglobulins. It is quite sensitive to the action of heat and protein-digesting enzymes. Several reports have linked IgD with antibody activity against hormones such as insulin and thyroxine, as well as milk proteins, penicillin, and cell nuclear antigens. While the main function of this class of immunoglobulins remains to be determined, it is known to be present in large quantities on the membranes of many circulating B lymphocytes, and it is believed that IgD molecules may trigger the differentiation of B lym-

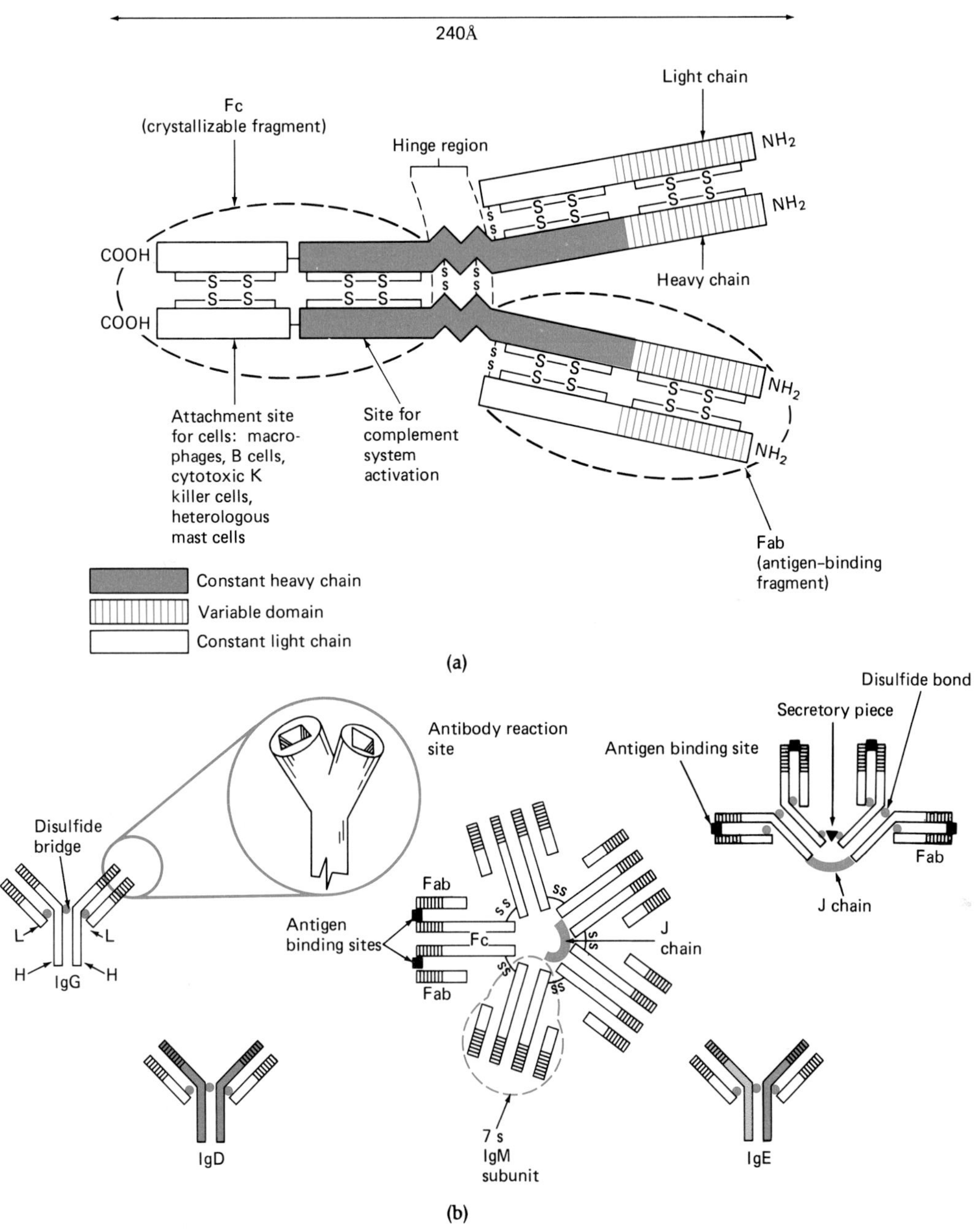

Figure 17–15 (a) The basic immunoglobulin structure (IgG). An immunoglobulin molecule contains four polypeptide chains, two heavy (H) and two light (L) chains. These components are joined by disulfide bonds (—S—S—). The N-terminal (NH_2) portions of a light chain and the adjacent heavy chain form an antibody reaction (antigen-binding) site. These sites are contained within the F_{ab} region of the molecule. Such immunoglobulin molecules are divalent; that is, each molecule has two antigen-binding sites. The F_c portion of the molecule contains the greater constant portion of the H chains. (b) A comparison of immunoglobulins showing their heavy- and light-chain components. The antigen-binding sites and distinctive components of certain immunoglobulins also are indicated. The unique appearance of the secretory IgA immunoglobulin molecule is shown in the upper right corner.

phocytes. Additional details of IgD involvement in the immune response are presented later in this chapter.

IgE. IgE molecules comprise only about 0.004% of the total serum immunoglobulins. However, they are heavily involved in allergic reactions and are known to have a strong attraction for mast cells. Interactions between IgE molecules and mast cells with allergens (antigens involved in allergies) trigger the release of the various chemical substances responsible for the well-known signs and symptoms exhibited by allergic individuals. The features of allergies are described in greater detail in Chapter 20. IgE is believed to play a role in defending the body against pathogenic protozoa and worms (helminths). [See mast cell reactions in Chapter 20.]

GENERAL STRUCTURE

Thousands of individual antibody molecules, differing in their primary structure, circulate in serum. Although there is great diversity in the sequence of amino acids in the molecule, all normal immunoglobulins contain at least one basic unit, or monomer, of four polypeptide chains (Figure 17–15). Two of these are identical H (heavy) chains and two are identical L (light) chains. The formula $(L_2H_2)_n$ can be used to represent the general basic composition of immunoglobulins. The chains of an immunoglobulin are covalently linked to each other by disulfide bonds (S—S). H chains are held together by one or more bonds, whereas each L chain is bound to an H chain by means of the disulfide bond. Immunoglobulins composed of more than one basic monomeric unit normally contain a J chain (Table 17–3), which is involved with the binding together of certain components (Figure 17–15).

Two different types of light chains, designated kappa (κ) and lambda (λ), occur in immunoglobulins. A given antibody molecule contains either two κ or two λ chains (or multiples of two), but never one of each. Each class of immunoglobulin has a different type of H chain structure. These chains are designated by a Greek letter corresponding to the Roman capital letters used for the immunoglobulin classes. Thus, the H chain designation for IgA is α (alpha); for IgD, δ (delta); for IgE, ϵ (epsilon); for IgG, γ (gamma); and for IgM, μ (mu). Figure 17–15b shows a comparison of the five classes of immunoglobulins.

Each chain of a basic immunoglobulin molecule (Figure 17–15a) may be divided into the two portions called the variable, or *V region*, and the constant, or *C region*. This division of the molecule emphasizes two basic roles of an immunoglobulin, namely, binding to an antigen and starting a process for the elimination or destruction of the antigen. The V region, which is the immunoglobulin portion exhibiting the greater variation in structure, binds antigens, while the C region performs the elimination/destruction function.

DETAILED IMMUNOGLOBULIN STRUCTURE

The finer details of immunoglobulin structure are collectively based on the studies of normal and abnormal proteins. The use of enzymes and other chemical treatments to digest and separate the immunoglobulin molecules into smaller parts has revealed the existence of functional subregions. Such different subregions include the F_{ab} and F_c fragments. The F_{ab} fragments contain the antigen-binding sites and are composed of the chemically variable portions of both H and L chains (Figure 17–15). A portion of the heavy chain in F_{ab} released by enzyme treatments is designated F_d. The F_c (crystallizable) fragment contains the chemically constant (c) portion of the H chain. This region participates in various biological activities or reactions, including complement fixation and sensitization. An unusual structure in the segment of the H chain, known as the *hinge region*, joins the F_d and F_c fragments (Figure 17–15a). It is found only in IgG, IgA, and IgD molecules.

Immunoglobulin Classification. Variations in immunoglobulin structure exist and may be conveniently divided into three general categories: *isotypes, allotypes,* and *idiotypes.* Isotypes are antigenic determinants (characteristic means of identification or markers) shared by all immunoglobulins within each of the five classes. The specificity of any immunoglobulin class is associated with the H chain. Allotypes are markers that reflect genetically determined antigenic differences among immunoglobulins. Thus, these antigenic determinants are not found in all individuals of a species. Idiotypes differ from the other two immunoglobulin antigenic determinants and are usually directed against a specific antigen. These unique determinants are located on the V region at or near the antigen-binding site (Figure 17–15) and are considered to be involved in regulating immune responses. The immunoglobulin fragment bearing the idiotype is designated F_v.

Genetic Aspects of Immunoglobulin Expression

Many genes and associated specific molecules have important roles in the immune system. Only the immunoglobulin gene system will be considered here. The histocompatibility complex system, which contains genes involved with distinct immunologic functions such as disease susceptibility; tissue transplantation interreactions between T cells, B cells, and macrophages; and allergic and autoimmune diseases is described in Chapter 20.

Most of the available knowledge concerning the arrangement and structure of immunoglobulin genes has come from the application of recombinant DNA techniques comparing DNA from cells either committed or not committed to the formation of immunoglobulins. Such studies have greatly clarified the process by which an individual can generate approximately 1 to 100 million different antibody types. Humans, as well as other mammals, have an elaborate system that not only uses gene segments on various chromosomes but also takes advantage of a flexible mechanism for genetic exchanges within or between chromosomes to reach the maximum limits of antibody diversity. [See protein synthesis in Chapter 6.]

HEAVY CHAINS

The genetic locations (loci) involved with heavy (H) chain formation include several variable (V_H) segment genes, short diversity (D_H) segments, four joining (J_H) segments, and eight constant (C_H) region genes arranged one behind the other and corresponding to the specific immunoglobulin molecule class to be formed (Figure 17–16). By the time a B cell reaches its mature or differentiated state, at least two types of genetic rearrangements can occur to produce an intact gene for the variable region. Such rearrangements can involve the deletion or duplication of genes or the shuffling of genes within or between chromosomes (translocation).

LIGHT CHAINS

The general design of the human gene complex for the two L chains classes varies considerably. The kappa chain complex is estimated to contain several hundred variable (V_κ) genes, five joining (J_κ) genes, and one constant (C_κ) gene region. The lambda (λ) L chain appears to consist of two variable (V_λ) genes, several duplicate constant (C_λ) gene regions, and one J_λ gene associated with each constant gene.

Antibody Production

Antibodies for a particular immunogenic substance are not detectable in the serum of an individual until exposure to the immunogen has occurred. The extent of the antibody response is known to be affected by various factors, including (1) the nature of the immunizing material, (2) the dosage received, (3) the number and frequency of exposures, (4) the particular animal species, and (5) the individual involved.

The effects of immunogenic stimulation on the body can be best studied by observing the response produced with a single injection of immunogen. This reaction to the first injection of an immunogen is called the *primary response.* The original description of the primary response noted a *latent* or *lag period* during which no increase in circulating antibody was detected. Improved techniques have shown that, in some instances, antibody synthesis does follow introduction of the stimulus almost immediately. After a period ranging from a few hours to several days, the antibody titer (level) reaches a peak or plateau (Figure 17–17). Peak concentration

Figure 17–16 A generalized view of the genetic and molecular events leading to the formation of an immunoglobulin light chain. Rearrangement of gene segments and aspects of transcription, RNA processing, translation, and the secretion of the kappa (κ) light chain are shown. Refer to the text for descriptions of the gene segments and to Chapter 6 for an explanation of protein synthesis.

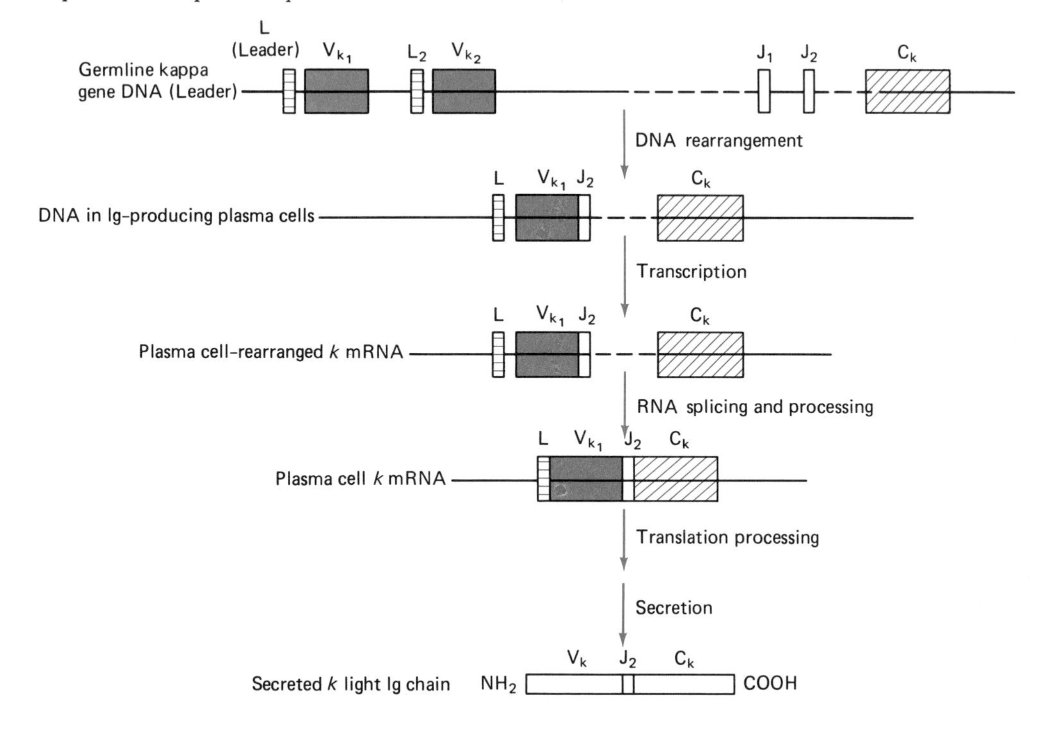

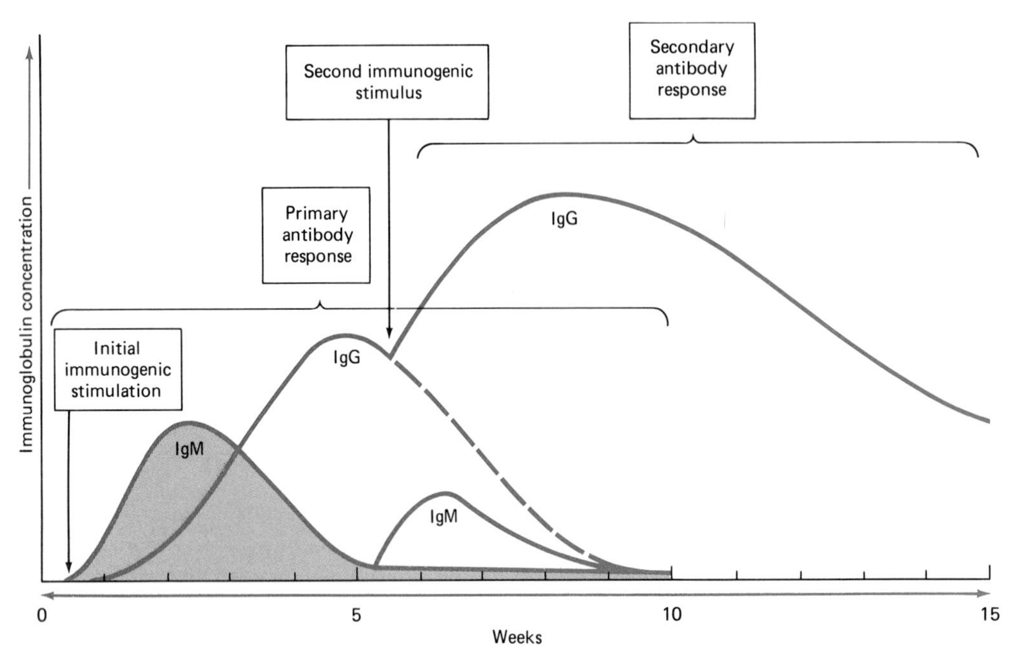

Figure 17–17 The primary and secondary antibody responses to an immunogenic (antigenic) stimulus.

is reached when the rates of antibody production and antibody breakdown are approximately the same. Such antibody levels may remain for several months or longer and then slowly begin to decline, as antibody breakdown exceeds production. The details depend, in part, upon the animal species and immunogenic material involved.

Some immunogens produce a sudden secondary rise in antibody titer when injected again some time after the first exposure. This effect is often called a specific **anamnestic response,** from the Greek term *anamnesis,* or "recall." The antibody titers associated with a specific anamnestic response generally are higher than those produced by the primary reaction, occur with little or no lag period, and remain for long periods.

Anamnestic reactions are produced by reimmunization with vaccines. The effectiveness of booster shots can be explained on this basis. **[See Chapter 18 for immunizations.]**

Antibody-associated *(in vivo)* Defense Mechanisms

There are a number of different kinds of antibodies involved in a host's resistance to infection and several mechanisms by which they protect against infection. These include neutralization of the poisonous effects of bacterial toxins; opsonization, involving IgG and IgM and complement, to promote the attachment of bacteria to phagocytes; the killing of bacteria by IgM in cooperation with complement (bacteriolysis); prevention of viral attachment to mammalian cell membranes (virus neutralization); and prevention of bacterial attachment to epithelial cells by IgA. **[Chapter 19 describes the diagnostic *in vitro* applications of antibodies.]**

The immune system specifically recognizes foreign substances with certain properties (antigens), forms immune products which include immunoglobulins, and—with the aid of specific lymphocytes known as memory cells (described later)—is ready to react quickly in the next encounter with the same type of foreign invader. We shall now examine the development of this system and the functions of its principal components, including the B and T lymphocytes.

Development of the Immunologic System

The Thymus Gland and Its Role

The thymus is located in the chest region, between the lungs and behind the sternum or breastbone. This gland normally consists of two lobes, each of which is divided into smaller regions called *lobules.* A lobule is made up of a *medulla* (center) composed

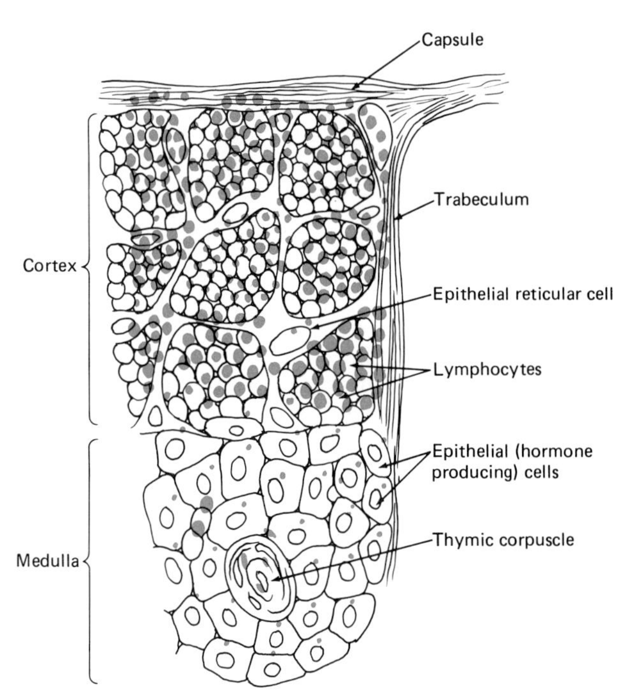

Figure 17–18 Thymus gland structure and organization. This portion shows the cortex and medulla and the location of lymphocytes and Hassall's (thymic) corpuscle.

of epithelial cells, lymphocytes, and Hassall's corpuscles and a surrounding *cortex* consisting primarily of lymphocytes (Figure 17–18).

At birth the thymus is the most fully developed of the peripheral lymphoid tissues with the exception of the bone marrow (the marrow is considered to be the central lymphoid tissue in mammals), and generally ranges in weight from 15 to 20 g. The structure increases in size until, at puberty, its weight approaches 40 g. Subsequently, the thymus gland atrophies and decreases in size, until at middle age it is relatively insignificant both physiologically and structurally.

The thymus gland is believed to be the central location for multiplication of *lymphocytes* (thymocytes) during embryonic development. It is from this region that certain stem (undifferentiated) cells acquire several new surface antigens (markers) that identify them as thymocytes. These cells then mature into T cells (thymus-derived lymphocytes) and spread peripherally to populate the lymph nodes, spleen, and related structures. Although the development of these cells is not completely understood, it is known that the epithelium (Figure 17–18) of the thymus synthesizes several substances that act as hormones and probably play a major role in T-cell regulation and differentiation. Such substances include thymosin, thymopoietin, and *facteur thymique sérique*.

The importance of the thymus is indicated by the obvious defects that occur on its removal from a newborn or very young individual. When the thymus is removed from an experimental animal such as a newborn mouse, no maturation of peripheral lymph nodes occurs. The number of lymphocytes in the peripheral blood decreases, and the antibody response is extremely poor. No antibodies are produced toward several antigenic substances. Such thymectomized animals accept foreign tissue grafts. These various defects can be corrected by grafting thymus tissue from a donor of the same inbred animal species to the thymectomized recipient.

Extracts of the thymus gland are under investigation in the treatment of different forms of cancer and for individuals suspected of having a congenital or acquired T-cell defect.

Immune Response of the Developing Individual

From conception to the time of birth, the human fetus develops in a highly protective environment. Beginning at approximately the third month of pregnancy, the mother begins to transfer large amounts of IgG antibodies to the fetus. The number of these maternal antibodies decreases gradually after birth (Figure 17–19). This method of passively equipping the infant with an immunologic "history" works well except in cases of Rh incompatability. Then passively transferred IgG can damage fetal blood cells, producing the condition known as the *Rh baby*. [See Microbiology Highlights, "Hemolytic Disease of the Newborn (The Rh Baby) and Its Control," in Chapter 19.]

It was long believed that the immune system of the newborn was immature and began to mature after birth. Currently available evidence suggests that this supposition is not true. It appears that the maturation of the human immune system begins *in utero* sometime during the second to third month of pregnancy, and involves the differentiation of cells that will carry out both specific and nonspecific immunologic activities. These cells appear to arise from a population of stem cells or hemocytoblasts located within the blood-forming tissues (Figure 17–20) of the developing embryo (bone marrow, fetal liver, etc.). Depending upon the type of environment the differentiated cells enter, they will develop into either hematopoietic or lymphopoietic tissue. The former will result in production of blood components, such as erythrocytes, granulocytes, monocytes, and platelets. The latter type of tissue can still undergo further differentiation and lead to the development and formation of two types of lymphocytes known as **B** and **T cells.** Both types of cells are involved in immune responses to antigens and originate as lymphocytic stem cells found in the human fetus and in the bone marrow after birth. The pathways to maturity followed by the descendants of stem cells are different (Figure 17–20).

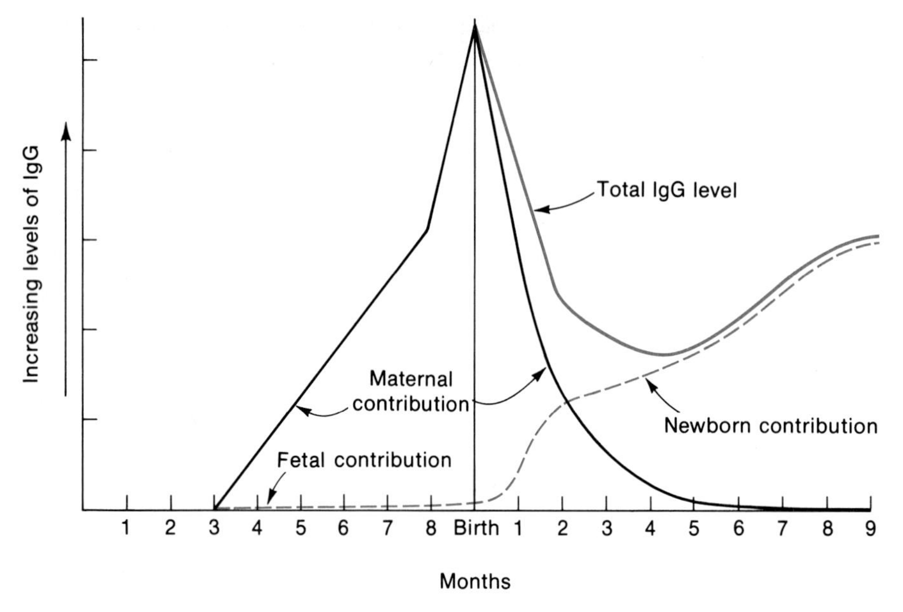

Figure 17–19
The development of IgG levels in the human fetus and newborn. The maternal immunoglobulin contribution is also shown. The immunoglobulins in the fetus and newborn are obtained from maternal sources. *[After M. R. Allensmith,* et al., J. Pediatr. ***75****:1231, 1969.]*

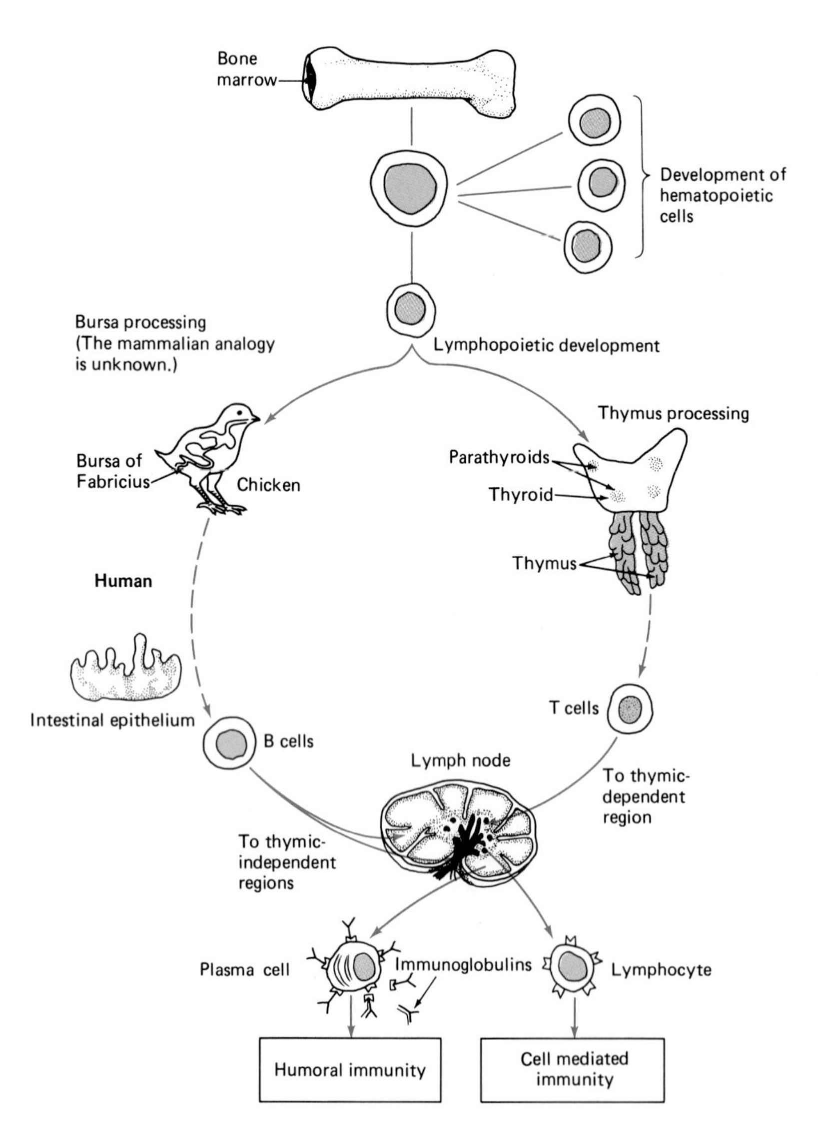

Figure 17–20 The development of the principal cells involved in the immune response. Parent cells are differentiated into hematopoietic (blood-forming) and immunocompetent T and B lymphocytes. Lymphopoietic development involves the processing, proliferation, and transformation of specific cell types to those of the lymphoblast and plasma cell series upon antigenic stimulation.

Both B and T cells circulate in the blood and lymph systems and are concentrated in the major lymphatic organs (Figure 17–6). These cells can live a long time. In humans they can persist for several years without dividing. However, upon exposure to an antigen, the lymphocytes enlarge greatly, divide rapidly, and secrete a number of protein substances that contribute to the elimination of foreign cells and materials.

B CELLS

The group of stem cells responsible for antibody-synthesizing cells, and *humoral immunity,* are believed to migrate to the bone marrow in mammals, where they differentiate into *bursal* or *B lymphocytes.* The name was given to these cells since the differentiation process is known with certainty to take place in birds in the *bursa of Fabricius,* a small pouch of lymphoid tissue attached to the animal's intestine. B cells leave the site of their differentiation to populate regional lymph nodes and the spleen. During its development, each B lymphocyte acquires a genetically determined surface immunoglobulin that is specific for a different antigen. Upon recognizing and reacting with its specific antigen, the B lymphocyte reproduces and undergoes further differentiation into specific antibody-secreting cells known as **plasma cells** (Figure 17–20).

B cells are distinguished from other lymphoid cells by the presence of easily detected immunoglobulins (mIg) on their membrane surfaces (Figure 17–21a). Such surface immunoglobulins function as specific antigen receptors and serve to mark or identify different B cell populations. The majority of B lymphocytes that populate the spleen and peripheral blood of mature animals exhibit surface IgM, IgD, and Ia (immune response) antigens. These surface molecules play critical roles in B cell activation. Specific antigens bind to and activate a particular B lymphocyte because its surface contains the immunoglobulin that the cell is capable of synthesizing. Both mIgM and mIgD bind antigen and may independently establish the basis for the recognition of a foreign (non-self) factor by the immune system. This immunoglobulin binding may result in the transmission of signals for B cell activation and further development into antibody-producing **plasma cells,** or for *tolerance induction,* in which no immune response leading to antibody production against the antigen (tolerogen) occurs. The plasma cells secrete antibodies that have the same antigen-binding properties as the antigen-binding (receptor) molecules on the surface of the parent B cell.

Once the antibodies are formed and released into the blood or lymph fluid, they attach to free antigen and mark it for elimination by other parts of the immune system. The general concept of how a B lymphocyte is selected by an antigen in order to give rise to specific antibody-producing cells is embodied in the clonal selection theory described later in this chapter.

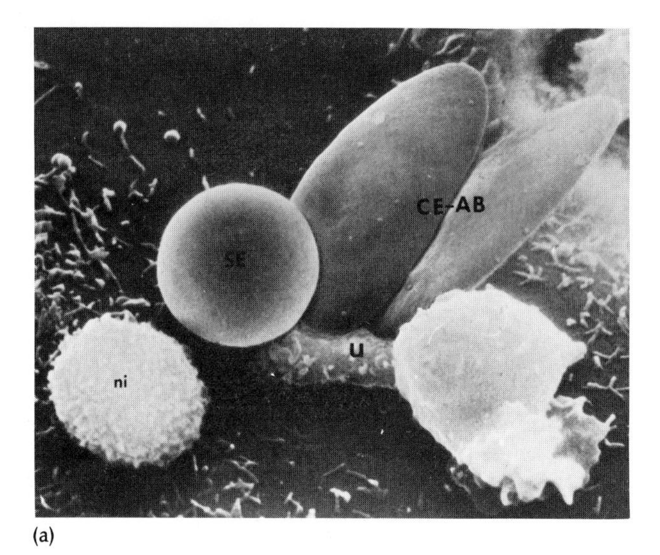

(a)

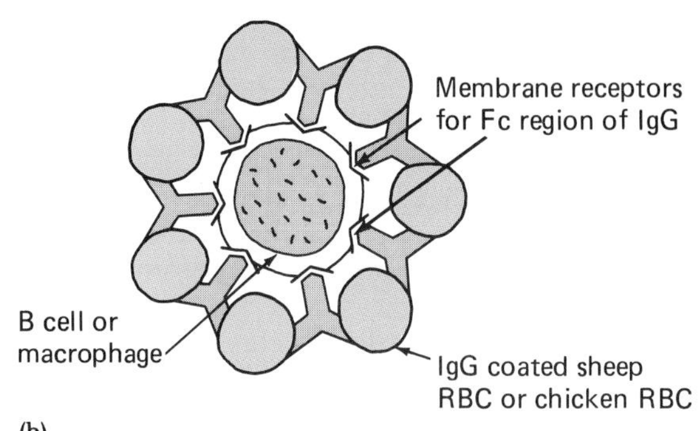

(b)

Figure 17–21 (a) The absorption of sheep red blood cells (SE) and chicken red cells and antibody complex (CE-AB) can be used to identify B and T cells. (b) The organization of EA rosette. The formation is called an *E (erythrocyte) A (antibody) rosette.* The B cell shown has an extending cell portion known as a *uropod (U),* (ni is a noninteracting lymphocyte). Original magnification 12,000×. *[B. Rentier, and W. C. Wallen,* Infect. Immun. ***30****:303–315, 1980.]*

The presence of surface immunoglobulins can be demonstrated *in vitro.* When a B cell bearing antibody contacts erythrocyte antigens, immune adherence occurs, resulting in an erythrocyte antibody (EA) rosette. Repetition of this exposure forms a rosette in which the B cell is surrounded by adhering erythrocytes (Figure 17–21b).

The bound surface immunoglobulin also serves as a means to attract and obtain help from specialized T (helper) cells for B cell activation. The immune response antigens function in situations of antigen activation involving both B and T lymphocytes. Additional details of antibody production are presented in a later section.

T CELLS

Under the influence of the thymus gland, a second group of lymphopoietic cells forms a population of lymphocytes responsible for **cell-mediated** immunity. These **thymic** or **T lymphocytes** participate in the tuberculin skin reaction, in the rejection or acceptance of certain tissue grafts, in the regulation of immunoglobulin production, and in an individual's defense against various microorganisms and cancers. T cells are involved in essentially all immune reactions, either as effector (directly attacking) cells or as regulators of both humoral and cellular responses.

CELLULAR MATURATION. The development and differentiation of T cells occurs in distinct stages. During such processing, a number of genetically determined T (thymus) cell-surface antigens or markers are acquired. These surface markers are associated with specific steps in the development and activation of mature T cell subpopulations or subsets (Figure 17–22). The T markers are designated by numbers 1 through 11 and can be detected by specific (monoclonal) antibodies.

Another important characteristic of developing T cells (thymocytes) is the ability to recognize the antigens (protein products) of closely liked genes in the region of DNA called the **major histocompatibility complex (MHC).** All mammalian species studied thus far have a single chromosome where the genes are located that encode the major tissue *(histo-)* antigens found on body cells. As a later section and Chapter 20 will describe, MHC antigens are recognized by different T cell types and are essential for immune recognition reactions.

The MHC gene products (antigens) are of three different kinds, *Class I MHC (host's self) protein,* a molecule present on the surface of all body nucleated cells, *Class II MHC proteins,* molecules found primarily on the surface of specialized cells known as antigen-presenting cells, and *Class III MHC proteins,* certain components of complement (C2, C4, and factor B). [See Chapter 20 for a description of complement activity.] Since most mature T cells cannot recognize free foreign (non-self) antigens circulating in the blood or lymph, they can respond to antigens on a cell surface only under particular conditions. Thus for an immune response involving T cells to occur, the foreign substance and the MHC protein must be displayed at the same time.

T-cell maturation also involves thymus hormones. Once differentiated, these cells leave the thymus gland and seed regional lymph nodes and the spleen (Figure 17–6). Each T cell has its own surface antigen receptor with which it can recognize foreign antigens and MHC proteins. The T cell receptor consists of two disulfide-linked polypeptides termed α (alpha) and β (beta).

When antigens enter the body, they will react only with the T cells bearing receptors specific for them (Figure 17–23). This type of reaction involves

Figure 17–22 T cell lines originate from stem cells. After differentiation in the thymus, thymocytes (developing T cells) acquire T antigens but are not yet immunocompetent. Later differentiation leads to various subpopulations (subsets) of immunocompetent peripheral T cells.

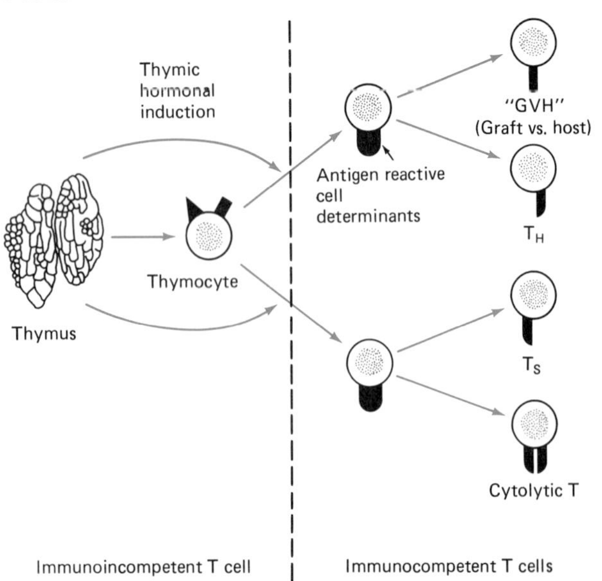

Figure 17–23 The activation of T cells in response to an infectious agent. A macrophage is activated and stimulated by the agent's antigens to secrete interleukin (IL-1). This protein, in turn, upon reaching resting T cells, stimulates them to secrete interleukin 2 (IL-2) and surface receptors for IL-2. The binding of IL-2 by T cells causes them to divide and assume active roles in the immune response.

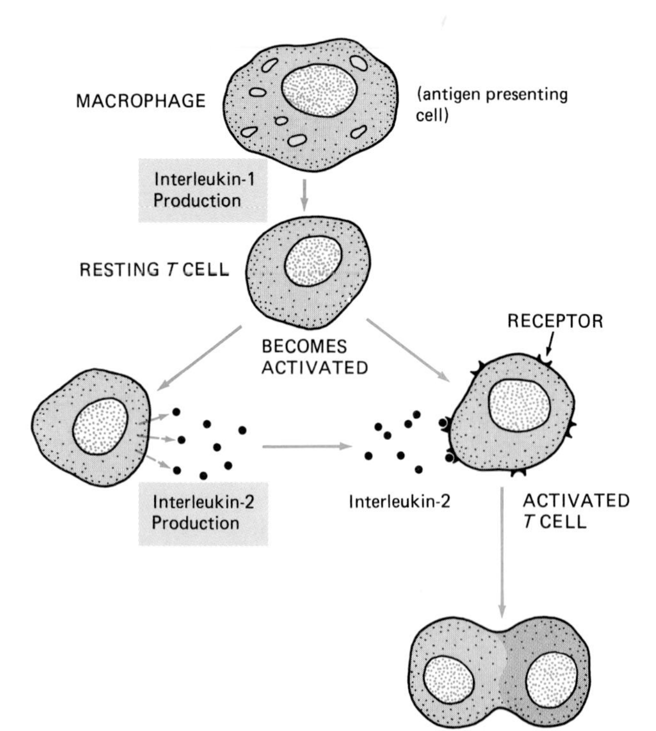

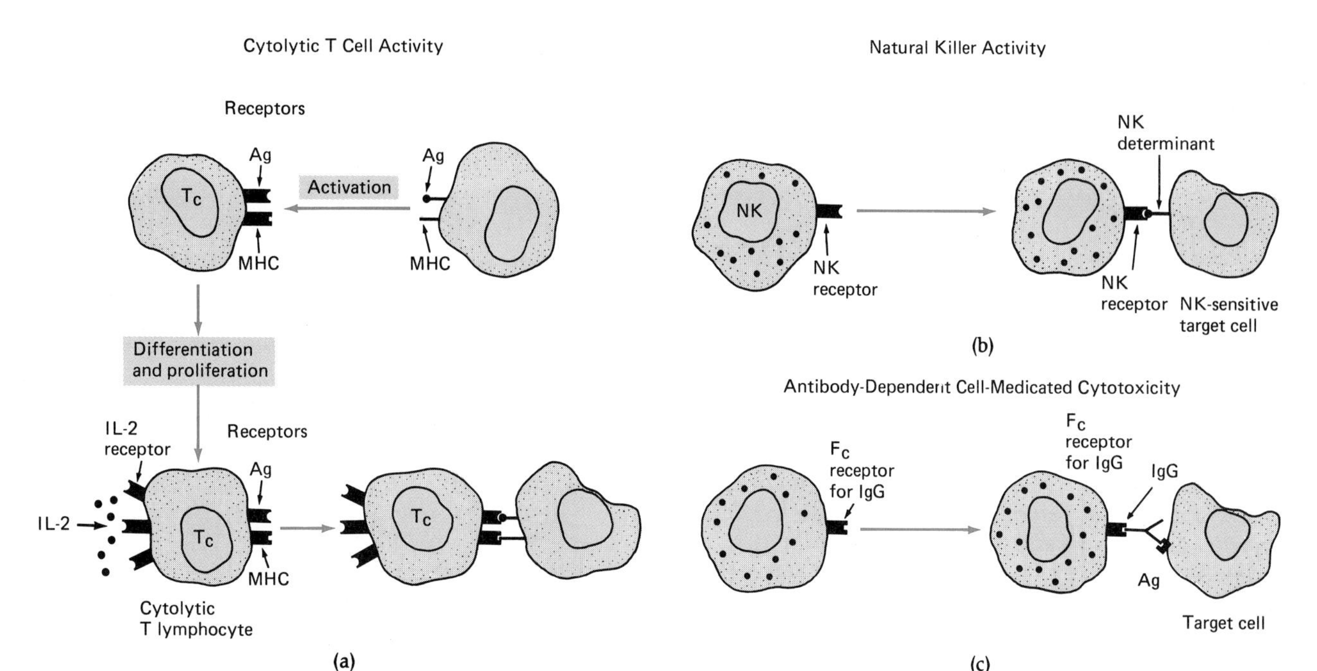

Figure 17–24 The cell-destructive (cytolytic) activity of T lymphocytes, natural killer (NK), and killer (K) cells. (a) Cytolytic T cells must be activated and differentiated before they can kill a target cell. (b) The activity exhibited by NK or K cells depends on the type of target. Certain microbially infected, or tumor, cells are among those susceptible to NK attack and destruction. (c) Cells coated with IgG molecules are vulnerable to NK attack. This type of situation is an antibody-dependent cell-mediated form of cytolytic activity (ADCC). Explanation of symbols: Tc, cytotoxic T cells; IL, interleukin; MHC, major histocompatibility complex; Ag, antigen; IgG, immunoglobulin G.

antigen-presenting cells (macrophages) that secrete interleukin 1, a type of lymphokine that activates T cells and causes them to divide, giving rise to individual cell lines or clones. Antigen-activated cells are referred to as being *sensitized.* Various types of lymphocytes and other cells are stimulated by the secretory products of lymphocytes known as **lymphokines.** These immunologically important substances affect the activation, reproduction, and chemical attraction (chemotaxis) of such cells as lymphocytes, macrophages, and mast cells. Lymphokines include macrophage chemotactic factor (MCF), which attracts macrophages to sites of inflammation, and cytotoxic factors that destroy microorganisms directly. **[See Chapter 20 for additional details of lymphokines.]**

T CELL SUBSETS

The functional T cell subsets, or subpopulations, are recognized and named according to their associated activities. Their designations and functions are as follows: (1) cytotoxic T (T_C) cells, which bind to their target cells bearing the appropriate MHC and antigenic determinants and disrupt such cells by direct cell-to-cell contact (Figure 17–24a); (2) natural killer (NK) cells, which have the ability to recognize and disrupt (lyse) a wide variety of target cells without having been exposed to relevant antigens; (3) T helper (T_H) cells, which work with B cells in immunoglobulin production and with other T cells in cytotoxic reactions; (4) T delayed-type hypersensitivity (T_{DTH}) cells, which bring about (mediate) inflammation and activate macrophages in delayed-type hypersensitivity reactions; and (5) T suppressor T_S cells, which act to regulate or suppress the activities of the other three types of T and B cells. The suppressive activities of T cells can be seen with newborn infants. Newborns essentially have adult levels of B cells and begin to form their own *antibody* immunoglobulins (Figure 17–19). However, their levels do not reach those of adults until they are nine to ten years of age. This situation occurs largely because newborns also have suppressor T cells that interfere with differentiation of B cells into plasma cells.

T cells contain surface antigens that can be detected by specific commercially available monoclonal antibody preparations. According to the nomenclature system of "cluster of differentiation" (CD) antigens, human helper and delayed hypersensitivity T cells are designated $CD4^+$ and $CD8^-$, while human suppressor T cells and cytotoxic T cells are indicated as $CD4^-$ and $CD8^+$. The expression of these CD4 and CD8 surface markers correlates with the specificity of T cells for either MHC class I or class II molecules and target cells.

T cells are also grouped, from a functional role

standpoint, into *effector* and *immunoregulator* categories. Effector cells, which include T_C and T_{DTH}, are sensitized and provide either cell-mediated immunity or account for delayed hypersensitivity reactions. Immunoregulator cells include T_H and T_S cells and control the activity of other immunocompetent B and T cells.

Cytotoxic T cells bind to target cells with appropriate surface antigens and the MHC class molecule (Figure 17–24). No antibody is needed for this type of attachment. The T cells involved are often $CD8^+$ and limited to reacting with MHC class I proteins (a phenomenon known as *MHC restriction*). After binding, swelling and disruption of the target cell follows, leaving the T_C cell intact and available to destroy more targets. The cytotoxic cell is the main type of T cell that defends the body by destroying infected, foreign, or cancerous cells. Much attention has been focused on this activity and the central role of T cells in the immune surveillance of the body, particularly in relation to cancer. Various studies have also pointed to the involvement of other effector cells including macrophages, monocytes, polymorphonuclear leukocytes, K (killer) cells, and NK cells (Table 17–4). Such cells and NK cells can also interact with and have considerable

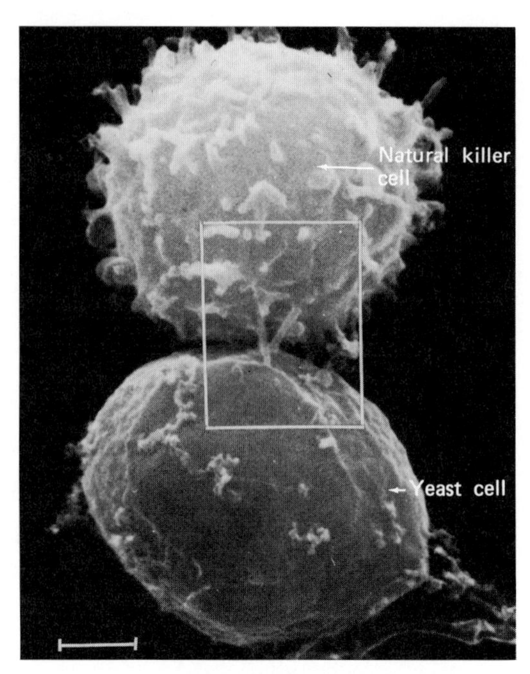

Figure 17–25 The attachment of a natural killer (effector) cell to the yeast (target) *Cryptococcus neoformans*. *[From N. Nabavi, and J. W. Murphy,* Infect. Immun. ***50**:50–57, 1985.]*

TABLE 17–4 Properties of Immunologic Effector Cells

	Cell Type			
Property	*Cytotoxic (T_C) Cells*	*Natural Killer (NK)*	*Monocyte or Macrophage*	*Polymorphonuclear Leukocyte*
Diameter (in μm)	9–12	12–15	16–20	12–18
Phagocytosis	No	No	Yes	Yes
Surface receptors for sheep erythrocytes	Yes	Yes (50%)	No	No
Surface receptors for IgG	Less than 10%	Yes	Yes	Yes
Spontaneous activity	No	Yes	Yes	Yes
Examples of activating factors	IL-1, IL-2, IFN, T cell helper factors[a]	IL-2, IFNs,[a] antibodies	MAF, IFNs,[a] a variety of foreign materials (bacterial endotoxins)	Contact and a variety of chemical substances
Cytotoxic reactivity against IgG antibody-coated targets (cells)	No	Yes	Yes	Yes
Possible mechanisms of cytotoxic effects	Protein-digesting enzymes, osmotic shock	Protein-digesting enzymes, cell toxins	Lysozyme, protein-digesting enzymes, reactive oxygen	Similar to macrophages
Examples of reactivity inhibitors	Specific and nonspecific T suppressor cells, IFN, PGE[b]	Nonspecific macrophages and other suppressor cells, PGE	Various enzyme inhibitors, PGE	Various enzyme inhibitors

[a] Lymphokine examples: IL-1 = interleukin-1; IL-2 = interleukin-2; MAF = macrophage activating factor; IFN = interferon. See Chapter 20 for additional details.
[b] PGE = prostaglandin E.

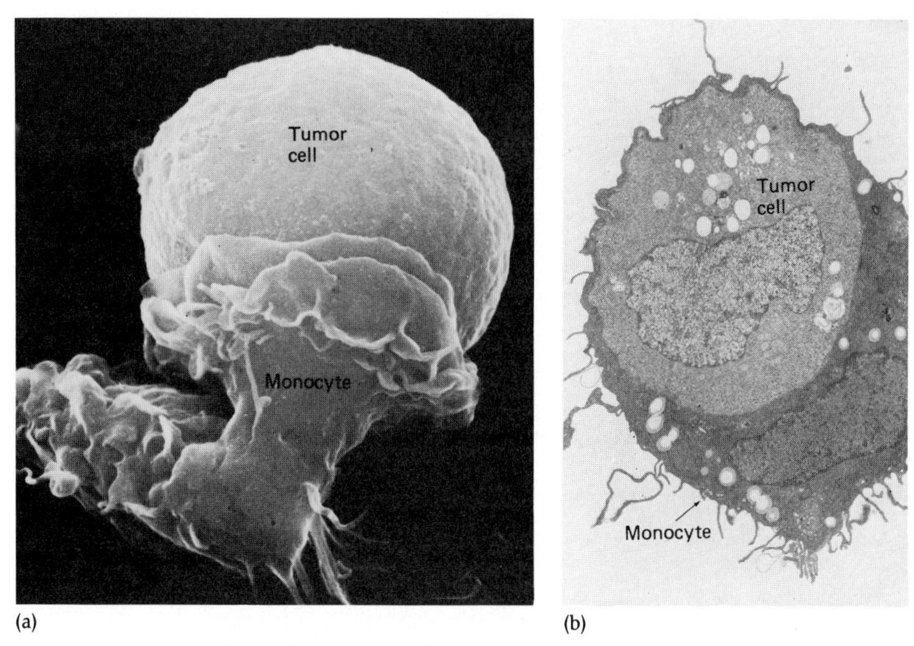

Figure 17–26 Antibody dependent cell-mediated cytoxicity (ADCC), mediated by human monocytes. (a) A monocyte establishing contact with an antibody-coated tumor cell. (b) An internal view of a monocyte that has completely engulfed an antibody-coated tumor cell. *[From T. Espevik, and J. Hammerstorm,* Acta Path. Microbiol. Immunol. Scand. *sect C.* ***91****:211–219, 1983.]*

cytotoxic activity against IgG-antibody coated tumor cells (Figure 17–26). Cells capable of participating in such antibody-dependent, cell-mediated cytotoxicity (ADCC) reactions possess F_C receptors for IgG (Figure 17–24c).

Natural killer (NK) cells (Figure 17–25) appear to be clearly different in several respects from other T_C cell types and effector cells (Table 17–4). In general, **NK** activity can be increased by exposure to various cells and substances including tumor cells, virus-infected cells, and **interferons (IFN).**

T helper cells bear the CD4 surface marker and respond to antigens that are associated with MHC class II proteins. In addition to cooperating with B cells to produce antibodies, these cells play several other vital roles in the host's immune response (Figure 17–27). Without their involvement exerted through the production of lymphokines (chemical products of lymphocytes) or through direct cellular contact, neither T cytotoxic or T suppressor cells could function. Furthermore, T helper cells produce interleukin-2 (a lymphokine), which stimulates natural killer cells, and gamma interferon, which stimulates macrophages in their role of engulfing viruses and presenting antigens.

The CD4 marker of these cells is of special interest and importance in the virus-causing disease, AIDS. The T helper cell is the major cell type attacked by the human immunodeficiency virus-III, and the CD4 surface marker appears to be the attachment site on the cell for the virus. **[See Chapter 27 for a description of AIDS.]**

Figure 17–27 The sphere of T4 cell importance. Reduction in the normal number of T4 cells seriously impairs the immune response and the level of host resistance.

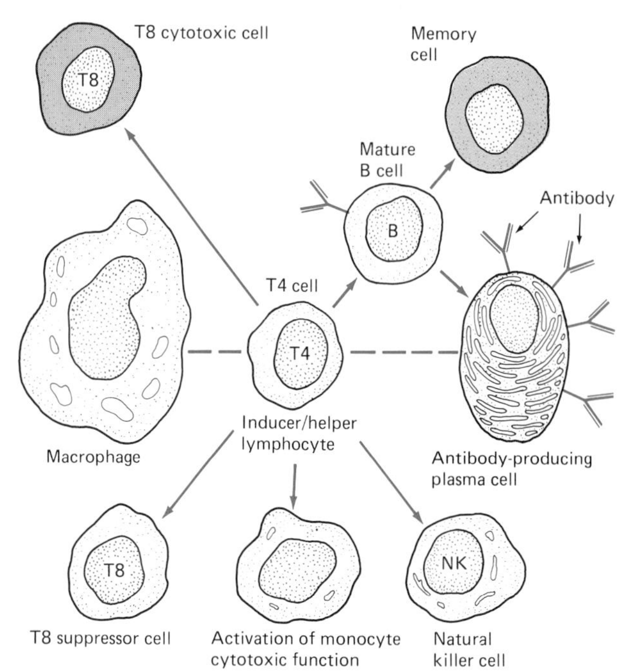

Microbiology Highlights

CELL SORTING

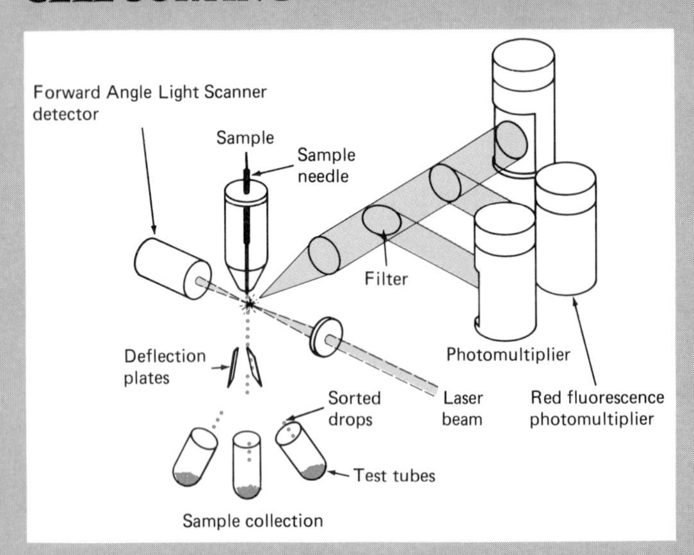

A typical flow cytometer.

Commercially available instruments used for the sorting or separation of cells and subcellular parts, such as chromosomes, have been available for several years. One sorter known as a fluorescent activated cell sorter (FACS) and others generally referred to as flow cytometers have numerous applications in immunology and protozoology. Immunological applications include identifying lymphocyte subsets, determining the phagocytic capacity of cells, and studying changes resulting from a response to infections, such as differences in the ratio of T helper to T suppressor cells, as in the case of AIDS. In protozoology, flow cytometry is used to separate and isolate pathogenic protozoa from tissue specimens.

In flow cytometry, stained cells flowing at high speeds in a solution are illuminated for a short time and produce optical signals detected by photosensors. Cells stained with a fluorescent dye emit a fluorescent signal when illuminated by the light beam of the instrument. The extent to which the light beam is scattered or dispersed determines the size, shape, and surface features of the cells. The fluorescent signals are analyzed by a computer, which can then control special plates that deflect and physically sort cells into collection tubes. The resulting sorted cell populations are very pure and thereby suitable for the previously listed applications. An argon-ion laser beam is the most commonly used source of illumination in flow cytometers.

Cells in the Immune Process

The B and T cells are the important responsive elements of an individual's immune system. While both cell types occupy different areas within the same lymphoid tissues (Figure 17–6), both are intimately associated with another form of white blood cell, the macrophage. Macrophages, which are also in lymphoid tissue, process and present antigens to T lymphocytes to enhance many immune responses (Figure 17–23). In addition, these cells synthesize and secrete chemical substances that regulate lymphocyte function. Other efficient antigen-presenting cells are found in the skin, blood, and lymph.

In the humoral form of immune response, T cells cooperate with B cells. While B cells are required for immunoglobulin production and may be stimulated by direct contact with strong immunogens, weaker

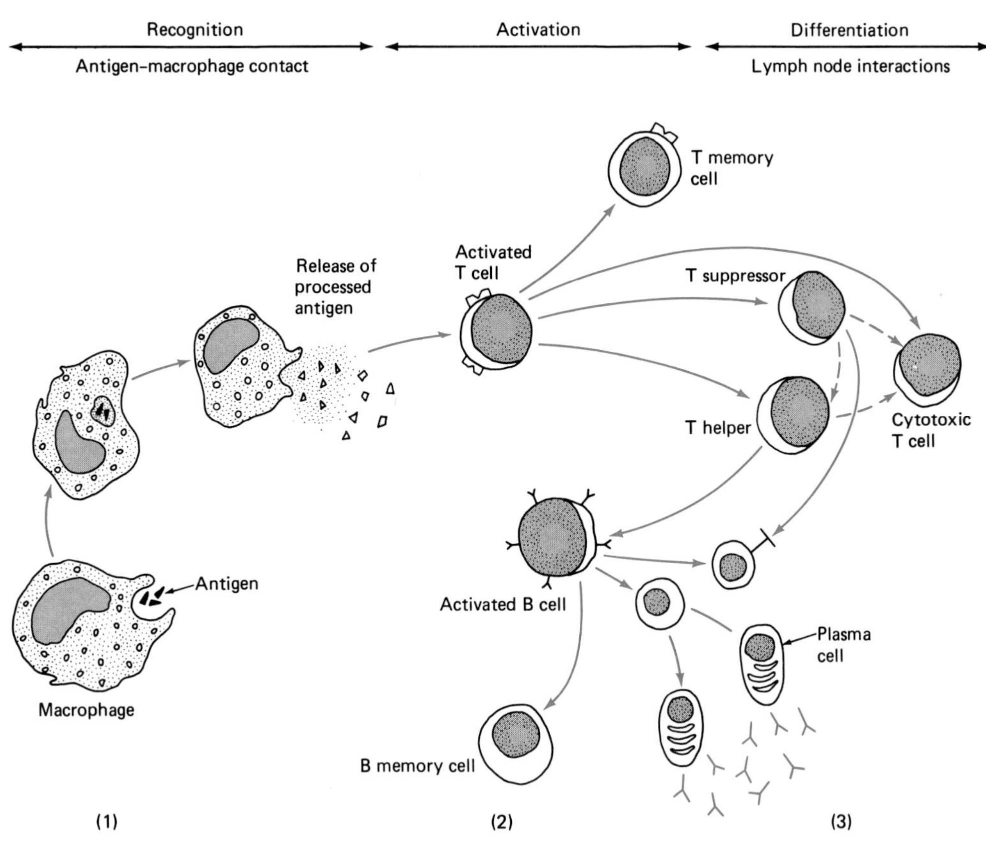

Figure 17–28 The steps in antibody production: recognition, activation, and differentiation. When an antigen enters the body, it comes into contact with a macrophage, is processed, and is then transported through lymphatic channels to a neighboring lymph node. There the cycle of regulatory interactions among macrophage, T cells and B cells in the immune response. (1) An antigen phagocytized and processed by macrophages is released along with cellular products. (2) Passing T cells with receptors for the processed antigen are activated and differentiate into helper (T_H), suppressor (T_s), or cytotoxic T cells. (3) B cells are stimulated to develop into mature antibody-producing plasma cells. The antibody molecules have receptor sites for the antigen that caused their production. Antigen T cells differentiate into memory cells and cells that control subsequent antigen recognition (B cell differentiation). The B cells also differentiate and proliferate into memory cells.

immunogens need the assistance of T cells to create an immune response by B cells. T helper cells, acting together with macrophages, may concentrate immunogens and present them to B cells, or T cells may release substances to activate B cells. Some of these activated cells enlarge, divide, and differentiate into a specific cell line (clone) of *plasma cells* (Figure 17–28). Plasma cells actively synthesize and secrete large quantities of one type of immunoglobulin for a limited time (four to five days) until cell death occurs. It should be noted that a B cell may switch from the production of one immunoglobulin class to another, but its specificity to bind a particular immunogen remains unchanged.

Activated B cells not undergoing differentiation into plasma cells remain as *memory cells* that can respond more rapidly to future exposures of the same immunogen. Once immunoglobulin production is under way, suppressor T cells are necessary to regulate the immune response. Their suppressor function prevents cells from differentiating into plasma cells.

The roles of macrophages, B and T cells, and plasma cells in antibody production are shown in Figure 17–28. The three basic events of the immune response are shown as they occur in sequence: *recognition,* the binding of immunogen to specific receptor sites; *activation,* the stimulation of a resting cell into an active cell; and *differentiation,* the production of the specific plasma or memory cells that partake in an immune response. The mechanism involved in this process supports the clonal selection hypothesis for antibody production discussed in the next section.

Microbiology Highlights

MECHANISM OF ANTIBODY FORMATION

Since the 1890s, much research has been aimed at finding the mechanism of antibody formation. As yet, no one hypothesis for antibody synthesis is universally accepted. The concepts developed to date fall in two general categories: *template hypotheses* and *selective hypotheses.*

Template Hypotheses

There are two types of template hypotheses, the *instructive* and the *directive.* The instructive concept suggests that the immunogen acts directly as a pattern to mold or form an antibody's active site. This modified portion of the antibody would have a structural image that is the reverse of the site on the immunogen that caused its formation.

The directive template mechanism assumes that protein synthesis (including antibody production) is under genetic control and that the specific pattern of the antibody is thereby dictated by the antibody-forming cell. The immunogen functions indirectly by causing the genetic mechanism to alter its antibody pattern, or template. This genetic modification is carried by all cells descending from the first one altered by the immunogen.

While the template hypotheses account for antibody specificity, they do not satisfactorily explain certain other known immunologic phenomena, including the primary and secondary responses to immunogens and the ability of the immune system to distinguish between self and not-self components. Both hypotheses have been discarded in favor of the selective concepts.

Selective Hypotheses

Selective hypotheses postulate that certain cells present in the body before the exposure to immunogens have the necessary information to synthesize a particular antibody. Such cells are capable of producing one antibody or a few of them. The number of cells necessary to meet the immunologic needs of the individual is in the thousands, because each cell or its descendants *(clones)* could produce, at most, a few types of antibody. According to the selection hypotheses, during embryonic development, normal (self) components of the body react with and destroy those cells that could produce antibodies against them. Only those cells that survive this period constitute the antibody-producing force of the adult. Immunogens here function by combining with such cells and stimulating their antibody-producing capability. The selective hypotheses appear to account better for various observed immunologic phenomena.

Conditions That Lower Host Resistance to Disease Agents

Preceding sections of this chapter have emphasized the importance of various factors and mechanisms to host resistance. Some consideration should be given to circumstances that can lower or impair resistance to infectious and other forms of disease agents. If we view the immune mechanisms normally available to a host collectively in the form of a protective umbrella to shield against harmful internal and external agents (Figure 17–29), it is quite obvious that defects in one or more protective factors could seriously impair host resistance. Disease-related defects contributing to impaired host resistance may be relatively subtle or quite severe and include (1) mononuclear phagocyte, neutrophil, or T-lymphocyte defects; (2) inability to produce immunoglobulins; (3) surgical removal of an organ such as the spleen; or (4) injury to mechanical defense barriers.

Individuals with certain forms of cancer, recipients of organ transplants, or patients receiving immunosuppressive drugs such as corticosteroids exhibit defects in T lymphocytes and mononuclear

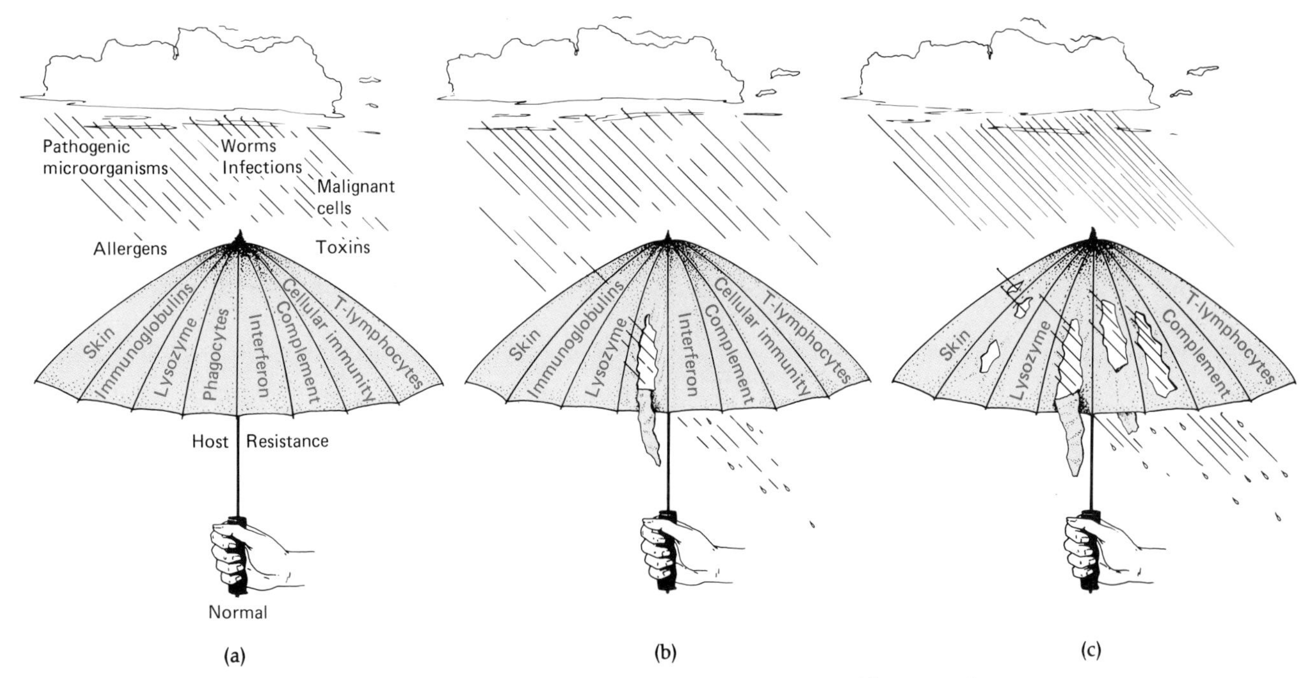

Figure 17–29 Collectively the various components of the immune system provide protection against both internal and external factors. (a) A high level of protection is evident in a host when exposure to pathogens, allergy-causing agents and even worms occurs. (b) With a reduction or deficiency in one immune factor such as normal phagocytic activity, repeated infections are quite likely to develop. (c) The more defects in the immune system or an underlying problem such as malnutrition the greater is the susceptibility to disease. *[After R. K. Chandra, "Malnutrition and Immunocompetence: An Overview," in Bellanti, J. A. (ed.),* Acute Diarrhea: Its Nutritional Consequences in Children. *New York: Raven Press, 1983.]*

phagocytes. A variety of microorganisms take advantage of such defects and cause disease. These disease agents include the bacteria *Streptococcus pneumoniae, Nocardia asteroides,* and *Mycobacterium avium,* the yeast *Cryptococcus neoformans,* the protozoon *Toxoplasma gondii,* and the herpesviruses.

Underlying diseases such as AIDS or acute leukemia and the use of cell-destructive drugs depress both the function and the number of neutrophils. Again, various microorganisms may take advantage of the lowered host resistance and cause disease. Such organisms include bacterial inhabitants of the individual's own gastrointestinal tract, as well as species of the yeast *Candida* and species of the molds *Aspergillus* and *Mucor.* **[See *Candida* infections in Chapter 24.]**

The inability to produce immunoglobulins in response to an antigenic challenge is characteristic of patients with AIDS and chronic forms of cancer. Major risks to such individuals are encapsulated bacteria including *S. pneumoniae, Haemophilus influenzae,* and *N. meningitidis,* as well as various viruses such as cytomegalovirus, Epstein-Barr virus, and hepatitis B virus. **[See viruses and cancer in Chapter 30.]**

Injury to the mechanical defense barriers against disease agents provides an opening for many types of microorganisms. Among the most commonly encountered agents are various gram-negative bacteria and the gram-positives *Staphylococcus aureus, S. epidermidis,* and *S. pneumoniae.*

A number of conditions and factors contributing to lowered host resistance are summarized in Table 17–5. In most cases of infection, more than one of these conditions is involved. The frequency of these conditions as direct or indirect causes varies with different segments of the population and with the nature of health care services provided. Developmental and genetic defects are also important.

An Introduction to the States of Immunity

Immunology is concerned with the specific responses of the immune system to cells and substances recognized by this system as being foreign. Many of the body systems—including those associ-

TABLE 17–5 Conditions or Factors That Lower Host Resistance to Infectious Disease Agents

Condition or Factor	*Selected Effects*
Acute radiation injury	Alteration of the cellular defenses of the host
Age	Decreased efficiency of antibody synthesis and cell-mediated immunity at extremes of age; decreased levels of certain complement components during first three months of gestation
Agranulocytosis	Reduction or absence of phagocytosis caused by neutrophils
Alcoholism	Nutritional deficiencies; possible depression of the inflammatory response to bacterial infection
Altered lysosomes	Extensive or limited inability of macrophages and neutrophils to destroy ingested microorganisms
Atmospheric pollutants	Depressed immunological function of PMNLs
Circulatory disturbances	Localized destruction of tissues; congestion; accumulation of fluid in tissue
Complement deficiencies and/or defects	Limited or extensive inability to inactivate and/or destroy certain infectious disease agents
Excessive or indiscriminate use of antibiotics	Elimination of natural flora that provide protection; overgrowth of resistant microbial forms; interference with digestive process and vitamin utilization
Immunological deficiency (includes AIDS)	Interference with immunoglobulin production and/or cell-mediated immunity
Immunosuppression	Impairment of cell-mediated immunity mechanisms
Mechanical obstruction of body drainage systems (urinary, tear, and respiratory mechanisms or systems)	Interference with the mobilization and functioning of phagocytic cells
Nutritional deficiencies	Interference with and/or changes in several immune mechanisms, including antibody production, phagocytic activity, and integrity of mucous membranes and skin
Traumatic injury	Direct access to body tissues for opportunists and pathogens; possible interference with immunity mechanisms; possible obstruction of body drainage systems

ated with the gastrointestinal tract and urinary organs, cell types such as macrophages, and B and T lymphocytes—are involved in the development of specific resistance. The next chapter describes the states of immunity (resistance) that can develop upon exposure to antigenic substances and the types of preparations used to provide protection against certain infectious disease agents or their products.

Summary

1. The immune system provides varying degrees of protection from or resistance to disease-causing microorganisms and cancer cells.
2. This system may impose barriers to invasion by microorganisms or selectively eliminate foreign invaders that do get into the body.
3. Immunological responses may be either nonspecific or specific.
4. *Specific mechanisms* provide protection against particular microorganisms and their products, while *nonspecific mechanisms* are used against any and all disease-causing agents. Factors contributing to nonspecific resistance include species or racial factors, mechanical and chemical barriers, phagocytosis, inflammation, and various chemical products of the body.

The Immune System

THE COMPONENTS OF NORMAL BLOOD AND THEIR ROLES IN HEALTH AND DISEASE

1. Depending on the circumstances, blood can play an important role in the transmission, production, diagnosis, cure, and prevention of many conditions caused by microbes.
2. In the body, blood consists of red blood cells, white blood cells (leukocytes), and platelets in a fluid called *plasma*.
3. When blood is removed from the body, the plasma can be separated from the cellular elements with the use of an anticoagulant. If the blood clots, the clear fluid formed is called *serum*.

4. *Erythrocytes,* or red blood cells, are nonnucleated; formed in the bone marrow; contain major, minor, and Rh blood factors; and appear as biconcave disks.
5. *Leukocytes,* or white cells, are nucleated and can be classified into two groups: *granulocytes,* which contain distinct cytoplasmic granules, and *agranulocytes,* which lack such granules.
6. Based on staining reactions, three types of granulocytes are found in blood smears: eosinophils (red granules); basophils (blue granules); and neutrophils or polymorphonuclear leukocytes, or PMNLs (orange granules).
7. Agranulocytes consist of two general types, the *lymphocyte,* with a rounded nucleus, and the larger *monocyte,* with a kidney-shaped nucleus.
8. Disease states can cause changes in the proportions of different types of leukocytes.

THE LYMPHATIC SYSTEM

1. The lymphatic system consists of lymphatic vessels, lymph fluid, lymph nodes, and lymphocytes.
2. The system participates in the transport of fatty acids, proteins, and white blood cells and in the removal of foreign cells and their products, white blood cells, and tissue debris that accumulate at sites of infection and injury.
3. The mononuclear phagocytic or reticuloendothelial system consists of a variety of cells that ingest and digest foreign and host substances.

Nonspecific Resistance

SPECIES RESISTANCE

1. Species or racial resistance to disease agents is determined by the interplay of many factors, including physiologic and anatomic properties of a particular animal species. This type of resistance is inheritable.
2. Changes in body temperature, diet, and stress can affect this kind of nonsusceptibility.

MECHANICAL, MICROBIAL, AND CHEMICAL BARRIERS: THE BODY'S FIRST LINES OF DEFENSE

1. The effectiveness of the body's barriers to potential disease-causing agents depends on the general health of the host.
2. Mechanical and chemical barriers include unbroken skin, mucous membranes of the respiratory and genitourinary systems, nasal hairs, coughing and sneezing reflexes, tears and their associated washing action, eyelashes, and the various secretions and microbial contents of the different portions of the gastrointestinal system.
3. The effectiveness of these barriers against disease can be reduced by conditions such as alcoholism, poor nutrition, the debilitating effects of aging, fatigue, and prolonged exposure to extreme temperatures and to immunosuppressive therapy.

INDIGENOUS MICROBIOTA OR NORMAL FLORA

1. The normal flora of the human include all microscopic forms of life normally found in or on the body and are referred to as microbiota or normal flora.
2. Certain normal inhabitants may, under certain conditions, cause disease or pose other problems. Pathogens that take advantage of a host's weakened defenses to cause disease are known as *opportunists.*
3. The major habitats for indigenous microorganisms include the skin, mucous membranes, portions of the upper respiratory tract, mouth, lower intestine, and the external and internal parts of the reproductive system. Different anatomic regions provide characteristic environments each favoring a different range of microorganisms.
4. Development of the indigenous microbiota begins with the normal birth process and changes with subsequent exposure to other environments throughout the lifetime of the individual.
5. Benefits provided by indigenous microorganisms include inhibiting foreign microorganisms, stimulating the production of antibodies and protective proteins, and synthesizing vitamins.

Antimicrobial Substances

1. Chemical substances capable of *in vitro* antimicrobial activity have been found in various animal fluids and tissue. These substances include complement, interleukins, interferons, lactoferrin, lysozyme, lymphokines, spermine, and transferrin.
2. One of these substances, *complement* (C), is a complex group of proteins. It is well known for its ability to react with a variety of antigen-antibody combinations to produce important physiological reactions, including the destruction of various tissue cells, the destruction of bacterial cells, and the enhancement of phagocytosis.
3. *Interferons* are proteins normally produced in response to certain virus infections, which have great potential for the treatment of virus infections. They interfere with the formation of viral proteins.

Other Lines of Defense in Nonspecific Resistance

PHAGOCYTOSIS

1. *Phagocytosis* is one of the most important defenses against invading matter. It is the ingestion and subsequent digestion of foreign matter by circulating granulocytes, monocytes, and fixed macrophages.
2. Some disease agents can escape the ingestion and intracellular destruction of phagocytosis.

INFLAMMATION

1. Inflammation is another defense mechanism of the body. It can be produced by infectious disease agents and by irritants such as chemicals, heat, and mechanical injury.
2. The characteristic or cardinal signs of inflammation are heat, pain, redness, swelling, and loss of normal function.
3. Pus formation may also be associated with inflammation. *Pus* is fluid formed by the remains of damaged tissue cells and dead phagocytes and microorganisms.

FEVER

1. Fever is an elevation of body temperature above normal.

2. Fever is a frequently treated symptom of many disease states; however, temperatures of 38.5° to 39°C are known to speed the destruction of disease agents by increasing immunoglobulin production and phagocytic activity.

Specific Immune Responses

1. Individuals who successfully recover from an infectious disease generally acquire some degree of resistance toward the disease agent. This resistance results from the ability of the body's immune system to recognize disease agents or their products as foreign.
2. Molecules that the body's immune system recognizes as foreign are called *antigens.*
3. One defense response of the body to antigens is the production of antibodies or immunoglobulins.

Antigens and Immunogens

1. *Antigens* are molecules that react with antibodies but do not necessarily cause their production.
2. *Immunogens* are substances that stimulate the formation of antibodies and can combine with them. Immunogenic substances play important roles in the prevention of disease and the preparation of vaccines.

Properties of Immunogens

1. Immunogenic substances include most free proteins, combinations of proteins and other organic substances, and certain polysaccharides.
2. Cellular components of bacteria and certain synthetic substances can also stimulate an immune response.
3. Certain drugs and molecules that by themselves cannot stimulate antibody production can become immunogenic by chemically bonding to larger protein molecules (carriers).
4. In general, the greater the incompatibility between the foreign substance and the recipient's tissues, the greater the immune response.

Factors That Determine Immunogenicity and Immune Responses

1. Immunogens generally have molecular weights greater than 5,000 and large molecular surfaces to accommodate many specific groups of atoms called *antigenic determinant sites,* or epitopes.
2. The number of antigenic determinant sites on the surface of an immunogen is known as the *valence.*
3. Other factors affect the immune response, including the species of animal involved, the animal's degree of immunological maturity, the route of inoculation, and the use of substances called *adjuvants,* which prolong and intensify the antibody-provoking stimulus.

Classes of Antigens

1. *Haptens* are single antigenic determinants that by themselves do not stimulate antibody production but can combine with antibodies that have already been formed.
2. Generally, antibodies are not produced in the body against its own substances or cells. The body's materials are recognized as *self,* in contrast to foreign matter, or *not-self.*
3. In exceptional cases, antibodies *(autoantibodies)* are produced against body components *(autoantigens).* This type of condition is known as *autoimmunization* or *autoallergy.*
4. *Isoantigens* are red-blood-cell immunogens that differ among individuals of the same species. The clumping reaction *(agglutination)* that is associated with these antigens serves as the basis of blood typing.
5. Antigens causing a state of immunological unresponsiveness are called *tolerogens.*
6. *Heterophile antigens* are substances present in the cells and tissues of some species that stimulate the production of antibodies effective against material from unrelated species.

Antibodies

Antibodies belong to a group of proteins known as *globulins.*

Electrophoresis and Immunoelectrophoresis (IEP)

1. The separation of charged molecules in an electrical field is the basis of electrophoresis.
2. IEP is used to identify antibodies or antigens in a variety of body fluids and other substances.
3. The technique combines the elements of electrophoresis and double gel diffusion.

Immunoglobulins: Activities and Occurrence

1. Known immunoglobulins are grouped into the five classes IgG, IgA, IgM, IgD, and IgE, based on differences in their immunogenicity to other animals and their physicochemical properties.
2. The most plentiful of these protein substances in humans is IgG; IgA and IgM are next in order of abundance.
3. All normal immunoglobulins share at least one basic molecular arrangement, which consists of two identical H (heavy) chains and two identical L (light) chains covalently linked to one another by disulfide bonds.
4. Subregions of an immunoglobulin include the F_{ab} fragments, which contain the antigen-binding sites, and the F_c fragment, which contains the chemically constant portion of the H chain.
5. Variations in immunoglobulin structure are divided into the three general categories: *isotype, allotype,* and *idiotype.*

Genetic Aspects of Immunoglobulin Expression

1. The human and other mammals have an elaborate genetically associated system for immunoglobulin production.
2. The immunoglobulin-generating process uses specific gene segments and takes advantage of a flexible mechanism of genetic exchange, deletion, or duplication to produce the maximum number of different antibody molecules.

Antibody Production

1. The extent of an antibody response is affected by several factors, including (1) the nature of the immunizing material, (2) the dosage received, (3) the number and frequency of exposures, (4) the particular animal species, and (5) the individual involved.

2. The reaction to the first injection of an immunogen is called the *primary response.* In time, the antibody level, or titer, reaches a plateau that remains for several months or longer and then slowly declines as antibody breakdown exceeds production.
3. A second injection of some immunogens will produce a secondary, or *anamnestic,* response that produces higher, longer-lasting antibody levels than those associated with the primary reaction.

Antibody-associated* (in vivo) *Defense Mechanisms

Antibodies contribute to a host's immunity by several mechanisms including neutralization of bacterial toxins, opsonization of bacteria, bacteriolysis, and prevention of viral and bacterial attachment to cell membranes.

Development of the Immunologic System

THE THYMUS GLAND AND ITS ROLE

1. The thymus gland, located in the chest region, is the most fully developed of the peripheral lymphoid tissues at birth, with the exception of the bone marrow.
2. After puberty, the gland begins to atrophy until at middle age it is relatively insignificant both physiologically and structurally.

IMMUNE RESPONSE OF THE DEVELOPING INDIVIDUAL

1. The maturation of the human immune system begins *in utero* sometime during the second to third month of pregnancy, with the differentiation of stem cells located within the embryonic blood-forming tissue.
2. Depending on environmental conditions, differentiated cells may develop into either lymphopoietic or hematopoietic tissue.
3. Lymphopoietic tissue forms either bursal or B lymphocytes associated with antibody-synthesizing cells (plasma cells and memory cells) and *humoral immunity,* or thymic or T lymphocytes associated with *cell-mediated immunity.*
4. B cells are distinguished from other lymphoid cells by the presence of easily detected surface-associated immunoglobulins (IgM, IgD, and Ia). These immunoglobulins bind antigens, resulting in B-cell activation or tolerance induction. Each specific antigen activates only those B cells capable of secreting antibody to it.
5. T cells are derived from the thymus; B cells are generated at sites other than the thymus.
6. T cells are differentiated from B cells by functions and by possessing several specific surface antigens *(markers).*
7. Five functional T cell subpopulations are recognized: *T helper cells, cytotoxic T cells, natural killer cells, T delayed hypersensitivity cells,* and *T suppressor cells.*
8. T cells are involved in essentially all immune reactions, either as effector cells or as regulators of both cellular and humoral responses. T cells, especially NK cells, have antitumor cell activity.

CELLS IN THE IMMUNE PROCESS

1. T cells, B cells, plasma cells, memory cells, and macrophages may act together in antibody production.
2. Macrophages are of central importance in the immune response of the host. They present antigens to T and B lymphocytes to initiate the immune response. The three basic stages of the immune response are *recognition* (the cellular binding of immunogen), *activation* (stimulation of a resting cell into an active one), and *differentiation* (production of the different types of cells required for an immune response).

Conditions That Lower Host Resistance to Disease Agents

1. Disease-related defects contributing to impaired host resistance include mononuclear phagocyte, neutrophil or T-lymphocyte defects; inability to produce immunoglobulins; surgical removal of a body organ; and injury to mechanical defense barriers.
2. Radiation injury, aging, alcoholism, atmospheric pollutants, circulatory disturbances, and complement deficiencies may lower host resistance to infectious disease agents.

Questions for Review

1. Distinguish between nonspecific and specific immunological response.

2. What are three functions served by immunological responses?

3. What parts or regions of the human body act as mechanical barriers to microorganisms? Explain how each functions in this capacity.

4. Compare the composition of whole blood and plasma; of plasma and serum.

5. Differentiate among the various types of formed elements in blood with regard to their morphological features and functions.

6. What is a differential count? What is its significance?

7. What is species resistance?

8. a. What significant roles do the indigenous microbiota (normal body flora) play in the control of infectious disease agents?
b. Are pathogens encountered as normal flora? If so, list several representative microorganisms in this category.
c. Are all types of microorganisms found as normal flora? Explain.

9. Explain the roles played by the following cells and processes in maintaining the defense mechanisms of the body:

a. B lymphocyte	i. properdin
b. interferons	j. lysozyme
c. monocyte	k. lactoferrin
d. phagocytosis	l. T lymphocyte
e. eosinophil	m. myeloperoxidase
f. neutrophil	n. wandering macrophages
g. complement	o. mast cell
h. inflammation	

10. Describe the stages of phagocytosis.

11. What is phagocytic dysfunction? Describe one type of phagocytic dysfunction.

12. Define or explain the following terms:

a. immunoglobulins	h. thymus gland
b. adjuvant	i. B and T cells
c. autoantigens	j. immunogen
d. antigen	k. allotype
e. isohemagglutination	l. thymic hormones
f. immunopoietic	m. bursa of Fabricius
g. immunoelectrophoresis	n. tolerogen

13. Which components of a bacterial cell are antigenic?

14. Discuss haptens.

15. What are heterophile antigens?

16. a. What are the classes of immunoglobulins?
b. How are they differentiated from one another?
c. Draw a representative figure of an immunoglobulin and indicate the region involved in biological activities, such as skin sensitization.

17. Describe the responses involved in antibody production. How do they differ?

18. Briefly describe one hypothesis for antibody production.

19. What are the sources of normal immunoglobulins?

20. List and describe the major categories (subsets) of T cells.

21. Figure 17–30 shows the development of the immune response in a newborn.
a. Identify the immunoglobulins indicated by the respective curves.
b. What is the response at the point of the arrow?
c. Indicate the specific cells and respective functions associated with the response specified in *b*.

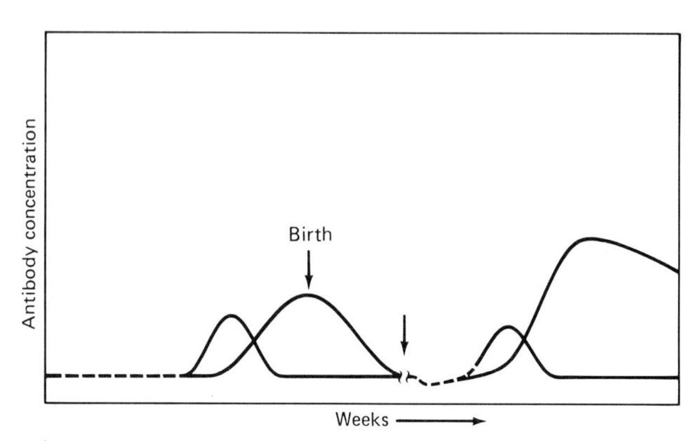

Figure 17–30

22. List and describe at least six conditions that lower host resistance to pathogens.

23. List and explain four disease-related defects contributing to impaired host resistance.

Suggested Readings

BELLANTI, J. A., *Immunology III.* Philadelphia: W. B. Saunders Company, 1985. *A highly comprehensive and detailed coverage of immunology.*

BRADFIELD, J. W. B., "Liver Sinusoidal Cells," *Journal of Pathology* **142:**5–6, 1984. *A short article presenting the characteristic properties of hepatic sinus-lining cells: fat-storing cells, endothelial cells, and Küpffer cells.*

BURNET, F. M. (ed.), *Reading from Scientific American: Immunology.* San Francisco: W. H. Freeman, 1976. *A collection of articles dealing with such topics as the immune response, cells associated with immunological reactions and immune deficiency, transplantation immunity and infectious disease, and autoimmunity. The articles are written on a general level.*

CAPRA, J. D., and A. B. EDMUNDSON, "The Antibody Combining Site," *Scientific American* **236:**50, 1977. *An article describing how an antigen combines with an antibody molecule at a site that fits like a lock and key. The nature of this site is discussed in great detail.*

EMERY, T., "Iron Metabolism in Human and Plants," *American Scientist* **70:**626–632, 1982. *This article describes how depriving microorganisms of the iron in animals and plants may act as a biological defense mechanism.*

GABIG, I. G., and B. M. BABIOR, "The Killing of Pathogens by Phagocytes," *Annual Review of Medicine* **32:**313, 1981. *A thorough consideration of the phagocytic mechanisms involved in the destruction of pathogens. Inherited abnormalities affecting microbicidal systems are also discussed.*

HERBERMAN, R. B., and J. R. ORTALDO, "Natural Killer Cells: Their Role in Defenses Against Disease," *Science* **214:**24–29, 1981. *A detailed article describing this newly discovered subpopulation of lymphoid cells. Comparisons are also made with macrophages, T cells, and polymorphonuclear leukocytes.*

HO, M., "Interferon for the Treatment of Infections," *Annual Review of Medicine* **38:**51–59, 1987. *An up-to-date report of both natural and recombinant DNA interferon effectiveness in the treatment of selected virus infections.*

JARET, P., "The Wars Within," *National Geographic,* June 1986. *A well-written and appropriately illustrated article describing the functions and the activities of the body's immune system.*

MANGAN, K. F., "Immune Disregulation of Hematopoiesis," *Annual Review of Medicine* **38:**61–70, 1987. *A report of experiments showing the potential role of various cells in the control of blood cell production.*

MILSTEIN, C., "Monoclonal Antibodies," *Scientific American* **243:**66–74, 1980. *A clear description of the processes involved with monoclonal antibody production.*

PAULING, L., *Vitamin C and the Common Cold.* San Francisco: W. H. Freeman, 1970. *A controversial study of the relationship between the common cold, general balanced health, and the use of vitamin C.*

THORBECKE, G. J., and G. A. LESLIE (eds.), *Immunoglobulin Structure and Function.* New York: New York Academy of Sciences, 1982. *This collection of papers presented at the New York Academy of Sciences focuses on the genetic aspects of complex structure and various roles of IgD.*

18

States of Immunity and the Control of Infectious Diseases

"An ounce of prevention is worth a pound of cure." — *Anonymous*

After reading this chapter, you should be able to:

1. Describe the general methods of preparing immunization material.
2. Distinguish between a toxoid and a toxin.
3. Compare the advantages and disadvantages of live versus inactivated microbial vaccines.
4. Outline the safety precautions and measures observed in preparation of the vaccine.
5. Explain the role of immunization in the control of disease.
6. Describe the general methods used for administering immunizing materials.
7. Differentiate between active and passive states of immunity.
8. List the different states of immunity and explain how each can be achieved.
9. Discuss the anamnestic response, emphasizing the differences between primary and secondary responses to immunogen.
10. Give the recommended schedules for the active immunization of infants, children, and adults.
11. List the preparations currently in use for both active and passive immunization.
12. List and discuss the types of complications associated with immunization.

How can the body acquire protection against infectious diseases? Chapter 18 explains the different states of resistance (immunity). It also discusses various immunizing preparations used to provide such protection and describes methods of vaccine preparation, including the develoment of synthetic vaccines by recombinant DNA technology. Methods of vaccine administration and some possible complications associated with immunization are also covered, along with recommended immunization programs for children, adults, and international travelers.

It has been recognized for centuries that individuals who recovered from certain diseases were protected from recurrences. The generally successful but hazardous introduction of small quantities of fluid from the blisters of smallpox victims into the skin of uninfected individuals was a human effort to imitate this natural phenomenon. Edward Jenner's approach to vaccination with vesicle fluid containing cowpox virus to protect against smallpox was the beginning of modern immunization (Figure 18–1). It was also an important step in the eventual eradication of this viral disease by the World Health Organization (WHO). Subsequent identification of the causative agents of common illnesses stimulated the development of other immunizing preparations. Thus, today it is quite clear that immunization with live or inactivated microorganisms and their toxic products is one of the most promising approaches to preventing and controlling microbial infections. Vaccines have been successful in reducing or eliminating the disease states of diphtheria, rubella or German measles, poliomyelitis, smallpox, and yellow fever from many regions of the world. Newer methods of vaccine production, incorporating recombinant deoxyribonucleic acid (DNA) techniques, are expected to provide more specific, more effective, and less expensive preparations. **[See recombinant DNA technology in Chapter 15.]**

Figure 18–1 The original vaccination procedure conducted by Edward Jenner (1749–1823). It was performed by removing some fluid from a lesion of cowpox on the hand of a dairymaid and injecting it into the arm of a small boy. The successful outcome of Jenner's vaccination against smallpox established a firm basis for the value of artificial immunization. *[Courtesy Fisher Scientific Co.]*

Immune Responses

Active immunization may not only result in the production of immunoglobulins directed against a disease agent or its toxic product but may also trigger cellular responses controlled by lymphocytes and macrophages. Both protective and nonprotective immunoglobulins are formed. The protective ones include *antitoxins,* which inactivate soluble toxic bacterial products; *opsonins,* which assist in the phagocytosis and intracellular digestion of bacteria; *lysins,* which interact with complement components to damage bacterial membranes; and *neutralizing antibodies,* which interfere with and prevent the replication of certain viruses.

Providing short-term protection against certain infections or changing their outcome may be accomplished through *passive immunization,* the administration of preformed immunoglobulins or immunoreactive cells. Although passive immunization is temporary, it is especially useful in the treatment and possible protection of virus-exposed individuals, some of whom are suffering from leukemia, other forms of cancer, or immunodeficiencies.

States of Immunity

The main function of the immune system is to provide protection, especially against the agents of infectious diseases. Special terms are used to indicate an individual's immune state and the type of existing protection.

The resistance to disease displayed by individuals varies considerably because it is greatly affected by many genetically determined (innate) or acquired factors (Figure 18–2). Generally speaking, two major categories are recognized, **innate** or **native immunities** and **acquired immunities.** Innate immunity includes species, race, and individual resistance to infection, discussed in Chapter 17. Acquired immunity may be either *natural* or *artificial,* depending on the processes involved in producing the im-

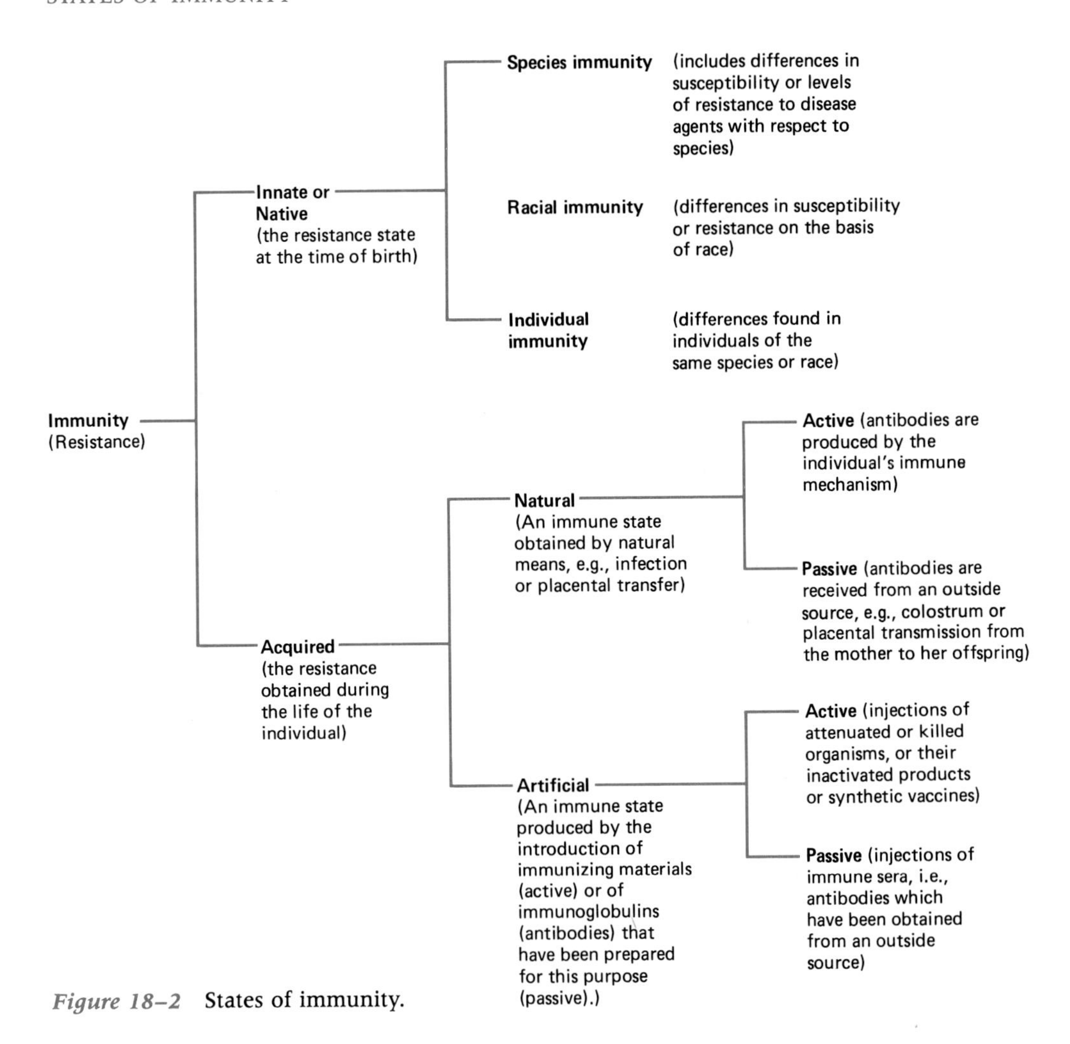

Figure 18–2 States of immunity.

munity. Immunization by the injection of a bacterial vaccine is an artificially produced contact with the organism, in contrast to a natural infection.

Both of these categories are further subdivided into active and passive types. In the *active state,* an individual makes antibody in response to an immunogenic stimulus; in the *passive state,* the antibody is acquired through transfer from an immunized individual. The acquired states of immunity are described in more detail below.

Naturally Acquired Active Immunity

An individual who recovers successfully from an infection usually acquires a specific resistance to the causative agent. This immunity is produced by the antibody-synthesizing mechanism, which was stimulated into action by the infecting organism. Depending on the immunogenic nature and dosage of such pathogens and related factors, this immunity may last for a period ranging from a few months to several years. Although immunity is never absolute, individuals having a naturally acquired active immunity are protected against ordinary attacks by infectious agents to which they have been previously exposed. Such resistance, however, is generally not substantial enough to overcome massive infections. Some of the diseases to which an individual can develop sufficient protection as a consequence of infection include chickenpox, classic measles and rubella (German measles), mumps, influenza, and typhoid fever. Immunity to reinfection by certain other pathogens is either minimal or nonexistent. This group of diseases includes gonorrhea, pneumonia, syphilis, and most protozoan infections.

SUBCLINICAL OR INAPPARENT INFECTION

Many persons experience attacks by pathogens that produce such mild symptoms that the disease goes undiagnosed. Nevertheless, the individual acquires a strong immunity through such repeated attacks. Diseases associated with this type of phenomenon include diphtheria, poliomyelitis, and scarlet fever.

Naturally Acquired Passive Immunity

Naturally acquired passive immunity is due to the natural transfer of antibodies from an immunized donor to a nonimmune recipient. This transfer can occur between mother and fetus, or via the *colostrum,* a protein-rich fluid produced by the mother before the appearance of true breast milk. During nursing, secretory IgA molecules and leukocytes are passed from the mother to her newborn infant in this breast fluid.

In humans, another form of naturally acquired passive immunity occurs when IgG molecules pass from the maternal circulation through the placenta into the fetal circulation through a single layer of cells separating these two systems. Thus, in the human species, an expectant mother having antibodies against diseases such as diphtheria, rubella, tetanus, poliomyelitis, and possibly salmonellosis imparts a share of these protective substances to her unborn child. These immunoglobulins are of the IgG class. While the duration of such immunity ranges from a few weeks to a few months, it is an important form of protection. Therefore, the importance of immunization with available vaccines for women of childbearing age can not be overemphasized.

In lower animals, since the placenta has several layers, the maternal and fetal circulatory systems are separated. Thus antibodies are prevented from passing into the fetal blood. Naturally-acquired passive immunity is achieved in these cases through the ingestion of colostrum by newborns. The immunoglobulins are absorbed, undigested, through the wall of the small intestine. This form of immunity is only temporary.

Artificially Acquired Active Immunity

Artificially acquired active immunity is induced by imitating nature: producing a mild infection. A carefully chosen immunogenic stimulus is provided without the severe effects of the actual disease. Preparations used to induce artificial active immunity include (1) killed or inactivated microorganisms; (2) inactivated bacterial exotoxins, known as *toxoids;* (3) living but attenuated (weakened) microbes; (4) parts of microorganisms; and (5) synthetic vaccines. The specific types of vaccines and related preparations, immunization schedules, and descriptions of the procedures used in their preparation are presented later in this chapter.

Any vaccine to be used against an infectious disease should possess the following five general properties listed by G. S. Wilson at the 1961 International Conference on Measles:

1. Vaccination should not be harmful to the individual receiving it.
2. The effects of the vaccine should not be greater than those associated with the disease itself.
3. The vaccine must be easy to administer.
4. The benefit from vaccination should serve the community as well as the individual.
5. The immunity conferred by the vaccine should be sufficient to eliminate the need for frequent revaccination.

In general, artificially active immune states require approximately one to two weeks to develop and are relatively long-lasting.

Artificially Acquired Passive Immunity

Protection against some diseases can be obtained only by provision of ample amounts of immunoglobulins. Individuals exposed to botulism, diphtheria, gas gangrene, rubella (especially in pregnant women), infectious hepatitis, mumps, rabies, and tetanus are treated in this way. In short, such persons must be given artificial passive immunization. Standardized doses of purified serum preparations obtained from immunized individuals are administered as soon as possible after exposure. Dosages of such protective sera are determined by the patient's body weight. The passive state of immunity that results is immediate but only temporary, because no active production of antibody toward the disease agent or its product occurs. If the immune serum used is obtained from the same animal species as the recipient, the rate of breakdown of injected immunoglobulin is approximately the same as that of normal globulins. However, if the immunoglobulins received come from another animal species, then their introduction into the individual provokes antibody production that rapidly eliminates them. Subsequent injections of the same preparation may cause severe allergic reactions, including anaphylactic shock and serum sickness. **[See immunologic disorders in Chapter 20.]**

ADOPTIVE IMMUNITY

This immune state is primarily an experimental phenomenon. At this time, it has no practical significance as a means of inducing antimicrobial protection. Adoptive immunity is a form of artificially acquired passive immune state; it is produced by transferring, from an immunized donor to a nonimmune recipient, cells capable of synthesizing immunoglobulins or of directly reacting with a specific antigen. Such cells are referred to as being "immunologically competent."

Immunization Preparations

The biological preparations used for immunization or diagnosis can be divided into three basic categories: (1) prophylactic agents for active immuni-

Microbiology Highlights

THE IMPACT OF IMMUNIZATION

Historically, many infectious diseases have commonly afflicted children between the ages of five and nine. One of the main reasons for this pattern of incidence was that susceptible children in this age range came together in large numbers at schools, where the introduction of any disease could lead to widespread transmission. The development and prevalent use of vaccines against the seven infectious diseases of diphtheria, measles, mumps, pertussis (whooping cough), poliomyelitis, rubella, and tetanus has had a remarkable impact on the occurrence of disease around the world and especially in the United States. The accompanying table summarizes the maximum number of cases of vaccine-preventable disease in or near the year when these conditions first became reportable to public health agencies. The table also shows the percentage decrease of reportable cases based on the number of cases in 1983. These dramatic reductions clearly emphasize immunization as an important tool in the prevention and control of infectious diseases.

Comparison of Maximum and Current Number of Vaccine-preventable Diseases

Disease	*Maximum no. of cases (year)*	*No. of cases in 1983*	*Percentage decrease*
Diphtheria	206,939 (1921)	5	99.99
Poliomyelitis (paralytic)	21,269 (1952)	15	99.93
Measles	894,134 (1941)	1,497	99.84
Congenital rubella syndrome†	20,000 (1964–1965)	22	99.09
Pertussis	265,269 (1934)	2,463	99.08
Rubella†	57,686 (1969)	970	98.32
Mumps*	152,209 (1968)	3,355	97.80
Tetanus‡	601 (1948)	91	84.86

* First reportable in 1968. † First reportable in 1966. ‡ First reportable in 1947.

zation, which include bacterial and viral vaccines, toxins, and toxoids (Table 18–1); (2) prophylactic preparations for passive immunization, mainly globulins (gamma globulins; Table 18–4); and (3) diagnostic reagents designed to demonstrate hyperimmune states or to detect susceptibility to disease agents. The last include purified protein derivative (PPD) and diluted diphtheria toxin. The applications of these diagnostic and related reagents are discussed mainly in the chapter dealing with respiratory bacterial and mycotic infections (Chapter 25).

Preparation of Representative Vaccines

Vaccines are preparations of either disease-causing microorganisms or certain of their components or products, such as toxins, which have been rendered unable to produce disease but are still antigenic. These products are used to produce an artificially acquired active immunity. **[See bacterial toxins in Chapter 22.]**

BACTERIAL VACCINES

Bacterial vaccines are generally suspensions of killed bacteria in an isotonic (physiologic) salt solution. [One exception is BCG (bacillus of Calmette and Guérin) vaccine, which is an attenuated, or weakened, preparation for immunization against tuberculosis.] Administering such vaccines is the most common means of inducing active immunity. In the preparation of these materials, the organisms are cultured, harvested after a suitable incubation period, and then killed with heat or chemical agents.

The use of heat to kill organisms has the disadvantage of reducing the immunizing potency of preparations even when the lethal temperature and the time of exposure are kept as low as possible to ensure sterilization. Chemical killing agents used for vaccine production include acetone, formalin, mer-

TABLE 18–1 Preparations Currently Used for Active Immunizations (Vaccines)

Bacterial Diseases	*Etiologic Agent*	*Means of Vaccine Preparation*
Cholera	*Vibrio cholerae*	Heat-killed suspension of *V. cholerae.*
Diphtheria	*Corynebacterium diphtheriae*	Alum-precipitated toxoid preparation of *C. diphtheriae* toxin. This antigenic material is combined with tetanus toxoid and killed *B. pertussis* in the DPT vaccine.
Epidemic typhus	*Rickettsia prowazekii*	Formalin-killed suspension of *R. prowazekii* cultivated in chick embryo yolk sacs.
Meningitis	*Haemophilus influenzae,* type b	Purified capsular polysaccharide.
Meningitis	*Neisseria meningitidis*	Purified capsular polysaccharides.
Plague	*Yersinia pestis*	Formalin-killed and alum-coated suspension of *Y. pestis.*
Pneumococcal pneumonia	*Streptococcus pneumoniae*	Purified capsular polysaccharides.
Rocky Mountain spotted fever	*Rickettsia rickettsii*	Formalin-killed suspension of *R. rickettsii* cultivated in chick embryo yolk sacs or obtained from infected ticks.
Tetanus	*Clostridium tetani*	Alum-precipitated toxoid preparation of *C. tetani* toxin. See section on diphtheria above.
Tuberculosis	*Mycobacterium tuberculosis*	Bacillus of Calmette and Guérin (BCG). Prepared from a strain of *M. tuberculosis* var. *bovis* attenuated by continuous subculture on glycerol-broth-bile-potato media.
Tularemia[a]	*Francisella tularensis*	Live attenuated strain of *F. tularensis.*
Typhoid and paratyphoid fevers	*Salmonella typhi, S. paratyphi, S. schottmuelleri*	Heat-killed, phenol-preserved, or acetone-dried preparations. Vaccines commonly contain *S. typhi* alone or in combination with *S. paratyphi* and *S. schottmuelleri (TAB).*
Whooping cough	*Bordetella pertussis*	Alum-precipitated or aluminum hydroxide- or aluminum phosphate-adsorbed, killed preparations of phase I *B. pertussis.* See section on diphtheria above.

Viral Diseases	*Means of Vaccine Preparation*
Hepatitis B	Formalin-treated, purified viral (subunits) antigen. Genetically engineered yeast-derived and mammalian-derived preparations.
Influenza	Formalin-killed preparation of prevalent viral strains. The viruses are usually grown in embryonated chick eggs.
Measles (rubeola)[a,b]	Two attenuated vaccines are in use. One preparation employs the Edmonston strain in combination with measles-immune gamma globulin. The other vaccine utilizes the Schwartz strain of measles virus. A new chicken embryo preparation is being evaluated.
Mumps[b]	Formalin-killed preparation of virus obtained from chick embryo cultivation. An attenuated vaccine is currently in use. It is prepared by cultivation of the virus in embryonated chicken eggs and then in chick embryo cell cultures.
Poliomyelitis	Two vaccines are in use. The Salk vaccine is a formalin or ultraviolet irradiation-inactivated preparation of the virus. The Sabin, or oral, vaccine is an attenuated preparation. Both vaccines contain types 1, 2, and 3 polio viruses (trivalent).
Rabies (hydrophobia)	Several vaccines are available. These include the human diploid cell vaccine (HDCV), and duck embryo vaccine (DEV). The human diploid cell vaccine is prepared from virus grown in tissue culture and inactivated by either tri (*n*)-butyl phosphate or beta-propiolactone. The duck-embryo vaccine (DEV) is a beta-propiolactone-inactivated preparation of fixed virus obtained from duck embryo cultivation. Production of DEV has been discontinued but stocks may still be available.
Rubella (German measles)[b]	Three attenuated vaccines are in use: (1) $HPV_{77}D_5$, obtained from tissue cultures of duck embryo cells; (2) $HPV_{77}DK_{12}$, prepared from dog kidney cells; and (3) Cendehill, obtained from rabbit kidney cells.
Yellow fever	Attenuated strain of the virus (17D) cultivated in chick embryos.

[a] Inactivated preparations do not provide adequate protection.
[b] Combined vaccines. Measles, mumps and rubella (MMR) virus vaccine is commonly used.

thiolate, phenol, and tricresol. The concentrations of these chemicals are adjusted to ensure the antigenic effectiveness of the bacterial preparations.

Rickettsial vaccines are formalin-killed preparations of the specific pathogen (Table 18–1). For the vaccines currently in use, the rickettsiae are grown in embryonated chicken eggs, specifically in the yolk sac. Because Cox, in 1938, was the first person to cultivate rickettsiae in this particular manner, vaccines so prepared are called *Cox vaccines.*

Two general types of bacterial preparations are utilized—*stock* and *autogenous vaccines.* Stock vaccine is prepared from laboratory stock cultures; autogenous vaccine is specially made from organisms freshly isolated from a patient. Autogenous vaccines are of significant value in certain staphylococcal infections. Before killed vaccines are released for human use, they are thoroughly tested for safety and potency.

TOXOIDS. The toxins of several bacteria can be converted into nontoxic but still immunogenic preparations called **toxoids.** Heat or formalin is used in the production of toxoids.

In most commercial preparations, 0.2% to 0.4% formalin is added to a bacterial toxin such as diphtheria or tetanus. The resulting mixture is incubated until detoxification is complete. Inert protein is then removed. The resulting preparation is called a *toxoid.*

Another type of toxoid is prepared through the incorporation of certain aluminum compounds, such as alum, aluminum hydroxide, or aluminum phosphate. When alum is added to the toxin, the immunogenic component of the toxin precipitates. This material is washed and suspended in sterile physiologic saline. A diphtheria toxoid prepared in this way would be called *diphtheria toxoid, alum precipitated.* When aluminum hydroxide or phosphate is added to a toxin, the immunogenic components adsorb onto the particles of these compounds. The resulting toxoid preparation is *aluminum hydroxide,* or *aluminum phosphate-adsorbed toxoid.*

Both types of toxoids, alum-precipitated and aluminum hydroxide-adsorbed, produce a prolonged and relatively continuous antigenic stimulus. The aluminum compounds function as adjuvants that slowly release the immunizing substances.

Currently available toxoids are shown in Table 18–1.

VIRAL VACCINES

Several viral vaccines are composed of viruses whose disease-producing capability (virulence) has been greatly reduced or eliminated. Such organisms are referred to as being **attenuated,** or weakened.

The virulence of viruses and other microorganisms can be reduced by several methods. These methods include (1) cultivation at temperatures above normal for the organism, (2) desiccation, (3) the use of unnatural hosts—for example, tissue culture or mice—for propagation of the microorganisms, and (4) continued and prolonged serial passages through laboratory animals. The first two techniques were discovered and used effectively by Louis Pasteur to combat anthrax and rabies. An example of the procedures involved in the preparation of an attenuated viral vaccine is the mumps, measles, and rubella **(MMR)** combined vaccine currently in use.

INACTIVATED VACCINES. Some viral vaccines consist of killed (inactivated) viruses. These include Salk poliomyelitis vaccine and influenza vaccine (Table 18–1). Formalin is usually the chemical killing agent.

The Safety of Live Versus Inactivated Preparations

The acceptability of a vaccine depends upon the general need for the preparation and the degree of protection it can provide. As severe and life-threatening bacterial diseases such as diphtheria and pneumococcal pneumonia progressively disappear, the public comes to expect greater freedom from viral diseases. Almost all the effective viral vaccines now used consist of living, attenuated viruses. Live vaccines produce a high level of immunity, usually in a single dose. However, although such immunizing preparations have obvious advantages when compared to inactivated ones, live vaccines are subject to certain problems. These disadvantages include the following possibilities: (1) infectious viruses will remain because of insufficient weakening; (2) the viruses will spread from the vaccinated individual to susceptible contacts; (3) contaminating viruses are present; (4) a genetic change will take place, resulting in a highly virulent viral strain capable of infecting various body areas; and (5) the vaccine will be inactivated by heat in facilities that lack adequate refrigeration. Inactivated vaccines, on the other hand, are free from most of these potential hazards. However, producing a level of immunity using inactivated vaccines comparable to that produced by live preparations requires larger amounts of antigen and several injections.

Vaccine Safety

The risk of accidents with vaccines has been considerably reduced by the development of elaborate series of tests of vaccine safety and effectiveness. For example, virus vaccines containing specific immunologically active virus components, both infectious (live) and inactivated, are checked for immunologic and pathogenic properties at all stages of preparation. The fate of viruses in the vaccinated individual is also studied. If the viruses are eventually shed, the possibility that they might act as pathogens for susceptible contacts is checked.

Antibody Production

The principal objective of immunization is to produce a high level of protection against infectious disease. Vaccine-conferred immunity should equal, if not exceed, that produced by an actual infection. Antibodies for a particular immunogenic substance are not detectable in the serum of an individual until exposure to the immunogen has occurred. The extent of the antibody response is known to be affected by various factors, including (1) the nature of the immunizing material, (2) the dosage received, (3) the number and frequency of exposures, (4) the particular animal species, and (5) the individual involved.

The effects of immunogenic stimulation on the body can be best studied by observing the response produced with a single injection of immunogen. This reaction to the first injection of an immunogen is called the *primary response.* The original description of the primary response noted a *latent* or *lag period* during which no increase in circulating antibody was detected. Improved techniques have shown that, in some instances, antibody synthesis does follow introduction of the stimulus almost immediately. After a period ranging from a few hours to several days, the antibody titer (level) reaches a peak or plateau (Figure 18–3). Peak concentration is reached when the rates of antibody production and antibody breakdown are approximately the same. Such antibody levels may remain for several months or longer and then slowly begin to decline, as antibody breakdown exceeds production. The details depend, in part, upon the animal species and the immunogenic material involved.

Some immunogens produce a sudden secondary rise in antibody titer when injected again some time after the first exposure. This effect is often called a specific **anamnestic response,** from the Greek term *anamnesis,* or "recall." The antibody titers associated with a specific anamnestic response generally are higher than those produced by the primary reaction, occur with little or no lag period, and remain for long periods.

Anamnestic reactions are produced by reimmunization with vaccines. The effectiveness of booster shots can be explained on this basis.

Vaccines Currently in Use

Immunization programs are largely directed at children and against a group of targeted diseases (Table 18–2). Even with the administration of effective

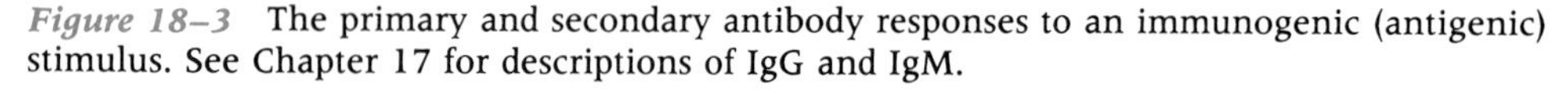

Figure 18–3 The primary and secondary antibody responses to an immunogenic (antigenic) stimulus. See Chapter 17 for descriptions of IgG and IgM.

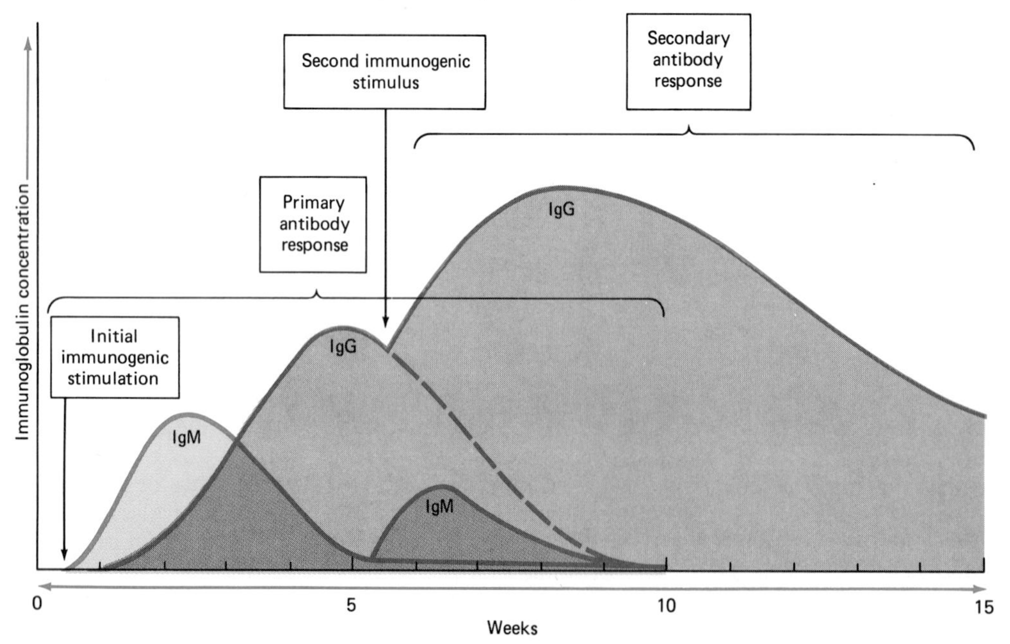

TABLE 18–2 A Recommended Schedule for Active Immunization and Tuberculin Testing[a]

Recommended Age	*Immunizing Agent*
2 months	Diphtheria toxoid
	Pertussis vaccine (adsorbed), Tetanus toxoid } DPT-1 (combined preparation)
	Oral polio vaccine-1 (TOPV or OPV-1)[b]
4 months	DPT-2
	OPV (trivalent preparation)
6 months	DPT-3
	TOPV (additional dose optional in areas with a high risk of polio exposure)
15 months	Measles, mumps, and rubella vaccines. These vaccines can be used in combined form (MMR) or as individual preparations.[c] Tuberculin testing is also included at this point.[d]
	DPT-4, TOPV-3 } Completion of primary series of DPT and TOPV
24 months	*Haemophilus influenzae b* polysaccharide vaccine
4–6 years	DPT-5 boosters, preferably given at or before school entry
	TOPV-4
14–16 years	Td (adult) booster[e]
Thereafter every 10 years (except in cases of injury)	Td (adult) booster[e]

[a] This schedule for active immunization is based in part on the recommendations of the *Red Book* of the American Academy of Pediatrics and the Centers for Disease Control, Atlanta, Georgia.
[b] TOPV, trivalent oral poliovirus vaccine.
[c] Children between the ages of 12 months and puberty may be vaccinated either with combined vaccines or with single-vaccine preparations.
[d] Initial testing is recommended at 1 year of age. The frequency of testing thereafter depends on the risk of exposure and the prevalence of tuberculosis in the community.
[e] Adult form of tetanus (T) and diphtheria (d) toxoids. The diphtheria toxoid dosage is reduced because of possible undesirable reactions in individuals with several previous inoculations.

viral vaccines and related immunizing agents during the childhood years, it is becoming evident that a portion of the adult population remains susceptible to one or more of the common childhood diseases. The reasons for this include the following: (1) the widespread use of vaccines creates a level of immunity within the general population that reduces the usual exposure to certain microbial disease agents; (2) not all individuals are immunized in childhood; (3) some persons receive ineffective or inactivated vaccines; or (4) maternal antibody is present in the infant at the time of immunization.

Because of the changing disease patterns (epidemiology) of some common bacterial and viral infections and a general lack of specific therapeutic agents, the use of vaccines for adults is increasing. A substantial number of adults born in the United States in the 1960s or earlier remain incompletely immunized against childhood diseases. Of most concern in this context are the viral diseases rubella and measles, which are more serious in adults than in children. Table 18–3 provides guidelines for adult immunization.

Combined Vaccines

The appearance of new vaccines makes the use of combined forms of immunizing preparations desirable. Combined forms are simpler to administer and less expensive to produce; they also require fewer visits by the subject to health-care facilities than single-immunogen preparations. Examples of combined preparations include DPT (diphtheria-pertussis-tetanus); MMR (measles-mumps-rubella viruses vaccine); TOPV (trivalent oral poliomyelitis vaccine), which contains all three immunologically distinct strains of attenuated polio viruses; and TAB (typhoid and paratyphoid A and B). Several studies have demonstrated excellent antibody responses to combined vaccine preparations. The combined vaccines provide simple, safe, and effective means of immunizing against important infectious disease. **[See measles and other skin diseases in Chapter 23.]**

BCG and Tuberculosis Prevention

The proper use of bacillus of Calmette and Guérin (BCG) in the prevention of tuberculosis has long been a controversial and emotionally charged issue in the United States. In contrast, BCG vaccination has been a major part of the World Health Organization's effort to control tuberculosis in countries with high rates of transmission. This group has been notably successful in reducing the incidence of tuberculosis. Control of the disease in the United States has met with some difficulty as evidenced by increasing numbers of cases. For this reason, there

TABLE 18–3 Guidelines for Adult Immunization

Vaccines and/or Toxoids	*Indications*	*Immunization Schedule*	*Special Considerations*
Diphtheria and Tetanus (Td) toxoids	All adults.	1. Two doses intramuscularly 4 to 8 weeks apart. 2. Third dose 6 to 12 months after second dose. 3. Boosters every 10 years.	
Polio vaccine[a]	For individuals traveling to areas where wild polio virus infection is epidemic or endemic; for health-care personnel at risk for exposure to wild virus.	1. Inactivated vaccine. (a) 3 doses, subcutaneously, 4 to 8 weeks apart. (b) fourth dose 6 to 12 months later. 2. Oral vaccine given for immediate protection.	Inactivated vaccine should be given to immunocompromised persons.
Rubella vaccine	All adults lacking (1) documentation of immunization on or after the first birthday or (2) laboratory evidence of immunity. Particularly important for women of childbearing age, health-care personnel, and travelers.	One dose subcutaneously; no booster.	Women pregnant when vaccinated or who become pregnant within 3 months should be advised of risks with live vaccine.
Measles vaccine	All adults born after 1956 (1) without documentation of live vaccine immunization or (2) without laboratory evidence of immunity.	One dose subcutaneously; no booster.	Vaccine should not be given in cases of pregnancy, immunodeficiency, or sensitivity to eggs.
Influenza vaccine	All elderly high risk persons; and health-care personnel.	Annual immunization.	Vaccine should not be given to persons with sensitivity to eggs.
Pneumococcal vaccine	All adults who are at risk for pneumococcal disease.	One dose; booster not recommended.	Vaccine should not be given during pregnancy.

[a] Inactivated polio vaccine is preferred for adults.

may be justification for instituting mass BCG vaccination in the United States. Under some circumstances, a BCG vaccination program might be of great value in certain areas where the usual surveillance and treatment programs have failed or cannot be readily applied. Such situations might involve (1) infants who are unavoidably exposed to family members with active cases of pulmonary tuberculosis; (2) persons such as ghetto dwellers, migrants, alcoholics, and others, who have no regular health care services and for whom the usual methods of disease detection and treatment are inadequate; (3) immigration of infected foreign persons; (4) populations with a known high frequency of tuberculosis; (5) health-care personnel working where an endemic incidence of tuberculosis is relatively high; and (6) individuals with negative tuberculin skin test results who are at high risk of repeated exposures to infected persons. [See tuberculosis in Chapter 25.]

Passive Immunization

Passive immunization is effective, acts immediately, and is temporary in its effect. The preparations used for passive immunization are of two basic types: **antitoxins** and **antimicrobial sera** (Table 18–4). Many of these preparations are produced in lower animals. They should be used only in clinical situations in which human sources of antitoxins and sera are not available, since these preparations present the possibility of hypersensitivity reactions. [See type III hypersensitivity in Chapter 20.]

Antitoxins

Antitoxins are antibodies capable of neutralizing the toxin that stimulated their production. These pro-

TABLE 18–4 Preparations Used for Passive Immunization

Disease or Condition	*Product*	*Source of Product*
Black widow spider bite[a]	Antivenin (antivenom) widow spider	Horse
Botulism	Antitoxin types A, B, and E	Horse
Diphtheria	Antitoxin	Horse
Hepatitis A	Immune globulin	Human
Hepatitis B	Immune globulin and hepatitis B immune globulin (HBIG)	Human
Hypogammaglobulinemia[a] (antibody deficient)	Immune globulin[b] IV (intravenous)	Human
Measles	Immune globulin	Human
Whooping cough (pertussis)	Pertussis immune globulin	Human
Rabies	Rabies immune globulin	Human
	Antirabies serum	Horse
Rh isoimmunization[a]	Rh_0 (D) immune globulin	Human
Snakebite[a]	Antivenin—coral snake	Horse
	Antivenin—rattlesnake, copperhead, water moccasin	Horse
Tetanus	Tetanus immune globulin	Human
	Antitoxins	Horse or cow
Vaccinia (cowpox)	Vaccinia immune globulin	Human
Varicella (chickenpox)	Varicella-zoster immune globulin	Human

[a] Passive immunization also is applied to noninfectious disease conditions.
[b] Product may cause allergic reactions in IgA deficient persons.

tein substances may be referred to as an *immune serum.* An antitoxin is specific in its action against its toxic counterpart, but it does not exert any effect on the microorganism that produced the toxin. Most commercial preparations used for the passive immunization of humans are produced in horses, although antisera from cows are available for use with individuals sensitive to horse products (Table 18–4). [See microbial toxins in Chapter 22.]

PREPARATION

The preparation of antitoxins, such as those of diphtheria and tetanus, involves injecting toxin solutions into a horse or other appropriate animal until sufficient antitoxin titers are produced. The animal is then bled and its serum or plasma processed to concentrate the antitoxin and eliminate a large proportion of native horse serum protein.

Removing as many of the horse protein components as possible reduces the likelihood of sensitizing individuals receiving antitoxin or of causing anaphylaxis or serum sickness in already sensitized persons. (Sensitization of an individual develops in response to the native horse serum protein, not to the antitoxins in the serum preparation.)

STANDARDIZATION

Before antitoxin preparations are released, they are sterilized, usually by filtration, treated with a chemical preservative to maintain the sterility of the preparation, and standardized. Three procedures are used for standardizing most antitoxins: animal *(in vivo)* protection procedures, tube flocculation tests, and skin challenge tests.

The first step in the *animal protection test* is the injection into test animals of different quantities of a particular toxin mixed with one unit of the corresponding standard antitoxin prepared by a central controlling agency, such as the National Institutes of Health in Bethesda, Maryland. This procedure determines the dose of toxin lethal to a laboratory animal of a certain weight after a specified length of time in the presence of one standard unit of antitoxin. This quantity is called L_+. For example, in the standardization of diphtheria and tetanus toxins, the L_+ dose is the quantity of toxin needed to kill a 250-g guinea pig on the fourth day following the injection of the toxin-antitoxin combination. Once the L_+ dose of the toxin has been obtained for the unit of standard antitoxin, the toxin is used to standardize the newly processed antitoxin. Standardization is achieved when the quantity of the new preparation producing the same results as one unit of the standard antitoxin is found.

In the *flocculation test,* certain proportions of the respective toxin and antitoxin preparations are mixed in test tubes. A flocculent precipitate generally forms in the tube that contains the proportions of toxin and antitoxin needed for neutralization.

The *skin challenge test* (or simply *skin test*) resembles the animal protection procedure except that the standard is not the death of the animal but the production of skin reactions by toxin-antitoxin combinations.

Commercial antitoxin preparations have been used principally in cases of diphtheria and tetanus. The designations for tetanus antitoxin preparations include ATS (antitetanic serum) and TAT (tetanus antitoxin). However, antitoxins have also been used to treat other diseases, including botulism and gas gangrene.

A physician should be present when antitoxin is administered, and the patient's blood pressure and pulse should be checked during and after the treatment to detect any possibility of a hypersensitive reaction.

IMMUNE GLOBULIN (IG)

Immune globulin (IG), also referred to as *gamma globulin,* is derived from human blood, plasma, or serum and contains most of the antibodies found in whole blood. The concentrations of specific antibodies vary among different preparations. Immune globulin contains primarily IgG immunoglobulins. Conspicuously absent are IgM-associated antibodies, which are important to the body's defense against bacterial pathogens, and secretory IgA immunoglobulins, which are important to the protective mechanism on mucous membranes. Nevertheless, IG preparations have advantages. These include (1) the absence of hepatitis B virus, (2) the presence of a large amount of antibodies in a small volume, and (3) stability during long-term storage. The value of human serum globulin has been shown unequivocally in cases of measles and certain congenital immune-deficiency disease states. IG can be useful in cases of posttransfusion hepatitis, rubella (German measles) during the first trimester of pregnancy, and exposure to chickenpox virus of patients receiving immunosuppressive drugs. The variable success rates observed with immune globulin in these situations are probably due to differences in the amounts of specific antibody molecules among the preparations used.

SPECIFIC IMMUNE GLOBULIN (SIG)

Specific immune globulins (SIG) have a higher concentration of specific antibody to the agent in question than IG preparations. The SIGs, which are used more frequently, include those of chickenpox (varicella), measles, mumps, whooping cough, tetanus, shingles, and the Rh_0 (D) blood factor. These globulin preparations are made from sera obtained from individuals recently recovered from an acute infection or from individuals who are hyperimmunized to a given material. SIGs are injected into persons with the disease or recently exposed to the causative agent.

Vaccines Under Development

A number of vaccines have been developed for experimental purposes or for individuals in high-risk occupations or situations. Several examples of such preparations, which are not yet available for general public use, include vaccines against adenoviruses (acute respiratory infections), cytomegaloviruses (birth defects), dental caries, gonorrhea, herpes viruses (genital herpes), *Legionella* (lee-jon-EL-lah) species (Legionnaires' disease), leprosy, malaria, rotaviruses (gastrointestinal infections), and syphilis.

Synthetic Vaccines

The content of most vaccines in common use is not completely known. In addition, such preparations are not totally specific, uniform in composition, or inexpensive to produce. The production of a natural virus vaccine, for example, requires first that the virus be obtained from humans or lower animals and maintained in tissue culture; then the various procedures that attenuate or inactivate the virus must be completed. Laboratory-synthesized vaccines may well provide functional solutions to these problems (Figure 18–3).

DNA sequencing techniques, briefly described in earlier chapters, have made it possible to determine the primary structure and amino acid composition of proteins transcribed from the triple combinations of DNA segments likely to function as actual genes. With such technology, the amino acid sequences can be determined and used to synthesize small protein or peptide fragments. These resulting molecules are the basis for synthetic vaccines to stimulate effective immunoglobulin production (Figure 18–4). Vaccines in various stages of development and testing have been made against several viruses, including those that cause AIDS, hepatitis B, influenza, rabies, the common cold, and mouse leukemia. Although most synthetic vaccines are still largely experimental, they offer great potential for the production of inexpensive, highly specific vaccines that have no side effects (Table 18–5). At least two synthetic vaccines for the prevention of lower animal diseases have been marketed. The preparations are used against foot-and-mouth disease and scours, a disease affecting piglets, caused by a toxin producing strain of *Escherichia coli* (esh-er-IK-ee-ah KOH-lee). **[Refer to Chapter 6 for a description of protein synthesis.]**

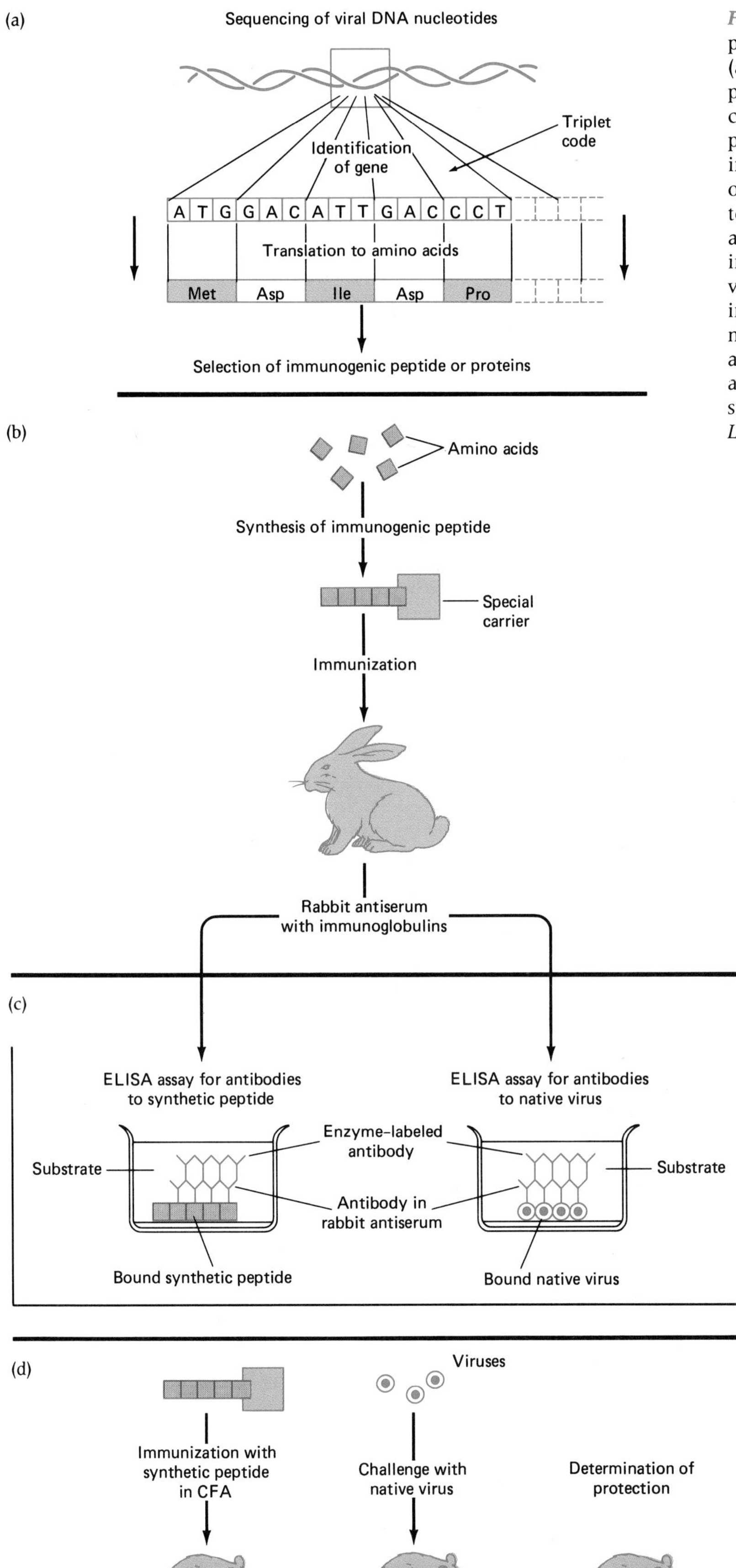

Figure 18–4 The various steps in the preparation of a synthetic virus vaccine. (a) DNA sequencing to determine the primary structure and amino acid composition of possible immunogenic peptides or proteins. (b) Synthesis of immunogenic material and immunization of laboratory animals. (c) Laboratory tests to determine the high activity against both the synthesized immunogenic peptide and the native virus. (d) Challenge of laboratory animals immunized with a synthetic vaccine material with the complete Freund's adjuvant (CFA). The use of CFA provides a prolonged period for exposure to the synthetic peptide. *[Modified from R. A. Lerner,* et al., Hosp. Pract. ***16****:55, 1981.]*

Microbiology Highlights

THE AIDS CHALLENGE

The development of a successful AIDS (acquired immunodeficiency syndrome) vaccine is a matter of great concern and importance. Unlike victims of many other viruses for which vaccines have been found, AIDS patients do not, at present, recover from the disease. The goals of investigations into vaccine development are twofold: first, prevention of the initial infection and, second, the elimination or control of the human immunodeficiency virus (HIV), the cause of AIDS.

Traditionally, viral vaccines have been made from whole viruses that have been killed or from an attenuated strain that stimulates antibody production but does not cause disease. Such vaccines have been effectively used to combat a number of viral diseases including measles, mumps, smallpox, and poliomyelitis. In the area of AIDS vaccines, investigators are hesitant to use whole viruses because they do not want to risk the possibility of live, disease-producing AIDS viruses finding their way into vaccine preparations. Moreover, they are not certain that truly harmless forms of HIV exist or whether it is possible to make an AIDS virus irreversibly harmless. An objection to the use of whole killed virus for a vaccine also has been raised by some researchers based on the concern for injecting viral nucleic acids that might integrate into the genetic material of the recipient's cells and cause problems such as the activation of harmful genes.

The potential sources of material for an AIDS vaccine are several and include purified components of the virus as well as synthetic viral subunits. **[See Chapter 10 for descriptions of viral parts.]** Among the most actively studied vaccine candidates are recombinant DNA products, such as those resulting from the insertion of HIV genes into vaccinia virus, the virus used to inoculate people against smallpox. The modified virus would be expected not only to produce the desired AIDS antigens but also to present the antibody-stimulating antigens to the immune system in a more natural manner, thereby possibly provoking a wider range of immune responses. Since vaccinia virus is a large virus, researchers can insert many genes. In 1987, Daniel Zagury of the Pierre and Marie Curie University in Paris made worldwide headlines when he disclosed that he had injected himself and several other volunteers in 1986 with this type of synthetic vaccine. As of yet there is no evidence that these subjects have taken the last step toward proving the vaccine's effectivensss, that is, exposing themselves to HIV.

Despite the view that the only hope for halting the spread of AIDS is widespread immunization, an effective vaccine is unlikely to become available until well into the 1990s. The extremely variable nature of the AIDS virus is one of the major obstacles to the making of a vaccine. While most viruses have only a few variant forms (*e.g.*, influenza viruses), HIV has many. This is quite apparent from the significant differences noted in the genetic makeup and the various parts of viruses isolated from different victims of the disease. Even viruses within the same individual keep changing. As a result, antibodies that neutralize one HIV variant form may not recognize the antigens of another.

Ever since the eighteenth-century physician Edward Jenner found an effective means of combating the dreaded disease of smallpox, vaccines have been indispensable, when available, in the battles against various infectious diseases. It is no small wonder, then, that researchers and people the world over are turning toward a vaccine in hopes of conquering the modern-day plague of AIDS.

The Role of Health-Care Personnel

All health-care personnel need a basic understanding of immunologic principles. Immunization against preventable diseases is one of the most important means of preventing or controlling communicable-disease epidemics. Nurses, for example, not only maintain and sterilize the instruments used in immunizing procedures but also frequently administer vaccines and related preparations. In addition, they often have the responsibility of maintaining the immunization records of patients and of informing parents about the importance to their children of certain immunizations and of completing immunization schedules (Table 18–2).

To perform these duties efficiently and intelligently, health-care personnel must be aware of the nature of the material to be administered, the correct method of administration, the anticipated results, and the contraindications and possible side effects (Table 18–5) of the material to be used.

Administering Vaccines

PREPARING FOR ADMINISTRATION

Aseptic precautions must be taken during any type of inoculation procedure. These procedures include the proper cleaning and sterilizing of all nondisposable equipment and the washing and drying of the inoculator's hands. The containers for the immunizing preparations should also be disinfected. The cap or any other part of the container to be used should be wiped with 70% alcohol or another appropriate agent.

ROUTES OF ADMINISTRATION

Injection. In order for antibody production to occur, the antigenic material must be introduced beneath the epithelial tissues. Among the possible injection routes for humans are the intradermal (intracutaneous), intramuscular (IM), and subcutaneous injection routes. For primary immunization of humans, the intramuscular route of injection is most commonly used. The intradermal procedure may be used in the revaccination of previously inoculated individuals. Other routes of injection, such as intraperitoneal, intrathecal, and intravenous (IV), have been used in certain situations requiring the administration of antisera and related materials, but such applications are rare. In the case of experimental animals used for the production of antibodies, the intraperitoneal and intravenous routes are generally employed. Both of these methods give excellent antisera. Immune globulins, depending on the preparation, are generally injected IM or IV.

Oral Administration. With the appearance of the Sabin live poliomyelitis vaccine, oral administration has come into widespread use in the mass immunization of whole community populations. In this technique the vaccine is usually given to the recipient in a small disposable cup, filled from a premeasured plastic packet or a calibrated dropper.

Intranasal Administration. To be effective, immunization procedures should stimulate the immune responses induced by natural infection. In the case of viral upper-respiratory disease, natural infection leads to the appearance of specific antibodies both in the serum and in the secretions at the local site of infection. This finding has given rise to the use of the nasal spray as a mode of vaccine delivery. In certain parts of the world, preparations for diseases such as rubella and influenza are administered intranasally.

Complications Associated with Vaccinations

Undesirable, although apparently inevitable, side reactions to immunization procedures have been reported ever since Jenner introduced vaccination in 1796. Before any vaccine is administered, patients should be questioned as to the history of previous reactions associated with vaccinations. Obtaining such information is especially important when immunizations involve vaccines produced in eggs. A full and careful documentation of the circumstances surrounding any adverse reaction is important. Data obtained should include the type of vaccine, its lot number and other pertinent features, the quantity of material administered, the route and site of injection, and as full a description as possible of any reaction from the time of onset to its termination. Details of the patient's health, age, and family history may also be significant. When an unfavorable reaction occurs after a specific immunization preparation has been administered, the fact must be fully described on the patient's record. Finally, the dangers involved with an additional exposure should be fully explained to the individual.

Representative reactions, their possible causes, and the complications that have been encountered with selected vaccines are summarized in Tables 18–5 and 18–6. Preventive measures and, in some cases, treatments are also listed.

TABLE 18–5 Representative Reactions, Possible Causes, and Prevention

Reaction	*Possible Cause*	*Prevention*
Anaphylaxis	1. Immunologic interaction between antigen and skin-sensitizing antibody causing the release of reaction-mediating substances, including histamine and serotonin (see Chapter 22)	1. Skin testing with diluted materials 2. Conspicuously labeling a patient's record regarding his or her allergic nature
Encephalomyelitis (inflammation of the central nervous system)	1. Immunologic basis of antigen-antibody reaction 2. Possible predisposition to reaction because of neurologic disorders, mental retardation, and so on	1. Avoiding revaccination to prevent recurrence of this reaction
Fetal injury (malformations)	1. Effects of live vaccines	1. Avoiding administration of live vaccines during pregnancy
High fever	1. Possible intolerance of vaccine 2. Reduction of tolerance to heat stress	1. Not exceeding recommended dosages 2. Avoiding administration of multiple toxic vaccines on the same day 3. Using fever-reducing medication, such as salicylates, after immunizations
Induction of disease	1. Existence of a subclinical infection with the disease agent against which immunization is being given 2. Susceptibility of individuals with immunologic deficiency states, such as hypogammaglobulinemias, or of patients under steroid or immunosuppressive therapy (see Chapter 20)	1. Avoiding inoculations of patients who appear to have an immunodeficiency or symptoms of the disease against which immunization is to be given 2. Avoiding administration of live vaccines to such persons
Serum-sickness-like reactions	1. Immunologic basis of antigen-antibody reaction	1. Using epinephrine and antihistamines in treatment
Severe local injury (abscess formation)	1. Bacterial contamination 2. Inadequate depth of vaccine administration	1. Proper cleansing of inoculation site 2. Proper injection technique
Sepsis (generalized infection)	1. Secondary abscess formation 2. Contaminated needles, syringes, or vaccines	1. Proper cleansing of inoculation site 2. Using disposable needle and syringes 3. Preventing vaccine contamination during storage and subsequent use
Toxic reactions	1. Sensitivity of the individual to vaccine	1. Reducing immunizing dosages

TABLE 18–6 Frequently Reported Reactions to Certain Vaccines

Vaccine	*Selected General Features of Side Reactions and Potential Dangers*	*Prevention or Treatment*
BCG	1. Regional lymph node enlargement followed by pus formation and perforation, with prolonged drainage in individuals under 1 month of age receiving dosages greater than 0.025 mL 2. Possible activation of quiescent (low-level) tuberculosis	1. Avoiding administration of vaccine to persons with positive tuberculin skin reactions
DPT and Td[a]	1. Toxic reactions 2. Encephalitis, when it occurs, is more commonly associated with the pertussis component	1. Not exceeding recommended dosages or reducing (tenfold) diphtheria toxoid in Td preparation 2. Most adverse effects averted through immunization of healthy children and administration of the final DPT dose at 4 to 7 years of age 3. Skin testing beginning with the tetanus toxoid

TABLE 18–6 (continued)

Vaccine	*Selected General Features of Side Reactions and Potential Dangers*	*Prevention or Treatment*
Influenza	1. Allergic reaction to immunizing material (anaphylaxis)	1. Avoiding use of the vaccine with persons who are pregnant or allergic to eggs
Measles (attenuated)	1. Fever, malaise, and regional lymph node enlargement 2. Possible infection in cases of pregnancy, and immunodeficiency 3. Allergic to the material given	1. Immunizing healthy children only 2. Administering adequate fluids and appropriate medications between the fifth and eighth days to children known to have histories of high fever reactions 3. Avoiding immunization in cases of pregnancy, immunodeficiency, or a history of sensitivity to eggs
Rubella	1. Fever, rash, and mild local reactions 2. Pain, swelling, and stiffness in joints 3. Numbness and tingling sensation	1. Avoiding administration of vaccine during pregnancy, or where a possibility exists of pregnancy within 3 months of vaccination or where evidence exists of a fever, respiratory disease, immunodeficiency, or cancerous state

[a] Td, adult dosages of tetanus and diphtheria toxoids.

The vaccines for cholera, typhoid fever, typhus, and yellow fever are not regularly used in the United States except by military personnel or persons intending to go into endemic regions where certain vaccines are required. Unfavorable reactions from these preparations include fever (cholera, typhoid, and yellow fever), anaphylaxis (rare with the typhus vaccine), and inflammation of the nervous system (yellow fever). The last reaction occurs most frequently in children less than one year old.

TABLE 18–7 Recommended Immunizations for American Travelers

Preparation	*Areas Involved*	*Qualifying Remarks*
Cholera vaccine	Countries requiring vaccination for entry	Not recommended during pregnancy
Diphtheria and tetanus (Td) Immune globulin	Recommended universally Africa, Asia, Central and South America	None Used to prevent infectious hepatitis
Measles vaccine	Areas where the disease is present	Non recommended during pregnancy; should be used for all susceptible travelers
Plague vaccine	Cambodia, Laos, Vietnam	Should be used where sylvatic (wildlife rodent) plague exists
Poliomyelitis vaccine	Other than usual tourist routes, remote sections of tropical areas, where the disease is common	None
Rabies (human diploid cell vaccine) (preexposure)	Africa, Asia, South America	Recommended only for personnel residing in these areas; not for short-term visitors
Rubella vaccine	Areas where the disease exists	Not recommended for cases of pregnancy, immunodeficiency or allergy to eggs
Typhoid fever vaccine	In endemic areas or regions where outbreaks are occurring	Not recommended during pregnancy
Typhus fever vaccine	Endemic areas for the disease include Asia, Africa, Central and South America, Europe	Should be used only for persons remaining in endemic regions for long periods of time
Yellow fever vaccine	Countries that require vaccination for entry, including parts of Africa and South America	Not recommended during pregnancy or for individuals with documented allergy to eggs

Immunization for International Travel

Several infectious diseases are considered quarantinable under WHO International Sanitary Regulations. These regulations require individuals traveling to, through, or from certain specified regions to be immunized against specific diseases. Verification of vaccination is required by some countries at the time of entry, and in certain situations it may be needed for reentry. For example, individuals traveling through regions considered to be yellow fever country are required to present verification of vaccination against that viral disease. Certain immunizations recommended for American travelers are listed in Table 18–7. Immunizations against poliomyelitis and typhoid fever are generally not needed if sanitary tourist facilities are available and used.

An Introduction to Immunodiagnosis

An important application of the products of the immune response, antibodies or immunoglobulins, and sensitized T lymphocytes is their use in disease-agent identification, disease diagnosis, and the monitoring of responses to vaccines. Since the interactions between antigens and immunoglobulins are quite specific, a great variety of *in vitro* (in containers) and *in vivo* (in the body) tests have been developed. These immunodiagnostic tests generally, but not exclusively, involve serum (the liquid portion of blood forming after coagulation). The next chapter describes many of these tests, their applications, and the principles on which they are based.

Summary

1. One current approach to the prevention of microbial infections involves the use of live or inactivated microbes or their products.
2. Such preparations are usually effective if they are used before a specific infection has begun.

Immune Responses

1. Active immunization may not only cause the production of specific immunoglobulins but may also trigger cellular processes.
2. Passive immunization provides a short-term protection against certain infections and may change the outcome of certain diseases.

States of Immunity

1. Two major states of resistance are recognized: *innate,* or native, immunities and *acquired* immunities.
2. Innate immunity includes species, race, and individual resistance.
3. Acquired immunity may be either natural or artificial, depending upon the processes involved in producing the immune state.
4. Immunities can also be subdivided into *active* and *passive* immunities. In the active state, the individual produces antibodies in response to an immunogenic stimulus; in the passive state, antibodies are introduced into the individual from an outside source.

ACQUIRED IMMUNITY

1. A *naturally acquired active immunity* is generally the result of successful recovery from an infection or series of infections.
2. A *naturally acquired passive immunity* is obtained through the natural transfer of antibodies from an immunized donor to a nonimmune recipient. The immunoglobulins transferred from the mother to her fetus through the placenta are an example.
3. *Artificially acquired active immunity* is produced through immunizations. Materials used for this purpose include killed organisms, inactivated toxins, attenuated organisms, and living organisms mixed with homologous antiserum. In general, this state of immunity is rarely long-lasting.
4. *Artificially acquired passive immunity* is produced by the injection of appropriate levels of specific immunoglobulins.

Immunization Preparations

Biological preparations for immunization or diagnosis can be divided into three categories: (1) prophylactic (preventive) agents for active immunization; (2) prophylactic (protective) preparations for passive immunization; and (3) diagnostic reagents.

PREPARATION OF REPRESENTATIVE VACCINES

1. Most preparations of bacterial vaccines consist of killed organisms in isotonic salt solution.
2. Organisms used for vaccines are killed with the aid of heat or chemical agents such as formalin.
3. Toxins converted by heat or chemical agents into nontoxic antigenic preparations are called *toxoids.*
4. Immunizing materials against virus disease contain either attenuated (weakened) or killed viruses.

The Safety of Live Versus Inactivated Preparations

1. The live vaccines now in use generally produce a high level of immunity and are usually given in single doses, but they are also more subject to problems than inactivated vaccines.
2. Problems encountered with live-virus vaccines include (a) insufficient weakening of the pathogen, (b) spread of viruses to susceptible contacts, (c) contamination, (d) genetic changes, and (e) inactivation caused by inadequate refrigeration.
3. Larger amounts of antigen and multiple injections are needed with killed vaccines to produce levels of immunity comparable to those of live preparations.
4. The risk of accidents has been reduced by the establishment of better safety controls.

Antibody Production

1. The extent of an antibody response is affected by several factors, including (1) the nature of the immunizing material, (2) the dosage received, (3) the number and frequency of exposures, (4) the particular animal species, and (5) the individual involved.
2. The reaction to the first injection of an immunogen is called the *primary response.* In time, the antibody level, or titer, reaches a plateau that remains for several months or longer and then slowly declines as antibody breakdown exceeds production.
3. A second injection of some immunogens will produce a secondary, or *anamnestic,* response that produces higher, longer-lasting antibody levels than those associated with the primary reaction.

Vaccines Currently in Use

1. Immunizations should produce a high level of protection against infectious diseases.
2. Preparations should be safe, free from unpleasant side effects, and relatively simple to use.
3. Several vaccines consist of combined forms of immunizing preparations. Combined forms are simpler to use, less expensive to produce, and require fewer visits to the health-care facility than single-vaccine preparations.

BCG AND TUBERCULOSIS PREVENTION

1. The proper role of the bacillus of Calmette and Guérin (BCG) in tuberculosis prevention has been controversial.
2. The preparation is considered appropriate in several situations, including those in which the usual surveillance procedures (skin testing, etc.) and treatment programs have failed or cannot be readily applied.

Passive Immunization

1. Passive immunization is effective, acts immediately, and is temporary in its effect.
2. Preparations used for passive immunization are of two basic types: antitoxins and antimicrobial sera (immune globulin).
3. *Antitoxins* are antibodies capable of neutralizing the toxin that stimulated their production.
4. *Immune globulin (IG)* is one product obtained from human blood, plasma, or serum. It contains most of the types of immunoglobulins (antibodies) found in whole blood.
5. *Specific immune globulin (SIG)* contains a higher specific antibody concentration to the disease agent than that found in IG preparations.

Vaccines Under Development

1. A number of vaccines have been developed for experimental purposes or for individuals in high-risk occupations or situations.
2. DNA recombinant technology is being applied to the synthesis of safe, highly specific, and inexpensive vaccines.

The Role of Health-Care Personnel

1. Immunization is one of the most important means available for preventing and controlling infectious diseases.
2. Individuals responsible for administering immunizing materials must be aware of the nature of the vaccine, the correct method of administration, anticipated results, and contraindications and possible side effects.

ADMINISTERING OF VACCINES

1. Aseptic precautions must be taken during any type of inoculation procedure.
2. Several injection routes can be used for immunizations; these include intradermal, intramuscular, and subcutaneous routes. Oral and intranasal administration are two other techniques for administering vaccines.

Complications Associated with Vaccinations

1. Unfavorable reactions to vaccinations occur in certain individuals.
2. When an unfavorable reaction occurs, it should be fully described in the medical records of the individual.

Immunization for International Travel

The immunization requirements for international travel vary according to the specific regions to be visited.

Questions for Review

1. Define or explain the following terms:
   a. attenuated viruses
   b. active immunity
   c. artificially acquired immunity
   d. toxoid
   e. gamma globulin
   f. adoptive immunity

2. Discuss the impact of immunization on childhood infectious diseases.

3. Distinguish between a primary and an anamnestic response.

4. List five factors known to influence an individual's antibody response.

5. Differentiate between native and naturally acquired immune states.

6. What host factors contribute to an individual's immunity? What accounts for differences in the immunity of individuals?

7. Distinguish between active and passive states of immunity. Which of these is longer lasting?

8. a. What types of preparations are used in producing artificially acquired active immune states?
b. What types of preparations are used in producing artificially acquired passive immune states?

9. What are subclinical infections? What role do they play in immunity?

10. What general types of immunizing agents are currently available?

11. What is the difference between a toxin and a toxoid?

12. What is the nature of the immunizing material used with each of the following infectious diseases?

a. typhoid fever
b. whooping cough
c. measles
d. yellow fever
e. tuberculosis
f. influenza
g. hepatitis B
h. tetanus
i. mumps
j. polio
k. rabies
l. diphtheria
m. meningitis

13. a. What is a synthetic vaccine?
b. What advantages does it offer in the control of diseases?

14. Describe at least three routes of administration for vaccines.

15. Does the administration of IGs have any value in combating the effects of disease agents? If so, with which diseases does this beneficial effect occur?

16. What complications may develop as a consequence of vaccinations? Describe several common examples.

Suggested Readings

ACIP, "New Recommended Schedule for Active Immunization of Normal Infants and Children," *Morbidity and Mortality Weekly Report* **35:**577–579, 1986.

BLOCH, A. B., *et al.*, "Health Impact of Measles Vaccination in the United States," *Pediatrics* **76:**524–532. 1985. *An article emphasizing the safety and effectiveness of the measles vaccine and the enormous reductions in illness, deaths, mental retardation and expenditures brought about by immunization.*

KRUGMAN, S., "The Conquest of Measles and Rubella." *Natural History* **92:**16–20, 1983. *A short but informative article describing the events leading to the elimination of viral diseases.*

LANGER, W. L., "Immunization Against Smallpox before Jenner," *Scientific American* **234:**64, 1976. *Before Jenner introduced inoculation with cowpox, smallpox was prevented by inoculation with material from smallpox victims. This article describes how this procedure was performed in the eighteenth century.*

LERNER, R. A., "Synthetic Vaccines," *Scientific American* **248:**64–74, 1982. *This article reflects the increasing interest in synthetic virus vaccines as effective means of immunization. Attention is also given to the use of computer graphics to locate the viral proteins on viral surfaces for vaccine production.*

WISHNOW, R. M., and J. L. STEINFELD, "The Conquest of the Major Infectious Diseases in the United States: A Bicentennial Retrospect," *Annual Review of Microbiology* **30:**427–450, 1976. *This article reviews past successes in controlling diseases such as smallpox, tuberculosis, cholera, typhoid fever, malaria, yellow fever, poliomyelitis, and diphtheria in the United States.*

19

Diagnostic Immunologic and Related Reactions

". . . It is of the highest importance, therefore, not to have useless facts elbowing out the useful one." — *Sir Arthur Conan Doyle*

After reading this chapter, you should be able to:

1. Describe the importance of immunologic procedures to the diagnosis of microbial diseases.
2. Describe the general principles and features of commonly used diagnostic immunologic procedures.
3. Recognize the appearance of negative and positive immunologic reactions.
4. Differentiate between hemagglutination and hemagglutination inhibition.
5. Describe the forms and applications of gel diffusion.
6. Explain fluorescent antibody techniques.
7. Outline the bases of the ELISA, the Western blot test, and immunoperoxidase procedures.
8. Describe the hybridoma production of monoclonal antibodies.
9. List five current and potential applications of monoclonal antibodies.
10. Discuss the importance and applications of investigative electron microscopy.
11. Define the ABO and Rh systems and their significance.
12. Discuss the relationship of Rh factors and disease.

Chapter 19 applies immunologic principles and techniques to the diagnosis of infectious diseases and the identification of various microorganisms and their products, blood cell types, and cancer cells. It also shows how certain immunologic tests are used to determine the level of an individual's resistance to a disease agent or to follow the course of recovery from an infection.

Knowing whether or not an individual has antibodies to a given antigen is valuable in establishing the identity of a disease agent, in charting a patient's recovery from infection, or in determining the effectiveness of immunization. Because of the specificity of the antigen-antibody reaction, if either the antigen or the antibody is known, it is possible to identify and measure the other by employing one of a variety of *in vitro* and *in vivo* techniques.

During the past two decades, immunologic research has yielded an overwhelming body of information on humoral (circulating) antibodies. The branch of immunology concerned with the use of serum in tests to determine the presence and behavior of antibodies to infectious disease agents is called *serology.* Because of space limitations, detailed descriptions of immunologic and serological reactions will not be given in this text. References in the Suggested Readings are sources of detailed descriptions of procedures and mechanisms concerned with specific antigen and antibody reactions.

This chapter surveys a number of current methods for the detection and measurement of antibodies in human serum. Many of these serological (immunodiagnostic) procedures, or antibody detection systems, are powerful tools not only in the diagnosis of disease states but also in the identification of microorganisms and blood types. They are based on antibodies produced *in vivo* in response to the antigenic components of microorganisms and other cells. Microorganisms contain a wide variety of different antigens. Some of these are *type-specific,* limited to a particular species, while others are *common group antigens,* antigenic to related groups of microorganisms. Certain important procedures, such as the serological tests for syphilis, have been placed in the respective chapters concerned with the specific disease agent or its product. A discussion of the human blood groups and associated problems is also included in this chapter. Table 19–1 summarizes the features of several commonly used immunologic procedures.

Production of Antisera

To obtain potent antisera (blood serum preparations containing antibodies) for use in diagnostic tests, an experimental animal is inoculated with suspensions of a particular antigen. Animals used for this purpose include chickens, mice, horses, rabbits, sheep, and even humans. In most laboratory situations, the course of immunization involves a series of inoculations. The immunogen—which may be introduced intraperitoneally, intravenously, or subcutaneously—usually presents a number of epitopes (minimum biochemical units capable of stimulating an antibody response).

Determinations of the antibody level *(trial titrations)* are performed periodically by taking a blood sample from the laboratory animal. The blood is allowed to clot and the serum is removed for testing. Once the antibody concentration reaches the desired level, the animal is bled. The immune serum obtained should contain the antibodies produced in response to the immunogenic stimulus. These antibodies are capable of binding in some manner with the antigenic determinant that caused their formation.

Diagnostic Significance of Rising Antibody Titers

Generally speaking, the titer, or concentration of antibody in serum, fluctuates as a consequence of immunizations and of subclinical as well as full-blown current infectious states. To distinguish the antibody production associated with an actual ongoing infection from the effects of immunization or from antibodies associated with a past infection, at least two specimens (acute and convalescent) of a patient's serum are necessary. The first is obtained soon after the onset of the disease (acute) and the other approximately 12 to 14 days later (convalescent). The sera from both specimens are tested to determine if a rise in concentration or titer of the suspected antibody has occurred. If the titer has risen, as indicated by a greater antibody activity in the later specimen, identification of the causative agent is possible. If little or no antibody is detected in either specimen, it can be assumed, barring any abnormalities, that an organism other than the one being tested for is the cause of the infection. Abnormalities that could cause a lack of antibody response include (1) the administration of immunosuppressing drugs, (2) exposure to excessive radiation, or (3) the presence of a congenital defect such as an inability to produce immunoglobulins (agammaglobulinemia). If antibody is present in both samples with no appreciable change noted in the titers, it can be assumed that these antibody levels were present before the onset of the current infection and bear no relationship to the disease state. **[See immunologic defects in Chapter 20.]**

TABLE 19–1 Immunologic Procedures Used in Diagnosis and/or Microbial Identification

Procedure	*Principle Involved*	*Positive Test Results*	*Applications*
Agglutination	Antibody clumps cells or other particulate antigen preparations (insoluble particles coated with antigens, e.g., latex particles, *Staphylococcus* protein A, etc.)	Aggregates (clumps) of antigens	1. Diagnosis of typhus, Rocky Mountain spotted fever (Weil-Felix test), typhoid fever (Widal test), and infectious mononucleosis 2. Identification of disease agents including *Haemophilus influenzae* type b, *Neisseria meningitidis,* and *Streptococcus pneumoniae*
Complement fixation	Antigen-antibody complex of test system binds complement, which thereby becomes unavailable for binding by sheep red blood cells and hemolysin of the indicator system	Cloudy red suspension	Diagnosis of various bacterial, mycotic, protozoan, viral, and helminth (worm) diseases
Countercurrent immuno-electrophoresis (one form of electro-immunodiffusion)	Antigen and antibody are placed in separate wells and driven toward each other with an electric current (electrophoresis)	Precipitation forms at a point intermediate between the two wells in the system	Antigen and/or antibody semiquantitative detection
Enzyme-linked immunoabsorbent assay (ELISA)	Antigen or antibody from specimens trapped by corresponding specific antibody or antigen coating a solid phase suppport combines with enzyme-labeled specific antibody. The formed complex reacts with an added enzyme substrate in proportion to the amount of antigen or antibody first bound by the coating antibody or antigen	Color changes occurring with the addition of an enzyme substrate are proportional to either antibody or antigen in specimens	1. Detection of IgM to rubella and influenza A 2. Identification and/or detection of herpes simplex viruses types 1 and 2, cytomegalovirus, measles, hepatitis B, and AIDS viruses 3. Detection of antibodies to bacterial toxins
Ferritin-conjugated antibodies	Antibody, to which ferritin (iron-containing) particles are attached, binds various types of antigens	Presence of localized dark spheres in electron micrographs	Locating bacterial, fungal, viral, and other biological antigens by electron microscopy
Hemagglutination	Homologous antibody causes (hemagglutinin) aggregates of red blood cells to form[a]	Aggregates of red blood cells	Blood typing
Hemagglutination inhibition (viral)	Antibody inhibits the agglutination of red blood cells by coating hemagglutinating virus	Formation of a circle of unagglutinated cells	1. Determining the immune status toward rubella (German measles) 2. Virus identification
Immunodiffusion	Antibody and soluble antigen diffuse toward one another through an agar gel and react where homologous antibody is in proper proportion to homologous antigen	Lines of precipitate form within the agar	Antigen and/or antibody identification
Immunofluorescent microscopy	Antibody (usually) or antigen is labeled with a fluorescent dye, which fluoresces on exposure to ultraviolet or blue light	Glowing on exposure to UV light	1. Detection of antigen or antibody 2. Identification of microbial pathogens of diseases such as rabies, syphilis, Legionnaires' disease, etc.

[a] Hemagglutination reactions caused by certain viruses and bacteria generally do not involve antibody.

TABLE 19–1 (continued)

Procedure	*Principle Involved*	*Positive Test Results*	*Applications*
Immunoperoxidase	Antibody is labeled (conjugated) with an enzyme, usually horseradish peroxidase, which is detected by a color reaction produced upon treatment with a peroxidase substrate	Color changes occurring with the addition of a peroxidase substrate	Detection and identification of several viruses, including cytomegaloviruses, rabies, and herpes viruses
Precipitation	Antibody and soluble antigen react where they are in proper proportion to one another	Lines of precipitate form	1. Diagnosis of microbial diseases 2. Detection of antigens
Radioimmunoassay	Antibody or antigen can be labeled with a radioactive element, and the resulting complex precipitated and monitored for radioactivity	Radioactivity counts	1. Detection of antigen and/or antibody 2. Detection of hepatitis antigen
Rocket electrophoresis (electroimmunodiffusion)	Antibodies to a specific antigen or antigens are incorporated into a solid supporting agar medium. The specimen is added to a small well and allowed to migrate in an electrical field (electrophoresed)	Precipitation pattern forms a rocket or spike shape. The amount of antigen is proportional to the length of the rocket	Quantitation of most antigens
Virus neutralization	Antibody neutralizes infectivity	Absence of virus-destructive effects	1. Determining neutralizing effects of antibody 2. Virus identification and diagnosis
Western blot	Proteins of antigen are separated by electrophoresis, transferred to and immobilized on nitrocellulose strips, and then exposed to serum specimens. Antigen-antibody reactions are detected by an added enzyme-linked anti-human immunoglobulin reagent	Formation of a black precipitate in the regions where enzyme-immunoglobulin reagent is bound	1. Diagnosis of infectious diseases such as AIDS 2. Detection of antibody against different antigenic components

The Agglutination Reaction

This classic serological reaction was formally described by Gruber and Durham in 1896. Agglutination involves the clumping of cellular or particlelike antigens by corresponding (homologous) antibodies (Figure 19–1). This phenomenon is widely used for the rapid diagnosis of several infectious diseases and for the determination of blood types (Color Photograph 76). Because blood typing procedures involve reactions between erythrocytes and corresponding antibiodies, the *hemagglutinins* (specifically *isohemagglutinins),* the phenomenon is often called **hemagglutination** (Color Photograph 76 and Figure 19–2).

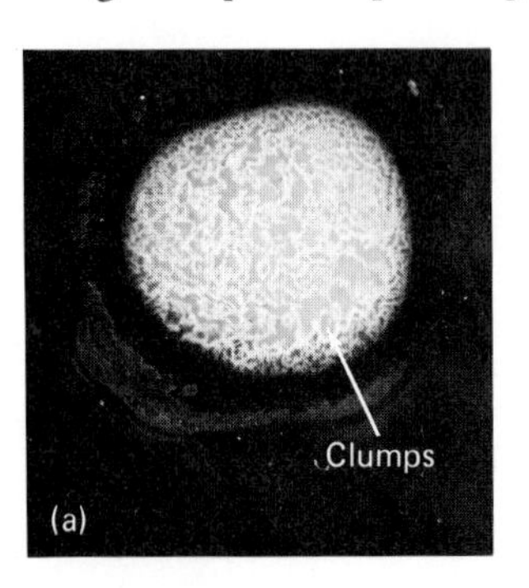

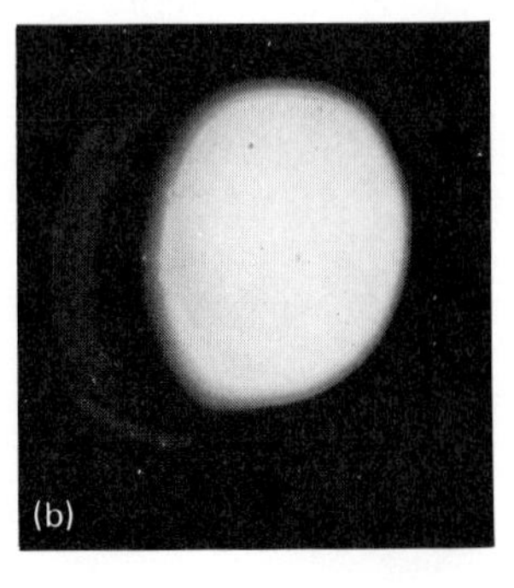

Figure 19–1 The agglutination test for the diagnosis of brucellosis in dairy cattle. This test is based on the presence of agglutinins in the blood of infected animals. When the antigen, a suspension of *Brucella* spp. cells, is added to serum, the agglutinins clump with the antigen. In the slide, or rapid, test these mixtures are spread into thin, even layers over a glass slide. Within a few minutes, clumping can be seen with the naked eye, as shown in (a). A negative result (absence of agglutinin) is shown in (b). *[Courtesy USDA.]*

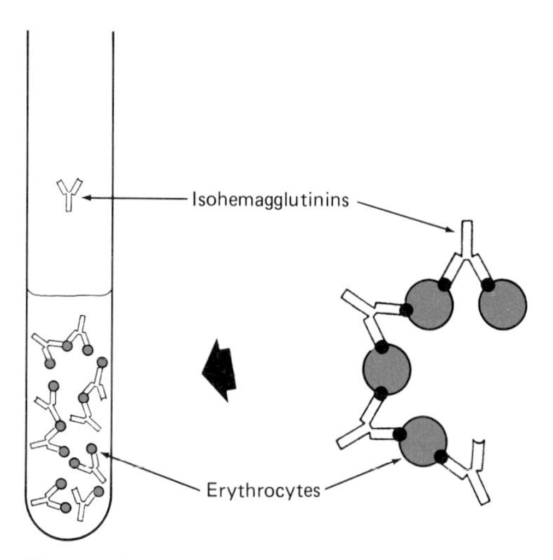

Figure 19–2 A diagram of the agglutination of erythrocytes by isohemagglutinins.

Examples of Agglutination and Related Reactions

COLD HEMAGGLUTINATION

The sera of patients with atypical primary pneumonia of mycoplasmal origin (Chapter 25) and protozoan infections, such as trypanosomiasis, contain antibodies that are capable of agglutinating erythrocytes from these patients at 2°C but not at 37°C. Such antibodies are called *cold agglutinins,* and the phenomenon is referred to as *cold hemagglutination.* These unusual antibodies are important to diagnosis in that they appear in association with few diseases.

HEMAGGLUTINATION

Other forms of hemagglutination exist, including the agglutination of red cells by viruses and mycoplasma. Several viruses, including influenza (Figure 19–3a and 3b), mumps, and cowpox or vaccinia are capable of binding to particular receptor sites of erythrocytes from suitable animal species. This activity forms a bridge between individual red cells, causing them to agglutinate.

WEIL-FELIX TEST

In 1916, E. Weil and A. Felix reported that a strain of the bacterium *Proteus* (PROH-tee-us), originally isolated from the urine of a typhus fever victim, was agglutinated by the sera of patients suffering from this disease. Serum from normal persons did not produce a similar result. Later studies showed that *Proteus* species (spp.) were not the cause of typhus fever and that antibodies against *Proteus* spp. normally occur quite commonly in humans. However, these bacteria can serve as an important tool in diagnosing various rickettsial infections.

The Weil-Felix reaction incorporates the somatic antigens (O) of *Proteus.* It is very important to use nonmotile organisms because motile strains have flagellar antigens that could interfere with the test. There are three strains that are used in the diagnosis of various rickettsial infections: OX2, OX19, and OX-K. [See rickettsial infections in Chapter 27.]

WIDAL TEST

The original Widal test was a microscopic procedure used for the laboratory diagnosis of typhoid fever. This method consisted of mixing drops of the patient's blood with a loopful of a 24-hour *Salmonella typhi* (sal-mon-EL-la TIE-fee) culture on a glass slide or other appropriate surface. After a 30- to 60-minute incubation period, the mixture was observed microscopically for the presence of clumping, which

Figure 19–3 (a) Viral hemagglutination. Virus particles attach to the receptor sites on the blood cells. Viral receptor sites here are different from those to which isohemagglutinins attach, in Figure 19–2. (b) A scanning micrograph showing agglutination of human erythrocytes by influenza viruses. The little dotlike structures on the blood cells are the viruses. *[Courtesy of Dr. L. F. Baker, Department of Microbiology, University of Southern California School of Medicine.]*

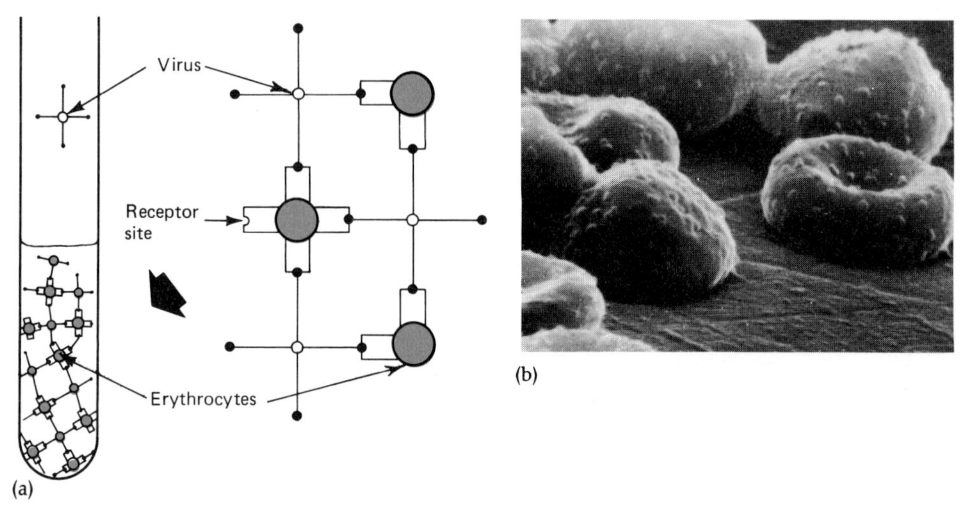

represented a positive test. A saline control consisting of saline and *S. typhi* was always included. As performed today, the Widal test has been modified to use more than one dilution of a patient's blood. **[See typhoid fever in Chapter 26.]**

PASSIVE AGGLUTINATION
In recent years, it has been possible to extend the agglutination reaction to a variety of soluble antigens by chemically attaching them to the surface of insoluble particles. In passive agglutination reactions, the types of particles used include bentonite (a mineral colloid), polystyrene latex spheres, red blood cells, and *Staphylococcus* (staff-il-oh-KOK-kuss) protein A. Adsorption of the soluble antigens usually is achieved by simply mixing them with the insoluble particles. Diagnostic tests based upon this technique are widely used for the detection and identification of a variety of disease states such as diphtheria and infections caused by the bacteria *Haemophilus influenzae* (hee-MAH-fi-lus in-floo-EN-zee) type B, *Neisseria meningitidis* (nye-SE-ree-ah men-in-jit-EE-diss), and several *Streptococcus* species. Examples of procedures include latex agglutination tests (Color Photograph 77), which are valuable in the diagnosis of several mycotic and other infections, and the bentonite flocculation test, which is used for the diagnosis of several helminth diseases. **[See helminth diseases in Chapter 31.]**

Hemagglutination Inhibition

In 1941, G. K. Hirst reported the phenomenon of viral hemagglutination and thereby provided a new picture and greater understanding of the relationship between viruses and host cells. His finding quickly led to the development of a new *in vitro* method, the hemagglutination-inhibition (HI) test, for the detection and titration of viral antibodies in a patient's serum and for the identification of specific viruses. Examples of viruses that clump erythrocytes and can, therefore, be tested by HI include influenza, mumps, vaccinia, Newcastle disease, and rubella (German measles). Viral hemagglutination is inhibited by specific antibodies against the virus in a reaction called **hemagglutination inhibition.** The mechanism involved in this reaction is quite simple. The antibody molecules attach to the viral particles, thus hindering adsorption of viral particles to erythrocytes. Failure of hemagglutination to occur constitutes a positive test for antibody. Unagglutinated red blood cells slide down the sides of the container that holds the reaction materials and settle directly to its bottom, forming a compact circular red "button" or "doughnut." Agglutinated cells stick to the sides and rounded portion of the chamber's bottom, forming a ragged-edged, filmlike deposit.

When the HI procedure is used to monitor the antibody response during and after suspected illness, two serum specimens *(paired sera)* are taken and tested, one shortly after the onset of symptoms *(acute phase)* and the other two to three weeks later *(convalescent phase).* Serial dilutions of the two specimens (e.g., 1:8, 1:16, 1:32, etc.) are prepared and then used to determine each specimen's antibody concentration or titer, which is the reciprocal of the highest serum dilution that completely prevents hemagglutination. Table 19–2 shows HI titers of a paired sera procedure. The 0 indicates hemagglutination inhibition, and the + represents hemagglutination. The titers are 8 and 256 for the acute and convalescent samples, respectively.

TABLE 19–2 Hemagglutination-Inhibition (HI) Titers of a Paired Sera Procedure

Serum Sample	*Serum Dilution Used*								*HI Titer*
	1:8	1:16	1:32	1:64	1:128	1:256	1:512	1:1024	
Acute	0	+	+	+	+	+	+	+	8
Convalescent	0	0	0	0	0	0	+	+	256

The Precipitin Reaction

Serological precipitation **(precipitin reaction)** is the reaction of a functional class of antibodies called *precipitins* with soluble antigens *(precipitinogens).* In experimental systems, a visible precipitate generally appears at the point where optimal proportions of antibody and antigen, as well as electrolytes, exist.

The main difference between precipitin reactions and agglutination reactions is the state of dispersion of the antigen used. In precipitin reactions, the molecules of precipitinogens are soluble so that their solutions appear clear, while the molecules of agglutinogens are particulate (resembling small particles) and produce cloudy solutions.

Ring or Interface Tests

In the course of an infection, precipitating antibody is often produced. This reaction is usually in response to soluble microbial substances released by a disintegrative process. The presence of these antibodies can be demonstrated by the *ring* or *interface* test based on a useful form of the precipitin reactions. In this procedure, antigenic material is carefully layered over an equal quantity of antisera in a narrow tube to form a sharp liquid interface. If the antigen and antibody are homologous, a ring of precipitation develops at the interface. In this test, the reactants diffuse into one another until the optimal, immunologically equivalent proportions for precipitation are achieved. In modern practice, agar gel bases (gel matrix) are usually incorporated to stabilize the precipitates that result. Examples of such gel diffusion tests include single-diffusion gel procedures, such as the *Oudin test, double diffusion procedure, radial immunodiffusion,* the *Ouchterlony test (double diffusion),* and *two-dimensional immunoelectrophoresis.*

OUDIN TEST (SINGLE DIFFUSION)

The Oudin (single diffusion) test is performed by introducing antiserum into a tube of melted agar at a temperature of about 45°C. This mixture, filling a narrow tube to one-third of its height, is allowed to solidify. An antigen-containing solution is then layered on the agar preparation, and the entire system is refrigerated. Diffusion of the antigen into the agar gel takes place during this refrigeration. Depending on the number of antigenic components in the test material, one or more precipitin bands form with homologous antibodies (Figure 19–4). One distinct band develops for each homologous soluble antigen-antibody system present. This feature of the Oudin method is a pronounced advantage.

RADIAL IMMUNODIFFUSION

In recent years, radial immunodiffusion has become widely used for the measurement of a variety of proteins in serum and other body fluids. In this technique, antiserum is incorporated at a relatively low concentration in a gel on a tray. Several wells are made in the gel, and various concentrations of a test antigen and controls are placed in the wells. As antigen diffuses into the gel, it forms a progressively widening circle of precipitate. The diameter of the precipitant ring is directly proportional to the antigen concentration in the well (Figure 19–5). This direct relationship between antigen concentration and ring diameter allows one to calculate the concentration of a known antigen.

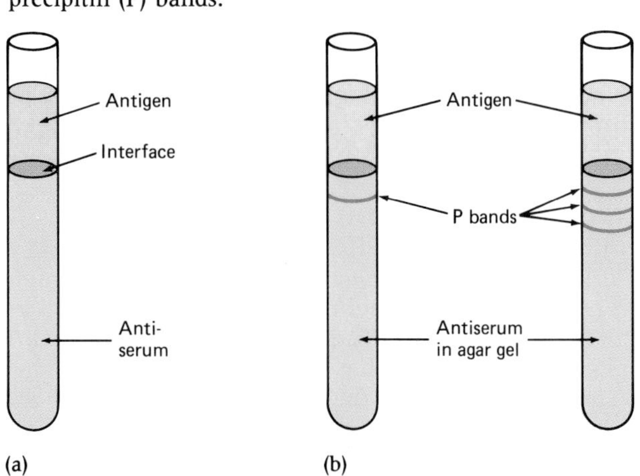

Figure 19–4 The precipitin test and its variations. (a) The ring test. (b) The Oudin technique. The left tube contains a simple system with one type of antigen and homologous antibody. The right tube contains a complex system of multiple antigens, as indicated by the multiple precipitin (P) bands.

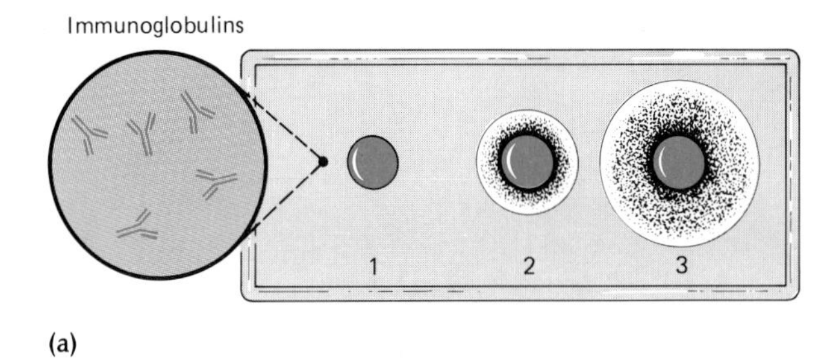

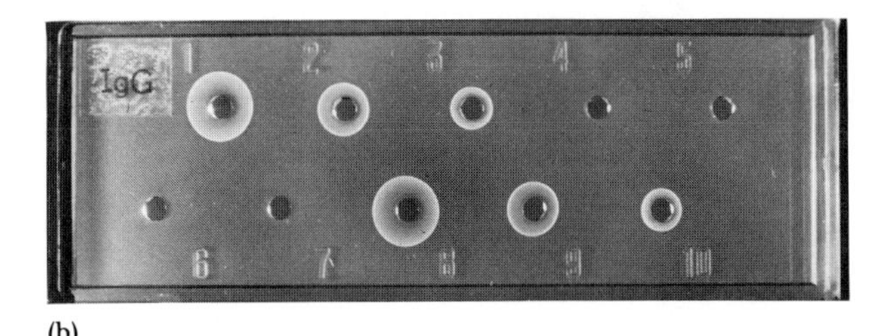

Figure 19–5 Single radial immunodiffusion using a gel containing immunoglobulins against the antigen under study. (a) Well 1 contains no antigen, well 2 contains a low concentration of antigen as indicated by the small surrounding precipitant ring, and well 3 contains a high antigen concentration as indicated by the diameter of the large surrounding ring. (b) The results of an actual immunodiffusion assay. This technique, which may take various forms, can be very accurate and is often the best method for measuring substances such as immunoglobulin classes or subclasses and serum complement components. *[Courtesy Helena Laboratories.]*

OUCHTERLONY TEST (DOUBLE DIFUSSION)

A variation of the single diffusion technique of Oudin is the Ouchterlony or double diffusion procedure. The Ouchterlony test also incorporates gel diffusion. In this test, an agar pour plate is made without antiserum. After the agar solidifies, circular or square holes or wells are made in the surface. Solutions containing antigens are added to certain wells while the antibody preparation is placed in another well (Figure 19–6). The plate is incubated to allow the various reactants to diffuse. Lines or bands of precipitate develop where antigenic components react with homologous antibodies from the antibody preparation. The results obtained by this technique depend upon several factors, including the relative

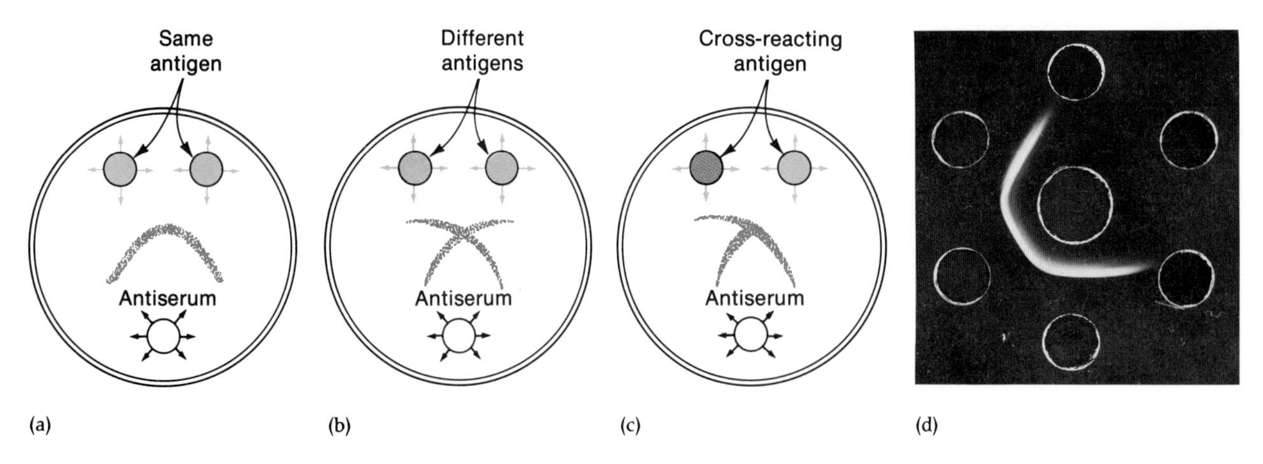

Figure 19–6 Selected geometrical patterns that indicate double-diffusion precipitin reactions. The small arrows around each well indicate the direction of diffusion. (a) Reaction of antigen identity. Both upper wells contain the same antigen, and the lower well contains homologous antiserum. (b) Reaction of antigen nonidentity. Each of the upper wells contains a different antigen, and the central well holds antisera to both antigens. (c) Reaction of partial identity or cross-reaction. The upper left well contains a cross-reacting antigen, and the upper right well is filled with an antigen that is homologous for the antiserum preparation in the lower well. (d) The actual appearance of an antigen identity reaction. *[D. Stickle, L. Kaufman, S. O. Blumer, and D. W. McLaughlin,* Appl. Microbiol. ***23****:490–499, 1972.]*

concentrations of reactants, the incubation period, the rate of diffusion, and the molecular weight of the antigens used. The arrangement shown in Figure 19–6c has a distinct advantage over the Oudin method in that several different antigen-containing solutions can be tested at one time.

The Ouchterlony double diffusion procedure can produce several different geometric patterns of precipitin reactions between the separate wells containing antigen and antibody solutions. Lines generally occur where the distance between the two reactants is at a minimum.

Much useful information can be obtained from this technique, including the detection of identical or cross-reacting components of various antigen or antibody preparations (gel diffusion analysis). Three patterns that represent possible reactions are shown in Figure 19–6. The precipitin band pattern of the *reaction of identity* develops when pure antigen preparations are placed in two wells adjacent to a centrally located well containing a homologous antibody solution (Figure 19–6a and d). Note that the bands fuse.

The pattern of Figure 19–6b shows the pattern formed by two unrelated antigens and an antiserum preparation that contains an antibody for each of the antigens. The arrangement of precipitin bands here is referred to as the *reaction of nonidentity.* The antigens and antibodies diffuse toward one another from their wells, forming two distinct precipitin bands that cross.

Antiserum preparations usually react with their homologous antigen counterparts. However, some antigen preparations are so similar to the homologous antigen substance that reactions between similar *heterologous* antigens and an antiserum can occur. These events are referred to as *cross-reactions.* Such reactions may be the result of several factors, including (1) the presence of several different antigenic molecules in an antigen preparation as laboratory contaminants; (2) the presence of numerous types of antigenic molecules as natural components; and (3) the limited specificity of the antiserum preparation.

The Ouchterlony *reaction of partial identity* or *cross-reaction* is shown in Figure 19–6c. In this situation, one of the antigens and the antiserum used form a homologous system; however, the second antigen present is a cross-reacting one. The precipitin bands that develop fuse, but one of them continues toward the cross-reacting antigen preparation. This *spur,* as it is sometimes called, is formed by those antibody molecules that have not combined with the cross-reacting antigen and have gone beyond the precipitin band facing it.

TWO-DIMENSIONAL IMMUNOELECTROPHORESIS

In two-dimensional electrophoresis, a serum sample is subjected to electrophoresis (movement caused by an electric current). The separated components are placed next to a gel that contains specific antibodies (Figure 19–7). Then the separated components are again subjected to an electric field, this one perpendicular to the original separating field. The compo-

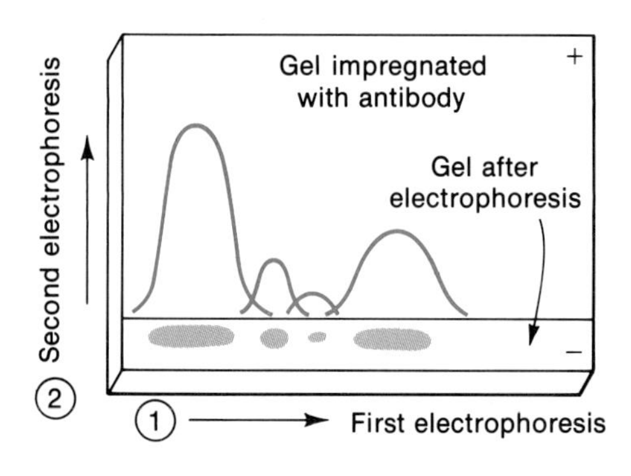

(a)

(b)

Figure 19–7 Two-dimensional electrophoresis. In this technique, components separated by normal electrophoresis are electrophoresed a second time in a gel containing specific immunoglobulins. The resulting reactions produce specific precipitant arcs. (a) The steps involved in two-dimensional electrophoresis. *[C. J. Smyth, A. E. Friedmans Kien, and M. R. J. Salton,* Infect. Immun. ***13****:1273–1288, 1978.]* (b) The appearance of actual precipitant arcs formed in the procedure.

nents move into the antibody-containing gel, where specific precipitant arcs are formed.

Electroimmunodiffusion

In the immunodiffusion techniques described earlier, antigen and antibody are allowed to move toward one another and to precipitate in an agar

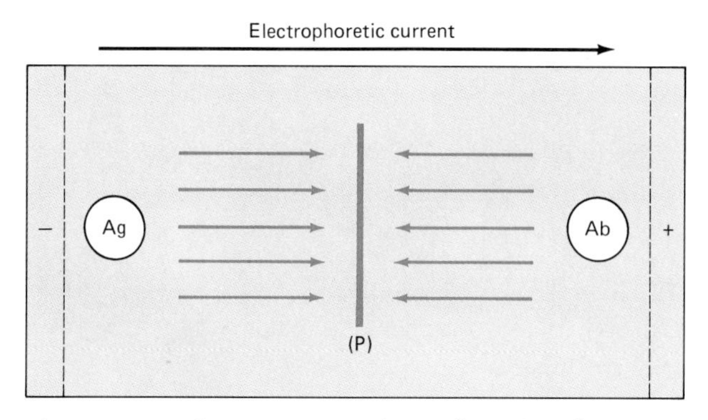

Figure 19–8 Countercurrent electrophoresis. When an electric current is applied across the gel, the antigen and antibody move towards each other. A precipitate (P) forms at the area where they are proportional to one another.

medium by diffusion. The technique of electroimmunodiffusion combines this form of immunoprecipitation with electrophoretic separation (movement in an electrical field) to identify as well as to approximate the concentration of specific proteins in serum, urine, or other body fluids. Although several variations of this combined approach have been described, only two have clinical applicability. These are *counterimmunoelectrophoresis* (one-dimensional double electroimmunodiffusion) and *Laurell's rocket electrophoresis* (one-dimensional single electroimmunodiffusion).

Counterimmunoelectrophoresis

This method is so named because antigen and antibody are made to move toward each other by an electric current (Figure 19–8). Visible precipitin lines develop at a point where they meet in optimal proportions. Counterimmunoelectrophoresis is at least ten times more sensitive than the double diffusion technique. It can be used for the detection of several fungal antigens and for certain antibodies used in the diagnosis of some cancers.

Rocket Electrophoresis

This method, also known as one-dimensional single electrophoresis, is used to determine the concentration of particular antigens, immunoglobulins, or complement components. Antiserum (antibodies) against the substance to be quantitated is incorporated into a gel-supporting medium on a glass slide. A specimen containing the unknown concentration is placed into a small well and electrophoresed. The resulting pattern of immunoprecipitation resembles the trail of a rocket (Figure 19–9). The total distance traveled by the unknown quantity of the substance is proportional to its concentration. Thus, a greater concentration will form a longer rocket trail.

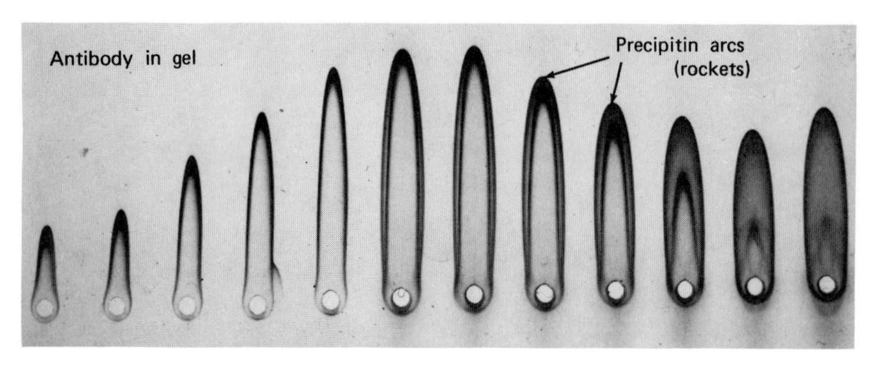

Figure 19–9 Rocket electrophoresis. Antigen standards in increasing amounts are placed in the first five wells on the left. The remaining wells contain unknown antigen specimens. The lengths of the individual rockets are proportional to the amount of antigen. *[From I. Brandslund, B. Teisner, P. Hole, J. G. Grudzinskas, and S-E. Svehag,* Acta Pathol. Microbiol. Immunol. Scand., *sect C.* ***91****:51–57, 1983.]*

Western Blot

Western blots can be used to detect and to quantitate antigen or antigen-antibody complexes (combinations). In the Western blotting technique, the components of an antigen or a mixture of proteins are electrophoretically separated and transferred to a nitrocellulose filter *(transblotting)*. The resulting transblots are cut into strips and incubated with test serum specimens to allow for antigen and antibody reactions to occur (Figure 19–10). After this step, the strips are washed to remove any unbound antibody molecules and treated with an enzyme-linked antihuman immunoglobulin (Ig) reagent, such as a horseradish peroxidase-conjugated antihuman immunoglobulin preparation. This reagent combines with the human antibodies in the test system. Antigen-antibody complexes formed in the blotting procedure are made visible by the addition of two other reagents, such as hydrogen peroxide and 3-diaminobenzidine, producing an insoluble black precipitate in the region of antihuman Ig-bound peroxidase (Figure 19–11).

The Western blot technique has been of major importance in the detection of antibody to the human immunodeficiency virus (HIV). Alternate approaches of the technique are available and include the use of radioactively labeled or fluorescent-dye-coupled antibody reagents.

In Vitro Hemolysis Tests

Complement Fixation

The normal serum component complement is found in a variety of animals. Historically, the contribu-

Figure 19–10 Detection of antibodies against specific antigen components by means of the Western blot technique (Western blotting). (a) Attachment of specific antibody to antigens immobilized on a nitrocellulose strip. (b) Detection of bound antibody (antigen-antibody complex) by the addition of peroxidase-labeled anti-human Ig. The enzyme reagent binds to the antibody of the complex. (c) The addition of the reagents hydrogen peroxide and 3-diaminobenzidine results in the formation of a black precipitate in the region of the anti-human Ig bound peroxidase.

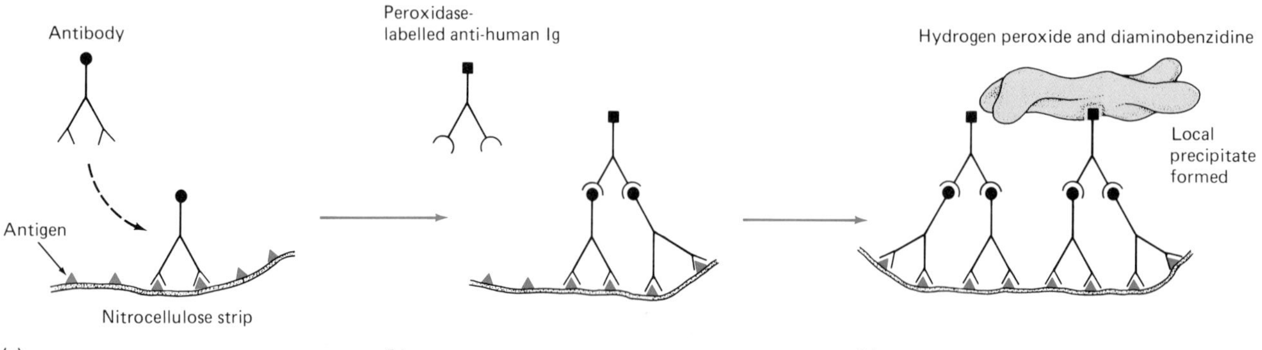

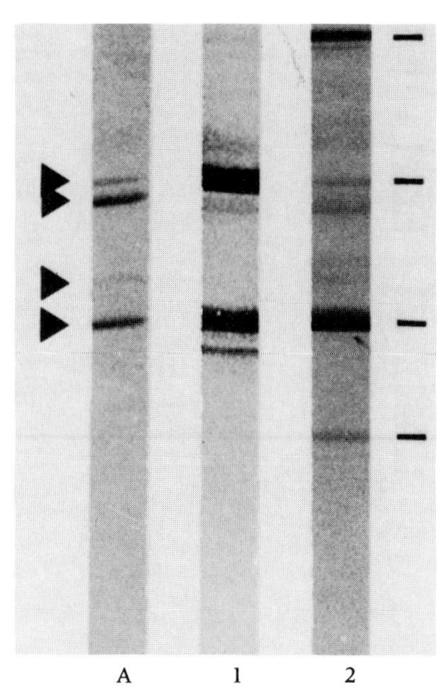

Figure 19–11 The appearance of immunoblots (Western blotting) showing the different patterns of IgG antibody binding to antigenic components. The antigenic components are indicated by triangles (►) in column A, and the reacting immunoglobulins by dashes (—) in columns 1 and 2. *[From E. Wedege, and L. O. Froholm,* Infect. Immun. ***51****:571–578, 1986.]*

tion of complement to the destruction of invading cells was first noted by Pfeiffer in 1896. His studies compared the fates of cholera-causing organisms (*Vibrio cholerae,* VIB-ree-oh KOL-er-ee) upon intraperitoneal injection into normal and immunized guinea pigs. The vibrios taken from normal animals at various intervals appeared normal. But, those organisms taken shortly after their initial injection into the immunized guinea pigs swelled, stained unevenly, and eventually burst. The immunized animals showed no sign of infection, and apparently the vibrios underwent *cytolysis,* or cellular disruption.

The fact that serum components participated in the phenomenon was shown *in vitro* by exposing a hanging drop preparation of the cholera-causing vibrios to (1) normal serum, (2) untreated immune serum, and (3) immune serum that had been heated to 56°C for 30 minutes, conditions known to inactivate complement. Only the untreated immune serum caused the lytic effect. Pfeiffer's experiments demonstrated the presence in immune serum of a thermolabile (heat-sensitive) substance that contributes to the destruction of bacteria.

The reaction was one of *immune cytolysis.* Normal sera had no effect. This type of reaction apparently is not limited to bacteria. In 1895 Bordet reported a somewhat similar phenomenon with antisera prepared against red blood cells. This observation led to the development of the complement fixation test (Color Photograph 78).

THE COMPLEMENT FIXATION PROCEDURE

Recognition of the fact that various antigen-antibody combinations have the ability to fix, or combine with, complement provides a means for detecting either antibodies or antigenic substances in unknown specimens. **[Refer to the discussion of the complement cascade in Chapter 20.]** Two systems are incorporated in this technique, the **test** and **indicator systems.** In a typical laboratory diagnostic situation, the *test system* consists of the patient's serum, an antigen preparation that may be commercially made or laboratory-produced, and complement, commonly obtained from guinea pig serum and commercially available. The components of the *indicator system* are sheep red blood cells and serum containing homologous antibodies against them. This antibody preparation is referred to as *hemolysin.*

All sera used in complement fixation tests are heated before use to 56°C for 30 minutes to inactivate any complement present. Reagents must be freshly and carefully prepared and must be combined in proper proportion to achieve a suitable balance between components. For example, enough hemolysin must be added to the indicator system to make the sheep erythrocytes susceptible to lysis by complement. The concentrations needed are determined before the complement fixation is performed.

A POSITIVE TEST. If a heated serum specimen is suspected of having specific antibodies, it is diluted and the dilutions are incorporated, together with antigen and complement, in the test system. Then, the components of the indicator system are added to the reaction mixture. If the serum sample contains antibody, the antibody will combine with antigen and fix the complement in the system so that no free complement remains (Figure 19–12). If the complement has been fixed, then no lysis (destruction) of the sensitized sheep erythrocytes occurs when the indicator system is added. Thus, a positive complement fixation test is represented by a cloudy red suspension (Color Photograph 78). This result indicates the presence of antibody toward the antigen used in the test system.

A NEGATIVE TEST. If a serum sample lacks antibody against the particular antigen used in the test system, complement fixation does not occur. Thus, complement present in solution is free to react with the components of the indicator system. The antigen-antibody complex formed by the sheep erythrocytes and hemolysin fixes complement, and lysis of the red blood cells occurs, producing a clear red solution (Color Photograph 78 and Figure 19–12). A negative test result (hemolysis) indicates the absence of antibodies toward the antigen used in the test system.

CONTROLS. Certain controls must be used in the complement fixation test because of the sensitive

Components in Complement Fixation

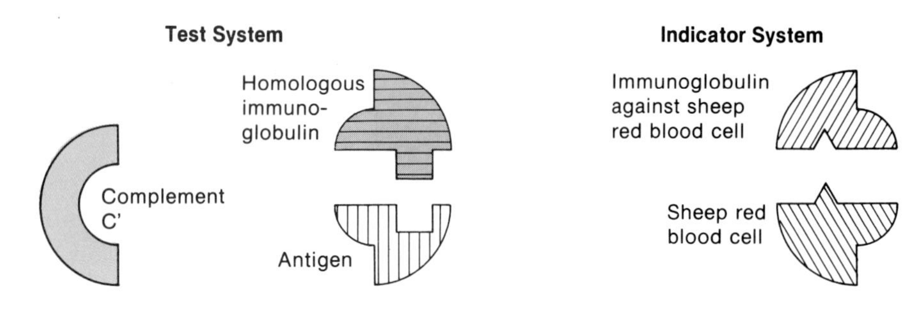

Final Combinations of Components

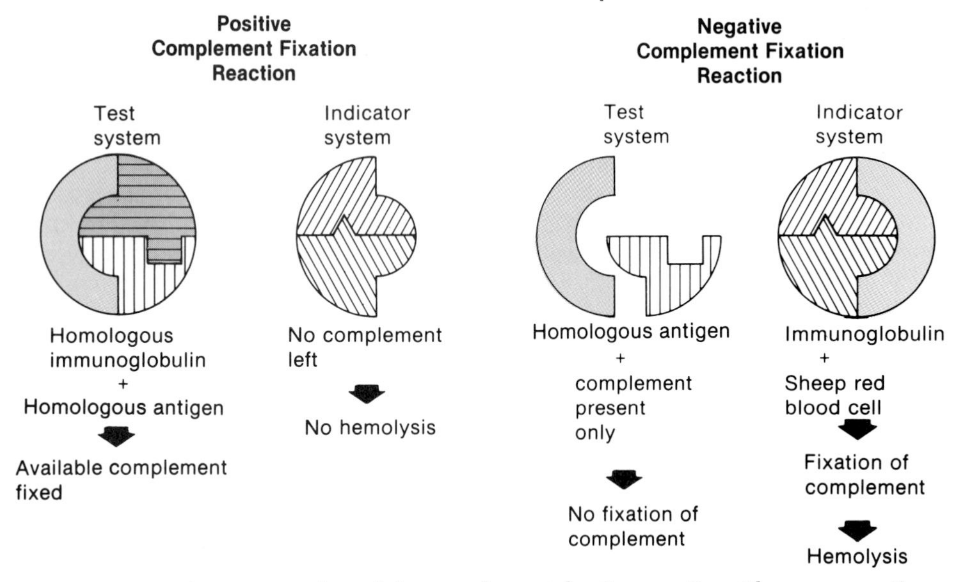

Figure 19–12 A representation of the complement-fixation reaction. The upper portion shows the individual reacting components (reagents) used. The lower section shows the combinations that occur in positive and negative tests.

nature of both the complement and the erythrocytes. Antigen and the patient's serum controls are used to detect any possible anticomplementary activity that might produce misleading test results. The reliability of the indicator system of sheep erythrocytes and hemolysin is also separately determined.

MECHANISM

The reaction of complement with sensitized erythrocytes has been studied extensively. The basic reaction occurs in several stages, the components of complement combining with sensitized cells in a precise manner. Once antibody (principally IgM and IgG) has attached to an erythrocyte's surface, the stage is set for the subsequent events. The entire sequence of reactions produces actual holes in the cell's membrane measuring about 10 nm. Through these circular lesions, small intracellular molecules escape and extracellular water enters rapidly, thus causing the cell to expand and to rupture (Figure 19–12b).

APPLICATIONS

Complement fixation procedures have been used in the laboratory diagnosis of a wide variety of microbial infections and certain helminthic (worm) diseases (Table 19–3). The most widely recognized application has been for the diagnosis of syphilis. The antibodylike material associated with this disease is the same as that involved with other flocculation tests, such as the Venereal Disease Research Laboratory (VDRL) and Kahn tests. [See serological tests for syphilis in Chapter 28.]

Antistreptolysin Test

The heat-sensitive hemolysin of streptococci, known as *streptolysin O,* is antigenic in humans and most laboratory animals. In various human streptococcal

TABLE 19–3 Representative Diseases for Which Diagnostic Complement Fixation Tests Can Be Used

Bacterial Disease	*Mycotic Disease*	*Protozoan Disease*	*Viral Agent or Disease*	*Helminthic Disease*[a]
Chlamydial infections	Aspergillosis	Amebic dysentery	ARD[b]	Echinococcosis
Gonorrhea	Blastomycosis	Chagas' disease	Cytomegalovirus	Paragonimiasis
Listeriosis	Coccidioidomycosis	Leishmaniasis	Enteroviruses	Schistosomiasis
Syphilis	Histoplasmosis	Malaria	Influenza	Trichinosis
Rickettsial infections	Sporotrichosis		Measles	
Tuberculosis			Mumps	
Whooping cough			Varicella (chickenpox)	
			Shingles	

[a] In several instances these tests are under development. Refer to Chapter 31 for a description of these worm infections.
[b] Acute respiratory diseases caused by adenoviruses.

diseases, antistreptolysin O titers (concentrations) increase during the times of infection and convalescence. In many diagnostic laboratories, the antistreptolysin test is routinely used to measure the levels of antibodies against this bacterial product and possibly predict the development of rheumatic fever and other streptococcus-associated disease states.

In Vitro Immunologic Procedures Incorporating Labeled Antibodies, Differential Staining, and Enzymatic Reactions

Fluorescent Antibody (Immunofluorescence) Techniques

Fluorescent antibody methodology is a combination of the techniques of immunology and cytology. In a broad sense, the procedure involves bringing a fluorescent dye (marker) into contact with serum proteins so that they combine together chemically (Figure 19–13). This process is referred to as *labeling* or *conjugation*. The resulting preparations generally remain active biologically. These fluorescent serum-protein conjugates, containing specific antibodies, can be used to detect homologous antigens in smears or tissue sections. The most commonly used of these methods are called *direct* and *indirect* methods (Figures 19–14, 19–15 and 19–16 and Color Photographs 17 and 79).

DIRECT FLUORESCENT ANTIBODY METHOD

The direct fluorescent antibody method is considered to be the simplest one because it requires a single staining reagent and because the manipulations involved are not complex. An antigen-containing specimen is fixed to a slide with acetone, ethanol, methanol, or heat. A few drops of a standardized labeled antibody preparation are applied to the specimen. This test system (the specimen and the labeled antibody) is incubated. Following incubation, the specimen is washed with saline and distilled water to remove excess antibody. The resulting preparation is usually dried, mounted in glycerol, and examined under a fluorescence microscope. If the ho-

Figure 19–13 The labeling of a serum protein with a fluorescein derivative by an isocyanurate amine linkage. *[Based on M. Goldman,* Fluorescent Antibody Methods. *New York: Academic Press, 1968.]*

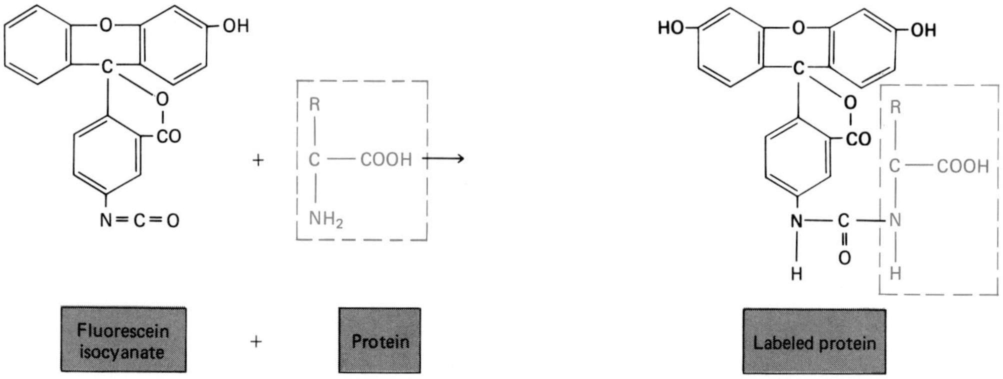

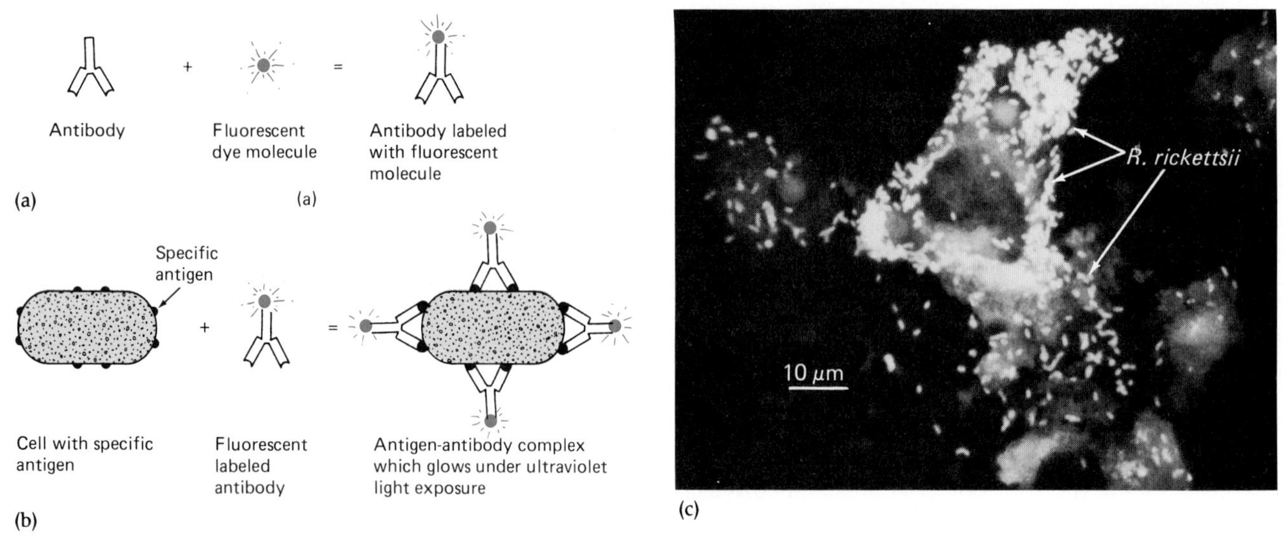

Figure 19–14 The direct fluorescent antibody technique. (a) The components used in the preparation of fluorescent antibodies are shown. (b) The direct fluorescent antibody technique schematically represented. Antigen and fluorescent-tagged antibody are combined to form the antigen-antibody complex, which will glow when exposed to ultraviolet light. (c) Numerous immunofluorescent *Rickettsia rickettsii* (the cause of Rocky Mountain spotted fever) mainly within the cytoplasm of infected human endothelial cells. The bar marker represents 10 μm. *[D. H. Walker, W. T. Firth, and C-J. S. Edgell,* Infect. Immun. ***37****:301–306, 1982.]*

mologous antigen is present in the specimen, the labeled antibody will combine with it and thus pinpoint its location (Figure 19–14).

CONTROLS. To verify the immunologic specificity of the fluorescence observed with the unknown antigen-containing specimen, certain controls must be applied. These may include:

1. Exposure of the antigen-containing specimen to labeled serum that does not contain antibody against the antigen in question. Fluorescence should not be observed.
2. Exposure of the antigen-containing specimen to homologous unlabeled antibody preparation. Here, too, fluorescence should be absent.

PURPOSE. Staining antigens by the direct method serves as a means of identifying unknown microbial agents or products and can be used to determine the distribution of antigens in tissue.

INDIRECT FLUORESCENT ANTIBODY METHOD

The indirect (antiglobulin) fluorescent antibody method involves formation of an "immunologic sandwich" in which nonfluorescent antibody (globulin) attached to primary antigen is, in turn, bound by fluorescent antiglobulin. The antigen itself is not directly rendered fluorescent, but rather the nonfluorescent antibody bound to the antigen. This indirect method utilizes both the antigen-binding capability of antibodies and their protein nature, which enables them to serve as antigens. The results of this method—the microscopic appearance of the stained antigen—are generally indistinguishable from those obtained with the direct fluorescent antibody techniques. Figure 19–15 is a schematic representation of this method.

PROCEDURE. Standardized reagents are used. After the specimen is fixed in a manner similar to that described for the direct test, unlabeled specific antiserum is applied to it. These reactants are incubated together for 15 to 60 minutes. The preparation is washed in saline, then in distilled water, and subsequently dried. Following this step, labeled antiglobulin is layered over the specimen, and the preparation is incubated again for 15 to 60 minutes. Rinsing and washing steps are carried out as in the direct test. The resulting antigen-globulin-antiglobulin complex is dried, mounted, and examined with a fluorescent microscope (Figure 19–16).

CONTROLS. This technique has much greater sensitivity than the direct method. For this reason, adequate controls are even more important to establish the specificity of the staining reaction. The possibility of nonspecific fluorescence in the test system must be ruled out. Suggested controls generally include incubating homologous antigen with combinations of the following reagents: (1) saline and labeled antiglobulin; (2) normal serum and labeled antiglobulin; (3) specific antiserum; and (4) labeled

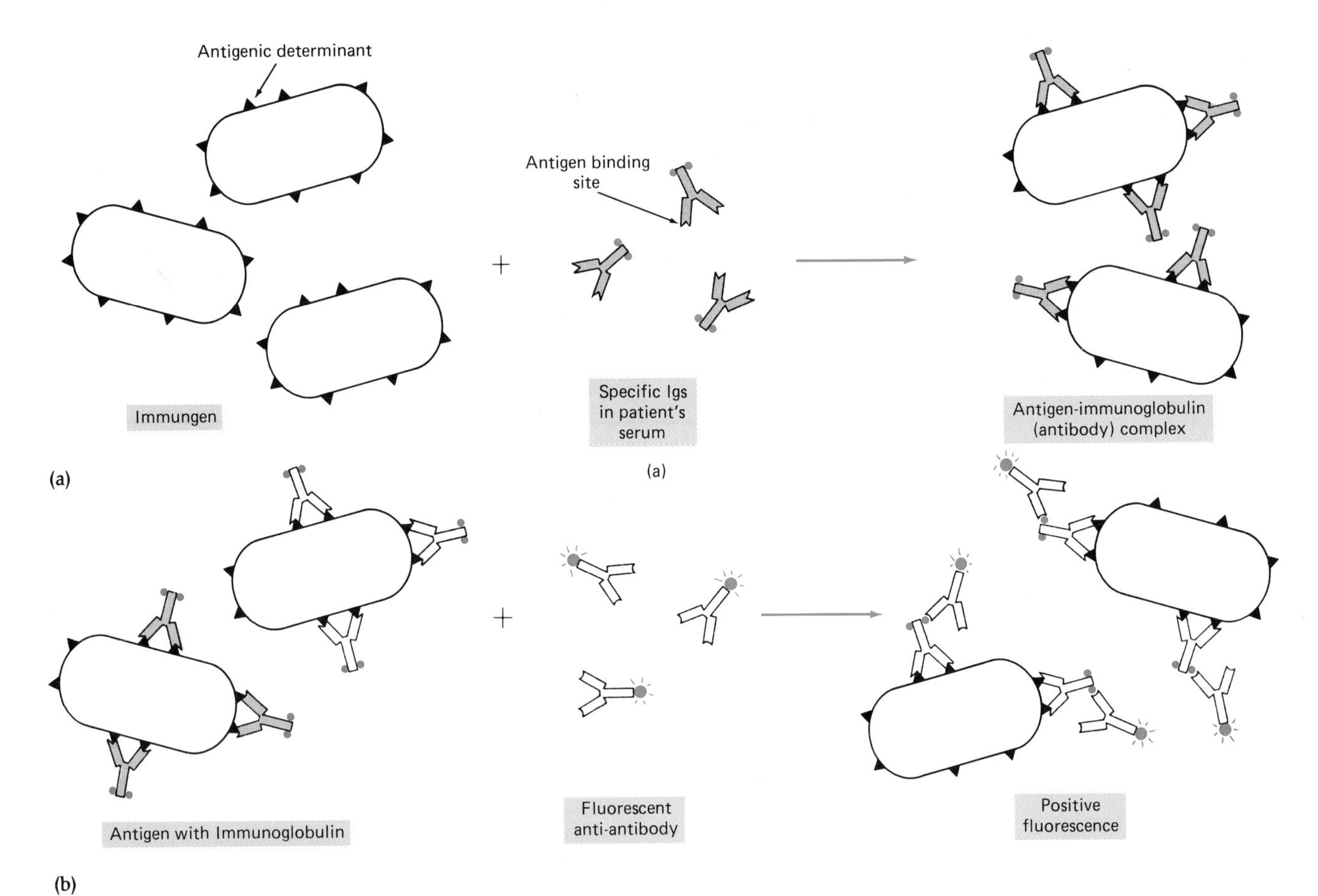

Figure 19–15 Demonstrating the presence of antigenic determinants on immunogens by the indirect fluorescent antibody technique. (a) Specific immunoglobulins react with antigenic determinants on the surface of immunogens. The reaction forms a specific antigen-antibody complex. (b) The addition of fluorescent anti-antibody to the complex results in the fluorescent antibody attaching to the specific immunoglobulin that is attached to the surface antigenic determinant. The resulting combination will glow (fluoresce) when examined under ultraviolet light.

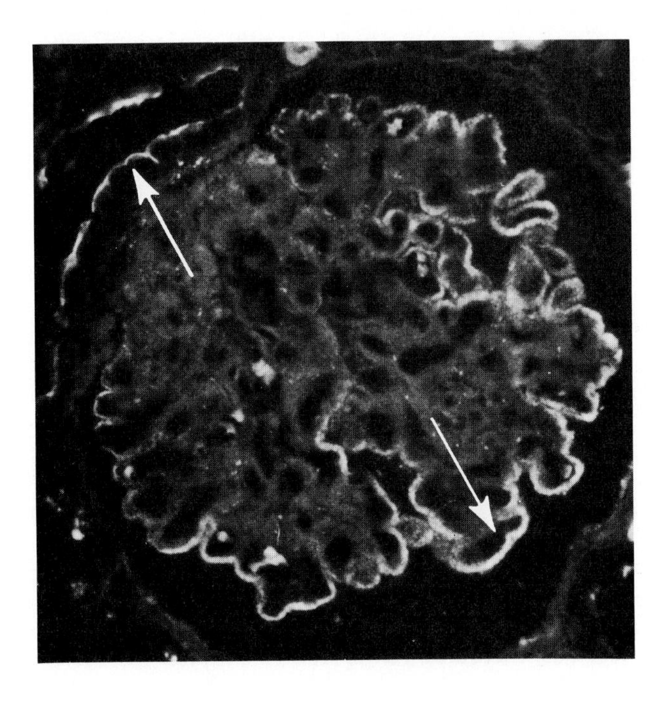

Figure 19–16 Immunofluorescent microscopy is an important laboratory method in diagnostic pathology, especially in diseases of the kidney and circulatory system. This micrograph shows the results obtained by the indirect method in an immunologic disorder of the kidney. The presence of immunoglobulins, IgG, is indicated by the glowing fringes and tufts in the kidney glomerulus. *[S. N. Huang, H. Minassian, and J. D. Moore,* Lab. Invest. ***35****:383, 1976.]*

normal serum. An additional control, which entails incubating heterologous antigen with a specific antiserum and labeled antiglobulin, could also be incorporated. If the fluorescence in the indirect test is specific, then fluorescence should not be observed with any of these controls.

Applications. The indirect fluorescent antibody technique has been used to demonstrate antibodies against various tissues (Figure 19–16), antigen-antibody complexes, and microbial pathogens, including *Treponema pallidum* (tre-poh-NEE-mah PAL-li-dum) the causative agent of syphilis, and herpes simplex virus (the causative agent of fever blisters). [See herpes simplex infections in Chapter 23.]

Variations of this basic technique exist. A test for the presence of antibody in serum, "fluorescent antibody serology," has been used in the diagnosis of syphilis and the detection of malaria antibodies.

In Vitro Immunologic Procedures Using Enzymatic Reactions

Two reliable diagnostic and sensitive techniques for the detection of viruses and certain cancer cells, the immunoperoxidase test and the enzyme-linked immunoabsorbent assay *(ELISA)*, incorporate enzymatic reactions.

Immunoperoxidase Test

In principle, the immunoperoxidase test is similar to fluorescent antibody techniques. It can be applied either directly or indirectly. Instead of a fluorescein-labeled conjugate, an enzyme, usually horseradish peroxidase, is conjugated (combined) with the specific antiserum (direct) or with an antispecies immunoglobulin (indirect). In this procedure, the conjugate is allowed to react with the specimen. If the test is positive, the peroxidate in the conjugate is connected to the antigen in the specimen by the antibody that is specific for the antigen. The preparation is then treated with a peroxidase substrate to detect the presence of the enzyme. A colored reaction is produced (Color Photograph 80). Several substrates give a colored reaction with peroxidase; alpha-naphthol and 3,3,-diaminobenzidine are two such substrates.

APPLICATIONS

The immunoperoxidase test offers several advantages over fluorescent antibody techniques since preparations are permanent, fewer nonspecific reactions occur, and results can be viewed under the ordinary light microscope. The immunoperoxidase test also can be used in electron microscopy. The test is of particular value in the identification and detection of chlamydia and viruses such as cytomegalovirus, herpes viruses (Color Photograph 80), and rabies virus. Detection of viruses in respiratory secretions is difficult because of the normal presence of peroxidase in such specimens. Modifications of the techniques are also applicable to the identification of certain cancerous states.

Enzyme-Linked Immunoabsorbent Assay (ELISA)

The enzyme-linked immunoabsorbent assay (ELISA) technique is similar in principle to the immunoperoxidase test. It is based on the ability of antigen or immunoglobulin (1) to adsorb to a solid-phase support surface of a tube or well in a plastic tray and (2) to be linked to an enzyme, forming a complex showing both detectable immunological and enzyme activity. Two methods are available, *direct* and *indirect* (Figure 19–17). In the direct assay method, a special immunoglobulin specific for a suspected pathogen first is used to coat the surface of a polystyrene well. The specimen containing the suspected microbial antigen is added to the treated well, followed by the enzyme-labeled specific immunoglobulin (conjugate). In a positive case, an invisible immunoglobulin-antigen complex is formed. The complex is made visible by the addition of a substrate, which the enzyme in the conjugate alters to produce a color change (Color Photograph 81). This reaction can be seen by the naked eye or measured by electronic means such as in a spectrophotometer.

In the indirect method, the solid surface of a well is first coated with antigen, then patient serum is added, and finally enzyme-labeled antihuman immunoglobulin is added to form a complex. Again, the appropriate substrate is added, and the procedure is virtually the same as in the direct technique.

APPLICATIONS

The ELISA technique offers important advantages over several other techniques used for the identification of viruses. These advantages include lower cost for supplies, greater sensitivity, and better reagent stability. ELISA is used for the identification of viral pathogens such as hepatitis B virus, rotaviruses, cytomegaloviruses, herpes viruses, rubella, bacteria such as species of *Legionella* (lee-jon-EL-lah), *Streptococcus, Neisseria, Staphylococcus,* and the protozoon *Toxoplasma gondii* (toks-oh-PLAZ-mah GON-dee-eye). The detection of IgM and IgG antibodies is also possible with this technique. The application of ELISA is reliable for the early detection of various viral infections, including AIDS, herpes simplex virus 2, in-

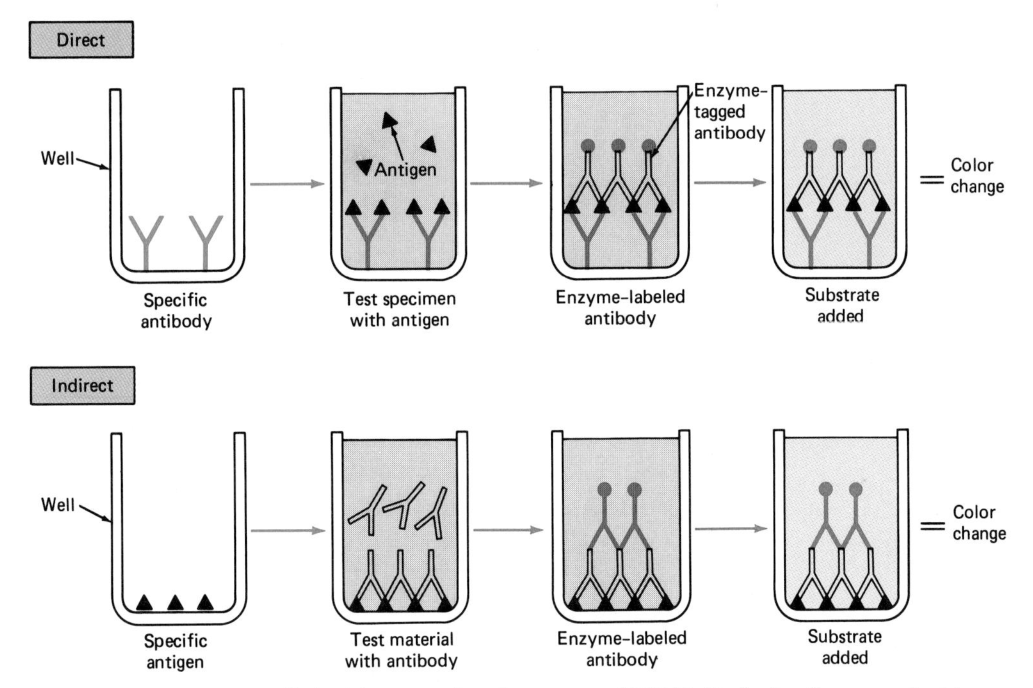

Figure 19–17 Enzyme-linked immunoabsorbent assay (ELISA). Both the direct, or double sandwich, technique and the indirect method are shown. Antibody to the antigen to be detected (capture antibody) is attached to the well. A specimen possibly containing antigen (triangles) is added and washed. Antibody to the antigen is then added. In the direct method, this antibody is tagged with enzyme so that when substrate is added a color change occurs. In the indirect method, the captured antigen is first treated with heterologous antibody, and an enzyme-labeled antibody to the heterologous antibody is added next, before the addition of substrate.

fectious mononucleosis, polio, hepatitis B, mumps, and measles.

Diagnostic and Investigative Electron Microscopy

Immunoelectron Microscopy

In recent years, the diagnostic potential and versatility of electron microscopy have been greatly extended by development of new techniques. One such technique, immunoelectron microscopy (IEM), was first used by Anderson and Stanley to observe the tobacco mosaic virus in the presence of a specific antibody. IEM has since been used to detect the presence of small amounts of antibodies (Figure 19–18), to show antigenic similarities and differences among the viruses, and to identify viral particles extracted directly from human tissues and feces. Feinstone, Kapikian, and Purceli found a viruslike particle in the feces of patients with hepatitis A (infectious hepatitis) using IEM. The discovery of the viruslike antigen by IEM provides the first technique for the diagnosis and study of hepatitis A infections. [See viral hepatitis in Chapter 26.]

Figure 19–18 Immunoelectron microscopy. Aggregates of regular tubular structures and capsids of human rotavirus. The clumping was caused by incubating fecal extracted specimens with rotavirus immunoglobulins. Original magnification 80,000×. *[T. Kimura,* Infect. Immun. *33:611–615, 1981.]*

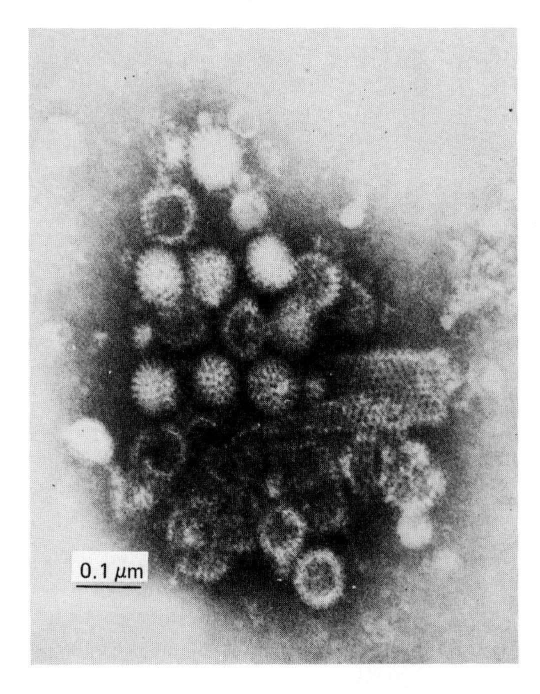

Microbiological Highlights

SINGLE-USE DIAGNOSTIC SYSTEM (SUDS)

The isolation and identification of infectious disease agents is generally the preferred method for diagnosing an infection. However, as this chapter shows, a wide variety of serological procedures are available for use as diagnostic tools. These indirect approaches to diagnosis rely on a host's response to an infectious agent during the course of a disease. Serological tests clearly are important tools for the detection of antibodies. As technological advances are made, improvements in the performance of these tests generally appear. An example of such an improvement is the single-use diagnostic system (SUDS), a form of enzyme immunoassay. With this test the immune status of an individual to the virus disease rubella, and to the protozoan disease toxoplasmosis, can be determined within ten minutes. Specific test cartridges are used for the mixing of a specimen and specific test reagent (see accompanying illustrations). In the case of positive tests, a distinct color (blue) forms within minutes. Similar types of assays are under development for rapidly determining the immune status of an individual to other disease agents.

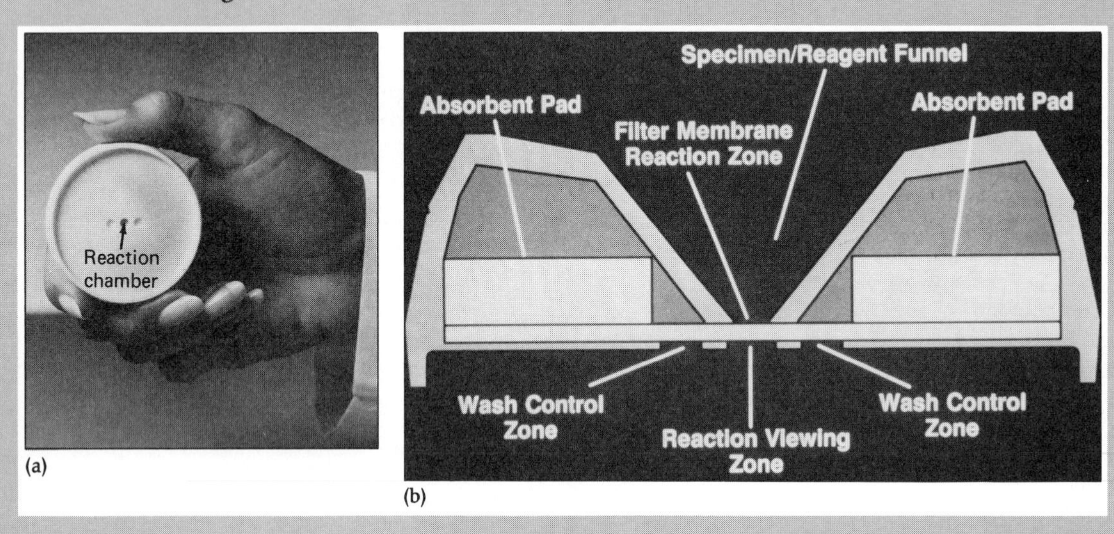

The SUDS cartridge and its parts. (a) The actual cartridge. (b) A diagram of the SUDS cartridge. (Courtesy of Murex Corporation.)

Ferritin-Conjugated Antibodies

Ferritin is a protein that has a molecular weight of 700,000 and contains about 23% iron, largely in the form of ferric hydroxide and phosphate. The iron is concentrated within the ferritin molecule in four particles called *micelles.* These units form a central core measuring 5.5 to 6.0 nm in diameter. Because of its composition and molecular arrangement, the ferritin molecule is electron dense and has a characteristic appearance in electron micrographs (Figure 19–19).

Immunogold Staining

Gold particles are also electron-dense and have a characteristic appearance in electron micrographs (Figure 19–20). When they are coupled to antibodies, the resulting gold probe can be used to locate surface as well as intracellular antigens. In addition, preparing additional probes by coupling different sized gold particles to different antibodies can be used to distinguish among antigens within the same specimen (Figure 19–20).

Monoclonal Antibodies

The production of pure or monoclonal (single-type) antibodies by fusing antibody-forming cells with

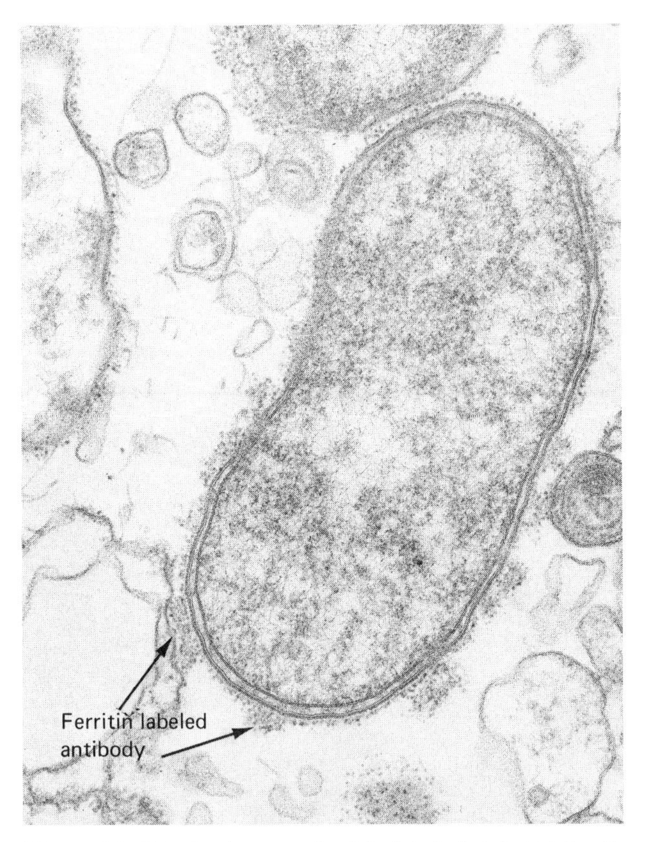

Figure 19–19 An immunoferritin-labeled rickettsial cell. The cell's outer membrane surface is evenly labeled with ferritin particles. *[Y. Rikihisa,* et al., Infect. Immun. ***26:*** *638–650, 1979.]*

continuously replicating cell lines has created a revolution in diagnostic immunology. The technique used, referred to as *somatic cell hybridization* or *hybridoma formation* (Figure 19–21), results in the formation of a hybrid cell usually containing two nuclei in a cytoplasm that is a mixture of the two parental cells. This hybridoma technique was developed in 1975 by C. Milstein and G. Köhler at the Medical Research Council's Laboratory of Molecular Biology in Cambridge, England. Both Drs. Milstein and Köhler were recognized for this contribution by being awarded the Nobel Prize in Physiology or Medicine in 1984. Monoclonal antibodies can be obtained by first injecting a mouse with an antigen and then obtaining and chemically fusing the antibody-making cells of its spleen with a cancerous type of mouse cell known as a *plasmacytoma.* The hybrid cell so formed produces the single-type antibody molecule of its spleen cell parent and continues to grow and divide like its plasmacytoma cell parent. Once the clone of cells producing the desired antibody has been selected from nonfunctional cells, it can be grown as a continuous cell line from which large amounts of the pure or monoclonal antibody can be harvested (Figure 19–21).

Another approach to hybridoma production using a low-intensity electrical field has been developed by U. Zimmerman (Figure 19–22). In this *electrofusion* technique, cells are placed between two electrodes, and a field of nonuniform alternating current is applied. The field causes the cells to separate on the basis of their electrical charges and to mi-

Figure 19–20 The double-label technique with an antibody-gold immunological probe. Small gold spheres (small arrows) are attached to pili, while larger spheres are attached to the surface of *Neisseria gonorrhoeae. [From E. N. Robinson, Jr.,* et al., Infect. Immun. ***46:*** *361–366, 1984.]*

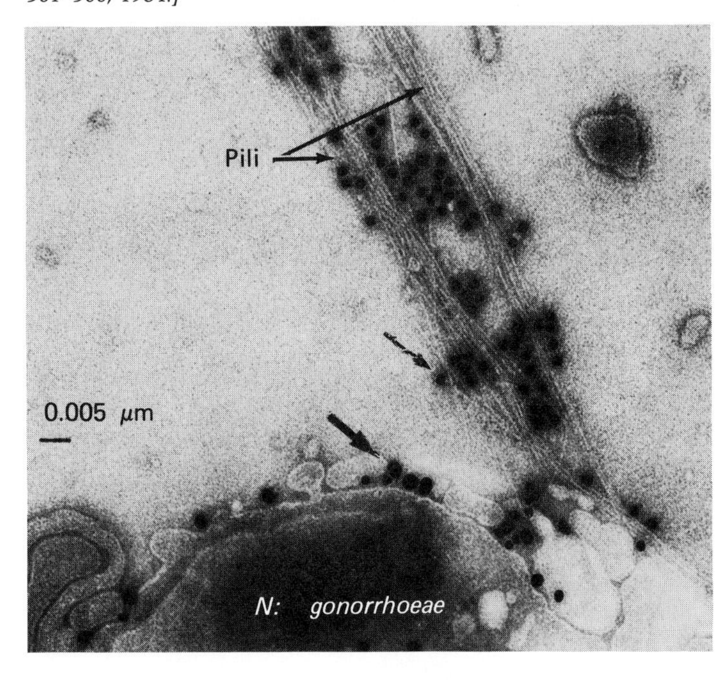

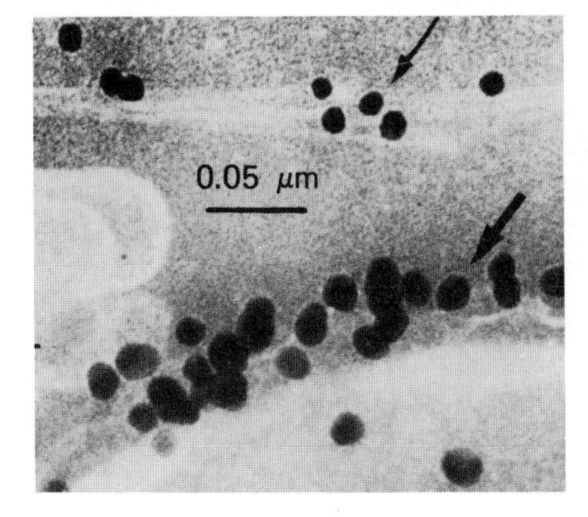

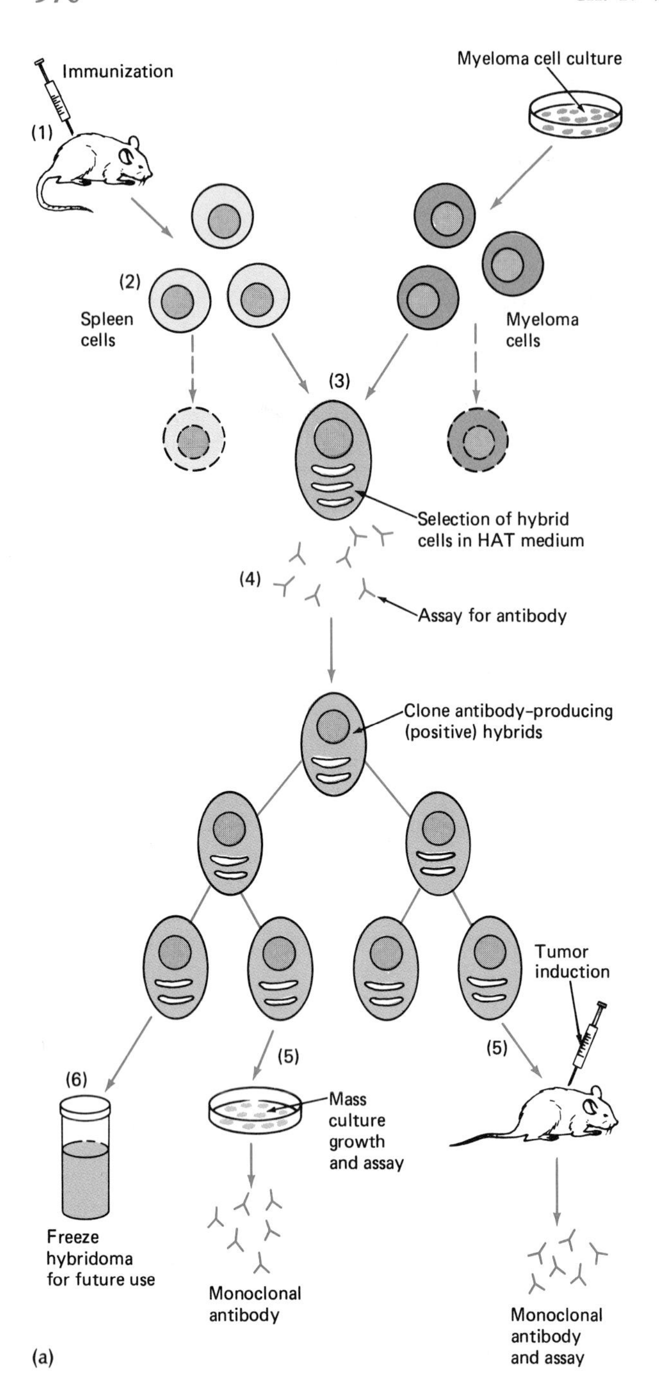

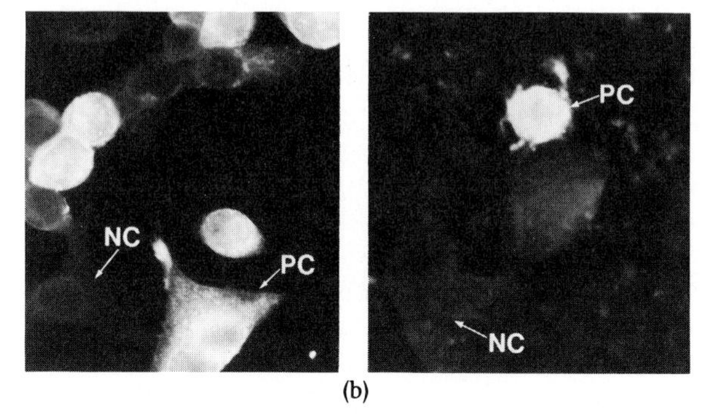

Figure 19–21 (a) Hybridoma formation and monoclonal antibody production. (1) The formation of hybridomas begins by immunizing a laboratory animal such as a mouse with a selected antigen. (2) The animal's spleen cells are harvested, mixed with mouse myeloma cells, and incubated (3) with polyethylene glycol to promote fusion. Cells are transferred to a selective growth medium to eliminate unfused myeloma cells and myeloma hybrids. (4) The remaining spleen cell-myeloma hybrids are assayed for antibody production against the immunogen, and positive hybrids are cloned. (5) These cells are again assayed for antibody production. (6) The positive cells are recloned and can be frozen, stored, and used to amplify antibody production in tissue culture or animals. (b) The use of monoclonal antibodies to distinguish between herpes simplex viruses 1 and 2. (left) Antibody staining of only herpes simplex virus type 1 infected cells produces a positive cell (PC) result. Noninfected cells (NC) produce a negative result. (right) Staining herpes simplex virus type 2 infected cells. Again, only infected cells produce a positive result. *[Courtesy of Drs. Lynn Goldstein, R. C. Nowinski, and associates.]*

grate toward one of the electrodes, which has a higher field intensity than the other. The cells form chains in parallel with the electrical field as their positive poles are attracted to the negative poles of adjacent cells. This step is called *dielectrophoresis.*

Once the chains of cells are formed, a high-intensity, short-duration pulse of direct current is applied. The pulse triggers a breakdown in the cell membranes along a narrow line within the cell chain at the points at which the cell membranes touch. This reaction causes the formation of a series of pores through which the cells in the chain exchange cytoplasm and fuse. The cytoplasm of the entire chain becomes continuous, and a giant cell is formed. Electrofusion is much less troublesome to perform and more efficient than other hybridoma-producing procedures.

The hybridoma technique provides a means by which a constant and uniform source of antibody can be produced. (The immune system usually produces a mixture of antibodies, with each type directed against a different feature of the antigen.) Monoclonal antibodies are chemically, physically, and immunologically uniform. These molecules, because of their specificity, are powerful research tools and can provide quick and accurate identification of

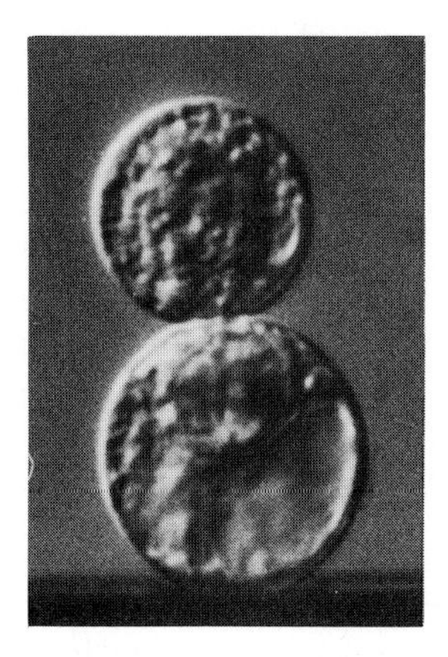
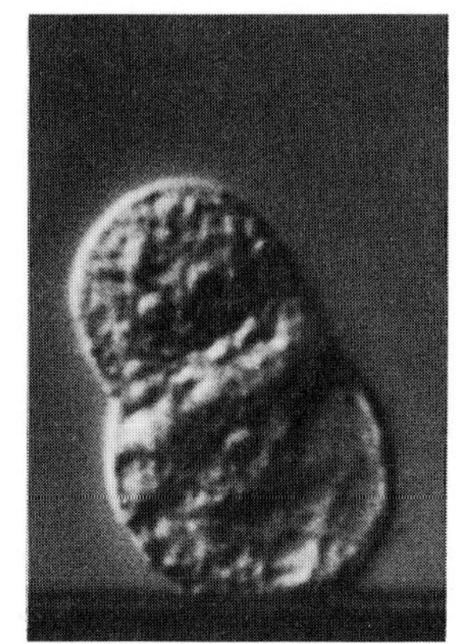
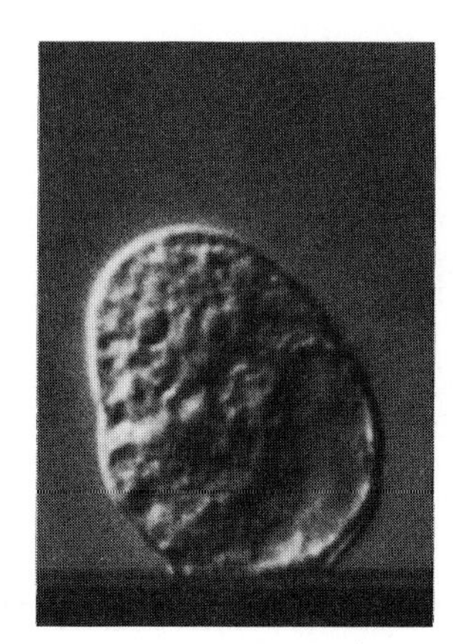
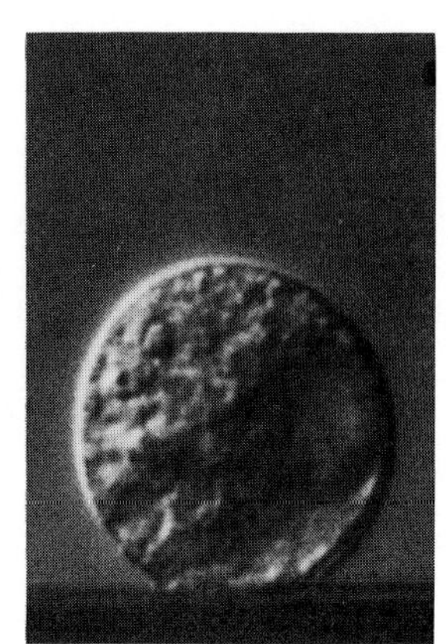

Figure 19–22 The results of electrofusion showing the joining of two cells. *[Courtesy of Drs. U. Zimmerman, and J. Vienken.]*

viruses (Color Photograph 79b), bacteria, and cancer cells. Monoclonal antibodies are used in various diagnostic procedures, including immunofluorescent antibody, radial immunodiffusion, and enzyme-linked immunoabsorbent assay (ELISA). The long-range promise of monoclonal antibodies lies in their therapeutic value as replacements for vaccines and as agents for the treatment of cancers. Several of the applications of monoclonal antibodies are given in Table 19–4.

In Vivo Testing Procedures

Virus Neutralization

Serological reactions of various kinds are used for identifying and classifying viruses, detecting antibodies against certain viral agents in the normal population, and studying the responses of individuals to prophylactic immunization. Among the procedures so used are animal protection, complement fixation, hemagglutination inhibition, and neutralization tests. Because *in vitro* methods are less expensive and quicker to perform, they are more commonly used than *in vivo* tests. Nevertheless, in many situations only the more complex *in vivo* techniques are applicable.

PROCEDURES

Viral neutralization tests may use either tissue culture systems or laboratory animals such as chick embryos, mice, or rats. The details of the procedure vary with the viral agent, the host, and related factors; however, there are two general procedures. In one of them, dilutions of a virus suspension are mixed with constant volumes of undiluted serum and incubated for one to two hours; a specific quantity of each mixture then is injected into separate groups of animals. For example, dilution mixtures ranging from 1/100 (10^{-2}) to 1/1,000,000 (10^{-6}) are each injected into four animals. The animals are observed for a specified length of time to determine any effects caused by the virus that were not inhibited by the antibodies present in the serum. Naturally, controls are also used; these include viral dilutions with serum lacking specific antibodies and without serum. In the other procedure, the viral concentration is held constant, and the serum is diluted. When tissue culture systems are used, cells are examined for any evidence of a specific inhibition or neutralization of cytopathic effects. **[Refer to Chapter 10 for virus cytopathic effects.]**

TABLE 19–4 **Monoclonal Antibody Applications**

Current Representative Uses	*Potential Uses*
Routine diagnostic and investigative serology	Passive immunization against infectious agents
Tissue typing identification and epidemiology of infectious disease agents such as viruses, bacteria, and worms	Elimination of drug toxicity
Identification of tumor antigens	Graft protection
Classification of leukemias and lymphomas	Control of tumor rejection
Identification of functional subpopulations of lymphoid cells	Manipulation of the immune response
	Targeting of diagnostic or therapeutic agents *in vivo* for detection of the delivery of cytotoxic agents to tumor cells

For general diagnostic purposes, two blood specimens should be obtained from the patient suspected of having a viral infection. One sample should be taken during the acute phase of the disease and the second during the recovery or convalescent period. The presence of an infection is generally indicated by a fourfold increase in antibody titer in the second blood specimen.

Diagnostic Skin Tests

The introduction of test antigens just under the skin can be used to determine cellular immunity and occasionally to establish a diagnosis (Table 19–5). Examples of such diagnostic skin tests are the Frei test for the venereal disease lymphogranuloma venereum and the tuberculin, Mantoux, tine, and patch tests for tuberculosis.

Positive responses of individuals to test antigens may be either *immediate* or *delayed.* An *immediate reaction,* which develops shortly after exposure to the test antigen, appears as an elevated, flat, pale, swollen area surrounded by a region of redness. It may be accompanied by intense itching around this **wheal and flare reaction.** Delayed hypersensitivity skin tests are of great value in determining overall immunocompetence and in epidemologic surveys. The inability to react to a number of common skin antigens is called *anergy.* This condition is found in several abnormal and disease states (Table 19–6). The *delayed reaction* appears several hours after the introduction of the antigen (Color Photograph 82). The area around the site of inoculation becomes reddened, firm, and swollen within 24 to 48 hours (Figure 19–23). Further aspects of these immunological reactions are discussed in Chapter 20.

Skin tests can also be used to determine an individual's susceptibility to a microbial toxin. Small quantities of toxin are injected into the skin, and the reaction is observed. Positive responses, those showing susceptibility, are indicated by reddened areas appearing within 24 hours. The Schick test for diphtheria and the Dick test for scarlet fever are examples of this type of procedure.

TABLE 19–5 Representative Skin Diagnostic Tests[a]

Disease State[b]	*Infective Disease Agent*	*Preparation Used*	*Type of Reaction*
Brucellosis	Bacterium	Brucellergin (extract of *Brucella* spp.)	Delayed
Leprosy	Bacterium	Lepromin (extract of lepromatous tissue)	Delayed
Lymphogranuloma venereum	Bacterium	Chorioallantoic membrane (extract from infected chick embryo)	Delayed
Psittacosis	Bacterium	Heat-killed organisms	Delayed
Tuberculosis	Bacterium	Purified protein derivative (PPD) or old tuberculin (OT)	Delayed
Blastomycosis	Fungus	Concentrated culture filtrate	Delayed
Coccidioidomycosis	Fungus	Coccidioidin (concentrated culture filtrate)	Delayed
Histoplasmosis	Fungus	Histoplasmin (concentrated culture filtrate)	Delayed
Leishmaniasis	Protozoon	Extract of cultured organisms	Delayed
Echinococcosis (sheep tapeworm)	Helminth	Hydatid fluid extract	Delayed
Trichinosis (pork roundworm)	Helminth	Extract of the causative agent	Immediate
Schistosomiasis	Helminth	Extract of the causative agent	Immediate
Contact dermatitis	Simple chemical compounds	Small quantities of suspected chemicals	Immediate or delayed

[a] NOTE: The information obtained from such tests also can be used in epidemiologic surveys.
[b] Refer to the individual chapters in which these disease states are discussed for further details.

TABLE 19–6 **Abnormal and Diseases Conditions Associated with Anergy**

Category	*Conditions*
Immunodeficiency (congenital)	Combined deficiencies of cellular and humoral immunity; cellular immunodeficiency alone (mucocutaneous candidiasis, yeast infection)
Immunodeficiency (acquired)	AIDS, cancers (carcinoma), Hodgkin's disease, immunosuppressive drugs, surgery, various forms of liver injury
Infections	Active tuberculosis, leprosy, scarlet fever, disseminated fungus infections, influenza, measles, mumps, and viral vaccines

Immunohematology

The ABO System

In 1900, Karl Landsteiner observed that mixing erythrocytes and sera from different human donors and incubating the mixture at body temperature resulted in the agglutination of the red cells. This phenomenon is known as *isohemagglutination* because it involves reactions between agglutinogens and agglutinins from the same *(iso-)* species. It enabled Landsteiner to demonstrate the existence of the A and B blood types. In addition, he found certain blood samples in which no visible reactions occurred between the reactants (corresponding agglutinogens and agglutinins). Blood cells exhibiting this negative result are designated O. The fourth and last of the major blood types of the ABO system, AB, was discovered by von Decastello and Sturli in 1902. AB blood is characterized by visible clumping, or positive results, with the reactants. The surfaces of red blood cells contain large numbers of antigenic determinants that are the direct or indirect products of genes. These antigenic determinants are arranged into blood groups.

In addition to having specific blood group antigens, certain humans possess antibodies that react with the red blood cells from individuals of other blood types, causing them either to agglutinate or to lyse (Figure 19–2). These immunoglobulins are called *isohemagglutinins* and *isohemolysins*, respectively. Lysis occurs when antibody molecules (isohemolysins) sensitize the blood cells, making them vulnerable to the hemolytic activity of complement. Usually, the antibodies in the blood serum or plasma of an individual are not directed against the blood factors present in his or her own blood. However, most individuals have antibodies against blood factors absent from their blood cells. For instance, a person of blood type O has antibodies against type A (anti-A) and type B (anti-B). Antibody concentrations can be increased as a result of transfusions or, in women, by bearing children of a different blood type. Table 19–8 lists the recognized blood groups with their respective agglutinogens and agglutinins.

THE UNIVERSAL DONOR AND RECIPIENT

A transfusion reaction can occur if the concentration of agglutinins in the recipient's plasma is high enough to cause agglutination or hemolysis of erythrocytes from the donor. It is obviously advisable to transfuse a patient with blood of his or her particular type. However, there may be circumstances in which this transfusion is impossible, because blood of the recipient's type is unavailable. Type O blood is used in such situations, because antibodies against it are very seldom encountered. Although the plasma portion of O blood contains agglutinins against both the A and B antigens (Table 19–7), these antibodies would be neutralized or diluted in the recipient's circulation.

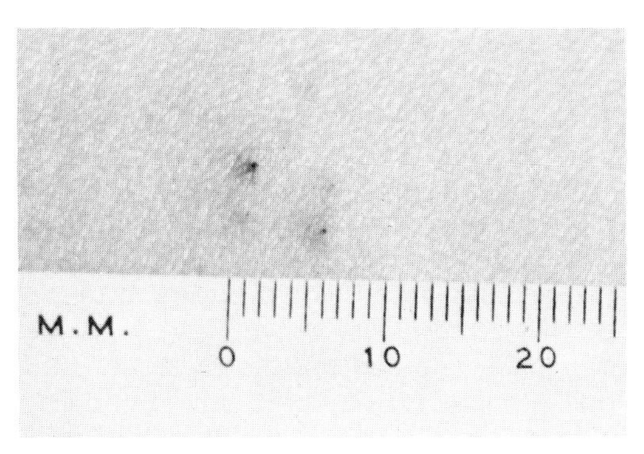

(a)

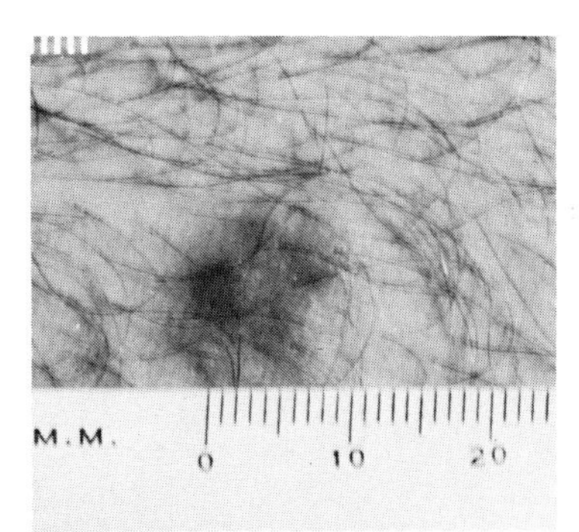

(b)

Figure 19–23
The results of a tuberculin tine test. The presence or absence of induration (hardened area) can be determined by visual observation or by stroking the region around the site of injection. (a) The response to the skin testing material 48 hours and (b) 96 hours after injection. *[Courtesy of Lederle Laboratories, Pearl River, N.Y.]*

TABLE 19–7 Selected Characteristics of the Major Human Blood Groups

International Designation	Agglutinogen Associated with Cells	anti-A	anti-B
A	A	−	+
B	B	+	−
AB	AB	−	−
O	O	+	+

NOTE: +, present; −, absent.

TABLE 19–8 Reaction Patterns in Blood Type Determinations

	Antiserum Used	
Blood Type	anti-A	anti-B
A	+	−
B	−	+
AB	+	+
O	−	−

NOTE: +, agglutination; −, no agglutination.

Individuals with type O blood are commonly referred to as "universal donors." Unfortunately, this designation is misleading. Transfusion of type O blood cannot be safely performed in all cases. Serious transfusion reactions can occur if these donors have antigen-antibody systems other than the ABO system. These may be minor blood group and Rh factors. In general, transfusion reactions are prevented by performing direct and indirect cross-matching procedures to determine the blood compatibilities of prospective donors and recipients. See the discussion of blood-testing techniques later in the chapter.

Persons with an AB blood type do not have agglutinins against the A or B factors. Consequently, such individuals can receive blood from donors belonging to any of the four major blood groups. They are, therefore, often called "universal recipients."

LABORATORY DETERMINATION OF BLOOD GROUPS (BLOOD TYPING)

Two diagnostic procedures, the tube and slide tests, are commonly used to determine blood types. The determination can be made either by testing red cells with standardized anti-A and anti-B sera or by testing the patient's serum with standard sensitive known A and B red cells. For reliable results, both tests should be employed; however, most often only the first is used.

The tube technique usually incorporates erythrocytes from the person with standard test sera of anti-A and anti-B reagents. Specific proportions of each antiserum and the erythrocyte suspension are mixed, incubated at 37°C for 30 to 60 minutes, and checked for agglutination. Table 19–8 shows the pattern of reactions observed in blood type determinations.

The slide test uses the same reagents. It is performed by placing one or two drops of each test antiserum on opposite portions of a glass slide and then adding a drop of the patient's cells to each of them. The combinations are mixed separately with the aid of applicator sticks and observed for agglutination. Color Photograph 76 shows the actual reactions characteristic of the major blood types. Reactions should develop within two to five minutes.

Suitable controls include (1) pretesting of sera for activity; (2) dilution of cells and sera in physiological saline if a reaction has not occurred after a sufficient length of time; and (3) incorporation of tests with known cells and antisera. The incubation temperature should also be checked, as variations in temperature can affect results.

THE ABO SUBGROUPS

Antigenic variations of the ABO system have been reported since 1935. The blood group A has been divided into the major subdivisions A_1 and A_2. The distinction between A_1 and A_2 is usually made by using extracts of certain plant seeds called *lectins.* Rare subtypes of A are also known and include A_3 and A_x. The AB group is subdivided into A_1B and A_2B. Subgroups of the B group include B_v, B_3, B_k, B_w, and B_x. The frequencies of these subgroups vary; A_1 and A_2 are more commonly encountered than the others.

INHERITANCE OF THE ABO BLOOD GROUPS

The ABO blood factors are one example of genetic characteristics determined by a multiple allelic series. An *allele* is a gene belonging to a group of alternate genes that occur on a specific locus (site) of a chromosome. Inheritance of A and B characteristics follows normal Mendelian principles. An individual's blood type is determined by receiving, from each parent, one of the four allelic genes A_1, A_2, B, or O. Blood factors A and B are carbohydrate antigens (Figure 19–24). On cell membranes these antigens are bound to glycoproteins, which are combinations of protein and carbohydrates, as well as to membrane lipids. The respective allelic genes produce enzymes called *transferases* that connect terminal sugars to the basic stem structure of the blood group substance H (Figure 19–24). These enzymes are specific in their reactions: transferase A connects *N*-acetylgalactosamine, and transferase B acts with galactose. The O gene does not produce a transferase that would change the blood group substance. Thus, blood type O individuals have only the blood group substance H. The specific blood group antigens become permanently established, a finding readily demonstrated in fetuses and newborns.

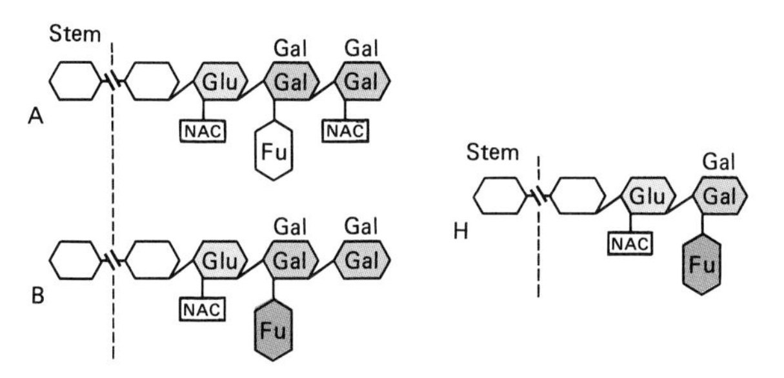

Figure 19–24 The structure of blood group antigens A, B, and H. The last sugars in the basic structure or stem chain form the H substance and are linked to either a peptide or a lipid. Gal = galactose; Glu = glucose; Fu = fucose; NAC = *N*-acetylglucosamine.

Blood group-associated isohemagglutinins, however, are not normally detectable at birth, but become so within three to six months.

The blood group substances A and B are not limited to red cells but are found in various body fluids and tissue cells including kidney, liver, lung, and muscle. They have been detected in amniotic fluid, gastric juices, ovarian cyst fluid, perspiration, saliva, semen, tears, and urine of approximately 80% of the population. The other 20% of the population does not secrete these substances into fluids and tissue.

The Rh System

Once the ABO system was understood, it was believed that transfusion problems could not develop if the donor and recipient belonged to the same blood group. Unfortunately, between 1921 and 1939, some hemolytic transfusion reactions were reported even when blood typing tests showed donor and recipient compatibility. No explanation for these reactions was offered. Moreover, prior to 1940, several reports appeared of newborns exhibiting a clinical state called *erythroblastosis fetalis* or hemolytic disease of the newborn, with swelling and marked anemia.

In 1937, Landsteiner and Wiener, immunizing guinea pigs and rabbits with rhesus monkey blood, discovered antisera that agglutinated red cells not only from the monkey but also from approximately 85% of their human blood samples. A hitherto unknown human blood agglutinogen had been found. The new factor was designated "Rh," indicating the source of the antigen, the rhesus monkey. Further research showed it to occur in all blood groups.

The antibody response of sensitized Rh-negative individuals to the administration of Rh-positive antigens clearly demonstrated the importance of the Rh factor. In 1939, Levine and Stetson observed a transfusion reaction in a woman shortly after the delivery of her stillborn fetus. Apparently in need of a transfusion, the woman received blood from her husband. The transfusion produced a pronounced hemolytic reaction. The mother's serum was found to contain an agglutinin against the Rh factor. Levine and Stetson theorized that the presence of this agglutinin was the direct result of *in utero* immunization of the mother by antigen that the fetus inherited from the father. Later studies by Levine and his colleagues proved this to be the cause and showed that the resulting maternal antibody could pass through the placenta and cause erythroblastosis fetalis, or Rh baby, which is discussed in the following Microbiology Highlights.

OTHER Rh AND Hr FACTORS

Since the original report of the Rh factor in human blood, more than 25 other blood factors that evidently belong to the Rh-Hr blood group system have been discovered. These blood factors have proved to be of great clinical significance, second only to the ABO system.

The Rh-Hr system is a complex one, and controversies have arisen over the nomenclature for the Rh agglutinogens. Two principal methods are currently used: the Wiener scheme (the original Rh-Hr nomenclature designations) and the Fisher-Race system (combination of the letters C, D, and E). Table 19–9 shows the comparison between these two systems of notation. The more recently discovered Rh factors are symbolized by various combinations of subscript or superscript letters and numerals, for example, D_u and C_w.

Blood Testing Techniques

Compatibility Testing (Cross Matching)

The primary purpose of compatibility testing is to prevent a transfusion reaction as a consequence of blood incompatibility. Specifically, the cross-matching technique is designed to detect any incompati-

TABLE 19–9 A Comparison of the Designations for the Rh Blood Factors (Wiener) and Rh Agglutinogens (Fisher-Race)

Wiener System	*Fisher-Race System*
Rh_0	D
rh′	C
rh″	E
hr′	c
hr″	e
hr	d

NOTE: Designations for the more recently discovered Rh factors or combinations are symbolized by various superscript letters and numerals, e.g., R^0, r^y.

HEMOLYTIC DISEASE OF THE NEWBORN (THE Rh BABY) AND ITS CONTROL

Hemolytic disease of the newborn (erythroblastosis fetalis) is the blood incompatibility normally occurring when certain differences in Rh factors exist between the mother and her child. Usually this condition develops in babies born to mothers who are negative for the Rh_0 factor and fathers who are positive for the same factor. (This specific Rh terminology is explained in Table 19–9.) The red cells of the child (fetus) carry the father's blood factor, which is foreign to the mother. If this antigen enters the maternal circulation, it immunizes the mother (isoimmunization) to the Rh factor. The resulting maternal isoantibodies pass from the mother's circulation into the fetal system and attack the baby's red cells. (see the accompanying diagram.)

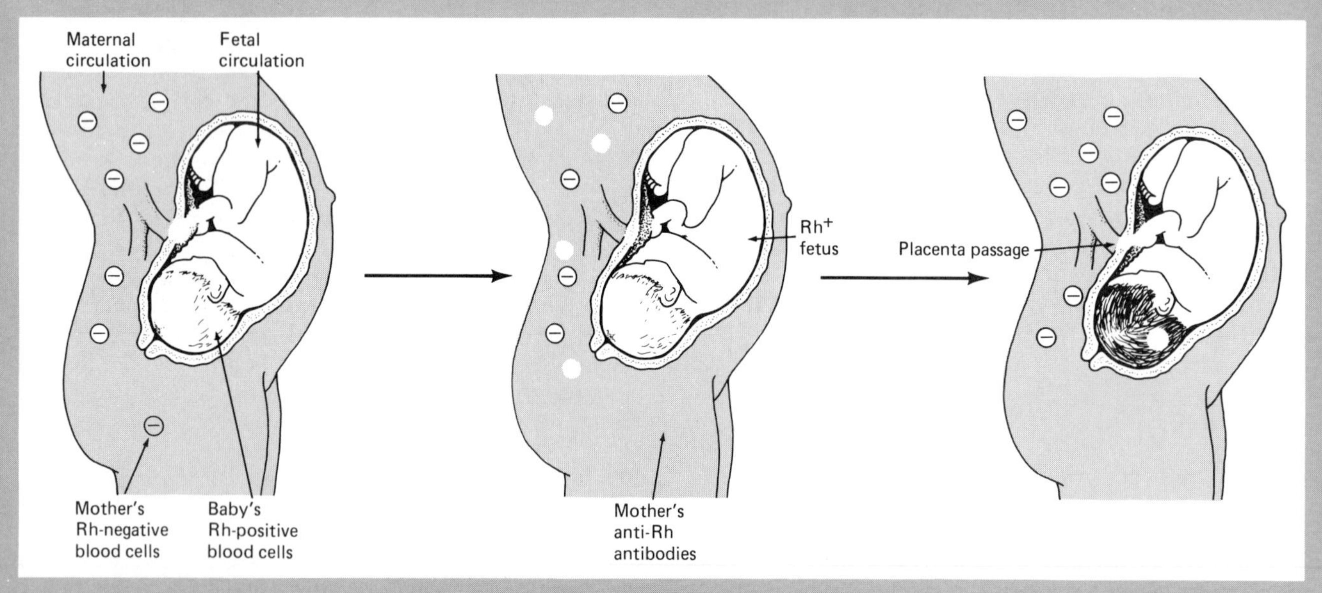

The sequence of events leading to the Rh baby condition. Late in the course of the first pregnancy or during the birth process, fetal red blood cells carrying the father's Rh antigens enter the maternal circulation.

bility between the recipient's serum and the donor's cells (known as the *major cross match* or forward typing) or between the recipient's cells and the donor's serum (referred to as the *minor cross match* or reverse typing). The cross-matching procedure is done after the respective blood specimens have been typed in regard to the ABO and Rh factors and any other factor that appears to be indicated.

Rh Typing. The Rh factors of red cells are detected by agglutination tests using the appropriate antisera, anti-Rh_0, -rh′, and -rh″. The classic and clinically most important factor is Rh_0 (D).

Agglutination or lysis of the red cells from either the donor or recipient in their tests indicates an incompatible situation. However, it should be stressed that occasionally an incompatible cross match may be due to an error in blood typing of the specimens used, or to the presence of atypical antibodies in the blood of either the donor or recipient. When these causes are suspected, it is customary to investigate the source of error.

Coombs' Tests

Two tests for the detection of incomplete antibodies, the direct and indirect Coombs tests, have great clinical importance in cases of hemolytic anemia and hemolytic disease of the newborn. Incomplete anti-

The best opportunities for fetal red cells to enter the maternal circulation are shortly before, during, or immediately after birth. Once a mother has been actively immunized against the Rh_0 (D) factor and has given birth to one child with erythroblastosis fetalis, the effects of the disease will be more severe with future Rh_0^+ offspring. If the father is heterozygous ($Rh_0^+Rh_0^-$), they may produce children lacking the Rh_0 factor. Almost all Rh_0 negative women without previous exposure to the Rh_0 factors via transfusions have given birth to one or more normal Rh_0-positive children before enough maternal antibodies were produced to cause trouble; if the father is heterozygous ($Rh_0^+Rh_0^-$), each fetus has a 50% chance of being Rh negative and, therefore, not subject to the disease.

Three clinical states can develop in a Rh baby. In the order of severity they are hydrofetalis (excessive outpouring of blood into the tissues), icterus gravis neonatorum (jaundice or yellowing), and congenital anemia (an abnormal reduction of red blood cells).

The Rh_0 (D) antigen is generally the main factor in hemolytic disease of the newborn. However, other blood antigens are capable of causing the condition. These include those of Kell, Kidd, and Duffy.

PREVENTION OF Rh ISOIMMUNIZATION

Fetal erythrocytes are commonly observed in the maternal circulation near the end of the pregnancy period, and their numbers generally increase after childbirth. Much clinical evidence indicates that the processes of labor and delivery cause fetal red cells to enter the maternal circulation. Some investigators hold that there is a continual leakage of the fetal cells into the mother's system throughout pregnancy. Possibly both processes contribute to the isoimmunization of certain Rh-negative women. The major threat of isoimmunization occurs at the end of the third trimester of pregnancy and immediately after childbirth.

The procedure for preventing maternal Rh problems utilizes passive immunization to suppress the antigenic stimulus provided by fetal cells during the postpartum period. An immunoglobulin G preparation that contains high titers of anti-D (anti-Rh_0) antibody is administered. Rh-negative unsensitized mothers receive this material within 72 hours after a delivery, abortion, or miscarriage involving an Rh problem. The antibody disappears after approximately six months, after producing a short-lived passive immunity and preventing a long-lasting active one that might endanger a later child. Failure to prevent Rh_0 sensitization with anti-D immunoglobulins can result from various causes, including maternal immunization during pregnancy, release of a large volume of fetal blood into the mother's circulation, incorrect typing of the newborn as Rh negative, and failure to administer the anti-D immunoglobulin.

bodies do not function in ordinary hemagglutination tests.

DIRECT COOMBS' TEST

The direct Coombs test is used to determine whether or not the patient's erythrocytes have been sensitized (coated) by antibodies *in vivo.* In this test, saline-washed blood cells suspected of being sensitized are mixed with antihuman globulin serum (Coombs' reagent) and observed for agglutination. A control of the patient's cells and saline is also observed. Clumping in the erythrocyte-Coombs' reagent system is a positive test and demonstrates the presence of incomplete antibodies, but it does not identify them specifically.

INDIRECT COOMBS' TEST

The indirect Coombs' test is a modification of the direct procedure and is used to detect either complete or incomplete antibodies in the serum of sensitized individuals. The procedure consists of the following three stages: (1) sensitization of erythrocytes (Rh positive) with Rh antiserum (patient's serum); (2) washing the sensitized cells to remove excess antibody; and (3) mixing the resulting preparation with the Coombs' reagent. Clumping of the Rh-antibody-coated red cells by the Coombs' reagent is a positive test.

The difference between these two tests is that, in the direct procedure, red cells are examined for the presence of antibody on cells (sensitization *in vivo*),

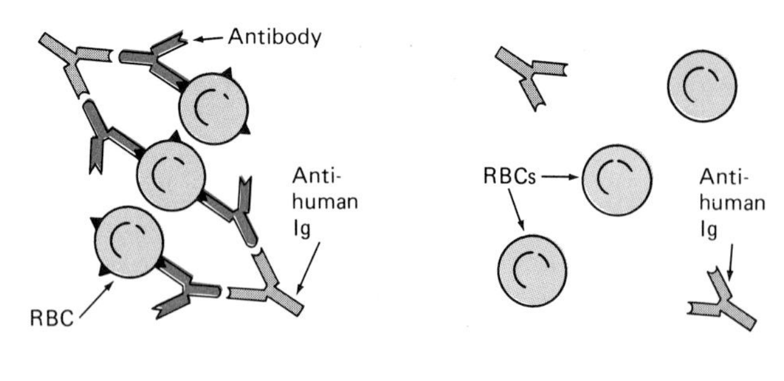

Figure 19–25 The indirect Coomb's test. This procedure is used to detect the presence of antibody on a patient's red blood cells (RBCs). Both a positive and negative situation are shown.

while in the indirect test, Rh-positive cells sensitized *in vitro* by the antibodies present in the patient's serum are tested for the presence of Rh antibodies on their surfaces. Antihuman immunoglobulins are used to detect the Rh antibodies in the indirect Coombs' test (Figure 19–25).

Disorders of the Immune System

Disorders and/or defects of the immune system may develop from genetic abnormalities, malignancies (cancers), infections, or drug usage. Portions of the system or even the entire system may fail in its function to protect the body. The next chapter will describe various disorders and the clinical diseases that result from them.

Summary

1. A variety of *in vitro* and *in vivo* techniques are available to detect and determine the levels of antibody toward antigens.
2. Such information is important in identifying a disease agent following recovery from an infection or in determining the effectiveness of immunizations.
3. Examples of *in vitro* immunologic procedures used in diagnosis or microbial identification include agglutination, complement fixation, ELISA, electroimmunodiffusion, ferritin-conjugated antibody, immunodiffusion, precipitin tests, radioimmunoassay, and virus neutralization.

Production of Antisera

1. Antigen-containing materials (immunogens) to produce potent antibody-containing serum (antiserum) for diagnostic tests may be introduced intraperitoneally, intravenously, or subcutaneously.
2. Determination of antibody levels is made from blood samples removed periodically from experimental animals.
3. Once the antibody concentration reaches the desired level, sufficient blood is removed to obtain the immune (antibody-containing) serum for diagnostic purposes.

Diagnostic Significance of Rising Antibody Titers

1. Two serum specimens of an individual are necessary to distinguish antibody production resulting from an actual ongoing infection from the effects of immunization or from antibodies associated with past infection. The first sample is taken soon after the appearance of disease symptoms and the other about two weeks later.
2. If, in testing such specimens for a particular disease agent, a greater degree of antibody activity occurs with the later specimen, identification of the causative agent is possible.
3. If little or no antibody activity is detected in either specimen, it can be generally assumed that an organism other than the one being tested for is the cause of the infection.
4. If antibody activity is present in both samples but no appreciable change is noted in antibody concentration, it can be assumed that such levels were present before the start of the current infection and have no relationship to the disease state.

The Agglutination Reaction

1. Agglutination involves the clumping of cellular or particulate (particlelike) antigen-containing materials by homologous antibodies.
2. Agglutination tests have been widely used for the rapid diagnosis of several infectious diseases, such as rickettsial infections and typhoid fever, and the identification of various bacterial pathogens.
3. Examples of agglutination and related reactions include cold hemagglutination, the Weil-Felix test, Widal test, latex agglutination tests, and viral hemagglutination.

HEMAGGLUTINATION INHIBITION

1. The hemagglutination inhibition (HI) test is used for the detection and titration of viral antibodies in a patient's serum and for viral identification.
2. Identification of specific hemagglutinating viruses is based on the fact that viral hemagglutination can be inhibited by specific antibodies against such viruses.
3. Adequate controls must be included in the procedure to ensure accuracy.

The Precipitin Reaction

1. Precipitin reactions involve interactions of antibodies called *precipitins* with soluble antigens known as *precipitinogens*.

2. A visible precipitate occurs at the point where optimal proportions of antibody and antigens and related factors exist.
3. Although the general form of this reaction is still functional today, it has undergone certain changes.
4. Modern procedures incorporate agar-gel bases to stabilize the precipitates that form. Examples of these procedures include the Oudin test (single diffusion), Ouchterlony test (double diffusion), radial immunodiffusion, and and two-dimensional immunoelectrophoresis.
5. Modern precipitin procedures are highly accurate and are used for the quantitation of a variety of specific proteins in serum and other fluids.

ELECTOIMMUNODIFFUSION
1. Electroimmunodiffusion combines agar diffusion and electrophoresis for the separation of proteins in body fluids.
2. Counterimmunoelectrophoresis and rocket electrophoresis are two clinically applicable techniques.

In Vitro *Hemolysis Tests*

COMPLEMENT FIXATION
1. Various antigen-antibody combinations have the ability to fix, or combine with, complement. This type of reaction can be used to detect either antibodies or antigen-containing substances in unknown specimens.
2. The complement fixation technique incorporates both a test and an indicator system.
3. The test system consists of a patient's serum, a commercially or laboratory-prepared antigen, and complement. All serum preparations must be heated at 56°C for 30 minutes to inactivate any complement in these sources of antibodies.
4. The indicator system consists of sheep red blood cells and antibody against sheep red blood cells known as *hemolysin.*
5. In a positive complement fixation test, antibody combines with antigen of the test system. Complement is fixed by this combination. Thus, complement is not available to react with the indicator system. The end result is a cloudy red suspension.
6. In a negative test, complement is fixed by the indicator system, resulting in a clear red solution. When complement is fixed by the sheep red blood cell-hemolysin combination, lysis of the blood cells occurs.
7. Specific controls must be used in the test to ensure accuracy.
8. Complement fixation procedures have been used in the laboratory diagnosis of a wide variety of microbial infections and helminthic diseases.

ANTISTREPTOLYSIN TEST
1. Determination of antibody levels against the thermolabile hemolysin of streptococci, streptolysin O, is known as the *antistreptolysin test.*
2. Levels of antibodies against this bacterial product increase during infection and convalescence.

In Vitro *Immunologic Procedures Incorporating Labeled Antibodies, Differential Staining and Enzymatic Reactions*

FLUORESCENT ANTIBODY (IMMUNOFLUORESCENCE) TECHNIQUES
1. Fluorescent antibody methodology involves chemically bonding a fluorescent dye (marker) with serum proteins.
2. Such preparations contain specific antibodies that can be used for the detection of homologous antigens in smears or tissue preparations.
3. The most commonly used fluorescent antibody procedures are the direct and the indirect methods.
4. The *direct fluorescent antibody method* can be used to identify unknown microbial agents and to determine the distribution of antigens in tissue.
5. The *indirect fluorescent antibody method* can be used to detect antibodies against various microbial pathogens.

In Vitro *Immunologic Procedures Using Enzymatic Reactions*

IMMUNOPEROXIDASE TEST
1. In this procedure an enzyme, usually horseradish peroxidase, is conjugated (combined) with specific antiserum (direct test) or with an antispecies immunoglobulin preparation (indirect test).
2. In positive tests, the peroxidase is joined to the antigen in a specific reaction by the antibody in the conjugate.
3. After treatment with a peroxidase substrate, the enzyme is detected by a specific color reaction.
4. The test is of value in the identification of cytomegalovirus, herpes viruses, and rabies virus, but it cannot be used in the detection of viruses in respiratory specimens.

ENZYME-LINKED IMMUNOABSORBENT ASSAY (ELISA)
1. This technique is based on the ability of antigen or immunoglobulin (a) to adsorb to a solid-phase surface of a tube or a well in a plastic tray and (b) to be linked to an enzyme, resulting in a complex exhibiting both detectable immunological and enzyme activity.
2. A positive reaction is detected by a color change visible to the naked eye or measurable by a spectrophotometer.
3. ELISA can be used for the identification of various viruses and/or the detection of IgM antibodies during the early phase of viral and other microbial diseases.

Diagnostic and Investigative Electron Microscopy

1. Immunoelectron microscopy (IEM) can be used to detect small amounts of antibody, show antigenic similarities and differences among viruses, and identify viral particles in various types of specimens.
2. Ferritin-conjugated antibody preparations are useful in finding various microbial antigens in tissues and in studying antigen-antibody interactions.
3. Immunogold probes are of value in locating surface as well as intracellular antigens.

Monoclonal Antibodies

1. The somatic cell hybridization or hybridoma formation technique is used to produce pure or monoclonal antibodies.

2. Monoclonal (single-type) antibodies are chemically, physically, and immunologically uniform.
3. These molecules can provide more rapid and accurate identification of viruses, bacteria, and cancer cells.

In Vivo *Testing Procedures*

VIRUS NEUTRALIZATION

1. Virus neutralization tests can use either tissue culture systems or laboratory animals, such as chick embryos, mice, or rats.
2. Tests such as these are used for identifying viruses, detecting antibodies against viruses, and studying responses to immunizations against viruses.
3. Details of the procedure vary with the viral agent, host used, and related factors.

DIAGNOSTIC SKIN TESTS

1. The intradermal introduction of test antigens can be used diagnostically, as well as to follow the course of recovery from a disease. Examples of such skin tests include the Frei test for lymphogranuloma venereum and the tuberculin test for tuberculosis.
2. Positive responses can occur shortly after exposure to the test antigen (immediately) or several hours later (delayed).
3. Skin tests also have value in determining an individual's susceptibility to microbial toxins. Examples of such tests include the Schick test for diphtheria and the Dick test for scarlet fever.

Immunohematology

THE ABO SYSTEM

1. The clumping of human red blood cells by homologous human immunoglobulins is known as *isohemagglutination.* Observing this phenomenon led to the discovery of the major blood types, A, B, AB, and O, frequently referred to as the *ABO system.*
2. Most individuals have antibodies against the major blood antigens that are absent from their blood cells. Thus, a person of blood type O has antibodies against types A and B, a person of blood type A has antibodies against type B, and a person of blood type B has antibodies against A. Individuals with blood type AB do not have antibodies against any of the other major blood types and are termed "universal recipients."
3. Tube and slide tests are commonly used to determine blood types.
4. These types of determinations can be made either by using an individual's red cells with standardized anti-A and anti-B sera or by testing the individual's serum with standard known A and B type red cells.
5. Clumping, or agglutination, resulting from cell serum combinations identifies the particular blood antigen present. Reactions establishing a particular blood type's identity are as follows:

	Antiserum Used	
Blood Type	*Anti-A*	*Anti-B*
A	Agglutination	No agglutination
B	No agglutination	Agglutination
AB	Agglutination	Agglutination
O	No agglutination	No agglutination

6. Antigenic subgroups have been found with both A and B blood types.
7. Inheritance of ABO blood factors follows normal Mendelian principles.
8. An individual's blood type is determined by receiving from each parent one of the four alternate (allelic) genes: A_1, A_2, B, or O.
9. Major blood-group antibodies become detectable within three to six months after birth.
10. Blood group substances are not limited to red cells but have been found in other cells and body fluids as well.

THE RH SYSTEM

1. The Rh system consists of blood factors that are also important.
2. *In utero* immunization of a mother by an Rh antigen inherited by a fetus from the father may bring about the condition known as the *Rh baby.*
3. Since the discovery of the Rh factor in human blood, other related blood factors have been discovered.
4. Determination of Rh factors in blood is made by agglutination tests using appropriate antisera.

Blood Testing Techniques

COMPATIBILITY TESTING (CROSS MATCHING)

1. The primary purpose of compatibility testing is to prevent transfusion reactions.
2. Cross matching includes mixing of the recipient's serum and the donor's cells *(major cross match)* and mixing of the recipient's cells with the donor's serum *(minor cross match).*
3. Agglutination (clumping) or lysis (destruction) occurring in either the major or minor cross matches indicates an incompatibility.

COOMBS' TESTS

1. The Coombs' tests are used to detect incomplete antibodies. Two forms of the procedure are used, the direct and the indirect.
2. The direct test is used to determine whether or not blood cells have been coated by antibodies *in vivo.*
3. The indirect test is used to detect either complete or incomplete antibodies in the serum of a sensitized individual.

Questions for Review

1. Differentiate between the following combinations:
   a. antigen and antibody
   b. serum and plasma
   c. toxin and antitoxin
   d. antitoxin and antibody
   e. agglutination and precipitation
   f. serology and immunology

2. What significance do serological tests have in the diagnosis of infectious diseases?

3. Construct a table listing:
   a. five representative serological tests
   b. the types of reagents used
   c. the appearance of both positive and negative reactions

4. What is viral neutralization?

5. a. What is complement? Where can this substance be found normally?
   b. Describe the complement fixation test. Include the components of the test and indication system, and the appearance of typical positive and negative tests.

6. a. Differentiate between the indirect and direct fluorescent antibody techniques.
   b. Distinguish between the immunoperoxidase and ELISA procedures.
   c. Distinguish between counterimmunoelectrophoresis and rocket electrophoresis.

7. a. How do monoclonal antibodies differ from those produced in a normal immune response?
   b. Of what diagnostic value are monoclonal antibodies?

8. Explain how electron microscopy can be used for diagnostic purposes.

9. a. Of what value are skin tests?
   b. List at least three that are of importance.
   c. What is anergy?

10. What are the major blood types, and how do they differ?

11. Distinguish between a universal recipient and a universal donor.

12. Differentiate between isoagglutinins and isohemolysins.

13. Describe the procedure used to detect ABO subgroups.

14. Discuss the major features of the Rh system and the possible complications of Rh incompatibility.

15. In Figure 19–26, the center well contains antigen. Each of the surrounding wells contains a different antiserum. With which of the wells is there antigen identity?

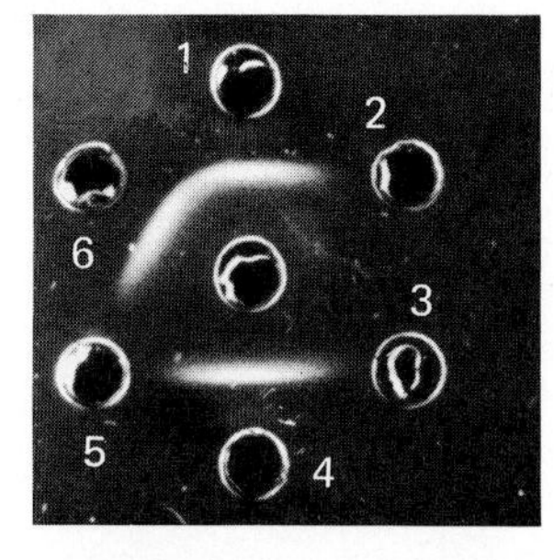

Figure 19–26 Double-diffusion precipitin reactions. *[D. L. Stickle, L. Kaufman, S. O. Blumer, and D. W. McLaughlin,* Appl. Microbiol. ***23****:490–499, 1972.]*

Suggested Readings

GARVEY, J. S., N. E. CREMER, and D. H. SUSSDORF, *Methods in Immunology,* 3rd ed. Reading, Mass.: Addison-Wesley, 1977. *A methodology text containing clear presentations of principles and techniques in "cookbook" fashion.*

LANDSTEINER, K., *The Specificity of Serological Reactions.* Cambridge, Mass.: Harvard University Press, 1945. *The basis of immunological specificity is described by the investigator most responsible for its definition.*

LENNETTE, E. H., A. BALOWS, W. J. HANSLER, JR., and H. J. SHADOMY (eds.), *Manual of Clinical Microbiology,* 4rd ed. Washington, D.C.: American Society for Microbiology, 1985. *A standard reference manual that lists and describes several important immunoserological tests, including tests for syphilis, antinuclear antibody, antistreptolysin O titer, streptococcal serology, and approaches to the identification of pathogenic protozoa and worms.*

MCMICHAEL, A. J., and J. W. FABRE, *Monoclonal Antibodies in Clinical Medicine.* New York: Academic Press, 1982. *A detailed presentation of monoclonal antibodies in diagnosis.*

MILSTEIN, C., "Monoclonal Antibodies," *Scientific American* **243**:66–74 1980. *A well-written, easy-to-understand article describing the production and applications of monoclonal antibodies.*

ROSE, R. R., H. FRIEDMAN, and J. L. FAHEY (eds.), *Manual of Clinical Laboratory Immunology.* Washington D.C.: American Society for Microbiology, 1986. *A highly functional reference dealing almost exclusively with techniques and methods used in laboratory settings.*

20

Immunologic Disorders

"The beginning of health is to know the disease." — Miguel de Cervantes

After reading this chapter, you should be able to:

1. Define *immunologic deficiency* and *hypersensitivity.*
2. List and describe the general categories of immunodeficiencies and hypersensitivity.
3. Describe the relationship of hypersensitivity to disease.
4. Identify the immunoglobulins associated with allergic reactions.
5. Outline the immunologic mechanisms and complications of anaphylaxis, atopy, cytotoxicity, allergy of infection, and graft rejection.
6. Discuss the roles of T cells, lymphokines, and related factors in delayed hypersensitivity and their potential for cancer immunotherapy.
7. Describe the general approaches used in the management of allergic states.
8. Explain the relationship and significance of the major histocompatibility complex (MHC) to histocompatibility antigens.
9. Discuss the basis for and outcomes of autoimmune diseases.
10. Describe the general features of the complement cascade.

The immune system is impressive in its high degree of specialization and its ability to protect the body from infection. In some situations, however, the immune system can fail. Worldwide interest and concern has been drawn to the existence of the new immune system defect called acquired immune deficiency syndrome (AIDS). Chapter 20 describes a number of other failures involving immune system deficiencies, exaggerated and damaging responses to immunogens, responses to drugs and various chemicals, and the transplantation of tissues.

The various components of the immune system are not only important in themselves but actually indispensable. To protect and preserve the body, the immune system must be able to recognize and destroy molecules that are *foreign*, or *non-self*, while conserving those body components that are *native*, or *self*. But what happens when this system fails and actually becomes the basis of disease or contributes to the lowering of the body's protective capabilities?

Immunodeficiencies

The immunodeficiency diseases, which include a wide range of disorders, are distinguished by the inability of the immune system to perform normally. Four major immune systems normally provide the defense against constant attacks by microbial and helminth (worm) agents and factors that have the potential to produce infection and disease. These systems are antibody-regulated (B cell) immunity, cell-mediated (T cell) immunity, phagocytosis, and complement. Each system may work in cooperation with one or more of the others, or it may function independently. The exact effects and symptoms of such conditions vary according to the immune system affected and the extent of the disorder. General features frequently associated with immunodeficiency disorders include recurrent infections, infections caused by unusual agents, long-lasting (chronic) infections, skin rash, diarrhea, growth failure, and recurrent skin abscesses and bone infections (osteomyelitis). Two general categories of these disorders are recognized—*congenital* or *primary* and *secondary immunodeficiencies*.

Primary Immunodeficiencies

Primary immunodeficiencies (Table 20–1) usually arise from an inherited failure of one or more immune system components to develop. The earliest and most devastating immune defect is *bone marrow stem cell deficiency*. In this condition, fatal infections develop within days after birth, since neither B nor T lymphocyte components are functional. In *DiGeorge's syndrome*, the thymus gland does not develop. Here T lymphocytes are lacking, and the infant is highly susceptible to the effects of various microorganisms, such as digestive system microbes, fungi, and atypical tubercle bacilli.

Individuals with X-linked agammaglobulinemia or deficiencies in all or selected classes of immunoglobulins represent another congenital abnormality. In this inborn condition, first described by Ogden C. Bruton in 1952, B-lymphocytes fail to develop. The disorder is transmitted as a sex-linked (X chromosome-linked) recessive trait and is therefore found primarily in males. In the past, infants with this disease died of infection. Today, most cases are recognized during the first or second year of life. In cases of severe bacterial infections, patients are saved with the aid of antibiotic therapy and injections of purified immunoglobulin preparations.

Electrophoretic examination of the sera from agammaglobulinemic persons shows that IgG is almost completely absent and IgA and IgM are lacking. Apparently, they are replaced by other nonfunctional proteins in the sera. It is important to demonstrate this severe and persistent immunoglobulin deficiency in order to distinguish it from the temporary hypogammaglobulinemia which can occur during infancy.

The basic defect in persons with this condition may be either an inability to form the necessary globulins or the activation of an abnormal mechanism that destroys the globulins shortly after they are formed. Among the major risks to such individuals are encapsulated bacteria such as *Haemophilus influenzae*, *Neisseria meningitidis*, and *Streptococcus pneumoniae*.

The absence or near absence of serum and secretory IgA is the most common immunodeficiency disorder. While this selective IgA deficiency has been found in apparently healthy individuals, it is commonly associated with poor health. Affected individuals experience infections that occur primarily in the respiratory, gastrointestinal, and urogenital tracts. The infecting bacteria are the same as those found in other immunoglobulin deficiencies. The defect that causes selective IgA deficiency is now known.

Severe combined immunodeficiency (SCID) disorders probably represent the most dangerous defects of the immune system. These defects are distinguished by a complete absence of immune function. Death usually occurs during the first year of life unless the affected individual is placed into a germ-free (sterile) environment and isolated from the outside world or is given a bone marrow transplant to establish a normally functioning immune system.

Phagocytic disorders result from a variety of external and internal defects (Table 20–1). Contributing factors in the external category include antibody and complement deficiencies, drug interference with

TABLE 20–1 **Immunodeficiency Disorders**

Disorder	*General Features*	*Diagnostic Tests*[a]
B CELL IMMUNODEFICIENCY		
X-linked agammaglobulinemia (X chromosome-linked)	B cells absent from peripheral blood; low levels of IgG; absence of IgA, IgM, IgD, and IgE; recurrent infections beginning early in life (4 to 5 months)	1. Protein electrophoresis 2. Radial immunodiffusion for immunoglobulins 3. B cell number determinations in peripheral blood
Selective IgA deficiency	Significantly low level of IgA; other immunoglobulin levels normal or increased; allergies, recurrent respiratory infections, and autoimmune disease commonly found	1. Radial immunodiffusion 2. Antinuclear and anti-DNA tests (Color Photograph 83)
T CELL IMMUNODEFICIENCY		
Chronic mucocutaneous candidiasis	Long-lasting infection of skin, nails, and mucous membranes with the yeast *Candida;* hormonal gland disorders (endocrinopathy)	1. Total lymphocyte count 2. T cell function tests 3. T cell rosette formation 4. Monoclonal antibody test 5. Detection of sensitized lymphocyte products (lymphokines)
DiGeorge syndrome	Decreased number of T cells; reduced T cell function; variable antibody production; symptoms appear shortly after birth; abnormal facial features and hormonal functions	Similar to those listed for chronic mucocutaneous candidiasis
PHAGOCYTIC DYSFUNCTION		
Chronic granulomatous disease (X chromosome-linked)	Features occur during first 2 years of life; males are mainly affected; susceptibility and frequent infections; enlarged lymph nodes	1. Nitroblue tetrazolium dye test to determine intracellular killing 2. Chemiluminescence to detect a normal level of radiation 3. Chemotaxis
Chédiak-Higashi syndrome	Susceptibility to infectious agents similar to that noted for chronic granulatomatous disease; both males and females affected; hemolytic anemia (destruction of red blood cells). Recurrent bacterial infections; enlarged liver and spleen; central nervous system abnormalities	Similar to those listed for chronic granulomatous disease
Glucose-6-phosphate dehydrogenase deficiency (X chromosome-linked)	Susceptibility to infectious agents similar to that noted for chronic granulatomatous disease; both males and females affected; hemolytic anemia (destruction of red blood cells). Recurrent bacterial infections; enlarged liver and spleen; central nervous system abnormalities	Similar to those listed for chronic granulomatous disease
Job's syndrome	Recurrent staphylococcal infections involving the skin, lymph nodes, middle ear, and respiratory system	Similar to those listed for chronic granulomatous disease
COMPLEMENT ABNORMALITIES		
C2 deficiency	Long-lasting kidney disease; antibody against DNA present; susceptibility to bacterial infections	Assay for specific complement component
C5 dysfunction	Susceptibility to bacterial infections; skin infections; diarrhea; large amounts of immunoglobulins produced	1. White blood cell count 2. Assay for specific complement component

[a] Several of these tests are described in Chapter 19.

phagocytic activities, suppression of circulating neutrophils by specific antibodies against neutrophil antigens, and suppression of phagocytic cells by drugs and other immunosuppressive agents. The internal factors are associated with enzymatic deficiencies needed by phagocytic cells to kill bacteria. **[See phagocytosis in Chapter 17.]**

The description of the complement system later in this chapter emphasizes its importance in phagocytosis, killing of bacteria, and attraction of neutrophils to sites of inflammation. Interference with such activities or deficiencies in complement components generally result in an increased susceptibility to microbial infections and to autoimmune disease states in which the body attacks its own tissues (Table 20–1).

Diagnosis of many immunodeficiency disorders can be made with the aid of tests that screen for the presence of respective immune system components. Table 20–1 lists examples of diagnostic approaches together with brief descriptions of representative disorders.

Secondary Immunodeficiencies

Secondary, or acquired, deficiencies (Color Photograph 84) occur much more frequently than primary immunodeficiencies. The causes of secondary deficiencies include malnutrition, malignancies (cancers), extensive exposure to radiation, burns, and various drugs that interfere with the lymphatic system. B lymphocyte malignancies such as Hodgkin's disease and chronic lymphocytic leukemia are known to produce immunoglobulin abnormalities. For example, in certain disease states, a single cell undergoes conversion to a cancerous state. The effect is a rapid and unregulated multiplication resulting in a large family of monoclonal cells (cells that are all descendants of one type). This formation, in turn, produces a single type of uniform immunoglobulin. This condition is called a *monoclonal gammopathy* (abnormality). These proteins are found in individuals suffering from multiple myeloma and other cancers involving the lymphatics. Immunoglobulin abnormalities may cause increases in the concentration of several or all immunoglobulins or abnormally large increases in the concentration of a single immunoglobulin type.

Secondary immunodeficiency also results in an increased susceptibility to opportunistic infection. One severe form of such a state is acquired immune deficiency syndrome, or AIDS.

Acquired Immune Deficiency Syndrome (AIDS)

AIDS is now recognized as an infection caused by human immunodeficiency virus (HIV). Among the primary targets of the virus are T4 (CD4) cells, the lymphocytes that play a central role in orchestrating the body's response to infections. Once the viral genes are integrated into the cell's DNA, they appear to remain inactive for an indefinite period without causing any ill effects. Problems arise when the viral genes are activated and new virus particles are formed, which then infect fresh T4 cells. The process of viral reproduction kills the infected cell. Eventually, the body's T4 cells are seriously depleted and the immune system collapses, thus making the victim susceptible to a variety of opportunistic infections. A more detailed description of AIDS and related immune defects can be found in Chapter 27.

Hypersensitivity (Allergy)

In 1906, von Pirquet introduced the term *hypersensitivity* to describe the increase in reactivity that can occur when persons vaccinated with vaccinia virus are reexposed to the virus. Later, he extended the limits of hypersensitivity to include any altered activity, or *allergy,* caused by contact with animate or inanimate substances. This state is generally described as an acquired exaggerated response toward a specific substance that does not produce similar reactions in the majority of previously unexposed members of the same species. The inciting, or inducing, agents are referred to as **allergens.**

Initial exposure to an allergen immunologically primes, or sensitizes, the individual. Subsequent contact with the same allergen can not only lead to a secondary boosting of the immune response but can also cause tissue-damaging reactions. Several factors influence the expression of hypersensitive responses, including the portal of entry, the cellular and tissue responses to the allergen, and the genetic makeup of the individual.

Categories of Hypersensitivity

Historically, hypersensitivity reactions were divided into two classes on the basis of the time required for an obvious physiological response to develop upon reexposure of a sensitized individual to the allergen. Reactions that developed within a few seconds to 24 hours following allergen exposure and involved circulating antibodies were usually classified as *immediate.* Those that developed after 24 hours but within 48 hours and involved the direct participation of sensitized lymphocytes were known as *delayed.* Although these designations are still used today, they are now understood to have different meanings. In 1968, R. A. Coombs and P. G. H. Gell defined four clinically important types of hypersensitivity (Table 20–2): type I (classic immediate), type II (cytotoxic), type III (immune-complex-mediated), and type IV (cell-mediated, or delayed). Types I through III are immunoglobulin-dependent hypersensitivities. Type IV reactions are mediated by antigen-sensitized T lymphocytes rather than by antibody molecules. In addition, specific interactions of sensitized T cells with antigens are involved in the rejection of transplanted tissues and organs and in protection against cancer.

Since the classification by Coombs and Gell, new techniques and results of molecular studies have uncovered additional forms of immunologic injury. One of these, which involves the actions of immunoglobulins against cellular receptors for various hormones, will also be described.

The immunoglobulin-dependent (immediate) hy-

TABLE 20–2 Comparison of Different Types of Hypersensitivity

	Immediate (Immunoglobulin dependent)			*Delayed*
Characteristic	*I (Classic Immediate)*	*II (Cytotoxic)*	*III (Immune-complex-mediated)*	*IV (Cell-Mediated)*
Maximum reaction time for clinical manifestation (response)	30 min	Variable	3–8 hr	24–48 hr
Immunoglobulins involved	IgE and IgG	IgG and IgM	IgE, IgG, and IgM	None
Cells involved	Mast cells, basophils	Red blood cells, lymphocytes, eosinophils, neutrophils, and body tissue cells	Body tissue cells	Body tissue cells, T lymphocytes, and macrophages (eosinophils in certain cases)
Reaction mediators (regulators)	Pharmacological substances including histamine, leukotriene D, ECF-A, kinins, and prostaglandins[a]	Complement	Complement, and lysosomal enzymes	Soluble mediators such as MIF and TF[b] (lymphokines)
Appearance of response to intradermal antigens	Wheal and flare		Erythema (redness) and edema (swelling)	Erythema and induration (hardened tissue)
Inhibition by antihistaminic drugs	Yes, in certain situations	No	No	No
Inhibition by cortisone	No, with normal dosages	No	No	Yes
Passive transfer with serum from sensitive donor	Generally yes	Yes	Yes	No
Passive transfer with lymphocytes or lymphocyte extracts from sensitive donor	No	No	No	Yes
Examples of allergic states	Anaphylaxis, asthma, serum sickness, and drug, food, and insect allergies	Transfusion reaction, hemolytic disease of the newborn, and drug-induced allergies	Arthus reaction, serum sickness, and certain autoimmune diseases	Infection allergies, autoimmune disease, graft rejection, contact dermatitis to drugs, and tumor immunity

[a] Leukotriene D was known as SRS-A, a slow-reacting substance of anaphylaxis; ECF-A is an eosinophilic chemotactic factor of anaphylaxis.
[b] MIF is a migration inhibitory factor; TF is a transfer factor.

persensitivities can be subdivided into two categories according to the agent primarily responsible for the immune response: the classic, heat-stable IgA, IgD, IgG, and IgM; immunoglobulins; or the heat-sensitive IgE antibodies. Heat-stable antibodies resist destruction by heat at 56° to 60°C for periods of 30 minutes to 4 hours. It should be noted that both immunoglobulin categories may be present at the same time.

From the characteristics listed in Table 20–2, it is obvious that hypersensitivity reactions involve several different types of physiological response by immunized cells or tissues. Nevertheless, one or more exposures are always required to establish hypersensitivity to the inciting allergenic substance. Exposure is followed by a so-called latent (inactive) period in which no symptoms are shown. The hypersensitive state is then triggered by another exposure to the allergen.

The particular type of hypersensitivity acquired is determined by several factors. These include (1) the chemical nature of the allergen; (2) the route involved in sensitization—for example, inhalation, ingestion, or injection; and (3) the physiological state of the individual. The form of the response depends on the type of contact with the precipitating allergenic dose and on the acquired hypersensitive state.

The four major types of hypersensitivity are distinguished by other differences as well. These distinctions are listed in Table 20–2 and will be discussed later.

Type I: Classic Immediate Hypersensitivity

Examples of type I allergic responses include certain forms of asthma, anaphylactic shock, hay fever *(allergic rhinitis)*, and hives *(urticaria)*. One of the distinguishing features of this hypersensitive state is the rapid liberation of physiologically active chemicals, such as histamine and heparin, from affected cells. Substances of this kind are normally released from cells as a consequence of antigen and antibody interaction. Initially, immunoglobulins (Ig) produced by plasma cells attach by means of their F_c portions (Figure 20–1) to basophils and mast cells. This attachment sensitizes the cells to antigens responsible for Ig production. Mast cells contain heparin and histamine in cytoplasmic granules and are found in

Figure 20–1 An immune system response to an allergen. Allergens come into contact with specific IgE antibody molecules on the surface of a mast cell. A subsequent membrane response causes the release of chemical mediators from granules in the basophil or mast cell.

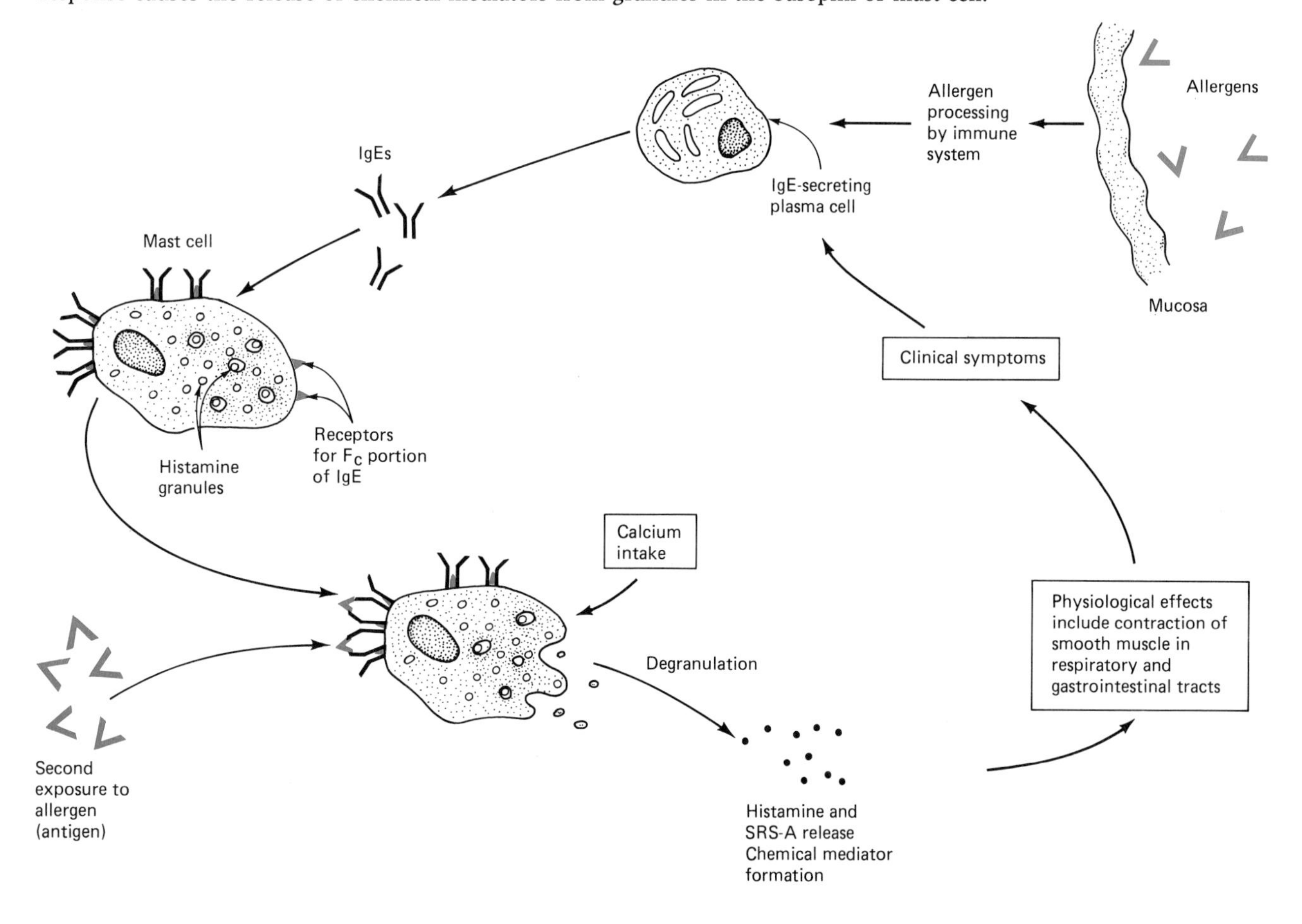

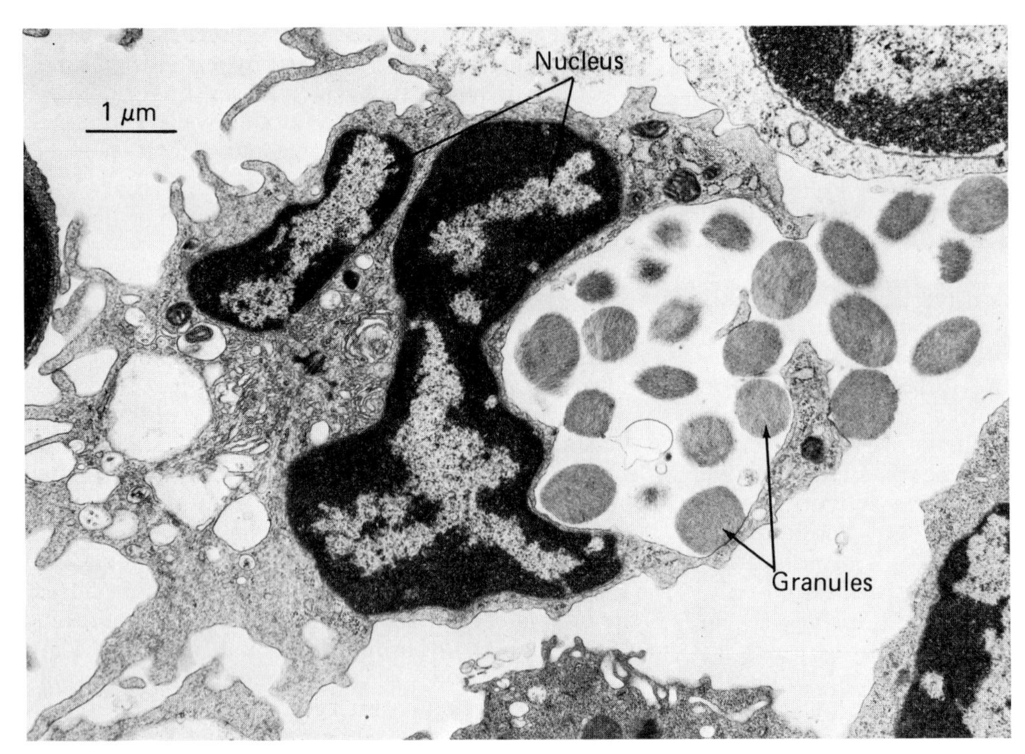

Figure 20–2 The appearance of degranulation. Note the release of granules from the cytoplasmic area. *[A. M. Dvorak,* et al., Lab. Invest. ***44**:174–190, 1981.]*

connective tissue. The F_{ab} regions of the attached immunoglobulins protrude from the cellular surfaces. When combined with the antigen, they alter the permeability of the mast cells. This reaction leads to the degranulation of mast cells and basophils (Figure 20–2) with the release of histamine and serotonin. [See immunoglobulin structure in Chapter 17.]

In humans, almost all type I reactions are associated with IgE. The liberated substances produce secondary involvement and responses in other cells. Among these effects are the destruction of blood cells, an increase in muscle activity, an increase in capillary permeability, and excessive mucus production.

The sensitivity response of type I hypersensitivity can be transferred to a normal, nonsensitive person simply by the injection of serum or the transfusion of blood from a sensitive individual.

Representative Type I Allergic States

ANAPHYLAXIS

Before approximately 1837, certain protein solutions, such as egg albumin, were used for the initial inoculation of laboratory animals and were considered to be largely harmless. In 1839, however, Magendie reported that repeated injections of such material often produced severe symptoms and even caused the sudden death of dogs. Instead of becoming immune to such foreign substances, the experimental animals became unusually sensitive to them. Moreover, death was caused by reinjection with dosages too small to affect normal laboratory animals. In essence, the effect was the opposite of protection *(prophylaxis)*. In 1902, Richet named this phenomenon *anaphylaxis*. The reaction has come to be known as *anaphylactic shock*.

Two general types of anaphylactic responses are recognized—*systemic*, or *generalized anaphylaxis*, and *cutaneous anaphylaxis*. Both states are temporary. If the individual does not die almost at once, recovery usually occurs within one hour.

The anaphylactic reaction results from the interaction between the specific allergen and the cell-fixed (or attached) anaphylactic antibodies. The cells specifically involved in this reaction include tissue mast cells and basophil leukocytes. Anaphylactic antibodies seem to have a special attraction for these cells. The interaction between the allergen and anaphylactic antibodies fixed to these cells triggers an increased intake of calcium, which is critical to a process inside the cells, ultimately leading to the release of both preformed and newly formed chemical mediators (Figure 20–1). The released mediators, in turn, act on certain target organs or tissues, such as smooth muscles in blood vessels, intestines, and portions of the respiratory passages.

GENERALIZED ANAPHYLAXIS. The generalized form of anaphylactic response is produced by the intravenous injection of specific soluble allergens, such as horse serum or egg albumin, to which an animal has been previously sensitized. The allergens, or sensitizers, that originally cause the hypersensitive state are called *inducers.* The anaphylactic response is brought about by substances similar to the inducers called *elicitors.* Generalized anaphylaxis produces abnormalities in different body systems. For example, nausea, vomiting, abdominal cramps, and diarrhea are common features of gastrointestinal system involvement. With the respiratory tract, suffocation and constriction and obstruction of passages are commonly found. Engorgement of blood vessels and failure of the peripheral circulation to function, leading to shock, are associated with the cardiovascular system. The features of anaphylactic shock vary depending on the species involved.

CUTANEOUS ANAPHYLAXIS IN HUMANS. Cutaneous anaphylaxis results from the intradermal injection of an elicitor into a sensitized individual. The reaction begins with an intense itching at the injection site. This symptom is followed shortly by the development of a pale, irregular wheal (a flat, elevated, swollen area) surrounded by a region of redness. The response, which usually subsides within approximately 30 minutes, is referred to as the *wheal and flare reaction.*

PRODUCTION OF ACTIVE ANAPHYLAXIS. The steps required to bring about anaphylactic responses include: (1) injection of the inducer substance (sensitizing allergen); (2) the passage of time, referred to as the *incubation,* or *latent, period,* during which IgE-class molecules sensitize mast cells and basophils; (3) reintroduction of the sensitizing allergen (elicitor), generally by injection or inhalation. The elicitor is frequently called the *shocking,* or *injection, dose.* The inducer and elicitor are closely related substances.

The mechanism governing an anaphylactic response generally involves a bridging process in which an allergen links antibody molecules. For this reason, the reaction requires (1) the presence of two or more antigenic determinants on the allergen, and (2) the presence of more than one antibody molecule per cell. (Actually hundreds or thousands of such molecules usually exist in a cell.) The intensity of the response depends on both the number of antibody-antigen complexes formed and the rate at which they are formed. The more rapidly such complexes develop, the stronger is the response. The release and rapid degradation of certain active substances play a vital role in anaphylaxis and related responses.

Several active participants, or *mediators,* are involved in producing the symptoms of anaphylaxis. Included are the preformed mediators (regulators) found existing in the granules of mast cells and basophils, histamine and heparin, the well-established secondary mediators such as the leukotrienes, and various chemical-attracting (chemotactic) factors and hydrolytic enzymes found in mast cell granules (Figure 20–1). The secondary mediators are produced by stimulated mast cells, basophils, and other cells recruited in the anaphylactic reaction. Table 20–3 lists these substances together with their associated functions. A major portion of secondary mediators are compounds derived metabolically from arachidonic acid, a component of membrane

TABLE 20–3 Representative Primary and Secondary Mediators of Anaphylaxis

Mediator	*Functions or Activities Include*
PRIMARY, PREFORMED	
Eosinophil chemotactic factors of anaphylaxis	Inactivate eosinophils and neutrophils; induce lysosomal enzyme release; increase expression of receptors on eosinophils for certain complement components
Heparin	Anticoagulant; anticomplement activity
Histamine	Increases blood vessel permeability; moves chemical substances (chemokinesis) and produces mucus; elevates levels of cyclic adenosine monophosphate and histamine suppressor factor
Kallikrein	Bradykinin synthesis
Neutrophil chemotactic factor	Inactivates neutrophils
Serotonin	Increases blood vessel permeability
Tryptase	Causes protein breakdown (proteolysis)
Vasoactive intestinal polypeptide	Relaxes smooth muscle
SECONDARY	
5-Hydroperoxy-eicosatetraenoic acid	Aids histamine release
Leukotriene C_4	Increases blood vessel permeability; powerful constrictor of bronchioles
Leukotriene D_4	Extremely powerful constrictor of respiratory passages; increases blood vessel permeability and dilatation (widening)
Leukotriene E_4	Same as leukotriene C_4
Prostaglandin D_2	Constricts bronchioles and increases effects of leukotrienes
Prostaglandin E_2	Inhibits histamine release and increases effects of leukotrienes

phospholipids. These substances are very powerful and specific in their activities and appear to be quite capable of stimulating or inhibiting one another's effects. The temporary nature of anaphylactic reactions is accounted for by the rapid degradation and excretion of these substances, which prevent them from accumulating.

The release of mediators from activated mast cells or basophils is initiated by the bridging of two IgE-specific receptors on these cells. This type of reaction can occur by the binding of a divalent antigen to IgE molecules that are themselves bound to two separate receptors (Figure 20–1).

The symptoms of laboratory-induced anaphylaxis differ in different laboratory animals. Such variations are explained partly by differences in the amount of active substances in the subjects' tissues and partly by differences in the response of specific tissues such as smooth muscle and blood vessels.

Anaphylaxis in Humans. Systemic, or generalized, anaphylactic shock can occur as a result of serum therapy or the administration of penicillin. Symptoms may include abdominal cramps, diarrhea, diffuse erythema, hives, intense itching, nausea, respiratory difficulties, and vomiting. Death may occur rapidly. The prompt intravenous or intracardial administration of epinephrine (adrenalin) will usually counter an attack of anaphylaxis.

Certain preventive measures against a fatal anaphylactic attack are available. One can skin-test individuals with minute amounts of substances known as *potential elicitors*—for example, serum, penicillin, and the like—before administering full doses of such substances. Also, one can keep a container of epinephrine close at hand to be administered as needed.

Several substances and activities have been implicated as causes of anaphylaxis. For example, sulfiting agents, such as gaseous sulfur dioxide and sulfite compounds, are added to foods and beverages as preservatives to prevent discoloration. Rapidly perishable foods—sush as vegetables, shellfish, wine, and beer—are likely candidates for such treatment. Strenuous exercise in susceptible individuals has also been reported as a likely cause.

ANAPHYLACTOID REACTIONS

Anaphylactoid reactions are not associated with IgE. However, they may result from exposure to various chemicals causing mast cell degranulation. Such chemicals include neuromuscular relaxing agents, various narcotics, and low-molecular-weight compounds.

ATOPY

At least 10% of the U.S. population is believed to have a natural hypersensitivity, in the absence of a deliberate exposure, to a large number of environmental allergenic substances. These materials include airborne pollens of grasses, ragweeds, and trees; animal danders; foods; certain fungi; and house dust components. Upon inhalation or ingestion of a specific allergen, a hypersensitive individual can develop various kinds of clinical syndromes, the most frequent of which are asthma, eczema, certain gastrointestinal disorders, hay fever, hives, and general cold symptoms. Coca applied the term *atopy* to these allergic disease states, from the Greek word meaning "out of place."

Apparently, hypersensitivity develops in susceptible persons because of repeated accidental absorption of the allergenic substances. Absorption can involve the mucous membranes of the gastrointestinal or respiratory tract or the skin. The various atopic allergic disease states are precipitated by later exposure to these allergens. The tendency to develop diseases of this type has been shown to be familial in distribution and is believed to be inheritable. Persons exhibiting atopic reactions are referred to as *atopics*. Such individuals have the genetic capacity to produce blood plasma concentrations of IgE ten times greater than those of nonatopics.

Skin Testing for Atopic States. Upon injection of an appropriate allergen, atopic individuals exhibit itching at the inoculation site followed by a wheal and flare response. The effect may persist for approximately 20 minutes. This type of response can be used to determine the substances to which the atopic is sensitive. To test an individual's susceptibility to a specific allergen, one introduces a sterile dilution into the skin either by intradermal injection or scarification (the making of a small number of superficial scratches). Sensitivity to the test substance is indicated by a wheal and flare response in the area of the inoculation. Intense reactions to such tests are usually considered to be associated with allergens responsible for the immediate difficulties of the subject.

Measurement of IgE. Since IgE levels are extremely low, highly sensitive tests are necessary to determine its concentration. The normal adult IgE level is about 250 ng/mL. In severely allergic individuals, IgE levels can reach 700 ng/mL. [A nanogram is one-billionth (10^{-9}) of a gram.] Although the technique of gel diffusion (see Chapter 19) can detect this amount of the immunoglobulin, more sensitive procedures are used. One such test is the radioallergosorbent test (RAST). In this procedure, purified allergen extract first is absorbed onto filter paper disks or cellulose particles (allergen-complex immunosorbent). The patient's serum, containing IgE or a control serum sample, is then reacted with the immunosorbent. The immunosorbent is then washed and exposed to a radioactive antibody preparation (^{125}I-labeled rabbit anti-IgE) against the patient's IgE. After further washing, the radioactivity on the sorbent is determined and is used as the measure of the amount of specific serum IgE antibodies

Microbiological Highlights

ALLERGY TO ANIMALS: AN OCCUPATIONAL DISEASE

Recent surveys have shown that allergies to laboratory animals are a common and important occupational health problem. Close contact with laboratory animals—their dander, secretions, or tissues—for animal technicians, veterinarians, and biomedical investigators is often unavoidable. Allergies to these animals can have serious consequences for such individuals, not only in terms of their comfort and health but also with regard to future career opportunities.

Animal allergies occur in 11.3% to 30% of the individuals who have direct contact with animals. Allergies usually appear within two years of the first exposure among the persons affected. Symptoms tend to develop within six months and include mild upper respiratory difficulties such as sneezing, runny nose, and conjunctivitis (an inflammation of the inner lining of the eyelid). Repeated exposures to airborne allergens may progress to more serious respiratory conditions sush as asthma. Direct skin contact during the handling of animals may produce reddening, raised areas (wheals), or a rash (eczema). These symptoms usually appear within five to ten minutes after the animals are handled.

The animals most commonly implicated in laboratory animal allergies are the rat, rabbit, guinea pig, mouse, cat, dog, and hamster. Allergies to more than one animal species are quite common.

Coping with this type of allergy may prove difficult. Avoiding the allergens is the best way to control it. However, this approach may not always be possible. Other means include the use of masks and protective clothing, improving ventilation, avoiding recirculation of animal-room air, increasing the frequency of cage cleaning, and using filtered and well-ventilated caging systems.

to that allergen. The RAST procedure measures 1 ng of specific IgE antibody.

Passive Transfer of Atopy (Prausnitz-Küstner, or P-K, Reaction). The presence of antibodies in the serum of atopics was discovered in 1921 by Prausnitz and Küstner. This discovery was directly related to Küstner's high sensitivity to fish. A small quantity of Küstner's serum was injected into Prausnitz's skin. After 24 hours, an extract of fish was introduced into both this site and an untreated region of Prausnitz's skin. The latter served as a control. Within 20 minutes of the fish extract injection, a typical wheal-and-flare response developed at the site of the serum injection. The control region showed no reaction.

The P-K procedure can be used for patients who are extremely sensitive to specific allergens and cannot undergo direct testing. In this test, several hours should elapse between the injection of serum from a sensitive individual and that of the suspected allergen. The two substances should not be mixed, as no reaction will develop when a mixture is injected.

Establishing the Atopic State. The IgE responses of allergic individuals to different allergens are not only sizable but may even increase upon reexposure to the allergen. Nonatopics, on the other hand, do not produce significant IgE responses against common allergens. Why?

Several factors are believed to be involved in the establishment of an atopic state. Recent studies have demonstrated the existence of circulating suppressive molecules that have the specific capacity to interfere with the activities of IgE. Although these suppressive substances are found predominantly in individuals who are not typically atopic, this does not mean that atopic individuals are unable to produce such materials. The exact nature of the suppressive molecules, the source of production, the conditions necessary for production, and the mechanism of their action have not been precisely determined.

Immunotherapy of Atopic Diseases. Several approaches to the treatment of allergy exist. The first step is to identify the specific allergen by obtaining a careful case history of the individual and the use of laboratory and skin tests. The next step consists of (1) pharmacologic intervention—the use of medications to block release of or to reduce the effects of the chemicals associated with allergic states (Figure

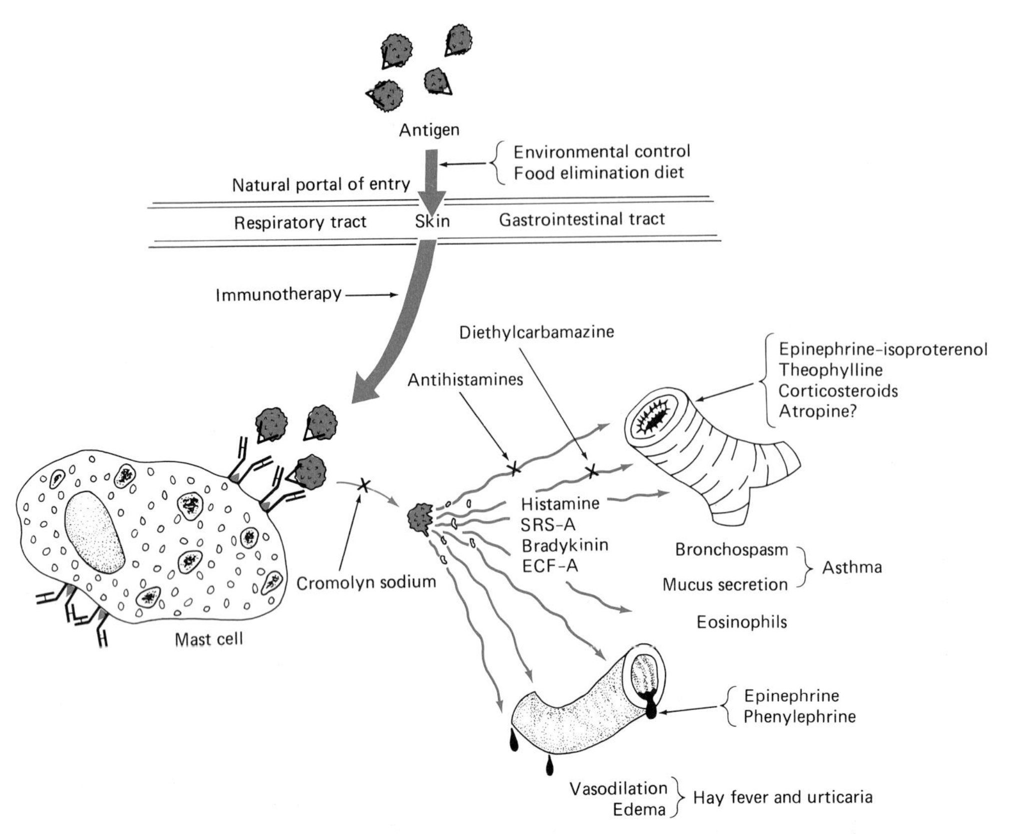

Figure 20–3 General therapeutic approaches to atopic reactions. The various substances, cells, and other factors, together with the neutralizing drugs and their actions, are shown. These approaches usually are successful in treating allergies due to animal dander, house dust, fungi, or foods. *[After D. P. Stites,* et al., Basic and Clinical Immunology. *Los Altos, Calif., Lange Medical Publications, 1982.]*

20–3)—and (2) immunotherapy—procedures to render an atopic individual immunologically unreactive to a particular allergen.

Several pharmacologic substances are used in allergy treatment. Among these are cromolyn sodium, which prevents the release of mediators from mast cells, probably by stabilizing lysosomal membranes; corticosteroids, which inhibit histidine decarboxylase, the enzyme necessary for histamine formation; and antihistamines that block histamine receptor sites on cells of shock organs. The last group of compounds include epinephrine and other beta-adrenergic substances. Figure 20–3 shows the sites of several pharmacologic agents used in allergy treatment.

Another immunotherapeutic approach, known as **desensitization** or **hyposensitization,** involves the controlled administration of specific allergen preparations. Injections of such material stimulate the formation of IgG-blocking antibodies, which remain in the circulation or tissues to interfere with the allergy (Figure 20–4). Upon subsequent natural exposure to allergen, these antibodies react with the antigenic material, forming a complex that is then removed by the monocytic-phagocytic (reticuloendothelial) system. Other promising approaches to the immunotherapy of allergic diseases include the stimulation of suppressor mechanisms such as those discussed earlier.

Figure 20–4 The results of immunotherapy against allergy. Injection of allergen produces IgG-blocking antibody. *[After D. P. Stites,* et al., Basic and Clinical Immunology. *Los Altos, Calif., Lange Medical Publications, 1984.]*

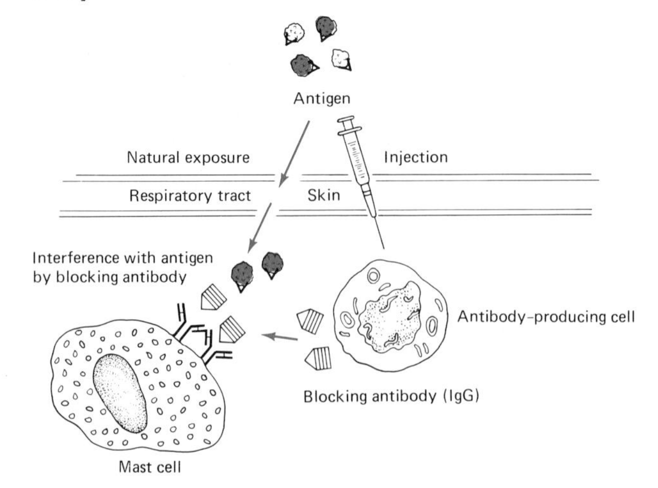

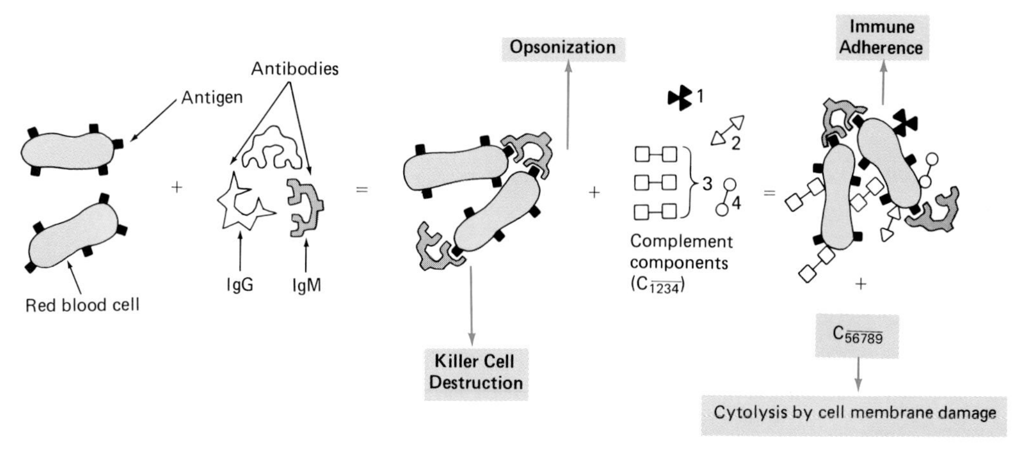

Figure 20–5 The mechanism of type II hypersensitivity with red blood cells. In most of these and related cytotoxic reactions, lesions are produced when fixation and activation of the complement system occur.

Type II: Cytotoxic Hypersensitivity

Cytotoxicity refers to reactions that disrupt or destroy cells. Examples of such type II responses in humans include (1) blood transfusion reactions, in which blood group antigens in red blood cell membranes are the inciting factors (Figure 20–5); (2) erythroblastosis fetalis (the Rh baby), in which Rh antigens on fetal or newborn red blood cells are the targets of antibodies formed by the mother; (3) drug-caused blood loss (anemias), in which a drug forms an antigenic complex with the surfaces of blood and other cells and brings about the production of antibodies that are destructive for the cell-drug complex; and (4) nephrotoxic nephritis, in which antibodies against antigen in the basement membrane of kidney glomeruli combine and produce a cytotoxic reaction.

Allergens create situations that cause cell disruption *(cytolysis)* or death. Two factors are necessary for lesions to occur in most type II reactions. These are (1) antibodies directed against target body cells or bacteria and (2) the **complement cascade system** (Figure 20–5), which is nonspecifically triggered by the combination of antibody and cell-associated antigenic determinants. [See Chapter 17 for a discussion of antigenic determinants.] The resulting type of reaction is known as antibody-dependent, complement-mediated cytotoxicity. The antibodies involved in type II hypersensitivity responses are IgG and IgM. Examples of responses not requiring the involvement of complement are discussed in Chapter 17 (Figure 17–26).

The Complement Cascade (Pathway)

The complement cascade system consists of a group of circulating plasma proteins that play a major role in host defenses. Complement proteins serve as mediators of inflammation, attracting factors for phagocytic cells, and as opsonins—substances that coat antigens to make them more susceptible to phagocytosis. These proteins also participate in the disruption (lysis) of bacterial and red blood cells. One of the important features of the complement cascade is that it increases antibody effectiveness. Complement activation leads to the generation of active enzymes that split (cleave), and thereby activate, later-forming members in the complement cascade (Figure 20–5).

Two different complement activation pathways are recognized: the classic and the alternate, or properdin, pathways. While the activation of these systems differs, the sequence of reactions forming the last portion is the same for both (Figure 20–6).

CLASSIC PATHWAY

The classic complement pathway consists of a series of sequential enzyme-substrate and protein-to-protein interactions leading to the formation of specific enzymes. The cascade of reactions begins with either the immunologic or nonimmunologic activation of the first component, C1. Immunologic activators include antigen-antibody complexes and clumps of immunoglobulins. Antigens include nucleated cells, red blood cells, or bacteria. With these substances, C1 binds to a site in the Fc region of interacting IgG and IgM molecules. Nonimmunologic activators include DNA, staphylococcal protein A, certain cellular membranes, and some protein-digesting enzymes such as trypsin. Activation occurs by the direct binding of C1 to these substances. The classic pathway consists of reactions involving the first four complement components, C1 to C4 (Figure 20–6), and leading to generation of the last enzyme, which activates the C5 component. This activation is the first step in the terminal, or last, portion of the complement sequence, referred to as the **membrane attack system.**

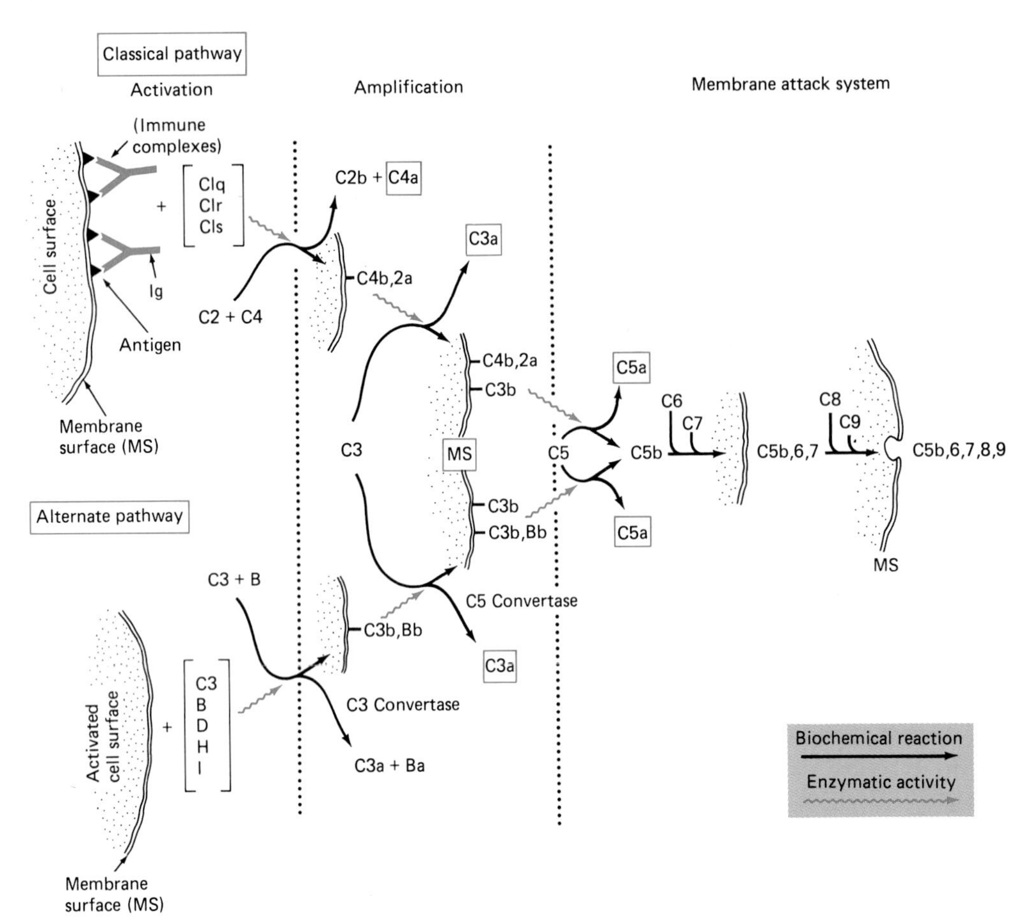

Figure 20–6 Summary of complement (C) activation in the classical (top) and alternate (bottom) pathways. The classical pathway is started when the components of complement (C1q, C1r, and C1s) are activated by immune complexes. The process continues with the subcomponents of $\overline{C1}$, enzymatically separating (cleaving) C4 and thereby releasing C4a. The major fragment C4b binds C2, which is in turn cleaved by the activated $\overline{C1}$ components, releasing fragment C2a. The major fragment C2b remains attached to C4b to form the enzyme C3 convertase, which converts C3 to C3a and C3b. In the alternate pathway, factors B and C3b combine to form the complex C3b, B, which is in turn converted into an active C3 convertase (C3b, Bb) by the loss of a small fragment Ba. This reaction is brought about by the enzyme factor D. Most of the remaining steps beyond this point in both pathways combine into a single cascade and include the following reactions: The C3b splits the C5 component into C5a and C5b. With this reaction, the C5b is fixed to biological membrane surfaces (MS) and is followed by the formation of the membrane attack or lytic complex (C6, C7, C8, and C9). The bioactive factors—C3a, C4a, and C5a—are anaphylatoxins.

ALTERNATE PATHWAY

The alternate pathway is recognized as an important mechanism involved with the destruction of certain bacteria, the neutralization of some viruses, and the disruption of red blood cells and other body cells from individuals with various physiological abnormalities.

This complement pathway can also be activated by either immunologic or nonimmunologic factors. Immunologically, IgA and IgG may activate the system, while nonimmunologic substances including complex polysaccharides, lipopolysaccharides, and protein-digesting enzymes may start the sequence of reactions. The *alternate pathway,* also originally referred to as the *properdin system,* contains several participating components including properdin; factor A, a high-molecular-weight protein similar in some respects to C4; factor B, another protein substance similar to but also different in some respects from C2; and other proteins designated as factors D, I, and H (Figure 20–6).

Activation of the alternate pathway requires the presence of a specific complement component, C3b, one of the fragments obtainable from C3. The levels

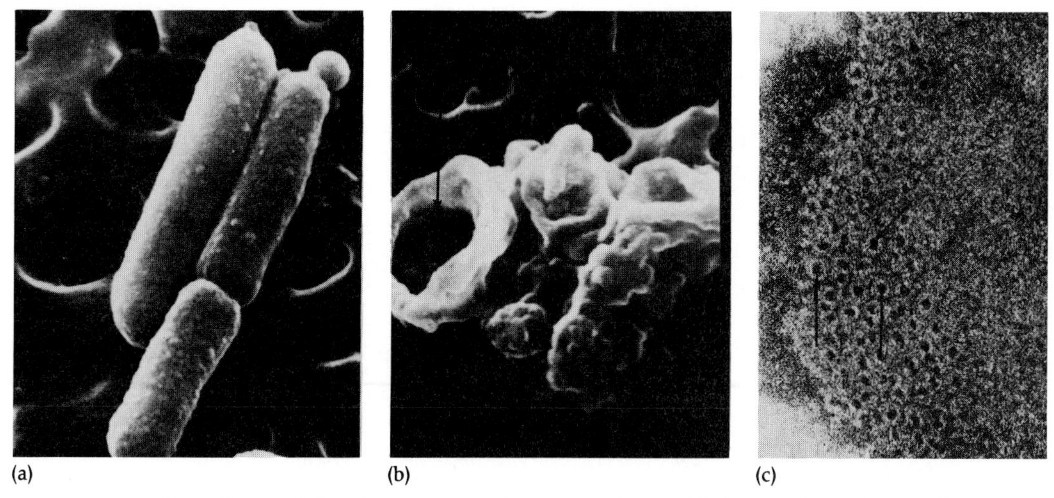

Figure 20–7 The effects of the alternate complement pathway. (a) A scanning micrograph showing *Escherichia coli* under normal conditions. (b) The bactericidal action of complement components of the alternate pathway. (c) A transmission micrograph showing the lytic effect on membranes caused by complement. *[R. D. Schreiber,* et al., J. Exp. Med. ***149****:870–882, 1979.]*

of newly formed C3b remain low, since the component is rapidly inactivated by the system's control factors, I and H. This condition continues until new sources of activators are introduced. These activators contribute substantially to an increased C3b concentration, which in turn eventually results in the formation of modified enzymes capable of activating the complement component C5, thereby triggering the membrane attack mechanism (Figure 20–6).

MEMBRANE ATTACK MECHANISM

The remaining portion of the complement cascade, the membrane attack mechanism, is common to both pathways and involves components C5 to C9. Following activation, the membrane attack system, components C5b to C9, must become attached to the surface of individual cells or viruses bearing the activating enzyme formed by the classic or alternate pathways. Attachment of the last component, C9, completes the cytolytic (cellular destructive) factor of the complement system and accelerates membrane changes and destruction (Figure 20–7).

CONTROL MECHANISMS

Uncontrolled activation of the complement cascade is prevented by the fragile nature of activated combining sites of certain complement components. Complement system factors I and H, mentioned earlier, and other serum enzymes also serve to regulate and limit complement activation. Uncontrolled complement activation and congenital (hereditary) deficiency of complement components may result in several diseases, including glomerulonephritis (a form of kidney disease), systemic lupus erythematosus, and repeated microbial infections.

Type III: Immune-complex-mediated Hypersensitivity

Immune complexes are formed by the interaction of one or more antibody molecules with one or more antigen molecules. Almost any foreign antigen that stimulates a detectable antibody response can lead to the development of an immune complex disorder. These antigens include a variety of bacterial and viral agents and therapeutic drugs. IgG and IgM are the antibody classes normally involved in immune complex formation. If complement is fixed by an antibody-antigen combination, biological molecules called *anaphylatoxins* are produced. These molecules ultimately cause the release of histamine and the production of other active substances, which in turn set in motion a chain reaction of events that damages tissues and intensifies the inflammatory response (Figure 20–8).

The effects of immune complexes in the body depend on the absolute concentrations of antigen and antibody, which determine the reaction's intensity, and on the relative proportions of these reactants, which determine the distribution of the complex within the body. In cases of antibody excess, the immune complexes are rapidly precipitated and tend to settle around the site of antigen introduction. In cases of antigen excess, soluble complexes are produced that can result in systemic reactions. Such complexes can be widely distributed throughout the body and deposited in such sites as the kidneys, joints, and skin. Many cases of severe kidney destruction (glomerulonephritis) are caused by immune complexes. Examples of type III responses include the Arthus reaction, serum sickness, and certain autoimmune reactions.

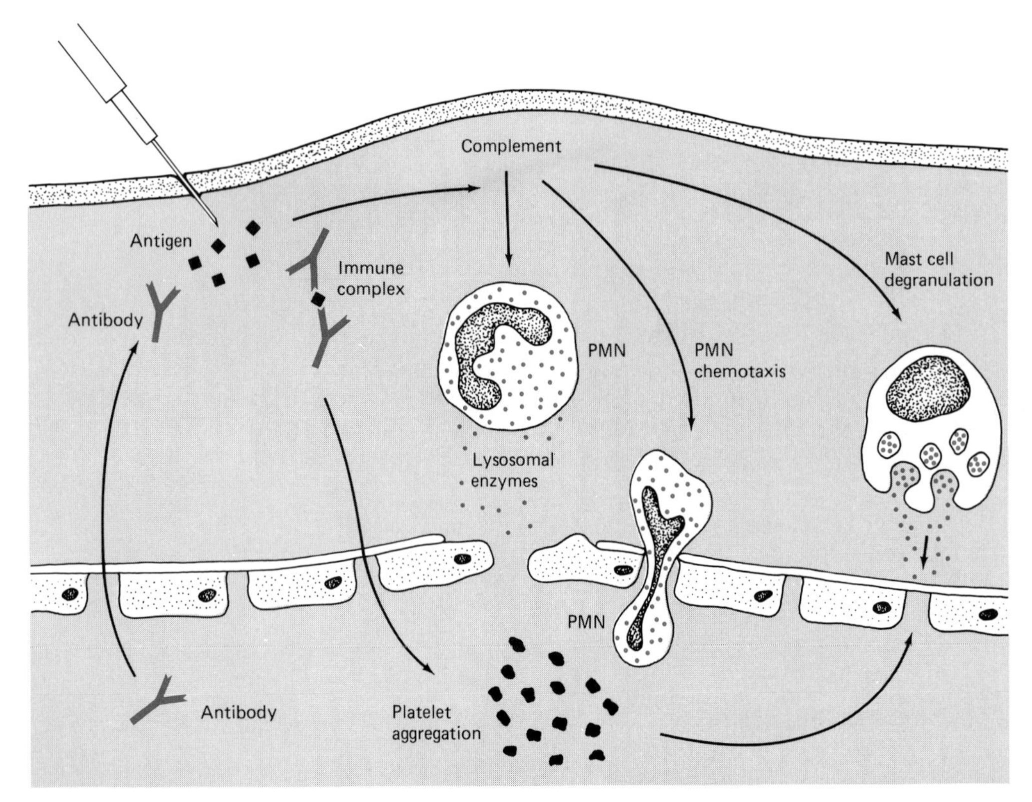

Figure 20–8 The mechanism involved with type III immune-complex-mediated hypersensitivity. PMN, polymorphonucleur leukocyte.

THE ARTHUS REACTION (ANTIBODY EXCESS)

In 1903, soon after the discovery of anaphylaxis, M. Arthus, a French physiologist, described the following antibody-dependent allergic reaction. When rabbits were given weekly injections of horse serum, no reaction was apparent. However, after several weeks, inoculation with the same type of material produced a localized inflammatory response. The same response was later recognized in humans and other animals.

Arthus reactions are not restricted to the skin. The phenomenon can involve most tissues in individuals having sufficiently high antibody levels. Regions exhibiting an Arthus response show the following tissue changes: (1) a marked reduction in blood flow, (2) the formation within small blood vessels of clots rich in platelets and leukocytes, (3) an escape of blood cells into neighboring connective tissue, (4) swelling, and (5) massive infiltration of the area by polymorphonuclear leukocytes. In short, the changes that take place are those of classic inflammation. Tissue destruction occurs in the later stages of the reaction. Certain of the components of complement play a role in the Arthus reaction.

SERUM SICKNESS (ANTIGEN EXCESS)

The original description of this anaphylactic condition was made by von Pirquet and Schick in 1905. Approximately 8 to 12 days after receiving an injection of large volumes of foreign protein such as antitoxin, individuals developed the distressing condition known as *serum sickness.*

Symptoms of serum sickness include fever, generalized swelling of lymph nodes, itching, hives, and swelling of the ankles, eyelids, and face. The severity of these symptoms is determined by the severity of the attack. The symptoms can subside in two days, but in some cases they persist for as long as two weeks.

The various effects of serum sickness are associated with complexes involving antibodies and the injected foreign protein that causes their formation. Together these substances form toxic complexes that not only injure blood vessels but also cause the various symptoms noted earlier. As the level of antibody increases, the level of foreign protein decreases. This sequence of events ultimately leads to the disappearance of the effects of serum sickness.

THE SHWARTZMAN REACTION

In 1927, G. Shwartzman observed that local bleeding and severe tissue destruction developed from intradermal injections of bacterial culture filtrates. The reaction could be made to occur quickly if the intradermal injection was followed in 30 minutes by a second so-called provocative intravenous injection. Some time later, Shwartzman proved that it was possible to induce reactions involving internal organs when both the initial injections and the provocative injections were given intravenously.

The Shwartzman reaction is caused by the endotoxins of gram-negative bacteria in the initial injection. The provocative injection can be either endotoxin or a variety of immunologically unrelated substances, including agar and starch. This reaction was shown not to be induced by an immune mechanism.

HYPERSENSITIVITY PNEUMONITIS

Hypersensitivity pneumonitis is another type III reaction. It is an inflammation of the respiratory passages and tissues, involving air sacs and bronchioles. Most commonly, the disorder is caused by repeated exposure and sensitization to inhaled organic dusts containing plant pollen, microorganisms, or animal proteins. Low-molecular-weight chemicals and drugs may also produce the disease (Table 20–4). With recurrent episodes, hypersensitivity pneumonitis may progress to irreversible lung damage.

To a large extent, this allergic state is one of several occupational lung diseases. Its association with the inhalation of dust dates as far back as the early eighteenth century. The first modern description of a hypersensitivity pneumonitis, farmer's lung, was given in 1932 by J. M. Campbell, who reported the condition in British farmers exposed to moldy hay. Since that time, many other similar diseases have been reported. Their exotic names reflect the epidemiologic conditions under which they develop (Table 20–4). These diseases can also result from environmental exposures that are neither strictly job- nor hobby-related.

Sensitized victims of these diseases present a similar clinical picture. Their symptoms include fever, chills, dry cough, general lack of energy, and difficulty and pain in breathing (dyspnea) four to eight hours after exposure to antigens. Commonly, acute hypersensitivity pneumonitis is mistaken for bacterial or viral pneumonia.

Once the condition has been correctly diagnosed, treatment is primarily directed toward relieving symptoms and avoiding the offending environment. The use of antifungal antibiotics is not recommended.

ANTIRECEPTOR ANTIBODIES: A POSSIBLE NEW TYPE OF HYPERSENSITIVITY

Certain abnormalities or disease states of unknown causes are associated with immunoglobulins that act against the cellular receptors for various hormones or reaction-controlling substances or mediators (Table 20–5). The best-known examples of these conditions are insulin-dependent diabetes mellitus (in which immunoglobulins interfere with insulin receptors, thus preventing the use of insulin by cells in their metabolism) and myasthenia gravis (in which antibodies are directed against the receptors for acetylcholine, the mediator of nerve impulses, resulting in great muscular weakness).

While the exact mechanisms of immunoglobulin-receptor interactions are not firmly established, several possible consequences have been demonstrated. The effects largely result in a disruption of cell function. The immunologic injury by the immunoglobulins against the receptors serves to activate target cells, resulting in disease.

Type IV: Cell-mediated (Delayed) Hypersensitivity

During the 80 years or so after the initial discovery of anaphylaxis, a variety of other allergic reactions were observed that appeared to be fundamentally

TABLE 20–4 **Selected Causes of Hypersensitivity Pneumonitis**

Disease	*Source of Antigen*	*Antigen*[a]
Air conditioner lung and humidifier lung	Contaminated air conditioner ducts and humidifiers	*Thermoactinomyces candidus* (B)
Bat lung	Bat fecal droppings	Bat serum protein (A)
Cheese washer's lung	Moldy cheese	*Aspergillus clavatus* (F)
Cheese worker's lung	Cheese mold	*Penicillium casei* (F)
Drug-induced hypersensitivity pneumonitis	Drugs, industrial materials	Nitrofurantoin, cromolyn sodium
Farmer's lung	Moldy hay	*Micropolyspora faeni* (B)[b]
Fog fever	Cattle	*Thermoactinomyces viridis* (B)[b]
Malt worker's lung	Moldy barley	*A. clavatus* (F)
Miller's lung	Contaminated grain	Wheat weevil (I)
Mushroom worker's lung	Moldy compost	*Thermoactinomyces sacchari* (F)[b]
Pigeon breeder's lung	Pigeon droppings	Pigeon serum proteins (A)
Wood pulp worker's lung	Moldy wood pulp	*Alternaria* (F)

[a] (B), bacteria; (F), fungus; (A), animal product; (I), insect products.
[b] These microorganisms can survive at higher temperatures; they are thermophilic.

TABLE 20–5 **Abnormalities or Diseases Possibly Caused by Antireceptor Immunoglobulins**

Target	*Hormone Involved*	*Abnormality or Disease*
Adrenal cortical cells	Adrenocorticotropic hormone (ACTH)	Adrenocortical insufficiency
Gastric parietal cells	Gastrin	Hypochlorhydria (decreased secretion of hydrochloric acid)
Graafian follicle; corpus luteum	Follicle-stimulating hormone (FSH); luteinizing hormone (LH)	Infertility; premature menopause
Melanocyte (skin pigment cell)	Melanocyte-stimulating hormone	Vitiligo (formation of white patches on the skin)
Motor endplate	Acetylcholine	Myasthenia gravis (great muscular weakness caused by failure of nerve impulses to induce muscular contractions)
Islets of Langerhans (pancreas)	Insulin	Diabetes (a disorder of carbohydrate metabolism)
Parathyroid chief cell	Parathormone	Primary hypoparathyroidism (interference with calcium and phosphorus metabolism)
Thyroid epithelial cells	Thyroxin	Thyrotoxicosis (toxic condition caused by excessive thyroid gland activity or hyperthyroidism), hypothyroidism

different from those associated with types I to III. Type IV responses do not involve globulins or complement or the release of histamine or chemically related substances. The sensitivity associated with this state cannot be transferred by the injection of serum from a sensitive individual to a nonsensitive one. Passive transfer of type IV responses to nonsensitive recipients can be accomplished by means of living lymphoid cells from sensitized donors or a nonantibody-active "transfer factor." Type IV responses are not inhibited by antihistamines but are inhibited by antiinflammatory agents such as the steroid compounds cortisone and hydrocortisone.

Type IV responses depend upon the interaction of a unique T lymphocyte subset (Tdh cells) with antigen. Several cell-mediated immune reactions are beneficial. They act, for example, to overcome certain types of infections, such as those caused by fungi, some protozoa, viruses, and various species of bacterial genera. They also prevent foreign tissue cells from becoming established in the body (tissue graft rejections) and stop the uncontrolled reproduction of cancer cells. When cell-mediated responses cause tissue damage, the condition is generally considered to be an allergic one. Such immunologic injury, also known as **delayed hypersensitivity,** involves Tdh cells bearing specific receptors on their surfaces that are stimulated by contact with antigens to release factors called **lymphokines** (Table 20–6). These factors cause inflammation and tissue injury. Whenever the Tdh receptors react with specific antigenic determinants, the T cells become sensitized. This exposure to antigens and sensitization is necessary for delayed hypersensitivity reactions to occur (Figure 20–9). **[Refer to Chapter 17 for descriptions of T cell subsets.]**

When a sensitized T lymphocyte interacts with

TABLE 20–6 **Representative Lymphokines and Associated Functions**

Lymphokine	*Functions or Activities*
Individual chemotactic factors for eosinophils, lymphocytes, macrophages, and neutrophils	Causes chemotaxis (chemical attraction) for specific types of white blood cells
Cloning inhibition factor (CLIF)	Blocks *in vitro* multiplication of certain cell types
Dialyzable transfer factors (TF)	Converts normal lymphocytes *in vivo* and *in vitro* to antigen-sensitized lymphocytes
Inhibitor RNA synthesis (IDS)	Reversibly inhibits mitosis of lymphocytes
Lymphotoxin (LT)	Destroys cells, including a variety of tumor cells, in a manner identical to that of sensitized lymphocytes
Macrophage activation factor (MAF)	Increases the motility and phagocytic activity of macrophages
Macrophage aggregation factor	Restricts macrophage movement and induces formation of giant cells
Macrophage inhibition factor (MIF)	Inhibits migration of macrophages
Mitogenic factor	Stimulates lymphocyte transformation
Proliferation inhibition factor (PIF)	Blocks *in vitro* multiplication of certain cell types
Skin reactive factor	Produces skin inflammation

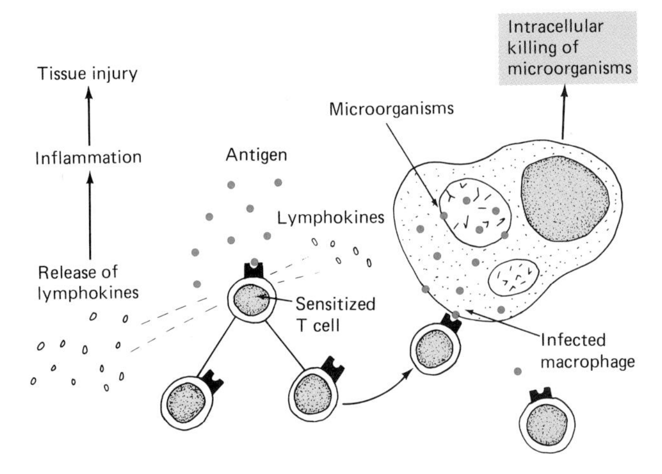

Figure 20–9 Type IV hypersensitivity reaction involving the release of lymphokines and intracellular killing of microorganisms by activated macrophages.

the specific antigenic determinant, a cell-mediated response is initiated. The T cell undergoes certain changes and divides, producing specifically sensitized cells and a corresponding increase in the number of reactive T cells. This activity takes place in blastogenetic (cell production) centers such as lymph nodes. From these areas, sensitized T cells can enter the circulation, seed other lymphatic tissues, or accumulate at sites where the antigen is introduced.

Lymphocytes and lymphokines are not the only factors involved in delayed hypersensitivity. Under the influence of lymphokines, activated macrophages show increased phagocytic, biosynthetic, and microbe-destroying properties. These macrophages are far more effective than normal macrophages in eliminating microorganisms (Figure 20–9).

Representative Type IV Allergic States

ALLERGY OF INFECTION

An allergy of infection brings about an accelerated and exaggerated tissue reaction to certain infectious agents. The most thoroughly studied example of this type of hypersensitive reaction is associated with tuberculosis. Hypersensitivity of this type was first demonstrated in 1891 by Robert Koch. He showed that guinea pigs infected with *Mycobacterium tuberculosis* two or more weeks earlier and uninfected (normal) guinea pigs reacted differently to subcutaneous injections of living, virulent tubercle bacilli. Within two days, the tuberculous animals developed massive inflammatory reactions at the inoculation site that gradually increased in intensity and produced areas of cellular death. The tissues of the infected animals reacted violently to the tubercle bacilli and tended to localize the infection by walling off the area of involvement. This sequence of events is known as the *Koch phenomenon.* Uninfected guinea pigs injected with similar material normally developed progressive tuberculosis.

Koch showed that the specific inflammatory reaction and associated tissue changes could be caused by dead as well as living tubercle bacilli. Furthermore, a bacteria-free protein-fraction extract prepared from these organisms, known as *tuberculin,* produced the same response.

In actual cases of tuberculosis infection, the allergic state becomes established early, generally before the obvious signs of the infections are apparent. In 1907, Koch proposed the use of tuberculin to detect tuberculous individuals. The tuberculin skin test and variations of it are widely used today for this purpose and for the standardization of skin-testing material. However, the significance of a positive delayed-type skin test (Color Photograph 82) to an infectious disease varies with the infection. *Reactions of this sort do not necessarily indicate current infection.*

There are a number of diseases in which delayed hypersensitivity develops. These are caused by pathogens that present a persistent, long-lasting antigenic stimulus. Examples include the bacterial infections, leprosy, tuberculosis, listeriosis, the protozoan infection leishmaniasis, the fungal infection coccidioidomycosis, and the worm infection schistosomiasis. Descriptions of these diseases are presented in Part VIII.

CONTACT DERMATITIS

The form of hypersensitivity known as **contact dermatitis** includes certain types of drug allergy. It is one of the most commonly encountered human allergic diseases. Sensitization and production of symptoms result from contact with various causative compounds, which combine with proteins in the skin. Included in the group of *incitants* are (1) simple chemicals, such as formaldehyde and picric acid; (2) metals, such as mercury and nickel; (3) various drugs and dyes; (4) certain cosmetics; (5) insecticides; and (6) the active components associated with poison ivy, poison oak, poison sumac, and other plants.

Two types of responses are generally recognized: *contact skin sensitivity* and *allergic contact dermatitis.* In contact skin sensitivity, exposure to the incitant causes local irritation. While these substances by themselves are not allergenic, they can become so by combining with proteins in the skin. In allergic contact dermatitis, both humans and various laboratory animals can be sensitized by skin contact with the incitant or by intradermal injection of the incitant. Approximately 5 to 20 days are needed for the development of the hypersensitivity reaction. To determine whether an individual has been sensitized, filter paper patches treated with the same sensitizing material are taped to the skin for about 24 hours. The test is generally read twice, once a few hours after the removal of the patch and again 24 hours later. A positive reaction is characterized by an erythematous region containing various-sized

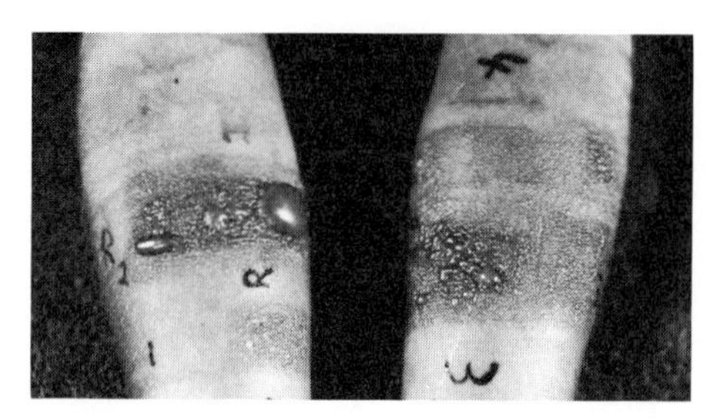

Figure 20–10 A positive patch test for contact dermatitis. *[Courtesy of Armed Forces Institute of Pathology, Washington, D.C., Neg. No. AFIP 57-15160-2.]*

blisters (Figure 20–10). In the diagnosis of certain cases of contact dermatitis, examinations may have to be continued for two or more weeks.

The treatment of contact dermatitis requires identification of the causative agent. Attempts to desensitize individuals have met with limited success. When the chemical involved has an oily nature, the chances of success are better. Total avoidance of the incitant is of course advisable, but it is not always practical.

Penicillin Allergy. Penicillin is considered to be among the least toxic drugs in current clinical use. Apparently, however, the antibiotic combines with certain protein derivatives in some manner to form stable inducers. These newly formed complexes can stimulate antibody production and bring about either type of hypersensitive state, immediate or delayed (Figure 20–11). Approximately 10% of the people who receive repeated doses of penicillin fall victim to this condition. The injection of 1 mg of the antibiotic can result in a fatal anaphylactic shock reaction in a sensitized individual.

Persons who are repeatedly exposed to penicillin in some form can develop type IV hypersensitivity to the antibiotic. Health care personnel and individuals involved in the preparation and packaging of penicillin fall into this category. The sensitivity develops as a drug contact dermatitis.

TISSUE TRANSPLANTATION REACTIONS

Experiments in the transplantation of animal kidneys began a chain of events that has led to an entirely new medical era. It is now possible surgically to replace or transfer a variety of tissues and organs, moving them from one site to another on the same individual or from one body to another. Transplantation of skin from one region to another in an individual is now a fairly common and usually successful operation. This procedure is used for burn victims and others who have experienced extensive skin destruction.

Since the first successful grafts of human kidneys between identical twins in the early 1960s, the surgical specialty known as **organ transplantation** has progressed from a limited and experimental

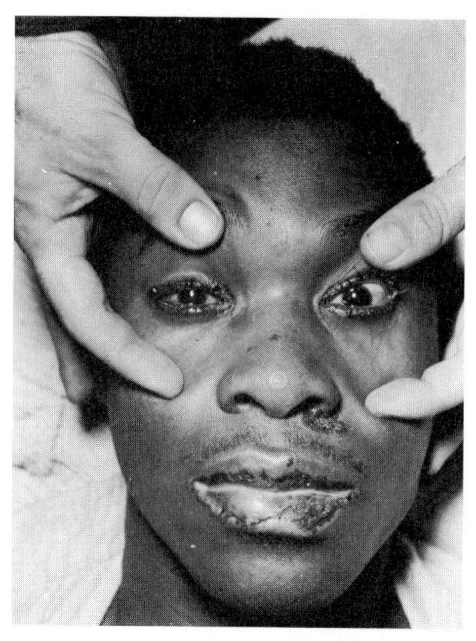

Figure 20–11 Penicillin sensitivity. This man shows a hemorrhagic reaction involving the eyelids and nasal and oral mucosae. *[Courtesy of Armed Forces Institute of Pathology, Washington, D.C., Neg. No. AFIP 54-1548-3.]*

practice to a mature and scientific technology. The transplantation of numerous organs can now be achieved with reasonable anticipation of success. Such organs include not only the kidney but also the heart, liver, pancreas, and lung, and the heart and lung in combination. Several thousand transplants are performed annually in the United States alone. In terms of frequency, heart transplantation is second to transplantation of the kidney. Liver transplantation comes next and is the most difficult of these surgical procedures to perform.

To some extent, the progress made in organ transplantation has resulted from the development of appropriate surgical techniques. However, without question, most gains have resulted from the ever-increasing ability to control and limit the major obstacle, namely, **immunologic rejection.**

The transplantation of tissues involves a complex collection of cells, each of which has a large variety of antigens controlled by DNA. Except in identical twins, each individual's chromosomal DNA is unique. This uniqueness of self (see Chapter 17) is accompanied by a highly sensitive mechanism for sensing and recognizing foreign substances. The recognition mechanism triggers the complex response that forms an individual's immunity. Tissue transplantation activates both humoral mechanisms (circulating antibodies) and cell-mediated mechanisms of immunity.

TRANSPLANT CATEGORIES

Transplants of tissues are classified in terms of genetic relationships (identity). For example, where

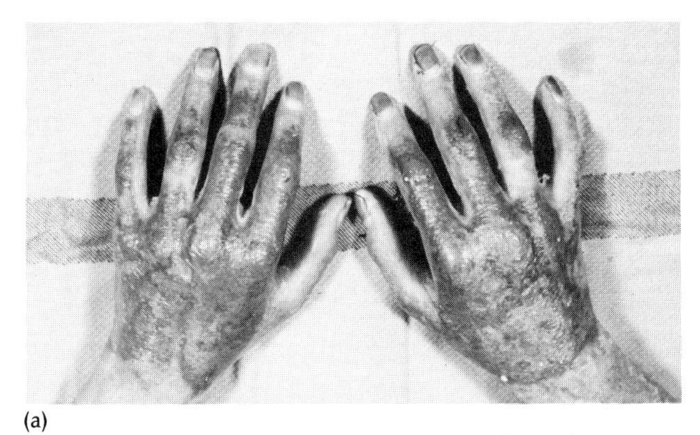

(a)

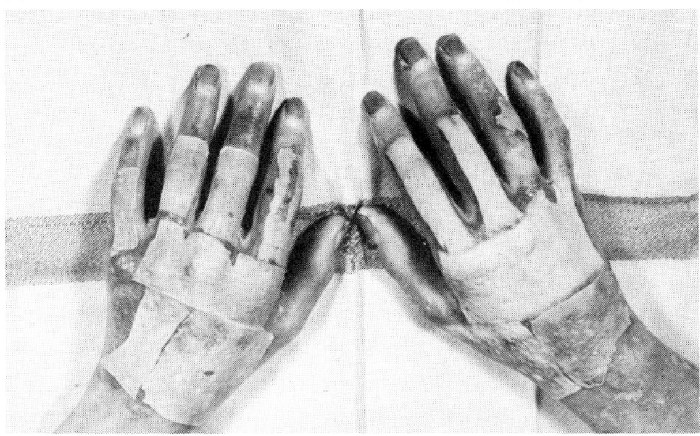

(b)

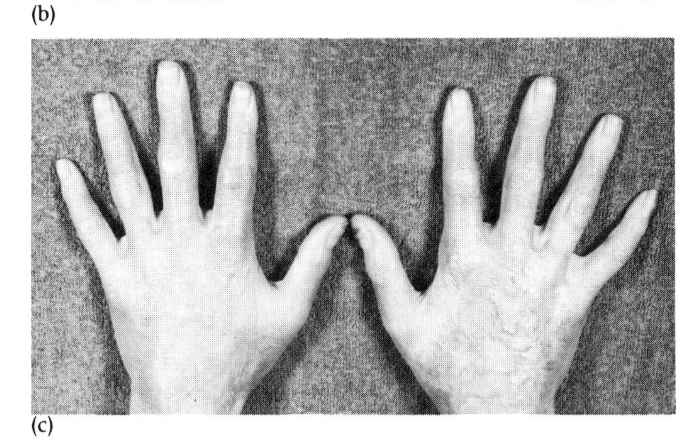

(c)

Figure 20–12 The successful transplantation of skin in a case of severe burn. (a) Deep dermal burns affecting both hands. (b) The application of autograft. (c) The result after three transplants. *[J. M. Shuck, B. A. Priutt, and J. A. Moncrief,* Arch. Surg. ***98:****472, 1969, © AMA.]*

the tissue of one individual is involved, the transplant is called an *autograft* (Figure 20–12). If a graft (the tissue to be transplanted) is taken from one animal and given to another of the same species (genetically dissimilar members of the same species), the tissue is referred to as an *allograft* (Figure 20–13). When both individuals are genetically identical—identical twins or mice from the same highly inbred line—the tissue is an *isograft.* Finally, a transplant involving two different species—for example, the transplantation of kidneys from a chimpanzee into a human—is referred to as a *xenograft.*

GENETIC CONTROL

The goal of transplantation is the long-term survival of the grafted tissue. The success of the procedure depends on several factors, including the degree of antigenic similarity between donor and recipient and the nature of the transplanted tissue itself.

One approach to preventing immune rejection of grafts is careful matching of donor tissues and cells with those of the recipient. Genetic differences between such individuals can also cause immunologic reactions in bone marrow transplant recipients and in graft-versus-host disease. Finding the best match involves the histocompatibility antigens of the *human leukocyte antigen (HLA) system.* This genetic system, also known in the human as the *major histocom-*

Figure 20–13 Allograft transplantation. (a) Preparation of the recipient animal. (b) The site of the graft. (c) The skin graft from an animal of the same species. (d) The allograft in place ten weeks after transplantation. *[A. A. van Es,* Lab. Anim. Sci. ***22:****404–406, 1972.]*

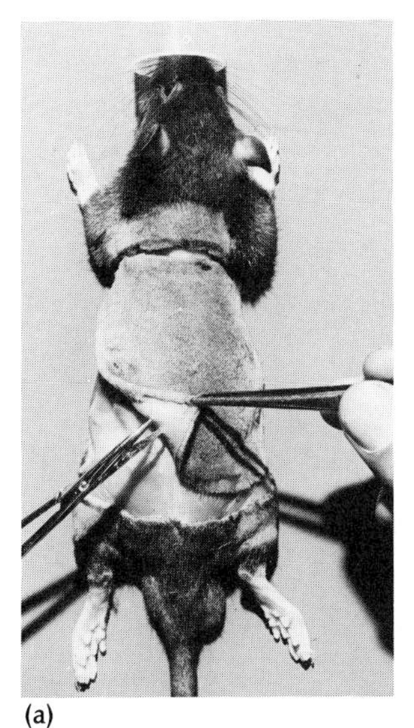

(a)

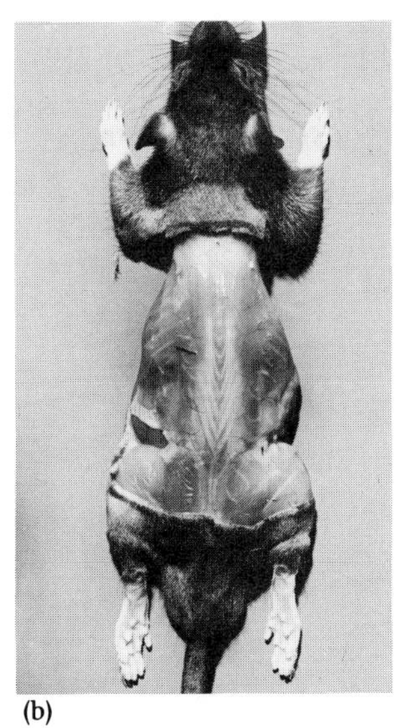

(b)

(c)

(d)

patibility complex (MHC), was discovered by J. Dausset in 1958. The MHC controls both the humoral immune and cellular immune responses of individuals.

THE HLA SYSTEM

The designations of the HLA system are devised by the HLA Nomenclature Committee of the World Health Organization. According to the committee's rules, the same term is used for the HLA gene, its alternate form (allele), and its product, the HLA antigen.

A great deal of attention has been focused on the histocompatibility systems of the mouse and the human. From a genetic point of view, the set of antigens in mice, known as the *H-2* system, provides by far the strongest barrier to transplantation. The system is under the control of a specific gene complex on the mouse's chromosome 17 (Figure 20–14b). Other parts of the H-2 gene complex play an important role in the functioning of the immune system and include the *I region* and the *S region.*

A series of genes within the I region known as *immune-response (Ir) genes* controls the manufacture of antibodies. Although the products of the *Ir* genes have not been identified, certain proteins designated *Ia (I-region-associated) antigens* have been found on the surface of lymphocytes involved in immunological reactions.

The S region has been shown to control the manufacture of a component of serum complement. This complement acts to destroy antigenic cells such as bacteria and grafted cells once they have reacted with specific antibodies.

The strong H-2 antigens in the mouse have their counterpart in the HLA system (Figure 20–14a). This major human system (MHC) is under the control of a cluster of genes on human chromosome 6, which controls the antigenic determinants forming the strongest antigenic barrier to transplantation between genetically nonidentical individuals. Other genes in the same chromosomal region play an important role in an individual's immune responsiveness to a wide variety of antigens.

On the basis of biochemical, serological, and related information, the MHC genes can be divided into three classes. Class I consists of genes that control the expression of the classic transplantation antigens of all cells of an individual. In humans these genes are HLA-A, -B, and -C. Class II contains the genes that control the expression of surface antigens (Ia) involved with the immune response on lymphocytes and macrophages. In humans these

Figure 20–14 The human leukocyte antigen (HLA) system. (a) The location of HLA genes on human chromosome 6: HLA-A, B, C, and D. (b) A comparison of the histocompatibility complexes of the mouse (H-2) and the human (HLA). Locations of genes for enzymes also are indicated: GLO = glyoxalase, 21 hydroxylase, PGM3 = phosphoglucomutase-3.

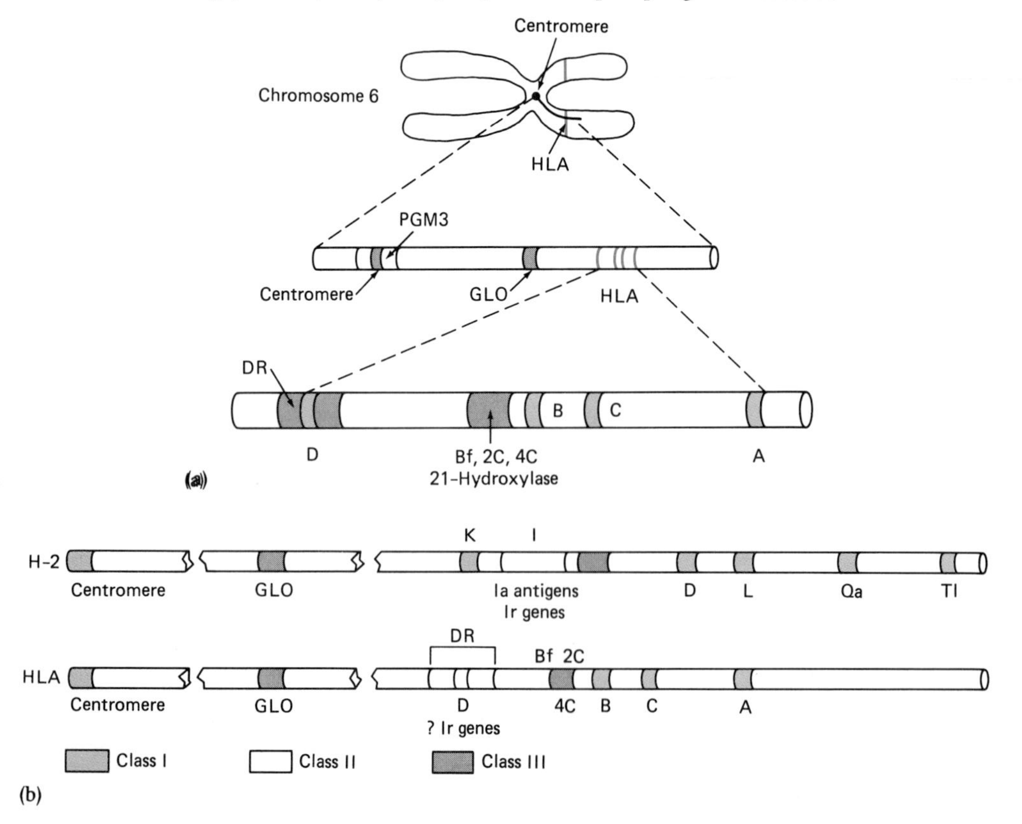

genes are found in the HLA-D region. Class III includes the genes that control a series of serum proteins that belong to the complement system. These are complement components C4, C2, and factor B of the properdin system. The specific human genes of this class have not been identified.

HLA-A, -B, and -C antigens are detected by a standard microcytotoxicity typing assay. Lymphocytes to be typed are incubated with human sera containing HLA antibodies of known specificity, exposed to complement, and then exposed to a vital dye. Cells bearing antigens recognized by the HLA test antibodies will exhibit membrane damage, which is detected by uptake of the dye (Figure 20–15).

A fourth HLA locus, HLA-D, marks a region that contains more than one histocompatibility gene. HLA-D is detected by the mixed lymphocyte culture assay.

Closely related to the HLA-D region is a series of serologically defined antigens that are distinct from HLA-A, -B, and -C antigens. These antigens, present on the surface of B lymphocytes and commonly known as *B-cell antigens,* are formally designated as *D-related,* or *DR, antigens.* HLA-DR antigens are found on B cells as well as on monocytes, macrophages, and some bone marrow cells. These antigens are not detectable on most T cells. DR types are determined by the microcytotoxicity typing assay, using the patient's B lymphocytes (separated from peripheral blood) and test sera of known DR specificity.

The HLA system is extremely variable, having several different alternate forms of genes (alleles) at each known locus. For example, there are at least 20 distinct alleles at the HLA-A locus, at least 42 distinct alleles at the HLA-B locus, 12 for the D locus, and 10 for DR.

HLA genes exhibit a close linkage. Because of this situation, the combination of alleles of each locus on a single chromosome is generally inherited as a unit. This unit is called the *haplotype.* One HLA haplotype is inherited from each parent.

Applications of HLA Typing. To date, the most fruitful use of HLA typing has been in selecting donors for tissue and organ transplantation. Another clear application of the typing procedure is in establishing paternity. The use of HLA typing in paternity cases in the last several years has increased significantly as a result of legislation that affects the responsibility of the biological father. HLA typing is also of value in anthropological studies, in genetic counseling, and in establishing associations of the HLA system with certain diseases.

HLA–Disease Associations. Two major groups of diseases have shown relationships with the HLA system. One group consists of the rheumatic diseases, which include arthritis of the spine and sacroiliac and axial joints. The second group of diseases is characterized by long-lasting inflammation and abnormal immunologic responses. Other HLA associations have been found with pernicious anemia, myasthenia gravis, psoriasis, and juvenile diabetes.

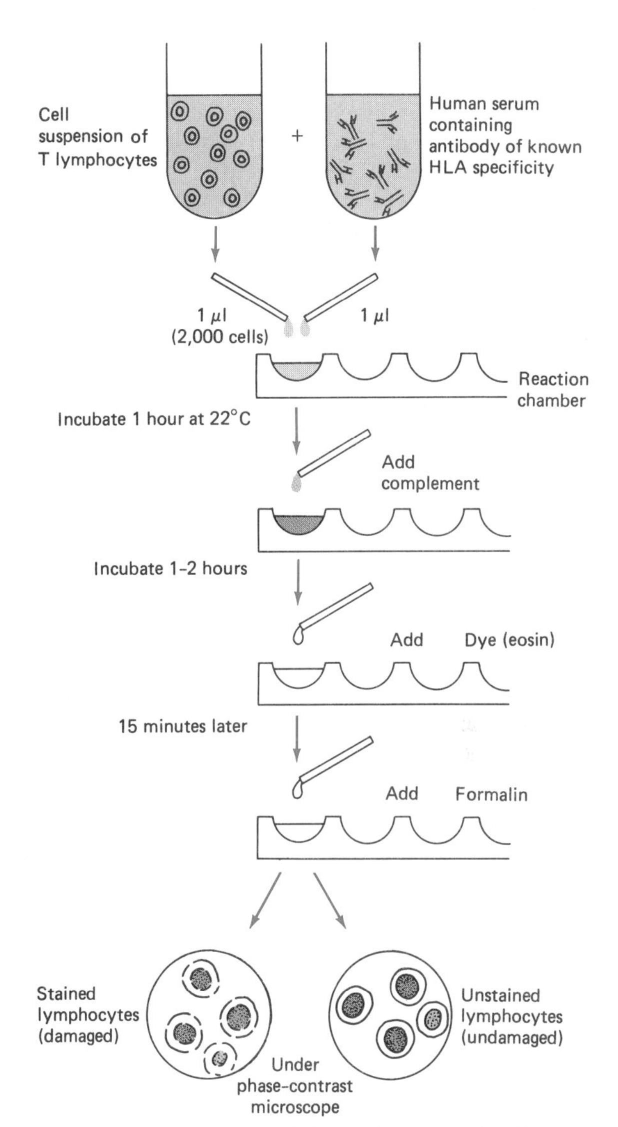

Figure 20–15 Microcytoxicity typing assay for the identification of HLA-A, B, and C antigens. Human lymphocytes are added to human serum containing HLA antibodies of known specificity and incubated in the presence of complement. Cells bearing antigens recognized by the HLA antibodies are damaged in the presence of complement and will be stained by vital dyes such as eosin. Formalin is added to the test to stop the reaction. The cells are examined by phase contrast microscopy. Stained cells show the end point of a cytotoxic reaction and indicate that the cells tested have the same HLA specificity as the typing serum used.

HLA antigen-associated diseases have several characteristics. These include an unknown cause or unknown underlying mechanism with a hereditary relationship, the presence of immunologic deficiency or abnormality, and little or no effect on reproduction.

GRAFT REJECTION

When tissue is transplanted from one individual to another, two distinct rejection processes can take place. The first, and traditionally the most studied, is the host rejection of grafted tissue, the *host-versus-graft reaction.* The second process is the reverse of this reaction, the *graft-versus-host response.* This reaction occurs when immunocompetent tissues are transferred to an immunologically handicapped host.

Determining the suitability of tissue for grafting (histocompatibility testing) is accomplished by the *lymphocyte toxicity test* or the *mixed-leukocyte (lymphocyte) culture assay.* The lymphocyte toxicity test is a widely applied method designed to measure the histocompatibility antigenic composition of the donor and recipient. The test is performed on lymphocytes of both the donor and recipient with a battery of antisera. The mixed-lymphocyte reaction test determines only the compatibility of donor and recipient and reflects incompatibilities of the major transplantation antigen.

Several methods are available for limiting or totally inhibiting graft rejection in certain situations. These include the injection of immunosuppressive drugs, such as corticosteroids or antilymphocytics, and total lymphoid irradiation. In several instances, a combination of these measures has been employed. It is important to note that, although this type of procedure can promote survival of allografts, it also lowers the individual's ability to produce immunoglobulins against pathogenic disease agents. Thus, antibiotics must be administered to prevent infection of the graft recipient.

Autoimmune (Autoallergy) Diseases

Normally, in the body of an immunocompetent individual, appropriate mechanisms exist to prevent the lymphoid system from responding to self components as antigens. Such unresponsiveness, referred to as *immunological tolerance,* is essential for human survival. Unfortunately, like all machinery, tolerance mechanisms occasionally falter and break down, thus creating autoimmune (autoallergic) disease states. In these disorders, self-antigens, or autoantigens, stimulate the production of circulating immunoglobulins (humoral response) or of specially sensitized lymphocytes (cell-mediated response), either of which will react with the autoantigen.

CAUSES OF AUTOIMMUNITY

Several factors may trigger an autoimmune response. These include the presence of (1) antigens that normally do not circulate in the blood; (2) the presence of an altered antigen (alterations can develop through exposure to chemicals, physical agents, or microorganisms); (3) the introduction of a foreign antigen similar to an autoantigen; (4) and the occurrence of a mutation in immunocompetent cells that results in a responsiveness to normal autoantigens. These agents may precipitate an autoimmunity following such events as tissue injury, administration of drugs, or certain microbial infections. Several human viruses are enveloped with membranes that are partly derived from nuclear, intracytoplasmic, or plasma membranes of the host but that also may include virus-specific antigens. Several of these viruses, together with bacteria, fungi, protozoa, and worms, cause temporary autoimmune symptoms such as arthritis as well as the formation of antinuclear antibodies (Color Photograph 83) and the rheumatoid factor. The rheumatoid factor is an antibody directed against denatured IgG present in the serum of individuals with rheumatoid diseases. Table 20–7 lists several examples of autoimmune symptoms or signs associated with infections.

Tissue lesions in autoimmune diseases can result from humoral, cellular, or combined mechanisms. In situations involving a humoral mechanism, antibody-antigen complexes are deposited in the tissue, especially if the immunoglobulin is IgG. The complement cascade is activated and inflammatory reactions produce a type II hypersensitivity reaction. When antigen is bound to cells, a type III hypersensitivity reaction, cell lysis, occurs. The direct action of cytotoxic T cells and antibody-dependent cell-

TABLE 20–7 Examples of Autoimmune Symptoms Associated with Infections

Symptom or Sign	*Infection or Disease Agent*[b]
Antinuclear antibodies[a]	Cytomegalovirus, Epstein-Barr virus (cause of infectious mononucleosis), leprosy, tuberculosis
Arthritis	Gonorrhea, plague, hepatitis B infection, tuberculosis
Immune-complex nephritis (kidney tissue destruction)	Chickenpox, elephantiasis (worm infection), hepatitis B infection, infectious mononucleosis, malaria, measles, schistosomiasis (blood fluke infection), toxoplasmosis (congenital), typhoid fever
Rheumatoid factor	Cytomegalovirus infection, hepatitis B infection, infectious mononucleosis, influenza, kala-azar (protozoan disease), leprosy, malaria, schistosomiasis, shingles, infective bacterial endocarditis, syphilis, tuberculosis

[a] Laboratory tests for these factors are used in diagnosis.
[b] These infections and disease agents are described in Part VIII.

TABLE 20–8 Autoimmune Diseases

Autoimmune Disease	*Autoantibody Directed Against*
ORGAN, CELL OR CELL PRODUCT, OR SPECIFIC DISEASE	
Addison's disease	Adrenal gland cells
Allergic rhinitis	Adrenal receptors on cells
Asthma	Adrenal receptors on cells
Autoimmune hemolytic anemia	Red blood cells
Bulbous pemphigoid	Basement membrane of skin and mucosa
Graves' disease	Thyroid receptors for thyroid-stimulating hormone
Hashimoto's thyroiditis	Thyroid hormone
Idiopathic hypoparathyroidism	Parathyroid cells
Idiopathic neutropenia	Neutrophils
Idiopathic thrombocytopenic purpura	Platelets
Insulin-resistant diabetes	Insulin receptors on cells
Juvenile insulin-dependent diabetes	Islets of Langerhans cells in pancreas
Myasthenia gravis	Acetylcholine (impulse transmission chemical at nerve connections)
Pemphigus	Substance between cells of skin and mucosa
Pernicious anemia	Gastric parietal cells and vitamin B_{12}
Premature ovarian failure	Corpus luteum cells and surrounding cells in the ovary
Primary biliary cirrhosis	Mitochondrial antigens
NON-ORGAN-SPECIFIC DISEASES	
Goodpasture's syndrome	Any basement membranes
Rheumatoid arthritis	Gamma globulin and Epstein-Barr virus-related antigens
Sjögren's syndrome	Gamma globulin and Epstein-Barr virus-related antigens
Systemic lupus erythematosus	Cell nuclei, double- and single-stranded DNA, soluble cell nuclear extracts, ribonucleic proteins, red blood cells, platelets, neuronal cells, gamma globulin

mediated cytotoxicity (ADCC) of killer cells and macrophages are examples of cell-mediated or type IV hypersensitivity reactions. (See Figure 17–26).

Autoimmune diseases comprise a broad range of disorders and a confusing array of overlapping symptoms, pathological lesions, and immunologic properties. Since the underlying mechanism of all these disorders is not clearly understood, autoimmune diseases may be classified according to the specific cell tissue or organ involved (Table 20–8). The specific autoimmune diseases are possibly related to the release of cellular components or modified cellular surfaces. Examples of such diseases include Hashimoto's thyroiditis and autoimmune hemolytic anemias. Nonspecific or systemic autoimmune diseases are also known and are probably caused by changes in the recognition mechanism of the central immune system (Table 20–8). In such disorders, neither lesions nor autoantibodies are limited to any one organ or tissue. Systemic lupus erythematosus is an example of this type of autoimmune disease. Autoimmunity is now accepted as being important in human disease, and it has created new approaches to a deeper understanding of the immune response itself.

The diagnosis and identification of autoimmune diseases involve the use of laboratory tests for autoantibodies. The use of such tests is a routine step in diagnosing suspected cases of systemic lupus erythematosus, thyroiditis, rheumatoid arthritis, some forms of progressive liver disease, and even pemphigus, a severe and formerly fatal skin disorder. Examples of tests include latex agglutination, complement fixation, radial immunodiffusion, fluorescence antibody, monoclonal antibody, and the specific antinuclear antibody procedure (Color Photograph 83). Antinuclear antibodies, which are directed against self-antigens such as cellular nucleoprotein, DNA, histones, RNA, and RNA nucleoli, represent a characteristic abnormality of patients with systemic lupus erythematosus.

Apart from the large number of currently recognized autoimmune disease states and the rather confusing signs and symptoms of these disorders, certain shared characteristics have become apparent. First, the incidence of autoimmune diseases and associated immunoglobulins tends to increase with age. Second, such diseases occur most commonly in persons with generalized immunological deficiencies. Third, autoimmunity occurs most often in fe-

Microbiology Highlights

COPING WITH IMMUNODEFICIENCIES

The development of immunotherapeutic agents, called biological response modifiers (BRMs), has generated much interest, especially in connection with the treatment of various immunodeficiency states, control of organ transplantation, and understanding of aging effects. Applications of BRMs fall into the field of immunopharmacology and include the immunosuppression and stimulation of various cells and activities to restore normal functioning of the immune system. The accompanying tables provide examples of agents in these two categories. **[See Chapter 30 for microbial BRMs.]**

TABLE 1 Immunosuppressive Agents

Agent	*Source(s)*	*Mechanism of Action*	*Application(s)*
Antilymphocyte globulin (ATGAM)	Immunoglobulins from various mammalian species	Covers T cell surfaces and disrupts B and T cells	Organ transplantation
Azathioprine	Synthetic chemical	Interferes with nucleic acid metabolism—selective for T cells	Kidney transplantation
Cyclosphosphamide	Synthetic chemical	Destroys reproducing lymphocytes	Autoimmunity; bone marrow transplants; cancer treatment
Cyclosporin A	Fungal product	Specific suppressor of T cells	Treatment of graft-vs.-host disease; organ transplantation
Glucocorticosteroids	Synthetic adrenal gland hormone	Interferes with macrophage and T cell functions	Allergic disorders; autoimmunity; organ transplantation

TABLE 2 Immunostimulating Agents[a]

		Experimental and Clinical Applications			
Agent	*Source(s)*	*Autoimmunity*	*Cancer*	*Immunodeficiency*	*Infection*
DRUGS					
Azimexon	Synthetic chemical	NT[b]	+	NT	NT
Bestatin	Fungal extract	NT	+	+	NT
Isoprinosine	Synthetic chemical	+	NT	+	2+
Levamisole	Synthetic chemical	2+	+	+	+
Interferon(s)	Lymphocyte product, and genetically engineered	−	2+	−	+
Lymphokine(s)	Lymphocyte products	+	+	+	+
Thymic hormones	Bovine thymus gland extracts gland extracts	+	+	+	+

[a] At the time of publication, most of the immunostimulatory agents listed were classified as experimental in the United States. However, many are licensed for use in Europe and Japan.
[b] NT = not tested; + = effective; 2+ = more effective; − = not effective.

males. In addition, it has been observed that any given autoimmune disease may produce destructive effects that resemble those associated with the other types of hypersensitivities, and high levels of immunoglobulins are generally present.

Therapy for autoimmune diseases varies widely and is influenced by several factors. These include the particular disease, the severity of the effects, and the associated hazards of prolonged therapy with anti-inflammatory drugs such as corticosteroids and other immunosuppressive agents and procedures.

Factors Involved in the Development of an Infectious Disease

Chapters 17 and 18 emphasize the important roles of various immune responses such as phagocytosis, inflammation, and immunoglobulin production in opposing microorganisms that threaten the welfare of a host. In this chapter, some examples of situations are described in which protective immune responses may also be harmful to the host, as in the cases of immune-complex disorders and type IV hypersensitivity. Thus immune responses to an infecting organism may eliminate the organism but, at the same time, may cause significant injury or death to the host.

As many of the chapters in Part VIII will emphasize, the effective treatment and eventual control of an infectious disease requires knowledge not only of host factors but also of facts about the disease agent. There are a great number and variety of infectious diseases. Microbial pathogens and helminths (worms) can cause infection in virtually any body tissue, organ, or system. Chapter 21 considers the development of an infectious disease (pathogenesis), with descriptions of the various factors involved with the distribution and transmission of infectious diseases in a population. Representative approaches to identifying causative agents of disease also are presented.

Summary

A properly functioning immune system protects the body through its capacity to recognize and destroy *foreign*, or *non-self*, molecules.

Immunodeficiencies

1. Immunodeficiency diseases result from the inability of the immune system to perform normally.
2. Four major immune systems normally provide the defense against agents of infection and disease: antibody-regulated (B cell) immunity, cell-mediated (T cell) immunity, phagocytosis, and complement. These systems may cooperate with one another or may function independently.
3. Two general categories of these disorders are recognized: *primary* and *secondary immunodeficiencies*.
4. Primary, or congenital, deficiencies result from an inherited lack of development of the immune system, and include the absence of T type lymphocytes, the inability to produce immunoglobulins, phagocytic disorders, and the absence of normal complement components. Several of these deficiencies result in an increased susceptibility to opportunistic infection.
5. Immunodeficiency disorders can be diagnosed with tests that screen for the presence of respective immune system components.
6. Secondary, or acquired, deficiencies occur much more frequently than primary conditions and result from an interference with lymphatic system activities. Factors causing such conditions include malnutrition, cancers, burns, radiation, and various drugs. These immunodeficiency disorders result in an increased susceptibilty to opportunistic infection.
7. One example of a severe form of increased susceptibility to opportunistic infection is acquired immunodeficiency syndrome (AIDS). Human immunodeficiency virus HIV causes the disease, primarily by attacking T4 lymphocytes.

Hypersensitivity (Allergy)

1. *Hypersensitivity*, or *allergy*, is an exaggerated response by individuals exposed to a substance that does not cause similar responses in previously unexposed persons. The inciting agents are referred to as *allergens*.
2. Initial exposure to an allergen sensitizes the individual to future contact with the same allergen. Various harmful reactions can result from such exposures.

Categories of Hypersensitivity

1. Historically, hypersensitivity reactions were divided into two categories based on the time required for responses to occur upon reexposure to the sensitizing allergen.
2. Categories today are based on several other factors. These include the presence or absence of substances regulating the reaction's *mediators*, circulating immunoglobulins, the actions of B and/or T lymphocytes, and the possibility of passive transfer of the hypersensitivity response.
3. Four categories of hypersensitivity are known: type I (classic immediate), type II (cytotoxic), type III (immune-complex-mediated), and type IV (cell-mediated, or delayed).
4. The particular type of hypersensitivity acquired is determined by several factors, including the chemical

makeup of the allergen, the route involved in sensitization, and the responsive and physiological condition of the individual.

Type I: Classic Immediate Hypersensitivity

1. A distinguishing feature of type I hypersensitivity is the rapid release of physiologically active chemicals, such as histamine, from cells affected by a reaction involving allergens and specific immunoglobulins (IgE).
2. Mediators involved in producing symptoms of anaphylaxis are of two types: *primary mediators* preformed and existing in the granules of mast cells and basophils, and *secondary mediators* produced by stimulated mast cells, basophils, and other cells recruited in the anaphylactic reaction.
3. Release of mediators from activated mast cells or basophils is started by the bridging of two IgE-specific receptors on these cells.
4. Examples of type I allergic states include certain forms of asthma, anaphylactic shock, hay fever, and hives.
5. Generalized anaphylaxis produces abnormalities in different body systems and produces symptoms such as nausea, vomiting, abdominal cramps, and diarrhea (gastrointestinal tract); suffocation and constriction and obstruction of passages (respiratory tract); and failure of peripheral circulation and shock (cardiovascular system).
6. Measurement of IgE levels may require highly sensitive laboratory procedures such as the radioallergosorbent test (RAST).
7. Anaphylactoid reactions resemble anaphylaxis but are not associated with IgE.
8. *Atopy* is an allergic response to a large variety of environmental allergic substances such as pollens of various types, foods, fungi, and house dust.
9. Several approaches to the treatment of allergy are possible, including the use of chemicals to reduce the effects of allergies and the use of specific allergen preparations to make allergic persons immunologically unreactive to such allergens (*desensitization* or *hyposensitization*).

Type II: Cytotoxic Hypersensitivity

1. Type II conditions result in cellular destruction and involve the immunoglobulins IgG and IgM, which can fix or react with complement.
2. Examples of type II allergies include blood transfusion reactions, the Rh baby, and certain forms of drug-caused tissue destruction and cell death.
3. Activation and fixation of the complement system are necessary for cellular lesions to develop.

The Complement Cascade (Pathway)

1. Two different mechanisms are recognized for the activation of complement, *classic* and *alternate pathways.*
2. Each pathway contains an initial set of different reactions, but at the midpoint of the system they share a common reaction sequence.
3. The common remaining portion of pathway, known as the *membrane attack mechanism,* requires the attachment of complement components C5b to 9 to cell and viral surfaces.
4. Control mechanisms exist to limit complement activation, including various serum proteins.
5. Uncontrolled complement activation or congenital complement deficiency may lead to various disease states.

Type III: Immune-complex-mediated Hypersensitivity

1. Type III allergic states result from reactions involving specific allergens, immunoglobulins, and complement.
2. Examples of type III responses include the Arthus reaction, serum sickness, and certain autoimmune states.

Antireceptor Antibodies: A Possible New Type of Hypersensitivity

1. Certain disease states are associated with the actions of immunoglobulins against the cellular receptors for various hormones or other reaction-controlling substances.
2. Examples of such disease states include insulin-dependent diabetes and myasthenia gravis.

Type IV: Cell-mediated (Delayed) Hypersensitivity

1. Type IV allergic responses do not involve immunoglobulins or the release of chemicals such as those associated with type I reactions. Type IV responses are dependent upon the interaction of Tdh cells with antigens.
2. Situations associated with immunologic injury are known as *delayed hypersensitivy* and involve factors released by certain T type cells known as *lymphokines.*

Representative Type IV Allergic States

1. Examples of type IV hypersensitivity include tissue destruction in certain infectious disease states such as tuberculosis tuberculin skin reaction, contact dermatitis, and tissue transplantation reaction.
2. The success of organ or tissue transplants involves a group of genes that control the formation of histocompatibility antigens. This genetic system, or *human leukocyte antigen (HLA),* also referred to in the human as the *major histocompatibility complex (MHC),* controls both humoral immune and cellular immune responses.
3. In humans, these genes, as well as others possibly responsible for immunologic reactions, are located on chromosome 6.
4. MHC genes can be divided into three classes. Class I genes (HLA-A, -B, and -C) control the expression of classic transplantation antigens. Class II genes (HLA-D) control the expression of surface antigens (Ia) involved with the immune response on lymphocytes and macrophages. Class III genes control certain complement system components.
5. The HLA system has several alternate genes (alleles) at each known locus. Closely linked allele combinations on a single chromosome form the unit called the *haplotype.* One HLA haplotype is inherited from each parent.
6. HLA typing is used in selecting donors for tissue and organ transplantation; establishing paternity; genetic counseling; and identifying HLA system-associated diseases.
7. The HLA system has been associated with rheumatic diseases, and with other conditions characterized by long-lasting inflammation and abnormal immunologic responses.

AUTOIMMUNE (AUTOALLERGY) DISEASES

1. The inability of the immune system to recognize normal parts of the body as self can result in disorders known as *autoallergies.*
2. Such conditions may develop from several factors, including exposure to chemicals, physical agents, or microorganisms, as well as aging and individual genetic defects.
3. Autoimmune diseases may be divided into cell, tissue, or organic-specific or nonspecific conditions.
4. Diagnosis and identification of autoimmune diseases involve the use of laboratory tests for autoantibodies against the antigens of various cellular components, tissues, or organs. Such tests include latex agglutination, complement fixation, radial immunodiffusion, fluorescent antibody, and the antinuclear antibody procedure.

Questions for Review

1. Distinguish between primary and secondary immunodeficiencies.
2. How are immunodeficiency disorders diagnosed? List three diagnostic approaches, together with the type of disorder with which each is associated.
3. What is AIDS?
4. List at least two differences among types of hypersensitivity. List and briefly describe two examples of the different states of sensitivity.
5. a. Distinguish between primary and secondary mediators of anaphylaxis.
   b. List four examples of each category, together with their respective function or activity.
6. a. What type of hypersensitivity state involves the complement cascade?
   b. Explain the mechanism involved in tissue destruction by the complement cascade.
   c. What is an anaphylactoid reaction?
7. a. What is an autoimmune disease?
   b. How are autoimmune diseases identified or diagnosed?
8. From what standpoint can type IV hypersensitivity be of diagnostic importance? To which diseases is the reaction applicable?
9. What is anaphylaxis? What accounts for the temporary nature of anaphylactic reactions? What are the symptoms of anaphylactic shock in humans?
10. a. What is atopy?
    b. Decribe the wheal and flare response.
    c. Discuss the Prausnitz-Küstner reaction. Does it have any medical importance?
11. What are lymphokines?
12. Does the Shwartzman reaction involve an immune mechanism of the host? Explain.
13. Describe the Koch phenomenon.
14. What is contact dermatitis? What types of substances act as incitants in this condition? How can an incitant be identified?
15. Define or explain and give examples of the following:
    a. xenograft
    b. rejection
    c. allograft
    d. H-2 system
    e. histocompatibility
    f. HLA system
    g. antireceptor antibodies
    h. BRM's
    i. immunosuppression
16. Differentiate between a humoral and a cell-mediated response.
17. What is desensitization?

Suggested Readings

"Acquired Immune Deficiency Syndrome (AIDS): Precautions for Clinical and Laboratory Staffs," *Morbidity and Mortality Weekly Report,* Vol. 31, No. 43. Atlanta, Ga.: Centers for Disease Control, 1982. *A presentation of appropriate precautions to be observed with persons and with specimens from individuals judged likely to have AIDS, or in studies involving experimental animals inoculated with tissues from suspected AIDS patients.*

BACH, M. K., "Mediators of Anaphylaxis and Inflammation," *Annual Review of Microbiology.* **36**:371–413, 1982. *A thorough and up-to-date presentation of the numerous preformed mast cell-derived and secondary chemical mediators involved with anaphylaxis and inflammation.*

BUISSERT, P. D., "Allergy," *Scientific American* **247**:86, 1982. *A clear discussion of why hay fever is a situation in which the immune system has gone wrong.*

DAVIS, T. F. (ed.), *Autoimmune Endocrine Disease.* New York: John Wiley & Sons, 1983. *A series of detailed presentations covering both established and new endocrine diseases having an autoimmune cause.*

LOCKLEY, R. F., AND S. C. BUKANTZ (eds.), *Fundamentals of Immunology and Allergy.* Philadelphia: W. B. Saunders Company, 1987. *A functional presentation of the cellular and molecular events underlying immunologic and allergic processes.*

ROSE, N. R., "Autoimmune Diseases," *Scientific American* **244**:80–103, 1981. *A well-illustrated and understandable article dealing with malfunctions of the immune system in which the body attacks its own tissues.*

SAMUELSSON, B., "Leukotrienes: Mediators of Immediate Hypersensitivity Reactions and Inflammation," *Science* **220**:568, 1983. *A detailed description of leukotrienes and their involvement in immediate hypersensitivity.*

STITES, D. P., J. D. STOBO, H. H. FUDENBERG, AND J. V. WELLS, *Basic and Clinical Immunology,* 4th ed. Los Altos, Calif.: Lange Medical Publications, 1984. *A thorough publication covering the major changes occurring in the field of immunology. Several chapters focus on primary immunologic diseases and disorders.*

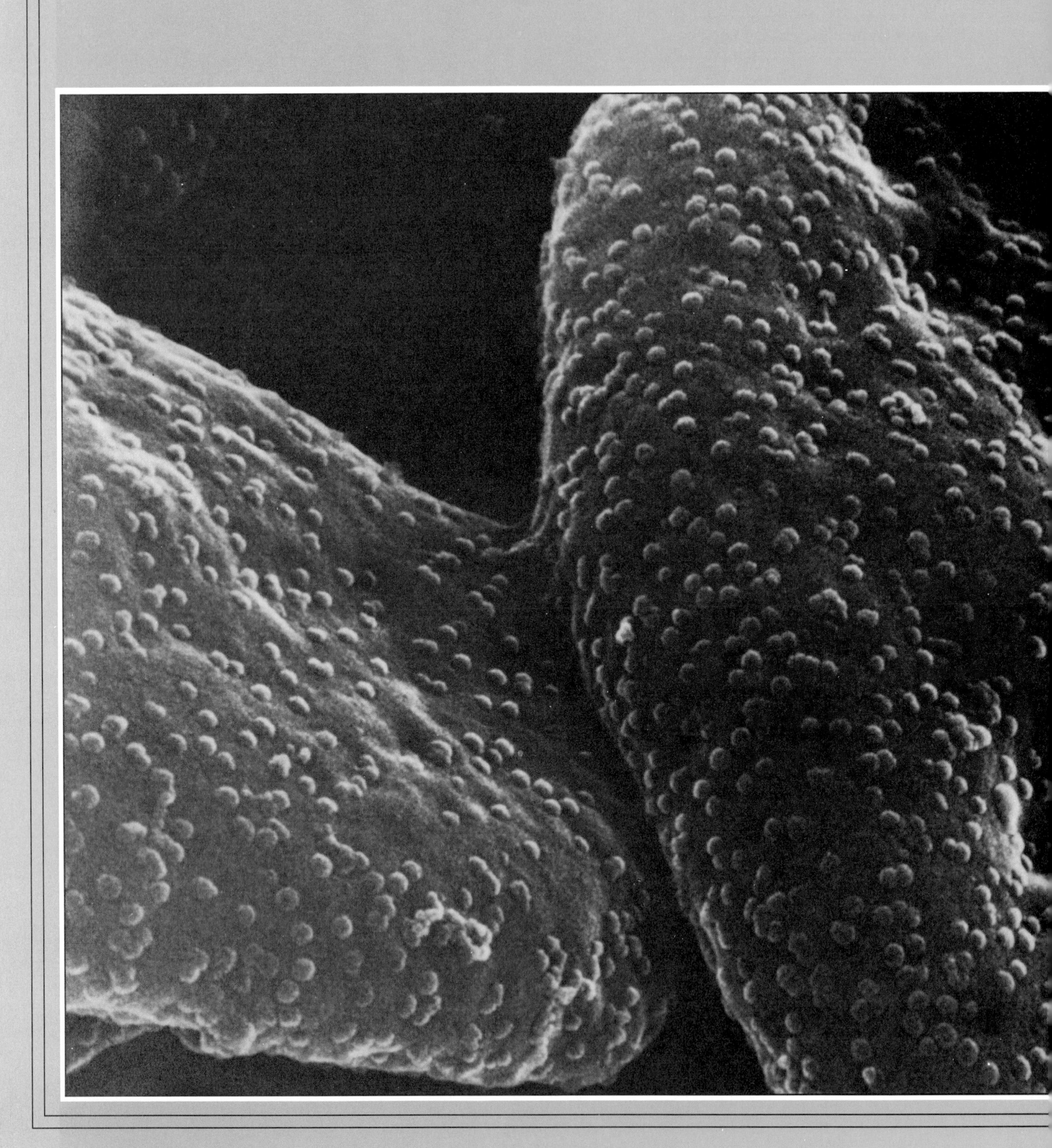

PART VIII

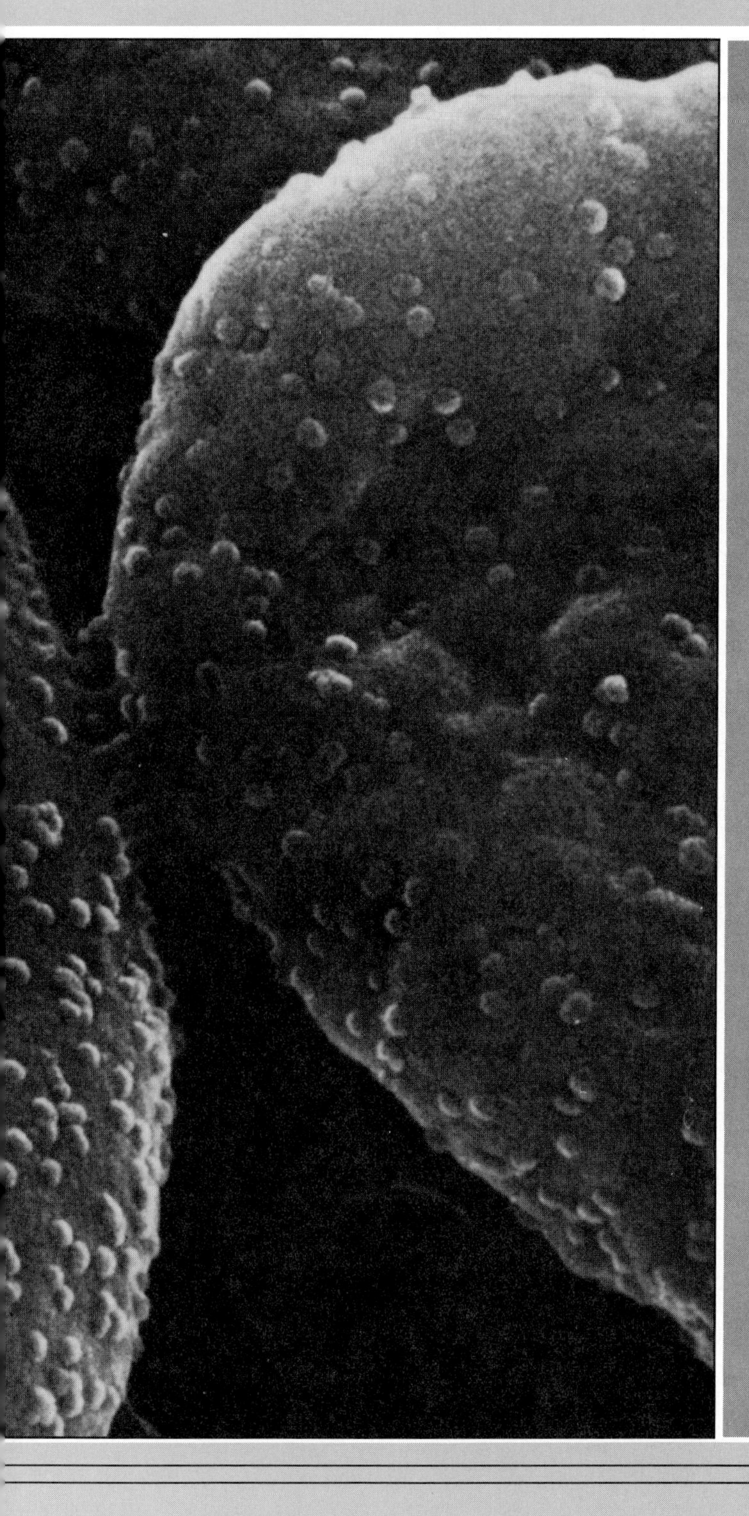

Micro-organisms and Infectious Diseases

Influenza virus particles on the surfaces of chicken red blood cells. [From J. Tawara, et al., J. Electron Microscopy 25:37–38, 1976.]

21

An Introduction to Epidemiology and the Identification of Disease Agents

"Epidemiology is more method than a body of knowledge. . . . it depends heavily on many other sciences. These include clinical medicine, microbiology, pathology, zoology, demography, anthropology, sociology, and almost universally statistics."
— J. P. Fox, C. E. Carrie, and L. R. Elveback

After reading this chapter, you should be able to:

1. Discuss the roles of signs, symptoms, and syndromes in recognizing a disease.
2. Describe the general methods of epidemiology.
3. Distinguish between an incubatory and a convalescent carrier.
4. Define zoonosis and give six examples.
5. List and explain five principal modes of transfer for infectious disease agents and describe appropriate control measures for each.
6. Describe the role played by arthropods in the transmission of infectious diseases.
7. Name a disease spread by each of the following arthropods: cockroaches, fleas, lice, mites, mosquitoes, and ticks.
8. Describe how plants become infected and the importance of these infections.
9. Outline some general approaches used to identify pathogenic bacteria, fungi, protozoa, viruses, and helminths.
10. Discuss the precautions necessary for the transport of microbe-containing specimens for diagnosis.
12. Discuss the growing impact of mechanization and automation in microbiology.
13. Discuss quality control and its importance in a diagnostic laboratory.
14. Distinguish between the following pairs of terms: "epidemic" and "pandemic"; "morbidity rate" and "mortality rates"; "fomite" and "vector"; "sign" and "symptom"; "epidemiology" and "etiology"; and "asepsis" and "nosocomial."

How is an infectious disease spread? What are the principles underlying the patterns of occurrence? How can the spread of pathogenic microorganisms be controlled in health-care facilities such as hospitals and in the population at large? Chapter 21 covers these and other topics related to the causes, distribution, and transmission of disease. Attention also is given to general approaches to the isolation and identification of various disease-causing microbes.

Humans, other animals, and plants can be successfully parasitized by a variety of microorganisms and by such forms of animal life as worms, mites, and ticks. The transmission of such agents from host to host is essential to both their survival and their biological success as pathogens. Pathogens are transmitted from the source of infection to susceptible hosts by many means and mechanisms. In this chapter, we describe some of the means of transfer as well as a variety of measures by which disease transmission can be controlled. The first step in disease control is the identification of pathogens. General and current approaches to the isolation and identification of various disease-causing microbes also are discussed.

Epidemiology is the study of the distribution and causes of disease prevalent in a population. Epidemiologists approach problems of a disease inductively. That is, they collect and analyze data from many individuals in order to reach conclusions about the presence or absence of a particular disease in a given group. For infectious diseases, various contributing factors are taken into consideration. These include the nature of the disease agent or pathogen, the host, and environmental conditions that together establish the disease in a specific population over a stated period of time. It should be noted that this area of investigation is not necessarily limited to the *communicable/infectious* type of disease. The so-called noninfectious diseases—for example, cancer, cardiovascular conditions, congenital defects, diabetes mellitus, emphysema, and those resulting from vitamin deficiencies (such as pellagra, rickets, and scurvy)—have been and continue to be studied epidemiologically. The purposes of such investigations include determining the effectiveness of methods used to control disease, describing the natural course or history of a disease, classifying diseases, and obtaining important information necessary for the planning and evaluation of health care.

Infection and Disease

An *infection* is caused by a disease-causing agent, or *pathogen,* and results from the ability of the agent to invade and multiply in the tissues of a host. A *disease,* on the other hand, is caused by a pathogen's interference with the normal functioning of body systems or organs. The mere presence of microorganisms, however, does not necessarily indicate that an infection or disease is occurring. Not only are many microorganisms present in and on the body of a host, but they can survive under a wide range of environmental and host conditions because of their versatile structural and metabolic mechanisms.

Some infectious diseases are communicable or transmittable, whereas others are not necessarily so. This type of situation is determined in part by the available sources of infection from which a disease may arise and in part on the ease and direction of the transfer of pathogens from such sources to susceptible hosts. Several of the following chapters will describe a wide range of infectious diseases.

Signs, Symptoms, and Syndromes

Recognition of the presence of disease is based to some extent on the obvious presence of objective **signs** or specific recognizable abnormalities known as **symptoms.** Groups of signs and symptoms occurring in a characteristic pattern are referred to as **syndromes,** and are of value in diagnosis and in determining the distribution as well as the cause, or **etiology,** of diseases. **[See specific disease symptoms in Chapters 23 through 31.]**

For centuries, the patterns of occurrence among various communicable/infectious diseases have differed noticeably. The following are the usual terms used to describe the occurrence of diseases:

1. **Endemic.** This term describes a disease that is constantly present but involves relatively few persons. Examples of such diseases, drawn from various localities, include the fungus infection coccidioidomycosis and the bacterial infections leprosy and tuberculosis.
2. **Epidemic.** An epidemic is an unusual occurrence of a disease involving large segments of a population for a limited period of time. Examples are herpes simplex virus infections and influenza and poliomyelitis epidemics.
3. **Pandemic.** A pandemic is a series of epidemics affecting several countries, or even major portions of the world. The influenza pandemic of 1918–1919 exhibited such a worldwide distribution. Currently, acquired immune deficiency syndrome (AIDS) is following a similar path.
4. **Sporadic.** Sporadic diseases are uncommon, occur irregularly, and affect only a relatively few persons. Infections such as the bacterially caused diphtheria, listeriosis, and whooping cough (pertussis) occur sporadically. These and other communicable/infectious diseases may ordinarily be sporadic or endemic, but, depending upon factors such as the immunity of the population and sanitation, they can, unfortunately, sometimes assume epidemic proportions.

Methods of Epidemiology

The design of an epidemiological study is generally determined by the need for specific information and may be **descriptive, analytic,** or **experimental** in nature. Descriptive studies are performed to determine the rate of occurrence of a disease, the kinds of individuals suffering from it, and where and when it occurs. Information about persons affected by the disease is analyzed to find the distribution of a particular characteristic, such as sex, or a variable factor, such as age. Similar types of data are gathered on the general population to deduce particular features of a disease. Descriptive studies yield information that is of immediate relevance to classifying diseases, to understanding the natural course of a disease, and to planning effective health care services.

Analytic studies are used to test hypotheses concerning factors believed to determine susceptibility to disease. Such studies are designed to show whether particular events or environmental conditions are responsible. Two basic types of epidemiological observations may be used to obtain information. One involves comparisons between persons with the disease and those without it. The other type of observation depends upon a comparison between individuals exposed to a suspected agent and those not exposed.

As described in Chapter 1, scientific investigation in many branches of science depends upon an orderly sequence of steps leading to the development of a hypothesis, followed by the design and performance of experiments to test the hypothesis. **[See Microbiology Highlights, "The Scientific Method: From Hypothesis to Law," in Chapter 1.]** Epidemiologists observe and study human populations and generally are limited in their ability to carry out actual experiments. However, epidemiologists can use an *experimental approach* in which their observations of changes and events are organized to formulate a hypothesis. An experimental model is then constructed in which one or more selected factors are manipulated. The result of the manipulation will either confirm or disprove the hypothesis. A major aid to such experiments has been the computer. As indicated in other chapters, computer technology has come of age as a useful tool in many areas of microbiology and related specialties. For epidemiologists, computers are of great value for the management of information, which includes storage and retrieval of data and the simulation of experimental models. In addition, computers serve as information aids with which to prepare reports more accurately, to update records easily, and to identify more rapidly, both qualitatively and quantitatively, factors that influence the causes and distribution of diseases.

Morbidity and Mortality Rates

When an outbreak of a communicable disease occurs, or even when a single case appears, epidemiologists gather many types of data. Some of the principal findings are frequently expressed in terms of morbidity and mortality rates. **Morbidity** is generally defined as the number of individuals having the disease per unit of the population within a given time period. Usually 100,000 is taken as the unit of population for such calculations. The **mortality rate** is the number of deaths attributable to a particular disease per unit of the population (usually 1,000) within a given time period. Reports may be compiled weekly, monthly, yearly, or for even longer periods, depending on the purpose of the study.

Occasionally, outbreaks of disease occur that are more or less limited to particular segments of a population. Consequently, morbidity and mortality rates may be calculated for that population segment alone. An *infant mortality rate* is an example. Figure 21–1 summarizes some of the factors involved in reducing the morbidity and mortality of infectious diseases.

Reporting Communicable/ Infectious Diseases

State administrative codes require that actual or even suspected cases of certain communicable/infectious diseases be reported to local health authorities. The number and kinds of such reportable diseases vary among the states. The specific reportable infections, as recommended by the U.S. Centers for Disease Control, are listed in Table 21–1; this list is fairly comprehensive. Individuals charged with the responsibility of notifying appropriate health authorities include physicians, coroners, directors of hospitals, clinics, and laboratories, and any persons knowing of the existence of a disease.

Sources and Reservoirs of Infection

The sources of infectious disease agents are many and varied (Figure 21–2). Generally speaking, however, the most disabling and most common infec-

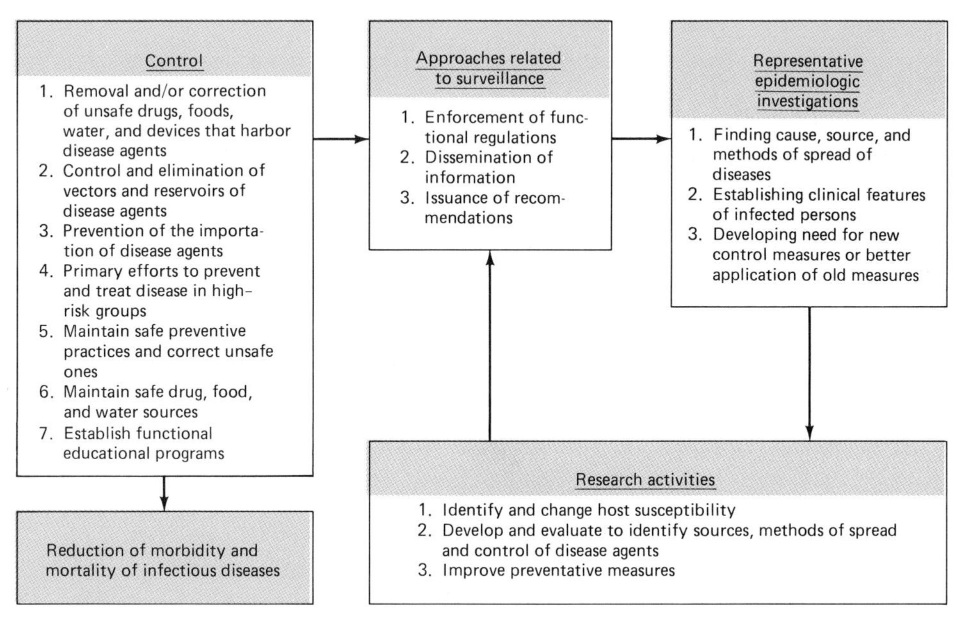

Figure 21–1
Factors involved in the reduction of morbidity and mortality of infectious diseases. The figure exhibits the interrelationships among control measures, epidemiologic investigations, and research. *[After J. V. Bennett,* Ann. Intern. Med. *89:761–763, 1978.]*

TABLE 21–1 Microbial and Related Diseases Recommended to Be Reported to the U.S. Centers for Disease Control

Disease	*Nature of Causative Agent*	*Disease*	*Nature of Causative Agent*
Acquired immune deficiency syndrome (AIDS)	V[b]	Meningitis (meningococcal or meningococcemia)	B
AIDS related complex (ARC)[a]	V	Mumps	V
Amebiasis	P[c]	Paratyphoid fever A, B, and C	B
Anthrax	B[d]	Pertussis (whooping cough)	B
Arbovirus infection	V	Plague	B
Aseptic meningitis	V	Poliomyelitis	V
Asiatic cholera	B	Psittacosis	B
Bacterial meningitis	B	Q fever	B
Botulism	B	Rabies (animal)	V
Brucellosis	B	Rabies (human)	V
Chancroid	B	Relapsing fever	B
Chickenpox	V	Rheumatic fever (acute)	B
Cholera	B	Rocky Mountain spotted fever	B
Diphtheria	B	Rubella	V
Dysentery (bacillary)	B	*Salmonella* infections (exclusive of typhoid fever)	B
Encephalitis (acute)	V	Scarlet fever	B
Food poisoning (excluding botulism)	B	*Shigella* infections	B
Gonorrhea	B	Syphilis (congenital)	B
Granuloma inguinale	B	Syphilis (primary and secondary)	B
Hepatitis: Type A	V	Tetanus	B
Type B	V	Toxic shock syndrome	B
Non-A, Non-B	V	Trachoma	B
Unspecified	—	Trichinosis	H[e]
Influenza	V	Tuberculosis	B
Legionnaire's disease	B	Tularemia	B
Leprosy	B	Typhoid fever (both actual cases and carriers)	B
Leptospirosis	B	Typhus fever	B
Lymphogranuloma venereum	B	Yellow fever	V
Malaria	P		
Measles (rubeola)	V		

[a] ARC may be reportable in some states.
[b] V = viral.
[c] P = protozoan.
[d] B = bacterial.
[e] H = helminthic.

Source: Adapted from *Morbidity and Mortality Weekly Report (MMWR)*, Centers for Disease Control, Atlanta, Georgia.

Microbiology Highlights

INFECTIOUS DISEASES AND DAY-CARE CENTERS

Infectious diseases, often occurring in epidemics, are not a surprising fact of life of day-care centers. Infant and toddlers have the highest age-specific attack rates for respiratory and enteric (intestinal) infections. Microorganisms are readily spread among groups of young children. Most endemic and epidemic infections occurring in day-care centers can be categorized as being respiratory, skin, gastrointestinal, or multisystem. The modes of spread for these diseases include airborne, fecal-oral, direct person-to-person contact, and indirect transmission by contaminated inanimate objects (fomites). Acute respiratory infections are the most frequent and universal illnesses suffered by children in day-care settings. Impetigo (a highly contagious bacterial infection) and lice infestation are the most commonly encountered skin problems. With respect to the gastrointestinal tract, acute infectious diarrhea, pinworm, and hepatitis A outbreaks have emerged as problems not only for the children attending day-care centers but for their families as well. The morbidity and economic costs of these and related diseases will reach immense proportions unless preventive and control measures are significantly improved.

tions among humans are caused by microorganisms capable of living and reproducing in human tissues. Where these organisms are present, human tissues and secretions serve as potential sources of pathogens. The sources of such infectious body fluids are referred to as *portals of exit.* They include: (1) the gastrointestinal tract, (2) the genitourinary system, (3) the mouth, (4) the respiratory tract, (5) the blood and blood products, and (6) lesions of the skin and other areas.

A host or a local environment that supports the survival and multiplication of pathogens is referred to as a **reservoir of infection.** Living reservoirs include infected (although not necessarily symptom-

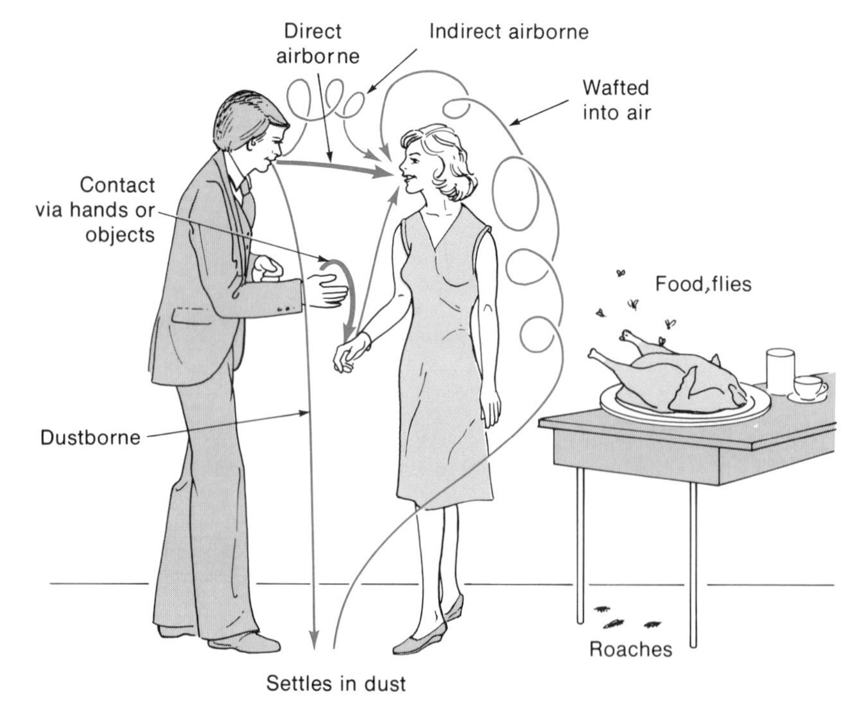

Figure 21–2 Some common sources of infection and portals of exit. Note that certain of these can also function as a means of disease transmission.

atic) humans and other animals, whereas inanimate reservoirs include air, food, soil, and water. Reservoirs of infection provide disease agents with a suitable environment for survival over a prolonged time period and also provide opportunities for their transmission to a new susceptible host. A newly infected host may, in turn, become a new reservoir capable of infecting others, thus extending the chain of infection. Individuals who harbor pathogens transmissible to others are called **carriers.** A carrier who apparently suffers no ill effects is called a *healthy carrier.* The individual who is in an incubating state, undergoing the initial stages of a disease but without exhibiting symptoms, is referred to as an *incubatory carrier.* Such persons may be infectious during the last stages of their incubation period. Another category of carrier is the *convalescent.* In certain situations, patients recovering from an infection may serve as sources of pathogens.

Zoonoses

Various warm-blooded animals are recognized as reservoirs of infectious disease agents for humans. Such animals include bats, birds, cattle, cats, dogs, horses, mice, monkeys, rabbits, rats, skunks, and various wild mammals. Rats, for example, are implicated in the transmission of plague, rat-bite fever, and certain tapeworms. Diseases that primarily affect lower animals but can also be transmitted to human beings by natural means are referred to as **zoonoses** (Table 21–2). Several animals also serve as sources of parasites that affect human beings. [See plague in Chapter 27.]

The number of zoonotic diseases is relatively large; over 150 are known worldwide, at least 40 of

TABLE 21–2 Representative Zoonoses Produced by Microorganisms[a]

Disease	*Associated Animals*	*Major Mode of Transmission*
BACTERIAL INFECTIONS		
Anthrax	Domestic livestock	Direct contact with infected and contaminated soil
Brucellosis (undulant fever)	Domestic livestock	Direct contact with infected tissues; ingestion of milk from diseased animals
Bubonic plague	Rodents	Fleas
Leptospirosis	Dogs, rodents, wild animals	Direct contact with infected tissues and urine
Relapsing fever	Various rodents	Lice and ticks
Rocky Mountain spotted fever	Dogs, rodents	Ticks
Salmonellosis	Dogs, poultry, rats	Ingestion of infected meat; contamination of water
Tularemia	Wild rabbits	Direct contact with infected tissues; deer flies, ticks
FUNGUS INFECTIONS		
Several forms of ringworm	Various domestic animals (e.g., cats, dogs)	Direct contact
PROTOZOAN INFECTIONS		
African sleeping sickness (trypanosomiasis)	Humans, wild game animals	Tsetse flies
Chagas' disease	Humans, wild animals	Kissing (assassin) bugs
Cryptosporidium species	Mice, cattle and other farm animals	Inhalation (aerosol)
Kala-azar (leishmaniasis)	Cats, dogs, rodents	Sandflies
Toxoplasmosis	Birds, wild rodents, domestic animals (e.g., cats)	Aerosols; possibly contamination of food and water; via the placenta
VIRAL INFECTIONS		
Eastern equine encephalitis	Birds; horses and related animals	Mosquitoes
Influenza	Humans, swine, horses	Direct contact with droplets
Jungle yellow fever	Various species of monkeys	Mosquitoes
Rabies	Bats, cats, dogs, humans, skunks, wolves, etc.	Bites, contamination of wounds with infectious saliva

[a] These diseases are described in the chapters in the last portion of the text.

these being important as occupational diseases in agriculture. In developing countries, zoonoses remain a major cause not only of short-term human illness but of long-term or chronic conditions as well. For example, blood fluke infection (schistosomiasis) and sheep tapeworm are associated with long-term physical disability and psychological stress. [See Chapter 31 for discussions of worm diseases.]

Certain infectious agents can be transmitted by the bite of warm-blooded animals. Perhaps the best known of these is rabies. The chief vectors of this viral infection include cats, coyotes, dogs, foxes, jackals, skunks, and a variety of bats. Since the turn of the century, vampire bats have been incriminated as vectors of rabies among cattle in various regions of the world, including Central and South America. When in a rabid state, these bats can bite one another as well as cattle and humans. The reports of bat- and skunk-associated rabies in humans have been steadily increasing.

Most species of bats are social animals, congregating together in buildings, caves, mines, or trees. In general, bats are considered beneficial to humans because they consume large quantities of insects and rodents. However, rabies has been reported to occur in more than 20 species of bats, including fruit-eating, insectivorous, and vampire varieties. Rabies may also be latent in these animals. Bats with latent rabies may serve as carriers, excreting the viral agents in their saliva and feces for several months.

Bat colonies may be removed by means of chemical repellents, batproofing procedures, or simple physical destruction. Batproofing, if feasible, is apparently the only truly satisfactory (but extremely difficult) method for the removal of bats. To batproof a building, one must determine the bats' actual roosting site and seal off the various means of access. A knowledge of the migratory habits of the colony is useful in this regard. It is necessary to guard against the possibility of trapping bats before they have had an opportunity to leave and of overlooking a possible entry to the site.

Categories of Zoonoses

Zoonotic diseases may be grouped in several ways. Generally they are classified according to the major reservoir of the infectious agent and the mode of transmission of the infectious agent among natural hosts. Zoonoses can also be less formally grouped as to the major human populations at risk for acquiring a zoonotic disease. Individuals at risk can include veterinarians, meat packers, and workers in the animal-hair and hide industries.

Table 21–3 defines the major categories of zoonotic disease and gives examples of each. Several of the diseases and pathogens listed are discussed in Chapters 23 to 31.

TABLE 21–3 Classification of Zoonotic Diseases

Category	*Brief Description*	*Example(s) of Diseases and/or Pathogens*
MAJOR RESERVOIR		
Zooanthroponosis	A zoonotic disease for which humans are the natural host of the infectious agent	A dairy farmer with tuberculosis who can transmit the infection to dairy cattle.
Anthropozoonosis	A zoonotic disease for an animal other than a human, which is the natural host	Domestic cattle, sheep, goats, and swine infected with brucellosis. Humans become infected by contact with diseased animals.
Amphixenosis	Zoonotic disease for which humans and other animals serve equally well as natural hosts	Pathogenic bacteria including *Escherichia coli, Salmonella, Staphylococcus,* and *Streptococcus.*
FORM OF TRANSMISSION		
Direct zoonosis	A disease requiring only one vertebrate (animal with a backbone) to maintain the pathogen	Rabies virus can be maintained in the wild skunk poplation by transmission from an infected skunk to a susceptible skunk.
Cyclozoonosis	A disease requiring two or three vertebrates to maintain the pathogen	Sheep tapeworm. This worm is maintained by a transmission cycle involving sheep and dogs.
Metazoonosis	A disease requiring both an arthropod (such as a mosquito) and a vertebrate	Equine encephalitis. Here a mosquito is needed to transmit the virus disease to a susceptible horse or human.
Saprozoonosis	A disease caused by a pathogen maintained in soil, water, or some other inanimate material (fomite)	Histoplasmosis, Legionnaires' disease.

Microbiology Highlights

THE HIGH COST OF ZOONOTIC DISEASES

In addition to causing human health problems, agricultural zoonoses are a major economic drain on the animal protein industry. For example, the bacterial diseases brucellosis, leptospirosis, and tuberculosis cost livestock producers millions of dollars annually. Such losses occur despite eradication programs that have dramatically reduced the incidence of these diseases in several countries in the past 30 years. Programs of this type are quite expensive. In 1978, the U.S. Department of Agriculture's brucellosis eradication program cost nearly $54 million. In many countries, millions of dollars are spent on inspection and related activities to prevent potentially hazardous red meat, poultry, and dairy products from reaching the public. Other major expenses include costs for the treatment and prevention of zoonoses in humans. For example, $15 million is spent annually in the United States to give the rabies vaccine, as a protective measure, to approximately 30,000 people who have been exposed to the deadly virus. In countries where rabies and other zoonoses are greater problems, the cost is much higher.

Although zoonoses are generally recognized as significant worldwide health problems, the actual prevalence and incidence of zoonotic infections is difficult to determine. Unfortunately, these infections are often not diagnosed or are misdiagnosed; even when they are correctly diagnosed, they are often inadequately treated.

Principal Modes of Transfer for Infectious Disease Agents

Infectious disease agents may be transmitted to susceptible individuals in a variety of ways. These include (1) direct contact with obviously infected persons or carriers; (2) indirect contact with inanimate objects, food, or water contaminated by infected individuals; (3) inhalation of airborne dust or droplet nuclei containing infectious agents; (4) inoculation; and (5) arthropods, which may either carry disease agents mechanically from one reservoir to another or serve as *biological vectors* by being both the host and the reservoir for such agents. *Mechanical transmission* refers to the situation in which an insect physically transports a pathogen from contaminated material such as food or water to other objects. Cockroaches and flies are good examples. In *biological transmission,* a portion of the pathogen's life cycle is carried out in the vector. Transmission of the malarial pathogen by the anopheline mosquito is an example of biological transmission.

The mechanical means of disease transmission include the five Fs: food, fingers, flies, feces, and fomites. The biological means of transmission include the injection of blood and blood products, the bites of warm-blooded animals, arthropod bites, and the introduction of arthropod feces into bites or wounds. It is important to note that many diseases can be spread in a variety of ways.

Direct Contact

Individuals who come into direct contact with infectious lesions such as open sores, boils, and draining abscesses (Color Photograph 4) obviously run the risk of acquiring the disease agent. Contagious diseases—from the Latin word *contagio,* meaning touch or contact—include anthrax, syphilis, herpes simplex virus infections (Color Photographs 118 and 120), and gonorrhea. Many of the disease states spread by direct contact gain access to the body through the nose and throat.

Pathogens can be transmitted through hand shaking or kissing. Examples of diseases spread in this manner include poliomyelitis, chickenpox, the common cold, bacillary dysentery, and streptococcal and herpes virus infections. Certainly, the washing of hands after blowing the nose, defecating, urinating, or working with infected persons helps to limit the spread of disease agents.

Sexual activities represent another form of direct contact. AIDS, syphilis, gonorrhea, and a host of other diseases can be spread in this manner. Descriptions of these conditions can be found in Chapters 27 and 28.

Indirect Contact

Various microorganisms can be transmitted by food, water, and, quite often, contaminated inanimate objects. Moist foods that are not highly acid can serve as excellent culture media for pathogenic microorganisms, including the causative agents of amebic dysentery, bacillary dysentery (shigellosis), cholera, and typhoid fever. These diseases can also be spread through contaminated water supplies.

Raw or inadequately cooked meat from infected animals is a well-known source of disease agents. In addition to microorganisms, such products can contain helminths (worms) capable of causing trichinosis (pork roundworm) and tapeworm infections. Proper sanitary measures and adequate meat inspection substantially reduce the possibility of transmission of these diseases. Many states have stringent sanitation requirements for the farms and ranches that supply meat for human consumption. Such measures are also important for bacterial disease control. Included in this category are infections such as undulant fever (brucellosis) and bovine tuberculosis, both of which can be transmitted by unpasteurized milk from infected cows.

The handling of food for human consumption by undiscovered carriers is always a serious hazard. The carriers may cough or sneeze on food or handle utensils or food without washing their hands after using the toilet or blowing the nose. Also, food handlers or dishwashers may have draining abscesses or boils, which serve as other sources of disease agents. Eating utensils and drinking glasses can also be important factors in the transmission of diseases. The thorough washing and proper disinfecting of such items is essential to good sanitation. Various types of commercial dishwashing equipment are available that clean and disinfect utensils mechanically. **[See microbial diseases of the G.I. tract in Chapter 26.]**

Regular inspection of restaurant personnel and equipment, including dishwashing machines, is necessary for the effective prevention of infections spread by food. Standardized methods exist for bacteriological examination of dishes and related items. Specimens can be taken directly by means of a sterile cotton swab or, if possible, the utensil can be introduced directly into a sterile medium. The American Public Health Association publishes information on appropriate methods, media, and other details for this purpose.

Maintenance of adequate sanitation in restaurants can be difficult, costly, and time-consuming. It is therefore not surprising that fast-food restaurants are increasing their use of disposable paper and plastic dishes and eating utensils. This practice greatly reduces operating costs and, more importantly, markedly increases sanitation levels. Pathogens are seldom found in bacterial counts of these plastic and paper products.

Air- and Dust-borne Infections

Particles bearing microorganisms are released into the general environment in two major ways. Some are produced during normal activities involving the respiratory tract—for example, talking, coughing, and sneezing. Significantly larger numbers of organisms are liberated by sneezing than by the other two activities. Microorganism-bearing particles from the skin, clothing, and even dressings covering wounds are also generated by normal body movements.

The second major means by which particles are introduced into the general environment is the redistribution of accumulated particles in room dust. Dust-borne infections include the fungal diseases of coccidioidomycosis and histoplasmosis. Once droplets are released into the air, they fall to the ground at a rate determined by their size. On the ground, these droplets stick to or become mixed with the variety of animal, plant, and mineral debris commonly known as "house dust." Evaporation, or drying, takes place next. The rate of evaporation depends on the size and composition of the droplets and the relative humidity of the atmosphere. The higher the humidity, the slower the rate of evaporation.

Depending on the types of microorganisms and the composition of droplets present, bacteria and related organisms may survive for long periods of time. This is especially true of droplets containing saliva, sputum, or other discharges. What remains of a droplet after evaporation is called the *droplet nucleus.* Droplet nuclei do not settle quickly after being disturbed, but remain suspended in the air for long periods of time, thus potentially giving rise to *droplet nucleus airborne infections.* This means of transmission contrasts to that of dust- and particle-borne microorganisms, which settle quickly. Droplet nuclei are reported to settle at a rate of 0.04 feet per minute, while dust particles settle at a rate of 1 to 5 feet per minute.

Obviously, coughs or sneezes, stifled or not, produce a microbial spray (Figure 21–3). In an effort to prevent hospital-acquired **(nosocomial)** infections, hospital personnel frequently wear masks to reduce the possibility of producing droplet nuclei. Unfortunately, preventing such sprays is impossible. Procedures and practices used to minimize disease transmission by such means include the application of bactericidal compounds to floors and the use of special floor coverings to help trap dust. **[See chemical control in Chapter 11.]**

Fomites

Inanimate objects or substances capable of absorbing and transferring infectious microorganisms are called **fomites.** A wide variety of materials can spread human diseases. These include clothing, eat-

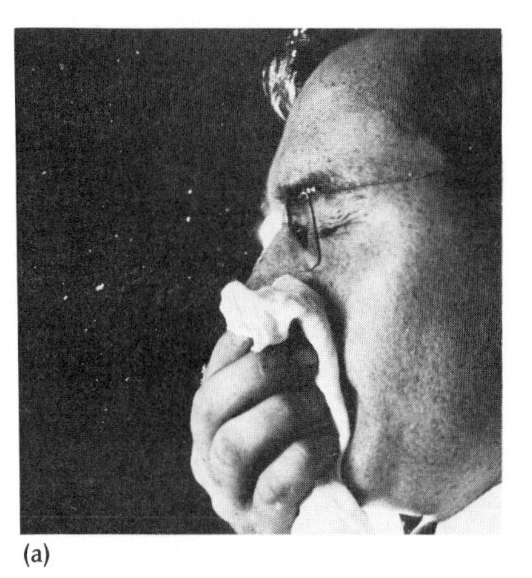
(a)

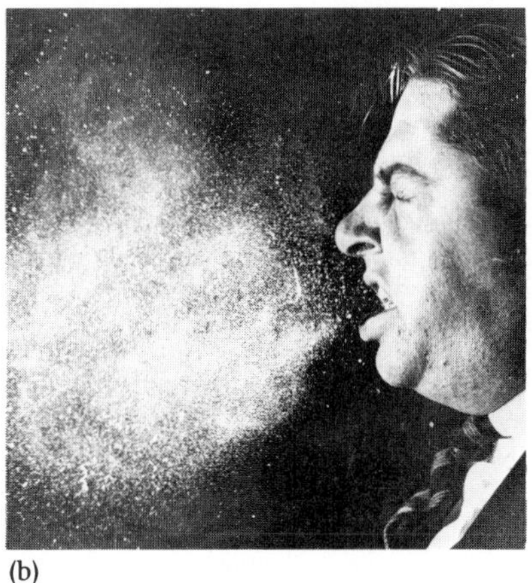
(b)

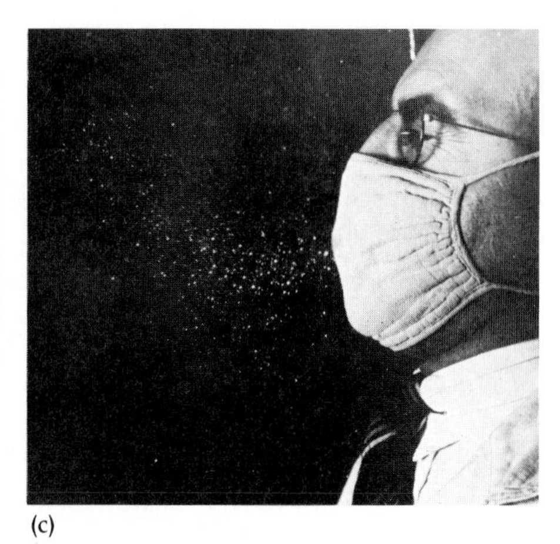
(c)

Figure 21–3 Sneezing. (a) Even this stifled sneeze produces many droplets. It is clear from the photo that the hands and arms can easily become contaminated with nasal secretions. (b) A full-blown, unstifled sneeze. Note the heavy cloud of material introduced into the air. (c) Despite the presence of a surgical mask, droplets from an unstifled sneeze are still propelled into the air. *[Courtesy of M. W. Jennison, Department of Bacteriology and Botany, Syracuse University, and the American Society of Microbiology, LS-5, LS-15.]*

ing utensils, instruments, bed linens, toys, and even fossils (Figure 21–4). Reports indicate that archeological relics and fossils from endemic areas pose a definite public health hazard. Dust and dirt that have accumulated on such objects may contain spores of infectious agents. Individuals can inhale this material while cleaning the relics.

Plant pathogens are transmitted by fomites such as gardening tools, gloves, and soil.

Figure 21–4 A fossilized sea shell taken from a dry creek bed in Simi Valley, California. This shell was incriminated as a source of the fungus *Coccidioides immitis,* the causative agent of coccidioidomycosis. A case of this disease apparently developed from the inhalation of fungal spores during the cleaning of the shell. *[P. E. Rothman,* et al., Am. J. Dis. Child. ***118****:792, 1969.]*

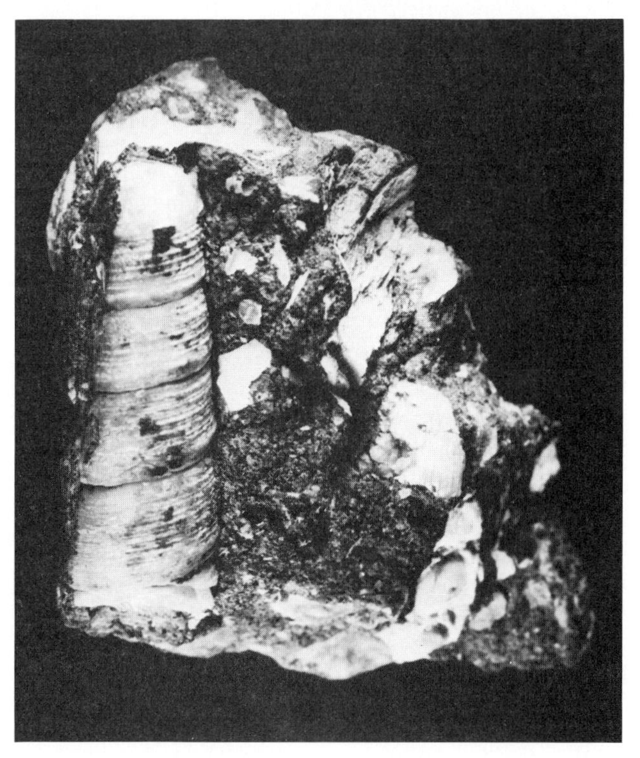

Accidental Inoculation

Infections can develop from the direct introduction of pathogens during surgery. Occasionally, individuals working with clinical specimens may introduce a pathogen into their bodies through a preexisting cut or an accidental wound with a contaminated hypodermic needle or inoculating loop. This is especially true for drug users. Contaminated hypodermic syringes and needles are major sources of virus infections such as AIDS and hepatitis B for such individuals. Any work with infectious materials should be done with great care. In the laboratories associated with microbiology, carelessness can lead to many possible hazards. Among the situations too often encountered are improperly sterilized inoculating needles, culture tubes or flasks left unplugged in incubators, partially opened culture flasks, inadequately disinfected microbial culture spills, and the practice of eating and drinking in laboratories.

Arthropods and Disease

Throughout the centuries, arthropods such as fleas and mosquitoes have interfered with human efforts to establish stable and safe environments. As new cities or agricultural communities developed, devastating diseases associated with arthropods appeared. During the sixth century B.C., malaria and plague flourished in newly established cities. One of the few recourses left to human beings was to escape from these centers of disease, returning long after the epidemic dangers had passed. [See malaria and plague in Chapter 27.]

Although various arthropods were known to live on the outer surfaces of mammals, until the nineteenth century little was known of the relationship between insects and disease agents. Toward the end of the nineteenth century, scientists launched an intensive, systematic study of infectious diseases. One product of this work was the demonstration in 1893 by T. Smith and F. L. Kilbourne that ticks transmit the protozoon *Babesia bigemina,* which causes Texas cattle, or red water, fever. This disease had been recognized in the United States since 1796, but its true etiology was masked until the investigations by Smith and Kilbourne.

The discovery of this arthropod-microbial connection provided a model that investigators could use to show the significance of other arthropods to disease epidemiology. Sir Ronald Ross applied the model in this way, demonstrating the importance of *Anopheles* mosquitoes to malaria. [See the life cycle of malaria in Chapter 27.]

In recent years, knowledge of arthropod-borne diseases has increased immeasurably. Along with several areas of related research, this knowledge forms the specialization referred to as *medical entomology.* This field of study is concerned with the recognition and description of arthropods, the distribution of arthropod-associated diseases, the effects of disease agents on the arthropod vector, the effects of the disease agent on the host, and the control of arthropods.

PROPERTIES OF THE ARTHROPODS

Arthropoda represent the largest group of animals. As of 1964, about 740,000 species were known. Many species are of medical and economic importance. Their significance lies partially in their ability to produce serious injuries or even sensitization, as in the case of centipedes, wasps, and spiders. Arthropoda are also important because several of them serve either as intermediate hosts for parasites or as vectors for pathogenic microorganisms. It is the latter aspect that concerns us in this section.

Although arthropods vary greatly, they all share several major characteristics: (1) a rigid or semirigid exoskeleton composed of chitin; (2) a complete digestive tract; (3) an open circulatory system (with or without a dorsally situated heart) that forms a body cavity (hemocoel); and (4) excretory, nervous, and respiratory systems. Insects, a type of arthropod, have segmented bodies and jointed legs.

A classic example of an arthropod that transmits disease agents mechanically is the common house fly, *Musca domestica* (MUSS-ka doe-MES-tik-a). The etiologic agents of such diseases as infectious hepatitis, polio, and salmonellosis may be picked up by flies during contact with fecal matter containing viable infectious microorganisms. These pathogens may contaminate various body parts of the fly and may be deposited on food when the fly rests or feeds on it. Factors affecting the survival of pathogens, the availability of arthropods, and the rapidity of transfer are all important in the mechanical transmission of pathogens by arthropods.

Most arthropod-borne diseases are transmitted biologically. Many animal and plant pathogens require an arthropod for purposes of development, multiplication, or both. Susceptible hosts become infected through the bites of such pathogen-carrying arthropod vectors.

TICKS AND MITES (ARACHNIDA)

Ticks and mites are important medically because they can (1) serve as reservoirs of infection, (2) transmit infectious disease agents, and (3) directly cause diseases such as tick paralysis. Two main groups of ticks are recognized, the hard- and soft-shelled ticks (Figure 21–5). Rocky Mountain spotted fever, deer fly fever (tularemia), Q fever, and relapsing fever are among the diseases spread by ticks.

The life cycle of both ticks and mites begins with an egg. Six-legged larvae hatch from the eggs and develop into eight-legged nymphs (Figure 21–6). These forms later become adults. Certain infectious disease agents are known to pass into the eggs of infected ticks (through *transovarian passage,* or *transmission*) and into the larvae of mites. In this way, the disease agents are transmitted to succeeding generations. Thus, these particular arthropods represent important reservoirs of infectious disease agents.

Distinguishing between mites and the two classes of ticks can be confusing. The following characteristics differentiate them: (1) ticks are larger than mites; (2) ticks have few or no body hairs, while long hairs are present on the membranous body of mites (when ticks do have hairs, the hairs are short); (3) the bodies of ticks are leathery in appearance; and (4) ticks have an exposed, tooth-bearing *hypostome* (a rodlike organ at the base of the beak) used for attachment purposes (Figure 21–6). [See rickettsial diseases in Chapter 27.]

Several species of mites infest humans and can thereby transmit certain diseases, including rickettsialpox and scrub typhus. Furthermore, sarcoptic acariasis, a noninfectious disease, is caused by the skin-burrowing mite *Sarcoptes scabies* (SAR-kop-tees scay-BEES). The term *acariasis* is used to denote a mite infestation (Figure 21–7).

THE INSECTS

LICE. Lice are parasitic on the hair and skin of birds and several mammalian species—cats, cattle, dogs, goats, horses, and humans. Generally, the mouth parts of lice are adapted either for sucking the blood and tissue fluids of mammals or for chewing epithelioid structures associated with the skin of their hosts.

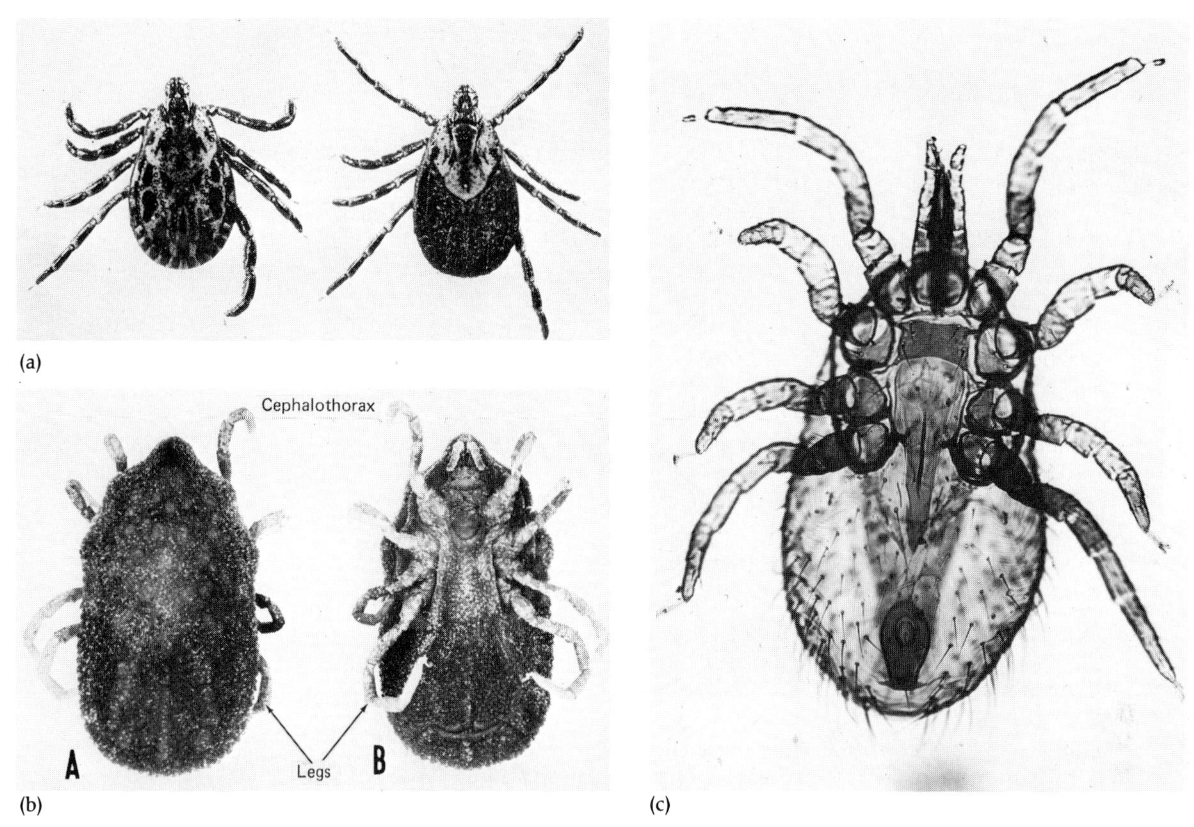

Figure 21–5 Ticks and a mite. (a) A dorsal (top) view of a male and a female hard-shelled dog tick. *(Dermacentor variabilis).* (b) Dorsal and ventral views of a female soft-shelled tick *(Ornithodoras concanensis).* (c) A female mite *(Liponyssus bacoti).* Mites are much smaller than ticks. *[Courtesy of the Rocky Mountain Laboratory, U.S. Public Health Service, Hamilton, Mont.]*

Figure 21–6 A larval form of the tick *Ornithodoros amblus,* magnified 150×. The rodlike hypostome used for attachment and body hairs of this arthropod can be easily seen. *[H. Clifford,* et al., J. Parasitol. ***66****:312–323, 1980.]*

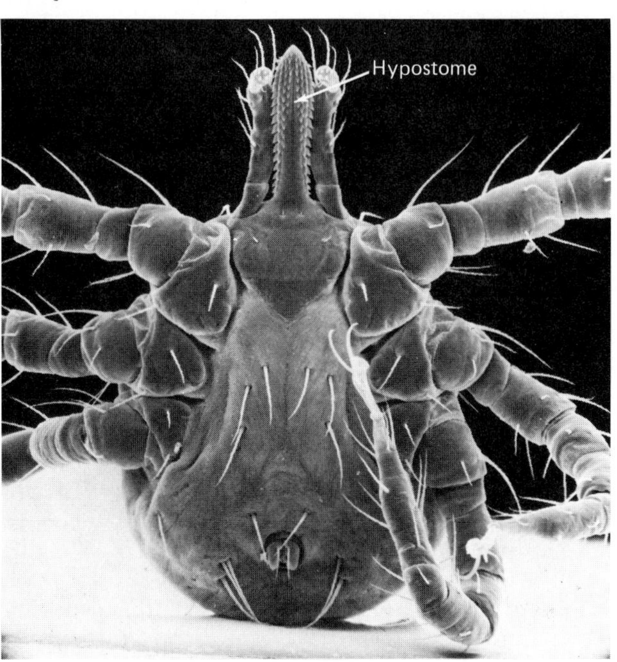

Two genera of lice (Figure 21–8) are associated with humans, *Pediculus* (pee-DIK-you-lus) and *Phthirus* (THIR-us). Members of the latter genus are known as the *pubic,* or *crab, lice.* The two recognized forms of *Pediculus* are *P. humanus var. corporis* (body louse) and *P. humanus var. capitis* (head louse). The two varieties are distinguished on the basis of size (head lice are slightly smaller), and the fact that body lice seldom infest the head region of the host. Head lice apparently roam over the entire body.

Several infectious diseases are transmitted by these lice. Diseases such as cholera, impetigo, and trachoma are spread mechanically. The rickettsial diseases of epidemic typhus and trench fevers, and the bacterial infection of relapsing fever, are biologically transmitted.

The life cycle of lice is composed of the egg (nit), nymph, and adult stages. Male and female organs are in separate insects. Blood meals can be taken by both the nymph and adult forms. Human lice can be found clinging to hair or clothing. A primary measure for controlling human lice is effective personal cleanliness. This precaution is especially important in crowded areas and during disasters such as earthquakes and floods. Control of these insects can also be accomplished through the judicious use of appropriate chemical agents.

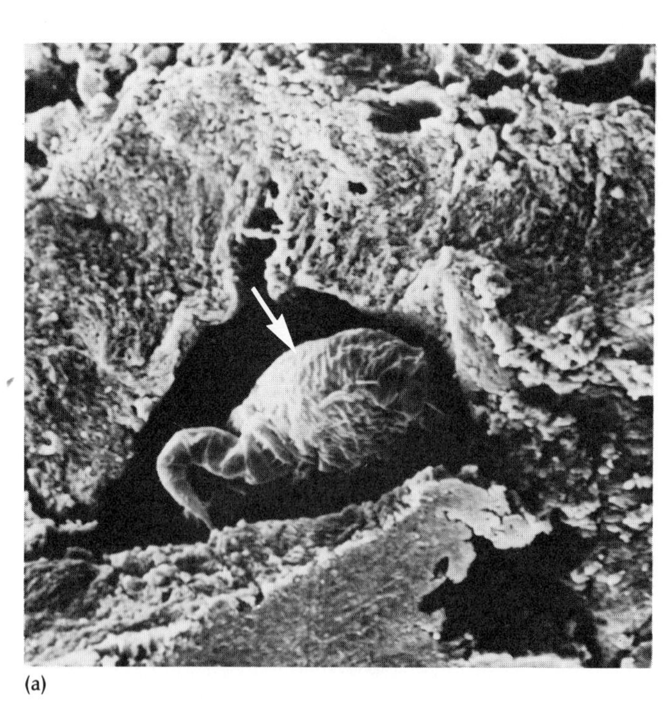

(a)

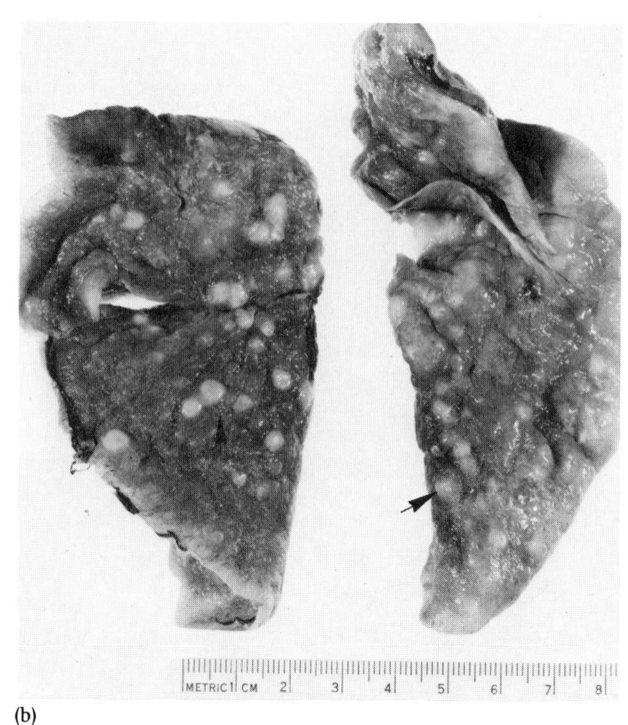

(b)

Figure 21–7 The mite. These arthropods are frequently considered to be disease agent vectors, but traditionally they are neglected as disease agents in their own right. (a) This scanning micrograph shows a mite (arrow) in lung tissue. *[J. C. S. Kim,* J. Med. Primato ***5:*** *3–12, 1976.]* (b) The lung damage in this rhesus monkey was caused by mite infestation. *[J. E. C. Kim,* et al., Infect. Immun. ***5:****137–142, 1972.]*

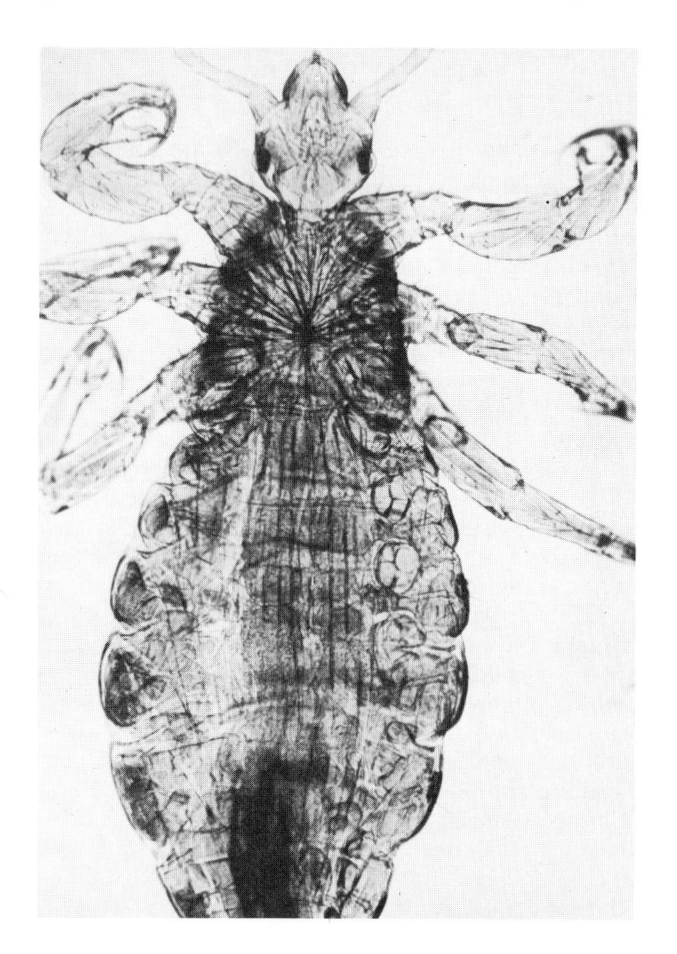

Figure 21–8 The body louse *(Pediculus humanus var. corporis).*

THE TRUE BUGS (HEMIPTERA)

Most of the true bugs are plant eaters. However, several species apparently have abandoned plant feeding and now eat other insects. Probably no other group of arthropods exerts as pronounced an effect on human welfare as these insects. Several species, such as aphids (Figure 21–9) and plant lice, cause extensive plant destruction. Others are vectors for viral plant disease agents and for protozoan pathogens of humans and animals. Certain reduviid insects, commonly referred to as "assassin or kissing bugs," attack higher forms of animals, including humans. Chagas' disease (American trypanosomiasis) is transmitted by species of the genera *Panstrongylus* (pan-stron-JEE-luss), *Rhodnius* (rod-NEE-us), and *Triatoma* (try-AT-oh-mah).

The bloodsucking insects vary in size. They are usually black or brown and occasionally have bright red or yellow markings (see Color Photograph 85). The life cycle of the reduviid bug comprises an egg, nymph, and adult stage. Male and female organs are in separate insects. **[See African sleeping sickness in Chapter 29.]**

Fleas. Fleas (Figure 21–10) are bloodsucking ectoparasites (outer-surface parasites) with long,

Figure 21–9 An aphid feeding on a juicy plant. This insect is a transmitter of a variety of plant pathogens. *[Courtesy of W. F. Rochow, Cornell University, Ithaca, N.Y.]*

compressed, wingless bodies ranging in size from 1.5 to 4.0 mm (males are generally smaller than females).

During their life cycle, fleas pass through the egg, larva, pupa (cocoon), and adult stages. The pattern of development is an example of complete metamorphosis: inside silky cocoons, the larvae are transformed into highly complex adult female and male fleas. When pupal development is complete, adult fleas emerge. The larvae feed mainly on nutritive debris, including blood-containing feces of adult fleas. Adults feed on their particular hosts.

Fleas are of medical interest primarily because they are involved in the transmission of plague and endemic typhus. *Xenopsylla cheopis* (the rat flea) is considered the most important vector for both of these diseases. *Nosopsyllus fasciatus,* as well as any other species of flea associated with rats, may also act as common vectors of endemic typhus.

Fleas serve as mechanical vectors for a number of helminths, such as the dog tapeworm *(Dipylidium caninum,* dip-ee-LID-ee-um kay-NINE-um*)* and the rat tapeworm *(Hymenolepis diminuta,* high-MEN-ol-ee-piss dim-IN-oo-tah*)*. Flea infestations also cause skin irritations. The chigger, or burrowing flea (*Tunga* spp.), burrows into the skin of mammals, including humans, and produces intense itching that can lead to ulceration.

Mosquitoes. Several species of mosquito are known to transmit helminthic, protozoan, and viral diseases of humans and lower animals. These insects are probably best known as the vectors for *Plasmodium* (plaz-MOH-dee-um) spp., which cause malaria. Other diseases transmitted to humans by mosquitoes include the virus infections, yellow fever, dengue fever, and Eastern equine encephalitis. Approximately 2,000 mosquito species have been described, many in almost every country.

Figure 21–10 The human flea *(Pulex irritans)*. *[Courtesy of the Bureau of Vector Control, California State Department of Public Health.]*

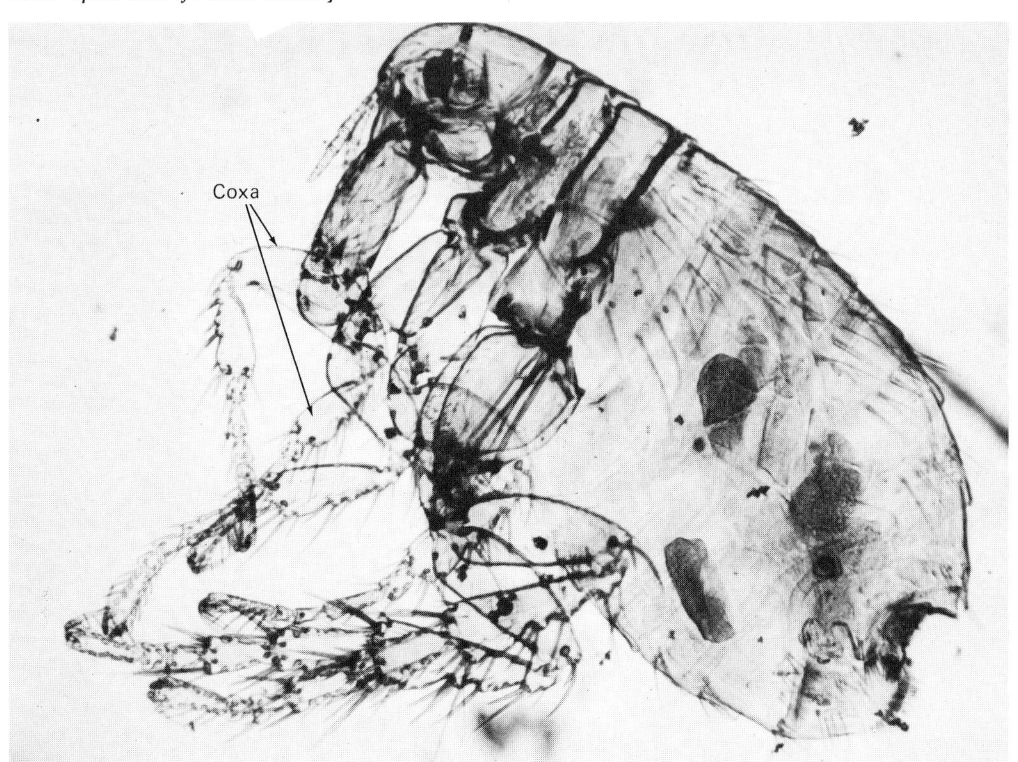

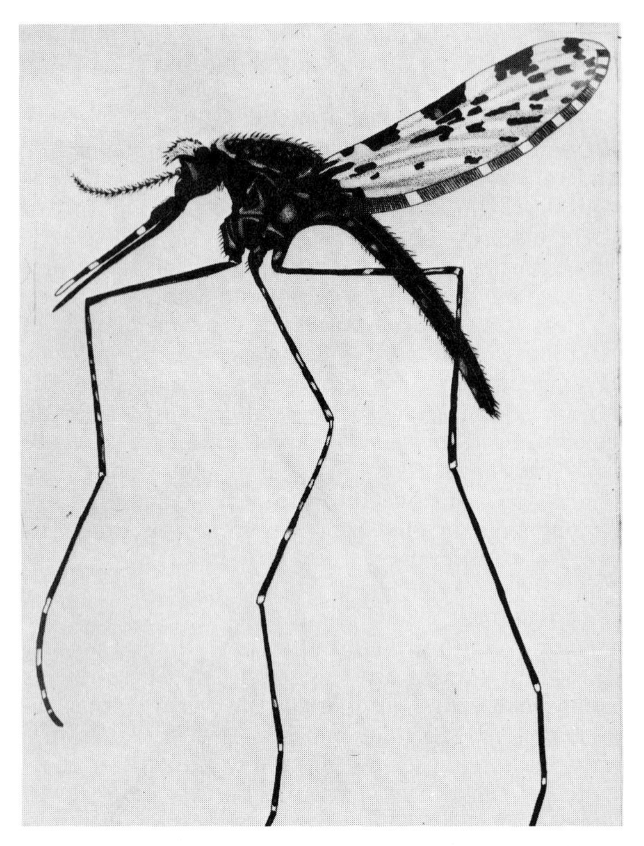

Figure 21–11 One vector of malaria, *Anopheles gambiae.* [*Courtesy of the World Health Organization, Geneva.*]

These "delicate flies of evil reputation" have long legs and slender bodies (Figure 21–11). Other distinguishing features of mosquitoes include (1) the elongated mouth parts (proboscis) of adult females, in most cases adapted for blood-sucking; (2) the bushier (plumose) antennae in the males (Figure 21–12); and (3) the characteristic wing veins and scales.

The life cycle of mosquitoes comprises the egg, larva, pupa, and adult stages. Moisture is a major factor in development of the larval form, or *instar.* Most mosquitoes live in fresh water, but species of certain genera—*Aëdes, Culex,* and *Mansonia,* for example—breed in brackish or salt water. These three genera, plus *Anopheles* and *Haemagogus,* are also the genera mainly involved with transmission of disease agents among humans and lower animals. Generally speaking, the most effective measures for controlling mosquitoes include elimination of breeding sites and the destruction of larval and adult mosquitoes.

COCKROACHES

The German cockroach is believed to be an important vector in the mechanical transmission of several infectious diseases. These include amebic dysentery, hepatitis A (infectious hepatitis), salmonellosis, and shigellosis. Cockroaches are widely distributed and survive under a great variety of conditions.

Figure 21–12 A comparison of anopheline mosquito antennae. Note the jointed nature of these structures. (a) Male antennae. (b) Female antennae.

(a)

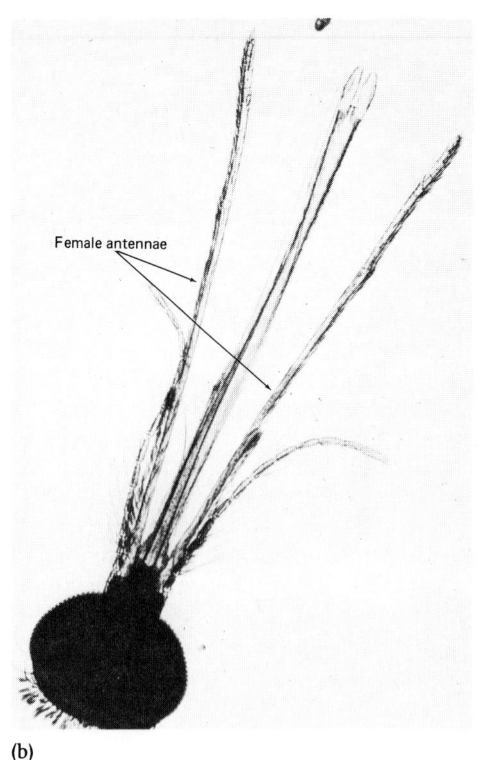

(b)

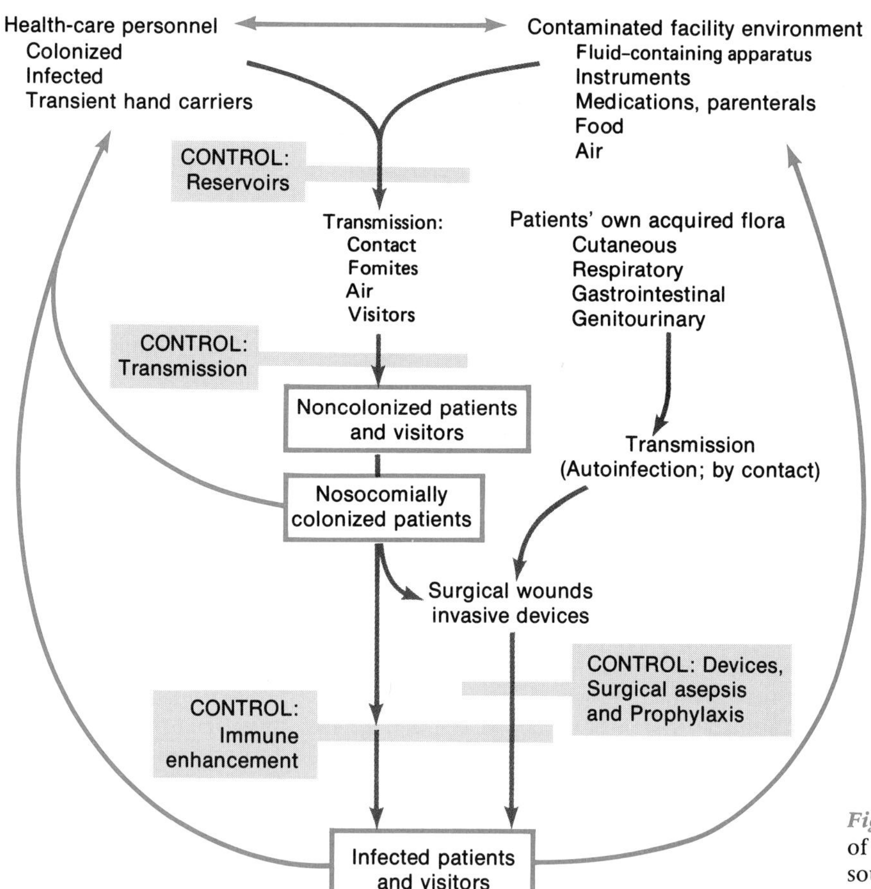

Figure 21–13 The epidemiology of nosocomial infection, including sources of infectious agents and some control measures.

The Hospital Environment

Hospital Infections

The hospital environment is a potential reservoir of infection, for it houses both patients with a variety of pathogenic microorganisms and a large number of susceptible individuals. Today, *nosocomial,* or hospital-acquired, infections pose serious and far-reaching problems. For example, since 1950 marked increases have been noted in bacteremia and deaths caused by staphylococci and gram-negative organisms such as *Escherichia coli* (esh-er-IK-ee-ah KOH-lee), *Enterobacter* (en-te-row-BACK-ter) spp., *Pseudomonas* (soo-doh-MOH-nass) spp., and *Proteus* (PROH-tee-us) spp.

Many factors contribute to the problem of nosocomial infections (Figure 21–13). They include (1) overcrowding and staff shortages in hospitals; (2) the closing of most communicable disease hospitals; (3) the indiscriminate, frequent, and prolonged use of broad-spectrum antibiotics; (4) the tendency toward longer, more complicated surgical procedures; (5) the design of health-care facilities; (6) a false sense of security that has fostered neglect of aseptic techniques; and (7) the use of immunosuppressing agents such as steroids, anticancer drugs, and irradiation. Such practices have provided fertile fields for previously harmless bacteria, which have emerged as the cause of more than 60% of all hospital-acquired infections. It is estimated that approximately 5% of all patients admitted to hospitals for reasons other than infectious states develop nosocomial infections.

Reported sources of contamination in outbreaks of nosocomial infections have included intravenous infusion products, respiratory therapy equipment, stethoscopes, medicinals and lotions, catheters, and shaving brushes used in the preoperative shaving of patients. A study has implicated rolls of adhesive tape, which are exposed to a variety of patients, as potential sources of nosocomial infections. This finding is not surprising, since adhesive tape is used in a variety of ways; used and unused portions of tape are exposed to patient secretions; and contaminated rolls of tape may, in turn, contaminate the hands of personnel.

Hospital personnel are considered to be important sources of infectious agents, as various parts of their bodies and clothes may serve to transport pathogens. In fact, nosocomial diseases may involve a major proportion of the individuals, equipment, and materials with which patients come into contact.

Certain areas of hospitals are considerd to involve an especially high risk regarding the transmission of diseases. Among these areas is the central supply unit, where most equipment used in patient care is cleansed and stored until needed again. Nurseries are another such area, because of the limited resistance of newborn infants to infection. Also, operating rooms and obstetric delivery rooms, burn units, trauma units, and intensive care units are considered to be high-risk areas in disease transmission, since broken skin is a common portal of entry for pathogenic organisms. [See skin infections in Chapter 23.]

MEDICAL DEVICES AND INFECTION

The use of various medical devices for diagnosis or treatment increases the risk of infection. These foreign objects, which are placed in contact with a patient's tissues either temporarily or semipermanently, represent one of the most important factors in the transmission of nosocomial infections. Interestingly enough, device-related infections appear to be the most preventable of the hospital-acquired infections.

Medical devices can bring about infections by (1) damaging or invading skin or membrane barriers to infection; (2) supporting the growth of microorganisms and thus serving as reservoirs of disease agents; (3) interfering with host defense mechanisms; and (4) when contaminated, directly infecting individuals. Table 21–4 lists several types of medical devices associated with infections. Many device-related infections are caused by gram-negative organisms.

Principles of Control

Health-care personnel (and perhaps nurses more than any other category) are responsible for patients 24 hours a day. To safeguard the well-being of patients and prevent the transmission of infectious diseases, health care personnel use the basic measures of medical asepsis and surgical asepsis.

MEDICAL ASEPSIS

The term *medical asepsis* refers to those techniques used to reduce the direct or indirect transmission of pathogenic microorganisms—both by reducing their number and by hindering their transfer from one person or place to another. A variety of techniques are used. These include washing, dusting, disinfection, isolation, and the wearing of gloves and gowns.

SURGICAL ASEPSIS

Surgical asepsis involves those practices that make and keep objects and areas sterile, that is, free from *all* microorganisms. Aseptic techniques are necessary in all surgical or other procedures involving the body's deeper tissues. During injections, for example, a break is made in the body tissues, rendering them more susceptible to infection. Surgical asepsis is also used throughout operating and delivery rooms and in nurseries to protect susceptible newborns. It is applied to the care of surgical wounds for several days following an operation, or until the injured tissues are healed sufficiently.

All materials used during these procedures are sterilized. Certain articles, such as linen or gauze, are obtained from the stock supply of sterile packages. The surgeon and all other personnel involved in the procedure limit the introduction of microorganisms into the operative area by wearing sterile gowns, caps, masks, and rubber gloves, and by washing their hands and associated areas adequately.

Surgical asepsis applies very high standards of sterility to materials used in the hospital. Any object not known to be sterile is assumed to be unsterile; this includes even the outside wrappings of supplies known to be sterile. In addition, sterile objects and packages must be kept dry, since moisture can carry bacteria to the sterile area.

TABLE 21–4 Nosocomial Infections Related to Common and Specialized Medical Devices

Infection	*Related Medical Devices*
Circulatory system, including inflammation of veins	Arterial pressure monitors; intravenous catheters and needles
Deep-wound infections	Artificial hip
Inflammation of heart tissue	Prosthetic heart valve
Eye infection	Prosthetic lens
Eyelid inflammation	Humidifier
Gastrointestinal inflammation	Suction machine
Hepatitis	Kidney machine equipment
Pneumonia	Respiratory equipment
Urinary tract infections	Examination equipment (cystoscope); urinary catheter

Microbiology Highlights

BACTERIAL BIOFILM ECOLOGY

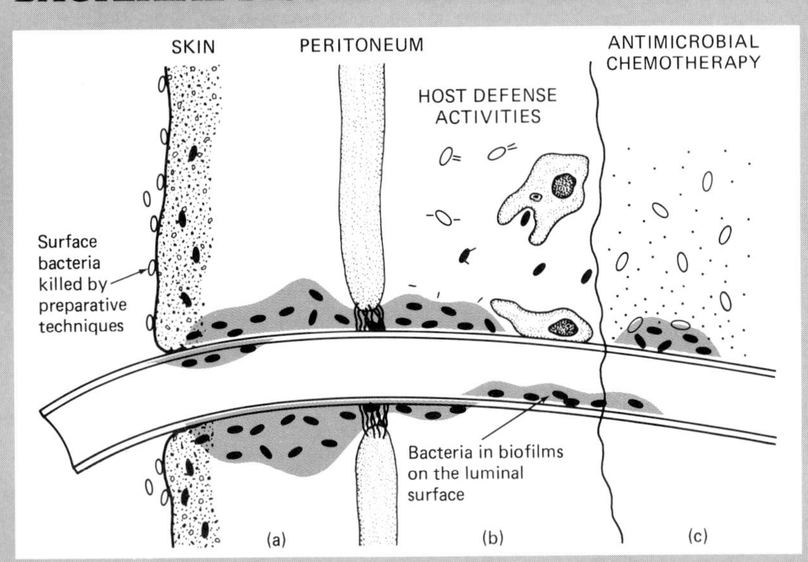

Skin bacteria colonize and are able to grow on implanted transcutaneous devices (a). Bacteria form biofilms and are protected from host defense mechanisms such as phagocytosis or actions of immunoglobulins (b). Antibiotics are effective in killing bacteria that are leaving the biofilms but ineffective in killing bacteria within the biofilms (c).

Several modern medical diagnostic or treatment techniques use devices to provide temporary as well as prolonged access to blood vessels or to body cavities. Since these devices pass through the skin, they are exposed to and, in some cases, eventually occupied by bacteria normally present on and in the layers of skin. The bacteria initiate the accumulation of large amounts of thick mucopolysaccharide material, resulting in the formation of protective *biofilms* (thin layers of metabolizing bacteria sticking to a surface). Such biofilms may develop within and on the surfaces of the transcutaneous devices, allowing bacteria to gather and form microcolonies. More than 240 different types of these devices have been found to be covered by adherent biofilms.

Biofilms, together with the tissue-surface environments, favor bacterial survival. Moreover, activated phagocytes, immunoglobulins, and antibiotics cannot kill established biofilm bacteria. These organisms can also serve as reservoirs of continuing infection. While biofilm bacteria are able to override host defense mechanisms in their protected environment, their protection decreases upon their separation from the biofilm.

In protecting patients from infection, nurses must understand and be able to apply the principles of sterile technique, since a breach in technique may become a threat to the patient's life. Even mild infections delay recovery and are expensive.

Measures of Control

HAND WASHING

Personnel should wash their hands before and after each patient contact, especially with patients considered to be potential sources of infectious agents. All areas of the hands should be well lathered and scrubbed, with special attention to the nails and nail beds. All rinsing should be performed under running water. The washing and rinsing steps should be repeated and followed by adequate drying. A sterile cream or lotion may be applied to prevent chapping. Soaps used for hand washing should contain a bacteriostatic agent, most commonly hexachlorophene. Although the value of this material has been questioned, it is still widely used to control staphylococcal infections.

ISOLATION

The purpose of isolating a patient is to contain an infectious agent within a presecribed area, thus preventing the spread of infection. A patient is placed in isolation for one of two reasons—to prevent the spread of a communicable disease to other persons or to protect an unusually susceptible patient from exposure to disease agents. The decision to isolate a patient may be based on a particular syndrome or microorganism or on the dictates of a specific hospital service, such as pediatrics or geriatrics. In addition, isolation practices are not without controversy. This is especially true in the case of AIDS patients.

Isolation by Microorganism. The more significant communicable diseases that should be considered for isolation are caused by such microorganisms as *Staphylococcus aureus* (staff-ill-oh-COK-us OH-re-us, Color Photograph 4), *Pseudomonas aeruginosa* (soo-doh-MOAN-us ah-roo-jin-OH-sah), group A beta-hemolytic streptococci, *Mycobacterium tuberculosis* (my-koh-back-TIR-ee-um too-ber-koo-LOW-sis), *Treponema pallidum* (tre-poh-NEE-mah PAL-li-dum), *Neisseria meningitidis, Salmonella,* and *Shigella* spp., cytomegalovirus, and hepatitis A and B viruses. Less common but still important organisms include *Bacillus anthracis, Vibrio cholerae* (VIB-ree-oh KOL-er-ee), *Corynebacterium diphtheriae* (ko-ri-nee-back-TI-ree-um dif-THI-ree-ah), *Yersinia pestis* (yer-SIN-ee-ah PES-tis), *Pseudomonas pseudomallei, Actinobacillus mallei* (ak-tin-oh-ba-SILL-us mall-EYE), *Leptospira* (lep-toe-SPY-rah) spp., and the agents of psittacosis (parrot fever) and rabies.

Isolation procedures for patients with communicable diseases are not rigid routines but depend in part on the microorganism involved and its virulence. The procedures are determined by the mode or route by which the organism is transmitted from one person to another, by the location of the microorganism within the host, and by its portal of exit (e.g., feces, wound drainage, or respiration secretions). The types of precautions taken also depend upon the usual portal of entry of the organism into the body (e.g., the skin, gastrointestinal tract, or respiratory tract).

Health-care personnel must explain to the patient, as well as immediate family members and visitors, the reasons for the isolation and the procedures to be followed.

Reverse Isolation. To shield highly susceptible patients from pathogens in the hospital environment, reverse isolation is employed. Such patients include premature infants, organ transplant patients, severely burned patients, leukemia patients, and individuals receiving radiation therapy. The person in reverse isolation is placed in a single room that has been thoroughly cleaned and disinfected prior to his or her admission. Everyone entering the room wears a gown to prevent pathogens from being carried into the room on clothes. No one with a known infection is allowed to enter the room.

If a more strict reverse isolation procedure is needed, the patient may be placed in an isolator, or plastic tent, which provides a germ-free environment. The isolator has a sterile air supply, and only sterile equipment is passed into it through special portholes. Attached to the sides of the plastic tent are rubber gloves with long sleeves. These features permit nursing and other personnel to render adequate care while protecting the patient from exposure to pathogens.

Total Protected Environment. Hospitalization is often necessary for patients whose host defenses are compromised by bone marrow failure, immunodeficiency, severe burns, or malignancy. Such individuals are treated for their underlying disorders as well as for the management of complications that pose a continued risk for the acquisition of infections. Of the methods developed, the *total protected environment (TPE)* has been shown to reduce the incidence of serious infections significantly in severely compromised patients. The goal of TPE is to eliminate the patient's endogenous microbial flora and to prevent colonization by new microorganisms. To achieve this goal, the method incorporates and involves physical patient isolation in a high efficiency particulate air (HEPA) filtered laminar airflow room, disinfection or sterilization of all objects that come into contact with the patient (including the wearing of sterile gowns, masks, gloves, caps, and boots by hospital personnel and patients, use of sterile water and at least semisterile food, in conjunction with decontamination of the gastrointestinal tract, body openings and skin surfaces.) TPE has certain disadvantages, two of which are the great expense it incurs and the heavy demands it places on hospital resources. **[See laminar airflow in Chapter 11.]**

Institutional Policies for Control

Health-care facilities vary greatly in their policies regarding the isolation and control (which includes prevention) of communicable diseases (Figure 21–14). Each institution should put its policies in writing, and complete details of care should be available to all staff members. Some of the generally accepted policies are as follows:

1. Correct hand washing is one of the most important measures in preventing the spread of infection.
2. Isolation gowns should be worn by persons giving direct nursing care to the patient or by persons whose uniforms are likely to come into contact with contaminated material.
3. There is no consistent policy on the use of masks, but it must be remembered that a wet or ill-fitting mask provides little or no protection.
4. Disposable equipment and supplies should be used whenever possible. Nondisposable equipment should be disinfected and/or sterilized when feasible. All used equipment should be disinfected as soon as possible and with a minimum of handling. Such items should be removed from the patient's room for sterilization or, if disposable, for incineration.

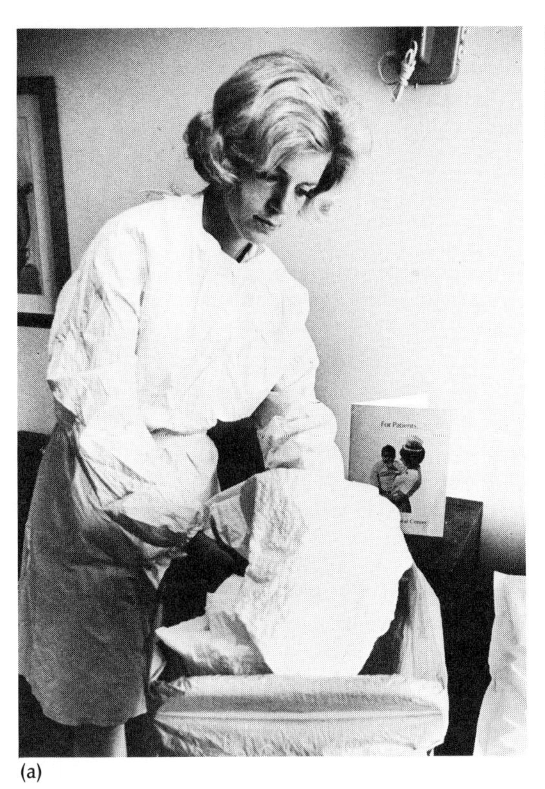

(a)

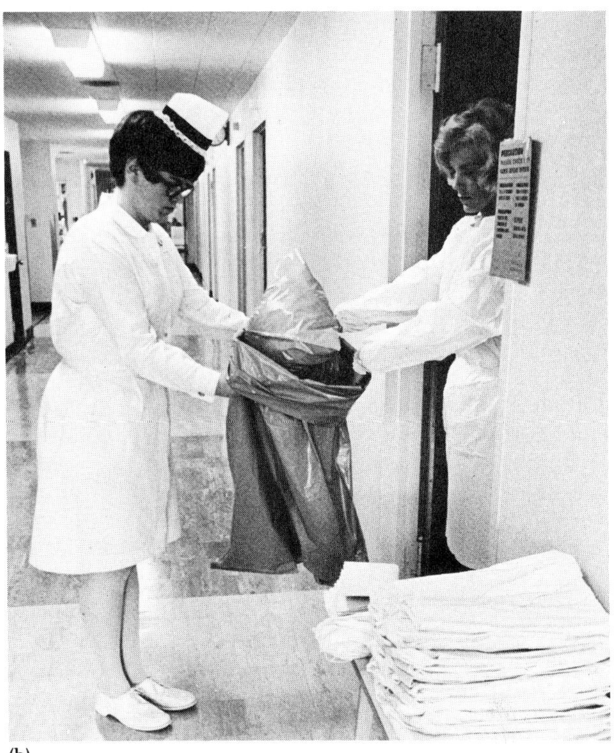

(b)

Figure 21–14 The double-bag technique. (a) A nurse discards disposable paper bed linen and a patient gown in a plastic hamper. The nurse also wears a disposable paper gown, which she deposits in the hamper before leaving the room. (b) The bag is secured at the top and placed in an uncontaminated bag at the unit door for transmittal to the incinerator. *[Courtesy of Andrew McGowan, St. Luke's Hospital Center, New York.]*

5. All contaminated material, including equipment, trash, linen, and specimens, should be removed from the patient's room by a double-bagging technique. In this method, contaminated material is placed in a paper or plastic bag, which is then passed into another clean bag held by someone outside the patient's room. The outside bag is then properly labeled "isolation" and identified as to contents (Figure 21–14b).
6. Terminal disinfection is performed when a patient has recovered, been transferred, or died. This procedure includes sterilizing or disinfecting all possibly contaminated material and equipment, such as mattresses, pillows, furniture, floors, and walls.

Diseases of Plants

The causative agents of plant infections and crop spoilage are for the most part of the same general types as those responsible for animal infections—bacteria, fungi, viruses, and parasitic nematodes (Color Photograph 8). Fortunately, however, none of the organisms affecting plants has thus far been shown to be capable of producing infections in humans or other animals. Plant disease agents are transmitted in a wide variety of ways—for example, by arthropods, through contaminated soil and tools, during grafting procedures, by mechanical inoculation (the rubbing of abrasives on the surfaces of leaves), and by seeds and even certain species of fungi and nematodes. Many factors—such as host resistance, temperature, moisture, and virulence of the disease agent—are important to the development of plant diseases. In many ways, the disease process in plants closely resembles that in animals.

Approaches to Identifying Pathogens

The isolation and identification of an unknown pathogen are extremely important, not only to adequate treatment but also to disease control. Laboratory diagnosis involves the collection and transport of appropriate specimens; prompt microscopic examination of such specimens (whenever practical); the selection and use of culture media in the isolation, identification, and determination of the antibiotic sensitivity of a pathogen; and the use of both specific and nonspecific diagnostic serological tests. In recent years, there has been a trend toward developing simple prepared systems for the rapid identification of microorganisms and toward the automation of various aspects of procedures to decrease the time required for diagnostic tests.

Upon recognizing the clinical symptoms of a particular infectious disease, a physician will request that specimens be taken and sent to the laboratory for processing and examination. With proper handling, the organisms in the specimen can be identified. Attention to details and good communication between members of the allied health-care team are necessary for the most rapid and accurate identification of disease agents. Once the pathogen has been identified, laboratory findings are quickly transmitted to the attending physician.

Protozoa and Helminths

The procedures required for identification of a disease agent vary significantly with the type of organism involved. For example, the examination of feces for helminth (worm) ova and other forms of parasites usually involves preparation of a wet mount of fresh or preserved material. A permanent, stained preparation may be made to assist in identification. The identification of protozoa such as the malarial parasites involves the examination of Wright- or Giemsa-stained blood smears. With extensive searching, a single smear of this type may be sufficient to locate various stages in the parasite's life cycle (Color Photograph 110). Thus, repeated blood samples and examinations are rendered unnecessary. **[See helminth diseases in Chapter 31; malaria in Chapter 27.]**

Fungi

Fungi are inoculated on a selective medium and incubated at room (25°C) and body (37°C) temperatures. Once growth appears, these organisms can usually be differentiated through microscopic examination. Structures such as hyphae and spores, plus the arrangement of spores and other structures, identify the pathogen. Yeasts may require one or several additional steps. For example, *Cryptococcus neoformans* (kryp-toe-KOK-kus nee-oh-FOR-manz) from clinical specimens has a broad capsule, and the presence of the enzyme urease distinguishes it from other yeast cells. This genus is the only one among the pathogenic yeasts that has the enzyme. Yeasts of *Candida* (KAN-did-ah) species, common causes of skin, mucous membrane, and nail infections, are generally differentiated from similar organisms by the formation of pseudomycelia. Here, in a culture of cornmeal Tween 80, cells remain together, forming a chain of yeasts. Thus, species in this genus are identified by the presence of pseudomycelia. Sugar fermentation and oxidation tests confirm the identification.

Bacteria

The isolation and identification of pathogenic bacteria may require relatively few or many different tests and media. For this reason, and also because many laboratories only occasionally encounter parasites and fungi, the differential identification of bacteria is stressed in this chapter. Selected aspects of the collection and handling of specimens, as well as the isolation and identification of microorganisms, are discussed in the following sections. The purpose of this chapter is to provide the reader with an overall view of diagnostic microbiology. Additional details are provided in the chapters that follow. **[See bacterial cultivation in Chapter 4.]**

Viruses

Rapid and accurate viral diagnosis is important in order to provide specific treatment or to take preventable measures to protect a community. Not all viral illnesses require laboratory confirmation. Some are of such a mild nature that the expense of diagnosis usually is not warranted. Other viral diseases are so characteristic in clinical appearance that the diagnosis is evident. Several viral skin diseases including chickenpox, herpes, shingles, measles, and hand-foot-and-mouth infection are of this type. For other virus infections, various methods for diagnosis are used. These include tissue cultures for virus isolation, electron microscopic examination of specimens, detection of viral antigens within infected cells using immunoglobulins tagged with a fluorescent dye molecule or an enzyme such as horseradish peroxidase (Color Photographs 55 and 80), or the detection of extracellular viral antigens in specimens by methods such as ELISA. Most of the possible methods for viral detection and identification are described in greater detail in Chapters 10 and 19.

Collecting and Handling Specimens

Initially, the physician must decide which type of specimen or specimens will provide confirmation of the clinical diagnosis. Next, these specimens are taken from the patient as ordered and sent to the

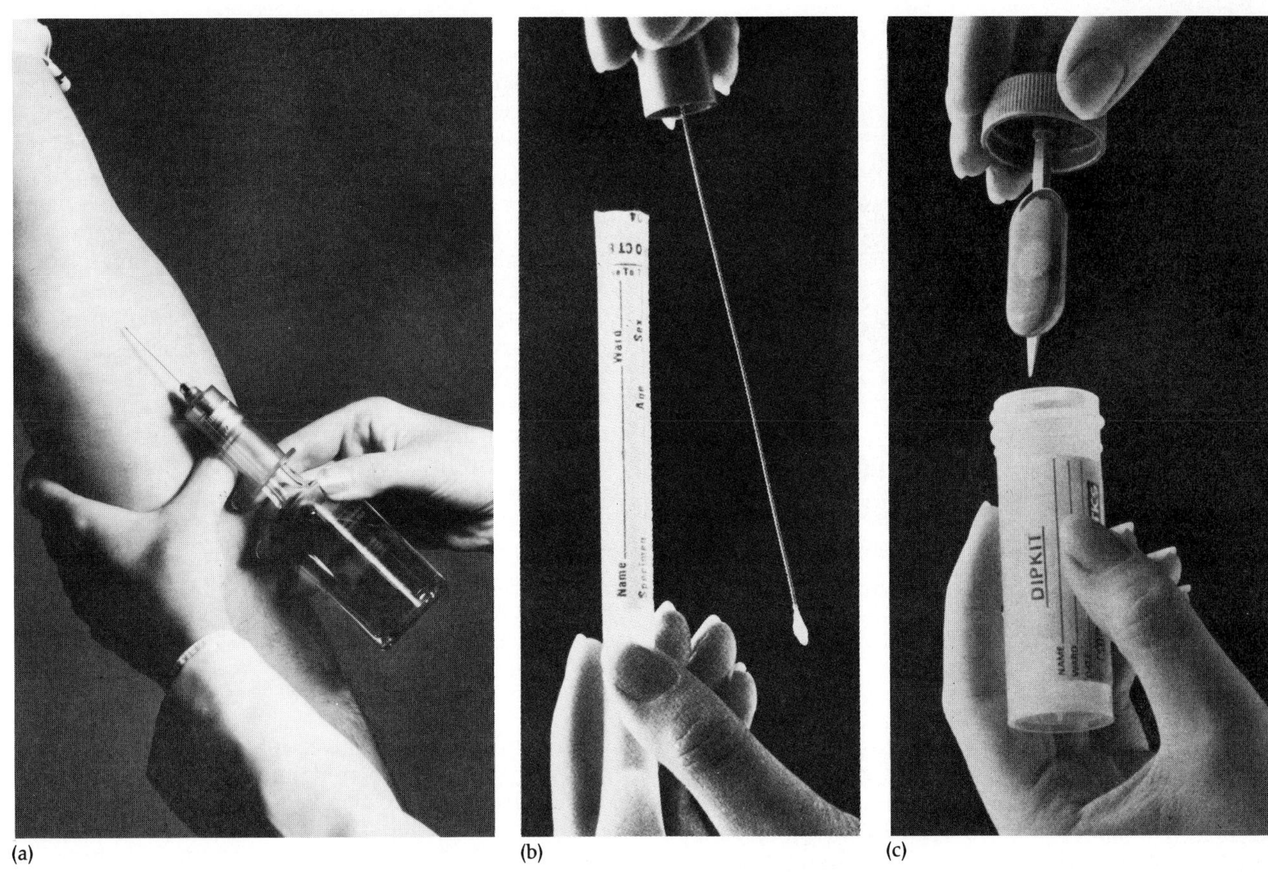

(a) (b) (c)

Figure 21–15 Methods and materials used to obtain specimens for microbial isolation and identification. (a) A Vacutainer blood culturing system. Blood is directly introduced in this device to a culture medium. *[Courtesy Becton Dickinson and Company.]* (b) A sterile cotton/Dacron swab used for the collection and transport of viral specimens. (c) A urine culture transport system. *[21–15b and c courtesy of MicroDiagnostics, Cleveland.]*

lab. Transport involves appropriate precautions to keep the specimen in good condition.

It is a good practice for physicians to indicate the reasons for the specimen on the laboratory tests order slip. Information of this sort can aid laboratory personnel in selecting the media and/or conditions that will make recovery of a suspected pathogen more likely. The proper media and proper conditions are particularly important where certain fastidious, unusual, or slow-growing organisms are involved.

Types of Specimens

In a hospital situation, most specimens are collected by physicians or nurses, but blood samples are generally obtained by medical technologists. Many different types of specimens are used in laboratory diagnosis. Representative specimens include blood, feces, sputum, urine, urethral and vaginal excretions, and cerebrospinal fluid.

The isolation and culture of bacteria from blood requires the aseptic collection of specimens by venous puncture. Current methods of collection and culture may include the direct inoculation of a medium with the specimen (Figure 21–15a).

Stool specimens are generally used for the isolation and detection of microbial pathogens of the intestinal tract. For bacteria, selective and differential media are used to isolate and separate disease agents from the large number of nonpathogens usually present in fecal material. **[See Chapter 4 for a discussion of selective and differential media.]**

Specimens from various parts of the body such as the throat and reproductive system are collected by means of sterile swabs (Figure 21–15b). The swabs may be used directly to inoculate appropriate media or transported to a laboratory for further processing. Sputum—an accumulation of fluid, cells, protein and related solids—is frequently the specimen of choice to isolate pathogens from the lower respiratory tract.

Urinary tract infections may involve all areas of the urinary system. The reliability of diagnosis and identification of the causative agent depends on the collection of clean-voided specimens. Various devices and techniques are used for specimen collection and preliminary diagnosis. **[See Chapter 28.]**

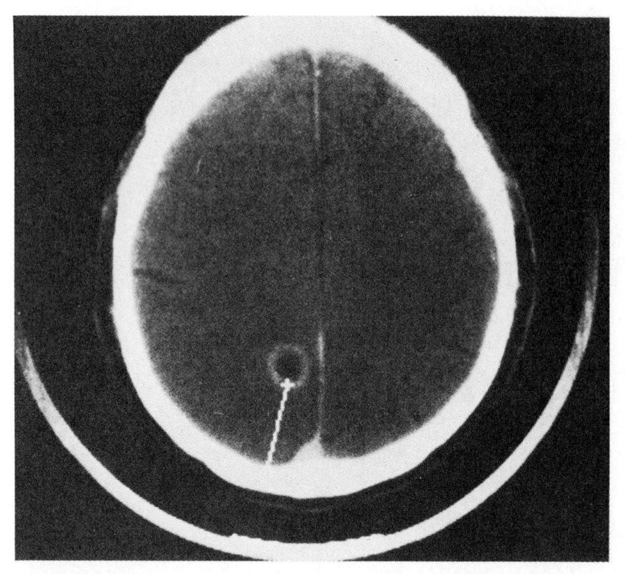

Figure 21–16 A computerized tomography-guided needle used to remove a specimen from a lesion. *[From W. H. Boom, and C. U. Tuazon,* Revs. Infect. Dis. *7:189–199, 1985.]*

In cases of suspected meningitis, the collection and examination of cerebrospinal fluid represent a major emergency and require immediate attention in a clinical laboratory. The reasons for the urgency are that bacterial meningitis is a rapidly fatal disease without adequate treatment, and appropriate treatment depends on the correct identification of the etiologic agent. Lumbar puncture is used for specimen collection.

The introduction of computerized tomograph (CT) scanning has dramatically changed diagnosis and specimen collection in a number of disease states, especially those involving the brain (Figure 21–16). CT scanning is an analytical process by which different levels of the body are swept by a pinpoint-sensitive X-ray beam. CT-guided needle removal (aspiration) of specimens from localized brain infections has greatly increased the quality of bacteriologic diagnosis and improved the delivery of specific antibiotic treatment. Additional information concerning specimens for other pathogens is provided in later chapters.

Transporting and Handling Microbe-containing Specimens

The rapid isolation and subsequent identification of microbial pathogens are important in determining the appropriate chemotherapy for the disease. Most microbial pathogens are fragile and short-lived. Because they deteriorate in transit or under unfavorable conditions, clinical specimens should be shipped to a laboratory by the most rapid means available.

Certain types of specimens are taken with swabs (Figure 21–15b). If these specimens dry out after collection or during transit to the laboratory, they may prove to be unsatisfactory. The introduction of Dacron swabs has greatly improved this situation, since many organisms survive well on dry Dacron. It is also advisable to transport a swab specimen in a tube containing a holding medium: any of several liquid or semisolid media that prevent drying of the pathogenic organisms without allowing overgrowth of normal flora or contaminants.

Most specimens remain suitable for culturing for several hours if they are refrigerated prior to inoculation. However, bacteria such as *Neisseria meningitidis* are sensitive to the cold, so a cerebrospinal fluid specimen suspected of containing this organism must be cultured immediately. Wound specimens should also be processed immediately, under anaerobic conditions, since any anaerobes present die rapidly upon exposure to oxygen. Refrigeration is generally adequate for urine, stools (feces), and sputum. Sputum collection is best done in an appropriate collection kit (Figure 21–17).

Figure 21–17 A sputum collection kit. A tight-fitting lid covers a funnel into which sputum is coughed. The funnel directs the specimen into a graduated, threaded plastic centrifuge tube. The screw cap is shown near the bottom of the tube. To process the sputum for culture, the clinician merely removes the bottom cap and detaches the centrifuge tube from the screw cap. *[Courtesy of Falcon Plastics, Division of BioQuest, Los Angeles.]*

Microbiology Highlights

FINE-NEEDLE-ASPIRATION BIOPSY

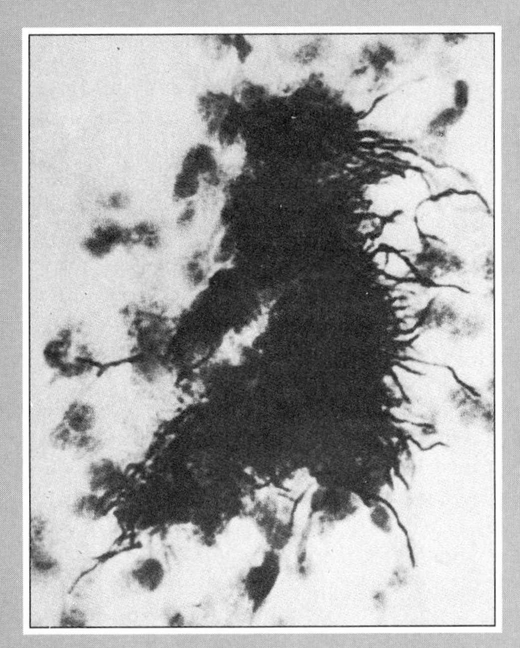

A gram-stained aspirate from an abscessed foot, showing the consistent features of *Actinomadura*. *[From K. Bottles,* et al., West J. Med. ***144****:695, 1986.]*

Fine-needle-aspiration biopsy (FNAB) is a method of obtaining specimens, without anesthesia, from inflamed masses in the body. The technique uses small-sized hypodermic needles and is useful for the diagnosis not only of various cancers but also of various infectious diseases, including tuberculosis and other mycobacterial infections, abscesses caused by staphylococci and *Actinomadura* (ak-tin-OH-ma-dur-ah) species, sheep tapeworm, and San Joaquin Valley fever (coccidioidomycosis). Diagnosis can be established from the results of a stained smear and/or culture of the aspirated biopsy material.

Samples of several types of clinical specimens (throat swabs, feces, blood serum) can be rapidly frozen and kept in this state for later viral examinations. Frozen samples may also be shipped in dry ice to a reference laboratory for further study. Serum for such viral or other serological testing must be separated from the whole blood and sent in a sterile tube. Whole blood, if frozen, proves unsatisfactory for serological tests. Clotted blood for examination should be refrigerated or transported in ice. Because some parasitic forms degenerate rapidly, stool specimens must be placed in preservative solutions to ensure that the organism will be maintained in a recognizable form.

Since specimens containing pathogens pose a clear danger to human life, they must be packaged properly whether they are to be transported by commercial aircraft or by other means. Adherence to all requirements specified by current federal guidelines is essential to the safeguarding of all persons.

This brief discussion has barely touched on the problems of proper collection and handling procedures. Because many microorganisms have unique requirements, laboratory personnel must be familiar with the specific organisms in order to find them in clinical specimens.

Laboratory Procedures

A small clinical laboratory that processes only a few specimens per day generally can be academic in its approach to identifying microorganisms. Each specimen can be examined soon after its arrival and, depending on the type of organism observed in the wet mount or stained smear, an appropriate isolation medium can be selected and inoculated. In examining the growth that appears, an experienced microbiologist may note several important clues to the identity of the suspected disease agent. Included in this group of properties are odor, colonial appearance, and staining reactions, among others (Color Photographs 11–16 and 19–26).

As the volume of specimens in a laboratory increases, the amount of time that can be spent on each decreases. In this case, the laboratory must adopt standard procedures. Often, specimen smears cannot be stained and observed soon enough to allow the selection of appropriate isolation media. In such situations, media must be chosen for the isolation of significant pathogens usually associated with a particular type of specimen (Color Photographs 23–26).

Media

A wide variety of media are available to assist the microbiologist. However, these media should not be used as a substitute for careful observation. Because of the personnel time required in making media and pouring plates, many laboratories either buy prepared media or use devices such as that shown in Figure 21–18. Such devices pour and stack Petri plates for use in a much shorter time than required by manual procedures. Computerized inoculating systems also are in use (Color Photograph 86a). The fact that a certain organism grows and produces a characteristic colony on a medium is only presumptive evidence of its identity. Biochemical testing and sometimes serotyping may be necessary for adequate identification.

MULTIPLE-TEST AND RAPID-METHOD SYSTEMS

The identification of bacterial species is based on schemes or keys that take into consideration a variety of characteristics, such as differential staining reactions and colonial properties. Biochemical tests determine the ability of a particular organism to utilize or attack certain substances and to produce chemical products that can be analyzed. Traditionally, the use of a variety of well-chosen biochemical tests has offered the best means of specific or near-specific identification of unknown bacterial cultures. References and identification schemes describe the results of key reactions obtained with known bacterial species.

In many cases, the biochemical tests used to identify unknown cultures are performed in separate Petri plates and/or test tubes (Color Photographs 25 through 28). These procedures are not only costly but also time-consuming. Efforts to minimize both the expense and the routine drudgery of microbiological methods have led to the development of several types of improved biochemical testing procedures and materials. These innovations include:

1. Combinations of several test substrates in one or two tubes. These substrates are inoculated and incubated in the conventional manner. Triple sugar iron agar (Color Photograph 28) is used for

Figure 21–18 Automation in microbiology. Various routine procedures and manual methods in microbiology can be automated. (a) Equipment used to help prepare plate media. This device pours and stacks filled Petri plates for immediate use in a much shorter time than manual preparation takes.

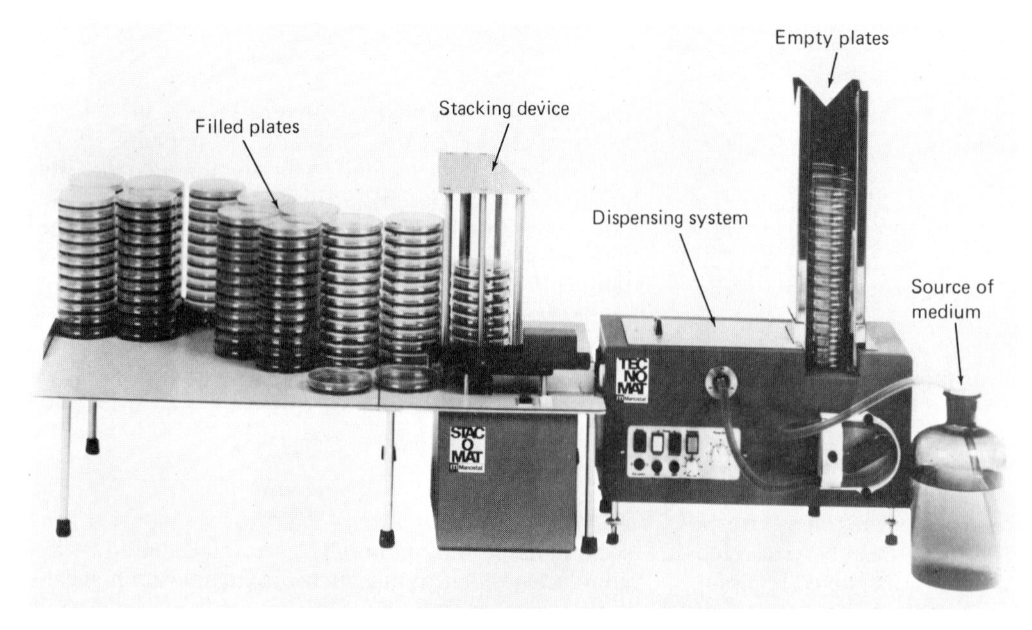

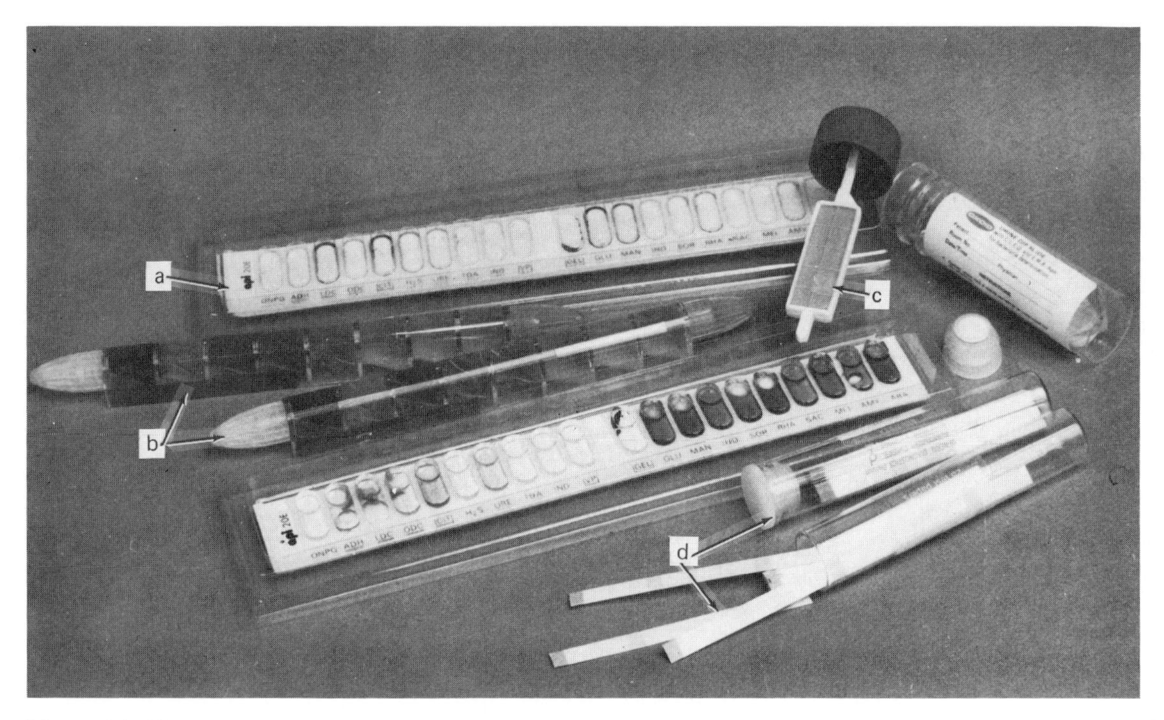

Figure 21–19 Representative examples of multiple rapid-test systems used in the identification of unknown bacterial cultures. (a) APl Enteric 20. *[Analytab Products Inc.]* (b) Enterotube. *[Roche Diagnostics.]* (c) Urine dipslide. *[Oxoid, Ltd.]* (d) Pathotec strips. *[Photo by C. Righter.]*

the differentiation of gram-negative enteric organisms by their ability to ferment dextrose (glucose), lactose, or sucrose and to reduce sulfites to sulfides. It is dispensed in the form of an agar slant.

2. Miniaturized, multicompartmental devices that perform separate biochemical tests. These devices are inoculated by unconventional methods but incubated according to standard practices (Color Photographs 86b and 88). Figure 21–19 shows several commercially available miniaturized multiple-test systems and devices. Complete instructions and identification keys are provided by the manufacturers. Several systems consist of a plastic strip that holds 20 or more miniaturized compartments, or capsules, each containing a dehydrated substrate for a different test (Color Photograph 86b). The dehydrated substrates are inoculated with a bacterial suspension and subsequently incubated according to a procedure described by the manufacturer.
3. Paper disks impregnated with biochemical substrates. These are inoculated by unconventional methods and produce reactions in significantly less time than do conventional tests.

The systems and techniques described here are intended to make the traditional biochemical approach for identifying unknown bacterial cultures more convenient. Each has certain advantages and disadvantages, and these should be weighed before a decision is made to adopt a particular system. To simplify identification, various keys and outlines have been devised that can involve the use of automation and computer analysis, such as the system shown in Figure 21–20. The identification of gram-negative enteric rods, fermenters and nonfermenters, anaerobes, and yeasts, as well as the determination of antibiotic sensitivities, can be obtained with the currently available automated and computerized systems. **[See antibiotic testing in Chapter 12.]**

Anaerobic Identification

Anaerobic bacteria are capable of causing every type of infection associated with aerobes or facultative (adaptable) anaerobes. Strict anaerobic bacteria may be overlooked unless appropriate isolation and anaerobic culture techniques are used. Various approaches to their identification can be found in clinical laboratories, ranging from the detection of metabolic products by gas-liquid chromatography to rapid biochemical test systems (Figure 21–21). Most anaerobes can be identified on the basis of cultural properties, cellular morphology, and a number of biochemical tests.

The Blood Culture

Bacteremia, the presence of bacteria in blood, may be of clinical importance. Various microorganisms

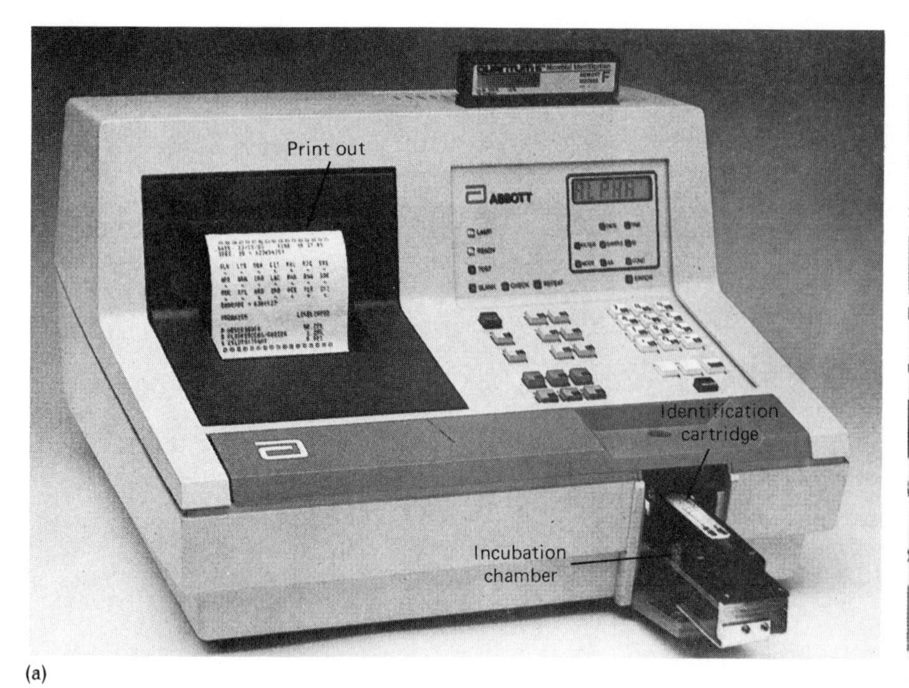

(a)

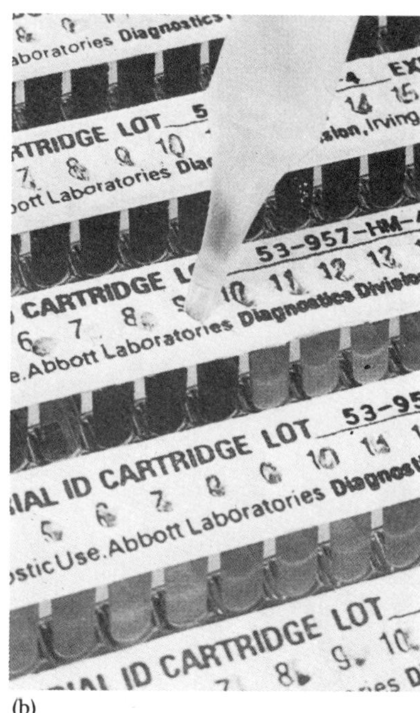

(b)

Figure 21–20 The Quantum II, one example of a functional microbial identification system. This system is capable of automatically interpreting a range of 20 biochemical tests and printing out the most likely identification. (a) The system and the location of the incubation chamber. (b) Individual compartments of an identification cartridge containing several biochemical substrates. In the procedure these are inoculated with the unknown bacterial specimen. The cartridge is then incubated for four to five hours and analyzed for test results and identification in the computerized system. *[After Abbott Laboratories Protocol Diagnostics Division.]*

gain access to the circulatory system, thus spreading from a diseased area by direct extension. Once the bloodstream is invaded by pathogens, potentially any or eventually all organ systems can become involved. The speed with which a positive blood culture is recognized and the attending physician notified is of critical importance to proper patient treatment. When a patient exhibits an elevated temperature that is unexplainable on a clinical basis, blood cultures are usually taken. Although tech-

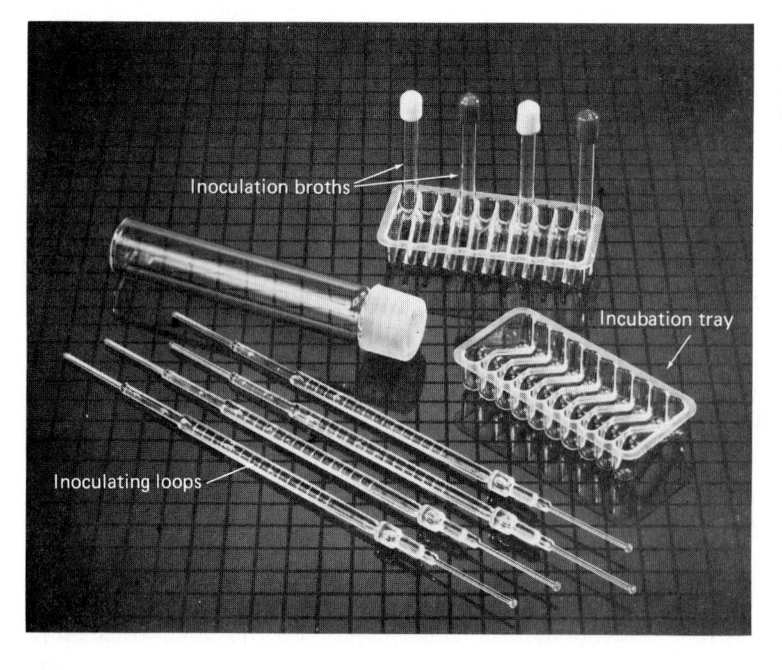

Figure 21–21 The components of the ABL rapid (two-hour) identification system for anaerobes. Such systems are accurate and cost-effective. *[Courtesy Curtin Matheson Scientific, Houston, Texas.]*

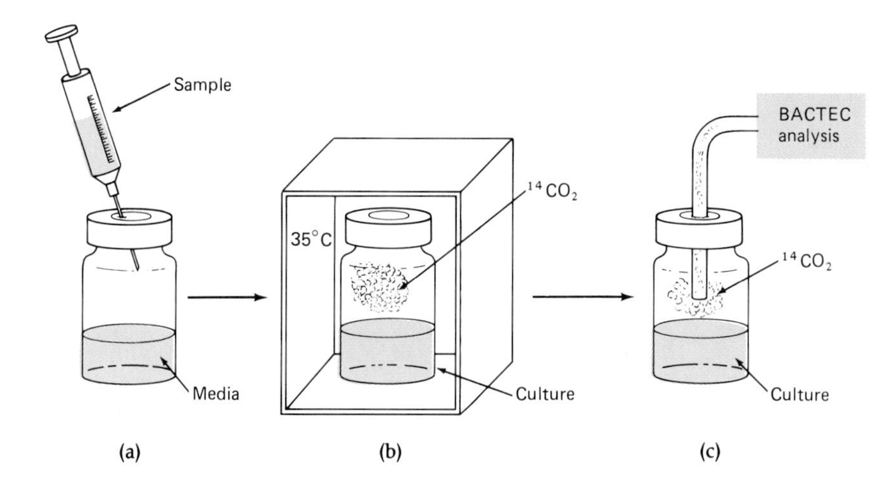

Figure 21–22
The BACTEC principle. (a) A specimen is inoculated into a culture medium containing ^{14}C substrate. (b) Bacterial metabolism produces ^{14}CO by exploiting the organisms' own use of the ^{14}C-containing substrate. (c) The BACTEC instrument (Figure 21–23) is used to measure the released $^{14}CO_2$.

niques vary, in one common practice three blood specimens are taken at approximately two-hour intervals and used to inoculate appropriate culture media. Careful preparation of the skin before obtaining the blood specimens is mandatory. Although many antiseptic preparations are available, the use of tincture of iodine is still probably the best choice for adequate skin disinfection. After this solution is applied, it is allowed to dry and then wiped off with isopropyl alcohol. Proper technique at this point minimizes the possibility of contamination of the blood specimen by skin bacteria.

MEDIA

A variety of media are available for blood culture; these include Trypticase Soy and Tryptic Soy broths, Thiol, and Liquoid. The growth-inhibiting effects of natural sera, various antimicrobial compounds, and antibiotics can be reduced or even eliminated through dilution of the blood specimen, accomplished when the blood is added to the broth medium in a ratio of 1:10. Some microbiologists believe that adding **anticoagulants** to the media enhances the isolation of organisms. Certain compounds may also be added to neutralize antibiotics that a patient may have received prior to collection of the blood specimen.

EXAMINATION

In the case of bacterial infections, blood cultures must be examined daily for the presence of growth or any indication of a microorganism's presence. If growth is detected, subcultures should be made with fresh media. These subcultures are incubated under aerobic and anaerobic conditions, and microscopic examinations of the positive cultures are made. At this point, the attending physician is notified so that he or she can evaluate the patient's treatment in light of this new information. Any additional information, including tentative identification of the isolated organism and antibiotic sensitivity, should be passed on to the physician as quickly as possible. The examination procedure outlined above is critical to the analysis of blood cultures.

Alternate methods for analyzing blood cultures are being used. Some involve the use of membrane filtration. In a more exotic procedure, liquid culture media are used that incorporate radioactive carbon (^{14}C). Bacterial metabolic activity releases radioactive carbon as $^{14}CO_2$. A special sensing device can detect the gas in this form. Figure 21–22 illustrates the principle of a radiometric BACTEC instrument available commercially, and Figure 21–23 shows the BACTEC system. This device is a fully automated model that analyzes multiple cultures in a controlled environment.

Figure 21–23 The BACTEC 225 automated system. *[Courtesy Johnston Laboratories, Cockeysville, Md.]*

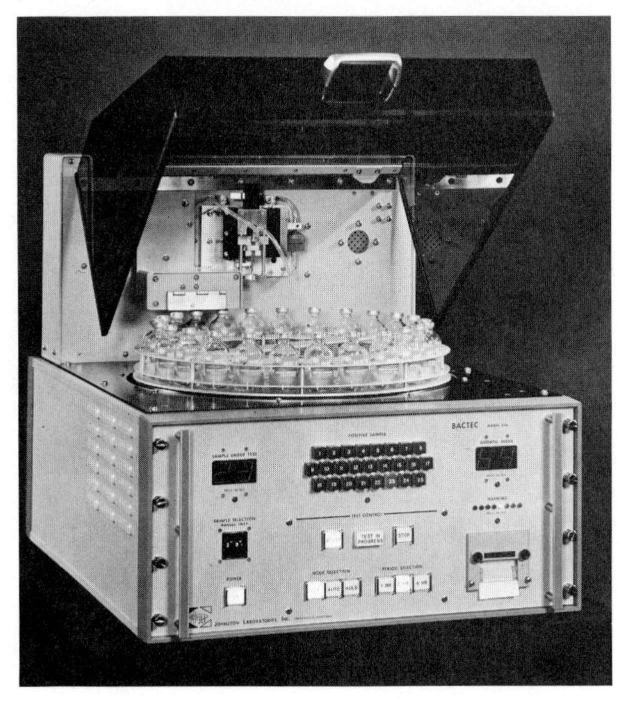

General Identification Procedures

Familiarity with the pathogens most likely to cause particular clinical symptoms and with the organisms most likely to be present in a certain specimen is important to good clinical microbiology. Other chapters of this book cover the various microorganisms encountered in the disease states and associated specimens. This chapter describes the acceptable techniques associated with various specimens.

Each specimen must be handled aseptically. After the specimen is used for the inoculation of appropriate media, one or more smears should be made of it for microscopic examination. Smears or wet mounts are usually omitted with fecal specimens, except when examination for worm ova or parasites is requested. However, some clinical microbiologists recommend the routine examination of a Gram stain of feces or rectal swabs. The purpose of this practice is to acquaint the medical technologist with the typical assortment of gram-negative and gram-positive organisms so that he or she will recognize an atypical assortment that might be diagnostic. For example, this would be important in cases of staphylococcal enterocolitis or *Clostridium perfringens* (klos-TRI-dee-um per-FRIN-jens) food poisoning. **[See Gram stain in Chapter 3.]**

A Gram stain of sputum is routinely prepared. A second smear for acid-fast staining may disclose an undiagnosed case of tuberculosis. Staining procedures in microbiology are becoming mechanized. Machines can perform any of the common procedures with considerable savings in personnel time. Once an unknown organism has been grown, the preliminary examination of colonies and/or the results of certain biochemical tests may suggest a particular pathogen. At this point, immunodiagnostic or serological testing can be used to save a great deal of time. Latex agglutination (Color Photograph 77) immunofluorescence tests and others may be of value. Commercially available rapid-identification kits are also in common use (Figure 21–24). Chapter 19 contains descriptions of several useful diagnostic tests.

Identification Keys

The accurate diagnosis of an infectious disease may require the isolation and identification of the pathogen, the finding of the disease agent's antigens in tissues, or the demonstration of increases in the levels of immunoglobulins specific for the pathogen. Most routine clinical procedures are designed to effectively screen and isolate suspected pathogens

Figure 21–24 One example of a rapid indentification system for streptococci. All the materials needed to begin identification are included in a small unit. *[Courtesy Marion Scientific.]*

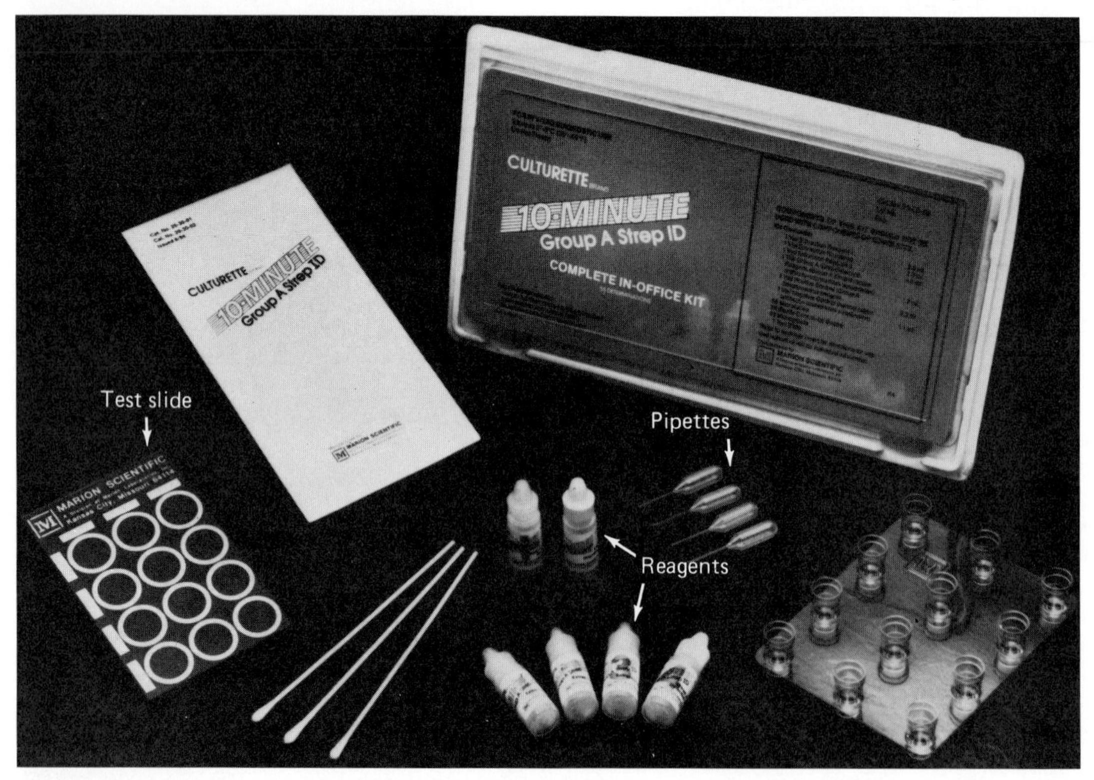

TABLE 21–5 Examples of Differential Tests Used in Bacterial Identification

Test	*Purpose*	*Positive Result*	*Application(s)*
Bacitracin sensitivity	Detects sensitivity to the antibiotic bacitracin	Inhibition of growth (Color Photograph 87)	Group A beta hemolytic streptococci show positive reaction
Blood agar	Detects enzymatic breakdown of hemoglobin	Alpha hemolysis—green zone; beta hemolysis—clear zone; nonhemolytic—no zone (Color Photographs 23a, 23b, and 23c)	Distinguishes among various bacterial species
CAMP (Christie, Atkins, and Munch-Peterson) factor	Detects heat-stable, extracellular streptococcal protein	Presence of factor intensifies beta hemolytic reaction of *Staphylococcus aureus* (Color Photograph 23b)	Group B streptococci produce a positive reaction
Catalase test	Detects the enzyme catalase	Formation of oxygen bubbles on addition of hydrogen peroxide to culture (Color Photograph 96)	Staphylococci and micrococci produce positive reaction; separates these organisms from streptococci
Coagulase test	Detects production of the enzyme coagulase	Coagulation of citrated plasma by culture (Color Photograph 88)	Pathogenic *S. aureus* produces positive reaction
Mannitol-salt agar	Detects mannitol fermentation	Acid (yellow color) production on the agar medium	*S. aureus* produces a positive reaction
Optochin test	Detects sensitivity to optochin reagent	Growth inhibition (Color Photograph 89)	*Streptococcus pneumoniae* is sensitive to reagent
Oxidase test	Detects presence of iron-containing enzyme that reduces oxygen	Color change of culture when exposed to oxidase reagent	Aids in identification of *Neisseria* and *Pseudomonas* species

from the appropriate specimens that are obtained from patients. Examples of enriched and differential media used for bacterial isolation together with various bacterial cultural characteristics are described in Chapter 4. Once the unknown organism is isolated, a number of tests can be performed to identify it. Certain characteristics, including Gram-stain results, colonial and cellular shapes, hemolytic reactions on blood agar, and biochemical activities, are useful in making a clinical identification. Most useful identification procedures are based on a process of elimination with each phase or step in the approach. The number of possibilities is reduced until only one possibility remains, thus revealing the pathogen's identity. Both microscopic observations and biochemical testing (Table 21–5) are major parts of conventional identification approaches. Test results for the unknown are compared with established test patterns for a large number of known organisms to find the closest fit. A match between the patterns determines the unknown organism's identification. Figure 21–25 shows the sequence of steps of an identification key or flowchart applicable to

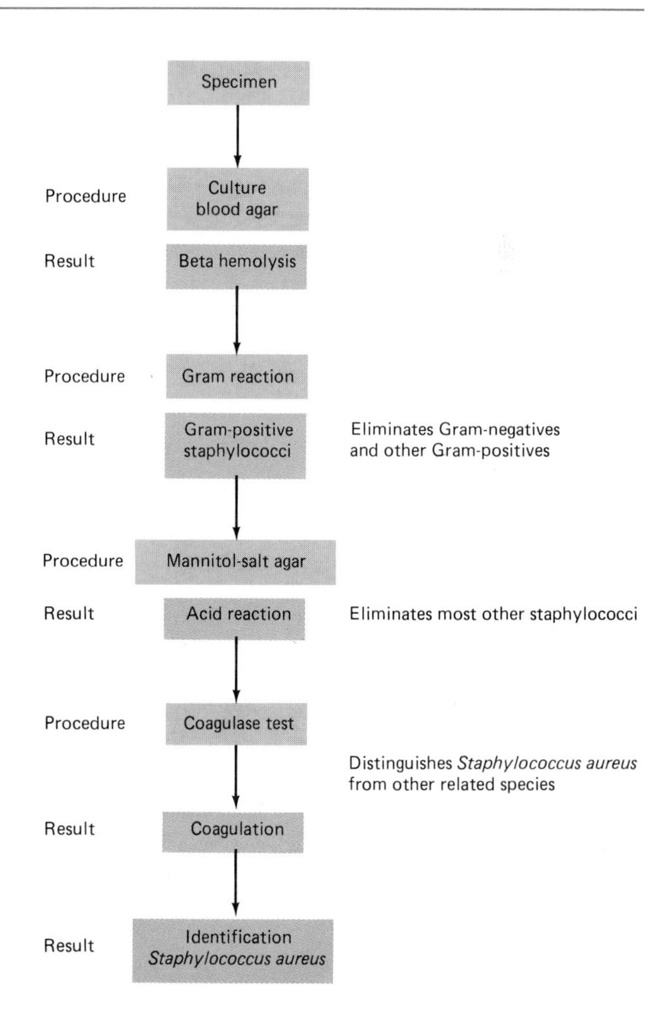

Figure 21–25 An identification key to uncover the causative agent of boils, a skin infection. With each step in the procedure, the possibilities narrow until eventually the identity of the unknown pathogen is established. [See Color Photographs 23b, 12a, 26, and 88 for the appearance of specific reactions.]

finding the bacterial etiologic agent of a case of boils. With additional testing, the pathogen's antibiotic susceptibilities can also be determined. **[See Chapter 12.]** Specific sections of Chapters 23 through 29 describe the approaches and tests used for the identification of other infectious disease agents.

Quality Control Considerations

The interpretation of any test result depends upon the investigator's confidence in the quality of the materials used and in the personnel who perform the test. To ensure high quality, all media reagents and staining solutions should be frequently checked under suitable control conditions. For example, one cannot rely upon the interpretation of a Gram stain unless one has performed the procedure with the same reagents on known gram-positive and gram-negative organisms. Prepared smears of a *Bacillus* (bah-SIL-lus) species and a *Neisseria* (nye-SE-ree-ah) species are good controls, since these organisms are particularly sensitive to deviations in technique. Federal regulations prescribe quality control tests, tests which all laboratories engaged in interstate commerce must perform (Table 21–6).

Periodically, laboratory personnel should be required to identify microbial unknowns to test their technical ability. The College of American Pathologists and the American Society for Clinical Pathologists have developed a survey program for this purpose, and individualized programs have been developed through the cooperative efforts of local and regional laboratories of universities and public health departments.

Periodic internal controls are also important. For example, routine specimens might be submitted to the clinical laboratory by the chief microbiologist in cooperation with hospital staff physicians. The specimen submitted would not differ in appearance from many others, but it would be specially designed to familiarize the technologists with unusual organisms or pathogens not often encountered—for example, *Pasteurella multocida* (pass-tour-EL-la mul-TOE-see-dah), *Brucella* (broo-SEL-lah) spp., or *Bordetella pertussis* (bor-de-TEL-lah per-TUSS-sis).

One important aspect of quality control is the preventive maintenance of equipment and the routine monitoring of equipment performance. Centrifuges must be checked periodically to see that operating speeds are accurate. Also, the minimum and maximum temperatures of all incubators, refrigerators, freezers, water baths, and thermal blocks should be monitored frequently. Significant deviations from the norm usually indicate a pending failure of the control mechanism. Such a failure could result in the cooking of cultures or spoilage of expensive frozen serological reagents.

Host Invaders

The human body can serve as a breeding ground for a wide variety of parasites—organisms that can live on or within such a host and derive nutrients from it. Chapter 22 describes the factors used by various organisms to invade a host, occupy sites within the host, and cause injury and sometimes death.

TABLE 21–6 Quality Control Required for Laboratories Engaged in Interstate Commerce

Item	*Particular Requirements*
Laboratory manual	1. Keep current. 2. Note dates on which procedures go into effect. 3. List complete references. 4. List criteria for quality control.
Records	1. Show any and all changes in procedure. 2. Give evidence of monitoring of materials and methods. 3. Indicate remedial action when necessary.
Stains	1. Incorporate procedures for control.
Media	1. Incorporate procedures for testing prior to or concurrent with use.
Serology	1. Test positive controls. 2. Test negative controls. 3. Test selected weak or variable controls.

Summary

1. Humans, other animals, and plants can be successfully parasitized by a variety of microorganisms, worms, mites, and ticks.
2. Essential to the survival of disease agents such as pathogens is their transmission from sources of infection to susceptible hosts.
3. *Epidemiology* is the study of the distribution and causes of disease, both infectious and noninfectious. Epidemiologists analyze data from many individuals in order to reach conclusions about the presence or absence of a particular disease in a given group.
4. The purposes of epidemiological investigations include determining the effectiveness of methods used to control disease, describing the natural course or history of a disease, classifying diseases, and obtaining information necessary for the planning and evaluation of health care.

Infection and Disease

Several terms are used to describe the prevalence of diseases: (1) *endemic,* which refers to a disease that is constantly present but which involves a relatively small number of victims; (2) *epidemic,* an unusual occurrence of a disease involving large segments of the population for a limited time period; (3) *pandemic,* a series of epidemics affecting several areas of the world; and (4) *sporadic,* which describes diseases that appear at unusual or irregular periods.

Methods of Epidemiology

1. Epidemiological studies may be either *descriptive, analytic,* or *experimental.*
2. Descriptive studies determine the rate of occurrence of a disease, the kinds of individuals suffering from it, and where and when it occurs.
3. Analytic studies are used to test hypotheses concerning factors believed to determine susceptibility to disease. Such studies are designed to show if particular events or environmental conditions are responsible.

Morbidity and Mortality Rates

1. *Morbidity* refers to the number of individuals per 100,000 population having a disease within a given time period.
2. The *mortality rate* is the number of deaths due to a particular disease per 1,000 individuals within a given time period.

Reporting Communicable/Infectious Diseases

State administrative codes require that actual and suspected cases of certain diseases be reported to local health authorities. Variations exist among states regarding specific diseases that should be reported.

Sources and Reservoirs of Infection

1. Sources of infectious agents are many and varied.
2. The most disabling and common human infections are caused by microbes capable of using human tissues for their own needs and thus making these tissues into sources of infectious material.
3. Individuals harboring disease agents without any ill effects are called *healthy carriers.*
4. Persons in an incubating state but still without symptoms are called *incubatory carriers.*
5. Individuals recovering from an infection and secreting disease agents in body fluids are referred to as *convalescent carriers.*

Zoonoses

1. Diseases that primarily affect lower animals but that can also be transmitted to humans by natural means are called *zoonoses.*
2. Zoonotic diseases may be grouped according to the major reservoir of the infectious agent and the mode of transmission among the infectious agent's natural hosts.

Principal Modes for Transfer of Infectious Disease Agents

1. Infectious disease agents can be transmitted to susceptible individuals by (1) direct contact, (2) indirect contact with contaminated food or inanimate objects (fomites), (3) inhalation of airborne dust or other particles, (4) inoculation, and (5) arthropods.
2. Arthropods, which include fleas, ticks, and mosquitoes, transmit disease agents either mechanically or biologically. Arthropods that serve as both host and reservoir for pathogens are called *biological vectors.*
3. *Mechanical transmission* is the simple physical transporting of a pathogen from one location to the next. In *biological transmission,* a portion of the pathogen's life cycle is carried out in the transporting body. These two forms of transmission are not limited to arthropods.
4. Disease agents can be mechanically transmitted by (a) fingers, (b) flies, (c) contaminated foods and water, (d) feces (waste materials), and (e) fomites.

DIRECT CONTACT

Individuals can acquire disease agents through direct contact with open sores and other lesions, through shaking hands, kissing, and sexual activities.

INDIRECT CONTACT

Various microorganisms can be transferred by contaminated food, water, and fomites.

AIR- AND DUST-BORNE INFECTIONS

Individuals can release particle-bearing microorganisms into the environment by normal activities involving the respiratory system such as talking, sneezing, and coughing. Another source of release is the redistribution of existing particles in room dust.

FOMITES

1. *Fomites* are inanimate objects or substances capable of absorbing and transferring pathogens.
2. Examples of potential fomites include eating utensils, toys, instruments, and soiled bed linens.

ACCIDENTAL INOCULATION

Infections can develop through the direct inoculation of pathogens into the body by means of contaminated needles and inoculating instruments.

ARTHROPODS AND DISEASE

1. Many arthropods can produce severe injuries to several body organs, cause allergies, and transmit disease agents.
2. Arthropods involved with the transmission of disease agents include ticks, mites, lice, fleas, mosquitoes, and cockroaches.

The Hospital Environment

HOSPITAL INFECTIONS

1. The hospital environment is a potential reservoir of infection.
2. Hospital-acquired, or *nosocomial,* infections are the result of several factors, including (a) patient overcrowding, (b) staff shortages, (c) indiscriminate and improper usage of antibiotics, (d) longer and more complicated surgical procedures, and (e) the use of such immunosuppressive agents as anticancer drugs and irradiation.
3. The use of medical equipment and devices, either for diagnosis or treatment, increases the risk of infection.
4. The *total protected environment (TPE)* is an effective but expensive approach to protecting compromised patients from infection.

PRINCIPLES OF CONTROL

Hospital-acquired infections can be controlled through the use of techniques that reduce the direct or indirect transmission of pathogens *(medical asepsis),* and practices that make and keep objects and areas involved in surgical procedures sterile *(surgical asepsis).*

Diseases of Plants

1. The causative agents of plant infections and crop spoilage are similar to agents of diseases in animals.
2. Plant pathogens can be transmitted by a variety of means, including arthropods, contaminated soil and tools, seeds, grafting procedures, and even fungi and roundworms.

Approaches to Identifying Pathogens

1. The isolation and/or identification of an unknown disease agent are extremely important to adequate treatment and disease control.
2. The numbers and types of procedures used in identifying disease agents vary significantly with the type of organism involved.

Collecting and Handling Specimens

1. The many different types of specimens used in laboratory diagnosis include blood, feces, sputum, stool, urine, urethral and vaginal secretions, and cerebrospinal fluids.
2. Specimens must be obtained with care and kept in good condition during delivery to the laboratory. Holding media (nutrient preparations) and special containers are used for the transport of some materials such as throat swabs.
3. Specimens should be sent to laboratories by the most rapid means available and according to federal guidelines. Proper handling of such materials is essential to protecting all persons.

Laboratory Procedures

The general procedures adopted by a diagnostic laboratory depend in part upon the volume of specimens it must handle.

MEDIA

1. Differential media are used to distinguish among organisms, but complete identifications generally involve further testing and other methods.
2. A variety of traditional or standardized biochemical tests have been combined in order to reduce the time and materials used in microbial identification.
3. Examples of these systems include the combination of several tests in one or two tubes, miniaturized, multicompartment devices, and biochemical substrates impregnated in paper strips.

THE BLOOD CULTURE

1. The detection of bacteria or other microbial types in blood may be of major importance.
2. Various media and devices are available for the rapid processing of specimens.

General Identification Procedures

1. All specimens must be handled aseptically.
2. Routine identification begins with staining procedures such as the Gram stain and acid-fast procedures.

IDENTIFICATION KEYS

1. Cellular and colonial appearance, hemolytic reactions, and various biochemical tests are important to the identification of pathogens.
2. Reaction patterns of biochemical tests for known bacterial species are used in the identification of isolated unknown pathogens.

Quality Control Considerations

Quality control measures consist of the monitoring of equipment, materials, and personnel involved in the performance of diagnostic tests.

Questions for Review

1. Define or explain the following terms:
   a. epidemic
   b. sporadic infection
   c. pandemic
   d. communicable disease
   e. endemic
   f. epidemiology
   g. sign
   h. syndrome
   i. vector
   j. reservoir of infection
2. Differentiate between morbidity and mortality rates.
3. What is a carrier? What types of diseases are associated with such individuals?
4. What are the five Fs, and why are they important?
5. What are fomites?

6. What types of fomites might one encounter in the following situations?
   a. a physician's office
   b. a dormitory
   c. a dentist's office
   d. a restaurant
   e. a hospital room
   f. a clinical laboratory
7. List at least six infectious diseases that can be transmitted by fomites.
8. Periodically, newspapers report the occurrence of both natural and human-caused disasters. Discuss the types of conditions favorable for the spread of diseases and the particular infectious diseases that might prevail in the following situations:
   a. an earthquake
   b. a war
   c. a flood
   d. a famine
   e. a fire
   f. a blizzard
9. Differentiate between biological and mechanical means of disease transmission.
10. a. What is a zoonosis?
    b. How are zoonoses classified?
11. What contributions did the following individuals make toward the understanding of disease transmission and processes?
    a. Sir Ronald Ross
    b. Theobald Smith
12. What dangers do bats pose for the well-being of human beings and the various domestic animals in their environment?
13. What arthropod vectors are associated with the following diseases?
    a. malaria
    b. Rocky Mountain spotted fever
    c. Western equine encephalitis
    d. plague
    e. relapsing fever
    f. Chagas' disease
    g. malaria
14. What methods are commonly available for the control of arthropods?
15. Discuss medical asepsis.
16. What practices are involved in surgical asepsis? How does it differ from medical asepsis?
17. What is reverse isolation?
18. Associate specific infectious diseases with the following potential means of disease transmission. (Mark your answers down on paper and keep for future reference.)
    a. flies
    b. ticks
    c. water (contaminated)
    d. mosquitoes
    e. hypodermic syringe
    f. cockroaches
    g. mouthpiece (used)
    h. nasal secretions
    i. milk
    j. kissing
    k. fleas
    l. sexual relations
    m. dogs
    n. soil
    o. mites
    p. lice
19. What importance do you attach to
    a. correct procedure in obtaining specimens?
    b. prompt delivery of specimens to the laboratory?
    c. proper inoculation of specimens?
    d. routine examination of smears and wet mounts?
    e. accurate identification of pathogenic organisms?
    f. prompt communication of information concerning a specimen?
20. Why is the careful transport of specimens containing microorganisms important?
21. Of what value are multiple-test and rapid-method systems?
22. Are there any disadvantages to using multiple-test media?
23. Of what value are computers to the identification of clinically isolated microorganisms?
24. How important is quality control to a laboratory? Describe three aspects of quality control in a microbiological laboratory.

Suggested Readings

BARKER, D. J. P., *Practical Epidemiology,* 3rd ed. New York: Churchill Livingstone, 1982. *A short, practical manual of epidemiology.*

BENENSON, A. L. (ed.), *Control of Communicable Diseases in Man.* Washington, D.C.: American Public Health Association, 1981. *A reference on the public health measures available to control the spread of infectious diseases.*

BURNET, M., and D. O. WHITE, *Natural History of Infectious Disease.* London: Cambridge University Press, 1972. *Interesting accounts of the history of various infectious agents—bacteria, protozoa, and viruses—and the spread and control of infections.*

CLIFF, A., and P. HAGGERT, "Epidemics," *Scientific American* **250**:138–147, 1984. *The spread of measles in Iceland is discussed as a model for understanding the epidemic spread of disease.*

DIXON, R. E. (ed.), *Nosocomial Infections.* New York: Yorke Medical Books, 1981. *The proceedings of an international conference dealing with important clinical and epidemiological features of nosocomial infections.*

HSIUNG, G. D., *Diagnostic Virology.* New Haven, Conn.: Yale University Press, 1982. *A guide and visual aid for the recognition of virus-caused cellular changes and virus structure. Particular attention is given to new diagnostic test procedures, virus isolation, and differential identification.*

KAPLAN, M. M., and R. G. WEBSTER, "The Epidemiology of Influenza," *Scientific American* **237**:88, 1977. *Describes how genetic recombination between human and animal strains of influenza virus may be responsible for the appearance of new subtypes of virus and thus new epidemics.*

LENNETTE, E. H., A. BALOWS, W. J. HAUSLER, JR., and J. SHADOMY (eds.), *Manual of Clinical Microbiology,* 4th ed. Washington, D.C.: American Society for Microbiology, 1985. *An excellent reference for several fields. Covers procedures for the collection, handling, and processing of a wide variety of specimens.*

THOMAS, G., and M. MORGAN-WITTS, *Anatomy of an Epidemic.* New York: Doubleday & Company, 1982. *The true story of one of the most puzzling medical mysteries of the century: Legionnaire's disease. This book presents examples of the problems faced by epidemiologists.*

22

Microbial Virulence

To talk of disease is a sort of Arabian Nights entertainment. — Sir William Osler

After reading this chapter, you should be able to:

1. Discuss virulence and list the virulence factors involved in the causation of an infectious disease.
2. Identify and describe the surface components of bacteria that contribute to their invasiveness.
3. Describe how bacterial components interfere with host defense mechanisms.
4. Distinguish between endotoxins and exotoxins.
5. Describe the properties of algal toxins, mycotoxins, and bacterial phytotoxins and their impact on human well-being.
6. Discuss the pathological effects of viruses.
7. Describe how antigenic variation favors the survival of a pathogen.
8. Explain the properties and importance of prion and slow virus infections.
9. List at least three factors that contribute to opportunistic infections.
10. Discuss the use of the Limulus test and other techniques for detecting microbial toxins.
11. Recognize the importance of laboratory animals in the study of infectious diseases.

How does a microorganism gain a foothold in the body and establish a disease process? Chapter 22 begins with a look at the bacterial structures, products, and mechanisms that assist disease agents in entering a host. Attention is then focused on the various types of microbial toxins and their effects. Finally, viruses and their destructive activities are discussed.

Infectious Disease and Virulence

Only a small percentage of the tens of thousands of known microorganisms are capable of overcoming the defense mechanisms of a host and causing disease in the host. Moreover, a large number of these disease-producing agents, or pathogens, can maintain themselves only within the systems of the animals and plants they invade. In short, their existence is dependent on the availability of a suitable host, a type of *obligatory parasitism.*

To cause disease, a pathogen must (1) attach to the surface of and gain entrance into a susceptible host (Figure 22–1); (2) multiply in host tissues; (3) resist, or not stimulate, host defenses; and (4) damage the host, often by the secretion of *toxins,* or poisons *(toxigenic).* An organism's capacity to establish a disease process in a specific animal or plant host depends on the means and mechanisms by which it carries out these steps.

An **infectious disease** may be described as an interference with the normal functioning of a host's physicochemical process caused by the activities of another organism living within its tissues or on its surface. The term **infection** refers to the multiplication of a disease-causing organism (pathogen) in or on the host, while the term **disease** represents the apparent response of the host to the infection.

Given suitable environmental conditions and a host, most microorganism have disease-producing potential. Table 22–1 lists mechanisms of action of several pathogens. Portions of this chapter and others in Part VIII describe these mechanisms, along with the pathogens exhibiting them.

The ability of an organism to cause disease is called **pathogenicity,** while **virulence** refers to the degree of pathogenicity. Virulence depends on certain features of the organism, including its ability to colonize body surfaces, its invasiveness, its ability to spread and reproduce in the face of the host's defensive mechanisms, and its production of toxins harmful to the host. Some pathogens may be both invasive and toxigenic. Experimentally, the *virulence* of a particular disease agent is measured by the number of such organisms required to kill a particular host

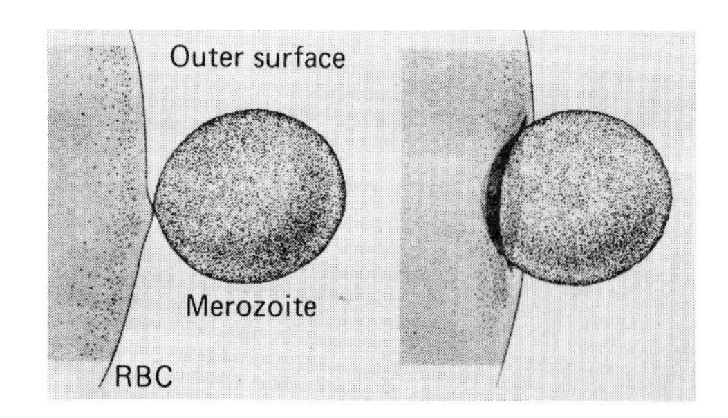

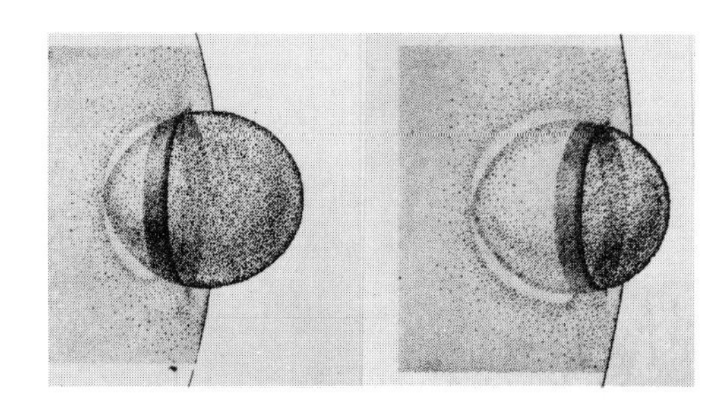

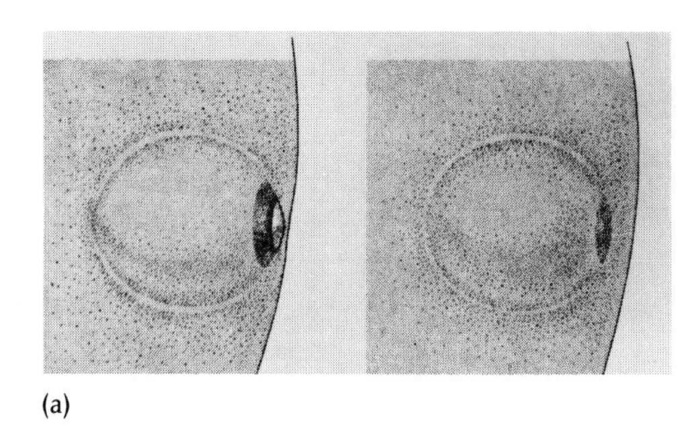

(a)

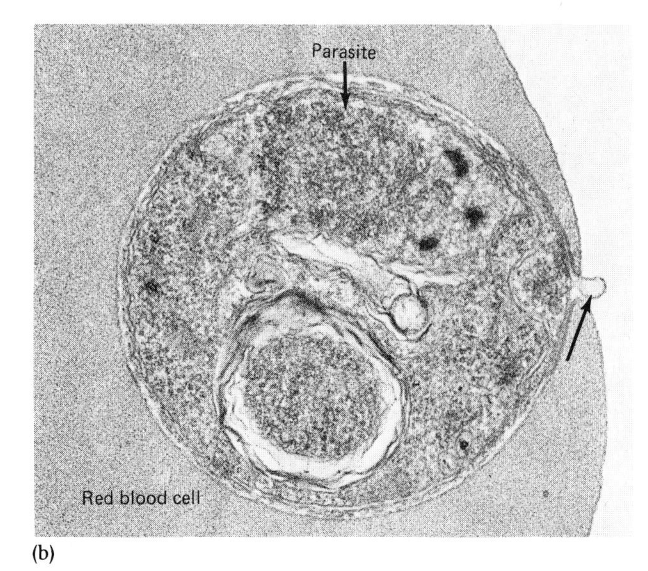

(b)

Figure 22–1 Adherence, penetration, and growth. (a) The sequence of steps in the invasion of a susceptible cell by a malarial parasite (merozoite). (b) An electron micrograph of the parasite beginning its growth and development. Note its point of entry (arrow). *[M. Aikawa, L. H. Miller, J. Johnson, and J. Rabbege,* J. Cell Biol. ***77****:72-82, 1978.]*

TABLE 22–1 Some Mechanisms of Action Exhibited by Infectious Disease Agents[a]

Mechanism	*Some Disease States and/or Pathogens Exhibiting the Mechanism*
Allergic reactions (e.g., delayed hypersensitivity)	Deep-seated (involving body tissues) fungus infections, fungus skin diseases, worm (helminthic) diseases, leprosy, syphilis, tuberculosis, and protozoan infections
Blood loss and/or utilization of vitamin B_{12}	Hookworm and fish tapeworm
Fusion of cellular and viral membranes; includes the formation of giant cells known as *syncytia*	Viruses including herpes simplex viruses 1 and 2, measles, parainfluenza, respiratory syncytial disease, and varicella-zoster agent (the cause of chickenpox and shingles)
Genetic integration (incorporation of nucleic acid of a plasmid or virus into that of the host)	The causative agents of botulism and diphtheria; cancerous states induced by oncogenic viruses
Immunodepression (interference with a host's immune responses)	Lepromatous leprosy, measles, syphilis, tuberculosis, virus-induced cancer and acquired immune deficiency syndrome (AIDS)[b]
Interference with essential body functions	Anthrax, botulism, cholera, diphtheria, plague, rickettsial infections, salmonellosis, and shigellosis
Interference with phagocytosis	Infections with bacterial pathogens such as anthrax bacilli, meningococci, pneumococci, staphylococci and group A streptococci, yeasts such as *Cryptococcus,* and influenza viruses
Interference with phagocytic killing	Bacterial diseases, including brucellosis, gonorrhea, leprosy, meningococcal meningitis, tuberculosis, and typhoid fever; fungus diseases, including histoplasmosis; and protozoan infections, such as leishmaniasis, pneumocystis pneumonia, and trypanosomiasis
Intracellular growth and cellular destruction	Bacterial diseases, such as brucellosis, leprosy, salmonellosis, shigellosis, tuberculosis, and rickettsial infections; most viral infections
Mechanical blockage of organs and/or associated vessels	Helminthic diseases, including ascariasis, filariasis, and schistosomiasis; fungus diseases, such as aspergillosis and candidiasis; the bacterial disease lymphogranuloma venereum; and the protozoan disease malaria
Migration through body tissues and/or organs	Helminthic diseases including ascariasis, fasciolopsiasis, hookworm, strongyloidiasis, trichinosis

[a] These include microorganisms as well as helminths (worms). The helminths are discussed in Chapter 31. The other disease states mentioned here are discussed in Chapters 23–30 (see the Index for page numbers).
[b] Human immunodeficiency virus (HIV) is the causative agent of AIDS; AIDS is discussed in Chapter 27.

under standardized conditions within a specified time. Naturally, the resistance of the host is an important factor. The relationship of virulence *(V)*, numbers of pathogens or dosage *(D)*, and the resistant state *(RS)* of the host to the establishment of an infectious disease can be shown by the frequently quoted formula:

$$\text{Infectious disease} = \frac{V \times D}{RS}$$

The outcome of an infectious disease is determined by several factors. These include the infecting dose of pathogens, the location or site of infection, the virulence of the pathogens involved, and the speed and effectiveness of the host's immune response.

Colonization

Certain bacteria have structures such as flagella or pili that enable them to *colonize,* or occupy, a site on a host's surface where they can reproduce without necessarily causing tissue invasion. Flagella not only propel organisms to areas containing nutritional factors that favor their growth and reproduction but also assist certain bacteria, such as *Vibrio cholerae* (VIB-ree-oh KOL-er-ee) and *Escherichia coli* (esh-er-IK-ee-ah KOH-lee), to establish close physical contact with specific surfaces such as those of the small intestine. In other situations, attachment of bacteria involves pili. Enterotoxin-producing *Escherichia coli* use these nonflagellar surface structures to attach to and invade intestinal linings (Figure 22–2). **[See Chapter 7 for a general description of pili.]**

The ability to attach to host surfaces increases the probability that a disease state will develop. This sit-

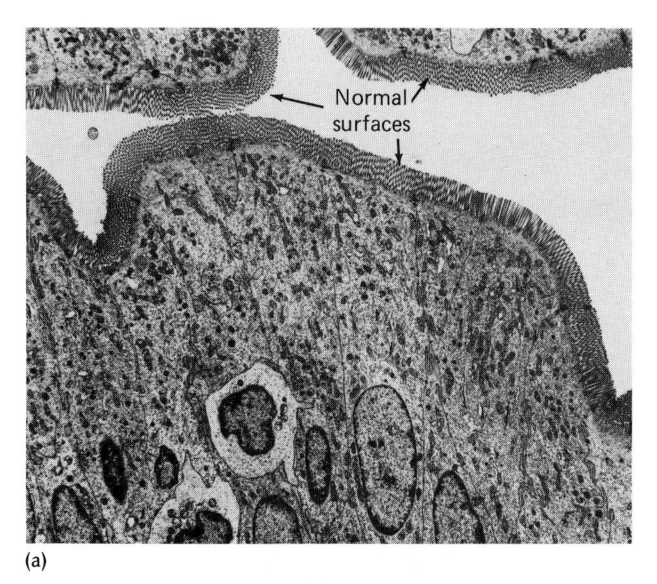

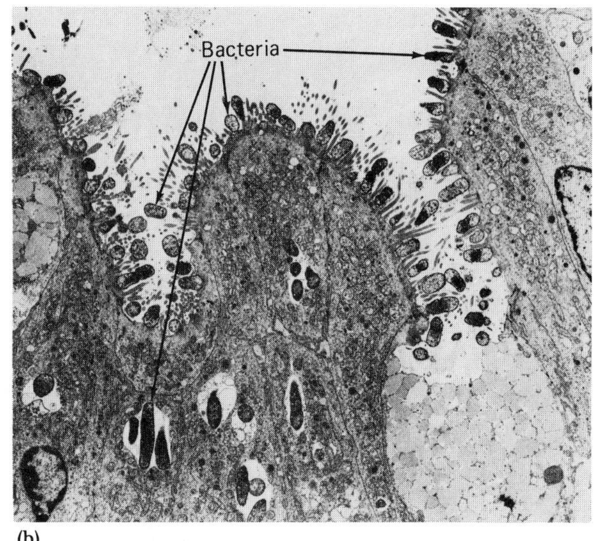

Figure 22–2 The use of pili for the attachment of enterotoxin producing *Escherichia coli.* (a) A normal portion of the small intestinal lining. (b) The effects of bacterial attachment and colonization. Bacterial cells not only attach and stick to surfaces, but also invade the deeper tissues. *[From Moon,* et al., Infect. Immun. ***41**:1340–1351, 1983.]*

uation is quite evident with certain protozoan pathogens, for example, that have specially constructed adhesive disks with which to attach to host tissues (Figure 22–3).

Invasiveness

Invasiveness refers to the ability of a parasite to survive and to establish itself in the tissues of the host. Thus, from a microorganism's point of view,

Figure 22–3 An electron micrograph of *Giardia* situated over the intestinal lining. The characteristic rigid adhesive disk used by this pathogen for attachment is shown. *[R. L. Owen, C. L. Allen, and D. P. Stevens,* Infect. Immun. ***33**:591–601, 1981.]*

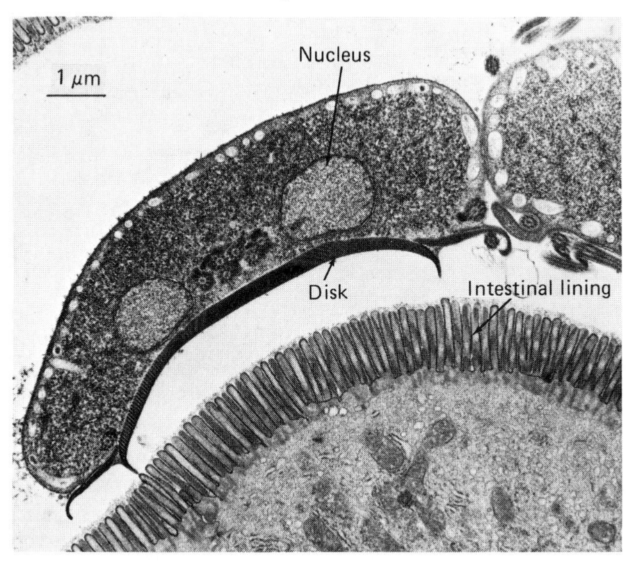

several obstacles must be overcome as it penetrates deeper into the tissues of the host. This group of barriers includes phagocytosis, the lytic (cellular destruction) action of serum in the case of gram-negative organisms, and the difficulty of spreading through tissues.

Unfortunately, the terms used to express the invasiveness of disease agents with respect to the bloodstream are sometimes applied loosely, leading to misunderstanding. Here we shall use these terms with the following definitions: The presence of bacteria in blood as detected either by means of bacteriological culture or by microscopic examination is referred to as *bacteremia.* This type of finding does not imply that the organisms are pathogenic, since blood may, from time to time, contain temporary invaders. The invasion of the bloodstream by virulent microorganisms from a local site of infection, accompanied by toxic or septic symptoms such as chills and fever, is referred to as *septicemia.* A disease state in which pathogens produce localized collections of pus in the tissues of the host is called *pyemia.*

Relatively Nontoxic Bacterial Structures and Products Contributing to Invasiveness

CAPSULES

Because phagocytic and other antimicrobial activities of the host involve the surfaces of invading organisms, it is not unusual to find bacterial and yeast pathogens equipped with certain protective substances. Early studies clearly described the presence of a halolike area surrounding various patho-

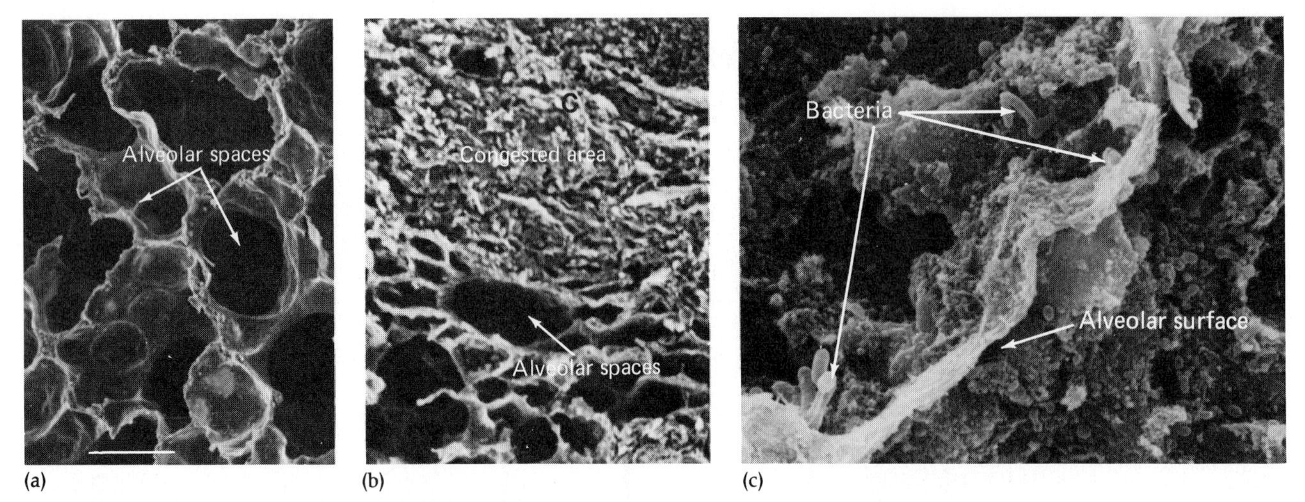

Figure 22–4 The effects of bacterial infection on lung tissue. (a) A scanning micrograph showing clear alveolar spaces. (b) Lung tissue from a chronic lung infection showing alveolar spaces and the massive congestion (upper region) caused by pathogenic bacteria. Both (a) and (b) are the same magnification. The bar marker in (a) represents 100 μm. (c) Bacteria embedded to various degrees in a shapeless mass of slime and discharge on an alveolar surface. *[I. Lam, R. Chan, K. Lam, and J. W. Costerton,* Infect. Immun. ***28****:546–556, 1980.]*

genic bacteria (Color Photograph 15). These regions, which later came to be known as *capsules,* are an extremely important class of surface components (Figure 22–4). Representative bacterial species with capsules include the anthrax bacillus, *Haemophilus influenzae* (he-MAH-fi-lus in-floo-EN-zee), *Klebsiella pneumoniae* (kleb-see-EL-la new-MOH-nee-ah), meningococci of groups A and C, the pneumococci, and certain strains of staphylococci. Loss of the ability to form capsules lowers the organism's virulence because it increases susceptibility to phagocytosis. **[See capsules in Chapter 7.]**

CELL WALL COMPONENTS

The cell walls of certain bacteria contain chemical surface components that obstruct phagocytosis and contribute to microbial invasiveness. Examples of antiphagocytic surface factors include hyaluronic acid and M proteins of group A streptococci such as *Streptococcus pyogenes* (strep-toe-KOK-kus pie-AH-gen-eez) and the A protein of *Staphylococcus aureus* (staff-il-oh-KOK-kuss OH-ree-us). **[See phagocytosis in Chapter 17.]**

PILI

As mentioned earlier, several bacterial pathogens use these nonflagellar surface appendages as their major means of attachment to a variety of cells. Enterotoxin-producing *Escherichia coli* (ETEC) and *Neisseria gonorrhoeae* (nye-SE-ree-ah go-nor-REE-ah) the cause of gonorrhea, are examples of such organisms. The gonococcal pili enable the organisms bearing them to attach to mucosal epithelial cells of reproductive organs, hitchhike on sperm cells, and interfere with phagocytosis. Pili from different bacterial strains of the same species appear to be antigenically quite different. Such differences may contribute to difficulties in preparing vaccines.

VIRULENCE PLASMIDS

Quite often, the extrachromosomal DNA, nonessential elements of bacteria known as *plasmids,* contain genes for some activity that allows the carrying bacterium to survive more efficiently in an unfavorable environment. Some plasmid-carrying bacteria are also able to compete more successfully with other microorganisms of the same or different species. From a medical standpoint, the best-known plasmids are those that specify resistance to various antibiotics. Other plasmids are also known that contribute to microbial pathogenicity. These *virulence plasmids* are encoded for the production of certain toxins, surface adherence proteins such as those of pili, and siderophores, which aid bacteria in acquiring growth-essential iron from the host. **[See plasmids in Chapters 7 and 12.]**

COAGULASE AND DNase (NUCLEASE)

Most pathogenic staphylococci are noted for their production of the extracellular enzyme *coagulase* or *staphylocoagulase,* which coagulates (clots) the blood protein, fibrinogen. The resulting fibrin clot may serve to protect pathogens from other host defense mechanisms. It should be noted that pathogenic staphylococci do not always produce coagulase, since an increasing number of coagulase-negative

Microbiology Highlights

THE A-LAYER AND DISEASE

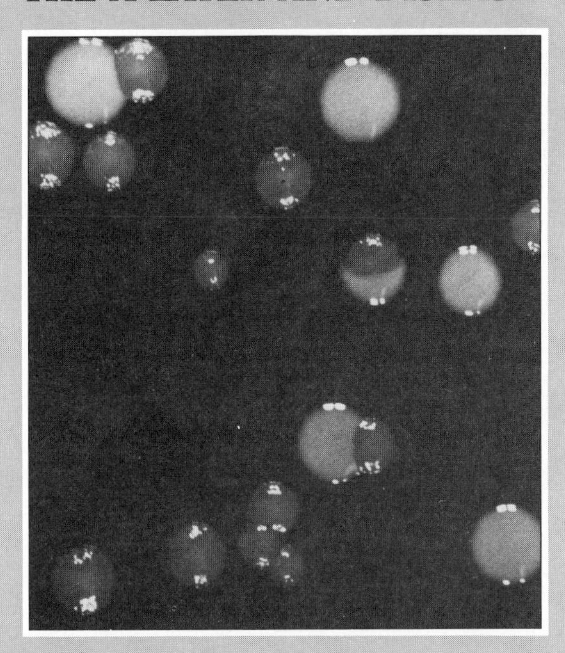

Colonies with the A-layer growing on Congo red agar are dark and smaller than the light colonies of cells lacking the layer. *[From E. E. Ishiguro,* et al., J. Bacteriol. *164:1233–1237, 1985.]*

Aeromonas salmonicida (air-oh-MOH-nass sal-mon-EE-sye-dah) is the cause of a fish disease known as furunculosis. Open sores on the skin surfaces of fish are usually symptoms of disease. If these sores are not treated, they eventually contribute to their deaths. *A. salmonicida* is a facultative intracellular pathogen capable of surviving within phagocytic cells called upon by the fish host to eliminate the invading bacteria.

What protects the disease agent and enables it to cause the death of fish? A cell surface protein called the *A-layer* has been found to be an absolute requirement for the bacterium to cause disease (virulence), since mutants lacking the layer are avirulent. It also appears that the layer provides *A. salmonicida* with a protective barrier against the defense mechanisms of fish hosts. Distinguishing between A-layer$^+$ cells and A-layer$^-$ cells can be done using bacterial viruses. However, a simpler approach is to use a differential medium of Congo red agar. Colonies of cells with the A-layer appear darker and smaller than colonies of cells lacking the layer. The A-protein, which is the major component of the layer, binds the Congo red in the medium, thus giving the colonies their characteristic appearance.

staphylococci are found in various human infections. Some of these organisms produce toxins, which react with the products of other bacteria to cause the destruction of red blood cells (Figure 22–5). Such synergistic hemolysis can be an important virulence factor. In the laboratory, human or rabbit blood plasma treated with citrate or oxalate to prevent its normal coagulation will be caused to clot by coagulase (Color Photograph 88). In order for the reaction to occur, an accessory factor, coagulase-reacting factor (CRF), must be present in the host's plasma. This serum substance is heat-labile, inactivated by heating to 56°C, and reacts slowly with the inactive coagulase to form the active enzyme. The CRF is a protein. Once activated, coagulase converts fibrinogen to fibrin, and a clot forms. Staphylocoagulase has antigenic properties. At least seven distinct varieties have been reported.

Virulent *S. aureus* strains also produce a heat-stable, calcium-activated enzyme called *deoxyribonuclease (DNase).* The particular contribution of this enzyme to microbial virulence is not well understood.

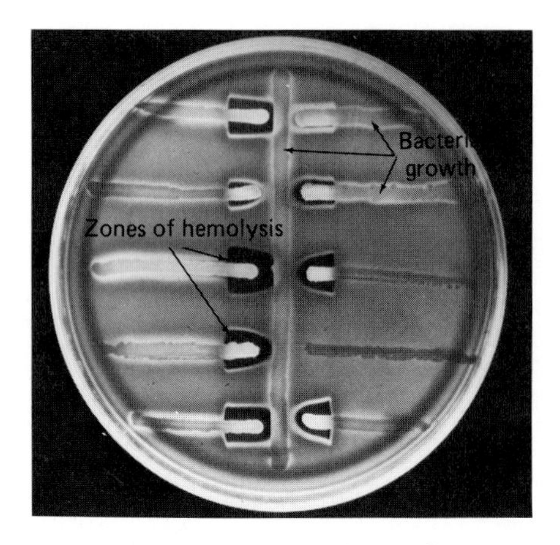

Figure 22–5 Synergistic associations. The pronounced hemolytic reactions (various arrow shaped regions) resulted from the interactions between a beta-lysin producing bacterial strain (vertical streak of growth) and exotoxin-producing coagulase negative strains (perpendicular streaks). *[G. A. Hébert, and G. A. Hancock,* J. Clin. Microbiol. ***22****:409–415, 1985.]*

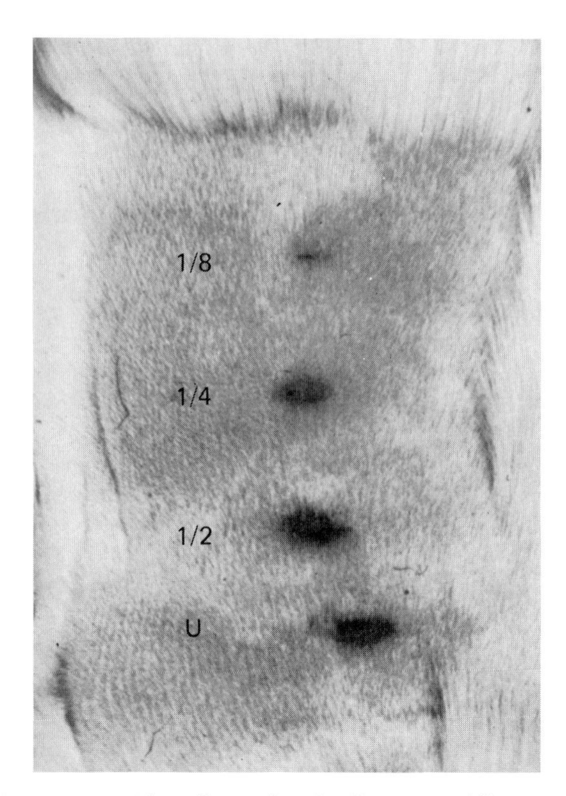

Figure 22–6 The effects of *Legionella pneumophila* on intradermal injection. Even though different amounts of purified protease were injected, tissue destruction always occurred. (U = undiluted; fractions represent dilutions of undiluted protease.) *[From A. Baskerville, J. W. Conlan, L. A. E. Ashworth, and A. B. Dowsett,* Br. J. Exp. Path. ***67****:527–536, 1986.]*

HYALURONIDASE

The ability of pathogens to spread among the tissues of their host has long been a subject for study. The existence of so-called spreading factors was suggested early in such investigations. One of the substances implicated was the enzyme *hyaluronidase,* which is produced by various organisms, including clostridia, pneumococci, and streptococci. Several parasitic worms, or *helminths,* also produce hyaluronidase. This enzyme hydrolyzes hyaluronic acid, a thick, high-molecular-weight polysaccharide that is an essential component of the intracellular ground substances of several tissues, and of clotted material that accumulates in host inflammation. The effect is to increase greatly tissue permeability and ease the pathogen's movement.

IMMUNOGLOBULIN A (IgA) PROTEASES

Certain bacterial pathogens, including *Haemophilus influenzae* and *Streptococcus pneumoniae,* known to cause respiratory infections, produce enzymes capable of cleaving human immunoglobulins of the IgA subclass. Other immunoglobulin subclasses appear to be resistant to these IgA proteases. The production of IgA proteases may be an important virulence factor, allowing certain bacteria to escape a host's mucosal defenses. The immunoglobulin molecule resulting after protease action is defective. Immunoglobulin A proteases may be involved in tooth decay and other diseases that occur in the mouth. **[See IgA structure in Chapter 17.]**

TISSUE-DESTROYING PROTEASES

The bacterium *Legionella pneumophila* (lee-jon-EL-lah new-MOH-fill-ah), the cause of Legionnaires' disease, has been found to have a broad range of extracellular enzymes that may act as virulence factors. A protease isolated from the bacterium is believed to be responsible for lung tissue damage and pneumonia. When injected intradermally in laboratory animals, it causes tissue injury (Figure 22–6). On inoculation into the respiratory tract, the protease produces rapid and severe lung damage. **[Chapter 25 describes Legionnaires' disease.]**

KINASES

Several gram-positive bacteria produce kinases. These enzymes dissolve fibrin clots formed by the host to isolate sites of infection. Among the better-known kinases are streptokinases, produced by streptococci. These enzymes are highly specific. For example, the kinases produced by human strains of streptococci dissolve only human fibrin, and those produced by canine strains of the microorganism liquefy only the fibrin of dogs. Purified streptokinase can be used therapeutically to dissolve blood clots. Other kinase producers are gas gangrene bacteria and staphylococci. **[See gas gangrene in Chapter 23.]**

SIDEROPHORES

All forms of life appear to have difficulty in acquiring enough iron for their well-being. Aerobic bacteria are no exception, since the inorganic element is an essential part of cytochromes, catalase, and nonheme iron-containing electron transport proteins. Iron is also needed to activate ribotide reductase, an essential enzyme for DNA synthesis. In various environments, iron forms insoluble ferric complexes that are largely unavailable to such bacteria. To be able to grow and reproduce, a bacterium must stabilize ferric ions and transport them into the cell, where they are reduced to the ferrous, usable form. The excretion of low-molecular-weight iron-binding proteins, called *siderophores,* enables the organism to acquire available iron. [See cytochromes in Chapter 5.]

Most of the human body's iron is packaged inside red blood cells in the form of hemoglobin and is unavailable to a bacterium infecting the bloodstream. For such an organism to grow and reproduce, it must compete for iron in the bloodstream with the body's iron-binding protein, *transferrin.* If the bacterium can excrete sufficient siderophores into the blood, it can compete for serum iron and survive to reproduce and cause disease. Bacteria that are unable to accumulate or produce siderophores are found to be avirulent (nonpathogenic).

Antigenic Variation

Several pathogens are able to evade or to repel a host's immune system by forming specific compounds or structures or by hiding in specific types of cells. However, trypanosomes, the protozoa that cause African sleeping sickness (Color Photograph 90), have a unique mechanism with which they can stay one jump ahead of the host's immune system. These organisms are able to change their outer antigenic coats during the course of an infection. Thus, while the host's immune system is working to eliminate the major population of trypanosomes with old coats, a subpopulation of pathogens begins to express a new antigenic surface coat. Under such conditions of antigen switching, the host is unable to cope with the infection. A somewhat similar situation has been noted to occur with the AIDS virus (human immunodeficiency virus, or HIV). One of the most important obstacles to making an AIDS vaccine appears to be the extreme variability of the virus and especially of its envelope proteins (antigens). While most viruses have only a few variants, HIV has many. There are major differences in the genetic makeup and envelope proteins of viruses isolated from different people; even within the same AIDS patient, these viruses keep changing. [See Chapter 10 for descriptions of viral organization.]

Microbial Toxin Production

Bacterial Toxins

Several normal components or products of bacterial cells are known to be toxic for higher forms of life. The toxins of bacteria are categorized as either *endotoxins* or *exotoxins.* Endotoxins are substances that are generally liberated only after the organism disintegrates by self-destruction or *autolysis.* Exotoxins are products generally released during the lifetime of an organism and, at times, by autolysis. Occasionally, enzymes released from cells are also considered exotoxins. Hyaluronidases and coagulase are two such enzymes. Others include fibrinolysins, such as streptokinases, which dissolve fibrin clots; proteinases, which dissolve proteins; and lecithinase, which decomposes lipids. Representative toxins and substances that aid microorganisms in their invasion of a host are discussed in the following sections.

ENDOTOXINS

Endotoxins are formed in the cell walls of gram-negative bacteria. Chemically, they are lipopolysaccharide-protein complexes. When the two components are separated, the lipopolysaccharide (LPS) fraction is toxic and pyrogenic *(fever-causing),* while the protein portion imparts antigenic properties to the entire complex identical to the cell wall or somatic (O) antigens of the intact bacterium.

Endotoxins from several pathogenic and nonpathogenic gram-negative organisms have been isolated and studied, including species of *Escherichia, Neisseria,* rickettsiae, *Salmonella, Serratia* (ser-RAY-sha), *Shigella* (she-GEL-lah), and *Veillonella* (vah-yon-ELL-ah). Experimentally the LPS of *Escherichia coli* was found to interact and cause changes in the shape of red blood cells (Figure 22–7). Several of the characteristics that distinguish endotoxins from exotoxins are listed in Table 22–2. [See the effects of endotoxins in Chapter 26.]

Modes of Endotoxin Action. Endotoxins are clearly involved in disease states caused by gram-negative bacteria, but their exact role is not understood. Various studies have shown their effects to be nonspecific. Apparently, they cause the release of a fever-inducing substance from polymorphonuclear leukocytes and other cells, which, in turn, interferes with the temperature regulatory centers in the brain. Endotoxins also cause blood-clotting disorders and shock.

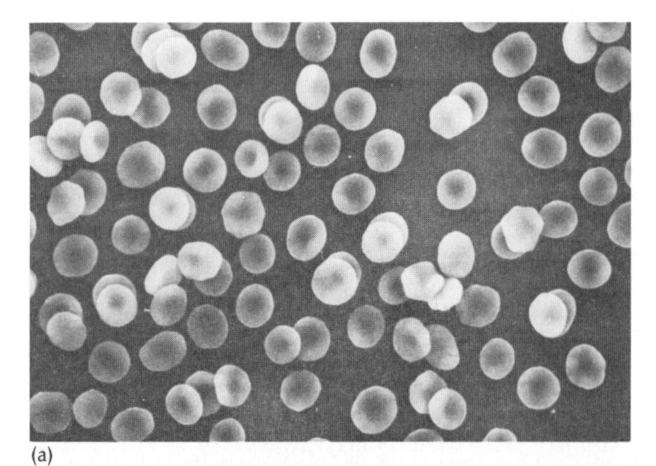

(a)

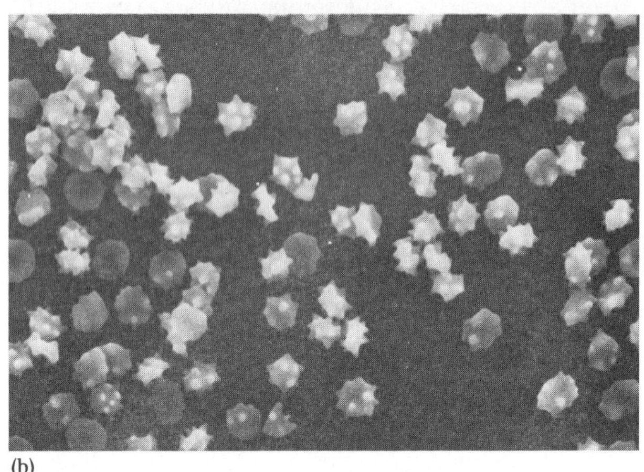

(b)

Figure 22–7 The interaction of *Escherichia coli* (LPS) with red blood cells. (a) Normal cells. (b) The formation of irregularly-shaped cells caused by incubation with *E. coli* LPS. *[Courtesy of Dr. J. R. Warren, Chief, Microbiology Laboratory, VA Lakeside Medical Center, Northwestern University.]*

ENDOTOXIN DETECTION. Even minute concentrations of endotoxins can be detected by the *Limulus* amoebocyte lysate test. This test is based on the fact that an aqueous extract of disrupted amoebocytes (lysate) from the blood of the horseshoe crab, *Limulus polyphemus* (Figure 22–8a), contains an enzyme that is activated in the presence of very small amounts of endotoxin and divalent cations such as calcium and magnesium. The end result of this combination is the formation of a cloudy gel.

The *Limulus* assay is performed by adding dilutions of specimen samples to equal volumes of Limulus lysate in glass test tubes. The reaction mixtures are incubated for 60 minutes at 37°C. After incubation, the presence of a solid gel or a marked increase in viscosity (Figure 22–8b) represents a positive test. Since its first description in 1968, the lysate procedure has been used for the detection of bacteriuria (bacteria in the urine), for the diagnosis of gram-negative spinal meningitis, and for the detection of pyrogens (fever-causing agents) in radiopharmaceuticals and biologicals. Because of its extreme sensitivity, it is quite possible that the test can be used routinely to determine pollution in natural bodies of water, such as lakes and streams, and to detect gram-negative bacterial contamination in various food products. The assay is unique in that it can be used to detect endotoxin in concentrations of as little as 5×10^{-4} μg/mL.

EXOTOXINS

Exotoxins may be secreted during the growth of bacteria or, as certain studies have shown, they may be released on the death and autolysis of cells. They are responsible for many of the disease symptoms and for the eventual disease state of the host. On the basis of their effects on host cells, exotoxins can be grouped into several categories, such as *cytotoxins, enterotoxins,* and *neurotoxins.* These poisonous substances are distinguished from endotoxins by several properties (Table 22–2). Chief among these are their protein nature and specificity of action. Several bacterial species produce a wide variety of exotoxins (Table 22–3).

TABLE 22–2 A Comparison of Selected Characteristics of Endotoxins and Exotoxins

Characteristic	*Endotoxin*	*Exotoxin*
Chemical composition	Lipopolysaccharide-protein complex	Protein
Source	Cell walls of gram-negative bacteria; released only on autolysis or artificial disruption of cells	Mostly from gram-positive bacteria; excretion products of growing cells or, in some cases, substances released upon autolysis and death
Effects on host	Nonspecific	Generally affects specific tissues
Thermostability	Relatively heat-stable (may resist 120°C for 1 hour)	Heat-labile; most are inactivated at 60° to 80°C
Toxoid[a] preparation for immunization possible	No	Yes

[a] Modified protein toxin that is not toxic but still causes the production of antibodies. See Chapter 18 for a description of toxoids.

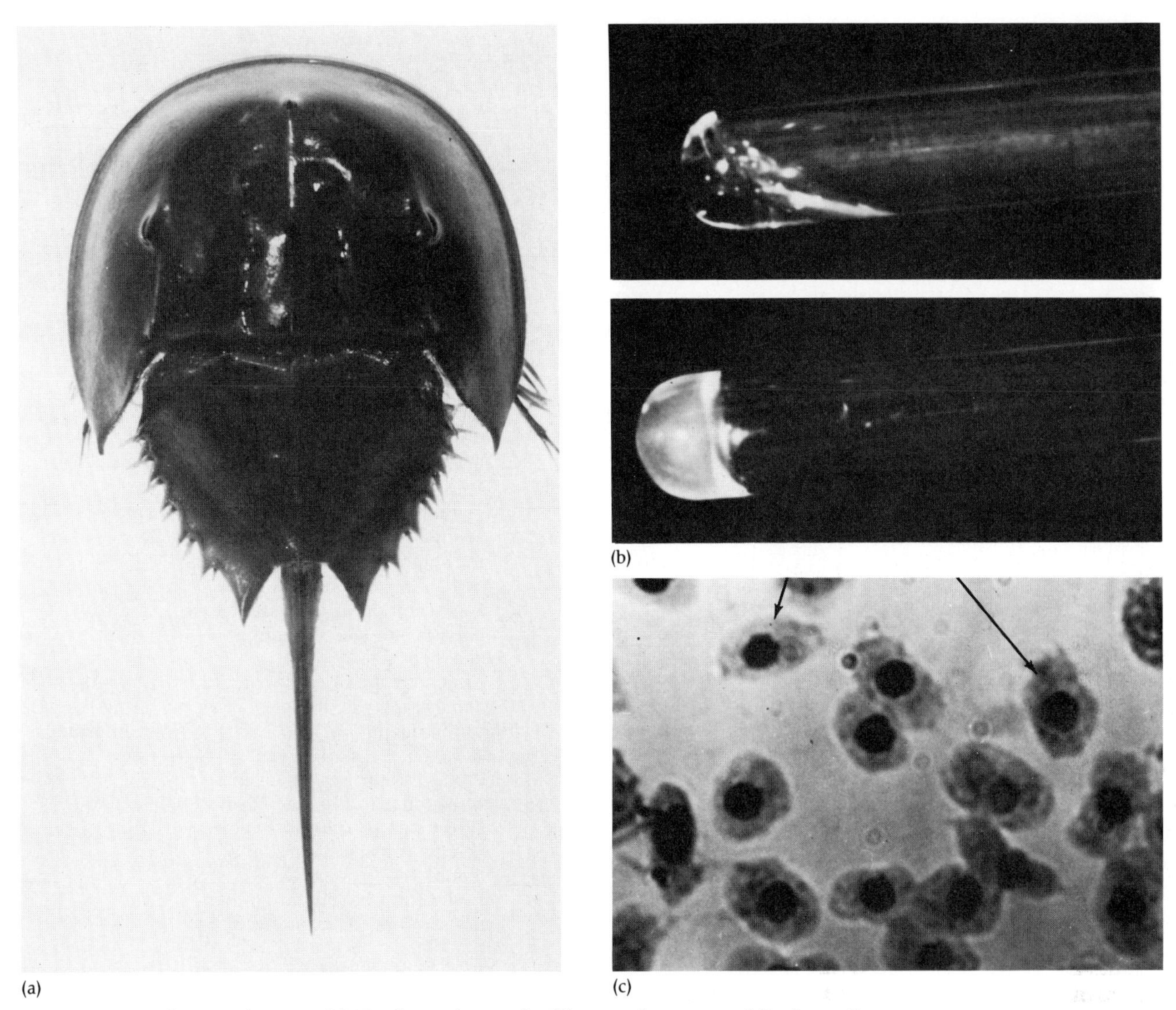

Figure 22–8 The Limulus test. (a) The horseshoe crab. *[Photographs courtesy of Dr. James H. Jorgensen, University of Texas Health Science Center at San Antonio, Texas.]* (b) A close-up view of a negative result (top tube) and a strongly positive result (bottom tube). (c) A microscopic view of the cells of the horseshoe crab, amoebocytes, involved with the test. *[J. H. Jorgensen, and R. F. Smith,* Appl. Microbiol. ***26**:43, 1973.]*

TABLE 22–3 Selected Representative Exotoxins Produced by Bacterial Pathogens

Bacterial Species	*Gram Reaction*	*Disease Produced*[a]	*Toxin Designation*	*Mechanism of Action*	*Tissue Invasion*
Bordetella pertussis	−	Whooping cough	Pertussis toxin	Damage to cilia of respiratory tract	Minimal
Clostridium botulinum	+	Botulism[b]	Seven types of specific neurotoxins: A, B, C_1, C_2, E, F, G[c]	Toxin binds at neuromuscular junction, blocks presynaptic release of acetylcholine, resulting in impaired breathing and swallowing, and paralysis	No
Clostridium difficile	+	Pseudomembranous colitis	Cytotoxin B	Direct cell destruction and suppression of cyclic adenosine monophosphate (AMP)	No
			Enterotoxin A	Not fully known	No

(continued)

TABLE 22–3 (continued)

Bacterial Species	*Gram Reaction*	*Disease Produced*[a]	*Toxin Designation*	*Mechanism of Action*	*Tissue Invasion*
Clostridium perfringens	+	Gas gangrene	Alpha toxin (a total of 11 toxins; may be active in different aspects of disease)	A lecithinase that causes massive red blood cell destruction	Minimal
		Perfringens-caused diarrhea	Enterotoxin	Causes elevation of cyclic adenosine triphosphate (ATP) levels	No
Clostridium tetani	+	Tetanus	Tetanospasmin	Blocks action of inhibiting neurons, which results in spasmodic muscle contractions	No
Corynebacterium diphtheriae	+	Diphtheria	Diphtheria toxin[c]	Blocks protein synthesis; toxin causes destructive effects on heart and cranial nerves	No
Escherichia coli	—	Enterotoxigenic[b] *E. coli* diarrhea	Heat-labile (LT) toxin	Toxin binds to gangliosides in epithelial target cells, activating	No
			Heat-stable (ST) toxin	adenylate cyclase and resulting in the production of cyclic AMP. The reaction causes excess secretion of chloride, bicarbonate, and water, leading to diarrhea	No
Shigella dysenteriae	—	Bacterial dysentery	Neurotoxin	Brain and spinal cord irritation; damage to cells lining capillaries	Epithelial cells lining the gastrointestinal tract
			Enterotoxin	Produces diarrhea by mechanism similar to those of *E. coli*	No
Streptococcus pyogenes	+	Scarlet fever[b]	Erythrogenic toxin	Direct skin toxicity or hypersensitivity, causing dilatation of blood vessels	Yes
Staphylococcus aureus	+	Staphylococcal scalded skin syndrome	Exfoliation	Destruction of cellular connections and cell membranes	No
		Staphylococcal food poisoning	Six types of enterotoxin	Toxin is absorbed, resulting in stimulation of vomiting center in the brain	No
		Toxic shock syndrome	TSS toxins	Unknown	No
Vibrio cholerae	—	Cholera	Choleragen	Toxin binds to gangliosides in epithelial target cells, activating adenylate cyclase and resulting in the production of cyclic AMP. The reaction causes excess secretions of chloride bicarbonate and water, leading to diarrhea	No

[a] Symptoms of specific states are described in later chapters of this section.
[b] Transmitted by virulence plasmids.
[c] Toxin produced by only tox^+ bacteriophage-infected cells.

Another aspect of exotoxin production is its association in certain cases with particular bacterial viruses. For example, diphtheria bacilli produce an exotoxin only when they are harboring a particular temperate bacteriophage. Only those bacteria infected by such specific viruses form the toxin. A similar situation exists with cells of *Clostridium botulinum* and *Streptococcus pyogenes,* which produce the erythrogenic toxin associated with scarlet fever only under the influence of certain temperate viruses. **[See lysogenic conversion in Chapter 6.]**

Modes of Exotoxin Action. The mechanisms and sites of action have been studied in several of the classic exotoxins, including botulism, diphtheria, and tetanus. In general, exotoxins function by destroying specific components of cells or by inhibiting certain cellular activities. Some of these substances work only on specific cell types. Further details are discussed in the chapters dealing with these toxin-producing agents.

EXOTOXINS OF MAJOR CLINICAL SIGNIFICANCE

Botulism. The exotoxin of *Clostridium botulinum* (klos-TRE-dee-um bot-you-LIE-num) causes a type of food poisoning that is fatal unless appropriate treatment is given. This powerful neurotoxin acts on motor neurons by preventing the release of the neurotransmitter acetylcholine, thereby preventing nerve impulse transmission and causing muscle paralysis.

Pseudomembranous Colitis. *Clostridium difficile* (C. diff-EE-seal) is a bacterium found infrequently and in small numbers in the large intestine. Treatment of patients with antibiotics—such as ampicillin, cephalosporins, lincomycin, and clindamycin—interferes with the ecological balance of the intestinal tract and permits increases in the growth and numbers of *C. difficile.* This species produces two toxins, A and B, both of which are cytotoxic and lethal to laboratory animals. [See Chapter 26 for details of this disease.]

Cholera. *Vibrio cholerae* produces an enterotoxin, choleragen, that stimulates excessive secretions of fluid and electrolytes from the lining of the surface of the small intestine. Enterotoxigenic strains of *Escherichia coli* also have a similar mechanism of action. Neither of these bacterial pathogens invades tissue to any great extent, so that penetration of the intestinal lining (mucosa) is not essential to virulence. In contrast, *Shigella dysenteriae,* the causative agent of bacterial dysentery, must penetrate the intestinal surface to establish the disease process. [Chapter 26 presents these diseases.]

Diphtheria. The diphtheria exotoxin is a cytotoxin. Although this substance can act on several cell types, its main mechanism of action is to interfere with protein synthesis. It is also noted for general destructive effects on various types of tissues (Figure 22–9).

Tetanus. The exotoxin formed by *Clostridium tetani* (C. TEH-tan-ee), the causative agent of lockjaw, causes its effects mainly on the Renshaw cells in the anterior horn of the spinal cord. It prevents the release of the neurotransmitter glycine, thereby inhibiting feedback inhibition. As a result, convulsions originating from the spinal cord develop.

OTHER EXOTOXINS OF CLINICAL SIGNIFICANCE

Several exotoxins normally play a lesser role in clinical disease states either because, under normal conditions, they are produced in small quantities or because they are not extremely toxic. However, under certain conditions, these exotoxins may cause serious illness. Examples of such exotoxins include dysentery bacillus neurotoxin, scarlet fever erythrogenic toxin, staphylococcal enterotoxin, and streptolysins O and S.

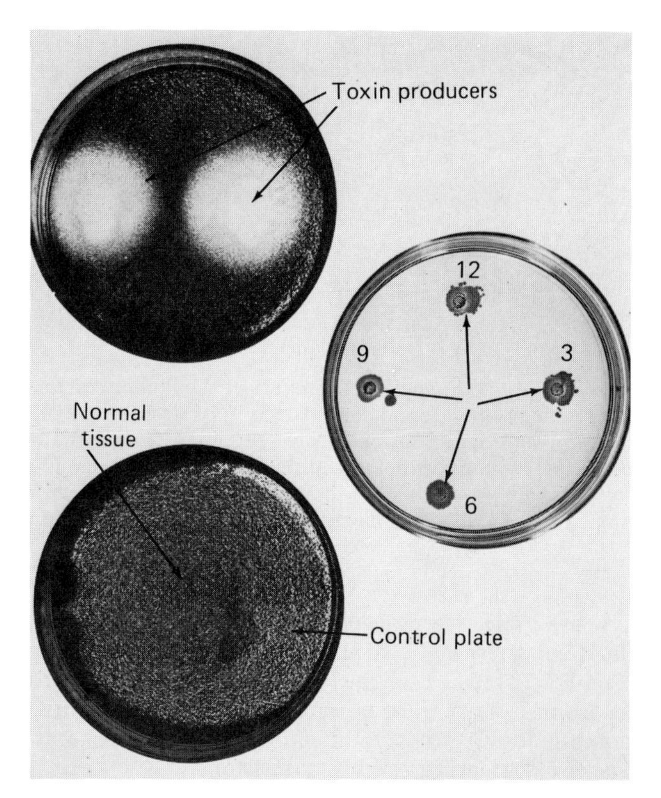

Figure 22–9 The colony overlay test (COT), a method for detecting toxin production (toxicogenicity) by *Corynebacterium diphtheriae* strains in tissue culture. The plate on the right shows growth of bacterial cultures 18 hours after inoculation. The cultures located at 3 and 9 o'clock are toxin producers, whereas the remaining two are not. This is evident in the top tissue culture plate, where toxin from two toxin-producing strains has destroyed a monolayer of tissue culture cells. The lower plate, a control, shows no tissue destruction.

The Erythrogenic Toxin of Scarlet Fever. A small number of the strains belonging to beta-hemolytic group A streptococci produce this minor toxin. These streptococci belong to one of the several immunological groups differentiated on the basis of their antigenic composition. Three toxins are recognized and labeled A, B, and C. The first of these toxins occurs most often. Erythrogenic toxins have a selective action on the skin and are neutralized by scarlet fever antitoxin. No effects are suffered by the streptococci multiplying at the time.

Leukocidins. Several strains of staphylococci and streptococci produce *leukocidins,* toxins that kill leukocytes and macrophages (cells active in phago-

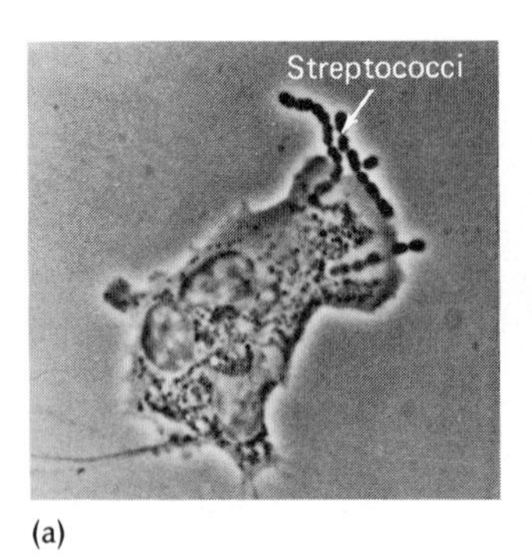

(a)

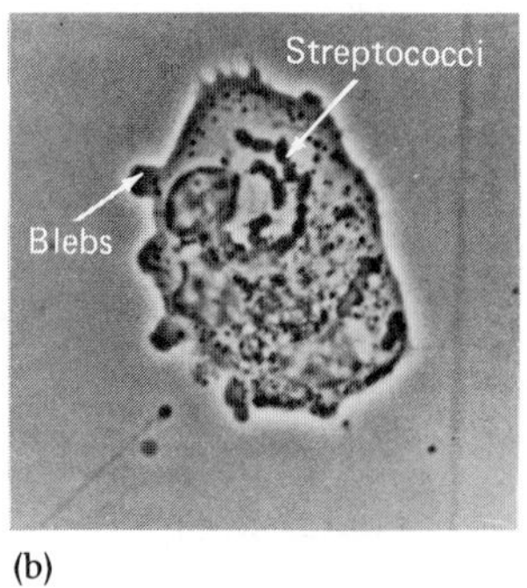

(b)

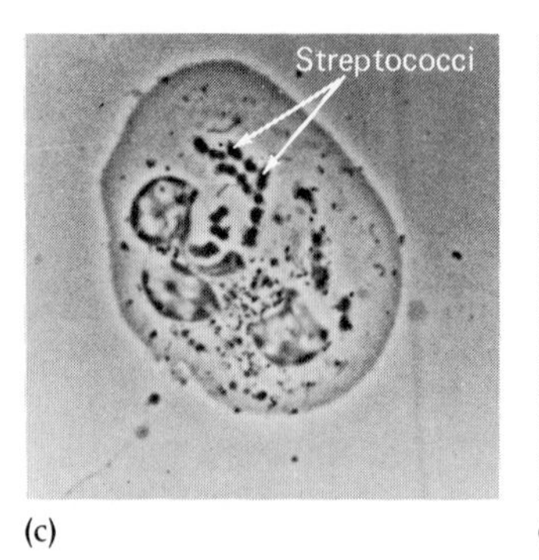

(c)

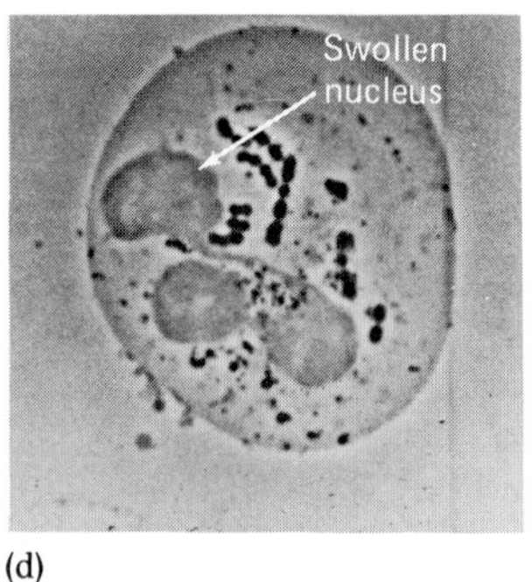

(d)

Figure 22–10 The destructive effect of the leukotoxic factor. These phase contrast micrographs show the leukotoxic action of streptococci on human polymorphonuclear leukocytes (PMNLs). (a) The PMNL engulfing several chains of leukotoxic streptococci. (b) The PMNL stops all movement, rounds up, and forms extended blebs. (c) The PMNL becomes swollen and immobile. (d) Most internal structures of PMNL have disappeared, and the nucleus swells. *[G. W. Sullivan, and G. L. Mandell,* Infect. Immun. ***30****:272–280, 1980.]*

cytosis). The streptococcal leukocidins cause the death of these cells by destroying their lysosomes (Figure 22–10). Enzymes released from damaged lysosomes may injure other cellular structures, thus making lesions of the infection more severe and thereby decreasing the host's resistance.

Staphylococcal Enterotoxins. Many coagulase-positive strains of *S. aureus* produce enterotoxins. These toxins are (1) protein in nature, (2) poor antigens, (3) resistant to boiling temperatures for approximately 30 minutes, and (4) not neutralized by antitoxins prepared against other staphylococcal toxins. Humans and monkeys are the only naturally susceptible victims of this enterotoxin. Affected individuals generally experience nausea and vomiting within a few hours after the toxin's ingestion. Fatalities are rare.

Streptolysin O. Certain streptococci produce two distinct soluble hemolysins, enzymes that destroy red blood cells. One of these bacterial products is called *O lysin* (SLO) because of its sensitivity to oxygen. The other hemolysin, designated *S lysin,* in addition to other properties, is noted for its extreme sensitivity to heat. Certain streptococci produce only O lysin, while others secrete only S lysin. Most members of group A produce both.

Streptolysin O is secreted by most streptococci in group A, those organisms found in group C, and certain members of group G. This streptolysin can be inactivated by heating at 37°C for two hours. It is an antigenic protein, as demonstrated during infections caused by streptococci. Streptolysin O is known for its toxic action on red blood cells and various other types of cells, including frog heart and mammalian kidney and heart (Figure 22–11). Several studies indicate that it may play a significant role in rheumatic heart disease. SLO molecules bind to cholesterol-containing portions of cells and form rings and arcs that penetrate membranes (Figure 22–12). These formations create slits and pores that lead to severe membrane damage similar to effects of the complement complex. **[Chapter 20 describes the action of complement.]**

Streptolysin S. Streptolysin S is produced by most strains of group A streptococci and probably members of other groups. Because streptolysin S has resisted attempts at purification, its chemical makeup remains in doubt. It is believed to be either polysaccharide or protein in nature.

The toxin is not antigenic. It is, however, toxic for the tissues of laboratory animals. Moreover, the hemolysis that develops around surface-located colonies on blood agar plates is caused by streptolysin S.

Bacterial and Fungal Phytotoxins

Humans are virtually dependent for survival on certain food crops. Among them are barley, cassava, corn, oats, potato, rice, soybean, sweet potato, sugar cane, and wheat. Unfortunately, all of these plants are hosts for certain pathogenic bacteria and fungi, many of which are known to produce toxins. Such **phytotoxins** (plant poisons) can cause plants to wilt or to lose their coloration, or they may bring about extensive destruction of leaves, stems, or flower parts. Species of the bacterial genera *Corynebacterium* (ko-ri-nee-back-TI-ree-um), *Erwinia* (er-WI-nee-ah), *Pseudomonas,* and *Xanthomonas* (zan-thoh-MOAN-us) are known to produce one or more phytotoxins. Chemically, these toxins are polysaccharides, peptides, and related compounds.

Phytotoxins are also products of fungi such as *Alternaria* (al-ter-NARE-ee-ah), *Ceratucystis, Fusicoccum, Helminthosporium, Rhynchosporium,* and *Stemphyllium.* Examples of plants known to be affected by their phytotoxins include almond trees, barley and other grasses, citrus, corn, cucumber, lettuce, potato, spinach, and strawberry.

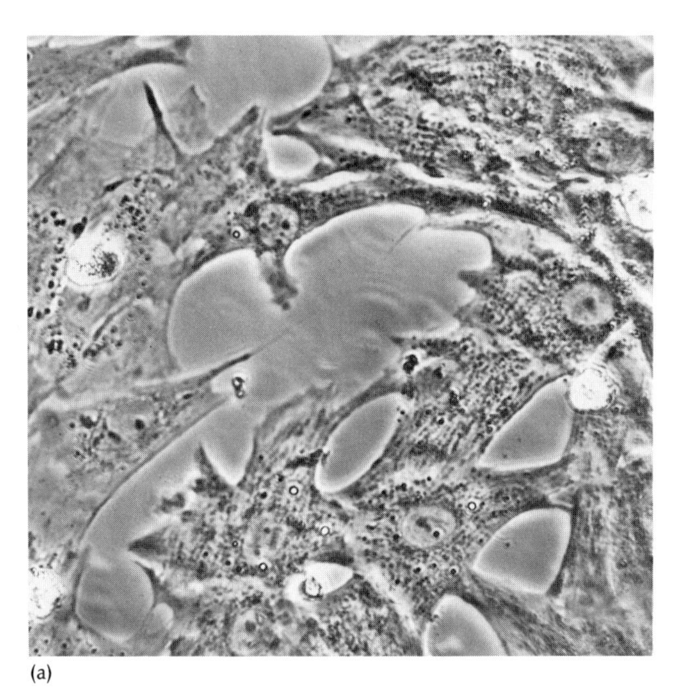

(a)

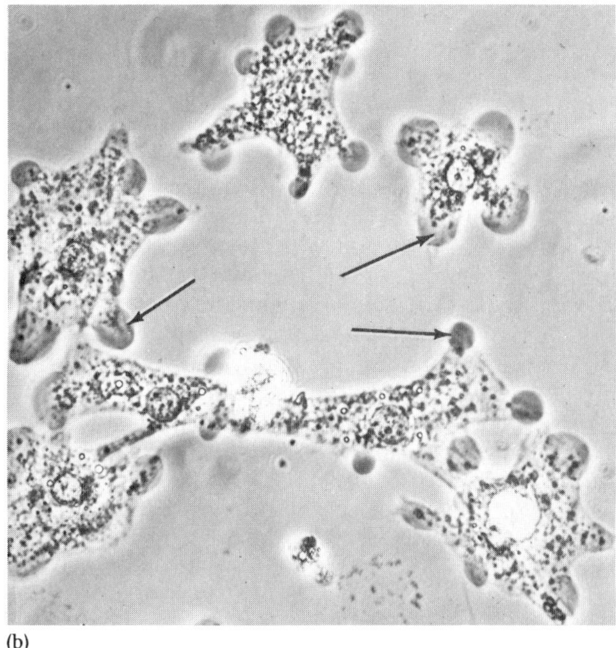

(b)

Figure 22–11 The toxic effect of streptolysin O on beating mammalian heart cells in tissue culture, as observed by phase-contrast microscopy. (a) The appearance of normal rat heart ventricle cells after 2 days of growth. Before exposure the myocardial cells on the right exhibited striations and were beating vigorously. (b) The same cells 3 min after exposure to group C streptolysin O. Note the granulation and numerous plasma membrane blebs (arrows) associated with killed myocardial cells. *[A. Thompson, S. P. Halbert, and U. Smith,* J. Exp. Med. ***131****:745–763, 1970.]*

Toxins of Cyanobacteria

Poisoning of lower animals and even of humans by cyanobacterial blooms is well documented. Toxic blooms have occurred in the lakes, ponds (Color Photograph 7a), and reservoirs of Africa, Asia, Australia, Europe, and North and South America. Toxic freshwater cyanobacteria are becoming an important factor in water quality, especially as more marginal water supplies are coming into public use. Unlike freshwater species, most marine cyanobacteria have not presented serious health or economic problems. However, swimmers in tropical waters have experienced rashes and blisters of the skin, lips, and genitals after exposure to disrupted cyanobacteria.

Figure 22–12 An electron micrograph of red blood cell disrupted (lyzed) with streptolysin O (SLO). The SLO lesions are seen as semicircular (c) and circular (o) forms measuring about 7 to 8 nm in diameter. *[From S. Bhakdi, J. Tranum-Jensen, and A. Sziegoleit,* Infect. Immun. ***47:*** *52–60, 1985.]*

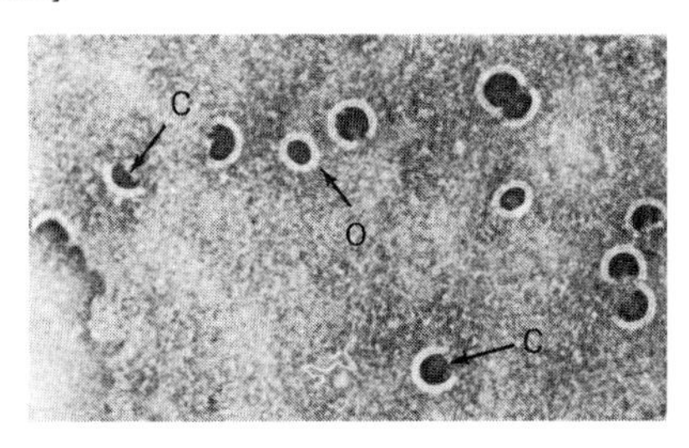

Algal Toxins

Algal toxins contribute a wide variety of poisonous substances to aquatic environments. Thus far, many algal toxins—which become poisonous to fish, waterfowl, mussels, and clams, and subsequently to humans who eat shellfish—have been studied in crude form only. One of the few that has been isolated in pure form is produced by the dinoflagellates associated with the toxic red tides. The toxin is a highly potent poison. It causes death in experimental animals in less than 30 minutes, and produces paralysis as well as death in humans.

Toxin-producing algae pose a serious threat to human well-being. Many algal species that may represent a potential food supply produce toxins, and the poisonous substances produced by marine algae threaten edible marine organisms. The problem could become more acute as the world becomes more dependent upon food from aquatic sources.

Mycotoxins

Mycotoxicosis is a disease state caused by the ingestion of foodstuffs contaminated with fungus toxins. This condition is to be distinguished from another fungus disease category, **mycosis,** which involves a general invasion of living tissue by actively growing fungi. Mold-caused deterioration of foods and feeds results in economic losses and poses a serious health hazard. Ergotism, the effect of the mycotoxin ergot, has been known for years. Historically, the unusual behavior of individuals, which included convulsions and incoherent speech, during the witch trials of the 1600s may have been due to the ingestion of this mycotoxin. There is also an increasing awareness of mycotoxins as potential natural environmental contaminants. **[See Chapter 8 for a discussion of fungal activities.]**

The distinctive characteristics of a mycotoxicosis include the following: (1) the disease is not transmissible; (2) drug or antibiotic treatments have little or no effect on the disease; (3) outbreaks in the field are often seasonal because certain climatic conditions affect mold development; (4) the outbreak is usually associated with a specific feed or foodstuff; and (5) examination of the suspected food or feed reveals signs of fungal activity.

Of the mycotoxins studied, those produced by species of *Aspergillus* (a-sper-JIL-lus), the *aflatoxins,* are the best known. These poisons are a unique group of low-molecular-weight compounds. Aflatoxins have been found in a wide variety of edible commodities, including beans, cereals, coconuts, milk, peanuts, sweet potatoes, and commercially prepared animal feeds.

In addition to their intoxicating effects, aflatoxins are known to be carcinogenic. This finding has reinforced but not confirmed the concept that naturally occurring mycotoxins may be a cause of human cancer. Studies during the 1960s demonstrated the involvement of aflatoxins with liver damage in several birds, fish, and mammals and with tumor formation in ducklings, ferrets, rats, and trout. Specific toxicological and biochemical studies with experimental animals have shown that mycotoxins cause such effects as ultrastructural cellular changes; changes in the synthesis of DNA, RNA, and protein; and mitochondrial activity. Despite the vast number of these studies, the precise mode of action of aflatoxin is not known. Like many carcinogens, aflatoxins act as nonspecific cell poisons that exert multiple effects on the structures and biochemistry of susceptible cells. Because most of these changes may be secondary to carcinogenic activity, they must be identified as such if the primary mode of action is to be defined.

Some mycotoxins also show antibacterial, antifungal, antiprotozoan, and antitumor effects. Plants are also susceptible, but although a great amount of data has been accumulated, few studies have been conducted on the *in vivo* effects of aflatoxin on plants. In these hosts, aflatoxin has been found to inhibit seed germination, growth, and chlorophyll development, and to induce chromosomal abnormalities and changes in cellular structures (Figure 22–13).

Other mycotoxins also exhibit a number of carcinogenic, mutagenic, and developmental, or teratogenic (Figure 22–14), effects in both procaryotic and eucaryotic systems. In animal systems, the biological action of mycotoxins is affected by the sex and species of the animal, environmental factors, and nutritional status. Thus, the biological effects of mycotoxins are as varied as their chemical structures.

Mushrooms are another source of toxic substances that, when ingested, may cause drowsiness, nausea, vomiting, hallucinations, convulsions, coma, and death. Mushroom poisoning involves many chemically unrelated toxins that cause cellular destruction and affect different body systems,

Figure 22–13 The effects of the fungal toxin aflatoxin on plants. Cellular changes after exposure to aflatoxin are obvious in the electron micrographs of root cells. (a) Control cells. (b) The effects of treatment with aflatoxin. (CW, cell wall; ER, endoplasmic reticulum; L, lipid bodies; M, mitochondria; N, nucleus; and arrow, light nucleolar cap.) *[E. V. Crisan,* Appl. Microbiol. ***12****:991–1000, 1973.]*

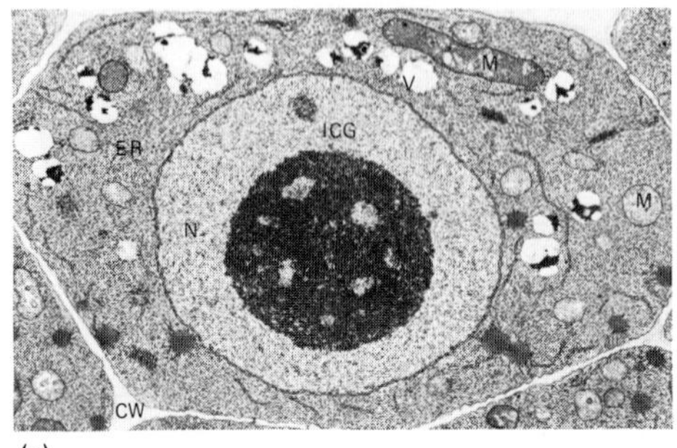

(a)

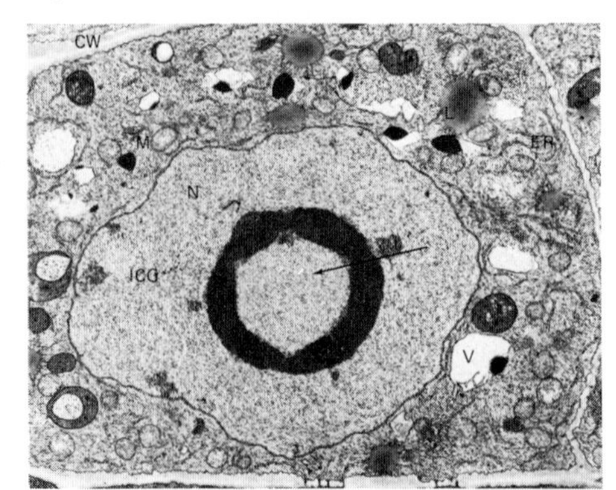

(b)

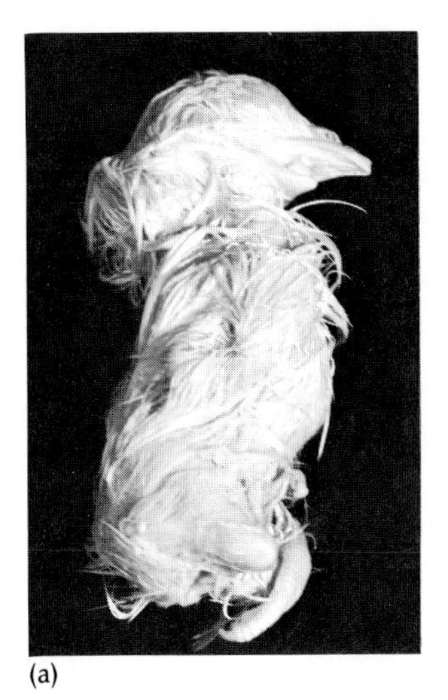
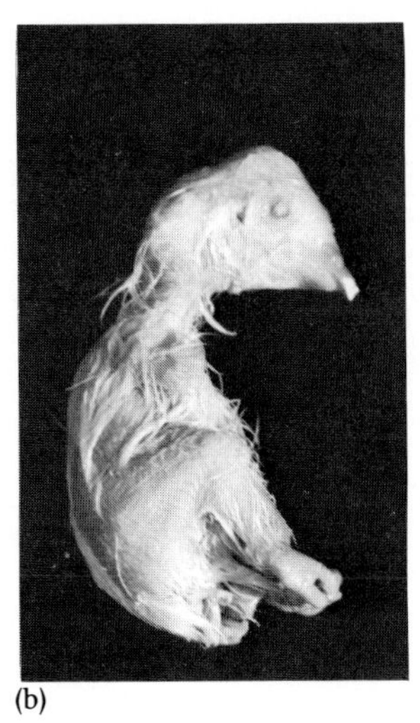
(a) (b)

Figure 22–14 Certain mycotoxins can interfere with the development of chicken embryos. (a) Control. (b) Embryo exposed to the mycotoxin citrinin. Note the twisted head. *[A. Ciegler, R. F. Vesonder, and L. K. Jackson,* Appl. Environ. Microbiol. ***33****:1004–1006, 1977.]*

such as the autonomic nervous system, central nervous system, and gastrointestinal tract. Several toxins interfere with both RNA and DNA transcription. Mushroom poisoning is associated with species of *Amanita* (Color Photograph 35), *Cortinarius*, and *Gyromitra*. More than 50% of all cases are caused by the *A. phalloides* group. **[See mushrooms in Chapter 9.]**

Viral Pathogenicity

Viruses are well-known intracellular parasites. Some of them enter the tissues of a host directly through some injury or insect bite. Most viral infections, however, start on the mucous membranes of the respiratory or gastroinestinal tract. To start an infectious process, virus *virions* must first establish themselves on these mucous membranes in the presence of other microorganisms (including normal flora). To replicate, they must enter susceptible host cells, either in the mucous membrane or in tissues distant from the point of entry. Replication of the virions that invade the mucous membrane can produce disease effects directly, as in the case of respiratory infections. Sometimes, however, it sets the stage for damaging replication in another part of the host. Poliovirus is a good example of this situation. The virus replicates first in alimentary tract cells and ultimately in specific sites in the central nervous system. Knowledge of the factors that affect the early stages of viral infections is incomplete.

Viruses break through host defenses to cause disease. As with bacteria, this process depends not only on the strength of the defenses and on the microorganism's capacity to counteract them but generally also on the number of invaders. A sufficiently large infecting dose can overwhelm the initial defenses of a susceptible host and cause severe injury before adequate defenses can be brought into action.

The Pathological Effects of Viruses

In studies on the effects and mechanisms of viral pathogenicity, two important questions arise: Which pathological effects are specific to virus attack, rather than being the host's nonspecific responses to general injury? How are the pathological effects produced? Our discussion will concern itself with the second question.

Cellular damage of animal tissues by virus attack has been recognized for many years. For example, brain cells are damaged by Newcastle disease virus, and respiratory epithelium is changed and injured by a variety of respiratory viruses (Figure 22–15).

CYTOPATHIC EFFECT

Excessive production of virus or viral components depletes and destroys cellular components essential for cell life or for the repair of mechanical injury. One process by which cellular damage occurs is virus cytotoxic activity (Color Photographs 54 and 80). There are two levels at which pathologically important cytotoxic activity can operate: biochemical damage without morphological damage and biochemical damage with morphological damage (cell lysis, fusion, or death). Morphological damage is usually referred to as a *cytopathic effect*. The general features of such reactions are described in Chapter 10.

LATENT INFECTIONS

Electron microscopy and immunofluorescent techniques have shown that viruses can exist in cells without causing damage. Clearly, not all viral infections result in the immediate death or destruction of the host cell. A good example of this situation is the herpes simplex virus, which causes cold sores or fever blisters. This virus can remain dormant or silent for months, years, or even decades. The viral genome (genetic material) lies latent within nerve cells of ganglia, producing no disease symptoms until stresses such as fever, exposure to extreme cold, emotional problems, or sunburn trigger it, causing small skin eruptions to develop. Yet, between outbreaks, the herpes virus apparently does not destroy the nerve cells in which it continues to exist. **[See herpes virus infections in Chapters 23 and 29.]**

SLOW INFECTIONS

In recent years, it has also become apparent that some slowly developing, persistent diseases that superficially do not appear to be infectious can be caused or triggered by unusual slow viruses or pro-

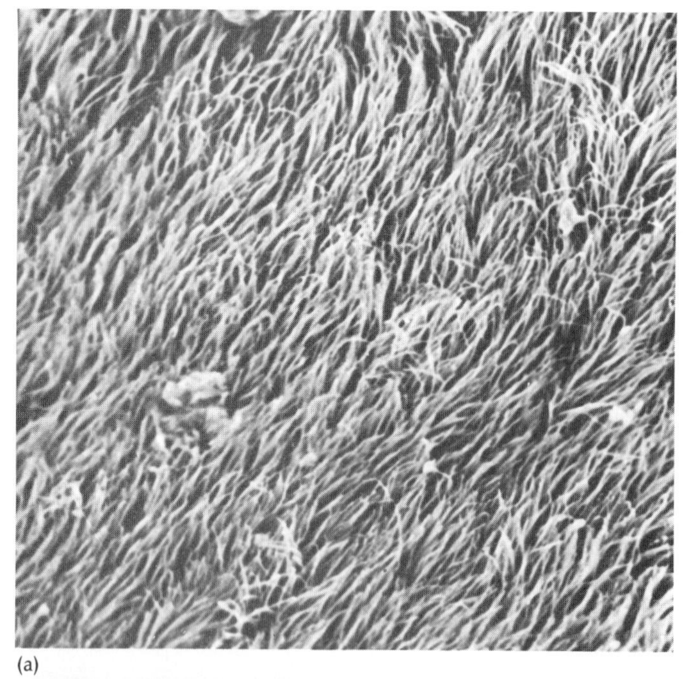

(a)

(b)

Figure 22–15 The destructive effects of respiratory viruses. (a) A scanning electron micrograph showing the appearance of normal tracheal tissue culture cells taken from a calf. Note the number of cilia. (b) The rapid destructive effects of a rhinovirus on the ciliated epithelium 6 days after inoculation of the tissue culture system. (c) Similar destruction by a parainfluenza virus 11 days after inoculation. *[S. E. Reed, and A. Boyde,* Infect. Immun. ***6***:*68–76, 1972.]*

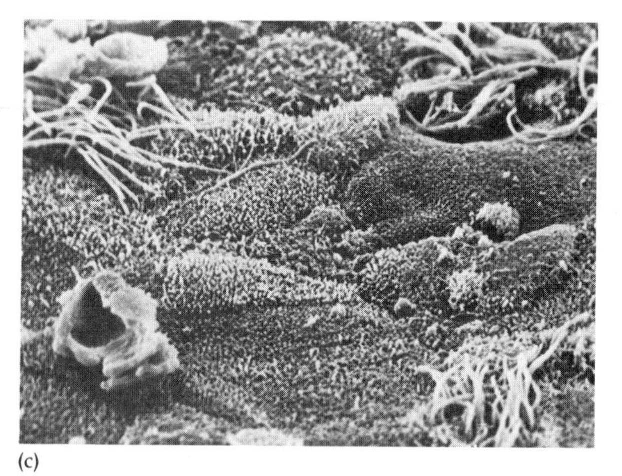

(c)

tein-containing agents called *prions.* Strong evidence has accumulated that several severe neurological diseases are caused by these protein-containing agents. There is also preliminary evidence, as yet inconclusive, that several common degenerative diseases, such as diabetes, leukemia, multiple sclerosis, and rheumatoid arthritis, may actually be the result of slowly developing viral or prion infection. **[See prions in Chapter 10.]**

Opportunists and True Pathogens

True pathogens are the relatively small fraction of those organisms harbored by most forms of animal and plant life that can directly invade tissues and cause obvious infections. *Opportunistic pathogens* are organisms that have the potential to produce infections if they accidentally gain access to the tissues of the host under special circumstances. An opportunistic microorganism takes advantage of weakened defense mechanisms to cause damage to a compromised host. Such microorganisms may or may not be members of a host's normal resident flora, and normally they may or may not be pathogenic to a normal host.

Knowledge of the mechanisms that open the way for invasion by opportunistic microbes is incomplete. Several situations have been identified as contributing to opportunistic infections: genetic defects, the use of antibiotics, and immunosuppressive therapy to limit the activity of the immune system as in cases of cancer or of tissue transplantation. Table 22–4 lists several of these factors and the bacteria to which they predispose the host.

The distinction between an opportunist and a true pathogen can be difficult to make. This is because, in a clinical setting, it is difficult to decide whether a patient's weakened defenses were or were not a necessary precondition for a given disease.

Animal Models

Experimental methods are not generally used to explore the responses of humans to infectious disease agents. However, since various studies have shown that the responses of lower animals appear to parallel those of humans, animal models are commonly used to gain an understanding of diseases. Appropriate laboratory animals such as chimpanzees, mice, rats, and rabbits are incorporated into experiments designed for various purposes, including (1) following the course of infection, (2) identifying factors involved in host resistance, (3) uncovering structural (Figure 22–16) and molecular abnormalities,

TABLE 22–4 Conditions and/or Factors Involved in Bacterial Opportunistic Infections

Predisposing Condition and/or Factor	*Bacteria*	*Infection or Disease State*
Abdominal and other forms of surgery	*Bacteroides fragilis*	Blood poisoning (septicemia)
	Streptococcus spp.	Blood poisoning and lung disease
Alcoholism	*Haemophilus* spp.	Infection of brain coverings (meningitis)
	Klebsiella pneumoniae	Lung disease (pneumonia)
	Mycobacterium tuberculosis	Tuberculosis
Antimicrobial therapy	*Escherichia coli, Clostridium difficile, Proteus vulgaris, Serratia marcescens, Staphylococcus aureus*	Urinary tract infection Pseudomembranous colitis Blood poisoning (septicemia)
Breaks in skin, wounds, burns, etc.	*Acinetobacter calcoaceticus, Bacteroides, Clostridium* spp.	Blood poisoning
	Pseudomonas spp., *Staphylococcus aureus, Streptococcus* spp.	Blood poisoning, lung infection
Diabetes mellitus	*Anaerobes*	Foot ulcers
	Haemophilus spp.	Infection of brain coverings
	Nocardia spp., *Streptococcus pneumoniae*	Lung infection
Immunosuppression	*Corynebacterium* spp.	Lung infection
	Escherichia coli	Urinary tract infection
	Staphylococcus aureus	Blood poisoning
	Mycobacterium avium complex (MAC)[a]	Acquired immune deficiency syndrome (AIDS)[a]
Malnutrition	*Mycobacterium tuberculosis*	Tuberculosis
Malignancies (various types of cancer)	*Aeromonas hydrophilia, Clostridium perfringens* and other clostridia, *Citrobacter* spp., *Escherichia coli,* and *Pseudomonas* spp.	Blood poisoning and lung infections
Sickle cell disease	*Haemophilus* and *Yersinia* spp.	Blood poisoning
Transplants	*Arizona* spp.	Blood poisoning
	Mycobacterium spp.	Lung infection
	Mycobacterium tuberculosis	Tuberculosis

[a] Several species of protozoa and fungi also are associated with AIDS.

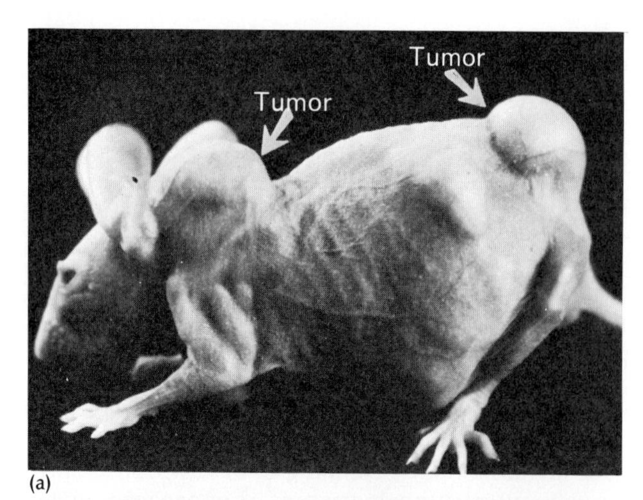

(a)

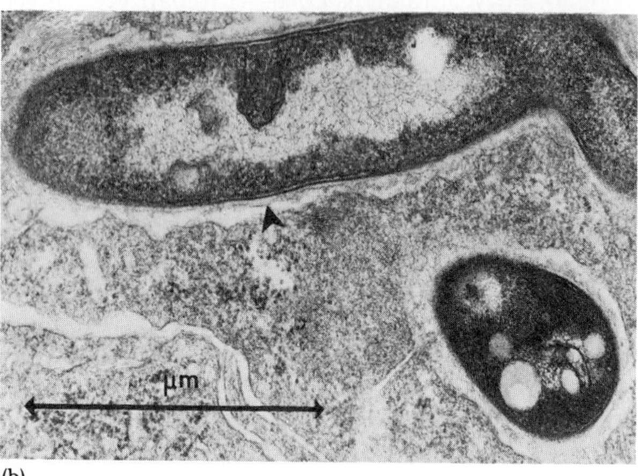

(b)

Figure 22–16 Experimental models for infection. (a) A nude mouse, one example of an animal model used in the study of disease processes. This mouse also is genetically immunodeficient (unable to overcome the effects of disease agents). It has been injected with *Nocardia caviae,* a known bacterial pathogen, in an effort to study the mechanism of host resistance. Note the presence of tumors. (b) An electron micrograph of material taken from one of the tumors. The micrograph shows the presence of nocardial cells. *[B. L. Beaman, and S. M. Scates,* Infect. Immun. ***33****:893–907, 1981.]*

(4) testing the effectiveness of drugs for treatment, (5) determining the effectiveness and safety of vaccines, and (6) identifying the role or virulence factors in disease processes. Much valuable data have been and continue to be gathered from such studies. **[See vaccine safety in Chapter 18.]**

Attacking the First Line of Defense

Microorganisms that can invade a host and cause disease possess a variety of mechanisms to avoid or modify the defense systems of the host. In addition, they have available to them a range of approaches for producing damaging injuries (lesions) and illnesses affecting the entire body (systemic infection). Chapter 23 begins the coverage of infectious diseases by taking a well-rounded look at the microbial diseases involving the first line of defense: the skin and related areas of the body.

Summary

Infectious Disease and Virulence

1. Only a small percentage of microorganisms cause disease.
2. To cause disease, a pathogen must be able to (1) attach to and gain entrance into a susceptible host; (2) multiply in host tissues; (3) resist or not provoke host defenses; and (4) damage the host, often by the secretion of toxins.
3. An infectious disease interferes with the normal function of a host's physiochemical process, because another organism is living within or attached to the host's tissues.
4. Given suitable environmental conditions and a susceptible host, most organisms will exhibit disease-producing potential.
5. The ability of an organism to cause disease is called *pathogenicity. Virulence* refers to the extent of pathogenicity, and depends on properties of the organism such as invasiveness, ability to survive and reproduce in the presence of the host's defense mechanisms, and toxin production.

COLONIZATION

1. Some microorganisms are able to colonize or occupy a site on a host's surface in order to reproduce without necessarily invading tissue.

2. Structures such as bacterial pili and protozoan attachment disks aid in colonization.

INVASIVENESS

1. *Invasiveness* refers to an organism's abilities to survive and to establish itself in host tissues.

2. Obstacles faced by pathogens include phagocytosis, the destructive action of serum upon gram-negative bacteria, and difficulty in spreading through tissues.

3. Three terms used to express the invasiveness of pathogens with respect to the bloodstream are *bacteremia,* the laboratory demonstration of bacteria in blood samples; *septicemia,* the presence of organisms in the blood and their association with toxic symptoms of the host; and *pyemia,* the production by pathogens of localized collections of pus in host tissues.

RELATIVELY NONTOXIC BACTERIAL STRUCTURES AND PRODUCTS CONTRIBUTING TO INVASIVENESS

Bacterial structures and products that can help pathogens to overcome host obstacles include capsules, pili, cell wall components, and enzymes such as coagulase, DNase, hyaluronidase, tissue-destroying proteases, and kinases.

ANTIGENIC VARIATION

Certain pathogens can escape the host's immune system by changing their antigenic surface parts.

Microbial Toxin Production

BACTERIAL TOXINS

1. Bacterial toxins are categorized as either *endotoxins* or *exotoxins.*

2. Endotoxins are associated with the cell walls of gram-negative bacteria. They are lipopolysaccharide-protein complexes involved with certain disease states caused by gram-negative organisms.

3. Minute amounts of endotoxins can be detected by means of the Limulus test, which uses cells from the horseshoe crab.

4. Exotoxins may be secreted during the growth of bacteria or upon the autolysis of cells. They are protein in nature and function by destroying specific parts of cells, or by inhibiting certain cellular activities. Some exotoxins are produced under the direction of bacterial viruses.

5. Exotoxins of clinical significance are associated with diseases such as botulism, pseudomembranous colitis, diphtheria, cholera, scarlet fever, staphylococcal food poisoning, and tetanus.

6. Certain bacterial and fungal species, called *phytotoxins,* produce toxins that affect important food crops.

7. Cyanobacteria are also sources of potentially dangerous toxins.

ALGAL TOXINS

Many algal toxins have been found to be poisonous to fish, waterfowl, mussels, and clams.

MYCOTOXINS

1. Ingestion of foods contaminated with fungus toxins can cause disease states called *mycotoxicoses.*

2. Distinctive properties of mycotoxicoses include non-transmissibility, ineffectiveness of drug or antibiotic treatment, seasonality of outbreaks, association of outbreaks with a specific food item, and demonstrated fungal activity in the suspected food.

3. Mycotoxins produced by *Aspergillus* species, the *aflatoxins,* are the best known.

4. Some mycotoxins also show antibacterial, antifungal, antiprotozoan, and antitumor effects. Some also are carcinogenic.

5. In animal systems, the biological action of mycotoxins is affected by host and environmental factors.

Viral Pathogenicity

1. Viral pathogens are intracellular parasites.

2. The establishment of a virus disease process depends not only on the host's defenses but on the pathogen's ability to counteract them and on the number of invading organisms.

THE PATHOLOGICAL EFFECTS OF VIRUSES

1. Cellular damage of animal tissues by virus attack is well known.

2. Viruses' cytotoxic activity can operate at two levels: biochemical injury without cellular damage and biochemical injury with cellular damage (e.g., cell destruction, fusion, or death). Cellular damage is usually referred to as a *cytopathic effect.*

3. Viral replication can occur in cells without significant damage. Not all virus infections result in cell death. An example of this type of situation is found with cold sores and fever blisters.

4. Some slowly developing, persistent diseases that do not appear to be infectious can be caused or triggered by unusual slow viruses or the protein disease agents known as *prions.* Certain severe neurological diseases may be caused by such agents.

Opportunists and True Pathogens

1. Opportunists are organisms that have the potential to produce infections but do not have the capacity to invade the tissues of a host directly.

2. Opportunists may or may not be members of a host's normal resident flora.

3. Several situations may contribute to the establishment of infections caused by opportunists. These factors include genetic host defects, use of antibiotics, and immunosuppressive therapy.

Animal Models

Experimental methods using laboratory animals are in wide use to study the various factors associated with the development and control of infectious diseases.

Questions for Review

1. Distinguish between the following:
  a. virulence and pathogenicity
  b. opportunist and pathogen
  c. endotoxins and exotoxins
  d. aflatoxin and carcinogen
  e. mycotoxin and algal toxin
  f. mycotoxicoses and mycosis
  g. colonization and invasiveness
  h. prion and virion

2. What microbial factors can influence the course of infection?

3. List five microbial structures or activities that contribute to virulence.

4. List at least four bacterial species that can produce powerful exotoxins together with the diseases each causes.

5. a. What factor or factors may cause a decrease in a pathogen's virulence?
  b. How might virulence be increased?

6. What mechanisms are involved in viral pathogenicity?

7. How important are algal toxins to human welfare?

8. Are all microorganisms pathogenic? Explain your answer.

9. a. What is the *Limulus* test?
  b. What other methods are available to detect microbial toxins? (Refer to Chapter 19.)

10. What effects do toxins cause in the human or other animals?

11. What factors or conditions contribute to opportunistic infections? List five.

12. Of what value are laboratory animals to the control of an infectious disease?

Suggested Readings

BURKE, J. F., and G. Y. YELDICK-SMITH, *The Infection-prone Hospital Patient.* Boston: Little, Brown and Company, 1978. *A short book containing a sufficient amount of information and references to serve as an introduction to the increasing problems of infection-susceptible patients.*

NEVILLE, D. M., and T. H. HUDSON, "Transmembrane Transport of Diphtheria Toxin, Related Toxins and Colicins," *Annual Review of Biochemistry* **55**:195, 1986. *An interesting presentation of the biochemical mechanisms of toxin action and mechanisms by which toxins reach their sites of attack.*

PETERSON, P. K., and P. G. QUIE, "Bacterial Surface Components and the Pathogenesis of Infectious Diseases," *Annual Review of Medicine* **32**:29–43, 1981. *A review article emphasizing the results of research studies over the past decade that have increased the general understanding of the biological significance of the bacterial surface.*

SCOTT, A., *Pirates of the Cell.* New York: Basil Blackwell, 1985. *A well-written examination of human viruses and their destructive effects. A glimpse into future uses of viruses for the treatment of bacterial diseases and genetic defects also is considered.*

VON GRAEVINITZ, A., "The Role of Opportunistic Bacteria in Human Disease," *Annual Review of Microbiology* **31**:447–471, 1977. *More and more attention has turned to infections that occur in individuals having generalized or local defects in the immune defense system. This article deals with bacteria that take advantage of such weakened or compromised hosts.*

WEISS, A. A., and E. L. HEWLETT, "Virulence Factors of *Bordetella pertussis*," *Annual Review of Microbiology* **40**:661–686, 1986. *A well-presented article identifying and describing the virulence factors of the cause of whooping cough.*

23

Microbial Diseases of the Integumentary and Musculoskeletal System

"Diseased nature oftentimes breaks forth in strange eruptions." — *William Shakespeare*

After reading this chapter, you should be able to:

1. Outline the general structure and organization of the skin and indicate the areas attacked by pathogenic microorganisms.
2. Distinguish among the general effects caused by pathogenic bacteria, fungi, and viruses associated with the skin.
3. Discuss at least two each of bacterial, mycotic, and viral diseases of the skin, including the causative agents, means of transmission, laboratory diagnosis, possible treatment, and control measures.
4. List the general differences in approaches used for the identification of microorganisms pathogenic for the skin, nails, and hair.
5. Discuss the roles of opportunistic fungi in causing disease.
6. Distinguish between superficial mycoses (fungus infections of the skin, nails, and hair) and deep-seated mycoses (infections of the deeper tissues and organs).
7. Explain the relationship of varicella-zoster virus to chickenpox and shingles and list possible complications of these diseases.
8. Describe the general types of complications that can be associated with microbial diseases of the skin, nails, and hair.
9. Describe the nature of more recently discovered viral diseases of the skin and their significance.
10. Describe the general organization of synovial joints and the sites most commonly infected.
11. Apply the clinical symptoms and laboratory findings to a possible diagnosis of a disease challenge.

Chapter 23 describes several microbial diseases associated with the skin, nails, and hair. The effects of bacteria, fungi, and viruses and the methods by which these microbes are spread, identified, and controlled are discussed. Attention is also given to the increasing number of infections associated with the musculoskeletal system.

Human beings are both the primary source and the target of various pathogenic or potentially pathogenic microorganisms. From birth to death, we live in an environment that is rarely free of such organisms. Moreover, many body regions such as the gastrointestinal tract, nose, skin, and throat serve as additional sources of pathogens. Diseases of the skin are caused by a variety of bacteria, fungi, protozoa, viruses, and worms. Many organisms capable of causing localized infections of the skin may also cause injury of internal tissues and organs (**systemic infection**). This chapter will begin with general descriptions of the skin's organization and the appearance of selected skin lesions (infected areas).

Organization of the Skin

The skin, together with hair, nails, and various glands, makes up the covering of the human body. As Figure 23–1 shows, the skin consists of two main structural parts: the *epidermis* (outer layer) and, underneath this layer, the *dermis.* A *subcutaneous tissue* layer composed of loose connective tissue is located below the dermis and serves to attach the skin to underlying structures.

The dermis region contains dense, irregularly arranged connective tissue and various types of cells, including fibroblasts, histiocytes (macrophages), T lymphocytes, mast cells, blood and lymphatic vessels, and nerves. Hair follicles, sweat glands, superficial sebaceous glands, and a variable amount of muscle are also found in this layer.

The subcutaneous layer contains a large number of components, including fat tissue, blood vessels, special nerve endings, nerve trunks, hair follicles, sebaceous glands, and sweat glands.

In addition to its functions of excretion, reception of external stimuli, secretion, and temperature regulation, intact skin serves as a natural protective barrier to the majority of infectious disease agents, as discussed in Chapter 17. However, hair follicles and the openings of secreting glands constitute potential portals of entry for pathogens (Figure 23–2). Certain physiologic factors are important in providing barriers to skin-invading microorganisms. These factors include the acidity of the skin, the presence of an indigenous microbial flora (microbiota), and a temperature that may prevent the growth of certain disease agents. Another extremely important factor, which applies not only to skin invasion but also to injuries of other body regions, is the inflammatory reaction. The outcome of an infection is largely determined by this local response. **[See Chapter 17 for a description of inflammation.]**

Skin Lesions

The breaks in the skin that occur during the course of normal living provide opportunities for infectious agents to enter the body. The nature and extent of injuries affect host–parasite relationships to varying degrees. So-called minor infections can develop into serious problems if they (1) spread and involve

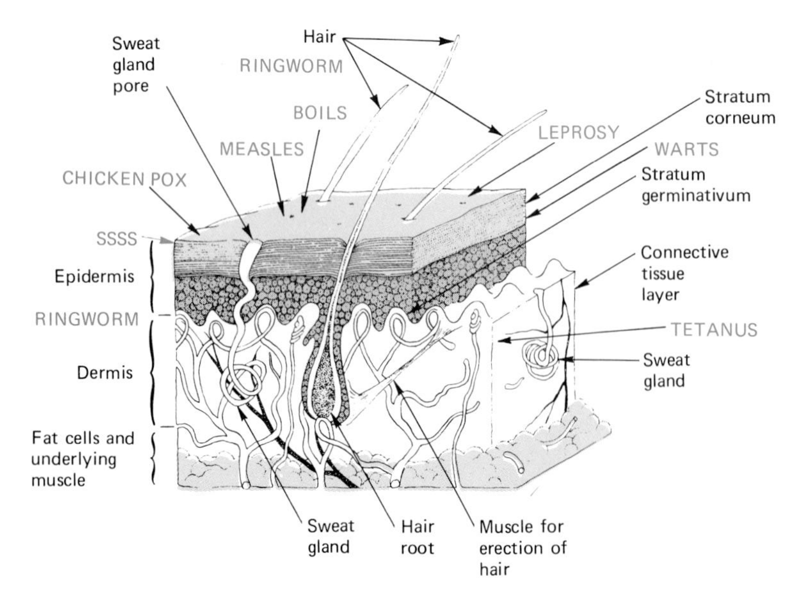

Figure 23–1 A vertical section of human skin, showing its structural arrangement. Sites of certain infections also are indicated. (SSSS, staphylococcal scalded skin syndrome.)

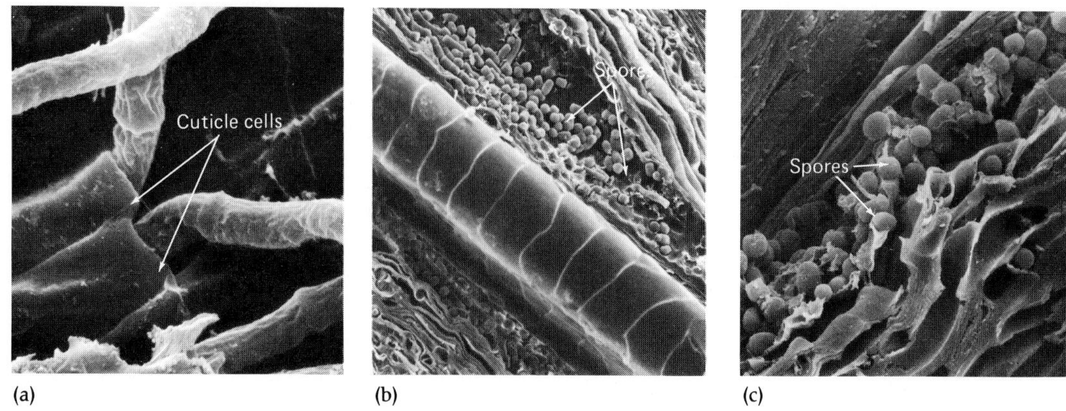

(a) (b) (c)

Figure 23–2 Scanning micrographs showing the intricate invasion pattern of skin- and hair-destroying fungi (dermatophytes). (a) Hyphae can be seen here wedging under the free edges of overlapping cuticle cells, which encase the hair shaft. (b) Spore formation from hyphae in progress next to a hair follicle. (c) Invasive hyphae carve tunnels in hair that resemble wormholes in wood. Note the presence of spores in the space between the hair shaft and follicle. *[R. D. Hutton, S. Kerbs, and K. Yee,* Infect. Immun. ***21****:247–253, 1978.]*

neighboring tissues, (2) cause bleeding, (3) produce local anemia because of stoppage of circulation, (4) cause the loss of skin, or (5) result in edema (swelling). Table 23–1 provides brief descriptions of selected lesions found with infectious diseases of the skin. (See Color Photographs 91a through 91d.)

This chapter will discuss some of the microbial skin infections in more detail.

TABLE 23–1 Brief Descriptions of Selected Skin Lesions

Type of Lesion	*Description*
Blebs, blisters	Thin-walled, rounded or irregularly shaped blisters containing serum or a combination of serum and pus
Carbuncle	A deep sore or ulcer lesion of the skin and subcutaneous tissue; usually a hardened border and draining of pus are evident
Crusts (crustae or scabs)	Dried accumulations of blood, pus, or serum combined with cellular and bacterial debris; detachment of thin crusts may leave dry or moist red bases; these generally heal, resulting in a smooth skin surface; scar formations are usually associated with thick crusts covering ulcers (See Color Photograph 91d)
Folliculitis	Inflammation of hair follicles characterized by a circumscribed, elevated, reddened lesion at the mouth of a follicle pierced by a hair; pus formation, crusting, and drying follow.
Furuncles	Localized inflamed regions that develop soft centers and eventually discharge pus
Macules (or spots)	Usually round, circumscribed changes in the color of the skin; the lesion is neither elevated nor depressed; the outline of a spot or macule may either be quite distinct or blend into the surrounding region (See Color Photograph 91b)
Maculopapules	Slightly raised macules
Papules (or pimples)	Circumscribed, solid, elevated lesions without visible fluid contents; the color, consistency, and size of papules vary
Pustules	Small elevated skin lesions containing pus; they may develop from papules; these lesions also vary in color, size, and contents (pus, blood, or both); pustules may consist of a single cavity or several compartments with fluid (See Color Photograph 91c)
Scars	Newly formed connective tissue that replaces tissue lost through injury or disease; these secondary lesions tend to be pink at first; then they assume a glistening appearance; scars are normal components of the healing process

(continues)

TABLE 23–1 (continued)

Type of Lesion	*Description*
Ulcers	Rounded or irregularly shaped depressions; these lesions vary in size (See Color Photograph 91d)
Vesicles (small blisters)	Elevations that may occur irregularly or in groups or rows; they may contain blood, pus, or serum; the color of the lesion depends upon its contents; vesicles may arise from a macule or papule and develop into pustules (See Color Photograph 91a)

Microbiology Highlights

SKIN CLUES IN THE DIAGNOSIS OF LIFE-THREATENING INFECTIONS

Early diagnosis and treatment are critical in life-threatening bacterial infections. The appearance and features of infected skin areas (lesions) may provide an important clue regarding the cause of such an infection. Also of value in diagnosis are smears of material oozing from lesions or scrapings from their bases, which are then gram-stained. This is especially the case in uncovering the cause of the presence of bacteria in blood (*septicemia*). Skin lesions that occur with septicemia may result from one or more of five main processes: (1) spreading of clots within blood vessels, (2) direct invasion through blood vessels and blockage by bacteria or fungi, (3) inflammation caused by an immunologic reaction, (4) blockage of blood vessels caused by heart tissue infection, and (5) the destructive effects of microbial toxins on blood vessels.

Bacterial Diseases

The signs, symptoms, and pathological features associated with skin infections vary from localized effects, such as those listed in Table 23–1, to extensive involvement and penetration of deeper tissues (Color Photograph 4) and even death. The identification of microbial pathogens of the skin frequently depends on the isolation and cultivation of the pathogen and the results of various biochemical tests of the types mentioned in Chapter 21. Skin tests or other immunological procedures may also be useful in the diagnosis. Table 23–2 describes the clinical features of selected bacterial skin infections. Laboratory diagnosis and treatment are included in the disease sections.

TABLE 23–2 Features of Selected Bacterial Diseases Associated with the Skin

Disease	*Causative Agent*	*Gram Reaction and Morphology*	*Incubation Period*	*General Features of the Disease*
Anthrax	*Bacillus anthracis*	+, Rod	Less than 1 week	A reddened, elevated, swollen pimple develops at site of infection; it may lead to bloodstream invasion (septicemia) and tissue death; oral lesions are reddened and swollen; pulmonary infection causes severe lung damage and death
Boils, carbuncles	*Staphylococcus aureus*	+, Coccus	4–10 days	Localized swollen areas of tissue destruction in deeper skin layers; may lead to bloodstream invasion; fever and general malaise

TABLE 23–2 (continued)

Disease	*Causative Agent*	*Gram Reaction and Morphology*	*Incubation Period*	*General Features of the Disease*
Erysipelas	Beta-hemolytic group A streptococci	+, Coccus	Unknown, probably 2 days	Fever, headache, stinging, or itching at site of infection, developing into widespread thickened, reddened areas
Folliculitis	*Pseudomonas aeruginosa*	−, Rod	1–2 days	Fever, swollen lymph nodes, general discomfort (malaise), reddened blisters, some with pus tips
	Staphylococcus aureus	+, Coccus	Same	Localized swollen areas that develop soft centers, some with pus
Furuncles	*Staphylococcus aureus*	+, Coccus	Several days	Localized swollen areas that develop soft centers and eventually discharge pus (Color Photograph 92)
Gas gangrene[b]	*Clostridium perfringens* and other clostridia	+, Rod	1–3 days	Usually affects muscle tissue; fever, fast heartbeat, severe pain; infected wounds smell foul, have a discharge, and accumulate gas within tissues
Green-nail syndrome and toe web infection	*Pseudomonas aeruginosa*	−, Rod	1 week or longer	Greenish discoloration of nail plate (Color Photograph 93); formation of thick, white scaling areas between toes
Impetigo contagiosa	*Staphylococcus aureus*	+, Coccus	4–10 days	Crust, scabs; localized pain and fever accompany the disease (Color Photograph 4)
	Beta-hemolytic group A streptococci	+, Coccus	Same	Less severe form than for *Staphylococcus aureus*
Leprosy (Hansen's disease)	*Mycobacterium leprae*	Not done; acid-fast Rod	1–5 years or longer	Four different types of the disease are recognized: *lepromatous*—round, nonelevated patches showing skin color changes (macules); *tuberculoid*—well-defined, reddened or nonpigmented areas, loss of sensation, and nerve destruction; *intermediate*—macules and nerve involvement; lepromin skin test may be positive; *borderline*—infectious form of the disease, showing the features of both lepromatous and tuberculoid leprosy
Pseudomonas pyoderma	*Pseudomonas aeruginosa*	−, Rod	1 to several days	Eroded and macerated skin surface, producing a bluish green pus and a grape odor
Scarlet fever	Beta-hemolytic group A streptococci	+, Coccus	4–10 days	Fever, headache, sore throat, vomiting, raised reddened rash, "strawberry tongue"; peeling of body surface and tongue may occur
Staphylococcal scalded skin syndrome (SSSS)	*Staphylococcus aureus*	+, Coccus	Less than 1 week	Starts with a distinctive faint, macular, yellow to brick-red rash following an eyelid or respiratory infection; skin tenderness involving central portions of the face, neck, armpits, and groin; spontaneous skin wrinkling and huge blisters develop, followed by skin peeling; skin drying and healing without scarring within 5–7 days after antibiotic therapy; malaise (general discomfort), irritability, and fever also accompany the disease
Tetanus (lockjaw)	*Clostridium tetani*	+, Rod	2–4 days or longer	Sudden and violent involuntary contractions of voluntary muscles, convulsions, locking of jaw muscles; fever and pain may be present
Wound botulism	*Clostridium botulinum*	+, Rod	Unknown, possibly 12–36 hr	Fever, double vision, difficulty in talking and swallowing, neck weakness

[a] Pathogenic *Staphylococcus aureus* produces acid from mannitol and a positive coagulase test (Color Photographs 26 and 88).
[b] *Staphylococcus aureus* and *Streptococcus pyogenes*, functioning together, can produce a form of this disease referred to as *synergistic gas gangrene*.
[c] Sulfuric acid rather than hydrochloric acid in combination with ethanol has been recommended as the decolorizing agent for staining.

Anthrax

Anthrax is primarily an infection of domestic animals such as cattle, goats, and sheep. It has also been reported in other warm-blooded animals, including camels, cats, chickens, elephants, horses, rodents, and wild deer. Although the human form of the disease (Figure 23–3) was first reported in the mid-eighteenth century, it was not recognized as related to the disease of animals. Robert Koch, applying the procedure now described in Koch's postulates, showed clearly that anthrax is caused by *Bacillus anthracis* (bah-SIL-lus an-THRAY-sis) (Figure 1–25). Two general forms of anthrax are recognized, cutaneous and pulmonary. Cutaneous anthrax is the more common and less severe form of the disease.

Transmission. Anthrax is found worldwide but is most prevalent in Africa, Asia, and central and southern Europe. Most human cases of the disease are acquired by handling infected animals and their products, such as meat, hides, hair, wool, bristles, bones, and manure. Oral forms of anthrax have developed from use of unsterilized toothbrushes made from bristles contaminated by *Bacillus anthracis* spores. Infection results from direct contact (Color Photograph 94), inhalation, or even possibly the ingestion of the etiologic agent. Insect bites have also been implicated as a means of transmission.

Control. In 1881, Louis Pasteur demonstrated the effectiveness of a vaccine against anthrax. The preparation contained organisms attenuated by cultivation at higher than optimal temperatures, 42° to 43°C. Today, similar vaccines are routinely used. Unfortunately, these preparations do not produce long-lasting immunity—protection ranges from 9 to 12 months—so vaccination affords only partial control.

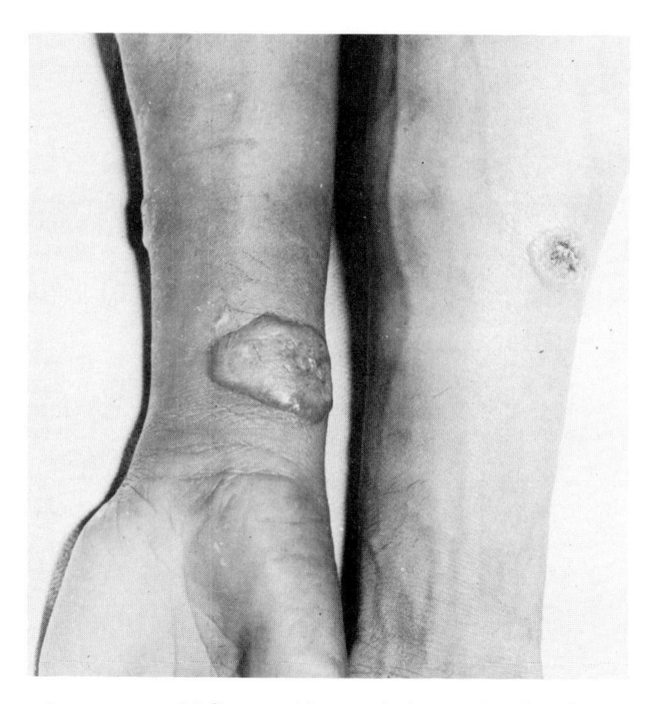

Figure 23–3 Malignant (destructive) pustule of anthrax. *[Courtesy of the Armed Forces Institute of Pathology, Washington, D.C., Neg. No. D-45409-10.]*

Other anthrax control methods include the disposal of infected carcasses, by cremation or deep burial with the addition of lime, and the disinfection of infected animal products by boiling or the use of formaldehyde.

Anthrax

Laboratory Diagnosis

B. anthracis can be microscopically demonstrated as well as cultured from tissue biopsy specimens. Characteristics for the identification of this pathogen should include the finding of "curled hair locks" or medusa-head colonies that are nonhemolytic on isolation; gram-positive rods, a capsule; susceptibility to penicillin, and the absence of motility.

Treatment

The administration of an appropriate antibiotic before *B. anthracis* gains access to the bloodstream (septicemia) is critical. Antibiotics are not effective against the pathogen's toxin. Penicillin is often the drug of choice. Chloramphenicol, erythromycin, and tetracycline also are used.

Gas Gangrene

LABORATORY DIAGNOSIS

The demonstration of gram-positive spore formers in a wound serves as a tentative diagnosis of gas gangrene. Further identification is based on the isolation of *C. perfringens* on blood agar, with colonies surrounded by an inner zone of complete hemolysis and an outer zone of incomplete hemolysis.

TREATMENT

Successful treatment of gas gangrene includes the immediate surgical removal of all dead tissue from the wound, the application of antitoxins against the major clostridial species causing the condition (polyvalent antitoxin preparation), and antibiotic therapy. Penicillin and tetracyclines are usually the drugs of choice. *Hyperbaric oxygen therapy,* the use of high concentrations of oxygen at elevated pressures, combined with the use of polyvalent antitoxins is also considered effective.

Gas Gangrene

Several species of the genus *Clostridium* (klos-TRE-dee-um) are known to cause gas gangrene. Among them are *C. novyi* (C. nov-EE-eye), *C. perfringens* (C. per-FRIN-jens), and *C. septicum* (C. sep-TE-kum). Most are not highly invasive, but they secrete highly injurious toxins during their growth in damaged tissues. Such toxic substances may be hemolytic, tissue-destroying, or lethal. The growth of clostridia in tissues often results in the accumulation of gas, mainly hydrogen, and of toxic breakdown products of tissues. The toxic products can cause extensive connective and muscle tissue destruction as well as serious systemic involvement.

TRANSMISSION. Infection occurs by the contamination of open wounds (incisions or lacerations) by clostridial spores. The process continues when the spores of the causative agent germinate in tissues that provide anaerobic conditions. Germinated cells undergo multiplication, grow, and secrete toxins. Infections have been associated with wounds resulting from attempted abortions, automobile accidents, frostbite, and military combat.

Leprosy (Hansen's Disease)

Leprosy is a chronic infectious disease of humans caused by the acid-fast, nonmotile rod *Mycobacterium leprae* (my-koh-back-TE-re-um LEP-ray) (Figure 28–4). This microorganism was first observed in the skin lesions of leprosy patients in 1873 by Armauer Hansen. Although *M. leprae* was one of the first microorganisms reported to cause a human disease, so far it has been difficult to culture *in vitro* on artificial laboratory media. Cultivation of mycobacteria isolated from leprous tissue and alleged to be *M. leprae* has been achieved in a specially prepared medium containing hyaluronic acid, which apparently plays a role in strengthening cell walls made leaky by the constant intracellular life of the bacteria. In addition the introduction of leprous specimens into experimental animals has only recently resulted in the development of a typical disease state. Inoculation of nine-banded armadillos and chimpanzees has resulted in experimental infections and the growth of bacteria that are metabolically similar to other cultured *M. leprae.* Armadillos and the mangabey monkey are excellent animal models for the study of leprosy and the evaluation of chemotherapeutic agents.

Several of the early descriptions of leprosylike diseases have made it very difficult to determine when the infection first appeared. Contrary to common opinion, the skin eruptions described in the Old Testament (Leviticus) were not what we now call leprosy. The word *lepra* was applied by early physicians to all scaly skin eruptions. The purpose of this practice was simply to avoid confusing such disease states with any of the better-recognized infections. Despite the vagueness of early descriptions, leprosy is though to have been present in various parts of Africa and the Far East before the Christian era. Leprosy subsequently spread throughout Europe as a consequence of military actions and an increase in migration. It appears likely that leprosy was introduced into the American continent from Europe and Africa. Today, the disease occurs mainly in tropical areas.

Microbiology Highlights

NATURALLY ACQUIRED LEPROSY IN A MONKEY

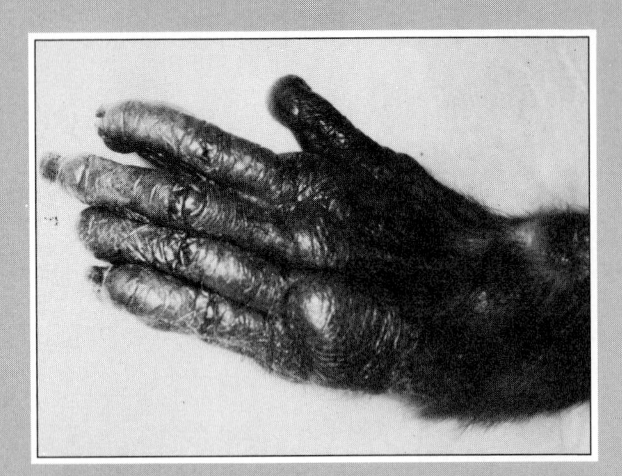

The appearance of the mangabey monkey with leprosy. Specimens from skin contained large numbers of acid-fast bacilli.

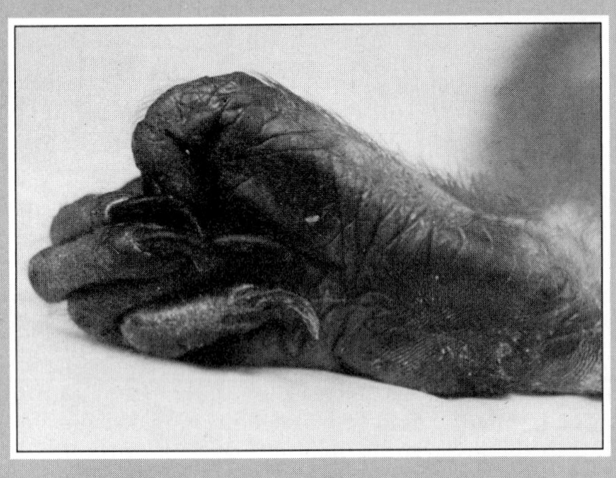

The clawing of the toes of the monkey's foot. *[From W. M. Meyers, and associates,* Internat. J. Leprosy *53:1, 1985.]*

The discovery of the leprosy bacillus in a lesion from a patient with lepromatous leprosy not only established that leprosy was infectious but also launched the search for a suitable animal model with which to follow the disease. The best animal model, of course, would duplicate a major number of features that were found with the disease in humans. Furthermore, in the words of Leader and Padgett in 1980, ". . . the discovery, recognition, and exploitation of an animal model to contribute ideally to human and animal welfare should accomplish the end of eliminating the disease in question."

In the case of human leprosy, the transmission of the disease agent to nine-banded armadillos was an important advance in the search for an animal model. However, the successful infection of one chimpanzee and the finding of naturally acquired infections in another as well as in a mangabey monkey offers greater promise as models with which to study and eventually find a means to eliminate the disease. The finding of naturally acquired leprosy infections raises a number of questions, including the following: What is the extent of leprosy in wild monkeys? Can the disease be transmitted from monkeys to humans? Or from humans to monkeys?

The leprosy infection in the mangabey monkey was believe to have been acquired from a human with an active form of the disease. Many of the laboratory and clinical features of the disease in the monkey are similar if not identical to leprosy in the human.

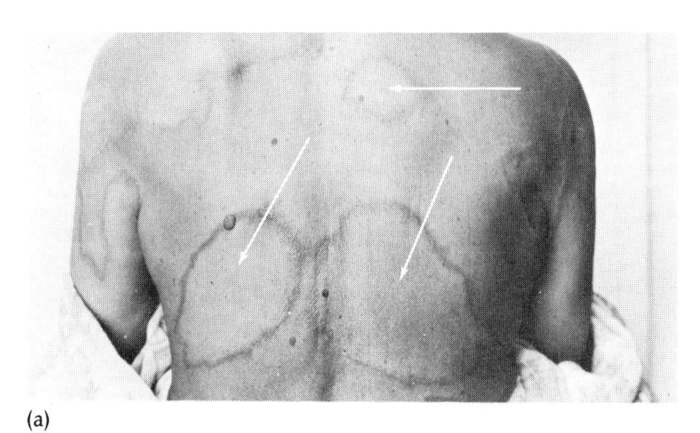

(a)

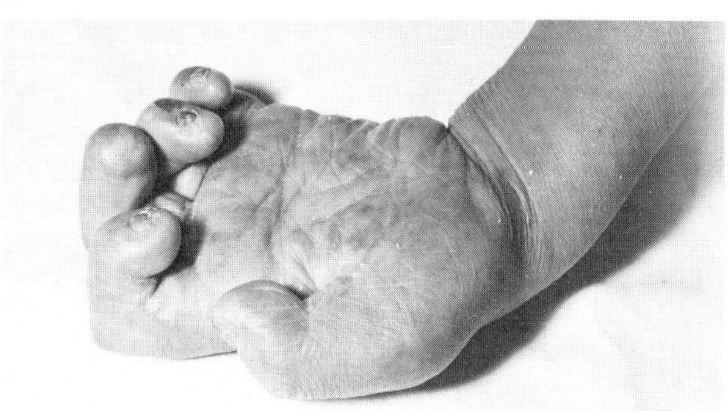

(b)

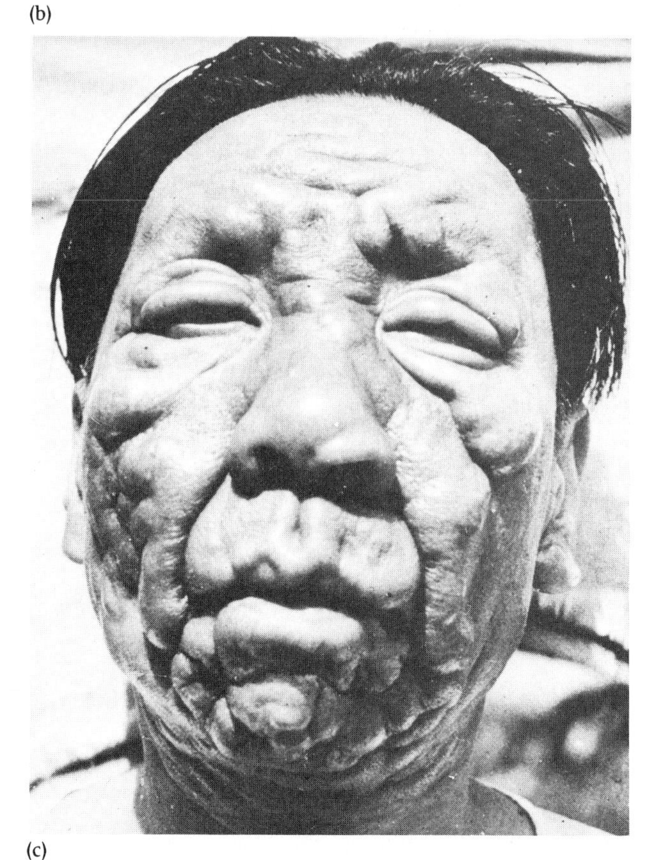

(c)

Figure 23–4 (a) In tuberculoid leprosy, the skin within the nodules (arrows) is completely without sensation. *[Courtesy of the Pathology Research Laboratory, Leonard Wood Memorial, Armed Forces Institute of Pathology, Washington, D.C.]* (b) The deformed "claw" hand associated with lepromatous leprosy. *[Courtesy of the Armed Forces Institute of Pathology, Washington, D.C., Neg. No. AFIP 56-14075-2.]* (c) The lion face (facies leprosa), one of the major deformities possible in lepromatous forms of leprosy. *[Courtesy of Dr. O. K. Skinsnes.]*

Leprosy affects predominantly the skin and peripheral nerves and results in a high prevalence of deformities involving these tissues in the face, hands, and feet (Figure 23–4b). Although several types of leprosy are known, the *tuberculoid* (nervous system involvement) and the *lepromatous* forms are more generally recognized (Table 23–2). Individuals with tuberculoid leprosy exhibit regions of the skin that have lost sensation and are surrounded by a border of small swellings or nodules (Figure 23–4a). Lepromatous leprosy is characterized by the formation of disfiguring nodules (masses) on various parts of the body, deformations of the hands and feet, destruction of tissue, and involvement of mucous membranes of the nose that may result in the "lion face" (Figure 23–4c). The *lepromin* skin test (Figure 23–5), which is patterned after the tuberculin skin test for tuberculosis, is positive in most cases of tuberculoid leprosy and negative in the lepromatous form. Although the lepromin test is not a diagnostic procedure, it does differentiate persons who are able to develop cell-mediated immunity to *M. leprae* from those who cannot do so. **[See type IV hypersensitivity in Chapter 20.]**

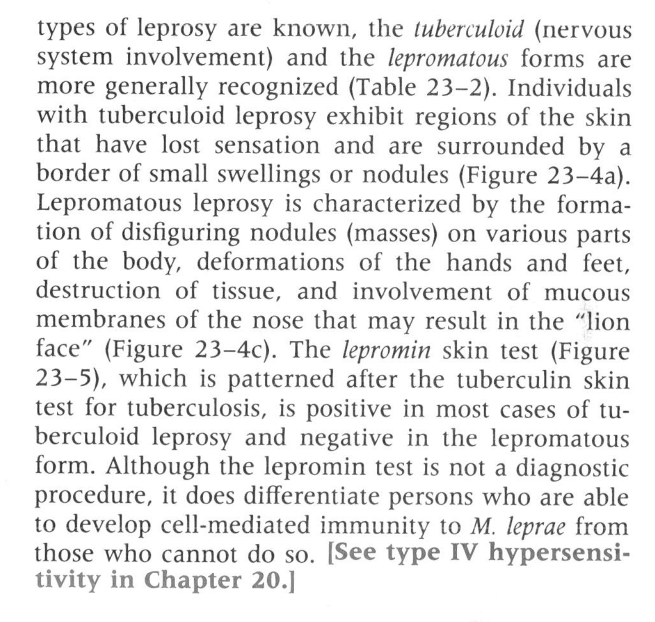

Predisposing Factors. Leprosy is usually acquired during childhood, although exposure and resultant infections are known to occur in adults. In general, the longer and more intimate the contact with infectious persons, the more likely it is that the infection will be transmitted. It is generally believed that the incubation period of leprosy is 3 to 5 years, but in some instances it may be as long as 15 to 20 years. Shorter periods are also known.

Sources of Infection and Mode of Transmission. Sources of human infection are human cases. This is no relationship between human and rat leprosy. Rat leprosy is caused by another microorganism, *M. lepraemurium* (M. LEP-ray-muir-ee-um).

Leprosy is usually transmitted by prolonged direct contact with infected (lepromatous) patients or by inhalation of organisms from sputum, nasal, or other types of discharges. Indirect contact via contaminated objects or various arthropods, including bedbugs, cockroaches, mosquitoes, and flies, has also been inferred. The portal of entry is open to question; the skin and the mucous membranes of the nose and throat have been suspected.

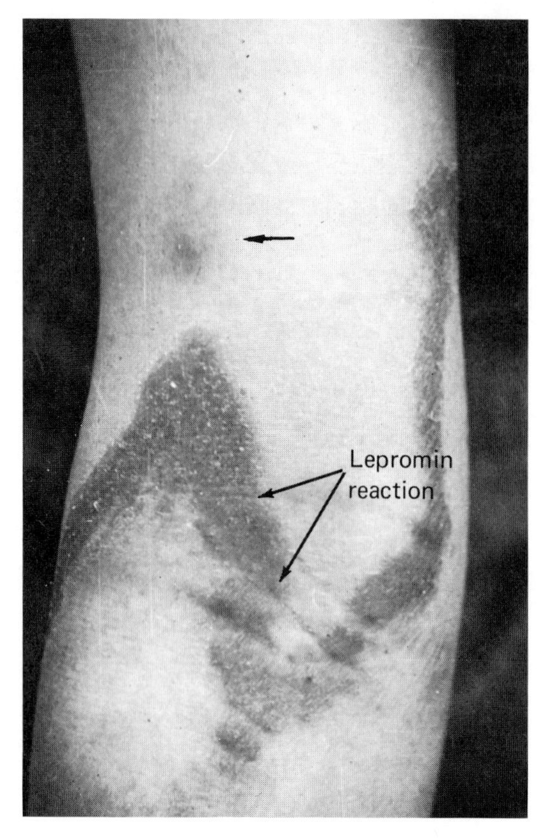

Figure 23–5 Immunologic responses in leprosy. A tuberculoid lesion (upper arrow) and an early lepromin reaction can be seen on this arm. *[Courtesy Dr. O. K. Skinsnes.]*

PREVENTION AND CORRECTION. Naturally, the reporting of leprosy cases and the hospitalization of patients are desirable. Hospitalization is especially important for all individuals with the lepromatous (progressive or infectious) form of the disease (Figure 23–4). One important aspect of hospitalization that is frequently overlooked is that it offers an opportunity to inform and educate patients, especially with respect to their treatment.

Contaminated articles should be adequately disinfected in all cases of suspected or known leprosy.

It has been estimated that approximately one-fourth of the 10 million people in the world who have leprosy need surgery for the correction of deformities. Deformity caused by leprosy represents one of the largest problems of reconstructive surgery in the world. Surgical procedures are being done in some cases to correct disfigurements and impaired functions of leprous deformities.

Several prophylactic measures against leprosy are also in use. These include bacillus of Calmette and Guérin (BCG) vaccination and the administration of small doses of dapsone (DDS) to children exposed to leprosy within their families. The vaccination results in a positive lepromin test, as well as a positive tuberculin test (Figure 23–5).

Pseudomonas Infections

Pseudomonas aeruginosa (soo-doh-MOH-nass ar-ra-jin-OH-sah) has been largely ignored as a causative agent of several specific skin diseases. This gram-negative rod has been reported to produce a variety of skin infections associated with epidermal destruction (Color Photograph 95), as with burns. *Pseudomonas* septicemia may develop from these cases or from changes in the normal human flora or immune mechanisms of the individual brought on by the use of antibiotics or drugs capable of suppressing antibody production. In healthy persons the pathogenic effects of this organism are limited. Examples of these dermatologic infections include folliculitis (Color Photograph 95), (inflammation of hair follicles), "green nail" syndrome (Color Photograph 93), and toe web infection (Table 23–2).

Leprosy

LABORATORY DIAGNOSIS

The demonstration of acid-fast rods in skin scrapings from characteristic lesions—along with clincal features of the disease—is generally sufficient for a diagnosis of leprosy. Additional types of specimens that can be used include nasal secretions and fluid obtained from swollen lymph nodes. Animal inoculations are sometimes used to confirm the diagnosis.

TREATMENT

Leprosy can be treated successfully with dapsone, a bacteriostatic sulfone drug. Rifampin is an alternate choice.

Pseudomonas Infection

Laboratory Diagnosis

Pseudomonas species can be cultured on standard media. These bacteria are noted for their production of green and purple pigments on appropriate media, and their inability to ferment carbohydrates. All strains are oxidase positive (Color Photograph 116b).

Treatment

Most infections are treated with colistin, gentamicin, and polymyxin B. Since these drugs are toxic, their use should be monitored. Severe *Pseudomonas* infections are treated with carbenicillin alone, or carbenicillin with gentamicin. Other antibiotics used include tobramycin together with ticarcillin.

Staphylococcal Infections

Staphylococcus aureus (staff-il-oh-KOK-kuss OH-rēe-us) is a gram-positive, spherical bacterium that usually appears in grapelike clusters (Color Photograph 12a). In humans, staphylococci produce a variety of infections that involve any and all tissues of the body. The disease states are characterized by pus formation. They include carbuncles, deep tissue abscesses, empyema, endocarditis, furuncles, boils, impetigo contagiosa (Color Photograph 4), meningitis, osteomyelitis, pneumonia, scalded skin syndrome, toxic shock syndrome, and wound infections. In recent years, staphylococcal infections have gained additional significance because of the appearance of drug-resistant mutants of *S. aureus.* [See toxic shock syndrome in Chapter 28.]

Transmission and Predisposing Factors. The exact mode of transmission involved in staphylococcal diseases is not always clear. Droplet nuclei, airborne organisms, carriers, and direct contact with individuals having open infected wounds have all been implicated as sources of disease agents. Hospital environments and hospital personnel in particular serve as sources of staphylococci in epidemic situations. Intimate contact with attendants, nurses, and physicians harboring these organisms in their nasopharyngeal regions poses serious problems in control. Carriers who happen to be food handlers and the improper disposal of contaminated inanimate objects (fomites) represent other difficult control situations. [See bacterial toxins in Chapter 22.]

In general, infections caused by *S. aureus* occur in persons whose local and general defense mechanisms are significantly lowered. Individuals with chronic debilitating diseases—such as cancer, diabetes mellitus, and cirrhosis of the liver—are vulnerable to staphylococcal infections. Furthermore, infections can result from exposure to skin irritants or as a complication of burns or of wounds produced either accidentally or from surgical operations.

Frequent, recurrent skin-limited infections often occur during the years of puberty. Examples of such diseases include inflammations of hair follicles (folliculitis) and acne vulgaris. The latter is a chronic condition that appears as inflammations of the sebaceous glands located on the back, chest, and face.

Prevention. Great care must be exercised in treating staphylococcal infections. The proper handling and disposal of contaminated objects must be observed in hospital as well as home environments. Individuals with infections of this type should be made aware of the potential danger they can pose to others if adequate hygienic habits are not observed. Infection of hair follicles by staphylococci frequently results in localized superficial abscesses or boils. Although these lesions generally heal spontaneously, they may spread to other tissues and produce furuncles.

FURUNCLES AND CARBUNCLES

Poor hygiene and nutrition, as well as the irritation produced by the rubbing of clothing, can contribute to the formation of both boils, or furuncles, and carbuncles (Figure 23–6).

IMPETIGO CONTAGIOSA

Impetigo contagiosa, a communicable infection of the superficial skin layers, is commonly found in areas where hygienic conditions are poor. Occasionally the disease is epidemic, but in most situations it is sporadic, with children under age ten the most

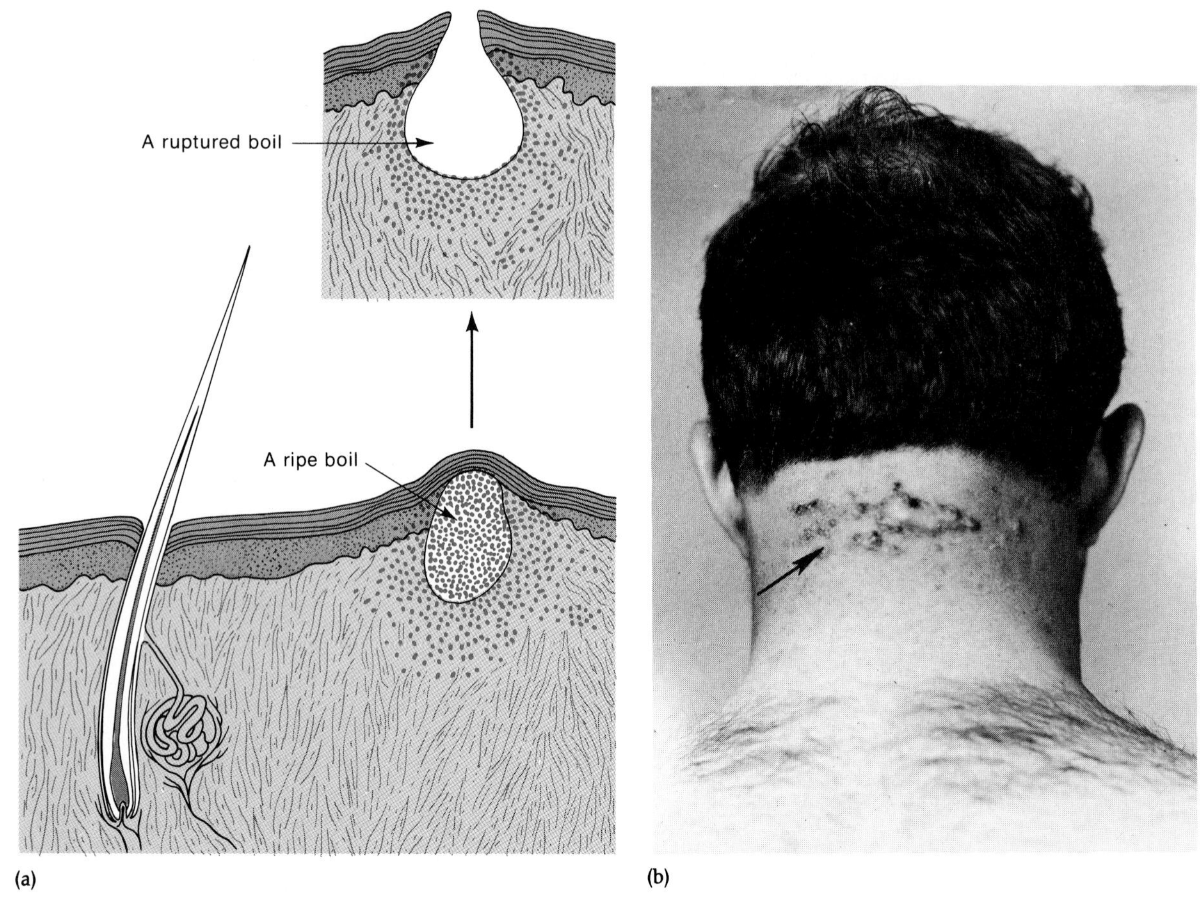

Figure 23–6 Staphylococcal infections of the skin. (a) A boil, its general formation and eventual rupture, with the release of bacteria. *[After T. D. Brock, and K. M. Brock,* Basic Microbiology. *Englewood Cliffs, N.J.: Prentice-Hall, 1978.]* (b) Chronic furunculosis. This patient also has inflammations of hair follicles (folliculitis) and the accumulation of tumorous growths called *keloids.* All of these conditions are due to staphylococcal infections. *[Courtesy of the Armed Forces Institute of Pathology, Washington, D.C., Neg. No. AFIP 53-12335.]*

common victims (Color Photograph 4). Orphanages and nurseries offer favorable environments for the spread of impetigo.

Transmission. Impetigo contagiosa is spread by direct contact with infected persons or indirectly by contact with fomites such as bed sheets, handkerchiefs, pencils, or towels. The crusts from lesions also serve as a source of *S. aureus.* Staphylococci are usually introduced into the body through some form of abrasion (skin scraping) or other lesion. The premature removal of scabs or crusts of viral skin infections, of which chickenpox is an example, may provide additional portals of entry.

SCALDED SKIN SYNDROME

Toxic epidermal scalded skin syndrome, often referred to as *scalded skin syndrome,* is one of the most rapidly skin-destroying diseases known. The disease exists in two distinct forms, both of which result in extreme skin tenderness and widespread skin loss (Table 23–2).

One type, *toxic epidermal necrolysis (TEN),* occurs chiefly in adults or older children and appears to be caused by hypersensitivity to drugs such as antibiotics, barbiturates, and sulfa compounds. TEN, which is also observed in patients undergoing graft-versus-host disease, resembles a massive second-degree burn and carries or shows high mortality in untreated cases.

The second form of the skin disease, *staphylococcal scalded skin syndrome (SSSS)* or *Ritter's disease,* is caused by a skin-destroying exotoxin produced by *S. aureus.* It is seen in infants and young children under the age of five years (Figure 23–7).

TEN may or may not have a staphylococcal cause, but when it does, it is referred to as *Ritter's-type toxic epidermal necrolysis.* Since localized lesions and the initial phases of both forms of the disease are indistinguishable, laboratory tests are necessary to differentiate one state from the other.

Patients are given semisynthetic, penicillinase-resistant penicillins to kill *S. aureus* at the earliest possible stage.

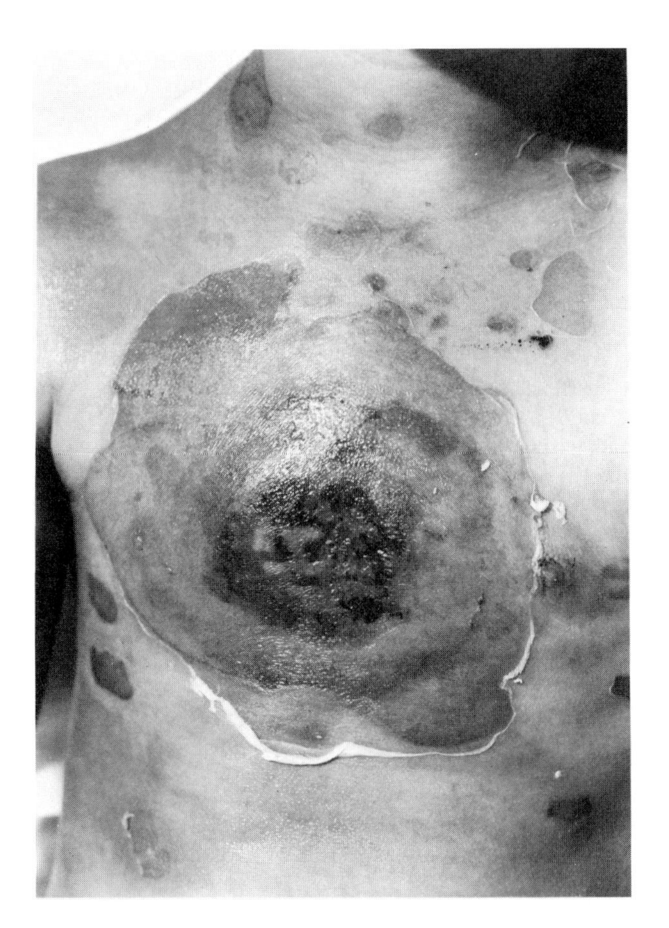

Figure 23–7 Early stages of generalized staphylococcal scalded skin syndrome (SSSS). *[P. E. Elias, et al., Arch. Dermat.* ***113****:207–218, 1977.]*

Streptococcal Infections

The streptococci include several organisms that, under the right conditions, are capable of infecting virtually all areas of the body. Primary infections can involve the respiratory, circulatory, or central nervous systems as well as the skin and genital and urinary tracts. Streptococci are known for their frequent secondary invasion of body tissues. Various strains of these organisms can cause the same disease state, and, conversely, a single given streptococcal species can produce several kinds of infections.

The virulence of streptococci (as in the case of many other parasitic microorganisms) seems to depend upon their cellular products, surface components, and related substances. These factors are important in the organism's ability to establish itself in the host. [See bacterial capsules and toxins in Chapter 22.]

In 1895, Mamorek first observed that streptococci were capable of causing *in vivo* as well as *in vitro* destruction (lysis) of red blood cells. In 1903, Schottmuller suggested that this hemolytic activity be utilized for classification purposes. This sug-

Staphylococcal Infection

Laboratory Diagnosis

Routine laboratory identification is generally straightforward. *S. aureus* produces beta hemolysis (Color Photograph 23b), grows on salt agar and ferments mannitol (Color Photograph 26), produces a positive catalase test (Color Photograph 96), and is usually coagulase-positive. [See Chapter 4 for selective and differential media and Chapter 22 for coagulase.] *S. aureus* is differentiated from streptococci on the basis of its positive catalase test.

Serological tests are of limited value in routine laboratory identification of *S. aureus*.

Antibiotic susceptibility testing is another important aspect of laboratory identification procedures.

Treatment

S. aureus can and does develop resistance to several antibiotics. The drug of choice for non-penicillinase-producing organisms is penicillin G or V. Infections caused by penicillinase-producing strains are treated with cloxacillin or dicloxacillin (for oral administration) and methicillin, nafcillin, or other penicillinase-resistant drugs. [See Chapter 12 for antibiotic
Alternative antibiotics, especially for individuals with a penicillin allergy, include cephalosporin, clindamycin, and vancomycin. [See Chapter 20 for a discussion of drug allergies.]

gestion was not fully explored until J. H. Brown undertook an intensive study in 1919. Based on his findings and those of others, streptococci were categorized into one of three groups on the basis of their hemolytic activity on blood agar plates. These categories were (1) *alpha-hemolytic,* generating a greenish zone surrounding the bacterial colony (Color Photograph 23a); (2) *beta-hemolytic,* generating a clear zone surrounding the colony (Color Photograph 23b), and (3) *gamma-hemolytic,* causing no obvious change on the medium (Color Photograph 23c).

In the early 1930s, Rebecca Lancefield further subdivided the hemolytic streptococci into groups based on immunological differences. These groups were designated by the capital letters A through O. The group A beta-hemolytic streptococci contain the greatest number of human pathogens.

Transmission. Humans are the ultimate sources of pathogenic streptococci. Factors involved in the transmission of these organisms to susceptible persons include climate, crowding, improper sanitation, and the creation of aerosols.

ERYSIPELAS

Erysipelas occurs worldwide, but with greater frequency in the temperate zone. It (Figure 23–8) is an acute infection arising as a complication of surgery or of accidental wounds. Victims of the disease usually show a history of minor injuries. Group A streptococci are most often found to cause this disease.

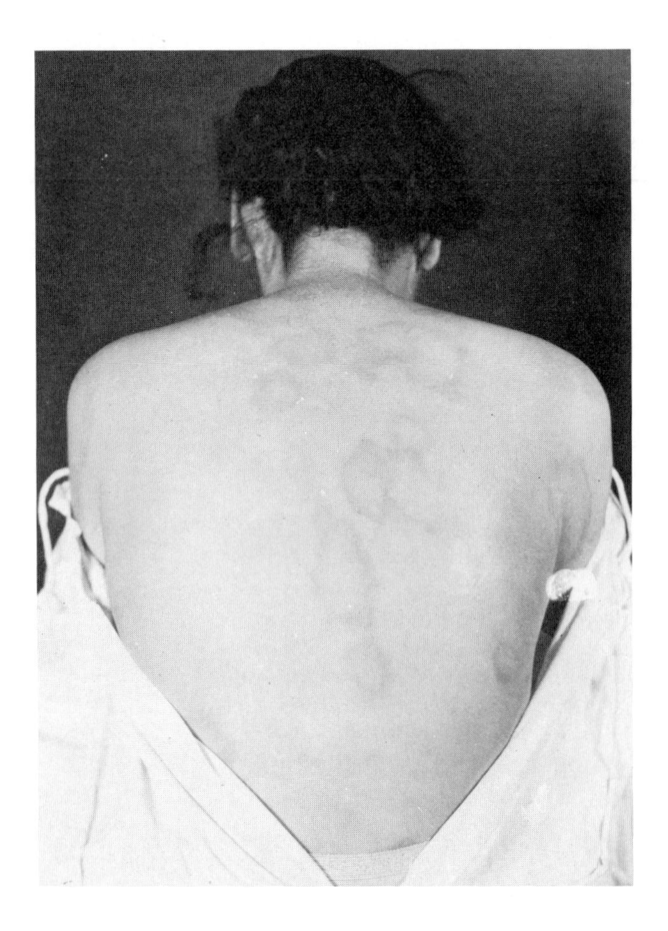

Figure 23–8 Erysipelas caused by group A streptococci. *[Courtesy of the Armed Forces Institute of Pathology, Washington, D.C., Neg. No. AF 18 58–6180.]*

SCARLET FEVER

Various strains of group A beta-hemolytic streptococci are known to cause scarlet fever. Many of these organisms also produce other infections, including septic sore throat, erysipelas, tonsillitis, puerperal fever, and wound abscesses. Typically, scarlet fever is an acute inflammation of the upper respiratory tract that may be accompanied by a generalized rash. In recent years this infection has appeared to be milder in its effects. The term *scarlatina* is often used to designate the milder forms of scarlet fever. The various effects of the infection are directly related to a toxin produced by streptococci infected by a temperate bacterial virus. The toxin usually spreads from the infected site to other parts of the body. [See lysogenic conversion in Chapter 6.]

Transmission. Scarlet fever infection is spread by means of (1) droplet nuclei, (2) aerosols, (3) contaminated food and water, and (4) direct contact with carriers, or individuals in the acute stage of the disease. Fomites have been implicated, but it is believed such contaminatd objects are not commonly involved.

Predisposing and Related Factors. This infection is worldwide. However, the development of a rash occurs more frequently in the temperate zones. Scarlet fever occurs most commonly during the fall, winter, and early spring months.

Studies indicate that this disease appears most frequently in Caucasians. Scarlet fever infection may occur in persons of any age. In general, however, it is most common in children under ten years of age.

One aid in the diagnosis of scarlet fever is the Schultz-Charlton, or blanching, test. This procedure involves the injection of either 0.1 mL of the scarlet fever antitoxin or 0.2 to 0.5 mL of convalescent serum into an area with a pronounced rash. A clearing or blanching will occur around the site of injection within four to eight hours if the rash is associated with scarlet fever infection.

IMPETIGO CONTAGIOSA

Although impetigo contagiosa is usually caused by staphylococci (Color Photograph 4), streptococci have also been associated with the condition. The streptococci are considered to act more as secondary invaders. Group A organisms appear to be the ones most often involved.

Streptococcal impetigo is a universal disease that usually occurs in areas where sanitation and personal hygiene are poor. The disease mainly affects young children, although when the abovementioned conditions exist during natural disasters and war, adults can also develop the infection.

The infectious agents can be transmitted by means of fomites (napkins, bedclothes, pencils) or by direct inoculation with infectious discharges from persons having the disease.

CONSEQUENCES OF GROUP A STREPTOCOCCAL INFECTIONS

Several consequences, or *sequelae,* can develop from streptococcal infections, especially throat involvement. Acute hemorrhagic glomerulonephritis and rheumatic fever are the most common complications. These conditions are discussed in later chapters.

Clostridrial Infections

TETANUS

Tetanus was described by Hippocrates some 24 centuries ago. Its effects, sudden contractions of voluntary muscles and convulsions, are caused by a neurotoxin known as *tetanospasmin,* which is among the most poisonous substances known, second only to botulinus type A toxin. Tetanospasmin is produced by the widely distributed anaerobic bacterium *Clostridium tetani* (C. TEH-tan-ee)—an organism present in dust, soil, and the feces of both domesticated and wild animals. **[See actions of bacterial toxins in Table 22–1.]**

Transmission. Tetanus is associated with wounds of all types. Any break in the skin, whether a superficial scratch or a puncture, is subject to contamination by the spores of *C. tetani* (Color Photograph 14b). Once these spores gain access to the injured area, if there is enough dead or dying tissue to provide reduced oxygen concentration, they may germinate into the growing cells that produce the tetanus-causing neurotoxin. The mechanism of action of the toxin is not entirely clear to researchers.

C. tetani is considered, for the most part, a relatively noninvasive organism. Foreign objects, including glass and slivers of metal or wood, can introduce spores into the deeper tissues. A newborn infant may develop *tetanus neonatorum* from infection of its severed umbilical cord. Tetanus can affect people of all ages and can develop from abortions, circumcisions, ear piercing, injections of drugs, and negligent surgical procedures.

Immunization is the only effective means of controlling tetanus. An initial course of three injections of tetanus toxoid is generally used, followed by a single booster injection after one year. Immunization schedules and the use of toxoid and antitoxin preparations are discussed in Chapter 18. **[See bacterial toxins in Chapter 22.]**

WOUND BOTULISM

Although wound botulism is a relatively rare disease, the marked increase in reported cases should serve as a signal of its potential danger. The causative agent of wound botulism, as well as of the more familiar food poisoning, is *Clostridium botulinum* (C. bot-choo-LIE-num). The spores of this bacterium are commonly found in the soil. Most infections are caused by contamination of wounds with soil. Although *C. botulinum* is noninvasive, it produces a potent exotoxin. The toxin binds at the neuromuscular junction and blocks the presynaptic release of acetylcholine. This reaction is responsible for the various symptoms, including impaired breathing and paralysis. **[Refer to Chapter 26 for additional features of botulism.]**

Streptococcal Infection

Laboratory Identification

Group B streptococci can be distinguished from other types because of their ability to produce the CAMP factor (Color Photograph 97) and to hydrolyze hippurate as well as because of their resistance to the antibiotic bactracin. **[See Chapter 21 for a description of the CAMP test.]**

Treatment

Penicillion is generally the drug of choice. Ampicillin alone or in combination with gentamicin is also used for group B streptococcal infections.

Tetanus

LABORATORY DIAGNOSIS

Anaerobic culturing of specimens is required for the isolation and identification of *C. tetani.* The organism is motile and produces characteristic terminal spores resembling tennis rackets (Color Photograph 14b). Definite identification of a neurotoxin-producing strain requires animal inoculation to show the effects of the toxin.

TREATMENT

An injection of tetanus antitoxin is administered to a nonimmunized individual and tetanus toxoid to an immunized individual to prevent the disease. Similar treatment is given to a suspected case of tetanus. Complications may develop in some individuals if the antitoxin is prepared in horses. Hypersensitivity to horses and their products can result in severe allergic reactions. In addition, antibiotics such as penicillin are given to eliminate *C. tetani.*

Mycotic Infections

Human skin, nails, and hair are particulary vulnerable to attack by certain pathogenic fungi. Of the 50,000 to 200,000 known species of fungi, about 50 are recognized as human pathogens. Most pathogens are of the class Deuteromycotina (fungi imperfecti). Several of these agents are capable of affecting the skin and related tissues that also contain the protein keratin (Color Photographs 41, 42, and 98a). These fungi are also called *dermatophytes,* and most are found worldwide.

Classification of Mycotic Infections

It is customary and useful to group the fungal diseases, or *mycoses,* according to the tissues and organs affected and the disease pattern or patterns. Our consideration of these diseases follows the terminology used in the *Ciba Foundation Symposium, Systemic Mycoses,* and their grouping according to the tissues and organs affected.

SUPERFICIAL MYCOSES

Fungi that attack mainly the epidermis, hair, nails, and mucosal surfaces are called *superficial fungi* or dermatophytes. The diseases caused by such agents include the various forms of ringworm, or tinea (from the Latin meaning "worm" or "larval") and yeast *(Candida)* infections of mucosal surfaces, such as oral thrush and vulvovaginitis. These infections are frequently referred to as the *superficial* or *surface mycoses.*

Superficial mycoses are further classified on the basis of the location of the effects produced by the causative fungus (Table 23–3). For example, ringworm of the scalp is *tinea capitis;* that of the feet *tinea pedis,* more commonly known as *athlete's foot.*

Representative Superficial Mycoses

TINEA BARBAE

The fungal infection known as *tinea barbae* is a chronic condition involving the bearded regions of the face and neck (Figure 23–9). Only men are affected.

Fungi that attack hair or hair follicles can produce *ectothrix* or *endothrix* infections. With the ecto-

Figure 23–9 Tinea barbae. Note the localized, boggy appearance of this infection. *[Courtesy of the Armed Forces Institute of Pathology, Washington, D.C., Neg. No. 56–4858.]*

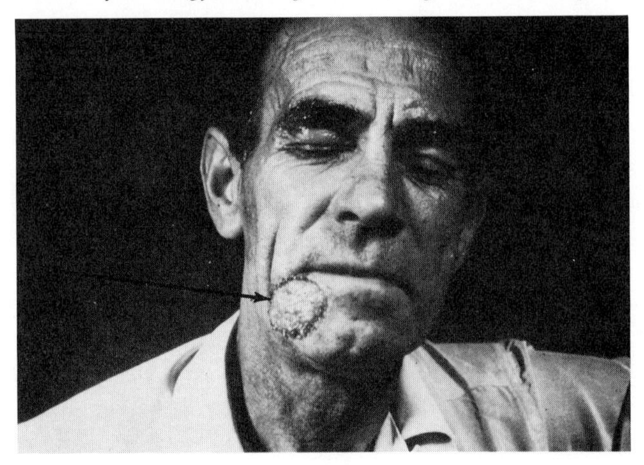

TABLE 23–3 Representative Superficial Mycoses

Disease	*Causative Agent*	*Source of Infection*	*Geographical Distribution*	*Possible Treatment*
Tinea barbae (ringworm of the beard)	*Microsporum canis* (rare); *Trichophyton mentagrophytes; T. rubrum, T. sabouraudi, T. verrucosum, T. violaceum*	Infected animals and children	Worldwide	Griseofulvin; application of warm saline compresses; antibiotics to prevent secondary bacterial infections
Tinea capitis (ringworm of the scalp)	*Microsporum audouini, M. canis, M. gypseum, T. mentagrophytes, T. sabouraudi, T. schoenleinii, T. sulfureum, T. tonsurans, T. violaceum*	Infected animals, people, and fomites	Worldwide	Griseofulvin; antibiotics to prevent secondary bacterial infections
Tinea corporis (ringworm of the body)	*M. audouini, M. canis, M. gypseum, T. concentricum, T. mentagrophytes, T. sabouraudi, T. schoenleinii, T. sulfureum, T. tonsurans, T. violaceum*	Infected animals and contaminated articles of clothing	Worldwide	Griseofulvin; for small lesions, fungicides such as Tinactin, Verdefam, or Whitfield's ointment
Tinea cruris (ringworm of the groin)	*Candida albicans, Epidermophyton floccosum, T. mentagrophytes, T. rubrum*	Contaminated articles of clothing or athletic supports	Worldwide	Whitfield's ointment
Tinea manuum and Tinea pedis (ringworm of the hands and feet)	*C. albicans, E. floccosum, M. canis, T. mentagrophytes, T. rubrum, T. schoenleinii*	Direct contact with fungi in moist environments including showers, swimming and wading pools	Worldwide	Aqueous potassium permanganate soaks; griseofulvin; antifungal ointments and powders
Tinea nigra (ringworm of the palms and soles)	*Exophiala (Cladosporium) werneckii*	Contaminated soil	Worldwide	Whitfield's ointment and tincture of iodine
Tinea unguium (ringworm of the nails)	*C. albicans, E. floccosum, T. mentagrophytes, T. rubrum, T. schoenleinii, T. violaceum*	Infected individuals or regions of the body	Worldwide	Griseofulvin, or nystatin, 1% clotrimazol
Tinea versicolor (branny scaling of the skin involving the face)	*Malassezia furfur*	Infected individuals	Worldwide	1% sodium hyposulfite
Black piedra	*Piedraia hortae*	Infected hair (beard, mustache, scalp)	Tropical countries	Shaving infected area, or adequate cleaning of hair followed by application of a mild fungicide
White piedra	*Trichosporon (beigelii) cutaneum*	Infected hair (beard, mustache, scalp)	Temperate and tropical regions	The treatment listed for black piedra may help

thrix condition, growth of the fungal agent (in the form of arthrospores) occurs in and on the hair shaft (Color Photograph 42). In an endothrix infection, the organism grows only within the hair shaft.

TINEA CAPITIS

An infectious mycotic condition, *tinea capitis* involves the scalp (Figure 23–10) and the follicle and shaft of the hair. Tinea capitis may be acquired by direct contact with infected animals, humans, or fomites. The disease is commonly found among individuals living in overcrowded areas and having poor hygiene. Children are most often affected.

PREVENTION. Preventive measures are important with tinea capitis. Parents should be made aware of its contagious nature. Shampooing after haircuts may help prevent the disease.

TINEA CORPORIS

The nonhairy skin of an individual's body is affected in the chronic mycotic disease known as *tinea corporis* (Color Photograph 98a). The infection is found

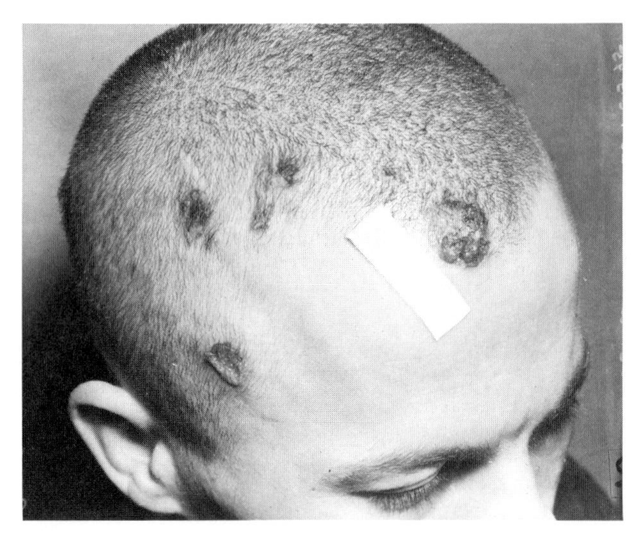

Figure 23–10 Favus, a severe form of tinea capitis, or ringworm of the head. *[Courtesy of the Armed Forces Institute of Pathology, Washington, D.C., Neg. No. 8–535–1.]*

in both sexes, with a greater frequency in moist, warm regions of the body. The disease agent in children is very often identified as *Microsporum canis* (my-krow-SPOH-rum KAY-niss); in adults several *Trichophyton* (trick-oh-FYE-ton) species produce the infection (Color Photograph 98c).

TINEA CRURIS

Tinea cruris is a chronic superficial infection usually confined to the inner surfaces of the groin. Perianal and armpit involvement can also occur.

TINEA MANUUM AND TINEA PEDIS

Tinea manuum and *tinea pedis* infections are long-lasting and usually develop when fungi acquired through contact with contaminated showers, swimming and wading pools, and wet tropical terrain are spread from the toe webs (Figure 23–11). Often the toenails and fingernails may become involved as a secondary effect (onychomycosis).

As fungal products are absorbed by the infected person, various nonfungal eruptions may occur on the extremities and trunk. These conditions indicate the so-called id or trichophytid reaction, an allergic reaction to the products of the causative fungi, *Trichophyton* spp. Reddened areas appear on the skin far from infected regions. No fungi are present in these eruptions.

PREVENTION. As with all tinea infections, preventive measures are important in the control of tinea pedis. These measures include keeping potential areas of infection clean and dry.

TINEA VERSICOLOR

Malassezia furfur (mal-ah-SEES-zee-ah FUR-fur) is the causative agent of the chronic fungus infection *tinea versicolor.* The condition is usually asymptomatic, although it may cause mild itching and loss of skin pigment (Figure 23–12). The disease most commonly affects young adults.

PIEDRA

Piedra is characterized by fungal growths securely attached to the hair's surface, forming firm black, brown, or white nodules. Two forms of the infection are recognized, black and white piedra (Table 23–3). All ages and both sexes appear to be vulnerable to the infection.

SUBCUTANEOUS MYCOSES

Subcutaneous fungal infections involve the skin and underlying subcutaneous tissue, generally without spreading to the internal organs of the body. Humans and lower animals serve as accidental hosts as a result of the inoculation of fungal spores into the skin and underlying tissue after an injury. Examples of subcutaneous mycoses include chromomycosis, sporotrichosis, and maduromycosis (Figure 23–13). Maduromycosis may progress to involve bone, muscle, and neighboring tissues, ultimately requiring amputation.

DEEP-SEATED, OR SYSTEMIC, MYCOSES

Infections in which the causative agents invade the subepithelial tissues (dermis and deeper regions) are known as *deep-seated, deep,* or *systemic mycoses* (Color Photograph 106a). Examples of these diseases, which include coccidioidomycosis and histoplasmosis, are described in Chapter 25.

Figure 23–11 Tinea pedis, or ringworm of the foot. *[Courtesy of the Armed Forces Institute of Pathology, Washington, D.C., Neg. No. 53–14665–1.]*

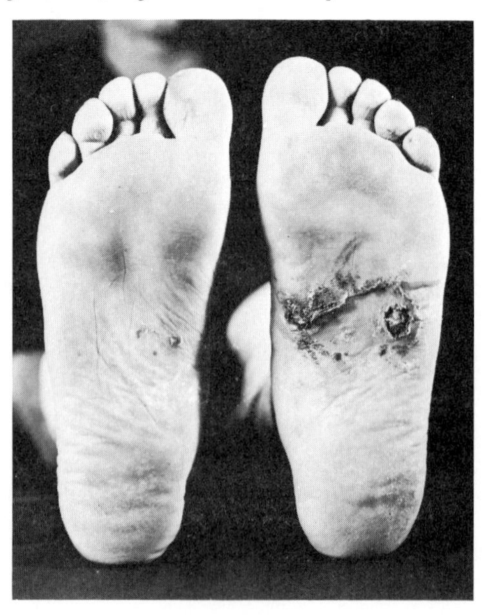

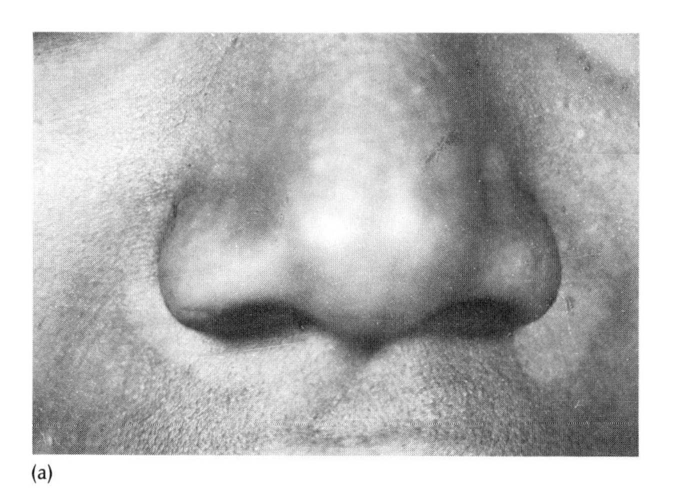

(a)

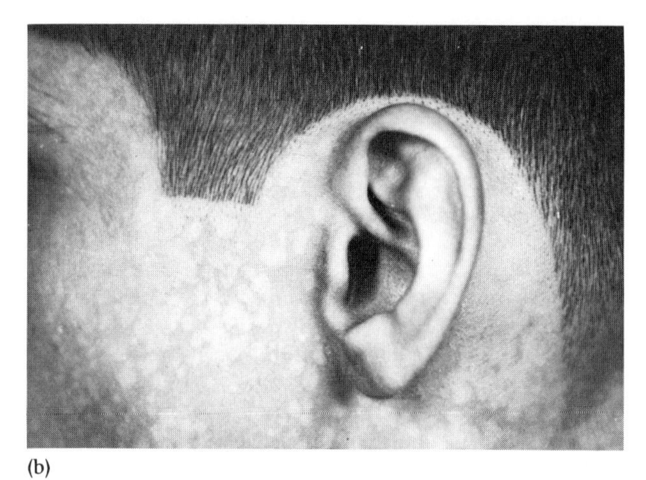

(b)

Figure 23–12 Tinea versicolor. The appearance of the fungus infection involving the face. Areas with a loss of pigment are quite obvious. *[A. Kamalam, and A. S. Thambiah,* Sabouraudia *14:129–148, 1976.]*

OPPORTUNISTIC FUNGI

Some fungi are *opportunistic* pathogens. They are not normally pathogenic to healthy persons, but under certain conditions they can produce severe infections. Included among these opportunistic agents are species of *Aspergillus* (a-sper-JIL-lus), *Candida* (KAN-did-ah), *Cryptococcus* (kryp-toe-KOK-kus), *Geotrichum* (gee-oh-TRICK-um), *Mucor* (MEW-kore), and *Rhizopus* (rye-ZOH-puss). Factors that have been found to predispose individuals to opportunistic infections include chronic anemia, leukemia, metabolic disorders (such as diabetes mellitus), and intensive treatment with broad-spectrum antibiotics and drugs that suppress antibody formation (immunosuppressive drugs).

Other Diseases Caused by Fungi

Several superficial fungi are capable of attacking tissues other than the skin.

Figure 23–13 Madura foot, an example of subcutaneous fungal infection. *[Courtesy Dr. A. Kamalam, Madras, India.]*

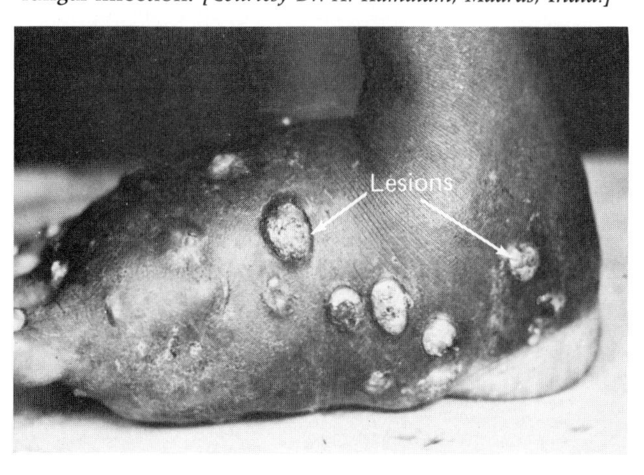

CANDIDIASIS

Records of candidiasis cases date back to the beginning of microbiology. Its effects range from simple, localized infections to uncontrollable, fatal septicemias. *Candida albicans* (C. al-BEH-kans) is considered the usual causative agent of bronchopulmonary candidiasis, skin and nail candidiasis (Figure 23–14), oral thrush, vaginitis, and paronychia (inflammation

Figure 23–14 *Candida*-caused onychomycosis. Note the brittle appearance of the infected toenail. *[Courtesy of the Armed Forces Institute of Pathology, Washington, D.C., Neg. No. 58–13966–4.]*

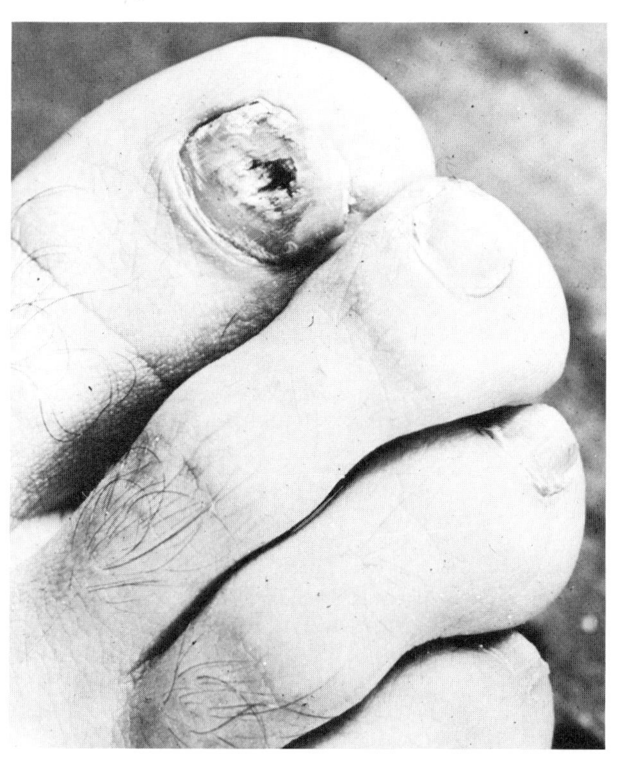

of the nail bed). [**Oral thrush is described in Chapter 24.**]

It should be noted that the mere presence of the fungus is not enough to cause disease, since *C. albicans* has often been isolated from the skin, oral cavity, and intestinal tract of healthy individuals. However, it can cause infection when host resistance is lowered, as in the case of AIDS, when there is some form of nutritional deficiency, or as a complication of bacterial or viral diseases. Candidiasis is commonly seen in both the very young and the very old, as well as in debilitated persons. Predisposing factors include diabetes mellitus, pregnancy, obesity, and vitamin deficiencies. The infection has shown an increase in recent years as a complication of therapy with antibiotics, corticosteroids, and cytotoxic drugs (used in cancer therapy). *Candida* species have often been grouped with the opportunistic fungi.

PREVENTION. Preventive measures are directed toward keeping susceptible skin areas as dry as possible. Rubber gloves and the avoidance of excessive exposure to detergents and related substances may help in controlling the disease.

Diagnostic and Related Features of the Dermatophytes

Unlike some systemic fungi, dermatophytes are not dimorphic; they do not have yeast and hyphal stages. Most dermatophytes look alike in skin lesions. Culturally, however, their properties are quite different (Color Photographs 98b and 98c). One of the most widely used media for the cultivation of fungi is Sabouraud's dextrose agar. This acidic medium (pH approximately 5.6) is especially suited for most fungi because they are able to survive in this environment while certain bacteria are prevented from growing. Media can be made more selective by adding antibiotics that discourage the growth of bacteria and saprophytic fungi.

The identification of several fungi is based on the presence and characteristics of hyphae and spores in culture.

Detection of fungi in specimens involves techniques for their isolation and cultivation and may also include direct microscopic examination of tissues. One of the procedures usually employed involves digesting specimens such as infected hairs in 10% potassium or sodium hydroxide with the aid of heat. This procedure is performed on a slide, which is then inspected under the microscope for the fungus.

Staining procedures, including the Gram stain and the periodic acid-Schiff (PAS) stain, are also used. Reactions obtained with Wood's light (a form of ultraviolet light) are also of diagnostic value with certain infections.

Viral Infections

Several viral infections are either limited to the skin or involve it in the course of disease development (Table 23–4). Some of the better-known viral diseases, including chickenpox, cold sores (herpes), measles, and warts, will be discussed.

TABLE 23–4 Representative Viral Diseases of Humans That Affect the Skin

Disease	*Causative Agent*	*Incubation Period*	*General Features of the Disease*	*Laboratory Diagnosis*	*Possible Treatment*
Chickenpox (varicella)	Varicella-zoster virus	2–3 weeks	General red rash leading to vesicles; different forms of rash appear in successive crops and are distributed mainly over the trunk and face; slight fever and itching are commonly experienced	1. Detection of giant cells in specimens[a] 2. Serological tests, such as fluorescent antibody test 3. Tissue culture for virus isolations	5-Iodo-2′-deoxyuridine (IUDR); symptomatic treatment; vitamin C; interferon and vidarabine are under consideration
Fever blister (cold sore, herpetic gingiostomatitis, herpes simplex virus infection)	Herpes simplex virus type 1	Unknown, possibly 4–7 days	Localized skin and/or mucous membrane lesions that appear as blisterlike eruptions on the lips, face, ears, etc.	1. Demonstration of typical inclusion bodies in specimens[a] 2. Viral isolations 3. Serological tests	5-Iodo-2′-deoxyuridine (IUDR); acyclovir

TABLE 23–4 (continued)

Disease	*Causative Agent*	*Incubation Period*	*General Features of the Disease*	*Laboratory Diagnosis*	*Possible Treatment*
Rubella (German measles)	Rubella virus	14–21 days	Slight fever, general discomfort, swollen lymph nodes, and macular rash	1. Tissue culture for viral isolation[a] 2. Serological tests, such as hemagglutination inhibition test	Immune globulin preparations
Measles (morbilli)[b]	Measles virus	14–21 days	Rash appears about 14 days after exposure; fever, cough, muscle pains, general discomfort, photophobia, redness of eyelids, characteristic lesions in mouth, Koplik's spots, which are pinpoint gray-white spots surrounded by a reddened area appearing on the mucous membranes opposite the molar teeth (Color Photograph 100).	1. Viral isolation in tissue culture[a] 2. Serological tests	Antibiotics to prevent secondary bacterial infections; symptomatic treatment; vitamin C
Molluscum contagiosum	Molluscum contagiosum virus (a poxvirus)	2–8 weeks	Small, pale, firm, pearllike masses (nodules) appear on the skin (Figure 23–20); a cheesy substance may be expressed from the center of each lesion; the condition clears within 2 to 12 months without complications or treatment	Demonstration of large red inclusions in skin cells	Generally none; however, individual nodules can be surgically removed
Shingles	Varicella-zoster virus	2–3 weeks	Blisters along nerve trunk, pain (sometimes extreme), slight fever, and general discomfort	Same as for chickenpox[a]	5-Iodo-2′-deoxyuridine (IUDR); zoster immunoglobulin; vitamin C; interferon and vidara bine are under con sideration
Warts (papilloma)	Polyomavirus	Unknown	Growths on the back of hands, palms, soles, and other body regions; generally no pain or fever	1. Electron microscopy[a] 2. Tissue culture for virus isolation	Surgical or chemical removal of warts; certain vaccines are under development

[a] The typical clinical features of the disease are usually sufficient for diagnosis.
[b] Rubeola is also used as a synonym for this disease. Unfortunately, the term has been used as a synonym for rubella (German measles).

The clinical features of viral skin diseases vary widely. Yet certain diseases may possess identical features, creating a problem in diagnosis.

Knowing the terminology of skin diseases presented in Table 23–1 is helpful, not only in diagnosis but also in following the development of a disease process. Table 23–4 lists general symptoms and approaches to laboratory diagnosis for several common viral diseases.

Although promising results have been obtained in treating some viral skin infections (Table 23–4), treatment is often limited to preventing secondary bacterial infections and relieving the victim's discomfort (symptomatic treatment). The high intake of vitamin C, with dosages based on the victim's body weight, is reported to be of value in the control of viral skin diseases, such as chickenpox, measles, and shingles.

The Herpesvirus Group

The viruses of the herpes group have several distinguishing properties, including (1) the possession of double-stranded DNA, (2) the presence of a nucleocapsid with cubical symmetry, (3) inactivation by chloroform and ether, (4) viral multiplication in the nucleus of infected cells, (5) the production of clinical effects such as blister eruptions of the skin (Figure 23–15) and mucous membranes, with occasional nerve involvement, and (6) the formation of intranuclear inclusions known as *Lipschütz* or *type A inclusions* (Figure 23–16).

According to the proposal suggested at the Third Conference on Oncogenesis and Herpesviruses on the basis of their biological properties, herpesviruses have been classified into three groups, alpha, beta, and gamma. Alpha herpesviruses are typified by herpes simplex 1, the beta group by human cytomegaloviruses (CMV), and gamma herpesviruses by Epstein-Barr virus (EBV).

HERPESVIRUS TYPE I (HERPES SIMPLEX VIRUS)

The most common form of herpes simplex virus infection is an inflammation of the gums *(herpetic gingiostomatitis)*. This condition occurs on the individual's first encounter with the viral agent (primary infection), chiefly in children between the ages of 1 and 5. Similar primary involvement in adults may be quite severe. After recovery, the virus assumes a latent form. However, infections involving the skin and mucous membranes producing blister-like lesions (Table 23–4) can recur quite readily as a consequence of precipitating causes, such as emotional disturbances, infections, lymphoma, menses, and sunburn. Although persons with recurrent herpes have circulating antibodies, they do not prevent the disease's effects (Figure 23–17).

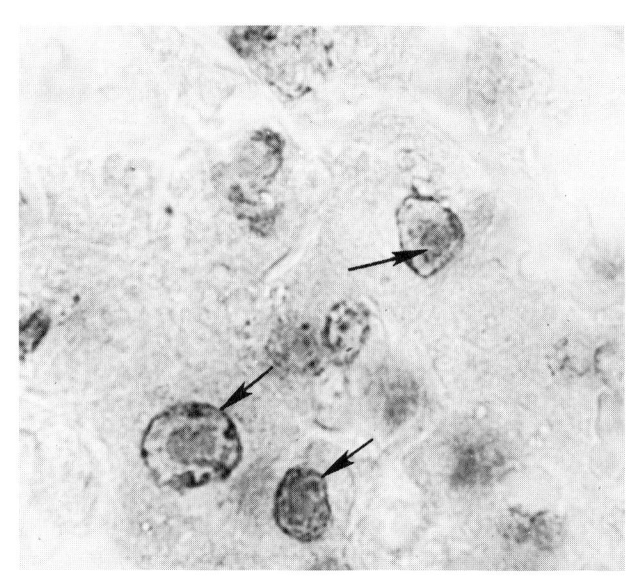

Figure 23–16 A liver specimen from a 10-day-old child, showing typical inclusions of herpes simplex virus (arrows). During the early stages of viral development, the inclusions stain blue when they contain viruses. The red state results after the viruses have been released. Thus the Lipschütz inclusion body is an empty shell, a "token" of viral infection. *[Courtesy of the Armed Forces Institute of Pathology, Washington, D.C., Neg. No. 56–2952.]*

Figure 23–15 Patient with generalized herpes simplex infection. The individual's immune state was altered by a cancerous condition. Identification of the viral agent was made by tissue culture studies of vesicle fluid. *[Hospital, Brooklyn. Photo by F. G. Hertling.]*

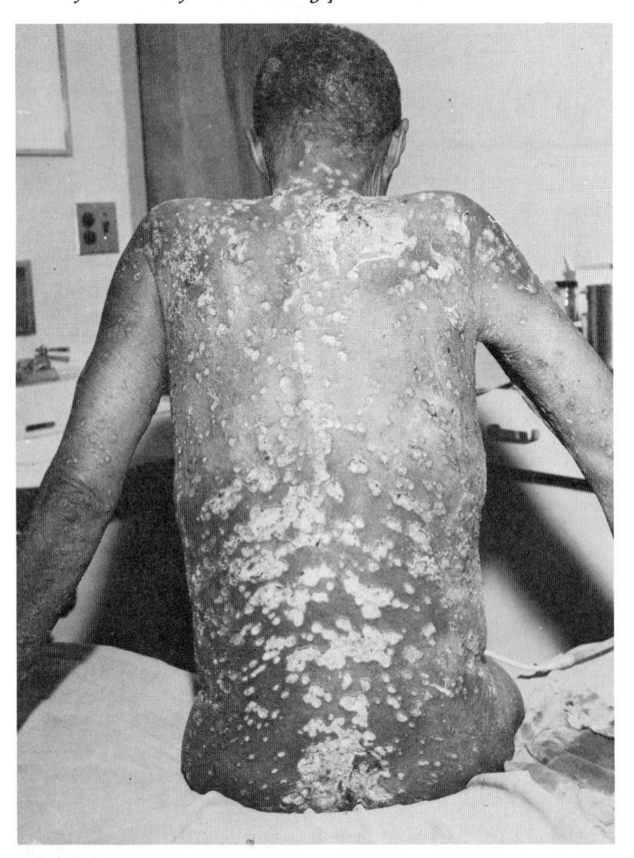

Herpes simplex infection of the finger, known as *herpetic whitlow* (Color Photograph 99), also can result from the inoculation of infectious secretions into a small cut in the skin. Such infections are considered occupational hazards for health-care personnel. Painful vesicular lesions are typical (Color Photograph 99).

Transmission of the virus can occur by direct contact, including kissing, hand touching, and sexual relations and by way of fomites.

VARICELLA (CHICKENPOX)-HERPES ZOSTER (SHINGLES) VIRUS

A variety of immunologic tests—agglutination, complement fixation, and viral neutralization—have shown the viruses that cause chickenpox (varicella)

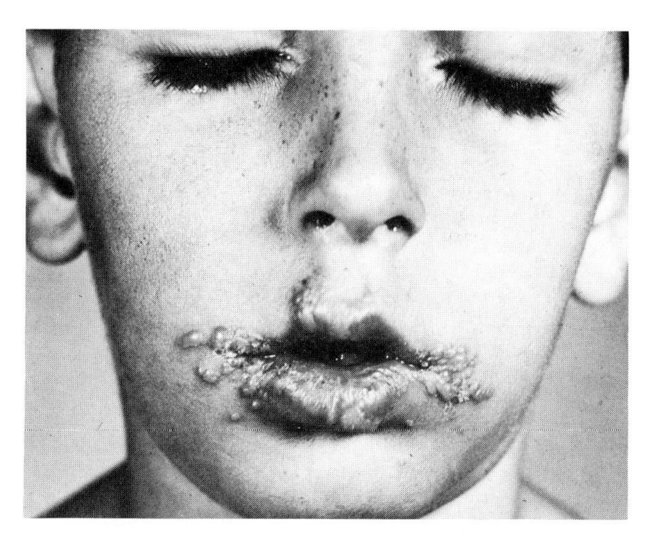

Figure 23–17 Common features of herpes simplex virus infection. Involvement of the lips. *[Courtesy of the Armed Forces Institute of Pathology, Washington, D.C., Neg. No. 55-11961-1.]*

and shingles (herpes zoster) to be immunologically identical. Furthermore, the morphological examination of the virions, inclusion bodies, and cytopathic effects associated with these agents found them to be physically indistinguishable. The identity of these agents has been further established by the production of typical cases of chickenpox in children following inoculation of shingles blister fluid. On the basis of these and related studies, the majority of investigators believe that varicella and herpes zoster represent different forms of infection with the same agent, the varicella-zoster virus.

VARICELLA (CHICKENPOX)

Chickenpox is one of the most common diseases in residential schools. It is primarily a childhood infection, but approximately 20% of cases are adults. Varicella is the primary disease state produced in an individual without immunity. Infections are transmitted via droplets of respiratory secretions and contact, either direct or indirect, with infectious skin surfaces. As noted above, an individual with shingles (herpes zoster infection) is also a source of infectious material for an outbreak of chickenpox.

Immunity. Recovery from an attack of chickenpox usually provides a relatively long-lasting active immune state. Recurrent infections are extremely rare.

Control. Control measures for this disease include isolation of infected persons and the administration of immunoglobulin to exposed individuals. Both of these means, however, are of limited effectiveness. An effective vaccine has been developed. High doses of vitamin C are reported to be of value in the control of chickenpox.

CONGENITAL VARICELLA SYNDROME

Women of childbearing age who are susceptible to varicella are an important high-risk group. During pregnancy, varicella infection has been associated with *congenital varicella syndrome.* The effects of fetal infection may include skin scars, defective development of the limbs, cataracts, and the formation of an abnormally small head. The most common maternal complications observed are pneumonia and premature labor. Although complications do not occur in the majority of women, the risk justifies the administration of varicella-zoster immune globulin to pregnant women who have no immunity to varicella and who have had a close exposure to varicella-zoster virus. The effectiveness of varicella-zoster immune globulin in preventing fetal complications is unkown. [See passive immunization in Chapter 18.]

HERPES ZOSTER (SHINGLES)

Herpes zoster is uncommon in children; most cases occur in adults (Figure 23–18). At least 70% of adults have had previous exposure to the varicella-zoster virus, as indicated by the presence of circulating antibodies. Shingles represents a reinvasion, or recurrence, of varicella-zoster virus in persons partially immune to varicella.

The natural means of transmission for shingles is not known. The virus may gain entrance to the body by way of the throat. From here it enters the bloodstream and then localizes in ganglionic nerve cells. In this location the virus remains inactive for varying periods. Activation may be caused by several factors, including cancer, trauma, and certain drugs, such as those containing antimony or arsenic.

Figure 23–18 Shingles in an adult patient, showing involvement of the ophthalmic nerve. *[Courtesy of the Armed Forces Institute of Pathology, Washington, D.C., Neg. No. 58-15409-4.]*

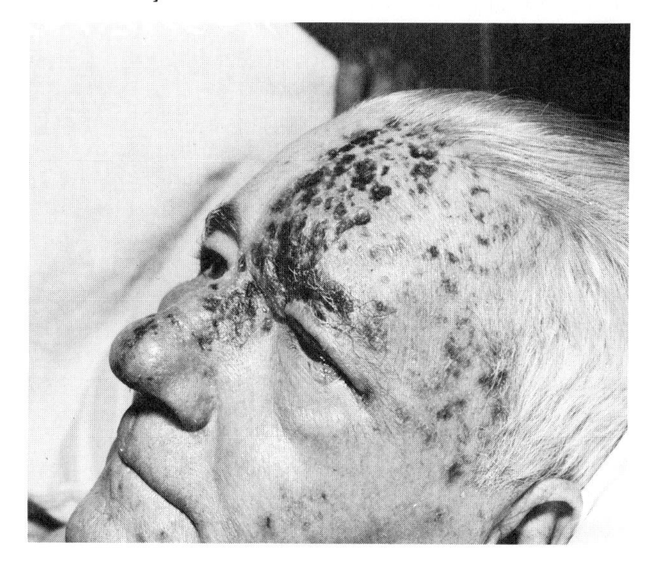

Warts (Papilloma Virus Infection)

HUMAN (PAPILLOMA) WARTS

Verruca vulgaris, otherwise known as the *common wart* or *condyloma* (Figure 23–19), is caused by human papilloma virus, which is a DNA virus.

Warts appear to be transferred by scratching. In addition, indirect spreading of viruses has been reported to occur by contact with contaminated bathroom and swimming pool floors, communal washroom facilities, and gymnastic equipment. Barbers, chiropodists, and masseurs have also been implicated in spreading the disease. Genital warts, or condylomata acuminata, are transmitted by sexual contact.

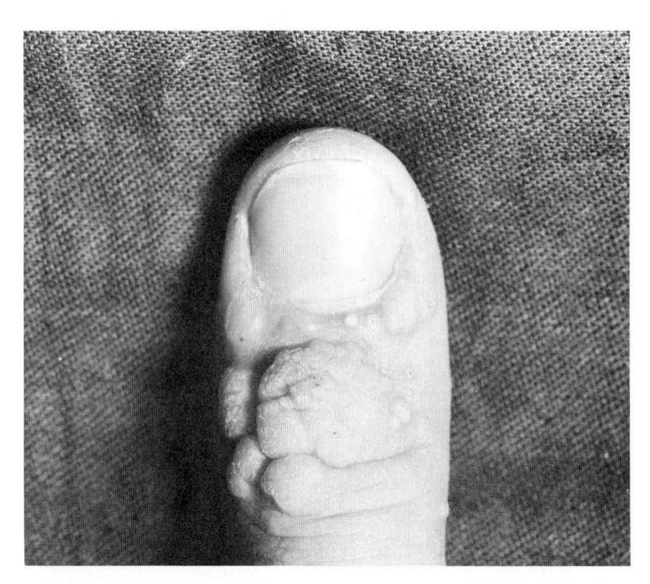

Figure 23–19 The common wart, *verruca vulgaris.* This infection is commonly found on the backs of hands, usually in the area of the nail folds. *[Courtesy of the Armed Forces Institute of Pathology, Washington, D.C., Neg. No. AMH 10737-2.]*

The Paramyxoviruses

The paramyxoviruses have several properties, including (1) the possession of one large molecule of single-stranded RNA, (2) an envelope with a lipid composition similar to that of the host's cell membrane, (3) replication in the cytoplasm, and (4) for most strains, hemagglutinating activity. The paramyxoviruses include the agents of classic measles (rubeola), mumps, Newcastle disease, and several human and lower animal respiratory infections. **[See virus and cancer in Chapter 30.]**

MEASLES (RUBEOLA)

Measles is a highly contagious disease usually contracted by exposure to respiratory secretions. As a rule, the virus causes disease in areas where large numbers of unimmunized young children reside. Outbreaks primarily affect children, as older persons have usually acquired some immunity from previous exposures to the virus. When susceptible adults are infected, symptoms are generally more severe than in children. Lesions known as Koplik's spots (Color Photograph 100), which give the appearance of salt grains on a red surface, typically appear in the mouth one to five days after the onset of the disease.

Complications may follow a measles virus infection, the most common being bacterial secondary infections such as pneumonia and middle ear infection. In general, it is advisable to administer antibiotics to protect infected children under three years of age and other persons with debilitating diseases, such as bronchitis and tuberculosis, against bacterial pathogens.

The virus can also produce serious complications such as infection of the central nervous system, sometimes leading to mental retardation. Fortunately, such cases are infrequent.

Pooled human immune globulin preparations can help modify the course of the disease. Timing and dosages are crucial.

Recovery from measles usually provides a long-lasting state of naturally acquired active immunity. Second bouts with the virus are uncommon.

PREVENTION. Several types of vaccines and programs have been developed for the express purpose of eradicating measles. One major approach involves immunization of all children at one year of age, administration of a vaccine to any unimmunized children upon their entry into school, and administration of measles immune globulin to all exposed susceptible persons. To further this program, many schools now refuse entry to children not immunized against measles (unless the failure to immunize is for religious or medical reasons).

A Togavirus Representative

Rubella virus is a typical togavirus in all respects except that it is not transmitted by arthropods. It contains a single-stranded RNA molecule.

RUBELLA (GERMAN MEASLES)

Rubella occurs in either epidemic or sporadic form worldwide. It is typically a mild disease with few and rare complications. However, among pregnant women the disease takes on an entirely different perspective because of the defects the viral agent can induce in fetuses during the early stages of pregnancy. The effects of the disease on the fetus, referred to collectively as the *rubella syndrome,* include cataract formation, congenital heart disease, permanent deafness, mental retardation, spontaneous abortion, and stillbirth. The consequences of rubella are not only varied but unpredictable. It appears that virtually any organ of a developing fetus may fall victim to the effects of the virus.

The mechanism or mechanisms contributing to these embryopathic effects are unknown. Studies of autopsied rubella infant victims show underdeveloped organs, and tissue cultures show pronounced chromosome breakage in lymphocytes from one-year-olds demonstrating the rubella syndrome. Based on these findings, several investigators have suggested that the virus may inhibit cell multiplication in the fetus.

Transmission of rubella usually occurs by direct contact with persons harboring inapparent infections. The nasal secretions of infected individuals are highly contagious. In addition, normal-appearing infants born to women who have had clinical rubella during the first three months of pregnancy excrete virus at birth. Such newborns come into close contact with a variety of people, such as health care personnel, expectant mothers, and other children.

CONTROL. Recovery from rubella generally imparts a long-lasting active immunity to the viral agent. Vaccines are also available. If teenage girls have not yet had rubella, immunization is recommended, since it is highly desirable for young girls to be exposed to the viral agent before reaching childbearing age.

Pooled preparations of immune globulin are administered to pregnant women as soon as possible after exposure if these women are in their first trimester. The effectiveness of this procedure is questionable, especially in relation to the prevention of congenital defects. Some physicians recommend a therapeutic abortion to their patients in the event of a rubella infection during the first trimester of pregnancy.

The Poxviruses

Agents of more than two dozen diseases in humans and lower animals belong to the poxvirus group. In the past, the two poxviruses of greatest significance for humans were the agents of smallpox and vaccinia, the virus used in smallpox vaccination. In May 1980, the World Health Organization declared that smallpox had been eradicated. The use of vaccinia virus for immunization purposes has largely been discontinued.

A relatively new human disease resembling smallpox has been identified. This viral disease, called monkey pox, was discovered in 1970. It is a zoonosis occurring in West and Central Africa (Color Photograph 101). Since smallpox, this new viral disease represents the most important human poxvirus infection. At the present time it does not appear to present a public health problem.

Most members of the poxvirus group (1) multiply and produce inflammatory lesions (pocks) after inoculation onto the chorioallantoic membranes of chick embryos (Color Photograph 53), (2) have double-stranded DNA, (3) replicate in the cytoplasmic regions of cells, (4) agglutinate (clump) fowl or mouse red blood cells, and (5) are resistant while in dried form to a wide variety of adverse environmental conditions.

MOLLUSCUM CONTAGIOSUM

Molluscum contagiosum, harmless in humans, is a viral skin disease (Table 23–4). It may be acquired through direct contact with contaminated objects such as towels and surfaces in gymnasium shower rooms and swimming pools. The disease also can be spread by sexual contact. Molluscum contagiosum is limited to the epidermis (Figure 23–20).

Figure 23–20 *Molluscum contagiosum.* Numerous papules of molluscum contagiosum on the back and buttocks of a patient. *[A. Ogino, and H. Ishida,* Acta Derm. Venereol. *(Stochh)* ***64****:83–86, 1984.]*

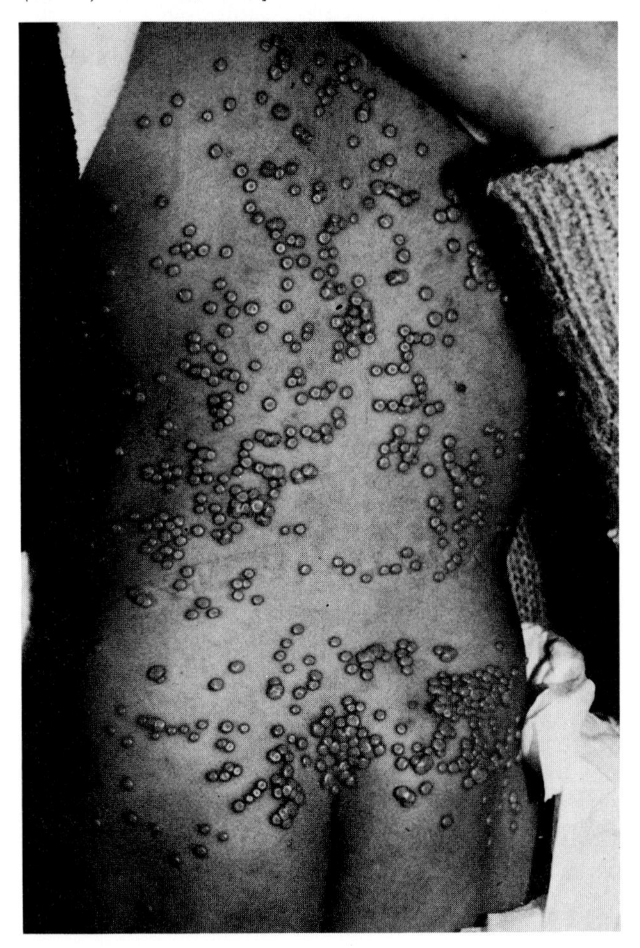

Infections of the Musculoskeletal System

The human skeletal system is composed of 206 individual bones that are held in position by strong fibrous connections called *ligaments.* The skeleton lies within the soft tissues of the body and performs several important functions, which include supporting soft tissues, providing points of attachment for most of the body muscles, determining the type and extent of body movement, protecting many of the vital organs from physical injury, and, after birth, producing the blood cells found in the circulatory system.

The skeleton contains a large number of joints that may be grouped according to the material that connects them or according to the movement allowed by the joints. Attention here will be focused on the most common *synovial (diarthrodial) joints.* These freely moving joints have several distinctive features, including an enclosing articular capsule, a thin layer of hyaline cartilage covering joint-associated bones, and a thin synovial membrane lining the joint (articular) capsule (Figure 23–21a). The synovial membrane secretes a clear, thick substance, synovial fluid, which lubricates joint surfaces and provides nourishment to other parts of the joints. Synovial membranes form two other structures that do not belong to synovial joints but are often associated with them. These are *bursae* (Figure 23–21) and *tendon sheaths.* Both reduce the friction that occurs during movement between a structure such as muscle, tendons, or skin and a bone. Bursae are small sacs distributed throughout the body, most of which are located between tendons and bones. Tendon sheaths are located where tendons cross joints, and generally are subjected to constant friction against bones.

Joints are subject to a variety of injuries, pathological changes, and microbial infections. The sites that may be infected include synovial joints (peripheral or central), central diarthrodial joints with or without synovial membranes, and surgically replaced joints; the tendon sheaths; the bursae; other soft tissues, including closed-space infections and muscle infections; the bones, including the spine; and the intervertebral disks. **[See inflammation in Chapter 17.]**

Arthritis, or infection of a joint, means involvement of the space within the joint capsule (Figure 23–21a). In most infections of a joint, the cardinal signs of inflammation are present along with a loss in the range of motion. With synovial joints, the most common form of infection is associated with a bacteremia caused by *Neisseria gonorrhoeae* (nye-SEH-ree-ah go-nor-REE-ah).

Septic Arthritis

The anatomical involvement in septic arthritis caused by microbial infection is highly dependent on the causative organism. Highly virulent microorganisms are likely to invade the blood, produce a septicemia, and invade bone tissue, causing an inflammation of the bone known as osteomyelitis. With some pathogens a specific or definite pattern of joint involvement is common. For example, mycobacteria and fungi tend to infect only one joint; gram-negative bacteria attack unusual joints and most commonly involve small joints in a symmetrical pattern.

Among the bacterial pathogens associated with septic arthritis, *S. aureus* is the most frequent cause of the disease in all age groups. *Haemophilus influenzae* (he-MAH-fi-lus in-floo-EN-zee) and *Streptococcus pneumoniae* (strep-toe-KOK-uss new-MOH-nee-eye) are also prominent causes of bacterial arthritis in young children and young adults, respectively.

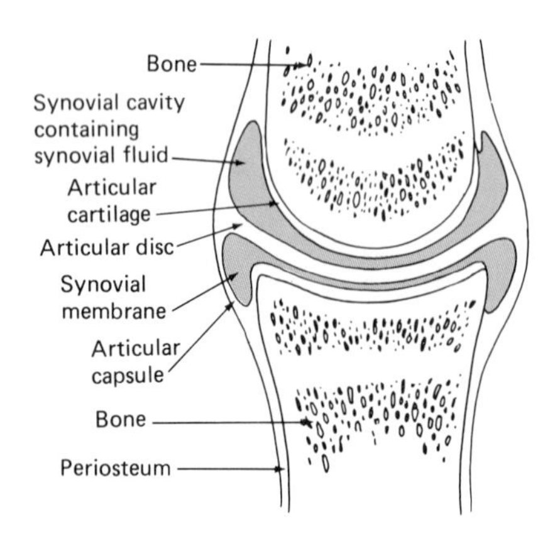

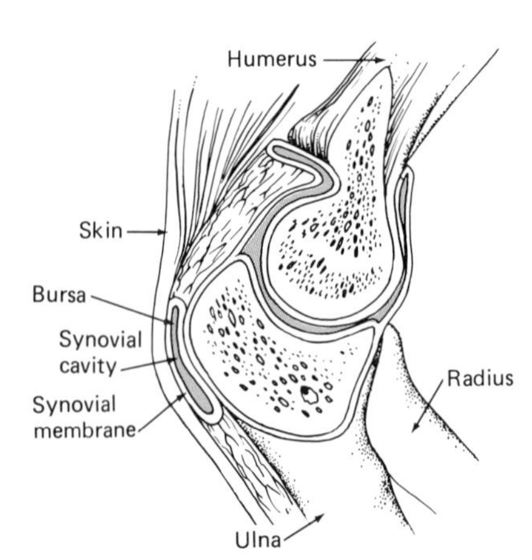

Figure 23–21 Synovial joints and bursae. (a) The structure of synovial joints. (b) Subcutaneous bursa of the human elbow.

Disease Challenge

The situation described has been taken from an actual case history. A review of treatment and epidemiological aspects is given at the end of the presentation. Answers to questions, laboratory findings, and interpretations are given immediately following a specific question. Test your skills and take the Disease Challenge.

Key Terms

folliculitis (foh-lick-you-LYE-tis): Inflammation of a hair follicle.
pseudohyphae (soo-doh-HIGH-fee): Yeast cells forming chains resembling the filaments of a mold.
pustule (PUS-tool): A small skin blister filled with pus or lymph.

Case

A twenty-five-year-old male with a five-year heroin addiction was admitted to the hospital because he had been suffering from painful scalp and beard folliculitis, which had been present for at least two weeks (See Figure). Three weeks earlier, the patient had injected, intravenously, brown heroin diluted with juice from a rotten lemon.

On admission, physical findings included normal blood pressure, pulse rate, and a slight fever. Additional painful pustules were found on his pubis and thighs.

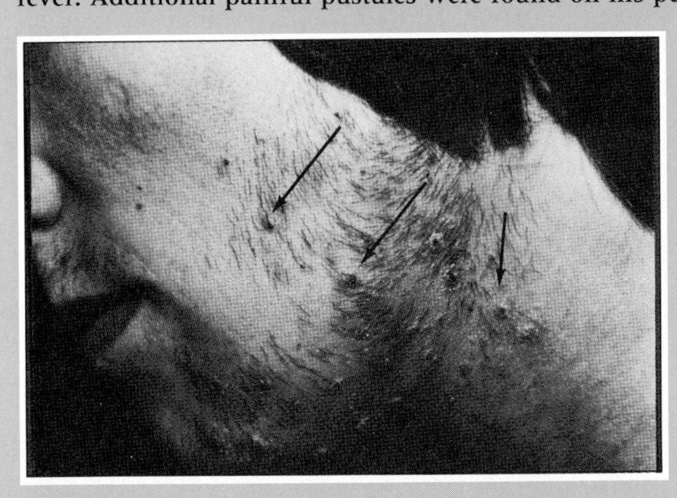

Follicular pustules and crusts on the beard. *[Courtesy Dr. Georgette Leclerc.]*

At this Point, What Type(s) of Specimen Should Be Taken?
Hairs from scalp and beard pustules for direct microscopic examination. In addition, pus and hair from the skin, sputum, urine, and blood for bacterial and fungal cultures.

Laboratory Results:
All hair preparations showed the presence of pseudomycelia. All cultures except the blood culture were positive for *Candida albicans.* Bacteria were not found.

What Is the Probable Diagnosis?
Candida folliculitis.

What Treatment Could Be Appropriate?
Ketoconazole.

Treatment:
Ketoconazole was administered. However, after one week, the lesions still persisted. The dosage was increased and continued for two weeks more. Skin lesions disappeared and were not evident at examination three months later.

Gonococcal arthritis develops in sexually active persons as a result of spreading via the bloodstream from a primary site of infection. A relatively new bacterial form of arthritis is *Lyme disease.* This illness is named after Lyme, Connecticut, where it was first studied in 1975. The infection occurs during the summer and autumn months in heavily wooded, tick-infested areas in several states in the Northeast, Midwest, and West. It is caused by a treponemalike spirochete belonging to the genus *Borrelia* (Figure 23–22) transmitted by *Ixodes* ticks. More than 500 cases have been reported since the original description of the disease.

The most common cause of viral arthritis is hepatitis B virus. Other viruses associated with arthritis include rubella virus and dengue fever virus.

Fungal involvement of synovial joints generally brings about a long-lasting form of arthritis. Most cases result from the spread of the pathogen from a remote infective site via the bloodstream. Fungus diseases in which septic arthritis occurs are infections involving body organs. These diseases, which are presented in other chapters, include coccidioidomycosis, histoplasmosis, cryptococcosis, and sporotrichosis.

Diagnosis involves laboratory examination of joint fluid or biopsy material. Such specimens are handled with care because they can serve as sources of infectious agents for laboratory personnel. Treatment is determined by the causative agent, and generally involves the use of antibiotics along with surgical removal of diseased tissue.

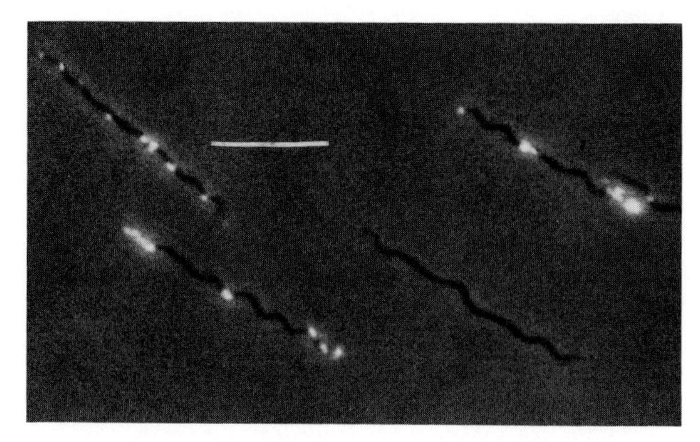

Figure 23–22 The spirochete causative agent of Lyme disease. These organisms have been treated with fluorescent monoclonal antibody to locate surface antigens. *[From A. G. Barbour, S. L. Tessier, and W. J. Todd,* Infect. Immun. ***45****:795–804, 1983.]*

Septic Bursitis

The bursae associated with the knee and shoulder are constantly at risk for injury with a resulting inflammation or infection *(septic bursitis).* These two areas are the major sites of septic bursitis, which is frequently seen in individuals engaged in gardening, carpentry, plumbing, golfing, and gymnastics. *S. aureus* is the causative agent in more than 90% of these bursal infections. Treatment usually includes drainage of the involved area and the use of antibiotics.

Summary

1. Diseases of the skin are caused by a variety of microorganisms and worms.
2. Organisms producing localized skin infections may also cause injury of internal organs.

Organization of the Skin

1. The skin, nails, and hair make up the covering of the human body.
2. The skin consists of the outer, *epidermis,* layer and the inner, *dermis,* layer.
3. Barriers to skin-invading microorganisms include skin acidity, indigenous microbial flora, temperature, and the inflammatory reaction.
4. Breaks in the skin during normal living can provide opportunities for microorganisms to enter the body.
5. The type and extent of injuries to the skin and the microorganisms involved influence the seriousness of the infections.

Bacterial Diseases

1. The signs, symptoms, and destructiveness of skin infections may vary from localized effects to rather extensive involvement and penetration into deeper tissues.
2. Identification of microbial pathogens of the skin frequently is dependent on the isolation, cultivation, and results of various biochemical tests.
3. Bacterial diseases associated with the skin include anthrax, erysipelas, gas gangrene, impetigo, green-nail syndrome, leprosy, scalded skin syndrome, scarlet fever, tetanus, and wound botulism.
4. *Scalded skin syndrome* exists in two forms. One form, *toxic epidermal necrolysis (TEN),* occurs mainly in adults and is associated with hypersensitivity to various chemicals. The other form, *staphylococcal scalded skin syndrome (SSSS),* is caused by a skin-destroying exotoxin of *Staphylococcus aureus* and occurs in infants under five years of age.

Mycotic Infections
Human skin, nails, and hair are particularly susceptible to attack by certain pathogenic fungi known as *dermatophytes.*

CLASSIFICATION OF MYCOTIC INFECTIONS
1. Fungus infections, or *mycoses,* are grouped according to the tissues or organs affected.
2. Fungi that mainly attack the skin, hair, nails, and mucosal surfaces are called *superficial* or *surface fungi.*
3. *Deep-seated,* or *systemic,* mycoses are associated with the deeper tissues.
4. Some fungi that normally are not pathogenic but take advantage of lower host resistance are called *opportunists.*

REPRESENTATIVE SUPERFICIAL MYCOSES
1. Fungal infections of the skin, hair, and nails are examples of ringworm (tinea).
2. Examples of such conditions include ringworm of the body (tinea corporis), feet (tinea pedis), head (tinea capitis), and groin area (tinea cruris).
3. Prevention of ringworm includes keeping potential or susceptible areas clean and dry.

OTHER DISEASES CAUSED BY FUNGI
The yeast infection thrush is caused by *Candida albicans.* This condition occurs in individuals with lowered resistance or suffering from complications of treatment or other microbial diseases.

DIAGNOSTIC AND RELATED FEATURES OF THE DERMATOPHYTES
1. The identification of several fungi is based on their growth characteristics and the presence and microscopic properties of their hyphae and spores.
2. The use of Wood's light (a form of ultraviolet light) is of diagnostic value with certain infections.

Viral Infections
1. Several viral infections are either limited to the skin proper or involve it in a disease process.
2. Examples of viral diseases of the skin include chickenpox, cold scores, measles, and warts.
3. The poxvirus disease smallpox has been eliminated as a human virus infection through immunization. However, a newly uncovered infection, monkey pox, is receiving considerable attention as a new human disease.

Infections of the Musculoskeletal System
1. Infections may involve bone tissue, synovial joints, surgically replaced joints, tendon sheaths, bursae, muscle, and the bones including the spine and vertebrae.
2. Infection of a joint involves the space within a joint capsule.
3. In most joint infections, inflammation and abnormal amounts of fluid are present.

SEPTIC ARTHRITIS
1. The causative agent determines the extent of infection.
2. The most frequent cause of septic arthritis is *Staphylococcus aureus. Haemophilus influenzae* and *Streptococcus pneumoniae* are prominent causes of arthritis in young children and adults. The gonococcus *(Neisseria gonorrhoeae)* is known to cause many cases of bacterial arthritis in sexually active persons.
3. Lyme disease is a relatively new tick-borne or -associated type of septic arthritis. The causative agent is a spirochete in the genus *Borrelia.*
4. Viruses and fungi cause septic arthritis involving synovial joints.

SEPTIC BURSITIS
1. Bursae associated with the knee and shoulder are at relatively great risk of infection *(septic bursitis).*
2. *S. aureus* is the causative agent in more than 90% of cases.

Questions for Review

1. a. What type of microorganisms make up the flora of the normal skin (see Chapter 20)?
   b. Are any of these organisms capable of invading normal skin? Explain.
   c. Does the microbial flora of the skin serve any useful function? Explain.

2. Select five bacterial species that are associated with diseases of the skin. Construct a table and compare these agents with respect to the following properties:
   a. morphology
   b. Gram reaction
   c. method of diagnosis
   d. determination of pathogenicity (if applicable)
   e. laboratory diagnosis
   f. preventive and control measures

3. Does there appear to be any form of immunity toward bacterial agents capable of causing skin infections? If so, toward which ones? Are vaccines available against these organisms?

4. a. What types of microorganisms are associated with wounds? Are any of these agents aerobic?
   b. With what types of wounds (e.g., abrasions, lacerations) is tetanus associated?

5. Compare the properties of an endotoxin with those of an exotoxin (see Chapter 22).

6. a. What is gas gangrene? How does it differ from an ordinary case of gangrene?
   b. How is gas gangrene treated?

7. a. What is leprosy?
   b. How is it transmitted?
   c. Where does this disease occur?

8. How does wound botulism differ from botulism associated with food? (See Chapter 26.)

9. Distinguish between staphylococcal scalded skin syndrome and toxic epidermal necrolysis.

10. Differentiate between subcutaneous and deep-seated mycoses.

11. What is a dermatophyte? Give six examples of disease states caused by dermatophytes.

12. How can one distinguish between endothrix and ectothrix infections?

13. How can the following diseases be contracted?
a. tinea capitis
b. tinea cruris
c. tinea pedis
d. tinea corporis
e. tinea versicolor
f. tinea unguium
g. white piedra

14. What methods are generally used in the diagnosis of superficial mycoses?

15. Discuss the diseases of *Candida albicans* in terms of their various forms and possible means of prevention.

16. a. If you were traveling in a tropical or subtropical area of the world, what mycotic infections might you find?
b. How could these diseases be prevented?

17. Differentiate between the following terms (refer to Table 23–1 and associated color photographs):
a. macule and pustule
b. vesicle and scar
c. maculopapule and crust

18. Describe the various disease states caused by herpes simplex virus 1. Are there any precipitating causes?

19. a. Compare the modes of transmission of the viral skin diseases discussed in this chapter.
b. What control measures could be employed to limit infection?

20. What is the relationship between varicella and shingles?

21. What types of complications are associated with viral skin disease?

22. Differentiate between rubella and rubeola.

23. For which of the viral skin diseases discussed in this chapter is a laboratory diagnosis essential?

24. Which of the infections discussed in this chapter have you experienced?

25. Against which viral diseases of the skin are vaccines currently available?

26. Define or explain the following terms:
a. vaccinia
b. Lipschütz body
c. rubella syndrome
d. verruca vulgaris
e. poxvirus
f. arthritis
g. bursitis
h. monkey pox

27. What is molluscum contagiosum?

Suggested Readings

BAKER, C. J., "Group B Streptococcal Infections," *Advances in Internal Medicine* **25:**475, 1980. *A general review of the diseases caused by this particular streptococcal group.*

CHERRY, J. D., "Viral Exanthems," *Disease-a-Month,* Vol. 28, No. 8. Chicago: Year Book Medical Publishers, 1982. *A publication covering several aspects of viral illnesses associated with the skin, including general descriptions of specific diseases, clinical features, clinical differentiation, and laboratory diagnosis.*

COOPER, L. Z., and S. KRUGMAN, "The Rubella Problem," *Disease-a-Month.* Chicago: Year Book Medical Publishers, 1969. *A detailed review of the various effects of rubella virus.*

COREY, L., and P. G. SPEAR, "Infections with Herpes Simplex Viruses," *New England Journal of Medicine* **314:**749–757, 1986. *Excellent descriptions of the clinical range of herpes simplex viruses are provided in this article.*

FINEGOLD, S. M., and W. J. B. MARTIN, *Bailey and Scott's Diagnostic Microbiology,* 6th ed. St. Louis: C. V. Mosby, 1982. *A functional reference to the isolation and identification of microbial pathogens.*

HENDERSON, D. A., "The Eradiction of Smallpox," *Scientific American* **235:**25–33 (1976). *A detailed description of the measures used to eliminate one of the historically important and serious diseases of humans. The article includes discussions of basic immunological and epidemiological topics.*

SHEAGREN, J. N., "*Staphylococcus aureus:* The Persistent Pathogen," *New England Journal of Medicine* **310:**1368, 1984. *A good review of the various diseases caused by this microbial pathogen.*

24

Oral Microbiology

He prayeth best who loveth best
All things great and small.
The Streptococcus *is the test—*
I love it least of all. — *Wallace Wilson*

After reading this chapter, you should be able to:

1. Describe the general organization and components of the human mouth and indicate the areas susceptible to attack by pathogenic microorganisms.
2. Discuss the relationship of the microorganisms in the mouth to the health of the individual.
3. List the general properties of four different microorganisms that make up the oral flora.
4. Describe the causative agent, means of transmission, and control measures for at least two each of bacterial, mycotic, and viral diseases affecting the oral regions.
5. Distinguish among the following disease states of the oral region in terms of general appearance or symptoms and treatment: gingivitis, periodontal disease, dental caries, necrotizing ulcerative gingivitis, and periodontitis.
6. Define or explain plaque formation and calculus.
7. Describe the cause, development, and control of dental caries.
8. Recognize the relationship of yeast infection to AIDS.
9. Solve a disease challenge involving a microbial agent.

Chapter 24 discusses specific microbial diseases affecting teeth, gums, and other supporting tissues. Attention is also given to the microbial flora of the oral region, the host immune response, and the complications that can develop from disease states associated with the mouth.

The mouth is continually exposed to organisms from the external environment, beginning with passage through the birth canal. In time, an ecological balance is reached that serves to establish a resident microbial flora that remains fairly stable throughout life. Oral infections result from disturbances in the relationship or balance between the oral resident flora and the host response. This imbalance may be caused by an allergic reaction or may be the result of an immunologic deficiency. In some other disease states, such as tooth decay or dental caries, specific pathogens may damage tissues directly regardless of a host's response.

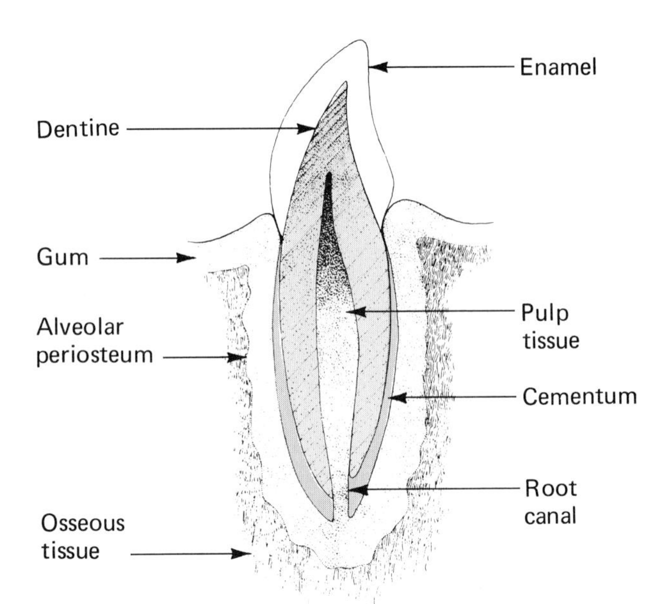

Figure 24–1 The structure of a human tooth.

Structure of the Mouth

The oral cavity, or mouth, is situated at the beginning of the gastrointestinal tract. This space is enclosed on the sides by the lips and cheeks, above by the hard and soft palates, and below by the floor of the mouth and the tongue. The lips are covered on the outside by skin and on the inside by mucous membrane. Small glands are present beneath the mucosa, and there are many muscle bundles within the lips.

The palate is divided into a hard palate at the front of the mouth and a soft palate at the back. The bones of the hard palate are covered by a thick layer of firm but soft tissue. The soft palate connects with the passageway from the mouth to the throat. It is continuous with the tissues encircling the opening to the pharynx.

The floor of the mouth lies in a horseshoe around the tongue and is continuous with the *gingiva* (the gum) and the tongue. Near the front end are the openings of the submandibular and sublingual *salivary glands*.

The human develops two sets of teeth, deciduous and permanent. There are 20 teeth in the first set, also called the *milk teeth;* the permanent set usually contains 32 teeth. Teeth are categorized into four groups: *incisors, canines, premolars* or *bicuspids*, and *molars*. Incisors are used for cutting food, canines for tearing food, and premolars and molars for grinding. Each tooth has three parts (Figure 24–1): the *crown*, the portion above the gum; the *root*, the structure embedded in the jaw; and the *neck*, which is the narrower region between the crown and the root.

A tooth's crown is coated with *enamel*, the hardest substance found in the body; the rest of the tooth is covered by a layer of modified bone called the *cementum*. Under the enamel coating is an ivorylike tissue called *dentine*, which comprises the bulk of the tooth. The dentine is quite hard and striated. Within this layer is a cavity, the *pulp chamber*, containing blood vessels, connective tissue, and nerve endings. The contents of this chamber are frequently referred to as the *dental pulp*.

Teeth tend to become covered with a gummy accumulation of salivary mucin and bacteria, called *dental plaque* (Figure 24–7), which can be seen by the naked eye. Microscopically, plaques can be seen as accumulations of bacteria and as various bacterial and host-produced units called "corncobs" (Figure 24–2). These highly specific microscopic units con-

Figure 24–2 Electron microscopic views of "corncobs." (a) A corncob consisting of *Fusobacterium* and adhering *Streptococcus sanguis*. (b) An ultrathin section showing the attachment of *S. sanguis* by means of its fimbriae (fm) or pili. *[From P. Lancy, Jr., J. M. Dirienzo, B. Appleman, B. Rosan, and S. C. Holt,* Infect. Immun. ***40****:303–309, 1983.]*

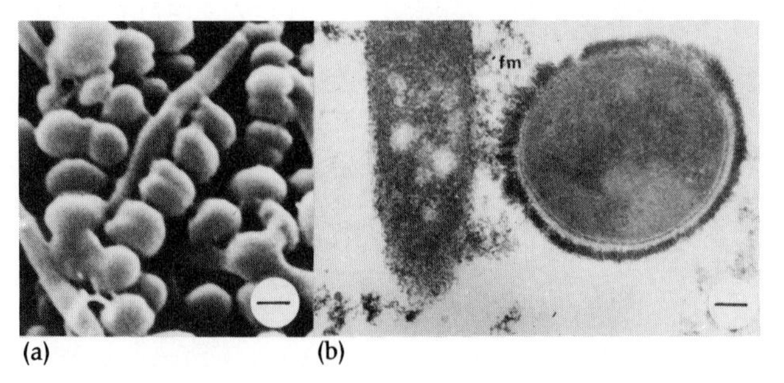

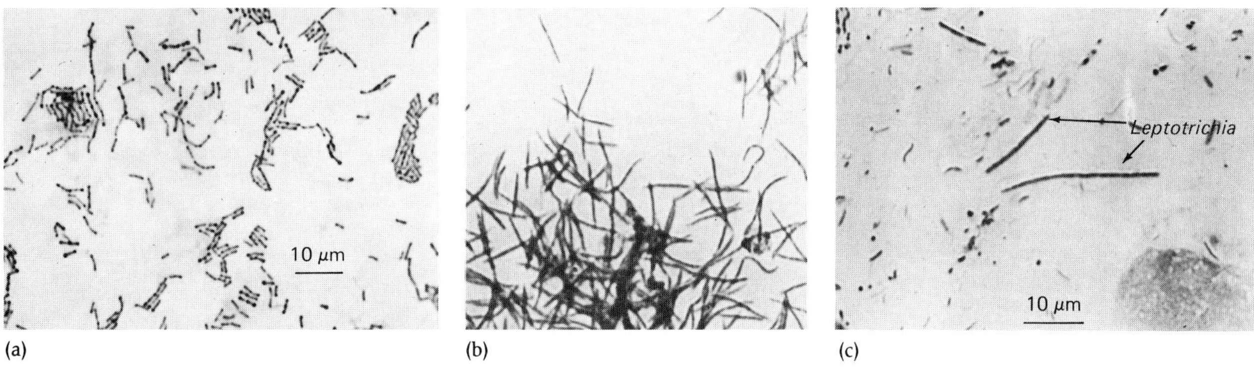

Figure 24–3 Representatives of oral flora. (a) A pure culture of lactobacilli. Note the polar staining in these organisms. (b) Fusobacteria, also rods, with tapered, pointed ends. *[Courtesy of Dr. Richard Parker.]* (c) Mixed oral flora. The long, granular organisms are *Leptotrichia*. A few rods, diplococci, and threadlike forms are also seen.

sist of a filamentous bacterium surrounded by adhering cocci and resemble an ear of corn. A study of the factors affecting corncob formation could be important to understanding plaque maturation and for developing strategies of prevention. It has been suggested that the early growth of aerobes and the buildup of plaque provide a suitable environment for the growth of the anaerobes that predominate in older plaque.

Oral Flora

Because the mouth is warm and moist and has a regular supply of fresh food, it makes an ideal growth environment for microorganisms. The study of *oral ecology* is both interesting and complex. Microscopic examination of plaque from a tooth's surface usually reveals a wide variety of bacteria (Figure 24–3). Many studies have been conducted to find which organisms, and in what concentration, predominate in the oral cavity.

The oral cavity of the fetus is essentially germ-free until it passes through the birth canal. At this time, lactobacilli, micrococci, alpha- and gamma-anaerobic streptococci, coliforms, corynebacteria, yeasts, viruses, and protozoa are obtained from the vagina and urethra. Staphylococci and pneumococci may be added from the air. Feeding and contact with people and new environments add many more organisms. *Streptococcus salivarius* (strep-toe-KOK-kuss sa-li-VA-ree-us) is one of the earliest colonizers of the oral cavity in infants. Oral streptococci attach to different sites. The associated attachment or adhesion process is highly selective and is believed to involve specific functional structures called fibrils (Figure 24–4). The fibrils differ ultrastructurally from pili, are densely packed, and carry either protein or glycoprotein surface-binding substances called adhesins. **[See pili in Chapter 7].**

The assortment of organisms in the oral flora is relatively stable, though subject to change with aging. Although other microorganisms may be introduced, perhaps with food, they are usually transient and seldom take up permanent residence.

The organisms present at any one time exist in balance with one another, and any change in this balance may result in disease. Attempts to study the oral microbiota in the laboratory have proved most difficult, since duplication of the oral environment is almost impossible. Mixed culture studies have yielded some data, but so far no one has developed a system able to duplicate the environment required by oral microorganisms.

The organisms of the human mouth fall into

Figure 24–4 Groups or tufts of long and short fibrils projecting from the surface of an adhesive strain of *Streptococcus salivarius. [P. S. Handley, P. L. Carter, and J. Fielding,* J. Bacteriol. ***157****:64–72. 1984.]*

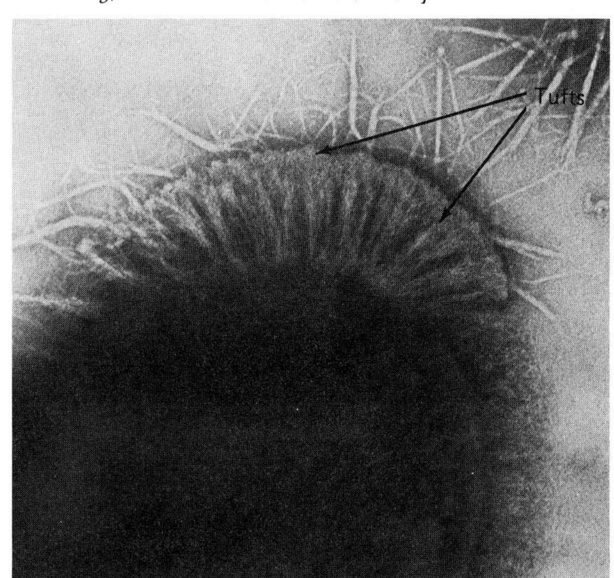

TABLE 24–1 Microorganisms in the Oral Flora[a]

Microorganism	*Gram Reaction*	*Morphology*	*Characteristics and/or Associated Activities*
Actinomyces	+	Rod to coccoid	Most are facultative anaerobes; organisms can be found between teeth and in gum grooves; certain species are pathogenic; major members of dental plaque
Bacterioides	–	Rods	Strict anaerobes; some species are pathogenic and are associated with gum disease (necrotizing ulcerative gingivitis)
Borrelia	–	Spirals	Strict anaerobes; some species may cause diseases involving supportive tissues in the mouth
Branhamella catarrhalis	–	Cocci	Aerobic; parasites of the human mucous membranes
Candida[b]	+	Large oval cells	Represent only a small percentage of the organisms in the total oral flora; known to cause oral infections in individuals with diabetes, cancer, immune defects or those who are receiving large doses of antibiotics
Capnocytophaga	–	Rods	Anaerobic rods with tapered ends found in gum region of molars; certain species are believed to be involved with periodontal disease
Corynebacterium	+	Rods	Aerobes and facultative anaerobes; picket fence arrangement of cells is a common feature; some pathogenic species *(C. diphtheriae)* produce exotoxins
Diphtheroids	+	Rods	Aerobes to microaerophilic; club-shaped cells arranged in patterns resembling Chinese characters; normal inhabitant of the mouth
Fusobacterium	–	Rods	Strict anaerobes; normally found in mouth and other human cavities; involved with corncob formation
Lactobacilli	+	Rods	Facultative organisms that produce large amounts of acid from carbohydrates; pathogenicity unusual
Leptotrichia	–	Rods	Highly anaerobic; found in recesses and crevices between teeth; appear as very thick, long, nonbranching rods with rounded ends (Figure 24–3c)
Mycoplasma	–	Variable shapes	Mostly facultative anaerobes; highly variable in shape (pleomorphic); certain species are parasitic as well as pathogenic
Neisseria	–	Cocci	Aerobes or facultative anaerobes
Nocardia	+	Coccoid to rods	Strict aerobes with branching; certain species are pathogenic
Streptococci	+	Cocci	These organisms make up the largest bacterial group in the oral cavity; streptococci are associated with plaque formation and the production of acids from carbohydrates; alpha-hemolytic streptococci (*viridans* group) pose danger in cases of tooth extraction and heart valve damage; beta-hemolytic streptococci are noted for diseases such as strep throat and scarlet fever
Treponema	–	Spirochete	Strict anaerobes; some species normally found in mouth and may cause dental plaques
Veillonella	–	Cocci	Parasitic anaerobes; commonly found in plaques

[a] Various protozoa and viruses are also found in the mouth.
[b] Yeast.

three groups with regard to their tolerance of or requirement for oxygen: strict anaerobes, strict aerobes, and facultatives. This last group includes everything between the two extremes—all those microorganisms that can tolerate some concentration of oxygen, from very high to very low.

Although bacteria are the most obvious inhabitants of the oral cavity (see Table 24–1), other microorganisms are often seen. These include several species of fungi, viruses, and protozoa.

Of the fungi, probably those most commonly found are of the genus *Candida.* It has been estimated that in approximately 40% of the normal population, these organisms can be cultivated from saliva. In a healthy mouth, they make up only a small percentage of the total oral flora. However, in children, the aged, and the debilitated, they are of major importance. Oral *Candida* infections often follow heavy dosages of antibiotics. These infections are undoubtedly caused by the change in bacterial population and the resulting imbalance in oral ecology. *Candida* (KAN-did-ah) infections also represent a feature of several disorders with an underlying immune defect. AIDS is an example here. [See AIDS in Chapter 27.]

Viruses, both human and bacterial, have been recovered from the oral cavity. The causative organisms of canker sores, herpes simplex fever blisters, and measles can be found in oral lesions during obvious disease. Little is known about the place of viruses in the normal ecology of the human mouth, but no doubt many types are present.

Microbiology Highlights

THE TEETHING VIRUS, OR NO MORE TOOTH FAIRY

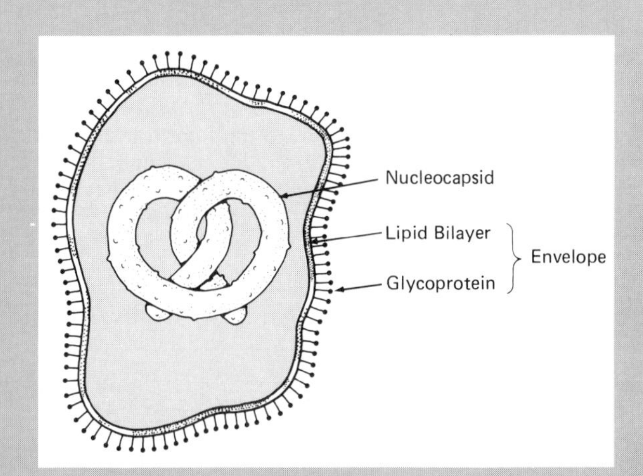

A schematic representation of the human teething virus. *[Modified from H. J. Bennett, and D. S. Brudino,* Pediant Inf. Dis. *5:399–401, 1986.]*

Teething has been a subject of major interest in the medical and nonmedical community for centuries. Arguments have resulted not only over the signs and symptoms associated with teething but also over the tooth fairy's influence on the family. The greatest controversy of all is whether or not teething causes fever.

Early in 1982, a sample of saliva from a soggy bagel chewed on by a teething infant was examined under an electron microscope. To everyone's amazement, a new, previously unknown virus was uncovered. The discovery of a human teething virus (HTV) opened the door to additional studies. In 1983 and 1984, during a "teething outbreak" in Washington, D.C., a study was begun in which 500 infants were followed from birth through 2½ years of age. These patients experienced repeated bouts of teething, and 84% became feverish (febrile) during the teething process. Specimens were collected on teething rings impregnated with tissue-cultured cells. The HT virus was isolated from 99% of the feverish patients. (The HT virus is a uniquely RNA-shaped particle with a diameter of 140 nm. Its helical neocapsid is surrounded by an envelope covered with spherical projections.) No virus was found with nonfebrile teething individuals. The virus is apparently dormant between teething periods.

TABLE 24–2 Features of Nonspecific Oral Foci of Infection

Infection and/or Condition	*Causes (C) and/or Contributing Factors (CF)*	*Brief Description*	*Signs and Symptoms*
Dental caries	*CF:* Climate, composition and amount of saliva, hormonal balance, nutritional state, oral hygiene, fluoride level in drinking water, diet, genetic makeup *C:* Interactions between the host tissues and specific caries-producing microorganisms, e.g., *Streptococcus* species	Loss of calcium salts (decalcification) of inorganic substances of teeth, followed by disintegration of organic portions	General pain, chronic irritation, headache, and complications resulting in infections of surrounding areas
Dry socket	*C:* Contamination of the extraction area, excessive injury, rinsing with hot fluids, dislodging of blood clot by vigorous rinsing, lowered host resistance, implanting bacteria or foreign material	Dislodging of a blood clot and exposure of bone from a tooth extraction site	Foul odor, swollen and inflamed gums, pain, and a mass of dead tissue (slough) along the margin nearest the socket
Gingivitis	*C:* Improper or inadequate oral hygiene, food impacting between poorly closing teeth or around teeth badly broken from decay	Inflammation of gums	Swollen, reddened, and bleeding gums; pus formation may occur
Necrotizing ulcerative gingivitis	*CF:* Fatigue, anxiety, and other forms of stress, debilitating illnesses, such as cancer or diabetes, severe vitamin deficiency diseases, local irritation of gums, calculus, and overhanging gums *C:* Implicated bacteria include *Borrelia vincentii* and *Fusobacterium fusiforme*	This disease is also known as *trench mouth* or *Vincent's disease;* destruction of gums and associated tissues	General pain in gums, slight fever, malaise, ulceration of gums, bleeding, loss of dead tissue, foul odor, and metallic taste; bone involvement may occur in untreated cases
Osteomyelitis	*C:* Infected pulp, residual infections, severe periodontal disease with extension into the bone, many forms of destructive injury, and specific infections such as actinomycosis, syphilis, and tuberculosis	Inflammation and eventual destruction of bone and surrounding tissues	Severe pain, elevated temperature, swollen lymph nodes associated with the area; loose teeth, and difficulty in eating
Pericoronitis	*C:* Contaminated instruments, infections following extractions, and specific bacteria including *Streptococcus* species	Inflammation around the crown of the tooth; the condition may spread to other surrounding tissues	Face swollen on the involved side, discoloration of tissue, draining of pus from involved area
Periodontitis	*C:* Untreated gingivitis *CF:* Various factors acting together are important considerations, including plaque formation, calculus, allergic responses to bacterial antigens, poor oral hygiene, injury by dental floss, genetic factors, hormonal balance, and poor closure of jaw (malocclusion)	This condition is also known as *pyorrhea.* It is an inflammation of the periodontium, the directly supporting tissue of the tooth	Inflamed gums, bleeding, loss of bone around the teeth, loose teeth; many cases exhibit few symptoms

Nonspecific Infections of the Oral Region

Infections of the face, oral cavity, and neck may be extremely serious, depending upon their location and the microorganisms involved. Specific infections caused by bacteria, fungi, or viruses occur here as well as in other body sites. More commonly, mixed bacterial infections occur in deep cavities in teeth or result from tissue injury.

Focal Infections

A localized area of infection anywhere on the body is called a *focus of infection.* When organisms or their toxic products spread from this focus to distant tissues, either to form another site of infection or to produce a hypersensitive reaction, the process is known as a **focal infection.**

The concept of focal infection has had a long and stormy history. Shortly after it was introduced, medical and dental practitioners enthusiastically condemned infected teeth and oral tissues as the cause of many unexplained conditions in the body, using the idea of focal infection to justify the extraction of teeth and related procedures. Countless thousands of teeth were sacrificed to the cause. Then the pendulum swung too far to the other side, and the oral regions were almost ignored as a source of focal infection. At present, certain conditions are known to be related to oral foci of infection. Good oral hygiene, along with elimination of infection, is considered an important part of restoring and maintaining good health.

Selected examples of nonspecific oral foci of infection in the periodontium, or supporting structures of the teeth, are presented in Table 24–2. Resultant changes in the tissues of infected teeth include abscesses and apical granulomas (a form of tumorous growth). Examples of disease states specifically related to oral foci are discussed in later sections.

DRY SOCKET

With dry socket, the blood clot that forms following the removal of a tooth is dislodged and lost from the extraction site, thus exposing the bone and allowing some degree of infection to develop. The factors listed in Table 24–2 have been implicated as causes of dry socket. The bony walls of the socket often show signs of tissue death and become infiltrated with bacteria of many types. A foul odor also develops.

OSTEOMYELITIS

Inflammation of the bone *(osteo)* marrow *(myelo)* may occur from the introduction of many different types of bacteria as either pure or mixed cultures (Figure 24–5). The resulting infection leads to inflammation, cellular degeneration, and necrosis of the tissues involved, often including the bone (osteitis) and the *periosteum,* or surrounding membrane (periostitis). Several different factors may cause osteomyelitis (Table 24–2).

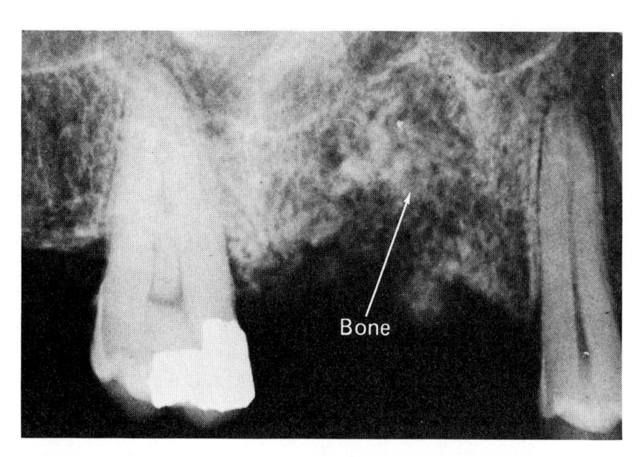

Figure 24–5 Chronic osteomyelitis involving the maxilla. This x-ray shows the moth-eaten appearance of the bone. *[Courtesy of Dr. N. H. Rickles, Pathology Department, University of Oregon Dental School.]*

The disease may be of short duration (acute) or long-lasting (chronic), depending upon many factors, such as the type and number of organisms involved, their virulence, and the age and resistance of the host. Since the teeth in the area may be loose, pain and difficulty in eating can occur. Thus, the patient's nutritional status can also be affected.

PERICORONITIS

Pericoronitis may begin in a flap of tissue overlying an erupting tooth or around an impacted or partially erupted third molar. Pericoronitis, or inflammation *(-itis)* around *(peri-)* the crown *(corona)* of the tooth, may spread into the surrounding tissues, resulting in a cellulitis, or diffuse inflammation of the soft tissues. The bacteria involved here produce large amounts of the enzymes hyaluronidase and fibrinolysins, which are capable of breaking down tissue cohesiveness, thus favoring the spread of the infection. As infection spreads, that side of the face begins to swell and the firm tissues may become discolored.

PERIODONTAL DISEASE

Periodontal disease, a worldwide affliction of humans, appears clinically as an inflammation of the soft tissues around the teeth (Figure 24–6). Depending on the severity of the disease, the destructive processes may involve both the gums (gingivitis) and the periodontal membrane and alveolar bone surrounding and supporting teeth (periodontitis). In advanced stages, destruction of cementum and periodontal membrane accompanied by loss of alve-

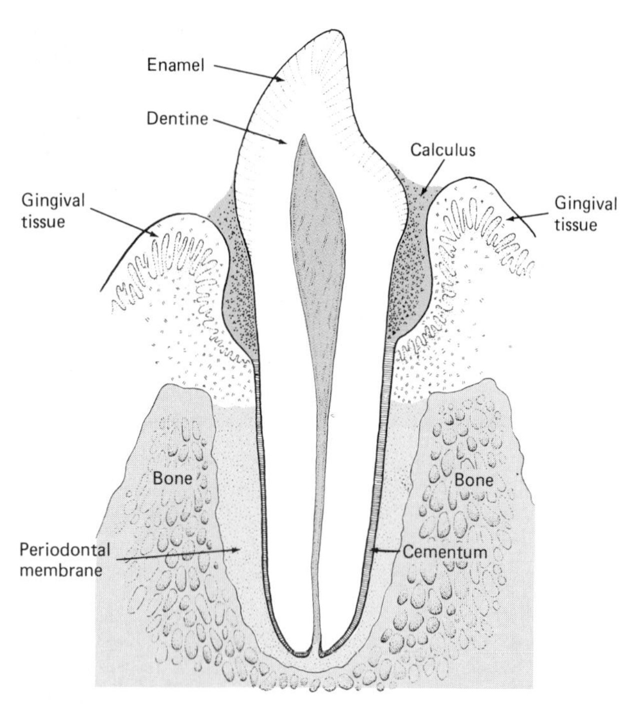

Figure 24–6 In periodontal disease, the gingival tissue shows inflammation, swelling, and loss of stippling, and plaque and calculus extend down the root, creating a marked inflammation reaction with resultant loss of bone. Pus may be expressed from the pocket. Also shown are tooth enamel, dentine, pulp, periodontal membrane, and cementum. *[Courtesy of Dr. N. H. Rickles, Pathology Department, University of Oregon Dental School.]*

olar bone occurs (Figure 24–7). The net result is the formation of a pocket between the root and the overlying soft tissue, usually with marked inflammation and pus formation. Unfortunately, there are few clinical symptoms, and the disease progresses relentlessly until the teeth are lost. In people over 35 years of age, this disease—not tooth decay—is the major cause of tooth loss.

It is quite possible that no single factor is responsible for the infection, but rather many factors acting in concert (Table 24–2). One current hypothesis is that periodontal disease is caused not by a particular bacterial species but rather by certain enzymatic and related activities of organisms in intimate contact with tissues surrounding the teeth. A large body of evidence already indicates that specific microorganisms must first colonize the tooth or epithelial surface as a prerequisite to periodontal disease. Taken as a whole, the *Capnocytophaga* (kup-NO-sigh-toe-fay-jah) have numerous properties that could place this genus in a central position in the development of periodontal disease. These thin, flexible bacteria could glide over the tooth surface in a direction pointed through narrow crevices, enter deep subgingival locations within a developing periodontal pocket, and ultimately colonize host tissue surfaces. The *Capnocytophaga* (Figure 24–8) also have the ability to transport nonmotile organisms from one location to another. This process may be critical to understanding the role of mixed anaerobic bacterial populations in establishing periodontal disease.

In recent years the roles of **dental plaque** and **calculus** (Figure 24–9) in the periodontal disease process have been studied. Dental plaque formation is of great importance, especially since it may well be the initiator of dental decay as well as periodontal disease. Plaque is a mixture of bacteria embedded in an accumulation of saliva and bacterial products sticking to the tooth surface (Figure 24–10). Several different microorganisms isolated from dental plaque are capable of producing alveo-

Figure 24–7 Periodontal disease has caused almost total destruction of the bone surrounding this lower molar. Clinically, the patient has a periodontal abscess with swelling and acute inflammation. *[Courtesy of Dr. N. H. Rickles, Pathology Department, University of Oregon Dental School.]*

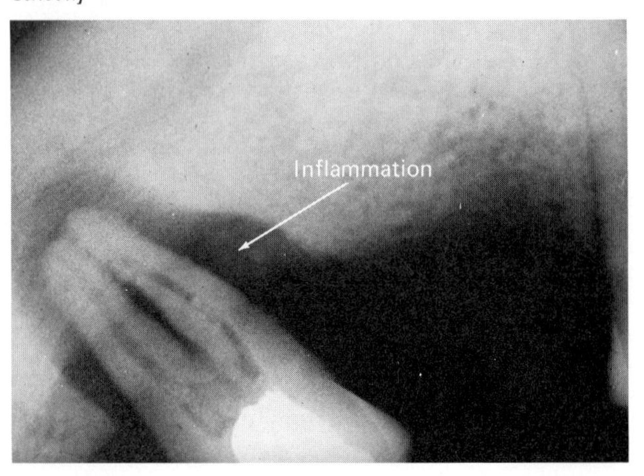

Figure 24–8 *Capnocytophaga.* These bacteria have been isolated from cases of rapidly progressing periodontal disease. This scanning micrograph shows pioneer colonies consisting of small accumulations of individual cells in an oriented, heaped, end-to-end arrangement. *[T. P. Poirier, S. J. Tonelli, and S. C. Holt,* Infect. Immun. ***26:****1146–1158, 1979.]*

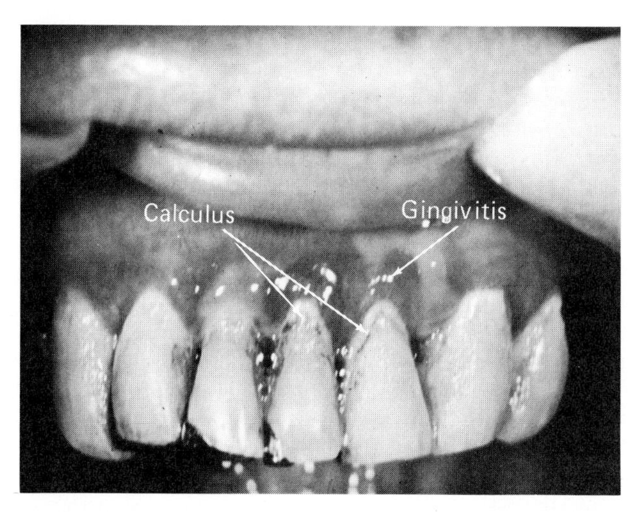

Figure 24–9 Gingivitis. Note the severe involvement of the gingiva around the teeth, where there is also heavy buildup of plaque and calculus. *[Courtesy of Dr. N. H. Rickles, Pathology Department, University of Oregon Dental School.]*

lar bone loss when inoculated directly into the oral cavity of germ-free rats. These bacteria include *Actinomyces naeslundii* (ak-tin-oh-MY-sees nees-LUN-dye), *A. viscosus* (A. viss-KOH-suss), *Bacteroides melaninogenicus* (back-te-ROY-dees mel-an-IN-oh-gen-ee-kuss),

Figure 24–10 An electron micrograph of *Bacteroides melaninogenicus* isolated from a case of periodontal disease. A large number of fibrils and several large, forming vesicles can be seen surrounding the cell. These fibrils differ in ultrastructure from pili. In addition, the forming vesicles may contain endotoxins. Original magnification 32,300×. *[Courtesy P. S. Handley, Department of Bacteriology and Virology, University of Manchester.]*

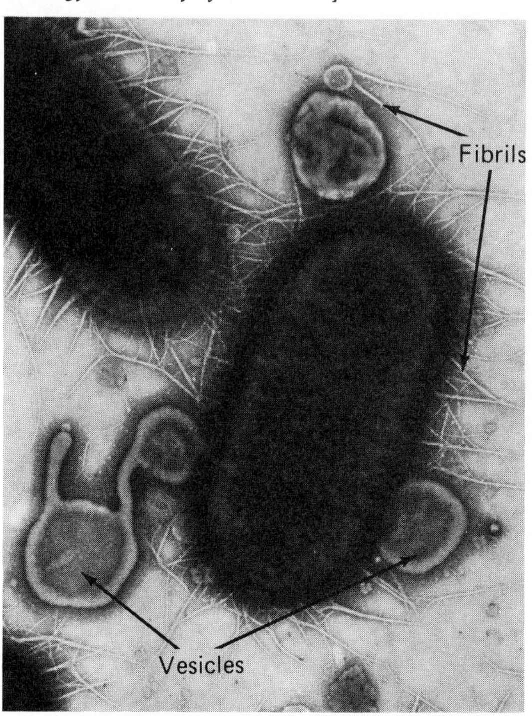

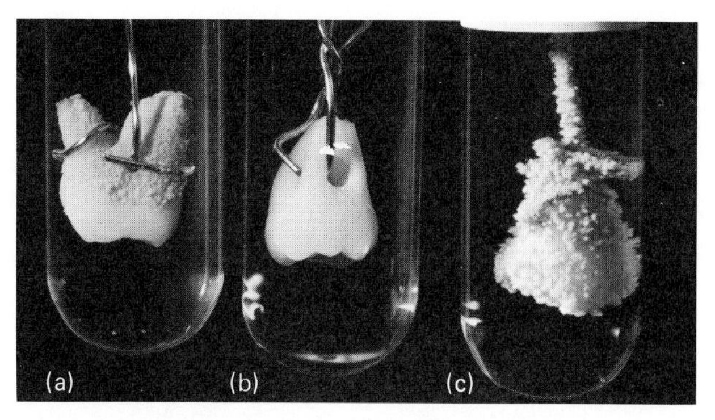

Figure 24–11 An *in vitro* plaque assay. Healthy extracted and cleaned teeth are used to determine the pattern of attachment and colonization of two plaque-associated oral bacteria. (a) A tooth showing the preference of *Cytophaga* (filamentous gram-negative bacteria) for the root surface. (b) Control tooth. (c) A tooth showing *Streptococcus mutans* adhering to the surfaces of the root and crown. *[R. Celesk, R. M. McCabe, and J. London,* Infect. Immun. ***26***:*15–18, 1979.]*

Eikenella corrodeus (ike-en-EL-lah korr-OH-dee-us), *Streptococcus mutans,* (S. MEW-tans), and *S. sanguis* (S. SAN-gwiss). *Cytophaga* exhibit a preference for colonizing the root surface of intact teeth (Figure 24–11). While these bacteria do not adhere well to smooth enamel surfaces, they can attach to the cementum surface of the root and produce acid from a limited number of sugars.

The contribution of *Cytophaga* to the progression of periodontal disease is still under study. However, experimental findings suggest that by binding to the root surface, these gram-negative microorganisms may provide a localized concentration of lipopolysaccharide on or near the alveolar bone-root juncture that could cause bone resorption by stimulating a host-regulated immune response. Dental plaque also contains immunologic agents that have the capacity to enhance or suppress immune responses. Among these immunoregulatory agents are lipopolysaccharides from gram-negative bacteria and carbohydrates such as dextrans and levans, which can control immune responses to antigens. Calculus is an almost constant companion of periodontal disease and is considered by many to be the major causative factor in its development (Figure 24–12).

Calculus is a hard, calcified substance that sticks firmly to the teeth or to bridges and other appliances worn in the mouth. The external environment, together with systemic factors of the host, determines whether the plaque will calcify or become associated with dental decay. Calcification of plaque begins in small areas that enlarge and combine to form large areas of calculus. The rough ragged calculus further increases the irritation to the supporting tissue surrounding the teeth (periodontium).

Immunologic mechanisms are also considered to be major factors in periodontal disease. Host immune responses may be involved in its initiation and development. These responses may be caused

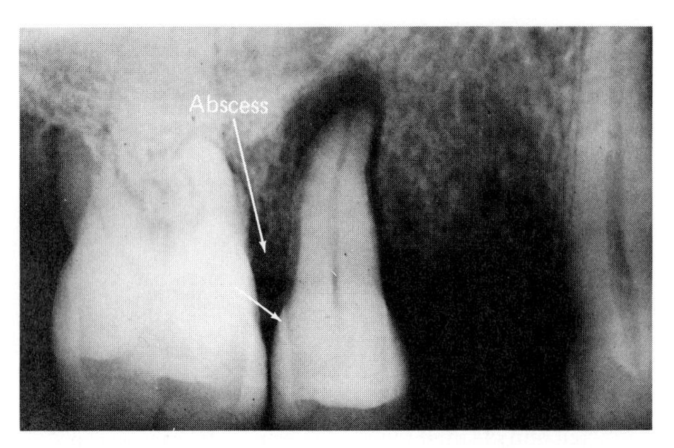

Figure 24–12 The very dark area at the apex and along the root surface of the tooth represents total destruction of the normal tissue. The area is an abscess that drains to the surface along the side of the tooth (arrows). Calculus is evident between the bicuspid and the molar. Note that the teeth are free from evidence of decay. *[Courtesy of Dr. N. H. Rickles, Pathology Department, University of Oregon Dental School.]*

nonspecifically by bacterial plaque, by a specific microorganism, or by a combination of microbial antigens. Moreover, it is becoming clear that host immune responses to bacterial plaque are extremely complex and may involve any one or a combination of type I, II, III, and IV immunological reactions. The tissue-destructive immunologic mechanisms thought to be involved in periodontal disease include complement-controlled damage initiated by both antibody and the alternate complement pathway, polymorphonuclear leukocyte-induced damage, and cell-mediated damage. **[See type I, II, III and complement reactions in Chapter 20.]**

GINGIVITIS

Gingivitis is an inflammation of the gingiva (that portion of the oral mucous membrane that surrounds a tooth). The inflammation is probably the result of an exaggerated response to large amounts of dental plaque. This condition is the most common disease affecting the soft tissues of the mouth (Color Photograph 102). In mild cases of gingivitis, an increase in the number of polymorphonuclear leukocytes and T lymphocytes occurs. With prolonged, severe forms of gingivitis, as well as with severe periodontal disease, significant numbers of lymphocytes and immunoglobulin G (IgG)-producing plasma cells infiltrate into the inflamed tissue. In gingivitis a serum discharge known as *crevicular fluid* is produced. This exudate, which flows around the tooth and is in contact with dental plaque, contains functional complement components and low levels of specific immunoglobulins to various microbial plaque antigens. The flow of the crevicular fluid is directly proportional to the severity of the disease. **[See IgG production in Chapter 17.]**

NECROTIZING ULCERATIVE GINGIVITIS

Necrotizing ulcerative gingivitis (NUG) is also called *Vincent's infection* or *trench mouth,* a name earned during World War I when it was common among soldiers (Figure 24–13). The outbreaks of NUG among soldiers and in other crowded groups under emotional or physical stress convinced many clinicians that it was an infectious and communicable disease. However, recent studies have shown that NUG is not communicable, and physicians and dentists are no longer required to report cases of the disease.

The infection is found in adolescents and young adults. Fatigue and anxiety evidently play a most important role in predisposing the mouth to this tissue-destroying (necrotizing) condition.

The infection may involve the gums as Vincent's infection, the oral mucosa as Vincent's stomatitis, or extend to the throat as Vincent's angina. Fusobacteria and *Bacteriodes* species are associated with NUG.

Figure 24–13 Trench mouth, or NUG. (a) The appearance of Vincent's infection with "punched out" necrotic projections, accompanied by swollen and inflamed gingiva. *[Courtesy of Dr. Francis Howell, Pathology Department, University of Oregon Dental School.]* (b) This smear was taken from a patient suffering from Vincent's infection. Note the large numbers of *Borrelia* and a variety of other cells. The large dark cell is a phagocytic inflammatory cell. *[Courtesy of Dr. N. H. Rickles, Pathology Department, University of Oregon Dental School.]*

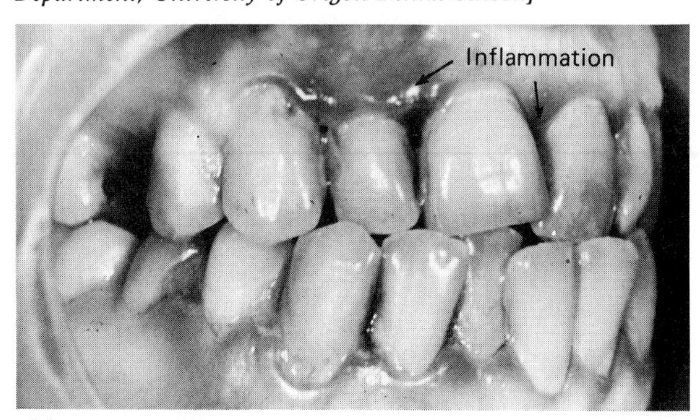

(a)

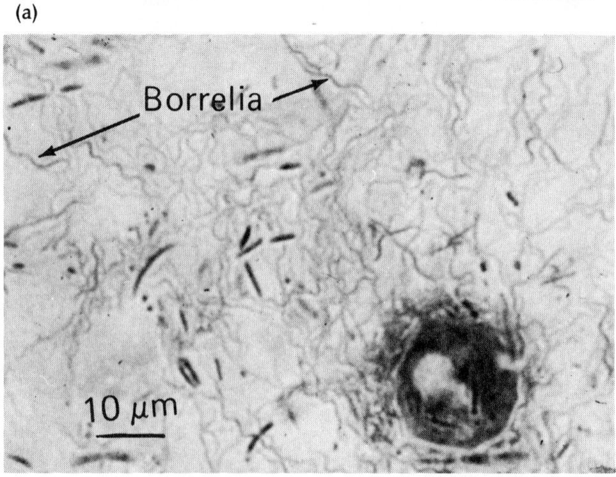

(b)

PERIODONTITIS

Untreated gingivitis leads to periodontitis, inflammation of the periodontium, and eventual periodontal bone loss. Another name for this condition is *pyorrhea.* In a small percentage of the population, bone loss occurs very rapidly, sometimes within two to five years. A characteristic gram-negative anaerobic bacterial flora is associated with this juvenile periodontitis. The flora is different from that found in the more slowly progressive form of the disease.

DENTAL CARIES

Tooth decay is a worldwide problem, although not all areas and peoples are affected equally. Dental caries is a disease of the calcified tissues of the teeth characterized by a decalcification, or loss of calcium salts, of the inorganic substance. This condition is either followed by or accompanied by disintegration of the organic portion of the tooth. The cause is complex, and considerable controversy exists over the exact mechanism of caries development. Dental decay in teeth is the result of an interaction between the host tissues and extremely specific *cariogenic* (caries-producing) microorganisms that utilize nutrients provided by the host's diet.

All bacteria known to cause enamel caries or decay on the smooth surfaces of the teeth are plaque formers, secreting the complex polysaccharides of plaque, which are derived chiefly from sucrose (table sugar). Three polysaccharide-producing streptococci are found in large numbers of humans: *Streptococcus mutans, S. sanguis,* and *S. salivarius.* The extracellular polysaccharides that these organisms produce from sucrose enable them to adhere to one another and thus form colonies on the tooth's surface. Although the streptococci are not the only bacteria known to synthesize polysaccharides, they are the major group of organisms that initiate plaque formation. Various experiments have shown that the principal causative microorganism of enamel caries is *S. mutans.* Although other organisms may play a role, *S. mutans* produces acids that are capable of dissolving calcium from the enamel surfaces. The process is shown in Figure 24–14.

Dental decay (Figure 24–15) and the defects produced by caries can be treated and the tooth restored to proper form and function in most cases. In recent years, greater effort has been expanded on measures to prevent this costly and often painful disease of humans. At present, plaque is most effectively eliminated by mechanical means such as brushing and flossing. An antiplaque chemotherapeutic agent contained in an easy-to-use, palatable vehicle would greatly simplify effective oral hygiene. Such a drug could drastically reduce periodontal disease while also reducing dental caries.

At this time, however, there is no chemotherapeutic agent that can be safely administered for long-term control of bacterial plaque. Antibiotics can be considered only for short-term plaque control because of their possible systemic side effects and limited spectrum of activity. All other compounds are limited in their usefulness by unpleasant or toxic side effects, by limited lack of effectiveness, or by inadequate experimental evidence.

Many substances have been tested in regard to their ability to eliminate or reduce the number of bacteria, change the flora to harmless non-acid formers, inhibit enzyme activity, and alter the tooth surface. One of the most popular is sodium fluoride. When used in proper amounts, this material has been shown to have significant effects on caries incidence without causing ill effects in the user.

Another preventative approach under consideration is the development of a caries vaccine. A basis for current studies is the observation that salivary immunoglobulin A (IgA) antibodies can inhibit the adherence of streptococci to epithelial cells in the mouth. The experimental use of streptococcal vaccine has been shown to influence the implantation of *S. mutans* in humans and lower animals. **[See IgA production in Chapter 17.]**

Diseases Related to Oral Foci of Infection

LUDWIG'S ANGINA

Certain serious complications can develop from infected teeth. One of these states, Ludwig's angina, involves beta-hemolytic streptococci (Figure 24–16). Microorganisms from such sites of infection pour into the surrounding tissues of the oral cavity. Ludwig's angina is a particularly dangerous infection of the throat and neck, since the swelling that develops may eventually block the air passages.

Bacterial Infections

Various bacterial pathogens involve the oral cavity at some time during their developmental cycles. Examples of such pathogens include *Actinomyces israelii, A. bovis, Mycobacterium leprae* (my-koh-back-TI-ree-um LEP-ray), *M. tuberculosis* (M. too-ber-koo-LOW-sis), *Treponema pallidum* (tre-poh-NEE-mah PAL-li-dum), and *T. pertenue* (T. per-TEN-you-ee). The properties of these microorganisms are discussed in other chapters of Part VIII.

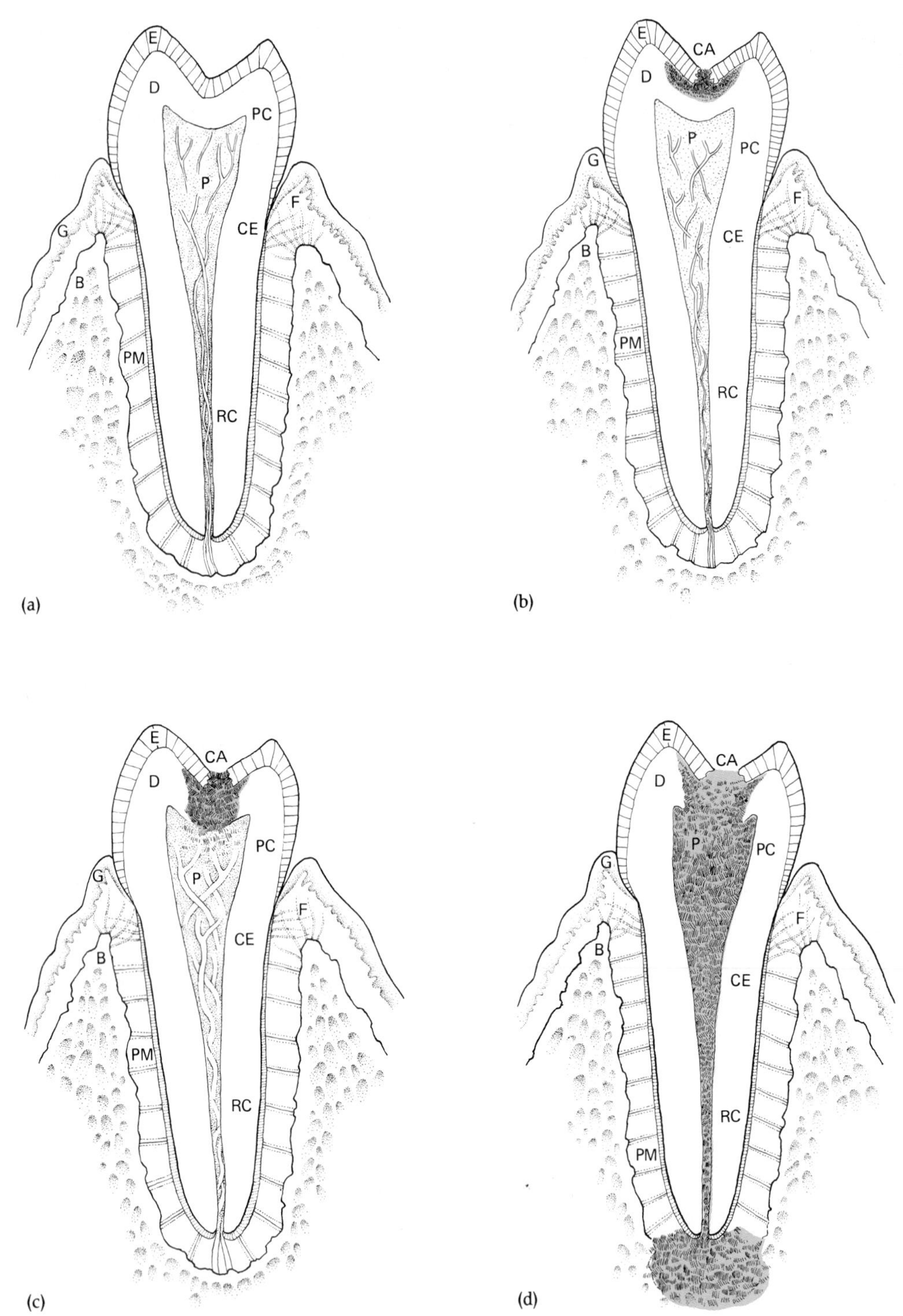

Figure 24–14 Diagrammatic representation of a section through a premolar tooth, showing the progress of decay on the chewing surface. (a) A normal tooth and its surrounding tissues. (b) A carious lesion. (c) The decay has passed through the dentine into the pulp, which has become inflamed, with many engorged vessels. (d) The decay has spread to involve the tissues around the end of the root. At this stage, there is some destruction of bone. Symbols: enamel barrier (E), dentine (D), gingiva (G), cementum (C), pulp tissue (P), pulp chamber and canal (PC), root canal (RC), alveolar bone (B), periodontal membrane (PM), fibers (F), cemento-enamel junction (CE), carious lesion (CA). *[Courtesy of Dr. William B. Wescott, Pathology Department, University of Oregon Dental School.]*

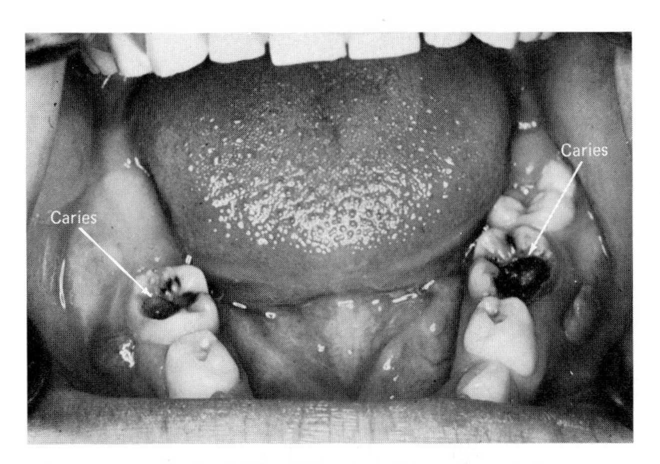

Figure 24–15 A child with extensive caries of the primary second molars. The dark areas of the teeth represent advanced caries, which have infected the pulp. Note the pus-containing swelling on the gingiva of these teeth. *[Courtesy of Dr. N. H. Rickles, Pathology Department, University of Oregon Dental School.]*

GONOCOCCAL STOMATITIS (ORAL GONORRHEA)

The urogenital system has traditionally been considered the major location for gonorrhea. However, it has become evident that the oral cavity and the throat may also be potential sites of *Neisseria gonorrhoeae* (nye-SE-ree-ah gon-or-REE-ah) infection. Individuals with oral infections associated with inflammation of the mouth, or stomatitis, present a fire-red, worn mucous lining of the mouth sometimes covered by a yellow or white film or patches. Severe sores measuring 2 to 3 cm in size may also be present. These lesions may be found on the tongue, gums, tonsils, and floor of the mouth. Oral gonorrheal infection may also cause destruction of the small nipplelike structures on a tongue's surface, producing a condition known as *central papillary atrophy* (Figure 24–17).

About 20% to 30% of those affected with gonococcal stomatitis have signs or symptoms of the disease. Diagnosis is generally based on positive culture results. Penicillin is usually the drug of choice for treatment.

Mycotic Infections

Fungus pathogens, as well as bacterial and viral agents of disease, involve the oral cavity either in a superficial manner or as a consequence of systemic disease. Representatives of such disease agents include *Candida albicans*, *Cryptococcus neoformans* (kryp-toe-KOK-kus nee-oh-FOR-mans), and *Geotrichum* (gee-oh-TRICK-um) species. One common mycotic

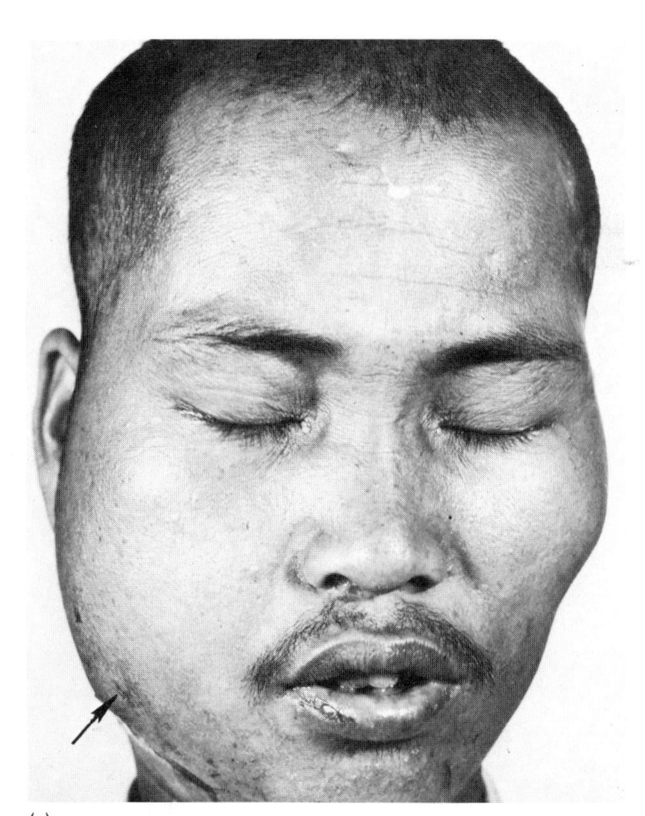

(a)

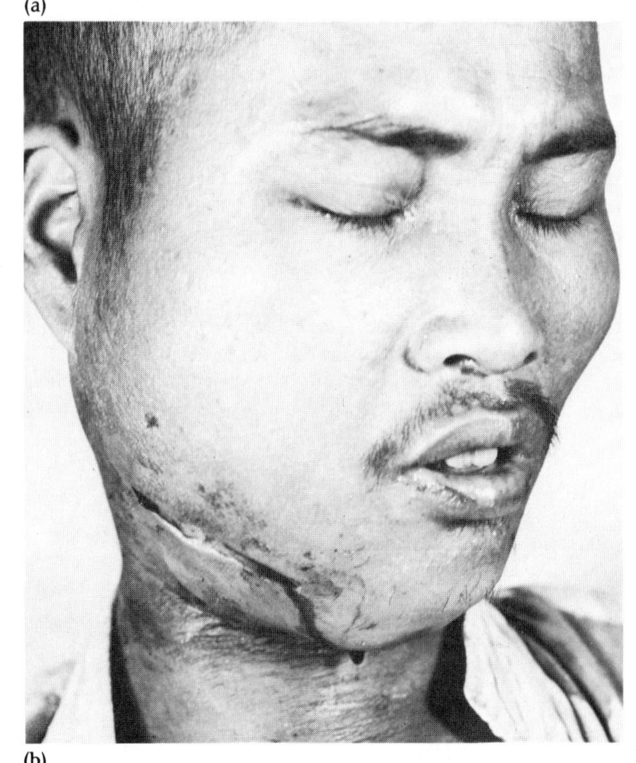

(b)

Figure 24–16 Ludwig's angina. (a) This photograph shows the typical swelling that can occur. The infection probably developed from an infected tooth on the side (arrow) of the individual's face. (b) The result of treatment, which included an incision to produce drainage. *[Courtesy of the Armed Forces Institute of Pathology, Washington, D.C., Neg. Nos. AFIP 44713-1 and 44713-2.]*

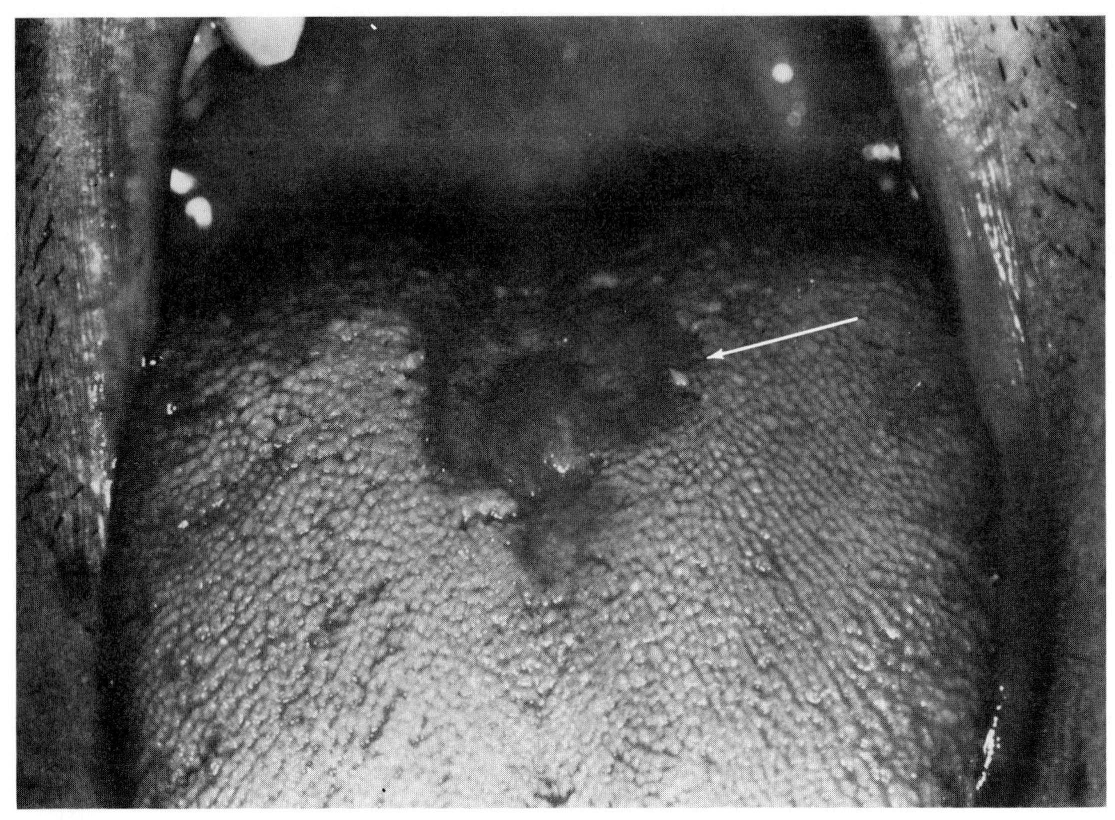

Figure 24–17 The tongue lesion (arrow) found in oral gonococcal infections. *[From V. Escobar,* et al., Int. J. Oral Surgery *13:549–554, 1984.]*

infection, caused by *C. albicans,* is discussed to illustrate fungal involvement in oral disease. Most of the other pathogens mentioned are described in the chapters dealing with diseases of the skin and the respiratory tract.

CANDIDIASIS (ORAL THRUSH)

In 1965, Cohen reported that vitamin deficiencies, iron deficiency anemia, pregnancy, and diabetes may predispose an individual to the development of monilial stomatitis, candidiasis (Color Photograph 103). Woods suggested in 1951 that the growth of *Candida* spp. appears to be restricted by the coexisting bacterial flora, but when this flora is suppressed by antibiotics, the *Candida* grow without restraint. As a result, patients may develop monilial infections of the oral cavity (Figure 24–18a) or a more severe, generalized *Candida* infection of the blood involving

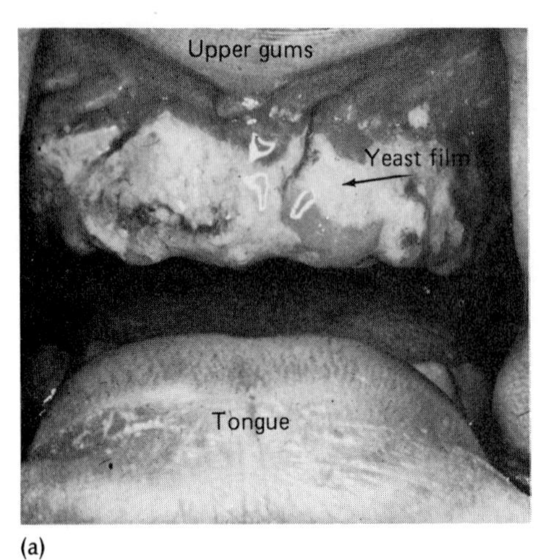

(a)

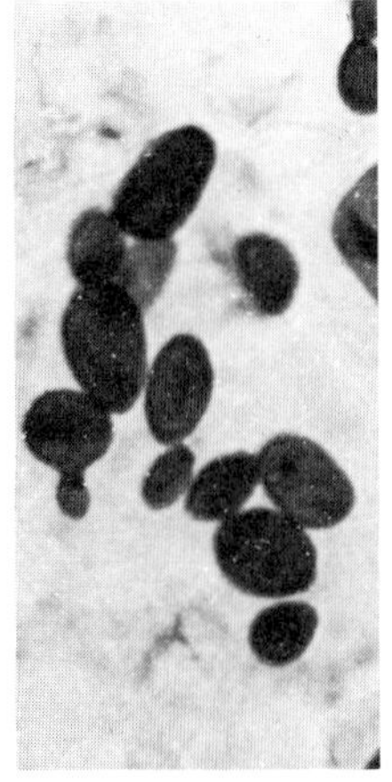

(b)

Figure 24–18 Candidiasis. (a) Forward view of the gums (gingiva) showing the extent of the grayish-white membranous growth. *[Courtesy of Dr. N. H. Rickles, Pathology Department, University of Oregon Dental School.]* (b) *Candida albicans* in a direct smear from sputum. *[P. J. Kozinn, and C. L. Taschdjian,* JAMA, ***198/2:****170–172, 1966.]*

the heart valves, respiratory system, gut, and brain. The oral lesions involve the tongue, palate, cheeks, and lips, and may further extend to the tonsils, pharynx, and larynx. *Candida* spp. may enter tissues at the time of tooth extractions.

Oral candidiasis or thrush also is a fairly common disease of newborns, who acquire *Candida* from passing through an infected birth canal. Older children may develop thrush because of hormonal disturbances or immune defects. *Candida* infection has gained particular attention as one of the signs of AIDS. **[See Chapter 27 for features of AIDS.]**

The diagnosis of candidiasis may include a direct examination of freshly obtained clinical specimens (Figure 24–18b) and specimen cultivation. *Candida* species are strict aerobes. While there is no specific treatment for oral candidiasis, topical application of sodium caprylate, Trichomycin, or amphotericin B are effective.

Viral Infections

A representative group of viral infections of the oral cavity are listed in Table 24–3 and discussed here. Other viruses involve the mouth either in a superficial way or as one phase or consequence of systemic disease. Diseases in this category include chickenpox (varicella), measles (rubeola), molluscum contagiosum, and shingles (herpes zoster). Most of these diseases and the viruses that cause them are described in Chapter 23.

Hairy Leukoplakia

"Leukoplakia" is currently a clinical descriptive term used for a wide range of nonspecific white patches

TABLE 24–3 Viral Infections Involving the Oral Cavity

Disease Entity	*Causative Agent*	*Symptoms and Clinical Appearance*	*Treatment*
Hairy leukoplakia	Human immunodeficiency virus (HIV)	Slightly raised, usually white areas on the surfaces of the tongue; some redness and swelling may occur; can be found with some AIDS patients	Symptomatic; and measures to prevent contamination
Hand, foot, and mouth disease	Coxsackie virus group A, types 16, 6, and 10	Vesicles and ulcerations involving the buccal mucosa, tongue, gingiva, and lips; ulcers are painful and interfere with eating; lesions also present on the hands and feet	Symptomatic
Herpangina (aphthous pharyngitis)	Coxsackie virus group A, types 2, 4, 5, 6, 8, and 10	Sudden onset of high fever, headache, and sore throat, accompanied by papules, vesicles, and later ulcers on inner throat surface, the uvula, and the soft palate	Symptomatic
Herpes simplex (fever blister)	Herpes simplex virus type 1	Primary lesions usually in the oropharyngeal mucosa as multiple, very small vesicles, which rupture and ulcerate; a bright red zone is present around the periphery; fever, malaise, loss of appetite, and swollen lymph glands are present; recurrent lesions of the mucosa or lips are common	Symptomatic; *Lactobacillus acidophilus* preparations; certain drugs that inhibit DNA synthesis (some are currently under study)
Hoof and mouth disease	Foot and mouth disease virus	Vesicles and ulcerations involving the lips, tongue, palate, and mucosa; heal within 2 weeks	Symptomatic
Mumps	Mumps virus	Painful, swollen salivary glands, usually the parotid gland near the ear	Symptomatic; convalescent serum; vitamin C supplements

appearing on the internal surfaces of the mouth. The condition known as hairy leukoplakia has been reported to occur on the tongue surfaces of some AIDS patients. Usually white, irregular, enlarged surface projections occur on the sides of the tongue (Color Photograph 104). Since viral particles can be found within such lesions, precautions such as the use of gloves and appropriate disinfection and sterilization of equipment should be observed by individuals treating or examining the mouths of infected persons.

Herpangina (Aphthous Pharyngitis)

Herpangina is specific, painful, and highly contagious disease transmitted by direct contact. Infected individuals experience high fever and a series of painful open sores on the internal surfaces of their throats. Sporadic outbreaks of the infection have been reported in many parts of the United States, usually during the summer months. The outbreaks begin during the warm weather and disappear with the first plant-killing frost. Children up to 15 years of age are commonly affected.

HERPES SIMPLEX VIRUS 1 INFECTION

The herpes simplex lesion has been called by many names, including *canker sore, fever blister, cold sore, aphthous ulcer,* and *herpes labialis.* This common viral disease affects the oral tissues (Figure 24–19), often remains localized, and produces considerable pain and discomfort for the patient suffering from an outbreak of lesions.

Many factors may stimulate the virus (herpes simplex virus 1) to produce clinical lesions. Included among these are excessive exposure to sun, fever, allergy, mechanical trauma, gastrointestinal upsets, immunological defects, and certain psychological influences.

Disease Challenge

The situation described has been taken from an actual case history. A review of treatment and epidemiological aspects is provided at the end of the presentation. Answers to questions, laboratory findings, and interpretations are given immediately following a specific question. Test your skill and take the Disease Challenge.

CASE

A 32-year-old male was examined in a dental clinic. He was complaining of sore, bleeding gums. In addition, the patient had been unable to sleep well for several weeks and indicated he had had a low but consistent fever for about three to four weeks.

Upon examination, the patient was found to have poorly fitting dentures and white patches covering portions of his tongue. Removal of some patches revealed raw and bleeding undersurfaces.

At This Point, What Disease(s), If Any, Would You Suspect?
Oral candidiasis (thrush) and oral gonorrhea appear to be the most likely diseases.

What Type(s) of Laboratory Specimens Should Be Taken?
Scrapings from the tongue and saliva specimens would be appropriate for microscopic examination, as well as yeast and bacterial cultures.

Laboratory Results:
Only the microscopic examination and cultures for yeast were found to be of diagnostic value. Creamy white colonies were found on culture and large numbers of yeasts were observed in the tongue scrapings and from cultures.

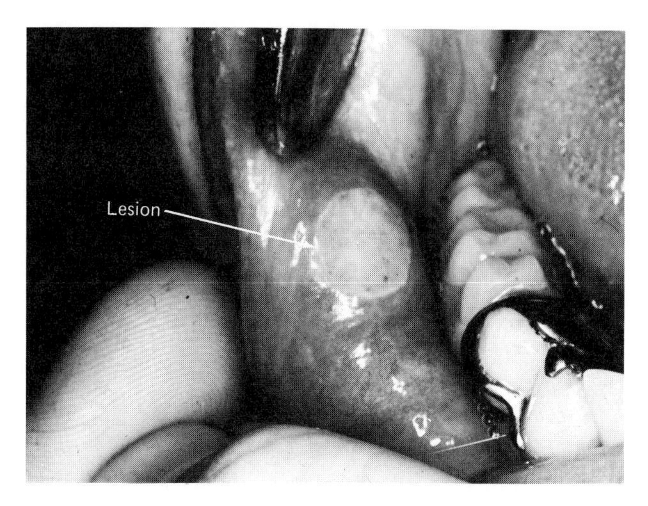

Figure 24–19 Herpes simplex virus type 1 involving the inner surface of the lip. This is an older lesion with secondary infection and a thin layer of necrotic tissue over the surface. *[Courtesy of Dr. N. H. Rickles, Pathology Department, University of Oregon Dental School.]*

MUMPS (EPIDEMIC PAROTITIS)

Mumps, a worldwide acute, communicable disease, is caused by a paramyxovirus. Humans are the only natural hosts for the causative agent. The virus apparently is spread as a droplet infection or by direct contact with saliva and respiratory secretions.

Mumps most frequently affect children between 8 and 15 years of age, with an overall incidence of at least 60%. Adults who develop mumps often suffer complications, including infections of the testes (*orchitis*) in young men and of the brain and its coverings (*meningoencephalitis*). The oral tissues involved are the salivary glands, most commonly the parotid glands. These are the largest of the salivary glands, found slightly below the ear. The immunity resulting from the disease is of long duration and may develop following subclinical infections. Subsequent attacks have rarely been reported. Swelling of a single salivary gland produces the same immunity as multiple gland involvement. **[See Chapter 18 for states of immunity.]**

What Is the Probable Diagnosis?
Oral candidiasis (also known as mucocutaneous candidiasis).

With What Other Disease(s) Could This Condition Be Associated?
Acquired immune deficiency syndrome (AIDS) and other immunological defects.

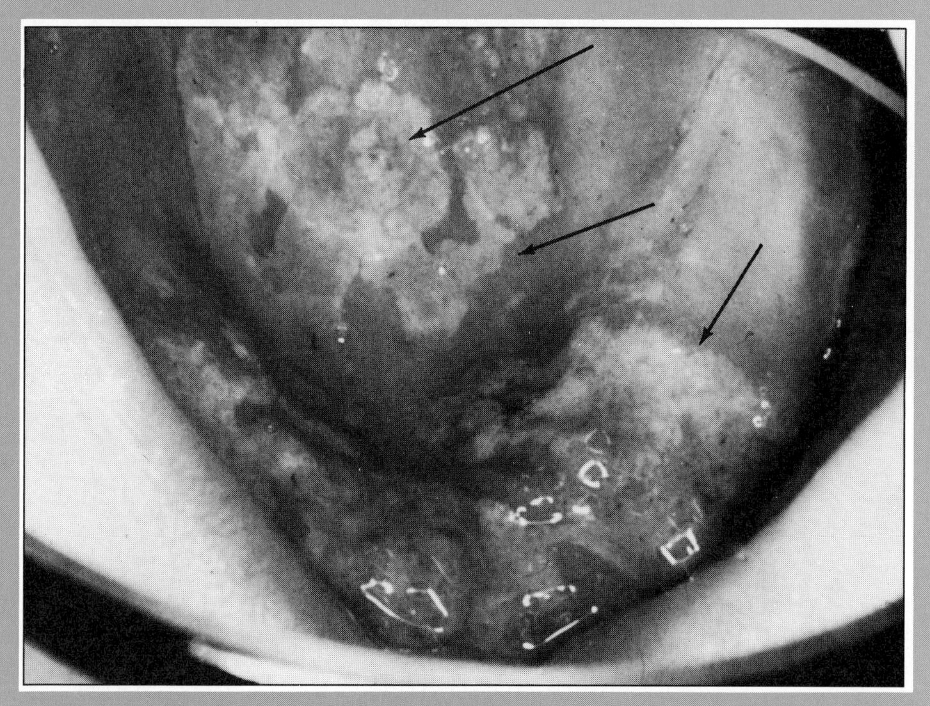

The appearance of the patient's tongue.

Temporary passive immunity to the virus may be obtained by giving injections of gamma globulin from serum known to contain mumps antibody in high concentrations. Active immunization by vaccination is also possible. High doses of vitamin C, based on body weight, is reported to be of value in the control of mumps infections.

Summary

Oral infections result from disturbances in the relationship of the individual's resident flora to the tissues of the mouth.

Structure of the Mouth

1. The human mouth has four groups of teeth: cutting *incisors,* tearing *canines,* and grinding *bicuspids* and *molars.*
2. Each tooth has three parts: the *crown,* the major exposed portion; the *root,* the portion below the gum and embedded in the jaw; and the *neck,* the constricted region between the crown and root.
3. A tooth's crown is coated with enamel; the rest is covered by a modified bone layer called the *cementum.* Dentine underlies the enamel.
4. The gummy accumulation of salivary mucin and various types of bacteria is called *dental plaque.* Microscopically, plaques are seen as units called "corncobs" because of their resemblance to an ear of corn.
5. In addition to containing microorganisms that can cause periodontal disease, dental plaque has immunologic agents than can enhance or suppress the immune responses. These responses may play a significant role in various diseases of the mouth.

Oral Flora

1. The microorganisms normally found in the human mouth can be divided into three groups on the basis of their oxygen requirements: strict anaerobes, strict aerobes, and facultative organisms.
2. The oral flora is relatively stable in terms of the types of organisms present. In addition to bacteria it includes several species of fungi, viruses, and protozoa.

Nonspecific Infections of the Oral Region

Infections of the face, oral cavity, and neck may be extremely serious, depending upon their location and the microorganisms involved.

FOCAL INFECTIONS

1. A localized area of infection anywhere on the body is called a *focus of infection.*
2. Oral foci of infection include those conditions that exist in the supporting structures of the teeth and the changes resulting from infected teeth. Examples include tooth decay (caries), infected gums (gingivitis), inflammation and destruction of the bone (osteomyelitis), and inflammation of the tissue directly supporting the teeth (periodontitis, or pyorrhea).
3. It is becoming clear that host immune responses to bacterial plaque and resulting disease states are extremely complex. They may involve any one or a combination of types I, II, III, and IV hypersensitivity immunological reactions.

Diseases Related to Oral Foci of Infection

Several serious complications can develop as a consequence of infected teeth or other portions of the mouth and various dental procedures. One such complication is Ludwig's angina.

Bacterial Infections

Various bacterial pathogens involve the mouth and related structures at some time during their developmental cycles. Examples of disease states of this kind include anthrax, syphilis, tuberculosis, and oral gonorrhea.

Mycotic Infections

Pathogenic fungi can also involve the mouth. Examples of associated diseases include the yeast infection candidiasis. This infection has particular significance in cases of immune defects such as AIDS.

Viral Infections

Viral infections of the mouth include chickenpox, fever blister (herpes simplex), hairy leukoplakia, herpangina, and mumps.

Questions for Review

1. List at least three structures of the oral cavity and a particular infection or disease state associated with each of them.
2. What is NUG?
3. What types of bacteria make up the largest microbial group of the oral cavity? Which one can cause diseases?
4. Discuss the following clinical states in terms of general features, causes, and treatment, if any:
   a. dry socket
   b. Vincent's infection
   c. osteomyelitis
   d. periodontitis
   e. pericoronitis
   f. dental caries
   g. periodontal disease
   h. subacute bacterial endocarditis
   i. oral thrush
   j. oral gonorrhea

5. What is the role of calculus in periodontal disease?
6. What mechanisms have been offered to explain how periodontal disease develops?
7. What is pyorrhea?
8. What is a focal infection? How important is it to the general health of the body?

Suggested Readings

DRIESSENS, F. C. M., and J. H. M. WOLTGENS (eds.), *Tooth Development and Caries.* Vol. II. Boca Raton, Florida: CRC Press, Inc., 1986. *A detailed treatment of tooth development together with the mechanisms associated with tooth decay caused by microorganisms.*

DUNLAP, C., and B. BARKER, *Oral Lesions.* Needham, Mass.: Hoyt Laboratories. *A highly illustrated reference to the diagnosis and treatment of a wide variety of oral lesions.*

GENCO, R. J., and S. E. MERGENHAGEN (eds.), *Host-Parasite Interactions in Periodontal Diseases.* Washington, D.C.: American Society for Microbiology, 1982. *A clearly written presentation of recent advances and concepts related to periodontal disease and other chronic oral infections.*

HOLT, S. C., and E. R. LEADBETTER, "Comparative Ultrastructure of Selected Oral Streptococci: Thin Sectioning and Freeze-etching Studies," *Canadian Journal of Microbiology.* **22**:475–485, 1976. *An electron microscopic study of the internal organization of streptococci associated with cariogenic activities.*

LONG, S. S., and R. M. SWENSON, "Determinants of the Developing Oral Flora in Normal Newborns," *Applied and Environmental Microbiology.* **32**:494–497, 1976. *A study of the factors that are important determinants in establishing the ecological place of bacteria in the oral microflora.*

25

Microbial Infections of the Respiratory Tract

It was the fashion to suffer from the lungs, everybody was consumptive [with tuberculosis], poets especially; it was good form to spit blood after each emotion that was at all sensational, and to die before reaching the age of thirty. — *Alexandre Dumas*

After reading this chapter, you should be able to:

1. Describe the areas of the human respiratory tract that are susceptible to attack by pathogenic microorganisms.
2. Discuss measures used to control and treat respiratory infections.
3. Discuss at least three bacterial, three fungal, and three viral diseases of the respiratory system in terms of the causative agent, means of transmission, and general symptoms.
4. List general approaches to the diagnosis of bacterial, fungal, and viral infections of the respiratory system.
5. Discuss the acute respiratory distress (ARD) syndrome.
6. Distinguish among the various diseases caused by species of *Legionella*.
7. Describe the effects of the protozoan disease agent *Pneumocystis carinii* and its relationship to AIDS.
8. Describe the types of complications associated with microbial diseases of the respiratory tract.
9. List and explain at least two approaches to the prevention of viral infections of the respiratory system.
10. Discuss the value of antibiotics in the treatment of virus-caused respiratory infections.
11. List factors that can predispose the individual to respiratory infections.

What are the common respiratory tract infections? How can they be prevented or controlled? Chapter 25 describes several major respiratory microbial diseases, including diphtheria, influenza, Legionnaires' disease, tuberculosis, whooping cough, the common cold, *Pneumocystis* infection, and the pneumonias.

The human respiratory tract starts at the nose, passes through the various parts of the respiratory tree, and ends in the air sacs, or alveoli. This entire system is adapted to making air containing oxygen available to the circulatory system and, by way of it, to the whole body. Unfortunately, the respiratory tract is a frequent portal of entry for various microorganisms. Different organisms gain access to different levels in the system, thus accounting in part for the differences in the types of infections occurring in the upper and lower portions of the respiratory tract (Figures 25–1 and 25–2). The system also provides a means of transmitting microbes to other individuals during speaking, coughing, and sneezing, when droplets of microbe-containing secretions are released. **[See disease transmission in Chapter 21.]**

The human respiratory tract provides an extensive area, approximately 60 square meters, of contact between the environment and the body. It includes the nose, pharynx (throat), larynx (voice box), trachea (windpipe), bronchi, and lungs (Figures 25–1 and 25–2). In addition to protecting against foreign particles and microorganisms, the respiratory system is designed for the efficient exchange of gases between the blood and the atmosphere.

Normal Flora of the Respiratory Tract

The normal flora of the human nose includes species of *Bacterioides* (back-te-ROY-dees), *Branhamella* (bran-ham-EL-la), *Corynebacterium* (ko-ri-nee-back-TI-ree-um), *Haemophilus* (he-MAH-fi-lus), *Micrococcus* (mi-krow-KOK-kuss), *Staphylococcus*, (staff-il-oh-KOK-kuss), and *Streptococcus* (strep-toe-KOK-kuss). Some of these microorganisms can pose serious problems and cause disease. At birth the throat, windpipe,

Figure 25–1 The human respiratory system, showing sites with which microbial infections are associated. The upper respiratory system and the passage of air (color arrows) are shown here.

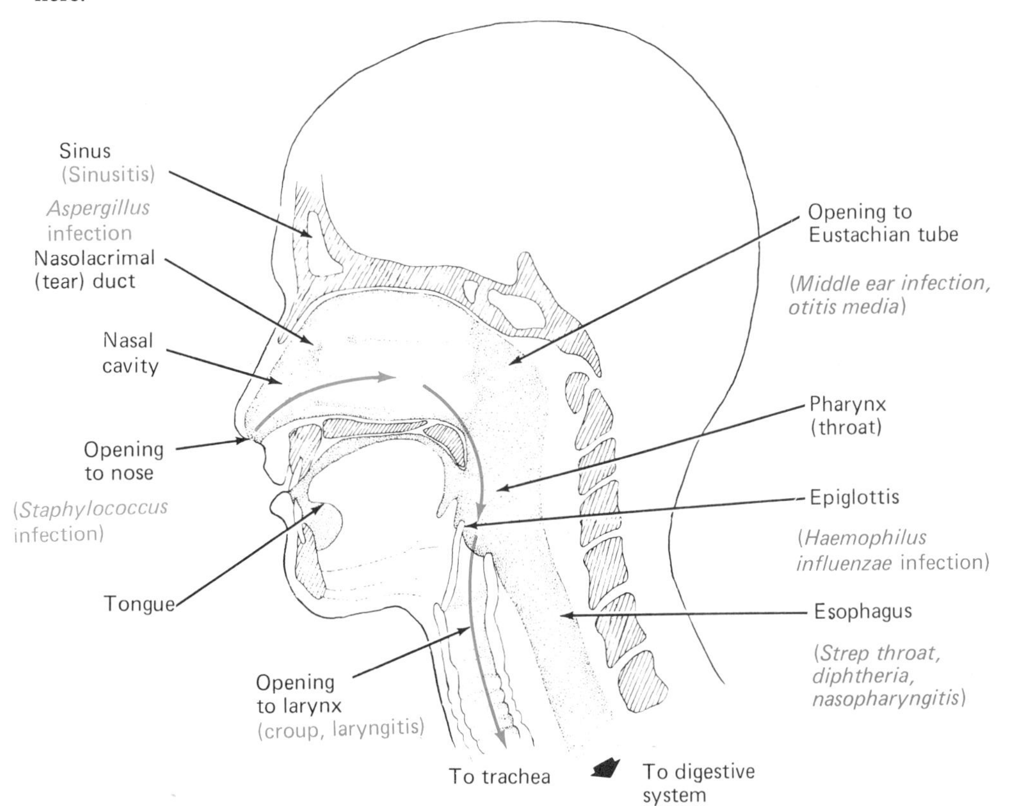

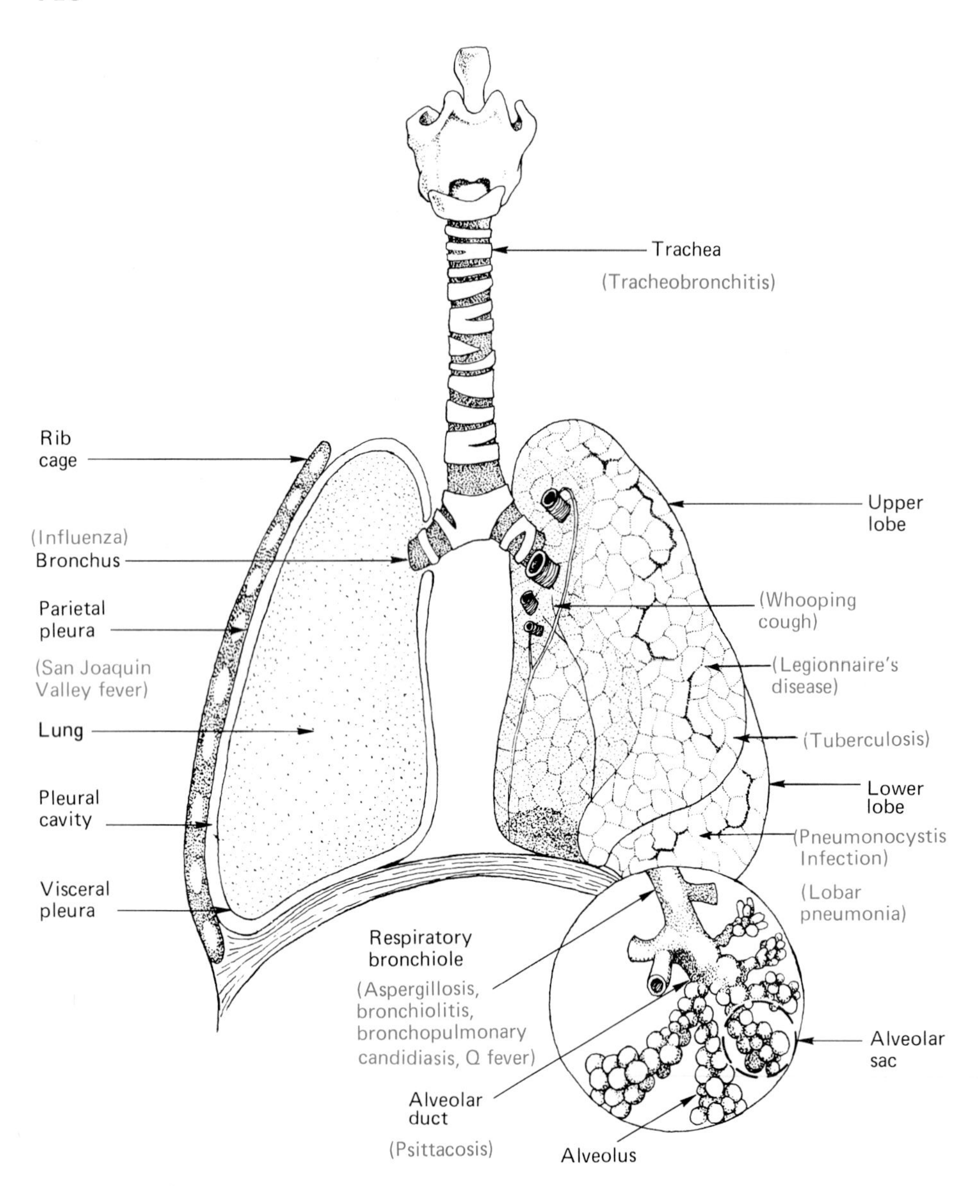

Figure 25–2 Portions of the respiratory tree. The relationship of the pleura, the membranes covering the lungs, is also shown. (a) The lower portion of the system. The alveoli (air sacs), which participate in gaseous exchanges, are emphasized. Specific sites associated with microbial infections are indicated. A depression on the middle surface of each lung is the region through which the primary bronchus, blood vessels, lymphatics, and nerves penetrate the structure. Compare this view with Figure 25–3.

and bronchi are sterile. However, within 24 hours after birth, these sites become colonized by streptococci and other bacteria. In the adult, the respiratory tract below the level of the epiglottis (Figure 25–1) is normally sterile. **[See acquiring microbial flora in Chapter 17.]**

Introduction to Microbial Infections of the Respiratory Tract

Several microorganisms can cause respiratory tract infections. Many of the resulting diseases are quite common and are among the most damaging of any that affect humans.

Various microbial pathogens—bacteria, fungi, and viruses—find suitable avenues for entry and sites for multiplication in the respiratory tract. Even though they use the respiratory tract as a portal of entry and at times a portal of exit, the resulting infectious process may easily extend to other regions of the body.

The various secretions associated with the respiratory system, such as sputum and droplets of mucus from sneezes and coughs, can be infectious. Relatively few microorganisms are introduced into the environment during normal breathing, even by an infected individual. However, an infected person or carrier who does not cover his or her nose or mouth while sneezing or coughing can easily contaminate the environment with disease agents. The communicability of infectious diseases is influenced by several different factors, including (1) the survival of respiratory pathogens on fomites or in the air, (2) the number of microorganisms inhaled, (3) the duration of contact, and (4) the anatomical site involved in the localization of the infectious agents. **[See reservoirs of infection in Chapter 21.]**

Control Measures

Control procedures are largely determined by the characteristics of the pathogen or the particular disease it causes. These factors may include (1) isolation of infected persons, (2) concurrent disinfection or sterilization of contaminated equipment—such as mouthpieces, thermometers, and rubber tubing—as well as any and all contaminated articles such as eating utensils and dishes; (3) the use and proper disposal of gowns following contact with infectious individuals; (4) the disinfection of rooms or associated facilities used by infected persons, and (5) the washing and disinfecting of hands before and after contact with patients or with any and all articles handled by patients, such as blankets, dishes, laundry, or pillows. **[See physical control in Chapter 11.]**

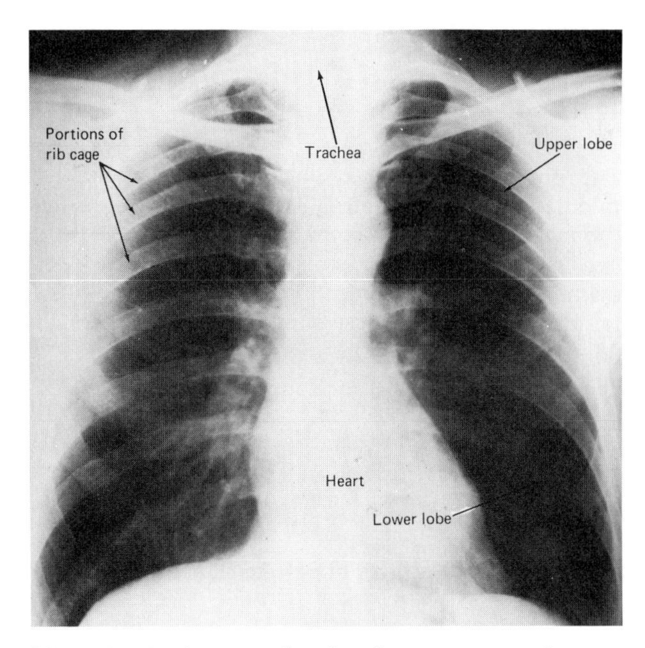

Figure 25–3 An x-ray showing the appearance of normal lungs. *[K. Kay,* Trans. N.Y. Acad. Sci. ***36****:511, 1974.]*

Figure 25–4 Lung abscess formation caused by *Staphylococcus aureus.* X-ray shows cloudy lung area (arrow) with abscess before treatment. Compare this view with Figure 25–3. *[Courtesy of Dr. Marcel Bilodeau, Hospital Laval, Quebec, Canada.]*

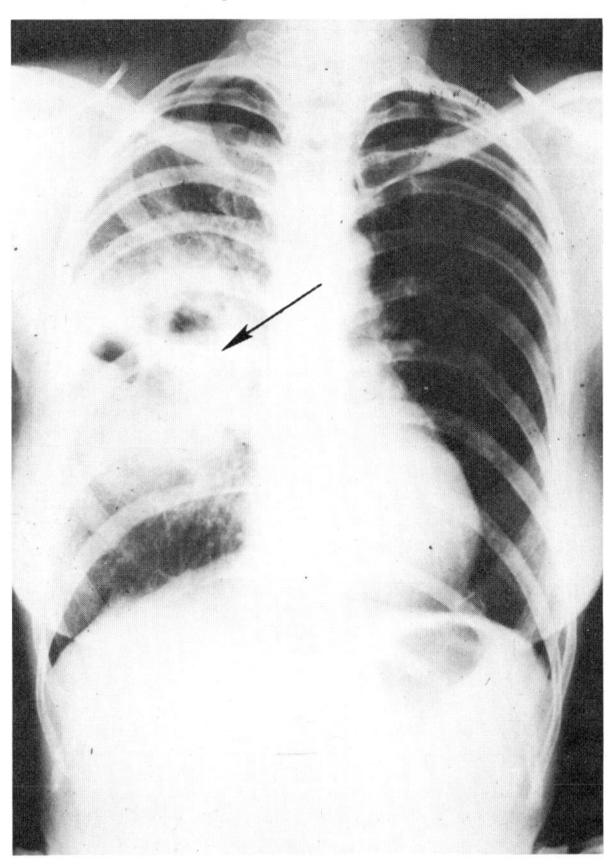

Representative Microbial Diseases

Several infections of the respiratory tract are of great concern because of the ease with which they are transmitted and contracted and the difficulty in eradicating their disease agents. A number of these microbial pathogens and the diseases they cause are described in Tables 25–1 through 25–9. Treatment, which includes the use of antibiotics in certain cases, is also indicated in several of these tables.

Diagnosis

Respiratory tract infections can often be diagnosed on the basis of x-ray films (Figures 25–3 and 25–4), the patient's history, and the examination. Confirmation is made by laboratory work, including microscopic examination of sputum, blood, and other specimens; isolation of pathogens from specimens; serological tests; and animal inoculations. Most of the sections dealing with specific pathogens in this chapter present the general diagnostic approaches used in respiratory tract infections as well as possible treatments.

Upper Respiratory Infections

The upper respiratory region—which includes the middle ear, the small air cells behind the ears called the *mastoids*, the sinuses, and the nasal corner of the eyes (Figure 25–1)—is exposed to a variety of pathogens when air is inhaled. These organisms can establish infections in these sites (Tables 25–1 through 25–3) and spread to other regions of the respiratory tract.

The majority of upper respiratory infections (URIs) result in relatively harmless and temporary illnesses. Many of them can be managed in the home with so-called household remedies and/or over-the-counter preparations.

Bacterial and Mycotic Diseases

DIPHTHERIA

Diphtheria, an acutely infectious communicable disease, has been known since ancient times, but its specific nature was not recognized until the nineteenth century. The bacterial cause of diphtheria was identified from stained smears by Edwin Klebs in 1883 and successfully cultivated and shown to

TABLE 25–1 Features of Selected Bacterial Upper Respiratory Tract Infections

Disease	*Causative Agent*	*Gram Reaction and Morphology*	*Incubation Period*	*General Features of the Disease*
Diphtheria	*Corynebacterium diptheriae* (Klebs-Löffler bacillus)	+, Rods appearing in aggregates resembling Xs, Ys, and Chinese characters	1–10 days	Symptoms appear suddenly and include fever, sore throat, general discomfort, formation of a diphtheritic pseudomembrane on tonsils, throat, or in the nasal cavity; complications such as heart and kidney failure can develop
Epiglottitis	*Haemophilus influenzae*, type b	−, Rod	Variable	Symptoms appear suddenly and include sore throat, difficulties in swallowing and breathing, hoarseness, and drooling
Middle ear infection (otitis media)	*Haemophilus influenzae*, type b *Staphylococcus aureus*, *Streptococcus pneumoniae*	−, Rod +, Coccus +, Coccus	1–2 days	Pain first limited to ear, then spreads to other head regions; other symptoms include varying degrees of deafness, dizziness, noises, feelings of revolving in space or rotation of surroundings, pus formation, and difficulty in swallowing
Sinus infection (sinusitis)	*Bacterioides* species, *Haemophilus influenzae*, *Streptococcus pneumoniae*, *Aspergillus* species[a]	−, Rod −, Rod +, Coccus Not done, Mold	1–2 days	Accumulation of mucus in the sinuses; pain, headache, general discomfort, difficulty in breathing
Streptococcal (strep throat)	*Streptococcus pyogenes*	+, Coccus	1–3 days	Chills, fever, headache, nausea or vomiting, rapid pulse, reddened and swollen throat and uvula, and presence of a gray or yellow-white material covering the throat; complications include middle ear infections, kidney infections, and blood poisoning (septicemia)

[a] Relatively few cases of sinusitis are caused by other fungi.

produce the disease soon after by Friedrich August Löffler.

Löffler, who first demonstrated that *Corynebacterium diphtheriae* (ko-ri-nee-back-TI-ree-um dif-THI-ree-ah) causes the disease, concluded from his studies that when the bacteria are injected into guinea pigs, they produce a powerful toxin. The disease state developed as a consequence of the transportation of poisonous material via the bloodstream to other parts of the animal's body. By separating organisms from their liquid environment, Pierre Paul Emile Roux and Alexandre Émile John Yersin in 1888 showed the presence of the *C. diphtheriae* toxin in the liquid phase of the medium.

In 1951, V. J. Freeman provided further insight into the course of diphtheria when he discovered that only those *C. diphtheriae* strains infected with a specific bacteriophage could produce the toxin that causes diphtheria. **[See lysogenic conversion in Chapter 6.]**

Toxin-producing *C. diphtheriae* produces a severe inflammation of the throat. A tough membranelike structure, pseudomembrane, forms on the tonsils and spreads to lower portions of the respiratory tract or upward into the nasal passages. This membrane may cause suffocation. The absorption of the toxin into the bloodstream can cause complications such as paralysis and cardiac arrest (Table 25–1). **[See mechanism of toxin action in Chapter 22.]**

The discovery of the diphtheria toxin had great immunological significance. By 1890, Emil von Behring and his associates were immunizing laboratory animals with a heat-attenuated toxin preparation and obtaining antitoxin with which to treat human victims of the disease. Unfortunately, their heat-attenuated form of diphtheria toxin proved too toxic for human immunization purposes. Not until 1923 was an effective and safe modified toxin for immunization prepared by treating the toxin with formalin. The resulting toxoid retains its antigenicity but is incapable of producing disease.

In the 1920s a mass immunization campaign was begun in the United States to reduce the extremely high attack rate of the disease. Within the period of 65 years, the number of cases were reduced from 151 per 100,000 in 1920 to less than 5 per year for the entire country. **[See immunizations in Chapter 18.]**

TRANSMISSION. Most instances of diphtheria result from direct contact with droplets. *C. diphtheriae* is a hardy organism, able to withstand cold, heat, and drying.

Diphtheria

LABORATORY DIAGNOSIS

Isolation of *Corynebacterium diphtheriae* from nasopharyngeal and related specimens is possible on Löeffler medium (coagulated serum) and the selective and differential medium potassium tellurite agar. Three characteristic biotypes of *C. diphtheriae* are recognized: *gravis*, *intermedius*, and *mitis*. Depending on the biotype of *C. diphtheriae*, colonies on tellurite agar generally vary from gray to black and from flat to convex. Smears of the isolated organisms are gram-positive, appear in parallel rows, and resemble Chinese characters. Stained with methylene blue, cells exhibit a beaded appearance caused by metachromatic granules (see Figure 7–31). Isolates can be tested for toxin production by means of animals, special tissue-culture systems, or gel diffusion.

TREATMENT

Diphtheria antitoxin should be given in any suspected case of the disease. Since the antitoxin is still prepared from horses, care should be taken to avoid allergic reactions. **[See Chapter 20.]**

Penicillin and erythromycin are the drugs of choice to eliminate *C. diphtheriae*. Erythromycin is used in the treatment of carriers.

Diphtheria occurs worldwide, mainly among individuals older than six months but not past middle age. Among the major sources of *C. diphtheriae*, carriers (individuals showing no outward signs of the disease) are believed to account for approximately one-fourth of the known cases. However, exposure to a recognizably infected person is far more likely to result in a case of diphtheria than exposure to a carrier. Infected persons may be a source of disease agents for as long as one to two weeks. **[See types of carriers in Chapter 21.]**

Predisposing factors associated with the disease include chilling, poor nutrition, overcrowded conditions, and operations involving the nose and throat. In the last case, diphtheria may result from the exposure of a noninfected individual to the pathogen or to a carrier of *C. diphtheriae.*

Immunity and Prevention. A relatively long-lasting acquired immunity results from one attack of diphtheria. Infants born to mothers immunized against *C. diphtheriae* infection usually have temporary protection. Because such immunity lasts for only a few weeks after birth, widespread immunization of infants is practiced. In the United States, the DPT vaccine (diphtheria, pertussis, and tetanus) is used to induce immunity. Later booster injections and exposure to infected persons contribute to the maintenance of adequate antitoxin levels in the individual. It is important to note that the antibody response of individuals varies widely. Therefore, it may be necessary to administer several booster shots during childhood and later in life to bring a person's degree of immunity up to a functional level. **[See DPT immunization in Chapter 18.]**

Schick Test. The Schick test, a skin test that determines susceptibility to diphtheria, is used in testing large groups of people. It is reliable in most situations. The procedure involves injection of 0.1 mL of a standardized preparation of diphtheria toxin just under the skin of the inside of the arm. Appearance of a reddened, swollen, tender area around the site of injection indicates susceptibility.

Hypersensitivity to toxin preparation has been reported. This reaction may result from sensitivity to other components of the skin test material. To distinguish such responses from a true indication of susceptibility, toxoid is injected into the opposite arm. Hypersensitive individuals will react to both types of preparations.

Among the measures adopted by health agencies to protect susceptible individuals are large-scale immunization programs. The quarantine of carriers and the daily use of appropriate antibiotics help prevent transmission of the disease until their systems

Haemophilus Influenzae Infection

Laboratory Diagnosis

H. influenzae is a gram-negative coccobacillus that can be isolated on chocolate agar or other types of enriched media from nasopharyngeal, blood, and related specimens. The organism grows best when placed on media containing the two nutritional factors X (hemin) and V (nicotinamide-adenine dinucleotide) and incubated in an atmosphere of 10% CO_2. Isolates can be identified with type-specific antiserum. Several serological tests are of value for detection and identification. These include the capsular swelling test, latex agglutination (Color Photograph 77), and coagglutination.

Treatment

Ampicillin, chloramphenicol, and sulfonamide are the drugs of choice for *H. influenzae* type b infections. For situations in which resistance to ampicillin or chloramphenicol occurs, the combination of erythromycin and sulfisoxazole may be effective. It is recommended that rifampin be used by members of a patient's household (except pregnant women) to prevent secondary cases of infection.

Sinusitis

Laboratory Identification

The processes for isolation and identification of *H. influenzae* type b are the same as those described earlier in this chapter. A preliminary identification of *S. pneumoniae* can be made by finding gram-positive streptococci in sputum and related specimens. Definitive diagnosis is generally based on isolation of alpha-hemolytic streptococci on blood agar. Once cultured, *S. pneumoniae* can be differentiated from other alpha-hemolytic streptococci by its lysis upon exposure to bile or bile salts and inhibition of growth by optochin, an antimicrobial drug obtained from quinine (Color Photograph 89). The capsule swelling or Quellung reaction with type-specific antisera is also a specific test used for identification.

As with all mycotic infections, diagnosis depends upon identification of the fungus, either by direct microscopic examination of specimens (Figure 25–5b) or by isolation and cultivation (Color Photograph 37). All aspergilli grow well on most laboratory media. They form identifiable mycelia and asexual spores (conida). [See Chapter 9 for the properties of fungi.]

Treatment

Several drugs are available for the treatment of bacterial sinusitis. These include a penicillin, amoxicillin-clavulanic acid combination, ampicillin, and tetracycline. *Aspergillus*-caused sinus infections are treated by the surgical removal of fungal masses (Figure 25–5a). Antimycotic creams may be applied to the involved areas as a follow-up treatment.

are free of *C. diphtheriae.* Carriers should be quarantined until bacterial cultures are negative.

EPIGLOTTITIS

Few diseases are as potentially dangerous as an acute inflammation of the epiglottis (Figure 15–1). A healthy person can die suddenly of suffocation resulting from blockage of the artery by swollen tissues. Epiglottitis was recognized as a clinical entity in the early part of this century. The condition is quite common among children, and it is also found among adults. Epiglottitis in adults appears to be different from the disease in children. On physical examination, the major difference observed in children is the cherry-red swelling of the soft tissues, whereas adults exhibit only a mild reddening of the same area.

Haemophilus influenzae (H. in-floo-EN-zee), type b is the predominant organism found in acute cases of the disease. Other bacteria, including *H. parinfluenzae* (H. par-rah-in-floo-EN-zee) and streptococci, have been found on occasion. Respiratory viruses may also play a role in the more severe form of epiglottitis.

SINUSITIS

Most episodes of acute sinusitis occur in asssociation with or following a viral upper respiratory tract illness. Some cases develop as a complication of an allergy or dental infection. In hospitals, sinusitis can arise from a blockage created by a nasotracheal or nasogastric tube. Because of the strategic location of the sinuses (Figure 25–1), infections of these structures can initiate life-threatening complications. The principal causative agents include *Haemophilus influenzae* type b, *Streptococcus pneumoniae* (strep-toh-KOK-kuss) and certain anaerobes. Chronic forms of sinusitis and its complications are often caused by fungi such as *Aspergillus* (a-sper-JIL-lus) species (Figure 25–5). Diagnosis is based on the clinical features of the patient (Table 25–1) and may be confirmed by x-ray or the newer diagnostic techniques, including ultrasound and computerized tomography (CT).

DISEASES OF THE EAR AND RELATED STRUCTURES

Various microorganisms can cause infections of the ear. This section is primarily concerned with those involving bacterial pathogens. Table 25–2 lists some

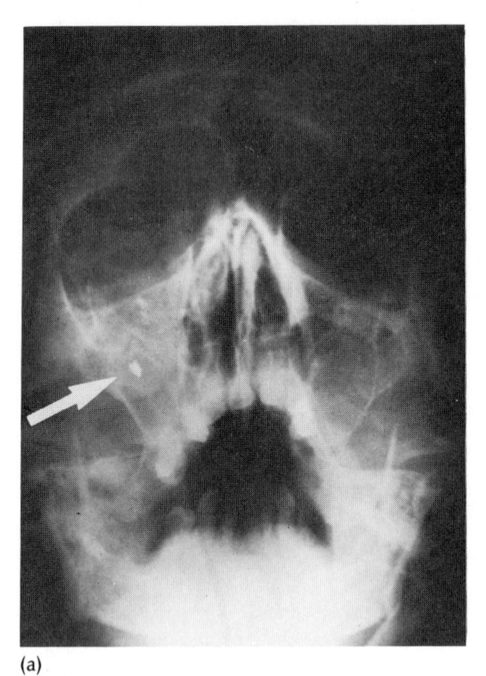
(a)

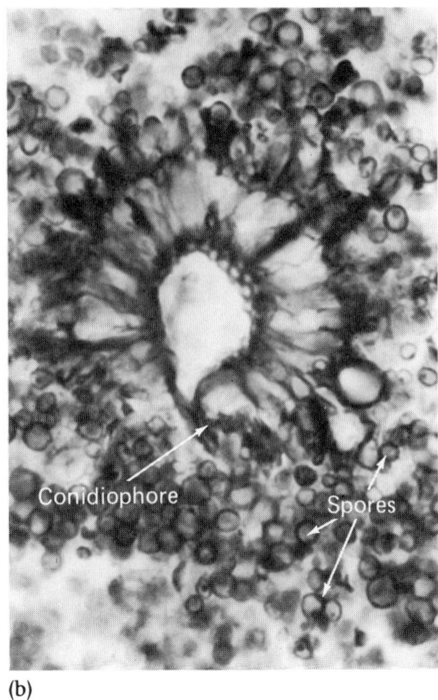

(b)

Figure 25–5 *Aspergillus* sinus infection. (a) An x-ray showing the location of a fungal mass. (b) The causative agent *Aspergillus*. [Courtesy Dr. H. Stammberger.]

of the diseases associated with such microorganisms.

It is well known that individuals with nasopharyngeal (nose and throat) diseases are often predisposed to ear infections. This is especially true in cases of adenoid growths, eustachian tube obstructions, and various forms of middle ear and sinus infections. Treatment must be directed not only to the ear disease but also to the predisposing condition.

TABLE 25–2 Representative Microbial Diseases of the Ear

Disease	*Causative Agent*	*Gram Reaction*	*Morphology (if applicable)*	*Possible Treatment*
Boils	*Staphylococcus aureus*	+	Coccus	Cephalothin, lincomycin, methicillin, nafcillin, vancomycin
Inflammation of the eardrum	Mixed infections involving beta-hemolytic streptococci and viruses	+	Cocci	Ampicillin, cephalothin, erythromycin, penicillin; symptomatic
Inflammation of the outer ear	*Escherichia coli*, *Proteus* spp., *Pseudomonas* spp.,	–	Rods	Chloramphenicol, gentamicin, kanamycin, penicillin, polymycin, tetracyclines, tobramycin
	Hemolytic streptococci, *Staphylococcus aureus*	+	Cocci	Ampicillin, cephalothin, erythromycin, penicillin, vancomycin
	Mixed infections			
Otitis media (middle ear infection)	*Streptococcus pneumoniae*, beta-hemolytic streptococci, *Staphylococcus aureus*	+	Cocci	Ampicillin, cephalothin, erythromycin, penicillin, vancomycin
	Haemophilus influenzae type b	–	Coccobacillus	Chloramphenicol, kanamycin, penicillin, streptomycin, tetracyclines
Mycotic infection of the external ear and ear canal	*Aspergillus niger*	Not useful	Mold	Amphotericin B, ketoconazole
	Candida albicans	+	Yeast	Amphotericin B, ketoconazole
Throat abscess	Beta-hemolytic streptococci, *Staphylococcus aureus*	+	Cocci	Ampicillin, cephalothin, erythromycin, penicillin, vancomycin

Pus-Producing Middle Ear Infection (Suppurative Otitis Media). Pus-forming ear infection is a common childhood disease, although it is quite frequently overlooked. The infection usually results from inflammatory conditions involving the upper air passages (Table 25–2). Contaminated swimming water entering the middle ear from the nose or nasopharynx, injuries to the eardrum, skull fractures associated with the temporal bone, and certain complications of epidemic cerebrospinal meningitis may all produce this condition. Children are predisposed to acute suppurative otitis media if they have adenoids in the nasopharynx and if they have a hereditary tendency toward nasal congestion (catarrh) and pus formation of the upper air passages.

Staphylococcus aureus, Haemophilus influenzae type b, and beta-hemolytic streptococci are the most common etiologic agents of this disease (Table 25–2). *Streptococcus pneumoniae* is also known for its ability to cause otitis media.

H. influenzae is a major cause of several important infections in young children. Encapsulated strains are divided serologically and chemically into six types designated by the letters a to f. Serious invasive infections are caused almost exclusively by serotype b. In addition to otitis media, as mentioned earlier, *H. influenzae* may cause an infection of the epiglottis (epiglottitis), which can produce sudden, unpredictable airway obstruction, a life-threatening emergency.

The laboratory diagnosis and treatment for suppurative otitis media are the same as those described for epiglottitis.

STREPTOCOCCAL SORE THROAT

Streptococcal sore throat is an acute, severe inflammation of both the tonsils (tonsillitis) and the throat (pharyngitis). Strep throat, one of the most important URIs, can be caused by several strains of Lancefield's group A beta-hemolytic streptococci (Table 25–1).

Transmission. Strep throat usually occurs in epidemic form and is associated with aerosols and with contaminated milk products and water. The nasopharynges of individuals suffering from the disease during the acute or convalescent stage serve as the source of infectious agents. There are also carriers.

Milk products may be contaminated by human handling or as a consequence of infection in the cows. In epidemics involving milk consumption, the source of contamination can usually be traced to a single dairy. Furthermore, strep throat is more common in regions where pasteurization of milk is not practiced. Epidemics may last from two to six weeks. The incidence of this disease is greater during the winter and spring months. **[See means of disease transmission in Chapter 21.]**

Prevention. Control measures for strep throat include pasteurization of milk, isolation of infected persons, and proper disinfection or disposal of objects contaminated by the discharges of infected persons. Such persons should not be allowed to come into contact with products for consumption during either the acute or convalescent stage of the disease.

Viral Diseases

This section deals with common viral agents for which the respiratory tract is both the principal site of replication and the site of their cytopathic effects

Strep Throat

Laboratory Identification

Accurate identification of group A hemolytic streptococci, (*Streptococcus pyogenes* S. pie-ah-GEN-eez) requires both clinical and laboratory findings. *S. pyogenes* can be isolated from throat swabs and related specimens on blood agar. The pathogen is beta-hemolytic and sensitive to bacitracin (Color Photograph 87). It can be further identified by a modified form of agglutination reaction in the Phadebact test (Color Photograph 105).

Treatment

Most strains of group A beta-hemolytic streptococci are susceptible to penicillin. Penicillin G benzathine is usually given because of its prolonged action.

TABLE 25–3 Viral Agents and Commonly Associated Upper Respiratory Tract Infections

Respiratory Tract Infection	*Virus by Generic Designation*[a]	*General Features of the Disease*	*Laboratory Diagosis*
Common cold	Coronavirus Rhinovirus	Cough, watery nasal discharge, head cold (coryza), headache, nasal obstruction, sneezing, sore throat	Isolation and identification of causative agent using tissue culture (not routinely done)
Croup (acute laryngotracheobronchitis)	Adenovirus Orthomyxovirus Paramyxoviruses Respiratory syncytial virus	Symptoms range from mild to severe and can be grouped into the following types: *Type 1*—cough, hoarseness, harsh breathing, high-pitched sounds *Type 2*—fever, toxic effects, vomiting *Type 3*—convulsions, bluish coloration (cyanosis), dehydration, and restlessness	1. Isolation and identification of causative agent using tissue culture 2. Serological tests
Minor respiratory illnesses	Adenoviruses Echoviruses Paramyxoviruses Reoviruses	Fever, sore throat, swollen lymph nodes in neck, persistent cough	1. Isolation and identification of causative agent using tissue culture 2. Serological tests

[a] Based on the results of isolations from cases of respiratory infections.

(Table 25–3). It is important to make this distinction because several viruses can grow in the respiratory tract or use the tract for entry into the body without producing their effects there. These viruses are not discussed here.

In 1968, A. J. Rhodes and C. E. Van Rooyen reviewed the major viral respiratory infections. They wrote that the number of viruses causing respiratory illness in humans and other animals is steadily increasing. In actual fact, the overwhelming majority of URIs are caused by viruses. Identification of the virus associated with a given disease is complicated by the fact that one viral agent may produce more than one set of symptoms. The antigenic characteristics of several respiratory viruses such as influenza change periodically, and there are no sharp distinctions in the symptoms of the various respiratory diseases; symptoms of many diseases overlap. In addition, susceptibility to respiratory illness is influenced by developmental, immunologic, and physiologic features of the host.

USE OF ANTIBIOTICS

In the treatment of various viral respiratory diseases, antibiotics are given primarily to prevent secondary bacterial infections. This procedure is done even though there is no clear evidence that these drugs reduce or eliminate the effects of the causative agents. Other types of treatment are directed toward relieving the discomfort of the patient.

THE COMMON COLD

According to many authorities, acute afebrile diseases (those without fever) of the upper respiratory tract are the most frequent human afflictions. As a group, they cause the loss of millions of hours of work each year and rank as the most frequent reason why patients of all ages seek medical attention. In most cases these infections are not serious, only extremely uncomfortable. Many studies have shown that several different viruses can cause the common cold. Many of these etiologic agents have been designated the *rhinoviruses* (from the Greek *rhino,* meaning nose). It is important to note that these viruses are not the sole causes of the common cold. About one-third of colds are caused by coronaviruses and other specific viruses.

The rhinovirus-caused common cold appears to have a wide geographic distribution, as judged by the presence of antibodies in persons throughout the world. The transmission of rhinovirus infections seems to require close and continued contact among individuals.

CONTROL. Attempts to develop vaccines against cold viruses have so far not been successful. Another approach to the problem of control has concentrated on *interferon,* the active inhibitor of viral activities. Here effective application has been limited. The intake of large quantities of vitamin C has been suggested as an effective means of providing resistance to respiratory illness under normal conditions. [See interferon in Chapters 10 and 12.]

Attempts to control the spread of common-cold agents by the use of disinfectant aerosols, irradiation with ultraviolet light, or the quarantine of infected persons are usually of little value.

CROUP

Croup, or acute laryngotracheobronchitis, is an acute infectious disease of children under three years of age. Boys are affected more often than girls. Croup may be mild or severe in its effects (Table 25–3). This disease should be distinguished from epiglottis infection from the standpoint of treatment. Epiglottitis, as indicated earlier, is caused almost entirely by the bacterium *Haemophilus influenzae* type b and can be controlled with antibiotic therapy.

Lower Respiratory Infections

The lungs inhale many pathogenic microorganisms that are normally eliminated efficiently by the defenses of the host. However, several diseases of the lower respiratory tract are life-threatening if not treated quickly and adequately (Tables 25–5 and 25–7 to 25–9). These may arise when microorganisms gain access to the lung in inhaled air or by means of the bloodstream.

Bacterial Pneumonias

Several bacterial species can cause pneumonia (Tables 25–1 and 25–4). One of these is associated with the condition known as Legionnaires' disease. This pneumonia earned its name and notoriety by striking many individuals, some fatally, who attended an American Legion convention in Philadelphia in 1976. Since that first report of Legionnaires' disease, other outbreaks have been noted. The general features of this disease and those of other types of pneumonia are given in Table 25–4. Selected pneumonias will be presented in greater detail.

PNEUMOCOCCAL PNEUMONIA

Streptococcus pneumoniae has caused the great majority of pneumonia cases in which a whole lobe or more than one lobe is involved, a condition referred to as *lobar pneumonia.*

More than 80 different serological types of *S. pneumoniae* have been reported. Differentiation among the various types is based on the existence of immunologically distinct polysaccharides, which form the capsules of these organisms (Color Photograph 15). The designation of each type is by number, such as type 3, type 7, and so on. The Quellung reaction, an immunologic laboratory test, is used to identify capsule types.

S. pneumoniae is also capable of causing diseases other than pneumonia (Figure 25–6). These diseases include infection of the middle ear (otitis media), pneumococcal meningitis, and inflammation of the sinuses (sinusitis). Antibiotic-resistant strains are being found, making treatment difficult. The development in 1977 of a polyvalent vaccine effective against the most common types of *S. pneumoniae* reduced the incidence of these diseases in high-risk groups and in persons with normal immune systems.

Other microorganisms that produce lobar pneumonia, although far less frequently, include *Klebsiella pneumoniae* (kleb-see-EL-lah new-MOH-nee-ah), *Mycobacterium tuberculosis* (my-koh-back-TI-ree-um too-ber-koo-LOW-sis), *Yersinia pestis* (yer-SIN-ee-ah PES-tis) (the cause of pneumonic plague), *Francisella tularensis* (fran-sis-EL-lah too-lah-REN-sis), (the cause of tularemia or rabbit fever) and *Serratia marcescens* (ser-RAY-shah mar-SES-sens).

TABLE 25–4 **Comparison of *Legionella pneumophila* and Related *Legionella* Organisms**

	Characteristics[a]				
Microorganism	*Growth on CYE Agar*	*Grown on F-G Agar*	*Beta-Lactamase*[b] *Production*	*Fluorescence under UVL*	*% DNA Relatedness to* L. pneumophila, *Philadelphia Strain*
Legionella pneumophila	+	+	+	Yellow	100
Legionella microdadeii	+	−	+	Yellow	5
Legionella bozemanii	+	−	+	Blue/white	15
Legionella dumoffii	+	+	+	Blue/white	12

[a] Explanation of symbols: CYE, charcoal yeast extract; F-G agar, Feeley-Gorman; UVL, longwave ultraviolet light.
[b] The presence of this enzyme is associated with penicillin resistance.

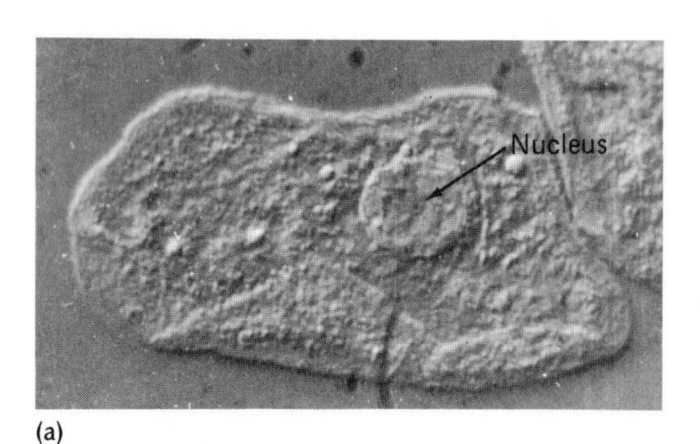

(a)

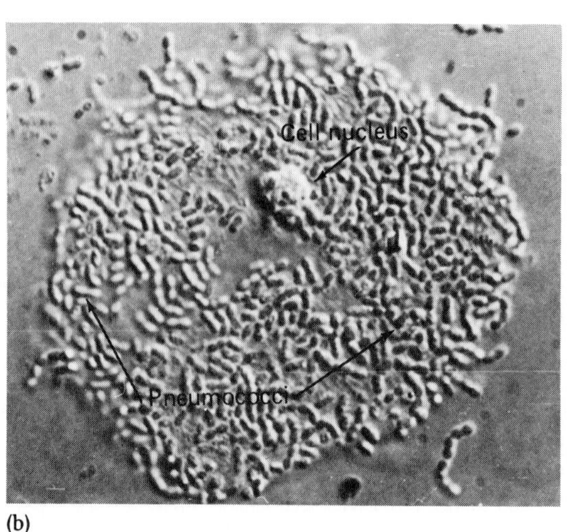

(b)

Figure 25–6 A virulence factor for pneumococci in causing middle ear infections. (a) A human pharyngeal surface cell without many adhering bacteria. (b) Pneumococci strongly sticking to a pharyngeal cell's surface. From surfaces such as these, pneumococci may move to the middle ear. *[From B. Anderson, B. Eriksson, E. Falsen, A. Fogh, L. A. Hanson, O. Nylen, H. Peterson, and C. S. Eden,* Infect. Immun. ***32****:311–317, 1981.]*

Microbiology Highlights

PSEUDOMONAS LUNG INFECTION

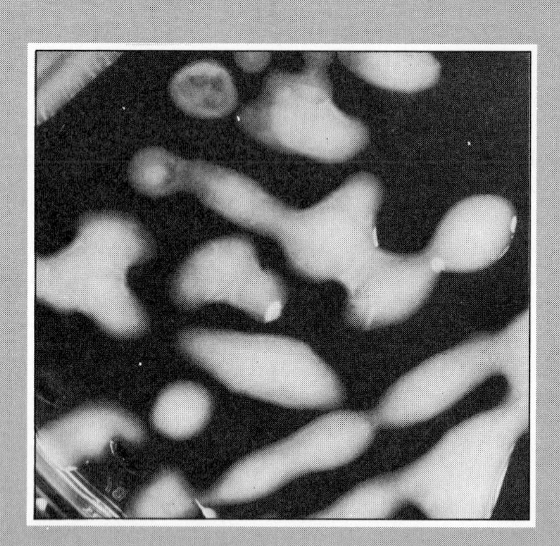

Mucoid colonies of *P. aeruginosa. [From R. Chan, J. S. Lam, K. Lam, and J. W. Costerton,* J. Clin. Microbiol. ***19****:8–16, 1984.]*

The mucoid type of growth of various pathogens protects them from opsonic antibodies and makes them more resistant to phagocytosis. In cystic fibrosis patients with *Pseudomonas* pneumonia, the causative agent *P. aeruginosa* (soo-doh-MOH-nass ar-roo-jin-OH-sah) grows in slimy polysaccharide-encloded colonies in the infected lung. A correlation appears to exist between the severity of these lung infections and the presence of mucoid isolates. In addition, the presence of these unusual isolates in patients may represent a response to extensive and sustained antibiotic therapy.

Streptococcus Pneumonia

LABORATORY IDENTIFICATION

The processes for isolation and identification of *S. pneumoniae* are the same as those described earlier in this chapter.

TREATMENT

Penicillin is the drug of choice for pneumococcal infections. Erythromycin or tetracycline are used in the treatment of penicillin-resistant strains, or individuals allergic to penicillin.

TRANSMISSION AND PREDISPOSING FACTORS. Pneumonia produced by *S. pneumoniae* occurs during the winter months in temperate zones, usually in persons between 15 and 45 years of age and in the elderly. Although this disease is not considered highly communicable, its incidence is greater where close personal contact exists, as in households, barracks, dormitories, hospital wards, prisons, and similar group living quarters. Mortality rates have been substantially reduced, especially in the young, by early and adequate administration of antibiotics.

Pneumococci are usually spread by droplet nuclei from nasal or pharyngeal secretions. Individuals exposed to such infectious material may contract the disease or become carriers. It is interesting to note that 40% to 70% of normal adults have *S. pneumoniae* as normal inhabitants of their throats.

Several factors have been reported to predispose individuals to pneumococcal and other bacterial pneumonias. Among these so-called secondary causes are viral infections of the upper respiratory tract and local or generalized pulmonary edema (fluid in the lungs). Pulmonary edema can be caused by irritating anesthetics, heart failure, and influenza. More general predisposing factors that have been implicated include increasing age, debilitating disease states (such as diabetes or malignancies), fatigue, and chilling. With many, if not all, of the factors mentioned, the underlying feature is impairment of the lung's defense mechanism, its phagocytic activity, by tissue macrophages.

COMPLICATIONS. Unfortunately, complications arise with at least 15% to 20% of the pathogen's victims. Commonly encountered complications include recurrence of the disease, inflammation of the pleura, otitis media, and sinusitis.

IMMUNITY AND PREVENTION. Individuals who have recovered from pneumococcal pneumonia often carry the causative agents in their upper respiratory tracts for long periods, despite the fact that they also have antibodies circulating in their bloodstreams. These antibodies remain for several months. If an individual suffers a second attack of pneumonia, it is probably caused by a serological type of *S. pneumoniae* other than the one that caused the first infection.

Preventive measures against pneumococcal diseases include immunization and the adequate treatment and isolation of infected persons. Indiscriminate use of antibiotics should be avoided because it can induce drug reactions and enhance the survival of drug-resistant organisms.

LEGIONNAIRES' DISEASE

In July 1976, the 58th Annual Convention of the Pennsylvania Department of the American Legion was held at the Bellevue-Stratford Hotel in Philadelphia. This meeting provided the background for the public's awareness of the now well-known Legionnaires' disease. The explosive outbreak of pneumonia, which affected 182 of the persons attending that convention and 72 other individuals who were also in Philadelphia during July, prompted the largest epidemic investigation ever conducted by the Centers for Disease Control (CDC). Interestingly enough, almost 30 years before the 1976 Philadelphia epidemic, a microorganism had been isolated from a patient suffering from a fever-producing (febrile) respiratory illness. The organism was described as "rickettsialike," partly because it would not grow on available artificial media, and it resembled several similar agents isolated from other patients in the 1940s and 1950s. This pathogen, which was never classified or studied further in 1947, later proved to be the causative agent of Legionnaires' disease, namely, the gram-negative rod *Legionella pneumophila* (lee-jon-EL-lah new-MOH-fill-ah). Between 1947, when the first isolated case occurred, and 1976, there were four epidemics of a similar respiratory disease investigated by the CDC. In each of these situations no causative agent was found.

From the various studies conducted since the 1976 episode, it is reasonably clear that two syndromes are caused by *L. pneumophila: Legionnaires' disease* and *Pontiac fever.* Legionnaires' disease is a severe disease involving several organs, but with pneumonia as a major feature (Table 25–5). The disease itself is an atypical pneumonia and can be confused with similar conditions such as psittacosis, Q fever, *Mycoplasma* (my-koh-PLAZ-mah) infections, and influenza. In addition, the clinical features vary from one case to another, thus making clinical diagnosis difficult. Legionnaires' disease can be diagnosed more completely in the laboratory. Culture and immunological tests are quite successful.

Pontiac fever is a mild, self-limited illness with a

TABLE 25–5 Features of Selected Bacterial Lower Respiratory Tract Infections

Disease	*Causative Agent*	*Gram Reaction and Morphology*	*Incubation Period*	*General Features of the Disease*
Atypical primary pneumonia	*Mycoplasma pneumoniae* (Eaton's agent)	[a]Pleomorphic	7–21 days	Cough, chills, headache, sore throat, general discomfort, thick sputum with pus
Legionnaires' disease	*Legionella pneumophila*	−, Rod	2–10 days	Chills, rapidly rising fever, abdominal pains, slight headache, muscle aches, nonproductive cough; complications leading to respiration failure can occur
Pontiac fever	*L. pneumophila*	−, Rod	36 hours	Fever, shaking chills, headache, severe muscle pains, some nausea and vomiting
Pneumonia	*Klebsiella pneumoniae* (Friedlander's bacillus)	−, Rod	1–2 days	Fever, chest pains, thick reddish-brown sputum
	Staphylococcus aureus	+, Coccus	1–2 days	High fever, blue coloration (cyanosis), frequent cough, pus-containing discharge from nose, eyes, and rapid breathing[b]; a complication of certain viral infections such as measles
	Streptococcus pneumoniae[c]	+, Coccus	1–3 days	Sudden onset of symptoms, including severe chills and shaking, high fever, chest pain, thick, rust-colored sputum, dry cough, and vomiting
	Streptococcus pyogenes (group A, beta-hemolytic)	+, Coccus	1–2 days	Chills, cough, difficulty in breathing, fever, general discomfort; complications include infections of the central nervous system and kidneys
Psittacosis	*Chlamydia psittaci*	−, Rod	6–15 days	Sudden onset of symptoms, including dry cough, difficulty in breathing, fever, pain, and headache
Q fever	*Coxiella burnetii*	−, Rod	14–28 days	Sudden onset of symptoms, including dry cough, fever, headache, and general stiffness

TABLE 25–5 (continued)

Disease	*Causative Agent*	*Gram Reaction and Morphology*	*Incubation Period*	*General Features of the Disease*
Tuberculosis	*Mycobacterium tuberculosis* and related species	[a], Rod	28–42 days for primary infection	Wide variety of symptoms, including fever, general discomfort, weight loss, productive blood, formation of tubercle (nodule in lung tissue); skin test eventually becomes positive
Whooping cough (pertussis)	*Bordetella pertussis* (Bordet-Gengou bacillus)	–, Rod	7–10 days	Disease occurs in three stages: *Catarrhal*—persistent dry cough, slight fever, poor appetite, excessive mucous secretions, tearing, and vomiting *Paroxysmal*—coughing attacks referred to as "whooping,"[e] a production of thick, stringy masses of mucus *Convalescent*–coughing attacks decrease in severity

[a] Gram reactions are of little value with this pathogen.
[b] This is a more severe form of pneumonia than that caused by *S. pneumoniae.*
[c] The Quellung (German, "swelling") test is diagnostic for infections caused by various encapsulated bacteria, including *Streptococcus pneumoniae, Haemophilus influenzae,* and *Klebsiella* spp. The procedure involves treating encapsulated cells with specific antisera that combine with the capsular polysaccharide, causing the impression of capsular swelling.
[d] The Gram stain reaction is not used diagnostically. The acid-fast staining procedure is used instead.
[e] The characteristic "whoop" is caused by rapidly inhaling air, which passes quickly over the vocal cords.

very high attack rate. Fatalities are extremely rare. Epidemics of Pontiac fever and legionnaires' disease tend to occur during summer months.

TRANSMISSION AND PREDISPOSING FACTORS. Transmission of *L. pneumophila* appears to be by air or by aerosol. In several outbreaks of the disease, various lines of evidence implicated cooling towers and evaporative condensers associated with air conditioning units as sources of the pathogen. *L. pneumophila* has also been isolated from sources that have not been associated with disease transmission. These sources include streams, lakes, showers, unrelated cooling towers, and soil.

Individuals above age 50, men, and smokers are more likely to contract Legionnaires' disease. No particular age group or sex appears to be especially susceptible to Pontiac fever.

Microbiology Highlights

THE AMEBA–LEGIONELLA CONNECTION

Amebas in natural water environments, especially species of *Acanthamoeba* (a-can-thah-MEE-bah), may act as sources of *Legionella pneumophila.* The trophozoites (active feeding stage) of *Acanthamoeba* have been shown to ingest the bacteria, which can then multiply within the amebas. *Legionella* trapped within amebic cysts (the resistant stage) would allow the bacterial contamination of water supplies to persist despite chlorine treatment, to which the cysts are resistant. Pontiac fever may be related to the presence—in aerosols breathed by patients—of *Legionella* enclosed within the remains of amebas in specific environments. Such aerosols may originate from the stored water of air conditioning systems.

OTHER *LEGIONELLA*. Several gram-negative rods, closely related to but distinguishable from *L. pneumophila,* have been implicated as causative agents of pneumonia. Twenty-two species and 33 serogroups have been identified and more are yet to be defined. The best known of these is *L. micdadeii,* or Pittsburgh agent. This organism has produced pneumonia in kidney transplant patients receiving steroid therapy. It differs from *L. pneumophila* by being susceptible to antibiotics, such as penicillins and cephalosporins. Some of the other related organisms reported as causes of human pneumonia include *L. bozemanii* (L. bozz-MAN-eye), *L. dumoffii* (L. dum-OFF-eye), and *Legionella*-like organisms. Various laboratory tests can be used to distinguish these pneumonia-causing agents from one another (Table 25–4). Such tests include growth on charcoal yeast extract (Color Photograph 2a) and Feeley-Gorman agar, beta-lactamase production, fluorescence under longwave ultraviolet light, direct fluorescent antibody staining to *L. pneumophila,* and the percentage of DNA relatedness to *L. pneumophila,* Philadelphia strain. **[See beta-lactam antibiotics in Chapter 12.] [See fluorescent antibody tests in Chapter 19; DNA relatedness in Chapters 6 and 13.]**

MYCOPLASMA INFECTIONS

Mycoplasma have been known for years to cause diseases of lower animals, but only recently have investigations demonstrated that one species is pathogenic for humans. This is *Mycoplasma pneumoniae,* the etiologic agent of primary atypical pneumonia (Table 25–5). Patients with the disease are known to develop high concentrations of immunoglobulin M (IgM) antibodies (cold agglutinins), which, when tested in the laboratory, cause red blood cells to clump when incubated together at 0° to 10°C. The reaction does not occur at 37°C.

PSITTACOSIS

Respiratory infections contracted from psittacine, usually exotic birds (canaries, cockatoos, lovebirds, parakeets, and parrots) are generally known as *psittacosis,* or *parrot fever.* The causative agent, *Chlamydia psittaci* (kla-MI-dee-ah SIT-tah-sigh), has been recovered from numerous other types of birds. In general, psittacosis is considered to be *zoonotic,* spread by lower animals through natural means to humans. However, some cases of human disease have resulted from direct contact with infected persons (Table 25–5). **[See zoonoses in Chapter 21.]**

TRANSMISSION. In birds, the infection is usually spread by direct contact, droplets, and droppings. Latent disease states are common with these animals, and under conditions of stress—chilling, dampness, or overcrowding—the inactive infection is known to develop into overt psittacosis. The symptoms of infected birds include diarrhea, weight loss, and a mucopurulent discharge from the mouth. Cases of human infection develop in pet owners, animal handlers, and pet-store attendants. The disease is an occupational hazard in the poultry industry, and individuals working in poultry processing plants

Legionnaires' Disease

LABORATORY DIAGNOSIS

The indirect fluorescent antibody test (Color Photograph 2b) is the most commonly used diagnostic procedure for *L. pneumophila,* while the direct form of the test is the most rapid approach to diagnosis. Isolation of the pathogen from sputum and related specimens is done on charcoal yeast extract agar or other special medium (Color Photograph 2a), biphasic blood culture broth, or other semiselective media that inhibit contaminants. Colonies appear within three to seven days. Final confirmation of *Legionella* species is made by DNA homology testing (gene probes). **[See Chapter 15 for gene probes.]**

TREATMENT

Erythromycin is the drug of choice for therapy of Legionnaires' disease and infections caused by *Legionella* species. Rifampin is given to patients failing to respond to erythromycin or who are gravely ill.

Mycoplasmal Pneumonia

Laboratory Diagnosis

Rapid and specific identification of mycoplasma in nasal or conjunctival (inner eyelid) specimens as well as in tissue can be achieved with the immunoperoxidase test (Color Photograph 80). Fluorescent antibody tests also are used. [See Chapter 19 for an explanation of these tests.]

Isolation of mycoplasmas from clinical specimens can be done using selective agar media containing penicillin and thallium acetate to inhibit the growth of contaminants. Colonies resembling fried eggs appear within 3 to 21 days (Figure 13–22). Specific identification of *Mycoplasma* species can be made by placing filter-paper disks containing growth-inhibiting antisera against specific species on inoculated papers while also using control disks with normal serum. Positive identification is made by the absence of growth (clear zone) around a growth-inhibiting disk as compared to the growth around the control. Distinctive properties of *M. pneumoniae* include an ability to utilize glucose and an inability to use arginine when grown in appropriate media.

Treatment

Erythromycin is the drug of choice in the treatment of mycoplasmal pneumonia. Tetracyclines also are effective.

are at particular risk. The birds associated with human infections may show obvious signs of illness, may have lesions apparent only at autopsy, or may have no signs of infection from the mouth.

The TWAR (Taiwan acute respiratory) strain of *C. psittaci* has been found to be an important cause of pneumonia in college students. Pneumonia associated with this strain is clinically similar to mycoplasmal pneumonia (Table 25–5).

Control. Control measures against psittacosis usually include the imposition of a six-month quarantine on shipments of psittacine birds, the use of spot laboratory examination of shipments of these

Psittacosis

Laboratory Diagnosis

Serological tests are used as methods of choice for diagnosis, since isolation of *C. psittaci* from clinical specimens can be hazardous and difficult. Complement fixation is the preferred procedure to shown increase in antibody levels.

Treatment

Tetracycline is the drug of choice for the management of *C. psittaci* infections.

animals, and the incorporation of antibiotics into the feed for psittacines. Chlortetracycline has proved extremely efficient in reducing the incidence of the disease in birds. It has been shown that inadequately treated animals are often released from quarantine stations and have been implicated as sources of human infection.

Q ("QUERY") FEVER

Q fever, a zoonotic rickettsial infection, was first described in 1933 by Derrick in Australia. The etiologic agent for the disease is *Coxiella burnetii* (Table 25–5). Arthropods do not play a major role in transmission of the disease to humans. However, ticks are considered to be important to the cycle of infection in wild rodents. Human infections, as well as those of other animals, occur through the inhalation of infectious discharges (droplet transmission) from diseased cattle, goats, and sheep. Milk and eggs have also been implicated in disease cycles.

TUBERCULOSIS AND OTHER MYCOBACTERIAL DISEASES

Tuberculosis has been a plague of humanity for centuries, especially in populations suffering from malnutrition and poor sanitary conditions. The clinical features as well as the communicable nature of this infectious disease have been known since at least 1000 B.C. Both Aristotle and Hippocrates described tuberculosis. The name of the infection probably arose from the postmortem observation of tubercles (small cellular masses) in the lungs of its victims.

In 1882 Robert Koch, fulfilling the dictates of his famous postulates, clearly proved *Mycobacterium tuberculosis* (Color Photograph 13a) to be the cause of this infection. Since Koch's discovery, a number of other species of the genus *Mycobacterium* have been reported to cause similar disease states in humans and other animals. Tuberculosis is still the preferred term for the mycobacterial diseases caused by *M. tuberculosis, M. bovis,* and *M. intracellulare-avium-complex* infections. However, increased recognition of the prevalence of pulmonary disease caused by mycobacteria other than *M. tuberculosis* has focused on the importance of differentiating among these mycobacterial pathogens. The National Tuberculosis and Respiratory Disease Association recommends that mycobacteria-caused diseases be reported in a way that specifies the causative agent (e.g., lung infection caused by *M. kansasii* or *M. intracellulare-avium-complex*). The approach to diagnosis in the case of *M. intracellulare* is more complex than that followed with other mycobacteria. This organism has been associated with several deaths of acquired immune deficiency syndrome (AIDS) victims. *M. tuberculosis* infection among AIDS victims has also increased substantially. **[See AIDS in Chapter 27.]**

Transmission and Predisposing Factors. Today, infection with *M. tuberculosis* occurs primarily through inhalation of droplet nuclei. Sputum, "coughing sprays," and droplets released by the sneezing of infected persons are common sources of the disease agent. So is contaminated dust.

Other ways by which mycobacteria may enter the susceptible individual include ingestion and direct inoculation. Tubercle organisms may be swallowed by children when they place contaminated objects in their mouths or consume food containing these bacteria. First infections are more common in children than in adults. The danger of acquiring tuberculosis from infected dairy products has been largely eliminated in many countries by the pasteurization of milk products and by the tuberculin testing of dairy cattle. (On rare occasions, butchers may be-

Q Fever

Laboratory Diagnosis

Serological testing is generally the most practical and efficient method of identification. The complement fixation and microagglutination tests are also of value. The Weil-Felix procedure is not applicable. **[See Chapter 19 for descriptions of these tests.]**

Treatment

Tetracycline is the preferred drug of choice. Chloramphenicol also is effective, but it is toxic.

come infected by handling diseased meats, or pathologists may become infected while examining infected tissues.) Congenital tuberculosis appears to be rare, as the placenta is usually an effective barrier to *M. tuberculosis.*

Predisposing factors in tuberculosis include advanced age, chronic alcoholism, poor diet, certain metabolic diseases, some occupations, race, and prolonged stress. Tuberculosis itself is not inherited. The question of whether air pollution and cigarette smoking predispose one to tuberculosis has not been settled.

Active pulmonary tuberculosis among infants and children has been substantially reduced in the United States. In general, however, the disease is severe in infants. Susceptibility to tuberculosis increases with age after adolescence. For some unknown reason, the typical victims of the disease in the United States are white males 40 years of age or older. However, the mortality with active cases of tuberculosis is substantially higher in the nonwhite population. This is true for both sexes of American Indians, Eskimos, and blacks.

Tuberculosis is a disease associated with poverty. Overcrowding, poorly ventilated rooms, and malnutrition favor the establishment of *M. tuberculosis* infection. Quite possibly, factors such as these may account for the greater mortality among nonwhites.

Of the metabolic diseases, diabetes mellitus appears to be the one that most frequently predisposes individuals to pulmonary tuberculosis.

Other significant predisposing factors include prolonged periods of fatigue or overexertion and chronic alcoholism, which is closely related to poor nutrition. Whether or not other respiratory diseases influence susceptibility to tuberculosis has not been firmly established. Frequent colds have been implicated, however.

The properties of several mycobacterial strains have been recorded for many years. From these records, it appears that the characteristics have remained relatively constant. Subsequently, several strains have been used as standards in studies and as material for human prophylactic immunization. The bovine tubercle bacillus or bacille Calmette-Guérin (BCG) is used for immunization. **[See vaccines in Chapter 19.]**

The Pathogenesis and Pathology of Tuberculosis. The pathogenesis and pathology of tuberculosis are relatively complex subjects. This discussion will emphasize the characteristic responses of the body to tubercle bacilli. Additional information can be obtained from the reference sources given at the end of the chapter.

The response of the individual to tubercle bacilli depends primarily upon the body's resistance and the organism's virulence. The size of the initial inoculum and the location of the infection are also important factors. In humans, a wide range of responses can occur. The tissue changes that develop are mainly those associated with inflammation and repair.

An inflammatory process is produced by the presence of *M. tuberculosis.* This tuberculosis inflammation is unusual in that it incorporates wandering tissue phagocytes known as *macrophages,* or histiocytes, and phagocytic pneumocytes. Most other inflammatory processes use polymorphonuclear leukocytes.

The initial and characteristic lesion of tuberculosis, called the *tubercle,* appears as a small tissue mass or nodule in the lung tissue. The tubercle represents an individual's hypersensitive response to invading organisms. Both B and T cells appear at the lesion, responding to the presence of the pathogens. Infection sites may be found in any portion of the lung (Figure 25–2). However, the membranes covering the lungs or the pleura appear to be frequently involved. In most cases, single localized lesions (single foci) form, although multiple foci are known to occur.

Pulmonary lesions develop in the most aerated regions of the lungs. The presence of a high oxygen concentration, which *M. tuberculosis* requires, may account for the development of lesions in the top parts of the upper lobes of human lungs. In cattle, the dorsal region of the lower lung lobes is involved. If large numbers of *M. tuberculosis* form within the tubercle, large numbers of neutrophils will accumulate. The release of lysomal enzymes by these leukocytes destroys both bacteria and host tissue, resulting in caseation necrosis (tissue with a cheesy consistency).

Healing. The healing of tuberculosis lesions may occur in several ways, including calcification, fibrosis, and resolution. Usually, the three are combined.

Calcification is the deposit of calcium within the semisolid (caseous) centers of older tuberculous lesions. Calcification usually occurs after two years.

Fibrosis, or *scarring,* accompanies the healing of most tuberculous lesions. Collagen is deposited during the process.

Resolution probably accompanies the healing of all tuberculous lesions. This process includes the disappearance of infiltrating macrophages and even of the tubercles described earlier.

When healing does not take place, the tuberculosis is considered to be progressive. Established lesions can extend into surrounding tissue simply by enlargement or can spread to other areas by means of the circulatory system (miliary tuberculosis).

Skin Tests. Robert Koch's discovery of the tuberculin reaction in 1890 provided one of the most valuable diagnostic procedures for the control of tuberculosis. In addition to its obvious importance from the standpoint of differential diagnosis, the test is an extremely important epidemiological tool.

The basis of the tuberculin reaction is the development during the course of an infection of a specific delayed hypersensitivity to certain products of *M. tuberculosis* and related mycobacteria. These products are contained in culture extracts and are referred to as *tuberculins.* The sensitivity that occurs in individuals may develop three to seven weeks after infection, usually remaining for several years or for an entire lifetime.

Depending on the diameter of the response *the tuberculin test reveals previous or current infections but does not prove the presence of an active disease state.* Confirmation of an active case of tuberculosis is done with x-ray examination and isolation of *M. tuberculosis.* [See skin tests in Chapter 19.]

Preparations Used for Testing. Two types of tuberculin, old tuberculin (OT) and purified protein derivative (PPD), are widely used for skin testing. The first material was originally described by Robert Koch and incorporates the heat sterilization of an *M. tuberculosis* culture. The active component in OT is a protein noted for its heat stability and retention of specificity for several years.

The second preparation, PPD, is a slightly more refined testing substance than OT. It is preferred because its strength lends itself to standardization of dosages; skin tests performed with the same dose are comparable. PPD contains an active protein obtained from filtrates of autoclaved tubercle bacilli cultures.

Techniques for Administration. Three procedures are currently used in the tuberculin test: (1) intradermal injection (Mantoux test), (2) jet injection, and (3) multiple puncture (Tine test). The first of these methods, intradermal injection, serves as the standard procedure for comparison with all other tests. Moreover, more accurate control of dosage is possible with the Mantoux test. The other two methods are utilized for epidemiological (survey and screening) purposes.

Interpretations of Skin Test Reactions. The following interpretations and recording of skin test results are in keeping with the current recommendations of the National Tuberculosis and Respiratory Disease Association.

The intradermal introduction of tuberculin into sensitized persons usually causes the formation of a hard area (Color Photograph 82) that may or may not be associated with a surrounding reddened area (erythema). The standard dose used is 5 tuberculin units. The intensity and size of the reaction vary according to the individual's sensitivity and the quantity of skin-testing material introduced. Those persons who exhibit tuberculin sensitivity are called *reactors.* The degree of sensitivity can be determined from the size and accompanying features of the skin reaction after 48 to 72 hours. The range of responses are: in duration of 1 to 5 mm, negative; 5 to 9 mm, doubtful; and 10 to 33 mm, previous or possible current infection. Reports from many parts of the world indicate a relationship between the size of the skin reaction and the risk of developing active tuberculosis: the larger the reaction, the greater the possibility of active disease. Skin reactivity to the introduction of skin-testing material may be suppressed by several factors or conditions. These include advanced age, terminal or severe acute dis-

Microbiology Highlights

DISTINGUISHING NONTUBERCULOSIS MYCOBACTERIA (NTM)

A number of mycobacterial species other than *Mycobacterium tuberculosis* cause pulmonary and other diseases in adults and children. Distinguishing mycobacterial from nontubercular mycobacterial diseases within the general population is a major clinical problem. Newer skin-testing materials that are under development may be the solution. In addition to the use of the international standard of purified protein derivative (PPD-S) with *M. tuberculosis,* other preparations are under consideration as skin-test antigens for the diagnosis of NTM infections. The preparations and their sources are PPD-B [*M. avium* (M. a-VEE-um) complex], PPD-G [*M. scrofulaceum* (M. scrowe-ful-ACE-ee-um)], PPD-platy [*M. marinum* (M. marr-I-num)], and PPD-Y [*M. kansasii* (M. kan-SASS-eye)]. Once these materials are perfected, they will also be helpful in distinguishing *M. tuberculosis* infections from nontuberculosis mycobacterial (NTM) illnesses and in separating NTM infections from one another.

Tuberculosis

LABORATORY DIAGNOSIS

A presumptive identification of *M. tuberculosis* can be made by finding acid-fast rods (Color Photograph 000) in sputum specimens or most other types of clinical material. Fluorescent microscopy and the use of an auramine-rhodamine dye combination is an alternative to the acid-fast procedure in several laboratories. [See Chapter 3 for fluorescent microscopy.]

Accurate identification of mycobacteria requires cultural methods and biochemical tests. Sputum specimens frequently require additional treatment before use because of their thick, slimy consistency and the presence of contaminants. Once prepared, specimens can be plated onto the selective Löwenstein-Jensen medium or 7H11 medium. Characteristic colonies generally appear four to eight weeks later. *M. tuberculosis* does not form pigmented colonies, while other pathogenic mycobacteria such as *M. kansasii* and *M. scrofulaceum* do. Biochemically *M. tuberculosis* produces niacin, reduces nitrate, degrades polysorbate 80 (Tween 80) in 10 to 20 days, and loses its ability to produce catalase if heated to 68°C for 20 minutes. Additional tests are needed to differentiate *M. tuberculosis* and other pathogenic members of the genus from nonpathogenic species.

TREATMENT

Several drugs are of value in tuberculosis treatment. These include isoniazid (INH), streptomycin, pyrazinamide, para-aminosalicylic acid (PAS), ethambutol, and rifampin. Certain chemotherapeutic agents are used in combination to prevent the development of drug-resistant forms. PAS and streptomycin, or INH, ethambutol, and rifampin are frequently used in such cases. [See Chapter 12 for the actions of these chemotherapeutic agents.]

INH is also used in preventative therapy. It is recommended for individuals exhibiting tuberculin skin test reactions greater than 5 mm in diameter.

eases (such as cancer), immune defects (AIDS and related conditions), and the administration of large quantities of cortisone, infectious diseases such as measles, and vaccination against this disease or a rapidly progressive case of tuberculosis may also cause suppression of the skin reaction.

The tuberculin test is practical not only because it can serve as a diagnostic tool but also because it can screen large groups of people for tuberculosis and detect infections and their sources. After initial infection with tubercle bacilli, the sensitivity to tuberculin develops in approximately two to ten weeks. Once this sensitivity is acquired, it usually persists.

PREVENTION. Preventive measures used for tuberculosis include case-finding and mass survey programs conducted among populations of apparently healthy individuals. Tuberculin testing and x-ray examinations are used for these purposes. Location of active cases can thereby be adequately determined.

The prophylactic use of chemotherapeutic agents (chemoprophylaxis) has been quite successful in reducing the incidence of active tuberculosis. This approach has been used with high-risk groups, such as recent contacts, infants, and tuberculin converters (individuals whose skin tests turn positive). BCG vaccination has also been used to protect uninfected persons in contact with known tuberculosis patients when normal measures of chemoprophylaxis cannot be performed. Because BCG vaccination influences the diagnostic value of tuberculin testing, its use is being questioned.

The prevalence of tuberculosis in many Western countries has been reduced. Unfortunately, however, despite the availability of treatment and early detection measures for tuberculosis, the infection is far from eradicated, especially in the tropics and in the Far East. This continued incidence is the result of several factors, including the nature of the disease itself (healthy-appearing persons may actually be infected and thereby serve as reservoirs for *M. tuberculosis*), limitations of detection and diagnosis, a general lack of interest in a community, an unwillingness of infected individuals to be isolated and adequately treated, and the development of drug-resistant infections. The antibiotic rifampin used in combination with other antituberculous agents has been successful in the treatment of drug-resistant infections.

WHOOPING COUGH (PERTUSSIS)

First described in 1578 by Baillou, pertussis is today found worldwide but appears to be more prevalent in colder regions. Children under five years of age, especially newborns, are the primary victims, with girls being affected more frequently than boys. However, aged and debilitated individuals are also quite susceptible.

TRANSMISSION. Whooping cough is characterized by spasmodic coughing attacks and the whooping that accompanies rapid air intake during the coughs. With this prominent feature of the disease, transmission is no doubt mainly by droplet infection. However, fomites should not be ruled out. Isolation of the causative agent is obtained by holding a Petri dish containing glycerin-potato-blood (Bordet-Gengou) agar in front of an ill child's mouth during a coughing attack.

Bordetella pertussis (bor-de-TEL-lah per-TUSS-is), the etiologic agent (formerly known as *Haemophilus pertussis*), was isolated by Jules Bordet and Octave Gengou in 1906. It is a small, nonmotile, gram-negative rod. Capsules are formed by virulent organisms.

Complications that may occur with whooping cough include collapsed lung, bronchopneumonia, interstitial emphysema, convulsions, and bleeding in various regions of the body (brain, conjunctivae, eyes, or skin).

PREVENTION. Active immunization is the most widely used practice for the control of pertussis. As noted earlier, control is accomplished by the use of the combined DPT (diphtheria, pertussis, and tetanus) vaccine.

Susceptible contacts are isolated and administered human hyperimmune serum. Exposed individuals who have been vaccinated one year or more previously should be given a booster. **[See immunizations in Chapter 19.]**

Fungus Diseases

Several fungus species are associated with respiratory diseases in humans (Table 25–6). Humans acquire these infections by inhaling spores from such reservoirs as dust, bird droppings, and soil. Certain fungi, *Coccidioides immitis,* for example, are not widely distributed in nature but appear to be found only in particular geographic regions. Other agents are more widely distributed (Table 25–7). Among these are the species of *Aspergillus.* These fungi are secondary invaders that cause disease in patients with immune defects or those whose immune response has been weakened by treatment with antibiotics or immunosuppressive agents. Allergic reactions to aspergillus spores also are common. Some of these pathogens are discussed in this chapter; others are presented elsewhere in the text. Diseases and treatments are presented in Tables 25–6 and 25–7. **[See fungi in Chapter 9.]**

Whooping Cough

LABORATORY DIAGNOSIS

Whooping cough is generally diagnosed on the basis of clinical features of the disease (Table 25–5). These, together with a finding of gram-negative coccobacilli in sputum, are sufficient for diagnosis.

The isolation of *B. pertussis* from nasopharyngeal specimens is done on the special medium known as the Bordet-Gengou plate. Antibiotics such as penicillin or methicillin can be added to inhibit contaminants. Small, smooth, pearllike colonies generally appear after 72 hours of incubation. Fluorescent antibody and other serological procedures are also used on isolates for identification.

TREATMENT

Erythromycin, tetracycline, and chloramphenicol are the antibiotics generally used for *B. pertussis* infections. To be effective, treatment must be started before the disease is well established. Penicillin also is used to reduce the possibility of secondary infections.

TABLE 25–6 Fungi Associated with Human Respiratory Diseases

Respiratory Disease	*Fungus Species*	*Possible Treatment*
Aspergillosis, asthma, bronchiectasis, rhinitis[a]	*Aspergillus fumigatus*	Amphotericin B, ketoconazole
Bronchopulmonary candidiasis	*Candida albicans*	Amphotericin B, flucytosine
Chronic pneumonitis	*Cryptococcus neoformans*	Amphotericin B, flucytosine
Coccidioidomycosis	*Coccidioides immitis*	For disseminated form, amphotericin B
Geotrichosis	*Geotrichum candidum*	Potassium iodide (orally), sodium iodide (intravenously)
Histoplasmosis	*Histoplasma capsulatum*	1. Amphotericin B 2. Surgical removal of oral lesions
North American blastomycosis (Gilchrist's disease)	*Blastomyces dermatitidis*	1. Amphotericin B 2. Hydroxystilbamidine 3. Surgical procedures together with chemotherapy
Sporotrichosis of the lungs (rare)	*Sporotrichum schenckii*	Amphotericin B

[a] Most common species causing human infections. Others include *A. flavus, A. nidulans,* and *A. niger.*

TABLE 25–7 Features of Selected Fungus-Caused Lower Respiratory Tract Infections

Disease	*Geographical Distribution (General)*	*Incubation Period*	*General Features of the Disease*	*Laboratory Diagnosis*[a]
Coccidioidomycosis	Southwestern United States (southern California), Central and South Americas	10–21 days	General flulike symptoms including chills, cough, fever, malaise, chest pain, and a pus-containing sputum in the case of pneumonia[b]	1. Demonstration of nonbudding spherules in clinical specimens 2. Culture and demonstration of arthrospores (Color Photograph 106c) 3. Serological testing, e.g., complement fixation, immunodiffusion (see Chapter 19) 4. Skin test
Histoplasmosis	Widespread in the United States, endemic in Missouri, Tennessee, Kentucky, Kansas, Iowa, Indiana, and Southern Illinois, Panama, Argentina, Brazil, Philippines, and Java	5–18 days	Either no detectable illness or mild effects; small calcified growths appear in several body organs upon recovery	1. Demonstration of typical yeast forms in specimens (Figure 25–10a) 2. Isolation and cultivation 3. Serological test, e.g., complement fixation, immunodiffusion, latex agglutination 4. Skin test
North American blastomycosis (Gilchrist's disease)	North America, more prevalent in South Central and mid-Atlantic United States and Ohio-Mississippi River valley	Several weeks	Symptoms are usually quite mild and self-healing and include cough, fever, and general discomfort; complications and spread to other body regions can occur	1. Microscopic demonstration of multinucleated, nonencapsulated yeast cells in specimens 2. Isolation and cultivation

[a] Tentative diagnosis is possible based on disease (clinical) symptoms.
[b] The spread of the disease throughout the body (disseminated form), usually starts from the respiratory system.

DIAGNOSIS

The laboratory identification of fungi-causing respiratory infections includes demonstration of the causative agent in various types of specimens and subsequent culture of the fungus. Skin tests and serological procedures as described in Chapter 19 are of value in some cases. Table 25–7 indicates approaches to diagnosis for several pathogenic fungi.

COCCIDIOIDOMYCOSIS

The fungus disease coccidioidomycosis (Figure 25–7) is also known as *desert fever, San Joaquin fever,* and *valley fever.* Three distinct forms are known with this infection: *acute, chronic pulmonary,* and *disseminated* (Color Photograph 106). The incidence of infection is related to climatologic conditions in the endemic areas (Table 25–7), the peak of infection occurring in the dry, dusty summer months, especially after a rather rainy winter season. At least 100,000 new infections are estimated to occur yearly in the United States. Sex and race often play a role in the frequency of severe coccidioidomycosis. Pregnant women, Filipinos, and black people show a greater incidence of this condition. Spreading of the disease agent also can occur via the blood in some patients. This type of situation appears to be intensely stimulated by certain sex hormones. In addition, allergic reactions to the causative agent, *C. immitis,* occur more frequently in adult white women than in any other group.

Transmission. Coccidioidomycosis is most often contracted by the inhalation of soil or dust containing the characteristic arthroconidia (arthrospores) of *C. immitis* (Color Photograph 106c). Domestic animals as well as rodents can develop the disease and may be partially responsible for its persistence in some areas.

Life Cycle. The cycle of this pathogen begins with the inhalation of arthroconidia (Figure 25–8). These spores bypass the host's defenses of the upper respiratory tract and become situated in the air sacs (alveoli). There the arthroconidia change to the round, endospore-containing spherules, and stimulate attacks by neutrophils and machophages. The spherules continue to grow, multiplying and eventually rupturing, whereupon they release large numbers of endospores (Figure 25–9).

Most infections are without symptoms or produce mild pulmonary symptoms that disappear within a few weeks provided a host cell-mediated response against the fungus occurs. Without this response, chronic pulmonary disease or the spreading or disseminated form of coccidioidomycosis develops. The disseminated disease state is associated with an overwhelming exposure to fungal endospores. All major organs and tissues may be affected.

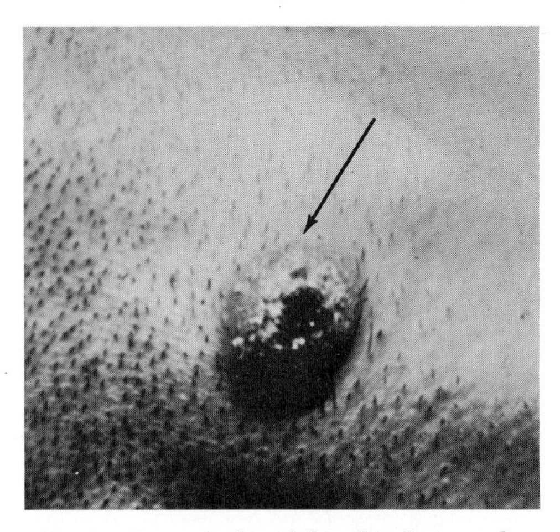

Figure 25–7 The typical nodular skin lesion of coccidioidomycosis. *[Courtesy of Dr. N. H. Rickles, Pathology Department, University of Oregon Dental School.]*

Immunity. Recovery from primary coccidioidomycosis confers a solid, permanent immunity to further infection.

HISTOPLASMOSIS

The fungus disease histoplasmosis is caused by *Histoplasma capsulatum.* (hiss-toe-PLAZ-mah cap-sul-

Figure 25–8 The life cycle of *Coccicidioides immitis. [After S. H. Sun, and M. Huppert,* Sabouraudia ***14****:185–198, 1976.]*

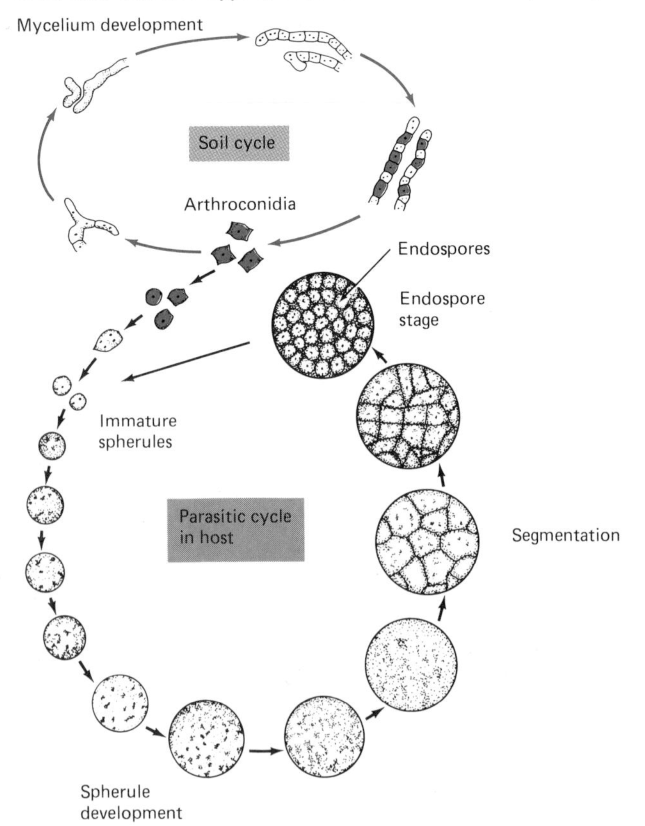

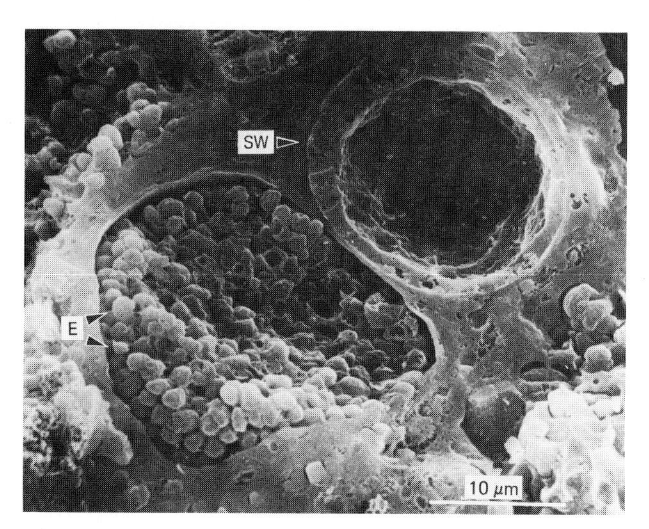

Figure 25–9 A scanning micrograph of *Coccidioides immitis* spores (arrows) in infected lung tissue. E, endospores, SW, spherule wall. *[From J. Drutz, and M. Huppert,* J. Infect. Dis. ***147****:372–390, 1983.]*

LAY-tum) (Figure 25–10). Since a sexual stage of the fungus has been discovered, it is also known as *Emmonsiella capsulatum* (eh-mon-see-EL-lah cap-sul-LAY-tum). The disease occurs throughout the world, with a relatively high incidence in certain countries (Table 25–7). Histoplasmosis is no longer thought of as a rare and fatal disease but rather as one that is widespread and generally mild. The organism appears localized within macrophages and reticuloendothelial cells (Figure 25–10a). The disease may affect one or several organs. It may be difficult to diagnose, since it presents a broad spectrum of effects ranging from an asymptomatic or mild infection to an acute, severe disease to a chronic pulmonary disease of long duration.

TRANSMISSION. Histoplasmosis is disseminated to human beings by spores (Figure 25–11b) from the fungus, which grows readily in acidic soil in areas with proper temperature and moisture. Spores are also found in bird droppings.

NORTH AMERICAN BLASTOMYCOSIS (GILCHRIST'S DISEASE)

A chronic systemic fungal disease, North American blastomycosis occurs only on the North American continent and is usually secondary to pulmonary involvement. Sporadic cases have been reported in all states. Males aged 20 to 40 are most often affected.

The causative agent, *Blastomyces dermatitidis* (blast-oh-MY-sees der-mah-tit-EE-diss), appears as yeast cells in infected tissues or in cultures at 37°C and as a mold in cultures at room temperatures. At least three clinical states of the disease are known: pulmonary and disseminated blastomycosis and a skin infection.

Influenza Virus Infection

Epidemics of influenza have plagued humans and other animals for centuries. Widespread epidemics, or pandemics, occurred in 1847–1848, 1889–1891, and 1918–1919. The last of these caused approximately 20 million deaths. Young adults between 20 and 40 were affected to a very large extent. Although a significant accumulation of data incriminated a virus as the causative agent of the pandemic of 1918–1919, *H. influenzae,* the so-called influenza bacillus, was still regarded as the cause of the disease. Subsequent research firmly established the viral nature of the disease. Moreover, it became clear that *H. influenzae*—as well as numerous other bacterial species, including *S. pneumoniae* and *S. aureus*—were secondary invaders in patients with influenza and quite often were responsible for the fatal forms of pneumonia that developed.

The original strains of human influenza viruses, isolated in 1933 and prevalent for about ten years, are referred to as *type A* or *A classic.* As additional strains were discovered on the basis of antigenicity studies, other designations were created (Table 25–8). Influenza viruses contain ribonucleic acid (RNA) and have a common virion structure (Figure 10–5) and mode of replication. Individual virus particles consist of a core with a matrix (M) protein

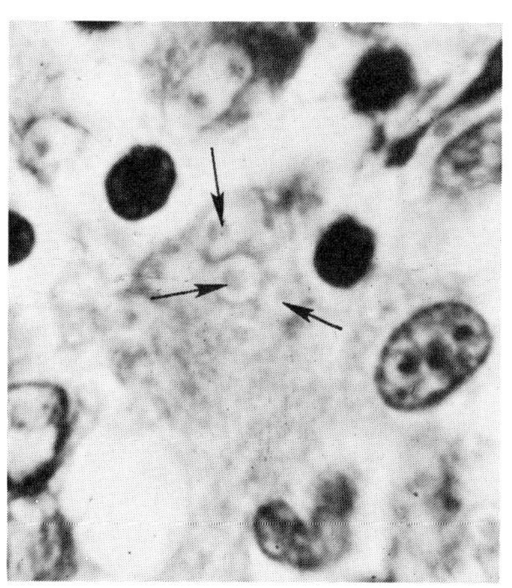
(a)

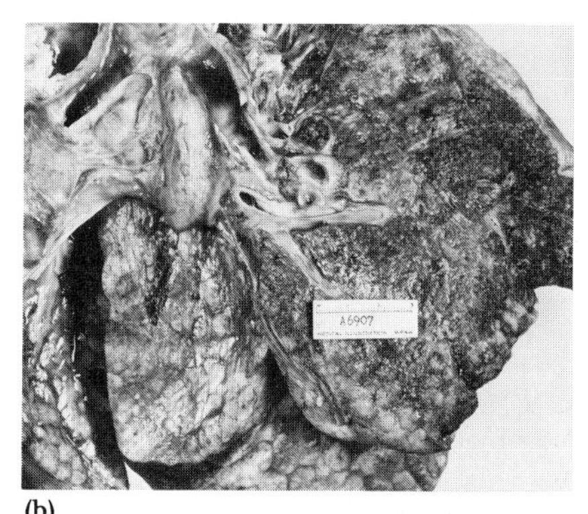
(b)

Figure 25–10 (a) *Histoplasma capsulatum* in the parasitic stage within tissue. Note the clear areas (arrows) that result from shrinkage of the cytoplasm away from the rigid walls. (b) The gross effect of *H. capsulatum* on the lung. *[Courtesy of Armed Forces Institute of Pathology, Washington, D.C., Neg. No. 58-1455-854965.]*

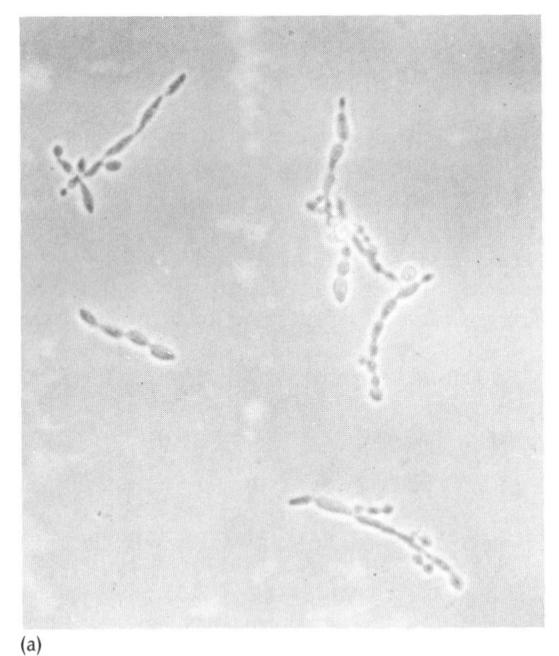
(a)

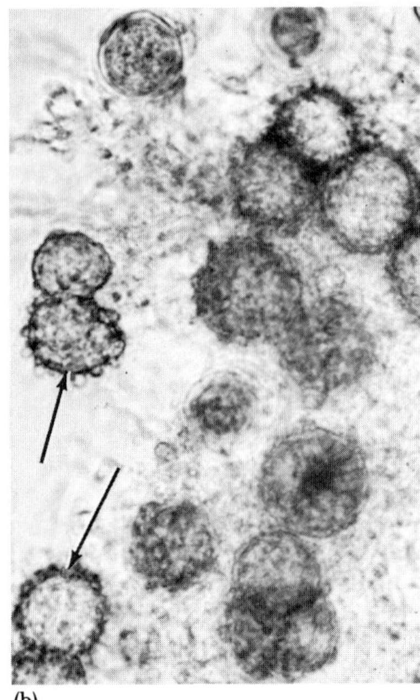
(b)

Figure 25–11
(a) Yeast phase of *Histoplasma capsulatum* strain A811 under dark phase. (b) Mycelial phase of *H. capsulatum* showing individual tuberculate macroconidia (arrows). *[L. Pine,* App. Microbiol. ***19****:413–420, 1970.]*

layer and a ribonucleoprotein complex, surrounded by a lipid bilayer membrane and a surface covering of glycoprotein molecules. Differences in the internal protein contents and the lengths of the RNA molecules serve as the basis for the classification of influenza viruses into types A, B, and C.

The capacity of influenza viruses in nature to vary rapidly and to undergo changes in antigenic structure have been the major limitations to controlling the disease. Such changes defeat the protective effects of an individual's immune response. In addition, the variability of influenza viruses is a major distinguishing feature when compared to the essentially unchanged antigenic structure of other viral agents of disease, such as measles and polio.

COMPLICATIONS. Unfortunately, influenza epidemics have been accompanied by a rise in mortality. The greatest proportional rise in deaths from complications has been caused by bronchitis and pneumonia. Complications have also been reported in organs other than the lungs, such as the components of the circulatory and central nervous systems. The development of chest complications during an influenza epidemic is determined by several factors, including the viral strain involved, the age and general health of the host, and the severity of the original attack of the disease.

Other complications found with influenza include Guillain-Barré syndrome (GBS) and Reye's syndrome (RS). GBS is a rapidly developing inflammation of several nerves resulting in a spreading muscular weakness of the extremities and possible paralysis. While in the majority of cases GBS follows recovery from viruslike illnesses associated with the respiratory and gastrointestinal tracts, it has also been linked to immunizations. In 1976, Guillain-Barré syndrome was directly related to the injections of a specific inactivated influenza vaccine. The subsequently prepared (1978–1979) influenza vaccines

TABLE 25–8 Features of Human Influenza—A Viral Disease of the Lower Respiratory Tract

Influenza Virus Type or Subtype	*General Features of the Disease*	*Laboratory Diagnosis*
A (classic) A_1 (A prime) A_2 (Asian)[a] B C	Uncomplicated influenza symptoms include backache, chills, fever, headache, general discomfort, nasal congestion, cough, dry and sore throat, loss of appetite, nausea, and vomiting; complications include pneumonia	1. Isolation of causative agent using chick embryo 2. Detection of viral antigens by immunofluorescence 3. Serological test, e.g., hemagglutination inhibition, complement fixation (see Chapter 22)

[a] Often incorrectly called *Asiatic.*

were not associated with an increased risk of the syndrome. Good recovery occurs in most individuals. However, it may not be complete for four to six months.

Reye's syndrome is another long-recognized complication of viral and bacterial infections and immunizations. Epidemiological evidence clearly shows a direct relationship between the condition and the viruses of chickenpox and influenza B. Severe vomiting is typical of the onset of RS. Individuals with RS experience a rapid accumulation of fluid in the brain (cerebral edema) without inflammation. While various organs, such as the liver, are affected during the course of the condition, death results from increased intracranial pressure. Several contributing factors have been suggested, including inborn metabolic error and certain medications used to relieve symptoms. Among medications, salicylate-containing preparations such as aspirin have been suspected and might prove to be a contributor.

IMMUNITY. Persons recovering from an infection usually acquire resistance to the particular antigenic strain responsible for the respiratory illness. Unfortunately, new strains with new antigens develop and consequently cause successive attacks of influenza.

PREVENTION. Immunization of certain key individuals, such as police officers, nurses, physicians, and other health care personnel, is recommended before an epidemic strikes. Pregnant women, the elderly, and persons with debilitating diseases, such as chronic heart or respiratory diseases, should also be immunized. Mortality rates in persons 45 years of age and over have been higher than in other age groups. As influenza can spread rapidly among residents of homes for the aged or nursing homes, the prophylactic use of vaccines is important there.

Pneumocystis carinii, a protozoan

The tissue parasite *Pneumocystis carinii* (new-moh-SIS-tis kar-I-nee-eye) (Figure 25–12a) has received considerable attention in recent years as the cause of a respiratory infection involving the pulmonary alveolar spaces and surrounding supporting structures (Figure 25–13). *P. carinii* infections have occurred almost exclusively in children with congenital immunologic disorders, in patients with certain generalized circulatory disorders, in debilitated infants, in newborn and premature infants, and in patients receiving immunosuppressive therapy and with acquired immune deficiency syndrome (AIDS). The causative agent, originally discovered in the guinea pig several years ago, has since been found in other animals, including cats, dogs, mice, monkeys, and sheep. Various reports, however, clearly indicate that *P. carinii* has a worldwide distribution. There

Figure 25–12 *Pneumocystis carinii* life cycle. There are three developmental stages: cyst, trophozoite, and sporozoite. (b) The penetration of respiratory tissue by the protozoon (arrow). *[From Y. Yoshida,* Zbl. Bkt. Hyg. ***256****:390–399, 1984.]*

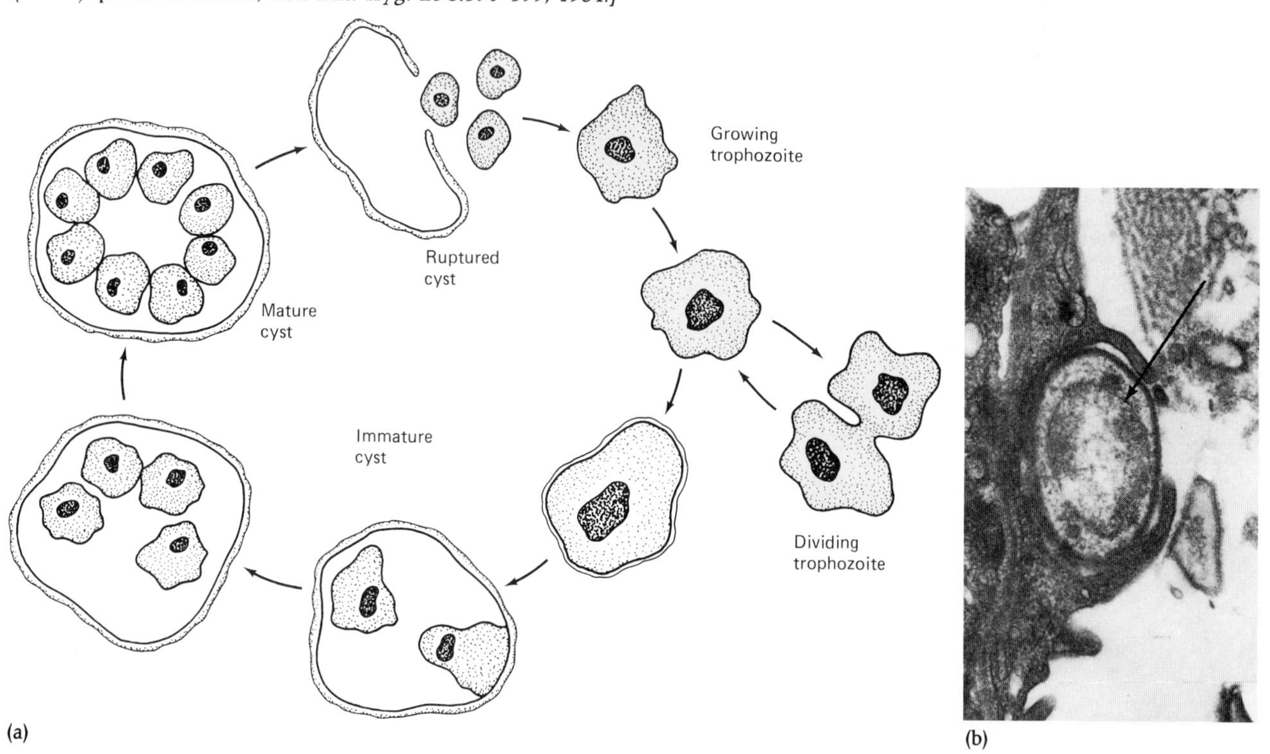

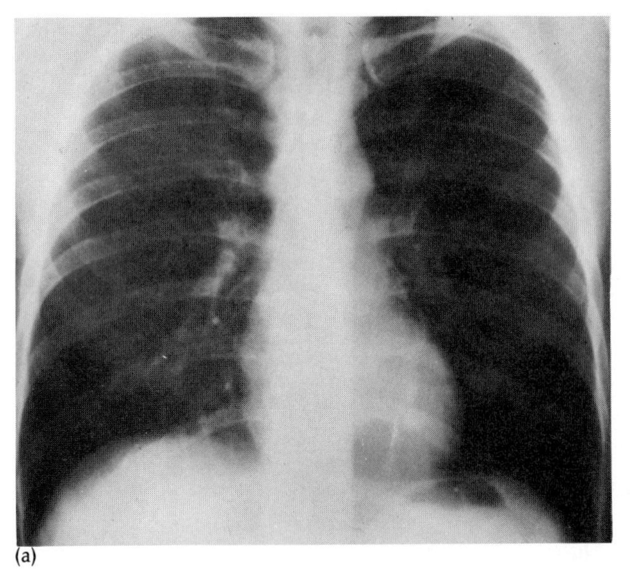

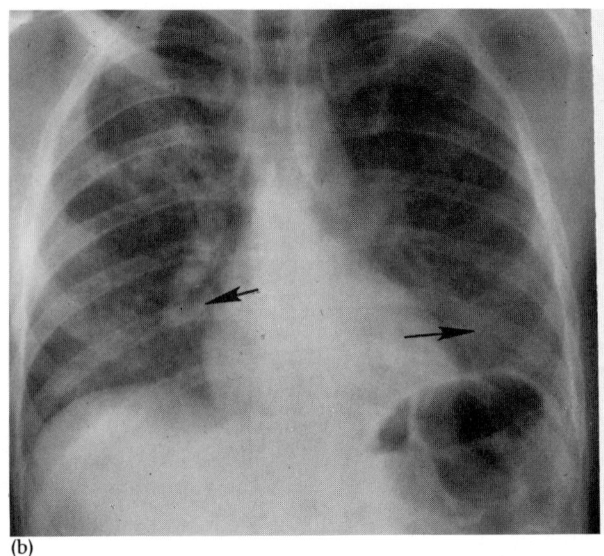

Figure 25–13 (a) Normal chest x-ray of a patient in the early stages of infection. (b) An x-ray film taken 4 days after the initial normal view, showing a rapid progression of pneumonia (arrows). *[D. W. Martin, Jr., D. G. Warnock, and L. H. Smith, Jr.,* West. J. Med. ***13**:400, 1982.]*

are several similarities between human and animal *P. carinii* pneumonitis with respect to the morphology and pathogenicity of the protozoon (Table 25–9) and histology of the diseased tissue. The organism infects the intra-alveolar space of the lung as an opportunist and may penetrate tissue.

There are three developmental stages of *P. carinii:* cyst, trophozoite, and sporozoite forms. Cysts are round, measure about 4 to 10 μm in diameter, and contain one to eight sporozoites (Color Photograph 107). The trophozoites vary in shape and range in size from 2 to 5 μm in diameter. A small amount of nuclear material in these forms, usually V-shaped, is generally located offcenter in the organism. The three stages of *P. carinii* can be demonstrated through the use of Giemsa, Wright, or polychrome methylene blue stains. However, one must be able to recognize *P. carinii* in each stage of its life cycle in order to use these stains. **[See protozoan structures in Chapter 9.]**

TRANSMISSION. Although the mode of transmission for *P. carinii* is not definitely established, inhalation of the parasite into the respiratory tract appears to be the most probable means.

TABLE 25–9 Features of *Pneumocystis carinii* Infection

Disease	*Incubation Period*	*General Features of the Disease*	*Laboratory Diagnosis*
Pneumonia	Unknown, possibly 7 days	Course of the disease ranges from 4 to 6 weeks; symptoms include cough, bluish coloration of the skin, fever, rapid breathing, and lung consolidation	1. X-ray 2. Demonstration of *P. carinii* cysts in specimens (Color Photograph 107); special stains

Disease Challenge

The situation described has been taken from an actual case history. It has been designed to show how clinical, laboratory, microbial, and related information is used in disease diagnosis. A review of treatment and epidemiological aspects is given at the end of the presentation. Answers to questions, laboratory findings, and interpretations are given immediately following a specific question. Test your skill and take the Disease Challenge.

Key Terms

AIDS: acquired immune deficiency syndrome, an immune system infection caused by human immunodeficiency virus (HIV).
ELISA test: enzyme-linked immunosorbent assay, a highly specific immunologic diagnostic test.
mediastinum (mee-di-as-TEE-num): a wall separating the two major portions of an organ.
opacification (oh-PAS-i-fe-kay-shun): a nontransparent area or structure.

Case

A three-year-old Hispanic female was admitted to a local medical center after a two-week history of fever, night sweats, cough, and dramatic weight loss. On physical examination, the patient appeared quite ill. She also exhibited breathing difficulties. A chest x-ray showed opacification of the lower left lung and most of the upper left lobe, moderate atelectasis, pneumonia of the right lung, and a shift of the heart mediastinum to the right.

At This Point What Disease(s) and/or Agents Would You Suspect Being Involved?
Several diseases can be suspected here, including tuberculosis, candidiasis, mycoplasma pneumonia, nontuberculous mycobacterial infection, and *Pneumocystis* infection.

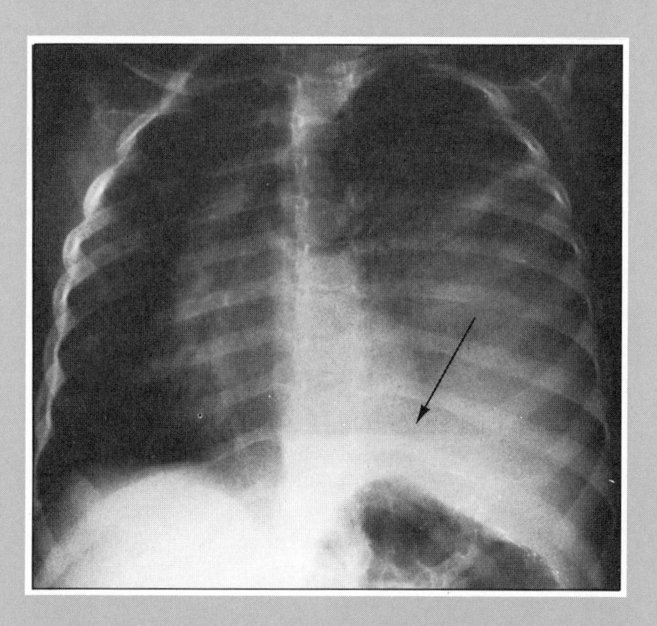

Admission chest x-ray showing opacification (arrow). *[From P. J. Krause,* et al., Pediatr. Infect. Dis. *5:269, 1986.]*

What Specimens Should be Taken and for What Purpose(s)?
Blood for bacterial, fungal, and viral cultures. Sputum for bacterial and fungal cultures, and Gram and acid-fast staining.

(continues)

Disease Challenge (continued)

Laboratory Results:
All cultures and the ELISA were negative after three days. [See Chapter 19 for ELISA procedures.]

Should Treatment Be Given At This Time?
Yes. Antibiotic administration is still appropriate. Isoniazid (INH), rifampin, and streptomycin were given. Unfortunately, no improvement was seen.

At This Point What Type(s) of Specimen(s) Should Be Taken?
A biopsy from the affected lung would be appropriate. Bacterial and fungal cultures and Gram and acid-fast staining should be performed.

Laboratory Results:
Numerous acid-fast rods were found on microscopic examination. Cultures were positive after one month only for *Mycobacterium intracellulare-avium* (*M. avium* complex).

What Should The Treatment Be Now?
Antibiotics effective against *M. intracellulare-avium* (*M. avium* complex).

Treatment Follow-up:
Two months after admission, the young patient became afebrile. Some clearing of the chest and movement of the heart and mediastinal structures occurred. Repeat cultures from appropriate specimens were negative. The patient was discharged after three months with appropriate antibiotic therapy.

Summary

1. The human respiratory tract starts at the nose, passes through various parts of the respiratory tree, and ends in the air sacs, or alveoli.
2. The system is not only susceptible to microbes that can cause infections at different levels but can also aid in the transmission of pathogens.

Normal Flora of the Respiratory Tract

1. The normal flora of the human nose includes a variety of bacterial species, some of which have the potential to cause disease.
2. In the adult human, the respiratory tract below the level of the epiglottis is normally sterile.

Introduction to Microbial Infections of the Respiratory Tract

1. Various microbial pathogens—bacteria, fungi, protozoa and viruses—can enter and multiply in the respiratory tract. Moreover, the various secretions associated with the respiratory system can be a source of disease agents.
2. The communicability of infectious diseases of the respiratory system is influenced by several factors such as survival of the pathogen on fomites or in the air, the number of microbes inhaled, the duration of contact, and the anatomical site involved in the infection.
3. Control measures for respiratory tract infections are similar, if not identical, to those of other types of diseases.

Representative Microbial Disease

1. Respiratory tract infections are of great importance because of the ease with which they are transmitted and contracted and the difficulty in eliminating disease agents.
2. Diagnosis is frequently based on the disease picture, which includes x-rays, the patient's history and physical examination, and on the laboratory isolation and

identification of the disease agent. Serological tests are important diagnostic tests.

Upper Respiratory Infections

1. The upper respiratory region, which includes the middle ear, mastoids, sinuses, and nasal corners of the eyes, is exposed to a variety of pathogens.
2. Infections in these sites can spread to other regions of the respiratory tract.
3. Examples of bacterial infections of this region include toxin-associated diphtheria, suppurative otitis media (pus-producing middle ear infection), sinusitis, and streptococcal sore throat. Certain fungi also cause sinusitis.
4. Viral diseases of the upper respiratory tract include the common cold, croup, and minor infections.

Lower Respiratory Infections

1. Several diseases of the lower respiratory tract are life-threatening if not treated quickly and adequately. These diseases are usually acquired through inhalation of the disease agent.
2. Bacterial infections of this portion of the respiratory system include the pneumonias caused by such organisms as *Streptococcus pneumoniae* and causative agents of Legionnaires' disease, mycoplasma infections, psittacosis (parrot fever), Q fever, tuberculosis, other mycobacterial diseases, and whooping cough.
3. *M. intracellulare* is a recognized problem in patients with poorly functioning or defective immune systems.
4. Two diseases are caused by *Legionella pneumophila:* Legionnaires' disease and Pontiac fever. The causative agent appears to be transmitted by aerosol or airborne means.
5. Several other gram-negative rods closely related to but distinguishable from *L. pneumophila* are causative agents of pneumonia. These microorganisms include *L. micdadeii, L. bozemanii, L. dumoffii,* and atypical *Legionella*-like organisms (ALLO). Various laboratory tests are used to distinguish among these microorganisms.
6. Fungus infections of the lower respiratory tract include coccidioidomycosis, histoplasmosis, and North American blastomycosis.
7. Influenza epidemics have caused severe infections in humans and other animals for centuries. These virus infections can be complicated by secondary bacterial invaders.
8. Influenza as well as other infections and certain immunizations may be associated with two specific complications: Guillain-Barré syndrome (GBS) and Reye's syndrome (RS).
9. The protozoon *Pneumocystis carinii* causes infections of the air sacs and supporting structures. The disease agent infects individuals who have an impairment of the immunologic system. This is especially true for AIDS victims.

Questions for Review

1. Compare five different bacterial respiratory infections. Construct a table for this purpose and include the following categories:
   a. specific causative agent
   b. distinguishing clinical features
   c. Gram reaction
   d. region of the tract involved
   e. means of transmission
   f. availability of diagnostic serological tests

2. a. What general measures are used in the prevention and control of respiratory diseases?
   b. Why are certain diseases of this type difficult to eradicate? Explain.

3. What significance does the acid-fast staining procedure have in the diagnosis of respiratory disease? Which, if any, respiratory system pathogens are noted for their acid-fast reaction?

4. List examples of complications that can develop from a bacterial respiratory infection. How can such problems be prevented?

5. What are the reservoirs and sources of infectious agents of the following diseases?
   a. tuberculosis
   b. psittacosis
   c. diphtheria
   d. whooping cough
   e. influenza
   f. atypical primary pneumonia
   g. lobar pneumonia
   h. the common cold
   i. Legionnaire's disease
   j. coccidioidomycosis
   k. sinusitis

6. What distinguishing property or properties does *Mycoplasma pneumoniae* have in comparison with other bacterial pathogens?

7. a. What fungus pathogens cause respiratory disease? Where can these diseases be found?
   b. What factors contribute to a person's susceptibility to fungal diseases?

8. List five viral respiratory tract infections together with their respective causative agents.

9. Discuss the measures used in the diagnosis, treatment, and prevention of viral respiratory system diseases.

10. Can complications develop from influenza? Explain.

11. If you were planning a world trip, including Arabia, southern California, Egypt, Greece, Japan, and India, to which respiratory diseases would you be exposed?

Suggested Readings

AUSTRIAN, R., *Life with the Pneumococcus, Notes from the Bedside, Laboratory, and Library.* Philadelphia: University of Pennsylvania Press, 1985. *A collection of lecture notes, reviews, and research reports covering a period of 25 years and emphasizing the great diversity and disease-producing capacity of the pneumococcus.*

KATZ, S. M. (ed.), *Legionellosis,* vol. I. Boca Raton, Florida: CRC Press, Inc., 1985. *A thorough review of the history of legionellosis, outbreaks, and clinical and pathological features of the disease and therapy.*

KILBOURNE, E. D., *Influenza.* New York; Plenum Publishing Corporaton, 1987. *A comprehensive and detailed review of influenza, covering the history of the disease and its development (pathogenesis), clinical symptoms, and pathology in humans and other animals.*

MCHENRY, M., "The Infectious Pneumonias," *Hospital Practice,* December 1980, pp. 41–52. *Important clinical aspects of microorganism-caused pneumonias are considered.*

PALESE, P., and J. F. YOUNG, "Variation of Influenza A, B, and C Viruses," *Science* **215:**1468–1474, 1982. *The general features of influenza viruses and the diseases they cause are detailed in this short article.*

PAPPENHEIMER, A. M., JR., "Diphtheria Toxin," *Annual Review of Biochemistry* **46:**69, 1977. *A detailed article dealing with the toxin of* Corynebacterium diphtheriae. *Its production, structure, and mode of action are discussed.*

RYDEN, I. C., and C. G. COBBS, "Diagnostic and Therapeutic Approaches in Respiratory Infections," *Medical Times* **109:**80–95, 1981. *A well-written, comprehensive article dealing with the major infections of the ear, nose, throat, and bronchi and their respective causes.*

26

Microbial Diseases of the Gastrointestinal Tract

As we have experienced during the last epidemic, the microscopic examination [of stools and vomit] is far more important than we were led to believe previously.
— Robert Koch

After reading this chapter, you should be able to:

1. Indicate the gastrointestinal areas attacked by microbial pathogens and/or their products.
2. Discuss the general features of the microbial ecology of this system, and the function of the microbiota.
3. Distinguish among the different forms of botulism.
4. Discuss at least three bacterial, three protozoan, and three viral diseases of the gastrointestinal system in terms of the causative agent, means of transmission, general symptoms, and control.
5. List and explain specific measures used to prevent or control bacterial food poisoning.
6. Distinguish among the various microorganisms associated with hepatitis and AIDS.
7. Explain the significance of bacterial spores and protozoan cysts to disease states of the gastrointestinal system.
8. Discuss the general approaches to the treatment of diseases associated with the gastrointestinal system.
9. Distinguish among the causative agents and features of the different types of viral hepatitis.

Food and water are important vehicles for the transmission of microbial disease agents that affect the gastrointestinal tract. Several infections and poisonings involving this human system are discussed in this chapter. Attention is also given to the identification, prevention, and control of disease states such as amebic dysentery, Asiatic cholera, typhoid fever, various forms of food poisoning, and several well-known as well as emerging virus infections.

Microbial diseases of the gastrointestinal tract usually result from the ingestion of food or water containing pathogenic microorganisms or their toxins. Despite the availability of functional control measures, diseases such as amebic dysentery, cholera, and salmonellosis still occur frequently in some parts of the world. The incidence of these and related diseases is greatly affected by various factors, including poverty, crowding, malnutrition, and natural reservoirs for pathogens.

The gastrointestinal tract (Figure 26–1), also called the *digestive tract,* is a canal consisting of the mouth, oropharynx, esophagus, stomach, small and large intestines, and the rectum and anus. It runs through the body from the mouth to the anus and is approximately 9.5 m long. The wall of the digestive tract contains absorptive surfaces, glands, and muscles. Variations in diameter and structures occur along the length of the tract. Certain glands and structures, including the liver and pancreas, which are outside the digestive tract, contribute secretions important to the digestive process. The function of the digestive system is to supply food and water to the body's internal environment. Once inside the body, basic substances such as amino acids, simple sugars, and fatty acids are carried to the cells via the circulatory system. The various portions of the digestive system that are vulnerable to direct attack by microbial pathogens or may be secondarily involved in a disease process are indicated in Figure 26–1. **[Chapter 31 describes the worm (helminth) infections of this system.]**

Gastrointestinal Microbial Ecology

The structure of the gastrointestinal tract determines the localization of the microbiota (microbial flora) and, to some extent, its composition as well. Microbial habitats may exist in any area from the esophagus to the anus. Each habitat provides a different kind of environmental or nutritional challenge.

The gastrointestinal tract is sterile in the normal fetus up to the time of birth. During normal birth, the baby picks up microorganisms from portions of the mother's reproductive tract and from any other environmental source to which it is exposed. Many

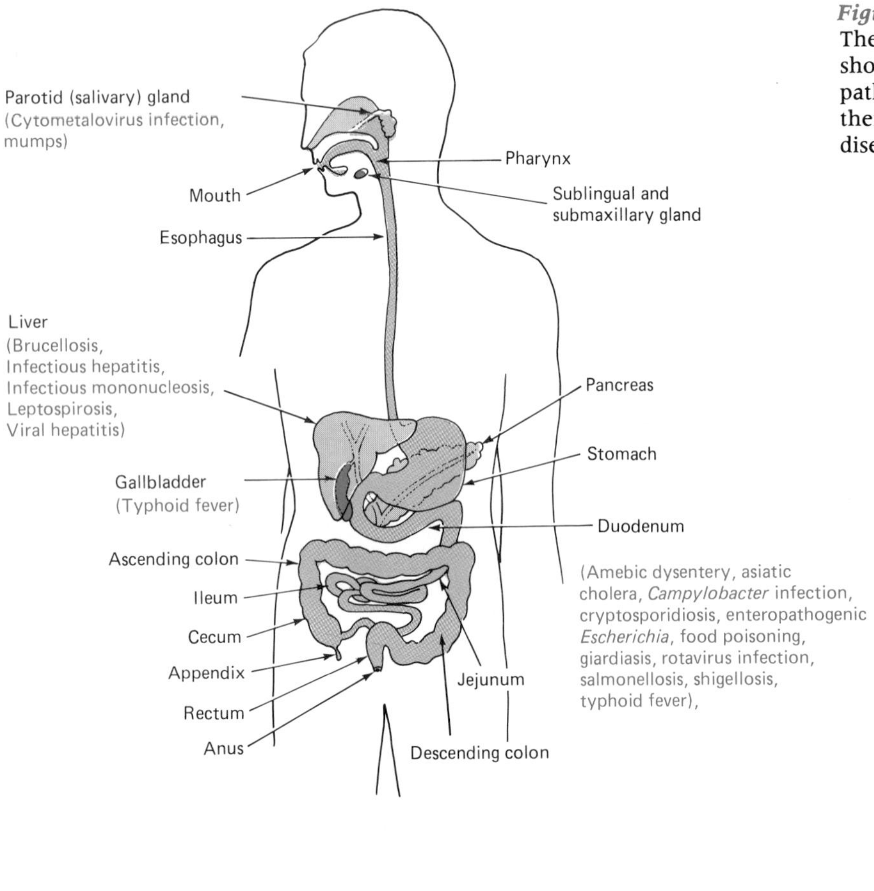

Figure 26–1
The human gastrointestinal tract showing specific locations at which pathogenic microorganisms and their infections and associated disease states can be found.

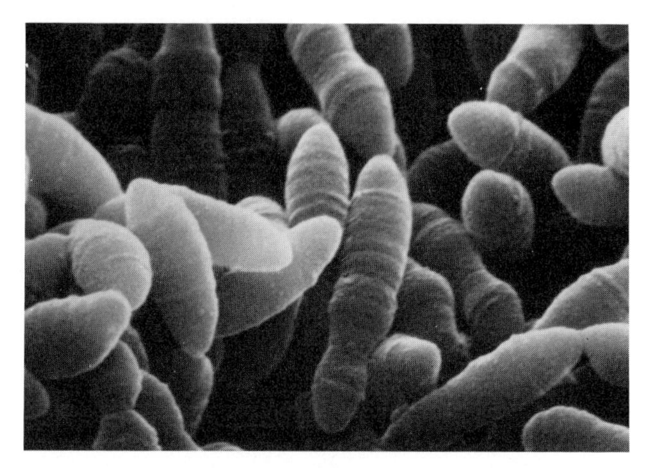

Figure 26–2 The bifidobacteria. These anaerobic organisms are important members of the natural microflora of the human gastrointestinal system. *[H. Bauer, and H. Sigarlakie,* Can. J. Microbiol. ***21:*** *1305–1316, 1975.]*

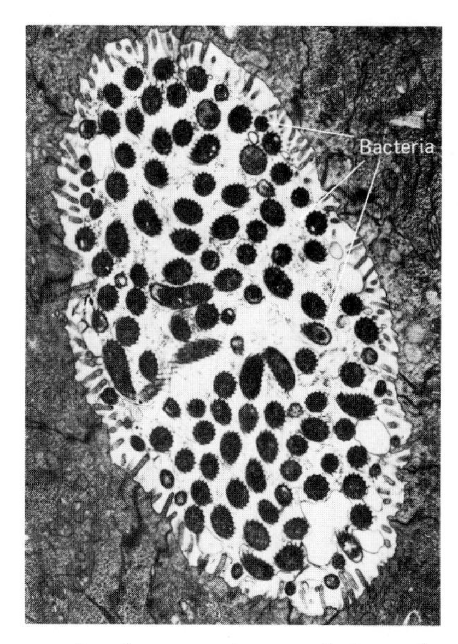

Figure 26–3 An electron micrograph through an intestinal cavity showing its colonization with bacteria. Original magnification 5000×. *[Courtesy A. Lee, University of New South Wales.]*

of these microbes are not able to establish themselves in the neonatal tract and disappear soon after birth. Other microbial types are pioneers, producing the offspring that eventually form established communities in the adult.

Soon after birth in most suckling infants, the microorganisms commonly found are primarily lactic acid bacteria. In breast-fed babies, species of *Bifidobacterium* (by-FEH-doh-back-TEH-ree-um) (Figure 26–2) predominate. With further development, a variety of complex interactions involving the individual, environment, diet, and even microorganisms themselves regulate the events that establish the microbiota of the gastrointestinal tract. While the stomach and upper small intestine do not have a permanent flora (microbiota), the lower small intestine contains an established flora of microorganisms that also are found in the large intestine (colon). The largest community of organisms of any body area is the colon. (Figure 26–3). Their number, representing more than 400 species, exceeds 10^{11} per gram of colon content. Most are primarily anaerobes. Table 26–1 lists the microorganisms isolated from various portions of this system. Interestingly enough, *Escherichia coli* is a minor rather than the chief inhabitant of most gastrointestinal ecosystems. A wide variety of microorganisms, such as strict anaerobes, normally outnumber *E. coli.*

TABLE 26–1 Microorganisms Isolated from Various Regions of the Human Gastrointestinal System[a]

Type of Microbe	*Gastrointestinal Region*		
	Stomach	*Small Intestine*	*Large Intestine*
Actinobacillus	+	+	−
Aeromonas	−	−	+
Bacillus species	−	−	+
Bacteroides	+	+	+
Bifidobacteria	+	+	−
Candida[b]	+	−	−
Clostridia	+	+	−
Coliforms	+	+	+
Lactobacilli	+	+	+
Peptostreptococcus	+	−	−
Staphylococcus	+	+	−
Streptococcus	+	+	+
Torulopsis[b]	+	−	−
Veillonella	+	+	−

[a] The microorganisms isolated and the numbers found vary to some extent, depending upon the diet, environmental factors, methods used for isolation, and geographical location of the subjects.
[b] Yeast.

One of the most remarkable features of the human intestinal microbiota is its stability. These microorganisms continually exert forces to maintain a stable environment. Only the most extreme stress situations, such as antibiotic administration, upset the intestine's ecological balance. Day-to-day changes in diet have little effect. The benefit from maintaining the microflora's stability is the exclusion of foreign organisms, including pathogens, that attempt to colonize the intestinal tract from time to time. It appears that the single most important function of the intestinal microbiota is to protect against infectious diseases originating in this body area. An important virulence factor for a pathogen is the ability to survive and multiply in its host. The successful enteric pathogen must be able to overcome the normal intestinal microbiota and to multiply unchecked in the gastrointestinal system.

Bacterial Diseases

The World Health Organization Expert Committee on Enteric Infections has adopted the designation *acute diarrheal disease* for disease conditions in which there is a disturbance of intestinal functions and which, once started, may produce dehydration and the passage of liquid stools. The specific bacterial pathogens that can cause such conditions are shown in Table 26–2, along with other agents of gastrointestinal disease. Among the organisms associated with acute diarrheal disease are the agents of cholera, gastroenteritis, bacillary dysentery, and traveler's diarrhea.

The intestinal microbiota may induce diarrhea in three ways: they may change dietary foodstuffs or host secretions into substances that affect gut fluid movement; they may penetrate the intestinal lining and damage the bowel wall; or they may produce exotoxins (Color Photograph 108) that cause an emptying of large amounts of water and electrolytes into the intestines without damaging the mucosa (intestinal lining).

Diarrheal illnesses occur when a patient's stools become loose, or watery, and significantly increase in number per day. Usually these conditions are accompanied by a loss of appetite, nausea, vomiting, abdominal pain, and fever. Infections of the gastrointestinal system result in some form of diarrheal illness. Such enteric infections can be grouped into three basic categories: (1) noninflammatory or water type, usually resulting from the action of an entero-

TABLE 26–2 Features of Bacterial Diseases of the Gastrointestinal Tract

Disease	*Causative Agent*	*Gram Reaction and Morphology*	*Incubation Period*	*General Features of the Disease*
Asiatic cholera	*Vibrio cholerae*	–, Vibrio	Usually 2–5 days	The disease ranges from mild to severe and symptoms include large amounts of mucus in stools (rice-water stools), sudden loss of water and electrolytes, and dehydration; collapse, shock and death can occur without treatment
Brucellosis (Malta fever, undulant fever)	*Brucella abortus, B. melitensis, B. suis*	–, Coccobacilli	1–3 weeks	Variable symptoms may include general discomfort, weakness, muscle aches and pains, elevated temperature late in the day, falling during the night, enlarged lymph nodes, spleen and liver involvement; disease may become chronic; residual tissue damage can occur
Gastroenteritis	*Campylobacter fetus,* subspecies *jejuni*	–, Curved rods	1–3 days, variable	Disease symptoms range from muted to severe and includ abdominal pain, diarrhea, fever, and bloody stools
	Vibrio parahaemolyticus	–, Vibrio	4–96 hours	Symptoms include abdominal cramps, nausea, watery diarrhea, and, in severe cases, dehydration

TABLE 26–2 (continued)

Disease	*Causative Agent*	*Gram Reaction and Morphology*	*Incubation Period*	*General Features of the Disease*
	Vibrio vulnificus[a]	−, Vibrio	Within 24 hours	Malaise, chills, diahea, vomiting, and rash in some cases
	Yersinia enterocolitica	−, Coccobacillus		Symptoms appear suddenly and include abdominal pain, diarrhea, dehydration, nausea, chills, vomiting, jaundice, possible convulsions, and weight loss; complications do not usually develop
Pseudo-membranous colitis	*Clostridium diffcile*	+, Rod	2 weeks following therapy	Fever, diarrhea
Salmonellosis (gastroenteritis)	*Salmonella typhimurium* and other *Salmonella* spp.	−, Rod	8–10 hours, possible up to 48 hours	Symptoms appear suddenly and include abdominal pain, diarrhea, dizziness, fever, headache, nausea, vomiting, and poor appetite
Shigellosis (bacillary dysentery)	*Shigella boydii, S. dysenteriae, S. flexneri, S. sonnei,* and other *Shigella* spp.	−, Rod	1–14 days	Symptoms appear suddenly and include abdominal pain, diarrhea, high fever, general discomfort, stools containing mucus, blood, and pus ("red currant jelly" appearance), rectal burning, and dehydration; complications include massive bleeding and perforation of the large intestine
Traveler's diarrhea	Enterotoxin-producing *Escherichia coli* strains and species of *Salmonella* and *Shigella*	−, Rod	Usually 2–5 days	Symptoms appear suddenly and include abdominal pain, diarrhea, dehydration, nausea, chills, vomiting, jaundice, possible convulsions, and weight loss; complications do not usually develop
Typhoid fever	*Salmonella typhi*	−, Rod	1–2 weeks or longer	Symptoms appear gradually and include abdominal distention, constipation, rising fever, headache, loss of appetite, nausea, vomiting, diarrhea, and appearance of a rash (rose spots) on abdomen; complications include inflammation of gall bladder, perforation of small intestine, intestinal bleeding, and pneumonia
Weil's disease (spirochetal jaundice, leptospirosis)	*Leptospira*[a] *interrogans*	Usually not done, Spiral	2–20 days	Symptoms appear suddenly and include lack of appetite, chest pains, head cold, difficulty in swallowing, swollen lymph nodes, fever, vomiting, and jaundice; complications are severe involvement of the skin, central nervous system, kidneys, and liver

[a] Infections of other tissues occur with this organism.

toxin or a process that interferes with the absorption or surface area of the small intestine; (2) inflammatory type, resulting from the invasion of the intestinal wall by the pathogen; and (3) penetrating type, resulting in invasion and the multiplication of pathogens in the cells of the intestine and causing involvement of other body organs. The following sections describe the wide range of enteric pathogens and the diseases they cause. The approaches to laboratory diagnosis, treatment, prevention, and control and the general features (signs and symptoms) of microbial diseases can be found in these sections and portions of Table 26–2. **[See Chapter 21 for additional features of laboratory diagnosis.]**

Asiatic Cholera

In the nineteenth century, pandemics of Asiatic cholera spread from the Far East to Africa, other parts of Asia, Europe, and North America. During the present century, the disease appears to have been more or less limited to India and surrounding areas, although epidemics have occurred in other parts of the world, including Egypt, Indonesia, Korea, and the Philippines.

Transmission. The causative agent of cholera is the gram-negative, slightly curved rod *Vibrio cholerae* (VIB-ree-oh KOL-er-ee) (Figure 26–4). Individuals acquire cholera through the ingestion of the causative organisms in sewage-contaminated food or water or by coming into direct contact with an infected person's feces. Houseflies may also spread the vibrios. Epidemics are usually associated with such sources of infection. There appear to be no long-term carriers.

Contamination of water supplies is usually caused by recent introduction of cholera organisms rather than the persistence of the agents in such an environment. Fish obtained from contaminated waters and eaten without sufficient cooking are also a source of disease agents.

Cholera epidemics have been described as either *protracted* or *explosive.* Explosive epidemics occur when the pathogenic agents are transmitted by means of contaminated food or water. Protracted forms occur when disease-causing organisms are spread by direct contact or by feces-contaminated objects (fomites). **[See disease transmission in Chapter 21.]**

Cholera is more frequent during the warmer months of the year. In the endemic areas of the Far East, cholera appears to be related to a warm cli-

Figure 26–4 Cholera is caused by a toxin produced by *Vibrio cholerae* in the human small intestine. The ability of the bacteria to adhere to the intestinal lining and establish themselves in this area (bound organisms) is thought to be an important aspect in producing the disease. (a) A portion of the intestine free of disease agents. (b) Large patches of *V. cholerae* on the adult villus. *[L. T. Nelson, J. D. Clements, and R. A. Finkelstein,* Infect. Immun. ***14****:527–547, 1976.]*

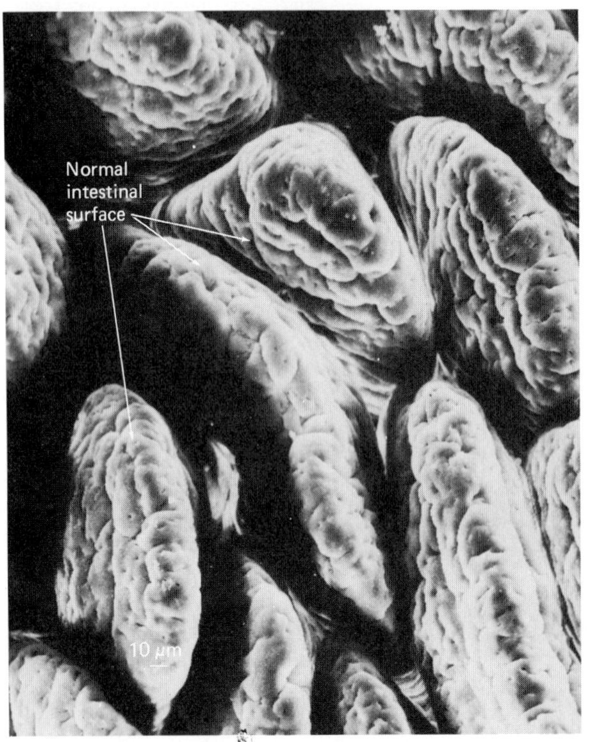

(a)

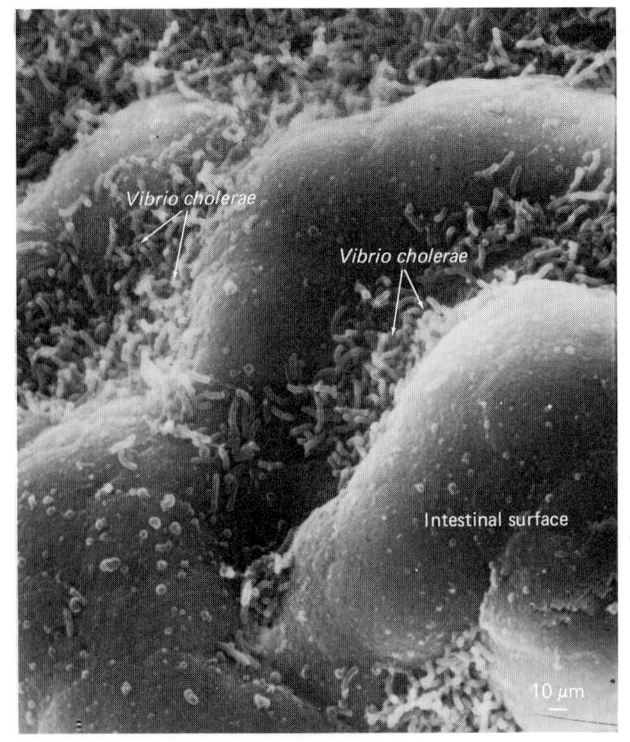

(b)

Cholera

LABORATORY DIAGNOSIS

Isolation of vibrios from fecal specimens is easily done. Selective media having a high pH (alkaline) are generally used. Once isolated and Gram stained, *Vibrio cholerae* is distinguished from other vibrios on the basis of biochemical reactions, and its cell wall or somatic 0 antigen. Serologically, *V. cholerae* isolates are divided into two groups, 01 and non-01. Strains of the 01 serogroup are associated with epidemic cholera, and they are subdivided into the Inaba and Ogawa serotypes. Determining the serotype is important because of the epidemic potential of 01 strains as opposed to non-01 strains.

TREATMENT

Cholera is a self-limiting condition which can be effectively controlled and cured if adequate water and electrolytes are administered. Fluid replacement by mouth, if started early, is sufficient for mild cases of the disease.

Antibiotics such as tetracyclines or chloramphenicol are given to shorten the duration of diarrhea, reduce the volume of fluid loss, and eliminate the vibro population in the intestinal tract.

mate, high absolute humidity, and large population. When the monsoons bring heavy rainfall to these regions, the incidence of cholera drops substantially because the heavy rains wash away contaminated matter.

It is fairly well accepted that the diarrhea of cholera is caused by the action of an exo-enterotoxin, choleragen, on an intact intestinal wall. It is believed that the ability of *V. cholerae* to adhere to the intestinal lining and to establish themselves in this area may be an important aspect in producing cholera (Figure 26–4b). [See mechanism of toxin action in Chapter 21.]

PREVENTION. A cholera vaccine consisting of heat-killed organisms can provide protection. But this immunity is not long-lasting. About four to six months of protection is believed possible. The vaccine is given in two 1-mL doses seven days apart. Newer and better immunizing preparations are under development. These include attentuated *V. cholerae* strains, combinations of toxoid and vibro organisms, and non-enterotoxin-producing strains prepared by recombinant DNA techniques. [See Chapters 6 and 15 for descriptions of recombinant DNA techniques.]

Probably the most effective means to protect against cholera infection is to establish and maintain purified water supplies. In areas where the disease prevails, this procedure would greatly help to reduce the incidence of infection. In addition, of course, the elimination of flies, suitable treatment of patients during recovery, and education of the general population regarding the means of transmission are effective preventive measures.

Other *Vibrio*-Caused Diseases

Several noncholera vibrios are known to be pathogenic for humans. Some of these organisms are of major clinical and ecological significance. The species receiving considerable attention include the halophile anaerobes *V. parahaemolyticus* (V. par-ah-hee-moh-LIT-ee-cuss), a cause of gastroenteritis, and *V. vulnificus* (V. vul-neh-FEE-cuss), a cause of bloodstream and wound infections. Disease states with both species are associated with the consumption of contaminated seafood or with exposure to marine environments.

Preventing infection is simple, regardless of *Vibrio* content, when seafoods are eaten cooked. Such foods will be safe if they are (1) cooked at high enough temperatures to kill any vibrios present, (2) protected from contamination after cooking, and (3) held until eaten at temperatures too low or too high to permit *Vibrio* reproduction. A typical problem is that seafoods considered cooked by traditional standards are not sterile. In addition, when seafoods are eaten raw, prevention of *Vibrio* infection is difficult. It is impossible to guarantee that any raw seafood is totally safe.

Noncholera *Vibrio* Infection

LABORATORY DIAGNOSIS

Isolation on selected media and biochemical testing are the major means of identification.

TREATMENT

Since the disease condition is usually self-limiting, antibiotics are not used. Replacement of lost fluids and electrolytes is the most important aspect of treatment.

Brucellosis

Brucellosis, commonly called *undulant* or *Malta fever,* occurs in countries bordering the Mediterranean Sea, the islands of Cyprus and Malta, the Scandinavian countries, Mexico, and the United States.

TRANSMISSION. Humans acquire brucellosis through contact with the tissues or excretions of infected animals or by the ingestion of contaminated meat or unpasteurized dairy products. Sixty percent of human infections in the United States in 1975 involved meat-processing workers. Accidental inoculations of brucellae also have been reported. In general, the disease is not readily transmitted from human to human.

CAUSATIVE AGENTS. The causative agents belong to the genus *Brucella* (broo-SEL-lah; Table 26–2), which was named after Sir David Bruce, who first isolated the causative agent of Malta fever. The genus contains the closely related species *B. abortus* (B. ah-BOR-tus), *B. canis* (B. KAY-niss), *B. neotomae* (B. nee-oh-TOME-eye), *B. ovis* (B. OH-viss), *B. suis* (B. SOO-iss), and *B. melitensis* B. mel-lee-TEN-sis). *B. canis* causes Bang's disease, or bovine brucellosis; *B. suis* causes swine (porcine) brucellosis; and *B. meli-*

Brucellosis

LABORATORY DIAGNOSIS

Isolation of brucellae can prove difficult, since brucellosis is generally long lasting and produces few symptoms. Organisms can be found more commonly in blood during the fever stage of the disease. Cultivation requires an enriched medium and an atmosphere of 2% to 10% CO_2. Such cultures may require two to four weeks of incubation. Isolation from contaminated milk is done on selective plating media containing bacteriostatic dyes and antibiotics.

Laboratory identification is often made serologically with tests such as agglutination, indirect fluorescent antibody, and ELISA. [See Chapter 19 for descriptions of these tests.]

TREATMENT

Because of the intracellular location of brucellae in the infected host, effective treatment is difficult. Tetracyclines are the drugs of choice and usually work best when given in the early or acute stage of the disease. Antibiotics are of little value in chronic or long-lasting cases.

tensis causes goat (caprine) brucellosis, or Malta fever. In some mammals, but not humans, the brucellae may infect mammary glands and be shed in milk. These organisms will also infect the placenta of domestic farm animals and cause abortions. Animals susceptible to abortion contain large amounts of the carbohydrate erythritol in their placentas. This sugar stimulates the growth of brucellae.

The brucellae are small, pleomorphic, gram-negative coccobacilli. These organisms are nonmotile and non-spore-forming. All species are obligate parasites capable of maintaining an intracellular existence.

Prevention. Prevention of brucellosis centers on the elimination of reservoirs of the disease. This process involves immunization of natural reservoirs—such as cattle, goats, and hogs—and the segregation and even destruction of infected animals. Vaccination of humans against this disease has been limited.

The routine pasteurization of dairy products is another effective means of prevention. This procedure has markedly reduced the number of human cases of brucellosis in the United States.

Campylobacter Gastroenteritis

It is now clear from numerous reports provided by a number of health care institutions throughout the world that *Campylobacter* (kam-pee-low-BACK-ter) is a major cause of diarrhea in adults and children. This bacterial genus contains the two human pathogens *C. fetus* subspecies (ss.) *intestinalis* (C. FEE-tuss ss in-tes-TIN-al-iss) and *C. fetus* ss. *jejuni.* Although *C. fetus* ss. *jejuni* (C. FEE-tuss ss jeh-JUNE-ee) may be involved as a cause of diarrhea at least as frequently as species of *Salmonella* (sal-mon-EL-lah) and *Shigella* (shi-GEL-lah), the *C. fetus* ss. *intestinalis* causes the vast majority of the more serious infections. Both contaminated food and water are important means of disease transmission. Infected puppies, kittens, and children with diarrhea have been implicated as sources of infection. The disease state tends to be self-limited, and most victims recover without treatment (Table 26–2). However, some individuals may require treatment. Relapses are also possible.

Leptospirosis

Leptospirosis is a zoonosis produced by distinct slender spirochetes of the genus *Leptospira* (lep-toe-SPY-rah; Table 26–2). Human leptospiral disease is a severe illness accompanied by a high fever, jaundice, some bleeding, and involvement of the kidney. The spirochetes may become established in the urinary system and thus serve as a major source of contamination of water. **[See zoonoses in Chapter 21.]**

Causative Agents. Strains of *Leptospira* can be found in various bodies of water, such as lakes, ponds, and rivers, on decaying matter, and sometimes even in tap water. Leptospires measure approximately 4 to 20 μm in length and 0.1 μm in width. These organisms are quite thin (*leptos* is from the Greek, meaning thin), elongated structures consisting of many small coils tightly set together. Quite often leptospires are shaped like the letters C, S, and J. Finely tapered or hooked ends are characteristic of these organisms.

Leptospires may be stained with a Giemsa stain preparation or with silver stains. These organisms can be grown and maintained in either broth or agar media.

Transmission. Several species of wild rodents and domestic animals, including cattle, cats, and dogs, serve as reservoirs of infection. The natural hosts may experience a mild infection, but they seldom die of the disease. A particular *Leptospira* strain

Campylobacter Gastroenteritis

Laboratory Diagnosis

Isolation of *Campylobacter* can be made by the inoculation of selective plating media with fecal specimens. Incubation at 42°C for 48 hours in an atmosphere of 10% CO_2 is required. Gram-stained preparations or dark-field examination is used to confirm the identification.

Treatment

When antibiotic treatment is necessary, erythromycin is the drug of choice.

is usually quite well adapted to its host and is capable of inhabiting portions of the animal's kidney without inflicting damage. Thus, the periodic release of large numbers of leptospires in the animal's urine is not unusual. Humans acquire the disease agent by coming into contact, either directly or indirectly, with this infected urine. *Leptospira* infections may result from bathing in or falling into stagnant bodies of water contaminated by the infectious urine of rodents. The spirochetes may gain access to the body tissues by penetrating the mucous membranes of the eyes and nasopharynx or by entering through skin abrasions or cuts. Cases involving ingestion of contaminated food and water have also been reported.

Several factors seem to affect the incidence of human infections, among them age, sex, occupation, and seasonal variations. In general, adults are more likely to acquire leptospirosis because of a greater possibility of exposure. For example, newborn infants are not exposed to contaminated bodies of water that might be used for swimming, boating, or fishing. The age group most commonly involved is between 20 and 30 years old. The higher percentage of cases among males is caused by the greater occupational exposure. Occupations that have previously been typically male—such as plumbing, meatpacking, and farming—provide a greater chance of exposure to infected animals and contaminated bodies of water. No difference in susceptibility between males and females has been reported.

In tropical areas, leptospirosis occurs throughout the year. In general, leptospires require moisture and warmth for survival. These conditions also increase the use of outdoor facilities for both pleasure and work, thereby creating greater exposure.

Recovery generally follows infection unless lesions are so far advanced that irreversible damage occurs in the kidney. When death occurs, it is caused by kidney failure.

IMMUNITY. Immunity appears to depend solely on immunoglobulins. There is no evidence of a role for T cell mechanisms or a cell-mediated immune response in resistance to initial infection or reinfection.

PREVENTION. Because the natural reservoirs of leptospirosis are varied and numerous, eradicating them would be quite difficult. An effective vaccine for this disease is not available for humans, and preventive measures are directed toward reducing human contact with contaminated water or urine from infected animals. Preventive measures include wearing protective clothing, such as gloves and rubber boots, while working with objects or water supplies that may be contaminated; avoiding bodies of water that might be contaminated by infected animals; and not using or ingesting water and food that may be contaminated. Because dogs may be exposed to infected animals or their urine, they are commonly immunized against the disease.

Salmonellosis

Salmonellosis, a food infection that bears no relationship to salmon, is primarily limited to the gastrointestinal tract. Outbreaks of salmonellosis are usually explosive and are associated with banquets, weddings, or other group meals. The suddenness of the disease tends to distinguish it from other infectious diseases involving the gastrointestinal system, such as amebic or bacillary dysentery (Tables 26–2 and 26–5).

Leptospirosis

LABORATORY DIAGNOSIS

Leptospires can be isolated in suitable media from blood and urine specimens. The organisms can also be detected in blood samples by dark-field microscopy during the first week of the infection. Serological tests including agglutination, complement fixation, and the enzyme-linked immunosorbent assay (ELISA) are also of value.

TREATMENT

Antibiotic therapy is effective only during the early stages of leptospirosis. Penicillin, streptomycin, or tetracycline are usually the drugs of choice.

Salmonellosis

LABORATORY DIAGNOSIS

Selective and differential media appropriate for species of both *Shigella* and *Salmonella* are used to isolate the organisms from fecal specimens. Examples include MacConkey, Salmonella-Shigella, and Hektoen enteric agars (Color Photographs 109a and 25). Further identification is based on biochemical tests. Serological tests are of little value.

TREATMENT

Replacing lost fluids and electrolytes is the most important aspect of treatment. Antibiotics are not indicated, since their use does not change the course of the infection significantly.

CAUSATIVE AGENTS. Several species of *Salmonella* are known to produce salmonellosis. Included in this group are *S. choleraesuis* (sal-mon-EL-lah KOL-er-ee-soo-iss), *S. enteritidis* (S. en-ter-IT-id-iss), and *S. typhi* (S. TIE-fee). These organisms are gram-negative, motile, non-spore-forming rods. At present, the members of the genus *Salmonella* are divided into serological groups based on their antigenic properties. All these organisms have the same somatic or O antigens, which are components of their cell walls. They are of several hundred serological types based on differences in their flagellar or H antigens. Another antigen found with certain salmonellae is the Vi or virulence antigen, another somatic antigen.

TRANSMISSION. Humans acquire salmonellosis through consumption of contaminated food or water. A variety of foods have been implicated as sources in outbreaks, including cream-containing bakery goods, ground meats, poultry, sausages, and eggs. Rodents such as mice and rats are often infected by salmonellae. Such animals, after recovery, may become carriers and by means of their excreta contaminate foods. This possibility should be guarded against in establishments where food is stored or prepared. Contamination by human carriers also occurs.

PREVENTION AND CONTROL. Measures that can be taken against salmonellosis include (1) proper cooking of foods obtained from animal sources, such as ground meat and sausages; (2) poultry refrigeration and covering of prepared foods; (3) protection of food from contamination by mice, rats, or flies and related insects; (4) the periodic inspection of food handlers; and (5) proper sanitation.

Salmonella food poisoning is a reportable disease. Reporting cases to public health authorities is important, so that suitable measures may be taken to prevent an epidemic. **[See reportable diseases in Chapter 21.]**

Shigellosis (Bacillary Dysentery)

The principal causes of shigellosis, an acute infectious disease, are the pathogenic but noninvasive shigellae. Shigellosis is distinct from amebic and viral dysentery. The various species of *Shigella* are widely distributed but are found primarily in human intestines and occasionally in the intestinal tracts of monkeys and other mammals. The most common causes of bacillary dysentery are *S. dysenteriae* (shi-GEL-lah dis-en-TEH-ree-eye) (discovered by Japanese bacteriologist Kiyoshi Shiga in 1896) and *S. flexneri* (S. FLEKS-ner-ee) In general, the shigellae are far less invasive than the salmonellae.

Classifying shigellae on the basis of fermentation reactions can be difficult in certain cases. The use of newer automated methods has simplified classification (Color Photograph 86). Separation into species and serological types on the basis of antigenic composition has been successful.

TRANSMISSION. Since the fourth century B.C., when it was first described, shigellosis has been reported during several major military campaigns. This disease seems to occur with greater frequency than other forms of dysentery.

The human is the sole reservoir of infection; no lower animal reservoir is known. Although all age groups are susceptible to infection, children and males between the ages of 20 and 30 are most commonly affected. Predisposing factors include lowered states of resistance, malnutrition, overcrowding, and poor sanitation.

Shigellosis

Laboratory Diagnosis

Confirmatory laboratory identification of *Shigella* requires the plating of fecal specimens on various selective and differential media. Species are differential on the basis of biochemical tests. *Shigella* are nonmotile, do not produce gas from carbohydrates, and do not form hydrogen sulfide.

Treatment

Ampicillin is generally the drug of choice. However, isolates should be tested for antibiotic susceptibilities.

Bacillary dysentery is usually contracted through the ingestion of contaminated food or water. The causative agents can be transmitted by feces, fingers, flies, or food. **[See disease transmission in Chapter 21.]**

Individuals recovering from bacillary dysentery may become carriers of shigellae. This is an important consideration in the control and prevention of shigellosis, as these persons serve as reservoirs between outbreaks.

Complications known to occur with shigellosis include the perforation of the large intestine and massive bleeding.

The particular factors responsible for the pathogenicity in shigellosis have been fully determined only for infections produced by *S. dysenteriae.* In addition to producing an endotoxin characteristic of all shigellae, *S. dysenteriae* is known to form an exotoxin. This soluble, heat-labile protein, called the *Shiga neurotoxin,* is one of the most powerful poisons known. When injected into experimental animals, the toxin causes bleeding, fever, diarrhea, paralysis, and death. **[See bacterial toxins in Chapter 22.]**

Prevention. The control of shigellosis involves far more than the appropriate use of antibiotics. Since human beings serve as the only source of infectious agents, preventive measures must be directed toward infected persons, carriers, and items that may have been contaminated. The elimination

Typhoid Fever

Laboratory Diagnosis

Identification of *S. typhi* depends on its isolation from blood, feces, urine, or contaminated food. Blood specimens are generally the first to yield organisms. Special selective and differential media, similar to those used for other salmonellae, are used. Serological tests are of major importance. Most patients with typhoid fever develop antibodies to the somatic (O) and flagellar (H) antigens of *S. typhi.* The organism also has the special heat-sensitive *Vi* (virulence) antigen.

Treatment

In addition to replacing lost fluid and electrolytes, the antibiotic trimethoprim is used in cases of *S. typhi* infections. Amicillin, amoxicillin, and chloramphenicol are alternative drugs.

of flies, the proper sanitary disposal of excreta, and the protection of food and water are also important.

An oral attentuated *Shigella* vaccine has been effective in reducing epidemics in certain situations. The preparation consists of streptomycin-dependent *Shigella* strains that cannot multiply in the intestine but can stimulate antibody production. In situations of widespread shigellosis, the use of mass chemoprophylaxis may be necessary.

Typhoid Fever

The various species of *Salmonella* are associated with at least three distinguishable human disease states: typhoid or enteric fever, blood poisoning *(septicemias)*, and the acute forms of infectious food poisoning discussed earlier.

Individuals suffering from blood infections (septicemia) exhibit a high fever in addition to the presence of bacteria in the blood. *S. choleraesuis* is a common causative agent. All of these conditions are collectively called *salmonellosis*. Typhoid fever is the classic example of enteric fever (Table 26–2). It is caused by the gram-negative, motile, non-spore-forming rod *S. typhi*.

Transmission. Humans acquire typhoid fever by ingesting contaminated food or water. Flies and fomites have also been implicated. In countries

Microbiology Highlights

THE SAGA OF TYPHOID MARY

During the sumer of 1906 an outbreak of typhoid fever occurred in the town of Oyster Bay in Long Island, New York. Six of eleven members of a family suddenly came down with the disease between August 27 and September 3. Having ruled out contaminated food and water as the possible sources of contamination, local health authorities were particularly puzzled as to where and how typhoid fever came into their community. In an effort to find the source and to prevent additional outbreaks, George Soper, a sanitary engineer from the New York City Health Department, was hired. Soper was well known for his investigative abilities. After a preliminary study of the situation, Soper's findings centered on Mary Mellon, the family cook, as the source of the problem. Unfortunately, she disappeared shortly after the outbreak of the disease. This turn of events did not discourage Soper, nor did it stop the investigation. Rather, it led to Soper's quiet search for Mary Mellon, who eventually became known as Typhoid Mary. From further investigations, which led him back to 1898, the year of Mary Mellon's first job as a family cook, Soper connected her to at least 22 cases of typhoid fever in the households for which she cooked from 1901 through 1906.

In 1907, Soper found Mary Mellon working under a false name and for a family in which typhoid fever had again surfaced. Because she was unwilling to be tested for the disease, she was forcibly detained by the New York Department of Health in the city hospital from March 19, 1907 until the time of her release in 1910. During this time, cultures from her stools, taken every few days, were generally found to contain enormous numbers of typhoid bacilli. There was no question as to her being a typhoid carrier.

Typhoid Mary's problems did not end with her release. In 1914 she was employed as a cook (again under an assumed name) at the Sloane Hospital for Women in New York. In January and February of the following year, an outbreak of typhoid involving 25 cases occured among health-care personnel. Eight people died. All avenues of suspicion pointed to the cook of the establishment, especially as she did not return after a short leave of absence and did not leave a home address. Typhoid Mary was later found again by the health department and again taken forcibly to the City Hospital, where she remained until her death in 1938. Mary Mellon, alias Typhoid Mary, died at age 70 from the effects of a stroke.

A subsequent study of Mary Mellon's involvement with other outbreaks of typhoid fever linked her to the well-known water outbreak in Ithaca, New York, in 1903. Over 1,300 cases of typhoid fever connected with her were recorded. Ironically, George Soper gained his reputation from his investigation of the incident.

where sanitation is adequate, typhoid fever appears either sporadically or in an endemic form. Often the source of infectious agents is traced to carriers. In regions with poor sanitation, impure water, improper waste disposal, and lack of pasteurization, typhoid epidemics are more likely to occur. All ages may be attacked.

S. typhi can gain access to various tissues and organs via the bloodstream. Thus, the bone marrow, gallbladder, and spleen can serve as future sources of reinfection, causing the relapses observed with typhoid fever when organisms gain access to the bloodstream from other foci of infection.

Prevention. The control of typhoid fever also involves far more than the appropriate use of antibiotics. Proper sewage disposal and the periodic examinaton of food handlers to determine that they are not carriers remain as the best methods of control. Convalescent patients may remain typhoid-fever carriers for long periods of time and ridding them of *S. typhi* may prove to be difficult.

A vaccine against *S. typhi* and two paratyphoid organisms is effective in reducing the incidence and severity of the disease. Immunization is generally limited to high-risk and military personnel and some international travelers. An oral vaccine is undergoing field trials.

Bacterial Disease States of Increasing Frequency

Several enteric or so-called coliform bacilli are normal, nonpathogenic inhabitants of the gastrointestinal tract (Table 26–1). However, other related strains or species are associated with pathological states involving the human urogenital and intestinal systems. Several of these diseases are caused by exotoxins produced during microbial growth in the intestinal tract. The toxins act directly on the intestinal tract (primarily in the small intestine) either by altering intestinal function or by causing structural damage to the intestinal lining (mucosa). Examples of these *infective intoxications* will be discussed here.

INFANT EPIDEMIC DIARRHEA

Epidemic diarrhea in newborn and young infants has often been reported. Most of these outbreaks have been associated with a specific group of enteropathogenic strains of *Escherichia coli* (EPEC). However, studies from India, Vietnam in 1971, and Japan in 1967 showed that EPEC can cause disease not only in children but in adults as well. There are at least two mechanisms of disease production: formation of a choleralike enterotoxin and invasion of intestinal epithelial lining.

At least two other major groups of *E. coli* associated with intestinal illness are recognized. These are **enterotoxigenic** *E. coli* (ETEC), noted for its ability to colonize the intestinal tract surfaces, and **enteroinvasive** *E. coli,* which invade the intestine and cause a dysentery characterized by bloody stools.

Transmission. Incidents of infant epidemic diarrhea pose a definite threat in hospital nurseries. Unfortunately, outbreaks of this kind may be accompanied by high mortality. Controlling the spread of enteropathogenic strains is a difficult problem because they develop resistance to commonly used antibiotics.

Epidemic diarrhea usually affects newborns and

Epidemic Diarrhea

Laboratory Diagnosis

E. coli can be isolated from various specimens on selective and differential media such as eosin-methylene blue (EMB). The species is a gram-negative lactose fermenter (Color Photograph 109b) and can be identified on the basis of biochemical tests. The IMViC reactions, a set of four tests, is commonly used for this purpose. *E. coli* metabolizes the amino acid tryptophan to form indole (I), produces acid in methyl red medium (MR), does not form acetyl-methyl carbinol (V), and does not use citrate (C) as a sole source of carbon in an appropriate medium. Slide agglutination tests are necessary to confirm the identification.

Treatment

If antibiotics are needed, isolated organisms should be tested for antibiotic susceptibility. Treatment includes replacement of water and electrolytes.

Clostridium difficile Infection

LABORATORY DIAGNOSIS

Identification of *C. difficile* includes microscopic examination of specimens, Gram stain, anaerobic culture, toxin detection, and serologic tests.

TREATMENT

Vancomycin given orally is used to prevent antibiotic-associated pseudomembranous colitis.

infants under two years of age. Premature babies are attacked most severely. Although the disease can occur in older children and adults, the effects are not serious.

Sources of enteropathogenic *E. coli* include convalescent infant carriers, cats, dogs, certain foods, and fomites.

PSEUDOMEMBRANOUS COLITIS

A serious side effect of treatment with certain antibiotics is pseudomembranous colitis (PMC). The majority of cases of this intestinal disease are caused by toxin-producing strains of the gram-positive rod *Clostridium difficile* (klos-TRE-dee-um diff-EE-seal). This organism is normally found in a significant number of newborn and young infants but in only a small percentage of adults. PMC does not totally depend on the presence of *C. difficile* alone. The disease can be precipitated by factors that disturb the ecology of the gastrointestinal tract, notably antibiotic therapy, surgery, or cancer chemotherapy. Unlike other toxigenic intestinal infections, PMC is accompanied by fever, severe diarrhea, and involvement of the colon.

Microscopically, inflammation is seen in the superficial or uppermost mucosal layers of the colon. It is accompanied by mucosal ulcerations and the formation of a pseudomembrane composed of fibrinogen, degenerating mucosal cells, and leukocytes.

TRAVELER'S DIARRHEA

Annually, more than one-quarter of a billion people travel from one country to another. Tourism has obviously become economically significant, especially to developing countries. Unfortunately, approximately a million travelers per year acquire a rapidly acting, dehydrating condition known as *traveler's diarrhea.* Laboratory findings have shown enterotoxigenic *E. coli* to be among the major causes of this illness.

A large number of bacteria are needed to initiate infection. For this reason, contaminated food is thought to be the major means by which the disease is spread. Species of *Salmonella* and *Shigella* have also been associated with cases of traveler's diarrhea. Recovery is usually without complications. Laboratory diagnosis and treatment are the same as described earlier for the causative agents.

YERSINIA ENTEROCOLITICA GASTROENTERITIS

Yersinia enterocolitica (yer-SIN-ee-ah en-ter-oh-koal-IT-ic-ah) is known to cause gastroenteritis in all age groups. The infection resembles conditions produced by enteropathogenic *E. coli* and species of *Salmonella* and *Shigella.* Infections are acquired through the ingestion of contaminated food or water. (Since *Y. enterocolitica* is found in domestic animals, many infections also result from human contact with animals.) A variety of additional clinical conditions may be associated with *Y. enterocolitica,* including abscesses, arthritis, and infections of the gallbladder, peritoneum, and blood. Infections in compromised hosts with lowered resistance produce high mortality. Generally, antibiotics are needed only for systemic infections.

Food Poisoning (Intoxications)

The clinical state of food poisoning discussed here refers to the symptoms resulting from the consumption of food or drink contaminated by pathogenic bacteria or their toxic products. Naturally poisonous foods and conditions resulting from the ingestion of foods sprayed with pesticides will not be considered here.

In the past, certain gastrointestinal upsets were generally labeled *ptomaine poisoning.* The term was used synonymously with food poisoning. This practice has declined as careful studies have shown the presence of pathogenic microorganisms or their products in foods consumed by stricken persons.

Yersinia enterocolitica Gastroenteritis

LABORATORY DIAGNOSIS

Isolation of *Y. enterocolitica* can be difficult, since very good selective media are not available. Identification is based on biochemical and serological tests.

TREATMENT

Antibiotics are generally not required for gastrointestinal infection. Emphasis is placed on the replacement of lost fluids and electrolytes. In the case of systemic disease, chloramphenicol, gentamicin, tobramycin, and the combination of trimethoprim-sulfamethoxazole are effective.

The agents of common bacterial food poisoning are well established. Among the conditions they produce are botulism, *perfringens* poisoning, salmonellosis, and staphylococcal poisoning. Salmonellosis (described earlier) is an example of an active infection (infectious food poisoning); the other three states are examples of poisonings or bacterial intoxications. In poisoning, it is the toxins alone that produce the symptoms (Table 26–3). [See microbial toxins in Chapter 22.]

Microorganisms have been implicated in causing other gastrointestinal upsets; these include *Bacillus cereus* (bah-SIL-lus SEE-ree us) certain *E. coli* strains, and members of the genus *Proteus* (PROH-tee-us). Moreover, viruses have been incriminated in certain cases of food-borne illnesses.

BOTULISM

The name *"botulism"* is derived from the Latin *botulus,* meaning sausage. Uncooked sausages were associated with disease for years. *Clostridium botulinum* (KLOS-TREH-dee-um bot-you-LIE-num), the causative agent of food poisoning, wound botulism, and infant botulism, produces a powerful neurotoxin when it grows under appropriate anaerobic conditions. Botulism is not considered an infectious disease. *C. botulinum* is not an invasive microorganism. However, persons who eat foods containing *C. botulinum* toxin develop an intoxication. Botulism has been associated with a variety of food products, including improperly preserved or prepared home-canned fruits and vegetables, smoked fish, and uncooked fish and meats. Outbreaks of the disease

TABLE 26–3 A Comparison of Common Bacterial Food and Other Poisonings

Disease	*Causative Agent*	*Gram Reaction and Morphology*	*Incubation Period*	*General Features of the Disease*
Botulism	*Clostridium botulinum*	+, Rod	12–96 hours	Difficulty in speaking, double vision, inability to swallow, nausea, vomiting, and paralysis of urinary bladder and all voluntary muscles; death caused by stoppage of heart action and/or breathing may occur
Perfringens poisoning	*Clostridium perfingens*	+, Rod	Within 18 hours	Abdominal cramps, chills, bluish coloration of the skin, diarrhea, headache, nausea, and vomiting
Staphylococcal intoxication	*Staphylococcus aureus*	+, Coccus	1–6 hours	Severity of symptoms depends on amount of enterotoxin ingested and include abdominal cramps, chills, bluish coloration of the skin, diarrhea, headache, nausea, and vomiting

have also occurred with commercial products. The neurotoxin causing botulism can usually be completely inactivated through heating at 100°C for ten minutes. This fact accounts for the relatively low incidence of the disease. It should be noted, however, that this procedure may not always be effective, as toxin inactivation also depends on the toxin's concentration.

It is important to note that foods contaminated by the toxin do not necessarily appear or smell any different from uncontaminated products. Furthermore, neither the gastric secretions of the stomach nor the protein-digesting enzymes of the duodenum inactivate the toxin. This poison is absorbed from both the stomach and the small intestine.

Of the eight immunologically distinct known types of exotoxins, five—A, B, E, F, and G—are associated with human diseases; types C_1, C_2, and D are reported to affect fowl and cattle, respectively. Most human cases and the highest mortality are caused by types A and E.

Botulinum toxin acts by becoming attached to the endings of efferent nerves. There it blocks the release of acetylcholine by nerve fibers when a nerve impulse passes through the peripheral nervous system. Antitoxin cannot neutralize the neurotoxin once it is attached. Thus, treatment should begin as soon as possible if botulism is suspected.

INFANT BOTULISM. A distinct form of infection with *C. botulinum*, infant botulism, is recognized. This condition should be distinguished from the two other better-known forms, food-borne botulism and wound botulism. The critical and frightening difference between the infant and adult forms of botulism is that children appear to become ill without having ingested toxin-containing foods. What is ingested are spores. Once in the intestine, *C. botulinum* spores germinate into vegetative cells, and then multiply and produce botulinal toxin *in vivo*. For as yet unknown reasons, the intestinal tracts of only some infants are susceptible to *C. botulinum* infection. Most infants, older children, and adults regularly ingest *C. botulinum* spores without ill effects. All cases of infant botulism recognized to date have been caused by either type A or type B botulinal toxin. Honey, house dust, and soil are considered to be potential environmental sources of spores. Most authorities do not recommend the feeding of honey to any child 12 months of age or less.

Because infants are unable to describe their symptoms, the onset of illness can be detected only by careful observation. Early signs of illness include constipation, increased sleepiness, loss of appetite, difficulty in swallowing, weak crying, and loss of head control. The exact number of infant botulism cases that occur is not known, largely because awareness of its existence remains limited and many cases remain incorrectly diagnosed.

Treatment for infant botulism, interestingly enough, generally does not involve the use of botulinal antitoxin. It appears that antitoxin is not needed for complete recovery. In addition, botulinal antitoxin is a horse serum product, which, if used in treatment of infants, may induce life-long hypersensitivity. Responding to symptoms and using antibiotics form the basis of treatment. **[See type III hypersensitivity in Chapter 20.]**

PERFRINGENS POISONING

Clostridium perfringens is more widespread than any other pathogenic bacterium. Its principal habitats are the soil and the intestinal contents of humans and animals. This organism has been recognized since the late 1800s, when reports linked it with food poisoning; however, it was not until 1945 that

Botulism

LABORATORY DIAGNOSIS

C. botulinum may be isolated from stools and suspected foods by anaerobic culture methods. The demonstration of botulism toxin in blood, intestinal contents, or food is shown by toxicity testing in mice.

TREATMENT

The administration of specific botulism antitoxin to neutralize any free toxin in the body is the major treatment approach. Antibiotics to prevent secondary infections and supportive measures to maintain normal breathing are also important aspects of treatment.

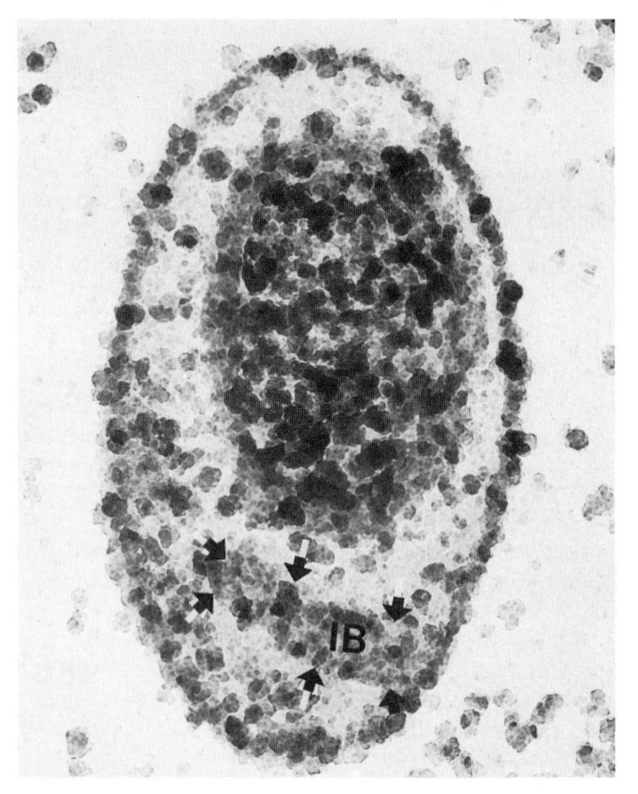

Figure 26–5 An electron micrograph showing the localization of enterotoxin in *Clostridium perfringens.* With special staining enterotoxin-containing inclusion bodies (IB) appear. *[From A. Loeffler, and R. Labbe,* J. Bacteriol. ***165****:542, 1986.]*

C. perfringens food-borne illness was reported. The illness received considerable attention in Great Britain in 1953 and has been recognized as a very important food poisoning organism in the United States since that time.

Strains of this bacterial species are divided into six types, A through F. The basis for this classification is the immunologically specific toxin produced by each strain (Figure 26–5). *C. perfringens* type A is known to cause a form of intoxication identical to that associated with staphylococcal enterotoxin. A more severe form of disease, enteritis necroticans, is produced by *C. perfringens* type F.

TRANSMISSION. Outbreaks of *perfringens* poisoning have been associated with several types of food, including cooked meats and poultry dishes.

PREVENTION. Preventive measures are similar to those described for staphylococcal poisoning.

STAPHYLOCOCCAL FOOD POISONING

As we have seen, staphylococci are normally found in various regions of the human body, including the nose, skin, and throat. These bacteria cause one of the most common types of food poisoning. The active agent of staphylococcal food poisoning is one of several enterotoxins produced by certain *Staphylococcus aureus* (staff-il-oh-KOK-kus ORE-ee-us) strains. At least six different enterotoxins have been reported. They are designated as enterotoxins A through F.

TRANSMISSION. Staphylococcal food poisoning occurs worldwide and is not a communicable disease. Age, race, and sex do not play a role in the occurrence of the condition. Humans show symptoms of the disease after consuming food in which these organisms have grown and produced a sufficient quantity of the toxin. Types of food that serve as growth media for staphylococci include bakery goods, especially those with custard or cream filling; cured, processed, or leftover meats; fish; and dairy products. Staphylococci grow in such foods when they are left unrefrigerated. In order for enterotoxin to be produced in sufficient concentration, food must remain at or above room temperature for several hours. Because the toxin is not inactivated by

Perfringens Poisoning

LABORATORY DIAGNOSIS

Isolation of a large number of *C. perfringens*, (10^5 cells/gram of ingested food), in the absence of other causes of food poisoning is sufficient for laboratory diagnosis.

TREATMENT

C. perfringens food poisoning does not require antibiotics since the disease is self-limiting and generally mild. Replacement of fluids and electrolytes is the major form of treatment used.

Staphylococcal Food Poisoning

LABORATORY DIAGNOSIS

Routine laboratory identification is generally straightforward. *S. aureus* produces beta hemolysis (Color Photograph 23b), ferments mannitol (Color Photograph 26), and in most cases causes positive catalase and coagulase tests (Color Photographs 88 and 96). [See Chapter 4 for selective and differential media and Chapter 22 for coagulase.] Enterotoxin may be detected in foods by means of animal toxicity tests.

TREATMENT

S. aureus food poisoning is generally self-limiting and does not require antibiotics. Replacement of fluids and electrolytes is the main course of treatment.

heat, cooking foods that have previously been left unrefrigerated provides no protection against staphylococcal food poisoning.

The sources of disease agents include individuals with staphylococcal infections, human carriers, infected animals, and milk or milk products contaminated by carriers.

Complications or deaths associated with this disease are rare.

PREVENTION. Preventive measures against food poisoning include the proper covering and refrigeration of all foods and the exclusion of persons with obvious skin infections or disorders from food handling during preparation or serving.

The true problems with staphylococcal food poisoning are not associated with food processing but rather with the mishandling of food in food service establishments and in the home.

Viral Infections

Certain viruses invade the gastrointestinal system and utilize its parts for purposes of replication only. Viral involvement may accompany other types of infections, but direct evidence showing the relationship of viruses to outbreaks of gastrointestinal (GI) upsets may be lacking. In addition symptoms of a disease process may be evident. When an infection does occur, the effects do not necessarily produce gastrointestinal disturbances but may cause reactions involving other body organs, such as those of the nervous and respiratory systems and the liver. Table 26–4 lists the general features, laboratory diagnosis, and treatment of the better-known GI viral infections.

The Liver

One of the largest organs or glands found in the body, the liver is vitally important to the well-being of the individual. It is located in the upper portion of the abdominal cavity just beneath the diaphragm.

Bile, or *gall,* is produced by all parts of the liver and stored in the gall bladder. Bile emulsifies fats so that they can pass through the intestines, prevents food from decaying, and stimulates the intestinal muscles. It enters the small intestine by way of the *bile duct.* If this duct becomes clogged or blocked, the condition known as *jaundice* (a yellowing of body tissues) develops. This condition has a variety of causes, including gallstones and various worm and microbial diseases, such as viral hepatitis.

Diseases of the liver cause destruction of its cells (hepatocytes), preventing them from carrying out important activities. Fortunately, the liver's immense capacity for regeneration makes it resistant to permanent damage. However, in cases of severe injury, the organ heals with the formation of nonfunctioning scar tissue, a condition known as *cirrhosis.*

LIVER FUNCTION TESTS

Since the general observable signs and symptoms of liver (hepatic) disease states are nonspecific, laboratory tests of liver function are required for a specific diagnosis. Such tests are used to detect causes, diagnose, follow recovery, and evaluate various forms of treatment in liver disease. Several viral infections of this organ also are considered here.

Picornavirus Infections

Coxsackie, echo, and polioviruses are capable of infecting the gastrointestinal system. Originally, these agents were named *enteroviruses* because of their obvious association with this system. It soon became apparent, however, that several viruses in the group could infect the respiratory tract and central nervous system as well. Another group of pathogens primarily affecting the respiratory system was consequently designated the *rhinoviruses.* All of these various viruses were classified in a more adequate group, the *picornaviruses. "Pico"* means small and *"rna"* comes from the type of nucleic acid found in this group. Coxsackieviruses were named after Coxsackie, New York, where the first isolations took place in 1948. The name *"echo"* was derived from certain of the properties of these viruses: E = enteric location; C = capable of causing cytopathic changes in tissue cells; H = human source; and O = orphan. At one time there were viral agents without diseases, hence the term *"orphan."*

Several pathogenic viruses of lower animals are also in the picornavirus group. These include the agents of encephalomyocarditis of mice, Teschen disease of pigs, and foot-and-mouth disease.

In general, picornaviruses enter the human body via the oral route. A few, however, enter by means of the respiratory tract. The disease states produced by this group differ in the tissues involved, the types of lesions resulting from infection, and the severity of the attack. Similar disease states (for example, aseptic meningitis) may be caused by different picornaviruses. Most of the infections produced by picornaviruses are discussed in greater detail elsewhere in the text. **[See aseptic meningitis in Chapter 29.]**

Cytomegalovirus Inclusion Disease

The cytomegaloviruses (CMVs) are a group of highly species-specific herpesviruses. They cause a distinctive cytopathic enlargement of host cells (cytomegaly) with intranuclear and cytoplasmic inclusion bodies (Figure 26–6). Humans, monkeys, and other animals can fall victim to this group of viruses. Human cytomegalovirus is found worldwide and causes severe, often fatal infections affecting the salivary glands, brain, kidney, liver, and lungs (Table 26–4). This is the leading cause of congenital viral infection in the United States. More than 95% of newborns with congenital CMV infection are free of abnormalities at birth. However, almost all newborns who exhibit symptoms at birth and about 10% of the larger asymptomatic newborn group will have lasting consequences ranging in severity from unilateral hearing loss to serious central nervous system problems. CMV also produces latent infections in healthy adolescents and adults; these include cytomegalovirus infectious mononucleosis. Such infections may be activated by pregnancy, multiple blood transfusions, or immunosuppressive therapy in the case of organ transplants. Victims of AIDS frequently have a history of CMV infection as well.

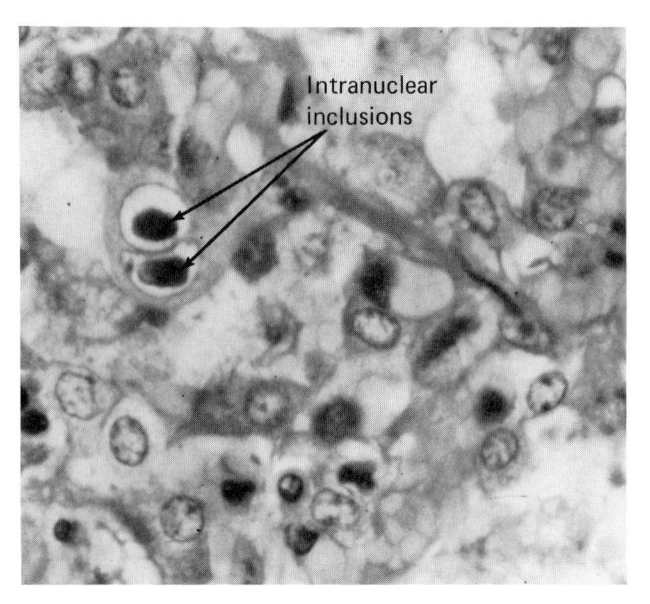

Figure 26–6 Cytomegalovirus infection, showing the presence of intranuclear inclusions in giant cells. CMV infections are of critical importance in newborns. *[A. Cangir, and M. P. Sullivan,* JAMA ***195:****1042, 1966. Courtesy of the University of Texas M. D. Anderson Hospital and Tumor Inst., Houston, Tex.]*

TRANSMISSION. The presence of CMV in urine, breast milk, cervical secretions, and semen provides the means for various routes of transmission. Congenital and neonatal infections (acquired before or at birth) from the mother occur in 5% to 7% of live births. The source of maternal CMV infection is rarely known. Acquisition of the virus through sexual contact has been proposed; some support for this hypothesis comes from studies involving homosexuals and of both heterosexual and homosexual patients seen in clinics for sexually transmitted diseases. Another possiblity is the shedding of CMV by young children. Such individuals may be a common source of maternal infections, as was true for rubella (German measles) before a vaccine was developed for the disease. No drug for prevention or treatment is currently available.

TABLE 26–4 General Features of Viral Diseases of the Gastrointestinal Tract

Disease	*Incubation Period*	*General Features of the Disease*	*Laboratory Diagnosis*	*Possible Treatment*
Cytomegalovirus inclusion disease (CMV)	Unknown	Symptoms of infected newborns include hepatitis, jaundice, increased size of liver and spleen, decreased number of blood platelets, and loss of sight; postnatally infected individuals may show no symptoms or may develop pneumonia; death can occur[a]	1. Demonstration of intranuclear inclusions (Figure 26–6) in specimens 2. Isolation of organism in tissue culture 3. Serological test, e.g., immunofluorescence, complement fixation, ELISA	No specific treatment; however, immune globulins and steroids have been used
Hepatitis A virus (HAV disease; infectious hepatitis)	15–40 days	Symptoms show a wide range and include abdominal discomfort, muscular pains, jaundice, dark urine, light-colored stools, fever, chills, and sore throat	1. Liver function tests 2. Serological tests to eliminate other disease states, e.g., ELISA 3. Demonstration of viral antigens and immunoglobulins	No specific therapy; supportive treatment including avoidance of physical stress, administration of vitamins and substances necessary to maintain caloric, fluid, and electrolyte balances; immune globulins have been used
Hepatitis B virus (HBV disease; serum hepatitis)	60–160 days	Symptoms are similar to those of hepatitis A disease; however, hepatitis B symptoms tend to be more severe	1. Liver function tests 2. Serological tests, e.g., complement fixation, hemagglutination inhibition, radio immunoassays, ELISA,[b] 3. Demonstration of viral antigens, e.g., (HBsAg) and immunoglobulins	Same as for infectious hepatitis A
Non-A, non-B hepatitis	50–70 days[c]	Similar to those listed for hepatitis B	1. Liver function tests 2. Serological tests to eliminate hepatitis A and B infection	1. Same as for hepatitis A 2. Alpha interferon has been successful in reducing liver injury
Infant diarrhea (rotavirus gastroenteritis)	48–72 hours	Abrupt onset of vomiting and diarrhea, fever, moderate dehydration, and normal white blood cell count	1. Demonstration of virus particles in stool specimens 2. Electron microscopic examination of stools 3. Serological tests, e.g., immunofluorescence, ELISA, complement fixation	No specific therapy, but fluid replacement is essential

[a] Several other infectious diseases may produce identical symptoms. These include the TORCH group (toxoplasmosis, rubella, cytomegalovirus inclusion disease, herpesvirus infection and syphilis).

[b] Radioimmunoassays are versatile, sensitive procedures that use antigens with radioactive labels for the measurement of antibody levels (titers) and/or the detection of antigens. Refer to Chapter 19.

[c] 10–40 days for infections other than those associated with past blood transfusions.

Viral Hepatitis

Viral hepatitis is an infectious enteric and systemic disease that characteristically involves the liver. Jaundice, a yellowing of tissues, is the most prominent symptom. The features of viral hepatitis are many (Table 26–4), ranging from viremia (viruses in the bloodstream) that does not seriously affect the liver to a destructive disease state ending in death within a few days.

It is customary to designate viral hepatitis in relation to a special causative agent, such as hepatitis A virus (HAV), formerly infectious hepatitis (IH); hepatitis B virus (HBV), formerly serum hepatitis (SH); and the non-A non-B viruses. The type A and type B microorganisms are separate viruses. Although they have several features in common, they differ in several respects. The most important of these include (1) the method of transmission, (2) the manner in which symptoms occur, (3) the incubation period, and (4) the chemical composition, size, and symmetry of the virus. Very little is known about the biological and physical properties of the non-A, non-B viruses. Studies have shown one of these agents to be sensitive to chloroform.

Humans are the only known natural hosts for viral hepatitis. Many cases of infectious hepatitis have been associated with closed communities such as camps, nurseries, housing tracts, and schools. Outbreaks have also occurred in hospitals. HAV is spread by fecal contamination of food or water. HBV, on the other hand, is usually transmitted by almost any type of injection, including those self-administered by drug addicts using contaminated needles and syringes. Moreover, doctors, nurses, technologists, and research technicians who handle blood or blood products are particularly vulnerable to this disease agent. Before the significance of the proper sterilization of syringes and needles in preventing serum hepatitis was realized, several outbreaks were reported in clinics giving routine injections. Unfortunately, even when suitable procedures are used, cases of the disease occur, especially in association with transfusions. Cases have also been traced to instruments used in oral surgery and to the practice of tattooing.

Non-A, non-B hepatitis accounts for 90% of post-transfusion cases of the disease. In addition to its transmission by blood transfusion, non-A, non-B hepatitis has occurred among drug abusers and patients undergoing kidney transplantation and dialysis. A large percentage of infections have also been shown to be unrelated to blood transfusions.

Experimental studies have shown mosquitoes to be potential transmitters of the disease agent.

Of those infected with HBV, 5% to 10% become chronic (long-term) carriers, and 15% to 25% of such carriers will develop significant chronic liver disease. Unfortunately, HBV carriers have a greater risk of developing a form of liver cancer (hepatic cell carcinoma).

PROPERTIES OF THE HEPATITIS VIRUSES A AND B AND HOST RESPONSES

Hepatitis A virus is one of the most stable viruses known to infect humans. It is a small (27-nm), nonenveloped, single-stranded RNA virus (Figure 26–7). The nucleocapsid of the virus, designated as the *hepatitis A antigen (HA Ag),* has a small protein component (VPg) that may aid viral attachment to host cell ribosomes. The capsid antigen of HAV consists of viral proteins 1 through 4 (VP1–VP4). These properties are typical of those found with picornaviruses such as polio and coxsackie.

During the course of a typical case of hepatitis A infection, the first serological indicators (markers) to

Figure 26–7 Hepatitis A virus. (a) An electron micrograph of purified negatively stained hepatitis A virus. *[Courtesy A. G. Coulepis.]* (b) Capsid organization. The protein capsid contains four specific virus proteins (VP1–VP4). The single-stranded (ss) RNA molecule of the particle has a genomic (genetically related) viral protein (VPg).

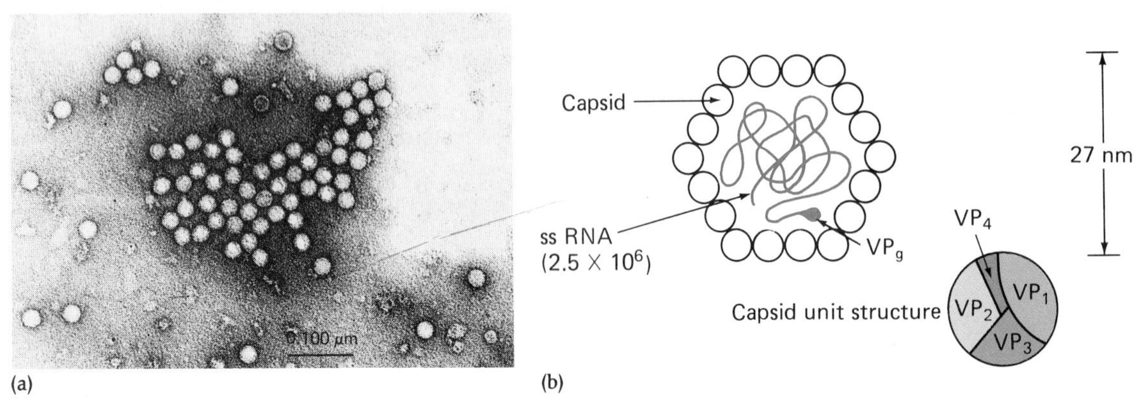

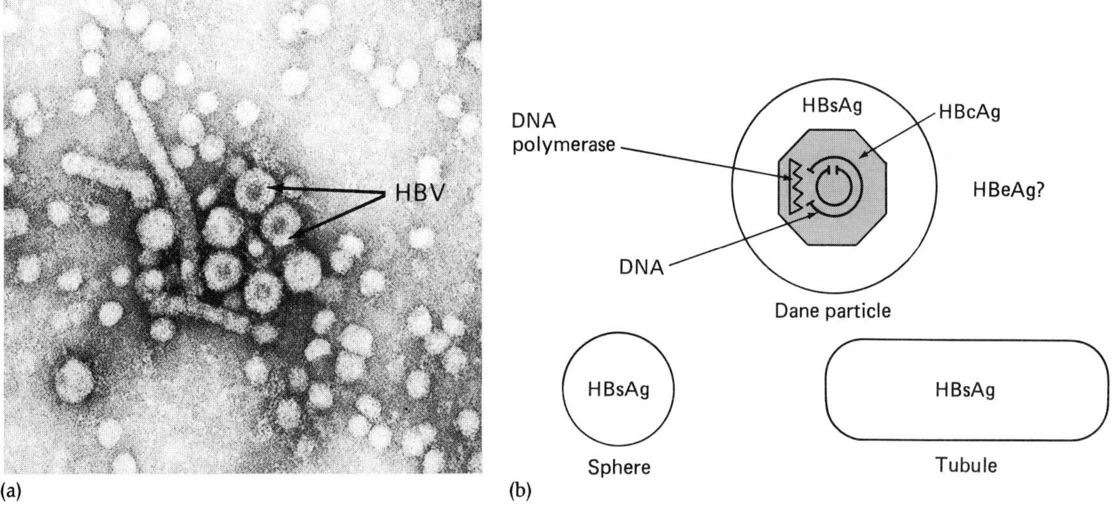

Figure 26–8 Hepatitis B virus structure. (a) An electron micrograph of human hepatitis B virus (HBV) and related parts. *[From C. R. Howard,* J. Gen. Virol. ***67****:1215–1235, 1986.]* (b) The structure of hepatitis B virus particles. The Dane particle, which is considered to be the hepatitis B virus, consists of an outer surface part (HBsAg) and an inner core, the hepatitis B core antigen (HBcAg). An enzyme, DNA polymerase, and a double-stranded, circular DNA molecule are found inside the core antigen. Incomplete virus particles have been found in specimens. These forms usually appear as spheres and tubules. They probably consist of ABsAg without HBcAg or the DNA molecule. The structural properties of the HBeAg component are not known at this time.

appear are hepatitis A antigen and virus in the stool (Figure 26–9a). The virus usually is not detectable in serum, but by the time symptoms appear, antibody to the virus (anti-HAV) can usually be found. The antibody titer rises rapidly and persists for a long time. The initial antibody response consists almost entirely of immunoglobulin M (IgM). However, in time the IgM titer decreases and IgG antibody levels increase. Diagnosis of HAV infection is based on one of the immunoglobulin responses (Table 26–4).

Through electron microscopy, several viruslike particles of hepatitis B have been demonstrated (Figure 26–8). Many of these components also serve as serological markers for sensitive and specific laboratory procedures with which to follow and determine the stage of infection, degree of infectivity, and immune status of a patient.

The hepatitis B virus is a 42-nm, double-shelled DNA virus (Figure 26–9b). The intact virion, referred to as the *Dane particle,* has an outer lipoprotein coat known as the *hepatitis B surface antigen (HBsAg)* and an inner core component known as the *hepatitis B core antigen (HBcAg).* The surface antigen, originally referred to as the *Australia antigen,* is found in the serum of many patients during both acute and chronic infections (Figure 26–9b). This situation occurs largely because the HBsAg circulates not only as the surface component of the Dane particle but also as an incomplete virus. While the second antigen, HBcAg, is not found in the serum, it can be demonstrated in the nuclei of liver cells from infected persons. HbsAg, in contrast, is found in the cytoplasm of infected liver cells.

The HBcAg replicates in the liver cell nucleus and there presumably causes the tissue damage associated with hepatitis B. Through some as yet unknown mechanism, the viral core migrates to the cell cytoplasm, where it is sheathed in the viral protein coat to become the Dane particle. This is believed to be the form in which the virus is transmitted. This mode of replication is similar to that of mouse leukemia and herpesviruses, whose protein coats are also synthesized in the cytoplasm.

A third antigen, *hepatitis Be antigen (HBeAg),* occurs in positive blood samples but does not appear to be associated with the Dane particle. Most evidence suggests that it is a subcomponent of the core antigen or its breakdown product.

Individual HAB particles have other distinctive properties, including a DNA polymerase which serves to complete a portion of the nucleocapsid, and a double-stranded, circular DNA, which may easily integrate into the DNA of host liver tissue.

OTHER TYPES OF VIRAL HEPATITIS

Other vital agents may also produce liver injury. Epstein-Barr virus, the cause of infectious mononucleosis, and cytomegalovirus are examples of recognized viral agents that produce hepatic damage, occasionally severe enough to be clinically recognized as hepatitis.

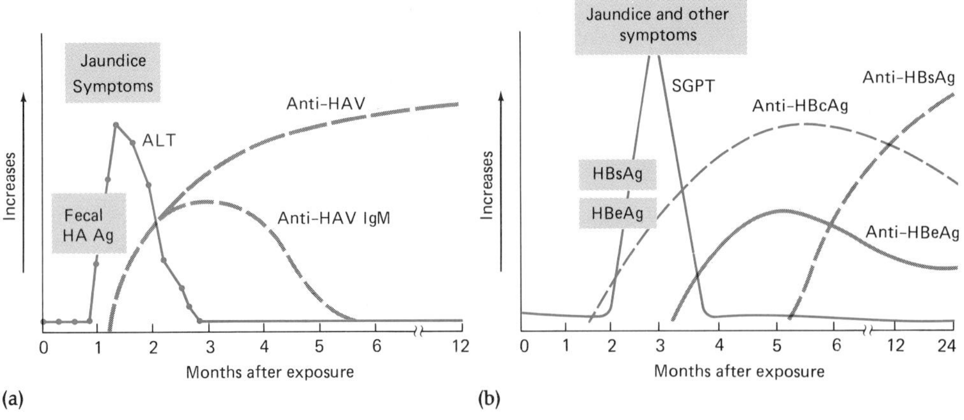

Figure 26–9 The serologic events associated with the typical course of (a) type A and (b) type B hepatitis acute infections. The appearance and relative concentrations of specific antigens and immunoglobulins and the onset of symptoms are shown. Abbreviations: HA Ag, hepatitis A antigen; HAV, hepatitis A virus; HBsAg, hepatitis B surface antigen; HBeAg, hepatitis B antigen; HBcAg, hepatitis B core antigen; anti-HBsAg, anti-HBeAg, and anti-HBcAg, immunoglobulins to respective hepatitis B viral antigens. Specific liver function tests indicated are ALT, alanine aminotransferase, and SGPT, serum glutamic-pyruvic transaminase.

A new agent distinct from the hepatitis B virus, called the **delta virus** (Figure 26–10), has been found in persons with a history of recent or past hepatitis B infection. The delta virus is found worldwide and exhibits an epidemiological pattern similar to that of hepatitis B infections. In general, a chronic or past hepatitis infection is required to establish a delta infection. The delta virus appears to piggyback on the hepatitis virus. Infected individuals may be asymptomatic or exhibit symptoms typical of viral hepatitis. Recombinant (gentically engineered) human alpha interferon is effective in reducing liver injury associated with the infection.

PREVENTION AND CONTROL

Since under most circumstances it is impossible to differentiate between the types of viral hepatitis, the preventive and control measures are applicable to both infectious and serum hepatitis.

In dealing with suspected or known cases of viral

Microbiology Highlights

ADMINISTRATION OF HEPATITIS B (HBV) VACCINE

Primary immunization for long-term protection against hepatitis B (HBV) infection includes three doses of HBV vaccine given intramuscularly. The second and third injections follow the first by respective intervals of one and six months. Immunization is appropriate for those individuals at significant risk for HBV infection. These include:

1. Contacts of HBV carriers (household members, sex partners, and institutional inmates)
2. Health-care workers (especially individuals handling blood specimens)
3. Users of illicit injectable drugs
4. Hemodialysis patients
5. Recipients of blood and related products that are possibly HBV-contaminated
6. Sexually active homosexual men
7. International travelers

Adverse reactions with the vaccine are few and consist mainly of localized pain at the injection site. Protection against disease appears to last beyond four years.

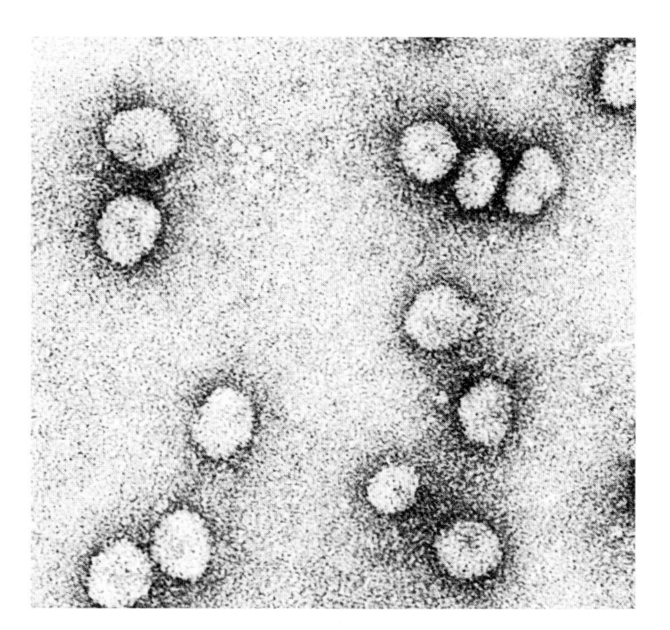

Figure 26–10 Delta virus particles. Original magnification 150,000×. *[From F. Bonino, B. Hoyer, J. W. Shih, M. Rizzetto, R. H. Purcell, and J. L. Gerin,* Infect. Immun. ***43**:1000–1015, 1984.]*

hepatitis, it is customary to observe enteric precautions. Such measures include (1) the hygienic disposal of feces, (2) the careful cleansing and disinfection of bedpans and toilet bowls, and (3) the proper sterilization of all dishes, eating utensils, bedclothes, and linen.

Preventing the transmission of viral hepatitis requires other important precautions as well. Nondisposable syringes, needles, tubing, and other types of equipment used for obtaining blood specimens or for the administration of therapeutic agents should be adequately sterilized before reuse. Most common means for chemical sterilization are not reliable for destroying hepatitis-causing viruses. Physical methods employed for this purpose are boiling in water for at least 20 minutes, heating in a drying oven at a temperature of 180°C for one hour, or autoclaving. Whenever possible, disposable equipment should be used for routine hospital and clinic procedures. *Disposable items should be properly sterilized before being discarded.*

The possibility of transmitting viral hepatitis by means of blood or plasma transfusions is yet another problem facing physicians and hospitals. An individual with a history of jaundice should not be used as a blood donor. Unfortunately, no method used today is effective in rendering whole blood containing hepatitis virus totally free of the agent. However, plasma exposed to the chemical viricide betapropiolactone and ultraviolet irradiation has been given to patients without adverse reactions.

When given shortly after exposure, or early during the incubation period, the use of immune globulin has proved effective in the prevention of hepatitis A. Specific recommendations for the administration of immune globulins (IGs) depend on the nature of the exposure to the virus. In general, IG is recommended for all household and sexual contacts of persons with the disease.

Two preparations, immune globulin (IG) and hepatitis B immune globulin (HBIG), are used for the prevention (prophylaxis) of hepatitis B infection. HBIG contains a much higher level of immune globulins and is more expensive than IG. HBIG is recommended for individuals experiencing a needle stick or mucous membrane exposure to blood containing HBsAg. The IG preparation is most effective if given immediately after exposure.

HEPATITIS VACCINES

One of the most important steps in producing a vaccine is the ability to grow the pathogen in question. Thus far, only hepatitis A has been grown (on a limited scale) in tissue culture. Such preparations are being used more as research tools than for diagnostic or preventive purposes. A vaccine for hepatitis B, however, that sidesteps this requirement has been prepared by using viral B antigens isolated from human carriers of hepatitis. The 22-nm viral component in the vaccine is HBsAg. It is readily found in and obtainable from the blood plasma of patients with hepatitis B. Filter-sterilized and formalin-treated vaccine preparations have been field-tested in clinical studies involving human volunteers. These preparations were found to stimulate antibody production and to provide protection to some individuals even after virus exposure. Rabies vaccine is the only preparation for which such a phenomenon has also been reported. Commercial vaccine preparation requires an extremely long production procedure and testing cycle. This elaborate approach is followed to confirm the elimination of any contaminating infectious agents, especially the AIDS virus and tissue components or proteins from blood donors that may be harmful to the liver.

In July 1986, a new, genetically engineered hepatitis B vaccine was licensed by the U.S. Food and Drug Administration. The recombinant vaccine is produced by the yeast *Saccharomyces cerevisiae* (sak-ah-row-MY-sees se-reh-VISS-ee-eye) into which a plasmid containing the gene for the hepatitis B surface antigen (HBsAg) has been inserted. The surface antigen is collected by disrupting the yeast cells and is separated from the yeast components by chemical and physical means. The immunogenicity of the recombinant vaccine is comparable to that of the plasma-derived product. Other recombinant vaccines derived from mammalian cells are currently under study and field testing.

Hepatitis B vaccine has great significance for humans in reducing the incidence of this most dangerous infection. The preceding "Microbiology Highlights" describes the leading candidates for immunization. No association has been reported between the vaccine and the virus of acquired immune deficiency syndrome (AIDS). The value of the hepatitis B vaccine to chronic carriers of the virus is highly questionable.

Rotavirus Gastroenteritis

Rapidly acting nonbacterial gastrointestinal infections are second only to acute respiratory infections as the cause of illness in families with young children (Table 26–4). Rotaviruses (originally known as *human reoviruslike agent*) have been identified as the major cause of sporadic acute gastrointestinal infections (enteritis) in infants and young children. Similar viruses are known to veterinary scientists as causes of diarrheal disease in calves, lambs, piglets, rabbits, antelope, and other lower animals.

Acute gastroenteritis caused by rotaviruses is a worldwide sporadic disease involving young children 6 to 24 months of age (Table 26–4). It is a leading cause of childhood deaths in developing countries. The disease agent is probably transmitted by the fecal-to-oral route. Infection is most common during the cooler months of the year, in contrast to bacteria-caused diarrheal disease. The infection can involve both adults and siblings within a family.

The human rotavirus (HRV) is spherical, contains double-stranded RNA, and exhibits a distinctive 70-nm double-capsid structure. In electron micrographs (Figure 26–11), its appearance suggests the wide hub, short spokes, and thin rim of a wheel, which provides the basis for the name *"rota,"* from the Latin word for wheel. HRV is a hardy virus, resisting temperatures of 56°C, a pH of 3, and most detergents and solvents. All rotaviruses thus far isolated belong to two immunologic groups, as demonstrated by complement fixation, ELISA, and immunoelectron microscopy. [See Chapter 19 for descriptions of these tests.]

Figure 26–11 Rotavirus particles from a filtered human stool specimen. Original magnification 224,840×.

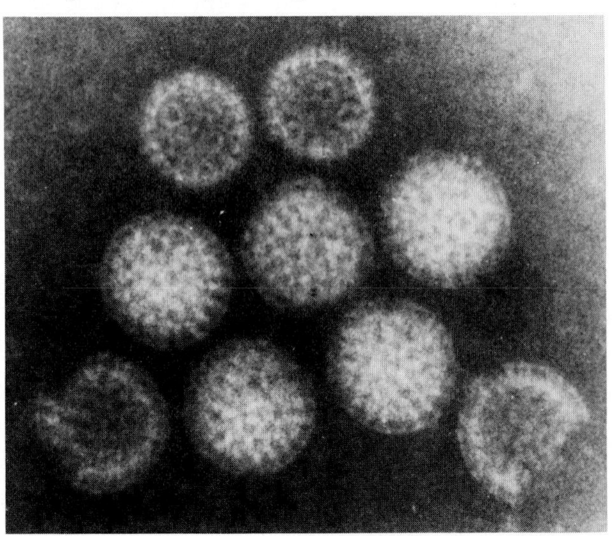

Prevention. Various studies have shown human colostrum and milk to contain HRV-specific IgA. Levels of immunoglobulin are high immediately after birth, and can be present for as long as nine months of lactation. Since human milk appears to provide some degree of protection against HRV, breast feeding of newborns generally is recommended. [See states of immunity in Chapter 18.]

A human vaccine against rotavirus gastroenteritis is currently under study. Studies are directed toward a live-attenuated HRV preparation that would produce both enteric (IgA) and serum (IgG) immunoglobulins after oral administration and in a manner similar to that obtained with the oral polio vaccine. [See vaccines in Chapter 18.]

Parasitic Protozoan Infections

Amebiasis (Amebic Dysentery)

Amebiasis, caused by *Entamoeba histolytica* (en-tah-ME-bah his-toe-LI-tee-kah; Table 26–5), occurs worldwide, with higher morbidity in warmer climates where opportunities for exposure to the protozoon are greater.

Transmission. Transmission of *E. histolytica* is most often by contaminated water supplies, flies, infected food handlers, and person-to-person contact. Communities with poor sanitary conditions provide opportunities for repeated exposure.

Life Cycle. Fully developed cysts (Figure 26–12 and Color Photograph 46) are ingested by humans. These forms undergo excystment, division, growth, and multiplication (Figure 26–13). The resulting *trophozoites* in the alimentary canal may penetrate the wall of the large intestine and multiply there. The parasites are known to feed well on red blood cells. Destruction of tissues may or may not produce the classic symptoms of amebic dysentery. Involvement of the liver and other organs may also occur as a result of bloodstream invasion by the protozoon. *Cyst* formation results when trophozoites enter the intestine and find unfavorable conditions. It should be noted that if trophozoites are rapidly discharged from this region, as in the acute diarrhea stage, these parasites die quickly. [See protozoan structures in Chapter 9.]

Prevention. Improvement in sanitation is the most important step toward disease prevention. Access to food by flies and cockroaches should be eliminated, and infected persons should not be allowed

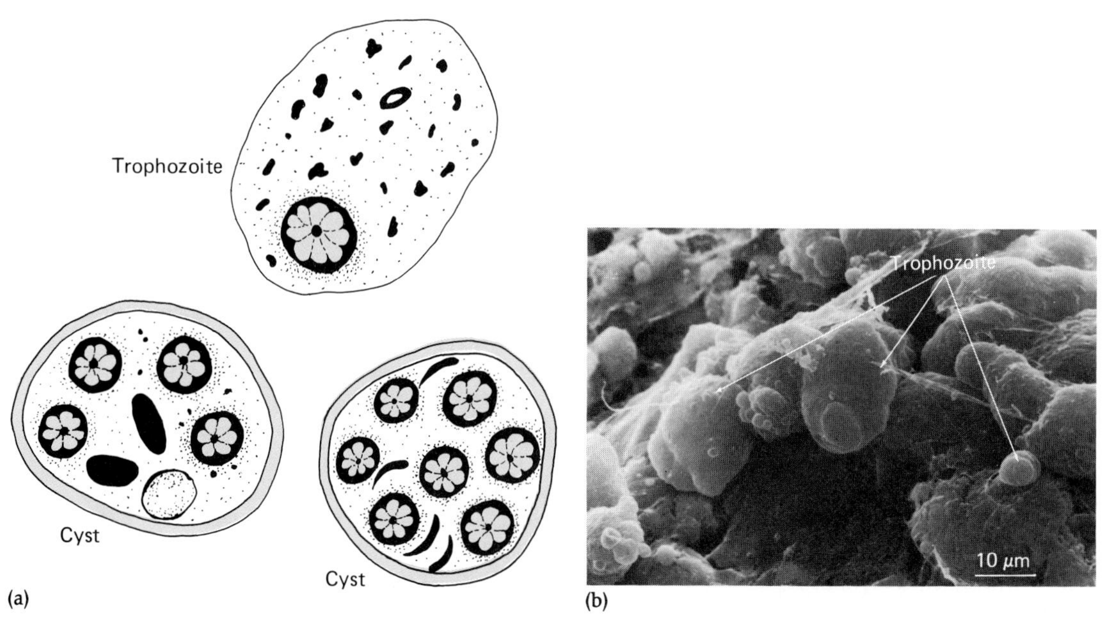

Figure 26–12 Intestinal protozoa. (a) *Entamoeba histolytica* trophozoite and cyst stages. An *E. coli* cyst is shown on the lower right for comparison purposes. (b) Amebic trophozoites sticking to the intestinal lining. *[From J. I. Ravdin,* et al., Infect. Immun. ***48****:292–297, 1985.]*

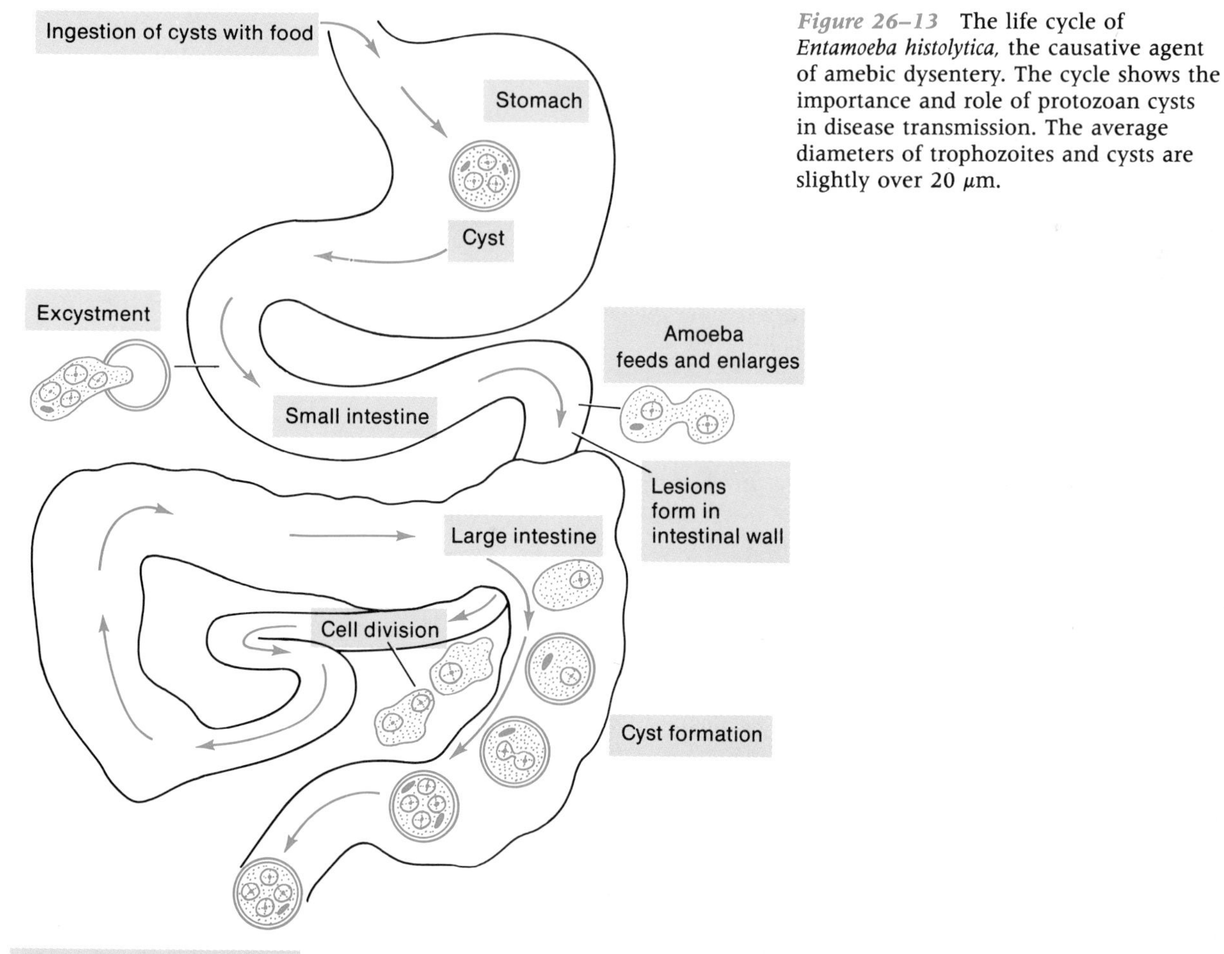

Figure 26–13 The life cycle of *Entamoeba histolytica,* the causative agent of amebic dysentery. The cycle shows the importance and role of protozoan cysts in disease transmission. The average diameters of trophozoites and cysts are slightly over 20 μm.

to handle food. Inadequate purification of water supplies must also be eliminated. Furthermore, education in personal hygienic habits and in the dangers of using as fertilizer untreated human fecal matter that may have come from infected persons should be expanded.

Balantidiasis

Balantidium coli (bal-an-TID-ee-um KOH-lee) (Color Photograph 48) is the only ciliate considered parasitic for humans. Geographically, *B. coli* is found practically worldwide. The parasite lives in the large intestine, where it can invade the mucosa and submucosa, causing ulceration and frequently a fatal form of dysentery.

Transmission. Humans acquire *B. coli* by ingesting viable cysts from contaminated food. Once *B. coli* becomes established in a human host, its transmission from person to person is greatly enhanced in areas where sanitation is poor.

Life Cycle. After cysts are ingested, the parasites in the host's intestine excyst, and the newly emerged trophozoites feed on various forms of organic matter, including starch grains, bacteria, and the host's cells. Invasion of the surrounding tissue may occur (Figure 26–14). Encystment of *B. coli* trophs takes place with the dehydration of fecal matter containing them, either before or after feces evacuation from the large intestine. Subsequent contamination of food or water with such material may start the cycle again.

Prevention of Infection. The most practical methods for the control of *B. coli* infection are better sanitary facilities, effective treatment of all infected persons, and increased education on the dangers of contaminated food and water.

Figure 26–14 *Balantidium coli* trophozoites in the intestine of an infected host. *[A. Westphal,* Z. Tropenmed. Parasit. ***22****:138–148, 1971.]*

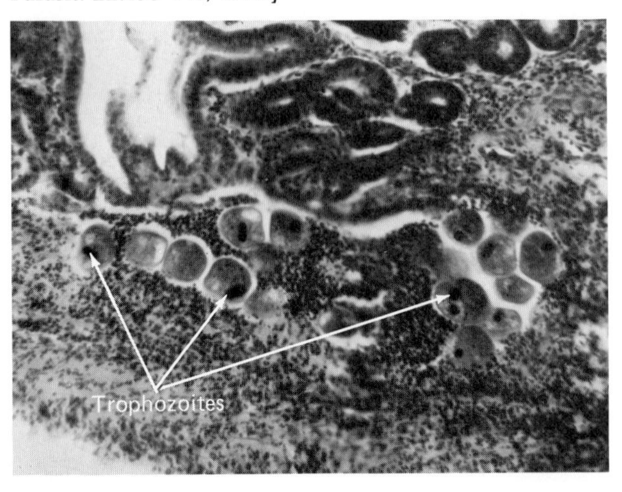

Cryptosporidiosis

Species of *Cryptosporidium* (crip-toe-spor-id-EE-um), a coccidial parasite in the intestinal tracts of several lower animals, causes intestinal disease in both immunosuppressed (primarily AIDS victims) and normal humans. Symptoms are generally mild (Table 26–5), not long-lasting, and characteristically include five to ten watery, frothing bowel movements per day, followed by constipation. *Cryptosporidium* is a common cause of diarrhea worldwide. Although 17 species of *Cryptosporidium* have been named, only two species, *C. parvum* (C. PAR-vum) and *C. muris* (C. MUIR-iss) are capable of infecting mammals. Transmission of the organism can occur from animal to human as well as from human to animal. The occurrence of infection in household contacts of infected individuals, children in day-care centers, healthy homosexual men, and people who have been hospitalized (hospital-acquired, or nosocomial infections) indicate that *Cryptosporidium* is highly infectious and transmissible from person to person. The disease agent is acquired through the ingestion of contaminated food and water. The development of *Cryptosporidium* (Figure 26–15) is similar in several aspects to the life cycles of protozoan agents of malaria and *Pneumocystis carinii* (new-mon-SIS-tis kar-I-nee-eye) pneumonia. **[See malaria in Chapter 27 for an explanation of life cycle stages.]**

Giardiasis

Giardia lamblia (jee-AR-dee-ah lam-BLEE-ah) is recognized as a cause of intestinal disease often acquired by travelers to foreign countries who ingest contaminated food or water, children in day-care nurseries, and homosexual males. Outbreaks of *Giardia* infections are being found with greater frequency in ski as well as summer resort areas in which sewage facilities are inadequate to destroy the causative agent in water supplies. Important additional sites of giardiasis are residential institutions for the mentally retarded, where hygienic practices may not be taught or enforced.

Life Cycle. Giardiasis generally results from the ingestion of *Giardia* cysts (Figure 26–10b). These organisms excyst in the stomach or upper small intestine, thereby releasing trophozoites that infect other parts of the small intestine and the body (Figure 26–16a). Encystment occurs later in the infection cycle. The infection can be either asymptomatic, mild, or severe in its effects (Table 26–5).

Prevention. Current studies indicate that well-operated conventional water treatment plants using processes including settling and filtration are necessary to prevent water-borne outbreaks. In addition, the approaches to prevention listed for amebic dysentery and balantidiasis are applicable.

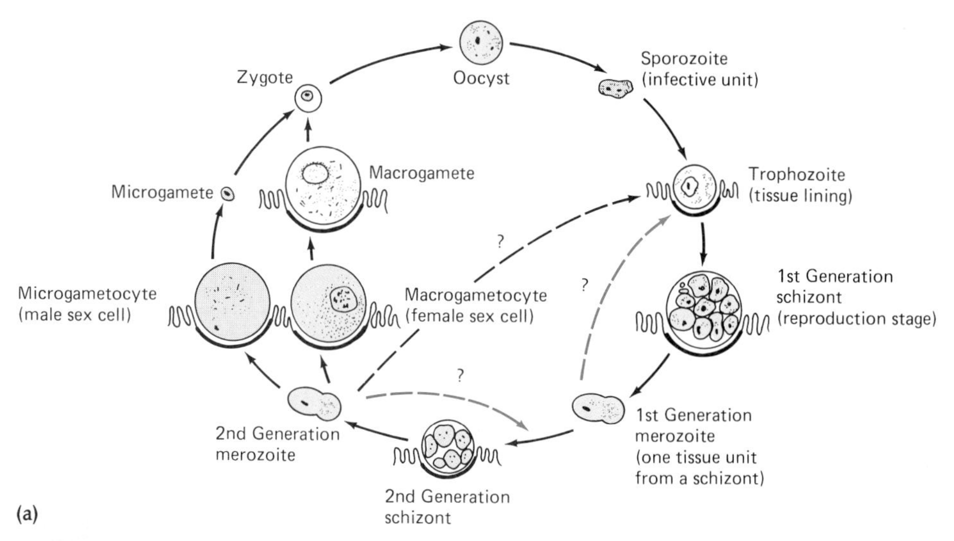

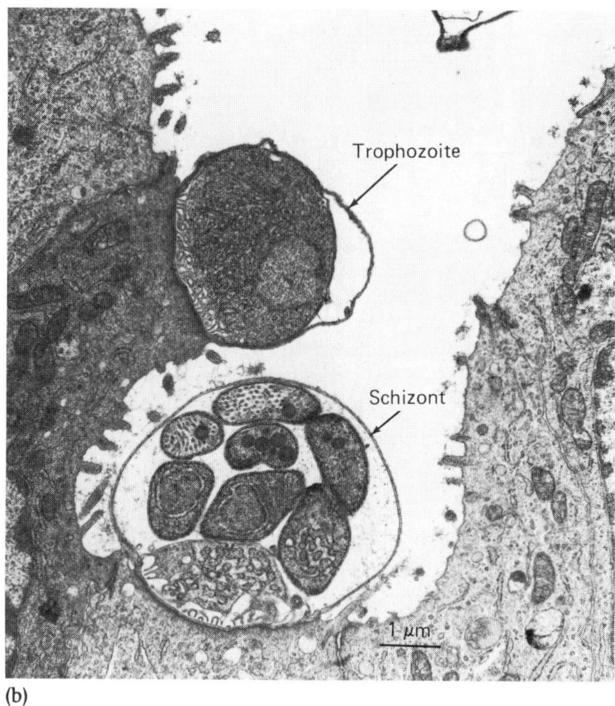

Figure 26–15
(a) The *Cryptosporium* life cycle. *[After T. R. Navin, and D. D. Juranek,* Rev. Infect. Dis. ***6****:313–327, 1984.]* (b) An electron micrograph showing the appearance of a trophozoite and reproductive stage (schizont) of *Cryptosporidium* species. *[From E. M. Liebler, J. F. Pohlenz, and D. B. Woodmansee,* Infect. Immun. ***54****:255–259, 1986.]*

TABLE 26–5 General Features of Protozoan Infections Associated with the Gastrointestinal System

Disease	*Causative Agent*	*Incubation Period*	*General Features of the Disease*	*Laboratory Diagnosis*	*Possible Treatment*
Amebiasis	*Entamoeba histolytica*	Usually 3–4 weeks	Abdominal cramps, flatulence, diarrhea, feces containing blood and mucus, weight loss, and general fatigue; complications such as invasion of the liver (amebic hepatitis) may develop	1. Demonstration of trophozoites and cysts in specimens (Color Photograph 46) 2. Serodiagnostic tests, e.g., immunodiffusion, fluorescent antibody (Chapter 19)	1. Chloroquine hydrochloride, metronidazole 2. Maintenance of fluid and electrolyte balance

(continues)

TABLE 26–5 (continued)

Disease	*Causative Agent*	*Incubation Period*	*General Features of the Disease*	*Laboratory Diagnosis*	*Possible Treatment*
Balantidiasis	*Balantidium coli*	Unknown	Intense abdominal pain, diarrhea, loss of weight, and vomiting; rapid destruction of tissue and death may occur	Demonstration of cysts in stool specimens (Color Photograph 48)	Antibiotics, including aureomycin and terramycin; carbarsone
Cryptosporidiosis	*Cryptosporidium* species	Unknown	Mild abdominal cramps, low-grade fever, nausea, loss of appetite, frequent watery, frothing stools followed by constipation	Histological examination of intestinal biopsy, demonstration of oocysts in stool specimens	1. Symptomatic 2. Stopping drug-induced immunosuppression, if possible
Giardiasis	*Giardia lamblia*	1–3 weeks	Generally no symptoms. In some cases, abdominal cramps, flatulence, nausea, and alternating constipation and diarrhea	Demonstration of cysts in specimens (Figure 26–16)	Quinacrine hydrochloride (Atabrine), metronidazole, furazolidone

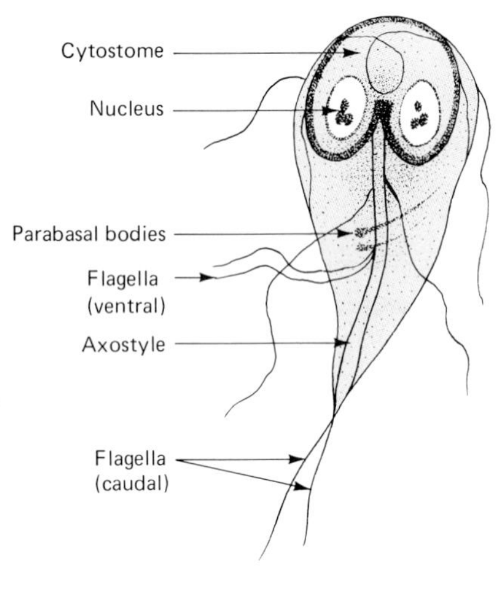

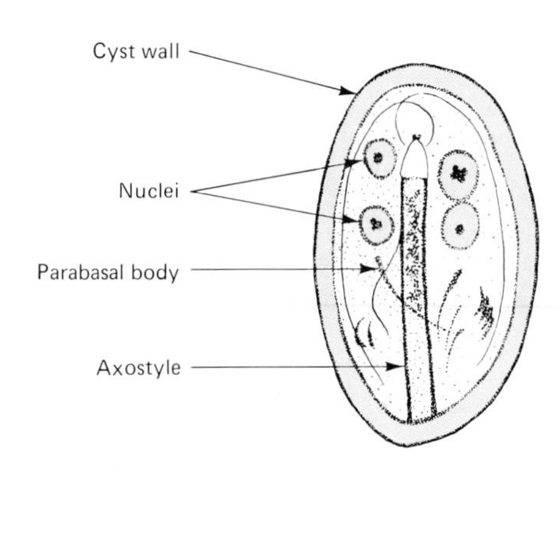

Figure 26–16 Composite sketches of the stages of *Giardia intestinalis.* (a) A typical trophozoite. The length of this form ranges from 9 to 21 μm, and the width can vary from 5 to 15 μm. (b) An ovoid cyst. These structures vary in length from 8 to 14 μm and in width from 7 to 10 μm.

Pediatric Gastroenteritis

Acute infectious diarrhea (gastroenteritis) is an extremely common illness in the pediatric population. It can be quite mild or severe and weakening. Most organisms that cause diarrhea are spread by the fecal-oral route, although certain pathogens, particularly those capable of causing infections in low doses, can be transmitted by direct person-to-person contact. In the last 12 to 18 years, remarkable progress has been made in recognizing "new" pediatric enteropathogens. For example, enteropathogenic *E. coli* and rotaviruses are recognized as important pathogens for newborns. With infants and toddlers, rotaviruses, *Giardia,* and *Shigella* species are known to be common enteropathogens. In adolescents, a unique group in the pediatric population, infections with enteric pathogens such as *Shigella* species, *Campylobacter, Giardia,* and *Entamoeba histolytica* are spread via sexual contact. Outbreaks of diarrhea in child day-care centers—with the pathogens mentioned here and others—are occurring with increasing frequency.

Disease Challenge

The situation described has been taken from an actual case history. It has been designed to show how clinical, laboratory, microbial, and related information is used in disease diagnosis. A review of treatment and epidemiological aspects is given at the end of the presentation. Answers to questions, laboratory findings, and interpretations, if appropriate, are given immediately following a specific question. Test your skill and take the Disease Challenge.

CASE

A 37-year-old woman was admitted to the hospital with a history of abdominal cramps, acute diarrhea, slight jaundice, and a low-grade fever. These symptoms had appeared five days earlier, approximately three days after her return from a river rafting trip in the Rocky Mountains. The patient indicated that she swam in and apparently swallowed some water from isolated streams.

At This Point, What Infectious Disease Would You Suspect?
Typhoid fever, infectious hepatitis, amebiasis, and giardiasis.

What Type(s) of Laboratory Specimens Should Be Taken?
Stool, small intestinal biopsy, and intestinal fluid should be taken for microscopic examination and bacterial and viral cultures. Blood for routine tests and immunological procedures should be taken.

Laboratory Results:
Bacterial and viral cultures failed to demonstrate the presence of any pathogens. However, microscopic examination of the stool and intestinal fluid specimens produced the results shown in the accompanying illustration. In addition, a large number of eosinophils was found on blood smear examination. Immunological tests were negative for typhoid fever and infectious hepatitis.

What Is the Probable Diagnosis?
Giardia lamblia infection.

What Treatment Should Be Given?
With patients exhibiting typical symptoms and upon laboratory confirmation, the drug of choice is currently quinacrine (Atabrine).

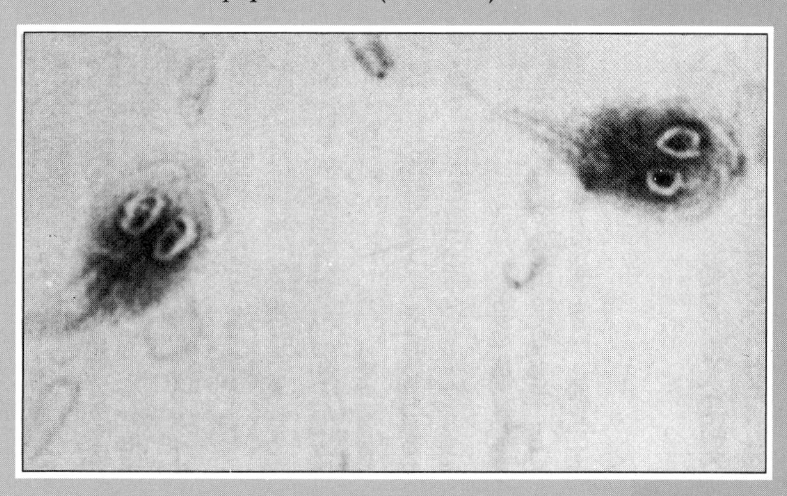

A stained preparation of intestinal fluid. *[Courtesy of Herbert L. DuPont, M.D.,* et al., *The University of Texas Health Science Center at Houston.]*

Summary

1. The human gastrointestinal system includes the mouth, oropharynx, esophagus, stomach, small and large intestines, the rectum, and several accessory organs such as the liver and pancreas.
2. The system serves as the means by which food and water are processed to meet the needs of the body.

Gastrointestinal Microbial Ecology

1. The structure of the gastrointestinal tract determines the location and composition of the microbial population (microbiota) of the system.
2. Prior to birth, the tract is sterile. During birth and thereafter, the system is exposed to a variety of microbes, some of which establish themselves (colonize) in specific locations.

Bacterial Diseases

1. Conditions in which there is a disturbance of intestinal functions accompanied by dehydration and the passage of liquid stools are referred to as *acute diarrheal disease.* Examples of diseases in which this condition exists are cholera, gastroenteritis, bacillary dysentery, and traveler's diarrhea.
2. The practical identification of bacterial pathogens includes extensive biochemical tests for the detection of specific enzymes.
3. Noncholera vibrios are also known to be human pathogens. These include *V. parahaemolyticus,* a cause of gastroenteritis, and *V. vulnificus,* a cause of bloodstream and wound infections. Important means of transmission include contaminated raw or improperly cooked seafood and exposure to contaminated seawater.
4. Several other bacteria can infect different regions of the system and cause serious illness. Examples of such disease are brucellosis, leptospirosis, salmonellosis and typhoid fever.

FOOD POISONING (INTOXICATIONS)

1. *Microbial food poisoning,* or *intoxication,* refers to symptoms resulting from the consumption of food or drink contaminated by pathogens or their toxins.
2. Examples of bacterial intoxications are botulism, *perfringens* poisoning, and staphylococcal food poisoning.
3. Three different forms of botulism are recognized, classic botulism, wound botulism, and infant botulism.

Viral Infections

1. Various viruses, as well as certain other microorganisms, invade the gastrointestinal system, use its parts for replication only, and then invade other body structures.
2. The pathogens that infect the gastrointestinal system include the coxsackie, echo, and polio viruses, cytomegalovirus, infectious hepatitis A and B and non-A, non-B viruses, and rotaviruses.
3. *Hepatitis A virus (HAV)* is a small, 27-nanometer, nonenveloped, single-stranded RNA virus. The nucleocapsid contains the *hepatitis A antigen (HA Ag)* and a specific small protein *(VPg)* that may aid viral attachment to host cell ribosomes.
4. *Hepatitis B virus (HBV)* is a 42-nanometer, double-shelled DNA virus. The intact virus *(Dane particle)* contains several components, including a *surface antigen (HBsAg)* and an *inner-core antigen (HBcAg).*
5. Hepatitis B infection is a particularly dangerous infection and can result in a chronic (long-lasting) state, with subsequent development of liver cancer. A new pathogen, the delta agent, requires previous or current B virus infection to cause disease.
6. Protection against hepatitis B can be provided with the use of hepatitis B immune serum for exposed individuals.
7. A new filter-sterilized, formalin-treated vaccine for HBV is available for general immunization.

Parasitic Protozoan Infections

1. Several protozoa can invade and infect the various parts of the gastrointestinal system. Examples of resulting protozoan infections are amebic dysentery, balantidiasis, cryptosporidiosis, and giardiasis.
2. These parasitic microbes exhibit two forms, the *cyst* (environmental resistant form) and the *trophozoite* (active feeding and invasive form).
3. Prevention of protozoan diseases and of diseases caused by other microbial types includes immunization (if available), improvements of sanitation, and adequate treatment of infected individuals.

Questions for Review

1. Could antibiotics taken by mouth affect an individual's microbiota (intestinal flora)? Explain.

2. Differentiate between wound botulism and infant botulism.

3. a. Would the heating of foods containing an enterotoxin inactivate the toxin? Explain.
b. What types of foods provide good growth conditions for staphylococci? For *Clostridium perfringens?*
c. List at least two measures to prevent staphylococcal poisoning.

4. Discuss ptomaine poisoning. Is it related to bacterial gastrointestinal diseases?

5. If you were going to tour Spain, Egypt, India, Peru, and Mexico, what diseases associated with the gastrointestinal system might you encounter? What precautionary measures would be advisable?

6. What disease outbreaks would you expect as a consequence of natural disasters such as earthquakes and floods?

7. Compare reservoirs for the following diseases:

a. Asiatic cholera	f. leptospirosis
b. bacillary dysentery	g. amebic dysentery
c. staphylococcal poisoning	h. rotavirus infection
	i. hepatitis A infection
d. EPEC	j. hepatitis B infection
e. brucellosis	k. delta agent

8. What protective mechanisms against viral and bacterial diseases are provided by the human gastrointestinal system?

9. Propose an approach to the identification of a bacterial pathogen of the gastrointestinal tract.

10. a. Distinguish between hepatitis A and hepatitis B infections.
b. What control measures can be used to prevent both diseases effectively?
c. Why do these diseases pose a particular problem as well as a danger to hospital personnel and patients?

11. Are vaccines or immunization procedures used for any gastrointestinal associated diseases? If so, list them.

12. What is cytomegalic inclusion disease? Are there are predisposing factors associated with it?

13. What is the relationship of the delta virus to hepatitis B virus?

14. What protozoon is an important factor in AIDS patients?

15. Discuss rotavirus infection with respect to the following topics:

a. causative agent	c. treatment and prevention
b. age group affected	

16. a. Is the liver vulnerable to infectious agents? Explain.
b. List other parts of the gastrointestinal system vulnerable to such attacks.

17. What general types of diagnostic methods are used for viral infections of the gastrointestinal tract?

18. What diseases are associated with the various members of the picornaviruses?

19. Unravel the alphabet mixture. Use the various letters shown to form meaningful abbreviations associated with tests and/or properties of hepatitis viruses. Note that the letters can be used more than once. Your answer should produce at least nine correct complete combinations.

D G virus s

A e Ig Ag

H B M c

delta A

Suggested Readings

CALLEA, F., M. ZORZI and V. J. DESMET (eds.), *Viral Hepatitis.* Secaucus, N.J.: Springer-Verlag New York, Inc. 1986. *A detailed presentation of the different forms of viral hepatitis, which include the features of hepatitis A, B, and delta virus infections.*

ERLANDSEN, S. L. (ed.), *Giardia and Giardiasis.* New York: Plenum Publishing Corporation, 1984. *This book provides an organized and comprehensive coverage of the major biological aspects of the protozoon* Giardia. *The information covered includes a correlation of structure and function, diagnosis and treatment of giardiasis, and host immunological responses.*

FEINSTONE, S. M., and R. H. PURCELL, "Non-A, Non-B Hepatitis," *Annual Review of Medicine* **29:**359, 1978. *Describes a new agent found to be one cause of human hepatitis. Its clinical characteristics, attempts to isolate a causative agent, and approaches to prevention are also discussed.*

HILL, M. J. (ed.), *Microbial Metabolism in the Digestive Tract.* Boca Raton, Fla: CRC Press, Inc., 1986. *A comprehensive coverage of the metabolic activities of the gut bacterial flora in relation to health and disease. Emphasis is given to topics such as hepatic disease, dental caries, acute enteric infection, and digestive tract cancer.*

IMMUNIZATION PRACTICES ADVISORY COMMITTEE (ACIP), CENTERS FOR DISEASE CONTROL, "Hepatitis B Vaccine: Evidence Confirm Lack of AIDS Transmission," *MMWR* **33:**685–687, 1984. *This report provides evidence for the safety of hepatitis B vaccine.*

MONDELLI, M. and A. L. W. F. EDDLESTON, "Mechanisms of Liver Cell Injury in Acute and Chronic Hepatitis B," *Seminars in Liver Disease* **4:**47–57, 1984. *This article highlights recent advances that to some extent have clarified the mechanisms of immune-mediated liver cell injury triggered by hepatitis B virus (HBV) infection.*

QUINN, T. C., B. S. BENDER, and J. G. BARTLETT, "New Developments in Infectious Diarrhea," *Disease-a-Month* **32:** 166–244, 1986. *An excellent review of infectious diarrhea, with emphasis on the more recent developments.*

27

Microbial Infections of the Cardiovascular and Lymphatic Systems

The Greeks, whenever doubt arose with regard to any disease, always thought it best to rely on nature and her doings, which were sure, in the last resort, to banish the disease. They based their opinion on the following grounds: Nature, being the servant and provider of all living things in healthy days, helps them also in disease. — Galen

After reading this chapter, you should be able to:

1. Describe the general organization and functions of the human circulatory system and indicate the areas attacked by microbial pathogens.
2. List and discuss at least six bacterial diseases associated with the circulatory system, specifying the causative agent, means of transmission, preventive measures, and methods of control for each disease.
3. Discuss the features of the viral diseases infectious mononucleosis and acquired immune deficiency syndrome (AIDS), and the protozoan disease malaria.
4. Describe diseases caused by retroviruses.
5. Distinguish among the general effects resulting from bacterial, protozoan, and viral infections of the circulatory system.
6. Describe the approaches to the diagnosis of rickettsial diseases and malaria.
7. Apply the clinical symptoms and laboratory findings to a possible disease diagnosis in solving a disease challenge.

Chapter 27 briefly describes the various components of the cardiovascular and immune systems and emphasizes the microbial diseases affecting them. These diseases include malaria, plague, and certain rickettsial infections. Special attention is given to the viral diseases infectious mononucleosis and AIDS.

The development of many infectious diseases follows a consistent pattern. Once pathogens enter the body and establish a local, or main, site of infection, they may spread to other regions of the body by means of the circulatory system, forming secondary sites of infection. Depending on the properties of the disease agent, the circulatory system may spread toxins released by the pathogen. In addition, certain microbial disease agents attack and destroy tissues of the cardiovascular system.

The circulatory system meets the need of humans and other vertebrates for internal transport and intercommunication between the various cells of the body. Blood, the transport medium, carries nutrients and hormones, removes cellular waste products, assists in the regulation of body temperature, and aids in the control and elimination of foreign organisms. The blood flows under pressure in a one-way path through the vessels (arteries, veins, and capillaries) and heart directly to the tissues and organs where it is needed. The blood is pumped from the heart into the *aorta,* the artery leading from the heart, then into other arteries and on to the smaller capillaries. Upon serving its purpose, the blood is returned to the heart through the veins.

The heart functions as a pump, providing the force necessary to maintain adequate blood flow throughout the body. Actually, it is a double pump: the right side receives blood from the body and pumps it to the lungs, where carbon dioxide in the blood is exchanged for oxygen, and the left side receives the freshly oxygenated blood from the lungs and pumps it to other body organs. Figure 27–1 shows the general direction, or path, followed by the blood through the human circulatory system.

As described in Chapter 17, T and B lymphocytes together are essential for the development and maintenance of immunity. Abnormalities in their structure and function produce a variety of specific immunodeficiency conditions as well as certain associated microbial diseases. Such conditions may be genetically determined, acquired, or associated with other diseases. The features of diseases caused by viruses, acquired immune deficiency syndrome (AIDS), and infectious mononucleosis are also considered in this chapter.

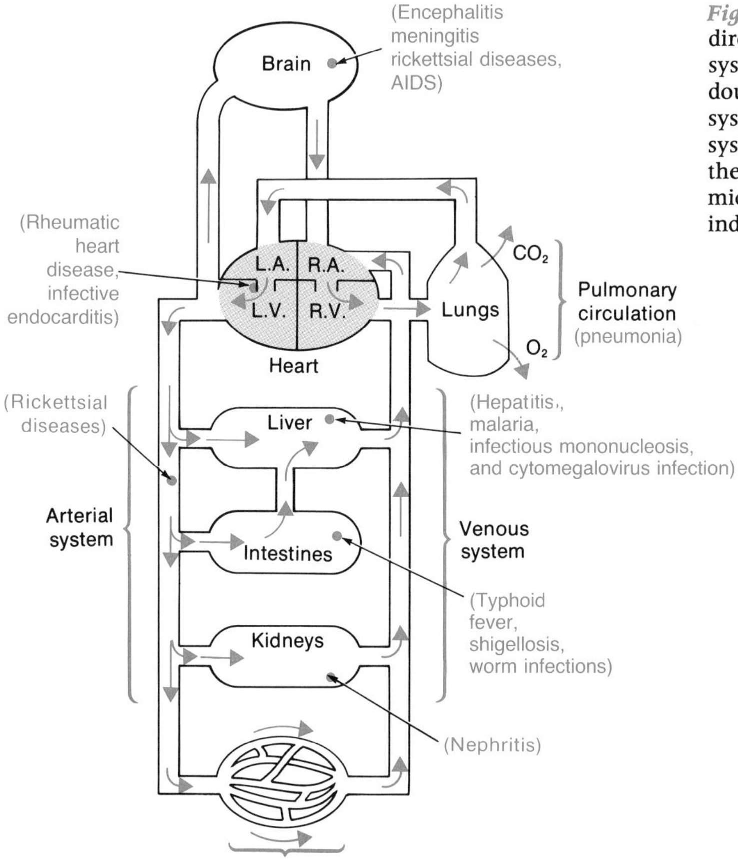

Figure 27–1 The structure and general direction of the human circulatory system. This closed system is really a double one consisting of the pulmonary system, which serves the lungs, and the systemic system, which serves the rest of the body. Specific organs with which microbial infections are associated are indicated.

Diseases of the Heart

In most developed countries, heart disease is the leading cause of death. The activities of the heart can be impaired in at least three ways: (1) **endocarditis,** including infection and damage of the heart's valves and associated tissues; (2) insufficient nourishment for cardiac muscle, due to a narrowing of the arteries of the heart sometimes associated with a blood clot (thrombus) within these vessels; and (3) overexertion of the heart, resulting in exhaustion. Any one of these situations can cause enlargement of the heart, one of the most important signs of heart disease.

The heart can suffer from infection just as any other part of the body can. Of the various infections of the heart, rheumatic heart disease and infective endocarditis are the most common.

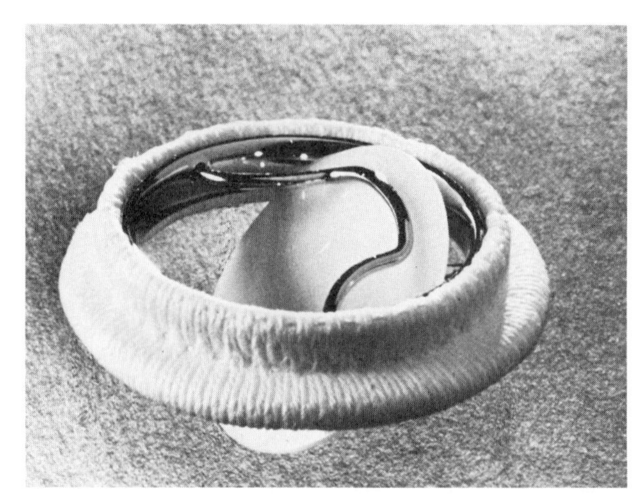

Figure 27–2 An artificial replacement valve. The device is surgically inserted into the area of the damaged valve. Note the centrally located movable valve.

Rheumatic Fever and Rheumatic Heart Disease

Many diseases, including those of the eye, skin, throat, gastrointestinal tract, bones, and joints, are known also to affect the heart. Rheumatic fever is one such disease. It is a hypersensitivity state that develops in a small percentage of individuals following such streptococcal infections as sore throat. Because proteins of group A hemolytic streptococci are antigenically similar to the proteins in the heart and other tissues of susceptible persons, the antibodies formed attack not only the bacteria but also the host's heart tissues. While the resulting inflammation is usually mild, it may produce rheumatic fever and serious involvement of the heart. The consequences of one or more episodes of rheumatic fever cause the heart valves, particularly the mitral valve, to become inflamed. Abnormal growths of connective tissue (fibrosis) form on portions of the valves, scarring and deforming them. This condition is known as *rheumatic heart disease.* The valve damage is produced by narrowing of the valve opening *(stenosis)* or failure of the valve to close completely *(valvular insufficiency).* In short, the valves do not function properly. If the damaged valve cannot be repaired surgically, it may be replaced with an artificial valve (Figure 27–2). Fibrosis of the mitral valve, the valve most commonly involved, usually leads to left heart failure.

Streptococcal Infection

LABORATORY DIAGNOSIS

Accurate identification of group A streptococci requires both clinical and bacteriological findings. Isolation of these organisms from throat swabs and related specimens can be made by streaking on sheep blood agar plates. Placing a bacitracin sensitivity disk over a streaked area is also helpful for a presumptive identification. Group A streptococci are beta hemolytic (Color Photograph 23b) and bacitracin-sensitive (Color Photograph 87). Highly specific and rapid serological methods such as those using Phadebact and Streptex are reliable tests for laboratory identification. Both tests are modified forms of an agglutination reaction.

TREATMENT

Penicillin is the drug of choice for infections caused by group A beta-hemolytic streptococci. Erythromycin is given to patients allergic to penicillin treatment.

Prevention. To prevent rheumatic heart disease, sore throats and other disease states caused by group A hemolytic streptococci *(Streptococcus pyogenes)* should be diagnosed and treated with antibiotics to minimize the development of antibodies to the streptococcal antigens.

The popular belief that rheumatic heart disease is restricted to young persons of low-income groups in cold, wet countries is incorrect. It is a disease of universal distribution, found in all climates, races, social strata, and ages.

Infective Endocarditis

Heart valves that have been scarred by rheumatic fever are readily attacked by bacteria, which can lodge in the irregular, roughened portions of these valves. The typical lesion in these situations is a vegetative meshwork of fibrin and blood platelets in which microorganisms are deeply embedded (Figure 27–3). These organisms are thereby protected from host defense mechanisms and the penetration of antimicrobial drugs. The left side of the heart is usually the main site of involvement except in the case of drug addicts, in whom the tricuspid valve tends to be affected. Until recently, this heart disease was called *bacterial endocarditis.* It is now referred to as *infective endocarditis.* The change in name reflects the fact that although several bacterial species—including specific *Rickettsia* (reh-KET-see-ah) and *Chlamydia* (kla-MEH-dee-ah)—account for most cases, fungi and viruses also may be causative agents. In addition, infective endocarditis is classified according to the causative agent and no longer according to acute or subacute categories.

Less commonly, infective endocarditis may develop in valves altered by congenital malformations, infections other than rheumatic fever, and roughened, irregular areas resulting from hardening of the arteries (arteriosclerosis).

Infective endocarditis may be a very serious complication of any dental procedure that allows bacteria to enter the bloodstream of a person with heart valve damage. Bacteria often enter the blood (producing *bacteremia,* or bacterial sepsis) from extractions, endodontics, surgical procedures associated with gums, deep scaling (removing infected material from the tooth's surface), and curettage (removal of material by scraping). Any manipulation of infected or inflammatory vascular tissues within the gums, even by chewing, can open small capillaries or force microorganisms into the bloodstream. The greater the injury, the greater the incidence of bacteremia.

Individuals with infective bacterial endocarditis experience prolonged fever, a changing heart murmur, and the growth of bulky bacterial vegetations on the heart valves (Figure 27–3). These accumulations consist of tangled masses of fibrin strands, platelets, and blood cell fragments along with the

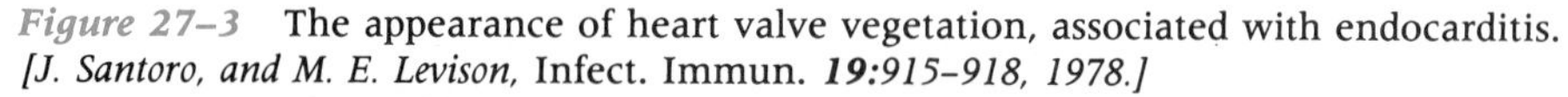
Figure 27–3 The appearance of heart valve vegetation, associated with endocarditis. *[J. Santoro, and M. E. Levison,* Infect. Immun. *19:915–918, 1978.]*

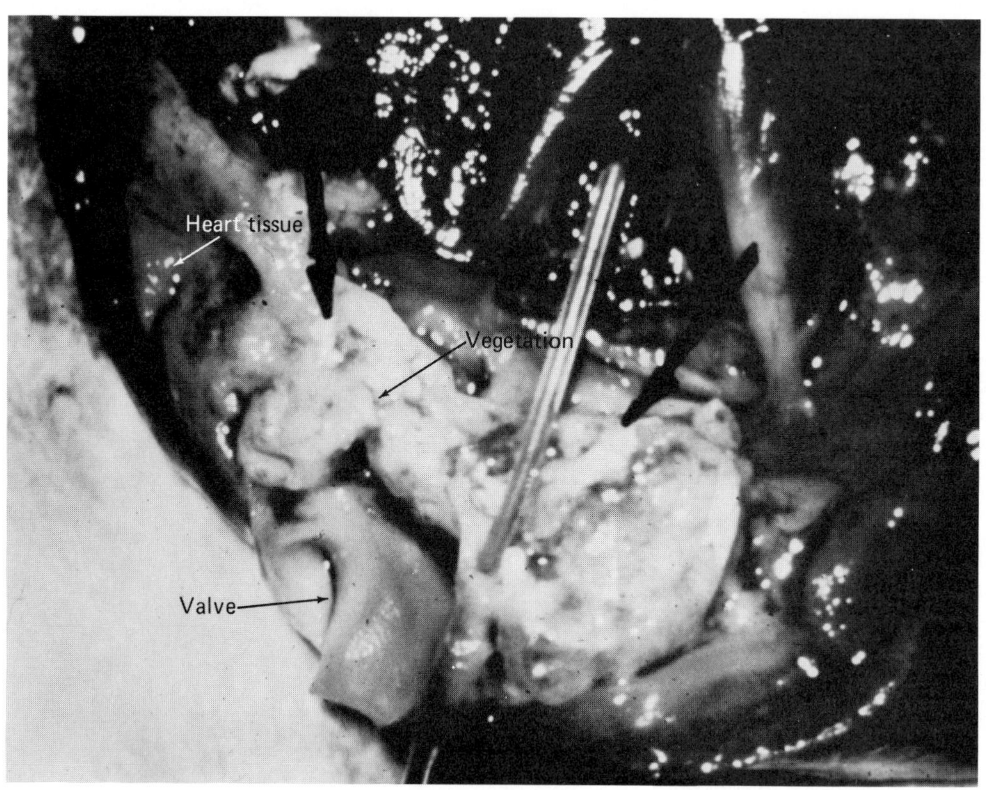

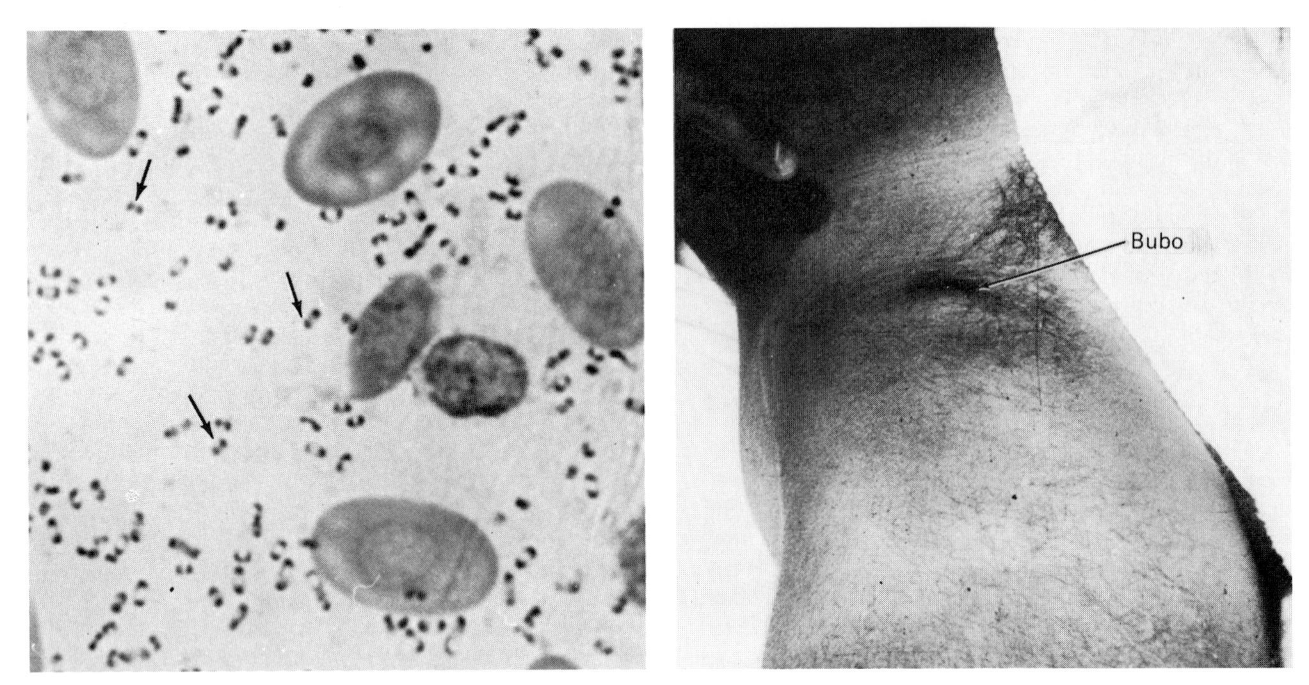

Figure 27–4 (a) The bipolar, or "safety pin," appearance of *Yersinia* and *Pasteurella* species are indicated by arrows. *[Courtesy of Drs. C. A. Manthei and K. L. Heddleston, USDA.]* (b) The appearance of a plague bubo on a victim's leg. Plague is a zoonosis and is caused by *Yersinia (pseudotuberculosis* subspecies*) pestis. [Courtesy of Mycology Section, Laboratory Division, Center for Disease Control, USPHS.]*

Infective Endocarditis

Laboratory Diagnosis

Laboratory identification of the pathogens listed in Table 27–1 is straightforward. Appropriate blood specimens streaked on blood agar generally yield colonies within 24 to 48 hours. In addition to Gram staining, biochemical and serological tests are required to differentiate the various causative agents.

General properties of these microorganisms include the following: *Cardiobacterium hominis* (kar-dee-oh-back-TEH-ree-um hom-EE-niss) is a nonhemolytic, catalase-negative, nonmotile fermentative rod. *Streptococcus pyogenes* is a beta-hemolytic gram-positive coccus; it is sensitive to bacitracin and can be serologically identified by a modified form of agglutination reaction in the Phadebact test (Color Photograph 105). *S. sanguis* and *S. faecalis* are alpha-hemolytic and nonhemolytic gram-positive cocci, respectively. Both species can be grouped by the Phadebact test. *S. aureus* is beta-hemolytic, produces catalase, and ferments mannitol (Color Photographs 26 and 96). Pathogenic strains may or may not be coagulase-positive. [See Chapter 22 for a discussion of coagulase.]

Treatment

Penicillin, tetracyclines, and aminoglycosides are the drugs of choice for *C. hominis* infections. A combination of penicillin G plus penicillin G procaine is appropriate for rheumatic heart patients and others with infections caused by most streptococci. Vancomycin may be substituted in cases of penicillin allergy.

bacterial masses. Infected individuals may also develop uncontrolled infections by antibiotic-resistant organisms, or organisms may be dislodged from damaged valves and travel to other organs, resulting in further destruction and even death.

CONTROL AND PREVENTION. The successful control of infective endocarditis includes prevention of rheumatic heart disease and prompt diagnosis and treatment.

Other Microbial Diseases of the Circulatory System

Signs and Symptoms of Infections

The signs and symptoms of various infectious diseases involving the circulatory system are nonspecific. However, in some cases the features of the disease are highly specific and suggestive of an infection. During the course of several infectious diseases, microorganisms such as bacteria invade the bloodstream, producing a bacteremia. Quite frequently, a bacteremia is associated with some form of underlying defect in the host's immune defenses. Other contributing factors include age (both extremes); underlying diseases such as solid tissue growths (tumors), leukemia, and other blood system-associated cancers; surgery or instrument manipulation of areas such as the urinary and gastrointestinal tracts; the use of certain antibiotics; and the administration of immunosuppressive drugs. In certain disease states, bacterial invasion and destruction of the circulatory system produce clinical signs and symptoms such as chills or fever. Tables 27–1 and 27–2 list the characteristic clinical signs and symptoms of certain microbial diseases. Approaches to laboratory diagnosis and treatment are included with the descriptions of the respective diseases in this chapter.

Bacterial Infections

PLAGUE

For centuries, the plague bacillus has caused pandemics that have ravaged Asia and Europe. The Great Plague, which began in A.D. 542, is believed to have been responsible for the deaths of more than 100 million people in 50 years. The pandemic that reached its height during the fourteenth century and became known as the Black Death has been considered the worst catastrophe to strike Europe and perhaps the world. An estimated one-third of the world's population died in it. Serious outbreaks of plague continued to appear in Europe and Asia from 1360 to 1400. The name *Black Death* was coined because of the severe cyanosis (blue or purple color of the skin) that developed in the terminal stages of the disease (Table 27–1). The last pandemic of the nineteenth century started in central Asia in 1871 and spread to other parts of the world. Epidemics continue to occur occasionally in many regions of Asia and Africa. There have also been reports of sporadic infections in South Africa, South America, and the southwestern United States. Plague was apparently spread to South Africa from South America in 1899.

The causative agent of plague, *Yersinia pestis* (yer-SIN-ee-ah PES-tis; Figure 27–4a), was identified in 1894, independently by Alexandre Yersin and Shibasaburo Kitasato.

This bacterial infection is primarily a disease of rodents, both domestic and wild. However, several other mammalian species, including cats, deer, kangaroos, and monkeys, can be infected. Recurrent outbreaks apparently occur among various species of wild rodents, such as pack rats, prairie dogs, rabbits, and squirrels. This form of the disease, known as *sylvatic plague,* poses a serious threat to human well-being, since such infected animals are a source of disease agents for future epidemics. **[See reservoirs of infection in Chapter 21.]**

TRANSMISSION. The major direct sources for humans are house rats, *Rattus rattus* and *R. r. dairdi,* and the ship's rat, *R. r. alexandrinus.* Plague is transmitted by the bite of rat fleas primarily belonging to the genera *Nosopsyllus* and *Xenopsylla.* Infected fleas regurgitate microorganisms together with aspirated blood into the wound caused by their bites and may deposit feces. *Y. pestis* does not reproduce within the tissues of the insect vector.

FORMS OF PLAGUE. Plague takes two main forms in human beings: *bubonic plague* and *primary pneumonic plague.* In untreated cases the mortality rate for bubonic plague varies from 70 to 90%. For primary pneumonic plague, it is 100%. Death occurs quickly if treatment is not administered.

Bubonic plague is acquired through the bite of an infected flea. The injected microorganisms gain entrance to the regional lymph nodes, usually in the groin. These lymph nodes become enlarged and extremely tender, thus resulting in *bubo formation* (Figure 27–4b). Without treatment, a large number of patients develop bacteremia and die of septic shock within hours or days after the appearance of the bubo. *Pneumonic plague* develops within 5% of infected patients. Symptoms include difficulty in breathing, fever, and the bluish to black skin discoloration that is associated, as indicated earlier, with the naming of the disease as the Black Death. Without specific treatment, death occurs on or about the second or third day after the appearance of symptoms.

TABLE 27–1 Selected Bacterial Diseases Associated with the Cardiovascular System

Disease	*Causative Agent*	*Gram Reaction and Morphology*	*Incubation Period*	*General Features of the Disease*
Infective endocarditis	*Cardiobacterium hominis* *Staphylococcus aureus* *Streptococcus faecalis* *S. faecium* *Streptococcus salivarius* *S. sanguis*	−, Rod +, Coccus +, Coccus +, Coccus +, Coccus +, Coccus	Days to weeks	Fever, general weakness, heart murmur
Plague	*Yersinia pestis*	−, Rod	2–6 days	*Bubonic plague:* formation of enlarged lymph nodes *(buboes)*, fever, chills, severe headache, and exhaustion *Pneumonic plague:* contagious, coughing, chest pains, difficulty in breathing, and bluish coloration of skin (cyanosis)
Relapsing fever	*Borrelia recurrentis*	Not done, Spirochetes	3–10 days	Sudden appearance of fever, which lasts for 4 days and ends suddenly, continued fever attacks (relapses), with each one milder than previous ones
Rheumatic fever[a]	Group A streptococci *(Streptococcus pyogenes)*	+, Cocci	1 to 5 weeks	Sore throat, fever, general discomfort, painful joints (arthritis)
Rickettsialpox	*Rickettsia akari*	−, Rod	10–24 days	Formation of a small, red, hard blister at site of mite bite, followed by sudden onset of backache, chills, fever 39.4°C and rash
Rocky Mountain spotted fever	*Rickettsia rickettsii*	−, Rod	3–10 days	Backache, chills, fever and rash, which develops on ankles, forehead, and wrists and spreads to trunk
Scrub typhus	*Rickettsia tsutsugamushi*	−, Rod	10–20 days	A localized sore site of mite bite followed by backache, chills, fever (38.5°–40.15°C), and rash; deafness and mental disturbances can be complications
Tularemia	*Francisella tularensis*	−, Rod	3–10 days	Fever, headache, general discomfort, and enlarged lymph nodes with pus formation
Typhus (endemic) fever	*Rickettsia typhi (mooseri)*	−, Rod	6–15 days	Symptoms similar to those of other typhus fevers
Typhus (epidemic) fever	*Rickettsia prowazekii*	−, Rod	10–15 days	Symptoms similar to those of other typhus fevers; rash begins on trunk and spreads to arms and legs

[a] Inflammation of heart tissue, injury to heart valves, and rheumatic heart disease can result from this disease.

Plague

Laboratory Identification

Sputum samples, specimens aspirated from buboes, and blood cultures should be examined for the presence of cells with typical bipolar staining (Figure 27–4). Wayson's stain (a mixture of methylene blue and basic fuchsin) is used for this purpose. Specific identification requires the fluorescent antibody technique with *Y. pestis* antiserum and bacteriophage typing.

Treatment

Because of the severe nature of plague, treatment should be started even before laboratory identification is complete. The drugs of choice include chloramphenicol, streptomycin, tetracycline, or a combination of trimethoprim and sulfamethoxazole.

Prevention. Vaccines, either of an attenuated or a killed type, have been used to produce an artificially acquired active immunity. **[See states of immunity in Chapter 18.]**

RELAPSING FEVER

Relapsing fever caused by the spirochetes of *Borrelia* (bore-ELL-ee-ah) (Figure 27–5a) are either louse- or tick-borne infections. Louse-borne disease is usually epidemic in nature, occurring in the crowded conditions that develop during times of war and natural disasters. *Pediculus humanus var. capitis* and *var. vestimenti* are the louse vectors involved. The tick-borne infections are endemic and are primarily limited to regions in which the vectors, *Ornithodorus* spp. (Figure 27–5b), live. Human infections with tick-borne borreliae are few and largely confined to field workers, hunters, soldiers, and tourists. Repeated attacks of fever, or *relapses,* give the disease its name (Table 27–1). Each successive episode tends to be less severe than the previous one, until the relapses subside altogether.

No natural animal reservoir of *Borrelia recurrentis* has been uncovered, although a variety of rodents have been implicated. A similar situation exists with *B. duttonii* (one of the tick-borne borreliae). A reservoir is created, however, when female ticks pass the borreliae to some of their eggs (transovarial passage).

Tick-borne fevers have been reported to be more intense and shorter in duration, involving the central nervous system more often, than the louse-borne variety. In addition, there are more relapses.

Relapsing Fever

Laboratory Identification

Borrelias can usually be easily observed in stained (Figure 27–5a) or unstained preparations. Laboratory animals such as white mice may be used for isolation of the pathogens. *Borrelia* usually appear within 48 hours after the specimens have been inoculated.

Treatment

Chloramphenicol and tetracyclines are the drugs of choice for the treatment of *Borrelia* infections.

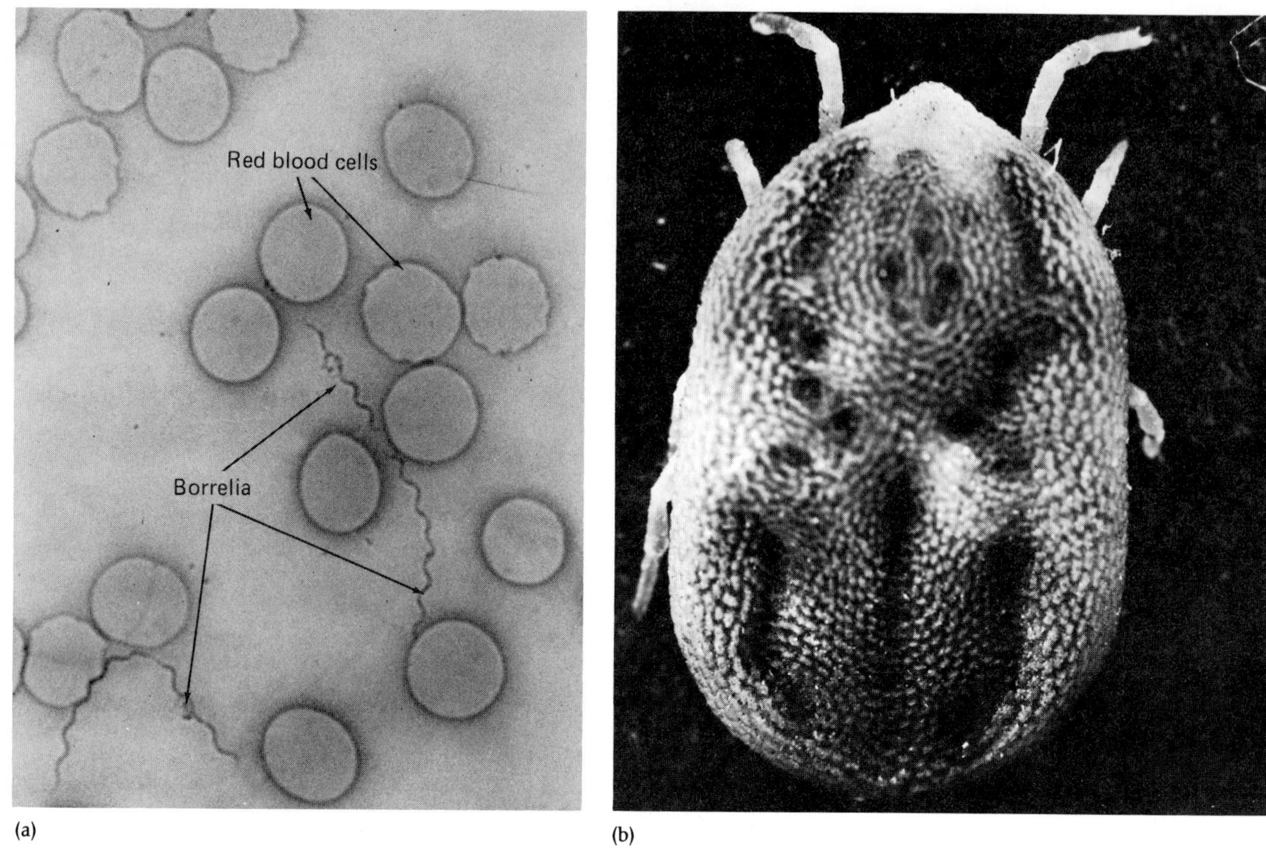

Figure 27–5 (a) Stained blood smear preparation of *Borrelia recurrentis.* This organism ranges from 8 to 40 μm in length and from 0.2 to 0.5 μm in width. (b) *Ornithodorus* sp., a vector for tick-borne relapsing fever. *[Courtesy of the Bureau of Vector Control, Berkeley, Calif.]*

RICKETTSIAL INFECTIONS

The first discovery of rickettsiae was largely a consequence of certain investigations by Howard Taylor Ricketts in 1909. While studying various aspects of the so-called Rocky Mountain spotted fever, he observed the presence of extremely small, distinct, rodlike formations in the blood of patients and certain associated ticks. This pathogen later was named *Rickettsia rickettsii* (reh-KET-see-ah reh-KET-see-eye), in honor of its discoverer, who contracted and in 1910 died of typhus fever.

Subsequent work has identified more than 30 varieties of rickettsiae that are pathogenic for humans or lower animals. This research has also provided techniques for the isolation and cultivation of rickettsiae as well as methods for their identification and control, including vaccines and insecticides against vectors.

Transmission. One distinctive feature of most rickettsiae is that their transmission, under natural conditions, is dependent upon a variety of arthropods, such as fleas, lice, mites, and ticks (Table 27–2). An exception to this mode of spread is Q fever, which is primarily a droplet-transmitted infection. Since the rickettsial agents multiply in the arthropod responsible for their transmission, the arthropods can be considered true natural vectors. Humans are the only vertebrate reservoirs for the rickettsiae causing Brill's relapsing typhus (Brill's disease), epidemic typhus, and trench fever. In all other rickettsioses, humans act as accidental hosts only (Table 27–2). Therefore, the rickettsial diseases, with the exceptions previously noted, are zoonoses. [See zoonoses in Chapter 21.]

General Clinical Features. Most rickettsial infections involve the circulatory system. Effects may extend to other tissues and organs as well (Table 27–1).

Epidemic Typhus Fever. The occurrence of *R. prowazekii* (R. prow-wah-zeh-KEE-eye) is traditionally linked to natural disasters, especially severe winters, famine, and war. In all of these situations, overcrowding makes maintenance of adequate personal hygiene difficult. The opportunities to bathe and wash clothes are limited, providing an environment in which body lice can flourish and typhus fever is easily transmitted.

TABLE 27–2 **Representative Human Rickettsioses**

Group Category and Disease	*Causative Agent*	*Principal Vector or Means of Transmission*	*Geographical Distribution*	*Mammals Concerned In Normal Cycle*
LOUSE-BORNE GROUP				
Epidemic typhus	*Rickettsia (R.) prowazekii*	*Pediculus capitis, P. corporis*	World	Humans
Brill's disease	*R. prowazekii*	None	Europe, North America	Humans
Trench fever	*Rochalimaea quintana*	*P. corporis*	Africa, Europe, North America	Humans
FLEA-BORNE GROUP				
Endemic (mu rine) typhus	*R. typhi (mooseri)*	*Xenopsyllus cheopis* (rat flea)	World	Rodents
TICK-BORNE GROUP				
Boutonneuse fever	*R. conorii*	*Amblyomma variegatum, Rhipicephalus sanguineus,* others	Africa, Europe, India, Middle East	Dogs, wild rodents
Queensland tick	*R. australis*	*Ixodes holocyclus*	Australia	Marsupials, wild rodents
Rocky Mountain spotted fever	*R. rickettsii*	*Amblyomma* spp., *Dermocentor* spp., *Rhipicephalus sanguineus*	Western Hemisphere	Dogs, wild rodents
MITE-BORNE GROUP				
Rickettsialpox	*R. akari*	*Allodermanyssus sanguineus*	Europe, North America	Wild rodents
Tsutsugamushi fever (scrub typhus)	*R. tsutsugamushi (R. orientalis)*	*Trombicula* spp.	Asia, Australia, Pacific Islands	Wild rodents
Q fever	*Coxiella burnetii (R. burnetii)*	Primarily droplet infection, although certain species of ticks have been implicated	World	Cattle, goats, sheep, wild rodents

Note: The specific rickettsial infections listed are arranged according to the arthropod vectors. This approach is commonly used.

The transmission of epidemic typhus involves a human-to-louse-to-human cycle. The disease can develop in persons of any age. However, the effects of the infection are more severe in older individuals (Figure 27–6). Under normal conditions, the rickettsiae are introduced into humans as a body or head louse obtains a blood meal. While the arthropod feeds, it defecates, releasing infective organisms. Since the bite of the louse causes extreme itching, the natural tendency is to scratch the affected area, leading to self-inoculation with the infectious feces. Although this is the way in which epidemic typhus is usually contracted, it is also possible to acquire an infection in one of two ways: by inhaling dried, infected louse fecal matter, or by having it blow into the eye, across the conjunctivae.

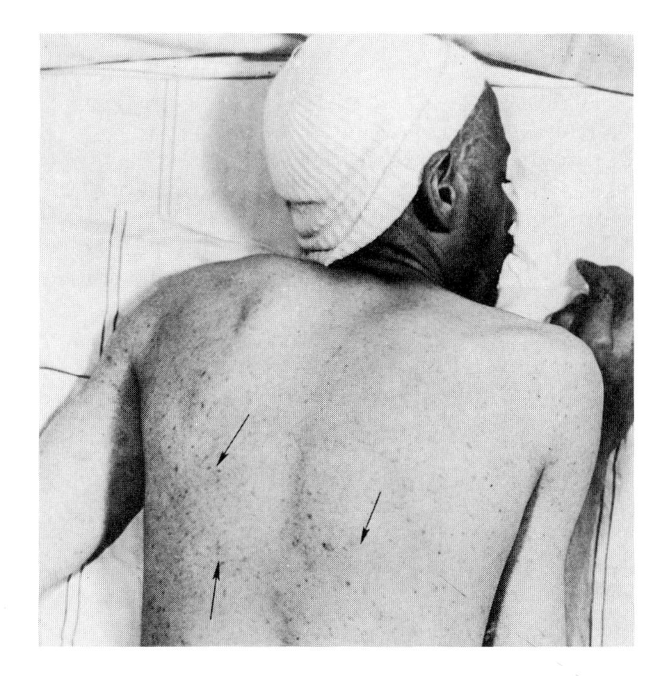

Figure 27–6 Rash on the back of a victim of epidemic typhus. *[Courtesy of the Armed Forces Institute of Pathology, Washington, D.C., Neg. No. 60–1011A.]*

TRENCH FEVER. The rickettsial disease trench fever, also known as *His-Werner disease, Polish-Russian intermittent fever,* and *shank* (or *shin*) *fever,* is also commonly associated with wartime conditions. Widespread during World Wars I and II, it has also been reported in Mexico and Tunisia. Transmission of the disease is similar to that of epidemic typhus.

FLEA-BORNE (MURINE) TYPHUS FEVER. Flea-borne typhus fever is a rickettsiosis found mainly in tropical and subtropical regions that is caused by *R. typhi.* Because this disease is common in rats, the infection can occur wherever these rodents and the tropical rat flea *Xenopsyllus cheopsis* exist. Localities such as granaries, storehouses, and waterfront areas are likely breeding sites.

The general transmission cycle of endemic typhus is from an infected rat to a flea to another rat. Humans acquire the infection accidentally; if a susceptible rat is not available, the infected rat flea uses the human as its new host. As in epidemic typhus, rickettsiae are introduced by the contamination of an infected flea bite with feces containing the infectious agents.

ROCKY MOUNTAIN SPOTTED FEVER. Rocky Mountain spotted fever, caused by *R. rickettsii,* was first recognized during the 1870s in Montana. Dogs, fieldmice, hares, rabbits, and squirrels all serve as reservoirs. Moreover, the tick vector itself serves as a reservoir, transmitting the rickettsiae to future generations of ticks through transovarian passage. Representative tick vectors (Figure 27–7a) and the geographic regions involved are identified in Table 27–2. Tick-borne rickettsial species (such as *R. rickettsii*) that cause spotted fevers (Figure 27–8) characteristically multiply in both the cytoplasmic and nuclear regions of infected cells (Figure 27–7b). Other rickettsiae that are pathogenic for humans reproduce only in the cytoplasm of cells.

SCRUB TYPHUS (TSUTSUGAMUSHI FEVER). Scrub typhus is an example of a **zoonosis** associated with rodents. The causative agent, *R. tsutsugamushi* (R. tsoo-tsoo-gam-OO-she), is also known as *R. orientalis* (R. or-ree-EN-tal-iss). Humans acquire the disease when bitten by infected mites, such as *Trombicula akamushi* and other *Trombicula* species (Figure 27–9).

RICKETTSIALPOX. *R. akari* (R. ah-KAR-ee) is the causative agent for rickettsialpox. The disease was first reported in 1946. Humans acquire the infectious agents through the bite of a common mouse mite, *Allodermanyssus sanguineus.* In the United States, the usual host for *A. sanguineus* is the common house mouse, *Mus musculus.* Although *R. akari* is not transmitted by ticks, it is grouped with other rickettsiae comprising the Rocky Mountain spotted fever category, because it has several properties in common with the group, including antigenicity and pattern of multiplication.

Figure 27–7 (a) An adult male *Dermacentor andersoni,* the Rocky Mountain wood tick. (b) *Rickettsia.* Note the appearance of these rickettsiae in both the cytoplasmic and nuclear regions of the cell. *[Courtesy of the Rocky Mountain Laboratory, USPHS, Hamilton, Mont.]*

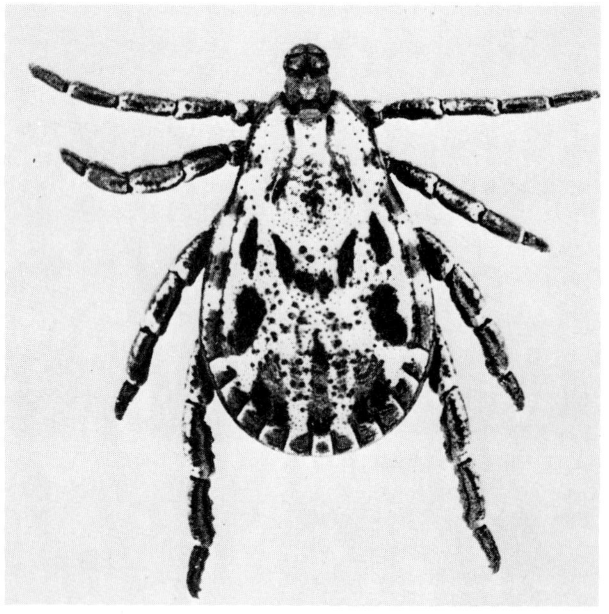

(a)

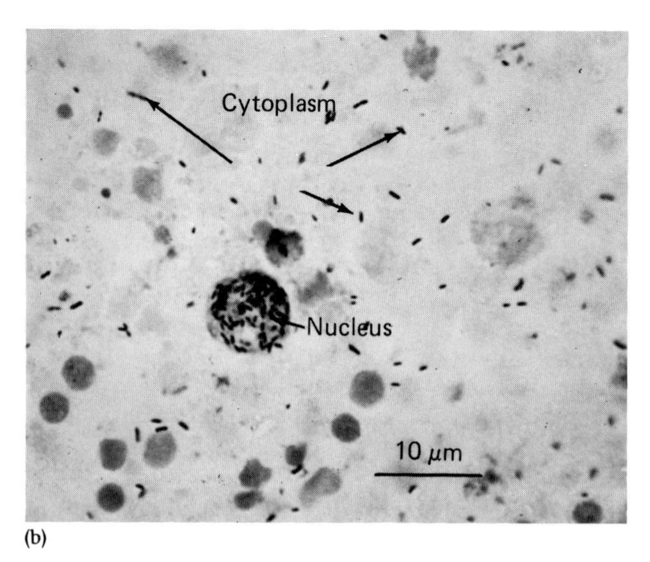

(b)

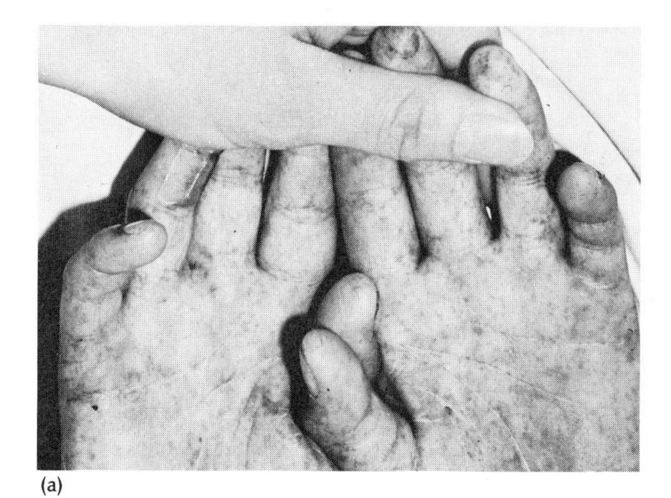

(a)

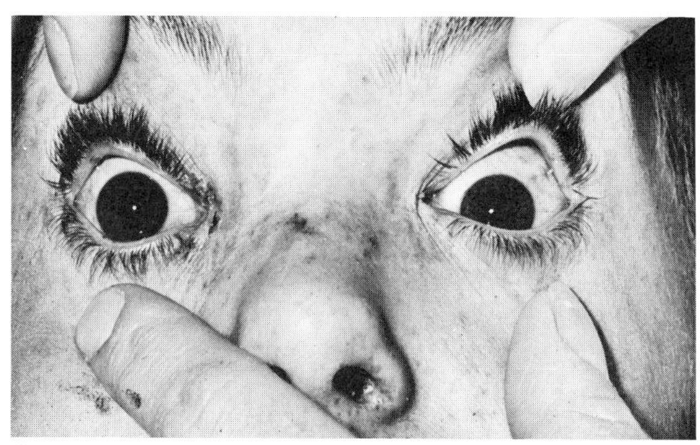

(b)

Figure 27–8 Clinical features of Rocky Mountain spotted fever in children. (a) A severe rash on a patient's hands. The skin on the fingertips of this child later became gangrenous. In general, the rash is most severe on the palms of the hands and soles of the feet. (b) Another feature of the disease is conjunctivitis (inflammation of eyelid lining), regularly noted in children. The photograph also shows the facial rash and hemorrhagic crusts in the nose. *[J. Haynes,* et al., J. Pediatr. ***76****:685–693, 1970.]*

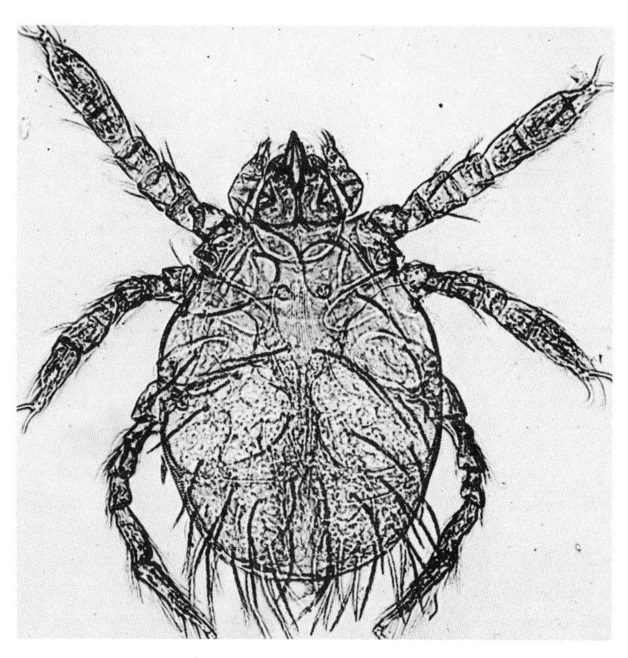

Figure 27–9 *Trombicula akamushi,* the mite vector for the rickettsial disease tsutsugamushi fever. *[Courtesy of the Rocky Mountain Laboratory, USPHS, Hamilton, Mont.]*

TULAREMIA

Tularemia, caused by *Francisella tularensis* (fran-sis-EL-lah too-lah-REN-sis), was first reported in ground squirrels from Tulare County, California (hence its name), in 1911. It is also known as *deer-fly fever, rabbit fever,* and *Ohara's disease.* The first human case of the disease was reported in 1914.

TRANSMISSION. Human beings can acquire tularemia by handling the carcasses of infected animals (Figure 27–10) or by contact with water contaminated by them. They may also be infected by the bites of infected flies and ticks (Figure 27–10b), *Am-*

Rickettsial Infection

LABORATORY DIAGNOSIS

The presence of rickettsia can be detected in infected tissues by staining. However, the use of serological tests is the most practical, efficient, and safe approach to rickettsial identification. The complement-fixation, microagglutination, and direct fluorescent antibody tests are generally used.

TREATMENT

Tetracycline is the drug of choice.

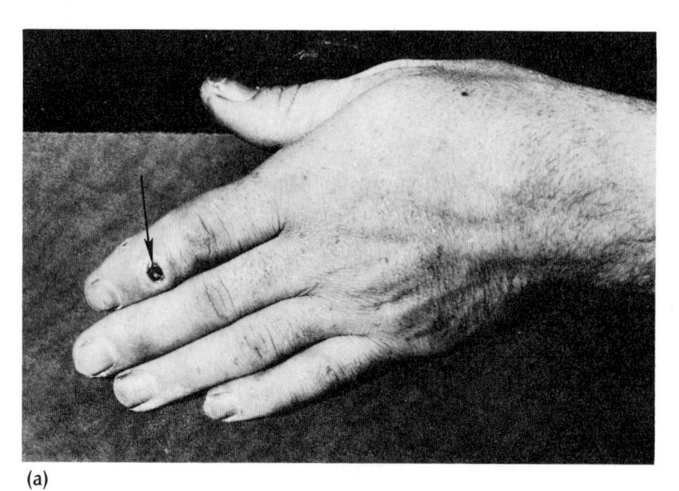

(a)

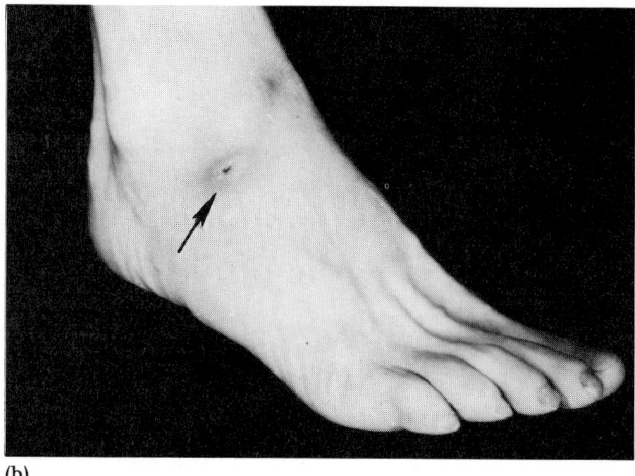

(b)

Figure 27–10 Tularemia. (a) An infection resulting from the handling of an infected rabbit carcass. (b) Site of primary involvement, probably resulting from an infected arthropod bite. *[Courtesy of the Armed Forces Institute of Pathology, Washington, D.C., Neg. Nos. 85387-2 and AN 1147-1A.]*

blyomma spp. and *Dermacentor* spp., or by inhalation of infectious aerosols. Infections also have resulted from the bites of animals that have fed on diseased rabbits.

IMMUNITY. Recovery from tularemia usually produces a relatively permanent, naturally acquired active state of immunity. However, second episodes of infection have been reported. Relapses may occur even though high antibody titers are present in the infected individual. This is probably caused by the fact that the organisms are normally inside the body cells, where they are safe from antibodies.

Protozoan Infections

The destructive effects of protozoa depend on their number, size, degree of activity, location, and toxic products. Consequently, symptoms may be absent, few, or severe (Table 27–3). [See protozoan structures in Chapter 9.]

KALA AZAR

The genus of flagellates that causes kala azar is named in honor of William Leishman, who discovered them in 1900. Morphologically, *Leishmania donovani* (lie-sh-may-NEE-ah don-oh-VON-ee) cannot be readily distinguished from other *Leishmania*. How-

Tularemia

LABORATORY DIAGNOSIS

Isolation of *Francisella tularensis* can be made with the use of cysteine-dextrose blood agar or other media enriched with the amino acid cystine. Small transparent colonies appear three to five days later. Cells from such colonies are stained by the fluorescent antibody technique or used in agglutination tests for definitive identification. The ELISA procedure also can be used for the diagnosis of tularemia. [See Chapter 19 for a description of these tests.]

TREATMENT

Streptomycin is generally the drug of choice. Chloramphenicol and tetracycline may also be used.

TABLE 27–3 Protozoan Diseases Associated with the Circulatory System

Disease	*Causative Agent*	*Geographical Distribution*	*Incubation Period*	*General Features of the Disease*	*Laboratory Diagnosis*	*Possible Treatment*
Kala azar (dum-dum fever)	*Leishmania donovani*	Central and South America, Europe, Asia, and Africa	From 10 days to several months	Fever, bleeding from gums, lips, and nose; enlarged liver and spleen; pneumonia can be a complication	1. Microscopic demonstration of causative agent in blood, bone marrow, and spleen specimens 2. Serological tests, e.g., complement fixation, fluorescent antibody test[a]	Sodium antimony gluconate, ethylstibamine, amphotercin B
Malignant tertian malaria	*Plasmodium falciparum*	Tropical and subtropical areas	12–30 days (depends on species)	Headache, nausea, abdominal pain, vomiting, chills,[b] prolonged fever stage, little sweating, convulsions, and coma; enlarged liver and spleen, heart failure, and bloody urine are complications	Demonstration of parasites in blood smears (Color Photograph 110)	Amodiacquine hydrochloride (acute infection), quinine dihydrochloride
Quartan malaria	*Plasmodium malariae*	Tropical and subtropical areas, including Africa and Ceylon	12–37 days (depends on species)	A relatively mild form of the disease with some chills, fever, headache, nausea, and abdominal pain	Demonstration of parasites in blood smear	Chloroquine phosphate (acute infection), quinine dihydrochloride
Tertian (ovale) malaria	*Plasmodium ovale* (rare)	East and west Africa, South America, United States	12–30 days (depends on species)	Periods of shaking chills, fever, headache, nausea, muscular pain, and vomiting	Demonstration of parasites in blood smear	Chloroquine phosphate (acute infection), quinine dihydrochloride
Tertian (vivax or benign) malaria	*Plasmodium vivax*	Reported to be established in central and western Africa and Philippines; also occurs sporadically in China, Greece, and Iran	12–30 days (depends on species)	Periods of shaking chills (cold stage), followed by fever (40°–41°C) with profuse sweating, nausea, vomiting, headache, and muscular pains	Demonstration of parasites in blood smear	Chloroquine phosphate (acute infection), quinine dihydrochloride

[a] Refer to Chapter 19 for a description of these procedures.
[b] The chill stage here is usually milder than that of other malarial infections.

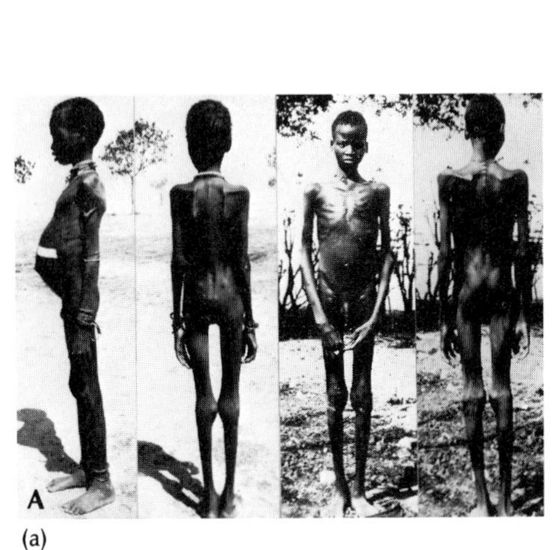

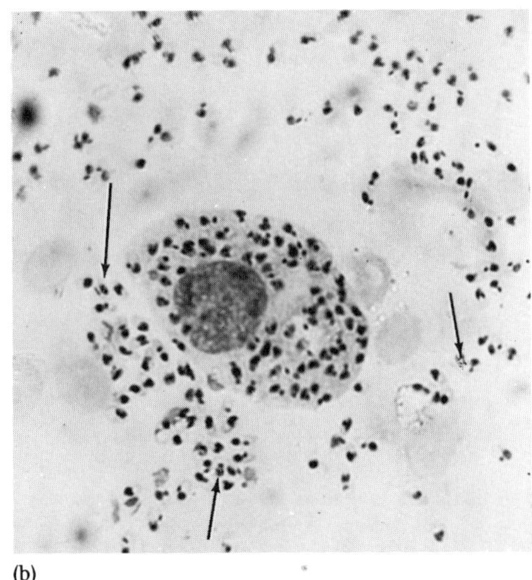

(a) (b)

Figure 27–11 (a) Two victims of kala azar. *[H. Hoogstraal, and D. Heyneman,* Amer. J. Trop. Med. Hyg. ***18****:1091, 1969.]* (b) *Leishmania donovani* in blood. *[Courtesy of the Armed Forces Institute of Pathology, Washington, D.C., Neg. No. AFIP 55–17580–3.]*

ever, differentiation can be made on the basis of the effects of the disease state (Figure 27–11b).

At present, closely related forms are described as either *dermotropic* or *viscerotropic* (Figure 27–11a), according to the part of the body they invade (Table 27–3). The dermotropic species are said to belong to the *L. tropica* species complex and are thought to cause the classical skin form of leishmaniasis called the *Oriental sore* (Color Photograph 49).

TRANSMISSION AND LIFE CYCLE. *L. donovani* is spread by the bite of various sandfly species of the genus *Phlebotomus.* The arthropod acquires the leishmanial form of the parasite by feeding on an infected host, which would be either a human or other suitable mammal.

PREVENTION. Control measures include (1) the use of insecticides to eliminate the arthropod vector, (2) treatment of infected individuals, and (3) the use of personal prophylactic devices such as insect repellents and sandfly nets. In addition, contaminated materials, such as dressings, clothing, and bedding, should be properly and thoroughly disinfected.

MALARIA

Nearly a century has passed since the discovery of the first malarial parasite. Considered to be among the greatest killers of the human race, malaria has been known from antiquity. Accounts of the disease's observable effects date back to at least 1500 B.C. While in Egypt, Hippocrates utilized the earlier records of malaria and noted the existence of differentiating fever cycles and other clinical features. With this information, he was able to divide the disease into quotidian, benign tertian, and quartan fevers. These designations denoted 24-, 48-, and 72-hour fever cycles, respectively. It is now known that they (and some other tertian fevers) are caused by infections with different species of the malarial parasite of the genus *Plasmodium* (plaz-MOH-dee-um). Human malaria may involve any area where *Anopheles* mosquitoes capable of supporting the parasite breed and are present in significant numbers, where reservoirs of the malarial parasite are available, and where control measures directed against mosquitoes are not adequate.

Four species of *Plasmodium* are known human pathogens: *P. falciparum* (fal-SIP-ar-um), *P. malariae* mal-AIR-ee-ah), *P. ovale,* and *P. vivax* (vye-VAX). Their respective distributions are shown in Table 27–3.

LIFE CYCLE. A typical life cycle of malaria (Figure 27–12) includes both an asexual phase, **schizogony,** which develops within the vertebrate host (human or other animal), and the sexual phase, **sporogony,** which develops within the invertebrate host (*Anopheles* mosquito).

Infective units **(sporozoites)** enter the human bloodstream through the bite of a parasite-harboring female mosquito and readily pass from the peripheral circulation into the parenchymal cells of the liver. There the parasites grow and undergo repeated divisions, eventually releasing *merozoites,* forms capable of invading red blood cells and thereby beginning the *erythrocytic* portion of the asexual cycle. Variations in this general cycle occur, depending upon the *Plasmodium* species (Color Photograph 110).

Schizogony continues with invasion of the red blood cells by merozoites. One such parasite assumes the shape of a signet ring in its new environment, grows, and develops into an amoeboid, single-

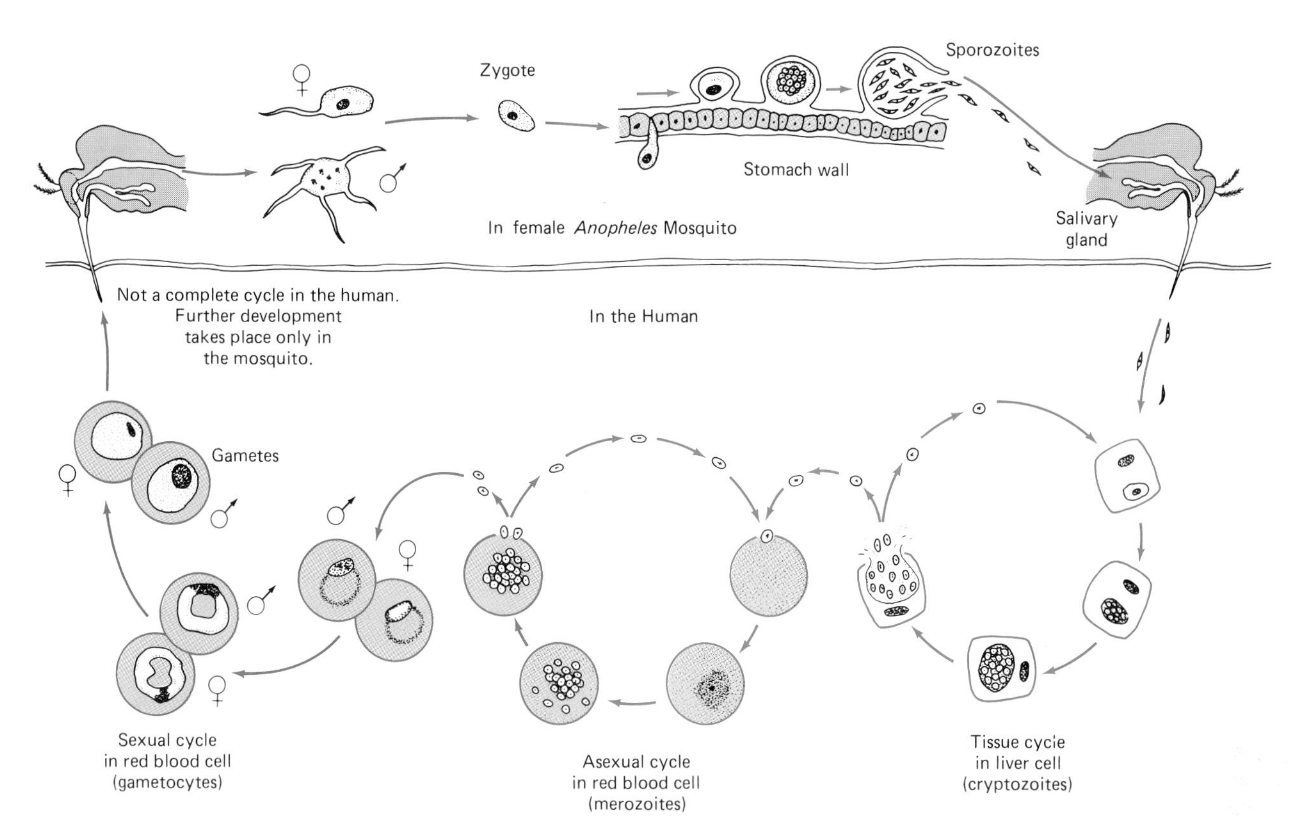

Figure 27–12 The life cycle of *Plasmodium* in a mosquito and in the human. The human host is infected by small elongated infective cells, the *sporozoites,* which are produced in the *Anopheles* mosquito and invade the salivary gland of the insect. The female mosquito introduces sporozoites into the human bloodstream when she takes a blood meal. The newly introduced parasites are removed from the blood by organs such as the liver and spleen, where they multiply and produce other infective forms that are released into the bloodstream to attack and carry out the asexual cycle in red blood cells. The outcome of this cycle is the formation of sex cells (gametocytes) and other infective units for red blood cells. If these sex cells are ingested by another female mosquito, they mature and participate in the sexual reproductive phase by forming a zygote. This fertilized cell undergoes developmental changes within the mosquito's stomach lining, where it enlarges and forms a large number of sporozoites. The cycle can begin again.

nucleus feeding form, known as a *trophozoite.* (See Color Photographs 110a, 110b, and 110c for a representation of these blood stages.) When the parasite reaches its full size, mitotic division accompanied by cytoplasmic segmentation occurs. Ultimately, a solid body composed of a varying number of merozoites forms. This structure is referred to as a *mature schizont.* The parasite-supporting red blood cell ruptures, releasing the merozoites, the toxic metabolic products of the parasite, and the remains of the host cell into the general circulation. This group of substances causes the chills and fever characteristic of the malarial syndrome (Table 27–3).

Certain merozoites, instead of continuing the asexual cycle, invade red cells and develop into female and male *gametocytes.* These forms circulate in the peripheral bloodstream, becoming available for ingestion by an anopheline mosquito. Under normal conditions, a large number of parasites are present when mosquitoes take their blood meals. In the mosquito, they undergo transformation into mature sex cells and participate in the sexual cycle, or **sporogony.** Fertilization of a mature female cell *(macrogamete)* by a male cell *(microgamete)* forms a *zygote.* This becomes an *oökinete,* which is elongated and capable of movement. The oökinete penetrates the mosquito's stomach wall, rounds out, and develops into a stationary form known as an *oöcyst.* A series of divisions within this structure produces many slender, threadlike, infective sporozoites. Rupture of the oöcyst releases the sporozoites, which migrate throughout the mosquito's entire body. Those infective units entering the salivary glands of the arthropod can be introduced into another human when the mosquito bites and obtains a blood meal. The malarial asexual cycle can then begin again. Another protozoan disease, babesiosis (Nantucket fever), resembles malaria both clinically and in the life cycle of the causative agent, *Babesia* (bah-BEE-zee-ah), species. The pathogenic protozoa reproduce in red blood cells and produce symptoms similar to those of malaria (Table 27–4).

RESISTANCE. Individuals carrying genes for sickle cell hemoglobin (HbS) exhibit resistance to *P. falciparum* infection. HbS-containing red cells are not suitable to support the growth of the blood stages of the parasite, especially when blood cells enter veins, where low oxygen tension exists. Under such conditions a distortion of cells, known as *sickling*, is increased and leads to a loss of potassium. With such a reduction, parasitic growth is inhibited and death of the protozoon eventually occurs. In some cells, needlelike hemoglobin molecules form that penetrate and disrupt the intracellular parasites (Figure 27–13).

TREATMENT. Several drugs are effective against malaria (Table 27–3). The choice of medication is determined by the particular situation, whether it is for ending an acute primary attack, suppressive or prophylactic therapy, or treatment of relapsing malaria. Intelligent therapy also depends upon an accurate diagnosis and a functional knowledge of the effects of antimalarial drugs.

PREVENTION. Control of malaria involves measures directed against the mosquito vector and against the spread of disease by the definitive hosts, human beings. Eradication of mosquitoes requires the destruction and elimination of breeding areas as well as larval and adult stages of the vector. Control measures in the case of humans include the use of suppressive and prophylactic drugs in endemic areas, prompt and adequate treatment of obvious clinical attacks, and avoidance of mosquito bites.

During late 1970 and early 1971, several cases of malaria in California were traced to heroin addicts who had had contact with one of the major sources of the malarial parasite, a Vietnam veteran. Since addicts frequently share unsterilized syringes and needles, the disease was easily transmitted. Suitable vectors are known to exist in California, but fortunately mosquitoes were not plentiful during these outbreaks of the disease. Had these cases appeared in the summer months, statewide epidemics might have developed.

Viral and Related Infections

ACQUIRED IMMUNE DEFICIENCY SYNDROME (AIDS)

Acquired immune deficiency syndrome, or AIDS (Table 27–4) has caused more fear and anxiety than any other disease of this decade. The panic associated with the disease is somewhat reminiscent of that experienced with bubonic plague centuries ago, even though there were considerably more cases and deaths over a much shorter time span then than there have been with AIDS. Unfortunately, great attention has been focused on the sensational aspects of AIDS, thereby alarming both the general public and the health-care community.

AIDS is a disease of the immune system that results in the development of life-threatening opportunistic infections (Table 27–5) or cancerous growths. One commonly associated cancer is Kaposi's sarcoma. This abnormality was rarely seen in North

Figure 27–13 Resistance to malarial infection in sickle cell anemia. (a) Normal intracellular location of *Plasmodium falciparum.* (b) The destructive effects of the internal environment of a sickle cell. The abnormal hemoglobin in such cells is believed to be involved with the process. [*M. J. Friedman,* J. Protozool. ***26****:195–199, 1979.*]

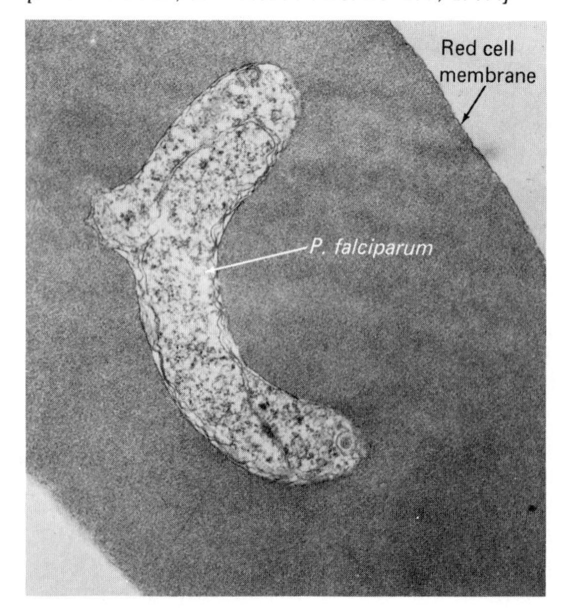

(a)

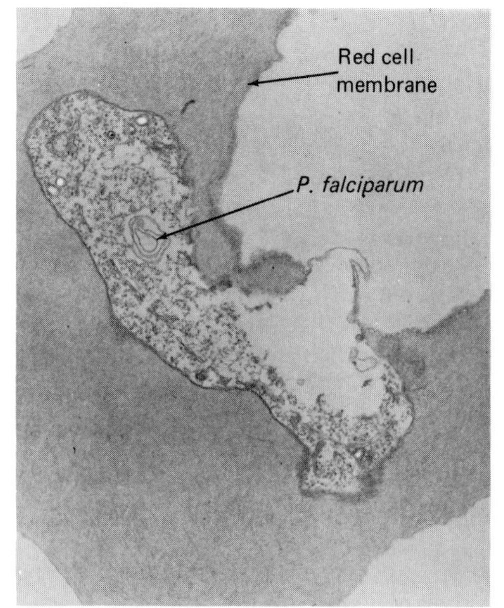

(b)

TABLE 27–4 Selected Viral and Other Microbial Agents Associated with the Circulatory and Immune Systems

Disease	*Incubation Period*	*Signs and Symptoms*	*Laboratory Diagnosis*	*Possible Treatment*
Acquired immune deficiency syndrome (AIDS)	Adult form: average 7 years (may be over 10 years) Pediatric form: 6 months to 1 year average	Refer to Table 27–6	Serological tests, ELISA, Western blot test (See Chapter 19); virus isolation	ATZ, possibly interferons; treatment of opportunistic infections; symptomatic
AIDS-related complex (ARC)[a]	Unknown	Symptoms include weight loss greater than 10%, night sweats, diarrhea, skin rash, fatigue, oral candidiasis (see Color Photograph 111d), swollen lymph glands, and long-lasting fever (37.8°C)	Positive HIV antibody test; decrease in T_4 (CD4) cells, loss of skin test sensitivity	Treatment of opportunistic infections; symptomatic
Infectious mononucleosis (acute form)[b]	19–49 days	Symptoms include mild jaundice, slight fever, enlarged and tender lymph nodes, sore throat, headache, and general weakness; complications can occur and include convulsions, anemia, and inflammation of heart tissue	1. Demonstration of atypical T lymphocytes 2. Liver function tests 3. Serological tests, e.g., specific agglutination tests, heterophile antibody	Bed rest; use of antiinflammatory agents in cases of hemolytic anemia; hospitalization of severe cases
Infectious mononucleosis (chronic form)	Unknown	Symptoms include extreme fatigue, fever, allergies or other environmental sensitivities, sore throat, muscle and joint pain, and enlarged lymph nodes	Possible findings include increased numbers of lymphocytes and elevated levels of EBV antibodies	No specific treatment other than relieving symptoms
Kawasaki syndrome	5–23 days	Fever (37.8°–40°C) for an average of 11 days, inflammation of conjunctivae (both eyes), red rash on palms and feet, enlargement of lymph nodes in the neck area; surface skin loss on tips of fingers and toes; hardened areas of hands and feet	Generally clinical nonspecific tests, since no specific causative agent is known	Supportive to relieve symptoms and prevent complications

[a] It should be noted that there is no standard definition of ARC. According to the Centers for Disease Control, defining this condition is the responsibility of individual states.
[b] Cytomegalovirus also can cause this disease.

America or Europe until it was found with several AIDS cases. Kaposi's sarcoma may be present on the skin, in the mouth, or in various body organs (Color Photographs 111a through 111d).

Epidemiology. The first cases of AIDS were reported in mid-1981; they were found in homosexual men and drug abusers in and around New York City and on the West Coast of the United States. While no underlying cause of the condition could be found, victims of AIDS suffered from two rare diseases, Kaposi's sarcoma and *Pneumocystis carinii* pneumonia, as well as from several opportunistic infections. These findings prompted the formation of a Centers for Disease Control task force to investigate the disease. The group established the following definition of AIDS to help in its diagnosis: "AIDS is characterized by the presence of Kaposi's sarcoma and/or life-threatening opportunistic infections in a previously healthy individual less than 60 years of age who has no underlying immunosuppressive disease and has not received immunosuppressive therapy." In August of 1987 the original case definition of AIDS was revised by the Centers for Disease Control in collaboration with other public health groups. The revised definition should simplify reporting, be consistent with current diagnostic approaches, and increase the sensitivity and specificity of detection through the greater use of laboratory evidence for human immunodeficiency virus (HIV) infection.

The current definition is organized into three sections that depend on the status of laboratory evidence of HIV infection. The major changes apply to persons with AIDS (PWA) showing laboratory evidence of infection and takes into account the presence of HIV encephalopathy (brain dysfunction), HIV wasting syndrome (dramatic weight loss), and a broader range of specific AIDS-indicator diseases. The complete revised case definition can be found in the 1987 *Morbidity and Mortality Weekly Report* listed at the end of this chapter. Table 27–6 lists additional features of AIDS.

As information concerning the disease accumulated, the possibility of a transmissible agent began to emerge. This hypothesis was supported by the finding of AIDS in hemophiliacs and recipients of blood transfusions in 1982 and 1983.

ETIOLOGY. Only three years after AIDS was first described, its cause was shown to be a third human retrovirus: human T lymphotropic virus or HTLV-III (Figure 27–14). The viral agent is also referred to as lymphadenopathy-associated virus (LAV), or AIDS-associated retrovirus (ARV; Figure 27–14). Currently the virus is called human immunodeficiency virus or HIV. Early in 1987 two distinct classes of AIDS virus, HIV-1 and HIV-2 were designated. The latter virus is endemic to West Africa.

As with other retroviruses, single-stranded RNA serves as the genetic material for the HIV virus. Upon entering a host cell, one enzyme called *reverse transcriptase* uses the HIV's RNA as a pattern to form a corresponding, temporary DNA molecule. The newly formed DNA then moves to the cell's nucleus and proceeds to insert itself among the host cell's chromosomes, from where it establishes the mechanism for viral replication. **[See Chapter 10 for descriptions of virus replication.]** Other retroviruses are described later in this chapter.

MECHANISM OF INFECTION. In AIDS, the host cell is often a T helper or (T4, CD4) lymphocyte. This cell is noted for its central role in regulating the various processes of the immune system (Figure

TABLE 27–5 Common Opportunistic Infections and Associated Conditions

Opportunist	*Resulting Condition*
BACTERIA	
Mycobacterium avium-intracellulare[a]	Spreading infection of blood, liver, spleen, and bone marrow; gastrointestinal infection
FUNGI	
Aspergillus species	Pneumonia (See Chapter 25)
Candida albicans	Oral thrush, infection of esophagus (See Chapter 24)
Cryptococcus neoformans	Meningitis (See Chapter 29), pneumonia; and skin lesions
PROTOZOA	
Crytosporidium muris	Gastrointestinal infection (See Chapter 26)
Isospora belli	Gastrointestinal infection
Pneumocystis carinii	Pneumonia (See Chapter 25)
Toxoplasma gondii	Brain abscess, encephalitis (See Chapter 29)
VIRUSES	
Cytomegalovirus	Hepatitis, pneumonia infection of the adrenal glands, and spreading infection throughout the body (See Chapter 26)
Epstein-Barr	Oral hairy leukoplakia, B-lymphocyte malignancy, pneumonia
Herpes simplex	Mucocutaneous lesions, especially perianal (See Chapter 23)
Varicella-zoster	Primary chickenpox or shingles (See Chapter 23)

[a]Also known as the *Mycobacterium avium* complex (MBC).

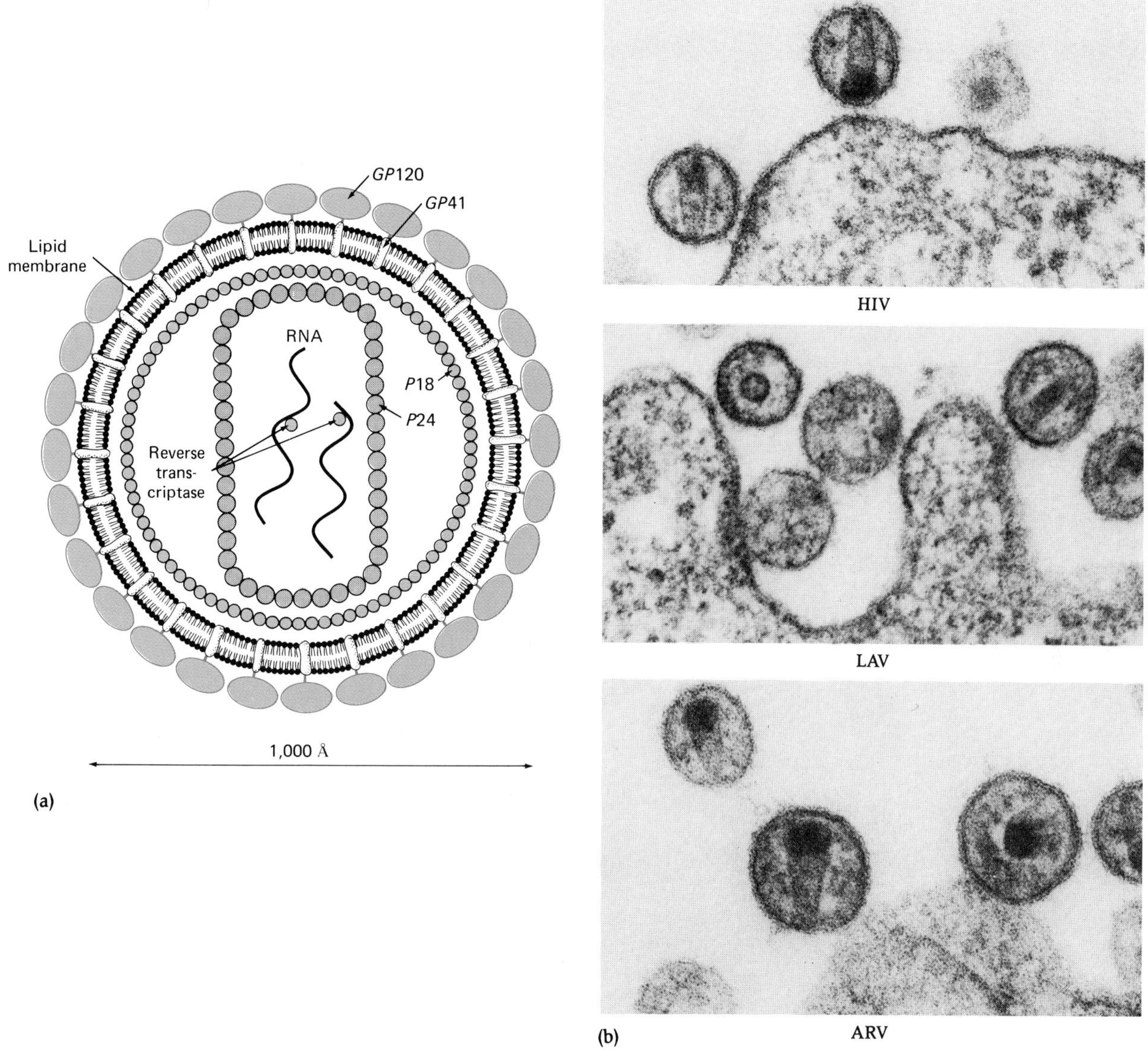

Figure 27–14 Portraits of the AIDS viruses. (a) The human immunodeficiency virus (HIV). The virus particle is covered by a membrane made up of two lipid layers derived from the host cell. Glycoproteins *(gp)*, or proteins with sugar chains are associated with these layers. Each glycoprotein has two components: *gp*41 spans the membrane and *gp*120 extends beyond it. The membrane-and-protein envelope covers a core made up of proteins designated *p*24 and *p*18. The viral RNA is carried in the core, along with several copies of the enzyme reverse transcriptase, which catalyzes the assembly of the viral DNA. (b) Electron micrographs of human immunodeficiency virus (HIV), lymphadenopathy virus (LAV) and AIDS-associated retrovirus (ARV). *[From R. J. Munn, P. A. Marx, J. K. Yamamoto, and M. B. Gardner,* Lab Invest. ***53****:194–199, 1985.]*

27–15). Once inside the lymphocyte, the virus may remain inactive until the lymphocyte is immunologically stimulated by another infectious agent. Then, the virus goes into action, replicating itself to such an extent that the cell is eventually depleted of its contents and dies. The resulting destruction and reduction of T4 cells, a distinguishing feature of AIDS (Table 27–6), leaves the AIDS victim open to opportunistic infections caused by microorganisms that normally would not create a problem for healthy individuals (Color Photograph 112).

While the crippling effects of HIV on the immune system are well known, it has become increasingly clear that the virus can also have a direct pathogenic effect in the brain and the spinal cord. AIDS victims so affected exhibit mental deterioration and physical degeneration of the nervous system.

In addition to the destructive actions of the virus on the central nervous system, HIV-infected individuals are at increased risk for developing at least three different types of cancers: Kaposi's sarcoma, cancerous growths associated with the skin and mucous membranes, and tumors originating in B lymphocytes.

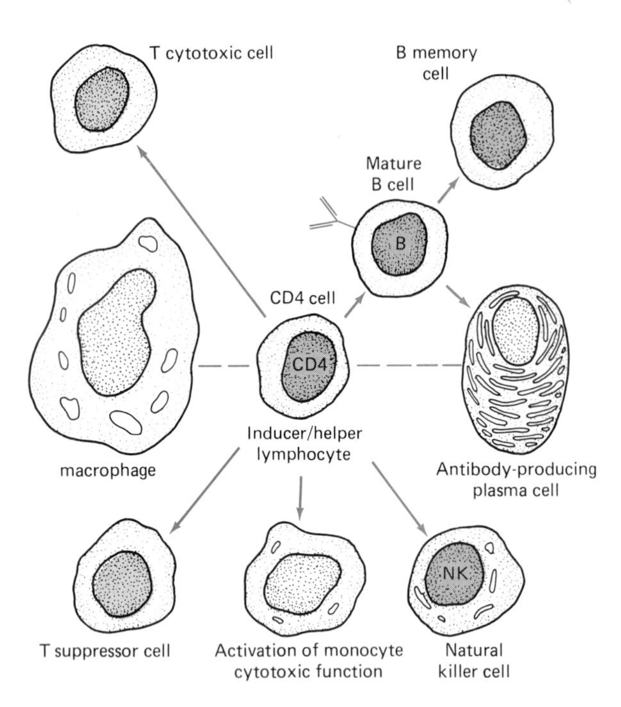

Figure 27–15 A diagrammatic representation of the important roles T helper cells play in the functioning of the immune system. Reduction or elimination of these cells results in a greater susceptibility to opportunistic organisms.

Transmission and Risk Factors. In adults, the HIV virus is transmitted primarily through sexual contact (homosexual or heterosexual) and through inoculation or transfusion with contaminated blood or commercial blood-product preparations such as those used for the treatment of hemophiliacs. The only other instances of adult transmission reported have involved artificial insemination or organ transplants from infected donors. In pediatric cases, the virus can be transmitted by infected women to their fetuses or offspring during pregnancy, during labor and delivery, and possibly shortly after birth. Even though the AIDS virus has been isolated from most body fluids, there is no evidence for its nonspecific

Microbiology Highlights

HBLV: A NEW HERPESVIRUS

A new herpesvirus that attacks B-type lymphocytes was found in several patients who had unusually high numbers of white blood cells. At first it was thought that the virus was associated in some way with AIDS. However, this possibility was eliminated after the HBLV was closely analyzed through the use of electron microscopy. The virus was clearly unlike the retrovirus that causes AIDS. Moreover, on the basis of its shape and size, it was found to be a herpesvirus. This new human B-lymphotropic virus (HBLV) may be responsible for one or more infectious diseases whose agents have not yet been identified.

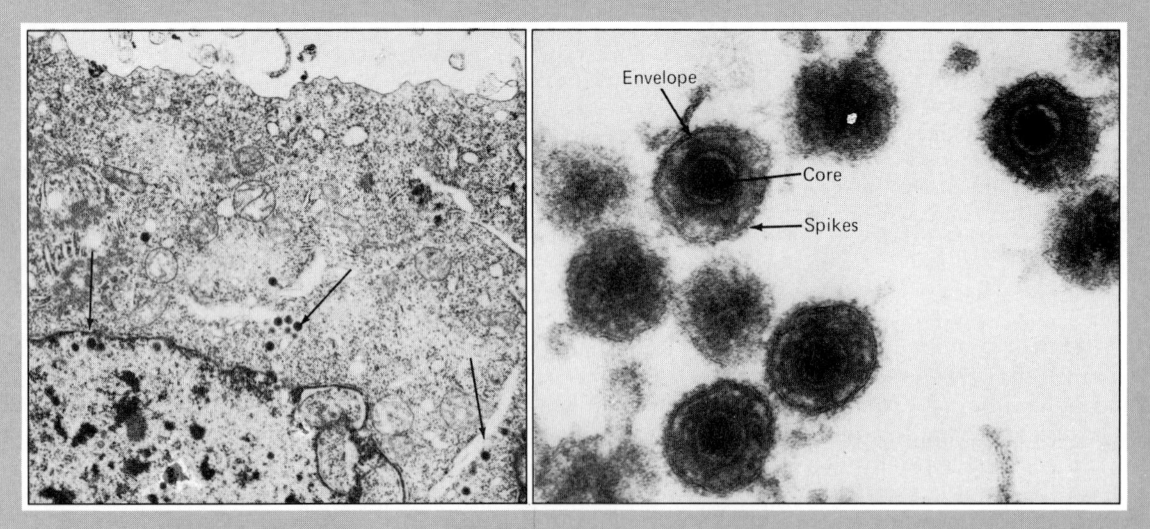

Electron micrographs of (a) an infected cell with several virus particles (arrows) and (b) an enlarged view of individual HBLVs. *[From J. F. Josephs,* et al., Science *234:601, 1986.]*

Microbiology Highlights

KAPOSI'S SARCOMA

Kaposi's sarcoma was first reported by Moritz Kaposi in 1872. Between 1868 and 1871, Kaposi studied five patients with these tumors. All were men between the ages of 40 and 68. This type of classic Kaposi's sarcoma was considered to be one of the least life-threatening forms of cancer. Today the picture has changed significantly.

Before the first cases of AIDS were recognized in the United States, Kaposi's sarcoma was a very rare disease. Annual cancer statistics indicated that 3 persons in 1 million might develop the condition.

The development of Kaposi's sarcoma and other forms of cancer in AIDS patients is one of the most frightening aspects of the immunodeficiency. The classic forms of Kaposi's sarcoma may appear in one of the following four ways:

1. First, the disease is seen as a relatively harmless, slowly progressing skin condition with highly pigmented red to violet surface patches or subcutaneous lumps (nodules) on the arms or legs. Heretofore, individuals showing these signs survived as long as 25 years. Death in such persons was usually due to unrelated problems.
2. Second, the disease is seen as progressively increasing skin tumors (growths) that tend to be invasive and to grow rapidly. Such tumors spread to the lymph nodes, bones, and internal organs, especially the intestines. Since they contain blood vessels, internal bleeding may occur and cause death. This form of Kaposi's sarcoma is seen with immunosuppression for kidney transplantation and in patients with treated or untreated lymphoid (lymph gland) forms of cancers.
3. Third, Kaposi's sarcoma is found in the lymph nodes of children in equatorial Africa or as a rapidly spreading and often fatal cancerous form in young adults. The latter condition starts in the skin and/or lymph nodes.
4. Fourth, the disease appears in epidemic numbers and resembles the spreading form seen in African young adults. Male homosexuals are the primary victims.

One of the most alarming features of Kaposi's sarcoma in AIDS victims is that more than one-third of them will eventually develop another kind of cancer, usually a leukemia or lymph gland (lymphoma) type.

Diagnosis of Kaposi's sarcoma (KS) is made by removing all or part of the tumor for pathological examination. The growth usually contains distinctive cancer cells called "spindle cells," large numbers of newly formed capillaries, and spaces filled with blood cells. The red to purple color of the typical skin lesions is caused by the capillaries and blood cell-filled spaces within these tumors (Color Photographs 111a–111c).

transmission through casual contact, blood-sucking arthropod bites, food, or water.

The populations at risk for HIV infection include homosexual/bisexual men with multiple sex partners, intravenous drug users, patients receiving multiple blood transfusions, hemophiliacs being treated with blood products (commercial sources), heterosexual contacts of individuals who have AIDS or are at increased risk for infection, and infants born to mothers in the described at-risk populations. Spread of infection does not involve household members who have not been sex partners or infants of infected mothers. The type of nonsexual person-to-person contact that generally occurs among health care personnel and clients or consumers in the working environment does not pose a risk for transmission of HIV provided precautions are taken to minimize the risk of exposure to the infectious virus. Infections among health-care workers have, however, occurred. As of June 1987, the Centers for Disease Control in Atlanta, Georgia, had recorded viral infections in nine persons who provided health care to patients with HIV infection. Four of these followed hypodermic needle-stick exposure. The others apparently arose in individuals who had extensive skin or mucous membrane contact with blood or body fluids of infected patients. In addition, precautions such as the wearing of gloves during the handling of blood specimens were not observed in these situations.

As of December 1986, approximately 30,000 cases of AIDS had been reported. If current trends continue, it is estimated that the number of AIDS victims will reach at least 270,000 by 1991.

FORMS OF HIV INFECTION. Although AIDS is the most visible stage, it is the last form of the entire range of HIV infection. In adults, the disease generally proceeds from an asymptomatic virus infection, continues with AIDS-related complex (ARC), and ends with AIDS (Figure 27–16). Patients with ARC, also known as pre-AIDS, generally have persistent swollen lymph glands and a significant reduction in T helper cells. Other signs of the condition are listed in Table 27–6. While there is no standard definition of ARC, according to the Centers for Disease Control, it has been defined by finding at least two clinical conditions and two laboratory abnormalities typical of AIDS (Table 27–4).

As indicated earlier, AIDS is the final stage of HIV infection. The natural course of the disease is not fully understood, but the majority of victims follow a slow but definite clinical pattern. In 1985, investigators at the Walter Reed Institute of Pathology in Washington, D.C., devised a system with which to follow and to possibly predict the development of AIDS. Each step or stage of the system is marked by one or more essential signs of infection (Figure 27–17).

The incubation period for adult AIDS is influenced by the means of transmission. This period for transfusion-related AIDS is longer than six years; in sexually acquired AIDS, it may range from three to ten years.

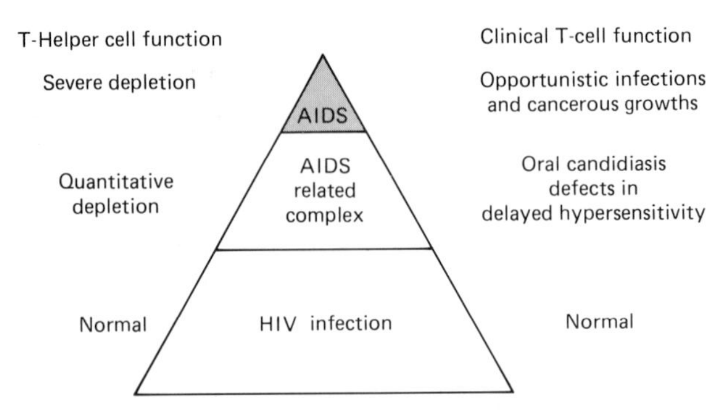

Figure 27–16 The course of AIDS, relating the loss of T helper cell (CD4) effectiveness, and ARC (AIDS-related complex).

Pediatric acquired immune deficiency syndrome (PAIDS) is a distinct disorder diagnosed on the basis of epidemiologic and virologic findings. In the United States, over 230 cases of the disease have been confirmed since July 1986. In addition, there are probably two or three times this number of infants who have AIDS-related disorders. The total number of AIDS cases in children continues to increase at about the same rate as in the adult population.

Several important risk factors can serve as the first indicators for PAIDS diagnosis. These include

TABLE 27–6 Features of AIDS in Adults and Children

Disease Signs	*Adult*	*Children*
MAJOR SIGNS[a]		
Weight loss	Greater than 10% of body weight	Abnormal development or weight loss
Chronic diarrhea	Longer than 1 month	Longer than 1 month
Prolonged fever	Longer than 1 month	Longer than 1 month
Kaposi's sarcoma	Present	May be present
MINOR SIGNS		
Persistent cough	Longer than 1 month	Longer than 1 month
General lymphadenopathy (enlarged lymph nodes)	Present and detectable	Present and detectable
Generalized dermatitis (skin rash)	Present with intense itching	Present
CANDIDA (yeast)		
Infection involving mouth and throat	Present	Present
Additional type(s) of infections	Shingles (herpes zoster), chronic and spreading herpes simplex infection	Repeated ear infection, sore throats
Confirmed maternal AIDS infection	Not applicable	Present[b]

[a] The presence of Kaposi's sarcoma or cryptococcal meningitis (yeast infection) is sufficient by itself for the diagnosis of AIDS.
[b] The presence of HIV antibody by itself, in a child *less* than 15 months of age is considered insufficient evidence of infection. This is largely because of the persistence of maternal antibody. (See Chapter 18 for passive immunization.)

Stage (Walter Reed, WR)	HIV Antibody and/or Virus Isolation	Chronic Lymphaden-opathy	T-Helper Cells/mm^3	Delayed Hyper-sensitivity	Oral Thrush	Opportunist Infection
WR 0	–	–	Greater than 400	Normal	–	–
WR 1	+	–	Greater than 400	Normal	–	–
WR 2	+	+	Greater than 400	Normal	–	–
WR 3	+	±	Less than 400	Normal	–	–
WR 4	+	±	Less than 400	Partial (P)	–	–
WR 5	+	±	–	Complete and/or + (C)		–
WR 6	+	±	–	Partial to complete ±		+

Figure 27–17
The Walter Reed Institute of Pathology classification staging system showing the features in the course of an AIDS infection. The hexagons indicate the critical feature of each stage. Explanation of symbols: +, present; ±, variable; –, absent; P, partial skin sensitivity to specific test antigens; C, complete absence of response to specific skin-test antigens.

mothers who use illicit injectable drugs or who are prostitutes, hemophiliac mothers with a history of transfusions with blood or blood concentrate products that were not adequately processed to inactivate the AIDS viruses, or mothers with bisexual husbands. An infant born into a family where the father is a hemophiliac who has used contaminated blood concentrate products is also at risk for PAIDS.

Infants with PAIDS exhibit symptoms similar to those found in patients with other immunodeficiencies (Table 27–6). These include recurrent and frequent infections, chronic diarrhea, enlarged lymph nodes, and abnormal development. Symptoms that appear unique to PAIDS are recurrent inflammation of the parotid glands (one of the pairs of salivary glands in the mouth) and chronic interstitial inflammation of the lungs. The age of onset may vary from the first month of life to as long as five years after exposure to the AIDS virus. Most infants show signs of the disease within the first year of infection. The finding of virus in the blood or tissues of infants or children and the demonstration of HIV antibodies in children older than 15 months of age also are indicative of infection.

Diagnosis and Treatment. Currently, two serological tests are used to detect HIV antibodies: the enzyme-linked immunosorbent assay (ELISA) and the Western blot test. ELISA detects antibodies to HIV with a high degree of sensitivity and specificity. A positive result may indicate virus infection. However, it should be noted that a small number of false-positive reactions occur. In addition, a positive ELISA does not indicate immunity or guarantee that AIDS will result from the virus infection. Negative results may not be conclusive either, since testing may be done too early and the infection may not yet have produced sufficient antibodies. The Western blot test is a highly sensitive procedure used to detect specific HIV proteins. Individuals with both a positive ELISA and Western blot test are likely to be infected even if clinical findings are lacking (Table 27–4). Virus isolation can also be done in suspected cases to provide conclusive proof of infection. [See Chapter 19 for explanations of ELISA and the Western blot.]

Antiviral chemotherapy for AIDS has been quite limited. Two drugs, azidothymidine (AZT, commercially known as Retrovir), and HPA-23, have been shown to stop the replication of HIV (Figure 27–18). Alpha and gamma interferons are also under study. Suramin, a drug used to treat protozoan infections such as African sleeping sickness, has been found to be unacceptable. Additional drugs are being tested.

Another approach being explored is the use of drugs that help to strengthen a deficient immune system. Since AIDS victims primarily suffer a defect in the functioning of T helper cells (Figure 27–15), efforts to restore immune function seem reasonable. Interleukin-2 and various drugs capable of activating the immune response or restoring immune activity are being tested.

Prevention and Control. While several viral diseases are controlled by immunization, AIDS presents a different and more difficult situation. This is largely because the AIDS virus is capable of regularly changing its antigenic components (Figure 27–19), thus making vaccine preparation a problem. As with influenza virus, several vaccine preparations

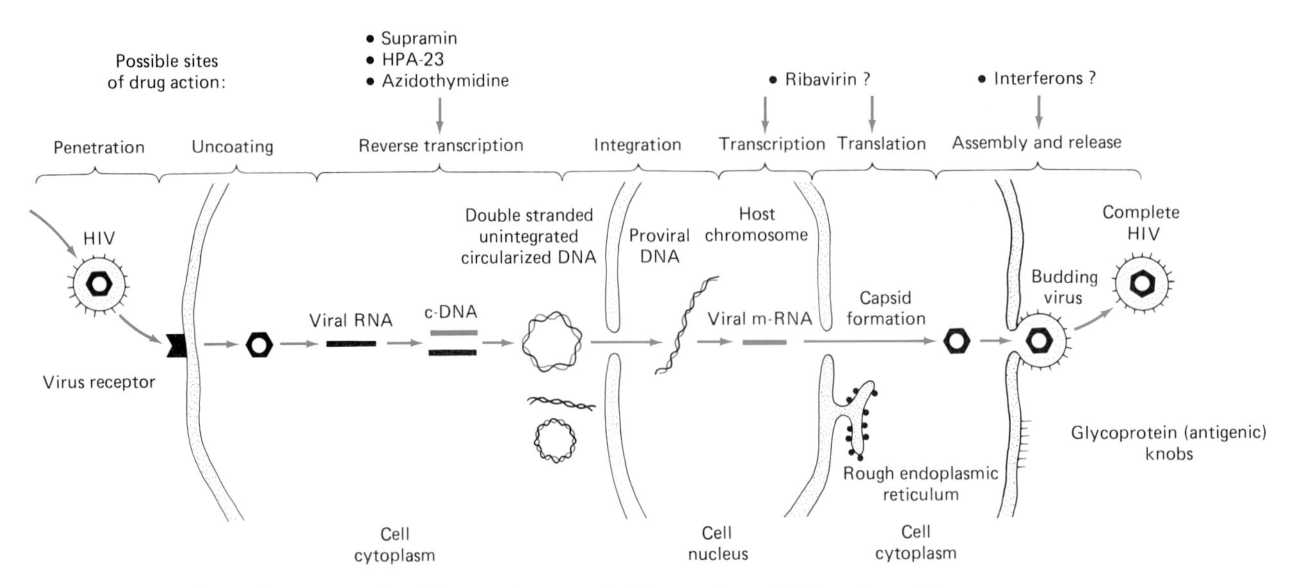

Figure 27–18 The replication cycle of human immunodeficiency virus (HIV) and possible sites of drug action. c-DNA, complementary DNA. *[After M. Vogt, and M. S. Hirsch,* Revs. Infect. Dis. ***8****:991, 1986.]*

may be needed. Human trials with a genetically engineered vaccine are expected to begin before the end of 1987.

Because vaccines and effective therapy are not yet available, preventative measures take on major importance in the control of HIV infection. These include (1) avoiding unsafe sexual activities with high-risk individuals (drug abusers, bisexuals, and infected, sexually active persons); (2) screening pregnant women at high risk to avoid transmission to their fetuses; and (3) screening of blood and blood products (if applicable) to prevent parenteral transmission. Education and nonjudgmental counseling programs for the public at large are also extremely important. Such programs should emphasize the means of transmission, encourage infected individuals to limit sexual contacts, and emphasize the use of barrier protective devices (condoms, diaphragms), during sexual contact. The use of condoms in conjunction with spermicides will reduce the risk of HIV transmission.

Sterilization and disinfection procedures currently recommended for use in health-care and dental facilities are adequate to sterilize and disinfect instruments, devices, and other items contaminated with blood or other body fluids from individuals infected with the HIV virus. The wearing of rubber gloves during the handling of contaminated items or body fluids is also very important. Masks, protective eyewear or face shields, and gowns or aprons should be worn during procedures in which droplets or splashes of blood are likely to occur.

In addition to hospital disinfectants, a freshly prepared solution of sodium hypochlorite (household bleach) is an inexpensive and very effective germicide. Concentrations of sodium hypochlorite ranging from 5000 ppm (a 1:10 dilution of household bleach) to 500 ppm (a 1:100 dilution) are effective for virus inactivation, depending on the amount of organic material (such as blood, mucus, etc.) present on the surface to be cleaned and disinfected. **[Refer to Chapter 11 for other methods and materials for physical and chemical control.]**

A View of the Retrovirus Group

In addition to human immunodeficiency virus (HIV)—also known as lymphadenopathy-associated virus (LAV), the cause of AIDS—an increasing number of related T-lymphotropic retroviruses are recognized. These include human T-cell lymphotropic virus I (HTLV-I), HTLV-II, HTLV-IV, bovine leukemia virus, and simian T-lymphotropic viruses I (STLV-I) and III (STLV-III), which infect certain nonhuman primates.

While the chief effect of HTLV-I is leukemia in adults, it is also known to cause a mild immune deficiency in some individuals and more recently a form of paralysis, called spastic paraparesis, in tropical areas of the world such as Jamaica and Martinique. The adult T-cell leukemia is a rapidly acting disease and is usually accompanied by *Pneumocystis carinii* infection. **[See Chapter 25.]** HTLV-I has also been implicated as a possible cause of multiple sclerosis. The main feature of tropical spastic paraparesis is the development of a spreading weakness of the legs and lower body. HTLV-II is associated with a form of nasopharyngeal cancer.

STLV-I naturally infects most species of Old World monkeys and great apes and has been linked to certain cancers in monkeys. STLV-III occurs in both normal and ill monkeys and is antigenically similar to the human AIDS virus. The antigenic similarity raises the possibility that a family of re-

TABLE 27–7 A Comparison of Human and Nonhuman Primate Retroviruses

	Viruses[a]			
Features	*STLV*	*HTLV-IV*	*LAV-2*[b]	*HIV-1*
Natural host	African green monkey	Human	Human	Human
Location(s) of infections	Africa	West Africa	West Africa and Europe	Worldwide
Virulence	No disease	No AIDS	AIDS	AIDS
Target cell in body	T_4^+(helper)[c]	T_4^+(helper)	T_4^+(helper)	T_4^+(helper) and macrophages

[a] STLV, simian T-cell lymphotropic virus; HTLV-IV, human T(cell)-lymphotropic virus; LAV, lymphadenopathy virus; HIV, human immune deficiency virus.
[b] Also may be known as HIV-2.
[c] Also known as CD4.

lated viruses may have existed in primates well before the AIDS epidemic began. Moreover, several investigators speculate that STLV-III may have been transmitted to humans at some time during the natural history of these viruses. AIDS cases may have been present in the mid 1970s, before the disease was recognized in the United States and Europe. The gap between STLV-III and HIV may soon be closed by the discovery of a group of intermediate viruses, human T-cell lymphotropic virus IV (HTLV-IV) and LAV-2. HTLV-IV is nonpathogenic, but it infects humans. LAV-2 is a closely related virus. Studies under way with these viruses should shed light on the origin and spread of the AIDS virus. Table 27–7 compares several members of the retrovirus group.

INFECTIOUS MONONUCLEOSIS

Infectious mononucleosis is an acute leukemialike infection generally caused by the Epstein-Barr virus. In 1968, Werner and Gertrude Henle detected this herpes virus in cell culture obtained from a cancerous state of children called *African Burkitt's lymphoma*

Microbiology Highlights

SIMIAN T-LYMPHOTROPIC VIRUSES

The first member of the human T-lymphotropic virus (HTLV) family, HTLV-I is known to be endemic in parts of Africa, the Caribbean, Japan, South America, and the southeastern United States. It has been found to be closely related to a retrovirus found in certain Old World monkeys called simian T-lymphotropic virus type I (STLV-I). Investigators propose that either HTLV-I represents a simian virus variation or both viruses originated from a common ancestor in Africa.

A simian virus similar to HIV, the cause of human AIDS, has also been identified and designated as STLV-III. The virus has been found in healthy, wild African green monkeys and in monkeys from macaque colonies in the United States. STLV-III causes simian acquired immunodeficiency syndrome (SAIDS), a condition which resembles human AIDS in most respects.

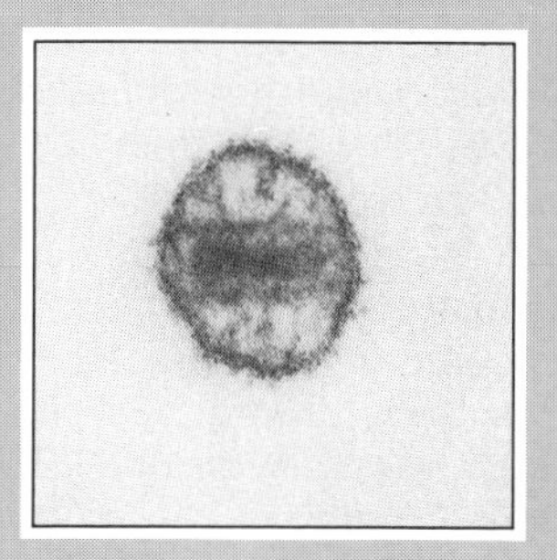

SAIDS (simian acquired immunodeficiency syndrome) virus. *[From P. A. Marx,* et al., Science *223*:1083–1086, 1984.]

(see Chapter 30). Infectious mononucleosis, which also can be caused by cytomegalovirus, has been found primarily in young adults, especially college students. It has often been called the "student's disease." The disease involves components of the mononuclear phagocyte (reticuloendothelial) system, including the lymph nodes and spleen. An increase in lymphocytes is also associated with the illness. These cells are abnormal and can be classified on the basis of certain cytoplasmic and nuclear features. Such atypical lymphocytes are called *Downey cells,* and at least three classes are recognized. In general, the leukocyte levels are related to the course of the disease. Epstein-Barr virus (EBV) infects B type lymphocytes (epithelial cells in tonsillar tissue), while T lymphocytes (helper and killer types) together with antibodies produced in response to the virus kill and destroy the infected cells. **[See Chapter 17 for description of B and T cells.]**

Interest in Epstein-Barr virus infection extends far beyond its causative role in infectious mononucleosis and its association with African Burkitt's lymphoma. The infection is one of the most common in humans. Up to 90% of children in developing nations have antibodies against the virus in the first years of life. Also, high concentrations of immunoglobulins and cell-mediated immune memory appear to persist for the life of the host.

Epstein-Barr virus has been associated with other cancerous states and infections, including Chinese nasopharyngeal carcinoma, rare tumors of the lymphatic systems in patients with immune deficiencies, and a chronic or relapsing form of infectious mononucleosis among adults. In the chronic condition, also referred to as sporadic neurasthenia (a generalized form of muscular weakness), few if any identifiable abnormalities are seen on physical examination or routine laboratory testing. About two-thirds of patients are women, the most common ages being between 25 and 45. Additional features of the condition are given in Table 27–4.

Transmission. Infectious mononucleosis is transmitted chiefly by kissing (direct oral means) or by drinking from a shared bottle or glass (indirect oral means) with the exchange of saliva and perhaps leukocytes. For Epstein-Barr virus infection to occur, an adequate quantity of virus must be introduced into the mouth of an antibody-negative host. Such factors as race, inherited characteristics of host defenses, titer of virus, and virus characteristics may determine whether or not symptoms occur (Table 27–4). Usually, once a person has had infectious mononucleosis an immunity is established. Definite recurrences are rare. However, certain neurological complications such as Guillain-Barré syndrome, aseptic meningitis, and encephalitis have been reported to occur.

Like other herpesviruses, Epstein-Barr virus can cause a true latent infection. The viral genome within the host B cell is arrested, or the production and release of new infectious viruses is suppressed. B lymphocytes may also be transformed. Here the viral genome is integrated into the host cell DNA.

Diagnosis and Treatment. In addition to the symptoms of this disease, certain laboratory findings are used for diagnostic purposes. These include the presence of atypical, large T lymphocytes and a characteristic positive test for heterophile antibodies—those that agglutinate red blood cells in sheep. Liver function tests may also be performed. While rarely needed in most cases of mononucleosis, there are diagnostic immunologic assays for a large number of antibodies that can be used to pinpoint, with great precision, the stage of infection in a given patient. These include determinations of (1) immunoglobulin M antibodies against viral capsid antigen; (2) antibodies against Epstein-Barr virus nuclear antigen; and (3) antibodies against membrane antigens of cells infected with Epstein-Barr virus. Treatment procedures usually include, among other things, bed rest and the hospitalization of those individuals with severe complications, such as convulsions and rupture of the spleen.

KAWASAKI SYNDROME

In 1967, Tomisaku Kawasaki of Tokyo's Red Cross Medical Center reported the first cases of an acute, fever-producing skin rash illness in children. Since that time, the clinical and epidemiologic experience with the disease has broadened considerably. Kawasaki syndrome is no longer considered rare and is now recognized as a disease of worldwide distribution, occurring in children of all racial groups. It is more prevalent in Japan and in children of Japanese ancestry. The majority of victims are under four years of age. While the lymph nodes are primarily involved, blood vessels and the heart are also damaged (Table 27–4). Clinical heart disease caused by inflammation of blood vessels and heart tissue occurs in at least 20% of patients. Recovery is frequently associated with some form of cardiovascular abnormality.

Transmission and Cause. At present, the transmission and cause of Kawasaki syndrome remain unknown. Several features of the disease suggest an infectious agent. Researchers believe a retrovirus to be the cause, since a reverse transcriptase was detected in several children with the disease. Reverse transcriptase, which is an enzyme peculiar to retroviruses, was not found in healthy children. Several investigators speculate that Kawasaki syndrome is triggered by a common agent or group of agents, which in certain susceptible children may stimulate an abnormal response and produce the damage in the blood vessels, heart, and other body tissues.

Disease Challenge

The situation described has been taken from an actual case history. It has been designed to show how clinical, laboratory, microbial, and related information is used in disease diagnosis. A review of treatment and epidemiological aspects is given at the end of the presentation. Answers to questions, laboratory findings, and interpretations are given immediately following a specific question. Test your skill and take the Disease Challenge.

KEY TERMS

myalgia (my-AL-ji-a): muscle pain

paroxysm (PAR-ok-sizm): A sudden, periodic occurrence of disease symptoms

CASE

A 36-year-old male was admitted to a local medical center after experiencing severe muscle stiffness (rigor) followed by a rapid temperature rise to 104°F. The fever lasted approximately six hours. Marked sweating, extreme fatigue, diarrhea, and chills followed. While waiting to be admitted, the patient complained of extreme nausea, headache, myalgia, and low back pain. A further history revealed that the patient had returned from a four-week tour of southern China approximately three months earlier.

On physical examination the patient was found to be normal. No rashes or bites were in evidence.

At This Point What Diseases Could Be Suspected?
Bacterial infections including typhoid fever and shigellosis, as well as the protozoan disease malaria. [Refer to Chapter 26 for descriptions of the bacterial infections.]

What Type(s) of Laboratory Specimens Should Be Taken?
Blood for smears, culture, serological tests, and standard laboratory tests. Stool samples for a bacterial culture should also be taken.

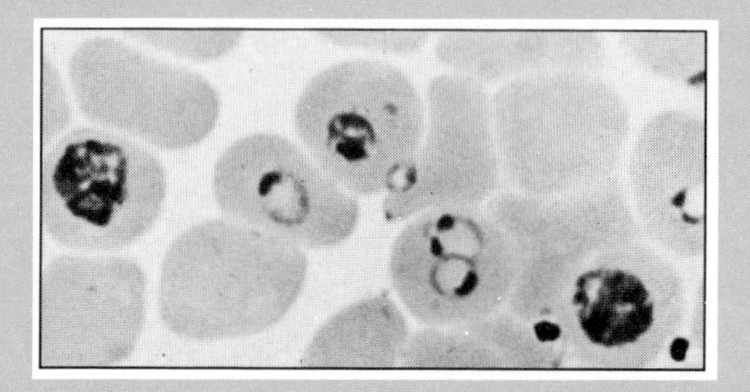

Results of blood smear examination.

Laboratory Results:
All bacterial cultures and serological tests were negative. Examination of blood smears revealed the results shown in the accompanying illustration. All standard blood tests were within normal ranges.

What Is the Probable Diagnosis?
A form of malaria.

Treatment:
The patient was given appropriate dosages of chloroquine phosphate and kept under observation during his stay.

Summary

1. The circulatory system, in addition to being susceptible to microbial diseases, can spread disease agents and their products to other regions of the body.

2. The circulatory system meets several needs of vertebrates, including internal transport of nutrients and hormones, removal of cellular wastes, regulation of body temperature, and control and elimination of foreign organisms.

3. Blood flows through a closed system that consists of the vessels—arteries, veins, and capillaries—and the heart.

Diseases of the Heart

1. Activities of the heart can be affected by endocarditis (which includes infection of its tissues), by insufficient nourishment for heart muscle, and by overexertion of the heart, resulting in exhaustion.

2. The heart can be subjected to infections, just as any other part of the body can.

3. Two infections of the heart are rheumatic heart disease and subacute bacterial endocarditis.

4. Rheumatic fever is a hypersensitivity state that develops in a small percentage of individuals who have a history of streptococcal infections.

5. Infective endocarditis results from inflammation of the endocardium by bacteria lodged in irregular or damaged heart valves. Infective endocarditis may be a complication of any dental procedure that allows bacteria to enter the bloodstream of a person with damaged heart valves. Accumulations of formless masses of fibrin strands, blood cell fragments, and bacterial masses called *vegetations* develop in damaged valves.

Other Microbial Diseases of the Circulatory System

BACTERIAL INFECTIONS

1. Several bacterial infections involve the circulatory system, including plague, relapsing fever, tularemia, and rickettsial diseases such as rickettsialpox, Rocky Mountain spotted fever, scrub typhus, and endemic and epidemic typhus fevers.

2. Many bacterial infections of this system are transmitted by arthropods such as fleas, lice, mites, or ticks.

3. Preventive measures include immunizations (if available) and elimination of the arthropod vectors.

PROTOZOAN INFECTIONS

1. As with other microbial disease agents, the destructive effects of pathogenic protozoa depend on their number, size, degree of activity, location, and toxic products.

2. Examples of protozoan diseases involving the circulatory system include kala azar (spread by sandflies) and malaria (spread by species of *Anopheles* mosquitoes).

3. Preventive measures are usually directed toward the elimination of the arthropod vector.

Viral and Related Infections

ACQUIRED IMMUNE DEFICIENCY SYNDROME (AIDS)

1. Acquired immune deficiency syndrome (AIDS) is an epidemic immunosuppressive disease that predisposes to life-threatening infections with opportunistic organisms, Kaposi's sarcoma, and, less commonly, other cancers.

2. AIDS is caused by a retrovirus known as human immunodeficiency virus (HIV).

3. Characteristically, the disorder is associated with a depletion of T4 or CD4 cells.

4. HIV infection has a wide range of effects, from asymptomatic disease states to AIDS-related complex (ARC), and finally full-blown AIDS, with or without central nervous system involvement.

5. Pediatric AIDS (PAIDS) is a distinct immune deficiency disease resulting from HIV infection transmitted by the mother to her fetus or by the use of contaminated blood or blood products.

INFECTIOUS MONONUCLEOSIS

1. Infectious mononucleosis is an acute leukemialike infection generally caused by the Epstein-Barr virus. The virus was originally detected in cell cultures from children with the cancerous condition called *African Burkitt's lymphoma.*

2. Epstein-Barr virus involves the lymph nodes and the spleen. It can cause a true latent infection or bring about the transformation of B lymphocytes. T lymphocytes together with specific antibodies produced in response to the virus kill infected cells. Cytomegalovirus also causes infectious mononucleosis.

3. A chronic or long-lasting form of infectious mononucleosis is being reported with increasing frequency among adults.

4. Transmission of the disease agent can occur through kissing or using a contaminated glass or other object that comes into contact with the mouth.

KAWASAKI SYNDROME

Kawasaki syndrome is a newly recognized clinical disease for which a specific cause has not been found. The disease involves several tissues, including lymph nodes, blood vessels, and the heart.

Questions for Review

1. What are the functions and activities of the human circulatory system?

2. How do infectious disease agents affect the various components of the circulatory system?

3. Give the general means of transmission for each of the following diseases:

a. relapsing fever	f. rheumatic heart disease
b. malaria	g. infectious mononucleosis
c. plague	h. AIDS
d. typhus fever	i. infective endocarditis
e. tularemia	j. ARC

4. Distinguish between bubonic and pneumonic plague.
5. What is infective endocarditis?
6. If you were assigned to speak to a group of representatives from developing countries on the transmission and control of arthropod-transmitted diseases, which diseases would you select? What major means of control would you recommend to this group?
7. What components and mechanisms of the circulatory system contribute to the elimination of disease agents? (Refer to Chapter 17 for assistance in answering this question.) Of what particular importance is infectious mononucleosis to college students?
8. Of what importance are T4 or CD4 cells in AIDS?
9. a. What is pediatric AIDS?
   b. Does it differ from the adult form of AIDS? If so, how?
10. Distinguish between AIDS and ARC.
11. Is Kawasaki syndrome an infectious disease?

Suggested Readings

Bartlett, R. C., P. D. Ellner, and J. A. Washington, II, *Cumitech I.* Washington, D.C.: American Society for Microbiology, 1974. *A short publication describing general procedures for the isolation of bacterial pathogens from blood (blood cultures).*

Centers for Disease Control, *Revision of the CDC Surveillance Case Definition for Acquired Immunodeficiency Syndrome.* Morbidity and Mortality Weekly (MMWR) **36**:3S-15S, 1987. *This supplement fully describes the new definition of AIDS that is to be used for reporting purposes.*

Cherry, J. D., "Viral Exanthems," *Disease-a-Month* **28**(8): 1–56, 1982. *A publication covering several aspects of viral illness and providing general descriptions of specific diseases, clinical features, clinical differentiation, and laboratory diagnosis.*

Gallo, R. C., "The AIDS Virus," *Scientific American* **256:** 46–56, 1987. *A functional review of the progress made in uncovering the viral cause, immunological dysfunction, and transmission and properties of the AIDS virus.*

Garcia, L. S., and M. Voge, "Diagnostic Clinical Parasitology: IV. Identification of the Blood Parasites," *American Journal of Medical Technology* **47**:21, 1981. *This article deals with the isolation and identification of human blood protozoan pathogens. Specific attention is given to the causative agents of malaria, babesiosis, leishmaniasis, and trypanosomiasis.*

Gregg, C. T., *Plague! The Shocking Story of a Dread Disease in America Today.* New York: Charles Scribner's Sons, 1978. *In addition to covering the properties of* Yersinia pestis, *the causative agent of plague, this book discusses in detail such topics as the biology of rats and other rodents, historical accounts of epidemics, control methods, and several unanswered questions about plague.*

Gupta, S. (ed.), *AIDS-Associated Syndromes.* New York: Plenum Publishing Corporation, 1985. *A functional series of articles dealing with research in the areas of the AIDS virus, the immunological status of patients, and approaches to treatment.*

McCue, J. D., "The Pathogenesis of Infectious Mononucleosis," *Hospital Practice* **17**:34, 1982. *A general description of the interactions between Epstein-Barr virus and human lymphocytes.*

Melish, M. E., R. V. Hicks, and V. Reddy, "Kawasaki Syndrome: An Update," *Hospital Practice* **17**:99, 1982. *A fairly detailed description of this new multisystemic disease of very young children.*

Mitscherlich, E., and E. H. Marth, *Microbial Survival in the Environment: Bacteria and Rickettsiae Important In Human and Animal Health.* New York: Springer-Verlag, 1984. *A comprehensive presentation of information emphasizing the resistance and environmental survival of rickettsia and other pathogenic bacteria responsible for contagious diseases or food-borne illness.*

Pankey, G. A., "Infective Endocarditis: Changing Concepts," *Hospital Practice* **21**:103–110, 1986. *A comprehensive description of the current features and approaches to control of the infection of the heart formerly known as bacterial endocarditis.*

Selwyn, P. A., "AIDS: What Is Now Known: II. Epidemiology," *Hospital Practice* **21**:127–164, 1986. *A well-written article outlining the features of AIDS epidemiologic patterns on a worldwide basis.*

28

Microbial Diseases of the Reproductive and Urinary Systems

There was a young man from Back Bay
Who thought syphilis just went away.
He believed that a chancre
was only a canker
That healed in a week and a day . . .
— Anonymous

After reading this chapter, you should be able to:

1. Describe the general organization of the human urinary system and indicate the areas attacked by microbial pathogens.
2. Describe the microbial flora of the human urinary tract.
3. Discuss predisposing factors and routes of transmission associated with urinary tract infections.
4. List and describe three infections of the urinary system, including the causative agents, means of transmission, and control measures.
5. Explain the importance of anaerobes to urinary tract infections.
6. Distinguish among the most common human sexually transmitted diseases (STDs) in terms of their causative agents, areas of the reproductive system attacked, general effects, complications, and measures for treatment and control.
7. Discuss the importance of STDs to pregnant women and newborn infants.
8. Compare the general approaches to diagnosis of urinary tract and reproductive system infections.
9. Explain the complications associated with diseases of the urinary tract and the reproductive system.

Urinary tract infections represent a major source of human discomfort. The most common diseases of this system are bacterial. Most of these infections remain limited to the urinary tract, but some can develop into extensive disease processes involving other organs and sometimes leading to death. Sexually transmitted diseases (STDs) represent major and growing public health problems. Some of the problems are caused by microorganisms unconnected with the traditional venereal diseases such as syphilis, gonorrhea, chancroid, lymphogranuloma venereum, and granuloma inguinale. Left untreated and unchecked, STDs pose serious threats to human well-being. Chapter 28 considers various genitourinary tract infections, their general features, and the complications they may cause.

The many functions of the urinary system include eliminating waste products, regulating the chemical compositions and acid-base balance of body tissues, and keeping the water content of the body constant. In general, the system is constructed and organized to prevent invasion by microorganisms. However, certain circumstances, such as obstruction of the flow of urine or the use of various medical devices, can promote infection. Many microbial agents are opportunists that are actually members of the body's normal flora.

The organs of the reproductive system are also subject to infection by a variety of agents. Of these, the *sexually transmitted diseases* (also referred to as *venereal diseases*) are of particular importance because of the discomfort and disability (including blindness, disfigurement, and sterility) that they can cause.

Anatomy of the Urinary System

The human urinary system includes two *kidneys,* two *ureters,* the *urethra,* and the *urinary bladder* (Figure 28–1). The bean-shaped kidneys are compound tubular glands that secrete urine. This excretory product goes by way of two muscular tubes, the ureters, to the bladder, which stores the product until its elimination. Urine is discharged by way of the urethra to the outside. The female urethra functions only in urination; the male counterpart serves for the passage of both urine and semen. As shown in Figure 28–1, several parts of the system are subject to infection.

A kidney consists of an outer cortex and an inner medulla (Figure 28–1). The basic functional unit of the kidney is the *nephron,* which is composed of a *glomerulus* (blood filter) and a long tubule (concentrator and urine collector). The glomerulus, a tuft or collection of capillaries, is surrounded by a spherically expanded portion of the nephron, the *Bowman's capsule* (Figure 28–1). Blood is filtered through the glomeruli and concentrated by the tubules,

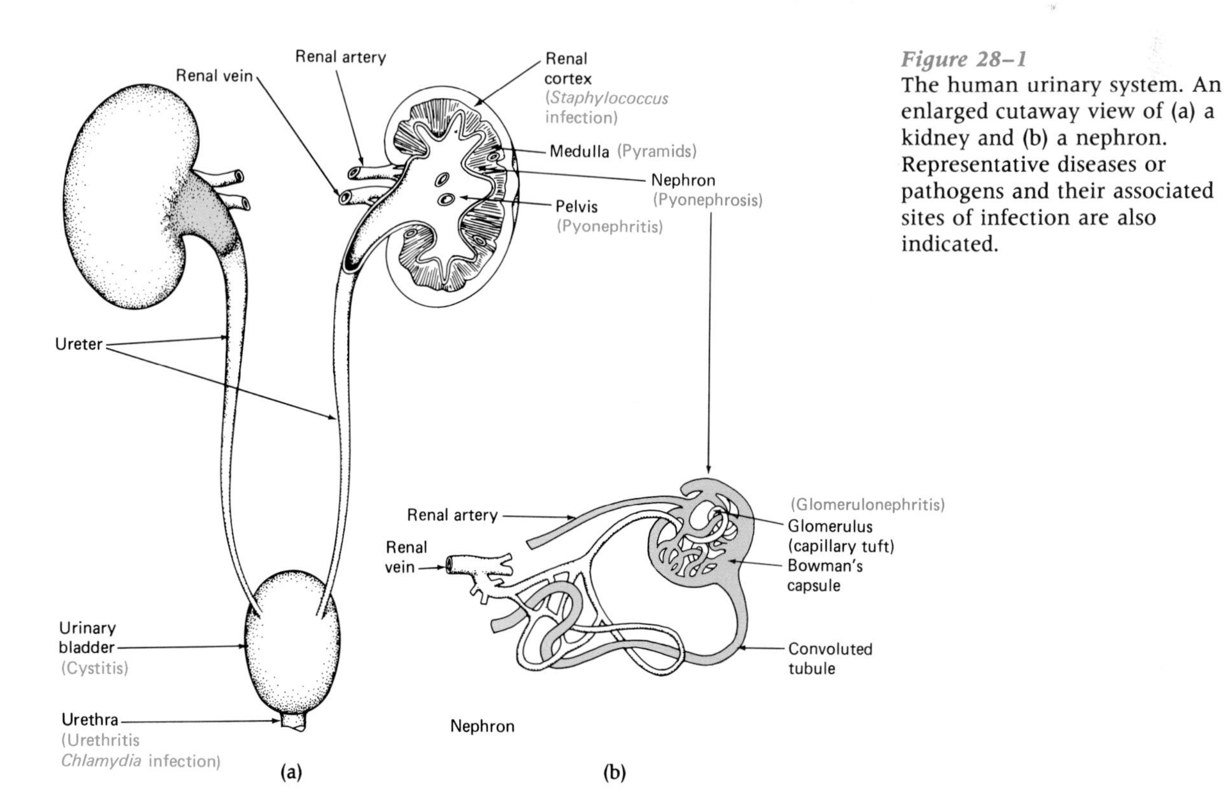

Figure 28–1
The human urinary system. An enlarged cutaway view of (a) a kidney and (b) a nephron. Representative diseases or pathogens and their associated sites of infection are also indicated.

which empty into the ureters (Figure 28–1). The renal capsule filters dissolved substances and some water out of the blood plasma; the renal tubule reabsorbs water, electrolytes, and other substances to help maintain the body's normal internal environment. The kidneys establish and help to maintain the concentration of electrolytes (for example, chloride, potassium, sodium, and bicarbonate) in the human body and excrete poisonous substances.

Flora of the Normal Urinary Tract

Experiments with various animals have shown that the normally sterile kidney is highly resistant to bacteria. Many well-known urinary tract pathogens are unable to invade the kidney and produce infection there. However, there are a few exceptions to this finding, including certain enterococcus strains, *Escherichia coli, Proteus* spp., and staphylococci.

In humans, bacteria are commonly present in the lower urethra. However, their numbers decrease in regions near the bladder. This decrease is apparently caused by an antibacterial secretion of the epithelial cells lining the urethra, which prevents bacteria from penetrating the organ, and the frequent flushing of the epithelial surface of the urethra by urine, which is normally sterile. Urinary tract infections are more common in females than in males. This difference in incidence may be due to the greater length of the male urethra and the fact that, by virtue of its location, it is less subject to fecal contamination.

Urinary tract infection is the most common bacterial infection affecting adult women. About 20% of women are affected at some time in their lives.

Diseases of the Urinary Tract

Introduction to Urinary Tract Infections

The features of common urinary tract infections (UTI) are summarized in Table 28–1. The bacteria commonly associated with urinary tract infections are intestinal flora called **coliforms.** These have been isolated from 80% of the cases that have developed in the absence of urinary obstructions or that have been unaffected by antibiotic therapy or urologic manipulations such as *catheterization* (the insertion of a surgical tubular device into organs or body cavities).

When infection follows catheterization, *Pseudomonas* spp. infections are found in greater numbers. Infections caused by enterococci, *Proteus* spp., and staphylococci are believed to result from direct contact with infected individuals or with contaminated articles, such as bedpans and catheters.

Anaerobic bacteria and several species of fungi are also known to cause urinary tract infections. The relationship of venereal disease to this category of disease will be discussed later.

PREDISPOSING CONDITIONS

Several clinical factors have been implicated as predisposing individuals to urinary tract infections. These factors include diabetes mellitus, neurologic diseases (for example, poliomyelitis and spinal cord injuries), toxemia of pregnancy, and lesions that interfere with urine outflow. Kidney stones, stricture, and tumors are examples of such urinary obstructions.

ROUTES OF INFECTION

Although a controversy exists regarding the pathways by which bacteria gain access to the kidney, three major routes of infection have been listed: *hematogenous* (via the blood), *lymphatic,* and *ascending urogenous* (via the ureter). Most clinical and experimental evidence shows that the last-mentioned route is the most common pathway in pyelonephritis (inflammation of the kidney and pelvis). The infecting organism determines the form and intensity of the resulting infection.

Sometimes it is possible to establish how causative agents are acquired. For example, infections that develop in women shortly after marriage may be the consequence of mechanical injury to the urethra during sexual intercourse. The onset of a typical acute infection is commonly associated with urinary tract instrumentation.

In the bladder cavity, several factors provide a favorable environment for bacterial persistence in urine. These include (1) the availability of adequate nutrients, (2) the absence of surface phagocytosis (which commonly occurs in other tissues), (3) an optimal temperature for growth, and (4) the fact that some bacteria remain in the bladder after urination. Apparently, the film composed of the remaining urine and bacteria, which coats the bladder surfaces, serves as an inoculum for newly formed urine. Evidence indicates that such contaminated urine can also be a source of infectious agents for other regions of the urinary tract. Some pathogens are capable of traveling in a direction opposite to normal urine flow.

ASPECTS OF DIAGNOSIS

The presence of bacteria in freshly voided urine may have little or no significance unless they happen to be microorganisms such as *Mycobacterium tuberculosis* (my-koh-back-TEH-ree-um too-ber-koo-LOW-sis) or

TABLE 28–1 Representative Diseases of the Urinary System

Disease	*Causative Agent*	*Gram Reaction and Morphology*	*General Features of the Disease*
Cystitis (inflammation of the urinary bladder)	*Escherichia coli, Proteus vulgaris, Pseudomonas aeruginosa*	−, Rods	Rapid onset of symptoms, which include painful urination (dysuria) and frequent urination (polyuria)
Glomerulonephritis (inflammation of kidney glomeruli)	*Streptococcus pyogenes*	+, Coccus	Blood and protein in urine, increased fluid in tissues (edema), high blood pressure (hypertension)
Kidney cortex infections	*Staphylococcus aureus*	+, Coccus	Painful urination, fever, and lower back pain
Nonspecific urethritis (inflammation of the urethra)	*Candida albicans*	+, Oval yeast	Painful urination, thick pus-containing discharge from urethral opening
	Chlamydia trachomatis[a]	−, Rods	
	Mycoplasma hominis and *Ureaplasma urealyticum*	−, Pleomorphic[b]	
	Staphylococci and streptococci	+, Cocci	
	Trichomonas vaginalis[a]	Not useful, Protozoan	
Pyelonephritis (inflammation of the kidney and pelvis)	*Bacteroides* spp., *Enterobacter aerogenes, Escherichia coli, Proteus* spp., *Pseudomonas aeruginosa*	−, Rods	Painful urination, frequent urination, fever, lower back pain, vomiting, diarrhea, and rapid heartbeat
	Staphylococci and streptococci	+, Cocci	
Pyonephrosis (inflammation of the kidney with pus formation)	Anaerobic streptococci	+, Cocci	Painful urination, fever, lower back pain, blood in urine
	Bacteroides spp.	−, Rods	

[a] Details of this pathogen are presented later, in the discussion of sexually transmitted diseases (STDs)
[b] These bacteria lack cell walls. See Chapter 13.

Salmonella (sal-mon-EL-lah) spp. Such urine, having passed through the urethra, will usually contain members of the region's normal flora, such as *Bacillus* (bah-SIL-lus) spp., coliforms, diphtheroids, *Proteus* (PROH-tee-us) spp., staphylococci, streptococci, and yeasts.

Catheterized specimens may also be contaminated, since the equipment used can come into contact with the urethra and consequently collect normal urethral flora. Since members of the normal flora can cause cystitis, prostatitis, pyelonephritis, and other severe urogenital infections, their presence in a urine specimen is not very helpful in identifying pathogens.

Standard Procedures. Several screening tests are available for use in suspected urinary tract infections. Included in these tests are dipstick devices for the detection of leukocytes, blood, and abnormal concentrations of proteins. Serological tests such as the fluorescent antibody technique are of diagnostic importance in chlamydial urethritis. [See Chapter 19 for fluorescent antibody techniques.]

The presence of more than 100,000 (10^5) bacteria per milliliter of urine is strongly indicative of acute disease. Diagnostic culturing systems are available not only to determine the concentrations of bacteria in specimens but also to detect the presence of specific medically important microorganisms (Figure 28–2), including the following: *Candida albicans* (a yeast), *Escherichia coli, Neisseria gonorrhoeae, Pseudomonas aeruginosa, Staphylococcus aureus, Streptococcus faecalis,* and *Trichomonas vaginalis* (trik-oh-MON-as va-jin-AL-iss) a protozoon. Several systems for screening large numbers of specimens are fully automated.

SYMPTOMS OF URINARY TRACT INFECTIONS

The following symptoms usually signal the presence of urinary tract disease: blood or pus in the urine, accumulation of fluid in tissues, pain, kidney enlargement, and blood loss, causing anemia. Such symptoms are usually found with infections of the

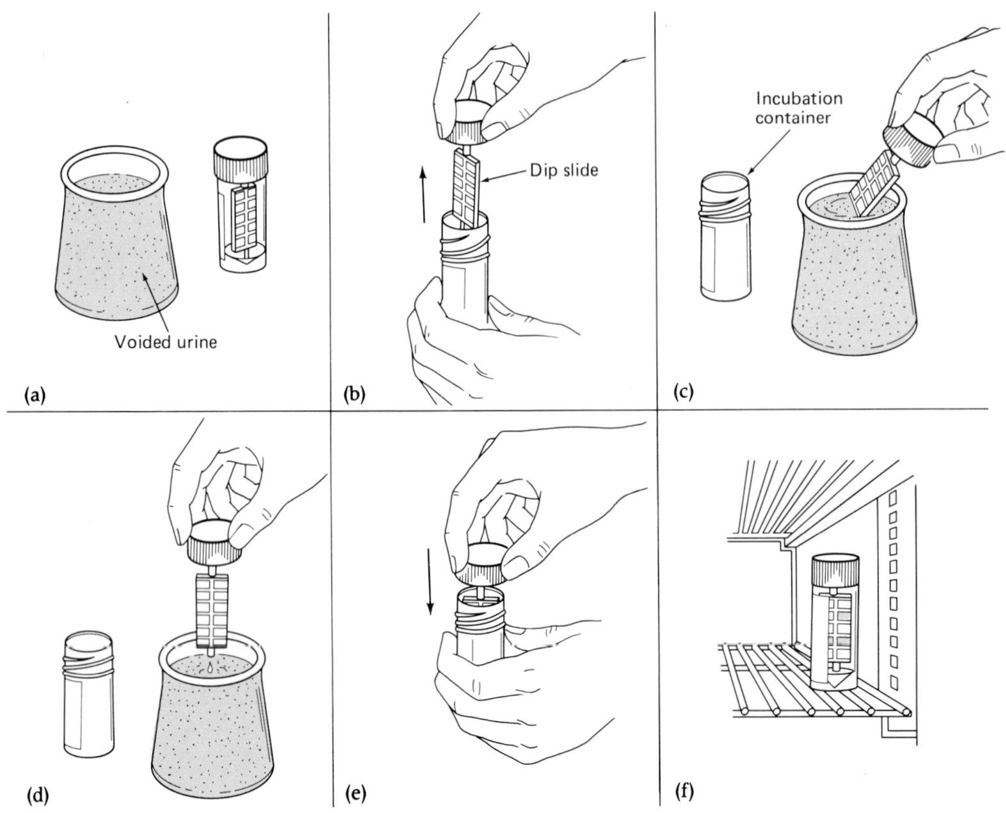

Figure 28–2 One type of diagnostic culture system, the urine dip-slide procedure. The components of the procedure, voided urine and a dip slide in its container. (a) The dip slide is removed from its container (b) and inserted into the urine specimen. (c) Excess urine is allowed to drip back into the container (d), and the dip slide is placed back into its container (e) for incubation (f). *[After G. J. Pazin, A. Wolinsky, and W. S. Lee,* Am. Fam. Phys. ***11****:85–96, 1975.]*

kidney, urinary bladder, ureter, and urethra (Table 28–1).

Kidney Diseases Caused by Bacteria

Two major types of infections affect the kidney—those that cause a diffuse inflammation of the tissues, *pyelonephritis,* and those that primarily involve the organ's cortex. Detailed discussion of the clinical and diagnostic features and treatment of these disease states can be found in the texts listed in the reference section at the end of this chapter.

ACUTE PYELONEPHRITIS

Acute pyelonephritis is probably the most common disease of the kidney. It is an inflammation of one or both kidneys, involving the tubules. It is generally not considered a primary infection but rather a complication brought on by an infectious process, such as respiratory disease or bloodstream infection (sepsis).

E. coli is the major cause of this condition. Some 60% to 80% of acute pyelonephritis is attributed to this organism. Other bacteria that have been associated with this infection include *Enterobacter aerogenes* (en-te-ro-BACT-ter a-RAH-jen-eez), Proteus spp., *Pseudomonas aeruginosa* (soo-doh-MOH-nass air-roo-jin-OH-sah), staphylococci, and *Streptococcus pyogenes* (strep-toh-KOK-kuss pie-AH-gen-eez). Most causes of acute pyelonephritis are blood-borne. However, some infections have been reported to develop after instrumentation of the urinary tract.

This kidney infection appears to occur more often in females than in males. In the female, acute pyelonephritis is a common complication of lower urinary tract infections. However, the majority of cases are harmless. Lesions usually heal spontaneously, leaving only small scars in the kidney tissue. Nevertheless, it is important not to underrate these infections, as they may recur or become chronic, eventually producing serious kidney damage.

KIDNEY CORTEX INFECTIONS

Infections of the cortical regions of the kidney are quite different from pyelonephritis in several ways. The former are localized infections of the cortex that are definitely spread via the blood. Pyelonephritis, on the other hand, is a diffuse infection that is not necessarily acquired from the bloodstream.

In cortical infections, the microorganism most

Urinary Tract Infection

LABORATORY IDENTIFICATION

Laboratory identification of the pathogens listed in Table 28–1 is straightforward. *E. coli, Proteus* species (spp.), *S. aureus,* and the streptococci can readily be isolated from specimens on selective and differential media such as eosin-methylene blue and MacConkey agars (Color Photographs 24 and 109). *E. coli* and *E. aerogenes* are lactose fermenters, whereas *Proteus* species are not. In addition, *Proteus* spp. form a spreading or swarming growth unless the agar concentration is increased in the isolation medium. The biochemical reactions of *E. coli* and *Proteus* spp. serve to separate them from other organisms. In the combination of tests known as the IMViC series, *E. coli* forms the metabolic breakdown product indole (I) from the amino acid tryptophane, produces acid from glucose in a methyl red (MR) broth containing the carbohydrate, does not produce acetylmethyl carbinol in the Voges-Proskauer (V) test, and is unable to use citrate (C) as the only source of carbon. *E. aerogenes* produces the reverse set of IMViC reactions, (−,−,+,+). *Proteus* species are distinguished biochemically by being non-lactose fermenters and producing urease. *P. aeruginosa* is an oxidase-positive non-lactose fermenter. Other distinguishing properties of the organism include the production of pyocyanin (blue pigment) and the ability to grow at 42°C.

Staphylococci and streptococci can be isolated from specimens on blood agar. Mannitol salt agar is also used for the isolation of *S. aureus.* The different species are differentiated on the bases of hemolytic activity (Color Photographs 23a and 23b) and biochemical tests (Color Photographs 26 and 86.) Commercial modified agglutination tests also are used for identification (Color Photograph 113).

Bacteroides (back-teh-ROY-deez) species require anaerobic cultivation techniques for isolation. Several species produce black pigment on blood agar (Color Photograph 21) and require vitamin K or related compounds for growth.

TREATMENT

Antibiotic susceptibility tests should be performed with isolated organisms. In general, chloramphenicol, tetracyclines, and nalidixic acid are used to treat gram-negative infections.

Penicillin G or V are the drugs of choice for non-penicillinase-producing staphylococci. Methicillin, nafcillin, or oxacillin are used with penicillinase (penicillin-inactivating enzyme)-producing strains. Alternate drugs for individuals with a penicillin allergy include cephalosporin, clindamycin, and vancomycin. Penicillin is generally the drug of choice for most streptococcal infections. Chloramphenicol, clindamycin, and metronidazole are commonly used for *Bacteroides* infections.

frequently present is *S. aureus. E. coli* may be a secondary invader.

Diseases of the Urinary Bladder

The bladder exhibits considerable resistance to infection. However, it, too, can succumb to disease.

CYSTITIS

The inflammation of the bladder that characterizes cystitis can be either acute or chronic. The acute form is one of the most common urinary tract lesions. However, it is a symptom rather than a specific disease. Thus, the underlying causative factor of acute cystitis must be uncovered if treatment is to be effective. Attacks are common among women.

Sources of bladder infection in the female include the gastrointestinal tract, the cervix (cervicitis), the uterus (endometritis), the urethra, and vaginal involvement by the protozoon *Trichomonas vaginalis.* In males, the condition may be associated with several body structures, including the gastrointestinal tract, the kidneys, and the urethra. A frequent source of cystitis is an infection of the prostate. Several other factors that can contribute to the development of acute cystitis are foreign bodies self-introduced, certain diseases involving the blood, and diabetes mellitus.

Kidney Cortex Infections

LABORATORY IDENTIFICATION

S. aureus can be isolated from clinical specimens on blood agar and selective and differential media. The pathogen is beta-hemolytic, produces catalase, and ferments mannitol (Color Photograph 26). Pathogenic strains generally produce coagulase. [See Chapter 22 for a description of coagulase.]

TREATMENT

Penicillin G or V are the drugs of choice for non-penicillinase-producing *S. aureus* strains. Methicillin, nafcillin, or oxacillin are used for penicillin-producing strains. Alternate drugs for patients with a penicillin allergy include cephalosporin, clindamycin, and vancomycin.

Microorganisms commonly associated with acute cystitis include *E. coli, P. vulgaris,* and *P. aeruginosa.*

Diseases of the Ureter

Infections of the ureter may develop independently or may be a consequence of disease states involving the bladder, intra-abdominal organs, or kidneys. Inflammatory lesions of the ureter are usually associated with disease and related states of the kidney. These conditions include acute and chronic pyelonephritis, tuberculosis, and various changes produced by kidney stones. Involvement occurs mostly through direct extension from other organs. Treatment usually includes the elimination of the original source of the problem. Therefore, depending on the clinical diagnosis, procedures such as antibiotic therapy or surgery may be necessary.

Nonspecific Urethritis

Inflammation of the urethra, nonspecific urethritis, is caused by microorganisms other than *Neisseria gonorrhoeae* (nye-SEH-ree-ah gon-or-REE-ah) and by other factors, including chemical agents (ingestion of alcoholic beverages or certain chemotherapeutic agents) and trauma, such as passage of a catheter. Nonspecific urethritis is a very common disorder among females. Microorganisms that have been associated with the condition include the yeast *Candida albicans* (KAN-did-ah AL-beh-kans), hemolytic staphylococci and streptococci, *Mycoplasma hominis* (my-koh-PLAZ-mah HOM-eh-niss), *Chlamydia trachomatis* (kla-MEH-dee-ah trah-koh-MAH-tiss), *Trichomonas vaginalis,* and *Ureaplasma urealyticum* (you-REE-ah-plaz-mah you-re-al-LIT-ee-kum). Mixed infections also occur, and several causative agents are sexually transmitted.

Cystitis

LABORATORY IDENTIFICATION

The approach to the laboratory identification of the pathogens listed in Table 28–1 is the same as described in the earlier section dealing with pyelonephritis.

TREATMENT

Therapy for cystitis is the same as described earlier for pyelonephritis.

Nonspecific Urethritis

Laboratory Identification

The bacterial pathogens causing urethritis can be isolated from clinical specimens on appropriate media. *M. hominis* grows rapidly on *Mycoplasma* agar. The pathogen can be distinguished from related organisms by its ability to use the amino acid arginine. *U. urealyticum* forms smaller circular colonies in *Ureaplasma* agar. The production of urease distinguishes this pathogen from related organisms.

C. albicans infections can readily be diagnosed on the basis of a microscopic examination of discharge material or epithelial scrapings. The finding of budding yeast cells and pseudohyphae is sufficient. If culture is needed, blood or Sabouraud's agars can be used.

T. vaginalis infection can generally be diagnosed microscopically by finding the protozoon in vaginal or urethral discharges or urine sediments.

Treatment

Tetracycline is the drug of choice for both *Ureaplasma* and *M. hominis* infections. Spectinomycin can be used for tetracycline-resistant strains of *Ureaplasma*. Superficial *Candida* infections can be treated with topical nystatin. Oral metronidazole (Flagyl) is the drug of choice for *Trichomonas* infections. Sex partners should always be treated simultaneously to minimize recurrent infections.

Urinary Tract Infections Caused by Anaerobes

Several anaerobic organisms are known for their ability to cause renal infections. These anaerobes include members of the genera *Actinomyces* (ak-tin-oh-MY-sees), *Bacteroides, Clostridium* (klos-TREH-dee-um), and *Streptococcus.* Diseases can come about from several causes, including ascending infection with bowel organisms, bacteremia, the introduction of organisms during instrumentation and surgical procedures, and lymphatic spread from the intestinal region.

Kidney stones are usually found in instances of anaerobic kidney infections. Although *Clostridium perfringens* (C. per-FRIN-jens) has been isolated from such stones, it is believed that the majority of infections develop in the dead tissue associated with the stones. Anaerobes may complicate various clinical states, such as congenital defects and renal or bladder tumors.

Immunological Kidney Injury

ACUTE GLOMERULONEPHRITIS

Kidney injury often develops following mild or moderate streptococcal infections. The intensity of the disease varies from a slight involvement of the kidney to severe injury ending in death. Most victims recover with varying degrees of kidney damage. This complication is a case of immunologic hypersensitivity. Two types of reactions can occur: (1) antibodies can form against the individual's kidney tissue; and (2) antigen-antibody complexes can become trapped in glomeruli and cause inflammation (glomerulonephritis). **[See hypersensitivity, types III and IV in Chapter 20.]**

Anatomy of the Reproductive Systems

The reproductive organs and associated structures are subject to a wide variety of inflammatory conditions and infectious diseases (Figures 28–3 and 28–4). The remainder of this chapter provides a brief orientation to the human reproductive system and describes a number of genitourinary infectious (nonvenereal and venereal) diseases (Tables 28–2, 28–4, and 28–5). Diseases that are sexually transmitted but do not cause injury to the reproductive system are described in other chapters. (See AIDS in Chapter 27.)

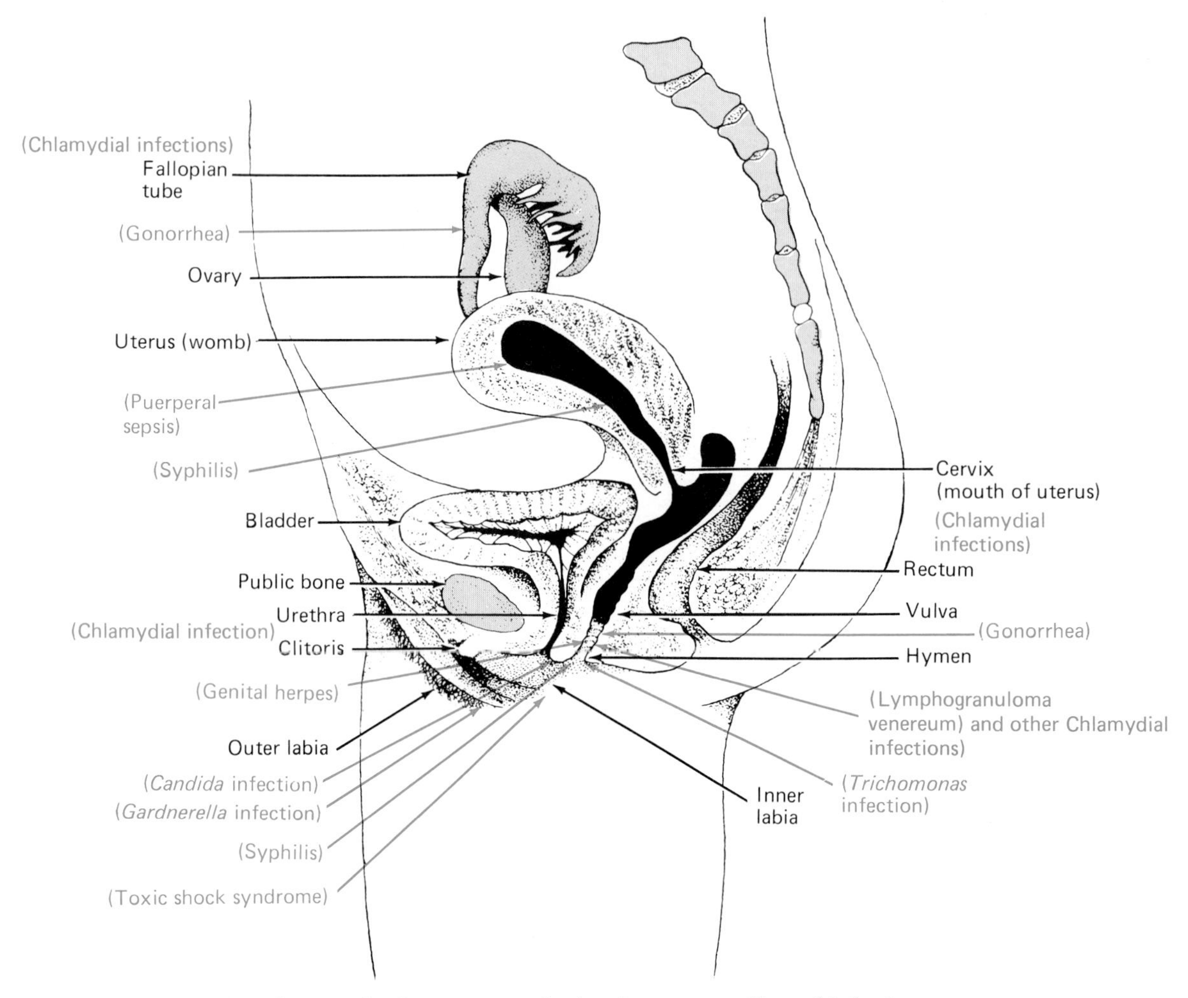

Figure 28–3 The female reproductive organs and related structures. Sites of infection, representative diseases and disease agents are also indicated.

Acute Glomerulonephritis

Laboratory Diagnosis

Accurate identification of group A streptococci requires both clinical (Table 28–1) and bacteriological findings. Isolation of these organisms from urine and related specimens can be made by streaking on sheep blood agar plates. For a presumptive identification, it is also helpful to place a bacitracin sensitivity disk over a streaked area. *S. pyogenes* and other group A streptococci are beta-hemolytic (Color Photograph 23b) and bacitracin-sensitive (Color Photograph 87). Highly specific and rapid serological methods such as Phadebact (Color Photograph 105) and Streptex are reliable tests for laboratory identification. Both tests are modified forms of an agglutination reaction.

Treatment

Penicillin is the drug of choice for infections caused by *S. pyogenes* and other group A beta-hemolytic streptococci. Erythromycin is given to patients sensitive to penicillin treatment.

The Female Reproductive System

The female reproductive system consists of two button-shaped organs called the *ovaries,* two *oviducts (Fallopian tubes),* the *uterus (womb),* and the *vagina* (Figure 28–3). The ovaries are situated near the kidneys on either side of the uterus.

Each ovary consists of a *medulla* (interior) and a *cortex* (outer region). Ova, or eggs, in various stages of development, are contained within the cortex. In addition to producing sex cells, the ovaries produce important hormones, such as estradiol and progesterone.

When an ovum is released *(ovulation),* the egg is transported to the opening of the oviducts. The oviducts convey the egg to the uterus. In case of fertilization, the developing embryo attaches itself in the uterine wall after about a week. The vagina is both a birth canal and a copulatory canal. A thin fold of mucous membrane, called the *hymen,* may be found covering the external opening of the vagina. As Figure 28–3 shows, the uterus projects into the upper portion of the vagina. This region is referred to as the *cervix,* or *neck,* of the uterus.

The external genitalia of the female, collectively referred to as the *vulva,* include the *clitoris, labia majora, labia minora,* and *vestibule.* The clitoris, a small, erectile organ homologous to the male penis, is located toward the front of the vulva. The labia majora and the labia minora are folds of skin that line the vaginal opening. The labia minora, which are situated between the labia majora, contain the *glands of Bartholin.* These glands are responsible for the secretion of an alkaline fluid that functions as a lubricant during copulation.

The Male Reproductive System

The male reproductive system includes two oval glandular *testes,* a system of *ducts,* auxiliary glands, and the *penis* (Figure 28–4). The testes are located outside of the abdominal cavity, suspended by the *spermatic cords* in a saclike structure. The testes are divided into lobes that contain several *seminiferous tubules.* They produce *spermatozoa* and *testosterone* (male sex hormone). Mature sex cells (sperm) go from the seminiferous tubules to the duct called the *vas deferens,* or *ductus deferens,* by way of the *epididymis.* This latter structure is a coiled duct located on the upper surface of each *testis.* The vas deferens carries the sperm to the *urethra.*

The three auxiliary glands, namely, the *seminal vesicles,* the *prostate gland,* and the *bulbourethral glands of Cowper,* are involved with the formation of *seminal fluid,* or *semen.* The secretions of the seminal vesicles empty into the vas deferens, while those of the prostate and bulbourethral glands empty into the urethra.

Diseases of the Reproductive System

Infections of the Female Genital Tract

VAGINAL MICROBIAL FLORA

Microbiologic and biochemical studies have shown that the vagina is a dynamic ecosystem, comprising

Figure 28–4 The male reproductive system. Sites of infection and structures attacked by microorganisms are also shown, as are specific diseases and disease sites.

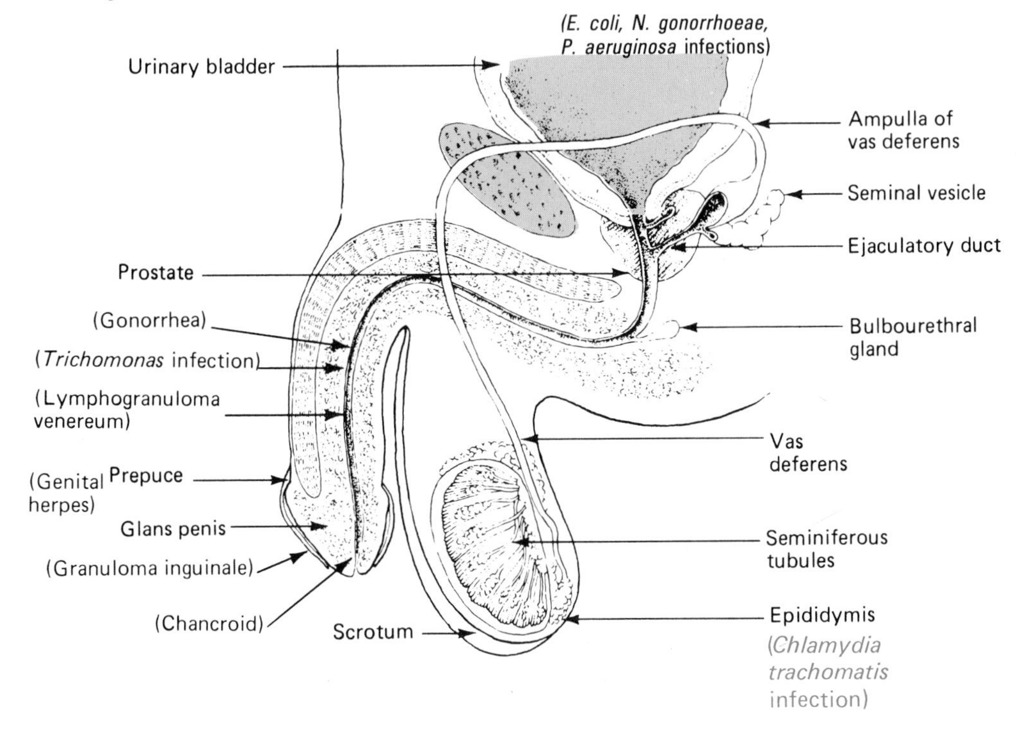

Puerperal Sepsis

LABORATORY IDENTIFICATION

The procedures for the isolation and identification of *E. coli, Proteus* spp., and anaerobes are the same as those described earlier for pyelonephritis. The laboratory identification of *N. gonorrhoeae* is discussed later under prostatitis.

TREATMENT

Therapy for the causative agents of puerperal sepsis are the same as presented in the section dealing with pyelonephritis. *N. gonorrhoeae* infections are treatable with several forms of penicillin and tetracyclines.

TABLE 28–2 Diseases of the Reproductive System

Disease	*Causative Agent*	*Gram Reaction and Morphology*	*General Features of the Disease*
Prostatitis (inflammation of the prostate gland)	*Escherichia coli, Pseudomonas aeruginosa*	−, Rods	Lower back pain, high fever, chills, painful urination (pain in testicles and perirectal area may also occur)
	Neisseria gonorrhoeae	−, Coccus	
	Staphylococci and streptococci	+, Cocci	
	Mixed infections of the above organisms		
Puerperal sepsis (childbed fever) (septicemia originating in uterus after delivery)	Anaerobic streptococci	+, Cocci	Fever, tenderness, and pain in the lower abdominal area and genital region
	Clostridia	+, Rods	
	Escherichia coli, Proteus spp.	−, Rods	
	Neisseria gonorrhoeae	−, Coccus	
Toxic shock	*Staphylococcus aureus (Streptococcus pyogenes)*[a]	+, Coccus	Low blood pressure, rash followed by shedding of skin, slight fever (38.9°C), and involvement of at least four organ systems: muscular, renal, hepatic, and gastrointestinal
Vaginitis	*Candida albicans*	+, Oval yeast	Refer to Table 28–3 for a comparison of vaginitis caused by these organisms
	Gardnerella vaginalis,	−, Rods	
	Trichomonas vaginalis	Not useful (Protozoon)[b]	

[a] A toxic shock-like syndrome has been reported for this organism.
[b] Details of this pathogen are given later, in the discussion of sexually transmitted diseases.

multiple microorganisms, that undergoes remarkable changes depending on both the menstrual cycle and the individual's age. Several factors influence the vaginal microbiota; these include pH, glucose concentration, hormonal support, pregnancy, sexual activity, and birth control methods. Lactobacilli are the most prevalent organisms found. Other bacteria commonly found in the vaginal microbiota of healthy women include *Staphylococcus epidermidis* (S. eh-pee-DER-meh-diss), nonhemolytic streptococci, and anaerobes such as *Bacteroides* species, clostridia, and peptococci.

Since the normal bacterial flora of the vaginal region can include several anaerobic species, these microorganisms as well as others may penetrate the uterine or pelvic cavity and, when suitable anaerobic conditions exist, cause severe infections. Most of these infections result from complications of the confinement period following labor, the *puerperium*, or from complications of septic abortion. Clostridia gain entrance to the uterus by means of unsterile instruments or other material used in the performance of the abortive procedure.

PUERPERAL (CHILDBIRTH) SEPSIS

Puerperal sepsis (Table 28–2) is an infection that develops from the invasion of open wounds which occur through injury or as the result of surgical procedures in the genital tracts of women who have just given birth. Postpartum (after-birth) hemorrhage, premature rupture of membranes, and prolonged labor predispose mothers to puerperal sepsis. Microorganisms reported as causative agents include anaerobic streptococci, clostridia, *E. coli*, *N. gonorrhoeae*, and *Proteus* spp.

Infections are acquired through direct contact with persons harboring pathogens in their upper respiratory passages. Even the patient herself may be the source of pathogens. Droplet nuclei and fomites also may serve as sources of infectious agents.

VAGINITIS

Vaginitis, or vulvovaginitis, and related infections of the female genital tract may be caused by a variety of microorganisms (Table 28–2). Among the most commonly encountered are the yeast *C. albicans (Monilia)*, the protozoon *T. vaginalis*, and the bacterium *Gardnerella vaginalis* (gard-ner-EL-lah va-jin-AL-iss).

There is some controversy regarding the pathogenic role of *G. vaginalis*, since it is present in large numbers along with various anaerobic bacteria in women with vaginitis and in low numbers in most normal, sexually active women. The possibility of a synergistic (combined) role for anaerobic bacteria has been suggested. The obvious physical signs of *G. vaginalis* vaginitis is a gray-white homogeneous discharge that coats the vulva, vagina, and cervix, without evidence of tissue infection.

Toxin-caused Disease

TOXIC SHOCK SYNDROME

Reports of a new disease, closely associated with women during menstruation, began to appear in the later 1970s mainly in the United States. In several cases, strains of *S. aureus* were isolated from the throat, the vagina, or a localized abscess, but never from the blood. The first reports of the disease,

Vaginitis

LABORATORY DIAGNOSIS

Pathogens frequently associated with vaginitis can be distinguished and diagnosed on the basis of clinical symptoms and laboratory findings (Table 28–3). On occasion in *Gardnerella* infection, a fishy odor develops when alkaline blood mixes with vaginal secretions. In the laboratory, the odor is released from specimens treated with potassium hydroxide. The microscopic finding of epithelial cells heavily stippled with *G. vaginalis*, known as *clue cells* (Color Photograph 114), in vaginal secretions confirms the diagnosis of vaginal disease.

TREATMENT

Candida infections are generally treated with intravaginal applications of nystatin. Imidazole derivatives are also effective. Metronidazole is used for *G. vaginalis* infections.

TABLE 28–3 Selected Clinical and Laboratory Features of Vaginitis

	Normal and Microbial Forms			
Feature	*Normal*	*Candida*[a] *(Yeast)*	*Trichomonas (Protozoon)*	*Gardnerella (Bacterial)*[b]
Appearance of discharge, consistency	White, gray, or clear	White, thick, and curdlike	Gray, green, creamy, frothy, and uniform	Gray, white, and uniform
Amount of discharge	Varies	Small amount	Large amount	Large amount
Fishy odor after addition of 10% potassium chloride to discharge	None	None	Present	Present
Microscopy	Epithelial cells; white blood cells rare	White blood cells, yeasts	White blood cells, motile protozoa	Clue cells[c] and bacteria (Color Photograph 114)

[a] *Candida* infections also are known by the older term moniliasis.
[b] Other bacterial forms of vaginitis produce a watery, pus-filled discharge.
[c] Epithelial cells with a ragged appearance caused by large numbers of adhering bacteria.

which became known as *toxic shock syndrome (TSS)*, linked the presence of a skin-destroying toxin to the condition. Later investigations of the disease in women during menses failed to implicate this toxin. Instead, a protein, fever-producing (pyrogenic) TSS toxin 1, produced by certain staphylococcal strains, has been found to cause the illness. *Streptococcus pyogenes* also has been reported to cause a toxic shock-like syndrome. Various reports have clearly linked the use of tampons with TSS, particularly if they are used continuously during menses. A nonmenstrual form of the syndrome has also been associated with staphylococcal enterotoxin B. TSS has been found in surgical patients with wound infections. While this condition is largely found in females, it can occur in males as well.

The pathogenesis of TSS is not fully understood, but its association with the use of tampons suggests that *S. aureus* is carried into the vagina during insertion of the tampon, with subsequent growth, colonization, and production of pyrogenic exotoxin(s). Toxin production appears to be linked to a specific site on the chromosome of TSS-associated *S. aureus* strains that is not found in non-TSS associated strains. The severity of TSS in some individuals may be caused by the reflux of menstrual blood from the vagina to the peritoneal cavity (via the host's Fallopian tubes), from which the toxin is rapidly absorbed. **[See bacterial toxins in Chapter 27.]**

Toxic shock syndrome as a clinical condition (Table 28–2) bears close comparison with both streptococcal scarlet fever and Kawasaki disease (mucocutaneous lymph node syndrome). **[See Kawasaki disease in Chapter 27.]**

Toxic Shock Syndrome

Laboratory Identification

The procedures of isolation and identification for *S. aureus* are the same as those described earlier in this chapter.

Treatment

Successful treatment of TSS involves early recognition of the condition and may require the removal of a tampon, drainage of an abscess, or surgical removal of damaged tissue. Antistaphylococcal antibiotics such as nafcillin and dicloxacillin may also be used.

Diseases of the Male Reproductive System

SURFACE INFECTIONS OF THE PENIS

Several nonvenereal infections are known to affect the penis. Inflammation of the glans penis *(balanitis)* and a similar condition of both the glans and the prepuce *(balanoposthitis)* are common examples (Color Photograph 115). A variety of microorganisms can produce such infections. The usual treatment is circumcision and the administration of appropriate antibiotics. The effects caused by venereal or sexually transmitted pathogens, such as herpes simplex viruses, are discussed elsewhere.

DISEASES OF THE PROSTATE

The prostate, which is composed of both glandular and smooth tissue, surrounds a portion of the male urethra. It is one of the most important male sex glands. Unfortunately, it is subject to various types of diseases. Two of these diseases associated with microorganisms are acute and chronic prostatitis (Table 28–2).

ACUTE PROSTATITIS. *N. gonorrhoeae* was once considered to be the chief cause of acute inflammations of the prostate gland. Now several other microorganisms, including *E. coli, P. aeruginosa,* staphylococci, and streptococci have also been identified as causative agents. Mixed infections have been commonly encountered.

Bacterial pathogens reach the prostate by routes similar to those listed for kidney infections. In cases of nonspecific urethral infections and gonorrhea, the prostate gland becomes involved as a consequence of an acute urethral inflammation. Blood-borne infections of the prostate may occur as complications of boils, carbuncles, osteomyelitis, and epidemics of acute respiratory tract infections. In some cases, the seminal vesicles may also be affected.

CHRONIC PROSTATITIS. Chronic prostatitis, an infection of the prostate, manifests itself most commonly in middle-aged males. It usually follows an acute infection of the gland. The bacteria producing this condition are similar to those associated with acute prostatitis.

The Sexually Transmitted Diseases (STDs)

Sexually transmitted diseases (STDs) are a major and growing public health problem. The well-known traditional venereal diseases—gonorrhea, syphilis,

Prostatitis

LABORATORY IDENTIFICATION

A presumptive diagnosis of *N. gonorrhoeae* infection may be made on the finding of gram-negative intracellular diplococci in smears of infectious material obtained from patients. The bacteria appear within white blood cells (Color Photograph 116a). Specimens used in the isolation and identification of *N. gonorrhoeae* include urethral discharges from males, endocervical exudates from females, and material from exposed sites such as the throat, cervix, and anal canal. *N. gonorrhoeae* can be isolated by placing specimens on appropriate media such as enriched chocolate agar and Thayer-Mayer medium and incubation in an environment with increased CO_2. Differentiation of *N. gonorrhoeae* from other species is based on its production of a positive oxidase test (Color Photograph 116b), acid from glucose, and a positive Superoxol test, which uses 30% hydrogen peroxide to show the presence of the enzyme catalase.

The procedures for the isolation and identification of *E. coli, P. aeruginosa,* staphylococci, and streptococci are the same as those described earlier in this chapter.

TREATMENT

Chemotherapy for *N. gonorrhoeae* infections varies somewhat according to the presence or absence of complications. Several forms of penicillin and tetracycline are used to treat uncomplicated infections. Spectinomycin and newer forms of cephalosporins are used in the treatment and possible prevention of infections caused by penicillinase-producing *N. gonorrhoeae* (commonly designated as PPNG).

TABLE 28–4 Agents of Sexually Transmitted Diseases (STD)

Bacteria	*Viruses*	*Protozoans*	*Ectoparasites*
Calymmatobacterium granulomatis	Cytomegalovirus[a]	*Entamoeba histolytica*[a]	*Phthirus pubis* (crab louse)
Campylobacter foetus	Genital warts	*Giardia lamblia* and other species[a]	*Sarcoptes scabiei* (scabies mite)
Chlamydia trachomatis	Hepatitis A[a]	*Trichomonas vaginalis*	
Gardnerella vaginalis[a]	Hepatitis B[a]		
Haemophilus ducreyi	Herpes simplex virus 1[c]		
Mycoplasma hominis	Herpes simplex virus 2		
Neisseria gonorrhoeae	Human immunodeficiency virus[d]		
Shigella species[a]	Molluscum contagiosum[c]		
Streptococcus, group B[b]			
Treponema pallidum			
Ureaplasma urealyticum			

[a] Refer to Chapter 26 for features of this pathogen.
[b] Doubtful status as an STD in several cases.
[c] Refer to Chapter 23 for a description of this pathogen.
[d] Refer to Chapter 27 for a description of the AIDS virus.

chancroid, lymphogranuloma venereum, and granuloma inguinale—account for some of the STDs in industrialized societies. Other problems are caused by an ever-increasing number of infections due to microorganisms other than those that produce the traditional venereal diseases. These include human immunodeficiency virus (HIV), *Chlamydia trachomatis*, hepatitis B virus, herpes simplex viruses, and several others (Table 28–4). These infectious agents have gained considerable importance both because they cause discomfort and disability and because some of them result in sterility or life-threatening situations. STDs are traditionally classified by cause (Table 28–4). Alternatively, they may be classified according to disease states, most of which may be caused by more than one pathogen. Pelvic inflammatory disease (PID) is one such example. The general features of this sexually acquired condition are also presented in this chapter.

Table 28–5 lists major STD states and complications, where appropriate, in the approximate order of their public health importance. General features of pathogens and the signs and symptoms of the commonly occurring sexually transmitted diseases that directly injure portions of the reproductive system are given in Table 28–6.

TABLE 28–5 Major STD States and Complications

STD State	*Examples of Associated Complication(s)*
Pelvic inflammatory disease (PID)	Abnormal bleeding, fever, painful urination, and abdominal pain
Infertility and abnormal pregnancy	
Fetal and perinatal infections	Congenital malformations; eye infections; inflammation of the central nervous system (encephalitis); deafness; pneumonia; stillbirth
Complications of pregnancy	Premature birth; spontaneous abortion; infection after delivery (postpartum endometritis)
Neoplasia (new growths)	Cervical cancer
Female lower genital tract infection	Infections of the cervix, urethra, and vagina
Male genitourinary tract infections	Urethritis and epididymitis
Genital ulcers	Pain and possible local infections
Intestinal involvement	Infections of the intestine and perianal area, hepatitis, and liver destruction
Arthritis	Pain on movement of some joints
Tertiary syphilis	Vascular and central nervous systems tissue destruction
Genital warts	Some pain
Infectious mononucleosis	General discomfort
Ectoparasitic infestation (lice, mites)	Pain, and possible localized infections

TABLE 28–6 **Features of the Established STDs**[a]

Disease	*Causative Agent*	*Gram Reaction and Morphology*	*Incubation Period*	*Type of Discharge*	*General Features of the Disease*
Chancroid	*Haemophilus ducreyi*	–, Small rod (coccobacillus)	1–3 days	Pus-containing and bloody	Symptoms include small, elevated lesions that form irregular soft-edged sores, soft chancres (Color Photograph 117), and some pain
Gonorrhea	*Neisseria gonorrhoeae*	–, Coccus	3–9 days or as late as 2 weeks	Mucus- and pus-containing	Symptoms in the male include uncomfortable sensation along the course of the urethra and painful, frequent urination; the majority of females do not show definite symptoms; skin lessions and a form of arthritis occur with disseminated gonococcal infection
Granuloma inguinale	*Calymmatobacterium granulomatis*	–, Small rod (coccobacillus)	Unknown	None	The appearance of a moist pimple (papule) on or in vicinity of external genitalia
Genital herpes	Herpes simplex virus type 2 (herpesvirus)	Not done[b], Virus	At least 36 hours	None	Itching and rash occur, followed by clusters of blister-like lesions that break and form sores
Lymphogranuloma venereum	*Chlamydia trachomatis*	– or variable (not useful), Small rod (coccobacillus)	3–21 days	Mucus discharge containing pus and blood	Swollen lymph nodes, which become filled with pus and eventually rupture and drain; these and other lesions occur in the vulvovaginal and/or rectal areas of females; leathery patches also occur
Monilial *(Candida)* vaginitis	*Candida albicans*	+ (usually not done), Yeast cell	Unknown	Thick, yellow, cheesy consistency	Severe itching in and around involved areas[c]
Syphilis	*Treponema pallidum*	Not useful, Spirochete	9–90 days	Generally none	*Primary:*[d] Formation of a hard sore (Hunterian chancre at infection site) (Color Photograph 118a); *Secondary:* Symptoms caused by generalized spreading of the disease include swollen lymph nodes, general discomfort, fever, headache, and skin rash (Color Photograph 118b); *Tertiary:* Soft, gummy, swollen areas, or tumors, the *gummas* (Color Photograph 118c) may form in any tissue *Congenital*[d]
Trichomoniasis	*Trichomonas vaginalis*	Not done, Flagellated protozoon	Unknown	Green-whitish, foamy, foul-smelling discharge	Profuse discharge and irritation

[a] AIDS is discussed in Chapter 27.
[b] Virus infection: the Gram stain is not done.
[c] Monilial vaginitis is common in pregnant women and has been associated with the use of oral contraceptives.
[d] Further details are provided in this chapter.

Bacterial Infections

CHANCROID (SOFT CHANCRE)

The gram-negative coccobacillus *Haemophilus ducreyi* (he-MAH-feh-lus do-KRAY-ee), which is occasionally called *Ducrey's bacillus*, is the causative agent of chancroid. The disease is highly contagious and specific and appears as a venereal ulcer on the genitals. Chancroid bears no relation to the primary lesion of syphilis (hard or Hunterian chancre). Generally, *H. ducreyi* is transmitted by sexual contact. However, some reports implicate dressings and surgical instruments as possible sources of infection. Poor sanitary habits are considered to be a predisposing factor.

CHLAMYDIAL INFECTIONS

Chlamydia trachomatis, an intracellular bacterial pathogen that requires tissue culture techniques for isolation, probably causes more STDs in industrialized countries than any other organism. *C. trachomatis* is considered to be a leading cause of nongonococcal urethritis and cervicitis in adults. Certain strains of the organism are involved with various disease conditions, including eye infections (trachoma) and pneumonia (Table 28–6). Another species, *C. psittaci* (kla-MEH-dee-ah SIT-tah-see), is the causative agent of psittacosis (parrot fever).

From an immunological standpoint, *C. trachomatis* can be divided into 15 immunotypes. Of these, types A, B, BA, and C are associated with trachoma; types D through K are involved in genital infections, the eye disease inclusion conjunctivitis of adults and newborns, and infant pneumonia; and types L_1, L_2, and L_3 are the agents of the venereal disease lymphogranuloma venereum.

TABLE 28–7 Chlamydial Groups and Diseases

Group	*Disease(s)*
Chlamydia psittaci	Psittacosis (parrot fever)
C. trachomatis[a]	
Group 1	
A, B, BA, C	Trachoma
Group 2	
D, G, F, G, H, I, J, K	Inclusion conjunctivitis, epididymitis, nongonorrhoeal urethritis, cervicitis, salpingitis, proctitis, pneumonia in newborns
Group 3	*Lymphogranuloma venereum*
L_1, L_2, L_3	(LGV)

[a] *Trachomatis* groups are organized into antigenic strains or *serovars*. Refer to the Microbiology Highlights entitled "Subdividing Subspecies," in Chapter 2, for an explanation of "serovar."

GENITAL INFECTIONS. In males, *C. trachomatis* is one of the most common causes of nongonococcal urethritis. In females, cervicitis is the main genital infection produced by this organism.

Chancroid

LABORATORY DIAGNOSIS

Chancroid is usually diagnosed on clinical grounds (Color Photograph 117). This process involves excluding other types of venereal infections and examining specimens microscopically. The absence of hard chancres and the failure to detect spirochetes by means of dark-field microscopic examination usually eliminate the possibility of syphilis. *H. ducreyi* may be detected in Gram-stained smears prepared from ulcer specimens. Isolation of *H. ducreyi* from genital ulcers can be obtained on clotted rabbit blood and specially enriched chocolate agar. Properties of *H. ducreyi* include hemin dependence (X factor), nitrate reduction, positive oxidase and negative carbohydrate fermentation, and urease and catalase tests. The fluorescent antibody can be used for confirmation purposes.

A skin test with a killed suspension of the causative agent is used in Europe to differentiate between chancroid and other venereal diseases. Usually 48 hours are needed to demonstrate a positive reaction. It is characterized by redness and hardening at the site of injection.

TREATMENT

Sulfisoxazole or a sulfonamide-trimethoprim combination are the drugs of choice.

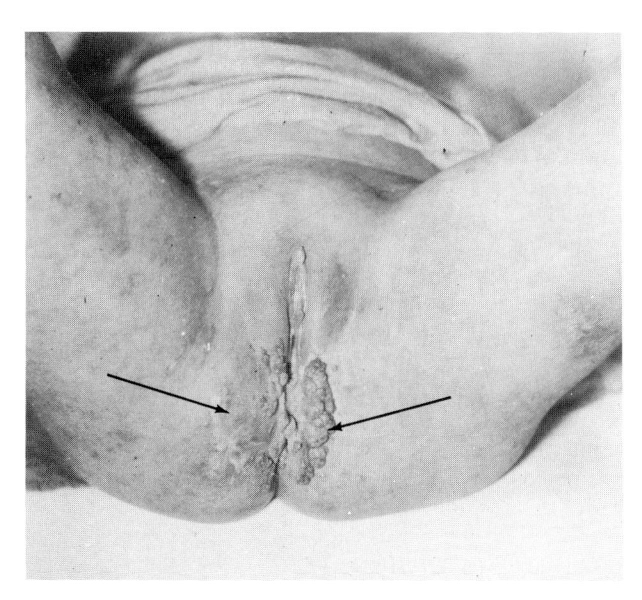

Figure 28–5 Perianal lesions (arrows) of lymphogranuloma venereum, a chlamydial infection. *[Courtesy of the Armed Forces Institute of Pathology, Washington, D.C., Neg. No. D4542–1.]*

Lymphogranuloma Venereum. Lymphogranuloma venereum (LGV) is the most severe of the genital chlamydial infections. Under normal conditions, this disease is sexually transmitted. However, infections can also be acquired through nonvenereal contacts, as by way of the hands. Dual infections of lymphogranuloma venereum and gonorrhea are common.

Lymphogranuloma venereum is also known by a variety of other names, including *Duran-Nicholas-Favre disease, lymphogranuloma inguinale, fifth venereal disease,* and *venereal bubo.* The last term refers to the enlarged regional lymph nodes that develop approximately one week to two months after the initial disease symptoms.

As healing proceeds, scars form and eventually obstruct lymph channels. Characteristic effects of this phase of LGV include elephantiasis (massive enlargement) of the external genitalia in males and rectal narrowing in females (Figure 28–5).

Neonatal Infections. Inflammation of the lining of the eye, inclusion conjunctivitis, and/or pneumonia are the most common types of neonatal infections. The neonate presumably acquires the infection during delivery, although the possibility of infection after birth has not been ruled out. Silver nitrate, which may be administered at birth to prevent gonococcal infection, does not protect against inclusion conjunctivitis.

GONORRHEA

Gonorrhea, a highly infectious, pus-producing disease, is often referred to as a *specific urethritis.* Gonorrhea is one of the most common communicable diseases in the United States today. It outnumbers syphilis by about 40 to 1. The disease is caused by the gram-negative, biscuit-shaped diplococcus *Neisseria gonorrhoeae* (Color Photographs 116a and 116b), which is usually situated intracellularly. The disease primarily involves the genitourinary tract. However, local infections involving the throat, lining of the eyelid, and anus occur. Gonococci can spread from original sites of infection by means of the bloodstream, resulting in disseminated gonococcal infection (DGI). Various complications can develop; included among them are endocarditis, inflammation of the coverings of the brain (meningitis), and pyelonephritis. Once the disease is acquired, it may persist for many years if treatment is not given. DGI can be caused by antibiotic-susceptible gonococci and penicillinase-producing *N. gonorrhoeae* (PPNG).

The most common means by which adults contract gonorrhea is through sexual intercourse. In the male, the disease initially remains limited to the lining of the front portion of the urethra (Figure 28–6). However, with additional sexual relations, improper instrumentation during examinations, or self-medication, the infectious agents can be introduced into the deeper, or posterior, portion of the urethra. From this focus, other parts of the urogenital system can be readily invaded.

Figure 28–6 The urethral discharge in an acute case of gonorrhea. *[Courtesy of the Armed Forces Institute of Pathology, Washington, D.C., Neg. No. 218663–5–96.]*

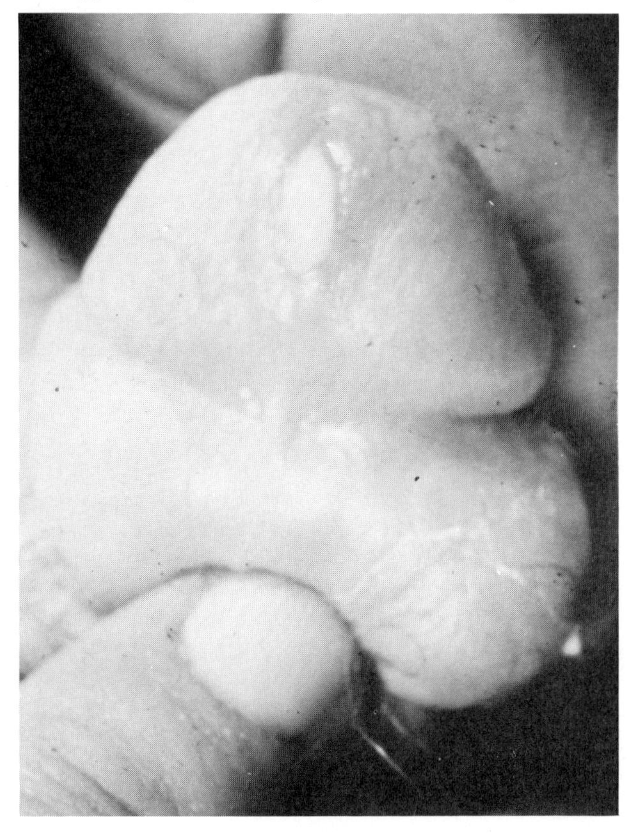

In the female, *N. gonorrhoeae* causes not only urethritis but also inflammations of the cervix and Fallopian tubes. In adult females, the label *external gonorrhea* is used when gonococcal involvement includes glands and related parts near the vagina and the urethra. If the infection spreads from these structures to the mucous membrane that lines the uterus, the membrane lining of the Fallopian tube, the ovaries, or the peritoneum, the term *internal gonorrhea* is used. In approximately 1% of patients, gonococci gain access to the bloodstream and produce disseminated gonococcal infection. Spreading skin lesions and inflammations of tendons and joints are common signs of the condition.

There are clear indications that the use of birth control pills increases the possibility of infection by establishing conditions more favorable to the survival of *N. gonorrhoeae.*

During childbirth, the eyes of newborns can become infected with *N. gonorrhoeae* as they pass through an infected birth canal. The resulting disease, known as *ophthalmia neonatorum,* was once the major cause of blindness in many parts of the world. The introduction of erythromycin or a 1% silver nitrate solution into the eyes of newborns greatly reduces the incidence of this infection.

Children can experience another manifestation of *N. gonorrhoeae, vulvovaginitis.* This type of gonococcal involvement is looked upon as an institutional disease, as it has been found to be transmitted by contaminated towels or similar articles in hospitals, orphanages, and schools. It appears that the conjunctivae and vaginas of children are quite vulnerable to gonococcal infections. In adults it is the internal genital organs and associated structures that have particular susceptibility. Vulvovaginitis is not caused exclusively by *N. gonorrhoeae.* Hemolytic staphylococci and streptococci, as well as other microorganisms, have also been isolated from cases of vulvovaginitis. Another problem is presented by isolation and treatment of *N. gonorrhoeae* L forms. **[See L forms in Chapter 7.]**

Prevention. Preventive measures for gonorrhea consist of detecting and treating the sexual contacts who transmitted the disease to the patients and educating the public as to the means of transmission, availability of treatment, and other pertinent details. Since 1976, penicillin-resistant strains of *N. gonorrhoeae* have been isolated from patients in several regions of the world. Such organisms make control of this disease more difficult. Effective vaccines are still under development.

GRANULOMA INGUINALE (GRANULOMA VENEREUM)

Granuloma inguinale is caused by the gram-negative bacterium *Calymmatobacterium (Donovania) granulomatis* (kal im mah toe back TER ee um gran-you-LOW-mah-tiss). Originally, the infection was believed to be limited to the tropics; however, several reports have proved it to be endemic in parts of the United States. The organism is presumed to be spread through sexual intercourse, although other means are thought to be possible. The disease itself is not highly contagious.

SYPHILIS AND THE OTHER TREPONEMATOSES

Most health experts regard the treponematoses as one of the world's worst communicable disease problems. During the late 1940s and early 1950s, a remarkable decrease in the incidence of venereal

Gonorrhea

Laboratory Identification

The procedures for the isolation and identification of *N. gonorrhoeae* are the same as those described earlier in this chapter for prostatitis.

Treatment

Several forms of penicillin and tetracycline are used to treat uncomplicated gonococcal infections. Spectinomycin and newer forms of cephalosporins are used to treat infections caused by penicillinase-producing gonococci.

As a preventative measure against ophthalmia neonatorum caused by *N. gonorrhoeae,* ointments containing either erythromycin or tetracycline are used. One percent silver nitrate solution (Crede's solution) has been replaced by antibiotics in most situations.

Granuloma Inguinale

Laboratory Identification

Examination of smears made from biopsy specimens are sufficient for diagnosis. Wright- or Giemsa-stained preparations will show clusters of encapsulated coccobacilli known as Donovan bodies in the cytoplasm of mononuclear cells.

Treatment

Tetracycline is the drug of choice.

syphilis was widely noted. This event coincided with the increased availability and use of penicillin. The general view was that syphilis was well on the road to eradication and would be relegated to the status of a historical disease.

Unfortunately, during the 1960s this downward trend reversed abruptly, and syphilis began to occur in near epidemic proportions. The number of cases in the United States increased from 6,399 in 1956 to 22,962 in 1964. These numbers are only minimum estimates because, according to many public health authorities, numerous cases go unreported. The number of reported syphilis cases in the United States remained relatively stable throughout the 1970s. Approximately 25,000 cases were reported in 1979. An extremely unfortunate aspect of the current status of syphilis is the increased involvement of teenagers and young adults 13 to 24 years of age.

In the United States during the first three months of 1987, after a five-year trend of decreasing incidence, a substantial increase in primary and secondary syphilis cases was reported. This increase in cases was 23% higher than during a comparable period in 1986 and was largely among heterosexuals. The increase is of great concern, since it may have a severe, unfavorable effect on efforts to control congenital syphilis and may be associated with an increased risk for infection with human immunodeficiency virus. [See Chapter 27 for a description of AIDS.]

The first recorded occurrence of syphilis in its present form was a great pandemic that erupted in Europe in 1497. One theory suggests that it was brought to Europe from Haiti by Columbus's crew and the six Indians whom they brought with them, but the origin of syphilis in the New World is by no means established. (See the references at the end of this chapter for a Public Health Service publication on the history of syphilis.)

Morphological and Cultural Properties. The causative agent of syphilis, *Treponema pallidum* (tre-poh-NEE-mah PAL-li-dum), was discovered in 1905 by Fritz Schaudinn and Erich Hoffman. This treponeme belongs to the order of Spirochaetales and the family of Treponemataceae.

The microorganism is one of three principal human pathogenic agents belonging to the genus *Treponema* (Color Photograph 62). The others are *T. pertenue* and *T. carateum,* the causative agents of yaws and pinta, respectively. The three species are morphologically and serologically indistinguishable. Selected epidemiologic features of these diseases are given in Table 28–8. *T. pallidum* is a thin, delicate, spiral bacterium ranging from 6 to 15 μm in length with a uniform cylindrical thickness of about 0.25 μm, arranged in a number of tight body coils (6 to 14 turns or spirals). Upon dark-field examination of clinical specimens from primary and secondary syphilitic lesions, the spiral appearance of the organism and its well-known corkscrewlike rotation are easily recognized.

Attempts to culture *in vitro* the treponemes pathogenic for humans have been successful (Figure 28–7). The spirochete can be grown in rabbit tissue culture cells under slightly aerobic conditions. *T. pallidum* generally is grown in the testes of rabbits. It has also been maintained in fluid media containing carbon dioxide, pyruvate, a reducing agent such as cysteine, serum albumin, and serum ultrafiltrate.

Transmission. Syphilis is transmitted through direct contact with infectious lesions in both the primary and secondary stages of the disease. These lesions may be either genital, oral, or epidermal (on the skin). *T. pallidum* is capable of passing through abraded skin and intact mucous membranes. From the lesions, the spirochetes gain access to the circulatory system of the individual and are carried to ev-

TABLE 28–8 **Comparison of Selected Epidemiologic Features of the Treponematoses**

				Mode of Transmission			
				Biological			*Mechanical*
Microorganism	*Disease*	*Geographical Distribution*	*Age Group Affected*	*Venereal*	*Nonvenereal*	*In Utero*	*Flies, Fomites, etc.*
Treponema pallidum	Venereal syphilis	World	Sexually mature individuals and some newborns	Common	Not usually	Not uncommon	Extremely rare
T. pallidum	Endemic syphilis (Bejel)	Rural regions, prevalent in Africa, Australia, Mediterranean countries	Children 2 to 10 years of age	Not usually	Common	Rare	Probably not uncommon
T. pertenue	Yaws	Rural regions, mostly tropical areas	Children 4 to 15 years of age	Not usually	Common	Questionable	Possibly not uncommon
T. carateum	Pinta	Rural regions, humid, tropical areas, only in the Americas	Individuals ranging in age from 10 to 30 years	Not reported	Common	Implicated	Implicated

ery organ. Outside the body, this microorganism is extremely vulnerable to the effects of both physical and chemical agents. These destructive forces include heat, drying, storage at refrigerator temperatures (approximately 3° to 4°C), and ordinary soap and water.

Figure 28–7 *Treponema pallidum* attached to the surface of tissue culture cells in which it replicates. *[From A. H. Fieldsteel, D. L. Cox, and R. A. Moecki.* Infect. Immun. ***32****:908–915, 1981.]*

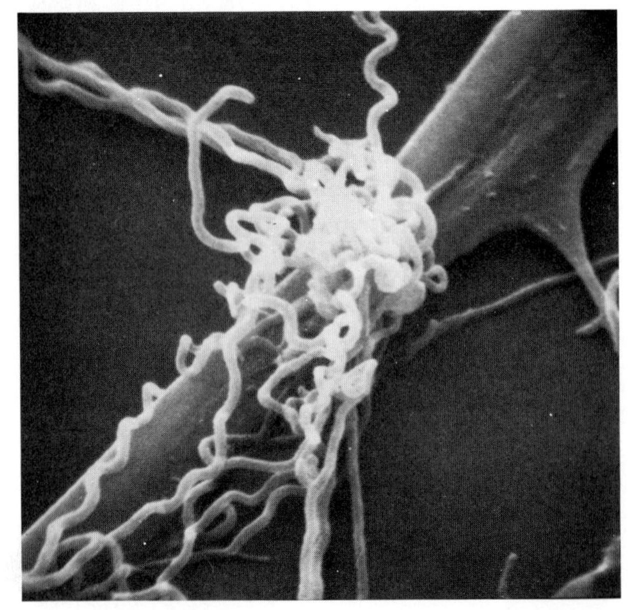

Clinical Features. Several clinical features of syphilis are given in Table 28–6.

The clinical manifestations of syphilis may conveniently be divided into three stages: primary, secondary, and tertiary.

Primary Syphilis. The primary lesion, or chancre, develops at the site of entry of *T. pallidum* within two to six weeks after exposure and lasts for approximately six weeks. During this stage, the treponemes invade lymph nodes, enter the bloodstream, and are distributed throughout the body, producing a *spirochetemia.* Extragenital chancres commonly involve the lip (Color Photograph 118a) but also may involve the tongue or other mucosal sites. Such lesions are relatively painless.

The organisms are transmitted directly from an individual with syphilitic lesions of this nature. The immediate, rapid spread of organisms throughout the body makes all attempts at local disinfection after contact ineffective and of no prophylactic value.

The primary chancre, which is also known as the *hard* or *Hunterian chancre,* begins as a single small, slightly elevated, round, red nodule on the tissue surface. It is usually painless. The central area breaks down, and the nodule ulcerates and discharges a fluid containing numerous treponemes.

These organisms can be demonstrated by darkfield examination and are fully capable of infecting others. The spirochetes cannot be demonstrated in tissue sections except with special stains such as silver impregnation. Diagnostic blood tests do not become positive until at least a week after the appearance of the chancre. The chancre heals spontaneously, leaving very little scarring unless there has been a superimposed secondary bacterial infection.

Secondary Syphilis. From one to two months following the appearance of the chancre, the generalized spread of the disease and widespread swelling of lymph nodes may become evident. Malaise accompanied by a rise in temperature, headaches, and various skin rashes are common symptoms (Color Photograph 118b). Swollen grayish-white areas may develop on the mucous membranes of the lips, soft palate, tongue, or other areas. The typical mucous patch is a slightly elevated and flattened grayish-white area that can be removed, leaving a red zone of erosion and ulceration. In the throat the chancres may have a creeping outline; they have been called "snail track ulcers." Such lesions contain large numbers of *T. pallidum* and are slow to heal.

In the secondary stage of syphilis, the mouth often contains many treponemes, a ready source of infection for dentists, their assistants, or anyone else who places an unprotected finger in the mouth of one of these infected individuals. Highly contagious skin lesions also occur.

Spotty eruptions may also occur in the mouth and on the skin. The oral lesions are often found on the palate, but they may involve the entire oral mucosa as nonelevated or slightly elevated reddish areas. Microscopically, the tissue shows a nonspecific chronic inflammation, often associated with capillaries having thickened walls.

The secondary stage of syphilis can be confirmed clinically by demonstrating *T. pallidum* from mucous patches or by diagnostic blood tests, which are positive in almost all cases. This stage can disappear, be latent, and then reappear several times before finally disappearing.

Tertiary Syphilis. The gummas, which are soft, gummy tumors (Color Photograph 118c), and nodules of tertiary syphilis usually take 5 to 20 years to appear, but they may occur, though rarely, within a short time after the secondary stage. Of the many changes that become apparent within tissues, two are most commonly seen: (1) *syphilitic arteritis,* in which the small arteries become narrowed as a result of a fibrous thickening of their walls; (2) *gummas,* lesions that are a type of tissue death followed by scarring and are assumed to be the result of a longstanding, progressive decrease in the blood supply to the tissue. Gummas may form in any tissue. Some involve the tongue, palate, and facial bones. These gummas ulcerate, leaving rounded, punched-out edges. When they occur in the palate they may eventually perforate, leaving an irregular opening into the nasal area, which will interfere with speech.

The third stage of the venereal disease includes conditions known as latent syphilis, neurosyphilis, late benign syphilis, and cardiovascular syphilis.

The lesion of *late benign syphilis,* which is the gumma, usually does not produce death or total physical incapacity. However, when lesions arise in vital organs, the complications that occur certainly are not benign. The gumma is believed to be a consequence of a hypersensitivity reaction to *T. pallidum.* It can involve any body organ. In addition, gummas may be found associated with the bones and skin of an infected individual. Serologic tests are usually highly reactive in this form of syphilis.

Congenital Syphilis. Congenital syphilis (CS) occurs when *T. pallidum* crosses the placenta and infects the fetus. This situation may arise after the eighteenth week of gestation, and reflects the presence of primary and secondary syphilis among women of childbearing age. Adequate treatment of an infected mother before this time usually prevents involvement of the fetus. If a woman conceives while she is in the primary or secondary stages of syphilis, her pregnancy in all probability will end in a stillbirth. When pregnancy occurs during the later stages of this infection, newborns may exhibit a variety of clinical effects ranging from fulminating fatal congenital infections to a normal, uninfected state. In the United States, the number of CS cases more than doubled during the period 1978–1985.

Congenital syphilis is divided into two principal stages, early and late congenital syphilis. A primary stage is not present, as *T. pallidum* is directly introduced into the fetus via the placenta.

The effects of early congenital syphilis may include skin (Figure 28–8) and mucous membrane lesions, hemolytic anemia, enlarged liver and spleen, and the involvement of teeth (Figure 28–9), bone,

Figure 28–8 Skin lesions of congenital syphilis appear soon after birth, frequently in the form of vesicles. These lesions can progress to superficial crusted erosions. *[Courtesy of the Armed Forces Institute of Pathology, Washington, D.C., Neg. No. 54-2488-7.]*

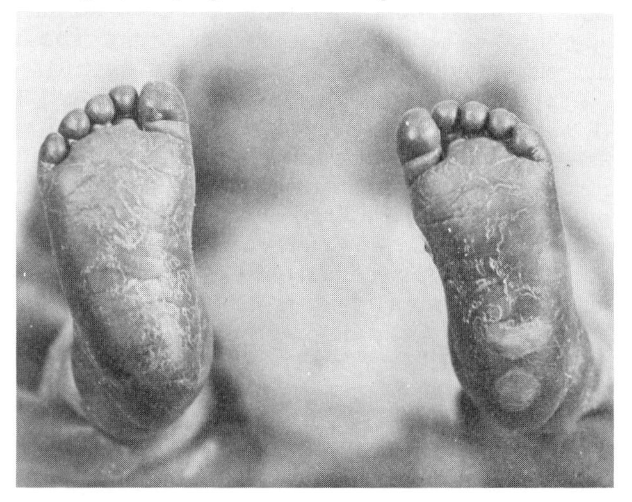

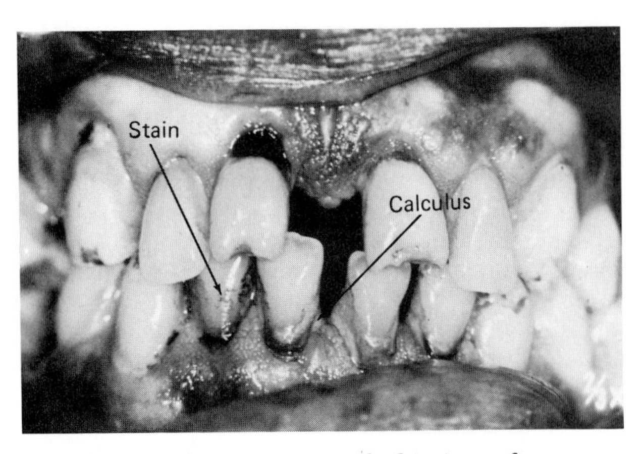

Figure 28–9 Characteristic notched incisors of congenital syphilis. Poor hygiene with considerable calculus and stain is also evident.

and the central nervous system. These symptoms appear before the child reaches two years of age.

The late form of congenital syphilis is defined by the persistence of these effects beyond the age of two. In approximately 60% of cases, the disease is latent and is characterized by reactive diagnostic immunologic reactions. However, several signs of this stage of syphilis may appear. These include rare cardiovascular lesions, Clutton's joints (a painless involvement of the joints, usually the knees), inflammation of the eyes, involvement of the bones and skin, Moon's molars (poor development of the cusps of the first molars), and neurosyphilis. Clutton's joints, eighth cranial nerve deafness, and eye inflammation usually occur together near the beginning of puberty and appear to be produced as a consequence of a hypersensitivity response rather than by the purely destructive effects of *T. pallidum.*

ENDEMIC SYPHILIS, PINTA, AND YAWS. Endemic syphilis is known by several other names, including *Bejel* and *Skerljevo.* The disease, found primarily in tropical areas, is believed to be transmitted through direct contact. The causative spirochete produces lesions during the early and late periods of the disease. Involvement of the cardiovascular and central nervous system as well as the skin usually occurs.

Pinta, a contagious disease caused by *T. carateum* (T. kar-ah-TEE-um), seems to be limited to tropical and subtropical regions, such as Central and South America. The infection is believed to be transmitted by direct contact. Flies have also been implicated as vectors. The lesions (which primary involve the feet, hands, and scalp) are described as dry, scaly, and exhibiting a variety of colors. The name of the disease is derived from the Spanish verb *pintar,* which means to paint. These lesions at first are highly pigmented and then, after several years, lose color. Other body tissues are not usually invaded. *T. carateum* is morphologically indistinguishable from *T. pallidum.* Wassermann antibodies are commonly detected, especially in the secondary and late stages of pinta. Penicillin is effective in the treatment of the disease.

Yaws is also a contagious disease, caused by *T. pertenue.* (T. per-TEN-you-ee). The infection is not transmitted by sexual contact, but by direct contact and possibly by flies. Yaws is found in tropical areas, especially those with heavy rainfall. The characteristic lesion resembles a raspberry and consequently has been referred to as a *framboise.* This disease is known by several other names, including *bouba* (Portugese), *buba* (Spanish), *framboesia* (Dutch and German), and *pian* (French). Ulceration of the lesion eventually occurs, follows by formation of a dry crust and healing. Secondary eruptions usually occur within two to four weeks. The late stage of the disease is characterized by the involvement of the skin and bones. Congenital yaws is apparently a rare phenomenon. Serologic tests for syphilis are positive in yaws infections. Penicillin is effective in the treatment of diseased persons. Unfortunately, *T. pertenue* infections occur in areas where medical services are extremely limited.

Virus Infection

GENITAL HERPES

Genital herpes (herpesvirus type 2) infections have caused genuine alarm among the public health-care experts—mainly because of the increasing numbers of cases, the ability of the causative agent to remain latent in the host and periodically cause recurrent infections, and the lack of effective treatment to eliminate recurrent herpes.

Two antigenic types of herpes simplex virus (HSV) can be distinguished. HSV type 1 usually causes oral lesions, infection of the cornea (keratitis) and encephalitis, while HSV type 2 causes most genital and neonatal infections. Neonatal herpes is acquired at birth as an infant passes through the vaginal canal of a woman experiencing active infection.

While initial infection with HSV may be asymptomatic, the more common situation is the appearance of moderately to severely painful lesions that begin as vesicles and progress to ulcers (Color Photographs 119 and 120). In males these appear on the penis, while in females they may appear on the vulva, vagina, or cervix. Lesions may also appear on the buttocks, thighs, and perianal area. The initial infection may also be accompanied by the enlargement of inguinal (groin) lymph nodes and other systemic symptoms. Recurrent infections are frequently less severe and may appear as only a few vesicles; in females, they may be completely asymptomatic. The disease is usually not incapacitating to adults.

HSV infection of the fetus or newborn infant is a severe disease with high mortality and serious resulting conditions. It is generally thought that intrauterine herpes infection, if it occurs, causes severe

Syphilis

Laboratory Identification

Syphilis was called the "great imitator" by Osler. Without an accurate history and physical examination of the patient, this disease may go undiagnosed. Therefore, a complete examination and a routine serological test for syphilis are of the utmost importance. While specimens taken during the primary and secondary stages of the disease can be used to identify *T. pallidum* with the aid of dark-field microscopy or fluorescence microscopy, serological tests are relied upon heavily. This situation is largely caused by the fact that most laboratories do not have the capability to perform the microscopy procedures properly.

Over the past 30 years, numerous serological tests for syphilis have been developed, including the original Wasserman complement fixation test, flocculation tests, and treponemal tests. All of the general tests used in the diagnosis of *T. pallidum* infection are referred to collectively as *serological tests for syphilis.* All of these procedures are dependent on an antibody-antigen reaction. The type of antigen used can serve as a criterion for partial classification. On this basis, two categories of tests are performed—procedures utilizing treponemes or extracts of treponemes *(treponemal tests)* and methods employing normal tissue or other substances *(nontreponemal or reagin tests).*

Two types of antibodies produced by the host in response to *T. pallidum* infection have been extensively studied. One of these substances is the reagin or Wasserman antibody. Its formation results from the interaction of the treponeme with the host's tissue. Antibodies of this nature react with a variety of nontreponemal substances. An example of such substances is cardiolipin, a highly purified preparation extracted from beef heart. The second type of antibody is produced in response to specific as well as group treponemal antigens. These substances are detected by the treponemal tests.

For many years, the standard screening procedure for syphilis was the nontreponemal antigen test VDRL (Venereal Disease Research Laboratory). Most laboratories currently perform the Rapid Plasma Reagin (RPR) card test for syphilis screening. Since nontreponemal tests are not entirely specific for *T. pallidum* infection, procedures using treponemal antigens such as the Fluorescent Treponema Antibody Absorption (FTA-ABS) test are used to confirm the disease. While a positive result confirms the presence of treponemal antibodies, it does not identify the specific stage of syphilis. The *T. pallidum* hemagglutination (TPHA) test is a popular recent addition. It is simple to perform, inexpensive, and detects the presence of treponemal antibodies by hemagglutination. The TPHA test compares favorably with the FTA-ABS except in the diagnosis of primary syphilis.

Treatment

Penicillin is generally the drug of choice. A number of other antibiotics including erythromycin, tetracycline, and chloramphenicol are used for the treatment of penicillin-allergic individuals.

damage to the fetus and is usually followed by spontaneous abortion. Since neonatal herpes is acquired at birth as the infant passes through the infected vaginal canal, cesarean section is recommended for a woman with active herpes infection at time of delivery. Two different clinical forms of neonatal herpes have been described: *localized* and *disseminated.* In localized herpes, the infection may be limited to the eye, mucous membranes, or central nervous system. With localized central nervous system infection, newborns are more likely to die or survive with severe sequelae. In disseminated herpes, the virus involves and may be isolated from a number of different organs.

Control and Preventive Measures. Although antiviral chemotherapy may be effective in reducing the number of cases and infant deaths resulting

Genital Herpes

Laboratory Identification

The techniques used for the identification of HSV type 2 infections are similar to those for other herpes viruses, including the immunoperoxidase test and the immunofluorescence test with monoclonal antibodies (Color Photographs 79 and 80).

Treatment

Therapy for HSV type 2 infection includes DNA synthesis-inhibiting drugs such as acyclovir for initial exposure and recurrent infections. Measures to reduce accompanying symptoms also are important considerations.

from HSV infections, preventive measures are currently the most important means of controlling neonatal infections.

The risk of infection to the newborn appears to be highest in primary genital infection of pregnant women, but it is also high in recurrent infection. Women with a history of recurrent genital HSV infection, those with active disease during a current pregnancy, and those whose sexual partners have proven genital HSV infection should be monitored by specific laboratory tests at least twice during the last six weeks of pregnancy. Clinical examination may not be sufficient to detect infection. If the partner has documented genital infection, sexual contact should be avoided in the last several months of pregnancy. Vaginal delivery can be performed when specific laboratory tests are negative on two successive examinations one week prior to delivery and if clinical lesions are not detected at the time of delivery.

An infant born by vaginal delivery to a woman with active genital HSV infection should be segregated from other infants and managed with proper isolation techniques. In addition, the infant should be monitored in the hospital for a period of up to two weeks. Appropriate antiviral drugs should be administered to the infant if HSV disease develops. Circumcision should be delayed in proved and suspected cases.

Since in this type of situation the newborn infant is separated from its mother, careful attention should be given to hygienic measures involving any type of contact. For example, breast-feeding is acceptable if there are no herpes lesions in the area and if exposed, active lesions are covered. In addition, when handling the infant in the hospital, the mother should wear a gown and observe proper hand-washing techniques; this practice should be continued at home. All precautions should be taken to prevent the exposure of the infant to any herpes lesion.

TRICHOMONAS VAGINALIS

Many flagellates capable of inhabiting humans have been found in human fecal matter, in the mouth, and in the genital tract. For the most part their pathogenicity is doubtful, and many authorities consider them harmless commensals. However, one species, *T. vaginalis*, is pathogenic (Figure 28–10). It is found only in the genitourinary system and has been acknowledged as a causative agent of genitourinary trichomoniasis. Several investigations have proved this disease to be a true venereal infection. The disease seems to be as frequent in adult males as in females. In women, the acute stage involves a noticeable discharge; the symptoms in males are generally latent and therefore not obvious. Prostate involvement occurs in the male.

Temporary and permanent sterility has been reported to result from long-lasting infections. However, in some cases successful therapy has enabled women to conceive and give birth to normal children.

Figure 28–10 Beta-hemolytic activity of *Trichomonas vaginalis*, a property that correlates well with virulence in patients. *[From J. N. Krieger, M. A. Poission, and M. F. Rein,* Infect. Immun. ***41****:1291–1295, 1983.]*

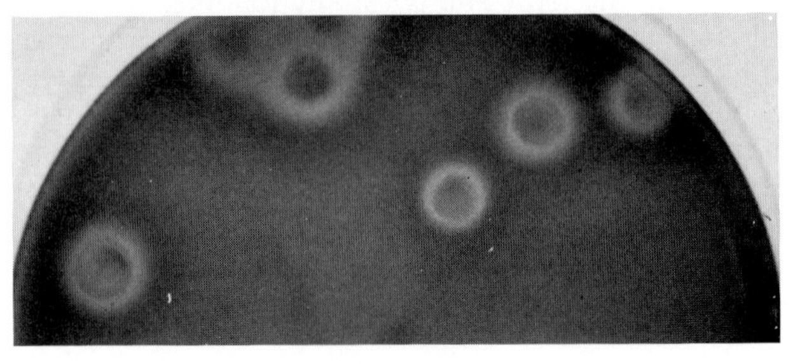

Protozoan Infection

LABORATORY IDENTIFICATION AND TREATMENT

The procedures for the identification of *T. vaginalis* and treatment of associated infections are the same as those described earlier for nonspecific urethritis.

T. vaginalis infections can be acquired by newborn babies by way of an infected birth canal or as a consequence of being exposed to highly unhygienic conditions. Infections in young girls have occasionally been reported.

Sexually Transmitted Enteric Disease Agents

The sexual transmission of enteric pathogens has been clearly recognized since 1977. There is both an increasing incidence in patients and an increasing awareness of the range of organisms that can be transmitted (Table 28–9). A major problem frequently found concerns the signs and symptoms of enteric disease that neither the patient nor the physician suspects to have been sexually transmitted. Understanding the epidemiologic factors, such as the mechanism of transmission and the range of sexually related enteric infections, is necessary for accurate diagnosis, prevention, and recognition of infection. The general mechanism of transmission is the fecal-oral passage of infectious agents.

THE TORCH COMPLEX

Infection of a fetus or a newborn during delivery with *Toxoplasma* (toks-oh-PLAZ-mah), a protozoon, rubella virus, cytomegalovirus, and herpesviruses may yield an inapparent disease. Even when the infections are clinically apparent, the associated signs and symptoms may be indistinguishable. Furthermore, all of the agents can produce long-term ill effects in the infected fetus or newborn, so that the prognosis must be guarded. Because the infections are often clinically inapparent, not only in the newborn but also in the mother, specific diagnosis is dependent on special laboratory testing. Because of the serious consequence of infection and the difficulty of identifying and distinguishing the disease state, the acronym TORCH was devised to focus attention on this group of microbial agents (T, *Toxoplasma;* R, rubella virus; C, cytomegalovirus; H, herpesviruses; and O, others). The specific features of these disease agents have been presented in other chapters. **[See cytomegalovirus in Chapter 26.]**

PELVIC INFLAMMATORY DISEASE

Pelvic inflammatory disease (PID) is a sexually acquired disease of young women; its greatest incidence is in the 15 to 24 age group. It is estimated that nearly 1 million women are treated for the condition annually in the United States, and that 25% of them require hospitalization. Most cases of PID occur following sexual contact and develop from pathogenic organisms spreading from the lower genital tract through the cervix. The uterus, fallopian tubes, and connecting reproductive system structures are primarily involved. Symptoms of PID

TABLE 28–9 Sexually Transmissible Enteric Diseases

Disease Agent	*Representative Diseases*
Bacterium	*Campylobacter foetus* proctitis[a], salmonellosis, shigellosis, streptococcal perianal cellulitis
Protozoon	Amebiasis, balantidiasis, and giardiasis[b]
Virus	Cytomegalovirus infection, hepatitis A and hepatitis B infections, herpes simplex type 1, non-A, non-B hepatitis[b]
Helminth (worm)	Pinworm[c], *Strongyloides stercoralis*

[a] Inflammation of the rectum and anus.
[b] Most of these diseases are discussed in Chapter 26.
[c] Worm infections are described in Chapter 31.

Disease Challenge

The situation described has been taken from an actual case history. A review of treatment and epidemiological aspects is given at the end of the presentation. Answers to questions, laboratory findings, and interpretations are given immediately following a specific question. Test your skills and take the Disease Challenge.

KEY TERMS

erythema (er-i-THEE-ma): redness over the skin

inguinal (ING-gwi-nal): pertaining to the groin

lymphadenopathy (lim-fad-ee-NOP-a-thee): swollen lymph nodes indicating an infection

purulent (PURR-you-lent): containing pus

CASE

A 23-year-old male from a nearby agricultural station was examined in a local free clinic by a volunteer physician. The patient complained of a single genital sore. Physical examination revealed the presence of a deep, invasive, pus-containing soft genital ulcer and painful inguinal lymphadenopathy. Erythema was also noted over the involved area. No purulent discharge was noted from the penis. According to the patient, the lesion first appeared about a month earlier.

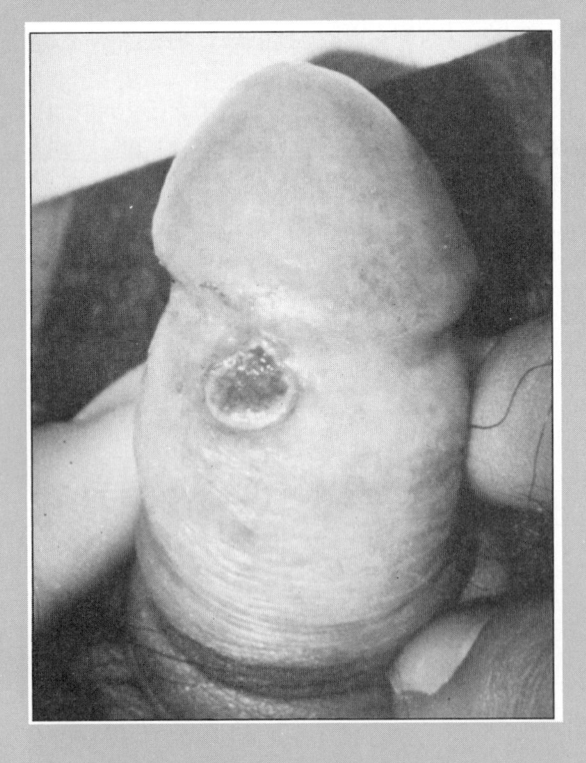

The appearance of the genital ulcer. *[Courtesy Centers for Disease Control.]*

At This Point, What Presumptive Diagnosis Could Be Made?
Syphilis, herpes simplex type 2 infection, chancroid, or lymphogranuloma venereum.

What Type(s) of Laboratory Specimens Should Be Taken?
Pus, scrapings and/or biopsies from lesions for staining, dark-field microscopic examination, and culture. A blood specimen for serological tests. Two specimens should be taken, one immediately and another approximately two to three weeks later to determine increases in specific antibody levels (titers).

What Specific Serological Tests Should Be Performed?
Standard serological tests for syphilis, complement fixation, fluorescent antibody, and ELISA for herpes simplex type 2; also, where applicable, tests for *Haemophilus ducreyi* and *Chlamydia trachomatis.*

Laboratory Results:
Gram staining showed the presence of gram-negative coccobacilli in several specimens. Dark-field examination was negative for spirochetes. After one week of incubation, extremely small colonies appeared on enriched chocolate agar plates.

All serological tests with the exception of the fluorescent antibody procedure for *H. ducreyi* were negative.

On the Basis of All Findings, What Disease Appears To Be the Most Probable?
Chancroid.

include abnormal bleeding, frequent and painful urination, fever, nausea and vomiting. The most frequently found pathogens are *Chlamydia trachomatis, Mycoplasma hominis,* and *Neisseria gonorrhoeae.* A diagnosis is generally based on Gram-stain results, the culturing and laboratory identification of the causative agent. Treatment of PID consists of the administration of the antibiotics cefoxitin or tetracycline.

Summary

The functions of the urinary system include eliminating waste products, regulating the chemical composition and the acid-base balance of body tissues, and maintaining the water content of the body at a constant level.

Anatomy of the Urinary System

1. The human urinary system consists of two kidneys, two ureters, the urethra, and the urinary bladder.
2. The basic unit of the kidney is the *nephron.* It is composed of a *glomerulus* (blood filter) and a long tubule (concentrator or urine collector).
3. Blood filters through the glomeruli and is concentrated by the tubules, which empty into the ureters.

Flora of the Normal Urinary Tract

The normal sterile kidney is highly resistant to bacteria.

Diseases of the Urinary Tract

1. Common urinary infections are caused by *coliforms,* intestinal organisms.
2. Predisposing conditions for urinary tract infections include diabetes mellitus, neurologic diseases, and lesions that interfere with urine flow.
3. The major routes of infection are by blood, lymphatics, and from the ureter.
4. Diagnostic culturing systems are used to determine the concentrations of bacteria in specimens and to detect the presence of specific medically important microbes. The presence of more than 100,000 (10^5) bacteria per milliliter of urine is strongly indicative of an infection.
5. Individuals with urinary tract infections usually exhibit symptoms such as blood or pus in the urine, accumulation of fluid in tissues, pain, kidney enlargement, and blood loss, causing anemia.

KIDNEY DISEASES CAUSED BY BACTERIA

1. The most common kidney infection is pyelonephritis, an inflammation of the organ's tissues. *Escherichia coli* is the major cause.
2. Other kidney infections involve the outer region, or cortex.

DISEASES OF THE URINARY BLADDER

Cystititis, or inflammation of the bladder, can be caused by various bacteria, including *E. coli, Proteus vulgaris, Pseudomonas aeruginosa,* and the protozoon *Trichomonas vaginalis.*

DISEASES OF THE URETER

Infections of the ureter may develop independently or may be a consequence of diseases involving other organs.

NONSPECIFIC URETHRITIS

1. Nonspecific urethritis, an inflammation of the urethra, can be caused by various microorganisms such as *Neisseria gonorrhoeae*, hemolytic staphylococci and streptococci, the yeast *Candida albicans*, and the protozoon *T. vaginalis*.
2. Chemicals and the insertion of medical devices such as catheters can also produce urethritis.

URINARY TRACT INFECTIONS CAUSED BY ANAEROBES

1. Anaerobes belonging to the genera *Actinomyces, Bacteroides, Clostridium*, and *Streptococcus* cause infections of the kidney.
2. Organisms are introduced from other sites in the body or by means of surgical procedures.
3. Anaerobes can complicate various clinical states.

IMMUNOLOGICAL KIDNEY INJURY

1. Immunologic hypersensitivity can cause severe kidney damage. Acute glomerulonephritis is an example of such injury.
2. Injury is caused either by antibodies that form against the individual's kidney tissue or by antigen-antibody complexes that are eventually trapped in glomeruli.

Anatomy of the Reproductive Systems

1. The female reproductive system includes two *ovaries*, two *oviducts*, the *uterus*, and the *vagina*. In addition to producing sex cells, the ovaries also produce several important hormones.
2. The male system includes the *testes*, a series of ducts or channels, auxiliary glands, and the *penis*. The testes also produce sex cells and male hormones.

Diseases of the Reproductive System

The reproductive organs and associated structures are susceptible to a variety of venereal and nonvenereal infectious agents.

INFECTIONS OF THE FEMALE GENITAL TRACT

1. Several microorganisms from the normal flora of the vaginal region, as well as others, can cause infections of this system.
2. Anaerobic streptococci and other bacteria can cause infections such as puerperal sepsis, which involves the genital tracts of women who have just given birth.
3. Vaginitis, or vulvovaginitis, and related infections of the female genital tract may be caused by a variety of microorganisms, including *Candida albicans, Gardnerella (Haemophilus) vaginalis*, and *Trichomonas vaginalis*.

TOXIN-CAUSED DISEASE

1. Toxic shock syndrome (TSS) is caused by certain *Staphylococcus aureus* exotoxin-producing strains.
2. TSS is linked to the use of tampons, particularly with their continuous use during menses. The disease also has been found in surgical patients.
3. Both females and males can acquire the disease.

DISEASES OF THE MALE REPRODUCTIVE SYSTEM

1. Several nonvenereal diseases are known to affect the penis and prostate gland.
2. *Neisseria gonorrhoeae, Escherichia coli*, and *Pseudomonas aeruginosa* are among the causes of prostatic infections.

THE SEXUALLY TRANSMITTED DISEASES (STDs)

1. Sexually transmitted or venereal diseases (STDs) are a major and growing public health problem.
2. STDs include the traditional venereal diseases, syphilis, gonorrhea, chancroid, lymphogranuloma venereum, and granuloma, as well as infections caused by *Chlamydia trachomatis*, herpes simplex viruses, and hepatitis B virus.
3. The agents of STDs include bacteria, protozoa, viruses, yeast, lice, and mites.
4. Chlamydiae are considered to be a leading cause of genital infections, including nongonococcal urethritis in males, cervicitis in females, and inclusion conjunctivitis and/or pneumonia in newborns.

THE TORCH COMPLEX

1. The TORCH complex is a set of certain infectious disease agents that produce long-term ill effects in an infected fetus or newborn infant. The infections are often clinically inapparent in both the newborn and the mother.
2. The complex consists of *Toxoplasma* (T), rubella virus (R), cytomegalovirus (C), herpesviruses (H), and others (O).

PELVIC INFLAMMATORY DISEASE

1. Pelvic inflammatory disease (PID) is a sexually acquired disease of young women.
2. The most frequently found pathogens include *C. trachomatis, M. hominis*, and *N. gonorrhoeae*.

Questions for Review

1. How do microorganisms gain access to the tissues of the kidney?
2. What types of urinary tract infections are associated with anaerobes?
3. Are there factors that can predispose an individual to some form of urinary tract infection? Explain.
4. Compare five STD (venereal diseases) discussed in this chapter as to causative agents, diagnosis, treatment, and prevention.
5. Can venereal diseases be transmitted by means other than sexual contact? Explain.
6. Which of the STDs are known to exert harmful effects during pregnancy? Describe these conditions and discuss preventive measures that can be taken.

7. a. What is TSS? b. How is the disease acquired?
8. Consider the professions of the individuals listed in answering this question: How can these persons acquire herpes simplex virus 2, syphilis, gonorrhea, or lymphogranuloma venereum? If there can be no professional involvement, say so:
   a. barber
   b. physical therapist
   c. nurse
   d. x-ray technician
   e. physician
   f. dietician
   g. dentist
9. Explain how STDs might be totally eliminated.
10. Define or explain
    a. TORCH complex
    b. PID
    c. toxic shock syndrome
    d. clue cells

Suggested Readings

CHARLES, D., and B. LARSEN, "Streptococcal Puerperal Sepsis and Obstetric Infections: A Historical Perspective," *Reviews of Infectious Diseases* **8**:411–422, 1986. *An article providing some historical insight into selected infectious disease problems during the last four centuries.*

NAHIAS, A. J., "The TORCH Complex," *Hospital Practice* **9**: 65, 1975. *An excellent summary of intrauterine infections of the fetus.*

OWEN, R. L., "Sexually Transmitted Enteric Disease," in *Current Clinical Topics in Infectious Diseases,* Remington, J. S., and M. N. Swartz (eds.). New York: McGraw-Hill, 1982, pp. 1–29. *An important description of the increasing range of infectious disease agents than can be sexually transmitted.*

PALAC, D. M., "Urinary Tract Infections in Women: A Physician's Perspective," *Laboratory Medicine* **17**:25–28, 1986. *An article summarizing recent developments in the areas of urinary tract infections and clinical laboratory diagnosis.*

PETERSON, E. M., and L. M. DELA MAZA, "*Chlamydia trachomatis:* A Genital Pathogen Worth Recognizing," *American Journal of Medical Technology* **48**:247, 1982. Chlamydia trachomatis *is rapidly becoming recognized as a leading cause of sexually transmitted diseases in adults, as well as pneumonia and conjunctivitis of the newborn. This article is concerned with specimen collection and approaches to the laboratory identification of* C. trachomatis.

PUBLIC HEALTH SERVICE, NATIONAL COMMUNICABLE DISEASE CENTER, ATLANTA, GEORGIA, Public Health Service Publication No. 1660. *An excellent, well-illustrated publication dealing with the historical and clinical features of syphilis. Properties of* Treponema pallidum, *together with diagnostic serological tests, also are discussed.*

"Sexually Transmitted Diseases Treatment Guidelines 1985," *Morbidity and Mortality Weekly Report* **31**:335 1985. Atlanta: U.S. Department of Health and Human Services. *A source of guidelines for the treatment of STDs.*

SMITH, C. B., and J. A. JACOBSON, "Toxic Shock Syndrome," *Disease-a-Month* **32**:77–118, 1986. *A thorough presentation of this severe staphylococcal infection.*

SPAGNA, V. A., and R. B. PRIOR (eds.), *Sexually Transmitted Diseases: Clinical Syndrome Approach (Reproductive Medicine Series),* vol. 7. New York: Marcel Dekker, Inc., 1985. *This publication provides a functional presentation of the clinical signs and symptoms of 25 of the more readily known sexually transmitted diseases (STDs).*

29

Microbial Diseases of the Central Nervous System and the Eye

Since the death of the child was almost certain, I decided in spite of my deep concern to try on Joseph Meister the method which had served me so well with dogs. . . . I decided to give a total of 13 inoculations in ten days. . . . Joseph Meister escaped not only the rabies that he might have received from his bites, but also the rabies which I inoculated into him. — Louis Pasteur

After reading this chapter, you should be able to:

1. Indicate the areas of the central nervous system attacked by pathogenic microorganisms.
2. Discuss the general routes of microbial invasion of the central nervous system.
3. List and describe one bacterial, one mycotic, one protozoan, and one viral infection of the central nervous system, giving their causative agents, means of transmission, and control measures.
4. Explain the nature of slow virus and prion diseases and their significance.
5. Compare the approaches used in the identification of bacterial, mycotic, protozoan, and viral infections of the central nervous system.
6. Describe commonly encountered infections of the eye.
7. Discuss the complications associated with infections of the eye.

Various microorganisms can invade the human central nervous system. Chapter 29 describes several nervous system diseases, including meningitis, rabies, slow virus diseases, and prion infections. Microbial infections of the eye are also discussed.

Organization of the Central Nervous System (CNS)

Most of the body's systems consist of several anatomically connected organs and structures so organized that together they perform important and specific functions for the entire body. The digestion of food, for example, would be of little value without a bloodstream to distribute the products. The working together of body systems is not haphazard; the time and location of one set of activities are closely related to those of others. The survival of any complex organism requires the *coordination* of vital processes taking place within it. The nervous system and the hormonal systems share this important body function.

The functions of the nervous system include (1) sensing and responding to changes that take place in the external environment; (2) regulating organ systems; and (3) maintaining several aspects of the organism's internal environment in response to conditions occurring within the organism. This system consists of a vast number of slender outgrowths of nerve cells, the *nerve fibers,* that connect cells sensitive to certain environmental changes *(receptors)* with those responsible for carrying out an organism's responses *(effectors).*

The CNS consists of the brain and spinal cord. The brain is encased within the bones of the skull. It is continuous with the spinal cord, which is surrounded by the segments of the vertebral column (Figure 29–1). The main regions of the brain are the *cerebrum, cerebellum,* and *brain stem.* It is the brain stem that serves as the relay between the brain and the spinal cord and controls heart rate, respiration, and several other body functions. Both the brain and spinal cord are hollow, contain *cerebrospinal fluid,* and are covered by the three *meninges* (membranes), called the *dura mater* (outer), *arachnoid* (middle), and *pia mater* (inner) *sheaths.* The sites for specific microbial infections are indicated on Figure 29–1.

The cerebrospinal fluid fills a space between the pia mater and arachnoid known as the *subarachnoid space.* The fluid provides nutrients and serves as a shock absorber for the brain and cord. Normally, the skull, vertebral column, and meninges provide considerable protection to nervous system components.

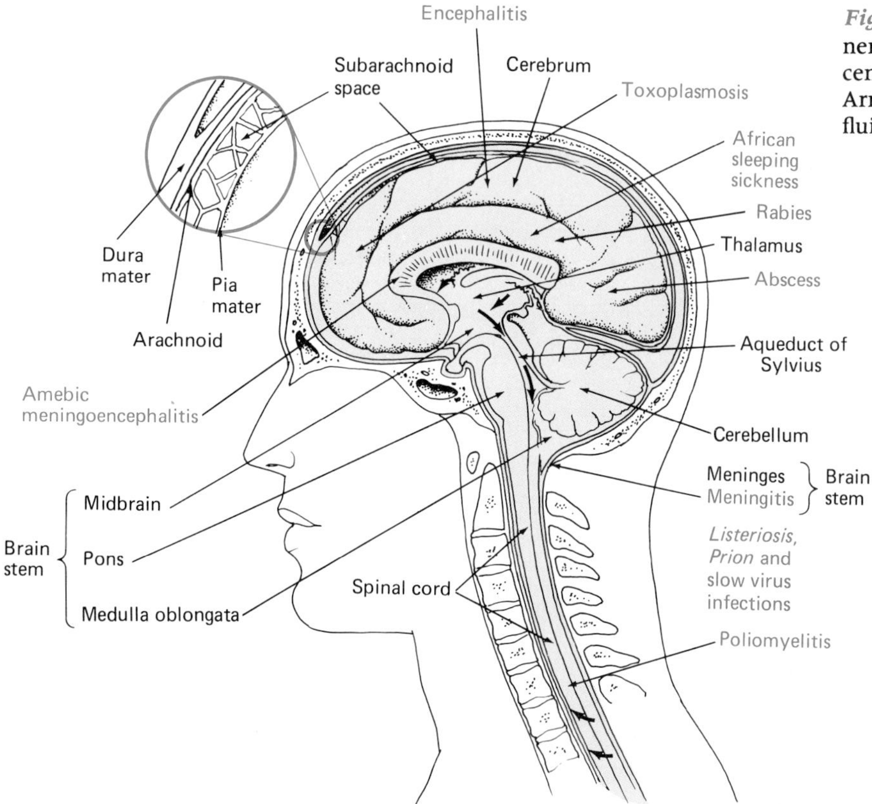

Figure 29–1 The human central nervous system. Diseases affecting the central nervous system are also indicated. Arrows show the flow of cerebrospinal fluid.

Diseases of the Central Nervous System

Injuries and disorders involving parts of the nervous system produce varied but definite effects. Symptoms of such conditions are particularly distressing. Among the most severe symptoms are paralysis, uncontrolled body movements, and loss of control over the body. Breathing or heartbeat is sometimes affected, and the disease may produce a variety of emotional disturbances. Injuries due to trauma, as from an automobile accident or a fall, are usually quite obvious. However, other causes may produce much the same effects but often some time later. Disorders of the system can also result from birth trauma, genetic defects, and tumors, as well as from microbial infections (Table 29–1).

Fever, general weakness, headache, and a stiff neck are typical signs of a CNS infection. Other

TABLE 29–1 Representative Microbial Diseases of the Nervous System

Disease	*Causative Agent*	*Gram Reaction and Morphology*	*Incubation Period*	*General Features of the Disease*
Aseptic meningitis	Enteroviruses, mumps virus	Not applicable	Unknown	Fever, irritation, stiffness of the neck, and general fatigue
Brain abscess	*Escherichia coil, Proteus* spp.	–, Rods	Unknown	Fever, headache, nausea, pus formation, and possible interference with vision, breathing, hearing and movement
	Staphylococcus aureus,		+, Cocci	
	Streptococcus pneumoniae,		+, Cocci	
	Streptococcus spp.		+, Cocci	
Haemophilus meningitis[a]	*Haemophilus influenzae* type b	–, Rods	Variable	Irritability, high fever, stiff neck, listlessness, headache, may be preceded by middle ear or upper respiratory infection
Listeriosis (meningitis)	*Listeria monocytogenes*	+, Rods	1–7 days	High fever, drowsiness, headache, confusion, seizures, tremors, uncoordinated movements, and coma
Meningococcal meningitis[a]	*Neisseria meningitidis*	–, Coccus	Unknown	Fever, headache, pus formation, and infections of the bones in the skull or ear
Rabies (hydrophobia)	Rabies virus	Not applicable	4–6 weeks or longer	Fever, general discomfort, headache, visual difficulties, painful throat spasms[b], convulsions, delirium, respiratory paralysis, and death
Viral encephalitis	Arboviruses including eastern equine, St. Louis, and western equine encephalitis viruses	Not applicable	4–21 days	Fever, chills, nausea, general fatigue, drowsiness, pain and stiffness of neck, and general disorientation; blindness, deafness, and paralysis may develop as consequence of infection

[a] Several other microorganisms can cause meningitis as a complication or secondary effect of disease states. These include the bacteria *Escherichia coli, Staphylococcus aureus, Streptococcus pneumoniae,* species of *Proteus* and *Pseudomonas,* and anaerobes such as *Bacteroides, Clostridium,* and *Streptococcus.* The yeast *Cryptococcus neoformans* is also associated with meningitis.

[b] This fear of painful swallowing of fluids, which can cause attacks of convulsive choking, is the reason for naming the disease *hydrophobia.*

signs and symptoms, as well as diagnostic features and treatment for some infections, are given in Tables 29–1 and 29–3.

Routes of CNS Infection

The majority of CNS infections appear to result from microbial invasion of the bloodstream at remote body locations. Once the pathogens gain access to the circulatory system, they may penetrate the blood-brain barrier. Other routes of infection include disease processes close to and continuous with the CNS. Examples of such conditions are middle ear infections *(otitis media),* infections of the mastoid bone *(mastoiditis),* sinus infections *(sinusitis),* and pus-producing infections of the skin or bone. Congenital anatomic defects of the CNS coverings (Figure 29–1) and surgical procedures may provide other routes of infection by which microorganisms invade and become established. **[See skin infections in Chapter 23.]**

Microbial Agents of Disease

Bacteria, fungi, viruses, and certain protozoa are capable of establishing CNS infections. The most common bacterial and mycotic pathogens have capsules that are necessary to establish a CNS infection. CNS diseases in newborns are caused by microorganisms that frequently are members of the flora in the maternal genital tract at the time of birth.

Several microbial CNS infections are rare in the normal host. Among these are fungal infections, including those caused by *Coccidioides immitis* and the yeast *Cryptococcus neoformans.* Only the more common diseases of the brain and spinal cord, such as *meningitis* (infection of the meninges), *encephalitis* (inflammation of the brain), and brain abscess (localized collections of pus) will be described.

Laboratory Diagnosis

In addition to culture methods, several noncultural approaches are used in the isolation and identification of microbial pathogens. Among the more specific and sensitive tests are the enzyme-linked immunosorbent assay (ELISA) and chemical procedures to determine lactic acid levels in cases of meningitis. Gas-liquid chromatography (GLC) and Monotest Lactate, the commercially available enzyme test, are examples of the chemical tests used to provide information concerning the cause of infectious meningitis. GLC is a highly complex procedure that uses a gas chromatograph and the extraction of cerebrospinal fluid specimens with organic solvents. The monotest utilizes a spectrophotometer. **[See ELISA in Chapter 19.]**

Bacterial Diseases of the Nervous System

BRAIN ABSCESS

Nervous tissue and the meninges respond to bacterial invasion as other tissues do. Pus-producing (pyogenic) organisms that invade brain tissue produce an inflammation with pus formation, which results in an abscess.

Several bacteria have been associated with this type of infection, including organisms such as *Escherichia coli* (esh-er-IK-ee-ah KOH-lee), *Streptococcus pneumoniae* (strept-toe-KOK-kus noo-MOH-nee-eye), *Proteus* (PROH-tee-us) spp., *Staphylococcus aureus* (staff-il-oh-KOK-kuss OR-ree-us), and other *Streptococcus* spp.

Anaerobic agents are believed to cause most, if not all, bacterial brain abscesses of a nontraumatic nature. The anaerobic diphtheroids and streptococci and *Bacteroides* (back-te-ROY-deez), *Clostridium* (klos-TREH-dee-um), and *Veillonella* (veh-yon-ELL-ah) are involved. The anaerobic streptococci are the most frequently found pathogens, followed by *Bacteroides.* Brain abscesses caused by *Clostridium* species are seldom found in civilian populations.

Most brain abscesses are complications that develop from chronic (old) pus-producing sites in other portions of the body, such as the lungs, middle ear, paranasal sinuses, pelvis, and pleura. Dissemination of organisms from these foci of infection can occur (1) by direct extension through bones, (2) via the covering of the olfactory nerves, and (3) by way of the venous system. Most brain abscesses are associated with mastoiditis, middle ear infections, or tumors of the middle ear. The cerebral hemispheres are the regions of the brain in which abscesses most commonly form. This infection rarely occurs in the spinal cord.

MENINGITIS

Microorganisms present in the circulatory system experience great difficulty in entering the CNS, largely because of the blood-brain barrier. Unless some form of injury or other condition occurs to alter the permeability of this barrier, organisms are unable to penetrate. However, once infectious agents gain entrance to the brain and the adjacent meninges, invasion and destruction of the nervous tissue can proceed rapidly. The outcome of the disease, of course, depends upon the initial treatment of patients.

Meningitis, inflammation of the membranes around the brain and spinal cord, can result from one of several mechanisms. These include the introduction of microorganisms (1) through penetrating injuries or primary infections involving the skull and spinal column; (2) by the direct extension of a disease process from primary foci of infection located in other parts of the body through bone via vascular channels, or along the covering of the ol-

Brain Abscess

Laboratory Identification

E. coli, Proteus species (spp.), *S. aureus,* and the streptococci can be readily isolated from specimens such as spinal fluid, on selective and differential media such as eosin-methylene blue and MacConkey agars. (Color Photograph 109). *E. coli* is a lactose fermenter, whereas *Proteus* spp. are not. In addition, *Proteus* spp. form a spreading or swarming growth unless the agar concentration is increased in the isolation medium. The biochemical reactions of *E. coli* and *Proteus* spp. serve to separate them from other organisms. In the combination of tests known as the IMViC series, *E. coli* forms the metabolic breakdown product indole (I) from the amino acid tryptophan, produces acid from glucose in a methyl red (MR) broth containing the carbohydrate, does not produce acetylemethyl carbinol in the Voges-Proskauer (V) test, and is unable to use citrate (C) as the only source of carbon. *Proteus* spp. are distinguished biochemically by being nonlactose fermenters and producing urease.

S. aureus is beta-hemolytic, produces catalase and ferments mannitol (Color Photographs 26 and 96). *S. pneumoniae* is alpha-hemolytic and can be differentiated from other streptococci by its sensitivity to bile salts and optochin (Color Photograph 89). Serological tests are also available for rapid identification.

Treatment

Antibiotic susceptibility tests should be performed with isolated organisms. In general, chloramphenicol, tetracyclines, and nalidixic acid are used to treat infections of gram-negatives.

Penicillin G or V are the drugs of choice for non-penicillinase-producing *S. aureus* strains. Methicillin, nafcillin, or oxacillin are used with penicillinase-producing strains. Alternate drugs for individuals with a penicillin allergy include cephalosporin, clindamycin, and vancomycin. Penicillin is generally the drug of choice for *S. pneumoniae.*

factory nerves; and (3) by means of the bloodstream (hematogenous route) during the course of a septicemia.

Meningeal inflammation can be classified into two types, inflammation of the dura mater and inflammation of the pia mater and the arachnoid of the brain and spinal cord. The first of these conditions is a localized infection and almost always results from a direct extension of an infection located in the surrounding tissues. Usually inflammation of the dura mater occurs as a complication of infected skull fractures or as a related effect. The dura appears to limit the infection and prevent involvement of the leptomeninges. Although the infection may be effectively localized by the dura, it may spread to the leptomeninges or the brain by means of infected veins.

At least 50 different organisms have been incriminated in these disease states. Three bacterial species, namely *S. pneumoniae, H. influenzae* (he-MAH-fi-lus in-floo-EN-zee), and *Neisseria meningitidis* (nye-SEH-ree-ah meh-nin-jit-EE-diss), are the most commonly encountered agents. Many of these pathogens reach the meninges by the routes discussed previously. However, organisms have sometimes been introduced through the injection of contaminated solutions, such as local anesthetics, into the cerebrospinal fluid. There are certain *Pseudomonas* (soo-doh-MOH-nass) meningitis cases that have been reported to occur in this manner. Contaminated foods and maternal transmission of the causative agents to fetuses are also known to occur with bacteria such as *Listeria monocytogenes (lis-TEH-ree-ah moh-no-SIGH-todge-ee-neez).*

The general clinical picture caused by many of these organisms is similar to that produced in meningococcal meningitis (Table 29–1).

Meningitis caused by *Haemophilus influenzae* type b occurs frequently in children under four years of age. For example, in the United States, the annual incidence is approximately 15,000 cases. Mortality in such cases, even with prompt antibiotic therapy, is estimated to be 5% to 10%. The number of cases is substantial, and unfortunately, survivors of meningitis frequently exhibit neurologic consequences. Table 29–1 lists several of the general symptoms of the disease. **[Refer to Chapter 25 for descriptions of other *H. influenzae,* type b-caused infections**

Listeriosis

LABORATORY DIAGNOSIS

L. monocytogenes may be isolated from blood, cerebrospinal fluid, or other specimens on blood agar. The bacterium is beta-hemolytic, grows well, and is motile at 25°C; it produces acid from several carbohydrates after a week of incubation at 35°C. The ocular inoculation of rabbits is used to demonstrate pathogenicity.

TREATMENT

Antibiotics, including ampicillin and tetracycline, are used to treat listeriosis.

and the approaches to laboratory diagnosis and treatment.]

LISTERIOSIS

Listeriosis is a bacterial disease that can develop into meningitis and sepsis, especially in immunocompromised hosts. The causative agent *L. monocytogenes* is primarily found in the intestinal tracts of birds, dairy cattle, and household pets. Human infections are acquired from contact with contaminated soil, and by the ingestion of contaminated foods.

In 1985 and 1986 a significant number of listeriosis cases were identified in Southern California. The first infections were found in pregnant Hispanic women, and all appeared to be associated with the consumption of Mexican-style cheeses. Several samples of suspected products were found to contain the bacterium *L. monocytogenes,* the cause of listeriosis. Unfortunately, the cheese products were distributed to at least 16 other states. Over 120 cases and 50 deaths from listeriosis occurred. To prevent the spread of the disease, the manufacturer instituted a voluntary recall of the cheeses and a major media campaign was instituted to warn the public.

TRANSMISSION. Infections may be acquired by the ingestion of contaminated food or contact with infected animals. Pregnant women may also transmit *L. monocytogenes* to their fetuses, resulting in abortion, stillbirth, or meningitis.

MENINGOCOCCAL MENINGITIS (CEREBROSPINAL FEVER)

N. meningitidis is capable of causing the death of a human faster than any other infectious agent. Death has been reported to occur in less than two hours after the appearance of the first symptoms. Infections with meningococci frequently develop in closed environments such as jails, military posts, schools, and ships. The causative agent is believed to be transmitted by droplets, as with other respiratory diseases. *N. meningitidis,* like certain other members of the genus, is unable to withstand unfavorable environments for any appreciable length of time and therefore can probably cause infections only by some form of direct contact. **[See disease transmission in Chapter 21.]**

Meningococcal meningitis has presented a serious problem in Africa since World War II. Group A organisms are predominant. The disease has been found in an area extending across Africa from the shores of the Atlantic Ocean to those of the Red Sea, and north of the Equator to south of the Sahara. This region is referred to as the "meningitis belt."

PREVENTION. Preventive measures against meningococcal infections are few. The isolation of carriers during epidemics would be extremely difficult in view of the sheer number of persons involved. Prophylactic mass treatment for substantial periods of time is dangerous since resistant strains would probably emerge. Vaccines appear to be one of the few possibilities to prevent epidemics. Preparations to immunize individuals actively against meningococcus have been developed. **[See vaccines in Chapter 18.]**

MENINGITIS CAUSED BY ANAEROBES

Meningitis is only rarely caused by anaerobic bacteria, although several of the species associated with brain abscesses have been implicated in this disease. Included in this group of causative agents are *Actinomyces* (ak-tin-oh-MY-sess), anaerobic streptococci, *Bacteroides,* and clostridia. Most meningitis caused by *Bacteroides* and anaerobic streptococci (Figure 29–2) has come from middle ear or sinus infections. Clostridia are usually considered to be the causative

Meningococcal Meningitis

LABORATORY DIAGNOSIS

N. meningitis may be isolated from a variety of specimens, including blood, nasopharyngeal swabs, cerebrospinal fluid (CSF), and skin lesions. Specimens are plated on enriched media such as chocolate agar or Thayer-Martin medium and incubated in an environment containing an elevated CO_2. The gram-negative diplococcus produces positive oxidase and superoxol tests and acid from both glucose and maltose. Serological tests including the Quellung reaction, fluorescent antibody, and counterimmunoelectrophoresis may be performed directly on CSF specimens. [See Chapter 19 for descriptions of these tests.]

TREATMENT

Because of the seriousness of meningococcal infection, therapy must be started before the completion of laboratory identification. Penicillin is the drug of choice. Chloramphenicol and a combination of ampicillin and moxalactam are appropriate alternates.

agents of anaerobic meningitis arising as a complication of head injuries or surgery.

Viral Infections of the Nervous System

Many viruses are associated with nervous system disease (Table 29–1). These pathogens are referred to as *neurotropic.* Brief descriptions of selected diseases and the viral agents associated with them are presented in Table 29–1. Viruses may gain access to the CNS by several means, including cellular blood components, cerebrospinal fluid, olfactory nerve fibers, and nerves. Isolation and identification of causative agents includes the use of tissue culture and laboratory animal inoculations with clinical specimens and immunodiagnostic or serologic tests. Examples of such tests are ELISA, immunoperoxidase, and fluorescent antibody techniques. [See Chapter 19 for descriptions of these tests.] Certain neurotropic viral infections are described in this section. AIDS, a disease also found to involve the CNS, is discussed in Chapter 27.

Figure 29–2 The effects of anaerobic streptococci on the human brain. *[Courtesy of the Armed Forces Institute of Pathology, Washington, D.C., Neg. No. 54-13308-1.]*

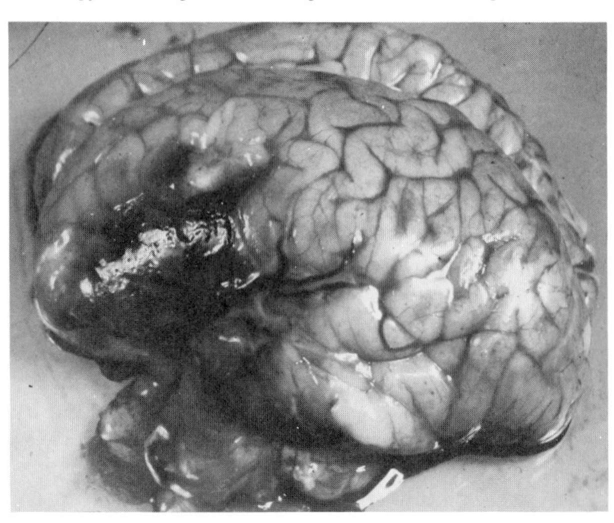

ASEPTIC (NONBACTERIAL) MENINGITIS

Aseptic meningitis is one of the most common syndromes and the easiest to recognize. Such infections occur worldwide in epidemic form and primarily involve children. Several viruses have been incriminated as causes. It should be noted that a wide spectrum of clinical conditions can be caused by these viruses.

PICORNAVIRUSES

The picornavirus group contains several viruses that can produce infections of the human nervous system, such as polioviruses, coxsackieviruses, and echoviruses.

POLIOMYELITIS. Poliomyelitis has been found in every country where it has been sought. However, since their introduction, the various types of polio vaccines have significantly reduced the incidence of the disease.

Three antigenic types are known. Poliovirus is generally (though not universally) believed to enter the body via the mouth and intestines. The organisms multiply primarily in the digestive system. The spreading of poliovirus to the nervous system occurs by way of the blood. Poliovirus is excreted in the feces before extensive involvement of the nervous system develops. Several authorities also strongly believe that poliovirus is transmitted to the CNS by way of the peripheral nerves (the neural route). Several effects, ranging from minor illness to paralysis, can occur as a result of poliovirus infection.

Several factors are known to either predispose to or aggravate paralytic poliomyelitis. These factors include physical exertion, especially during the early phase of the disease, pregnancy near term, routine intramuscular injections, and tonsillectomy.

The effective control of poliomyelitis can be achieved by (1) adequate diagnosis, classification, and reporting of polio cases, (2) routine immunizations, (3) control of outbreaks, and (4) mass immunization programs. [See Chapter 18.]

Individuals recovering from polio infection usually acquire an active immunity to the viral type causing the disease.

The Coxsackieviruses. The first coxsackie viral agent was isolated by Dalldorf and Sickles in 1948. The virus was recovered from suckling (unweaned) mice that were inoculated with fecal matter of children from Coxsackie, New York. Two main groups of these viruses are recognized, A and B. Coxsackieviruses exhibit the general characteristics described for the picornavirus group. However, their pathogenicity for suckling mice rather than for adult animals is a feature that distinguishes them from the other picornaviruses, echoviruses, and polioviruses.

Diseases caused by group A coxsackieviruses include aseptic meningitis, herpangina (an ulcerative condition of the throat), minor respiratory illness, myocarditis, paralytic illness, and rubelliform rashes accompanied by fever. The diseases associated with the group B coxsackieviruses include aseptic meningitis, epidemic pleurodynia, myocarditis, neonatal

Microbiology Highlights

THE PERSISTENCE OF VIRAL INFECTIONS

Viral infections can produce disease long after an individual has been initially exposed to the virus. When such conditions become persistent, they produce a variety of disease patterns and are often described as long-lasting or chronic, latent or slow virus infections. For a virus to establish a persistent infection, it must be able not only to replicate over a long time period (which means that the virus cannot destroy its host) but also to escape from the actions of the host's immune system.

In chronic infections, virus particles are shed over a long period and can be isolated and identified. Congenital rubella and cytomegalovirus infections are examples. These viruses can be isolated from infected infants for years after birth.

In latent viral infections, viruses exist in a form in which no identifiable viral particles are present. This state occurs with some herpesvirus and retrovirus infections. When viruses are latent, detection by the host's immune system or by standard laboratory isolation procedures or electron microscopic techniques is not possible.

Slow virus infections generally refers to disease states with inperiods that last for months or years and that are followed by see disability and usually death.

Regardless of the type of persistent infection, both host and virus factors contribute to the survival of the viral agent. In many persistent viral infections, the initial infection is conacquired or occurs at a very early age, even though the disease may not appear until later in life. The maturity of the immune sys of infants plays a significant role in the establishment of pertent viral infections. Some viruses are able to integrate their genomes into the host's DNA, thus providing the posof pass the infection through sex cells.

Persistent viral infections are often difficult to identify by the usual methods of viral isolation and cultivation. With the increasing number of new approaches to viral identhis group of viral infections may be recognized more accurately and rapidly.

encephalomyocarditis, and paralytic illness. A virus related to this group has been implicated as a diabetes-causing agent. It was isolated from the pancreas of a 10-year-old boy who had rapidly developed symptoms of diabetes and died after a week's hospitalization.

Most often, these diseases agents are spread either directly or indirectly by contact with contaminated articles (fomites) and aerosols.

The Echoviruses. The echoviruses are commonly found in the human gastrointestinal tract. They are considered to be among the most common cause of aseptic meningitis, as nearly all strains comprising the group have been associated with the illness. Other diseases caused by echoviruses include diarrhea, fever, and mild respiratory illness.

RABIES (HYDROPHOBIA)

The viral infection rabies has been a dreaded disease since the time of the ancient civilizations of Egypt, Greece, and Rome. According to the fifth report of the World Health Organization (WHO) Expert Committee on Rabies, issued in 1966, rabies exists in two epidemiologic varieties: (1) the urban form, characteristically associated with dogs, and (2) the wildlife variety, which occurs in animals such as bats, coyotes, foxes, jackals, mongooses, skunks, and weasels.

The principal source of rabies virus is believed to be wild mammals. Humans and domesticated animals—such as cows, goats, horses, and sheep—acquire this disease accidentally. Saliva containing the rabies virus introduced into humans by the bite of a rabid animal is the principal means of transmission. Infection through minor scratches and via the respiratory tract has also been reported. The virus responsible for the disease is called *street virus,* or virus *de rage de rue.* Strains of this street virus passed through rabbits are known as *fixed virus,* or *virus fixé.*

Rabies in Lower Animals. The incubation period in dogs ranges from two to eight weeks. Infected animals exhibit changes in their behavioral patterns, such as stumbling and gnawing on sticks and stones. In the "furious" type of rabies, dogs characteristically snap and bite; in "dumb" rabies, they exhibit paralysis. The infection in dogs can be easily transmitted to other animals by biting. Often the saliva of infected dogs contains virus shortly before the disease symptoms appear.

Bat-transmitted rabies has been found to involve fruit-eating, insectivorous, and vampire bats. When these animals become rabid, they bite one another as well as other animals in the area. Rabid bats have been reported from every state in the United States except Hawaii.

Other important reservoirs of rabies are the wildlife populations of several countries. The infection can occur in either the furious or dumb varieties. The tendency of rabid animals, such as skunks and wolves, to enter villages, cities, and even homes poses a continuous threat to the well-being of humans and domestic animals.

Laboratory diagnosis depends on the demonstration of rabies virus in brain tissue. The fluorescent antibody test is the procedure of choice. Mouse inoculations with infected brain tissue; demonstration of viral inclusions, or Negri bodies; and diagnostic electron microscopy also are used.

Treatment. Various procedures have been used to treat individuals bitten by rabid animals. The WHO Expert Committee on Rabies recommends the following:

1. Animals suspected of being rabid should be confined, if possible, observed, and examined for at least ten days. Treatment should not await the confirmation of rabies in the animal. However, the failure to detect the virus by a thorough laboratory investigation is a sufficient basis for stopping the course of vaccine injections.
2. Local treatment of bites should be carried out. Procedures include thorough washing of the wound with soap and water or other appropriate materials known for their viricidal effects (e.g., quaternary ammonium compounds). The topical administration of rabies antiserum or specific immune globulin (rabies immune globulin) preparations should be considered and should be conducted or supervised by a physician.
3. The injection of rabies vaccine and antiserum should be instituted.
4. Preventive measures against tetanus and other bacterial infections also should be taken.

The vaccines used in antirabies treatment have been of several types. The first of these, originally developed by Pasteur, utilized infected rabbit spinal cord material, which was suspended over a drying agent in a closed jar. The purpose of this procedure was to inactivate the viral agent. Today most vaccines are inactivated or attenuated, and prepared in human diploid cell cultures or duck embryos. [See rabies vaccines in Table 18–2.]

The treatment of individuals exposed to rabies virus first involves extensive cleansing of the wound, then passive immunization by injections with hyperimmune serum (rabies immune globulin), and finally active immunization with a vaccine preparation. The use of interferon may be a possibility in the near future. Prompt cleansing of the wound reduces the number of virus particles but does not eliminate the possibility of infection. The original procedure for the administration of vaccine preparations developed by Pasteur consisted of 14 to 21 subcutaneous injections of inactivated virus material over the abdominal area. A period of two to three weeks was necessary to complete the series of injections. With the development of newer vaccines, fewer injections are required. With human diploid cell vaccine, five intramuscular injections are given at one site in the thigh. A sixth injection is recommended two months later. In the case of duck embryo vaccine, a large number of subcutaneous injections still are necessary. In addition, serological tests are performed to determine if sufficient antibody has been produced. If the antibody level is not adequate, booster doses are given as needed.

Preexposure active immunization is available and advisable for persons at high risk; such individuals include veterinarians, cave explorers, laboratory personnel, and animal handlers. Three intramuscular injections are given at weekly intervals for this purpose.

CONTROL. The control of rabies presents several problems. However, certain measures recommended by the fifth report of the WHO Expert Committee on Rabies, if properly implemented, could significantly reduce the incidence of the disease. Included in their recommendations were the following: (1) prophylactic immunization not only of dogs but also of cats; (2) establishment of clinics in localities where rabies prevails; (3) elimination of stray dogs and quarantine of dogs imported into island communities; and (4) incorporation of programs to eliminate or significantly reduce proven wildlife vectors.

SLOW VIRUS AND PRION DISEASES

The concept of slow virus infections was first introduced by Sigurdsson in 1954 in his classic descriptions of several chronic diseases of Icelandic sheep. Since then, additional slow virus infections have been described in humans and domestic animals. These disorders are characterized by a slow, almost imperceptible onset and a long-lasting and progressive course leading eventually to death. *Slow infections* may be defined as a group of disease states caused by viruses in which the incubation periods are extremely long and in which the clinical expressions or course of the infections are relatively slow to appear. The outstanding feature of these infections is the persistence of a viral agent or its genetic components in a host who ultimately experiences cellular and tissue injury from the activities of the virus.

The reason for the long incubation period of these diseases is unknown. The viral agents may be masked or hidden as a result of either immunoglobulin activity or an integrated effect of the virus that is not fully expressed.

The agents associated with this group of disease states are classified into two major categories, *conventional* and *nonconventional,* based on the recognized characteristics of the viruses and the less well-defined transmissible agents called *prions* (see Chapter 10). Usually included are the agents of lymphocytic choriomeningitis, herpesvirus-associated persistent infections, rabies, progressive multifocal choriomeningitis, and subacute sclerosing panencephalitis. Subacute sclerosing panencephalitis is a rare consequence of measles virus chronic infection. The virus persists in the body despite the fact that the host has high levels of antimeasles antibody. This situation suggests an underlying defect in the host's immune system. Progressive multifocal leukoencephalopathy is another rare disease found in adults and is associated with polyomalike tumor viruses. The disease occurs most frequently in patients with debilitating diseases, most of which produce some form of immunosuppression. These diseases include various cancers and autoimmune diseases such as systemic lupus erythematosus. All conventional agents show evidence of complete viral structural or substructural components. In addition, they provoke immune responses in the infected host and cause either degenerative or inflammatory changes in a variety of tissues. The nonconventional group includes the agents of two rare human diseases, kuru and Creutzfeldt-Jakob disease, as well as the lower animal diseases, a transmissible brain disease of mink, and scrapie (Figure 29–3), the well-known disease that kills sheep and goats. Another disease possibly caused by prions is Alzheimer's. This condition usually affects the elderly and is accompanied by irreversible memory loss, disorientation, and speech difficulties.

Prions contain protein, but unlike the members of the conventional group, they do not exhibit a virion structure or substructure, are not antigenic, and produce degenerative changes confined to the CNS. The transmission of Creutzfeldt-Jakob disease has been associated with corneal explants and the implantation of electrodes in brain wave determinations. George Balanchine, the once athletic choreographer, died after several months of suffering from this rare disease in 1983. **[See autoimmune diseases in Chapter 20; prions in Chapter 10.]**

Figure 29–3 An electron micrograph of scrapie-associated protein fibrils from an infected hamster brain. *[From H. J. Cho,* J. Gen. Virol. ***67****:243–253, 1986.]*

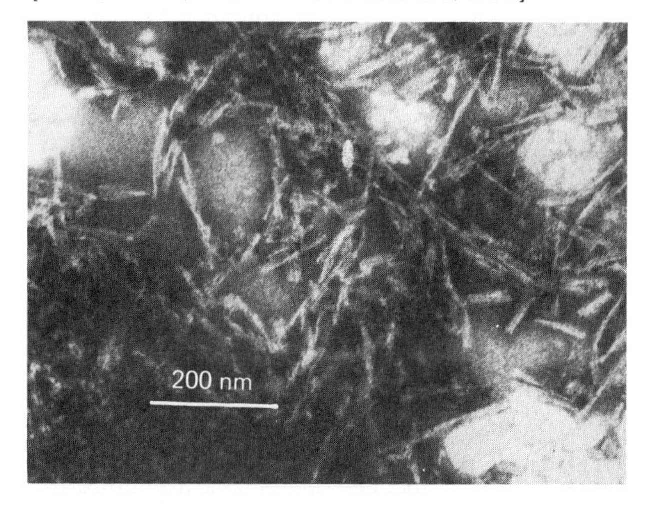

Microbiology Highlights

HUMAN GROWTH HORMONE AND CREUTZFELDT-JAKOB DISEASE

Growth hormone (GH) deficiency in children produces the congenital condition known as pituitary dwarfism. This is treated by the administration of human growth hormone prepared from human pituitary glands obtained at autopsy. Despite the success of this treatment, it was suspended in April 1985 in both the United States and the United Kingdom due to reports of several associated cases of the fatal, degenerative neurological Creutzfeldt-Jakob disease. The problem of neurological disease transmission through the use of human GH has been solved by recombinant DNA technology. In 1985 the Food and Drug Administration approved a synthetic form of GH which produces effects identical to those of pituitary-derived human GH. In addition to correcting congenital disorders, the synthetic product might prove useful in the treatment of delayed wound or fracture healing and in the metabolic problems associated with aging. It might also be used to increase the height and muscle mass of athletes.

THE ARBOVIRUSES (ARTHROPOD-BORNE)

To date, well over 300 arboviruses have been reported. Only about 100 of these are capable of causing human infection. The first arbovirus to be identified was the causative agent of yellow fever. In 1901, Major Walter Reed and his colleagues not only clearly demonstrated the relationship of the mosquito *Aëdes aegypti* to the transmission of yellow fever but also established the existence of the first human viral pathogen.

The arboviruses have the unique capability of multiplying within the tissues of vertebrates and of certain blood-sucking arthropods. Following their inoculation into the tissues of a susceptible vertebrate, viruses of this group multiply rapidly. In cases of human and certain lower animal infections, the virus may localize in the CNS of the host, causing extensive viral multiplication, tissue injury, encephalitis, and eventual death.

Most arboviruses are either togaviruses or bunyaviruses. The togaviruses are divided into two groups, *Alphavirus* (previously known as *group A arboviruses*) and *Flavivirus* (previously known as *group B arboviruses*). The bunyaviruses represent about 100 arboviruses. Table 29–2 lists features of some arboviruses.

ALPHAVIRUSES. Among the alphaviruses are several pathogens that can cause severe encephalitis. A wide range of vertebrates and arthropods can be involved in their life cycles, including domestic fowl, horses, humans, rodents, snakes, and wild birds. Humans are considered to be only incidental hosts. Horses do not appear to be significant natural reservoirs either. The primary hosts are birds, while snakes and certain rodents are probably secondary reservoirs for some alphaviruses. The majority of alphaviruses are transmitted by mosquitoes. Hibernating mosquitoes have been reported to harbor viruses for several months.

Two general types of clinical states have been observed with these viruses. One form, which is found in eastern, western and Venezuelan equine encephalitis, consists of both a systemic phase with chills, fever, headache, and vomiting, and an encephalitic phase with stupor or coma, perhaps followed by slight paralysis.

Certain alphaviruses, including Chikungunya and Sindbis, are also known to produce arthritis in humans. The most common symptoms of infected persons are severe joint pain (arthralgia), fever, and rash. Treatment is directed toward relieving the symptoms. No vaccines are presently available.

FLAVIVIRUSES. The flaviviruses include at least 36 prototype strains. Most of these arboviruses are transmitted by culicine *(Culex)* mosquitoes, exhibit a host range and initial stages of pathogenesis similar to those of alphaviruses, and produce variable clinical syndromes. Flavivirus infections may produce three types of symptomatology: (1) acute CNS involvement resulting in encephalitis and death; (2) severe systemic illness affecting important visceral organs, including the kidneys and liver; and (3) a milder form of systemic involvement characterized

TABLE 29–2 Some Groups of Arboviruses

Group Designation	Disease	Vector[a]	Geographic Distribution
Alphaviruses	Chikungunya	M	India, most of Africa south of the Sahara, Philippines, Southeast Asia
	Eastern equine encephalitis (EEE)	M	Argentina, Brazil, Dominican Republic, Guyana, Panama, Trinidad, United States
	Sindbis	M	Africa, Asia, Australia, Europe, Philippines
	Venezuelan equine encephalitis (VEE)	M	Brazil, Colombia, Ecuador, Panama, Trinidad, Venezuela
	Western equine encephalitis (WEE)	M	Argentina, Brazil, Canada, Guyana, Mexico, United States
Flaviviruses	Dengue fever	M	Australia, Greece, New Guinea, Pacific Islands, Southeast Asia
	Japanese B encephalitis	M	Eastern Asian mainland, Guam, India, Japan, Malaya
	Russian spring-summer encephalitis	T	Northern European Russia, Siberia
	St. Louis encephalitis	M	United States, Jamaica
	Yellow fever	M	Africa, Central and South America, Trinidad
Bunyavirus	Bunyamwera	M	South Africa, Uganda
	Germiston	M	South Africa
	Guaroa	M	Brazil, Colombia

[a] M = mosquito; T = tick.

by fever, a rash, and severe muscle pains. Subclinical infections can also be produced by these viruses.

Bunyaviruses. The bunyaviruses are immunologically distinct from the other arbovirus groups. Animals implicated as natural reservoirs on the basis of epidemiological investigations include monkeys, opossums, rats, and sloths. The full range of mosquito vectors has not been fully determined.

Control of Arthropod-transmitted Diseases. The control measures used for arbovirus infections are directed against the arthropod vectors (lice, mites, mosquitoes, sandflies, and ticks). This process usually involves the use of insecticides, natural predators, and other functional means. Eradication programs designed to destroy the breeding sites or structures that provide protection are necessary to reduce the reproduction rates of vectors. Measures of this type will decrease successive generations.

Control measures also involve attacks on the reservoirs of infectious agents, which can be done through the use of traps and professional hunters. Furthermore, effective vaccines can also be used as prophylactic measures against such diseases as certain rickettsial infections and yellow fever.

Protozoan Infections of the CNS

TOXOPLASMA GONDII

The causative agent of toxoplasmosis was originally discovered in 1908 by Nicolle and Manceaux in a small North African rodent, the gondi *(Tenodactylus gondi)*. The organism's name is derived in part from this animal. *Toxoplasma gondii* (toks-oh-PLAZ-mah GON-dee-eye) (Color Photographs 47 and 121) exhibits a wide host range. In the late 1930s it was recognized that toxoplasma are associated with a type of infection that results in brain damage of the newborn. Infants may show evidence of the infection at birth or shortly thereafter. Toxoplasmosis has gained increasing importance because of its involvement with AIDS patients. As of 1987 *T. gondii* has been implicated in over 100 cases of AIDS-associated meningoencephalitis (inflammation of the brain and its coverings). Features of the disease, laboratory diagnosis, and possible treatment are given in Table 29–3.

Transmission. *T. gondii* can be transmitted through the placenta (transplacental route). Other modes of transmission have created interest among investigators, particularly because the domesticated cat was implicated as a carrier of the disease agent.

TABLE 29–3 Selected Features of Protozoan Diseases of the Central Nervous System

Disease	*Causative Agent*	*Incubation Period*	*General Features of the Disease*	*Laboratory Diagnosis*	*Possible Treatment*
Toxoplasmosis	*Toxoplasma gondii*	Unknown	When symptoms occur in adults they include: chills, fever, headache, extreme discomfort, and muscle pain; symptoms resemble infectious mononucleosis; symptoms of newborns who contract the disease *in utero* include fever, convulsions, enlarged spleen, and serious central nervous system defects causing blindness and mental retardation	Serodiagnostic tests, e.g., ELISA, fluorescent antibody procedures (Color Photograph 79) and Sabin-Feldman dye test[a]; demonstration of the pathogen in host tissues (Color Photographs 47 and 121)	Pyrimethamine, sulfonamides
Primary amebic meningoencephalitis	*Naegleria fowleri*	5–23 days	In order of frequency of occurrence, headache and stiff neck	Microscopic identification of amebae in specimens	No satisfactory treatment
Trypanosomiasis (African sleeping sickness)	*Trypanosoma brucei* variety (v.) *gambiense*, *T. brucei* (v.) *rhodesiense*	6–14 days	Disease may continue for years; symptoms include enlargement of lymph nodes, spleen, and liver; fever, chills, disturbed vision, general weakness, headache, loss of appetite, occasional rash, nausea, vomiting, and serious defects of the CNS, ending in death	1. Demonstration of organisms in specimens (Color Photograph 90) 2. Serodiagnostic tests, e.g., fluorescent antibody procedures[a]	For early stages, aromatic diamidine suramin; for later stages, Med b (a phenylarsonate derivative)

[a] Refer to Chapter 19 for a discussion of these techniques.

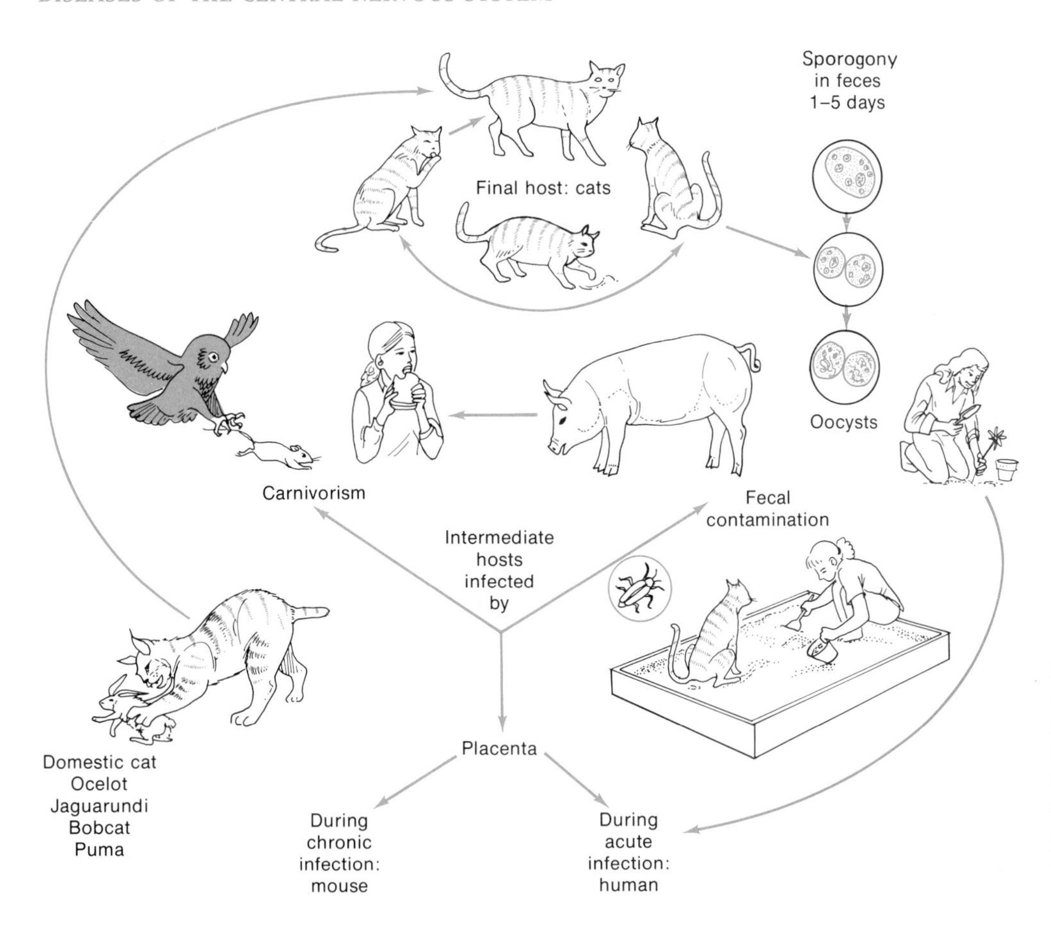

Figure 29–4 Life cycle of *Toxoplasma gondii.* The means of transmission for intermediate hosts are shown in the center of the figure. *[After J. K. Frenkel, and J. P. Dubey,* J. Infect. Dis. ***126:****664–673, 1972.]*

Evidence has indicated that cat feces contain one stage of the life cycle (Figure 29–4) that has a high resistance to the environment as long as it is in the presence of moisture. Because domesticated cats bury their excreta, usually in moist soil, they are a potential link in disease transmission. Cat litter and sandboxes, in particular, may be a source of infection for individuals who clean the box and inhale or ingest the cysts. Because toxoplasmosis is a disease of rodents, cats that eat infected mice can become infected in turn (Figure 29–5). A cat that is not permitted to roam or to eat raw meat is unlikely to acquire the infection.

While the cat is the main animal whose feces

Figure 29–5 *Toxoplasma gondii* infection within the small intestine of a cat. (a) *T. gondii* on the intestinal surface. (b) Portion of a villus showing many rupture sites through which the protozoan parasites have escaped. *[W. M. Hutchinson, R. M. Pittilo, S. J. Ball, and J. Chr. Siim,* Ann. Trop. Med. Physiol. ***74:****427–437, 1980.]*

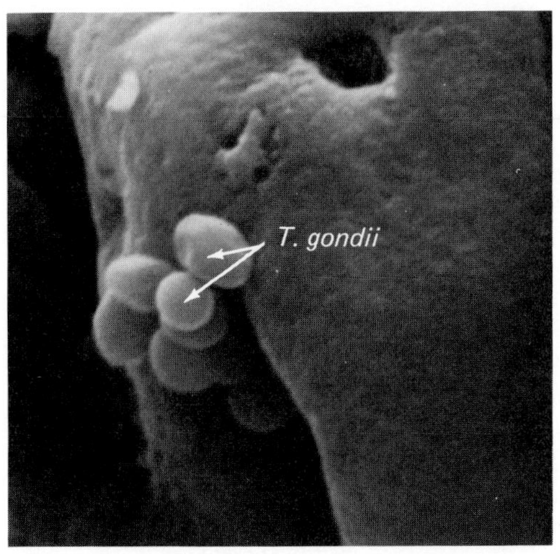

(a)

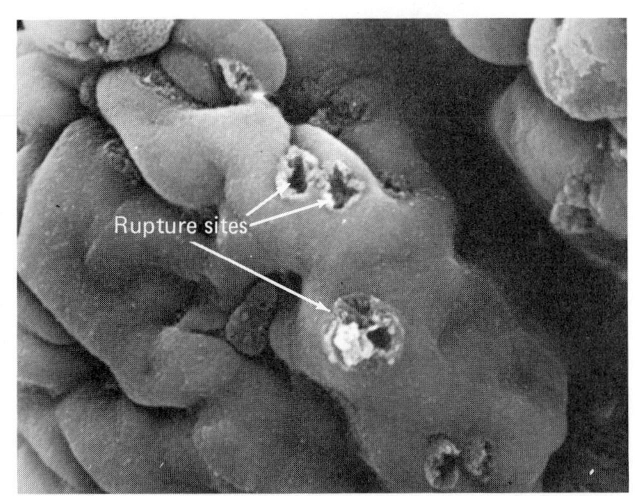

(b)

have been implicated in transmission, toxoplasma may be found in practically every mammal and in some birds. Therefore, ingestion of raw meat by humans or other animals can contribute to the spread of the disease agent (Figure 29–4). *T. gondii* is also found in soil and may be acquired through inhalation. Although the actual incidence is uncertain, it is now believed that many retarded persons as well as individuals with immune deficiences are victims of toxoplasmosis. In addition, subclinical cases of the disease occur in various parts of the United States.

PREVENTION. Infection in a pregnant woman represents the greatest public health problem by virtue of transmission to the developing fetus. Experimental studies have resulted in the finding of a destructive *toxofactor* (Figure 29–6). This substance interferes with fetal development.

The following suggestions are of particular importance to pregnant women as well as individuals with immune defects.

1. Avoid eating undercooked or raw meat.
2. If a cat is already a family pet, and if a litterbox or sandbox is used, have another person clean and handle it.
3. If a cat is to be adopted and kept as a family pet, keep it indoors and do not feed it raw meat products.
4. Avoid contact with wildlife, such as squirrels, exhibiting abnormal movements or behavior.

It should be noted that if serological testing shows that a woman had antibodies against *T. gondii* prior to becoming pregnant, the danger to the fetus is low.

AFRICAN SLEEPING SICKNESS

The effects of African sleeping sickness (Figure 29–7) were originally recognized in two separate geographical regions and were therefore considered to be two distinctly different diseases (Table 29–3). *Trypanosoma brucei* v. *gambiense* (try-pan-oh-SOH-mah BROO-see v. gam-bee-EN-zee), the causative agent of Gambian trypanosomiasis, was first observed in 1901. The discovery of *T. brucei* v. *rhodesiense,* the causative agent of Rhodesian trypanosomiasis, was reported some eight years later. *T. brucei* v. *gambiense* is widely distributed in the western and central portions of Africa, while *T. brucei* v. *rhodesiense* appears to be restricted almost entirely to the southeastern regions of the Africa continent. Neither form should be confused with the virus-induced sleeping sickness infections eastern and western equine encephalitis, which are found in the United States and other parts of the world.

TRANSMISSION AND LIFE CYCLE. Gambian trypanosomiasis is contracted by humans through the bite of the tsetse fly, *Glossina palpalis.* The insect vectors for *T. rhodesiense* (T. row-deez-ee-EN-zee), specifically *G. morsitans* and *G. pallidipes,* are closely related tsetse fly species (Figure 29–8).

Trypanosomes are taken into the tsetse fly during a blood meal from an infected host. The parasites pass into the insect's intestine and other regions, where they develop, mature, and eventually transform into infective units called *metacyclic trypanosomes.* This last form is similar to the stage of the parasite as it appears in blood (Color Photograph 90). The infected tsetse fly bites a susceptible host, and the cycle is thereby perpetuated.

Rhodesian trypanosomiasis presents very much

Figure 29–6 The effects of the toxofactor in *Toxoplasma gondii* infection. The normal animal is shown on the left, while the remaining animals show the interference on development caused by the toxofactor. *[From B. Grimwood, G. O'Connor, and H. A. Gaafar,* Infect. Immun. ***42**:1126–1135, 1983.]*

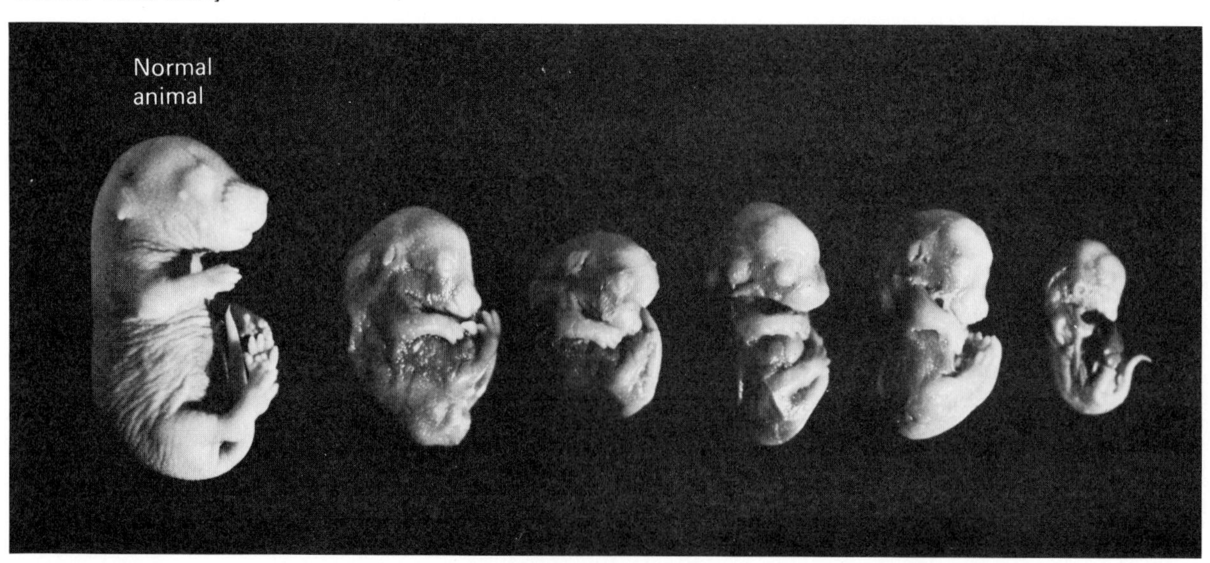

Figure 29–7 The ravages of African sleeping sickness. *[Courtesy of the World Health Organization, Geneva.]*

the same clinical picture as Gambian sleeping sickness (Table 29–3). However, the infection has a much shorter incubation period and a more rapid course, with death occurring within one year.

Prevention. Practical control measures for reducing the incidence of African sleeping sickness include (1) destruction of breeding sites of the tsetse flies, (2) diagnosis and treatment of the disease in patients, (3) quarantine of infected individuals, (4) wearing protective clothing against the tsetse flies, and (5) prophylactic drug administration, especially in areas where the risk of infection is great.

CHAGAS' DISEASE (AMERICAN TRYPANOSOMIASIS)

Chagas' disease, once considered rare and exotic, is now known to be a dangerous plague of the Americas. The most frequent signs of the disease involve the cardiovascular system. Involvement of the brain, liver, spleen, and spinal cord also occurs.

Trypanosoma cruzi (try-pan-oh-SOH-mah KRUZ-ee), the causative agent of Chagas' disease, was first discovered in 1908 by Carlos Chagas in the intestines of blood-sucking, winged reduviid bugs (Color Photograph 85) common in the huts of the Brazilian hinterland. Later, the parasites were found in the blood of domestic animals and of several hundred patients apparently suffering from the manifestations of *T. cruzi*. Unfortunately, it was not until much later that Chagas' disease was rediscovered

Figure 29–8 A tsetse fly, the vector of African sleeping sickness, perched on a fingertip. *[Courtesy of the World Health Organization, Geneva.]*

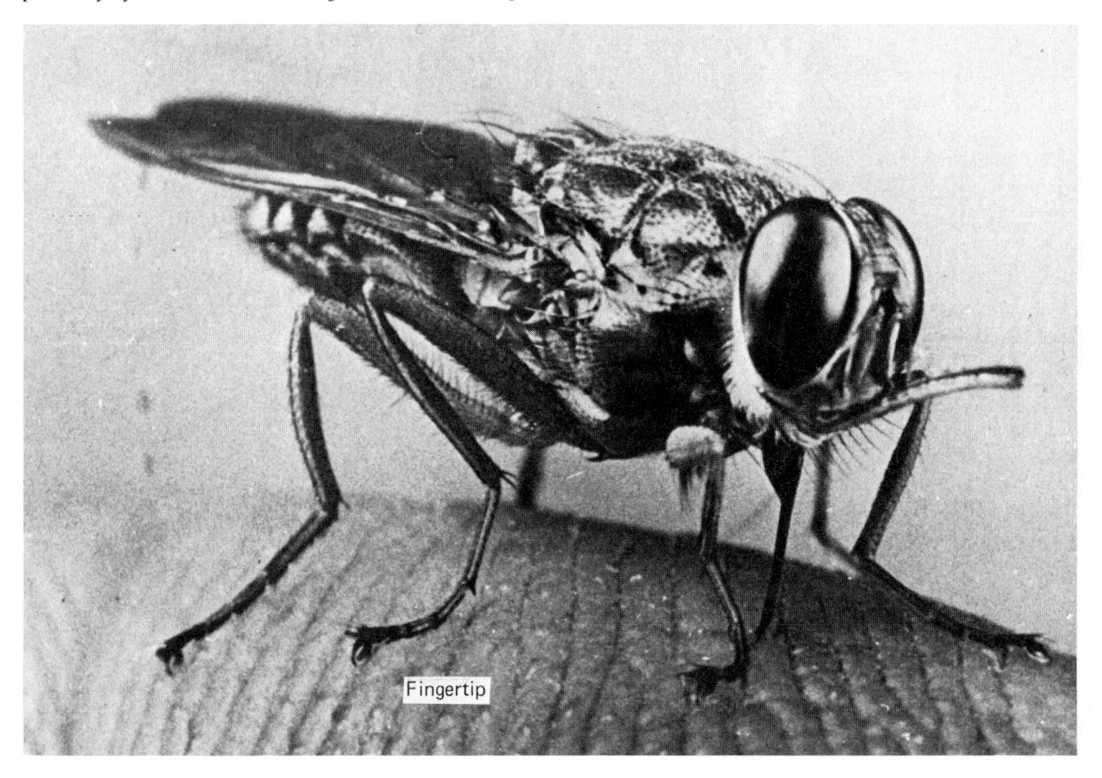

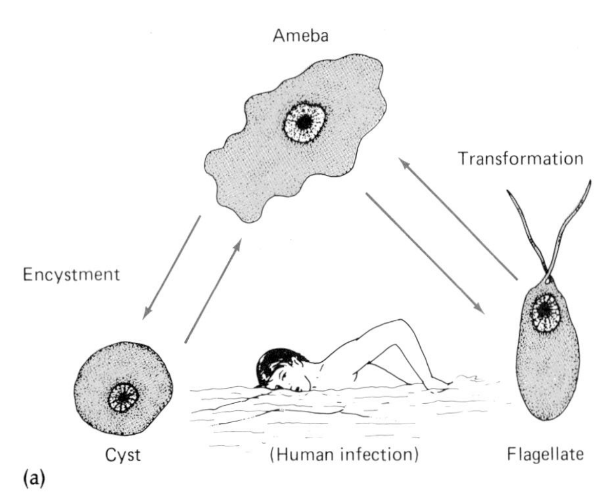

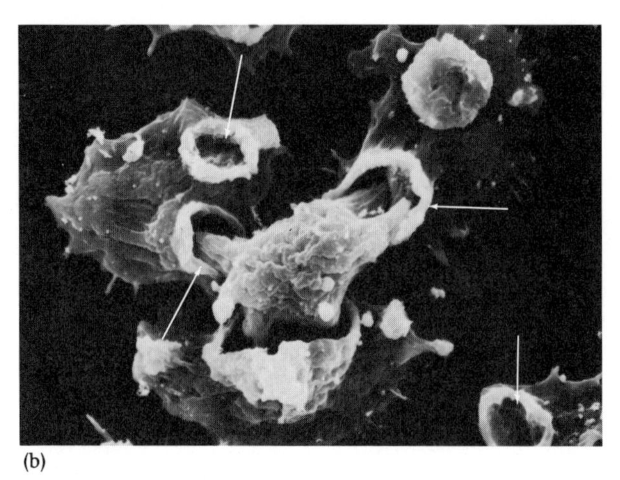

Figure 29–9 *Naegleria fowleri.* (a) The life cycle of the free-living amoeboflagellate. *[After D. T. John, in E. Markell, M. Voge, and D. T. John,* Medical Parasitology. *Philadelphia: W. B. Saunders, Inc., 1986.]* (b) The ameba stage showing the ameba's suckerlike amebastomes (arrows), which are used for engulfing other cells (phagocytosis). *[From D. T. John, T. B. Cole, Jr., and R. A. Bruner,* J. Protozool. ***32****:12–19, 1985.]*

and its importance recognized by Mazza and his colleagues, who recorded more than 1,000 acute cases by 1944. Chagas' disease is emphasized in the World Health Organization (WHO) Report of 1962, which referred to "an estimated minimum of 7 million infected individuals."

T. cruzi is known to have a geographic distribution extending from the southern United States through Mexico, Central America, and into South America to Argentina. Various wild rodents, opossums, and armadillos can harbor the parasite.

TRANSMISSION. The disease is acquired by contaminated feces of a reduviid bug (Color Photograph 85) dropped into a bite wound caused by the arthropod. After the bug has taken its blood meal, the involved area itches intensely, and scratching it moves the feces into the wound.

The parasite can also penetrate through the ocular conjunctiva. Blood transfusions from infected individuals are also an important means of disease transmission in endemic areas.

Infections are most common among the very young. Both sexes are equally involved.

PRIMARY AMEBIC MENINGOENCEPHALITIS (PAM)

Amebic infections of the CNS may be caused by the parasitic ameba *Entamoeba histolytica* or by the opportunistic free-living amebas *Acanthamoeba* (a-can-tha-ME-bah) species or *Naegleria fowleri* (nye-GLEH-ree-ah FAU-ler-eye). *E. histolytica,* known for its enteric infections, may produce a brain abscess after penetrating the gastrointestinal tract and spreading to the nervous system via the circulatory system. *Acanthamoeba* produces an amebic encephalitis usually in chronically ill or debilitated individuals, some of whom may be undergoing immunosuppressive therapy. *N. fowleri,* an ameba flagellate, is the only species known to cause *primary amebic meningoencephalitis* (Figure 29–9).

PAM, first detected in 1965 in Australia, is a very rapidly acting fatal human infection. Death usually occurs within 72 hours after the onset of symptoms (Table 29–3). Although PAM is a relatively rare disease, it has been reported worldwide. It is acquired from freshwater lakes or ponds, and occurs in healthy children or young adults. Infection follows ingestion of water containing ameba or flagellate stages of *N. fowleri* (Figure 29–9a). This protozoon has been isolated from a variety of environmental sources. However, there is no evidence for the existence of animal reservoirs or carriers. Currently, there is no satisfactory treatment. Because of the relationship between swimming and *Naegleria* infection, measures for control and prevention are directed toward educating the public, increasing the awareness of the medical community, and providing adequate chlorination (10 parts per million) for public swimming facilities.

The Eye

The eye consists of the eyeball and accessory structures, including eyebrows, eyelids, conjunctiva, and the *lacrimal* apparatus, which produces tears (Figure 29–10).

The eyeball consists of three concentric coverings that enclose the various transparent media through which light must pass in order to reach the photosensitive retina. The outermost covering protects the inner regions and gives form to the eyeball. This coat is made up of two regions, the transparent outer *cornea* and the *sclera* (white of the eye). Light rays pass through the cornea and enter the eyeball. The sclera, behind the cornea, is opaque and mainly protective in function.

The middle vascular covering, behind the cornea, is primarily nutritive in function. It is made up of the *chorioid proper,* the *ciliary body,* and the *iris.*

The third and innermost coat, the *retina,* lines the vascular covering. The retina contains several cell layers, including *visual cells (rods* and *cones), ganglionic cells* (portions of which form the optic nerve), and *bipolar cells,* which are involved in the visual pathway from the eye to the brain.

Figure 29–10 shows other parts of the eye, including the *anterior chamber* (which contains the *aqueous humor*), the *posterior chamber* (which holds the *vitreous body* or *humor*), and the crystalline, circular, transparent *lens.* The infections of the eye also are indicated on this figure.

The conjunctiva, a mucous membrane that lines the inner part of the eyelid, provides a thin cover for the forepart of the sclera, the inner surfaces of the eyelids, and the cornea. Tears lubricate the conjunctiva and keep it free from particulate matter.

The *lacrimal glands* are located above and to the side of the eyeball. Tears secreted by these glands lubricate the front of the eye and thereby prevent drying of the cornea and the development of friction between the eyelids and the eyeball. Tears have a protective function by virtue of the presence of the enzyme lysozyme. Lysozyme is bactericidal for certain saprophytic and pathogenic bacterial species.

The discussions of microbial-caused eye diseases will be limited mainly to bacterial and viral states, since other microorganisms are rarely encountered. Tables 29–4 and 29–5 list causative agents, general disease features, laboratory diagnostic approaches, and possible treatment for microbial diseases of the eye. For discussions of treatment and of diagnostic aspects, the reader is referred to the textbooks listed in the reference section of this chapter.

Flora of the Normal Conjunctiva

A small number of microorganisms can be isolated from the normal conjunctiva, including *Corynebacterium* spp., *Branhamella catarrhalis,* staphylococci, and

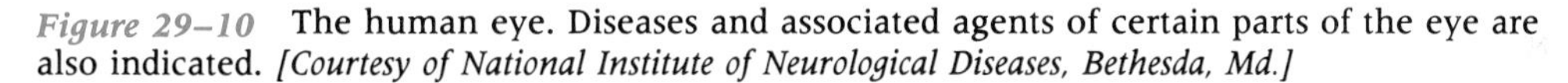

Figure 29–10 The human eye. Diseases and associated agents of certain parts of the eye are also indicated. *[Courtesy of National Institute of Neurological Diseases, Bethesda, Md.]*

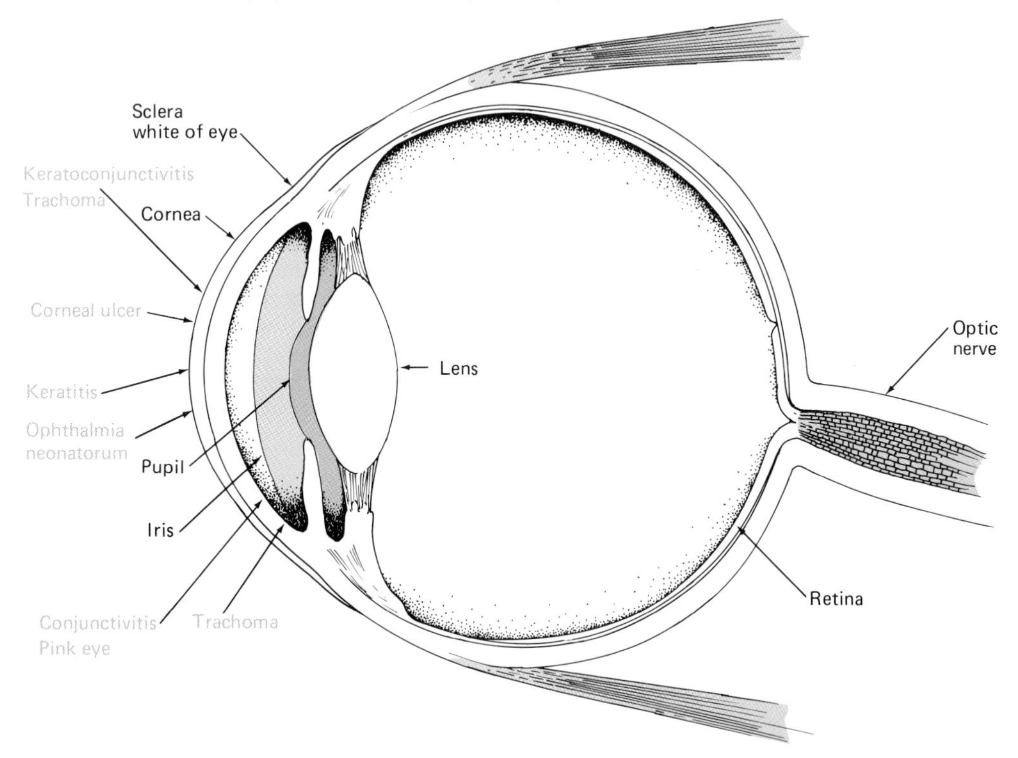

TABLE 29–4 Diseases of the Eye Caused by Bacteria

Disease	Causative Agent	Gram Reaction and Morphology	Incubation Period	General Features of the Disease
Conjunctivitis	*Neisseria gonorrhoeae* *Staphylococcus aureus* *Streptococcus pneumoniae* *Streptococcus* spp.	−, Coccus +, Coccus +, Coccus +, Coccus	24–27 hours	Irritation around eyelid, tearing, swelling, redness, and discharge from the involved eye (Color Photo graph 122)
Corneal ulcer	*Streptococcus pneumoniae*	+, Coccus	Unknown	Destruction of cornea, leading to scarring and cataract (formation of opaque area)
Inclusion conjunctivitis (newborn)	*Chlamydia trachomatis*	−, Rod	Within 36 hours after birth	Reddening of eyelid and pussy discharge; no scarring
Keratitis (inflammation of the cornea)[a]	*Moraxella lacunata* *Neisseria meningitidis* *Staphylococcus aureus* *Streptococcus mitis* *Streptococcus pneumoniae*	−, Rod −, Coccus +, Coccus +, Coccus +, Coccus	Unknown	Fever, headache, and swelling of tissues around eyes; pus formation and scarring can occur as complications; this condition can appear in combination with conjunctivitis
Ophthalmia neonatorum (conjunctivitis of the newborn)	*Neisseria gonorrhoeae* *Staphylococcus aureus*	−, Coccus +, Coccus	1–3 days	Effects include the involvement of both eyes, discharge consisting of blood and pus, swollen eyelids
Pink eye (acute mucopurulent conjunctivitis)	*Haemophilus aegyptius*[b]	−, Rod	1–4 days	Abundant discharge, redness, and extreme swelling of the eyelids; bleeding within the conjunctiva can occur
Trachoma	*Chlamydia trachomatis*	−, Rod	5–12 days	Typical effects include accumulation of blood vessels on the surface and penetration of cornea, tumor formation, scarring, and blindness

[a] The protozoon *Acanthamoeba* has been reported to cause keratitis through the contamination of contact lenses.
[b] This organism also causes a serious invasive infection.

streptococci. Although almost any pathogenic bacterial species can cause ocular infection, such organisms may be present without producing any destructive effects. Several factors play prominent roles in protecting the human eye from infection (Color Photograph 122), including the washing and bactericidal action of tears and the mechanical barrier of an intact mucous membrane.

Microbial Diseases of the Eye

Bacterial Diseases

Several bacteria are known for their specific attraction to certain ocular structures (Table 29–4). Some of them are limited only to certain regions of the eye (Color Photograph 122). *Neisseria gonorrhoeae* (nye-SEH-ree-ah gon-or-REE-ah) attacks the conjunctiva, but not the lacrimal apparatus; *Haemophilus aegyptius* (he-MAH-fi-lus eye-JIP-tee-us) and *H. influenzae,* (biotype III), is associated with the lining of the eyelid (conjunctiva) but not with the cornea. In 1984 the first cases of Brazilian purpuric fever (BPF) were recognized. A pus-associated conjunctivitis caused by *H. aegyptius* preceded BPF.

Furthermore, several bacterial species that do not cause primary invasion of the eye or associated appendages do produce secondary infections. Examples of such pathogens include *Brucella* (broo-SEL-lah) spp., *Escherichia coli,* and *Proteus* spp. Several organisms, such as *Bacillus anthracis* (bah-SIL-lus an-THRAY-sis), *Clostridium* spp., and *Mycobacterium tuberculosis,* which can produce infections of the eye, rarely do so.

In several types of ocular diseases, the practice of local eye hygiene may be totally effective without the use of any other type of treatment. Local hygiene involves the removal of crusts and scales from the margins of the eyelids. The use of hot water compresses is most helpful for this purpose. Before the application of ointments, crusts and scales should always be removed.

GONOCOCCAL CONJUNCTIVITIS

Neisseria gonorrhoeae may cause two types of conjunctival involvement—ophthalmia neonatorum (gonococcal conjunctivitis of the newborn) and conjunctivitis of the adult (Table 29–4). Complications of these infections include corneal ulceration (keratitis), scarring, and blindness.

OPHTHALMIA NEONATORUM. Ophthalmia neonatorum develops as a consequence of the passage of the fetus through the birth canal of an infected mother. Usually, both eyes of the newborn infant are involved (Table 29–4).

GONOCOCCAL CONJUNCTIVITIS IN ADULTS. In adults, gonococcal conjunctivitis is similar to that of newborns, but there is likely to be greater involvement of the conjunctiva and cornea. Adults usually contract the disease by carrying organisms to the eye from a genitourinary infection. [See gonorrhea in Chapter 28.]

PINKEYE

Pinkeye is a bacterial disease caused by *Haemophilus aegyptius,* also known as *Koch-Weeks bacillus.* Pinkeye is a highly contagious variety of acute catarrhal conjunctivitis. The infection occurs more commonly during the warmer months of the year.

Symptoms of pinkeye infection include a copious discharge, redness and extreme swelling of the eyelids, and subconjunctival hemorrhages.

Several reports have implicated *H. aegyptius* as the cause of the serious illness Brazilian purpuric fever. The illness begins with a purulent (pus-associated) conjunctivitis and progresses to fever and muscle aches and pains. In untreated cases, skin hemorrhages, shock, and an overwhelming toxic state

Gonococcal Conjunctivitis

LABORATORY IDENTIFICATION

N. gonorrhoeae can be isolated by placing specimens on enriched media such as chocolate agar and Thayer-Martin medium, and incubation in an environment with increased CO_2. All members of *Neisseria* are cytochrome-oxidase-positive. Differentiation of *N. gonorrhoeae* from other species is based on its production of acid from glucose and a positive Superoxol test, which uses 30% hydrogen peroxide to show the presence of the enzyme catalase.

TREATMENT

Chemotherapy for *N. gonorrhoeae* infections varies somewhat depending on the presence or absence of complications.

As noted in Chapter 28, preventive measures consist of the administration of erythromycin or of a 1% silver nitrate solution (Crede's solution) into each thoroughly cleansed eye of the newborn infant. This procedure is applied to every infant regardless of whether or not gonorrhea infection of the mother is suspected. Erythromycin also is effective against ophthalmia neonatorum caused by *Chlamydia trachomatis.*

Pinkeye

Laboratory Identification

H. aegyptius can be isolated from specimens on blood agar enriched with X (hemin) and V (nicotinamide adesine dinucleotide) factors. This species is nonhemolytic, susceptible to the lytic action of bile salts, and reduces nitrates to nitrites.

Treatment

Several antibiotics are used for *H. aegyptius* infections. These include ampicillin and chloramphenicol. For ampicillin-resistant strains, a streptomycin-sulfisoxazole combination may be used.

ending in death occur. The clinical features of the disease are similar to those of meningococcal meningitis.

MORAXELLA LACUNATA (MORAX-AXENFELD DIPLOBACILLUS) INFECTIONS

Moraxella lacunata (more-ax-EL-ah la-KOON-at-ah), a gram-negative rod, has particular importance for the ophthalmologist, as it seems to cause eye disease involving only the cornea and conjunctiva. Infections primarily affect adults, although there have been reports of the involvement of newborns. Hot, dusty environments are closely associated with these disease states.

STREPTOCOCCAL INFECTIONS

Streptococcus pneumoniae and other species, in addition to causing upper respiratory tract infections, are known for their ability to produce severe diseases of the eye, including conjunctivitis, corneal ulcer, keratitis, and postoperative infections of the eye.

STAPHYLOCOCCAL INFECTIONS

Staphylococci are considered to be the most common causative agents of eye infections. Of the various eye diseases caused by staphylococci, inflammation of the eyelids and glands (blepharitis) is the most common.

In addition to blepharitis, these gram-positive cocci cause several other diseases, including conjunctivitis in adults and newborns (staphylococcal ophthalmia neonatorum), sties, keratitis, and postoperative infections.

Moraxella Infection

Laboratory Identification

Moraxella species can be isolated from specimens on enriched media. These organisms are oxidase-positive and biochemically inactive.

Treatment

Several *Moraxella* species are generally susceptible to antibiotics such as penicillin, tetracycline, aminoglycosides, and erythromycin.

Streptococcal Infection

LABORATORY IDENTIFICATION

A definitive identification of *S. pneumoniae* is generally based on the isolation of gram-positive, alpha-hemolytic streptococci on blood agar (Color Photograph 23a). *S. pneumoniae* can be differentiated from other alpha-hemolytic streptococci by its destruction upon exposure to bile or bile salts and inhibition of growth by optochin (Color Photograph 89). The capsule swelling, or Quellung reaction, with specific antisera and other serological tests is available for specific identification.

S. mitis (S. MY-tis) is alpha-hemolytic, and resistant to the actions of bile salts and optochin (Color Photograph 89). Commercial, modified agglutination tests also are used for identification.

TREATMENT

Penicillin is generally the drug of choice.

Chlamydial Infections

TRACHOMA

The contagious keratoconjunctivitis of trachoma is prevalent in various semitropical and tropical parts of the world, including Africa, the Far East, the Middle East, and South America. The disease also occurs in certain regions of North America (such as the southwestern United States), as well as in certain European countries. Although trachoma does not kill, it is the largest single cause of preventable blindness (Figure 29–11). WHO has estimated that nearly 400 million people suffer from this disease. In certain rural portions of northern India, for example, the infection rate may be as high as 90% of the population. The incidence of trachoma is closely associated with poor sanitation and high population density. Transmission of the disease agents occurs by direct contact. Flies are important mechanical vectors for *Chlamydia trachomatis*.

IMMUNITY. Successful recovery from trachoma does not provide a guarantee against additional infections. Apparently, immunity toward trachoma does not develop as a consequence of a clinical in-

Staphylococcal Infection

LABORATORY IDENTIFICATION

The procedures for isolating and identifying *S. aureus* are the same as those described earlier in the Brain Abscess section in this chapter.

TREATMENT

Several forms of penicillin and tetracycline are used for nonpenicillinase-producing *S. aureus* strains. Alternate antibiotics, especially for patients with penicillin allergy, include cephalosporin, clindamycin, and vancomycin.

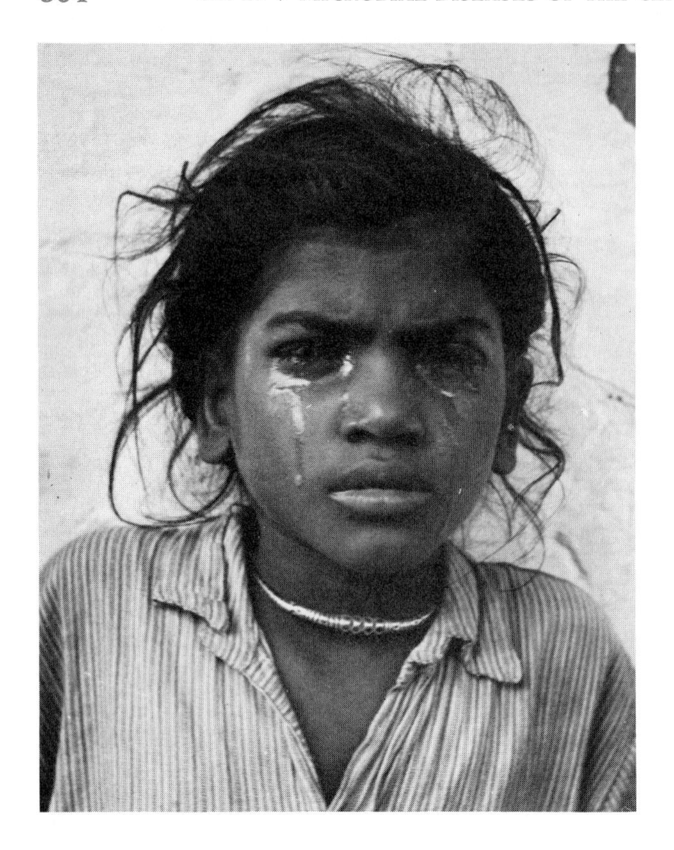

Figure 29–11 A victim of trachoma. A secondary bacterial infection is also present. *[Courtesy of the World Health Organization, Geneva. Photo by Homer Page.]*

fection. The development of an effective vaccine to provide some degree of protection is now being sought.

INCLUSION CONJUNCTIVITIS

Strains of *C. trachomatis* (kla-MEH-dee-ah trah-KOH-mah-tiss) also cause clinical diseases in the genital tract and in conjunctiva. The designation *inclusion blennorrhea* applies to genital infections such as infections of the cervix and urethra, while *inclusion conjunctivitis* designates conjunctival inflammations caused by this organism.

Females harboring *C. trachomatis* may not exhibit any signs of infection. Mild vaginal discharges may occasionally be observed. Males acquiring the pathogenic agent usually produce a discharge, which may be a chronic or intermittent manifestation.

Inclusion conjunctivitis infection is contracted by babies on passage through an infected cervix. In the case of older individuals, the disease can be acquired if an infectious genital tract discharge reaches the conjunctiva or by direct or indirect contact with swimming pool patrons. This type of infection, which is frequently referred to as *swimming pool conjunctivitis,* is quite mild and may heal without treatment. Scarring usually does not occur.

Eye Infections Caused by Viruses

Viruses cause some of the most common and destructive eye diseases of humans (Table 29–5). Selected features of some of these pathogens and the infections they cause are included here.

EPIDEMIC KERATOCONJUNCTIVITIS (EKC)

Human adenoviruses have a worldwide distribution. In general, these agents produce mild infections. Occasionally, involvement of the conjunctiva occurs as a consequence of direct contact with respiratory and ocular secretions. Reports have also implicated swimming pools in epidemics of conjunctivitis and sore throats with fever. EKC is commonly caused by adenovirus type 8, and appears to be the result of traumatic injury associated with dirt and dust in factories and shipyards. In addition, ophthalmologists,

C. trachomatis Infection

Laboratory Identification

Cell culture systems are available for the isolation of *C. trachomatis* from clinical specimens. The finding of intracytoplasmic inclusions in infected cells or demonstration of the pathogen by fluorescent antibody are important aids in laboratory diagnosis.

Treatment

Tetracylines are the drugs of choice for *C. trachomatis* infections. Erythromycin is used as an alternate (Color Photograph 17).

TABLE 29–5 Common Eye Diseases Caused by Viruses

Disease[a]	*Causative Agent (Common Designation)*	*Possible Treatment*
Conjunctivitis	Newcastle disease virus	Symptomatic
Conjunctivitis, keratitis, or a combination of both conditions	Molluscum contagiosum virus *(Molluscovirus hominis)*	Symptomatic
Keratitis (associated with recurrent herpes); herpetic conjunctivitis (commonly found in children); purulent conjunctivitis; herpetic keratitis (corneal herpes)	Herpes simplex virus type 1	5-iodo-2-deoxyuridine and other DNA synthesis inhibitors
Epidemic keratoconjunctivitis (EKC)	Adenovirus, types 7 and 8	Symptomatic
Non-pus-forming conjunctivitis	Measles virus, Rhinovirus	Symptomatic

[a] Refer to Table 29–4 for descriptions of these conditions.

optometrists, and nurses may also transmit the disease agents with contaminated or improperly sterilized instruments.

Prevention. Control of epidemic keratoconjunctivitis (EKC) mainly involves improvement of working conditions in factories and the incorporation of suitable measures to prevent the spread of the disease during the examination and treatment of patients. An inactivated type 8 vaccine has been effectively used in Japan to prevent this adenovirus infection.

Disease Challenge

The situation described has been taken from an actual case history. A review of treatment and epidemiological aspects is provided at the end of the presentation. Answers to questions, laboratory findings, and interpretations are given immediately following a specific question. Test your skill and take the Disease Challenge.

Key Terms

abscess **(AB-ses): A localized collection of pus.**

aspiration **(as-pi-RAY-shun): Removal by suction.**

computerized axial tomography (CAT) scan: **A noninvasive technique using a radiographic beam to sweep the body in a crosswise direction to show the relationship of body structures and to detect abnormalities.**

parietal **(pa-RYE-e-tal): Refers to the sides of the brain under the bones forming the top of the skull.**

pyogenic **(pie-oh-JEN-ik): Producing pus.**

(continues)

Disease Challenge (continued)

CASE

A 27-year-old male who had been behaving quite abnormally was brought to the hospital by a family friend. The patient had a long history of drug abuse, which included the use of cocaine. A computerized axial tomography (CAT) scan of the brain showed a left parietal mass, which was suspected of being a pyogenic abscess.

At This Point, What Type(s) of Laboratory Specimens Should Be Taken?
Material from the abscess should be taken, generally by aspiration for direct microscopic examination and bacterial and yeast culture.

Laboratory Results:
Microscopic examination of India-ink preparations revealed the view shown.

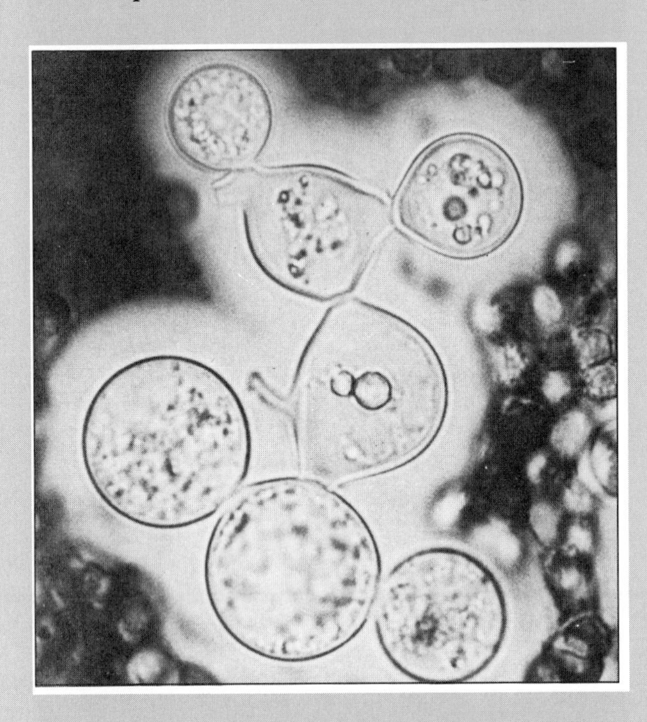

India-ink preparation of specimen. *[From G. I. Love, G. D. Boyd, and D. L. Greer,* J. Clin. Microbiol. *22:1068–1070, 1985.]*

On the Basis of This Finding What Disease or Agent Would You Suspect?
The yeast *Cryptococcus* (kryp-toe-KOK-kuss) species.

Additional Laboratory Findings
Cultures of brain-aspirate fluid were negative for both aerobic and anaerobic bacteria. Seventy-two-hour blood agar cultures incubated at room temperature showed the presence of yeast colonies and cells typical of *Cryptococcus* species.

What Treatment Would Be Appropriate?
Amphotericin B.

Case Follow-up:
Unfortunately, the patient's condition worsened and he died on the thirty-sixth day of hospitalization.

HERPES CORNEAE

Herpes simplex virus type 1 is well known for its ability to cause a wide variety of lesions in humans. One particular group of clinical manifestations associated with the skin or mucous membranes includes a unilateral ulcer on the conjunctiva or cornea. This ocular infection, referred to as *corneal herpes,* may constitute an initial (primary) infection. A recurrent type is the most commonly encountered type of eye infection caused by herpes simplex virus type 1; this particular infection develops in the presence of circulating antibodies. **[See herpes simplex virus type 1 infections in Chapter 23.]**

Summary

Organization of the Central Nervous System (CNS)

1. The coordination of body processes is an important function of the CNS.
2. The CNS consists of the brain and spinal cord. The brain is encased within the skull and is continuous with the spinal cord, which is surrounded by vertebrae.
3. The brain and spinal cord are hollow, contain cerebrospinal fluid, and are covered by three meninges, or membranes.

Diseases of the Central Nervous System

ROUTES OF CNS INFECTION

1. The majority of CNS infections result from microbial invasion of the bloodstream.
2. Other routes include disease processes close to or continuous with the CNS. Congenital anatomic defects and surgical manipulations also provide access to the CNS.

MICROBIAL AGENTS OF DISEASE

1. Infections and disorders of the CNS are caused by a variety of microorganisms. Typical symptoms of such infections are fever, general weakness, headache, and a stiff neck.
2. Major microbial diseases of the brain and spinal cord are abscesses, encephalitis, and meningitis.

LABORATORY DIAGNOSIS

1. Both culture and nonculture methods are used to identify causative agents.
2. ELISA and chemical methods are among the newer approaches.

BACTERIAL DISEASES OF THE NERVOUS SYSTEM

1. *Brain abscesses* are inflammations of nervous tissues caused by pus-forming (pyogenic) aerobic and anaerobic microbes. Most brain abscesses are complications that develop from chronic (old) locations of microorganisms in other regions of the body.
2. *Meningitis* is an inflammation of the various protective membranes around the brain and spinal cord. The most rapidly acting form of the disease is caused by *Neisseria meningitidis.* Infections with these meningococci frequently develop in closed environments such as jails, military posts, schools, and ships.
3. *Listeria monocytogenes,* another cause of meningitis, occurs in immunocompromised individuals. Infections may also be acquired through contact with contaminated soil and through the ingestion of contaminated foods.
4. Control measures include immunization (if available) and adequate treatment.

VIRAL INFECTIONS OF THE NERVOUS SYSTEM

1. Viruses causing infections of the CNS are called *neurotropic.* Examples of such infections include aseptic (nonbacterial) meningitis caused by pathogens including picornaviruses (coxsackieviruses, echoviruses, and polioviruses); rabies (hydrophobia); and slow virus diseases.
2. Slow chronic and latent virus diseases are associated with conventional viruses and less well-defined transmissible agents called *prions.*
3. Examples of slow virus infections include lymphocytic choriomeningitis, progressive multifocal choriomeningitis, and subacute sclerosing panencephalitis. Prion-associated conditions include the human diseases kuru and Creutzfeldt-Jakob disease, as well as the lower animal diseases, a brain disease of mink and scrapie of sheep.
4. Arboviruses, which replicate in and are spread by various types of arthropods, can cause extensive nervous tissue injury and encephalitis. The majority of these pathogens are known as *togaviruses* or *bunyaviruses.*

PROTOZOAN INFECTIONS OF THE CNS

1. Protozoan infections of the CNS include toxoplasmosis, African sleeping sickness, Chagas' disease, and primary amebic meningoencephalitis.
2. Toxoplasmosis, caused by *Toxoplasma gondii,* can result in brain damage in newborns that were infected while in the uterus. Other effects include blindness and mental retardation. Sources of the disease agent include infected cats and contaminated raw meat and soil.
3. African sleeping sickness should not be confused with viral forms of encephalitis. This disease is caused by *Trypanosoma brucei gambiense* and *T. brucei rhodesiense,* which are spread by tsetse flies.
4. Chagas' disease is caused by another trypanosome, *T. cruzi.* Its geographic range extends from the southern portions of the United States through Mexico and into South America. Chagas' disease frequently affects part of the cardiovascular system and the CNS and is transmitted by reduviid bugs.
5. Amebic infections of the CNS include brain abscess caused by *Entamoeba histolytica,* amebic encephalitis caused by *Acanthamoeba* species, and the rapidly fatal primary amebic meningoencephalitis (PAM), caused by *Naegleria fowleri.* PAM is acquired by swimming in freshwater lakes or ponds containing *N. fowleri.*

The Eye

The eye consists of the eyeball and accessory structures, including the eyebrows, eyelids, conjunctiva (inner part of the eyelid), and the lacrimal apparatus, which produces tears.

Flora of the Normal Conjunctiva

A small number of microorganisms can be isolated from the normal conjunctiva. The bactericidal action of lysozyme in tears, the flushing mechanism of tears, and the intact membranes and coverings of the eye protect against a large number of diseases.

Microbial Diseases of the Eye

BACTERIAL DISEASES

1. Bacterial diseases of the eye include conjunctivitis, keratitis, pinkeye, and trachoma. Brazilian purpuric fever is associated with *Haemophilus aegyptius*, the cause of pinkeye.
2. The destructive effects of these diseases can be prevented by adequate treatment with antibiotics.

EYE INFECTIONS CAUSED BY VIRUSES

Examples of virus infections of the eye include epidemic keratoconjunctivitis (EKC) (caused by adenovirus type 8) and herpes corneae (caused by herpes simplex virus type 1).

Questions for Review

1. Discuss five viral diseases of the CNS, including the causative agent, mode of transmission, specific structures of the system involved, diagnostic procedures used, and treatment.
2. What measures are currently available for the control of viral agents affecting the CNS?
3. What are slow virus infections? What significance do they have?
4. a. What are prions?
   b. What is the relationship of prions to viruses? (Refer to Chapter 10).
5. List the causative agents and the means of disease transmission for the infections given below:
   a. trachoma
   b. adult gonococcal conjunctivitis
   c. EKC
   d. angular blepharoconjunctivitis
   e. pinkeye
   f. central corneal ulcers
   g. inclusion conjunctivitis
   h. sty
   i. trachoma
   j. western equine encephalitis
   k. listeriosis
6. What is Brazilian purpuric fever?
7. What types of organisms comprise the normal flora of the conjunctiva? Are any of these normally pathogenic?
8. What are the distinguishing features of the arboviruses?
9. Which of the arthropod-borne diseases mentioned in this chapter are found in your immediate geographical area? What measures are utilized to control mosquito populations in your area?
10. Sleeping sickness (encephalitis) can be caused by several microbial types. Name and compare at least six causative agents in terms of (a) transmission, (b) geographical distribution, (c) preventive and control measures, and (d) significance as a world problem.
11. If you were planning a trip to Arabia, Egypt, Greece, Japan, and India, to which viral, bacterial, and protozoan diseases affecting the CNS would you be exposed? Refer to other chapters in this part.
12. Which of the protozoan infections mentioned in this chapter are known to occur in the United States? In your state or region of the country?
13. What is toxoplasmosis? Discuss this disease from the standpoint of occurrence, life cycle, and the types of individuals susceptible to its effects.

Suggested Readings

GAJDUSEK, C. D., "Unconventional Viruses and the Origin and Disappearance of Kuru," *Science* **197**:943–960, 1977. *An interesting description of a slow virus infection and the discovery of its mechanism of action and transmission. The article is written by a Nobel Prize recipient.*

GERSTER, G., "Tsetse—The Deadly Fly," *National Geographic* **170**:814–832, 1986. *A well-illustrated article showing the general features of this most important transmitter of African sleeping sickness.*

JONES, D. B., T. J. LIESEGANG, and N. M. ROBINSON, "Laboratory Diagnosis of Ocular Lesions," *Cumitech 13.* Washington, D.C.: American Society for Microbiology, 1981. *A functional, short publication dealing with specimen collection and the use and interpretation of various laboratory procedures for infectious diseases of the eye.*

KRISTENSON, K, and E. NORRBY, "Persistence of RNA Viruses in the Central Nervous System," *Annual Review of Microbiology* **40**:159–185, 1986. *A well-written article describing the central nervous system as one target for several persistent virus infections.*

KURSTAK, E., Z. J. LIPOWSKI, and P. V. MOROZOV (eds.), *Viruses, Immunity, and Mental Disorders.* New York: Plenum Publishing Corporation, 1987. *An interesting publication that not only deals with viral infections of the central nervous system but also explores the role of viruses in mental disorders such as schizophrenia.*

ROSEN, L., "The Natural History of Japanese Encephalitis Virus," *Annual Review of Microbiology* **40**:395–414, 1986. *An interesting article describing the features of one of the most widespread forms of viral encephalitis in humans.*

THOMPSON, R. A., and J. R. GREEN, *Advances in Neurology,* Vol. 6: *Infectious Diseases of the Central Nervous System.* New York: Raven Press, 1974. *A functional summary of infectious diseases of the fetus, children, and adults.*

30

Microbiology and Cancer

What checks the natural tendency of each species to increase in number is most obscure.
— Charles Darwin

After reading this chapter, you should be able to:

1. Distinguish between tumors and malignancies.
2. Describe the four different forms of human cancer.
3. Explain viral transformation.
4. Summarize the properties of oncogenic RNA viruses and oncogenic DNA viruses.
5. Discuss the role of microorganisms and their products in causing cancer.
6. Describe the use of microorganisms and immunological procedures in the detection of carcinogens.
7. Outline and explain the current cancer virus hypothesis.
8. Discuss the use of microorganisms or their products and monoclonal antibodies in the treatment of cancerous states.
9. Define or explain: tumor antigen, benign, malignant, topoinhibition, BRMs, HTLV-1, fetal proteins, Epstein-Barr virus, Ames test, transformation, LAK cells, and oncogenes.

Of all the diseases that afflict us, those labeled "cancer" produce the strongest emotions. This disease state is greatly and justifiably feared by many. What are cancers? Do microorganisms play a role in cancer causation? Chapter 30 describes the different types of malignancies and outlines the relationship of microorganisms and their products to cancer. The use of microorganisms, their products, and monoclonal antibodies for the detection and control of cancer-causing substances are considered as well.

Although some animal cancers are known to be virus caused, it is widely assumed that cancers are caused by environmental factors such as chemicals and radiation, as well as certain microbial products. Despite extensive and increasingly sophisticated studies dealing with both experimentally and naturally occurring human cancers, the various causes of cancer, the role of microorganisms and their products, and the mechanisms operating in these disease states are not well understood. However, advances in immunology, such as the discovery of monoclonal antibodies, have provided valuable investigative and diagnostic tools. Possible roles for such antibodies in direct cancer treatment are under serious consideration; these may be the long-sought "magic bullets." **[See monoclonal antibodies in Chapter 19.]**

General Characteristics of Cancerous States

All the tissues of the human body are composed of individual cells of varying microscopic size. These cells must be able to divide and reproduce in an exact and orderly manner for the well-being of the body to be maintained. Normal cell growth is inhibited by crowding. When sufficient cells have been produced in a given area, intercellular controls operate to stop further growth, a property called *contact inhibition. Topoinhibition* (from the Greek word meaning "place") has been proposed as a replacement for this term, since it is not clear if the inhibition of cellular division is caused by intercellular contact or simply by the closeness of cells to one another. If cellular reproduction becomes uncontrolled and topoinhibition is lost, then normal layering is replaced by growths or swellings, called *neoplasms* or *tumors.* Tumors are classified by their general pattern of growth. Those that form in a localized area and do not spread to other parts of the body are referred to as *benign.* Tumors that reproduce progressively and unrestrictedly, spreading to other body regions, are called *malignant* or *cancerous.* Cancer cells spread from one part of the body to others either by invading and destroying normal tissue around the malignancy (a direct extension process) or by *metastasis,* the separation of cells from the major portion of a tumor and the movement of these cells to new sites. Tumors, both benign and malignant, are usually named by adding the suffix *-oma* to a term characteristic of their microscopic appearance and tissue origin. For example, *hepatomas* develop in liver cells or hepatocytes.

Cancers have been recognized by physicians for several centuries. In ancient times they were relatively rare, and they remained a fairly insignificant cause of death until the present century. Today, however, cancers are the cause of approximately 16% of all deaths in the United States and more in many European countries. This great increase, especially in the past few decades, is the result of improved detection procedures, environmental causes, and the increase in the human life span. The effects of environmental agents on human health is currently a subject of considerable interest. Of special concern to scientists are the long-term effects of chemical, physical, and microbial agents and, more specifically, their possible *oncogenic* (tumor-causing) and *carcinogenic* (cancer-producing) effects. Many factors have been identified as modifying cellular nucleic acids and possessing carcinogenic potential.

A given tumor arises from certain normal cells of one type (its tissue of origin) that have undergone a series of changes influenced by intrinsic or extrinsic factors (Figure 30–1). *Intrinsic* or internal factors are associated with those events that subject a cell or a body containing such cells to change. Among these are age, heredity, sex, hormonal influences, and a natural predisposition to tissue overgrowth. *Extrinsic,* or external, factors are agents that originate outside the body, such as chemicals, irradiation, and possibly viruses.

Three major characteristics seem to define cancerous states: **anaplasia, hyperplasia,** and **metastasis.** *Anaplasia* is a structural abnormality in which involved cells do not mature and therefore resemble primitive or embryonic cells. *Hyperplasia* is an uncontrolled reproduction of cells. Because malignant cells do not respond to a host's signal to stop dividing, they form a localized accumulation of tissue. In *metastasis,* a malignant cell detaches itself from a tumor and moves to another anatomical site, where it establishes a new tumor.

It is interesting to note that cancers are not in themselves fatal. In general, cancerous growths do not produce toxins or otherwise kill hosts directly. Cancers create a condition of malnutrition by using nutrients needed by the tissues of the host. The malnutrition state produces a generalized loss of weight and poor state of health called *cachexia.*

Microbiology has had a great impact on several aspects of present-day cancer research. For example, certain microbial products have been shown to in-

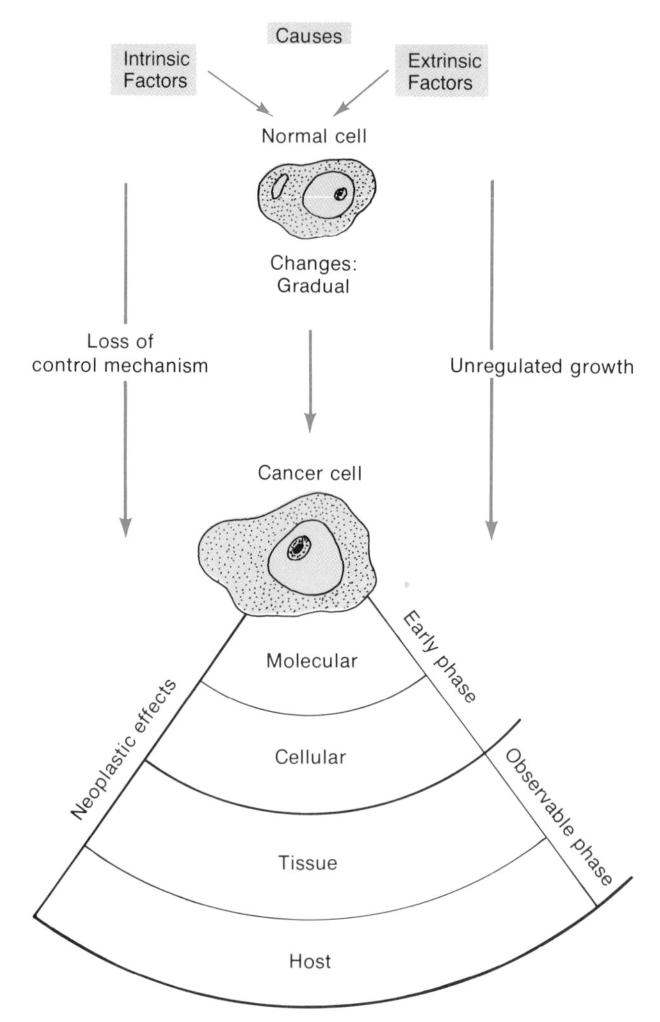

Figure 30–1 Stages believed to take place in carcinogenesis. A normal cell interacts with the causative factors in the environment. Through a gradual process, the normal cell is transformed into a cancerous one. With the establishment of the cancerous state, the stage is set for the development of tumor cell masses and the appearance of symptoms in the host.

hibit cancer. Others have helped to stimulate the immune response to a limited degree. Moreover, recent progress in our understanding of tumor growth and prevention has come from work with microorganisms and their interactions with their hosts. This chapter describes some aspects of tumor production and malignancy with which microorganisms have been associated.

Forms of Human Cancer

It is not yet clear whether the condition referred to as *cancer* consists of several diseases having a common pattern of general symptoms or whether it is a single disease that occurs in many forms, depending upon the tissue from which it evolves. At present, the first view is more widely accepted. In any case, more than 100 clinically distinct types of cancer are recognized, each having a unique set of specific symptoms and requiring a specific course of therapy.

Several cancers are named according to the cell types in which they develop. For example, *melanomas* (Figure 30–3a) arise in skin-pigment-producing cells, or melanocytes, and *lymphomas* develop in lymphocytes. Most cancers can be grouped into four major categories: carcinomas, leukemias, lymphomas, and sarcomas.

Carcinomas

Carcinomas are solid tumors formed by layers of cells and derived from epithelial tissues such as those found in breasts, glands, skin, nerves, and the linings of the gastrointestinal, genital, respiratory, and urinary systems.

Leukemias

Leukemia, also called *cancer of the blood,* is characterized by the uncontrolled proliferation and accumulation of leukocytes, most of which do not mature into functional cells. Just as there are many different types of white cells, there are many different types of leukemia. These include (1) acute and chronic lymphocytic leukemias (lymphoblastic leukemias; Figure 30–2), which are malignancies of lymphocytes, and (2) acute and chronic myelocytic

Figure 30–2 A scanning micrograph of leukemic lymphoblasts. *[U. Haemmerli, and H. Felix,* Blood Cells *2:415–430, 1976.]*

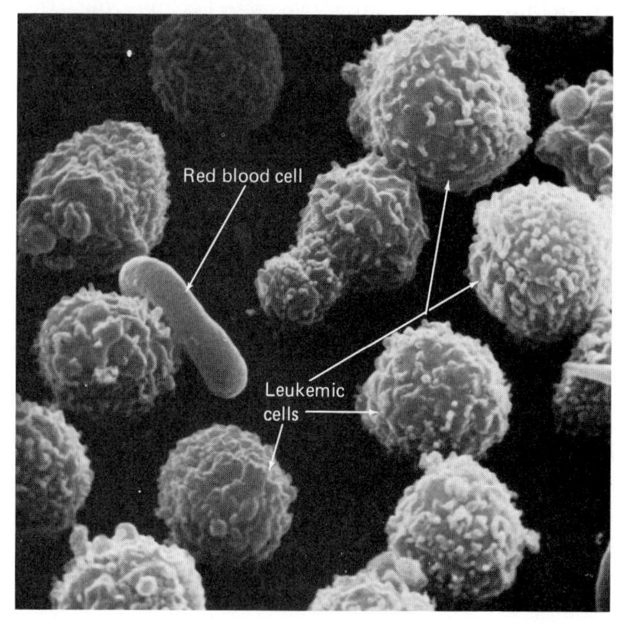

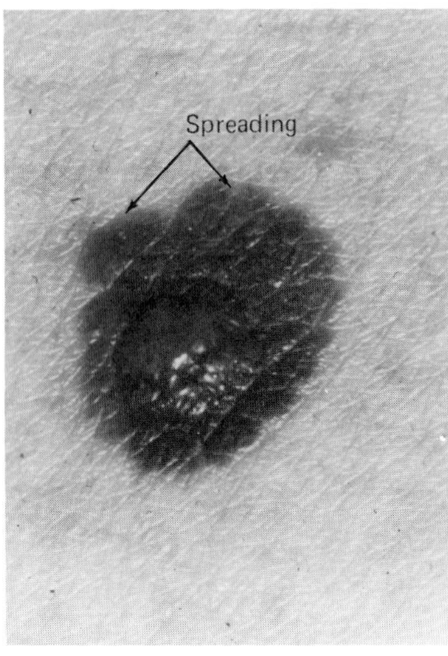

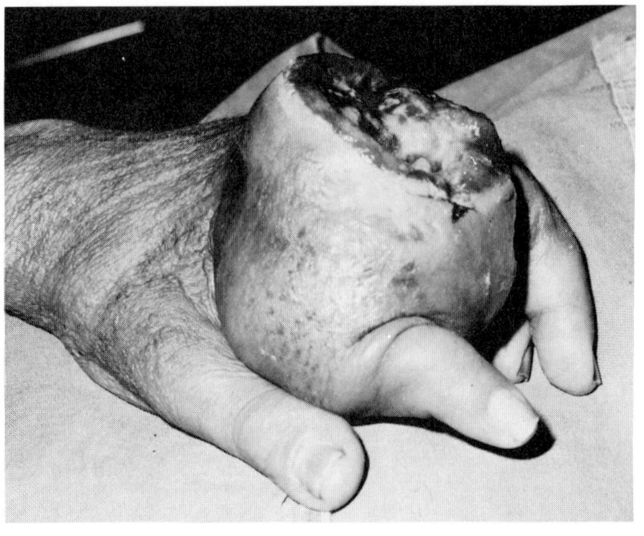

(a) (b)

Figure 30–3 Examples of cancers. (a) The appearance of a melanoma, showing a spreading pattern. *[From J. S. Lederman, T. B. Fitzpatrick, and A. J. Sober,* Arch. Dermatol. ***120****:1449, 1984.]* (b) Tumor in the bones of the hand. Because the bones of the hand are so near the surface, certain tumors of the hand extending beyond the bone tend to break through the skin. *[L. M. Vistnes, and W. J. Vermuelen,* J. Bone Joint Surg. ***57A****:865–867, 1975.]*

leukemias (granulocytic or myelogenous leukemias), which are disorders of granulocytes. **[See leukocytes in Chapter 17.]**

Acute leukemias usually appear suddenly, with symptoms similar to those of a cold, and progress rapidly. The lymph nodes, spleen, and liver may become infiltrated with leukocytes and enlarged. Symptoms frequently include bone pain, paleness, a tendency to bleed easily, and a high susceptibility to infectious diseases. The most common causes of death, which occurs if the leukemia is not treated, are bleeding and uncontrolled infections.

The chronic or long-lasting leukemias begin much more slowly, and several years may pass before significant symptoms appear. The symptoms are similar to those of the acute leukemias. The life expectancy of an individual who goes without treatment, however, is about three years after the onset of the condition.

Although leukemias represent one of the most common malignancies of children, persons of any age can be victims.

Lymphomas

Lymphomas consist of abnormal numbers of lymphocytes in various stages of development. They are produced by the spleen and lymph nodes. These diseases are similar to leukemia, but in some lymphomas the immature lymphocytes aggregate in the lymphoid tissues. Hodgkin's disease is one of the best-known forms of lymphoma.

Sarcomas

Sarcomas are characterized by tumors growing from bone, cartilage, connective tissue, fat, and muscle cells (Figure 30–3).

Viral Transformation

Viral transformation is the process by which normal cells are altered by viral infection to become malignant. Transformed cells often undergo many changes in morphology (Figure 30–4), metabolic functions, and antigenicity. One of the important conclusions that some scientists have reached from animal experiments is that a virus must be integrated into the host cell's genome before it can transform the cell.

Certain cancer-inducing viruses readily produce cell transformation in tissue culture systems. The viral agent is often no longer recognizable in the cultures by its best-known properties, such as infectivity and antigenicity. Traces of the virus can be detected in the form of new antigens and viral DNA or RNA. Findings such as these suggest that standard procedures for the isolation and identification of viral pathogens may not be adequate for human viral cancer agents. The ultimate test to show the true relationship of any suspected cancer virus and the actual cancer is to cause a cancerous process to

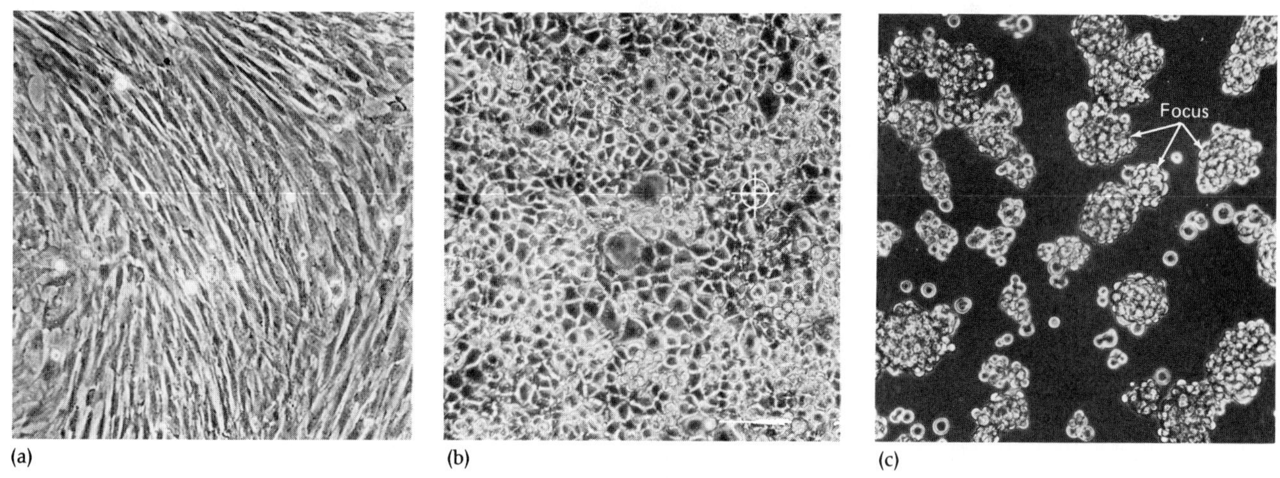

Figure 30–4 Most viruses capable of causing tumor formation in experimental animals also transform cells to the malignant state in an *in vitro* system. Transformation results in altered cell shapes and a capacity for unregulated growth. (a) Untransformed cells. (b) The unregulated growth of transformed cells. (c) The piling up of cells over one another instead of formation of a layer such as the one shown in (a). *[N. Auersperg,* Cancer Res. ***38**:1872, 1978.]*

develop in an experimentally infected individual. [See Koch's postulates in Chapter 1.]

In the laboratory, tissue culture assays are used for the detection of transformation by tumor viruses. For such purposes, several normal animal cell types are used, including the *fibroblast,* which is an unspecialized component of all connective tissues in the body. The most obvious signs of neoplastic transformation are changes in cellular shape and the loss of surface attachment or place (topoinhibition). Transforming viruses added to a fibroblast culture cause infected cells to form small piles, which resemble miniature tumors, on normal cells (Figure 30–4c). Each little tumorlike mass is called a *focus* of transformed cells and is the basis of the procedure called the *focus assay* for tumor viruses.

Early Discoveries Relating Viruses and Cancer

Acceptance of the concept that cancer can be caused by viruses has been slow despite the unequivocal and long-standing evidence from animal experiments. In 1908, V. Ellerman and O. Bang showed that leukemia in chickens could be transmitted by injecting bacteria-free filtrates from infected chickens into healthy chickens. Three years later, in 1911, Francis Peyton Rous demonstrated a similar transfer of a chicken sarcoma. But the occurrence of virus-induced tumors in chickens was generally regarded as a biological curiosity of domestic fowl. Over the next 40 years, accumulating evidence pointed to viral induction of malignancies of various types in rabbits, frogs, and mice (Figure 30–5).

The first mouse cancer viruses discovered were found to contain RNA. Later discoveries have shown that several DNA-containing viruses also cause cancer in mice and other rodents. Virus-induced cancers in hamsters, mice, and rats by human adenoviruses were also reported. In addition, many new leukemia viruses were discovered in the 1960s.

Sarcoma viruses are leukemia-causing agents containing an additional cancer causing gene (oncogene), called *sarc.* This gene enables the sarcoma viruses to transform not only white blood cells but other cells having a similar embryonic origin. Since

Figure 30–5 The virion of mouse mammary tumor virus has an envelope surface covered with spikes. The bar marker represents 0.1 μm. *[J. B. Sheffield,* J. Virol. ***12**: 616–624, 1973.]*

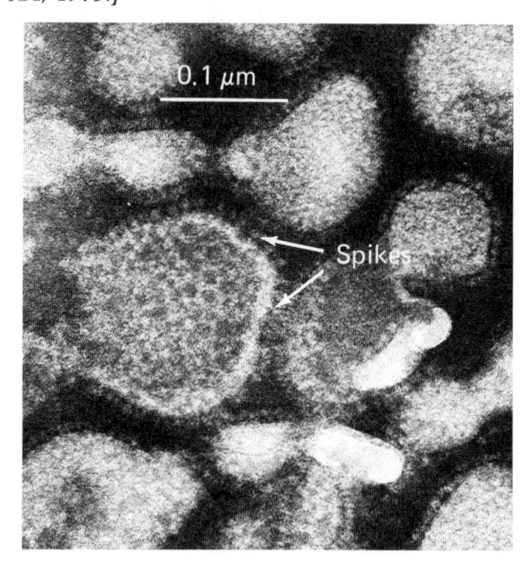

1967, leukemia viruses carrying their own respective oncogenes have been found in an increasing number of animals, including chickens, cats, cattle, and the gibbon ape, a nonhuman primate. These viruses are members of the RNA-containing retrovirus group. [See AIDS, another retrovirus infection, in Chapter 27.]

The transmission of leukemia viruses under natural conditions occurs in various ways. For example, in the domestic cat and dairy cattle, leukemia is a highly infectious disease and causative agents are spread by contact with diseased animals. In mice, transmission of the disease occurs through genetic mechanisms, while in the domestic chicken, both contact and genetic transmission of leukemia are involved. Because there is a close physical association between humans and some animals in which leukemia is transmitted through contact, studies have been undertaken to determine if these viruses may be harmful to humans. The results of such studies show no correlation between particular lower animal leukemias and leukemia in humans.

Human T Cell Leukemia Virus and AIDS

In 1983, the results of several studies showed the cause of acquired immune deficiency syndrome (AIDS), described in Chapter 27, to be a human retrovirus related to the human T cell lymphotrophic virus (HTLV-I), the cause of some types of adult T cell leukemia. Retrovirus infection, known to cause leukemia, lymphomas, and tumors in several animal species and T cell human malignancies, has also been shown to result in immune suppression in some cats infected with feline leukemia virus. More cats are killed as a result of feline leukemia virus causing immunosuppression than by the virus causing leukemia.

Characteristics of Oncogenic RNA Viruses

Oncogenic RNA viruses are found in two of the four designated classes, A, B, C, and D. Type A RNA viruses, which are not infectious, are a very small group of viruslike particles that have not been found outside the confines of cells and have not been shown to be oncogenic. Type B RNA viruses have been associated primarily with certain carcinomas of the breast. Type C RNA viruses, the most important class, have been shown to infect a large number of animal species. Most type C RNA viruses are oncogenic, mainly causing leukemias, lymphomas, and sarcomas—all tumors arising in tissues such as bone, cartilage, connective tissue, and lymph nodes. Type D RNA viruses have been isolated from rhesus monkeys and certain tumor cell cultures. Table 30–1 lists a number of RNA tumor viruses, together with their hosts and the conditions produced. [See RNA viruses in Chapter 10.]

TABLE 30–1 RNA Tumor Viruses

Virus	*Type*	*Host*	*Condition Produced*
Lymphomatosis	C	Birds	Leukemias in chickens
Rous sarcoma	C	Birds, monkeys, rodents	Sarcomas
Feline leukemia	C	Cats	Leukemia
Gibbon lymphosarcoma	C	Monkeys	Tumors
Leukemia (MLV)	C	Mice, other rodents	Leukemia in mice (hamsters)
Mammary tumor (MTV)	B	Mice	Mammary tumors
Mason-Pfizer	C	Monkeys	Tumors
Wooley fibrosarcoma	C	Monkeys	Tumors
Simian T cell lymphotropic virus III[a]	C	Old World monkeys	AIDS in monkeys
T cell lymphotropic virus I[b]	C	Humans	Adult T cell leukemia and a form of paralysis
T cell lymphotropic virus II	C	Humans	Lymphoma

[a] See Microbiology Highlights "Simian T Cell Lymphotropic Virus" in Chapter 27.
[b] See Chapter 27 for a view of the retrovirus group.

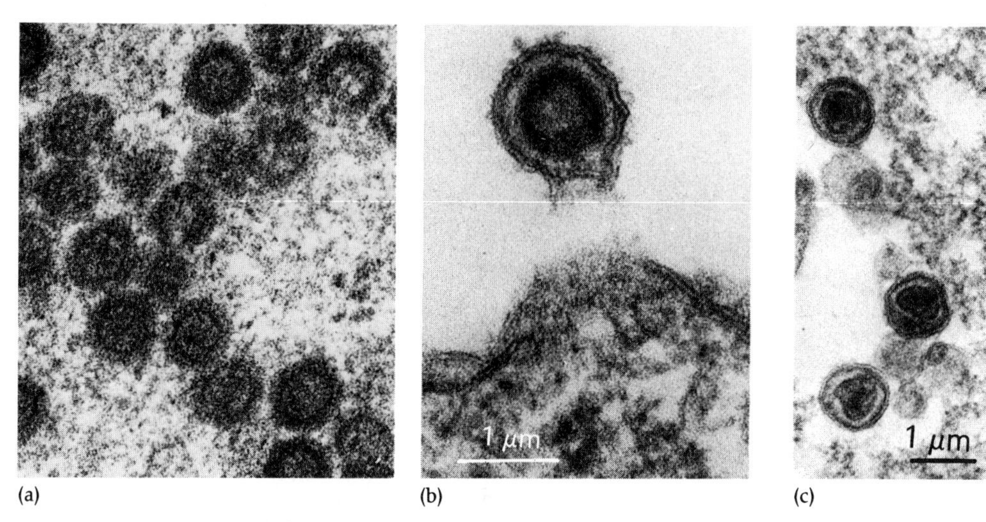

Figure 30–6 Types B and C RNA production. Type C viruses have been implicated in causing leukemia, lymphomas, and sarcomas in a variety of avian and mammalian systems. (a) Mouse mammary tumor virus in tissue culture. This virus is a type B particle. *[Courtesy of Dr. T. M. Murad.]* (b) A budding particle from a rat tissue culture preparation. (c) Free and budding particles from a pig cell line. Note the similarities in particle appearance. *[M. M. Liever,* et al., Science ***182****:56–59, 1973.]*

Differences Among Oncogenic RNA Viruses

Distinctions among the various types of RNA tumor viruses have traditionally been based on morphology, although they can also be made on the basis of immunological differences and modes of maturation.

The type C RNA viruses consist of a roughly spherical, compact RNA core and associated proteins (nucleoid) surrounded by a lipid layer partially transparent to electrons, which gives electron micrographs of the virus a targetlike appearance (Figure 30–6).

The nucleoid of type B viruses is more eccentric in shape, apparently because its major internal protein is about two-thirds larger than that of the type C viruses. The glycoprotein surface spikes of the type B viruses are larger and more regularly spaced than those of the type C viruses.

The type A particles occur in two subtypes, one found in cellular cytoplasm and one found in the lymph and other extracellular body fluids. Those found in the cytoplasm are believed to be immature forms of type B viruses, to which they are immunologically similar. The morphology of type A particles is similar to that of the other viruses, but the type A particles are encapsulated by a protein shell rather than by a lipid-containing membrane. Type D particles are intermediate in size, shape, and appearance to B and C type particles.

A principal difference between oncogenic RNA viruses and other animal RNA viruses lies in the size of the complete set of hereditary information contained in their genome. RNA tumor virus genomes have a mass of about 12×10^6 daltons, as compared to about 6×10^6 daltons for the paramyxoviruses and about 2×10^6 daltons for the poliomyelitis virus.

The Role of DNA Viruses in Carcinogenesis

Interest in the DNA viruses as possible human cancer viruses centered at first on the adenoviruses and papovaviruses. Adenoviruses, which cause respiratory infections in humans, and papova (papilloma-polyoma-vacuolating) viruses, which may be responsible for a variety of tumors, including those found in the mouth (Figure 30–7) and warts produce new growths in experimental animals and transform animal cells in culture. Despite the fact that these viruses are widespread, there is little evidence implicating them in human cancers.

During the mid-1960s, however, the herpesviruses, a group of complex DNA viruses, attracted the attention of investigators seeking to establish a link between cancer and viruses. Several herpesviruses may be involved in human and lower animal cancers and may thus serve as models for the study of human disease. Table 30–2 lists the DNA viruses that are associated with cancerous states.

One of the virus diseases of lower animals, a malignant tumorous growth in the lymphoid system of chickens known as *Marek's disease,* has provided the first unequivocal proof that a herpesvirus is the cause of a cancer. First described by Joseph Marek in 1901, Marek's disease remained almost unknown

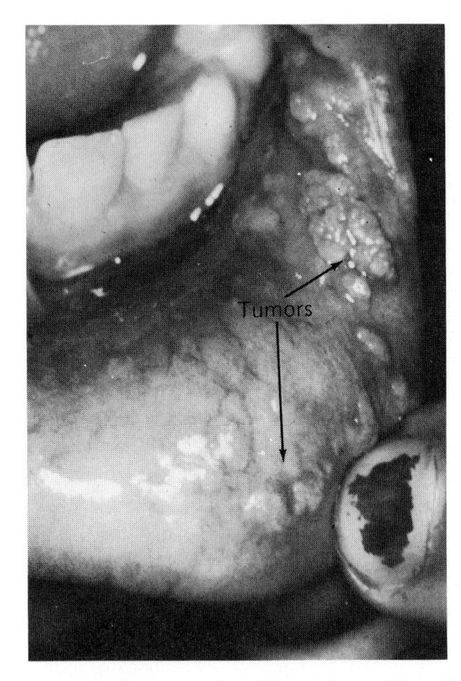

Figure 30–7 Oral tumors (papillomas) resulting from human papillomavirus infection. *[Courtesy Drs. E. M. de Villiers and Christina Neumann.]*

except among academic veterinary pathologists until the late 1960s, when it became the scourge of U.S. commercial chicken flocks. In Marek's disease, cells of the lymphoid system become cancerous and invade the nerve cells, causing paralysis. As the disease progresses, other organs are invaded and the bird eventually dies. The virus that causes Marek's disease is extremely resistant to drying and may remain infectious for a long time after it has been liberated in bits of sloughed-off dead chicken skin (dander). In Marek's disease, the tumor virus is not produced in the tumor cells. The virus transforms normal cells into tumor cells without necessarily producing a new generation of viruses. The replication of the herpesvirus of Marek's disease takes place in nonmalignant skin cells that normally die in any case.

An attenuated (weakened) strain of the Marek's disease virus has been used for live-virus vaccines that have effectively eradicated the disease in treated flocks. This suggests that cancer vaccines for humans may be possible.

The herpesvirus of Marek's disease is immunologically related to a herpesvirus closely associated with a human cancer: Burkitt's lymphoma (Figure 30–8). In 1958, Denis Burkitt, a missionary surgeon working in Uganda, reported that a large number of African children between the ages of 4 and 16 were suffering from tumors in the connective tissue of the jaw. These jaw sarcomas were thought to be very rare in Europe and in the United States, but children of white missionaries living in Uganda sometimes acquired the disease. Although the lower jaw is the most frequent site of the tumor in Burkitt's lymphoma, tumors also arise at other sites, such as the

TABLE 30–2 DNA Viruses and Cancerous States

Virus Group	*Virus*	*Host*	*Conditions Produced*
Adenovirus	Adenovirus, type 3	Cow	Sarcomas in newborn hamsters, mice, rats
	Adenovirus (certain types)	Human	Sarcomas in newborn hamsters, mice, rats
	Adenovirus (certain types)	Monkey	Type 7 produces malignant lymphoma
Herpesvirus	Marek's	Chicken	Lymphoma
	Lucké	Frog	Adenocarcinoma frog kidney
	Herpes simplex virus type 2	Human	Suspected of producing cervical cancer
	Epstein-Barr	Human	Asiatic nasopharyngeal carcinoma, Burkitt's lymphoma, infectious mononucleosis, carcinoma of the thymus, brain lymphoma
	Herpesvirus saimiri	Monkey	Lymphomas
	Cytomegalovirus	Human	Infectious mononucleosis, jaundice, and possibly associated with Kaposi's sarcoma
	BK, JC[a]	Human	Benign wartlike tumors, brain tumors
	Hepatitis B	Human	Liver cancer
Papovavirus	Papillomavirus	Cow, dog, human, rabbit	Warts or papillomas in their host of origin; suspected cause of cervical cancer
	SV40	Monkey	Sarcomas in hamsters
	Polyomavirus	Mouse	Tumors in newborn mice and other rodents
Poxvirus	Poxvirus-molluscum contagiosum	Human	Wartlike tumors on skin
	Yaba	Monkey	Tissue tumors
	Fibromamyxoma	Rabbit	Sarcomas

[a] The initials of patients from whom the viruses were isolated.

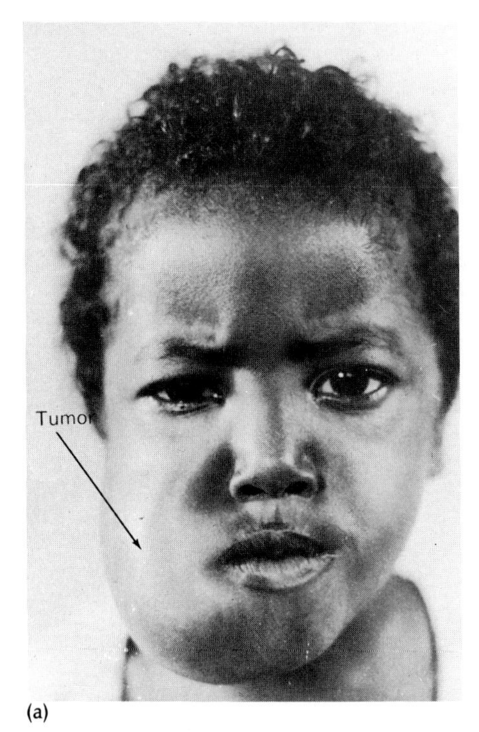

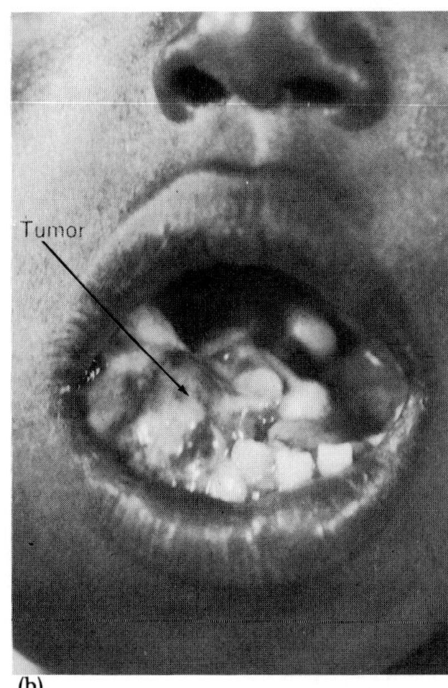

Figure 30–8 A case of Burkitt's lymphoma caused by Epstein-Barr virus (EBV). (a) An external view with right jaw involvement. (b) The same patient showing the extensive internal involvement of the left jaw. *[Courtesy S. R. Prabhn, and K. I. Yagi.]*

upper jaw, the thyroid, the ovaries, the liver, and the kidneys. Intensive study of Burkitt's lymphoma began immediately after Burkitt's report was published.

Electron microscopic examinations of tumors for virus particles were fruitless, as were efforts to infect laboratory animals and tissue culture with extracts of tumors. Efforts to establish tissue cultures with the cells from tumors also met with little success. Then M. A. Epstein and Y. M. Barr, as well as R. J. V. Pulvertaft, who worked independently of the other two investigators, undertook a series of tissue culture experiments using fresh tumor tissue flown in from Uganda. Electron microscopic examinations of some cells from these cultures clearly demonstrated the presence of herpesvirus particles (Figure 30–9). These particles have since been named *Epstein-Barr virus (EBV)*, and associated with the other forms of human cancer (Table 30–2). **[See infectious mononucleosis in Chapter 27.]**

The identification of a herpesvirus in Burkitt's lymphoma was a significant finding, since it is well known that certain herpesviruses, such as herpes simplex virus types 1 and 2, are widespread in the human population. In 1968, Werner and Gertrude Henle and V. Diehl reported that infectious mononucleosis is caused not only by a herpesvirus but also by an agent that cannot be distinguished from the EBV. One question raised by this finding is: Do people who have had infectious mononucleosis have a greater probability of contracting cancer in later life? Thus far, it appears that a history of infectious mononucleosis is not associated with the incidence of cancer. Another interesting finding in the relationship between infectious mononucleosis and EBV is that antibodies to EBV are associated with protection against infectious mononucleosis; conversely, infectious mononucleosis occurs only in individuals who do not have antibodies against EBV.

Several investigators have subsequently shown that EBV can induce lymphoid tumors in monkeys. Since such experiments are too hazardous to be carried out using humans, this is as far as experimentation can go in demonstrating that a virus from human cancer is in fact a cancer-causing agent. Demonstrating that a virus can induce a tumor is a different matter from finding a virus in a tumor. EBV will also transform healthy human tissue cultures into cancerlike cells.

It seems possible that several human cancers are caused by RNA viruses (Table 30–1), but the matter still remains to be established. More direct evidence is needed to prove their relationship conclusively. The most convincing evidence that some kinds of human cancers are caused by viruses comes from studies of DNA viruses, including herpes simplex viruses 1 and 2, EBV, cytomegalovirus, and hepatitis B virus.

The study of and search for a viral cause of cancer, while extremely complex, continues at an intensive pace. The knowledge that has accumulated more than hints at a fundamental role for microorganisms, particularly viruses, in various cancerous processes. However, if and when human cancers are shown to be caused by viruses, the mechanism of action will still need to be determined in order to develop effective measures for their prevention and control.

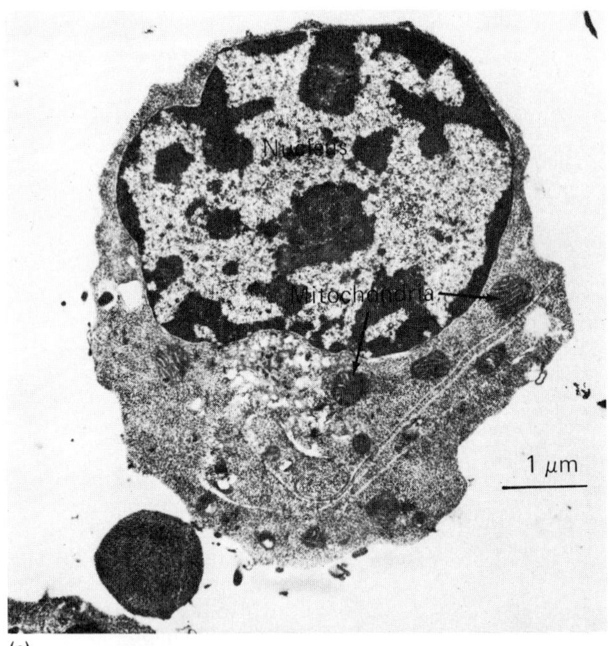

(a)

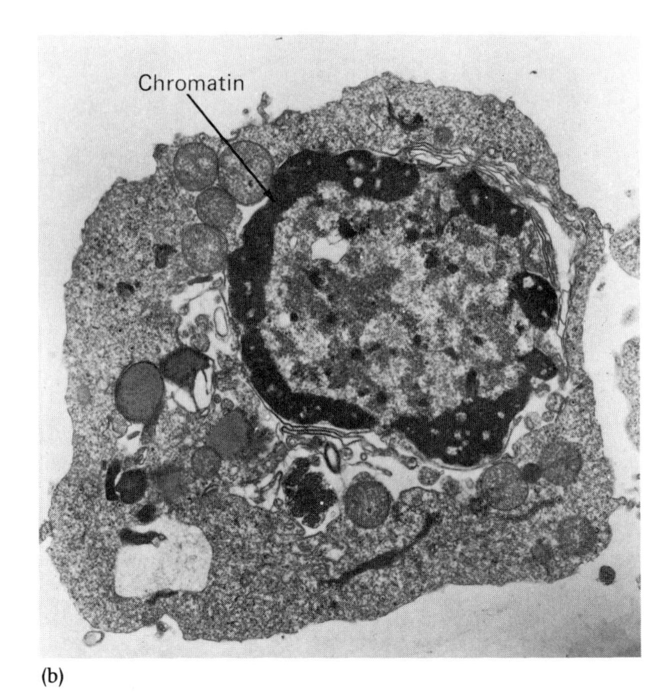

(b)

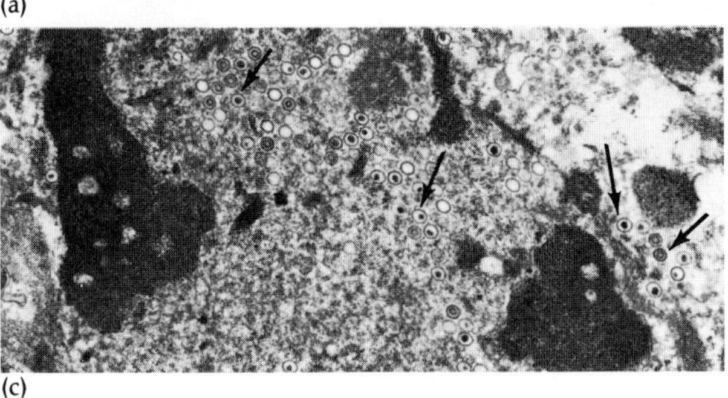

(c)

Figure 30–9 Epstein-Barr virus (EBV) infection. (a) An ultrathin section of a non-virus-producing Burkitt lymphoma cell. The nucleus, cytoplasm, and organelles such as mitochondria appear normal. (b) An EBV-producing cell reveals ultrastructural changes in the nucleus, margination of chromatin, and production of excess nuclear membrane. The cytoplasm contains swollen mitochondria. (c) Virus capsids in various stages of maturation (arrows).

Cancer Virus Hypotheses

Reverse Transcriptase

In 1970 the surprising report was first made by Temin and Mizutani, and then by Baltimore, of an enzyme that catalyzes the flow of genetic information from RNA to DNA, a reversal of the usual DNA-to-RNA direction of genetic expression. The discovery was of particular significance to cancer virus studies, because some viruses known to cause cancer in animals have an RNA core. The enzyme, known as *reverse transcriptase,* provided, for the first time, a mechanism by which genetic material in the RNA of a virus might be incorporated into the DNA of a cell, where it could function like any gene (Figure 30–10). Temin gave reverse transcriptase the name *RNA-directed DNA-polymerase,* which explained the enzyme's function. It appears that the transcriptase brings about the formation of a DNA copy of the viral RNA in the cytoplasm of affected cells. The complementary DNA *(cDNA)* thus formed can then integrate into the chromosomal DNA of the host cell. Such cell-integrated viral information, referred to as *proviral DNA,* is transmitted to daughter (descendant) cells in much the same way as are other cellular genes. The integration of proviral DNA would explain the maintenance of a cancerous state and of viral genetic information over numerous cell generations without the formation of fully formed virus particles. The finding of reverse transcriptase explains the genetic transmission of retroviruses and others that can be transmitted genetically through DNA obtained from viral RNA. In addition, incomplete proviral information can also be genetically transmitted. Such limited viral DNA may be sufficient to transform normal cells into malignant ones, even without the presence of a complete virus. Confirming evidence for such transformations has been obtained from various experiments, including the infection of laboratory animals with retroviruses. Two *in vitro* mechanisms involved in retroviral cell transformation are shown in Figure 30–10. These are *contact,* or *horizontal,* and *genetic* (Figure 30–10a) or *vertical* (Figure 30–10b and 10c), modes of transmission. The former process involves cells in an existing population, while the genetic mode of transmission includes future generations.

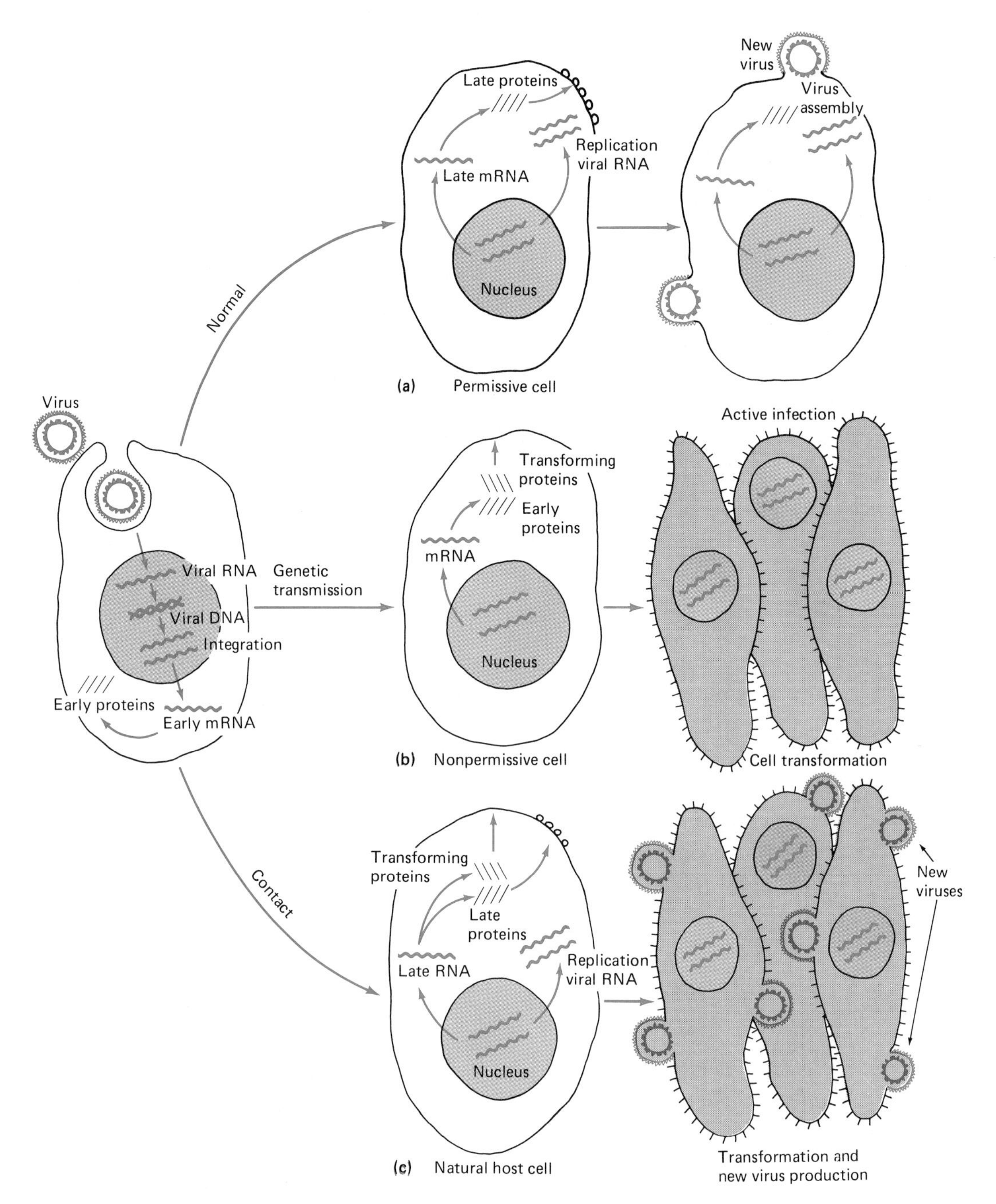

Figure 30–10 Cell transformation by retroviruses. All leukemia and sarcoma viruses belong to the retrovirus group and are often C-type viruses. (a) Invading viruses can use susceptible (permissive) cells for new virion production. (b) In other situations, such as those involving the genetic mechanism of nonpermissive cells, viral DNA is integrated into the cell's nucleus, causing changes (transformation) in cell growth but no production of new virus particles. (c) Situations such as those involving the contact mechanism also occur in which transformation and new virions result.

In the contact mechanism, a normal cell with virus receptor sites may become infected by a retrovirus virion. The virus enters the cell's cytoplasm, where it sheds its protein covering and uses its reverse transcriptase to bring about the formation of cDNA by the infected cell. The cDNA integrates into the cell's chromosomal DNA, becomes proviral DNA, and eventually causes a malignant transformation. Cells of this type may, in turn, produce virus particles or, possibly because of the proviral DNA, may remain transformed without producing virus particles.

The genetic transmission mechanism occurs only

when dormant proviral DNA is present in a normal-appearing cell. In such situations, exposure to chemicals or radiation can trigger the inactive DNA, causing chromosomal changes and the malignant transformation of the cell. New virus particles will eventually be produced by such a cell if the dormant proviral DNA was complete; if it was incomplete, no virus particles will be formed but the cell will remain transformed.

RNA Cancer Hypotheses

There are three RNA virus cancer hypotheses based on the assumption that viruses do, in fact, cause human cancer. These hypotheses have been developed to account for the way in which genetic information associated with a cancerous process is expressed in cells. We should note again that cancer appears to be a number of different diseases. The demonstration of viral association with the causation of one type of cancer or, for that matter, several types should not be interpreted to mean that the problem of cancer will be solved. Various studies suggest that most cancers may be chemically induced. Carcinogenic chemicals are powerful mutation-causing agents in a variety of organisms. These chemicals presumably transform cells by inducing mutations in cellular genes.

PROVIRUS HYPOTHESIS

Of the three proposed RNA cancer virus hypotheses, the first was the provirus hypothesis, formulated by H. M. Temin in the early 1960s. According to the provirus concept, after infection of a cell by an RNA tumor virus, the cell makes a DNA copy of the viral RNA and incorporates this genetic information into its own DNA. This reaction gives the cell the capacity to produce oncogenic viruses and transforms it from a normal cell to a neoplastic one (Figure 30–10).

It should be emphasized that the integration of viral genetic material takes place after a viral reverse transcriptase makes a DNA copy of the viral RNA. Once this step is performed, the reverse transcriptase is no longer needed to establish the virus in the cell (Figure 30–10b).

Many researchers accept the provirus hypothesis as an accurate model of oncogenesis in lower animals. The possibility also exists that a similar set of reactions can develop in humans.

ONCOGENE-TRANSFORMING HYPOTHESIS

According to the oncogene hypothesis, advanced in 1969 by R. J. Huebner and G. J. Todaro, the genetic information for cancer is present in every cell and is transmitted from parent to child. According to this model, infection of cells by type C RNA viruses occurred millions of years ago during the course of evolution. Every cell is assumed to contain an oncogene, a region of DNA that is normally *repressed* (prevented from functioning). When the oncogene becomes derepressed, possibly by a virus, a chemical carcinogen, or radiation, it expresses itself by bringing about the formation of a transforming protein (Figure 30–11). A transforming protein of this type could change a normal cell into a malignant one even though no viruses could be recovered from it.

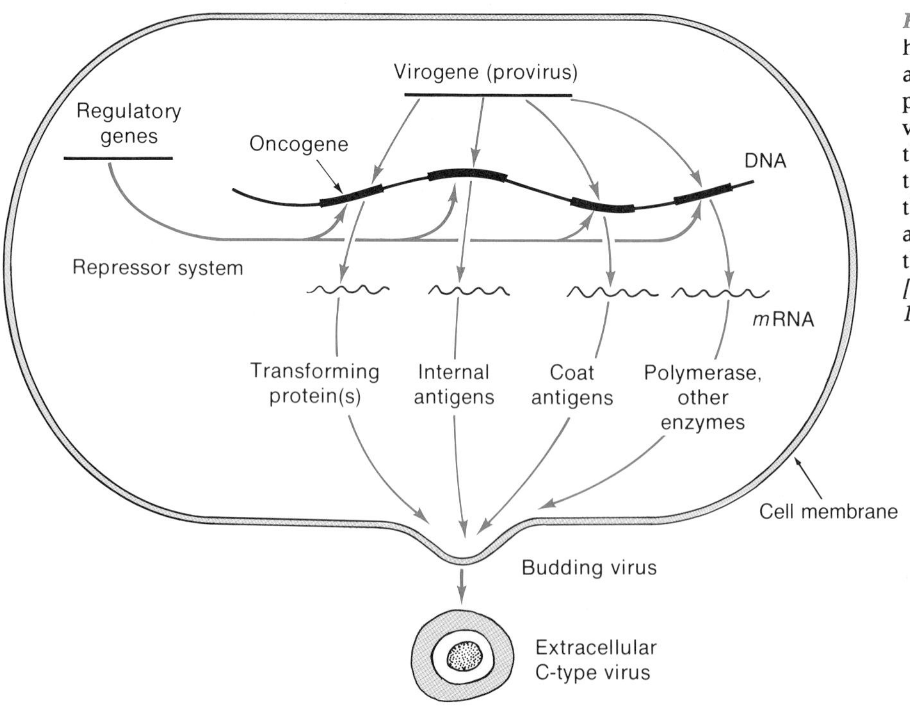

Figure 30–11 The oncogene hypothesis proposes that tumors are induced by transforming proteins, which are the products of virogene activity. A virogene has the capacity to produce a complete tumor virus. According to this theory, in normal cells oncogenes are turned off by regulatory genes that code for a repressor system. *[B. J. Culliton,* Science **177:***44–47, 1972.]*

Microbiology Highlights

THE RECESSIVE ONCOGENE

According to the concept of an oncogene, a tumor or cancer occurs through a change in a gene that activates a normally silent portion of DNA to begin a chain of cancer-causing events. The existence of another type of oncogene, the *recessive oncogene,* has also been hypothesized. According to this hypothesis, this gene's cancer-causing effects result from its being inactivated rather than abnormally activated. S. H. Friend and his colleagues isolated the first recessive oncogene, which causes rare tumors in the eyes *(retinoblastomas)* and in bone *(osteosarcomas).* Retinoblastomas generally appear within the first year of life and are treatable. However, if they occur at the time of puberty, control is more difficult.

The oncogene associated with these tumors was found on human chromosome 13. How common recessive oncogenes are is not known. However, tumors and related cancerous conditions known to have a pattern of inheritance may turn out to be caused by this newly uncovered form of oncogene.

The oncogene is pictured as only one portion of a larger structure, the *virogene* (Figure 30–11). The virogene consists of several segments of genetic information, all of which must be activated for complete viruses to be made. This means, then, that the virogene contains the necessary genetic information for the transforming protein, for the various components of the virus, and for the enzymes needed for making a complete virus. Thus, it would not be necessary for the complete virogene to be expressed in order to produce a transformed cell. In fact, such complete expression could work against a virus-associated cancerous state. The body might recognize and destroy the antigens of the whole virus, while it would be unable to act against the transforming protein of the oncogene.

PROTOVIRUS HYPOTHESIS

The protovirus hypothesis, proposed by Temin in 1970, is similar in many ways to the oncogene concept. However, there are some distinguishing points. The protovirus hypothesis holds that cancer viruses arise from segments of genetic information randomly brought together by a variety of cellular and genetic events. These segments form the protovirus. An important distinction between this and the oncogene hypothesis is that cells do not come into being with all the genetic information necessary for the development of a malignancy. What they do have is the potential for assembling such information.

As more and more information about the molecular processes of normal and malignant cells accumulates, evidence in favor of each of the cancer virus hypotheses correspondingly increases. However, while evidence is increasing in support of the presumption that viruses cause human cancers, there is still reason to be skeptical. It is possible, for example, that a virus may be a necessary, but not a sufficient contributor to the development of a cancerous state. Other factors, such as genetic disposition, immunological deficiency, or exposure to chemicals or radiation, may also be required. The need for two or more causes acting together in the proper sequence may help to explain why—if, in fact, viruses or portions of viruses are involved—there is little evidence that cancer is contagious.

Microbial Carcinogens

Although it is known that some fungi are harmful or toxic to plants and animals, there is a common tendency to regard molds as harmless. However, in addition to the side effects of treatment with fungal antibiotics, both humans and other animals can be affected by contamination of foodstuffs with mycotoxins. Evidence from field and storage studies confirms contamination of human and animal foods with carcinogenic mycotoxins such as aflatoxins. Extensive studies have shown that mycotoxins, especially aflatoxins, produce carcinogenic effects on organs such as the kidney, liver, intestines, stomach, and trachea. [See Chapter 22 for aflatoxins.]

Carcinogens are also known to be synthesized or activated by bacteria. Among the chemicals in this category are nitrites and nitrosamines.

Cancer Detection

The treatment of cancer can be undertaken only after the condition is detected. There are several methods that can be used, but no single test will find all cases. A basic method used today involves routine screening tests, such as physical examinations, cervical smears (the Papanicolaou test, or Pap test) for women of childbearing age or older, x-rays, and proctoscopic examination of the rectum and lower bowel.

Fetal Antigen

Biochemical tests for cancer are also important. Researchers have known for several years that a protein normally present only in human fetuses shows up in cancer patients. At first the protein, named *carcino-embryonic antigen (CEA),* was thought to occur only in patients with cancer of the colon. Then it was discovered in patients with other types of cancers as well. Later, small amounts of CEA were found in individuals who seemed to be free of cancer. It now appears that the amount of CEA in a person's bloodstream is related to the presence or absence of cancer.

Although CEA level does not signal any specific type of cancer (and may not, by itself, indicate cancer at all), it is already proving to be a valuable aid in monitoring treatment and in diagnosing a tumor's recurrence. Antigens such as CEA may eventually be used to indicate specific types of cancer, just as different microorganisms indicate specific diseases.

Other fetal antigens have been associated with certain human cancers and are detected by immunological tests. These include alpha-fetoprotein, which is found in cases of liver and embryonic tumors and cancers of the reproductive organs (Figure 30–12).

Figure 30–12 A malignant tumor of the uterus responsible for the production of alpha-fetoprotein. *[From K. Kawagoe,* Jpn. J. Clin. Oncol. ***15:****577–583, 1985.]*

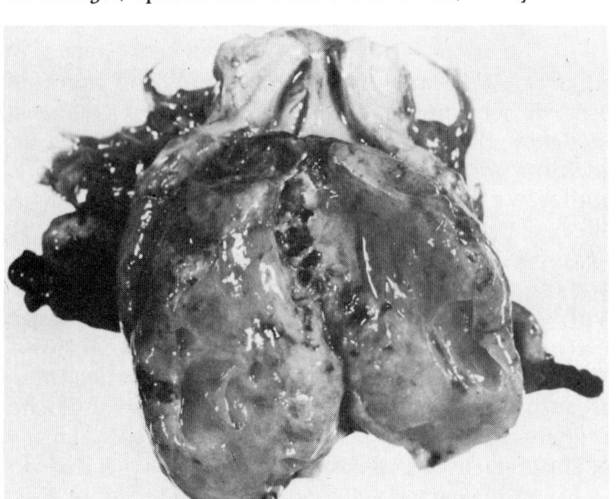

Tumor-specific Antigens

Immune reactions of humans to their cancers depend on tumor-specific antigens (TSAs) that are absent from or hidden in normal cells. Because these antigens are usually present in the surface membranes of tumor cells, the host can react against the tumor as if it were a transplant involving genetically different members of the same species. Immunological surveillance by the host's immune system probably eliminates many potentially neoplastic cells during a normal lifetime. However, individuals with certain impairments of the immune system have an increased probability of developing a tumor or related condition. Direct evidence for TSAs on human cancer cells have been provided by immunofluorescent procedures (immunoperoxidase reactions; Figure 30–13) and other tests. Increasing evidence for the presence of these antigens in a wide variety of tumors and for host-associated immune responses to the TSAs has raised hopes for the development of diagnostic allergic skin tests or serological procedures for the early detection of cancers as well as for some form of immunization against certain tumors.

Figure 30–13 An immunoperoxidase reaction showing the presence of an antigen found in cancerous tumors. This T antigen may be responsible for delayed hypersensitivity in patients with breast carcinoma and can be seen as blackened areas along cell membranes. (See Color Photograph 80.) *[D. R. Howard, and J. G. Batsakis,* Science ***210:****201–203, 1980.]*

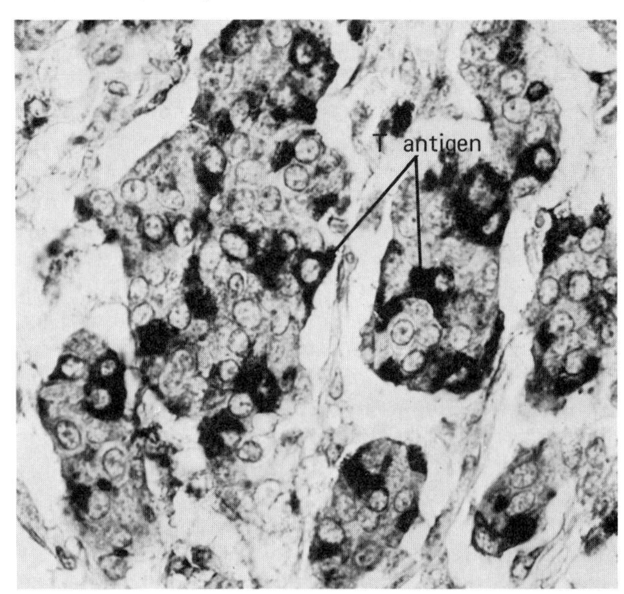

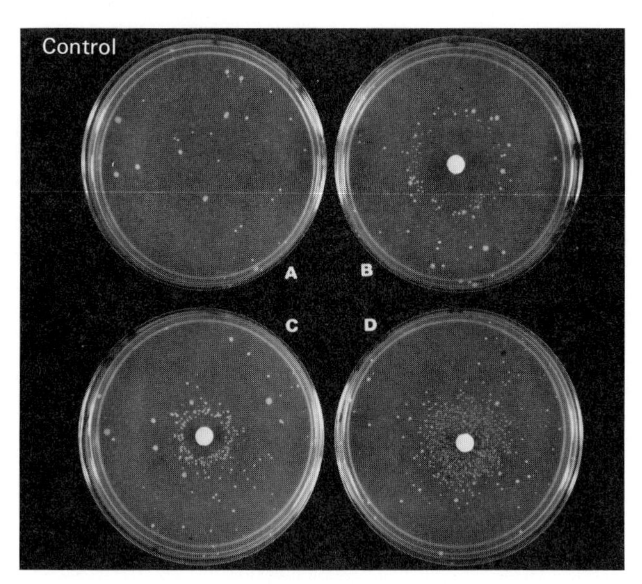

Figure 30–14 The Ames test for the detection of mutagenic or carcinogenic activity. In this "spot test," the bacterial colonies develop if mutations occur. In (B), (C), and (D), a chemical mutagenic agent was applied to the filter paper disc seen in the center of each plate. (A) is a control and shows the amount of bacterial growth that occurred due to spontaneous mutation. *[B. N. Ames, J. McCann, and E. Yamasaki,* Mutation Res. ***31****:347, 1975.]*

Microbial Detection of Carcinogens and Mutagens

In the past, the use of laboratory animals was the only acceptable means of demonstrating the carcinogenicity and mutagenicity of chemicals. A rapid, accurate, and inexpensive *in vitro* procedure, the Ames/*Salmonella* microsome mutagenicity test, was developed for the detection of potential chemical mutagens and carcinogens (Figure 30–14). This test, which was developed by Bruce N. Ames, has been widely used as an early warning system to identify and reduce human exposure to dangerous chemicals. For example, prior to the commercial production of a new food additive or pharmaceutical product, manufacturers can utilize the Ames/*Salmonella* bioassay to determine the carcinogenic potential of the new substance. With the test results, which are obtainable in a matter of days, hazardous chemicals can be eliminated before they are introduced to the public.

The Ames test detects compounds that are capable of inducing permanent genetic changes in DNA molecules of specially developed histidine-deficient strains of *Salmonella typhimurium* (sal-mon-EL-lah tie-fee-MUR-ee-um) and *Escherichia coli* (esh-er-IK-ee-ah KOH-lee). Since these organisms cannot produce the required amino acid histidine, they cannot grow unless the nutrient is provided or a genetic change occurs that restores their ability to produce histidine. Chemicals can quickly be detected if they cause mutations at levels greater than those occurring normally (spontaneously).

The Use of Microorganisms and Monoclonal Antibodies in Cancer Treatment

The major forms of cancer treatment used today are surgery, radiation, chemotherapy, or a combination of the three. However, for neoplastic disease, there are limitations to these conventional methods—a realization that has resulted in the search for new measures of treatment. One approach is to stimulate the host's immune system to kill new abnormal (neoplastic) cells. This method involves the use of nonspecific immunological stimulators (NISs) or biological response modifiers (BRMs), which are a series of microbial preparations that can modify immunological or, in a broader sense, biological responses (Table 30–3). The activity of BRMs depends on many factors that are mainly related to the immunological status of the individual to be treated. In general, two types of these immunomodulators are recognized: (1) those that require a functional immune system to work and (2) those that appear capable of restoring immune competence in individuals with defective immune systems. Components of many bacteria and their metabolic products appear to be among the most powerful immunomodulating agents. Many of these BRMs are of low molecular weight and are produced by industrial laboratories. Nonspecific immunological stimulators appear to offer several advantages over conventional approaches. Radiotherapy, hormonal therapy, and chemotherapy all reduce immunological responsiveness; they can therefore severely impair normal immunological function, particularly resistance to microorganisms and their products. We have seen in preceding chapters many cases in which use of these methods or of immunosuppressive drugs predisposed individuals to serious disease. Nonspecific stimulators can prevent or reverse such immunosuppression.

The tumor-destructive effectiveness of BCG (Figure 30–15) serves as an example of NIS use. BCG is a live, attenuated organism derived from a bovine strain of *Mycobacterium tuberculosis.* There is a general consensus that infection with BCG bacilli causes a change in the immunological apparatus of the host. One expression of this altered state is an increased immunoglobulin response to unrelated antigens. Tumor inhibition requires the host to develop and ex-

TABLE 30–3 Bacterial Products as Immunomodulating Agents

Microorganism	*Preparation*	*Uses*
Brucella abortus	Extracts	Stimulates interferon production, has some anticancer activity in laboratory animals
Propionibacterium acnes (Corynebacterium parvum)	Heat-killed and formaldehyde-treated cells	Treatment of lung cancer, breast cancer, melanomas (pigment cell cancers), and sarcomas; stimulates antibody production
Gram-negatives	Endotoxins	Nonspecific stimulation of immune system; increases resistance to tumors in mice
Group A streptococci	Cellular extracts	Some antitumor activity; enhances host resistance to tumors
Klebsiella pneumoniae	Extract	Restores skin test hypersensitivity in cancer patients
Mycobacterium bovis	Bacillus of Calmette and Guérin (BCG)	Increases the responsiveness of immune system; some antitumoral activity
Nocardia	*Nocardia* water-soluble mitogen	Stimulation of both humoral and cell-mediated immunity in mice

press a delayed hypersensitivity reaction to BCG antigens. **[See type IV hypersensitivity in Chapter 20.]**

The current understanding of the cellular and molecular basis for BCG-mediated tumor destruction can be summarized as follows. After injection of the host with BCG, sensitized T lymphocytes develop that are able to recognize distinctive tubercle bacillus antigens. These lymphocytes react with the bacillus and produce a variety of potent molecules, the lymphokines. Some of these molecules immobilize or activate macrophages; others may be directly cytotoxic. If tumor cells are located close to the bacteria, they are killed by activated macrophages and lymphotoxins. As a result, immunity to specific tumor antigens may develop and, in turn, cause additional tumor cell death. **[See T cells in Chapter 17.]**

Figure 30–15 A skin tumor (left) in guinea pigs at the site of injection of BCG (bacille *Calmette* and Guérin). The photograph on the right shows the tumor 26 days after administration of one million (10^6) tumor cells only. *[A. Bekierkunst,* et. al., Infect. Immun. ***10**:1044, 1974.]*

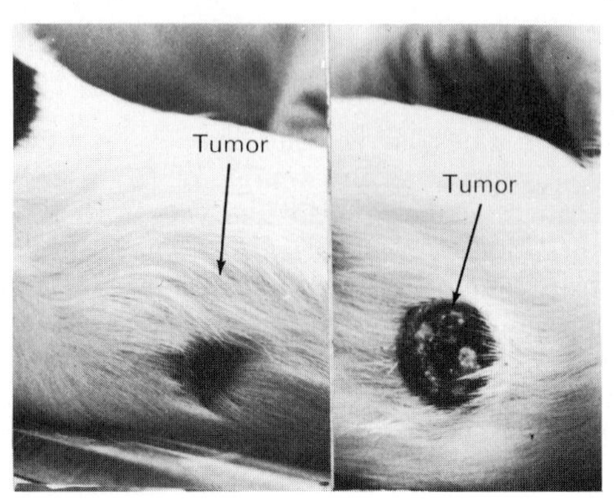

Living BCG has occasionally been used effectively in humans. Preparations have also proved effective in lengthening remission in acute leukemia. Unfortunately, living BCG can cause disseminated disease in an immunosuppressed host. Although local reactions at the site of injections may be severe, they usually disappear. It seems that these adverse effects of using living BCG can be avoided by using lyophilized (freeze-dried), killed BCG alone or with other preparations having antitumor effects. One of these is the cord factor (trehalose-6,6′-dimycolate) obtained from mycobacterial cell walls.

Other bacterial sources of antitumor agents include bacterial endotoxins from *Salmonella enteritidis* and *Serratia marcescens,* both of which have been shown to inhibit tumor growth in laboratory animals. However, such preparations may have side effects.

Monoclonal Antibodies

Monclonal antibodies, in addition to being highly effective detectors of cancer cells, are under consideration as a potential means of tumor therapy. The possible roles for monoclonal antibodies include the delivery of high concentrations of cell-killing radioactive chemicals, drugs, or toxins to cancerous tissues, as well as the use of these antibodies to locate and attack tumor cells.

For effective delivery of anticancer agents linked to monoclonal antibodies, the combinations (conjugates) should penetrate and destroy the parts of a tumor that contribute to its growth and then bind

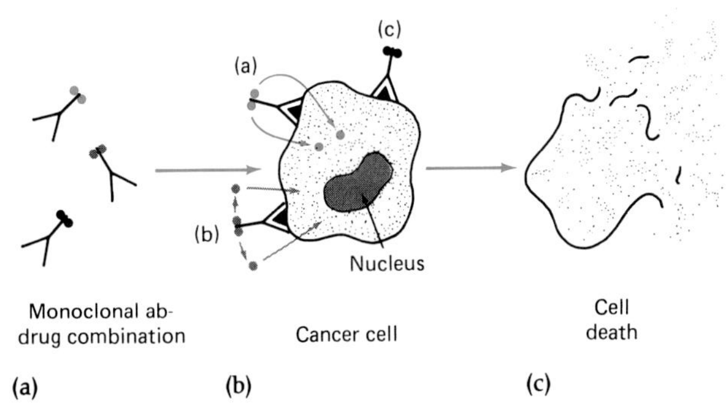

Figure 30–16 Pathways to deliver tumor cell-destroying agents (drugs) linked to monoclonal antitumor antibodies. (a) After binding to the tumor cell, the drug may enter the cell to produce its cytotoxic reactions. (b) The drug may be released at the tumor cell surface and then function as a free agent to affect the tumor. (c) Drug-antibody complexes may attach, remain at the tumor cell surface, and cause their effects.

directly to most, if not all, individual tumor cells. Additionally, anticancer drugs may be released around the tumor cells (extracellularly) following monoclonal antibody attachment to tumor cell surfaces. Figure 30–16 shows several pathways that can be used for the delivery of anticancer drugs via monoclonal antibodies.

Future Outlook

It has been difficult to prove that a virus causes human cancer because both Koch's and Rivers' postulates, which have served for years as the criteria for establishing that a disease is caused by a given infectious agent, cannot be fulfilled. The first postulate requires the isolation of the causative agent from all infected organisms. But infectious particles cannot be recovered from most fresh human tumor cells. Only after careful manipulations have been performed on tumor cells (for example, culturing them, sometimes with other kinds of cells) can infectious virus be demonstrated. Thus, it is possible that the virus is a contaminant and is not involved in the development of the tumor itself.

Another postulate requires induction of the disease state in a suitable host by a pure preparation of the suspected causative agent. However, such experiments are not normally performed on humans.

Thus, investigators must rely on indirect, circumstantial evidence to prove the case for a suspected oncogenic virus. Current strategies include (1) epidemiological studies, usually done in conjunction with immunological studies, to determine whether the virus has left traces of its presence in the form of antibodies against it in the patient's blood; (2) study of tumor cells to detect the presence of viral DNA or RNA or of virus-associated antigens; (3) comparison with virus-induced animal tumors; and (4) study of the oncogenic potential of the virus both in cultured cells and in living animals, especially nonhuman primates.

If human oncogenic viruses can be isolated and obtained in sufficient quantities in pure form, there is hope that the associated forms of cancer can be eliminated, or at least reduced in frequency, through the development of effective vaccines. The discovery of ways in which to interrupt the primary routes of infection would also be of great importance. For example, contrary to the normal preventive practice of giving vaccines before the onset of a disease, it is quite likely that an antileukemic vaccine would be administered to patients who already had the disease. Most patients under treatment experience a period of remission (relaxation of the disease process). The vaccine would be given at this time in an attempt to stimulate the host's immune

Figure 30–17 Application of adoptive immunotherapy in patients receiving lymphokine (interleukin-2) activated killer cells (LAKs). (a) A patient with metastatic melanoma exhibiting cutaneous disease (arrows) before treatment. (b) The effects of treatment on the cutaneous condition. *[From S. A. Rosenberg,* et al., New England J. Med., ***316****:889–897, 1987.]*

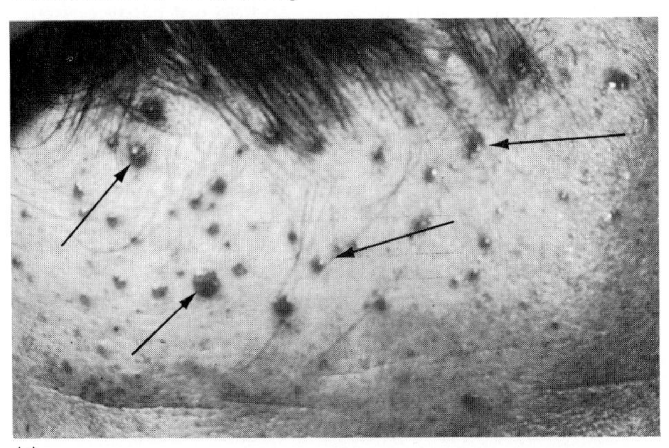

(a)

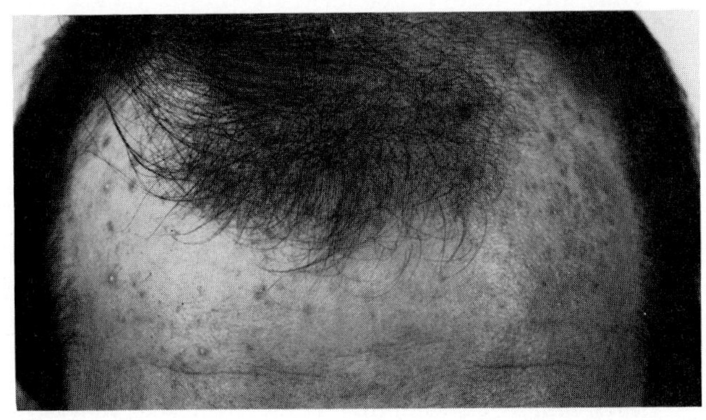

(b)

system so as to prevent the development and reproduction of undetectable malignant cells. This type of vaccine would be expected only to prolong the remission period and thereby the survival of leukemia patients for a limited time.

LYMPHOKINE ACTIVATED KILLER (LAK) CELLS
The treatment of humans with advanced forms of spreading cancers represents a major challenge. A relatively recent approach to treating such conditions is *adoptive immunotherapy.* In this treatment, lymphocytes taken from a patient are incubated with recombinant interleukin-2 (rIL-2), a lymphokine (Figure 30–17). The rIL-2 activates these cells to destroy tumor cells that are resistant to attack by natural killer cells. **[See Chapters 17 and 20 for descriptions of lymphokines and killer cells.]** The interleukin-stimulated cells are referred to as *lymphokine activated killer (LAK) cells.* Potential LAK cells are widely distributed in the blood, lymph nodes, and bone marrow. When the LAK cells are reintroduced into the cancer patient, tumor cells are destroyed without interfering with the activities of the patient's normal cells. The antitumor actions of LAK cells are related to their production of *perforin,* a pore-forming protein that perforates membranes, and a *tumor necrosis factor* that also is active against tumor cells. Applications of adoptive immunotherapy have been successful in a number of patients and appear to hold great promise for the future.

Summary

General Characteristics of Cancerous States

1. Normal cells must be able to divide and reproduce in an exact and controlled manner for the well-being of the body.
2. Uncontrolled cellular reproduction leads to the formation of new, abnormal swellings called *neoplasms* or *tumors.*
3. Tumors that remain localized and do not cause destruction and death are called *benign,* while those that spread to other body areas and endanger the life of the individual are called *malignant* or *cancerous.*
4. The process by which cells separate from tumors and spread to other sites is called *metastasis.*
5. Cancers are believed to be caused by extrinsic, or environmental, factors including chemicals, radiation, certain microbial products, and viruses. Cancers are also influenced by intrinsic factors such as age, heredity, sex, and hormones.
6. Cancerous states have several major properties, including uncontrolled cellular reproduction and an inability of the cells to mature into recognized structures.
7. While some microbial products are known to have cancer-causing (carcinogenic) capabilities, other products have been shown to inhibit the progress of cancer.

Forms of Human Cancer

1. More than 100 clinically distinct types of cancer are recognized, each with a unique set of symptoms.
2. Most cancerous conditions can be grouped into one of four general types: *carcinomas,* solid tumors from tissues covering the internal and external portions of the body; *leukemias,* uncontrolled production and accumulation of different white blood cells; *lymphomas,* abnormal production of immature lymphocytes; and *sarcomas,* tumors growing from bone, cartilage, connective tissue, and muscle.

Viral Transformation

1. Viral transformation is the process by which normal cells are changed into malignant ones.
2. Such altered cells show changes in appearance, metabolism, and antigenicity.
3. The formation of tumorlike masses in tissue culture systems is the basis for the focus assay for tumor viruses.

Early Discoveries Relating Viruses and Cancer

1. Leukemia viruses are found in a large number of lower animals, including cats, chickens, dairy cattle, and the gibbon ape.
2. Transmission of these viruses includes genetic inheritance and direct contact.
3. Lower animal leukemia viruses do not appear to cause disease in humans.

Characteristics of Oncogenic RNA Viruses

1. Cancer-producing (oncogenic) RNA viruses are found in two of the four main classes, A, B, C, and D.
2. Type A viruses are not infectious or oncogenic and are found in the cytoplasm of cells.
3. Type B viruses are associated with carcinomas of the breast.
4. Type C viruses, the most important class, cause a variety of cancerous states, including leukemias, lymphomas, and sarcomas.
5. Type D viruses are intermediate to types B and C in appearance and have been isolated from rhesus monkeys and certain tumor cell cultures.

The Role of DNA Viruses in Carcinogenesis
Several DNA viruses are considered to be possible causes of human cancer. These include adenoviruses, papovaviruses, and herpesviruses.

Cancer Virus Hypotheses

REVERSE TRANSCRIPTASE

1. The enzyme *reverse transcriptase,* found in RNA cancer viruses, promotes the incorporation of viral genetic material into a susceptible host cell by transcribing complementary DNA (cDNA) from the viral RNA host cell. The cDNA can integrate into the host cell chromosomal DNA. Such integrated viral information (proviral DNA) is transmitted to future generations in the same way as are other genes.
2. The finding of reverse transcriptase explains the genetic transmission and transformations of retroviruses and related viruses.
3. Two *in vitro* mechanisms are involved in retroviral cell transformation, contact and genetic.

RNA CANCER HYPOTHESES

1. Three current RNA virus cancer hypotheses are recognized: the *provirus, oncogene,* and *protovirus* hypotheses.
2. According to the provirus hypothesis, normal cells are changed into cancerous ones by infection with an RNA tumor virus. These transformed cells produce new oncogenic viruses.
3. The oncogene hypothesis assumes that the genetic information, or *oncogene,* for cancer exists in every cell in an inactive or repressed form. Viruses or environmental factors activate the oncogene, which changes the normal cell into a malignant one.
4. The protovirus hypothesis is similar to the oncogene concept. It differs in several ways, however, including the process for cancer virus production and the assumption that cells do *not* come into being with all of the information necessary to become malignant.

Microbial Carcinogens

Several products of microorganisms are harmful to various forms of life. Mycotoxins, which include aflatoxins, have been shown to be carcinogenic.

Cancer Detection

1. Methods of cancer detection include physical examinations and various laboratory tests for the detection of tumors and other cancer-related antigens.
2. A rapid, accurate, and inexpensive procedure using bacteria known as the *Ames*/Salmonella *microsome mutagenicity test* exists for the detection of potential chemical mutagens and carcinogens.

The Use of Microorganisms and Monoclonal Antibodies in Cancer Treatment

1. The major forms of cancer treatment used today are surgery, radiation, chemotherapy, or a combination of these treatments.
2. Biological response modifiers (BRMs) are microbial preparations that can modify biological responses and may restore immune competency in individuals with defective systems.
3. Monoclonal antibodies are under consideration to deliver anticancer drugs to tumor locations in the body.

Future Outlook

If researchers can succeed in proving that a virus or viruses cause human cancer, there is hope that cancer can be eliminated or treated by the development of vaccines or the interruption of the primary routes of infection. Treatment of cancers with immune cells having antitumor reactivity is among new approaches having great promise for cancer victims.

Questions for Review

1. Distinguish between the following:
   a. carcinoma and leukemia
   b. sarcoma and lymphoma

2. What is reverse transcriptase? Of what significance is it to cancer virus studies?

3. Summarize the current RNA virus cancer hypotheses.

4. What is a retrovirus?

5. a. What is an oncogene?
   b. What is a provirus?

6. Do oncogenic RNA viruses differ structurally from other RNA viruses? How?

7. Distinguish between a type B and a type C virus particle.

8. What role do DNA viruses have in carcinogenesis?

9. What promise do microorganisms or monoclonal antibodies hold for the treatment of tumors?

10. Of what value are biological response modifiers?

11. How are microorganisms used to detect potential carcinogens?

12. What is the relationship of CEA and alpha-fetoprotein to cancers?

Suggested Readings

BISHOP, J. M., "Oncogenes," *Scientific American,* March: 82–92, 1982. *The story of the sarc (sarcoma) gene and how it causes cancer.*

CAIRNS, J., "The Cancer Problem," *Scientific American* **233**:64, 1975. *A general review of the biological aspects of cancer. Emphasis is given to possible environmental causes of this dreaded disease state.*

FURMANSKI, P., J. C. HAGER, and M. A. RICH (eds.), *RNA Tumors, Viruses, Oncogenes, Human Cancer and AIDS: On the Frontiers of Understanding.* Dordrecht, The Netherlands: Martinus Nijhoff Publishers, 1985. *This publication contains a wealth of information concerning the molecular and related levels of interactions between retroviruses and various disease states.*

GALLO, R. C., "The First Human Retrovirus," *Scientific American* **256**:88–98, 1986. *An interesting article describing the features of retroviruses and the events leading to the uncovering of the cause of adult T-cell leukemia.*

KARPAS, A., "Viruses and Leukemia," *American Scientist* **70**: 277–295, 1982. *A short article presenting the evidence establishing certain viruses as the causative agents of leukemia in numerous lower animals.*

LYNCH, P. J., "Warts and Cancer," *American Journal of Dermatology* **4**:55–58, 1982. *The potential of human papilloma (wart) virus to cause malignant growths is discussed.*

NAGAO, M., T. SUGIMURA, and T. MATSUSHIMA, "Environmental Mutagens and Carcinogens," *Annual Review of Genetics* **12**:117–159, 1978. *A rather complete, detailed review of the effects environmental mutagens and carcinogens have on humans. Attention is given to screening methods and specific industrial and natural sources of dangerous substances.*

NICOLSON, G. L., "Cancer Metastasis," *Scientific American* **240**:66–76, 1979. *A view of current knowledge of how and which cancer cells spread throughout a tumor system (metastasis). The article stresses that metastasis, not the initial or primary tumor, is the threat to life.*

PETO, R., et al. (eds.), *Viral Etiology of Cervical Cancer:* Banbury Report 21. Cold Spring Harbor, N.Y.: Cold Spring Harbor Laboratory, 1986. *A comprehensive consideration of the viral cause of a most commonly encountered disease state.*

REIF, A. E., "The Causes of Cancer," *American Scientist* **69**: 437–447, 1981. *A well-written article describing the essential differences between normal and tumor cells and emphasizing the importance of environmental factors in causing cancer.*

TAYLOR-PAPADIMITRIOV, J. (ed.), *Interferons: Their Impact in Biology and Medicine.* Oxford, England: Oxford University Press, 1985. *A series of presentations examining the influence of interferons in immunity and on the cancer.*

31

Helminths and Diseases

Lines on the antiquity of parasites:
"Adam . . . had 'em." — *Gillian Strickland*

After reading this chapter, you should be able to:

1. List five structural, metabolic, or other modifications that favor the parasitic existence of worms.
2. Outline the life cycles of at least two different parasitic nematodes, cestodes, and trematodes.
3. Describe three characteristics of tapeworms that distinguish them from roundworms.
4. List three characteristics of flukes that distinguish them from roundworms and tapeworms.
5. List three preventative and control measures used against nematodes, cestodes, and trematodes.
6. Identify the basic structures of cestodes and flukes.
7. List two methods for the diagnosis of three nematode, three cestode, and three trematode infections.
8. Discuss the importance of parasitic worms to human welfare.
9. Solve a disease challenge involving a parasitic helminth.

Helminths or worms are among the major causes of human misery and death in the world today. Since the survival of these parasites typically depends on the other forms of life, the resulting diseases frequently pose important medical, socioeconomic, and political obstacles to human well-being. Chapter 31 deals with helminthic diseases from the standpoints of transmission and control and discusses those factors that determine the severity of helminthic diseases.

Worms are among the most abundant animals on earth. Most are free-living and do not cause serious problems. However, some worms are parasitic, causing diseases of great importance to humans and to domestic (Color Photographs 123, 124 and 125) and wild animals and plants. Parasitic worms live in varying environments and must adapt to conditions in order to survive a host body's cellular and chemical defenses. These multicelled forms differ in several ways from all other infectious agents, including bacteria, fungi, protozoa, and viruses. Examples of such differences include much longer generation times and the absence or low rate of direct multiplication within the host.

Figure 31–2 A cotton bollworm is shown with its damage done. This parasite is one of the serious pests of cotton. *[USDA photo by Jim Strawser.]*

Figure 31–1 Rotifers. A common rotifer, *Philodina,* shows various primitive organ systems. The bilateral symmetry, cephalization (definition of the head region), and organ systems familiar throughout the animal kingdom appear at this level of life. These organisms are free-living and commonly found in natural bodies of water such as ponds and streams.

One of the major subdivisions of parasitology is the study of **helminths.** This term, derived from the Greek word *helmins,* meaning "worm," designates both parasitic and free-living species of worms. The helminths are a large group of animals, including some of the most notorious forms of parasitic life, such as leeches and tapeworms. There are also, however, a great many free-living (Figure 31–1) and beneficial forms as well.

The worms of medical and agricultural importance discussed in this chapter include representative parasites from the phyla Nematoda (roundworms) and Platyhelminthes (flatworms, or flukes, and tapeworms). Several nematode species, as well as other parasites, are known to be damaging to plants, causing spoilage and destruction of agricultural products (Figure 31–2). These forms will also be discussed.

Equipment of Parasitic Worms

Adaptation of a helminth to a parasitic existence is in large part determined by the development of certain structural and metabolic modifications. Many intestinal worms (and certain others) have an espe-

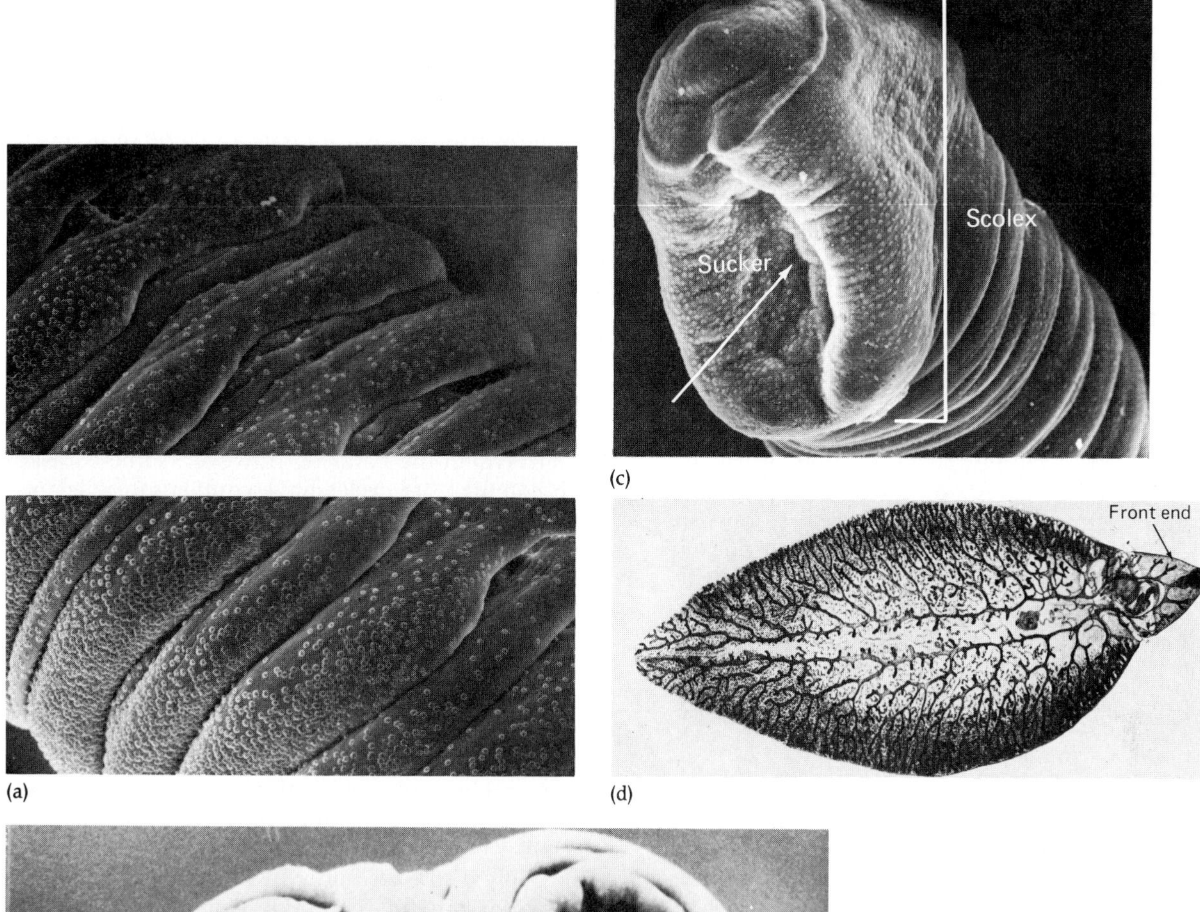

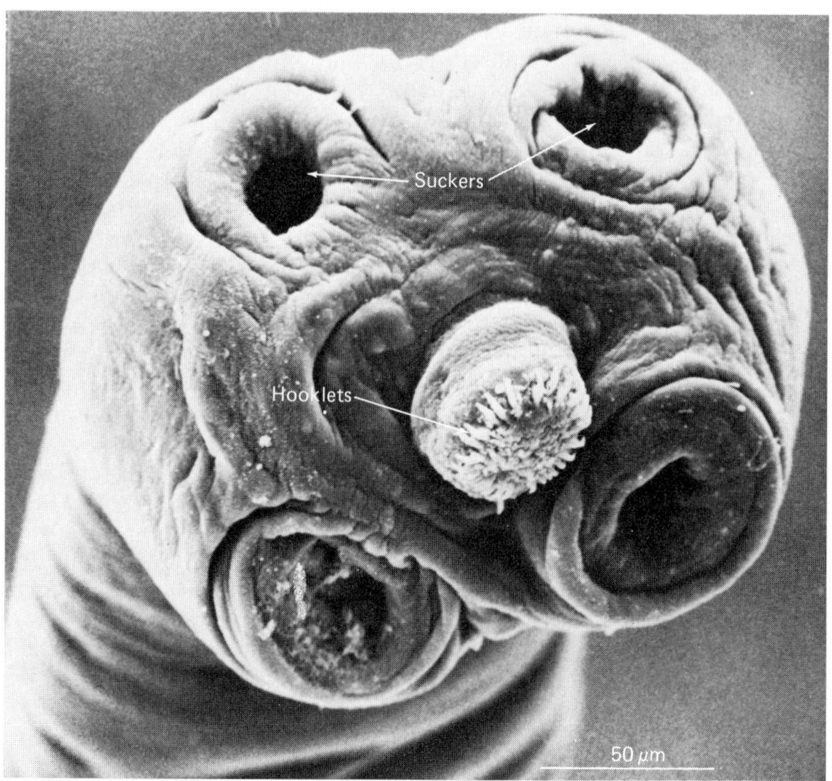

Figure 31–3
The equipment of parasitic worms. (a) A scanning micrograph showing the surface of a tapeworm's body covered with small structures that probably function in the uptake of nutrients. *[(a and c) Reproduced by permission of the National Council of Canada, from N. P. Boyce,* Can. J. Zool. ***54:****610–613, 1976.]* (b) The head (scolex) of the mouse bile duct tapeworm *(Hymenolepis microstoma).* The hooklets and suckers shown are used for attachment. *[P. W. Pappas, and H. R. Gamble,* J. Parasitol. ***64:****760–762, 1978.]* (c) A sucker of the cestode *Eubothrium salvelini.* (d) The reproductive network of the fluke *Fasciola hepatica.* The fluke is another type of flatworm. *[G. H. Sahla,* et al., J. Parasitol. ***58:*** *712–716, 1972.]*

cially hard outer covering, or integument (Figures 31–3a and 31–3c), enabling them to resist digestion by the host. Other modifications include (1) the possession of hooks (Figure 31–3b), spines, cutting plates, suckers (Figure 31–3c), various enzyme secretions, and additional weapons for purposes of attachment or penetration and (2) the development of elaborate reproductive systems (Figure 36–3d). The latter feature is represented by hermaphroditism in cestodes (tapeworms) and a large number of trematodes (flukes), whereby the reproductive organs of both sexes occur in the same animal. Most worm species do not multiply within the final, or **definitive, host.** More detailed descriptions of specific helminths follow.

Several helminths also are known to produce toxins that poison their hosts and contribute to their pathogenic effects. The filarial roundworm, *Wuchereria bancrofti* (woo-cheh-rer-EE-ah ban-CROF-tee) (Color Photograph 125), which lives in the lymphatic system after it invades the human body, produces numerous larvae. Many of these young forms move into the surrounding tissues, where they eventually die and release toxins that stimulate a powerful immune response from the infected host. These toxins, together with the host's reaction to them, result in swelling and thickening of the involved body area, sometimes causing dramatic enlargement (Figure 31–9). Other examples showing how helminths damage their hosts through poisoning include the effects of tapeworm waste products, which cause a condition known as *verminous intoxication* that produces nausea, vomiting, and dizziness; and the dropping down (prolapse) of the rectum in whipworm infections caused by a toxin that affects the nervous control of intestinal muscles. **[See effects of toxins in Chapter 22.]**

Life Cycles and Control of Parasitic Worms

Hosts

Some worms are capable of either a parasitic or a free-living existence. Such organisms are referred to as **facultative** parasites. Other forms, the **obligate** parasites, cannot complete their life cycles without the participation of the required host or hosts.

Many other terms denote the special functions, types of parasites, or states of parasitoses encountered. The term **infection** indicates the relationship of a parasite to its host; it applies to animal species found internally **(endoparasites).** The term **infestation,** however, is used for parasites that attach to the skin or temporarily invade its superficial layers **(ectoparasites).** Thus, a state of parasitosis is caused by either an infection or infestation by an animal parasite. Figure 31–4 shows the human body sites associated with worm infections.

Temporary parasites invade a host intermittently during their life cycles only to obtain nutrients. *Incidental* parasites may establish an infection in a host that ordinarily is not parasitized. Other types of diagnosed parasites include *coprozoic,* or *spurious,* forms (foreign organisms that pass through the intestinal tract without causing an infection) and *pseudoparasites* (particles misdiagnosed as parasites).

Several different types of hosts are distinguished. For example, the *definitive,* or *final,* host harbors the adult or sexually mature parasite; the *intermediate* host provides the environment for some or all of the immature or larval stages. Although human beings may be the only definitive host for several parasites, occasionally humans are accidental victims. When other animal species act as hosts for species that are parasitic for humans, such animals are called *reservoir* hosts. **[See importance of reservoirs in Chapter 21.]**

The life cycles of parasites can be complex, sometimes involving several hosts. Apparently, the more complicated life cycle greatly decreases the chance for survival of the parasite. However, some parasites with complex life cycles have compensating adaptations, including parthenogenesis (reproduction without a male) and overdeveloped reproductive organs.

Distribution and Transmission

The distribution of parasites such as worms is influenced by the habits of suitable definitive hosts and by favorable environmental conditions such as appropriate temperature and moisture and the availability of intermediate hosts. Many parasitic species are widely distributed only in the tropical regions of the world, but this condition appears to be disappearing; because of the revolutionary changes in the international transportation of cargo and passengers, opportunities for the distribution of parasites and their vectors have increased substantially.

Economic and social conditions are also important in the distribution of parasites. A low standard of living, inadequate sanitation, and ignorance of the means to control parasitic diseases favor the establishment of parasitoses.

The sources of parasitic diseases and their means of transmission include (1) domestic or wild animals in which parasites can live, domestic and sylvatic reservoirs, respectively (Color Photograph 123); (2) bloodsucking insects such as mosquitoes; (3) various foods containing immature infective parasites; (4) contaminated soil or water; and (5) humans and any portion of their environment that has been contaminated. Even the individual harboring a parasite can cause his or her own reexposure with the same species of parasite, a process known as *autoinfection.* **[See arthropod disease transmission in Chapter 21.]**

Worm Reproduction in Humans

Most helminth parasites of humans are unable to multiply within the human host. Thus, the number of worms to which the individual is exposed deter-

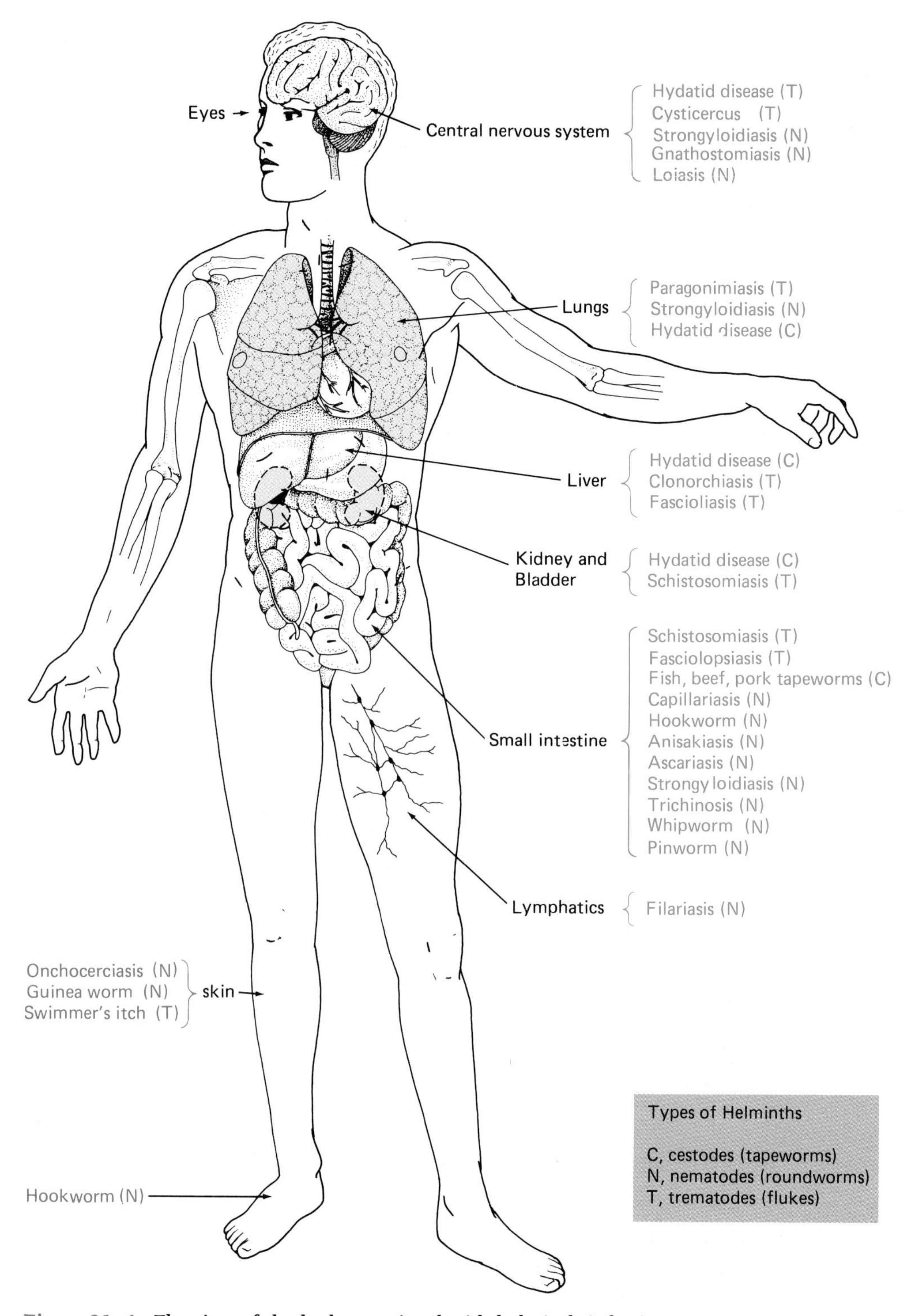

Figure 31–4 The sites of the body associated with helminth infection.

mines the course and severity of the disease. However, for some helminths the ability to multiply within humans has a dramatic effect on the development and length of infection. Take, for example, the roundworm *Strongyloides stercoralis* (stron-gee-LOY-deez ster-core-AL-liss). This parasite can reproduce int the soil as well as in an infected host.

In an infected human, all the parasitic adult worms are female. Reproduction in the host takes place by parthenogenesis (fertilization does not require males). Eggs apparently hatch rapidly within the large intestine, thus giving rise to large numbers of young forms (larvae) that penetrate the intestinal lining, establishing and extending the infection.

Microbiological Highlights

LUNG FLUKE INFECTION AND ENVIRONMENTAL FACTORS

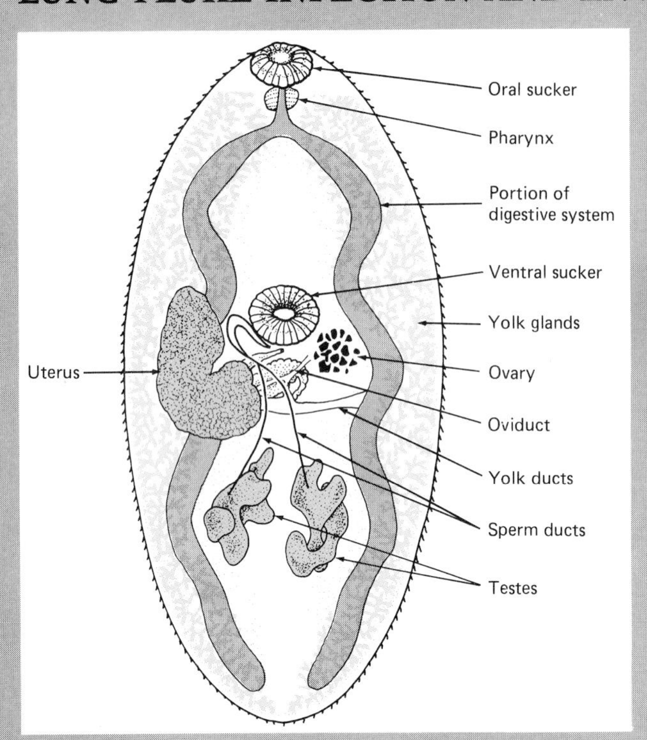

The lung fluke *Paragonimus westermani.*

Oriental lung fluke infection, or paragonimiasis, is caused by *Paragonimus westermani* (par-ah-gone-EE-mus wes-ter-MAN-ee) and occurs in areas of the world where freshwater crab, crayfish, or shrimp may be eaten raw, pickled, or undercooked. Although most reported cases are from countries in the Orient, the disease also is endemic in Africa and South America. Several cases caused by another species, *P. kelicotti* (P. kell-EE-cott-ee), have been reported from the province of Quebec, Canada, as well.

Recently lung fluke infection has been found to be present in a large number of immigrants from Indochina. In addition to their usual means of transmission in Southeast Asia, two factors have contributed to the increased number of cases. First, the political turmoil and hardships in that geographical area over the past 11 years forced many individuals to leave their homes and to alter their dietary habits. For example, the group known as the Hmongs fled from the highlands and mountain areas and were frequently forced to eat raw or uncooked foods for fear that open cooking fires might reveal their location to their political enemy. Secondly, many of the refugee camps for the Indochinese were situated along the Mekong River, a known habitat for the lungworm and a site for recent outbreaks.

Symptoms

The clinical effects observed in helminth infections usually depend upon the number and location of the parasites in the body and upon the general health and resistance of the host. Many infections are asymptomatic. If symptoms do occur, they may include anemia, excess numbers of eosinophils (Figure 31–5), fever, intestinal obstruction, muscular pains, and respiratory difficulties. Tables 31–2, 31–4, and 31–6 list clinical features of the more common helminth infections.

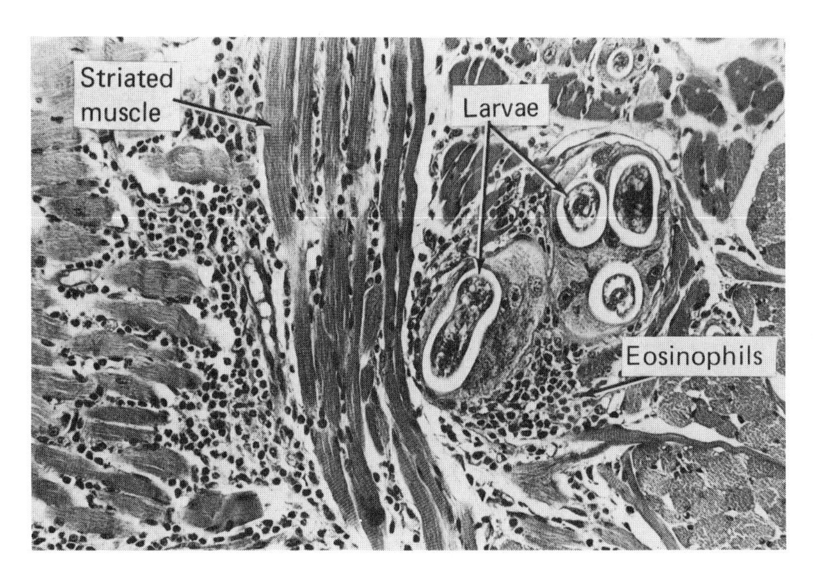

Figure 31–5 Host responses to worm infections such as pork roundworm infection, trichinosis, include an increased number of eosinophils. The encysted *Trichinella spiralis* larvae here have stimulated a moderate eosinophilic reaction. Original magnification 200×. *[L. Gustowska, E. J. Ruitenberg, and A. Elgersma,* Parasite Immunol. ***2:****133–154, 1980.]*

Severity of Helminth Infections

Three major factors determine how severe worm infections will be: the number of worms (*worm burden*), the location of the worms, and the reaction of the infected host to the parasite. Of these factors, the worm burden is the most critical. Thus, the degree of damage to the liver and spleen in schistosomiasis or the extent of an iron-deficiency anemia in hookworm infection depend on the number of worms. In some infections, the location of the worm is important. For example, a hydatid cyst (sheep tapeworm) in the liver may cause no significant problem; however, the same worm settled at the base of the brain may prove to be life-threatening (Color Photograph 127). Finally, the interaction between a worm and its host may influence the development of disease and/or complications. Thus, a victim of hookworm infection who is also poorly nourished is likely to develop anemia, even with a low number of worms. For all infections, the integrity of the host's immune system is always of major importance in determining the course and duration of helminthic diseases.

Laboratory Diagnosis

The finding and identification of ova (eggs), larvae (Figure 31–6), or adult worms are sufficient in most cases for the diagnosis of specific helminth infections. Skin tests and other immunological procedures are used in certain cases to identify susceptible individuals or to distinguish one type of infection from another. Most tests do not provide a reliable indication of the state of the helminth disease, the effects of treatment, or the future outlook for recovery **(prognosis).** The results of serologic procedures or tests, however, do confirm past infection. Among the serologic procedures used for diagnosis are complement fixation, hemagglutination, immunofluorescence, enzyme-linked immunosorbent assay (ELISA), and radioimmunodiffusion. Tables 31–2, 31–4, and 31–6 list the types of specimens and method of identification for the more common helminths. **[See serologic tests in Chapter 19.]**

Figure 31–6 The presence of a young worm (larval form) of the roundworm *Strongyloides stercoralis* (arrows) in a vaginal smear. Original magnification 400×. *[From E. Avram, M. Yakovlevitz, and A. Schachter. Acta Cytologica* ***28:****468–470, 1984.]*

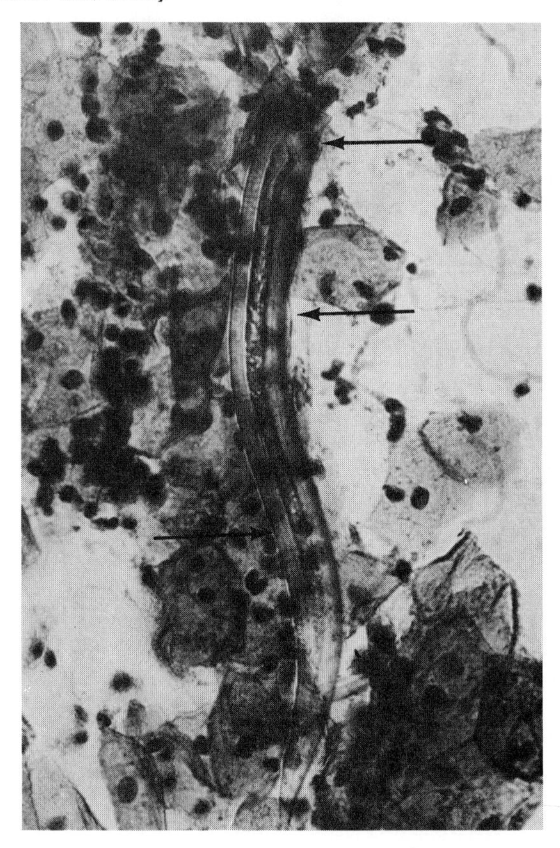

Microbiological Highlights

Approaches to Diagnosis

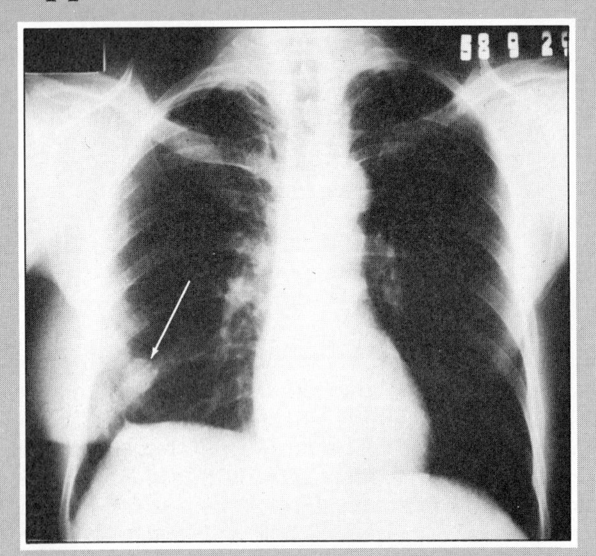

A chest x-ray showing a coin lesion in the right lobe of the lung.

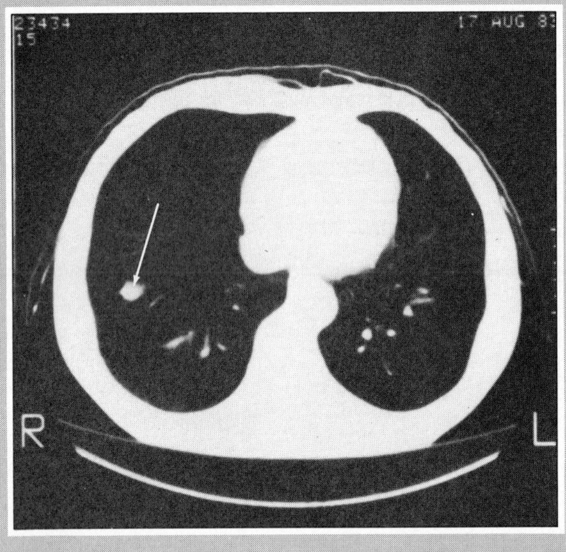

A CT scan showing the coin lesion found in the x-ray. *[From H. Yoshimura, and N. Akao,* Int. J. Zoon. *12:53–60, 1985.]*

A number of suspected helminth infections are detected clinically through applications of x-ray and computerized tomography (CT) scans. Human infection with *Dirofilaria immitis,* the dog heartworm, may involve the lungs, skin, and subcutaneous tissues. The causative worm is transmitted by mosquitoes from infected dogs. Infections appear to be increasing in various parts of the world, especially Japan. In pulmonary dirofilariasis, coin lesions, which are tumorlike masses measuring about 3 × 3 cm in diameter, are detectable as abnormal shadows on x-ray film and CT scans. Since other helminths affect the lungs, specific follow-up immunologic procedures such as intradermal testing, gel diffusion tests, and immunoelectrophoresis are necessary. Considering the prevalence of dog *Dirofilaria* infection and the degree of human exposure to infected mosquitoes, x-ray and CT scans will prove useful in uncovering pulmonary and related forms of helminth infections.

Treatment of Parasitic Infections

Several antiparasitic medications have been developed. Some of them are unlicensed, and their use is limited in various countries, including the United States. The increase in the number of Americans traveling or working in tropical areas where parasitic disease agents abide, and the significant number of new arrivals from such areas, seem to be adding to U.S. infection rates. Therefore, it may become essential to have appropriate medications licensed for use and available to physicians.

This chapter emphasizes the parasitic helminth infections that have attracted most attention in recent years. Classification is based upon a system now commonly used (see Appendix F).

Immunological Control

Approaches to the immunological control of helminth infections have involved the use of three types of immunogens: irradiated-attenuated live helminths, extracts of helminths' bodies, and metabolic or excretory/secretory substances produced by the *in vitro* culture of helminths. Of these immunizing materials, secreted antigens appear on the whole to be most effective and to induce protection against reinfection with cestodes and nematodes in lower animals. [See properties of immunogens in Chapter 17.]

Progress in the isolation and identification of functional immunogens has been limited. In addition, the only vaccines currently produced on a commercial scale utilize infective larvae irradiated to a level at which such forms are still capable of surviving, migrating, and producing functional immunogens, but possessing greatly reduced virulence. With the technical advances in genetic engineering and the laboratory *in vitro* cultivation of helminths, it is expected that large quantities of functional immunogens will be provided for immunological control. The applications of such immunizing material on a broad scale will have considerable economic value for farmers and will improve public health.

The Nematodes

Parasitosis caused by roundworms, or nematodes, can involve the skin and organs of the circulatory (Color Photograph 8), digestive, nervous, and respiratory systems (Table 31–1). Practically every tissue of the body is vulnerable to attack by certain nematode species. True nematodes are unsegmented, typically cylindrical and long, tapered at both ends, fundamentally bilaterally symmetrical, and without any appendages. These parasites all have digestive, nervous, and reproductive systems. The digestive system is complete. With relatively few exceptions, the sexes are in separate worms. Excretory systems are not found in all nematodes; adults of certain species lack such a system.

The worm's body is covered by a cuticle, which may have ridges, striations, or wartlike structures. These structures are considered in classification. The sizes of roundworms vary widely. Some species are microscopic (Color Photograph 125); others are several centimeters in length (Figure 31–7).

Figure 31–7 *Ascaris lumbricoides.* Female worms range in size from 20 to 35 cm in length; male specimens measure approximately 30 cm (see Table 31–1). *[Courtesy of the Dow Chemical Company.]*

TABLE 31–1 Representative Medically Important Nematodes

Organism	*Disease*	*Host Range*	*Location of Adult Forms in the Body*	*Geographical Distribution*
Ancylostoma duodenale	Old World hookworm infestation	Primarily humans	Duodenum and jejunum	Chiefly found in Africa, Europe, and the Orient
Angiostrongylus cantonensis	Angiostrongyliasis	Humans, rodents	Central nervous system, meninges, and eyes	Thailand
Anisakis species	Anisakiasis	Humans and marine animals	Small intestine	Europe and the Orient
Ascaris lumbricoides	Ascariasis	Humans and other vertebrates	Small intestine	World
Brugia malayi	Malayan filariasis	Humans, other primates, cats	Lymphatics	Far East
Capillaria philippinensis	Capillariasis	Humans, monkeys, and birds	Small intestine	Philippines, Thailand
Dracunculus medinensis	Dracunculiasis (guinea worm infestation)	Humans, dogs, cats, several wild mammals	Skin, connective tissue	Certain areas of Africa and Asia, and rarely in South America
Enterobius vermicularis	Enterobiasis or oxyuriasis (pin- or seatworm infestation)	Humans, especially children	Large quantities occur in cecum and appendix; the female worm is especially found in the rectum	Widespread
Gnathostoma hispidum, G. spinigerum, G. vivarina	Gnathostomiasis	Humans and other meat-eating animals	Eyes, central nervous system	Thailand
Loa loa	Loiasis (eye worm infection)	Humans, monkeys	Connective tissue, eyes	Central and West Africa
Necator americanus	New World or American hookworm	Humans	Small intestine	Generally in southern United States, Central and South America
Onchocerca volvulus	Onchocerciasis	Humans	Skin, subcutaneous connective tissue, eyes	Africa, tropical America
Strongyloides stercoralis	Strongyloidiasis	Humans, dogs, cats	Intestinal mucosa, lungs	World, but more commonly encountered in tropical areas
Trichinella spiralis	Trichinosis	Humans and several other mammals, including rats, rabbits, dogs, and wolves	Small intestine	World, encountered in areas where pork is eaten
Trichuris trichiura	Whipworm infestation	Humans	Cecum	World
Wuchereria bancrofti	Bancroft's filariasis (elephantiasis)	Humans	Lymphatics	Australia, Eastern Europe, Near East, Orient, Central and South America, Mediterranean, and central Africa

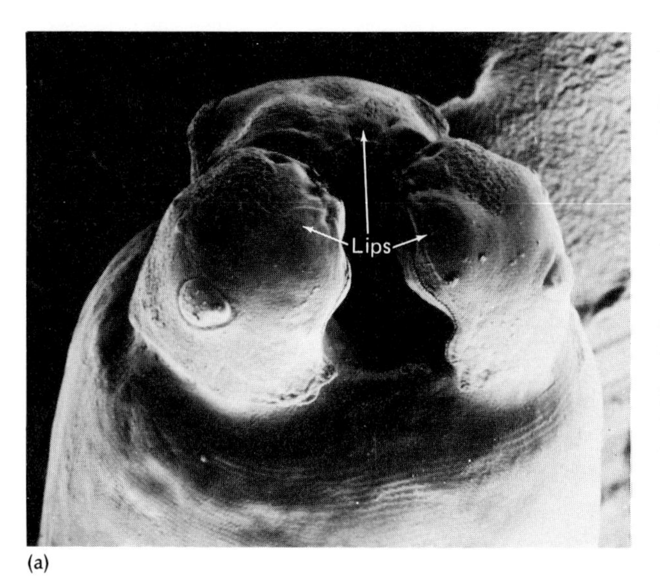

(a)

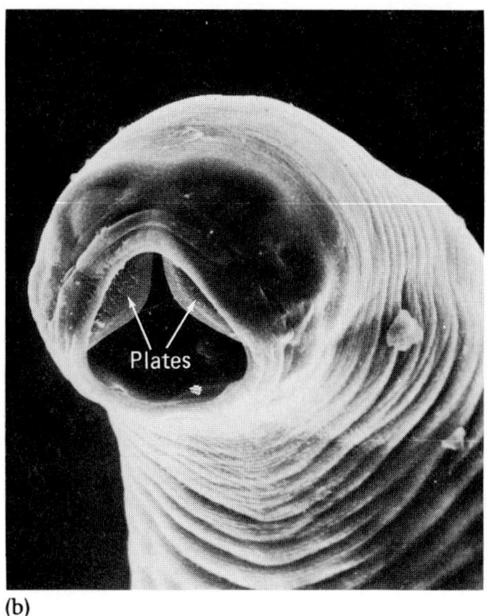

(b)

Figure 31–8 (a) *Ascaris suum.* A scanning electron micrograph showing the attachment surface of the nematode's head with three lips. This worm infects both humans and pigs. *[Dr. R. W. Weise, Department of Pathology, Wayne State University School of Medicine.]* (b) The head of an adult hookworm, *Necator americanus,* shown by scanning electron microscopy. Note the internal biting plates. *N. americanus* is one cause of hookworm. *[D. J. McLaren,* Int. J. Parasitol. ***4****:25–37, 1974.]*

The mouths of primitive nematodes are surrounded by three lips (Figure 31–8a). In *Ancylostoma duodenale* (an-sil-oss TOE-mah doo-oh-den-AL-le) and *Necator americanus* (nee-KAY-tore ah-mer-ee-CANE-uss), the Old and New World hookworms, respectively, these regions are equipped with cutting plates or teeth (Figure 31–8b). Such modifications, as well as others that will be mentioned, are important in nematode identification. **[See medically important nematodes in Appendix F.]**

Life Cycle

The life cycles of nematodes vary. However, there are fundamental stages involving an adult worm, egg, and larvae. The eggs of some roundworms are discharged in the host's feces, mature on the ground, and may either be swallowed by a host or hatch out larvae that undergo further development and consequently become infective. With infective larvae, penetration of the host's skin leads to parasitosis. In still other nematode life cycles, larvae are discharged in the fecal matter, and the infective forms that develop are capable of penetrating the host.

Some parasites utilize two hosts. *Trichinella spiralis* (trick-a-NEL-ah speh-RAH-liss), the causative agent of pork roundworm infection, or trichinosis, is one such worm. All the stages of its life cycle can be completed in the pig. Human infection usually depends on ingestion of encapsulated larvae in the striated muscles of the pig as inadequately cooked pork (Color Photograph 126).

These and other examples of life cycles will be described, along with the features of specific roundworms and the infections they cause.

Larval Forms

Frequently, in describing the life cycles of these worms, references are made to *filariform* and *rhabditiform* larvae. These terms reflect the type of esophagus of the larval forms. The esophagus of filariform larvae is a tube with a uniform diameter throughout. The esophagus of rhabditiform larvae expands toward the rear of the worm, forming a swollen structure called a *bulb* that has a valve mechanism.

The geographical distribution of medically important roundworms is shown in detail in Table 31–1.

Representative Animal Roundworm Infections

The effects of roundworm infections, and their laboratory diagnoses and treatment, are listed in Table 31–2.

ANISAKIASIS

Many species of *Anisakis* (an-ee-SAK-iss) are parasites in the stomachs of birds, marine fishes, and mammals. The larvae of these nematodes are widely distributed. Most cases of human infection have been reported from Japan, Europe, portions of the eastern United States, and Scandinavia, where raw fish is commonly eaten and relished.

TABLE 31–2 **Characteristics of Representative Nematode (Roundworm) Infections**

Disease	*Clinical Features*	*Location of Adult Worm*	*Specimen of Choice*	*Laboratory Diagnosis*	*Treatment*
Anisakiasis	Intestinal blockage, sharp abdominal pain, abscesses, and tumorlike growths at site of worm attachment	Stomach	Biopsy	Microscopic identification of larvae	Mebendazole (Vermox)
Filariasis (elephantiasis)	Lymphatic inflammation, chills, fever, inflammation of testes with sudden enlargement, obstruction of lymphatics, thickened and cracked skin[a]	Blood, lymphatics	Blood	Microscopic identification of larvae	Diethylcarbamazine citrate
Hookworm	Iron deficiency anemia, abdominal pain, protein deficiency, delayed puberty, lowered antibody response, "potbelly" in children[b]	Small intestine	Feces	Microscopic identification of ova or adult worms	Mebendazole, pyrantel pamoate, or thiabendazole
Pinworm	Tickling or intense itching in perianal area	Intestine, anus	Cellophane-tape swab[c]	Microscopic identification of ova	Improved hygiene and pyrivinium pamoate, piperazine-citrate, or mebendazole
Trichinosis	Diarrhea, muscular pain, nervous disorders, eosinophilia, respiratory complications[b]	Intestine	Muscle	Microscopic identification of larvae; serodiagnostic tests, latex agglutination, immunofluorescence	Mebendazole, pyrantel pamoate

[a] Most symptoms here are caused by wandering larvae.
[b] Most light infections are asymptomatic.
[c] A short piece of cellophane tape, held against a flat, wooden applicator, with the sticky surface out. This device is pressed against the perianal area to obtain a specimen. The tape is reversed on a slide for microscopic examination.

Transmission and Life Cycle. The adults of *Anisakis* species live in the stomach of marine mammals such as dolphins, porpoises, and whales. These parasites use shellfish and other marine forms as first and second intermediate hosts, respectively. Humans, as well as the definitive host, become infected by eating raw, salted, or pickled marine fish containing the infective larvae. Serious effects occur upon infection (Table 31–2). Fatalities have also been reported.

While most cases of anisakiasis are caused by *Anisakis* larvae, a few cases have been reported where the cause is the larvae of another roundworm, *Terranova* (terr-a-NOV-ah).

Prevention. The proper cooking of fish dishes and related products, and the deep-freezing of such foods, are relatively effective procedures in preventing anisakiasis.

FILARIASIS

The filaria are long, slender nematodes that invade portions of the lymphatic system and various other tissues of the human body (Figure 31–9). As a group, these helminths undergo a unique stage in their life cycle, the production of motile larval forms called **microfilaria** (Color Photograph 125). Although there are at least six species of filaria for which human beings are believed to be the defini-

Figure 31–9 A victim of elephantiasis. *[Courtesy of the World Health Organization, Geneva.]*

tive host, the discussion here is limited to the two most important forms of infection, Bancroftian and Malayan filariasis.

Bancroftian filariasis has been known since 600 B.C. The causative agent, *Wuchereria bancrofti,* is transmitted by the bite of mosquitoes, specifically *Culex fatigans. Brugia malayi* (broo-GEE-ah may-LAY-eye) is the causative agent of Malayan filariasis. Selected characteristics of other filarian worms—e.g., *Loa loa* (LOW-ah LOW-ah) and *Onchocerca volvulus* (ON-CHOH-SIR-kah vol-VOO-lus)—can be found in Table 31–1.

Transmission and Life Cycle. The filaria capable of infecting humans have similar life cycles. Mosquito vectors include species of *Aëdes, Anopheles, Culex,* and *Mansonia.* Microfilaria are introduced into humans by the bite of an infected mosquito. The human host provides a suitable environment for the further development of the parasite.

The microfilaria of *W. bancrofti* (Color Photograph 125) appear in the peripheral circulation of the human host only at night (nocturnal periodicity). They are ingested by the appropriate species of mosquito during the course of a blood meal from an infected individual.

Once in the body, the microfilaria usually pass into the lymphatics and occasionally into subcutaneous tissues, where they mature into adult male and female worms. After mating, the female releases her offspring into the lymphatic system of the host. The larvae become widely distributed in the bloodstream and exhibit the nocturnal pattern described. If suitable vectors do not ingest the microfilaria, the parasites die and disappear from the circulation. If ingested by mosquitoes, the larvae develop into active, elongated, infective forms. These larvae migrate to the mosquito's mouth parts, and from there they can be introduced into the human host at the time of the next blood meal.

The life cycle of *B. malayi*—also known as *Wuchereria malayi* or *Filaria malayi*—is similar to that of *W. bancrofti,* except that these filaria are present in the peripheral blood for approximately 20 hours a day.

Prevention. Measures taken to prevent or control filariasis are similar to those for other arthropod-borne diseases.

HOOKWORM INFECTION

The term "hookworm" was coined either on the basis of the curved or bent front end of the worm or on the basis of the hooklike supporting structures found in the posterior extremity of male worms, the bursa (Figure 31-10c shows a copulatory structure of the male).

Transmission and Life Cycle. The two hookworm species most important for humans are *Ancylostoma duodenale* and *Necator americanus.* Both have the same general life cycle. The ova of hookworms (Figure 31–10a) are passed in the fecal matter of an infected person. These eggs develop in the soil and give rise to rhabditiform larvae, which develop further into infective filariform larvae (Figure 31–10b). Under favorable conditions, the larvae may live for several months in the soil. A hookworm infection is acquired when these larvae penetrate the skin, often through the soles of the feet. Once in the body, the parasites are carried to the heart and lungs by the bloodstream. After further growth and development, the larvae penetrate the lung air sacs, ascend the respiratory tree, and are swallowed. Thus, the parasites reach the small intestine, where they mature into adult male and female worms. Mating takes place in this area, the female lays eggs (which are passed out with the host's feces), and the cycle begins again.

Prevention. Measures to prevent hookworm infections include (1) the use of footwear in endemic areas, (2) improvement of general sanitation (especially disposal of human fecal matter), (3) mass

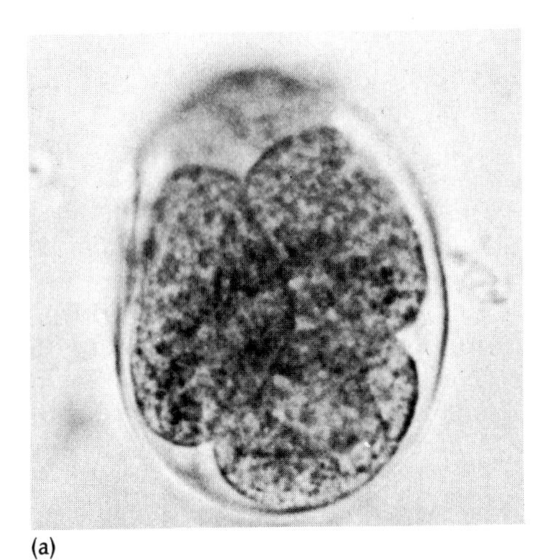

(a)

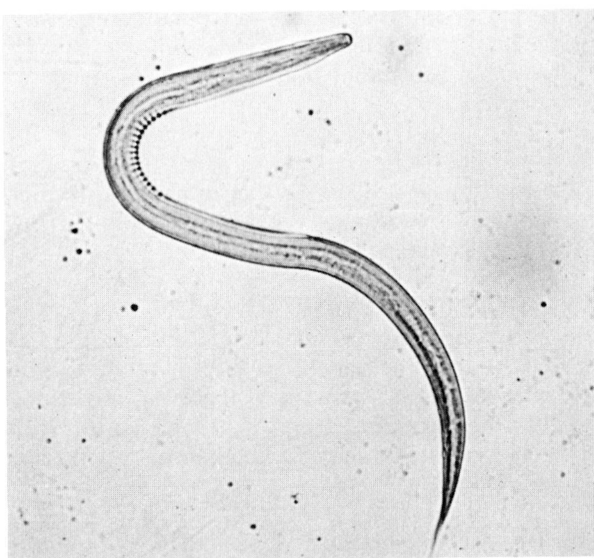

(b)

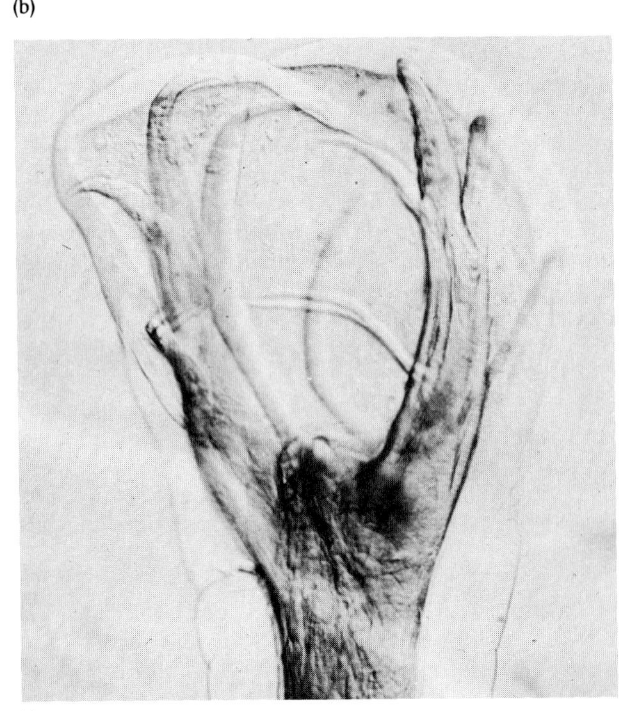

(c)

treatment of infected persons, and (4) public education about the parasite and its effects.

ENTEROBIUS VERMICULARIS INFECTION (PINWORM, SEATWORM)

Enterobiasis (oxyuriasis) has been known since ancient times. Its incidence is higher in temperate and colder climates than in tropical regions. Infections are more prevalent where large groups congregate, as in schools, mental institutions, and even large families. Children appear to have higher rates of enterobiasis than adults.

TRANSMISSION AND LIFE CYCLE. Infection results from the ingestion of ova containing infective larvae. Such infections can be acquired through the handling of fomites, inhalation of eggs (Figure 31–11), or a transfer of ova to the mouth from the perianal region via the fingers.

Ingested ova hatch in the small intestine, forming infective larvae that migrate to the first portion of the large intestine. Here maturation and mating of the parasites occur. The female worm, carrying fertilized eggs, migrates to the perianal region, releasing masses of eggs as it crawls. The deposited eggs are fully embryonated and infective. Eventually, the female's body ruptures (Figure 31–11).

The migration of worms produces a tickling or intense itching, which causes the individual to scratch the area, often contaminating the fingers. Infection of other individuals by means of contaminated bedding or clothing is frequently encountered. Larvae that hatch in the perianal region eventually may migrate back into the large bowel. This particular type of infection is referred to as *retrofection.*

PREVENTION. Preventive measures include treatment of all infected individuals and the development of good personal hygiene habits.

TRICHINOSIS

Trichinella spiralis was first observed in encapsulated larval form in the early 1800s. The pathogenic significance of trichinosis in humans was not truly recognized until 1860, when Friedrich Albert von Zenker showed the serious and often fatal consequences of the disease.

T. spiralis infection was first found to be a public health problem in the pork-consuming populations of Europe. However, trichinosis has a much wider distribution today, especially in the United States.

Figure 31–10 Stages in the life cycle of hookworms. (a) The ovum of *Necator americanus.* Hookworm ova range from 56 to 76 μm in length and 31 to 49 μm in width. (b) An infective (filariform) larva of *N. americanus.* These forms average 700 μm in length. (c) The expanded posterior end or bursa of the male worm.

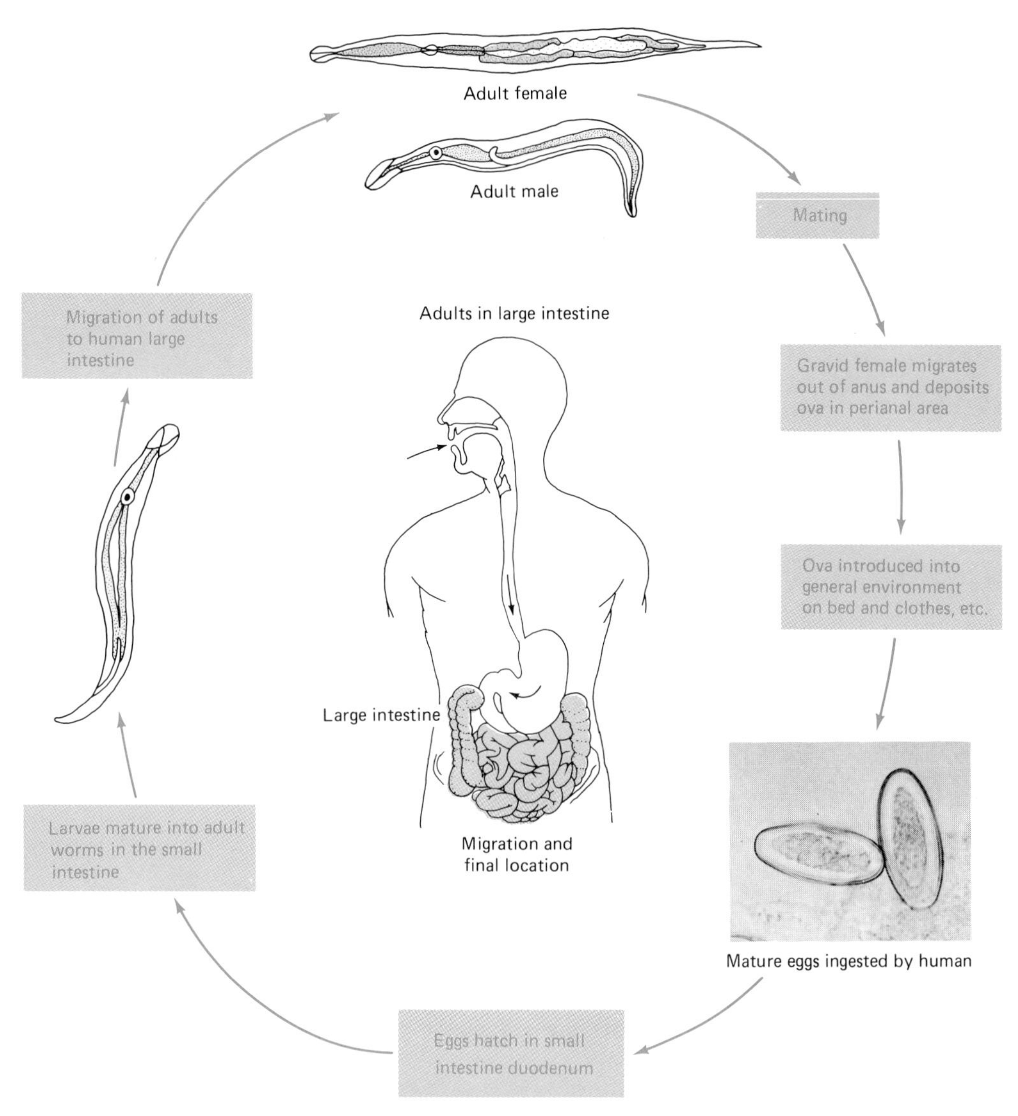

Figure 31–11 Life cycle of the pinworm *Enterobius vermicularis.*

Transmission and Life Cycle. *T. spiralis* infection generally results from consumption of improperly cooked or inadequately processed pork (Color Photograph 126). The parasite exhibits very little host specificity, as evidenced by trichinal infections associated with bear meat and walrus flesh. The life cycle of this parasite is shown in Figure 31–12.

Prevention. The proper cooking of pork and related products and the freezing of pork at $-10°C$ or less for more than 24 hours are relatively effective procedures in preventing *T. spiralis* infection.

Parasitic Nematodes of Plants

The roundworm parasites of plants are often called *eelworms.* The enormous damage they cause cultivated plants is responsible for agricultural losses amounting to millions of dollars per year. Free-living and parasitic plant nematodes occur in great numbers in all types of soil capable of supporting plant growth. Nematodes may either damage susceptible plants directly or introduce microorganisms such as viruses, which in turn cause plant destruction and death.

Damage to plants attacked by nematodes is caused primarily by the feeding of the nematodes on the plant tissues (Figure 31–13a). The survival of the worms depends on the availability of suitable hosts. Some nematodes attack the root surfaces of plants as well as those parts that are above ground. These worms are *ectoparasitic* (Figure 31–13b). Other worms enter and attack the internal parts of plants. Such roundworms are referred to as *endoparasitic.* Formation of a **gall** (a type of tumor) is one of the plant's responses to endoparasitic activity.

Parasitic plant roundworms have a special feeding organ known as a *stylet* or *spear.* Its hollow form enables the worm to pierce plant tissue or cell walls

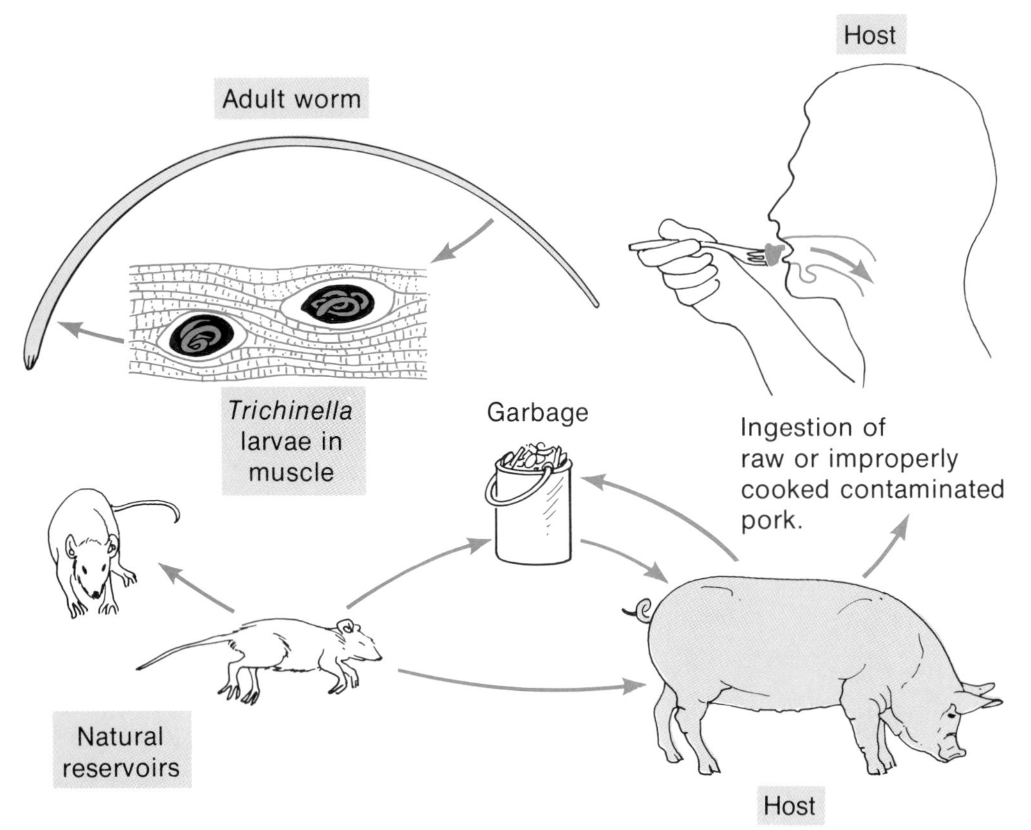

Figure 31–12 The life cycle of *Trichinella spiralis.* In the case of human trichinosis infections, the full life cycle of the parasite usually involves two omnivorous (capable of eating every kind of food) mammals. *T. spiralis* is commonly found in rats and in swine raised on uncooked garbage. The ingestion of meat containing encysted larvae begins the parasite's life cycle in the host. The ingested larvae excyst, invade the intestinal mucosa, and soon develop into adult female and male worms. Mating occurs and the female releases larvae, which eventually enter the host's bloodstream and are distributed throughout the body. These parasites are often filtered out in striated muscle where they reach full size, embed in the host's muscle tissue and eventually become calcified.

Figure 31–13 A potential biological control agent for weeds. (a) The nematode *Nothanguina phyllobia* causes lethal infections of the weed silverleaf nightshade *(Solanum elaegnifolum).* This weed interferes with cotton production in the southwestern United States. The leaf-damaging effects caused by the nematode are evident. (b) Larvae of the roundworm on the weed's leaf surface before penetration. *[C. C. Orr, J. R. Abernathy, and E. B. Hudsepth,* Plant. Dis. Rep. ***59****:416–418, 1975.]*

(a)

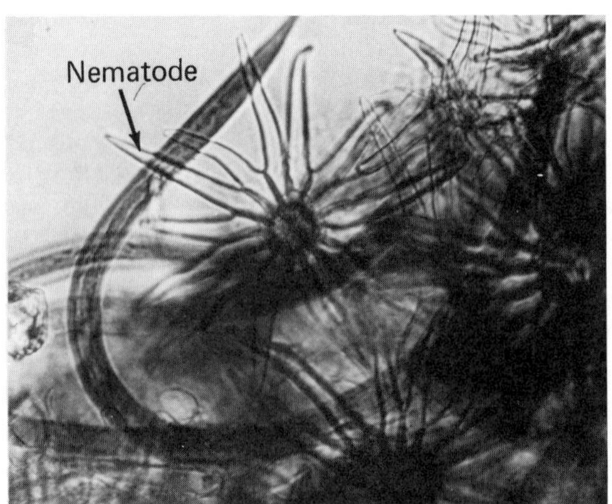

(b)

and to suck out cellular contents. Nematodes may enter a plant to feed, may feed from the outside, or may be only partially embedded. The most common effects of roundworm infections are the rotting of the attacked plant parts and adjacent tissue and the development of *galls* (knots) and other abnormal growths. Either of these effects can interfere with the development of a plant and cause shortening of roots and stems, twisting, crinkling, or death of parts of stems and leaves, and other abnormalities.

Root knot galls distort plant tissue that conducts nutrients to the upper portions of the plant. Injuries to stems and leaves also interfere with normal growth. The effects of plant nematode parasites may not be evident soon after the organisms are introduced into a field. However, once the parasites have had sufficient time to multiply, their effects become obvious. For example, crops attacked by root-damaging nematodes give the impression of a deficiency of water and fertilizer even though abundant quantities of these materials are available in the soil. Heavily infected plants may also exhibit lighter coloration and die prematurely.

In physical appearance, plant nematodes are similar to the ascarid worms associated with vertebrates (Figure 31–7), except that they are much smaller. The life cycles of the worms consist of eggs, larval stages, and adults. Details of the life cycle vary considerably according to species and environment.

Plant parasitic nematodes have several enemies in the soil. At any stage of their life cycle they may be captured and eaten by other soil animals, such as insects or predatory, free-living roundworms. Certain soil fungi almost seem designed to catch nematodes. Some of these traps are hyphal loops (Figure 8–1) that close when roundworms start to crawl through them. Other fungal traps have surfaces to which nematodes stick. In either case, the fungus penetrates the body of the worm and subsequently kills it.

The Platyhelminths

The platyhelminths, or flatworms, are among the most primitive groups of animals to have bilateral symmetry, the basic body plan exhibited by all higher forms of animal life. Flatworms range in size from less than a millimeter to several meters in length. Tapeworms have been reported to measure 75 m. The flattened bodies of these worms may be slender, broadly leaflike, or long and resembling a ribbon.

The platyhelminths include both free-living and parasitic forms. Most free-living worms are bottom dwellers in marine or fresh water or live in moist areas on land; they are found under stones and other hard objects in freshwater streams. The common planaria are found in such environments.

The true parasitic flatworms include the classes Cestoda (tapeworms) and Trematoda (flukes). The members of both groups have undergone several adaptative changes that make them well equipped for their parasitic existence. For example, they possess adhesive organs such as suckers, excessive reproductive capacity, and no digestive system. (Additional details are given in the following sections and in Table 31–4.)

The Cestodes

The cestodes are endoparasites, which as adults live in the intestines of vertebrate hosts. Several species are capable of parasitizing humans (Table 31–3). [See medically important cestodes in Appendix F.]

The bodies of cestodes are flattened front-to-back and consist of (1) a head or attachment organ (**scolex,** plural *scolices*), which, depending on the particular species, may or may not have hooks or suckers (Figures 31–3b and 31–3c); (2) a neck region (the region where new body segments, or **proglottids,** grow); and (3) the body, properly called the *strobila.* Figure 31–14 shows the relationship of these tapeworm parts to one another.

The hooks of the armed scolices of tapeworms are attached to a fleshy elevation region, or rostellum (Figure 31–3b), which may or may not be retractable (Color Photograph 127). The armed scolex is characteristic of the pork tapeworm *Taenia solium* (TEE-nee-ah SOH-lee-um); the unarmed structure is found in the beef tapeworm, *T. saginata* (TEE-nee-ah sah-jin-AH-tah). Still other tapeworm species, such as the fish tapeworm, *Dibothriocephalus latus* (dye-both-ree-oh-SEFF-al-us LAY-tus), have heads in the form of a trenchlike groove (Figure 31–15). Usually scolices of this type have weak suction power.

The strobila is composed of a series of proglottids, which, proceeding from the neck of the tapeworm, include first sexually immature units, then sexually

Figure 31–14 A typical tapeworm and its components.

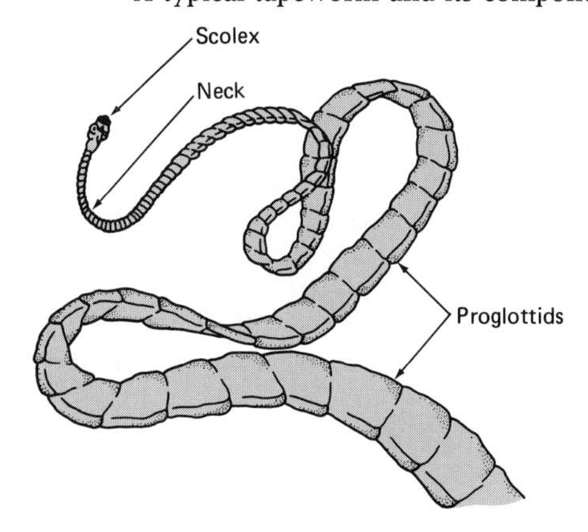

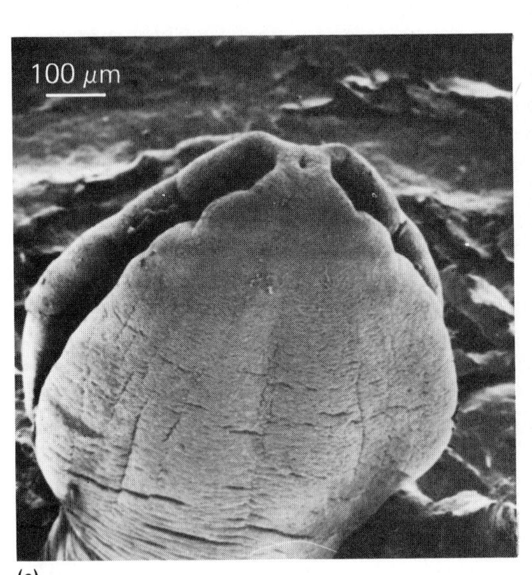

(a) (b)

Figure 31–15 The fish tapeworm *Diphyllobothrium ditremum.* (a) The scolex. (b) The parasite attached to the intestinal lining of a host (arrows). *[K. Anderson,* Int. J. Parasitol. *5:487–493, 1975.]*

mature segments (Figure 31–16), and finally gravid proglottids filled with eggs. These body segments account for the ribbonlike appearance of tapeworms (Figure 31–14). Usually each proglottid has a complete set of female and male reproductive organs, making tapeworms **hermaphroditic.** Fertilization may occur between proglottids of the same or different worms *(reciprocal fertilization)* or between the sexual organs of a single body segment *(self-fertilization).*

Cestodes do not have circulatory, digestive, respiratory, or skeletal organs. They have no special sense organs but do have free sensory nerve endings.

The life cycles of cestodes are complex. Most utilize a nonhuman host for the development of infective larvae. The eggs of the cestodes, with one exception (Figure 31–17b) do contain lids (they are operculated; Figure 31–17a). They vary in appearance, as well as in the protective coverings of embryos (embryonic membranes). The embryo is referred to as the *oncosphere.* Cestodes, with few exceptions, involve at least two hosts of different species. The adult worms are found only in vertebrates. Other stages may be found in either invertebrates or vertebrates. The life cycle of the beef tapeworm is shown in Figure 31–18. Table 31–4 lists the signs and symptoms, features of laboratory diagnosis, and treatment of several tapeworms that affect humans.

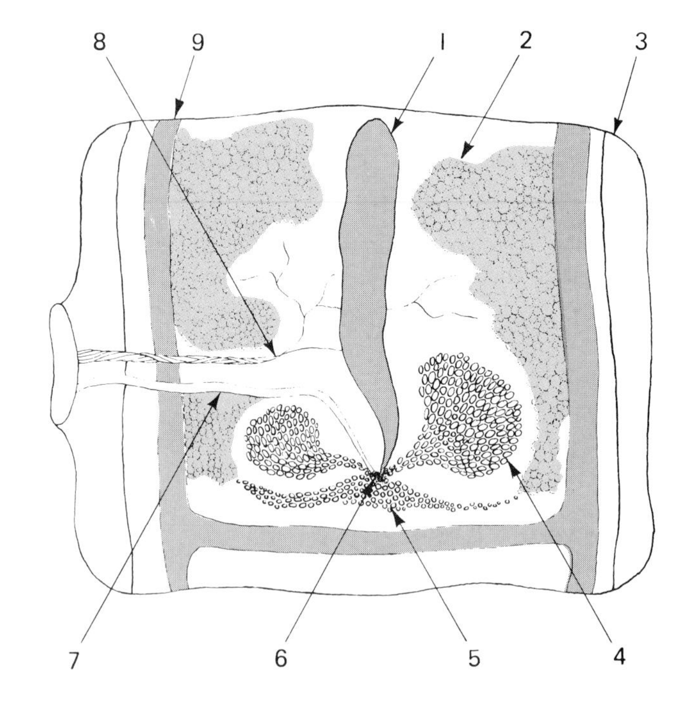

Figure 31–16 A mature proglottid from *Taenia solium,* the pork tapeworm.

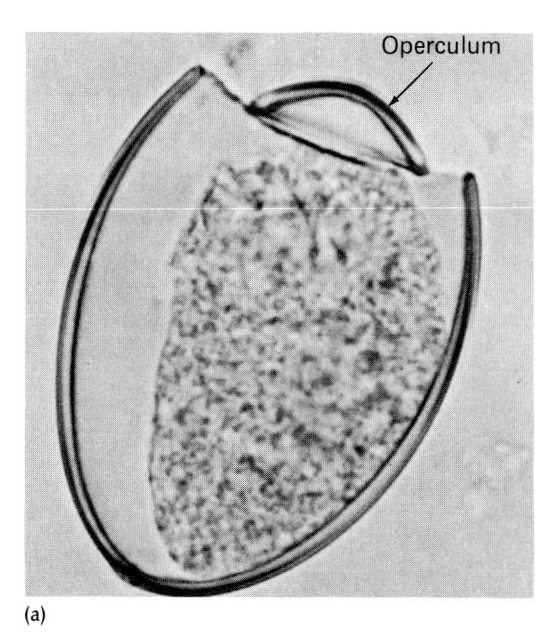

(a)

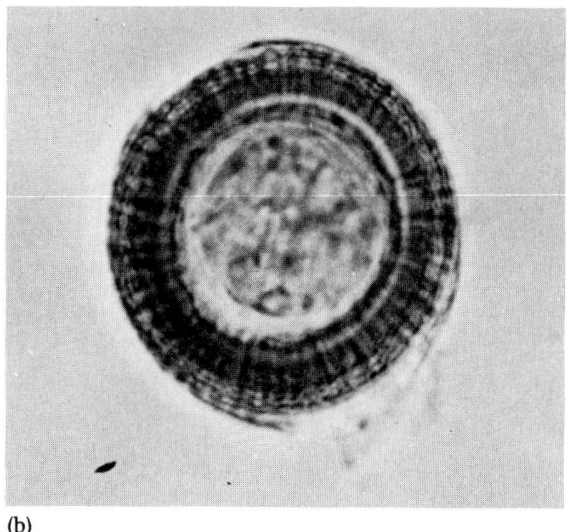
(b)

Figure 31–17 Representative cestode eggs. (a) An operculated (with lid) egg of *Diphyllobothrium latum,* the fish tapeworm. (b) A nonoperculated egg of *Taenia solium.*

Figure 31–18 The beef tapeworm *Taenia saginata* and its life cycle. Several important structures of the helminth are shown, including those of a typical adult tapeworm and a mature proglottid.

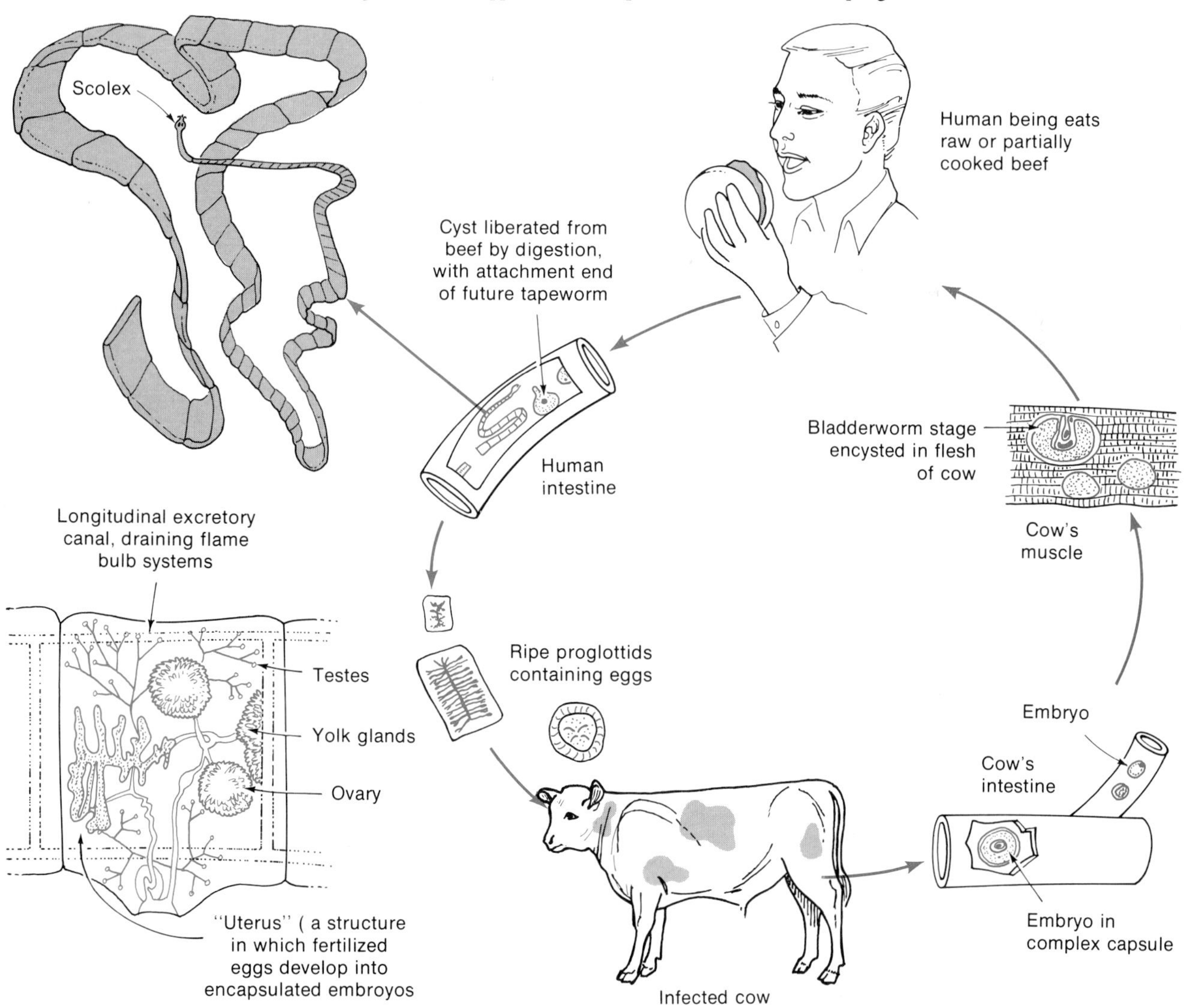

TABLE 31–3 Representative Medically Important Cestodes

Organism	*Disease*	*Host Range*	*Location in the Body*	*Geographical Distribution*
Cysticercus solium	Cysticercosis	Humans and other meat-consuming mammals	Small intestine, eyes, and central nervous system	Areas in which pork is consumed
Dibothriocephalus latus (also known as *Diphyllobothrium latum*)	Dibothriocephaliasis (Diphyllobothriasis)	Humans, dogs, cats, bears, other fish-consuming mammals	Intestine	Orient, Europe, United States, most of the great lakes of the world
Echinococcus granulosis	Hydatid disease	Cattle, deer, dogs, foxes, horses, humans, jackals, pigs, rabbits, sheep, wolves	Bone, brain, heart, kidney, liver, lungs, spleen	Australia, Middle East, portions of South America, United States
Hymenolepsis diminuta	Hymenolepiasis diminuta	Rats, mice, humans, dogs	Intestine	Widespread; sporadic in most instances
Hymenolepsis nana	Hymenolepiasis nana	Rats, mice, humans	Intestine	United States, Asia, Europe; sporadic in many instances
Taenia saginata	Taeniasis saginata	Cysticercus in cows, adult worms in humans	Small intestine	Areas in which beef is consumed
Taenia solium	Taeniasis solium	Cysticercus in pigs, adult worms in humans	Small intestine	Areas in which pork is consumed

TABLE 31–4 Characteristics of Representative Tapeworm Infections

Disease	*Clinical Features*	*Location of Adult Worm*	*Specimen of Choice*	*Laboratory Diagnosis*	*Treatment*
Diphyllobothriasis (fish tapeworm)	Abdominal discomfort, diarrhea, nausea, vitamin B_{12} deficiency (pernicious anemia), general weakness	Intestines	Feces	Microscopic demonstration of ova or proglottids	Nicolsamide, praziquantel
Hydatid disease (sheep tapeworm)	Dependent on body location, and includes blockage and/or interference with organ function	Various locations, including nervous system and bone marrow	Surgical specimens	1. Intradermal skin test (Casoni) 2. Medical x-ray examination 3. Complement fixation, fluorescent antibody, etc.	Mebendazole, surgical removal of cysts
Beef tapeworm	Abdominal pain, headache, loss of appetite, localized sensitivity to touch, nausea[a]	Intestines	Feces	Microscopic identification of scolex or gravid proglottid	Nicolsamide, praziquantel
Pork tapeworm	Mild symptoms, if any[a]	Intestines	Feces	Same as for beef tapeworm	Nicolsamide, praziquantel

[a] Most cases are asymptomatic or have poorly defined distinctive symptoms.

The Trematodes

The Trematoda, or flukes, are another parasitic class of platyhelminths. Although thousands of species are known today, not all flukes are of medical significance. Table 31–5 lists some of the parasites belonging to this group that cause disease in humans (Figure 31–19). **[See medically important trematodes in Appendix F.]**

Generally speaking, flukes are flattened (somewhat in the form of a leaf or pear), lack a distinct and segmented body, and are covered by an external cuticular layer. They range in size from less than one millimeter to several centimeters, depending on the species.

As shown in Figure 31–20, these worms possess suckers and excretory, digestive, and reproductive systems. The digestive systems of the medically im-

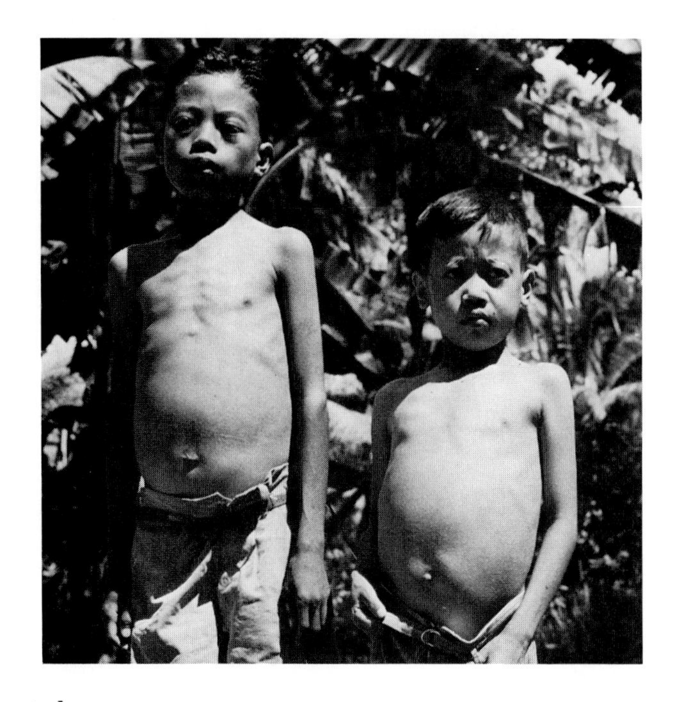

Figure 31–19 Victims of schistosomiasis, blood fluke infection. Note the enlarged abdomens and puffy faces. *[Courtesy of the World Health Organization, Geneva.]*

TABLE 31–5 Representative Medically Important Trematodes

Organism	*Disease*	*Host Range*	*Location in the Body*	*Geographical Distribution*
Opisthorchis (Clonorchis) sinensis (Chinese liver fluke)	Clonorchiasis	Humans, dogs, cats	Liver, bile ducts	China, Japan, Korea, Indochina
Fasciola hepatica (sheep liver fluke)	Fascioliasis	Humans, sheep, goats, cattle	Liver, bile ducts	Sheep-raising regions
Fasciolopsis buski	Fasciolopsiasis	Humans, pigs	Small intestine (duodenum, jejunum)	Orient
Heterophyes heterophyes	Heterophyiasis	Humans, cats, dogs	Small intestine	Near East, Far East
Metagonimus yokogawai	Metagonimiasis	Humans, cats, dogs, pigs	Small intestine	Far East, Siberia, Balkan states
Opisthorchis felineus	Opisthorchiasis	Cats, occasionally humans	Biliary and pancreatic ducts	Mainly in Central and Eastern Europe, USSR
Paragonimus westermani	Paragonimiasis	Humans, cats	Lungs	Far East, Nigeria, Belgian Congo, Central America
Paragonimus kallkoti	Paragonimiasis	Mink, humans	Lungs	Midwest United States
Schistosoma haematobium	Schistosomiasis	Humans, monkeys	Blood vessels, urinary bladder	Mainly Africa and Madagascar
S. intercalatum	Schistosomiasis	Humans	Blood vessels, intestines	Central Africa
S. japonicum	Schistosomiasis	Humans, domestic animals	Blood vessels	Japan, China, Formosa, Philippines
S. mansoni	Schistosomiasis	Humans	Blood vessels, intestines	Africa, South America, including Puerto Rico and the Lesser Antilles
S. mekongi	Mekong Schistosomiasis	Humans	Blood vessels, intestines	Laos and Cambodia

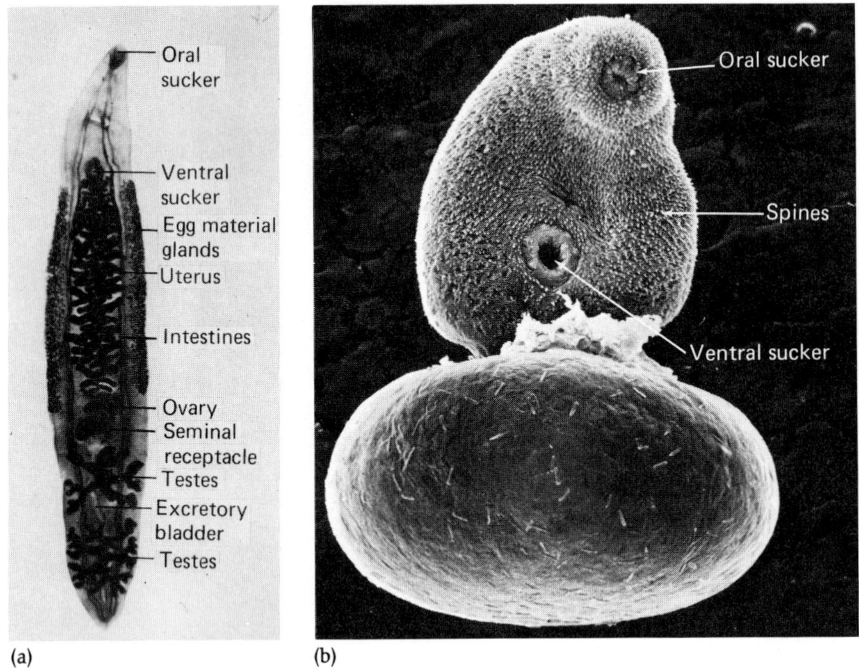

Figure 31–20 Characteristics of flukes. (a) The anatomy of a sexually mature *Opisthorchis (Clonorchis) sinensis* (Chinese liver fluke). This morphological type the *distome* (possessing two suckers), is one of the most frequently encountered. *[Courtesy of Dr. Y. Komiya, director, National Institute of Health, Tokyo.]* (b) A newly emerging fluke, *Fasciola hepatica*, showing its oral sucker, ventral sucker, and an outer covering of spines. *[C. E. Bennet,* J. Parasitol. ***61****:886–891, 1975.]*

portant parasites are incomplete. With one exception—the schistosomes, or blood flukes—all trematodes are hermaphroditic (one worm contains both male and female sex organs). Sexual reproduction is by self-fertilization or cross-fertilization between two individual parasites.

LIFE CYCLES

Most human-associated flukes produce operculated eggs (with lids) and pass through four larval stages. The one exception is the schistosomes, which have nonoperculated (without lids) eggs and only three larval stages (Figure 31–22).

The eggs of trematodes can be found in sputum, urine, or feces, depending upon the location of the parasite in the host such as *Paragonimus westermani* (par-ah-GONE-rr-mus wes-ter-MAN-ee) in the lungs and *Schistosoma haematobium* (shis-toe-SOH-mah he-mah-toh-BEE-um) in the urinary bladder. By contrast, in human-associated cestodes, the eggs of adult worms are encountered only in the feces.

The individual larval stages of flukes are (1) *miracidium,* (2) *sporocyst,* (3) *redia,* and (4) *cercaria* (Figure 31–21). The last three stages are found in freshwater snails. In certain trematode species, such as *Opisthorchis (Clonorchis) sinensis* (oh-pis-THOR-kis seh-NEN-sis), *Fasciola hepatica* (fas-see-OH-lah hep-AT-ick-ah), and *Paragonimus westermani,* the last infective stage (known as a *metacercaria*) can be found in crustaceans, in fishes, or on vegetation. The life cycle of a blood fluke shows the relationship of these larval stages (Figure 31–22).

The life cycle of a schistosome begins when ova containing fully developed miracidia are passed by human infected feces. If the fecal matter comes into contact with water, the miracidium becomes active, ruptures the egg, and escapes into the watery surroundings. The miracidium swims about in search of an appropriate snail host. Upon penetrating the intermediate host, these larval forms lose their ciliated covering, develop into sporocysts, and migrate to the visceral mass of the mollusc. The sporocysts multiply rapidly, and within two weeks large numbers of minute daughter sporocysts develop.

Unlike other flukes of medical importance, schistosomes do not produce rediae at any time during their life cycles. The final larval stage, cercaria, arises from the sporocyst. Once formed, these fork-tailed cercariae (Figure 31–21a) escape from the snail and swim freely in a water environment, looking for an appropriate host. In penetrating the unbroken skin of a mammalian host, their forked tails are left on the host's outer skin surface (Figure 31–21b). These motile immature schistosomes, *schistosomules,* migrate into blood vessels and eventually develop into male and female flukes (Figure 31–21c), mate, and deposit spined eggs within the tissues of the new host (Color Photograph 128). The eggs, containing the larval forms, miracidia, pierce

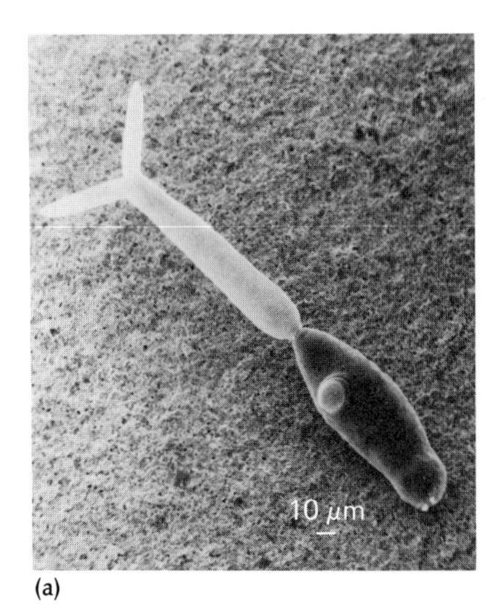

(a)

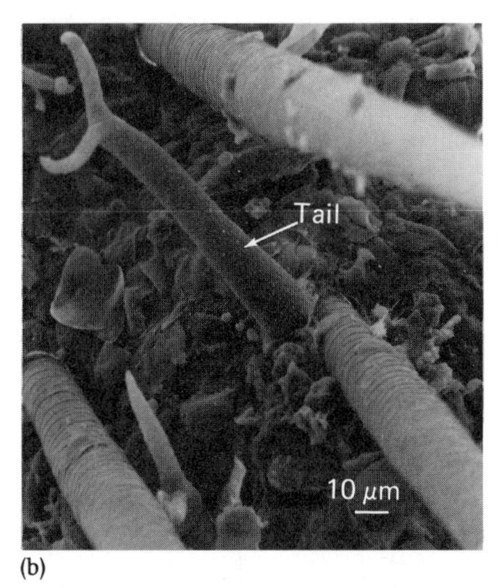

(b)

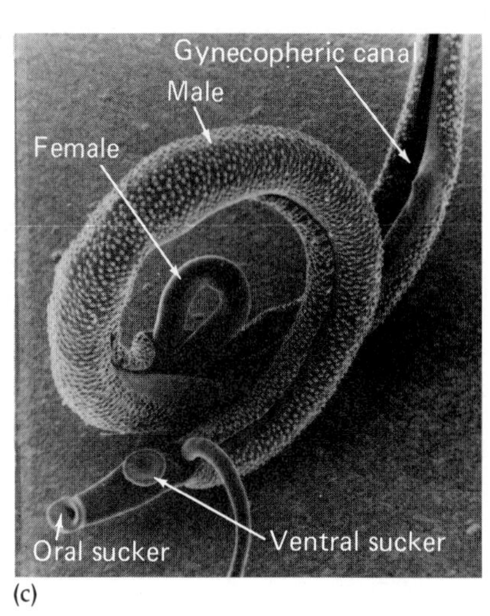

(c)

Figure 31–21 Stages of flukes as shown by scanning electron microscopy. (a) The fork-tailed cercarium of the schistosome. This form penetrates the skin surfaces of a susceptible host immersed in contaminated water. (b) The appearance of an infected skin surface showing the attachment of cercarial tails. *[Courtesy of Dr. H. D. Blankespoor, the University of Michigan.]* (c) Male and female adult *Schistosoma mansoni*. Note the following anatomical parts: gynecopheric canal, oral sucker, and ventral sucker. *[Courtesy of Dr. H. D. Blankespoor, the University of Michigan.]*

their way through blood vessels and surrounding tissue and enter the intestinal space or, in the case of *S. haematobium*, the urinary bladder. The soluble egg antigens stimulate inflammatory reactions and tumorlike growths. In the liver, such reactions give rise to hepatitis and other forms of liver disease.

The effects and laboratory diagnosis of fluke infections are given in Table 31–6.

TABLE 31–6 Characteristics of Fluke Infections

Disease	*Clinical Features*	*Location of Adult Worm*	*Specimen of Choice*	*Laboratory Diagnosis*	*Treatment*
Clonorchiasis	Abdominal pains, diarrhea, enlarged spleen and liver, increase in eosinophils	Liver and bile duct	Feces	Microscopic demonstration of ova	Praziquantel
Paragonimiasis (lung fluke)	Chronic cough, fever, respiratory difficulties, sputum with brownish streaks	Lungs	Sputum, surgical specimens	1. Microscopic identification of adults in surgical specimens or ova in sputum 2. Skin test and immunological laboratory tests	Bithionol, praziquantel
Schistosomiasis (bilharziasis)	Abdominal pain, fever, allergic reactions, inflammation of urinary bladder *(Schistosoma haematobium)*, neurological destruction, spleen and liver enlargement	Urinary bladder, liver, mesenteric vein in large intestine	Feces, urine,[a] biopsy	1. Microscopic demonstration of ova 2. Intradermal tests 3. Complement fixation 4. Circumoval precipitin test[b] 5. Immunofluorescence	Oxamiquine *(S. masoni)* Metrifonate *(S. haematobium)* Praziquantel *(S. japonicum)* Praziquantel *(S. intercalatum)* Praziquantel *(S. mekongi)*

[a] *S. haematobium* only.

[b] Test depends on the formation of a precipitate around preserved eggs incubated in immune serum.

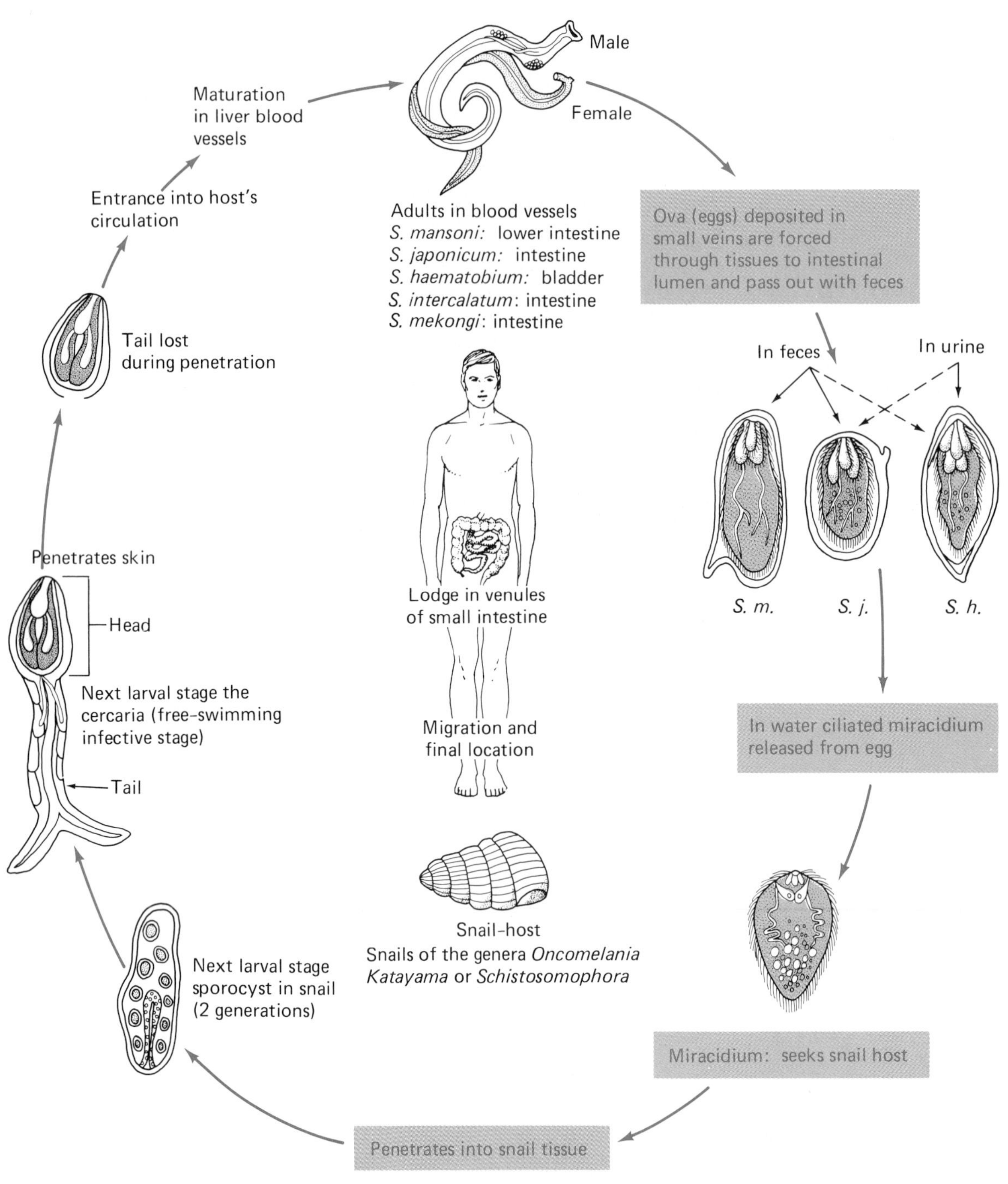

Figure 31–22 The life cycle of schistosomes. Specific stages in both the human and intermediate snail hosts are shown. The arrangement of the cycle suggests the specific phases that are possible to control. *S. mekongi* produces eggs similar to those of *S. japonicum* and *S. intercalatum* eggs resemble those of *S. haematobium.*

PREVENTION

Preventive measures for many helminthic infestations involve interrupting the life cycle of the parasite, the destruction of the adult flukes and ova, and the control of intermediate hosts. More public health education is needed to improve sanitation and persuade individuals in endemic areas to stop eating raw or improperly cooked, feces-contaminated foods.

Trematode Larval Infections: Schistosome Dermatitis or Swimmer's Itch

This condition results from coming into contact with the schistosome cercariae of birds or mammals in fresh or brackish waters. The infection usually occurs during the warm months of the year and is caused when schistosome cercariae penetrate the

MICROBIOLOGY HIGHLIGHTS

A SCHISTOSOMIASIS VACCINE?

Schistosomiasis is an ancient infectious disease that afflicts 200 to 300 million people today. Actually, 5% of the world's population is infected by a worm that can live as long as 20 years in a human host. Schistosomiasis is a disease of poverty and is associated with conditions such as a lack of adequate living facilities and poor sanitation. Refer to Table 31–6 for disease symptoms. Theoretically, if living conditions were improved in the major areas of infection—which include Africa, China, Central and South America, the Middle East, the Philippines, and Malaysia—eradication of the diseases would be possible. A further complication is the high cost of praziquantel, a very effective drug. Because its expense limits its use, the disease continues to flourish. Thus, the problems with eradicating schistosomiasis are not only medical but also economic and probably political as well.

However, what seemed to be a hopeless situation now is showing a positive side. The recent advances in recombinant-DNA technology and related areas indicate that the development of a schistosomiasis vaccine is feasible. All too often, victims of the disease who are cured after taking praziquantel develop a new infection within a year's time. The use of a vaccine would eliminate this situation and, more importantly, the vaccine would have to be given only once.

skin. Certain local snail species serve as intermediate hosts for these larval forms, the adults of which are usually parasitic for aquatic migratory birds. Schistosome dermatitis is also caused by the penetration of the human skin with the cercariae associated with cattle, dogs, raccoons, rodents, and water buffalo. The disease is found in many areas of the world, including the United States, Europe, Latin America, and Australia.

From a clinical standpoint, individuals with the dermatitis exhibit an itching, flat, red rash. Without treatment, the rash becomes elevated into papules. (Color Photograph 129).

Treatment for swimmer's itch includes the oral administration of trimeprazine and the application of lotions to relieve the itching. Control measures usually include the elimination of snail hosts.

Disease Challenge

The situation described has been taken from an actual case history. It has been designed to show how clinical, laboratory, microbial, and related information is used in disease diagnosis. A review of treatment and epidemiological aspects is given at the end of the presentation. Answers to questions, laboratory findings, and interpretations are given immediately following a specific question. Test your skill and take the Disease Challenge.

CASE

An 8-year-old female was taken to the family doctor with complaints of intense itching around the vaginal and perianal areas of her body. The child's case history indicated that she had been well until three weeks earlier, when she became quite irritable and began to scratch herself to the point of drawing blood. The child lost her appetite for favorite foods and was unable to sleep through the night. The patient's younger sister, who shares her bed, had developed similar symptoms two days earlier, which prompted a visit to the family doctor.

(continues)

Disease Challenge (continued)

Microscopic view of a vaginal smear. Original magnification, 100×. *[From E. Avram, M. Yakorelitz, and A. Schacter,* Acta Cytologica *28:468, 1984.]*

At This Point, What Disease(s) Could Be Suspected?
Several conditions may be possible, including pubic lice (commonly referred to as "crabs"), vaginal thrush (*Candida* infection), ringworm, and pinworm.

What Type(s) of Laboratory Specimens Should Be Taken?
Vaginal smears and swabs for microscopic examination and fungal cultures, respectively. A cellophane-tape specimen from the perianal area.

Laboratory Results
Cultures for potential fungal agents (*Candida* and ringworm species) were all negative. Microscopic examination of vaginal smears showed the results seen in the accompanying photo. Cellophane-tape examination yielded similar results.

What Is the Possible Diagnosis?
The patient's history and the finding of *Enterobius vermicularis* ova in both vaginal smear and perianal specimens clearly point to the diagnosis of pinworm.

Treatment and Control:
Several drugs can be used to treat pinworm, including mebendazole and piperazine hexahydrate. The patient's sister should also be treated. In addition, since reinfection is possible, treatment may be required every five to six weeks.

Summary

1. Most helminths, or worms, are free-living, but some are parasitic and can cause diseases in humans and in domestic and wild animals and plants.
2. Helminths of medical and economic importance include roundworms (nematodes), tapeworms (cestodes), and flukes (trematodes).

Equipment of Parasitic Worms

Various worms have several modifications that favor their survival as parasites. These include hardened outer surface coverings, the possession of hooks, spines, cutting plates, suckers, and enzymes for attachment to host tissues, and elaborate reproductive systems.

Life Cycles and Control of Parasitic Worms

1. Worms capable of either a parasitic or a free-living existence are *facultative* parasites. Those incapable of completing their life cycles without a host are *obligate* parasites.
2. *Endoparasites* are found internally in hosts, *ectoparasites* are found attached to the skin or in the outer layers of skin.
3. *Definitive hosts* harbor the adult, or sexually mature, form of the parasite; *intermediate hosts* provide the environment for the development of some or all larval stages.

Distribution and Transmission

1. The distribution of parasitic worms is influenced by the habits of suitable hosts and by environmental conditions such as appropriate temperature and moisture.
2. The sources and transmission of parasitic diseases involve susceptible domestic and wild hosts, bloodsucking insects, foods containing immature infective parasites, and contaminated soil, water, or other portions of the environment.
3. Destructive effects of parasitic worms depend on their number, size, location, degree of activity, and toxic products.

Worm Reproduction in Humans

1. Most helminths are unable to multiply within the human.
2. The course and severity of the disease is determined by the number of worms to which the individual is exposed.

Symptoms

Symptoms of helminth infections may be absent, few, or severe.

Severity of Helminth Infections

The destructive effects of parasties depend on their number, their location, and the reaction of the infected host.

Laboratory Diagnosis

1. The finding and identification of ova, larvae, and adult worms are sufficient for diagnosis.
2. Skin tests and other immunological procedures are used to identify susceptible individuals, to distinguish one type of infection from another, or in a limited number of cases to follow the progress of recovery.

Immunological Control

1. Three types of immunogens are involved in approaches to controlling helminth infections: irradiated-attenuated live helminths, extracts of helminths, and helminth metabolic or excretory-secretory substances.
2. Progress in the isolation and identification of functional immunogens has been limited.

The Nematodes

1. Practically every tissue of the human body is vulnerable to attack by certain roundworms.
2. Roundworms are unsegmented, cylindrical, tapered at both ends, bilaterally symmetrical, and do not have any appendages.
3. These worms have digestive, nervous, and reproductive systems. The sexes, with few exceptions, are in separate worms.
4. The life cycles of nematodes vary and usually include the adult worm, egg, and larvae.

Representative Animal Roundworm Infections

1. Examples of human nematode infections include anisakiasis, filariasis (elephantiasis), hookworm, pinworm, and trichinosis (pork roundworm).
2. Examples of measures for the prevention and control of these and other helminth infections are mass treatment of infected persons, improved general sanitation, and education of individuals regarding the parasite, its mode of transmission, and its effects.

Parasitic Nematodes of Plants

1. Roundworm plant parasites cause enormous damage to cultivated plants and are responsible for major agricultural losses.
2. Both internal and external plant parts are subject to attack. Examples of the most common effects include tumors *(galls)*, rotting of plant tissue, and interference with normal plant growth and development.
3. Several species of soil fungi are natural enemies of nematodes.

The Platyhelminths

1. The platyhelminths, or flatworms, include both free-living and parasitic forms.
2. The true parasitic flatworms are the tapeworms (cestodes) and the flukes (trematodes). Both of these groups have undergone several adaptive changes that equip them for a parasitic existence. These changes include the possession of structures for attachment, excessive reproductive capacity, and the loss of a digestive system.

The Cestodes

1. Adult cestodes, or tapeworms, live in the intestines of vertebrate hosts.
2. These worms consist of a head, or *scolex,* a neck region, and a main body composed of segments called *proglottids.* The proglottids contain the reproductive organs and are responsible for the ribbonlike appearance of tapeworms.
3. The life cycles of cestodes are complex and involve eggs, larval forms, and adult worms.
4. Human tapeworm infections can be acquired by ingestion of contaminated, inadequately cooked fish, beef, pork, or lamb.

The Trematodes

1. Most flukes are flat, are covered by an external resistant layer, have suckers for attachment, and have excretory, digestive, and reproductive systems. These worms lack a head and a segmented body.

2. Except for the schistosomes, all trematodes are hermaphroditic.
3. The life cycles of most flukes include eggs, four larval stages, and adults.
4. Flukes can be acquired through the ingestion of inadequately cooked, contaminated fish, vegetation, or crayfish, as well as by swimming or working in contaminated waters.
5. Preventive and control measures are similar to those for other worm diseases.

TREMATODE LARVAL INFECTIONS

Contact with schistosome cercariae of birds and mammals in fresh and brackish water can result in schistosome dermatitis or swimmer's itch.

Questions for Review

1. Define or explain the following terms:
   a. scolex
   b. proglottid
   c. cercaria
   d. ova
   e. hermaphrodite
   f. trematode
   g. gall
   h. operculum
   i. worm burden
   j. stylet
   k. stobila
2. What factors determine the severity of worm infections?
3. What types of structural and functional modifications favor the parasitic existence of helminths? List at least four.
4. List the types of parasitic worms one might find if the following food sources were contaminated.
   a. beef and beef products
   b. pork
   c. fish
   d. crayfish
   e. lamb
   f. vegetation
5. What role do arthropods play in the transmission of parasitic helminths?
6. By what methods are helminthic infections controlled?
7. How are the following worm infections acquired?
   a. hookworm
   b. pinworm
   c. elephantiasis
   d. pork roundworm
8. Identify and give the functions of each labeled structure in Figure 31–23*a* through *e*.
9. What is swimmer's itch?

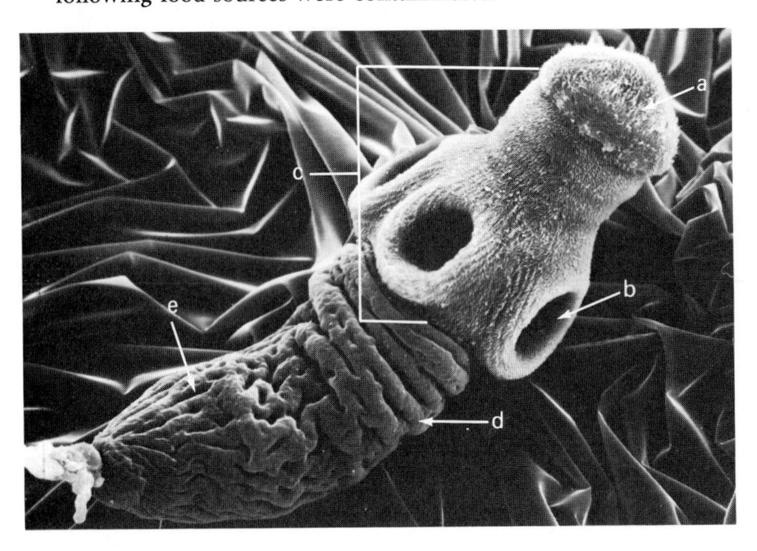

Figure 31–23 A scanning micrograph showing the head or scolex of the tapeworm *Echinococcus multiocularis*. *[From A. A. Marchiondo, and F. L. Andersen,* J. Parasitol. ***69****:709–718, 1983.]*

Suggested Readings

BEAVER, P. C., and R. C. JUNG, *Animal Agents and Vectors of Human Disease,* 5th ed. Philadelphia: Lea & Febiger, 1985. *Parasitic diseases, epidemiology, means of treatment, identification, and life cycles are presented in a readable manner.*

CAMPBELL, W. C. (ed.), *Trichinella and Trichinosis.* New York: Plenum Publishing Corporation, 1983. *This publication presents an organized overview of recent biological information on the helminth* Trichinella *as parasite and pathogen. Leading international authorities provide a critical review of both old and new formation, making this book a comprehensive account of the many aspects of the pathogen.*

LOBEL, H. O., and I. G. KAGAN, "Seroepidemiology of Parasitic Diseases," *Annual Review of Microbiology* **32**:329, 1978. *This review describes the use of immunological tests in investigations concerned with measuring the intensity and geographic distribution of infection. Disease surveillance and control are also discussed.*

OLSEN, O. W., *Animal Parasites: Their Life Cycles and Ecology,* 3rd ed. Baltimore: University Park Press, 1974. *An excellent atlas of tapeworm life cycles.*

SCHMIDT, G. D., *Handbook of Tapeworm Identification.* Boca Raton, Fla.: CRC Press Inc., 1986. *An up-to-date publication providing valuable sections on tapeworm identification, geographical localities, host range, and methodology.*

Appendices

Appendix A
Basic Chemistry and Biochemistry

The major large biological molecules—also referred to as macromolecules, such as lipids, carbohydrates, proteins, and nucleic acids—enter into the topics discussed in various parts of this text. For students already familiar with these molecular groups, this appendix will serve as a brief review. For others, the information presented will provide a foundation with which to follow molecular functions and interactions described in the various chapters of the text.

Basic Chemistry Review

Atoms: The Particles of Matter

In describing the basic units of matter, the early Greek philosopher Democritus (460–370 B.C.) used the term "atom" for the ultimate, smallest particle of matter. His usage of the word implied "that which cannot be further divided." Modern research, however, has demonstrated that atoms are divisible into smaller particles that are arranged as a central nuclear core surrounded by moving units (Figure A–1). The three most important types of atomic particles are *protons* (positively charged), *electrons* (negatively charged), and *neutrons* (electrically neutral). Positively and negatively charged particles attract one another. Alternatively, two similarly charged particles repel one another. The **nucleus,** or center of an atom, which accounts for most of its total mass, contains protons and neutrons. Electrons whirl around the nucleus along predictable paths called **orbitals.** An atom is electrically neutral because the number of protons it contains is equal to the number of its electrons. What distinguishes one type of atom from another is its characteristic number and arrangement of electrons, neutrons, and protons (Figure A–1).

Elements

A substance consisting of atoms that all have the same numbers of protons and electrons is called an *element.* More than 100 such elements are known. These elements have particular chemical and physical characteristics that account for their different activities. Carbon, hydrogen, nitrogen, oxygen, and phosphorus are all elements that are present in living systems and participate in processes vital to life. The letters C, H, N, O, and P are the atomic symbols for these elements. These letters are essentially a means of chemical shorthand that is handy when discussing or writing the composition of a substance containing many different atoms. The number of protons in a single atom's nucleus is known as the *atomic number* of an element, while the total number of protons and neutrons represents its *atomic weight.*

Electrons in various orbitals form a series of concentric electron shells around the nucleus. The chemical characteristics of the atom, including its degree of reactivity with other atoms, are determined by the distribution of these electrons. Hydrogen, for example, is chemically reactive because it has only one electron (Figure A–1c), but its electron shell is most stable if occupied by two electrons. Thus, hydrogen will tend to share or capture an electron from some other atom. Heavier elements are most often stable if their outermost shell contains eight electrons. In an attempt to reach this stable condition, atoms may either capture, yield, or share electrons. Oxygen (Figure A–1b), which has six electrons in its outermost shell, will readily share electrons with two atoms of hydrogen to produce one molecule of water, H_2O. Examples of other chemical bonds involving complete transfer of electrons are discussed later.

Compounds, Molecules, and Macromolecules

The physical-chemical joining of two or more different kinds of atoms results in the formation of a *compound.* The smallest unit of a compound that is representative of its composition and properties is called a *molecule.* There are two kinds of compounds:

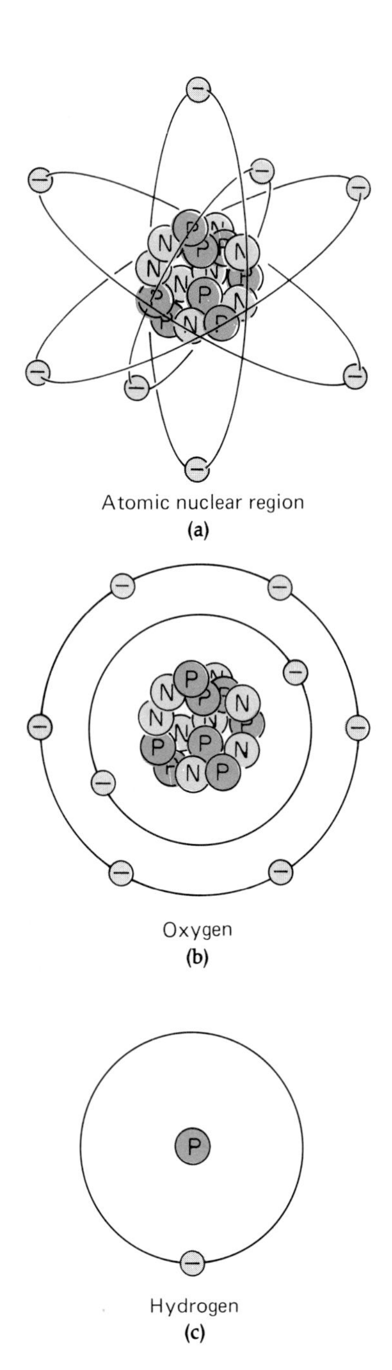

Figure A–1 Three ways of showing the components of atoms. (a) A generalized scheme illustrating a central atomic nuclear region consisting of neutrons (N) and protons (P), which is circled by electrons (−) at varying distances from the center. The electrons move in orbital paths. (b) The oxygen atom. Note the eight protons, eight neutrons, and eight electrons. The electrons form shells around the atom's nucleus. (c) The hydrogen atom. How many protons, neutrons, and electrons are there in this chemical element's atom?

organic compounds, which contain carbon and are found in all organisms, forming their structures and important components; and **inorganic** compounds, which lack carbon in the form described.

The number and types of atoms in a molecule are indicated by its *chemical formula.* For example, the formula for sodium chloride, ordinary table salt, is NaCl. It is made up of sodium (Na) and chlorine (Cl) atoms held together by a *chemical bond.* The outer orbit of sodium contains one electron, whereas that of chlorine has seven electrons. In this case chlorine needs one electron to stabilize its orbit or energy level at eight electrons. Thus, as sodium loses one and chlorine gains one, the transfer of electrons from one atom to another causes electrically neutral atoms to become electrically charged atoms or *ions* (Figure A–2). The force of attraction between ions is called an *ionic bond* and results from the transfer of an electron or electrons from one atom to another.

In the reaction, the sodium atom lost an electron and became a positive ion or *cation,* while the chlorine atom gained an electron and became a negative ion or *anion.* These ions are written as Na^+ and Cl^-. Ionically bonded compounds are held together because of the opposite charges of their components. Ionic bonds dissociate (break) easily in water. This type of reaction has important biological consequences.

One type of chemical bond commonly found in cells and related matter is the *covalent bond.* This type of bond occurs when electrons are *shared* by the

Figure A–2 In the formation of sodium chloride, a sodium atom will lose one electron to a chloride atom, producing the positive sodium ion (Na^+), and the negative chloride ion (Cl^-).

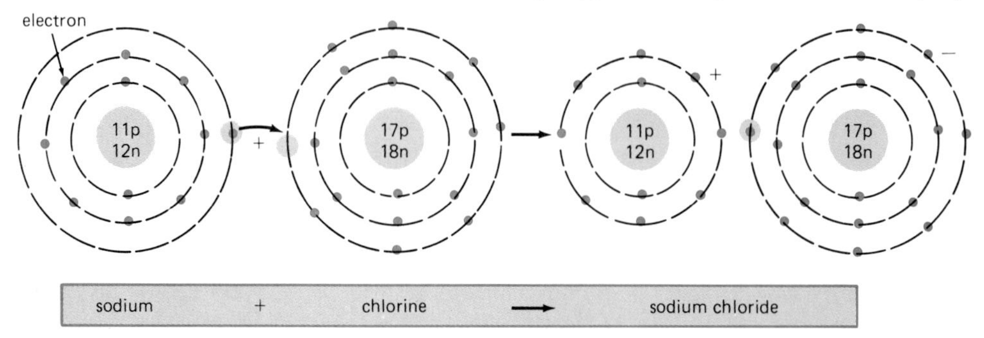

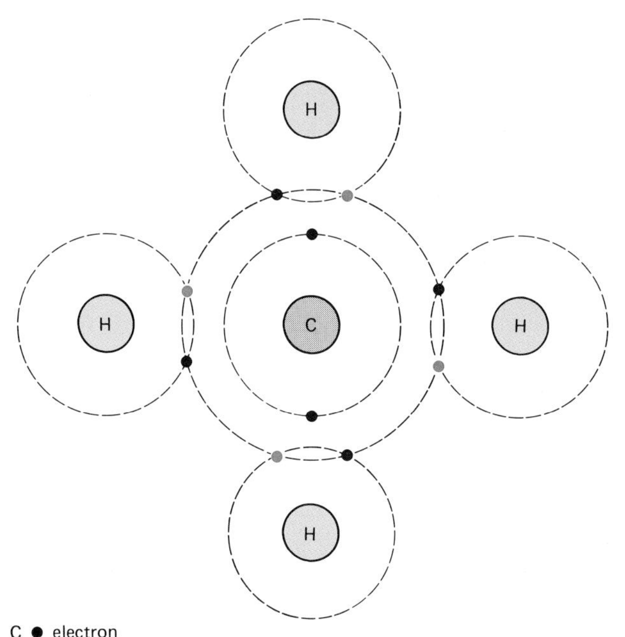

Figure A–3 The chemical formula for methane. Methane, or natural gas, is an example of a molecule with covalent bonding.

atoms making the bond. Therefore, ions are not generated and these compounds do not dissociate readily in water. Natural gas, or methane, is a good example of covalent bonding. Hydrogen has one electron to share, and carbon needs four electrons to stabilize its orbit. Four hydrogen atoms, therefore, share their electrons with one carbon atom, producing methane, chemically noted as CH_4 or as shown in Figure A–3.

Many molecules involve the sharing of four or even six electrons between two atoms. Carbon dioxide (CO_2) is one such molecule. Each oxygen atom shares a total of four electrons with carbon, forming *double bonds.* These are symbolized by two dashes: O═C═O.

Covalent bonds allow the formation of large molecules, or *macromolecules,* that are relatively stable. Many cellular activities involve the synthesis, or manufacture, of their macromolecules, while other processes break them down.

Another common bond found in biological molecules is the *hydrogen bond.* This results from the nature of hydrogen to form *polar molecules.* In a polar molecule there are regions of positive and negative charge in an overall neutral molecule. An excellent example is water, in which the oxygen atom is able to hold the shared electrons more strongly than the two hydrogen atoms. The hydrogen bonding here is intermolecular, that is, between molecules (Figure A–4).

In this case oxygen has a partial negative charge. The hydrogen atoms, as a result, will develop a partial positive charge. Thus, in water, the oxygen

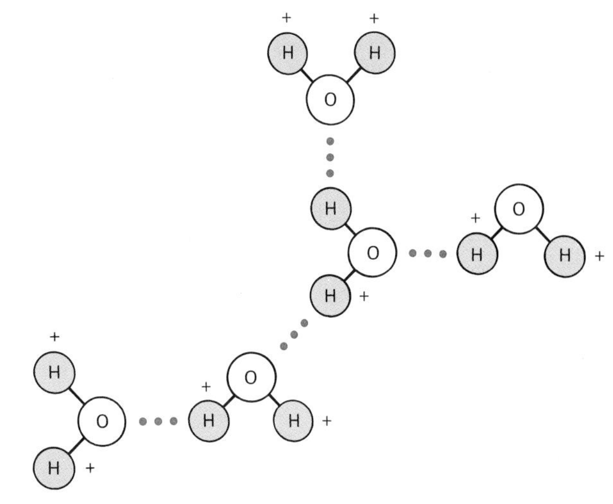

Figure A–4 Hydrogen bonds in water. The hydrogen bonds (· · ·) form between the positively charged hydrogen atoms of one molecule of water and the negatively charged oxygen atom of another.

atom of one molecule is attracted and loosely bonded to a hydrogen atom of another molecule (Figure A–4). This, in part, gives water its cohesiveness and accounts for its surface tension and unusually high boiling point. The same phenomenon occurs in many biological molecules and accounts for some of the characteristic structures of large molecules such as proteins and nucleic acids. In contrast to ionic and covalent bonds, hydrogen bonds are relatively weak and easily broken.

Functional Groups

A typical cell is a complex array of chemical molecules. Moreover, cells manufacture a wide variety of organic compounds. The main types of compounds contained in and synthesized by cells are **carbohydrates, lipids, nucleic acids,** and **proteins.** Some of these are used as components of cell parts; others provide energy for cellular activities; and still others are of particular importance to the regulation of cellular chemical activities **(metabolism).** *Such organic molecules contain certain recurring assemblies of atoms called functional groups. Each group has its own characteristics, including solubility in aqueous or nonaqueous materials and ability to react with other functional groups. It is the arrangement of these groups in a biological molecule that gives that molecule its specific and characteristic chemical properties.* Some common functional groups are shown in Table A–1. Two particularly common bonding arrangements are peptide bonds and ester formation, shown in Table A–2.

TABLE A–1 Common Functional Groups

Functional Group	*Formula*
Aldehyde	O ‖ —C—H
Amino	—N(H)(H)
Ketone (Carbonyl)	O ‖ —C—
Carboxyl (acid)	O ‖ —C—O—H
Hydroxyl	—O—H
Methyl	H | —C—H | H
Phosphate	O | —P—O | O
Sulfhydryl (thiol)	—S—H

Oxidation–Reduction

Whenever an atom, molecule, or ion loses one or more electrons, the process is called **oxidation** and the particle is said to have been *oxidized.* The name suggests that oxygen is involved in the process, but this is not necessarily so.

The electrons that have been lost by an oxidized molecule do not float randomly. They are reactive and are readily picked up by another molecule. The resulting gain of one or more electrons is referred to as **reduction.** The second molecule has been reduced. In these so-called redox reactions, oxidation is always accompanied by an equal reduction.

Certain molecules readily lose or gain electrons. Those that lose electrons supply them to molecules to be reduced and are therefore called *reducing agents.* In the same way, those that gain electrons are obtaining them from molecules being oxidized and are therefore called *oxidizing agents.*

Oxidation and reduction in biological systems are intimately associated with the many metabolic reactions necessary for life. In particular, the passage of electrons from one molecule to another is a major form of energy exchange in all living cells. This can

TABLE A–2 Two Common Bonding Arrangements

Reaction Type	*Functional Groups*	*Reaction*[a]	*Biological Significance*
Peptide bond	—N—H (amino), with H on N —C—O—H, with ‖O (carboxyl)	R—C(H)(NH₂)—C(=O)OH (amino acid) + H—N(H)—C(H)(R)—C(=O)OH (amino acid) ↓ R—C(H)(NH₂)—C(=O)—N(H)—C(H)(R)—C(=O)—OH (peptide) + H_2O	Peptide bonds serve to form proteins by linking various amino acids together.
Ester formation	—O—H (hydroxyl) —C—OH, with ‖O (carboxyl)	R—OH (alcohol) + HO—C(=O)—R (acid) ↓ R—O—C(=O)—R (ester) + H_2O	Ester formation is the means by which components of fats, oils, and other lipids are bonded. A comparable reaction between hydroxyl groups and phosphate is the basis for the linking of nucleic acids.

[a] The symbol R in each organic chemical represents a general chemical group that varies among molecules of the related chemicals.

TABLE A–3 Representative Saturated and Unsaturated Fatty Acids

Saturated Fatty Acid		*Unsaturated Fatty Acid*	
Compound Designation	*Chemical Formula*	*Compound Designation*	*Chemical Formula*
Butyric acid (found in butter)	$CH_3(CH_2)_2$—COOH	Oleic acid (olive oil)	$CH_3(CH_2)_7$—CH = CH$(CH_2)_7$COOH
Caproic acid (found in butter)	$CH_3(CH_2)_4$—COOH	Linoleic acid (linseed oil)	$CH_3(CH_2)_7$—CH = CH—CH_2CH = CH$(CH_2)_4$COOH
Stearic acid (found in animal and plant fats)	$CH_3(CH_2)_{16}$—COOH		

be shown with the oxidation of hydrogen sulfide (H_2S) by oxygen, resulting in the formation of elemental sulfur (S^0) and water.

$$H_2S + \frac{1}{2} O_2 \longrightarrow S^\circ + H_20$$

In this reaction hydrogen sulfide serves as the electron donor and oxygen functions as the electron acceptor. Additional aspects of oxidation and reduction can be found in Chapter 4 and 5.

The Organic Compounds

Lipids

The term **lipid** usually refers to any substance that is soluble in organic solvents such as acetone, benzene, carbon tetrachloride, chloroform, and ether. Lipids include fatty acids, fats, oils, waxes, phospholipids, and steroids.

These molecules serve as (1) important cellular structural components, particularly in membranes; (2) sources of energy; (3) insulation and padding; (4) lubricants for mammalian gastrointestinal tracts; and (5) certain hormones and other cell function regulators such as prostaglandins.

FATTY ACIDS

Naturally occurring **fatty acids** are even-numbered carbon chains (of 4 to 30 carbon atoms) with an acid group (—COOH) at one end. They may be *saturated* or *unsaturated.* Saturated fatty acids do not contain double bonds in their chains, whereas unsaturated fatty acids have one or more double bonds (see Table A–3).

FATS, OILS, AND WAXES

Fats and *oils* are *esters* formed by a reaction between the carboxyl group of fatty acids and the hydroxyl (OH) group of glycerol. The basic difference between fats and oils is that fats are solid at normal room temperature and oils are liquid. This characteristic is determined by the degree of saturation. The compound containing unsaturated fatty acids is the liquid. Waxes are esters of fatty acids with long-chain alcohols.

A fat may consist of glycerol, a carbohydrate esterified with one, two, or three different fatty acids. Tristearin is an example of a simple triglyceride (Figure A–5). Linolenyl-oleylstearin is an example of a mixed triglyceride. As shown in Figure A–6, this compound is formed from glycerol (a), linolenic acid (b), oleic acid (c), and stearic acid (d). Saturated fatty acids such as stearic acid have no double bonds (=) between their carbon atoms. Unsaturated fatty acids, including linolenic and oleic acids, have one or more double bonds. Those fatty acids with more than one double bond are also called *polyunsaturated* fatty acids.

Glycerol + 3 stearic acid (fatty acid) $C_{17}H_{35}COOH$ ⟶ Tristearin + 3 water molecules

H—C—OH HO—C(=O)—$(CH_2)_{16}$—CH_3 ⟶ H—C—O—C(=O)—$(CH_2)_{16}$—CH_3 + 3 H_2O

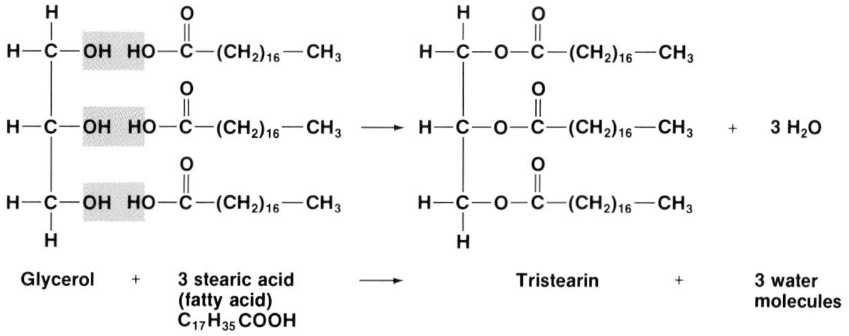

Figure A–5 The formation of the simple triglyceride tristearin.

Part (a)
Part (b)
Part (c)
Part (d)

Figure A–6 The components of the mixed triglyceride linolenyloleylstearin.

Waxes occur in a wide variety of materials, including beeswax, carnauba wax, lanolin, and the cutin on surfaces of leaves and fruits.

PHOSPHOLIPIDS

Phospholipids play an important role in the cell because they contain groups that generally do not dissolve in water (*hydrophobic* groups), like those in lipids, as well as functional groups that do dissolve in water (*hydrophilic* groups). This dual nature appears to be a key factor in the function of the cell membrane (Figure A–7).

STEROIDS

Steroids are important forms of lipids that do not contain fatty acids. Many are physiologically important substances and include bile acids, cholesterol, cortisone, ergosterol, and several hormones. The basic structure of steroids is the four-ring structure shown in Figure A–8a.

The wide variety of activities ascribed to steroids depends upon the various side groups that can be attached to the nucleus. Two examples of this are cholesterol (Figure A–8b), which is an important component of animal and certain microbial cell membranes, and cortisone (Figure A–8c), which is a hormone used as a drug to control inflammation and certain allergic reactions.

Carbohydrates

The group of organic compounds known as **carbohydrates** consists of a wide variety of simple sugars and their derivatives and *polymers* (high-molecular-weight compounds formed by combinations of smaller units) such as starch, glycogen, and cellulose. For the sake of brevity, only a few representative carbohydrates will be discussed.

Both the Fischer and the Haworth systems can be used to present the structure of these compounds. In the Fischer system, the molecule is drawn with the aldehyde (HC═O) or ketone (C═O) group at the top of the carbon skeleton, the primary alcohol (CH_2OH) at the bottom, and hydrogen (H) and hydroxyl (OH) groups at right angles to the carbon backbone of the compound. As examples, four individual aldo sugars (sugars containing an aldehyde group) and one keto (ketone-containing sugar) are shown in Figure A–9. These compounds are referred to as triose, tetrose, pentose, and hexose, or as three-, four-, five-, and six-carbon sugars, respectively.

Figure A–7 The organization of a phospholipid, showing the polar and nonpolar regions. In water, polar regions become oriented toward polar water molecules and away from hydrogen bonds. Nonpolar regions arrange themselves to form contact with neighboring nonpolar regions.

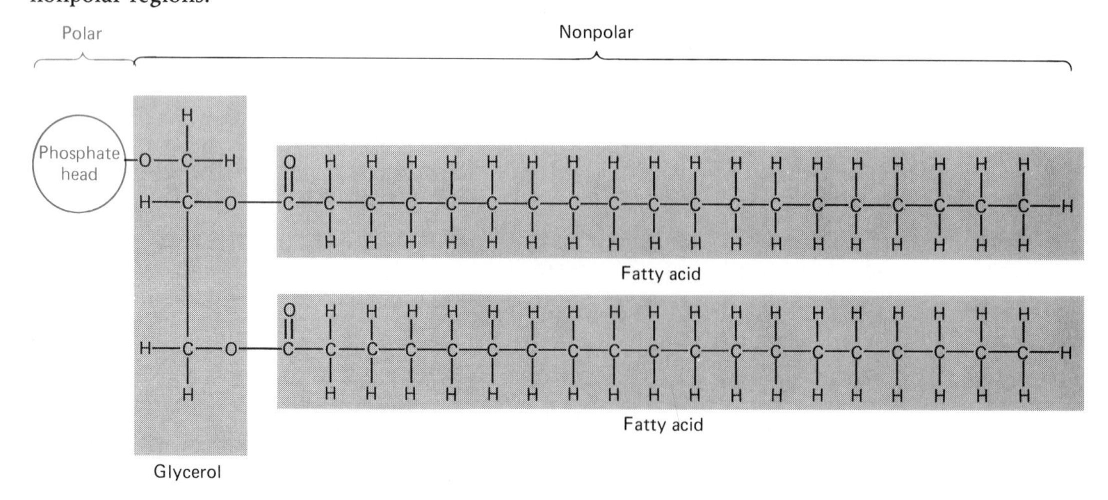

Figure A–8 (a) The basic four-ring structure of steroids. (b) Cholesterol. (c) Cortisone.

Figure A–9 Structural formulas of representative aldo sugars and a keto sugar (e). (a) Glyceraldehyde. (b) Erythrose. (c) Ribose. (d) Glucose. (e) Fructose (a keto sugar).

Figure A–10 The cyclic arrangements of ribose, a five-carbon sugar as represented in the Haworth system. (a) The pyran ring. (b) The furan ring.

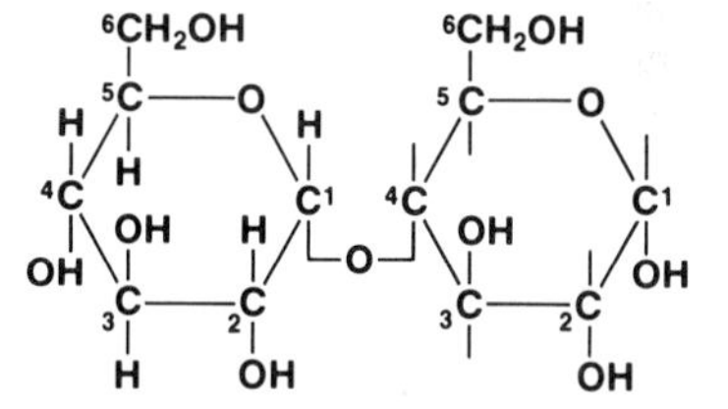

Figure A–12 The disaccharide maltose.

With the Haworth system pentoses and hexoses are shown in cyclic form, and different cyclic arrangements can be drawn. Figure A–10 shows the cyclic arrangements of ribose, a five-carbon sugar that is a significant structural component of coenzymes, nucleic acids, and adenosine triphosphate.

GLYCOSIDES

Glycosides are formed by the reaction of a sugar with another compound containing a hydroxyl group. If the sugar happens to be glucose, the compound then is called a **glucoside.**

The glycosidic linkage is an essential feature of carbohydrates composed of more than one sugar unit: disaccharides, trisaccharides, and polysaccharides. In chained polysaccharides, the linkage between the sugar monomers is commonly expressed, as 1,4 or 1,6, based upon the attachment points between the sugar units. Numbering the ring carbons clockwise starting at the ring oxygen of the sugar molecule produces the diagram shown in Figure A–11. The disaccharide maltose is formed by 1,4 glycosidic linkage of two molecules of glucose (Figure A–12). In this structure we have used a convention common in organic chemistry. Where no substituent is shown, as in the right-hand glucose molecule, hydrogen is assumed. We shall use this convention later with a variety of ring compounds.

Figure A–11 The clockwise numbering system for sugar molecules.

MONOSACCHARIDES

Important **monosaccharides** (simple sugars) include the six-carbon hexoses. Among the common hexoses are fructose, galactose, and glucose. Glucose is the primary energy source for nutrition, but mammals metabolize all four hexoses similarly.

Other important monosaccharides include amino sugars, sugar acids, and sugar alcohols. Representative monosaccharides of these types are shown in Figure A–13. The amino sugars are found in mucoproteins, in chitin, and as components of some antibiotics. Glucuronic acid represents one type of sugar acid found in certain polysaccharides. The sugar alcohols occur in polysaccharides and phospholipids.

H—C=O
H—C—OH
HO—C—H
H—C—OH
H—C—OH
CH_2OH
Glucose
(a)

H—C=O
H—C—NH_2
HO—C—H
H—C—OH
H—C—OH
CH_2OH
Glucosamine
(b)

H—C=O
H—C—OH
HO—C—H
H—C—OH
H—C—OH
HCOOH
Glucuronic Acid
(c)

CH_2OH
H—C—OH
HO—C—H
H—C—OH
H—C—OH
CH_2OH
Glucitol (sorbitol)
(d)

Figure A–13 Structural formulas of different forms of the monosaccharide glucose. (a) General formula for glucose. (b) The amino form, glucosamine. (c) The uronic acid form, glucuronic acid. (d) The alcohol form, glucitol or sorbitol.

DISACCHARIDES

Disaccharides are molecules composed of two monosaccharides. Usually they can be split to yield simple sugars. The three most common disaccharides are lactose, maltose, and sucrose. Lactose, or milk sugar, a prime component of milk, consists of glucose and galactose molecules. Maltose is made up of two glucose molecules, as noted earlier (Figure A–12). It is produced by the breakdown of starch molecules. Sucrose, composed of glucose and fructose, is the natural sugar found in most fruits and vegetables; it is the most commonly used sweetener.

POLYSACCHARIDES

Polysaccharides, the long-chain polymers (combinations) of sugar, may consist of only one type of sugar molecule or of different ones. *Starch,* which is the main food storage product in plants, is composed entirely of glucose units. It is of two basic types, *amylose* and *amylopectin.* The amylose molecule has a linear arrangement of units; amylopectin is a branched molecule with some side chains. The linear structures of starch are essentially chains of maltose units with 1,4 linkages. Amylopectin contains linear 1,4 linked chains that are attached by 1,6 linkages to the backbone polymer at intervals along the main chain.

The chief animal polysaccharide is *glycogen.* It is structurally similar to amylopectin in that it contains 1,4 linked glucose units cross-connected with 1,6 linkages.

Cellulose, the main structural component of wood, is yet another polysaccharide of glucose units connected by 1,4 linkages. This component of plants has a different 1,4 linkage from that found in starch, which makes it resistant to the effects of mammalian enzymes. It is therefore indigestible to humans. However, various microorganisms present in certain animal digestive systems are able to cleave the molecule into utilizable sugar units. It is because of these microbes that ruminants (such as cattle, sheep, and goats) and termites are able to digest cellulose. [See Chapter 14 for microbial interactions.]

Nucleic Acids

Nucleic acids are found in all microorganisms, plants, and animals with the exception of mature mammalian red blood cells. They contain the cell's storehouse of information that enables the cell to govern activities such as protein synthesis and reproduction.

Nucleic acids are polymers of subunits called **nucleotides.** These smaller components, in turn, are made up of nitrogen-containing bases **(purines** and **pyrimidines),** phosphate, and a five-carbon sugar or pentose. The pentose classifies nucleic acids into two main classes: ribonucleic and deoxyribonucleic acids. **Ribonucleic acid (RNA)** contains ribose, while **deoxyribonucleic acid (DNA)** has deoxyribose (Figure A–14).

The primary nitrogenous bases found in RNA are adenine, guanine, cytosine, and uracil. Adenine and guanine are purines; cytosine and uracil are pyrimidines. DNA (Figure A–14) also contains adenine, guanine, and cytosine, but the pyrimidine thymine replaces uracil.

When a purine or pyrimidine is linked with ribose or deoxyribose, the resulting molecule is called a **nucleoside.** With the attachment of a phosphate group, the molecule becomes a **nucleotide.** An example of a ribose nucleoside is adenosine (adenine + pentose), shown in Figure A–15. The corresponding nucleotide would be adenylic acid (adenine + pentose + phosphate), also known as **adenosine monophosphate (AMP).** This compound is involved in energy metabolism and protein synthesis. The addition of a second phosphate yields **adenosine diphosphate (ADP).** Adding a third phosphate forms **adenosine triphosphate (ATP),** the high-energy-containing compound in the cell. The names of var-

TABLE A–4 Common Nomenclature of Nucleosides and Nucleotides

Nitrogen Base	*Nucleoside*	*Nucleotide*
Adenine	Adenosine	Adenylic acid
Guanine	Guanosine	Guanylic acid
Cytosine	Cytidine	Cytidylic acid
Uracil	Uridine	Uridylic acid
Thymine	Thymidine	Thymidylic acid

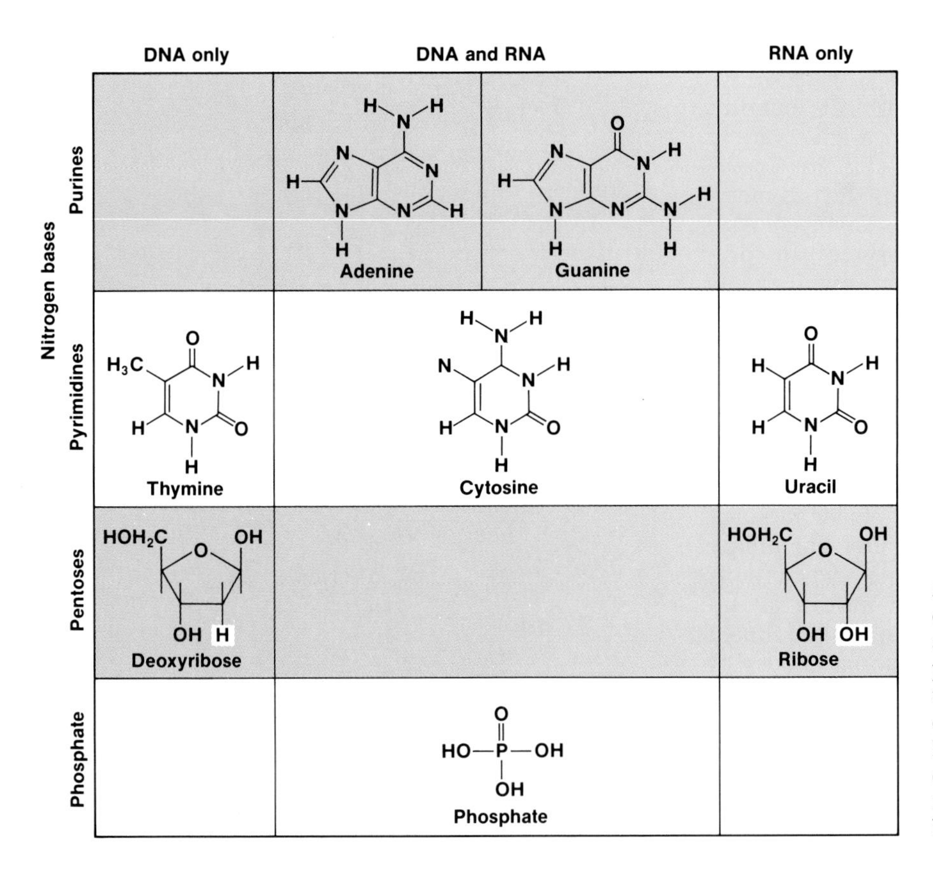

Figure A–14 The components of nucleic acids. Both purines are found in DNA and RNA. The pyrimidines cytosine and thymine are found in DNA; cytosine and uracil are found in RNA. DNA contains the pentose deoxyribose, and RNA contains ribose. Phosphate is found in both DNA and RNA.

ious nucleosides and nucleotides are presented in Table A–4.

Other nucleoside triphosphates are important in some metabolic reactions. For example, the triphosphates of uridine, cytidine, and guanosine are essential to the synthesis of polysaccharides. Various nucleotides are also involved in other cellular activities. Four molecules responsible for energy transfer in oxidation-reduction reactions are nicotinamide adenine dinucleotide (NAD), nicotinamide adenine dinucleotide phosphate (NADP), flavin adenine dinucleotide (FAD), and flavin mononucleotide (FMN). They contain the vitamins niacin and riboflavin in their active forms. Coenzyme A (CoA), important to the catabolism (breakdown) of carbohydrates and lipids, consists of the nucleotide AMP, the vitamin pantothenic acid, and mercaptoethylamine. Further details concerning the biochemistry of these structures and the nucleic acids may be found in Chapters 4 and 5, respectively.

Figure A–15 The building blocks of nucleic acids. (a) Using the purine adenine, the construction of a nucleotide (purine + five-carbon sugar + PO_4) and related compounds can be visualized. Note that adenine is common to both DNA and RNA. A nucleoside consists of a purine or a pyrimidine and a five-carbon sugar.

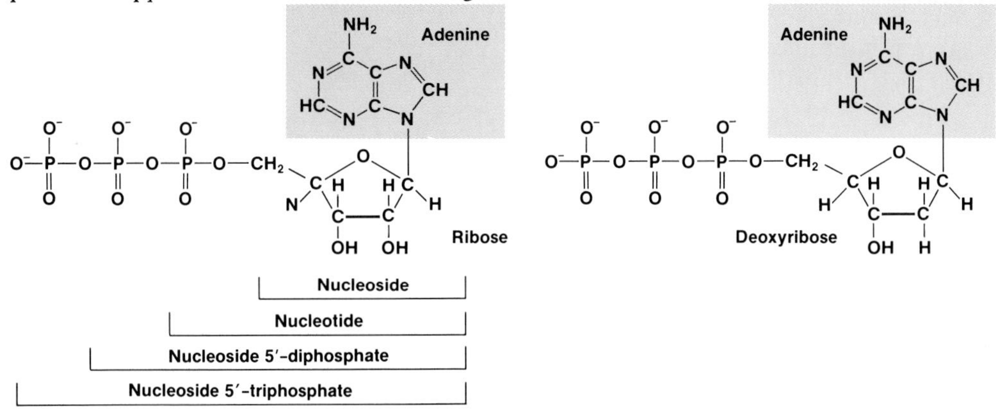

Proteins

Proteins include large molecules. They are found in all forms of life and are involved in a wide variety of cell functions. For example, proteins are important constituents of cellular structures such as membranes, cilia, flagella (the structures involved with movement), and ribosomes. All enzymes are proteins, as are certain hormones and protective agents in the body, namely antibodies (immunoglobulins). [See Chapter 17.]

THE STRUCTURE OF AMINO ACIDS

Proteins are polymers of smaller functional units called **amino acids.** Carbon, hydrogen, oxygen, and nitrogen are the major elements found in amino acids and proteins. Other elements appear in small amounts, sulfur being the one most commonly found. As its name suggests, an amino acid contains a carboxylic acid group (—COOH) and an amino group ($—NH_2$). Both of these components are linked or bonded to a carbon skeleton. The remainder of the amino acid is the radical, or R, group, as shown in Figure A–16. Examples can be found in Table A–5. The R group may be any of the following:

1. Hydrogen (H), as in glycine, or a methyl group (CH_3), as in alanine
2. Larger carbon chains, as in isoleucine, leucine, and valine
3. An alcohol group (CH_2OH), as in serine and threonine
4. Sulfur (S), as in cysteine, cystine, and methionine
5. Acid or basic in nature, as in glutamic acid or lysine, respectively
6. Heterocyclic, as in the amino acids histidine, hydroxyproline, proline, and tryptophan, which have a ring structure in which at least one atom is not a carbon atom (Figure A–17a)
7. A benzene ring, as in the aromatic amino acids phenylalanine and tyrosine (Figure A–17b)

Structural formulas of 10 amino acids with R groups are shown in Table A–5.

Cystine and cysteine are especially significant amino acids in that their structures permit the formation of disulfide linkages (S—S) between thiol groups (—SH) in proteins. Such bonds contribute to the ultimate shape of the protein molecules and in this way are important to the activity of enzymes.

NH_2
R—C—COOH
H

Figure A–16 The typical structure of an amino acid, the basic unit of proteins. Note the presence of an amino group (NH_2), a carboxyl group (COOH), and the radical, or R, group. The latter is a carbon chain to which the amino and carboxyl groups are linked.

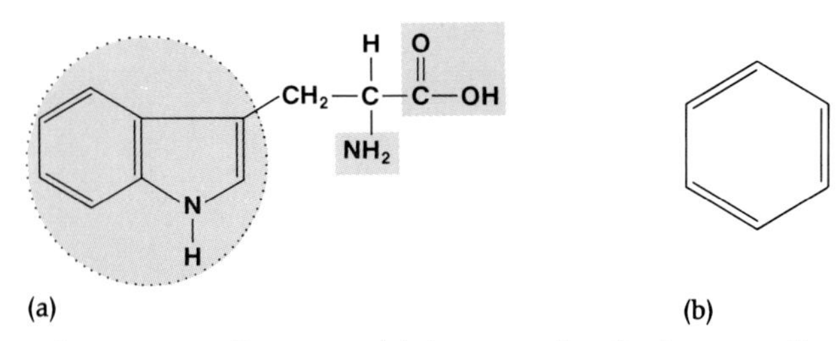

Figure A–17 R groups. (a) An example of a heterocyclic R group in tryptophan. (b) An example of the benzene ring.

TABLE A–5 Structural Formulas of Representative Amino Acids

Amino Acid	*Name*
H—C(H)(NH$_2$)—C(=O)—OH	Glycine
CH$_3$—C(H)(NH$_2$)—C(=O)—OH	Alanine
(CH$_3$)$_2$CH—C(H)(NH$_2$)—C(=O)—OH	Valine
HO—CH$_2$—C(H)(NH$_2$)—C(=O)—OH	Serine
HS—CH$_2$—C(H)(NH$_2$)—C(=O)—OH	Cysteine
S—CH$_2$—C(H)(NH$_2$)—C(=O)—OH | S—CH$_2$—C(H)(NH$_2$)—C(=O)—OH	Cystine
HO—C(=O)—CH$_2$—CH$_2$—C(H)(NH$_2$)—C(=O)—OH	Glutamic acid
H$_2$N—CH$_2$—CH$_2$—CH$_2$—CH$_2$—C(H)(NH$_2$)—C(=O)—OH	Lysine
HC=C—CH$_2$—C(H)(NH$_2$)—C(=O)—OH (ring: HC=C, N, NH, C—H)	Histidine
C$_6$H$_5$—CH$_2$—C(H)(NH$_2$)—C(=O)—OH	Phenylalanine

Note: The R components are shown in a different color.

The disulfide bridge is shown in the cystine molecule in Table A–5. Other features of amino acids are described in various chapters.

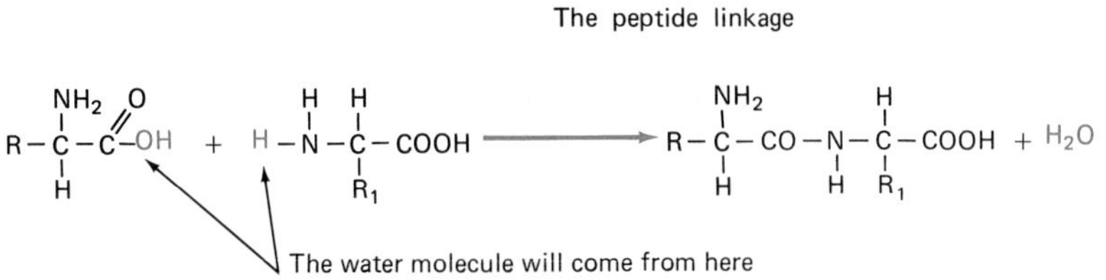

The peptide linkage

$$R-\underset{H}{\overset{NH_2}{C}}-\overset{O}{\overset{/\!/}{C}}-OH + H-\underset{H}{N}-\underset{R_1}{\overset{H}{C}}-COOH \longrightarrow R-\underset{H}{\overset{NH_2}{C}}-CO-\underset{H}{N}-\underset{R_1}{\overset{H}{C}}-COOH + H_2O$$

The water molecule will come from here

Figure A–18 The formation of a peptide. Two amino acids are joined by means of a newly formed peptide linkage. In this reaction, one molecule of water (H_2O) is removed.

PROTEIN STRUCTURE AND FORMATION

The formation of proteins from amino acids involves the bonding of amino acids by a type of linkage referred to as the **peptide bond.** When such a bond is formed, the acid group of one amino acid reacts with the amino group of another, with the removal of one molecule of water (H_2O) (Figure A-18). The protein molecule is constructed from this general type of reaction repeated many times. When two amino acids are joined, they form a dipeptide; three, a tripeptide; several, a polypeptide. Finally, the linking together of large numbers of polypeptides and larger amino acid combinations produces a protein molecule. It should be noted that in living organisms, this process of protein synthesis involves several enzymes and the expenditure of energy. Additional types of bonding both within and between peptide chains include disulfide bonds, hydrogen bonds, and ionic linkages. All three types of bonds result in the coiling or folding of a protein molecule containing them.

As mentioned earlier, sulfhydryl groups (—SH) in sulfur-containing amino acids permit formation of disulfide bonds that can hold peptide chains together in a larger molecule.

Hydrogen bonding often occurs between the hydrogen atoms bonded to nitrogen and adjacent oxygen (Figure A–19). The amino and carboxyl groups of peptide bonds are often involved in hydrogen bonding.

The structures of protein are usually described as either primary, secondary, tertiary, or quaternary. The *primary structure* is the sequence of amino acids held together by peptide bonds in the macromolecule. The *secondary structure* is the coiled, or twisted, chain shape, often helical (springlike), that results when hydrogen bonds form between adjacent parts of the molecule. The secondary structure may appear as either an *alpha helix* or a *beta pleated sheet* (Figure A–20). The folding or bending of the long amino acid chain results in a three-dimensional form called the *tertiary structure.* Several globular proteins, such as hemoglobin (the respiratory pigment), and some enzymes consist of more than one polypeptide chain. The resulting arrangement is called a *quaternary structure.* Figure A–20 illustrates the variations in protein structure. It should be noted that three-dimensional surfaces of proteins are quite flexible and capable of large position changes depending on environmental conditions.

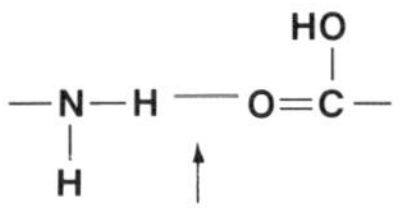

$$-\underset{H}{N}-H \;\text{---}\; O=\overset{HO}{C}-$$

Figure A–19 An example of hydrogen bonding.

The primary structure of a protein determines the manner in which the protein will fold. Changes in the pattern of protein folds can alter chemical reactions. Thus, a change in sequence may produce an overall change in the shape of the protein and a change in the spatial arrangement (configuration) of the protruding R groups on each of the amino acid residues. The sequence of the amino acids and the manner in which these sequences are folded produce an almost infinite variety of proteins.

Since the bonds associated with the secondary and tertiary structures of globular proteins are not very strong chemically, they are easily ruptured, thereby causing a change in the compound's shape and often destroying its original properties. This process, called **denaturation,** may be produced by several factors including heat and results in the loss of structural properties. The cooking of an egg and the sterilizing of surgical instruments both involve denaturation. The spatial arrangement of protein is severely altered, causing the coagulation of the egg protein and the destruction of any infectious material on the instruments. Additional properties of proteins are presented throughout the book.

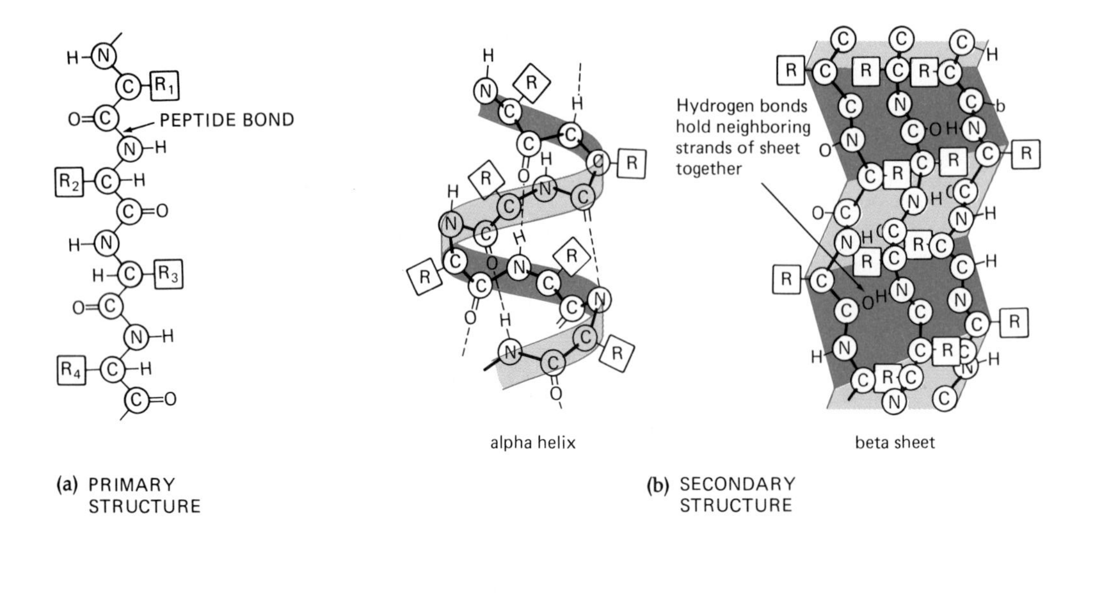

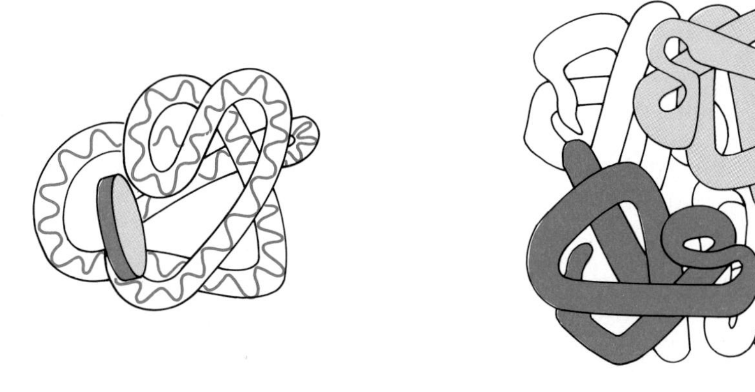

(c) TERTIARY STRUCTURE

(d) QUATERNARY STRUCTURE

Figure A–20 A diagrammatic representation of protein structure. The connecting link or bond between amino acids is the *peptide bond.* The arrangement of chains of amino acids determines protein structure. (a) The primary structure. (b) The secondary structures of proteins. In the alpha helix, hydrogen bonds form between the hydrogen and oxygen atoms that lie next to peptide bonds. In a beta pleated sheet, the atoms next to peptide bonds form hydrogen bonds between different polypeptide strands that zigzag next to each other. (c) The tertiary structure. (d) The quaternary structure.

Appendix B
Classification Outline of Bacteria, Kingdom *Procaryotae*

The classification scheme presented here is based on the information contained and/or to be provided in the four volumes of *Bergey's Manual of Systematic Bacteriology* (Baltimore: Williams and Wilkins). The respective editors and dates of the publication are as follows: Volume 1, R. Krieg, 1984; Volume 2, P. H. A. Sneath (ed.), S. N. Mair, and M. E. Sharpe, (associate eds.), 1986; Volume 3, J. T. Stanley, 1988; and Volume 4, S. T. Williams, 1988. The arrangements clearly follow a scheme based on genetic relatedness (similarity) and are reflections of the biotechnological influence on the reshaping of bacterial classification. Brief descriptions of the respective procaryotic divisions are presented first, before listing the members of each division, so as to familiarize the reader with the characteristics of the group.

Division I—Gracilicutes

Members of this division have a complex cell wall, consisting of an outer membrane and an inner thin peptidoglycan layer, and a variety of other components between or outside of these layers. The procaryotes are usually gram-negative and exhibit cell shapes including spheres, ovals, rods (straight or curved), springlike helices, and filaments. Sheaths or capsules also may be present. Reproduction generally is by binary fission; however, budding and multiple fission occur rarely. Endospores are not formed. The division includes motile and nonmotile species; aerobes, anaerobes, and facultative anaerobes; phototrophs and nonphototrophs; fruiting body and myxospore formers; and obligate intracellular parasites.

APPENDIX B **Classification Outline of Bacteria, Kingdom Procaryotae**

Group	*Order*	*Family*	*Genus*
The spirochetes	*Spirochaetales*	*Spirochaetaceae*	*Spirochaeta*
			Cristispira
			Treponema
			Borrelia
		Leptospiraceae	*Leptospira*
	Other organisms: Hindgut spirochetes of termites and *Cryptocercus punctulatus*		
Aerobic/microaerophilic, motile, helical/vibrioid gram-negative bacteria			*Aquaspirillum*
			Spirillum
			Azospirillum
			Oceanospirillum
			Campylobacter
			Bdellovibrio
			Vampirovibrio
Nonmotile (or rarely motile), gram-negative curved bacteria		*Spirosomaceae*	*Spirosoma*
			Runella
			Flectobacillus
		Other genera	*Microcyclus*
			Meniscus
			Brachyarcus
			Pelosigma
Gram-negative aerobic rods and cocci		*Pseudomonadaceae*	*Pseudomonas*
			Xanthomonas
			Frateuria
			Zoogloea
		Azotobacteriaceae	*Azotobacter*
			Azomonas
		Rhizobiaceae	*Rhizobium*
			Bradyrhizobium
			Agrobacterium
			Phyllobacterium
		Methylococcaceae	*Methylococcus*
			Methylomonas

continues

APPENDIX B Classification Outline of Bacteria, Kingdom Procaryotae

Group	*Order*	*Family*	*Genus*
		Acetobacteriaceae	*Acetobacter*
			Gluconobacter
		Legionellaceae	*Legionella*
		Neisseriaceae	*Neisseria*
			Moraxella
			Acinetobacter
			Kingella
		Other genera	*Beijerinckia*
			Derxia
			Xanthobacter
			Thermus
			Thermomicrobium
			Halomonas
			Alteromonas
			Flavobacterium
			Alcaligenes
			Serpens
			Janthinobacterium
			Brucella
			Bordetella
			Francisella
			Paracoccus
			Lampropedia
Facultatively anaerobic gram-negative rods		*Enterobacteriaceae*	*Escherichia*
			Shigella
			Salmonella
			Citrobacter
			Klebsiella
			Enterobacter
			Erwinia
			Serratia
			Hafnia
			Edwardsiella
			Proteus
			Providencia
			Morganella
			Yersinia
			Obesumbacterium
			Xenorhabdus
			Kluyvera
			Rahnella
			Cedecea
			Tatumella
		Vibrionaceae	*Vibrio*
			Photobacterium
			Aeromonas
			Plesiomonas
		Pasteurellaceae	*Pasteurella*
			Haemophilus
			Actinobacillus
		Other genera	*Zymomonas*
			Chromobacterium
			Cardiobacterium
			Calymmatobacterium
			Gardnerella
			Eikenella
			Streptobacillus

APPENDIX B Classification Outline of Bacteria, Kingdom Procaryotae

Group	*Order*	*Family*	*Genus*
Anaerobic gram-negative straight, curved, and helical rods		*Bacteroidaceae*	*Bacteroides* *Fusobacterium* *Leptotrichia* *Butyrivibrio* *Succinimonas* *Succinivibrio* *Anaerobiospirillum* *Wolinella* *Selenomonas* *Anaerovibrio* *Pectinatus* *Acetivibrio* *Lachnospira*
Dissimilatory sulfate- or sulfur-reducing bacteria			*Desulfuromonas* *Desulfovibrio* *Desulfomonas* *Desulfococcus* *Desulfobacter* *Desulfobulbus* *Desulfosarcina*
Anaerobic gram-negative cocci		*Veillonellaceae*	*Veillonella* *Acidaminococcus* *Megasphaera*
Chemolithotrophic bacteria		*Nitrobacteraceae* (Nitrifiers)	*Nitrobacter* *Nitrospina* *Nitrococcus* *Nitrosomonas* *Nitrosospira* *Nitrosococcus* *Nitrosolobus*
		Sulfur oxidizers	*Thiobacillus* *Thiomicrospira* *Thiobacterium* *Thiospira* *Macromonas*
		Obligate hydrogen oxidizers	*Hydrogenbacter*
		Siderocapsaceae (Metal oxidizers and depositers)	*Siderocapsa* *Naummaniella* *Ochrobium* *Siderococcus*
		Other magnetotactic bacteria	
Anoxygenic photosynthetic bacteria	*Purple Bacteria*	*Chromatiaceae*	*Erythrobacter* *Chromatium* *Thiocystis* *Thiospirillum* *Thiocapsa* *Amoebobacter* *Lamprobacter* *Lamprocystis* *Thiodictyon* *Thiopedia*
		Ectothiorhodospiraceae	*Ectothiorhodospira*
		Rhodospirillaceae (Purple nonsulfur bacteria)	*Rhodospirillum* *Rhodopseudomonas* *Rhodobacter* *Rhodomicrobium* *Rhodopila* *Rhodocyclus*

APPENDIX B Classification Outline of Bacteria, Kingdom Procaryotae

Group	*Order*	*Family*	*Genus*
	Green Bacteria	*Chlorobiaceae* (Green sulfur bacteria)	*Chlorobium* *Prosthecochloris* *Ancalochloris* *Pelodictyon* *Chloroherpeton* *Symbiotic consortia*
		Chloroflexaceae (Multicellular filamentous green bacteria)	*Chloroflexus* *Heliothrix* *Oscillochloris* *Chloronema*
		Others	*Heliobacterium*
Cyanobacteria			*Anabaena*, *Chroococcus*, etc.
Others	*Prochlorales*	*Prochloraceae*	*Prochloron*
Budding and/or appendaged bacteria	*Prosthecate Bacteria*	(Budding bacteria)	*Hyphomicrobium* *Hyphomonas* *Pedomicrobium* *Filomicrobium* *Dicotomicrobium* *Tetramicrobium* *Stella* *Ancalomicrobium* *Prosthecomicrobium*
		(Nonbudding bacteria)	*Caulobacter* *Asticcacaulis* *Prosthecobacter* *Thiodendron*
	Nonprosthecate Bacteria	(Budding bacteria)	*Planctomyces* *Pasteuria* *Blastobacter* *Angulomicrobium* *Gemmiger* *Ensifer* *Isosphaera*
		(Nonbudding bacteria)	*Gallionella* *Nevskia*
		Morphologically unusual budding bacteria involved in iron and manganese deposition	*Seliberia* *Metallogenium* *Caulococcus* *Kuznezovia*
		Others	Spinate bacteria
Sheathed bacteria			*Sphaerotilus* *Leptothrix* *Haliscominobacter* *Lieskeella* *Phragmidiothrix* *Crenothrix* *Clonothrix*
Gliding, fruiting bacteria	*Myxobacteriales*	*Myxococcaceae* *Archangiaceae* *Cystobacteriaceae*	*Myxococcus* *Archangium* *Cystobacter* *Melittangium* *Stigmatella*

APPENDIX B Classification Outline of Bacteria, Kingdom Procaryotae

Group	*Order*	*Family*	*Genus*
		Polyangiaceae	*Polyangium*
			Nannocystis
			Chondromyces
		Others	*Angiococcus*
Gliding, nonfruiting bacteria	*Cytophagales*	*Cytophagaceae*	*Cytophaga*
			Sporocytophaga
			Capnocytophaga
			Flexithrix
			Flexibacter Microscilla
			Saprospira
			Herpetosiphon
	Lysobacteriales	*Lysobacteriaceae*	*Lysobacteriaceae*
			Lysobacter
	Beggiatoales	*Beggiatoaceae*	*Beggiatoa*
			Thioploca
			Thiospirillopsis
			Thiothrix
			Achromatium
		Simonsiellaceae	*Simonsiella*
			Alysiella
		Leucothricaceae	*Leucothrix*
		Pelonemataceae	*Pelonema*
			Achroonema
			Peloploca
			Desmanthus
		Families and genera of uncertain affiliation	*Toxothrix*
			Vitreoscilla
			Chitinophagen
			Desulfonema
The Rickettsias and Chlamydias	*Rickettsiales*	*Rickettsiaceae*	*Rickettsia*[a]
			Rochalimaea[a]
			Coxiella[a]
			Ehrlichia[b]
			Cowdria[b]
			Neorickettsia[b]
			Wolbachia[c]
			Rickettsiella[c]
		Bartonellaceae	*Bartonella*
			Grahamella
		Anaplasmataceae	*Anaplasma*
			Aegyptianella
			Haemobartonella
			Eperythrozoon
	Chlamydiales	*Chlamydiaceae*	*Chlamydia*

[a] Tribe Rickettsieae [b] Tribe Ehrlichieae [c] Tribe Wolbachieae

Division II—Firmicutes

Members of this division have a cell wall with a thick peptidoglycan layer that is characteristic of gram-positive organisms. These procaryotes generally stain gram-positive and exhibit shapes including spheres, rods, and filaments. Branching of the rods and filaments does occur. Reproduction is generally by binary fission. Endospores or spores on hyphae are formed by some species. Most organisms are chemosynthetic heterotrophs and not photosynthetic. The division includes aerobes, anaerobes, facultative anaerobes, asporogenous and sporogenous species, and the actinomycetes together with related forms.

APPENDIX B Classification Outline of Bacteria, Kingdom Procaryotae

Group	*Order*	*Family*	*Genus*
Gram-positive cocci		*Micrococcaceae*	*Micrococcus* *Stomatococcus* *Planococcus* *Staphylococcus*
		Deinoccaceae	*Deinococcus* "Pyogenic" streptococci "Oral" streptococci "Lactic" streptococci and enterococci *Leuconostoc* *Pediococcus* *Aerococcus* *Gemella* *Peptococcus* *Peptostreptococcus* *Reminococcus* *Coprococcus* *Sarcina*
Endospore-forming gram-positive rods and cocci			*Bacillus* *Sporolactobacillus* *Clostridium* *Desulfotomaculum* *Sporosarcina* *Oscillospira*
Regular, nonsporing, gram-positive rods			*Lactobacillus* *Listeria* *Erysipelothrix* *Brochothrix* *Renibacterium* *Kurthia* *Caryophanon*
Irregular, nonsporing, gram-positive rods			Animal and saprophytic Corynebacteria Plant-pathogenic Corynebacteria *Gardnerella* *Arcanobacterium* *Arthrobacter* *Brevibacterium* *Curtobacterium* *Caseobacter* *Microbacterium* *Aureobacterium* *Cellulomonas* *Agromyces* *Arachnia* *Rothia* *Propionibacterium* *Eubacterium* *Acetobacterium* *Lachnospira* *Butyrivibrio* *Thermoanaerobacter* *Actinomyces* *Bifidobacterium*
Mycobacteria		*Mycobacteriaceae*	*Mycobacterium*

APPENDIX B Classification Outline of Bacteria, Kingdom Procaryotae

Group	*Order*	*Family*	*Genus*
Nocardioforms			*Nocardia* *Rhodococcus* *Nocardioides* *Pseudonocardia* *Oerskovia* *Saccharopolyspora* *Micropolyspora* *Promicromonospora* *Intrasporangium*
Actinomycetes dividing in more than one plane			*Geodermatophilus* *Dermatophilus* *Frankia* *Tonsilophilus*
Sporangiate actinomycetes			*Actinoplanes* (including *Amorphosporangium*) *Streptosporangium* *Ampullariella* *Spirillospora* *Pilimelia* *Dactylosporangium* *Planomonospora*
Streptomycetes and their allies			*Streptomyces* *Streptoverticillium* *Actinopycnidium*[d] *Actinosporangium*[d] *Chainia*[d] *Elytrosporangium*[d] *Microellobosporia*[d]
Other conidiate genera			*Actinopolyspora* *Actinosynnema* *Kineospora* *Kitasatospora* *Microbispora* *Micromonospora* *Microtetrospora* *Saccharomonospora* *Sporichthya* *Streptoalloteichus* *Thermomonospora* *Actinomadura* *Nocardiopsis* *Excellospora* *Thermoactinomyces*

[d] The last five genera may be merged with the Streptomycetes

Division III—Tenericutes

Members of this division lack a cell wall and do not synthesize peptidoglycan precursors. These procaryotes are enclosed by a unit membrane—the plasma membrane—and stain gram-negative. The cells are highly pleomorphic and may appear as large irregular vesicles to filamentous forms with branching projections. Many resemble naked L-forms that can be generated from various bacterial species in other divisions. Reproduction may occur by budding, fragmentation, and/or binary fission. Most species are nonmotile, but some exhibit a type of gliding movement. The organisms in the division generally require complex media for growth and the inclusion of cholesterol and long-chain fatty acids. Growth on solid media is characterized by the formation of "fried egg" colonies. The division includes saprophytes and pathogens of animals, plants, and tissue cultures. All members are completely resistant to beta-lactam antibiotics.

APPENDIX B Classification Outline of Bacteria, Kingdom Procaryotae

Group	*Order*	*Family*	*Genus*
The mycoplasmas	*Mycoplasmatales*	*Mycoplasmataceae*	*Mycoplasma* *Ureaplasma*
		Acholeplasmataceae	*Acholeplasma*
		Spiroplasmataceae	*Spiroplasma*
		Other Genera	*Anaeroplasma* *Thermoplasma* Mycoplasmalike organisms of plants and invertebrates

Division IV—Mendosicutes

Most members of this division have some type of cell wall that does not contain typical peptidoglycan components. The composition of cell walls may include large protein molecules or a variety of polysaccharides. The species in the division are either gram-positive or gram-negative and exhibit cell shapes including cocci, rods, filaments, and those resembling the mycoplasma (cells lacking cell walls). Many organisms are motile. Endospores or other resting forms have not been reported. Most species are strict anaerobes. Members of the division include organisms that are ecologically and metabolically diverse and capable of living in extreme environments.

APPENDIX B Classification Outline of Bacteria, Kingdom Procaryotae

Group	*Order*	*Family*	*Genus*
Archaeobacteria	*Methanobacteriales*	*Methanobacteriaceae*	*Methanobacterium* *Methanobrevibacter*
		Methanothermaceae	*Methanothermus*
	Methanococcales	*Methanococcaceae*	*Methanococcus*
	Methanomicrobiales	*Methanomicrobiaceae*	*Methanogenium* *Methanomicrobium* *Methanospirillum*
		Methanosarcinaceae	*Methanosarcina* *Methanococcoides* *Methanolobus*
		Methanoplanaceae	*Methanoplanus*
		Other genera	*Methanothrix*
	Halobacteriales	*Halobacteriaceae*	*Halobacterium* *Halococcus*
	Thermoplasmales	*Theroplasmaceae*	*Thermoplasma*
	Sulfolobales	*Sulfolobaceae*	*Sulfolobus*
	Thermoproteales	*Thermoproteaceae*	*Thermoproteus* *Thermofilum*
		Desulfurococcaceae	*Thermococcus* *Desulfurococcus* *Thermodiscus* *Pyrodictium*
Unassigned Organisms			
Endosymbionts			

APPENDIX B Classification Outline of Bacteria, Kingdom Procaryotae

Group	Order	Family	Genus
Endosymbionts of protozoa			*Holospora* *Caedibacter* *Pseudocaedibacter* *Lyticum* *Tectibacter*
Endosymbionts of insects			*Blattabacterium*
Endosymbionts of fungi and invertebrates other than arthropods			

Appendix C
Classification Outline of the True Fungi (Myceteae)[a]

Division	Subdivision	Class	Order	Family	Representative Genera
Division Mastigomycota		Chytridiomycetes[b]	Chytridiales	Megachytriaceae	*Nowakowskiella*
				Olipidiaceae	*Entophyctis*
					Olpidium
				Phylctidiaceae	*Rhizophydium*
				Synchytriaceae	*Synchytrium*
			Blastocladiales	Blastocladiaceae	*Allomyces*
					Blastocladia
					Blastocladiella
				Coelomonycetaeae	*Coelomomyces*
			Monoblepharidales		*Gonopodya*
					Monoblepharis
					Monoblepharella
			Harpochytriales		*Harpochytrium*
					Oedogoniomyces
		Hyphochytridiomycetes[b]	Hyphochytriales	Rhizidiomycetaceae	*Hyphochytrium*
					Rhizidiomyces
		Plasmodiophoromycetes	Plasmodiophorales		*Plasmodiophora*
					Spongospora
		Oomycetes	Saprolegniales	Saprolegniaceae	*Saprolegnia*
				Leptomitaceae	*Sapromyces*
			Lagenidiales		*Lagenidium*
					Olipidiopsis
			Peronosporales		*Albugo*
					Pythium
					Phyotophthora
Division Amastigomycota	Zygomycotina	Zygomycetes	Mucorales		*Absidia*
					Blakeslea
					Cunninghamella
					Endogone
					Mucor
					Phycomyces
					Rhizopus
			Entomophthorales		*Basidiobolus*
					Entomophthora
					Massospora
			Zoopagales		*Cochlonema*
					Endocochlus
		Trichomycetes[b]	Harpellales		*Genistellospora*
					Smittium
					Trichozygospora
			Amoebidiales		*Amoebidium*

APPENDIX C (continued)

	Subdivision	*Class*	*Order*	*Family*	*Representative Genera*
Division Amastigomycota	Ascomycotina	Hemiascomycetes	Endomycetales	Ascoideaceae	*Dipodascus*
				Endomycetaceae	*Eremascus*
				Saccharomycetaceae	*Cryptococcus*
					Hansenula
					Saccharomyces
			Protomycetales		—
			Taphrinales		*Taphrina*
		Plectomycetes	Eurotiales	Ascophaeriaceae	*Ascophaera*
				Eurotiaceae	*Eurotium*
					Histoplasma
					Microsporum
					Trichophyton
				Gymnoascaceae	*Arthroderma*
					Ctenomyces
		Pyrenomycetes	Erysiphales	Erysiphaceae	*Sphaerotheca*
			Meliolales		*Asterina*
					Meliola
			Coronophorales		*Coronophora*
			Sphaeriales	Clavicipitaceae	*Claviceps*
				Diaporthaceae	*Diaporthe*
				Hypocraeceae	*Gibberella*
					Hypomyces
					Nectria
				Sordariaceae	
					Neurospora
					Sordaria
		Discomycetes	Tuberales		*Elaphomyces*
					Tuber
			Phacidiales		*Lophodermium*
					Rhytisma
			Ostropales		*Stictis*
					Vibrissea
			Pezizales		*Morchella*
					Peziza
					Pyronema
			Helotiales	Phacidiaceae	
				Scleratiniaceae	*Bisporella*
					Botryotina
					Monilinia
					Sclerotinia
		Laboulbeniomycetes	Laboulbeniales		*Ceratomyces*
					Rhizomyces
					Stigmatomyces
		Loculoascomycetes	Dothideales	Dothideaceae	*Dothidea*
	Basidiomycotina	Teliomycetes	Uredinales	Melampsoraceae	*Conartium*
				Pucciniaceae	*Puccinia*
			Ustilaginales	Tilletiaceae	*Tilletia*
				Ustilaginaceae	*Ustilago*
		Hymenomycetes[b]	Auriculariales	Auriculariaceae	*Auricularia*
					Eocronartium
					Herpobasidium
			Septobasidiales	Septobasidiaceae	*Septobasidium*
			Agaricales	Agaricaceae	*Agaricus*
					Mycena
				Amanitaceae	*Amanita*
				Boletaceae	*Boletus*
				Coprinaceae	*Coprinus*
				Russulaceae	*Russula*
			Aphyllophorales		*Clavaria*
					Fomes
					Stereum

APPENDIX C (continued)

	Subdivision	*Class*	*Order*	*Family*	*Representative Genera*
Division Amastigomycota			Dacrymycetales		*Calocera*
			Tulasnellales		—
		Gasteromycetes	Hymenogastrales		*Hymenogaster*
					Trunocolumella
			Phallales		*Phallogaster*
			Lycoperdales		*Lycoperdon*
			Tulostomatales		*Tylostoma*
			Nidulariales		*Cyathus*
	Deuteromycotina[c]	Blastomycetes			*Candida*
					Cryptococcus
		Hyphomycetes			*Aspergillus*
					Penicillium
		Coelomycetes			*Macrophoma*
					Septuria
					Colletotrichum
		Mycelia Sterilia			*Rhizoctonia*

[a] This classification is based on information contained in the following publications: E. Moore-Landecker, *Fundamentals of the Fungi* (Englewood Cliffs, N.J.: Prentice-Hall, 1982) and G. C. Ainsworth, "Introduction and Keys to Higher Taxa," in G. C. Ainsworth, F. K. Sparrow, and A. S. Sussman, eds., *The Fungi—An Advanced Treatise* (New York: Academic Press, 1973, 4A:1–7). Only representative families, orders, and genera are listed in the scheme shown.
[b] Subdivisions into orders, families, or genera were not always specified.
[c] The majority of the fungi in this subdivision are the asexual stages, or *anamorphs*, of sexually reproducing members of the Ascomycotina or Basidiomycotina. This subdivision is also referred to as *Deuteromycetes*, or *Fungi Imperfecti*. The subdivision is also an artificial arrangement of species and is not intended to group closely related organisms.

Appendix D
Classification Outline of Medically and Agriculturally Important Protozoa[a]

Phylum	*Subphylum*	*Class*	*Order*	*Representative Genera*
Sarcomastigophora	Mastigophora	Phytomastigophorea	Dinoflagellida	*Ceratium*
				Gymnodinium
			Euglenida	*Astasia*
				Euglena
			Chrysomonadida	*Ochromonas*
				Synura
			Volvocida	*Chlamydomonas*
				Volvox
		Zoomastigophorea	Kinetoplastida	*Blastocrithidia*
				Bodo
				Cryptobia
				Leishmania
				Trypanosoma
			Retortamonadida	*Chilomastix*
				Retortamonas
			Diplomonadida	*Enteromonas*
				Giardia
				Hexamita
				Trepomonas
			Trichomonadida	*Dientamoeba*
				Histomonas
				Monocercomonas
				Trichomonas
			Hypermastigida	*Lophomonas*
				Trichonympha

APPENDIX D (continued)

Phylum	Subphylum	Class	Order	Representative Genera
	Opalinata	Opalinatea	Opalinida	*Opalina*
	Sarcodina	Lobosea	Amoebida	*Acanthamoeba*
				Amoeba
				Entamoeba
			Schizopyrenida	*Naegleria*
				Tetramitus
				Vahlkampfia
			Arcellinida	*Arcella*
				Difflugia
		Acrasea	Acrasida	*Acrasis*
		Eummycetozoea	Dictyosteliida	*Dictyostelium*
			Physarida	*Physarum*
		Heliozoea	Actinophrydia	*Actinophrys*
Labyrinthomorpha		Labyrinthulea	Labyrinthulida	*Labyrinthula*
Apicomplexa		Sporozoea	Eugregarinida	*Gregarina*
			Eucoccidiida	*Eimeria*
				Isopora
				Sarcocystis
				Toxoplasma
				Haemoproteus
				Leucocytozoon
				Plasmodium
			Piroplasmida	*Babesia*
				Nosema
				Theileria
Microspora		Microsporea	Microsporida	*Thelophania*
Ciliophora		Kinetofragminophorea	Prostamitida	*Didinium*
			Trichostomatida	*Balantidium*
				Blepharocorys
				Isotrichia
				Sonderia
			Suctorida	*Tokophrya*
		Oligohymenophorea	Hymenostomatida	*Tetrahymena*
				Paramecium
			Peritrichida	*Vorticella*
		Polymenophorea	Heterotrichida	*Blepharisma*
				Stentor
			Hypotrichida	*Euplotes*

[a] This 1980 classification incorporates a great deal of new taxonomic information, much of it obtained by electron microscopy. It is a major revision of the 1964 scheme from the Society of Protozoologists, and can be found in its entirety in N. E. Levine, *et al.*, "A Newly Revised Classification of the Protozoa," *J. Protozool.* **27** (1):37–58, 1980.

Appendix E
Classification Outline of the Algae[a]

Division	Class	Representative Genera
Chlorophycophyta	Chlorophyceae	*Chlamydomonas, Chlorococcum, Chlorogonium, Closterium, Cosmarium, Eudorina, Gonium, Haematococcus, Oedogonium, Scenedesmus, Spirogyra, Ulothrix, Volvox*
	Charophyceae	*Chara, Nitella, Tolypella*
Cryptophycophyta	Cryptophyceae	*Chilomonas, Chroomonas, Cryptomonas, Tetragonidium*
Chrysophycophyta[b]	Bacillariophyceae	*Asterionella, Cymbella, Diatoma, Navicula, Pinnularia, Tabellaria, Thalassiothrix*
	Chrysophyceae	*Chrysamoeba, Chrysococcus, Dinobryon, Lagynion, Mallomonas, Synura*
	Eustigmatophyceae	*Ellipsoion, Pleurochloris, Polyedriella, Vischeria*

APPENDIX E (continued)

Division	*Class*	*Representative Genera*
	Haptophyceae	*Apistonema, Chrysochromulina, Coccolithus, Hymenomonas, Phaeocyctis, Prymnesium*
	Xanthophyceae	*Botrydium, Tribonema, Vaucheria*
Euglenophycophyta[b]	Euglenophyceae	*Astasia, Euglena, Phacus, Trachelomonas*
Phaeophycophyta	Phaeophyceae	*Agarum, Desmarestia, Dictyota, Ectocarpus, Elachistea, Fucus, Giffordia, Laminaria, Scytosiphon*
Pyrrophycophyta[b]	Dinophyceae	*Ceratium, Gonyaulax, Gymnodinium, Ornithocecus, Peridinium, Procentrum, Pyrocystis*
Rhodophycophyta	Rhodophyceae	*Ceramium, Polysiphonia, Porhyridium, Rhodella, Rhodosorus*

[a] Developed from information contained in F. R. Trainor, *Introductory Phycology* (New York: John Wiley & Sons, 1978), and H. C. Bold and M. J. Wynne, *Introduction to the Algae* (Englewood Cliffs, N.J.: Prentice-Hall, Inc., 1978).
[b] A recognized protist division.

Appendix F
Classification Outline of the Pathogenic Helminths[a]

Phylum	*Class*	*Order*	*Family*	*Genus*	*Species*
Platyhelminthes (flatworms)	Trematoda (flukes)	Strigeata	Schistosomatidae	*Schistosoma*	*haematobium*
				S.	*japonicum*
				S.	*mansoni*
				S.	*intercalatum*
				S.	*mekongi*
		Echinostomata	Fasciolidae	*Fasciola*	*hepatica*
				F.	*gigantica*
				Fasciolopsis	*buski*
		Plagiorchiata	Troglotrematidae	*Paragonimus*	*westermani*
		Opisthorchioidea	Opisthorchiidae	*Opisthorchis*	*sinensis*
				O.	*felineus*
	Cestoidea (tapeworms)	Cyclophyllidea	Taeniidae	*Taenia*	*saginata*
				T.	*solium*
				Echinococcus	*granulosus*
				E.	*multiocularis*
				Multiceps	*multiceps*
			Hymenolepididae	*Hymenolepsis*	*diminuta*
				H.	*nana*
			Dilepididae	*Dipylidium*	*caninum*
		Pseudophyllidea	Diphyllobothriidae	*Diphyllobothrium*	*latum*
Nematoda (roundworms)	Aphasmidia	Trichurata	Trichinellidae	*Trichinella*	*spiralis*
			Trichuridae	*Trichuris*	*trichiura*
			Capillariidae	*Capillaria*	*philippinensis*
	Phasmidia	Rhabditida	Ascaridae	*Ascaris*	*lumbricoides*
			Aniskidae	*Anisakis*	*species*
			Ancylostomatidae	*Ancylostoma*	*duodenale*
				Necator	*americanus*
			Strongyloididae	*Strongyloides*	*stercoralis*
			Oxyuridae	*Enterobius*	*vermicularis*
		Spirurida	Dipetalonematidae	*Brugia*	*malayi*
				Wuchereria	*bancrofti*
				Dirofilaria	*immitis*
				Onchocerca	*volvulus*
				Loa	*loa*
			Dracunuculidae	*Dracunculus*	*medinensis*
		Spirulata	Gnathostomatidae	*Gnathostoma*	*hispidum*
				G.	*spinigerum*
				G.	*vivarina*
		Strongylata	Strongylidae	*Angiostrongylus*	*cantonensis*
				A.	*costaricensis*

[a] Modified in part from T. Sun, *Pathology and Clinical Features of Parasitic Diseases* (New York: Masson Publishing USA, 1982).

Word Roots and Combining Forms, Prefixes and Suffixes of Medical Importance

Understanding and using medicine-related terms is a major challenge to beginning students of microbiology. The first step in building a working medical vocabulary is to break words down into their respective word elements and then to determine the meanings of those. This section contains the word elements—*word roots, prefixes,* and *suffixes*—generally used in biological and medical terms. Examples, as well as phonetic pronunciations are given.

Word roots and combining forms

acanth-o- thorn or spine; *acanthocyte* (a-KAN-thoh-site) an abnormal erythrocyte exhibiting projections that microscopically appear as thorns

aco-o-, acous-o- hearing; *acoustic* (a-KOOS-stik) pertaining to sound or hearing

acr-o- extremity; *acropathy* (ak-ROW-pah-thee) any disease of the extremities

acu-o- severe, sudden, sharp; *acute* (a-KYOOT) having sudden onset

aden-o- gland; *adenosis* (ad-e-NO-sis) any disease of a gland, especially of a lymphatic gland

aer-o- air or gas; *aerobe* (AIR-ob) a microorganism able to live and metabolize in an environment of free oxygen or air

agglutin-o- clumping, sticking together; *agglutination* (a-gloo-teh-NAY-shun) a reaction in which particles or related materials are clumped together

alb-o- white; *albicans* (al-BEH-kanz) white or whitish

amyl-o- starch; *amylogenic* (am-i-loh-JEN-ik) producing starch

angi-o- vessel; *angiography* (an-ji-OG-ra-fee) an x-ray or roentgenogram of blood vessels after injection of an x-ray opaque substance

aque-o- water; *aqueous* (AH-kwee-us) watery

arteri-o- related to artery; *arteriosclerosis* (ar-tee-ri-oh-skle-ROH-sis) thickening, hardening and loss of elasticity of artery walls

arthr-o- joint; *arthropathy* (ar-THROP-a-thee) disease of a joint

articul-o- joint; *articulation* (ar-TIK-you-lay-shun) interface or union between two or more bones

audi-o- ear; *auditory* (AW-di-tor-ee) pertaining to the sense of hearing

axill-o- armpit; *axilla* (aks-IL-a) the armpit

azot-o- nitrogen; *Azotobacter* (a-zoh-toh-BAK-ter) a bacterial genus that fixes atmospheric nitrogen

bacteri-o- bacteria; *bacteriostatic* (bak-tee-ri-oh-STAT-ic) inhibiting or retarding the growth of bacteria

balan-o- pertaining to the glans penis or glans clitoridis; *balanitis* (bal-a-NYE-tis) inflammation of the glans penis and mucous membrane beneath it

bi-o- life or living; *biopsy* (BY-op-see) tissue specimen removed from living body for examination purposes

blenn-o- mucus or pertaining to mucus; *blennorrhea* (blen-oh-REE-ah) any discharge from mucous membranes

blephar-o- eyelid; *blepharitis* (blef-a-RYE-tis) inflammation of the eyelids

bronch-o- windpipe (trachea); *bronchoscopy* (bron-KOS-koh-pee) direct examination of the bronchi (in the trachea)

bucc-o- cheek; *buccal* (buk-KAL) pertaining to cheek or mouth

carcin-o- cancer; *carcinogenic* (kar-sin-oh-JEN-ik) cancer causing

cardi-o- heart; *cardiomegaly* (kar-di-oh-MEG-a-lee) enlargement of the heart

cauter-o- burn, heat; *cauterize* (KOH-ter-eyes) to destroy [unhealthy] tissue by means of heat, corrosive chemicals, electricity, or extreme cold

cephal-o- head; *cephalomeningitis* (sef-a-loh-men-in-JEYE-tis) inflammation of the cerebral meninges (covering of the brain)

cerebr-o- brain; *cerebrospinal* (se-ree-broh-SPY-nal) *fluid* fluid contained within the cranium (skull) and the spinal canal

cervic-o- neck; *cervix* (neck of the uterus); *cervicovaginitis* (ser-vik-oh-vaj-in-EYE-tis) inflammation of the cervix of the uterus and of the vagina

chem-o- chemical, drug; *chemotherapy* (key-moh-THER-a-pee) the use of chemicals to treat disease

chromat-o-, chrom-o- color; *hyperchromic* (hi-per-KROH-mik) excessive pigmentation.

conjunctiv-o- conjunctiva; *conjunctivitis* (kon-junk-teh-VYE-tis) inflammation of the conjunctiva (lining of the eyelids)

crani-o- skull; *cranium* (KRAY-nee-um) portion of the skull enclosing the brain

cry-o- cold; *cryotherapy* (kry-oh-THER-a-pee) the use of cold for treatment

cyan-o- blue; *cyanobacteria* (sye-an-oh-bak-TEE-ree-ah) a large group of procaryotes that perform photosynthesis, generally with the release of oxygen

cyst-i-, cyst-o- sac, or bladder; *cystitis* (sis-TIE-tis) inflammation of the urinary bladder

cyt-o- cell; *cytoplasm* (SYE-toe-plasm) the cellular substance outside of the nucleus

dacry-o- tear duct; *dacryopyorrhea* (dak-ri-oh-pie-oh-REE-ah) pus discharge from lacrimal (tear) duct

dent-i- tooth; *dentifrice* (DEN-teh-fris) a powder or other material used for cleaning the teeth

dermat-o-, derm-o- skin; *dermatitis* (der-ma-TIE-tis) inflammation of the skin

dors-i-, dors-o- back (of a body); *dorsal* (DOOR-sal) pertaining to the back

electr-o- electricity; *electrophoresis* (ee-lek-troh-for-EE-sis) a technique used to separate charged particles moving in an electrical field

encephal-o- brain; *encephalitis* (en-sef-a-LIE-tis) inflammation of the brain

enter-o- intestine; *enteritis* (ent-e-RYE-tis) inflammation of the intestine

eosin-o- red; *eosinophilia* (ee-oh-sin-oh-FIL-i-a) accumulation of an unusual number of eosinophils

eryth-er-o- flush, redness; *erythema* (er-i-THE-mah) a reddened surface area

erythr-o- red; *erythrocyte* (e-RITH-roh-site) red blood cell

esophag-o- esophagus; *esophagitis* (ee-sof-a-JEYE-tis) inflammation of the esophagus

eti-o- cause; *etiology* (ee-tee-OL-oh-jee) the study of the cause of disease

fluor-o- luminous; *fluorescence* (floo-oh-RES-ents) the property of emitting light when exposed to certain types of irradiation, such as ultraviolet light

follicul-o- follicle, small sac; *folliculitis* (foh-lik-you-LIE-tis) inflammation of a follicle or follicles (of hairs)

galact-o- milk; *galactose* (gah-LAK-tose) a component of milk sugar

gangli-o- ganglion; *ganglionitis* (gang-li-on-EYE-tis) inflammation of a ganglion

gastr-o- stomach; *gastrointestinal* (gas-troh-in-TES-tin-al) pertaining to the stomach and intestine

gingiv-o- gum; *gingivitis* (jin-ji-VYE-tis) inflammation of the gums

glomerul-o- glomerulus; *glomerulonephritis* (glow-mer-you-low-ne-FRY-tis) a form of nephritis (kidney disease) in which the glomeruli are primarily involved

gloss-o- tongue; *glossopharyngeal* (glos-oh-fa-RIN-ji-al) pertaining to the tongue and palate (roof of the mouth)

glyc-o- glucose, sugar; *glycosuria* (gly-koh-SHUR-ee-ah) sugar present in the urine

granul-o- granule(s); *granulocytopenia* (gran-you-low-sye-toe-PEE-nee-ah) an abnormal reduction of granulocytic white blood cells

hema- blood; *hemagglutination* (hem-a-gloo-teh-NAY-shun) clumping of red blood cells

hemat-o- blood; *hematocrit* (he-MAT-oh-krit) the volume of red blood cells packed by centrifugation in a given volume of blood

hepat-o- liver; *hepatitis* (hep-a-TIE-tis) inflammation of the liver

histi-o- tissue; *histiocyte* (his-tee-OH-site) a cell present in all loose connective tissue

home-o- constant; *homeostasis* (hoh-me-oh-STAY-sis) state of equilibrium of the internal body environment maintained by regulatory processes

hydr-o- water; *hydrophobia* (hi-droh-FOE-bi-a) fear of water; common name for the viral disease rabies

hyster-o- uterus; *hysteropathy* (his-ter-OP-ah-thee) any disorder of the uterus

immun-o- immune, protection; *immunogen* (i-MU-no-jen) a substance that stimulates immunoglobulin (antibody) production

kerat-o- cornea; *keratitis* (ker-a-TIE-tis) inflammation of the cornea

lacrim-o- tear; *lacrimal* (LAK-rim-al) pertaining to tears

laryng-o- *larynx* (LAR-rinks) voice box; *laryngitis* (lar-in-JEYE-tis) inflammation of the larynx

leuk-o- white; *leukocytic* (loo-koh-SIT-ik) pertaining to white blood cells (leukocytes)

lip-o- fat; *lipoid* (LIP-oyd) resembling fat

lymph-o- lymph; *lymphatic* (lim-FAT-ik) pertaining to a lymph vessel

lymphadenen-o- lymph gland; *lymphadenopathy* (lim-fad-ee-NOP-a-thee) disease of the lymph nodes

lymphangi-o- lymph vessel; *lymphangitis* (lim-fan-JEYE-tis) inflammation of lymphatic channels

mast-o- breast; *mastitis* (ma-STY-tis) inflammation of a mammary gland

mening-o- membrane; *meningitis* (men-in-JEYE-tis) inflammation of the meninges (membranes covering the brain and spinal cord)

metr-o- uterus; *endometrium* (en-doh-ME-tree-um) lining of the uterus

morph-o- form or shape; *morphology* (mor-FOL-oh-jee) the study of form and structure

myc-o- fungus; *mycosis* (my-KOH-sis) any disease caused by a fungus

myel-o- marrow, spinal cord; *myelitis* (my-eh-LIE-tis) inflammation of the spinal cord

my-o- muscle; *myocardium* (my-oh-KARD-ee-um) heart muscle

necr-o- corpse, dead; *necrosis* (neh-KROH-sis) area of dead tissue surrounded by healthy tissue

nephr-o- kidney; *nephrosis* (neh-FROH-sis) degeneration of kidney tissue

neur-o- nerve; *neuritis* (noo-RYE-tis) inflammation of a nerve or nerves

neutr-o- neutral; *neutrophil* (NOO-troh-fil) a leukocyte that stains with neutral dyes

nucle-o- nucleus; *nucleotoxin* (noo-klee-oh-TOKS-in) a poison acting upon or produced by cell nuclei

ocul-o- eye; *ocular* (OK-you-lar) eyepiece of a microscope

odont-o- tooth; *odontitis* (oh-don-TIE-tis) inflammation of a tooth

onc-o- tumor; *oncogenesis* (ong-koh-JEN-ee-sis) tumor formation

onych-o- nail; *onychia* (oh-NIK-ee-ah) inflammation of the nail bed

opthalm-o- eye; *opthalmia* (off-THAL-me-ah) severe inflammation of the eye

orch-o- testis (testicle); *orchitis* (or-KYE-tis) inflammation of a testis

oste-o- bone; *osteomyelitis* (os-tee-oh-my-ee-LIE-tis) inflammation of bone

ot-o- ear; *otodynia* (oh-toe-DIN-ee-ah) pain in the ear

path-o- disease; *pathogenic* (path-oh-JEN-ik) causing disease

phag-o- to eat; *phagocytosis* (fag-oh-sye-TOE-sis) the ingestive process of phagocytes

pharyng-o- throat; *pharyngitis* (far-in-JEYE-tis) inflammation of the throat

phil-o-, -philic to like, having an affinity for; *thermophilic* (ther-moe-FIL-ik) preferring high temperatures

phleb-o- vein; *phlebitis* (fli-BY-tis) inflammation of the veins

phot-o- light; *photophilic* (foh-toe-FIL-ik) seeking light

pneum-o- lung, air; *pneumonia* (noo-MOE-nee-ah) inflammation of the lungs

py-o- pus; *pyogenic* (pie-oh-JEN-ik) producing pus

pyr-o- fever; *pyrogen* (PIE-roh-jen) any fever-producing substance

rhin-o- nose; *rhinitis* (rye-NYE-tis) inflammation of the nose lining

staphyl-o- clusters, grapes; *staphylococcus* (staf-il-oh-KOK-us) spherical bacteria in clusters

stomat-o- mouth; *stomatitis* (stoh-ma-TIE-tis) inflammation of the mouth

therm-o- heat; *thermoduric* (ther-moh-DO-rik) able to live at high temperatures

tonsill-o- tonsil; *tonsillar* (TON-si-lar) pertaining to a tonsil

tox-o- poison: *toxemia* (toks-EE-mi-a) distribution of poisonous substances throughout the body

ureter-o- ureter; *ureteritis* (you-ree-ter-EYE-tis) inflammation of the ureter

urethr-o urethra; *urethritis* (you-ree-THRI-tis) inflammation of urethra

vagin-o- vagina; *vaginitis* (vaj-in-EYE-tis) inflammation of the vagina

zo-o- animal life; *zoonosis* (zoh-NO-sis) a disease communicable from other animals to humans under natural conditions

Prefixes

a- no, not, without; *acaryote* (a-KAR-ee-ot) without a nucleus

ab- away from; *abnormal* (ab-NOR-mal) away from normal

ad- to, toward; *adneural* (ad-NOO-ral) toward a nerve

an- no, not, without; *anaerobe* (an-AIR-obe) a form of life able to live and develop in the absence of oxygen

ante- before; *antefebrile* (an-te-FE-bril) before the development of fever

anti- against; *antibiosis* (an-tee-by-OH-sis) association between two forms of life, wherein one is harmful or destructive to the other

apo- separation from; *apoenzyme* (ap-oh-EN-zime) the protein portion of a conjugated enzyme

arch-, arche-, archi- first, chief, beginning, original; *archetype* (AR-kee-type) original type from which other forms have developed through differentiation

auto- self; *autolysis* (aw-TOL-ee-sis) cellular destruction caused by cell's own enzymes

bi- two, double; *bicuspid* (by-KUS-pid) having two projections, cusps, or leaflets

blasto- germ, or bud; *blastospore* (BLAS-toe-sporr) a spore formed by a budding process, as in yeast

brady- slow; *bradycardia* (brad-ee-KAR-dee-ah) slow heartbeat

cata- down, downward, against; *catabolism* (ka-TAB-oh-lism) metabolic decomposition of a complex substance into simpler ones

circum- around; *circumscribe* (sur-kum-SCRIYB) to limit or enclose within certain boundaries

con- together, with; *congenital* (kon-JEN-ee-tal) existing or present at birth

contra- opposite or against; *contraindication* (kon-tra-in-di-KAY-shun) inappropriateness of a form of treatment normally advisable

de- lack of, down; *deficiency* (dee-FISH-en-see) lack

di-, diplo- two; *diplococci* (dip-loh-KOK-si) spherical bacteria occurring in pairs

dis- separation, away from; *disinfect* (dis-in-FECT) to free from sources of infection by physical or chemical measures

dys- painful, difficult, bad; *dysphagia* (dis-FAH-jee-ah) difficulty in swallowing

e- out; *emigration* (EM-ee-gray-shun) passage of white blood cells out through capillary and venous walls during inflammation

ecto- outside, out; *ectoparasite* (ek-toh-PAR-a-site) a parasite that lives on the outer surface of a host

em- in; *empyema* (em-pie-EE-mah) pus in a body cavity

en- in, within; *encephalitis* (en-sef-a-LIE-tis) inflammation within the brain

endo- in, within; *endobiotic* (en-doh-by-OT-ik) pertaining to an organism living inside another organism

epi- above, upon, on; *epidermis* (ep-ee-DER-mis) the outer layer of skin

ex- out, away from; *excise* (ek-SIZE) to cut out or remove surgically

exo- out, away from; *exoenzyme* (ek-soh-EN-zime) enzyme functioning outside of the cell that secretes it

extra- outside, in addition to; *extranuclear* (eks-tra-NOO-klee-ar) outside of the nucleus

hemi- half; *hemisphere* (HEM-ee-sfeer) either half of the cerebrum or cerebellum (in the brain)

hetero- other, different; *heterophil* (HET-er-oh-fil) pertaining to an antibody that reacts with a substance other than the antigen that caused its formation

hyper- above, excessive, beyond; *hypertonic* (hi-per-ton-IK) pertaining to a solution having a higher osmotic pressure than another

hypo- less, below, deficient; *hypocalcemia* (hi-poh-kal-SEE-mi-a) below normal levels of calcium in the blood

idio- individual, separate, distinct; *idiopathic* (id-i-oh-PATH-ik) pertaining to a condition (disease state) without a recognizable cause

in- in, inside, within; *inclusion* (in-KLOO-zhun) being enclosed or included

inter- between; *intercellular* (in-ter-SEL-you-lar) between cells

intra- within, inside; *intramuscular* (in-tra-MUSS-kyuh-lar) within a muscle

macro- large; *macrogametocyte* (mac-row-ga-ME-toh-site) a large nonmotile reproductive cell found among certain protozoa

mal- bad, abnormal; *malformation* (mal-for-MAY-shun) abnormal shape or structure

mega- great, large; *megakaryocyte* (meg-a-KAR-oh-site) giant cell of the bone marrow

meta- change; *metachromatic* (met-a-krow-MAT-ik) *granules* granules that stain differently from the dye used originally

micro- small; *microtome* (MY-kroh-tomm) an instrument used to obtain thin slices for microscopic examination

multi- many; *multicellular* (mul-tee-SEL-you-lar) consisting of many cells

neo- new; *neoplasm* (NEE-oh-plazm) a new and abnormal growth

oliga- small, deficient; *oligodynamic* (ol-ee-go-die-NAM-ik) effective in a small quantity

pan- all; *pandemic* (pan-DEM-ik) a disease affecting the majority of the population of a large region

para- near, beside, abnormal; *paracolitis* (par-a-koh-LIE-tis) inflammation of the tissue surrounding the large intestine

per- through; *percolate* (PUR-koh-late) to allow a liquid to pass through a powdered or coarse substance

peri- surrounding; *periarthritis* (per-ee-ar-THRI-tis) inflammation of an area around a joint

polio- gray matter of the brain and spinal cord; *poliomyelitis* (pol-ee-oh-my-el-EYE-tis) inflammation of the gray matter of the spinal cord

poly- many; *polyhedral* (pol-ee-HE-dral) having many surfaces

post- after, behind; *postfebrile* (post-FEE-bril) occurring after a fever

pre- before, in front of; *preclinical* (pree-KLIN-ee-kal) occurring before diagnosis of a disease state is possible

pro- before; *prodromal* (proh-DROH-mal) pertaining to the initial stage of a disease

re- back, again; *reactivate* (ree-AK-teh-vate) to make active again

retro- behind, back, backward; *retrograde* (RET-roh-grade) moving backward

semi- half; *semicircular* (sem-ee-SUR-koo-lar) in the form of a half circle

sub- under, below, less than normal; *subcutaneous* (sub-koo-TAY-nee-us) beneath the skin

super- above, beyond; *superficial* (soo-per-FISH-al) confined or limited to the surface

sym- together, with; *symbiosis* (sim-bi-OH-sis) the living together of two different forms of life

syn- together with; *syndrome* (SIN-drome) a group of signs or symptoms characteristic of a disease or abnormal state

tachy- rapid; *tachycardia* (tack-ee-KAR-dee-ah) rapid heart activity

tetra- four; *tetracoccus* (tet-rah-KOK-us) spherical bacteria arranged in boxlike groups of four

tri- three; *tricuspid* (try-KUS-pid) having three points or cusps

uni- one; *unilateral* (you-nee-LAT-er-al) affecting or occurring on one side only

Suffixes

-ac pertaining to; *cardiac* (KARD-ee-ak) pertaining to the heart

-al pertaining to; *abdominal* (ab-DOM-ee-nal) pertaining to the abdomen

-algia pain; *abdominalgia* (ab-dom-in-AL-jee-ah) abdominal pain

-ar pertaining to; *glandular* (GLAN-dyou-lar) pertaining to or of a gland

-ary pertaining to; *coronary* (KOR-oh-na-ree) pertaining to the blood vessels encircling the heart

-ase enzyme; *lactase* (LAK-tayse) enzyme that breaks down lactose or milk sugar

-asthenia lack of strength; *myasthenia* (my-as-THEE-nee-ah) muscular weakness

-cele hernia, swelling; *meningocele* (men-IN-go-seel) enlargement of the meninges (coverings of the brain)

-centesis surgical puncture used to remove fluid; *paracentesis* (par-a-sen-TEE-sis) puncture of a cavity with the removal of fluid

-cidal killing; *bactericidal* (bak-ter-ee-oh-SYE-dal) pertaining to substances that kill bacteria

-clasis break; *bacterioclasis* (bak-teh-ree-OK-lay-sis) fragmenting of bacteria

-coccus spherical or berry-shaped bacterium; *Streptococcus* (strep-toh-KOK-us) spherical bacteria appearing in chains

-crine secrete, separate; *endocrine* (EN-doh-krin) pertaining to an internal secretion

-crit separate; *hematocrit* (he-MAT-oh-krit) the volume of erythrocytes, packed by the centrifugation of blood

-cyte cell; *leukocyte* (LOO-koh-site) white blood cell

-cytosis condition of numerous cells; *leukocytosis* (loo-koh-sye-TOE-sis) increase in the number of white blood cells

-dynia pain; *gastrodynia* (gas-troh-DIN-ee-ah) stomach pain

-eal pertaining to; *laryngeal* (lar-IN-jee-al) pertaining to the larynx (voice box)

-ectasis stretching, dilation; *bronchiectasis* (brong-kee-EK-tah-sis) expansion of the bronchus

-ectomy excision, surgical removal; *tonsillectomy* (ton-sill-EK-toh-me) surgical removal of the tonsils

-emia blood condition; *glycemia* (gly-SEE-me-ah) sugar in the blood

-fusion to pour; *infusion* (in-FEW-zhun) introducing a liquid into the body via a vein; the fluid resulting from steeping a substance in hot or cold water

-genesis producing, forming; *carcinogenesis* (kar-si-no-JEN-ee-sis) producing cancer

-genic produced by or in; *bacteriogenic* (bak-ter-ee-oh-JEN-ik) caused by bacteria

-globulin protein; *immunoglobulin* (im-myoo-no-GLOB-you-lin) a group of closely related but not identical proteins formed in response to immunogens

-gram record; *electroencephalogram* (ee-lek-troh-en-SEF-a-low-gram) a tracing or record of the brain's electrical activity

-graphy process of recording; *electrocardiography* (ee-lek-troh-KAR-dee-oh-grah-fee) process of recording the electrical activity of the heart

-ia condition; *leukemia* (loo-KEE-mi-a) a form of cancer of the blood

-iasis abnormal condition; *amebiasis* (am-ee-BY-ah-sis) infection with amebae (amebic dysentery)

-ic pertaining to; *chronic* (KRON-ik) of long duration

-ism process; *rheumatism* (ROO-ma-tizm) condition with inflammation and pain of muscles and joints

-itis inflammation; *tonsillitis* (ton-sil-EYE-tis) inflammation of the tonsil(s)

-ium structure, tissue; *cranium* (kray-nee-UM) portion of the skull that covers the brain

-logy study of; *microbiology* (my-krow-by-OL-oh-jee) study of microscopic forms of life

-lysis breakdown, destruction; *hemolysis* (he-MOL-ee-sis) destruction of red blood cells

-lytic reducing, destroying; *autolytic* (oh-toe-LIT-ik) self-digestive or self-destructive

-malacia softening; *osteomalacia* (os-tee-mal-A-shee-ah) softening of the bones

-megaly enlargement; *hepatomegaly* (hep-a-toe-MEG-a-lee) enlargement of the liver

-meter measure; *nanometer* (NAH-no-me-ter) one billionth of a meter

-oid resembling; *viroid* (VYE-royd) resembling a virus

-ole little, small; *arteriole* (ar-TEE-ree-ol) small artery

-oma tumor, mass; *adenoma* (ad-eh-NO-mah) tumor of a gland

-opsy view of; *biopsy* (BY-op-see) removal of a small piece of tissue for examination

-ose pertaining to; *adipose* (AD-eh-poz) pertaining to fat

-osis abnormal condition; *dermatosis* (der-ma-TOE-sis) skin condition

-ous pertaining to; *mucous* (MYOO-kus) pertaining to mucus (secretion)

-partum birth; *postpartum* (post-PAR-tum) occurring after birth

-pathy disease; *myelopathy* (my-ee-LOP-ah-thee) disease state of the spinal cord

-penia deficiency; *leukopenia* (loo-koh-PEE-nee-ah) abnormal decrease of white blood cells

-phagia eating, swallowing; *dysphagia* (dis-FAY-jee-ah) inability or difficulty in swallowing

-philia attraction for; *eosinophilia* (ee-oh-sin-oh-FIL-ee-ah) accumulation of a large number of eosinophils; attraction for eosin stain

-phobia distaste for; *photophobia* (fo-toe-FO-bee-ah) dislike of light

-phoresis carrying; *electrophoresis* (ee-lek-troh-for-EE-sis) movement of charged particles in an electrical field

-phylaxis protection; *prophylaxis* (pro-feh-LAK-sis) use of procedures to prevent disease

-phyte plant; *dermatophyte* (DER-ma-toe-fyte) a fungal pathogen that attacks the skin, nails, or hair

-plasia formation, development; *hyperplasia* (hi-per-PLAY-zee-ah) excessive increase in the number of normal cells

-plasm formation; *cytoplasm* (SYE-toe-plazm) cellular material outside of the nucleus

-plasty surgical repair; *osteoplasty* (OS-tee-oh-plas-tee) surgical repair of bone

-plegia paralysis; *paraplegia* (par-a-PLEE-jee-ah) paralysis of the lower portion of the body

-pnea breathing; *apnea* (ap-NEE-ah) temporary stoppage of breathing

-poiesis formation; *hematopoiesis* (hem-a-toh-poy-EE-sis) the production and development of blood cells

-porosis passage; *osteoporosis* (os-tee-oh-por-OH-sis) increased porosity of bone

-rrhage flowing or bursting forth; *hemorrhage* (HEM-or-rij) abnormal discharge of blood

-rrhea flow, discharge; *diarrhea* (di-a-REE-ah) frequent passage of watery bowel movements

-sclerosis hardening; *arteriosclerosis* (ar-tee-ri-oh-skle-ROW-sis) hardening and loss of elasticity of arteries

-scope instrument for viewing; *microscope* (MY-kroh-skope) optical instrument that gives a magnified image of minute objects

-scopy visual examination; *cystoscopy* (sis-TOS-koh-pee) examination of the urinary bladder

-sis state of, condition; *parasitosis* (par-ah-seh-TOE-sis) a disease or condition caused by a parasite

-stasis stopping; *bacteriostasis* (bak-tee-ri-oh-STAY-sis) inhibition of bacterial growth

-stomy new opening; *tracheostomy* (tray-kee-OS-toe-me) surgical cutting into the trachea

-therapy treatment; *chemotherapy* (kee-moh-THER-a-pee) the use of chemicals in the treatment of disease

-tic pertaining to; *necrotic* (ne-KROT-ik) related to dead tissue

-tome cutting instrument; *microtome* (MY-krow-tomm) instrument used to obtain thin slices of a specimen for microscopic examination

-tomy cutting process; *neurotomy* (new-ROT-oh-mee) dissection or division of a nerve

-trophy nourishment; *heterotrophy* (HET-er-oh-troff-ee) a form of nutrition requiring complex organic food for growth and development

-ule little; *venule* (VEN-yule) small vein

-um structure, object; *pericardium* (per-ee-KAR-de-um) membranous sac surrounding the heart

-uria urination, urine; *polyuria* (pol-ee-YOOR-ee-a) excessive urine secretion

-y process or condition; *atrophy* (AT-roh-fee) a decrease in the size of a tissue or organ

Organism Pronunciation Guide

This section contains the phonetic pronunciations for the majority of organisms described in this text. Taking time to sound out new and unfamiliar terms and to say them aloud several times will help you to learn and to use a specialized vocabulary. Organisms have been arranged according to their respective groups: algae, bacteria, fungi, protozoa, and helminths.

Algae

Acetabularia (A-say-tab-you-LAIR-ee-ah)
Alaria (A-LAIR-ee-ah)
Arachnoidiscus (A-rack-NOI-dis-kuss)
Chlorella (klor-ELL-ah)
C. pyrenoidosa (C. pie-re-NOI-doh-sah)
C. stigmatophora (C. stig-meh-TOFF-for-rah)
Colacium mucronatum (coh-lay-SEE-um myoo-KROW-nah-tum)
Ectocarpus (ek-toh-KAR-pus)
Euglena (you-GLEE-nah)
Fucus (FEW-kus)
Gambierodiscus (GAM-beer-oh-diss-kus)
Gonyaulax catanella (gon-ee-AW-lax cat-ah-NEL-lah)
Gymnodium (jim-no-DEE-um)
Mastogloia (mass-toh-glow-EE-ah)
Melosira (mel-LOW-ser-ah)
Navicula (na-VIK-you-lah)
Nereocystis (neer-ee-oh-SIS-tiss)
Oedegonium (ee-dee-GO-nee-um)
Pinnularia (pin-you-lair-EE-ah)
Platoma abbotiana (pla-TOE-ma a-boh-TEE-an-na)
Scenedesmus quadricauda (sin-EE-des-mus quad-REE-kaw-dah)
Sorastrum (soh-RASS-trum)
Stephanopyxis turris (steff-an-oh-PICKS-iss TURR-iss)
Vacuolaria virescens (vacu-oh-LAIR-ee-ah vire-ESS-sens)

Bacteria

Acetobacter aceti (a-see-toh-BACK-ter a-SEE-tee)
Achromobacter (a-KROME-oh-back-ter)
Acidaminococcus (a-sid-a-ME-no-KOK-kuss)
Acinetobacter (a-sin-et-OH-back-ter)
A. calcoaceticus (A. kal-coh-a-see-TEE-kus)
Actinomyces israelii (ak-tin-oh-MY-sees is-RAY-lee-eye)
Agrobacterium tumefaciens (a-grow-back-TIR-ee-um too-me-FAYSH-ee-enz)
Alcaligenes (al-ka-LIJ-en-eez)
A. eutrophus ((A. you-TROH-fus)
Anabaena (an-ah-BE-nah)
Aphanizomenon (a-fan-ee-zoh-MEN-on)
Aquaspirillum (a-kwah-spy-RIL-lum)
Arthrobacter vinelandii (are-THROW-back-ter vin-LAN-dee-eye)
Azomonas (a-zoh-MOH-nas)
Azospirillum lipoferrum (a-zoh-spee-RIL-lum lip-oh-FER-rum)
A. agilis (A. aj-EE-liss)
A. chroococcum (A. krow-oh-COH-kum)
Bacillus anthracis (bah-SIL-lus ann-THRAY-sis)
B. cereus (B. SEE-ree-us)
B. circulans (B. sir-KOO-lanz)
B. coagulans (B. koh-AG-you-lanz)
B. colistinus (B. coh-LIST-in-us)
B. globisporus (B. glob-EE-spor-us)
B. pasteurii (B. pas-TUER-ee-eye)
B. polymyxa (B. poh-lee-MIKS-ah)
B. popillae (B. poh-PILL-ee)
B. stearothermophilus (B. ste-row-ther-MAH-fil-us)
B. subtilis (B. SAH-til-us)
B. subtilis variety *niger* (B. SAH-til-us var-EYE-eh-tee NYE-jer)
B. thuringiensis (B. thur-in-jee-EN-sis)
Bacteroides fragilis (back-teh-ROY-deez FRAH-jill-us)
Bdellovibrio bacteriovorus (dell-o-VIB-ree-oh back-tir-ee-OH-vor-us)
Beggiatoa (beg-ee-ah-TOW-ah)
Bordetella pertussis (bor-de-TEL-lah per-TUSS-sis)
Branhamella catarrhalis (bran-ham-EL-a ca-tah-RAL-is)
Brevibacterium (brev-ee-back-TI-ree-um)
Brucella abortus (broo-SEL-la ah-BORE-tus)
B. canis (B. KAY-niss)
B. melitensis (B. mel-lee-TEN-sis)
Calymmatobacterium granulomatis (kal-im-mah-toe-back-TIR-ee-um gran-you-LOW-mah-tiss)
Campylobacter foetus (kam-peh-low-BACK-ter FEE-tus)
Carboxydomonas (kar-box-EE-doh-moh-nass)
Caulobacter (koh-loe-BACK-ter)
Cellulomonas (sell-you-LOW-moh-nass)
Chlamydia psittaci (kla-MEH-dee-ah SIT-tah-sye)
C. trachomatis (C. trah-KOH-mah-tiss)

Chlorobium (kloh-ROW-bee-um)
Chromatium (krow-MAH-tee-um)
Chroococcus (krow-oh-KOK-kuss)
Chromobacterium (krow-moh-back-TIR-ee-um)
Clostridium bifermentans (klos-TREH-dee-um by-fer-MEN-tans)
C. botulinum (C. bo-tyoo-LIE-num)
C. nigrificans (C. nye-GREH-feh-cans)
C. perfringens (C. per-FRIN-jens)
C. sporogenes (C. spoh-RAH-jen-eez)
C. tetani (C. TEH-tan-ee)
Corynebacterium diphtheriae (koh-rye-nee-back-TIR-ee-um dif-THEH-ree-ah)
C. pseudodiphtheriticum (C. soo-doh-dif-theh-RIT-ee-cum)
C. renale (C. REN-al)
C. xerosis (C. zeh-ROW-sis)
Cowdria (kow-DREE-ah)
Coxiella burnetii (kocks-ee-EL-la bur-NEH-tee-eye)
Crenothrix (kren-OH-thriks)
Cristispira (kris-tee-SPY-rah)
Cytophaga (sye-TOF-aj-ah)
Desulfotomaculatum (dee-sul-foh-toe-MAC-you-lay-tum)
Desulfovibrio (dee-sul-foh-VIB-ree-oh)
D. desulfuricans (D. dee-sul-fur-EE-kans)
Desulfuromonas (dee-sul-fur-OH-moh-nass)
Ectothiorhodospira mobilis (ek-toe-thigh-oh-row-doh-SPY-rah MOH-bill-iss)
Ehrlichia (er-LICK-ee-ah)
Enterobacter aerogenes (en-te-roh-BACK-ter ah-RAH-jen-eez)
E. cloacae (E. klow-EH-kah)
Erwinia (er-WEH-nee-ah)
Escherichia coli (esh-er-IK-ee-ah KOH-lee)
Flavobacterium (flay-voh-back-TIR-ee-um)
Flexibacter (flex-ee-BAK-ter)
Francisella tularensis (fran-sis-EL-lah too-lah-REN-sis)
Frankia (FRANK-ee-ah)
Fusobacterium (few-zoh-back-TIR-ee-um)
Gallionella (gal-ee-oh-NELL-ah)
Gardnerella vaginalis (gard-NEH-rel-lah vah-JIN-al-iss)
Gluconobacter (gloo-kon-oh-BAK-ter)
Haemophilus aegyptius (he-MOF-il-us ee-jip-TEE-us)
Haemophilus ducreyi (he-MOF-il-us do-KRAY-eye)
H. influenzae (H. in-floo-EN-zee)
Halobacterium (hale-oh-back-TIR-ee-um)
Hydrogenomonas (hi-droh-jeh-no-MOH-nas)
Hyphomicrobium (hi-foh-my-KROW-be-um)
Klebsiella pneumoniae (kleb-see-EL-lah new-MOH-nee-ah)
Lactobacillus acidophilus (lack-to-bah-SIL-lus ay-seh-DAH-fill-us)
L. bulgaricus (L. bul-GAH-reh-kuss)
L. casei (L. KAY-see-eye)
L. delbruckii (L. dell-BROOK-ee-eye)
L. plantarum (L. plan-TAR-um)
Legionella pneumophila (lee-jon-EL-lah new-MOH-fill-ah)
Leptospira interogans (lep-toe-SPY-rah in-TER-oh-ganz)
Leptothrix (lep-TOE-thricks)
Leptotrichia (lep-TOE-trick-ee-ah)
Leuconostoc cremoris (loo-koh-NOS-tok krem-OH-riss)
L. mesenteroides (L. mes-en-ter-OY-deez)
Listeria monocytogenes (L. moh-no-sye-TAH-jen-neez)
Methanobacterium (meh-THAN-oh-back-TIR-ee-um)
M. bryantii (M. bry-ANN-tie)
Methyloccus (meth-ee-low-KOK-kuss)
Methylomonas methanica (meth-ee-low-MOH-nass meth-an-EE-kah)
Micrococcus citreus (my-krow-KOK-kuss SIT-ree-us)
M. luteus (M. LOO-tee-us)
M. radiodurans (M. ray-dee-oh-DUR-anz)
Microcyclus marinus (my-krow-SYE-klus mar-EE-nus)
Mycobacterium avium-intracellulare (my-koh-back-TIR-ee-um ey-VEE-um-in-tra-cell-you-lare)
M. bovis (M. BOH-viss)
M. leprae (M. LEP-ree)
M. smegmatis (M. smeg-MEH-tiss)
M. tuberculosis (M. too-ber-koo-LOW-sis)
Mycoplasma pneumoniae (my-koh-PLAZ-mah new-MON-nee-eye)
Neisseria gonorrhoeae (nye-SEH-ree-ah go-nor-REE-ah)
N. meningitidis (N. meh-nin-jit-EE-diss)
Neorickettsia (nee-oh-reh-KET-see-ah)
Nitrobacter (nye-troh-BACK-ter)
Nitrococcus (nye-troh-KOK-kuss)
Nitrosococcus (nye-troh-so-KOK-kuss)
Nitrosomonas (nye-troh-so-MOH-nass)
Nocardia asteroides (no-KAR-dee-ah as-ter-OI-deez)
Nostoc (nos-TOCK)
Oscillatoria (os-sill-a-toe-REE-ah)
Pasteurella (pass-tur-EL-lah)
Pedicoccus cerevisiae (ped-ee-KOK-kuss seh-ri-VISS-ee-eye)
Photobacterium (foto-back-TIR-ee-um)
Propionibacterium acnes (pro-pee-on-ee-back-TIR-ee-um AK-neez)
Proteus mirabilis (PRO-tee-us meh-RA-bill-iss)
P. vulgaris (P. vul-GA-riss)
Pseudomonas aeruginosa (soo-doh-MOH-nass a-ruh-jin-OH-sah)
Rhizobium (rye-ZOH-be-um)
Rhodomicrobium vannielii (roh-doh-my-KROW-be-um van-neel-ee-eye)
Rhodophyta (roh-doh-FYE-tah)
Rhodopseudomonas (row-doh-soo-doh-MOH-nass)
Rhodospirillum rubrum (roh-doh-spy-RIL-lum RUBE-rum)
Rickettsia prowazekii (ri-KET-see-ah prow-wah-ZEH-kee-eye)
R. rickettsii (R. ri-KET-see-eye)
R. typhi (R. TIE-fee)
Rochalimaea quintana (roh-sha-LIM-ee-ah quin-TAN-ah)
Salmonella enteritidis (sal-mon-EL-lah en-ter-IT-id-iss)
S. typhi (S. TIE-fee)
S. typhimurium (S. tie-fee-MUIR-ee-um)
Sarcina (sar-SEE-nah)
Serratia marcescens (ser-RAY-sha mar-SES-sens)

Shigella dysenteriae (shi-GEL-la dis-en-TEH-ree-eye)
S. flexneri (S. FLEKS-ner-eye)
S. sonnei (S. SONN-nee-eye)
Simonsiella (sye-mon-see-EL-lah)
Sphaerotilus natans (sfeh-RAH-teh-lus NAY-tans)
Spirillum (spy-RIL-lum)
Spiroplasma (spy-row-PLAZ-mah)
Spirosoma (spy-row-SO-mah)
Spirulina (spy-roo-LIE-nah)
Sporosarcina (spoh-row-sar-SEE-nah)
Staphylococcus aureus (staff-il-oh-KOK-kuss ORE-ee-us)
S. epidermidis (S. e-pee-DER-meh-diss)
Stigmatella (stig-MAH-tel-lah)
Streptobacillus moniliformis
(strep-toe-bah-SILL-uss moh-nil-ee-FOR-miss)
Streptococcus cremoris (strep-toe-KOK-kuss krem-OR-riss)
S. faecalis (S. fee-KAL-iss)
S. lactis (S. LACK-tiss)
S. mitis (S. MY-tiss)
S. mutans (S. MYOO-tans)
S. pneumoniae (S. new-MOH-nee-eye)
S. pyogenes (S. pie-ah-GEN-eez)
S. salivarius (S. sa-li-vah-REE-uss)
S. sanguis (S. SAN-gwiss)
Streptomyces (strep-toe-MY-sees)
Sulfolobus (sul-foh-LOW-bus)
Symbiotes (sim-BY-oh-tees)
Thermoactinomyces (ther-moh-ak-tin-oh-MY-sees)
Thermoplasma (ther-moh-PLAZ-mah)
Thiobacillus (thigh-oh-bah-SIL-lus)
Thiothrix (thigh-oh-TRICKS)
Treponema pallidum (tre-poh-NEE-mah PAL-li-dum)
T. pertenue (T. per-TEN-you-ee)
Ureaplasma urealyticum
(you-ree-ah-PLAZ-mah you-ree-ah-LIT-eh-kum)
Veillonella (vah-yon-ELL-ah)
Vibrio cholerae (VIB-ree-oh KOL-er-eye)
V. parahaemolyticus (V. par-ah-he-moh-LIT-ee-cuss)
Xanthomonas oryzae (zan-thoh-MOAN-us or-EE-zay)
Yersinia enterocolitica (yer-SIN-ee-ah en-ter-oh-koal-IT-ic-ah)
Y. pestis (Y. PES-tis)
Zoogloea (zoh-oh-GLEE-ah)

Fungi

Absidia (ab-sid-EE-ah)
Agaricus (a-GAR-i-kuss)
Allescheria (al-lesh-ER-ee-ah)
Alternaria (al-ter-NARE-ee-ah)
Amanita (am-an-I-tah)
Arthrobotrys (ar-throw-BOT-ris)
Aspergillus flavus (a-sper-JIL-lus FLAY-vus)
A. niger (A. NYE-jer)
A. oryzae (A. or-EE-zay)
A. soyae (A. SOY-ee)
Aureobasidium pullulans
(o-ree-oh-bah-SEH-dee-um pull-YOU-lans)
Candida (KAN-did-ah)
C. albicans (C. AL-beh-kans)
C. intermedia (C. in-ter-ME-dee-ah)
C. tropicalis (C. trop-EE-kal-iss)
C. utilis (C. YOU-til-iss)
Cephalosporium (seff-a-low-SPOR-ee-um)
Claviceps purpurea (KLA-vee-seps pur-pooh-REE-ah)
Coccidioides immitis (kok-sid-ee-OI-deez IM-mi-tiss)
Coprinus (ko-PRIN-us)
Cryptococcus neoformans
(kryp-toe-KOK-kus nee-oh-FOR-manz)
Dictyostelium discoideum
(dic-tee-oh-STEL-lee-um dis-KOY-dee-um)
Epidermophyton (ep-ee-der-moh-FYE-ton)
Fusarium (foo-sar-EE-um)
Geotrichum (gee-oh-TRICK-um)
Hansenula (han-sen-YOU-la)
Histoplasma capsulatum
(hiss-toe-PLAZ-mah kap-soo-LAY-tum)
Kluyveromyces (kli-ver-oh-MY-sees)
Microsporum (my-krow-SPO-rum)
Mucor (MYOO-kore)
Neurospora crassa (new-RAH-spor-ah KRASS-ah)
Penicillium (pen-ee-SIL-lee-um)
P. camemberti (P. cam-em-BER-tie)
P. notatum (P. no-TAY-tum)
P. roqueforti (P. rok-FOR-tie)
Philalophora verruscosa
(fee-lah-LOW-fo-rah ver-roo-SKOH-sah)
Pichia (PIK-ee-ah)
Pityrosporum (pit-ee-ROS-poh-rum)
Rhizopus nigricans (rye-ZOH-pus NYE-gri-kans)
Rhodotorula (row-doh-TOUR-you-la)
Saccharomyces cerevisiae
(sak-a-row-MY-sees se-ri-VISS-ee-eye)
Saprolegnia (sap-row-LEG-nee-ah)
Sporothricum schenkii
(spor-oh-TREE-kum SHEN-kee-eye)
Trichoderma (trik-oh-DER-mah)
Trichophyton mentagrophytes
(trik-oh-FYE-ton men-tag-row-FYE-teez)
Uromyces (ur-oh-MY-seez)

Protozoa

Acanthamoeba (a-can-tha-ME-bah)
Amoeba (a-ME-bah)
Balantidium coli (bal-an-TID-ee-um KOH-lee)
Cryptobia (crip-toe-BEE-ah)
Cryptosporidium (krip-toe-spoh-REH-dee-um)
Didinium (die-din-EE-um)
Difflugia (dif-floo-GEE-ah)
Eimeria (EYE-meer-ee-ah)
Entamoeba histolytica (en-tah-ME-bah his-toe-LI-tee-kah)

Euplotes (you-PLOH-teez)
Giardia lamblia (jee-AR-dee-ah LAM-lee-ah)
Leishmania donovani (lysh-may-NEE-ah don-oh-VAY-nee)
Naegleria fowleri (nye-GLI-ree-ah FOU-ler-eye)
Opalina (oh-pah-LINE-ah)
Paramecium (par-ah-ME-see-um)
Plasmodium falciparum (plaz-MOH-dee-um fal-SIP-ar-um)
P. malariae (P. mah-LAR-ee-eye)
P. ovale (P. oh-VAH-lee)
P. vivax (P. VYE-vaks)
Pneumocystis carinii (new-moh-SIS-tiss kar-IN-nee-eye)
Tetrahymena (tet-rah-hi-MEN-ah)
Tokophyra (toe-KAH-fir-ah)
Toxoplasma gondii (toks-oh-PLAZ-mah GON-dee-eye)
Trichomonas vaginalis (trik-oh-MON-as va-jin-AL-iss)
Trypanosoma brucei variety *gambiense* (tri-pan-oh-SO-ma broo-SEE-eye var-EYE-ee-tee gam-bee-ENS-zee)
T.b.v. rhodesiense (T.b.v. row-dees-ee-ENS-zee)
T. cruzi (T. KRUS-ee)
Vorticella (vor-tee-SEL-lah)

Helminths

Ancylostoma duodenale (an-sil-oss-TOE-mah doo-ah-den-AL-leh)
Anisakis (an-ee-SAK-iss)
Ascaris lumbricoides (AS-kar-iss lum-bre-KOY-deez)
Brugia malayi (broo-GEE-ah may-LAY-eye)
Clonorchis sinensis (klo-NOR-kiss seh-NEN-sis)
Diphyllobothrium latum (die-fil-low-BOW-three-um LAY-tum)
Dirofilaria immitis (die-row-fil-AIR-ree-ah IM-mit-tis)
Echinococcus granulosus (e-ky-no-KOK-kus gra-new-LOH-suss)
Enterobius vermicularis (en-ter-OH-bee-us ver-mi-kyoo-LAR-iss)
Fasciola (fas-see-OH-la)
Hymenolepsis (hi-men-ol-EP-sis)
Loa loa (LOH-ah LOH-ah)
Necator americanus (nee-KAY-tor ah-mer-ee-CAN-uss)
Onchocerca (on-choh-CER-ka)
Paragonimus westermanni (par-ah-GON-ee-mus wes-ter-MAN-nye)
Schistosoma haematobium (shis-toe-so-mah he-mah-toe-BE-um)
S. intercalatum (S. in-ter-kal-A-tum)
S. japonicum (S. ja-PON-ee-cum)
S. mansoni (S. man-SO-nee)
S. mekongi (S. may-KONG-eye)
Strongyloides stercoralis (stron-gee-LOY-deez ster-cor-AL-liss)
Taenia saginata (TEE-nee-ah sa-jin-AH-tah)
T. solium (T. SO-lee-um)
Trichinella spiralis (trick-a-NEL-lah spi-RA-liss)
Trichuris trichiura (tri-CURE-iss tri-CURE-ah)
Wuchereria bancrofti (woo-cheh-rer-EE-ah ban-KROF-tee)

Glossary

Over 600 important and widely used terms and concepts are included in this glossary. Other terms defined in the text are referred to by page number in the index. Phonetic pronunciations of selected terms are included.

Abscess *(ab-sess)*: A localized collection of pus.

Acid: A compound that releases hydrogen (H^+) ions when dissolved in water.

Acid-fast staining technique: One type of differential stain procedure. It is used to determine a particular property of certain bacteria to retain the primary stain, carbol fuchsin, and resist decolorization with acid alcohol.

Acquired immunity: Resistance obtained during development and after birth.

Activated sludge *(slujh)*: A method of secondary sewage treatment for the rapid handling of large volumes of sewage.

Active immunity: The state of resistance following exposure to an antigenic stimulus; this immune state may be naturally or artificially acquired.

Active site: The location at which a substrate binds an enzyme.

Active transport: A process occurring at the cell membrane in which a cell expends energy to move materials through the membrane, often against a concentration gradient.

Acute *(a-KYOOT)*: The term used to indicate the rapid appearance of the signs and symptoms of an illness or disease.

Acute necrotizing ulcerative gingivitis *(a-KYOOT ne-krow-TIE-zing UL-sir-a-tiv jin-jeh-VYE-tis)*: The rapidly developing form of periodontal disease also known as Vincent's angina *(an-JIE-nah)* or trench mouth.

Acyclovir *(a-SIGH-kloh-vir)*: A chemotherapeutic purine derivative that acts as a competitive inhibitor of viral nucleic acid synthesis.

Adenosine triphosphate (ATP): The macromolecule that functions as an energy carrier in cells. The energy is stored in a high-energy bond between the second and third phosphates.

Adhesins *(ad-HE-zuhns)*: Surface-associated parts of microorganisms that are used for attachment to host tissues.

Adjuvant: A compound capable of increasing an immune response.

Aerobe: A microorganism whose growth requires free oxygen.

Aerobic respiration: The process by which a cell uses oxygen and releases the energy in glucose, producing adenosine triphosphate (ATP). Aerobic respiration includes glycolysis, the Krebs (citric acid) cycle, and electron and hydrogen transport.

Aerosol: A fine suspension of particles or liquid droplets sprayed into the air.

Aflatoxin: One type of mycotoxin produced by some strains of the fungi *Aspergillus flavus* and *A. parasiticus.*

Agar *(AH-garr)*: A dried polysaccharide (galactan), extract of red algae, used as the solidifying agent in various microbiological media.

Agglutination (immunology): The visible clumping of cellular or particlelike antigens by corresponding antibodies; e.g., blood typing.

AIDS (acquired immune deficiency syndrome): A viral condition due to a retrovirus (human immunodeficiency virus) causing a defect in the host's cell-mediated immunity so that there are a greater number of T suppressor cells than T helper cells; opportunistic infections and/or Kaposi's sarcoma usually cause the death of AIDS victims.

Antimetabolite *(an-tie-meh-TAB-oh-light)*: A chemical that interferes with the microbial metabolism of an essential compound by acting as a competitive inhibitor.

Aldehyde *(AL-de-hide)*: Any one of a large group of substances derived from the primary alcohols by oxidation, and containing CHO.

Alkaline: A condition in which hydroxyl (OH^-) ions are in abundance. Solutions with a pH of 7.1 or higher are alkaline or basic.

Allergen *(al-ER-gen)*: A substance capable of bringing about an allergic state when introduced into the body.

Allergy: Any altered activity of an individual caused by contact with animate or inanimate substances that normally do not cause a reaction (allergens), e.g. pollen, fur, feathers, certain chemicals, etc.

Allograft (also homograft): A tissue or organ graft between two genetically dissimilar members of the same species.

Allotype *(AL-low-type)*: The genetically determined antigenic difference among immunoglobulins in different members of the same species.

Alternative complement pathway (also properdin pathway): The system of activation of the complement pathway through the involvement of properdin factor D, properdin factor B, and C3b, finally activating C3 and following the classic pathway.

Ames test: A test used to determine the mutagenic or carcinogenic potential of chemical substances. This test measures the degree to which the test chemical promotes mutation in a particular bacterial sample.

Amino acid: A nitrogenous organic compound that serves as a basic unit of a protein molecule.

Aminoglycoside *(ah-ME-no-GLY-koh-side)*: A group of bacterial antibiotics (e.g., streptomycin and gentamycin) that contain amino groups (NH_2) bonded to carbohydrate groups and derived from various species of *Streptomyces;* aminoglycosides interfere with the protein synthesizing function of bacterial ribosomes.

Ammonification: The microbial decomposition of organic nitrogen-containing compounds, such as proteins, with the formation of ammonia.

Anabolism: The phase of metabolism involving the formation of organic compounds; usually an energy-utilizing process.

Anaerobe *(AN-air-robe)*: A microorganism that grows only or best in the absence of free oxygen. Organisms utilize bound oxygen.

Anamnestic *(an-am-NES-tick)* **response:** The sudden secondary rise in immunoglobulin (antibody) concentration produced by a second exposure to an immunogen some time after the initial exposure.

Anemia *(ah-NEE-me-ah)*: A deficiency of red blood cells, hemoglobin or both.

Angstrom unit (Å): A unit of length measuring 10^{-8} cm (1/100,000,000 cm).

Animal reservoir: A wild or domestic animal that serves as a host for an infective agent, and from which the agent may periodically be transmitted to humans or other forms of life.

Antibiotic: A microbial metabolic or laboratory-synthesized product with the capacity to inhibit or kill bacteria and other microorganisms. Antibiotics are used in the treatment and control of various infectious diseases.

Antibody (immunoglobulin): A protein molecule formed in response to a foreign substance (immunogen). Antibodies react with and/or inactivate the foreign matter causing their formation.

Antibody combining site: The region on an antibody molecule that links with a corresponding antigenic determinant.

Antibody-dependent cell-mediated cytotoxicity (ADCC): A form of lymphocyte-mediated cellular injury (cytotoxicity) in which an activated cell kills an antibody-coated target cell.

Anticodon: A sequence of three nucleotides in transfer RNA (tRNA) that is complementary to and combines with the three-nucleotide codon of messenger RNA (mRNA). This reaction binds the activated amino acid-tRNA combination to mRNA.

Antigenic determinant (see also epitope): The specific portion of an antigen (immunogen) that interacts with the corresponding antigenic combining site of an immunoglobulin.

Antiseptic: Against or opposing sepsis (infection), putrefaction, or decay. An antimicrobial agent used to prevent or arrest microbial growth.

Antiserum: Serum containing specific antibodies.

Antitoxin: An antibody (immunoglobulin) capable of neutralizing the toxin or toxoid that stimulated its production.

Archaeobacteria *(ar-kee-bak-TEE-ree-ah)*: A group of procaryotic microorganisms that is more closely related to the earliest of life forms than to present-day bacteria; these microbes differ from eubacteria in several properties including the absence of peptidoglycan from cell walls and chemical differences in cell membrane and ribosomes.

Arthropod: An animal lacking a backbone (invertebrate) and having jointed legs; e.g., insects, crustaceans.

Arthrospore (also arthroconidium): An asexual spore of fungi formed by the fragmentation of hyphae.

Arthus *(ar-TOOS)* **reaction:** A localized form of anaphylaxis in which immune complexes are deposited in small blood vessels, and tissue destruction occurs.

Artificially acquired active immunity: The state of resistance resulting from immunizations and in which immunoglobulins are produced.

Artificially acquired passive immunity: The state of resistance usually resulting from the injection of immunoglobulin-containing preparations.

Ascomycotina (Ascomycetes): The group of fungi known for the formation of a saclike structure, the ascus.

Ascospore: A sexual spore found in the Ascomycotina.

Ascus: A saclike structure found in Ascomycotina fungi within which ascospores are formed.

Aseptic *(a-SEP-tick)* **technique:** A procedure in which precautions against microbial contamination are taken.

Asexual reproduction: A form of reproduction not involving sex cells or fusion of their nuclei.

Asymptomatic *(a-simp-toe-MAT-ik)*: Without symptoms.

Attenuated: Weakened; reduced in virulence.

Autoantigens: Body components that provoke antibody production against themselves.

Autoclave: An apparatus utilizing pressurized steam for sterilization.

Autogamy: A modified form of conjugation that occurs within one protozoon.

Autograft: A tissue graft between genetically identical members of the same species.

Autoimmunity *(oh-toe-eh-MU-neh-tee)*: The condition in which the body's immune system reacts to the body's own tissues.

Autoinfection: An infection of one part of the body caused by organisms derived from another region of the body.

Autolysis: Cellular disintegration caused by the organism's own enzymes.

Autotroph: An organism that can synthesize all of its organic components from inorganic sources.

Auxotrophic *(ox-oh-TRO-fic)* **mutant:** An organism that requires one or more specific growth factors not needed by the parental (wild-type) strain.

Avirulent *(a-VEER-you-lent)*: Unable to produce disease.

Axial *(AKS-i-al)* **filaments:** Flagellalike structures found in spirochetes, between the main body of cells and their outer coverings.

Axenic *(aks-EN-ik)*: Refers to a pure culture.

Azidothymidine (AZT): A drug used in the treatment of AIDS and other human immunodeficiency virus infections.

Bacteremia *(bak-teh-REE-mee-uh))*: The presence of bacteria in the blood.

Bacterial conjugation: A form of plasmid-regulated transfer of some or all of a donor cell's chromosomal DNA to a recipient cell.

Bactericidal *(back-tier-ee-SI-dal)*: Lethal to bacteria.

Bacteriophage *(bak-TEE-reh-oh-fayj)*: A virus that infects bacteria.

Bacteriorhodopsin *(bak-TEE-reh-oh-roh-DOP-sin)*: A light-reactive chemical related to human retinal pigment; is used by certain bacteria for the capture of light energy.

Bacteriostatic *(back-tier-ee-oh-STAT-ic)*: Inhibiting bacterial growth without killing organisms.

Bacteriocin: A protein produced by bacteria that kills other bacteria; some bacteriocins appear to be defective viruses or viral components.

Bacteriuria *(bak-TEE-re-you-ri-ya)*: The presence of bacteria in urine.

Baeocysts *(bay-OH-sists)*: Reproductive cells formed by certain cyanobacteria.

Barophile *(BA-roh-file)*: A bacterium adapted to intense atmospheric pressure.

Basidiomycotina (basidiomycetes): A subdivision of fungi that includes molds, rusts, mushrooms, and yeasts. Members of the group form a special structure called a basidium, which bears basidiospores.

Basidiospore *(bah-SID-ee-oh-spor)*: The spore type found on a basidium.

B cell (also B lymphocyte): The precursors of antibody producing plasma cells.

Benign *(be-NINE)*: Harmless.

Beta-lactamase *(LACK-tam-ace)*: The general name for a group of enzymes, such as penicillinase, that attack the beta-lactam ring in the penicillin-class antimicrobials.

Beta-lactam ring: The portion of the penicillin molecule that contains its antimicrobial activity.

Binary fission *(FISH-shun)*: An asexual reproductive process in which one cell splits into two independent daughter cells.

Binomial system of nomenclature: The method of naming organisms devised by Carolus Linnaeus. Each organism is given two names designating genus and species, such as *Staphylococcus aureus.*

Bioassay: Determination of the activity or amount of a biologically active material by measuring its effect on living organisms.

Biochemical oxygen demand (BOD): The quantity of oxygen needed by microorganisms in a body of water to decompose the organic matter present. An index of water pollution.

Bioconcentration: The reactions during natural cycles that lead to the storage of particular materials, e.g., sulfur, nitrate.

Bioconversion: The enzymatic modification of chemicals.

Biosphere: The portion of the earth in which all life exists, including the upper layers of soil and water and the lower atmosphere.

Brownian movement: A type of jiggling motion of particles and bacteria in suspension caused by the bombardment of molecules in the suspending fluid.

Bubo *(BOO-boh)*: A swollen, painful lymph node usually seen in the bubonic form of the bacterial disease plague.

Budding: The production of a new cell from a small outgrowth from another cell.

Buffer: Any substance present in a preparation or body fluid that tends to control the change in pH when either an acid or base is added.

Bursa of Fabricius *(fab-REE-see-us)*: A cloacal organ in fowl from which the immunoglobulin-forming B lymphocytes originate.

Cancer *(KAN-ser)*: A broad group of disease states in which harmful or life-threatening growths (tumors) generally develop.

Candidiasis *(kan-deh-DYE-a-sis)*: Infection of the skin or mucous membranes with any species of the yeast *Candida.*

Capneic *(KAP-nay-ick)*: Refers to microorganisms having a growth requirement for carbon dioxide above the level normally found in air; requirements range from 2 to 10 percent.

Capsid *(CAP-sid)*: Regular, shell-like structure, composed of protein subunits, that encloses the nucleic acid of individual virus particles with them.

Capsomere *(cap-SOH-mer)*: Specific protein molecules that represent the building block of a viral capsid.

Capsule: A form of glycocalyx that is a thick, sticky structure surrounding the cell walls of certain bacterial and yeast cells. Usually composed of polysaccharide or polypeptide.

Carbohydrate: A class of organic compounds made of carbon, hydrogen, and oxygen, with the latter two elements in a ratio of 2 to 1; e.g., sugars, starches, cellulose.

Carbon cycle: A geochemical cycle in which carbon is recycled by natural processes. Carbon in the form of a gas (CO_2) is converted into food by producers; consumer organisms obtain their carbon by eating producers; carbon leaves producers and consumers in the form of wastes and dead material, which are broken down by decomposers.

Carbuncle *(KAR-bung-kel)*: A localized inflammation of the skin and deeper tissues accompanied by pus formation.

Carcinoma *(car-sin-OH-mah)*: Cancer of epithelial tissue.

Carcinogen: A cancer-causing agent.

Carditis: Inflammation of the heart.

Caries: Gradual decay of a bone or tooth associated with the inflammation of bone or progressive decalcification of teeth.

Carrier (biological): An individual harboring a disease agent without apparent symptoms.

Carrier (immunologic): An immunogenic substance that, when linked to a hapten, makes the hapten immunogenic.

Caseation *(kay-zee-A-shun)*: A form of necrosis (tissue destruction) in which the tissue is changed into a dry, shapeless mass resembling cheese.

Casein *(KAY-seen)*: The major protein in milk.

Catabolism: The chemical reactions by which food materials or nutrients are converted into simpler substances for the production of energy and cell materials.

Catalyst: A substance that can speed up a reaction or cause a reaction to occur without itself being altered permanently.

Catheter: A tubular surgical device used for the withdrawal of fluids from a body cavity or structure.

Cell fusion: The formation of a hybrid cell with nuclei and cytoplasm from different cells.

Cell wall: The cell structure exterior to the cell membrane of typical plants, algae, bacteria, and fungi. It give cells form and shape.

Cementum: The bonelike connective tissue covering the root of a tooth and assisting in tooth support.

Centers for Disease Control: A branch of the United States Public Health Service responsible for the collection and dissemination of epidemiological information and involved with the identification of the causes of major public health problems.

Cercaria *(sir-CARE-e-ah)*: The major invasive larval (young) form of flukes (trematodes).

Cestode *(SES-tode)*: A tapeworm.

Chemoautrotroph: An nonphotosynthesizing organism that obtains its energy from the oxidation of inorganic chemical compounds in order to fix CO_2 as the only source of carbon.

Chemoheterotroph: An organism that obtains its carbon and energy from the oxidation of organic chemical compounds.

Chemolithotroph: An organism that utilizes inorganic compounds as energy sources.

Chemotherapy: The treatment of disease by the use of chemicals that inhibit or kill the causative agents but ideally do not injure the cells or tissues of the host.

Chitin: A complex polysaccharide that is the main component of the exoskeleton of certain shellfish and arthropods and of the cell walls of some fungi.

Chloroplast: A chlorophyll-containing intracellular organelle of eucaryotic organisms such as algae, plants, and certain protozoa that is the site of photosynthesis.

Chromatophore *(kroh-MAT-oh-four)*: A photosynthetic pigment contained in folded membranes or sacs in photosynthetic bacteria.

Chronic: Recurrent or of long duration.

Cilium *(SILL-ee-uhm)*: A short, hairlike, eucaryotic cellular structure used for movement.

Class: A major taxonomic subdivision of a phylum. Each class is composed of one or more related orders.

Classic complement pathway: A series of enzyme–substrate and protein–protein interactions that ultimately leads to biologically active complement enzymes. It proceeds sequentially from C1 through other C′ components 4,2,3,5,6,7,8,9.

Clone: A group of identical cells, all of which are the descendants (progeny) of a single cell.

Cloning: A term frequently and loosely used for the entire process of isolating, identifying, and manipulating single genes in recombinant DNA procedures.

Coagulase: An enzyme produced by disease-causing strains of *Staphylococcus aureus;* the enzyme catalyzes the formation of a fibrin clot in blood plasma.

Coccus *(KOCK-us)*: A spherical bacterium.

Codon *(KOH-don)*: Nucleotide triplet of messenger RNA containing the information needed to specify a particular amino acid to be used in the formation of proteins and related compounds.

Coenzyme: The simpler portion of a conjugated enzyme, which is necessary for the enzyme's activation and reaction with a substrate.

Cold agglutinins *(ah-GLUE-tin-ins)*: Antibodies that agglutinate (clump) bacteria or erythrocytes more efficiently at temperatures below 37°C than at 37°C. The test is used to detect mycoplasma pneumonia.

Coliform: Gram-negative, lactose fermenting rods, including *Escherichia coli* and similar species that normally inhabit the colon (large intestine). Commonly included in the coliform group are *Enterobacter aerogenes, Klebsiella* species, and other related bacteria.

Colonization: The establishment of a reproduction site for microorganisms without necessarily resulting in tissue invasion or injury.

Colostrum *(ko-LOS-trum)*: The watery, yellow, milky fluid secreted by the mammary glands a few days before or after delivery.

Commensalism: The symbiotic association between two organisms in which one is benefited and the other is neither harmed nor benefited.

Community: All organisms that occupy the same habitat and interact with one another.

Competent cell: A living recipient bacterial cell capable of accepting DNA during transformation.

Compound: A chemical containing two or more kinds of atoms.

Competition: An association between two species, both of which need some limited environmental factor such as a nutrient for survival.

Complement: A complex system of serum proteins that is the primary humoral mediator (regulator) of antigen-antibody reactions. The complex is found in the blood of most warm-blooded animals, and participates in the destruction of bacteria and certain other cells when it is combined with an antigen-antibody complex.

Conjugation (protozoan): A form of sexual reproduction in which there is a temporary fusion of mating partners for the exchange of nuclear material.

Constant (C) region: The portion of an immunoglobulin chain in which there is little or no variation in the amino acid sequence; this region determines the immunoglobulin class. The C region is the carboxyl terminal (COOH) portion of the heavy or light chain that is identical in immunoglobulin molecules of a given class.

Contact dermatitis: A particular form of delayed hypersensitivity (type IV), which may be associated with simple chemicals, metals, certain drugs, cosmetics, insecticides, or the active components of plants such as poison ivy and poison oak.

Contractile vacuole: A pulsating vacuole found in certain protozoa and used for the excretion of wastes and the maintenance of proper osmotic balance.

Coombs (KOOMS) test: The immunological procedure used to detect the presence of incomplete (nonagglutinating) antibodies.

Crossing over: Exchange of genetic material between corresponding (homologous) chromosomes.

Crustose: Flat, crustlike.

Culture: Any growth or cultivation of microorganisms.

Cyclic adenosine monophosphate (cAMP): A nucleotide associated with starting the formation of catabolic enzymes.

Cyclosporine *(sye-kloh-SPOR-in)*: A fungus-produced peptide used in the control of graft tissue exchanged between two nonidentical members of the same species (allograft).

Cyst *(sist)*: A walled sac or pouch that contains fluid, semisolid, or solid material. Also, a resting structure formed by certain microorganisms.

Cystitis *(sis-TIE-tis)*: Infection of the urinary bladder.

Cytopathic effect (CPE): Morphological changes in tissue culture cells caused by a pathogen such as a virus.

Dalton: The unit of mass equal to that of a single hydrogen atom.

Dark-field microscopy: A microscopy technique in which no direct lighting of the microscope field occurs. An object is made visible only by light rays reflected by it.

Deamination: The removal of an amino (NH_2) group from a compound.

Debridement: The removal of foreign material and devitalized or contaminated tissue from an injured or infected lesion.

Degranulation: The process of losing granules.

Delayed hypersensitivity: A form of allergy characterized by several factors including the absence of circulating antibodies (immunoglobulins), reactions taking place from 24 to 48 hours after exposure to antigen, possible transfer in humans by a nonantibody "active" transfer factor, and lack of an inhibition of the reaction by antihistamines.

Denaturation: A change in the secondary or tertiary structure of a macromolecule, such as protein or nucleic acid, that affects solubility and various biological activities.

Dendrogram: A branching diagram constructed to show relationships between organisms or clusters (groups) of organisms.

Denitrification: The microbial process whereby nitrates (NO_3) are transformed into nitrogen (N_2) or ammonia (NH_3), which then reenters the atmosphere as a gas.

Deoxyribonucleic acid (DNA): A macromolecule that contains genetic information coded in specific sequences of its constituent nucleotides.

Dermatophyte: A pathogenic fungus that attacks the skin, nails, or hair.

Desensitization: The elimination or reduction of an allergic sensitivity, usually by means of a programmed course of allergen injections.

Desquamation: The shedding of chiefly skin in scales or small sheets of cells; exfoliation.

Deuteromycotina *(DEW-tier-oh-my-co-teen-ah)*: An artificially established class of fungi that have not been observed to form sexual reproductive structures. This group is also known as the Fungi Imperfecti.

Dextrans: Polysaccharides composed of a single sugar.

Diaminopimelic acid: A compound found in nature only in procaryotic organisms, particularly in the cell wall mucocomplex of most bacteria.

Diapedesis: The outward movement of cells through intact blood vessel (capillary) walls.

Diatomaceous *(dye-ah-toe-MAY-shus)* **earth:** The remains of diatoms, often used as filtering material.

Differential medium: A growth medium for some bacteria on which the appearance of growth exhibited by certain organisms is sufficient to distinguish them from others growing on the same medium.

Differential staining methods: Procedures used to distinguish among bacterial species and in some cases between the bacterial cell and its parts; e.g., Gram stain, acid-fast stain, spore stain.

Diluting: The procedure of increasing the proportion of liquid diluent to particular matter or other material being diluted.

Dimer: The molecular structure resulting from covalent bonding or aggregation of two identical subunits.

Dimorphic *(dye-MORE-fick)*: Exhibiting two forms in two different environments.

Dipicolinic *(dye-pick-oh-LIN-ic)* **acid:** A compound of bacterial spores that contributes to their heat resistance.

Direct count: The microscopic analysis of the sample that leads to determining the number of microorganisms present per specified volume of sample.

Disaccharide: A sugar consisting of two monosaccharides.

Disinfection: The chemical or physical treatment used to remove or kill microorganisms on inanimate surfaces.

Disseminated: Spread to neighboring or distant areas.

Disulfide bonds: Chemical S—S bonds between sulfhydryl-containing amino acids that contribute to the three-dimensional structure of proteins.

Division: A major taxonomic subdivision, usually of the plant kingdom. Each division is composed of one or more classes. This subdivision is equivalent to the taxonomic rank of phylum.

DNase: Deoxyribonuclease; applies to a group of differing enzymes all of which hydrolyze DNA in various ways.

DNA base ratio: The ratio in mole percent of adenine plus thymine to guanine plus cytosine.

DNA ligase *(LYE-gace)*: An enzyme that binds DNA fragments together.

DNA polymerase *(POL-eh-mer-ace)*: A DNA-forming enzyme that combines nucleotide fragments during DNA replication.

DNA probe: A highly specific genetic engineering tool consisting of nucleic acid sequences that will match and combine with genetic sequences of material extracted from or left in place in cancer cells, specific microorganisms, or cells from individuals with specific genetic defects.

Dormancy: A state of inactivity.

EAC rosette: A cluster of red blood cells sensitized with antibody and complement surrounding human B lymphocytes.

Early proteins: The proteins of viral origin that appear early in the viral replication cycle.

Early versus late genes: Genes transcribed early and late after virus infection of cells.

Ecology: The study of the interrelationship of an organism and its environment.

Ecosystem: All of the organisms in a habitat plus all of the factors in the environment with which the organisms interact.

Ectoparasite: A parasite that lives or feeds on the outer surface of the host's body.

Ectosymbiont: An organism living on the surface of a host.

Edema *(eh-DEE-mah)*: Swelling caused by an accumulation of fluid in tissues.

Electron transport system: A series of oxidation-reduction reactions in which electrons are transported from a substrate to a final acceptor, usually O_2, and ATP is formed.

Electrophoresis: A laboratory procedure for the separation of charged molecules in an electrical field.

Encephalitis *(en-seff-ah-LYE-tiss)*: Inflammation of the brain.

Endemic: Present more or less continuously in a community.

Endocarditis: Inflammation of the inner portions of the heart.

Endoparasite: A parasite found within host tissues and cells.

Endoplasmic reticulum: Interconnected membranes within the cytoplasm of a eucaryotic cell.

Endospore: A resistant body formed within certain bacterial cells.

Endotoxins: Lipopolysaccharides that are derived from the cell walls of gram-negative microorganisms and have toxic and fever-causing (pyrogenic) effects when injected *in vivo.*

Envelope (viral): The outer lipid-containing layer possessed by some virions, obtained from modified host cell membranes upon the release of the virus particle.

Enzyme: An organic (protein) catalyst that causes changes in other substances without undergoing any alteration itself.

Enzyme-linked immunosorbent assay (ELISA): A highly specific diagnostic test that includes the reaction involving an antigen-antibody complex with an antiglobulin linked to an enzyme. The presence of the complex can be revealed by the formation of a colored product, upon the addition of a specific substrate.

Enzyme-substrate complex: The combination of an enzyme and its substrate that initiates an enzymatic reaction.

Eosinophilia *(ee-oh-sin-oh-PHIL-ee-ah)*: An increase in the number of eosinophils.

Epidemic: An outbreak of a disease that appears rapidly and attacks a large number of persons in a community at approximately the same time.

Epidemiology *(eh-pee-dee-me-OL-oh-gee)*: The investigative science concerned with the occurrence and distribution of various diseases.

Epithelial tissue: Cells making up an exterior and interior body surface, including the liver and other organs.

Epitope *(eh-PEE-toep)*: The simplest form of an antigenic determinant present on a complex antigenic molecule.

Epstein-Barr virus (EBV): The human herpesvirus that causes acute and chronic forms of infectious mononucleosis and the cancerous condition Burkitt's lymphoma.

E rosette: A formation of a cluster (rosette) of cells consisting of sheep erythrocytes and human lymphocytes.

Erythema *(er-eh-THEE-mah)*: A reddened area of the skin caused by an increased accumulation of blood.

Etiology *(ee-tee-OL-oh-gee)*: Cause(s) of a disease.

Eucaryotic *(u-care-ee-OH-tick)* **cells:** Cells of higher organisms that have nuclear membranes, membrane-bound organelles, 80S ribosomes, and biochemistry differentiating them from procaryotes (bacteria and cyanobacteria).

Eutrophication: Nutrient increase and enrichment in lakes and other bodies of waters, leading to an overproduction of algae and a decrease in dissolved oxygen levels.

Exfoliation: A falling off in scales or layers.

Exocytosis: The cellular elimination of a vacuole's contents, the reverse of pinocytosis.

Exoenzyme: An enzyme secreted by the cell to the environment.

Exon *(EKS-on)*: A region coding for proteins in the mRNA of eucaryotic cells.

Exotoxins: Diffusible toxins produced by certain gram-positive and gram-negative microorganisms.

Exudate *(EKS-u-date)*: Material consisting of or containing pus that has escaped from blood vessels and has been deposited in tissues or on tissue surfaces, usually as a result of inflammation.

Fab: An antigen-binding fragment produced by enzymatic digestion of an IgG molecule.

Facultative: Adjustable.

Fastidious organism: An organism that has complex nutritional and environmental requirements.

Fat: One of the several classes of lipids, composed of glycerol and fatty acids.

Fatty acid: A straight chain of carbon atoms with a COOH at one end, in which most of the carbons are attached to hydrogen atoms.

F^+ cell: A bacterial cell that can transfer the extrachromosomal F (fertility) particle to a recipient (F^-) cell.

F^- cell: A bacterial cell that does not contain an F particle but can act as a recipient and receive one from an F^+ cell.

F′ cell: A bacterial cell containing a plasmid that is a combination of the F factor and a portion of its chromosome.

Fc fragment: A crystallizable fragment obtained by enzyme digestion of IgG molecules. It consists of the C-terminal half of two H chains linked by disulfide bonds. It contains no antigen-binding capability but determines important biologic characteristics of the intact molecule.

Fc receptor: A receptor present on members of various subclasses of lymphocytes for the Fc fragment of immunoglobulins.

Feedback inhibition: Inhibition of enzyme activity resulting from the effect of the concentration of the end product in a biosynthetic pathway. The enzyme is crucial to the formation of an intermediate metabolite. This inhibition is a self-regulatory mechanism by which the rate of synthesis of a metabolite is controlled.

Fermentation: The enzymatic breakdown of complex organic compounds under anaerobic conditions in which the final hydrogen acceptor is an organic compound.

Ferritin: A liver protein containing 20% iron, which is useful as a marker in electron microscopy.

Fetus: The unborn offspring after it has largely completed its embryonic development.

Fibroblast: Connective tissue cell type.

Filtration: Passage of a liquid through a porous membrane for the removal of particles. Also, the use of certain types of glass to remove particular wavelengths of light.

Fission: A form of splitting asexual reproduction in which a unicellular organism divides into two new cells.

Five-kingdom system: The arrangement of eucaryotic and procaryotic forms of life into the five kingdoms of Animalia, Plantae, Fungi, Protista, and Procaryotae (Monera).

Fixation: The process of preserving a specimen.

Flagellum: A long, whiplike organelle used in the movement of certain organisms and cells.

Floc: A mass of microorganisms caught together in a slime produced by certain bacteria; usually found in waste treatment plants.

Fluorescence *(floor-ESS-sense)*: Unused potential energy released as visible light by a substance that has absorbed radiation from another source.

Fluorescent antibody technique: Detection of a specific antigen in cells by staining with a specific antibody linked (conjugated) with a fluorescent dye.

Focal infection: A local infection from which microorganisms continuously or intermittently spread to other areas.

Foliose *(FOE-lee-ohss)*: Flat, leaflike.

Fomite *(FOE-might)*: Inanimate contaminated object.

Food chain: A sequence of organisms by which food energy is transferred from autotrophs to heterotrophs; each organism consumes the preceding one and in turn is eaten by the following member of the sequence.

Frameshift mutation: A mutation resulting in a shift in the reading frame of nucleic acid molecules; it is caused by the addition or deletion of one or more nucleotide bases in DNA.

Freeze-fracture: A special preparative procedure used in electron microscopy in which the exposed surface of a supercooled specimen is used for a metal impression (replica).

Frustule *(FRUSS-tool)*: The characteristic cell wall of diatoms.

Functional groups: The arrangement of atoms in organic molecules that is responsible for most of the chemical characteristics of such molecules.

Fungus: Eucaryotic unicellular and sometimes multicellular organisms with rigid cell walls and an absorptive type of nutrition; e.g., molds, mushrooms, puffballs, and yeasts.

Furuncle *(FUR-un-kel)*: A localized bacterial skin infection resulting from the blockage of hair follicles or sebaceous glands.

Gall (plant): A tumor produced on various plant parts due to infections caused by certain microorganisms or insects.

Gene splicing: A term used to describe the formation of recombinant DNA.

Generation time: The length of time required by bacterial species to complete their division cycle.

Genetic code: The composition and sequence of all the sets of three nucleotides that code for the amino acids in a protein molecule. The sequence of three nucleotides that codes for one amino acid is referred to as a code word.

Genome *(GEE-nome)*: The entire set of genetic information in an organism.

Genotype: Genetic makeup of an organism, as distinguished from its physical appearance or phenotype.

Genus (pl. genera): A taxonomic category of related organisms, usually containing several species; the first name of an organism in the binomial system of nomenclature.

Germination: The sprouting of a spore.

Giemsa stain: A stain (composed of the dyes azure and eosin) used for the demonstration of chlamydia, rickettsia, and protozoa.

Glucoside: A compound resulting from the combination of glucose and another compound containing an alcohol group; a glycoside.

Glycocalyx *(gli-co-KAY-licks)*: The polysaccharide-containing structure of bacteria lying outside of the cell wall.

Glycolysis *(gly-KAHL-eh-sis)*: The main metabolic pathway for the oxidation of glucose to pyruvic acid.

Gnotobiotics *(no-toe-by-OH-ticks)*: The study of animals that have been born and raised in a germ-free environment.

Golgi apparatus: A cytoplasmic organelle of eucaryotic cells consisting of unit membrane foldings. The Golgi apparatus serves as a collecting and processing center for lipids and proteins, changing these macromolecules to meet the needs of the cell.

Gram stain: A differential staining procedure by which bacteria are categorized as either gram-positive or gram-negative based on their retention of a primary dye (usually crystal violet) after treatment with a decolorizing agent such as acetone alcohol.

Granulocytes: One group of white blood cells containing cytoplasmic granules that stain differently; e.g., basophil, eosinophil, neutrophil.

Granuloma *(gran-you-LOW-mah)*: A tumor or growth consisting of lymphocytes, macrophages, and giant cells; develops in various infectious diseases including leprosy, syphilis, and the protozoan disease cutaneous leishmaniasis.

Growth curve: The curve obtained by plotting increases in the numbers of microorganisms against the elapsed time.

Guillain-Barré *(GIL-ane-BAR-ray)* **syndrome:** Acute febrile inflammation of nerves (polyneuritis).

Gumma *(GOO-mah)*: A tumorlike fleshy mass of tissue (granuloma) found in the late stages of syphilis.

Habitat: The specific location where a particular organism lives.

Halophile *(HA-lo-fil)*: An organism that grows in environments containing high salt concentrations.

Hapten *(HAP-ten)*: A substance that is not immunogenic but can react with an antibody of appropriate specificity.

HBcAg (also core antigen): The 27-nanometer core of hepatitis B virus that has been identified in the nuclei of hepatocytes.

HBsAg: The surface coat or envelope of hepatitis B virus.

H chain (heavy chain): One pair of identical polypeptide chains making up an immunoglobulin molecule. The heavy chain is twice the molecular weight of the light chain.

Helminth: Worm.

Helper T cells: A subtype (subpopulation) of T lymphocytes that function with B cells in antibody formation.

Hemagglutination *(hee-mah-GLOO-tin-a-shun)*: The clumping of red blood cells.

Hemagglutination inhibition: The prevention of hemagglutination, usually by means of specific immunoglobulins or specific enzymes.

Hemolysis *(hee-MOLL-ee-sis)*: The disruption of red blood cells with the leakage of hemoglobin.

Hemorrhage: The escape of blood from the vessels.

Hepatitis: Inflammation of the liver.

Hermaphroditic: Having both female and male sex cell-producing organs.

Heterophile antigen: A substance present in various tissues that causes the production of immunoglobulins, which react with the tissues of several mammals, fish, and plants.

Heterotroph: An organism incapable of utilizing CO_2 as its sole source of carbon and requiring one or more organic compounds for its nutrition.

Hfr cell (high frequency of recombination): A rare bacterial cell in the F^+ population that can transfer its chromosomes to an F^- cell with a greater frequency than other cell types.

Hinge region: The area of the H chains in the C region between the first and second C region of an immunoglobulin.

Histamine *(HISS-tah-mean)*: An active chemical released from mast cells that causes smooth muscle contraction of human bronchioles and small blood vessels, increased permeability of capillaries, and increased secretion by nasal and bronchial mucous glands.

Histocompatible: Sharing transplantation antigens.

HLA (human leukocyte antigen): The major histocompatibility genetic region in humans.

H-2 locus: The major genetic histocompatibility region in the mouse.

Holozoic: A form of nutrition exhibited by an organism capable of existing only by ingesting entire organisms or solid or particulate matter.

Homo: A prefix meaning "same."

Homologous: Derived from the same system or species.

Homologous antigen: An antigen that stimulates production of an antibody and reacts specifically with it.

Hormogonium: A motile section of a filamentous cyanobacterium, usually located between heterocysts.

Hormone: A substance produced in minute amounts in one part of the body and transported to another region where it produces its effects; e.g., insulin, gonadotropin.

Host: The particular animal, plant, or microorganism harboring another form of life as a parasite.

Humoral immunity: Antibody-mediated immunity.

Humus: A dark mass of decayed animal and plant matter remaining after microbial decomposition, which gives soil a loose texture and brown or black color.

Hybrid *(HIGH-brid)* **cells:** The result when unrelated cells, including their nuclei, become fused through the action of certain viruses or other agents.

Hybridoma *(HIGH-brid-oh-mah)*: A cell formed from the fusion of two cells, usually one a cancerous (myeloma) cell and the other an antibody-producing cell; produces monoclonal antibodies.

Hydrogen bonds: A weak chemical bond that forms between protons (H^+) and pairs of electrons on adjacent atoms.

Hydrolysis: Breaking of a molecule into smaller molecules by the addition of a water molecule.

Hydrophilic: Chemical groups with the ability to form hydrogen bonds, thus having affinity for water and similar groups (—OH, =O, —NH, etc.).

Hydrophobic: Chemical groups unable to form hydrogen bonds, thus lacking affinity for water.

Hyphae *(HIGH-fee)*: The filaments or threads that form a mycelium.

Hypotroph: An obligate intracellular parasite, including viruses and other forms of microorganisms requiring living cells for their nutrition.

Idiotype: A unique antigenic determinant present on homogeneous antibody or myeloma protein. The idiotype represents the antigenicity of the antigen-binding site of an antibody and is located in the V (variable) region.

IgA: The major immunoglobulin class present in secretions.

IgD: The major immunoglobulin class present on human B lymphocytes.

IgE: An antibody involved in immediate hypersensitivity reactions.

IgG: The major immunoglobulin class present in human serum.

IgM: A pentameric (five-unit) immunoglobulin comprising approximately 10% of normal human serum immunoglobulins. It is the largest immunoglobulin, having a molecular weight of 900,000.

Immune complexes: Antigen-antibody complexes (combinations) that may be deposited in tissues.

Immunofluorescence: Fluorescence resulting from a reaction between a substance and specific immunoglobulins that are bound to a fluorescent dye.

Immunogen: A substance that provokes immunoglobulin (antibody) production.

Immunogenicity: The capacity to stimulate the formation of specific immunoglobulins (antibodies).

Immunoglobulin: A glycoprotein composed of heavy (H) and light (L) chains that functions as antibody. All antibodies are immunoglobulins.

Immunoglobulin class: A subdivision of immunoglobulin molecules based on differences in antigenic determinants in the Fc region of the H chains. In humans there are five classes of immunoglobulins designated IgG, IgA, IgM, IgD, and IgE.

Immunopoietic: Immunoglobulin (antibody)-forming.

Impetigo *(im-peh-TIE-go)*: A bacterial skin infection, usually caused by either staphylococci or streptococci.

Infestation: The attachment to or temporary invasion of the superficial skin layers by a parasite.

Inflammation: Tissue reaction to injury; a defensive response to irritation.

Innate *(in-NATE)* **immunity:** The various factors providing resistance to disease determined by the genetic makeup of the individual.

Inoculum: Usually the microorganism-containing specimen used to start cultures.

Inorganic compound: Any substance that either does not contain carbon atoms or is oxidized in the form of carbon monoxide (CO), carbon dioxide (CO_2), or bicarbonate (HCO_3^-).

Interferon *(in-ter-FEER-on)*: A special class of proteins normally produced in response to viruses, endotoxins, and certain chemicals or synthesized through genetic engineering techniques; associated primarily with anticancer and antiviral activities.

Interleukins *(in-ter-LOO-kins)*: small, immunologically nonspecific substances that act on leukocytes; they transmit information related to growth and differentiation among leukocytes.

Intoxication: Poisoning.

Intron: A noncoding region for proteins in the mRNA of eucaryotic cells: separates exons, the protein coding regions.

In-use tests: Procedures used to test the effectiveness of a disinfectant under conditions similar to or identical to actual applications.

In vitro **(Latin: *in glass*):** Pertains to experiments done outside of the natural environment of living cells or organisms.

In vivo **(Latin: *in life*):** Pertains to experiments done in living cells or organisms.

Ion *(eye-uhn)*: An atom that has either lost or gained one or more orbital electrons, and thus becomes an electrically charged particle.

I region: That portion of the major histocompatibility complex which contains genes that control immune responses.

Isoantigens: Antigens (immunogenic substances) of the same species.

Isotype: Antigenic characteristics of a given class or subclass of immunoglobulin H and L chains.

Jaundice *(john-diss)*: Yellowing of body tissues.

J chain: A glycopeptide chain that is normally found in polymeric immunoglobulins, particularly IgA and IgM.

Kaposi's *(KAP-oh-shee)* **sarcoma:** A type of skin cancer that has several forms; the condition can assume an aggressive, fatal form in AIDS patients.

Kappa (κ) chains: One of two major types of L chains in immunoglobulins.

K cell: A killer cell (lymphocyte) responsible for antibody-dependent cell-mediated cytotoxicity.

Kingdom: A major taxonomic category, consisting of several phyla or divisions.

Kleinschmidt technique: The method used to spread nucleic acid molecules for good electron microscopic visualization (usually using proteins as agents).

Koch's postulates: A definite sequence of experimental steps that shows the causal relationship between a specific organism and a specific disease.

Koplik's *(KOPE-liks)* **spots:** A characteristic sign of measles; usually bluish spots surrounded by a reddened area in the mucous membranes on the sides of the mouth.

Küpffer cells: Fixed mononuclear phagocytes that are present within the spaces or sinusoids of the liver.

Lactoferrin *(lack-toe-FERR-in)*: An iron-containing compound that produces a slight antimicrobial action by binding the iron necessary for microbial growth.

Lambda (λ) chain: One of two major types of L chains in immunoglobulins.

Larva: An immature form of certain animals that differs morphologically from the adult form.

Leukotrienes *(LOO-koh-treens)*: A class of immunological mediators that appear when allergens combine with IgE on mast cell surfaces; leukotrienes attract neutrophils, contract smooth muscles, and increase blood vessel permeability.

L-forms: A procaryotic form lacking a cell wall, developing from a normally wall-containing organism.

Lichen *(LIE-kin)*: A mutualistic association of a fungus and either algae or cyanobacteria, resulting in a new structural form. The new formation is morphologically distinct from either of its microbial components and possesses separate chemical and physiological properties.

Light chain (L chain): A polypeptide chain present in all immunoglobulin molecules. Two types exist in most species and are termed kappa (κ) and lambda (λ).

Lipids: A group of organic compounds composed of carbon and hydrogen; e.g., fats, phospholipids, waxes, steroids.

Lipopolysaccharides (LPS): Cell wall components of gram-negative bacteria.

Locus: The specific site of a gene on a chromosome.

Lymphokines *(LIMF-foe-kines)*: Soluble products of lymphocytes that are responsible for the multiple effects of a cellular immune reaction.

Lymphoma *(LIMF-foe-mah)*: A solid tumor of lymphatic tissue.

Lyophilization *(lie-oh-FIL-eh-zay-shun)*: The process of freeze-drying; a substance and/or cells are rapidly frozen at a low temperature and then dehydrated in a vacuum.

Lysis: Bursting of a cell by the destruction of its cell membrane.

Lysogenic conversion: The occurrence of particular genetic changes when a virus integrates itself into the host cell's DNA.

Lysogeny *(lie-SOJ-eh-nee)*: A state in which a bacteriophage's genome is integrated or closely associated with the host cell's chromosome.

Macromolecule: A very large molecule formed by a linear or chainlike structure consisting of simpler repeating molecules or monomers. Examples of macromolecules include proteins, polysaccharides, and other synthetic and natural polymers.

Macrophages: Phagocytic mononuclear cells that originate from bone marrow monocytes and serve accessory roles in cellular immunity.

Macule *(MAC-yuhl)*: A pink-red, not elevated spot, often associated with a skin rash.

Major histocompatibility complex (MHC): A number of genes located in close proximity to one another that determine histocompatibility antigens of members of a species; MHC codes for the majority of cell-surface glycoproteins that guide antigen recognition and cell interactions, especially by T cells.

Malaise: General discomfort, uneasiness.

Mast cell: A tissue cell that resembles a peripheral blood basophil and contains granules with serotonin and histamine.

Medium: A liquid or solid nutrient preparation used for the cultivation of microorganisms.

Meiosis: Nuclear division, characteristic of sex cell formation, in which the number of chromosomes in daughter (newly formed) cells is half the number in the original cell.

Meninges *(men-IN-gees)*: Protective coverings of the brain and spinal cord.

Meningitis: Inflammation of the membranes covering the brain and spinal cord.

Mesophile *(ME-zoh-fil)*: An organism that grows in the temperature range between 20–45°C.

Mesosome *(MEEZ-oh-some)*: A membranous infolding (involution) of the cytoplasmic membrane of procaryotes found mainly with gram-positive cells.

Messenger RNA (mRNA): Ribonucleic acid (RNA) that serves as a pattern, or template, for protein synthesis.

Metabolism: The sum total of cellular chemical reactions by which energy is provided for vital processes and new cell substances are assimilated.

Metastasis *(meh-TAST-ah-sis)*: Spreading of a disease.

Methanogen: A methane-producing procaryote.

Metric system: Standardized decimal measurement based upon the meter as a unit of length, the gram as a unit of weight, and the liter as a unit of liquid measure.

Microaerophilic: Growing best in the presence of low concentrations of oxygen.

Microfibril: Threadlike structure found in the cell walls of filamentous fungi. Chemically, microfibrils consist of chitin.

Microfilaria *(my-crow-fill-AIR-ee-ah)*: Microscopic larvae of certain nematodes, e.g., *Wuchereria bancrofti*, the causative agent of elephantiasis.

Micrometer (μm): One millionth (10^{-6}) part of a meter, or 10^{-3} of a millimeter.

Microorganism: A microscopic or submicroscopic form of life.

Microtubule: A microscopic structural protein unit of eucaryotic cilia and flagella. Microtubules aid in maintaining cell shape and serve as spindle fibers in mitosis.

Minimal inhibitory concentration (MIC): The lowest concentration of a drug that will prevent growth of a standard microbial suspension.

Mitochondrion: A eucaryotic cellular organelle consisting of an outer membrane and an inner one folded into cristae.

Mitogens: Substances that cause DNA synthesis, primitive or blast cell transformation, and ultimately, division of lymphocytes.

Mitosis: The asexual process in which genetic material is duplicated and equally distributed in newly formed cells.

Mixed lymphocyte culture (mixed leukocyte culture) (MLC): An *in vitro* test for cellular immunity in which lymphocytes or leukocytes from genetically dissimilar individuals are mixed and mutually stimulate DNA synthesis.

Monera: The kingdom of procaryotic organisms with unicellular and simple colonial organization; the term Procaryotae has replaced Monera.

Monoclonal immunoglobulin (antibody) molecules: Identical copies of antibody that consist of one H chain class and one L chain type.

Monomer *(MON-o-mer)*: A small molecule; monomers are the units that combine to form polymers.

Monosaccharide: A category of six-carbon sugars having the molecular formula $C_nH_{2n}O_n$; e.g., glucose.

Morbidity rate: The number of cases of a disease per 1,000 of the population within a given time period.

Mordant *(MOR-dant)*: A chemical added to a dye to make it stain more intensely.

Morphology *(more-FOL-oh-gee)*: The study of shape and structure.

Mortality rate: The percentage of individuals dying from a specific disease.

Must: Pressed grape juice used for wine making.

Mutation: A sudden change in the genetic code of an organism, resulting in a hereditable (permanent) property differing from the parent cell.

Mutualism: A symbiotic association in which two different organisms are obligately dependent upon each other.

Myalgia *(my-AL-jee-ah)*: Muscle pain.

Mycelium (pl. mycelia): An interwoven mat of hyphae, i.e., mass of fungal filaments.

Mycology: The study of fungi.

Mycoplasma *(my-co-PLAZ-mah)*: A group of bacteria composed of cells lacking cell walls and exhibiting a variety of shapes (pleomorphism).

Mycorrhiza: A growth occurring as a result of the symbiotic relationship between certain fungi and the roots of plants and trees.

Mycosis: A fungus infection of animals.

Mycotoxicosis: Poisoning caused by fungal toxins (mycotoxins).

Mycotoxin: A poison (toxin) produced by and associated with fungi.

Myeloma *(my-ee-LOH-mah)*: A tumor composed of cells of the type normally found in the bone marrow.

Nanometer (nm): 10^{-6} millimeter or 10^{-9} meter. This designation has replaced the millimicron.

Native immunity: A state of resistance conferred by one's genetic makeup, not acquired by exposure to infectious disease agents.

Necrosis *(neh-KROW-sis)*: Death or destruction of tissue.

Neonate *(KNEE-oh-nate)*: A newborn infant.

Neoplasm: Any new and abnormal growth.

Niche: The biological role of an organism in its community, which includes its location and its activities.

Nitrogen cycle: A geochemical cycle in which atmospheric nitrogen gas is converted into a usable form for plants by microbial enzymatic reactions, and back to nitrogen gas again.

Nitrogen fixation: The formation of nitrogen compounds (NH_3, organic nitrogen) from free atmospheric nitrogen (N_2).

NK cells (natural killer cells): Cytotoxic cells belonging to the T-cell subpopulation, responsible for cellular injury or cytotoxicity without prior sensitization.

Nonself: A term referring to an antigen normally not found in the body.

Nonsense mutation: A mutation that results in a codon's not being translated into an amino acid in protein synthesis.

Nosocomial *(no-so-CO-me-al)* **infection:** Hospital-acquired infection.

Nucleic acid: One of a class of macromolecules consisting of a series of nucleotides; e.g., deoxyribonucleic acid (DNA), ribonucleic acid (RNA).

Nucleocapsid: The capsid and the enclosed nucleic acid core of a virus particle.

Nucleolus: A dense oval to round organelle found in the nucleus of eucaryotic cells. It is largely composed of RNA and thought to be the site for RNA formation.

Nucleoside *(NEW-klee-oh-side)*: A chemical compound consisting of a five-carbon sugar and a purine or pyrimidine base.

Nucleotide *(NEW-klee-oh-tide)*: A chemical compound consisting of a five-carbon sugar with a phosphate group and a purine or pyrimidine attached. Nucleotides are building blocks of nucleic acids.

Nucleus: The eucaryotic cellular organelle enclosed by a nuclear membrane and containing the cell's chromosomes. Also, the center, or core, of an atom.

Numerical taxonomy: A technique for determining the relationships among organisms by determining the number of properties the organisms have in common.

Nutrient broth: A standard bacteriological growth medium consisting of beef extract, peptones, and distilled water; the addition of agar converts this medium to nutrient agar.

Okazaki *(oh-ka-ZOCK-ee)* **fragments:** Short segments of DNA that are enzymatically combined with one another to form a DNA molecule during DNA replication.

Oncogene: The gene of cancer-causing viruses responsible for inducing transformation of cells.

Operator: Genome region capable of interacting with a specific repressor, thereby controlling the functioning of an adjacent operon.

Operon *(OH-per-on)*: A group of adjacent genes that are under the joint control of an operator and a repressor.

Opportunist: A microorganism that causes infection only under especially favorable conditions (e.g., when defense host mechanisms are not fully functioning).

Opsonization: A form of phagocytosis enhanced by specific immunoglobulins and complement.

Order: In taxonomy, a major subdivision of a class. Each order consists of one or more related families.

Organelle: Particular structures (little organs) in eucaryotic cells, such as the nucleus, mitochondria, chloroplasts, Golgi body, cilia, and so on.

Organic compounds: Any compound that contains carbon (C) with the exception of those substances oxidized in the form of carbon monoxide (CO), carbon dioxide (CO_2) or carbonate (HCO_3^-).

Otitis media *(oh-TIE-tis ME-dee-ah)*: Middle ear infection.

Oxidase test: A test for the presence of the cytochrome oxidase complex, widely used in the presumptive identification of the *Neisseria* species.

Oxidation: The loss of electrons; the opposite of reduction.

Oxidative phosphorylation: The generation of energy in the form of adenosine triphosphate (ATP) that results from the passage of electrons through the electron transport chain to the final electron acceptor, oxygen.

Oxygen cycle: The general cycle by which oxygen is exchanged and distributed within the biosphere.

Pandemic: Affecting the majority of the population of a large region, or epidemic concurrently in many different parts of the world.

Papule *(PAH-pule)*: A small, firm, generally round, elevated skin lesion.

Parasite: An organism that lives within or upon another form of life.

Passive immunization: The transfer of antibodies from an immunized donor to a nonimmune recipient.

Pasteurization: The processes of heating food or other substances under controlled conditions of time and temperature, such as 63°C for 30 minutes, in order to kill pathogenic microorganisms.

Pathogenic: Capable of producing disease.

Pathogenicity: The ability of an organism to produce disease.

Pellicle: A thin, membranelike film formed on the surface of microbial broth culture. Also, the thickened, outer surface of certain protozoa.

Pelvic inflammatory disease (PID): Inflammation of the female pelvic organs, especially the ovaries, fallopian tubes, and uterus.

Penicillinase: An enzyme produced by penicillin-resistant bacteria that destroys the beta-lactam ring of penicillin.

Peptide bond: The covalent bond that joins an amino group of one amino acid to the carboxyl group of another amino acid, with the formation of water.

Periplasmic space: In gram-negative bacteria, the area between the plasma membrane and the cell wall, containing various enzymes some of which are involved in nutrition.

Peroxisome: The microbody in which photorespiration takes place.

Petri *(PEE-tree)* **dish:** A commonly used container in the cultivation of various microorganisms.

pH: The term used to indicate the hydrogen-ion concentration, which reflects the relative acidity or alkalinity of a solution.

Phage: A bacterial virus (bacteriophage).

Phagocytosis: Cellular engulfment of foreign particles.

Phosphorus cycle: A geochemical cycle by which phosphorus is exchanged and distributed within the biosphere.

Phosphorylation: The introduction of a phosphate (PO_4^{3-}) group into an organic molecule.

Phylum (pl. phyla): A large taxonomic rank of related families in the classification of living organisms. The classification system is subdivided progressively from world → kingdoms → subkingdoms → phyla → classes → orders → families → genera → species.

Phytotoxin: Bacterial toxins that affect plants.

Pilus (pl. pili): Surface-associated structures found on certain bacterial cells and some yeasts that are not organelles of locomotion. Depending on the pilus type, activities such as attachment and transfer of genetic material are associated with them.

Pinocytosis *(pee-no-sigh-TOE-sis)*: The process performed by a cell in which liquid or small particles are enveloped in a vacuole and brought into the cell.

Plankton: Microscopic, floating animal, microbial, and plant life distributed throughout various bodies of water.

Plantae: The kingdom of plants.

Plaque *(plack)* **(dental):** A collection of bacteria tightly adhering to a tooth surface, involved in the production of dental caries.

Plaque (viral): A clear area in a single layer or monolayer of cells created by viral lysis of infected cells.

Plasma: The liquid portion of lymph and blood.

Plasma cells: Fully differentiated antibody-forming cells that are derived from B lymphocytes.

Plasmid: A small amount of DNA that exists in procaryotes independently of the chromosome. Plasmids are coded for nonessential information.

Plasmolysis: Contraction or shrinking of a cell's cytoplasm due to the loss of water by osmotic action; i.e., a loss of water molecules.

PMNL: Polymorphonuclear leukocyte.

Polyhedron: Many-sided figure.

Polymer *(POL-ee-mer)*: A compound, usually of high molecular weight, formed by the linear combination of simpler repeating molecules, or monomers.

Polymerase: Any enzyme that catalyzes the process of polymerization.

Polymerization: The process of forming a compound (polymer), usually of high molecular weight, by the combination of simpler molecules.

Polysaccharide: A carbohydrate composed of more than three molecules of monosaccharides.

Polysome: A chain of ribosomes held together temporarily by messenger RNA.

Porin: Specific clusters of protein molecules that form pores in the outer membranes of gram-negative cell walls.

Prion *(PRE-ahn)*: An infectious disease agent that is neither virus nor viroid and consists only of protein.

Procaryotae: The kingdom of procaryotic organisms, includes the eubacteria, rickettsias, chlamydias, and archaeobacteria. See Appendix B.

Proglottid: Body segment of a tapeworm.

Proton gradient: The difference in concentration of H^+ between the interior of a cell and the surrounding medium.

Pseudopod: A temporary cytoplasmic protrusion of amebae, or ameboid cells, which functions in cellular feeding and movement.

Psychrophile: Microorganisms that grow well between 0° and 20°C. The optimum is usually above 4°C.

Purified protein derivative (PPD): Substance obtained from cultures of *Mycobacterium tuberculosis* or related organisms and used in skin testing.

Purines: Organic bases with carbon and nitrogen atoms in two interlocking rings; components of nucleic acids, adenosine triphosphate (ATP), and other biologically active substances.

Pustule: Small elevated skin lesion containing pus.

Putrefaction *(pew-treh-FAC-shun)*: The anaerobic decomposition of protein resulting in foul odors.

Pyogenic *(pie-oh-GEN-ic)*: Pus-producing.

Pyrimidines: Nitrogenous bases consisting of a single ring of carbon and nitrogen; components of nucleic acids and certain antimicrobial drugs.

Pyrogens: Substances that are released either from within leukocytes or from bacteria, and that produce fever in susceptible hosts.

Quellung reaction: A laboratory test in which swelling of the capsules of bacteria occurs when the organisms are exposed to specific antibodies.

Radioallergosorbent test (RAST): A test using radioactive materials capable of detecting IgE antibody directed at specific allergens.

Reactant: A starting substance entering into a chemical reaction.

Reaction: The interaction of two or more chemical substances resulting in their chemical change.

Recombinant: The new cell that results from genetic recombination.

Recombinant DNA: DNA that results from the combining of DNA molecules from two or more sources.

Recombination: Interaction of two double-stranded DNA molecules through crossing-over of strands (breakage and reunion) that leads to a molecular rearrangement.

Reduction: The addition of electrons to an atom or molecule; opposite of oxidation.

Replication fork: The site of the copying and separation of DNA, and where new DNA strands are formed.

Reservoir: The source or sources of an infectious agent in nature.

Resistance transfer factors (RTF): A set of genes that is associated with antibiotic resistance and that carries the genetic information for the transfer of such resistant (R) factors.

Respiration: Any biochemical process in which energy is released; two forms are known, aerobic respiration (in the presence of oxygen) and anaerobic respiration (in the absence of oxygen).

Reverse transcriptase: An enzyme involved in the formation of DNA complementary to an RNA pattern or template.

R factor: A transferable plasmid found in many enteric bacteria, which carries genetic information for resistance to one or more chemotherapeutic agents; often associated with a transfer factor that is responsible for conjugation with another bacterium and for transfer of the R factor.

Ringworm: The common term for fungus infections of the skin, nails, and hair; tinea.

Rumen: The first stomach of a ruminant, or cud-chewing animal.

Saprophyte: Any organism that derives its nutrition from dead or decaying organic matter.

Sarcoma *(sar-CO-mah)*: A solid tumor developing from connective tissue, bone, muscle, and fat.

Satellite viruses: Small viruses that are replicated only in the presence of a specific full-sized helper virus.

Schizogony: The asexual phase of sporozoans, e.g., malarial parasites.

Scolex *(SCOH-lex)*: The head of a tapeworm.

Secondary immunodeficiency: Immunodeficiency caused by such factors as malnutrition, malignancies, radiation exposure, burns, and various drugs that interfere with the functions of the lymphatic system.

Secretory IgA: A dimer of IgA molecules linked by J chain and secretory component.

Selective medium: A growth medium containing a component that will inhibit the growth of certain microorganisms and either enhance or not affect the growth of other organisms.

Sequelae *(SEE-quell-ee)*: Conditions following or resulting from a disease.

Serology: Literally, the study of serum. Refers to the laboratory determination of antibodies to infectious agents important in clinical medicine.

Serum: The light yellow fluid left after the clotting of blood has occurred.

Sexual reproduction: The formation of an organism from the union of two different sex cells (gametes).

Sheath: A tubular structure surrounding certain microbial cells.

Simple staining: A procedure using only one dye.

Sinus: Any cavity having a relatively narrow opening.

SI system: Système International d'Unités. An international system of units for the measurement of length, volume, and mass. It is based on the metric system.

Sludge: The precipitated solid matter produced by water and sewage treatment.

Smear: A thin film of material spread on a clean glass.

Sodium hypochlorite (NaOCl): Bleach; a chlorine derivative used in disinfection.

Species: The unit of taxonomic classification for most forms of life; a population of similar individuals, alike in their structural and functional properties and able to reproduce sexually with each other but with no other types of organisms.

Spontaneous generation: A theory that certain forms of life arose spontaneously from nonliving matter.

Spontaneous mutation: A mutation for which no immediate cause is known.

Sporadic *(spore-AD-ic)*: Irregular in occurrence.

Spore: A resistant structure or stage formed by procaryotes. Also, the reproductive cell of certain types of eucaryotic organisms.

Sporogony: The sexual phase of sporozoans such as *Plasmodium* spp., which cause malaria.

Sporozoite *(spore-oh-ZOH-ite)*: An elongated, sickle-shaped body formed in an oöcyst as the result of the sexual reproduction phase of sporozoans such as *Plasmodium* spp.

Sputum *(SPUE-tum)*: The thick mucus secreted by irritated tissues in the lower respiratory tract and expelled by coughing; specimen of choice in several respiratory infections.

Starter culture: A pure or mixed culture of microorganisms used to initiate a desired fermentation process, as in cheese manufacture.

Sterile *(STER-il)*: Free from all living organisms.

Sterilization: The destruction or removal of all living forms.

Steroid: A complex macromolecule containing carbon atoms arranged in four interlocking rings, three of which contain six carbon atoms each and the fourth of which contains five.

STS: Serological tests for syphilis.

Subclinical infection: A state in which the individual does not experience all of the characteristic symptoms of a particular disease, or such effects are less severe.

Substrate *(SUB-straight)*: A substance acted upon, as by an enzyme.

Sulfur cycle: A geochemical cycle by which sulfur is utilized and distributed through the biosphere.

Suppressor T cells: A subset of T lymphocytes that suppress antibody synthesis by B cells or inhibit other cellular activities such as those of T helper cells.

Suppurative *(sup-you-RAY-tiv)*: pus forming.

S value: Svedberg unit. Indicates the sedimentation coefficient (*S*) of a protein; usually determined by ultracentrifugation.

Swimmer's itch: A mild inflammation of the skin caused by the cercariae of bird schistosomes (flukes).

Symbiosis: The living together of two dissimilar organisms; such associations may be commensal, mutualistic, or parasitic.

Symptomatic treatment: Treatment of existing or expected symptoms.

Syngamy *(sin-GAM-ee)*: Sexual reproduction, the union of the two haploid chromosome complements in fertilization.

Synthetic medium: A medium of known composition.

Syntropism: Feeding together.

Systemic: Pertaining to or affecting the body as a whole.

T antigens: Tumor antigens, protein products of the viral genome present only on infected neoplastic cells.

Taxonomy: The description, classification, and naming of organisms.

T cell (T lymphocyte): A thymus-derived cell that participates in a variety of cell-mediated (controlled) immune reactions; four subsets of T cells are known.

Temperate viruses: Viruses that can become stably established within the host, either by integration into the host's genome or by plasmid formation, and establish immunity to reinfection by the same virus.

Template: The molecule that serves as the pattern for processes such as the formation of another molecule; a pattern or mold that guides the formation of a duplicate.

Test: An outer protective covering or shell formed by certain protozoa, e.g., foraminifera.

Thallus: The vegetative body of certain microorganisms and plants that does not show differentiation into roots, stems, or leaves.

Thermophile: An organism that grows best at high temperatures, e.g., 45° to 75°+C.

Thylakoid: An organelle consisting of layers in the cytoplasm; the site of photosynthesis.

Thymosin: A thymic hormone protein of MW 12,000 that can restore T-cell immunity in animals having their thymus glands removed.

Thymus: The central lymphoid organ, which is present in the thorax and controls the development (ontogeny) of T lymphocytes.

Tissue matching: The process of determining the histocompatibility of donor and recipient tissues prior to grafting.

Titer: The potency of a biological reactant expressed in units per milliliter. The level of immunoglobulins is frequently given in this way.

Toxin: A poisonous substance.

Toxoid: A converted toxin; the resulting preparation is nontoxic but still capable of provoking immunoglobulin production.

Transcriptase: An enzyme that catalyzes the flow of genetic information from DNA to RNA.

Transcription: The synthesis of a complementary copy of a DNA or RNA template strand.

Transduction: Transfer of bacterial genes from one bacterium to another within the DNA of a bacteriophage.

Transformation *(bacterial)*: A form of recombination in which DNA fragments from disintegrated donor cells are taken up by living competent recipient cells and incorporated into their chromosomes.

Transformation *(mammalian cells)*: The changing of a normal cell into a cancer cell.

Translation: A process whereby the genetic information present in an mRNA molecule directs the order and position of the specific amino acids during protein synthesis.

Transposons *(trans-POE-zons)*: Genes that are capable of moving from one DNA molecule to another within the same cell; jumping genes.

Trichocyst *(TRICK-oh-sist)*: An organelle in the cytoplasm of ciliated protozoa, such as *Paramecium* spp., which can discharge a filament for defense or for trapping prey.

Trichome: A single row of cells of a multicellular organism in which the multicellular nature is evident without staining.

Trophozoite *(tro-foe-ZOH-ite)*: The vegetative cell of a protozoon, i.e., active feeding stage.

T lymphocyte: A thymus-derived lymphocyte that is responsible for cell-mediated immunity (CMI).

Tubercle *(TOO-burr-kel)*: A nodule, a specific lesion in certain diseases such as tuberculosis.

Tumor: An accumulation of cells resulting in the appearance of an observable lump.

Turbidity: Cloudiness.

Ulcer: A rounded or irregularly shaped area of inflammatory tissue destruction in the epithelial lining of a surface.

Ulceration: The process of ulcer formation.

Ultracentrifugation: A high-speed centrifugation technique that can be used for the analytic identification of proteins of various sedimentation coefficients or as a preparative technique for separating proteins of different shapes and densities.

Ultrahigh temperature (UHT) treatment: The process of milk at a temperature higher than that used in regular pasteurization; the processed milk can be stored for several months at room temperature in unopened containers.

Vacuole: A clear area in a cell's cytoplasm.

VDRL (Venereal Disease Research Laboratory) test: A precipitin test for the diagnosis of syphilis.

Vector: A carrier of pathogenic agents, especially an arthropod.

Venereal *(ven-EAR-re-al)* **disease:** A disease, such as gonorrhea or syphilis, which is spread by sexual contact and involves parts of the genitourinary system.

Viremia: Presence of viruses in the circulatory system.

Virion *(VEER-ee-on)*: A complete, fully infectious virus particle.

Viroid: Short strands of RNA capable of causing diseases in plants.

Virulence *(VEER-u-lenz)*: The capacity to produce disease; it is a function of microbial invasiveness and toxigenicity and is measured with reference to a particular host.

V (variable) region: The amino (NH_2) terminal portion of the H or L chain of an immunoglobulin molecule, containing considerable heterogeneity in the amino acid residues compared to the constant region.

Water activity: The ratio of the water pressure of a food substance to the vapor pressure of pure water at the same temperature.

Wheal and flare: A circumscribed, reddened elevation of the skin.

Yeast: A type of unicellular fungus that characteristically does not form typical mycelia.

Zoonosis *(zoo-NO-sis)*: An animal disease that is transmitted to humans by natural means.

Zoospore: An asexual, flagellated, motile spore.

Index

Explanation of symbols: *f*, figure; *t*, table; *cp*, color photograph

A

C

D

E

F

G

J

K

L

M

N

O

T

U

V

W

X

Y

Z